RAND McNALLY COSMOPOLITAN WORLD ATLAS

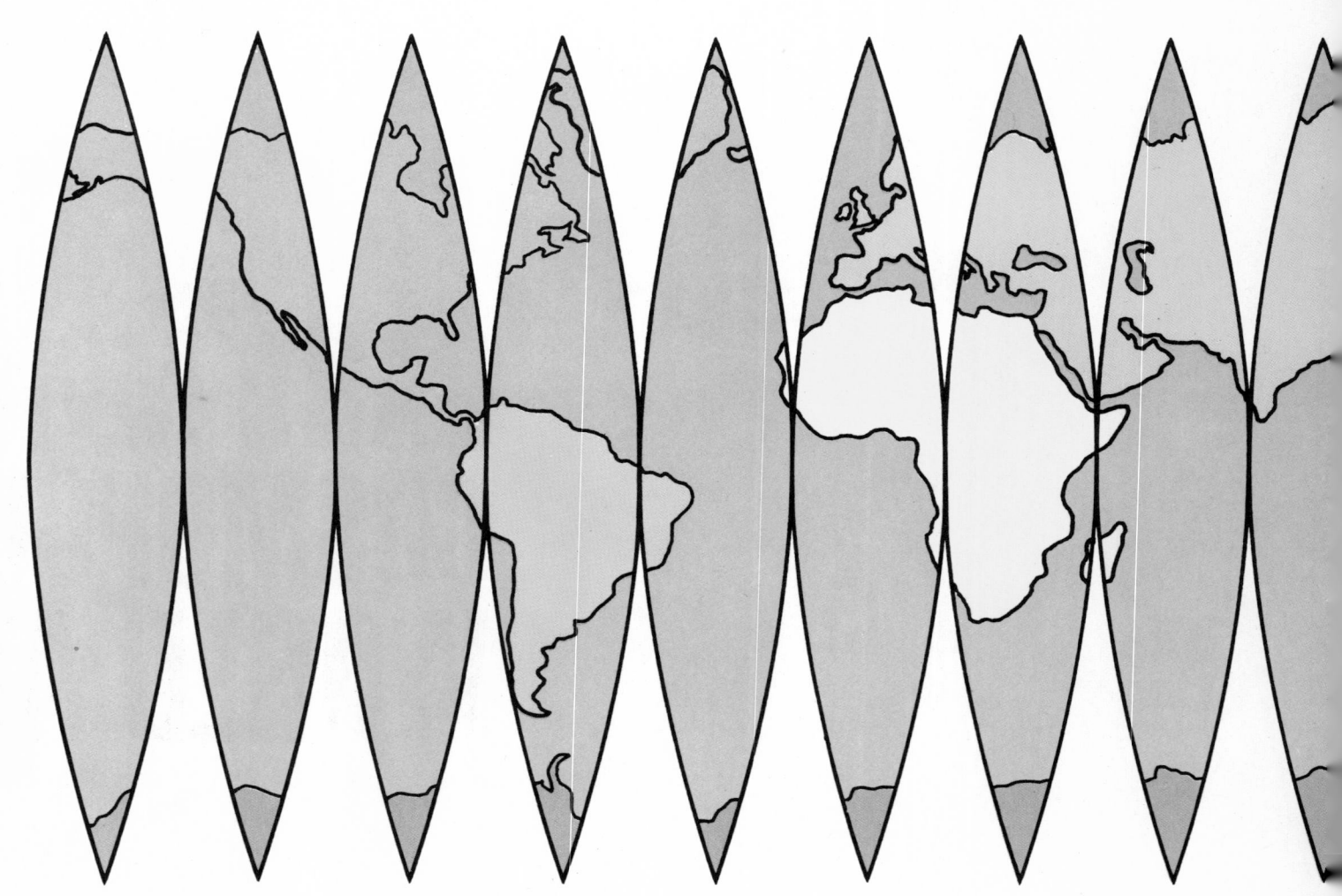

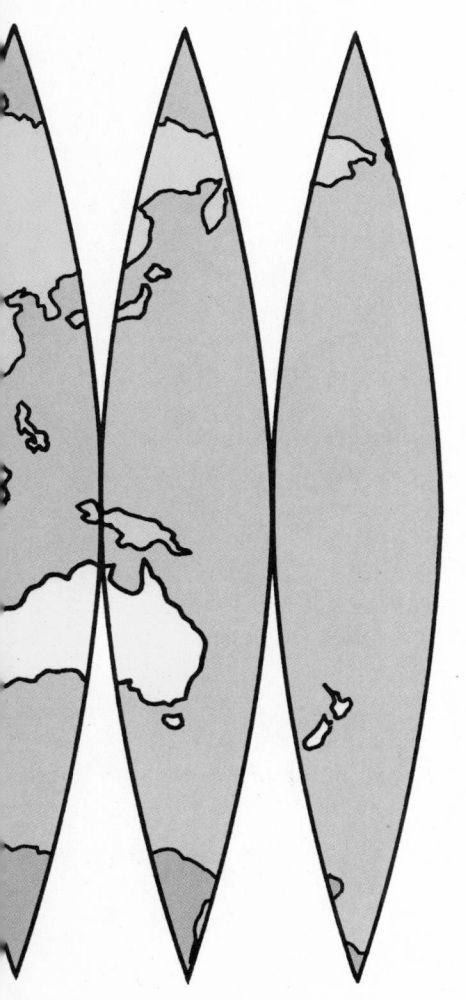

RAND McNALLY
COSMOPOLITAN
WORLD ATLAS

RAND McNALLY & COMPANY

CHICAGO · NEW YORK · SAN FRANCISCO

SPECIAL FEATURES OF THE ATLAS

In design, scope, and organization the Cosmopolitan World Atlas is quite different from any reference atlas previously published in the United States. The world has been mapped on a broad regional basis, not country by country; but each regional map is centered around a major country or a significant grouping of countries.

Each of the great land masses of the earth is shown as a whole and then broken down into major continental regions. All the regional maps for each continent are drawn on the same scale. Thus it is possible to make direct visual comparisons of the sizes of countries and the distances between places by turning from one map to another. North America is mapped in four sections, all on the scale of 1:12,000,000. South America is shown in seven sections on the scale of 1:8,000,000. The continents of Eurasia and Australia are divided into five major regions, and are mapped on the scale of 1:16,000,000. Africa is mapped in five sections on the scale of 1:11,400,000. Smaller areas are shown in greater detail on larger scale maps, which are even multiples of these regional scales.

Small marginal maps show the location of areas covered by the sectional maps. No detail has been neglected that will help the user to compare areas or see the world relationships of the area in which he is interested.

The value of any atlas depends almost as much upon the quality of its index as upon the quality of its maps. For this reason the publishers of the Cosmopolitan World Atlas have given to the index the same careful attention they have given to the maps.

The index includes in one alphabetical listing all the names that appear in the Atlas. Often the form of a name commonly used in English-speaking countries differs from the official name. Whenever the anglicized form of the name of a capital city differs from the official spelling, the official name is included on the maps in parentheses. Both forms appear in the index. For other cities, the local, official form is given first on the map, with the customary English form in parentheses. Wherever two forms of a name are in common use because of a recent change in sovereignty, an official change in name, or for any other reason, both forms are given on the map and in the index. This policy has been followed to facilitate the use of the Atlas by English-speaking readers.

In general, spellings follow the recommendations of the United States Board of Geographical Names of the Department of the Interior, which determines the official spelling of foreign geographical names. For some areas and languages, however, more anglicized names are used than the Board recommends for official maps. For the spelling of place names in the United States, the United States Postal Guide is the authority that is followed.

The index of the Cosmopolitan World Atlas includes the names of the political subdivisions of all major countries. The names of political divisions are distinguished from the names of cities by descriptive terms such as "county," "country," "province," "region," "state," or their abbreviations, which follow the names.

The maps carry as many political subdivisions as space will permit. Counties are shown on all state maps in the United States and on most of the maps of Canadian provinces. Other countries may not be mapped on a large enough scale to permit showing present administrative subdivisions. For some of these countries, the names of larger administrative subdivisions appear without the boundaries. In others, regions with historical significance are shown instead of present subdivisions.

The index includes the names of all administrative divisions of all major countries, whether or not these names appear on the maps. By means of the index, the location of the political subdivision can be found on the map. When a name does not appear on the maps, an asterisk precedes the locational key in the index.

On the maps and in the index, the names of physical features are distinguished by italic type. A descriptive term, such as "mountain," "river," "bay," or "gulf," follows each name of a physical feature in the index, whenever the descriptive term is not part of the name itself.

The index also includes the population of cities, countries, and administrative divisions. Population figures are based on the latest available sources. As a result of many mid-century census reports becoming available, new population figures have been entered for most cities of the United States, England, France, Germany, Italy, Argentina, Japan, and many other countries. All other statistical facts have likewise been brought down to date.

Three special alphabetical lists increase the usefulness of the maps and index. These are: (1) a gazetteer of historical names that do not usually appear on modern maps; (2) a glossary of English equivalents of foreign geographic terms which appear in the names of physical features; and (3) an explanation of geographic terms frequently found on maps or needed for their interpretation.

The supplementary materials include tables of areas and populations, climatic and economic data, and the capitals and largest cities. It also includes tables of steamship, railroad, and air distances, as well as historical facts, and much other useful information.

A special color section is devoted to the Saga of Space which reviews the space activities of the last decade and the plans established for the future. The new section of maps in full color depict major periods of world history.

CONTENTS

Map Projections . . . . viii–ix
A Study in Map Scale x
Explanation of the Index Reference System and Map Symbols xi

SAGA OF SPACE – *pages xii to xxvi*

A New Frontier xii
The First Steps Upward xiii
Orbiting The Earth – Project Mercury . . . . xiv–xv
Atmosphere: Physical Nature xvi
Atmosphere: Man's Activity xvii
The Moon . . . . xviii–xix
To The Moon – Project Apollo xx

To the Planets xxi
Principal Planets xxii
The Dangers of Outer Space xxiii
Planet Information Table xxiv
Star Chart for the Northern Skies xxv
Members of the Universe xxvi

PHYSICAL WORLD IN MAPS – *pages xxvii to xl*

Introduction xxvii

Physical Map of the World . . . . . xxviii–xxix
Europe and Western Asia . . . . xxx–xxxi
Eurasia, Palestine Region, Malaya . . . . xxxii–xxxiii
Africa xxxiv

South America xxxv
United States of America . . . . xxxvi–xxxvii
Canada . . . . xxxviii–xxxix
North America xl

THE POLITICAL WORLD IN MAPS – *pages 1 to 125*

THE WORLD AND SPECIAL REGIONS – *pages 1 to 7*   PAGES
Polar Map of the World 1
The World 2–3
North Polar Regions 4
South Polar Regions 5
Atlantic Ocean 6
Pacific Ocean 7

EUROPE – *pages 8 to 31*
Europe and Western Asia 8–9
British Isles 10
Ireland 11
England and Wales 12
Scotland and Northern England 13
France and the Low Countries 14
Belgium, Netherlands, Luxembourg and Western Germany . 15
Germany, Austria and Switzerland 16
Central Germany 17
Alpine Regions 18
Switzerland 19
Spain and Portugal 20
Italy 21
Yugoslavia, Hungary, Romania and Bulgaria 22
Greece, Albania and Western Turkey 23
Denmark and Northern Germany 24
Norway, Sweden and Finland 25
Poland and Czechoslovakia 26
Western Soviet Union 27
Soviet Union 28
South Central Soviet Union 29
Western Mediterranean 30
Eastern Mediterranean 31

ASIA – *pages 32 to 41*
Israel and Northern Egypt 32
Eurasia 33
China, Mongolia, Korea, Japan and Taiwan 34

Indonesia and the Philippines 35
Northeastern China 36
Japan and Korea 37
Cambodia, Laos, Vietnam, Thailand and Malaysia . . . 38
India, Pakistan, Burma and Ceylon 39
Central India and Nepal 40
Iran and Afghanistan 41

AFRICA – *pages 42 to 49*
Africa 42
Northeast Africa 43
Northwest Africa 44
West Africa 45
Equatorial Africa 46
East Africa 47
Central Africa 48
Southern Africa 49

AUSTRALIA AND OCEANIA – *pages 50 to 52*
Australia 50
Southeastern Australia and New Zealand 51
Pacific Islands 52

SOUTH AMERICA – *pages 53 to 60*
South America 53
Central and Southern Argentina and Chile 54
Bolivia, Paraguay, Northern Argentina and Chile . . . 55
Uruguay and Southern Brazil 56
Eastern Brazil 57
Peru and Ecuador 58
The Guianas and Northern Brazil 59
Colombia and Venezuela 60

NORTH AMERICA – *pages 61 to 125*
North America 61
Central America 62
Mexico 63

CONTENTS – *continued*

NORTH AMERICA – *continued*

PAGES

West Indies 64
Puerto Rico and Virgin Islands 65
Canada 66–67
    British Columbia 68
    Alberta 69
    Saskatchewan 70
    Manitoba 71
    Ontario 72
    Quebec 73
    Maritime Provinces 74
    Newfoundland 75

United States 76–77
    Alabama 78
    Alaska 79
    Arizona 80
    Arkansas 81
    California 82
    Colorado 83
    Connecticut and
      Rhode Island . . . 84
    Delaware and
      Maryland 85
    Florida 86
    Georgia 87
    Hawaii 88
    Idaho 89
    Illinois 90
    Indiana 91

PAGES

Iowa 92
Kansas 93
Kentucky 94
Louisiana 95
Maine 96
Maryland 85
Massachusetts 97
Michigan 98
Minnesota 99
Mississippi 100
Missouri 101
Montana 102
Nebraska 103
Nevada 104
New Hampshire . . . 105
New Jersey 106
New Mexico 107
New York 108
North Carolina . . . 109

PAGES

North Dakota . . . 110
Ohio 111
Oklahoma 112
Oregon 113
Pennsylvania 114
Rhode Island 84
South Carolina . . . 115
South Dakota . . . 116
Tennessee 117
Texas 118
Utah 119
Vermont 120
Virginia 121
Washington 122
West Virginia 123
Wisconsin 124
Wyoming 125

COMPARATIVE WORLD MAPS – *pages 126 to 133*

PAGES

CLIMATES 126
NATURAL VEGETATION 127
POPULATION 128
RACES 129

PAGES

LANGUAGES 130
RELIGIONS 131
THE WESTERN HEMISPHERE 132
THE EASTERN HEMISPHERE 133

MAPS OF THE WORLD THROUGH THE AGES – *pages 134 to 160*

PAGES

ANCIENT MAPS 134
EURASIA – THE GROWTH OF CIVILIZATION TO 200 A.D. 135
ALEXANDER'S EMPIRE 136
ROMAN EMPIRE ABOUT 120 A.D. 137
EUROPE AND THE CRUSADER STATES ABOUT 1140 138
ASIA AT THE DEATH OF KUBLAI KHAN, 1294 139
EUROPE ABOUT 1360 140
EUROPEAN CIVILIZATION DURING THE RENAISSANCE 141
THE AGE OF DISCOVERY 142
EUROPE IN 1721 143
EASTERN AND SOUTHERN ASIA ABOUT 1775 144
REVOLUTIONS IN THE ATLANTIC WORLD, 1776–1826 145
BRITISH NORTH AMERICA AFTER THE SEVEN YEARS' WAR . . 146
REVOLUTIONARY WAR 146
STATE CLAIMS TO WESTERN LANDS 146
THE UNITED STATES, 1775–1800 146
WESTWARD EXPANSION, 1800–50 147
THE CIVIL WAR 148
CAMPAIGNS OF THE CIVIL WAR 148
    THE VICKSBURG CAMPAIGN, 1863 148

PAGES

THE CHATTANOOGA AND ATLANTA CAMPAIGNS, 1863–64 . . 148
THE EASTERN THEATER, 1862–63 148
THE EASTERN THEATER, 1864–65 148
LATIN AMERICA ABOUT 1790 149
LATIN AMERICA AFTER INDEPENDENCE 150
CANADA – 1792–1840 151
DOMINION OF CANADA FORMED 1867 151
EUROPE IN 1810 AT THE HEIGHT OF NAPOLEON'S POWER . . 152
EUROPE IN 1815 AFTER THE TREATY OF VIENNA 152
EXPANSION OF RUSSIA IN EUROPE 153
UNIFICATION OF GERMANY 154
UNIFICATION OF ITALY 154
LANGUAGES OF EUROPE IN THE 19TH CENTURY 155
THE PARTITION OF AFRICA 156
BALKAN PENINSULA TO 1914 157
EUROPE IN 1914 157
THE WORLD ABOUT 1900 158
EUROPE, 1922–40 159
EUROPE AFTER WORLD WAR II SHOWING CHANGES TO 1950 . . 160
ASIA AFTER WORLD WAR II SHOWING CHANGES TO 1950 . . . 160

CONTENTS – *continued*

SELECTED WORLD INFORMATION – *pages 161 to 188*

Introduction to the World and United States Information 161

PAGES

WORLD POLITICAL INFORMATION TABLE . . . 162–166

WORLD FACTS AND COMPARISONS 167–169
    Movements, Measurements, Inhabitants, Surface and Extremes of Temperature and Rainfall of the Earth; The Continents; Approximate Population of the World, 1650–1962; Largest Countries of the World in Population; Largest Countries of the World in Area 167
    Principal Mountains of the World 168
    Great Oceans and Seas of the World 169
    Principal Lakes of the World 169
    Principal Rivers of the World 169
    Principal Islands of the World 169

LARGEST METROPOLITAN AREAS AND CITIES OF THE WORLD, 1962 170

PAGES

PRINCIPAL WORLD CITIES AND POPULATIONS 170–171

HISTORICAL GAZETTEER 172–181

PRINCIPAL DISCOVERIES AND EXPLORATIONS 182–185

WORLD AIR DISTANCE TABLE 186–187
    Showing Great Circle Distances Between Principal Cities of the World in Statute Miles.

WORLD STEAMSHIP DISTANCE TABLE 186–187
    Showing Steamship Distances Between Principal Ports of the World in Statute Miles.

ALTITUDES OF SELECTED WORLD CITIES 188

SELECTED UNITED STATES INFORMATION – *pages 189 to 232*

PAGES

GEOGRAPHICAL FACTS ABOUT THE UNITED STATES 189
    Elevation, Extremities, Length of Boundaries, Geographic Centers, Extremes of Temperature, Precipitation.

HISTORICAL FACTS ABOUT THE UNITED STATES . . 189
    Territorial Acquisitions; Westward Movement of Center of Population.

STATE AREAS AND POPULATIONS 189
    Giving Land, Water, and Total Area; Rank in Area; Population in 1960; Population Per Square Mile in 1960; Rank in Population in 1960; Population in 1950; Rank in Population in 1950; Population in 1940, for each state in the United States.

U.S. STATE CLIMATIC AND ECONOMIC TABLE 190–191
    Giving Topography, Climate Information, Principal Mineral Products, Principal Agricultural Products, Principal Forest and Fishery Products, and Principal Manufactures for Each State in the United States.

NATIONAL PARKS AND MONUMENTS MAP OF THE UNITED STATES 192

U.S. NATIONAL PARK SYSTEM 193
    National Parks; National Monuments; National Historical Parks, National Historic Parks, Sites, and Memorials; National Military Parks and Cemeteries; National Battlefield Parks and Sites; National Recreational Areas.

PAGES

RAILROAD MAP OF THE UNITED STATES 194

U.S. RAILROAD DISTANCE TABLE 195
    Showing Travel Distances (Short Line) Between Railroad Centers of the United States in Statute Miles.

U.S. AIR DISTANCE TABLE 196
    Showing Great Circle Distances between Principal Cities of the United States in Statute Miles.

U.S. STATE GENERAL INFORMATION TABLE 197
    Giving Capital, Largest City, Entry Into Union as State, Extreme Length, Extreme Width, Highest Point, Flower, Bird and Nickname for Each State in the United States.

U.S. POPULATION BY STATE OR COLONY 198
    Giving the Population by State or Colony from 1650 to 1960.

U.S. METROPOLITAN AREAS OF 100,000 OR MORE, 1962 199–200
    Listed in Order of Population.

NUMBER OF U.S. COUNTIES, CITIES, TOWNS, AND LOCALITIES BY STATES, 1962 200

PLACES OF INTEREST IN THE UNITED STATES . 201–215

UNIVERSITIES AND COLLEGES OF THE U.S. . . 216–232

PAGES

GLOSSARY OF FOREIGN GEOGRAPHICAL TERMS 233

GLOSSARY OF MAP TERMINOLOGY 234–236

EXPLANATION OF THE INDEX AND ABBREVIATIONS 1A

INDEX TO POLITICAL-PHYSICAL MAPS 2A–123A

MAP PROJECTIONS

A map projection is merely an orderly system of parallels and meridians on which a flat map can be drawn. There are hundreds of projections, but no one represents the earth's spherical surface without some distortion. The distortion is relatively small for most practical purposes when a small part of the sphere is projected. For larger areas, a sacrifice of some property is necessary.

Most projections are designed to preserve on the flat map some particular property of the sphere. By varying the systematic arrangement or spacing of the latitude and longitude lines, a projection may be made either equal-area or conformal. Although most projections are derived from mathematical formulas, some are easier to visualize if thought of as projected upon a plane, or upon a cone or cylinder which is then unrolled into a plane surface. Thus, many projections are classified as plane (azimuthal), conic, or cylindrical.

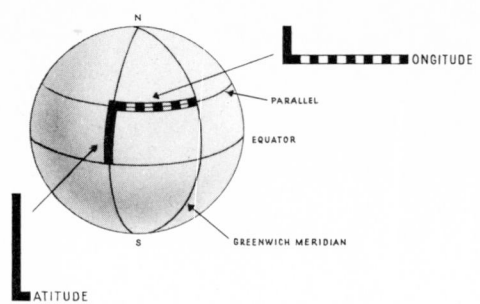

SIMPLE CONIC PROJECTIONS

A perspective projection on a tangent cone with the origin point at the center of the globe. At the parallel of tangency, all elements of the map are true angles, distances, shapes, areas. Away from the tangent parallel, distances increase rapidly, giving bad distortion of shapes and areas.

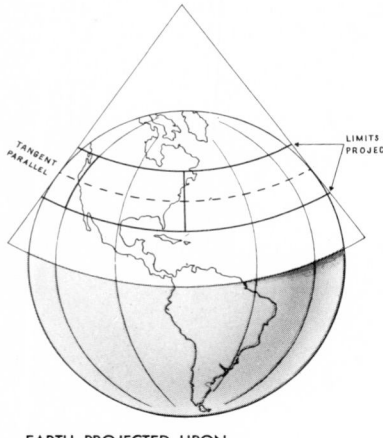

EARTH PROJECTED UPON
A TANGENT CONE

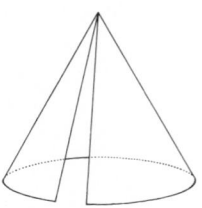

CONE CUT FROM BASE TO APEX

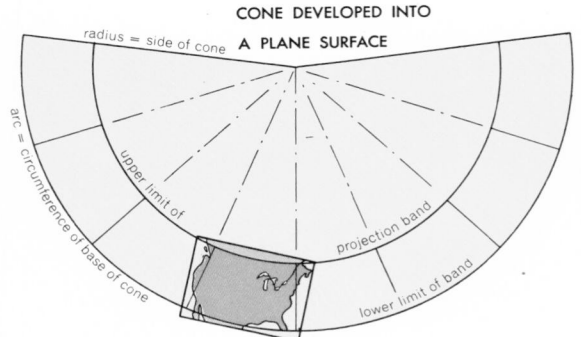

CONE DEVELOPED INTO
A PLANE SURFACE

MODIFIED CONIC PROJECTION

EARTH PROJECTED UPON AN INTERSECTING CONE

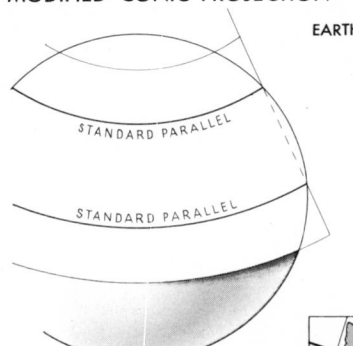

This modification of the conic has two standard parallels, or lines of intersection. It is not an equal-area projection, the space being reduced in size between the standard parallels and progressively enlarged beyond the standard parallels. Careful selection of the standard parallels provides however, good representation for areas of limited latitudinal extent.

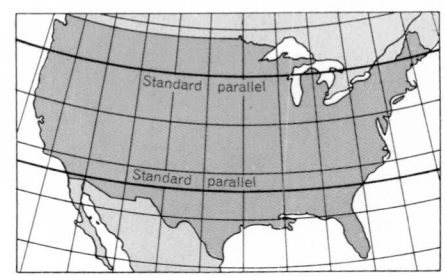

CONIC PROJECTION WITH TWO STANDARD PARALLELS

BONNE PROJECTION

An equal-area modification of the conic principle. Distances are true along all parallels and the central meridian; but away from it, increasing obliqueness of intersections and longitudinal distances, with their attendant distortion of shapes, limits the satisfactory area.

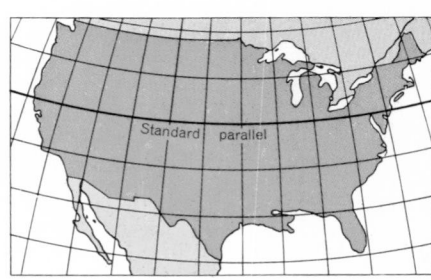

POLYCONIC PROJECTION

This variation is not equal-area. Parallels are nonconcentric circles truly divided. Distances along the straight central meridian are also true, but along the curving meridians are increasingly exaggerated. Representation is good near the central meridian, but away from it there is marked distortion.

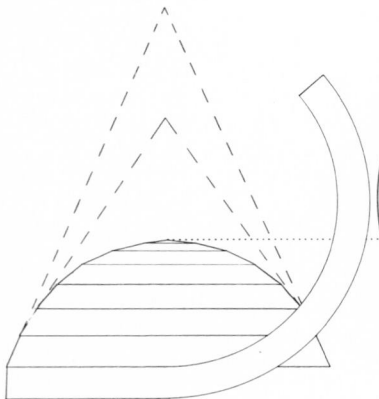

EARTH CONSIDERED AS FORMED
BY BASES OF CONES

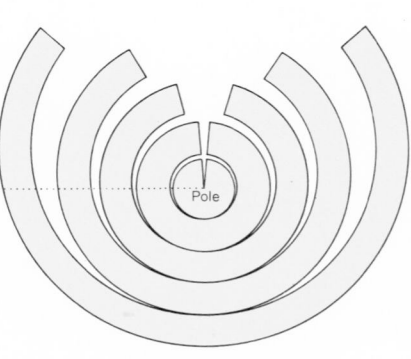

DEVELOPMENT OF THE CONICAL BASES

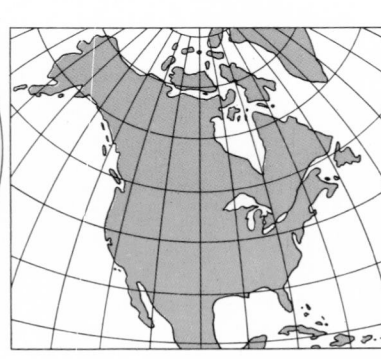

POLYCONIC PROJECTION

TYPICAL PLANE PROJECTIONS

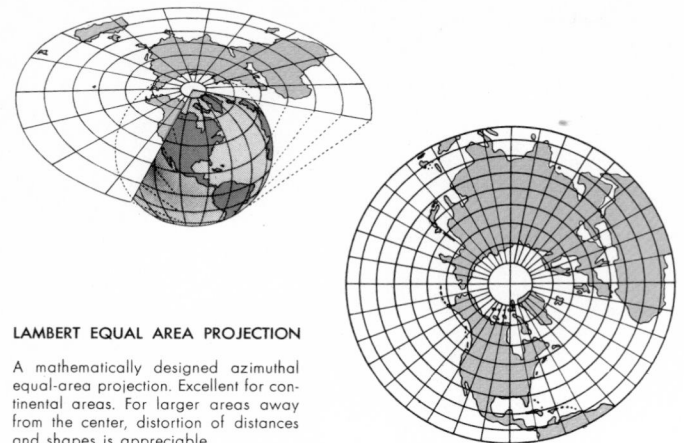

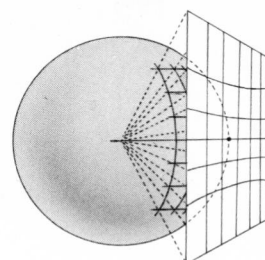

LAMBERT EQUAL AREA PROJECTION

A mathematically designed azimuthal equal-area projection. Excellent for continental areas. For larger areas away from the center, distortion of distances and shapes is appreciable.

GNOMONIC PROJECTION

A geometric or perspective projection on a tangent plane with the origin point at the center of the globe. Shapes and distances rapidly become increasingly distorted away from the center of the projection. Important in navigation, because all straight lines are great circles.

CYLINDRICAL PROJECTIONS

EARTH PROJECTED UPON A CYLINDER

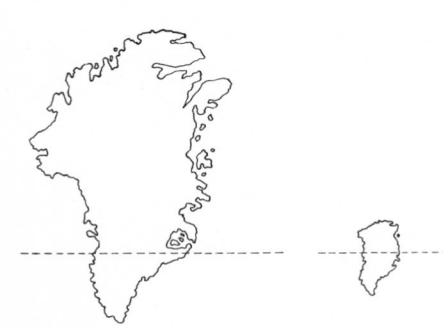

PERSPECTIVE PROJECTION

A perspective projection on a tangent cylinder. Because of rapidly increasing distortion away from the line of tangency and the lack of any special advantage, it is rarely used.

Note the increasing distortion of Greenland (above left) compared to an equal area projection (above right).

MERCATOR CONFORMAL PROJECTION

Mercator's modification increases the longitudinal distances in the same proportion as latitudinal distances are increased. Thus, at any point shapes are true, but areas become increasingly exaggerated. Of value in navigation, because a line connecting any two points gives the true direction between them.

MILLER PROJECTION

This recent modification is neither conformal nor equal-area. Whereas shapes are less accurate than on the Mercator, the exaggeration of areas has been reduced somewhat.

EQUAL AREA PROJECTIONS OF THE WORLD

The earth's surface peeled like the skin from an orange.

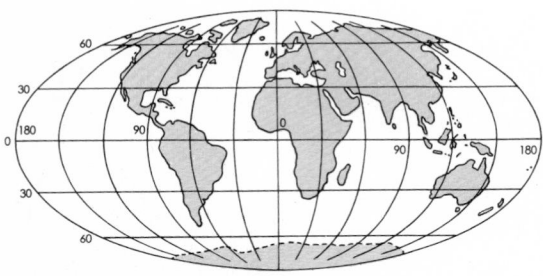

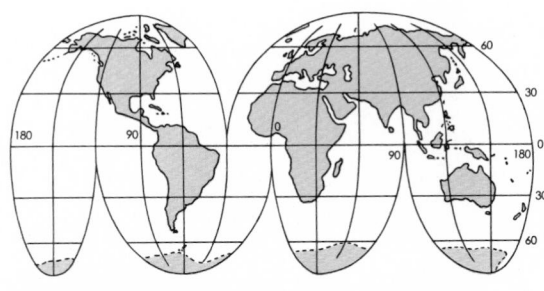

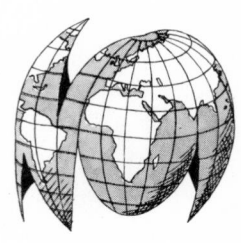

MOLLWEIDE'S HOMOLOGRAPHIC PROJECTION

GOODE'S INTERRUPTED HOMOLOGRAPHIC PROJECTION

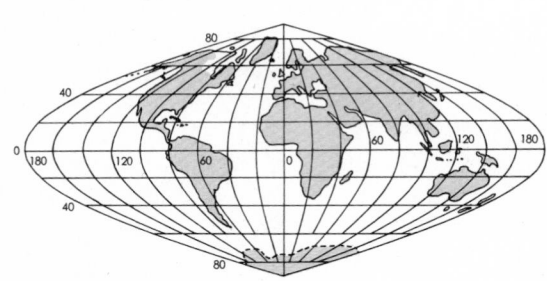

SINUSOIDAL PROJECTION

GOODE'S INTERRUPTED HOMOLOSINE PROJECTION

Although each of these projections is equal-area, differences in the spacing and arrangement of latitude and longitude lines result in differences in the distribution and relative degree of the shape and distance distortion within each grid. On the homolographic, there is no uniformity in scale. It is different on each parallel and each meridian. On the sinusoidal, only distances along all latitudes and the central meridian are true. The homolosine combines the homolographic, for areas poleward of 40°, with the sinusoidal. The principle of interruption permits each continent in turn the advantage of being in the center of the projection, resulting in better shapes.

A STUDY IN MAP SCALE

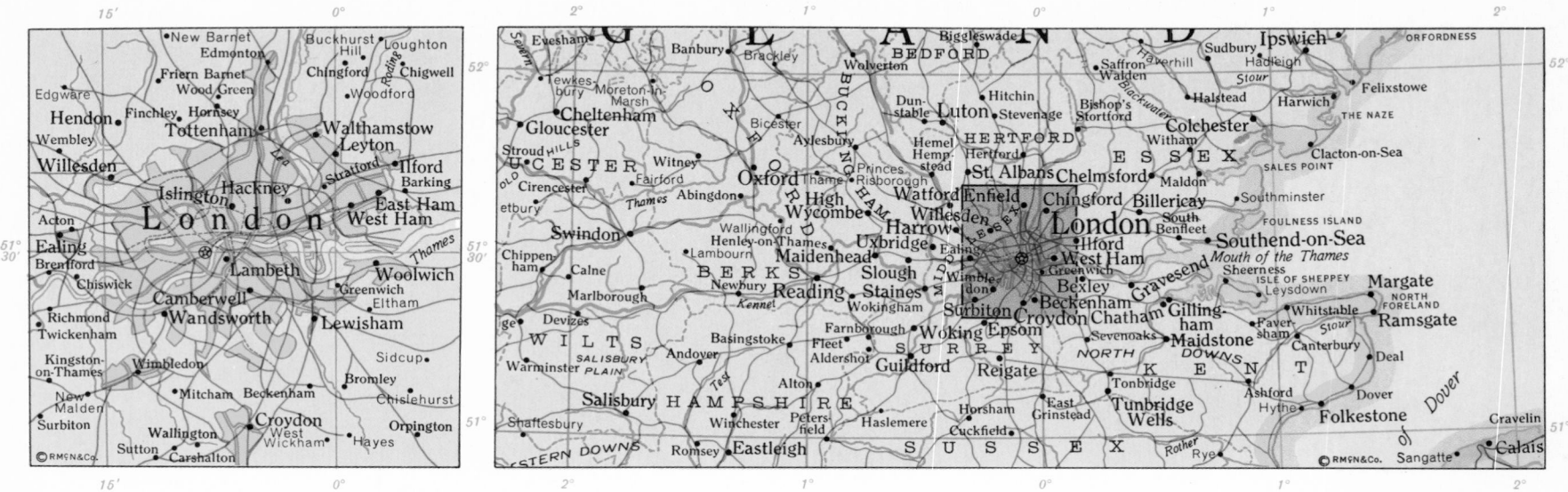

LONDON   **1 : 500,000**

1 inch = approx. 8 statute miles.

LONDON AND VICINITY   **1 : 2,000,000**

Statute Miles 5 0 5 10 20 30   1 inch = approx. 32 statute miles, or distances are ¼ as great as on LONDON map.

SOUTHERN ENGLAND   **1 : 4,000,000**

Statute Miles 25 0 25 50 75

1 inch = approx. 64 statute miles, or distances are ½ as great as on LONDON & VICINITY map;
or distances are ⅛ as great as on LONDON map.

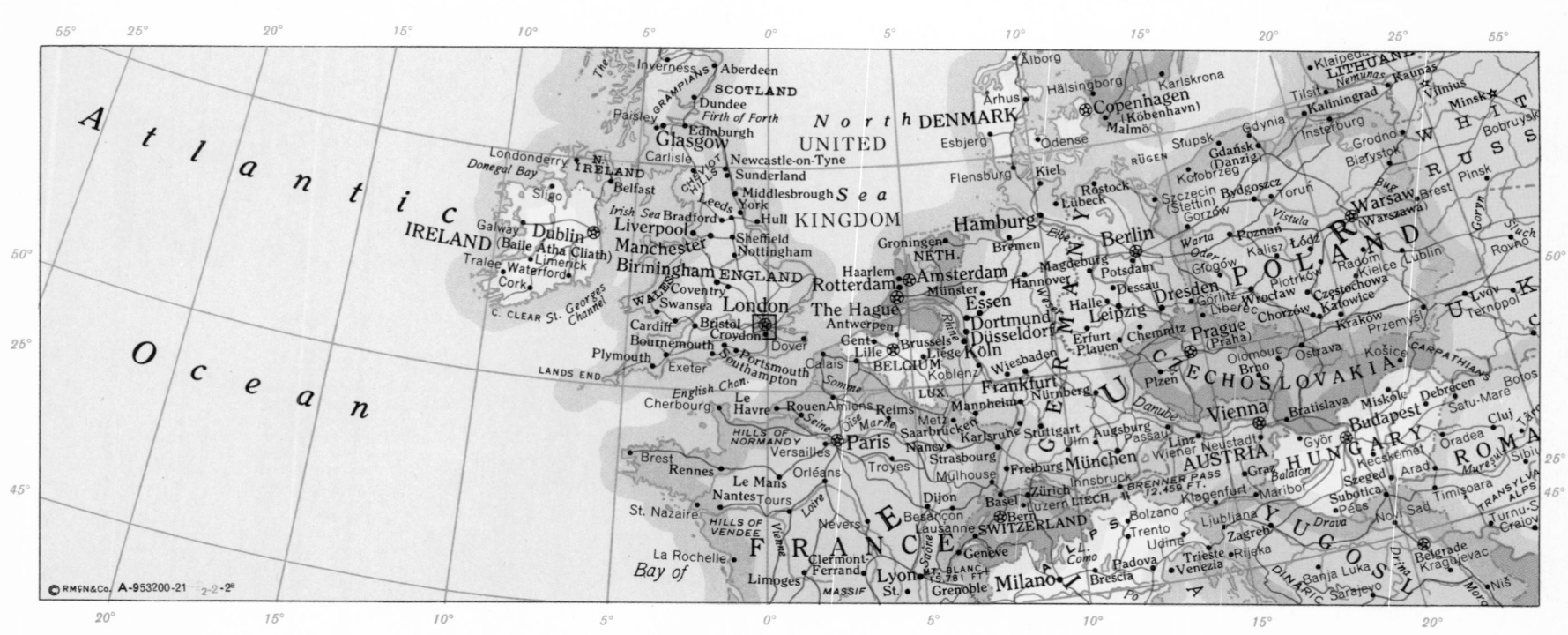

CENTRAL EUROPE   **1 : 16,000,000**

Statute Miles 100 0 100 200 300

1 inch = approx. 252 statute miles, or distances are ¼ as great as on SOUTHERN ENGLAND map;
or distances are ⅛ as great as on LONDON & VICINITY map;
or distances are 1/32 as great as on LONDON map.

X

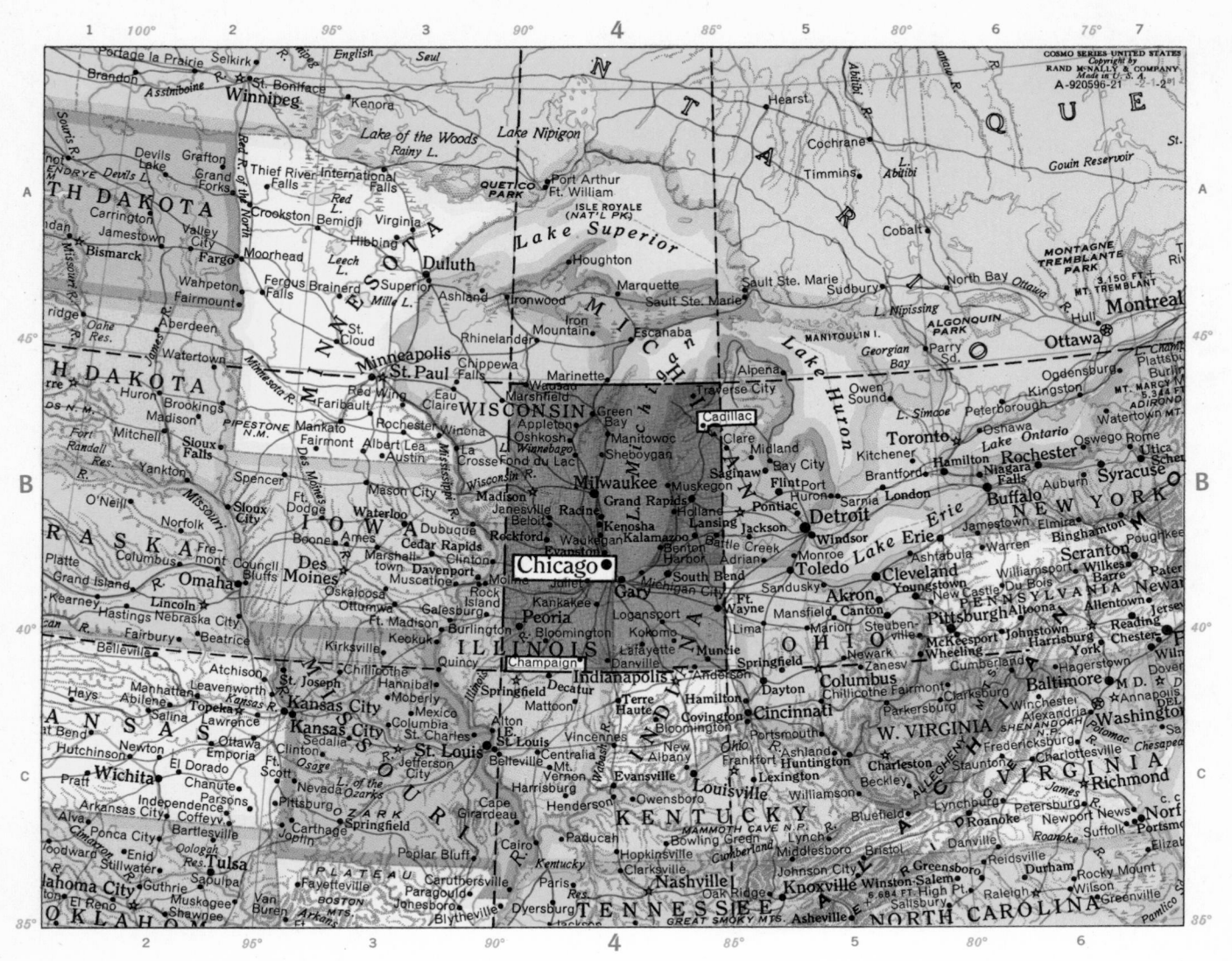

Place	Location	Pop.	Index Key	Page
Cades, Gibson, Tenn., 15			B3	117
➡Cadillac, Wexford, Mich., 10,112			B4	98
Cádiz, Sp., 117,871			D2	20
Cádiz, prov., Sp., 818,847			*D2	20
Calabogie, lake, Ont., Can.			B8	72
Calais, Washington, Vt., 150 (684▲)			C4	120
Calchaqui, riv., Arg.			E2	55
Calhoun, co., Ala., 95,878			B4	78
Caltanissetta, It., 48,500			F5	21
Cambodia, country, Asia, 3,800,000			F6	38
Campechuela, Cuba, 2,782			E5	64
Campton, Okaloosa, Fla., 100			G2	86
Caney, fork, Tenn.			C8	117
Cannon Falls, Goodhue, Minn., 2,055			F6	99
Capilano, B.C., Can., 800			A9	68
Carbon, Clay, Ind., 409			E3	91
Cardiff, Jefferson, Ala., 202			E4	78
Carlsbad Caverns, nat. park, N. Mex			E5	107
Caroleen, Rutherford, N.C., 1,168			B2	109
Carouge, Switz., 12,760			D1	19
Casey, co., Ky., 14,327			C5	94
Castleton, Rutland, Vt., 375 (1,902▲)			D2	120
Catacombs, tombs, It.			h9	21
Catharine, Ellis, Kans., 225			D4	93
Caudéran, Fr., 28,715			E3	14
Cavite, Phil., 54,900			o13	35
Cayo Largo, isl., Cuba			E3	64
Cedar Rapids, Linn, Iowa, 92,035			C6	92
Ceduna, Austl., 1,292			F5	50
Centertown, Cole, Mo., 190			C5	101
Ceska Lipa, Czech., 14,000			C3	26
➡Champaign, Champaign, Ill., 49,583			B4	90
Chandpur, Pak., 32,048			D9	39
Charente-Maritime, dept., Fr., 470,897			*D3	14
Charleston, harbor, S.C.			F8	115
Château-Thierry, Fr., 10,006			C5	14
Chellala, Alg., 2,899			G8	30
Chernovtsy, Sov. Un., 147,000			A7 22, G5	27
Cheviot, N.Z., 440			O14	51
Cheyenne, Laramie, Wyo., 43,505			D8	125
Chiapa de Corzo, Mex., 6,972			D6	63
➡Chicago, Cook, Ill., 3,550,404 (*6,517,600)			B4	90
Chiefland, Levy, Fla., 1,459			C4	86

EXPLANATION OF THE INDEX REFERENCE SYSTEM

The indexing system used in this atlas is based upon the conventional pattern of parallels and meridians used to indicate latitude and longitude. The index sample beside the map indicates that the cities of *Chicago, Cadillac,* and *Champaign* are all located in *B4.* Each index key letter, *in this case "B,"* is placed between corresponding degree numbers of latitude in the vertical borders of the map. Each index key number, *in this case "4,"* is placed between corresponding degree numbers of longitude in the horizontal borders of the map. Crossing of the parallels above and below the index letter with the meridians on each side of the index number forms a confining "box" in which the given place is certain to be located. It is important to note that location of the place may be anywhere in this confining "box."

Small insets on many foreign maps are indexed independently of the main maps by separate index key letters and figures. All places indexed to these insets are identified by the lower case reference letter in the index key.

Place names are indexed to the location of the city symbol. Political divisions and physical features are indexed to the names.

MAP SYMBOLS

CULTURE

Political Boundaries

International
State and Provincial
County

Cities, Towns, and Villages

Principal Cities
Other cities, towns, and villages are indicated by size of type and symbol according to relative population.
County Seats are indicated by dot-centered symbol.
⊛ Major Capital Cities
☆ Minor Capital Cities

Miscellaneous

National Parks
National Monuments
Indian Reservations
△ Points of Interest
Railroads (Initialed in U.S. and Canada)
Tunnels
Underground or Subway
∴ Ruins
Dikes
Bridges
Dams
Race Tracks, Buildings, etc.

TOPOGRAPHY

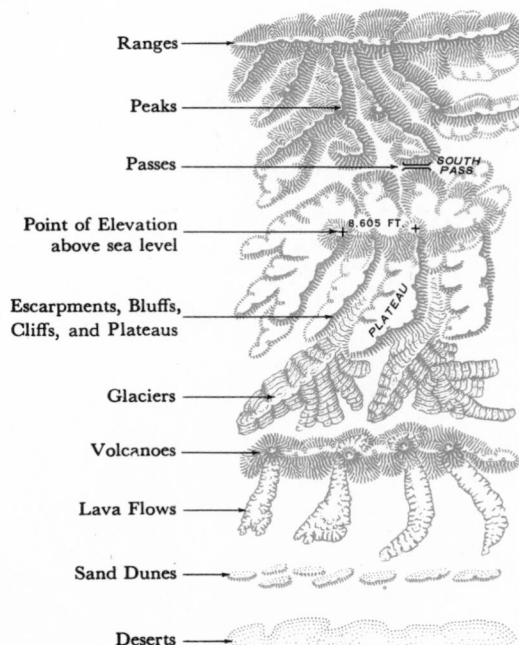

Ranges
Peaks
Passes — SOUTH PASS
Point of Elevation above sea level — 8,605 FT.
Escarpments, Bluffs, Cliffs, and Plateaus — PLATEAU
Glaciers
Volcanoes
Lava Flows
Sand Dunes
Deserts

HYDROGRAPHY

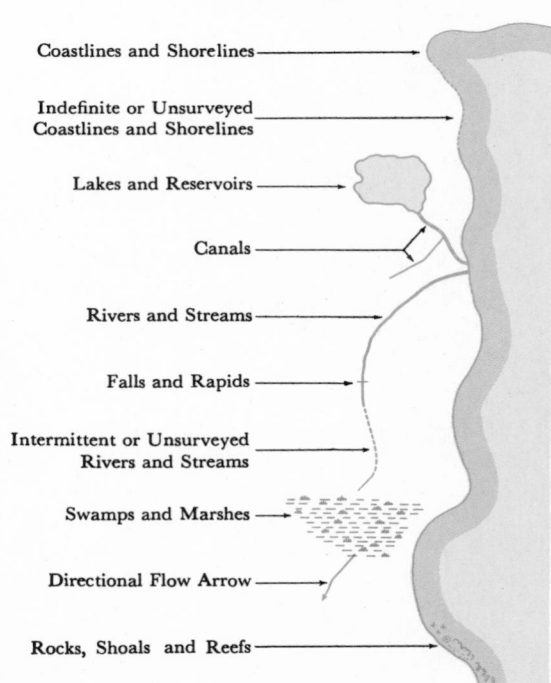

Coastlines and Shorelines
Indefinite or Unsurveyed Coastlines and Shorelines
Lakes and Reservoirs
Canals
Rivers and Streams
Falls and Rapids
Intermittent or Unsurveyed Rivers and Streams
Swamps and Marshes
Directional Flow Arrow
Rocks, Shoals and Reefs

The Saga of Space

A NEW FRONTIER

For more than a decade men of science have been attempting to penetrate the void of space. Their approach has been methodical; their weapons have been varied but always increasingly effective. At first only "robot"-like devices were used—large rockets carrying hundreds of pounds of data-gathering instruments. These scientific tools probed, studied, and returned by radio or by television the invaluable data about the nature of the physical environment which lies just beyond our protective mantle—our atmosphere. Thanks to these reports we know something about cosmic radiation intensity, temperature extremes, electromagnetic storms, meteor showers, and other factors which have enabled the scientists to realize how bitterly hostile that outer environment really is. They know what to expect. They know what tremendous effort must be made to send a man into space and return him alive.

In the early days our reach above the earth was limited. Our rockets lacked sufficient power to raise great pay-loads to hundreds of miles. The fantastic energy required to launch large earth satellites lay beyond our technological means. Furthermore, the needed support of the people was not yet forthcoming; there was still a belief that much of rocketry and space travel was a faraway dream reserved for future generations. Through this ATLAS Rand McNally & Company wishes to contribute to the greater understanding of space science.

Man's accomplishments to date have been noteworthy. The launching of the first two Soviet satellites on October 4 and November 3, 1957 (See page xvii) revealed dramatically a new era in space exploration. It also demonstrated that man was capable of producing the enormous brute force required to penetrate the realm of space. Later the American satellites, such as the Explorer series, Vanguard series, and Discoverer series, set the pace for refinement in the field of instrumented exploration in the fringes of space. Within the next fifteen years the heavens will be literally glittering with a large variety of scientific satellites to be hurled into orbits at varying distances from the Earth's surface.

In recent months we have seen in the newspapers photographs of portions of the Earth taken from an orbited satellite, Tiros-1, the forerunner of a series of geodetic satellites which will map the entire Earth in great detail. No doubt such geodetic satellites will be used to map the Moon for a better knowledge of it before man actually sets foot upon it. Yet the study of satellites is not the only thing that occupies men of science. Presently our scientists are attacking the problems of space along a dual front. First, of course, is the technological equipment: enormous rocket engines, new exotic rocket fuels, elaborate electronic controls (even now capable of directing a rocket to the Moon with high accuracy), and above all, high reliability of performance. But the mastering of space

exploration is not for machines alone. The second goal is to test and condition man himself to enable him to be sent into orbits around the Earth, on journeys to the Moon and, eventually, to other planets.

Our scientists, in their attack against the problems of space exploration, left man out of the rocket originally. Because of the uncertainty of rocket firings, the lack of needed power, and the lack of information along the whole vast frontier of space, they sent the missiles alone, far above the earth. Simultaneously with these "robot" experiments they slowly and carefully raised man himself to new heights. By using balloons and rocket-powered aircraft for actual flights, and by developing a new line of laboratory equipment, scientists subjected pilots to many of the conditions to be encountered beyond our atmosphere, such as weightlessness, isolation, and the effects of high acceleration and sudden deceleration.

Actually, man has already begun to penetrate space. The first person to do so was Captain Iven C. Kincheloe, Jr., who flew the X-2 rocket plane (See page xvii) straight up in a howling climb. At 103,000 feet (19.3 miles) and a speed of 2,000 miles per hour his fuel ran out. Nevertheless he reached an altitude of 126,200 feet (24.7 miles) by coasting along on momentum only. He had left behind more than 99 per cent of the Earth's atmosphere. Physiologically, Captain Kincheloe was in the vacuum of space; without his special pressure suit, without special heating and cabin pressurization, he would have died horribly in fifteen seconds. Today a similar experiment is about to occur. A new rocket plane, the X-15, is scheduled to be flown to over one hundred miles above the surface of the Earth. At such altitudes environmental conditions are even more variable than Captain Kincheloe found them. Consequently this forthcoming trip will contribute further to an understanding of the conditions man must overcome to survive in space.

A different type of experiment has been the usage of manned balloons in data gathering ascensions, such as *Project Manhigh.* In 1957 Major David G. Simons reached 102,000 feet (19 miles) in a Project Manhigh balloon and stayed at extreme altitudes in a flight that lasted some thirty-two hours. During this period he was subjected to many conditions of space flight, including utter loneliness, a feeling of absolute isolation, and strange psychological reactions that greatly concern doctors of space medicine. Also Major Simons' body was exposed to intense floods of cosmic radiation.

Many men have been subjected to the effects of space flight. Volunteers in centrifuges have suffered, not without pain and discomfort, the crushing effects of high acceleration, simulating the terrible forces of spaceship take-off. These men have withstood a force of 17 gravities (at 200 pounds normal weight), thus causing them to weigh the fantastic total of 3,400 pounds! Even rapid deceleration—or slowing down—is a problem, because a return at high speed to the Earth's atmosphere produces the same effect as plunging into a mass of heavy fluid. But these tests proved beyond doubt that man *can* withstand such tremendous forces, and that man *can* endure flight at enormous altitudes. Isolation experiments simulating cabin conditions also provide valuable data. Airman Donald Farrell proved that a man can live for a week or longer in a sealed, cramped, environment, rebreathing his own air, functioning properly, and working technical equipment. Subsequently groups of men have been similarly

Internal structure of Tiros-I Satellite.

National Aeronautics and Space Administration

The Red Sea area; a photographic view taken by Tiros-I. Major geographic features have been labelled for reference.

National Aeronautics and Space Administration

Geographic map portraying same area photographed by Tiros-I.

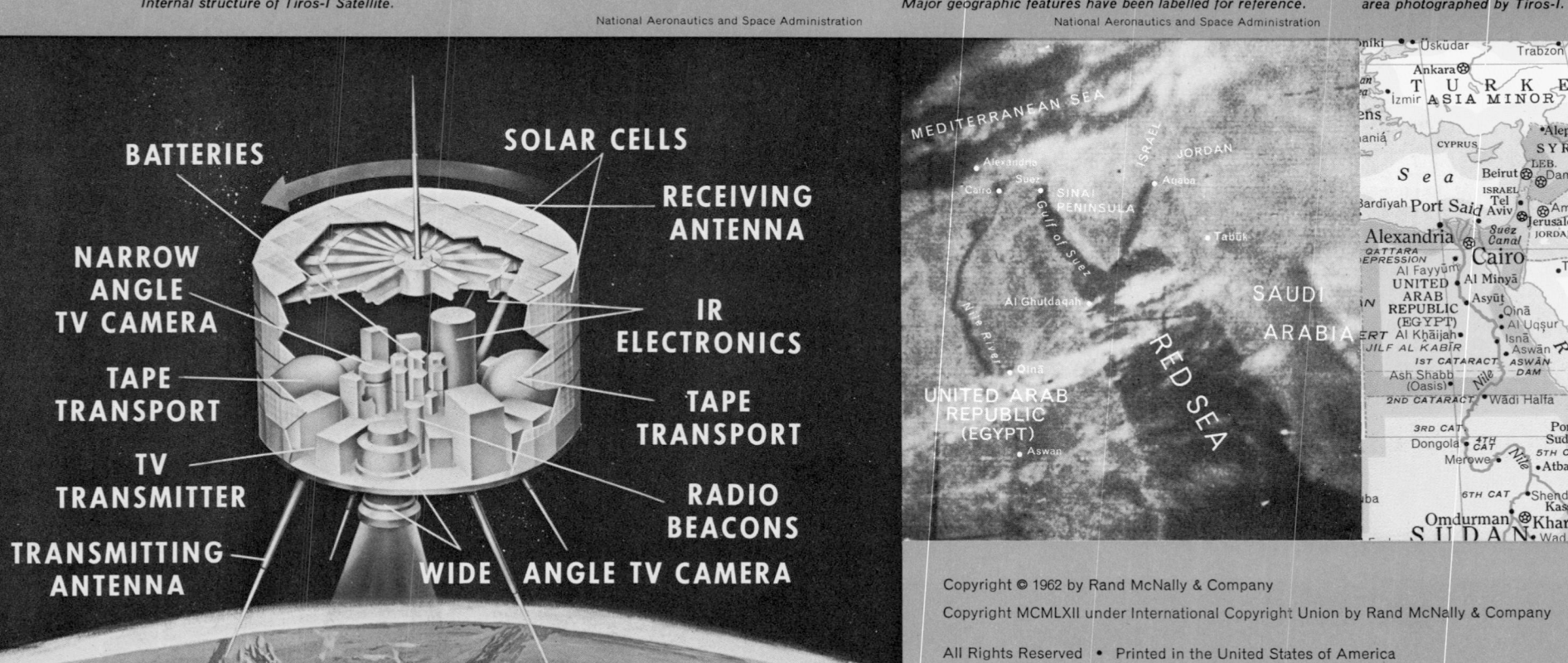

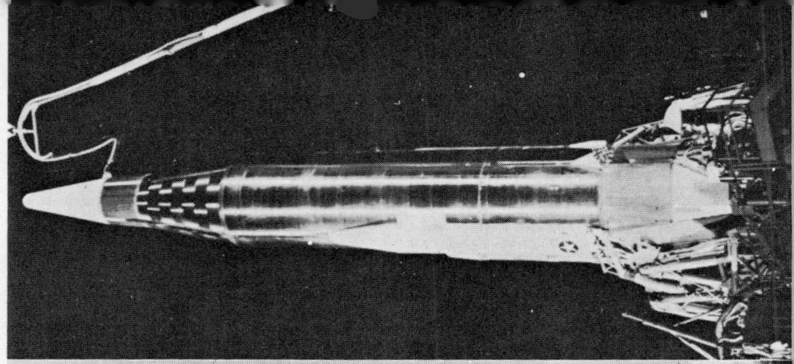

U.S. Air Force Photo

ATLAS

Type / Intercontinental Ballistic Missile.
Range / 5,500 nautical miles.
Weight / 262,000 lbs. with fuel.
Propulsion / 3 motors.
Fuel: liquid oxygen.

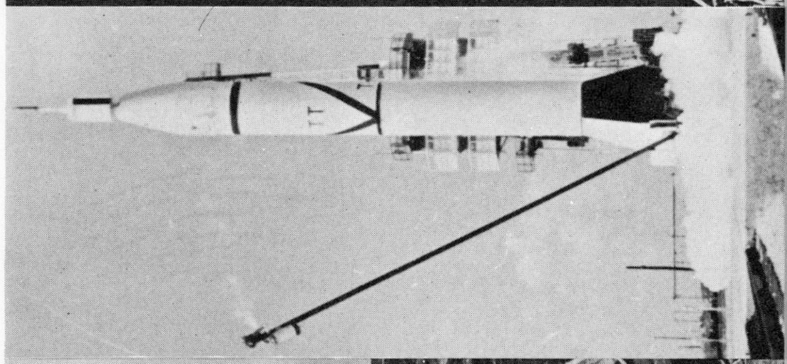

U.S. Army Photo

JUPITER-C

Type / Intercontinental Ballistic Missile and Spacecraft.
Range / 3,300 nautical miles.
Weight / Restricted information.
Propulsion / One rocket motor.
Fuel: liquid oxygen.

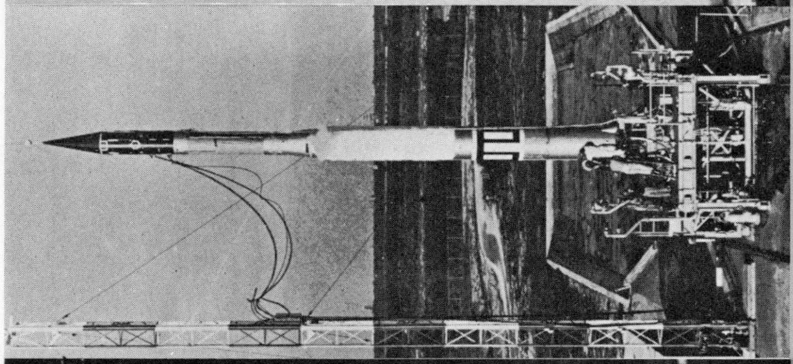

Official U.S. Navy Photo

VANGUARD

Type / Spacecraft.
Range / Restricted information
Weight / 22,600 lbs. (without fuel).
Propulsion / 3 stages (one rocket motor each). Fuel: First and second stages—liquid oxygen; third stage—solid.

National Aeronautics and Space Administration

JUNO—II

Type/Spacecraft
Range/Restricted information.
Weight/120,000 lbs. with fuel.
Propulsion/4 stages: first stage—liquid fuel; upper stages solid.

REDSTONE

Type/Intercontinental Ballistic Missile and Spacecraft.
Range/Restricted information.
Weight/Restricted information.
Propulsion/One motor. Fuel: Liquid oxygen and ethyl alcohol.

Official U.S. Navy Photo

POLARIS

Type / Intermediate Ballistic Missile.
Range / 1,200 — 1,500 nautical miles.
Weight / Restricted information
Propulsion / 2 stages: (one rocket motor each). Fuel: solid.

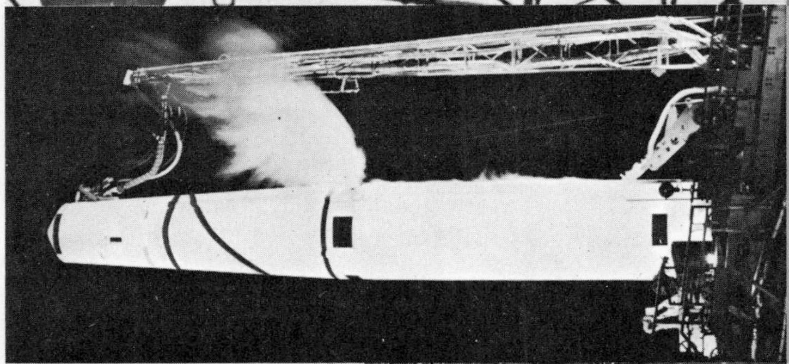

U.S. Air Force Photo

THOR

Type / Intermediate Ballistic Missile
Range / 1,500 nautical miles
Weight / 100,000 lbs. with fuel
Propulsion / One rocket motor.
Fuel: liquid oxygen.

TITAN

Type / Intercontinental Ballistic Missile.
Range / 5,500 nautical miles
Weight / Approximately 300,000 lbs. with fuel.
Propulsion / 2 stages (one rocket motor each). Fuel: liquid oxygen.

tested, and scientists have found great encouragement in these experiments. It is now clearly understood that not all men are suited to the rigors of space flight, but from among the young, the talented, and the highly trained there are those select individuals to whom space has become a challenge eagerly accepted. Such is the case of the men who have already begun participating in Project Mercury (See p. xv), in an effort to place a manned capsule into orbit around the Earth for twenty-four hours. To that end, in the spring of 1961 the United States launched two sub-orbital manned capsules, followed by the successful three-orbit voyages of J. H. Glenn and M. S. Carpenter, the six-orbit flight of W. M. Schirra in 1962, and the twenty two-orbit voyage of L. G. Cooper in 1963. Such experiments are a major step toward the inevitable exploration of space by American astronauts. The Soviet Union has also achieved noteworthy success by orbiting four manned capsules whose maximum achieved height exceeded 115 miles.

All these complex efforts, attained accomplishments, failures and experiments, constitute the beginning . . . only the beginning!

THE FIRST STEPS UPWARD

The rocket is, at this time, the only usable means of providing propulsion in the vacuum of space. It needs neither ground, nor water, nor air to push against, as do other vehicles that move by such resistance (such as wheels turning on a roadway or propeller blades churning air or water). The rocket's movement forward is an illustration of the law of reaction—in this case, reaction to the gases which are expelled from the rocket's exhaust. Out in space the rocket is even more efficient than inside a planet's atmosphere.

Men traveling in a rocket or in a "space station satellite" will live a confining and peculiar existence. A rocket engine operates only during part of its journey; the rest of the time the space vehicle keeps moving because it has achieved momentum and there is no air resistance to slow it down. When in this stage of flight, the rocket and its contents are in a state of "free fall" and weightlessness. This latter condition cannot be produced on the earth except for extremely brief periods, therefore we can only imagine what physical and psychological effects such a trip might have on human beings. In 1957, however, the dog Laika seemed to survive weightlessness in Sputnik II (See page xvii), and remained healthy as long as the air supply lasted. More recently, in 1960, the dogs Strelka and Belka climbed to very high altitudes inside a rocket and were safely returned with no evidence of ill effects.

The accumulative results of these experiments using live animals, including human beings, point out that the next step forward after Project Mercury is a manned space station satellite such as the ones shown on this page. These manned satellites would move in an orbit around the Earth and would rotate their central hubs, inducing a centrifugal force that would cause objects and people to "fall" toward the outside rim, thus providing an artificial gravity.

Traveling through space, man must carry with him absolutely everything he needs—even the air he breathes. He will also need a vehicle with sufficient fuel for his eventual return to earth, though he will not, ordinarily, need fuel to continue on his way—his rocket engine provided the initial push, and, because

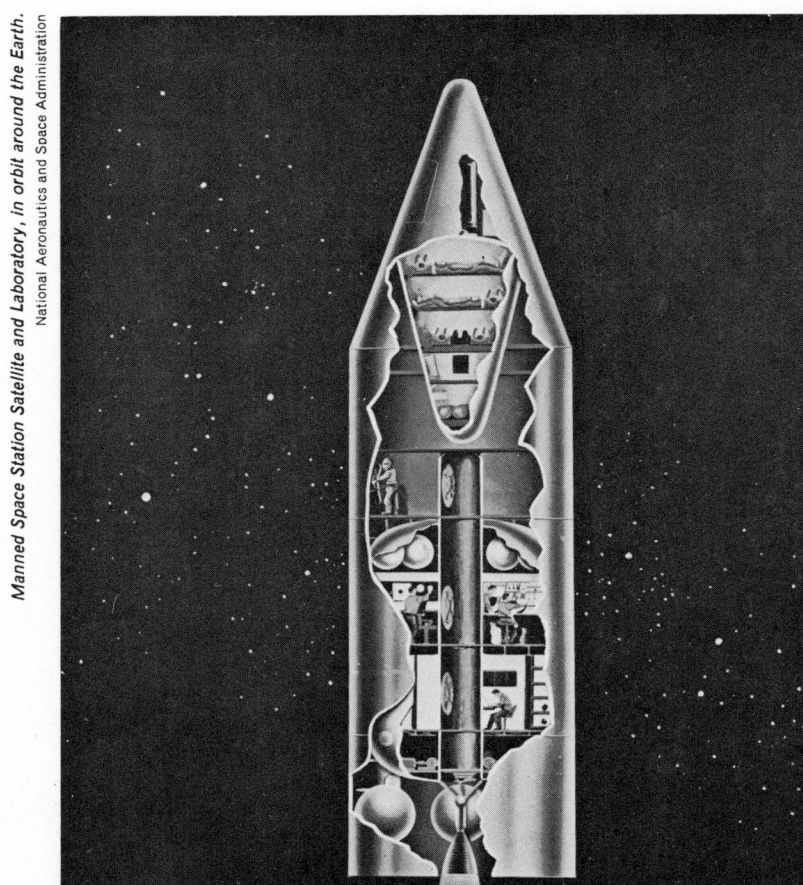

Manned Space Station Satellite and Laboratory, in orbit around the Earth.
National Aeronautics and Space Administration

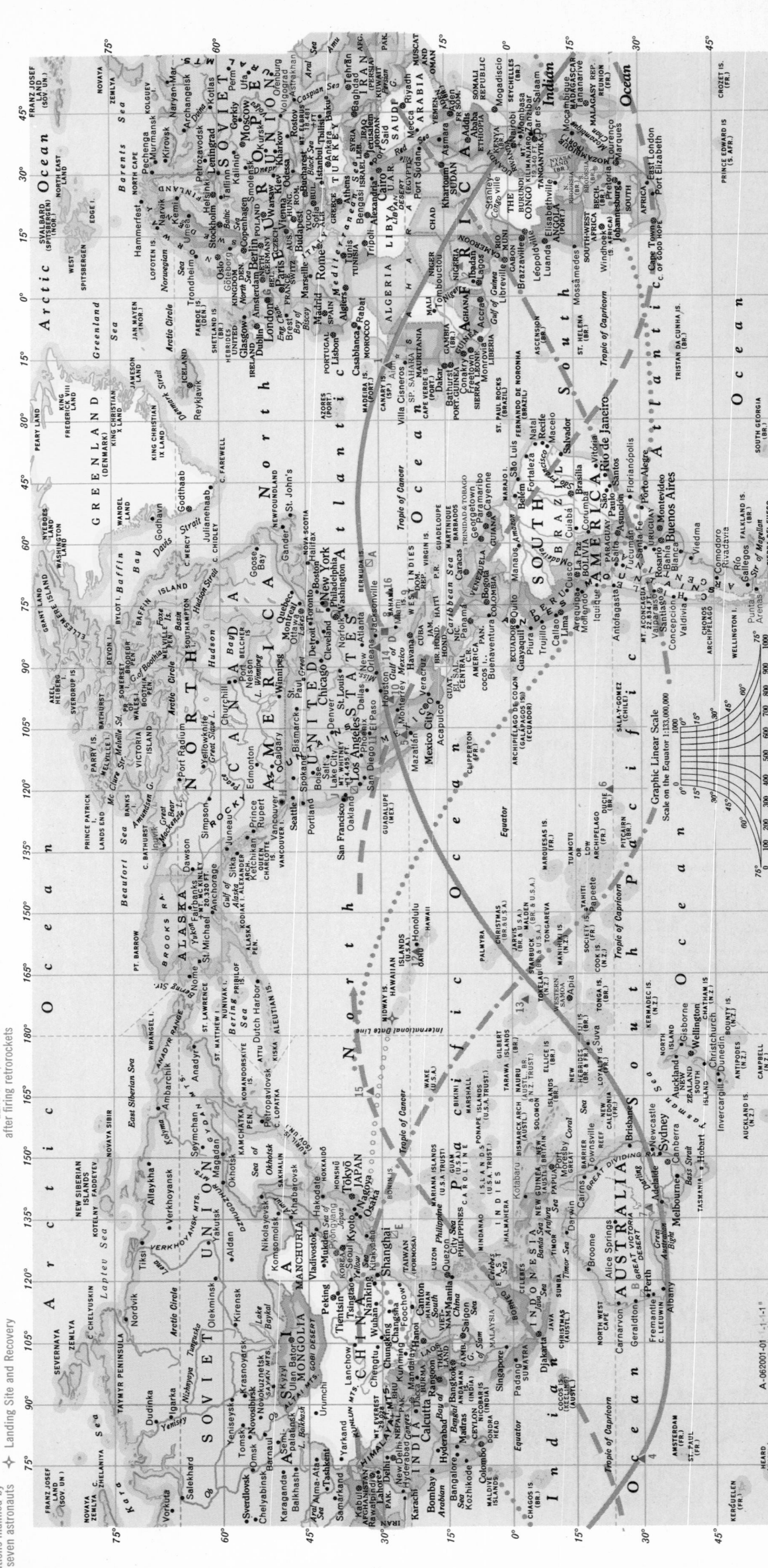

"FAITH 7" IN ORBIT

Maj. L. G. Cooper, aboard spacecraft "Faith 7," was launched into orbit on May 15, 1963, by an Atlas 130-D rocket, generating 360,000 pounds of thrust.

The principal orbital characteristics of "Faith 7" were: perigee, 100 miles; apogee, 168 miles; average velocity, 17,544 m.p.h. and inclination, 32.5°.

Astronaut Cooper circled the Earth twenty-two times in 34 hours, 20 minutes, and 30 seconds. During that time he performed numerous pre-assigned experiments, took photographs of subjects in space as well as of the Earth below him, ate an adequate low residue diet, drank water, and slept soundly through a normal period of over eight hours.

Some of the experiments performed were intended to collect data that will help design equipment to be used in the future Gemini and Apollo missions. Other experiments were of an aeromedical nature, designed to gather information that will allow an evaluation of man's physiological reactions under conditions of extended orbital flight environment. Still other experiments were related to radiation measurements, infrared weather photography, television system operation, temperature measurements, and dim light phenomenon photography.

Unlike all previous orbited American astronauts, Maj. Cooper's voyage took him over both Southwest and Southeast Asia, the People's Republic of China, and southern South America.

During the 22nd and last orbit, when "Faith 7" reached a point 170 miles southeast of Kyushu, Japan, Astronaut Cooper fired his retrorockets and manually piloted his spacecraft to a safe and spectacularly close splash-down near its recovery ship. It was the most successful Mercury flight on record.

TRACKING STATIONS FOR "FAITH 7"

1. Gran Canaria (Canary Is.)
2. Kano (Nigeria)
3. Zanzibar
4. Ship: "Coastal Survey" (Indian Ocean)
5. (Near) Guaymas (Mexico)
6. Ship: "Rose Knot" (S. E. Pacific Ocean)
7. (Near) Woomera (Australia)
8. Grand Bahama (Bahama Is.)
9. Grand Turk I. (West Indies Federation)
10. Corpus Christi (Tex., U.S.A.)
11. White Sands (N.M., U.S.A.)
12. Kauai I. (Hawaii, U.S.A.)
13. Canton I. (Pacific Ocean)
14. Eglin (Fla., U.S.A.)
15. Ship: "Range Tracker"
16. Ship: "Twin Falls Victory"

CONTROL STATIONS

A. Bermuda I. (Atlantic Ocean)
B. (Near) Muchea (Australia)
C. Point Arguello (Calif., U.S.A.)
D. Cape Canaveral
E. Ship: "Coastal Sentry Quebec"

PUBLIC COMMUNICATIONS CENTERS

F. New York (U.S.A.)   H. Vancouver (Canada)
G. London (U.K.)   J. San Francisco (U.S.A.)

▲ Tracking Stations

■ Mercury Control Center:
  Cape Canaveral

■ Control Stations manned by
  one of the seven astronauts

Ⓢ Goddard Space Flight Center
  (computing nucleus)

• Public Communications Center used in relaying
  part of the data to Goddard Space Flight Center

✧ Landing Site and Recovery

ORBITS OF "FAITH 7"

— — — 1st Orbit
— — — 5th Orbit
— — — 9th Orbit
· · · · · 22nd orbit; descent
          after firing retrorockets

A-062001-01   -1--1'-1"
Copyright by
RAND McNALLY & COMPANY
Made in U.S.A.

Courtesy of NASA

Spacecraft being brought along side recovery ship

TOP—View of a sunset, taken by J.H. Glenn, from "Friendship 7."

BOTTOM—View of the Florida coast taken by J.H. Glenn, at the beginning of his second orbit

The beginning of the voyage: lift-off!

TOP—Astronaut J.H. Glenn boards "Friendship 7."

BOTTOM—Instrument panel of the Mercury capsule

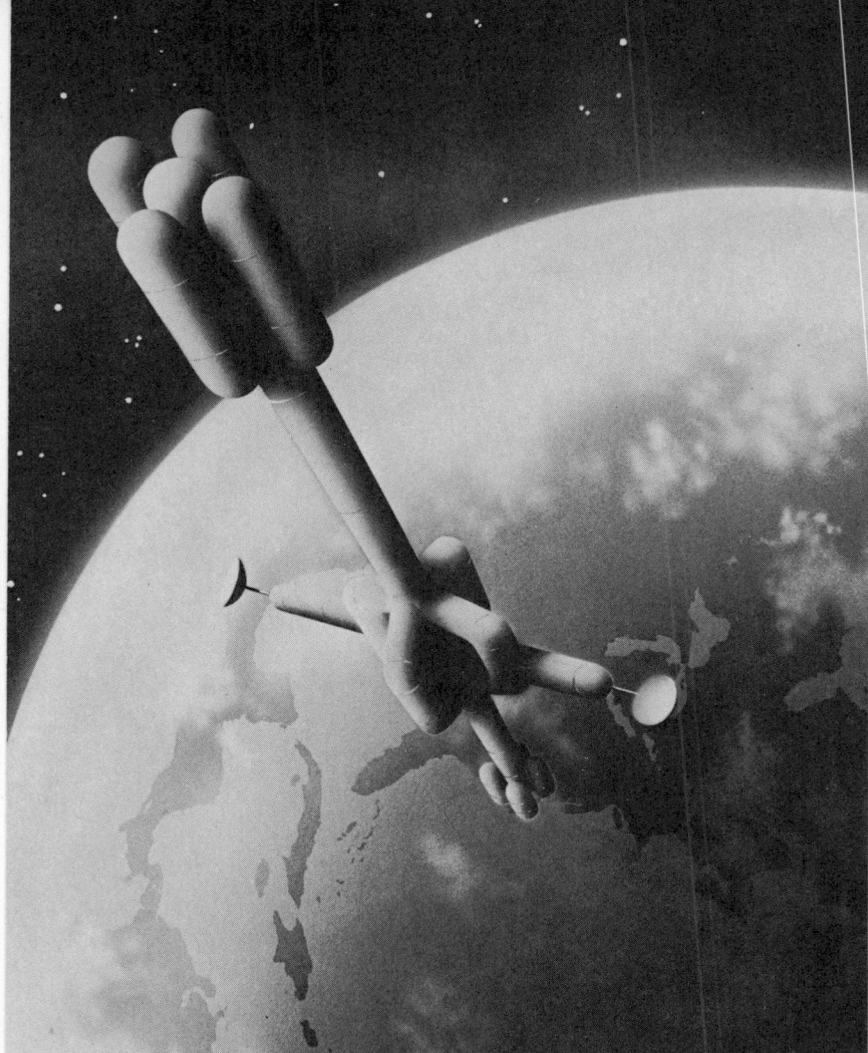

Space Station Satellite orbiting the Earth, shown over the eastern coast of North America.

there is no air to slow the rocket down, it will continue in a predetermined path. An artificial satellite, given the proper shove to put it in orbit, keeps going for similar reasons and will follow an elliptical path around the Earth. Needless to say, the satellite must be pushed hard enough and in the right direction to stay aloft.

From a space station satellite or an orbiting rocket plane passengers will see the earth as it has never before been seen by man. Weather conditions can be studied over vast areas; beginning storms can be spotted and warnings radioed to Earth. In darkness our great cities will seem vast whirlpools of light glowing beneath the shimmering atmosphere.

Scientifically, a manned space station satellite represents an unprecedented opportunity to gather valuable astronomical data by taking advantage of the absence of atmospheric turbulance which now interferes with the work of all ground observatories. At the same time, extensive information will become available about the characteristics of the upper atmosphere of the Earth, and data can be collected on the Van Allen radiation belts and their effect on our atmosphere's behavior.

Technically, the space station satellite will serve as a testing laboratory, not only for the construction of various machines or experimental models of exploratory equipment, but for gathering data on phenomena that can best be evaluated while in orbit, through controlled experimentation. One of the most valuable contributions will be the testing of propulsion and guidance systems under actual space conditions. It is essential that such systems be extremely reliable before it is possible to enter into manned space exploration in regions beyond the Earth's vicinity. The vastness of space demands full accuracy in navigation and propulsion. A small mistake coupled with an erring guidance system would be sufficient to render a group of explorers helpless; perhaps, even, condemned to wander through the solar system forever. Eventually, the space station satellite will also serve as an engineering center for the assembly of space ships, for subsequent launching into uncharted portions of the planetary system.

Just as we used instrumented rockets to probe ahead into space we have also gone farther in probing the space around the Moon with lunar probes, and further yet in probing interplanetary space by means of the three solar artificial satellites now in orbit around the sun. So from our first steps outward we shall probe onward, carrying our curiosity to the vicinity of other planets. Plans for these future undertakings are already under way. Thus man will, in the age just dawning, make his way step by step through space.

PROJECT MERCURY

Project Mercury is the program to be carried out by National Aeronautics and Space Administration with the objective of placing manned satellites into orbit around the Earth. Other objectives are: to study the effects of space environment upon a human being by means of medical instrumentation, to collect data concerning the capsule's external and internal environments, to make scientific observations, and, finally, to recover successfully the capsule and its occupant.

LT. COL. J. H. GLENN: FIRST AMERICAN IN ORBIT

An Atlas-D rocket (360,000 pounds of thrust) launched Lt. Col. J. H. Glenn, aboard spacecraft "Friendship 7," into orbit on February 20, 1962. The principal orbital characteristics of the manned capsule were: perigee, 97.6 miles; apogee, 159.5 miles; average velocity, 17,545 m.p.h.; and inclination, 32.5°.

The capsule afforded a maximum degree of protection for its astronaut against the effects of heat, acceleration changes, and various aerodynamic forces. Within the 9 ft. 6 in. long capsule the temperature averaged about 90°F while traversing over the "night" regions of the Earth and over 100°F while traversing over the "daylight" regions. The separately controlled suit, however, kept astronaut Glenn at a more comfortable 67°F, approximately, while he made valuable observations, performed preassigned experiments on human reaction and adaptation, and took photographs through the porthole. One of the most interesting observations was the phenomenon of numerous small particles of metal-like texture moving at approximately the same velocity as the spacecraft which Glenn encountered in those regions undergoing the process of sunrise.

The re-entry of the capsule began during the end of the third orbit, over the Pacific Ocean, while approaching the West Coast of the United States. This was achieved with the help of retro-rockets which slowed down the spacecraft. During parts of the re-entry, the outer temperature of the capsule rose to about 3,000°F. Subsequently, the spacecraft deployed a parachute which enabled it to execute a gentle splashing near Grand Turk Island. Thus, a three-orbit voyage had been successfully completed.

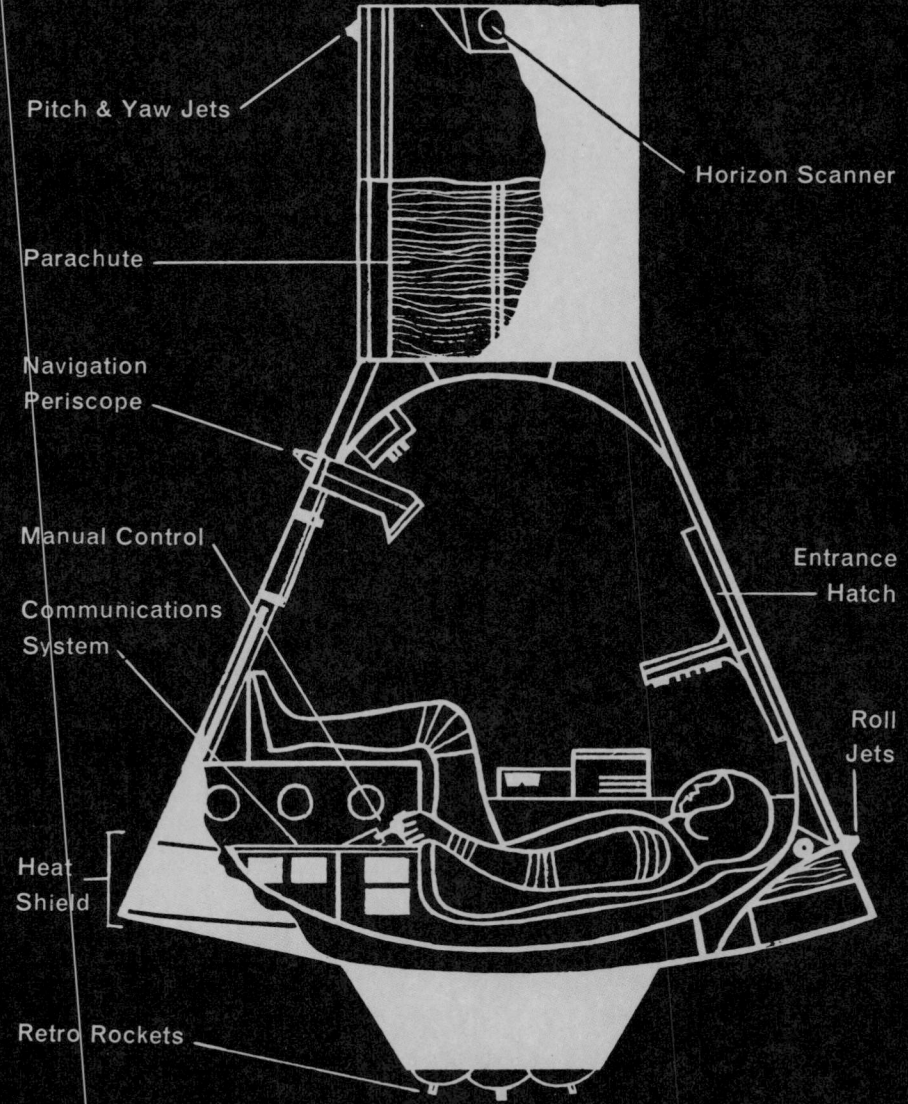

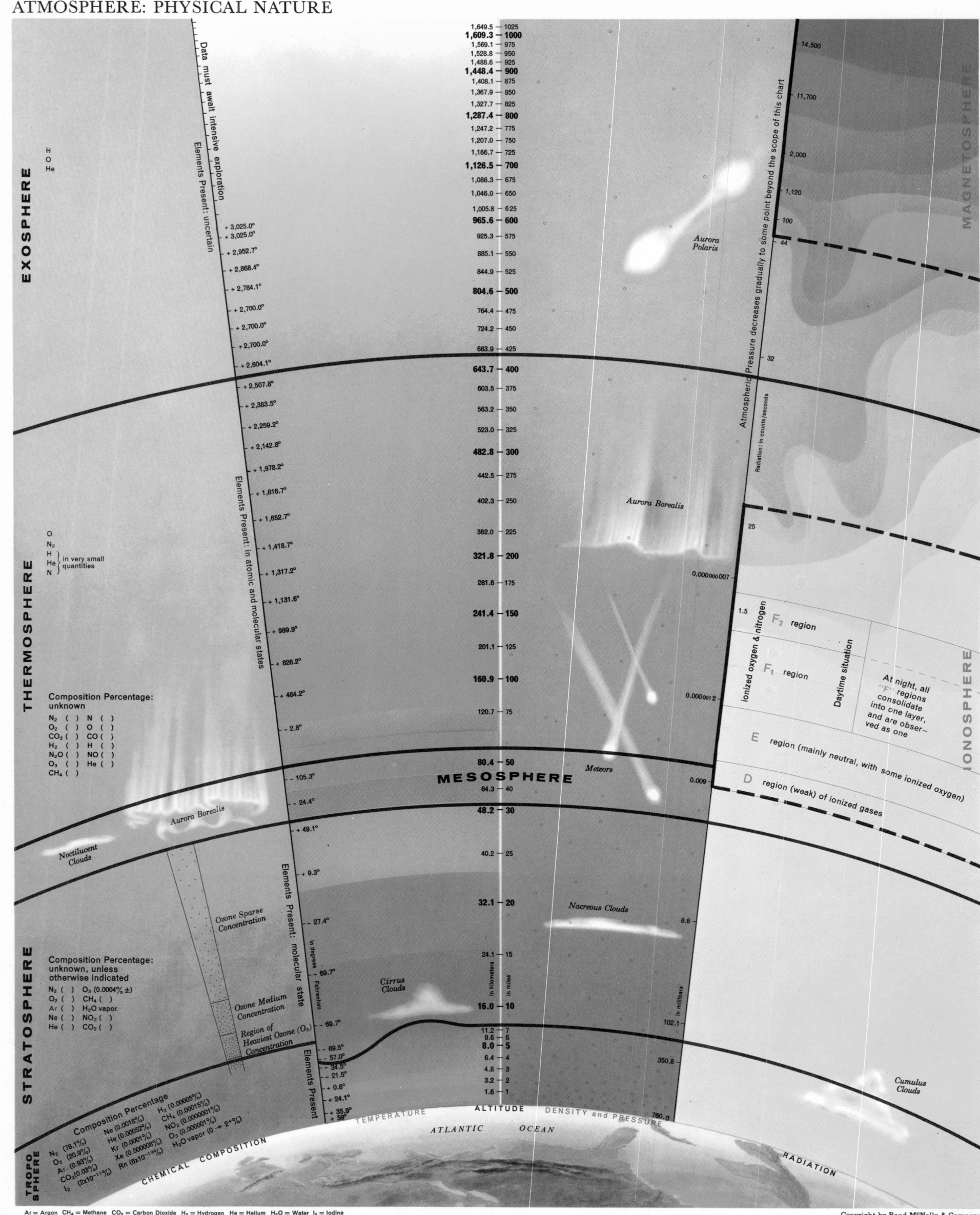

Ar = Argon CH₄ = Methane CO₂ = Carbon Dioxide H₂ = Hydrogen He = Helium H₂O = Water I₂ = Iodine

Kr = Krypton N₂ = Nitrogen Ne = Neon NO₂ = Nitrogen Dioxide O₂ = Oxygen O₃ = Ozone Rn = Radon

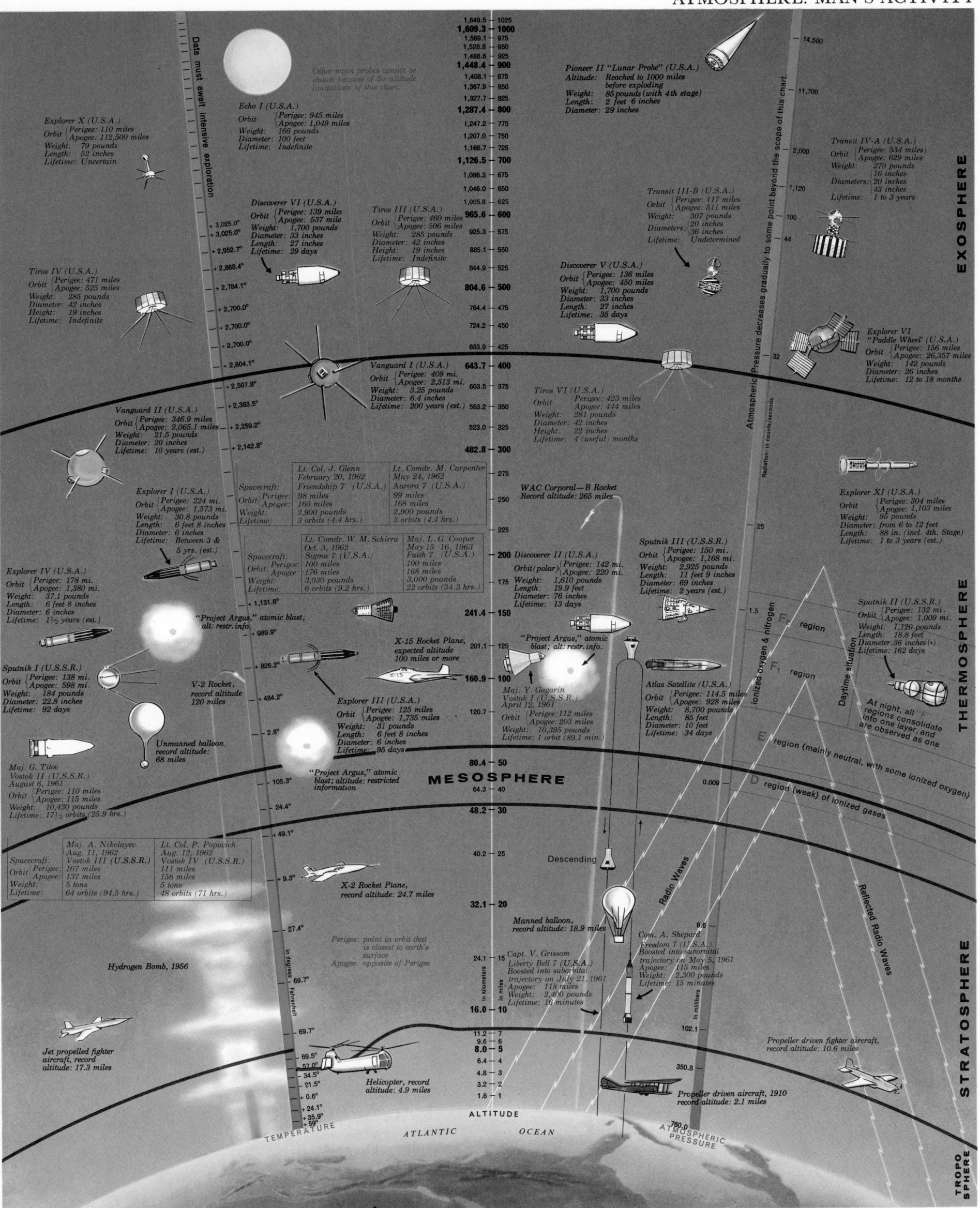

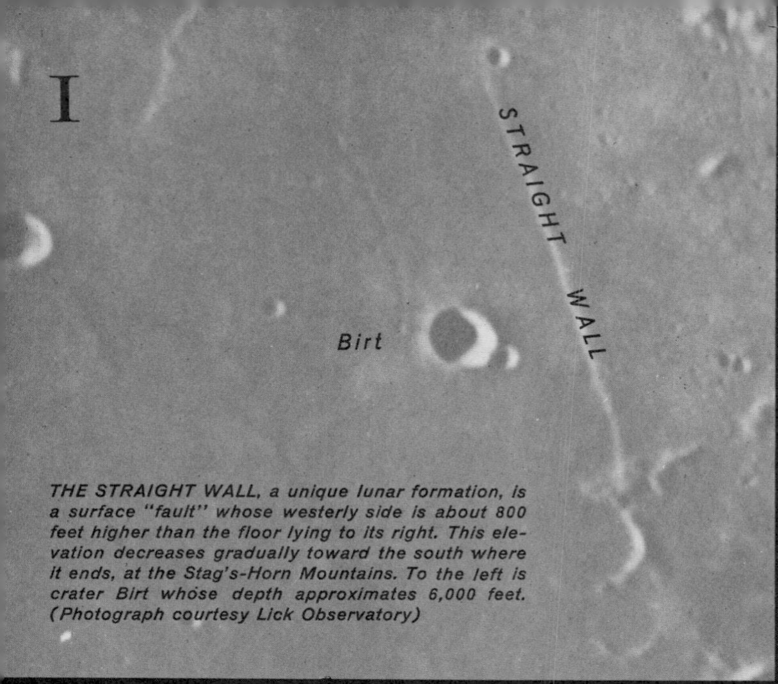

I

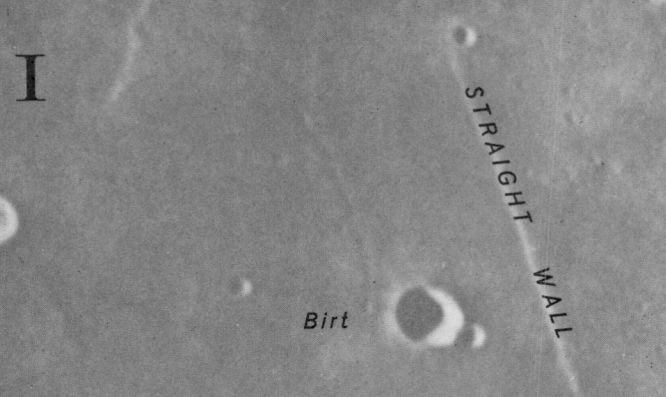

THE STRAIGHT WALL, a unique lunar formation, is a surface "fault" whose westerly side is about 800 feet higher than the floor lying to its right. This elevation decreases gradually toward the south where it ends, at the Stag's-Horn Mountains. To the left is crater Birt whose depth approximates 6,000 feet. (Photograph courtesy Lick Observatory)

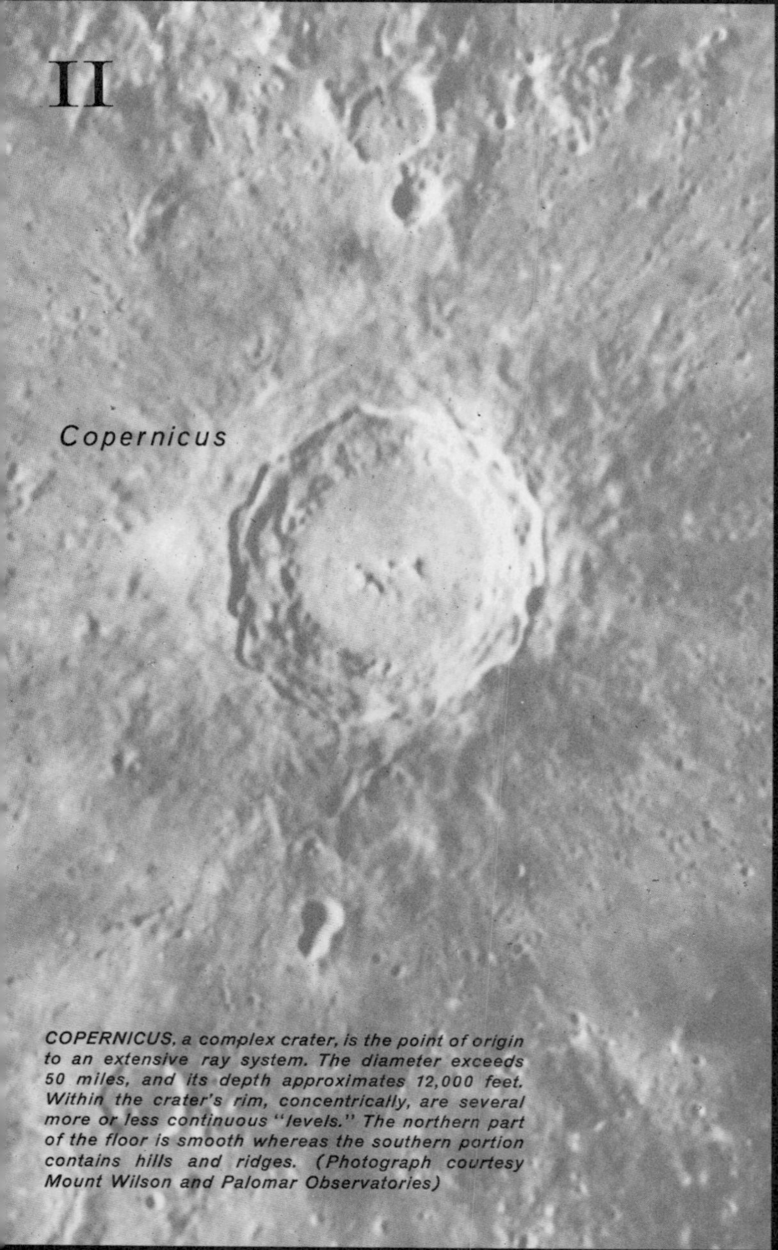

II

Copernicus

COPERNICUS, a complex crater, is the point of origin to an extensive ray system. The diameter exceeds 50 miles, and its depth approximates 12,000 feet. Within the crater's rim, concentrically, are several more or less continuous "levels." The northern part of the floor is smooth whereas the southern portion contains hills and ridges. (Photograph courtesy Mount Wilson and Palomar Observatories)

III

Tycho

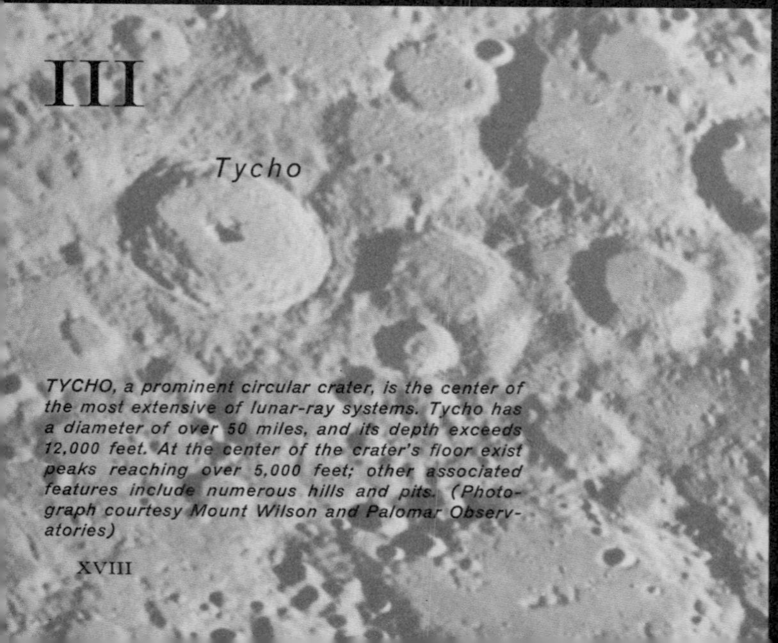

TYCHO, a prominent circular crater, is the center of the most extensive of lunar-ray systems. Tycho has a diameter of over 50 miles, and its depth exceeds 12,000 feet. At the center of the crater's floor exist peaks reaching over 5,000 feet; other associated features include numerous hills and pits. (Photograph courtesy Mount Wilson and Palomar Observatories)

NORTH POLE

II

I

III

SOUTH POLE

Last Quarter

The photograph above was taken during the last quarter of the Moon. The relative positions of the Earth, Moon, and Sun during this time are shown in the diagram below.

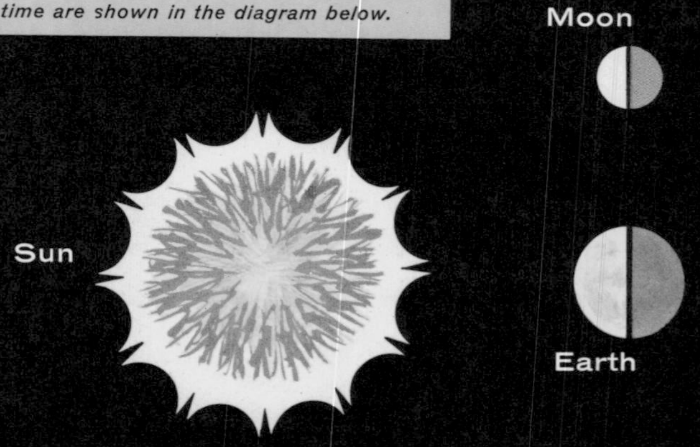

Sun

Moon

Earth

NORTH

Philolaus

MARE FRIGORIS
(Sea of Cold)

SINUS RORIS
(Bay of Dew)

Harpalus

JURA MTS.

STRAIGHT
RANGE

TENERIFE MTS.

Plato

ALPS MTS.

Alpine Valley

LAPLACE
PROM.

PICO

PITON

Cassini

Helicon

Leverrier

ALPS MTS.

Aristillus

MARE
IMBRIUM
(Sea of Rains)

Autolycus

Archimedes

HARBINGER
MTS.

Herodotus

Aristarchus

Lambert

Timocharis

Euler

Pytheas

APENNINE MTS.

OCEANUS

CARPATHIAN MTS.

Eratosthenes

SINUS
AESTUUM
(Bay of Billows)

Reiner

Kepler

Copernicus

Encke

Pallas

PROCELLARUM
(Ocean of Storms)

Reinhold

SINUS
MEDII
(Central Bay)

Landsberg C

Grimaldi

Mösting

Lalande

Herschel

RIPHAEUS MTS.

Parry

Ptolemaeus

Billy

Guericke

Alphonsus

Albategnius

Gassendi

Bullialdus

MARE
NUBIUM
(Sea of Clouds)

Arzachel

MARE
HUMORUM
(Sea of Moisture)

Campanus

STRAIGHT
WALL

Purbach

Vitello

Mercator

Pitatus

Aliacensis

Walter

Wurzelbauer

Gauricus

Hainzel

Wilhelm I

Tycho

Schickard

Longomontanus

Schiller

Maginus

Scheiner

Clavius

Bettinus

Moretus

SOUTH

The Moon

POLE

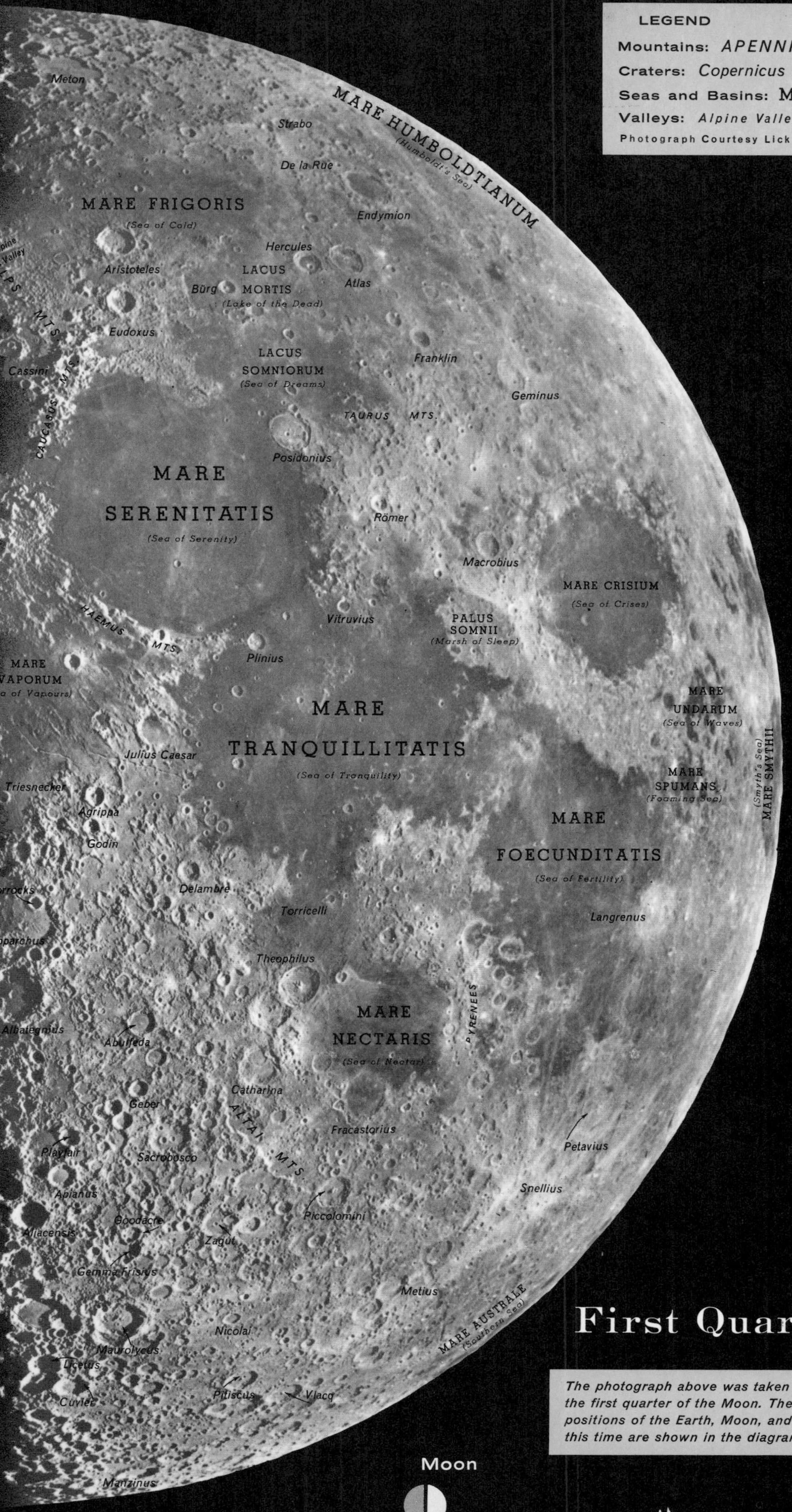

Meton

MARE HUMBOLDTIANUM
(Humboldt's Sea)

Strabo

De la Rue

MARE FRIGORIS
(Sea of Cold)

Endymion

Hercules

Aristoteles

LACUS
MORTIS
Burg *(Lake of the Dead)*

Atlas

Eudoxus

Franklin

LACUS
SOMNIORUM
(Sea of Dreams)

Geminus

TAURUS MTS.

Posidonius

MARE
SERENITATIS

Römer

(Sea of Serenity)

Macrobius

MARE CRISIUM
(Sea of Crises)

HAEMUS MTS.

Vitruvius

PALUS
SOMNII
(Marsh of Sleep)

MARE
VAPORUM
(Sea of Vapours)

Plinius

MARE
UNDARUM
(Sea of Waves)

Julius Caesar

MARE
TRANQUILLITATIS

MARE
SPUMANS
(Foaming Sea)

MARE SMYTHII

(Smyth's Sea)

Triesnecker

(Sea of Tranquility)

Agrippa

Godin

MARE
FOECUNDITATIS

(Sea of Fertility)

Delambre

Langrenus

Torricelli

Theophilus

MARE NECTARIS

PYRENEES

(Sea of Nectar)

Catharina

Fracastorius

Petavius

ALTAI MTS.

Snellius

Playfair

Sacrobosco

Piccolomini

Apianus

Boodacus

Metius

Gemma Frisius

Zagut

Nicolai

MARE AUSTRALE
(Southern Sea)

Maurolycus

First Quarter

Cuvier

Pitiscus

Vlacq

Manzinus

POLE

The photograph above was taken during the first quarter of the Moon. The relative positions of the Earth, Moon, and Sun during this time are shown in the diagram below.

Moon

Sun

Earth

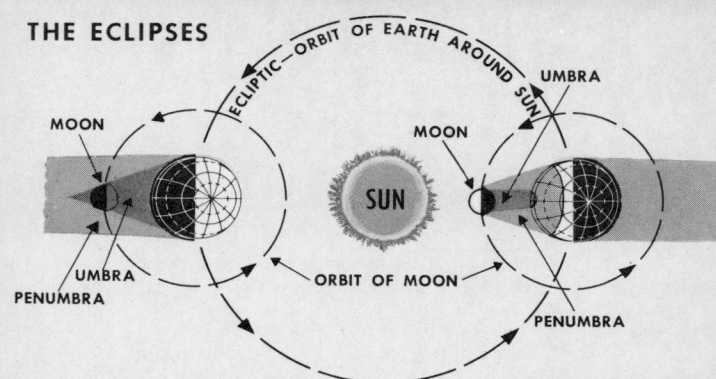

THE ECLIPSES

ECLIPTIC—ORBIT OF EARTH AROUND SUN

MOON

UMBRA

MOON

SUN

ORBIT OF MOON

UMBRA

PENUMBRA

PENUMBRA

TOTAL ECLIPSE OF THE MOON

Eclipse of the moon occurs only when moon is full. Moon usually appears dull red during the eclipse due to the refraction of the red rays of the sun by the atmosphere of the earth.

TOTAL ECLIPSE OF THE SUN

Eclipse of the sun occurs only during new moon. Sun is invisible in *umbra* and partly invisible in *penumbra*. Total eclipse is visible only in portion of earth touched by shadow of moon (umbra).

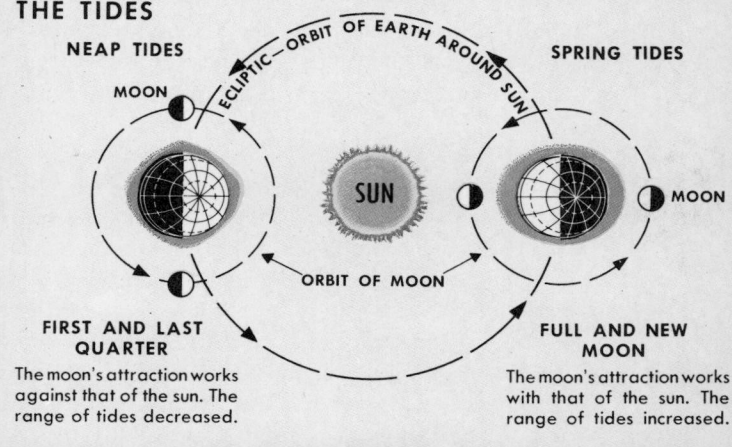

THE TIDES

NEAP TIDES

ECLIPTIC—ORBIT OF EARTH AROUND SUN

SPRING TIDES

MOON

SUN

MOON

ORBIT OF MOON

FIRST AND LAST QUARTER

The moon's attraction works against that of the sun. The range of tides decreased.

FULL AND NEW MOON

The moon's attraction works with that of the sun. The range of tides increased.

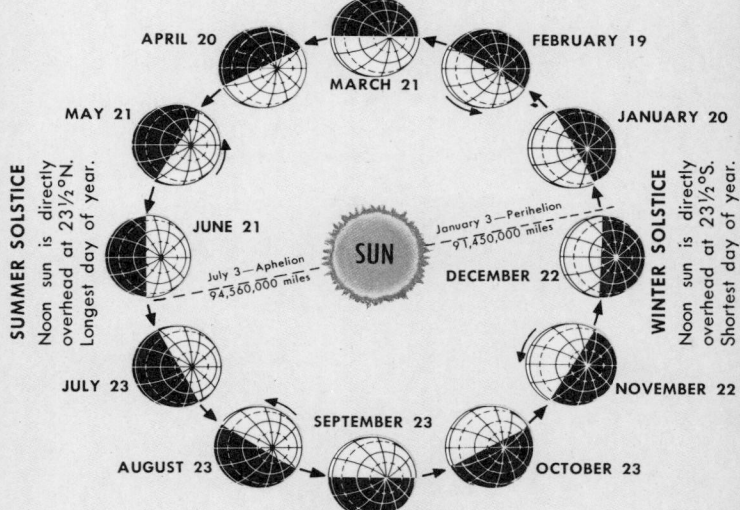

VERNAL EQUINOX
Noon sun is directly overhead at the Equator on its apparent migration north. Day and night are equal.

APRIL 20

FEBRUARY 19

MARCH 21

MAY 21

JANUARY 20

SUMMER SOLSTICE
Noon sun is directly overhead at 23½°N. Longest day of year.

JUNE 21

January 3—Perihelion
91,450,000 miles

July 3—Aphelion
94,360,000 miles

SUN

DECEMBER 22

WINTER SOLSTICE
Noon sun is directly overhead at 23½°S. Shortest day of year.

JULY 23

NOVEMBER 22

AUGUST 23

SEPTEMBER 23

OCTOBER 23

AUTUMNAL EQUINOX
Noon sun is directly overhead at the Equator on its apparent migration south. Day and night are equal.

THE SEASONS

Northern Hemisphere

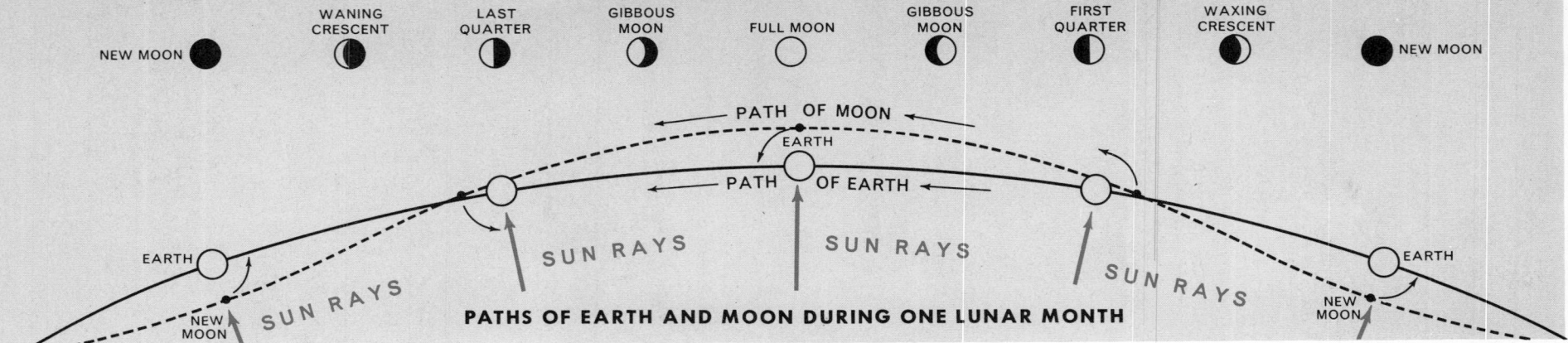

PATH OF MOON
EARTH
PATH OF EARTH
SUN RAYS — SUN RAYS — SUN RAYS
EARTH — NEW MOON — SUN RAYS — EARTH — NEW MOON

PATHS OF EARTH AND MOON DURING ONE LUNAR MONTH

FIRST STEP—THE MOON

Nearly airless, this dead world of steep mountains, huge craters and barren plains is the object nearest the Earth. It is the logical step outward from the Earth's surface or from a space station satellite orbiting the Earth. No doubt, in the future the Moon will also become a way station for space voyages to other planets. (See pages xviii-xix.)

Our beautiful "lantern of the night"—the Moon—is a convenient natural satellite which is 2,160 miles in diameter. As it travels majestically and dependably around the Earth each 27⅓ days, at a mean distance of 238,862 miles, it provides proof—undisputed for millions of years—of the laws of motion and gravitation.

In some respects we know the Moon well. All of the side which faces the Earth has been carefully examined and charted, and about 30,000 craters on its surface have been labeled. These craters range in size from walled plains over one hundred miles in diameter to the smallest objects that can be photographed. Copernicus (See page xviii), the well-formed crater near the center of the picture on the moon page, is a classic example. It has a diameter of fifty-six miles. A system of rays extend outward across its great, dry Mare Imbrium, "The Sea of Showers." Selenographers are not certain about the nature of these rays but think they are light material from beneath the Moon's surface splashed outward by the impact of the meteoroid that caused the crater.

Presumably the Moon is a dead world—it has a very thin atmosphere consisting of rare gases such as argon and krypton, and therefore has no weather. Sound does not carry on the moon. Its surface features remain unchanged except for the damage done by meteoroids falling on its surface. Life as we know it, plant or animal, cannot be expected without abundant air to sustain it. The means of existence for a group of explorers must be transported, almost intact, from the Earth. Later some of the needed materials may be extracted from the rocks at the surface, but at first the visitors will have to rely on their space ships and space suits for all the necessities of life.

It may be possible to enclose enough lunar space inside a plastic dome or bubble for a small city on the moon. Another alternative is to find or construct a cave underground—this would provide better protection against falling meteoroids. Furthermore, it would protect man against the extremes of temperature—over 200 degrees F. to below zero by night.

To the adventurer the Moon is a goal for varied reasons. To the astronomer, however, it is a potential super-market of astronomical data. The near absence of an atmosphere, which will prove troublesome to colonizers, will be a welcome blessing to those who wish an undistorted, unimpaired look at the heavens. Up to the present time, man has only been able to view the stars from the surface of the Earth through its atmosphere which impresses its characteristics on the light received from celestial bodies. When telescopes are installed on the Moon a new era of astronomical observation will be opened.

The far side of the Moon is hidden from our direct view, because tidal forces hold the Moon to one rotation on its axis which takes place in the same amount of time that the Moon revolves around the Earth. Efforts have already been made to discover the nature of the "Far Side of the Moon." Lunik III, a Soviet lunar probe, is claimed to have photographed a major portion of the hidden side, using two cameras, as it moved around the Moon. These photographs were successfully transmitted to Earth a few hours later.

Several experiments similar to that of Lunik III, with improved instrumentation and camera equipment, will permit selenographers to chart the far side of the Moon in the same degree of detail as the visible side or "Near Side." The visible side was touched physically by an instrumented payload on September 13, 1959, when Lunik II made a fast impact somewhere along the Haemus Mountains, south of the Sea of Serenity.

PROJECT APOLLO

The N.A.S.A. program that will lead to a successful manned spacecraft flight to the Moon, and its return to Earth, is called Project Apollo. The project will consist of three phases, and each of these will enable the scientists to learn more about the space environment between the Earth and the Moon. From this information, the lunar-landing spacecraft Apollo will be designed and constructed.

The Apollo will be manned by a crew of three, who will not wear space suits except in emergencies. The interior of the spacecraft will be approximately 13 feet wide and 12 feet high. No actual designs of its construction exist as yet,

Far Side of the Moon, taken by Lunik III.

United Press International Photo

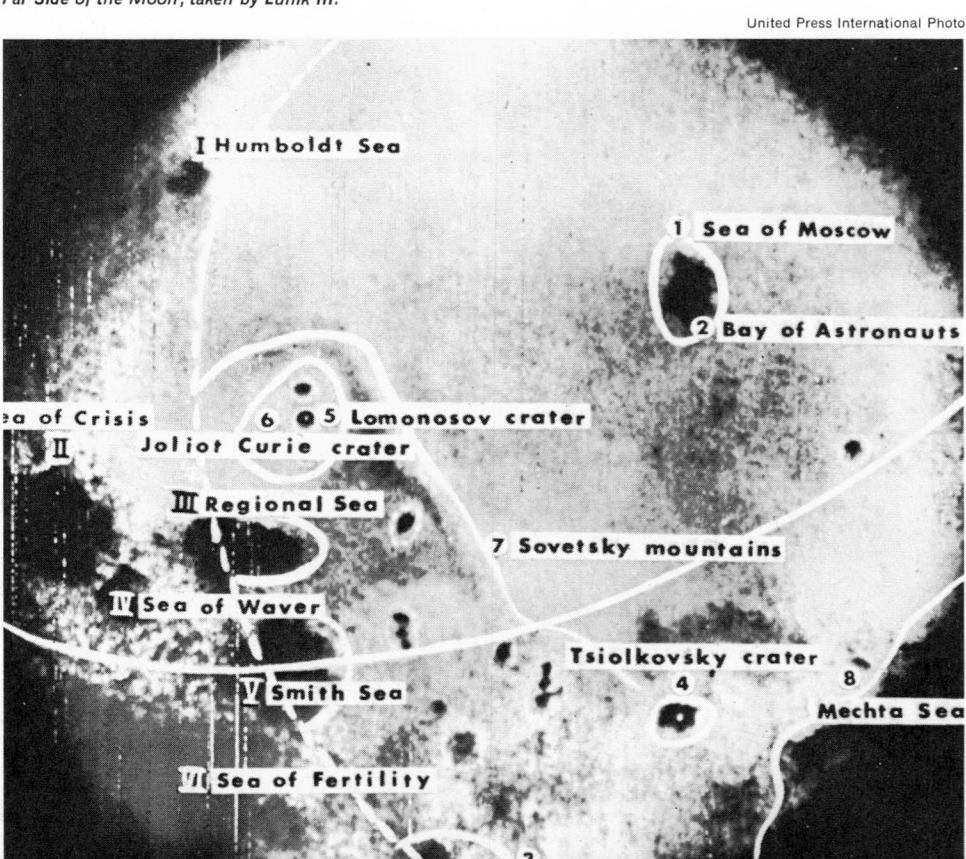

I Humboldt Sea
1 Sea of Moscow
2 Bay of Astronauts
Sea of Crisis
6 5 Lomonosov crater
Joliot Curie crater
II
III Regional Sea
7 Sovetsky mountains
IV Sea of Waver
Tsiolkovsky crater
4
8
V Smith Sea
Mechta Sea
VI Sea of Fertility
3

Circumlunar Apollo spacecraft
N.A.S.A.

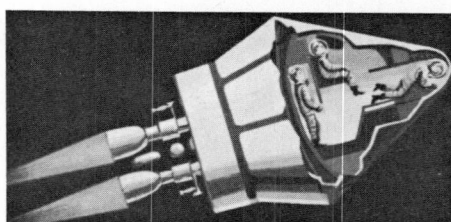

Lunik III photographing the Far Side of the Moon.

Sovfoto

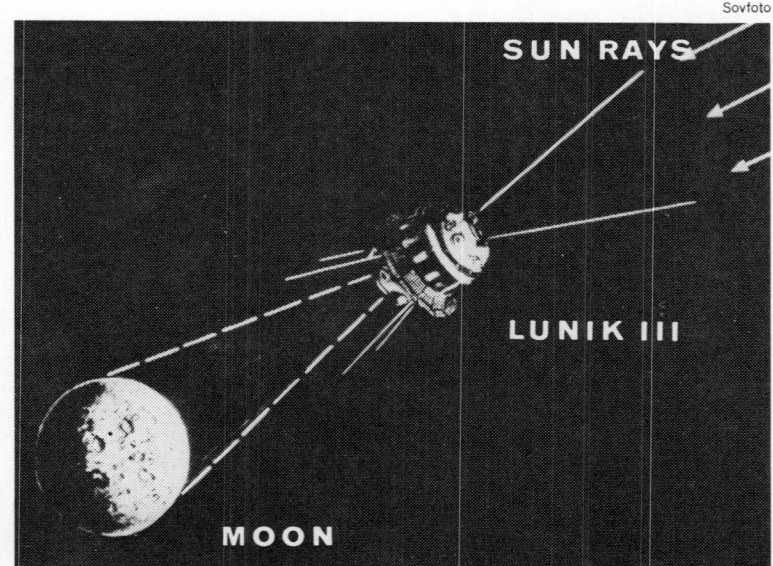

SUN RAYS
LUNIK III
MOON

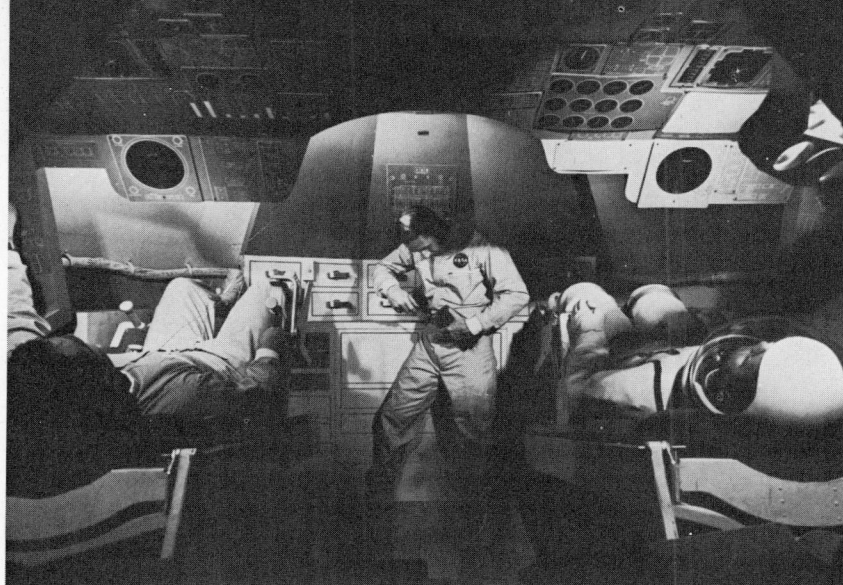

Interior of Apollo spacecraft

N.A.S.A.

N.A.S.A.

Apollo spacecraft after lunar landing

N.A.S.A.

Take-off from Moon of Apollo spacecraft

but the illustrations will give an idea of how the Apollo will look when it is ready for its moon flight.

The first Project Apollo phase is a laboratory stage. This will consist of a series of Earth-orbital flights of varying durations designed to test the spacecraft, improve its guidance and control systems, train and condition the crew, and gather data on human behavior and body reactions under space conditions.

The second phase will consist of a series of lunar-approach trips that will take the Apollo to progressively greater distances from Earth, gathering data on the physical environment being encountered and further testing the navigational systems. This phase will culminate with a two week circumlunar voyage, during which the crew will view and photograph the Far Side of the Moon and gather information on lunar features.

The third and most important phase is the manned lunar-landing mission. The launch-vehicle will boost the Apollo spacecraft to an initial velocity of 25,000 m.p.h., which will progressively decrease under the influence of the Earth's gravitational pull. Velocity, however, will increase again when the spacecraft comes into the Moon's gravitational field. The spacecraft must follow an extremely accurate course, making only a minimum of course-corrections to avoid wasting the limited fuel supply that must be used for a carefully controlled soft-landing at a pre-selected spot on the Moon's surface.

The crew, wearing moon-suits, will be able to leave the Apollo to observe and explore the lunar environment and terrain within its immediate vicinity. They will gather mineral samples, make astronomical observations, and radio their findings to Earth.

For the return voyage the lunar take-off is the most crucial part of the mission. The time of take-off must be precisely planned. After take-off the spacecraft must still have enough fuel for course-correction, and for braking during re-entry into the Earth's atmosphere.

SECOND STEP—THE PLANETS

After the Moon has been reached and studied man will turn his attention to other worlds nearby. The mind of every space explorer will immediately turn to Mars and Venus. As shown on p. xxii these are the two planets nearest the Earth. We know a great deal about them, but many mysteries are yet to be learned by the first visitors from Earth.

Mars (See p. xxii) is a reddish-colored wanderer of the night sky that has greatly attracted man's imagination. Probably more has been written about it than any other planet. Every two years it swings close to the Earth in a part of the sky opposite the Sun — a fine position for observation — and occasionally as near as 35 million miles.

"Voyager" spacecraft, designed to orbit around **another** planet on data-gathering mission.

National Aeronautics and Space Administration

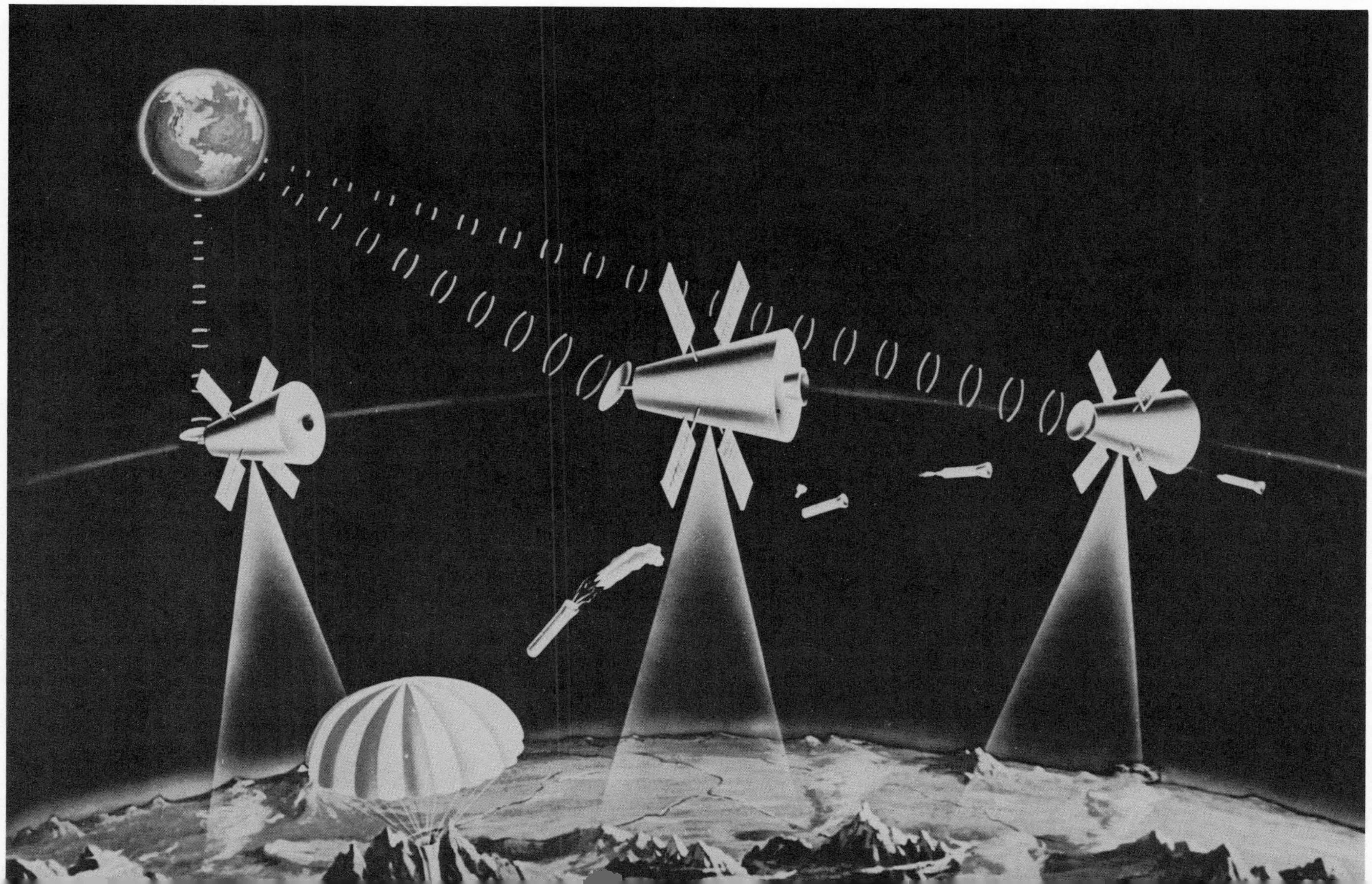

VENUS

The surface of Venus has never been seen due to the obstructive nature of its upper atmosphere. However, it is believed to have a rugged topography, mostly covered with water. Its upper atmosphere is rich in carbon dioxide, with the occurrence of frequent underlying clouds of formaldehyde droplets. The possibility of organic life is extremely limited. Perhaps there might exist some forms of primitive organic life in association with its extensive water bodies. The Venusian year is composed of 224.7 days. The planet has an orbital velocity of 21.78 miles per second, and any given object would require an escape velocity of 6.4 miles per second to leave the planet as compared with the Earth's 18.5 miles per second and 7.0 miles per second respectively. Venus has no moons.

MARS

The surface of Mars has been the object of extensive studies which seem to indicate the existence of vegetation in both hemispheres. Mars has similar climatic seasons to those of Earth. The Polar caps are easily distinguishable as are their seasonal changes. The surface configurations can be summarized as "debatable in nature." These include a large network of visible lines which at one time were thought to be canals, and regions of four different colors which vary with the seasons. The atmosphere is thinner than that of Earth, nevertheless, it is sufficiently dense to eliminate the need for pressurized suits, although a breathing helmet would still be required. The Martian year consists of 687 days. The planet has an orbital velocity of 15 miles per second and an escape velocity of 3.1 miles per second. This planet has two very small moons.

JUPITER

The surface of this planet has never been seen. Consequently, its visible upper atmosphere has been the object of intensive study. That region of the atmosphere is rich in ammonia, methane and hydrogen. Jupiter rotates more rapidly than Earth and, therefore, its day is only 9.8 hours in length. As a contrast, it takes 11.9 Earth years for Jupiter to travel once around the Sun, with an orbital velocity of about 8 miles per second. The planet requires an escape velocity of 37 miles per second, or over 5 times that required to escape Earth. Jupiter has 12 moons.

SATURN

Saturn's most distinguishing characteristic consists of the three concentric rings that revolve around it. The nearest of these is located a little over 6,000 miles above the planet itself. The rings are extremely thin and are made up of drifting material of varying sizes. Saturn has nine moons. The atmosphere is deep and shows evidence of some turbulence; it does not permit the observation of the actual surface of the planet. It revolves around the Sun once in 29.5 Earth years, with an orbital velocity of 6 miles per second. The escape velocity is 22 miles per second.

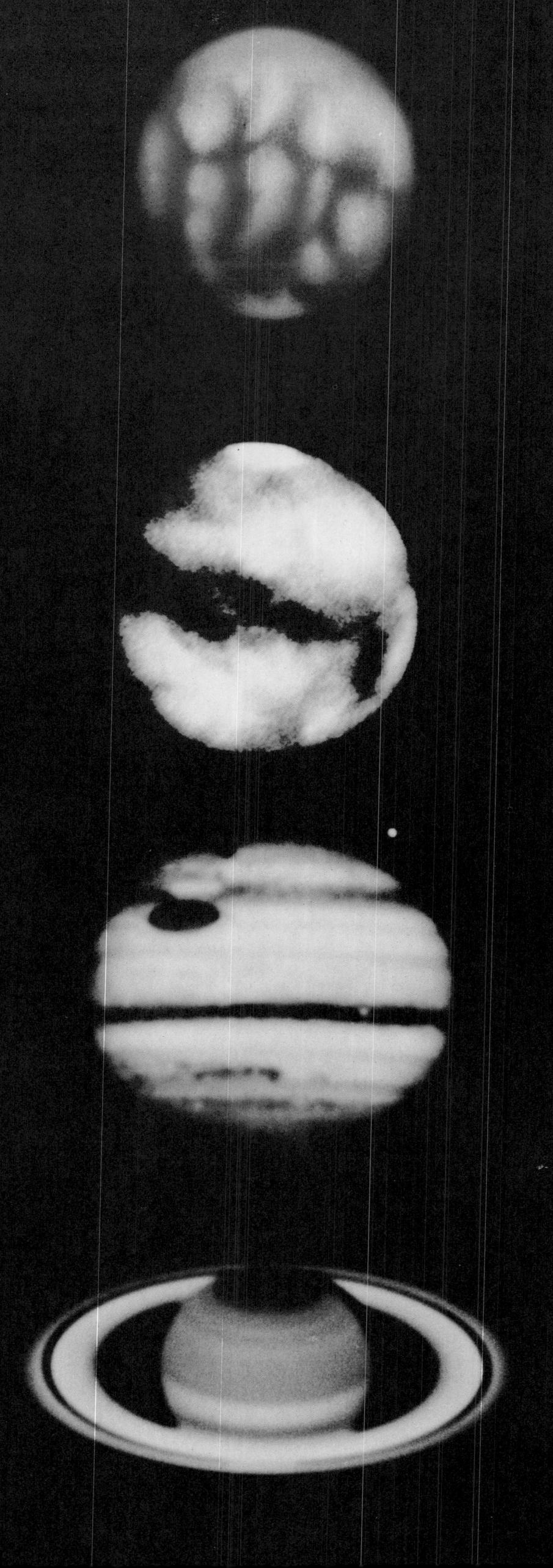

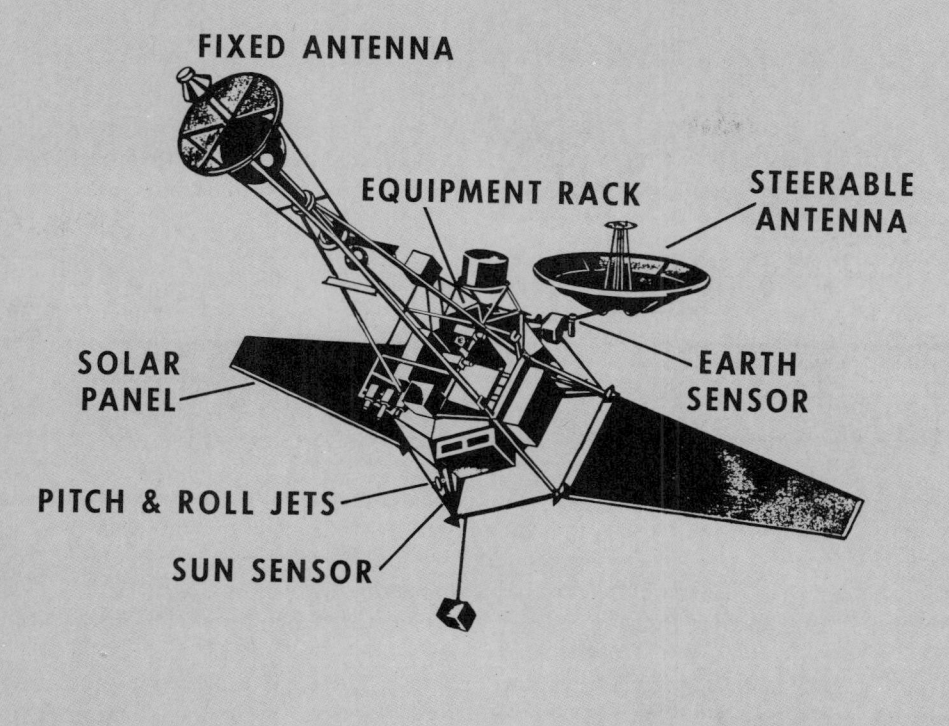

"Mariner" spacecraft designed for data-gathering voyages on early planetary missions to Mars and Venus.

Dr. Charles B. Moore at work during balloon ascension of November, 1959.

Mars has Earth-like features: an axis tilted 25° 10′ to provide the seasons; a year equal to 687 Earth days; a day consisting of 24 hours 37 minutes, and temperatures moderated by an atmosphere. The presence of polar caps and the periodic occurrence of changes in color suggest that the surface of Mars experiences seasonal cycles. There may even be life, such as lichens or mosses, which is dependent on the carbon dioxide (which is known to exist in Mars' atmosphere) and the water (which is believed to be present in moderate quantities).

By earthly standards, Mars is largely barren. The atmosphere is very thin, perhaps less than 10 per cent as dense as our own. Most of its water is probably locked in the ice and snow of the polar caps but may also be found in the clouds that are occasionally detected. The "canals," never well photographed but sketched by many observers after watching Mars through the telescope, may be no more than optical illusions. It has been suggested, however, that these "canals" are natural waterways carrying the water from the melting caps toward the Martian equator in springtime, causing the primitive plant life to flourish and turning the normally barren surface temporarily green. Most of the time the surface has the appearance of a rusty-red desert area, and gives the planet its characteristic color.

Mars has two small satellites, Phobos and Deimos, each less than ten miles in diameter, and both quite near the planet. Since space scientists point to definite advantages in exploratory orbiting before approaching the surface of a strange world, one of these small moons may provide a temporary way station en route.

Venus (See p. xxii) also offers a logical target for the early years of the space era. At times it comes within 26 million miles of the Earth—closer than any other regular body excluding the Moon. In many respects Venus is like the Earth and has been called our "twin sister" because it is so nearly the same size, but the clouds covering it are so dense that no telescope has ever penetrated them.

By careful analysis of sunlight reflected by the atmosphere of Venus, man has learned that there is an abundance of carbon dioxide. There is also definite evidence that the atmosphere has water vapor. This fact was established through the observations made by meteorologist Dr. Charles B. Moore during a manned balloon ascending over 81,000 feet in November, 1959. Some scientists believe that Venus may have oceans and swamps, but other astronomers maintain that the surface of Venus might be dry and wind-beaten. The evidence for either view is rather slim.

Venus does not have a moon (natural satellite), therefore an artificial satellite—with men or with instruments—placed into orbit around the planet becomes the logical approach for a close study of Venus before attempting a landing. Then, when safety is assured, man will descend to the surface.

POSSIBLE THIRD STEP—
THE DANGEROUS WORLDS

Not all of the planets in the Sun's family are likely to be visited by man in the early years following the first exploration of space. On the contrary, all the worlds except Mars and Venus are extremely unattractive and in some respects much more dangerous than these comparatively nearby planets. What man would brave the extreme temperatures of Mercury, or the deadly gases of Jupiter's atmosphere, or the super-coldness of beautiful Saturn.

Mercury is the planet nearest the Sun. It has a diameter of 3,100 miles, and because of its small size and low surface gravity its atmosphere—if it ever had any—escaped into space eons ago. At a distance averaging about 36 million miles from the Sun, this little world is bathed in intense solar radiation. As a result, it is arid, airless, and sun-drenched—hotter, indeed, than a baking oven. A view of its surface as we conceive it today is characterized by a volcanic and extremely rugged appearance.

The close pull of the Sun's tremendous gravitational force has bestowed another distinction on Mercury: its period of revolution around the Sun is precisely equal to the period of its rotation on its axis—88 days. As a result, the same side of the planet is always toward the Sun, much as the Moon keeps the same side to the Earth. Therefore there is no day and night. The sunlit side is always hot; the dark side is always cold, receiving no light or heat from the Sun. It is likely that Mercury has the most extreme variations in heat and cold of any of the planets.

Since there is a cold side and a hot side, neither ever changing, we might reason that there must be a dividing line where temperatures are moderate. If there were an atmosphere, that would be true, but its complete absence eliminates this possible zone of moderation. Objects are either cold if in shadow or hot if in sunlight, with no in-between condition. Mercury must top our list of dangerous worlds.

Turning from Mercury, so dangerously close to the Sun, let us look out in the opposite direction. Outward from the Sun we find that the next orbital path beyond Mars is that of mighty Jupiter (See p. xxii), the largest planet, which is more massive than all the others put together. Its mean diameter is about 88,700 miles, and although it never comes closer to Earth than 367 million miles it usually outshines everything in our night sky except the Moon, Venus, and, very occasionally, Mars.

We admire it, glowing so elegantly—yet how deceiving is its beauty. In reality Jupiter is a world of turmoil and change, rotating on its axis so rapidly—once each 9 hours and 50 minutes—that the gases at its surface are forced into colored bands of turbulent motion. Its relatively small rocky interior is surrounded by a huge shell of ice, which is in turn surrounded by solid, then liquid,

then gaseous hydrogen topped with clouds of deadly methane and ammonia. This is science's present-day image of the planet. The average temperature, to be expected at this distance from the Sun, is 200° F. below zero. No space traveler, in this or any foreseeable age, will venture into this maelstrom of uncertainty and danger.

Fortunately Jupiter possesses a family of twelve natural satellites, affording a possible haven to the traveler. Io, Ganymede, and Callisto are all larger than our Moon, and Europa is almost as large. It is conceivable that a space ship may some day successfully navigate the millions of miles to Jupiter's vicinity and land on one of these large satellites to establish a scientific outpost for the study of this titanic world, though we need to know much more about the satellites before such a plan is formulated.

Beyond Jupiter is another giant—Saturn (See p. xxii)—with a family of nine satellites and a ring system that distinguishes it from anything else we can see in the skies. Physically Saturn is quite similar to its huge neighbor Jupiter. Somewhat colder because of the greater distance from the Sun, Saturn nevertheless has an atmosphere agitated by rapid rotation, containing quantities of ice, hydrogen, methane, and ammonia. The ring system is made up of millions of small solid particles revolving like satellites around the planet, each in its own orbit. The outside diameter of all the rings is 175,000 miles.

Uranus and Neptune complete the list of four giant worlds. They, too, are danger points, with more of the methane and ammonia clouds (here frozen because of the increasingly lower temperatures) in atmospheres of hydrogen. Neptune cannot be seen without telescopic assistance. Occasionally an observer who knows where to look may be able to spot Uranus. They are greenish-colored globes much larger than the Earth but much smaller than Jupiter and Saturn. For more data consult the Planet Information Table.

The remotest known member of the solar family is little Pluto—so distant that it is visible only through large telescopes. Physical information concerning it is very sketchy. It was discovered in 1930 at Lowell Observatory in Flagstaff, Arizona. Pluto can be observed for only a small part of one of its trips around the Sun—each trip taking 248.4 Earth-years. The planet must at all times be horri-

fyingly cold. From its surface the Sun would seem weak and small and far away, as shown in the corresponding illustration on this page.

Besides the planets mentioned—these "dangerous worlds"—there are other objects in our system that might cause trouble to space travelers. Between Mars and Jupiter, for example, there is an entire family of hazards—the "planetoids" or asteroids. These are tiny planet-like objects ranging in diameter from a few miles to 480 miles for Ceres, the largest known. Nearly 3,000 asteroids have been discovered, and there are probably thousands still undetected.

The asteroids travel around the Sun in ellipses that vary tremendously, but on the average their orbits are beyond that of Mars and about a third of the distance from Mars to Jupiter. All the asteroids are airless because of their small size and low surface gravity. Their principal interest to the space traveler is the nuisance of their presence.

Other nuisances will be the meteoroids—bits of celestial debris ranging in size upward from specks of dust. These objects are known as meteors when they reach the Earth's atmosphere. Millions of meteors collide with the Earth each day, but few are large enough to attract any attention. However, collision between a space ship and a sizable meteoroid would be deadly.

Still another danger in space is the comet. Although we see relatively few of these strange objects, one never knows when a comet will turn up—or where. The typical comet orbit is highly eccentric. The comet may go billions of miles into space and then return to the heart of the Solar System for perhaps its first time since the dawn of history. It may glow brilliantly for a few nights, exciting a great deal of comment, and then return to its limbo in the extreme limits of the space controlled by the Sun's gravitation.

Comets consist of fragments of ice, or rock, or both, held loosely together by mutual gravitational forces. When near the Sun these materials are heated and made to glow; in addition, the reflection of sunlight is great.

The tail of a comet adds special interest. It is always spread out away from the Sun—even when the comet is receding, in which case the tail precedes the comet. Pressure of the Sun's radiation seems to be forcing tiny pieces of material from the head of the comet and then leaving them in space.

PLANET INFORMATION TABLE

PLANET	Mercury	Venus	Earth*	Mars	Jupiter†	Saturn	Uranus	Neptune	Pluto
Number of Natural Satellites per Planet	0	0	1	2	12	9	5	2	0
Mean Diameter (in Miles)	3,100	7,700	7,918	4,220	88,700	71,600	32,000	31,000	3,600
Mean Distance to the Sun (in millions of miles)	36.0	67.25	93.0	141.7	484.0	887.0	1,787.0	2,797.0	3,675.0
Comparative Volume (Earth = 1.00)	0.06	0.92	1.00	0.15	1,318	736	64	60	0.09
Comparative Mass (Earth = 1.00)	0.04	0.81	1.00	0.11	316.94	94.9	14.7	17.2	0.1
Necessary Escape Velocity	2.66	6.38	6.95	3.16	37.0	22.10	13.70	15.40	3.30(?)
Mean Surface Gravity (Earth = 1.00)	0.29	0.86	1.00	0.37	2.64	1.17	0.91	1.12	<0.5
Weight of a Human Being (in pounds)	38	88	100	39	265	117	105	123	55
Rotation on Planet's Own Axis	88.0 days	Unknown	23ʰ56ᵐ	24ʰ37ᵐ	9ʰ50ᵐ	10ʰ14ᵐ	10ʰ45ᵐ	15ʰ48ᵐ	6.39 days
Revolution Around the Sun	88.0 days	224.7 days	365.2 days	687.0 days	11.9 years	29.5 years	84.0 years	164.8 years	248.4 years
Mean Orbital Velocity (in miles per second)	29.76	21.78	18.52	15.00	8.12	6.00	4.23	3.37	2.95
Inclination of Planet's Orbit to the Ecliptic	7°00′	3°24′	0°00′	1°51′	1°18′	2°29′	0°46′	1°47′	17°09′
Inclination of Planet's Equator	0°(?)	32°(?)	23°27′	25°10′	3°07′	26°45′	97°53′	29°	Unknown

*EARTH   †LARGEST PLANET

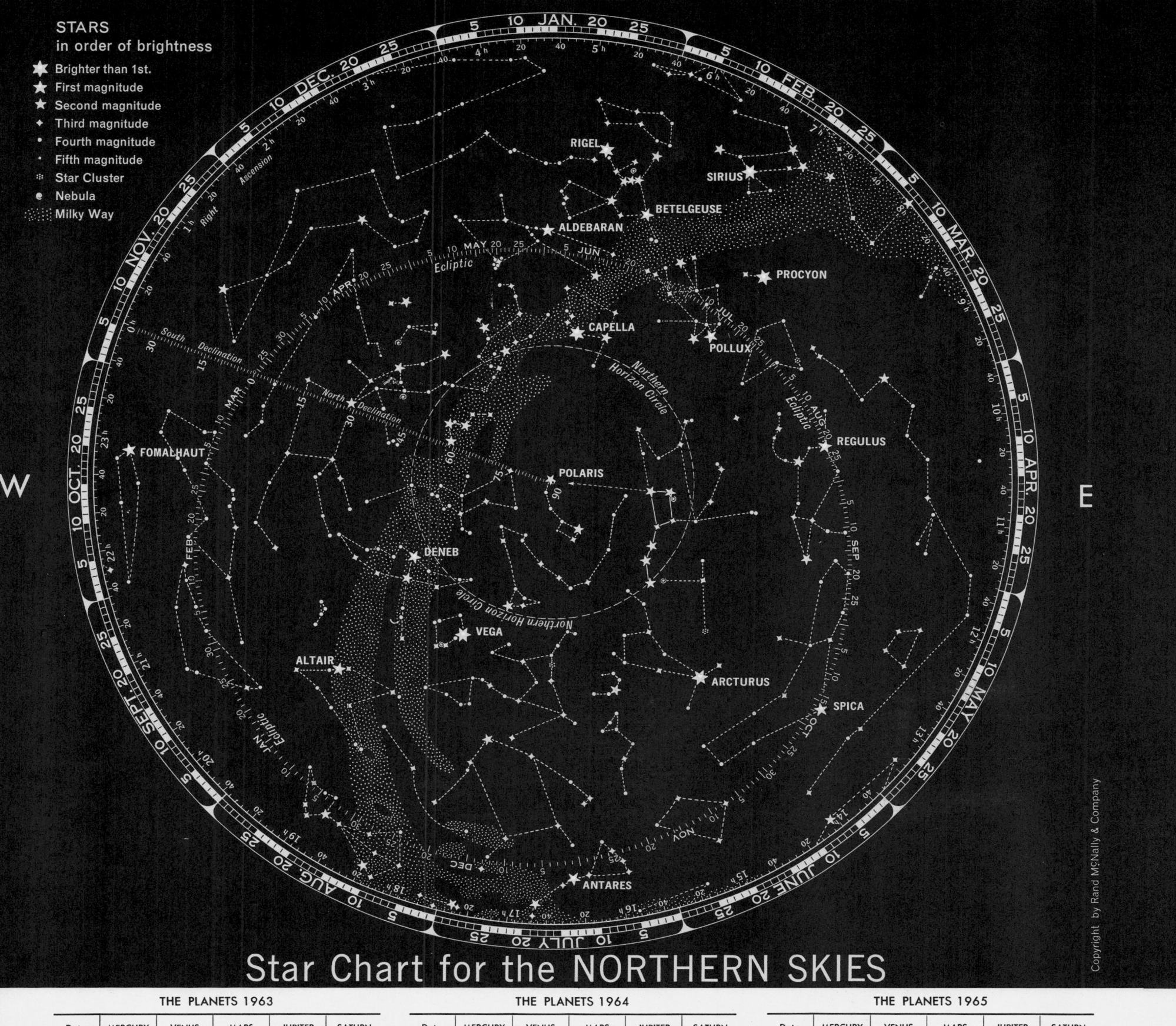

STARS in order of brightness

- Brighter than 1st.
- First magnitude
- Second magnitude
- Third magnitude
- Fourth magnitude
- Fifth magnitude
- Star Cluster
- Nebula
- Milky Way

Star Chart for the NORTHERN SKIES

THE PLANETS 1963

Date	MERCURY R.A.	Decl.	VENUS R.A.	Decl.	MARS R.A.	Decl.	JUPITER R.A.	Decl.	SATURN R.A.	Decl.
	h m	°	h m	°	h m	°	h m	°	h m	°
Jan. 1	20 05	22 S	15 35	15 S	9 53	17 N	22 44	9 S	20 50	18 S
15	20 30	17	16 26	18	9 44	18 N	22 54	8	20 57	18 S
Feb. 1	19 28	19	17 38	20	9 21	20 N	23 07	7	21 05	18 S
15	20 07	20	18 43	21	8 59	22	23 18	6	21 12	17 S
Mar. 1	21 22	17	19 51	20	8 42	23	23 31	4	21 18	17 S
15	22 48	10	20 58	17	8 35	22	23 44	3	21 24	16 S
Apr. 1	0 44	4 N	22 18	11	8 39	22	23 59	1	21 31	16 S
15	2 27	16	23 22	6	8 52	20	0 11	0	21 36	15 S
May 1	3 45	23 N	0 33	2 N	9 13	18 N	0 25	1 N	21 40	15 S
15	3 44	20	1 36	8 N	9 36	16 N	0 36	3 N	21 42	15 S
June 1	3 21	15 N	2 55	15 N	10 06	13 N	0 48	4 N	21 43	15 S
15	3 54	17	4 03	20	10 57	0 57	5 N	21 43	15 S
July 1	5 33	23	5 26	23 N	11 05	7 N	1 05	6 N	21 41	15 S
15	7 41	23	6 40	23	11 34	3 N	1 10	6 N	21 38	15 S
Aug. 1	9 55	14 N	8 10	21	12 11	1 S	1 14	6	21 34	16 S
15	11 12	5 N	9 21	17	12 43	4 S	1 14	6	21 30	16 S
Sept. 1	12 08	4 S	10 42	10 N	13 23	9 S	1 11	5 N	21 25	17 S
15	11 60	4	11 47	3 N	13 58	12	1 07	5	21 21	17 S
Oct. 1	11 27	13	12 59	5 S	14 40	16	0 59	4 N	21 19	17 S
15	12 28	1	14 05	12	15 19	19	0 52	4	21 17	17 S
Nov. 1	14 13	13	15 28	19	16 10	22 S	0 45	3 N	21 18	17 S
15	15 41	20	16 41	23	16 53	23	0 40	3	21 19	17 S
Dec. 1	17 26	25	18 07	25	17 45	24	0 37	2 N	21 23	17 S
15	18 53	25	19 24	24	18 32	24 S	0 38	3 N	21 27	16 S

THE PLANETS 1964

Date	MERCURY R.A.	Decl.	VENUS R.A.	Decl.	MARS R.A.	Decl.	JUPITER R.A.	Decl.	SATURN R.A.	Decl.
	h m	°	h m	°	h m	°	h m	°	h m	°
Jan. 1	19 16	21	20 52	19	19 29	23 S	0 42	3 N	21 33	16 S
15	18 19	20	22 01	14	20 15	21	0 47	4 N	21 39	15 S
Feb. 1	19 13	22	23 18	6	21 10	17 S	0 56	5 N	21 46	15 S
15	20 35	20	0 19	2 N	21 54	14 S	1 05	6 N	21 53	14 S
Mar. 1	22 13	13	1 22	9 N	22 39	10 S	1 17	7 N	22 00	14 S
15	23 48	3	2 22	16 N	23 20	5 S	1 28	8 N	22 06	13 S
Apr. 1	1 43	12 N	3 35	22 N	0 09	0	1 43	10 N	22 13	12 S
15	2 32	18	4 34	26 N	0 48	4 N	1 56	11	22 18	12 S
May 1	2 10	13 N	5 35	27 N	1 34	9 N	2 10	12	22 23	12 S
15	2 04	9 N	6 14	27	2 14	13 N	2 23	13 N	22 27	11 S
June 1	3 01	14	6 30	25 N	3 03	17 N	2 38	14 N	22 29	11 S
15	4 33	21	6 07	23 N	3 44	20 N	2 50	15	22 30	11 S
July 1	7 00	24 N	5 28	19 N	4 32	22 N	3 03	16 N	22 29	11 S
15	8 55	19 N	5 21	17 N	5 13	24 N	3 13	17	22 27	11 S
Aug. 1	10 29	9 N	5 48	18 N	6 04	24 N	3 24	17	22 24	12 S
15	11 07	2 N	6 31	19 N	6 44	24 N	3 30	18 N	22 20	12 S
Sept. 1	10 44	4 N	7 36	19 N	7 32	23 N	3 35	18 N	22 15	13 S
15	10 28	19	8 36	17 N	8 09	21 N	3 36	18	22 11	13 S
Oct. 1	11 50	9	9 47	13 N	8 49	19 N	3 35	18 N	22 08	14 S
15	13 19	7	10 50	8 N	9 23	17	3 30	18	22 05	14 S
Nov. 1	15 04	18	12 05	1 N	10 00	14 N	3 23	17 N	22 04	14 S
15	16 30	24	13 08	5 S	10 28	12 N	3 15	17	22 05	14 S
Dec. 1	18 01	26	14 22	12 S	10 58	9 N	3 07	16 N	22 07	13 S
15	18 08	23	15 31	17 S	11 20	7 N	3 01	16 N	22 10	13 S

THE PLANETS 1965

Date	MERCURY R.A.	Decl.	VENUS R.A.	Decl.	MARS R.A.	Decl.	JUPITER R.A.	Decl.	SATURN R.A.	Decl.
	h m	°	h m	°	h m	°	h m	°	h m	°
Jan. 1	17 14	20 S	16 59	22 S	11 42	5 N	2 56	16 N	22 16	12 S
15	18 10	23	18 14	23	11 54	4 N	2 56	16 N	22 21	12 S
Feb. 1	19 54	22	19 46	23	11 58	4 N	2 59	16 N	22 28	11 S
15	21 28	17	20 59	18	11 52	5 N	3 04	16 N	22 34	11 S
Mar. 1	23 05	7	22 09	13	11 37	7 N	3 11	17 N	22 42	10 S
15	0 37	5 N	23 15	6	11 17	9 N	3 20	18 N	22 47	10 S
Apr. 1	1 20	12 N	0 32	2 N	10 55	11 N	3 33	18 N	22 55	9 S
15	0 50	6 N	1 36	9 N	10 46	11 N	3 45	19 N	23 00	8 S
May 1	0 59	3 N	2 52	16 N	10 47	10 N	4 00	20 N	23 06	8 S
15	1 53	8 N	4 02	21 N	10 57	8 N	4 13	21 N	23 10	7 S
June 1	3 43	19 N	5 31	24 N	11 16	6 N	4 30	21 N	23 14	7 S
15	5 50	25	6 46	24 N	11 37	4 N	4 44	22 N	23 15	7 S
July 1	8 06	22 N	8 11	22 N	12 04	0	4 59	22 N	23 16	7 S
15	9 25	15 N	9 20	17 N	12 31	3 S	5 12	22 N	23 15	7 S
Aug. 1	10 04	8 N	10 40	10 N	13 06	7 S	5 27	23 N	23 13	7 S
15	9 37	9 N	11 42	3 N	13 38	11 S	5 38	23 N	23 10	8 S
Sept. 1	9 30	14 N	12 55	6 S	14 19	15 S	5 50	23 N	23 05	8 S
15	10 52	9	13 56	13	14 58	17 S	5 57	23 N	23 01	9 S
Oct. 1	12 39	3	15 15	19	15 41	21	6 03	23 N	22 57	9 S
15	14 04	13	16 15	24	16 23	23	6 06	23 N	22 54	9 S
Nov. 1	15 42	22 S	17 37	27 S	17 16	24 S	6 05	23 N	22 52	10 S
15	16 53	25	18 43	26	18 02	25	6 01	23 N	22 51	10 S
Dec. 1	16 50	20	19 49	24	18 56	24	5 54	23 N	22 52	9 S
15	16 07	18	20 34	21	19 43	23 S	5 46	23 N	22 54	9 S

Members of the Universe

The Universe consists of a vast Space, within which are contained untold numbers of galaxies, billions of stars, planets, asteroids, and other celestial objects of varying sizes, shapes and characteristics.

Among the important members of the Universe are the galaxies, known also as "Universe Islands."

FIGURE 1

FIGURE 2

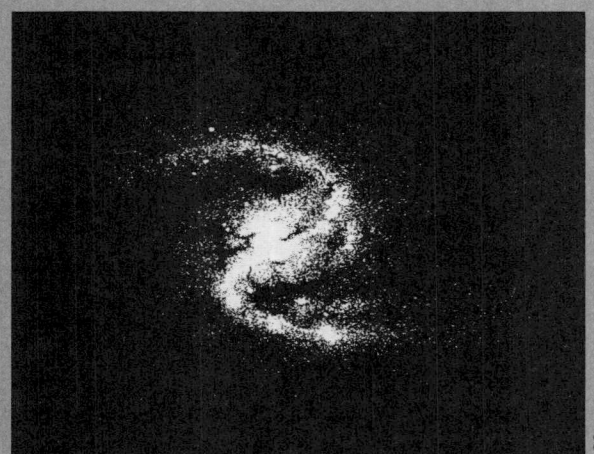

FIGURE 3

FIGURE 4

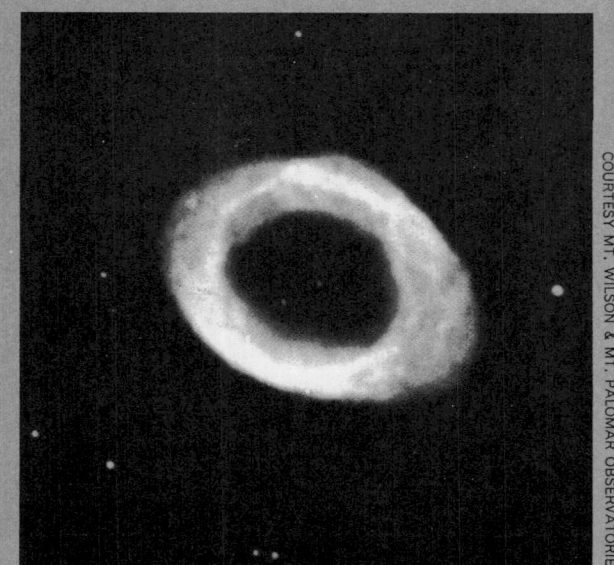

FIGURE 5

A Galaxy is a large community of stars. Several types of Galaxies can be classified by their shape and structure. Thus, our "Milky Way" is a Spiral Galaxy (Figure No. 1). The Sun is a star in the Milky Way and is located over half way out from the center of the Galaxy. In Figure No. 2 the Milky Way is shown as a side view; its diameter approximates 100,000 light years.

There are other types of Galaxies, among them, Barred Spirals (Fig. 3), which have straighter spiral arms than the normal spirals. Also Elliptical galaxies, such as M32, which is located in the region of Andromeda. And, Irregular galaxies, such as the Magellanic Clouds, characterized by having no particular shape. Stars are also grouped close together in Star Clusters (Fig. 4), such as that of Hercules.

In Space, there are also clouds of gases and dust. Any of these clouds are termed Nebula. There are two kinds of them: Diffuse Nebulae and Dark Nebulae. The former are located near a star, which is their source of light (Fig. 5), such as the Ring Nebula in Lyra. Dark Nebulae are those which have no star nearby from which to draw light, (Fig. 6), such as the Horsehead Nebula in Orion.

Our Sun is a smaller than average star. It is completely gaseous and rotates irregularly. The two major portions of the Sun are: its Atmosphere composed of 3 layers, and the Nucleus, or Sun proper. The atmosphere is separated from the Nucleus by the Photosphere, an opaque layer, which does not permit the observer to see under it, (Fig. 7). The effective temperature of the Sun approximates 10,000° F. The amount of energy received from it at the earth's surface is of 1.94 calories per square centimeter per minute.

The upper layer of the Sun's atmosphere registers some very turbulent phenomena known as solar prominences, which resemble flame tongues, streaking upward to tremendous heights dwarfing the size of earth to a mere dot, (Fig. 8).

FIGURE 6

FIGURE 7

FIGURE 8

This fourteen page section of the atlas provides a fascinating graphic review of the most significant geographic features of the World, the Continents, Canada, and the United States. These relief maps are strikingly new in cartographic design and technique, giving the visual effect of a three-dimensional model. Multicolor layer tints, enclosed by a generalized contour line, give the general elevation above sea level. Shades of green are used for the lowlands below 1,000 feet, hues of light tan, buff, and yellow are used for successively higher elevations, and areas higher than 10,000 feet are left white. Each of the colors increases in value with increasing elevation, and thus higher regions visually appear closer to the observer.

A shaded mountain overprint has been used to show the local relief and the general slope of the land. The three-dimensional effect is more noticeable where it is important—in the higher, mountainous areas where the slopes are steepest— because the shadow contrast of the mountain overprint is greatest in the same areas where the color values are highest.

These new maps will give immediate and lasting impressions of the great natural regions of each continent—the lowlands and valleys, the highlands and mountain barriers, the vast plateaus and high plains.

Superimposed on this physical background are the chief political divisions and their most important cities. Such information on a single map makes it easy to see how some countries have good communications across lowland plains or inland seas, while others are separated by highland barriers or long ocean routes.

THE
PHYSICAL
WORLD
IN MAPS

A GRAPHIC REVIEW OF WORLD GEOGRAPHY

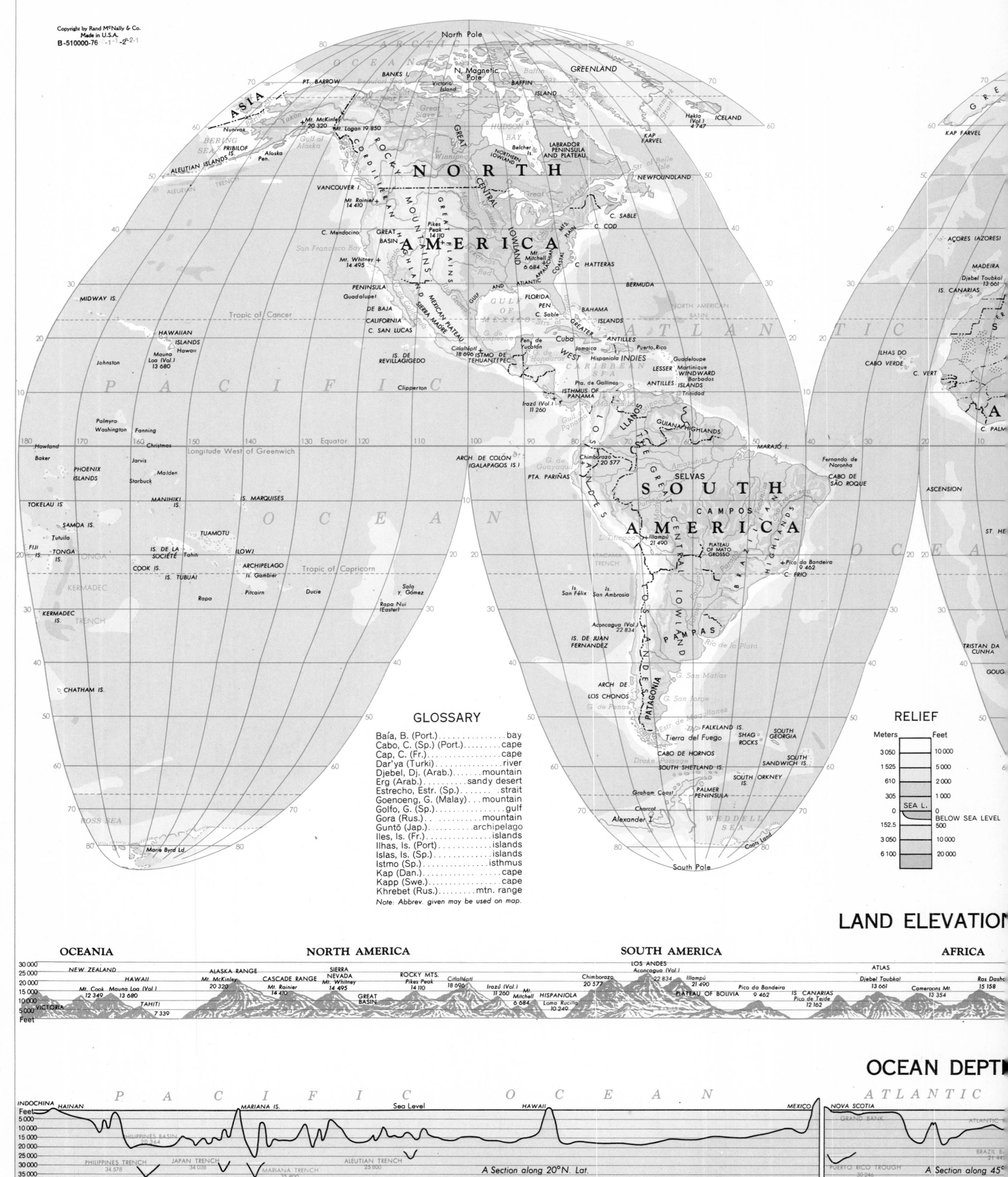

GLOSSARY

Baía, B. (Port.) . . . . bay
Cabo, C. (Sp.) (Port.) . . . . cape
Cap, C. (Fr.) . . . . cape
Dar'ya (Turki) . . . . river
Djebel, Dj. (Arab.) . . . . mountain
Erg (Arab.) . . . . sandy desert
Estrecho, Estr. (Sp.) . . . . strait
Goenoeng, G. (Malay) . . . mountain
Golfo, G. (Sp.) . . . . gulf
Gora (Rus.) . . . . mountain
Guntô (Jap.) . . . . archipelago
Iles, Is. (Fr.) . . . . islands
Ilhas, Is. (Port.) . . . . islands
Islas, Is. (Sp.) . . . . islands
Istmo (Sp.) . . . . isthmus
Kap (Dan.) . . . . cape
Kapp (Swe.) . . . . cape
Khrebet (Rus.) . . . . mtn. range
Note: Abbrev. given may be used on map.

RELIEF

Meters	Feet
3 050	10 000
1 525	5 000
610	2 000
305	1 000
0 SEA L.	0
	BELOW SEA LEVEL
152.5	500
3 050	10 000
6 100	20 000

LAND ELEVATION

OCEAN DEPTH

Elevations and depress

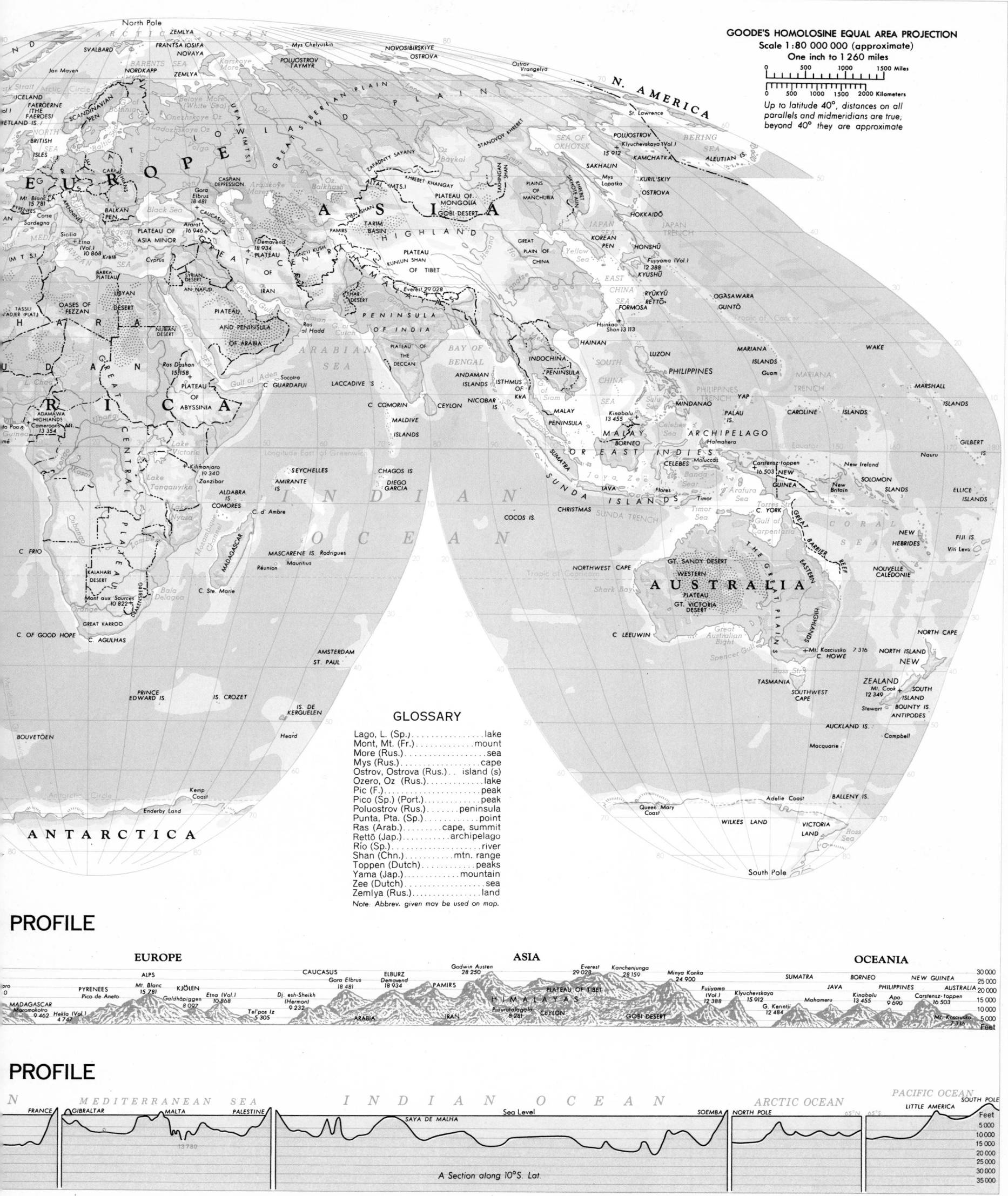

GOODE'S HOMOLOSINE EQUAL AREA PROJECTION

Scale 1:80 000 000 (approximate)
One inch to 1 260 miles

PROFILE

PROFILE

GLOSSARY

Lago, L. (Sp.)	lake
Mont, Mt. (Fr.)	mount
More (Rus.)	sea
Mys (Rus.)	cape
Ostrov, Ostrova (Rus.)	island (s)
Ozero, Oz (Rus.)	lake
Pic (F.)	peak
Pico (Sp.) (Port.)	peak
Poluostrov (Rus.)	peninsula
Punta, Pta. (Sp.)	point
Ras (Arab.)	cape, summit
Rettō (Jap.)	archipelago
Río (Sp.)	river
Shan (Chn.)	mtn. range
Toppen (Dutch)	peaks
Yama (Jap.)	mountain
Zee (Dutch)	sea
Zemlya (Rus.)	land

Note: Abbrev. given may be used on map.

EUROPE

Relief

Meters		Feet
3050		10 000
1525		5000
610		2000
305		1000
152.5		500
0	Sea Level	0
152.5	500	Below
1525	5000	Sea Level
3050	10 000	

Scale 1: 16 000 000; one inch to 250 miles. Conic Projection

Elevations and depressions are given in feet

Longitude West of Greenwich   Longitude East of Greenwich

xxx

EURASIA

Relief

Meters	Feet
3050	10 000
1525	5000
610	2000
305	1000
0 Sea Level	0 Sea Level
152.5	500
1525	5000
3050	10 000
6100	20 000
	Below Sea Level

B-519695-76 -2-2-2
COPYRIGHT BY
RAND MCNALLY & COMPANY
MADE IN U.S.A.

Scale 1:31 933 000; one inch to 504 miles. Lambert's Azimuthal, Equal Area Projection
Elevations and depressions are given in feet

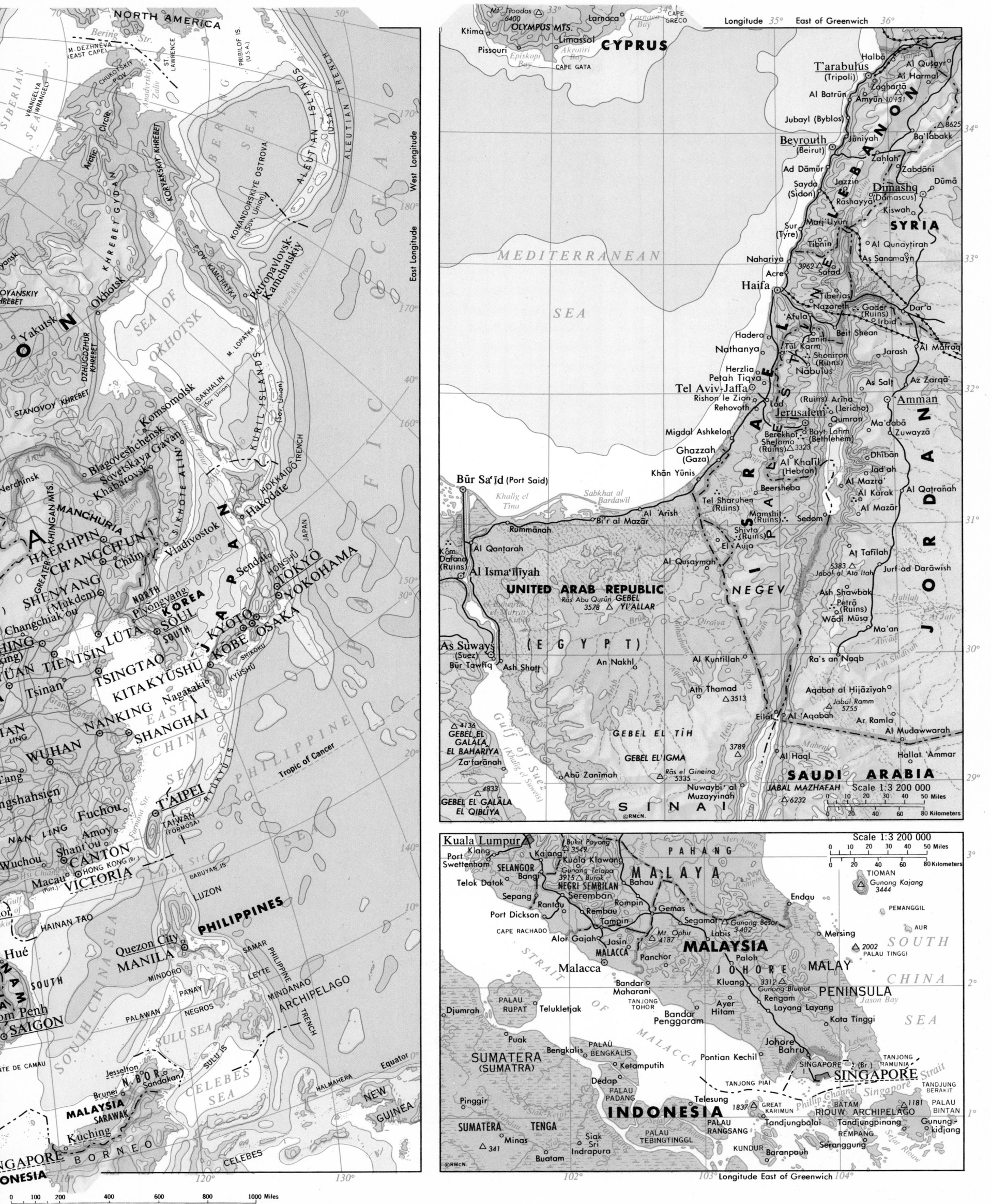

North America / East Asia map

NORTH AMERICA

Bering Str.
M. DEZHNEVA
(EAST CAPE)
CHUKOTSKIY PLOV.
Arctic Circle
ST. LAWRENCE I. (U.S.A.)
PRIBILOF IS. (U.S.A.)
ALEUTIAN ISLANDS
ALEUTIAN TRENCH

SIBERIAN SEA
WRANGELYA (WRANGEL I.)
M. SCHMIDTA

West Longitude
East Longitude

OYANSKIY KHREBET
Kolyma
DZHUGDZHUR KHREBET
Okhotsk
KORYAKSKIY KHREBET
KHREBET GYDAN
SOV. KAMCHATKA
KOMANDORSKIYE OSTROVA (Sov. Union)
Petropavlovsk-Kamchatskiy

Yakutsk
STANOVOY KHREBET
Nerchinsk
Blagoveshchensk
Komsomolsk
Sovetskaya Gavan
Khabarovsk
SAKHALIN (Sov. Union)
M. LOPATKA
KURIL ISLANDS (Sov. Union)
HOKKAIDO TRENCH

SEA OF OKHOTSK

MANCHURIA
HAERHPIN
CH'ANGCH'UN
SHENYANG (Mukden)
Chilin
Vladivostok
SIKHOTE ALIN'
Hakodate
Sendai
HONSHU
JAPAN

GREATER KHINGAN MTS.

KOREA
P'yongyang
SOUL
NORTH
SOUTH
KYOTO
TOKYO
YOKOHAMA

YÜAN TIENTSIN LÜTA
TSINGTAO
KITAKYUSHU KOBE OSAKA
KYUSHU
Shikoku
Nagasaki

Tsinan
NANKING
SHANGHAI
CHINA

WUHAN
EAST CHINA SEA
RYUKYUS
Tropic of Cancer

T'AIPEI
TAIWAN (FORMOSA)
Fuchou
NAN LING
Amoy
Formosa Str.
Shant'ou
CANTON
Macau
HONG KONG (Br.)
VICTORIA (Port.)

HAINAN TAO

PHILIPPINE SEA
PHILIPPINE TRENCH

LUZON
Quezon City
MANILA
MINDORO
SAMAR
LEYTE
PANAY
NEGROS
PALAWAN
MINDANAO
ARCHIPELAGO

SOUTH CHINA SEA
Hué
SAIGON

PHILIPPINES

SULU SEA
SULU IS.
Jesselton
Brunei
Sandakan
N. BORNEO
CELEBES SEA
HALMAHERA
NEW GUINEA

MALAYSIA
SARAWAK
Kuching
BORNEO
SINGAPORE
CELEBES
INDONESIA

0 100 200 400 600 800 1000 Miles
0 200 400 800 1200 1600 Kilometers

Eastern Mediterranean map

CYPRUS
Mt. Troodos 6400
OLYMPUS MTS.
Ktima
Pissouri
Episkopi Bay
CAPE GATA
Limassol
Akrotiri Bay
Larnaca
Larnaca Bay
CAPE GRECO

Longitude 35° East of Greenwich 36°

MEDITERRANEAN SEA

LEBANON
T'arabulus (Tripoli)
Al Qusayr
Al Harmal
Zaghartā
Al Batrūn
Amyūn 10,431
Jubayl (Byblos)
Jūniyah
Ba'labakk 8625
Beyrouth (Beirut)
Zahlah
Zabdānī
Ad Dāmūr
Jazzīn
Rāshayyā
Dimashq (Damascus)
Dūmā
Sayda (Sidon)
Kiswah
SYRIA
Marj 'Uyūn
Sur (Tyre)
Al Qunaytirah
As Sanamayn
Nahariya
Tibnin
Acre
Safad 3962
Haifa
Tiberias
Gader (Ruins)
Dar'a
Nazareth
Afula
Irbid
Janin
Beit Shean
Jarash
Hadera
Tūl Karm
Shomron (Ruins)
Al Mafraq
Nathanya
Nabulus
Herzlia
Petah Tiqva
As Salt
Tel Aviv-Jaffa
Az Zarqā
Rishon le Zion
Amman
Lod
Ariha (Jericho)
Rehovoth
Jerusalem
Qumran
Ma'dabā
Zuwayzā
Migdal Ashkelon
Berekhot Shelomo (Ruins) 3323
Bayt Lahm (Bethlehem)
Dhibān
Ghazzah (Gaza)
Al Khalil (Hebron)
Jad'ah
Khān Yūnis
Al Mazra
Al Qatrānah
Būr Sa'īd (Port Said)
Beersheba
Al Karak
Al Mazar
Khalig el Tina
Sabkhat al Bardawil
Tel Sharuhen (Ruins)
Sedom
Bi'r al Mazār
El 'Arish
Mamshit (Ruins)
Rummānah
Shivta (Ruins)
El Auja
At Tafilah
Kôm Defana (Ruins)
Al Qantarah
Al Qusaymah
5383 Jabal al 'Ata Itah
Jurf ad Darāwish
Al Isma'īlīyah
UNITED ARAB REPUBLIC
NEGEV
Ash Shawbak
Petrā (Ruins)
Rās Abu Qurūn GEBEL
3578 YI'ALLAR
Wādi Mūsa
Ma'ān
(EGYPT)
Qiraiya
'Abda
As Sūways (Suez)
Būr Tawfiq
An Nakhl
Al Kuntillah
Ra's an Naqb
Ash Shawbak
Ash Shidiyah
Ash Shatt
Ath Thamad 3513
Aqabat al Hijāziyah
GEBEL EL GALALA EL BAHARIYA 4136
Za'farānah
Jabal Ramm 5755
GEBEL EL 'IGMA
3789
Eilat
Al 'Aqabah
Ar Ramla
Al Mudawwarah
Abū Zanimah
Rās el Gineina 5335
Al Haql
Hallat 'Ammar
4833
Nuwaybi' al Muzayyinah
SINAI
JABAL MAZHAFAH 6232
Scale 1:3 200 000
0 10 20 30 40 50 Miles
0 20 40 60 80 Kilometers
SAUDI ARABIA
GEBEL EL GALALA EL QIBLIYA
ISRAEL
JORDAN
Gulf of Suez (Khalig es Suweis)
Gulf of Aqaba
©RMCN.

Malaysia / Singapore map

Scale 1:3 200 000
0 10 20 30 40 50 Miles
0 20 40 60 80 Kilometers

Kuala Lumpur
Port Swettenham
Klang
Kajang
SELANGOR
Bangi
Bukit Payong 3549
Kuala Klawang
Gunong Telapa
NEGRI SEMBILAN 3915 Burok
Telok Datok
Bahau
PAHANG
MALAYA
Sepang
Rantau
Seremban
Rompin
Gemas
Port Dickson
Rembau
Tampin
Segamat
Gunong Besar 3402
Endau
TIOMAN
Gunong Kajang 3444
Alor Gajah
Jasin
Mt Ophir 4187
Labis
Mersing
MALACCA
Panchor
MALAYSIA
PEMANGGIL
Malacca
JOHORE
Paloh
2002 PALAU TINGGI
MALAY
AUR
Bandar Maharani
Kluang
Gunong Blumut 3312
Rengam
PENINSULA
CAPE RACHADO
TANJONG TOHOR
Ayer Hitam
Layang Layang
Jason Bay
Djumrah
Telukletjak
Bandar Penggaram
Pontian Kechil
Kota Tinggi
SOUTH CHINA SEA
Puak
PALAU RUPAT
PALAU BENGKALIS
Johore Bahru
TANJONG RAMUNIA
Bengkalis
Ketamputih
SINGAPORE (Br.)
TANJONG BERAKIT
SUMATERA (SUMATRA)
Dedap
SINGAPORE
TANJONG PIAI
Singapore Strait
PALAU RANGSANG
GREAT KARIMUN
BATAM
PHILLIP CHANNEL
PALAU BINTAN
Pinggir
Telesung
1837
Gunung-kidjang
RIOUW ARCHIPELAGO
INDONESIA
SUMATERA TENGA
Siak Sri Indrapura
PALAU TEBINGTINGGI
Tandjungbalai
Tandjungpinang
REMPANG
341
Minas
KUNDUR
Buatam
Baranpauh
Seranggung
Selat Riouw

Longitude East of Greenwich
102° 103° 104°
©RMCN.

XXXIII

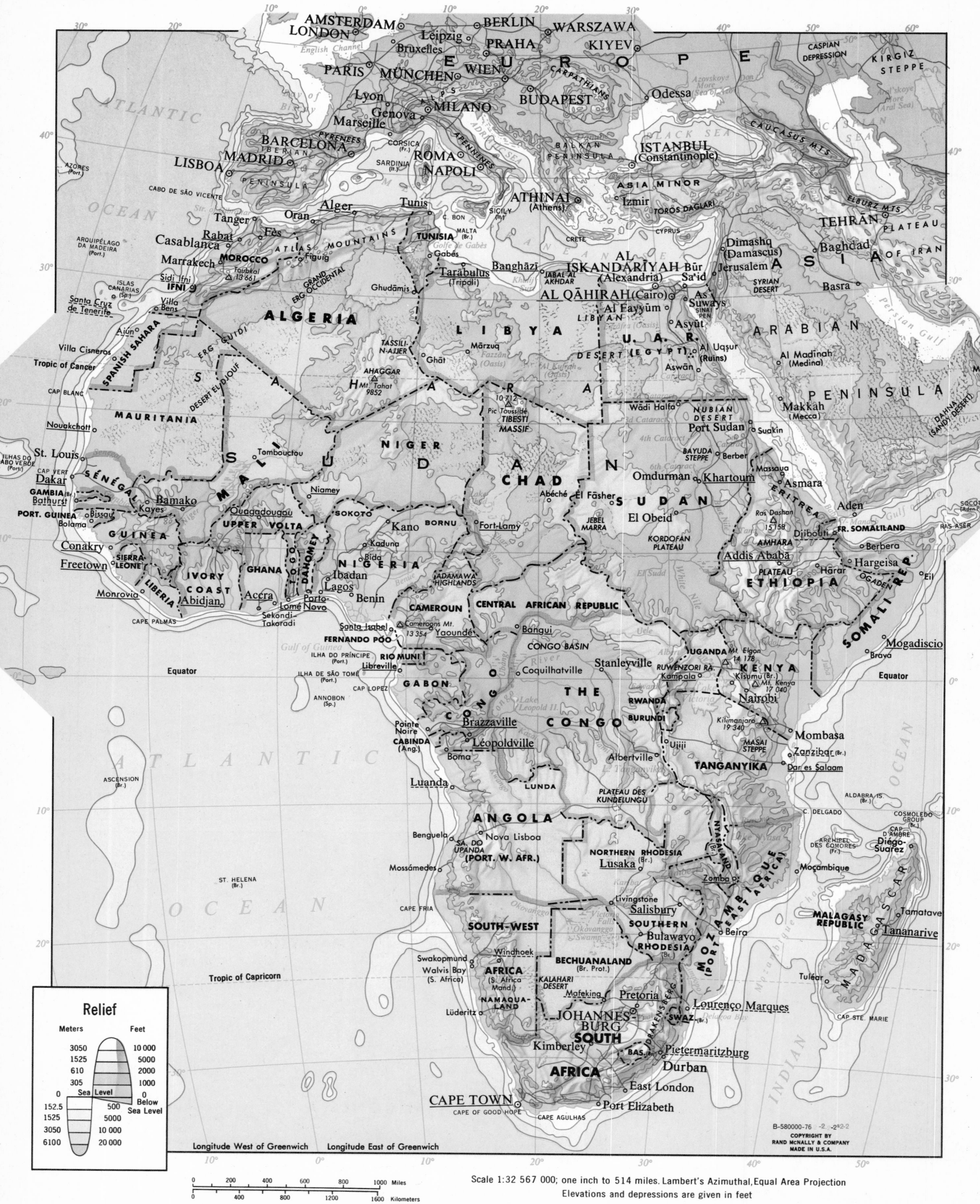

AFRICA

Relief

Meters		Feet
3050		10 000
1525		5000
610		2000
305		1000
0	Sea Level	0
		Below
152.5		500 Sea Level
1525		5000
3050		10 000
6100		20 000

Longitude West of Greenwich   Longitude East of Greenwich

0   200   400   600   800   1000 Miles
0   400   800   1200   1600 Kilometers

Scale 1:32 567 000; one inch to 514 miles. Lambert's Azimuthal, Equal Area Projection
Elevations and depressions are given in feet

B-580000-76 -2 -2*2-2
COPYRIGHT BY
RAND McNALLY & COMPANY
MADE IN U.S.A.

LA HABANA
CUBA
PEN DE YUCATÁN
Gulf de Campeche
HISPANIOLA
San Juan
JAMAICA
PUERTO RICO (U.S.A.)
GUADELOUPE (Fr.)
MARTINIQUE (Fr.)
BARBADOS (Br.)
CARIBBEAN SEA
WEST INDIES
NORTH AMERICAN BASIN
ATLANTIC OCEAN

Tropic of Cancer

PUNTA DE GALLINAS
Barranquilla
Cartagena
Maracaibo
Valencia
La Guaira
CARACAS
TRINIDAD AND TOBAGO
Port-of-Spain
CENTRAL AMÉRICA
Panamá
ISTMO DE PANAMÁ
Medellín
Bogotá
Nevado del Tolima 17,110
Mérida
Ciudad Bolívar
VENEZUELA
Cerro Icutú 7800
Georgetown
Paramaribo
Cayenne

ISLA DEL COCO (Costa Rica)
MALPELO (Colombia)
COLOMBIA
Boa Vista do Rio Branco
GUIANA (Br.) (Neth.) (Fr.)
GUIANA HIGHLANDS

Quito
Cotopaxi 19,344
ARCHIPIÉLAGO DE COLÓN (GALÁPAGOS ISLANDS) (Ec.)
ECUADOR
Chimborazo 20,577
Guayaquil
Golfo de Guayaquil
Iquitos
Leticia
Manaus (Manáos)
Belém (Pará)
São Luís (Maranhão)
ILHA DE MARAJÓ
Equator
ROCEDOS SÃO PEDRO E SÃO PAULO (Brazil)

Chiclayo
Trujillo
Nevs. Huascarán 22,205
PERU
Pôrto Velho
Rio Branco
BRAZIL
Fortaleza (Ceará)
Teresina
Natal
João Pessoa (Paraíba)
Recife (Pernambuco)
Maceió
CABO DE SÃO ROQUE
ARQUIPÉLAGO FERNANDO DE NORONHA (Brazil)

LIMA
Callao
Cuzco
El Misti 19,144
Arequipa
Mollendo
La Paz
Nev. Illimani 21,151
BOLIVIA
Sucre
Potosí
CHAPADA DE MATO GROSSO
Cuiabá
Brasília
Diamantina
Salvador (Bahia)

Iquique
DESIERTO DE ATACAMA
Antofagasta
Salta
GRAN CHACO
PARAGUAY
Asunción
SÃO PAULO
Belo Horizonte
Pico da Bandeira 9462
Vitória
ILHA DE TRINIDADE (Brazil)

ISLA DE SAN FÉLIX ISLA DE SAN AMBROSIO (Chile)
Cerro Azufre (Copiapó) Vol. 19,947
Copiapó
Tucumán
Corrientes
Santos
RIO DE JANEIRO
CABO FRIO

Coquimbo
Cerro Aconcagua 22,834
Córdoba
Santa Fe
Salto
Rosario
URUGUAY
Rio Grande
Florianópolis
Pôrto Alegre

Valparaíso
SANTIAGO
Mendoza
BUENOS AIRES
Montevideo
La Plata
ISLAS DE JUAN FERNÁNDEZ (Chile)

Concepción
PAMPAS
Valdivia
Bahía Blanca
Viedma
Golfo San Matías
Puerto Montt
ISLA DE CHILOÉ
ARCHIPIÉLAGO DE LOS CHONOS
Comodoro Rivadavia
Golfo San Jorge

Monte San Valentín 13,314
ARGENTINA
WELLINGTON
HANOVER
Río Gallegos
FALKLAND IS. (Br.)
Port Stanley

Punta Arenas
DESOLACIÓN
Mt. Sarmiento 7546
Estrecho de Magallanes
TIERRA DEL FUEGO
ISLA DE LOS ESTADOS
CABO DE HORNOS (CAPE HORN)

SOUTH GEORGIA (Br.)
SOUTH SANDWICH ISLANDS (Br.)

Drake Passage

SOUTH SHETLAND ISLANDS
SOUTH ORKNEY IS. (Br.)
JOINVILLE
PALMER PENINSULA
JAMES ROSS
Antarctic Circle

PACIFIC OCEAN
ATLANTIC OCEAN

Tropic of Capricorn

Longitude West of Greenwich

Relief

Meters		Feet
3050		10 000
1525		5000
610		2000
305		1000
0	Sea Level	0
152.5		500
1525		5000
3050		10 000
6100		20 000

0   200   400   600   800   1000 Miles
0   400   800   1200   1600 Kilometers

Scale 1:32 567 000; one inch to 514 miles. Lambert's Azimuthal, Equal Area Projection
Elevations and depressions are given in feet

CANADA

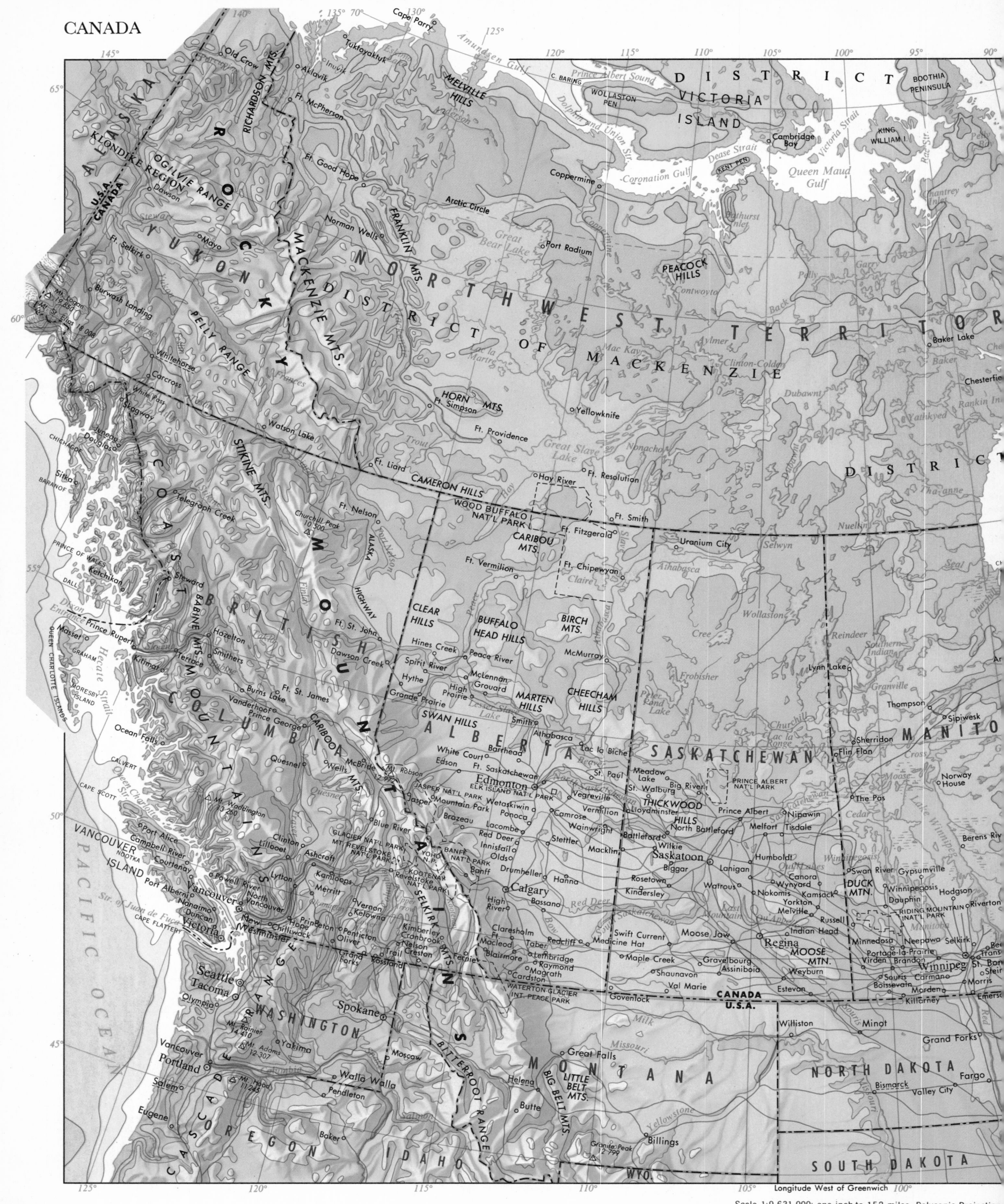

Scale 1:9 631 000; one inch to 152 miles. Polyconic Projection
Elevations and depressions are given in feet

Longitude West of Greenwich

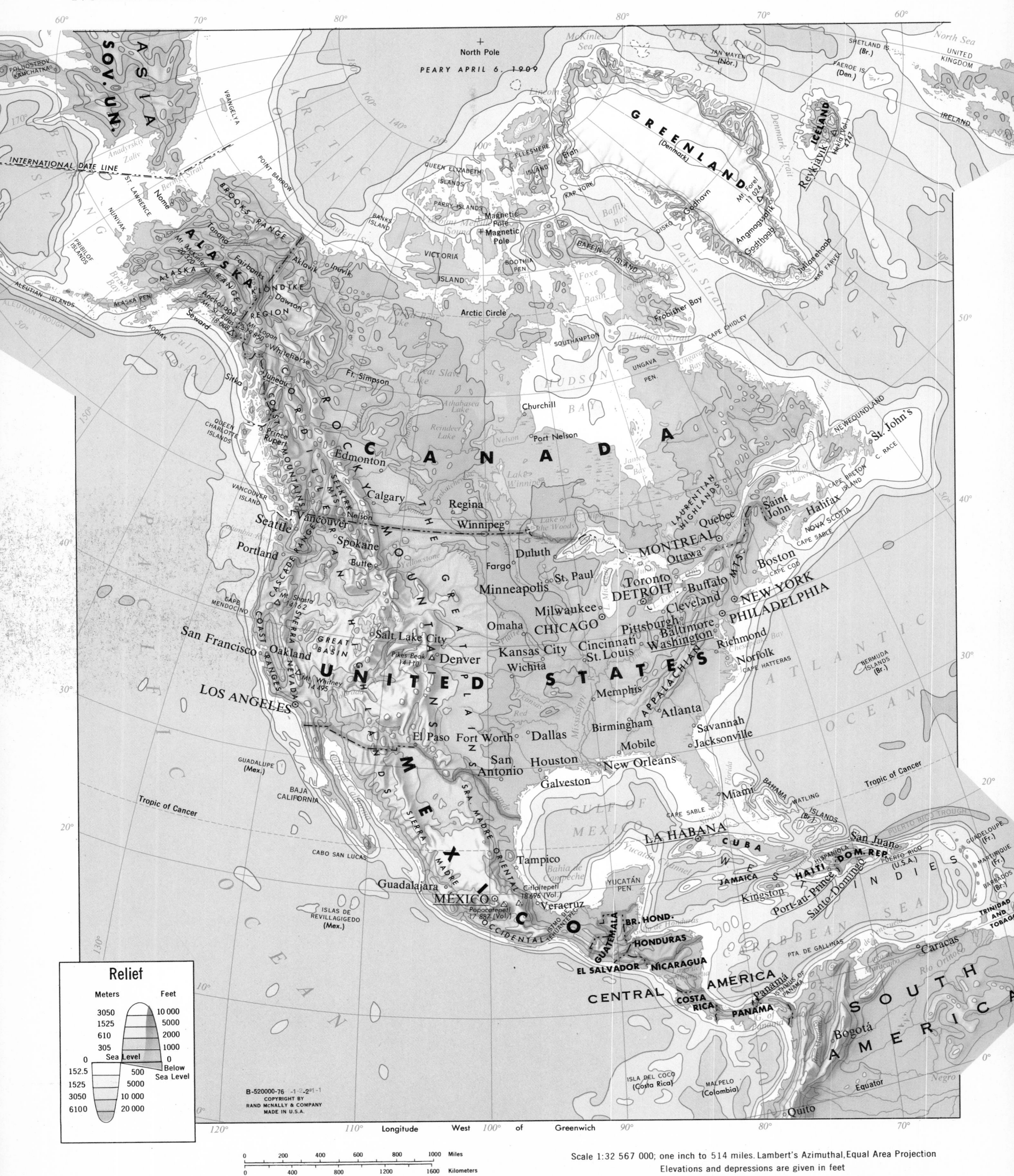

Relief

Meters | Feet
3050 | 10 000
1525 | 5000
610 | 2000
305 | 1000
Sea Level | 0
| Below
152.5 | Sea Level
| 500
1525 | 5000
3050 | 10 000
6100 | 20 000

B-520000-76   -1-2-2¹-1
COPYRIGHT BY
RAND McNALLY & COMPANY
MADE IN U.S.A.

XL

Longitude   West   100°   of   Greenwich

0   200   400   600   800   1000 Miles
0   400   800   1200   1600 Kilometers

Scale 1:32 567 000; one inch to 514 miles. Lambert's Azimuthal, Equal Area Projection
Elevations and depressions are given in feet

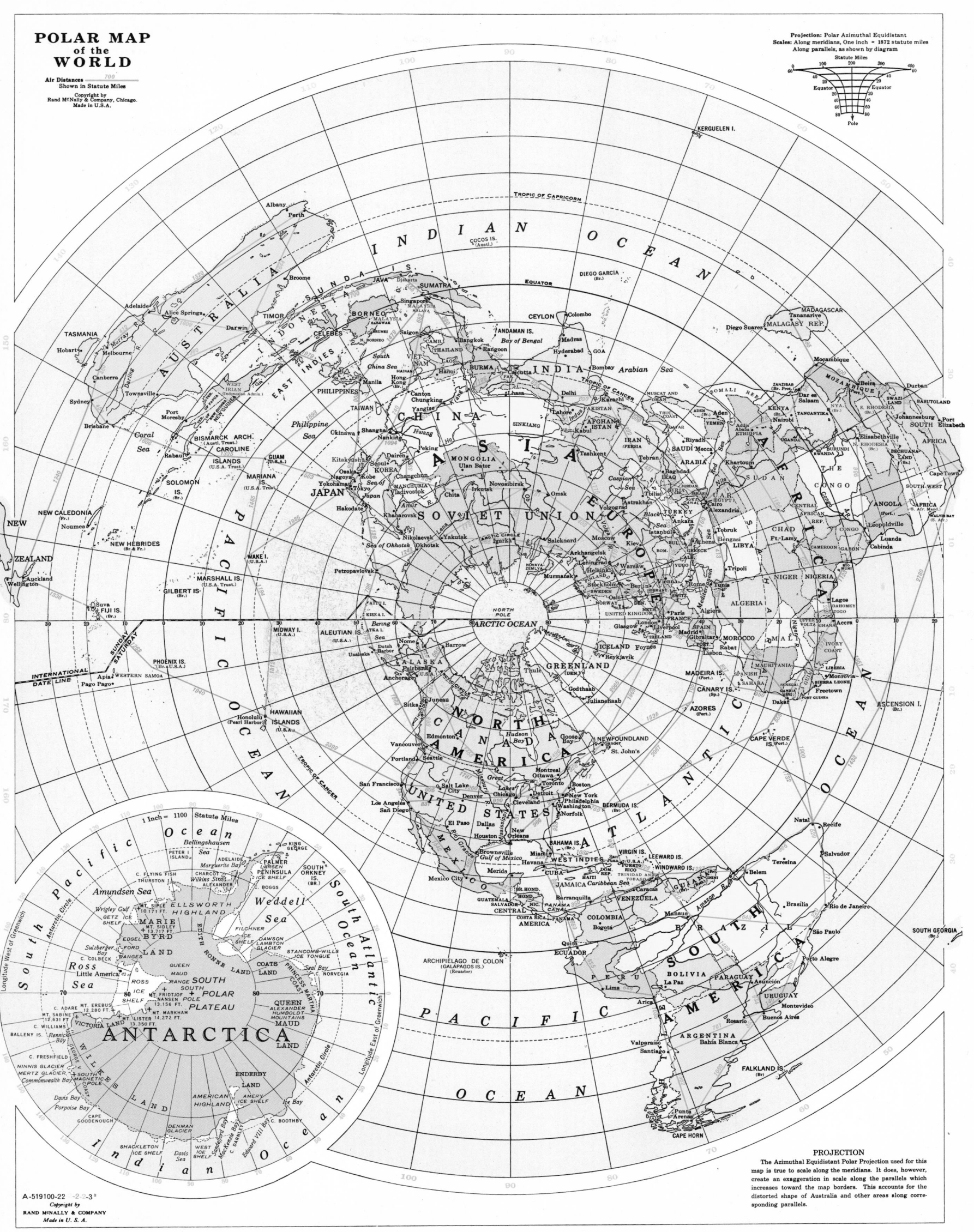

Polar Map of the WORLD

Air Distances
Shown in Statute Miles

Copyright by
Rand McNally & Company, Chicago.
Made in U.S.A.

Projection: Polar Azimuthal Equidistant
Scales: Along meridians, One inch = 1872 statute miles
Along parallels, as shown by diagram

Statute Miles

PROJECTION

The Azimuthal Equidistant Polar Projection used for this map is true to scale along the meridians. It does, however, create an exaggeration in scale along the parallels which increases toward the map borders. This accounts for the distorted shape of Australia and other areas along corresponding parallels.

A-519100-22 -2-2-3 ³
Copyright by
RAND McNALLY & COMPANY
Made in U.S.A.

1

COMPARATIVE WORLD TIME
(Legal Clock Time)

In comparing the time of one zone with another, consider the zone numbers as hours, then by subtracting find the difference in time. The lower zone number represents the earlier hour and the higher zone number the later hour. (If the difference is greater than 12 hours, subtract this difference from 24 hours to find the nearest time difference.)

The following areas have no legal time: Antarctica, interior of Greenland, northern Canadian Islands, Mongolia, Svalbard.

Cook Islands
Zone 1 + 30 min.

Tonga Is.
Zone 24 + 40 min.

Chatham Is.
Zone 24 + 15 min.

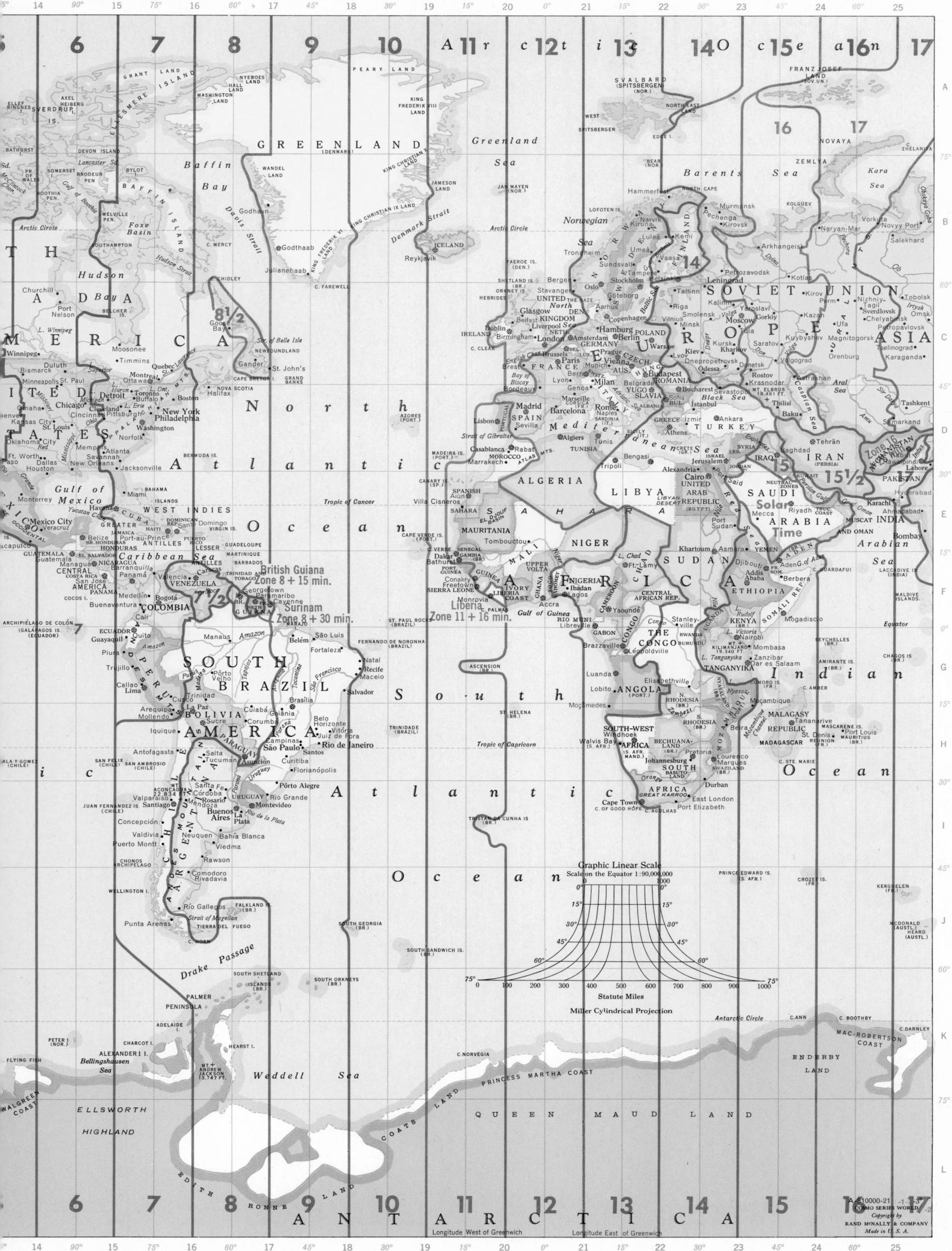

NORTH POLAR REGIONS

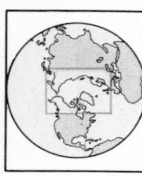

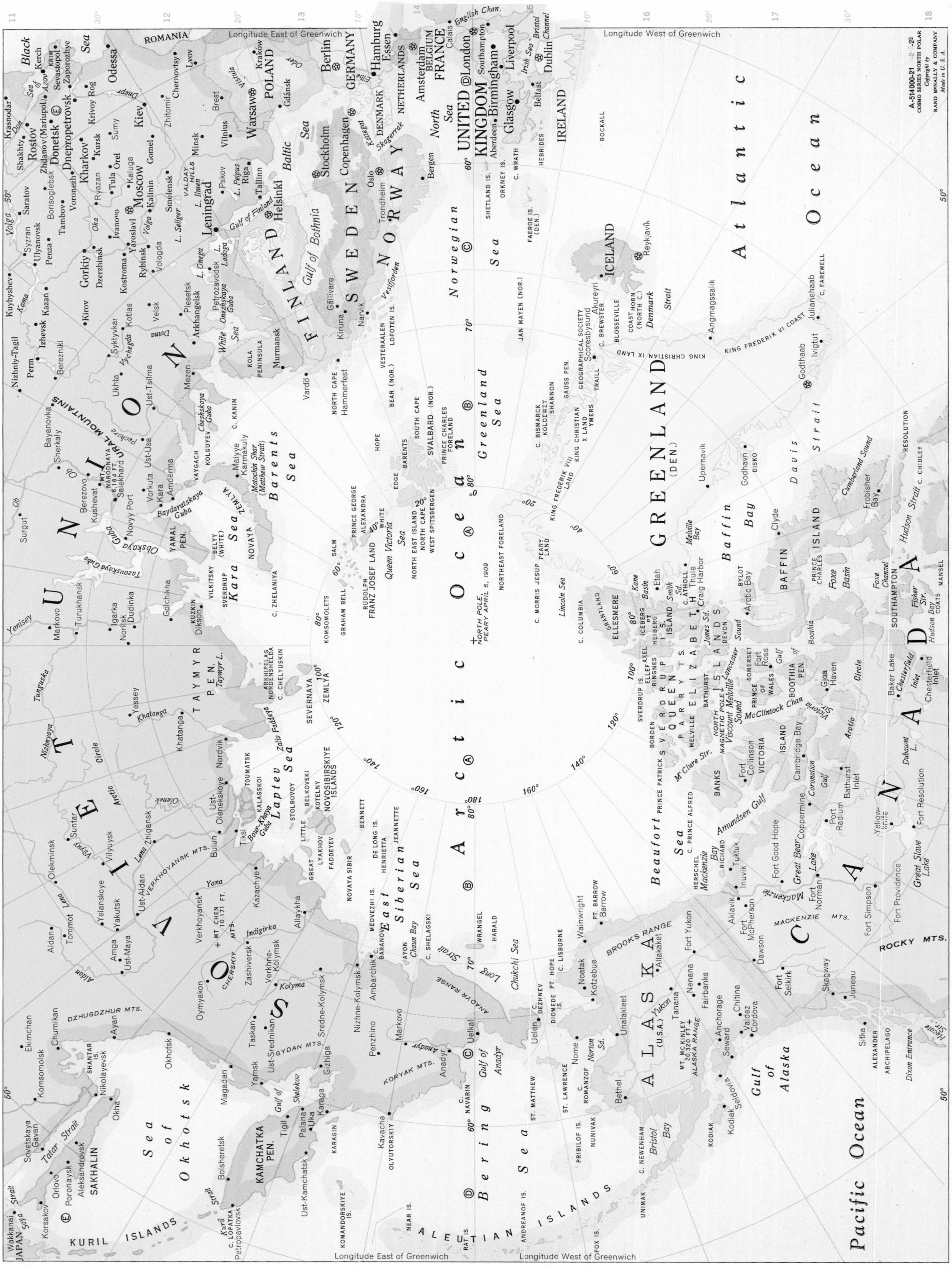

Longitude East of Greenwich
Longitude West of Greenwich
Arctic Ocean
Atlantic Ocean
Pacific Ocean
Norwegian Sea
Greenland Sea
Barents Sea
Kara Sea
Laptev Sea
East Siberian Sea
Chukchi Sea
Beaufort Sea
Bering Sea
Sea of Okhotsk
Baffin Bay
Davis Strait
Denmark Strait
North Sea
Baltic Sea
Black Sea
Hudson Bay
Gulf of Alaska
SOVIET UNION
CANADA
ALASKA (U.S.A.)
GREENLAND (DEN.)
ICELAND
NORWAY
SWEDEN
FINLAND
UNITED KINGDOM
IRELAND
SVALBARD (NOR.)
FRANZ JOSEF LAND
QUEEN ELIZABETH ISLANDS
ALEUTIAN ISLANDS
KURIL ISLANDS
URAL MOUNTAINS
BROOKS RANGE
ROCKY MTS.
NORTH POLE PEARY APRIL 6, 1909
Moscow
Leningrad
Reykjavík
Godthaab
Fairbanks
Anchorage
A-514000-21 COSMO SERIES NORTH POLAR Copyright by RAND MCNALLY & COMPANY Made in U. S. A.
Statute Miles 100 0 100 200 300 400 500
Kilometers 100 0 100 300 500 700

Lambert Azimuthal Equal Area Projection
SCALE 1:28,000,000 1 Inch = 442 Statute Miles

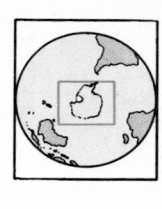

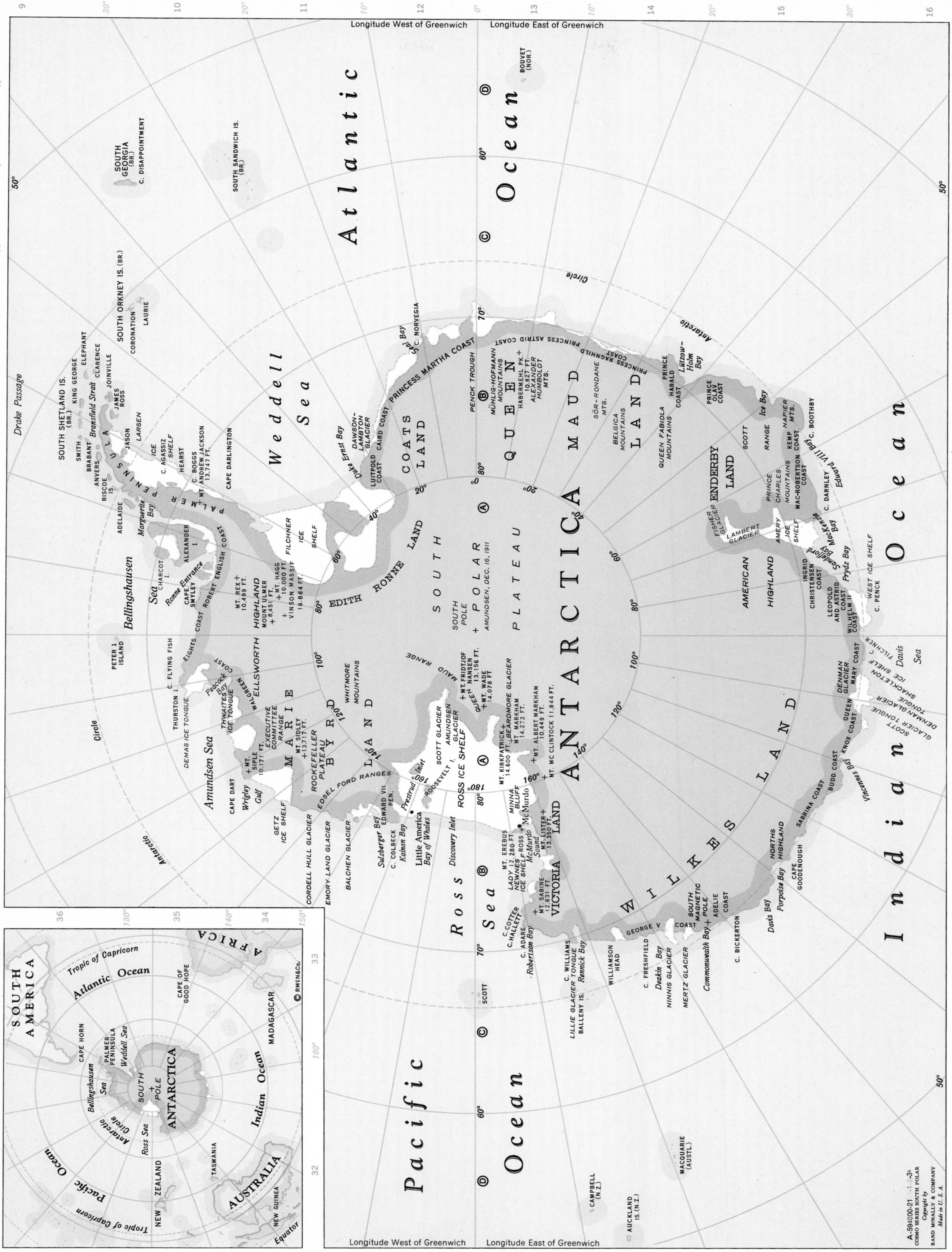

Longitude West of Greenwich Longitude East of Greenwich

BOUVET (NOR.)

A t l a n t i c   O c e a n

SOUTH GEORGIA (BR.)
C. DISAPPOINTMENT
SOUTH SANDWICH IS. (BR.)

Antarctic   Circle

SOUTH ORKNEY IS. (BR.)
CORONATION
LAURIE
ELEPHANT
CLARENCE
JOINVILLE
JAMES ROSS
SMITH
KING GEORGE
SOUTH SHETLAND IS. (BR.)
BRANSFIELD STRAIT
JASON
BRABANT
ANVERS
BISCOE IS.
LARSEN
ICE SHELF
C. AGASSIZ
HEARST
MT. ANDREW JACKSON 13,747 FT.
CAPE DARLINGTON

W e d d e l l   S e a

Drake Passage

Bellingshausen Sea

ADELAIDE
Marguerite Bay
CHARCOT
ALEXANDER I.
CAPE ROBERT ENGLISH COAST
SMYLEY
Ronne Entrance

FILCHNER ICE SHELF
RONNE LAND
EDITH

HIGHLAND
MOUNT ULMER +8,451 FT.
MT. HAGG +10,000 FT.
VINSON MASSIF 16,864 FT.
MT. REX + 10,499 FT.

C. FLYING FISH
PETER I ISLAND
Antarctic Circle
EIGHTS
THURSTON I.
Demas Ice Tongue
WALGREEN COAST
ELLSWORTH

CAPE DART
Wrigley Gulf
GETZ ICE SHELF
+ MT. SIPLE 10,171 FT.
EXECUTIVE COMMITTEE RANGE
MT. SIDLEY 13,717 FT.

Amundsen Sea

M A R I E   B Y R D   L A N D

WHITMORE MOUNTAINS
ROCKEFELLER PLATEAU
EDSEL FORD RANGES

CORDELL HULL GLACIER
BALCHEN GLACIER
C. COLBECK
SULZBERGER BAY
EDWARD VII PEN.
Kainan Bay
Little America
Bay of Whales

R o s s   S e a

Discovery Inlet
Prestrud Inlet
ROOSEVELT I.
SCOTT GLACIER
AMUNDSEN GLACIER
+ MT. FRIDTJOF NANSEN 13,156 FT.
QUEEN MAUD RANGE
MT. WADE 14,078 FT.
MT. KIRKPATRICK 14,600 FT.
BEARDMORE GLACIER
MT. MARKHAM 14,272 FT.
MT. ALBERT MARKHAM 10,449 FT.
MT. MC CLINTOCK 11,844 FT.

ROSS ICE SHELF

MINNA BLUFF
McMurdo Sound
MT. EREBUS 12,280 FT.
McMurdo
LADY NEWNES BAY
MT. LISTER 13,350 FT.
MT. SABINE 12,631 FT.
C. COTTER
C. HALLETT
C. ADARE
Robertson Bay

V I C T O R I A   L A N D

A N T A R C T I C A

S O U T H   P O L E   &   P O L A R   P L A T E A U
SOUTH POLE
+ AMUNDSEN, DEC. 16, 1911

PENCK TROUGH
MÜHLIG-HOFMANN MOUNTAINS
HABERMEHL PK. + 10,827 FT.
ALEXANDER HUMBOLDT MTS.
SÖR-RONDANE MTS.
BELGICA MOUNTAINS
QUEEN FABIOLA MOUNTAINS

Q U E E N   M A U D   L A N D

PRINCESS RAGNHILD COAST
PRINCESS ASTRID COAST
PRINCE HARALD COAST
PRINCE OLAV COAST
Lützow-Holm Bay
SCOTT
RANGE
NAPIER MTS.
C. BOOTHBY
C. DARNLEY

E N D E R B Y   L A N D

PRINCE CHARLES MOUNTAINS
FISHER GLACIER
LAMBERT GLACIER
AMERY ICE SHELF
MAC-ROBERTSON COAST
INGRID CHRISTENSEN COAST
Prydz Bay
LEOPOLD AND ASTRID COAST
WILHELM II COAST
C. PENCK
WEST ICE SHELF

AMERICAN HIGHLAND

D a v i s   S e a
SHACKLETON ICE SHELF
DENMAN GLACIER
QUEEN MARY COAST
KNOX COAST
Vincennes Bay
BUDD COAST
SABRINA COAST
SCOTT GLACIER TONGUE
Davis Bay

W I L K E S   L A N D

NORTHS HIGHLAND
CAPE GOODENOUGH
SOUTH MAGNETIC POLE
George V COAST
ADÉLIE COAST
C. BICKERTON
Commonwealth Bay
Porpoise Bay
WILLIAMSON HEAD
C. FRESHFIELD
Deakin Bay
NINNIS GLACIER
MERTZ GLACIER
LILLIE GLACIER TONGUE
BALLENY IS.
C. WILLIAMS
Remnick Bay

I n d i a n   O c e a n

P a c i f i c   O c e a n

CAMPBELL (N.Z.)
AUCKLAND IS. (N.Z.)
SCOTT
MACQUARIE (AUSTL.)

COATS LAND
PRINCESS MARTHA COAST
CAIRD COAST
LUITPOLD COAST
DAWSON-LAMBTON GLACIER
Duke Ernst Bay
Card Bay
C. NORVEGIA

SOUTH POLAR REGIONS

Lambert Azimuthal Equal Area Projection
SCALE 1:28,000,000   1 Inch = 442 Statute Miles

Statute Miles 100 0 100 200 300 400 500
Kilometers 100 0 100 300 500 700

SOUTH AMERICA
Atlantic Ocean
Tropic of Capricorn
CAPE OF GOOD HOPE
AFRICA
MADAGASCAR
CAPE HORN
PALMER PENINSULA
Bellingshausen Sea
Weddell Sea
ANTARCTICA
SOUTH POLE +
Ross Sea
Antarctic Circle
Indian Ocean
NEW ZEALAND
TASMANIA
AUSTRALIA
NEW GUINEA
Equator
Tropic of Capricorn
Pacific Ocean

© RM N & Co.

A-594000-21 -1.-3ᵇ
COSMO SERIES SOUTH POLAR
Copyright by
RAND McNALLY & COMPANY
Made in U.S.A.

Longitude West of Greenwich Longitude East of Greenwich

5

6

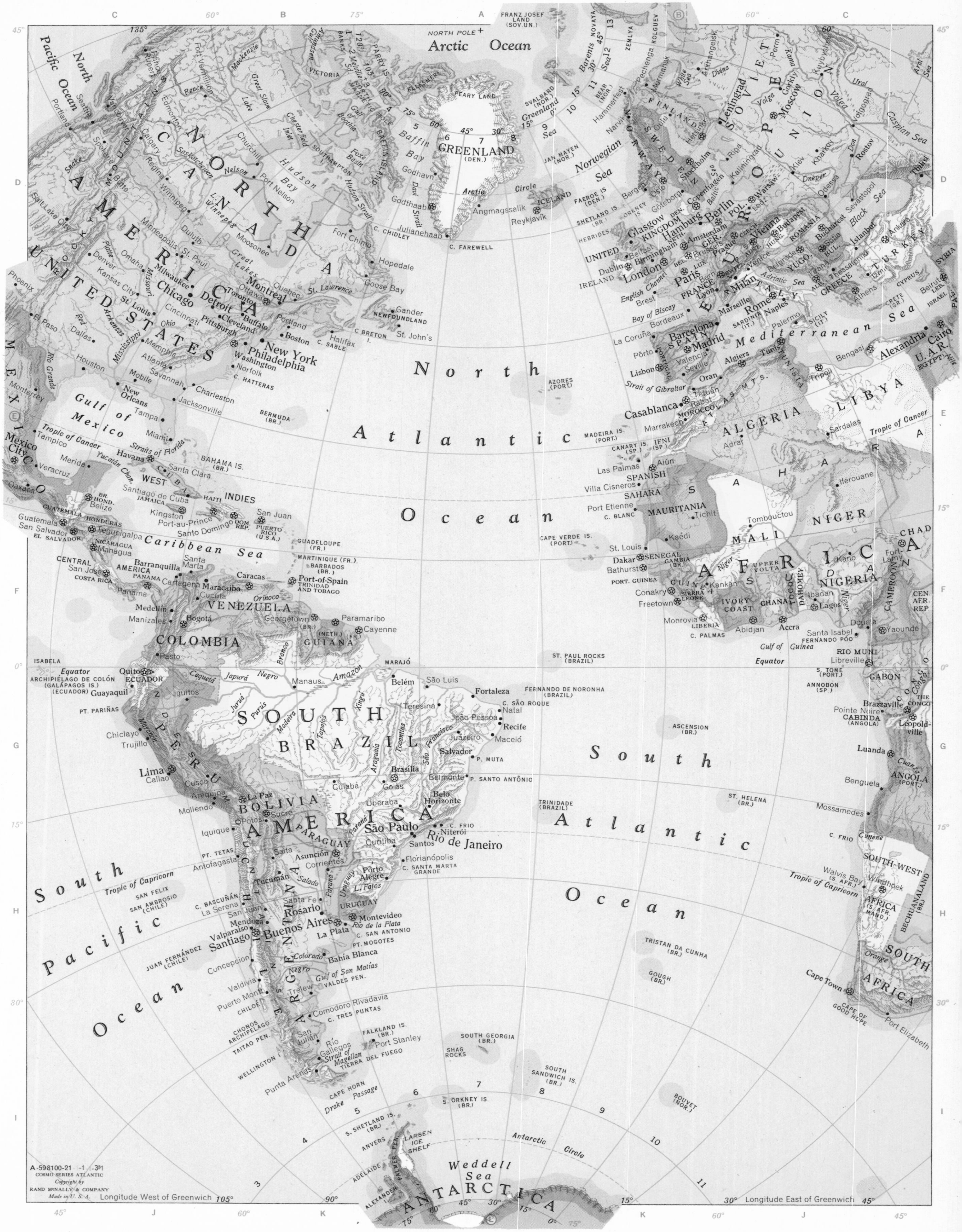

Statute Miles 200 0 200 600 1000 1400
Kilometers 200 0 200 600 1000 1400 1800 2200

Modified Polyconic Projection
SCALE 1:60,728,000 1 Inch = 960 Statute Miles

A-598100-21 -1 -3³¹
COSMO-SERIES ATLANTIC
Copyright by
RAND McNALLY & COMPANY
Made in U.S.A.

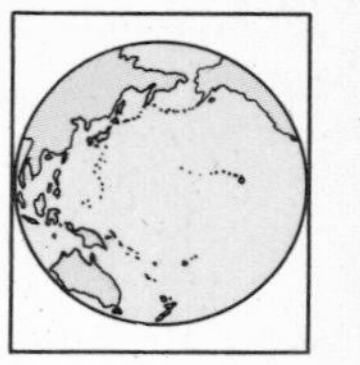

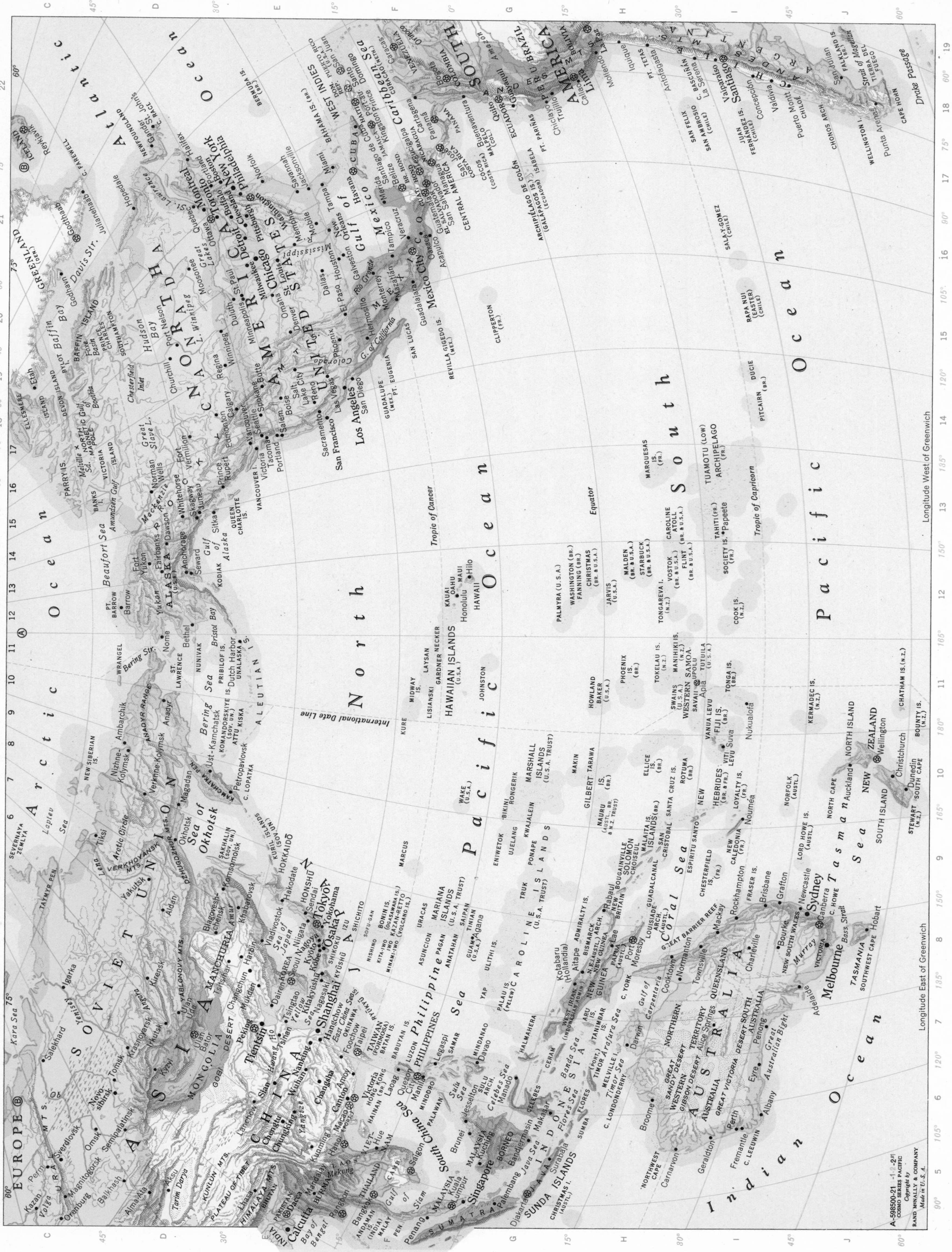

Modified Secant Conic Projection
SCALE 1:66,800,000   1 Inch = 1,040 Statute Miles

Statute Miles 200 0   200   600   1000   1400
Kilometers   200 0   200   600 1000 1400 1800 2200

A-598500-21   -1-: 2P1
COSMO SERIES PACIFIC
Copyright by
RAND M9NALLY COMPANY
Made in U.S.A.

Ocean Sea
Kara Sea
NOVAYA
ZEMLYA
VAYGACH
Baydaratskaya Guba
YAMAL PEN.
Obskaya Guba
Longitude East of Greenwich
CENTRAL SIBERIAN UPLANDS

C. KANIN
Cheshskaya Indiga Guba
Pechorskaya Guba
KOLGUEV ISLAND
Novyy Port
Arctic Circle
Nyda
Tarko-Sale
Taz
Krasnoselkup
Yenisey
Tunguska
Turukhansk

KOLA PENINSULA
C. KANIN
Mezenskaya Guba
White Sea
Duinskaya Guba
Arkhangelsk
Naryan-Mar
Amderma
Kara
Vorkuta
Salekhard
Muzhi
Kushevat
Nadym
Yenisey

Onega
Koslan
Mezen
Ust-Tsilma
Kozhva
Ukhta
Berezovo
Khanty-Mansiysk
Surgut
Ob
Yelizarovo
Sherkaly
Yakh

WEST SIBERIAN SOCIALIST REPUBLIC
SOVIET FEDERATED
RUSSIAN
SOVIET

Vologda
Plesetsk
Nyandoma
Velsk
Syktyvkar
Kerchemya
Krasnovishersk
Ural
Ivdel
Serov
Severouralsk
Karpinsk
Tobolsk
Irtysh
Tomsk
Yurga

Rybinsk Reservoir
Buy
Yaroslavl
Kostroma
Kirov
Glazov
Perm
Sverdlovsk
Tyumen
Tara
Omsk
Novosibirsk

MOSCOW
Moskva
Gorkiy
Kazan
Izhevsk
Ufa
Chelyabinsk
Kurgan
Petropavlovsk
Kokchetav

Tula
Ryazan
Penza
Ulyanovsk
Kuybyshev
Orenburg (Chkalov)
Orsk
Kustanay

Orel
Lipetsk
Tambov
Saratov
Uralsk
Aktyubinsk
Turgay

Kursk
Voronezh
Borisoglebsk
Balashov

Kharkov
Volgograd
Astrakhan
Guryev
Aralsk
Aral Sea

KAZAKHSTAN S.S.R.
Karaganda
Lake Balkhash
Alma-Ata
Frunze

Dnepropetrovsk
Lugansk
Donetsk
Rostov
Sea of Azov
Kerch

Caspian Sea
UST-URT PLATEAU
Nukus
KYZYL-KUM (DESERT)
Tashkent
UZBEK S.S.R.
KIRGHIZ S.S.R.

Krasnodar
Stavropol
Grozny
Ordzhonikidze
Baku
Krasnovodsk
TURKMEN S.S.R.
KARA-KUM (DESERT)
Ashkhabad
Samarkand
TADZHIK S.S.R.

GEORGIA
Tbilisi
Batumi
ARMENIA
Yerevan
AZERBAIDZHAN
Lenkoran
Atrek
Mashhad
HINDU KUSH
Kabul
Rawalpindi

TURKEY
ASIA MINOR
Erzurum
MT. ARARAT
Tabriz
L. Urmia
Rasht
Tehrān
ELBURZ
Semnan
Herāt
AFGHANISTAN
Kandahar
Quetta
PAKISTAN

SYRIA
Aleppo (Halab)
Mosul
Kirkuk
Hamadān
IRAN (PERSIA)
PLATEAU
DASHT-I-KAVIR (DESERT)
DASHT-I-LUT (DESERT)
Kermān
Zāhedān

LEBANON
Damascus (Dimashq)
Baghdad
IRAQ
Karbala
Esfahān
Yazd
Shīrāz
Bam
Hyderabad

JORDAN
Amman
SYRIAN DESERT
SAUDI ARABIA
NEUTRAL ZONE
KUWAIT
Kuwait
Basra
Abadan
Bandar Abbas
Karachi
INDIA

Conic Projection
SCALE 1:16,000,000   1 Inch = 252 Statute Miles

Statute Miles 25 0 25 50 75
Kilometers 25 0 25 50 100

Conic Projection
SCALE 1:4,000,000   1 Inch = 63 Statute Miles

1 Inch = 8 Statute Miles

COSMO SERIES BRITISH ISLES
Copyright by
RAND McNALLY & COMPANY
Made in U.S.A.
A-553600-21

A-551700-21 -1 -2"-1
COSMO SERIES IRELAND
Copyright by
RAND McNALLY & COMPANY
Made in U.S.A.

Longitude West of Greenwich

Lambert Conformal Conic Projection
SCALE 1 : 2,000,000 1 Inch = 32 Statute Miles

Statute Miles 5 0 5 10 20 30 40 50
Kilometers 5 0 5 10 20 30 40 50 60

Statute Miles 5 0 5 10 20 30 40 50
Kilometers 5 0 5 10 20 30 40 50 60

Lambert Conformal Conic Projection
SCALE 1:2,000,000 1 Inch = 32 Statute Miles

COSMO SERIES ENGLAND
Copyright by
RAND McNALLY & COMPANY
Made in U.S.A.
A-553292-21 -1 -2½

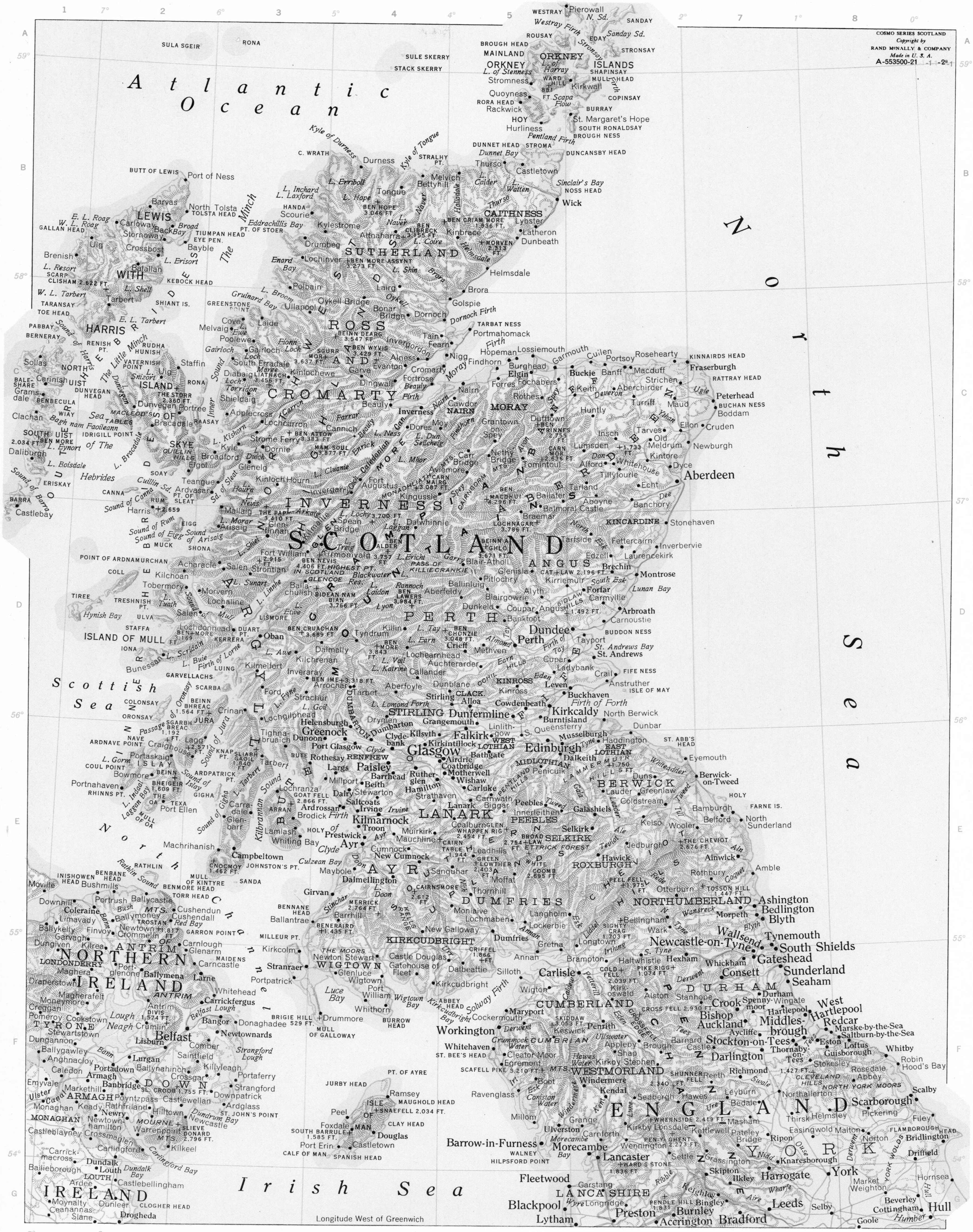

SCOTLAND AND NORTHERN ENGLAND

COSMO SERIES SCOTLAND
Copyright by
RAND McNALLY & COMPANY
Made in U.S.A.
A-553500-21 -1 -2ᴿ

Lambert Conformal Conic Projection
SCALE 1 : 2,000,000   1 Inch = 32 Statute Miles

Statute Miles

Kilometers

Longitude West of Greenwich

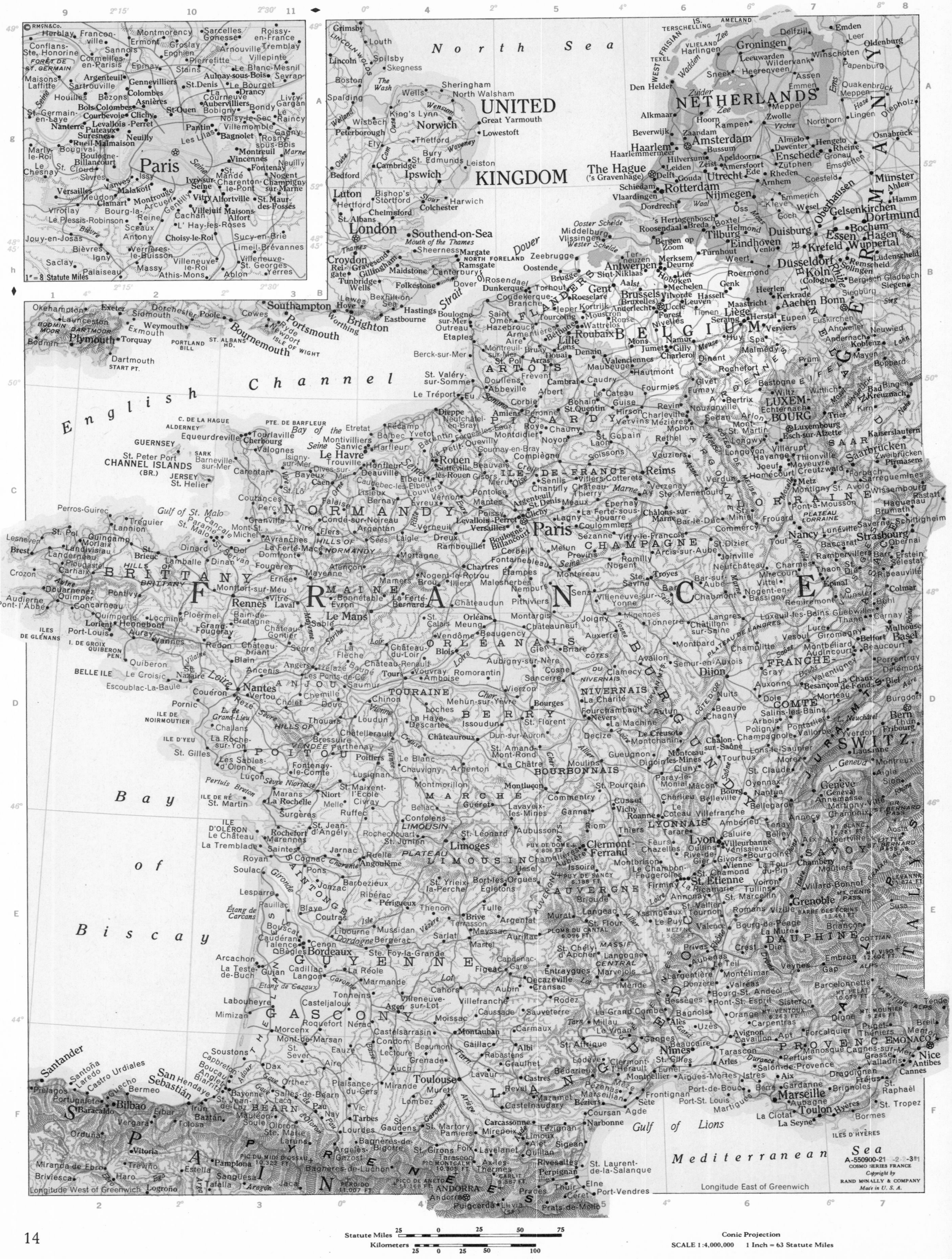

FRANCE AND THE LOW COUNTRIES

Statute Miles

Kilometers

Conic Projection

SCALE 1 : 4,000,000   1 Inch = 63 Statute Miles

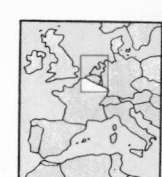

GERMANY, AUSTRIA AND SWITZERLAND

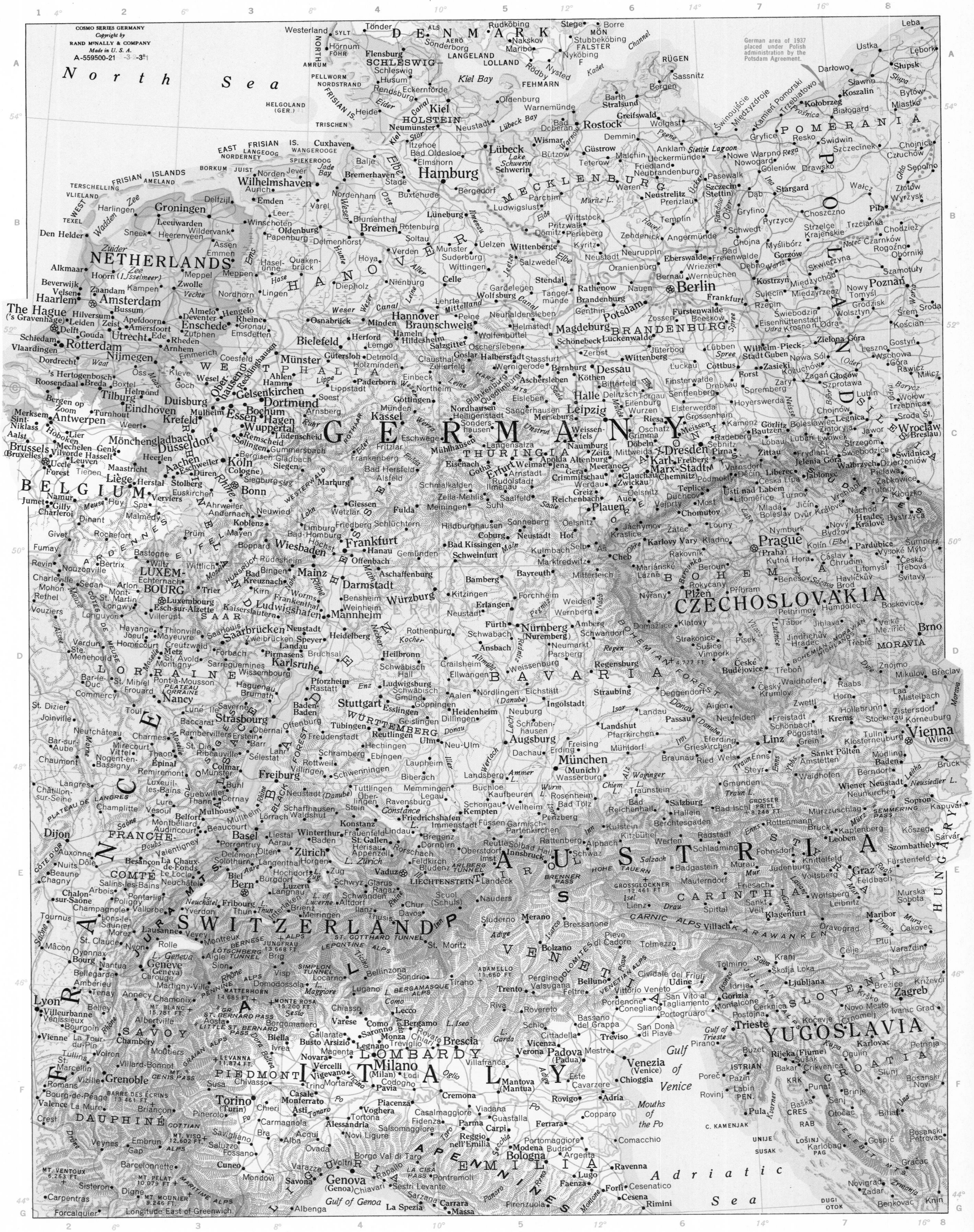

COSMO SERIES GERMANY
Copyright by
RAND McNALLY & COMPANY
Made in U.S.A.
A-559500-21 -3 -3⁵1

Lambert Conformal Conic Projection
SCALE 1:2,000,000   1 Inch = 32 Statute Miles

Statute Miles 5 0 5 10 20 30 40 50

Kilometers 5 0 5 10 20 30 40 50 60

Statute Miles 5 0 5 10 20 30 40 50
Kilometers 5 0 5 10 20 30 40 50 60

Lambert Conformal Conic Projection
SCALE 1:2,000,000   1 Inch = 32 Statute Miles

Lambert Conformal Conic Projection
SCALE 1: 1,100,000   1 Inch = 17 Statute Miles

Statute Miles
Kilometers

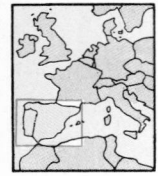

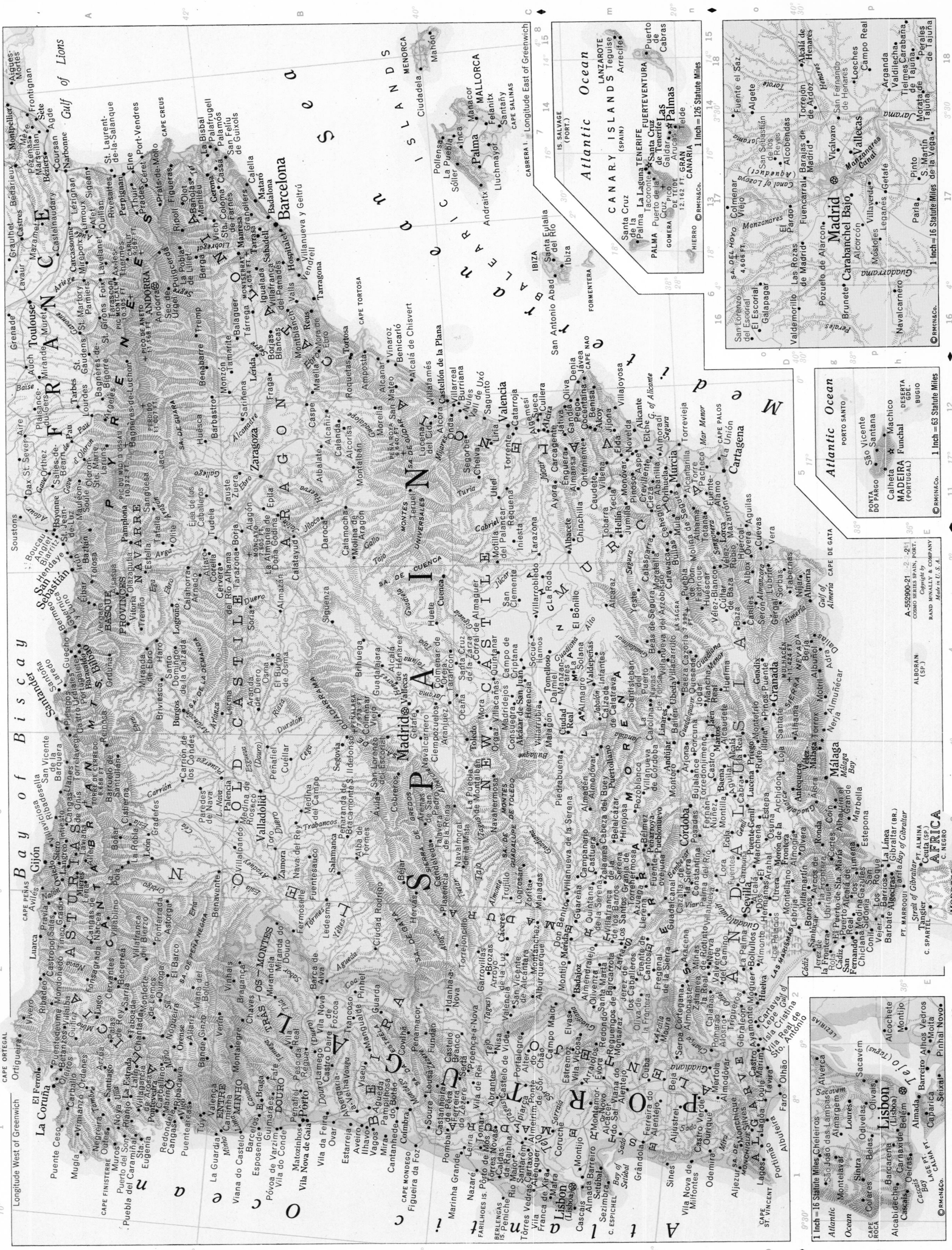

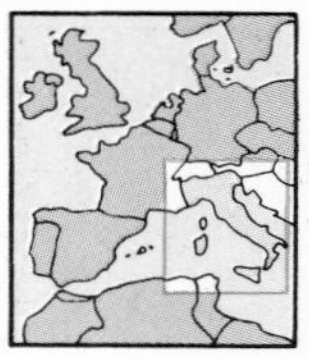

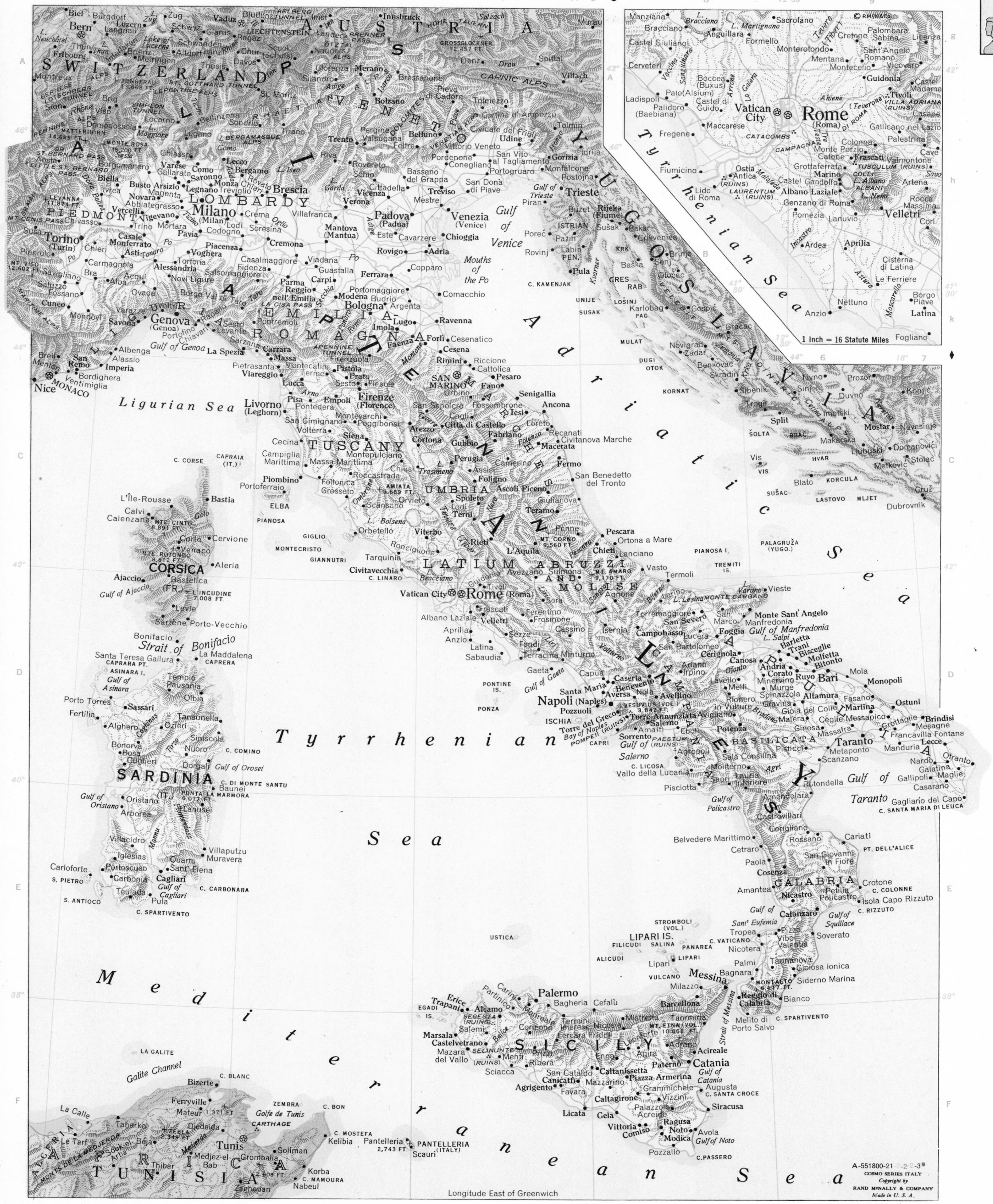
SWITZERLAND
AUSTRIA
LIECHTENSTEIN
YUGOSLAVIA
ALPS
APENNINES
LOMBARDY
PIEDMONT
EMILIA-ROMAGNA
TUSCANY
UMBRIA
LATIUM
ABRUZZI AND MOLISE
CAMPANIA
APULIA
BASILICATA
CALABRIA
SICILY
SARDINIA
CORSICA (FR.)
TUNISIA
ALGERIA
SAN MARINO
MONACO
Vatican City
Rome (Roma)
Milano (Milan)
Torino (Turin)
Genova (Genoa)
Venezia (Venice)
Firenze (Florence)
Napoli (Naples)
Palermo
Bari
Bologna
Catania
Messina
Trieste
Tunis
Adriatic Sea
Tyrrhenian Sea
Ligurian Sea
Mediterranean Sea
Gulf of Venice
Gulf of Genoa
Gulf of Taranto
Strait of Bonifacio
Strait of Messina
Galite Channel
Gulf of Manfredonia
LIPARI IS.
Tyrrhenian Sea
Vatican City
Rome (Roma)
1 Inch = 16 Statute Miles
© RMcN&Co.
Longitude East of Greenwich
A-551800-21
COSMO SERIES ITALY
Copyright by RAND McNALLY & COMPANY
Made in U. S. A.

Conic Projection
SCALE 1:4,000,000 1 Inch = 63 Statute Miles

Statute Miles 25 0 25 50 75
Kilometers 25 0 25 50 100

YUGOSLAVIA, HUNGARY, ROMANIA AND BULGARIA

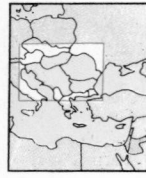

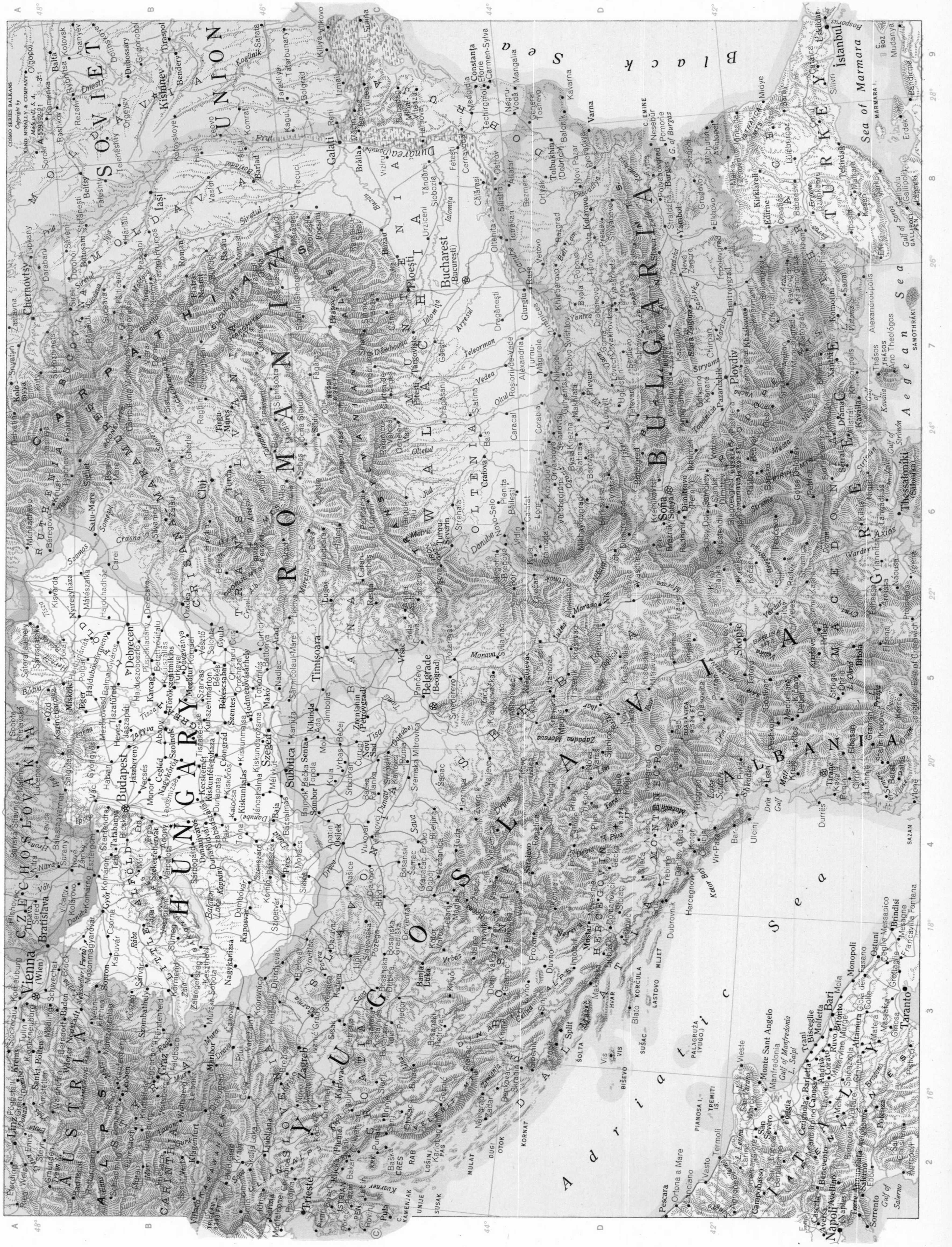

Statute Miles 25 0 25 50 75

Kilometers 25 0 25 50 100

Conic Projection

SCALE 1:4,000,000 1 Inch = 63 Statute Miles

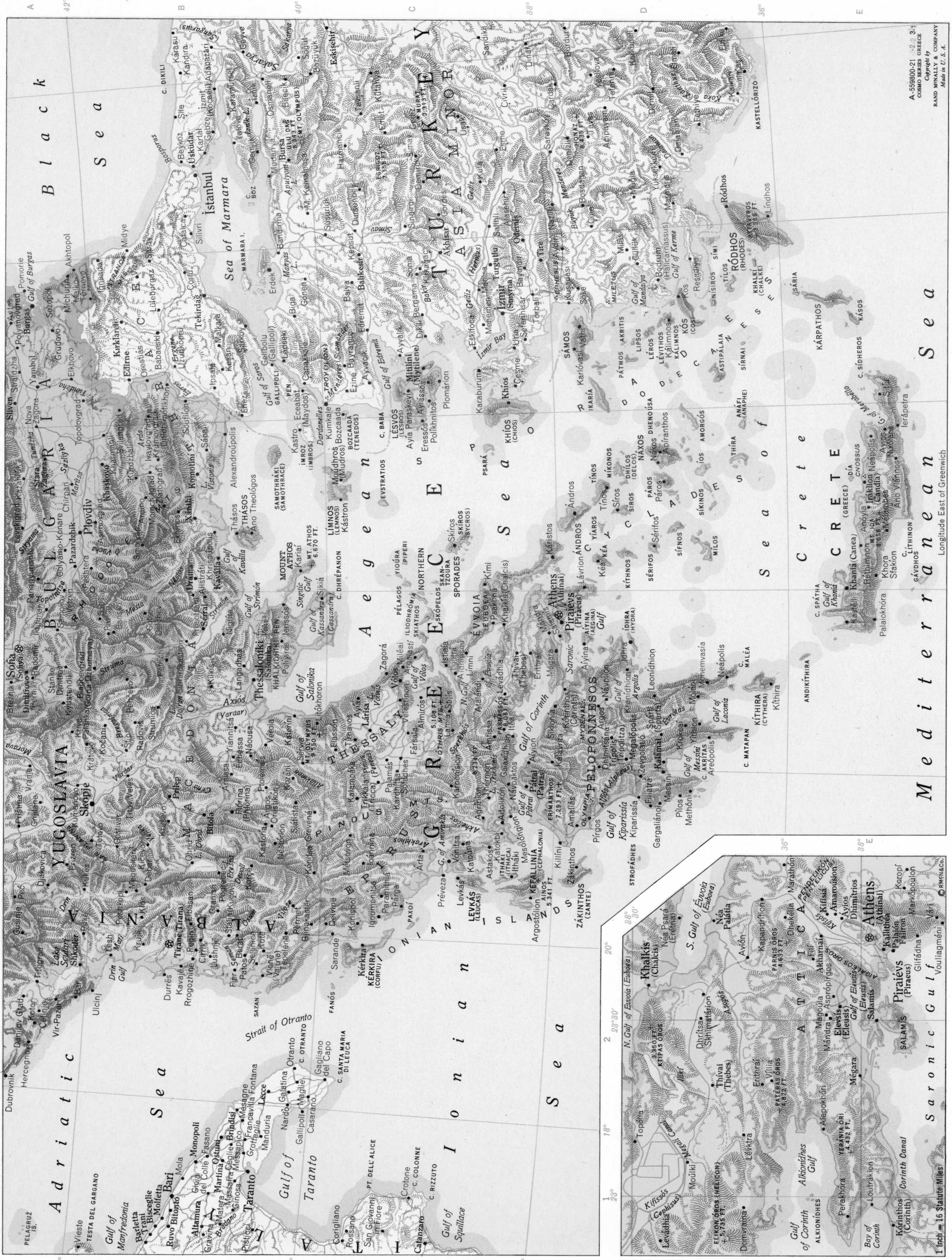

A-559800-21 -22.) 3.7
COSMO SERIES GREECE
Copyright by
RAND M°NALLY & COMPANY
Made in U.S.A.

Conic Projection
SCALE 1:4,000,000   1 Inch = 63 Statute Miles

Statute Miles
Kilometers

23

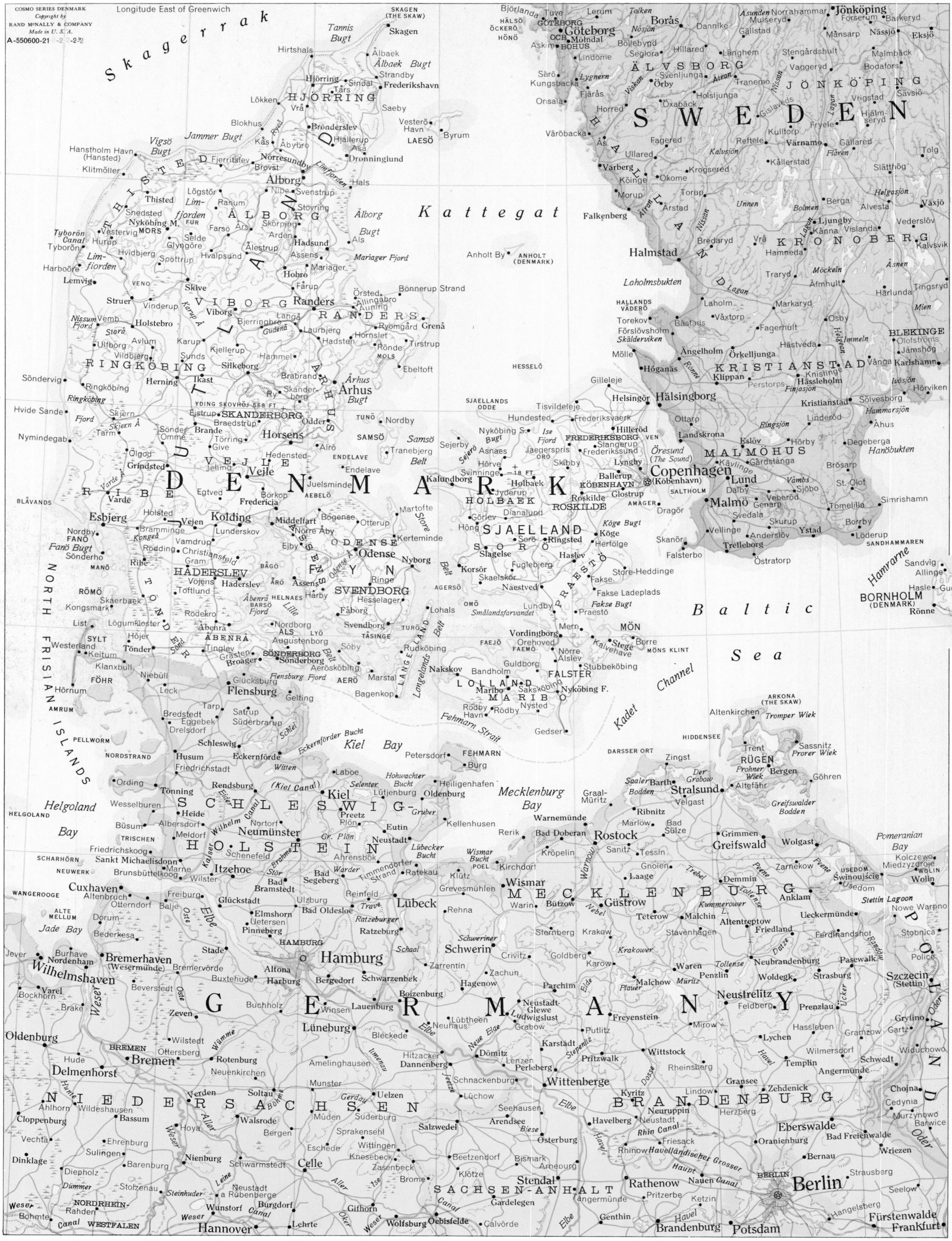

COSMO SERIES DENMARK
Copyright by
RAND McNALLY & COMPANY
Made in U.S.A.
A-550600-21

Longitude East of Greenwich

Skagerrak

Kattegat

SWEDEN

JÖNKÖPING

ÄLVSBORG

KRONOBERG

KRISTIANSTAD

BLEKINGE

MALMÖHUS

DENMARK

JYLLAND

ÅLBORG

VIBORG

RANDERS

RINGKÖBING

ÅRHUS

SKANDERBORG

VEJLE

RIBE

HADERSLEV

FYN

ODENSE

SVENDBORG

SJAELLAND

HOLBAEK

ROSKILDE

SORÖ

Copenhagen (København)

FREDERIKSBORG

LOLLAND

MARIBO

FALSTER

MÖN

BORNHOLM
(DENMARK)

Baltic Sea

NORTH FRISIAN ISLANDS

SCHLESWIG-HOLSTEIN

MECKLENBURG

RÜGEN

Kiel Bay

Lübeck

HAMBURG

BREMEN

GERMANY

NIEDERSACHSEN

NORDRHEIN-WESTFALEN

SACHSEN-ANHALT

BRANDENBURG

POLAND

Hamburg

Bremen

Hannover

Berlin

24

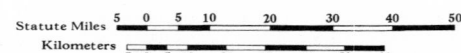

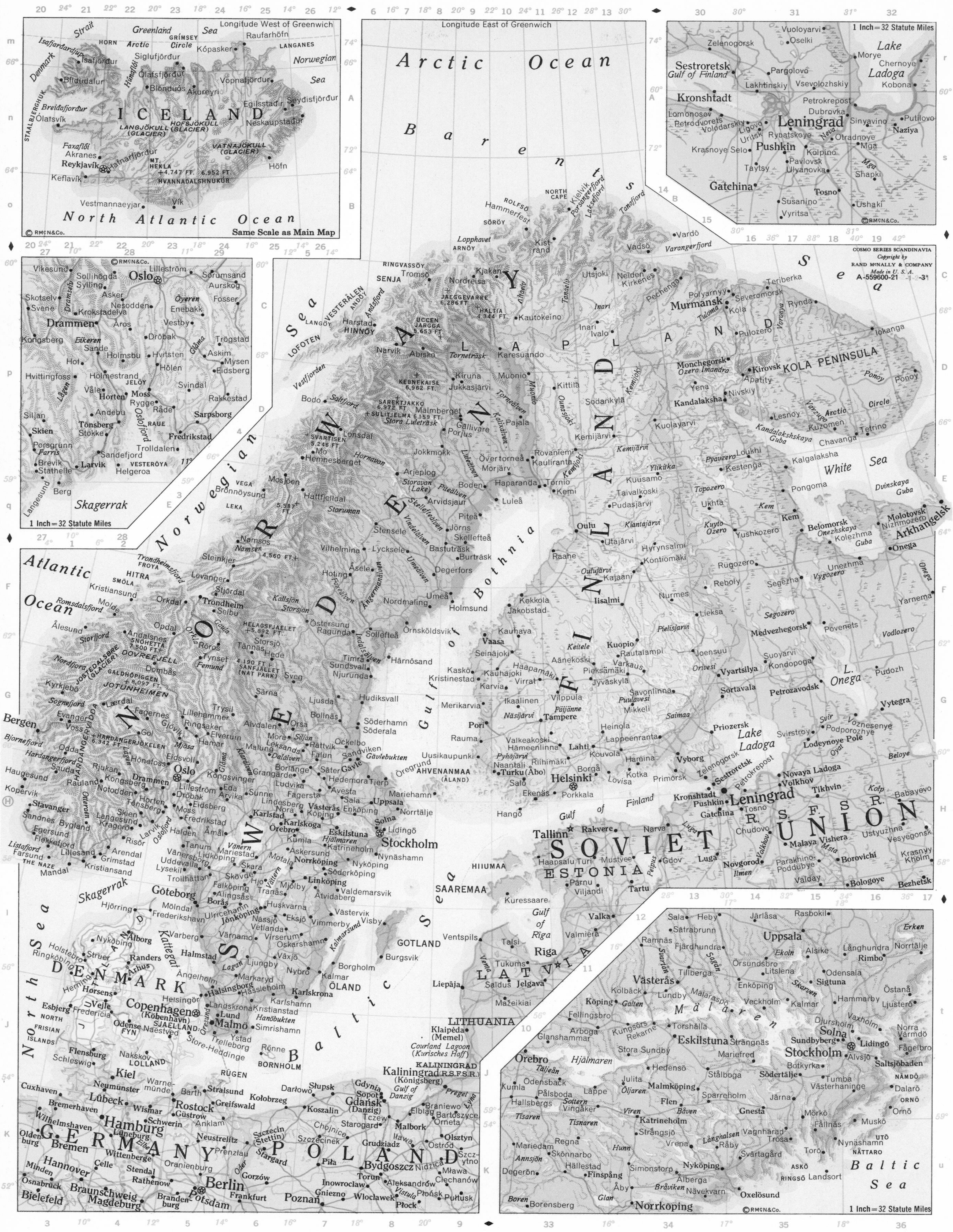

Zawiercie

Chorzów · Bytom · Katowice · Gliwice

SILESIA

Bielsko-Biała · Oświęcim

Warsaw (Warszawa) · Okęcie · Pruszków · Grodzisk Mazowiecki · Mińsk Mazowiecki

Praga · Zižkov · Prague (Praha) · Smíchov

1 Inch = 16 Statute Miles

1937 Boundaries Defining Danzig and
German Areas placed under Polish or Russian
Administration by the Potsdam Agreement

COSMO SERIES POLAND, CZECH.
Copyright by
RAND MCNALLY & COMPANY
Made in U.S.A.
A-559391-21 -3- -3⁰1

Baltic Sea

DENMARK · SWEDEN · BORNHOLM (DENMARK)

LITHUANIA · KALININGRAD (R.S.F.S.R.) (Königsberg) · MASURIA

SOVIET UNION · BYELO-RUSSIA

POMERANIA · Gulf of Danzig · Gdańsk (Danzig) · Gdynia · Sopot

POLAND

Warsaw (Warszawa) · Łódź · Lublin

Berlin · GERMANY · BRANDENBURG · MECKLENBURG

Copenhagen (København) · Rostock · Magdeburg · Leipzig · Dresden · Prague (Praha)

CZECHOSLOVAKIA · BOHEMIA · MORAVIA · SLOVAKIA · CARPATHIANS · RUTHENIA

Kraków · Katowice · Ostrava · Brno · Bratislava

Vienna (Wien) · AUSTRIA · BAVARIA · HUNGARY · Budapest · ROMANIA · Debrecen

26

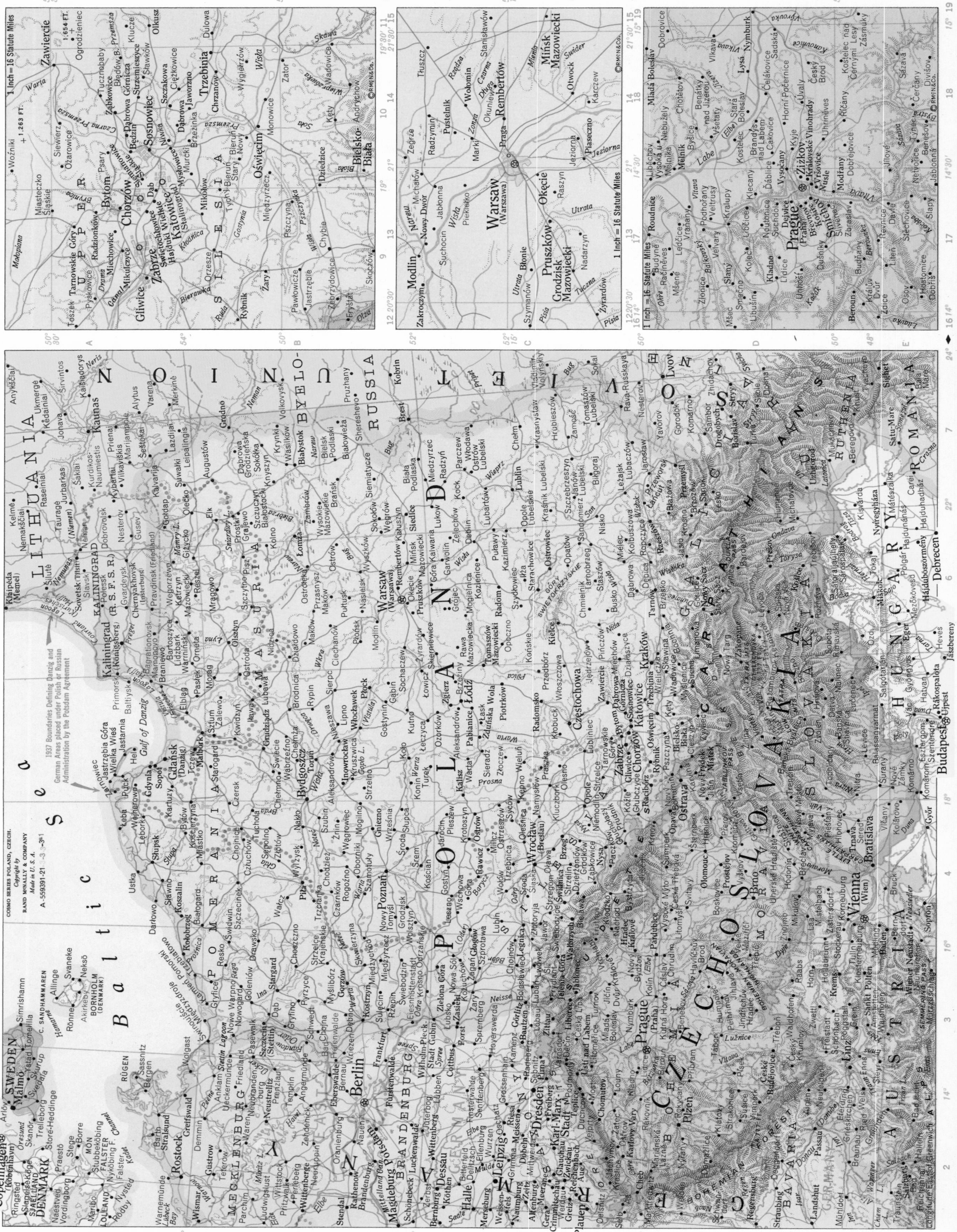

Statute Miles   25   0   25   50   75
Kilometers   25   0   25   50   100

Conic Projection
SCALE 1:4,000,000   1 Inch = 63 Statute Miles

Conic Projection
SCALE 1:8,000,000   1 Inch = 126 Statute Miles

Statute Miles
Kilometers

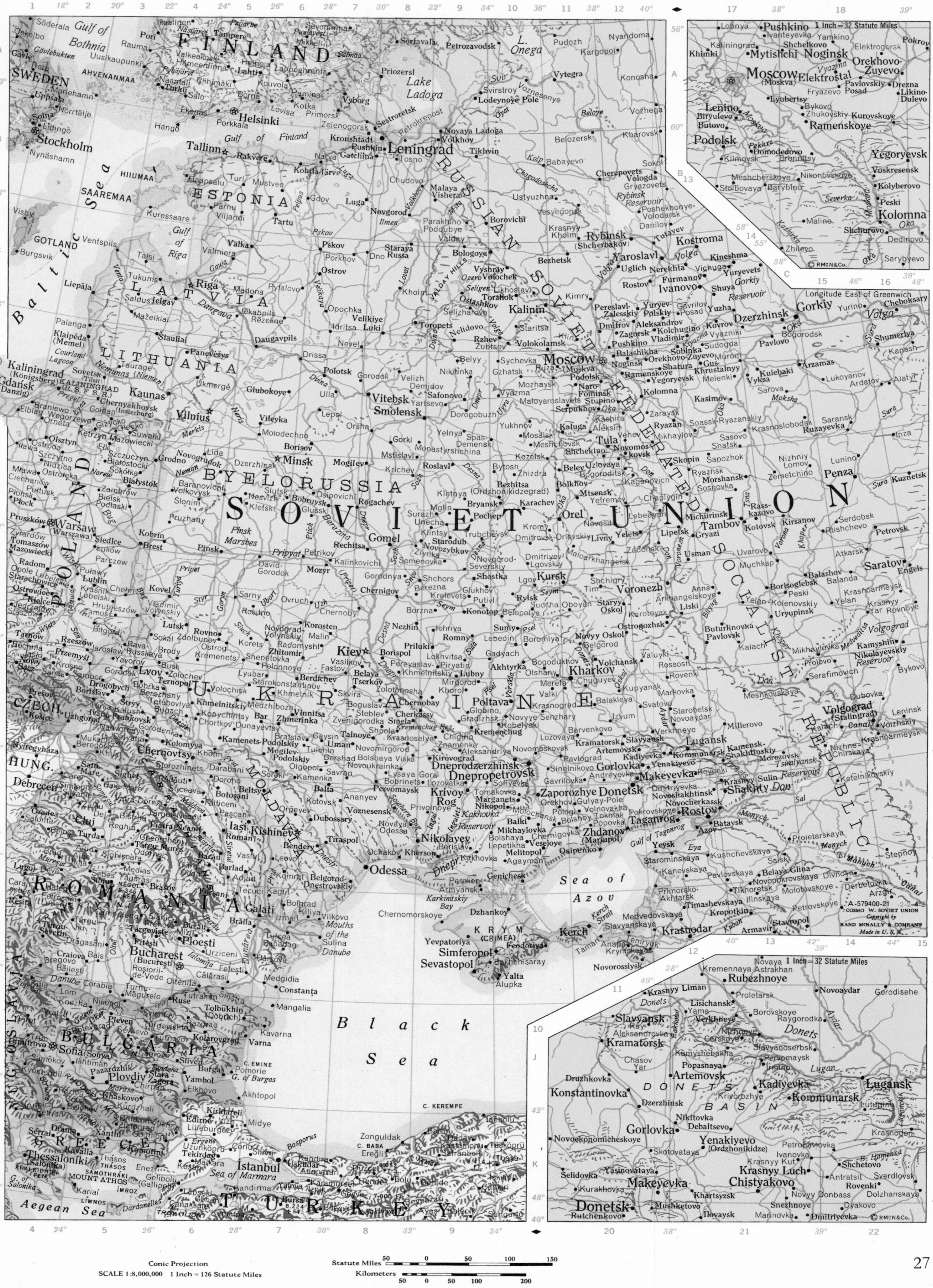

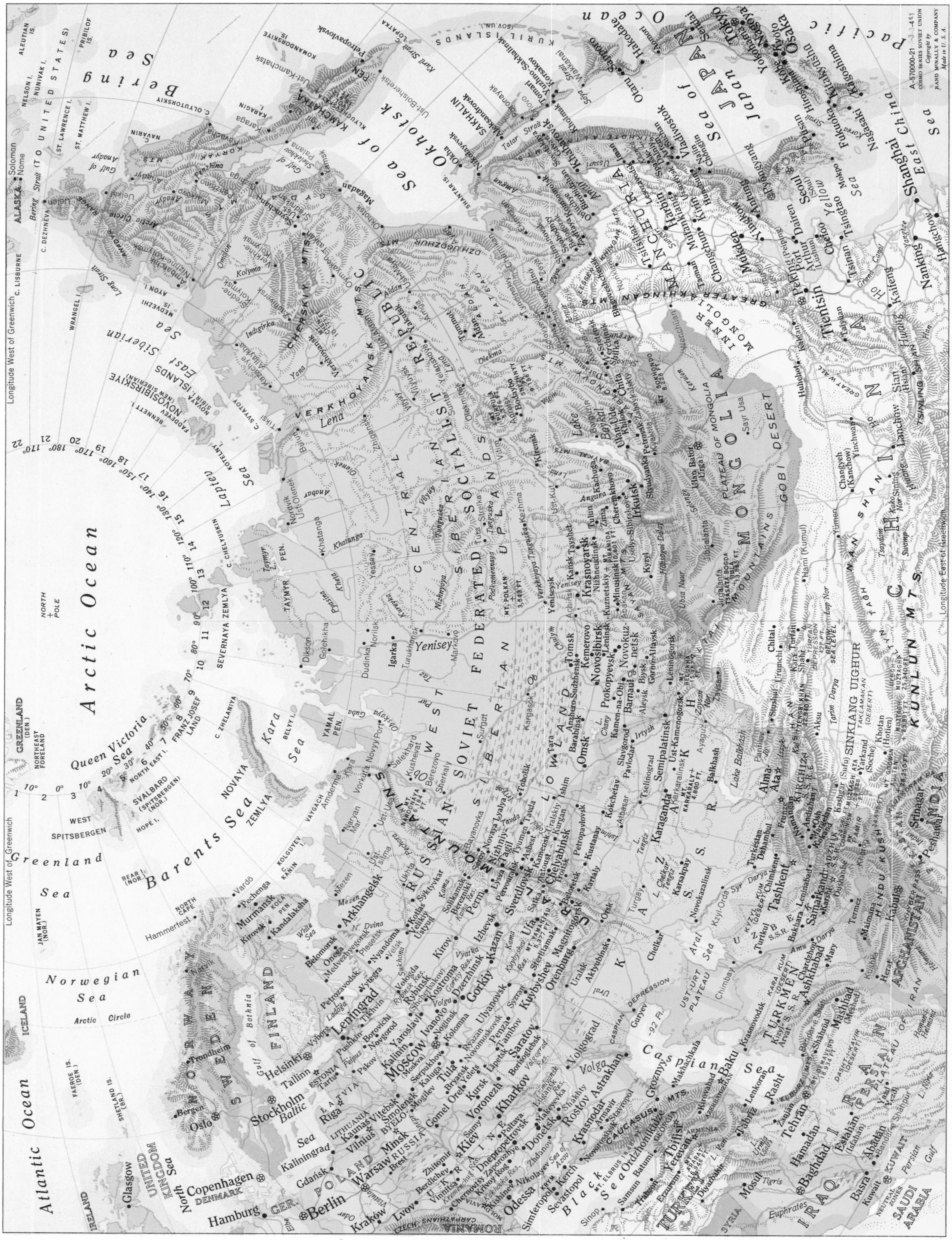

SOVIET UNION

Statute Miles   100 0 100 200 300 400 500

Kilometers   100 0 100 300 500 700

Lambert Azimuthal Equal Area Projection
SCALE 1:28,000,000   1 Inch = 442 Statute Miles

A-570000-21 -3-461
Copyright by
RAND McNALLY & COMPANY
Made in U.S.A.

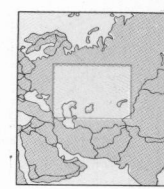

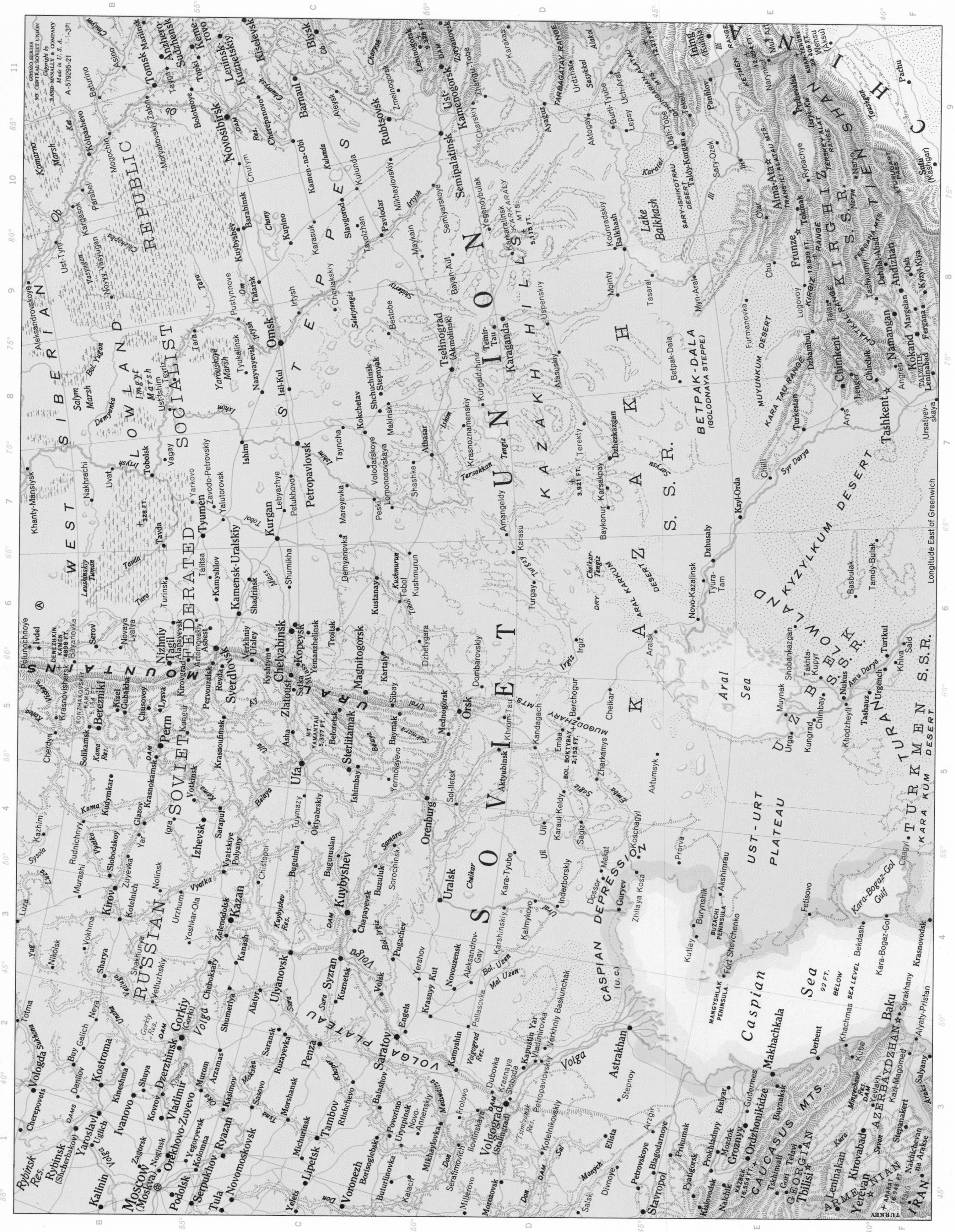

Sinusoidal Projection

SCALE 1: 11,400,000 1 Inch = 180 Statute Miles

Statute Miles 50 0 50 100 150 200 250

Kilometers 50 0 50 100 150 200 250 300

Statute Miles 50 0 50 100 150
Kilometers 50 0 50 100 200

Lambert Conformal Conic Projection
SCALE 1 : 8,000,000   1 Inch = 126 Statute Miles

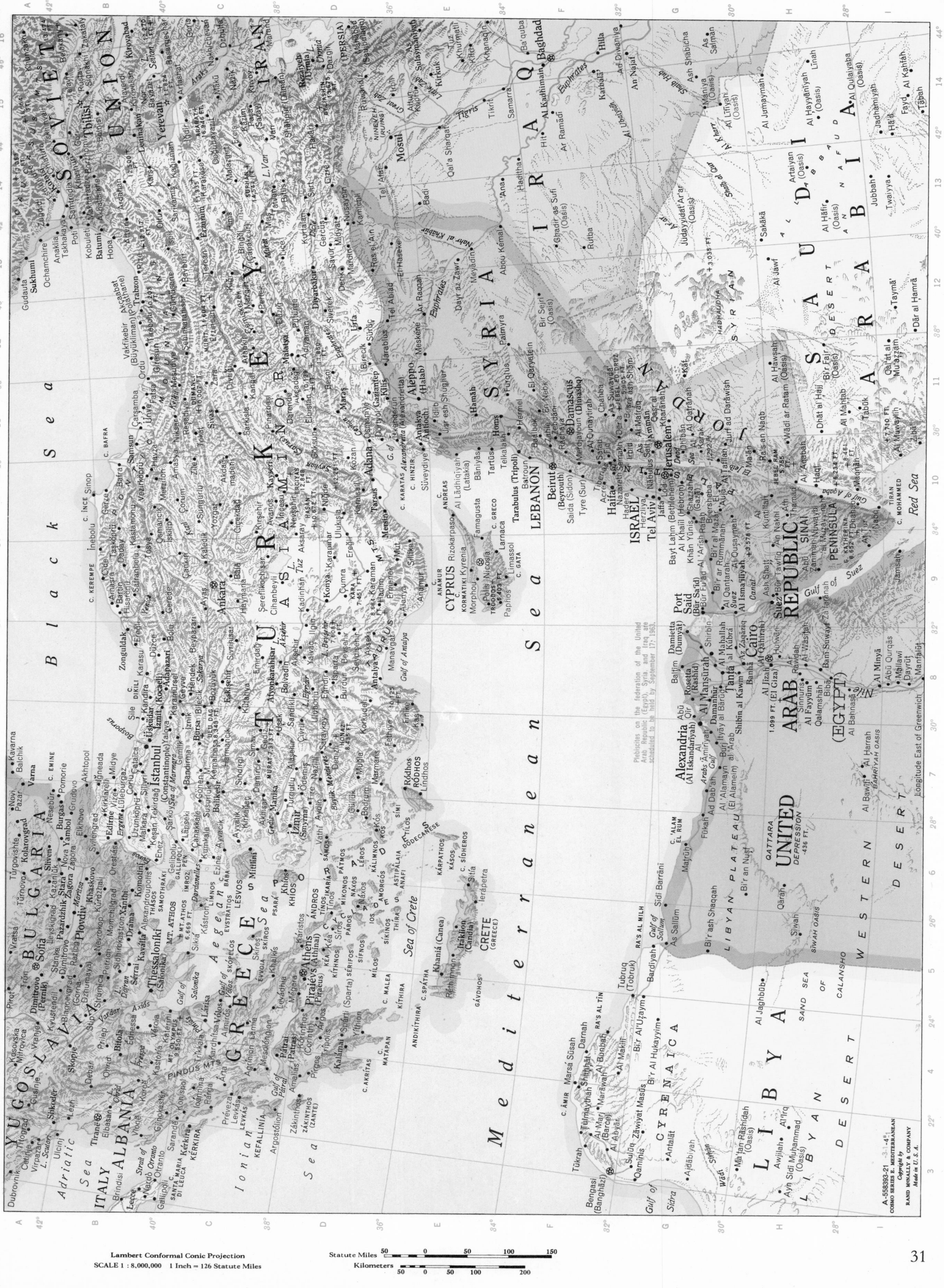

Lambert Conformal Conic Projection
SCALE 1 : 8,000,000   1 Inch = 126 Statute Miles

Statute Miles
50   0   50   100   150

Kilometers
50   0   50   100   200

A-558363-21

COSMO SERIES E. MEDITERRANEAN
Copyright by
RAND McNALLY & COMPANY
Made in U.S.A.

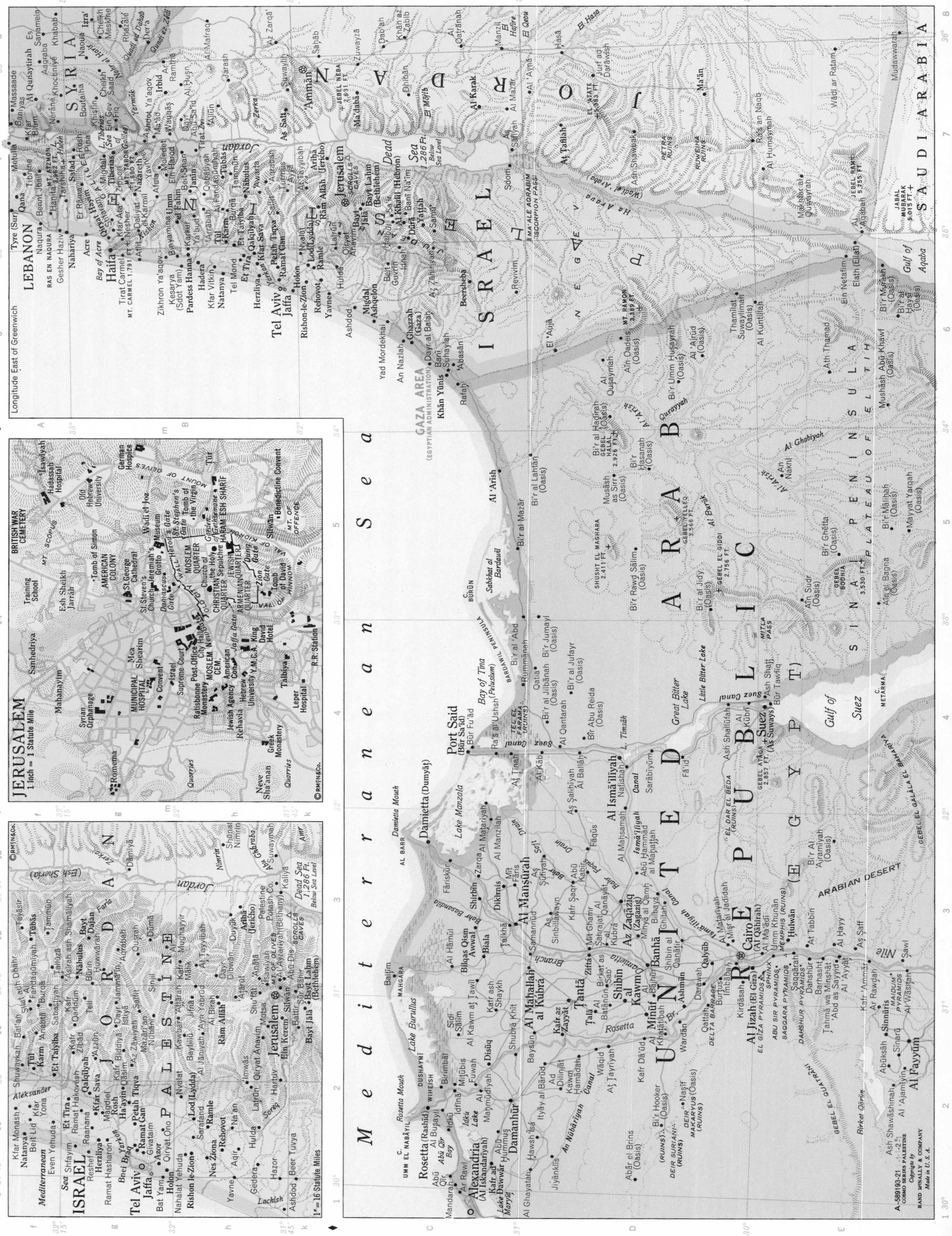

JERUSALEM
1 inch = 1 Statute Mile

Statute Miles 5 0 5 10 20 30 40 50
Kilometers 5 0 5 10 20 30 40 50 60

Lambert Conformal Conic Projection
SCALE 1:2,000,000   1 Inch = 32 Statute Miles

A-589193-21

Copyright by
RAND McNALLY & COMPANY
Made in U.S.A.

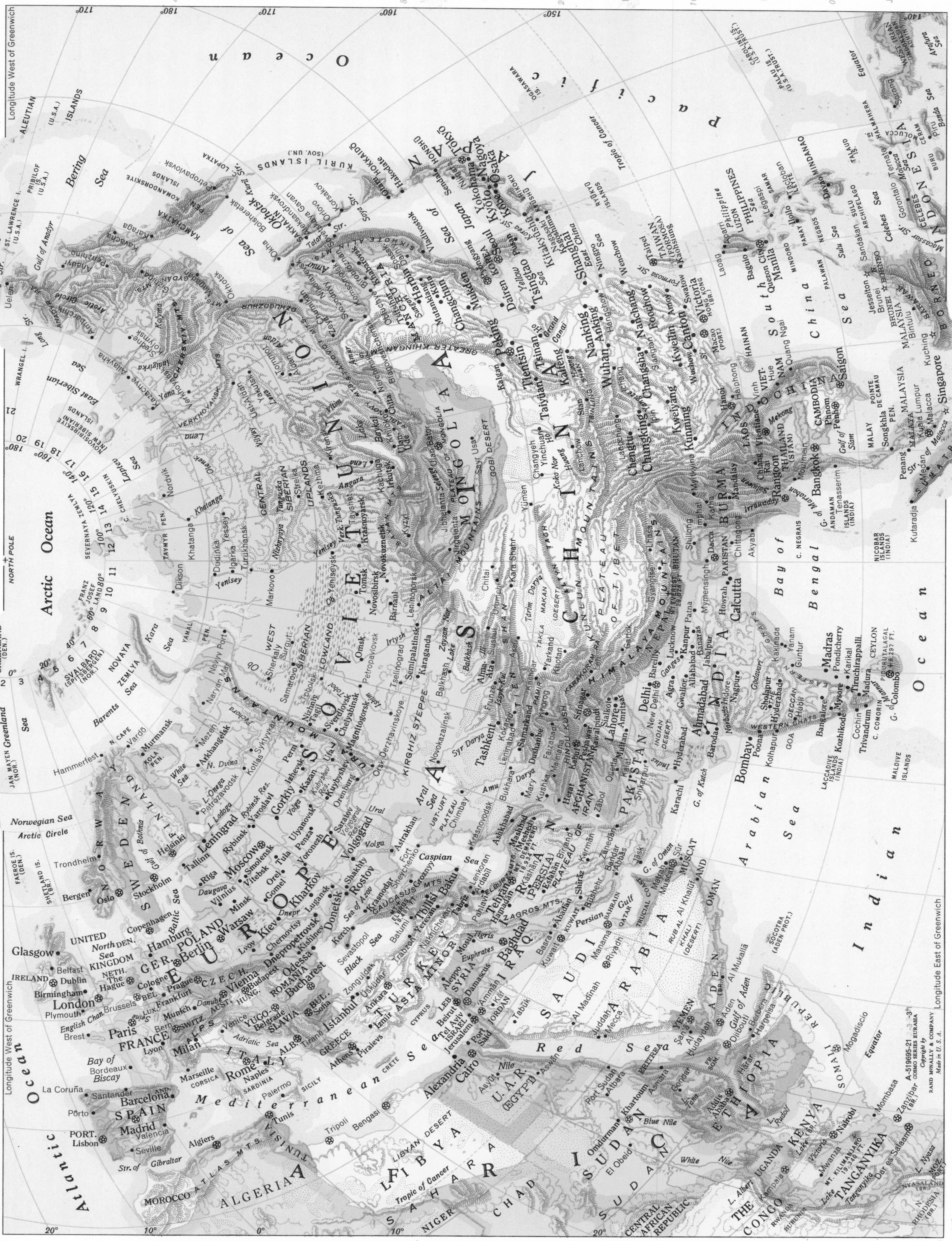

Lambert Azimuthal Equal Area Projection
SCALE 1:42,000,000   1 Inch = 663 Statute Miles

Statute Miles
100 0 100   300   500   700   900

Kilometers
100 0 100   300   700   1100

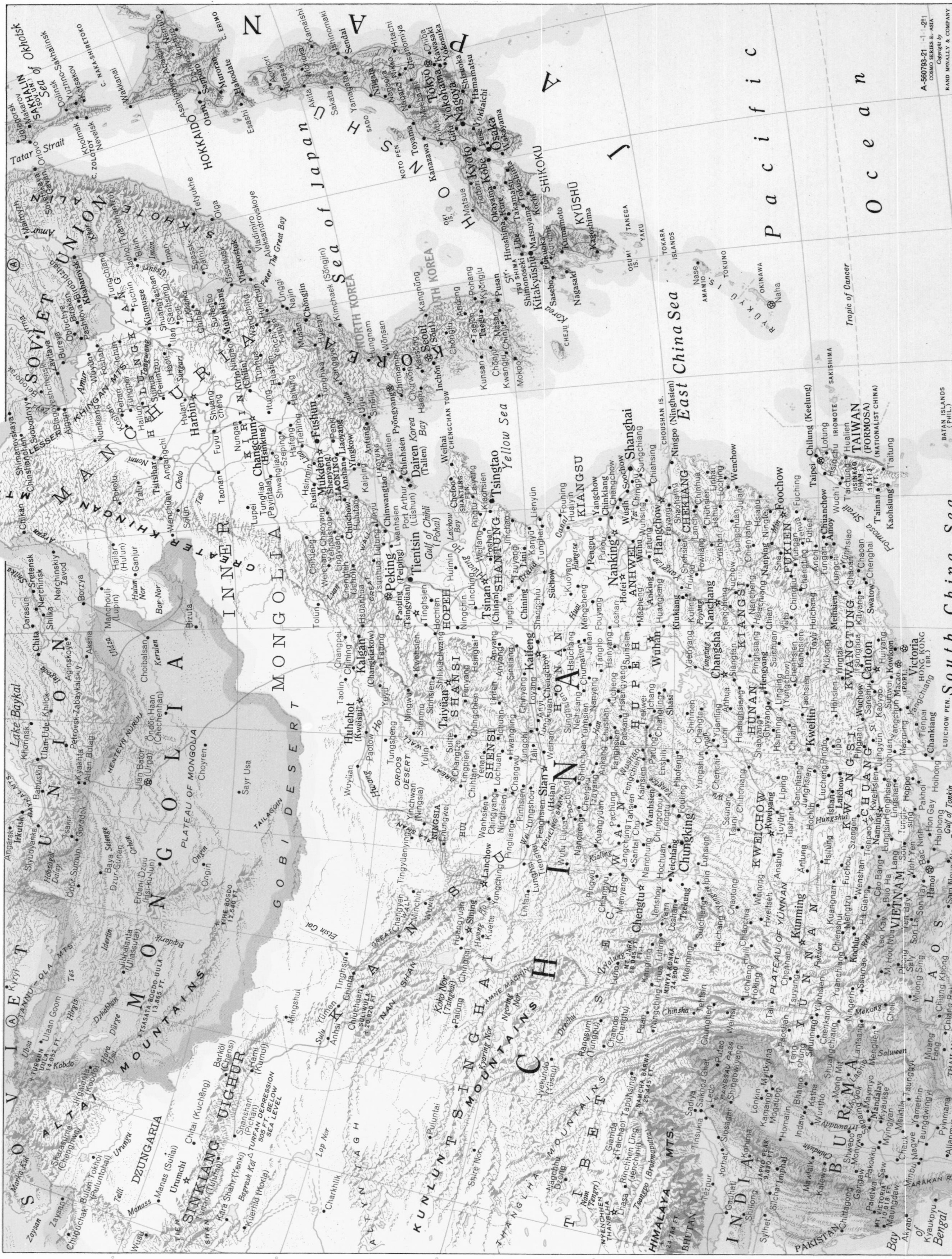

CHINA, MONGOLIA, KOREA, JAPAN and TAIWAN

Statute Miles 100 0 100 200 300

Kilometers 100 0 100 200 300 400

Polyconic Projection

SCALE 1:16,000,000   1 Inch = 252 Statute Miles

A-560793-21 -1 -2!!
COSMO SERIES 8: ASIA
Copyright by
RAND MCNALLY & COMPANY
Made in U.S.A.

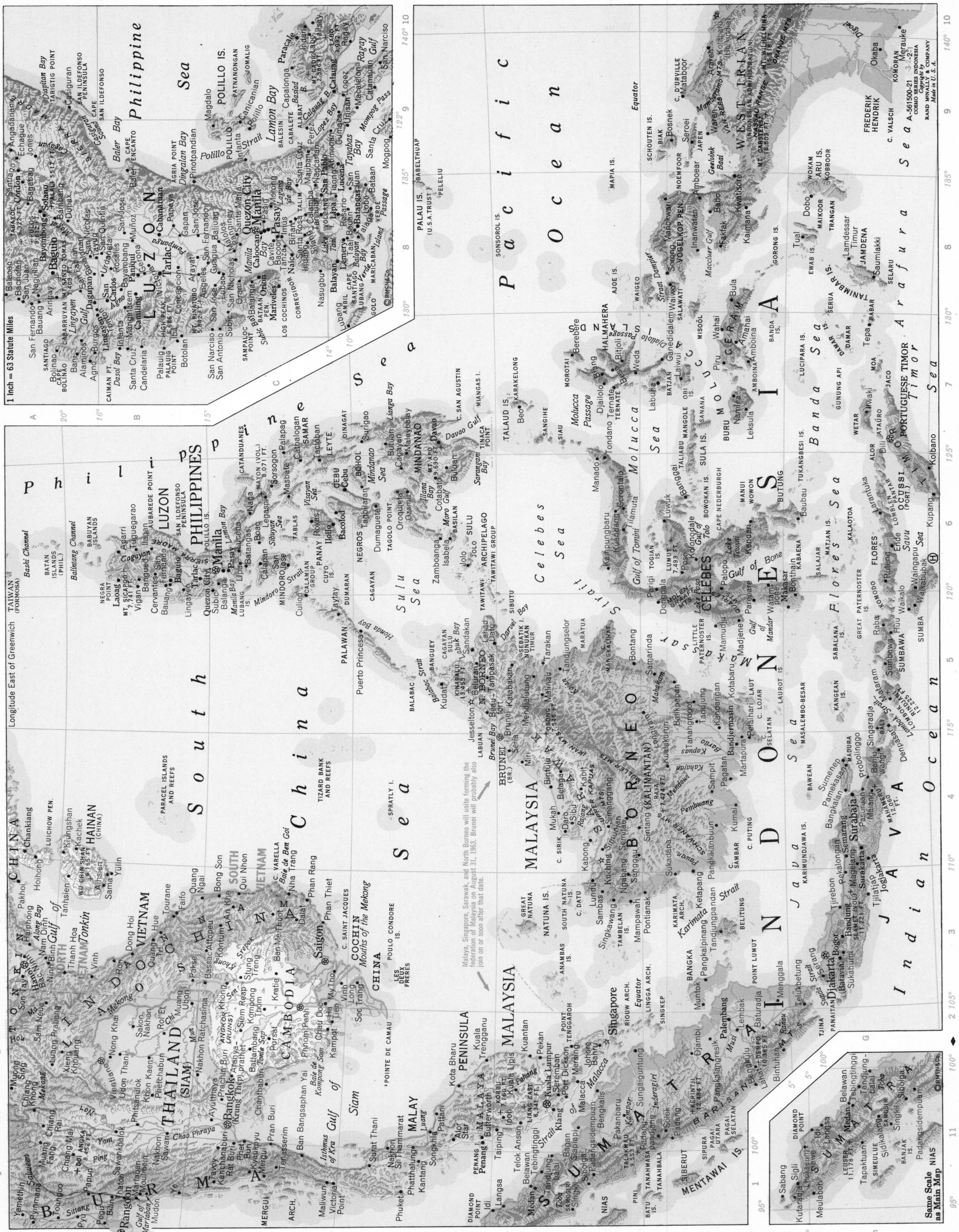

Philippine Sea

PHILIPPINES

LUZON

Manila

South China Sea

TAIWAN (FORMOSA)

HAINAN (CHINA)

NORTH VIETNAM

SOUTH VIETNAM

CAMBODIA

THAILAND (SIAM)

Bangkok

BURMA

MALAY PENINSULA

MALAYSIA

Singapore

SUMATRA

BORNEO (KALIMANTAN)

BRUNEI (BR.)

SARAWAK

NORTH BORNEO

I N D O N E S I A

J A V A

Djakarta

Surabaja

CELEBES

Celebes Sea

Molucca Sea

HALMAHERA

WEST IRIAN

Banda Sea

Flores Sea

Java Sea

Timor Sea

Arafura Sea

PORTUGUESE TIMOR

Pacific Ocean

Indian Ocean

Sulu Sea

Mindanao

Davao

Celebes Strait

MALAYSIA

Malaya, Singapore, Sarawak, and North Borneo will unite forming the
Federation of Malaysia on August 31, 1963. Brunei will probably also
join on or soon after that date.

Longitude East of Greenwich

Polyconic Projection
SCALE 1:16,000,000   1 Inch = 252 Statute Miles

Statute Miles
Kilometers

1 Inch = 63 Statute Miles

Same Scale as Main Map

A-561500-21
CORSO SERIES INDONESIA
Copyright by
RAND McNALLY & COMPANY
Made in U.S.A.

35

36

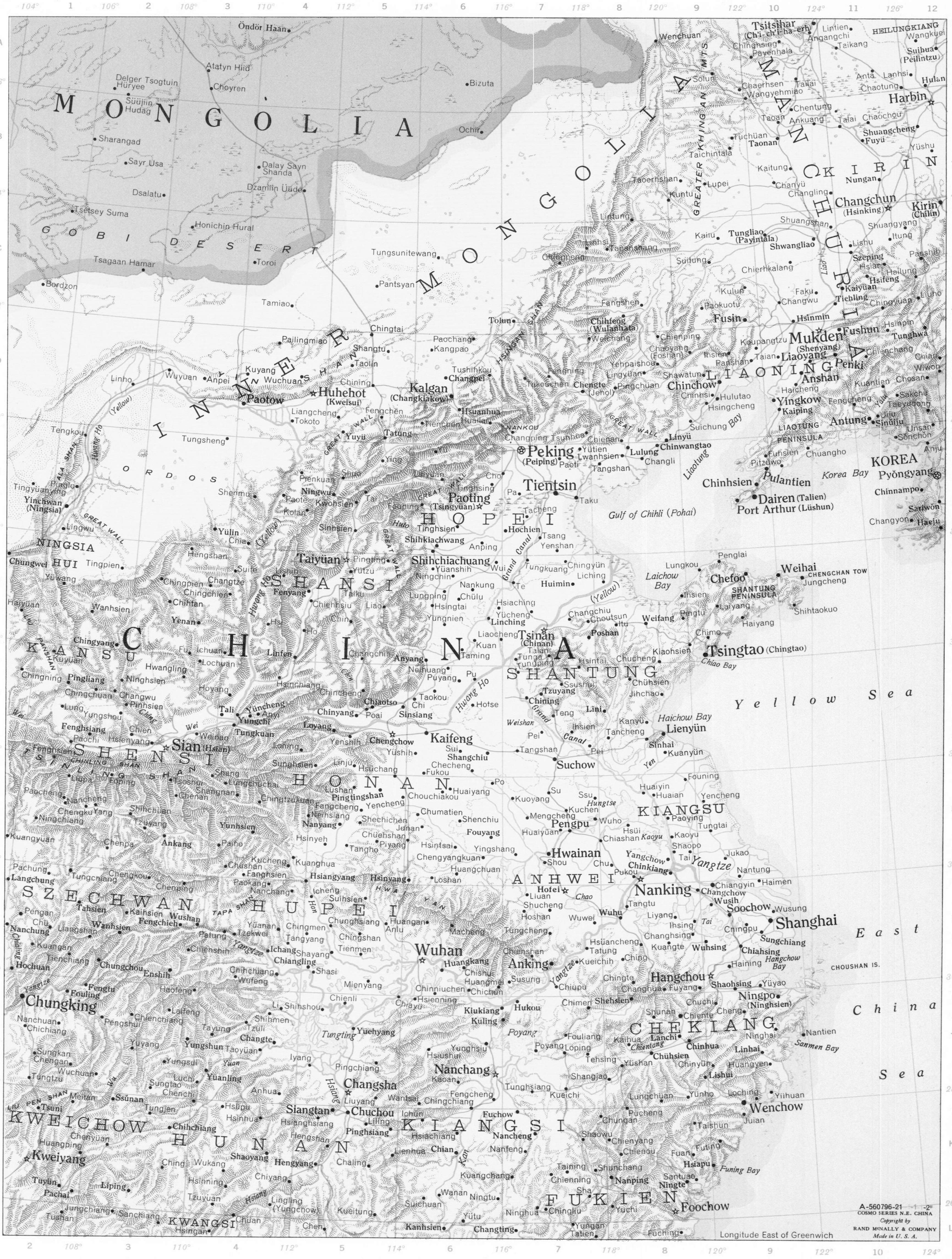

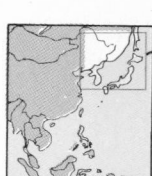

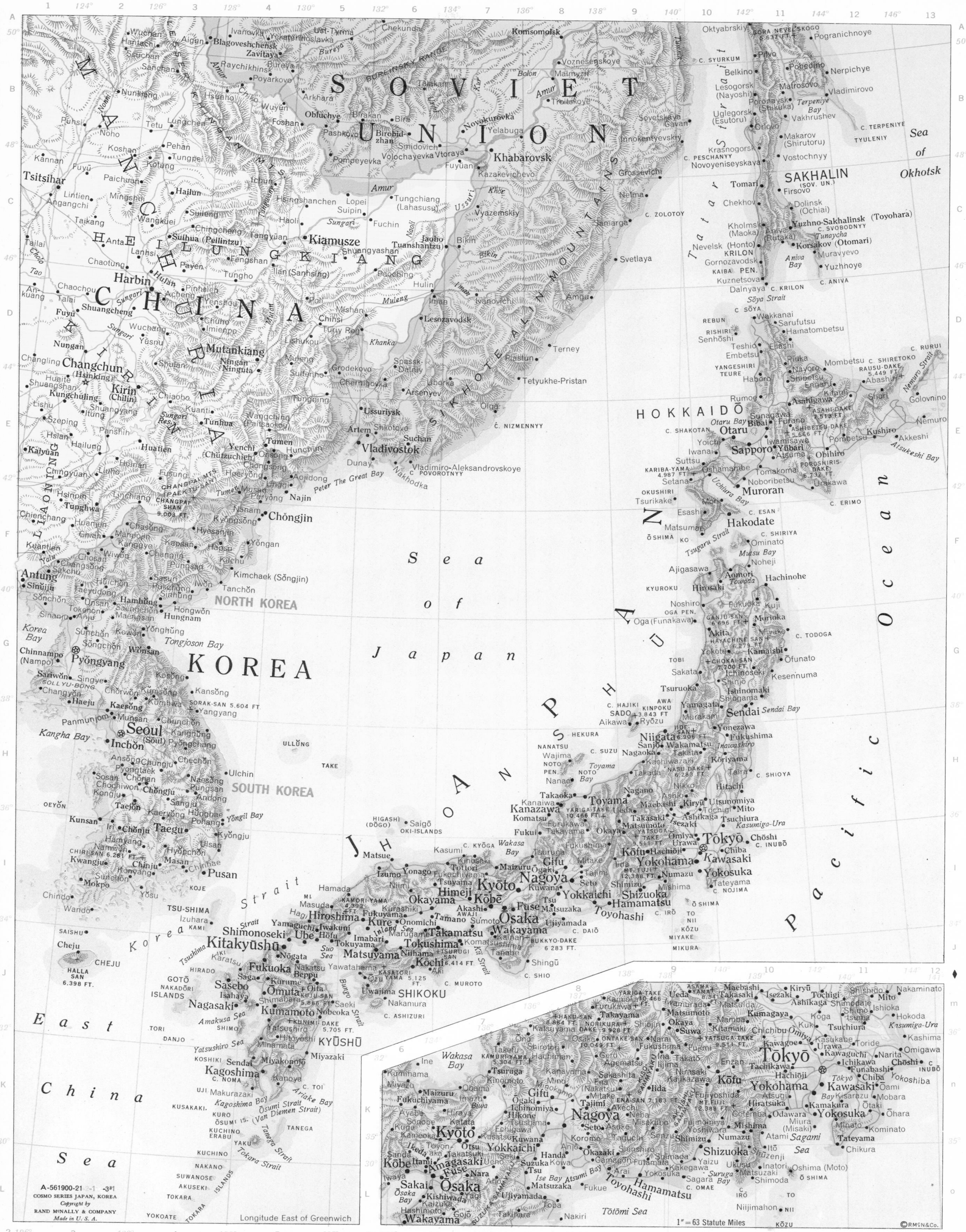

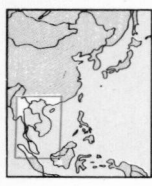

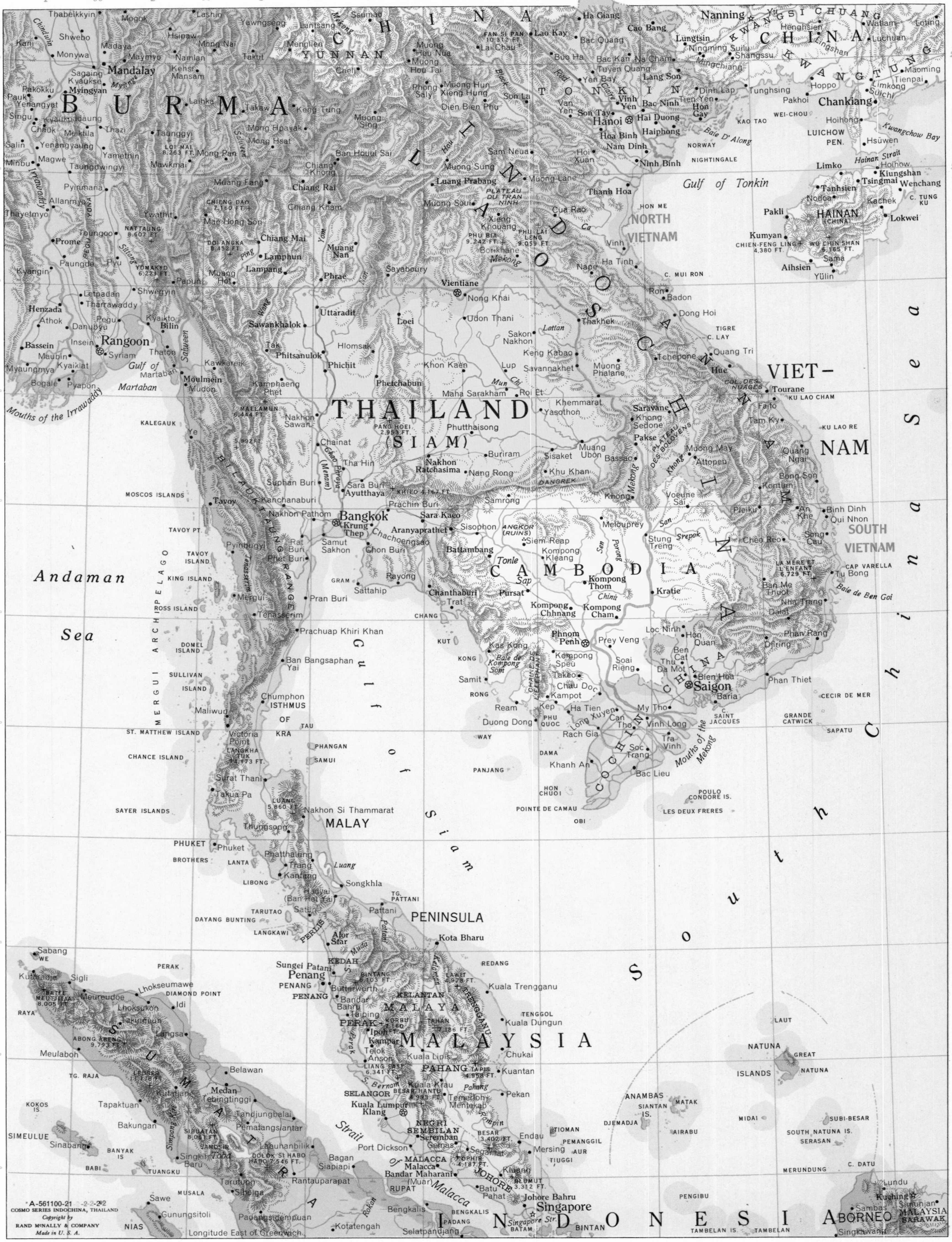

Statute Miles 50 0 50 100 150

Kilometers 50 0 50 100 200

Lambert Conformal Conic Projection

SCALE 1:8,000,000 1 Inch = 126 Statute Miles

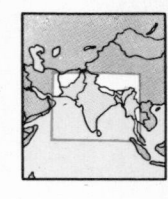

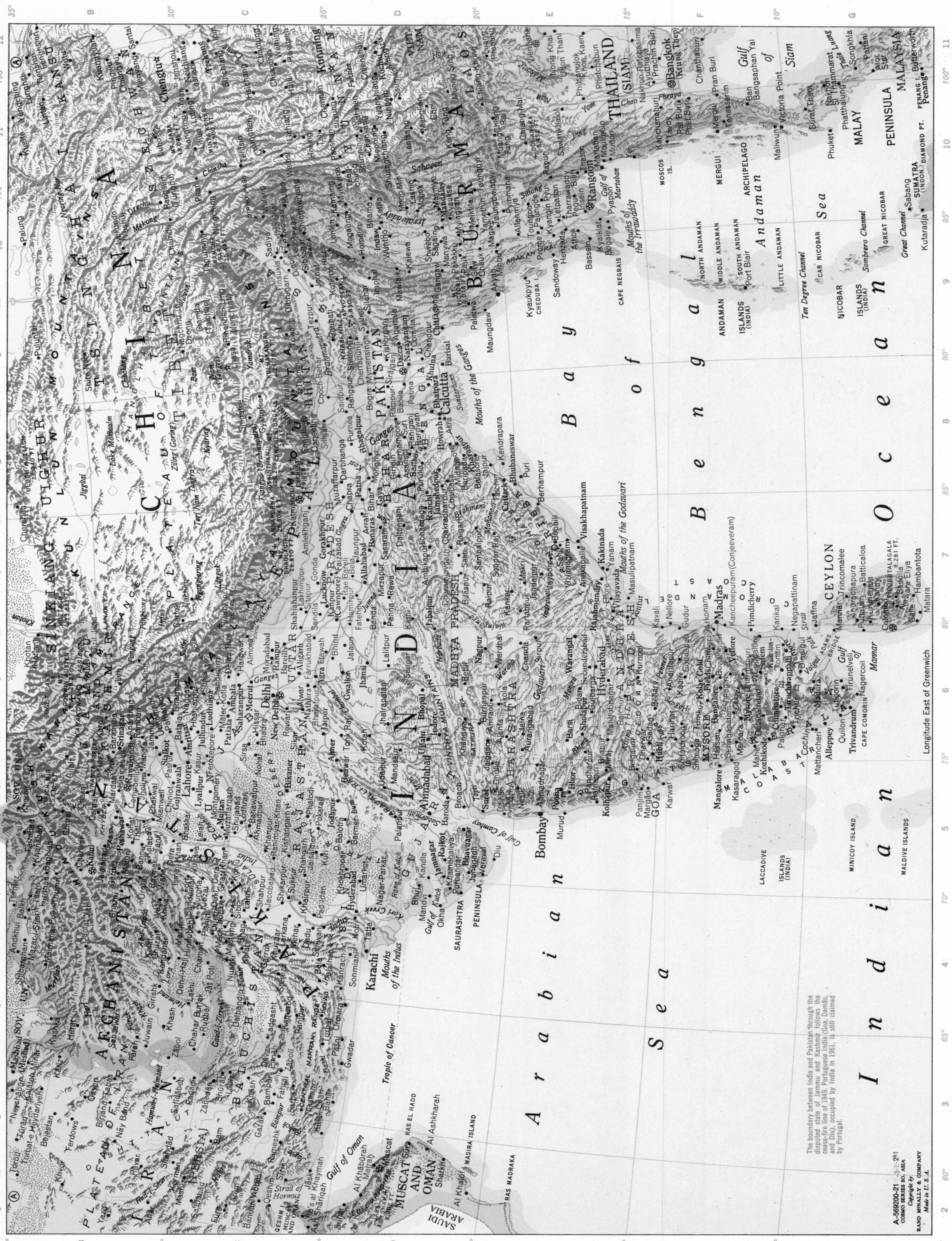

The boundary between India and Pakistan through the disputed state of Jammu and Kashmir follows the cease-fire line of 1949. Portuguese India (Goa, Damão and Diu), occupied by India in 1961, is still claimed by Portugal.

A-569200-21
COSMO SERIES NO. 39A
Copyright by
RAND McNALLY & COMPANY
Made in U.S.A.

Polyconic Projection
SCALE 1:16,000,000 1 inch = 252 Statute Miles

Statute Miles 100 0 100 200 300
Kilometers 100 0 100 200 300 400

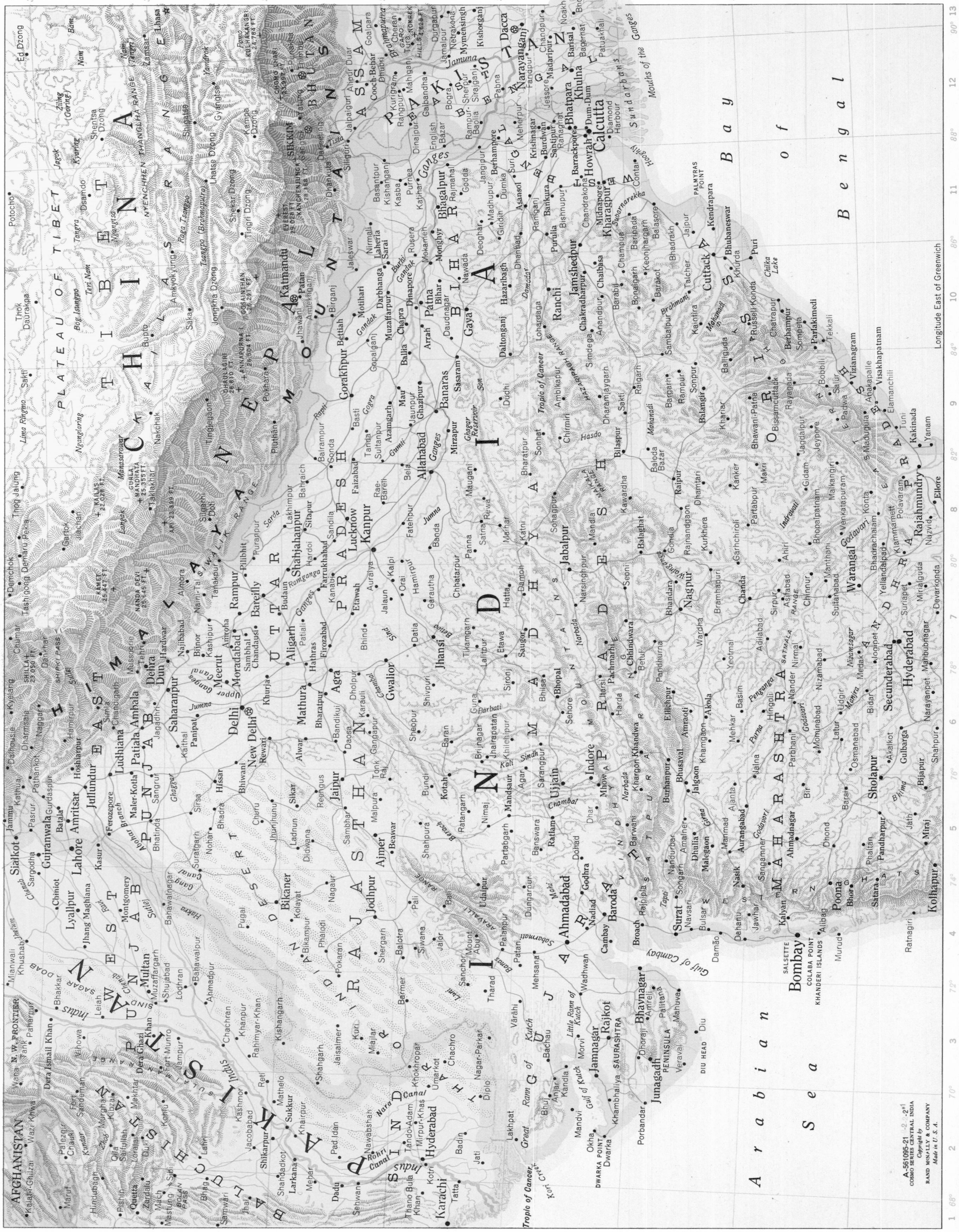

40

Statute Miles 50 0 50 100 150

Kilometers 50 0 50 100 200

Lambert Conformal Conic Projection

SCALE 1 : 8,000,000   1 Inch = 126 Statute Miles

A-561095-21
COSMO SERIES CENTRAL INDIA
Copyright by
RAND McNALLY & COMPANY
Made in U.S.A.

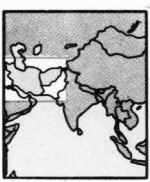

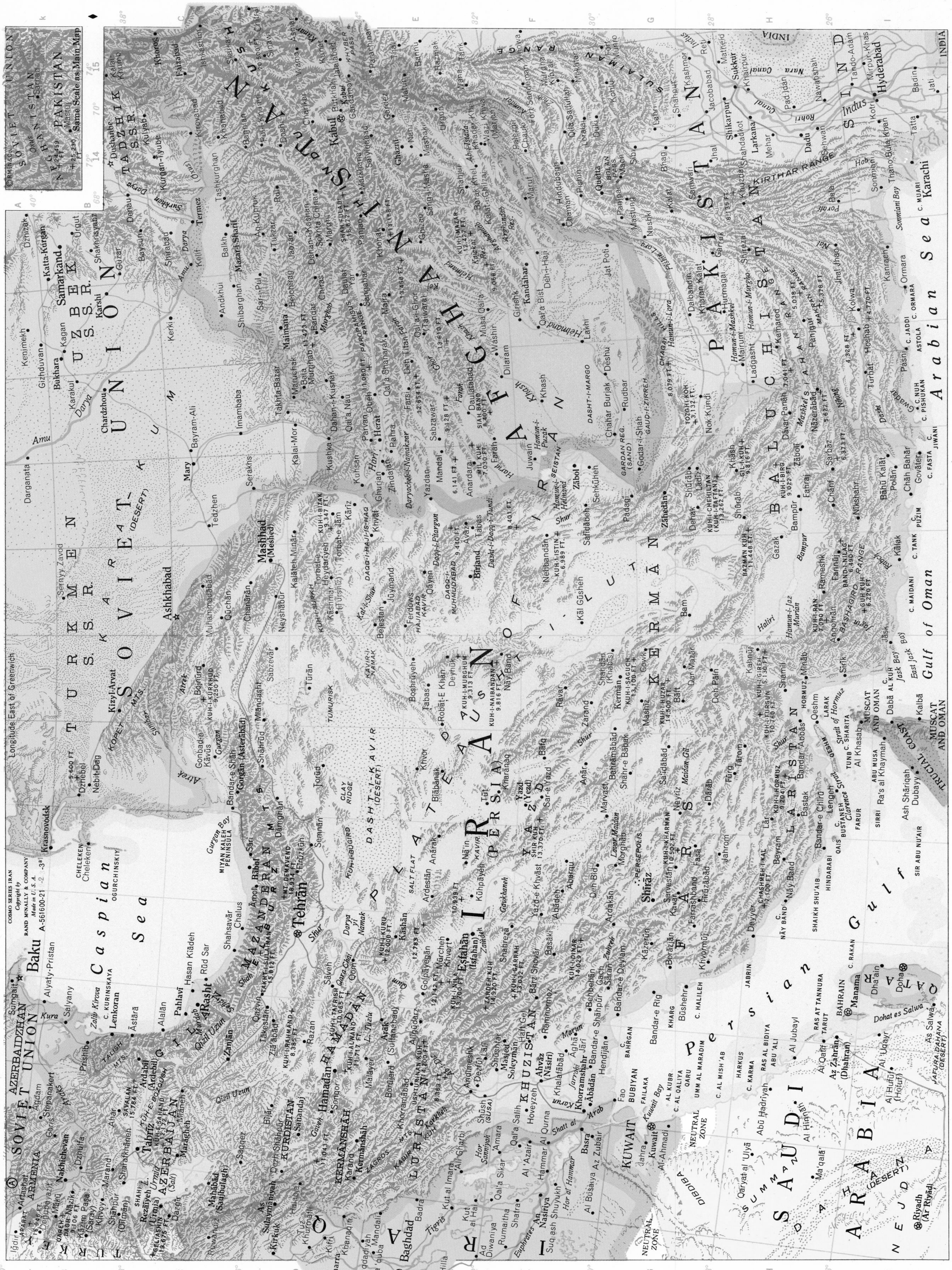

Lambert Conformal Conic Projection

SCALE 1 : 8,000,000 1 Inch = 126 Statute Miles

Statute Miles 50 0 50 100 150

Kilometers 50 0 50 100 200

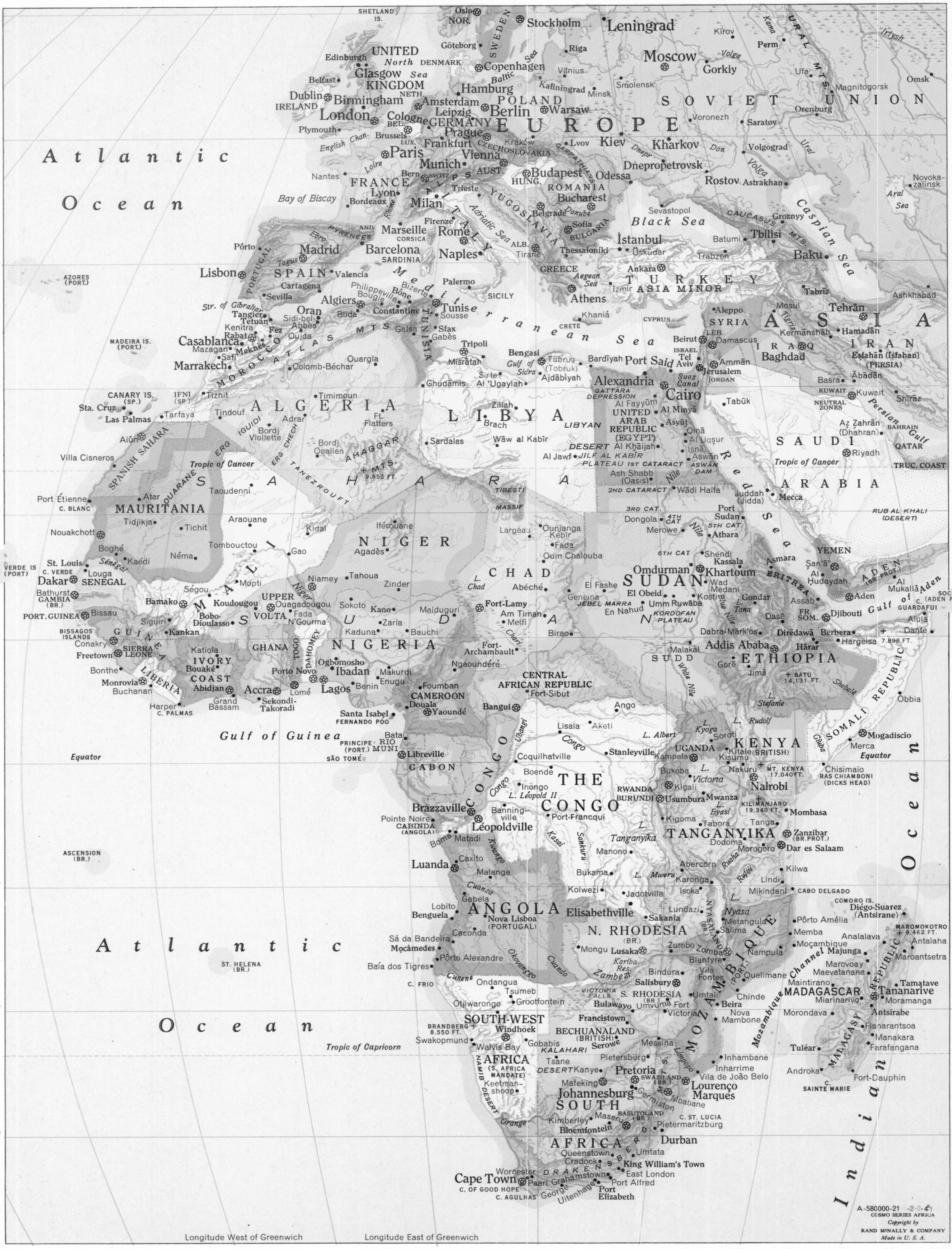

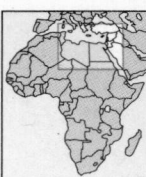

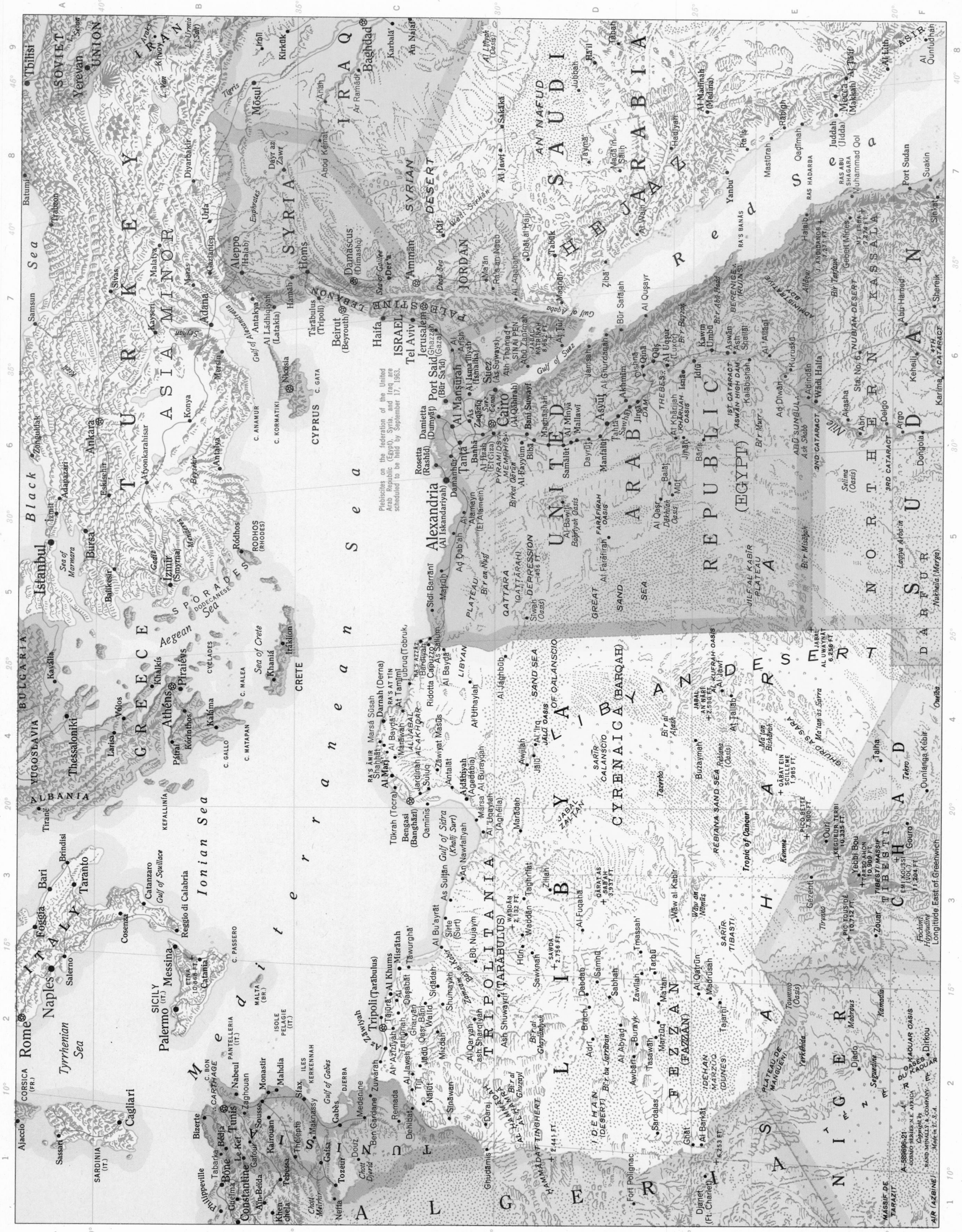

Sinusoidal Projection
SCALE 1: 11,400,000 1 Inch = 180 Statute Miles

Statute Miles 50 0 50 100 150 200 250

50 0 50 100 150 200 250 300

NORTHWEST AFRICA

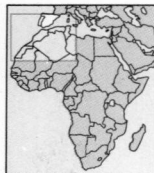

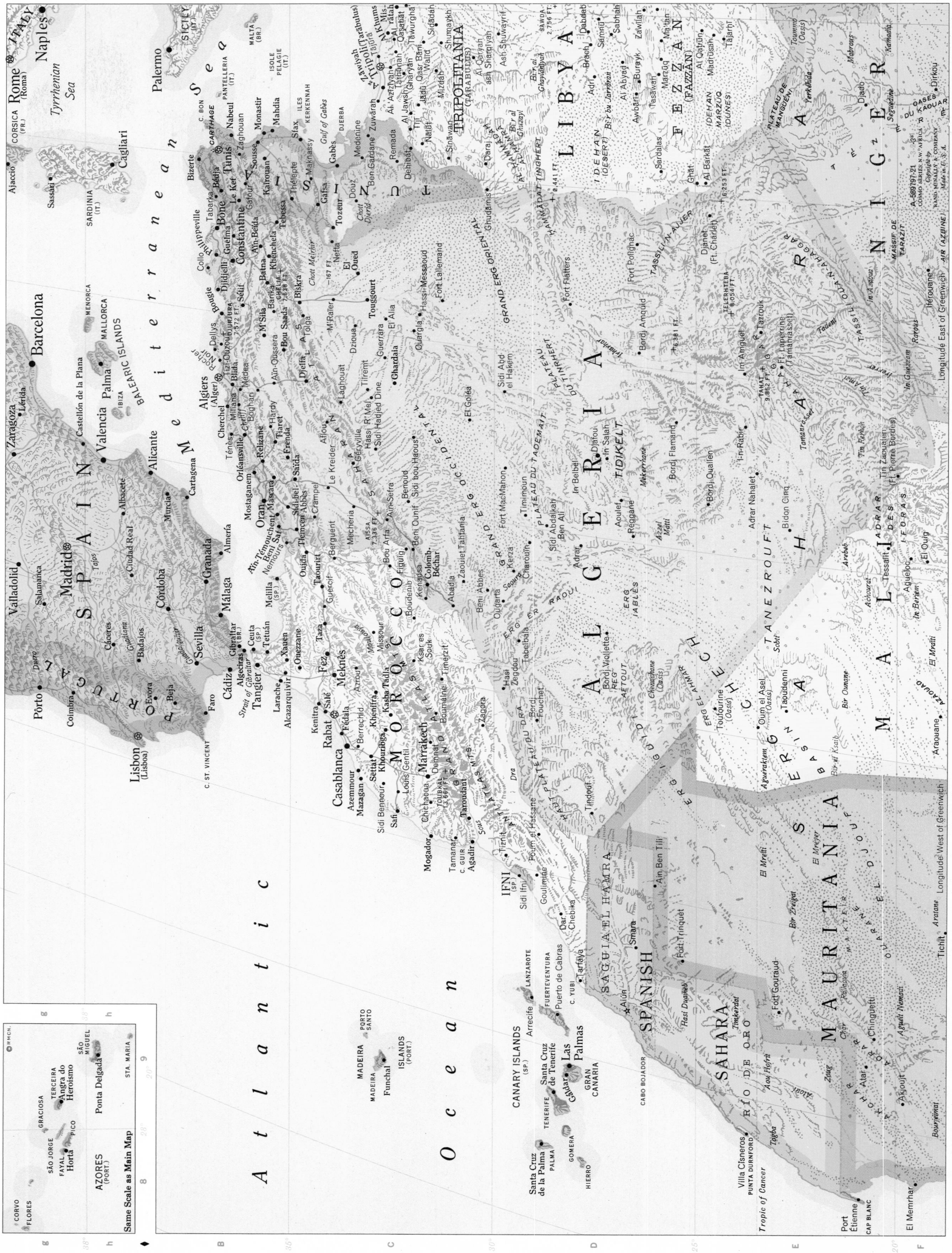

Statute Miles 50 25 0 50 100 150 200 250

50 0 50 100 150 200 250 300

Sinusoidal Projection

SCALE 1: 11,400,000 1 Inch = 180 Statute Miles

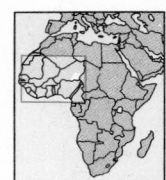

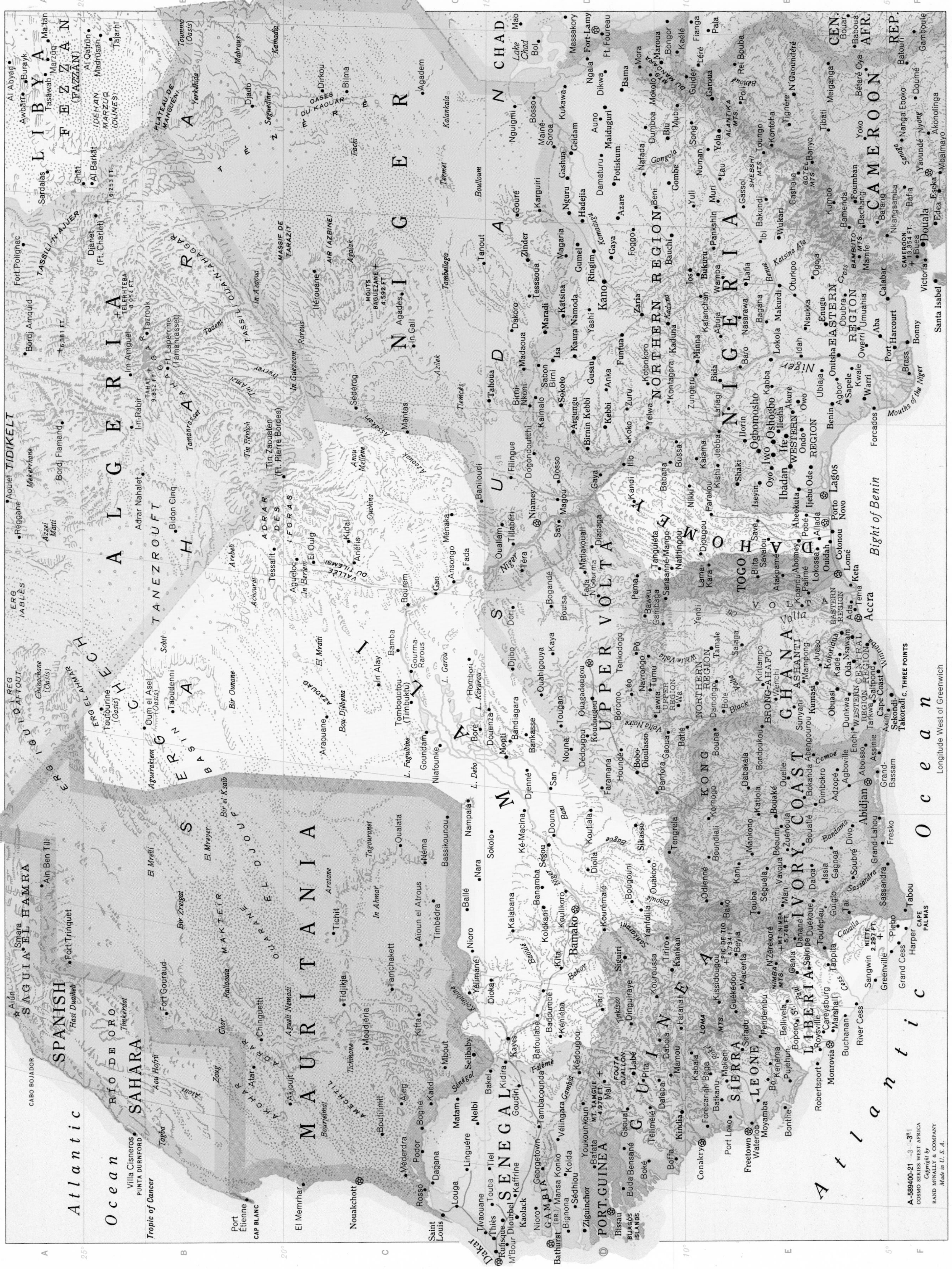

Sinusoidal Projection
SCALE 1 : 11,400,000   1 Inch = 180 Statute Miles

Statute Miles

Kilometers

A-589400-21 ·-3-·-3¹¹
COSMO SERIES WEST AFRICA
Copyright by
RAND M¢NALLY & COMPANY
Made in U.S.A.

EQUATORIAL AFRICA

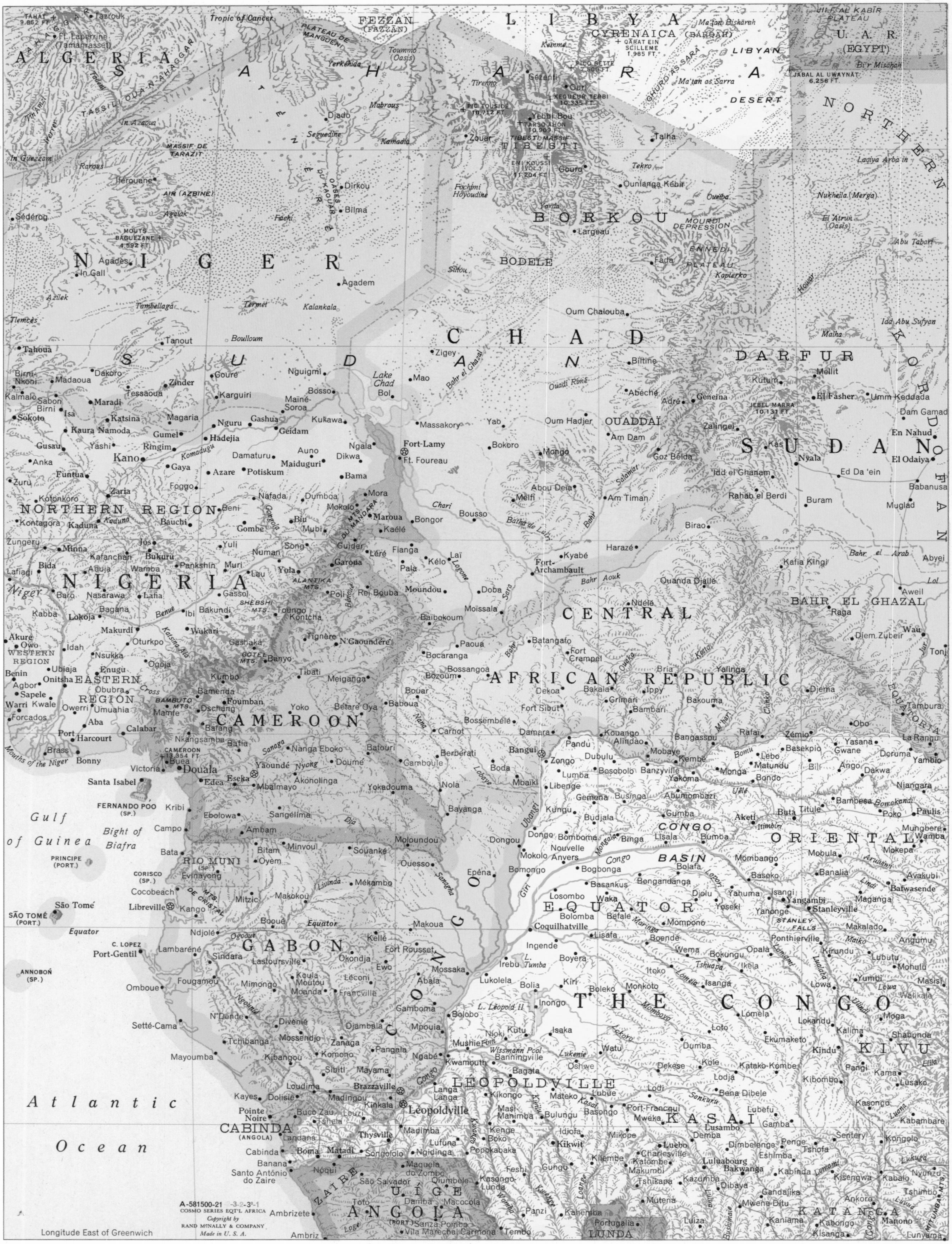

Longitude East of Greenwich

A-581500-21 -3-2-3²-1
COSMO SERIES EQ'T AFRICA
Copyright by
RAND McNALLY & COMPANY
Made in U.S.A.

Statute Miles 50 25 0   50   100   150   200   250

50 0 50 100 150 200 250 300

Sinusoidal Projection
SCALE 1: 11,400,000   1 Inch = 180 Statute Miles

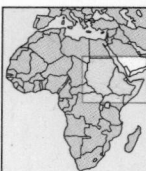

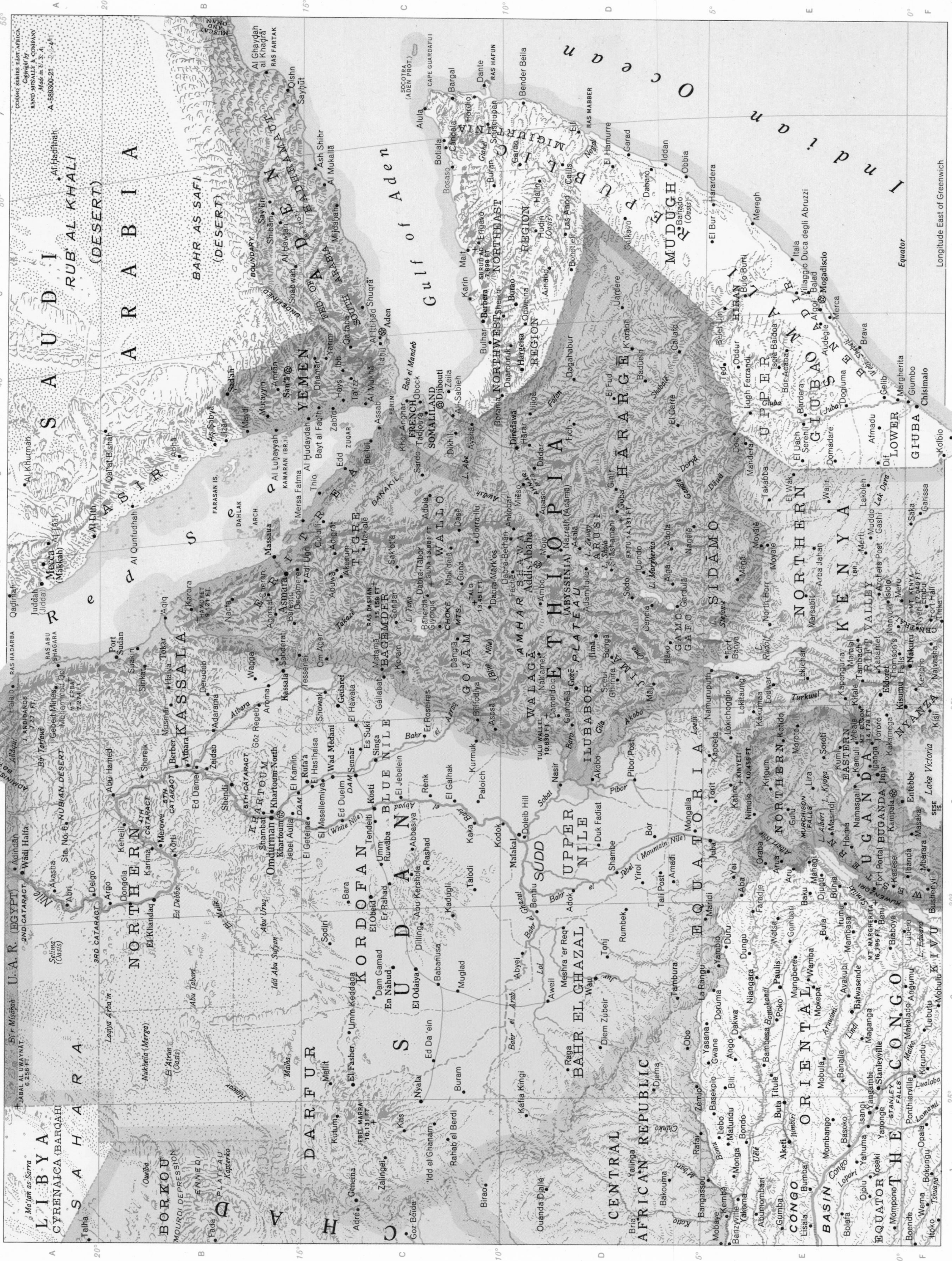

Sinusoidal Projection
SCALE 1 : 11,400,000 1 Inch = 180 Statute Miles

Statute Miles 50 25 0 50 100 150 200 250
Kilometers 50 0 50 100 150 200 250 300

CENTRAL AFRICA

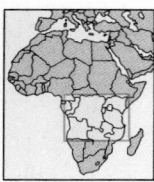

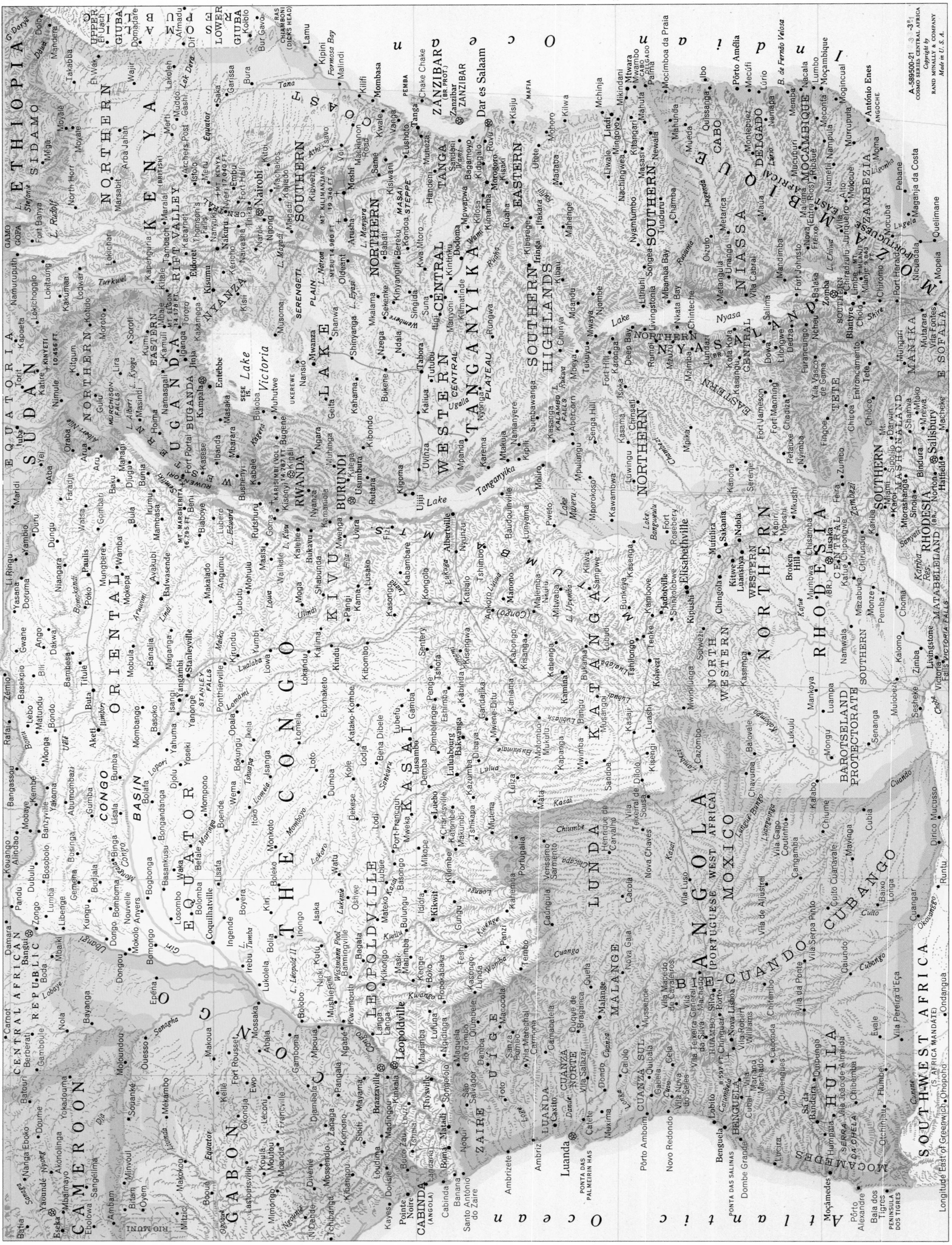

Statute Miles 50 0 50 100 150 200 250

50 0 50 100 150 200 250 300

Sinusoidal Projection

SCALE 1: 11,400,000   1 Inch = 180 Statute Miles

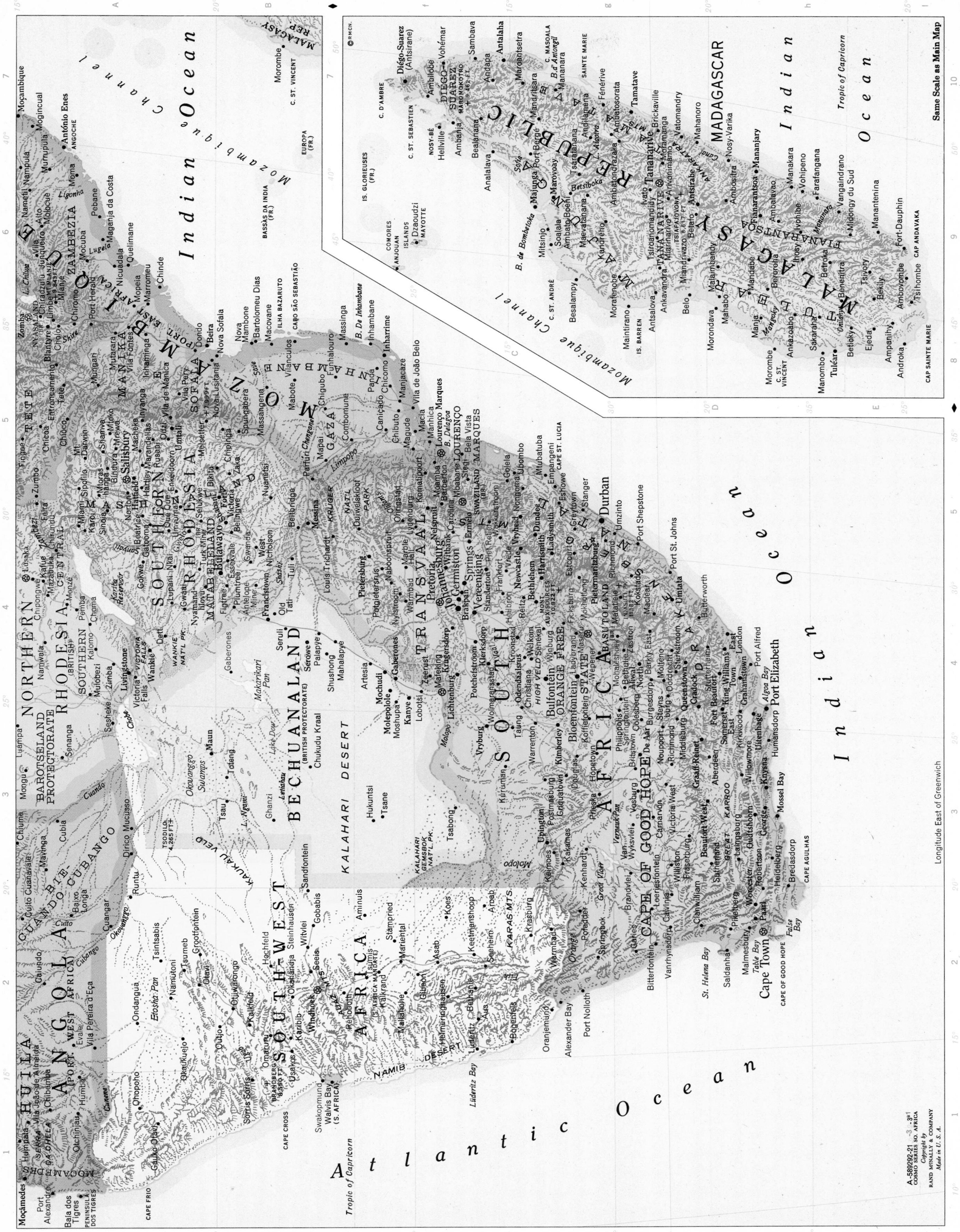

Same Scale as Main Map

Sinusoidal Projection
SCALE 1:11,400,000   1 Inch = 180 Statute Miles

Statute Miles
Kilometers

A-589292-21 -3- 3⁵¹
COSMO SERIES NO. AFRICA
Copyright by
RAND McNALLY & COMPANY
Made in U. S. A.

Longitude East of Greenwich

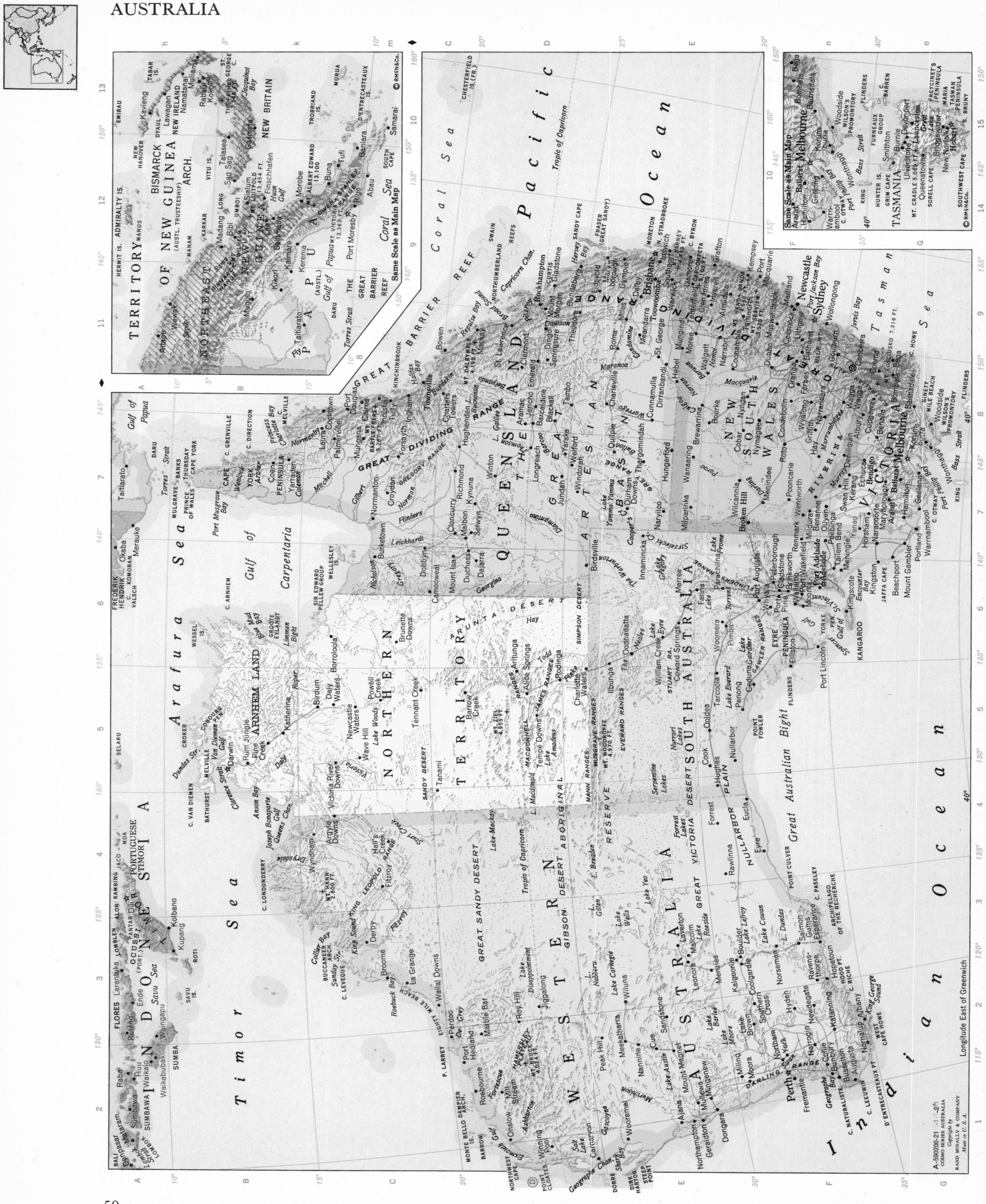

AUSTRALIA

Statute Miles
Kilometers

Lambert Azimuthal Equal Area Projection
SCALE 1:16,000,000   1 Inch = 252 Statute Miles

50

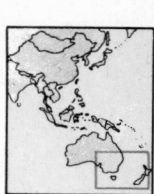

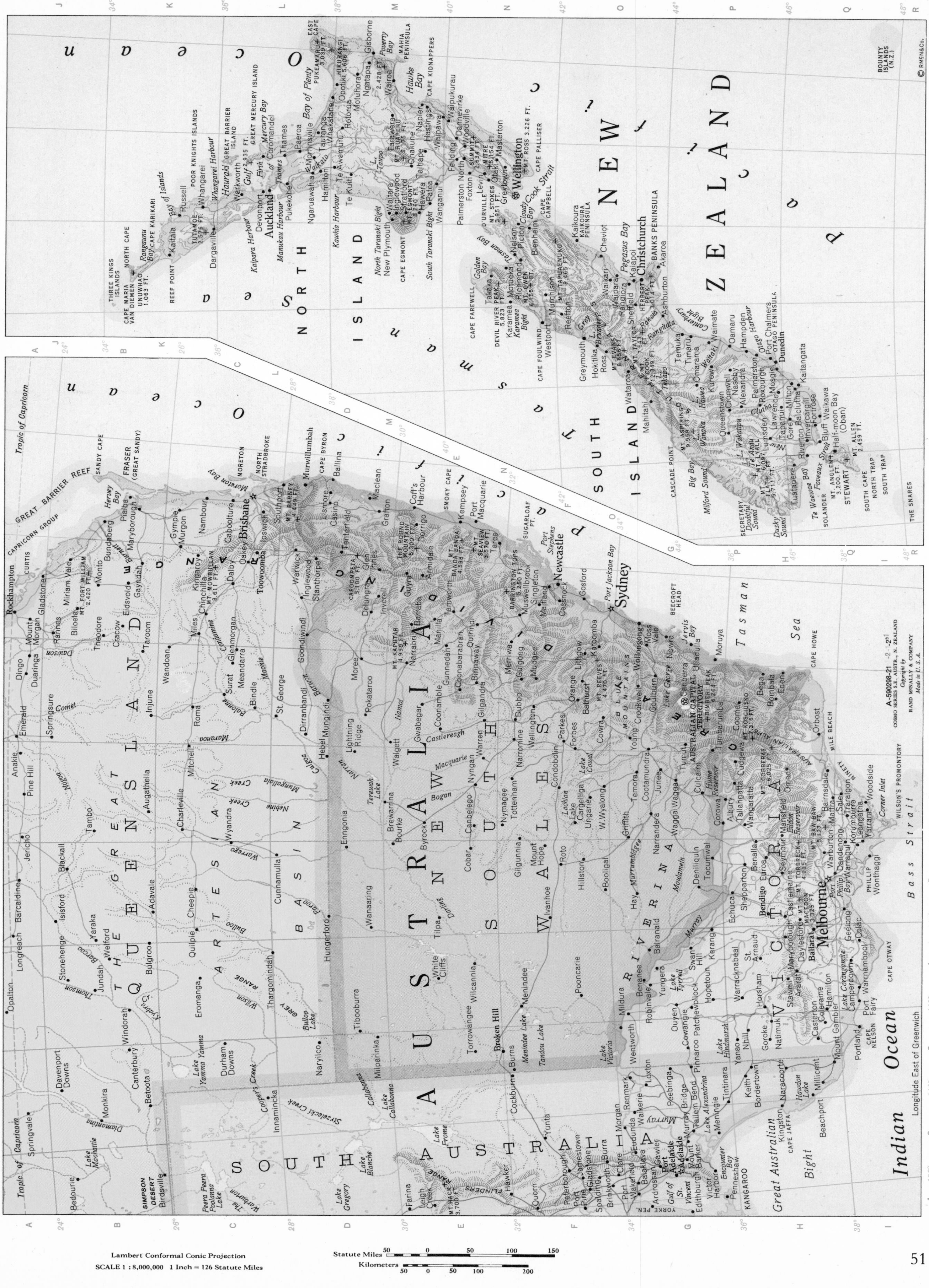

Lambert Conformal Conic Projection
SCALE 1 : 8,000,000 1 Inch = 126 Statute Miles

Statute Miles 50 0 50 100 150
Kilometers 50 0 50 100 200

PACIFIC ISLANDS

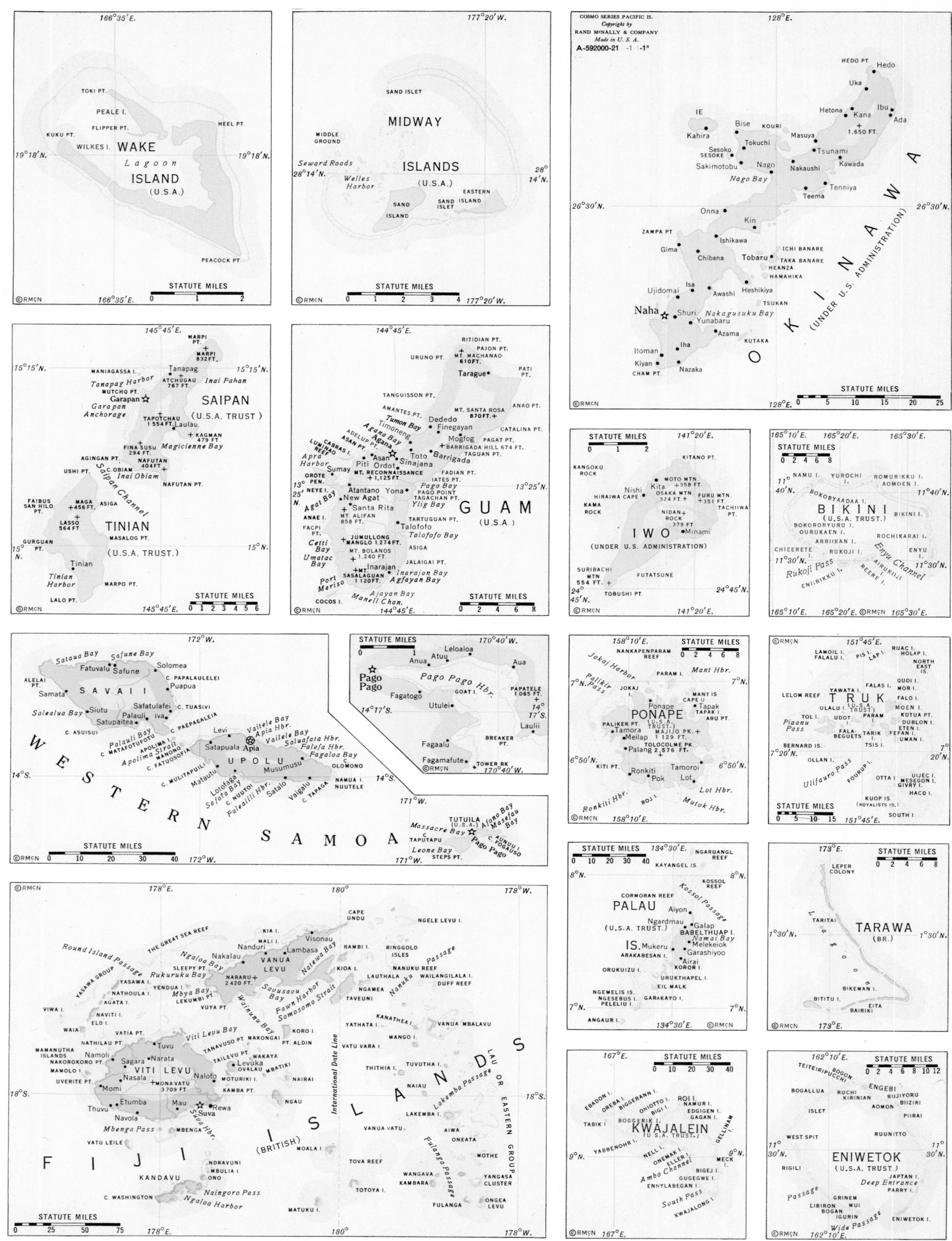

CENTRAL AND SOUTHERN ARGENTINA AND CHILE

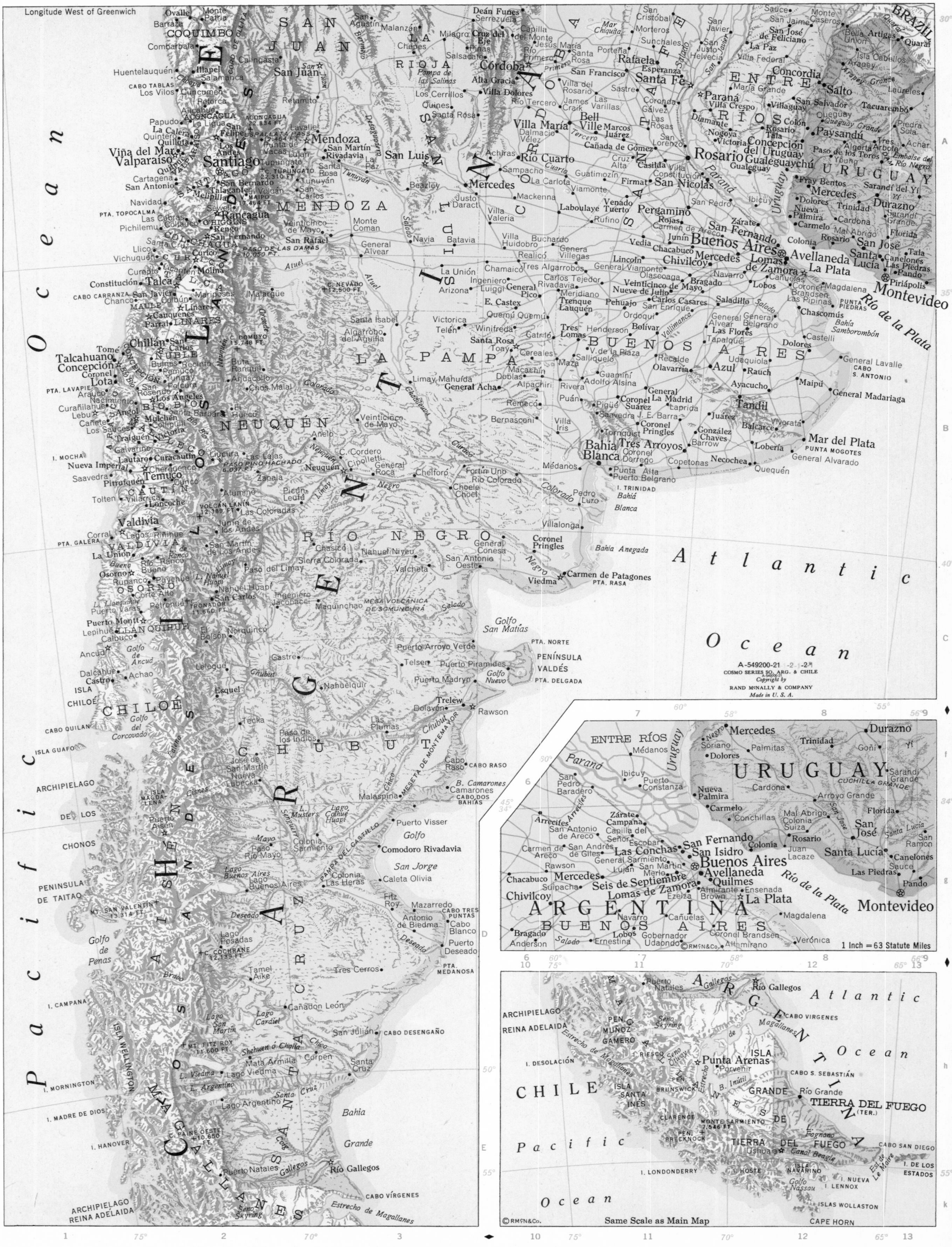

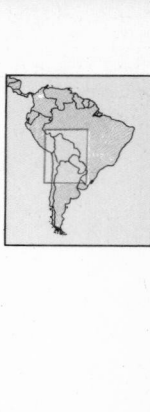

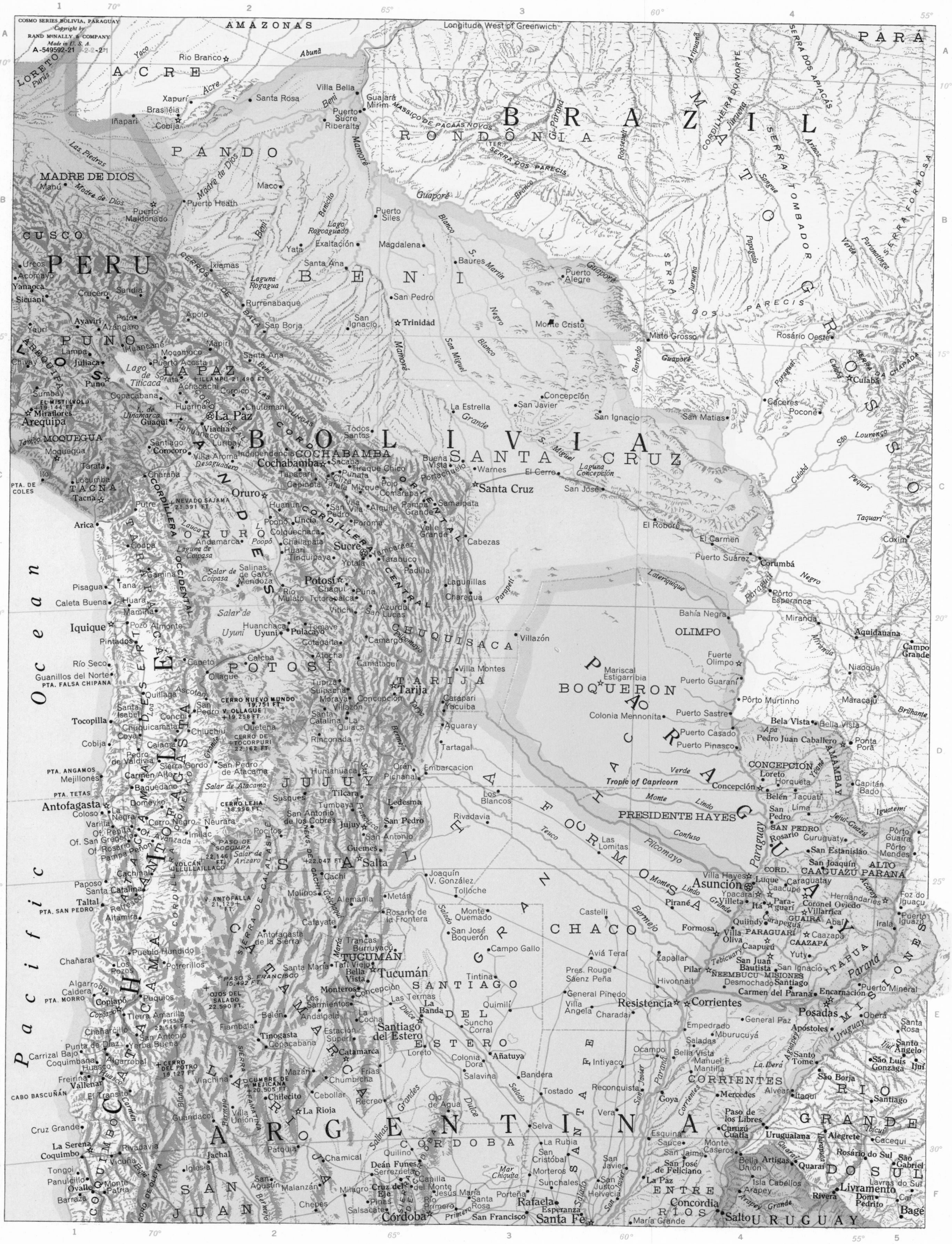

Oblique Conic Conformal Projection
SCALE 1:8,000,000 1 Inch = 126 Statute Miles

Statute Miles 50 0 50 100 150
Kilometers 50 0 50 100 150 200

MATO GROSSO

BRAZIL

MINAS GERAIS

BAHIA

ESPÍRITO SANTO

PARAGUAY

Asunción

ARGENTINA

SÃO PAULO

São Paulo

PARANÁ

Curitiba

SANTA CATARINA

Florianópolis

RIO GRANDE DO SUL

Porto Alegre

URUGUAY

Montevideo

Buenos Aires

La Plata

Atlantic Ocean

Rio de Janeiro

Tropic of Capricorn

A-540392-21 -2 -3°
COSMO SERIES URUGUAY
Copyright by
RAND McNALLY & COMPANY
Made in U.S.A.

1 Inch = 63 Statute Miles

MINAS GERAIS

RIO DE JANEIRO

Rio de Janeiro

Atlantic Ocean

MINAS GERAIS

SÃO PAULO

São Paulo

Santos

Atlantic Ocean

1 Inch = 63 Statute Miles

Statute Miles   50   0   50   100   150
Kilometers   50   0   50   100   150   200

Oblique Conic Conformal Projection
SCALE 1:8,000,000   1 Inch = 126 Statute Miles

Longitude West of Greenwich

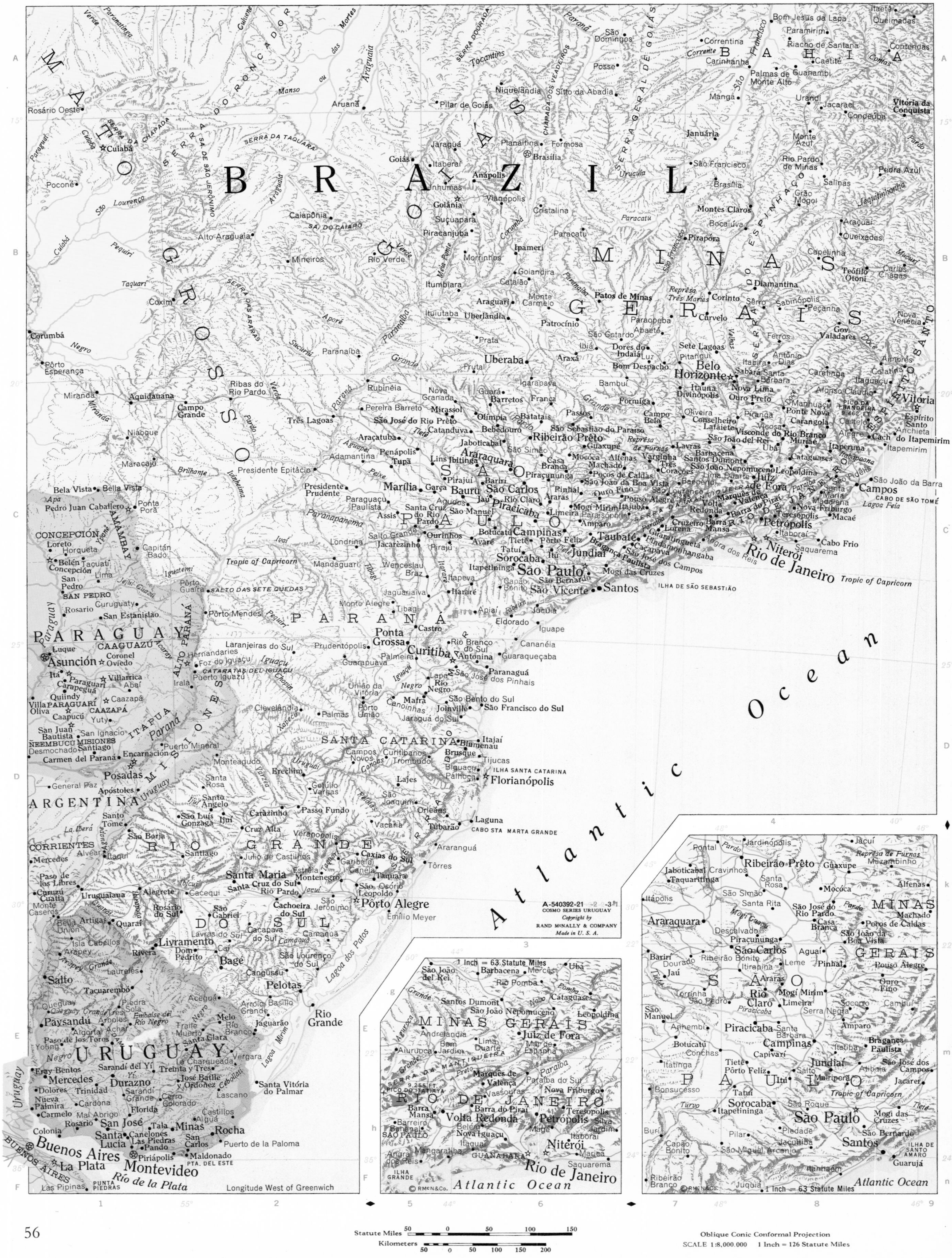

Longitude West of Greenwich

Atlantic Ocean

Oblique Conic Conformal Projection
SCALE 1:8,000,000   1 Inch = 126 Statute Miles

Statute Miles
Kilometers

1 Inch = 63 Statute Miles

A-540393-21
COSMO SERIES E. BRAZIL
Copyright by
RAND McNALLY & COMPANY
Made in U.S.A.

PERU AND ECUADOR

COLOMBIA

ECUADOR

PERU

BRAZIL

BOLIVIA

CHILE

VEN.

Pacific Ocean

Quito

Guayaquil

Lima

Callao

Cusco

Arequipa

Iquitos

La Paz

Sucre

Cochabamba

Potosí

Oruro

Pacific Ocean

ARCHIPIELAGO DE COLÓN (GALÁPAGOS IS.) (ECUADOR)

Same Scale as Main Map

I. DARWIN
I. WOLF
I. PINTA
I. MARCHENA
PTA. ALBEMARLE
I. GENOVESA
Equator
C. BERKELEY
B. de Banks
I. SAN SALVADOR (SANTIAGO)
I. FERNANDINA (NARBOROUGH)
I. PINZÓN
I. BALTRA
I. STA. CRUZ (INDEFATIGABLE)
B. Elizabeth
ISLA ISABELA (ALBEMARLE I.)
I. SAN CRISTÓBAL (CHATHAM)
PTA. ESSEX
Villamil
PTA. STA. FÉ
I. STA. FÉ
Puerto Baquerizo
I. STA. MARÍA
I. ESPANOLA

A-549400-2   -1-2-2-1
COSMO SERIES PERU, ECUADOR
Copyright by
RAND McNALLY & COMPANY
Made in U.S.A.

Longitude West of Greenwich

Statute Miles 50 0 50 100 150
Kilometers 50 0 50 100 150 200

Oblique Conic Conformal Projection
SCALE 1:8,000,000   1 Inch = 126 Statute Miles

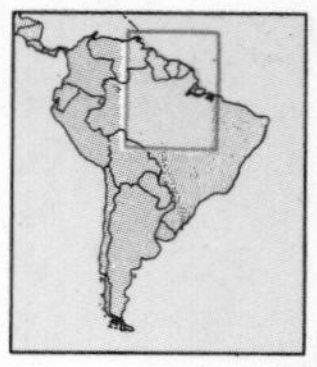

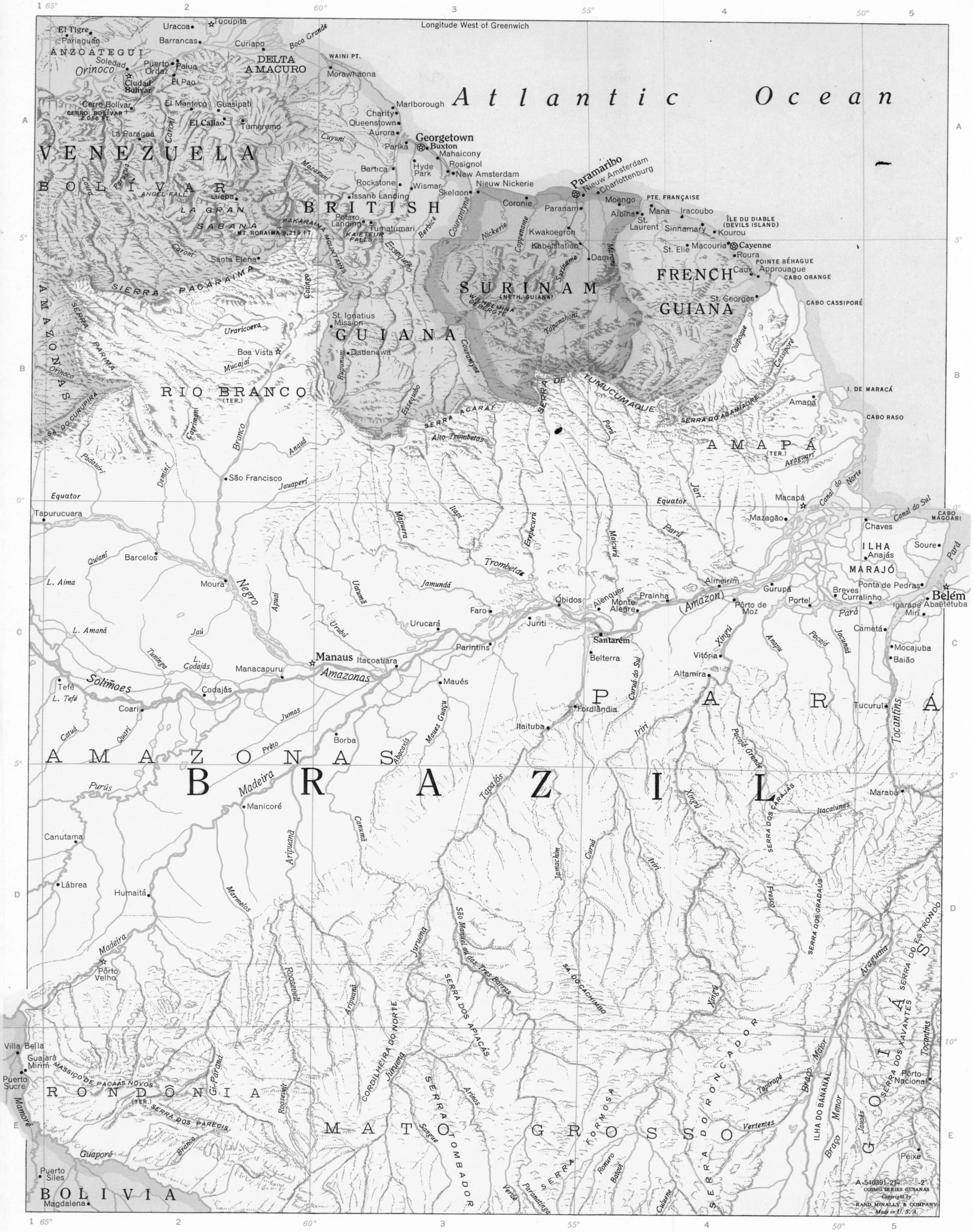

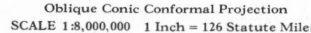

Oblique Conic Conformal Projection
SCALE 1:8,000,000 1 Inch = 126 Statute Miles

Statute Miles 50 0 50 100 150
Kilometers 50 0 50 100 150 200

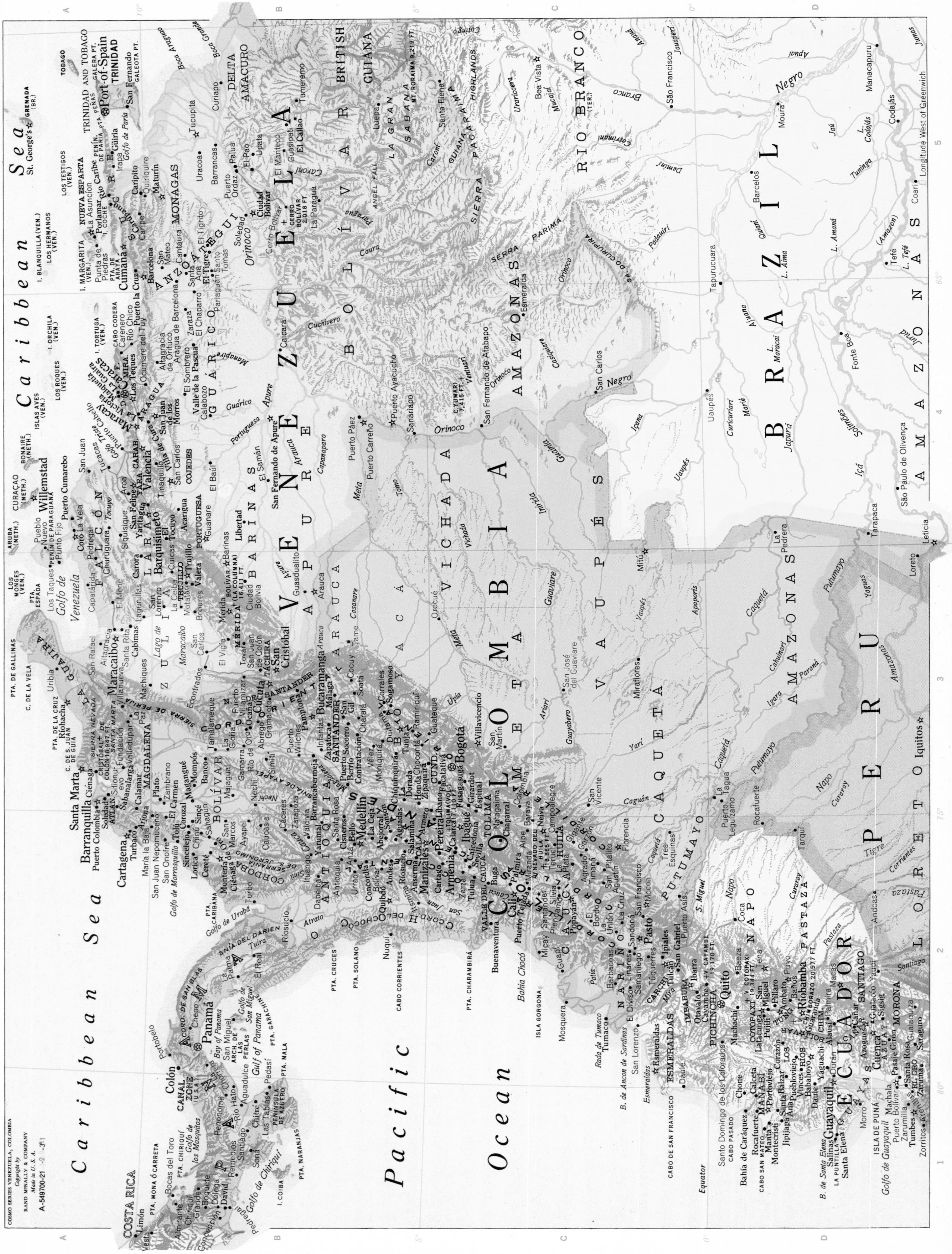

Caribbean Sea

Pacific Ocean

VENEZUELA

COLOMBIA

BRAZIL

PERU

ECUADOR

COSTA RICA

BRITISH GUIANA

TRINIDAD AND TOBAGO

Gulf of Venezuela

Lago de Maracaibo

CANAL ZONE

Bay of Panama

Gulf of Panama

Bogotá ⋆

Caracas ⋆

Quito ⋆

MT. RORAIMA 9,219 FT.

ANGEL FALL

SERRA PARIMA

GUIANA HIGHLANDS

LA GRAN SABANA

Orinoco

Rio Negro

Amazonas

ZULIA · LARA · FALCÓN · TRUJILLO · MÉRIDA · TÁCHIRA · APURE · BARINAS · PORTUGUESA · COJEDES · CARABOBO · ARAGUA · GUÁRICO · ANZOÁTEGUI · MONAGAS · SUCRE · DELTA AMACURO · BOLÍVAR · AMAZONAS · APURE

LA GUAJIRA · MAGDALENA · BOLÍVAR · CÓRDOBA · ANTIOQUIA · SANTANDER · NORTE DE SANTANDER · BOYACÁ · CUNDINAMARCA · CHOCÓ · CALDAS · TOLIMA · VALLE DEL CAUCA · CAUCA · HUILA · META · NARIÑO · PUTUMAYO · CAQUETÁ · VICHADA · GUAINÍA · VAUPÉS · AMAZONAS

ARUBA (NETH.) · CURAÇAO (NETH.) · BONAIRE (NETH.)

I. MARGARITA (VEN.) · LOS ROQUES (VEN.) · LA ORCHILA (VEN.) · LA TORTUGA (VEN.) · LOS TESTIGOS (VEN.)

GRENADA (BR.) · St. George's ⋆

Statute Miles 50 0 50 100 150
Kilometers 50 0 50 100 150 200

Oblique Conic Conformal Projection
SCALE 1:8,000,000 1 Inch = 126 Statute Miles

COSMO SERIES VENEZUELA, COLOMBIA
Copyright by
RAND McNALLY & COMPANY
Made in U.S.A.
A-549700-21

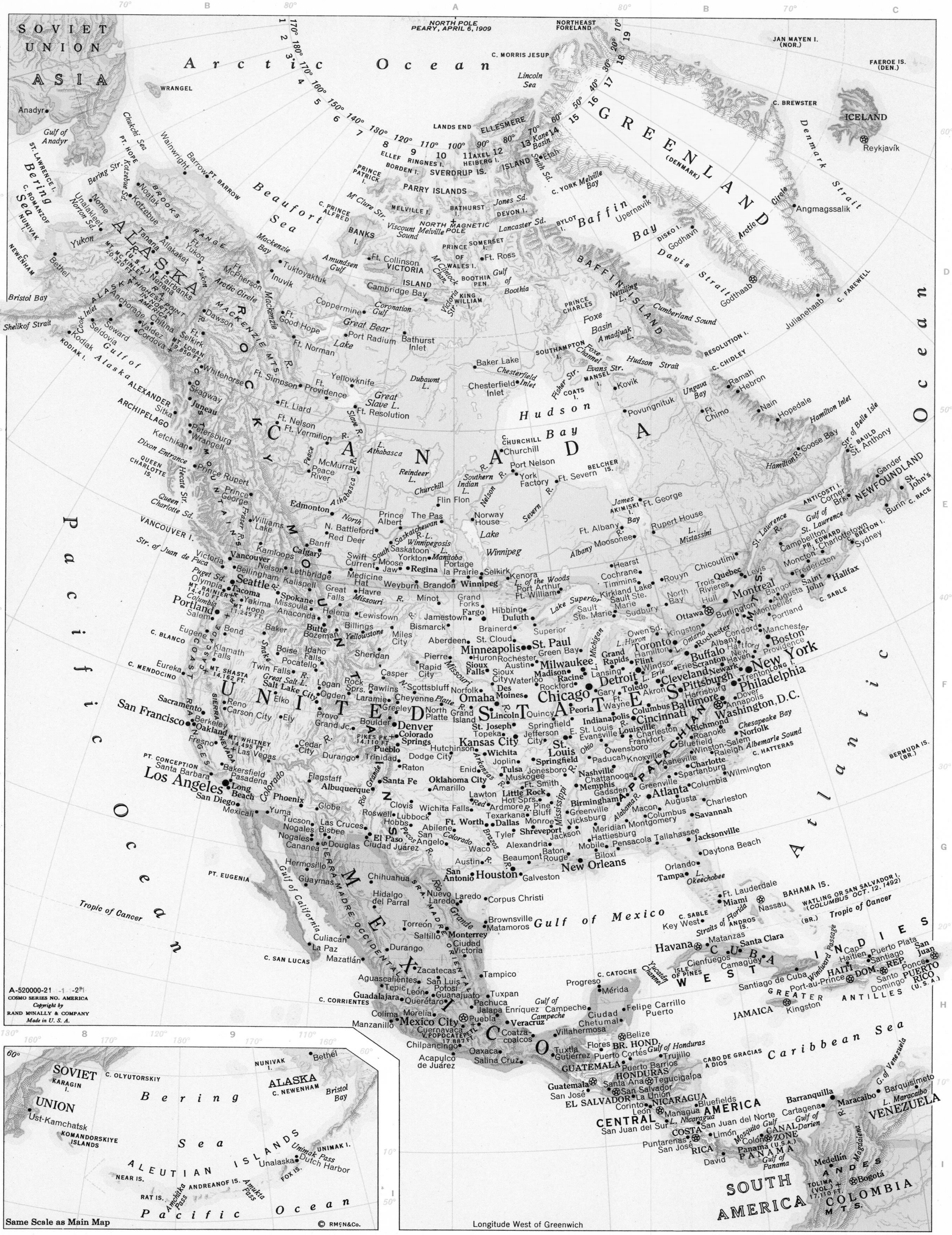

SOVIET UNION ASIA

Arctic Ocean

NORTH POLE
PEARY, APRIL 6, 1909

NORTHEAST FORELAND

JAN MAYEN I. (NOR.)

FAEROE IS. (DEN.)

ICELAND
Reykjavík

GREENLAND (DENMARK)

Beaufort Sea

Baffin Bay

Davis Strait

Denmark Strait

C. Farewell

BAFFIN ISLAND

ALASKA

Nome
Fairbanks
Anchorage

Bering Sea

Gulf of Alaska

MACKENZIE

R O C K Y   M O U N T A I N S

C A N A D A

Hudson Bay

Edmonton
Calgary
Vancouver
Victoria
Seattle
Tacoma
Portland

Regina
Winnipeg

NEWFOUNDLAND
St. John's

Pacific Ocean

Atlantic Ocean

U N I T E D   S T A T E S

San Francisco
Oakland
Los Angeles
Long Beach
San Diego
Sacramento
Fresno

Minneapolis
St. Paul
Milwaukee
Chicago
Detroit
Cleveland
Buffalo
Toronto
Montreal
Ottawa
Quebec
Boston
New York
Philadelphia
Baltimore
Washington, D.C.
Pittsburgh

Denver
Colorado Springs
Kansas City
St. Louis
Wichita
Oklahoma City
Dallas
Ft. Worth
Houston
San Antonio
New Orleans

Phoenix
Albuquerque
El Paso

A P P A L A C H I A N

Atlanta
Birmingham
Memphis
Nashville
Charlotte

Miami
Tampa
Jacksonville

Gulf of Mexico

M E X I C O

SIERRA MADRE OCCIDENTAL

Monterrey
Mexico City
Guadalajara
Veracruz
Puebla
Acapulco

Tropic of Cancer

HAVANA
CUBA
W E S T   I N D I E S
GREATER ANTILLES
HAITI
DOM. REP.
PUERTO RICO (U.S.A.)
JAMAICA

Caribbean Sea

CENTRAL AMERICA
GUATEMALA
BR. HOND.
HONDURAS
EL SALVADOR
NICARAGUA
COSTA RICA
PANAMA
CANAL ZONE

VENEZUELA
Caracas
COLOMBIA
Bogotá

SOUTH AMERICA
ANDES MTS.

Inset map (lower left):

SOVIET UNION

Bering Sea

ALASKA

ALEUTIAN ISLANDS

NEAR IS.
RAT IS.
ANDREANOF IS.
FOX IS.
Unalaska
Dutch Harbor

Pacific Ocean

Same Scale as Main Map

Longitude West of Greenwich

Lambert Azimuthal Equal Area Projection
SCALE 1:31,977,000   1 Inch = 490 Statute Miles

Statute Miles
100 0 100 200 300 400 500 600 700 800

Kilometers
100 0 100 200 400 600 800 1000

A-520000-21
COSMO SERIES NO. AMERICA
Copyright by
RAND McNALLY & COMPANY
Made in U.S.A.

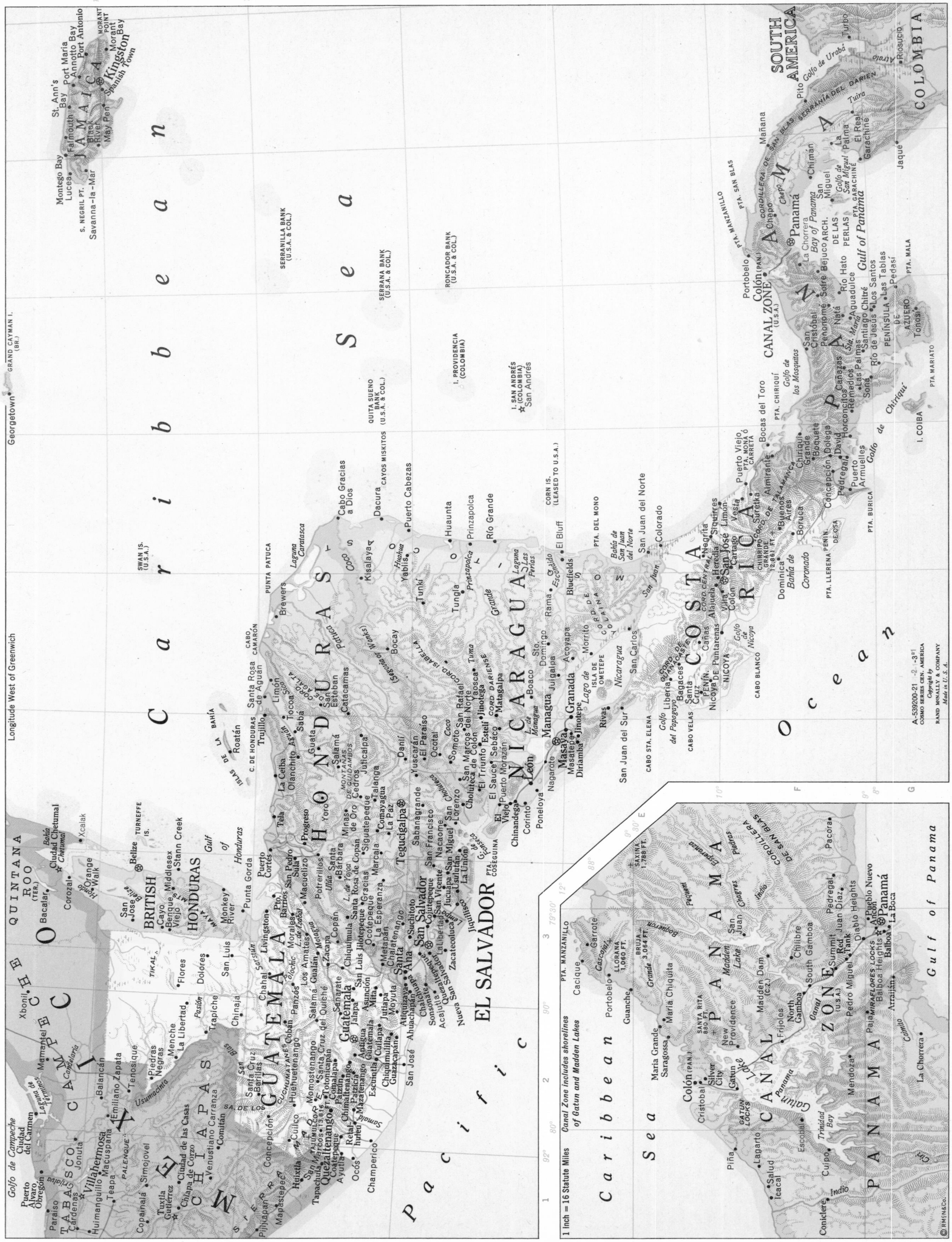

SOUTH AMERICA

COLOMBIA

C a r i b b e a n   S e a

P a c i f i c   O c e a n

Gulf of Panama

QUINTANA ROO

CAMPECHE

TABASCO

CHIAPAS

M E X I C O

GUATEMALA

BRITISH HONDURAS

EL SALVADOR

H O N D U R A S

N I C A R A G U A

C O S T A   R I C A

P A N A M A

CANAL ZONE

JAMAICA

GRAND CAYMAN I. (BR.)

SWAN IS. (U.S.A.)

SERRANILLA BANK (U.S.A. & COL.)

SERRANA BANK (U.S.A. & COL.)

RONCADOR BANK (U.S.A. & COL.)

I. PROVIDENCIA (COLOMBIA)

I. SAN ANDRÉS (COLOMBIA) / San Andrés

CORN IS. (LEASED TO U.S.A.)

CAYOS MISKITOS

I. COIBA

Longitude West of Greenwich

A-539200-21 -2,-3**
COSMO SERIES CEN. AMERICA
RAND MCNALLY & COMPANY
Made in U.S.A.

1 Inch = 16 Statute Miles

Canal Zone includes shorelines of Gatun and Madden Lakes

Statute Miles   25   0   25   75   125

Kilometers   25   0   25   75   125   175

Oblique Conic Conformal Projection
SCALE 1:6,000,000 1 Inch = 95 Statute Miles

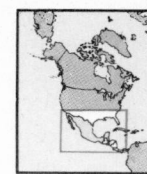

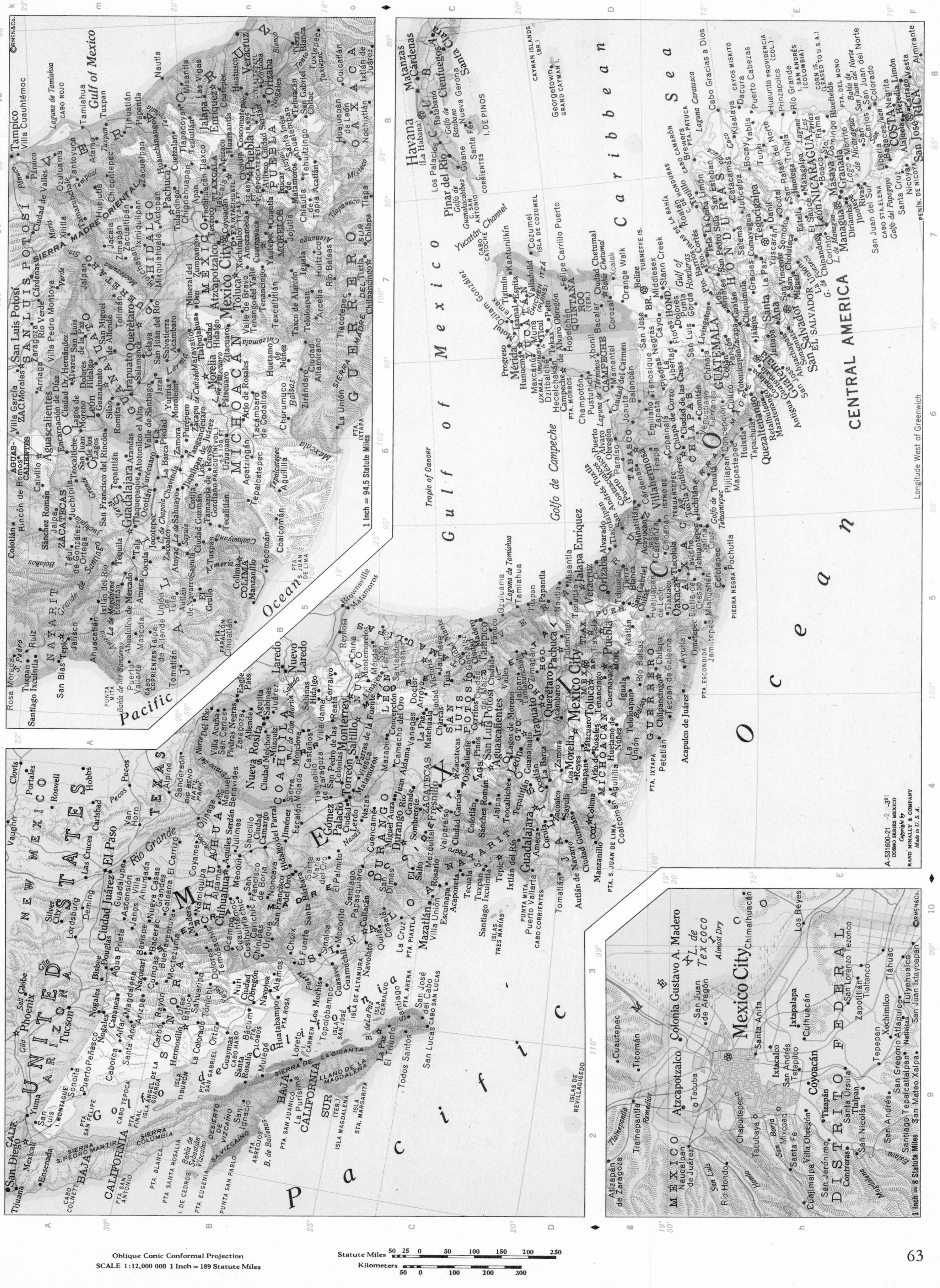

Oblique Conic Conformal Projection
SCALE 1:12,000 000 1 Inch = 189 Statute Miles

Statute Miles 50 25 0 50 100 150 200 250
Kilometers 50 0 100 200 300

WEST INDIES

A t l a n t i c   O c e a n

C a r i b b e a n   S e a

Gulf of Mexico

FLORIDA

BAHAMA ISLANDS

LEEWARD ISLANDS

WINDWARD ISLANDS

HISPANIOLA

HAITI

DOMINICAN REPUBLIC

PUERTO RICO

CUBA

JAMAICA

Same Scale as Main Map

A-533200-21 -1 -2
COMBO SERIES W. INDIES
Copyright by
RAND McNALLY & COMPANY
Made in U.S.A.

Statute Miles 25 0 25 75 125
Kilometers 25 0 25 75 125 175

Oblique Conic Conformal Projection
SCALE 1:6,000,000 1 Inch = 95 Statute Miles

1 Inch = 4 Statute Miles

HAVANA (La Habana)

©RM&N&Co.

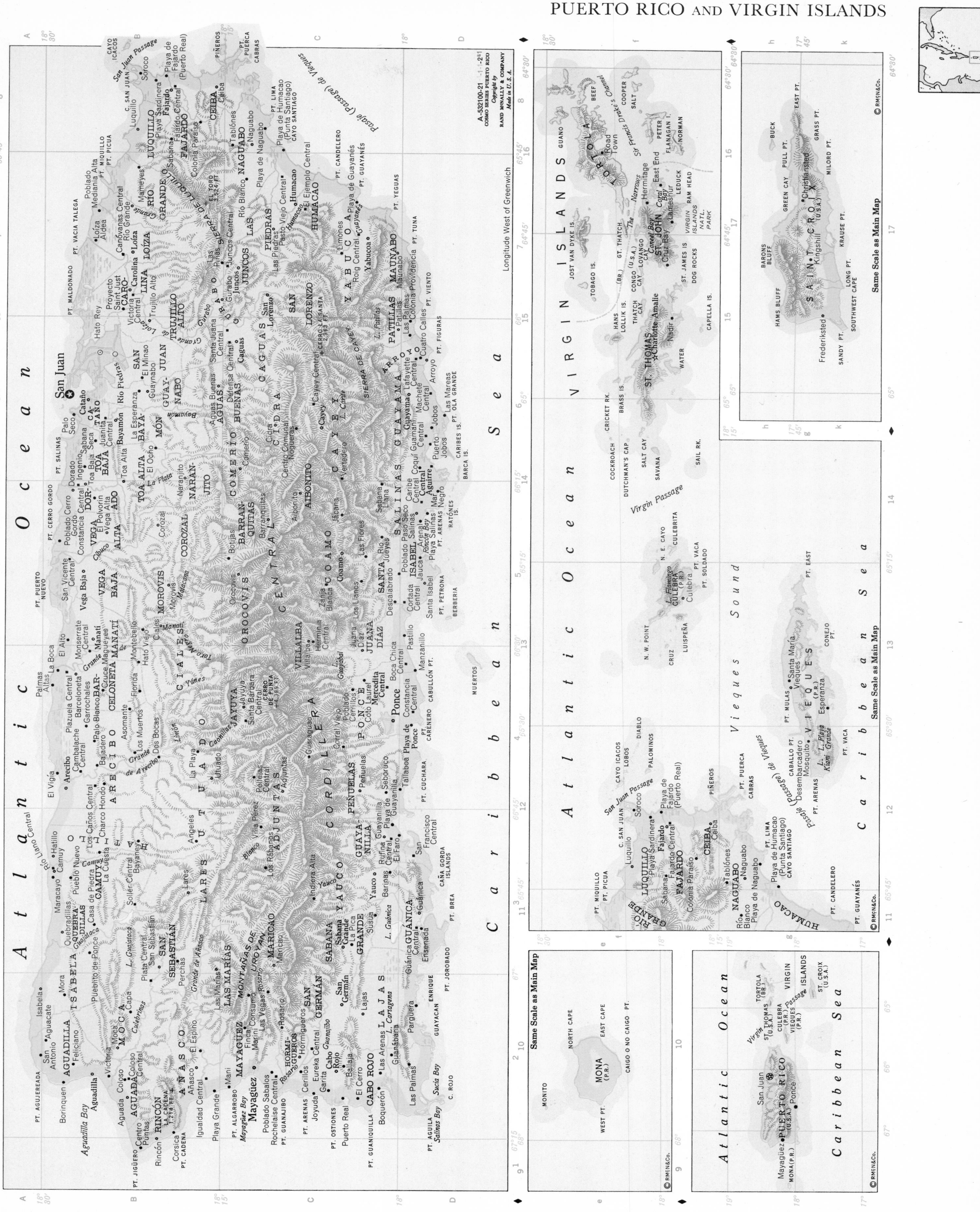

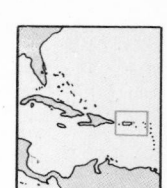

Atlantic Ocean
Caribbean Sea
San Juan
Ponce
Mayagüez
Arecibo
CORDILLERA CENTRAL
Pasaje (Passage) de Vieques
VIRGIN ISLANDS
ST. THOMAS
ST. JOHN
TORTOLA
SAINT CROIX (U.S.A.)
VIEQUES
CULEBRA (P.R.)
MONA (P.R.)
Virgin Passage
Vieques Sound
Same Scale as Main Map
Longitude West of Greenwich
A-532100-21 COSMO SERIES PUERTO RICO Copyright by RAND McNALLY & COMPANY Made in U. S. A.

Lambert Conformal Conic Projection
SCALE 1 : 600,000 1 Inch = 9.5 Statute Miles

Statute Miles 5 0 5 10
Kilometers 5 0 5 10 15

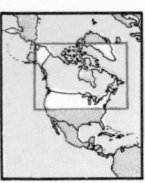

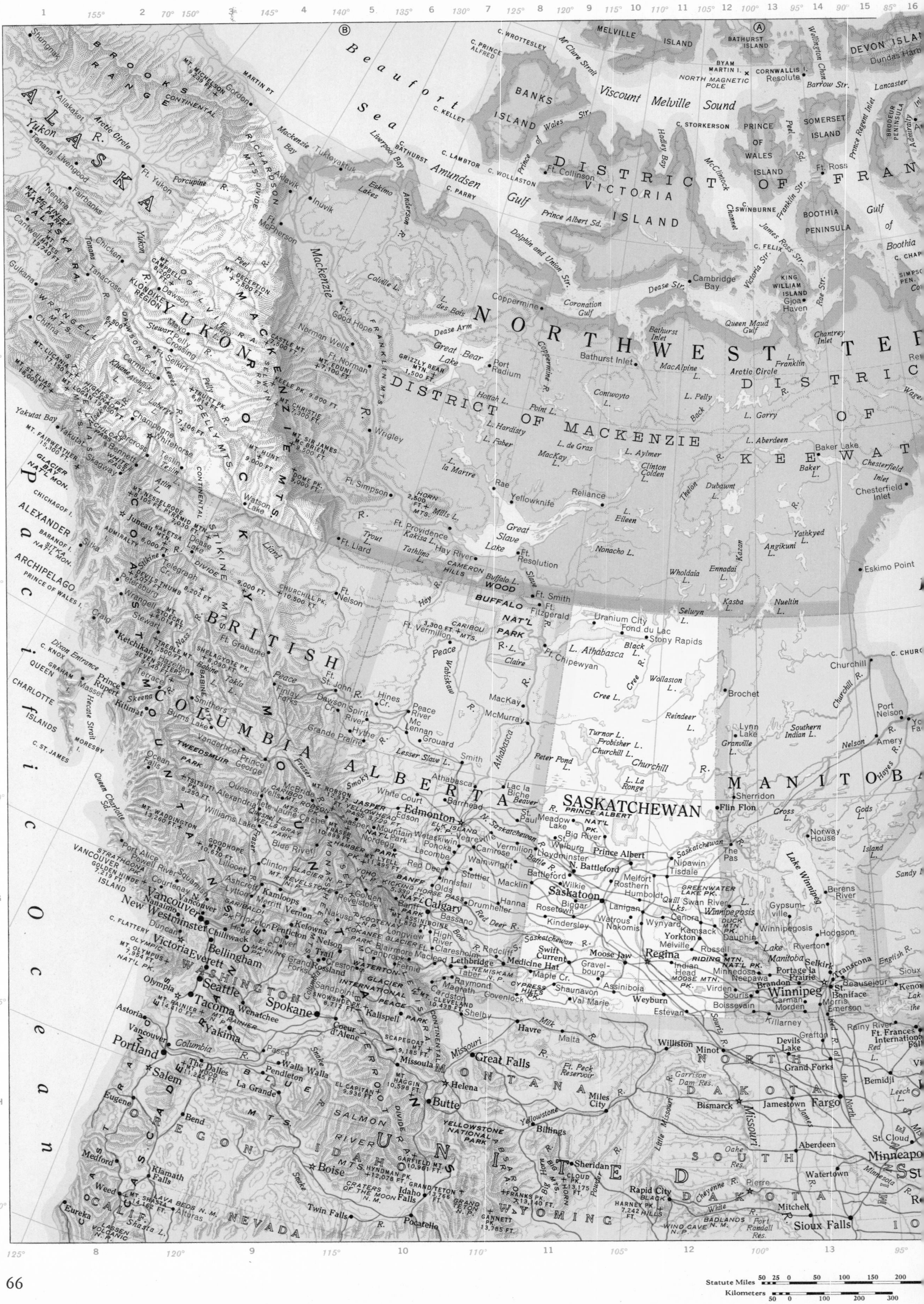
Beaufort Sea
Pacific Ocean
ALASKA
YUKON
BRITISH COLUMBIA
ALBERTA
SASKATCHEWAN
MANITOBA
NORTHWEST TERRITORIES
DISTRICT OF MACKENZIE
DISTRICT OF KEEWATIN
DISTRICT OF FRANKLIN
WASHINGTON
OREGON
IDAHO
MONTANA
WYOMING
NORTH DAKOTA
SOUTH DAKOTA
NEVADA
CALIF.
UNITED STATES
BROOKS RANGE
RICHARDSON MTS.
MACKENZIE MTS.
ROCKY MTS.
COAST MTS.
CASCADE RA.
BANKS ISLAND
VICTORIA ISLAND
MELVILLE ISLAND
PRINCE OF WALES ISLAND
SOMERSET ISLAND
DEVON ISLAND
BOOTHIA PENINSULA
KING WILLIAM ISLAND
QUEEN CHARLOTTE ISLANDS
VANCOUVER ISLAND
ALEXANDER ARCHIPELAGO
Viscount Melville Sound
Amundsen Gulf
Coronation Gulf
Queen Maud Gulf
Gulf of Boothia
Great Bear Lake
Great Slave Lake
L. Athabasca
Lake Winnipeg
Lake Winnipegosis
Dixon Entrance
Hecate Strait
Yukon R.
Mackenzie R.
Peace R.
Fraser R.
Columbia R.
Saskatchewan R.
Missouri R.
WOOD BUFFALO NAT'L PARK
JASPER NAT'L PK.
BANFF NAT'L PARK
YELLOWSTONE NATIONAL PARK
GLACIER NAT'L PARK
Whitehorse
Dawson
Fairbanks
Inuvik
Aklavik
Tuktoyaktuk
Yellowknife
Ft. Smith
Hay River
Uranium City
Edmonton
Calgary
Lethbridge
Medicine Hat
Red Deer
Vancouver
Victoria
New Westminster
Prince Rupert
Prince George
Kamloops
Kelowna
Penticton
Nelson
Saskatoon
Regina
Prince Albert
Moose Jaw
Swift Current
Flin Flon
The Pas
Brandon
Winnipeg
Churchill
Seattle
Tacoma
Spokane
Portland
Salem
Eugene
Boise
Helena
Butte
Great Falls
Billings
Bismarck
Fargo
Minot
Rapid City
Sioux Falls
Eskimo Point
Baker Lake
Chesterfield Inlet
Cambridge Bay
Coppermine
NORTH MAGNETIC POLE
Statute Miles 50 25 0 50 100 150 200
Kilometers 50 0 100 200 300

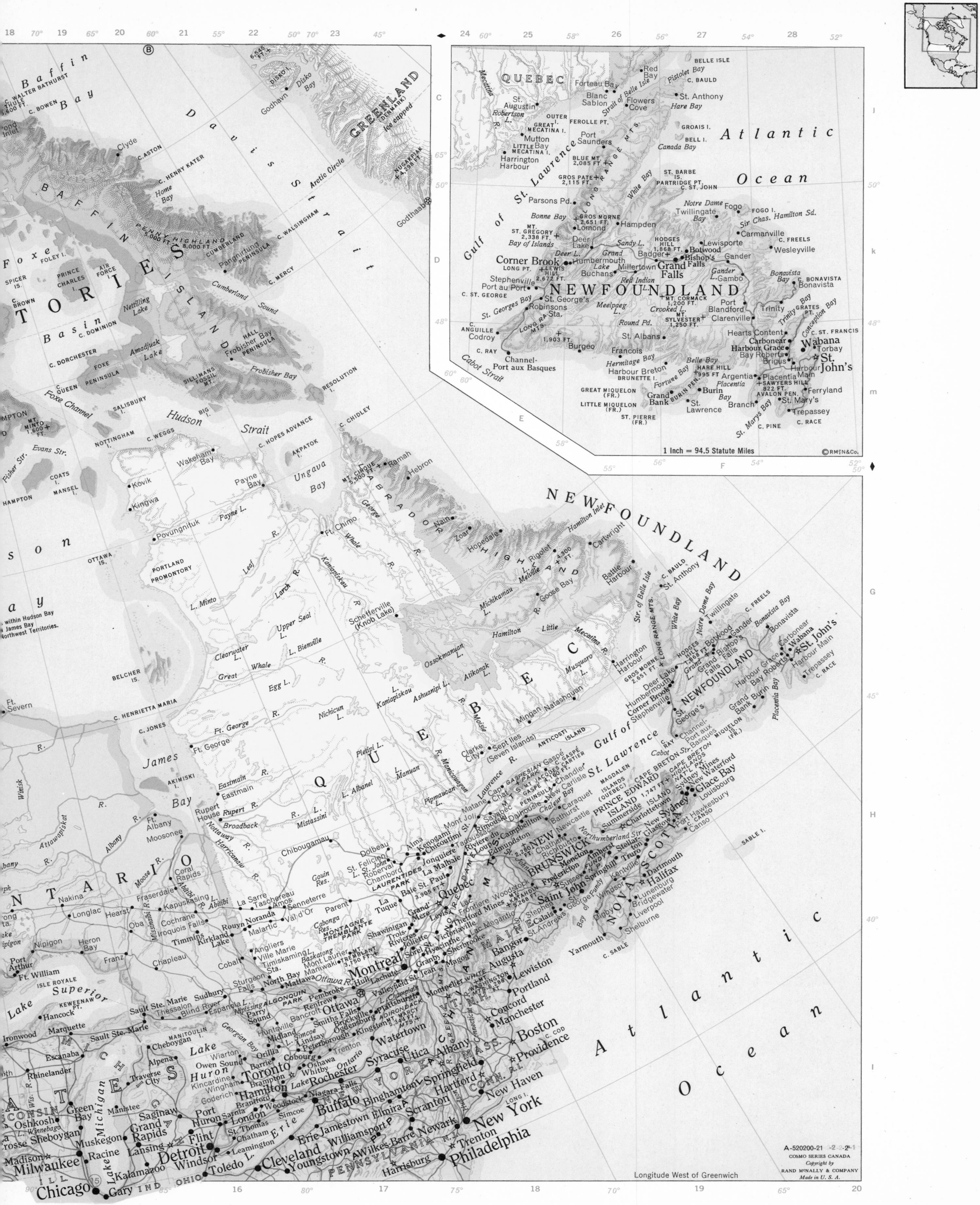

Lambert Conformal Conic Projection
SCALE 1:12,000,000   1 Inch = 189 Statute Miles

67

BRITISH COLUMBIA

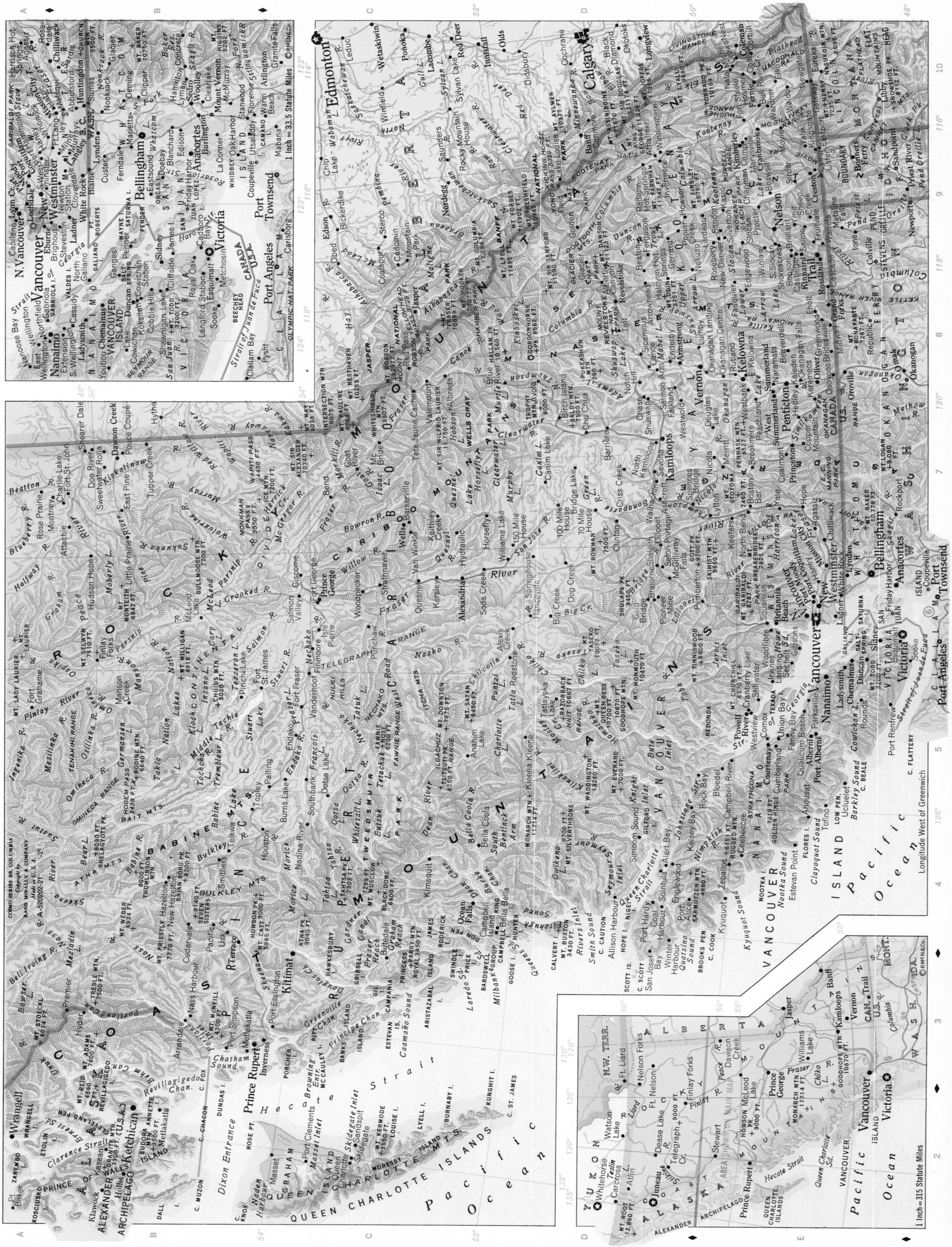

Statute Miles 10 0 10 20 30 40 50 60 70 80 90 100

Kilometers 10 0 10 20 30 40 50 60 70 80 90 100 120 140

Oblique Cylindrical Projection
SCALE 1:4,255,000 1 Inch = 67 Statute Miles

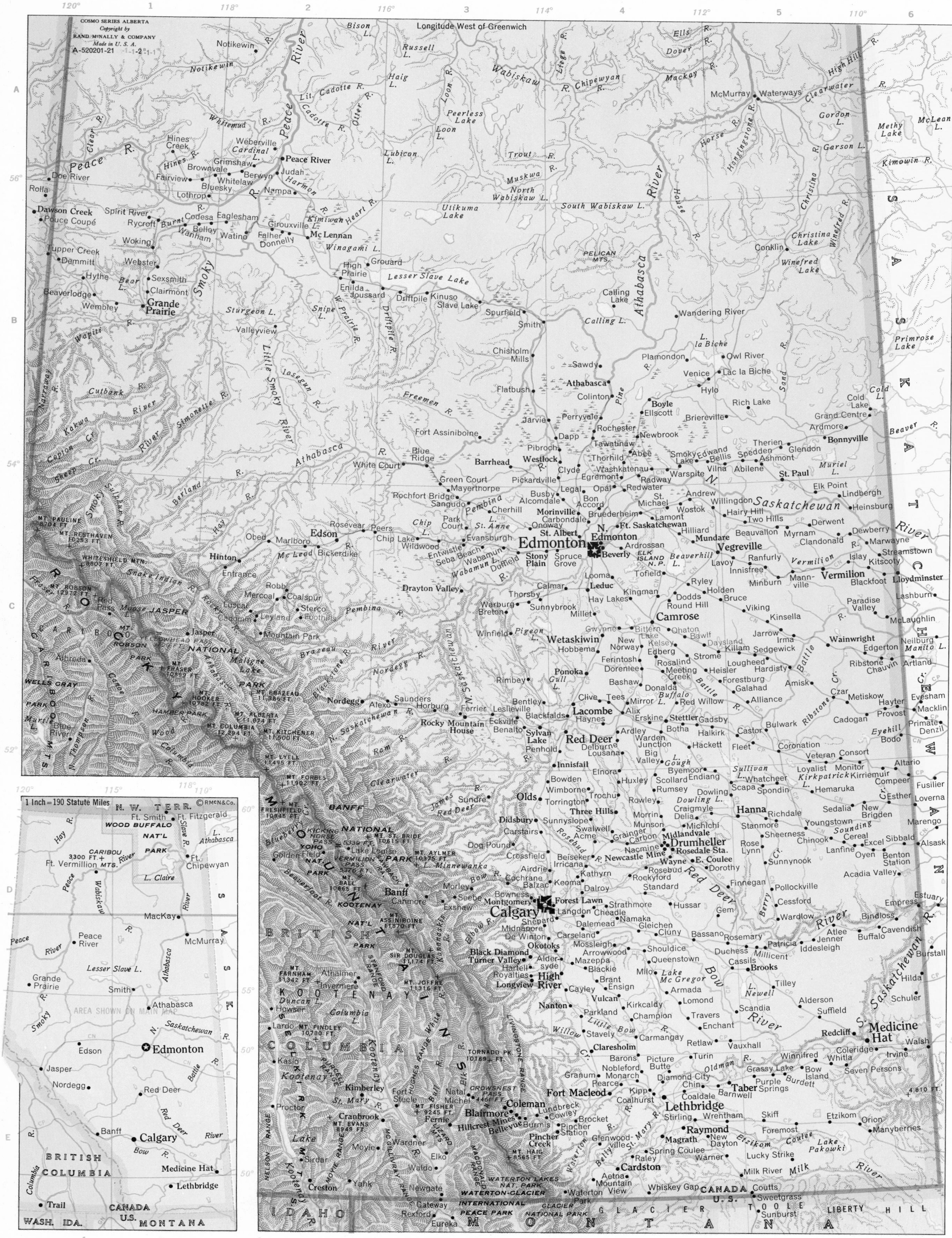

COSMO SERIES SASKATCHEWAN
Copyright by
RAND McNALLY & COMPANY
Made in U.S.A.
A-520209-21

1 Inch = 210 Statute Miles

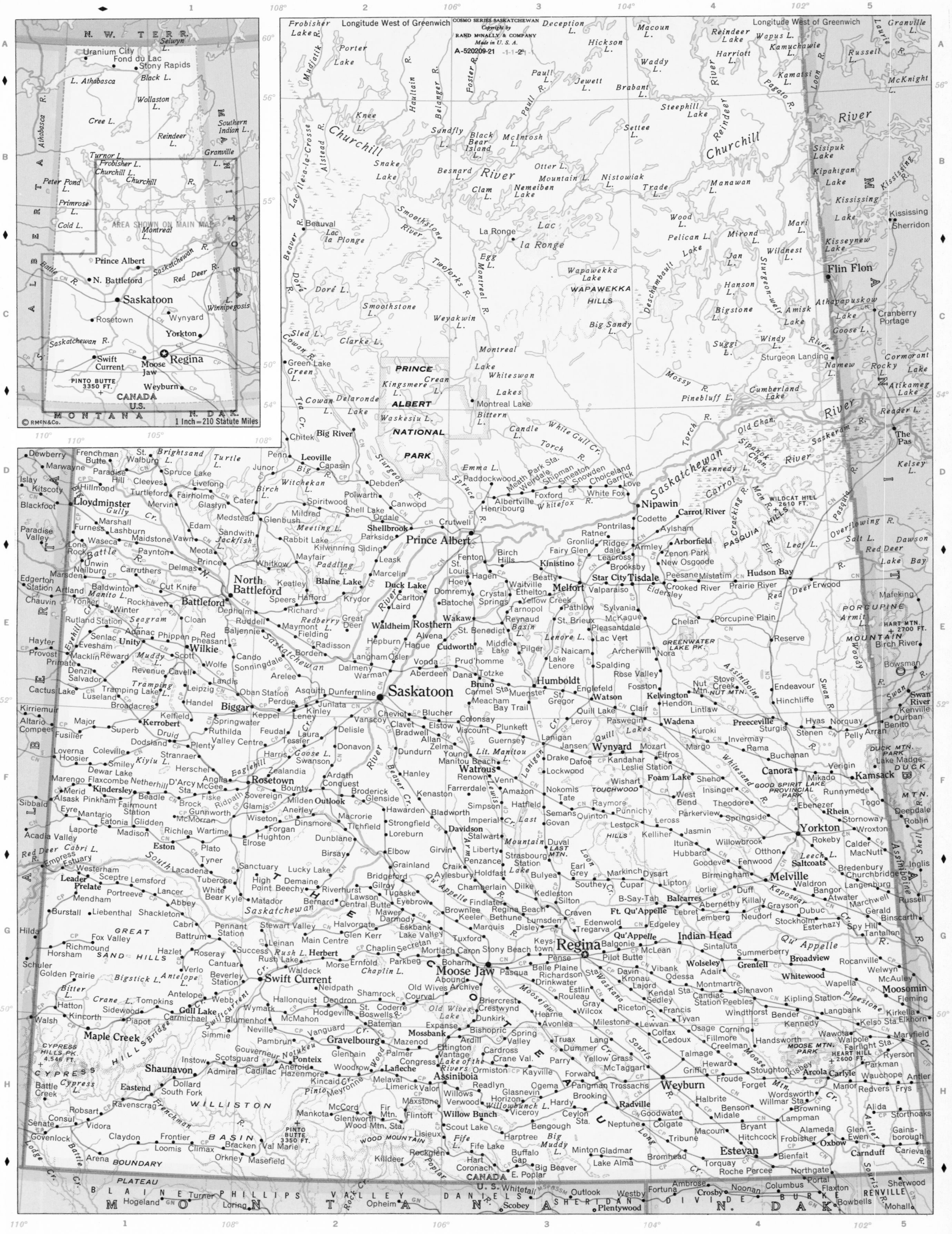

Statute Miles 10 0 10 20 30 40 50 60
Kilometers 10 0 10 20 40 60 80

Oblique Cylindrical Projection
SCALE 1:2,827,000 1 Inch = 44 Statute Miles

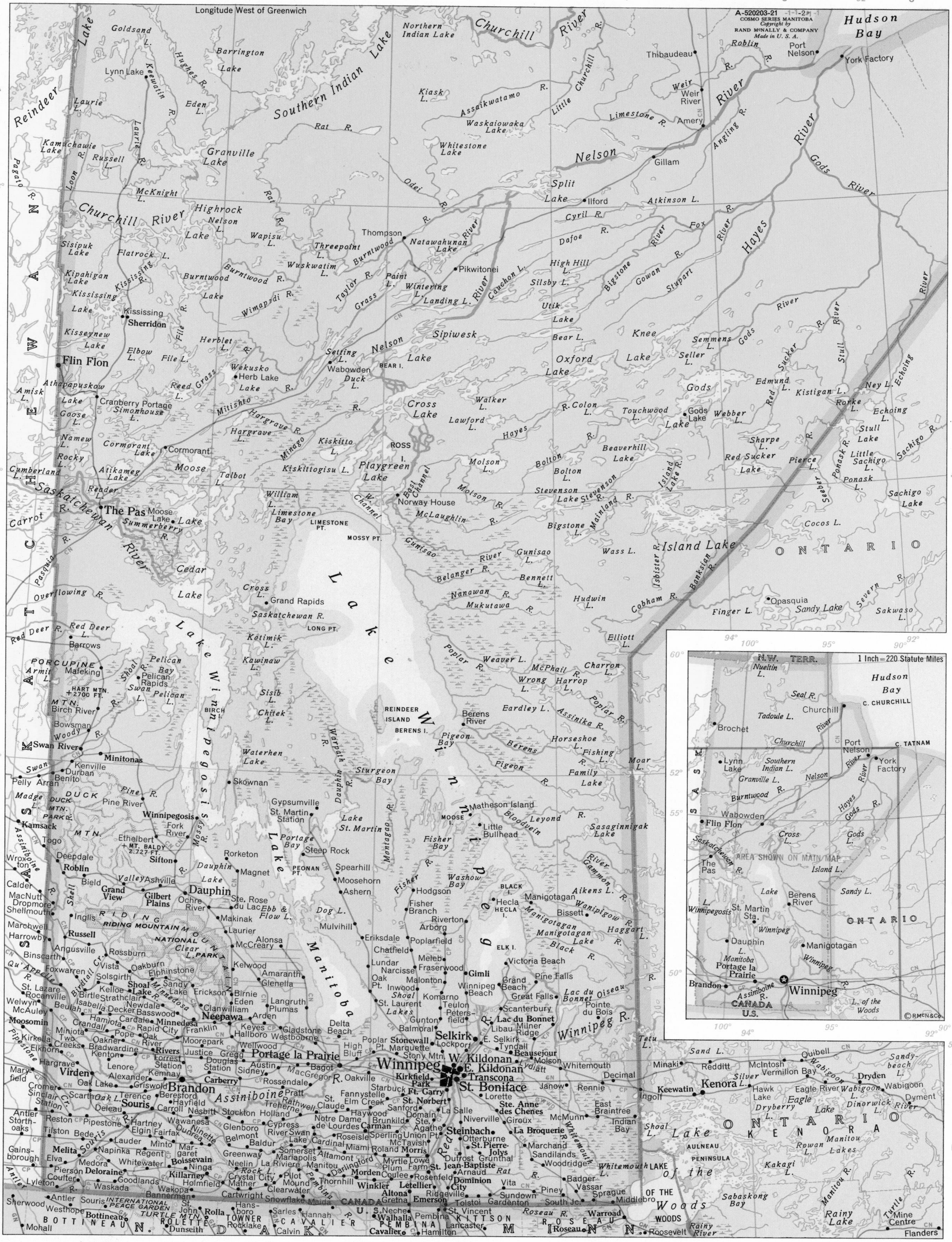

A-520203-21
COSMO SERIES MANITOBA
Copyright by
RAND McNALLY & COMPANY
Made in U.S.A.

Longitude West of Greenwich

Oblique Cylindrical Projection
SCALE 1:3,167,000 1 Inch = 50 Statute Miles

Statute Miles
10 0 10 20 30 40 50 60 70

Kilometers
10 0 10 20 40 60 80 100

1 Inch = 220 Statute Miles

AREA SHOWN ON MAIN MAP

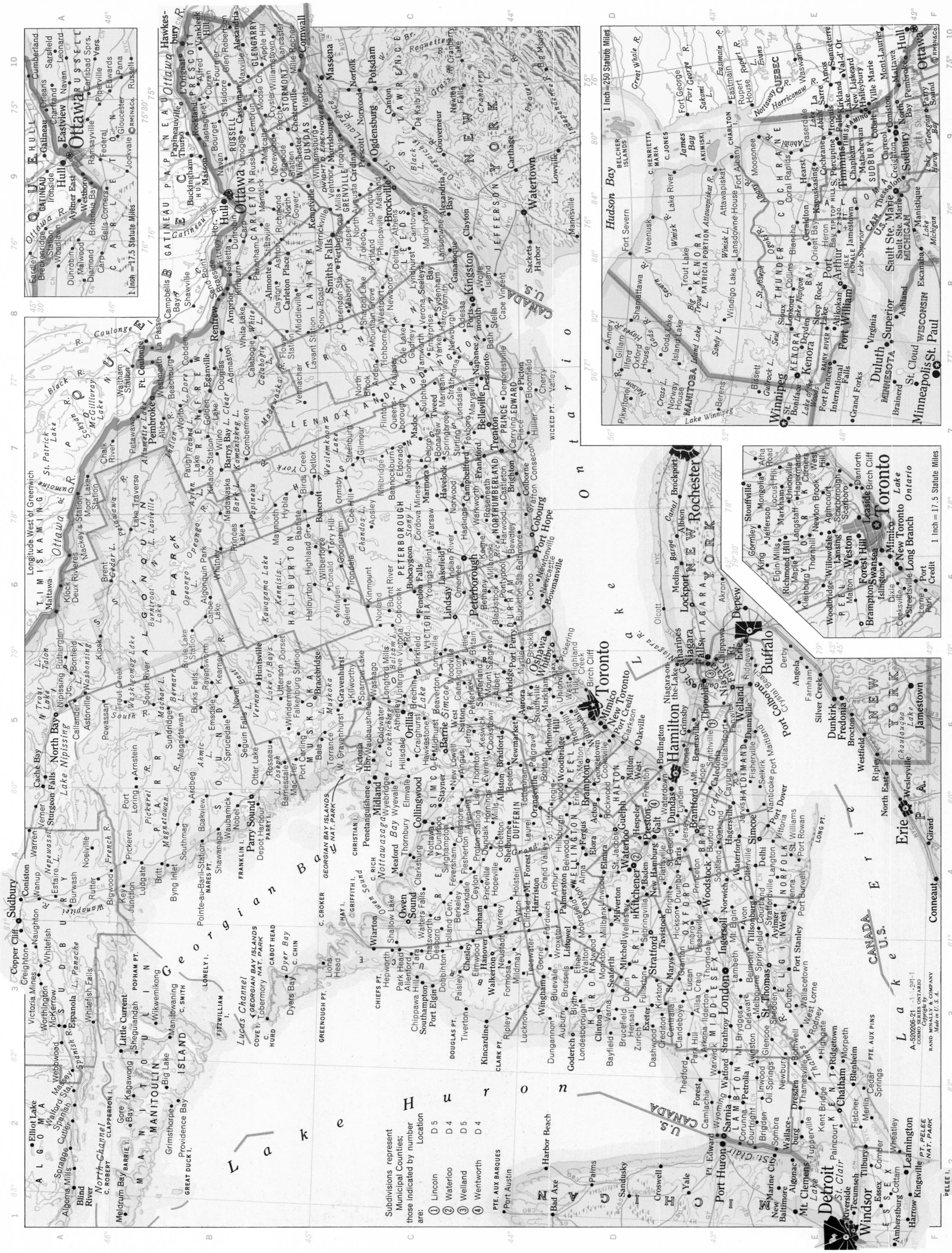

ONTARIO

Oblique Cylindrical Projection
SCALE 1:2,226,000 1 Inch = 35 Statute Miles

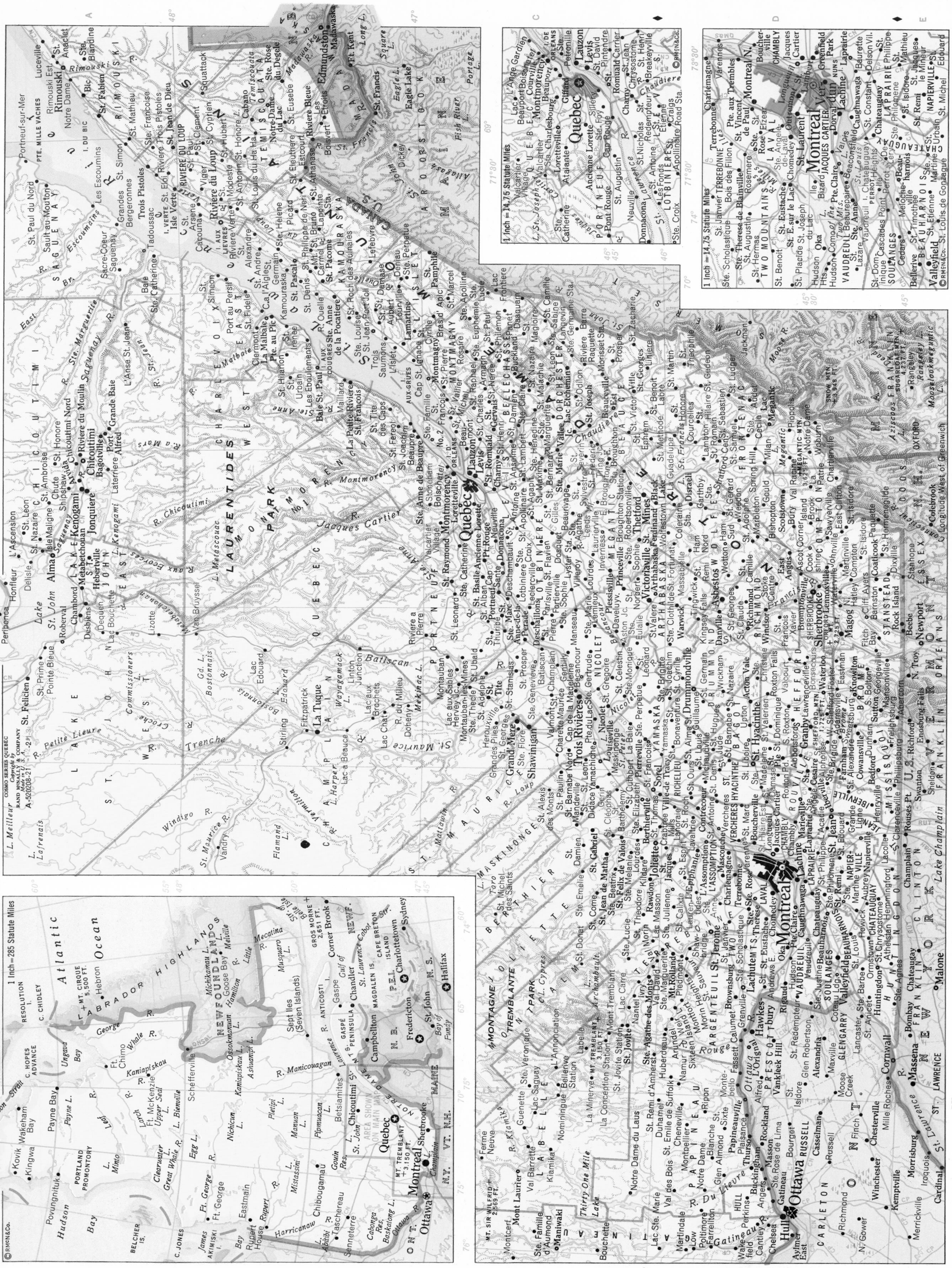

COSMO SERIES QUEBEC
Copyright by
RAND MCNALLY & COMPANY
A-520008-21

Oblique Cylindrical Projection
SCALE 1:1,929,000   1 Inch = 30.5 Statute Miles

Statute Miles
Kilometers

1 Inch = 14.75 Statute Miles

1 Inch = 14.75 Statute Miles

1 Inch = 285 Statute Miles

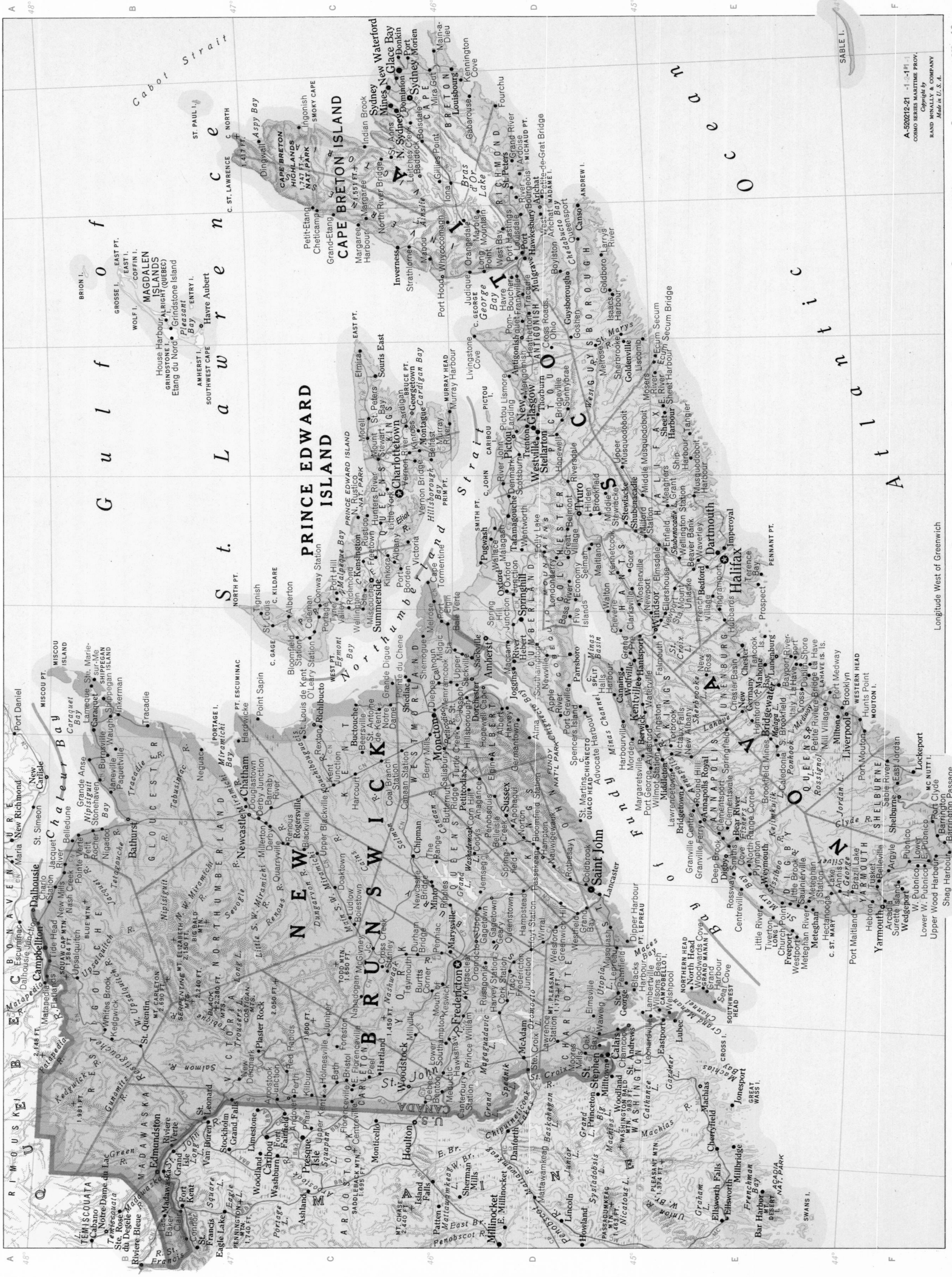

MARITIME PROVINCES

Gulf of St. Lawrence

Cabot Strait

MAGDALEN ISLANDS

PRINCE EDWARD ISLAND

CAPE BRETON ISLAND

NEW BRUNSWICK

NOVA SCOTIA

Atlantic Ocean

Bay of Fundy

Chaleur Bay

Northumberland Strait

CANADA
U.S.

Charlottetown

Fredericton

Saint John

Moncton

Halifax

Dartmouth

Sydney

Yarmouth

Oblique Cylindrical Projection
SCALE 1:2,312,000   1 Inch = 36.5 Statute Miles

Statute Miles   5 0 5 10   20   30   40   50
Kilometers   5 0 5   15   25   35   45   55   75

Oblique Cylindrical Projection
SCALE 1:2,312,000   1 Inch = 36.5 Statute Miles

Longitude West of Greenwich

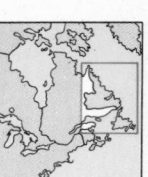

Districts indicated by number are:
Location
① Carbonear-Bay de Verde E-5
② Harbour Grace E-5
③ Harbour Main-Bell Island E-5
④ Port de Grave E-5
⑤ St. John's East E-5
⑥ St. John's West E-5

COSMO SERIES NEWFOUNDLAND
Copyright by
RAND McNALLY & COMPANY
Made in U.S.A.
A-520204-21 -1 -2

Lambert Conformal Conic Projection
SCALE 1 : 3,000,000   1 Inch = 47 Statute Miles

Statute Miles
Kilometers

1 Inch = 126 Statute Miles

©RMcN&Co.

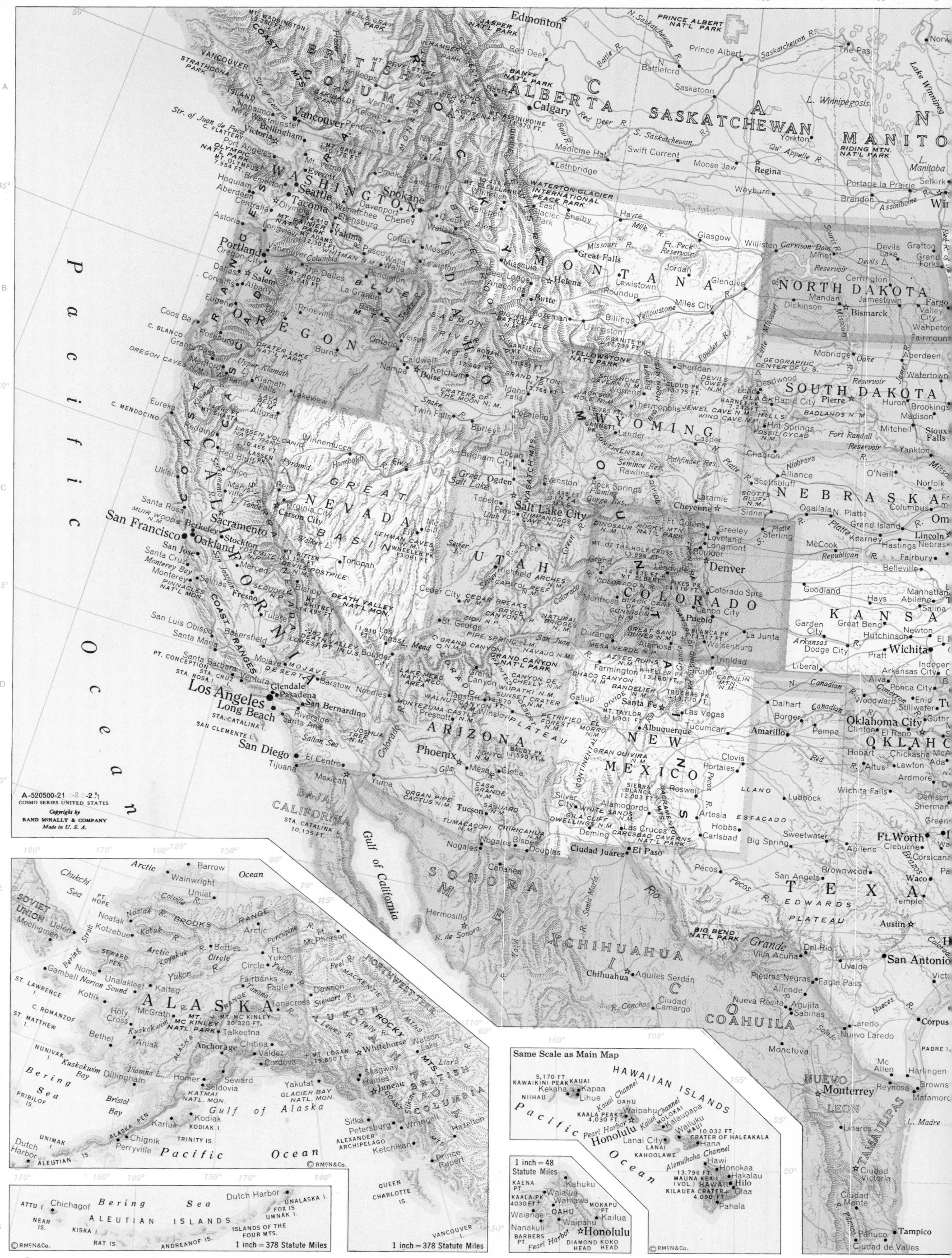

Lambert Conformal Conic Projection
ALE 1:12,000,000 1 Inch = 189 Statute Miles

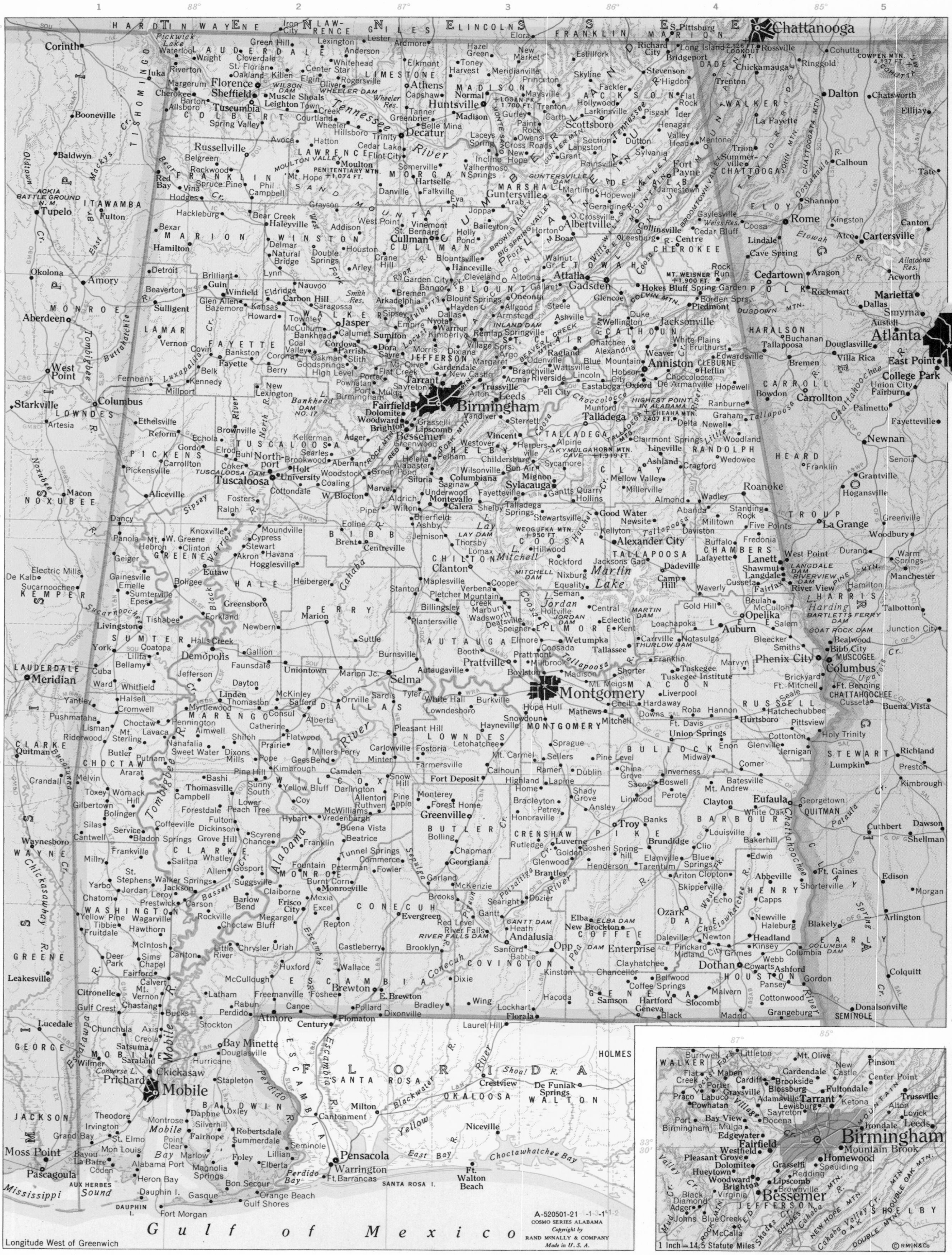

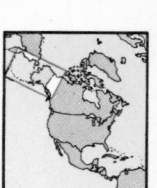

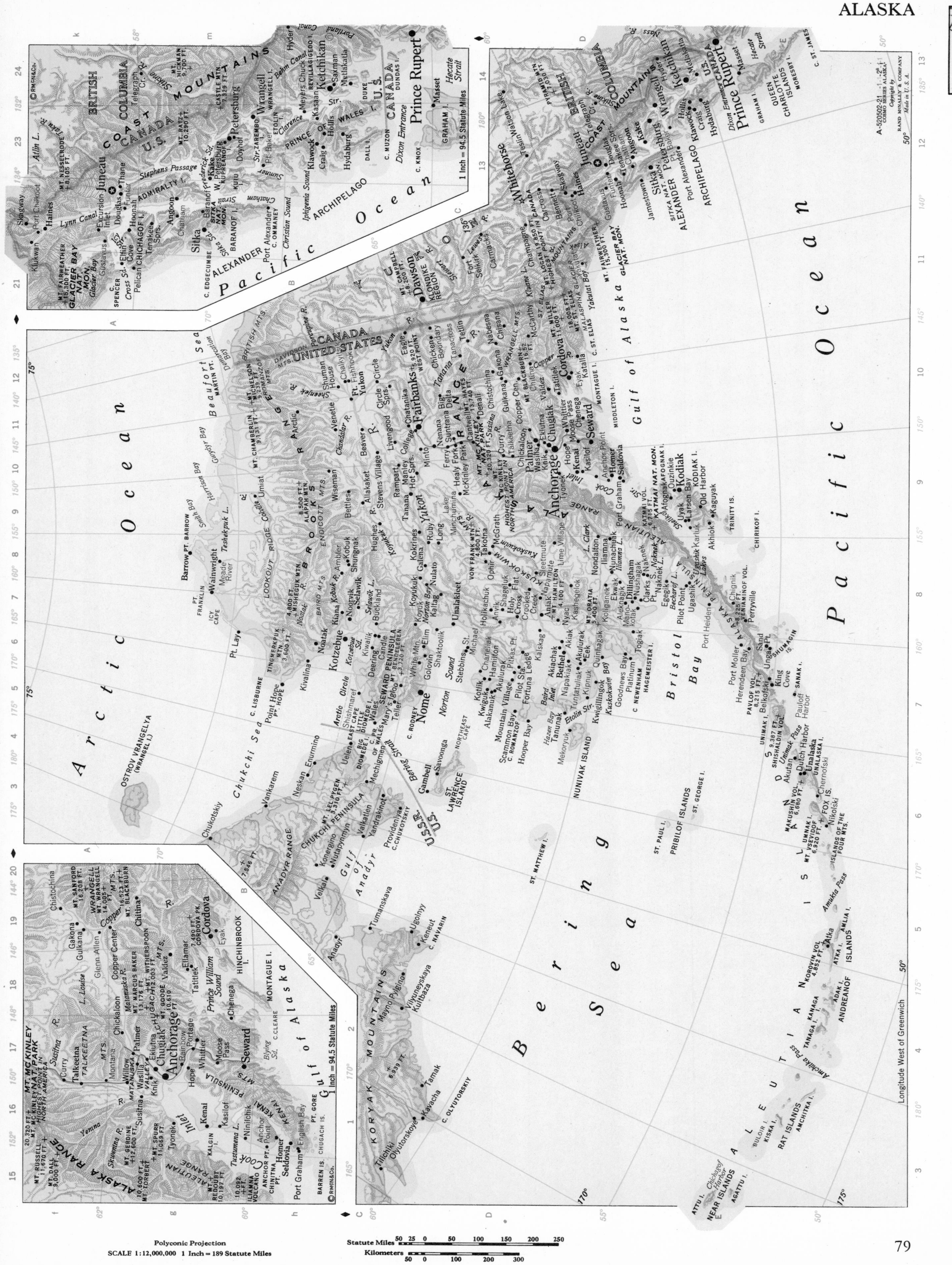

Arctic Ocean
Pacific Ocean
Bering Sea
Gulf of Alaska
Bristol Bay
Beaufort Sea
Chukchi Sea
Norton Sound
Gulf of Anadyr
BROOKS RANGE
ALASKA RANGE
ALEUTIAN ISLANDS
COAST MOUNTAINS
KORYAK MOUNTAINS
ANADYR RANGE
CANADA
UNITED STATES
U.S.S.R.
U.S.
YUKON
BRITISH COLUMBIA
ALEXANDER ARCHIPELAGO
Anchorage
Fairbanks
Juneau
Nome
Kodiak
Sitka
Ketchikan
Prince Rupert
Whitehorse
Dawson
Seward
Cordova
Kotzebue
Barrow
MT. McKINLEY
Longitude West of Greenwich
1 Inch = 94.5 Statute Miles
Polyconic Projection
SCALE 1:12,000,000 1 Inch = 189 Statute Miles
Statute Miles 50 25 0 50 100 150 200 250
Kilometers 50 0 100 200 300
A-520502-21 -1-1-2-1
COSMO SERIES ALASKA
Copyright by RAND McNALLY & COMPANY
Made in U. S. A.

Arizona

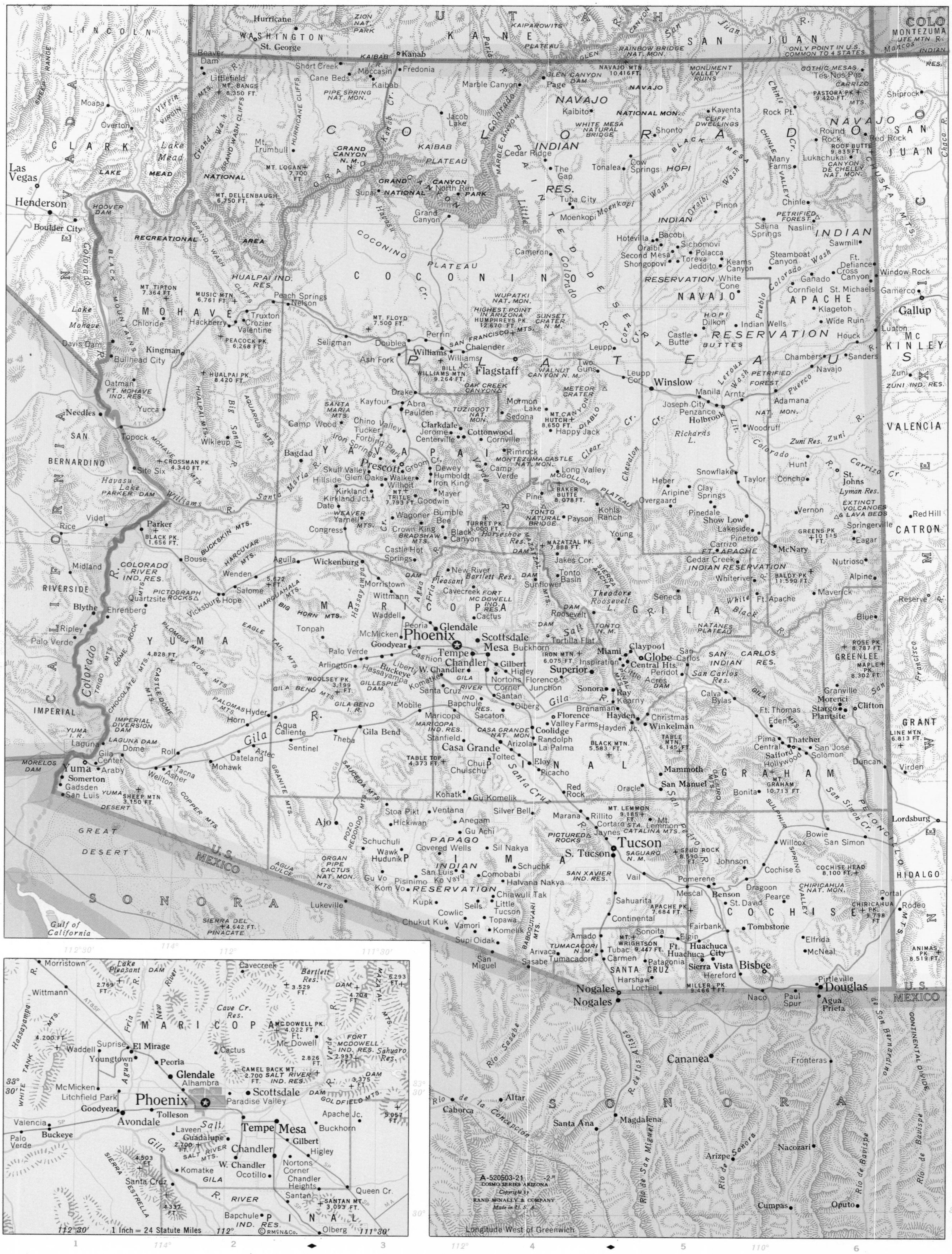

Statute Miles

Kilometers

Lambert Conformal Conic Projection
SCALE 1:2,725,000   1 Inch = 43 Statute Miles

1 Inch = 24 Statute Miles

A-520503-21

COSMO SERIES ARIZONA
Copyright by
RAND M\cNALLY & COMPANY
Made in U.S.A.

Longitude West of Greenwich

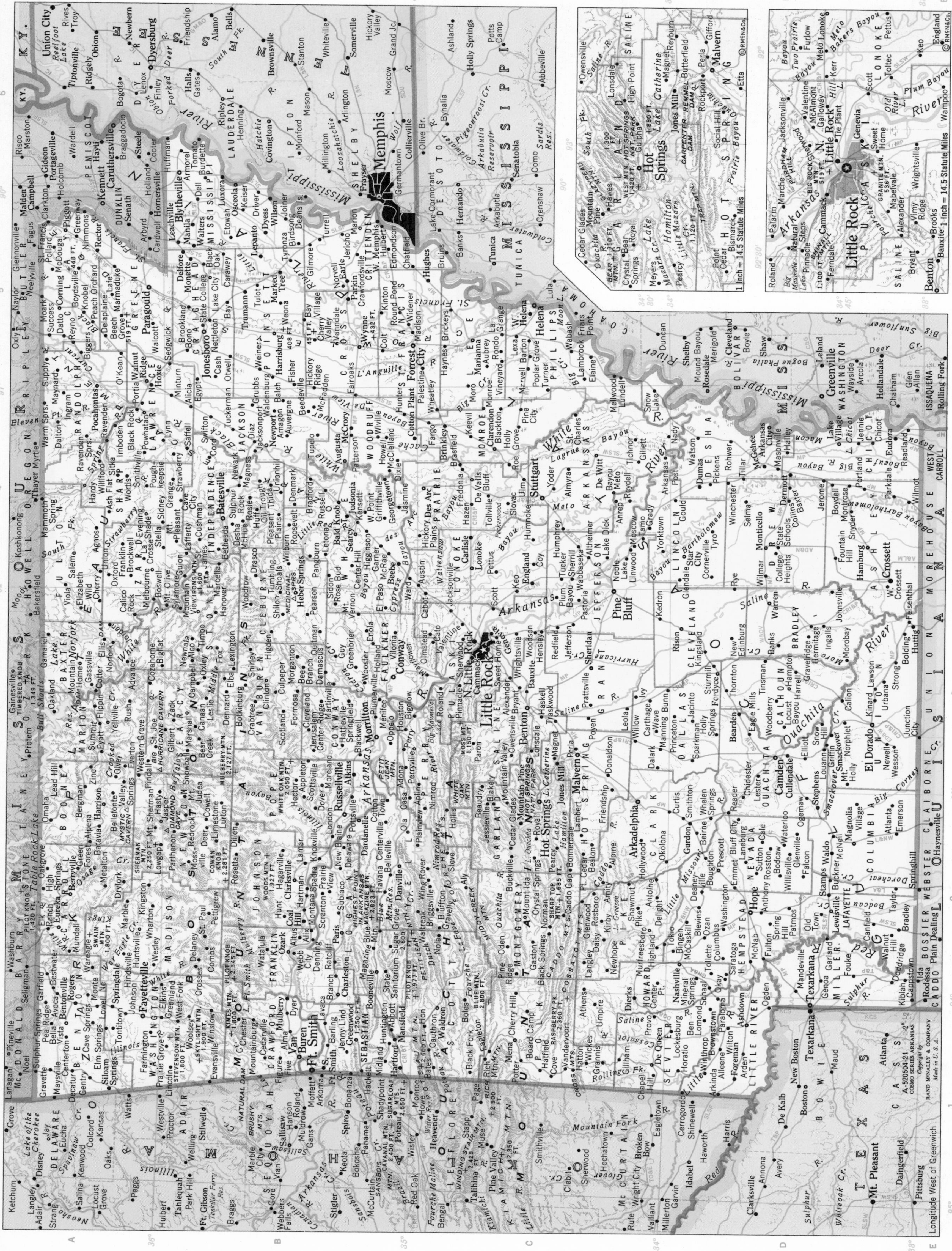

Lambert Conformal Conic Projection
SCALE 1:1,832,000 1 Inch = 29 Statute Miles

Statute Miles 5 0 5 10 20 30 40
Kilometers 5 0 5 15 25 35 45 55

CALIFORNIA

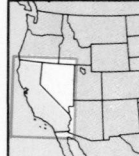

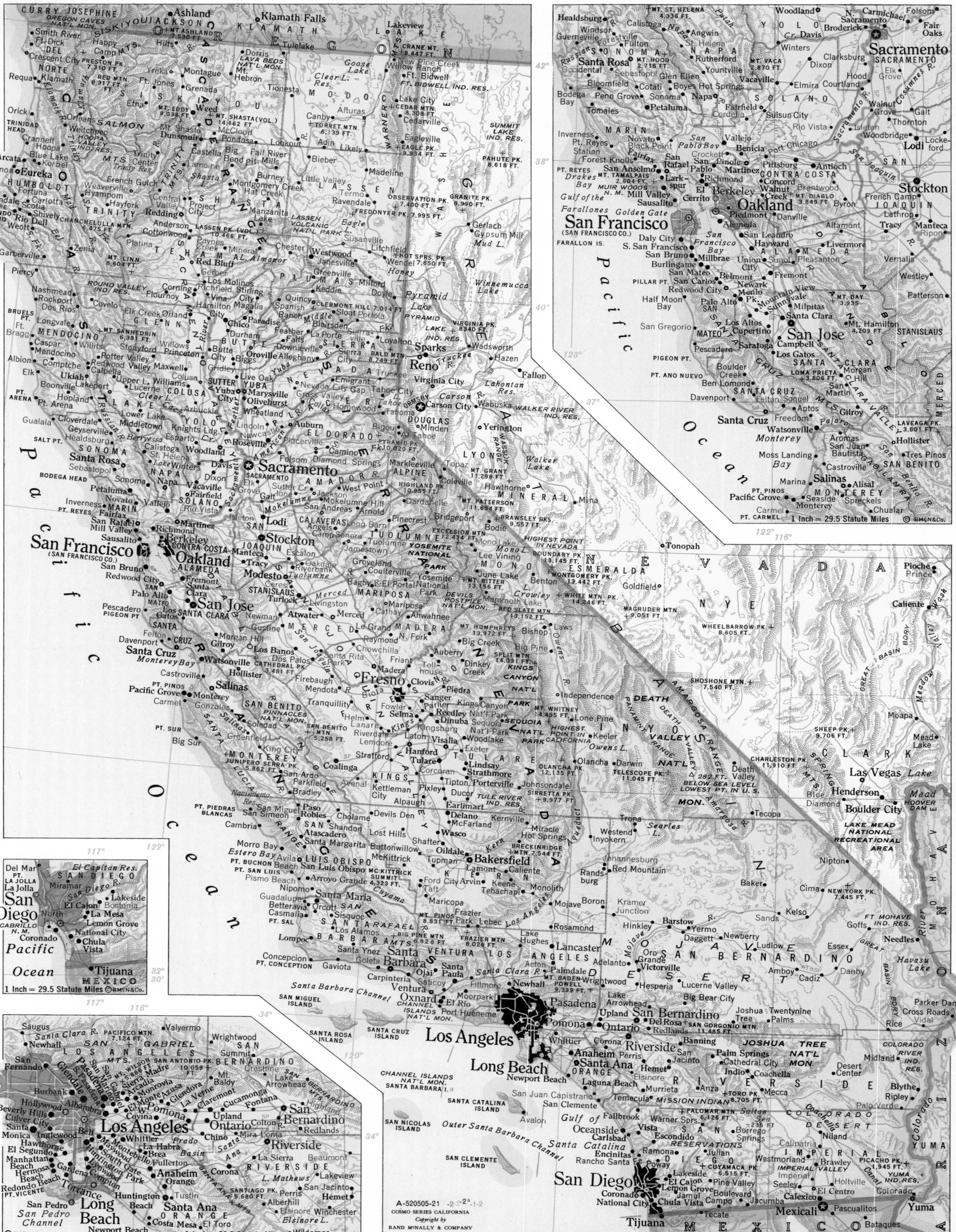

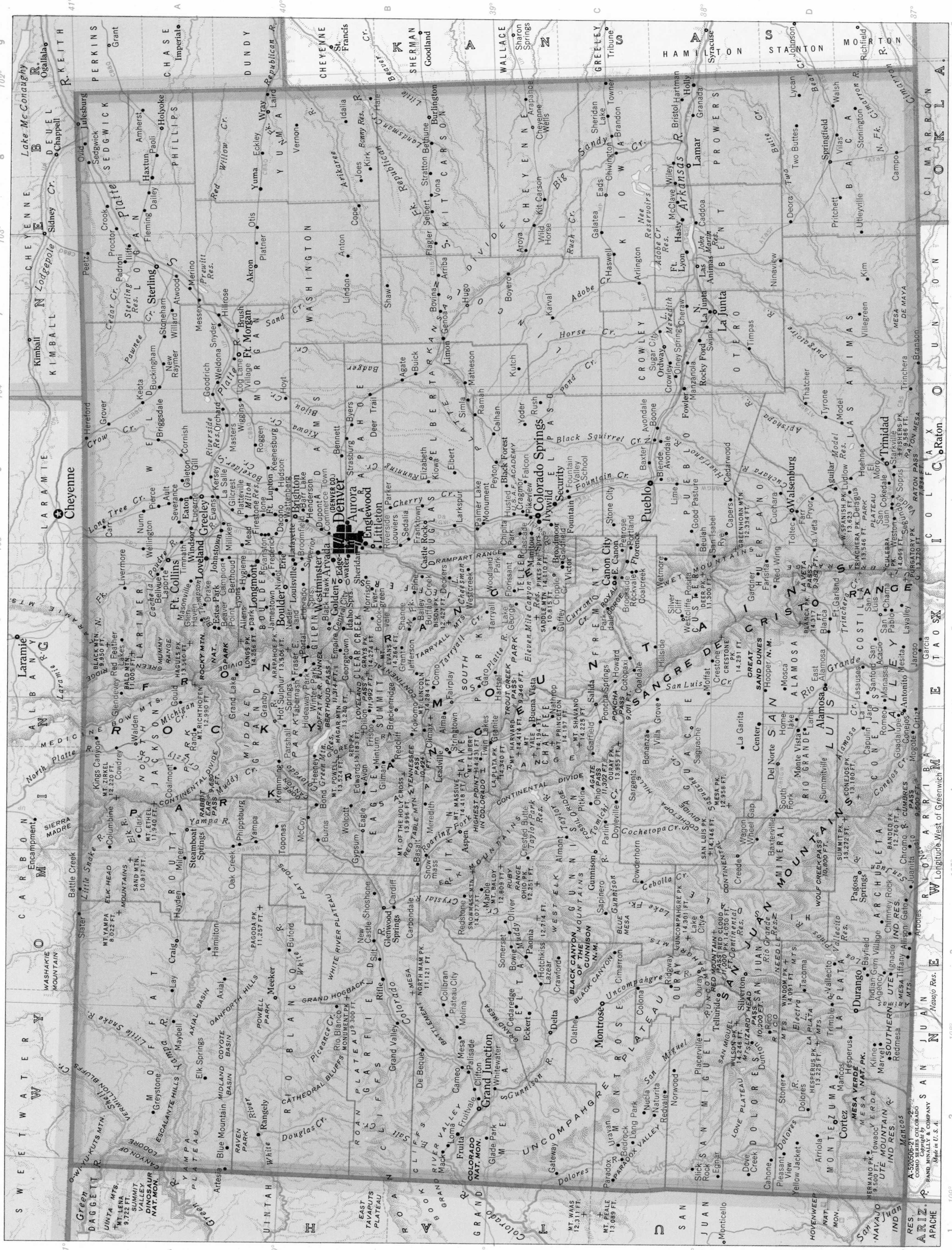

COLORADO

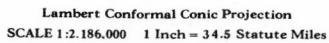

Lambert Conformal Conic Projection
SCALE 1:2,186,000   1 Inch = 34.5 Statute Miles

Statute Miles 5 0 5 10 20 30 40 50
Kilometers 5 0 5 15 25 35 45 55 65 75

83

CONNECTICUT AND RHODE ISLAND

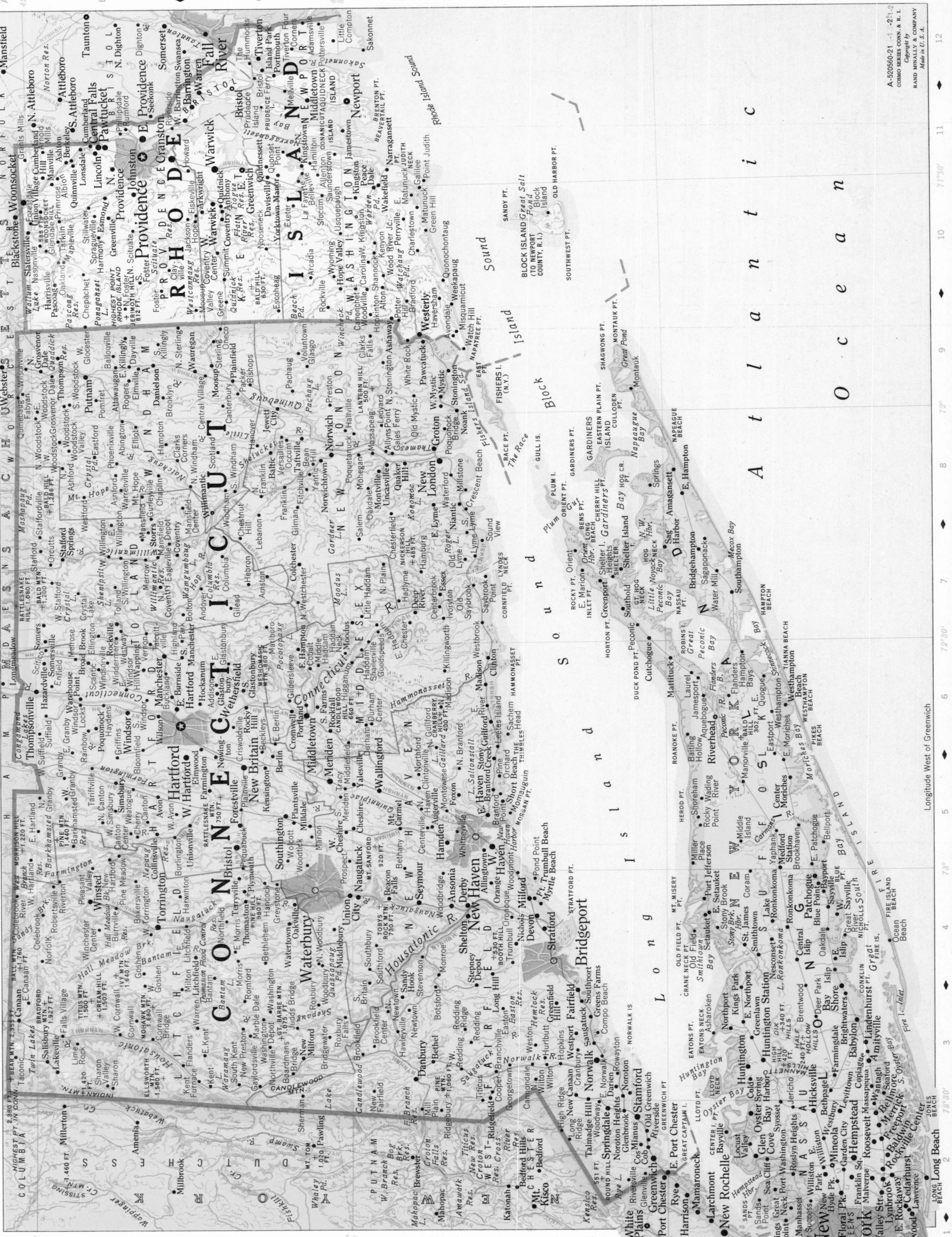

Statute Miles

Kilometers

Lambert Conformal Conic Projection
SCALE 1:731,000 1 Inch = 11.5 Statute Miles

A-520560-21 -1 -2²-2
COSMO SERIES CONN. & R.I.
Copyright by
RAND McNALLY COMPANY
Made in U.S.A.

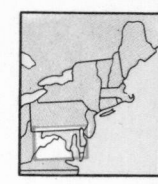

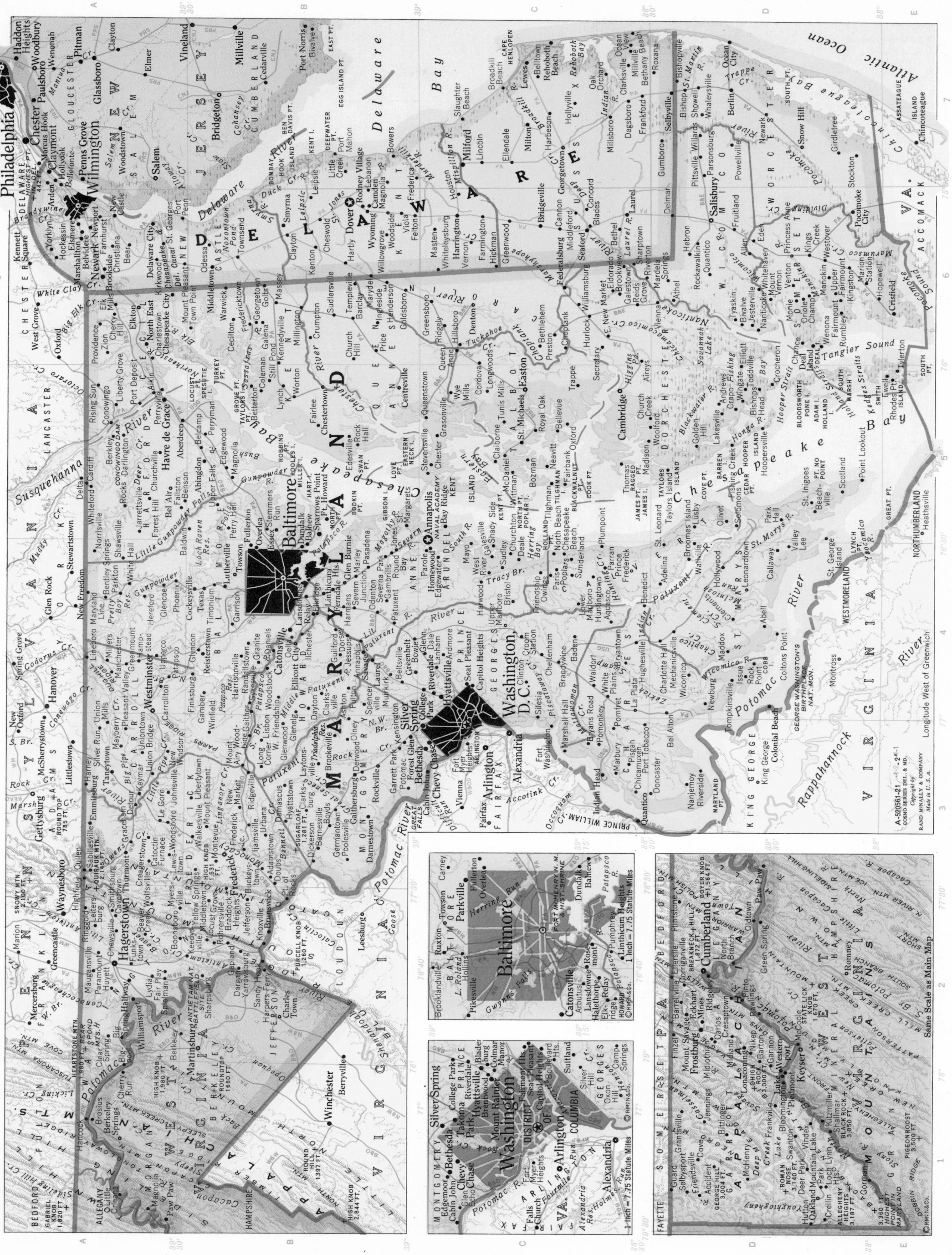

Lambert Conformal Conic Projection
SCALE 1:985,000 1 Inch = 15.5 Statute Miles

Statute Miles
5 0 5 10 15 20

Kilometers
5 0 5 10 15 20 25 30

FLORIDA

Gulf of Mexico

Atlantic Ocean

Straits of Florida

Tallahassee · Jacksonville · St. Augustine · Gainesville · Ocala · Orlando · Winter Park · Tampa · St. Petersburg · Clearwater · Lakeland · Winter Haven · Sarasota · Bradenton · Fort Myers · Naples · Fort Lauderdale · Hollywood · Miami · Miami Beach · Coral Gables · Key West · Pensacola · Panama City · Daytona Beach · De Land · Melbourne · Ft. Pierce · West Palm Beach · Lake Worth · Okeechobee

Jacksonville (inset) — 1 Inch = 19 Statute Miles

Tampa · St. Petersburg · Sarasota (inset) — 1 Inch = 19 Statute Miles

Miami · Fort Lauderdale (inset) — 1 Inch = 19 Statute Miles

Pensacola · Panama City (inset) — Same Scale as Main Map

Statute Miles 5 0 5 10 20 30 40 50
Kilometers 5 0 5 15 25 35 45 55 65

Lambert Conformal Conic Projection
SCALE 1:2,425,000 1 Inch = 38 Statute Miles

A-520510-21

Copyright by
RAND McNALLY & COMPANY
Made in U.S.A.

Longitude West of Greenwich

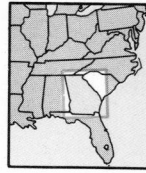

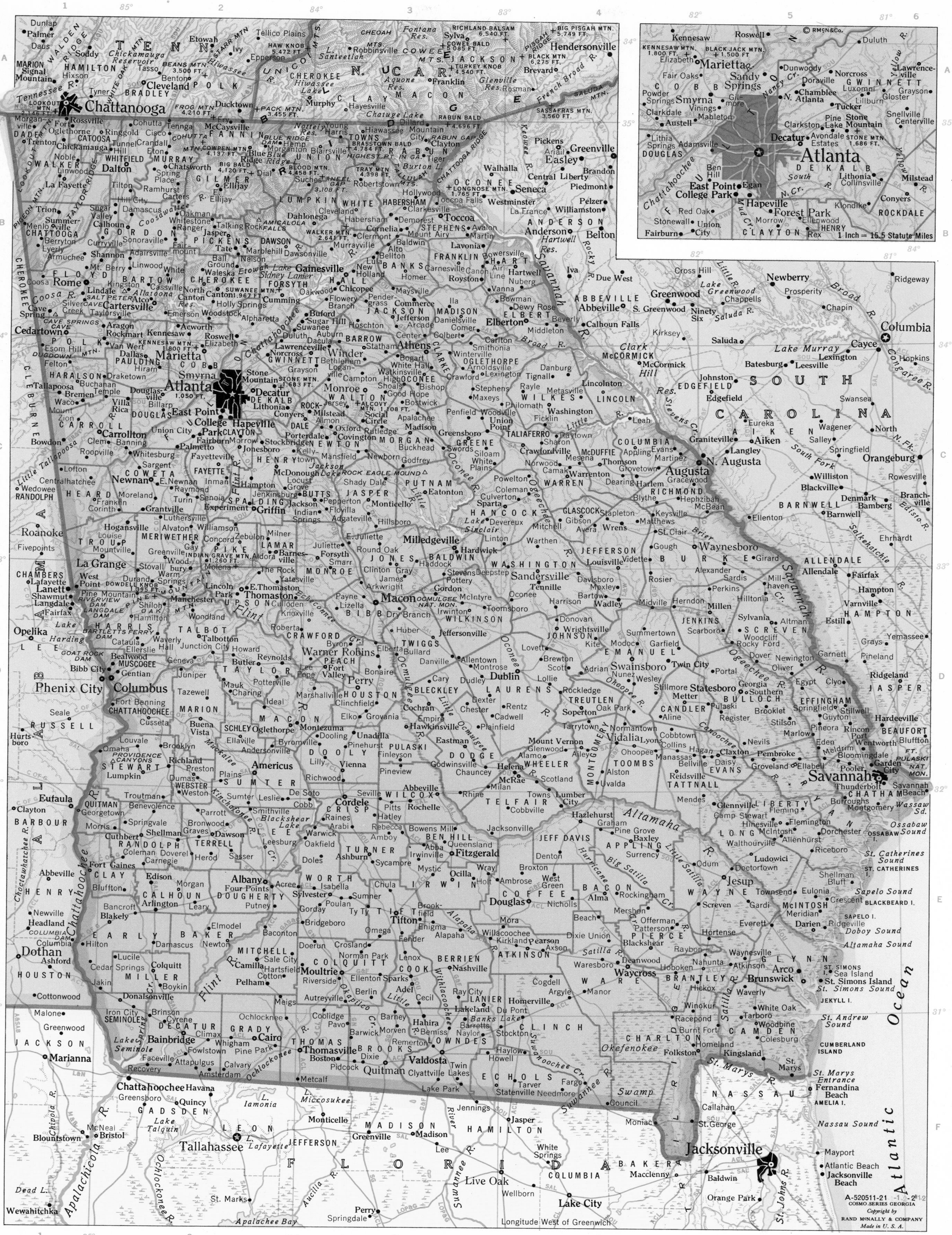

Lambert Conformal Conic Projection
SCALE 1:1,962,000   1 Inch = 31 Statute Miles

Statute Miles
Kilometers

A-520511-21
COSMO SERIES GEORGIA
Copyright by
RAND MCNALLY & COMPANY
Made in U.S.A.

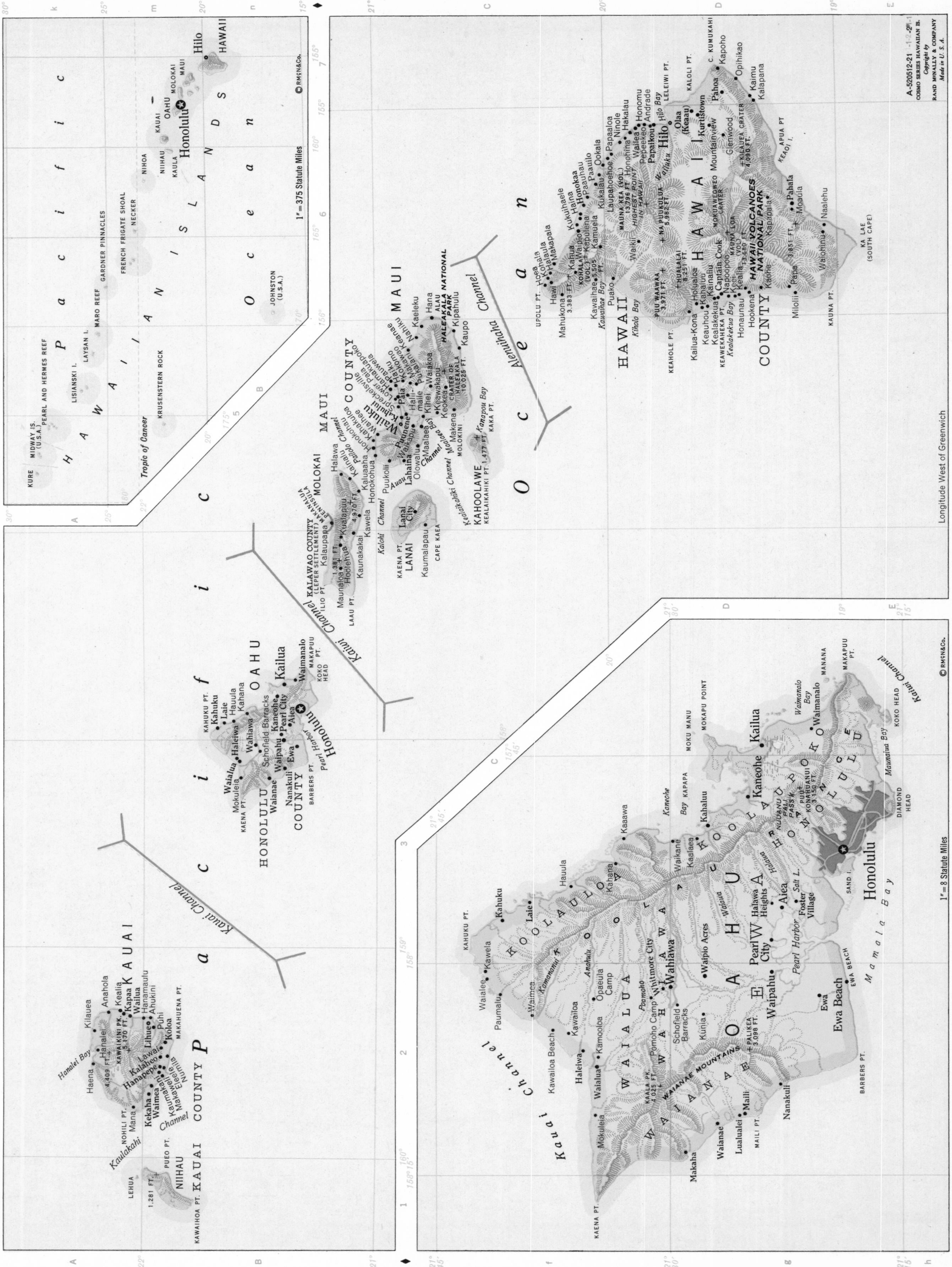

HAWAII

Statute Miles

Kilometers

Lambert Conformal Conic Projection
SCALE 1:2,000,000   1 Inch = 32 Statute Miles

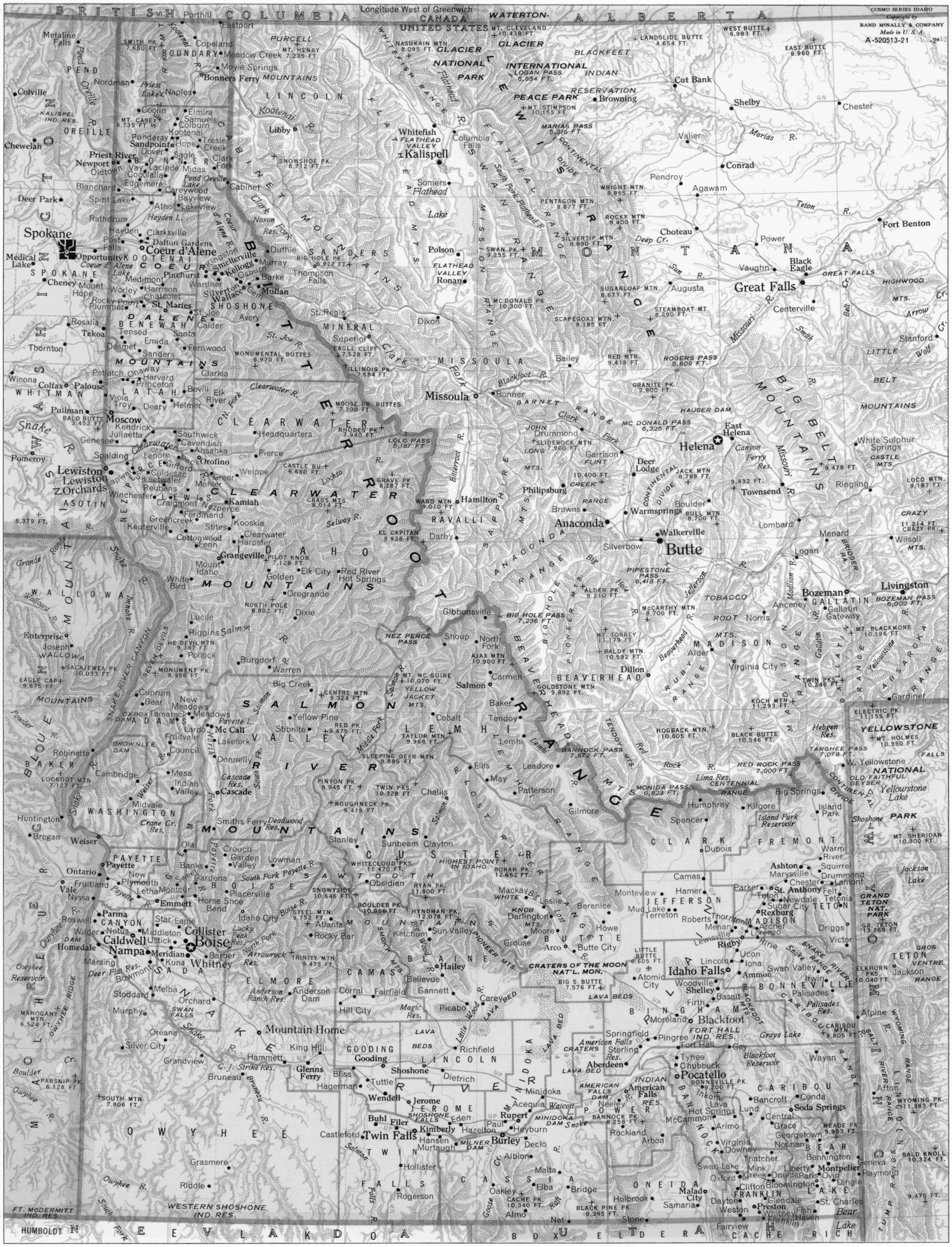

Longitude West of Greenwich

Cosmo Series Idaho
Copyright by
RAND McNALLY & COMPANY
Made in U.S.A.
A-520513-21

Lambert Conformal Conic Projection
SCALE 1:2,633,000 1 Inch = 41.5 Statute Miles

Statute Miles
5 0 5 10   20   30   40   50   60

Kilometers
5 0 5 15 25 35 45 55 65 75

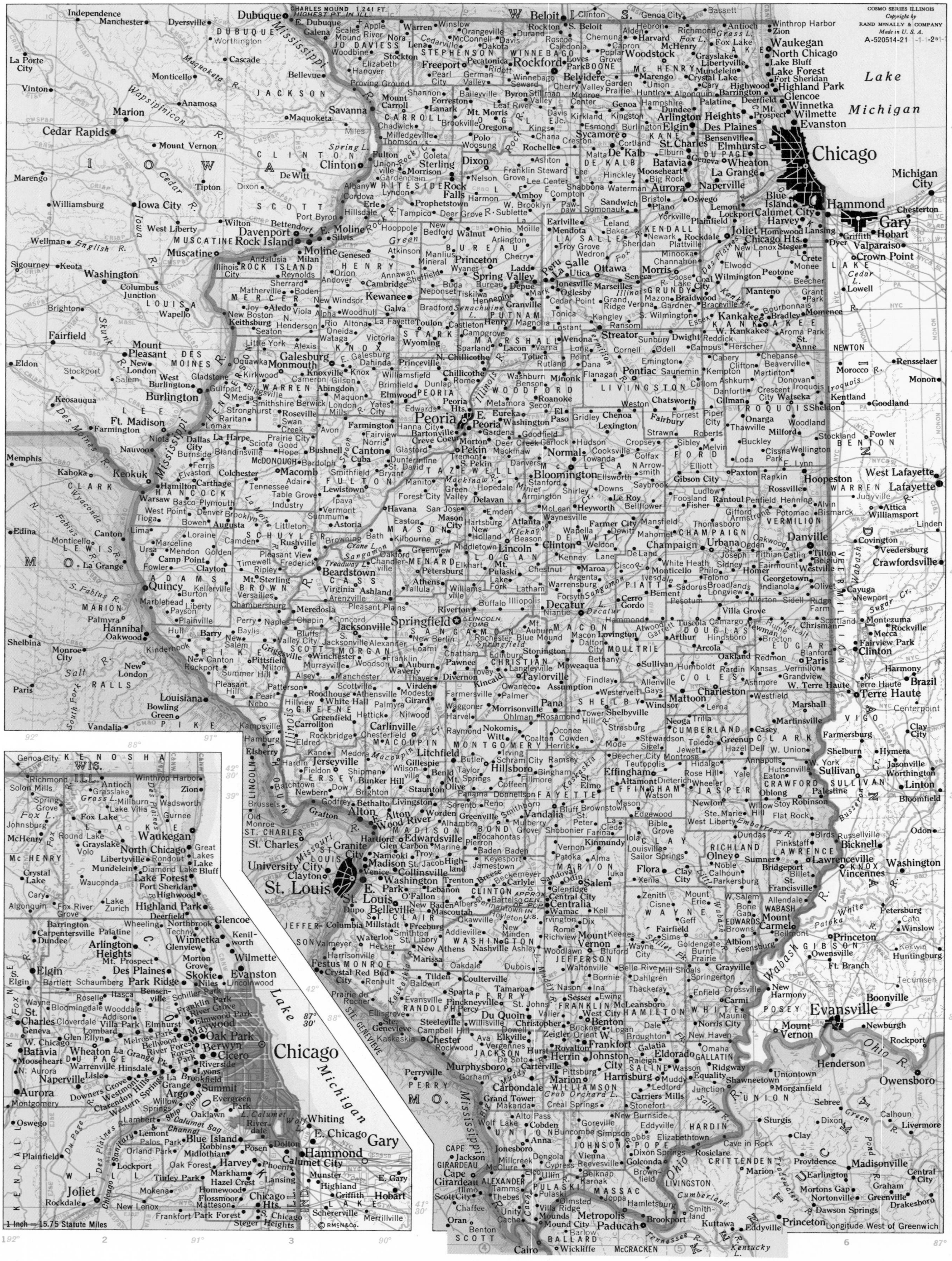

ILLINOIS

COSMO SERIES ILLINOIS
Copyright by
RAND McNALLY & COMPANY
Made in U.S.A.
A-520514-21

Statute Miles
Kilometers

Lambert Conformal Conic Projection
SCALE 1:1,997,000   1 Inch = 31.5 Statute Miles

1 Inch = 15.75 Statute Miles

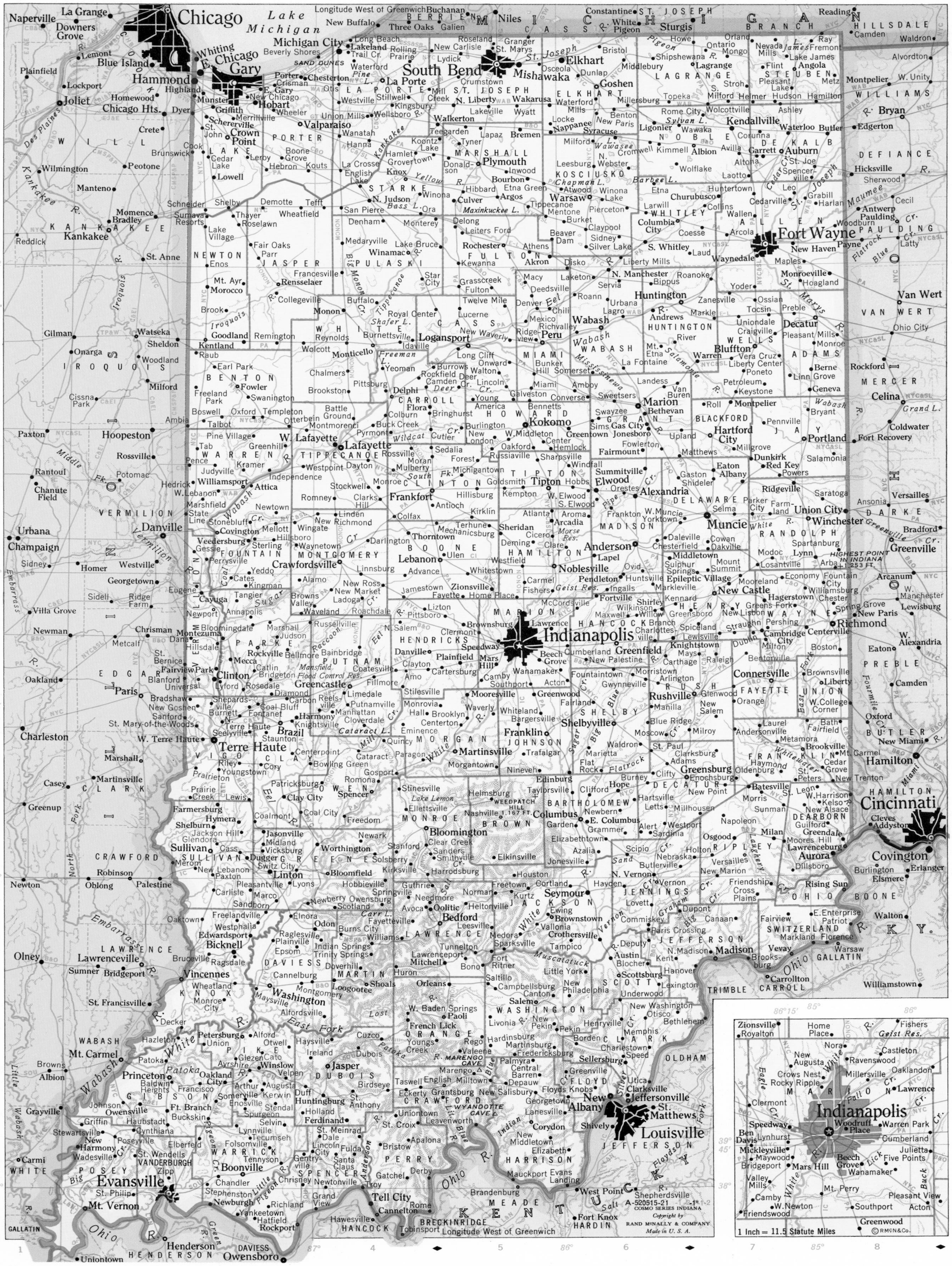

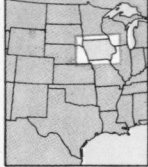

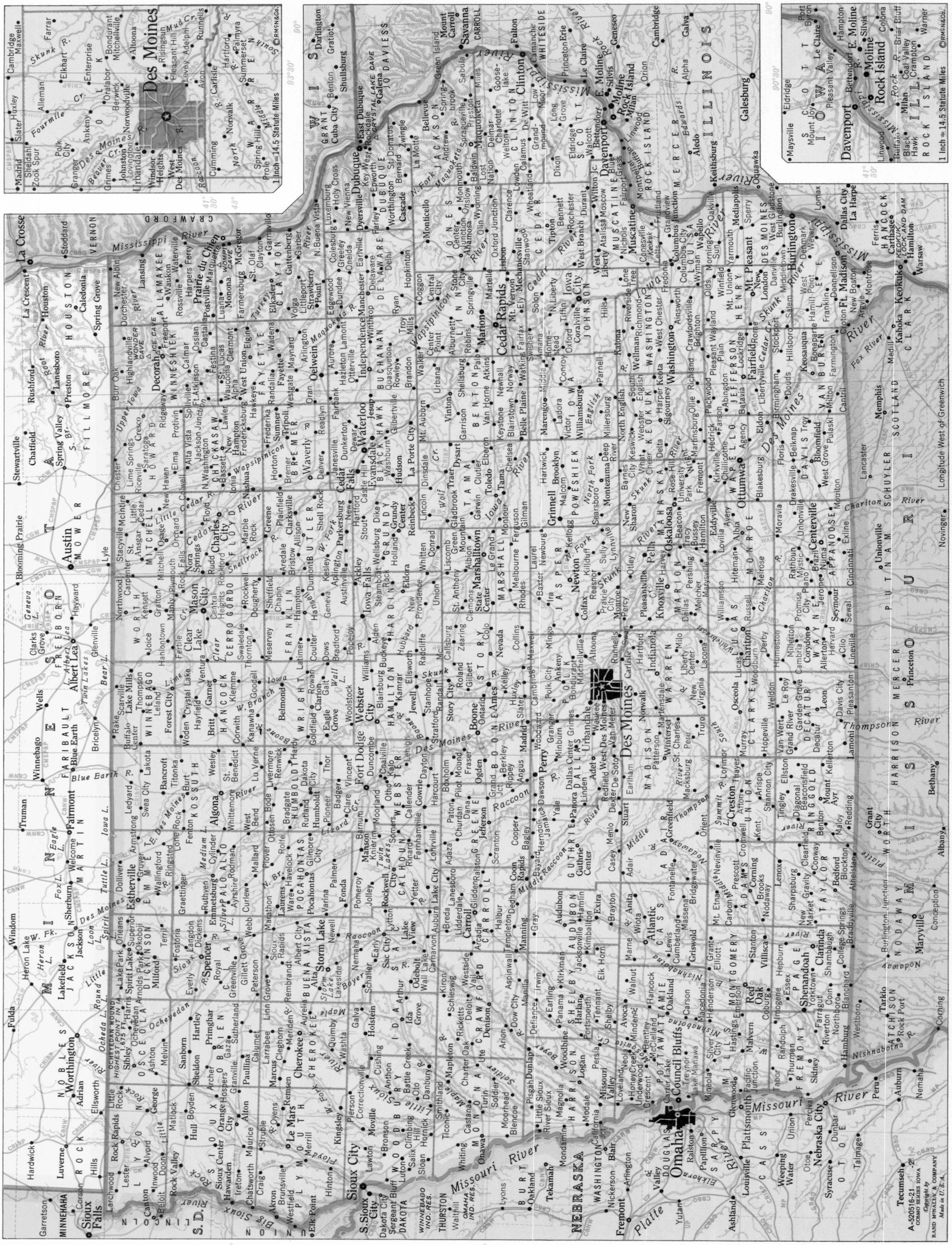

Statute Miles 5 0 5 10 20 30 40

Kilometers 5 0 5 15 25 35 45 55

Lambert Conformal Conic Projection

SCALE 1:1,834,000 1 Inch = 29 Statute Miles

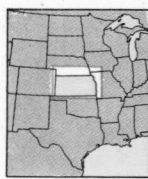

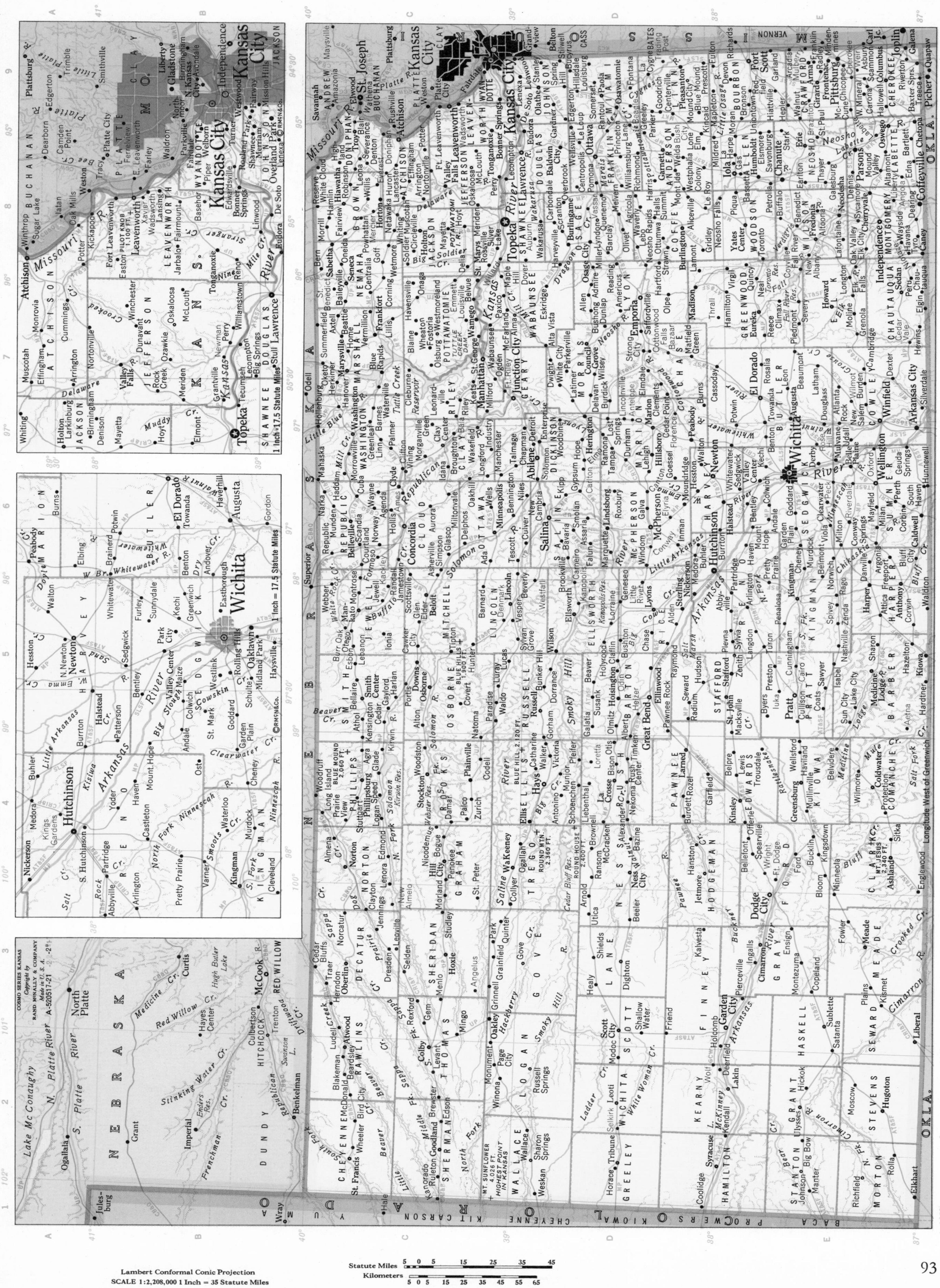
Wichita
1 Inch = 17.5 Statute Miles
Topeka
Kansas City
Lambert Conformal Conic Projection
SCALE 1:2,208,000 1 Inch = 35 Statute Miles
Statute Miles 5 0 5 15 25 35 45
Kilometers 5 0 5 15 25 35 45 55 65

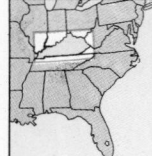

KENTUCKY

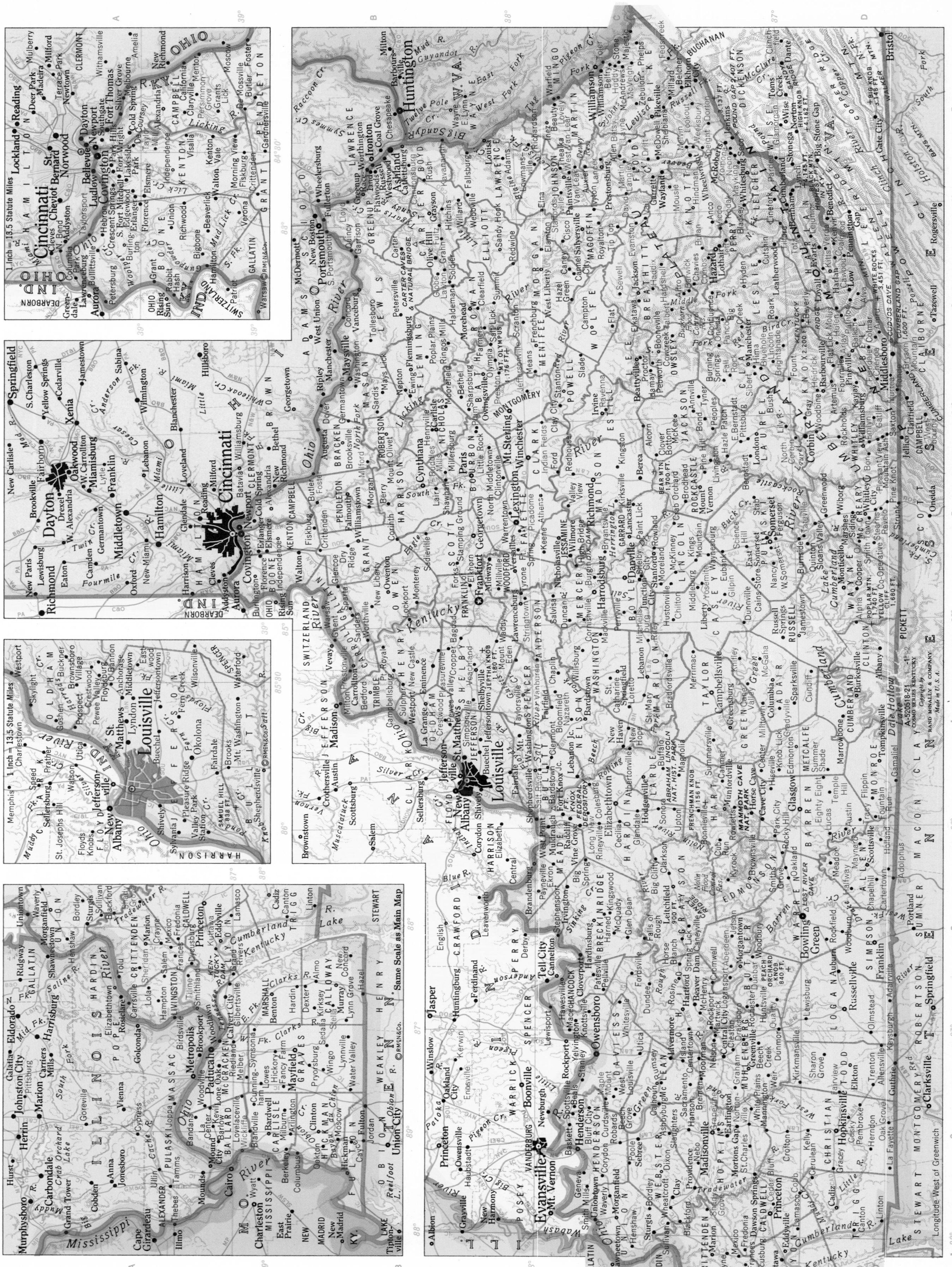

Statute Miles 5 0 5 10 20 30 40

Kilometers 5 0 5 10 20 30 40 50 60

Lambert Conformal Conic Projection

SCALE 1:1,738,000 1 Inch = 27 Statute Miles

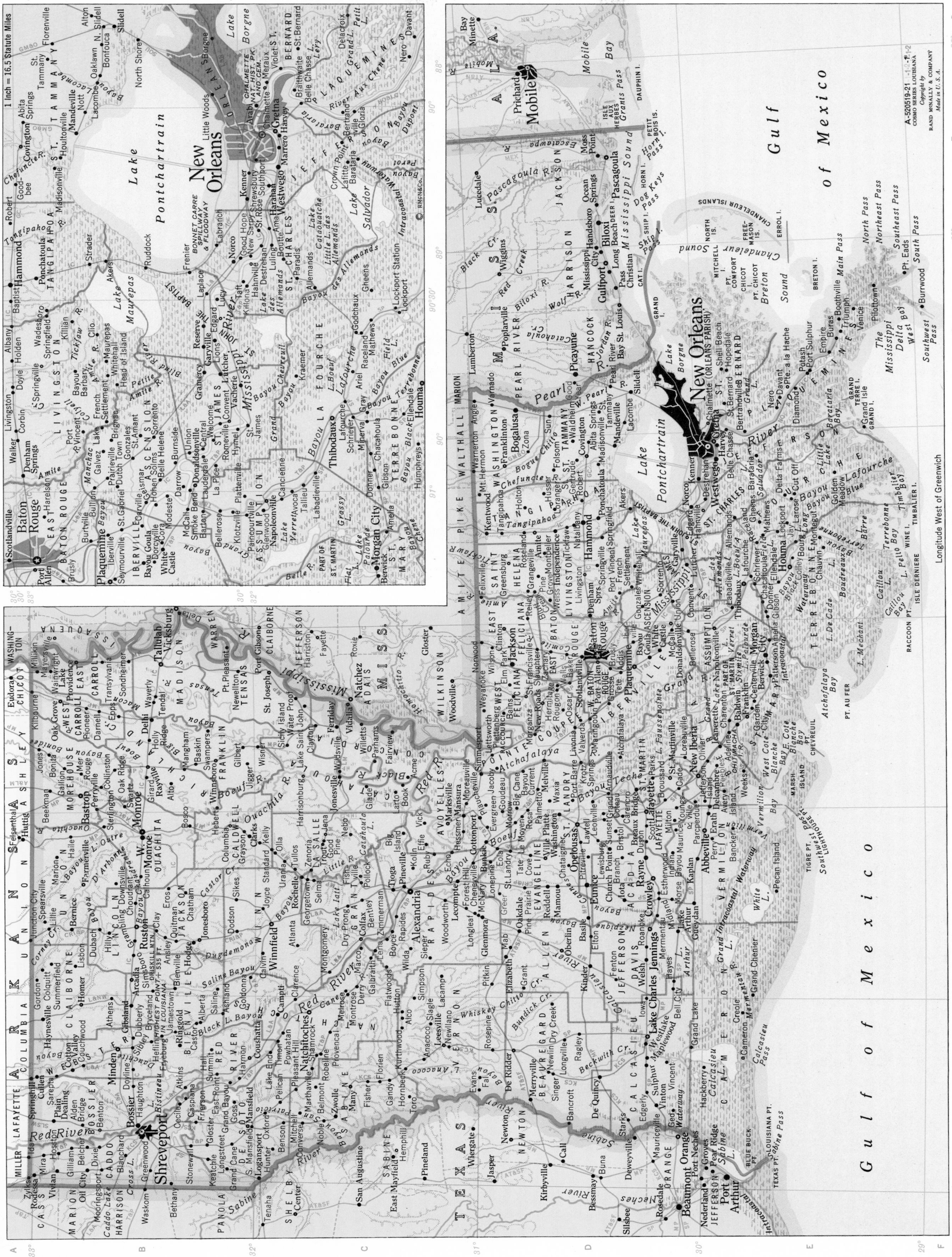

Gulf of Mexico

Lake Pontchartrain

New Orleans

Baton Rouge

Shreveport

Mobile

Lambert Conformal Conic Projection
SCALE 1:2,083,000   1 Inch = 33 Statute Miles

Statute Miles 5 0 5 10   20   30   40
Kilometers 5 0 5   15   25   35   45 55

A-520520-21

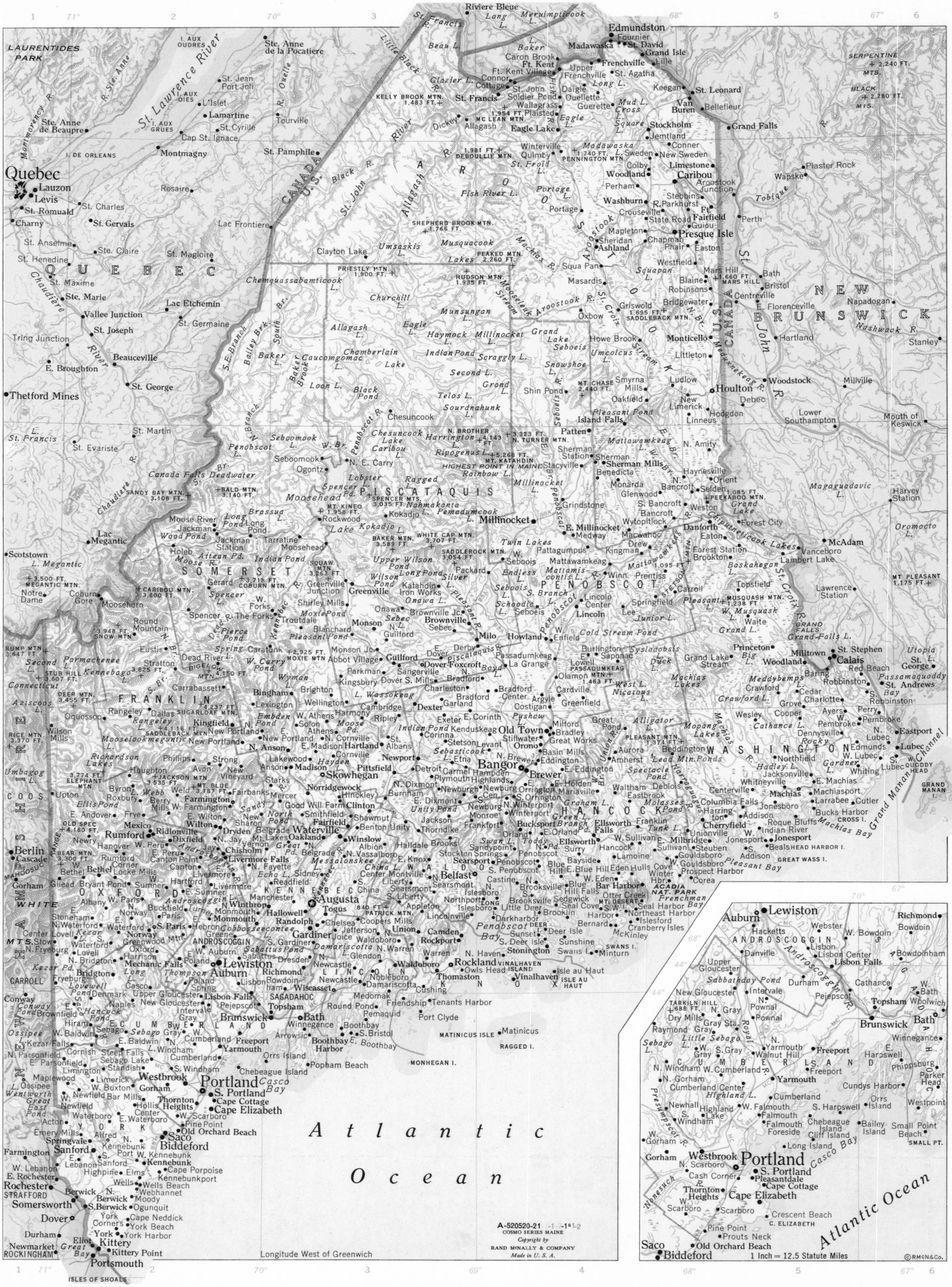

COSMO SERIES MAINE
Copyright by
RAND McNALLY & COMPANY
Made in U.S.A.

Longitude West of Greenwich

Statute Miles

Kilometers

Lambert Conformal Conic Projection
SCALE 1:1,581,000 1 Inch = 25 Statute Miles

1 Inch = 12.5 Statute Miles

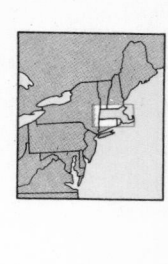

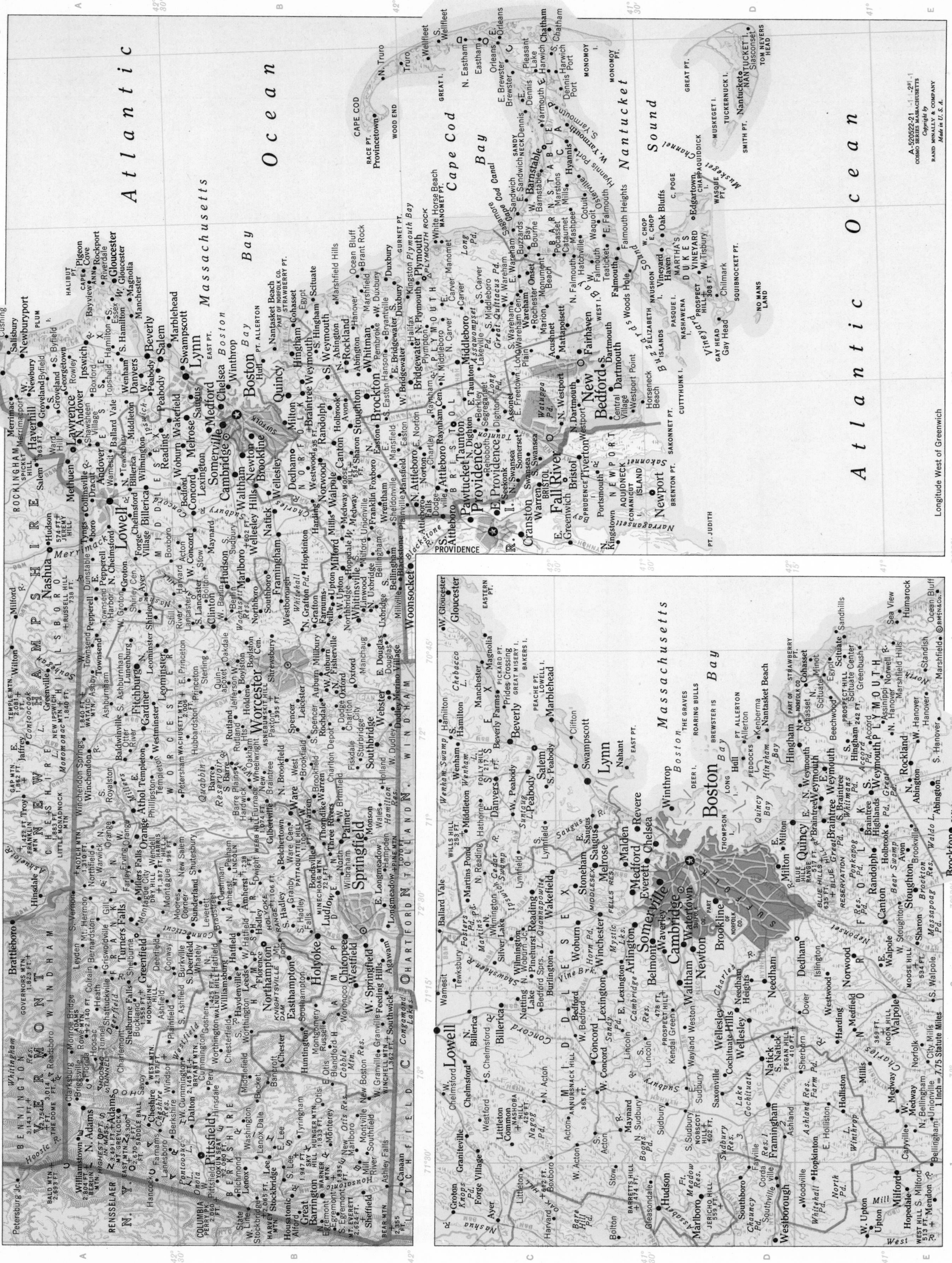

Atlantic Ocean

Massachusetts Bay

Cape Cod Bay

Nantucket Sound

Longitude West of Greenwich

Lambert Conformal Conic Projection
SCALE 1:978,000 1 Inch = 15.5 Statute Miles

Statute Miles

Kilometers

1 Inch = 7.75 Statute Miles

A-520522-21 -2A-1
COSMO SERIES MASSACHUSETTS
Copyright by
RAND McNALLY & COMPANY
Made in U.S.A.

MICHIGAN

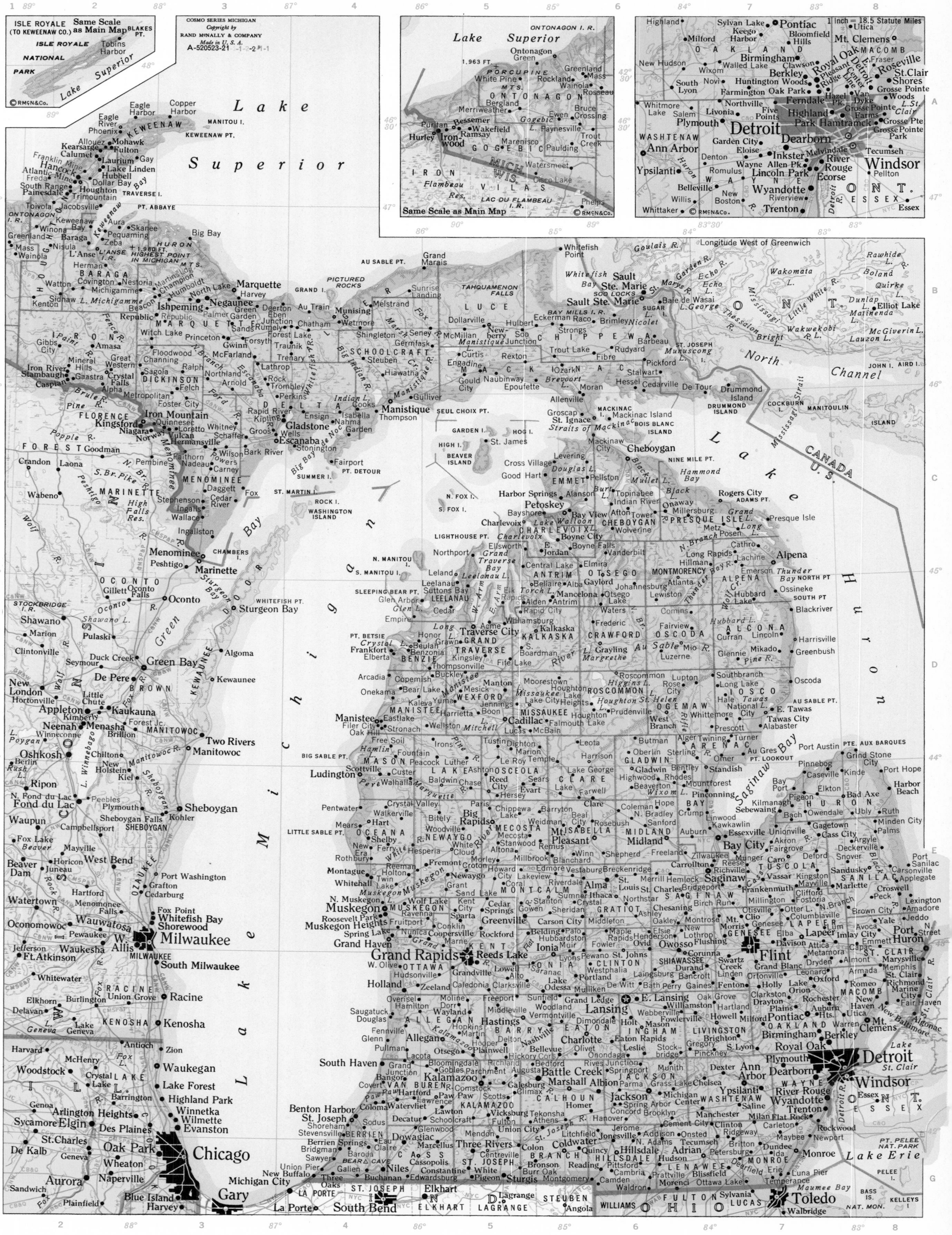

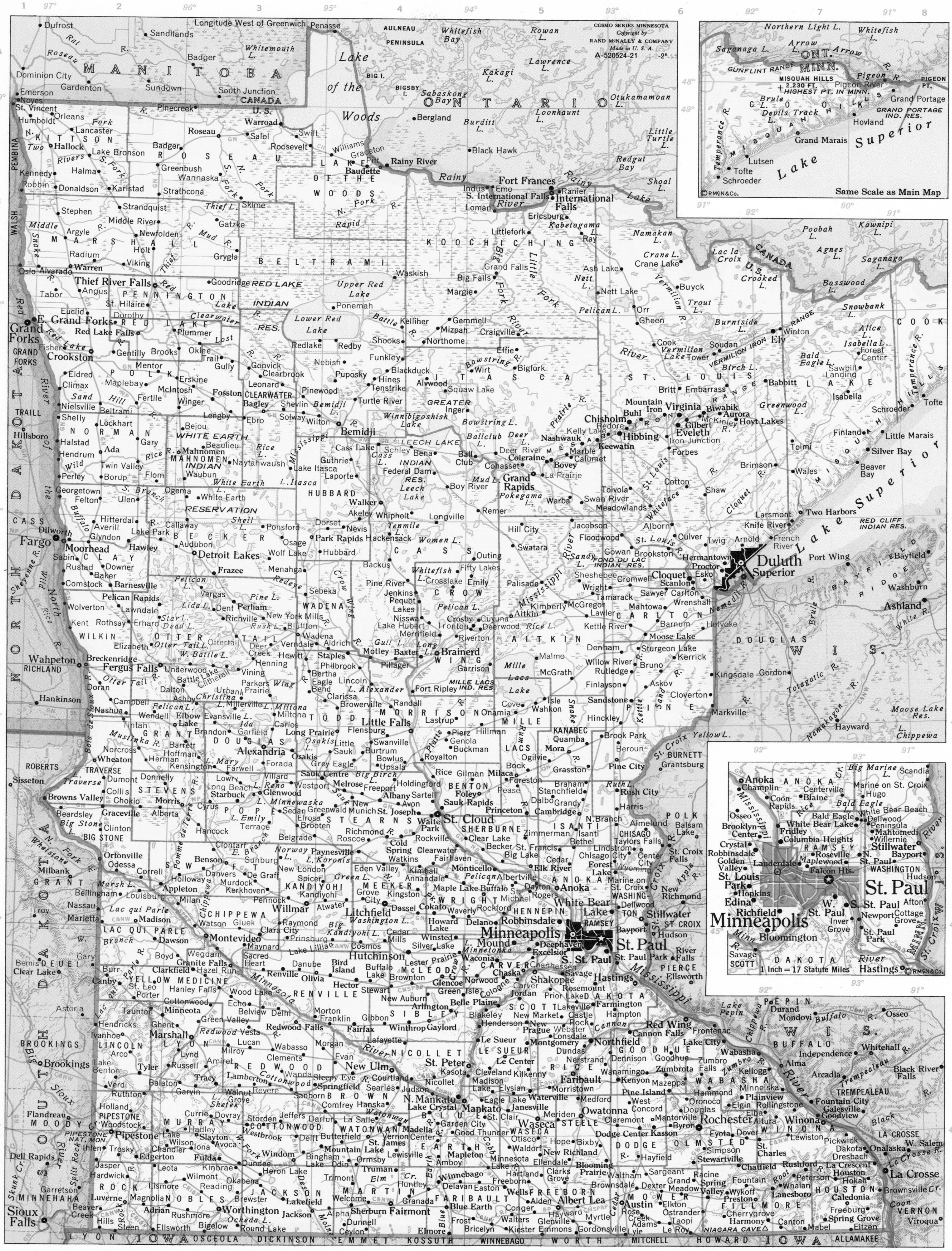

MINNESOTA

99

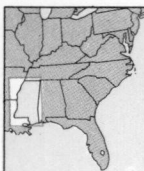

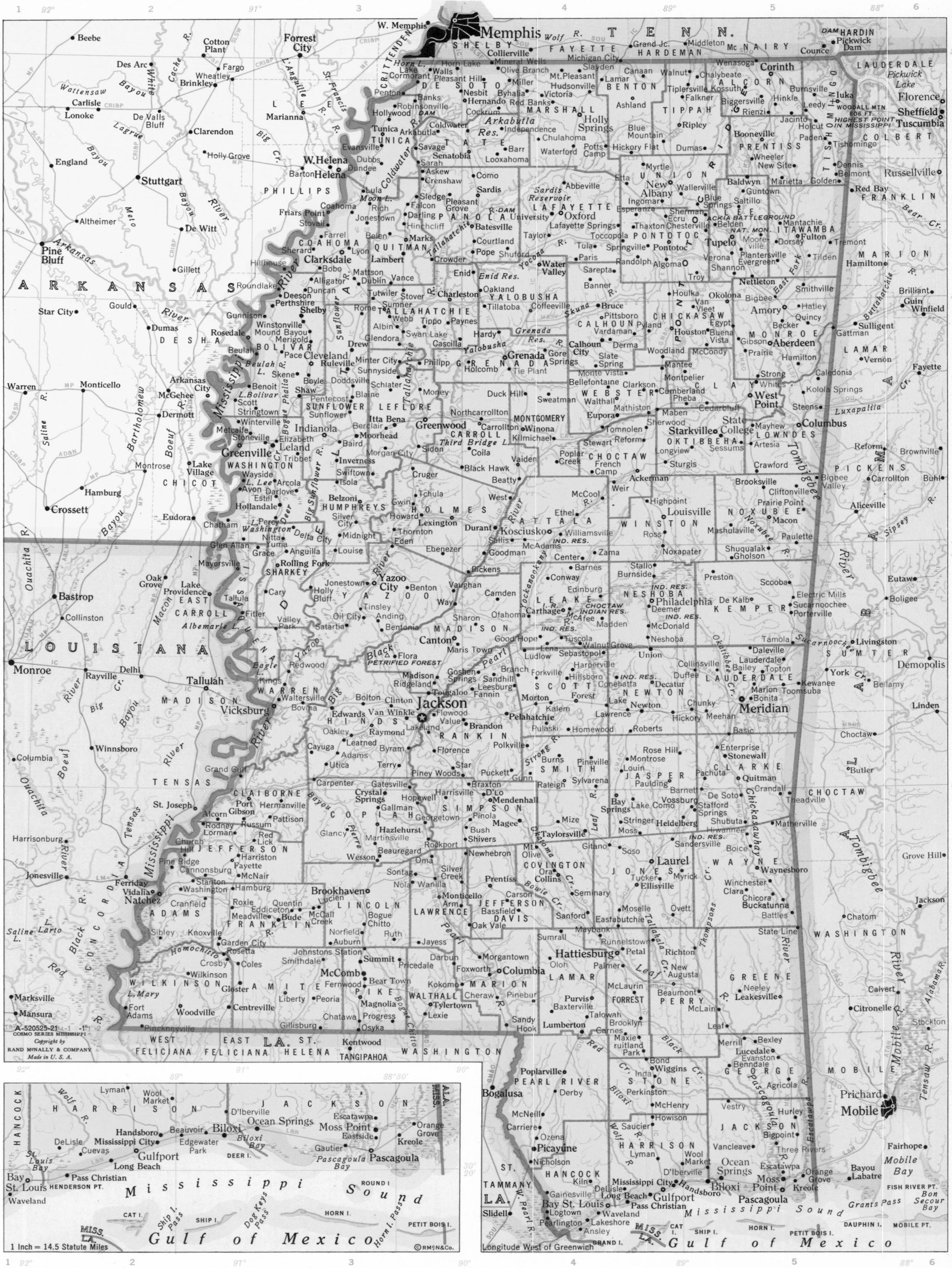
Memphis
T E N N.
A R K A N S A S
L O U I S I A N A
Jackson
Vicksburg
Meridian
Hattiesburg
Greenville
Natchez
Biloxi
Gulfport
Mobile
Mississippi Sound
Gulf of Mexico
1 Inch = 14.5 Statute Miles
Longitude West of Greenwich
A-520525-21
COSMO SERIES MISSISSIPPI
Copyright by RAND McNALLY & COMPANY Made in U. S. A.

Statute Miles 5 0 5 10 20 30 40

Kilometers 5 0 5 15 25 35 45 55

Lambert Conformal Conic Projection

SCALE 1:1,837,000 1 Inch = 29 Statute Miles

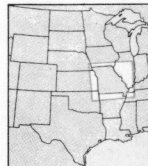

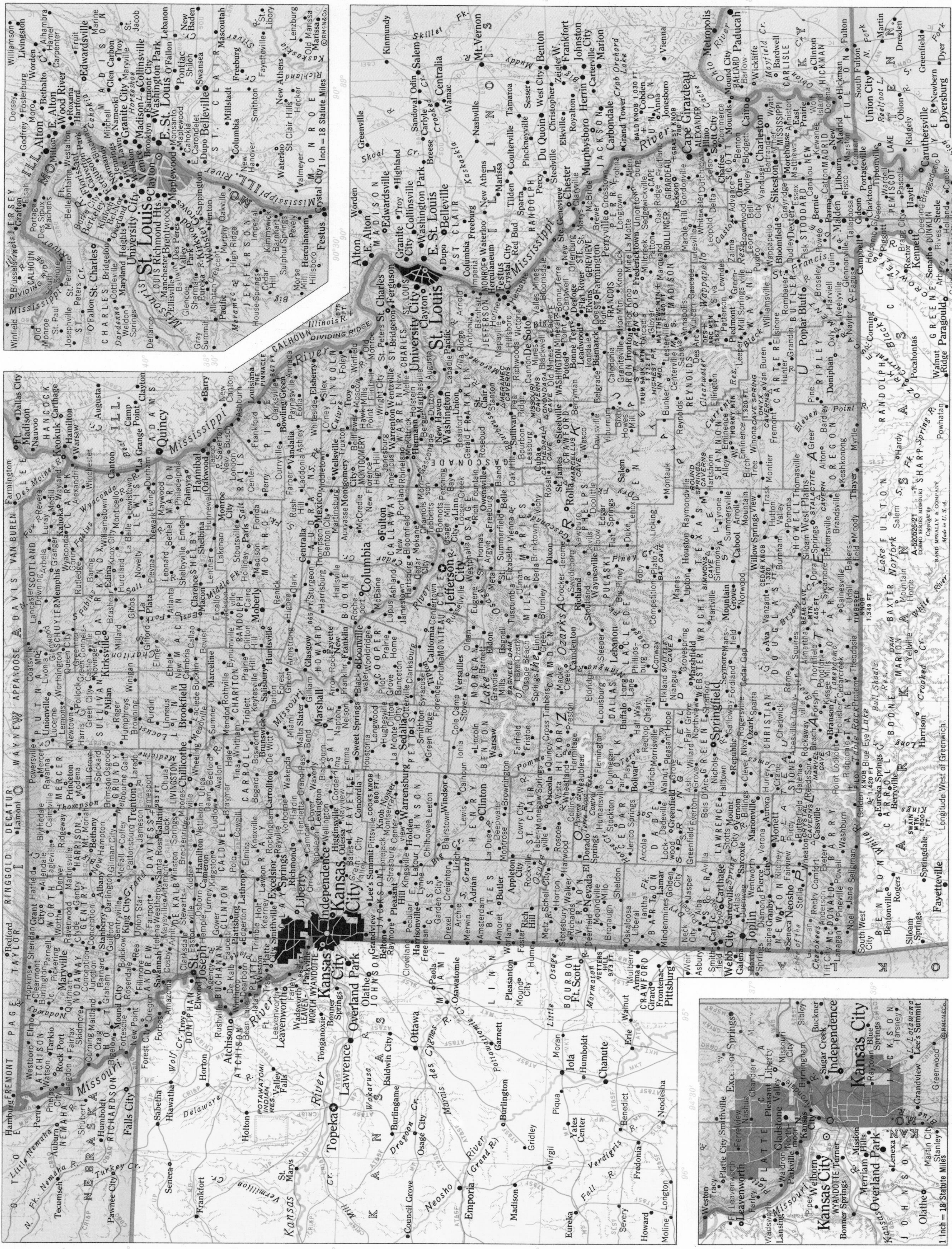

Lambert Conformal Conic Projection
SCALE 1:2,283,000 1 Inch = 36 Statute Miles

Statute Miles 5 0 5 15 25 35 45
Kilometers 5 0 5 15 25 35 45 55 65

MONTANA

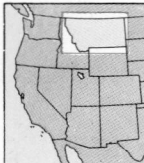

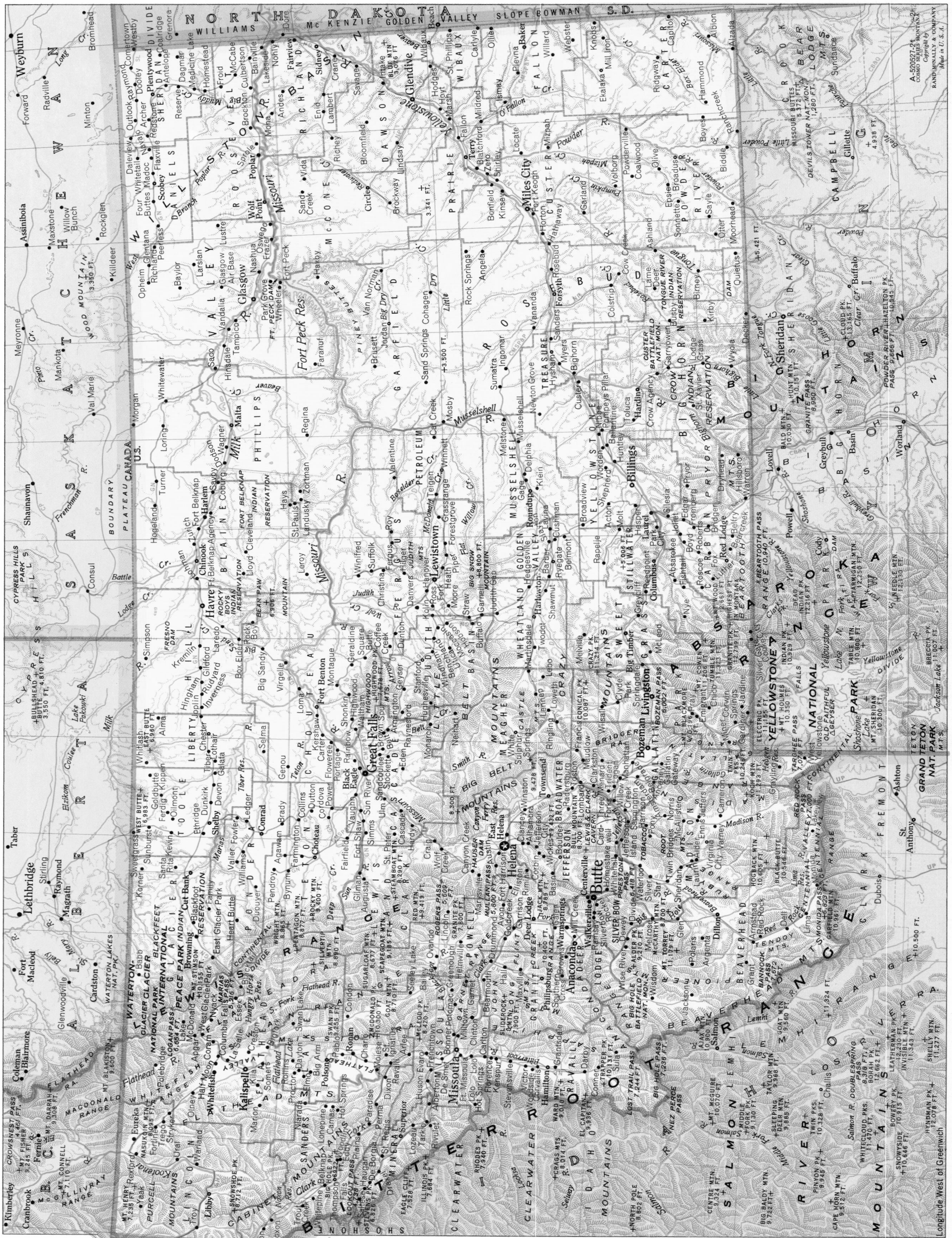

Statute Miles 10 0 10 20 30 40 50 60 70

Kilometers 10 0 10 30 50 70 90

Lambert Conformal Conic Projection

SCALE 1:2,999,000 1 Inch = 47.5 Statute Miles

NEBRASKA

Lambert Conformal Conic Projection
SCALE 1:2,460,000 1 Inch = 39 Statute Miles

Statute Miles
5 0 5 10   20   30   40   50   60
Kilometers
5 0 5 15   35   55   75   95

1 Inch = 19.5 Statute Miles

NEVADA

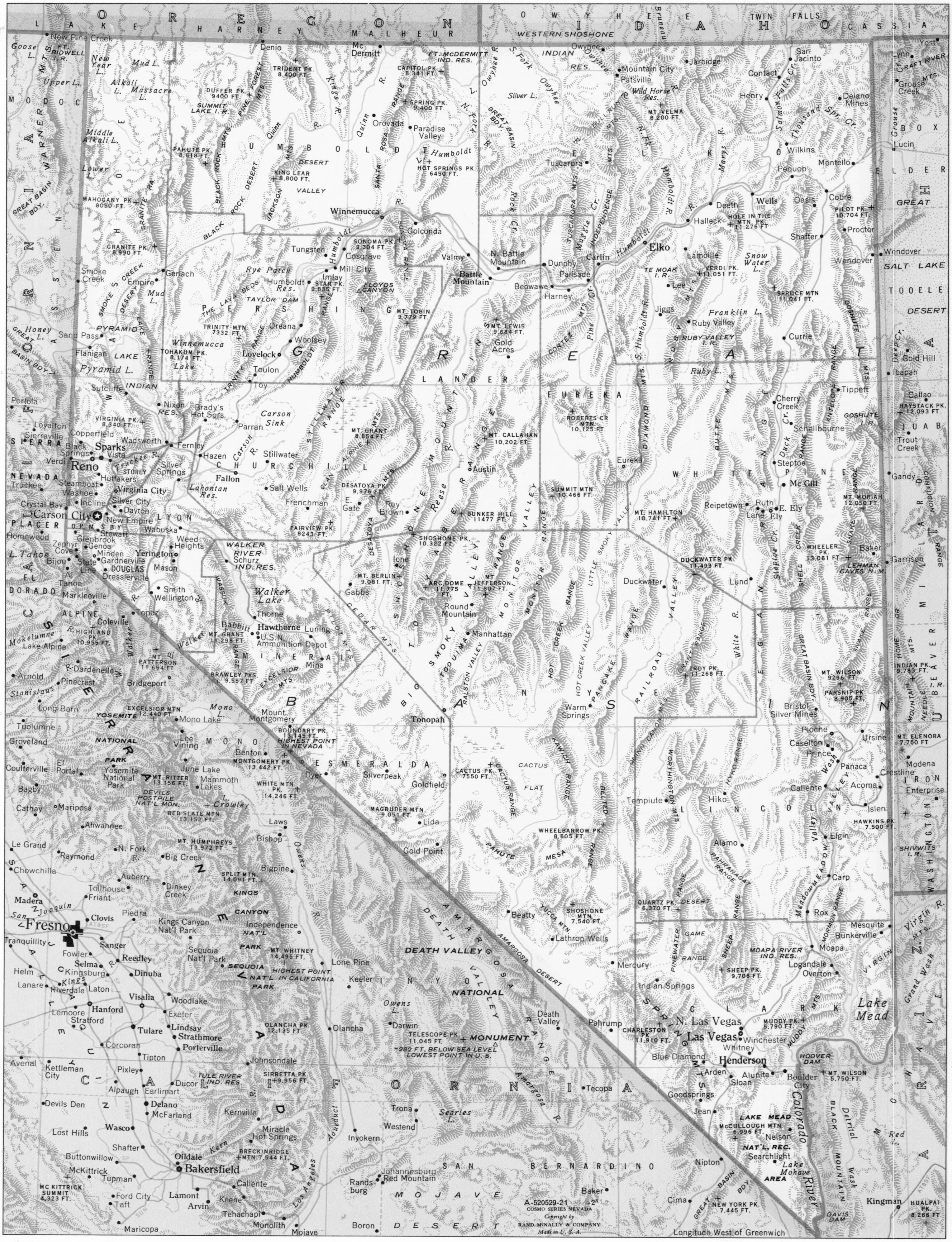

Statute Miles 5 0 5 10 20 30 40 50 60 70 80
Kilometers 5 0 10 20 40 60 80 100 120

A-520529-21
COSMO SERIES NEVADA
Copyright by
RAND M?NALLY & COMPANY
Made in U.S.A.

Longitude West of Greenwich

Lambert Conformal Conic Projection
SCALE 1:2,630,000 1 Inch = 41.5 Statute Miles

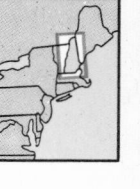

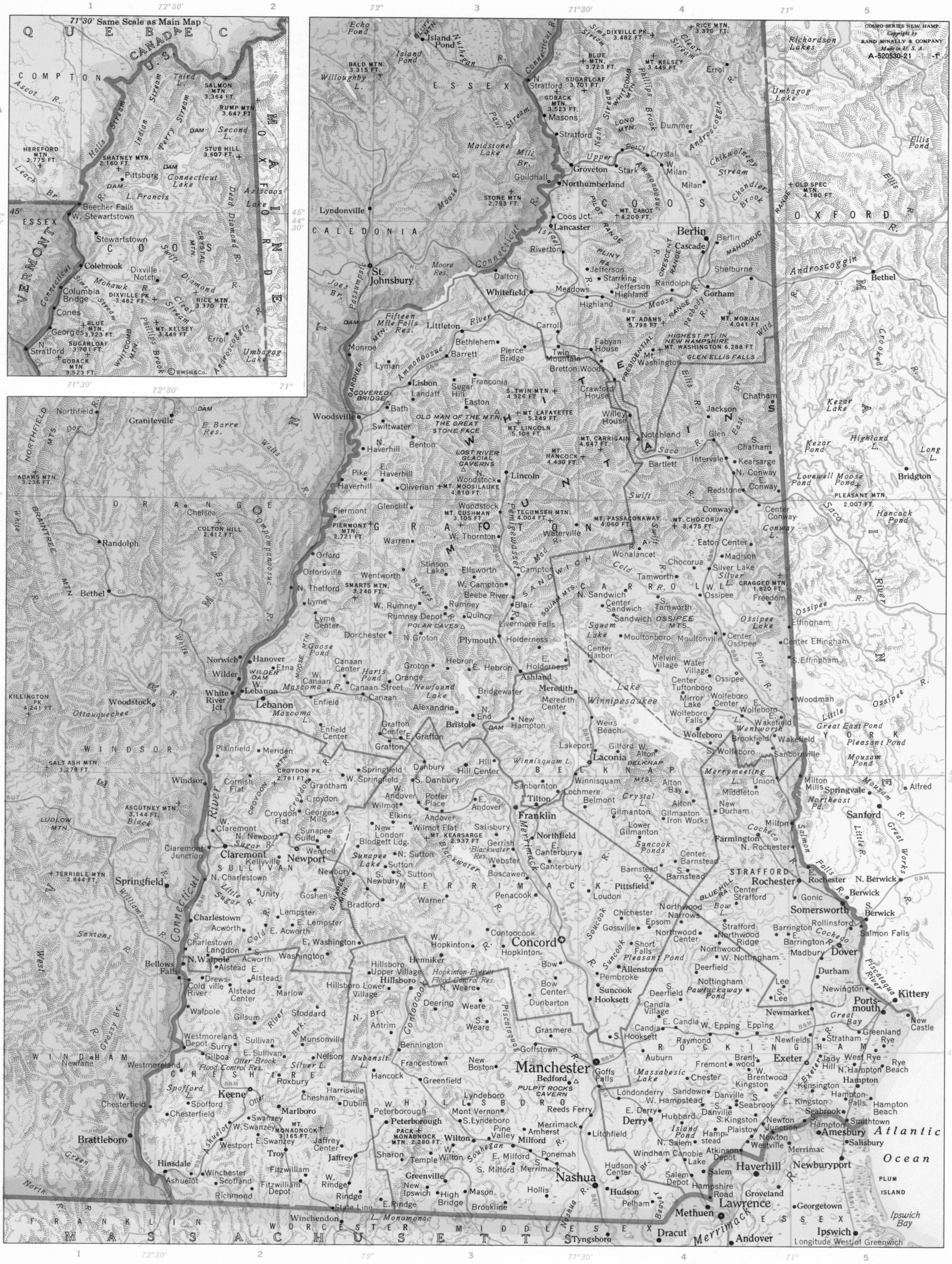

COSMO SERIES NEW HAMP.
Copyright by
RAND McNALLY & COMPANY
Made in U. S. A.
A-520530-21

71°30' Same Scale as Main Map

Lambert Conformal Conic Projection
SCALE 1:792,000   1 Inch = 12.75 Statute Miles

Statute Miles
Kilometers

Longitude West of Greenwich

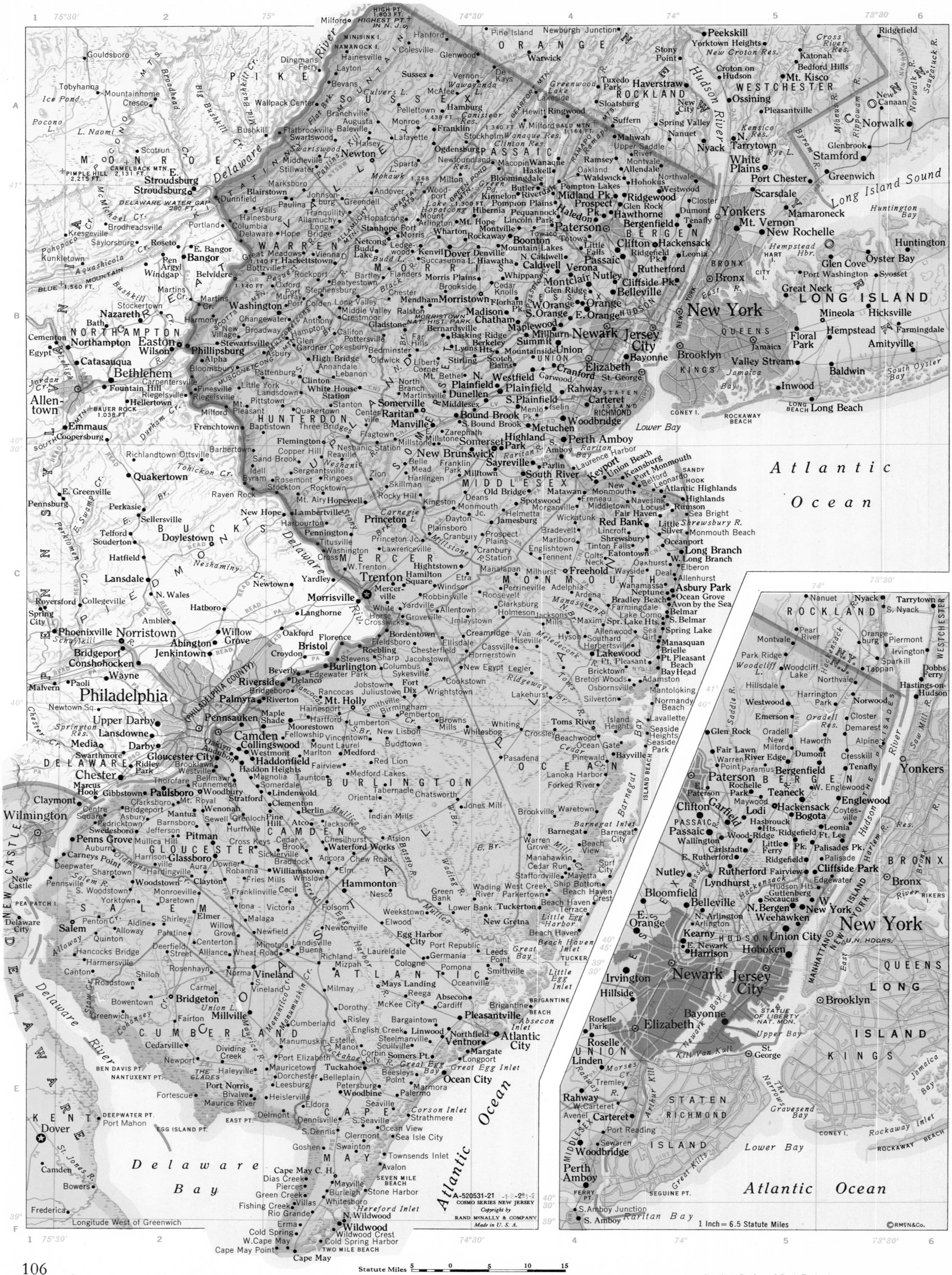

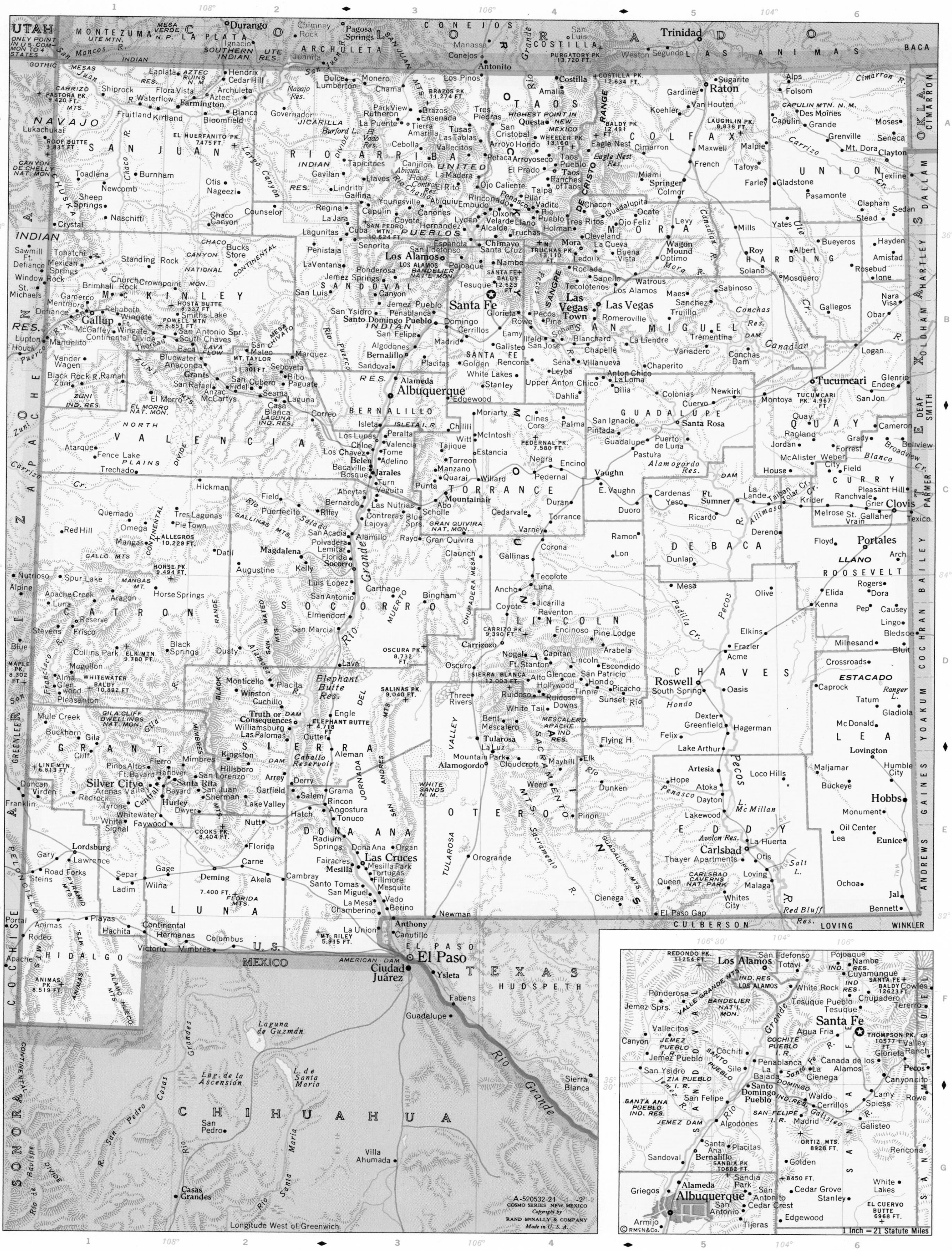

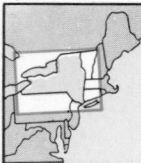

NEW YORK

Statute Miles 0 5 10 20 30 40

Kilometers 5 0 5 15 25 35 45 55

Lambert Conformal Conic Projection
SCALE 1:1,862,000 1 Inch = 29 Statute Miles

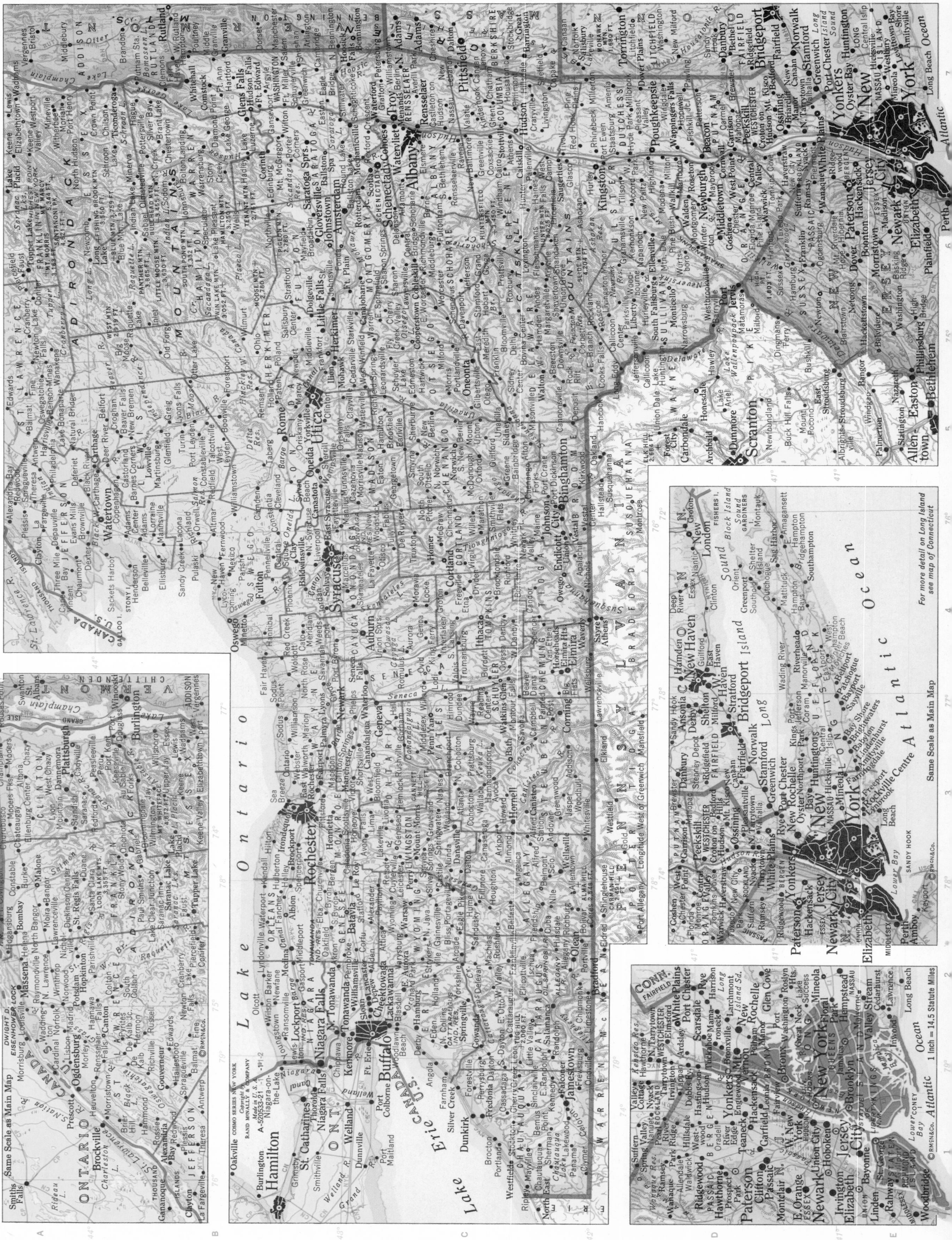

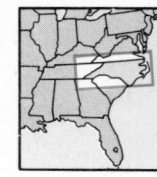

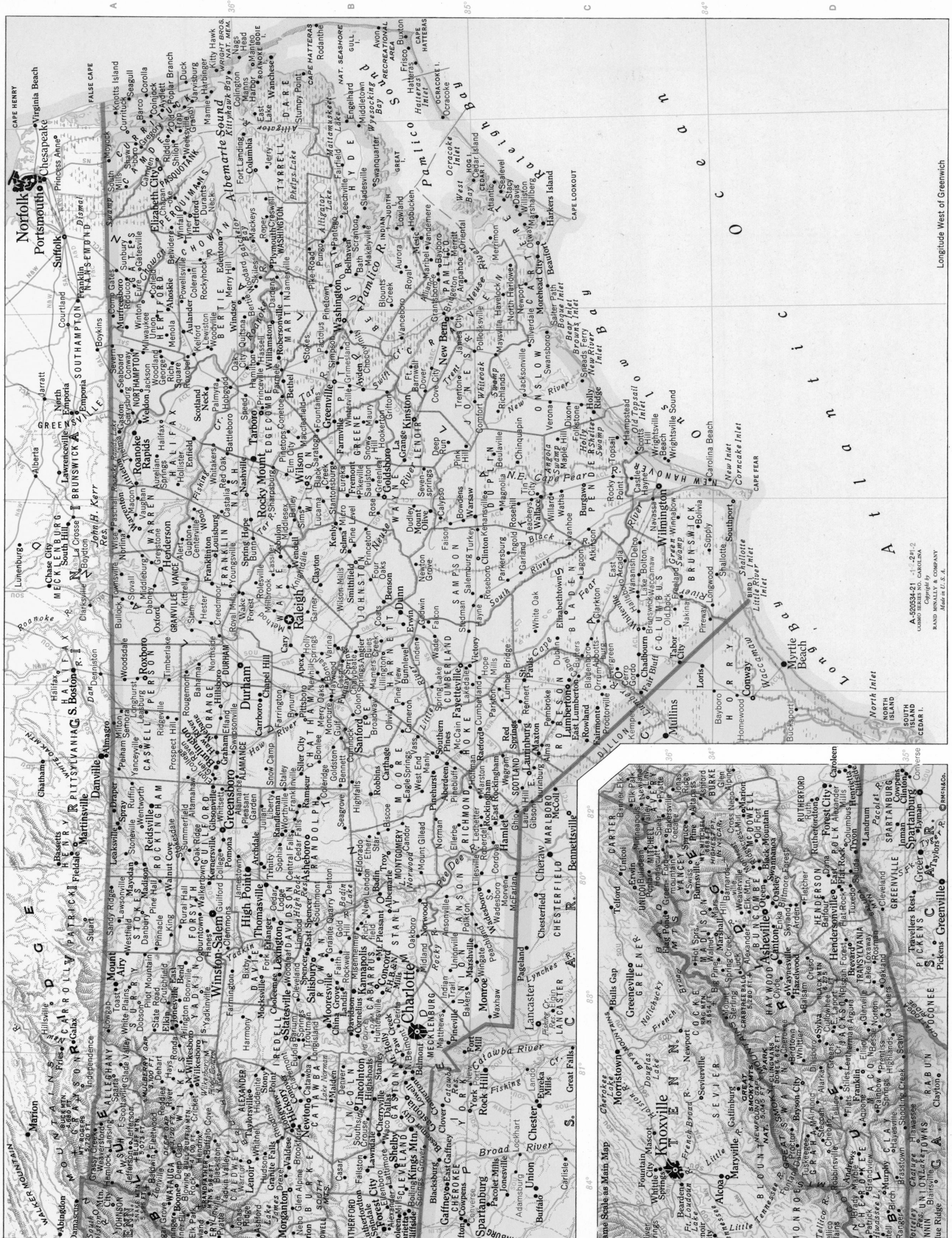

Longitude West of Greenwich

A-50053A-21 -1 -2½-2
COSMO AL2 NORTH CAROLINA
Copyright by
RAND M⁹NALLY & COMPANY
Made in U.S.A.

Lambert Conformal Conic Projection
SCALE 1:1,950,000 1 Inch = 31 Statute Miles

Statute Miles
Kilometers

109

NORTH DAKOTA

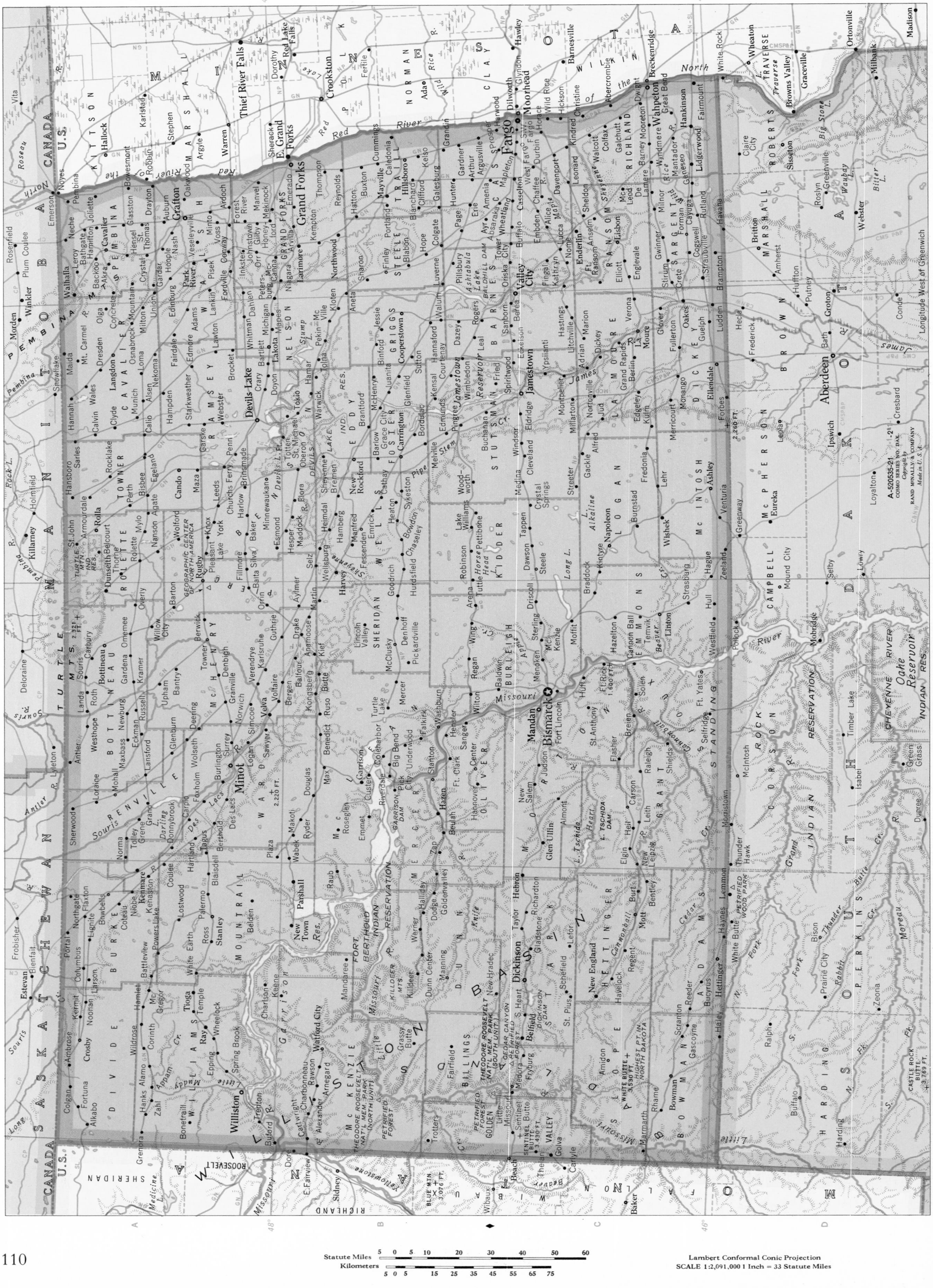

Statute Miles 5 0 5 10 20 30 40 50 60
Kilometers 5 0 5 15 25 35 45 55 65 75

Lambert Conformal Conic Projection
SCALE 1:2,091,000 1 Inch = 33 Statute Miles

A-520535-21
COSMO SERIES WO. DAK.
RAND McNALLY & COMPANY
Made in U.S.A.

Lambert Conformal Conic Projection
SCALE 1:1,714,000 1 Inch = 27 Statute Miles

Statute Miles 5 0 5 10 20 30 40
Kilometers 5 0 5 15 25 35 45 55

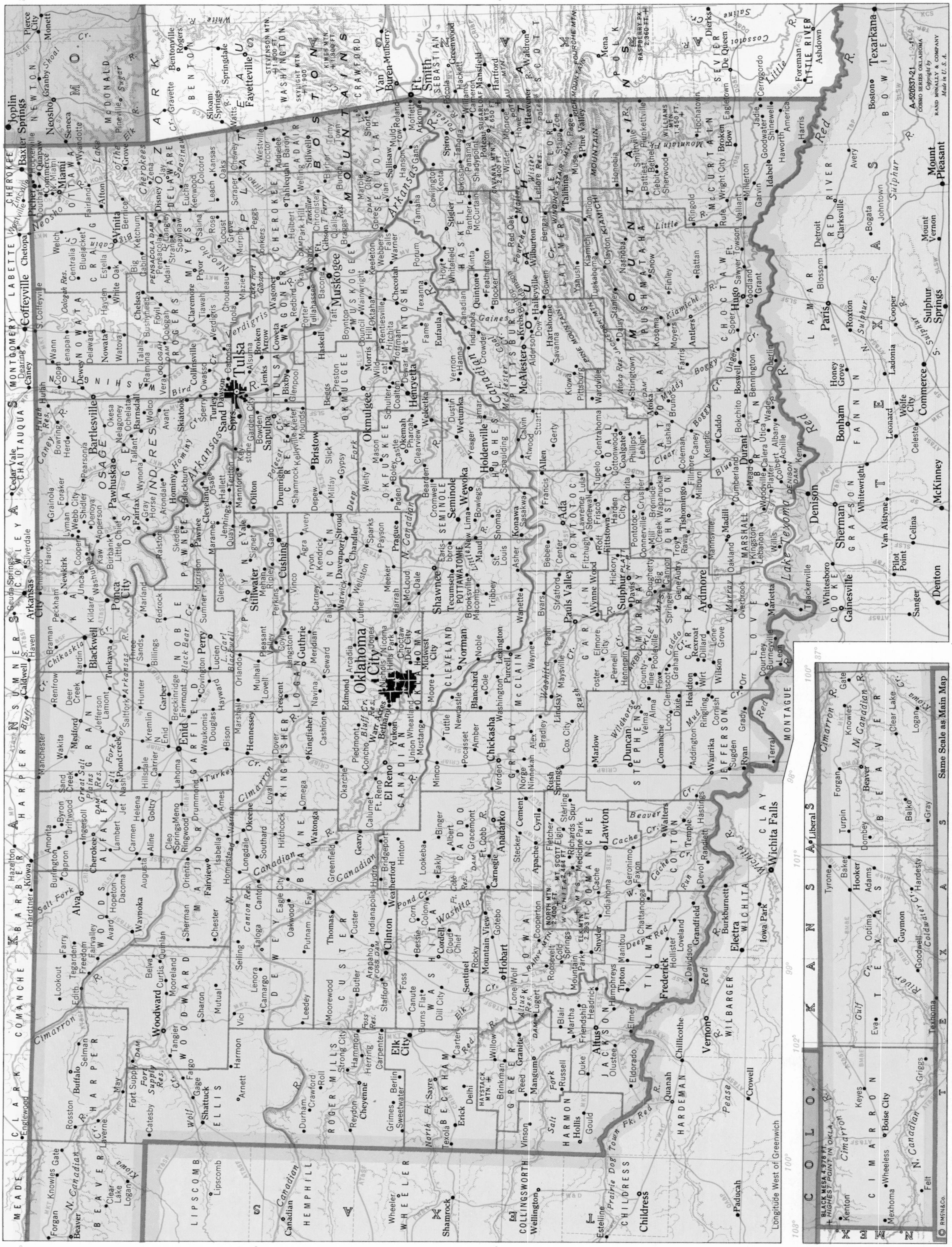

OKLAHOMA

Statute Miles 5 0 5 10 20 30 40
Kilometers 5 0 5 15 25 35 45 55

Lambert Conformal Conic Projection
SCALE 1:1,957,000 1 Inch = 31 Statute Miles

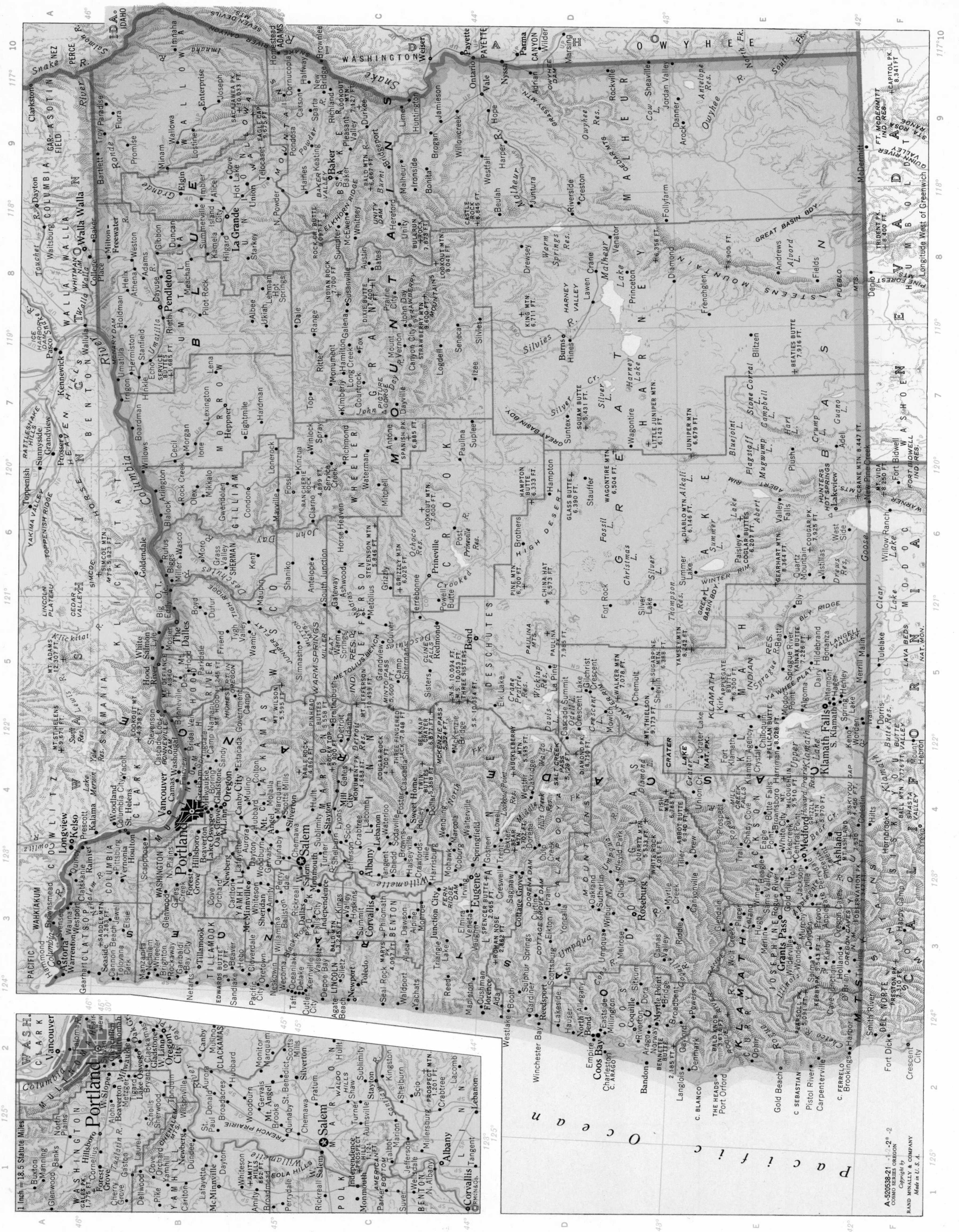

Lambert Conformal Conic Projection
SCALE 1:2,329,000   1 Inch = 37 Statute Miles

Statute Miles
Kilometers

1 Inch = 18.5 Statute Miles

Portland
Salem
Albany
Corvallis

Longitude West of Greenwich

A-50053B-21 ... -2°-2
COMBINED SERIES—OREGON
RAND M?NALLY & COMPANY
Made in U.S.A.

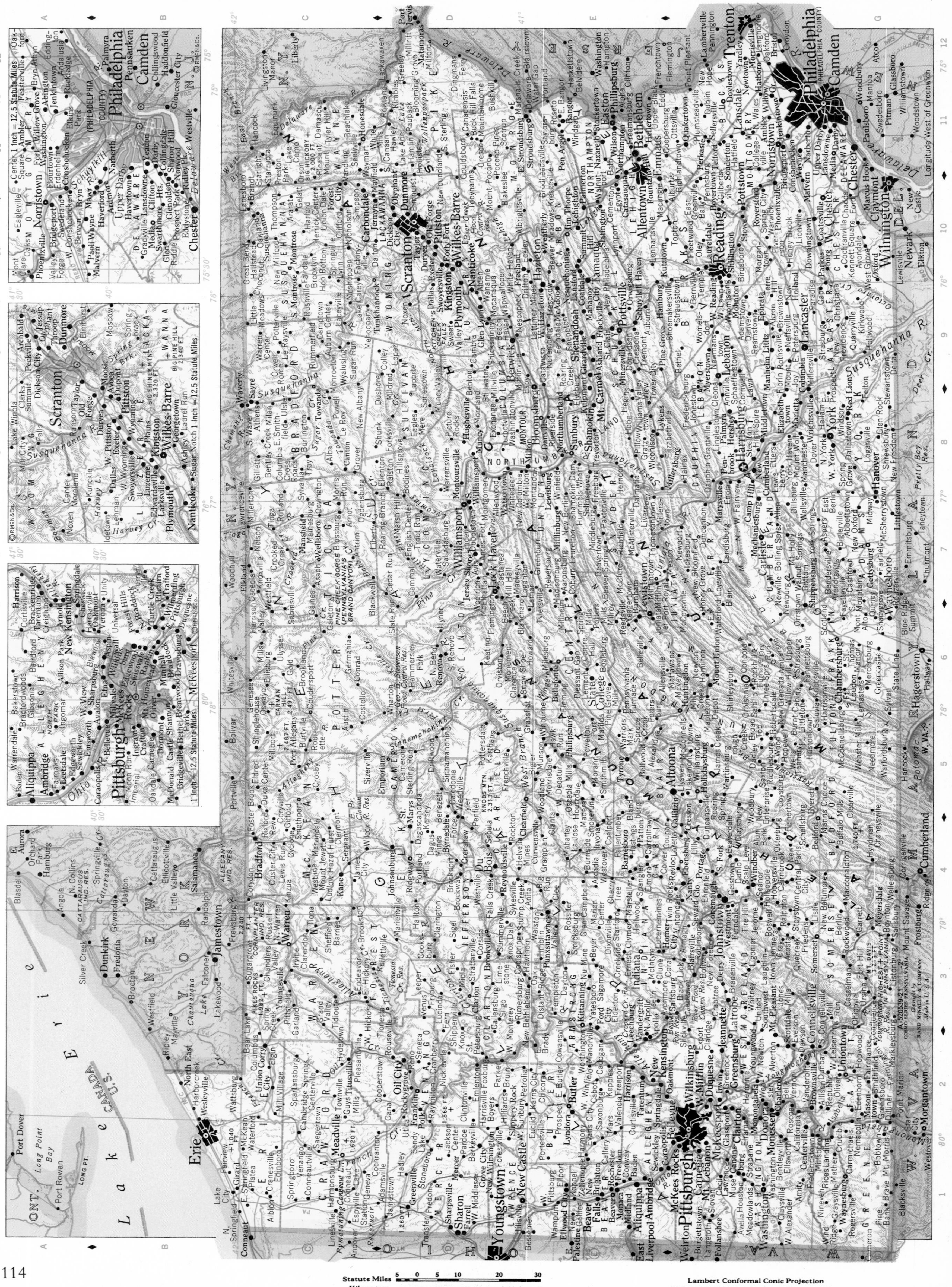

PENNSYLVANIA

Statute Miles
Kilometers

Lambert Conformal Conic Projection
SCALE 1:1,593,000   1 Inch = 25 Statute Miles

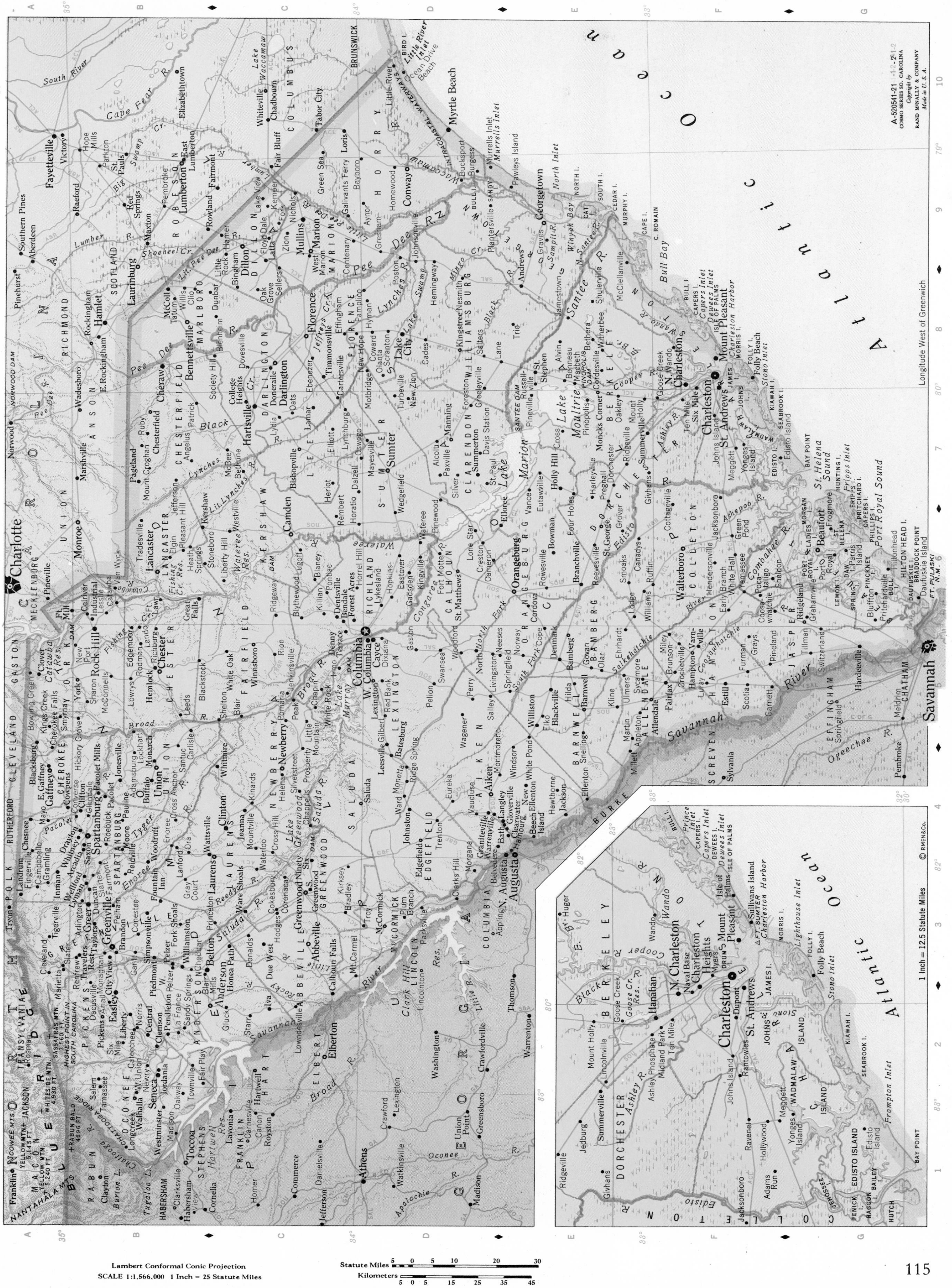

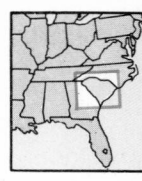

Atlantic Ocean

Lambert Conformal Conic Projection
SCALE 1:1,566,000 1 Inch = 25 Statute Miles

Statute Miles
Kilometers

1 Inch = 12.5 Statute Miles

A-50541-21
COSMO SERIES SO. CAROLINA
Copyright by
RAND McNALLY & COMPANY
Made in U.S.A.

Statute Miles
Kilometers

Lambert Conformal Conic Projection
SCALE 1:2,091,000   1 Inch = 33 Statute Miles

A-520542-21 · 1 · 21
COSMO SERIES SO. DAK.
Copyright by
RAND M\cNALLY & COMPANY
Made in U.S.A.

Longitude West of Greenwich

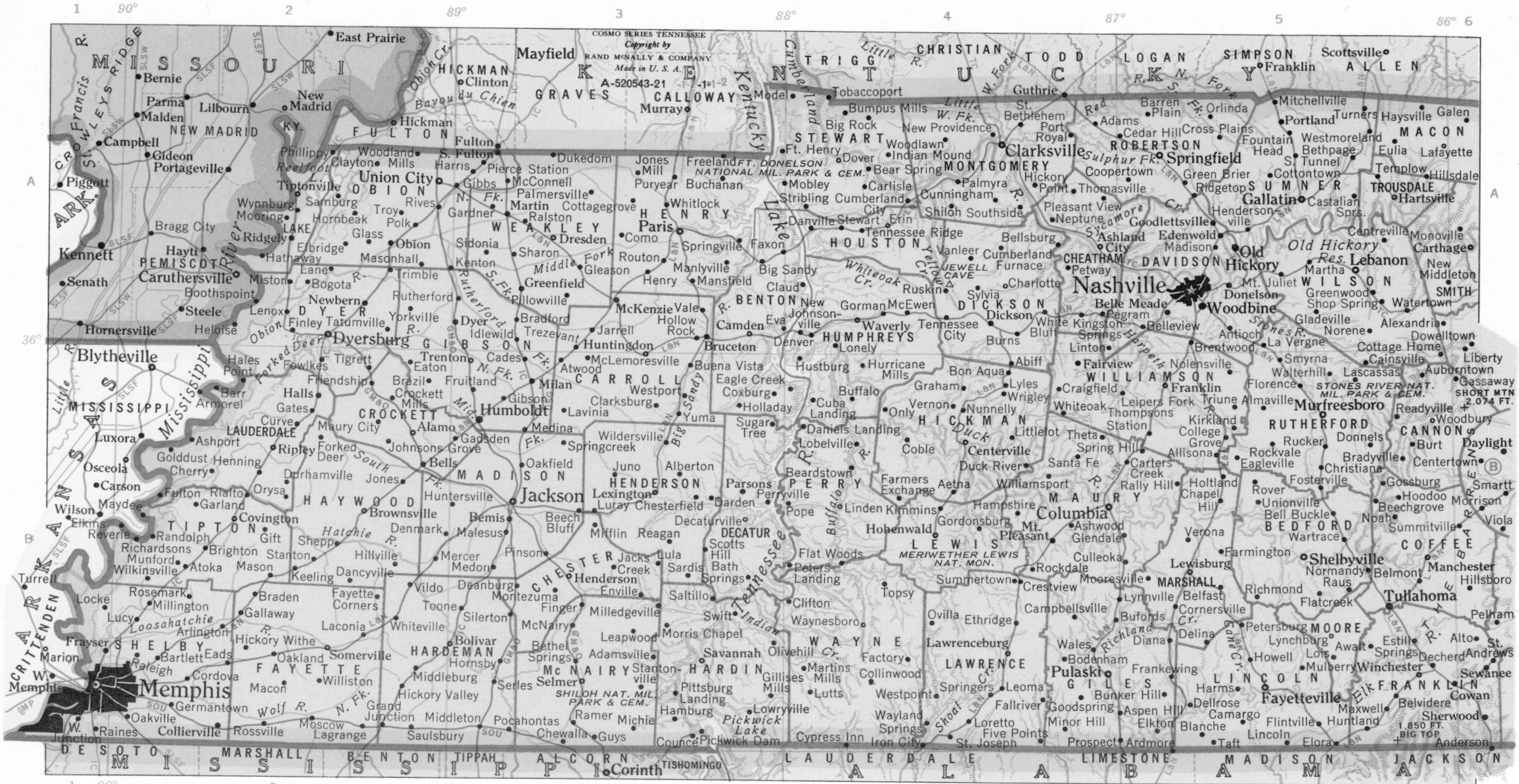

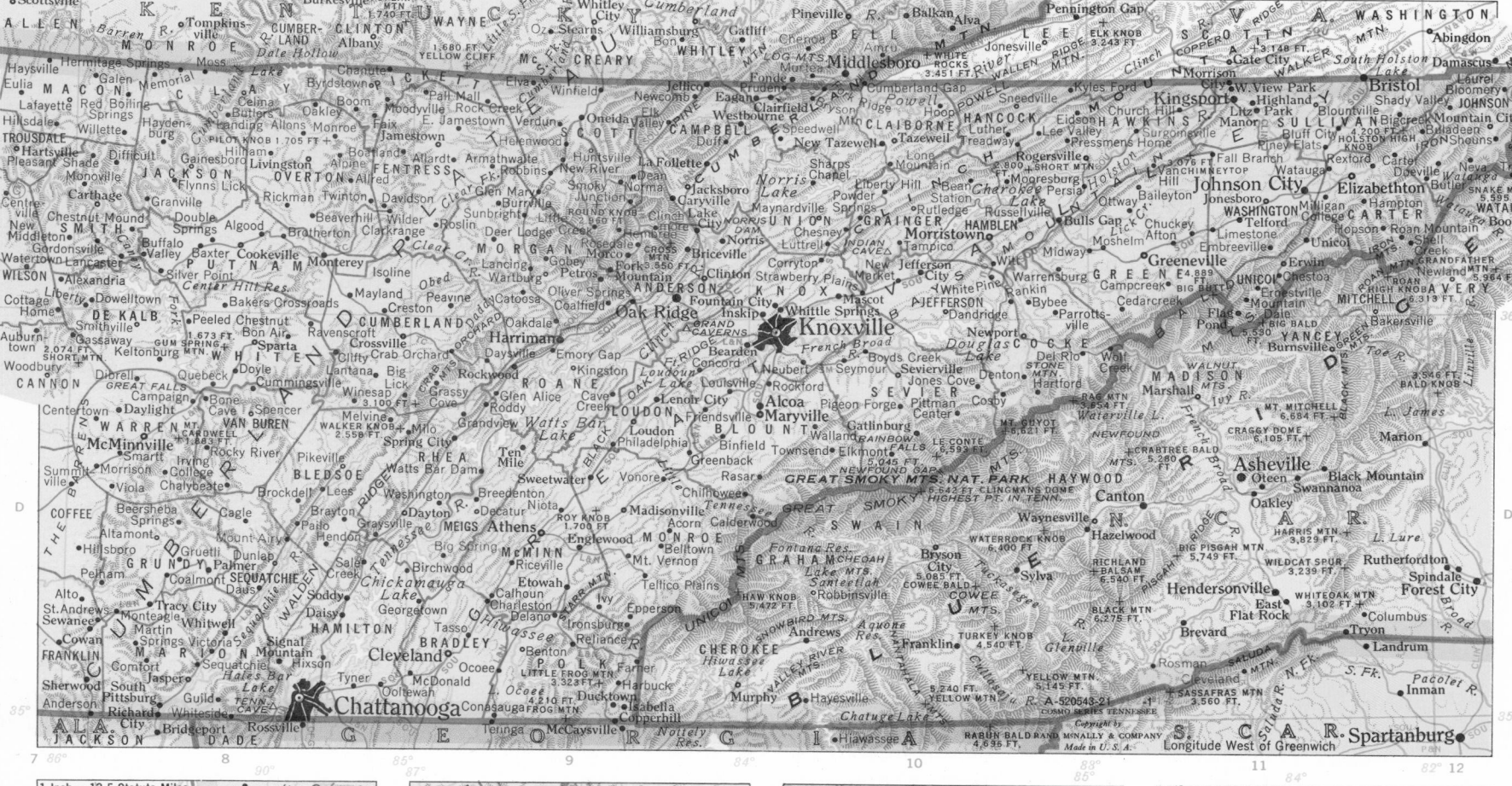

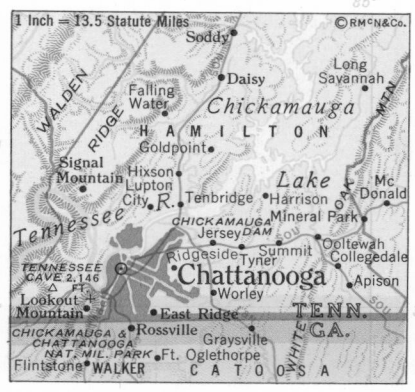

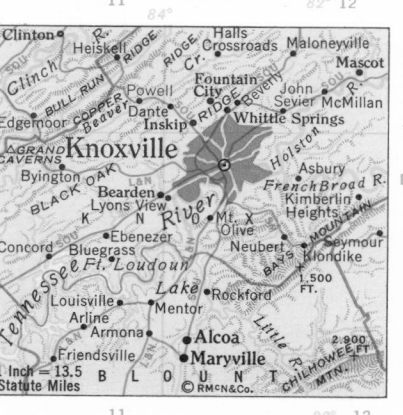

Lambert Conformal Conic Projection
SCALE 1:1,713,000 1 Inch = 27 Statute Miles

Statute Miles
5 0 5 10   20   30   40

Kilometers
5 0 5 15 25 35 45 55

117

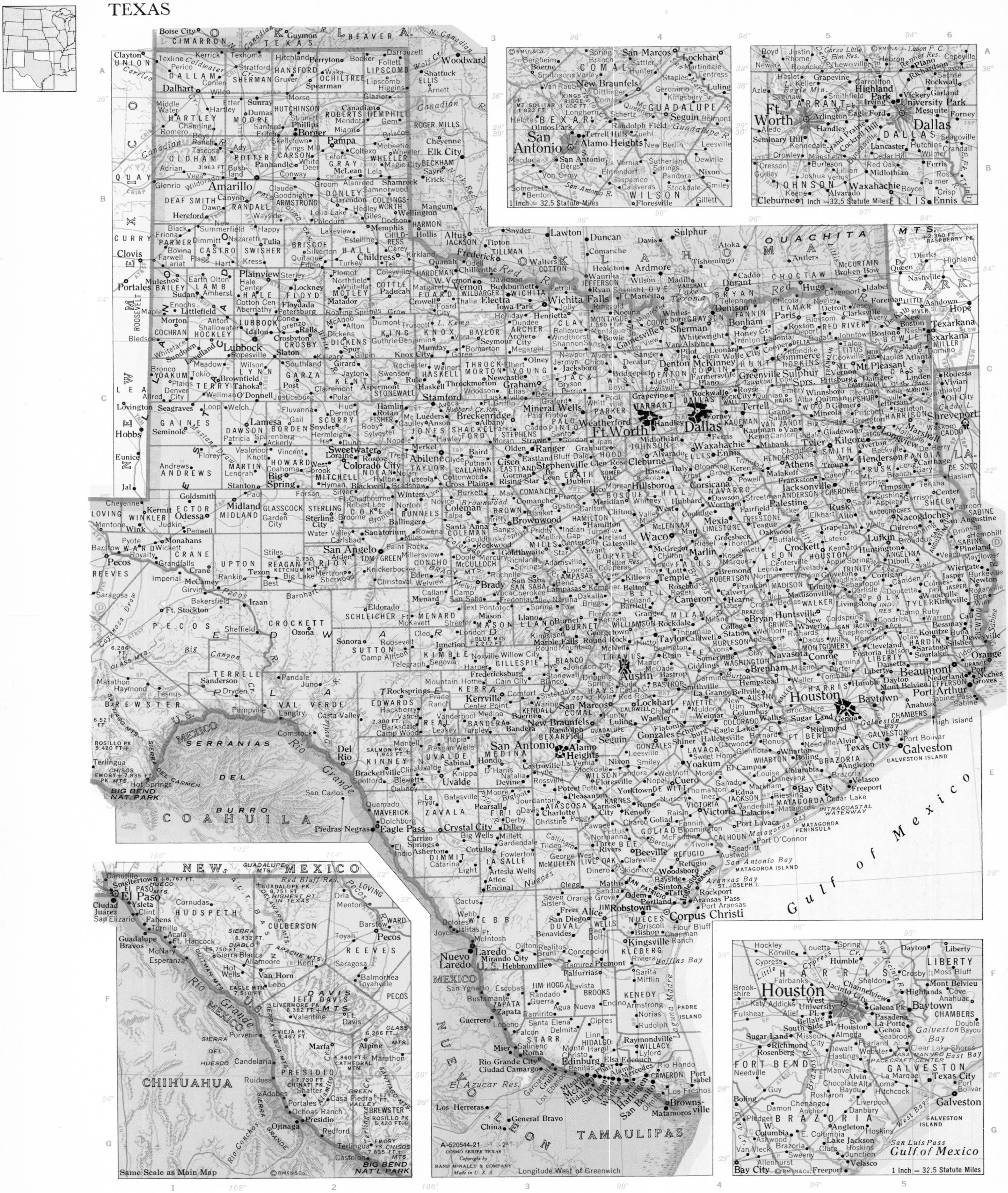

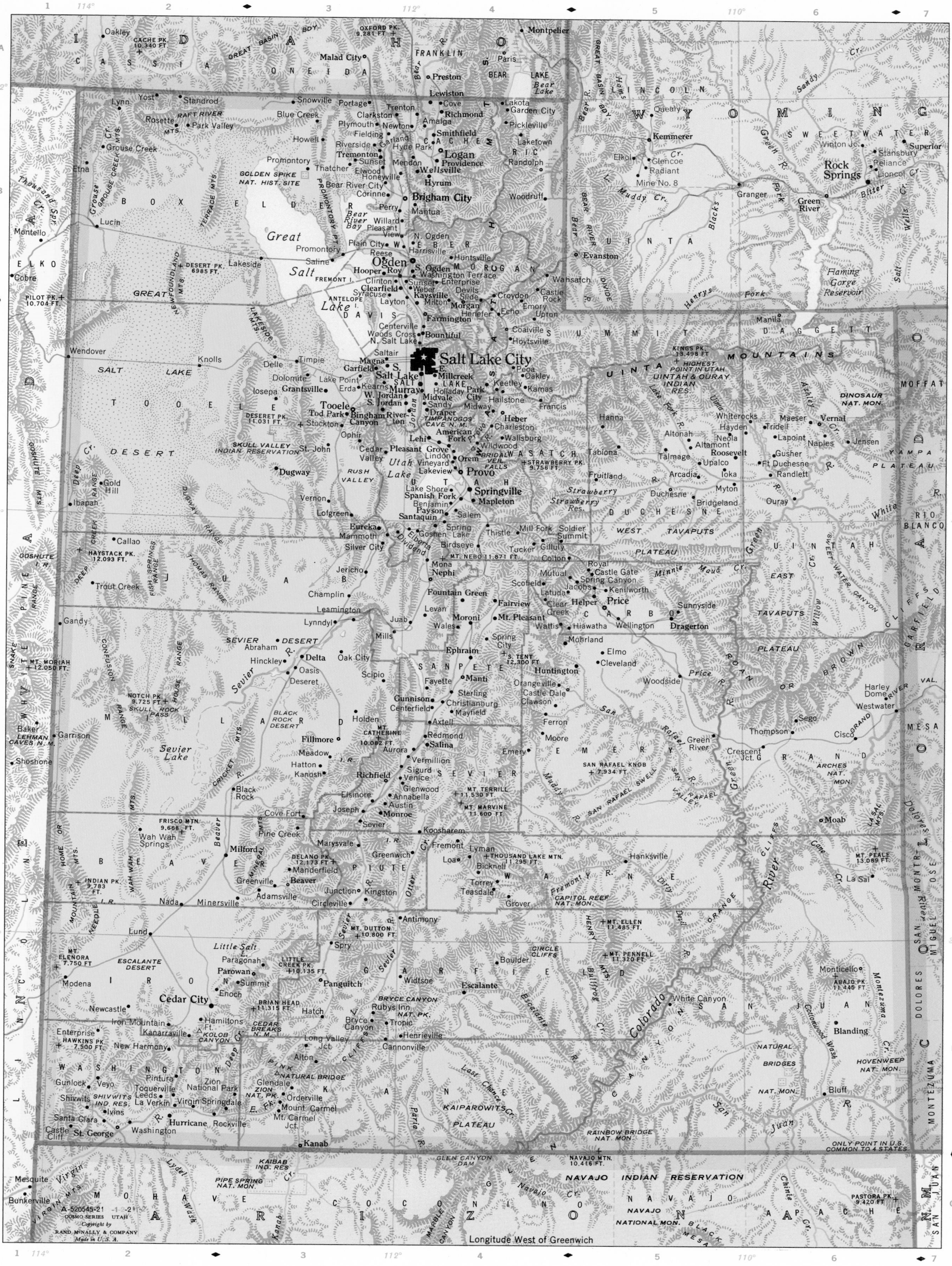

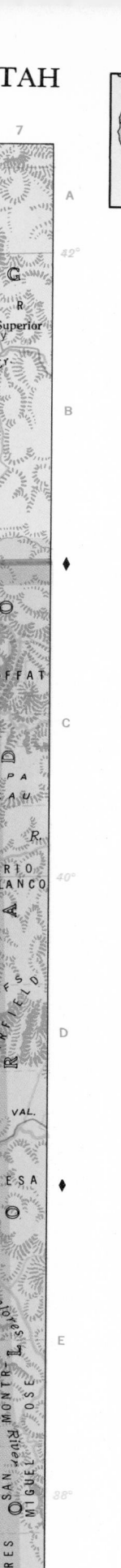

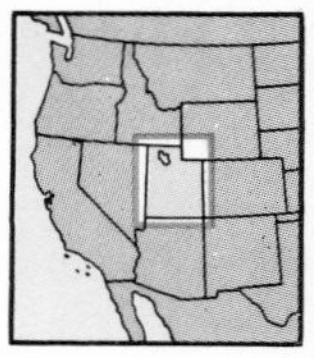

Lambert Conformal Conic Projection
SCALE 1:2,100,000   1 Inch = 33 Statute Miles

Statute Miles
Kilometers

A-520545-21   -1 -2+1
COSMO SERIES UTAH
Copyright by
RAND McNALLY & COMPANY
Made in U.S.A.

Longitude West of Greenwich

VERMONT

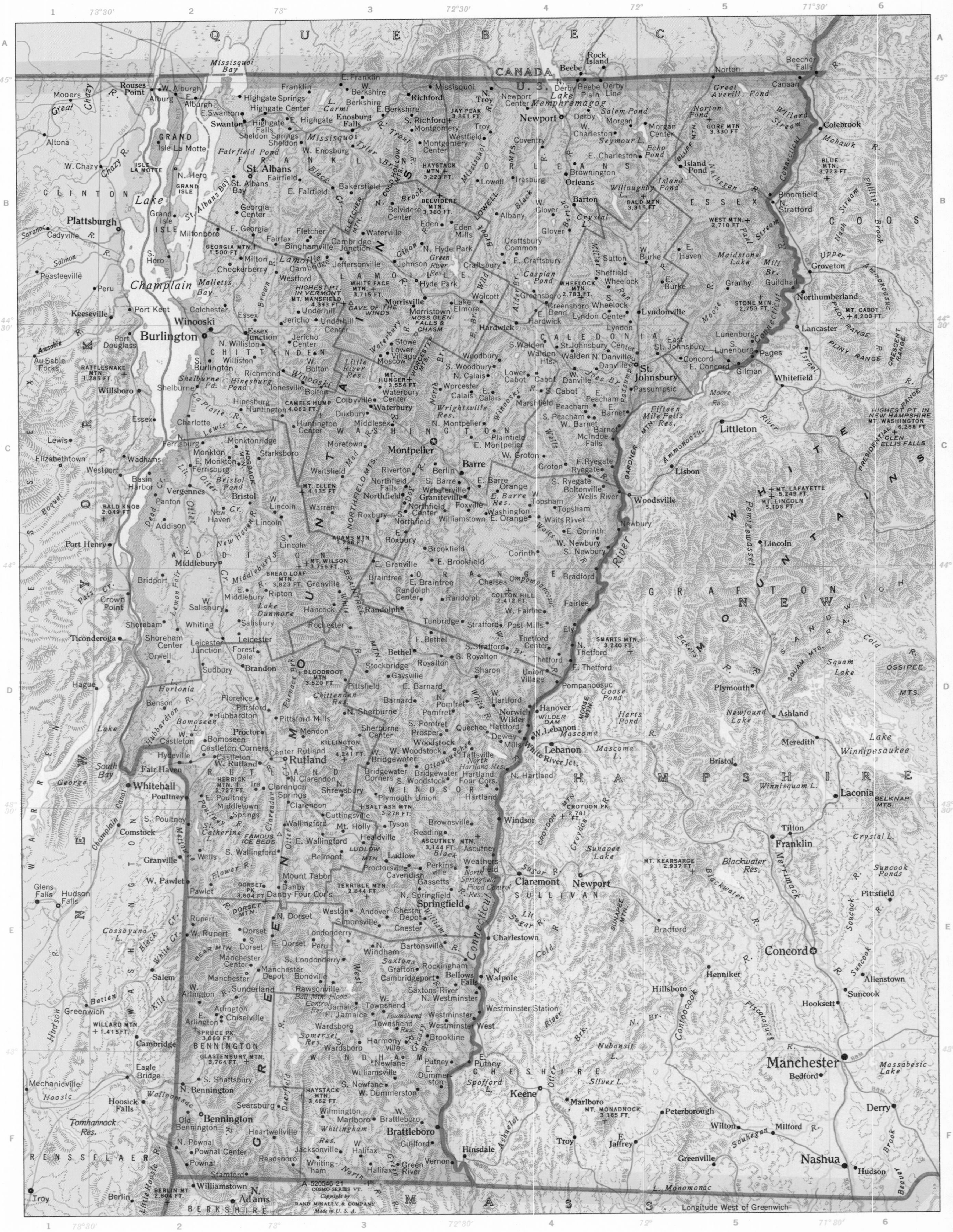

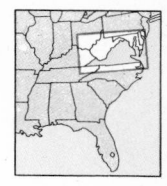

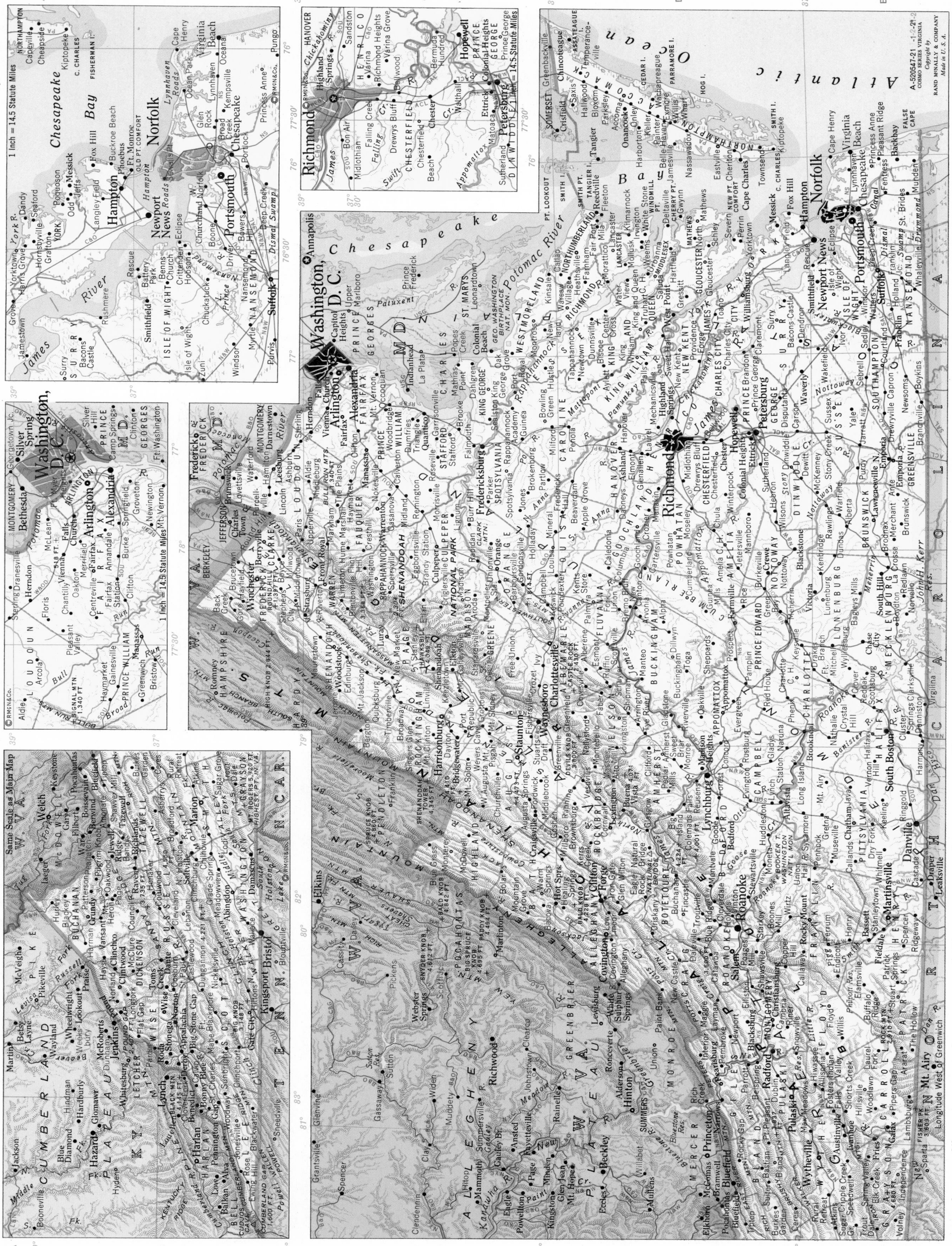

Lambert Conformal Conic Projection
SCALE 1:1,822,000 1 Inch = 29 Statute Miles

Statute Miles 5 0 5 10 20 30 40
Kilometers 5 0 5 15 25 35 45 55

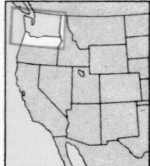

WASHINGTON

Statute Miles 0 5 10 20 30 40 50

Kilometers 5 0 5 15 25 35 45 55 65

Lambert Conformal Conic Projection
SCALE 1:2,068,000 Inch = 33 Statute Miles

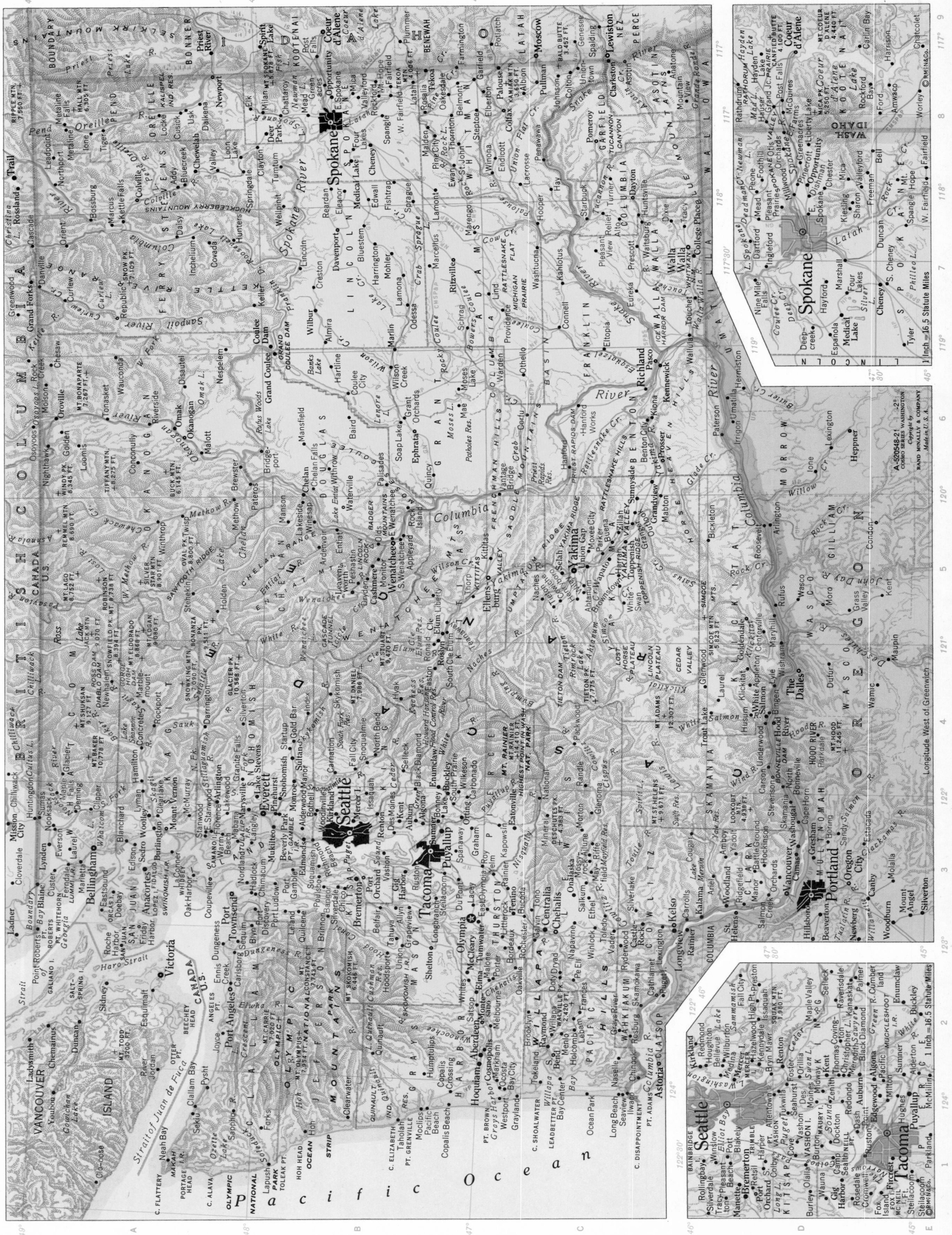

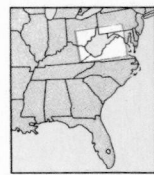

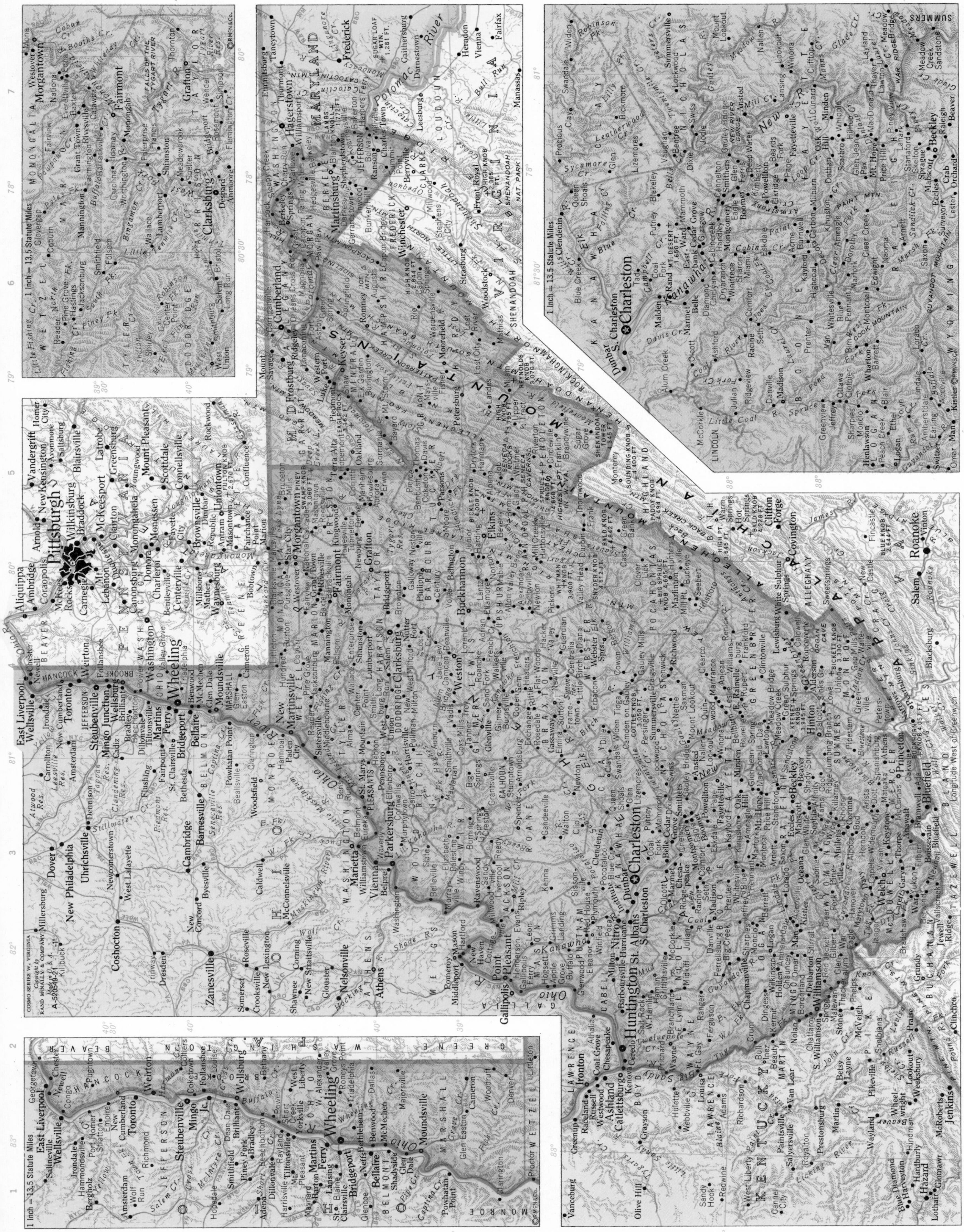

Lambert Conformal Conic Projection
SCALE 1:1,704,000 1 Inch = 27 Statute Miles

Statute Miles 5 0 5 10 20 30 40
Kilometers 5 0 5 15 25 35 45 55

WISCONSIN

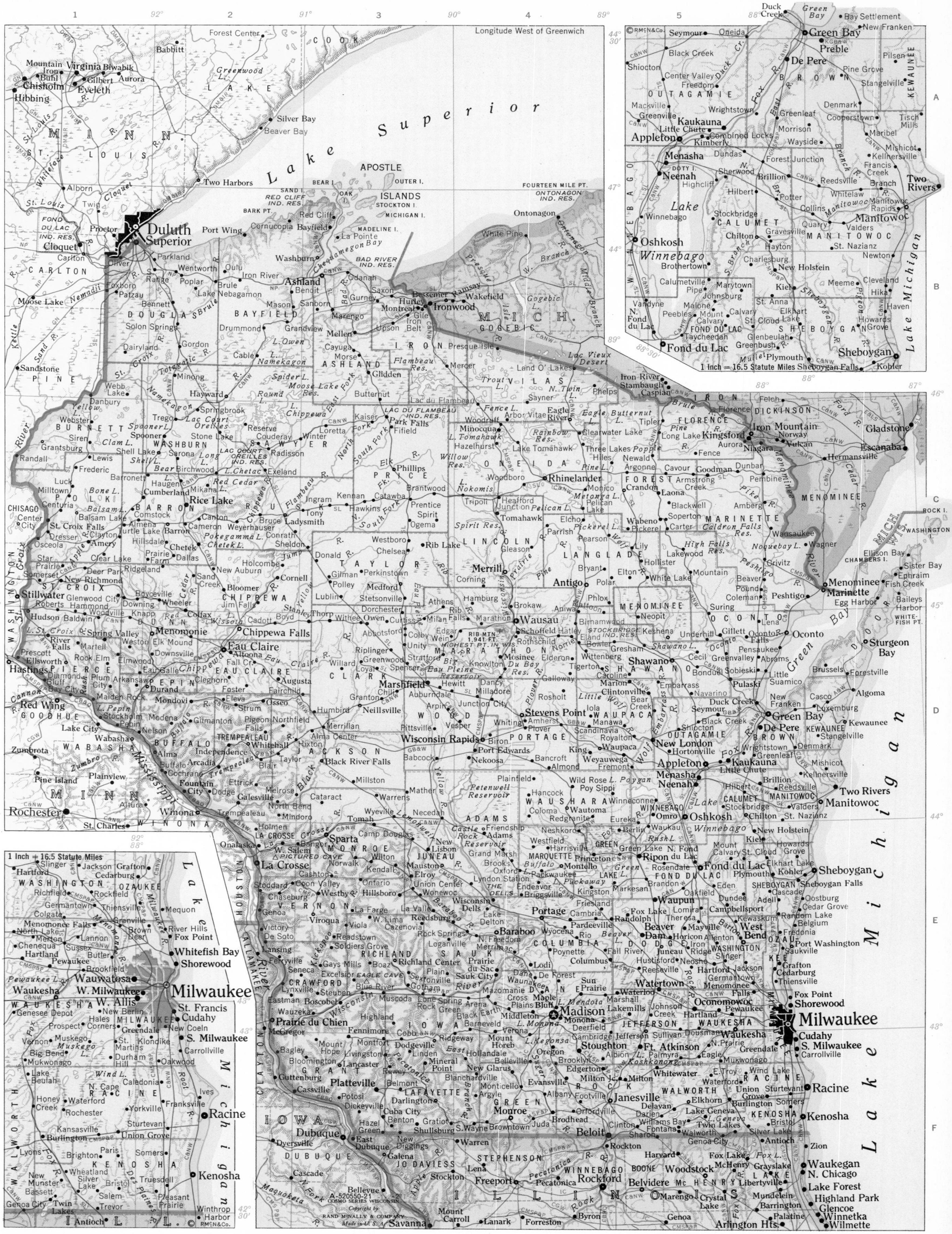

Statute Miles

Kilometers

Lambert Conformal Conic Projection
SCALE 1:2,088,000   1 Inch = 33 Statute Miles

1 Inch = 16.5 Statute Miles

Copyright by
RAND McNALLY & COMPANY
Made in U.S.A.

Lambert Conformal Conic Projection
SCALE 1:2,186,000   1 Inch = 34.5 Statute Miles

Statute Miles

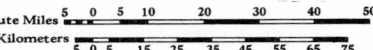
Kilometers

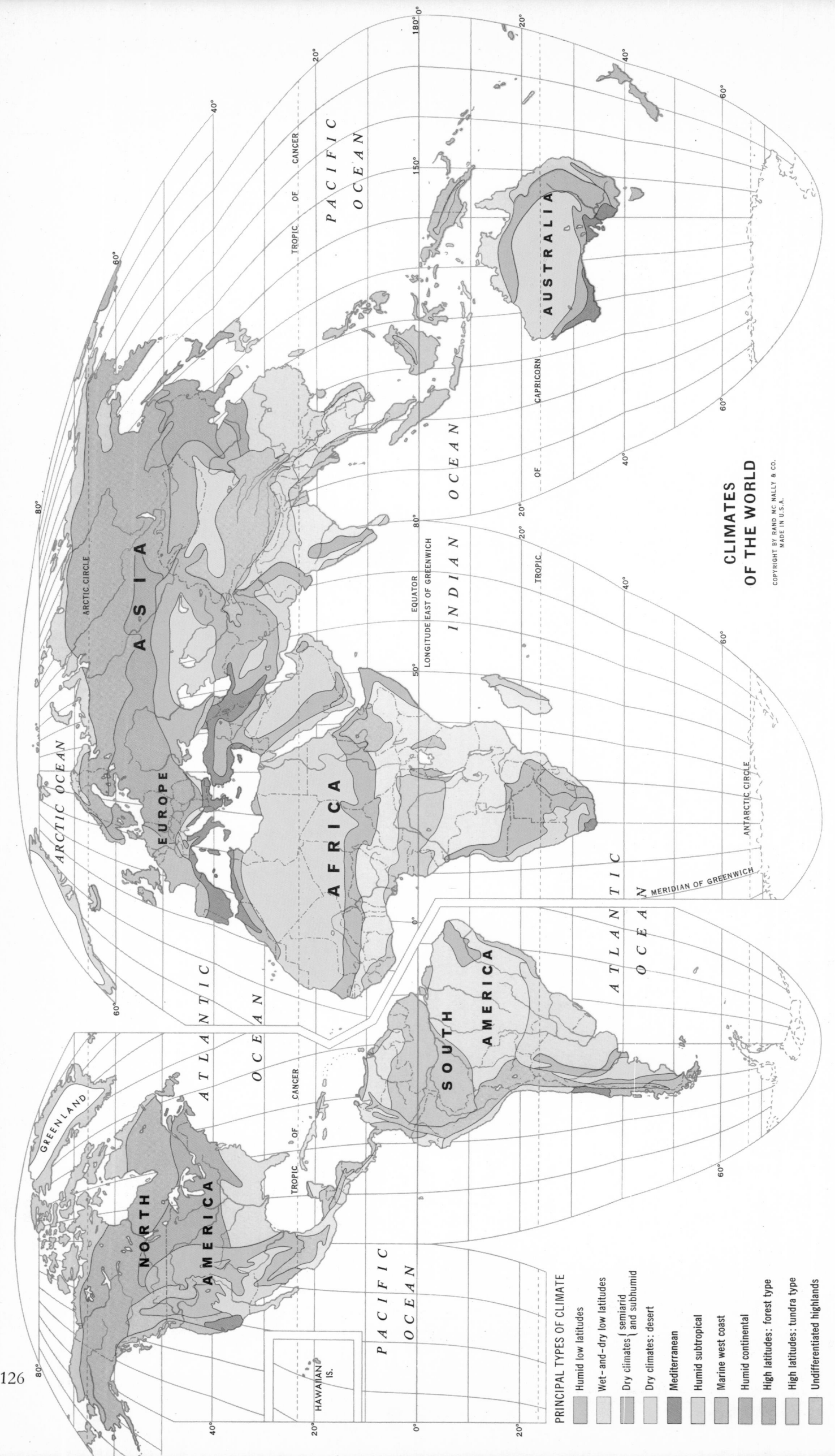

PRINCIPAL TYPES OF CLIMATE

Humid low latitudes

Wet-and-dry low latitudes

Dry climates { semiarid
                and subhumid

Dry climates: desert

Mediterranean

Humid subtropical

Marine west coast

Humid continental

High latitudes: forest type

High latitudes: tundra type

Undifferentiated highlands

CLIMATES
OF THE WORLD

126

The principal types of climate shown on the map above indicate the average weather conditions of a given region over a long period of time. Climate is one of the most important natural factors, for upon it depend the kind of crops man can grow, the length of the growing season and the amount of food produced, the type of homes built, and how life on the land is organized.

On this map, the earth's surface has been divided into a small number of large regions. There are numerous local variations within these regions, depending upon the lay of the land, exposure to the winds, supply of moisture, and the

vegetative cover. The characteristics of the natural vegetation and the soil are also greatly influenced by climate.

The easiest of the climate types to distinguish is that of the humid low latitudes, sometimes referred to as the equatorial rain forest. In these areas temperatures are uniformly high throughout the year, and there are no distinct seasons. Rain occurs in all months; one can distinguish only between wet months and very wet months.

Near the equator and in areas where the marine west coast influence is well developed, the variation in temperature from season to season is relatively small, but on the east coasts

in the middle latitudes and in the continental interiors the seasonal temperatures may vary greatly.

The wet and dry climate of low latitudes is much easier on man than the humid equatorial climate. Also the soils are better and more suited to the cultivation of crops.

The majority of the world's population lives in the middle latitudes. Here climate varies from the Mediterranean, with its alternating wet and dry seasons, to the warm summers and rainy winters of the humid subtropics, and the seldom cold, but seldom hot, marine west coasts, to the humid continental climate with its short hot summers and cold snowy winters.

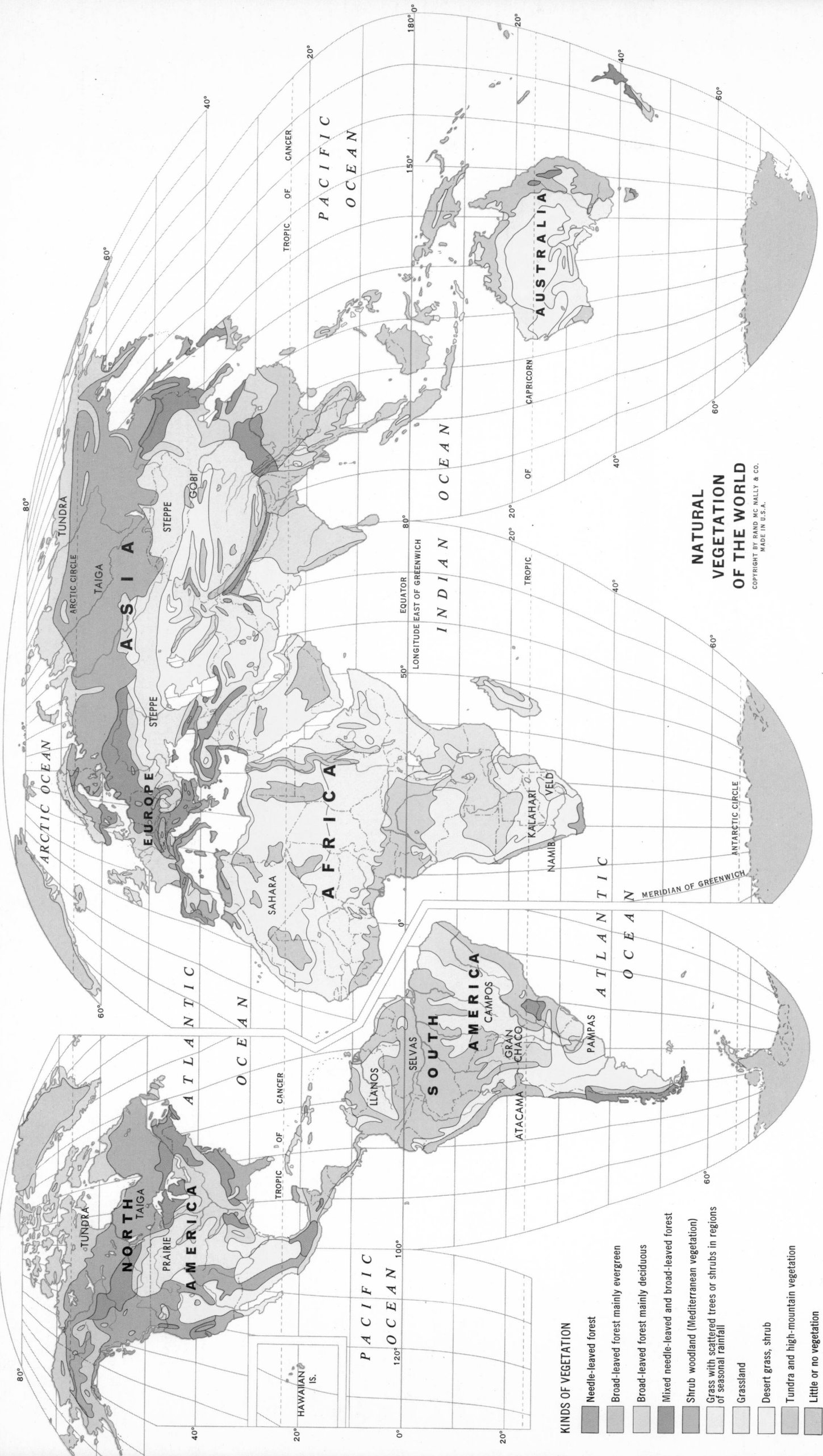

NATURAL VEGETATION OF THE WORLD

COPYRIGHT BY RAND MC NALLY & CO.
MADE IN U.S.A.

KINDS OF VEGETATION

Needle-leaved forest

Broad-leaved forest mainly evergreen

Broad-leaved forest mainly deciduous

Mixed needle-leaved and broad-leaved forest

Shrub woodland (Mediterranean vegetation)

Grass with scattered trees or shrubs in regions of seasonal rainfall

Grassland

Desert grass, shrub

Tundra and high-mountain vegetation

Little or no vegetation

Nearly everywhere on earth some vegetation grows; only the ice caps and the most barren deserts lack vegetation. The most important single control of vegetation is climate, but soil and the altitude of the land also are significant in determining vegetative cover.

The pattern of natural vegetation resembles closely the pattern of world climates. In the rainy lands forests extend for hundreds of miles. Where the climate is too dry or too cold for forests, shrub, grasses, and lichens are found.

In general there are three main forest types, which correspond with low, middle, and high latitude lands. In the humid low latitudes luxuriant evergreen forests of broad-leaved trees predominate. The humid middle latitudes have forests which consist of broad-leaved trees which are deciduous, losing their leaves in winter. Much of the middle latitude forest has been cleared for farm land, and few regions of these hardwood forests remain. High latitude forests occur only in the Northern Hemisphere, and they all but girdle the earth. The northern forest, or taiga, is almost entirely composed of needle-leaved, cone-bearing softwood trees. The southern parts of this great forest region are today the suppliers of the world's lumber and paper.

The moister parts of the semiarid lands where the natural vegetation was grass have in places been plowed to raise small grain. The drier semiarid lands remain in grass and are the world's great grazing regions. Tundra areas, covered with mosses and lichens, are found in polar regions.

127

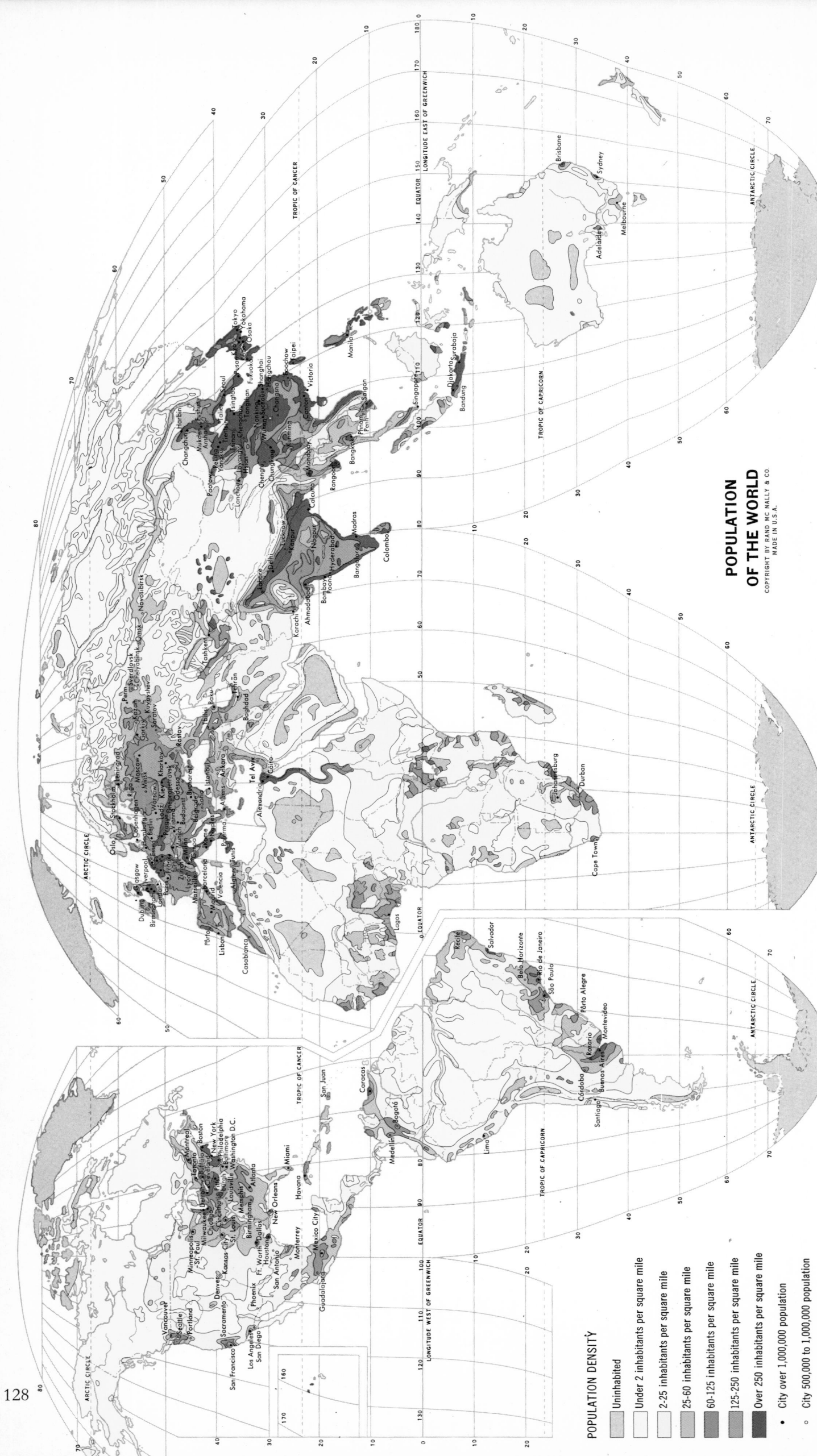

The population of the world is nearly three billion. One hundred and fifty years ago there were only about one-quarter as many people. Yet, today, there is still a square mile for every fifty people in the world. Approximately one-fourth of mankind lives, or at least works, in cities. This population map shows the densest population categories as more than 250 persons per square mile, but this is not the actual extreme. Population densities range from no people at all in parts of desert and polar regions to over 1,000 per square mile in the congested lands of India, China, Egypt, or the Island of Java. The highest densities anywhere, of course, are in the great cities.

The map above shows three large areas of very dense population, along with numerous small ones. The densest areas are in central eastern Asia, the Indian subcontinent, and Europe. Other than in these three large areas, the rest of the world is very much less densely populated.

The United States shows clusters of dense population in the Northeast and Midwest, but most of the land has no more than the average density and much of the land has a great deal less. In South America nine-tenths of the population lives around the margin of the continent.

Compare the population map with the climate and vegetation map. Why do people live where they do? Three-fourths of the world's population lives and works on the land producing food, not only for its own needs, but for the rest of humanity. The remaining one-quarter lives mainly in cities and is employed in manufacturing, transportation, and service industries. The cities named on this map are classified by their metropolitan area populations.

POPULATION OF THE WORLD

COPYRIGHT BY RAND MC NALLY & CO
MADE IN U.S.A.

POPULATION DENSITY

- Uninhabited
- Under 2 inhabitants per square mile
- 2-25 inhabitants per square mile
- 25-60 inhabitants per square mile
- 60-125 inhabitants per square mile
- 125-250 inhabitants per square mile
- Over 250 inhabitants per square mile
- • City over 1,000,000 population
- ○ City 500,000 to 1,000,000 population

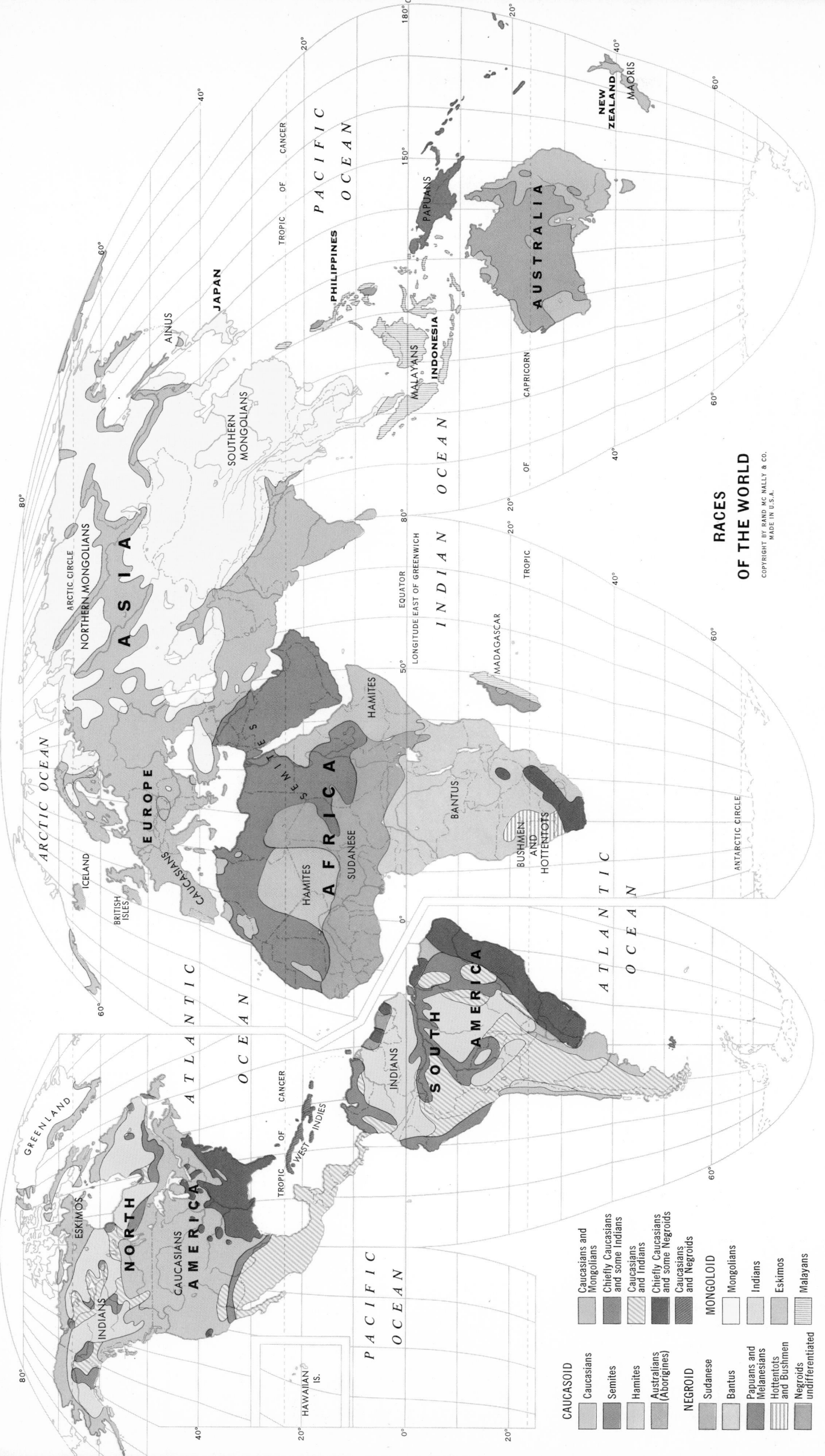

RACES OF THE WORLD

COPYRIGHT BY RAND MC NALLY & CO.
MADE IN U.S.A.

CAUCASOID
- Caucasians
- Semites
- Hamites
- Australians (Aborigines)

NEGROID
- Sudanese
- Bantus
- Papuans and Melanesians
- Hottentots and Bushmen
- Negroids undifferentiated

- Caucasians and Mongolians
- Chiefly Caucasians and some Indians
- Caucasians and Indians
- Chiefly Caucasians and some Negroids
- Caucasians and Negroids

MONGOLOID
- Mongolians
- Indians
- Eskimos
- Malayans

The term race denotes a group of people who have inherited certain physiological characteristics such as shape of the skull, the nose, and chin, skin color, and texture of the hair. Most conspicuous of racial attributes is skin color, although it may cover a wide range even within a single race. There is no "typical person" in any race.

During the course of history the major races of the world have so intermixed and have given rise to so many subgroups that a precise schematic diagram of the human race cannot be achieved. The map above shows what the anthropologists recognize as the predominant racial types in each area. Every racial group in the world is well represented in the Americas. There are Mongoloids, represented by descendants of the original Indian peoples, and also Negroids, who make up important groups in parts of Brazil, the United States, and especially in the West Indies. In Latin America the Caucasoids, mainly from Spain and Portugal, are mixed in varying degrees with the Negroids and Mongoloids.

The people of Europe are almost all Caucasoid. Today there is only slight evidence in a few areas of the Mongoloid or Negroid races. In Africa south of the Sahara most of the people are Negroid; northward the majority are Caucasoid. The chief race of southwest and west central Asia is Caucasoid with a wide variance of characteristics, depending upon the area. In India there are many people who have hair and facial features of the Caucasoid but are likely to have brown skins, dark brown or black hair, and dark eyes. The rest of Asia is mainly Mongoloid. Nearly all of these people have dark straight hair and dark eyes. Skin color varies from medium brown to very pale yellow.

129

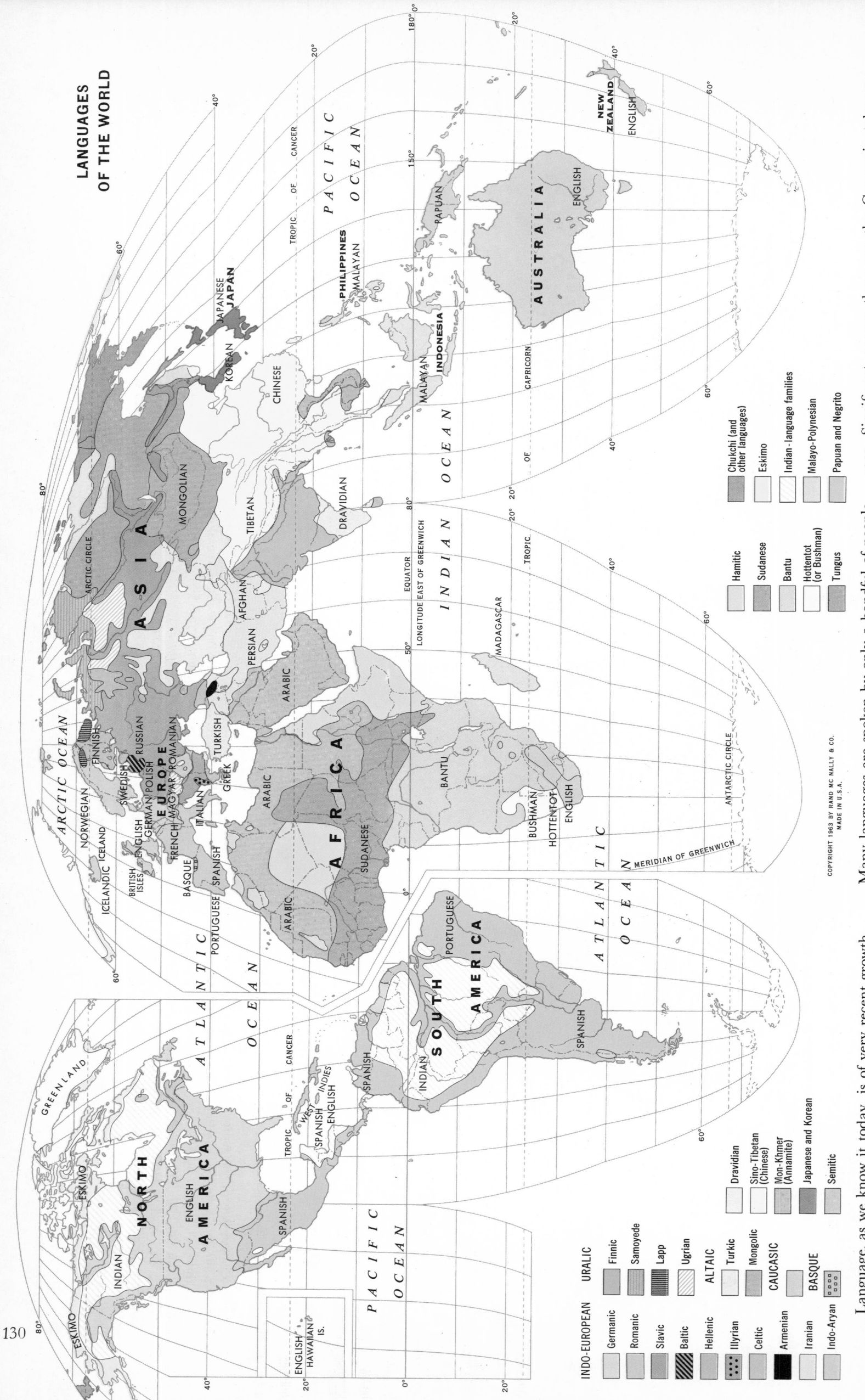

LANGUAGES OF THE WORLD

Language, as we know it today, is of very recent growth and is continually changing, absorbing new words and simplifying older grammatical forms. A world map cannot attempt to show individual areas in which a single language is spoken. This would be an impossible task for it is claimed by linguists that there are some 4,000 distinct languages spoken in the world today and, within them, an almost infinite number of dialects. The languages on the map above are grouped according to relationships determined by language specialists.

Many languages are spoken by only a handful of people while another language, Chinese, if we include all its dialect forms, is spoken by almost one-quarter of the human race. In the area or areas shown on the map in a single color, several different languages may be spoken, and a person speaking one of these languages might understand little or none of the other languages in the group, but since they are related, it would be easier for him to learn them.

The Indo-European languages are by far the most important and are spoken by almost two-thirds of the human race. Significant among these are the Germanic subgroup which includes English, the Romanic with Spanish, French Italian, and others, and Slavic, especially Russian.

Next in importance are the Sino-Tibetan group, mainly Chinese and related languages; the Uralic and Altaic groups which include Finnish, Hungarian, Turkish, and Mongolian; Japanese and Korean; the Semitic languages of the Middle East; and the Sudanese and Bantu languages of Africa. The distribution of these language groups can be related to the spread of cultures in the world.

COPYRIGHT 1963 BY RAND MC NALLY & CO.
MADE IN U.S.A.

INDO-EUROPEAN
Germanic
Romanic
Slavic
Baltic
Hellenic
Illyrian
Celtic
Armenian
Iranian
Indo-Aryan

URALIC
Finnic
Samoyede
Lapp
Ugrian

ALTAIC
Turkic
Mongolic

CAUCASIC

BASQUE

Dravidian
Sino-Tibetan (Chinese)
Mon-Khmer (Annamite)
Japanese and Korean
Semitic

Hamitic
Sudanese
Bantu
Hottentot (or Bushman)
Tungus

Chukchi (and other languages)
Eskimo
Indian-language families
Malayo-Polynesian
Papuan and Negrito

130

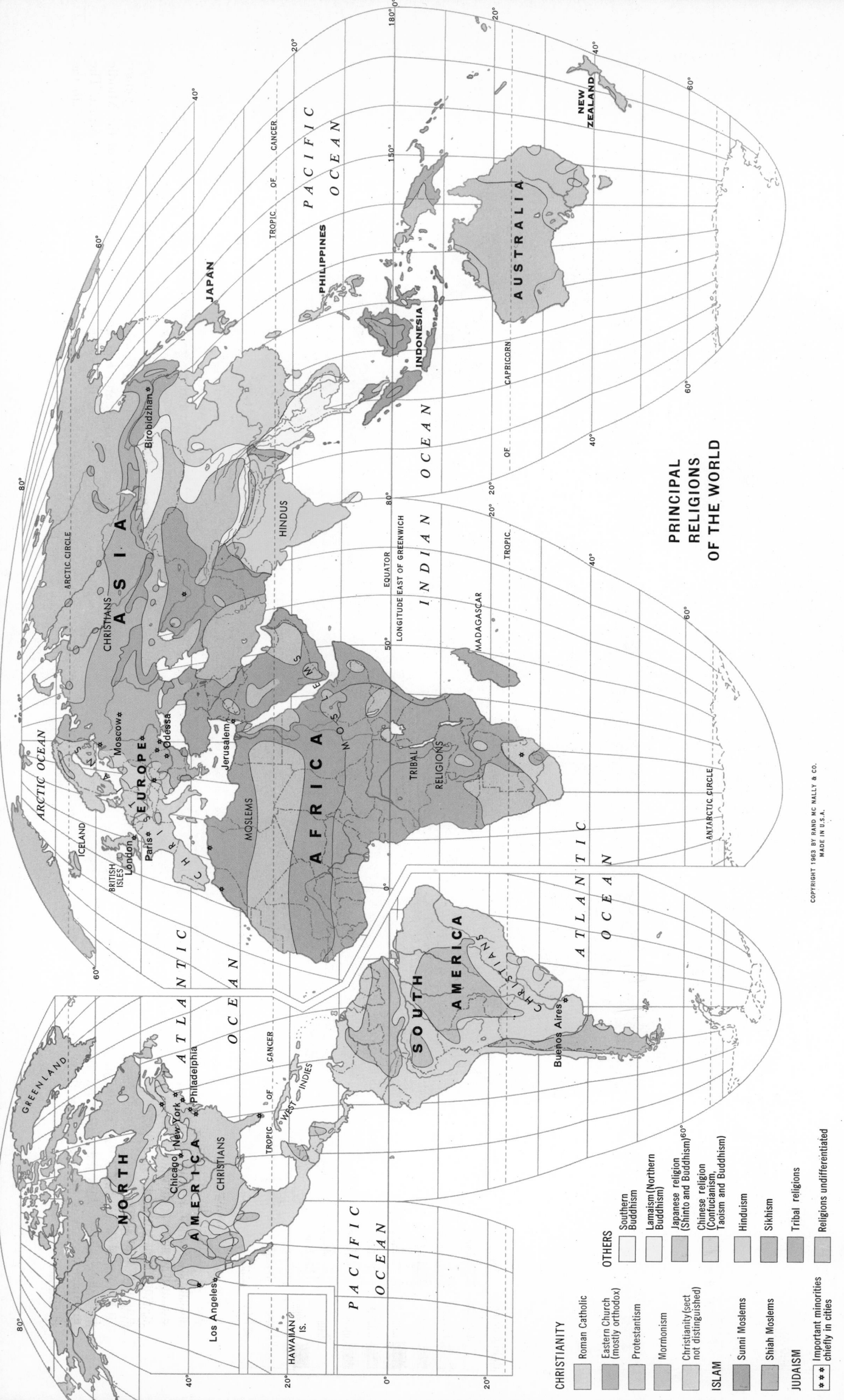

PRINCIPAL
RELIGIONS
OF THE WORLD

COPYRIGHT 1963 BY RAND MC NALLY & CO.
MADE IN U.S.A.

CHRISTIANITY
- Roman Catholic
- Eastern Church (mostly orthodox)
- Protestantism
- Mormonism
- Christianity (sect not distinguished)

ISLAM
- Sunni Moslems
- Shiah Moslems

JUDAISM
✳✳✳ Important minorities chiefly in cities

OTHERS
- Southern Buddhism
- Lamaism (Northern Buddhism)
- Japanese religion (Shinto and Buddhism)
- Chinese religion (Confucianism, Taoism and Buddhism)
- Hinduism
- Sikhism
- Tribal religions
- Religions undifferentiated

The distribution of the main religions of the world shows a certain similarity to the language map although in comparison there are fewer religions. The Romanic-speaking world with certain notable exceptions, like Romania, is Roman Catholic; the Arabic world is predominantly Islamic; the Germanic world is heavily Protestant; the Slavic world is chiefly Orthodox or Eastern, and the Chinese world is a blend of Buddhism, Confucianism, and Taoism. Western Christianity was propagated from Rome, the source also of the Latin language from which the Romance languages derived. Islam was spread by the Arab invasions, and missionaries of the Eastern or Orthodox church played an important role in shaping the Slavic languages and providing them with an alphabet. Protestantism, derived chiefly from the German Reformation, is important in the countries where languages of Germanic origin are widely spoken.

Common religious beliefs have often brought peoples into closer association, but religious differences have at times kept people apart. Two or more religions may flourish in the same area but on a world map it is often only possible to show the religion of the majority. Because, other than in Israel, the Jewish religion is not concentrated in an area large enough to show by color, symbols have been used to show other areas where there is a large concentration of Jews. In some places people may be members of two or more religious groups, as in Japan. Here the joint practice of Shinto and Buddhism is common and the term Japanese religion is used.

131

THE WESTERN HEMISPHERE

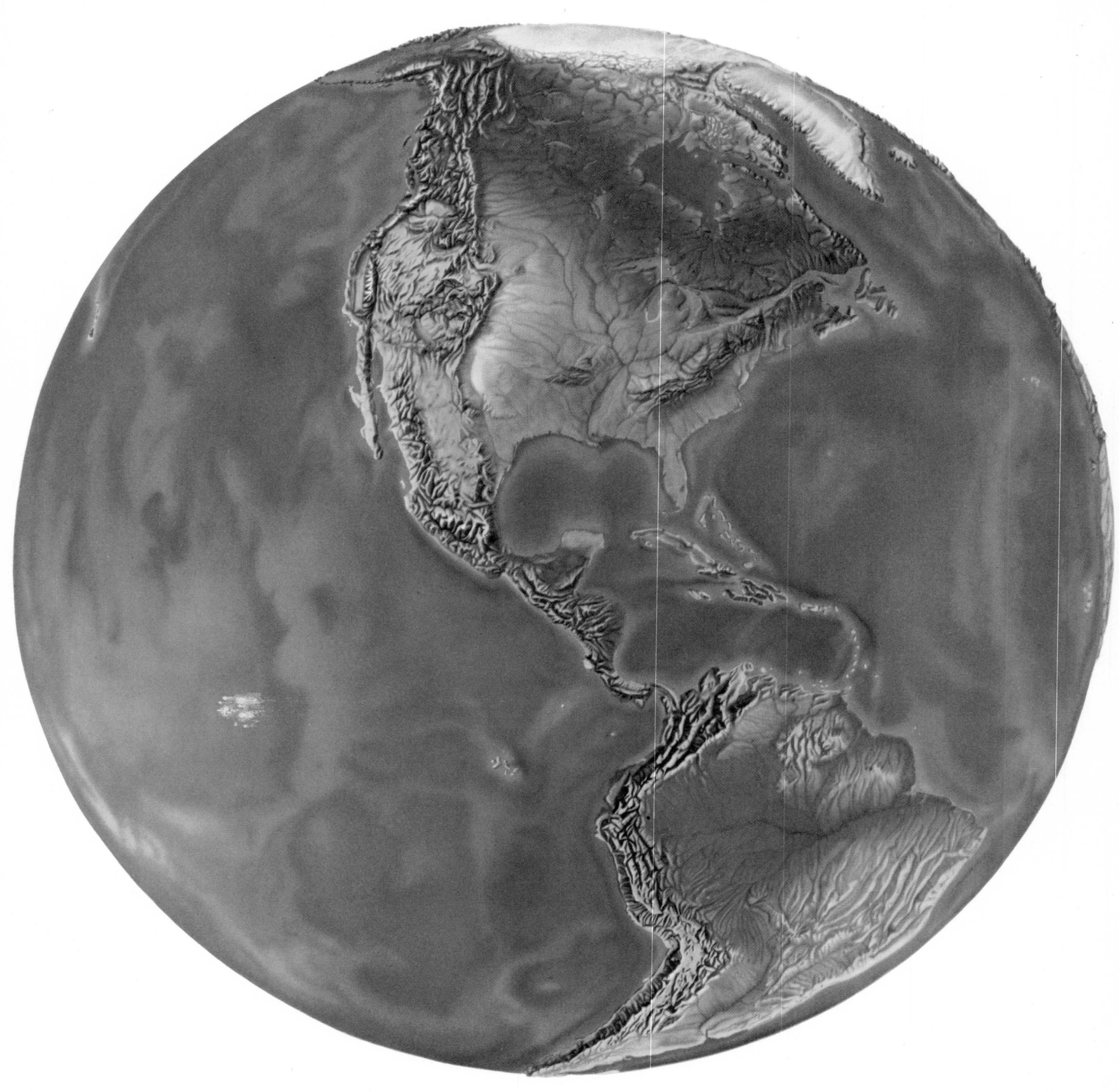

This distinctive relief interpretation shows the New World as it might be seen from outer space if all hindrances to vision were removed. You can see how the mountain backbone of the Rockies and the Andes stretches nearly from pole to pole, and how the major river systems of the Mississippi-Missouri and the Amazon each drain half a continent.

The face of the land is shown from the ice pack of Greenland and the tundra region of Northern Canada to the great plains of the Midwest and the deserts of southwestern United States and Mexico. The vegetation of South America is depicted from the highlands of Venezuela and the tropical rain forest of the Amazon Valley to the Pampas and Patagonia.

132

A realistic picture of the Old World is afforded by precise and accurate shaded relief combined with subtly blended colors portraying the face of the land. You can see how the mighty Sahara and the deserts of Arabia confined early civilizations in the Nile and Tigris-Euphrates river valleys.

The topography and vegetation of the earth as shown here reveal something of how such great barriers as the Himalayas, the dense jungles of Africa, and the vast deserts have affected and controlled man's efforts in transportation, communication, and broad cultural interchange.

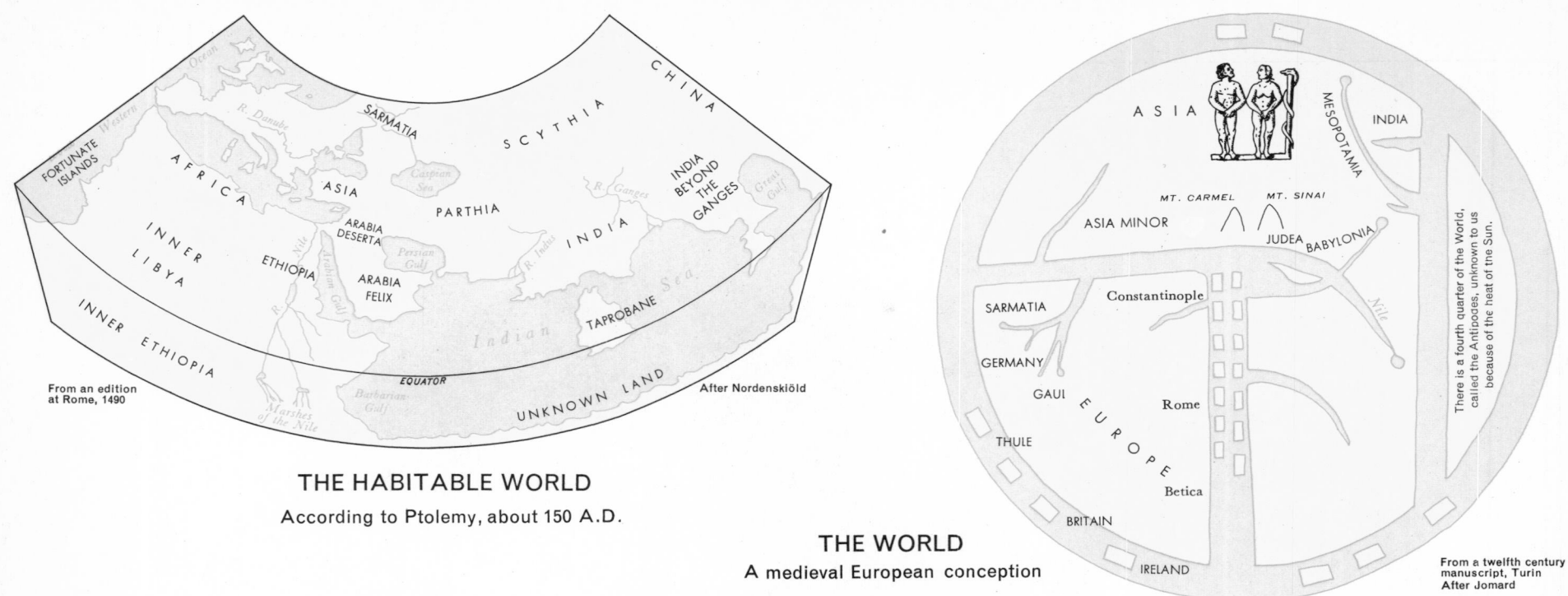

THE HABITABLE WORLD
According to Ptolemy, about 150 A.D.

From an edition at Rome, 1490

After Nordenskiöld

THE WORLD
A medieval European conception

From a twelfth century manuscript, Turin
After Jomard

WORLD HISTORY IN MAPS

This world history section of 27 pages is a collection of modern maps illustrating past periods of time, not a collection of old maps made in times past. The maps on this page, however, are drawings of old maps, reproduced here only to show how men at three points in history conceived the world in which they lived.

Centuries of Greek and Near Eastern thought are summarized in the map made by Ptolemy at Alexandria about 150 A.D. He knew that the earth is a sphere, but he believed about only one-third of the Northern Hemisphere was habitable. His map represents this portion of the globe. He understood the principles of map projection, that is of representing a curved surface on a flat page, and he located places according to longitude and latitude, defining longitude by distance east of the

Fortunate Islands (now the Canaries) and latitude by the length of the longest day of the year. His map was defective not in conception but by lack of information.

The sample of a medieval map represents a common form of map in use at this period of time—a diagram rather than a true map. The outer circle represents the ocean, the vertical radius containing oblong islands in the Mediterranean Sea. The East is at the top and the Holy Land is in the center. A person using this map could get a rough idea of direction, but he would have no idea of distance, size, or proportion.

The Behaim globe, below, represents man's conception of the world at the time of Columbus.

There are 48 maps in this section, including insets. Asia is represented at successive periods of its history, and we have given attention to Africa and Latin America, but since this book is mainly for American use, the treatment of North American and European history is given better coverage. The maps both illustrate general ideas and supply particular information. Modern maps differ from old maps in being more exact in projection and in scale.

THE WORLD
According to Behaim, 1492

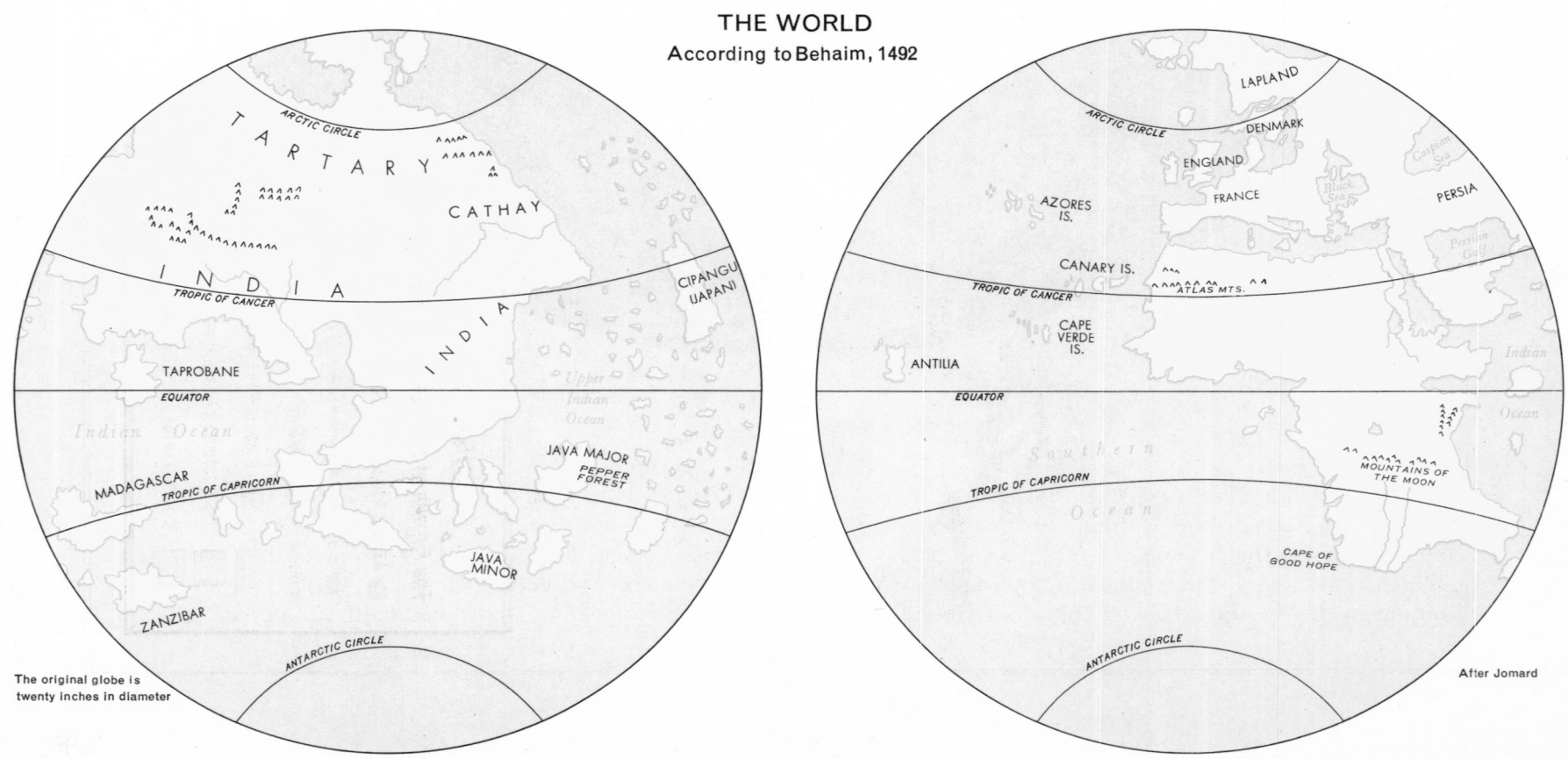

The original globe is twenty inches in diameter

After Jomard

134

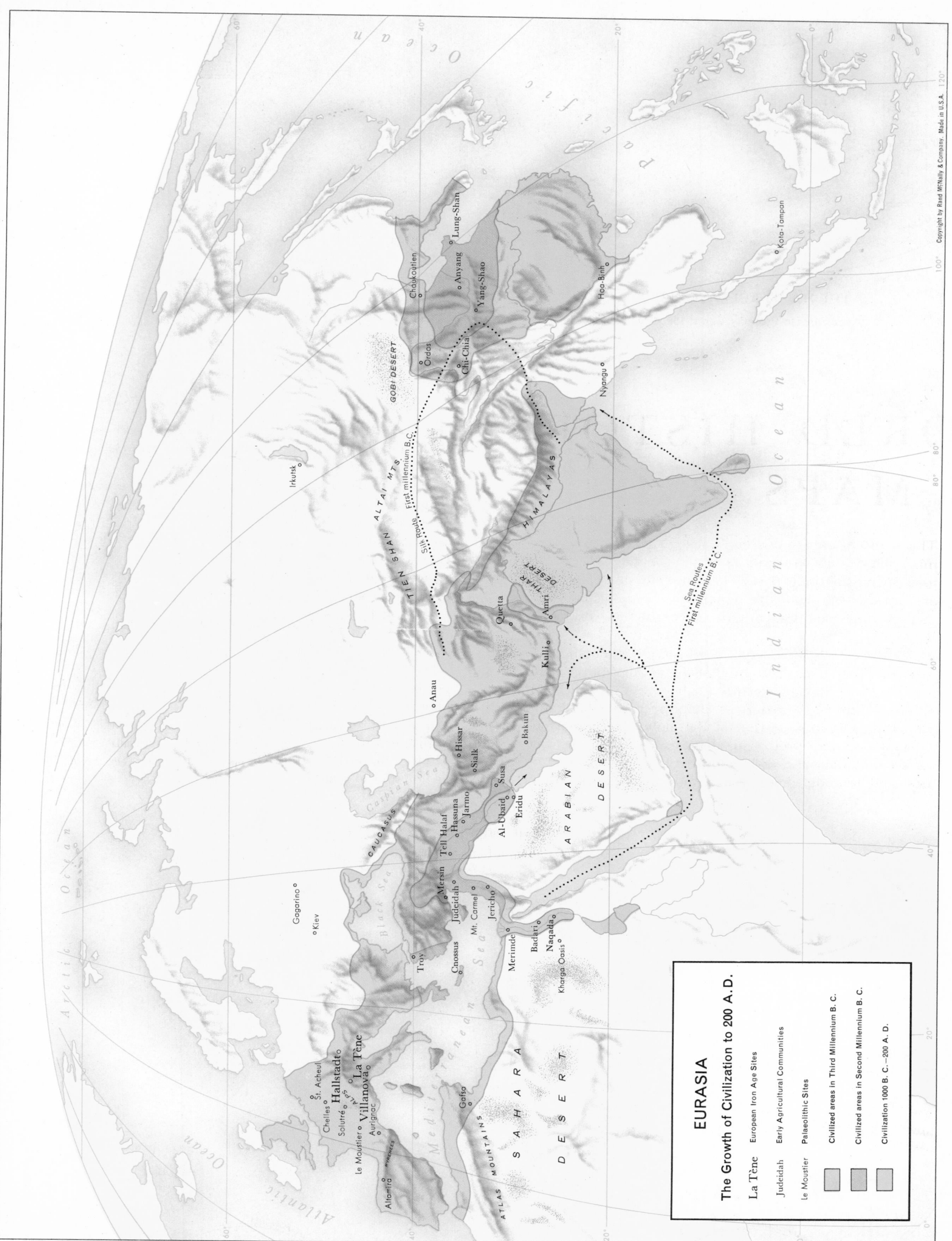

La Tène
European Iron Age Sites

Judeidah
Early Agricultural Communities

le Moustier
Palaeolithic Sites

EURASIA

The Growth of Civilization to 200 A. D.

Civilized areas in Third Millennium B. C.

Civilized areas in Second Millennium B. C.

Civilization 1000 B. C. —200 A. D.

Arctic Ocean

Pacific Ocean

Indian Ocean

Atlantic Ocean

Mediterranean Sea

Black Sea

Caspian Sea

GOBI DESERT

ALTAI MTS

TIEN SHAN

HIMALAYAS

THAR DESERT

ARABIAN DESERT

SAHARA DESERT

ATLAS MOUNTAINS

CAUCASUS

Silk Route, First millennium B.C.

Sea Routes First millennium B. C.

Irkutsk

Gagarino
Kiev

Choukoutien
Lung-Shan
Anyang
Yang-Shao
Ordos
Chi-Chia
Hoa-Binh
Nyangu
Koto-Tampan

Anau

Quetta
Kulli
Amri

Hissar
Sialk
Bakun

Susa
Al-Ubaid
Eridu

Tell Halaf
Hassuna
Jarmo

Mersin
Judeidah
Mt. Carmel
Jericho

Troy
Cnossus
Merinde
Badari
Naqada
Kharga Oasis

St. Acheul
Chelles
Solutré Hallstadt La Tène
Aurignac Villanova La Tène
le Moustier
Gafsa
Altamira

135

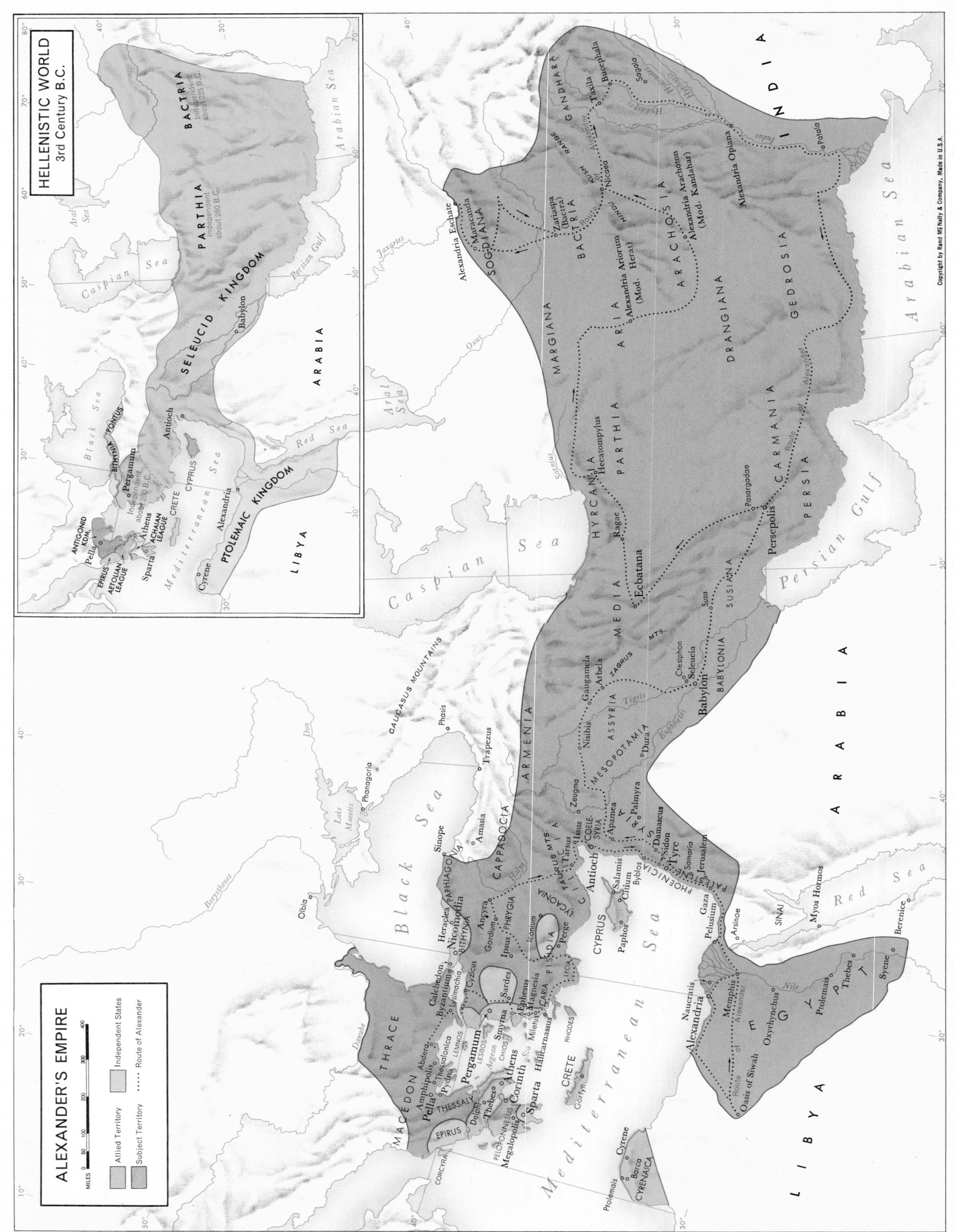

ALEXANDER'S EMPIRE

MILES
0   50 100   200   300   400

Allied Territory
Subject Territory
Independent States
Route of Alexander

HELLENISTIC WORLD
3rd Century B.C.

BACTRIA
Independent
about 225 B.C.

PARTHIA
Independent
about 250 B.C.

SELEUCID
KINGDOM

Babylon

Antioch

PONTUS
BITHYNIA
Pergamum
Independent
about 250 B.C.

ANTIGONID
KDM.
Pella
EPIRUS
AETOLIAN
LEAGUE
Athens
Sparta
ACHAIAN
LEAGUE

CRETE
CYPRUS

Alexandria

PTOLEMAIC
KINGDOM

Cyrene

LIBYA

ARABIA

Red Sea

Aral
Sea

Caspian
Sea

Black Sea

Persian Gulf

Arabian
Sea

Mediterranean
Sea

INDIA

SOGDIANA
Alexandria Eschate
Maracanda
Zariaspa
(Bactra)
BACTRIA
GANDHARA
Taxila
Bucephala
Nicaea
Sangala
Patala

MARGIANA
ARIA
Alexandria Ariorum
(Mod. Herat)
ARACHOSIA
Alexandria Arachoton
(Mod. Kandahar)
Alexandria Opiana

HYRCANIA
PARTHIA
Hecatompylus
DRANGIANA
GEDROSIA

MEDIA
Ragae
Ecbatana
Pasargadae
Persepolis
CARMANIA
PERSIA

Gaugamela
Arbela
ZAGRUS MTS.
ASSYRIA
Nisibis
MESOPOTAMIA
Ctesiphon
Seleucia
Babylon
BABYLONIA
SUSIANA
Susa
Dura
Zeugma

ARMENIA
Trapezus
CAUCASUS MOUNTAINS
Phasis
Phanagoria

CAPPADOCIA
Amasia
Sinope
Heraclea
PAPHLAGONIA
Ancyra
PHRYGIA
Gordium
BITHYNIA
Nicomedia

THRACE
MACEDON
Byzantium
Calchedon
Cyzicus
Lysimachia
Sardes
Ephesus
Miletus
Halicarnassus
CARIA
LYDIA
Pergamum
Smyrna
LESBOS
CHIOS
RHODES
Athens
Corinth
Sparta
THESSALY
Amphipolis
Thessalonica
Pydna
Abdera
Pella
Thebes
Delphi
EPIRUS
PELOPONNESUS
Megalopolis
CORCYRA

PISIDIA
PAMPHYLIA
LYCIA
Perge
LYCAONIA
Iconium
CILICIA
Tarsus
Issus
Iaisos
TAURUS MTS.
GALATIA

COELE
SYRIA
Antioch
Apamea
Palmyra
SYRIA
Damascus
PHOENICIA
Sidon
Tyre
Samaria
Jerusalem
PALESTINE
Gaza
Pelusium
Arsinoe
SINAI

CYPRUS
Salamis
Citium
Paphos
Byblos

Naucratis
Alexandria
Memphis
(Mod. Cairo)
Oasis of Siwah
Oxyrhynchus
Ptolemais
Thebes
Syene
EGYPT
Nile
Berenice
Myos Hormos

ARABIA

Cyrene
Barca
CYRENAICA
Ptolemais
LIBYA

Black Sea
Olbia
Borysthenes
Don
Lake
Maeotis

Caspian Sea

Jaxartes
Oxus

Tigris
Euphrates

136

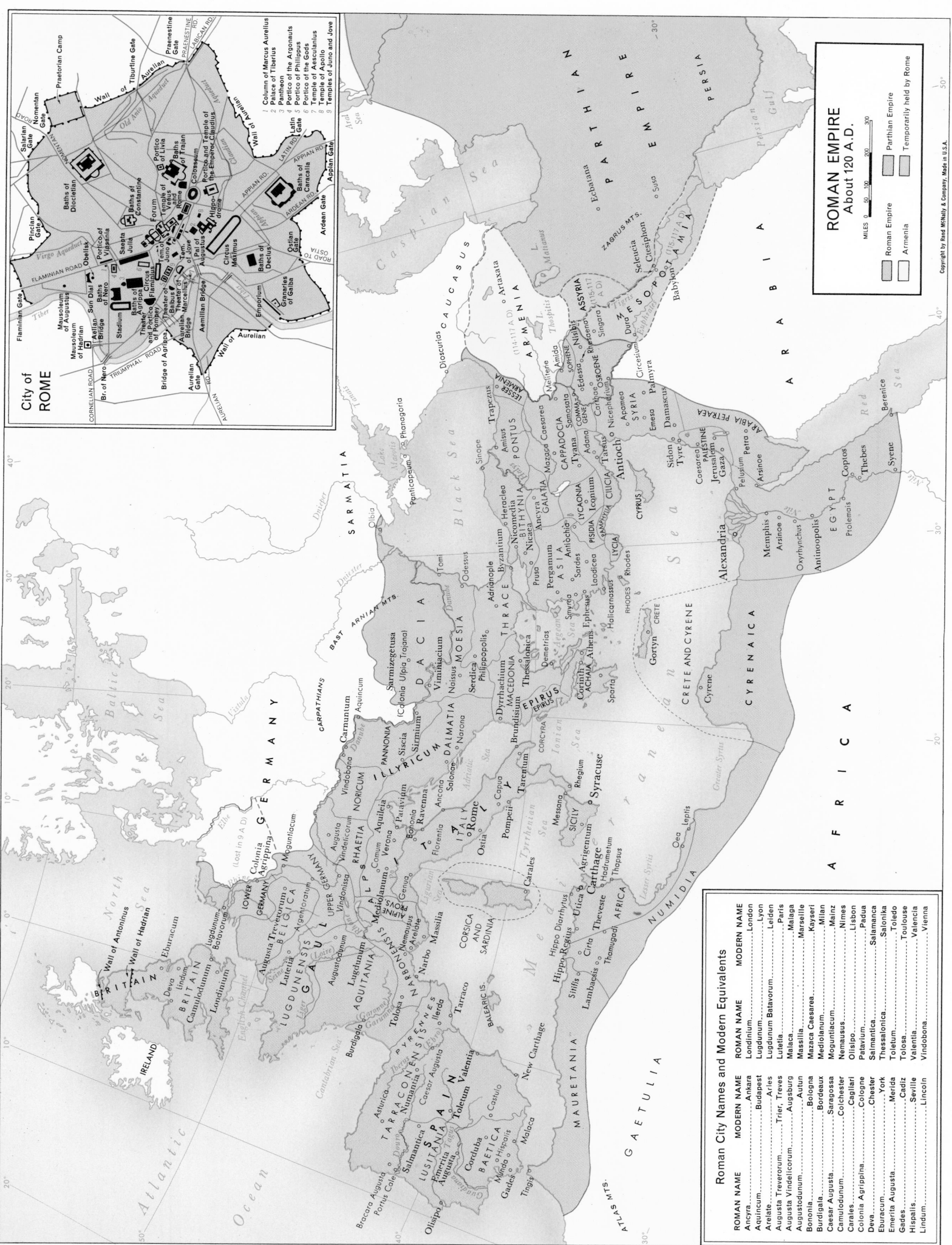

ROMAN EMPIRE
About 120 A.D.

MILES 0   50   100   200   300

Roman Empire
Armenia
Parthian Empire
Temporarily held by Rome

Copyright by Rand McNally & Company, Made in U.S.A.

City of ROME

1 Column of Marcus Aurelius
2 Palace of Tiberius
3 Pantheon
4 Portico of the Argonauts
5 Portico of Philippus
6 Portico of the Gods
7 Temple of Apollo
8 Temple of Aesculanius
9 Temples of Juno and Jove

Roman City Names and Modern Equivalents

ROMAN NAME	MODERN NAME	ROMAN NAME	MODERN NAME
Ancyra	Ankara	Londinium	London
Aquincum	Budapest	Lugdunum	Lyon
Arelate	Arles	Lugdunum Batavorum	Leiden
Augusta Treverorum	Trier, Treves	Lutetia	Paris
Augusta Vindelicorum	Augsburg	Malaca	Malaga
Augustodunum	Autun	Massilia	Marseille
Bononia	Bologna	Mazaca Caesarea	Kayseri
Burdigala	Bordeaux	Mediolanum	Milan
Caesar Augusta	Saragossa	Moguntiacum	Mainz
Camulodunum	Colchester	Nemausus	Nimes
Carales	Cagliari	Olisipo	Lisbon
Colonia Agrippina	Cologne	Patavium	Padua
Deva	Chester	Salmantica	Salamanca
Eburacum	York	Thessalonica	Salonika
Emerita Augusta	Merida	Toletum	Toledo
Gades	Cadiz	Tolosa	Toulouse
Hispalis	Seville	Valentia	Valencia
Lindum	Lincoln	Vindobona	Vienna

137

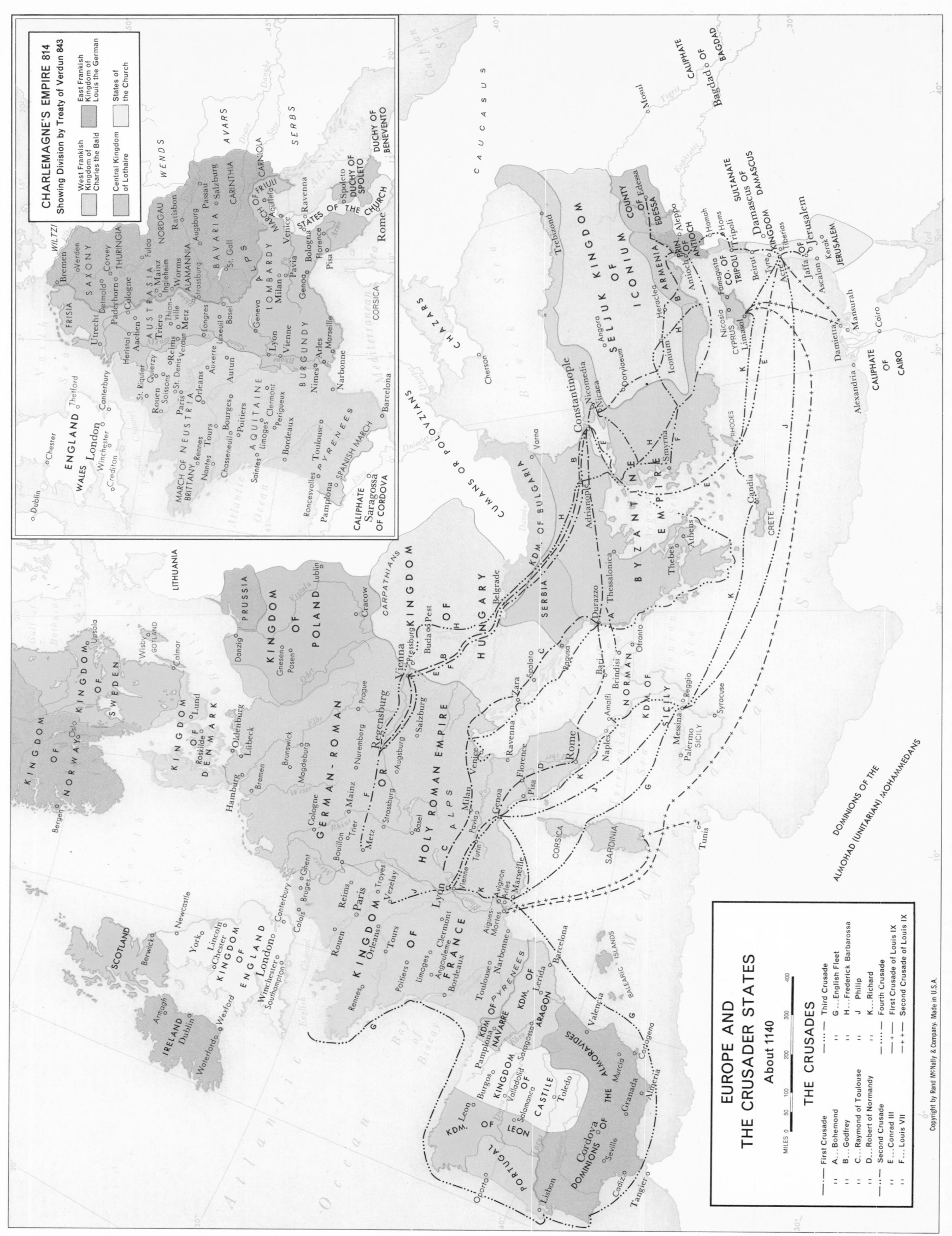

CHARLEMAGNE'S EMPIRE 814
Showing Division by Treaty of Verdun 843

West Frankish Kingdom of Charles the Bald

East Frankish Kingdom of Louis the German

Central Kingdom of Lothaire

States of the Church

**EUROPE AND
THE CRUSADER STATES**
About 1140

MILES 0 50 100 200 300 400

THE CRUSADES

First Crusade
A...Bohemond
B...Godfrey
C...Raymond of Toulouse
D...Robert of Normandy

Second Crusade
E...Conrad III
F...Louis VII

Third Crusade
G...English Fleet
H...Frederick Barbarossa
J...Philip
K...Richard

Fourth Crusade
First Crusade of Louis IX
Second Crusade of Louis IX

Copyright by Rand McNally & Company. Made in U.S.A.

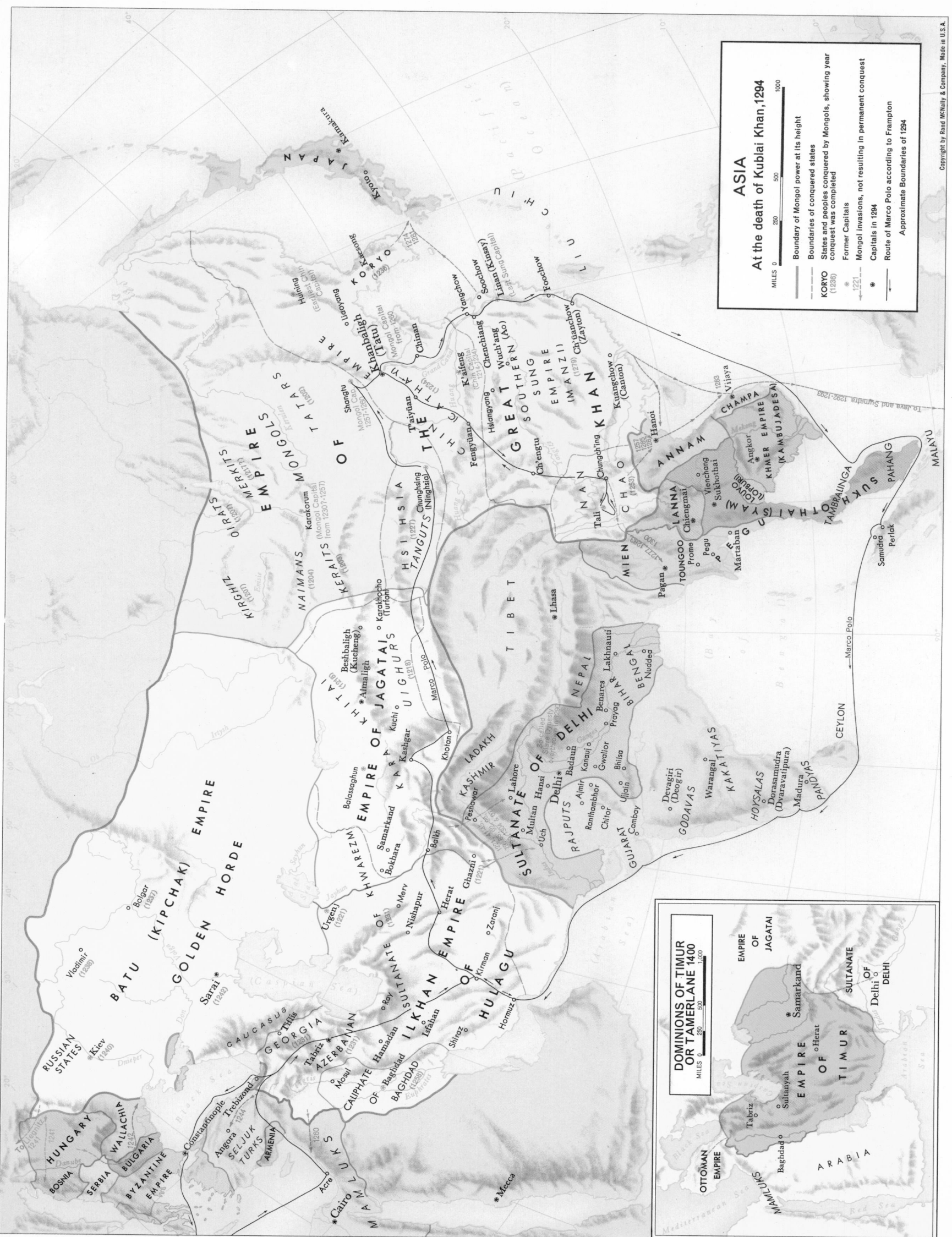

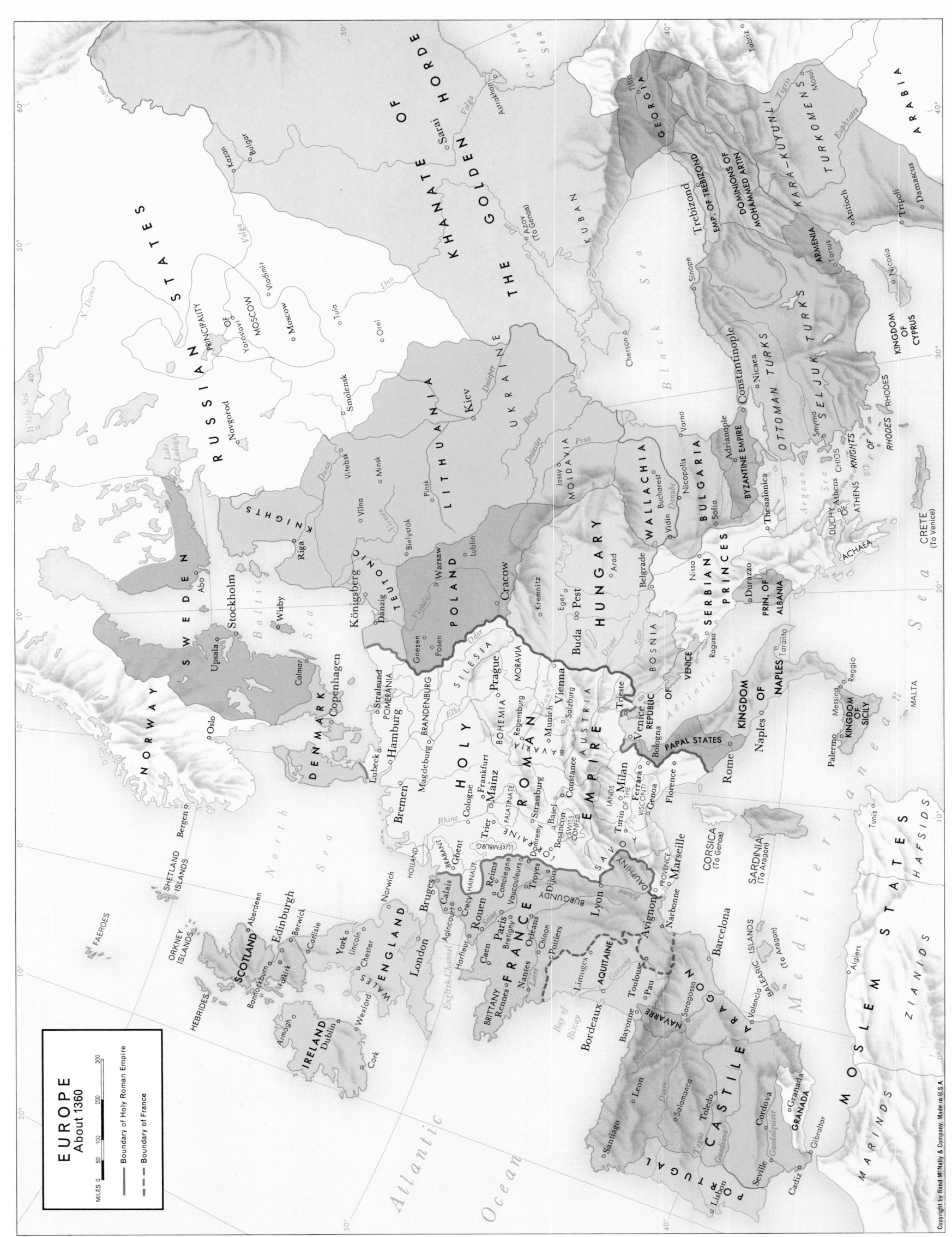

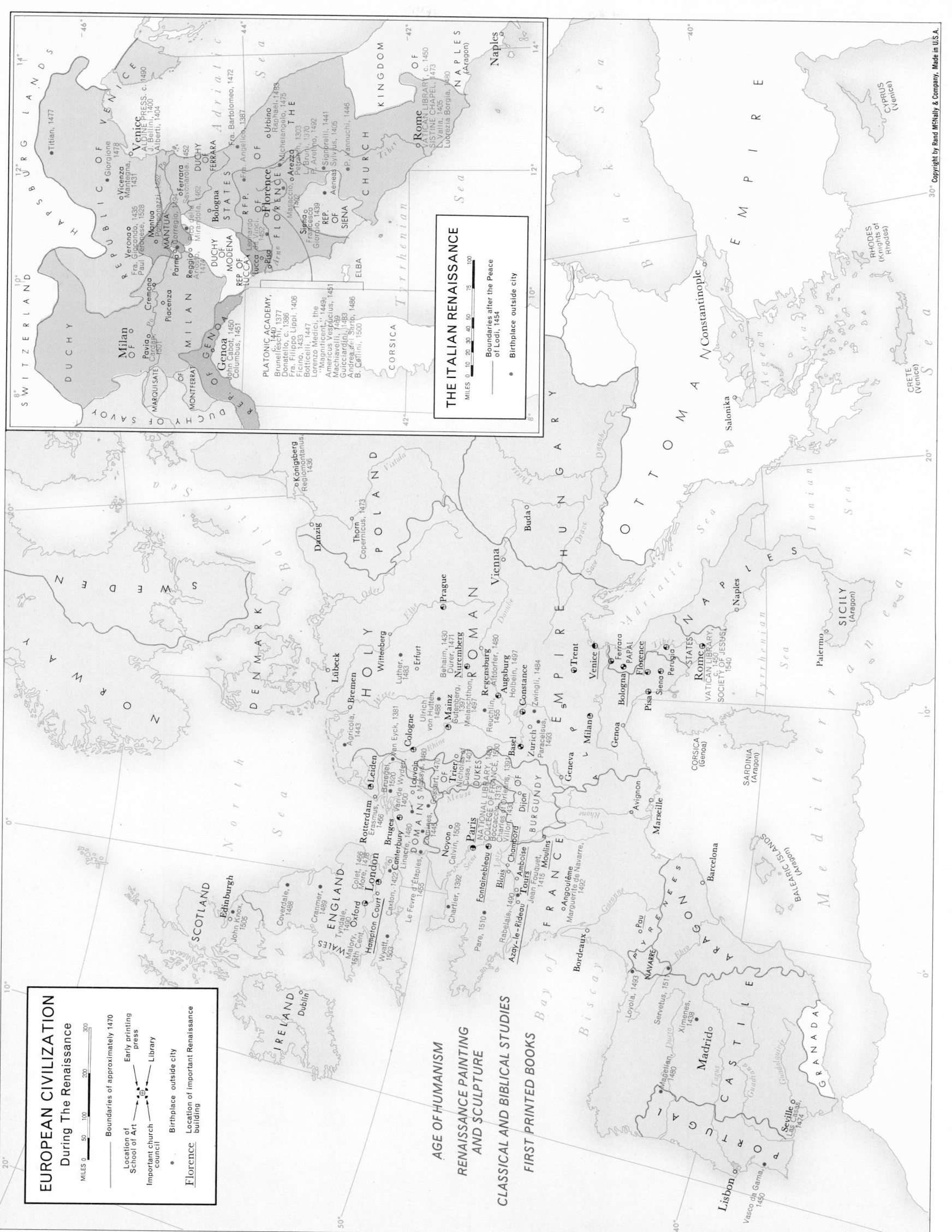

EUROPEAN CIVILIZATION
During The Renaissance

THE ITALIAN RENAISSANCE

141

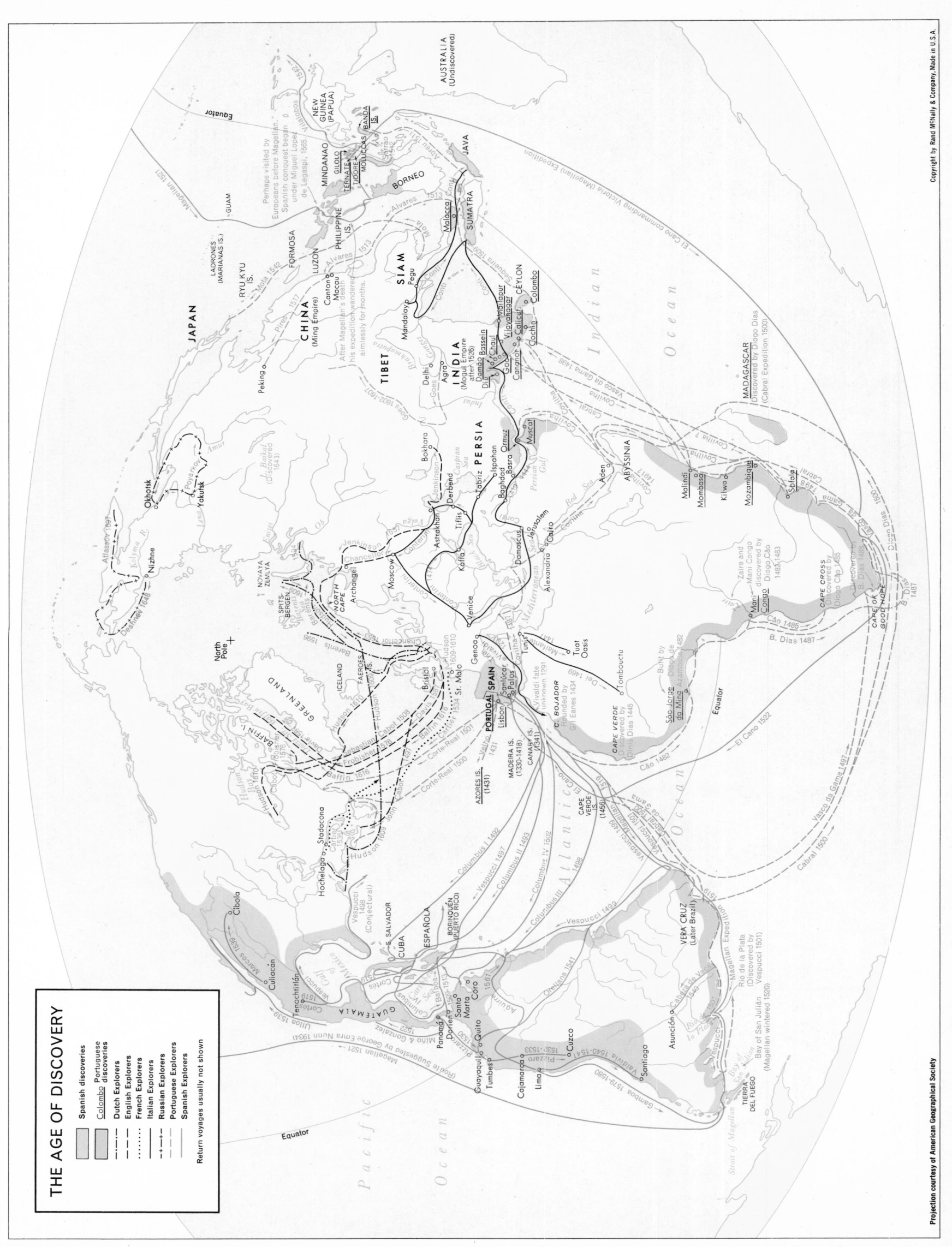

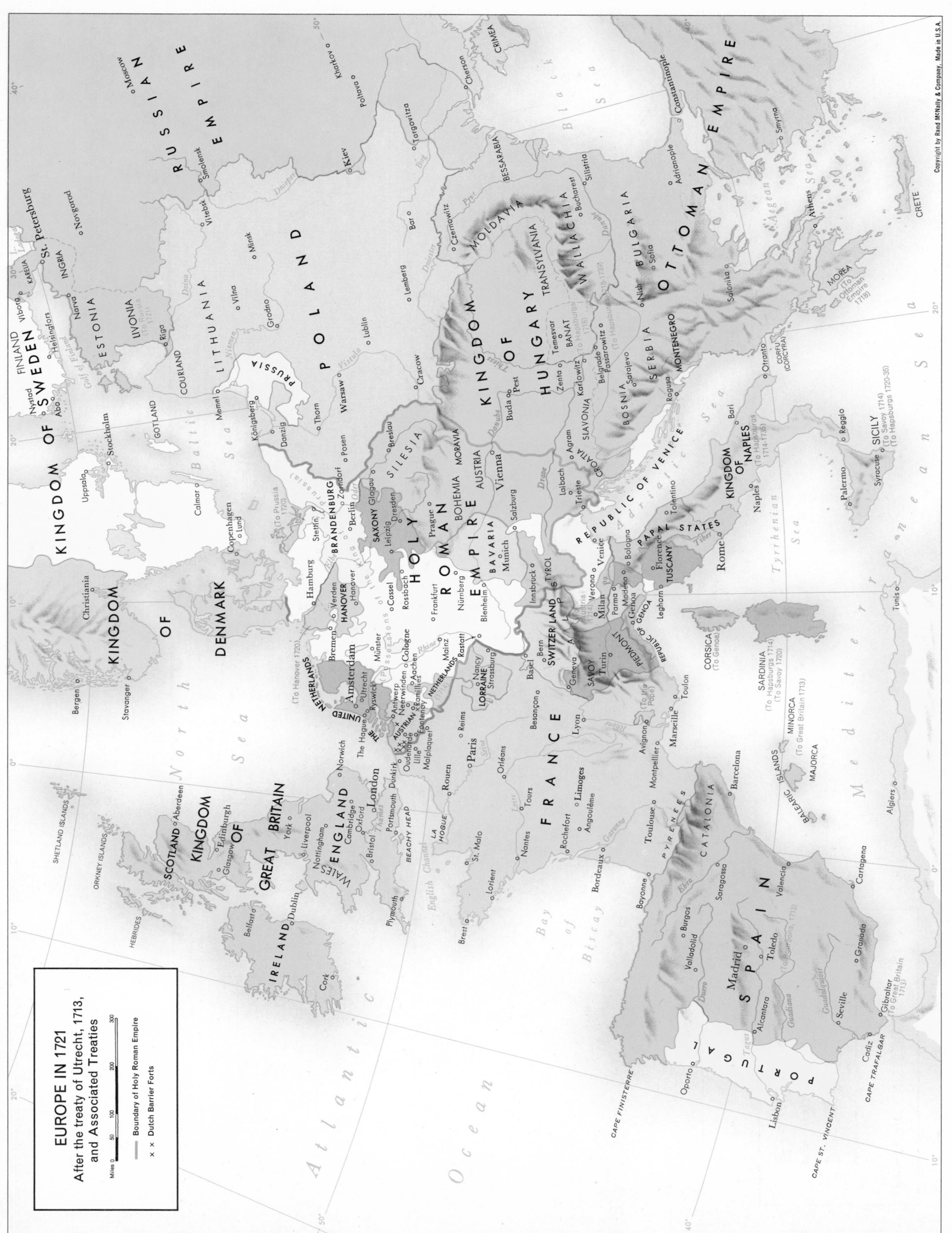

EUROPE IN 1721
After the treaty of Utrecht, 1713,
and Associated Treaties

Miles 0 50 100 200 300

—— Boundary of Holy Roman Empire

x x Dutch Barrier Forts

KINGDOM OF SWEDEN

RUSSIAN EMPIRE

POLAND

KINGDOM OF DENMARK

KINGDOM OF GREAT BRITAIN

SCOTLAND

IRELAND

ENGLAND

WALES

HOLY ROMAN EMPIRE

KINGDOM OF HUNGARY

OTTOMAN EMPIRE

FRANCE

SPAIN

PORTUGAL

THE UNITED NETHERLANDS

AUSTRIAN NETHERLANDS

BRANDENBURG

HANOVER

SAXONY

SILESIA

MORAVIA

BOHEMIA

AUSTRIA

BAVARIA

TYROL

SWITZERLAND

LORRAINE

SAVOY

PIEDMONT

REPUBLIC OF VENICE

PAPAL STATES

TUSCANY

KINGDOM OF NAPLES

REPUBLIC OF GENOA

MOLDAVIA

WALLACHIA

TRANSYLVANIA

BESSARABIA

BULGARIA

SERBIA

BOSNIA

CROATIA

SLAVONIA

MONTENEGRO

BANAT

CRIMEA

FINLAND

KARELIA

INGRIA

ESTONIA

LIVONIA

COURLAND

LITHUANIA

PRUSSIA

GOTLAND

SHETLAND ISLANDS

ORKNEY ISLANDS

HEBRIDES

CATALONIA

BALEARIC ISLANDS

MINORCA

MAJORCA

CORSICA (To Genoa)

SARDINIA
(To Hapsburgs 1714)
(To Savoy 1720)

SICILY
(To Savoy 1714)
(To Hapsburgs 1720-35)

MOREA
(To Ottoman 1718)

CRETE

CORFU
(CORCYRA)

PYRENEES

Atlantic Ocean

North Sea

Baltic Sea

Black Sea

Mediterranean Sea

Aegean Sea

Adriatic Sea

Tyrrhenian Sea

Bay of Biscay

English Channel

Gulf of Finland

Moscow

Novgorod

St. Petersburg

Viborg

Nystad

Åbo

Helsingfors

Narva

Reval

Riga

Memel

Königsberg

Danzig

Stettin

Berlin

Warsaw

Thorn

Posen

Cracow

Lublin

Lemberg

Grodno

Vilna

Minsk

Bar

Kiev

Vitebsk

Smolensk

Mogilev

Poltava

Kharkov

Targovitza

Cherson

Czernowitz

Bucharest

Silistria

Niš

Sofia

Adrianople

Salonica

Constantinople

Smyrna

Athens

Ragusa

Sarajevo

Belgrade

Passarowitz

Karlowitz

Zenta

Temesvar

Pest

Buda

Agram

Laibach

Trieste

Venice

Verona

Milan

Turin

Genoa

Parma

Modena

Bologna

Florence

Leghorn

Rome

Naples

Bari

Brindisi

Otranto

Reggio

Syracuse

Palermo

Tunis

Algiers

Cartagena

Cadiz

Seville

Granada

Cordova

Madrid

Toledo

Alcántara

Valencia

Barcelona

Saragossa

Burgos

Valladolid

Oporto

Lisbon

Bayonne

Bordeaux

Toulouse

Montpellier

Marseille

Toulon

Avignon

Lyon

Besançon

Limoges

Angoulême

Rochefort

Nantes

Lorient

Brest

St. Malo

Rouen

Orléans

Paris

Reims

Nancy

Strassburg

Rastatt

Mainz

Aachen

Cologne

Münster

Frankfurt

Nürnberg

Rossbach

Leipzig

Dresden

Zorndorf

Glogau

Breslau

Prague

Brünn

Vienna

Salzburg

Munich

Innsbruck

Basel

Bern

Geneva

Blenheim

Cassel

Hanover

Verden

Bremen

Hamburg

Amsterdam

Utrecht

The Hague

Antwerp

Ryswick

Oudenarde

Malplaquet

Neerwinden

Ramillies

Lille

Tournay

Dunkirk

Calmar

Upsala

Stockholm

Christiania

Stavanger

Bergen

Copenhagen

Lund

London

Oxford

Cambridge

Norwich

Portsmouth

Plymouth

Bristol

Liverpool

Nottingham

York

Edinburgh

Glasgow

Aberdeen

Dublin

Belfast

Cork

Beachy Head

La Hogue

Gibraltar (To Great Britain 1713)

Cape Finisterre

Cape St. Vincent

Cape Trafalgar

(To Prussia 1720)

(To Hanover 1720)

(To Russia 1721)

Rhine

Danube

Elbe

Oder

Vistula

Dnieper

Don

Dniester

Volga

Ebro

Tagus

Guadiana

Guadalquivir

Seine

Loire

Garonne

Rhône

Po

Tiber

Drave

Save

Theiss

Pruth

Niemen

Düna

143

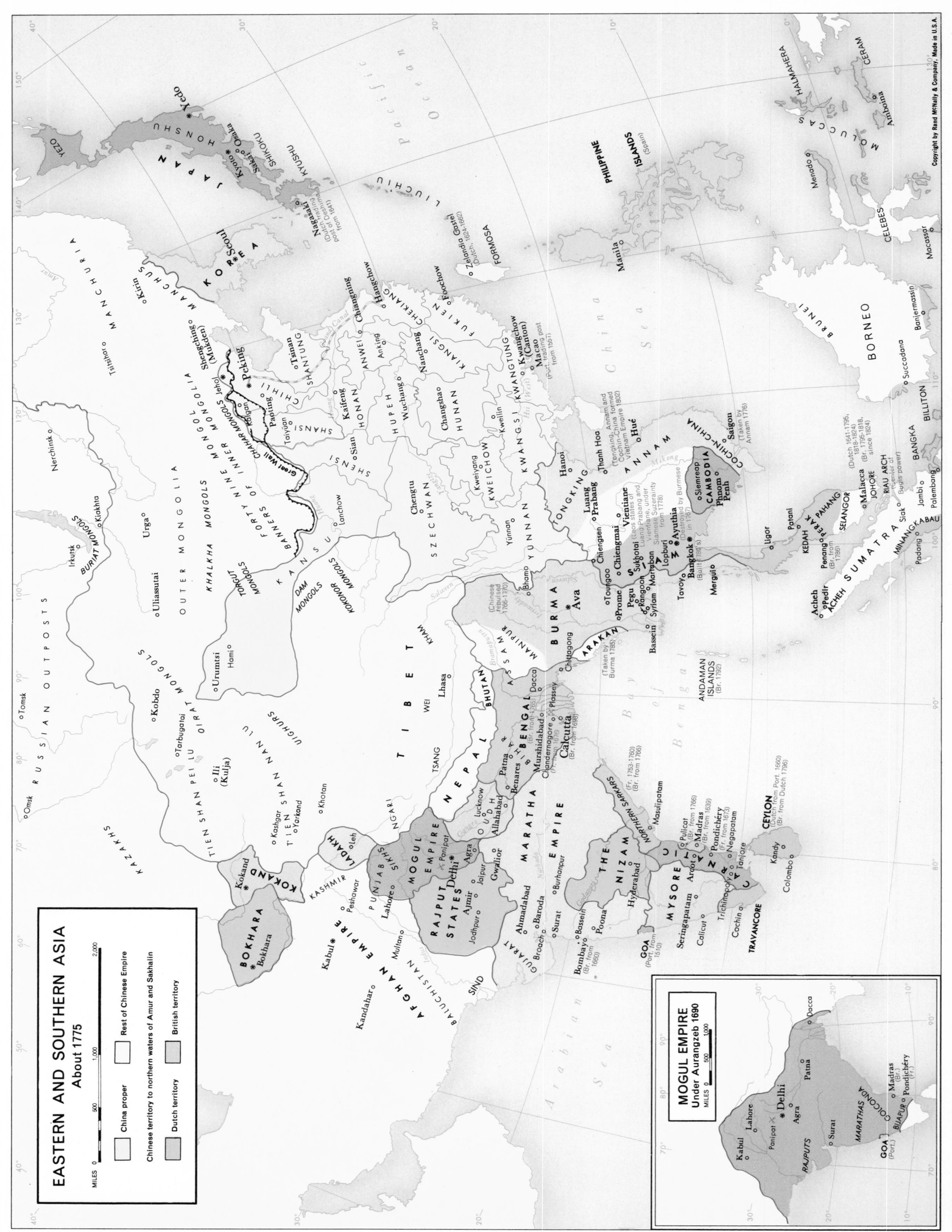

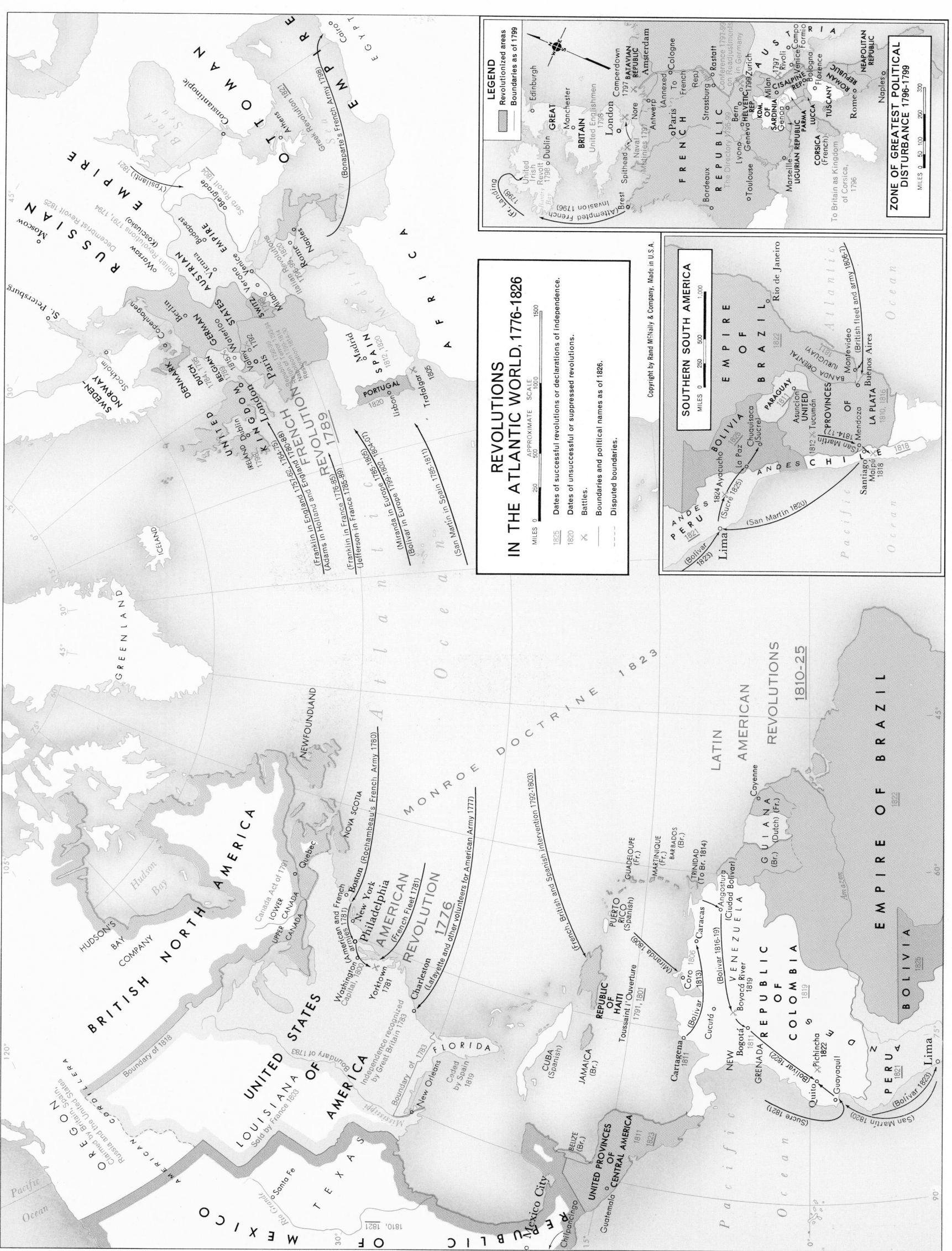

REVOLUTIONS
IN THE ATLANTIC WORLD, 1776-1826

MILES 0   250   500
        APPROXIMATE SCALE
        500   1000   1500

1825   Dates of successful revolutions or declarations of independence.
1820   Dates of unsuccessful or suppressed revolutions.
X   Battles.
        Boundaries and political names as of 1826.
------ Disputed boundaries.

Copyright by Rand McNally & Company, Made in U.S.A.

SOUTHERN SOUTH AMERICA

MILES 0   250   500   1,000

LEGEND
        Revolutionized areas
        Boundaries as of 1799

ZONE OF GREATEST POLITICAL
DISTURBANCE 1796-1799

MILES 0 50 100 200 300

145

Map 1 (upper left)

HUDSON'S BAY COMPANY

CREE
OJIBWA
SAC & FOX
Ft. Mackinac
POTAWATOMI
Detroit
KICKAPOO
ILLINOIS
MIAMI
ERIE Ft. Pitt
St. Louis 1764
Vincennes
Kaskaskia
SHAWNEE
CHICKASAW
CHOCTAW
CREEK
WEST FLORIDA 1767
WEST FLORIDA 1763
Mobile 1702
Pensacola 1698
New Orleans 1718
EAST FLORIDA
St. Augustine 1565

ALGONKIN QUEBEC Quebec 1608
MICMAC
Montreal 1642
MAINE DIST. (MASS.) ABNAKI
Port Royal 1605
NOVA SCOTIA Halifax
ADIRONDACK MTS. GREEN MTS. WHITE MTS. N.H.
NEW YORK
IROQUOIS CATSKILL MTS.
Salem 1626
MASS. Boston 1630
Plymouth 1620
CONN. Hartford
Providence
R.I.
PENNSYLVANIA N.J. New York 1626-64 (Nieu Amsterdam)
Philadelphia 1682
Baltimore 1745
MD. DEL.
VIRGINIA
Richmond
Jamestown 1607 / 1609
ALLEGHENY MTS. CUMBERLAND MTS.
CHEROKEE GREAT SMOKY MTS. BLUE RIDGE MTS.
NORTH CAROLINA
SOUTH CAROLINA
GEORGIA
Charleston 1672
Savannah 1733

PRESERVED FOR INDIANS
LOUISIANA (To Spain)

Gulf of Mexico
Atlantic Ocean

BRITISH NORTH AMERICA
After the Seven Years' War
MILES 0 50 100 200 300

Approximate extent of settlement, 1690
Approximate extent of settlement 1760
Boston 1630 Town, with date of first settlement
Proclamation Line of 1763
Limit of British territory

Map 2 (upper right)

REVOLUTIONARY WAR
MILES 0 50 100 200

British routes
American routes
Major battles

Quebec
Montreal
MAINE DIST. (MASS.)
Ft. Niagara
Oriskany 1777
Saratoga 1777
N.H.
NEW YORK
Bennington 1777
Lexington
Concord
MASS. Breed's Hill 1775
Boston
Detroit
CONN. R.I.
Washington
White Plains
New York
PENNSYLVANIA
Ft. Pitt
Valley Forge 1777-78
Princeton
Trenton
Brandywine
Philadelphia
N.J.
MD. DEL.
VIRGINIA
Bedford
Williamsburg
Petersburg
Yorktown 1781
Chesapeake Capes 1781
NORTH CAROLINA
Guilford Courthouse 1781
Cowpens 1781
Kings Mt. 1780
Camden 1780
Marion
Wilmington
Georgetown
SOUTH CAROLINA
Augusta 1779
Lincoln 1779
Charleston
GEORGIA
Savannah 1778

Copyright by Rand McNally & Company, Made in U.S.A.

Map 3 (lower left)

Quebec
Montreal
VIRGINIA Area north
MASS. CLAIM
CONN. CLAIM Ceded 1786
Ceded 1785
CONN. WESTERN RESERVE Ceded 1800
Not Ceded Admitted as State of Kentucky 1792
VIRGINIA CLAIM
NORTH CAROLINA CLAIM Ceded 1790
S. CAROLINA CLAIM 1787
GEORGIA CLAIM Ceded 1802
Ceded 1784
Ceded To George

MASS. & N.Y. CLAIM To New York
VT. N.H.
NEW YORK
MASSACHUSETTS
CONN. R.I.
Boston
PENNSYLVANIA
Philadelphia
N.J. New York
MD. DEL.
VIRGINIA
Norfolk
NORTH CAROLINA
SOUTH CAROLINA
Charleston
GEORGIA
Mobile Pensacola
New Orleans
St. Augustine
Gulf of Mexico
Atlantic Ocean

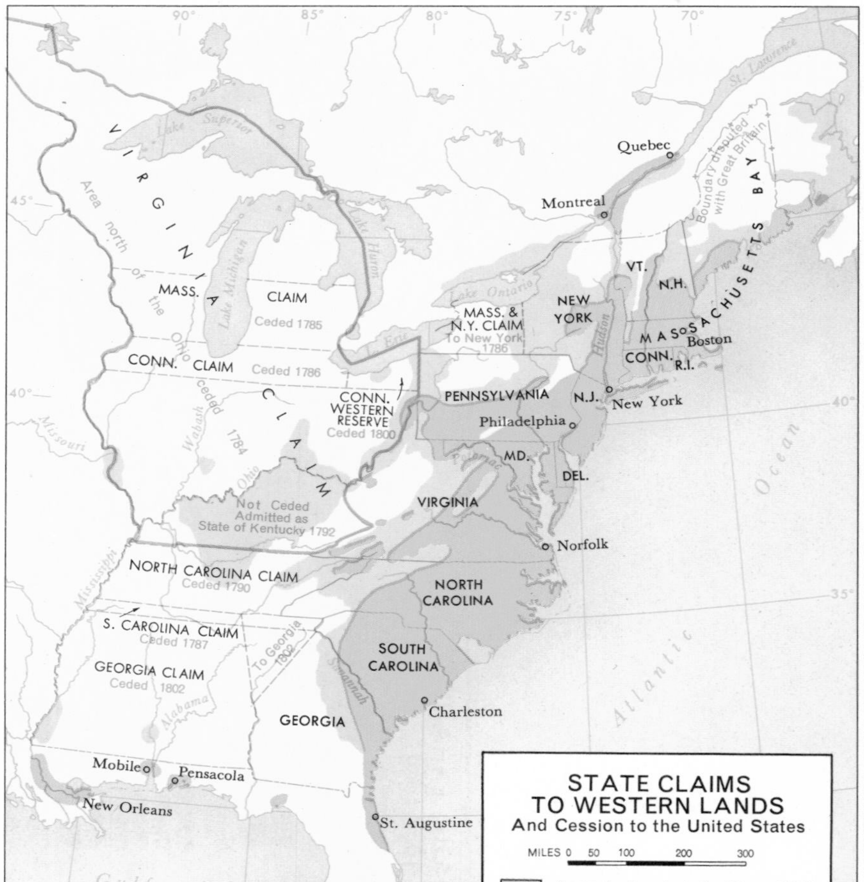

STATE CLAIMS TO WESTERN LANDS
And Cession to the United States
MILES 0 50 100 200 300

Approximate extent of settlement 1775
Approximate extent of settlement 1800
Boundaries of thirteen original states
Boundaries of western land claimed
Boundary of territory claimed by Virginia; Ceded 1784

Copyright by Rand McNally & Company, Made in U.S.A.

Map 4 (lower right)

HUDSON'S BAY COMPANY
NORTHWEST TERRITORY
Ft. Mackinac
Detroit
Fallen Timbers
Ft. Niagara
Ft. Oswegatchie
Ft. Ontario
Quebec
Montreal
NEW BRUNSWICK
DISTRICT OF MAINE (MASS.)
NOVA SCOTIA
VT. 1791
N.H.
NEW YORK
MASS. Boston
CONN. Providence Newport
R.I.
Hamilton Br. 1778
Cincinnati 1788
Marietta 1788
Vincennes
Clark 1778
St. Louis 1764
Frankfort
Louisville Lexington 1779
KENTUCKY
TENNESSEE
Nashville
TERRITORY OF THE OHIO Organized 1790
TERRITORY OF THE SOUTH Organized 1790
PENNSYLVANIA
Pittsburgh
APPALACHIAN MTS.
CUMBERLAND MTS.
N.J. Philadelphia
MD. DEL.
Baltimore
VIRGINIA
Richmond
NORTH CAROLINA
SOUTH CAROLINA
Charleston
Savannah
GEORGIA
Nogales
Ft. Confederación
MISSISSIPPI TERRITORY
Baton Rouge Mobile
New Orleans
WEST FLORIDA
Pensacola
EAST FLORIDA
St. Augustine
BAHAMA IS. (Br.)
LOUISIANA
Gulf of Mexico
Atlantic Ocean

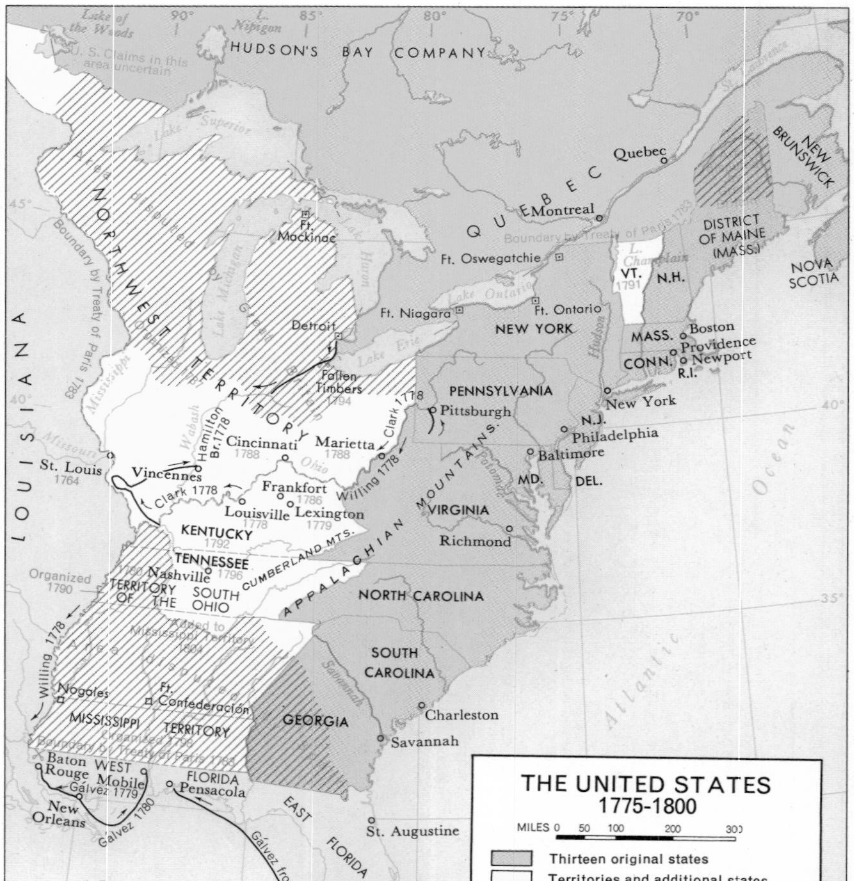

THE UNITED STATES
1775-1800
MILES 0 50 100 200 300

Thirteen original states
Territories and additional states
British possessions after 1783
Spanish possessions after 1783
Disputed territory
Posts retained by the British 1783-96
Revolutionary War routes
Areas colored as of 1783

Copyright by Rand McNally & Company, Made in U.S.A.

146

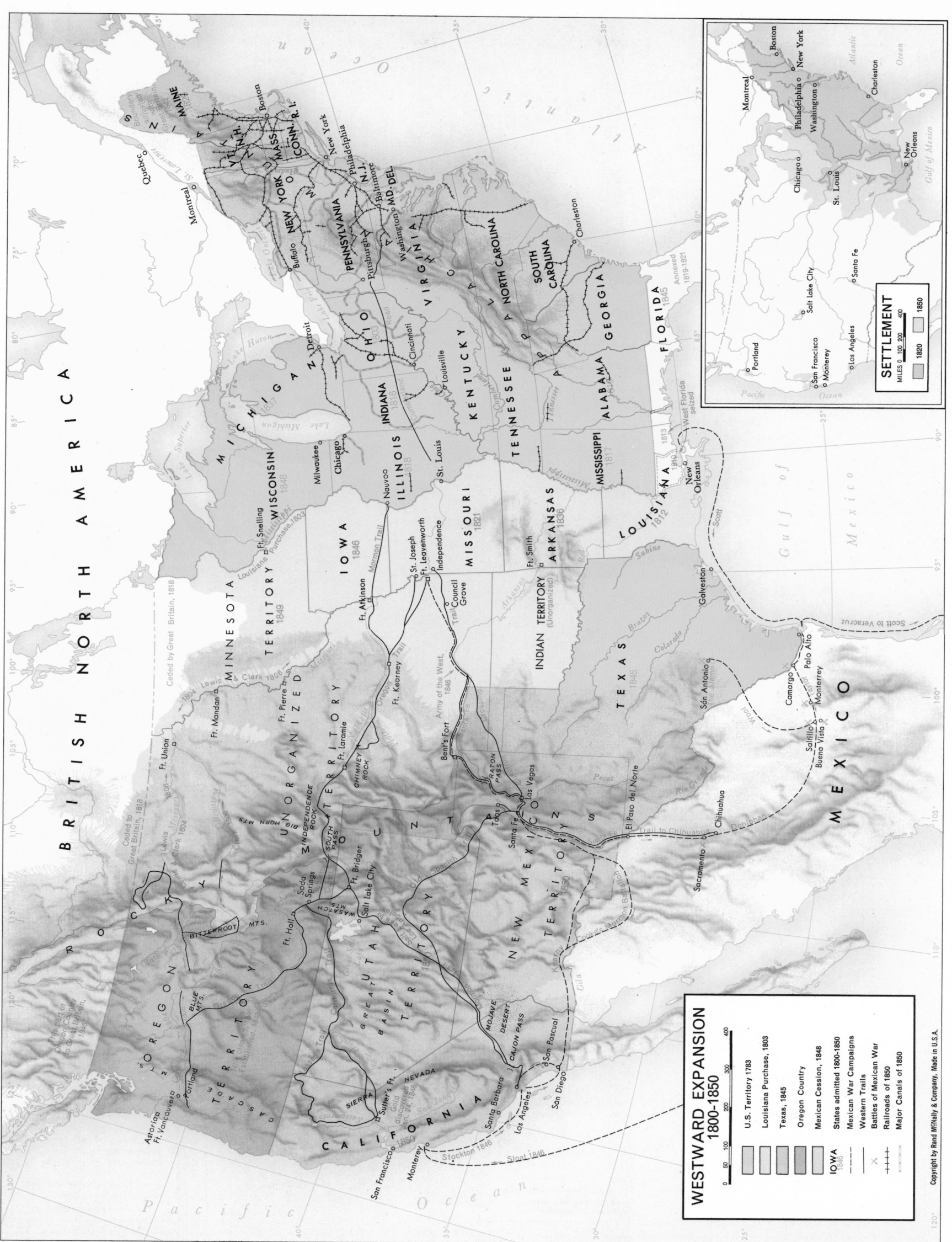

WESTWARD EXPANSION 1800–1850

U.S. Territory 1783
Louisiana Purchase, 1803
Texas, 1845
Oregon Country
Mexican Cession, 1848
States admitted 1800–1850

IOWA 1846

Mexican War Campaigns
Western Trails
Battles of Mexican War
Railroads of 1850
Major Canals of 1850

MILES 0 100 200 300 400

Copyright by Rand McNally & Company. Made in U.S.A.

SETTLEMENT

MILES 0 100 200 400

1820
1850

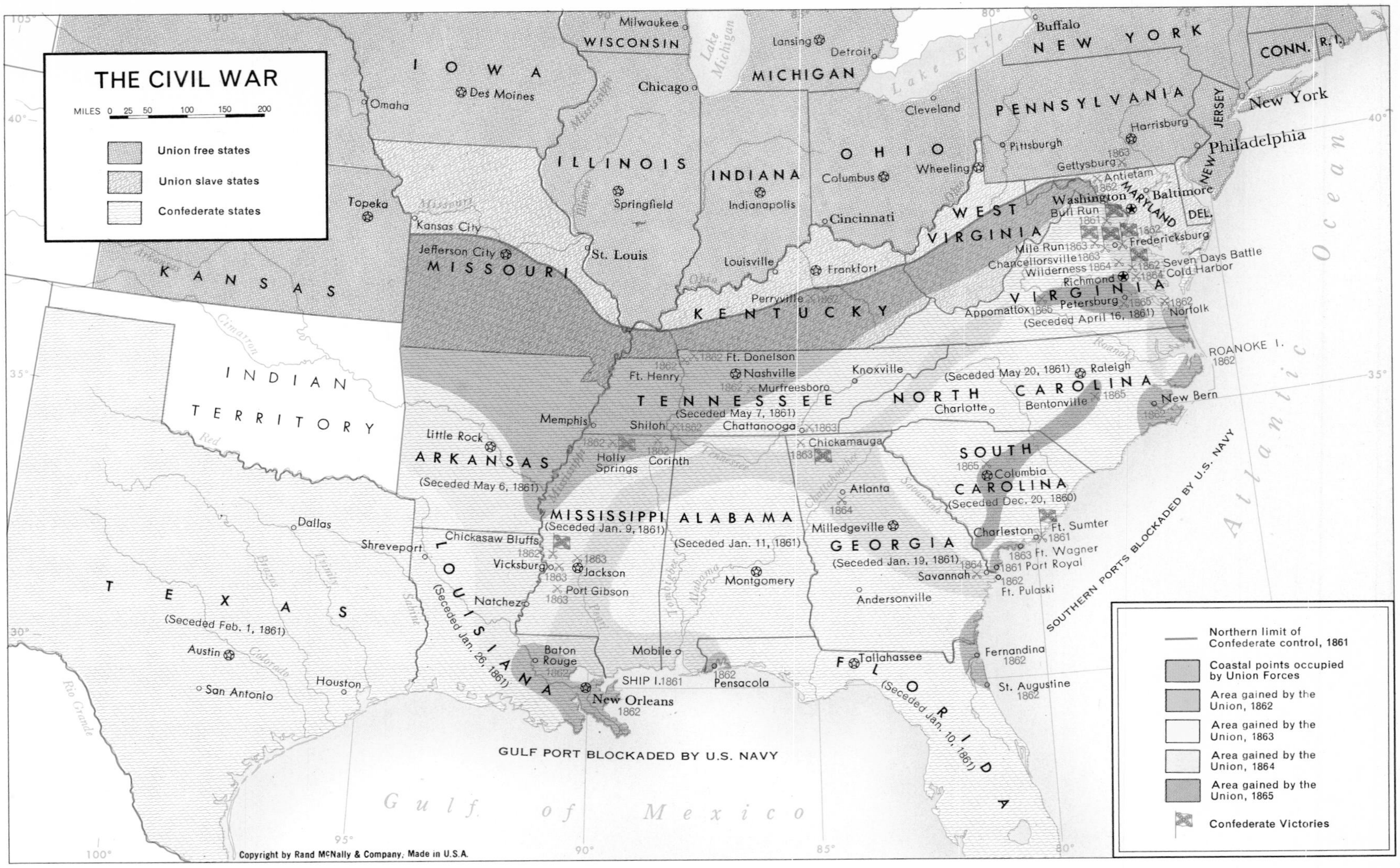

THE CIVIL WAR

MILES 0 25 50 100 150 200

Union free states
Union slave states
Confederate states

IOWA · WISCONSIN · Milwaukee · Lansing · Detroit · MICHIGAN · Lake Michigan · Lake Erie · Buffalo · NEW YORK · CONN. R.I.
Omaha · Des Moines · Chicago · Cleveland · PENNSYLVANIA · New York · JERSEY
ILLINOIS · INDIANA · OHIO · Columbus · Pittsburgh · Harrisburg · Philadelphia
Topeka · Springfield · Indianapolis · Cincinnati · Wheeling · WEST VIRGINIA · Gettysburg · Antietam · DEL.
Kansas City · Louisville · Frankfort · Washington · Bull Run 1861 · Baltimore · MARYLAND
KANSAS · Jefferson City · St. Louis · MISSOURI · Mile Run 1863 · Fredericksburg
Chancellorsville 1863 · Wilderness 1864 · Seven Days Battle 1862 · Cold Harbor
Richmond · VIRGINIA · Petersburg 1865 · 1862
Perryville 1862 · KENTUCKY · Appomattox 1865 · Norfolk · (Seceded April 16, 1861)
1862 · Ft. Donelson · ROANOKE I. 1862
Ft. Henry · Nashville · Knoxville · NORTH · Raleigh · (Seceded May 20, 1861)
INDIAN · Murfreesboro · CAROLINA
TERRITORY · Memphis · TENNESSEE · Chattanooga · New Bern
(Seceded May 7, 1861) · Shiloh 1862 · Chickamauga 1863 · Bentonville 1865
Little Rock · Charlotte · SOUTH · Columbia 1865
ARKANSAS · Holly Springs · Corinth · CAROLINA
(Seceded May 6, 1861) · Atlanta · (Seceded Dec. 20, 1860)
MISSISSIPPI · ALABAMA · GEORGIA · Charleston · Ft. Sumter 1861
Chickasaw Bluffs · (Seceded Jan. 9, 1861) · (Seceded Jan. 11, 1861) · Milledgeville · Ft. Wagner
DALLAS · Vicksburg · Jackson · Montgomery · Savannah · 1864 · 1861 Port Royal
Shreveport · Port Gibson · Ft. Pulaski
TEXAS · LOUISIANA · Natchez · Andersonville
(Seceded Feb. 1, 1861) · (Seceded Jan. 26, 1861) · Fernandina
Austin · Baton Rouge · Mobile · Tallahassee · St. Augustine 1862
San Antonio · Houston · New Orleans 1862 · SHIP I. 1861 · Pensacola · FLORIDA · (Seceded Jan. 10, 1861)
Rio Grande

GULF PORT BLOCKADED BY U.S. NAVY

Gulf of Mexico

Copyright by Rand McNally & Company, Made in U.S.A.

Northern limit of Confederate control, 1861
Coastal points occupied by Union Forces
Area gained by the Union, 1862
Area gained by the Union, 1863
Area gained by the Union, 1864
Area gained by the Union, 1865
Confederate Victories

SOUTHERN PORTS BLOCKADED BY U.S. NAVY

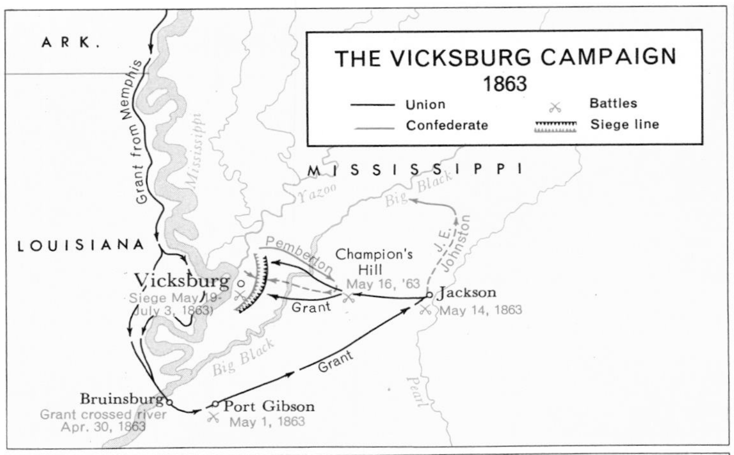

THE VICKSBURG CAMPAIGN 1863

Union · Confederate · Battles · Siege line

ARK. · MISSISSIPPI · LOUISIANA · Yazoo · Big Black
Grant from Memphis
Vicksburg · Siege May 19 – July 3, 1863 · Pemberton · Champion's Hill May 16, '63 · J.E. Johnston · Jackson May 14, 1863
Grant · Bruinsburg · Port Gibson May 1, 1863
Grant crossed river Apr. 30, 1863 · Big Black · Pearl

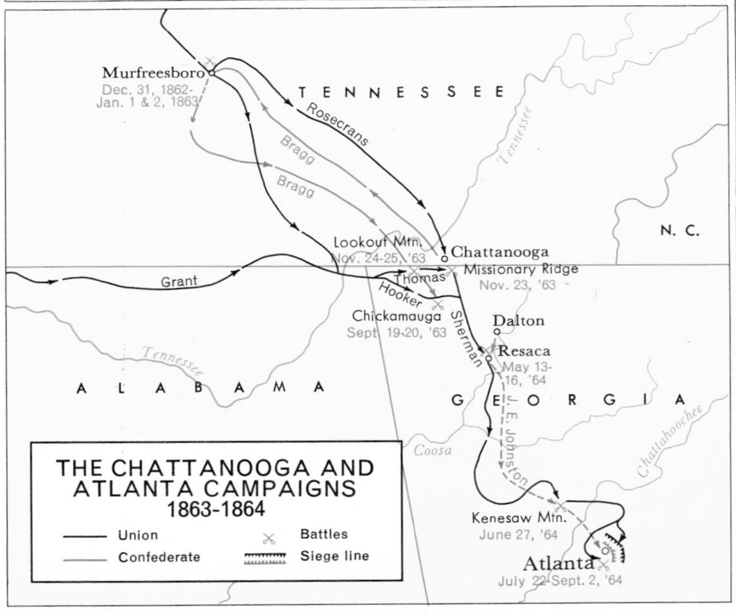

THE CHATTANOOGA AND ATLANTA CAMPAIGNS 1863-1864

Union · Confederate · Battles · Siege line

TENNESSEE · Murfreesboro Dec. 31, 1862 – Jan. 1 & 2, 1863 · Rosecrans · Bragg
N.C.
Lookout Mtn. Nov. 24-25, '63 · Chattanooga · Missionary Ridge Nov. 23, '63 · Thomas · Hooker
Grant · Chickamauga Sept. 19-20, '63 · Dalton · Resaca May 13-16, '64
ALABAMA · GEORGIA · Johnston · Sherman · Coosa · Chattahoochee
Kenesaw Mtn. June 27, '64 · Atlanta July 22-Sept. 2, '64

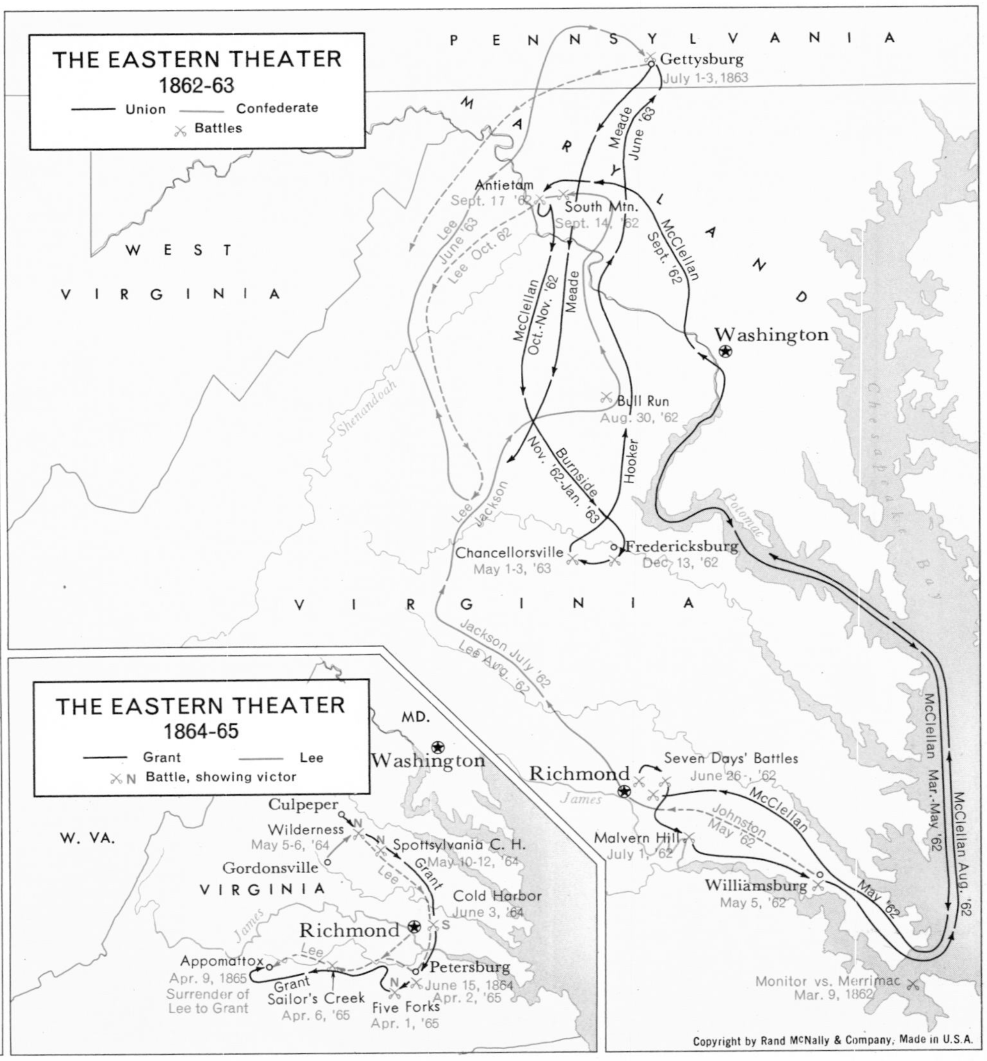

THE EASTERN THEATER 1862-63

Union · Confederate · Battles

PENNSYLVANIA · Gettysburg July 1-3, 1863 · Meade June '63
MARYLAND · Antietam Sept. 17 '62 · South Mtn. Sept. 14, '62 · McClellan Sept. '62
WEST VIRGINIA · Lee June '63 · Lee Oct. '62 · McClellan Oct.-Nov. '62 · Meade
Washington
Jackson · Bull Run Aug. 30, '62 · Burnside Nov. '62-Jan. '63 · Hooker
Shenandoah
Chancellorsville May 1-3, '63 · Fredericksburg Dec. 13, '62
VIRGINIA · Lee Aug. '62

THE EASTERN THEATER 1864-65

Grant · Lee · ✕N Battle, showing victor

MD. · Washington · Richmond · Seven Days' Battles June 26-, '62
W. VA. · Culpeper · Wilderness May 5-6, '64 · Spottsylvania C.H. May 10-12, '64 · Johnston May '62 · McClellan Mar.-May '62
Gordonsville · Cold Harbor June 3, '64 · Malvern Hill July 1, '62 · Williamsburg May 5, '62
VIRGINIA · Richmond · Petersburg June 15, 1864 – Apr. 2, '65
Appomattox Apr. 9, 1865 · Surrender of Lee to Grant · Sailor's Creek Apr. 6, '65 · Five Forks Apr. 1, '65 · Grant
Monitor vs. Merrimac Mar. 9, 1862
James · Chesapeake Bay

Copyright by Rand McNally & Company, Made in U.S.A.

BRITISH NORTH
AMERICA

UNITED STATES
OF
AMERICA

INTENDANCY OF NUEVA
CALIFORNIA
San Francisco 1776
Monterey
1770
Santa
Barbara
San Luis
Obispo
1772
Los Angeles
1781
San Juan
San Diego
1769
Capistrano
San Juan
INTENDANCY OF
NUEVO MEXICO
PRESIDENCY
OF
El Paso
INTENDANCY OF
SONORA (AUDIENCIA)
Chihuahua
INTENDANCY
OF DURANGO
La Paz
1535
INTENDANCY
OF
ZACATECAS
VICEROYALTY
OF NEW
Culiacán
1531
INTENDANCY
OF VIEJA CALIFONIA
SPAIN
Guadalajara
Querétaro
Mexico City
1325
INTENDANCY OF
VERA CRUZ
Vera Cruz
1519
INTENDANCY OF VALLADOLID
INTENDANCY
OF
OAXACA
INTENDANCY
OF
CHIAPAS
CAPTAINCY-
GENERAL (AUDIENCIA)
OF GUATEMALA
San Salvador
1525
Granada
1524
León
San José
1735
Cartago
1564
Portobelo
1584

St. Louis
1764
CAPTAINCY-
GENERAL
OF
LOUISIANA
Disputed with
U.S. 1783-1795
WEST FLORIDA
Pensacola
1698
New
Orleans
1718
EAST FLORIDA
St. Augustine
1565

WESTERN
INTERIOR PROVINCES
EASTERN
INTERIOR PROVINCES
Santa Fé
1609
San Juan
del Norte
1550
San Antonio
1718
Laredo
Saltillo
INTENDANCY OF
SAN LUIS POTOSI

Habana
1515
CAPTAINCY-GENERAL OF CUBA

INTENDANCY
OF
YUCATAN
Belice
CAPTAINCY
GENERAL (AUDIENCIA)
OF GUATEMALA
Loosely joined Spain

Santiago
1514
JAMAICA
Br. 1655
Port au
Prince
1749
Santo
Domingo
1496
CAPTAINCY-GENERAL OF SANTO DOMINGO
Ceded to France 1795
PUERTO
RICO
San Juan
1511

Santa Marta
Cartagena
1533
La Guaira
1588
Caracas
CAPTAINCY-GENERAL
OF CARACAS
TRINIDAD
Ceded to Great Britain,
1802

Stabroek (Georgetown)
Approx. 1740
Paramaribo
1640
Cayenne
1664
DUTCH
GUIANA
Dutch
in
1790
FRENCH
GUIANA

VICEROYALTY
OF
NEW GRANADA
Bogotá
Santa Marta
Santa Fé
SANTA FÉ
Established 1717, Restored 1740

Quito
PRESIDENCY
(AUDIENCIA)
OF QUITO
Guayaquil
1535
Tabatinga
1780
Barcelos
1658
Barra do
Rio Negro
1660
CAPTAINCY
OF
RIO NEGRO
CAPTAINCY
OF
PARÁ
Belem
1616
São Luis
1612
Fortaleza
1609

GALAPAGOS IS.
Claimed by Spain,
but unoccupied

AUDIENCIA
OF
Trujillo
1535
LIMA
VICEROYALTY
OF PERU
Callao
1537
Lima
1535
Cuzco
1534
PRESIDENCY
(AUDIENCIA)
OF
CUZCO
PRESIDENCY
(AUDIENCIA)
OF
CHARCAS
La Paz
1549
Chuquisaca
1538
Potosí
1545
Principe
da Beira
1760
CAPTAINCY
OF
MATO GROSSO
Villa Bella
(Mato Grosso)
1752
Santa Anna
(Goiaz)
1736
CAPTAINCY
OF
GOIAZ
CAPTAINCY
OF
MARANHÃO
CAPTAINCY
OF
PIAUÍ
Recife
(Pernambuco)
1561
CAPTAINCY
OF
PERNAMBUCO
CAPTAINCY
OF
SERGIPE
CAPTAINCY
OF
BAÍA
Salvador
(Baía)
1549
VICEROYALTY
OF
BRAZIL
Definitively established
1714
CAPTAINCY
OF
Tijuca
(Diamantina)
MINAS
GERAIS
Ouro Preto
CAPTAINCY OF
RIO DE JANEIRO
CAPTAINCY OF
ESPIRITO SANTO

VICEROYALTY
OF
LA PLATA
Established 1776
Salta
Tucumán
1565
Asunción
1537
PARAGUAY
São Paulo
1554
CAPTAINCY OF
SÃO PAULO
Santos
1536
Rio de Janeiro
1567
La Serena
1544
CAPTAINCY-GENERAL (AUDIENCIA) OF CHILE
Loosely joined to Peru
Valparaíso
1544
Santiago
1541
Concepción
1550
Valdivia
1552
Mendoza
1561
Córdoba
1573
Santa Fé
1573
AUDIENCIA OF
BUENOS AIRES
Buenos Aires
1580
Colonia
Montevideo
1724
BANDA
ORIENTAL
CAPTAINCY OF
SANTA CATARINA
CAPTAINCY OF
RIO GRANDE DO SUL
Porto Alegre
Rio Grande
1737
Abandoned
Jesuit Missions

CHILOE I.
PATAGONIA

TIERRA DEL
FUEGO
MALVINAS
(FALKLAND
ISLANDS)
CAPE HORN
Drake Passage

Atlantic
Ocean
Tropic of Cancer
Caribbean Sea
Pacific
Ocean
Tropic of Capricorn
Ocean

Gulf of Mexico

LATIN AMERICA ABOUT 1790

MILES 0   250   500   1,000

European Colonies

	Spain		Portugal
	Great Britain		France
	Netherlands	*	Seat of Government

1535 Lima   Dates indicate year of founding

Copyright by Rand McNally & Company, Made in U.S.A.

149

UNITED STATES

42nd Parallel
Ceded to U.S. 1848
San Francisco
Monterey
San Diego
LOWER
CALIFORNIA
Columbia
Santa Fé
TEXAS
Independent 1836
Annexed to U.S. 1845
Chihuahua
Gila
Mesilla Strip
Sold to U.S. 1853
Grande de Santiago
Monterrey
Tampico
MEXICO
Independent 1821
Monarchy 1822-23
Republic 1824
Jalapa
Mexico City
Puebla
Vera Cruz
Acapulco
YUCATAN
Independent 1839-43
CHIAPAS
To Mexico 1824
Guatemala
GUATEMALA
San Salvador
SALVADOR
HONDURAS
BRITISH
HONDURAS
Belice
Tegucigalpa
NICARAGUA
Managua
CENTRAL AMERICA
Independent 1821
United with Mexico 1821
Independent Confederation 1823
Divided into five states 1838
San José
COSTA RICA
PANAMA ISTHMUS
To Colombia 1821-1903
Gulf of Mexico
New Orleans
Tropic of Cancer
Habana
CUBA
Sp. until 1898
Santiago
HAITI
Port au Prince
DOMINICAN REPUBLIC
United with Haiti until 1844
Santo Domingo
JAMAICA
(British)
PUERTO RICO
Sp. until 1898
VIRGIN ISLANDS
(Den.)
Caribbean Sea
CURACAO (Dutch)
MOSQUITO COAST
British Protectorate
1841-50
Panama
La Guaira
Caracas
VENEZUELA
GREAT
COLOMBIA
(1819-1830)
TRINIDAD
(British)
BRITISH GUIANA
DUTCH GUIANA
FRENCH GUIANA
Bogotá
GALAPAGOS IS
Ecuador since 1832
Quito
ECUADOR
Guayaquil
Paita
Ceded by Ecuador to Brazil 1904
MARAJO I.
Belem
São Luiz
Fortaleza
Amazon
Atlantic Ocean

PERU
(1824)
Trujillo
Callao
Lima
Cuzco
Arequipa
Mollendo
CHINCHA IS.
(Peru)
Ceded by Bol. to Braz. 186
Claim relinquished
by Peru 1909
Peru and Bolivia
Confederated 1836-1839
Republic of Bolívar 1825,
Later Bolivia
La Paz
BOLIVIA
Sucre
BRAZIL
Empire of Brazil, Monarchy 1822-1889
United States of Brazil since 1889
Recife
(Pernambuco)
São Salvador
(Baia)
Ceded to
Brazil 1907
Belo Horizonte
To Chile 1883
To Peru 1929 TACNA
To Chile 1883 ARICA
TARAPACÁ
To Chile 1883
ATACAMA
To Chile 1884
Antofagasta
Iquique
CHACO
Claimed by
Bolivia and
Paraguay
PARAGUAY
Asunción
São Paulo
To Brazil 1895
Santos
Rio de Janeiro
Tropic of Capricorn
JUAN FERNANDEZ
ISLANDS
Chile since 1818
Salta
Tucumán
ARGENTINA
United Provinces of
Rio de la Plata 1816
Argentine Confederation
1835
Córdoba
Santa Fé
Argentine Republic
1853
Mendoza
Rosario
URUGUAY
Cisplatine Province
Spanish expelled 1814
To Portugal 1817
To Brazil 1822
Republic of Uruguay 1828
Valparaíso
Santiago
Buenos Aires
Montevideo
Argentine
Nation
1860
PROVINCE OF
BUENOS AIRES
Independent
1853-1859
Federal District
since 1880
PAMPAS
Bahia Blanca
Pacific Ocean
Original Republic of Chile 1818
CHILE
CHILOE
Boundary fixed by treaty 188
PATAGONIA
Conquered by Argentina 1876-1879
Strait of Magellan
FALKLAND IS.
Held by Great Britain
since 1833
Claimed by Argentina
TIERRA DEL FUEGO
Disputed between
Argentina and Chile
Divided 1902

LATIN AMERICA AFTER
INDEPENDENCE

MILES 0   250   500   1,000

Copyright by Rand McNally & Company, Made in U.S.A.

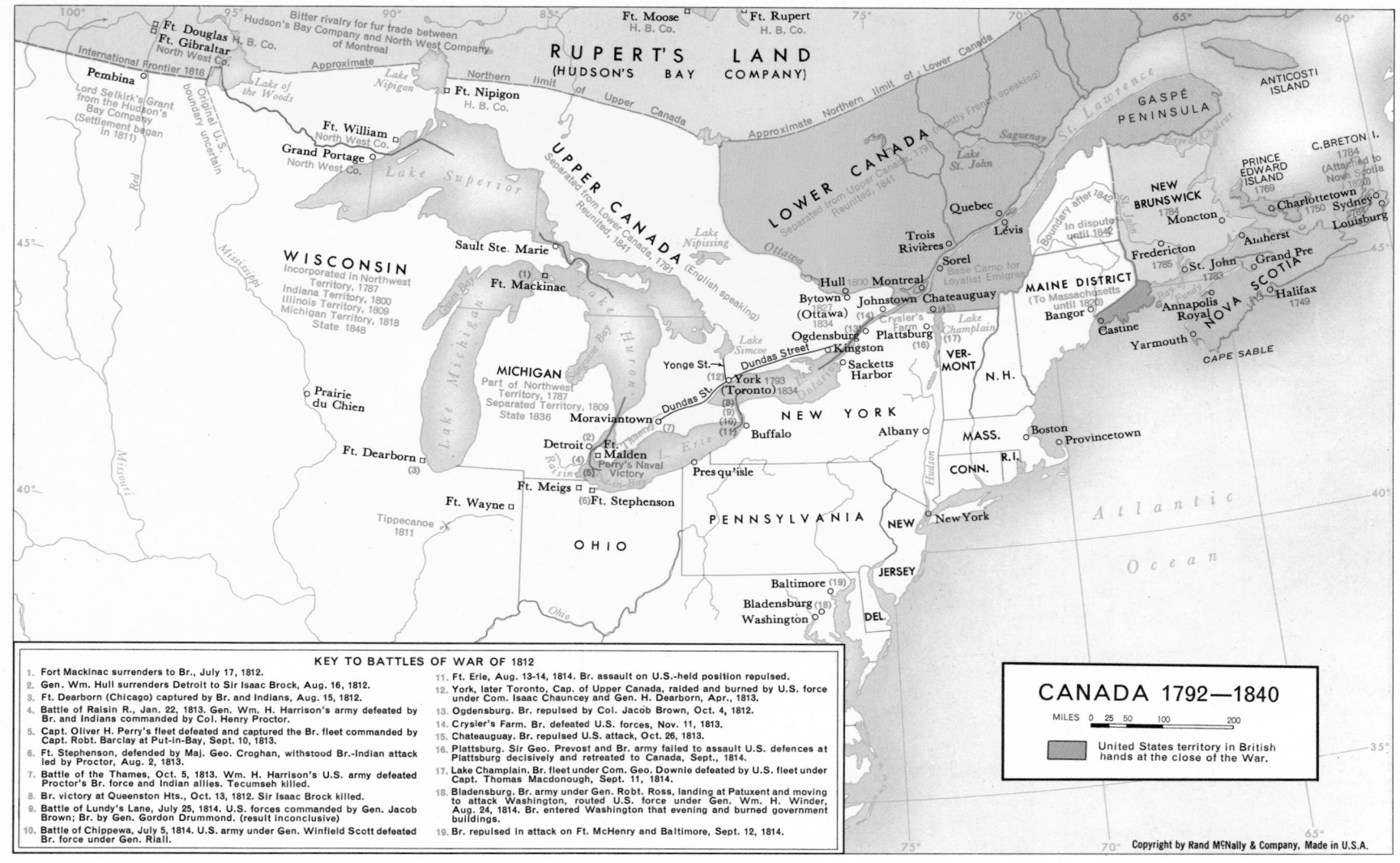

CANADA 1792—1840
MILES 0 25 50 100 200
United States territory in British hands at the close of the War.
RUPERT'S LAND (HUDSON'S BAY COMPANY)
Bitter rivalry for fur trade between Hudson's Bay Company and North West Company of Montreal
Ft. Moose H. B. Co.
Ft. Rupert H. B. Co.
Ft. Douglas H. B. Co.
Ft. Gibraltar North West Co.
International Frontier 1818
Pembina
Lord Selkirk's Grant from the Hudson's Bay Company (Settlement began in 1811)
Original U.S. boundary uncertain
Ft. Nipigon H. B. Co.
Ft. William North West Co.
Grand Portage North West Co.
Lake Superior
Approximate Northern limit of Upper Canada
Approximate Northern limit of Lower Canada
UPPER CANADA Separated from Lower Canada, 1791 Reunited, 1841 (English speaking)
LOWER CANADA Separated from Upper Canada, 1791 Reunited, 1841 (Mostly French speaking)
GASPÉ PENINSULA
ANTICOSTI ISLAND
NEW BRUNSWICK 1784
PRINCE EDWARD ISLAND 1769
C. BRETON I. 1784 (Attached to Nova Scotia 1820)
NOVA SCOTIA
WISCONSIN Incorporated in Northwest Territory, 1787 Indiana Territory, 1800 Illinois Territory, 1809 Michigan Territory, 1818 State 1848
MICHIGAN Part of Northwest Territory, 1787 Separated Territory, 1809 State 1836
MAINE DISTRICT (To Massachusetts until 1820)
Boundary after 1842
In dispute until 1842
Base Camp for Loyalist Emigres
Sault Ste. Marie
Ft. Mackinac
Quebec
Levis
Trois Rivières
Sorel
Hull 1800
Montreal
Bytown 1827 (Ottawa) 1834
Johnstown
Chateauguay
Crysler's Farm
Ogdensburg
Plattsburg
Kingston
Sackets Harbor
Yonge St.
Dundas Street
York 1793 (Toronto) 1834
Moraviantown
Detroit
Ft. Malden
Perry's Naval Victory
Ft. Meigs
Ft. Stephenson
Buffalo
Pres qu'isle
Prairie du Chien
Ft. Dearborn
Ft. Wayne
Tippecanoe 1811
Albany
Boston
Provincetown
New York
Baltimore
Bladensburg
Washington
Bangor
Castine
Fredericton 1785
St. John 1783
Moncton
Amherst
Grand Pre
Charlottetown 1750
Sydney
Louisburg
Annapolis Royal
Halifax 1749
Yarmouth
CAPE SABLE
VER-MONT
N.H.
MASS.
CONN.
R.I.
NEW YORK
PENNSYLVANIA
OHIO
NEW JERSEY
DEL.
Atlantic Ocean
KEY TO BATTLES OF WAR OF 1812
1. Fort Mackinac surrenders to Br., July 17, 1812.
2. Gen. Wm. Hull surrenders Detroit to Sir Isaac Brock, Aug. 16, 1812.
3. Ft. Dearborn (Chicago) captured by Br. and Indians, Aug. 15, 1812.
4. Battle of Raisin R., Jan. 22, 1813. Gen. Wm. H. Harrison's army defeated by Br. and Indians commanded by Col. Henry Proctor.
5. Capt. Oliver H. Perry's fleet defeated and captured the Br. fleet commanded by Capt. Robt. Barclay at Put-in-Bay, Sept. 10, 1813.
6. Ft. Stephenson, defended by Maj. Geo. Croghan, withstood Br.-Indian attack led by Proctor, Aug. 2, 1813.
7. Battle of the Thames, Oct. 5, 1813. Wm. H. Harrison's U.S. army defeated Proctor's Br. force and Indian allies. Tecumseh killed.
8. Br. victory at Queenston Hts., Oct. 13, 1812. Sir Isaac Brock killed.
9. Battle of Lundy's Lane, July 25, 1814. U.S. forces commanded by Gen. Jacob Brown; Br. by Gen. Gordon Drummond. (result inconclusive)
10. Battle of Chippewa, July 5, 1814. U.S. army under Gen. Winfield Scott defeated Br. force under Gen. Riall.
11. Ft. Erie, Aug. 13-14, 1814. Br. assault on U.S.-held position repulsed.
12. York, later Toronto, Cap. of Upper Canada, raided and burned by U.S. force under Com. Isaac Chauncey and Gen. H. Dearborn, Apr., 1813.
13. Ogdensburg. Br. repulsed by Col. Jacob Brown, Oct. 4, 1812.
14. Crysler's Farm. Br. defeated U.S. forces, Nov. 11, 1813.
15. Chateauguay. Br. repulsed U.S. attack, Oct. 26, 1813.
16. Plattsburg. Sir Geo. Prevost and Br. army failed to assault U.S. defences at Plattsburg decisively and retreated to Canada, Sept., 1814.
17. Lake Champlain. Br. fleet under Com. Geo. Downie defeated by U.S. fleet under Capt. Thomas Macdonough, Sept. 11, 1814.
18. Bladensburg. Br. army under Gen. Robt. Ross, landing at Patuxent and moving to attack Washington, routed U.S. force under Gen. Wm. H. Winder, Aug. 24, 1814. Br. entered Washington that evening and burned government buildings.
19. Br. repulsed in attack on Ft. McHenry and Baltimore, Sept. 12, 1814.
Copyright by Rand McNally & Company, Made in U.S.A.

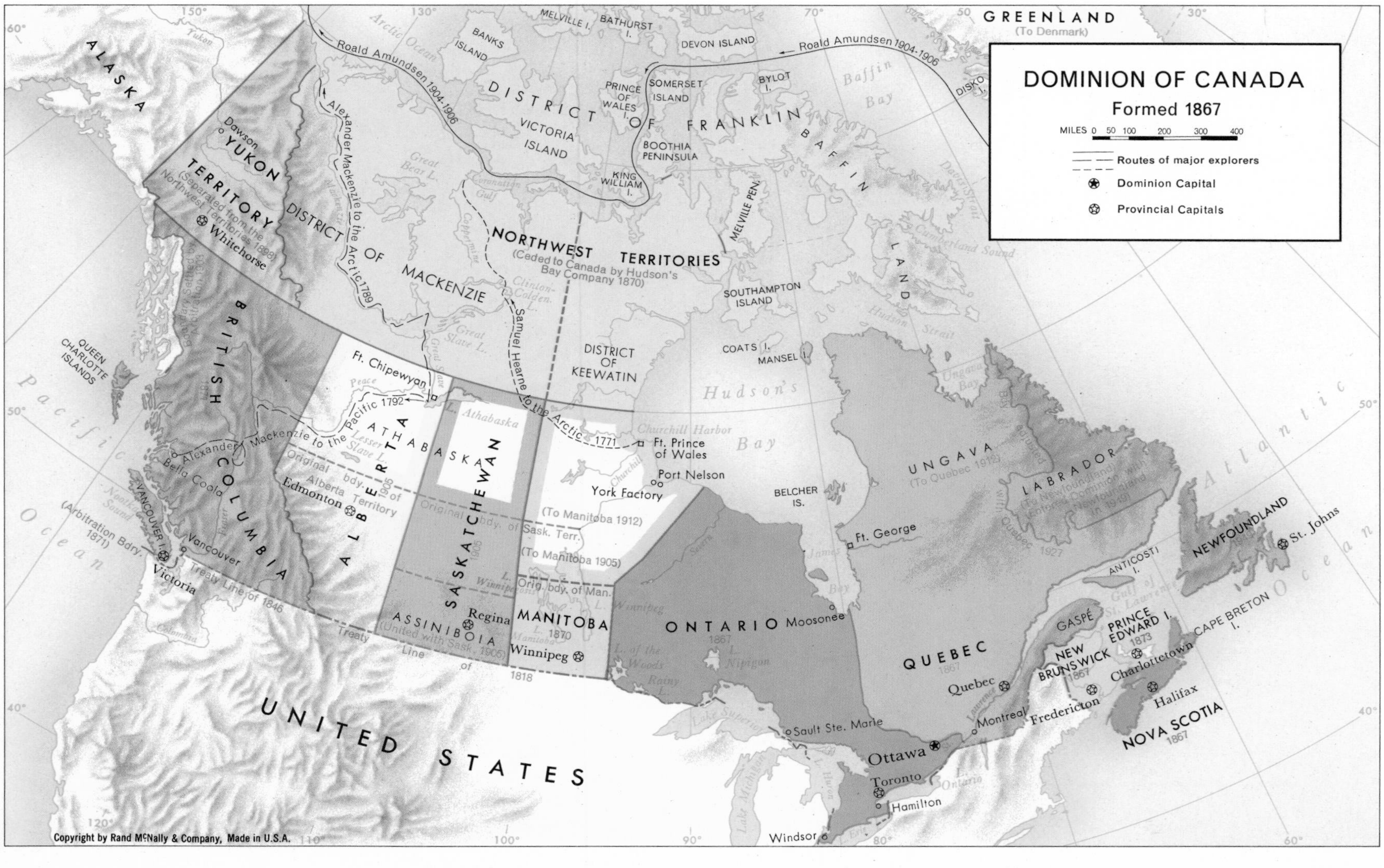

DOMINION OF CANADA
Formed 1867
MILES 0 50 100 200 300 400
Routes of major explorers
Dominion Capital
Provincial Capitals
GREENLAND (To Denmark)
ALASKA
YUKON TERRITORY (Separated from the Northwest Territories 1898)
Dawson
Whitehorse
DISTRICT OF FRANKLIN
DISTRICT OF MACKENZIE
DISTRICT OF KEEWATIN
NORTHWEST TERRITORIES (Ceded to Canada by Hudson's Bay Company 1870)
Roald Amundsen 1904-1906
Alexander Mackenzie to the Arctic 1789
Mackenzie to the Pacific 1792
Samuel Hearne to the Arctic 1771
BANKS ISLAND
MELVILLE I.
BATHURST I.
DEVON ISLAND
SOMERSET ISLAND
BYLOT I.
PRINCE OF WALES I.
VICTORIA ISLAND
BOOTHIA PENINSULA
KING WILLIAM I.
MELVILLE PEN.
BAFFIN LAND
Baffin Bay
DISKO I.
SOUTHAMPTON ISLAND
COATS I.
MANSEL I.
BELCHER IS.
Hudson's Bay
BRITISH COLUMBIA
QUEEN CHARLOTTE ISLANDS
VANCOUVER I.
(Arbitration Bdry. 1871)
Bella Coola
Vancouver
Victoria
Treaty Line of 1846
Treaty Line of 1818
ALBERTA
Edmonton
Original bdy. of Alberta Territory
ATHABASKA
Ft. Chipewyan
SASKATCHEWAN
Original bdy. of Sask. Terr.
ASSINIBOIA (United with Sask. 1905)
Regina
MANITOBA 1870
Orig. bdy. of Man.
(To Manitoba 1912)
(To Manitoba 1905)
Winnipeg
Ft. Prince of Wales
Churchill Harbor
Port Nelson
York Factory
ONTARIO 1867
Moosonee
Ft. George
UNGAVA (To Quebec 1912)
LABRADOR
NEWFOUNDLAND
St. Johns
ANTICOSTI I.
GASPÉ
QUEBEC 1867
Quebec
Montreal
Ottawa
Toronto
Hamilton
Windsor
Sault Ste. Marie
NEW BRUNSWICK 1867
Fredericton
PRINCE EDWARD I. 1873
Charlottetown
CAPE BRETON I.
NOVA SCOTIA 1867
Halifax
UNITED STATES
Pacific Ocean
Atlantic Ocean
Copyright by Rand McNally & Company, Made in U.S.A.

EUROPE IN 1810
At the Height of Napoleon's Power

MILES 0 50 100 200 300

- French Empire ★
- "Greater Empire," subject to Napoleon, undergoing internal reform. ★
- Nominal Allies of Napoleon. ★
- Openly hostile to Napoleon; protected by British fleet.
- Hostile to Napoleon.
- ✕ Battles
- ★ Continental System, boycotting British trade.

NORWAY AND DENMARK
SWEDEN
GRAND DUCHY OF FINLAND (To Russia 1808)
Christiania
Helsingfors
St. Petersburg
Stockholm
RUSSIAN EMPIRE
Moscow 1812
Napoleon's advance and retreat 1812
Borodino
Smolensk
Riga
Düna
SCOTLAND
Edinburgh
Copenhagen
Vilna
Tilsit
UNITED KINGDOM OF GREAT BRITAIN AND IRELAND (Act of Union, 1801)
Dublin
IRELAND
Liverpool
ENGLAND
Hartwell
London
Plymouth
Portsmouth
HELGOLAND (Britain)
Hamburg
King of Prussia in refuge from Berlin
Königsberg
Friedland 1807
Niemen
Bourbon King of France in refuge
Amsterdam
Brussels
Berlin
PRUSSIA
Posen
GRAND DUCHY OF WARSAW (1808-13)
Lemberg
Kiev
UKRAINE
Dnieper
Bug
English Channel
Waterloo 1815
Cologne
WESTPHALIA
CONFEDERATION OF THE
Leipzig
SAXONY
Dresden 1813
Odra
Vistula
Dniester
Frankfort
Jena 1806
Prague
BOHEMIA
AUSTRIAN
Austerlitz 1805
BESSARABIA
MOLDAVIA
Yassy
Paris
Strasburg
RHINE (1806-13)
WÜRT.
BADEN
Ulm
Hohenlinden
1800
Vienna
Wagram 1809
EMPIRE
CARPATHIANS
Pruth
Brest
Seine
FRENCH
Munich
BAVARIA
Buda Pest (Proclaimed 1804)
HUNGARY
WALLACHIA
Bucharest
Valençay
Bourbon King of Spain in captivity
Loire
EMPIRE (1804-14)
SWITZERLAND
Geneva
Milan
KDM. OF
ILLYRIAN
Trieste
PROVINCES
Venice (1810-13)
Save
Belgrade
Danube
Sofia
Black Sea
Bordeaux
Lyon
Turin
Marengo 1800
ITALY
LUCCA (1804-13)
Florence
Ancona
MONTENEGRO
Adrianople
Constantinople
Bay of Biscay
PYRENEES
Marseille
Savona
Genoa
Pope in captivity
CORSICA (France)
ELBA
Rome
KDM. OF Benevento
NAPLES (1806-13)
CORFU (France)
IONIAN ISLANDS (Britain)
Saloniki
OTTOMAN
Smyrna
Coruña
Oporto
KDM. OF PORTUGAL
Madrid
KINGDOM OF SPAIN
Valencia (1808-13)
Bailén
French surrender 1808
Barcelona
Balearic Is.
KDM. OF SARDINIA
King of Savoy in refuge from Turin
Naples
Otranto
EMPIRE
CYPRUS
Torres Vedras
Lisbon
Regent of Portugal in refuge in Brazil
KDM. OF PORTUGAL
Duero
Tagus
Guadiana
Seville
Guadalquivir
Cape Trafalgar 1805
Gibraltar (Britain)
Mediterranean
Palermo
KDM. OF SICILY
Bourbon King of Naples in refuge
CRETE
Sea
MALTA (Britain)
Copyright by Rand McNally & Company. Made in U.S.A.

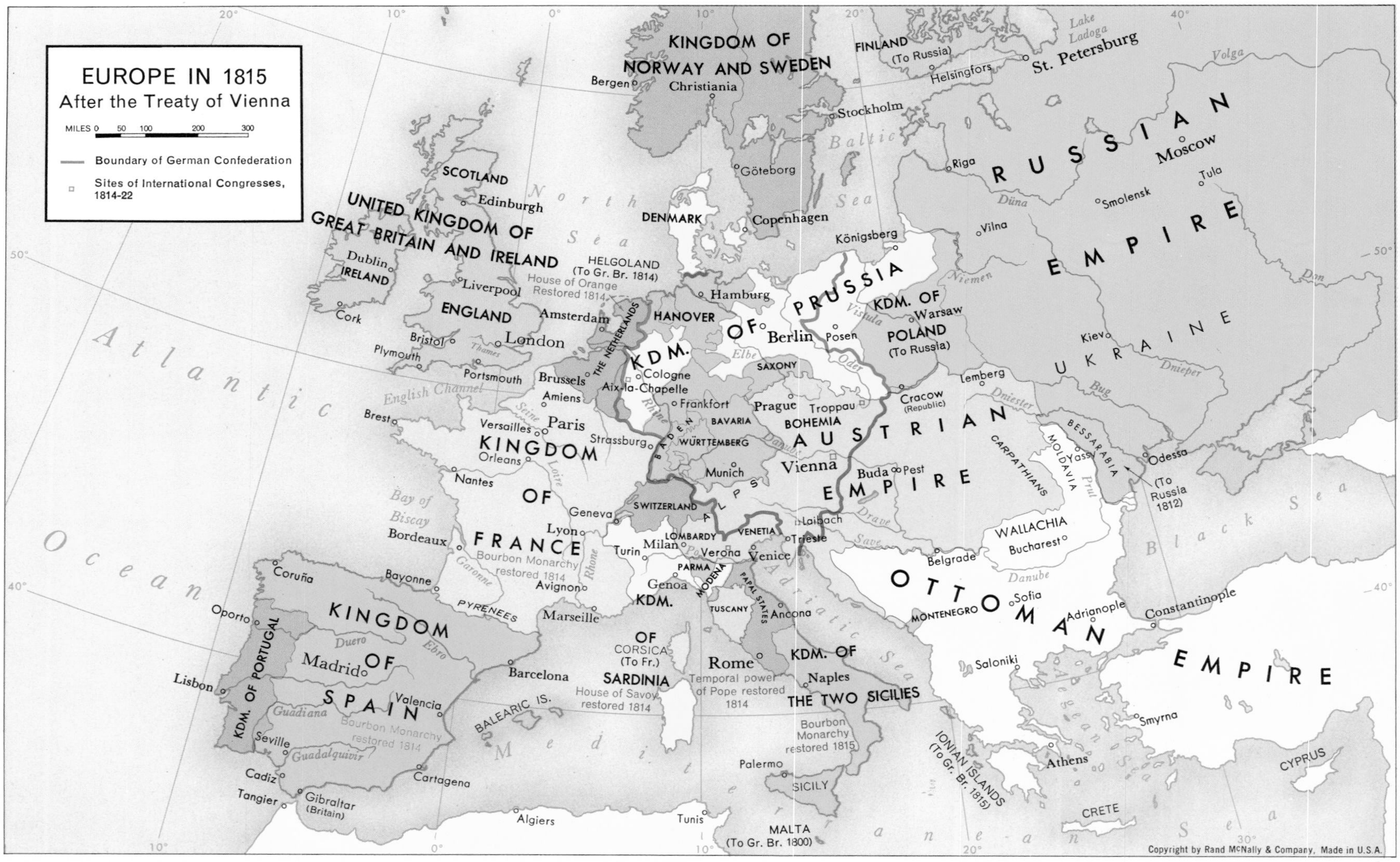

EUROPE IN 1815
After the Treaty of Vienna

MILES 0 50 100 200 300

- Boundary of German Confederation
- ◻ Sites of International Congresses, 1814-22

KINGDOM OF NORWAY AND SWEDEN
FINLAND (To Russia)
Bergen
Christiania
Göteborg
Stockholm
Helsingfors
St. Petersburg
Volga
Lake Ladoga
RUSSIAN EMPIRE
Moscow
Tula
SCOTLAND
Edinburgh
DENMARK
Copenhagen
Königsberg
Smolensk
Riga
Düna
Vilna
Kiev
UNITED KINGDOM OF GREAT BRITAIN AND IRELAND
Dublin
IRELAND
Cork
Liverpool
Bristol
ENGLAND
London
Plymouth
Portsmouth
HELGOLAND (To Gr. Br. 1814)
House of Orange restored 1814
Hamburg
HANOVER
THE NETHERLANDS
Berlin
PRUSSIA
Posen
KDM. OF POLAND (To Russia)
Lemberg
UKRAINE
Dnieper
Bug
English Channel
Thames
Amsterdam
Brussels
Amiens
KDM. OF
Cologne
Aix-la-Chapelle
SAXONY
Elbe
Frankfort
Prague
BOHEMIA
Troppau
Cracow (Republic)
Dniester
Versailles
Paris
Orleans
Seine
Strassburg
WÜRTTEMBERG
BADEN
BAVARIA
Vienna
AUSTRIAN
Laibach
CARPATHIANS
Pruth
BESSARABIA
MOLDAVIA
Yassy
Brest
KINGDOM
Nantes
Loire
OF FRANCE
Bourbon Monarchy restored 1814
Geneva
SWITZERLAND
Munich
EMPIRE
Buda Pest
Troppau
WALLACHIA
Bucharest
(To Russia 1812)
Bay of Biscay
Bordeaux
Garonne
Lyon
Turin
Milan
LOMBARDY
VENETIA
Verona
Trieste
Save
Belgrade
Danube
Sofia
Black Sea
Avignon
Marseille
Genoa
PARMA
MODENA
Venice
PAPAL STATES
TUSCANY
Ancona
MONTENEGRO
Adrianople
Constantinople
PYRENEES
KDM. OF CORSICA (To Fr.)
Rome
Temporal power of Pope restored 1814
KDM. OF Naples
THE TWO SICILIES
Saloniki
Smyrna
Coruña
Oporto
Lisbon
KDM. OF PORTUGAL
Duero
Madrid
KINGDOM OF SPAIN
Valencia
Barcelona
Balearic Is.
SARDINIA
House of Savoy restored 1814
Bourbon Monarchy restored 1815
IONIAN ISLANDS (To Gr. Br. 1815)
Athens
OTTOMAN EMPIRE
CYPRUS
Guadiana
Seville
Guadalquivir
Bourbon Monarchy restored 1814
Cadiz
Tangier
Gibraltar (Britain)
Cartagena
Mediterranean Sea
Algiers
Tunis
Palermo
SICILY
MALTA (To Gr. Br. 1800)
CRETE
Copyright by Rand McNally & Company. Made in U.S.A.

EXPANSION OF RUSSIA IN EUROPE

MILES 0 50 100   200   300   400

Russia 1533

Acquired to 1598

Acquired to 1914

Held at other times

Dates indicate time area held or gained by Russia.

Copyright by Rand McNally & Company, Made in U.S.A.

153

UNIFICATION OF GERMANY
Bismarck's Empire

MILES 0 50 100 200

– – – Boundary of the German Confederation of 1815.

——— Boundary of the German Empire, 1871—1918

1866 Absorbed by Prussia

1867 Entered North German Confederation, as a member state

1871 Entered German Empire, with preceding, as a member state.
Alsace-Lorraine annexed

Copyright by Rand McNally & Company, Made in U.S.A.

DENMARK SWEDEN

Baltic Sea

BORNHOLM

Copenhagen Malmö

RÜGEN

(To Oldenburg)

SCHLESWIG
1866

Flensborg

Kiel

HOLSTEIN

(To Hamburg)
Hamburg

LÜBECK
LAUENBURG
1867

MECKLENBURG
SCHWERIN
1867

POMERANIA

MECKLENBURG
STRELITZ
1867

Stettin

Danzig

Thorn

WEST PRUSSIA

EAST PRUSSIA

Königsberg

Tilsit

Bielostock

EAST FRIESLAND

Bremen
1867

OLDEN-BURG
1866

KINGDOM OF HANOVER
1866

Hanover

BRANDENBURG

Berlin

Magdeburg

POSEN
Poseno

Warsaw

Lodz

RUSSIAN EMPIRE

POLAND

Kalisz

Lublin

NETHERLANDS

Amsterdam

Rotterdam

LIPPE
1867

BRUNSWICK

HANOVER

ANHALT

SAXONY

Leipzig

KINGDOM OF SAXONY
1867

Dresden

Breslau

SILESIA

Cracow

GALICIA

(Republic of Cracow 1815)
(To Austria 1846)

ENGLAND

London

Cleves

Munster

WALDECK

Kassel

Weimar

SAXON DUCHIES
1867

REUSS

BELGIUM

Ghent

Antwerp

Brussels

Liége

Mons

Namur

Lille

Essen

RHINE PROVINCE

Düsseldorf

Cologne
Aachen

OF Bonn
Wetzlar

PRUSSIA

K.DM.
WESTPHALIA

Ruhr

NASSAU
1866

Coblenz

HESSE-KASSEL

HESSE-DARMSTADT

Frankfurt

Ems

Sadowa

Karlsbad

Eger

Prague

Pilsen

BOHEMIA

MORAVIA

Olmütz

Brünn

SILESIA

(Neutralized 1867)

Sedan

Luxemburg

(To Prussia 1834)

Mainz

Darmstadt

Mannheim

BAVARIAN
PALATINATE

Baireuth

Würzburg

KINGDOM OF
Nuremberg

Regensburg

AUSTRIAN

FRANCE

Reims

Verdun

Paris

Metz

Nancy

LORRAINE

Strassburg
1871

ALSACE

GRAND DUCHY OF BADEN

Karlsruhe

Stuttgart

KINGDOM OF
WÜRTTEMBERG

HOHENZOLLERN
(To Prussia 1849)

Augsburg

Munich
1871

BAVARIA

Danube

Inn

Vienna

AUSTRIA

EMPIRE

Belfort

Basel

Zurich

Constance

SWITZERLAND

Innsbruck

TYROL

North Sea

Rhine

Meuse

Moselle

Rhine

Main

Elbe

Oder

Warta

Vistula

Neisse

Spree

GERMAN TARIFF UNITY
The Zollverein

Showing years of adherence of various states to the tariff union initiated by Prussia. The old free cities of Hamburg and Bremen were not brought under the national tariff until long after political unification.

Hamburg and Bremen
1888

1854

1819
1828
1835
1836
1834

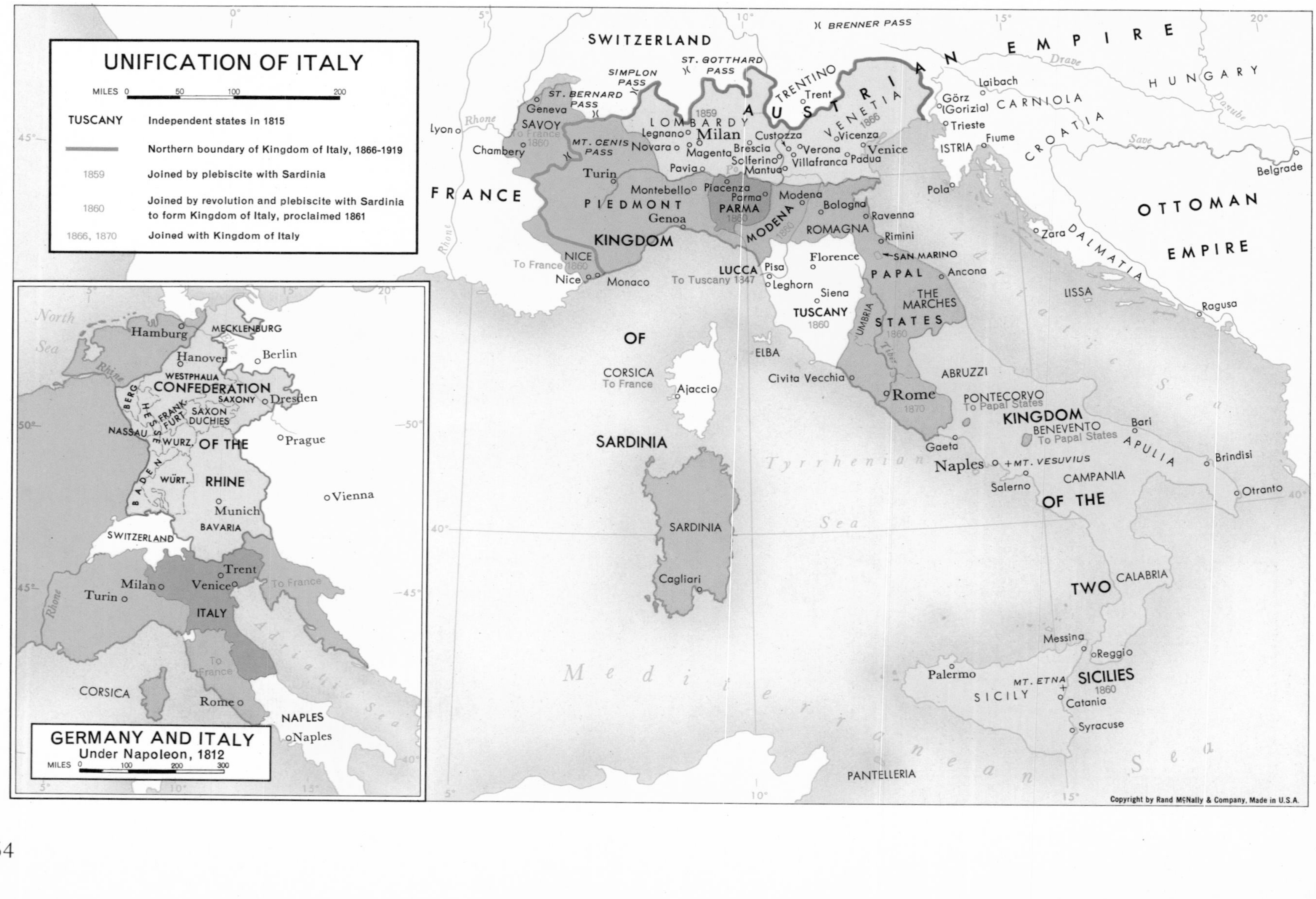

UNIFICATION OF ITALY

MILES 0 50 100 200

TUSCANY Independent states in 1815

——— Northern boundary of Kingdom of Italy, 1866-1919

1859 Joined by plebiscite with Sardinia

1860 Joined by revolution and plebiscite with Sardinia to form Kingdom of Italy, proclaimed 1861

1866, 1870 Joined with Kingdom of Italy

SWITZERLAND

BRENNER PASS

SIMPLON PASS

ST. GOTTHARD PASS

ST. BERNARD PASS

TRENTINO

Trent

AUSTRIAN EMPIRE

Laibach

Görz
(Gorizia)

CARNIOLA

HUNGARY

Geneva

SAVOY
1860

Chambery

ST. CENIS PASS

LOMBARDY
1859

Legnano

Milan

Novara

Magenta

Brescia

Custozza

VENETIA
1866

Vicenza

Verona

Villafranca

Padua

Venice

Trieste

ISTRIA

Fiume

CROATIA

DALMATIA

Belgrade

Lyon

Rhone

FRANCE

Turin

PIEDMONT

Pavia

Montebello

Genoa

Piacenza

Parma

PARMA

Solferino

Mantua

MODENA

Modena

Bologna

ROMAGNA

Ravenna

Rimini

Pola

Zara

Lissa

Ragusa

OTTOMAN EMPIRE

KINGDOM

NICE
1860

Nice

Monaco

To France

LUCCA
To Tuscany 1847

Pisa

Leghorn

Siena

Florence

TUSCANY
1860

UMBRIA
1860

San Marino

PAPAL

THE MARCHES
1860

Ancona

Adriatic Sea

OF

CORSICA
To France

Ajaccio

ELBA

Civita Vecchia

Rome
1870

PONTECORVO
To Papal States

ABRUZZI

STATES

SARDINIA

Tyrrhenian Sea

KINGDOM

BENEVENTO
To Papal States

Gaeta

Naples

MT. VESUVIUS

APULIA

Bari

Brindisi

CAMPANIA

Salerno

Otranto

Cagliari

Mediterranean Sea

OF

THE

CALABRIA

TWO

Messina

Reggio

Palermo

MT. ETNA

SICILIES

SICILY

Catania

Syracuse

PANTELLERIA

Copyright by Rand McNally & Company, Made in U.S.A.

GERMANY AND ITALY
Under Napoleon, 1812

MILES 0 100 200 300

North Sea

Hamburg

MECKLENBURG

Hanover

Berlin

BERG

WESTPHALIA

CONFEDERATION

SAXONY

Dresden

HESSE

FRANKFURT

SAXON DUCHIES

Prague

NASSAU

WURZ.

OF THE

BADEN

WÜRT.

RHINE

Munich

BAVARIA

Vienna

SWITZERLAND

Rhine

Turin

Milano

Trent

Venice

ITALY

To France

Adriatic Sea

CORSICA

Rome

NAPLES

Naples

To France

154

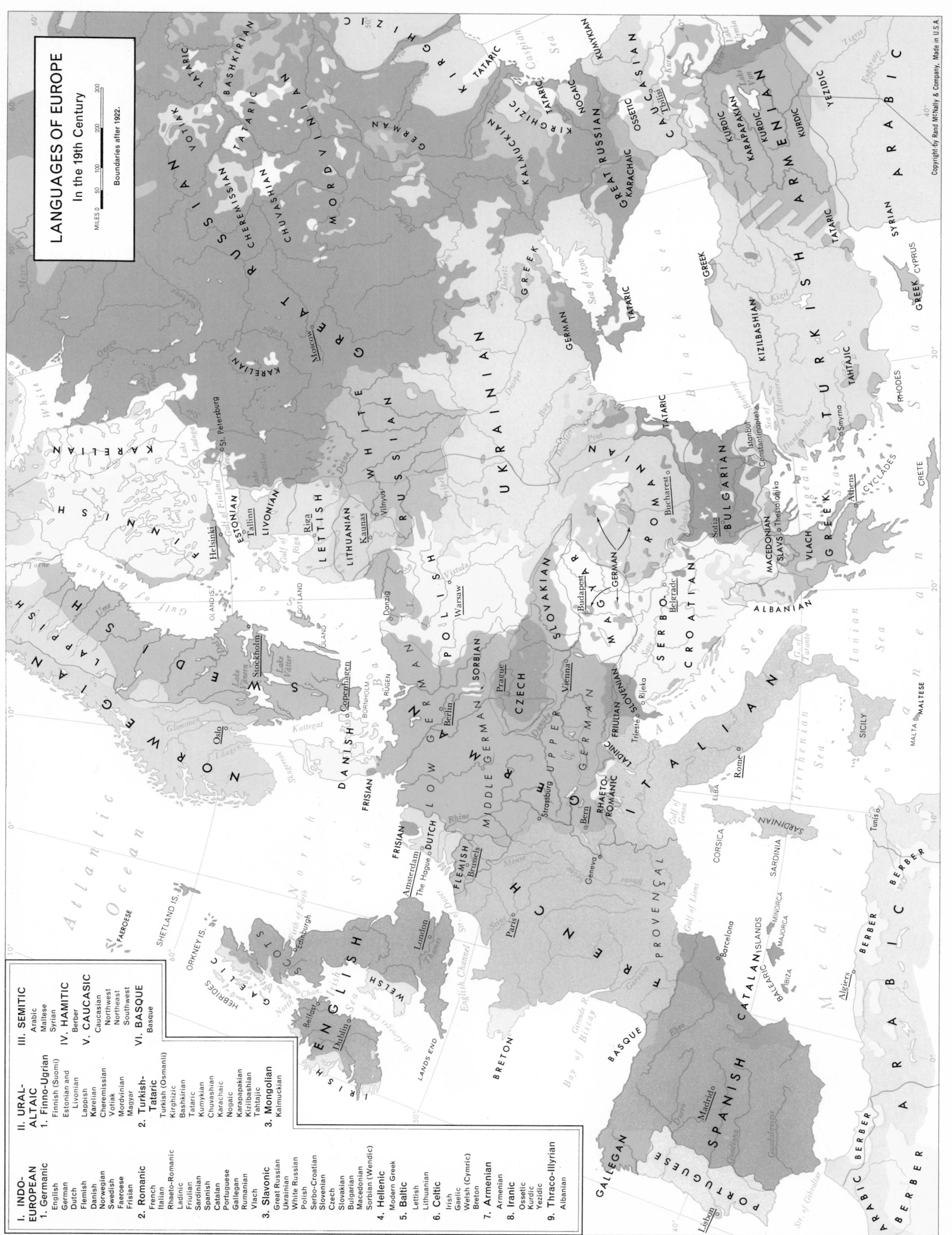

LANGUAGES OF EUROPE
In the 19th Century

MILES 0 50 100 200 300

Boundaries after 1922.

Copyright by Rand McNally & Company. Made in U.S.A.

I. INDO-EUROPEAN

1. Germanic
English
German
Dutch
Flemish
Danish
Norwegian
Swedish
Faeroese
Frisian

2. Romanic
French
Italian
Rhaeto-Romanic
Ladinic
Friulian
Sardinian
Spanish
Catalan
Portuguese
Gallegan
Rumanian
Vlach

3. Slavonic
Great Russian
Ukrainian
White Russian
Polish
Serbo-Croatian
Slovenian
Czech
Slovakian
Bulgarian
Macedonian
Sorbian (Wendic)

4. Hellenic
Modern Greek

5. Baltic
Lettish
Lithuanian

6. Celtic
Irish
Gaelic
Welsh (Cymric)
Breton

7. Armenian
Armenian

8. Iranic
Ossetic
Kurdic
Yezidic

9. Thraco-Illyrian
Albanian

II. URAL-ALTAIC

1. Finno-Ugrian
Finnish (Suomi)
Estonian and
Livonian
Lappish
Karelian
Cheremissian
Votiak
Mordvinian
Magyar

2. Turkish-Tataric
Turkish (Osmanli)
Kirghizic
Bashkirian
Tataric
Kumykian
Chuvashian
Karachaic
Nogaic
Karapapakian
Kizilbashian
Tahtajic

3. Mongolian
Kalmuckian

III. SEMITIC
Arabic
Maltese
Syrian

IV. HAMITIC
Berber

V. CAUCASIC
Caucasian
Northwest
Northeast
Southwest

VI. BASQUE
Basque

155

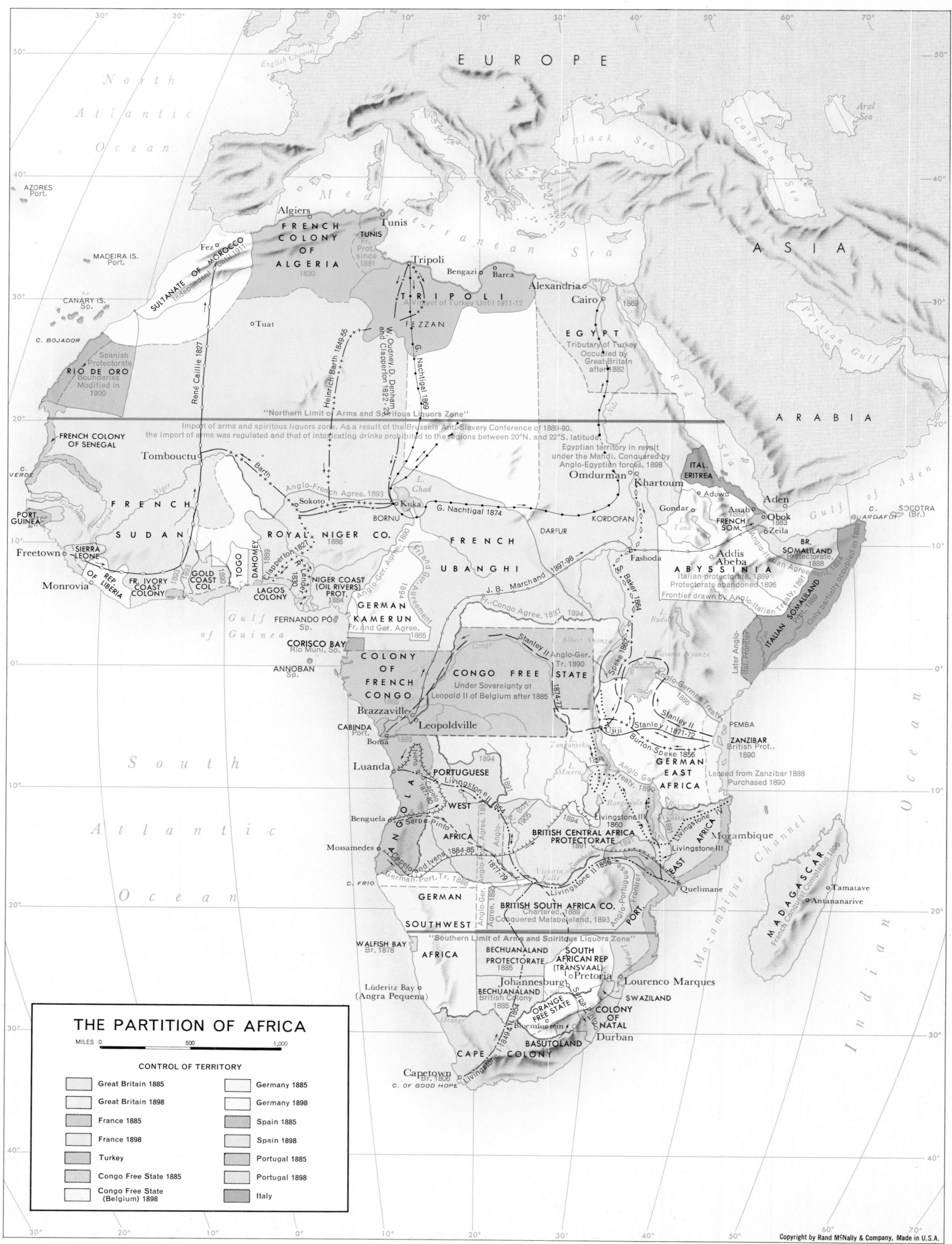

THE PARTITION OF AFRICA

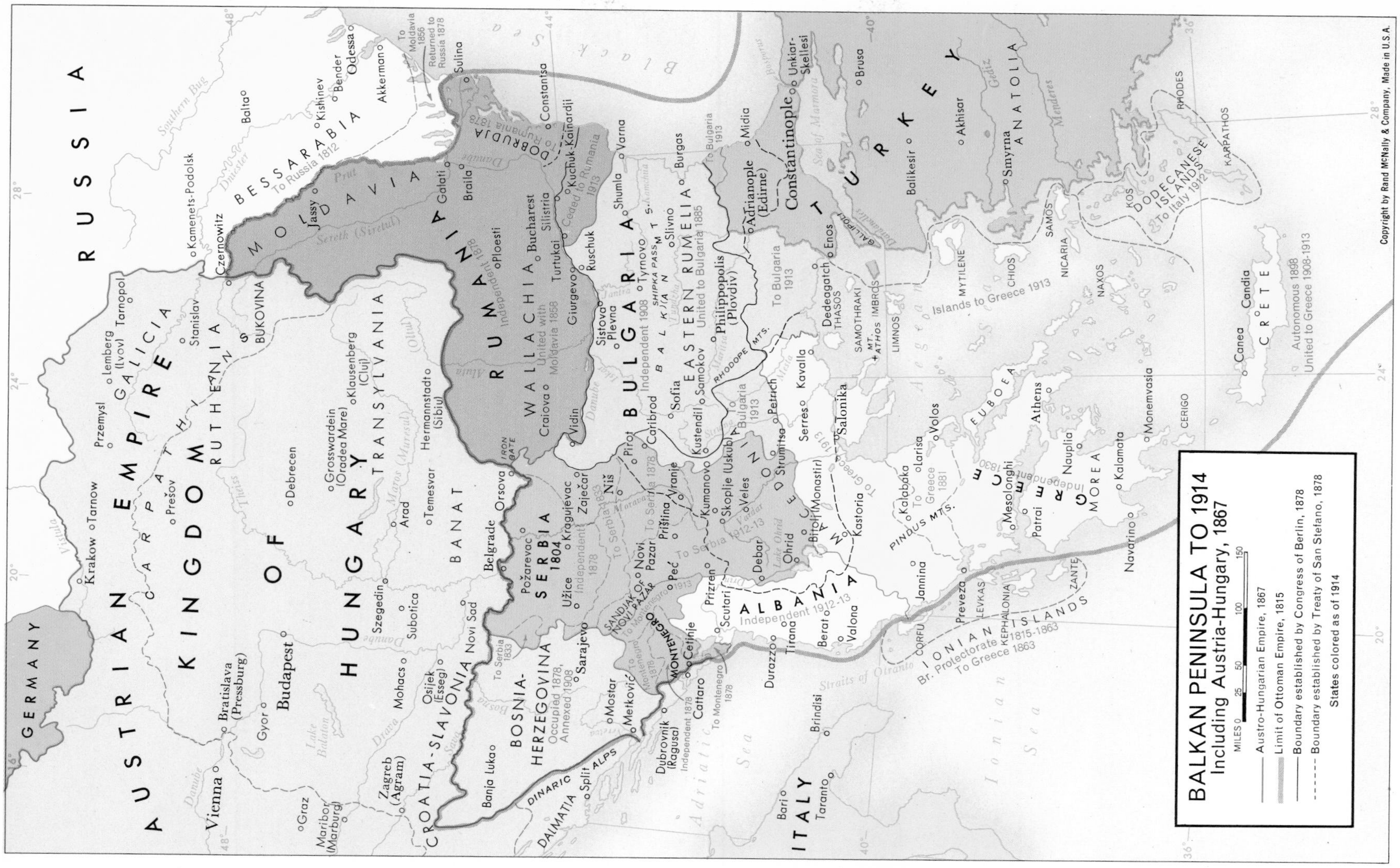
BALKAN PENINSULA TO 1914
Including Austria-Hungary, 1867
MILES 0 25 50 100 150
Austro-Hungarian Empire, 1867
Limit of Ottoman Empire, 1815
Boundary established by Congress of Berlin, 1878
Boundary established by Treaty of San Stefano, 1878
States colored as of 1914
Copyright by Rand McNally & Company, Made in U.S.A.

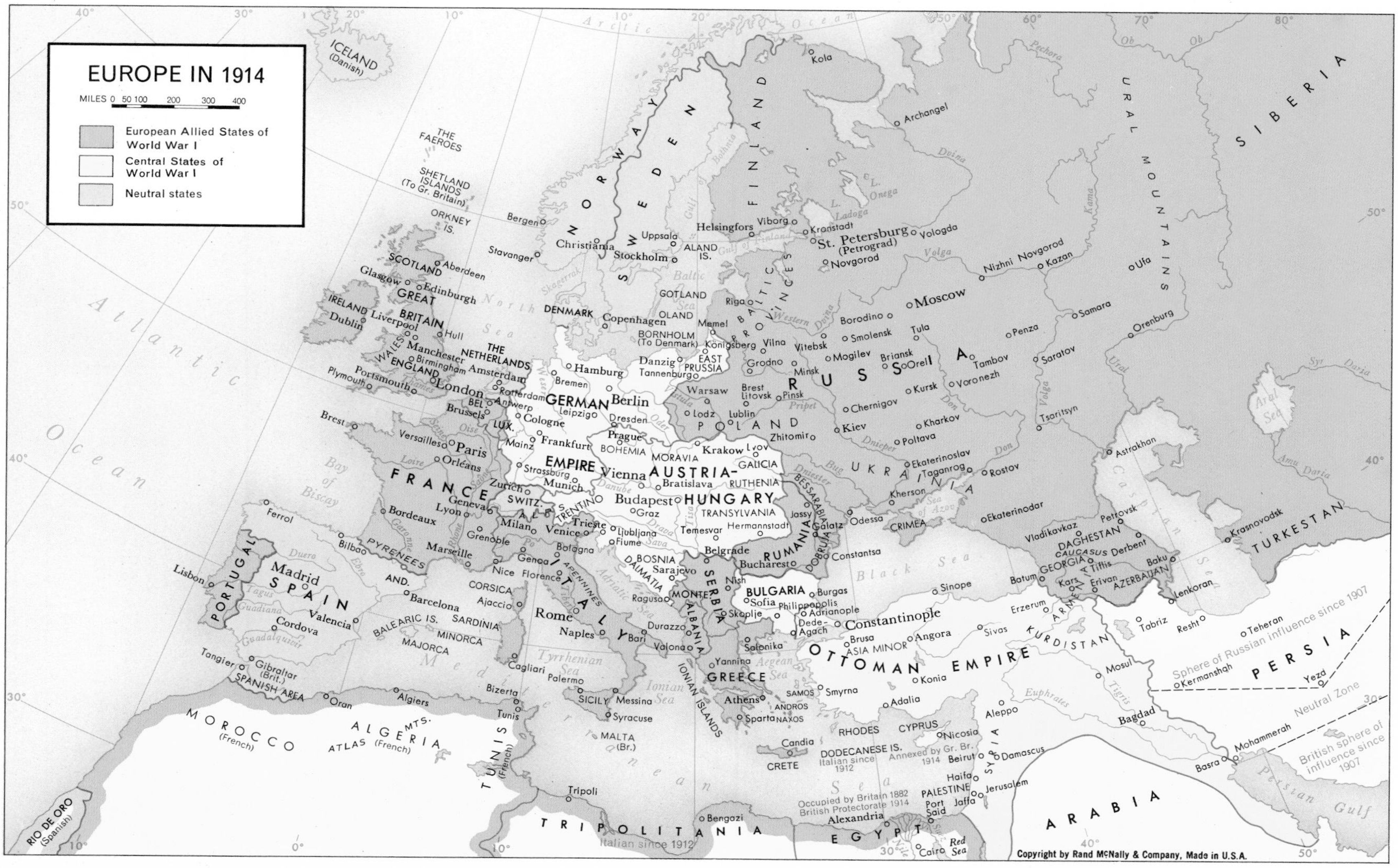
EUROPE IN 1914
MILES 0 50 100 200 300 400
European Allied States of World War I
Central States of World War I
Neutral states
Copyright by Rand McNally & Company, Made in U.S.A.

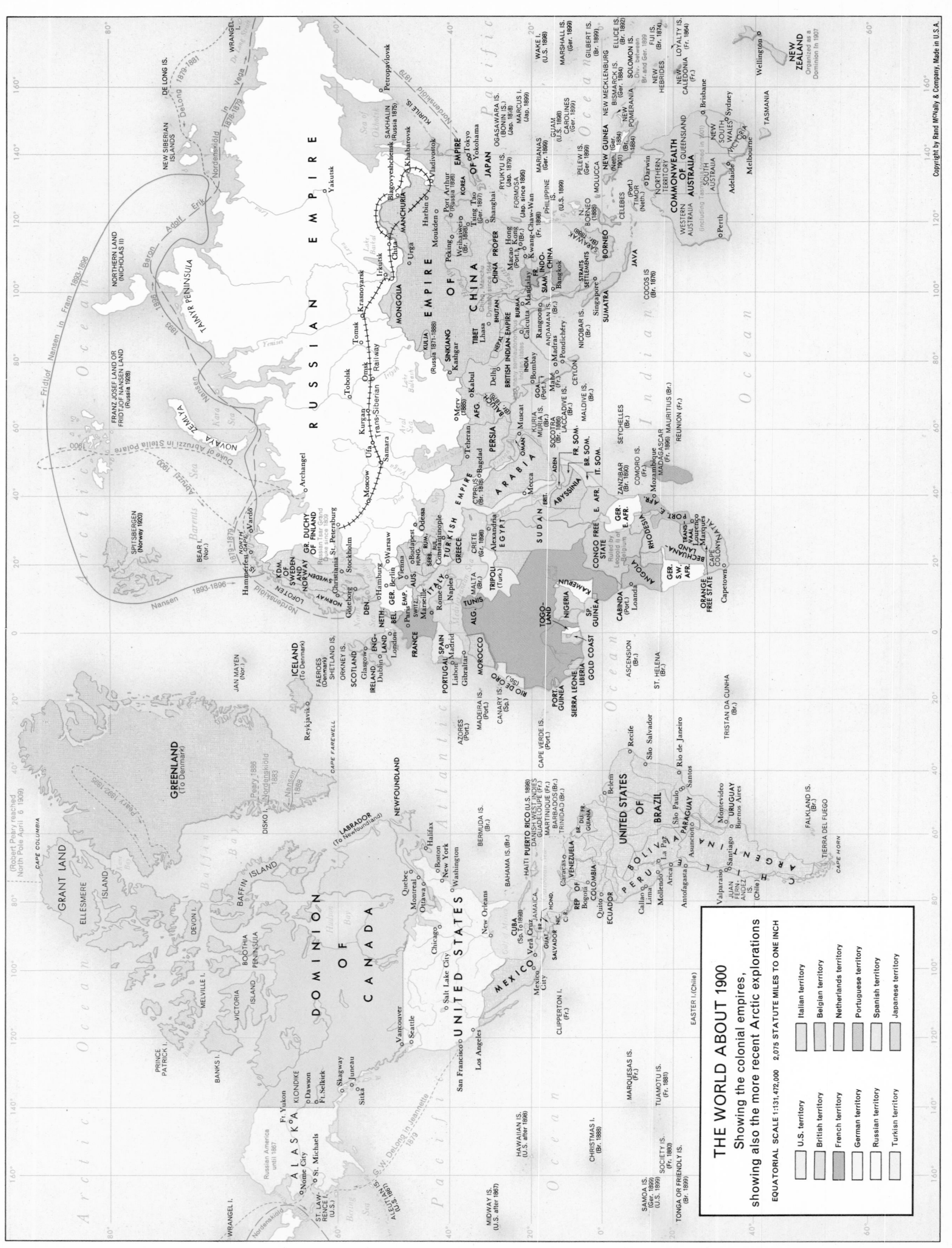

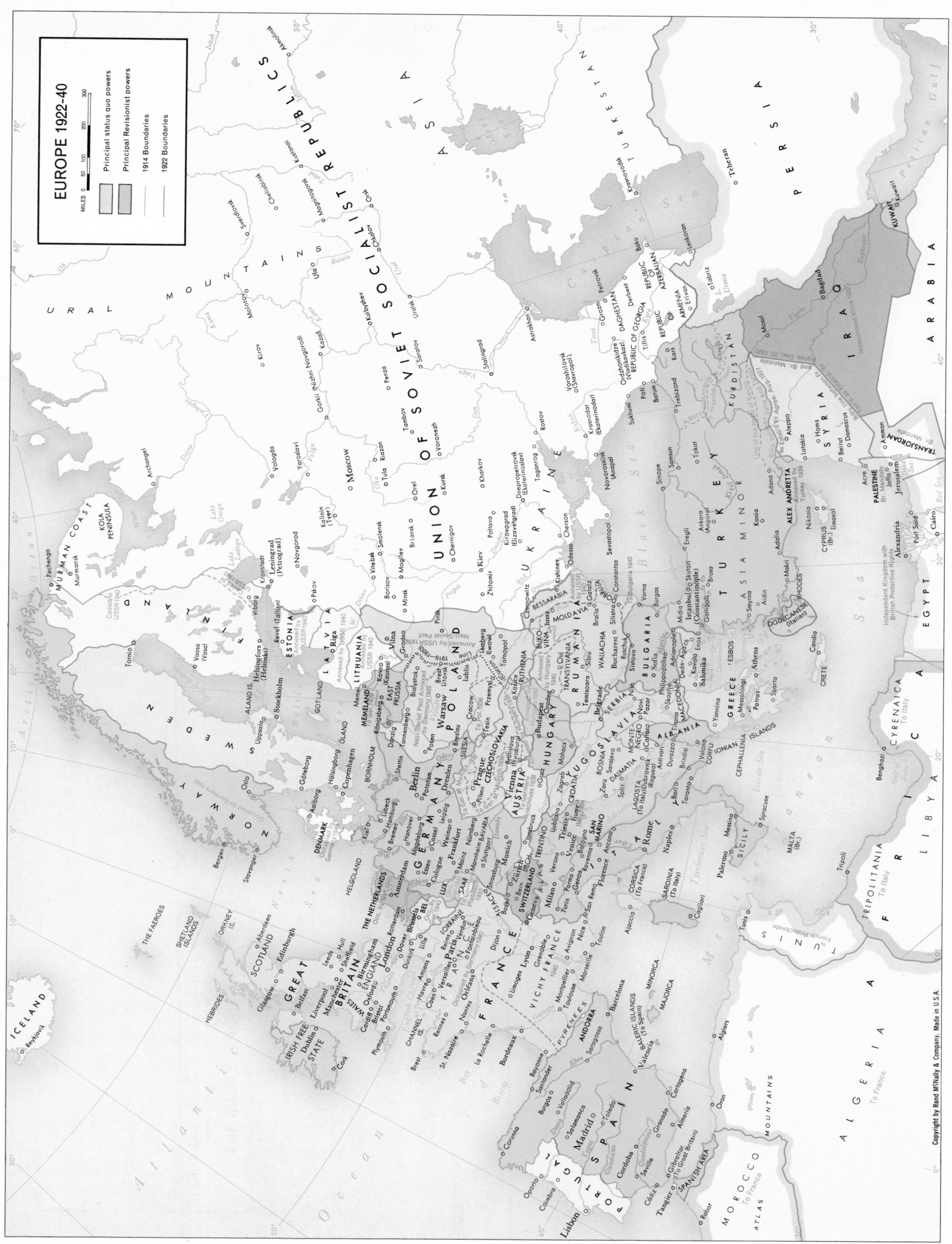

EUROPE 1922-40

Principal status quo powers
Principal Revisionist powers

1914 Boundaries
1922 Boundaries

MILES 0 50 100 200 300

Copyright by Rand McNally & Company. Made in U.S.A.

159

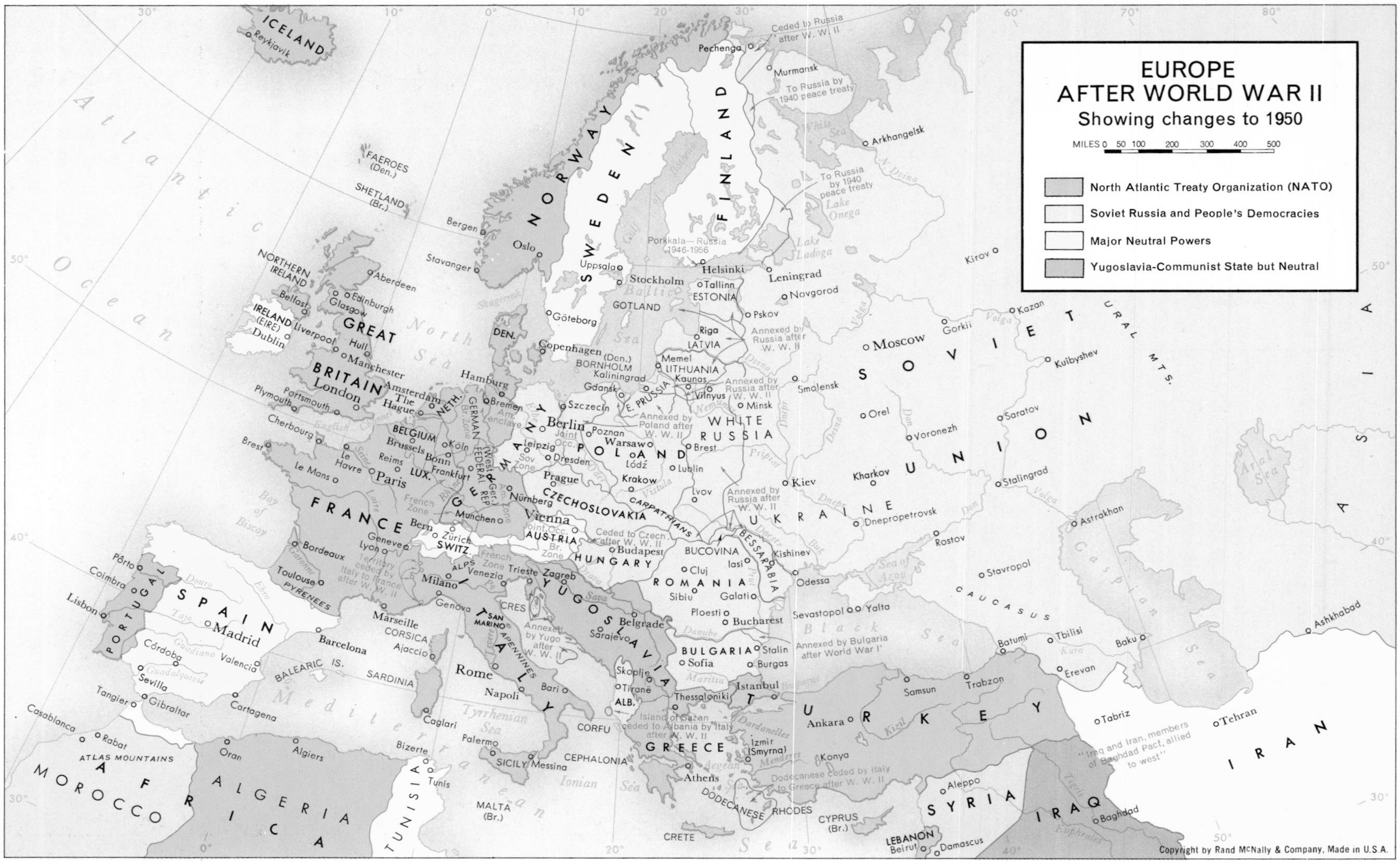
EUROPE
AFTER WORLD WAR II
Showing changes to 1950
MILES 0 50 100 200 300 400 500
North Atlantic Treaty Organization (NATO)
Soviet Russia and People's Democracies
Major Neutral Powers
Yugoslavia-Communist State but Neutral
ICELAND
Reykjavik
Atlantic Ocean
FAEROES (Den.)
SHETLAND (Br.)
NORWAY
Bergen
Oslo
Stavanger
SWEDEN
Uppsala
Stockholm
Göteborg
FINLAND
Helsinki
Pechenga
Ceded to Russia after W. W. II
Murmansk
To Russia by 1940 peace treaty
Arkhangelsk
Porkkala—Russia 1946-1956
Leningrad
Tallinn
ESTONIA
Novgorod
Pskov
Riga
LATVIA
LITHUANIA
Kaunas
Vilnyus
Memel
Kaliningrad
E. PRUSSIA
Annexed by Russia after W. W. II
Minsk
WHITE RUSSIA
Smolensk
Moscow
Gorkii
Kazan
Kirov
Kuibyshev
URAL MTS.
SOVIET UNION
Orel
Voronezh
Saratov
Stalingrad
Kharkov
Kiev
UKRAINE
Dnepropetrovsk
Rostov
Astrakhan
Caspian Sea
Aral Sea
CAUCASUS
Stavropol
Odessa
Sevastopol
Yalta
Black Sea
Tbilisi
Baku
Batumi
Erevan
Ashkhabad
NORTHERN IRELAND
Belfast
IRELAND (EIRE)
Dublin
Aberdeen
Edinburgh
Glasgow
GREAT BRITAIN
Liverpool
Hull
Manchester
London
Plymouth
Portsmouth
North Sea
DEN.
Copenhagen
BORNHOLM (Den.)
GOTLAND
Gdansk
Szczecin
Hamburg
Bremen
Amsterdam
The Hague
NETH.
BELGIUM
Brussels
Köln
Bonn
Frankfurt
LUX.
GERMAN FEDERAL REP.
GERMANY
Berlin
Leipzig
Dresden
Poznan
Warsaw
Lodz
Lublin
Brest
POLAND
Annexed by Poland after W. W. II
Prague
Krakow
Lvov
CZECHOSLOVAKIA
CARPATHIANS
Nürnberg
München
Vienna
AUSTRIA
Budapest
HUNGARY
Ceded to Czech. after W. W. II
Cherbourg
Brest
Le Havre
Reims
Paris
Le Mans
FRANCE
Bay of Biscay
Bordeaux
Lyon
Geneve
Bern
Zurich
SWITZ.
Toulouse
PYRENEES
Marseille
Ceded by Italy to France after W. W. II
Milano
Venezia
Trieste
Zagreb
Genova
ALPS
ITALY
SAN MARINO
CORSICA
Ajaccio
Rome
Napoli
Bari
SARDINIA
Cagliari
Tyrrhenian Sea
Palermo
SICILY
Messina
Annexed by Yugo. after W. W. II
YUGOSLAVIA
Belgrade
Sarajevo
Skoplje
ALB.
Tirane
CORFU
CEPHALONIA
Ionian Sea
BUCOVINA
Cluj
Iasi
Kishinev
BESSARABIA
ROMANIA
Sibiu
Galati
Ploesti
Bucharest
BULGARIA
Sofia
Stalin
Burgas
Annexed by Bulgaria after World War I
Thessaloniki
Istanbul
GREECE
Athens
Izmir (Smyrna)
Island of Sazan ceded to Albania by Italy after W. W. II
Dodecanese ceded by Italy to Greece after W. W. II
DODECANESE
RHODES
CRETE
TURKEY
Ankara
Samsun
Trabzon
Konya
Aleppo
SYRIA
CYPRUS (Br.)
LEBANON
Beirut
Damascus
IRAQ
Baghdad
Tabriz
Tehran
IRAN
"Iraq and Iran, members of Baghdad Pact, allied to west"
Pôrto
Coimbra
Lisbon
PORTUGAL
SPAIN
Madrid
Valencia
Barcelona
Córdoba
Sevilla
Gibraltar
Cartagena
BALEARIC IS.
Tangier
Casablanca
Rabat
ATLAS MOUNTAINS
MOROCCO
AFRICA
Oran
Algiers
ALGERIA
Bizerte
Tunis
TUNISIA
MALTA (Br.)
Mediterranean Sea
Copyright by Rand McNally & Company, Made in U.S.A.

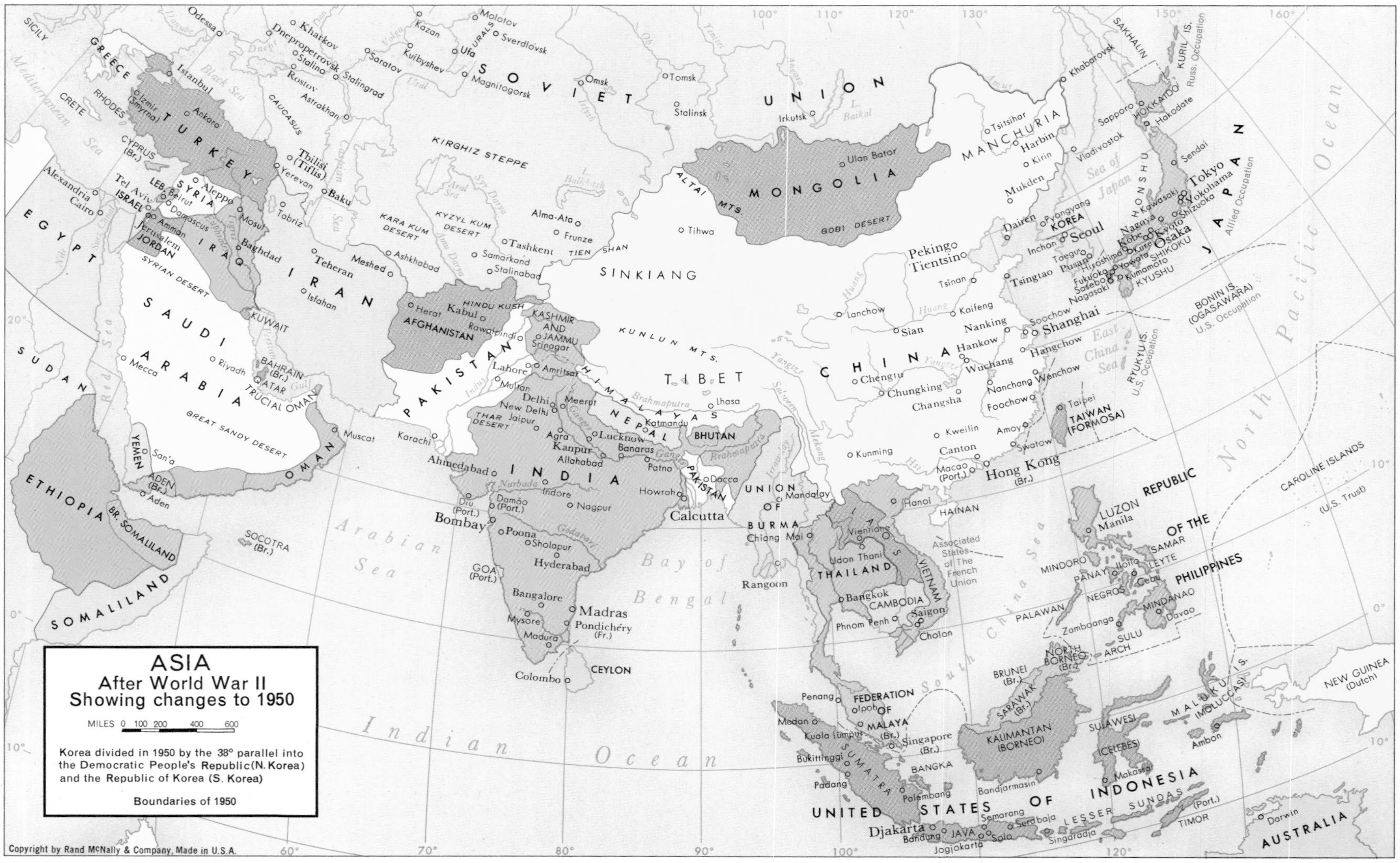
ASIA
After World War II
Showing changes to 1950
MILES 0 100 200 400 600
Korea divided in 1950 by the 38° parallel into the Democratic People's Republic (N. Korea) and the Republic of Korea (S. Korea)
Boundaries of 1950
SOVIET UNION
Odessa
Kharkov
Dnepropetrovsk
Stalino
Rostov
Astrakhan
Stalingrad
Saratov
Kazan
Kuibyshev
Molotov
Sverdlovsk
Ufa
Magnitogorsk
Omsk
Tomsk
Stalinsk
Irkutsk
Khabarovsk
SAKHALIN
KURIL IS. Russ. Occupation
KIRGHIZ STEPPE
GREECE
CRETE
RHODES
Istanbul
Ankara
Izmir (Smyrna)
TURKEY
CYPRUS (Br.)
Tbilisi (Tiflis)
Yerevan
Baku
Tabriz
Teheran
Meshed
Ashkhabad
IRAN
SYRIA
Aleppo
LEB.
Tel Aviv
ISRAEL
Damascus
Amman
Jerusalem
JORDAN
Mosul
Baghdad
IRAQ
Alexandria
Cairo
EGYPT
SYRIAN DESERT
KUWAIT
Isfahan
SAUDI ARABIA
Riyadh
Mecca
BAHRAIN (Br.)
QATAR
TRUCIAL OMAN
OMAN
Muscat
GREAT SANDY DESERT
YEMEN
San'a
ADEN (Br.)
Aden
SUDAN
ETHIOPIA
BR. SOMALILAND
SOMALILAND
SOCOTRA (Br.)
Arabian Sea
KARA KUM DESERT
KYZYL KUM DESERT
Alma-Ata
Frunze
Tashkent
Samarkand
Stalinabad
TIEN SHAN
SINKIANG
Herat
Kabul
HINDU KUSH
AFGHANISTAN
KASHMIR AND JAMMU
Srinagar
Rawalpindi
Lahore
Amritsar
Multan
PAKISTAN
Karachi
THAR DESERT
Delhi
New Delhi
Jaipur
Meerut
Agra
Lucknow
Kanpur
Allahabad
Banaras
Patna
HIMALAYAS
NEPAL
Katmandu
BHUTAN
KUNLUN MTS.
TIBET
Lhasa
Ahmedabad
Diu (Port.)
Daman (Port.)
Indore
Nagpur
INDIA
Bombay
Poona
Sholapur
Hyderabad
GOA (Port.)
Bangalore
Mysore
Madras
Pondichéry (Fr.)
Madura
Colombo
CEYLON
Calcutta
Howrah
Dacca
PAKISTAN
Bay of Bengal
UNION OF BURMA
Mandalay
Chiang Mai
Rangoon
MONGOLIA
Ulan Bator
ALTAI MTS.
GOBI DESERT
Tihwa
MANCHURIA
Tsitsihar
Harbin
Kirin
Mukden
Vladivostok
Dairen
Pyongyang
KOREA
Seoul
Inchon
Taegu
Pusan
Peking
Tientsin
Tsinan
Tsingtao
Lanchow
Sian
Kaifeng
Nanking
Hankow
Wuchang
Shanghai
Soochow
Hangchow
Wenchow
Chengtu
Chungking
Changsha
Nanchang
Foochow
CHINA
Kweilin
Amoy
Canton
Swatow
Kunming
Macao (Port.)
Hong Kong (Br.)
TAIPEI
TAIWAN (FORMOSA)
East China Sea
Sea of Japan
JAPAN
HOKKAIDO
Sapporo
Hakodate
Sendai
HONSHU
Tokyo
Yokohama
Shizuoka
Nagoya
Kyoto
Osaka
Kobe
Kawasaki
SHIKOKU
KYUSHU
Hiroshima
Fukuoka
Nagasaki
Kumamoto
Allied Occupation
RYUKYU IS. U.S. Occupation
BONIN IS. (OGASAWARA) U.S. Occupation
North Pacific Ocean
HAINAN
Hanoi
Vientiane
LAOS
Udon Thani
THAILAND
VIETNAM
Associated States of The French Union
Bangkok
CAMBODIA
Phnom Penh
Saigon
Cholon
South China Sea
REPUBLIC OF THE PHILIPPINES
LUZON
Manila
MINDORO
SAMAR
LEYTE
PANAY
NEGROS
Cebu
Iloilo
PALAWAN
MINDANAO
Zamboanga
Davao
SULU ARCH.
CAROLINE ISLANDS (U.S. Trust)
NORTH BORNEO (Br.)
BRUNEI (Br.)
SARAWAK (Br.)
KALIMANTAN (BORNEO)
Penang
Ipoh
FEDERATION OF MALAYA (Br.)
Medan
Kuala Lumpur
Singapore (Br.)
Bukittinggi
Padang
SUMATRA
BANGKA
Palembang
Bandjarmasin
SULAWESI (CELEBES)
Makasar
MALUKU (MOLUCCAS)
Ambon
NEW GUINEA (Dutch)
UNITED STATES OF INDONESIA
Djakarta
Bandung
JAVA
Semarang
Surabaja
Jogjakarta
Solo
Singaradja
LESSER SUNDAS
TIMOR (Port.)
Darwin
AUSTRALIA
Indian Ocean
Copyright by Rand McNally & Company, Made in U.S.A.

Geographical Facts, Figures, and Information about the World and the United States

In the seventy-five pages which follow, the editors of the atlas have provided factual information of geographic interest on the world, the continents, individual foreign countries, and the fifty United States. Presented mainly in tabular form, these pages are designed to supplement the maps with data not readily available from the maps themselves. Here will be found answers to many of the questions raised by atlas users, particularly questions that ask "how large?" "how many?" and "when?"

The first of the tables, and in many ways the most useful of all, is the World Political Information Table. For each political unit listed, this table specifies the 1962 estimated population, area in square miles, population density, capital, largest city, and principal languages. In addition, the table states the precise political or administrative status of the units listed, and classifies them into major types.

Under the heading of World Facts and Comparisons appear the answers to many frequently asked questions. Here are the basic facts about the earth's movements and measurements, as well as information on the physiographic and temperature extremes found in each of the continents. The population growth of the continents is summarized for the period since the year 1650, and the countries with the largest areas and populations are listed.

Following the World Facts and Comparisons come listings of the major physical features of the world, including mountains, oceans and seas, lakes, rivers, and islands. Each list includes the outstanding features in each category and provides a ready answer to questions about which of two mountains is higher, which of two rivers is longer, and many similar queries.

Next are two tables giving current population figures for the world's major metropolitan areas and cities. The data for metropolitan areas have been especially prepared by Rand McNally to make possible accurate comparisons of size among the world's great urban centers.

Atlas users frequently encounter a geographical name in their reading that is of historical significance but does not appear on most contemporary maps. The Historical Gazetteer identifies a large number of these places and references, and gives modern names or descriptions to aid the user in locating the feature on today's maps.

Many atlas users will find the list of Principal Discoveries and Explorations among the most fascinating in the entire series. Here are traced the contributions of the long procession of explorers, traders, and soldiers whose voyages successively opened up to civilized man the continents and oceans familiar to us today. That the Age of Discovery has not entirely passed is confirmed by the listing of several recent events in the exploration of Antarctica.

The summary of discoveries is followed by a convenient pair of tables giving air and steamship distances between important world cities and ports. The group of tables relating to the world as a whole closes with a listing of major world cities and their altitudes.

The series of tables relating primarily to the United States begins with a summary of geographical and historical facts about the nation, including extremes of elevation, distance, temperature, and rainfall. The historical growth of the country's territory is charted; so is the course of settlement, as reflected in the steady westward migration of the center of population. The land and water areas and 1960, 1950, and 1940 populations of the fifty States are conveniently summarized, and each State's rankings in area and population are included.

Concise sketches of the topography, climate, and principal products of each of the States comprise the State Climatic and Economic Table, condensing a great volume of information into convenient and readily utilized form. The special characteristics of the different regions of the country are shown in another way in the list of National Parks, Monuments, and Historical Sites, accompanied by a convenient map. This list and map will, of course, be of special interest to tourists, vacation planners, and armchair travelers.

The Railroad Map of the United States serves to present a bird's-eye view of the entire railroad network, shown in detail on the individual color maps of the States. Supplementing the map is a table of U.S. railroad distances, and immediately following, a table of air distances between principal American cities.

The U.S. State General Information Table is another that will provide the answers to many questions of atlas users. Besides the capital, largest city, and official State bird, flower, and nickname, the table gives the extremes of length and width, the highest point, and information on the date of entry into the Union. The next tables trace the growth of the population of the nation and the individual States from the small Colonial communities of 1650 down to the present.

Next, U.S. metropolitan areas of 100,000 or more are listed ranked according to 1962 population, specifying both city and suburban population. The table following provides a detailed breakdown of the more than 100,000 named localities in the United States by population size groups, for the nation as a whole and for each State.

Armchair travelers may find the summary of Places of Interest in the United States the most interesting section of the atlas after the color maps themselves. Here the historical and scenic highlights of each State are presented in convenient alphabetical form, and supplemented with a number of illustrations.

The detailed listing of principal Universities and Colleges will prove useful to many, including students presently in college and prospective college students and their parents.

Closing this section and introducing the detailed index to the maps are a Glossary of Foreign Geographical Terms and a Glossary of Map Terminology. The first will aid the atlas user in understanding and translating the foreign geographical terms found on the maps and in the index. The Glossary of Map Terminology offers convenient definitions and descriptions of the more frequently found geographical and map terms.

WORLD POLITICAL INFORMATION TABLE

This table lists all countries and dependencies in the world, U.S. States, Canadian provinces, and other important regions and political subdivisions. Besides specifying the form of government for all political areas, the table classifies them into six groups according to their political status. Units labeled **A** are independent sovereign nations. (Several of these are designated as members of the British Commonwealth of Nations.) Units labeled **B** are independent as regards internal affairs, but for purposes of foreign affairs they are under the protection of another country. Areas under military government are also labeled **B**. Units labeled **C** are colonies, overseas territories, dependencies, etc., of other countries. Together the **A**, **B**, and **C** areas comprise practically the entire inhabited area of the world. The areas labeled **D** are physically separate units, such as groups of islands, which are *not* separate countries, but form part of a nation or dependency. Units labeled **E** are States, provinces, Soviet Republics, or similar major administrative subdivisions of important countries. Units in the table with no letter designation are regions or other areas that do not constitute separate political units by themselves.

Region or Political Division	Area in sq. miles	Estimated Population 1/1/1962	Pop. per sq. mi.	Form of Government and Ruling Power	Capital; Largest City (unless same)	Predominant Languages	Map Index Key	Map Page
Aden [Protectorate] (incl. former Aden Colony)	111,100	932,000	8.4	Two Protectorates (U.K.)...........C	Aden; Al Mukallā	Arabic, English	C6	47
Afghanistan†	251,000	14,000,000	56	Monarchy...........A	Kabul	Pushtu (Afghan), Persian	E11	41
Africa	11,685,000	261,300,000	22		........; Cairo	....	.....	42
Alabama	51,609	3,303,000	64	State (U.S.)...........E	Montgomery; Birmingham	....	.....	78
Alaska	586,400	239,000	0.4	State (U.S.)...........E	Juneau; Anchorage	English, Indian, Eskimo	.....	79
Albania†	11,099	1,680,000	151	People's Republic...........A	Tiranë	Albanian	B3	23
Alberta	255,285	1,355,000	5.3	Province (Canada)...........E	Edmonton	English	.....	69
Algeria†	917,537	11,150,000	12	Republic...........A	Algiers (Alger)	Arabic, French	B5	44
American Samoa	76	21,000	276	Unincorporated Territory (U.S.)...........C	Pago Pago	Polynesian, English	G11	7
Andaman & Nicobar Is.	3,215	66,000	21	Territory of India...........D	Port Blair	Andaman, Nicobar Malay	F9	39
Andorra	175	9,000	51	Principality...........A	Andorra	Catalan	A6	20
Angola	481,351	4,900,000	10	Overseas Province (Portugal)...........C	Luanda	Bantu languages	D2	48
Antarctica	5,100,000	....	.....		....	....	.....	5
Antigua (incl. Barbuda)	171	56,000	327	Colony (U.K.)...........C	St. John's	English	n16	64
Arabian Peninsula	933,231	13,241,000	14		........; Mecca	Arabic	G7	33
Argentina†	1,072,070	[20,500,000	19	Federal Republic...........A	Buenos Aires	Spanish	.....	54, 55
Arizona	113,909	1,412,000	12	State (U.S.)...........E	Phoenix	....	.....	80
Arkansas	53,104	1,791,000	34	State (U.S.)...........E	Little Rock	....	.....	81
Armenia (S.S.R.)	11,500	1,960,000	170	Soviet Socialist Republic (Sov. Un.)...........E	Yerevan	Armenian	E7	28
Aruba	68	60,000	882	Division of Netherlands Antilles (Neth.)...........D	Oranjestad	Dutch, Spanish, English, Papiamento	A4	60
Ascension I.	34	500	15	Dependency of St. Helena (U.K.)...........D	Georgetown	English	G9	6
Asia	17,085,000	1,792,000,000	105		........; Tōkyō	....	.....	33
Australia†	2,971,081	10,740,000	3.6	Monarchy (Federal) (Br. Commonwealth of Nations)...........A	Canberra; Sydney	English	.....	50
Australian Capital Territory	939	62,000	66	Federal Territory (Australia)...........E	Canberra	see Australia	G7	51
Austria†	32,374	7,090,000	219	Federal Republic...........A	Vienna (Wien)	German	E6	16
Azerbaidzhan (S.S.R.)	33,450	4,120,000	123	Soviet Socialist Republic (Sov. Un.)...........E	Baku	Turkic languages, Russian, Armenian	E7	28
Azores Is.	894	339,000	379	Part of Portugal (3 Districts)...........D	........; Ponta Delgada	Portuguese	h8	44
Baden-Wurttemberg	13,803	7,865,000	570	State (Germany, West)...........E	Stuttgart	German	E4	17
Bahama Is.	4,375	108,000	25	Colony (U.K.)...........C	Nassau	English	B5	64
Bahrain	231	153,000	662	Sheikdom (U.K. protection)...........B	Manama	Arabic	H5	41
Balearic Is.	1,936	446,000	230	Part of Spain (Baleares Province)...........D	Palma de Mallorca	Catalan	C6	20
Baltic Republics	67,200	6,255,000	93	Soviet Union...........E	........; Riga	Lithuanian, Latvian, Estonian, Russian	C5	27
Barbados	166	240,000	1,446	Colony (U.K.)...........C	Bridgetown	English	p17	64
Basutoland	11,716	700,000	60	Territory (Protectorate) (U.K.)...........C	Maseru	Kaffir, other Bantu languages	C4	49
Bavaria (Bayern)	27,239	9,635,000	354	State (Germany, West)...........E	München (Munich)	German	D6	17
Bechuanaland	275,000	345,000	1.3	Protectorate (U.K.)...........C	Mafeking, S. Afr.; Kanye	Bechuana, other Bantu languages	B3	49
Belgium†	11,778	9,230,000	784	Monarchy...........A	Brussels (Bruxelles)	Flemish, French	D4	15
Benelux	25,754	21,257,000	825		Brussels (Bruxelles)	Dutch, Flemish, French, Luxembourgeois	.....	15
Berlin, West	186	2,175,000	11,693	State (Germany, West)...........E	Berlin (West)	German	F7	24
Bermuda	21	53,000	2,524	Colony (U.K.)...........C	Hamilton	English	E13	77
Bhutan	19,300	675,000	35	Monarchy (Indian protection)...........B	Thimbu and Paro	Tibetan dialects	C9	39
Bismarck Archipelago	20,415	175,000	8.6	Part of Australian Trust Ter. of New Guinea (3 Districts)...........D	........; Rabaul	Melanesian, Papuan	h12	50
Bolivia†	424,163	3,530,000	8.3	Republic...........A	Sucre; La Paz	Spanish, Quechua, Aymará, Guaraní	C2	55
Bonin Islands	40	200	5.0	U.S. Military Administration...........B	....	English	E8	7
Borneo, Indonesian (Kalimantan)	208,286	4,100,000	20	Part of Indonesia (4 Provinces)...........D	........; Bandjermasin	Bahasa Indonesia	E4	35
Brazil†	3,286,478	73,400,000	22	Federal Republic...........A	Brasília; Rio de Janeiro	Portuguese	.....	55-59
Bremen	156	718,000	4,603	State (Germany, West)...........E	Bremen	German	E2	24
British Columbia	366,255	1,651,000	4.5	Province (Canada)...........E	Victoria; Vancouver	English	.....	68
British Commonwealth of Nations	11,179,061	737,974,000	66		London	....	.....	3
British Guiana	83,000	592,000	7.1	Colony (U.K.)...........C	Georgetown	English	B3	59
British Honduras	8,866	93,000	10	Colony (U.K.)...........C	Belize	English, Spanish, Indian languages	B3	62
Brunei	2,226	90,000	40	Part of Malaysia...........E	........; Brunei	Malay-Polynesian languages	E4	35
Bulgaria†	42,729	7,975,000	187	People's Republic...........A	Sofia (Sofiya)	Bulgarian	D7	22
Burma†	261,789	21,650,000	83	Federal Republic...........A	Rangoon	Burmese, English	D10	39
Burundi (Urundi)†	10,747	2,300,000	214	Monarchy...........A	Usumbura	Bantu and Hamitic languages	B4	48
Byelorussia (S.S.R.)†	80,150	8,325,000	104	Soviet Socialist Republic (Sov. Un.)...........E	Minsk	Byelorussian, Polish	E7	27
California	158,693	16,821,000	106	State (U.S.)...........E	Sacramento; Los Angeles	....	.....	82
Cambodia†	66,606	5,100,000	77	Monarchy...........A	Phnom Penh	Cambodian (Khmer), French	F6	38
Cameroon†	183,569	4,150,000	23	Federal Republic...........A	Yaoundé; Douala	Native languages, French	D2	46
Canada†	3,851,809	18,460,000	4.8	Monarchy (Federal) (Br. Commonwealth of Nations)...........A	Ottawa; Montreal	English, French	.....	66, 67
Canal Zone	553	42,000	76	Under U.S. Jurisdiction...........C	Balboa Heights; Rainbow City	Spanish, English	k11	62
Canary Is.	2,808	960,000	342	Part of Spain (2 Provinces)...........D	........; Las Palmas	Spanish	D1	44
Canton & Enderbury	27	300	11	U.K.-U.S. Administration...........C	Canton Island	Malay-Polynesian languages, English	.....	.....
Cape of Good Hope	277,543	5,460,000	20	Province (South Africa)...........E	Cape Town	see South Africa	D3	49
Cape Verde Is.	1,552	208,000	134	Overseas Province (Portugal)...........C	Praia	Portuguese	E3	42
Caroline Is.	457	54,000	118	Part of U.S. Pacific Is. Trust Ter. (4 Districts)...........D	....	Malay-Polynesian languages	F8	7
Cayman Is.	100	7,700	77	Colony (U.K.)...........C	Georgetown	English	F3	64
Celebes (Sulawesi)	72,987	6,620,000	91	Part of Indonesia (2 Provinces)...........D	........; Makassar	Malay-Polynesian languages	F6	35
Central African Republic†	238,200	1,275,000	5.4	Republic...........A	Bangui	Bantu languages, French	D4	46
Central America	200,412	12,444,000	62		........; Guatemala	Spanish, Indian languages	.....	62
Central Asia, Soviet	478,150	15,115,000	32	Soviet Union...........E	........; Tashkent	Uzbek, Russian, Kirghiz, Turkoman, Tadzhik	F9	28
Ceylon†	25,332	10,300,000	407	Commonwealth (Br. Commonwealth of Nations)...........A	Colombo	Sinhalese, Tamil, English	G7	39
Chad†	495,800	2,750,000	5.5	Republic...........A	Fort-Lamy	Hamitic languages, Arabic, French	B3	46
Channel Is. (Guernsey, Jersey, etc.)	75	109,000	1,453		........; St. Helier	English, French	F5	10

†*Member of the United Nations (1963).*　　*Not shown on map; index key denotes approximate location.*

WORLD POLITICAL INFORMATION TABLE

Region or Political Division	Area in sq. miles	Estimated Population 1/1/1962	Pop. per sq. mi.	Form of Government and Ruling Power	Capital; Largest City (unless same)	Predominant Languages	Map Index Key	Map Page
Chile†	286,397	7,880,000	28	Republic … A	Santiago	Spanish	.....	54, 55
China (excl. Taiwan)	3,691,500	700,000,000	190	People's Republic … A	Peking (Peiching); Shanghai	Chinese, Mongolian, Turkish, Tungus	.....	34
China (Nationalist), see Taiwan	.....	.....	.....	.....	.....	.....	.....	....
Christmas I. (Indian Ocean)	55	3,100	56	External Territory (Australia) … C	.....	Chinese, Malay, English	G6	7
Christmas I. (Pacific Ocean)	222	400	1.8	Part of Gilbert & Ellice Is. (U.K.); also claimed by U.S.	.....	Malay-Polynesian languages, English	F12	7
Cocos (Keeling) Is.	5	600	120	External Territory (Australia) … C	.....	Malay, English	G3	2
Colombia†	439,513	14,625,000	33	Republic … A	Bogotá	Spanish	C3	60
Colorado	104,247	1,824,000	17	State (U.S.) … E	Denver	.....	.....	83
Commonwealth of Nations, see Br. Commonwealth of Nations	.....	.....	.....	.....	.....	.....	.....	....
Comoro Is.	838	187,000	223	Overseas Territory (France) … C	Dzaoudzi	Malagasy, French	G23	3
Congo (Rep. of Congo; Capital: Brazzaville)†	132,000	850,000	6.4	Republic … A	Brazzaville	Bantu languages, French	B2	48
Congo, The (Rep. of The Congo; Capital: Léopoldville)†	905,565	14,500,000	16	Republic … A	Léopoldville	Bantu languages, French	B3	48
Connecticut	5,009	2,629,000	525	State (U.S.) … E	Hartford	.....	.....	84
Cook Is.	90	18,000	200	Island Territory (New Zealand) … C	Avarua	Malay-Polynesian (Maori)	H12	7
Corsica	3,368	161,000	48	Part of France (Corse Department) … D	Ajaccio; Bastia	French, Italian	C2	21
Costa Rica†	19,600	1,254,000	64	Republic … A	San José	Spanish	E5	62
Crete	3,219	484,000	150	Part of Greece (4 Prefectures) … D	........; Iraklion	Greek	E5	23
Cuba†	44,217	7,000,000	158	Republic … A	Havana (Habana)	Spanish	.....	64
Curaçao	174	127,000	730	Division of Netherlands Antilles (Neth.) … D	Willemstad	Dutch, Spanish, English, Papiamento	A4	60
Cyprus†	3,572	586,000	164	Republic (Br. Commonwealth of Nations) … A	Nicosia	Greek, Turkish, English	E9	31
Czechoslovakia†	49,367	13,840,000	280	People's Republic … A	Prague (Praha)	Czech, Slovak	D4	26
Dahomey†	44,696	2,060,000	46	Republic … A	Porto Novo	Native languages, French	E5	45
Delaware	2,057	467,000	227	State (U.S.) … E	Dover; Wilmington	.....	.....	85
Denmark†	16,619	4,635,000	279	Monarchy … A	Copenhagen (København)	Danish	C3	24
Denmark and Possessions	857,159	4,702,000	5.5	.....	Copenhagen (København)	Danish, Faeroese, Greenlandic	.....	3
District of Columbia	69	755,000	10,942	District (U.S.) … E	Washington	.....	C1	85
Dominica	305	61,000	200	Colony (U.K.) … C	Roseau	English, French	o16	64
Dominican Republic†	18,704	3,150,000	168	Republic … A	Santo Domingo	Spanish	F8	64
Ecuador†	104,506	4,500,000	43	Republic … A	Quito; Guayaquil	Spanish, Quechua	B2	58
Egypt, see United Arab Republic	.....	.....	.....	.....	.....	.....	.....	....
El Salvador†	8,260	2,550,000	309	Republic … A	San Salvador	Spanish	D3	62
England (excl. Monmouthshire)	50,327	43,605,000	866	United Kingdom …	......; London	English	D6	10
England & Wales	58,344	46,250,000	793	Administrative division of United Kingdom … E	London	English, Welsh	.....	12
Estonia (S.S.R.)	17,400	1,235,000	71	Soviet Socialist Republic (Sov. Un.) … E	Tallinn	Estonian, Russian	B5	27
Ethiopia (incl. Eritrea)†	457,267	20,000,000	44	Monarchy … A	Addis Ababa	Amharic and other Semitic languages, English, various Hamitic languages	D4	47
Eurasia	20,910,000	2,375,000,000	114	.....	........; Tōkyō	.....	.....	33
Europe	3,825,000	583,600,000	153	.....	........; London	.....	.....	8, 9
Faeroe Is.	540	35,000	65	Self-Governing Territory (Denmark) … C	Thorshavn	Danish, Faeroese	C7	8
Falkland Is. (excl. Deps.)	4,618	2,200	0.5	Colony (U.K.) … C	Port Stanley	English	I5	53
Fernando Poo	785	65,000	83	African Province (Spain) … C	Santa Isabel	Bantu languages, Spanish	E1	46
Fiji	7,040	415,000	59	Colony (U.K.) … C	Suva	Malay-Polynesian languages, English, Hindi	H11	7
Finland†	130,119	4,517,000	35	Republic … A	Helsinki	Finnish, Swedish	F11	25
Florida	58,560	5,341,000	91	State (U.S.) … E	Tallahassee; Miami	.....	.....	86
France†	212,822	46,200,000	217	Republic … A	Paris	French	.....	14
France and Possessions	276,238	47,639,000	172	.....	Paris	.....	.....	3
Franklin	549,253	5,800	0.01	District of Northwest Territories, Canada … E	.......; Cambridge Bay	see Canada	B13	66
French Guiana	35,100	32,000	0.9	Overseas Department (France) … C	Cayenne	French	B4	59
French Polynesia	1,550	79,000	51	Overseas Territory (France) … C	Papeete	Malay-Polynesian languages, French	H13	7
French Somaliland	8,500	69,000	8.1	Overseas Territory (France) … C	Djibouti	Somali, French	C5	47
French Southern & Antarctic Ter. (excl. Adélie Coast)	2,917	100	0.03	Overseas Territory (France) … C	.....	French	J25	3
French West Indies	1,112	565,000	508	.....	......; Fort-de-France	French	o16	64
Gabon†	103,100	450,000	4.4	Republic … A	Libreville	Bantu languages, French	F2	46
Galápagos Is.	3,028	1,800	0.6	Province of Ecuador … D	Puerto Baquerizo	Spanish	g5	58
Gambia	4,008	315,000	79	Colony & Protectorate (U.K.) … C	Bathurst	Mandingo, Fula, English	D1	45
Georgia (S.S.R.)	26,900	4,280,000	159	Soviet Socialist Republic (Sov. Un.) … E	Tbilisi	Georgic, Armenian, Russian	E7	28
Georgia	58,876	4,025,000	68	State (U.S.) … E	Atlanta	.....	.....	87
Germany (Entire)	137,743	73,850,000	536	.....	........; Berlin	German	.....	16
Germany, East	41,815	17,170,000	411	People's Republic … A	Berlin (East)	German	.....	16
Germany, West (incl. West Berlin)	95,928	56,680,000	591	Federal Republic … A	Bonn; Berlin (West)	German	.....	16
Ghana†	91,843	7,050,000	77	Republic (Br. Commonwealth of Nations) … A	Accra	Twi, Fanti, Ewe-Fon, English	E4	45
Gibraltar	2	26,000	13,000	Colony (U.K.) … C	Gibraltar	Spanish, English	D3	20
Gilbert & Ellice Is.	369	47,000	127	Colony (U.K.) … C	Tarawa	Malay-Polynesian languages	G10	7
Goa, Damão, and Diu	1,615	650,000	402	Territory (India) … E	Panjim (Nova Goa)	Indo-Aryan languages, Dravidian languages	E5	39
Great Britain & Northern Ireland, see United Kingdom	.....	.....	.....	.....	.....	.....	.....	....
Greece†	50,547	8,435,000	167	Monarchy … A	Athens (Athinai)	Greek	C4	23
Greenland	840,000	32,000	0.04	Overseas Territory (Denmark) … C	Godthaab	Greenlandic, Danish, Eskimo	B19	4
Grenada	133	90,000	677	Colony (U.K.) … C	St. George's	English	p16	64
Guadeloupe (incl. Dependencies)	687	280,000	408	Overseas Department (France) … C	Basse-Terre; Pointe-à-Pitre	French	n16	64
Guam	212	70,000	330	Unincorporated Territory (U.S.) … C	Agana	English, Chamorro	.....	52
Guatemala†	42,042	3,930,000	93	Republic … A	Guatemala	Spanish, Indian languages	C2	62
Guernsey (incl. Dependencies)	30	47,000	1,567	Bailiwick (U.K.) … C	St. Peter Port	English, French	F5	10
Guinea†	94,925	3,000,000	32	Republic … A	Conakry	Native languages, French	D2	45
Haiti†	10,714	3,570,000	333	Republic … A	Port-au-Prince	Creole, French	F7	64
Hamburg	288	1,850,000	6,424	State (Germany, West) … E	Hamburg	German	E4	24
Hawaii	6,424	662,000	103	State (U.S.) … E	Honolulu	English, Japanese, Hawaiian	.....	88
Hesse (Hessen)	8,150	4,850,000	595	State (Germany, West) … E	Wiesbaden; Frankfurt am Main	German	C3	17
Hispaniola	29,530	6,720,000	228	.....	.......; Santo Domingo	French, Spanish	E8	64
Holland, see Netherlands	.....	.....	.....	.....	.....	.....	.....	....
Honduras†	43,277	1,925,000	44	Republic … A	Tegucigalpa	Spanish	C4	62
Hong Kong	398	3,225,000	8,103	Colony (U.K.) … C	Victoria	Chinese, English	G7	34
Hungary†	35,919	10,065,000	280	People's Republic … A	Budapest	Hungarian	B4	22
Iceland†	39,800	182,000	4.6	Republic … A	Reykjavík	Icelandic	n23	25
Idaho	83,557	682,000	8.2	State (U.S.) … E	Boise (Boise City)	.....	.....	89
Ifni	580	51,000	88	African Province (Spain) … C	Sidi Ifni	Spanish, Arabic	D2	44
Illinois	56,400	10,346,000	183	State (U.S.) … E	Springfield; Chicago	.....	.....	90
India† (incl. part of Kashmir)	1,227,275	443,400,000	361	Republic (Br. Commonwealth of Nations) … A	New Delhi; Calcutta	Hindi and other Indo-Aryan languages, Dravidian languages, English	D6	39
Indiana	36,291	4,746,000	131	State (U.S.) … E	Indianapolis	.....	.....	91
Indonesia†	574,670	95,800,000	167	Republic … A	Djakarta	Chinese, Bahasa Indonesia, English	.....	35
Iowa	56,290	2,785,000	49	State (U.S.) … E	Des Moines	.....	.....	92

†Member of the United Nations (1963). **Not shown on map; index key denotes approximate location.*

Region or Political Division	Area in sq. miles	Estimated Population 1/1/1962	Pop. per sq. mi.	Form of Government and Ruling Power	Capital; Largest City (unless same)	Predominant Languages	Map Index Key	Map Page
Iran (Persia)†	636,300	20,925,000	33	Monarchy...A	Tehrān	Persian, Turkish dialects, Kurdish	.....	41
Iraq†	171,599	7,350,000	43	Republic...A	Baghdad	Arabic, Kurdish	I17	9
Ireland†	27,136	2,805,000	103	Republic...A	Dublin	English, Irish	.....	11
Isle of Man	227	48,000	211	Possession (U.K.)...C	Douglas	English	C4	10
Israel†	7,815	2,232,000	286	Republic...A	Jerusalem; Tel Aviv	Hebrew, Arabic	.....	32
Italy†	116,303	49,800,000	428	Republic...A	Rome (Roma)	Italian	.....	21
Ivory Coast†	124,503	3,340,000	27	Republic...A	Abidjan	French, native languages	E3	45
Jamaica†	4,411	1,650,000	374	Self-Governing Member (Br. Commonwealth of Nations)...A	Kingston	English	F5	64
Japan†	142,726	94,500,000	662	Monarchy...A	Tōkyō	Japanese	.....	37
Java (Djawa) (incl. Madura)	51,040	63,100,000	1,236	Part of Indonesia (5 Provinces)...D	......; Djakarta	Bahasa Indonesia, English	G4	35
Jersey	45	62,000	1,378	Bailiwick (U.K.)...C	St. Helier	English, French	F5	10
Jordan†	37,301	1,800,000	48	Monarchy...A	Amman	Arabic	G11	31
Kansas	82,264	2,218,000	27	State (U.S.)...E	Topeka; Wichita		.....	93
Kashmir, Jammu &	86,024	4,600,000	53	In dispute (India & Pakistan)	Srinagar	Kashmiri, Punjabi	B6	39
Katanga	191,879	1,750,000	9.0	Province (Congo: Léopoldville)...E	Élizabethville	see Congo: Léopoldville	C4	48
Kazakh S.S.R.	1,050,000	10,900,000	10	Soviet Socialist Republic (Sov. Un.)...E	Alma-Ata	Turkic languages, Russian	E9	28
Keewatin	228,160	2,400	0.01	District of Northwest Territories, Canada...E	......; Chesterfield Inlet	see Canada	C14	66
Kentucky	40,395	3,074,000	76	State (U.S.)...E	Frankfort; Louisville		.....	94
Kenya	224,960	7,400,000	33	Self-Governing Colony & Protectorate (U.K.)...C	Nairobi	Swahili and other Bantu languages, English	A6	48
Kerguélen	2,700	100	0.03	Part of French Southern & Antarctic Ter. (Fr.)...D	....	French	J1	2
Kirghiz S.S.R.	76,650	2,305,000	30	Soviet Socialist Republic (Sov. Un.)...E	Frunze	Turkic languages, Persian	E10	28
Korea (Entire)	85,255	33,900,000	398		Seoul (Soul)	Korean	.....	37
Korea, North	47,255	8,200,000	174	People's Republic...A	Pyongyang	Korean	.....	37
Korea, South	38,000	25,700,000	676	Republic...A	Seoul (Soul)	Korean	.....	37
Kuwait†	6,000	232,000	39	Sheikdom...A	Kuwait	Arabic	J18	9
Labrador	112,826	14,000	0.1	Part of Newfoundland Province, Canada...D	......; Goose Bay	English, Eskimo	g8	75
Laos†	91,400	1,875,000	21	Monarchy...A	Vientiane	Lao, French	C5	38
Latin America	7,923,124	217,950,000	28		......; Buenos Aires		.....	3
Latvia (S.S.R.)	24,600	2,170,000	88	Soviet Socialist Republic (Sov. Un.)...E	Riga	Latvian, Russian	C5	27
Lebanon†	4,000	1,850,000	463	Republic...A	Beirut	Arabic, French, English	E10	31
Liberia†	43,000	1,300,000	30	Republic...A	Monrovia	Native languages, English	E2	45
Libya†	679,360	1,225,000	1.8	Monarchy...A	Tripoli and Bengasi; Tripoli	Arabic	B3	43
Liechtenstein	61	17,000	279	Principality...A	Vaduz	German	B5	18
Lithuania (S.S.R.)	25,150	2,850,000	113	Soviet Socialist Republic (Sov. Un.)...E	Vilnius	Lithuanian, Polish, Russian	D4	27
Louisiana	48,523	3,357,000	69	State (U.S.)...E	Baton Rouge; New Orleans		.....	95
Lower Saxony (Niedersachsen)	18,294	6,630,000	362	State (Germany, West)...E	Hannover (Hanover)	German	A4	17
Luxembourg†	998	317,000	318	Grand Duchy...E	Luxembourg	Luxembourgeois, French	E6	15
Macao	6	225,000	37,500	Overseas Province (Portugal)...C	Macao	Chinese, Portuguese	G7	34
Mackenzie	527,490	15,000	0.03	District of Northwest Territories, Canada...E	......; Yellowknife	see Canada	D10	66
Madeira Is.	308	284,000	922	Part of Portugal (Funchal District)...D	Funchal	Portuguese	h11	20
Maine	33,215	981,000	30	State (U.S.)...E	Augusta; Portland		.....	96
Malagasy Republic (Madagascar)†	227,800	5,625,000	25	Republic...A	Tananarive	French, Malagasy	h9	49
Malaya†	50,700	7,225,000	143	Part of Malaysia...E	......; Kuala Lumpur	Malay, Chinese, English	J4	38
Malaysia‡	130,788	10,310,000	79	Self-Governing Member (Br. Commonwealth of Nations)...A	Kuala Lumpur; Singapore	Malay, Chinese, English	.....	35
Maldive Is.	115	89,000	774	Sultanate (U.K. protection)...B	Malé	Arabic	G5	39
Mali†	464,874	4,150,000	8.9	Republic...A	Bamako	Native languages, French, Arabic	C4	45
Malta	122	330,000	2,705	Self-Governing Colony (U.K.)...C	Valletta	English, Maltese	G14	39
Manchuria	309,498	57,000,000	184	China	......; Mukden (Shenyang)	Chinese, Tungus, Mongolian	B10	34
Manitoba	251,000	930,000	3.7	Province (Canada)...E	Winnipeg	see Canada	.....	71
Mariana Is. (excl. Guam)	154	10,000	65	Part of U.S. Pacific Is. Trust Ter. (2 Districts)...D	....	Malay-Polynesian languages	E8	7
Maritime Provinces (excl. Newfoundland)	51,963	1,449,000	28	Canada	......; Halifax	see Canada	.....	74
Marshall Is.	61	16,000	262	District of U.S. Pacific Is. Trust Ter...D	Majuro	Malay-Polynesian languages	F10	7
Martinique	425	285,000	671	Overseas Department (France)...C	Fort-de-France	French	o16	64
Maryland	10,577	3,237,000	306	State (U.S.)...E	Annapolis; Baltimore		.....	86
Massachusetts	8,257	5,285,000	640	State (U.S.)...E	Boston		.....	97
Mauritania†	419,230	740,000	1.8	Republic...A	Nouakchott; Kaédi	Arabic, French	C2	45
Mauritius (incl. Dependencies)	808	685,000	848	Colony (U.K.)...C	Port Louis	Indo-Aryan languages, French, Creole	H24	3
Mexico†	761,602	36,650,000	48	Federal Republic...A	Mexico City	Spanish	.....	63
Michigan	58,216	7,943,000	136	State (U.S.)...E	Lansing; Detroit		.....	98
Middle America	1,053,124	69,350,000	66		......; Mexico City		H11	61
Midway Is.	2	3,000	1,500	Possession (U.S.)...C		English	E11	7
Minnesota	84,068	3,498,000	42	State (U.S.)...E	St. Paul; Minneapolis		.....	99
Mississippi	47,716	2,200,000	46	State (U.S.)...E	Jackson		.....	100
Missouri	69,686	4,387,000	63	State (U.S.)...E	Jefferson City; St. Louis		.....	101
Moldavia (S.S.R.)	13,000	3,120,000	240	Soviet Socialist Republic (Sov. Un.)...E	Kishinev	Moldavian, Russian, Ukrainian	H7	27
Monaco	0.8	23,000	28,750	Principality...A	Monaco	French, Italian	F7	14
Mongolia†	592,700	950,000	1.6	People's Republic...A	Ulan Bator	Mongolian	B5	34
Montana	147,138	687,000	4.7	State (U.S.)...E	Helena; Great Falls		.....	102
Montserrat	32	12,000	375	Colony (U.K.)...C	Plymouth	English	n15	64
Morocco†	171,305	11,950,000	70	Monarchy...A	Rabat; Casablanca	Arabic, Berber, French	C3	44
Mozambique	297,846	6,500,000	22	Overseas Province (Portugal)...C	Lourenço Marques	Bantu languages, Portuguese	B5	49
Muscat & Oman	82,000	575,000	7.0	Sultanate...A	Muscat; Maṭraḥ	Arabic	G8	33
Natal	35,284	3,010,000	85	Province (South Africa)...E	Pietermaritzburg; Durban	English, Afrikaans, Bantu languages	C5	49
Nauru	8	4,600	575	Trust Territory (Austl.-U.K.-N.Z.)...C	....	Malay-Polynesian languages, Chinese, English	G10	7
Nebraska	77,227	1,429,000	19	State (U.S.)...E	Lincoln; Omaha		.....	103
Nepal†	54,362	9,480,000	174	Monarchy...A	Katmandu	Nepali, Tibeto-Burman languages	D10	40
Netherlands†	12,978	11,710,000	902	Monarchy...A	The Hague (s' Gravenhage) and Amsterdam; Amsterdam	Dutch	.....	15
Netherlands and Possessions	68,492	12,231,000	179		The Hague and Amsterdam; Amsterdam	Dutch	.....	3
Netherlands Antilles	371	197,000	531	Self-Governing Territory (Netherlands)...C	Willemstad	Dutch, Spanish, English, Papiamento	A4	60
Netherlands Guiana, see Surinam	....	....	.....	....			...	...
Netherlands New Guinea, see West Irian	....	....	.....	....			...	...
Nevada	110,540	306,000	2.8	State (U.S.)...E	Carson City; Las Vegas		.....	104
New Brunswick	28,354	603,000	21	Province (Canada)...E	Fredericton; Saint John	see Canada	H9	74
New Caledonia (incl. Deps.)	6,531	80,000	12.2	Overseas Territory (France)...C	Nouméa	Malay-Polynesian languages, French	B13	7
New England	66,608	10,792,000	162	United States	......; Boston	English	.....	77
Newfoundland	156,185	463,000	3.0	Province (Canada)...E	St. John's	see Canada	.....	75
Newfoundland (excl. Labrador)	43,359	449,000	10		......; St. John's	see Canada	k26	67
New Guinea, North-East	69,695	1,220,000	18	Part of Australian Trust Ter. of New Guinea (3 Districts)...D	......; Lae	Papuan and Negrito languages	k12	50
New Guinea, Ter. of	94,430	1,450,000	15	Trust Territory (Austl.; administered from Papua)...C	Port Moresby, Papua; Rabaul	Papuan and Negrito languages, English	k12	50
New Hampshire	9,304	622,000	67	State (U.S.)...E	Concord; Manchester		.....	105
New Hebrides	5,700	61,000	11	Condominium (France-U.K.)...C	Vila	Malay-Polynesian languages, French	H10	7

†Member of the United Nations (1963). *Not shown on map; index key denotes approximate location. ‡ Malaya, Singapore, Sarawak, Brunei, and North Borneo are scheduled to unite forming the Federation of Malaysia by August 31, 1963.

WORLD POLITICAL INFORMATION TABLE

Region or Political Division	Area in sq. miles	Estimated Population 1/1/1962	Pop. per sq. mi.	Form of Government and Ruling Power	Capital; Largest City (unless same)	Predominant Languages	Map Index Key	Map Page
New Jersey	7,836	6,320,000	807	State (U.S.) ... E	Trenton; Newark		.....	106
New Mexico	121,666	978,000	8.0	State (U.S.) ... E	Santa Fe; Albuquerque		.....	107
New South Wales	309,433	3,975,000	13	State (Australia) ... E	Sydney	see Australia	F6	51
New York	49,576	17,123,000	345	State (U.S.) ... E	Albany; New York		.....	108
New Zealand†	103,736	2,466,000	24	Monarchy (Br. Commonwealth of Nations) ... A	Wellington; Auckland	English	.....	51
Nicaragua†	48,600	1,550,000	32	Republic ... A	Managua	Spanish	D5	62
Niedersachsen, see Lower Saxony								
Niger†	458,995	2,900,000	6.3	Republic ... A	Niamey	Hausa, Arabic, French	C6	45
Nigeria†	356,669	36,100,000	98	Federation (Br. Commonwealth of Nations) ... A	Lagos; Ibadan	Hausa, Ibo, Yoruba, English	E6	45
Niue	100	5,000	50	Island Territory (New Zealand) ... C	Alofi	Malay-Polynesian languages, English	*H11	7
Norfolk Island	13	800	62	External Territory (Australia) ... C	Kingston	English	H10	7
North America	9,420,000	273,100,000	29		.........; New York		.....	61
North Borneo	29,388	475,000	16	Part of Malaysia ... E	.........; Sandakan	Malay, Chinese	D5	35
North Carolina	52,712	4,645,000	88	State (U.S.) ... E	Raleigh; Charlotte		.....	109
North Dakota	70,665	641,000	9.1	State (U.S.) ... E	Bismarck; Fargo		.....	110
Northern Ireland	5,459	1,425,000	261	Administrative division of United Kingdom ... E	Belfast	English	C3	10
Northern Rhodesia	288,130	2,525,000	8.8	Protectorate (U.K.) ... E	Lusaka	Bantu languages, English	E3	48
Northern Territory	523,620	43,000	0.08	Territory (Australia) ... E	Darwin	English, Aboriginal languages	D5	50
North Polar Regions								4
North Rhine-Westphalia (Nordrhein-Westfalen)	13,119	16,100,000	1,227	State (Germany, West) ... E	Düsseldorf; Köln	German	B2	17
Northwest Territories	1,304,903	23,000	0.02	Territory (Canada) ... E	Ottawa, Ontario; Yellowknife	see Canada	C12	66
Norway†	125,064	3,629,000	29	Monarchy ... A	Oslo	Norwegian (Riksmål and Landsmål)	E5	25
Nova Scotia	21,425	741,000	35	Province (Canada) ... E	Halifax	see Canada	D6	74
Nyasaland	45,747	2,920,000	64	Protectorate (U.K.) ... C	Zomba; Blantyre-Limbe	Bantu languages	D5	48
Oceania (incl. Australia)	3,295,000	16,400,000	5.0		.........; Sydney			7
Ohio	41,222	9,943,000	241	State (U.S.) ... E	Columbus; Cleveland		.....	111
Oklahoma	69,919	2,360,000	34	State (U.S.) ... E	Oklahoma City		.....	112
Ontario	412,582	6,315,000	15	Province (Canada) ... E	Toronto	see Canada	.....	72
Orange Free State	49,649	1,430,000	29	Province (South Africa) ... E	Bloemfontein	see South Africa	C4	49
Oregon	96,981	1,811,000	19	State (U.S.) ... E	Salem; Portland		.....	113
Orkney Is.	376	18,500	49	County of Scotland, U.K. ... D	Kirkwall	English	A5	13
Pacific Islands Trust Territory	672	80,000	119	Trust Territory (U.S.) ...	Saipan	Malay-Polynesian languages, English	.....	7
Pakistan† (incl. part of Kashmir)	399,373	96,600,000	242	Federal Republic (Br. Comm. of Nations) ... A	Rawalpindi and Dacca; Karachi	Urdu, Bengali, English	C4, D9	39
Pakistan, East	55,134	51,700,000	938	Province of Pakistan ... D	Dacca	Bengali, English	D9	39
Pakistan, West (incl. Karachi and part of Kashmir)	344,239	44,900,000	145	Pakistan	.........; Karachi	Urdu, English	C4	39
Palestine (Gaza Area)	78	390,000	5,000	Military Government (U.A.R.) ... B	Gaza	Arabic	C6	32
Panamá†	29,290	1,100,000	38	Republic ... A	Panamá	Spanish	F7	62
Papua (excl. New Guinea Ter.)	90,600	530,000	5.8	External Territory (Australia) ... C	Port Moresby	Papuan and Negrito languages, English	k11	50
Paraguay†	157,047	1,835,000	12	Republic ... A	Asunción	Spanish, Guaraní	D4	55
Pennsylvania	45,333	11,521,000	254	State (U.S.) ... E	Harrisburg; Philadelphia		.....	114
Persia, see Iran								
Peru†	496,223	11,400,000	23	Republic ... A	Lima	Spanish, Quechua	.....	58
Philippines†	115,707	29,200,000	252	Republic ... A	Quezon City; Manila	Tagalog and other Malay-Polynesian languages, English	B6	35
Pitcairn (excl. Dependencies)	2	150	75	Colony (U.K.) ... C	Adamstown	English	H14	7
Poland†	120,348	30,265,000	251	People's Republic ... A	Warsaw (Warszawa)	Polish	C5	26
Portugal†	35,340	9,000,000	255	Republic ... A	Lisbon (Lisboa)	Portuguese	C1	20
Portugal and Possessions	837,733	21,991,000	26		Lisbon (Lisboa)		.....	3
Portuguese Guinea	13,948	580,000	42	Overseas Province (Portugal) ... C	Bissau	Native languages, Portuguese	D1	45
Portuguese India, see Goa, Damão, and Diu								
Portuguese Timor	7,332	510,000	70	Overseas Province (Portugal) ... C	Dili	Malay, Papuan languages, Portuguese	G7	35
Prairie Provinces	757,985	3,214,000	4.2	Canada ... E	.........; Winnipeg	see Canada	F12	66
Prince Edward Island	2,184	105,000	48	Province (Canada) ... E	Charlottetown	see Canada	C6	74
Puerto Rico	3,435	2,410,000	702	Commonwealth (U.S.) ... C	San Juan	Spanish, English	.....	65
Qatar	8,500	59,000	6.9	Sheikdom (U.K. protection) ... B	Doha	Arabic	I5	41
Quebec	594,860	5,330,000	9.0	Province (Canada) ... E	Quebec; Montreal	French, English	.....	73
Queensland	667,000	1,563,000	2.3	State (Australia) ... E	Brisbane	see Australia	D7	50
Reunion	969	352,000	363	Overseas Department (France) ... C	St. Denis	French	H24	3
Rhineland-Palatinate (Rheinland-Pfalz)	7,657	3,440,000	449	State (Germany, West) ... E	Mainz; Ludwigshafen am Rhein	German	D2	17
Rhode Island	1,214	868,000	715	State (U.S.) ... E	Providence		.....	84
Rhodesia & Nyasaland, Fed. of, see Northern Rhodesia, Southern Rhodesia, and Nyasaland.								
Rio Muni	10,045	185,000	18	African Province (Spain) ... C	Bata; Evinayong	Bantu languages, Spanish	E2	46
Rodrigues	42	18,000	428	Dependency of Mauritius (U.K.) ... D	Port Mathurin	English, French	*H25	3
Romania†	91,698	18,665,000	204	People's Republic ... A	Bucharest (Bucureşti)	Romanian, Hungarian	B7	22
Ruanda-Urundi, see Rwanda and Burundi								
Russian Soviet Federated Socialist Republic	6,592,850	122,200,000	19	Soviet Federated Socialist Republic (Sov. Un.) ... E	Moscow (Moskva)	Russian, Finno-Ugric languages, various Turkic, Iranian, and Mongol languages	C11	28
Russian S.F.S.R. in Europe	1,919,750	89,975,000	47	Soviet Union	.........; Moscow	Russian, Finno-Ugric	.....	27
Rwanda†	10,169	2,725,000	268	Republic ... A	Kigali	Bantu and Hamitic languages	B4	48
Ryukyu Is. (Southern)	848	900,000	1,061	U.S. Military Administration ... B	Naha	Japanese	F10	34
Saar (Saarland)	991	1,085,000	1,095	State (Germany, West) ... E	Saarbrücken	German	E6	15
St. Helena (incl. Dependencies)	160	5,900	37	Colony (U.K.) ... C	Jamestown	English	H9	6
St. Kitts-Nevis-Anguilla	153	58,000	379	Colony (U.K.) ... C	Basseterre	English	n15	64
St. Lucia	238	88,000	370	Colony (U.K.) ... C	Castries	English	p16	64
St. Pierre & Miquelon	93	5,000	54	Overseas Territory (France) ... C	St. Pierre	French	m26	67
St. Vincent	150	82,000	547	Colony (U.K.) ... C	Kingstown	English	p16	64
Samoa (Entire)	1,206	137,000	114		.........; Apia	Malay-Polynesian languages, English	C4	52
San Marino	23	17,000	739	Republic ... A	San Marino	Italian		21
Sao Tome & Principe	372	68,000	183	Overseas Province (Portugal) ... C	São Tomé	Bantu languages, Portuguese	E1	46
Sarawak	48,250	800,000	17	Part of Malaysia ... E	.........; Kuching	Malay, Chinese, English	E4	35
Sardinia	9,301	1,400,000	151	Part of Italy (3 Provinces) ... D	.........; Cagliari	Italian	D2	21
Saskatchewan	251,700	929,000	3.7	Province (Canada) ... E	Regina	English	.....	70
Saudi Arabia†	617,800	6,200,000	10	Monarchy ... A	Riyadh; Mecca	Arabic	G7	33
Scandinavia (incl. Finland and Iceland)	509,909	20,538,000	40		.........; Copenhagen (København)	Swedish, Danish, Norwegian, Finnish, Icelandic		25
Schleswig-Holstein	6,045	2,332,000	386	State (Germany, West) ... E	Kiel	German	D4	24
Scotland	30,411	5,185,000	170	Administrative division of United Kingdom ... E	Edinburgh; Glasgow	English, Gaelic	D2	13
Senegal†	76,124	3,050,000	40	Republic ... A	Dakar	Wolof, Poular, French	G24	45
Seychelles	156	43,000	276	Colony (U.K.) ... C	Victoria	French, Creole, English		3
Shetland Is.	550	17,700	32	Part of Scotland, U.K. (Zetland County) ... D	Lerwick	English	C8	8
Siam, see Thailand								
Sicily	9,926	4,650,000	468	Part of Italy (Sicilia Autonomous Region) ... D	Palermo	Italian	F5	21
Sierra Leone†	27,925	2,500,000	90	Monarchy (Br. Commonwealth of Nations) ... A	Freetown	Temne, Mende, English	E2	45
Sikkim	2,744	162,000	59	Monarchy (Indian protection) ... B	Gangtok	Tibeto-Burman languages	D12	40

†Member of the United Nations (1963).   *Not shown on map; index key denotes approximate location.

Region or Political Division	Area in sq. miles	Estimated Population 1/1/1962	Pop. per sq. mi.	Form of Government and Ruling Power	Capital; Largest City (unless same)	Predominant Languages	Map Index Key	Map Page
Singapore	224	1,720,000	7,679	Part of Malaysia...............E	........; Singapore	Chinese, Malay, English	L5	38
Solomon Is. (Austl. Trust)	4,320	55,000	13	Part of Australian Trust Ter. of New Guinea (Bougainville District)..............D	Sohano; Kieta	Malay-Polynesian languages	G9	7
Solomon Is., British	11,500	130,000	11	Protectorate (U.K.)..............C	Honiara	Malay-Polynesian languages	G9	7
Somali Republic†	246,202	2,035,000	8.3	Republic..............A	Mogadiscio	Somali	E5	47
South Africa†	472,926	16,350,000	35	Federal Republic..............A	Pretoria and Cape Town; Johannesburg	Afrikaans, English, Bantu languages	C3	49
South America	6,870,000	148,600,000	22		........; Buenos Aires			53
South Australia	380,070	992,000	2.6	State (Australia)..............E	Adelaide	see Australia	E6	50
South Carolina	31,055	2,421,000	78	State (U.S.)..............E	Columbia			115
South Dakota	77,047	689,000	8.9	State (U.S.)..............E	Pierre; Sioux Falls			116
Southern Rhodesia	150,333	3,180,000	21	Self-Governing Colony (U.K.)..............C	Salisbury	Bantu languages, English	A4	49
South Georgia	1,450	400	0.3	Dependency of Falkland Is. (U.K.)..............D	Grytviken	English, Norwegian	J7	6
South Polar Regions								5
South-West Africa	317,887	535,000	1.7	Mandate (South Africa)..............C	Windhoek	Bantu languages, Hottentot, Bushman, Afrikaans, English	B2	49
Soviet Union (Union of Soviet Socialist Republics)†	8,599,300	220,000,000	26	Federal Soviet Republic..............A	Moscow (Moskva)	Russian and other Slavic languages, various Finno-Ugric, Turkic and Mongol languages, Caucasian languages, Persian		28
Soviet Union in Europe	1,919,750	151,400,000	79	Soviet Union	........; Moscow (Moskva)	Russian, Ruthenian, Finno-Ugric and Caucasian languages		27
Spain†	194,345	30,700,000	158	Monarchy (Regency)..............A	Madrid	Spanish, Catalan, Galician, Basque		20
Spain and Possessions	308,540	31,180,000	101		Madrid			3
Spanish Possessions in North Africa	82	154,000	1,878	Five Possessions (no central government) (Spain)..............C	........; Melilla	Spanish, Arabic, Berber	*G4	30
Spanish Sahara	102,703	25,000	0.2	African Province (Spain)..............C	Aiún	Arabic, Spanish	E2	44
Spitsbergen, see Svalbard								
Sudan†	967,500	12,300,000	13	Republic..............A	Khartoum	Arabic, native languages, English	C2	47
Sumatra (Sumatera)	182,860	15,550,000	85	Part of Indonesia (6 Provinces)..............D	........; Medan	Bahasa Indonesia, English, Chinese	F2	35
Surinam (Neth. Guiana)	55,143	324,000	5.9	Self-Governing Territory (Netherlands)..............C	Paramaribo	Dutch, Indo-Aryan, Indian	B3	59
Svalbard (Spitsbergen)	24,101	4,000	0.2	Dependency (Norway)..............C	Longyearbyen	Norwegian, Russian	B12	4
Swaziland	6,705	270,000	40	Territory (Protectorate) (U.K.)..............C	Mbabane	Swazi and other Bantu languages, English	C5	49
Sweden†	173,666	7,540,000	43	Monarchy..............A	Stockholm	Swedish	F7	25
Switzerland	15,941	5,520,000	346	Federal Republic..............A	Bern (Berne); Zürich	German, French, Italian	C4	18
Syria†	71,498	4,825,000	67	Republic..............A	Damascus (Esh Sham)	Arabic	E12	31
Tadzhik S.S.R.	55,250	2,170,000	39	Soviet Socialist Republic (Sov. Un.)..............E	Dushanbe	Tadzhik, Turkic languages, Russian	F10	26
Taiwan (Formosa) (Nationalist China)†	13,884	11,150,000	803	Republic..............A	Taipei	Chinese	G9	34
Tanganyika†	361,800	9,500,000	26	Republic (Br. Commonwealth of Nations)..............A	Dar es Salaam	Swahili and other Bantu languages, English	C5	48
Tasmania	26,215	355,000	14	State (Australia)..............E	Hobart	see Australia	o15	50
Tennessee	42,244	3,632,000	86	State (U.S.)..............E	Nashville; Memphis			117
Texas	267,339	9,928,000	37	State (U.S.)..............E	Austin; Houston			118
Thailand (Siam)†	198,500	27,000,000	136	Monarchy..............A	Bangkok (Krung Thep)	Thai, Chinese	E5	38
Tibet	471,660	1,300,000	2.8	Autonomous Region (China)..............E	Lhasa	Tibetan	B8	39
Togo†	21,850	1,475,000	68	Republic..............A	Lomé	Negro languages, French	E5	45
Tokelau (Union) Is.	4	2,000	500	Island Territory (New Zealand)..............C	........; Fakaofo	Malay-Polynesian languages	G11	7
Tonga	270	67,000	248	Protected Monarchy (U.K.)..............C	Nukualofa	Malay-Polynesian languages, English	H11	7
Transcaucasia	71,850	10,360,000	144	Soviet Union..............E	........; Baku		E7	28
Transvaal	110,450	6,450,000	58	Province (South Africa)..............E	Pretoria; Johannesburg	see South Africa	B4	49
Trinidad & Tobago†	1,980	864,000	436	Self-Governing Member (Br. Commonwealth of Nations)..............A	Port-of-Spain	English, Spanish	A5	60
Tristan da Cunha	40	....	.....	Dependency of St. Helena (U.K.)..............D	Edinburgh		I9	6
Trucial Coast	32,300	88,000	2.7	Seven Sheikdoms (no central government) (U.K. protection)..............B	........; Dubayy	Arabic	I7	41
Tunisia†	48,332	4,260,000	88	Republic..............A	Tunis	Arabic, French	B6	44
Turkey†	301,381	28,675,000	95	Republic..............A	Ankara; Istanbul	Turkish	C10	31
Turkey in Europe	9,121	2,375,000	260	Turkey..............E	........; Istanbul	Turkish	B6	31
Turkmen S.S.R.	188,450	1,690,000	9.0	Soviet Socialist Republic (Sov. Un.)..............E	Ashkhabad	Turkic languages, Russian	F8	28
Turks & Caicos Is.	166	5,700	34	Colony (U.K.)..............C	Grand Turk	English	E8	64
Uganda†	92,525	6,925,000	75	Self-Governing Member (Br. Commonwealth of Nations)..............A	Entebbe; Kampala	Bantu languages	A5	48
Ukraine (S.S.R.)†	232,050	43,725,000	188	Soviet Socialist Republic (Sov. Un.)..............E	Kiev	Ukrainian, Russian	G8	27
Union of Soviet Socialist Republics, see Soviet Union								
United Arab Republic (Egypt)†	386,000	27,000,000	70	Republic..............A	Cairo (Al Qāhirah)	Arabic	D5	43
United Kingdom of Great Britain & Northern Ireland†	94,214	52,860,000	561	Monarchy (Br. Commonwealth of Nations)..............A	London	English, Gaelic, Welsh		10
United Kingdom & Possessions	1,342,338	78,570,000	59		London			3
United States†	3,675,633	185,200,000	50	Federal Republic..............A	Washington; New York	English		76, 77
United States and Possessions	3,680,757	187,860,000	51		Washington; New York	English, Spanish		3
Upper Volta†	105,869	3,735,000	35	Republic..............A	Ouagadougou	Voltaic and Mande languages, French	D4	45
Uruguay†	72,172	2,380,000	33	Republic..............A	Montevideo	Spanish	E1	56
Utah	84,916	929,000	11	State (U.S.)..............E	Salt Lake City			119
Uzbek S.S.R.	171,900	8,950,000	52	Soviet Socialist Republic (Sov. Un.)..............E	Tashkent	Turkic languages, Sart, Russian	E9	28
Vatican City (Holy See)	0.2	1,000	5,000	Ecclesiastical State..............A	Vatican City	Italian, Latin	D4	21
Venezuela†	352,143	7,600,000	22	Federal Republic..............A	Caracas	Spanish	B4	60
Vermont	9,609	393,000	41	State (U.S.)..............E	Montpelier; Burlington			120
Victoria	87,884	2,990,000	34	State (Australia)..............E	Melbourne	see Australia	H5	51
Vietnam (Entire)	125,881	31,500,000	250		........; Saigon	Annamese, Chinese	D8	38
Vietnam, North	59,933	16,900,000	282	People's Republic..............A	Hanoi	Annamese, Chinese	B6	38
Vietnam, South	65,948	14,600,000	221	Republic..............A	Saigon	Annamese, Chinese	G7	38
Virgin Is., British	59	7,500	127	Colony (U.K.)..............C	Road Town	English	m14	64
Virgin Is. of the U.S.	133	33,000	248	Unincorporated Territory (U.S.)..............C	Charlotte Amalie	English	f15	65
Virginia	40,815	4,044,000	99	State (U.S.)..............E	Richmond; Norfolk			121
Wake I.	3	1,200	400	Possession (U.S.)..............C		English		53
Wales (incl. Monmouthshire)	8,017	2,645,000	330	United Kingdom..............E	Cardiff	English, Welsh	B4	12
Wallis & Futuna	106	9,000	85	Overseas Territory (France)..............C	Mata-Utu	Malay-Polynesian languages	*G11	7
Washington	68,192	2,928,000	43	State (U.S.)..............E	Olympia; Seattle			122
Western Australia	975,920	760,000	0.8	State (Australia)..............E	Perth	see Australia	E3	50
Western Samoa	1,130	110,000	97	Self-Governing Member (Br. Commonwealth of Nations)..............A	Apia	Malay-Polynesian languages, English	*G11	7
West Indies	91,110	20,256,000	222		........; Havana			64
West Irian	160,600	750,000	4.7	Under Indonesian Administration..............C	Kotabaru (Hollandia)	Various Papuan languages	F9	35
West Virginia	24,181	1,831,000	76	State (U.S.)..............E	Charleston; Huntington			123
White Russia, see Byelorussia								
Wisconsin	56,154	4,047,000	72	State (U.S.)..............E	Madison; Milwaukee			124
World	57,280,000	3,075,000,000	54		........; New York			2, 3
Wyoming	97,914	338,000	3.5	State (U.S.)..............E	Cheyenne			125
Yemen†	75,300	5,000,000	66	Republic..............A	San'ā'	Arabic	C5	47
Yugoslavia†	98,766	18,700,000	189	Socialist Federal Republic..............A	Belgrade (Beograd)	Serbo-Croatian-Slovenian, Macedonian	C3	22
Yukon Territory	207,076	15,000	0.07	Territory (Canada)..............E	Whitehorse	English	D5	66
Zanzibar	1,020	313,000	307	Protectorate (U.K.)..............C	Zanzibar	Arabic, English	C6	48

†Member of the United Nations (1963).   *Not shown on map; index key denotes approximate location.

WORLD FACTS AND COMPARISONS

MOVEMENTS OF THE EARTH

The earth makes one complete revolution around the sun every 365 days, 5 hours, 48 minutes, and 46 seconds.

The earth makes one complete rotation on its axis in 23 hours and 56 minutes.

The earth revolves in its orbit around the sun at a speed of 66,700 miles per hour.

The earth rotates on its axis at an equatorial speed of more than 1,000 miles per hour.

MEASUREMENTS OF THE EARTH

Estimated age of the earth, at least 3 billion years.
Equatorial diameter of the earth, 7,926.68 miles.
Polar diameter of the earth, 7,899.99 miles.
Mean diameter of the earth, 7,918.78 miles.
Equatorial circumference of the earth, 24,902.45 miles.
Polar circumference of the earth, 24,818.60 miles.
Difference between equatorial and polar circumference of the earth, 83.85 miles.

Weight of the earth, 6,600,000,000,000,000,000,000 tons, or 6,600 billion billion tons.

Total area of the earth, 196,940,400 square miles.

Total land area of the earth (including inland water and Antarctica), 57,295,000 square miles.

THE EARTH'S INHABITANTS

Total population of the earth is estimated to be 3,075,000,000 (January 1, 1962).

Estimated population density of the earth, 54 per square mile.

THE EARTH'S SURFACE

Highest point on the earth's surface, Mount Everest, China (Tibet)–Nepal, 29,028 feet.

Lowest point on the earth's land surface, shores of the Dead Sea, Israel-Jordan, 1,286 feet below sea level.

Greatest ocean depth, the Marianas Trench, south of Guam, Pacific Ocean, 36,198 feet.

EXTREMES OF TEMPERATURE AND RAINFALL OF THE EARTH

Highest temperature ever recorded, 136.4°F. at Azizia, Libya, Africa, on September 13, 1922.

Lowest temperature ever recorded, −126.9°F. at Vostok, Antarctica, on August 24, 1960.

Highest mean annual temperature, 88°F. at Lugh Ferrandi, Somali Republic.

Lowest mean annual temperature, −67°F at Vostok, Antarctica.

At Baguio, Luzon, in the Philippines, 46 inches of rainfall was reported in a 24-hour period, July 14–15, 1911. This is believed to be the world's record for a 24-hour rainfall.

An authenticated rainfall of 366 inches in 1 month—July, 1861—was reported at Cherrapunji, India. More than 131 inches fell in a period of 7 consecutive days in June, 1931. Average annual rainfall at Cherrapunji is 450 inches.

THE CONTINENTS

CONTINENT	Area (sq. mi.)	Population Estimated Jan. 1, 1962	Population per sq. mi.	Mean Elevation (feet)	Highest Elevation (Feet)	Lowest Elevation (Feet)	Highest Recorded Temperature	Lowest Recorded Temperature
North America	9,420,000	273,100,000	29	2,000	Mt. McKinley, United States (Alaska), 20,320	Death Valley, California, 282 below sea level	Death Valley, California, 134°F.	Snag, Yukon, Canada, −81°F.
South America	6,870,000	148,600,000	22	1,800	Mt. Aconcagua, Argentina, 22,834	Salinas Grandes, Península Valdés, Argentina, 131 below sea level	Rivadavia, Argentina, 120°F.	Sarmiento, Argentina, −27.4°F.
Europe	3,825,000	583,600,000	153	980	Mt. Elbrus, Soviet Union, 18,481	Caspian Sea, Soviet Union—Iran, 92 below sea level	Sevilla (Seville), Spain, 122°F.	Ust-Shchugor, Soviet Union, −67°F.
Asia	17,085,000	1,792,000,000	105	3,000	Mt. Everest, China (Tibet)-Nepal, 29,028	Dead Sea, Israel-Jordan, 1,286 below sea level	Tirat Zvi, Israel, 129.2°F.	Oymyakon, Soviet Union, −89.9°F.
Africa	11,685,000	261,300,000	22	1,900	Mt. Kilimanjaro, Tanganyika, 19,340	Qattara Depression, U.A.R. (Egypt), 436 below sea level	Azizia, Libya, 136.4°F.	Ifrane, Morocco, −11.2°F.
Oceania, incl. Australia	3,295,000	16,400,000	5	.....	Mt. Wilhelm, New Guinea Ter., 15,400	Lake Eyre, South Australia, 39 below sea level	Cloncurry, Queensland, Australia, 127.5°F.	Charlotte Pass, New South Wales, Australia, −8°F.
Australia	2,971,081	10,740,000	4	1,000	Mt. Kosciusko, New South Wales, 7,316	Lake Eyre, South Australia, 39 below sea level	Cloncurry, Queensland, 127.5°F.	Charlotte Pass, New South Wales, −8°F.
Antarctica	5,100,000	Uninhabited	...	6,000	Vinson Massif, 16,864	Sea level	Esperanza (Palmer Peninsula), 58.3°F.	Vostok, −126.9°F.
World	57,280,000	3,075,000,000	54	.....	Mt. Everest, China (Tibet)-Nepal, 29,028	Dead Sea, Israel-Jordan, 1,286 below sea level	Azizia, Libya, 136.4°F.	Vostok, −126.9°F.

APPROXIMATE POPULATION OF THE WORLD, 1650–1962*

AREA	1650	1750	1800	1850	1900	1914	1920	1939	1950	1962
North America	5,000,000	5,000,000	13,000,000	39,000,000	106,000,000	141,000,000	147,000,000	186,000,000	219,000,000	273,100,000
South America	8,000,000	7,000,000	12,000,000	20,000,000	38,000,000	55,000,000	61,000,000	90,000,000	111,000,000	148,600,000
Europe	100,000,000	140,000,000	190,000,000	265,000,000	400,000,000	470,000,000	453,000,000	526,000,000	530,000,000	583,600,000
Asia	335,000,000	476,000,000	593,000,000	754,000,000	932,000,000	1,006,000,000	1,000,000,000	1,247,000,000	1,418,000,000	1,792,000,000
Africa	100,000,000	95,000,000	90,000,000	95,000,000	118,000,000	130,000,000	140,000,000	170,000,000	199,000,000	261,300,000
Oceania, incl. Australia	2,000,000	2,000,000	2,000,000	2,000,000	6,000,000	8,000,000	9,000,000	11,000,000	13,000,000	16,400,000
Australia					4,000,000	5,000,000	6,000,000	7,000,000	8,000,000	10,740,000
World	550,000,000	725,000,000	900,000,000	1,175,000,000	1,600,000,000	1,810,000,000	1,810,000,000	2,230,000,000	2,490,000,000	3,075,000,000

Figures prior to 1962 are rounded to the nearest million. Figures in italics represent very rough estimates.

LARGEST COUNTRIES OF THE WORLD IN POPULATION

	Population 1/1/1962
1 China (excl. Taiwan)	700,000,000
2 India (incl. part of Kashmir)	443,400,000
3 Soviet Union	220,000,000
4 United States	185,200,000
5 Pakistan (incl. part of Kashmir)	96,600,000
6 Indonesia	95,800,000
7 Japan	94,500,000
8 Brazil	73,400,000
9 Germany, West (incl. West Berlin)	56,680,000
10 United Kingdom of Great Britain & Northern Ireland	52,860,000
11 Italy	49,800,000
12 France	46,200,000
13 Mexico	36,650,000
14 Nigeria	36,100,000
15 Spain	30,700,000
16 Poland	30,265,000
17 Philippines	29,200,000
18 Turkey	28,675,000
19 Thailand	27,000,000
20 United Arab Republic (Egypt)	27,000,000
21 Korea, South	25,700,000
22 Burma	21,650,000
23 Iran	20,925,000
24 Argentina	20,500,000
25 Ethiopia	20,000,000

LARGEST COUNTRIES OF THE WORLD IN AREA

	Area (sq. mi.)
1 Soviet Union	8,599,300
2 Canada	3,851,809
3 China (excl. Taiwan)	3,691,500
4 United States	3,675,633
5 Brazil	3,286,478
6 Australia	2,971,081
7 India (incl. part of Kashmir)	1,227,275
8 Argentina	1,072,070
9 Sudan	967,500
10 Algeria	917,537
11 Congo, The (Léopoldville)	905,565
12 Greenland (Den.)	840,000
13 Mexico	761,602
14 Libya	679,360
15 Iran	636,300
16 Saudi Arabia	617,800
17 Mongolia	592,700
18 Indonesia	574,670
19 Peru	496,223
20 Chad	495,800
21 Angola (Port.)	481,351
22 South Africa	472,926
23 Mali	464,874
24 Niger	458,995
25 Ethiopia	457,267

PRINCIPAL MOUNTAINS OF THE WORLD

Height (feet)

NORTH AMERICA

McKinley, △Alaska (△United States;
  △North America)...................20,320
Logan, △Canada (△St. Elias Mts.)..........19,850
Citlaltepetl (Orizaba), △Mexico..........18,696
St. Elias, Alaska–Canada...............18,008
Popocatepetl, Mexico.................17,887
Foraker, Alaska....................17,395
Ixtacihuatl, Mexico.................17,343
Lucania, Yukon, Canada...............17,150
Whitney, △California.................14,495
Elbert, △Colorado (△Rocky Mts.).........14,431
Massive, Colorado..................14,418
Harvard, Colorado..................14,414
Rainier, △Washington (△Cascade Range)......14,410
Williamson, California...............14,384
Blanca Pk., Colorado
  (△Sangre de Cristo Range)............14,317
Uncompahgre Pk., Colorado
  (△San Juan Mts.)..................14,301
Grays Pk., Colorado (△Front Range)........14,274
Evans, Colorado....................14,264
Longs Pk., Colorado.................14,256
Colima, Nevado de, Mexico.............14,235
Shasta, California..................14,162
Pikes Peak, Colorado................14,110
Wrangell, Alaska...................14,005
Tajumulco, △Guatemala (△Central America)...13,816
Mauna Kea, △Hawaii (△Hawaii I.).........13,796
Gannett Pk., △Wyoming...............13,785
Grand Teton, Wyoming...............13,766
Mauna Loa, Hawaii.................13,680
Kings Pk., △Utah...................13,498
Waddington, Canada (△Coast Mts.)........13,260
Cloud Pk., Wyoming (△Big Horn Mts.)......13,175
Wheeler Pk., △New Mexico.............13,160
Boundary Pk., △Nevada...............13,145
Robson, Canada (△Canadian Rockies).......12,972
Chirripó Grande, △Costa Rica...........12,861
Granite Pk., △Montana...............12,799
Humphreys Pk., △Arizona.............12,670
Borah Pk., △Idaho..................12,662
Adams, Washington.................12,307
Gunnbjörn, △Greenland...............12,139
San Gorgonio, California
  (△Southern California)..............11,485
Chiriquí, △Panama..................11,410
Hood, △Oregon....................11,245
Lassen Pk., California...............10,466
Duarte, Pico, △Dominican Rep. (△West Indies)..10,417
Haleakala, Hawaii (△Maui)............10,025
Paricutín, Mexico...................9,100
Selle, Massif de la, △Haiti.............8,793
Guadalupe Pk., △Texas...............8,751
Olympus, Washington (△Olympic Mts.)......7,954
Santa Ana, △El Salvador..............7,828
Blue Mountain Pk., △Jamaica...........7,520
Harney Pk., △South Dakota (△Black Hills)....7,242
Mitchell, △North Carolina (△Appalachian Mts.)..6,684
Clingmans Dome, North Carolina–
  △Tennessee (△Great Smoky Mts.).......6,642
Turquino, Pico de, △Cuba..............6,496
Washington, △New Hampshire (△White Mts.)...6,288
Rogers, △Virginia...................5,720
Marcy, △New York (△Adirondack Mts.).....5,344
Katahdin, △Maine...................5,268
Kawaikini, Hawaii (△Kauai)............5,170
Spruce Knob, △West Virginia...........4,860
Pelée, △Martinique..................4,800
Mansfield, △Vermont (△Green Mts.).......4,393
Punta, Cerro de, △Puerto Rico..........4,389
Black Mtn., △Kentucky...............4,145
Kilauea, Hawaii (Hawaii I.)............4,090
Kaala Pk., Hawaii (△Oahu)............4,025

SOUTH AMERICA

Aconcagua, △Argentina (△Andes Mts.;
  △South America)..................22,834
Ojos del Salado, Nudos, Argentina–△Chile....22,590
Pissis, Argentina...................22,546
Tupungato, Argentina–Chile............22,310
Mercedario, Argentina................22,211
Huascarán, △Peru..................22,205
Llullaillaco, Argentina–Chile...........22,146
Yerupaja, Peru....................21,758
Incahuasi, Argentina–Chile............21,719
Illampu, △Bolivia..................21,490
Ancohuma, Bolivia.................21,489
Sajama, Nevado, Bolivia..............21,391
Illimani, Bolivia...................21,151
Chimborazo, △Ecuador...............20,577
Cotopaxi, Ecuador.................19,344
Misti, El, Peru....................19,144
Cristóbal Colón, △Colombia............18,947

△ *Highest mountain in state, country, range, or region named.*

168

Height (feet)

Huila, Colombia (△Cordillera Central).......18,865
Bolívar (La Columna), △Venezuela.........16,411
Fitz Roy, Argentina.................11,600
Bandeira, Pico da, △Brazil.............9,462

EUROPE

Elbrus, Soviet Union (△Caucasus Mts.,
  △Europe)......................18,481
Shkhara, Soviet Union...............17,059
Dykh-Tau, Soviet Union..............17,054
Kazbek, Soviet Union...............16,554
Blanc, Mont, △France (△Alps)..........15,781
Rosa, Monte (Dufourspitze) (△Switzerland)...15,200
Rosa, Monte (Grenzgipfel) (△Italy–
  Switzerland)....................15,194
Weisshorn, Switzerland..............14,803
Matterhorn, Switzerland.............14,685
Finsteraarhorn, Switzerland...........14,026
Jungfrau, Switzerland...............13,668
Gross Glockner, △Austria............12,461
Teide, Pico de, △Spain (△Canary Is.)......12,162
Mulhacén, △Spain (continental).........11,424
Aneto, Pico de, Spain (△Pyrenees)........11,168
Perdido (Perdu), Spain..............11,007
Etna, Italy (△Sicily)................10,868
Clapier, France-Italy (△Maritime Alps).....10,817
Zugspitze, △Germany...............9,721
Coma Pedrosa, Andorra..............9,665
Musala, △Bulgaria.................9,592
Corno, Italy (△Apennines)............9,560
Olympus, △Greece.................9,550
Triglav, △Yugoslavia...............9,393
Korab, △Albania..................9,068
Cinto, France (△Corsica).............8,891
Gerlachovka, △Czechoslovakia
  (△Carpathian Mts.)...............8,737
Galdhöpiggen, △Norway (△Scandinavia)....8,400
Negoi, △Romania.................8,346
Rysy Pk., Czechoslovakia............8,212
Parnassós, Greece.................8,061
Idhi (Ida), Greece (△Crete)...........8,058
Pico, △Portugal (△Azores Is.).........7,615
Kebnekaise, △Sweden..............6,962
Hvannadalshnúkur, △Iceland..........6,952
Malhão, △Portugal (continental).......6,532
Narodnaya, Soviet Union (△Ural Mts.).....6,184
Marmora, Punta La, Italy (△Sardinia).....6,017
Hekla, Iceland...................4,747
Nevis, Ben, △United Kingdom (△Scotland)...4,406
Haltia, △Finland.................4,344
Vesuvius, Italy..................3,842
Snowdon, △Wales................3,560
Carrantuohill, △Ireland.............3,414
Kekes, △Hungary.................3,330
Scafell Pike, △England.............3,210
Stromboli, Italy.................3,038

ASIA

Everest, △China (△Tibet)–△Nepal (△Himalaya
  Mts.; △Asia; △World)..............29,028
Godwin Austen (K²), △Pakistan (△Kashmir)
  (△Karakoram Range)...............28,250
Kanchenjunga, Nepal–△Sikkim...........28,168
Makalu, China (Tibet)–Nepal...........27,790
Dhaulagiri, Nepal.................26,810
Nanga Parbat, Pakistan (Kashmir)........26,660
Annapurna, Nepal.................26,504
Gasherbrum, Pakistan (Kashmir).........26,470
Gosainthan, (Tibet) China............26,291
Nanda Devi, △India................25,645
Rakaposhi, Pakistan (Kashmir)..........25,551
Kamet, India...................25,447
Namcha Barwa, China (Tibet)..........25,445
Gurla Mandhata, China (Tibet).........25,355
Ulugh Muztagh, China (△Kunlun Mts.).....25,340
Tirich Mir, Pakistan (△Hindu Kush).......25,230
Minya Konka, China...............24,900
Kangri, △Bhutan.................24,740
Communism Pk., △Soviet Union
  (△Pamir-Alay Mts.)..............24,590
Pobeda Pk., China–Soviet Union (△Tien Shan)..24,409
Muztagh Ata, China...............24,388
Api, Nepal....................23,398
Lenin Pk., Soviet Union.............23,382
Tengri Khan, Soviet Union...........22,940
Kailas, China (Tibet)..............22,028
Hkakabo Razi, △Burma.............19,296
Demavend, △Iran................18,934
Ararat, △Turkey.................16,946
Carstensz, △West Irian (△New Guinea)....16,503
Klyuchevskaya, Soviet Union (△Kamchatka)...15,912

Height (feet)

Wilhelmina, West Irian..............15,518
Tabun Bogdo (Khuitun), △Mongolia (△Altai
  Mts.)........................15,266
Belukha, Soviet Union..............15,157
Turgun Uula, Mongolia.............13,996
Kinabalu, △Malaysia (△Borneo).........13,455
Hsinkao, China (△Formosa)...........13,113
Erciyas, Turkey..................12,848
Munku-Sardyk, Mongolia–Soviet Union
  (△Sayan Mts.)..................12,821
Kerintji, △Indonesia (△Sumatra)........12,484
Fuji, △Japan (△Honshu).............12,388
Hadūr Shu'ayb, △Yemen
  (△Arabian Peninsula)..............12,336
Rindjani, Indonesia (△Lombok).........12,225
Mahameru, Indonesia (△Java).........12,060
Qalate Qarrāde, △Iraq.............12,000
Razih, Jabal, △Saudi Arabia..........11,999
Rantemario, Indonesia (△Celebes)......11,286
Qurnet es Sa'uda, △Lebanon.........10,131
Shām, Jabal ash, △Muscat and Oman....9,902
Apo, △Philippines (△Mindanao)........9,690
Pulog, Philippines (△Luzon)..........9,612
Phu Bia, △Laos.................9,242
Hermon, Lebanon–△Syria...........9,232
Anai Mudi, △India (peninsular)........8,841
Angka, Doi, △Thailand............8,452
Kwanmo, △Korea................8,336
Mayon, Philippines (Luzon).........8,284
Pidurutalagala, △Ceylon...........8,281
Asahi, Japan (△Hokkaido)..........7,513
Tahan, Gunong, Malaysia (△Malaya)....7,186
Troodos, △Cyprus...............6,403
Kuju-San, Japan (△Kyushu).........5,866
Atzmon, △Israel................3,962
Krakatoa (Rakata), Indonesia........2,667
Carmel, Israel.................1,791

AFRICA

Kilimanjaro (Kibo), △Tanganyika (△Africa)....19,340
Kenya, △Kenya..................17,040
Margherita, Mt., △Congo L.–△Uganda......16,795
Ras Dashan, △Ethiopia..............15,158
Meru, Tanganyika................14,979
Elgon, Kenya–Uganda..............14,178
Toubkal, Djebal, △Morocco (△Atlas Mts.)....13,661
Cameroon, △Cameroon.............13,354
Thabantshonyana, △Basutoland (△Southern
  Africa)......................11,425
Emi Koussi, △Chad (△Tibesti Mts.).......11,204
Injasuti, △South Africa.............11,182
Neiges, Piton des, △Reunion..........10,069
Tahat, △Algeria (△Ahaggar Mts.).......9,852
Maromokotro, △Malagasy Republic......9,462
Santa Isabel, △Fernando Poo..........9,350
Cano, △Cape Verde Is.............9,760
Katrīnah, Jabal, △United Arab Republic
  (Egypt).......................8,652
São Tomé, Pico de, △Sao Tome........6,640

OCEANIA

Wilhelm, △New Guinea Ter...........15,400
Bangeta, New Guinea Ter............13,434
Giluwe, △Papua.................13,660
Victoria, Papua (△Owen Stanley Range)....13,363
Cook, △New Zealand (△South Island).....12,349
Balbi, △Solomon Is. (△Bougainville).....10,170
Ruapehu, New Zealand (△North Island)....9,175
Egmont, New Zealand.............8,260
Mauga Silisili, △Western Samoa........8,000
Orohena, △Fr. Polynesia (△Tahiti)......7,618
The Father, New Guinea Ter.
  (△Bismarck Archipelago)...........7,546
Kosciusko, △Australia (△New South Wales)...7,316
Hombolt, △New Caledonia...........5,380
Panié, New Caledonia.............5,348
Ossa, △Australia (△Tasmania).........5,305
Bartle Frere, Australia (△Queensland).....5,287
Woodroffe, Australia (△South Australia)....4,970
Victoria, △Fiji (△Viti Levu)..........4,341
Bruce, Australia (△Western Australia)....4,024

ANTARCTICA

Vinson Massif (△Antarctica)..........16,864
Kirkpatrick....................14,600
Markham.....................14,272
Wade.......................14,078
Andrew Jackson................13,747
Sidley......................13,717

GREAT OCEANS AND SEAS OF THE WORLD

OCEANS AND SEAS	Area (sq. mi.)	Average Depth (feet)	Greatest Depth (feet)	OCEANS AND SEAS	Area (sq. mi.)	Average Depth (feet)	Greatest Depth (feet)	OCEANS AND SEAS	Area (sq. mi.)	Average Depth (feet)	Greatest Depth (feet)
Pacific Ocean	63,855,000	14,050	36,198	Bering Sea	876,000	4,710	16,800	Hudson Bay	476,000	420	850
Atlantic Ocean	31,744,000	12,690	27,498	Caribbean Sea	750,000	7,310	24,580	Japan, Sea of	389,000	4,490	12,280
Indian Ocean	28,371,000	13,000	26,400	Gulf of Mexico	596,000	4,960	14,360	North Sea	222,000	310	2,170
Arctic Ocean	5,427,000	5,010	17,880	Okhotsk, Sea of	590,000	2,760	11,400	Black Sea	178,000	3,610	7,360
Mediterranean Sea	967,000	4,780	15,900	East China Sea	482,000	620	9,840	Red Sea	169,000	1,610	7,370
South China Sea	895,000	5,420	18,090	Yellow Sea	480,000	150	300	Baltic Sea	163,000	180	1,440

PRINCIPAL LAKES OF THE WORLD

LAKES	Area (sq. mi.)	LAKES	Area (sq. mi.)	LAKES	Area (sq. mi.)
Caspian, Soviet Union–Iran (salt)	152,084	Winnipeg, Canada	9,465	Torrens, Australia (salt)	△2,200
Superior, United States–Canada	31,820	Ontario, United States–Canada	7,540	Albert, Uganda–Congo L.	2,162
Victoria, Kenya–Uganda–Tanganyika	26,828	Ladoga, Soviet Union	7,092	Vänern, Sweden	2,156
Aral, Soviet Union (salt)	26,518	Balkhash, Soviet Union	6,678	Winnipegosis, Canada	2,103
Huron, United States–Canada	23,010	Chad, Chad–Nigeria–Cameroon	△6,300	Bangweulu, Northern Rhodesia	△1,900
Michigan, United States	22,400	Onega, Soviet Union	3,821	Nipigon, Canada	1,870
Great Bear, Canada	12,275	Eyre, Australia (salt)	△3,700	Manitoba, Canada	1,817
Baykal, Soviet Union	12,159	Titicaca, Peru–Bolivia	3,500	Great Salt, United States (salt)	1,700
Great Slave, Canada	10,980	Athabasca, Canada	3,120	Koko Nor, China	1,650
Tanganyika, Congo L.–Tanganyika–Burundi– Northern Rhodesia	10,965	Nicaragua, Nicaragua	2,972	Dubawnt, Canada	1,600
Nyasa, Nyasaland–Tanganyika–Mozambique	10,900	Rudolf, Kenya–Ethiopia (salt)	2,473	Gairdner, Australia (salt)	△1,500
Erie, United States–Canada	9,940	Reindeer, Canada	2,467	Lake of the Woods, United States–Canada	1,485
		Issyk-Kul, Soviet Union	2,393	Van, Turkey (salt)	1,470
		Urmia, Iran (salt)	△2,229		

△ Due to seasonal fluctuations in water level, areas of these lakes vary considerably.

PRINCIPAL RIVERS OF THE WORLD

River	Length (miles)	River	Length (miles)	River	Length (miles)
Nile, Africa	4,132	Amu Darya, Asia	1,628	Si, Asia	930
Amazon, South America	3,900	Kolyma, Asia	1,615	Oka, Europe	920
Mississippi-Missouri–Red Rock, North America	3,860	Murray, Australia	1,600	Canadian, North America	906
Ob-Irtysh, Asia	3,461	Ganges, Asia	1,550	Dnestr, Europe	876
Yangtze, Asia	3,430	Pilcomayo, South America	1,550	Brazos, North America	870
Hwang Ho, Asia	2,903	Angara, Asia	1,549	Salado, South America	870
Congo, Africa	2,900	Ural, Asia	1,522	Fraser, North America	850
Amur, Asia	2,802	Vilyuy, Asia	1,513	Parnaíba, South America	850
Irtysh, Asia	2,747	Arkansas, North America	1,450	Colorado, North America	840
Lena, Asia	2,653	Colorado, North America	1,450	Rhine, Europe	820
Mackenzie, North America	2,635	Irrawaddy, Asia	1,425	Narbada, Asia	800
Mekong, Asia	2,600	Dnepr, Europe	1,420	Athabasca, North America	765
Niger, Africa	2,590	Aldan, Asia	1,392	Donets, Europe	735
Yenisey, Asia	2,566	Negro, South America	1,305	Pecos, North America	735
Missouri, North America	2,466	Paraguay, South America	1,290	Green, North America	730
Paraná, South America	2,450	Kama, Europe	1,261	Elbe, Europe	720
Mississippi, North America	2,348	Juruá, South America	1,250	James, North America	710
Plata-Paraguay, South America	2,300	Xingú, South America	1,230	Ottawa, North America	696
Volga, Europe	2,293	Don, Europe	1,224	White, North America	690
Madeira, South America	2,060	Ucayali, South America	1,220	Cumberland, North America	687
Indus, Asia	1,980	Columbia, North America	1,214	Gambia, Africa	680
Purús, South America	1,900	Saskatchewan, North America	1,205	Yellowstone, North America	671
St. Lawrence, North America	1,900	Peace, North America	1,195	Tennessee, North America	652
Rio Grande, North America	1,885	Orange, Africa	1,155	Gila, North America	630
Brahmaputra, Asia	1,800	Tigris, Asia	1,150	Vistula, Europe	630
Orinoco, South America	1,800	Sungari, Asia	1,140	Loire, Europe	625
São Francisco, South America	1,800	Pechora, Europe	1,118	Tagus, Europe	625
Yukon, North America	1,800	Tobol, Asia	1,093	North Platte, North America	618
Danube, Europe	1,770	Snake, North America	1,038	Albany, North America	610
Darling, Australia	1,750	Uruguay, South America	1,025	Tisza, Europe	607
Salween, Asia	1,730	Red, North America	1,018	Back, North America	605
Euphrates, Asia	1,675	Churchill, North America	1,000	Ouachita, North America	605
Syr Darya, Asia	1,653	Marañón, South America	1,000	Cimarron, North America	600
Zambezi, Africa	1,650	Ohio, North America	981	Sava, Europe	585
Tocantins, South America	1,640	Magdalena, South America	950	Nemunas (Niemen), Europe	582
Araguaia, South America	1,630	Roosevelt (River of Doubt), South America	950	Branco, South America	580
		Godavari, Asia	930	Oder, Europe	565

PRINCIPAL ISLANDS OF THE WORLD

Island	Area (sq. mi.)	Island	Area (sq. mi.)	Island	Area (sq. mi.)
Greenland, Arctic Region	840,000	Hispaniola, West Indies	29,530	Ceram, Indonesia	6,046
New Guinea, Oceania	316,856	Sakhalin, Soviet Union	29,344	New Caledonia, Oceania	5,671
Borneo, Indonesia	286,967	Banks, Canadian Arctic	23,230	Flores, Indonesia	5,513
Madagascar, Indian Ocean	227,800	Tasmania, Australia	26,215	Samar, Philippines	5,124
Baffin, Canadian Arctic	183,810	Ceylon, Indian Ocean	25,332	Negros, Philippines	4,903
Sumatra, Indonesia	182,859	Devon, Canadian Arctic	20,861	Palawan, Philippines	4,500
Honshū, Japan	88,930	Tierra del Fuego, Argentina-Chile	18,600	Panay, Philippines	4,448
Great Britain, North Atlantic Ocean	88,756	Kyūshū, Japan	16,215	Jamaica, West Indies	4,411
Ellesmere, Canadian Arctic	82,119	Melville, Canadian Arctic	16,141	Hawaii, Oceania	4,030
Victoria, Canadian Arctic	81,930	Southampton, Hudson Bay, Canada	15,700	Cape Breton, Canada	3,970
Celebes, Indonesia	72,986	West Spitsbergen, Arctic Region	15,260	Bougainville, Oceania	3,880
South Island, New Zealand	58,093	New Britain, Oceania	14,592	Mindoro, Philippines	3,794
Java, Indonesia	50,745	Formosa, China Sea	13,885	Cyprus, Mediterranean Sea	3,572
North Island, New Zealand	44,281	Hainan, South China Sea	13,127	Kodiak, Gulf of Alaska	3,569
Cuba, West Indies	44,217	Timor, Timor Sea	13,094	Puerto Rico, West Indies	3,435
Newfoundland, North Atlantic Ocean	42,734	Prince of Wales, Canadian Arctic	12,830	Corsica, Mediterranean Sea	3,367
Luzon, Philippines	40,814	Vancouver, Canada	12,408	Crete, Mediterranean Sea	3,238
Iceland, North Atlantic Ocean	39,800	Sicily, Mediterranean Sea	9,925	New Ireland, Oceania	3,205
Mindanao, Philippines	36,906	Somerset, Canadian Arctic	9,370	Leyte, Philippines	3,090
Ireland, North Atlantic Ocean	32,596	Sardinia, Mediterranean Sea	9,301	Wrangel, Soviet Arctic	2,819
Novaya Zemlya, Soviet Arctic	31,390	Shikoku, Japan	7,245	Guadalcanal, Oceania	2,500
Hokkaidō, Japan	29,950	North East Land, Svalbard Group	6,350	Long Island, United States	1,620

LARGEST METROPOLITAN AREAS AND CITIES OF THE WORLD, 1962

This table lists all metropolitan areas in the world with 1,000,000 or more population. For ease of comparison, each metropolitan area has been defined by Rand McNally & Company according to consistent rules. A metropolitan area includes a central city, neighboring communities linked to it by continuous built-up areas, and more distant communities if the bulk of their population is supported by commuters to the central city. All populations are estimates for January 1, 1962. The "city proper" figures refer to the area locally considered to be the city, provided it is under a single municipal government. Some metropolitan areas, such as Tōkyō–Yokohama, have more than one central city; in such cases the "city proper" figure is for the first-named city only.

Rank 1962		Estimated Population, 1/1/1962 Metropolitan Area	City Proper
1	New York, New York	15,775,000	7,775,000
2	Tōkyō–Yokohama, Japan	14,700,000	8,600,000
3	London, England	10,900,000	3,190,000
4	Ōsaka–Kōbe, Japan	8,350,000	3,125,000
5	Moscow (Moskva), Soviet Union	8,200,000	6,275,000
6	Shanghai, China	7,800,000	10,700,000▲
7	Paris, France	7,750,000	3,050,000
8	Buenos Aires, Argentina	7,175,000	2,965,000
9	Los Angeles, California	6,955,000	2,565,000
10	Chicago, Illinois	6,735,000	3,540,000
11	Calcutta, India	6,450,000	2,950,000
12	Mexico City, Mexico	5,150,000	2,775,000
13	São Paulo, Brazil	4,900,000	4,100,000
14	Rio de Janeiro, Brazil	4,900,000	3,425,000
15	Essen–Dortmund–Duisburg, Germany (West)	4,825,000	733,000
16	Bombay, India	4,475,000	4,250,000
17	Detroit–Windsor, Michigan–Canada	4,110,000	1,645,000
18	Philadelphia, Pennsylvania	4,100,000	1,995,000
19	Peking, China	4,000,000	6,800,000▲
20	Cairo (Al Qāhirah), United Arab Republic (Egypt)	4,000,000	3,400,000
21	Berlin, Germany	4,000,000	2,175,000
22	Leningrad, Soviet Union	3,875,000	3,050,000
23	San Francisco–Oakland–San Jose, California	3,500,000	745,000
24	Tientsin, China	3,200,000	3,600,000▲
25	Boston, Massachusetts	3,000,000	680,000
26	Djakarta, Indonesia	2,950,000	2,400,000
27	Hong Kong (Victoria), Hong Kong	2,850,000	1,150,000
28	Manchester, England	2,815,000	658,000
29	Delhi–New Delhi, India	2,750,000	2,150,000
30	Seoul, Korea (South)	2,700,000	2,550,000
31	Manila, Philippines	2,650,000	1,170,000
32	Mukden (Shenyang), China	2,600,000	2,600,000
33	Birmingham, England	2,590,000	1,112,000
34	Milano (Milan), Italy	2,575,000	1,590,000
35	Wuhan, China	2,500,000	2,500,000
36	Madrid, Spain	2,430,000	2,325,000
37	Rome (Roma), Italy	2,325,000	2,175,000
38	Hamburg, Germany (West)	2,280,000	1,850,000
39	Sydney, Australia	2,265,000	171,000
40	Bangkok (Krung Thep), Thailand	2,250,000	1,500,000
41	Santiago, Chile	2,225,000	645,000
42	Montreal, Canada	2,150,000	1,200,000
43	Cleveland, Ohio	2,150,000	870,000
44	Budapest, Hungary	2,150,000	1,835,000
45	Washington, D.C	2,140,000	755,000
46	St. Louis, Missouri	2,115,000	730,000
47	Johannesburg–Germiston, South Africa	2,100,000	580,000
48	Barcelona, Spain	2,070,000	1,585,000
49	Chungking, China	2,050,000	2,400,000▲
50	Madras, India	2,050,000	1,750,000
51	Lima, Peru	2,025,000	1,750,000
52	Nagoya, Japan	2,025,000	1,655,000
53	Canton, China	2,025,000	2,025,000
54	Vienna (Wien), Austria	2,005,000	1,630,000
55	Karachi, Pakistan	2,000,000	1,675,000
56	Pittsburgh, Pennsylvania	1,990,000	595,000
57	Melbourne, Australia	1,960,000	76,000
58	Brussels (Bruxelles), Belgium	1,930,000	170,000
59	Katowice–Zabrze–Bytom, Poland	1,925,000	273,000
60	Istanbul, Turkey	1,900,000	1,510,000
61	Athens (Athinai), Greece	1,890,000	635,000
62	Glasgow, Scotland	1,885,000	1,055,000
63	Toronto, Canada	1,860,000	673,000
64	Tehrān, Iran	1,850,000	1,750,000
65	Harbin, China	1,800,000	1,800,000
66	Singapore, Malaysia	1,720,000	1,040,000
67	Napoli (Naples), Italy	1,700,000	1,183,000
68	Saigon, Vietnam (South)	1,700,000	1,500,000
69	Baltimore, Maryland	1,695,000	936,000
70	Donetsk–Makeyevka, Soviet Union	1,675,000	775,000
71	Liverpool, England	1,660,000	744,000
72	Amsterdam, Netherlands	1,650,000	865,000
73	Sian (Hsian), China	1,600,000	1,600,000
74	Nanking, China	1,600,000	1,600,000
75	Caracas, Venezuela	1,550,000	800,000
76	Havana (Habana), Cuba	1,550,000	885,000
77	Alexandria (Al Iskandarīya), United Arab Republic (Egypt)	1,525,000	1,460,000
78	Warsaw (Warszawa), Poland	1,515,000	1,170,000
79	Kyōto, Japan	1,510,000	1,300,000
80	Minneapolis–St. Paul, Minnesota	1,505,000	480,000
81	Köln (Cologne), Germany (West)	1,500,000	820,000
82	München (Munich), Germany (West)	1,450,000	1,135,000
83	Frankfurt am Main, Germany (West)	1,405,000	690,000
84	Gorkiy, Soviet Union	1,380,000	1,035,000
85	Lahore, Pakistan	1,375,000	1,250,000
86	Stuttgart, Germany (West)	1,375,000	645,000
87	Bucharest (Bucureşti), Romania	1,375,000	1,245,000
88	Kitakyūshū–Shimonoseki, Japan	1,370,000	910,000
89	Buffalo–Niagara Falls, New York	1,370,000	525,000
90	Copenhagen (Köbenhavn), Denmark	1,365,000	722,000
91	Lisbon (Lisboa), Portugal	1,350,000	820,000
92	Houston, Texas	1,335,000	970,000
93	Leeds–Bradford, England	1,335,000	512,000
94	Miami–Fort Lauderdale, Florida	1,330,000	296,000
95	Ahmadabad, India	1,325,000	1,175,000
96	Hyderabad, India	1,325,000	940,000
97	Kiev, Soviet Union	1,300,000	1,210,000
98	Milwaukee, Wisconsin	1,285,000	750,000
99	Chengtu, China	1,275,000	1,275,000
100	Tsingtao (Chingtao), China	1,275,000	1,275,000
101	Torino (Turin), Italy	1,255,000	1,030,000
102	Bangalore, India	1,250,000	915,000
103	Cincinnati, Ohio	1,250,000	503,000
104	Taiyüan, China	1,200,000	1,200,000
105	Bogotá, Colombia	1,200,000	1,050,000
106	Dairen (Talien), China	1,200,000	1,150,000
107	Kharkov, Soviet Union	1,200,000	998,000
108	Baku, Soviet Union	1,185,000	685,000
109	Pusan, Korea (South)	1,185,000	1,185,000
110	Montevideo, Uruguay	1,175,000	1,070,000
111	San Diego–Tijuana, California–Mexico	1,165,000	610,000
112	Fushun, China	1,150,000	1,150,000
113	Newcastle-on-Tyne, England	1,145,000	266,000
114	Mannheim–Ludwigshafen–Heidelberg, Germany (West)	1,140,000	318,000
115	Stockholm, Sweden	1,140,000	806,000
116	Tashkent, Soviet Union	1,135,000	1,000,000
117	Changchun (Hsinking), China	1,125,000	1,125,000
118	Recife, Brazil	1,100,000	830,000
119	Dallas, Texas	1,090,000	710,000
120	Casablanca, Morocco	1,080,000	1,015,000
121	Taipei, Taiwan	1,075,000	925,000
122	Surabaja, Indonesia	1,075,000	1,000,000
123	Kansas City, Missouri	1,070,000	510,000
124	Novosibirsk, Soviet Union	1,065,000	1,005,000
125	Atlanta, Georgia	1,060,000	500,000
126	Prague, Czechoslovakia	1,060,000	1,005,000
127	Düsseldorf, Germany (West)	1,035,000	705,000
128	Kunming, China	1,025,000	1,025,000
129	Bandung, Indonesia	1,025,000	975,000
130	Kanpur, India	1,025,000	910,000
131	Kuybyshev, Soviet Union	1,010,000	890,000
132	Sverdlovsk, Soviet Union	1,000,000	860,000
133	Rotterdam, Netherlands	1,000,000	729,000
134	Antwerpen (Antwerp), Belgium	1,000,000	255,000

▲ *Municipal boundaries of Shanghai, Peking, Tientsin, and Chungking now include extensive rural zones, which have been excluded in estimating their metropolitan populations.*

PRINCIPAL WORLD CITIES AND POPULATIONS

This table includes all cities with 400,000 or more population, as well as many smaller cities of importance. The populations for all United States cities and for foreign cities listed in the table of World Metropolitan Areas above are estimates for January 1, 1962. For other cities, the populations are recent census figures or official estimates. Metropolitan populations are given for as many cities as possible, and identified by a star symbol (*). Some metropolitan areas, such as Minneapolis-St. Paul, include more than one large city. In such cases, the entry for the first-named city carries the entire metropolitan population, and other cities in the metropolitan area carry a reference to the first-named city with a star symbol.

A

Aachen, Germany (West) (*430,000)	169,800
Abadan, Iran	226,103
Abidjan, Ivory Coast	180,000
Accra, Ghana (491,060▲)	388,231
Addis Ababa, Ethiopia	500,000
Adelaide, Australia (*630,000)	23,051
Aden, Aden (*138,441)	99,285
Agra, India (*508,680)	462,020
Ahmadabad, India (*1,325,000)	1,175,000
Akron, Ohio (*590,000)	293,000
Albany, New York (*600,000)	129,000
Aleppo (Halab), Syria	425,500
Alexandria (Al Iskandarīyah), U.A.R. (*1,525,000)	1,490,000
Algiers (Alger), Algeria (*587,570)	361,285
Allahabad, India (*430,730)	411,955
Alma-Ata, Soviet Union	508,000
Amagasaki, Japan (*Ōsaka)	405,955
'Ammān, Jordan	108,304
Amoy (Hsiamen), China	224,300
Amritsar, India (*398,047)	376,295
Amsterdam, Netherlands (*1,650,000)	865,000
Ankara (Angora), Turkey	646,200
Anshan, China	805,000
Antwerpen (Antwerp), Belgium (*1,000,000)	255,000
Apia, Western Samoa	16,000
Arkhangelsk (Archangel), Soviet Union	271,000
Ashkhabad, Soviet Union	187,000
Astrakhan, Soviet Union	313,000
Asunción, Paraguay	205,605
Athens (Athinai), Greece (*1,890,000)	635,000
Atlanta, Georgia (*1,060,000)	500,000
Auckland, New Zealand (*448,365)	143,583

B

Baghdad, Iraq (*650,000)	355,958
Baku, Soviet Union (*1,185,000)	685,000
Baltimore, Maryland (*1,695,000)	936,000
Bamako, Mali	68,200
Banaras (Benares), India (*489,864)	471,258
Bandung, Indonesia (*1,025,000)	975,000
Bangalore, India (*1,250,000)	915,000
Bangkok (Krung Thep), Thailand (*2,250,000)	1,500,000
Bangui, Central African Republic	80,000
Barcelona, Spain (*2,070,000)	1,585,000
Barnaul, Soviet Union	338,000
Barranquilla, Colombia	356,920
Basel (Bâle), Switzerland (*460,000)	206,746
Basra, Iraq	164,623
Bathurst, Gambia	21,022
Beirut (Beyrouth), Lebanon	400,000
Belém (Pará), Brazil (*405,000)	359,988
Belfast, Northern Ireland (*575,000)	416,094
Belgrade (Beograd), Yugoslavia	587,899
Belo Horizonte, Brazil (*775,000)	642,912
Bengasi (Banghāzī), Libya	69,718
Bergen, Norway (*205,000)	115,900
Berlin, East, Germany (*Berlin)	1,055,300
Berlin, West, Germany (*4,000,000)	2,175,000
Bern (Berne), Switzerland (*238,000)	163,172

PRINCIPAL WORLD CITIES AND POPULATIONS *(Continued)*

Bilbao, Spain (*565,000)...297,942
Birmingham, Alabama (*638,000)..344,000
Birmingham, England (*2,590,000)...1,112,000
Bochum, Germany (West) (*Essen).361,400
Bogotá, Colombia (*1,200,000)...1,050,000
Bologna, Italy...444,872
Bombay, India (*4,475,000)...4,250,000
Bonn, Germany (West) (*280,000).143,900
Bordeaux, France (*480,000)...249,688
Boston, Massachusetts (*3,000,000).680,000
Bradford, Eng and (*Leeds)...295,768
Brasília, Brazil...150,000
Bratislava, Czechoslovakia...242,100
Brazzaville, Congo...99,000
Bremen, Germany (West)...564,500
Bridgetown, Barbados (*110,000)...11,289
Brighton, England (*400,000)...162,757
Brisbane, Australia (*625,000)...593,668
Bristol, England (*595,000)...436,440
Brno, Czechoslovakia...314,400
Brussels (Bruxelles), Belgium (*1,930,000)...170,000
Bucharest (Bucureşti), Romania (*1,375,000)...1,245,000
Budapest, Hungary (*2,150,000).1,835,000
Buenos Aires, Argentina (*7,175,000)...2,965,000
Buffalo, New York (*1,370,000)...525,000
Bulawayo, Southern Rhodesia (*195,500)...160,000

C

Cairo (Al Qāhirah), U.A.R. (*4,000,000)...3,400,000
Calcutta, India (*6,450,000)...2,950,000
Calgary, Canada (*279,062)...249,641
Cali, Colombia...429,170
Canberra, Australia...56,449
Canton, China (*2,025,000)...2,025,000
Cape Town, South Africa (*807,211)...508,341
Caracas, Venezuela (*1,550,000)...800,000
Cardiff, Wales (*600,000)...256,270
Casablanca, Morocco (*1,080,000)...1,015,000
Catania, Italy...363,928
Cebu, Philippines...259,200
Changchun (Hsinking), China (*1,125,000)...1,125,000
Changsha, China...703,000
Charleroi, Belgium (*375,000)...26,175
Chelyabinsk, Soviet Union (*910,000)...733,000
Chengchow, China...766,000
Chengtu, China (*1,275,000)...1,275,000
Chicago, Illinois (*6,735,000)...3,540,000
Chittagong, Pakistan (*364,205)...180,000
Christchurch, New Zealand (*220,510)...151,671
Chungking (Chungching), China (2,400,000▲)...*2,050,000
Cincinnati, Ohio (*1,250,000)...503,000
Cleveland, Ohio (*2,150,000)...870,000
Colombo, Ceylon (*725,000)...424,816
Columbus, Ohio (*745,000)...483,000
Conakry, Guinea (*109,500)...43,000
Concepción, Chile (*295,000)...150,000
Copenhagen (København), Denmark (*1,365,000)...722,000
Córdoba, Argentina...580,000
Cork, Ireland (*118,000)...77,980
Coventry, England (*560,000)...305,060

D

Dacca, Pakistan (*720,000)...556,712
Dairen (Talien), China (*1,200,000)...1,150,000
Dakar, Senegal (*270,000)...231,000
Dallas, Texas (*1,090,000)...710,000
Damascus (Dimashq), Syria...530,000
Dar es Salaam, Tanganyika...128,742
Dayton, Ohio (*675,000)...264,000
Delhi, India (*2,750,000)...2,150,000
Denver, Colorado (*905,000)...505,000
Derby, England (*260,000)...132,325
Detroit, Michigan (*4,110,000)...1,645,000
Djakarta (Batavia), Indonesia (*2,950,000)...2,400,000
Dnepropetrovsk, Soviet Union (*795,000)...707,000
Donetsk, (Stalino), Soviet Union (*1,675,000)...775,000
Dortmund, Germany (West) (*Essen)...641,500
Dresden, Germany (East) (*700,000)...491,700
Dublin (Baile Átha Cliath), Ireland (*690,000)...537,448
Duisburg, Germany (West) (*Essen)...503,000
Durban, South Africa (*659,934)...560,010
Dushanbe, Soviet Union...224,000
Düsseldorf, Germany (West) (*1,035,000)...705,000

E

Edinburgh, Scotland (*600,000)...468,378
Edmonton, Canada (*337,568)...281,027
Elisabethville, The Congo...183,000
El Paso, Texas (*615,000)...294,000
Entebbe, Uganda...10,941
Eşfahān (Isfahan), Iran...254,876
Essen, Germany (West) (*4,825,000)...733,000

F

Firenze (Florence), Italy (*535,000).436,516
Flint, Michigan (*392,000)...198,000
Foochow, China...616,000
Fortaleza, Brazil (*525,000)...354,942
Fort-Lamy, Chad...23,470
Fort Worth, Texas (*535,000)...365,000
Frankfurt [am Main], Germany (West) (*1,405,000)...690,000
Freetown, Sierra Leone...100,000
Frunze, Soviet Union...252,000
Fukuoka, Japan (*790,000)...647,122
Fushun, China (*1,150,000)...1,150,000

G

Gdańsk (Danzig), Poland (*500,000)...286,000
Gelsenkirchen, Germany (West) (*Essen)...382,700
Genève (Geneva), Switzerland (*270,000)...176,183
Genova (Genoa), Italy (*825,000)..784,194
Gent (Ghent), Belgium (*330,000)..157,811
Georgetown, British Guiana...92,000
Glasgow, Scotland (*1,885,000)..1,055,000
Gorkiy (Gorki), Soviet Union (*1,380,000)...1,035,000
Göteborg, Sweden (*505,000)...408,300
Grand Rapids, Michigan (*380,000)...204,000
Guadalajara, Mexico (*635,000)...580,617
Guatemala, Guatemala...284,276
Guayaquil, Ecuador...258,966
Gwalior, India...300,587

H

Haifa, Israel (*250,000)...182,007
Halle [an der Saale], Germany (East) (*425,000)...276,200
Hamburg, Germany (West) (*2,280,000)...1,850,000
Hamilton, Canada (*395,189)...273,991
Hangchow, China...784,000
Hannover (Hanover), Germany (West) (*710,000)...573,100
Hanoi, Vietnam (North)...237,500
Harbin, China (*1,800,000)...1,800,000
Hartford, Connecticut (*800,000)..160,000
Havana (Habana), Cuba (*1,550,000)...885,000
Heidelberg, Germany (West) (*Mannheim)...125,300
Helsinki, Finland (*545,000)...452,800
Hiroshima, Japan (*560,000)...431,336
Honolulu, Hawaii (*515,000)...306,000
Houston, Texas (*1,335,000)...970,000
Howrah, India (*Calcutta)...512,598
Huhehot (Kweisui), China...314,000
Hull (Kingston-upon-Hull), England (*360,000)...303,268
Hyderabad, India (*1,325,000)...940,000
Hyderabad, Pakistan (*434,537)...420,000

I

Ibadan, Nigeria...459,196
Inchŏn, Korea (South)...402,000
Indianapolis, Indiana (*830,000)...480,000
Indore, India...394,941
Irkutsk, Soviet Union...380,000
Istanbul, Turkey (*1,900,000)...1,510,000
Ivanovo, Soviet Union...352,000
Izmir (Smyrna), Turkey...370,900

J

Jabalpur (Jubbulpore), India (*367,014)...295,375
Jacksonville, Florida (*475,000)...201,000
Jaipur, India...403,444
Jamshedpur, India (*328,044)...291,791
Jersey City, New Jersey (*New York)...273,000
Jerusalem, Israel (*240,000)...166,301
Jerusalem, Jordan (*Jerusalem)...46,713
Johannesburg, South Africa (*2,100,000)...580,000
Juddah, Saudi Arabia...160,000

K

Kabul, Afghanistan...206,208
Kaliningrad (Königsberg), Soviet Union...226,000
Kampala, Uganda (*123,332)...46,735
Kano, Nigeria (*130,173)...93,016
Kanpur (Cawnpore), India (*1,025,000)...910,000
Kansas City, Missouri (*1,070,000)...510,000
Karachi, Pakistan (*2,000,000)...1,675,000
Karaganda, Soviet Union...441,000
Karl-Marx-Stadt (Chemnitz), Germany (East) (*400,000)...286,100
Karlsruhe, Germany (West) (*335,000)...241,900
Katmandu, Nepal...106,579
Katowice, Poland (*1,925,000)...273,000
Kaunas, Soviet Union...232,000
Kawasaki, Japan (*Tōkyō)...632,975
Kazan, Soviet Union...693,000
Khabarovsk, Soviet Union...349,000
Kharkov, Soviet Union (*1,200,000)...998,000
Khartoum, Sudan (*260,000)...93,103
Kiev (Kiyev), Soviet Union (*1,300,000)...1,210,000
Kigali, Rwanda...4,000
Kingston, Jamaica...421,718
Kirin, China...568,000
Kishinev, Soviet Union...236,000
Kitakyūshū, Japan (*1,370,000)...910,000
Kōbe, Japan (*Ōsaka)...1,113,977
Köln (Cologne), Germany (West) (*1,500,000)...820,000
Kowloon (*Victoria), Hong Kong...726,926
Krakow (Cracow), Poland...479,000
Krasnodar, Soviet Union...343,000
Krasnoyarsk, Soviet Union...468,000
Krivoy Rog, Soviet Union...436,000
Kuala Lumpur, Malaysia (*400,000)...316,230
Kumamoto, Japan...373,922
Kunming, China (*1,025,000)...1,025,000
Kuwait, Kuwait (*125,929)...104,551
Kuybyshev, Soviet Union (*1,010,000)...890,000
Kweiyang, China...504,000
Kyoto, Japan (*1,510,000)...1,300,000

L

Lagos, Nigeria...364,000
Lahore, Pakistan (*1,375,000)...1,250 000
Lanchow, China...699,000
La Paz, Bolivia...321,073
La Plata, Argentina (*410,000)...295,000
Leeds, England (*1,335,000)...512,000
Le Havre, France (*220,000)...183,776
Leicester, England (*420,000)...273,298
Leipzig, Germany (East) (*800,000)...585,300
Leningrad, Soviet Union (*3,875,000)...3,050,000
Léopoldville, The Congo...390,000
Lhasa, China...50,000
Libreville, Gabon...19,700
Liège, Belgium (*550,000)...153,240
Lille, France (*865,000)...193,096
Lima, Peru (*2,025,000)...1,750,000
Lisbon (Lisboa), Portugal (*1,350,000)...820,000
Liverpool, England (*1,660,000)...744,000
Łódź, Poland (*850,000)...708,000
Lomé, Togo...69,448
London, England (*10,900,000)..3,190,000
Long Beach, California (*Los Angeles)...350,000
Los Angeles, California (*6,955,000)...2,565,000
Louisville, Kentucky, (*765,000)...390,000
Lourenço Marques, Mozambique...99,000
Luanda, Angola...189,590
Lucknow, India (*675,000)...595,440
Lusaka, Northern Rhodesia (*85,800)...75,000
Luxembourg, Luxembourg...71,653
Lvov, Soviet Union...436,000
Lyon (Lyons), France (*920,000)...528,535

M

Macao, Macao (*169,299)...153,630
Madras, India (*2,050,000)...1,750,000
Madrid, Spain (*2,430,000)...2,325,000
Madura, India...361,781
Magdeburg, Germany (East) (*360,000)...262,400
Magnitogorsk, Soviet Union...328,000
Makasar (Macassar), Indonesia...367,882
Makeyevka, Soviet Union (*Donetsk)...381,000
Managua, Nicaragua...226,300
Manchester, England (*2,815,000).658,000
Manila, Philippines (*2,650,000).1,170,000
Mannheim, Germany (West) (*1,140,000)...318,000
Maracaibo, Venezuela...432,902
Marrakech, Morocco...243,134
Marseille (Marseilles), France (*870,000)...778,071
Mecca (Makkah), Saudi Arabia...200,000
Medan, Indonesia...466,370
Medellín, Colombia (*700,000)...485,250
Melbourne, Australia (*1,960,000). 76,000
Memphis, Tennessee (*658,000)...515,000
Mexico City, Mexico (*5,150,000)...2,775,000
Miami, Florida (*1,330,000)...296,000
Middlesbrough, England (*530,000)...157,308
Milano (Milan), Italy (*2,575,000)...1,590,000
Milwaukee, Wisconsin (*1,285,000)...750,000
Minneapolis, Minnesota (*1,505,000)...480,000
Minsk, Soviet Union...570,000
Mogadiscio, Somali Rep...77,000
Mombasa, Kenya...189,800
Monrovia, Liberia...41,391
Monterrey, Mexico (*665,000)...596,993
Montevideo, Uruguay (*1,175,000)...1,070,000
Montreal, Canada (*2,150,000)..1,200,000
Moscow (Moskva), Soviet Union (*8,200,000)...6,275,000
Mosul, Iraq...179,646
Mukden (Shenyang), China (*2,600,000)...2,600,000
München (Munich), Germany (West) (*1,450,000)...1,135,000
Murmansk, Soviet Union...237,000
Muscat, Muscat & Oman...5,000

N

Nagasaki, Japan...344,153
Nagoya, Japan (*2,025,000)...1,655,000
Nagpur, India (*700,000)...643,659
Nairobi, Kenya...297,000
Nanchang, China...508,000
Nanking, China (*1,600,000)...1,600,000
Nantes, France (*335,000)...240,028
Napoli (Naples), Italy (*1,700,000)...1,183,000
Nashville, Tennessee (*428,000)...260,000
Newark, New Jersey (*New York)..400,000
Newcastle-on-Tyne, England (*1,145,000)...266,000
New Delhi, India (*Delhi)...261,545
New Orleans, Louisiana (*920,000)...635,000
New York, New York (*15,775,000)...7,775,000
Niamey, Niger...30,000
Nice, France...292,958
Nicosia, Cyprus (*81,744)...48,864
Norfolk, Virginia (*593,000)...310,000
Nottingham, England (*615,000)...311,645
Novokuznetsk (Stalinsk), Soviet Union...405,000
Novosibirsk, Soviet Union (*1,065,000)...1,005,000
Nürnberg (Nuremberg), Germany (West) (*655,000)...454,500

O

Oakland, California (*San Francisco)...365,000
Odessa, Soviet Union...696,000
Oklahoma City, Oklahoma (*475,000)...345,000
Omaha, Nebraska (*455,000)...320,000
Omsk, Soviet Union...630,000
Ōsaka, Japan (*8,350,000)...3,125,000
Oslo, Norway (*635,000)...477,100
Ottawa, Canada (*429,750)...268,206
Ouagadougou, Upper Volta...63,000

P

Palembang, Indonesia...458,661
Palermo, Italy...587,985
Panamá, Panama...273,440
Paotow, China...400,000
Papeete, French Polynesia...12,428
Paramaribo, Surinam...75,233
Paris, France (*7,750,000)...3,050,000
Patna, India (*450,000)...363,700
Peking (Peiping), China (6,800,000▲)...*4,000,000
Penang (George Town), Malaysia (*325,000)...234,903
Perm, Soviet Union...678,000
Perth, Australia (*450,000)...94,508
Philadelphia, Pennsylvania (*4,100,000)...1,995,000
Phnom Penh, Cambodia...420,000
Phoenix, Arizona (*685,000)...475,000
Pittsburgh, Pennsylvania (*1,990,000)...595,000
Plymouth, England (*250,000)...204,279
Poona, India (*800,000)...597,562
Port-au-Prince, Haiti...134,117
Port Elizabeth, South Africa (*274,180)...249,211
Portland, Oregon (*750,000)...371,000
Port Louis, Mauritius...60,500
Pôrto (Oporto), Portugal (*750,000)...303,400
Pôrto Alegre, Brazil (*850,000)...617,629
Port-of-Spain, Trinidad & Tobago (*170,000)...94,100
Porto Novo, Dahomey...30,800
Port Said (Būr Sa'īd), U.A.R...212,973
Portsmouth, England (*415,000)...215,198
Poznan, Poland...408,000
Prague (Praha), Czechoslovakia (*1,060,000)...1,005,000
Pretoria, South Africa (*422,590)..303,684
Providence, Rhode Island (*810,000)...200,000
Pusan, Korea (South) (*1,185,000)...1,185,000
Pyŏngyang, Korea (North)...653,100

Q

Quebec, Canada (*357,568)...171,979
Quezon City, Philippines (*Manila)...397,400
Quito, Ecuador...209,932

R

Rabat, Morocco (*310,000)...227,445
Rangoon, Burma...752,000
Rawalpindi, Pakistan (*340,175)...250,000
Recife (Pernambuco), Brazil (*1,100,000)...830,000
Reykjavík, Iceland (*90,000)...73,388
Richmond, Virginia (*420,000)...218,000
Riga, Soviet Union...607,000
Rio de Janeiro, Brazil (*4,900,000)...3,425,000
Riyadh (Ar Riyāḍ), Saudi Arabia..150,000
Rochester, New York (*615,000)...316,000
Rome (Roma), Italy (*2,325,000).2,175,000
Rosario, Argentina...595,000
Rostov [-na-Donu], Soviet Union (*735,000)...645,000
Rotterdam, Netherlands (*1,000,000)...729,000
Rouen, France (*325,000)...120,857

S

Saarbrücken, Germany (West) (*350,000)...130,700
Sacramento, California (*595,000).205,000
Saigon, Vietnam (South) (*1,700,000)...1,500,000
St. Louis, Missouri (*2,115,000)...730,000
St. Paul, Minnesota (*Minneapolis)...313,500
St. Petersburg, Florida (*385,000)..193,000
Salisbury, Southern Rhodesia (*299,200)...190,000
Salt Lake City, Utah (*435,000)...190,000
Salvador, Brazil...630,878
Samarkand, Soviet Union...209,000
San'ā', Yemen...89,000
San Antonio, Texas (*725,000)...606,000
San Bernardino, California (*485,000)...94,500
San Diego, California (*1,165,000).610,000
San Francisco, California (*3,500,000)...745,000
San Jose, California (*San Francisco)...245,000
San José, Costa Rica (*260,000)...115,700
San Juan, Puerto Rico (*588,805)..432,377
San Salvador, Salvador (*350,000)...248,100
Santiago, Chile (*2,225,000)...645,000
Santo Domingo, Dominican Republic...367,053
Santos, Brazil (*400,000)...262,048
São Paulo, Brazil (*4,900,000)...4,100,000
Sapporo, Japan (*615,000)...523,839
Saratov, Soviet Union (*740,000)..622,000
Seattle, Washington (*975,000)...635,000
Semarang, Indonesia...487,006
Sendai, Japan (*515,000)...425,272
Seoul, Korea (South) (*2,700,000)...2,550,000
Sevastopol, Soviet Union...163,000
Sevilla (Seville), Spain...442,300
Shanghai, China (10,700,000▲)..*7,800,000
Sheffield, England (*725,000)...493,954
Shihchiachuang, China...598,000
Shizuoka, Japan (*485,000)...323,819
Sian (Hsian), China (*1,600,000).1,600,000
Singapore, Malaysia (*1,720,000)...1,040,000
Sofia (Sofiya), Bulgaria (*434,888).366,925
Soochow (Suchou), China...663,000
Southampton, England (*340,000).204,707
Springfield, Massachusetts (*450,000)...178,000
Srinagar, India (*295,084)...285,257
Stockholm, Sweden (*1,140,000)...806,000
Stoke-on-Trent, England (*430,000)...265,506
Strasbourg, France (*320,000)...228,971
Stuttgart, Germany (West) (*1,375,000)...645,000
Suchow, China...676,000
Sucre, Bolivia...40,128
Suez, U.A.R...162,826
Surabaja (Soerabaja), Indonesia (*1,075,000)...1,000,000
Suva, Fiji...37,371
Sverdlovsk, Soviet Union (*1,000,000)...860,000
Sydney, Australia (*2,265,000)...171,000
Syracuse, New York (*460,000)...216,000
Szczecin (Stettin), Poland...269,000

T

Tabrīz, Iran...290,195
Taegu, Korea (South)...678,300
Taipei (Taihoku), Taiwan (*1,075,000)...925,000
Taiyüan, China (*1,200,000)...1,200,000
Tallinn, Soviet Union...298,000
Tampa, Florida (*370,000)...288,000
Tananarive, Malagasy Republic...248,000
Tangier, Morocco...141,714
Tangshan, China...800,000
Tashkent, Soviet Union (*1,135,000)...1,000,000
Tbilisi, Soviet Union (*815,000)...724,000
Tegucigalpa, Honduras...133,887
Tehrān, Iran (*1,850,000)...1,750,000
Tel Aviv [-Jaffa], Israel (*640,000).386,612
The Hague ('s Gravenhage), Netherlands (*820,000)...605,900
Thessaloniki (Salonika), Greece (*373,635)...250,920
Tientsin, China (3,600,000▲)...*3,200,000
Tiranë, Albania...119,000
Tōkyō, Japan (*14,700,000)...8,600,000
Toledo, Ohio (*525,000)...324,000
Torino (Turin), Italy (*1,255,000)...1,030,000
Toronto, Canada (*1,860,000)...673,000
Torreón, Mexico (*300,000)...179,955
Trieste, Italy...272,723
Tripoli (Tarābulus), Libya...170,000
Tsinan (Chinan), China...862,000
Tsingtao (Chingtao), China (*1,275,000)...1,275,000
Tsitsihar, China...668,000
Tula, Soviet Union...333,000
Tulsa, Oklahoma (*382,000)...278,000
Tunis, Tunisia...410,000

U

Ufa, Soviet Union...588,000
Ulan Bator, Mongolia...70,000
Usumbura, Burundi...50,000
Utrecht, Netherlands (*395,000)...256,300

V

Valencia, Spain (*660,000)...505,066
Valletta, Malta (*181,414)...18,666
Valparaíso, Chile (*440,000)...255,000
Vancouver, Canada (*790,165)...384,522
Venezia (Venice), Italy...347,347
Victoria, Hong Kong (*2,850,000)...1,150,000
Vienna (Wien), Austria (*2,005,000)...1,630,000
Vientiane, Laos...100,000
Vilnius, Soviet Union...255,000
Vladivostok, Soviet Union...317,000
Volgograd (Stalingrad), Soviet Union (*725,000)...632,000
Voronezh, Soviet Union...496,000

W

Warsaw (Warszawa), Poland (*1,515,000)...1,170,000
Washington, D.C. (*2,140,000)...755,000
Wellington, New Zealand (*249,532)...123,969
Wichita, Kansas (*360,000)...260,000
Wiesbaden, Germany (West) (*500,000)...253,300
Willemstad, Netherlands Antilles (*90,000)...44,062
Winnipeg, Canada (*475,989)...265,429
Wroclaw (Breslau), Poland...429,000
Wuhan, China (*2,500,000)...2,500,000
Wuppertal, Germany (West) (*880,000)...420,700
Wusih, China...613,000

Y

Yaoundé, Cameroon...61,750
Yaroslavl, Soviet Union...433,000
Yerevan, Soviet Union...558,000
Yokohama, Japan (*Tōkyō)...1,375,710
Youngstown, Ohio (*475,000)...166,000

Z

Zagreb, Yugoslavia...427,319
Zanzibar, Zanzibar...45,284
Zaporozhye, Soviet Union...475,000
Zürich, Switzerland (*665,000)...440,170

* *Population of metropolitan area, including suburbs. See headnote.*
▲ *Population of entire municipality or district, including rural area. Starred population in these entries refers to urban portion of municipality only.*

HISTORICAL GAZETTEER

A

A B C Countries. Term applied to three South American countries: Argentina, Brazil, and Chile.

Abraham, Plains of. Battlefield near the city of Quebec where the English under Wolfe defeated the French under Montcalm, 1759.

Abydos. Ancient town on the Hellespont, site of the Bridge of Xerxes. Also an ancient Egyptian town on the Nile below Thebes.

Abyssinia. Former name of Ethiopia.

Acadia. Old French colonial territory bounded by the Atlantic, the river and gulf of St. Lawrence, and a line running north from Penobscot Bay.

Achaia. Separate regions of ancient Greece in southern Thessaly and northern Peloponnesus. Later, a Roman province embracing all but the northern part of modern Greece.

Acropolis. Hill in Athens, Greece, where some of the finest monuments of antiquity now stand.

Actium. Off this promontory at the entrance to the Gulf of Amvrakia in northwestern Greece, Octavius won a naval battle against Antony and Cleopatra, 31 B.C.

Aegospotami. Off the mouth of this river in ancient Thrace, the Spartans crushed the Athenian fleet in 405 B.C., in the final battle of the Peloponnesian War.

Aelia Capitolina. Jerusalem in late Roman times.

Aeolian Islands. Ancient name of the Lipari Islands, off the coast of Sicily.

Aetolia. District of ancient Greece along the north shore of the Gulf of Corinth.

Africa. Roman province corresponding with modern Tunisia. Name was later applied to entire continent.

Agassiz, Lake. Prehistoric lake temporarily created by the withdrawal of glaciers. Covered an area which includes parts of present-day Manitoba, Saskatchewan, Ontario, Minnesota, and North Dakota.

Agincourt. Village near Boulogne, France, where outnumbered English forces under Henry V defeated the French, 1415.

Agrigentum. Prosperous commercial center on the southern coast of Sicily in the fifth century, B.C. The modern Agrigento.

Ai. Old city of the Canaanites near Bethel. It was destroyed by Joshua.

Aix-la-Chapelle. Northern capital and residence of Charlemagne, and coronation site of later German emperors. Now Aachen, Germany.

Akkad. Very ancient land of the Akkadians in north Babylonia at the closest approach of the Tigris and Euphrates. The capital was Agade.

Albania. Ancient country west of the Caspian Sea in the territory roughly corresponding to Soviet Azerbaidzhan. Also the small Adriatic republic.

Albion. Ancient name of Britain.

Alsace-Lorraine. Region in northeastern France seized by Germany in 1871, and returned to France following World War I. The capital was at Strasbourg.

America. Inclusive name of the continents and adjacent islands of North and South America. It is also commonly used when referring to the United States.

Amphipolis. City of ancient Macedonia near the mouth of the Struma River, site of a Spartan victory over Athens in 422 B.C.

Anáhuac. Old Indian term, now applied to the high plateau containing Mexico City.

Anatolia. See Asia Minor.

Anau. Ruins of an ancient city in the desert near Ashkhabad, Soviet Union.

Ancyra. Important city of Phrygia, Asia Minor, and later capital of Galatia. The modern Ankara or Angora, the capital of Turkey.

Andalusia. Region comprising the eight provinces of southern Spain in basin of Guadalquivir River.

Angkor. Extensive ruins of magnificent old city in northern Cambodia.

Antarctic Ocean. Obsolete name for southern part of the Pacific, Atlantic, and Indian Oceans bordering the continent of Antarctica.

Antilles. Collective name for the islands of the West Indies which enclose the Caribbean Sea. They consist of the Greater Antilles (Cuba, Jamaica, Hispaniola, and Puerto Rico) to the north and the Lesser Antilles (which include a number of islands and groups) to the east.

Antioch. Capital of ancient Syria and one of the greatest commercial centers of the time, Antioch later became a stronghold of Christianity. The Roman Antiochia, it is now Antakya, Turkey. It was also a city of Pisidia, Asia Minor, prominent in the journeys of Paul.

Antipodes. Name occasionally applied to New Zealand and Australia, because of their location on the globe diametrically opposite the British Isles.

Apulia. Division of southeastern Italy, containing the important cities of Bari, Taranto, and Foggia.

Aquitania. Originally a region in Transalpine Gaul between the Pyrenees and the Garonne, later extending northward to the Loire.

Arabia. Large Asiatic peninsula extending south of Iraq and Jordan between the Red Sea, the Persian Gulf, and the Arabian Sea on the south.

Arab League. A group of Arab states organized in 1945, with headquarters in Cairo. Original members were Egypt, Iraq, Syria, Lebanon, Transjordan (Jordan), Saudi Arabia, and Yemen. The purpose of the League is to coordinate the foreign policy of the member states for the mutual benefit of all Arab countries.

Aragon. Medieval kingdom in northeastern Spain, whose conquests included the Balearic Islands, Sardinia, and Sicily. Its union with Castile in 1479 created the Spanish kingdom.

Aram. Biblical name of Syria. Also called Aramea.

Ararat. Hebrew name of an ancient kingdom in eastern Armenia. Noah's Ark reputedly came to rest on one of its mountains. The Assyrian Urartu.

Araucania. Old name for the land of the Araucanian Indians in southern Chile, between Concepción and Puerto Montt.

Araxes. Ancient name of the Aras River on the border of Armenia and Media.

Arbela. The modern town of Arbil, Iraq, east of Mozul, where Alexander the Great defeated a huge Persian army, 331 B.C.

Argolis. Region of ancient Greece on the east coast of the Peloponnesus. Its principal cities were Argos and Mycenae.

Ariel. Poetic name of Jerusalem.

Armageddon. Greek name of "The Hill of Megid-do," near an Israelite battlefield. In present usage, the name refers to a final battle between the powers of good and evil.

Armenia. Region of western Asia now roughly comprising northeastern Turkey, Soviet Armenia, and adjacent parts of Soviet and Iranian territory.

Arnon. Biblical name of the Wady Mojib, which flows from Jordan into the Dead Sea.

Ascalon. Also called Ashkelon, an ancient Philistine city on the Mediterranean north of Gaza, Palestine. The present village of Askalan is on its site.

Asculum. Capital of ancient Picenum and site of the original "Pyrrhic victory." Here, in 279 B.C., Pyrrhus of Epirus defeated the Romans, suffering very heavy losses. Asculum is modern Ascoli Piceno, Italy.

Ashdod. Ancient city of the Philistines, later called Azotus. Now it is a village in Israel.

Asia. Roman province comprising western part of Asia Minor. Name was later given to entire continent.

Asia Minor. Name of the peninsula in western Asia bounded by the Black, Aegean, and Mediterranean Seas. The area is nearly identical with Anatolia and contains a large part of Asiatic Turkey.

Asshur. First capital and original name of Assyria. Also Assur and Asur.

Assyria. Very ancient empire which developed on the right bank of the upper Tigris. The empire eventually extended from Elam in the east to Egypt and eastern Asia Minor in the west. Its power ended with the fall of the capital, Nineveh, 612 B.C.

Asturias. Located in northwestern Spain in the early Middle Ages, Asturias was the first Christian kingdom to be established on the Iberian peninsula.

Athos, Mount. Mountain southeast of Salonika (Thessaloniki), Greece, where a community of monks has maintained almost complete autonomy since the early Middle Ages.

Atlantis. Legendary island of great size in the Atlantic Ocean west of the Pillars of Hercules. After reaching an advanced state of civilization, it was supposedly destroyed by a subterranean cataclysm.

Attica. Ancient Greek state southeast of Boeotia. Its capital and chief city was Athens.

Augusta Treverorum. Capital of ancient Belgica; the modern Trier, Germany.

Austerlitz. Town near Brno, Czechoslovakia, where Napoleon defeated armies of Russia and Austria, 1805.

Australasia. That part of the southwest Pacific containing Australia, New Zealand, and the islands of Melanesia. Term is often used to include all Oceania.

Austria-Hungary. Dual monarchy which ruled an empire extending from Bohemia to Transylvania and from Galicia to the Adriatic between 1867 and 1918. Its capital was Vienna.

Axis. Term used to denote first the early understanding between Rome and Berlin and later the military alliance that developed in World War II between Germany, Italy, and their allies.

B

Baalbek. Also Baalbec and the Greek Heliopolis, an ancient city of Syria near headwaters of Litani River, now in Lebanon.

Babel. Biblical name of the city, probably Babylon, where the notorious tower of Babel was located.

Babylon. Capital city of ancient Babylonia on both sides of the lower Euphrates.

Babylonia. Very ancient and powerful kingdom—also called Shinar and Chaldea—in the lower valley of the Tigris and Euphrates. Its capital was Babylon. The Chaldean Empire eventually extended westward to the Mediterranean.

Bactria. Ancient country of central Asia between the Oxus River and the Hindu Kush Mountains. Its capital Bactra is the modern Balkh, Afghanistan.

Baden. Old grand duchy and state of Germany along the right bank of the upper Rhine. Karlsruhe was its capital.

Bad Lands. Name applied to barren, badly eroded areas in the western United States. The best known region is located in western South Dakota.

Balaklava. Town in the Crimea, southeast of Sevastopol. Near here the courageous "Charge of the Light Brigade" took place, 1854. Also Balaclava.

Balkan Peninsula. That part of southeastern Europe south of the Sava and Danube Rivers.

Balkan States. Countries located partly or entirely within the Balkan Peninsula: Yugoslavia, Albania, Greece, Romania, Bulgaria, and European Turkey.

Baltic States. Term used for the former republics of Estonia, Latvia, and Lithuania, bordering Baltic Sea. They were absorbed into Soviet Union in 1940.

Banat, The. Region in Central Europe bounded by the Transylvanian Alps and the Danube, Tisza, and Muresul rivers.

Banda Oriental. Old Spanish name for the "eastern shore" of Plata River, now the country of Uruguay.

Bannockburn. Village south of Stirling, Scotland, where Bruce defeated the English under Edward II, 1314, and assured the independence of Scotland.

Bantam. Seaport, formerly of great commercial importance, west of Djakarta, Java.

Barbary States. Old collective term for the countries of Africa along the "Barbary Coast" of the Mediterranean, between Egypt and the Atlantic Ocean.

Basque Provinces. Region along the Bay of Biscay extending from the areas around Bilbao and Vitoria, Spain, across the Pyrenees into southwestern France.

Batavia. Latin name for island home of Batavi on lower Rhine, later applied to all the Netherlands.

Behistun. Place near Kermanshah, Iran, where a famous rock carries ancient Assyrian inscriptions on its precipitous face.

Belgica. Northerly region of Transalpine Gaul between the Rhine and Seine. It eventually extended southward to the Rhône-Saône confluence and eastward to include most of Switzerland.

Benelux. Collective term for Belgium, the Netherlands, and Luxembourg.

Beneventum. The modern Benevento, Italy; in antiquity, the battleground for the decisive Roman victory over Pyrrhus of Epirus, 275 B.C.

Bengal. Former province in eastern British India between the Himalayas and the mouths of the Ganges. In 1947, most of the province, except the western part containing Calcutta, was absorbed into the new state of Pakistan.

Berea. Ancient Macedonian town near Mount Olympus, where Paul found willing converts to Christianity. It is now Verria, Greece.

Bessarabia. Region in southeastern Europe between the Prut and the Dnestr rivers. After its cession by Romania, in 1940, it became a part of the Soviet Union.

Bethabara. Site of the baptism of Jesus by John the Baptist. It may have been located at a ford of the Jordan 13 miles south of the Sea of Galilee.

Bethany. Famous village that once stood at the foot of the Mount of Olives, east of Jerusalem. The ascension of Christ took place near by.

Bethel. Shrine city of ancient Israel north of Jerusalem. The modern Beitin.

Beth-horon. Site of the great victory of Joshua over the Canaanites. The battle took place between the present villages of Beit Ur el Foka and Beit Ur el Tahta, midway between Jerusalem and Lydda.

Bethsaida. Ancient town at the north end of the Sea of Galilee where Jesus fed the five thousand.

Bimini. Island or region of West Indian legend where the Fountain of Youth was supposed to be located. The name has been given to an island group in the western Bahamas.

Black Country, The. Mining and manufacturing area in the vicinity of Birmingham, England.

Blarney. Village near Cork, Ireland, where the castle containing the famous Blarney Stone is located.

Boeotia. Ancient district in central Greece northwest of Attica. The chief city and capital was Thebes.

Bohemia. Medieval duchy and kingdom of varying frontiers and later a crownland of Austria-Hungary. After World War I, Bohemia became the western section of the new republic of Czechoslovakia. Its capital is Prague.

Bonneville, Lake. Extinct lake that covered northwestern Utah during the glacial period.

Borodino. Village 70 miles west of Moscow where Napoleon defeated the Russians, 1812.

Bosnia. Mountainous region south of the Sava River in Yugoslavia. It was formerly a medieval kingdom and part of the Austro-Hungarian Empire.

Bosporus. Name of an ancient Greek kingdom encircling the Sea of Azov, Soviet Union. This name was derived from the Cimmerian Bosporus, now called Kerch, or Yenikale Strait. Also name of famous strait at Istanbul, ancient Thracian Bosporus.

Bourbon. Medieval duchy in central France, the early home of the famous royal house. Also the county and province of Bourbonnais. Capital was Moulins.

Brabant. Old duchy containing the cities of Antwerp, Brussels, and Louvain. The territory is now divided between Belgium and the Netherlands.

Brandenburg. Former Prussia, now divided between Germany and Poland. Its capital was Potsdam.

Brandywine. Creek near Philadelphia where British defeated Americans commanded by Washington, 1777.

Britannia. Roman name for island of Great Britain, specifically, the southern part. Now used poetically.

British America. Name usually restricted to Canada.

British East Africa. Old descriptive name for the British territories of Kenya, Uganda, and Zanzibar. Also applied to Tanganyika.

British India. That part of the subcontinent of India before August, 1947, under the direct control of Great Britain. The relatively independent princely states were not considered in this category.

British West Africa. Name occasionally used collectively for the former British colonies in western Africa—Nigeria, Gold Coast, now Ghana, Sierra Leone, Gambia, Togo, and Cameroon.

Brundisium. Ancient Roman port on Adriatic and terminus of famed Appian Way. Now Brindisi, Italy.

Bukovina. Region on the eastern slope of the Carpathians, formerly a part of Austria-Hungary, now divided between the Soviet Ukraine in the north and Romania in the south. Also Bukowina.

Bull Run. Small river east of Manassas, in northern Virginia. Two battles were fought in this vicinity (1861 and 1862) between the armies of the Union and the Confederacy.

Bunker Hill. Famous hill in Charlestown, Mass. The battle of that name was actually fought on nearby Breed's Hill in 1775.

Burgenland. Fertile region along the Austro-Hungarian border. Its chief city is Sopron, Hungary.

Burgundy. Early medieval kingdom in southeastern France largely east of the Rhône and Saône and west of the Alps, including the western half of present-day Switzerland. Later, a duchy in the Seine, Loire, and Saône river basins of northeastern France.

Byblos. Important port in ancient Phoenicia north of Beirut, Lebanon. Called Gebal in Old Testament.

Bytown. Early name of Ottawa, Canada.

Byzantine Empire. Name for the Eastern Roman Empire, established in 330 A.D. Included in the Empire were the territories of modern U.A.R., Israel, Jordan, Syria, Turkey, Greece, and Bulgaria.

Byzantium. Ancient Greek city on the Hellespont. Constantinople was built on its site in 330 A.D. and became the capital of the Byzantine Empire.

C

Caesarea Philippi. Ancient town of northern Palestine at the foot of Mt. Hermon, near the modern Banias, Syria.

Calabria. Roman name for the "heel" of Italy, now applied to the "toe."

Caledonia. Roman name for that part of Scotland lying north of the Firths of Clyde and Forth. The name is now used poetically to include all Scotland.

Calvary. Latin word for "skull," an unidentified place outside Jerusalem where Christ was crucified. Also Golgotha.

Campagna di Roma. Extensive reclaimed lowlands around Rome, Italy, comprising roughly the territory of ancient Latium.

Campania. Division of Italy containing the cities of Naples and Salerno. Roman cities of importance were Capua and Nola.

Camulodunum. Early Roman city in Britannia; now Colchester, England.

Cana. Town where Jesus performed his first miracle. Probably the modern Kefr Kenna, Israel.

Canaan. The biblical land of Canaan lay west of the Jordan and the Dead Sea, and extended from Mount Lebanon to the southern deserts of Palestine.

Cannae. North of this ancient town in Apulia, a Roman army was almost completely destroyed by Hannibal, 216 B.C.

Capernaum. Biblical town on the west shore of the Sea of Galilee. It was often visited by Christ.

Cappadocia. Ancient country in eastern Asia Minor extending originally from Cilicia to the Euxine. Under the Romans, the region comprised the provinces of Cappadocia to the south and Pontus to the north.

Carchemish. Old capital of the Hittites on the Euphrates, at its closest approach to the Mediterranean. It was the site of a decisive Babylonian victory over Egypt, 605 B.C.

Caria. Ancient country south of the Meander River in southwest Asia Minor. The principal cities were Halicarnassus and Miletus.

Caribbees. General name for the Lesser Antilles.

Carinthia. Mountainous province in southern Austria, a former crownland of Austria-Hungary. The capital is Klagenfurt.

Carmel, Mount. Mountain ridge south of the Bay of Acre, prominent in the Old Testament.

Carniola. Medieval duchy, later a crownland of Austria-Hungary, located in what is now the northwestern part of Yugoslavia. Laibach, now called Ljubljana, was the principal city.

Carthage. Ancient Phoenician city in northern Africa near Tunis. In 146 B.C., it was destroyed by its great commercial rival, Rome.

Castile. Medieval kingdom in northern and central Spain which united with Aragon in 1479 to form the Spanish monarchy. Old Castile was in the northern part of kingdom and New Castile was in southern part.

Catalonia. Originally an independent state in the northeast corner of the Iberian Peninsula, Catalonia united with Aragon in the twelfth century. Barcelona is chief city of region. Spanish name is Cataluña.

Cathay. Old name for China, used by Marco Polo.

Caucasia. General name for that part of the Soviet Union between the Black and Caspian seas and traversed by the Caucasus Mountains.

Celestial Empire. Once a popular name of China.

Central America. Name applied to that portion of North America south of Mexico and often including Panama. Specifically, the area between the isthmuses of Tehuantepec and Panama.

Central Powers. Collective term formerly used for Germany and Austria-Hungary due to their location in central Europe. In World War I the definition was expanded to include Bulgaria and Turkey, the allies of Germany.

Chaco. Low-lying, swampy region in South America between Paraguay River and the Andes, in Argentina, Paraguay, and Bolivia. Also called Gran Chaco.

Chaeronea. Ancient town in western Boeotia, the site of the victory of Philip over the forces of Thebes and Athens, 338 B.C., which introduced the era of Macedonian supremacy.

Chalcedon. Ancient Greek town on the Bosporus opposite Byzantium.

Chalcidice. Three-armed peninsula of ancient Macedonia containing the cities of Olynthus and Potidaea. It was colonized and named by Euboeans from Chalcis—the modern Khalkis, Greece.

Chaldea. Ancient name for the lowlands at the head of the Persian Gulf, but commonly applied to all of Babylonia. The second and final empire of Babylonia was called the Chaldean Empire.

Champagne. Old province of northeastern France whose capital was at Troyes.

Chancellorsville. Village near Spotsylvania, Virginia, the scene of a victory gained over the Union army by the outnumbered Confederates, 1863.

Chapultepec. Rocky hill southwest of Mexico City, the site of Aztec fortifications and the summer residence of the viceroys and presidents of Mexico.

Charcas. Old Spanish colonial province, a part successively of the viceroyalties of Peru and La Plata. It corresponded roughly to modern Bolivia and extended westward to the Pacific.

Chichén Itzá. Principal city of the Mayas. Its ruins lie in the jungles of eastern Yucatán.

Chinnereth. Sea of Galilee in the Old Testament.

Chosen. Native name of Korea.

Christiania. Former name of Oslo, Norway.

Cibola. Old name given to a region of New Mexico containing seven cities, supposedly rich and powerful. Coronado identified them in 1540 as commonplace Zuni Indian villages.

Cimmeria. Ancient country of the Cimmerians, probably located along northern shores of Black Sea.

Cinnamon, Land of. Old name of the Napo River region in Peru, first explored in the 16th century.

Cinque Ports. In medieval England, a group of coastal towns in Kent and Sussex carrying special royal privileges and responsibilities. Included were the cities of Hastings and Dover.

Cipango, or Cipangu. Name given by Marco Polo to an island group east of Asia, probably Japan.

Circassia. Region in the southern Soviet Union, bordering on Kuban River, Black Sea, and Caucasus Mountains.

Cisalpine Republic. Short-lived state in the valley of the Po between Piedmont and Venetia. In 1802, it was reconstituted as the Italian Republic.

Cisleithania. Old name for the Austrian portion of Austria-Hungary to the west and north of Transleithania. Besides Austria, it included what is now western Czechoslovakia, Galicia, northern Bukovina, and Slovenia. Name is derived from Leitha River, which formed part of the boundary south of Vienna.

Colchis. Ancient country on the eastern shores of the Euxine south of the Caucasus, now part of Soviet Georgia. The legendary land of the Golden Fleece.

Colonia Agrippina. Roman colony on the Rhine; the modern Köln (Cologne), Germany.

Colossae. Early Christian stronghold of Phrygia, Asia Minor, whose people were recipients of Paul's Epistle to the Colossians. Also Colosse.

Columbia. Poetic name of the New World or, more specifically, the United States, in honor of Christopher Columbus.

Constantinople. Old name of Istanbul, Turkey, the ancient Byzantium.

Copais. Ancient lake northwest of Thebes, Boeotia. Recent drainage of the lake and its surrounding marshes has altered the map of central Greece.

Courland. Old Baltic duchy with the capital at Mitau. Most of its territory is now contained in the southern part of Soviet Latvia. Also called Kurland.

Crécy. Small town near Abbeville, France, where an English army badly defeated the French, 1346.

Crimea. Large peninsula of the Soviet Union on the northern coast of the Black Sea. Its chief cities are Simferopol, Sevastopol, and Yalta.

Croatia. Important state on the northwestern coast of the Adriatic in early Middle Ages. More recently it became, with Slavonia, a crownland of Austro-Hungarian Empire. Capital is Zagreb, in Yugoslavia.

Cumae. Ancient city, reputedly the earliest Greek settlement in Italy, 10 miles west of Naples.

Cunaxa. Ancient town of north Babylonia, the site of the Persian victory which precipitated the retreat of the Ten Thousand to Trepezus, 401 B.C.

Cush. Biblical name applied to ancient Ethiopia and also to a land in Mesopotamia. Also called Kush.

Cynoscephalae. Heights in southeastern Thessaly, Greece, the site of a Roman victory over the Macedonians, 197 B.C.

Cyzicus. Ancient name for a city and peninsula on the south coast of the Propontis in Asia Minor. In 410 B.C., near here, the Athenian navy won a victory over Sparta.

D

Dacia. Ancient territory of Rome which lay north of the Danube in an area now occupied in large part by Romania.

Dalmatia. Former crownland of Austria-Hungary along the Adriatic, now part of Yugoslavia.

Dalriada. Kingdom of Scots on southwest coast of Scotland in early Middle Ages. They united with the Picts in ninth century, and combined territories eventually formed the greater part of Scotland.

Dan. Biblical town, originally Laish, in the extreme north of Israel. It is noted for the saying, "from Dan to Beersheba."

Danelaw. That part of northeastern England, from the Tyne to the Thames, under Danish law and control in the early Middle Ages. Also called Danelagh.

Darién. Old name of eastern Panama. The name "Isthmus of Darién," formerly applied to the Isthmus of Panama, now pertains particularly to the isthmus between the gulfs of Darién and San Miguel.

Dark Continent. Name popularly applied to Africa.

Dartmoor. Desolate moorland in south Devonshire, England. Site of the famous prison of the same name.

Dauphiné. A medieval province between the Rhône and the Alps, with its capital at Grenoble. From it was derived the French royal title, "Dauphin."

Deccan. In its broadest sense, the peninsula of India south of the Narbada River. In particular, that part of the peninsula north of the Kistna River.

Decelea. Village in northern Attica used as a Spartan base and giving its name to the final stage of the Peloponnesian War between Sparta and Athens.

Delos. Sacred island of the ancient Greeks, in the Cyclades southwest of Mykonos Island.

Delphi. Ancient city of Phocis, Greece, the site of the Delphic oracle. It is now the village of Kastri.

Deseret. Early Mormon name of Utah.

Dixie. Popular term for the southern states of the United States. It may have stemmed from the ten dollar bills or "Dixies" issued in Louisiana before the Civil War. These bills bore the French word "dix," meaning "ten."

Dobruja. Region bounded on the west and north by the Danube, on the east by the Black Sea, and extending south into Bulgaria. Its chief port and town is Constanta, Romania.

Dodecanese. In Greek, literally "twelve islands." Actually there are about fourteen main islands, with numerous smaller ones belonging to this group. The largest of the islands is Rhodes.

Dogger Bank. Shallow sand bank covering a wide area in central North Sea; valuable fishing ground.

Doris. Small mountainous territory in the central part of ancient Greece; also, a maritime region of Caria colonized by Dorians.

Down Under. A British term for the lands south of the equator; specifically, in the Antipodes.

Drogio. Unidentified island to the south of Estotiland on late medieval maps. Modern Cape Breton Island most nearly answers the description.

Dur Sharrukin. Ancient Assyrian city north of Nineveh. For a short time it was a royal residence.

E

East Anglia. Early medieval kingdom comprising present English counties of Norfolk and Suffolk.

Eastern Archipelago. Name for Malaysia or the Malay Archipelago.

Eastern Empire. Name of the Eastern Roman or Byzantine Empire.

East Indies. Old collective name for southeast Asia including India, Indochina, and Malaysia.

Ebal, Mount. In Old Testament, mountain of the curse, overlooking Shechem (now Nābulus, Jordan).

Eburacum. Old Roman town and military base on site of modern York, England. Also Eboracum.

Ecbatana. Capital of ancient Media; it is now the city of Hamadan, Iran.

Eden. In the Bible, first home of man. Some scholars place it in the valley of the Tigris and Euphrates.

Edessa. Capital of an ancient kingdom in Mesopotamia. The modern Urfa, Turkey.

Edom. In the Old Testament, a mountainous district south of Moab and the Dead Sea. The Roman Idumaea extended westward to include the Negev.

Eire. Official name of Ireland.

Ekron. Important city of the ancient Philistines. The modern Akir, Israel.

Elam. Very ancient kingdom east of the Tigris and north of Persian Gulf, later absorbed into Persian Empire as province of Susiana. Capital city was Susa.

El Dorado. Legendary land of opulence sought by early Spanish explorers in northern South America. The name of the country was taken from its fabled king, El Dorado (the gilded one), who ruled in the golden city of Manoa.

Emerald Isle. Popular name of Ireland.

Emilia. Division of northern Italy between the Po River and the Apennines. Its chief city is Bologna.

Ephesus. Metropolis of ancient Ionia, Asia Minor, important as a center of commerce and Christianity through much of the Roman period. Its ruins are located south of Izmir, Turkey.

Ephraim, Mount. In the Old Testament, the land of the tribe of Ephraim in central Palestine. It contained the village of Ephraim (the Old Testament Ophrah) northeast of Bethel, in Jordan.

Epirus. Ancient country along Ionian Sea in northwestern Greece and southern Albania. Capital was Ambracia, modern Arta.

Eridu. Very ancient city of Sumeria. Originally built on the Euphrates near the Persian Gulf, it has now been placed over one hundred miles inland by the growth of the delta.

Erin. Popular name of Ireland.

Esdraelon. Great plain of northern Israel, famous as a battlefield and surrounded by Mounts Carmel, Gilboa, and Tabor.

Estotiland. Unidentified island on late medieval maps, in the general location of Newfoundland.

Ethiopia. Ancient kingdom of the upper Nile extending south from Nubia. It was known in the Bible as the land of Cush, with successive capitals at Napata and Meroe. Also used in antiquity for all of known Africa, name is now restricted to Empire of Ethiopia.

Etruria. Land of ancient Etruscans, extending from outskirts of Rome northward to include Tuscany.

Eurasia. Continental land mass of Europe and Asia.

Euxine, The. Name for the Black Sea, known in Roman times as Pontus Euxinus.

Eylau. Town southeast of Königsberg, East Prussia (now Kaliningrad, Soviet Union), where the Russians and Prussians met Napoleon in a bloody, inconclusive battle, 1807.

F

Faesulae. Ancient Roman city in northern Etruria; the modern Fiesole, near Florence.

Far East. Collective name for all Asiatic countries east of Iran and Afghanistan, including eastern Siberia, Japan, and the islands of Malaysia.

Farther India. Old name for peninsula of Indochina.

Fertile Crescent. Name for the fertile lands of antiquity, which extended in a great crescent around the Syrian desert from Israel and Jordan through the valley of the Tigris and the Euphrates.

Finger Lakes. Beautiful group of long, narrow lakes in the western part of New York state.

Flanders. Medieval country along the North Sea from the Straits of Dover to the mouth of the Schelde River, containing the cities of Ypres, Ghent, and Bruges. It is now divided between Belgium and France.

Flodden. Field southwest of Berwick, England, where the Scots under James IV were defeated by the English, 1513.

Fontenoy. Village near Tournai, Belgium, the site of the French victory over the English and their allies in the War of the Austrian Succession (1745).

Fortunate Isles. Legendary islands in the western seas, also called the Isles of the Blest and the Happy Isles. When the Canary and Madeira Islands were discovered the name was attached to them.

Francia. Early medieval land of the Franks embracing the kingdoms of Neustria and Austrasia. The name was applied later to a region of varying boundaries in the basins of the Seine and Loire. Also the Italian and Spanish names for France.

Franklin. Short-lived state government organized in eastern Tennessee in 1784. Capital was Jonesboro.

French North Africa. Collective name sometimes used when referring to former possessions of France in North Africa—Algeria, French Morocco, and Tunisia.

French Shore. The western coastal regions of Newfoundland where the French for many years exercised exclusive fishing rights.

Friedland. Former German town southeast of Kaliningrad, Soviet Union, where Napoleon won a great victory over the Russians, 1807. A temporary peace with Russia and Prussia was the consequence.

Friesland. Medieval state which included, at its greatest extent, most of the Netherlands north of the Schelde and Meuse. Now a province of northern Netherlands with its capital at Leeuwarden.

G

Gadara. Ancient city southeast of the Sea of Galilee, which gave its name to the region along the eastern shore. The name has been perpetuated by the biblical account of the Gadarene swine.

Gades. Westernmost colony of the Phoenicians, also called Gadeira. On site is modern city of Cadiz, Spain.

Galatia. Region and later Roman province in central Asia Minor settled by, and named after, the Gauls. Chief city was Ancyra, now Ankara, Turkey.

Galicia. Former Austro-Hungarian crownland to the north of the western Carpathians. Its chief cities were Lvov, now in the Soviet Ukraine, and Krakow, now in Poland. Also a coastal region of northwestern Spain.

Galilee. In the Roman period, the northern part of Palestine containing the towns of Nazareth, Capernaum, and Ptolemais. It was located west of the Jordan between Phoenicia and Samaria.

Gascony. Old duchy and province of southwestern France bordering the Bay of Biscay. Auch was the chief city of the region, which later became a part of the province of Guyenne and Gascogne with its capital at Bordeaux.

Gath. Ancient capital of the Philistines, birthplace of giant Goliath. The modern Tell es Safi, Israel.

Gaul. Ancient name for the land of the Gauls, the Roman Gallia. The greater part, located west of the Rhine and Alps and north of the Pyrenees, was called Transalpine Gaul. That part in the Po Basin of northern Italy was called Cisalpine Gaul.

Gedrosia. Country of ancient Persia along the northern coast of the Arabian Sea. Now part of Pakistan and Iran.

Gehenna. New Testament name for Tophet.

Gennesaret, Lake of. The Sea of Galilee.

Germania. Roman name for the region east of the Rhine, north of the Danube, and west of the Vistula.

German Ocean. Former name of the North Sea.

Germantown. Site of a British victory over the Americans under Washington, 1777. Now a part of residential Philadelphia.

Giant's Causeway. Peninsula northeast of Coleraine, Northern Ireland. Its unusual columns of basalt have given rise to the legend that it was constructed by giants as a causeway to Scotland.

Gilead. Mountainous biblical region east of the Jordan River, now the northwestern part of modern Jordan, between the Parmak and the Mojib rivers.

Gilgal. Camp city of the Israelites near Jericho. It was their base during the conquest of Canaan.

Golgotha. Hebrew word for "skull," and the site of the crucifixion of Christ. Also Calvary.

Goshen. Region of ancient Egypt colonized by the early Israelites. It was located between the Nile delta and the present Suez Canal.

Grand Banks. Extensive submarine plateau east of Newfoundland, noted as a fishing ground.

Great Basin. Extensive region of the western United States between Wasatch Mountains and the Sierra Nevadas. It contains no drainage outlet to the sea except at the northern and southern extremities.

Great Britain. The name of the island containing England, Scotland, and Wales.

Great Lakes. Collective name for Lakes Superior, Michigan, Huron, Erie, and Ontario.

Great Plains. A vast elevated region of North America between the Rockies and the central prairies, and extending from the Mackenzie River south to the Rio Grande. Included in the area are western parts of the Dakotas, Nebraska, Kansas, Oklahoma, and Texas, and the eastern sections of Montana, Wyoming, Colorado, and New Mexico.

Guano Islands. Islands off the Peruvian coast famed for their guano deposits. The most important are the Lobos and Chincha Islands.

Guinea. All lands on west coast of Africa between Sénégal River and the southern boundary of Angola.

H

Halicarnassus. Important seaport of ancient Caria whose site is now occupied by Budrum, Turkey.

Hamelin. German city, commonly Hameln, noted for the legend of the Pied Piper. It is located on the Weser southwest of Hanover.

Haran. Ancient city of Mesopotamia, southeast of Urfa, Turkey. It later became the Roman Carrhae, site of a Parthian victory over the Romans, 53 B.C.

Hatay. District around Iskenderon (Alexandretta), Turkey. It was formerly the Syrian mandated Sanjak of Alexandretta and the short-lived republic of Hatay.

Heartland. The vast center of the Eurasian continent, protected on almost every side by great mountains and seas. The largest portion is the Soviet Union.

Heliopolis. Important city of ancient Egypt whose ruins lie a few miles north of Cairo. The biblical On.

Hellas. Ancient Greece, land of the Hellenes.

Hellespont, The. Ancient name of the Dardanelles.

Helluland. Early Norse name for a desolate land southwest of Greenland, possibly Labrador.

Helvetia. That part of Gaul comprising the western part of Switzerland. The word is used poetically with reference to the entire country.

Heraclea. Name of numerous Greek cities of antiquity, the most important of which was located on the Gulf of Taranto, Italy. In the vicinity, Pyrrhus of Epirus defeated the Romans, 280 B.C., in the first known battle between Greeks and Romans.

Hercegovina. Old duchy in the mountains south of Sarajevo, Bosnia, now absorbed into the republic of Yugoslavia. Also Herzegovina.

Herculaneum. Roman city at the foot of Mt. Vesuvius. Destroyed with Pompeii, 79 A.D.

Hermopolis. City of ancient Egypt midway between Memphis and Thebes. Also called Hermopolis Magna.

Heshbon. Biblical capital of the Amorites near Mt. Nebo, now Hesban in Jordan.

Hesperia. Name given by the Greeks to lesser–known lands in the extreme western Mediterranean.

Hesse. Former grand duchy and state in western Germany with capital at Darmstadt. German Hessen.

Hibernia. Ancient name of Ireland.

Hiddekel. Biblical name of the Tigris River.

Hieromax. Old name of the Wady Varmuk, which flows into the Jordan south of the Sea of Galilee.

Hierosolyma. Greek name of Jerusalem.

Himera. Important seaport founded by the ancient Greeks near the Sicilian city of Termini. It was destroyed by Carthage about 408 B.C.

Hindustan. Old name for predominantly Hindu India north of the Vindhya Mountains, including the upper Ganges Basin and much of the Punjab. In a looser sense, the entire subcontinent.

Hispania. Roman name of the Iberian Peninsula.

Hispaniola. Name of the second largest island in the West Indies, a corruption of the Spanish Española. The island contains Haiti and the Dominican Republic.

Holland. Name of a district along the west coast of the Netherlands; now often applied to entire country.

Holy Alliance. Loose agreement signed in 1815 under which the signatories, Russia, Prussia, and Austria, proposed to conduct their affairs in a Christian manner. The alliance, to which all European states except Great Britain, Turkey, and the Papal States adhered, became reactionary and had lost all its significance by 1848.

Holy Land. Familiar term for Palestine, now a region divided between Israel and Jordan.

Holy Roman Empire. Medieval confederation of Germanic peoples of central Europe ruled by emperors claiming succession to the emperors of Rome. Although the empire lost northern Italy at an early date, it retained close connection with the papacy until the Reformation. The last monarch abdicated in 1806 after a long period of imperial decline.

Hondo. Island of Honshu, Japan.

Hump, The. Nickname given the eastern end of the lofty Himalayas, which lay astride the route of the India-China air transport service in World War II.

Hyrcania. Ancient region on the southeastern shores of the Caspian Sea; now part of northern Iran.

I

Iberia. That part of ancient Europe south of the Pyrenees, now called the Iberian Peninsula. Also an ancient name of Georgia, Soviet Union.

Île-de-France. Old French province whose capital was Paris.

Ilium. Roman name of ancient Troy, Asia Minor.

Illyria. Very ancient region along the Adriatic corresponding to western Yugoslavia. The name Illyricum was later applied by the Romans to western Yugoslavia, northern Albania, Hungary, and eastern Austria. The former Austro-Hungarian kingdom of Illyria included Carinthia, Carniola, and Kustenland.

Indies. Name assigned by early geographers to the newly discovered islands and coasts of America, which they believed to be the Indies of Asia. The islands later became known as the West Indies. The Netherlands East Indies, on the other hand, assumed the official title of Indonesia in 1948.

Indochina. That part of southeast Asia between India and China (i.e., Burma, Thailand, Laos, Cambodia, Vietnam, and Malaya), formerly called Farther India.

Ingermanland. Old name of the region of the Soviet Union south of the Neva and the Gulf of Finland, the ancient Ingria.

Inland Empire. Name of the great white–pine lumber region in the interior basin of the Columbia River. It encompasses northern Idaho, western Montana, and eastern Washington and Oregon.

Ionia. Maritime region of western Asia Minor colonized by the Ionian Greeks. Its chief cities were Ephesus, Miletus, and Smyrna.

Ipsus. Famous Phrygian town where successors of Alexander the Great fought over division of his empire, 301 B.C. It was north of Pisidian Antioch.

Isabela. Name of the first known European settlement in New World. Located on north coast of Hispaniola in 1493 by Columbus, it was soon abandoned.

Israel. Northern kingdom of the Jews. Its territory included Samaria, Galilee, Bashan, and Gilead, and was bounded on the south by Judah. The capital was successively at Shechem, Tirzah, and Samaria. Also, the name of the modern Jewish republic.

Issus. Famous town in ancient Cilicia where Alexander the Great defeated the Persians in battle, 333 B.C. It was located on Turkish Gulf of Iskenderun.

Ithaca. The modern island of Ithake, Greece, the legendary home of Ulysses.

Ivry. Village of Ivry-la-Bataille, south of Rouen, France. Here, in 1590, Henry IV led the Huguenots to victory over forces of the Catholic League.

J

Jamestown. First permanent English settlement in the United States, established on a peninsula on the left bank of the James in 1607. The site is now on Jamestown Island, near Williamsburg, Virginia.

Jaxartes. Ancient name of the Syr Darya River.

Jebus. Early name of Jerusalem—also called Salem, Ariel, Hierosolyma, and Aelia Capitolina.

Jericho. Walled city of biblical fame which once stood in the valley of the Jordan northwest of the Dead Sea. A small village remains.

Judah. Southern kingdom of the Jews, to the south of Israel between the Dead Sea and the Mediterranean. Its chief cities were Hebron, Beersheba, and the capital, Jerusalem.

Judea. Also Judaea, the Roman division of Palestine south of Samaria and west of the Dead Sea. At its greatest extent, the province included Idumaea, Samaria, Galilee, and Peraea.

Jutland. The Danish Jylland, the continental peninsula of Denmark.

K

Kadesh. Southernmost capital of the ancient Hittites near the source of the Orontes.

Kadesh-Barnea. Desert headquarters of Moses and his wandering Israelites near the present border of the U.A.R. and Israel.

Karafuto. Japanese name of the southern half of Sakhalin Island, transferred to Soviet Union in 1945.

Karakorum. Medieval capital of Genghis Khan whose ruins lie along the Orkhon River west of Urga, Mongolia.

Karelia. Old name of district between Lake Ladoga and Gulf of Finland, now part of the Soviet Union.

Kaskaskia. Early French settlement in Illinois on the right bank of the Kaskaskia River near its confluence with the Mississippi.

Kedesh. The present village of Kades, town of Old Testament prominence in northern Palestine near the Waters of Merom.

Kingsmill Islands. The Gilbert Islands.

Kiptchak. Medieval khanate or kingdom ruled by the descendants of the Mongol leader, Genghis Khan. Also called the Kingdom of the Golden Horde, it extended from central Asia to the Black Sea. Its capital was at Sarai on the lower Volga.

Kishon. Famous biblical river, watering plain of Esdraelon in northern Israel.

Kittim. Biblical name of the island of Cyprus, probably derived from its Phoenician port of Citium. The name Chittim also referred to Cyprus, as well as to islands and shores of the eastern Mediterranean.

Klondike. Name of a stream which flows westward into the Yukon River at Dawson, Yukon. Since the gold rush of 1897, name has been applied to entire area of gold fields extending westward into Alaska.

Kurdistan. Land of the Kurds, comprising part of Turkey, Iraq, Iran, and Soviet Armenia.

Kustenland. Old province of Austria-Hungary containing Trieste and the Istrian Peninsula.

L

Laconia. In antiquity, the southeastern division of the Peloponnesus, with its capital at Sparta. Also called Lacedaemon.

Ladrone Islands. Old Spanish name of the Mariana or Marianne Islands. Also called the Ladrones.

Lake District. Picturesque, lake-studded mountain region in England north of Morecambe Bay.

Landes, The. Extensive marshy plain in France along the Bay of Biscay south of the Gironde.

Land of the Midnight Sun. Norway.

Languedoc. Medieval government and province of southern France bordering the Rhône, Mediterranean, and eastern Pyrenees. Its capital was at Toulouse.

Lanka. Ancient name of Ceylon.

Laodicea. Capital of ancient Phrygia and site of one of Paul's seven Asiatic churches. Its ruins now lie near Denizli, Turkey.

Lapland. Country of the Lapps in arctic Europe. This region extends westward from the Kola Peninsula in the Soviet Union through northern Finland, Sweden, and Norway.

Latin America. Collective term for all countries and islands south of the United States where people of Spanish, Portuguese, and French ancestry predominate. It includes Mexico, Central and South America, and most of the West Indies.

Latium. Ancient land of the Latins along the Tyrrhenian Sea. Rome lay in the extreme north.

Laurentian Lakes. Lakes of the glacial period that occupied largely the same area as the present Great Lakes. Huge Lake Algonquin (now divided into lakes Superior, Michigan, and Huron) emptied into Lake Iroquois (now Lake Ontario), which in turn emptied into the Atlantic through the Mohawk and Hudson valleys.

Leeward Islands. General name of the northern group of islands in the Lesser Antilles, extending southeastward from Puerto Rico to include Guadeloupe.

León. Kingdom in northwestern Spain in the early Middle Ages. It absorbed the kingdom of Asturias.

Lepanto. Greek strait between the gulfs of Patras and Corinth, the site of an Italo-Spanish naval victory over Turkey, 1571.

Leuctra. Ancient Greek village near Thebes, Boeotia, where the Thebans destroyed the military power of Sparta, 371 B.C.

Levant, The. Collective name for the lands on the eastern shores of the Mediterranean, including the U.A.R. and Asia Minor. The name was derived from the Latin "levare" (to raise), meaning "the rising

176

sun" or "the east." It was therefore originally applied to all the Mediterranean lands east of Italy.

Lidice. Small village west of Prague, Czechoslovakia, whose inhabitants were put to death by the Germans in World War II.

Lilybaeum. Fortress city on the western tip of ancient Sicily. The modern Marsala.

Little America. Admiral Byrd's base in Antarctica at 78° 40′ S. and 164° 03′ W.

Little Russia. Name given to an old division of Russia, now in the northern part of the Soviet Ukraine. The region contains the cities of Kharkov, Kiev, Poltava, and Chernigov.

Livonia. Medieval country on the Baltic Sea with its capital at Riga. Its territory is now divided between the Soviet republics of Latvia and Estonia.

Llano Estacado. Vast arid plateau in western Texas and eastern New Mexico, called in English "The Staked Plains."

Lombardy. Early medieval kingdom of the Lombards extending originally from the Alps to southern Italy. It was reduced later to the Po Valley region of northern Italy.

Loochoo Islands. The Ryukyu Islands.

Lorraine. Former kingdom, duchy, and province of varying boundaries in the Meuse and Moselle river basins between France and Germany. The chief city is Metz, France.

Lotharingia. Latin name of the kingdom of Lothaire which extended in the early Middle Ages from the Alps along the left bank of the Rhine to its mouth. It dwindled in size and became duchy of Lorraine.

Louisbourg. Famous French fortress south of Sydney, Nova Scotia, destroyed by the British in 1758. The small port of Louisburg survives near its site.

Louisiana Territory. Name given to the vast frontier region which the United States purchased from France in 1803. It extended roughly from the Mississippi to the Rockies and from Texas to Canada.

Low Countries. Term originally applied to the medieval Netherlands, which then included the present Netherlands, Belgium, and Luxembourg. The term still applies to all three countries.

Lugdunensis. Roman province in central Gaul lying largely between the Seine and Loire rivers and extending westward to the ocean. Its capital city of Lugdunum, the modern Lyon, France, lay in its southeastern extremity.

Lutetia. Roman name of Paris. Also called Lutetia Parisiorum.

Lydia. Powerful kingdom of antiquity extending from the Aegean to the Halys River in Asia Minor. It was conquered by Persia in 546 B.C. and reduced to a small Aegean coastal province. The chief cities were Sardes and Smyrna (now Izmir, Turkey).

M

Macedonia. Also called Macedon. An ancient kingdom, empire, and Roman province of varying boundaries in northern Greece around the head of the Gulf of Salonika. Under Alexander the Great the empire extended from Greece eastward to Egypt and India. The northern part of modern Macedonia is contained within Yugoslavia and Bulgaria.

Maeander. Old name of Menderes River, which flows into the Aegean south of Izmir, Turkey. Its winding course has given rise to the modern word "meander."

Maelström. Also Malström. A strait south of Moskenäsö island in the Lofoten Islands, noted for its dangerous current and whirlpool.

Magna Graecia. Latin name of the southern part of ancient Italy colonized by the Greeks. It extended from the Bay of Naples to the eastern shore of the Gulf of Taranto.

Magnesia. Coast district of ancient Thessaly. It has given its name to magnetic ore which may have been first discovered there.

Maine. Old French province south of Normandy.

Maipo. Chilean river which reaches the Pacific south of Santiago. San Martín's victory over the Spanish here, at the town of Maipú in 1818, assured the independence of Chile.

Malacca. Old name occasionally applied to the Malay Peninsula.

Malaysia. General name for all the islands in the East Indies except New Guinea and including the Philippines. Also called the Malay Archipelago.

Mancha, La. Former province of central Spain southeast of Madrid. Famous in fiction as the home of Don Quixote.

Manche, La. French name of the English Channel.

Manchukuo. State created in Manchuria and part of North China through Japanese intervention in 1932. Its existence ended with the fall of Japan in 1945.

Mantinea. Ancient Arcadian city in the Peloponnesus west of Argos, Greece. It was the site of several ancient battles. Also Mantineia.

Marathon. Famous coastal village and plain of ancient Greece northeast of Athens. Here, in 490 B.C., the Athenian army defeated the Persian invaders.

Mareotis, Lake. Ancient name of the Birket-el-Mariut, a large lake southeast of Alexandria, Egypt.

Maritime Provinces. Historic term for the Canadian provinces of New Brunswick, Nova Scotia, and Prince Edward Island.

Markland. Early Norse name of a region southwest of Greenland, probably Newfoundland.

Mason and Dixon Line. Popular name for the dividing line between North and South prior to the Civil War. Originally, it was the boundary between Maryland and Pennsylvania as surveyed by two English astronomers, 1763-1767.

Massilia. Roman name of Marseilles, France.

Mauretania. Ancient name of northwestern Africa, embracing modern Morocco and part of Algeria. Distinguished from Mauritania, the former French protectorate north of Senegal.

Mayapán. Old capital of the Mayas. Its ruins are located south of Merida, Yucatan.

Mazaca. Capital of ancient Cappadocia. The Roman Caesarea and the modern Kayseri, Turkey.

Mecklenburg. Medieval duchy and later a province of Germany along the Baltic, west of Pomerania. Its chief cities are Schwerin and Rostock.

Media. Ancient country of northwestern Iran with its capital at Ecbatana. Once a powerful empire itself it became in the sixth century, B.C., a part of the empire of Persia.

Megaris. District of ancient Greece on the eastern end of the Isthmus of Corinth. The capital was Megara.

Megiddo. Ancient city on the plain of Esdraelon which gave its name to Armageddon, "the mountain of Megiddo."

Melanesia. Collective name for certain islands in the Southwest Pacific whose inhabitants have similar racial characteristics. They comprise the entire chain of islands from New Guinea to the Fijis, including the Admiralties, the Bismarck Archipelago, the Solomons, New Caledonia, and the New Hebrides.

Melita. Ancient name of Malta.

Memphis. Very ancient capital of Egypt on the left bank of the Nile below Cairo.

Mercia. Early medieval kingdom of central England.

Meroe. Capital of the ancient kingdom of Ethiopia on the Nile below Khartoum.

Merom, Waters of. Biblical name of the small lake in the Jordan valley between Mt. Hermon and the Sea of Galilee. Its modern name is Hula.

Mesopotamia. Old name for the land between the Tigris and Euphrates, the biblical Aram Naharaim. Also a Roman province. It is now a part of Iraq.

Messenia. Ancient country in the southwestern Peloponnesus, Greece. Its chief city was Messene. Now a province of Greece.

Michmash. Site of early Israelite victory over Philistines, now Mukhmas, Jordan, northeast of Jerusalem.

Micronesia. Name applied to an extensive chain of islands in the western Pacific whose races are related. The chief Micronesian groups are the Marianas and Palaus (volcanic), and the Carolines, Marshalls, and Gilberts (coral).

Middle East. Originally, a collective name for the lands between the Near and Far East, including Iraq, Iran, Afghanistan, and the countries of the Arabian peninsula. The name has recently been applied also to the countries of the eastern Mediterranean, from the U.A.R. to Turkey, and east to Iran and is replacing "Near East" in official U.S. State Department usage.

Middle West. Collective name for the lowland states of the north-central United States. West of the Mississippi River it includes Missouri and Kansas and all the states to the north; east of the Mississippi it includes all the states north of the Ohio River.

Miletus. Important seaport of ancient Ionia, Asia Minor, south of Ephesus.

Mizpeh. Mountain meeting–place of the ancient Israelites, northwest of Jerusalem. Also Mizpah.

Mizraim. Hebrew name of Egypt.

Moab. Ancient kingdom between Gilead and Edom on the eastern shores of the Dead Sea. Now it is a part of Jordan.

Moldau. The Vltava River which flows through Prague, Czechoslovakia.

Moldavia. Former principality along the right bank of the Prut River, with its capital at Iasi. In 1861 it united with Walachia to create the kingdom of Romania. Also, the name given to the new Soviet republic in previously Romanian Bessarabia.

Mongolia. Vast land of the Mongols in central Asia between China and the Soviet Union. Outer Mongolia is now the Soviet-dominated People's Republic of Mongolia with its capital at Urga. Inner Mongolia comprises an extensive plateau area in north and northeast China. In the Middle Ages the Mongols ruled an empire which reached to western Russia.

Montenegro. Former principality and kingdom in the mountains northwest of Albania. Now part of Yugoslavia.

Mont-Saint-Michel. Famous island and village west of Avranches, France, containing an historic monastery. It is connected to the mainland by a causeway.

Moon, Mountains of the. Mountain range believed by ancient geographers to lie across central Africa, and to give rise to the headwaters of the Nile.

Moravia. Former Austro-Hungarian crownland in the Morava River valley; now the central province of Czechoslovakia. The capital is Brno.

Morgarten. Mountain slope near Lucerne, Switzerland, where an Austrian army was routed, 1315, and the independence of Switzerland assured.

Moriah, Mount. Hill on the east side of Jerusalem, the site of the Temple of Solomon.

Mount of Olives. This mountain, east of Jerusalem, is frequently mentioned in scripture. It is 2,665 ft. high. Also called Mount Olivet.

Muscovy. Old name of Russia derived from Moscow, the capital of a strong medieval grand duchy.

Mutina. Early Roman colony, the modern Italian city of Modena. It was the site of the last resistance of the Cisalpine Gauls to the growing power of Rome, 193 B.C.

Mycale. Promontory east of island of Samos where Greeks destroyed a Persian naval force, 479 B.C.

Mycenae. Very ancient Greek city whose ruins have been of great archaeological interest. It was located in Argolis northeast of Argos.

N

Narbonensis. Roman province extending along the Mediterranean coast of Gaul from the Alps to the Pyrenees. Its important cities were Massilia, Tolosa, and the capital, Narbo Martius (now Narbonne).

Naseby. Village near Northampton in central England; site of the decisive victory of the Parliamentarians under Cromwell over Charles I, 1645.

Naucratis. Ancient Greek trading center on the Nile delta south of Alexandria.

Navarre. Early medieval kingdom in the western Pyrenees. Its capital was at Pamplona, Spain. Most of the kingdom later united with Aragon, while the remainder, on the northern slopes of the Pyrenees, was absorbed by France.

Navigators' Islands. Old name of Samoan Islands.

Neanderthal. Valley between Düsseldorf and Wuppertal, Germany, where the skull of the famed prehistoric man was found.

Near East. Name formerly associated with the Ottoman Empire, and more recently applied to those countries making up the "Middle East."

Nebo (Pisgah), Mount. Mountain of Gilead near the north end of the Dead Sea, with an altitude of 2,631 ft. Now Jebel Neba in Jordan.

Negev. Vast desert region of southern Israel.

New Albion. Name given by Drake to part of the Pacific Coast north of San Francisco Bay.

New Amsterdam. Old name of New York City under the Dutch regime.

New Carthage. Important colony of the Carthaginians; now the city of Cartagena, Spain.

New England. Collective name since the early colonial days for the northeastern part of the United States. It includes the states of Maine, New Hampshire, Vermont, Massachusetts, Connecticut, and Rhode Island.

New France. Old name applied to the French colonies in North America, specifically, those in eastern Canada which were yielded to the British after the capture of Quebec City in 1763.

New Georgia. Old name of Vancouver Island, Canada, and adjacent mainland territory.

New Granada. Old Spanish viceroyalty in that part of Latin America now occupied by the republics of Colombia and Panama.

New Holland. Old name of Australia.

New Netherland. Originally an English land grant to the Duke of York, for lands between the Connecticut and Delaware Rivers. After the English defeat by the Dutch, it became a Dutch colony, roughly comprising much of New Jersey, New York, and Delaware.

New Spain. Early Spanish colony and viceroyalty of varying boundaries, now the republic of Mexico.

New Sweden. Old Swedish colony along the lower Delaware River.

New World. Name first applied in the 15th century to the newly discovered Western Hemisphere.

Nineveh. Capital of the Assyrian Empire and one of the most splendid cities of antiquity. Its ruins lie across the Tigris from Mosul, Iraq. Also called Ninus.

Nippon. Name of the old Japanese Empire, now reduced to the main and adjacent islands of Japan. Name sometimes restricted to Honshu, largest island.

Nod. Unknown land east of Eden to which Cain fled for refuge.

Normandy. Old French duchy and province along the English Channel containing the cities of Rouen (the capital), Le Havre, Caen, and Cherbourg.

Northeast Passage. Maritime passage from the Atlantic to the Pacific Ocean along north coast of Europe and Asia. It is navigable only in summer months.

North Polar Sea. The Arctic Ocean.

North River. Name of Hudson River near mouth.

Northumbria. Powerful kingdom of the early Middle Ages, located in northeastern England between the Humber and the Firth of Forth. Also called the kingdom of Northumberland, it was created by the union of the kingdoms of Bernicia and Deira.

Northwest Passage. A navigable passage from the Atlantic to the Pacific and the Orient, sought for centuries along the northern coast of North America. The Arctic passage finally found proved to be impassable most of the year.

Northwest Territory. Old name of the region west of Pennsylvania bounded by the Mississippi and Ohio rivers and the Great Lakes. It was ceded by England to the United States at close of Revolution, 1783.

North Woods. Popular name for the forest regions of northern Michigan, Wisconsin, and Minnesota.

Norumbega. In early maps of North America, a mysterious city and region on the Atlantic coast. Its existence has never been proven, though speculation has placed it in Massachusetts and Maine.

Notium. Ancient port west of Ephesus in Asia Minor, noted for the Spartan naval victory over Athens in 407 B.C.

Nubia. Name applied since the dawn of history to a region of the upper Nile now lying in the northern Sudan. It was part of the ancient kingdom of Ethiopia.

Numantia. Ancient town near Soria, Spain, famed for resistance to Roman conquest, 143-133 B.C.

Numidia. Ancient kingdom and later Roman province of northern Africa, in the northeastern part of modern Algeria.

O

Oberammergau. Famous resort in the Bavarian Alps, Germany, noted for its decennial "Passion Play."

Occident. The lands toward the setting sun, a general name for Western Europe and the New World in contradistinction to "Orient."

Oceania. Term used for the Pacific island divisions of Melanesia, Micronesia, and Polynesia. Occasionally it includes the islands of Australasia and Malaysia as well. Also called Oceanica.

Oceanus. According to the ancients, a swift and boundless stream flowing around all the known seas and continents.

Old World. Name applied to Eastern Hemisphere, particularly Europe, since discovery of America.

Olivet. Mount of Olives, east of Jerusalem, in Jordan.

Olympia. Valley of ancient Elis, Greece, site of the Olympic games. It was located along the Alfios River in the western Peloponnesus.

Olynthus. City of ancient Chalcidice, the subject of the famed Olynthiac orations of Demosthenes. It was destroyed in 347 B.C.

On. Biblical name of ancient Heliopolis, Egypt.

Ophir. In the Old Testament, a country which supplied Solomon with gold, silver, and other luxuries. It may have been located in India, southern Arabia, or eastern Africa.

Orange. Medieval principality of France, eventually absorbed by the house of Nassau. It has given its name to the royal house of the Netherlands. The capital, Orange, is located north of Avignon.

Orange River Free State. Former independent republic in South Africa with its capital at Bloemfontein. It was defeated by the British in the Boer War (1899-1902) and later became a province of South Africa.

Orient. The lands toward the rising sun; a western term for the countries of Asia and the Far East.

Ormuz. Famous medieval trade emporium on an island at the mouth of the Persian Gulf. Also called Ormus and Hormuz.

Orontes. Chief river of western Syria, the Arabic Nahr el Asi.

Ottoman Empire. Old Turkish Empire which, at its height in the 16th century, embraced parts of three continents: Asia, Europe, and Africa. Turkish power in Europe advanced to the gates of Vienna and nearly encircled the Black Sea. Turkey also ruled Syria and Mesopotamia in Asia and the entire African coast from Egypt to Algeria.

Oxus. Ancient name of the Amu Darya River in central Asia, which flows into the Aral Sea.

P

Paestum. Ancient city of Magna Graecia whose ruins lie on the southern shore of Italy's Gulf of Salerno. An earlier name was Posidonia.

Palatine Hill. One of the "seven hills" of Rome reputedly selected by Romulus as original site of city.

Palmyra. Desert metropolis of ancient Syria which flourished particularly in Roman times. It was the Tadmor of the Bible, and is now a small Syrian village.

Pampas, The. A vast fertile plain of central Argentina extending from the Atlantic and the Paraná in the east to the Andes and from the Gran Chaco in the north to Patagonia.

Panhandle. Name applied to any strip of territory projecting from the main body of a state or territory, for example, northern Texas, Idaho, and West Virginia; western Oklahoma; southeastern Alaska.

Papal States. Formerly an independent country of central Italy ruled by the Papal See. Stretching from coast to coast, it included the cities of Rome, Bologna, and Ancona. The papacy also possessed Venaissin, an enclave at Avignon, France.

Papua. Former name sometimes applied to the entire island of New Guinea, but now restricted to the Australian-owned territory in the southeast.

Paran. Desert wilderness north of Mt. Sinai where the Israelites wandered before reaching Canaan.

Parima. Legendary lake which early explorers vainly sought in northern South America. The name has since been given to a mountain range at the source of the Orinoco.

Parthia. Ancient country southeast of the Caspian Sea containing the city of Hecatompylos. The Parthian Empire extended from the Indus to the Euphrates before its downfall in 226 A.D.

Pas-de-Calais. French name for the Strait of Dover.

Patagonia. That part of South America south of the Argentine Río Negro between Andes and Atlantic.

Patmos. Small Greek island in the Sporades where John the Divine is believed to have experienced the visions of the Apocalypse.

Pella. Capital of ancient Macedonia, west of Salonika (Thessaloniki), Greece.

Peloponnesus. Ancient name of the large peninsula in southern Greece. The modern Morea.

Peraea. In the New Testament, a region east of the Jordan corresponding closely to Gilead.

Pergamum. Also Pergamon and Pergamus, the name of an ancient Greek kingdom and its capital in Mysia, Asia Minor. The city is identified with the biblical Pergamos, site of one of Paul's seven Asiatic churches. The modern Bergama, Turkey.

Persepolis. Capital of the Persian Empire. Its famous ruins are located northeast of Shiraz, Iran.

Persia. Old name of Iran. The original Persia or Persis was a mountainous country on the northeastern shores of the Persian Gulf. It later grew into an empire extending from the Aegean to the Indus.

Petra. The biblical Sela, midway between the Dead Sea and the Gulf of Aqaba; capital of ancient Edom and, later, of Roman province of Arabia Petraea.

Petrograd. The name of Leningrad between 1914 and 1924. Prior to 1914 city was called St. Petersburg.

Pharos. Peninsula near Alexandria, U.A.R. In ancient times it was an island and the site of a great lighthouse that was considered one of the seven wonders of the world.

Pharsalus. City in southern Thessaly near which Caesar defeated Pompey, 48 B.C. The modern Farsala, Greece.

Philadelphia. Ancient Lydian city where Paul established one of his seven Asiatic churches. It is now the town of Alasehir, Turkey. Also an old name of Amman, the capital of Jordan.

Philippi. City of ancient Greece, inland from the present port of Kavalla. Near here, in 42 B.C., Augustus and Antony defeated the Roman Republicans led by Brutus and Cassius. Here also Paul founded the first Christian church in Europe.

Philistia. Ancient country of the Philistines in southwest Palestine. Its chief cities were Gaza, Ashkelon, Ashdod, Ekron, and Gath. The name Palestine was derived from "Philistine."

Phoenicia. The first nation to engage in large-scale Mediterranean commerce and colonization. Its chief cities were Sidon and Tyre, the present ports of Saida and Tyre (Sur), Lebanon.

Phrygia. In the time of Persian power a large country in west-central Asia Minor. Its important cities were Celaenae and, later, Apamea.

Picardy. Old province of northern France with its capital at Amiens.

Pichincha. Lofty volcano west of Quito, Ecuador, in Pichincha province. It is the site of one of the world's highest battlefields. Here, in 1822, Sucre defeated the Spanish and insured the independence of Ecuador.

Piedmont. Originally, a division of northern Italy in the upper basin of the Po, a part of the old duchy of Savoy and kingdom of Sardinia. Now, a name applied to any foothill region, for example, the plateau between the Appalachians and the Atlantic Coastal Plain, extending from Virginia to Alabama.

Pillars of Hercules. Ancient name of the two promontories forming the Strait of Gibraltar.

Pisgah, Mount. The spur of Mount Nebo from which Moses beheld the Promised Land.

Plassey. Indian village north of Calcutta where Clive defeated Bengal army, 1757, assuring British control of lower Ganges basin and, eventually, of all India.

Plataea. Town south of ancient Thebes, Greece, where the Greeks, principally Spartans, decisively defeated an invading Persian army, 479 B.C.

Polish Corridor. Narrow corridor of former German territory assigned to Poland in 1919 to give that country access to the Baltic Sea. As a result of its World War II acquisitions from Germany east and west of corridor, Poland has extensive Baltic frontage.

Polynesia. Collective name of a myriad of islands in the Central and South Pacific inhabited by closely related races. Important in this category are the Hawaiian, Marquesas, Tuamotu, Society, Samoa, Tonga, Ellice, and Phoenix Islands.

Pomerania. Formerly, a Prussian province along the coast of the Baltic Sea. Most of the old province including Stettin, its capital, is now Polish territory. The German Pommern.

Pompeii. Famous Roman city southeast of Naples. Destroyed by the eruption of Mt. Vesuvius in 79 A.D.

Pontine Marshes. Also called Pomptine Marshes. Extensive areas of reclaimed marshland along the southern coast of the Campagna di Roma, Italy.

Pontus. Old kingdom of northeastern Asia Minor, later a Roman province. It took its name from the Euxine, or Pontus Euxinus, on whose shores it lay.

Porto Bello. Famous port for gold shipment in Spanish colonial days. It is now a small Panamanian village (Portobelo) on the Caribbean east of Colón. Also Puerto Bello.

Port Royal. Old French name of Annapolis Royal, Nova Scotia, founded in 1604.

Prairie Provinces. Collective name for the Canadian provinces of Manitoba, Saskatchewan, and Alberta.

Prussia. Medieval duchy on the Baltic around the lower Vistula and Niemen. Under the later kingdom of Prussia it was divided into provinces; East Prussia (capital, Königsberg) and West Prussia (capital, Danzig). The kingdom, around which greater Germany was constructed, extended westward to Rhenish Prussia (capital, Coblenz).

Pteria. Old capital of the Hittites, believed to be the site of a Persian victory over Lydia. Its ruins are near the village of Boghaz-Keui, Turkey.

Punjab. Former province in the lowlands of northwestern British India. In 1947 much of the Punjab including the capital at Lahore became part of the new state of Pakistan.

Puteoli. A leading seaport of ancient Italy; the modern Pozzuoli, west of Naples.

Pydna. Town in ancient Macedonia where the Romans won a final decisive victory over the Macedonians, 168 B.C.

R

Rabbah. Biblical capital of the Ammonites. Later Philadelphia, it is now Amman, capital of Jordan.

Ragae. First capital of ancient Media, on site of modern Tehrān, Iran. Also called Rhagae and Rages.

Ramah. Ancient home of Samuel, north of Jerusalem.

Ramoth-Gilead. Place of biblical importance east of the Jordan River, possibly near Es Salt in Jordan. May be identical with Ramoth-Mizpah.

Ratisbon. A Bavarian city on the Danube northeast of Munich; now called Regensburg. The story of its capture by Napoleon in 1809 is told in Browning's poem "Ratisbon."

Reval. Also Revel, the former name of Tallinn, capital of Soviet Estonia.

Rhaetia. Alpine province of Rome which has given its name to the Rhaetian Alps between Austria and Switzerland. Also called Raetia.

Rhineland. Also called Rhenish Prussia or Rhine Province, in the basin of the German lower Rhine. Important cities are Köln (Cologne) and Düsseldorf.

Rimland. That part of the World Island adjacent to the Heartland but close enough to the temperate seas and oceans to be influenced by sea power. Included are Western Europe, the Middle East, the Far East, and Africa.

Riviera. Popular name for a narrow strip of territory between the mountains and the Mediterranean extending from Hyères, France to La Spezia, Italy. It includes the cities of Nice and Genova (Genoa) and many smaller resorts.

Roman Empire. The greatest empire of ancient times, administered from the city of Rome. The lands under Roman control included entire Mediterranean littoral and extended from Britain to Mesopotamia.

Roncesvalles. Modern French, Roncevaux, a Spanish village near a pass in the Pyrenees northeast of Pamplona; scene of the death of Roland in 778.

Rubicon. Small stream north of Rimini marking the boundary between Cisalpine Gaul and ancient Italy. Caesar crossed the stream in 49 B.C., precipitating a Roman civil war.

Ruhr. Name applied to industrial district in Ruhr River valley of western Germany. Chief city is Essen.

Rumelia. Also Roumelia. Name, loosely used, for the European territory of the Ottoman Empire along the northern shore of the Aegean. It corresponded roughly with ancient Thrace and Macedonia. Eastern Rumelia was annexed by Bulgaria in 1885.

Runnymede. Also Runnimede. A meadow on the right bank of the Thames west of London, where King John signed the Magna Carta, 1215.

Rupert's Land. Also Rupert Land, an early name for the vast area in the basin of Hudson Bay, later known as the Hudson's Bay Territory.

Russia. Popular name of the Union of Soviet Socialist Republics (Soviet Union), formerly the Russian Empire. The name is now applied more properly to two Soviet republics, the Russian Socialist Federated Soviet Republic and White Russia. Soviet Russia is also a popular name of the Soviet Union.

Ruthenia. Land of the Ruthenians in the central Carpathian Mountains. Parts of Ruthenia have at various times been under the control of Poland, Austria-Hungary, Czechoslovakia, and Russia. At the close of World War II, Czechoslovakia ceded that part of Ruthenia south of the Carpathians to the Soviet Union.

Ryswick. Village near the Hague, the Netherlands, where the French signed a treaty with England, the Netherlands, and Spain, 1697; now Rijswijk.

S

Saguntum. Ancient city noted for its valiant resistance to Hannibal, 219 B.C. Later called Murviedro, it is now the town of Sagunto, Spain.

Saikio. Name occasionally given to Kyoto, the old "western capital" of Japan, to distinguish it from Tokio or Tokyo, the present "eastern capital."

Saint Brendan's Island. According to medieval legend, one of several islands in the Atlantic visited by Irish monks and their leader, Saint Brendan.

Saint Petersburg. Name of the former capital of Russia. The name was changed to Petrograd in 1914 and Leningrad in 1924. In 1918, the capital was transferred to Moskva (Moscow).

Salamis. Chief seaport of ancient Cyprus, now a heap of ruins north of Famagusta. Also, an island near Peiraieus (Piraeus), Greece, noted for Athenian naval victory over Persia off its shores, 480 B.C.

Salem. Original name of Jerusalem, still used poetically. Also Shalem.

Salt Sea. One of the biblical names for the Dead Sea, the others being the Sea of the Plain and the East Sea. Known to Romans as Lacus Asphaltites.

Samaria. Name for the capital of the ancient kingdom of Israel, later applied to the entire region between Galilee and Judea. The city of Samaria became the Roman Sebaste and the present village of Sebustieh, northwest of Nābulus, Jordan.

Sandwich Islands. Old name of Hawaiian Islands.

San Juan Hill. Celebrated elevation southeast of Santiago de Cuba, taken by storm by the Americans, 1898, in an engagement in the Spanish-American War. The famous Rough Riders were prominent in the assault.

San Stefano. Small port west of Constantinople, now Istanbul. The treaty signed there, 1878, ended Russo-Turkish War.

179

Sarai. Capital of the medieval khanate of Kiptchak. Soviet city of Leninsk, east of Volgograd, formerly Stalingrad, now on site.

Sardes. Also Sardis, capital of the early kingdom of Lydia and seat of one of the seven Asiatic churches of Paul. It was located east of what is now Izmir, Turkey.

Sardinia. Old kingdom created in 1720 out of Savoy, Piedmont, Genoa, and the island of Sardinia. The kings of Sardinia passed the royal succession from the dukes of Savoy to the later kings of Italy.

Sargasso Sea. An extensive ocean area northeast of the West Indies, laden with drifting weeds, chiefly Sargassum or gulf weed. Since it is the center of a great elliptical gyration of ocean currents in the North Atlantic, the region is relatively calm.

Satsuma. Formerly a province on Kyushu Island, Japan; famed for its fine porcelain.

Savoy. Old county and duchy in the French and Italian Alps. At its height, the duchy controlled the Piedmont and an outlet to the sea at Nice, France. In 1860, the Nice district and that part of Savoy on the western slopes of the Alps were ceded to France. The early counts of Savoy founded the reigning house of the kingdom of Italy.

Saxony. Early medieval duchy extending from the North Sea almost to Leipzig in the southeast and the Rhine in the southwest. It later became an electorate and a kingdom north of the Erzgebirge (Ore Mountains), with Dresden and Leipzig as the chief cities.

Scandinavia. Sweden, Norway, and Denmark.

Scapa Flow. Large protected bay in southern Orkney Islands. Main anchorage of the British Home (Grand) Fleet. Chief British naval base in World War I and World War II.

Schleswig-Holstein. Old duchy and province north of the Elbe between the North and Baltic Seas, for whose possession Germany and Denmark have contended for centuries. Northern Schleswig (the Danish Slesvig) is now in Danish territory

Scotia. Early medieval name of Ireland, the original home of the Scots. It was later applied poetically to Scotland, the land to which many Scots emigrated.

Scylla. Prominent rock on the Italian side of the Strait of Messina. It was the legendary home of the sea monster, Scylla, across the strait from the whirlpool, Charybdis.

Sea of the Plain. Biblical name of the Dead Sea, also called in scripture the Salt Sea and the East Sea. The Roman name was Lacus Asphaltites.

Sedgemoor. Moor near Bridgwater, England, where the Royalists of James II defeated the Duke of Monmouth in 1685.

Seleucia. Capital of the Seleucid Empire on the Tigris south of Baghdad. This empire extended from the Mediterranean to the Indus before being absorbed by Rome. Also, a Syrian city, the port for ancient Antioch.

Senlac. Hill near Hastings, England, where the invading Normans, led by William the Conqueror, won their great victory over the Saxons, 1066.

Sepharvaim. Biblical name of Sippar, a city north of Babylon.

Serbia. Formerly an independent kingdom, now a part of eastern Yugoslavia. Also called Servia. The capital was Belgrade.

Sharon. Coastal plain in Israel between Jaffa and Mt. Carmel.

Sheba. Biblical name of Saba or Sabea, a wealthy kingdom in southwestern Arabia.

Shechem. Also Sichem, the earliest capital of the kingdom of Israel. The Sychar of the New Testament and the Roman Flavia Neapolis, it is now the city of Nābulus, Jordan.

Shiloh. Early religious center of Israelites between Nābulus and Jerusalem. Also a great battlefield near Pittsburg Landing, Tennessee, in the Civil War.

Shinar. Biblical name of Babylonia or its southern part; also called Chaldea or Sumer.

Shire. In England, a land division or county.

Shushan. Biblical name of Susa.

Siberia. Vast expanse of Soviet territory in Asia extending from the Urals to the Pacific. It is bounded on the south by the Soviet republics of central Asia, and by Sinkiang, Mongolia, and Manchuria, and on the north by the Arctic Ocean.

Sick Man of Europe. Term given to the Ottoman or Turkish Empire in the years prior to World War I, when its control over southeastern Europe and its Arabic and North African empire was beginning to weaken.

Sicyon. Important city of ancient Greece northwest of Corinth. The area surrounding the city was called Sicyonia.

Sidon. Principal port of ancient Phoenicia, also called Zidon. Saida, Lebanon, now occupies its site.

Silesia. Industrial region, largely in the valley of the upper Oder. Formerly a province of Prussia and Germany, it now forms the southwestern part of Poland. The chief city is Wroclaw, formerly Breslau, Germany. Austrian Silesia, in the Moravska Ostrava region, was divided between Czechoslovakia and Poland following World War I. German Schlesin.

Sinai, Mount. Famous biblical mountain where Moses received the Ten Commandments. Many identify it with Ras es Sufsafeh or Jebel Musa, near Jebel Katherina on the U.A.R.'s Sinai Peninsula.

Slavonia. Region between Drava and Sava rivers in north-central Yugoslavia, formerly part of Austro-Hungarian crownland of Croatia and Slavonia.

Slot, The. Name given in World War II to the sound which bisects the Solomon Island chain to the northeast and southwest.

Slovakia. Mountainous province of eastern Czechoslovakia. Capital is Bratislava, the German Pressburg.

Slovenia. Region in northwestern Yugoslavia. The chief city is Ljubljana, formerly called Laibach.

Smyrna. Former Greek name of Izmir, Turkey. One of Paul's seven churches of Asia was located here.

Sofala. Seaport of Mozambique, southwest of Beira, taken from the Arabs by Portugal in 1505.

Southern Ocean. Name formerly applied to the ocean waters between 40° South Latitude and the Antarctic Circle.

South Sea. Old name of the Pacific Ocean. It is now applied to the South Pacific, commonly in the plural form, "South Seas."

Spanish America. Those portions of the New World settled by the Spaniards in the early Colonial period.

Spanish Main. Old name of the mainland of Spanish America; particularly the Caribbean coast of South and Central America, the Spanish Tierra Firme. Popularly, the name has become associated with the entire Caribbean area during the colorful pirate period.

Stonehenge. This mysterious roofless monument stands on Salisbury Plain, near Wiltshire, England. It is formed by huge standing stones, apparently the ruins of a prehistoric sacred enclosure. It is believed to have been erected by men of the Bronze Age.

Straits Settlements. Former crown colony of Great Britain on the Malay Peninsula. The principal settlements were at Singapore (the capital), Penang Island, and Malacca.

Straits, The. Name used collectively in referring to the strategic waterway between the Black and the Aegean Seas. It includes the Bosporus, the Sea of Marmara, and the Dardanelles.

Strathclyde. Old Celtic kingdom of northwestern England south of the Clyde, including the districts of Cumbria and Galloway.

Stresa. Resort town on the southwestern shore of Lake Maggiore in northern Italy. It was the site of an important international conference in 1935.

Styria. Crownland of Austria–Hungary located north of the Drava River, largely within the southeastern boundaries of modern Austria.

Sudan. An extensive region south of the Sahara stretching from the horn of Africa on the Atlantic to the southern shores of the Red Sea. Includes most of the independent nation of the same name.

Sudetenland. Strip of territory in northwestern Czechoslovakia across the Erzgebirge (Ore Mountains) from Germany. It contained a majority of Sudeten Germans prior to World War II.

Sumer. That part of ancient Babylonia south of Akkad at the head of the Persian Gulf. Also called Sumeria, Shumer, and the Land of Shinar.

Susa. Capital of ancient Elam or Susiana and royal residence of kings of Persia. Called Shushan in Bible.

Swanee. Musical adaptation by Stephen Foster of the name of the Suwannee River in Georgia and northern Florida.

Sybaris. Wealthy and luxury-loving city of ancient Magna Graecia. Its ruins are now located in northern Calabria near the Gulf of Taranto.

Sychar. The New Testament name of Shechem; now Nābulus, Jordan.

Syrtis. Ancient name applied to two North African gulfs. Syrtis Major, southwest of Bengazi, is now called the Gulf of Sidra. Syrtis Minor, in Tunisia, is now called the Gulf of Gabes.

T

Tabor, Mount. A mountain east of Nazareth, Palestine, prominent in the Old Testament. Its altitude is 1,880 ft.

Tadmor. Biblical name of Palmyra, Syria.

Tanagra. Ancient town in eastern Boeotia, Greece, the site of a Spartan victory over Athens, 457 B.C. Famed also for its terra-cotta figurines.

Tannenberg. Small village south of Ostróda, Poland, formerly within the boundaries of East Prussia. Here, in 1914, a German army surrounded and partially destroyed a large Russian force.

Taprobane. Ancient name of Ceylon. Also, in the domain of the legendary Prester John somewhere in Asia or Africa, a remarkable island containing deposits of pure gold.

Tarraconensis. Largest of the three Roman provinces in Spain, located west of Baetica and north and east of Lusitania. The capital was Tarraco, the modern city of Tarragona.

Tarshish. Biblical name for a region remote from Palestine. It probably refers to the area around Tartessus, a Phoenician city near the mouth of the Guadalquivir in southern Spain.

Tarsus. Town of southeastern Asia Minor, southwest of what is now Adana, Turkey. It is chiefly famous as the birthplace of the Apostle Paul.

Tatary. Old name (commonly, but less correctly, Tartary) of the vast center of the Eurasian continent inhabited by the Tatars. In the Middle Ages, they extended their control from Mongolia westward to the Dnepr River in European Tatary and eastward to the Pacific. An autonomous Tatar republic now exists with its capital at Kazan, Soviet Union.

Tenochtitlán. Capital of the old Aztec Empire, captured by the Spanish in 1521. Modern Mexico City has been built on the same site.

Thapsus. Town southeast of Sousse, Tunisia, where Caesar defeated the followers of Pompey, 46 B.C.

Thebes. Very ancient capital of Egypt on the right bank of the Nile near the present town of Luxor. Also, the chief city of Boeotia, Greece.

Therma. Ancient name for the modern Thessaloniki, or Salonika, Greece.

Thermopylae. Famous pass between the sea marshes and the mountains southeast of Lamia, Greece. It is the site of the heroic stand made by several hundred Spartans against the invading Persians, 480 B.C.

Thessaly. Division of ancient Greece bordering the Aegean Sea and completely surrounded by mountains, including the famous Olympus. The chief city of the region today is Larisa.

Thrace. Region of antiquity bounded by the Danube and the Black and Aegean seas. The subsequent Roman province, Thracia, included only territory south of the Balkan Mountains, what is now southern Bulgaria, European Turkey, and northeastern Greece.

Thule. Ancient name of the most northerly part of the known world, probably Iceland, or Norway, or one of the Shetland Islands. The expression "Ultima Thule" refers to any distant human objective, geographical or otherwise.

Thyatira. City of ancient Lydia, the site of one of Paul's seven churches of Asia. Its ruins lie near Akhisar, Turkey.

Tierra del Fuego. Desolate archipelago at the southern tip of South America, separated from the continent proper by the Strait of Magellan.

Timbuktu. Remote town on the upper Niger in the interior of Mali. The French form is Tombouctou.

Tiphsah. Biblical name of ancient Thapsacus, Syria.

Tippecanoe. River in northwestern Indiana noted for Harrison's victory over the Indians, near its confluence with the Wabash, 1811.

Tipperary. County and city in south-central Ireland, prominent in a popular song of World War I.

Tiryns. Very old city southeast of Argos, Greece. Its ruins are of great archaeological significance.

Tirzah. Early capital of the kingdom of Israel, east of modern Sebustieh in Jordan.

Tolosa. Roman name of Toulouse, France.

Tophet. That part of the Valley of Hinnom south of Jerusalem where the idolatrous Jews of antiquity worshiped their god of fire. Also called Gehenna.

Touraine. Medieval duchy and province of France with its capital at Tours.

Trafalgar. Cape in southern Spain between Cadiz and Gibraltar. Off this promontory, the British fleet under Lord Nelson defeated the combined navies of France and Spain, 1805.

Transcaucasia. That part of the Soviet Union south of the Caucasus Mountains, comprising the republics of Georgia, Azerbaidzhan, and Armenia.

Transleithania. Old name for the Hungarian portion of Austria-Hungary, to the east and south of Cisleithania. It included all of modern Hungary and parts of Yugoslavia, Czechoslovakia, and Romania.

Transylvania. Region in northwestern Romania separated from the rest of the country by the Carpathians and the Transylvanian Alps. A former part of Austria-Hungary, the area has long been a source of conflict between Hungary and Romania. The capital is Cluj.

Trapezus. Ancient name of Trabzon, Turkey.

Trebizond. Medieval empire of the Byzantines along the southeastern coast of the Black Sea. The capital was Trebizond, which is now the city of Trabzon, Turkey.

Tremont. Also Trimontaine, original name of Boston, derived from the three summits of Beacon Hill.

Trinacria. Ancient name of Sicily, so called because of the island's three large promontories to the northeast, southeast, and west.

Troas. Also called the Troad, a region of ancient Asia Minor bordering the Aegean and the Hellespont. Located here was the famous city of Troy or Ilium and the ancient seaport Troas or Alexandria Troas.

Troy. Ancient city of Troas, Asia Minor, destroyed by the Greeks in the Trojan War. The Greek Ilion and the Roman Ilium.

Turkestan. Extensive region in central Asia bounded by the Caspian to the west, Iran and Afghanistan to the south, the Tien Shan Mtns. to the east, and Siberia to the north. It is divided among constituent republics of the Soviet Union.

Tuscany. Medieval duchy in western Italy containing the old cities of Florence and Pisa.

Two Sicilies. United medieval kingdom of Sicily and southern Italy.

Tyrol. Mountainous region which formed, with Vorarlberg, the westernmost crownland in Austria-Hungary. After World War I, the southern part became the Italian Trentino. The capital of the Tyrol, also called Tirol, was Innsbrück, Austria.

U

Ukraine. Formerly, a general name for the lands in the basin of the Dnepr (Dnieper) River of southwestern Russia. In its broadest sense, the name included all of Russia westward to Polish Galicia, and is now applied to a Soviet republic in that area. It has often been called "Little Russia."

Ulster. Northern division of Ireland largely contained within the boundaries of Northern Ireland, a unit of the United Kingdom. The other divisions, Connaught, Leinster, and Munster, are in Eire (Republic of Ireland).

Umbria. Ancient land of the Umbrians in central Italy, probably extending at one time from the Adriatic to the Tyrrhenian Sea. Modern Umbria lies landlocked in the basin of the upper Tiber. Its capital is Perugia.

United Kingdom. Collective name now applied to Great Britain and Northern Ireland. Before 1931, all of Ireland was included.

Ur (Ur of the Chaldees). Capital of a very ancient kingdom in Chaldea. Its ruins, called Mugheir by the Arabs, are located near the left bank of the lower Euphrates.

Urartu. Assyrian name of Ararat.

Utica. After the fall of Carthage, 146 B.C., the capital of the Roman province of Africa. Utica was located northwest of Carthage.

Uxmal. Ruined Maya city in southern Yucatan.

Uz. In the Old Testament, the home of Job, somewhere in northern Arabia.

V

Van Diemen's Land. Former name of Tasmania.

Veii. Ancient Etruscan city which fell to the Romans in 396 B.C. after a ten-year siege. Generally thought to be site of modern Isola Farnese, Italy.

Veld (Veldt). Open plateau country in South Africa. The name is taken from the Dutch word for "field," or "grassland."

Venaissin. Old county and papal possession whose chief city was Avignon. It was ceded to France in 1791.

Venetia. Province of ancient Italy north of the Adriatic and the lower Po. Modern Venetia is divided into three parts: Venetia proper or Veneto (capital, Venice); Venezia Tridentina (capital, Trento); and Venezia Giulia (capital, Trieste), now ceded in part to Yugoslavia.

Vindelicia. Ancient region of north Rhaetia between the Danube and the Alps. Its chief city was Augusta Vindelicorum, now Augsburg, Germany.

Vinland. Name given by Norsemen to an unidentified region, possibly New England, on northeastern coast of North America. In English, Wineland.

W

Wagram. Village northeast of Vienna. It was the site, in 1809, of a Napoleonic victory over Austria.

Wahlstatt. Village in Silesia where the Germans fought the Mongols and checked their westward advance through Europe, 1241.

Walachia. Also Wallachia. That part of southern Romania between the Danube and the Transylvanian Alps. It united with Moldavia in 1861 to form the kingdom of Romania. Chief city is Bucharest.

Waterloo. Belgian village south of Brussels where Napoleon suffered his final defeat, 1815.

Western Reserve. Early name of a vast tract of land on Lake Erie which was claimed by the state of Connecticut. Incorporated in Northwest Territory in 1800, it is now the northeastern corner of Ohio.

Westphalia. Province of northwestern Germany, formerly duchy and kingdom. Capital is Münster.

White Russia. General name of the land of the White Russians in the western Soviet Union and adjacent parts of Poland and Lithuania. The name applies specifically to the Byelorussian (White Russian) Soviet Republic, whose capital is located at Minsk.

Wilderness, The. In the Civil War, a battlefield south of the Rapidan River, Virginia.

Windward Islands. General name of the southern group of islands in the Lesser Antilles, extending from Dominica to Grenada.

World Island. Term used when referring to the combined land mass of Eurasia and Africa.

Wurtemberg. Old kingdom of southern Germany comprising much of medieval Swabia and almost completely encircled by Baden and Bavaria. German name was Württemberg. Capital was Stuttgart.

X

Xanthus. Famous city of western Lycia, destroyed by the Persians in 545 B.C. and the Romans in 43 B.C. Only ruins remain.

Y

Yedo. Also Yeddo, an old name of Tokyo, Japan.

Yellow River. In Chinese, the Hwang Ho, second of the great rivers of China. Rising in Tibet, it flows through Mongolia and China to the Gulf of Pohai, a distance of 2,903 miles. Its periodic floods have earned it the name of "China's sorrow."

Yezo. Former name of island of Hokkaido, Japan.

Z

Zealand. Largest island in Denmark, the Danish Sjaelland. Also, low-lying province of islands in southwestern Netherlands, commonly called Zeeland.

Zela. Ancient town on the site of Zile, Turkey. After defeating the king of Pontus there in 47 B.C., Caesar sent his famous message, "Veni, vidi, vici."

Zidon. Biblical name of Sidon, Phoenicia; the modern Saida, Lebanon.

Zion, Mount. The highest hill in Jerusalem, where the ancient City of David stood. The name "Zion" was later applied to Jerusalem itself and to the Zionist movement. Also called Sion.

Zoar. In early biblical history, the sole Canaanite city on the plain near the Dead Sea which was spared destruction. It stood near Sodom.

Zuider Zee. Inlet of the North Sea which projects into the Netherlands for a distance of 80 miles. The seaward end is enclosed by a dike. It was once a lake but in the 13th century was joined to the sea by a great flood. Since 1920, site of major land reclamation project. Also IJsselmeer.

Zululand. Former native kingdom north of Durban, Natal. Now part of South Africa.

PRINCIPAL DISCOVERIES AND EXPLORATIONS

Ancient and Medieval (to the Discovery of America)

DATE	EXPLORER	COUNTRY REPRESENTED	DESCRIPTION
600 B.C.	Phoenician Sailors	Egypt	Reported by Herodotus to have sailed around Africa from east to west in three years, under orders of King Necho.
500–450 B.C.	Himilco	Carthage	Said to have explored the west coast of Europe, possibly reaching Britain.
500 B.C.	Hanno	Carthage	Explored west coast of Africa to Sierra Leone or about 5°N.
450 B.C.	Herodotus	Greece	Visited Black Sea, eastern Mediterranean, and Egypt, and described the world of his time.
334–323 B.C.	Alexander the Great	Macedonia	Explored and conquered all of southwestern Asia from Egypt to the Jaxartes and Indus rivers.
320 B.C.	Pytheas	Marseilles	Visited Britain and northwestern Europe and, possibly, either Iceland or Norway, which he called Thule.
59–44 B.C.	Julius Caesar	Rome	Added information about Gaul, Britain, and Germany to current geographical knowledge.
20 B.C.	Strabo	Rome	Traveled widely throughout Mediterranean lands; compiled most complete geography of ancient times.
570 A.D.	St. Brendan	Ireland	Alleged to have sailed the western seas for seven years in search of tropical islands; may have reached Madeira or West Indies.
690 A.D.	Bishop Arculf	France	Visited Jerusalem and other holy places; described Egypt.
721–31 A.D.	Willibard	England	Visited and described the Holy Land, Constantinople, and Rome.
890 A.D.	Othere	Norway	Sailed around North Cape, along the Lapland coast, and discovered the White Sea.
925–950 A.D.	Al Masudi	Baghdad	Traveled in India, Ceylon, China, Russia, Persia, and Egypt.
982 A.D.	Eric the Red	Norway	Discovered and colonized southern Greenland.
1000 A.D.	Leif Ericson	Norway	Discovered Labrador, Newfoundland, and nearby coasts.
1003–06 A.D.	Thorfinn Karlsefni	Iceland	Explored and attempted to colonize northeast coast of North America.
1099–1154 A.D.	Idrisi	Spain and Sicily	Traveled in north Africa and Asia Minor; compiled a description and map of the world.
1106 A.D.	Daniel of Kiev	Russia	Visited Jaffa, Jerusalem, the Jordan, and Damascus on pilgrimage to the Holy Land.
1160–73 A.D.	Benjamin of Tudela	Spain	Traveled through Egypt, Assyria, Persia, and central Asia, visiting Jewish centers.
1245–47 A.D.	John de Plano Carpini	Italy	Traveled through Poland, Russia, and central Asia to Karakoram, in Mongolia, as legate of the pope.
1253–55 A.D.	William of Rubruck	France	Visited Karakoram, in Mongolia, by way of southern Russia and Turkestan.
1270 A.D.	Lancelot Malocello	Italy	Rediscovered the Fortunate or Canary Islands.
1271–95 A.D.	Marco Polo	Italy	Journeyed to China by way of central Asia; returned by sea by way of Sumatra, Ceylon, India, and Persia; reported existence of Japan and Madagascar.
1281–91 A.D.	Vivaldi Brothers	Italy	Attempted voyage to India by sea along west coast of Africa, but never returned.
1323–28 A.D.	Friar Odoric	Italy	Traveled to China by way of India and Malaya; returned through central Asia.
1325–54 A.D.	Ibn Battuta		Visited every Islamic country from Spain to India; traveled widely in Far East, Arabia, and western Africa.
1346 A.D.	Jayme Ferrer	Catalonia	Credited by 14th-century maps with having rounded Cape Bojador on west coast of Africa.
1427–31 A.D.	Diogo de Seville	Portugal	Discovered some of the Azores Islands.
1433–35 A.D.	Gil Eannes	Portugal	Rounded Cape Bojador in exploration of west coast of Africa.

America (1492–1850) continued

DATE	EXPLORER	COUNTRY REPRESENTED	DESCRIPTION
1577–80	Francis Drake	England	Explored west coast of North America to 46° or 48° N. and named it New Albion; circumnavigated the earth.
1583	Humphrey Gilbert	England	Made first effort to establish an English colony in North America; ship lost returning to England.
1585–87	John Davis	England	Made several voyages in search of Northwest Passage; discovered Davis Strait and Baffin Bay.
1602–03	Sebastian Vizcaino and Martin Aguilar	Spain	Sailed along west coast of California to about 42° or 43° N.; discovered Monterey Bay but missed that of San Francisco; Aguilar reported large river near 43° N.
1603–15	Samuel de Champlain	France	Explored and mapped St. Lawrence R., New England coast, Ottawa R., Lake Huron, Lake Ontario; discovered Lake Champlain (1609).
1607–14	John Smith	England	Explored and mapped vicinity of Jamestown, Virginia (1608) and coast of New England (1614).
1609–11	Henry Hudson	Holland and England	Explored Hudson R. to Albany for Holland (1609); discovered and explored Hudson Bay for England (1610–1611).
1612–13	Thomas Button	England	Explored Hudson Bay in search of strait to the Western Ocean.
1615–16	William Baffin	England	Made two voyages in search of Northwest Passage; explored Baffin Bay to 78° N.
1631–32	Luke Foxe and William James	England	Explored northern and southern extensions of Hudson Bay without finding passage westwards.
1634	Jean Nicollet	France	Crossed Lake Huron to Mackinac Strait and Green Bay; reported "Western Sea" three days distant.
1658–59	Pierre Radisson and Sieur des Groseillers	France	Explored upper Mississippi R. and western shores of Lake Superior.
1669–70	John Lederer	England	Crossed the Blue Ridge and explored the Shenandoah Valley.
1669–87	Robert Cavalier, Sieur de la Salle	France	Explored Lake Ontario and upper Ohio R. (1669) and the Great Lakes to head of Lake Michigan (1679); descended Illinois and Mississippi rivers to Gulf of Mexico (1681–82); killed in Texas after failing to locate Mississippi R. by sea (1684–87).
1673	Jacques Marquette and Louis Joliet	France	Descended Mississippi R. from the Wisconsin R. to the Arkansas and returned to the Great Lakes via the Illinois-Chicago portage.
1680	Louis Hennepin	France	Explored upper Mississippi R. from the Illinois R. to the Minnesota.
1688	Louis de la Hontan	France	Explored upper Mississippi region; spread reports of fictitious "Long River" leading to Western Sea.
1699	Pierre le Moyne, Sieur d' Iberville	France	Entered mouth of Mississippi from Gulf of Mexico and explored delta.
1701–02	Eusebio Francisco Kino	Spain	Explored the Gila and lower Colorado rivers; proved that California was not an island.
1718–19	Bernard de la Harpe	France	Explored the Red and Arkansas rivers.
1721	Pierre François Xavier de Charlevoix	France	Visited French settlements in North America from Quebec to New Orleans.
1728–41	Vitus Bering	Russia	Confirmed existence of strait between Asia and America (1728); discovered northwest coast and named Mt. St. Elias (1741).
1730–43	Sieur de La Vérendrye and sons	France	Explored territory northwest of Lake Superior; discovered Lake Winnipeg; sons may have seen Rocky Mountains.
1742	Christopher Middleton	England	Discovered Repulse Bay in search of passage to Western Sea.
1749	Celoron de Bienville	France	Buried plates along the Ohio R., claiming formal possession for France.

Date	Explorer	Country	Description
…	… Dias Diaz	Portugal	Rounded Cape Verde on west coast of Africa.
1455–57 A.D.	Alvise da Cadamosto	Portugal	Explored the Sénégal and Gambia rivers; discovered Cape Verde Islands.
1472 A.D.	Fernando Póo	Portugal	Discovered island bearing his name in Gulf of Guinea.
1482–86 A.D.	Diogo Cao	Portugal	Discovered mouth of Congo R. (1482), reached Cape Negro at 16° S. (1486).
1487 A.D.	Pedro de Covilhã	Portugal	Traveled to India via Egypt and Arabia; visited east coast of Africa, south to Zambezi R.
1487–88 A.D.	Bartolomeu Dias	Portugal	Discovered Cape of Good Hope; explored coast east to Mossel Bay.

America (1492–1850)

Date	Explorer	Country	Description
1492–1502	Christopher Columbus	Spain	Discovered the West Indies (1492); in three later voyages explored coasts of northern South America and Central America.
1497–98	John and Sebastian Cabot	England	Discovered shores of Nova Scotia and Newfoundland, and visited southern Greenland.
1499–1500	Amerigo Vespucci, Juan de la Cosa and Alonso de Ojeda	Spain	Discovered and explored northeastern coast of South America.
1499–1500	Vincente Yáñez Pinzón	Spain	Discovered mouth of Amazon R.
1500	Pedro Álvares Cabral	Portugal	Discovered or visited coast of Brazil on voyage to India.
1500–01	Gaspar Corte Real	Portugal	Made two voyages to northeastern North America, but never returned.
1501–02	Amerigo Vespucci	Portugal	Explored coast of Brazil to 30° S. or farther.
1513	Juan Ponce de León	Spain	Discovered and explored coasts of Florida.
1513	Vasco Núñez de Balboa	Spain	Crossed Isthmus of Panama and discovered the South Sea (Pacific Ocean).
1515	Juan Diaz de Solis	Spain	Explored mouth of Río de la Plata.
1517	Francisco Fernández de Córdoba	Spain	Discovered Yucatán and evidence of Mayan culture.
1518	Juan de Grijalva	Spain	Explored east coast of Mexico north of Yucatán.
1519	Álvárez Pineda	Spain	Explored Gulf of Mexico and may have discovered mouth of Mississippi R.
1519–22	Ferdinand Magellan	Spain	Discovered Strait of Magellan (1520) during first circumnavigation of the earth.
1519–27	Hernando Cortes	Spain	Explored and conquered Mexico.
1523–41	Francisco Pizarro	Spain	Explored northwestern South America and conquered Peru.
1524	Giovanni da Verrazano	France	Discovered New York Bay and explored coast northward.
1524–25	Estéban Gomez	Spain	Sailed along east coast of North America from Nova Scotia to Florida.
1527–37	Cabeza de Vaca	Spain	Wandered for nine years along and near coast of Gulf of Mexico from Florida to Mexico.
1534–41	Jacques Cartier	France	Explored Gulf of St. Lawrence (1534) and river as far as sites of Quebec and Montreal (1536).
1535–36	Diego de Almagro	Spain	Explored and conquered Chile.
1536–38	Gonzalo Jiménez de Quesada	Spain	Explored and conquered New Granada, and founded Bogotá.
1539	Francisco de Ulloa	Spain	Explored Gulf of California to its head.
1539–43	Hernando de Soto	Spain	Explored southeastern United States from Florida to Tennessee; discovered Mississippi R. (1541).
1540	Hernando de Alarcón	Spain	Sailed up Gulf of California and entered Colorado R.
1540–42	Francisco Vasquez de Coronado	Spain	Led expedition into southwestern United States; explored Great Plains northward to Kansas; Grand Canyon of Colorado R. discovered by one of his party.
1541	Francisco de Orellana	Spain	Crossed the Andes and descended Amazon R. to its mouth.
1542–43	Bartolomé Ferrelo and Juan Rodriguez Cabrillo	Spain	Discovered San Diego Bay and explored California coast to about 42° N. or Cape Mendocino.
1562–65	René de Laudonnière and Jean de Ribaut	France	Failed in effort to establish a permanent colony on coast of South Carolina.
…	…	…	…cuascue of the Oregon or River of the West.
1769–75	Daniel Boone	England	Explored eastern Kentucky (1769–71) and blazed the famous Wilderness Road (1775).
1769	José Ortega	Spain	Discovered San Francisco Bay during overland expedition into upper California.
1770–72	Samuel Hearne	England	Traced the Coppermine R. to the Northern Ocean and discovered Great Slave Lake.
1774–75	Juan Pérez and Bruno Heceta	Spain	Sent to explore northwest coast, reaching 55° N.; Heceta observed entrance to Columbia R.; Pérez discovered Nootka Sound.
1778–79	James Cook	England	Rediscovered Hawaiian Islands; explored and charted northwest coast from 45° N. to Arctic Ocean.
1788–92	Robert Gray	United States	Explored northwest coast; discovered Grays Harbor; entered and named the Columbia R. (1792).
1789–93	Alexander Mackenzie	England	Traced Mackenzie R. to its mouth (1789); crossed Rocky Mountains via Peace R. and reached Pacific Ocean.
1792–94	George Vancouver	England	Explored and mapped Puget Sound; charted inside passage and inlets along northwest coast.
1804–06	Meriwether Lewis and William Clark	United States	Ascended Missouri R. to its source, crossed Rocky Mountains and descended Columbia R. to Pacific Ocean.
1805–07	Zebulon M. Pike	United States	Explored and mapped upper Mississippi R. (1805); and southwestern section of Louisiana Territory (1806–07).
1807–08	Manuel Lisa and John Colter	United States	Explored Northern Rockies (Yellowstone-Big Horn region) as trappers and fur traders.
1811–12	Wilson Price Hunt (Astorians)	United States	Discovered overland route to Pacific via the Snake and Columbia rivers.
1819–20	Stephen H. Long	United States	Explored the high plains between Platte and Arkansas rivers; called the Great Plains the "Great American Desert."
1821	William Becknell	United States	Opened trade route between Missouri R. and Santa Fe.
1823–29	Jedediah Smith	United States	Located famous South Pass across Rocky Mts.; crossed desert between Colorado R. and California.
1824–28	Peter Skene Ogden	England	Explored upper Snake R. and northern Great Basin; discovered Humboldt R. and Great Salt Lake.
1829–30	Ewing Young and party	United States	Opened up Spanish Trail between Santa Fe and Los Angeles.
1832–33	Nathaniel J. Wyeth	United States	Led first expedition along Oregon Trail to Columbia R.
1833	Joseph E. Walker	United States	Crossed Great Basin between Great Salt Lake and California.
1841	Charles Wilkes	United States	Visited Oregon country and California during official Pacific exploring expedition by sea.
1842–45	John C. Frémont	United States	First official government explorer to re-trace explorations of fur trappers in the Far West.

Africa

Date	Explorer	Country	Description
1520–27	Francisco Alvarez	Portugal	Visited Ethiopia and described it in detail.
1541	Christopher da Gama	Portugal	Led expedition into Ethiopia.
1578–89	Duarte López	Portugal	Visited the Kingdom of Congo; his reports a chief source of information until 19th century.
1604–22	Pedro Páez	Portugal	First European to visit Ethiopian sources of the Nile R.
1616	Gaspar Boccaro	Portugal	Explored interior from upper Zambezi R. to west coast.
1618–19	George Thompson	England	Explored the Gambia R.
1625–35	Jerome Lobo	Portugal	Lived in Ethiopia as a missionary.
1698–1700	C. J. Poncet	France	Traveled as a physician into Ethiopia to treat the Emperor.
1768–73	James Bruce	England	Explored Ethiopia, especially source of the Blue Nile R.
1777–79	William Patterson	England	Made several trips into the Kaffir country as a naturalist.
1795–1805	Mungo Park	England	Explored the Gambia R. and was first modern European to reach the Niger R.
1797–98	John Barrow	England	Journeyed from Cape of Good Hope to upper Orange R.

PRINCIPAL DISCOVERIES AND EXPLORATIONS (Continued)

Arctic Regions continued

DATE	EXPLORER	COUNTRY REPRESENTED	DESCRIPTION
1871-74	Julius Payer and Carl Weyprecht	Austria	Discovered Franz Josef Land Archipelago.
1876	Albert H. Markham	England	Reached 83° 20' on northwest coast of Greenland.
1878-79	Nils A. E. Nordenskjöld	Sweden	Completed the Northeast Passage in two seasons in ship *Vega*.
1879-81	George W. DeLong	United States	Explored Arctic Ocean northwest of Bering Strait; ship *Jeannette* and most of party lost.
1881-83	Adolphus W. Greely	United States	Explored northern Greenland and Ellesmere Island; party established new record of 83° 24' N.
1888-96	Fridtjof Nansen	Norway	Made first crossing of Greenland (1888); reached record of 86° 14' during drift of ship *Fram* (1895).
1897	Salomon A. Andrée	Sweden	Attempted balloon flight to North Pole from Spitsbergen (Svalbard); remains of party found in 1930 on White Island.
1898-1902	Otto Sverdrup	Norway	Explored northern Ellesmere Island and discovered Axel Heiberg Island.
1899-1900	Umberto Cagni	Italy	Reached new record at 86° 34' N. by sledge from Franz Josef Land; member of Abruzzi expedition.
1900-09	Robert E. Peary	United States	Made repeated efforts to reach North Pole, succeeding (April 6, 1909) by sledge from Grant Land.
1903-6	Roald Amundsen	Norway	Completed first trip through Northwest Passage from east to west.
1907-9	Frederick A. Cook	United States	Claimed to have reached North Pole on April 20, 1908.
1925-26	Lincoln Ellsworth	United States	Made flight with Amundsen from Spitsbergen (Svalbard) to 87° 43' N. and return; co-leader of dirigible flight over North Pole (1926).
1926	Richard Byrd and Floyd Bennett	United States	Made successful flight from Spitsbergen (Svalbard) to North Pole and return.
1926-28	Umberto Nobile	Italy	Made numerous dirigible flights across arctic region; rescued after *Italia* crashed on ice in 1928.
1937-38	Otto Schmidt	Russia	Spent nine months with scientific expedition near North Pole.

Antarctic Regions

DATE	EXPLORER	COUNTRY REPRESENTED	DESCRIPTION
1738-39	J.B.C. Bouvet de Lozier	France	Discovered Bouvet Island south of Africa in latitude 54° S.
1768-75	James Cook	England	Established non-existence of southern continent in habitable latitudes; reached record of 71° 10' S.
1771-73	Yves Joseph de Kerguélen-Trémarec	France	Discovered and explored Kerguélen Island in latitude 49° 50' S., longitude 69° 30' E.
1819	William Smith	England	Discovered South Shetland Islands.
1819-21	Fabian von Bellingshausen	Russia	Circumnavigated Antarctica; discovered Alexander I Land.
1821	Nathaniel Palmer	United States	Discovered Palmer Peninsula on sealing expedition.
1823	James Weddell	England	Discovered Weddell Sea; reached 74° 15' S.
1837-40	Jules Dumont d'Urville	France	Discovered Adélie Land south of Tasmania.
1839-40	Charles Wilkes	United States	Sighted Antarctic coast between 108° and 148° E.
1840-43	James Ross	England	Charted coast in neighborhood of Ross Sea; reached record of 78° 9' S.
1902-04	Robert F. Scott	England	Explored coast of Edward VII Land; reached 82° 17' S.
1903-05	Jean B. Charcot	France	Explored Palmer Peninsula; discovered Loubet Coast.
1908-09	Ernest Shackleton	England	Explored head of Ross Sea; reached 88° 23' S.
1910-12	Roald Amundsen	Norway	Discovered Queen Maud Range; reached South Pole Dec. 16, 1911.
1910-12	Robert Scott	England	Reached South Pole January 18, 1912; entire party perished during return.
1911-13	Douglas Mawson	England and Australia	Explored coast from King George V Land to Enderby Land in two expeditions.
1929-31	Ernest Shackleton	England	Discovered Coats coast; ship lost in Weddell Sea.

Africa continued

DATE	EXPLORER	COUNTRY REPRESENTED	DESCRIPTION
1797-1800	Frederick Hornemann	England	Traveled from Egypt to Marzūq and the Niger R., disguised as an Arab.
1798-99	Francisco de Lacerda	Portugal	Explored southeastern interior north of Zambezi R.
1801	John Trutter and William Somerville	England	Explored Bechuanaland, north of Orange R.
1802-06	Pedro Baptista and A. Jose	Portugal	Made first recorded crossing of continent eastward from Angola.
1812-14	Johann L. Burckhardt	Switzerland	Traveled up the Nile R. and across to the Red Sea.
1822-25	Dixon Denham and Hugh Clapperton	England	Crossed desert from Tripoli to Lake Chad and westward to the Niger R.
1825-26	Alexander G. Laing	England	Reached Timbuktu from Tripoli, but was murdered on return trip.
1827-28	René Caillé	France	Traveled from Guinea Coast to 'Fez and Tangier by way of Timbuktu.
1830-34	Richard Lander	England	Explored the lower Niger R. and located its mouth.
1849-73	David Livingstone	England	Discovered Zambezi R. (1851), Victoria Falls (1855), and Lake Nyasa (1859); explored upper Congo tributaries; found by Stanley on Lake Tanganyika in 1871.
1856-59	Richard Burton	England	Discovered Lake Tanganyika and explored surrounding area.
1858-63	John Speke	England	Discovered Victoria Nyanza as source of the Nile R.
1861-69	Samuel Baker	England	Explored upper Nile R.; discovered Lake Albert.
1863-71	Georg A. Schweinfurth	Germany	Explored extensively in the Sudan and equatorial Africa.
1871-90	Henry Stanley	United States	Continued Livingstone's explorations in the lakes region; descended Congo R. to Atlantic Ocean (1877); discovered Stanley Pool and Lake Edward.
1877-86	Serpa Pinto	Portugal	Crossed the continent from Angola to Mozambique.
1879-90	Joseph Thomson	England	Explored new areas in Tanganyika, Kenya, and Uganda.
1888	Samuel Teleki	Hungary	Discovered lakes Rudolph and Stephanie.

Asia

DATE	EXPLORER	COUNTRY REPRESENTED	DESCRIPTION
1497-99	Vasco da Gama	Portugal	Discovered sea route to India by way of South Africa and Indian Ocean.
1502-07	Ludovici di Varthema	Portugal	Traveled as convert to Islam in Arabia, Persia, India, and East Indies.
1511	Mathias Albuquerque	Portugal	Conquered Malacca, East Indian spice center.
1520-21	Thomé Pires	Portugal	Sent to Peking as commercial envoy.
1537-58	Fernão Mendes Pinto	Portugal	Described travels in India, China, and Japan.
1549-51	Francis Xavier	Portugal	Introduced Christianity into Japan.
1561-63	Anthony Jenkinson	England	Visited Persia by overland route from Russia.
1578-1610	Matteo Ricci	Portugal	Established first Christian missions in China.
1603-05	Benedict de Goez	Portugal	Made first overland trip to China after Marco Polo.
1632-68	Jean B. Tavernier	France	Traveled as commercial trader in Persia, India, and East Indies.
1656	Pieter van Goyer and Jacob von Keyser	Holland	Visited Peking by overland route from Canton.
1665-77	John Chardin	France	Described extensive travels in Persia and India.
1683-93	Engelbert Kaempfer	Holland	As physician with Dutch embassy, visited and described Thailand (Siam) and Japan.
1715-47	John Bell	Russia	Traveled as physician with Russian embassies to Persia and through Siberia to China.
1716-21	Ipolito Desideri	Italy	Reached Tibetan city of Lhasa from Kashmir.
1761-64	Carsten Niebuhr	Denmark	Explored Yemen, reaching cities of Şanʻā and Mocha (Al Mukhā); also visited Oman, Syria, and Palestine.

Dates	Explorer	Country	Description
1935–36	Lincoln Ellsworth	United States	flight over South Pole (1929); second expedition remained through winter of 1934; third expedition (1939–40) made extensive aerial explorations; fourth expedition concentrated on scientific work.
1947–48	Finn Ronne	United States	Explored by air between Palmer Peninsula and Little America.
1955–58	Vivian Fuchs and Edmund Hillary	United Kingdom and New Zealand	Commonwealth Trans-Antarctic Expedition crossed the Continent through the South Pole from Weddell Sea to McMurdo Sound.
1957–58	I. G. Y. (International Geophysical Year)	Arg.; Austl.; Bel.; Chile; Fr.; Jap.; N.Z.; Nor.; S. Afr.;Sov. Un.; U.K.; U.S.	Established research stations; field expeditions led to new discoveries of physical features, as well as new information on ice conditions; extensive oceanographic surveys and mapping conducted.
1959–61	Australian National Antarctic Expedition	Australia	Gathered data on weather, cosmic rays, geomagnetism, seismology; field explorations and mapping; extensive aerial surveys.

Pacific Ocean and Australia

Dates	Explorer	Country	Description
1520–21	Ferdinand Magellan	Spain	Crossed the Pacific from South America to the Philippines during first circumnavigation of the earth.
1542	Lopez de Villalobos	Spain	Sailed from Mexico to the Philippines; discovered Caroline and Palau Islands.
1565	Andrés de Urdaneta	Spain	Discovered northern sailing route from Philippines to Mexico in latitude of the Forties.
1567–95	Alvaro de Mendana	Spain	Discovered Solomon, Marshall, and Ellice Islands (1567); also Marquesas and Santa Cruz (1595).
1578	Francis Drake	England	Crossed the Pacific from California to the East Indies on first English circumnavigation.
1606	Pedro de Quiros	Spain	Discovered Tahiti and New Hebrides Islands.
1606	Luis de Torres	Spain	Sailed through Torres Strait between Australia and New Guinea.
1616	Dirk Hartog	Holland	Explored section of west coast of Australia.
1616	William Van Schouten and Jacob Lemaire	Holland	Rounded Cape Horn and crossed Pacific; discovered Bismarck Archipelago.
1642–44	Abel Tasman	Holland	Discovered Tasmania and part of New Zealand; explored the north coast of Australia.
1699	William Dampier	England	Explored west and northwest coasts of Australia.
1721	Jacob Roggeveen	Holland	Discovered Easter Island and Samoa.
1767–69	Louis de Bougainville	France	Explored South Pacific islands, including Tahiti, Samoa, and the New Hebrides.
1768–79	James Cook	England	Made three voyages into the Pacific; explored coasts of New Zealand and eastern Australia (1769–70); proved non-existence of continental land north of Antarctic Circle (1772–75); discovered Hawaiian Islands and explored northwest coast of North America (1776–79).
1785–88	Jean de La Pérouse	France	Explored North Pacific Ocean, especially coasts of Siberia and Japan; lost at sea.
1798	George Bass	England	Discovered strait separating Tasmania from Australia.
1802–03	Matthew Flinders	England	Explored south coast of Australia and sailed completely around the continent.
1816–22	John Oxley	England	Explored the interior of New South Wales, Australia.
1828–45	Charles Sturt	England	Discovered the Darling R.; descended Murray R. to its mouth; reached center of continent (1845).
1833–35	Charles Darwin	England	Explored South Pacific islands as a naturalist.
1839–41	Edward Eyre	England	Crossed southern Australia from Spencer Gulf to King George Sound.
1844–48	Ludwig Leichhardt	Germany	Explored interior of northern Queensland and Arnhem Land.
1858–62	John Stuart	England	Explored interior of South Australia and made unsuccessful attempt to cross the continent (1860); succeeded (1862).
1860–61	Robert Burke and W. J. Wills	England	Succeeded in crossing Australia from Melbourne to Gulf of Carpentaria.
1873	Peter E. Warburton	England	Crossed western Australia from Alice Springs to the coast, using camels.
1874	John Forrest	England	Crossed desert region of Australia from Perth to Adelaide.
1875–76	Ernest Giles	England	Made trip across desert from Port Augusta to Perth and return.

Dates	Explorer	Country	Description
1851–54	Matthew C. Perry	United States	Opened Japan to foreign trade.
1862–67	Peter Kropotkin	Russia	Made geographical surveys of North Manchuria.
1867–88	Nikolai Przhevalsky	Russia	Led expeditions into Central Asia, Mongolia, and Tibet; rediscovered Lop Nor.
1868–72	Ferdinand Richthofen	Germany	Explored and described most of Chinese Empire.
1869–70	Joseph Halévy	France	Explored interior of southwestern Arabia.
1873	Jean Dupuis	France	Explored Tonkin route into China.
1885–1908	Sven Hedin	Sweden	Traveled extensively in Persia, Turkestan, China, and Tibet.
1886–1904	Francis Younghusband	England	Explored and surveyed in Kashmir, Central Asia, and Tibet.
1889–92	William W. Rockhill	United States	Explored eastern Tibet.
1899–1914	Gertrude Bell	England	Traveled widely in Palestine, Mesopotamia, and inner Arabia.
1899–1926	Aurel Stein	England	Made archaeological explorations in India, Persia, and central Asia.
1901–06	Ellsworth Huntington	United States	Explored upper Euphrates R. and Chinese Turkestan.
1914–29	Roy Chapman Andrews	United States	Explored western China and Mongolia as a naturalist, discovering many animal fossils.
1917–32	St. John Philby	England	Crossed Arabia from sea to sea; explored oases of Nejd.

Arctic Regions

Dates	Explorer	Country	Description
1553–54	Hugh Willoughby and Richard Chancellor	England	Attempted exploration of Northeast Passage; Willoughby lost, but Chancellor reached Archangel and opened trade with Russia.
1576–78	Martin Frobisher	England	Made three voyages in search of Northwest Passage; discovered Frobisher Bay.
1580	Arthur Pet and Charles Jackman	England	Reached the Kara Sea, exploring Northeast Passage.
1585–87	John Davis	England	Reached latitude 73° N. in Baffin Bay, exploring Northwest Passage.
1594–97	Willem Barents	Holland	Discovered Spitsbergen (Svalbard) and reached Novaya Zemlya along Northeast Passage.
1607–11	Henry Hudson	England and Holland	Made several voyages in search of both Northeast and Northwest Passages to India; reached 73° N. on east Greenland coast.
1615–16	William Baffin and Robert Bylot	England	Explored Baffin Bay; reached 78° N.
1648	Simon Dezhnev	Russia	Explored northeastern Siberian coast from the Kolyma to Anadyr rivers.
1728	Vitus Bering	Russia	Discovered Bering Strait and the St. Lawrence and Diomede Islands.
1737–42	Dimitri Laptev	Russia	Explored north Siberian coast from Lena R. to Cape Baranov.
1742	T. Chelyuskin	Russia	Discovered northernmost point of Asia by land.
1773	C. J. Phipps	England	Reached 80° 48' north of Spitsbergen (Svalbard).
1818–27	William E. Parry	England	Explored Canadian arctic and Spitsbergen (Svalbard) areas; reached 82° 45' (1827).
1820–22	William Scoresby	England	Explored Scoresby Sound in eastern Greenland; published standard description of Arctic regions.
1825–28	Frederick W. Beechey	England	Explored arctic coast of North America from Bering Strait to Point Barrow.
1829–49	John and James Ross	England	Discovered Boothia Peninsula and Gulf; James located North Magnetic Pole (1831); both participated in search for Franklin (1848–49).
1845–48	John Franklin	England	Lost two ships and 129 men in attempt to sail through Northwest Passage; reached King William Island.
1850–54	Richard Collinson and Robert McClure	England	Reached Melville Sound from Bering Strait and proved existence of northwest waterway passage.
1853–55	Elisha K. Kane	United States	Explored Smith Sound and Kane Basin; reached 80° 10' N.
1857–58	Francis L. McClintock	England	Discovered McClintock Channel and relics of Franklin expedition on King William Island.
1860–71	Charles F. Hall	United States	On third expedition, explored northern shores of Ellesmere Island and Greenland, reaching 82° 26' N.

	Apia	Azores Islands	Berlin	Bombay	Buenos Aires	Calcutta	Cape Town	Cape Verde Islands	Chicago	Darwin	Denver	Gibraltar	Hong Kong	Honolulu	Istanbul	Juneau	London	Los Angeles	Manila	Melbourne	Mexico City
Apia, Western Samoa		9644	9743	8154	6931	7183	9064	10246	6557	3843	5653	10676	5591	2604	10175	5415	9789	4828	4993	3113	5449
Azores Islands	9644		2185	5967	5417	6549	5854	1499	3093	10209	3991	1249	7572	7180	2975	4526	1527	4794	8250	12101	4385
Berlin, Germany	9743	2185		3910	7376	4376	5977	3194	4402	8036	5077	1453	5500	7305	1078	4560	574	5782	6128	9919	6037
Bombay, India	8154	5967	3910		9273	1041	5134	6297	8054	4503	8383	4814	2673	8020	2991	6866	4462	8701	3148	6097	9722
Buenos Aires, Argentina	6931	5417	7376	9273		10242	4270	4208	5596	9127	5928	5963	11463	7558	7568	7759	6918	6118	11042	7234	4633
Calcutta, India	7183	6549	4376	1041	10242		6026	7148	7981	3744	8050	5521	1534	7037	3646	6326	4954	8148	2189	5547	9495
Cape Town, South Africa	9064	5854	5977	5134	4270	6026		4509	8449	6947	9327	5076	7372	11532	5219	10330	6005	9969	7525	6412	8511
Cape Verde Islands	10246	1499	3194	6297	4208	7148	4509		4066	10664	4975	1762	8539	8311	3507	5911	2731	5772	9221	10856	4857
Chicago, U.S.A.	6557	3093	4402	8054	5596	7981	8449	4066		9346	920	4258	7790	4244	5476	2305	3950	1745	8128	9668	1673
Darwin, Australia	3843	10209	8036	4503	9127	3744	6947	10664	9346		8557	9265	2642	5355	7390	7105	8598	7835	1979	1964	9081
Denver, U.S.A.	5653	3991	5077	8383	5928	8050	9327	4975	920	8557		5122	7465	3338	6154	1831	4688	831	7661	8759	1434
Gibraltar, Gibraltar	10676	1249	1453	4814	5963	5521	5076	1762	4258	9265	5122		6828	8075	1874	5273	1094	5936	7483	10798	5629
Hong Kong, Asia	5591	7572	5500	2673	11463	1534	7372	8539	7790	2642	7465	6828		5537	4980	5634	5981	7240	693	4607	8776
Honolulu, Hawaii, U.S.A.	2604	7180	7305	8020	7558	7037	11532	8311	4244	5355	3338	8075	5537		8104	2815	7226	2557	5296	5513	3781
Istanbul (Constantinople), Turkey	10175	2975	1078	2991	7568	3646	5219	3507	5476	7390	6154	1874	4980	8104		5498	1551	6843	5659	9088	7102
Juneau, Alaska, U.S.A.	5415	4526	4560	6866	7759	6326	10330	5911	2305	7105	1831	5273	5634	2815	5498		4418	1842	5869	8035	3219
London, United Kingdom	9789	1527	574	4462	6918	4954	6005	2731	3950	8598	4688	1094	5981	7226	1551	4418		5439	6667	10501	5541
Los Angeles, U.S.A.	4828	4794	5782	8701	6118	8148	9969	5772	1745	7835	831	5936	7240	2557	6843	1842	5439		7269	7931	1542
Manila, Philippines	4993	8250	6128	3148	11042	2189	7525	9221	8128	1979	7661	7483	693	5296	5659	5869	6667	7269		3941	8829
Melbourne, Australia	3113	12101	9919	6097	7234	5547	6412	10856	9668	1964	8759	10798	4607	5513	9088	8035	10501	7931	3941		8422
Mexico City, Mexico	5449	4385	6037	9722	4633	9495	8511	4857	1673	9081	1434	5629	8776	3781	7102	3219	5541	1542	8829	8422	
Moscow, Soviet Union	9116	3165	996	3131	8375	3447	6294	3982	4984	7046	5485	2413	4439	7033	1088	4534	1549	6068	5130	8963	6688
New Orleans, U.S.A.	6085	3524	5116	8865	4916	8803	8316	4194	833	9545	1082	4757	8480	4207	6171	2905	4627	1673	8724	9275	934
New York, U.S.A.	7242	2422	3961	7794	5297	7921	7801	3355	713	9959	1631	3627	8051	4959	5009	2854	3459	2451	8493	10355	2085
Nome, Alaska, U.S.A.	5438	4954	4342	5901	8848	5271	10107	6438	3314	6235	2925	5398	4547	3004	5101	1094	4381	2876	4817	7558	4309
Oslo, Norway	9247	2234	515	4130	7613	4459	6494	3444	4040	8022	4653	1791	5337	6784	1518	4045	714	5325	6016	9926	5706
Panamá, Panama	6514	3778	5849	9742		10114	7014	3734	2325	10352	2636	4926	10084	5245	6750	4460	5278	3001	10283	9022	1495
Paris, France	9990	1659	542	4359	6877	4889	5841	2666	4133	8575	4885	964	5956	7434	1401	4628	213	5601	6673	10396	5706
Peking (Peiping), China	5903	6565	4567	2964	11974	2024	8045	7763	6592	3728	6348	6009	1226	5067	4379	4522	5054	6250	1770	5667	7733
Port Said, U.A.R.	10485	3391	1747	2659	7362	3506	4590	3672	6103	7159	6819	2179	4975	8738	693	6215	2154	7528	5619	8658	7671
Quebec, Canada	7406	2240	3583	7371	5680	7481	7857	3355	878	9724	1752	3383	7650	5000	4644	2660	3101	2579	8124	10497	2454
Reykjavik, Iceland	8678	1777	1479	5191	7099	5409	7111	3248	2954	8631	3596	2047	6031	6084	2558	3268	1171	4306	6651	10544	4622
Rio de Janeiro, Brazil	8120	4428	6144	8257	1218	9376	3769	3040	5296	9960	5871	4775	10995	8190	6395	7598	5772	6296	11254	8186	4770
Rome, Italy	10475	2125	734	3843	6929	4496	5249	2772	4808	8190	5561	1034	5768	8022	854	5247	887	6326	6457	9934	6353
San Francisco, U.S.A.	4786	4872	5657	8392	6474	7809	10241	5921	1858	7637	949	5936	6894	2392	6700	1525	5355	347	6963	7854	1885
Seattle, U.S.A.	5222	4501	5041	7741	6913	7224	10199	5714	1737	7619	1021	5462	6471	2678	6063	899	4782	959	6641	8186	2337
Shanghai, China	5399	7229	5215	3133	12197	2112	8059	8443	7053	3142	6698	6646	772	4934	4959	4869	5710	6477	1152	5005	8039
Singapore, Singapore	5850	8326	6166	2429	9864	1791	6016	8700	9365	2075	9063	7231	1652	6710	5373	7235	6744	8767	1479	3761	10307
Tokyo, Japan	4656	7247	5538	4188	11400	3186	9071	8589	6303	3367	5795	6988	1796	3850	5556	4011	5938	5470	1863	5089	7035
Valparaiso, Chile	6267	5678	7795	10037	761	10993	4998	4649	5268	8961	5452	6408	11607	6793	8172	7271	7263	5527	10930	6998	4053
Washington, D.C., U.S.A.	7066	2667	4167	7988	5216	8088	7894	3486	591	9923	1494	3822	8148	4829	5216	2834	3665	2300	8560	10173	1878
Wellington, New Zealand	2062	11269	11265	7677	6260	7042	7019	10363	8349	3310	7516	12060	5853	4708	10663	7475	11682	6714	5162	1595	6899
Vienna, Austria	10010	2291	328	3718	7368	4259	5671	3147	4694	7974	5383	1386	5429	7626	783	4895	772	6108	6120	9792	6306
Winnipeg, Canada	6283	3389	4286	7644	6297	7424	9054	4556	714	8684	798	4435	7096	3806	5361	1597	3918	1525	7414	9319	2097
Zanzibar, Africa	9892	5323	4309	2855	6421	3859	2346	4635	8358	6409	9221	4103	5414	10869	3312	8795	4604	10021	5763	6802	9484

	Bombay	Buenos Aires	Cape Town	Colombo	Gibraltar	Halifax	Hamburg	Honolulu	Istanbul	Le Havre	Lisbon	Liverpool	Manila	Melbourne	New Orleans	New York	Panama Roads	Port Said	Rio de Janeiro	San Francisco	Shanghai	Singapore	Valparaiso	Wellington
Bombay, India		9601	5469	1042	5639	8760	7552	9631	4412	7024	6036	7156	4361	6365	10927	9413	14921	3511	8998	11247	5328	2824	11356	7961
Buenos Aires, Argentina	9601		4345	9415	6074	6600	7622	8744	8488	7074	6148	7178	12128	8477	7233	6761	8259	1325	10062	13087	3181	10782	6956	
Cape Town, South Africa	5469	4345		5070	5982	7386	7388	11948	7058	6861	5912	7001	7821	6998	9382	7814	7417	6148	3769	11154	8787	6511	6977	7531
Colombo, Ceylon	1042	9415	5070		6227	9278	8090	8594	4920	7563	6577	7717	3399	5380	11489	9941	13919	4010	8839	10289	4370	1825	11073	7058
Gibraltar, Gibraltar	5639	6074	5982	6227		3051	1863	10433	2099	1336	350	1490	9641	11257	5271	3714	5038	2217	4816	8775	10553	8008	9006	12847
Halifax, Canada	8760	6600	7386	9278	3051		3480	8152	5147	3082	2792	2891	12591	11876	2517	686	2718	5257	5332	6456	12707	11047	5731	10196
Hamburg, Germany	7552	7622	7388	8090	1863	3480		11283	3939	573	1083	1083	16678	13066	5935	4166	5888	4058	6354	9625	12349	9838	8900	13758
Honolulu, Hawaii, U.S.A.	9631	8744	11948	8594	10433	8152	11283		12510	10757	10363	10682	5571	5691	7046	7718	5395	12604	9875	2408	4986	6772	6816	4736
Istanbul, Turkey	4412	8488	7058	4920	2099	5147	3939	12510		3421	2430	3543	8245	9928	7384	5788	7115	910	6897	10884	9210	6700	11020	11540
Le Havre, France	7024	7074	6861	7563	1336	3082	573	10757	3421		1017	578	10856	12540	5315	3640	5363	3521	5820	9095	11822	9312	8347	12801
Lisbon, Portugal	6036	6148	5912	6577	350	2792	1083	10363	2430	1017		1148	9867	11551	5377	3403	4968	2532	4858	8737	10833	8323	7975	12459
Liverpool, United Kingdom	7156	7178	7001	7717	1490	2891	1083	10682	3543	578	1148		11111	12764	5266	3539	5287	3652	5932	9024	12201	9490	8299	12778
Manila, Philippines	4361	12128	7821	3399	9641	12591	16678	5571	8245	10856	9867	11111		5214	12414	13086	10764	7335	1524	7164	1338	1578	11967	5647

...HOWING GREAT CIRCLE DISTANCES BETWEEN PRINCIPAL CITIES OF THE WORLD IN STATUTE MILES

New Orleans	New York	Nome	Oslo	Panamá	Paris	Peking (Peiping)	Port Said	Quebec	Reykjavik	Rio de Janeiro	Rome	San Francisco	Seattle	Shanghai	Singapore	Tokyo	Valparaiso	Washington, D.C.	Wellington	Vienna	Winnipeg	Zanzibar	
6085	7242	5438	9247	6514	9990	5903	10485	7406	8678	8120	10475	4786	5222	5399	5850	4656	6267	7066	2062	10010	6283	9892	Apia
3524	2422	4954	2234	3778	1659	6565	3391	2240	1777	4428	2125	4872	4501	7229	8326	7247	5678	2667	11269	2291	3389	5323	Azores Islands
5116	3961	4342	515	5849	542	4567	1747	3583	1479	6114	734	5657	5041	5215	6166	5538	7795	4167	11265	328	4286	4309	Berlin
8865	7794	5901	4130	9742	4359	2964	2659	7371	5191	8257	3843	8392	7741	3133	2429	4188	10037	7988	7677	3718	7644	2855	Bombay
4916	5297	8848	7613	3381	6877	11974	7362	5680	7099	1218	6929	6474	6913	12197	9864	11400	761	5216	6260	7368	6297	6421	Buenos Aires
8803	7921	5271	4459	10114	4889	2024	3506	7481	5409	9376	4496	7809	7224	2112	1791	3186	10993	8088	7042	4259	7424	3859	Calcutta
8316	7801	10107	6494	7014	5841	8045	4590	7857	7111	3769	5249	10241	10199	8059	6016	9071	4998	7894	7019	5671	9054	2346	Cape Town
4194	3355	6438	3444	3734	2666	7763	3672	3355	3248	3040	2772	5921	5714	8443	8700	8589	4649	3486	10363	3147	4556	4635	Cape Verde Islands
833	713	3314	4040	2325	4133	6592	6103	878	2954	5296	4808	1858	1737	7053	9365	6303	5268	591	8349	4694	714	8358	Chicago
9545	9959	6235	8022	10352	8575	3728	7159	9724	8631	9960	8190	7637	7619	3142	2075	3367	8961	9923	3310	7974	8684	6409	Darwin
1082	1631	2925	4653	2636	4885	6348	6819	1752	3596	5871	5561	949	1021	6698	9063	5795	5452	1494	7516	5383	798	9221	Denver
4757	3627	5398	1791	4926	964	6009	2179	3383	2047	4775	1034	5936	5462	6646	7231	6988	6408	3822	12060	1386	4435	4103	Gibraltar
8480	8051	4547	5337	10084	5956	1226	4975	7650	6031	10995	5768	6894	6471	772	1652	1796	11607	8148	5853	5429	7096	5414	Hong Kong
4207	4959	3004	6784	5245	7434	5067	8738	5000	6084	8190	8022	2392	2678	4934	6710	3850	6793	4829	4708	7626	3806	10869	Honolulu
6171	5009	5101	1518	6750	1401	4379	693	4644	2558	6395	854	6700	6063	4959	5373	5556	8172	5216	10663	783	5361	3312	Istanbul
2905	2854	1094	4045	4460	4628	4522	6215	2660	3268	7598	5247	1525	899	4869	7235	4011	7271	2834	7475	4895	1597	8795	Juneau
4627	3459	4381	714	5278	213	5054	2154	3101	1171	5772	887	5355	4782	5710	6744	5938	7263	3665	11682	772	3918	4604	London
1673	2451	2876	5325	3001	5601	6250	7528	2579	4306	6296	6326	347	959	6477	8767	5470	5527	2300	6714	6108	1525	10021	Los Angeles
8724	8493	4817	6016	10283	6673	1770	5619	8124	6651	11254	6457	6963	6641	1152	1479	1863	10930	8560	5162	6120	7414	5763	Manila
9275	10355	7558	9926	9022	10396	5667	8658	10497	10544	8186	9934	7854	8186	5005	3761	5089	6998	10173	1595	9792	9319	6802	Melbourne
934	2085	4309	5706	1495	5706	7733	7671	2454	4622	4770	6353	1885	2337	8039	10307	7035	4053	1878	6899	6306	2097	9484	Mexico City
5756	4662	4036	1016	6711	1541	3597	1710	4242	2056	7179	1474	5868	5199	4235	5238	4650	8792	4883	10279	1044	4687	4270	Moscow
	1171	3937	4795	1603	4788	7314	6756	1534	3711	4796	5439	1926	2101	7720	10082	6858	4514	966	7794	5385	1418	8754	New Orleans
1171		3769	3672	2231	3622	6823	5590	439	2576	4820	4273	2571	2408	7357	9630	6735	5094	205	8946	4224	1281	7698	New York
3937	3769		3836	5541	4574	3428	5745	3489	3366	8586	5082	2547	1976	3784	6148	2983	8360	3792	7383	4657	2599	8209	Nome
4795	3672	3836		5691	832	4360	2211	3263	1083	6482	1243	5181	4591	5020	6246	5221	7914	3870	10974	850	3854	4803	Oslo
1603	2231	5541	5691		5382	8906	7146	2659	4706	3294	5903	3322	3651	9324	11687	8423	2943	2080	7433	6026	2998	8245	Panamá
4788	3622	4574	832	5382		5101	1975	3235	1380	5703	682	5441	4993	5752	6671	6033	7251	3828	11791	644	4118	4396	Paris
7314	6823	3428	4360	8906	5101		4584	6423	4903	10768	5047	5902	5396	662	2774	1307	11774	6922	6698	4639	5907	5803	Peking (Peiping)
6756	5590	5745	2211	7146	1975	4584		5250	3227	6244	1317	7394	6759	5132	5088	5842	8088	5796	10249	1429	6032	2729	Port Said
1534	439	3489	3263	2659	3235	6423	5250		2189	5125	3943	2642	2353	6981	9097	6417	5504	610	9228	3858	1199	7443	Quebec
3711	2576	3366	1083	4706	1380	4903	3227	2189		6118	2044	4199	3614	5559	7160	5472	7225	2800	10724	1805	2804	5757	Reykjavik
4796	4820	8586	6482	3294	5703	10768	6244	5125	6118		5684	6619	6891	11340	9774	11535	1855	4797	7349	6136	6010	5589	Rio de Janeiro
5439	4273	5082	1243	5903	682	5047	1317	3943	2044	5684		6240	5659	5677	6232	6124	7420	4435	11524	463	4803	3712	Rome
1926	2571	2547	5181	3322	5441	5902	7394	2642	4199	6619	6240		678	6132	8479	5131	5876	2442	6739	5988	1504	9958	San Francisco
2101	2408	1976	4591	3651	4993	5396	6759	2353	3614	6891	5659	678		5703	8057	4777	6230	2329	7242	5376	1150	9359	Seattle
7720	7357	3784	5020	9324	5752	662	5132	6981	5559	11340	5677	6132	5703		2377	1094	11650	7442	6054	5270	6350	5971	Shanghai
10082	9630	6148	6246	11687	6671	2774	5088	9097	7160	9774	6232	8479	8057	2377		3304	10226	9834	5292	6036	8685	4480	Singapore
6858	6735	2983	5221	8423	6033	1307	5842	6417	5472	11535	6124	5131	4777	1094	3304		10635	6769	5760	5679	5575	7040	Tokyo
4514	5094	8360	7914	2943	7251	11774	8088	5504	7225	1855	7420	5876	6230	11650	10226	10635		4977	5785	7783	5931	7184	Valparaiso
966	205	3792	3870	2080	3828	6922	5796	610	2800	4797	4435	2442	2329	7442	9834	6769	4977		8745	4429	1243	7884	Washington, D.C.
7794	8946	7383	10974	7433	11791	6698	10249	9228	10724	7349	11524	6739	7242	6054	5292	5760	5785	8745		11278	8230	8122	Wellington
5385	4224	4657	850	6026	644	4639	1429	3858	1805	6136	463	5988	5376	5270	6036	5679	7783	4429	11278		4604	3983	Vienna
1418	1281	2599	3854	2998	4118	5907	6032	1199	2804	6010	4803	1504	1150	6350	8685	5575	5931	1243	8230	4604		8416	Winnipeg
8754	7698	8209	4803	8245	4396	5803	2729	7443	5757	5589	3712	9958	9359	5971	4480	7040	7184	7884	8122	3983	8416		Zanzibar

...HOWING STEAMSHIP DISTANCES BETWEEN PRINCIPAL PORTS OF THE WORLD IN STATUTE MILES

	Bombay	Buenos Aires	Cape Town	Colombo	Gibraltar	Halifax	Hamburg	Honolulu	Istanbul	Le Havre	Lisbon	Liverpool	Manila	Melbourne	New Orleans	New York	Panama Roads	Port Said	Rio de Janeiro	San Francisco	Shanghai	Singapore	Valparaiso	Wellington	Yokohama
...lbourne, Australia............	6365	8477	6998	5380	11257	11876	13066	5691	9928	12540	11551	12764	5214		10780	11452	9130	9040	9416	8011	6012	4396	7222	1737	5606
...w Orleans, U.S.A.............	10927	7233	9382	11489	5271	2517	5935	7046	7384	5315	5377	5266	12414	10780		1970	1650	7498	5965	5287	11495	13207	4663	9133	10489
...w York, U.S.A..............	9413	6761	7814	9941	3714	686	4166	7718	5788	3640	3403	3539	13086	11452	1970		2323	5895	5493	6059	12176	11693	5335	9814	11169
...ama Roads, Canal Zone......	14921	6311	7417	13919	5038	2718	5888	5395	7115	5363	4968	5287	10764	9130	1650	2323		7217	5058	3737	9853	12097	3013	7491	8846
...t Said, U.A.R..............	3511	8259	6148	4010	2217	5257	4058	12604	910	3521	2532	3652	7335	9040	7498	5895	7217		7006	10986	8301	5791	10225	10630	9128
...de Janeiro, Brazil..........	8998	1325	3769	8839	4816	5332	6354	9875	6897	5820	4858	5932	11524	9416	5965	5493	5058	7006		8794	12490	10179	4191	7915	13317
...Francisco, U.S.A............	11247	10062	11154	10289	8775	6456	9625	2408	10884	9095	8737	9024	7164	8011	5287	6059	3737	10986	8794		6339	8467	5919	6800	5223
...nghai, China..............	5328	13087	8787	4370	10553	12707	12349	4986	9210	11822	10833	12201	1338	6012	11495	12176	9853	8301	12490	6339		2545	11806	6184	1199
...gapore, Singapore..........	2824	10782	6511	1825	8008	11047	9838	6772	6700	9312	8323	9490	1578	4396	13207	11693	12097	5791	10179	8467	2545		12534	5992	3345
...paraiso, Chile..............	11356	3181	6977	11073	9006	5731	8900	6816	11020	8347	7975	8299	11967	7222	4663	5335	3013	10225	4191	5919	11806	12534		5799	10740
...llington, New Zealand.......	7961	6956	7531	7058	12847	10196	13758	4736	11540	12801	12459	12778	5647	1737	9133	9814	7491	10630	7915	6800	6184	5992	5799		5736
...ohama, Japan..............	6155	13921	9614	5151	11353	11592	14734	3908	10037	12649	11660	13399	2023	5606	10489	11169	8846	9128	13317	5223	1199	3345	10740	5736	

ALTITUDES OF SELECTED WORLD CITIES

Abbreviations

Abbreviation	Meaning
Afg.	Afghanistan
Ala.	Alabama
Arg.	Argentina
Ariz.	Arizona
Ark.	Arkansas
Aus.	Austria
Austl.	Australia
Bel.	Belgium
Calif.	California
Colo.	Colorado
Conn.	Connecticut
Con. B.	Congo; Capital: Brazzaville
Con. L.	Congo, The; Capital: Léopoldville
Czech.	Czechoslovakia
D.C.	District of Columbia
Del.	Delaware
Den.	Denmark
Eng.	England
Eth.	Ethiopia
Fla.	Florida
Ga.	Georgia
Ger.	Germany
Guat.	Guatemala
Ill.	Illinois
Ind.	Indiana
Ire.	Ireland
Kans.	Kansas
Ky.	Kentucky
La.	Louisiana
Leb.	Lebanon
Lux.	Luxembourg
Malag. Rep.	Malagasy Republic
Mass.	Massachusetts
Md.	Maryland
Mex.	Mexico
Mich.	Michigan
Minn.	Minnesota
Mo.	Missouri
N. Car.	North Carolina
Nebr.	Nebraska
Neth.	Netherlands
N. H.	New Hampshire
N. J.	New Jersey
N. Mex.	New Mexico
N. Y.	New York
N. Z.	New Zealand
Okla.	Oklahoma
Oreg.	Oregon
Pa.	Pennsylvania
Par.	Paraguay
R. I.	Rhode Island
S. Afr.	South Africa
Sov. Un.	Soviet Union
Switz.	Switzerland
Tenn.	Tennessee
U.A.R.	United Arab Republic (Egypt)
Ur.	Uruguay
U. S.	United States
Va.	Virginia
Ven.	Venezuela
Wash.	Washington
Wis.	Wisconsin
Wyo.	Wyoming

A

City	Height in Feet
Aachen, Ger.	580
Addis Ababa, Eth.	7,749
Adelaide, Austl.	40
Aden, Aden Protectorate	20
Agra, India	545
Aguascalientes, Mex.	6,258
Ahmadabad, India	180
Ajmer, India	1,685
Akron (Ohio), U. S.	875
Albany (N. Y.), U. S.	20
Albuquerque (N. Mex.), U. S.	4,950
Aleppo, Syria	1,290
Alexandria, U.A.R.	25
Algiers (Alger), Algeria	200
Amman, Jordan	2,548
Amritsar, India	760
Amsterdam, Neth.	8
Andorra, Andorra	3,376
Ankara, Turkey	2,250
Antwerpen (Antwerp), Belgium	30
Asheville (N. Car.), U. S.	1,985
Aspen (Colo.), U. S.	7,930
Astrakhan, Sov. Un.	—50
Asunción, Par.	246
Athens (Athinai), Greece	300
Atlanta (Ga.), U. S.	1,050
Augsburg, Ger.	1,623

B

City	Height in Feet
Baden-Baden, Ger.	594
Badgastein, Aus.	3,323
Baghdad, Iraq	112
Baguio, Philippines	4,640
Baku, Sov. Un.	—40
Baltimore (Md.), U. S.	20
Banff, Canada	4,534
Banaras (Benares), India	250
Bangalore, India	2,950
Bangkok, Thailand	10
Barcelona, Spain	43
Basel, Switz.	1,050
Batavia, see Djakarta, Indonesia.	
Beirut (Beyrouth), Leb.	200
Belém, Brazil	30
Belfast, N. Ireland	20
Belgrade (Beograd), Yugoslavia	433
Belo Horizonte, Brazil	2,490
Benares, see Banaras, India.	
Beograd, see Belgrade, Yugoslavia.	
Berlin, Ger.	115
Bern, Switz.	1,876
Beyrouth, see Beirut, Leb.	
Bilbao, Spain	30
Birmingham, Eng.	452
Birmingham (Ala.), U. S.	600
Bogor (Buitenzorg), Indonesia	570
Bogotá, Colombia	8,659
Bologna, Italy	180
Bolton, Eng.	480
Bombay, India	25
Bonn, Ger.	197
Bordeaux, France	49
Boston (Mass.), U. S.	21
Bradford, Eng.	635
Brasília, Brazil	3,474
Braunschweig, Ger.	230
Brawley (Calif.), U. S.	—109
Brazzaville, Con. B.	1,055
Bremen, Ger.	10
Breslau, see Wroclaw, Poland.	
Brisbane, Austl.	35
Bristol, Eng.	220
Brno, Czech.	750
Brunswick, see Braunschweig, Ger.	
Brussels (Bruxelles), Bel.	190
Bucharest (Bucuresti), Romania	276
Budapest, Hungary	370
Buenos Aires, Arg.	45
Buffalo (N. Y.), U. S.	585
Bydgoszcz, Poland	233

C

City	Height in Feet
Cairo, U.A.R.	98
Calcutta, India	20
Campinas, Brazil	2,220
Canberra, Austl.	1,875
Canton, China	33
Canton (Ohio), U. S.	1,030
Cape Town, S. Afr.	36
Caracas, Ven.	3,164
Cardiff, Wales	60
Cawnpore, see Kanpur, India.	
Cerro de Pasco, Peru	14,385
Chamonix [-Mont-Blanc], France	3,402
Changchun (Hsinking), China	735
Changsha, China	200
Chattanooga (Tenn.), U. S.	675
Chelyabinsk, Sov. Un.	700
Chemnitz, see Karl-Marx-Stadt, Ger.	
Chengteh (Jehol), China	1,630
Chengtu, China	1,195
Chernovtsy, Sov. Un.	1,040
Chicago (Ill.), U. S.	595
Chihuahua, Mex.	4,443
Chita, Sov. Un.	2,300
Chungking, China	787
Cincinnati (Ohio), U. S.	550
Clermont-Ferrand, France	1,316
Cleveland (Ohio), U. S.	580
Cochabamba, Bolivia	8,435
Cody (Wyo.), U. S.	4,980
Cologne, see Köln, Ger.	
Colombo, Ceylon	15
Colorado Springs (Colo.), U. S.	5,980
Columbus (Ohio), U. S.	780
Constantine, Algeria	1,670
Constantinople, see Istanbul, Turkey.	
Conway (N. H.), U. S.	465
Copenhagen (København), Den.	45
Córdoba, Arg.	1,240
Cortina d'Ampezzo, Italy	4,003
Croydon, Eng.	325
Cuiabá, Brazil	771
Cusco, Peru	11,440

D

City	Height in Feet
Dacca, Pakistan	26
Dairen, China	80
Dallas (Texas), U. S.	435
Damascus (Dimashq), Syria	2,250
Danzig, see Gdansk, Poland.	
Darjeeling, India	7,000
Davos, Switz.	5,062
Dayton (Ohio), U. S.	745
Dehra Dun, India	2,282
Delhi, India	770
Denver (Colo.), U. S.	5,280
Des Moines (Iowa), U. S.	805
Detroit (Mich.), U. S.	585
Djakarta (Batavia), Indonesia	16
Dnepropetrovsk, Sov. Un.	250
Dortmund, Ger.	249
Dresden, Ger.	375
Dublin (Baile Átha Cliath), Ire.	35
Duisburg, Ger.	108
Duluth (Minn.), U. S.	610
Durango, Mex.	6,196
Durban, S. Afr.	25
Düsseldorf, Ger.	118

E

City	Height in Feet
Edinburgh, Scotland	195
Elisabethville, Con. L.	4,101
El Paso (Texas), U. S.	3,695
Ely (Minn.), U. S.	1,415
Erfurt, Ger.	656
Erie (Pa.), U. S.	671
Essen, Ger.	269
Evansville (Ind.), U. S.	385

F

City	Height in Feet
Fairbanks, Alaska	512
Fez, Morocco	1,020
Firenze (Florence), Italy	164
Flagstaff (Ariz.), U. S.	6,890
Flint (Mich.), U. S.	715
Florence, see Firenze, Italy.	
Fort Wayne (Ind.), U. S.	790
Fort Worth (Texas), U. S.	620
Frankfurt [am Main], Ger.	312
Fukuoka, Japan	25
Fusan, see Pusan, Korea	

G

City	Height in Feet
Garmisch-Partenkirchen, Ger.	2,330
Gartok, Tibet, China	14,240
Gdańsk (Danzig), Poland	49
Genève (Geneva), Switz.	1,329
Genova (Genoa), Italy	62
Glasgow, Scotland	45
Gorkiy, Sov. Un.	230
Göteborg, Sweden	45
Granada, Spain	2,265
Grand Rapids (Mich.), U. S.	610
Graz, Aus.	1,170
Greenwich, Eng.	235
Grenoble, France	770
Groningen, Neth.	5
Guadalajara, Mex.	5,180
Guatemala, Guat.	4,850
Guayaquil, Ecuador	30

H

City	Height in Feet
Halle, Ger.	328
Hamburg, Ger.	20
Hamilton, Canada	323
Hangchow, China	49
Hannover, Ger.	180
Hanoi, Vietnam	30
Harbin, China	460
Hartford (Conn.), U. S.	40
Havana (Habana), Cuba	35
Helsinki, Finland	25
Hiroshima, Japan	75
Hong Kong, see Victoria, Hong Kong.	
Honolulu, Hawaii	25
Hot Springs (Ark.), U. S.	607
Houston (Texas), U. S.	40
Hull, Eng.	25
Huntington (W. Va.), U. S.	565
Hyderabad, India	1,750

I

City	Height in Feet
Ibadan, Nigeria	768
Indianapolis (Ind.), U. S.	710
Innsbruck, Aus.	1,985
Interlaken, Switz.	1,850
Iquitos, Peru	350
Irkutsk, Sov. Un.	1,400
Istanbul, Turkey	30
Izmir (Smyrna), Turkey	35

J

City	Height in Feet
Jacksonville (Fla.), U. S.	20
Jaipur, India	1,258
Jersey City (N. J.), U. S.	20
Jerusalem, Israel–Jordan	2,618
Jiachan, Tibet, China	15,870
Johannesburg, S. Afr.	5,689
Juf, Switz.	6,975

K

City	Height in Feet
Kabul, Afg.	5,890
Kaifeng, China	670
Kaliningrad (Königsberg), Sov. Un.	75
Kandy, Ceylon	1,602
Kanpur (Cawnpore), India	410
Kansas City (Kans.), U. S.	750
Kansas City (Mo.), U. S.	750
Karachi, Pakistan	60
Karaganda, Sov. Un.	2,000
Karl-Marx-Stadt (Chemnitz), Ger.	1,013
Karlsruhe, Ger.	377
Kassel, Ger.	476
Katmandu, Nepal	4,223
Kaunas, Sov. Un.	255
Kazan, Sov. Un.	290
Kharkov, Sov. Un.	450
Khartoum, Sudan	1,252
Kiel, Ger.	46
Kiev (Kiyev), Sov. Un.	450
Kingston, Jamaica	25
Kishinev, Sov. Un.	130
Knoxville (Tenn.), U. S.	890
Kobe, Japan	50
Köln (Cologne), Ger.	174
Königsberg, see Kaliningrad, Sov. Un.	
Krakow, Poland	700
Kumamoto, Japan	200
Kunming, China	6,080
Kure, Japan	275
Kuybyshev, Sov. Un.	570
Kyoto, Japan	360

L

City	Height in Feet
Lahore, Pakistan	3,270
Lake Placid (N.Y.), U. S.	1,740
Lanchow, China	5,185
La Paz, Bolivia	12,795
La Plata, Arg.	40
Leadville (Colo.), U. S.	10,152
Leeds, Eng.	245
Leh, India	11,253
Le Havre, France	200
Leipzig, Ger.	387
Leningrad, Sov. Un.	33
Léopoldville, Con. L.	1,045
Lhasa, Tibet, China	11,800
Lille, France	69
Lima, Peru	501
Lisbon (Lisboa), Portugal	285
Liverpool, Eng.	130
Lodz, Poland	660
London, Eng.	80
Long Beach (Calif.), U. S.	35
Los Angeles (Calif.), U. S.	340
Louisville (Ky.), U. S.	450
Lourdes, France	1,345
Luanda, Angola	243
Lucknow, India	425
Ludwigshafen, Ger.	308
Luxembourg, Lux.	1,096
Luzern, Switz.	1,634
Lvov, Sov. Un.	980
Lyon, France	555

M

City	Height in Feet
Madras, India	30
Madrid, Spain	2,150
Magdeburg, Ger.	164
Mainz, Ger.	269
Managua, Nicaragua	195
Manaus, Brazil	141
Manchester, Eng.	275
Manila, Philippines	30
Mannheim, Ger.	318
Maracaibo, Ven.	30
Marrakech, Morocco	1,535
Marseille, France	150
Mecca, Saudi Arabia	919
Medellín, Colombia	5,044
Melbourne, Austl.	30
Memphis (Tenn.), U. S.	275
Merano, Italy	1,062
Mérida, Mex.	65
Mérida, Ven.	5,415
Mexico City, Mex.	7,349
Miami (Fla.), U. S.	10
Milano (Milan), Italy	397
Milwaukee (Wis.), U. S.	635
Minneapolis (Minn.), U. S.	815
Minsk, Sov. Un.	690
Monterrey, Mex.	1,765
Montevideo, Ur.	80
Montreal, Canada	104
Mont Tremblant, Canada	745
Moscow (Moskva), Sov. Un.	425
Mosul, Iraq	800
Mukden, China	560
München (Munich), Ger.	1,699
Münster, Ger.	203

N

City	Height in Feet
Nagasaki, Japan	210
Nagoya, Japan	50
Nagpur, India	1,020
Nairobi, Kenya	5,452
Nancy, France	675
Nanking, China	90
Nantes, France	26
Napoli (Naples), Italy	33
Nashville (Tenn.), U. S.	546
Newark (N. J.), U. S.	55
Newcastle-on-Tyne, Eng.	175
New Orleans (La.), U. S.	5
New York (N. Y.), U. S.	55
Nice, France	94
Niigata, Japan	50
Nikko, Japan	1,746
Norfolk (Va.), U. S.	10
Northampton, Eng.	235
Novosibirsk, Sov. Un.	490
Nürnberg, Ger.	1,110

O

City	Height in Feet
Oakland (Calif.), U. S.	25
Odessa, Sov. Un.	165
Oklahoma City (Okla.), U. S.	1,195
Omaha (Neb.), U. S.	1,040
Omsk, Sov. Un.	286
Oran, Algeria	35
Orizaba, Mex.	4,028
Oruro, Bolivia	12,149
Osaka, Japan	40
Oslo, Norway	55
Ottawa, Canada	286

P

City	Height in Feet
Palermo, Italy	46
Panamá, Panama	40
Paris, France	250
Patna, India	170
Peking (Peiping), China	165
Peoria (Ill.), U. S.	470
Perm, Sov. Un.	250
Perth, Austl.	55
Petropavlovsk, Sov. Un.	313
Petrópolis, Brazil	2,800
Philadelphia (Pa.), U. S.	100
Phnom Penh, Cambodia	33
Phoenix (Ariz.), U. S.	1,090
Piraievs (Piraeus), Greece	30
Pittsburgh (Pa.), U. S.	745
Plauen, Ger.	1,335
Plzen, Czech.	995
Poona, India	1,720
Port-au-Prince, Haiti	25
Portland (Oreg.), U. S.	75
Pôrto, Portugal	300
Port Said, Egypt	30
Portsmouth, Eng.	20
Potosí, Bolivia	13,600
Poznan (Posen), Poland	292
Prague (Praha), Czech.	575
Pretoria, S. Afr.	4,472
Providence (R. I.), U. S.	80
Puebla, Mex.	7,091
Puno, Peru	12,648
Pusan (Fusan), Korea	25
Pyongyang, Korea	60

Q

City	Height in Feet
Quebec, Canada	19; 305
Quito, Ecuador	9,320

R

City	Height in Feet
Rangoon, Burma	20
Rawalpindi, Pak.	1,726
Recife, Brazil	10
Regensburg, Ger.	1,075
Reims, France	280
Reno (Nevada), U. S.	4,490
Richmond (Va.), U. S.	164
Riga, Sov. Un.	30
Rio de Janeiro, Brazil	30
Riyadh, Saudi Arabia	1,897
Rochester (N. Y.), U. S.	510
Rome (Roma), Italy	66
Rosario, Arg.	86
Rostov, Sov. Un.	100
Rotterdam, Neth.	—8; 15
Rouen, France	90

S

City	Height in Feet
Saarbrücken, Saar	625
Sacramento (Calif.), U. S.	30
Saigon, Vietnam	26
St. Étienne, France	1,800
St. Louis (Mo.), U.S.	455
St. Paul (Minn.), U.S.	780
Salisbury, S. Rhodesia	4,780
Salt Lake City (Utah), U. S.	4,390
Salvador, Brazil	135
Salzburg, Aus.	1,391
Şan'ā', Yemen	7,700
San Antonio (Texas), U. S.	650
San Francisco (Calif.), U. S.	65
San José, Costa Rica	3,021
San Juan, Puerto Rico	10
Sankt Moritz, Switz.	6,037
San Salvador, Salvador	2,178
Santa Fe (N. Mex.), U. S.	6,950
Santiago, Chile	1,795
Santos, Brazil	30
São Paulo, Brazil	2,545
Sapporo, Japan	245
Saratov, Sov. Un.	250
Scranton (Pa.), U. S.	725
Seattle (Wash.), U. S.	75
Sendai, Japan	250
Seoul, Korea	75
Sevilla (Seville), Spain	39
's Gravenhage, see The Hague, Neth.	
Shanghai, China	20
Sheffield, Eng.	325
Sholapur, India	1,435
Sian (Siking), China	1,364
Simla, India	7,186
Singapore, Malaysia	35
Smyrna, see Izmir, Turkey.	
Sodom, Israel	—1,286
Sofia (Sofiya), Bulgaria	1,700
Soochow, see Wuhsien, China.	
Southampton, Eng.	45
Springfield (Mo.), U. S.	1,200
Spokane (Wash.), U. S.	1,890
Srinagar, India	5,130
Stalingrad, see Volgograd, Sov. Un.	
Stettin, see Szczecin, Pol.	
Stockholm, Sweden	46
Strasbourg, France	450
Stuttgart, Ger.	853
Sucre, Bolivia	8,950
Sun Valley (Idaho), U. S.	6,000
Surabaja, Indonesia	25
Sverdlovsk, Sov. Un.	860
Sydney, Austl.	75
Syracuse (N. Y.), U. S.	400
Szczecin (Stettin), Pol.	50

T

City	Height in Feet
Tabriz, Iran	5,250
Tacoma (Wash.), U. S.	110
Taipei, Taiwan (Formosa)	26
Tallinn, Sov. Un.	25
Tampa (Fla.), U. S.	15
Tananarive, Malag. Rep.	4,200
Taos (N. Mex.), U. S.	6,985
Tashkent, Sov. Un.	1,500
Taxco [de Alarcón], Mex.	5,756
Tbilisi, Sov. Un.	1,450
Tegucigalpa, Honduras	3,070
Tehran, Iran	3,865
The Hague ('s Gravenhage), Neth.	—4; 25
Tientsin, China	15
Tiranë, Albania	374
Tokyo, Japan	15
Toledo (Ohio), U. S.	1,799
Tombouctou, Mali	938
Tomsk, Sov. Un.	303
Torino (Turin), Italy	784
Toronto, Canada	356
Toulouse, France	490
Tours, France	160
Trieste, Italy	7
Tripoli, Libya	115
Tsinan, China	125
Tucson (Ariz.), U. S.	2,390
Tucumán, Arg.	1,385
Tulsa (Okla.), U. S.	804
Tunis, Tunisia	30
Turin, see Torino, Italy.	

U

City	Height in Feet
Ulan Bator, Mongolia	4,160

V

City	Height in Feet
Valencia, Spain	49
Valparaíso, Chile	35
Vancouver, Canada	38
Venezia (Venice), Italy	5
Verona, Italy	194
Versailles, France	445
Victoria, Hong Kong	50
Vienna (Wien), Aus.	550
Vilnius, Sov. Un.	500
Vladivostok, Sov. Un.	65
Volgograd (Stalingrad), Sov. Un.	180

W

City	Height in Feet
Wanchuan (Kalgan), China	2,550
Warsaw (Warszawa), Poland	344
Washington (D. C.), U. S.	25
Wellington, N. Z.	415
Whitehorse, Canada	2,083
White Sulphur Springs (W. Va.), U. S.	2,000
Wichita (Kans.), U. S.	1,290
Wien, see Vienna, Aus.	
Wiesbaden, Ger.	361
Wilmington (Del.), U. S.	135
Winnipeg, Canada	764
Worcester (Mass.), U. S.	475
Wroclaw (Breslau), Poland	390
Wuhan, China	55
Wuhsien (Soochow), China	30
Wuppertal, Ger.	1,100

Y

City	Height in Feet
Yakutsk, Sov. Un.	210
Yokohama, Japan	110
Youngstown (Ohio), U. S.	863

Z

City	Height in Feet
Zacatecas, Mex.	8,010
Zagreb, Yugoslavia	515
Zakopane, Poland	2,733
Zaragoza, Spain	820
Zermatt, Switz.	5,279
Zürich, Switz.	1,867
Zwickau, Ger.	855

GEOGRAPHICAL FACTS ABOUT THE UNITED STATES

ELEVATION

The highest elevation in the United States is Mount McKinley, Alaska, 20,320 feet.

The lowest elevation in the United States is in Death Valley, California, 282 feet below sea level.

The average elevation of the United States is 2,500 feet.

EXTREMITIES

Direction	Location	Latitude	Longitude
North	Point Barrow, Alaska	71°23′N.	156°29′W.
South	South Cape, Hawaii	18°56′N.	155°41′W.
East	West Quoddy Head, Maine	44°49′N.	66°57′W.
West	Cape Wrangell, Alaska	52°55′N.	172°27′E.

The two places in the United States separated by the greatest distance are Kure Island, Hawaii, and Mangrove Point, Florida. These points are 5,848 miles apart.

LENGTH OF BOUNDARIES

The total length of the Canadian boundary of the United States is 5,525 miles.

The total length of the Mexican boundary of the United States is 2,013 miles.

The total length of the Atlantic coastline of the United States is 2,069 miles.

The total length of the Pacific and Arctic coastline of the United States is 8,683 miles.

The total length of the Gulf of Mexico coastline of the United States is 1,631 miles.

The total length of all coastlines and land boundaries of the United States is 19,921 miles.

The total length of the tidal shoreline and land boundaries of the United States is 96,171 miles.

GEOGRAPHIC CENTERS

The geographic center of the United States (including Alaska and Hawaii) is in Butte County, South Dakota at 44°58′N., 103°46′W.

The geographic center of North America is in North Dakota, a few miles west of Devils Lake, at 48°10′N., 100°10′W.

EXTREMES OF TEMPERATURE

The highest temperature ever recorded in the United States was 134°F., at Greenland Ranch, Death Valley, California, on July 10, 1913.

The lowest temperature ever recorded in the United States was —76°F., at Tanana, Alaska, in January, 1886.

PRECIPITATION

The average annual precipitation for the United States is approximately 29 inches.

Hawaii is the wettest state, with an average annual rainfall of 82.48 inches. Nevada, with an average annual rainfall of 8.81 inches, is the driest state.

The greatest local average annual rainfall in the United States is at Mt. Waialeale, Kauai, Hawaii, 460 inches.

Greatest 24-hour rainfall in the United States, 23.22 inches at New Smyrna, Florida, October 10–11, 1924.

Extreme minimum rainfall records in the United States include a total fall of only 3.93 inches at Bagdad, California, for a period of 5 years, 1909–13, and an annual average of 1.78 inches at Death Valley, California.

Heavy snowfall records include 60 inches at Giant Forest, California, in 1 day; 42 inches at Angola, New York, in 2 days; 87 inches at Giant Forest, California, in 3 days; and 108 inches at Tahoe, California, in 4 days.

Greatest seasonal snowfall, 1,000.3 inches, more than 83 feet, at Paradise Ranger Station, Washington, during the winter of 1955–56.

HISTORICAL FACTS ABOUT THE UNITED STATES

TERRITORIAL ACQUISITIONS

Accession	Date	Area (sq. mi.)	Cost in Dollars
Original territory of the Thirteen States	1790	888,811	....
Purchase of Louisiana Territory, from France	1803	827,192	$11,250,000.00
By treaty with Spain: Florida	1819	58,560 }	$ 5,000,000.00
Other areas	1819	13,443 }	
Annexation of Texas	1845	390,144	....
Oregon Territory, by treaty with Great Britain	1846	285,580	....
Mexican Cession	1848	529,017	$15,000,000.00
Gadsden Purchase, from Mexico	1853	29,640	$10,000,000.00
Purchase of Alaska, from Russia	1867	586,400	7,200,000.00
Annexation of Hawaiian Islands	1898	6,454	....
Puerto Rico, by treaty with Spain	1898	3,435	....
Guam, by treaty with Spain	1898	212	....
American Samoa, by treaty with Great Britain and Germany	1900	76	....
Panama Canal Zone, by treaty with Panama	1904	553	*$10,000,000.00
Virgin Islands, by purchase from Denmark	1917	133	$25,000,000.00
Total		3,619,650	$83,450,000.00

Note: The Philippines, ceded by Spain in 1898 for $20,000,000.00, were a territorial possession of the United States from 1898 to 1946. On July 4, 1946 they became the independent republic of the Philippines.

• $25,000,000.00 was also paid to the republic of Colombia, out of whose territory the republic of Panama was created. In addition, an annual payment of $430,000.00 is made to the republic of Panama.

WESTWARD MOVEMENT OF CENTER OF POPULATION

Year	U.S. Population Total at Census	Approximate Location
1790	3,929,214	23 miles east of Baltimore, Md.
1800	5,308,483	18 miles west of Baltimore, Md.
1810	7,239,881	40 miles northwest of Washington, D.C.
1820	9,638,453	16 miles east of Moorefield, W. Va.
1830	12,866,020	19 miles southwest of Moorefield, W. Va.
1840	17,069,453	16 miles south of Clarksburg, W. Va.
1850	23,191,876	23 miles southeast of Parkersburg, W. Va.
1860	31,443,321	20 miles southeast of Chillicothe, Ohio
1870	39,818,449	48 miles northeast of Cincinnati, Ohio
1880	50,155,783	8 miles southwest of Cincinnati, Ohio
1890	62,947,714	20 miles east of Columbus, Ind.
1900	75,994,575	6 miles southeast of Columbus, Ind.
1910	91,972,266	Bloomington, Ind.
1920	105,710,620	8 miles southeast of Spencer, Ind.
1930	122,775,046	3 miles northeast of Linton, Ind.
1940	131,669,275	2 miles southeast of Carlisle, Ind.
1950	150,697,361	8 miles northwest of Olney, Ill.
1960	179,323,175	6 miles northwest of Centralia, Ill.

STATE AREAS AND POPULATIONS

STATE	Land Area (square miles) in 1960	Water Area (square miles) in 1960	Total Area (square miles) in 1960	Rank in Area	Population in 1960	Population Per Square Mile in 1960	Rank in Population in 1960	Population in 1950	Rank in Population in 1950	Population 1940
Alabama	51,060	549	51,609	29	3,266,740	63	19	3,061,743	17	2,832,961
Alaska	571,065	15,335	586,400	1	226,167	0.4	50	128,643	50	72,524‡
Arizona	113,575	334	113,909	6	1,302,161	11	35	749,587	37	499,261
Arkansas	52,499	605	53,104	27	1,786,272	34	31	1,909,511	30	1,949,387
California	156,573	2,120	158,693	3	15,717,204	99	2	10,586,223	2	6,907,387
Colorado	103,884	363	104,247	8	1,753,947	17	33	1,325,089	34	1,123,296
Connecticut	4,899	110	5,009	48	2,535,234	506	25	2,007,280	28	1,709,242
Delaware	1,978	79	2,057	49	446,292	217	46	318,085	47	266,505
District of Columbia†	61	8	69		763,956	11,072	..	802,178	..	663,091
Florida	54,252	4,308	58,560	22	4,951,560	85	10	2,771,305	20	1,897,414
Georgia	58,274	602	58,876	21	3,943,116	67	16	3,444,578	13	3,123,723
Hawaii	6,415	9	6,424	47	632,772	99	43	499,794	45	423,330
Idaho	82,708	849	83,557	13	667,191	8.0	42	588,637	43	524,873
Illinois	55,930	470	56,400	24	10,081,158	179	4	8,712,176	4	7,897,241
Indiana	36,185	106	36,291	38	4,662,498	128	11	3,934,224	12	3,427,796
Iowa	56,032	258	56,290	25	2,757,537	49	24	2,621,073	22	2,538,268
Kansas	82,048	216	82,264	14	2,178,611	26	28	1,905,299	31	1,801,028
Kentucky	39,863	532	40,395	37	3,038,156	75	22	2,944,806	19	2,845,627
Louisiana	45,106	3,417	48,523	31	3,257,022	67	20	2,683,516	21	2,363,880
Maine	31,012	2,203	33,215	39	969,265	29	36	913,774	35	847,226
Maryland	9,874	703	10,577	42	3,100,689	293	21	2,343,001	23	1,821,244
Massachusetts	7,867	390	8,257	45	5,148,578	624	9	4,690,514	9	4,316,721
Michigan	57,019	1,197	58,216	23	7,823,194	134	7	6,371,766	7	5,256,106
Minnesota	80,009	4,059	84,068	12	3,413,864	41	18	2,982,483	18	2,792,300
Mississippi	47,223	493	47,716	32	2,178,141	46	29	2,178,914	26	2,183,796
Missouri	69,138	548	69,686	19	4,319,813	62	13	3,954,653	11	3,784,664
Montana	145,736	1,402	147,138	4	674,767	4.6	41	591,024	42	559,456
Nebraska	76,612	615	77,227	15	1,411,330	18	34	1,325,510	33	1,315,834
Nevada	109,788	752	110,540	7	285,278	2.6	49	160,083	49	110,247
New Hampshire	9,014	290	9,304	44	606,921	65	45	533,242	44	491,524
New Jersey	7,521	315	7,836	46	6,066,782	774	8	4,835,329	8	4,160,165
New Mexico	121,510	156	121,666	5	951,023	7.8	37	681,187	39	531,818
New York	47,939	1,637	49,576	30	16,782,304	339	1	14,830,192	1	13,479,142
North Carolina	49,067	3,645	52,712	28	4,556,155	86	12	4,061,929	10	3,571,623
North Dakota	69,457	1,208	70,665	17	632,446	8.9	44	619,636	41	641,935
Ohio	40,972	250	41,222	35	9,706,397	235	5	7,946,627	5	6,907,612
Oklahoma	68,887	1,032	69,919	18	2,328,284	33	27	2,233,351	25	2,336,434
Oregon	96,248	733	96,981	10	1,768,687	18	32	1,521,341	32	1,089,684
Pennsylvania	45,007	326	45,333	33	11,319,366	250	3	10,498,012	3	9,900,180
Rhode Island	1,058	156	1,214	50	859,488	708	39	791,896	36	713,346
South Carolina	30,272	783	31,055	40	2,382,594	77	26	2,117,027	27	1,899,804
South Dakota	76,378	669	77,047	16	680,514	8.8	40	652,740	40	642,961
Tennessee	41,762	482	42,244	34	3,567,089	84	17	3,291,718	16	2,915,841
Texas	262,840	4,499	267,339	2	9,579,677	36	6	7,711,194	6	6,414,824
Utah	82,339	2,577	84,916	11	890,627	10	38	688,862	38	550,310
Vermont	9,276	333	9,609	43	389,881	41	47	377,747	46	359,231
Virginia	39,838	977	40,815	36	3,966,949	97	14	3,318,680	15	2,677,773
Washington	66,709	1,483	68,192	20	2,853,214	42	23	2,378,963	24	1,901,974
West Virginia	24,079	102	24,181	41	1,860,421	77	30	2,005,552	29	1,901,974
Wisconsin	54,705	1,449	56,154	26	3,951,777	70	15	3,434,575	14	3,137,587
Wyoming	97,411	503	97,914	9	330,066	3.4	48	290,529	48	250,742
United States	3,548,974	66,237	3,675,633*		179,323,175	49		151,325,798		132,165,129

† District. * Includes the United States parts of the Great Lakes (60,422 square miles). These are not included in state figures. ‡ Census taken in 1939.

U.S. STATE CLIMATIC AND ECONOMIC TABLE

State	Topography	Climate Information: Weather Station	Annual Rainfall	January Mean Temp.	July Mean Temp.	Principal Mineral Products	Principal Agricultural Products	Principal Forest and Fishery Products	Principal Manufactures
ALABAMA	Mountainous in north and northeast; southward the land gradually slopes to sea level.	Birmingham Mobile	53.52 in., 67.57 in.,	45.2°F., 52.7°F.,	79.6°F. 80.7°F.	Coal, cement, iron ore, stone, lime, sand & gravel, clays, natural gas.	Poultry and eggs, cotton, cattle, dairy products, hogs, peanuts, corn, cottonseed, soybeans, potatoes.	Shortleaf and loblolly pine, longleaf and slash pine, oak, gum, naval stores; shrimp, red snapper.	Steel rolling, textiles, paper products, rubber & plastics, iron & steel products, lumber & wood products.
ALASKA	Very high elevations in E. Broad, rolling, central plateaus, wide river valleys; NE.–W., mountain range sloping gradually to Arctic.	Anchorage Fairbanks	14.27 in., 11.92 in.,	13.0°F., –9.8°F.,	57.3°F. 60.9°F.	Coal, gold, sand & gravel, petroleum, mercury, stone, natural gas, copper, silver, clay.	Dairy products, field crops, poultry & poultry products, vegetables, cattle, sheep, hogs, horses.	Western hemlock, spruce lodgepole pine; salmon, halibut, herring, crabs.	Canned and frozen salmon & other fish products, lumber & wood products, newspapers.
ARIZONA	South—high plains with scattered mountains; north—plateaus, rough mountains, Grand Canyon of Colorado River.	Flagstaff Tucson	18.47 in., 10.66 in.,	25.3°F., 49.7°F.,	65.2°F. 86.2°F.	Copper, sand & gravel, zinc, uranium, molybdenum, stone, gold, silver, lime, lead.	Cotton, cattle, lettuce, dairy products, cottonseed, hay, cantaloups, barley, poultry & eggs.	Ponderosa pine, Douglas fir, true firs, spruce.	Printing & publishing, concrete products, dairy products, lumber & wood products, machinery.
ARKANSAS	Boston Mountains and Ouachita Mountains in northwest, separated by Arkansas River Valley; rest of state slopes southeast to Mississippi River.	Little Rock El Dorado	47.38 in., 51.58 in.,	41.8°F., 47.1°F.,	81.9°F. 82.9°F.	Petroleum, bauxite, stone, sand & gravel, natural gas, natural gas liquids, coal, barite, clays, gypsum, gem stones.	Cotton, poultry & eggs, soybeans, cattle, rice, dairy products, cottonseed, hogs, turkeys, wheat, peaches.	Shortleaf and loblolly pine, oak, gum, hickory, cypress, walnut, cottonwood; buffalofish, catfish.	Lumber & wood products, paper products, chemicals, shoes, primary metals, electrical machinery, clothing, furniture.
CALIFORNIA	Coast Ranges along western edge, bisected by Central Valley; Sierra Nevada inland; low areas in southeast at Death Valley and Imperial Valley.	Los Angeles San Francisco San Diego Sacramento Eureka	14.54 in., 17.43 in., 10.86 in., 16.32 in., 36.15 in.,	55.0°F., 47.9°F., 54.9°F., 45.2°F., 47.2°F.,	72.5°F. 60.4°F. 69.3°F. 75.3°F. 56.4°F.	Petroleum, natural gas, cement, sand & gravel, natural gas liquids, stone.	Cattle, dairy products, cotton, poultry & eggs, grapes, oranges, tomatoes, hay, lettuce, potatoes, turkeys,	Douglas fir, true firs, Ponderosa pine, redwood, sugar pine, lodgepole pine; tuna, sardines, crabs, salmon.	Metal products, electrical machinery, aircraft, machinery, chemicals, printing & publishing, canned & frozen foods.
COLORADO	Great Plains in east rise abruptly westward to high ranges of Rocky Mountains; Colorado Plateau in west central part.	Denver Sterling Pueblo Grand Junction	13.43 in., 14.00 in., 11.87 in., 9.06 in.,	31.4°F., 24.8°F., 29.4°F., 24.0°F.,	73.7°F. 74.2°F. 74.9°F. 78.2°F.	Petroleum, molybdenum, uranium ore, coal, sand & gravel, natural gas, natural gas liquids, zinc.	Cattle, wheat, sugar beets, dairy products, sheep, potatoes, dry edible beans, hay, poultry & eggs, hogs.	Spruce, lodgepole pine, Ponderosa pine, true firs, cottonwood, aspen, Douglas fir.	Machinery, metal products, meat products, dairy products, beverages, newspapers.
CONNECTICUT	Wide central Connecticut Valley; coastal plain rises inland to hills in east, low mountains in west.	Hartford New Haven	40.48 in., 44.99 in.,	27.0°F., 29.1°F.,	73.8°F. 71.2°F.	Stone, sand & gravel, lime, clays, peat, beryllium concentrate, gem stones.	Poultry & eggs, dairy products, tobacco, cattle, potatoes, apples, tomatoes, sweet corn, turkeys.	Oysters, clams, flounders, lobsters, scup or porgy, scallops, shad, cod, swordfish, butterfish, sea bass.	Electrical machinery, rubber & plastics, chemicals, precision instruments, copper rolling, hardware.
DELAWARE	Low plain, swampy along coast; low rolling hills in north.	Dover	46.40 in.,	36.8°F.,	77.2°F.	Sand & gravel, clays, stone.	Poultry & eggs, corn, soybeans, dairy products, potatoes, cattle, lima beans, hogs.	Menhaden, oysters, clams, crabs, sea trout or gray weakfish, flounders, white perch, shad.	Periodicals, chemicals, clothing, metal products, primary metals, rubber & plastics, animal feeds.
FLORIDA	Generally low and flat; many swamps, most extensive in south (The Everglades); many lakes in central part.	Jacksonville Pensacola Tampa Miami	52.08 in., 61.60 in., 49.94 in., 47.20 in.,	55.9°F., 54.0°F., 61.5°F., 68.5°F.,	82.1°F. 81.0°F. 81.7°F. 81.6°F.	Phosphate rock, stone, cement, titanium concentrate, clays, sand & gravel, lime.	Oranges, dairy products, cattle, tomatoes, poultry & eggs, grapefruit, tobacco, potatoes, snap beans.	Naval stores, longleaf and slash pine, cypress, gum; shrimp, mullet, red snapper, catfish and bullheads.	Paper products, canned & frozen food, metal products, concrete & plaster, agricultural chemicals.
GEORGIA	Ridges of the southern Appalachian Mountains in northwest, separated from wide coastal plain by Piedmont.	Atlanta Macon Savannah	49.16 in., 46.31 in., 45.75 in.,	44.6°F., 49.5°F., 51.6°F.,	79.5°F. 82.4°F. 81.2°F.	Clays, stone, cement, sand & gravel, iron ore, mica sheets, talc & soapstone, peat, coal.	Poultry & eggs, cotton, tobacco, hogs, peanuts, cattle, dairy products, corn, pecans, peaches.	Longleaf and slash pine, shortleaf and loblolly pine, gum, oak, cypress, poplars; shrimp, crabs, shad.	Textiles, transportation equipment, paper, clothing, lumber & wood, machinery.
HAWAII	Rugged with young volcanoes, lava slopes, small plains areas.	Honolulu	21.70 in.,	72.2°F.,	78.8°F.	Stone, sand & gravel, pumice, cement.	Pineapples, sugar, poultry & eggs, cattle, coffee, vegetables, hogs.	Sandalwood, ohia, lehua, kukui, koa; tuna, snapper, marlin, big-eye scad.	Sugar products, canned & frozen food, printing, clothing, bakery products.
IDAHO	High Snake River plains in south and west; Bitterroot and other rugged mountain ranges to north, east, and south.	Boise Idaho Falls Lewiston	11.48 in., 7.69 in., 13.12 in.,	27.3°F., 15.7°F., 30.8°F.,	74.8°F. 69.2°F. 75.2°F.	Silver, phosphate rock, lead, zinc, sand & gravel, copper, stone, mercury.	Cattle, potatoes, wheat, dairy products, sugar beets, dry edible beans, sheep, hay, poultry & eggs.	Douglas fir, Ponderosa pine, true firs, sugar and white pines, spruce, lodgepole pines, hemlock.	Lumber & wood products, chemicals, dairy products, frozen fruits & vegetables, printing, machinery.
ILLINOIS	Broad, plain, undulating and ridged in north, hilly in extreme south.	Chicago Rockford Springfield	33.28 in., 36.36 in., 36.65 in.,	27.4°F., 24.0°F., 27.4°F.,	75.2°F. 75.3°F. 76.3°F.	Petroleum, coal, stone, sand & gravel, cement, natural gas, natural gas liquids.	Cattle, hogs, corn, soybeans, dairy products, wheat, poultry & eggs, oats, sheep, hay, turkeys, sweet corn, tomatoes.	Oak, hickory, maple; catfish, buffalofish, carp, chubs.	Electrical machinery, metal products, printing & publishing, primary metals, chemicals, heavy machinery, precision instruments.
INDIANA	Undulating glaciated plains with many lakes in north; sand dunes along Lake Michigan; rocky hill lands in south.	Indianapolis South Bend Evansville	39.69 in., 35.59 in., 41.37 in.,	28.8°F., 24.6°F., 34.7°F.,	76.0°F. 73.4°F. 78.2°F.	Coal, cement, stone, petroleum, sand & gravel, clays.	Hogs, cattle, corn, soybeans, dairy products, poultry & eggs, wheat, turkeys, tomatoes, oats.	Oak, soft maple, hickory, walnut, sugar maple, poplar, gum; buffalofish, carp, catfish.	Steel products, electrical machinery, automobiles & parts, chemicals, machinery, metal products.
IOWA	Land slopes gradually eastward to Mississippi River; undulating to rolling surface.	Des Moines Sioux City Davenport	30.74 in., 24.90 in., 34.27 in.,	22.1°F., 19.1°F., 24.3°F.,	76.2°F. 76.3°F. 77.3°F.	Cement, stone, sand & gravel, gypsum, coal, clays.	Cattle, hogs, corn, dairy products, poultry & eggs, soybeans, turkeys, sheep, oats, hay, wheat.	Oak; catfish, buffalofish, sheepshead.	Meat products, farm machinery, electrical machinery, grain mill products, printing & publishing.
KANSAS	Rolling, valley-cut plain in east; high, gently undulating plain in west. Elevations increase gradually westward.	Topeka Wichita Dodge City	33.28 in., 30.70 in., 20.58 in.,	29.9°F., 32.0°F., 30.3°F.,	80.6°F. 80.9°F. 79.9°F.	Petroleum, natural gas, cement, stone, salt, natural gas liquids, sand & gravel, coal, clays, zinc.	Cattle, wheat, sorghum, dairy products, hogs, poultry & eggs, corn, soybeans, hay, sheep, barley, turkeys.		Aircraft, chemicals, meat products, petroleum refining, machinery, grain mill products.
KENTUCKY	Southeast—Appalachian ridges and valleys; south central—highland rim; north central—Bluegrass plain.	Frankfort Bowling Green	43.55 in., 48.67 in.,	36.6°F., 38.7°F.,	77.7°F. 79.4°F.	Coal, petroleum, stone, natural gas, sand & gravel, fluorspar.	Tobacco, cattle, dairy products, hogs, poultry & eggs, corn, soybeans, wheat, sheep, hay.	Oak, hickory, beech, poplar, shortleaf and loblolly pine, maple, ash, walnut; catfish, buffalofish.	Tobacco, electrical machinery, chemicals, distilled liquor, machinery, metal products.
LOUISIANA	Broad Mississippi Valley on eastern border, with delta at southeast; west undulating to rolling; coastlands marshy.	New Orleans Shreveport	63.54 in., 45.10 in.,	55.9°F., 47.8°F.,	83.1°F. 85.5°F.	Petroleum, natural gas, natural gas liquids, sulfur, salt, sand & gravel, stone.	Cotton, cattle, rice, dairy products, sugarcane, poultry & eggs, soybeans, sweet potatoes.	Shortleaf and loblolly pine, oak, gum, tupelo, hickory, walnut, naval stores; shrimp, menhaden, oysters.	Chemicals, petroleum products, paper, primary metals, lumber & wood products, sugar.
MAINE	Rugged, dissected by numerous stream valleys; coast fringed with promontories and rocky islands.	Portland Farmington	41.78 in., 44.47 in.,	20.7°F., 18.3°F.,	67.8°F. 69.3°F.	Sand & gravel, stone, cement, clay, gem stones.	Poultry & eggs, potatoes, dairy products, cattle, apples, blueberries, hay, oats, hogs, lettuce.	Spruce and balsam fir, red and white pine, birch; shellfish, lobsters, perch, herring, clams, haddock.	Paper, leather, textiles, lumber & wood products, transportation equipment, canned & frozen food.
MARYLAND	Parallel mountain ridges in west; rolling hill lands in central part; coastland lowlands in East, penetrated by Chesapeake Bay.	Baltimore Frederick	42.59 in., 40.23 in.,	34.2°F., 32.7°F.,	76.3°F. 76.7°F.	Stone, sand & gravel, coal, natural gas, clays, gem stones.	Poultry & eggs, dairy products, cattle, tobacco, corn, soybeans, hogs, wheat, tomatoes, apples.	Oak, poplar, gum, pine; shellfish, oysters, crabs, clams, bass, shad, fluke.	Primary metals, chemicals, clothing, communication equipment, machinery, ships, beverages.
MASSACHUSETTS	West—hilly to mountainous, split by Connecticut Valley; Berkshire Hills in far west; rolling lowlands on the coast, with Cape Cod extension.	Boston Springfield New Bedford	38.76 in., 44.87 in., 41.43 in.,	29.1°F., 28.8°F., 31.8°F.,	72.2°F. 74.2°F. 72.0°F.	Sand & gravel, stone, lime, clay, gem stones.	Dairy products, poultry & eggs, tobacco, cattle, apples, hogs, potatoes, sweet corn, turkeys, hay, tomatoes, carrots, lettuce.	Pine, oak; shellfish, haddock, perch, yellowtail, cod, whiting, lobsters, pollock, clams, blackback, fluke, sole, alewives.	Electrical machinery, machinery, leather products, textiles, printing, paper, clothing, rubber & plastics, transportation equipment.
MICHIGAN	Northern peninsula—hilly, with mountain ranges in west; southern peninsula—rolling, glaciated surface with many moraines.	Detroit Grand Rapids Sault Ste. Marie	31.03 in., 31.50 in., 30.19 in.,	26.2°F., 23.5°F., 13.8°F.,	73.1°F. 71.5°F. 63.9°F.	Iron ore, cement, petroleum, sand & gravel, gypsum, salt, stone, clays, natural gas, peat, copper.	Dairy products, cattle, wheat, poultry & eggs, corn, dry edible beans, hogs, apples, cherries.	Maple, hemlock, pine, fir; chubs, herring, trout, pike, perch, whitefish, smelt.	Automobiles & parts, food products, chemicals, metal working machinery, steel products, paper.
MINNESOTA	Glaciated surface dotted with lakes and swamps; Mesabi and other ranges in northeast.	Minneapolis Duluth Bemidji	24.71 in., 29.72 in., 21.95 in.,	14.6°F., 8.3°F., 4.9°F.,	74.1°F. 66.4°F. 68.5°F.	Iron ore, granite, dolomite, sandstone, sand & gravel, clay, manganiferous ore.	Cattle, dairy products, hogs, poultry & eggs, corn, soybeans, turkeys, wheat, oats, barley.	Cottonwood and aspen, oak, red and white pine, jack pine, spruce and balsam fir; herring, pike.	Machinery, meat products, printing, paper, chemicals, electrical machinery.
MISSISSIPPI	Rolling to hilly, sloping gently to south and west; Yazoo-Mississippi Delta in west.	Jackson Greenville Biloxi	50.86 in., 51.69 in., 57.59 in.,	48.3°F., 47.0°F., 54.2°F.,	82.1°F. 82.4°F. 81.8°F.	Petroleum, natural gas, sand & gravel, clays, natural gas liquids, stone.	Cotton, poultry & eggs, cattle, dairy products, soybeans, cottonseed, hogs, corn, rice, pecans.	Shortleaf and loblolly pine, oak, gum, longleaf and slash pine, hickory, naval stores; shellfish, shrimp.	Forest products, paper, clothing, transportation equipment, lumber & wood products, chemicals.
MISSOURI	Highly dissected Ozark Plateau in south; Missouri River traverses northern hill lands from west to east.	St. Louis Kansas City Springfield	37.86 in., 35.31 in., 41.51 in.,	33.3°F., 30.0°F., 32.7°F.,	80.6°F. 80.9°F. 77.5°F.	Cement, stone, lead, lime, coal, sand & gravel, clays, barite, zinc, copper, natural gas, silver.	Cattle, hogs, dairy products, soybeans, corn, poultry & eggs, cotton, wheat, turkeys, sheep.	Oak, hickory, shortleaf and loblolly pine, walnut, cottonwood, aspen, maple; buffalofish, catfish, carp.	Automobiles, chemicals, printing, leather, machinery, electrical machinery, clothing.
MONTANA	Rockies cover western third; remainder consists of plateaus and undulating plains.	Butte Great Falls Billings	12.67 in., 14.03 in., 13.10 in.,	14.2°F., 22.7°F., 22.9°F.,	62.4°F. 69.6°F. 73.3°F.	Petroleum, copper, sand & gravel, chromium ore & concentrate, silver, zinc, natural gas, manganese ore.	Cattle, wheat, barley, dairy products, sheep, sugar beets, wool, hay, hogs, poultry & eggs.	Douglas fir, Ponderosa pine, lodgepole pine, spruce, sugar and western white pine.	Lumber & wood products, petroleum, printing & publishing, dairy products, sugar, beverages.

State	Topography	Climate Information: Weather Station	Annual Rainfall	January Mean Temp.	July Mean Temp.	Principal Mineral Products	Principal Agricultural Products	Principal Forest and Fishery Products	Principal Manufactures
NEBRASKA	Platte River flows eastward through undulating sand-and-loess-covered plains; foothills of the Rockies in far west.	Omaha Grand Island Scottsbluff	25.90 in., 22.70 in., 15.00 in.,	23.0°F., 23.0°F., 23.5°F.,	78.5°F. 78.5°F. 74.6°F.	Petroleum, sand & gravel, stone, natural gas liquids, natural gas, clays.	Cattle, corn, wheat, hogs, dairy products, poultry & eggs, sorghum grain, sheep, sugar beets, hay, soybeans.		Packed meat, grain mill products, printing & publishing, metal products, dairy products, machinery.
NEVADA	Broken series of roughly parallel ranges and basins with a north to south orientation. High Sierra Nevada on west.	Carson City Las Vegas Elko	11.50 in., 4.35 in., 9.13 in.,	31.7°F., 44.2°F., 21.9°F.,	69.6°F. 90.5°F. 70.2°F.	Copper, sand & gravel, iron ore, manganese ore, gypsum, gold.	Cattle, dairy products, hay, sheep, wool, cotton, wheat, onions, potatoes, hogs, barley, cottonseed.	Ponderosa pine, true firs.	Concrete & plaster products, newspapers, lumber & wood products, dairy products.
NEW HAMPSHIRE	North central—rugged, culminating in White Mountains; high ridges separate Connecticut and Merrimac river valleys; low coastal plain in southeast.	Concord Berlin	37.23 in., 39.13 in.,	20.1°F., 16.0°F.,	69.0°F. 66.6°F.	Sand & gravel, mica, stone, clays, gem stones, feldspar.	Dairy products, poultry & eggs, apples, cattle, hay, potatoes, turkeys, hogs, maple, snap beans.	Red and white pine, birch, maple, spruce and balsam fir, hemlock; shellfish, lobster, smelt.	Leather footwear, textiles, machinery, paper, electrical machinery, lumber & wood products, printing & publishing, food products.
NEW JERSEY	Southern half—low coastal plain, stream-indented coastline with protecting sand bars; northern half—parallel ridges and valleys.	Trenton Atlantic City	40.06 in., 41.77 in.,	32.6°F., 35.8°F.,	75.3°F. 73.6°F.	Stone, sand & gravel, clays, lime, iron ore, uranium, peat.	Poultry & eggs, dairy products, tomatoes, cattle, asparagus, peaches, potatoes, blueberries, apples, corn, hogs.	Oak, beech, maple, gum, poplar; shellfish, menhaden, clams, flounders, oysters, scup or porgy, bass, shad, perch.	Chemicals, electrical machinery, transportation equipment, machinery, metal products, clothing, primary metals.
NEW MEXICO	Rolling plains & plateaus; scattered mountain ranges running north to south; Rio Grande & Pecos valleys drain southward.	Albuquerque Roswell	8.68 in., 12.07 in.,	33.7°F., 39.6°F.,	79.0°F. 79.0°F.	Petroleum, natural gas, potassium salts, uranium ore, natural gas liquids, copper, sand & gravel.	Cattle, cotton, dairy products, wheat, hay, sorghum grain, sheep, poultry & eggs, cottonseed.	Ponderosa pine, Douglas fir, spruce, true firs, cottonwood, aspen.	Food products, lumber & wood products, petroleum & coal products, newspapers.
NEW YORK	Rolling plateau; Adirondacks in northeast, Catskills in southeast, separated by Mohawk Valley; Hudson-Champlain lowland along eastern border.	New York Buffalo Albany	42.03 in., 32.29 in., 35.81 in.,	32.9°F., 25.5°F., 25.2°F.,	74.6°F. 70.6°F. 73.1°F.	Stone, sand & gravel, iron ore, salt, zinc, abrasive garnet, petroleum, gypsum, clays, natural gas.	Dairy products, poultry & eggs, cattle, potatoes, apples, grapes, wheat, snap beans, dry edible beans, hay, onions, tomatoes.	Maples, beech, white and red pine, hemlock, firs; shellfish, clams, oysters, lobster, whiting, bass.	Clothing, printing & publishing, electrical machinery, machinery, chemicals, transportation equipment, precision instruments.
NORTH CAROLINA	Coastal plain in eastern third; Piedmont Uplands in central; Blue Ridge and Smoky Mountains along western border.	Asheville Raleigh Winston-Salem	37.22 in., 45.05 in., 43.15 in.,	39.4°F., 41.4°F., 39.3°F.,	73.8°F. 78.5°F. 77.5°F.	Stone, sand & gravel, feldspar, mica, clays, asbestos, talc & pyrophyllite, silver.	Tobacco, poultry & eggs, dairy products, hogs, corn, cotton, cattle, peanuts, soybeans, wheat, turkeys.	Shortleaf and loblolly pine, oak, gum, cypress, poplar, hickory, hemlock; menhaden, shellfish, crabs.	Textiles, cigarettes, furniture, chemicals, electrical machinery, clothing, paper products.
NORTH DAKOTA	Red River Valley on eastern border; central section, glaciated plains; west, Missouri Plateau cut by Missouri River system.	Bismarck Grand Forks Williston	15.40 in., 20.05 in., 14.66 in.,	9.2°F., 4.4°F., 10.0°F.,	72.1°F. 70.7°F. 70.9°F.	Petroleum, sand & gravel, coal, natural gas, clays, stone, natural gas liquids.	Wheat, cattle, barley, flaxseed, dairy products, hogs, potatoes, poultry & eggs, oats, sheep, sugar beets.	Bullheads, carp, catfish, suckers, buffalofish, perch, burbot, shovelnose.	Dairy products, newspapers, meat products, concrete products.
OHIO	Nearly level plains in north; rolling glacial plains in west; dissected Allegheny Plateau in southeast.	Cleveland Columbus Cincinnati	32.08 in., 34.36 in., 39.34 in.,	28.5°F., 31.1°F., 34.6°F.,	73.7°F. 75.8°F. 78.1°F.	Coal, cement, stone, sand & gravel, lime, salt, petroleum, clays, natural gas.	Dairy products, hogs, cattle, corn, poultry & eggs, wheat, soybeans, tomatoes, oats, turkeys, sheep, tobacco, apples.	Oak, hickory, maple, beech; pike, perch, catfish, bass, carp, sheepshead, whitefish, suckers, bullheads.	Iron & steel products, transportation equipment, machinery, electrical machinery, metal products, rubber products, chemicals.
OKLAHOMA	Ouachita Mountains in southeast; rolling prairies and high plains rising westward.	Oklahoma City Tulsa Lawton	30.22 in., 37.68 in., 29.68 in.,	37.1°F., 37.4°F., 41.1°F.,	82.1°F. 82.1°F. 83.9°F.	Petroleum, natural gas, natural gas liquids, stone, coal, sand & gravel, helium.	Cattle, wheat, cotton, dairy products, poultry & eggs, hogs, peanuts, pecans, hay, sorghum grain.	Shortleaf and loblolly pine, oak, hickory.	Petroleum refining, aircraft, metal products, oil-field machinery, meat products, glass.
OREGON	Broad Cascade Range separates wide Willamette Valley and Coast Range from lava plateaus and basins in southeast, and from mountains in northeast.	Portland Klamath Falls	39.91 in., 13.94 in.,	39.5°F., 29.2°F.,	68.5°F. 68.7°F.	Stone, sand & gravel, nickel, cement, clays, mercury, gold, pumice, gem stones, copper.	Cattle, wheat, dairy products, poultry & eggs, potatoes, barley, hay, beans, pears, strawberries, sheep, hogs, turkeys.	Douglas fir, Ponderosa pine, true firs, hemlock, sugar and western white pine; salmon, tuna, shellfish, crabs, flounders.	Lumber & wood products, canned & frozen fruits & vegetables, paper, iron & steel products, dairy products, machinery.
PENNSYLVANIA	Rolling piedmont in southeast; folded ranges of Appalachians from northeast to southwest; Allegheny Plateau in north and west.	Pittsburgh Philadelphia Scranton	36.23 in., 41.13 in., 40.49 in.,	33.0°F., 33.2°F., 26.9°F.,	75.4°F. 76.3°F. 72.2°F.	Coal, cement, stone, natural gas, petroleum, sand & gravel, clays, lime, zinc.	Dairy products, poultry & eggs, cattle, corn, hogs, wheat, potatoes, tobacco, apples, turkeys, hay, peaches, tomatoes.	Oak, maple, hemlock, white and red pine.	Iron & steel products, heavy machinery, metal products, electrical machinery, chemicals, clothing, glass.
RHODE ISLAND	Glaciated highland in west; Narragansett Bay penetrates rolling eastern lowlands; Block Island lies 10 miles off shore.	Providence	39.63 in.,	28.7°F.,	71.0°F.	Stone, sand & gravel.	Dairy products, poultry & eggs, potatoes, cattle, apples, hogs, tomatoes, sweet corn, turkeys, hay.	Shellfish, clams, scup or porgy, yellowtail, butterfish, fluke, lobster, blackback, menhaden.	Textiles, jewelry, machinery, printing & publishing, metal products, silverware.
SOUTH CAROLINA	Coastal plain occupies three-fifths of state; in remainder, Piedmont Upland rises to Blue Ridge Mountains on northwest border.	Charleston Columbia Greenville	45.99 in., 46.15 in., 47.65 in.,	51.4°F., 47.0°F., 43.0°F.,	81.5°F. 81.4°F. 78.4°F.	Stone, clay, sand & gravel, mica.	Tobacco, cotton, poultry & eggs, dairy products, soybeans, peaches, corn, cottonseed, wheat, oats.	Shortleaf and loblolly pine, gum, oak, cypress, poplar, maple; shellfish, shrimp, oysters, crabs, menhaden.	Textiles, chemicals, clothing, paper, lumber & wood products.
SOUTH DAKOTA	Black Hills and Badlands in southwest; Missouri River Valley separates glaciated plains in east from high, dissected plains of west.	Sioux Falls Aberdeen Rapid City	25.24 in., 19.63 in., 17.10 in.,	14.2°F., 11.0°F., 21.1°F.,	74.8°F. 73.9°F. 72.3°F.	Gold, sand & gravel, stone, cement, uranium ore, feldspar, clays, mica, silver.	Cattle, hogs, wheat, poultry & eggs, dairy products, corn, sheep, oats, flaxseed, wool, barley.	Ponderosa pine, cottonwood and aspen; carp, bullheads, buffalofish, suckers.	Meat products, dairy products, newspapers, wood products.
TENNESSEE	Regions from east to west; Smoky Mountains; parallel valleys; eroded Cumberland Plateau; Nashville Basin; sloping plains to Mississippi River Valley.	Memphis Nashville Chattanooga	46.81 in., 45.03 in., 53.60 in.,	41.9°F., 39.9°F., 41.6°F.,	81.3°F. 80.0°F. 78.3°F.	Stone, cement, zinc, coal, phosphate rock, copper, sand & gravel, clays.	Cotton, cattle, dairy products, tobacco, hogs, poultry & eggs, soybeans, corn, cottonseed, wheat, apples, strawberries, hay.	Oak, hickory, poplar, shortleaf and loblolly pine, gum; catfish, buffalofish, carp, paddlefish.	Chemicals, metal products, plastics, textiles, clothing, paper, printing & publishing, wood products, footwear.
TEXAS	Gulf coastal plain in east; broad, rolling, central plains; high, rolling plateaus in west; mountains extreme southwest.	Houston Dallas Amarillo El Paso San Antonio	45.37 in., 34.42 in., 21.12 in., 7.83 in., 27.93 in.,	53.8°F., 45.7°F., 35.3°F., 43.4°F., 50.6°F.,	83.8°F. 85.5°F. 77.8°F. 81.3°F. 84.2°F.	Petroleum, natural gas, natural gas liquids, cement, sulfur, stone, sand & gravel.	Cotton, cattle, sorghum grain, wheat, cottonseed, rice, hogs, sheep, mohair, wool, peanuts, turkeys.	Shortleaf and loblolly pine, oak, gum, longleaf and slash pine; shellfish, shrimp, menhaden, snapper, trout.	Chemicals, petroleum refining, aircraft & parts, metal products, construction equipment, oilfield machines.
UTAH	High Colorado Plateau in east; basins and ranges in west; Great Salt Lake Plain in northwest; high mountains in northeast.	Salt Lake City Logan Richfield	14.74 in., 17.03 in., 8.19 in.,	26.5°F., 23.6°F., 28.0°F.,	76.6°F. 73.5°F. 71.8°F.	Copper, petroleum, coal, uranium ore, iron ore, gold, asphalt, lead, natural gas.	Cattle, dairy products, poultry & eggs, turkeys, sheep, sugar beets, wheat, hay, wool, potatoes, hogs.	Spruce, lodgepole pine, Ponderosa pine, Douglas fir, true fir, cottonwood and aspen.	Primary metals, petroleum refining, construction machinery, metal products, concrete & plaster products.
VERMONT	The Green Mountains are the main feature; Champlain lowlands in northwest; Connecticut River forms eastern border.	Burlington Rutland	32.22 in., 38.67 in.,	17.9°F., 21.5°F.,	70.4°F. 69.5°F.	Stone, sand & gravel, asbestos, clays, lime, talc, gem stones.	Dairy products, cattle, poultry & eggs, apples, hay, maple, potatoes, hogs, turkeys.	Birch and maple, spruce and balsam, fir, beech, hemlock, white and red pine, oak, walnut.	Machinery & machine tools, paper products, cut stone & stone products, lumber & wood products.
VIRGINIA	Coastal plain merges with Piedmont Upland; Great Valley lies between Blue Ridge and other Appalachian ranges in west.	Richmond Norfolk Roanoke	42.89 in., 43.26 in., 41.58 in.,	38.3°F., 41.5°F., 37.9°F.,	77.5°F. 77.5°F. 75.9°F.	Coal, stone, sand & gravel, lime, zinc, clays, cement, gypsum, gem stones, iron ore.	Dairy products, tobacco, poultry & eggs, cattle, peanuts, hogs, apples, soybeans, turkeys, corn.	Oak, shortleaf and loblolly pine, poplar, gum, hickory, maple and beech, ash, walnut; shellfish, oysters.	Fibers, plastics & rubber, cigarettes, textiles, paper, furniture, chemicals, lumber & wood products.
WASHINGTON	Coast Ranges and Cascade Range, separated by Puget Sound lowland, parallel the west coast; rolling plateau in southeast; Rockies in northeast.	Seattle Spokane Walla Walla	31.92 in., 14.92 in., 15.07 in.,	40.7°F., 24.9°F., 32.0°F.,	65.6°F. 69.6°F. 76.2°F.	Sand & gravel, stone, zinc, uranium ore, lead, coal, barite, pumice, diatomite, clay, peat, gypsum.	Wheat, dairy products, cattle, apples, poultry & eggs, barley, potatoes, hay, hops, sugar beets, green peas, dry field peas.	Douglas fir, western hemlock, true firs, Ponderosa pine, spruce, red alder; salmon, halibut, shellfish, crabs, oysters.	Aircraft, lumber & wood products, pulp & paper, primary metals, chemicals, canned & frozen foods, printing & publishing.
WEST VIRGINIA	Greater portion in Allegheny Plateau; Appalachian ranges and valleys in extreme east.	Charleston Clarksburg	45.00 in., 41.82 in.,	36.4°F., 33.0°F.,	75.4°F. 73.6°F.	Coal, natural gas, natural gas liquids, stone, sand & gravel, salt, clays.	Poultry & eggs, dairy products, cattle, apples, turkeys, sheep, hogs, corn, tobacco.	Oak, maple, birch, beech, yellow poplar, hickory, ash, basswood, walnut, gum, hemlock.	Chemicals, steel rolling & finishing, metal products, electrical machinery, glassware, machinery.
WISCONSIN	Southwest, rough and dissected; remainder, rolling to level glaciated plateau with many lakes and moraines.	Milwaukee Madison Green Bay	27.57 in., 30.71 in., 26.51 in.,	21.9°F., 19.3°F., 16.1°F.,	71.3°F. 73.1°F. 69.9°F.	Sand & gravel, stone, zinc, cement, lime, iron ore, clays.	Dairy products, cattle, hogs, poultry & eggs, corn, potatoes, turkeys, hay, green peas, tobacco, oats, sweet corn, soybeans.	Oak, birch, maple, red and white pine, hemlock, ash, basswood, walnut, cottonwood, aspen; chubs, perch, herring, trout.	Heavy machinery, paper, electrical machinery, automobiles & parts, metal products, dairy products, malt liquors, leather.
WYOMING	Numerous large, high, basins surrounded by high ranges of the Rockies; high plains in northeast quarter.	Cheyenne Sheridan Saratoga Moran	16.25 in., 16.75 in., 9.53 in., 21.21 in.,	25.5°F., 20.1°F., 20.4°F., 10.3°F.,	69.7°F. 70.6°F. 65.8°F. 57.6°F.	Petroleum, uranium ore, natural gas, clays, natural gas liquids, coal, sand & gravel, stone.	Cattle, sheep, wool, wheat, sugar beets, dairy products, dry edible beans, hay, poultry & eggs, hogs.	Lodgepole pine, spruce, Ponderosa pine, Douglas fir.	Petroleum & coal products, food products, stone, clay & glass products, lumber & wood products.

NATIONAL PARKS AND MONUMENTS MAP OF THE UNITED STATES

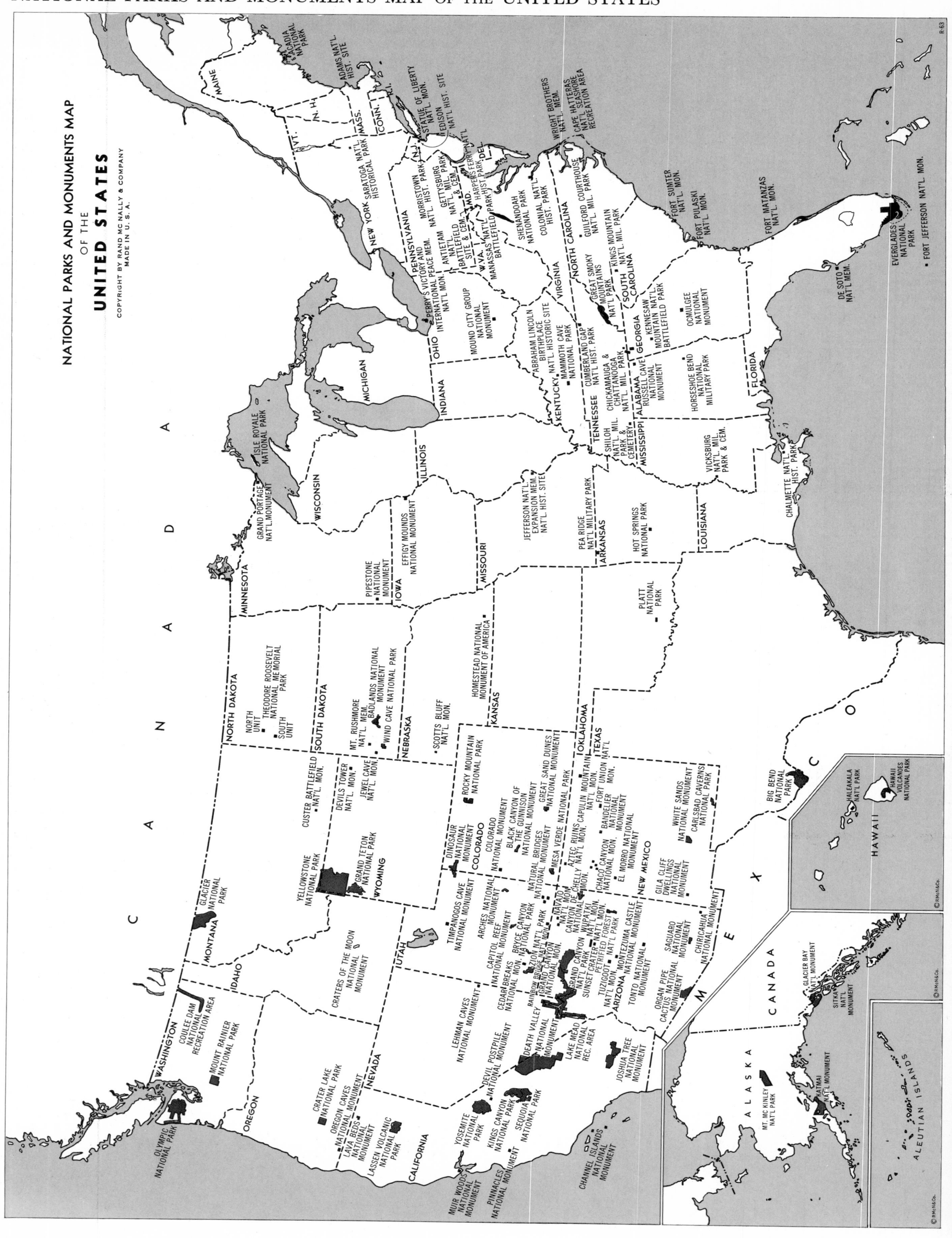

NATIONAL PARKS AND MONUMENTS MAP
OF THE
UNITED STATES

COPYRIGHT BY RAND McNALLY & COMPANY
MADE IN U. S. A.

R-63

U.S. NATIONAL PARK SYSTEM

Name	Year Established	Location	Gross Acreage	Description
PARKS				
Acadia	1919	Maine	41,634	Wilderness area on Mount Desert Island, Schoodic Point, and Isle au Haut
Big Bend	1944	Texas	708,221	Spectacular mountains and deserts in bend of Rio Grande
Bryce Canyon	1928	Utah	36,010	Wonderland of colorful rocks and pinnacles
Carlsbad Caverns	1930	N.Mex.	49,448	Immense limestone caverns with brilliantly colored formations
Crater Lake	1902	Oreg.	160,290	High cliffs encircle deep-blue lake in extinct volcano crater
Everglades	1947	Fla.	1,406,248	Subtropical area; mangrove swamps; rare birds and plants
Glacier	1910	Mont.	1,013,129	Part of Waterton-Glacier International Peace Park
Grand Canyon	1919	Ariz.	673,575	Tremendous gorge; ever-changing colors; fantastic rock shapes
Grand Teton	1929	Wyo.	310,350	Majestic, snow-capped peaks, lakes, and evergreen forests
Great Smoky Mountains	1930	N.C.-Tenn.	511,714	Highest eastern mountains; primeval hardwood forests
Haleakala	1961	Hawaii	26,403	One of world's largest dormant volcanoes
Hawaii Volcanoes	1916	Hawaii	220,345	Active volcanoes; tropical forests; tree ferns
Hot Springs	1921	Ark.	989	Forty-seven historically famous mineral hot springs
Isle Royale	1940	Mich.	539,338	Forested island in Lake Superior; large moose herd
Kings Canyon	1940	Calif.	454,650	Imposing peaks and canyons; giant sequoias
Lassen Volcanic	1916	Calif.	106,934	Recently active volcano; many lakes and volcanic exhibits
Mammoth Cave	1936	Ky.	51,354	Beautiful limestone caverns, underground rivers, and lakes
Mesa Verde	1906	Colo.	51,334	Well-preserved prehistoric cliff dwellings and pueblo houses
Mount McKinley	1917	Alaska	1,939,493	Highest mountain in N.A.; glaciers; interesting wildlife
Mount Rainier	1899	Wash.	241,782	Glaciers radiating from snow-capped peak; dense forests
Olympic	1938	Wash.	896,599	Mountain wilderness; luxuriant forests of huge evergreens
Petrified Forest	1962	Ariz.	94,161	Spectacular display of petrified wood; part of Painted Desert
Platt	1906	Okla.	912	Cold mineral springs, some of which contain bromides
Rocky Mountain	1915	Colo.	260,018	Magnificent mountain scenery; interesting wildlife
Sequoia	1890	Calif.	386,551	Immense groves of sequoias; Mt. Whitney and other peaks
Shenandoah	1935	Va.	212,304	Blue Ridge Mountains; Skyline Drive
Theodore Roosevelt National Memorial	1947	N.Dak.	70,374	Section of badlands and part of Roosevelt's Elkhorn Ranch
Virgin Islands	1956	Virgin Is.	9,500	A tropical area having historic and prehistoric interest
Wind Cave	1903	S.Dak.	28,059	"Boxwork" limestone caverns in Black Hills; buffalo herd
Yellowstone	1872	Wyo-Mont.-Idaho	2,221,773	World's greatest geyser area; bubbling hot springs and colorful pools; beautiful Yellowstone Falls; wildlife sanctuary
Yosemite	1890	Calif.	760,951	Spectacular gorges, domes, and waterfalls; giant sequoias
Zion	1919	Utah	147,035	Colorful canyon, displaying picturesque rock formations
MONUMENTS				
Andrew Johnson	1942	Tenn.	16	President Johnson's tailor shop, home, and grave
Arches	1929	Utah	34,010	Giant arches, bridges, windows, spires eroded from sandstone
Aztec Ruins	1923	N.Mex.	27	Ruins of 12th-century prehistoric Indian town
Badlands	1939	S.Dak.	111,530	Deeply eroded hills containing prehistoric animal fossils
Bandelier	1916	N.Mex.	30,703	Prehistoric Indian ruins of the later Pueblo period
Black Canyon of the Gunnison	1933	Colo.	13,548	Remarkably deep, narrow gorge of great geologic interest
Booker T. Washington	1957	Virginia	200	Memorial to the famous educator and reformer
Cabrillo	1913	Calif.	81	Memorial to Juan Cabrillo, discoverer of San Diego Bay, 1542
Canyon de Chelly	1931	Ariz.	83,840	Prehistoric Indian ruins built in caves; modern Navajo homes
Capitol Reef	1937	Utah	39,173	Twenty-mile-long buttressed cliff of colored sandstone
Capulin Mountain	1916	N.Mex.	680	Large cinder cone of a recently extinct volcano
Casa Grande	1918	Ariz.	473	Adobe tower built by the Indians 600 years ago
Castillo de San Marcos	1924	Fla.	22	Oldest masonry fort in Continental U.S.—built by Spanish
Castle Clinton	1950	N.Y.	1	Through its door, 1855–90, 7,500,000 people entered America
Cedar Breaks	1933	Utah	6,155	Brilliantly colored amphitheater 2000 feet deep; eroded cliffs
Chaco Canyon	1907	N.Mex.	21,509	Most extensive Indian ruins in the U.S.; shows Pueblo culture
Channel Islands	1938	Calif.	18,167	Sea lion rookery; other interesting wildlife, fossil beds
Chesapeake and Ohio Canal	1961	Md.-W.Va.	4,475	One of the least altered of the older American canals
Chiricahua	1924	Ariz.	10,646	Strange rock shapes depict billion years of geologic history
Colorado	1911	Colo.	17,693	Deep canyons with weirdly eroded red-sandstone walls
Craters of the Moon	1924	Idaho	53,545	Craters, lava flows, caves and other volcanic Phenomena
Custer Battlefield	1946	Mont.	765	Site of battle in which Custer and all his men were killed
Death Valley	1933	Calif.-Nev.	1,907,760	Colorful desert area; lowest point in Western Hemisphere
Devils Postpile	1911	Calif.	798	Sixty-foot basaltic cliff, composed of blue-gray columns
Devils Tower	1906	Wyo.	1,347	Fluted tower of volcanic rock, 865 feet high
Dinosaur	1915	Utah-Colo.	205,136	Well-preserved fossils of many dinosaurs; spectacular canyons
Effigy Mounds	1949	Iowa	1,476	Indian mounds shaped like birds and other animals
El Morro	1906	N.Mex.	1,279	Monolith inscribed by early Spaniards and Americans
Fort Frederica	1945	Ga.	250	Built by Oglethorpe, 1736, as defense against the Spaniards
Fort Jefferson	1935	Fla.	47,125	Huge masonry fortification; notable bird refuge; marine life
Fort McHenry National Monument and Historic Shrine	1939	Md.	43	Its defense, 1814, inspired writing of "Star Spangled Banner"
Fort Matanzas	1924	Fla.	228	Historical Spanish fort built, 1737, to protect St. Augustine
Fort Pulaski	1924	Ga.	5,517	Southern fortress important in the Civil War
Fort Sumter	1948	S.C.	2.4	Scene of first battle of the Civil War
Fort Union	1956	N.Mex.	721	Site of an old fort erecteed to protect the Santa Fe Trail
Geo. Washington Birthplace	1930	Va.	394	Memorial house and gardens on site of Washington's birthplace
Geo. Washington Carver	1951	Mo.	210	Site of birthplace and childhood home of famous scientist
Gila Cliff Dwellings	1907	N.Mex.	160	Remains of dwellings built in face of an overhanging cliff
Glacier Bay	1925	Alaska	2,274,595	Area of receding glaciers and post-glacial forests
Grand Canyon	1932	Ariz.	198,280	Portion of the Grand Canyon, including Toroweap Point
Grand Portage	1960	Minn.	770	Portage on principal route of Indians, explorers, missionaries and fur traders into the Northwest
Gran Quivira	1909	N.Mex.	611	Ruins of 17th-century Spanish mission and Indian pueblo
Great Sand Dunes	1932	Colo.	36,740	Very large and high shifting sand dunes
Homestead	1939	Nebr.	163	First claim under the Homestead Act of 1862
Hovenweep	1923	Utah-Colo.	505	Several groups of cliff dwellings, pueblos, and towers
Jewel Cave	1908	S.Dak.	1,275	Series of limestone rooms incrusted with calcite crystals
Joshua Tree	1936	Calif.	557,992	Desert area containing a fine stand of the rare Joshua tree
Katmai	1918	Alaska	2,697,590	Crater of Katmai Volcano; Valley of Ten Thousand Smokes
Lava Beds	1925	Calif.	46,239	Volcanic formations; scene of Modoc Indian War, 1873
Lehman Caves	1922	Nev.	640	Limestone caverns containing stalactites and other formations
Montezuma Castle	1906	Ariz.	842	Unusually well-preserved, 5-story, 20-room cliff dwelling
Mound City Group	1923	Ohio	68	Prehistoric Indian mounds
Muir Woods	1908	Calif.	504	Virgin forest of Coast redwoods and other interesting plants
Natural Bridges	1908	Utah	2,650	Three bridges carved from sandstone walls by running water
Navajo	1909	Ariz.	360	Three large and elaborate cliff dwellings
Ocmulgee	1936	Ga.	683	Mounds and other remains constructed by Southern mound-builders
Oregon Caves	1909	Oreg.	480	Caverns containing unique limestone formations
Organ Pipe Cactus	1937	Ariz.	330,874	Desert area containing rare cactus species
MONUMENTS *(Continued)*				
Perry's Victory and International Peace Memorial	1936	Ohio	14	Monument to Perry's naval victory and amity with Canada
Pinnacles	1908	Calif.	14,498	Rock spires 500–1200 feet high; caves of volcanic origin
Pipe Spring	1923	Ariz.	40	Historic fort and other Mormon structures
Pipestone	1937	Minn.	283	Quarry from which Indians obtained materials for peace pipes
Rainbow Bridge	1910	Utah	160	Greatest known natural bridge—a rainbow in stone
Russell Cave	1961	Ala.	310	Believed to have been continuously inhabited by stoneage men from 7000 B.C. to 1650 A.D.
Saguaro	1933	Ariz.	78,644	Forest of giant saguaro cactus and other rare desert plants
Scotts Bluff	1919	Nebr.	3,452	High bluff on Oregon Trail—landmark for early wagon trains
Sitka	1910	Alaska	54	Scene of last stand of Kik-Siti Indians against Russians
Statue of Liberty	1924	N.Y.	10	Famous statue on Liberty Island—gift of the French people
Sunset Crater	1930	Ariz.	3,040	Volcanic cone with highly colored rim; ice caves; lava flows
Timpanogos Cave	1922	Utah	250	Brilliantly colored caves on the slope of Mt. Timpanogos
Tonto	1907	Ariz.	1,120	Two 14th-century Pueblo Indian cliff dwellings
Tumacacori	1908	Ariz.	10	Site of historic Spanish mission dating from 1691
Tuzigoot	1939	Ariz.	43	Excavated ruins of large pueblo dating from 1000–1400
Walnut Canyon	1915	Ariz.	1,879	Cliff dwellings built in shallow caves along canyon walls
White Sands	1933	N.Mex.	146,535	Sand dunes 10 to 60 feet high, composed of snow-white Gypsum
Wupatki	1924	Ariz.	35,545	Red-sandstone pueblo believed built by ancestors of Hopis
HISTORICAL PARKS, SITES, AND MEMORIALS				
Abraham Lincoln Birthplace	1939	Ky.	117	Log cabin thought to have been Lincoln's birthplace
Adams	1946	Mass.	5	Home of John Adams and John Quincy Adams, U.S. Presidents
Appomattox Court House Park	1954	Va.	972	Scene of Robert E. Lee's surrender to Grant in Civil War
Chalmette Park	1939	La.	136	Part of site of the Battle of New Orleans, War of 1812
City of Refuge	1961	Hawaii	182	An area preserving the history of the Polynesian people and early Hawaiian culture
Colonial Park	1936	Va.	9,430	Jamestown Isl., Cape Henry Mem., Yorktown, Williamsburg Parkway
Coronado Mem.	1952	Ariz.	2,745	Place where Coronado entered U.S. (1540–42) to explore southwest
Cumberland Gap	1955	Ky.-Tenn.-Va.	20,193	Gap through which passed the main artery of the great trans-Allegheny migration
Custis-Lee Mansion	1925	Va.	3	Home of the Confederate General Robert E. Lee
De Soto Mem.	1948	Fla.	30	Commemorates landing of De Soto in Florida, 1539
Dutch Reformed (Sleepy Hollow) Church	1962	N.Y.	. . .	Dutch Colonial church made famous by Washington Irving
Edison	1955	N.J.	15	Glenmont, home of Thomas Alva Edison from 1886 until his death, 1931; also includes his laboratory, stock room and library
Federal Hall Mem.	1939	N.Y.	.45	Subtreasury building; site of first seat of U.S. Government
Fort Caroline Mem.	1950	Fla.	120	Overlooks site of French colony of 1564
Fort Clatsop	1958	Ore.	125	Commemorates Lewis & Clark's winter camp
Fort Davis	1962	Texas	454	Ruins of Army post used from 1854 to 1891
Fort Laramie	1938	Wyo.	564	Military post on Oregon Trail, famous in covered-wagon days
Fort Raleigh	1941	N.C.	144	Scene of Sir Walter Raleigh's "Lost Colony" settlement
Fort Vancouver	1954	Wash.	90	Site of stockaded fur trading post from 1824 to 1846
General Grant	1959	N.Y.	1	Memorial to the commander of the Union armies
Hampton	1948	Md.	45	Eighteenth-century Georgian mansion
Harpers Ferry	1962	W. Va.-Md.	1,500	Scene of John Brown's famous raid in 1859
Home of Franklin D. Roosevelt	1944	N.Y.	94	Birthplace and home of President F. D. Roosevelt
Hopewell Village	1938	Pa.	848	Ruins of an 18th- and early 19th-century iron-making village
House Where Lincoln Died	1896	D.C.	.05	Petersen House, where Lincoln was taken after he was shot
Independence Park	·1956	Pa.	22	Structures in Philadelphia associated with the founding of the U.S.
Jefferson National Expansion Mem.	1935	Mo.	85	Riverfront area in St. Louis commemorating westward expansion
Lincoln Boyhood Mem	1962	Ind.	200	Site of Lincoln's residence from ages 7 to 21
Lincoln Mem.	1911	D.C.	.61	White-marble structure enclosing seated figure of Lincoln
Lincoln Museum	1866	D.C.	.18	Site of Ford's Theatre, where Abraham Lincoln was shot
Minute Man	1959	Mass.	750	Site of Revolutionary War skirmish
Morristown Park	1933	N.J.	958	Washington's Headquarters, 1779–80, and other historical sites
Mount Rushmore Mem.	1925	S.Dak.	1,278	Features of four great Presidents carved on face of mountain
Sagamore Hill	1962	N.Y.	. . .	Victorian-style home of Theodore Roosevelt on Long Island
Salem Maritime	1938	Mass.	11	Several buildings important in New England maritime history
San Juan	1949	P.R.	40	16th Century fortifications guarding San Juan Harbor
Saratoga Park	1948	N.Y.	5,500	Scene of the defeat of Burgoyne, British general, in 1777
Theodore Roosevelt Birthplace	1962	N.Y.	. . .	New York City birthplace of President Roosevelt
Thomas Jefferson Mem.	1934	D.C.	1	Structure in classic style introduced in America by Jefferson
Vanderbilt Mansion	1940	N.Y.	212	Palatial residence of the "Gay Nineties"
Washington Monument Mem.	1848	D.C.	.37	An obelisk 555 feet high—memorial to George Washington
Whitman Mission	1962	Wash.	96	Site of mission where Indians massacred Whitman family
Wright Brothers Mem.	1927	N.C.	425	Commemorates Wright brothers' airplane flight at Kitty Hawk
RECREATION AREAS				
Cape Cod National Seashore	1961	Mass.	26,666	An immense area of historical interest, geological significance, and beaches and dunes for recreation
Cape Hatteras Seashore	Project	N.C.	28,500	Extensive beaches with surf fishing and waterfowl refuge
Coulee Dam	1946	Wash.	98,500	Dam forms Franklin D. Roosevelt Lake extending to Canada
Glen Canyon	1958	Ariz.-Utah	1,429,007	Third highest dam in world—forms Lake Powell
Lake Mead	1936	Ariz.-Nev.	1,951,928	Formed by Hoover Dam across the gorge of the Colorado River
Padre Island Seashore	Project	Texas	. . .	80 miles of subtropical offshore bar; excellent fishing
Point Reyes Seashore	Project	Calif.	53,000	Peninsular area containing beaches, meadows and timberland
Shadow Mountain	1952	Colo.	18,240	Included is Shadow Mountain Lake and Granby Reservoir

MILITARY PARKS, BATTLEFIELDS, AND CEMETERIES

Name	Location
Antietam (Battlefield and Cemetery)	Md.
Battleground (Cemetery)	D.C.
Big Hole (Battlefield)	Mont.
Brices Cross Roads (Battlefield)	Miss.
Chickamauga and Chattanooga (Battlefields)	Ga.-Tenn.
Cowpens (Battlefield)	S.C.
Fort Donelson (Park and Cemetery)	Tenn.
Fort Necessity (Battlefield)	Pa.
Fredricksburg and Spotsylvania Co. Battlefields Mem.	Va.
Fredricksburg (Cemetery)	Va.
Gettysburg (Battlefield and Cemetery)	Pa.
Guilford Courthouse (Battlefield)	N.C.
Horseshoe Bend (Battlefield)	Ala.
Kennesaw Mountain Park	Ga.
Kings Mountain (Battlefield)	S.C.
Manassas (Battlefield Park)	Va.
Moores Creek (Battlefield)	N.C.
Pea Ridge (Battlefield)	Ark.
Petersburg (Military Park)	Va.
Poplar Grove (Cemetery)	Va.
Richmond (Battlefield Park)	Va.
Shiloh (Battlefield)	Tenn.
Stones River (Battlefield and Cemetery)	Tenn
Tupelo (Battlefield)	Miss.
Vicksburg (Park and Cemetery)	Miss.
Yorktown (Cemetery)	Va.

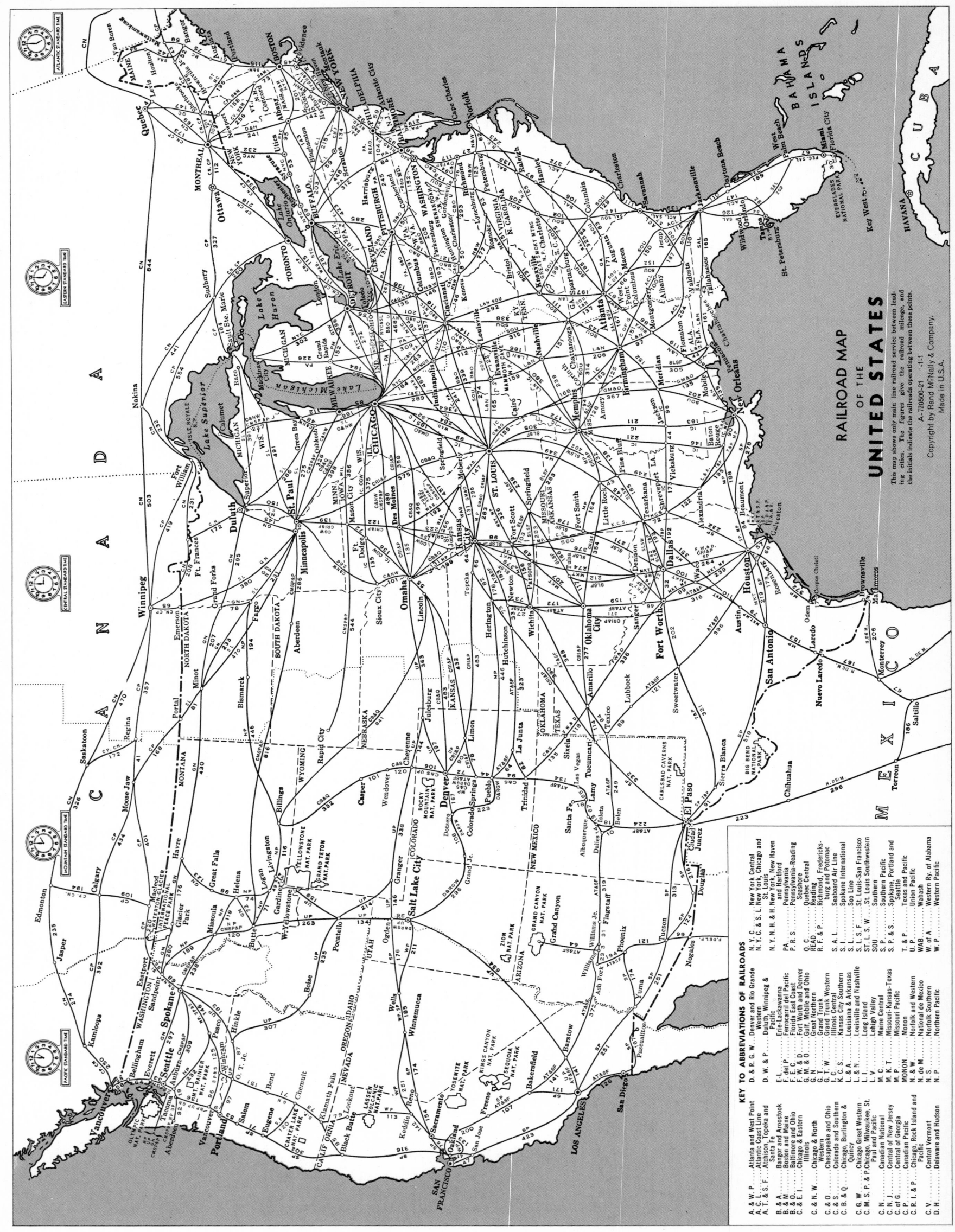

RAILROAD MAP
OF THE
UNITED STATES

This map shows only main line railroad service between leading cities. The figures give the railroad mileage, and the initials indicate the railroads operating between these points.

Copyright by Rand McNally & Company.
Made in U.S.A.
A-7205OO-21   -1:1

KEY TO ABBREVIATIONS OF RAILROADS

A. & W. P.	Atlanta and West Point	N. Y. C.	New York Central
A. C. L.	Atlantic Coast Line	N. Y. C. & S. L.	New York, Chicago and St. Louis
A. T. & S. F.	Atchison, Topeka and Santa Fe	N. Y. N. H. & H.	New York, New Haven and Hartford
B. & A.	Bangor and Aroostook	P. R. S.	Pennsylvania
B. & M.	Boston and Maine		Pennsylvania-Reading
B. & O.	Baltimore and Ohio		Seashore
C. & E. I.	Chicago and Eastern Illinois	Q. C.	Quebec Central
C. & N. W.	Chicago & North Western	READ.	Reading
		R. F. & P.	Richmond, Fredericksburg and Potomac
C. & O.	Chesapeake and Ohio	S. A. L.	Seaboard Air Line
C. & S.	Colorado and Southern	S. I.	Spokane International
C. B. & Q.	Chicago, Burlington & Quincy	S. L.	Soo Line
		St. L. S. W.	St. Louis Southwestern
C. G. W.	Chicago Great Western	SOU.	Southern
C. M. S. P. & P.	Chicago, Milwaukee, St. Paul and Pacific	S. P.	Southern Pacific
C. N.	Canadian National	S. P. & S.	Spokane, Portland and Seattle
C. N. J.	Central of New Jersey	T. & P.	Texas and Pacific
C. of G.	Central of Georgia	U. P.	Union Pacific
C. P.	Canadian Pacific	WAB.	Wabash
C. R. I. & P.	Chicago, Rock Island and Pacific	W. of A.	Western Ry. of Alabama
C. V.	Central Vermont	W. P.	Western Pacific
D. H.	Delaware and Hudson		
D. & R. G. W.	Denver and Rio Grande Western		
D. W. & P.	Duluth, Winnipeg & Pacific		
E. L.	Erie-Lackawanna		
F. del P.	Ferrocarril del Pacifico		
F. E. C.	Florida East Coast		
F. W. & D.	Fort Worth and Denver		
G. M. & O.	Gulf, Mobile and Ohio		
G. N.	Great Northern		
G. T.	Grand Trunk		
G. T. W.	Grand Trunk Western		
I. C.	Illinois Central		
K. C. S.	Kansas City Southern		
L. & A.	Louisiana & Arkansas		
L. & N.	Louisville and Nashville		
L. V.	Lehigh Valley		
M. C.	Maine Central		
M. K. T.	Missouri-Kansas-Texas		
M. P.	Missouri Pacific		
MONON	Monon		
N. & W.	Norfolk and Western		
N. de M.	National de Mexico		
N. S.	Norfolk Southern		
N. P.	Northern Pacific		

U.S. RAILROAD DISTANCE TABLE

SHOWING TRAVEL DISTANCES (SHORT LINE) BETWEEN RAILROAD CENTERS OF THE UNITED STATES IN STATUTE MILES

The table is a symmetric mileage matrix. Column headings (left to right) and row headings (top to bottom) are the same 49 cities:

Albuquerque, N. Mex.; Amarillo, Tex.; Atlanta, Ga.; Baltimore, Md.; Billings, Mont.; Birmingham, Ala.; Boston, Mass.; Buffalo, N.Y.; Butte, Mont.; Cheyenne, Wyo.; Chicago, Ill.; Cincinnati, Ohio; Cleveland, Ohio; Columbia, S.C.; Dallas, Tex.; Denver, Colo.; Des Moines, Iowa; Detroit, Mich.; Duluth, Minn.; El Paso, Tex.; Fargo, N. Dak.; Houston, Tex.; Indianapolis, Ind.; Jacksonville, Fla.; Kansas City, Mo.; Knoxville, Tenn.; Los Angeles, Calif.; Memphis, Tenn.; Miami, Fla.; Mobile, Ala.; Nashville, Tenn.; New Orleans, La.; New York, N.Y.; Oklahoma City, Okla.; Omaha, Nebr.; Philadelphia, Pa.; Pittsburgh, Pa.; Portland, Oreg.; Richmond, Va.; St. Louis, Mo.; St. Paul, Minn.; Salt Lake City, Utah; San Antonio, Tex.; San Francisco, Calif.; Seattle, Wash.; Spokane, Wash.; Tucson, Ariz.; Washington, D.C.; Wichita, Kans.

Selected legible distances (in statute miles):

From \ To	Amarillo	Atlanta	Baltimore	Billings	Birmingham	Dallas	Denver	El Paso	Los Angeles	Wichita
Albuquerque, N. Mex.	374	1554	2102	1133	1388	723	477	253	889	721
Amarillo, Tex.	—	1181	1728	1221	1014					347
Atlanta, Ga.		—	676		167					986
Baltimore, Md.			—							1401
Billings, Mont.				—						1236

U.S. AIR DISTANCE TABLE

Showing Great Circle Distances Between Principal Cities of the United States in Statute Miles

	Albuquerque, N. Mex.	Amarillo, Tex.	Atlanta, Ga.	Billings, Mont.	Birmingham, Ala.	Boston, Mass.	Buffalo, N.Y.	Burlington, Vt.	Charleston, S.C.	Charlotte, N.C.	Cheyenne, Wyo.	Chicago, Ill.	Cincinnati, Ohio	Cleveland, Ohio	Dallas, Tex.	Denver, Colo.	Des Moines, Iowa	Detroit, Mich.	El Paso, Tex.	Fargo, N. Dak.	Houston, Tex.	Indianapolis, Ind.	Jacksonville, Fla.	Kansas City, Mo.	Knoxville, Tenn.
Albuquerque, N. Mex.		273	1272	744	1138	1972	1580	1878	1539	1457	429	1129	1251	1421	588	334	837	1364	229	961	754	1169	1488	720	1280
Amarillo, Tex.	273		999	809	866	1722	1338	1640	1266	1185	440	894	992	1173	334	358	626	1124	358	847	533	915	1219	481	1009
Atlanta, Ga.	1272	999		1519	140	937	697	951	267	227	1229	587	369	554	721	1212	739	596	1291	1114	701	426	285	676	155
Billings, Mont.	744	809	1519		1425	1861	1473	1713	1761	1617	370	1073	1304	1369	1092	453	798	1283	973	565	1315	1204	1796	846	1447
Birmingham, Ala.	1138	866	140	1425		1052	776	1049	402	361	1119	578	406	618	581	1095	670	641	1152	1060	567	433	374	579	235
Boston, Mass.	1972	1722	937	1861	1052		400	182	820	721	1735	851	740	551	1551	1769	1159	613	2072	1300	1605	807	1017	1251	818
Buffalo, N.Y.	1580	1338	697	1473	776	400		304	699	538	1335	454	393	173	1198	1370	760	216	1692	919	1286	435	879	861	548
Burlington, Vt.	1878	1640	951	1713	1049	182	304		884	755	1612	749	690	476	1501	1654	1049	516	1995	1149	1580	739	1079	1161	815
Charleston, S.C.	1539	1266	267	1761	402	820	699	884		177	1486	757	506	609	981	1474	967	681	1552	1317	936	594	197	928	316
Charlotte, N.C.	1457	1185	227	1617	361	721	538	755	177		1362	587	335	435	930	1358	819	504	1496	1153	927	428	341	803	180
Cheyenne, Wyo.	429	440	1229	370	1119	1735	1335	1612	1486	1362		891	1082	1199	726	96	583	1125	653	563	947	986	1493	560	1183
Chicago, Ill.	1129	894	587	1073	578	851	454	749	757	587	891		252	308	803	920	309	238	1252	569	940	165	863	414	454
Cincinnati, Ohio	1251	992	369	1304	406	740	393	690	506	335	1082	252		222	814	1094	510	235	1335	820	892	100	626	541	219
Cleveland, Ohio	1421	1173	554	1369	618	551	173	476	609	435	1199	308	222		1025	1227	617	90	1525	835	1114	263	770	700	400
Dallas, Tex.	588	334	721	1092	581	1551	1198	1501	981	930	726	803	814	1025		663	632	999	572	972	225	763	908	451	767
Denver, Colo.	334	358	1212	453	1095	1769	1370	1654	1474	1358	96	920	1094	1227	663		610	1156	557	642	879	1000	1467	558	1178
Des Moines, Iowa	837	626	739	798	670	1159	760	1049	967	819	583	309	510	617	632	610		546	983	397	821	411	1023	180	651
Detroit, Mich.	1364	1124	596	1283	641	613	216	516	681	504	1125	238	235	90	999	1156	546		1479	745	1105	240	831	645	442
El Paso, Tex.	229	358	1291	973	1152	2072	1692	1995	1552	1496	653	1252	1335	1525	572	557	983	1479		1163	676	1264	1473	839	1326
Fargo, N. Dak.	961	847	1114	565	1060	1300	919	1149	1317	1153	563	569	820	835	972	642	397	745	1163		1183	725	1399	549	1004
Houston, Tex.	754	533	701	1315	567	1605	1286	1580	936	927	947	940	892	1114	225	879	821	1105	676	1183		865	821	644	790
Indianapolis, Ind.	1169	915	426	1204	433	807	435	739	594	428	986	165	100	263	763	1000	411	240	1264	725	865		699	453	290
Jacksonville, Fla.	1488	1219	285	1796	374	1017	879	1079	197	341	1493	863	626	770	908	1467	1023	831	1473	1399	821	699		950	410
Kansas City, Mo.	720	481	676	846	579	1251	861	1161	928	803	560	414	541	700	451	558	180	645	839	549	644	453	950		624
Knoxville, Tenn.	1280	1009	155	1447	235	818	548	815	316	180	1183	454	219	400	767	1178	651	442	1326	1004	790	290	410	624	
Little Rock, Ark.	816	543	456	1143	325	1259	913	1214	723	649	813	552	524	740	293	780	478	723	847	869	388	483	690	325	479
Los Angeles, Calif.	664	937	1936	959	1802	2596	2198	2485	2203	2119	882	1745	1897	2049	1240	831	1438	1983	701	1427	1374	1809	2147	1356	1941
Louisville, Ky.	1178	915	319	1275	331	826	483	780	500	343	1033	269	90	311	726	1038	476	316	1254	818	803	107	594	480	188
Memphis, Tenn.	939	667	337	1213	217	1137	803	1100	604	521	902	482	410	630	420	879	485	623	976	882	484	384	590	369	350
Miami, Fla.	1698	1441	604	2085	665	1255	1181	1347	482	652	1763	1188	952	1087	1111	1726	1333	1152	1643	1716	968	1024	326	1241	736
Minneapolis, Minn.	983	812	907	742	862	1123	731	985	1104	939	642	355	605	630	862	700	235	543	1157	214	1056	511	1191	413	792
Nashville, Tenn.	1119	848	214	1309	182	943	627	916	455	340	1032	397	238	459	617	1023	525	470	1169	902	665	251	499	473	161
New Orleans, La.	1029	776	424	1479	312	1359	1086	1361	630	649	1131	833	706	924	443	1082	827	939	983	1222	318	712	504	680	547
New York, N.Y.	1815	1560	748	1760	864	188	292	260	641	533	1604	713	570	405	1374	1631	1022	482	1905	1210	1420	646	838	1097	632
Omaha, Nebr.	721	526	817	703	732	1282	883	1171	1058	918	463	432	622	739	586	488	123	669	878	390	794	525	1098	166	745
Philadelphia, Pa.	1753	1494	666	1727	783	271	279	328	562	451	1556	666	503	360	1299	1579	973	443	1836	1184	1341	585	758	1038	552
Phoenix, Ariz.	330	598	1592	872	1456	2300	1906	2202	1857	1783	663	1453	1581	1749	887	586	1155	1690	346	1225	1017	1499	1794	1049	1607
Pittsburgh, Pa.	1499	1244	521	1479	608	483	178	445	528	362	1298	410	257	115	1070	1320	715	205	1590	949	1137	330	703	781	375
Portland, Oreg.	1107	1304	2172	686	2066	2540	2156	2385	2425	2290	947	1758	1985	2055	1633	982	1475	1969	1286	1239	1836	1885	2439	1497	2115
Raleigh, N.C.	1576	1306	356	1698	491	609	490	665	220	130	1461	642	396	428	1057	1463	902	510	1621	1210	1056	495	414	905	296
St. Louis, Mo.	942	685	467	1057	400	1038	662	966	704	568	795	262	309	492	547	796	273	455	1034	660	679	231	751	238	392
Salt Lake City, Utah	484	668	1583	387	1466	2099	1699	1969	1845	1727	371	1260	1453	1568	999	371	953	1492	689	863	1200	1356	1837	925	1547
San Antonio, Tex.	617	444	882	1252	744	1766	1430	1729	1122	1105	882	1051	1039	1256	252	802	882	1238	503	1207	189	999	1011	702	959
San Francisco, Calif.	896	1157	2139	904	2013	2699	2300	2568	2405	2301	967	1858	2043	2166	1483	949	1550	2091	995	1446	1645	1949	2374	1506	2121
Seattle, Wash.	1184	1359	2182	668	2082	2493	2117	2333	2428	2285	973	1737	1972	2026	1681	1021	1467	1938	1376	1197	1891	1872	2455	1506	2114
Spokane, Wash.	1030	1176	1961	443	1865	2266	1888	2108	2204	2059	768	1508	1744	1796	1489	826	1240	1709	1239	969	1704	1644	2237	1287	1890
Syracuse, N.Y.	1718	1475	781	1600	875	264	138	177	738	595	1472	592	514	303	1326	1508	898	354	1828	1042	1403	567	928	998	641
Tulsa, Okla.	604	335	678	930	552	1398	1023	1327	945	853	588	598	661	853	236	550	396	813	674	741	442	591	921	216	676
Washington, D.C.	1653	1391	543	1669	661	393	292	432	453	330	1477	597	404	306	1185	1494	896	396	1728	1140	1220	494	647	945	430
Wichita, Kans.	549	304	776	801	658	1424	1036	1337	1039	933	465	591	702	873	340	437	334	821	661	634	559	620	1031	177	753

	Little Rock, Ark.	Los Angeles, Calif.	Louisville, Ky.	Memphis, Tenn.	Miami, Fla.	Minneapolis, Minn.	Nashville, Tenn.	New Orleans, La.	New York, N.Y.	Omaha, Nebr.	Philadelphia, Pa.	Phoenix, Ariz.	Pittsburgh, Pa.	Portland, Oreg.	Raleigh, N.C.	St. Louis, Mo.	Salt Lake City, Utah	San Antonio, Tex.	San Francisco, Calif.	Seattle, Wash.	Spokane, Wash.	Syracuse, N.Y.	Tulsa, Okla.	Washington, D.C.	Wichita, Kans.
Albuquerque, N. Mex.	816	664	1178	939	1698	983	1119	1029	1815	721	1753	330	1499	1107	1576	942	484	617	896	1184	1030	1718	604	1653	549
Amarillo, Tex.	543	937	915	667	1441	812	848	776	1560	526	1494	598	1244	1304	1306	685	668	444	1157	1359	1176	1475	335	1391	304
Atlanta, Ga.	456	1936	319	337	604	907	214	424	748	817	666	1592	521	2172	356	467	1583	882	2139	2182	1961	781	678	543	776
Billings, Mont.	1143	959	1275	1213	2085	742	1309	1479	1760	703	1727	872	1479	686	1698	1057	387	1252	904	668	443	1600	930	1669	801
Birmingham, Ala.	325	1802	331	217	665	862	182	312	864	732	783	1456	608	2066	491	400	1466	744	2013	2082	1865	875	552	661	658
Boston, Mass.	1259	2596	826	1137	1255	1123	943	1359	188	1282	271	2300	483	2540	609	1038	2099	1766	2699	2493	2266	264	1398	393	1424
Buffalo, N.Y.	913	2198	483	803	1181	731	627	1086	292	883	279	1906	178	2156	490	662	1699	1430	2300	2117	1888	138	1023	292	1036
Burlington, Vt.	1214	2485	780	1100	1347	985	916	1361	260	1171	328	2202	445	2385	665	966	1969	1729	2568	2333	2108	177	1327	432	1337
Charleston, S.C.	723	2203	500	604	482	1104	455	630	641	1058	562	1857	528	2425	220	704	1845	1122	2405	2428	2204	738	945	453	1039
Charlotte, N.C.	649	2119	343	521	652	939	340	649	533	918	451	1783	362	2290	130	568	1727	1105	2301	2285	2059	595	853	330	933
Cheyenne, Wyo.	813	882	1033	902	1763	642	1032	1131	1604	463	1556	663	1298	947	1461	795	371	882	967	973	768	1472	588	1477	465
Chicago, Ill.	552	1745	269	482	1188	355	397	833	713	432	666	1453	410	1758	642	262	1260	1051	1858	1737	1508	592	598	597	591
Cincinnati, Ohio	524	1897	90	410	952	605	238	706	570	622	503	1581	257	1985	396	309	1453	1039	2043	1972	1744	514	661	404	702
Cleveland, Ohio	740	2049	311	630	1087	630	459	924	405	739	360	1749	115	2055	428	492	1568	1256	2166	2026	1796	303	853	306	873
Dallas, Tex.	293	1240	726	420	1111	862	617	443	1374	586	1299	887	1070	1633	1057	547	999	252	1483	1681	1489	1326	236	1185	340
Denver, Colo.	780	831	1038	879	1726	700	1023	1082	1631	488	1579	586	1320	982	1463	796	371	802	949	1021	826	1508	550	1494	437
Des Moines, Iowa	478	1438	476	485	1333	235	525	827	1022	123	973	1155	715	1475	902	273	953	882	1550	1467	1240	898	396	896	334
Detroit, Mich.	723	1983	316	623	1152	543	470	939	482	669	443	1690	205	1969	510	455	1492	1238	2091	1938	1709	354	813	396	821
El Paso, Tex.	847	701	1254	976	1643	1157	1169	983	1905	878	1836	346	1590	1286	1621	1034	689	503	995	1376	1239	1828	674	1728	661
Fargo, N. Dak.	869	1427	818	882	1716	214	902	1222	1210	390	1184	1225	949	1239	1210	660	863	1207	1446	1197	969	1042	741	1140	634
Houston, Tex.	388	1374	803	484	968	1056	665	318	1420	794	1341	1017	1137	1836	1056	679	1200	189	1645	1891	1704	1403	442	1220	559
Indianapolis, Ind.	483	1809	107	384	1024	511	251	712	646	525	585	1499	330	1885	495	231	1356	999	1949	1872	1644	567	591	494	620
Jacksonville, Fla.	690	2147	594	590	326	1191	499	504	838	1098	758	1794	703	2439	414	751	1837	1011	2374	2455	2237	928	921	647	1031
Kansas City, Mo.	325	1356	480	369	1241	413	473	680	1097	166	1038	1049	781	1497	905	238	925	702	1506	1506	1287	998	216	945	177
Knoxville, Tenn.	479	1941	188	350	736	792	161	547	632	745	552	1607	375	2115	296	392	1547	959	2121	2114	1890	641	676	430	753
Little Rock, Ark.		1480	435	129	949	708	325	355	1081	492	1007	1137	779	1759	774	291	1148	516	1688	1785	1573	1038	231	892	348
Los Angeles, Calif.	1480		1829	1603	2339	1524	1780	1673	2451	1315	2394	357	2136	825	2237	1589	579	1204	347	959	940	2336	1266	2300	1197
Louisville, Ky.	435	1829		320	919	605	154	623	652	580	582	1508	344	1950	429	242	1402	949	1986	1943	1717	603	582	476	633
Memphis, Tenn.	129	1603	320		872	699	197	358	957	529	881	1263	660	1849	645	240	1250	631	1802	1867	1650	923	341	765	442.
Miami, Fla.	949	2339	919	872		1511	815	669	1092	1397	1019	1982	1010	2708	695	1061	2089	1148	2594	2734	2520	1212	1176	923	1297
Minneapolis, Minn.	708	1524	605	699	1511		697	1051	1018	290	985	1280	743	1427	996	466	987	1110	1584	1395	1166	861	626	934	546
Nashville, Tenn.	325	1780	154	197	815	697		469	761	607	685	1446	472	1969	457	254	1393	823	1963	1975	1752	739	515	569	594
New Orleans, La.	355	1673	623	358	669	1051	469		1171	847	1089	1316	919	2063	776	598	1434	507	1926	2101	1898	1187	548	966	677
New York, N.Y.	1081	2451	652	957	1092	1018	761	1171		1144	83	2145	317	2445	426	875	1972	1584	2571	2408	2179	194	1231	205	1266
Omaha, Nebr.	492	1315	580	529	1397	290	607	847	1144		1094	1036	836	1371	1008	354	833	828	1429	1369	1146	1021	352	1014	257
Philadelphia, Pa.	1007	2394	582	881	1019	985	685	1089	83	1094		2083	259	2412	345	811	1925	1507	2523	2380	2151	220	1163	123	1204
Phoenix, Ariz.	1137	357	1508	1263	1982	1280	1446	1316	2145	1036	2083		1828	1005	1903	1272	504	849	653	1114	1019	2044	932	1983	879
Pittsburgh, Pa.	779	2136	344	660	1010	743	472	919	317	836	259	1828		2165	330	559	1668	1291	2264	2138	1908	268	917	192	950
Portland, Oreg.	1759	825	1950	1849	2708	1427	1969	2063	2445	1371	2412	1005	2165		2377	1723	636	1720	534	145	290	2281	1531	2354	1411
Raleigh, N.C.	774	2237	429	645	695	996	457	776	426	1008	345	1903	330	2377		667	1829	1235	2410	2367	2139	519	972	233	1044
St. Louis, Mo.	291	1589	242	240	1061	466	254	598	875	354	811	1272	559	1723	667		1162	792	1744	1724	1500	796	361	712	394
Salt Lake City, Utah	1148	579	1402	1250	2089	987	1393	1434	1972	833	1925	504	1668	636	1829	1162		1087	600	701	550	1835	917	1848	808
San Antonio, Tex.	516	1204	949	631	1148	1110	823	507	1584	828	1507	849	1291	1720	1235	792	1087		1490	1787	1614	1553	486	1388	573
San Francisco, Calif.	1688	347	1986	1802	2594	1584	1963	1926	2571	1429	2523	653	2264	534	2410	1744	600	1490		678	727	2435	1461	2442	1369
Seattle, Wash.	1785	959	1943	1867	2734	1395	1975	2101	2408	1369	2380	1114	2138	145	2367	1724	701	1787	678		229	2238	1560	2329	1437
Spokane, Wash.	1573	940	1717	1650	2520	1166	1752	1898	2179	1146	2151	1019	1908	290	2139	1500	550	1614	727	229		2010	1353	2100	1227
Syracuse, N.Y.	1038	2336	603	923	1212	861	739	1187	194	1021	220	2044	268	2281	519	796	1835	1553	2435	2238	2010		1157	290	1173
Tulsa, Okla.	231	1266	582	341	1176	626	515	548	1231	352	1163	932	917	1531	972	361	917	486	1461	1560	1353	1157		1058	130
Washington, D.C.	892	2300	476	765	923	934	569	966	205	1014	123	1983	192	2354	233	712	1848	1388	2442	2329	2100	290	1058		1106
Wichita, Kans.	348	1197	633	442	1297	546	594	677	1266	257	1204	879	950	1411	1044	394	808	573	1369	1437	1227	1173	130	1106	

U.S. STATE GENERAL INFORMATION TABLE

STATE	CAPITAL	LARGEST CITY	Date of Entry	Rank of Entry	Greatest N-S Measurement (miles)	Greatest E-W Measurement (miles)	Highest Point Location	Altitude (feet)	STATE FLOWER	STATE BIRD	STATE NICKNAME
Alabama	Montgomery	Birmingham	Dec. 14, 1819	22	330	200	Cheaha Mountain	2,407	Camellia	Yellowhammer	Cotton
Alaska	Juneau	Anchorage	Jan. 3, 1959	49	1,332	2,250	Mt. McKinley	20,320	Forget-me-not	Willow Ptarmigan	Last Frontier
Arizona	Phoenix	Phoenix	Feb. 14, 1912	48	390	335	Humphreys Peak	12,670	Saguaro Cactus	Cactus Wren	Grand Canyon
Arkansas	Little Rock	Little Rock	June 15, 1836	25	240	275	Magazine Mtn.	2,823	Apple Blossom	Mockingbird	Land of Opportunity
California	Sacramento	Los Angeles	Sept. 9, 1850	31	800	375	Mt. Whitney	14,495	Golden Poppy	California Valley Quail	Golden
Colorado	Denver	Denver	Aug. 1, 1876	38	270	380	Mt. Elbert	14,431	Rocky Mountain Columbine	Lark Bunting	Centennial
Connecticut*	Hartford	Hartford	Jan. 9, 1788	5	75	90	S. slope of Mt. Frissell	2,380	Mountain Laurel	Robin	Constitution
Delaware*	Dover	Wilmington	Dec. 7, 1787	1	95	35	Ebright Road, New Castle Co.	442	Peach Blossom	Blue Hen Chicken	Diamond
District of Columbia†	Washington	Washington	March 3, 1791	..	15	15	Tenleytown	410	American Beauty Rose	....	....
Florida	Tallahassee	Miami	March 3, 1845	27	460	400	N. boundary, Walton Co.	345	Orange Blossom	Mockingbird	Sunshine
Georgia*	Atlanta	Atlanta	Jan. 2, 1788	4	315	250	Brasstown Bald (mtn.)	4,784	Cherokee Rose	Brown Thrasher	Peach State
Hawaii	Honolulu	Honolulu	Aug. 21, 1959	50	...	1,600	Mauna Kea	13,796	Red Hibiscus	Nene (Hawaiian Goose)	The Aloha
Idaho	Boise	Boise	July 3, 1890	43	480	305	Borah Peak	12,662	Syringa	Mountain Bluebird	Gem
Illinois	Springfield	Chicago	Dec. 3, 1818	21	380	205	Charles Mound	1,241	Native Violet	Cardinal	Prairie
Indiana	Indianapolis	Indianapolis	Dec. 11, 1816	19	265	160	Near Spartanburg	1,253	Peony	Cardinal	Hoosier
Iowa	Des Moines	Des Moines	Dec. 28, 1846	29	205	310	Ocheyedan Mound	1,675	Wild Rose	Eastern Goldfinch	Hawkeye
Kansas	Topeka	Wichita	Jan. 29, 1861	34	205	410	Mt. Sunflower	4,026	Sunflower	Western Meadowlark	Sunflower
Kentucky	Frankfort	Louisville	June 1, 1792	15	175	350	Black Mountain	4,145	Goldenrod	Kentucky Cardinal	Bluegrass
Louisiana	Baton Rouge	New Orleans	April 30, 1812	18	275	300	Driskill Mountain	535	Magnolia	Eastern Brown Pelican**	Pelican
Maine	Augusta	Portland	March 15, 1820	23	310	210	Mt. Katahdin	5,268	White Pine Cone and Tassel	Chickadee	Pine Tree
Maryland*	Annapolis	Baltimore	April 28, 1788	7	120	200	Backbone Mountain	3,360	Black-eyed Susan	Baltimore Oriole	Old Line
Massachusetts*	Boston	Boston	Feb. 6, 1788	6	110	190	Mt. Greylock	3,491	Mayflower	Chickadee	Bay
Michigan	Lansing	Detroit	Jan. 26, 1837	26	400	310	N.E. Baraga Co.	1,980	Apple Blossom	Robin	Wolverine
Minnesota	St. Paul	Minneapolis	May 11, 1858	32	400	350	Misquah Hills	2,023	Showy Lady's-slipper	Loon	Gopher
Mississippi	Jackson	Jackson	Dec. 10, 1817	20	340	180	Woodall Mountain	806	Magnolia	Mockingbird	Magnolia
Missouri	Jefferson City	St. Louis	Aug. 10, 1821	24	280	300	Taum Sauk Mountain	1,772	Hawthorne	Bluebird	Show Me
Montana	Helena	Great Falls	Nov. 8, 1889	41	315	570	Granite Peak	12,799	Bitterroot	Western Meadowlark	Treasure
Nebraska	Lincoln	Omaha	March 1, 1867	37	210	415	S.W. corner Kimball Co.	5,424	Goldenrod	Western Meadowlark	Cornhusker
Nevada	Carson City	Las Vegas	Oct. 31, 1864	36	485	315	Boundary Peak	13,145	Sagebrush	Mountain Bluebird**	Silver
New Hampshire*	Concord	Manchester	June 21, 1788	9	185	90	Mt. Washington	6,288	Purple Lilac	Purple Finch	Granite
New Jersey*	Trenton	Newark	Dec. 18, 1787	3	166	70	High Point	1,803	Purple Violet	Eastern Goldfinch	Garden
New Mexico	Santa Fe	Albuquerque	Jan. 6, 1912	47	390	350	Wheeler Peak	13,160	Yucca	Road Runner	Land of Enchantment
New York*	Albany	New York	July 26, 1788	11	310	330	Mt. Marcy	5,344	Rose	Bluebird	Empire
North Carolina*	Raleigh	Charlotte	Nov. 21, 1789	12	200	520	Mt. Mitchell	6,684	Dogwood	Cardinal	Tar Heel
North Dakota	Bismarck	Fargo	Nov. 2, 1889	39	210	360	White Butte	3,530	Wild Prairie Rose	Western Meadowlark	Flickertail
Ohio	Columbus	Cleveland	March 1, 1803	17	230	205	Campbell Hill	1,550	Scarlet Carnation	Cardinal	Buckeye
Oklahoma	Oklahoma City	Oklahoma City	Nov. 16, 1907	46	210	460	Black Mesa	4,978	Mistletoe	Scissor-tailed Flycatcher	Sooner
Oregon	Salem	Portland	Feb. 14, 1859	33	290	375	Mt. Hood	11,245	Oregon Grape	Western Meadowlark	Beaver
Pennsylvania*	Harrisburg	Philadelphia	Dec. 12, 1787	2	180	310	Mt. Davis	3,213	Mountain Laurel	Ruffed Grouse	Keystone
Rhode Island*	Providence	Providence	May 29, 1790	13	50	35	Jerimoth Hill	812	Violet**	Rhode Island Red	Little Rhody
South Carolina*	Columbia	Columbia	May 23, 1788	8	215	285	Sassafras Mountain	3,560	Yellow Jessamine	Carolina Wren	Palmetto
South Dakota	Pierre	Sioux Falls	Nov. 2, 1889	40	240	360	Harney Peak	7,242	American Pasque Flower	Ringnecked Pheasant	Coyote
Tennessee	Nashville	Memphis	June 1, 1796	16	120	430	Clingmans Dome	6,642	Iris	Mockingbird	Volunteer
Texas	Austin	Houston	Dec. 29, 1845	28	710	760	Guadalupe Peak	8,751	Bluebonnet	Mockingbird	Lone Star
Utah	Salt Lake City	Salt Lake City	Jan. 4, 1896	45	345	275	Kings Peak	13,498	Sego Lily	Sea Gull	Beehive
Vermont	Montpelier	Burlington	March 4, 1791	14	155	90	Mt. Mansfield	4,393	Red Clover	Hermit Thrush	Green Mountain
Virginia*	Richmond	Norfolk	June 25, 1788	10	205	425	Mt. Rogers	5,720	American Dogwood	Cardinal	Old Dominion
Washington	Olympia	Seattle	Nov. 11, 1889	42	230	340	Mt. Rainier	14,410	Western Rhododendron	Willow Goldfinch	Evergreen
West Virginia	Charleston	Charleston	June 20, 1863	35	200	225	Spruce Knob	4,860	Big Rhododendron	Cardinal	Mountain
Wisconsin	Madison	Milwaukee	May 29, 1848	30	300	290	Rib Mountain	1,941	Wood Violet	Robin	Badger
Wyoming	Cheyenne	Cheyenne	July 10, 1890	44	275	365	Gannett Peak	13,785	Indian Paint Brush	Meadowlark	Equality
United States	Washington, D.C.	New York	....	..		...	Mt. McKinley, Alaska	20,320		Bald Eagle	

*One of the Thirteen Original States. †District. **Unofficial.

197

U.S. POPULATION BY STATE OR COLONY
1650 to 1960

STATES	1650	1700	1750	1770	1790	1800	1820	1840	1860	1880	1900	1910	1920	1930	1940	1950	1960
Alabama							127,901	590,756	964,201	1,262,505	1,828,697	2,138,093	2,348,174	2,646,248	2,832,961	3,061,743	3,266,740
Alaska										33,426	63,592	64,356	55,036	59,278	72,524	128,643	226,167
Arizona										40,440	122,931	204,354	334,162	435,573	499,261	749,587	1,302,161
Arkansas							14,273	97,574	435,450	802,525	1,311,564	1,574,449	1,752,204	1,854,482	1,949,387	1,909,511	1,786,272
California									379,994	864,694	1,485,053	2,377,549	3,426,861	5,677,251	6,907,387	10,586,223	15,717,204
Colorado	4,139								34,277	194,327	539,700	799,024	939,629	1,035,791	1,123,296	1,325,089	1,753,947
Connecticut	185	25,970	111,280	183,881	237,946	251,002	275,248	309,978	460,147	622,700	908,420	1,114,756	1,380,631	1,606,903	1,709,242	2,007,280	2,535,234
Delaware		2,470	28,704	35,496	59,096	64,273	72,749	78,085	112,216	146,608	184,735	202,322	223,003	238,380	266,505	318,085	446,292
District of Columbia						8,144	23,336	33,745	75,080	177,624	278,718	331,069	437,571	486,869	663,091	802,178	763,956
Florida								54,477	140,424	269,493	528,542	752,619	968,470	1,468,211	1,897,414	2,771,305	4,951,560
Georgia			5,200	23,375	82,548	162,686	340,989	691,392	1,057,286	1,542,180	2,216,331	2,609,121	2,895,832	2,908,506	3,123,723	3,444,578	3,943,116
Hawaii											154,001	191,874	255,881	368,300	422,770	499,794	632,772
Idaho										32,610	161,772	325,594	431,866	445,032	524,873	588,637	667,191
Illinois							55,211	476,183	1,711,951	3,077,871	4,821,550	5,638,591	6,485,280	7,630,654	7,897,241	8,712,176	10,081,158
Indiana						5,641	147,178	685,866	1,350,428	1,978,301	2,516,462	2,700,876	2,930,390	3,238,503	3,427,796	3,934,224	4,662,498
Iowa								43,112	674,913	1,624,615	2,231,853	2,224,771	2,404,021	2,470,939	2,538,268	2,621,073	2,757,537
Kansas									107,206	996,096	1,470,495	1,690,949	1,769,257	1,880,999	1,801,028	1,905,299	2,178,611
Kentucky				15,700	73,677	220,955	564,317	779,828	1,155,684	1,648,690	2,147,174	2,289,905	2,416,630	2,614,589	2,845,627	2,944,806	3,038,156
Louisiana							153,407	352,411	708,002	939,946	1,381,625	1,656,388	1,798,509	2,101,593	2,363,880	2,683,516	3,257,022
Maine[3]				31,257	96,540	151,719	298,335	501,793	628,279	648,936	694,466	742,371	768,014	797,423	847,226	913,774	969,265
Maryland	4,504	29,604	141,073	202,599	319,728	341,548	407,350	470,019	687,049	934,943	1,188,044	1,295,346	1,449,661	1,631,526	1,821,244	2,343,001	3,100,689
Massachusetts[3]	16,603	55,941	188,000	235,308	378,787	422,845	523,287	737,699	1,231,066	1,783,085	2,805,346	3,366,416	3,852,356	4,249,614	4,316,721	4,690,514	5,148,578
Michigan							8,896	212,267	749,113	1,636,937	2,420,982	2,810,173	3,668,412	4,842,325	5,256,106	6,371,766	7,823,194
Minnesota									172,023	780,773	1,751,394	2,075,708	2,387,125	2,563,953	2,792,300	2,982,483	3,413,864
Mississippi						8,850	75,448	375,651	791,305	1,131,597	1,551,270	1,797,114	1,790,618	2,009,821	2,183,796	2,178,914	2,178,141
Missouri							66,586	383,702	1,182,012	2,168,380	3,106,665	3,293,335	3,404,055	3,629,367	3,784,664	3,954,653	4,319,813
Montana										39,159	243,329	376,053	548,889	537,606	559,456	591,024	674,767
Nebraska									28,841	452,402	1,066,300	1,192,214	1,296,372	1,377,963	1,315,834	1,325,510	1,411,330
Nevada									6,857	62,266	42,335	81,875	77,407	91,058	110,247	160,083	285,278
New Hampshire	1,305	4,958	27,505	62,396	141,885	183,858	244,161	284,574	326,073	346,991	411,588	430,572	443,083	465,293	491,524	533,242	606,921
New Jersey		14,010	71,393	117,431	184,139	211,149	277,575	373,306	672,035	1,131,116	1,883,669	2,537,167	3,155,900	4,041,334	4,160,165	4,835,329	6,066,782
New Mexico									93,516	119,565	195,310	327,301	360,350	423,317	531,818	681,187	951,023
New York[4]	4,116	19,107	76,696	162,920	340,120	589,051	1,372,812	2,428,921	3,880,735	5,082,871	7,268,894	9,113,614	10,385,227	12,588,066	13,479,142	14,830,192	16,782,304
North Carolina		10,720	72,984	197,200	393,751	478,103	638,829	753,419	992,622	1,399,750	1,893,810	2,206,287	2,559,123	3,170,276	3,571,623	4,061,929	4,556,155
North Dakota[4]										36,909	319,146	577,056	646,872	680,845	641,935	619,636	632,446
Ohio						45,365	581,434	1,519,467	2,339,511	3,198,062	4,157,545	4,767,121	5,759,394	6,646,697	6,907,612	7,946,627	9,706,397
Oklahoma[5]											790,391	1,657,155	2,028,283	2,396,040	2,336,434	2,233,351	2,328,284
Oregon[6]									52,465	174,768	413,536	672,765	783,389	953,786	1,089,684	1,521,341	1,768,687
Pennsylvania		17,950	119,666	240,057	434,373	602,365	1,049,458	1,724,033	2,906,215	4,282,891	6,302,115	7,665,111	8,720,017	9,631,350	9,900,180	10,498,012	11,319,366
Rhode Island	785	5,894	33,226	58,196	68,825	69,122	83,059	108,830	174,620	276,531	428,556	542,610	604,397	687,497	713,346	791,896	859,488
South Carolina		5,704	64,000	124,244	249,073	345,591	502,741	594,398	703,708	995,577	1,340,316	1,515,400	1,683,724	1,738,765	1,899,804	2,117,027	2,382,594
South Dakota[4]									4,837	98,268	401,570	583,888	636,547	692,849	642,961	652,740	680,514
Tennessee				1,000	35,691	105,602	422,823	829,210	1,109,801	1,542,359	2,020,616	2,184,789	2,337,885	2,616,556	2,915,841	3,291,718	3,567,089
Utah									40,273	143,963	276,749	373,351	449,396	507,847	550,310	688,862	890,627
Vermont[6]				10,000	85,425	154,465	235,981	291,948	315,098	332,286	343,641	355,956	352,428	359,611	359,231	377,747	389,881
Virginia[6]	18,731	58,560	231,033	447,016	691,737	807,557	938,261	1,025,227	1,219,630	1,512,565	1,854,184	2,061,612	2,309,187	2,421,851	2,677,773	3,318,680	3,966,949
Washington[6]									11,594	75,116	518,103	1,141,990	1,356,621	1,563,396	1,736,191	2,378,963	2,853,214
West Virginia[6]					55,873	78,592	136,808	224,537	376,688	618,457	958,800	1,221,119	1,463,701	1,729,205	1,901,974	2,005,552	1,860,421
Wisconsin								30,945	775,881	1,315,497	2,069,042	2,333,860	2,632,067	2,939,006	3,137,587	3,434,575	3,951,777
Wyoming										20,789	92,531	145,965	194,402	225,565	250,742	290,529	330,066
Total[1]	50,368	250,888	1,170,760	2,148,076	3,929,214	5,308,483	9,638,453	17,069,453[2]	31,443,321	50,189,209	76,212,168	92,228,496	106,021,537	123,202,624	132,164,569	151,325,798	179,323,175

[1] All figures exclude uncivilized Indians. Figures for 1650 through 1770 include only the British colonies that later became the United States. No areas are included prior to their annexation to the United States. However, many of the figures refer to territories prior to their admission as States. U.S. total includes Alaska from 1880 through 1960 and Hawaii from 1900 through 1960.

[2] U.S. total for 1840 includes 6,100 persons on public ships in service of the United States, not credited to any State.

[3] Maine figures for 1770 through 1800 are for that area of Massachusetts which became the State of Maine in 1820. Massachusetts figures exclude Maine from 1770 through 1800, but include it from 1650 through 1750. Massachusetts figure for 1650 also includes Plymouth, a separate colony until 1691.

[4] South Dakota figure for 1860 represents entire Dakota Territory. North and South Dakota figures for 1880 are for the parts of Dakota Territory which later constituted the respective States.

[5] Oklahoma figure for 1900 includes population of Indian Territory (392,060).

[6] West Virginia figures for 1790 through 1860 are for that area of Virginia which became West Virginia in 1863. These figures are excluded from the figures for Virginia from 1790 through 1860.

U.S. METRO. AREAS OF 100,000 OR MORE, 1962

To facilitate comparisons, Rand McNally Metro. Areas are defined according to consistent rules, so as to include with each city those neighboring communities linked to it by continuous built-up areas, and more distant communities if the bulk of their population is supported by commuters to the central city.

Rank 1962		Metro. Area	Central City	Suburbs
1	New York, N.Y.	15,775,000	**7,775,000	7,456,000
2	Los Angeles, Calif.	6,955,000	2,565,000	4,390,000
3	Chicago, Ill.	6,735,000	3,540,000	3,195,000
4	Philadelphia, Pa.	4,100,000	1,995,000	2,105,000
5	Detroit, Mich.	3,915,000	1,645,000	2,270,000
	incl. part in Canada	4,110,000		
6	San Francisco–Oakland–San Jose, Calif.	3,500,000	745,000 / 365,000 / 245,000	2,145,000
7	Boston, Mass.*	3,000,000	**680,000	2,244,000
8	Cleveland, Ohio	2,150,000	870,000	1,280,000
9	Washington, D.C.*	2,140,000	755,000	1,385,000
10	St. Louis, Mo.	2,115,000	730,000	1,385,000
11	Pittsburgh, Pa.	1,990,000	595,000	1,395,000
12	Baltimore, Md.	1,695,000	936,000	759,000
13	Minneapolis–St. Paul*, Minn.	1,505,000	480,000 / 313,500	711,500
14	Houston, Tex.	1,335,000	970,000	365,000
15	Miami–Fort Lauderdale, Fla.	1,330,000	296,000 / 88,000	946,000
16	Milwaukee, Wis.	1,285,000	750,000	535,000
17	Buffalo–Niagara Falls, N.Y.	1,285,000	525,000 / 103,500	656,500
	incl. part in Canada	1,370,000		
18	Cincinnati, Ohio	1,250,000	502,500	747,500
19	Dallas, Tex.	1,090,000	710,000	380,000
20	Kansas City, Mo.	1,070,000	510,000	560,000
21	Atlanta, Ga.*	1,060,000	500,000	560,000
22	San Diego, Calif.	985,000	610,000	375,000
	incl. part in Mexico	1,165,000		
23	Seattle, Wash.	975,000	559,000	416,000
24	New Orleans, La.	920,000	635,000	285,000
25	Denver, Colo.*	905,000	505,000	400,000
26	Indianapolis, Ind.*	830,000	480,000	350,000
27	Providence*–Pawtucket–Woonsocket, R.I.	810,000	200,000 / 81,000 / 46,000	483,000
28	Hartford*–New Britain, Conn.	800,000	160,000 / 83,500	556,500
29	Louisville, Ky.	765,000	390,000	375,000
30	Portland, Oreg.	750,000	371,000	379,000
31	Columbus, Ohio*	745,000	483,000	262,000
32	San Antonio, Tex.	725,000	606,000	119,000
33	Phoenix, Ariz.*	685,000	475,000	210,000
34	Dayton, Ohio	675,000	264,000	411,000
35	Memphis, Tenn.	658,000	515,000	143,000
36	Birmingham, Ala.	638,000	344,000	294,000
37	Rochester, N.Y.	615,000	316,000	299,000
38	Albany*–Schenectady–Troy, N.Y.	600,000	129,000 / 80,000 / 66,500	324,500
39	Sacramento, Calif.*	595,000	205,000	390,000
40	Norfolk–Portsmouth, Va.	593,000	310,000 / 117,000	166,000
41	Akron, Ohio	590,000	293,000	297,000
42	Fort Worth, Tex.	535,000	365,000	170,000
43	Toledo, Ohio	525,000	324,000	201,000
44	Honolulu, Hawaii*	515,000	306,000	209,000
45	San Bernardino–Riverside, Calif.	485,000	94,500 / 97,500	293,000
46	Oklahoma City, Okla.*	475,000	345,000	130,000
47	Jacksonville, Fla.	475,000	201,000	274,000
48	Youngstown–Warren, Ohio	475,000	166,000 / 61,000	248,000
49	Syracuse, N.Y.	460,000	216,000	244,000
50	Omaha, Nebr.–Council Bluffs, Iowa	455,000	320,000 / 57,000	78,000
51	Springfield–Holyoke, Mass.	450,000	178,000 / 52,500	219,500
52	Salt Lake City, Utah*	435,000	190,000	245,000
53	Nashville, Tenn.*	428,000	260,000	168,000
54	Richmond, Va.*	420,000	218,000	202,000
55	Flint, Mich.	392,000	198,000	194,000
56	St. Petersburg–Clearwater, Fla.	385,000	193,000 / 37,000	155,000
57	Tulsa, Okla.	382,000	278,000	104,000
58	Grand Rapids, Mich.	380,000	204,000	176,000
59	Tampa, Fla.	370,000	288,000	82,000
60	Wichita, Kans.	360,000	260,000	100,000
61	Bridgeport, Conn.	335,000	156,500	178,500
62	Wilmington, Del.	334,000	93,500	240,500
63	New Haven, Conn.	330,000	150,000	180,000
64	El Paso, Tex.	327,500	294,000	33,500
	incl. part in Mexico	615,000		
65	Mobile, Ala.	318,000	208,000	110,000
66	Worcester, Mass.	318,000	183,000	135,000
67	Allentown–Bethlehem, Pa.	308,000	108,500 / 76,000	123,500
68	Tacoma, Wash.	305,000	150,000	155,000
69	Chattanooga, Tenn.	295,000	130,000	165,000
70	Knoxville–Oak Ridge, Tenn.	292,000	110,000 / 28,000	154,000
71	Trenton, N.J.*	290,000	111,500	178,500
72	Charlotte, N.C.	288,000	207,000	81,000
73	Canton–Massillon, Ohio	288,000	113,000 / 31,500	143,500
74	Albuquerque, N. Mex.	285,000	220,000	65,000
75	Orlando, Fla.	280,000	90,000	190,000
76	Beaumont–Port Arthur, Tex.	276,000	122,000 / 66,500	87,500
77	Peoria, Ill.	273,000	102,000	171,000
78	Des Moines, Iowa*	270,000	214,000	56,000
79	South Bend, Ind.	270,000	133,000	137,000
80	Davenport, Iowa–Rock Island–Moline, Ill.	267,000	91,000 / 52,500 / 43,500	80,000
81	Tucson, Ariz.	266,000	230,000	36,000
82	Harrisburg, Pa.*	265,000	78,500	186,500
83	Baton Rouge, La.*	262,000	158,000	104,000
84	Spokane, Wash.	260,000	184,000	76,000
85	Shreveport, La.	258,000	170,000	88,000
86	Little Rock, Ark.*	250,000	109,000	141,000
87	Wilkes-Barre, Pa.	248,000	61,500	186,500
88	Fresno, Calif.	242,000	143,000	99,000
89	Huntington, W. Va.–Ashland, Ky.	234,000	83,000 / 31,300	119,700
90	Pomona–Ontario, Calif.	230,000	72,000 / 49,500	108,500
91	Newport News–Hampton, Va.	227,000	117,500 / 92,500	17,000
92	Columbia, S.C.*	225,000	98,000	127,000
93	Fort Wayne, Ind.	220,000	164,000	56,000
94	Erie, Pa.	219,000	139,000	80,000
95	Binghamton, N.Y.	219,000	75,000	144,000
96	Austin, Tex.*	218,000	194,000	24,000
97	Charleston, W. Va.*	218,000	86,500	131,500
98	Lansing, Mich.*	215,000	115,000	100,000
99	Scranton, Pa.	214,000	109,500	104,500
100	Charleston, S.C.	210,000	76,000	134,000
101	Columbus, Ga.	209,000	121,000	88,000
102	Corpus Christi, Tex.	206,000	174,000	32,000
103	Jackson, Miss.*	202,000	152,000	50,000
104	Evansville, Ind.	202,000	142,000	60,000
105	Rockford, Ill.	201,000	131,000	70,000
106	Lawrence–Haverhill, Mass.	199,000	69,000 / 46,500	83,500
107	Savannah, Ga.	196,000	153,000	43,000
108	Waterbury, Conn.	196,000	107,500	88,500
109	Reading, Pa.	195,000	96,000	99,000
110	Winston-Salem, N.C.	194,000	115,000	79,000
111	Madison, Wis.*	188,000	138,000	50,000
112	Pensacola, Fla.	178,000	57,500	120,500
113	Macon, Ga.	178,000	126,000	52,000
114	Kalamazoo, Mich.	177,000	83,500	93,500
115	Muskegon, Mich.	173,000	46,000	127,000
116	West Palm Beach, Fla.	172,000	57,500	114,500
117	Stockton, Calif.	170,000	88,000	82,000
118	Greenville, S.C.	170,000	67,000	103,000
119	Bakersfield, Calif.	168,000	60,000	108,000
120	Augusta, Ga.	168,000	70,500	97,500
121	Duluth, Minn.–Superior, Wis.	167,000	107,500 / 33,300	26,200
122	Saginaw, Mich.	166,000	99,000	67,000
123	Utica, N.Y.	165,000	100,000	65,000
124	Greensboro, N.C.	164,000	123,000	41,000
125	Roanoke, Va.	164,000	97,500	66,500
126	Montgomery, Ala.*	160,000	138,000	22,000
127	Amarillo, Tex.	154,000	148,500	5,500
128	Lubbock, Tex.	153,000	136,000	17,000
129	Lincoln, Nebr.*	152,000	133,000	19,000
130	York, Pa.	152,000	53,500	98,500
131	Lowell, Mass.	152,000	91,500	60,500
132	Colorado Springs, Colo.	151,500	75,000	76,500
133	Ogden, Utah	149,000	72,500	76,500
134	New Bedford, Mass.	147,000	101,000	46,000
135	Portland, Maine	144,000	72,000	72,000
136	Galveston–Texas City, Tex.	143,000	66,000 / 33,500	43,500
137	Topeka, Kans.*	142,000	123,500	18,500
138	Ventura–Oxnard, Calif.	140,000	31,000 / 43,000	66,000
139	Fayetteville, N.C.	140,000	49,000	91,000
140	Fall River, Mass.	139,500	97,500	42,000

* National or State capital.
** Newark (1962 population 400,000) and Paterson (144,000), N.J., are also central cities of the New York Metro. Area; and Brockton, Mass. (76,000), is also a central city of the Boston Metro. Area.

U.S. METRO. AREAS OF 100,000 OR MORE, 1962 *(Continued)*

Rank 1962		Metro. Area	Central City	Suburbs
		Estimated Population, 1/1/1962		
141	Joliet, Ill.	139,000	67,500	71,500
142	Lake Charles, La.	136,000	67,000	69,000
143	Raleigh, N.C.*	136,000	97,000	39,000
144	Atlantic City, N.J.	134,000	59,500	74,500
145	Waco, Tex.	133,000	104,000	29,000
146	Las Vegas, Nev.	132,000	71,000	61,000
147	Gulfport— Biloxi, Miss.	132,000	{ 31,500 / 45,000 }	} 55,500
148	Poughkeepsie, N.Y.	131,000	38,000	93,000
149	Lexington, Ky.	130,000	64,000	66,000
150	Lancaster, Pa.	129,500	60,500	69,000
151	Appleton, Wis.	128,500	50,000	78,500
152	Eugene, Oreg.	128,000	53,000	75,000
153	Johnstown, Pa.	127,000	52,000	75,000
154	Wheeling, W. Va.	126,000	52,500	73,500
155	Springfield, Ill.*	125,500	84,000	41,500
156	Cedar Rapids, Iowa	125,000	95,000	30,000
157	Jackson, Mich.	124,000	50,500	73,500
158	Steubenville, Ohio— Weirton, W. Va.	123,000	{ 34,000 / 28,500 }	} 60,500
159	Wichita Falls, Tex.	121,500	107,000	14,500
160	Waterloo, Iowa	118,500	73,000	45,500
161	Racine, Wis.	118,000	92,000	26,000
162	Pueblo, Colo.	115,500	95,000	20,500
163	Decatur, Ill.	115,000	80,000	35,000
164	Manchester, N.H.	114,500	89,500	25,000
165	Springfield, Ohio	114,000	83,500	30,500
166	Springfield, Mo.	113,000	98,000	15,000
167	New London, Conn.	111,000	34,500	76,500
168	Green Bay, Wis.	110,000	64,500	45,500
169	Battle Creek, Mich.	109,500	48,500	61,000
170	Newburgh, N.Y.	108,500	30,500	78,000
171	Durham, N.C.	108,000	79,500	28,500
172	Hamilton, Ohio	107,000	73,000	34,000
173	Petersburg— Hopewell, Va.	106,000	{ 37,000 / 18,500 }	} 50,500
174	Provo, Utah	106,000	37,000	69,000
175	Asheville, N.C.	105,500	60,500	45,000
176	High Point, N.C.	104,500	64,000	40,500
177	Abilene, Tex.	104,000	96,000	8,000
178	Altoona, Pa.	103,500	68,500	35,000
179	Sioux City, Iowa	102,500	90,000	12,500
180	Santa Barbara, Calif.	102,000	63,000	39,000
181	Huntsville, Ala.	102,000	78,000	24,000
182	Muncie, Ind.	102,000	69,500	32,500
	Port Huron, Mich.	64,000	36,000	28,000
	incl. part in Canada	127,000		
	Laredo, Tex.	63,500	63,000	500
	incl. part in Mexico	163,000		
	Brownsville, Tex.	52,000	49,500	2,500
	incl. part in Mexico	152,000		

* National or State capital.

NUMBER OF U.S. COUNTIES, CITIES, TOWNS, AND LOCALITIES BY STATES, 1962

STATES	Number of Counties*	Number of Places Under 1,000 Population	Number of Places 1,000-2,500 Population	Number of Places 2,500-5,000 Population	Number of Places 5,000-10,000 Population	Number of Places 10,000-25,000 Population	Number of Places 25,000-50,000 Population	Number of Places 50,000-100,000 Population	Number of Places Over 100,000 Population	Total Places
Alabama	67	2,568	82	57	28	15	8	3	3	2,764
Alaska†	4	713	16	2	3	1	1	.....	.....	736
Arizona	14	732	37	16	10	6	2	.....	2	805
Arkansas	75	2,342	61	29	19	13	3	2	1	2,470
California	58	3,921	210	143	97	122	57	31	14	4,595
Colorado	63	1,306	54	23	15	11	5	3	1	1,418
Connecticut	8	685	39	31	17	21	13	6	4	816
Delaware	3	319	24	8	5	2	.....	1	.....	359
District of Columbia	1*	96	.....	.....	.....	.....	.....	.....	1	97
Florida	67	1,880	128	100	61	35	14	7	4	2,229
Georgia	159	2,268	130	69	32	24	6	2	4	2,535
Hawaii	4	648	25	13	3	3	2	.....	1	695
Idaho	45*	856	34	16	7	8	3	.....	.....	924
Illinois	102	3,111	272	136	91	76	25	14	3	3,728
Indiana	92	2,222	149	58	39	29	10	3	6	2,516
Iowa	99	1,622	138	46	33	11	7	6	1	1,864
Kansas	105	1,436	100	46	19	22	5	.....	3	1,631
Kentucky	120	3,953	101	45	27	13	5	2	1	4,147
Louisiana†	64	1,977	97	46	29	23	5	3	3	2,183
Maine	16	1,610	43	24	14	11	2	1	.....	1,705
Maryland	24*	2,085	94	40	22	20	6	4	1	2,272
Massachusetts	14	1,330	89	50	29	47	27	17	5	1,594
Michigan	83	2,506	185	86	53	37	21	13	6	2,907
Minnesota	87	1,898	141	50	38	28	7	1	3	2,166
Mississippi	82	1,822	71	31	22	10	6	1	1	1,964
Missouri	115*	2,596	136	67	44	26	8	4	2	2,883
Montana	57*	1,141	39	16	6	5	2	2	.....	1,211
Nebraska	93	790	71	86	13	8	1	.....	2	971
Nevada	17	340	10	5	2	3	.....	2	.....	362
New Hampshire	10	668	31	11	5	6	3	1	.....	725
New Jersey	21	1,717	104	119	93	87	28	10	6	2,164
New Mexico	32	915	24	13	12	5	7	.....	1	977
New York	62	5,064	280	150	113	91	29	13	8	5,748
North Carolina	100	2,877	162	60	29	20	9	4	3	3,164
North Dakota	53	803	49	3	5	3	4	.....	.....	867
Ohio	88	3,238	230	127	103	78	24	10	8	3,818
Oklahoma	77	1,446	84	38	26	17	6	1	2	1,620
Oregon	36	1,499	64	40	24	15	1	2	1	1,646
Pennsylvania	67	7,631	467	203	171	93	18	11	5	8,599
Rhode Island	5	292	13	12	7	6	3	3	1	337
South Carolina	46	1,399	112	37	30	11	4	3	.....	1,596
South Dakota	67	679	37	13	4	6	1	1	.....	741
Tennessee	95	3,157	97	46	25	23	4	.....	4	3,356
Texas	254	3,835	270	142	83	61	19	10	12	4,432
Utah	29	709	47	15	16	6	2	1	.....	797
Vermont	14	655	23	5	9	3	1	.....	.....	696
Virginia	131*	3,792	99	40	25	21	3	4	5	3,989
Washington	39	1,862	93	54	24	23	6	.....	3	2,065
West Virginia	55	3,096	102	31	14	8	4	3	.....	3,258
Wisconsin	72	1,980	134	53	38	26	11	6	2	2,250
Wyoming	24*	470	16	11	3	3	2	.....	.....	505
Total	3,115*	96,557	5,114	2,562	1,637	1,242	440	211	134	107,897

* Includes 3,076 counties, parishes, and judicial divisions; the District of Columbia; Baltimore city, Md.; St. Louis city, Mo.; 33 independent cities in Virginia; and the parts of Yellowstone National Park in Idaho. Montana, and Wyoming.
† The divisions of Louisiana are known as parishes. Those shown for Alaska are the former judicial divisions.

Arizona's Grand Canyon—Mighty Gorge of the Colorado

PLACES OF INTEREST IN THE UNITED STATES

For the state capital, largest city, highest point, and other general information about each state, see the U.S. State General Information Table on page 198. Also see the U.S. National Park System Table and Map, pages 192-193, for additional information about the national parks, monuments, historical parks, and other units under the National Park Service.

ALABAMA

Birmingham. Largest city in Alabama and steel center of the South. On top of Red Mountain, overlooking the city, stands the 55-foot statue of Vulcan on a pedestal 124 feet high. Cast in a Birmingham foundry, from pig iron mined in the area, the statue is second in size only to the Statue of Liberty. Other points of interest are Woodrow Wilson Park, in the city, and Vestavia, on nearby Shades Mountain. Built as a circular residence and patterned after the Roman Temple of Vesta, Vestavia now houses a museum and is surrounded by gardens.

Mobile. Important southern seaport and Alabama's first permanent white settlement, the city is famed for its historic old-colonial homes, a five-day Mardi Gras, and a thirty-five mile Azalea Trail. On this Trail, southwest of the city, are beautiful Bellingrath Gardens, containing many kinds of flowers and shrubs. On Dauphin Island, Alabama's annual Deep Sea Fishing Rodeo is held.

Montgomery. Capital of Alabama and first capital of the southern Confederacy. The Capitol and its grounds, situated on Goat Hill, contain many historic relics and monuments. The first White House of the Confederacy, the Montgomery home of Jefferson Davis, has many war relics, as well as some of Davis' personal belongings.

Mound State Monument. Thirty-four prehistoric Indian mounds. The largest is 58½ feet high and covers one and a quarter acres. A museum near the center of the park contains burial exhibits, tools, pottery, beads, and other relics excavated on the site. The monument is near Moundville.

Muscle Shoals. Site of Wilson Dam and two nitrate plants, construction of which was begun in 1916 as a defense measure and which later was expanded into the vast Tennessee Valley Authority project. Wilson Dam, 137 feet high and nearly a mile long, has two navigation locks that lift boats 89 feet in less than an hour. Fifteen miles up the Tennessee is Wheeler Dam.

ALASKA

Aleutian Islands. A 1,000-mile chain of twenty volcanic, treeless islands stretching out from the Alaska Peninsula. Scattered along the archipelago are small native villages peopled by Aleuts.

Anchorage. Largest city in the state and gateway to a fabulous hunting area for bear, mountain sheep and goats, caribou, and moose. Ski areas are nearby. Short trips from the city lead to Portage Glacier and to Matanuska Valley where giant vegetables and fruits are produced.

Barrow. Located on Point Barrow, the northernmost point in the United States. From here the Distant Early Warning (DEW) Line sweeps across the Arctic to the Atlantic Ocean. Here Eskimos hunt polar bear on the ice and harpoon whale.

Fairbanks. Situated on a loop of the Chena River, Alaska's second largest city lies in the heart of a great gold mining region where giant dredges have replaced the early miner's pan. The University of Alaska is at the nearby town of College. Sightseeing trips on the Tanana River can be taken in a "paddlewheeler," and there are air trips to Chena Hot Springs for hunting and fishing.

Glaciers. The best known glaciers are the Columbia, at the head of Columbia Bay, near Valdez; the Mendenhall, the most photographed glacier, available from Juneau; Portage, available from Anchorage; and Glacier Bay National Monument at the uppermost portion of the Panhandle.

Juneau. The State capital nestles on a narrow ledge of land at the foot of the steep slope of Mt. Juneau. A lumbering and fishing center, its early glory is preserved in the Alaska-Juneau Gold Mine.

Ketchikan. Salmon fishing center, noted for its picturesque harbor. Interesting collection of totem poles at Saxman Indian Village.

Mt. McKinley National Park. Twin peaks of Mt. McKinley overlook more than 3,000 square miles, in which ice-capped domes rise above sphagnum growth, and glaciers wait to be explored.

National Monuments. Glacier Bay, west of Juneau; Katmai, on the Alaska Peninsula; Sitka. See U.S. NATIONAL PARK SYSTEM, pages 228–29.

Nome. Metropolis of Seward Peninsula, where gold, silver, and other valuable minerals are mined.

More than 350 varieties of wildflowers bloom in season. Also see King Island ivory carvers.

Pribilof Islands. Breeding grounds for government protected fur seals which arrive each May.

Seward. Lumbering center on terminus of Alaska Railroad; gateway to the interior.

Sitka. Settled in 1799, it became the Russian capital of Alaska in 1802. Points of interest include Alaska Pioneer's Home, St. Michael's Cathedral, the Old Russian Blockhouse, and Sitka National Monument with its eighteen totem poles.

Skagway. Port of entry to Canada at the terminus of the Yukon Railway. White Pass and Chilkoot Pass are points of interest.

Wrangell. Early Russian fur trading post, now lumber and fishing center. Headquarters for the lovely Stikine River trip through a cut in the Coast Range Mountains. Also see the Tlingit Indian Community House and the remarkable totem poles.

ARIZONA

Apache Trail. Now a graveled road, this famous trail winds through rugged mountain scenery.

Bisbee. Center of a rich copper district. Located on the steep upper slopes of Mule Pass Gulch, Bisbee is built in tiers, and many of the houses are reached by steep flights of wooden and stone steps. Sacramento Pit, 435 feet deep and covering 35 acres, is at Bisbee.

Flagstaff. Home of the Lowell Observatory and Arizona State College. Situated near Humphreys Peak, highest point in the state. The city is popular with summer tourists because of its cool climate.

Grand Canyon National Park. One of nature's most magnificent spectacles, 5,000 feet deep Grand Canyon of the Colorado is lined with rocky towers and pinnacles and is filled with constantly changing colors. Many observation points along the North and South Rims afford excellent views of the Canyon from above, and mule trails lead to the bottom of the gorge. Grand Canyon is from four to eighteen miles wide and is more than two hundred miles long.

Hoover (Boulder) Dam and Lake Mead Recreation Area. See NEVADA

Alaska Vis. Ass'n.   *Hole-in-the-Wall Glacier, Near Juneau*

Union Pacific R.R.   *Los Angeles' Fabulous Freeway System*

Redwood Empire Ass'n.   *Historic Golden Gate Bridge, San Francisco*

202

Indian Reservations. Located in the northeastern part of the state and completely surrounded by NAVAJO reservations, the HOPI reservation is one of the most interesting of the numerous Indian reservations. Most of its eleven villages are built on the tops of mesas. Oraibi has existed since 1400.

The HAVASUPAI reservation is on the bottom of Cataract Canyon in Grand Canyon National Park, and the HUALPAI is south of Grand Canyon. FORT MOJAVE and COLORADO RIVER reservations, near the California border, are the homes of the Mojave and Southern Paiute Indians. Other reservations are the MARICOPA, south of Phoenix, the PAPAGO, west of Tucson, and the Apache reserves—SAN CARLOS and FORT APACHE—near the New Mexico border.

Meteor Crater. This crater, 600 feet deep and a mile wide, was formed centuries ago by a meteor believed to have displaced between five and six million tons of earth; west of Winslow.

Monument Valley. Colorful valley in the northeastern corner of the state. Its red sandstone pillars and spires resemble huge temple ruins.

National Monuments. There are 16 national monuments in Arizona. CANYON DE CHELLY is in the Navajo Reservation and contains more than 300 cliff-dwelling ruins in an area of nearly 84,000 acres.

PETRIFIED FOREST MONUMENT includes six forests of petrified wood, much of it in the form of huge logs. The Administration Building contains many fine specimens of the wood, as well as Indian relics and fern fossils. For other National Monuments, see the U.S. NATIONAL PARK SYSTEM, pages 192-193.

Painted Desert. Three hundred miles of sand, shale, and sandstone formations splashed with brilliant shades of red, amethyst, yellow, and purple that change to other hues in the shifting light. Part of it is in the Petrified Forest National Monument.

Phoenix. Capital of Arizona and, because of its semi-tropical climate, a popular winter tourist resort. The city is built on the site of a prehistoric community; not far from the business district is La Ciudad (the city), an excavation of an ancient pueblo. The Heard Museum contains relics from La Ciudad, relics of other Indians, including a collection of blankets. The Arizona Museum is also in Phoenix, as is the South Mountain Park.

San Xavier del Bac. Founded in 1700 by a Jesuit missionary and completed in 1797 by the Franciscans, this mission, one of the best preserved in the Southwest, is still in use. It is near Tucson.

Southwestern Arboretum. Located near Superior, this 120-acre tract of land contains 10,000 varieties of plants, brought from all over the world. Founded by the late William Boyce Thompson.

Tombstone. Roaring mining camp of the 1880's, noted for its gambling houses and gun fights. In Boothill Graveyard are buried many desperadoes and others who died violent deaths. Some of the old buildings still in existence are the Bird Cage Theater, Crystal Palace Saloon, Oriental Bar, Russ House, and office of the *Epitaph*, Tombstone newspaper.

Tucson. Resort and site of the University of Arizona. In the southern city, modern metropolitan sections rub elbows with the remnants of the old Spanish town. Businessmen and cowboys, students and Indians, mingle on the streets and in the stores.

ARKANSAS

Arkansas Post State Park. Scene of one of the first white settlements in the Mississippi River Valley, made by the French in 1686; also the first capital of Arkansas Territory. The park contains Confederate fortifications and other historical ruins.

Bauxite. Site of numerous bauxite mines and several aluminum refineries. The mines are worked largely by a stripping process.

Devil's Den State Park. A heavily wooded area of more than 4,000 acres in the Boston Mountains, southern section of the Ozarks. Sandstone formations with deep cracks and crevices are a feature.

Diamond Cave. Large, brilliantly colored cave near Jasper, in the Ozark Mountains.

Eureka Springs. A health resort in the Ozarks, northeast of Fayetteville. On steep mountain slopes, town contains many springs and limestone caves.

Fort Smith. Second largest town in Arkansas. The old fort, erected to protect the settlers against the Indians, is still standing.

Hot Springs. Famous tourist resort and site of Hot Springs National Park. Containing 989 acres the Park includes the mountains around the city.

Little Rock. State capital and largest city. A fascinating blend of ante-bellum homes and modern metropolitan buildings. The restored Territorial Capitol was the scene of the last territorial legislature in 1835. The War Memorial Building, one of the most beautiful buildings in Arkansas, was the state Capitol for three-quarters of a century. The present Capitol faces Capitol Avenue and contains, in addition to the legislative chambers and other state offices, the State History Museum. In Mac-Arthur Park, birthplace of General Douglas MacArthur, is the Old Arsenal, last building of the old army post. It now contains the Arkansas Museum of Natural History and Antiquities.

Mammoth Spring. One of the largest springs in the world and the source of Spring River; located in Fulton County, near the Missouri Line.

Murfreesboro. Scene of the only diamond mine in the United States.

Ozark Mountains. Highlands extending from southern Missouri into northern Arkansas and Oklahoma, famous for rugged scenery.

Petit Jean State Park. Situated on Petit Jean Mountain, a wooded plateau on the edge of the Arkansas Valley, near Morrilton. Attractions include Bear Cave and Cedar Falls.

CALIFORNIA

Catalina Island. Famous resort island where fishing and boating are prime attractions.

Coloma. The site of Sutter's Sawmill, where discovery of gold in 1848 precipitated the California gold rush. Placerville, a few miles south, was a typical "rip-roaring" camp and a station for Pony Express, Overland Mail, and freight. Gold is still mined in the area; dredging is in operation at Chico, to the north.

Death Valley National Monument. Almost 3,000 square miles of desert—salt flats, volcanic craters, hills and valleys and jagged peaks weirdly eroded. Its normal colorings of mauve to red and white to nearly black take on sunset shades as the sunlight changes, with deep shadows of purple and blue. Near Badwater, in the sink of the Armargosa River, is the lowest spot in North America, 282 feet below sea level.

Donner Pass. Six miles above Donner Lake and the spot where the hapless Donner party camped through the winter of 1846.

Hollywood. Part of Los Angeles and world motion picture capital.

Kings Canyon National Park. Established in 1940 and incorporating the old General Grant National Park, as well as Redwood Mountain and Redwood Canyon. Much of this park is a roadless wilderness of mountains, canyons, and sequoia forests. In the southern section a paved automobile road traverses Kings Canyon, whose peaks and crags include Sentinel Dome, North Dome, and Lookout Peak. Famous General Grant Grove Section, containing the General Grant Tree, is at the southwest corner.

Lake Tahoe. This colorful mountain lake is shared by Nevada through most of its length.

Lassen Volcanic National Park. Lassen Peak, a dormant volcano, centers the western half of this region of boiling mudpots, steaming sinks, and weird volcanic formations.

Los Angeles. Spreading for miles inland and along the ocean, this great city is the center of far-flung

orange groves and other agriculture and of an important oil-producing area. It is rich in manufactures, with a fine harbor, and its superb beaches are an unexcelled resort attraction. In the Olvera Street district, the point of the city's origin, Mexican pottery and other wares are made and sold in quaint old shops. Fine residential sections, handsome gardens and parks, flowers in exotic masses of color—for these Los Angeles is famous. Here are the Bernheimer Oriental Gardens and a fine botanic garden at the southern branch of the University of California. (Los Angeles is also home-city for the University of Southern California.) Numerous museums include the Southwest, showing early art, handicraft, and history of the American Indians; the Los Angeles County Museum of History, Science, and Art, containing fossils from La Brea Tar Pits, also here; and Lyons Pony Express Museum, with a fine collection of western relics. The Olympic Stadium and the race track at Santa Anita Park are sports attractions; Great Western Livestock Show is held in December. See HOLLYWOOD, PASADENA, and CATALINA ISLAND.

Missions. In the earliest days of the white man in California, the Spanish missions were the outposts of civilization. A day's journey apart, they made a chain that still reaches from San Diego to Sonoma. Soundly and beautifully built, they contain today many examples of the work of skilled artisans. SAN LUIS REY DE FRANCIA MISSION has changed little since 1798; at SAN JUAN CAPISTRANO beautiful grounds are background for swallows whose regularity of migration is legend; at MISSION SANTA BARBARA, the altar fire has never died since it was first set. The Mission Festival is held in September, at SAN GABRIEL ARCHANGEL.

Monterey and Carmel. South of Monterey Bay, the two towns are at the base of Monterey Peninsula, whose jutting, rocky coast is a series of superb crags, reefs, and white surf. Here are the famous Monterey cypress trees, and the pines. Monterey was the early key-city of the state and its first capital.

Mount Shasta. Famed for its beauty the peak is a favorite of mountain climbers. The region south of it contains Castle Crags State Park, and, farther south, Shasta Lake, with gigantic Shasta Dam.

National Monuments. See DEATH VALLEY NATIONAL MONUMENT, and U. S. NATIONAL PARK SYSTEM, pages 192-193.

Palomar Observatory. Northwest of San Diego. Home of the world's largest reflector telescope.

Pasadena. Home of the famed Rose Bowl. Here each January 1 the Tournament of Roses takes place. Nearby is the Mount Wilson Observatory; and the Huntington Library, Art Gallery, and Gardens are in San Marino to the south.

Redwoods. All along the northern half of California's coast, the redwood forests are spectacular. From Mill Creek Redwoods State Park in the north to the Big Basin Redwoods State Park south of San Francisco, the coastal area is a series of state parks and groves.

Sacramento. California's capital, on the Sacramento River. Sutter's Fort is a faithful restoration of the ranch and fort where Captain John A. Sutter first settled in 1839, and where scores of starving, exhausted pioneers were given succor. The Crocker Art Gallery, partly housed in the Old Crocker Home, is one of the finest in the country.

San Diego. At the southwestern tip of California. San Diego has a huge training station and military base for the United States Marine Corps and a United States Naval Training Center. The ocean drive from Long Beach ends here. Numerous missions are near the highway. Cabrillo National Monument, the Old Spanish Lighthouse, is a half-acre monument. Tijuana is over the line in Old Mexico.

San Francisco. "City of the world, of the hills, and of the sea." The Golden Gate Bridge links it with the mainland to the north; the Bay Bridge connects it with Oakland, across San Francisco Bay; by a narrow neck of land its peninsula joins the mainland. Twin Peaks Scenic Drive offers a superb view of the city, bay, waterfront, and suburban cities across the bay. A scenic highway skirts the ocean along the peninsula's length. Sea-going ships in the harbor, Fisherman's Wharf, the Latin Quarter, and Chinatown's temples and bazaars lend an air of adventure. Beauty spots are the Civic Center and the all-year parks. Other features of interest are the San Francisco Symphony Orchestra, the Opera, the M. H. DeYoung Memorial Museum (art), the Wells Fargo Historical Collection, and the Steinhart Aquarium. Across the bay are Oakland and Berkeley, site of the University of California; Stanford University is at nearby Palo Alto.

San Jose. Key-city of California's prune and apricot industry and world's largest canning center. Lick Observatory is on nearby Mount Hamilton.

Sequoia National Park. Home of the Big Trees. Grove after grove of these gigantic sequoias are the park's greatest attraction. (See also KINGS CANYON and YOSEMITE NATIONAL PARK.) Chief among them is the Giant Forest, where the sequoias are intermingled with white firs, ponderosa pines, sugar pines, and incense cedars. The park rises sharply to the crest of the High Sierras at its eastern border. Here is Mt. Whitney, at 14,495 feet the highest peak in conterminous United States.

Sonoma and Napa Counties. The vineyard of California, one of the greatest grape and wine areas in the world. Luther Burbank conducted his famous experiments here, and his experimental gardens still may be seen at Santa Rosa and Sebastopol. Calistoga, at the northern end of Napa Valley, is a region of mineral springs and geysers.

Yosemite National Park. Yosemite, in California's High Sierras, is a scenic park of mountains, canyons, lakes, streams, waterfalls and mountain parks. Its sights include Yosemite Valley, walled by towering granite cliffs and monoliths such as El Capitan and Half Dome, accented by waterfalls—Yosemite, Bridalveil, Nevada, Vernal; the Mariposa grove of giant sequoia trees (there are two others in the park); and the mule deer and black bear.

COLORADO

Aspen. West of Continental Divide on Roaring Fork River, this early mining camp is now a popular ski center and winter resort, and a summer cultural center. Beautiful Snowmass Mountain, streams, and lakes are in the vicinity.

Black Canyon of the Gunnison National Monument. State road leads to rim of canyon where rock walls drop 2200 feet to Gunnison river. An old trail takes skilled climbers down to the river.

Boulder. Home of University of Colorado and site of Colorado's first schoolhouse. Boulder Canyon, Roosevelt National Forest, and Rocky Mountain National Park are nearby.

Central City. First town on Gregory Gulch, scene of the first great gold rush to the Rockies and called "the richest square mile on earth." Now the town is partly "ghost."

Colorado National Monument. Near Grand Junction, 17,693 acres of canyons and desert hills are filled with red and yellow sandstone monoliths and columns of fantastic size and shape. There are good views from Serpent's Trail, Rimrock Drive, and Cold Shivers Point. Dinosaur beds are numerous.

Colorado Springs. Famed health and pleasure resort. Here are Pikes peak, Garden of the Gods, containing many fantastic formations of red and yellow sandstone, Manitou Springs, Cave of the Winds, Cheyenne Canyon, Seven Falls, Will Rogers Shrine of the Sun on Cheyenne Mountain.

Cripple Creek. In the mining era, one of the richest gold fields in the world.

Denver. The state capital's beautiful Civic Center and Capitol Hill are marked by fine architecture, murals, and statuary. The city park has a color-lighted electric fountain in the center of its lake. Rare western history collections are in the Public Library and the State Museum. Denver is the home

Cliff Palace, Mesa Verde National Park

Spectacular Royal Gorge, Colorado D.&R.G.W. R.R.

Longs Peak, from Bear Lake, Colorado D.L. Hopwood

203

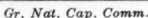
Gr. Nat. Cap. Comm.

Capitol of the United States, Washington, D.C.

Miami N.B.

Miami, Florida, City of Sunshine

of the first Juvenile Court, the Emily Griffith Opportunity School and the University of Denver. A U.S. Mint is here, and the Denver Union Stockyards. The National Western Livestock and Rodeo Show is an annual attraction in January. The Mountain Parks System includes Red Rocks, with its outdoor theater, Echo Lake, Mount Evans, and Lookout Mountain, with Buffalo Bill's grave.

Dinosaur National Monument. Nearly 205,000 acres of dinosaur beds in Colorado and Utah.

Estes Park. Gateway to Rocky Mountain National Park and headquarters for many vacationers.

Glenwood Springs. Hot mineral springs resort at mouth of Glenwood Canyon of the Colorado River.

Grand Mesa. Flat-topped, high-altitude mountain (10,500 feet) in Grand Mesa National Forest—a land of lakes, fishing, camping, and winter sports. Land's End is a vantage point on the Mesa's rim that offers a panoramic view of hundreds of miles.

Leadville. This highest city in the United States (10,200 feet) has around it the storied gold and silver mines of early-day Colorado. It was the home of the fabulous H. A. W. Tabor and the locale of his mines; it is still the heart of a rich mining region. Nearby Mt. Elbert is the highest peak in Colorado.

Mesa Verde National Park. A great plateau carved by deep canyons, in whose walls are hundreds of prehistoric Indian cliff dwellings. Cliff Palace alone contains more than 200 rooms and a score of kivas. The region is reached by the spectacular Million Dollar Highway, through the most rugged mountains and canyons of the Rockies.

National Monuments. See BLACK CANYON OF THE GUNNISON; COLORADO NATIONAL MONUMENT; DINOSAUR NATIONAL MONUMENT; and U. S. NATIONAL PARK SYSTEM, pages 192-193.

Ouray. Mountain mining town, famous in gold-mining history as the home of Thomas F. Walsh and his Camp Bird Mine, which is still in operation, as are others in the region. Footpaths and horseback trails to mountain viewpoints, towering peaks and mountain cataracts, Box Canyon with its falls and cavern, are attractive to tourists. This is the northern terminal of the Million Dollar Highway to the southwestern tip of the state.

Pagosa Springs. Mineral springs and health resort. It is just west of the Continental Divide, crossed by spectacular Wolf Creek Pass.

Pueblo. Colorado's second largest city, notable for its huge steel and smeltering plants. The Mineral Palace has a fine collection of ore samples.

Rocky Mountain National Park. Astride the Continental Divide, here is a land of snow-capped mountain ranges, sparkling lakes, pine and fir and quaking aspen, elk, deer, and bighorn sheep, and much other wildlife. Trail Ridge Road, a wide and beautiful highway, crosses the park and drops down

the west side of the Divide to Grand Lake. Among many other sights, the Mummy Range, the Never Summer Range, Iceberg Lake, and Specimen Mountain are seen from the Trail Ridge Road.

Royal Gorge. Deep and colorful canyon of the Arkansas River, near Canon City. An 880-foot suspension bridge swings across the top of the Gorge, 1,053 feet above the river.

San Luis Valley. Largest mountain park in Colorado, near the head of the Rio Grande, and a fertile and important agricultural area. The state's earliest-settled region, it still retains the atmosphere of the first Spanish and Mexican settlements. Among many scenic attractions, it contains the Great Sand Dunes National Monument.

Uravan. Site of one of the most extensive uranium deposits in the world; in Montrose county.

CONNECTICUT

Bridgeport. Industrial city, famous as the home of the circusman, P. T. Barnum. Among the points of interest are the Tom Thumb House, P. T. Barnum Museum, and Seaside Park, which contains Barnum's statue.

East Haddam. At the cemetery, above the Connecticut River, stands the schoolhouse where Nathan Hale taught. Uniquely furnished Gillette Castle, once the country estate of William Gillette, the famous actor, has become a state park.

Guilford. More than 150 old houses still remain in this well-preserved village, settled in 1639 by Reverend Henry Whitfield. His stone home is now a historical museum.

Hartford. State capital and famous insurance center. The Capitol, on a swelling crest of land, overlooks the other State buildings to the south and Bushnell Park to the north. Several blocks to the northeast stands the Old State House, designed by the colonial architect, Bulfinch, in 1796. Between Main and Prospect Streets are grouped the Wadsworth Athenaeum, Morgan Memorial, Colt Memorial, and Avery Memorial.

Marine Historical Museum. At Mystic, former clipper shipbuilding center. The museum houses an outstanding collection of clipper models.

Newgate Prison. In the countryside near East Granby. The dungeons and leg chains serve as grim reminders of punishments used in Colonial days.

New Haven. Graceful elms edge the streets of the home of Yale University. Noteworthy structures connected with the University are Connecticut Hall (built in 1752), Payne Whitney Gymnasium, Harkness Memorial Tower, Sterling Memorial Library, Gallery of Fine Arts, Sprague Memorial Hall, and the Yale Bowl.

New London. Former whaling town. Among its educational institutions are the U. S. Coast Guard

Academy and Connecticut College. Across the river, north of Groton, lies the U. S. Navy Submarine Base.

DELAWARE

Dover. State capital since 1777. The Georgian-Colonial-style State House, first completed in 1792, has undergone several remodelings and additions. Christ Church dates from 1734.

Lewes. Here the first white men of the Delaware River Region settled in 1630. Early Colonial antiques are exhibited in Zwaanendael House, an adapted model of Town Hall at Hoorn, Holland. Cannons used in War of 1812 stand in Memorial Park. Lewes Beach stretches along Delaware Bay.

Newark. University of Delaware campus cuts through the town. Of special interest are Old College, Elliott Hall, and Mitchell Hall.

New Castle. Colonial appearance is preserved by stately 17th-century homes and buildings. Market and Courthouse Squares comprise The Green, plotted by Peter Stuyvesant. The Old Court House was the State Capitol until 1777. Its central portion was completed in 1704, and its east wing was begun before 1698. Immanuel Church dates from 1710. Town Common, a common land grant of the Dutch era, lies west of the city; this farming tract still yields revenue to the city.

Rehoboth Beach. White sand beaches, fronting Atlantic Ocean, draw vacationers throughout the state and give town the title "Delaware's Summer Capital." Annually, art exhibits line the boardwalk near Virginia Avenue.

Wilmington. Largest city in the state and heavily industrialized. A stone monument, at The Rocks, marks the site of Fort Christina, the original Swedish settlement of 1638. Old Swedes (Holy Trinity) Church, built in 1698, presents a gray-stone facade with a hooded gable roof. The approach from Church Street leads through the cemetery where the earliest legible stone marker bears the date 1719. The Historical Society of Delaware, housed in Old Town Hall (1798), includes in its collection a portion of the eastern terminal stone of the Mason and Dixon Line.

DISTRICT OF COLUMBIA

Washington. Considered by many the most beautiful capital in the world. Washington was planned by George Washington, Thomas Jefferson, and the French engineer, Major Pierre Charles L'Enfant.

Among the Government buildings are the Capitol; White House; Library of Congress, largest in the world; the State Treasury, and other department buildings; the Pentagon, home of the Department of Defense; and the United States Supreme Court Building. The Smithsonian Institution administers ten bureaus, including the United States National Museum, the National Zoological Park, the Bureau of American Ethnology, the Freer Gallery of Art, and the National Gallery of Art, home of the Andrew Mellon Collection and others.

The Grecian charm of the Lincoln Memorial, the clean, straight shaft of the 555-foot Washington Monument, and the classic dignity of the Thomas Jefferson Memorial are beauty spots in the capital, which is at its loveliest when the cherry trees blossom in the spring. The Lincoln Museum (Ford's Theatre), containing a great collection of Lincolniana, and the House Where Lincoln Died have been made National Memorials. At Arlington National Cemetery, on the Virginia side of the Potomac, a sentry always stands guard at the Tomb of the Unknown Soldier. The Custis-Lee Mansion, on a hillside in the cemetery, is now a National Memorial.

FLORIDA

Cape Canaveral. Home of the astronauts. From here on May 5, 1961, Commander Alan Shepard became our first man in space.

Daytona Beach. World-famous speedway beach 23 miles long, whose hard-packed white sands are used by motorists, sand sailboaters, and bathers.

Everglades National Park. Including part of the Florida Keys, the Ten Thousand Islands, and Big

Cypress Swamp, the 2,000 square miles of tropical and sub-tropical wilderness, marsh, and semi-aquatic grasslands comprise America's third largest national park. Here are alligator and otter, raccoon, bobcat, and cougar; snowy egret and white and wood ibis; and tropical trees and other plants found nowhere else in the United States.

Jacksonville. Third largest city in Florida and most important commercially on the south Atlantic seaboard. Near the mouth of the St. Johns River, it is a great ship port and yacht harbor. The Oriental Gardens offer an orange grove and lush plantings of Florida flowers.

Key West. At the tip of Florida's long chain of keys, Key West is a tropical city with Latin atmosphere; its most modern note is the superb Overseas Highway that reaches it by traversing the whole chain of keys. Turtle, sponge, and deep-sea fishing are of interest, as are also the quaint old houses built by ships' carpenters. A United States Naval Station has been based there since 1823. Fort Jefferson National Monument is reached by boat or plane.

Lake Okeechobee. Florida's largest fresh-water lake, claiming the country's best black bass fishing.

Miami and Miami Beach. Cosmopolitan Miami, largest city in Florida, is important as a financial, transportation, and recreation center. Sights include the Hialeah Park with its race track, its avenues of royal palms, and its flamingos; the city docks where the fishing fleets unload, and the steamship piers; the International Airport; the Seminole Indian Village. At nearby Coral Gables is the University of Miami. MacArthur and other causeways cross Biscayne Bay to Miami Beach, fashionable resort.

Ocala. Hub of Florida's thriving Brahman cattle industry, center of an important citrus fruit area, and resort town in the north-central lake district, which is studded with lakes and springs, parks and recreational areas. At Silver Springs, the largest of the state's many springs, the flow of water is estimated at from 500,000,000 to 800,000,000 gallons a day; glass-bottomed boats, a jungle tour, and a fine bathing beach are attractions.

Orlando. Citrus center of lake area. Lake Eola Park and Orlando Zoo are natural homes for Florida's native flowers, trees, and animals.

Palm Beach and West Palm Beach. Luxury resorts popular among the socially elect.

Pensacola. Old and historically interesting city, with a definitely Spanish air. Plaza Ferdinand VII is on the site of the city's first fort; many old forts are around the harbor and elsewhere. Old Christ Church was built in 1835. Modern points of interest are the huge Naval Air Station nearby, the picturesque red snapper fleet, and the fine beaches across the bay on Santa Rosa Island.

St. Augustine. Oldest city in the United States. Founded in 1565. The Plaza de la Constitucion was the parade grounds of the first settlement; other landmarks include America's oldest house and oldest schoolhouse; the Spanish Treasury; Zero Milestone where the Spanish trail to California began; the old city gates; Castillo de San Marcos, the oldest masonry fort in the nation, now a National Monument. The Fountain of Youth Gardens have a central fountain where Ponce de León supposedly drank the waters of eternal life.

St. Petersburg. The west coast's important resort, lying on a climatically favorable and sunny peninsula between Tampa Bay and the Gulf of Mexico; numerous causeways and bridges.

Suwannee River. Beautiful river made romantic by the song, "Old Folks at Home." At White Springs is the Stephen Foster Memorial; dioramas in the museum tell his songs' stories.

Tallahassee. The state's capital, of which the grounds are a beautifully wooded park. There are many points of interest, among them Walker Memorial Library, containing Indian relics and pieces of armor dating from De Soto.

Tampa. South Florida's most important commercial city. An early Spanish trading post, it was

Hawaii Vis. Bur.   *Hawaii's Capitol—Iolani Palace*

later a base for José Gaspar and other west coast pirates. Among important exports, its handmade cigars are famous.

GEORGIA

Andersonville Prison Park. Site of famous Confederate Military Prison. Providence Spring, said to have bubbled forth in answer to prayers of prisoners, is still to be seen.

Athens. Home of the University of Georgia, chartered in 1785 and oldest state university. Overlooking Oconee River, Athens retains much of the atmosphere of the old South.

Atlanta. Capital of the state and one of the great railroad and banking centers of the United States. Here is the "Wren's Nest," home of Joel Chandler Harris. In Grant Park stands Fort Walker, historic breastworks of the Battle of Atlanta.

Chickamauga and Chattanooga National Military Park. Battlefields where Confederate and Union armies met in one of the great campaigns of the war.

Fort Benning. Largest infantry school in the United States, located nine miles from Columbus.

Macon. Industrial city in the State's great agricultural belt. A stately city, it is rich in historic tradition. Nearby, at Ocmulgee National Monument, 683 acres of prehistoric mounds disclose homes and relics of six successive Indian cultures.

Okefenokee Swamp. More than 600 square miles of swampland wilderness in southern Georgia and northern Florida. A wonderland of flowers, canopied by forests of gum, pine, bay, and cypress, it is inhabited by deer, bear, panthers, alligators, birds, and great varieties of fish.

Savannah. "Birthplace of Georgia," founded in 1733. Famous resort and great export city, Savannah's traditions are earliest in America.

Stone Mountain. Near Atlanta; immense granite dome, rises 800 feet above the surroundings.

HAWAII

Island of Oahu

Honolulu. This romantic south-seas tourist center, home of fabled Waikiki Beach and its sentinel Diamond Head, an extinct volcano, is the capital of the newest state. Iolani Palace, the present capitol, served as the royal palace from the 1880's; its Throne Room is kept as it was during the reign of King Kalakaua. The world-famous Bishop Museum houses Polynesian and royal Hawaiian objects; other museums include Queen Emma Museum and the Royal Mausoleum. Foster Park Botanical Gardens with its orchid displays and Kapiolani Park with both an aquarium of Pacific marine life and a zoo with a tropical bird aviary are well worth a visit. Also see the National Memorial Cemetery of the Pacific and the University of Hawaii.

Hawaii Vis. Bur.   *Iao Needle—Maui*

Kahana Beach. Here breadfruit, bamboo, and mango trees mark the site of a once populous old-Hawaiian settlement.

Kaneohe. Fantastic coral formations may be viewed from glass-bottomed boats.

Koko Head Park. Named for the volcanic mountain which dominates the head of land that juts into the sea. A feature of the area is lovely Hanauma Bay, created by volcanic action 10,000 years ago. At the Blow Hole the mighty sea forces its way through a tiny hole in the lava ledge to form miniature geysers rising high in the air.

Laie. A quiet Samoan village where visitors are charmed by the splendid white Mormon Temple, often called the Taj Mahal of Hawaii.

Nuuanu Pali. Seven miles from Honolulu, at the head of the Nuuanu Valley, lies this scenic mountain pass which separates the windward and leeward sides of the island. Here in 1795, during the wars to unite the islands, Kamehameha the Great forced defeated Oahuan warriors over the precipice to death on the rocks below.

Pearl Harbor. United States Naval base. In the harbor lie the sunken remains of the battleships *Arizona* and *Utah*, which entomb the heroes who went down with their ships on December 7, 1941. A monument was erected over the sunken battleship *Arizona* and was dedicated on Memorial Day in 1962.

Sacred Falls. Off the highway near Hauula a crystal stream leaps from the sheer cliffs to a beautiful pool below.

Island of Hawaii

Hawaii Volcanoes National Park. The Kilauea-Mauna Loa section of the park features volcanoes, lava flows, and luxuriant tropical forest with giant fern trees. Mauna Loa, largest single mountain mass in the world, rises 13,680 feet above sea level. A trip to the rim of Kilauea is rewarded with a view of lava activity in the firepit. Kilauea last erupted in 1961 and enlarged the Puna Lava Flow, an area of raw cinder cones and steaming lava.

Hilo. Famous as "Orchid Capital," its daily export of blooms numbers in the thousands. The lovely Japanese gardens and a banyan-tree grove in Liluokalani Park delight visitors, as do the ever changing colors in Rainbow Falls, where the Wailuku River cascades over a volcanic ledge. Another spectacular waterfall, Akaka Falls, lies north of the city. The Lyman Memorial Museum features a collection of ancient Hawaiian relics.

Honaunau. The City of Refuge National Historical Park was established in July 1961. It has 180 acres and contains temple foundations dating from the twelfth century.

Kailua-Kona. Site of the old Hulihee Palace, a reminder that the Islands' monarchy sprang from this island. Here also is the First Christian Church, established by missionaries in 1806.

Kalapana Black Sand Beach. A tremendous coco palm grove surrounds a jet-black beach, where the volcanic sands are washed by creamy surf.

Kohala. Birthplace of Kamehameha I, marked by the original of the statue of the king which stands in the Palace Square in Honolulu. Nearby is the scenic Pululu Valley once the site of ancient Hawaiian temples.

Mauna Kea. Its snow-capped majesty marks the highest point in the Islands; 13,796 feet

Puna Warm Springs. Near Kapoho. Lush tropical vegetation, including green ferns, surround waters warmed by volcanic heat.

Island of Maui
Haleakala National Park. The Park features the enormous Haleakala Crater; its floor measures twenty-five square miles, and the rim extends twenty-one miles around. Along the walls, and in the crater itself, bloom the rare silversword plant. Rain clouds, borne by the trade winds, find their way into this crater to form world-famous cloud effects.

Iao Valley. A dense green gorge west of Wauluku, best known for the Iao Needle. This freak formation, a solid mass of volcanic stone, rises 1,200 feet.

Lahaina. Romantic and easy-going, this town was a favorite spot of ancient Hawaiians. Here Kamehameha II established the royal capital. Nineteenth-century whalers, also delighted with the locale, spent their winters here.

Maui Ditch. On the winding highway between Wailuku and Hana, through dense forests of bamboo and ape plant, here and there are glimpses of the water gates and tunnels of this famous ditch, important in the battles between the warriors of Maui and Hawaii.

Island of Kauai
Hanalei Valley. This majestic valley can be viewed from a look-out on the highway which skirts the northern coast of the island.

Waimea Canyon. Volcanic rock, sculptured and plunging to great depths, reflects varied colors as mist and sunshine alternately sweep the great expanse which stretches miles into the interior of the island from the south shore highway.

Wet and Dry Caves of Haena. Situated at the base of a volcanic cliff, these large caverns were the gathering place of Hawaiian chiefs. The dry cave is large enough to drive in an automobile.

IDAHO

American Falls. Dam and reservoir on the Snake River, adjacent to Fort Hall Indian Reservation.

Boise. Idaho's capital and largest city is the center of a rich agricultural and mining region.

Of interest are the Municipal Art Gallery, the Capitol, the Veterans' Hospital, and the Julia Davis Park. East of Boise is Arrowrock Dam; northwest is Black Canyon Dam.

Coeur d'Alene. Lumber, agricultural, and tourist city in noted Coeur d'Alene mining area. In this region are Coeur d'Alene and Pend Oreille lakes.

Craters of the Moon National Monument. Volcanic area of craters, cinder cones, lava flows, and stalactite caves; resembles craters on the moon.

Hell's Canyon of the Snake River. Spectacular canyon along the Snake at Oregon boundary. From four to nine miles wide and almost 8,000 feet deep at some points; its walls are red, purple, yellow.

Idaho Falls. Center of rich farming region. Idaho Falls of the Snake and the Lavas are nearby.

Primitive Area. In central Idaho, 1,500,000 acres in Boise, Challis, and Sawtooth National Forests; access is by pack-train only. Rugged mountains, gigantic precipices, and sheer canyons.

Sun Valley. World-famous winter sports resort, popular also with summer tourists. To the northwest is the Lost River Range that includes Mt. Borah, Idaho's highest mountain (12,652 feet).

Twin Falls. Center of great irrigated agricultural region. Nearby are the Blue Lakes, the Twin Falls-Jerome Bridge, Twin and Shoshone Falls, and the Thousand Springs.

ILLINOIS

Brookfield Zoo. Chicago Zoological Park, in Brookfield, near Chicago. Most of the lairs are of the natural habitat type, without bars. The zoo contains everything, from insects to elephants.

Cahokia Mounds State Park. Site of Monks Mound, largest Indian mound in United States. A museum exhibits many Indian relics taken from the Cahokia Mounds. The park is near East St. Louis.

Champaign-Urbana. These two cities are the home of the University of Illinois.

Charleston. Scene of fourth Lincoln-Douglas debate and site of the grave of Dennis Hanks. Near town is the Lincoln Log Cabin State Park, containing reconstructed cabin of Lincoln's father, built on the original foundation.

Chicago. Vibrant and noisy, the great metropolis of the Midwest sprawls over a 200-square-mile area at the lower western end of Lake Michigan. It is the greatest railway center in the world and the heart of an arterial system of steel rails, concrete roads, waterways, and airways, radiating in all directions. It is called the Convention City. Outstanding points of interest include the Art Institute,

Chicago Natural History Museum, parks and boulevards, Navy Pier, University of Chicago, Planetarium, Aquarium, Soldier Field, Museum of Science and Industry, Board of Trade, Prudential Building, McCormick Place, Stock Yards, airports, and Chicago Zoological Park in Brookfield, where most of the lairs are of the natural habitat type.

Dickson Mounds State Park. Here 230 Indian remains have been exhumed and left lying in their original postures; located near Havana.

Galena. Scene of famous "lead rush" in first half of the 19th century. Built on terraces cut by the Galena River, Galena's houses are placed like chalets in an Alpine village. The first home of Ulysses S. Grant and the one presented to him after the Civil War are still standing.

Great Lakes Naval Training Center. Only major naval unit in Middle West and one of three naval training centers in the United States.

Lincoln's New Salem State Park. Authentic reproduction of the village, near Springfield, where Lincoln spent six of his early Illinois years.

Peoria. Third largest city of state, on the northwest bank of the Illinois River. Located in the Corn Belt, Peoria distilleries draw a considerable portion of the corn crop of the area.

Rock Island Arsenal. Located on an island in the Mississippi River. Here are stored a portion of the nation's war supplies; War Museum.

Springfield. Capital of Illinois. The Capitol and the Lincoln Home are attractions. The Lincoln Tomb in Oak Ridge Cemetery contains the bodies of Abraham Lincoln, his wife, and three of his children.

Starved Rock State Park. On Starved Rock, rising 140 feet above the Illinois River, La Salle built Fort St. Louis in 1683. Legend has it that a band of Indians starved to death on the rock.

INDIANA

Bedford and Bloomington. In limestone region from which comes stone for world's finest buildings. Visiting parties are taken through quarries.

Brown County State Park. In the heart of Indiana's rolling hills and dense woods, Brown County's scenic beauty is world famous. Nearby Nashville has a world-famous art colony.

Gary. In Gary and nearby cities are northern Indiana's steel and associated industries; most of the large plants have tours for visitors.

Greenfield. Birthplace of James Whitcomb Riley. Riley Memorial Park contains the original "Old Swimmin' Hole," and the old Riley Homestead is an authentic and notable Riley museum.

Indiana Dunes State Park. East of Gary. Three miles of beach and sand dunes on Lake Michigan; wooded inland.

Indianapolis. State capital, Indiana's largest city, and center of a highly developed agricultural area. The city is large and spacious; four great diagonals reach its center, Monument Circle, holding the impressive Soldiers and Sailors Monument. West of the Circle are the Capitol and Indiana University Medical Center for Children, where murals and stained-glass windows illustrate the Riley poems. East of the Circle is the Riley Home. (See also Greenfield.) North are the World War Memorial, Scottish Rite Cathedral with its carillon and gigantic pipe organ, and the Benjamin Harrison Home. At the famous Indianapolis Motor Speedway, Memorial Day races are an annual event.

International Friendship Gardens. A hundred acres of beautifully landscaped gardens of all nations; colorful outdoor theaters. Near Michigan City.

Nancy Hanks Lincoln Memorial. In Lincoln State Park, near Gentryville, are the grave of Nancy Hanks Lincoln and site of the Lincoln cabin, the fireplace reconstructed from the original stones. At nearby Rockport is the Lincoln Pioneer Village.

Ky. Dept. of Cons.
Cumberland Falls, Ky. — Scene of the "Moonbow"

Ewing Galloway
New Orleans' Imposing Jackson Square

New Harmony. Wabash village founded in 1815 by the Rappites, a religious communal sect. Many of the original stone buildings still remain: Old Rappite Fort, the Community Houses, the Rapp-Maclure Home, the Workingmen's Institute.

South Bend. Home of University of Notre Dame and automobile manufacturing city.

Spring Mill State Park. A pioneer village has been restored around an old stone grist mill, powered by water-wheel. Meal is ground for visitors.

Turkey Run State Park. On picturesque Sugar Creek, famed for its lovely gorges cut deep in sandstone. There is much virgin forest.

Vincennes. Mellow old city on the Wabash River, in the midst of rich orchards and farmlands. First settlement in Indiana, a key city of the old Northwest Territory, and Indiana Territory's capital, Vincennes is rich in history. One of its many fine monuments is the George Rogers Clark Memorial.

Wyandotte Cave. Spectacular caverns, with 20 miles of passages on five levels; near Wyandotte.

IOWA

Amana Colonies. Seven former communal villages, the first settled in 1855.

Davenport. See nation's largest roller gate dam which operates even when the Mississippi is frozen. Wild Cat Den with old dam and grist mill nearby.

Des Moines. State capital; atop Capitol Hill, the gold-domed Capitol dominates other State buildings clustered on the slopes.

Dubuque. In lead and zinc region, one of the oldest towns in Iowa. Trappist Monastery nearby.

Effigy Mounds National Monument. Near McGregor. Great Indian mounds, some 300 feet long, trace shapes of eagles, bears, and wolves. Nearby are McGregor State Park and Painted Rocks.

Fort Dodge. Vast gypsum fields surround town. In Dolliver Memorial Park, sandstone bluffs edge the Des Moines River.

Iowa City. The University of Iowa campus includes the Old Capitol, of Territorial days.

Maquoketa Caves State Monument. Natural bridge, balanced rock, and many limestone caves.

Okoboji Region. Fine resort area including East and West Okoboji and Spirit lakes.

Sioux City. Grand View Park offers a natural amphitheater with outdoor pavilion. At War Eagle Park is the grave of the famous Sioux chief.

KANSAS

Abilene. Famous frontier cow town. Site of the Eisenhower family home and the new Dwight D. Eisenhower Museum.

Council Grove. Historic town on the site of an old Santa Fe Trail campground.

Dodge City. On the Arkansas River, the town was famous as a watering place and shipping point for Texas longhorns. "Boot Hill Cemetery" contains early-day memorials.

Emporia. Birthplace and home of the late William Allen White, noted Kansas journalist. His residence is a show place of the city.

Fort Riley. Military reservation and cavalry post established in 1853 to protect travelers on the Santa Fe Trail from the Indians. The Old Territorial Capitol is nearby.

Fort Scott. On the site of an old army post, of which Carroll Plaza was the parade ground.

Hanover. Near town stands the only original unaltered Pony Express station in the country.

Kaufmann & Fabry

Chicago, Giant of the Midwest

John Brown Memorial State Park. This park commemorates the Battle of Osawatomie and includes a museum built around John Brown's cabin, moved from its original site.

Kansas City. Important meat-packing and railroad center. Nearby is Wyandotte County Park and Lake, largest in the state, and also the Old Shawnee Mission.

Lawrence. Scene of the Quantrell raid in 1863; now home of the University of Kansas and Haskell Institute, government Indian school.

Leavenworth. First settlement in Kansas Territory. Nearby Fort Leavenworth, military reservation and officers' training school, was established in 1827. A Federal penitentiary is also at Leavenworth.

Topeka. Capital of Kansas and third largest city. In the Capitol are the "John Brown" murals, by John Steuart Curry. Some of the historical buildings are Constitution Hall, Underground Railroad Station, and the Old Settlers' Memorial Cabin in Gage Park.

Wichita. Largest city in Kansas and important airplane manufacturing and oil refinery center. Points of interest are the Art Museum and the airplane wind tunnel at the University of Wichita campus.

KENTUCKY

Abraham Lincoln Birthplace National Historic Site. At Hodgenville, the birthplace of Abraham Lincoln. Some of the original Lincoln farm is contained in the park. A granite memorial room encloses the birthplace cabin.

Audubon State Park. Near Henderson, once the home of John James Audubon. The museum has superb collection of his bird prints and other Audubonia.

Berea College. Oldest and largest of Kentucky's mountain schools, at Berea; weaving produced by the school is famous throughout America.

Cumberland Falls State Park. Woodland park on the Cumberland River, where a semi-circular cataract makes a "moonbow" under the moon's light.

Cumberland Gap National Historical Park. Daniel Boone blazed a trail into Kentucky through this pass in the Cumberland Mountains. The park is in three states—Kentucky, Tennessee, and Virginia.

Frankfort. Capital of Kentucky. The Capitol and the Governor's Mansion are characteristic of the stately south. The old Capitol now houses the Kentucky Historical Society. Daniel and Rebecca Boone are buried at Frankfort.

Harrodsburg. First white settlement in Kentucky. In Pioneer Memorial State Park is a replica of old Fort Harrod, complete with stockade and blockhouse. The Lincoln Marriage Temple contains the cabin that was the first home of Lincoln's parents.

Kentucky Lake. Man-made lake, formed by Kentucky Dam across the Tennessee River.

Lexington. Home of the University of Kentucky and "Heart of the Bluegrass." Here some of the world's finest race horses have been raised.

Louisville. Kentucky's largest and most important commercial and industrial city; also the famous home of the Kentucky Derby. Points of interest include the Public Library and the Filson Club, with collections of Kentucky history; Iroquois Park; the grave of George Rogers Clark.

Mammoth Cave National Park. In 150 miles of charted passages there are three underground rivers, eight waterfalls, and two lakes, as well as stalactite, stalagmite, and drapery formations.

My Old Kentucky Home State Shrine. The stately old mansion where Stephen Foster was a guest when he wrote the beloved song.

LOUISIANA

Avery Island. Of special interest are the McIlhenny Jungle Gardens and the bird refuges.

Baton Rouge. Winding drives, scenic lakes and bayous, and ante-bellum homes typify the state capital. The present Capitol was built in 1932. The Old Capitol overlooks the river.

Grand Isle. Semitropical island in Gulf of Mexico. Once headquarters for the pirate, Lafitte.

New Iberia and St. Martinville. Both on Bayou Teche and settled by French colonists in 18th century. In cemetery at St. Martinville Church the Evangeline Monument marks grave of Emmeline Labiche, supposedly Longfellow's "Evangeline."

New Orleans. Founded in 1718, the "Crescent City" was nearly 100 years old when it became part of the United States. The flavor of its glamorous past has been preserved in the Vieux Carre, or French Quarter. A tour through the Quarter leads along narrow streets, lined with two- and three-story French and Spanish buildings, many with balconies of exquisite iron-lace grill. Views through doorways and porte-cocheres reveal lovely patios and oleander, camellia, and wistaria gardens. Surrounding Jackson Square are St. Louis Cathedral, the Pontalba buildings, the Presbytere, and the Cabildo, home of the Spanish governors. At the French Market foodstuffs are vended in the open. The Mardi Gras, preceding Lent, is one of the most famous annual events.

MAINE

Augusta. State capital on the Kennebec River, which bisects the town. The State House, started in 1829, retains only the graceful portico of Bulfinch's design; the 1911 enlargement demolished the major portion. The James G. Blaine House serves as the Executive Mansion. Fort Western has been restored to its appearance in 1754.

Warren—Md. Dept. Econ. Develop.
Old Senate Chamber, State House, Annapolis

Massie—Mo. Res. Div.
Interesting Old Mill in Shannon County, Mo.

Bar Harbor. Fashionable resort on Mt. Desert Island, gateway to Acadia National Park. The park, a 41,634-acre tract, preserves a section of the magnificent wilderness that was once part of the French grant to Sieur de Monts.

Baxter State Park. More than 160,000 acres of spectacular wilderness. Within area is Mt. Katahdin, highest point in Maine (5,268 feet).

Brunswick. Seat of Bowdoin College. Longfellow stayed at Emmons House while teaching at the college. Harriet Beecher Stowe wrote her famous book "Uncle Tom's Cabin" at the house bearing her name.

Moosehead Lake. Largest lake in state. It is popular with trout and salmon fishermen.

Portland. State's largest city and principal port since colonial days. A monument in Longfellow Square honors Portland's noted citizen. The Wadsworth-Longfellow House, the writer's boyhood home, is open. The Museum of the Maine Historical Society and the Portland Museum of Art, including Sweat Mansion, are among the cultural institutions in the city. Near South Portland, the Portland Head Light, built in 1791, marks the entrance to the Portland Harbor.

MARYLAND

Annapolis. State capital, famous as the seat of the U. S. Naval Academy and St. John's College. In the Old Senate Chamber of the State House, Washington surrendered his commission and (1784) the peace treaty with Great Britain was ratified.

Baltimore. Maryland's largest city, and an important port and industrial center. Products of its diversified industries leave the city through its great harbor on the Patapsco River. Of the many monuments marking the parks and squares, Washington Monument is best known. Among educational and cultural institutions are Johns Hopkins University, the Walters Art Gallery, Maryland Historical Society, and Enoch Pratt Free Library. Outside the city, on a point reaching into the harbor, lies Fort McHenry National Monument, site of the terrible bombardment during the War of 1812, which inspired Francis Scott Key to write "The Star-Spangled Banner."

Frederick. Barbara Fritchie House, reconstructed, contains relics supposedly used by the heroine of Whittier's poem. The home of Roger Brooke Taney, Chief Justice who handed down the Dred Scott decision, preserves his personal effects and those of his brother-in-law, Francis Scott Key.

Hagerstown. In a city park the house of the founder, Jonathan Hager, is maintained. It dates from 1739. Nearby is the Antietam National Battlefield Site.

Ocean City. Important seashore and fishing resort, with marlin ranking highest.

MASSACHUSETTS

Adams National Historic Site. Home of four generations of the John Adams and John Quincy Adams family; near Quincy.

Amherst. Home of Emily Dickinson, Noah Webster, and Robert Frost, and seat of two famous institutions of learning, Amherst College, founded in 1821, and the University of Massachusetts.

Boston. "Hub of the Universe," state capital, leading seaport, industrial and cultural center. Greater Boston, at the mouths of the Mystic and Charles Rivers, includes East Boston, Charlestown, South Boston, Roxbury, Dorchester, Brighton, Hyde Park, and other sections. In the crooked streets and narrow alleys of old Boston it is almost impossible to proceed in any direction without passing a spot of historic significance. The first free public school in America was established at Boston in 1635, followed next year by the founding of Harvard University at Cambridge. In Boston Navy Yard is the reconstructed U.S. frigate *Constitution*, revered by the nation as "Old Ironsides." A tablet marks the spot of the Boston Tea Party. Other points of interest are the Paul Revere House, Old North Church, Battle of Bunker Hill Monument, Boston Athenaeum, Faneuil Hall, King's Chapel, Massachusetts Historical Society Museum, Old South Meeting House, Old State House, The Common and Trinity Church.

Cambridge. Home of Harvard University, Massachusetts Institute of Technology, and Radcliffe College. Among the points of interest are the Botanic Garden, Fogg Museum, the noted glass flower collection at University Museum, and the homes of Longfellow and Lowell.

Concord. Noted for its historic and literary associations, it shares with Lexington the honor of being the "Birthplace of the American Revolution." It was the home of Emerson, Hawthorne, Louisa M. Alcott, Thoreau, and Channing. Hawthorne lived at Old Manse and Wayside, both standing; at Orchard House Miss Alcott wrote *Little Women*.

Gloucester. On the peninsula of Cape Ann, this great fishing port has been dominated by the seafaring tradition for more than 300 years.

Lexington. "Birthplace of American Liberty." It was here that the Minute Men faced the British redcoats and fired "the shot heard 'round the world." In the Common stands a bronze statue of a Minute Man, commemorating the event.

Martha's Vineyard. A large triangular island between Buzzard's Bay and Nantucket Sound, a few miles from the mainland. Its variety of scenery and lively villages have made it a popular summer resort.

Nantucket. Last seaward outpost of New England. With its crooked cobbled streets, quaint comfortable homes, yacht club, artists' colony, beaches, it still retains an atmosphere of the great whaling days. Places of interest are its Whaling Museum; Friends

Meetinghouse; Old Mill, dating from 1746; and Oldest House, built in 1686.

Plymouth. Here is enshrined "the cornerstone of the nation," Plymouth Rock, on which the Pilgrims landed in 1620. There are many fine 17th-century houses and other reminders of the Pilgrim era.

Salem. Witches, clipper ships, privateers, exotic wares from the Indies and China, wharves and docks laden with incoming and outgoing merchandise—all these are conjured up by Salem, historic treasure chest of New England. Nathaniel Hawthorne was born in an old gambrel-roofed house built in 1692; the House of Seven Gables, said to be the setting for his novel of that name, is one of the celebrated structures of Salem. Another is the Witch House.

Woods Hole. Cape Cod port; site of Oceanographic Institute and Marine Biological Laboratory.

MICHIGAN

Ann Arbor. Campus of the University of Michigan extends over half the town. The famous Nichol Arboretum features lilacs and peonies.

Benton Harbor and St. Joseph. Central points in the state's southern fruit belt. Benton Harbor holds an annual Blossom Festival; fruit is marketed wholesale at the Municipal Fruit Market.

Dearborn. Home of Henry Ford. Greenfield Village, model of early American town, exhibits historic buildings moved to this site—Henry Ford's Birthplace; Logan County Courthouse, where Lincoln practiced law; Edison Buildings, including Menlo Park Laboratory; and Luther Burbank's office.

Detroit. "Motor Capital of the World." The Rivera Murals at the Institute of Arts depicts its industrial greatness. In the channel of the Detroit River lies the great amusement park, Belle Isle. In the Detroit Zoological Park, at Royal Oak, animals roam in open areas resembling their natural habitat. Grosse Point, residential section, fronts on Lake St. Clair.

Holland. Dutch settled here in 1847. Now famous for the Annual Tulip Festival, the Netherlands Museum, and the Wooden Shoe Factory.

Isle Royale National Park. This island wilderness lies in Lake Superior fifty miles off the Upper Peninsula. Moose, coyote, mink, beaver, and snowshoe rabbit find a haven here. Wildflowers grow in profusion among the hardwoods and conifers.

Lansing. State capital and important automobile manufacturing center.

Mackinac Island. Boats ferry from Mackinaw City, where automobiles are left behind. Horse-drawn buggies and bicycles provide transportation to Ft. Mackinac, Astor Fur Post, and Arch Rock. Father Marquette is buried at St. Ignace, across the Straits of Mackinac.

Sault Ste. Marie. Michigan's first permanent white settlement. Through the famous "Soo" locks and St. Marys Ship Canal passes America's greatest marine commercial tonnage.

Traverse City. Deep-sea fishing resort in the heart of cherry land. Each May the Blessing of the Cherry Blossoms is held and there is an annual Cherry Festival in mid-July.

MINNESOTA

Alexandria. Resort town in western part of state. The much discussed Kensington Runestone is on exhibit here. Runic inscription on stone tells of visit to this area by Norsemen in year 1362.

Bemidji. Named for Chief Bemidji, leader of a band of Chippewa, whose settlement was located at the present site of the city on the southern end of Lake Bemidji. The legendary stories of Paul Bunyan add to the romantic folklore of Bemidji.

Duluth. Third of Minnesota's cities in size. From the western tip of Lake Superior, the city is 800 feet above the lake level. The Duluth-Superior port is fifth only to New York City in annual tonnage.

Grand Marais. Gateway to the Arrowhead country and the Superior-Quetico canoe country.

Hibbing-Virginia. Two important mining towns in the great Mesabi Iron Range. The largest open pit iron mine in the world is at Hibbing.

Indian Reservations. The WHITE EARTH INDIAN RESERVATION is the largest; authentic Indian customs and language are maintained at the RED LAKE RESERVATION. Chippewas are predominant.

Lake of the Woods. On the Canadian border, this region is a paradise for fishers and hunters.

Minneapolis. Largest city in the state, and one of the largest flour-producing cities in the world. It has a wonderful system of boulevards, lakes, and parks, including the Falls of the Minnehaha, immortalized by Longfellow's "Hiawatha." Minneapolis is the home of the University of Minnesota and the well known Minneapolis Symphony Orchestra.

Pipestone National Monument. Famed quarries of red stone, used by the Plains Indians for material for their ceremonial pipes.

Rochester. Home of famed Mayo Clinic. Health seekers descend on Rochester from every state in the Union and from nearly every country in the world. Transient guests double the official population; because of this unique turnover of inhabitants, the town has an amazing number of apartment buildings, rooming houses, and restaurants.

St. Paul. Capitol and second largest city. The industrial, social, and educational life of the entire state revolves around the hub of St. Paul and its Twin City, Minneapolis, just across the Mississippi River.

Superior National Forest. The nation's largest wilderness park. Only by water, foot, or seaplane can the camper or fisherman traverse much of this area. Hunting is prohibited within the greater portion of its boundaries.

MISSISSIPPI

Biloxi. First permanent settlement in the lower Mississippi River Valley. At nearby Ocean Springs, now a resort, d'Iberville founded Ft. Maurepas in 1699. The Old Lighthouse, built in 1848, still blinks its warning to sailors. Beauvoir, Jefferson Davis' last home, is near the city.

Gulfport. Coastal resort. Fort Massachusetts, used as a Federal prison during the Civil War, is on Ship Island, nearby.

Jackson. State capital. At the Old Capitol the Mississippi Ordinance of Secession was passed and Jefferson Davis made his last public appearance. The State Hall of Fame and the Department of Archives and History are housed in the modern War Memorial Building. Mississippi governors since 1842 have resided at the charming old Governor's Mansion. Manship Home was the Confederate headquarters during the siege of the city.

Natchez. Famous for two annual garden pilgrimages, the town is a virtual museum of ante-bellum mansions. The old Natchez Trail, which terminated at Nashville, Tennessee, started at Natchez.

Pascagoula. Modern shipbuilding center on Pascagoula Bay. Old Spanish Fort, constructed of oyster shells and moss, was built in 1718. Horn Island Light, built on stilts over the water, affords an excellent point for study of marine life. It is eight miles from town.

Vicksburg. Site of the famous Civil-War siege which lasted forty-seven days. The battleground of this campaign is well marked in the Vicksburg National Military Park.

Washington. Second Territorial capital and first capital of the state. Under "Burr Oaks," on the campus of Jefferson Military College, Aaron Burr stood his preliminary trial for treason.

MISSOURI

Arrow Rock State Park. On the Missouri River in the central part of the state. Here is the historic Old Tavern, built in 1834, and the Arrow Rock Academy, furnished with ante-bellum objects.

Hannibal. Important industrial city, and home of Mark Twain. Huck Finn and Tom Sawyer roamed the banks of the Mississippi River in this section. The Mark Twain Museum and Home contains much Mark Twain and Hannibal memorabilia.

Independence. Historic frontier town on the Missouri River, eastern terminal of Santa Fe and Oregon Trails; the Overland Stage started here. Permanent settlers came in 1825; the Jackson County Courthouse, still standing, was erected two years later. Independence is the home of former President Harry S. Truman and here is Truman Museum and Library.

Jefferson City. State capital. In the Capitol, of Carthage marble, are the murals of Thomas Hart Benton—a crowded design of Frankie and Johnny, Huck Finn and Nigger Jim, the James Boys, a political rally, and other subjects.

Kansas City. Second largest city, railroad center, and market place. Kansas City grew out of two frontier towns—Kansas, on the Missouri River, and Westport, four miles south, on the Santa Fe Trail. Incorporated as a city in 1853, it is noted for its packing plants and as one of the nation's largest horse and mule markets. Among the points of interest are William Rockhill Nelson Gallery of Art and Atkins Museum of Fine Arts; the $2,000,000 Liberty Memorial; Union Station, designed by Jarvis Hunt of Chicago; Swope Park, with its Zoological Gardens; and the Kansas City Museum.

Lake of the Ozarks. Large artificial lake, formed by impounding the waters of the Osage River behind Bagnell Dam. The lake is about 130 miles long and has an irregular shoreline of 1,300 miles.

Ste. Genevieve. First permanent settlement in state; noted for 18th-century French buildings.

St. Joseph. Important as a manufacturing, grain, and meat-packing center. Situated on the bluffs above the Missouri River, it was once the focal point of the trade lanes westward. This was the eastern end of the Pony Express route, and the Pony Express stables are still standing.

St. Louis. Largest city in Missouri, situated on the Mississippi River just below its junction with the Missouri. Founded by French fur traders, St. Louis is still an important market for raw furs, as well as a center for grain and the production of stoves, machinery, and other manufactures. Points of interest are the Old Courthouse, historically associated with the Dred Scott case; Wainwright Building, one of the first skyscrapers, completed in 1891; seven bridges across the Mississippi, Missouri Botanical (Shaw's) Garden, containing more than 12,000 species of trees and plants; Forest Park, site of the Louisiana Purchase Exposition in 1904; St. Louis Municipal Opera, an open-air theater seating 10,000; Jefferson Memorial, containing pioneer and Indian relics, trophies and gifts to Charles A. Lindbergh after his 1927 trans-Atlantic flight; St. Louis Zoological Garden; and the Jefferson Expansion National Memorial, commemorating westward expansion.

MONTANA

Anaconda. Smelting point for the ores of copper and zinc that are mined at Butte; the Anaconda smelter has the world's biggest smokestack.

Bear Tooth Mountains. Spectacular high plateau, (12,799 feet altitude), carved by deep canyons and jutting precipices, in southern Montana and northern Wyoming. A scenic highway travels through the area from Red Lodge to Cooke City, at the northeastern entrance to Yellowstone National Park. One of the many peaks is Granite, highest point in Montana (12,850 feet). Nearby is Grasshopper Glacier, where grasshopper hordes are frozen in the ice.

Butte. Second largest city and center of the state's great copper-producing region. Numerous old homes and other structures compete in interest with the nearby mines.

Custer Battlefield National Monument. Within today's Crow Indian Reservation, General George A. Custer, with a command of about 262 men, made his ill-fated stand against overwhelmingly large numbers of Sioux and Cheyenne Indians. Marble slabs mark the spots where the men fell, and the cemetery also holds the graves of many soldiers and civilians removed from other forts of the Northwest.

Flathead Lake. Largest body of fresh water west of the Great Lakes, surrounded by magnificent mountain scenery. To the south is the Flathead Indian Reservation and the National Bison Range.

Fort Peck Dam. The huge earth-fill dam across the Missouri forms a 175-mile-long lake.

Glacier National Park. Almost 1,600 square miles, high at Montana's Canadian border, this park is a glacier-carved Rocky Mountain wonderland. The park has sixty small glaciers, in the process of disappearing, and nearly 200 glacier-formed lakes. Alpine flowers and other vegetation are of interest, as well as animal life, including the white mountain goat, elk, moose, bighorn sheep, deer, bear, and many others. Park is part of Waterton-Glacier International Peace Park.

Helena. Capital of Montana. Its streets still follow the line of gulches, as they did when this mining town was first laid out. Helena's oldest building still stands; the State Historical Library has a collection of Custer and Indian relics; a feature of the State Capitol is the paintings depicting the state's history.

Lewis and Clark Cavern State Park. Largest limestone cavern in the Pacific Northwest.

Virginia City. This colorful early gold mining camp has been restored to the 1865 period.

Minneapolis—Nation's Flour-Milling Center

Minneapolis C. of C.

Busy Soo Locks, at Sault Ste. Marie, Michigan

Mich. Tour. Council

Glacier Nat. Park

Grinnell Glacier, in Glacier National Park

Union Pacific R.R.

Immense Hoover Dam at Night

Yellowstone and Powder Rivers. Heading in Wyoming, but traveling most of its length through Montana and draining the whole southeastern corner of the state, the Yellowstone and its tributaries are closely associated with the state's present agricultural economy and with its eventful Indian and cow-country history. On the Yellowstone are Miles City and Billings, rip-roaring cow towns in the day of cattle empires and still holding the flavor now that the range has taken on the character of irrigated ranches. Here, emptying into the Yellowstone near Miles, is the famous Powder River; from this country still comes some of Montana's best beef.

NEBRASKA

Arbor Lodge State Park. At Nebraska City, home of J. Sterling Morton, sponsor of Arbor Day and a distinguished Secretary of Agriculture. The 52-room mansion has fine pioneer and art collections; the arboretum is a mass of color from spring to fall.

Boys Town. A thousand-acre city, 11 miles from Omaha, founded by Father Flanagan for homeless boys, and managed largely by the boys themselves.

Homestead National Monument of America. Site of first land claimed under the Homestead Act of 1862. Includes Daniel Freeman's pioneer cabin.

Kearney. Near Fort Kearney State Park, site of historic Fort Kearney on the Oregon Trail. Farther west is Gothenburg, where a fur-trading post and Pony Express cabin are preserved; lower 96 Ranch, another Pony Express station, is near Gothenburg.

Kingsley Dam. This earthen dam forms the 23-mile-long Lake McConaughy, which affords excellent boating and fishing; near Ogallala.

Lincoln. Capital of the state and home of the University of Nebraska. From far across Nebraska's prairie, the towering 400-foot shaft of the modern Capitol can be seen.

Omaha. Railroad center of Nebraska and the state's largest city. From here the first transcontinental railway was built across the prairies. Points of interest include the Union Passenger Terminal, the Union Stockyards and meat-packing plants. The Joslyn Memorial, tribute to the founder of the Nebraska Western Newspaper Union, houses an art gallery and a concert hall.

Oregon Trail. Crosses the whole state of Nebraska. All along the course of the wide, shallow Platte River, choked with sand, are markers and monuments showing the path of the white-topped wagons, the Pony Express, and the Overland Stage. Later, the first transcontinental railroad was built along this path. Famous Chimney Rock may be seen along this trail.

Pioneer Village. In Minden a two-block area of twelve museum buildings houses pioneer items.

Scotts Bluff National Monument. Scotts Bluff, a great butte of sandstone and clay in the western

part, was an important landmark of the Oregon Trail. Mitchell Pass, a defile through the Bluff, was made passable for wagons in 1852 and used thereafter; the Pony Express and the first transcontinental telegraph went through it. Today's Oregon Trail Museum at its foot contains relics, paintings, and dioramas of the Trail; an automobile road leads to the summit for a view of distant Trail landmarks.

Valentine. Near the junction of the Minnechaduza and Niobrara rivers. Here are dozens of small lakes offering boating, fishing, and camping.

NEVADA

Carson City. Smallest state capital; grown from a Pony Express station and mining-boom town named for Kit Carson. Besides the Capitol, points of interest include the Supreme Court and Library Building, the office of the Carson City *Daily Appeal*, where a newspaper has been published since 1865, the old firehouse, the old mint, the Abe Curry House, and other old homes. The Carson Indian Agency, School, and Museum are nearby.

Hoover (Boulder) Dam. Gigantic dam across the Colorado's Black Canyon, vital to water supply of the whole Colorado River Valley; reservoir is enormous Lake Mead, where water backs up for more than a hundred miles and spreads into tributary canyons. Lake Mead National Recreational Area, surrounding almost entire lake, makes it a vast play center.

Las Vegas. Glamorous winter-summer resort featuring gambling and spectacular shows.

Lehman Caves National Monument. Near Baker; delicate, beautiful stalactites and stalagmites.

Reno. Notorious for gambling and easy divorces, Reno is still a mining town and increasingly an all-year sports center. On the Truckee River in the Sierras, with Lake Tahoe to the south and Pyramid Lake to the north, it has great scenic attraction. The University of Nevada is at Reno.

Ruth. Site of large open-cut copper pits.

Tonopah. Famous old mining town and supply center for Nevada's south-central mining country.

Virginia City. Site of the Comstock Lode, colossal gold and silver lode that made endless millions. Largely "ghost" today, the town preserves many famous old buildings—the Crystal Bar, Piper's Opera House, International Hotel, *Territorial Enterprise* office, where Mark Twain reported and edited.

NEW HAMPSHIRE

Berlin. Northern manufacturing city, split by rushing waters of Androscoggin River, which form Berlin Falls; important ski center.

Concord. State capital. Points of interest include granite quarries, the State House, Historical Society and Museum, Kent House, State Library.

Dixville Notch. Sheer rock prominences rise on each side of a two-mile section of highway.

Franconia Notch. Deep cut between the Kinsman and Franconia ranges. Here is the 40-foot granite profile of the Old Man of the Mountain, made famous by Hawthorne's *The Great Stone Face*. In the vicinity are Indian Head, Echo Lake, Cannon Mountain Aerial Tramway, and the Flume, a deep rift with high granite walls.

Hanover. Seat of Dartmouth College, founded as an Indian school in 1769.

Lake Winnipesaukee. The state's largest lake and well-known New England resort area.

Manchester. Largest city and great industrial center. Located on Merrimack River at Amoskeag Falls, Manchester's industrial growth dates from the development of the Amoskeag Mills in 1810. Points of interest include the Manchester Historic Association Building, Home of General John Stark, Stark Park, St. Anselm's College, the Currier Art Gallery, Weston Observatory, and Massabesic Lake.

Mount Washington. Highest point in New Hampshire; a bald, rocky peak of the Presidential Range, 6,288 feet above sea level. The three-mile cog railway and a motor road lead to the summit.

Portsmouth. Port city for New Hampshire since before the Revolutionary War. Today it is a shipping and service center for the United States Navy Yard situated in the harbor. John Paul Jones' *Ranger* was built here; so was the 74-gun ship *Washington* (1815), as well as the *Kearsage* of Civil War fame. There are many notable structures in the vicinity, including Old State House, Pitt Tavern, Fort Constitution, Fort Stark, the John Paul Jones House, and the Jackson House.

NEW JERSEY

Atlantic City. Warmed by the Gulf Stream and protected by the New Jersey pine belt, this pleasure resort on the Atlantic Ocean is visited by some 16,000,000 persons annually. It is famous for its boardwalk and five great piers. The Miss America Pageant is held here every year.

Burlington. Settled by Quakers about 1677 and one time capital of the Province of West New Jersey, Burlington is most noted for its historic associations. Among its interesting old buildings are the James Fenimore Cooper House, Thomas Revel House, Friends Meeting-House, and General Grant House.

Delaware Water Gap. On the Pennsylvania border, where the Delaware River cuts through the rocky ridge of the Kittatinny Mountains, the Gap affords a pass for railroads and highways. It is surrounded by a picturesque resort section.

Edison Laboratory National Monument. At West Orange; the prototype for industrial laboratories of today. Established in 1887.

Elizabeth. Settled in 1664, Elizabeth is a residential suburb of the New York and New Jersey metropolitan area, and an important industrial center. Of interest are the Statue of the Minute Man, Boudinot House, Galloping Hill Monument, Nathaniel Bonnell House, Belcher Mansion, and others.

Highlands. Fishing village and summer resort on the Atlantic Ocean. Nearby is the Navesink Lighthouse, which flashes its beam 22 miles at sea. North of Highlands, on Sandy Hook, is the Sandy Hook Lighthouse, 85 feet high and built in 1763.

Jersey City. Situated on the Hudson River, at the western end of the Holland Tunnel, Jersey City is a shipping, manufacturing, and industrial center. Settled about 1630, the city has many reminders of its Dutch forebears, such as the Old Bergen Church and the Statue of Peter Stuyvesant.

Lakehurst. Famous for many years for the U. S. Naval Air Station for lighter-than-air ships.

Morristown National Historical Park. Headquarters for Washington and his men during the winters of 1777 and 1779-80.

Newark. Largest city in the state, active shipping port, and one of the leading manufacturing centers of the country. Newark's airport is one of the busiest in the world. Among the interesting relics of Newark's past are Military Park, used in colonial days as a drill ground; Trinity Episcopal Church built in 1743; John Plume Home, probably erected in 1710; Newark Academy, founded in 1774 and used as barracks by Revolutionary troops.

Paterson. Known as the "Silk City," because the manufacture of silk is one of its leading industries, Paterson is situated on the Passaic River, which supplies power for many manufacturing plants. Paterson Museum contains an excellent mineral exhibit and the first submarine built by John Holland in 1878.

Princeton. Seat of Princeton University, situated on a 1,300-acre campus. Most famous of the many fine buildings is Nassau Hall, built in 1756. The Chapel is the largest university chapel in America.

Trenton. Situated at the headwaters of the Delaware. Important historically, Trenton was the scene of one of the most decisive battles of the Revolution. See the Trenton Battle Monument. The State House is composed of a part of the original structure of 1792; the Annex houses the state museum.

Washington Crossing State Park. Here Washington made his famous crossing of the Delaware. McKonkey Ferry House, restored to its appearance in colonial days, is maintained as a museum.

NEW MEXICO

Acoma Pueblo. On Acoma Rock 60 miles west of the Rio Grande, this fortress-like Indian village is thought to be the oldest community in the United States. Enchanted Mesa is seen from the pueblo.

Albuquerque. Largest city and important financial and transportation center. Old Town section has much of the picturesque Spanish period; the new town is a modern business and residence section. Buildings of the University of New Mexico are in Spanish-Pueblo architecture, as are many fine residences. Of interest are Old Town Plaza, with the Church of San Felipe de Neri (built in 1706); the Sandia Mountain Rim Drive; and Isleta Indian Pueblo.

Bandelier National Monument. In the beautiful Frijoles Canyon where a northern wall is lined with cliff houses and cave rooms. On the floor of the canyon is the Tyuonyi Pueblo, with its ceremonial kivas, excavated 1908–10.

Carlsbad Caverns National Park. Vast limestone caverns, world-famous for size and beauty, are central feature of 49,448-acre park. Many miles have been explored, on three levels; but the actual size of the caverns is still unknown. In the vast chambers open to the public, huge stalactites and stalagmites and massive draperies form King's Palace, Queen's Chamber, Big Room, Giant Dome, Rock of Ages, and many others. In summer several million bats fly from the caverns every evening.

Chaco Canyon National Monument. Prehistoric Indian ruins, the Pueblo Bonito dates back more than 1,000 years and is believed to have housed in four stories 1,200 people.

Fort Sumner. Ruins of Old Fort Sumner where Kit Carson held 7,000 Navajos prisoner for four years. Billy the Kid is buried nearby.

Fort Union National Monument. Ruins of an important outpost, and guardian, of the Santa Fe Trail; near Watrous.

Gallup. Coal mining center, once an Overland Stage station; today it attracts nation-wide interest with Intertribal Ceremonial each August, when Navajos, Apaches, and Pueblo Indians gather for tribal dances. The Navajo Reservation and the Zuñi Pueblo Reservation are nearby.

Lincoln. Theater of the famous Lincoln County War of the late 1870's. Lincoln County Courthouse State Monument, branch of the Museum of New Mexico, displays local art and traveling exhibits, and houses much historical and archaeological material.

National Monuments. See Bandelier, Chaco Canyon, Fort Union and White Sands and U.S. National Park System, pages 228-29.

Pecos. Ranch town and outfitting point for hunting and fishing on Pecos River. Nearby are ruins of Pecos Pueblo (about 1350) and Apache Canyon where Geronimo, great Apache chief, met defeat.

Santa Fe. Capital of New Mexico and a capital city for 300 years under the flags of Spain, Mexico, the Confederacy, and the United States. Oldest city in the West, it was the terminus of the Santa Fe Trail and was trade and supply center for the whole southwest for many years. Points of interest are the Plaza; the adobe Palace of the Governors, dating from the city's earliest years and housing some of the West's finest historical and archaeological collections; ancient San Miguel Church; Cathedral of St. Francis; art and historical museums; Seton Village.

Taos. Famed artists' and writers' colony. There are three settlements: Taos (Don Fernando de Taos), Spanish town, with the home and grave of Kit Carson; Taos Pueblo, with the ruins of the Mission of San Geronimo de Taos, founded in 1600; and Ranchos de Taos, an old Indian farming center, whose massive adobe church, built in 1772, is one of the most impressive in the southwest. Taos Canyon is nearby.

White Sands National Monument. Desert of glistening white gypsum near Alamogordo. In this vicinity the atom bomb was first tested.

NEW YORK

Adirondack Forest Preserve. Covering almost two-thirds of the state north of the Mohawk River, between Lakes Champlain and Ontario, this park consists of several million acres of private and public land, comprising lakes, mountains, and valley regions. Included is Adirondack Forest Preserve, more than 2,000,000 acres of primitive forest land. Famous Ausable Chasm, three miles from Lake Champlain, is in the preserve; also Lake Placid and Saranac Lake.

Albany. State capital, industrial city, and freight center. Founded in 1614 by the Dutch, it is the second oldest permanent white settlement in the thirteen colonies. Points of interest include the Capitol, costing $25,000,000, and Schuyler Mansion.

Buffalo. On the eastern end of Lake Erie, Buffalo is one of the nation's great ports and grain distributing centers. Among the points of interest are the International Peace Bridge, McKinley Monument, and Albright Art Gallery.

Catskills. Scenic wooded mountain area west of the Hudson. Kingston, the first state capital, and Rip Van Winkle country are nearby.

Finger Lakes Region. In central part of state, it includes the lakes Cayuga, Keuka, Seneca, and Watkins Glen State Park.

Lake George. This beautiful body of water lies between majestic mountains in Adirondack Forest Preserve. At its southern end are the ruins of Fort William Henry; at the northern tip is Fort Ticonderoga, captured from the British in 1775 by Ethan Allen.

Long Island. Extends from the mouth of the Hudson River 125 miles along the Connecticut coast, from which it is separated by Long Island Sound. Considered today as one of America's most important playgrounds, with great country estates, beaches, and resorts, Long Island is rich in historic tradition. The Battle of Long Island was fought here; Captain Kidd is said to have buried his gold on nearby Gardiner's Island; the Whaling Museum and Whaler's Church at Sag Harbor are interesting reminders of whaling days. At the extreme tip is Montauk Point Light.

New York City. Greatest metropolis in the world today, New York City embraces five boroughs—Manhattan, business and financial district; Queens, Brooklyn, and the Bronx, industrial and residential areas; and Richmond (Staten Island), reached by bridge. Flags of all nations may be seen on vessels in the harbor; people of all nationalities are a part of its heterogeneous population. Major among its attractions are Wall Street and the New York Stock Exchange; Central Park and other parks, boulevards, highways, and tunnels; Bronx Zoo; Aquarium; Greenwich Village; the Bowery; Chinatown, Harlem; famous cathedrals and churches; Broadway and Times Square; Metropolitan Opera; Empire State, United Nations Building, and other buildings and skyscrapers; City Hall; great museums and art galleries; General Grant's National Memorial; Statue of Liberty National Monument; Radio City; Columbia University; Brooklyn Navy Yard; Governor's Island; Coney Island.

Niagara Falls. Acclaimed one of the seven natural wonders of the world.

Poughkeepsie. Home of Vassar College, founded in 1861 as one of the first institutions in the world to offer women the same educational advantages that universities offered men. Near the city is Hyde Park, estate of the late Franklin D. Roosevelt.

Syracuse. Home of Syracuse University. Once known as "Salt City," because of its production of salt, Syracuse is rich in historic, geologic, and scenic features; Onondaga Indian Reservation nearby.

Thousand Islands. Vacation area of beautiful islands, on the St. Lawrence River.

West Point. Overlooking the Hudson River, West Point is the home of the U. S. Military Academy, established in 1802.

NORTH CAROLINA

Asheville. Mountain resort city and North Carolina's gateway to the Great Smoky Mountains

Niagara Falls, Spectacular Source of Power
Bell Aircraft Corp.

World-famous Brooklyn Bridge, New York
"NYSPIX"

Oreg. Highway Comm.
Mount Hood, Oregon's Highest Mountain

Pittsburgh C. of C.
Pittsburgh's "Golden Triangle"

National Park. The magnificent Biltmore House displays a fortune in art objects. A drive over Elk Mountain Scenic Highway affords an excellent view of the plateau on which the city is located.

Cape Hatteras National Seashore Recreational Area. A 28,500-acre area preserving the flora and fauna, with beach recreational areas. Site of the old lighthouse (built in 1793), is part of the area. Wright Brothers National Memorial is also here.

Fayetteville. Market Hall, once a slave market, then the town hall, now houses the public library. Fort Bragg, a large military reservation is nearby.

Fontana Dam. The highest dam in the TVA system has formed the 30-mile-long Fontana Lake, which is a paradise for fishermen. Fontana Village is the center for this resort area.

Fort Raleigh National Historic Site. Scene, on Roanoke Island, of Sir Walter Raleigh's "lost colony." A facsimile of the setting has been made.

Great Smoky Mountains National Park. Overlaps the Tennessee-North Carolina border. This spur of the Appalachian Range contains many peaks 6,000 feet high. Virgin forest and frontier mountain communities lie in the park. See TENNESSEE.

Raleigh. State capital. The Capitol, a noteworthy example of Greek Revival architecture, is surrounded by statuary of state and national heroes. In the Hall of History, the state historical museum, all phases of North Carolina's history, from the Roanoke Colony to modern transportation, are represented.

Winston-Salem. In 1766 a Moravian group settled Salem. The Moravian Brothers' House, dating from 1769, the Home Moravian Church, and the Moravian Graveyard are reminders of the religious settlers and their influence. In 1913 Winston and Salem were united. At nearby Bethabara (Oldtown), the church and graveyard are all that remain of the original communal settlement of 1753.

NORTH DAKOTA

Badlands. A strange and fantastic region along the Little Missouri River, where the land is carved into a jumble of weirdly colored buttes, domes, ridges, pyramids, and other shapes. The Theodore Roosevelt Memorial National Park encloses the heart of the Badlands area.

Bismarck. State capital; located in the south-central part of the state, on the Missouri River. The 19-story Capitol, replacing the former building which burned in 1930, is an outstanding example of modern architecture. There are interesting Indian and pioneer relics in the Historical Society Museum. The log cabin from Theodore Roosevelt's Elkhorn Ranch stands nearby.

Crow Flies High Butte. From this vantage point the French explorer Verendrye first looked across the Missouri River.

Fargo. Largest city in North Dakota, and the

chief shipping and distributing center. Because of the location there of the North Dakota State University, it is also state agricultural headquarters.

Fort Abraham Lincoln State Park. Lying on the west bluffs of the Missouri River, near Mandan, this park encloses the site of an old Mandan Indian village, now partially restored, and two frontier forts, also restored to show their original appearance.

Garrison Dam. Huge earth-fill dam at Riverdale forms 200-mile-long Garrison Lake.

International Peace Garden. A formal garden, one mile square, on the international border between North Dakota and Manitoba.

OHIO

Akron. The rubber capital of the world. Points of interest include the rubber plants, Goodyear Air Dock, Baptist Temple, and the parks.

Cincinnati. Metropolis of southern Ohio, the city is a cultural center supported by diversified industry.

Cleveland. Largest city, Lake Erie port, iron and steel center. Severance Hall is the home of the symphony, and the white marble Museum of Art houses a collection of Byzantine and European art. The Cultural Gardens are divided into sections in which each nationality presents a representative display.

Columbus. State capital, situated at the confluence of the Scioto and Olentangy rivers. The Doric-style Capitol and the beautiful Civic Center are the show places of the city.

Dayton. Industrial city and home of Wright-Patterson Air Force Base.

Fort Ancient. Prehistoric earthworks enclosing an area of 100 acres; in Warren County.

Marietta. Peaceful trading center of a large farm area. The Campus Martius Museum is built around Rufus Putnam's Block House, where early pioneers took refuge during Indian attacks.

Mound City Group National Monument. Near Chillicothe. Twenty-three prehistoric burial mounds.

Put-In-Bay. Here Commodore Perry defeated the British in the Battle of Lake Erie (1813). The Perry's Victory and International Peace Memorial National Monument commemorates this event and the subsequent peace between Canada and the United States.

Schoenbrunn Village State Memorial. At New Philadelphia. Schoolhouse, church, and cabins of early Moravian village (1772) have been restored.

Serpent Mound State Memorial. In Adams County. Mound extends for a quarter of a mile in the shape of a snake holding an egg in its mouth.

Toledo. Diversified industry supports this city, originally Anthony Wayne's Fort Industry.

Arbuckle Mountains. Located in the south-central part of the state, these aged mountains, worn down to a height of only 700 feet above the surrounding plains, contain a great variety of geological formations of limestone, sandstone, shale, and granite.

Claremore. Rogers County seat, named for Clem Rogers, father of Oklahoma's well-known citizen, Will Rogers. The state erected the Will Rogers Memorial in Claremore on land donated by his widow.

Lake Texoma and Murray. Denison Dam on the Red River, impounds waters to form Lake Texoma, which nudges the southeastern tip of Lake Murray. Both lakes offer water sports.

Muskogee. Once capital of the Five Civilized Tribes —Cherokee, Choctaw, Chickasaw, Creek and Seminole. The Bacone University for Indians is located here, and Fort Gibson National Cemetery is nearby.

Oklahoma City. State capital and largest city. Settled in the afternoon of April 22, 1889, the date of the "Historic Run." Some of country's largest oil companies have their headquarters here, and there are many oil derricks in the city.

Platt National Park. Well-known for its sulphur, iron, bromide, and fresh water springs. Wild flowers, birds, and small game animals are also abundant.

Sequoyah Shrine. Log cabin of the famous Cherokee Indian statesman and educator. Sequoyah invented the Cherokee syllabry after being intrigued by the white man's language. The Shrine is located near Hulbert and Wagoner.

Tulsa. The state's largest oil refining center. There is a large population of Indians and white people.

Wichita Mountains Wildlife Refuge. Mountainous area lying entirely within Comanche County. Bison, elk, Texas longhorns, wild turkeys and other birds are found in this wildlife sanctuary.

OREGON

Bonneville Dam. This huge dam spans the Columbia River about 42 miles northeast of Portland, supplying electric power over a wide area, and making the Columbia River navigable for seagoing vessels as far as The Dalles. Fish ladders allow the salmon to ascend the Columbia past the dam to their spawning grounds.

Crater Lake National Park. One of the scenic wonders of America, the waters of the lake, 1,996 feet deep, are crystal clear and intensely blue. Hemmed in by steep mountain walls and towering forests, the lake lies 6,000 feet above sea level, in a sunken volcanic crater.

Mount Hood. Oregon's highest mountain, rising 11,245 feet in the Cascade Range east of Portland. Its impressive snow-capped pyramidal peak can be seen from a great distance. Comparatively easy to ascend, by tramway and chairlift, it is a popular recreational area, offering magnificent views from its summit.

Oregon Caves National Monument. A series of caverns in Elijah Mountain in the Siskiyou Range, known as "the Marble Halls of Oregon." The mountain is a labyrinth of corridors and caverns of weird beauty, filled with strange and intricate marble formations. The region around the caves has been set aside as a game refuge.

Portland. The largest city of Oregon and one of the leading lumber-shipping ports in the world; also known as "the City of Roses." The annual Rose Festival is held in June. Points of interest include the Portland Art Museum; the Civic Auditorium, which houses the Oregon Historical Society Collection; the Rough Rider Statue of Theodore Roosevelt; the statue of Sacajawea, the "Bird Woman," in Washington Park; the sunken rose gardens in Peninsula Park; and the Forestry Building.

Snake River Canyon. See IDAHO.

PENNSYLVANIA

Bethlehem. Steel city and music center; home of the Bach Festival, and of the Moravian College (1742). The log Gemein Haus (1741), is the oldest structure in the city.

Chester. Second oldest city in state, settled by the Swedes in 1644. The Caleb Pusey House, in nearby Upland, is the oldest English-built house in the state (1683). Other old-time houses are the Washington House, where General Washington stayed after the Battle of Brandywine, and the Friends Meeting-house, erected in 1736.

Delaware Water Gap. A scenic three-mile gorge in the Kittatinny Mountains near Stroudsburg. Northwest are the beautiful Pocono Mountains.

Erie. State's only port on the Great Lakes. Of historic interest are the Wayne Memorial, a reproduction of the blockhouse where "Mad Anthony" Wayne died in 1796; the Old Custom House; and Perry's flagship, the *Niagara*.

Fort Necessity National Battlefield. Reconstructed stockade, scene of opening of French and Indian Wars.

Gettysburg. Scene of the Battle of Gettysburg and site of the Soldiers' National Monument, where Lincoln delivered his immortal Gettysburg Address.

Harrisburg. State capital, on the Susquehanna River. There are six buildings in the Capitol group, including the Capitol, flanked by the George Gray Barnard sculptured group.

Lancaster County. Pennsylvania Dutch country.

Philadelphia. Fourth largest American city, important port, and great industrial center. Here starts the great Pennsylvania Turnpike, a fast four-lane highway that goes to the Ohio border, via Pittsburgh. Founded in 1682 and first capital of the United States, Philadelphia is second to no other city in historic associations. Among its many points of interest are Independence Hall, where the Liberty Bell is exhibited; Carpenters' Hall, where the first Continental Congress assembled in 1774; Benjamin Franklin's grave in the burial ground of Christ Church; Betsy Ross House; William Penn's House; the United States Mint; Bartram's Gardens; Philadelphia Zoological Gardens; Philadelphia Navy Yard; University of Pennsylvania; Philadelphia Museum of Art, and others.

Pittsburgh. Second largest city in the state and one of the greatest steel centers in the world. Pittsburgh is on the Ohio River, at the junction of the Monongahela and Allegheny rivers. The city began as the French Fort Duquesne (1754), which later became the English Fort Pitt. Points of interest are Carnegie Institute and Library; Carnegie Institute of Technology; University of Pittsburgh's Cathedral of Learning, 42-story skyscraper; Schenley Park; Liberty Tunnels, which pierce the bluff of Mount Washington; Fort Pitt Blockhouse; Stephen Collins Foster Memorial Building, Mellon Institute, and many others.

Valley Forge State Park. Here Washington camped in the winter 1777–78. Remains of entrenchments are to be seen and commemorative buildings.

RHODE ISLAND

Block Island. Pear-shaped and studded with low hills and ponds, it lies in the mouth of Long Island Sound. The Great Salt Pond which covers more than 100 acres almost bisects the island.

Charlestown. The Indian Burial Ground and Fort Neck Lot are nearby.

Narragansett. Fine beach; the old casino, the Towers, at the pier is a landmark.

Newport. Famed as an elite summer capital, the city's historic associations are often overlooked. Washington Square is the center of the old city. Nearby are Old Colony House, once the statehouse, and Friends Meeting House. Touro Synagogue, now

a National Historic Site, dates from 1763. The U. S. Naval Base and Goat Island, the Navy's Torpedo Station, are in Newport. The Breakers, Vanderbilt's palatial residence, is open to the public.

Pawtucket. Home of Samuel Slater, the man who built, from memory, the first spinning frame in America, thereby founding the textile industry. The Old Slater Mill, now a memorial, contains some of the original machinery and early relics.

Providence. State capital. Built on three hills, the city was founded by Roger Williams, "Father of Rhode Island," in 1636. Old State House was seat of the General Assembly until 1901. The new Capitol, made of white marble, boasts of a dome which rivals that of St. Peter's in Rome. Rolling woodlands, notable rose gardens, and many lagoons extend throughout Roger Williams Park. Noteworthy collections are found in the Providence Athenaeum and in the libraries at Brown University.

SOUTH CAROLINA

Beaufort. Historic port town on the Intracoastal Waterway. Mellowed residences, surrounded by live oaks, front on narrow, crooked streets. Points of interest are the Oldest House, Tabby Manse, the Beaufort National Cemetery, and Fort Frederick. Parris Island, U. S. Marine base, is nearby.

Charleston. Important city and deep-sea port on the Intracoastal Waterway, six and one-half miles from the ocean. The beautiful harbor with its historic forts spreads before the palmetto-fringed waterfront—"the Battery." In the harbor is Fort Sumter, where the first shots of the Civil War were fired. Charleston gardens and those of nearby plantations have won world acclaim for more than a century. North of the city are the tropical Cypress Gardens, whose moss-laden trees grow in a freshwater lake.

Columbia. State capital. The walls of the gray-granite State House bear scars of Sherman's bombardment. The World War Memorial Building, at the University of South Carolina, houses a historic collection and a shrine to the state's soldiers.

Georgetown. Paper pulp and lumbering center. At the nearby Brookgreen Gardens an outdoor museum of statuary is set among boxwoods, oaks, and flowers. Azaleas and camellias bloom among live oaks at Belle Isle Gardens, also nearby. Myrtle Beach is a little farther up the coast.

Kings Mountain National Military Park. Third largest battlefield park in the United States. In 1780 a Whig force defeated a larger British force entrenched on the summit of the mountain, thus breaking British resistance in the South.

SOUTH DAKOTA

Angostura Dam. This dam impounds the waters of the Cheyenne River, one of the tributaries of the Missouri; near Hot Springs in the southern Black Hills. A reservoir and recreation area are nearby.

Badlands National Monument. Fantastically eroded sedimentary rock formations in many unusual pastel shades. Story book castles and temples seem very real in pink and lavender hues. A weird beauty characterizes this strange land just east of the Black Hills. Numerous fossils and rocks are found among the prickly cactus and wild flowers.

Belle Fourche. Trading center for cattle and sheep empire extending into Montana and Wyoming.

Black Hills. National forest region on the west embracing the highest mountains between the Rockies and the Appalachians. Rising dark against the surrounding colorless plains, the "black" is in reality the brown and green of lofty pines. Popular as a vacation land.

Custer State Park. Features some of the most spectacular scenery in the Black Hills—Needles highway, Sylvan Lake, and Harney Peak (7,242 feet), highest point east of the Rockies. The park covers 128,000 acres and includes a wildlife sanctuary.

Deadwood. Historic mining town and setting for Wild Bill Hickok, Calamity Jane, and other frontier characters. Deadwood Gulch was the center of gold rush of 1876. Adams Memorial Museum houses an excellent collection of pioneer mining devices. In No. 10 Saloon, Wild Bill Hickok was shot in the back holding a poker hand of aces and eights.

Lead. Mile-high city built around the fabulous Homestake Mine, largest gold mine in the Western Hemisphere. Underground shafts go down a mile.

Mt. Rushmore National Memorial. Faces of Washington, Jefferson, Lincoln, and Theodore Roosevelt, carved in granite by Gutzon Borglum, are proportioned to men 465 feet tall. Each head is 60 feet from chin to forehead, twice the height of the Great Sphinx of Egypt. A 500 word history of the United States in letters three feet high is also shown. Three tunnels through adjoining mountains are lined up to focus on Mt. Rushmore.

Pierre. Second smallest capital city in the United States. Nearby Ft. Pierre, built in 1817, is the oldest continuous settlement in the state.

Rapid City. Gateway to the Black Hills, it was the supply point for mining camps during the gold rush of 1876. Stratosphere Bowl was the takeoff point for the 72,395-foot balloon flight by General Orvil Anderson and Albert Stevens on Nov. 11, 1935. South Dakota School of Mines and Technology features fossil and mineral exhibits. Dinosaur Park has five prehistoric monsters, molded of cement.

Sioux Falls. Largest city, located on the Big Sioux River. An important processing center for farm products, it is also a cultural and medical center for a large area. Extensive use is made of locally quarried pink quartzite.

Wind Cave National Park. Located in the southern Black Hills, it is a limestone cavern featuring unusual boxwork formations; several miles of lighted routes. The temperature is always 47°F. There is a buffalo herd in the park area.

Great White Throne, Zion National Park   Union Pacific R.R.   *Mount Vernon—American Shrine*   Va. Dept. of C.&D.

Wash. Adv. Comm.

Grand Coulee Dam, Washington

TENNESSEE

Andrew Johnson National Monument. Former president's home and tailor shop; at Greeneville.

Chattanooga. Manufacturing city on the Tennessee River near the Georgia border. Trips from Chattanooga lead to Lookout Mountain (which also may be reached by the Incline Railway) and to Chickamauga and Chattanooga National Military Park. Atop the mountain are the unusual rock formations of Rock City, and Point Park, vantage point for a breath-taking view of Moccasin Bend.

Great Smoky Mountains National Park. Smoky haze hovering about the multi-green mountain tops gives this scenic wonderland its name. The highway from Gatlinburg, headquarters for the Park, passes Mt. Le Conte, Mountaineer Museum, Chimney Tops, and Newfound Gap. At the Gap, trails for mountain climbers wind to the top of Clingman's Dome, highest point in Tennessee. See NORTH CAROLINA.

Knoxville. Industrial center and farm market, especially for tobacco. About 20 miles north is Norris Dam. Oak Ridge, a few miles northwest, features the American Museum of Atomic Energy.

Memphis. Largest city, inland port, and cotton market. The Cotton Exchange is the hub of the city's economy. The annual Cotton Carnival, in May, draws thousands of visitors.

Meriwether Lewis Monument. Honors co-leader of famous western expedition and marks his place of death at Grinder's Inn and his grave.

Nashville. Capital and second largest city. The grounds of the Capitol exhibit statues of Tennessee's prominent people. A replica of Fort Nashborough (1780) is built on its original site; "the Parthenon," a reproduction of the Athenian structure, stands in Centennial Park. Near the city, Andrew Jackson's home, "The Hermitage," remains much as it was when he lived there.

National Military Parks. See U. S. NATIONAL PARK SYSTEM, page 192.

TEXAS

Amarillo. Commercial center of the Panhandle section, Amarillo has developed from a barren prairie to a thriving metropolis. Helium gas is produced in large quantities here. Southeast of Amarillo is the spectacular Palo Duro Canyon.

Austin. State capital, commercial city, and home of the University of Texas. Planned by the founders of the Republic of Texas, Austin was the national

capital until Texas became a state in the United States. Points of interest are the Capitol and grounds, Governor's Mansion, former French Legation, O. Henry Museum, and others.

Big Bend National Park. Vast frontier area located in a picturesque section of Mexican-border wilderness, in the big bend of the Rio Grande. Spectacular mountains, desert, and canyon areas contain a variety of unusual geological formations.

Dallas. Industrial and commercial city, cotton market and oil center. The city is modern in every way, with beautiful parks and highways, theaters and museums. Points of interest are the John Neely Bryan Cabin, reconstructed on the site where the founder of the city built his home in 1843; Dallas Cotton Exchange; Southern Methodist University; Museum of Fine Arts, and others.

El Paso. Opposite Juarez, Mexico, on the Rio Grande, El Paso dates from the time of the Spanish conquistadores. Today it is a thriving commercial center and important port of entry. Noteworthy are Texas Western College; Fort Bliss, once the largest cavalry post in the United States; and International Bridge.

Galveston. One of the largest cotton ports in the world, year-round resort, and commercial city. Galveston lies on the eastern extremity of Galveston Island, which parallels the Texas mainland. Interesting are the sea wall, built to prevent a disaster similar to the Galveston Flood of 1900; the landlocked harbor; the coast-defense forts.

Houston. Largest city and important port, especially for the export of cotton and oil. Although 50 miles inland, Houston is connected with the Gulf Coast by the Houston Ship Channel. Points of interest are the Hermann Park Zoo, Rice University, and museums. The San Jacinto Battlefield and State Park, where the *Texas* is berthed, is nearby.

Kingsville. Home of Texas College of Arts and Industries; headquarters of million-acre King Ranch.

San Antonio. Historic old city and busy industrial community. Among its many interesting places are the Alamo, where 187 Americans made their last stand; San Fernando Cathedral; Spanish Governor's Palace; Mexican quarter; Franciscan Missions; La Villita; Randolph and Kelly fields.

UTAH

Bingham Canyon. Largest open cut copper mine in world; highway leads to observation points from which operations can be seen in the many-tiered pit. Town is a single street, three miles long.

Bryce Canyon National Park. Vast amphitheater filled with pink and white sandstone eroded into myriads of spires, shafts, minarets, and like formations; sharply defined stratification heightens the color. The highway winds along the west rim with frequent short side roads to viewpoints at the very edge, from which are seen such groups as Fairyland, Queen's Gardens, Fairy Castle, Wall of Windows.

Natural Bridges National Monument. Three massive natural bridges have been cut through red sandstone by the eroding action of the river; Sipapu, largest and most beautifully proportioned, is 220 feet above the canyon floor. Similar natural bridges and arches are found throughout southern Utah. For other National Monuments, see U. S. NATIONAL PARKS SYSTEM, pages 192-193.

Salt Lake City. Capital of the state, home of the University of Utah, and one of the most beautiful cities in America. Wide and shaded streets, green lawns of beautiful homes, spacious grounds of the Capitol, are in sharp contrast to the encroaching desert. The Capitol itself holds a masterful exhibit of the resources and the means by which this green empire has been wrested from an arid wasteland. Famous all over the world are the beautiful Mormon Temple and the Tabernacle with its great dome put together with wooden pegs; in its auditorium a 10,000-pipe, hand-built organ is played every noon in public concert. Nearby is Great Salt Lake, where a swimmer can't sink because the water is so salty. The Great Salt Lake Desert holds the Bonneville Salt Flats that supply much of the world's salt.

Zion National Park. In all the world there is no more awe-inspiring sight than the great shafts and temples and sheer walls that rise from the bottom of Zion Canyon, where the road winds. Vermilion toward the base, they blend to rose, to pink, and finally to white, adding the glow of red and white fire to majestic line and towering height. Some of the formations are the Temple of Sinawava, Angel's Landing, the Great White Throne, the Great Arch.

VERMONT

Barre. Granite center. Cutting and polishing of granite can be watched at the Granite Sheds, which stretch along the floor of the valley.

Bennington. Textile mills and furniture plants in new section. Old First Church, Walloomsac Inn, the Ethan Allen House and the Tichenor Mansion in old section. On a nearby hill is the 300-foot Bennington Battle Monument.

Burlington. Largest city, lake port, and home of the University of Vermont. Historic points are Battery Park, site of big guns in the War of 1812; Green Mountain Cemetery, where Ethan Allen and other soldiers are buried; Ethan Allen Park.

Grand Isle. Several islands and a peninsula, all in Lake Champlain, comprise this resort isle.

Green Mountains. This magnificent chain of rounded, verdant peaks bisects the state and extends from the Massachusetts line to the Canadian border. The highest point in the range, Mt. Mansfield, is also the highest point in the state. Scenic points near here are Smuggler's Notch, Bingham Falls.

Montpelier. From every approach to the city the golden dome of the State Capitol can be viewed. Museum of Natural History contains the Daye Press.

Mount Equinox. Highest mountain overlooking the Vermont valley; has skyline drive to summit.

Rutland. Industrial center surrounded by great marble fields. Methods of cutting and finishing marble and exhibits are shown at nearby Proctor.

VIRGINIA

Appomattox Court House National Historical Park. Commemorates Lee's surrender to Grant on April 9, 1865. For National Military Parks, Battlefield Parks, etc., see U. S. NATIONAL PARK SYSTEM, page 193.

Arlington National Cemetery. Burial ground of the nation's heroic dead. The Tomb of the Unknown Soldier is here, as is the Custis-Lee Mansion.

Colonial National Historical Park. Park includes Jamestown, first permanent English settlement, which has reconstructed Glasshouse and Jamestown Festival Park; Yorktown, where Cornwallis surrendered; and a parkway between these two places and Williamsburg.

Fredericksburg. Washington's mother's home and Kenmore, his sister's home, are here, as is James Monroe's law office, and the home of John Paul Jones. East of the city, reached by State Route 3, is Stratford Hall, birthplace of Robert E. Lee, and built by his family 1725-30.

Monticello. Home of Thomas Jefferson, third president of the United States. His tomb is here.

Mount Vernon. Famous country place of George Washington, standing since 1743 and filled with Washington treasures.

Natural Bridge. Higher than Niagara, it spans the Cedar Creek; near Lexington.

Norfolk and Portsmouth. Location of Naval Station and Navy Yard, respectively. In vicinity are Newport News, site of immense shipbuilding plant, Hampton Roads, and Virginia Beach, popular resort.

Richmond. State capital since 1779 and Confederate capital from 1861 to 1865. The Capitol, designed by Thomas Jefferson, has Houdon's marble figure of Washington. Other points of interest are the Edgar Allan Poe Shrine, St. John's Church,

Chief Justice John Marshall's Home, Confederate Museum, Hall of Delegates.

Shenandoah National Park. Embraces a portion of the Blue Ridge Mountains. Scenic Skyline Drive traverses the Park. Luray Caverns are nearby.

Williamsburg. Home of the College of William and Mary and subject of extensive restoration to the 18th-century days when Williamsburg was the colonial capital of Virginia.

WASHINGTON

Grand Coulee. A 50-mile dry gorge cut 800 feet deep during glacial times by an ice-diverted Columbia River. At mid point is Dry Falls, once the greatest waterfall known to have existed in North America. Grand Coulee Dam, 550 feet high and 4,173 feet long, furnishes electric power to all parts of the state. Surplus power pumps water 300 feet higher into Grand Coulee, where retaining dams create a huge reservoir. From here it spreads south and east into the Columbia Basin, to irrigate, ultimately, a million acres.

Mt. Rainier National Park. Rising from near sea level to over 14,410 feet, Mt. Rainier, dormant if not extinct volcano, is the dominant natural feature of the state. Park has permanent snow fields and glaciers, high meadows and crystal clear lakes, great variety of trees and flowers. Paradise Valley on south side has hotel and camping; Yakima Park on north, cabin camp and picnic grounds. Summit climb of two days starts from Paradise Inn.

Olympia. New capitol buildings on impressive site overlooking Budd Inlet emphasize character of city as state capital. Suburb of Tumwater, first American settlement on Puget Sound (1845). Gateway to famed Olympic Peninsula recreation area.

Olympic National Park. Occupies the heart of rugged Olympic Peninsula between Pacific Ocean and Puget Sound; preserves primeval forests of giant trees and native animal life, zones ranging from temperate rain forests to mountain glaciers.

Puget Sound Islands. These include beautiful Vashon and Bainbridge, island suburbs of Tacoma and Seattle; Whidbey, the second largest in the United States and connected to the mainland by a bridge across tide-ripped Deception Pass; the San Juan group of nearly 200 habitable islands and hundreds of barren rocks; and scores of others. Protected waters and sheltered coves are a paradise for yachtsman, fisherman, picnicker, and camper.

Seattle. Metropolis of Washington and the Pacific Northwest. Salt water harbor of Elliott Bay rimmed by port facilities for largest ocean going vessels; fresh water harbor via government locks a haven for fishing and pleasure boats. Scenic boulevards link waterfronts with hilltops and provide views of distant mountains. University of Washington, farmers' Public Market, Frozen Fish Museum, Volunteer Park Conservatory, Smith Cove docks, and Lake Washington Bridge are of interest. Industries are varied—lumbering, airplane manufacturing, aluminum fabrication, shipbuilding, and others. Snoqualmie Falls are to the east.

Spokane. Metropolis of the so-called "Inland Empire"—portions of four states between Cascade and Rocky Mountains. Flour mills, lumber mills, and light metal industries are based upon agricultural, timber, and power resources. Spokane Falls in heart of city supplied water-power for original settlement (1871). Nearby lakes and Mt. Spokane provide summer and winter recreation.

Tacoma. Important lumber manufacturing center. Large smelter handles copper ores from all parts of the world. In Point Defiance Park is full size replica of Hudson's Bay Co. Fort Nisqually, first white settlement in the area. Nearby Fort Lewis is one of largest permanent army posts.

Vancouver. River fort and industrial center; dates from Hudson's Bay Co. post of 1825 (Fort Vancouver National Historic Site) when fur-trading and trapping brigades fanned out in all directions. Aluminum industry, based on low-cost power from Bonneville Dam, and wood-processing plants now are major activities.

Walla Walla. Old Fort Walla Walla of fur-trading days was at Wallula; Waiilatpu Mission of Marcus Whitman is now in Whitman National Monument.

WEST VIRGINIA

Berkeley Springs. Historic spa, willed to the state by Lord Fairfax. The Washington Elm, Lovers' Leap, and the Castle are of interest.

Charleston. State capital and state's largest city, near coal, natural gas, and oil fields. The gold-leaf dome of the State Capitol rises 300 feet above the ground. "Stonewall" Jackson Monument and Pioneer Monument are on the Capitol grounds.

Charles Town. In the Eastern Panhandle. The town is named for Charles Washington, brother of George; his home "Harewood" is preserved. John Brown stood trial at the Jefferson County Courthouse and was hanged at Site of John Brown Gallows.

Grave Creek Mound. At Moundsville. It is the largest conical Indian mound in the country.

Harpers Ferry. John Brown Monument marks site of the engine house where the abolitionist made his last stand; "John Brown's Fort," part replica and part reconstruction of the engine house, is on the campus of Storer College. Harpers Ferry National Monument includes Federal hillsides surrendered to Jackson, thus permitting his quick juncture with Lee at Antietam.

Huntington. Picturesquely situated on the Ohio, the state's second largest city is an industrial center primarily known for its huge nickel plant.

Monongahela National Forest. A vacation land along the eastern boundary. Points of interest include Spruce Knob, highest point in the state; Blackwater Falls, which drop 63 feet and rush through a 1,000-foot gorge; "Pictured Rocks" at Petersburg Gap, and nearby Smoke Hole Caverns; the eroded, castle-like white sandstone mountain, Seneca Rock, near Mouth of Seneca; and the Sinks, near Osceola, where the Gandy River goes underground.

Wheeling. Industrial center, leading city of the Northern Panhandle. Oglebay Park contains Mansion Museum, Nature Museum, Greenhouse and Conservatory, and recreational facilities.

White Sulphur Springs. A popular spa since the 1800's. The President's Cottage has housed several Chief Executives since the time of Andrew Jackson, and the Robert E. Lee Cottage remains unchanged.

WISCONSIN

Ashland. Center of hunting area; nearby is beautiful Copper Falls State Park. A side trip leads to Bayfield and from here the offshore Apostle Islands in Lake Superior can be visited.

Big Manitou Falls. Highest waterfall in the state. In Pattison State Park, south of Superior, Black River cascades 165 feet down a narrow, rocky gorge.

Devil's Lake State Park. Unusual rock formations and geological curiosities. Of glacial origin, Devil's Lake is hemmed in by towering cliffs 400 to 500 feet high; near West Baraboo.

Green Bay. State's oldest settlement and gateway to Door Peninsula, famed for cherry blossoms.

Lake Geneva. Fashionable summer resort; many fine estates. Sailboat races are popular on lake. World-famous Yerkes observatory, of the University of Chicago, is at Williams Bay.

Lake Winnebago. Largest lake in state and extensive summer-winter vacation area.

Madison. Capital of Wisconsin and site of the University of Wisconsin; on narrow isthmus between Lake Monona and Lake Mendota, with three other lakes close by. The Capitol, second in height only to the National capitol in Washington, has the only granite dome in the United States.

Milwaukee. Largest city in Wisconsin and important center for the manufacture of heavy machinery. It is also noted for its many large breweries, some of which conduct tours for visitors. Has a Lake Michigan harbor and many fine parks. Among the points of interest are the botanical gardens of Mitchell Park; Court of Honor, with its war memorials; Milwaukee Public Library and Museum.

Prairie du Chien. The second oldest settlement in the state, this picturesque little city has numerous reminders of the old fur-trading days.

Rib Mountain. Near Wausau; 1,941 feet above sea level, highest point in Wisconsin; popular winter sports area, with majestic scenic views from summit.

Superior. Sister city to Duluth. *See* MINN.

Wisconsin Dells. Spectacular rock formations formed by the Wisconsin River as it dashes through eight miles of rocky gorges. Tribal ceremonies by the Winnebago Indians are a summer feature.

WYOMING

Casper. Trading center for rich oil fields (Teapot Dome is to the north) and lively cattle country; on Oregon Trail and rich in frontier history.

Cheyenne. Wyoming's capital; founded as a construction terminus for the building of the first transcontinental railroad, today an important transportation and trade center. Drawing much of its wealth from the cattle industry, it is world-famous for its Cheyenne Frontier Days celebration (July). Points of interest include the Capitol, Ft. Warren, Frontier Park, the State Historical Museum.

Cody. Near entrance of Yellowstone National Park and Shoshone Canyon and Dam; founded by Buffalo Bill Cody; it contains statue and museum of his effects.

Devils Tower National Monument. The first National Monument; of columnar basalt, it rises 865 feet high; northwest of Sundance.

Fort Bridger. Historic trading post established by the famous frontiersman, Jim Bridger.

Fort Laramie National Historic Site. Trade, military, and supply outpost for many years before and during wagon travel on the Oregon Trail; called "Cradle of American civilization in the west." Some original buildings are still standing.

Fort Washakie. Agency for Shoshone Indian Reservation; Sacajawea and Chief Washakie buried here.

Grand Teton National Park. Vistas of shining gray-blue peaks, snow-frosted, reflected in sky-blue lakes amid close ranks of spruce and fir—these are the Tetons. No less typical is the little log Church of the Transfiguration, its altar window framing Grand Teton Peak. Jackson and Jenny are two of several beautiful lakes; adjacent is the famous Jackson Hole country.

Hot Springs State Park. Hot mineral springs. Big Horn Spring is world's largest, flowing 18,000,000 gallons (135°) a day. At Thermopolis.

Independence Rock. On the Sweetwater; famed landmark of Oregon Trail. Called "Register of the Desert" by Father De Smet (1840), it bears 5,000 names scratched by explorers, traders, and emigrants.

Sheridan. Gateway to the Big Horn Mountains; outfitting point for dude and cattle ranches. To the south are sites of Ft. Phil Kearney and Wagon Box Fight, famous turning-point of the Red Cloud War.

Yellowstone National Park. Probably the most famous of all the national parks, the Yellowstone is a fabulous combination of mountains and canyons, lakes and waterfalls, spouting geysers and wildlife that includes moose, buffalo, grizzly bear, and trumpeter swan. Old Faithful Geyser, erupting regularly every hour, is hallmark to the Park. Others that play at more or less regular intervals are Jewel, Daisy, Rocket, and Grotto; the spectacular Grand Geyser performs about every 16 to 20 hours. Many pools, like sparkling blue Morning Glory, terraces like Mammoth Hot Springs, and boiling mud pots are of interest. Yellowstone Canyon and Falls are among the world's most beautiful; from Fishing Bridge almost anyone can catch a trout.

UNIVERSITIES AND COLLEGES
IN THE UNITED STATES

This listing, divided by states, includes those accredited schools which have 500 or more students and offer a degree only after the completion of at least four years of college-level work.

Immediately following the name of the institution in bold face type is the city or town where it is located unless this is evident from its title.

Following the name and address of the school, and in this order, are given the date of founding, the type of legal control, a description of the student body and total enrollment, the number of faculty members*, and

the calendar system under which it operates, whether semester or quarter.

Following this information, a description of the institution's program is given if it offers more than a liberal arts and general course; for the latter category no description of program is given. For those schools which offer liberal arts and general courses and have one or more professional schools, a detailed description of their program is given.

Those schools which grant degrees beyond the bachelor's or first professional degree are designated by the phrase "advanced degrees conferred."

ALABAMA

Alabama Agricultural and Mechanical College. Normal. Founded 1875. State control. Coeducational, 1,200 students. Faculty, 68. Quarter system. Liberal arts, terminal-occupational, teacher preparatory.

Alabama College. Montevallo. Founded 1896. State control. Coeducational, 1,200 students. Faculty, 65. Semester system. Liberal arts, home economics, music. Advanced degrees conferred.

Alabama State College. Montgomery; branch at Mobile. Founded 1874. State control. Coeducational, 2,200 students. Faculty, 40. Quarter system. Liberal arts, teacher preparatory. Advanced degrees conferred.

Athens College. Athens. Founded 1822. Methodist control. Coeducational, 500 students. Faculty, 28. Quarter system. Liberal arts, teacher preparatory.

Auburn University. Auburn. Founded 1872. State control. Coeducational, 8,800 students. Faculty, 572. Quarter system. Liberal arts, agriculture, architecture, chemistry, education, engineering, home economics, pharmacy, veterinary medicine. Advanced degrees conferred.

Birmingham-Southern College. Birmingham. Founded 1856. Methodist control. Coeducational, 1,000 students. Faculty, 74. Quarter system. Liberal arts and teacher preparatory. Advanced degrees conferred.

Florence State College. Florence. Founded 1872. State control. Coeducational, 1,900 students. Faculty, 90. Semester system. Liberal arts, terminal-occupational, teacher preparatory. Advanced degrees conferred.

Howard College. Birmingham. Founded 1842. Southern Baptist control. Coeducational, 2,100 students. Faculty, 98. Semester system. Liberal arts and pharmacy.

Huntingdon College. Montgomery. Founded 1854. Methodist control. Coeducational, 700 students. Faculty, 50. Semester system. Liberal arts and teacher preparatory.

Jacksonville State College. Jacksonville. Founded 1883. State control. Coeducational, 2,300 students. Faculty, 134. Semester system. Liberal arts and teacher preparatory. Advanced degrees conferred.

Livingston State College. Livingston. Founded 1840. State control. Coeducational, 700 students. Faculty, 45. Quarter system. Liberal arts and teacher preparatory. Advanced degrees conferred.

St. Bernard College. St. Bernard. Founded 1892. Roman Catholic control. Men only, 550. Faculty, 10. Semester system. Liberal arts and theology.

Spring Hill College. Spring Hill. Founded 1830. Roman Catholic control. Coeducational, 1,400 students. Faculty, 79. Semester system. Liberal arts and teacher preparatory.

Stillman College. Tuscaloosa. Founded 1876. Presbyterian control. Coeducational, 500 students. Faculty, 40. Semester system. Liberal arts, terminal-occupational, teacher preparatory.

Troy State College. Troy. Founded 1887. State control. Coeducational, 1,800 students. Faculty, 57. Quarter system. Primarily teacher preparatory. Advanced degrees conferred.

Tuskegee Institute. Tuskegee Institute. Founded 1881. Private control. Coeducational, 2,100 students. Faculty, 183. Semester system. Liberal arts, terminal-occupational, teacher preparatory. Advanced degrees conferred.

University of Alabama. University; Medical Center at Birmingham. Founded 1831. State control. Coeducational, 13,500 students. Faculty, 872. Semester system. Liberal arts, chemistry, commerce, education, engineering and mines, home economics, law, nursing; at Birmingham, medicine and dentistry. Advanced degrees conferred.

ALASKA

University of Alaska. College. Founded 1922. State control. Coeducational, 2,300 students. Faculty, 76. Semester system. Liberal arts, civil engineering, mines. Advanced degrees conferred.

ARIZONA

Arizona State College. Flagstaff. Founded 1899. State control. Coeducational, 2,000 students. Faculty, 79. Semester system. Liberal arts and teacher preparatory. Advanced degrees conferred.

Arizona State University. Tempe. Founded 1885. State control. Coeducational, 12,000 students. Faculty, 434. Semester system. Liberal arts, applied arts and sciences, business administration, education. Advanced degrees conferred.

University of Arizona. Tucson. Founded 1885. State control. Coeducational, 13,500 students. Faculty, 759. Semester system. Liberal arts, agriculture, business and public administration, education, engineering, fine arts, home economics, law, mines, music, nursing, pharmacy. Advanced degrees.

ARKANSAS

Agricultural, Mechanical, and Normal College. Pine Bluff. Founded 1875. State control. Coeducational, 1,600 students. Faculty, 88. Semester system. Liberal arts and terminal-occupational.

Arkansas Agricultural and Mechanical College. College Heights. Founded 1909. State control. Coeducational, 1,000 students. Faculty, 69. Semester system. Liberal arts, terminal-occupational, teacher preparatory.

Arkansas Polytechnic College. Russellville. Founded 1909. State control. Coeducational, 1,500 students. Faculty, 71. Semester system. Liberal arts and teacher preparatory.

Arkansas State College. Jonesboro. Founded 1909. State control. Coeducational, 3,000 students. Faculty, 97. Semester system. Liberal arts and teacher preparatory. Advanced degrees conferred.

Arkansas State Teachers College. Conway. Founded 1907. State control. Coeducational, 1,800 students. Faculty, 81. Semester system. Liberal arts, terminal-occupational, teacher preparatory. Advanced degrees conferred.

Harding College. Searcy. Founded 1924. Church of Christ control. Coeducational, 1,200 students. Faculty, 63. Semester system. Liberal arts and teacher preparatory. Advanced degrees.

Henderson State Teachers College. Arkadelphia. Founded 1929. State control. Coeducational, 1,500 students. Faculty, 71. Semester system. Liberal arts and teacher preparatory. Advanced degrees conferred.

Hendrix College. Conway. Founded 1884. Methodist control. Coeducational, 600 students. Faculty, 41. Semester system. Liberal arts and teacher preparatory.

Little Rock University. Little Rock. Founded 1927. Private control. Coeducational, 1,600 students. Faculty, 88. Semester system. Liberal arts, terminal-occupational, teacher preparatory.

Ouachita Baptist College. Arkadelphia. Founded 1886, Southern Baptist control. Coeducational, 1,200 students. Faculty, 52. Semester system. Liberal arts and teacher preparatory. Advanced degrees conferred.

Philander Smith College. Little Rock. Founded 1868. Methodist control. Coeducational, 700 students. Faculty, 44. Semester system. Liberal arts, terminal-occupational, teacher preparatory.

Southern State College. Magnolia. Founded 1909. State control. Coeducational, 1,000 students. Faculty, 60. Semester system. Liberal arts, terminal-occupational, teacher preparatory.

University of Arkansas. Fayetteville; campus at Little Rock. Founded 1871. State control. Coeducational, 6,500 students. Faculty, 746. Semester system. Liberal arts, agriculture and home economics, business administration, education, engineering, law; at Little Rock, medicine, nursing, pharmacy. Advanced degrees conferred.

CALIFORNIA

Alameda County State College. Hayward. Founded 1957. State control. Coeducational, 1,450 students. Faculty, 80. Quarter system. Liberal arts, terminal-occupational, teacher preparatory.

Art Center School, The. Los Angeles. Founded 1930. Private control. Coeducational, 1,100 students. Faculty, 76. Semester system. Professional and terminal-occupational. Advanced degrees conferred.

California College of Arts and Crafts. Oakland. Founded 1907. Private control. Coeducational, 650 students. Faculty, 38. Semester system. Professional and teacher preparatory. Advanced degrees conferred.

California Institute of Technology. Pasadena. Founded 1891. Private control. Men only, 1,300. Faculty, 407. Quarter system. Technical only. Advanced degrees conferred.

California State Polytechnic College. San Luis Obispo; Kellogg-Voorhis Campus at Pomona-San Dimas. Founded 1901. State control. Coeducational, 6,900 students. Faculty, 391. Quarter system. Liberal arts, terminal-occupational, teacher preparatory. Advanced degrees conferred.

California Western University. San Diego. Founded 1924. Methodist control. Coeducational, 1,500 students. Faculty, 83. Semester system. Liberal arts, business, law. Advanced degrees conferred.

Chapman College. Orange. Founded 1861. Disciples of Christ control. Coeducational, 800 students. Faculty, 38. Semester system. Liberal arts and teacher preparatory. Advanced degrees conferred.

Chico State College. Chico. Founded 1889. State control. Coeducational, 3,700 students. Faculty, 206. Semester system. Liberal arts, terminal-

* Statistics of total faculty come from *American Universities and Colleges*, 1960 Edition, with the kind permission of the American Council on Education, the publishers.

occupational, teacher preparatory. Advanced degrees conferred.

Chouinard Art Institute. Los Angeles. Founded 1921. Private control. Coeducational, 600 students. Faculty, 62. Semester system. Professional only.

Claremont University College. Claremont. Founded 1925. Private control. Coeducational, 600 students. Faculty, 172. Semester system. Liberal arts and teacher preparatory. Advanced degrees only.

College of the Holy Names. Oakland. Founded 1868. Roman Catholic control. Women only, 900. Faculty, 62. Semester system. Liberal arts and teacher preparatory. Advanced degrees conferred.

Dominican College of San Rafael. San Rafael. Founded 1890. Roman Catholic control. Women only, 700. Faculty, 62. Semester system. Liberal arts and teacher preparatory. Advanced degrees conferred.

Fresno State College. Fresno; center at Bakersfield. Founded 1911. State control. Coeducational, 7,200 students. Faculty, 325. Semester system. Liberal arts, terminal-occupational, teacher preparatory. Advanced degrees conferred.

Golden Gate College. San Francisco. Founded 1901. Private control. Coeducational, 1,600 students. Faculty, 123. Three 16-week terms. Liberal arts and law. Advanced degrees conferred.

Hastings College. (*See* University of California, San Francisco campus.)

Hebrew Union College—Jewish Institute of Religion, California school. Los Angeles. (*See* Ohio.)

Humboldt State College. Arcata. Founded 1913. State control. Coeducational, 2,100 students. Faculty, 135. Semester system. Liberal arts, terminal-occupational, teacher preparatory. Advanced degrees conferred.

Immaculate Heart College. Los Angeles. Founded 1916. Roman Catholic control. Women only, 1,700. Faculty, 91. Semester system. Liberal arts and teacher preparatory. Advanced degrees conferred.

La Sierra College. Arlington. Founded 1922. Seventh Day Adventist control. Coeducational, 1,000 students. Faculty, 57. Semester system. Liberal arts and teacher preparatory.

La Verne College. La Verne. Founded 1891. Church of Brethren control. Coeducational, 500 students. Faculty, 40. Semester system. Liberal arts and teacher preparatory.

Loma Linda University. Loma Linda. Founded 1910. Seventh Day Adventist control. Coeducational, 1,000 students. Faculty, 161. Quarter and semester systems. Professional only. Advanced degrees conferred.

Long Beach State College. Long Beach. Founded 1949. State control. Coeducational, 10,100 students. Faculty, 421. Semester system. Liberal arts and teacher preparatory. Advanced degrees.

Los Angeles State College of Applied Arts and Sciences. Los Angeles. Founded 1947. State control. Coeducational, 14,000 students. Faculty, 571. Semester system. Liberal arts and teacher preparatory. Advanced degrees conferred.

Loyola University of Los Angeles. Los Angeles. Founded 1870. Roman Catholic control. Men only, 1,800. Faculty, 99. Semester system. Liberal arts, business, engineering, law. Advanced degrees conferred.

Mills College. Oakland. Founded 1852. Private control. Women only, 750. Faculty, 86. Semester system. Liberal arts and teacher preparatory. Advanced degrees conferred.

Mount St. Mary's College. Los Angeles. Founded 1925. Roman Catholic control. Women only, 1,200. Faculty, 82. Semester system. Liberal arts, terminal-occupational, teacher preparatory. Advanced degrees conferred.

Northrop Institute of Technology. Inglewood. Founded 1942. Private control. Men only, 1,400. Faculty, 71. Quarter system. Technical and terminal-occupational.

Occidental College. Los Angeles. Founded 1887. Private control. Coeducational, 1,500 students. Faculty, 122. Semester system. Liberal arts and teacher preparatory. Advanced degrees conferred.

Orange County State College. Fullerton. Founded 1960. State control. Coeducational, 1,650 students. Faculty, 95. Semester system. Liberal arts and teacher preparatory. Advanced degrees conferred.

Pacific Union College. Angwin. Founded 1882. Seventh Day Adventist control. Coeducational, 1,000 students. Faculty, 71. Quarter system. Liberal arts and teacher preparatory. Advanced degrees.

Pasadena College. Pasadena. Founded 1902. Nazarene control. Coeducational, 1,000 students. Faculty, 56. Semester system. Liberal arts and graduate school of religion. Advanced degrees conferred.

Pepperdine College. Los Angeles. Founded 1937. Private control. Coeducational, 1,100 students. Faculty, 100. Semester system. Liberal arts and teacher preparatory. Advanced degrees conferred.

Pomona College. Claremont. Founded 1887. Private control. Coeducational, 1,100 students. Faculty, 116. Semester system.

Sacramento State College. Sacramento. Founded 1947. State control. Coeducational, 7,500 students. Faculty, 294. Semester system. Liberal arts and teacher preparatory. Advanced degrees.

St. Mary's College of California. St. Mary's College. Founded 1863. Roman Catholic control. Men only, 850. Faculty, 59. Semester system.

San Diego College for Women. San Diego. Founded 1952. Roman Catholic control. Women only, 600. Faculty, 39. Semester system. Liberal arts and teacher preparatory. Advanced degrees conferred.

San Diego State College. San Diego. Founded 1897. State control. Coeducational, 11,800 students. Faculty, 491. Semester system. Liberal arts and teacher preparatory. Advanced degrees conferred.

San Fernando Valley State College. Northridge. Founded 1958. State control. Coeducational, 6,200 students. Faculty, 158. Semester system. Liberal arts and teacher preparatory. Advanced degrees conferred.

San Francisco College for Women. San Francisco. Founded 1930. Roman Catholic control. Women only, 500. Faculty, 40. Semester system. Liberal arts and teacher preparatory. Advanced degrees conferred.

San Francisco State College. San Francisco. Founded 1899. State control. Coeducational, 17,000 students. Faculty, 520. Semester system. Liberal arts, terminal-occupational, teacher preparatory. Advanced degrees conferred.

San Jose State College. San Jose. Founded 1857. State control. Coeducational, 17,300 students. Faculty, 724. Semester system. Liberal arts, terminal-occupational, teacher preparatory. Advanced degrees conferred.

Stanford University. Palo Alto. Founded 1891. Private control. Coeducational, 9,300 students. Faculty, 1,820. Quarter system. Liberal arts, business, education, engineering, law, medicine, mineral sciences. Advanced degrees conferred.

United States Naval Postgraduate School. Monterey. Founded 1909. Federal control. Men only, 1,400. Faculty, 128. Quarter system. Professional only. Advanced degrees conferred.

University of California. Berkeley; campuses also at Los Angeles, San Francisco, Santa Barbara, Davis, Riverside, La Jolla (Scripps Institution of Oceanography). Founded 1868. State control. Coeducational, 49,100 students. Faculty, 3,540. Semester system. Liberal arts at all campuses except San Francisco and La Jolla; *in addition at Berkeley*, agriculture, business administration, chemistry, criminology, education, engineering, environmental design, forestry, law, librarianship, optometry, public health, social welfare; *in addition at Los Angeles*, agriculture, applied arts, business administration, education, engineering, law, medicine, nursing, social welfare; *in addition at Santa Barbara*, applied arts; *in addition at Davis*, agriculture, veterinary medicine; *in addition at Riverside*, Citrus Experiment Station. At San Francisco: (Medical Center), medicine, dentistry, nursing, pharmacy; Hastings College (law). Advanced degrees conferred.

University of Judaism. Los Angeles. (*See* New York, Jewish Theological Seminary of America.)

University of the Pacific. Stockton. Founded 1851. Methodist control. Coeducational, 2,200 students. Faculty, 113. Semester system. Liberal arts, education, engineering, music, pharmacy. Advanced degrees conferred.

University of Redlands. Redlands. Founded 1909. Baptist control. Coeducational, 1,600 students. Faculty, 119. Semester system. Liberal arts and music. Advanced degrees conferred.

University of San Diego, College for Men. San Diego. Founded 1954. Roman Catholic control. Men only, 600. Faculty, 33. Semester system.

University of San Francisco. San Francisco. Founded 1855. Roman Catholic control. Men only, 4,100. Faculty, 236. Semester system. Liberal arts, business administration, law, nursing. Advanced degrees conferred.

University of Santa Clara. Santa Clara. Founded 1851. Roman Catholic control. Men only, 2,300. Faculty, 138. Semester system. Liberal arts, business administration, education, engineering, law. Advanced degrees conferred.

University of Southern California. Los Angeles. Founded 1880. Private control. Coeducational, 17,300 students. Faculty, 709. Semester system. Liberal arts, architecture, business administration, dentistry, education, engineering, law, library science, medicine, music, pharmacy, public administration, social work. Advanced degrees conferred.

Westmont College. Santa Barbara. Founded 1940. Private control. Coeducational, 500 students. Faculty, 33. Semester system. Liberal arts and teacher preparatory.

Whittier College. Whittier. Founded 1901. Private control. Coeducational, 1,500 students. Faculty, 84. Semester system. Liberal arts and teacher preparatory. Advanced degrees conferred.

Woodbury College. Los Angeles. Founded 1884. Proprietary control. Coeducational, 1,500 students. Faculty, 52. Quarter system. Technical and terminal-occupational.

COLORADO

Adams State College. Alamosa. Founded 1921. State control. Coeducational, 1,500 students. Faculty, 69. Quarter system. Liberal arts, terminal-occupational, teacher preparatory. Advanced degrees conferred.

Colorado College. Colorado Springs. Founded 1874. Private control. Coeducational, 1,300 students. Faculty, 102. Semester system. Liberal arts, terminal-occupational, teacher preparatory. Advanced degrees conferred.

Colorado School of Mines. Golden. Founded 1874. State control. Men only, 1,000. Faculty 105. Semester system. Technical only. Advanced degrees conferred.

Colorado State College. Greeley. Founded 1890. State control. Coeducational, 4,600 students. Faculty, 186. Quarter system. Primarily teacher preparatory. Advanced degrees conferred.

Colorado State University. Fort Collins. Founded 1879. State control. Coeducational, 6,100 students. Faculty, 389. Quarter system. Liberal arts, agriculture, engineering, forestry, home economics, veterinary medicine. Advanced degrees conferred.

Colorado Woman's College. Denver. Founded 1889. American Baptist control. Women only, 650. Faculty, 50. Semester system. Liberal arts, terminal-occupational, teacher preparatory.

Loretto Heights College. Loretto. Founded 1918. Roman Catholic control. Women only, 850. Faculty, 70. Semester system. Liberal arts, terminal-occupational, teacher preparatory.

Regis College. Denver. Founded 1888. Roman Catholic control. Men only, 1,000. Faculty, 71. Semester system. Liberal arts and teacher prep.

Union College. Denver. Clinical Division Campus. (*See* Nebraska.)

United States Air Force Academy. Colorado Springs. Founded 1955. Federal control. Men only, 1,900. Faculty, 450. Semester system.

University of Colorado. Boulder; Campus at Denver. Founded 1877. State control. Coeducational, 17,900 students. Faculty, 1,148. Semester system. Liberal arts, business, education, engineering, journalism, law, music, pharmacy; at Denver, medicine, nursing. Advanced degrees conferred.

University of Denver. Denver. Founded 1864. Methodist control. Coeducational, 5,900 students. Faculty, 317. Quarter system. Liberal arts, art, business administration, communication arts, education, engineering, hotel and restaurant management, law, librarianship, music, nursing, public administration, social work, speech. Advanced degrees conferred.

Western State College of Colorado. Gunnison. Founded 1911. State control. Coeducational, 1,600 students. Faculty, 56. Quarter system. Liberal arts and teacher preparatory. Advanced degrees.

CONNECTICUT

Central Connecticut State College. New Britain. Founded 1849. State control. Coeducational, 3,200 students. Faculty, 87. Semester system. Primarily teacher preparatory. Advanced degrees conferred.

Connecticut College. New London. Founded

1911. Private control. Women only, 1,100. Faculty, 106. Semester system. Advanced degrees conferred.

Danbury State College. Danbury. Founded 1904. State control. Coeducational, 1,100 students. Faculty, 39. Semester system. Primarily teacher preparatory. Advanced degrees conferred.

Fairfield University. Fairfield. Founded 1942. Roman Catholic control. Men only, 1,800. Faculty, 66. Semester system. Liberal arts and teacher preparatory. Advanced degrees conferred.

Hillyer College. (*See* University of Hartford.)

New Haven College. New Haven. Founded 1920. Private control. Coeducational, 1,200 students. Faculty, 130. Semester system. Professional, technical, and terminal-occupational.

Quinnipiac College. Hamden. Founded 1929. Private control. Coeducational, 1,400 students. Faculty, 68. Semester system. Liberal arts and terminal-occupational.

Rensselaer Polytechnic Institute. Hartford Graduate Division, East Windsor Hill. (*See* New York.)

St. Joseph College. West Hartford. Founded 1932. Roman Catholic control. Women only, 650. Faculty, 53. Quarter system. Liberal arts and teacher preparatory. Advanced degrees conferred.

Southern Connecticut State College. New Haven. Founded 1893. State control. Coeducational, 3,600 students. Faculty, 134. Semester system. Liberal arts and teacher preparatory. Advanced degrees conferred.

Trinity College. Hartford. Founded 1823. Private control. Men only, 1,400. Faculty, 107. Semester system. Advanced degrees conferred.

United States Coast Guard Academy. New London. Founded 1876. Federal control. Men only, 600. Faculty, 60. Semester system. Professional only.

University of Bridgeport. Bridgeport. Founded 1927. Private control. Coeducational, 5,600 students. Faculty, 218. Semester system. Liberal arts, business administration, education, engineering, nursing. Advanced degrees conferred.

University of Connecticut. Storrs; branches at Hartford, Stamford, Torrington, Waterbury. Founded 1881. State control. Coeducational, 11,100 students. Faculty, 799. Semester system. Liberal arts, agriculture, business administration, education, engineering, home economics, insurance, law, nursing, pharmacy, physical education, physical therapy, social work. Advanced degrees conferred.

University of Hartford, Hillyer College. Hartford. Founded 1879. Private control. Coeducational, 6,000 students. Faculty, 316. Semester system. Liberal arts, terminal-occupational, teacher preparatory. Advanced degrees conferred.

Wesleyan University. Middletown. Founded 1831. Private control. Men only, 950. Faculty, 134. Semester system. Liberal arts and teacher preparatory. Advanced degrees conferred.

Willimantic State College. Willimantic. Founded 1889. State control. Coeducational, 650 students. Faculty, 23. Semester system. Primarily teacher preparatory. Advanced degrees conferred.

Yale University. New Haven. Founded 1701. Private control. Undergraduate schools, men only; 8,200 students. Faculty, 1,815. Semester system. Liberal arts, art and architecture, divinity, drama, engineering, forestry, law, medicine, music, nursing. Advanced degrees conferred.

DELAWARE

University of Delaware. Newark. Founded 1833. State control. Coeducational, 5,800 students. Faculty, 220. Semester system. Liberal arts, agriculture, education, engineering, home economics. Advanced degrees conferred.

DISTRICT OF COLUMBIA

American University, The. Founded 1893. Methodist control. Coeducational, 8,300 students. Faculty, 404. Semester system. Liberal arts, business administration, government and public administration, international service, law. Advanced degrees conferred.

Catholic University of America. Founded 1889. Roman Catholic control. Coeducational, except theology and canon Law, men only, and nursing, women only; 4,400 students. Faculty, 445. Semester system. Liberal arts, canon law, engineering and architecture, law, nursing, philosophy, sacred theology, social science, social service. Advanced degrees conferred.

Columbian College. (*See* George Washington University.)

District of Columbia Teachers College. Founded 1851. Municipal control. Coeducational, 1,400 students. Faculty, 125. Semester system. Primarily teacher preparatory.

George Washington University. Founded 1821. Private control. Coeducational, 14,000 students. Faculty, 917. Semester system. Liberal arts (Columbian College), education, engineering, government, law, medicine, pharmacy. Advanced degrees conferred.

Georgetown University. Founded 1789. Roman Catholic control. Men only in college, women only in nursing, coeducational in graduate schools; 6,100 students. Faculty, 1,000. Semester system. Liberal arts, business administration, dentistry, foreign service, languages and linguistics, law, medicine, nursing. Advanced degrees conferred.

Howard University. Founded 1867. Federal and private control. Coeducational, 5,200 students. Faculty, 563. Semester system. Liberal arts, dentistry, engineering, law, medicine, music, pharmacy, social work, theology. Advanced degrees conferred.

School of Advanced International Studies of the Johns Hopkins University. (*See* Maryland, Johns Hopkins University.)

Trinity College. Founded 1897. Roman Catholic control. Women only, 900. Faculty, 64. Semester system. Liberal arts, and teacher preparatory. Advanced degrees conferred.

FLORIDA

Barry College. Miami. Founded 1940. Roman Catholic control. Women only, 800. Faculty, 56. Semester system. Liberal arts and nursing. Advanced degrees conferred.

Bethune-Cookman College. Daytona Beach. Founded 1872. Private control. Coeducational, 650 students. Faculty, 41. Semester system. Liberal arts, terminal-occupational and teacher preparatory.

Florida Agricultural and Mechanical University. Tallahassee. Founded 1887. State control. Coeducational, 3,300 students. Faculty, 217. Semester system. Liberal arts, agriculture, education, law, nursing education, pharmacy, vocational-technical. Advanced degrees conferred.

Florida Southern College. Lakeland. Founded 1885. Methodist control. Coeducational, 2,400 students. Faculty, 123. Semester system. Liberal arts and teacher preparatory.

Florida State University. Tallahassee. Founded 1857. State control. Coeducational, 9,000 students. Faculty, 550. Semester system. Liberal arts, business, education, home economics, library science, music, nursing, public affairs, social work. Advanced degrees conferred.

Rollins College. Winter Park. Founded 1885. Private control. Coeducational, 1,400 students. Faculty, 77. Quarter system. Liberal arts and teacher preparatory.

Stetson University. De Land. Founded 1883. Southern Baptist control. Coeducational, 1,900 students. Faculty, 124. Semester system. Liberal arts, business, law, music. Advanced degrees conferred.

University of Florida. Gainesville. Founded 1853. State control. Coeducational, 14,400 students. Faculty, 938. Semester system. Liberal arts, agriculture, architecture and fine arts, business administration, education, engineering, forestry, health related service, journalism, law, medicine, nursing, pharmacy, physical education and health. Advanced degrees conferred.

University of Miami. Coral Gables. Founded 1925. Private control. Coeducational, 12,800 students. Faculty, 1,236. Semester system. Liberal arts, business administration, education, engineering, law, medicine, music. Advanced degrees conferred.

University of Tampa. Tampa. Founded 1931. Private control. Coeducational, 1,900 students. Faculty, 72. Semester system. Liberal arts, terminal-occupational, teacher preparatory.

GEORGIA

Agnes Scott College. Decatur. Founded 1889. Private control. Women only, 650. Faculty, 70. Quarter system.

Albany State College. Albany. Founded 1903. State control. Coeducational, 950 students. Faculty, 40. Quarter system. Liberal arts, terminal-occupational, teacher preparatory.

Atlanta University System. Atlanta. *Atlanta University.* Founded 1865. Private control. Coeducational, 600 students. Faculty, 69. Semester system. Liberal arts, business administration, education, library service, social work. Advanced degrees conferred. *Morehouse College.* Founded 1867. Private control. Men only, 800. Faculty, 44. Semester system. *Spelman College.* Founded 1881. Baptist control. Women only, 500. Faculty, 47. Semester system. Liberal arts and teacher preparatory.

Berry College. Mount Berry. Founded 1926. Private control. Coeducational, 600 students. Faculty, 54. Semester system. Liberal arts, terminal-occupational, teacher preparatory.

Clark College. Atlanta. Founded 1869. Methodist control. Coeducational, 850 students. Faculty, 48. Semester system. Liberal arts and teacher preparatory.

Emory University. Atlanta; branch at Oxford. Founded 1836. Methodist control. Coeducational, 4,500 students. Faculty, 476. Quarter system. Liberal arts, business administration, dentistry, law, medicine, nursing, theology. Advanced degrees conferred.

Fort Valley State College. Fort Valley. Founded 1895. State control. Coeducational, 1,000 students. Faculty, 68. Quarter system. Liberal arts, and teacher preparatory. Advanced degrees.

Georgia Institute of Technology. Atlanta; branch, The Southern Technical Institute, at Chamblee. Founded 1885. State control. Coeducational, 6,600 students. Faculty, 464. Quarter system. Technical and terminal-occupational. Advanced degrees conferred.

Georgia Southern College. Collegeboro (Statesboro). Founded 1908. State control. Coeducational, 1,400 students. Faculty, 55. Quarter system. Liberal arts and teacher preparatory. Advanced degrees conferred.

Georgia State College of Business Administration. Atlanta. Founded 1914. State control. Coeducational, 3,600 students. Faculty, 217. Quarter system. Liberal arts, business administration. Advanced degrees conferred.

Mercer University. Macon. Founded 1833. Southern Baptist control. Coeducational, 1,500 students. Faculty, 99. Quarter system. Liberal arts, law, pharmacy. Advanced degrees conferred.

Morehouse College. Atlanta. (*See* Atlanta University System.)

Morris Brown College. Atlanta. Founded 1881. African Methodist Episcopal control. Coeducational, 900 students. Faculty, 52. Semester system. Liberal arts and teacher preparatory.

North Georgia College. Dahlonega. Founded 1873. State control. Coeducational, 800 students. Faculty, 46. Quarter system. Liberal arts and teacher preparatory.

Oglethorpe University. Atlanta. Founded 1835. Private control. Coeducational, 400 students. Faculty, 30. Quarter system. Liberal arts and teacher preparatory.

Savannah State College. Savannah. Founded 1891. State control. Coeducational, 1,100 students. Faculty, 71. Quarter system. Liberal arts and teacher preparatory.

Shorter College. Rome. Founded 1873. Southern Baptist control. Coeducational, 500 students. Faculty, 32. Semester system. Liberal arts and teacher preparatory.

Southern Technical Institute. Chamblee. Founded 1948. (*See* Georgia Institute of Technology.)

Spelman College. Atlanta. (*See* Atlanta University System.)

Tift College. Forsyth. Founded 1847. Baptist control. Women only, 550. Faculty, 30. Quarter system. Liberal arts and teacher preparatory.

University of Georgia. Athens. Founded 1785. State control. Coeducational, 9,600 students. Faculty, 484. Quarter system. Liberal arts, agriculture, business administration, education, forestry, home economics, journalism, law, pharmacy, veterinary medicine. Advanced degrees conferred.

Valdosta State College. Valdosta. Founded 1913. State control. Coeducational, 800 students. Faculty, 40. Quarter system. Liberal arts and teacher preparatory.

Wesleyan College. Macon. Founded 1836. Methodist control. Women only, 500. Faculty, 57. Semester system.

Woman's College of Georgia, The. Milledgeville. Founded 1889. State control. Women only, 850. Faculty, 57. Quarter system. Liberal arts and teacher preparatory. Advanced degrees conferred.

HAWAII

University of Hawaii. Honolulu. Founded 1907. State control. Coeducational, 9,400 students. Faculty, 468. Semester system. Liberal arts, business administration, education, engineering, nursing, social work, tropical agriculture. Advanced degrees conferred.

IDAHO

College of Idaho. Caldwell. Founded 1891. Presbyterian control. Coeducational, 1,000 students. Faculty, 57. Semester system. Liberal arts and teacher preparatory. Advanced degrees conferred.

Idaho State College. Pocatello. Founded 1901. State control. Coeducational, 2,300 students. Faculty, 145. Semester system. Liberal arts, education, pharmacy, trade and technical education. Advanced degrees conferred.

Lewis-Clark Normal School. Lewiston. (*See* University of Idaho.)

Northwest Nazarene College. Nampa. Founded 1913. Nazarene control. Coeducational, 600 students. Faculty, 38. Semester system. Liberal arts and teacher preparatory.

University of Idaho. Moscow; branch, Lewis-Clark Normal School, Lewiston. Founded 1899. State control. Coeducational, 4,000 students. Faculty, 329. Semester system. Liberal arts, agriculture, business administration, education, engineering, forestry, law, mines. Advanced degrees conferred.

ILLINOIS

Augustana College. Rock Island. Founded 1860. Lutheran control. Coeducational, 1,500 students. Faculty, 96. Semester system. Liberal arts and teacher preparatory.

Aurora College. Aurora. Founded 1893. Adventist Christian control. Coeducational, 1,100 students. Faculty, 55. Quarter system. Liberal arts and teacher preparatory.

Bradley University. Peoria. Founded 1897. Private control. Coeducational, 4,700 students. Faculty, 252. Semester system. Liberal arts, business administration, education, engineering, horology, industrial education, music. Advanced degrees conferred.

Carthage College. Carthage. Founded 1870. Lutheran control. Coeducational, 550 students. Faculty, 52. Semester system. Liberal arts and teacher preparatory.

Chicago Teachers College. Chicago. Founded 1869. Municipal control. Coeducational, 5,200 students. Faculty, 208. Semester system. Primarily teacher preparatory. Advanced degrees conferred.

College of St. Francis. Joliet. Founded 1925. Roman Catholic control. Women only, 550. Faculty, 42. Semester system. Liberal arts and teacher preparatory.

Concordia Teachers College. River Forest. Founded 1864. Lutheran control. Coeducational, 950 students. Faculty, 67. Quarter system. Primarily teacher preparatory. Advanced degrees conferred.

DePaul University. Chicago. Founded 1898. Roman Catholic control. Coeducational, 8,700 students. Faculty, 365. Semester system. Liberal arts, commerce, law, music. Advanced degrees conferred.

Eastern Illinois University. Charleston. Founded 1899. State control. Coeducational, 2,900 students. Faculty, 197. Quarter system. Liberal arts and teacher preparatory. Advanced degrees conferred.

Elmhurst College. Elmhurst. Founded 1871. Evangelical and Reformed control. Coeducational, 1,400 students. Faculty, 63. Semester system. Liberal arts, terminal-occupational, teacher preparatory.

Garrett Biblical Institute. Evanston. Founded 1855. Methodist control. Coeducational, 500 students. Faculty, 28. Semester system. Professional only. Advanced degrees conferred jointly with Northwestern University.

Greenville College. Greenville. Founded 1892. Methodist control. Coeducational, 600 students. Faculty, 35. Semester system. Liberal arts and teacher preparatory.

Illinois College. Jacksonville. Founded 1829. Presbyterian and Congregational Christian control. Coeducational, 500 students. Faculty, 34. Semester system.

Illinois Institute of Technology. Chicago. Founded 1892. Private control. Coeducational, 7,300 students. Faculty, 407. Semester system. Liberal arts and engineering. Advanced degrees conferred.

Illinois State Normal University. Normal. Founded 1857. State control. Coeducational, 5,000 students. Faculty, 313. Semester system. Primarily teacher preparatory. Advanced degrees conferred.

Illinois Wesleyan University. Bloomington. Founded 1850. Methodist control. Coeducational, 1,200 students. Faculty, 100. Semester system. Liberal arts, art, dramatics, music. Advanced degrees conferred.

John Marshall Law School. Chicago. Founded 1899. Private control. Coeducational, 600 students. Faculty, 71. Semester system. Professional only. Advanced degrees conferred.

Knox College. Galesburg. Founded 1837. Private control. Coeducational, 950 students. Faculty, 73. Semester system. Liberal arts and teacher preparatory.

Lake Forest College. Lake Forest. Founded 1857. Presbyterian control. Coeducational, 1,300 students. Faculty, 59. Semester system. Liberal arts and teacher preparatory.

Loyola University. Chicago; West Baden College at West Baden, Indiana. Founded 1870. Roman Catholic control. Coeducational, 8,900 students. Faculty, 940. Semester system. Liberal arts, commerce, dentistry, law, medicine, nursing, social work; at West Baden College, theology. Advanced degrees conferred.

MacMurray College. Jacksonville. Founded 1846. Methodist control. Men, women, 1,000. Faculty, 59. Semester system. Liberal arts and teacher preparatory. Advanced degrees conferred.

Millikin University. Decatur. Founded 1903. Presbyterian control. Coeducational, 1,700 students. Faculty, 90. Semester system. Liberal arts, business and industry, music. Advanced degrees conferred.

Monmouth College. Monmouth. Founded 1853. United Presbyterian control. Coeducational, 850 students. Faculty, 51. Semester system. Liberal arts and teacher preparatory.

Mundelein College. Chicago. Founded 1930. Roman Catholic control. Women only, 1,200. Faculty, 87. Semester system. Liberal arts and teacher preparatory.

National College of Education. Evanston. Founded 1886. Private control. Coeducational, 850 students. Faculty, 54. Semester system. Primarily teacher preparatory. Advanced degrees conferred.

North Central College. Naperville. Founded 1861. Evangelical United Brethren control. Coeducational, 900 students. Faculty, 54. Semester system. Liberal arts and music.

Northern Illinois University. De Kalb. Founded 1895. State control. Coeducational, 7,600 students. Faculty, 300. Semester system. Liberal arts and teacher preparatory. Advanced degrees.

Northwestern University. Evanston; Campus also at Chicago. Founded 1851. Private control. Coeducational, 15,500 students. Faculty, 1,907. Quarter system. At Evanston, liberal arts, business, education, journalism, music, speech, technological institute; at Chicago, dentistry, law, medicine. Advanced degrees conferred.

Olivet Nazarene College. Kankakee. Founded 1907. Nazarene control. Coeducational, 1,100 students. Faculty, 55. Semester system. Liberal arts and teacher preparatory.

Parks College of Aeronautical Technology. East St. Louis, a part of St. Louis University. (*See* Missouri.)

Principia College. Elsah. Founded 1910. Private control. Coeducational, 550 students. Faculty, 45. Quarter system. Liberal arts and teacher preparatory.

Quincy College. Quincy. Founded 1860. Roman Catholic control. Coeducational, 1,100 students. Faculty, 60. Semester system. Liberal arts and teacher preparatory.

Rockford College. Rockford. Founded 1847. Private control. Coeducational, 1,300 students. Faculty, 93. Semester system. Liberal arts and teacher preparatory. Advanced degrees conferred.

Roosevelt University. Chicago. Founded 1945. Private control. Coeducational, 5,100 students. Faculty, 248. Semester system. Liberal arts, business

administration, labor education, music. Advanced degrees conferred.

Rosary College. River Forest. Founded 1848. Roman Catholic control. Women only, 1,000. Faculty, 76. Semester system. Liberal arts and teacher preparatory. Advanced degrees conferred.

St. Procopius College. Lisle. Founded 1887. Roman Catholic control. Men only, 550. Faculty, 42. Semester system. Liberal arts, teacher prep.

St. Xavier College. Chicago. Founded 1846. Roman Catholic control. Women only, 1,100. Faculty, 70. Semester system. Liberal arts and teacher preparatory. Advanced degrees conferred.

School of the Art Institute of Chicago. Chicago. Founded 1879. Private control. Coeducational, 2,300 students. Faculty, 85. Quarter system. Professional and teacher preparatory. Advanced degrees conferred.

Southern Illinois University. Carbondale; campus also at Edwardsville. Founded 1869. State control. Coeducational, 13,800 students. Faculty, 768. Quarter system. Liberal arts at both campuses; *in addition at Carbondale*, agriculture, applied science, business, communications, education, fine arts, home economics, nursing, technical and adult education; *in addition at Edwardsville*, business, education, fine arts, social studies, science. Advanced degrees conferred.

University of Chicago. Chicago. Founded 1892. Private control. Coeducational, 9,000. Faculty, 808. Quarter system. Liberal arts, biological sciences, business, law, library school, medicine, physical sciences, social sciences, social service administration, theology. Advanced degrees conferred.

University of Illinois. Urbana; undergraduate division, Navy Pier, Chicago; medical courses at Chicago. Founded 1868. State control. Coeducational, 30,800 students. Faculty, 4,212. Semester system; quarter system for medical courses. Liberal arts, agriculture, aviation, commerce and business administration, dentistry, education, engineering, fine and applied arts, government and public affairs, journalism and communications, labor and industrial relations, law, library science, medicine, nursing, pharmacy, physical education, social work, veterinary medicine. Advanced degrees conferred.

Western Illinois University. Macomb. Founded 1899. State control. Coeducational, 3,000 students. Faculty, 167. Quarter system. Liberal arts and education. Advanced degrees conferred.

Wheaton College. Wheaton. Founded 1860. Private control. Coeducational, 1,800 students. Faculty, 132. Semester system. Liberal arts and teacher preparatory. Advanced· degrees conferred.

INDIANA

Anderson College and Theological Seminary. Anderson. Founded 1917. Church of God control. Coeducational, 1,100 students. Faculty, 80. Semester system. Liberal arts and theology.

Ball State Teachers College. Muncie. Founded 1918. State control. Coeducational, 7,100 students. Faculty, 282. Quarter system. Liberal arts and teacher preparatory. Advanced degrees conferred.

Butler University. Indianapolis. Founded 1855. Disciples of Christ control. Coeducational, 3,900 students. Faculty, 235. Semester system. Liberal arts, business administration, education, music, pharmacy. Advanced degrees conferred.

DePauw University. Greencastle. Founded 1837. Private control. Coeducational, 2,300 students. Faculty, 184. Semester system. Liberal arts and music. Advanced degrees conferred.

Earlham College. Richmond. Founded 1847. Society of Friends control. Coeducational, 900 students. Faculty, 79. Semester system. Liberal arts and teacher preparatory. Advanced degrees conferred.

Evansville College. Evansville. Founded 1854. Private control. Coeducational, 2,700 students. Faculty, 93. Quarter system. Liberal arts, terminal-occupational, teacher preparatory.

Franklin College of Indiana. Franklin. Founded 1834. Baptist control. Coeducational, 600 students. Faculty, 38. Semester system. Liberal arts and teacher preparatory.

Goshen College. Goshen. Founded 1894. Mennonite control. Coeducational, 1,100 students. Faculty, 61. Semester system. Liberal arts, biblical seminary, nursing.

Hanover College. Hanover. Founded 1827. Presbyterian control. Coeducational, 850 students.

219

Faculty, 48. Semester system. Liberal arts and teacher preparatory.

Indiana Central College. Indianapolis. Founded 1902. Evangelical United Brethren control. Coeducational, 1,500 students. Faculty, 48. Semester system. Liberal arts and teacher preparatory.

Indiana State College. Terre Haute. Founded 1870. State control. Coeducational, 4,900 students. Faculty, 183. Quarter system. Primarily teacher preparatory. Advanced degrees conferred.

Indiana University. Bloomington; campus also at Indianapolis. Founded 1820. State control. Coeducational, 26,800 students. Faculty, 2,555. Semester system. Liberal arts, business, education, health, physical education and recreation, law, music, optometry; at Indianapolis, dentistry, law, medicine, nursing, physical education, social service. Advanced degrees conferred.

Manchester College. North Manchester. Founded 1889. Church of Brethren control. Coeducational, 1,000 students. Faculty, 61. Quarter system. Liberal arts and teacher preparatory.

Marian College. Indianapolis. Founded 1937. Roman Catholic control. Coeducational, 700 students. Faculty, 48. Semester system. Liberal arts and teacher preparatory.

Purdue University. Lafayette. Founded 1874. State control. Coeducational, 19,200 students. Faculty, 1,204. Semester system. Agriculture; engineering; home economics; industrial management; pharmacy; science, education and humanities; veterinary science and medicine. Advanced degrees conferred.

Saint Francis College. Fort Wayne. Founded 1940. Roman Catholic control. Women only, 500. Faculty, 37. Semester system. Liberal arts and teacher preparatory.

St. Joseph's College. Rensselaer. Founded 1889. Roman Catholic control. Men only, 1,600. Faculty, 73. Semester system. Liberal arts and teacher preparatory.

St. Mary-of-the-Woods College. St. Mary-of-the-Woods. Founded 1840. Roman Catholic control. Women only, 600. Faculty, 47. Semester system. Liberal arts and teacher preparatory.

St. Mary's College. Notre Dame. Founded 1844. Roman Catholic control. Women only, 1,100. Faculty, 95. Semester system. Liberal arts and teacher preparatory. Advanced degrees conferred.

Taylor University. Upland. Founded 1846. Private control. Coeducational, 800 students. Faculty, 42. Semester system. Liberal arts and teacher preparatory.

University of Notre Dame. Notre Dame (South Bend). Founded 1843. Roman Catholic control. Men only, 6,500. Faculty, 449. Semester system. Liberal arts, commerce, engineering, law, science. Advanced degrees conferred.

Valparaiso University. Valparaiso. Founded 1859. Lutheran control. Coeducational, 3,200 students. Faculty, 186. Semester system. Liberal arts, engineering, law.

Wabash College. Crawfordsville. Founded 1832. Private control. Men only, 650. Faculty, 56. Semester system.

West Baden College. (*See* Illinois, Loyola University.)

IOWA

Briar Cliff College. Sioux City. Founded 1930. Roman Catholic control. Women only, 500. Faculty, 41. Semester system. Liberal arts and teacher preparatory.

Buena Vista College. Storm Lake. Founded 1891. Presbyterian control. Coeducational, 700 students. Faculty, 34. Semester system. Liberal arts and teacher preparatory.

Central College. Pella. Founded 1854. Reformed control. Coeducational, 500 students. Faculty, 42. Semester system. Liberal arts, terminal-occupational, teacher preparatory.

Clarke College. Dubuque. Founded 1843. Roman Catholic control. Women only, 900. Faculty, 67. Semester system. Liberal arts and teacher preparatory.

Coe College. Cedar Rapids. Founded 1851. Private control. Coeducational, 1,000 students. Faculty, 63. Semester system. Liberal arts and teacher preparatory.

Cornell College. Mount Vernon. Founded 1853. Private control. Coeducational, 750 students. Faculty, 72. Semester system. Liberal arts, terminal-occupational, teacher preparatory.

Drake University. Des Moines. Founded 1881. Private control. Coeducational, 6,700 students. Faculty, 230. Semester system. Liberal arts, business administration, divinity, education, fine arts, law, pharmacy. Advanced degrees conferred.

Graceland College. Lamoni. Founded 1895. Reorganized Latter Day Saints control. Coeducational, 800 students. Faculty, 55. Semester system. Liberal arts, terminal-occupational, teacher preparatory.

Grinnell College. Grinnell. Founded 1846. Private control. Coeducational, 1,100 students. Faculty, 103. Semester system. Liberal arts and teacher preparatory.

Iowa State University of Science and Technology. Ames. Founded 1858. State control. Coeducational, 9,700 students. Faculty, 855. Quarter system. Agriculture, engineering, home economics, science, veterinary medicine. Advanced degrees conferred.

Iowa Wesleyan College. Mount Pleasant. Founded 1842. Methodist control. Coeducational, 750 students. Faculty, 40. Semester system. Liberal arts and teacher preparatory.

Loras College. Dubuque. Founded 1839. Roman Catholic control. Men only, 1,300. Faculty, 74. Semester system. Liberal arts and teacher preparatory.

Luther College. Decorah. Founded 1861. American Lutheran control. Coeducational, 1,300 students. Faculty, 77. Semester system. Liberal arts and teacher preparatory.

Marycrest College. Davenport. Founded 1939. Roman Catholic control. Women only, 750. Faculty, 44. Semester system. Liberal arts and teacher preparatory.

Morningside College. Sioux City. Founded 1889. Methodist control. Coeducational, 1,300 students. Faculty, 67. Semester system. Liberal arts and music.

Parsons College. Fairfield. Founded 1875. Presbyterian control. Coeducational, 1,700 students. Faculty, 68. Trimester system. Liberal arts and teacher preparatory.

St. Ambrose College. Davenport. Founded 1882. Roman Catholic control. Coordinated, 1,300 students. Faculty, 76. Semester system.

Simpson College. Indianola. Founded 1860. Methodist control. Coeducational, 700 students. Faculty, 45. Quarter system. Liberal arts and teacher preparatory.

State College of Iowa. Cedar Falls. Founded 1876. State control. Coeducational, 4,600 students. Faculty, 271. Semester system. Primarily teacher preparatory. Advanced degrees conferred.

State University of Iowa. Iowa City. Founded 1847. State control. Coeducational, 11,100 students. Faculty, 804. Semester system. Liberal arts, commerce, dentistry, education, engineering, fine arts, journalism, law, medicine, nursing, pharmacy, religion, social work. Advanced degrees conferred.

University of Dubuque. Dubuque. Founded 1852. Presbyterian control. Coeducational, 800 students. Faculty, 51. Semester system. Liberal arts and theology.

Upper Iowa University. Fayette. Founded 1857. Private control. Coeducational, 1,200 students. Faculty, 32. Semester system. Liberal arts and teacher preparatory.

Wartburg College. Waverly. Founded 1852. Lutheran control. Coeducational, 1,100 students. Faculty, 60. Semester system. Liberal arts and teacher preparatory.

Westmar College. Le Mars. Founded 1890. Evangelical United Brethren control. Coeducational, 600 students. Faculty, 45. Semester system. Liberal arts and teacher preparatory.

KANSAS

Baker University. Baldwin City. Founded 1858. Methodist control. Coeducational, 600 students. Faculty, 42. Semester system. Liberal arts and teacher preparatory.

Bethany College. Lindsborg. Founded 1881. Lutheran control. Coeducational, 900 students. Faculty, 46. Semester system. Liberal arts and fine arts.

Bethel College. North Newton. Founded 1887. Mennonite control. Coeducational, 550 students. Faculty, 45. Quarter system. Liberal arts and teacher preparatory.

Fort Hays Kansas State College. Hays. Founded 1902. State control. Coeducational, 2,900 stu-

dents. Faculty, 141. Semester system. Liberal arts and teacher preparatory. Advanced degrees.

Friends University. Wichita. Founded 1898. Society of Friends control. Coeducational, 650 students. Faculty, 49. Semester system. Liberal arts and teacher preparatory.

Kansas State College of Pittsburg. Pittsburg. Founded 1903. State control. Coeducational, 3200 students. Faculty, 169. Semester system. Liberal arts, terminal-occupational, teacher preparatory. Advanced degrees conferred.

Kansas State Teachers College. Emporia. Founded 1863. State control. Coeducational, 4,200 students. Faculty, 166. Semester system. Liberal arts, terminal-occupational, teacher preparatory. Advanced degrees conferred.

Kansas State University of Agriculture and Applied Science. Manhattan. Founded 1863. State control. Coeducational, 7,800 students. Faculty, 672. Semester system. Liberal arts, agriculture, engineering and architecture, home economics, veterinary medicine. Advanced degrees conferred.

Marymount College. Salina. Founded 1922. Roman Catholic control. Women only, 500. Faculty, 43. Semester system. Liberal arts and teacher preparatory.

McPherson College. McPherson. Founded 1887. Church of Brethren control. Coeducational, 500 students. Faculty, 33. Semester system. Liberal arts and teacher preparatory.

Mount St. Scholastica College. Atchison. Founded 1863. Roman Catholic control. Women only, 500. Faculty, 46. Semester system. Liberal arts and teacher preparatory.

Ottawa University. Ottawa. Founded 1865. Baptist control. Coeducational, 600 students. Faculty, 37. Semester system. Liberal arts and teacher preparatory.

St. Benedict's College. Atchison. Founded 1858. Roman Catholic control. Men only, 650. Faculty, 55. Semester system. Liberal arts and theology.

Saint Mary College. Xavier. Founded 1923. Roman Catholic control. Women only, 550. Faculty, 53. Semester system. Liberal arts and teacher preparatory. Advanced degrees conferred.

St. Mary's College. St. Mary's. (*See* Missouri, St. Louis University.)

Southwestern College. Winfield. Founded 1885. Methodist control. Coeducational, 600 students. Faculty, 38. Semester system. Liberal arts and teacher preparatory.

University of Kansas. Lawrence; medical courses at Kansas City. Founded 1866. State control. Coeducational, 10,000 students. Faculty, 790. Semester system. Liberal arts, business, education, engineering and architecture, fine arts, journalism and public information, law, medicine, pharmacy. Advanced degrees conferred.

University of Wichita. Wichita. Founded 1895. Municipal control. Coeducational, 5,700 students. Faculty, 223. Semester system. Liberal arts, business administration and industry, education, engineering, music. Advanced degrees conferred.

Washburn University of Topeka. Topeka. Founded 1865. Municipal control. Coeducational, 3,000 students. Faculty, 154. Semester system. Liberal arts and law. Advanced degrees conferred.

KENTUCKY

Asbury College. Wilmore. Founded 1890. Private control. Coeducational, 900 students. Faculty, 61. Quarter system. Liberal arts and teacher preparatory.

Bellarmine College. Louisville. Founded 1950. Roman Catholic control. Men only, 1,300. Faculty, 75. Semester system. Liberal arts, terminal-occupational, teacher preparatory.

Berea College. Berea. Founded 1855. Private control. Coeducational, 1,300 students. Faculty, 116. Semester system. Liberal arts and teacher preparatory.

Bowling Green College of Commerce. Bowling Green. Founded 1922. Proprietary control. Coeducational, 500 students. Faculty, 26. Semester system. Technical and teacher preparatory.

Brescia College. Owensboro. Founded 1925. Roman Catholic control. Coeducational, 750 students. Faculty, 42. Semester system. Liberal arts, terminal-occupational, teacher preparatory.

Centre College of Kentucky. Danville. Founded 1819. Private control. Coeducational, 500 students. Faculty, 39. Semester system. Liberal arts and teacher preparatory.

Eastern Kentucky State College. Richmond. Founded 1906. State control. Coeducational, 3,500 students. Faculty, 124. Semester system. Liberal arts and teacher preparatory. Advanced degrees conferred.

Georgetown College. Georgetown. Founded 1829. Southern Baptist control. Coeducational, 1,300 students. Faculty, 54. Semester system. Liberal arts and teacher preparatory. Advanced degrees conferred.

Kentucky State College. Frankfort. Founded 1886. State control. Coeducational, 600 students. Faculty, 51. Semester system. Liberal arts, terminal-occupational, teacher preparatory.

Kentucky Wesleyan College. Owensboro. Founded 1860. Methodist control. Coeducational, 600 students. Faculty, 42. Semester system. Liberal arts, terminal-occupational, teacher preparatory.

Morehead State College. Morehead. Founded 1923. State control. Coeducational, 2,600 students. Faculty, 92. Semester system. Liberal arts, terminal-occupational, teacher preparatory. Advanced degrees conferred.

Murray State College. Murray. Founded 1923. State control. Coeducational, 3,000 students. Faculty, 135. Semester system. Liberal arts and teacher preparatory. Advanced degrees conferred.

Nazareth College. Louisville. Founded 1920. Roman Catholic control. Women only, 1,300. Faculty, 82. Semester system. Liberal arts and teacher preparatory. Advanced degrees conferred.

Southern Baptist Theological Seminary. Louisville. Founded 1859. Southern Baptist control. Coeducational, 800 students. Faculty, 77. Semester system. Church music, religious education, theology. Advanced degrees conferred.

Transylvania College. Lexington. Founded 1780. Private control. Coeducational, 550 students. Faculty, 40. Quarter system. Liberal arts and teacher preparatory.

Union College. Barbourville. Founded 1879. Methodist control. Coeducational, 700 students. Faculty, 39. Semester system. Liberal arts and teacher preparatory.

University of Kentucky. Lexington. Founded 1865. State control. Coeducational, 10,200 students. Faculty, 535. Semester system. Liberal arts, agriculture and home economics, commerce, education, engineering, law, nursing, pharmacy. Advanced degrees conferred.

University of Louisville. Louisville. Founded 1798. Municipal control. Coeducational, 6,200 students. Faculty, 906. Semester system. Liberal arts, business, dentistry, engineering, law, medicine, music, social work. Advanced degrees conferred.

Ursuline College. Louisville. Founded 1938. Roman Catholic control. Women only, 550. Faculty, 42. Semester system. Liberal arts, terminal-occupational, teacher preparatory.

Villa Madonna College. Covington. Founded 1921. Roman Catholic control. Coeducational, 1,300 students. Faculty, 64. Semester system. Liberal arts and teacher preparatory.

Western Kentucky State College. Bowling Green. Founded 1906. State control. Coeducational, 3,800 students. Faculty, 113. Semester system. Liberal arts and teacher preparatory. Advanced degrees conferred.

LOUISIANA

Centenary College. Shreveport. Founded 1825. Methodist control. Coeducational, 1,600 students. Faculty, 114. Semester system. Liberal arts and teacher preparatory.

Dillard University. New Orleans. Founded 1935. Private control. Coeducational, 900 students. Faculty, 72. Semester system. Liberal arts and teacher preparatory.

Grambling College. Grambling. Founded 1901. State control. Coeducational, 2,700 students. Faculty, 148. Semester system. Liberal arts and teacher preparatory.

Louisiana College. Pineville. Founded 1906. Southern Baptist control. Coeducational, 1,000 students. Faculty, 61. Semester system. Liberal arts and teacher preparatory.

Louisiana Polytechnic Institute. Ruston. Founded 1894. State control. Coeducational, 3,700 students. Faculty, 221. Semester system. Liberal arts, agriculture and forestry, business administration, education, engineering, home economics. Advanced degrees conferred.

Louisiana State University and Agricultural and Mechanical College. Baton Rouge; campuses also at New Orleans and Alexandria. Founded 1860. State control. Coeducational, 13,900 students. Faculty, 1,379. Semester system. Liberal arts, agriculture, chemistry and physics, commerce, education, engineering, law, library science, medicine, music, social welfare. Advanced degrees conferred.

Loyola University. New Orleans. Founded 1912. Roman Catholic control. Men; women in professional departments only; 2,500 students. Faculty, 226. Semester system. Liberal arts, business administration, dentistry, law, music, pharmacy. Advanced degrees conferred.

McNeese State College. Lake Charles. Founded 1939. State control. Coeducational, 2,700 students. Faculty, 117. Semester system. Liberal arts and teacher preparatory.

New Orleans Baptist Theological Seminary. New Orleans. Founded 1917. Southern Baptist control. Coeducational, 750 students. Faculty, 40. Semester system. Religious education, sacred music, theology. Advanced degrees conferred.

Newcomb College. New Orleans. (*See* Tulane University.)

Northeast Louisiana State College. Monroe. Founded 1931. State control. Coeducational, 2,600 students. Faculty, 119. Semester system. Liberal arts, agriculture and home economics, business administration, education, fine arts, pharmacy.

Northwestern State College of Louisiana. Natchitoches. Founded 1884. State control. Coeducational, 3,000 students. Faculty, 148. Semester system. Liberal arts, applied arts and sciences, education, nursing. Advanced degrees conferred.

St. Mary's Dominican College. New Orleans. Founded 1910. Roman Catholic control. Women only, 500. Faculty, 43. Semester system. Liberal arts and teacher preparatory.

Southeastern Louisiana College. Hammond. Founded 1925. State control. Coeducational, 2,100 students. Faculty, 126. Semester system. Liberal arts, terminal-occupational, teacher preparatory.

Southern University and Agricultural and Mechanical College. Baton Rouge. Founded 1880. State control. Coeducational, 5,400 students. Faculty, 286. Semester system. Liberal arts, agriculture, education, engineering, law. Advanced degrees conferred.

Tulane University of Louisiana. New Orleans. Founded 1834. Private control. Coordinated, 6,800 students. Faculty, 1,324. Semester system. Liberal arts (men only), Newcomb College (women only), architecture, business administration, engineering, graduate medicine, law, medicine, social work. Advanced degrees conferred.

University of Southwestern Louisiana, The. Lafayette. Founded 1898. State control. Coeducational, 4,900 students. Faculty, 281. Semester system. Liberal arts, agriculture, commerce, education, engineering, nursing. Advanced degrees conferred.

Xavier University. New Orleans. Founded 1925. Roman Catholic control. Coeducational, 800 students. Faculty, 90. Semester system. Liberal arts and pharmacy. Advanced degrees conferred.

MAINE

Bates College. Lewiston. Founded 1864. Private control. Coeducational, 850 students. Faculty, 54. Semester system.

Bowdoin College. Brunswick. Founded 1794. Private control. Men only, 800. Faculty, 77. Semester system.

Colby College. Waterville. Founded 1813. Private control. Men, women, 1,200 students. Faculty, 103. Semester system.

Farmington State Teachers College. Farmington. Founded 1864. State control. Coeducational, 500 students. Faculty, 45. Semester system. Primarily teacher preparatory.

Gorham State Teachers College. Gorham. Founded 1878. State control. Coeducational, 750 students. Faculty, 36. Semester system. Primarily teacher preparatory.

University of Maine. Orono. Founded 1865. State control. Coeducational, 5,900 students. Faculty, 319. Semester system. Liberal arts, agriculture, education, technology. Advanced degrees conferred.

MARYLAND

College of Notre Dame of Maryland, Inc. Baltimore. Founded 1873. Roman Catholic control. Women only, 950. Faculty, 58. Semester system. Liberal arts and teacher preparatory.

Columbia Union College. Takoma Park, Maryland. Founded 1904. Seventh Day Adventist control. Coeducational, 500 students. Faculty, 63. Semester system. Liberal arts, terminal-occupational, teacher preparatory.

Goucher College. Towson, (Baltimore). Founded 1885. Private control. Women only, 750. Faculty, 86. Quarter system. Liberal arts and teacher preparatory. Advanced degrees conferred.

Hood College. Frederick. Founded 1893. Private control. Women only, 650. Faculty, 54. Semester system. Liberal arts and teacher preparatory.

Johns Hopkins University. Baltimore; branch, School of Advanced International Studies, Washington, D.C. Founded 1876. Private control. Men only in undergraduate schools, coeducational in graduate schools; 8,000 students. Faculty, 1,407. Semester system. Liberal arts, engineering, hygiene and public health, medicine, philosophy. Advanced degrees conferred.

Loyola College. Baltimore. Founded 1852. Roman Catholic control. Men only in undergraduate schools, coeducational in evening and graduate schools; 1,600 students. Faculty, 101. Semester system. Liberal arts and teacher preparatory. Advanced degrees conferred.

Maryland State College. Princess Anne. (*See* University of Maryland.)

Morgan State College. Baltimore. Founded 1867. State control. Coeducational, 2,500 students. Faculty, 150. Semester system. Liberal arts, terminal-occupational, teacher preparatory.

Mount St. Mary's College. Emmitsburg. Founded 1808. Roman Catholic control. Men only, 750. Faculty, 44. Semester system.

St. Charles College. Catonsville. (*See* St. Mary's Seminary and University.)

St. Joseph College. Emmitsburg. Founded 1809. Roman Catholic control. Women only, 550. Faculty, 51. Semester system. Liberal arts, terminal-occupational, teacher preparatory.

St. Mary's Seminary and University. Baltimore; branch, St. Charles College, Catonsville. Founded 1791. Roman Catholic control. Men only, 750. Faculty, 65. Semester system. Liberal arts and theology. Advanced degrees conferred.

State Teachers College. Frostburg. Founded 1902. State control. Coeducational, 1,100 students. Faculty, 42. Semester system. Primarily teacher preparatory. Advanced degrees conferred.

State Teachers College at Towson. Baltimore. Founded 1866. State control. Coeducational, 1,500 students. Faculty, 91. Semester system. Liberal arts and teacher preparatory. Advanced degrees conferred.

United States Naval Academy. Annapolis. Founded 1845. Federal control. Men only, 3,900. Faculty, 505. Semester system. Professional and general.

University of Maryland. College Park; campuses also at Baltimore and Princess Anne. Founded 1807. State control. Coeducational, 19,900 students. Faculty, 879. Semester system. Liberal arts, agriculture, business and public administration, education, engineering, home economics, physical education, recreation and health; at Baltimore, dentistry, law, medicine, nursing, pharmacy. Advanced degrees conferred. At Maryland State College division, Princess Anne, liberal arts, terminal-occupational, teacher preparatory.

Western Maryland College. Westminster. Founded 1867. Methodist control. Coeducational, 850 students. Faculty, 65. Semester system. Liberal arts and teacher preparatory. Advanced degrees.

MASSACHUSETTS

American International College. Springfield. Founded 1885. Private control. Coeducational, 1,700 students. Faculty, 73. Semester system. Liberal arts, terminal-occupational, teacher preparatory. Advanced degrees conferred.

Amherst College. Amherst. Founded 1821. Private control. Men only, 1,000. Faculty, 126. Semester system. Advanced degrees conferred.

Atlantic Union College. South Lancaster. Founded 1882. Seventh Day Adventist control. Coeducational, 600 students. Faculty, 43. Semester system. Liberal arts and theology.

Babson Institute of Business Administration. Babson Park. Founded 1919. Private control. Men only, 700. Faculty, 33. Quarter system. Professional only. Advanced degrees conferred.

Boston College. Chestnut Hill. Founded 1863.

Roman Catholic control. Coeducational, 8,000 students. Faculty, 564. Semester system. Liberal arts, business, education, law, nursing, philosophy, social work, theology. Advanced degrees conferred.

Boston University. Boston. Founded 1839. Private control. Coeducational, 19,000 students. Faculty, 1,444. Semester system. Liberal arts, business administration, education, fine and applied arts, industrial technology, law, medicine, nursing, physical education and physical therapy (Sargent College), public relations and communications, social work, theology. Advanced degrees conferred.

Brandeis University. Waltham. Founded 1948. Private control. Coeducational, 1,600 students. Faculty, 164. Semester system. Liberal arts, social welfare. Advanced degrees conferred.

Clark University. Worcester. Founded 1887. Private control. Coeducational, 1,900 students. Faculty, 86. Semester system. Advanced degrees conferred.

College of the Holy Cross. Worcester. Founded 1843. Roman Catholic control. Men only, 1,800. Faculty, 133. Semester system. Advanced degrees conferred.

College of Our Lady of the Elms. Chicopee. Founded 1928. Roman Catholic control. Women only, 650. Faculty, 32. Semester system. Liberal arts and teacher preparatory.

Crane Theological School. Medford. (*See* Tufts University.)

Eastern Nazarene College. Wollaston. Founded 1918. Nazarene control. Coeducational, 650 students. Faculty, 41. Semester system. Liberal arts and teacher preparatory.

Emerson College. Boston. Founded 1880. Private control. Coeducational, 700 students. Faculty, 52. Semester system. Liberal arts, terminal-occupational, teacher preparatory. Advanced degrees conferred.

Emmanuel College. Boston. Founded 1919. Roman Catholic control. Women only, 900. Faculty, 75. Semester system. Liberal arts and teacher preparatory.

Fletcher School of Law and Diplomacy. Medford. (*See* Tufts University.)

Gordon College. Beverly Farms. Founded 1889. Private control. Coeducational, 500 students. Faculty, 28. Semester system. Liberal arts and divinity.

Harvard University. Cambridge; branch at Boston. Founded 1636. Private control. Men only, 11,400. Faculty, 3,042. Semester system. Liberal arts, design, divinity, education, law, public administration; at Boston, business administration, dental medicine, medicine, public health. Advanced degrees conferred.

Jackson College. Medford. (*See* Tufts University.)

Lesley College. Cambridge. Founded 1910. Private control. Women only, 500. Faculty, 31. Semester system. Primarily teacher preparatory. Advanced degrees conferred.

Lowell Technological Institute. Lowell. Founded 1895. State control. Coeducational, 1,200 students. Faculty, 111. Semester system. Technical only. Advanced degrees conferred.

Massachusetts College of Art. Boston. Founded 1873. State control. Coeducational, 500 students. Faculty, 35. Semester system. Professional and teacher preparatory.

Massachusetts College of Pharmacy. Boston. Founded 1823. Private control. Coeducational, 550 students. Faculty, 44. Semester system. Professional only. Advanced degrees conferred.

Massachusetts Institute of Technology. Cambridge. Founded 1861. Private control. Coeducational, 6,300 students. Faculty, 1,093. Semester system. Liberal arts, architecture, engineering, industrial management, science. Advanced degrees conferred.

Merrimack College. North Andover. Founded 1947. Roman Catholic control. Coeducational, 1,900 students. Faculty, 85. Semester system. Liberal arts, business, engineering.

Mount Holyoke College. South Hadley. Founded 1837. Private control. Women only, 1,500. Faculty, 157. Semester system. Liberal arts and teacher preparatory. Advanced degrees conferred.

Newton College of the Sacred Heart. Newton. Founded 1946. Roman Catholic control. Women only, 600. Faculty, 38. Semester system.

Northeastern University. Boston. Founded 1898. Private control. Coeducational, 17,300 students. Faculty, 633. 10-week terms. Liberal arts, business administration, education, engineering. Advanced degrees conferred.

Radcliffe College. Cambridge. Founded 1879. Private control. Women only, 1,800. Faculty, arts and science faculty of Harvard University. Semester system. Advanced degrees conferred.

Regis College. Weston. Founded 1927. Roman Catholic control. Women only, 700. Faculty, 74. Semester system. Liberal arts and teacher preparatory.

Sargent College (*See* Boston University.)

Simmons College. Boston. Founded 1899. Private control. Women only, 1,400. Faculty, 208. Semester system. Liberal arts, business, education, home economics, library, nursing, Prince School of Retailing, publication, science, social science, social work. Advanced degrees conferred.

Smith College. Northampton. Founded 1871. Private control. Women only, 2,400. Faculty, 242. Semester system. Liberal arts and social work. Advanced degrees conferred.

Springfield College. Springfield. Founded 1885. Private control. Coeducational, 1,500 students. Faculty, 102. Quarter system. Professional and teacher preparatory. Advanced degrees conferred.

State College at Bridgewater. Founded 1840. State control. Coeducational, 1,300 students. Faculty, 68. Semester system. Primarily teacher preparatory. Advanced degrees conferred.

State College at Fitchburg. Founded 1894. State control. Coeducational, 1,000 students. Faculty, 68. Semester system. Primarily teacher preparatory. Advanced degrees conferred.

State College at Framingham. Founded 1838. State control. Women only, 950. Faculty, 45. Semester system. Primarily teacher preparatory.

State College at Lowell. Founded 1894. State control. Coeducational, 700 students. Faculty, 53. Semester system. Primarily teacher preparatory.

State College at North Adams. Founded 1894. State control. Coeducational, 550 students. Faculty, 30. Semester system. Primarily teacher preparatory. Advanced degrees conferred.

State College at Salem. Founded 1854. State control. Coeducational, 2,000 students. Faculty, 69. Quarter system. Primarily teacher preparatory. Advanced degrees conferred.

State College at Westfield. Founded 1839. State control. Coeducational, 950 students. Faculty, 45. Semester system. Primarily teacher preparatory.

State College at Worcester. Founded 1871. State control. Coeducational, 1,500 students. Faculty, 40. Semester system. Primarily teacher preparatory. Advanced degrees conferred.

Stonehill College. North Easton. Founded 1948. Roman Catholic control. Coeducational, 650 students. Faculty, 45. Semester system. Liberal arts and teacher preparatory.

Suffolk University. Boston. Founded 1906. Private control. Coeducational, 1,700 students. Faculty, 100. Semester system. Liberal arts and law. Advanced degrees conferred.

Tufts University. Medford. Founded 1852. Private control. Coeducational, 4,300 students. Faculty, 498. Semester system. Liberal arts (men), engineering, dental medicine, medicine, special studies; Crane Theological School; Fletcher School of Law and Diplomacy; Jackson College (women). Advanced degrees conferred.

University of Massachusetts. Amherst. Founded 1863. State control. Coeducational, 6,200 students. Faculty, 453. Semester system. Liberal arts, agriculture, business administration, education, engineering, home economics, nursing, physical education. Advanced degrees conferred.

Wellesley College. Wellesley. Founded 1875. Private control. Women only, 1,700. Faculty, 184. Semester system. Advanced degrees conferred.

Wheaton College. Norton. Founded 1834. Private control. Women only, 800. Faculty, 73. Semester system.

Wheelock College. Boston. Founded 1888. Private control. Women only, 500. Faculty, 26. Semester system. Primarily teacher preparatory. Advanced degrees conferred.

Williams College. Williamstown. Founded 1793. Private control. Men only, 1,100. Faculty, 129. Semester system. Advanced degrees conferred.

Worcester Polytechnic Institute. Worcester. Founded 1865. Private control. Men only, 1,200. Faculty, 150. Semester system. Technical only. Advanced degrees conferred.

MICHIGAN

Adrian College. Adrian. Founded 1845. Methodist control. Coeducational, 850 students. Faculty, 37. Semester system. Liberal arts, teacher prep.

Albion College. Albion. Founded 1835. Methodist control. Coeducational, 1,400 students. Faculty, 88. Semester system. Liberal arts and teacher preparatory. Advanced degrees conferred.

Alma College. Alma. Founded 1887. Presbyterian control. Coeducational, 750 students. Faculty, 57. Semester system. Liberal arts and teacher preparatory.

Andrews University. Berrien Springs. Founded 1874. Seventh Day Adventist control. Coeducational, 1,200 students. Faculty, 68. Semester system. Liberal arts and teacher preparatory.

Aquinas College. Grand Rapids. Founded 1922. Roman Catholic control. Coeducational, 1,000 students. Faculty, 61. Semester system. Liberal arts and teacher preparatory.

Calvin College. Grand Rapids. Founded 1876. Christian Reformed control. Coeducational, 2,100 students. Faculty, 108. Semester system. Liberal arts and teacher preparatory.

Central Michigan University. Mount Pleasant. Founded 1892. State control. Coeducational, 6,900 students. Faculty, 220. Semester system. Liberal arts, business administration, education, fine and applied arts, health and physical education. Advanced degrees conferred.

Detroit College of Law. Detroit. Founded 1891. Private control. Coeducational, 500 students. Faculty, 22. Semester system. Professional only.

Eastern Michigan University. Ypsilanti. Founded 1849. State control. Coeducational, 6,700 students. Faculty, 331. Semester system. Liberal arts and teacher preparatory. Advanced degrees.

Ferris Institute. Big Rapids. Founded 1884. State control. Coeducational, 2,600 students. Faculty, 150. Quarter system. Professional and terminal-occupational.

Hillsdale College. Hillsdale. Founded 1844. Private control. Coeducational, 700 students. Faculty, 46. Semester system. Liberal arts and teacher preparatory.

Hope College. Holland. Founded 1851. Reformed control. Coeducational, 1,500 students. Faculty, 88. Semester system. Liberal arts and teacher preparatory.

Kalamazoo College. Kalamazoo. Founded 1833. Baptist control. Coeducational, 700 students. Faculty, 52. Semester system. Liberal arts and teacher preparatory. Advanced degrees conferred.

Marygrove College. Detroit. Founded 1910. Roman Catholic control. Women only, 950. Faculty, 97. Semester system. Liberal arts and teacher preparatory.

Mercy College. Detroit. Founded 1941. Roman Catholic control. Women only, 700. Faculty, 62. Semester system. Liberal arts and teacher prep.

Michigan College of Mining and Technology. Houghton. Founded 1885. State control. Coeducational, 3,200 students. Faculty, 193. Quarter system. Technical only. Advanced degrees.

Michigan State University of Agriculture and Applied Science. East Lansing; branch at Oakland. Founded 1855. State control. Coeducational, 24,300 students. Faculty, 1,170. Quarter system. Liberal arts, agriculture, business and public service, communication arts, education, engineering, home economics, international programs, veterinary medicine. Advanced degrees conferred.

Northern Michigan College. Marquette. Founded 1899. State control. Coeducational, 2,600 students. Faculty, 91. Semester system. Liberal arts, terminal-occupational, teacher preparatory.

Olivet College. Olivet. Founded 1844. Private control. Coeducational, 600 students. Faculty, 30. Semester system. Liberal arts and teacher preparatory.

Siena Heights College. Adrian. Founded 1919. Roman Catholic control. Women only, 550. Faculty, 35. Semester system. Liberal arts, terminal-occupational, teacher preparatory. Advanced degrees conferred.

University of Detroit. Detroit. Founded 1877. Roman Catholic control. Coeducational, 10,800 students. Faculty, 528. Semester system. Liberal arts, commerce and finance, dentistry, engineering, law. Advanced degrees conferred.

University of Michigan, The. Ann Arbor; branches at Flint and Dearborn. Founded 1817. State control. Coeducational, 27,600 students. Faculty, 1,666. Semester system. Liberal arts, architecture and design, business administration, dentistry, education, engineering, law, medicine, music,

natural resources, nursing, pharmacy, public health, social work. Advanced degrees conferred.

Wayne State University. Detroit. Founded 1868. State control. Coeducational, 21,400. Faculty, 1,139. Semester system. Liberal arts, business administration, education, engineering, law, medicine, nursing, pharmacy, social work. Advanced degrees conferred.

Western Michigan University. Kalamazoo. Founded 1903. State control. Coeducational, 10,300 students. Faculty, 325. Semester system. Liberal arts, applied arts and sciences, business, education. Advanced degrees conferred.

MINNESOTA

Augsburg College and Theological Seminary. Minneapolis. Founded 1874. Lutheran control. Coeducational, 1,000 students. Faculty, 69. Semester system. Liberal arts, seminary.

Bemidji State College. Bemidji. Founded 1919. State control. Coeducational, 1,700 students. Faculty, 69. Quarter system. Liberal arts and teacher preparatory. Advanced degrees conferred.

Bethel College and Seminary. St. Paul. Founded 1871. Baptist control. Coeducational, 700 students. Faculty, 55. Semester-quarter system. Liberal arts and theology.

Carleton College. Northfield. Founded 1866. Private control. Coeducational, 1,200 students. Faculty, 109. Semester system. Liberal arts and teacher preparatory.

College of St. Catherine. St. Paul. Founded 1905. Roman Catholic control. Women only, 1,400. Faculty, 95. Semester system. Liberal arts and teacher preparatory.

College of St. Teresa. Winona. Founded 1907. Roman Catholic control. Women only, 800. Faculty, 89. Semester system. Liberal arts and teacher preparatory.

College of St. Thomas. St. Paul. Founded 1885. Roman Catholic control. Men only, 1,900. Faculty, 103. Semester system. Liberal arts and teacher preparatory. Advanced degrees conferred.

Concordia College. Moorhead. Founded 1891. Lutheran control. Coeducational, 1,700 students. Faculty, 111. Semester system. Liberal arts and teacher preparatory.

Gustavus Adolphus College. St. Peter. Founded 1862. Lutheran control. Coeducational, 1,100 students. Faculty, 85. Semester system. Liberal arts and teacher preparatory.

Hamline University. St. Paul. Founded 1854. Methodist control. Coeducational, 1,100 students. Faculty, 80. Semester system. Liberal arts and nursing. Advanced degrees conferred.

Luther Theological Seminary. St. Paul. Founded 1876. American Lutheran control. Men only, 550. Faculty, 30. Semester system. Professional only. Advanced degrees conferred.

Macalester College. St. Paul. Founded 1885. Private control. Coeducational, 2,100 students. Faculty, 130. Semester system. Liberal arts and teacher preparatory. Advanced degrees conferred.

Mankato State College. Mankato. Founded 1867. State control. Coeducational, 5,600 students. Faculty, 234. Quarter system. Liberal arts, terminal-occupational, teacher preparatory. Advanced degrees conferred.

Minneapolis School of Art. Minneapolis. Founded 1886. Private control. Coeducational, 550 students. Faculty, 35. Semester system. Professional only.

Moorhead State College. Moorhead. Founded 1887. State control. Coeducational, 1,700 students. Faculty, 80. Quarter system. Liberal arts and education. Advanced degrees conferred.

St. Cloud State College. St. Cloud. Founded 1869. State control. Coeducational, 3,100 students. Faculty, 150. Quarter system. Liberal arts and teacher preparatory. Advanced degrees conferred.

St. John's University. Collegeville. Founded 1857. Roman Catholic control. Men only, 1,200. Faculty, 106. Semester system. Liberal arts and theology.

St. Mary's College. Winona. Founded 1913. Roman Catholic control. Men only, 900. Faculty, 52. Semester system. Liberal arts and teacher preparatory. Advanced degrees conferred.

St. Olaf College. Northfield. Founded 1874. Lutheran control. Coeducational, 1,800 students. Faculty, 137. Semester system. Liberal arts and teacher preparatory.

University of Minnesota. Minneapolis; branches at St. Paul and Duluth. Founded 1851. State control. Coeducational, 37,900 students. Faculty, 2,081. Quarter system. Liberal arts, architecture, business administration, dentistry, education, journalism, law, medicine, nursing, pharmacy, public health, social work; at St. Paul, agriculture, forestry and home economics, veterinary medicine. Advanced degrees conferred.

Winona State College. Winona. Founded 1860. State control. Coeducational, 1,600 students. Faculty, 70. Quarter system. Liberal arts and teacher preparatory. Advanced degrees conferred.

MISSISSIPPI

Alcorn Agricultural and Mechanical College. Lorman. Founded 1871. State control. Coeducational, 850 students. Faculty, 57. Semester system. Liberal arts, terminal-occupational, teacher preparatory.

Delta State College. Cleveland. Founded 1925. State control. Coeducational, 1,000 students. Faculty, 50. Semester system. Liberal arts and teacher preparatory.

Jackson State College. Jackson. Founded 1877. State control. Coeducational, 1,500 students. Faculty, 72. Quarter system. Liberal arts and teacher preparatory.

Millsaps College. Jackson. Founded 1892. Methodist control. Coeducational, 900 students. Faculty, 57. Semester system.

Mississippi College. Clinton. Founded 1826. Southern Baptist control. Coeducational, 1,800 students. Faculty, 75. Semester system. Liberal arts and teacher preparatory. Advanced degrees conferred.

Mississippi Southern College. Hattiesburg. Founded 1912. State control. Coeducational, 4,800 students. Faculty, 203. Quarter system. Liberal arts and teacher preparatory. Advanced degrees conferred.

Mississippi State College for Women. Columbus. Founded 1884. State control. Women only, 1,500. Faculty, 84. Semester system. Liberal arts and teacher preparatory.

Mississippi State University. State College. Founded 1878. State control. Coeducational, 5,200 students. Faculty, 267. Semester system. Agriculture, business and industry, education, engineering, science. Advanced degrees conferred.

Tougaloo Southern Christian College. Tougaloo. Founded 1869. American Missionary Association and United Christian Mission Society control. Coeducational, 500 students. Faculty, 32. Semester system. Liberal arts and teacher prep.

University of Mississippi. University; campus also at Jackson. Founded 1848. State control. Coeducational, 5,100 students. Faculty, 306. Semester system. Liberal arts, commerce, education, engineering, law, pharmacy; at Jackson, medicine, nursing. Advanced degrees conferred.

William Carey College. Hattiesburg. Founded 1906. Southern Baptist control. Coeducational, 550 students. Faculty, 31. Semester system. Liberal arts and teacher preparatory.

MISSOURI

Central Methodist College. Fayette. Founded 1854. Methodist control. Coeducational, 750 students. Faculty, 52. Semester system. Liberal arts, teacher preparatory.

Central Missouri State College. Warrensburg. Founded 1871. State control. Coeducational, 4,000 students. Faculty, 119. Quarter system. Liberal arts, terminal-occupational, teacher preparatory. Advanced degrees conferred.

College of St. Teresa. Kansas City. Founded 1917. Roman Catholic control. Women only, 500. Faculty, 54. Semester system. Liberal arts and teacher preparatory.

Culver-Stockton College. Canton. Founded 1853. Disciples of Christ control. Coeducational, 550 students. Faculty, 39. Semester system. Liberal arts and teacher preparatory.

Drury College. Springfield. Founded 1873. Affiliated with Congregational Christian Churches. Coeducational, 1,000 students. Faculty, 104. Semester system. Liberal arts and teacher preparatory. Advanced degrees conferred.

Fontbonne College. St. Louis. Founded 1923. Roman Catholic control. Women only, 650. Faculty, 62. Semester system. Liberal arts and teacher preparatory.

Harris Teachers College. St. Louis. Founded 1857. Municipal control. Coeducational, 1,400 students. Faculty, 74. Semester system. Liberal arts, terminal-occupational, teacher preparatory.

Lincoln University. Jefferson City. Founded 1866. State control. Coeducational, 1,500 students. Faculty, 97. Semester system. Liberal arts and teacher preparatory. Advanced degrees conferred.

Lindenwood College for Women. St. Charles. Founded 1827. Presbyterian control. Women only, 550. Faculty, 55. Semester system. Liberal arts and teacher preparatory.

Missouri Valley College. Marshall. Founded 1888. Presbyterian control. Coeducational, 500 students. Faculty, 34. Semester system. Liberal arts and teacher preparatory.

Northeast Missouri State Teachers College. Kirksville. Founded 1867. State control. Coeducational, 2,800 students. Faculty, 107. Quarter system. Liberal arts, terminal-occupational, teacher preparatory. Advanced degrees conferred.

Northwest Missouri State College. Maryville. Founded 1905. State control. Coeducational, 2,100 students. Faculty, 98. Semester system. Liberal arts and teacher preparatory. Advanced degrees conferred.

Park College. Parkville. Founded 1875. Presbyterian control. Coeducational, 550 students. Faculty, 42. Semester system.

Rockhurst College. Kansas City. Founded 1910. Roman Catholic control. Men only, 1,900. Faculty, 99. Semester system.

St. Louis College of Pharmacy and Allied Sciences. St. Louis. Founded 1865. Private control. Coeducational, 500 students. Faculty, 28. Semester system. Professional only. Advanced degrees conferred.

St. Louis University. St. Louis; branches at East St. Louis, Illinois, and St. Mary's, Kansas. Founded 1818. Roman Catholic control. Coeducational, 8,400 students. Faculty, 1,219. Semester system. Liberal arts, commerce, dentistry, law, medicine, nursing and health service, philosophy and letters, social order, social service, technology; at St. Mary's College, Kansas, theology; at Parks College, East St. Louis, aeronautical technology. Advanced degrees conferred.

Southeast Missouri State College. Cape Girardeau. Founded 1873. State control. Coeducational, 2,700 students. Faculty, 113. Quarter system. Liberal arts and teacher preparatory.

Southwest Missouri State College. Springfield. Founded 1906. State control. Coeducational, 3,200 students. Faculty, 117. Semester system. Liberal arts and teacher preparatory.

University of Kansas City. Kansas City. Founded 1933. Private control. Coeducational, 3,500 students. Faculty, 391. Semester system. Liberal arts, business administration, dentistry, education, law, music, pharmacy. Advanced degrees conferred.

University of Missouri. Columbia; campus also at Rolla. Founded 1839. State control. Coeducational, 15,300 students. Faculty, 1,083. Semester system. Liberal arts, agriculture, business and public administration, education, engineering, journalism, law, medicine, veterinary medicine. Advanced degrees conferred.

Washington University. St. Louis. Founded 1853. Private control. Coeducational, 13,300 students. Faculty, 1,660. Semester system. Liberal arts, architecture, business and public administration, dentistry, engineering, fine arts, law, medicine, nursing, social work. Advanced degrees conferred.

Webster College. Webster Groves. Founded 1915. Roman Catholic control. Women only, 700. Faculty, 47. Semester system. Liberal arts and teacher preparatory.

Westminster College. Fulton. Founded 1851. Presbyterian control. Men only, 600. Faculty, 44. Semester system.

William Jewell College. Liberty. Founded 1850. Baptist control. Coeducational, 950 students. Faculty, 52. Semester system. Liberal arts and teacher preparatory.

MONTANA

Carroll College. Helena. Founded 1909. Roman Catholic control. Coeducational, 700 students. Faculty, 65. Semester system. Liberal arts, terminal-occupational, teacher preparatory.

College of Great Falls. Great Falls. Founded 1932. Roman Catholic control. Coeducational, 950

students. Faculty, 31. Semester system. Liberal arts and teacher preparatory.

Eastern Montana College of Education. Billings. Founded 1927. State control. Coeducational, 1,400 students. Faculty, 66. Quarter system. Liberal arts, terminal-occupational, teacher preparatory. Advanced degrees conferred.

Montana State College. Bozeman. Founded 1893. State control. Coeducational, 3,900 students. Faculty, 300. Quarter system. Agriculture, architecture, art, education, engineering, home economics, letters and science, nursing. Advanced degrees conferred.

Montana State University. Missoula. Founded 1893. State control. Coeducational, 3,900 students. Faculty, 263. Quarter system. Liberal arts, commerce, education, forestry, journalism, law, music, pharmacy. Advanced degrees conferred.

Northern Montana College. Havre. Founded 1929. State control. Coeducational, 650 students. Faculty, 50. Quarter system. Liberal arts, terminal-occupational, teacher preparatory.

Western Montana College of Education. Dillon. Founded 1897. State control. Coeducational, 600 students. Faculty, 29. Quarter system. Primarily teacher preparatory. Advanced degrees.

NEBRASKA

College of Saint Mary. Omaha. Founded 1923. Roman Catholic control. Women only, 550. Faculty, 34. Semester system. Liberal arts and teacher preparatory.

Concordia Teachers College. Seward. Founded 1894. Lutheran control. Coeducational, 700 students. Faculty, 37. Semester system. Primarily teacher preparatory.

Creighton University. Omaha. Founded 1878. Roman Catholic control. Coeducational, 3,000 students. Faculty, 450. Semester system. Liberal arts, business administration, dentistry, law, medicine, pharmacy. Advanced degrees conferred.

Dana College. Blair. Founded 1884. United Evangelical Lutheran control. Coeducational, 550 students. Faculty, 29. Semester system. Liberal arts, terminal-occupational, teacher preparatory.

Hastings College. Hastings. Founded 1882. Presbyterian control. Coeducational, 800 students. Faculty, 52. Semester system.

Midland College. Fremont. Founded 1887. United Lutheran control. Coeducational, 850 students. Faculty, 36. Semester system. Liberal arts and teacher preparatory.

Municipal University of Omaha. Omaha. Founded 1908. Municipal control. Coeducational, 6,800 students. Faculty, 127. Semester system. Liberal arts, applied arts and sciences, business administration, education. Advanced degrees conferred.

Nebraska State Teachers College. Chadron. Founded 1911. State control. Coeducational, 1,100 students. Faculty, 60. Semester system. Liberal arts, terminal-occupational, teacher preparatory. Advanced degrees conferred.

Nebraska State Teachers College. Kearney. Founded 1905. State control. Coeducational, 2,200 students. Faculty, 83. Semester system. Liberal arts and teacher preparatory. Advanced degrees conferred.

Nebraska State Teachers College. Peru. Founded 1867. State control. Coeducational, 700 students. Faculty, 53. Semester system. Liberal arts, terminal-occupational, teacher preparatory. Advanced degrees conferred.

Nebraska State Teachers College. Wayne. Founded 1910. State control. Coeducational, 1,400 students. Faculty, 81. Semester system. Liberal arts and teacher preparatory. Advanced degrees conferred.

Nebraska Wesleyan University. Lincoln. Founded 1887. Methodist control. Coeducational, 1,100 students. Faculty, 62. Semester system. Liberal arts and teacher preparatory.

Union College. Lincoln; branch at Denver, Colorado. Founded 1891. Seventh Day Adventist control. Coeducational, 800 students. Faculty, 72. Semester system. Liberal arts and teacher preparatory.

University of Nebraska. Lincoln. Founded 1869. State control. Coeducational, 10,600 students. Faculty, 1,021. Semester system. Liberal arts, agriculture, business administration, dentistry, education, engineering, fine arts, journalism, law, medicine, nursing, pharmacy, social work. Advanced degrees conferred.

NEVADA

University of Nevada. Reno. Founded 1887. State control. Coeducational, 4,100 students. Faculty, 196. Semester system. Liberal arts, agriculture, business administration, education, engineering, mining engineering, nursing. Advanced degrees conferred.

NEW HAMPSHIRE

Dartmouth College. Hanover. Founded 1769. Private control. Men only, 3,200. Faculty, 402. Semester system. Liberal arts, business administration, engineering, medical science. Advanced degrees conferred.

Keene Teachers College. Keene. Founded 1909. State control. Coeducational, 1,000 students. Faculty, 54. Semester system. Primarily teacher preparatory. Advanced degrees conferred.

Plymouth Teachers College. Plymouth. Founded 1871. State control. Coeducational, 800 students. Faculty, 45. Semester system. Primarily teacher preparatory. Advanced degrees conferred.

St. Anselm's College. Manchester. Founded 1889. Roman Catholic control. Men only, 1,200. Faculty, 79. Semester system. Liberal arts and teacher preparatory.

University of New Hampshire. Durham. Founded 1866. State control. Coeducational, 5,000 students. Faculty, 283. Semester system. Liberal arts, agriculture, business and economics, technology. Advanced degrees conferred.

NEW JERSEY

Bloomfield College. Bloomfield. Founded 1926. Presbyterian control. Coeducational, 750 students. Faculty, 32. Semester system. Liberal arts and teacher preparatory.

Caldwell College for Women. Caldwell. Founded 1939. Roman Catholic control. Women only, 650. Faculty, 36. Semester system. Liberal arts, terminal-occupational, teacher preparatory.

College of St. Elizabeth. Convent Station. Founded 1899. Roman Catholic control. Women only, 950. Faculty, 85. Semester system. Liberal arts and teacher preparatory.

College of South Jersey. (*See* Rutgers.)

Douglass College. (*See* Rutgers.)

Drew University. Madison. Founded 1867. Methodist control. Coeducational, 950 students. Faculty, 76. Semester system. Liberal arts and theology. Advanced degrees conferred.

Fairleigh Dickinson University. Rutherford; campuses also at Teaneck and Florham-Madison. Founded 1941. Private control. Coeducational, 14,300 students. Faculty, 330. Semester system. Liberal arts, business administration, dentistry, education, engineering and science. Advanced degrees conferred.

Glassboro State College. Glassboro. Founded 1923. State control. Coeducational, 2,800 students. Faculty, 74. Semester system. Primarily teacher preparatory. Advanced degrees conferred.

Jersey City State College. Jersey City. Founded 1929. State control. Coeducational, 2,000 students. Faculty, 78. Semester system. Primarily teacher preparatory.

Monmouth College. West Long Branch. Founded 1933. Private control. Coeducational, 2,500 students. Faculty, 94. Semester system. Liberal arts and terminal-occupational.

Montclair State College. Upper Montclair. Founded 1908. State control. Coeducational, 3,400 students. Faculty, 139. Semester system. Primarily teacher preparatory. Advanced degrees conferred.

Newark College of Engineering. Newark. Founded 1881. State and municipal control. Coeducational, 3,600 students. Faculty, 163. Semester system. Technical only. Advanced degrees.

Newark State College. Union. Founded 1855. State control. Coeducational, 3,900 students. Faculty, 267. Semester system. Primarily teacher preparatory. Advanced degrees conferred.

Paterson State College. Wayne. Founded 1855. State control. Coeducational, 3,000 students. Faculty, 66. Semester system. Primarily teacher preparatory. Advanced degrees conferred.

Princeton Theological Seminary. Princeton. Founded 1812. Presbyterian control. Coeducational, 400 students. Faculty, 43. Quarter system. Professional and teacher preparatory. Advanced degrees conferred.

Princeton University. Princeton. Founded 1746. Private control. Men only, 3,900. Faculty, 563. Semester system. Liberal arts, architecture, engineering, public and international affairs. Advanced degrees conferred.

Rider College. Trenton. Founded 1865. Private control. Coeducational, 3,400 students. Faculty, 131. Semester system. Liberal arts and teacher preparatory. Advanced degrees conferred.

Rutgers, The State University. New Brunswick; campuses also at Camden, Newark. Founded 1766. State control. Coordinated, 17,867 students. Faculty, 1,714. Semester system. At New Brunswick (men), liberal arts, agriculture; at Douglass College, New Brunswick (women), education, engineering, journalism, library service, microbiology, social work; at Camden, College of South Jersey, law; at Newark, liberal arts, business, law, nursing, pharmacy. Advanced degrees conferred.

St. Peter's College. Jersey City. Founded 1872. Roman Catholic control. Men only, 2,000. Faculty, 131. Semester system. Liberal arts and business administration.

Seton Hall University. South Orange; branches at Newark, Jersey City, Paterson. Founded 1856. Roman Catholic control. Men, women, 10,300. Faculty, 442. Semester system. At South Orange (men), liberal arts, business administration, divinity, education; *in addition at Newark*, law, nursing. At Jersey City, liberal arts, business administration, dentistry, education, medicine. At Paterson, liberal arts, business administration, education. Advanced degrees conferred.

Stevens Institute of Technology. Hoboken. Founded 1870. Private control. Men only, 2,100. Faculty, 168. Semester system. Technical only. Advanced degrees conferred.

Trenton State College. Trenton. Founded 1855. State control. Coeducational, 3,600 students. Faculty, 102. Semester system. Primarily teacher preparatory. Advanced degrees conferred.

Upsala College. East Orange. Founded 1893. Lutheran control. Coeducational, 1,900 students. Faculty, 108. Semester system. Liberal arts and teacher preparatory.

NEW MEXICO

Eastern New Mexico University. Portales. Founded 1934. State control. Coeducational, 2,400 students. Faculty, 90. Semester system. Liberal arts, terminal-occupational, teacher preparatory. Advanced degrees conferred.

New Mexico Highlands University. Las Vegas. Founded 1893. State control. Coeducational, 1,000 students. Faculty, 58. Quarter system. Liberal arts, terminal-occupational, teacher preparatory. Advanced degrees conferred.

New Mexico State University of Agriculture, Engineering and Science. University Park. Founded 1889. State control. Coeducational, 4,100 students. Faculty, 167. Semester system. Liberal arts, agriculture and home economics, engineering, teacher education. Advanced degrees conferred.

New Mexico Western College. Silver City. Founded 1893. State control. Coeducational, 750 students. Faculty, 45. Semester system. Liberal arts, terminal-occupational, teacher preparatory. Advanced degrees conferred.

University of New Mexico. Albuquerque. Founded 1889. State control. Coeducational, 7,700 students. Faculty, 260. Semester system. Liberal arts, business administration, education, engineering, fine arts, law, nursing, pharmacy. Advanced degrees conferred.

NEW YORK

Adelphi College. Garden City. Founded 1896. Private control. Coeducational, 6,100 students. Faculty, 274. Semester system. Liberal arts; institute of health, education and welfare; nursing; social work. Advanced degrees conferred.

Alfred University. Alfred. Founded 1836. State and private control. Coeducational, 1,500 students. Faculty, 115. Semester system. Liberal arts, engineering, nursing, theology. Advanced degrees conferred.

Auburn Theological Seminary. (*See* Union Theological Seminary.)

Barnard College. New York. (*See* Columbia University.)

Brooklyn College. New York. (*See* City University of New York, The.)

224

Brooklyn College of Pharmacy. New York (*See* Long Island University.)

Brooklyn Law School. Brooklyn. Founded 1901. Private control. Coeducational, 1,100 students. Faculty, 36. Semester system. Professional only. Advanced degrees conferred.

C. W. Post College. Brookville. (*See* Long Island University.)

Canisius College. Buffalo. Founded 1870. Roman Catholic control. Coeducational, 2,300 students. Faculty, 118. Semester system. Liberal arts and teacher preparatory. Advanced degrees conferred.

City University of New York, The. New York. Municipal control. *Brooklyn College.* Brooklyn. Founded 1930. Coeducational, 18,100 students. Faculty, 773. Semester system. Liberal arts, terminal-occupational, teacher preparatory. Advanced degrees conferred. *City College.* New York. Founded 1847. Men, women, 29,800. Faculty, 720. Semester system. Liberal arts, business and public administration, education, engineering. Advanced degrees conferred. *Hunter College.* New York. Founded 1870. Coeducational, 16,900 students. Faculty, 545. Semester system. Liberal arts and teacher preparatory. Advanced degrees conferred. *Queens College.* Flushing. Founded 1937. Coeducational, 10,600 students. Faculty, 305. Semester system. Liberal arts and teacher preparatory. Advanced degrees conferred.

Clarkson College of Technology. Potsdam. Founded 1896. Private control. Men only, 1,600. Faculty, 97. Semester system. Technical only. Advanced degrees conferred.

Colgate University. Hamilton. Founded 1819. Private control. Men only, 1,400. Faculty, 119. Semester system. Liberal arts and teacher preparatory. Advanced degrees conferred.

College of Mount St. Vincent. New York. Founded 1847. Roman Catholic control. Women only, 900. Faculty, 62. Semester system.

College of New Rochelle. New Rochelle. Founded 1904. Roman Catholic control. Women only, 950. Faculty, 70. Semester system. Liberal arts and teacher preparatory.

College of Saint Rose, The. Albany. Founded 1920. Roman Catholic control. Women only, 1,300. Faculty, 84. Semester system. Liberal arts and teacher preparatory. Advanced degrees conferred.

Columbia University. New York. Founded 1754. Private control. Coeducational, 23,600 students. Faculty, 2,473. Semester system. *Columbia College* (men), liberal arts, architecture, business, dentistry, engineering, international affairs, journalism, law, library service, medicine, nursing, pharmacy, public health. Advanced degrees conferred. *Barnard College* (women). Founded 1889. Faculty, 158. *New York School of Social Work.* Founded 1898. *Teachers College.* Founded 1889. Coeducational. Professional only. Advanced degrees conferred.

Cooper Union. New York. Founded 1859. Private control. Coeducational, 1,200 students. Faculty, 161. Semester system. Professional only.

Cornell University. Ithaca; medical courses at New York. Founded 1865. Private control and state control for courses in State University division. Coeducational, 11,700 students. Faculty, 1,595. Semester system. Liberal arts, architecture, business and public administration, education, engineering, hotel administration, law, medicine, nursing, nutrition. (*See* State University of New York for agriculture, home economics, industrial and labor relations, veterinary.) Advanced degrees conferred.

D'Youville College. Buffalo. Founded 1908. Roman Catholic control. Women only, 700 students. Faculty, 63. Semester system. Liberal arts and nursing.

Elmira College. Elmira. Founded 1855. Private control. Women only, 1,400. Faculty, 58. Semester system. Liberal arts and teacher preparatory.

Fordham University. New York; branches, Shrub Oak; Novitiate of St. Andrew-on-Hudson, Poughkeepsie. Founded 1841. Roman Catholic control. Coeducational, 9,100 students. Faculty, 524. Semester system. Liberal arts (men), business, education, law, pharmacy, philosophy, social service. Advanced degrees conferred.

Good Counsel College. White Plains. Founded 1923. Roman Catholic control. Women only, 550. Faculty, 36. Semester system. Liberal arts and teacher preparatory.

Hamilton College. Clinton. Founded 1812. Private control. Men only, 750. Faculty, 68. Semester system.

Harpur College. Binghamton. (*See* State University of New York.)

Hartwick College. Oneonta. Founded 1928. Lutheran control. Coeducational, 700 students. Faculty, 42. Semester system. Liberal arts and teacher preparatory.

Hebrew Union College—Jewish Institute of Religion, New York School. New York. (*See* Ohio.)

Hobart and William Smith Colleges. Geneva. Founded 1822. Private control. Hobart College (men), William Smith College (women); 1,200 students. Faculty, 90. Semester system. Liberal arts and teacher preparatory. Advanced degrees conferred.

Hofstra College. Hempstead. Founded 1935. Private control. Coeducational, 8,600 students. Faculty, 410. Semester system. Liberal arts, terminal-occupational, teacher preparatory. Advanced degrees conferred.

Houghton College. Houghton. Founded 1883. Wesleyan Methodist control. Coeducational, 800 students. Faculty, 44. Semester system. Liberal arts and teacher preparatory.

Hunter College. New York. (*See* City University of New York, The.)

Iona College. New Rochelle. Founded 1940. Roman Catholic control. Men only, 2,000. Faculty, 97. Semester system. Liberal arts and teacher preparatory.

Ithaca College. Ithaca. Founded 1892. Private control. Coeducational, 1,800 students. Faculty, 91. Semester system. Liberal arts, health and physical education, music.

Jewish Theological Seminary of America. New York; branch, University of Judaism, at Los Angeles, Calif. Founded 1887. Jewish control. Coeducational, 450 students. Faculty, 89. Semester system. Professional and teacher preparatory. Advanced degrees conferred.

Juilliard School of Music. New York. Founded 1905. Private control. Coeducational, 850 students. Faculty, 132. Semester system. Professional only. Advanced degrees conferred.

Keuka College. Keuka Park. Founded 1890. Baptist control. Women only, 550. Faculty, 42. Quarter system. Liberal arts, terminal-occupational, teacher preparatory.

LeMoyne College. Syracuse. Founded 1946. Roman Catholic control. Coeducational, 1,300 students. Faculty, 69. Semester system. Liberal arts and teacher preparatory.

Long Island University. Brooklyn; branches—Brooklyn College of Pharmacy; C. W. Post College, Brookville; Mitchel College, Mitchel Air Force Base. Founded 1926. Private control. Coeducational, 7,900 students. Faculty, 293. Semester system. At Brooklyn Center, liberal arts, business administration. Advanced degrees conferred.

Manhattan College. New York. Founded 1853. Roman Catholic control. Men only, 3,200. Faculty, 206. Semester system. Liberal arts, business administration, engineering. Advanced degrees conferred.

Manhattan School of Music. New York. Founded 1917. Private control. Coeducational, 600 students. Faculty, 118. Semester system. Professional and teacher preparatory. Advanced degrees conferred.

Manhattanville College of the Sacred Heart. Purchase. Founded 1841. Roman Catholic control. Women only, 800. Faculty, 90. Semester system. Liberal arts, Pius X School of Liturgical Music (coeducational). Advanced degrees conferred.

Marymount College. Tarrytown. Founded 1907. Roman Catholic control. Women only, 1,200. Faculty, 117. Semester system. Liberal arts and teacher preparatory.

Marymount Manhattan College. New York City. Founded 1947. Roman Catholic control. Women only, 550. Faculty, 36. Semester system. Liberal arts and teacher preparatory.

Millard Filmore College. Buffalo. (*See* University of Buffalo.)

Mitchel College. Mitchel Air Force Base. (*See* Long Island University.)

Nazareth College. Rochester. Founded 1924. Roman Catholic control. Women only, 950. Faculty, 57. Semester system. Liberal arts and teacher preparatory. Advanced degrees conferred.

New School for Social Research. New York. Founded 1919. Private control. Coeducational, 900 students. Faculty, 52. Semester system. Liberal arts, philosophy, politics and social studies. Advanced degrees conferred.

New York Law School. New York. Founded 1891. Private control. Coeducational, 800 students. Faculty, 53. Semester system. Professional only. Advanced degrees conferred.

New York Medical College. New York. Founded 1860. Private control. Coeducational, 500 students. Faculty, 1,000. Semester system. Professional only. Advanced degrees conferred.

New York School of Social Work. New York. (*See* Columbia University.)

New York University. New York. Founded 1831. Private control. Coeducational, 32,500 students. Faculty, 4,096. Semester system. Liberal arts, commerce, dentistry, education, engineering, law, medicine, public administration, retailing. Advanced degrees conferred.

Niagara University. Niagara University. Founded 1856. Roman Catholic control. Coeducational, 1,600 students. Faculty, 108. Semester system. Liberal arts, business administration, education, nursing, seminary. Advanced degrees conferred.

Novitiate of St. Andrew-on-Hudson. Poughkeepsie. (*See* Fordham University.)

Nyack Missionary College. Nyack. Founded 1882. Christian and Missionary Alliance control. Coeducational, 550 students. Faculty, 44. Semester system. Professional only.

Pace College. New York. Founded 1906. Private control. Coeducational, 4,500 students. Faculty, 198. Semester system. Advanced degrees conferred.

Polytechnic Institute of Brooklyn. Brooklyn. Founded 1854. Private control. Men only, 5,500. Faculty, 455. Semester system. Technical only. Advanced degrees conferred.

Pratt Institute. Brooklyn. Founded 1887. Private control. Coeducational, 4,400 students. Faculty, 236. Semester system. Technical and teacher preparatory. Advanced degrees conferred.

Queens College. Flushing. (*See* City University of New York, The.)

Rensselaer Polytechnic Institute. Troy; Hartford Graduate Division, East Windsor Hill, Connecticut. Founded 1824. Private control. Men only, 4,600. Faculty, 350. Semester system. Architecture, engineering, humanities and social sciences, science. Advanced degrees conferred.

Rochester Institute of Technology. Rochester. Founded 1829. Private control. Coeducational, 6,300 students. Faculty, 156. Quarter system. Technical only. Advanced degrees conferred.

Rosary Hill College. Buffalo. Founded 1948. Roman Catholic control. Women only, 700. Faculty, 43. Semester system. Liberal arts and teacher preparatory.

Russell Sage College. Troy. Founded 1916. Private control. Women only, 2,400. Faculty, 72. Semester system. Liberal arts, business, home economics, nursing, physical education. Advanced degrees conferred.

St. Bernardine of Siena College. Loudonville. Founded 1937. Roman Catholic control. Men only, day sessions; coeducational, evening sessions; 1,800 students. Faculty, 97. Semester system. Liberal arts and teacher preparatory. Advanced degrees conferred.

St. Bonaventure University. St. Bonaventure. Founded 1859. Roman Catholic control. Coeducational, 2,000 students. Faculty, 147. Semester system. Liberal arts, business administration, education, theology (men only); St. Elizabeth's Teachers College. Advanced degrees conferred.

St. Elizabeth's Teachers College. (*See* St. Bonaventure University.)

St. Francis College. Brooklyn. Founded 1884. Roman Catholic control. Men only, 1,500. Faculty, 58. Semester system.

St. John Fisher College, Inc. Rochester. Founded 1951. Roman Catholic control. Men only, 550. Faculty, 40. Semester system. Liberal arts and teacher preparatory.

St. John's University. Jamaica. Founded 1870. Roman Catholic control. Men, women, 10,000. Faculty, 387. Semester system. Liberal arts, commerce, education, law, nursing, pharmacy. Advanced degrees conferred.

St. Joseph's College for Women. Brooklyn. Founded 1916. Roman Catholic control. Women only, 600. Faculty, 47. Semester system. Liberal arts and teacher preparatory.

St. Lawrence University. Canton. Founded 1856. Private control. Coeducational, 1,700 students. Faculty, 102. Semester system. Liberal arts and theology. Advanced degrees conferred.

Sarah Lawrence College. Bronxville. Founded 1926. Private control. Women only, 450. Faculty,

70. Quarter system. Liberal arts and teacher preparatory. Advanced degrees conferred.

Skidmore College. Saratoga Springs. Founded 1911. Private control. Women only, 1,300. Faculty, 118. Semester system. Liberal arts and teacher preparatory.

State University of New York. Albany. In this state controlled system there are 22 colleges offering a degree after 4 years' work. (Also included in the system are 6 institutions which require less than 4 years' work for a degree and which are not listed here.) *College of Agriculture at Cornell University.* Founded 1904. Coeducational, 1,600. Faculty, 378. Semester system. Technical and teacher preparatory. Advanced degrees conferred. *College of Ceramics at Alfred University.* Founded 1900. Coeducational. Faculty, 34. Semester system. Technical only. Advanced degrees conferred. *College of Education at Albany.* Founded 1844. Coeducational, 3,300 students. Faculty, 177. Semester system. Primarily teacher preparatory. Advanced degrees conferred. *College of Education at Brockport.* Founded 1866. Coeducational, 2,000 students. Faculty, 102. Semester system. Primarily teacher preparatory. Advanced degrees conferred. *College of Education at Buffalo.* Founded 1867. Coeducational, 4,300 students. Faculty, 211. Semester system. Primarily teacher preparatory. Advanced degrees conferred. *College of Education at Cortland.* Founded 1863. Coeducational, 3,200 students. Faculty, 158. Semester system. Primarily teacher preparatory. Advanced degrees conferred. *College of Education at Fredonia.* Founded 1866. Coeducational, 1,500 students. Faculty, 69. Semester system. Primarily teacher preparatory. Advanced degrees conferred. *College of Education at Geneseo.* Founded 1867. Coeducational, 1,800 students. Faculty, 90. Semester system. Primarily teacher preparatory. Advanced degrees conferred. *College of Education at New Paltz.* Founded 1886. Coeducational, 2,800 students. Faculty, 113. Semester system. Primarily teacher preparatory. Advanced degrees conferred. *College of Education at Oneonta.* Founded 1889. Coeducational, 2,400 students. Faculty, 105. Semester system. Primarily teacher preparatory. Advanced degrees conferred. *College of Education at Oswego.* Founded 1861. Coeducational, 2,800 students. Faculty, 139. Semester system. Primarily teacher preparatory. Advanced degrees conferred. *College of Education at Plattsburgh.* Founded 1890. Coeducational, 1,800 students. Faculty, 95. Semester system. Primarily teacher preparatory. Advanced degrees conferred. *College of Education at Potsdam.* Founded 1866. Coeducational, 1,600 students. Faculty, 99. Semester system. Primarily teacher preparatory. Advanced degrees conferred. *College of Forestry at Syracuse University.* Founded 1911. Coeducational, 700 students. Faculty, 95. Semester system. Professional only. Advanced degrees conferred. *College of Home Economics at Cornell University.* Founded 1925. Coeducational. 700 students. Faculty, 104. Semester system. Professional and teacher preparatory. Advanced degrees conferred. *Downstate Medical Center.* Brooklyn. Founded 1860. Coeducational, 600 students. Faculty, 441. Quarter system. Professional only. Advanced degrees conferred. *Harpur College.* Binghamton. Founded 1946. Coeducational, 1,400 students. Faculty, 77. Semester system. *Long Island Center.* Oyster Bay. Founded 1957. Coeducational, 500 students. Faculty, 44. Semester system. Liberal arts and teacher preparatory. Advanced degrees conferred. *Maritime College.* Bronx. Founded 1874. Men only, 600. Faculty, 42. Semester system. Professional only. *School of Industrial and Labor Relations at Cornell University.* Founded 1945. Coeducational, 350 students. Faculty, 60. Semester system. Professional only. Advanced degrees conferred. *Upstate Medical Center.* Syracuse. Founded 1872. Coeducational, 400 students. Faculty, 179. Annual system. Professional only. Advanced degrees conferred. *Veterinary College at Cornell University.* Founded 1894. Coeducational, 220 students. Faculty, 40. Semester system. Professional only.

Syracuse University. Syracuse; branch, Utica College, Utica. Founded 1870. Private control. Coeducational, 17,500 students. Faculty, 1,074. Semester system. Liberal arts, architecture, art, business administration, citizenship, education, engineering, forestry, home economics, journalism, law, library science, music, nursing, social work, speech and dramatic art. Advanced degrees conferred.

Teachers College. New York. (*See* Columbia University.)

Union College and University. Schenectady (men only); campus also at Albany (coeducational). Founded 1795. Private control. 2,700 students. Faculty, 117. Semester system. Liberal arts, law, medicine, pharmacy, Dudley Observatory. Advanced degrees conferred.

Union Theological Seminary. New York; branch, Auburn Theological Seminary, Auburn. Founded 1836. Interdenominational control. Coeducational, 650 students. Faculty, 80. Semester system Professional only. Advanced degrees.

United States Merchant Marine Academy. Kings Point. Founded 1938. Federal control. Men only, 1,000. Faculty, 90. Quarter system. Professional only.

United States Military Academy. West Point. Founded 1802. Federal control. Men only, 2,600. Faculty, 375. Semester system. Professional only.

University of Buffalo. Buffalo; branch, Millard Filmore College. Founded 1846. Private control. Coeducational, 11,600 students. Faculty, 1,344. Semester system. Liberal arts, business administration, dentistry, education, law, medicine, pharmacy; at Millard Filmore College, engineering, nursing, social work. Advanced degrees conferred.

University of Rochester. Rochester. Founded 1850. Private control. Coeducational, 6,800 students. Faculty, 1,107. Semester system. Liberal arts, business administration, education, engineering, medicine, music, nursing. Advanced degrees.

Utica College. Utica. (*See* Syracuse University.)

Vassar College. Poughkeepsie. Founded 1861. Private control. Women only, 1,500. Faculty, 190. Semester system. Advanced degrees conferred.

Wagner College. Staten Island. Founded 1883. Lutheran control. Coeducational, 1,800 students. Faculty, 151. Semester system. Liberal arts and nursing. Advanced degrees conferred.

William Smith College. Geneva. (*See* Hobart and William Smith Colleges.)

Yeshiva University. New York. Founded 1886. Jewish Congregations Control. Coordinated, Yeshiva College (men), Stern College (women); 3,000 students. Faculty, 855. Semester system. Liberal arts, cantorial training, education, higher Jewish studies, mathematical sciences, medical sciences, medicine, social work, theology. Advanced degrees.

NORTH CAROLINA

Agricultural and Technical College of North Carolina. Greensboro. Founded 1891. State control. Coeducational, 2,100 students. Faculty, 175. Quarter system. Liberal arts, agriculture, education, engineering, nursing, technical institute. Advanced degrees conferred.

Appalachian State Teachers College. Boone. Founded 1903. State control. Coeducational, 2,600 students. Faculty, 107. Quarter system. Primarily teacher preparatory. Advanced degrees conferred.

Atlantic Christian College. Wilson. Founded 1902. Disciples of Christ control. Coeducational, 1,100 students. Faculty, 61. Semester system. Liberal arts and teacher preparatory.

Belmont Abbey College. Belmont. Founded 1878. Roman Catholic control. Men only, 550. Faculty, 50. Semester system. Liberal arts and teacher preparatory.

Bennett College. Greensboro. Founded 1873. Methodist control. Women only, 550. Faculty, 39. Semester system. Liberal arts and teacher preparatory.

Catawba College. Salisbury. Founded 1851. Evangelical Reformed control. Coeducational, 800 students. Faculty, 52. Semester system. Liberal arts, terminal-occupational, teacher preparatory.

Davidson College. Davidson. Founded 1836. Presbyterian control. Men only, 950. Faculty, 71. Semester system.

Duke University. Durham. Founded 1838. Private control. Coordinated, Trinity College (men), Woman's College (women); 5,500 students. Faculty, 795. Semester system. Liberal arts, divinity, engineering, forestry, law, medicine, nursing. Advanced degrees conferred.

East Carolina College. Greenville. Founded 1907. State control. Coeducational, 6,500 students. Faculty, 200. Quarter system. Liberal arts, terminal-occupational, teacher preparatory. Advanced degrees conferred.

Elizabeth City State Teachers College. Elizabeth City. Founded 1891. State control. Coeducational, 750 students. Faculty, 31. Quarter system. Liberal arts, terminal-occupational, teacher preparatory.

Elon College. Elon College. Founded 1889. Congregational-Christian control. Coeducational, 1,200 students. Faculty, 63. Semester system. Liberal arts and teacher preparatory.

Fayetteville State Teachers College. Fayetteville. Founded 1877. State control. Coeducational, 750 students. Faculty, 35. Quarter system. Primarily teacher preparatory.

Gaston Technical Institute. Gastonia. (*See* University of North Carolina.)

Greensboro College. Greensboro. Founded 1838. Methodist control. Coeducational, 500 students. Faculty, 48. Semester system. Liberal arts and music.

Guilford College. Guilford College. Founded 1837. Society of Friends control. Coeducational, 1,300 students. Faculty, 46. Semester system. Liberal arts and teacher preparatory. Advanced degrees conferred.

High Point College. High Point. Founded 1924. Methodist control. Coeducational, 1,200 students. Faculty, 60. Semester system. Liberal arts and teacher preparatory.

Johnson C. Smith University. Charlotte. Founded 1867. Presbyterian control. Coeducational, 800 students. Faculty, 48. Semester system. Liberal arts and theology.

Lenoir-Rhyne College. Hickory. Founded 1891. Lutheran control. Coeducational, 950 students. Faculty, 57. Semester system. Liberal arts and teacher preparatory.

Livingstone College. Salisbury. Founded 1879. African Methodist Episcopal Zion control. Coeducational, 550 students. Faculty, 43. Semester system. Liberal arts and theology.

Meredith College. Raleigh. Founded 1899. Southern Baptist control. Women only, 700. Faculty, 47. Semester system. Liberal arts and teacher preparatory.

North Carolina College at Durham. Durham. Founded 1910. State control. Coeducational, 2,100 students. Faculty, 132. Semester system. Liberal arts, law, library sciences. Advanced degrees.

Pembroke State College. Pembroke. Founded 1887. State control. Coeducational, 600 students. Faculty, 29. Semester system. Liberal arts and teacher preparatory.

Pfeiffer College. Misenheimer. Founded 1885. Methodist control. Coeducational, 850 students. Faculty, 54. Semester system. Liberal arts, terminal-occupational, teacher preparatory.

Queens College. Charlotte. Founded 1857. Presbyterian control. Women only, 650. Faculty, 43. Semester system. Liberal arts and teacher preparatory.

St. Andrews Presbyterian College. Laurinburg. Founded 1958. Southern Presbyterian control. Coeducational, 850 students. Faculty, 50. Semester system. Liberal arts, terminal-occupational, teacher preparatory.

St. Augustine's College. Raleigh. Founded 1867. Protestant Episcopal control. Coeducational, 500 students. Faculty, 36. Semester system. Liberal arts and teacher preparatory.

Shaw University. Raleigh. Founded 1865. Baptist control. Coeducational, 600 students. Faculty, 39. Semester system. Liberal arts and teacher preparatory.

Southeastern Baptist Theological Seminary. Wake Forest. Founded 1951. Baptist control. Coeducational, 700 students. Faculty, 32. Semester system. Professional and teacher preparatory. Advanced degrees conferred.

University of North Carolina. Consolidated Office, Chapel Hill. State control. *State College of Agriculture and Engineering.* Raleigh; branch, Gaston Technical Institute, Gastonia. Founded 1889. Coeducational, 7,400 students. Faculty, 562. Semester system. Agriculture, design, education, engineering, forestry, general studies, physical sciences and applied mathematics, textiles. Advanced degrees conferred. *University of North Carolina at Chapel Hill.* Chapel Hill. Founded 1795. Coeducational, 8,600 students. Faculty, 1,176. Semester system. Liberal arts, business administration, dentistry, education, journalism, law, library science, medicine, nursing, pharmacy, public health, social work. Advanced degrees conferred. *Woman's College.* Greensboro. Founded 1892. Women only, 2,900. Faculty, 184. Semester system. Liberal arts, education, home economics, music. Advanced degrees conferred.

Wake Forest College. Winston-Salem. Founded 1834. Baptist control. Men, women, 2,600. Faculty, 311. Semester system. Liberal arts, business administration, law, medicine.

Western Carolina College. Cullowhee. Founded 1889. State control. Coeducational, 1,900 students. Faculty, 72. Quarter system. Primarily teacher preparatory. Advanced degrees conferred.

Winston-Salem Teachers College. Winston-Salem. Founded 1892. State control. Coeducational, 1,000 students. Faculty, 57. Quarter system. Professional and teacher preparatory.

NORTH DAKOTA

Jamestown College. Jamestown. Founded 1884. Presbyterian control. Coeducational, 500 students. Faculty, 44. Semester system. Liberal arts, terminal-occupational, teacher preparatory.

North Dakota State University. Fargo. Founded 1889. State control. Coeducational, 3,400 students. Faculty, 254. Quarter system. Liberal arts, agriculture, applied arts and sciences, chemistry, engineering, home economics, pharmacy. Advanced degrees conferred.

State Teachers College. Dickinson. Founded 1918. State control. Coeducational, 600 students. Faculty, 34. Quarter system. Liberal arts and teacher preparatory.

State Teachers College. Mayville. Founded 1889. State control. Coeducational, 650 students. Faculty, 38. Quarter system. Primarily teacher preparatory.

State Teachers College. Minot. Founded 1913. State control. Coeducational, 1,600 students. Faculty, 80. Quarter system. Liberal arts, terminal-occupational, teacher preparatory.

State Teachers College. Valley City. Founded 1890. State control. Coeducational, 800 students. Faculty, 51. Quarter system. Liberal arts and teacher preparatory.

University of North Dakota. Grand Forks. Founded 1883. State control. Coeducational, 4,400 students. Faculty, 280. Semester system. Liberal arts, business and public administration, education, engineering, law, medical sciences. Advanced degrees conferred.

OHIO

Adelbert College. Cleveland. (*See* Western Reserve University.)

Antioch College. Yellow Springs. Founded 1852. Private control. Coeducational, 1,400 students. Faculty, 91. Quarter system. Liberal arts and teacher preparatory. Advanced degrees conferred.

Ashland College. Ashland. Founded 1878. Church of the Brethren control. Coeducational, 1,100 students. Faculty, 58. Semester system. Liberal arts and theology.

Athenaeum of Ohio, The. Cincinnati. Founded 1829. Roman Catholic control. Men only, 500. Faculty, 43. Semester system. Liberal arts and teacher preparatory. Advanced degrees conferred.

Baldwin-Wallace College. Berea. Founded 1845. Methodist control. Coeducational, 2,400 students. Faculty, 141. Quarter system. Liberal arts and teacher preparatory.

Bowling Green State University. Bowling Green. Founded 1910. State control. Coeducational, 7,100 students. Faculty, 231. Semester system. Liberal arts, business administration, education. Advanced degrees conferred.

Capital University. Columbus. Founded 1850. Lutheran control. Coeducational, 1,500 students. Faculty, 100. Semester system. Liberal arts, music, terminal-occupational, teacher preparatory.

Case Institute of Technology. Cleveland. Founded 1880. Private control. Men only, 2,300. Faculty, 237. Semester system. Technical only. Advanced degrees conferred.

Central State College. Wilberforce. Founded 1887. State control. Coeducational, 1,700 students. Faculty, 76. Semester system. Liberal arts, terminal-occupational, teacher preparatory.

Cleveland College. (*See* Western Reserve University.)

Cleveland-Marshall Law School. Cleveland. Founded 1897. Private control. Coeducational, 550 students. Faculty, 24. Semester system. Professional only. Advanced degrees conferred.

College of Mount St. Joseph-on-the-Ohio. Mount St. Joseph. Founded 1920. Roman Catholic control. Women only, 750. Faculty, 65. Semester system. Liberal arts and teacher preparatory.

College of St. Mary of the Springs. Columbus. Founded 1911. Roman Catholic control. Women

only, 500. Faculty, 42. Semester system. Liberal arts and teacher preparatory.

College of Steubenville, The. Steubenville. Founded 1946. Private control. Coeducational, 750 students. Faculty, 37. Semester system. Liberal arts, terminal-occupational, teacher preparatory.

College of Wooster. Wooster. Founded 1866. Presbyterian control. Coeducational, 1,200 students. Faculty, 97. Semester system. Liberal arts and teacher preparatory.

College-Conservatory of Music of Cincinnati. Cincinnati. Founded 1867. Private control. Coeducational, 500 students. Faculty, 71. Semester system. Professional only. Advanced degrees conferred.

Defiance College, The. Defiance. Founded 1886. Congregational-Christian control. Coeducational, 750 students. Faculty, 32. Semester system. Liberal arts, terminal-occupational, teacher preparatory.

Denison University. Granville. Founded 1831. Baptist control. Coeducational, 1,500 students. Faculty, 125. Semester system. Liberal arts and music.

Fenn College. Cleveland. Founded 1923. Private control. Coeducational, 5,800 students. Faculty, 231. Quarter system. Liberal arts, business administration, engineering.

Flora Stone Mather College. Cleveland. (*See* Western Reserve University.)

Franklin University. Columbus. Founded 1902. Young Men's Christian Association control. Coeducational, 550 students. Faculty, 51. Semester system. Professional and terminal-occupational.

Hebrew Union College—Jewish Institute of Religion. Cincinnati; branches also at Los Angeles and New York. Founded 1875. Jewish Congregations control. Men only, 800 (Cincinnati, 180, New York, 500, Los Angeles, 120). Faculty, 46 (Cincinnati, 23, New York, 13, Los Angeles 10). Semester system. Professional only. Advanced degrees conferred.

Heidelberg College. Tiffin. Founded 1850. Evangelical-Reformed control. Coeducational, 950 students. Faculty, 67. Semester system. Liberal arts and music.

Hiram College. Hiram. Founded 1850. Private control. Coeducational, 650 students. Faculty, 50. 11-week terms. Liberal arts and teacher preparatory.

John Carroll University. Cleveland. Founded 1886. Roman Catholic control. Men only, 4,000. Faculty, 220. Semester system. Liberal arts and business. Advanced degrees conferred.

Kent State University. Kent; centers at Cleveland and Ashtabula. Founded 1910. State control. Coeducational, 11,600 students. Faculty, 410. Quarter system. Liberal arts, business administration, education, fine and professional arts. Advanced degrees conferred.

Kenyon College. Gambier. Founded 1824. Private control. Men only, 600. Faculty, 63. Semester system. Liberal arts and divinity.

Lake Erie College. Painesville. Founded 1856. Private control. Women only, 750. Faculty, 48. 11-week terms. Liberal arts and teacher preparatory.

Marietta College. Marietta. Founded 1835. Private control. Coeducational, 1,500 students. Faculty, 64. Semester system. Liberal arts and teacher preparatory.

Mary Manse College. Toledo. Founded 1922. Roman Catholic control. Women only, 1,000. Faculty, 37. Semester system. Liberal arts and teacher preparatory.

Miami University. Oxford. Founded 1809. State control. Coeducational, 10,400 students. Faculty, 402. Semester system. Liberal arts, applied sciences, business administration, education, fine arts. Advanced degrees conferred.

Mount Union College. Alliance. Founded 1846. Methodist control. Coeducational, 950 students. Faculty, 63. Semester system. Liberal arts and teacher preparatory.

Muskingum College. New Concord. Founded 1837. United Presbyterian control. Coeducational, 1,400 students. Faculty, 76. Semester system. Liberal arts and teacher preparatory.

Oberlin College. Oberlin. Founded 1833. Private control. Coeducational, 2,400 students. Faculty, 215. Semester system. Liberal arts, music, theology. Advanced degrees conferred.

Ohio Northern University. Ada. Founded 1871. Methodist control. Coeducational, 1,700 students. Faculty, 94. Quarter system. Liberal arts, engineering, law, pharmacy.

Ohio State University, The. Columbus; branches at Lima, Mansfield, Marion, Newark. Founded 1870. State control. Coeducational, 25,200 students. Faculty, 1,819. Quarter system. Liberal arts, agriculture and home economics, commerce and administration, dentistry, education, engineering, law, medicine, pharmacy, veterinary medicine. Advanced degrees conferred.

Ohio University. Athens. Founded 1804. State control. Coeducational, 11,100 students. Faculty, 409. Semester system. Liberal arts, applied science, commerce and journalism, education, fine arts and architecture. Advanced degrees conferred.

Ohio Wesleyan University. Delaware. Founded 1841. Methodist control. Coeducational, 2,100 students. Faculty, 170. Semester system. Liberal arts and teacher preparatory. Advanced degrees conferred.

Otterbein College. Westerville. Founded 1847. Evangelical-United Brethren control. Coeducational, 1,000 students. Faculty, 70. Semester system. Liberal arts, terminal-occupational, teacher preparatory.

Our Lady of Carey Seminary. Carey. (*See* University of Dayton.)

Our Lady of Cincinnati College. Cincinnati. Founded 1935. Roman Catholic control. Women only, 900. Faculty, 53. Semester system. Liberal arts and teacher preparatory.

St. Charles Seminary. Carthagena. (*See* University of Dayton.)

St. John College of Cleveland. Cleveland. Founded 1928. Roman Catholic control. Women only, 1,200. Faculty, 69. Semester system. Professional and teacher preparatory. Advanced degrees conferred.

Salmon P. Chase College. Cincinnati. Founded 1885. Young Men's Christian Association control. Coeducational, 550 students. Faculty, 20. Semester system. Commerce and law.

University of Akron, The. Akron. Founded 1870. Municipal control. Coeducational, 5,800 students. Faculty, 189. Semester system. Liberal arts, business administration, education, engineering, law. Advanced degrees conferred.

University of Cincinnati. Cincinnati. Founded 1819. Municipal control. Coeducational, 17,500 students. Faculty, 1,270. Semester system. Liberal arts, applied arts, business administration, education and home economics, engineering, law, medicine, nursing, pharmacy. Advanced degrees.

University of Dayton. Dayton; branches, St. Charles Seminary, Carthagena, Our Lady of Carey Seminary, Carey. Founded 1850. Roman Catholic control. Coeducational, 6,000 students. Faculty, 401. Semester system. Liberal arts, business administration, education, engineering, technical institute. Advanced degrees conferred.

University of Toledo. Toledo. Founded 1872. Municipal control. Coeducational, 5,800 students. Faculty, 316. Semester system. Liberal arts, business, education, engineering, law, pharmacy. Advanced degrees conferred.

Western Reserve University. Cleveland. Founded 1826. Private control. Adelbert College (men), Flora Stone Mather College (women), Cleveland College (coeducational); 7,700 students. Faculty, 1,045. Semester system. Liberal arts, applied social sciences, business, dentistry, law, library science, medicine, nursing. Advanced degrees conferred.

Wilmington College. Wilmington. Founded 1870. Society of Friends control. Coeducational, 700 students. Faculty, 55. Semester system.

Wittenberg University. Springfield. Founded 1845. United Lutheran control. Coeducational, 2,700 students. Faculty, 119. Semester system. Liberal arts, community education, music, theology. Advanced degrees conferred.

Xavier University. Cincinnati. Founded 1831. Roman Catholic control. Men only, day sessions; coeducational other sessions; 4,000 students. Faculty, 214. Semester system. Liberal arts and teacher preparatory. Advanced degrees conferred.

Youngstown University, The. Youngstown. Founded 1908. Private control. Coeducational, 6,700 students. Faculty, 246. Semester system. Liberal arts, business, educ., engin., music.

OKLAHOMA

Bethany-Nazarene College. Bethany. Founded 1899. Nazarene control. Coeducational, 1,000 students. Faculty, 48. Semester system. Liberal arts and teacher preparatory.

Central State College. Edmond. Founded 1890. State control. Coeducational, 4,000 students. Faculty, 114. Semester system. Liberal arts and teacher preparatory. Advanced degrees conferred.

East Central State College. Ada. Founded 1909. State control. Coeducational, 1,800 students. Faculty, 73. Semester system. Liberal arts and teacher preparatory. Advanced degrees conferred.

Langston University. Langston. Founded 1897. State control. Coeducational, 650 students. Faculty, 50. Semester system. Liberal arts, terminal-occupational, teacher preparatory.

Northeastern State College. Tahlequah. Founded 1909. State control. Coeducational, 2,800 students. Faculty, 70. Semester system. Liberal arts and teacher preparatory. Advanced degrees conferred.

Northwestern State College. Alva. Founded 1897. State control. Coeducational, 1,100 students. Faculty, 46. Semester system. Liberal arts and teacher preparatory. Advanced degrees conferred.

Oklahoma Baptist University. Shawnee. Founded 1910. Southern Baptist control. Coeducational, 1,500 students. Faculty, 93. Semester system. Liberal arts and fine arts.

Oklahoma City University. Oklahoma City. Founded 1904. Methodist control. Coeducational, 2,800 students. Faculty, 152. Semester system. Liberal arts, business, fine arts, industrial arts, law.

Oklahoma College for Women. Chickasha. Founded 1908. State control. Women only, 600. Faculty, 45. Semester system. Liberal arts and teacher preparatory.

Oklahoma State University of Agriculture and Applied Science. Stillwater. Founded 1891. State control. Coeducational, 11,800 students. Faculty, 559. Semester system. Liberal arts, agriculture, business, education, engineering, home economics, veterinary medicine. Advanced degrees conferred.

Panhandle Agricultural and Mechanical College. Goodwell. Founded 1909. State control. Coeducational, 950 students. Faculty, 50. Semester system.

Phillips University. Enid. Founded 1907. Disciples of Christ control. Coeducational, 1,100 students. Faculty, 72. Semester system. Liberal arts, bible, fine arts. Advanced degrees conferred.

Southeastern State College. Durant. Founded 1909. State control. Coeducational, 1,600 students. Faculty, 64. Semester system. Liberal arts and teacher preparatory. Advanced degrees conferred.

Southwestern State College. Weatherford. Founded 1901. State control. Coeducational, 1,800 students. Faculty, 72. Semester system. Liberal arts and pharmacy. Advanced degrees conferred.

University of Oklahoma. Norman; medical branch at Oklahoma City. Founded 1890. State control. Coeducational, 12,400 students. Faculty, 716. Semester system. Liberal arts, business administration, education, engineering, fine arts, law, medicine, nursing, pharmacy. Advanced degrees conferred.

University of Tulsa. Tulsa. Founded 1894. Private control. Coeducational, 5,000 students. Faculty, 277. Semester system. Liberal arts, business administration, law, music, petroleum engineering. Advanced degrees conferred.

OREGON

Eastern Oregon College. La Grande. Founded 1929. State control. Coeducational, 800 students. Faculty, 57. Quarter system. Liberal arts, terminal-occupational, teacher preparatory. Advanced degrees conferred.

Lewis and Clark College. Portland. Founded 1867. Presbyterian control. Coeducational, 1,100 students. Faculty, 86. Semester system. Liberal arts and teacher preparatory. Advanced degrees conferred.

Linfield College. McMinnville. Founded 1849. Baptist control. Coeducational, 950 students. Faculty, 56. Semester system. Liberal arts and teacher preparatory. Advanced degrees conferred.

Marylhurst College. Marylhurst. Founded 1930. Roman Catholic control. Women only, 600. Faculty, 44. Semester system. Liberal arts, terminal-occupational, teacher preparatory.

Oregon College of Education. Monmouth. Founded 1856. State control. Coeducational, 1,100 students. Faculty, 82. Quarter system. Liberal arts and teacher prep. Advanced degrees conferred.

Oregon State University. Corvallis. Founded 1865. State control. Coeducational, 7,900 students. Faculty, 730. Quarter system. Agriculture, business and technology, education, engineering, forestry, home economics, humanities and social sciences, pharmacy, science. Advanced degrees conferred.

Pacific University. Forest Grove. Founded 1849. Private control. Coeducational, 800 students. Faculty, 57. Semester system. Liberal arts, music, optometry. Advanced degrees conferred.

Portland State College. Portland. Founded 1955. State control. Coeducational, 4,500 students. Faculty, 214. Quarter system. Liberal arts and teacher preparatory.

Reed College. Portland. Founded 1911. Private control. Coeducational, 800 students. Faculty, 76. Semester system. Liberal arts and teacher preparatory. Advanced degrees conferred.

Southern Oregon College. Ashland. Founded 1926. State control. Coeducational, 1,400 students. Faculty, 93. Quarter system. Liberal arts, terminal-occupational, teacher preparatory. Advanced degrees conferred.

University of Oregon. Eugene; medical courses at Portland. Founded 1872. State control. Coeducational, 8,600 students. Faculty, 648. Quarter system. Liberal arts, architecture and allied arts, business administration, dental, education, health and physical education, journalism, law, medicine, music. Advanced degrees conferred.

University of Portland. Portland. Founded 1901. Roman Catholic control. Coeducational, 1,700 students. Faculty, 162. Semester system. Liberal arts, business, engineering, music, nursing, science. Advanced degrees conferred.

Willamette University. Salem. Founded 1842. Methodist control. Coeducational, 1,200 students Faculty, 92. Semester system. Liberal arts, law. music. Advanced degrees conferred.

PENNSYLVANIA

Albright College. Reading. Founded 1856. Evangelical-United Brethren control. Coeducational, 1,100 students. Faculty, 59. Semester system. Liberal arts and teacher preparatory.

Allegheny College. Meadville. Founded 1815. Private control. Coeducational, 1,300 students. Faculty, 83. Semester system. Liberal arts and teacher preparatory. Advanced degrees conferred.

Beaver College. Jenkintown. Founded 1853. Presbyterian control. Women only, 700. Faculty, 58. Semester system. Liberal arts and teacher preparatory.

Bloomsburg State College. Bloomsburg. Founded 1839. State control. Coeducational, 1,800 students. Faculty, 123. Semester system. Primarily teacher preparatory.

Bryn Mawr College. Bryn Mawr. Founded 1885. Private control. Women only, 1,000. Faculty, 144. Semester system. Liberal arts and social work. Advanced degrees conferred.

Bucknell University. Lewisburg. Founded 1846. Private control. Coeducational, 2,600 students. Faculty, 172. Semester system. Liberal arts and engineering. Advanced degrees conferred.

California State College. California. Founded 1852. State control. Coeducational, 2,400 students. Faculty, 74. Semester system. Primarily teacher preparatory.

Carnegie Institute of Technology. Pittsburgh. Founded 1900. Private control. Coeducational; also Margaret Morrison Carnegie College, women; 5,100 students. Faculty, 352. Semester system. Engineering and science, fine arts (architecture), humanistic and social studies, industrial administration (business), library school, printing management. Advanced degrees conferred.

Cedar Crest College. Allentown. Founded 1867. Evangelical Reformed control. Women only, 450. Faculty, 40. Semester system. Liberal arts and teacher preparatory.

Chatham College. Pittsburgh. Founded 1869. Private control. Women only, 500. Faculty, 54. Semester system. Liberal arts and teacher preparatory.

Chestnut Hill College. Philadelphia. Founded 1871. Roman Catholic control. Women only, 950. Faculty, 51. Semester system. Liberal arts and teacher preparatory.

Cheyney State College. Cheyney. Founded 1837. State control. Coeducational, 850 students. Faculty, 40. Semester system. Primarily teacher preparatory.

Clarion State College. Clarion. Founded 1867. State control. Coeducational, 1,400 students. Faculty, 52. Semester system. Primarily teacher preparatory.

College Misericordia. Dallas. Founded 1924. Roman Catholic control. Women only, 1,200. Faculty, 89. Semester system. Liberal arts, terminal-occupational, teacher preparatory.

Dickinson College. Carlisle. Founded 1773. Private control. Coeducational, 1,100 students. Faculty, 102. Semester system. Liberal arts and teacher preparatory.

Drexel Institute of Technology. Philadelphia. Founded 1891. Private control. Coeducational, 5,900 students. Faculty, 301. Quarter system. Business administration, engineering, home economics, library science. Advanced degrees conferred.

Duquesne University. Pittsburgh. Founded 1878. Roman Catholic control. Coeducational, 5,400 students. Faculty, 288. Semester system. Liberal arts, business administration, education, law, music, nursing, pharmacy. Advanced degrees conferred.

East Stroudsburg State College. East Stroudsburg. Founded 1893. State control. Coeducational, 1,400 students. Faculty, 62. Semester system. Primarily teacher preparatory.

Edinboro State College. Edinboro. Founded 1861. State control. Coeducational, 1,500 students. Faculty, 56. Semester system. Primarily teacher preparatory. Advanced degrees conferred.

Elizabethtown College. Elizabethtown. Founded 1899. Church of the Brethren control. Coeducational, 1,100 students. Faculty, 47. Semester system. Liberal arts and teacher preparatory.

Franklin and Marshall College. Lancaster. Founded 1787. Private control. Men only, 1,900. Faculty, 101. Semester system. Liberal arts and teacher preparatory. Advanced degrees conferred.

Gannon College. Erie. Founded 1944. Roman Catholic control. Men only, 1,700. Faculty, 83. Semester system. Liberal arts, terminal-occupational, teacher preparatory.

Geneva College. Beaver Falls. Founded 1848. Reformed Presbyterian control. Coeducational, 1,600 students. Faculty, 104. Semester system. Liberal arts and teacher preparatory.

Gettysburg College. Gettysburg. Founded 1832. Lutheran control. Coeducational, 1,700 students. Faculty, 119. Semester system. Liberal arts and teacher preparatory.

Grove City College. Grove City. Founded 1876. Private control. Coeducational, 1,500 students. Faculty, 79. Semester system. Liberal arts and teacher preparatory.

Immaculata College. Immaculata. Founded 1921. Roman Catholic control. Women only, 950. Faculty, 66. Semester system. Liberal arts and teacher preparatory.

Indiana State College. Indiana. Founded 1875. State control. Coeducational, 3,600 students. Faculty, 153. Semester system. Primarily teacher preparatory. Advanced degrees conferred.

Jefferson Medical College of Philadelphia. Philadelphia. Founded 1825. Private and state control. Men only, 700. Faculty, 760. Special term system. Professional only. Advanced degrees.

Juniata College. Huntingdon. Founded 1876. Church of the Brethren control. Coeducational, 800 students. Faculty, 53. Semester system. Liberal arts and teacher preparatory.

King's College. Wilkes-Barre. Founded 1946. Roman Catholic control. Men only, 1,100. Faculty, 64. Semester system. Liberal arts and teacher preparatory.

Kutztown State College. Kutztown. Founded 1866. State control. Coeducational, 1,700 students. Faculty, 66. Semester system. Primarily teacher preparatory. Advanced degrees conferred.

Lafayette College. Easton. Founded 1826. Presbyterian control. Men only, 1,800. Faculty, 150. Semester system. Liberal arts and teacher preparatory.

LaSalle College. Philadelphia. Founded 1863. Roman Catholic control. Men only, 4,500. Faculty, 197. Semester system. Liberal arts, terminal-occupational, teacher preparatory.

Lebanon Valley College. Annville. Founded 1866. Evangelical United Brethren control. Coeducational, 1,200 students. Faculty, 56. Semester system. Liberal arts and teacher preparatory.

Lehigh University. Bethlehem. Founded 1865. Private control. Men only, 3,500. Faculty, 358. Semester system. Liberal arts, business administration, engineering. Advanced degrees conferred.

Lock Haven State College. Lock Haven. Founded 1870. State control. Coeducational, 1,100 students. Faculty, 54. Semester system. Primarily teacher preparatory.

Lycoming College. Williamsport. Founded 1812. Methodist control. Coeducational, 1,100 students. Faculty, 77. Semester system. Liberal arts and teacher preparatory.

Mansfield State College. Mansfield. Founded 1857. State control. Coeducational, 1,000 students. Faculty, 70. Semester system. Primarily teacher preparatory.

Margaret Morrison Carnegie College. Pittsburgh. (See Carnegie Institute of Technology.)

Marywood College. Scranton. Founded 1915. Roman Catholic control. Women only, 1,300. Faculty, 82. Semester system. Liberal arts and teacher preparatory. Advanced degrees conferred.

Mercyhurst College. Erie. Founded 1926. Roman Catholic control. Women only, 450. Faculty, 46. Semester system. Liberal arts and teacher preparatory.

Millersville State College. Millersville. Founded 1855. State control. Coeducational, 2,000 students. Faculty, 91. Semester system. Primarily teacher preparatory. Advanced degrees conferred.

Moravian College. Bethlehem. Founded 1742. Moravian control. Coeducational, 850 students. Faculty, 61. Semester system. Liberal arts and theology.

Mount Mercy College. Pittsburgh. Founded 1929. Roman Catholic control. Women only, 1,200. Faculty, 68. Semester system. Liberal arts and teacher preparatory.

Muhlenberg College. Allentown. Founded 1848. Lutheran control. Coeducational, 1,400 students. Faculty, 71. Semester system. Liberal arts and teacher preparatory.

Pennsylvania Military College. Chester. Founded 1821. Private control. Men only, 1,700. Faculty, 100. Semester system.

Pennsylvania State University, The. University Park; campuses also at Allentown, Altoona, Behrend (Erie), DuBois, Hazleton, McKeesport, New Castle, New Kensington, Ogontz (Philadelphia), Pottsville, Scranton, Wilkes-Barre, Wyomissing, York. Founded 1855. State and private control. Coeducational, 20,200 students. Faculty, 1,793. Semester system. Liberal arts, agriculture, business administration, chemistry and physics, education, engineering and architecture, home economics, mineral industries, physical education and athletics. Advanced degrees conferred.

Philadelphia College of Pharmacy and Science. Philadelphia. Founded 1821. Private control. Coeducational, 750 students. Faculty, 72. Semester system. Professional only. Advanced degrees conferred.

Philadelphia College of Textiles and Science. Philadelphia. Founded 1884. Private control. Men, women, 700. Faculty, 42. Semester system. Professional and technical only.

Philadelphia Museum College of Art. Philadelphia. Founded 1876. Private control. Coeducational, 650 students. Faculty, 116. Semester system. Professional and teacher preparatory.

Rosemont College. Rosemont. Founded 1921. Roman Catholic control. Women only, 550. Faculty, 59. Semester system. Liberal arts and teacher preparatory.

St. Francis College. Loretto. Founded 1847. Roman Catholic control. Coeducational, 1,100 students. Faculty, 61. Semester system. Liberal arts and teacher preparatory.

St. Joseph's College. Philadelphia. Founded 1851. Roman Catholic control. Men only, 4,000. Faculty, 82. Semester system. Liberal arts, terminal-occupational, teacher preparatory. Advanced degrees conferred.

St. Vincent College. Latrobe. Founded 1846. Roman Catholic control. Men only, 950. Faculty, 79. Semester system. Liberal arts and theology. Advanced degrees conferred.

Seton Hill College. Greensburg. Founded 1883. Private control. Women only, 800. Faculty, 64. Semester system. Liberal arts and teacher preparatory.

Shippensburg State College. Shippensburg. Founded 1871. State control. Coeducational, 1,300 students. Faculty, 54. Semester system. Primarily teacher preparatory. Advanced degrees conferred.

Slippery Rock State College. Slippery Rock. Founded 1889. State control. Coeducational, 1,300 students. Faculty, 68. Semester system. Primarily teacher preparatory.

Susquehanna University. Selinsgrove. Founded 1858. Lutheran control. Coeducational, 650 students. Faculty, 40. Semester system. Liberal arts and teacher preparatory.

Swarthmore College. Swarthmore. Founded 1864. Private control. Coeducational, 950 students. Faculty, 117. Semester system. Advanced degrees conferred.

Temple University. Philadelphia. Founded 1884. Private control. Coeducational, 18,600 students. Faculty, 1,119. Semester system. Liberal arts, business and public administration, dentistry, education, fine arts, law, medicine, nursing; Community College and Technical Institute. Advanced degrees conferred.

Thiel College. Greenville. Founded 1866. United Lutheran control. Coeducational, 900 students. Faculty, 49. Semester system. Liberal arts, terminal-occupational, teacher preparatory.

University of Pennsylvania. Philadelphia. Founded 1740. Private control. Men, women, 17,900. Faculty, 3,263. Semester system. Liberal arts (men), liberal arts (women), allied medical professions, communications, dentistry, education, engineering (chemical, civil and mechanical, metallurgical, Moore School of Electrical Engineering), finance and commerce (Wharton School), fine arts, law, medicine, nursing, social work, veterinary medicine. Advanced degrees conferred.

University of Pittsburgh. Pittsburgh. Founded 1787. Private control. Coeducational, 13,400 students. Faculty, 1,175. Quarter system. Liberal arts, business administration, dentistry, education, engineering and mines, law, medicine, natural sciences, nursing, pharmacy, public health, public and international affairs, retailing, social sciences, social work. Advanced degrees conferred.

University of Scranton. Scranton. Founded 1888. Roman Catholic control. Men only, 2,300. Faculty, 96. Semester system. Liberal arts and teacher preparatory. Advanced degrees conferred.

Ursinus College. Collegeville. Founded 1869. Private control. Coeducational, 1,300 students. Faculty, 56. Semester system. Liberal arts and teacher preparatory.

Villa Maria College. Erie. Founded 1925. Roman Catholic control. Women only, 700. Faculty, 49. Semester system. Liberal arts and teacher preparatory.

Villanova University. Villanova. Founded 1842. Roman Catholic control. Men only, 6,700. Faculty, 336. Semester system. Liberal arts, commerce and finance, engineering, law, nursing. Advanced degrees conferred.

Washington and Jefferson College. Washington. Founded 1781. Private control. Men only, 800. Faculty, 60. Semester system. Liberal arts and teacher preparatory. Advanced degrees conferred.

Waynesburg College. Waynesburg. Founded 1849. Presbyterian control. Coeducational, 1,200 students. Faculty, 59. Semester system. Liberal arts and teacher preparatory.

West Chester State College. West Chester. Founded 1871. State control. Coeducational, 3,000 students. Faculty, 112. Semester system. Primarily teacher preparatory. Advanced degrees conferred.

Westminster College. New Wilmington. Founded 1852. United Presbyterian control. Coeducational, 1,300 students. Faculty, 78. Semester system. Liberal arts and teacher preparatory. Advanced degrees conferred.

Wilkes College. Wilkes-Barre. Founded 1933. Private control. Coeducational, 1,700 students. Faculty, 86. Semester system. Liberal arts, terminal-occupational, teacher preparatory.

Wilson College. Chambersburg. Founded 1869. Private control. Women only, 550. Faculty, 50. Semester system. Liberal arts and teacher prep.

PUERTO RICO

Catholic University of Puerto Rico. Ponce. Founded 1948. Roman Catholic control. Coeducational, 3,000 students. Faculty, 128. Semester system. Liberal arts, terminal-occupational, teacher preparatory.

Inter American University of Puerto Rico. San German. Founded 1912. Private control. Coeducational, 3,200. Faculty, 103. Semester system. Liberal arts and teacher preparatory.

University of Puerto Rico. Río Piedras; campuses also at Mayagüez and San Juan. Founded 1903. Territorial control. Coeducational, 18,600 students. Faculty, 1,305. Semester system. At Río Piedras, liberal arts, commerce, education, law, natural sciences, pharmacy, social sciences; at Mayagüez, agriculture, engineering, natural sciences; at San Juan, dentistry, medicine-tropical medicine. Advanced degrees conferred.

RHODE ISLAND

Brown University. Providence. Founded 1764. Private control. Coordinated, The College (men), Pembroke College (women); 4,100 students. Faculty, 302. Semester system. Liberal arts and teacher preparatory. Advanced degrees conferred.

Pembroke College. Providence (See Brown University.)

Providence College. Providence. Founded 1917. Roman Catholic control. Men only, 3,000. Faculty, 112. Semester system. Liberal arts and teacher preparatory. Advanced degrees conferred.

Rhode Island College. Providence. Founded 1854. State control. Coeducational, 2,200 students. Faculty, 64. Semester system. Primarily teacher preparatory. Advanced degrees conferred.

Rhode Island School of Design. Providence. Founded 1877. Private control. Coeducational, 800 students. Faculty, 92. Semester system. Professional and teacher preparatory. Advanced degrees conferred.

Salve Regina College. Newport. Founded 1947. Roman Catholic control. Women only, 600. Faculty, 41. Semester system. Liberal arts and teacher preparatory.

University of Rhode Island. Kingston. Founded 1892. State control. Coeducational, 5,600 students. Faculty, 267. Semester system. Liberal arts, agriculture, business administration, engineering, home economics, nursing, pharmacy. Advanced degrees conferred.

SOUTH CAROLINA

Allen University. Columbia. Founded 1870. African Methodist Episcopal control. Coeducational, 750 students. Faculty, 52. Semester system. Liberal arts and teacher preparatory.

Benedict College. Columbia. Founded 1870. Baptist control. Coeducational, 800 students. Faculty, 42. Semester system. Liberal arts and theology.

Citadel, The—The Military College of South Carolina. Charleston. Founded 1842. State control. Men only, 2,000. Faculty, 131. Semester system. Liberal arts and teacher preparatory.

Clemson Agricultural College. Clemson. Founded 1889. State control. Coeducational, 4,000 students. Faculty, 294. Semester system. Liberal arts, agriculture, architecture, engineering, textiles. Advanced degrees conferred.

Columbia College. Columbia. Founded 1854. Methodist control. Women only, 1,100. Faculty, 40. Semester system.

Converse College. Spartanburg. Founded 1889. Private control. Women only, 600. Faculty, 58. Semester system. Liberal arts and music. Advanced degrees conferred.

Erskine College. Due West. Founded 1839. Reformed Presbyterian control. Coeducational, 600 students. Faculty, 42. Semester system. Liberal arts and theology.

Furman University. Greenville. Founded 1826. Southern Baptist control. Coeducational, 1,500 students. Faculty, 96. Semester system. Liberal arts and teacher preparatory. Advanced degrees conferred.

Medical College of South Carolina. Charleston. Founded 1823. State control. Coeducational, 650 students. Faculty, 210. Quarter system. Professional only. Advanced degrees conferred.

Newberry College. Newberry. Founded 1856. United Lutheran control. Coeducational, 700 students. Faculty, 37. Semester system. Liberal arts and teacher preparatory.

Presbyterian College. Clinton. Founded 1880. Presbyterian control. Coeducational, 500 students. Faculty, 34. Semester system. Liberal arts and teacher preparatory.

South Carolina State College. Orangeburg. Founded 1896. State control. Coeducational, 2,000 students. Faculty, 115. Semester system. Liberal arts, agriculture and home economics, education, industrial education, law. Advanced degrees conferred.

University of South Carolina. Columbia. Founded 1801. State control. Coeducational, 6,800 students. Faculty, 271. Semester system. Liberal arts, business administration, education, engineering, journalism, law, pharmacy. Advanced degrees conferred.

Winthrop College. Rock Hill. Founded 1886. State control. Women only, 1,800. Faculty, 93. Semester system. Liberal arts and teacher preparatory. Advanced degrees conferred.

Wofford College. Spartanburg. Founded 1854. Methodist control. Men only, 700. Faculty, 46. Semester system.

SOUTH DAKOTA

Augustana College. Sioux Falls. Founded 1860. Lutheran control. Coeducational, 1,500 students. Faculty, 75. Semester system. Liberal arts and teacher preparatory.

Black Hills Teachers College. Spearfish. Founded 1883. State control. Coeducational, 1,000 students. Faculty, 44. Quarter system. Liberal arts and teacher preparatory.

Dakota Wesleyan University. Mitchell. Founded 1885. Methodist control. Coeducational, 600 students. Faculty, 27. Semester system. Liberal arts and teacher preparatory.

Huron College. Huron. Founded 1883. Presbyterian control. Coeducational, 550 students. Faculty, 30. Semester system. Liberal arts and teacher preparatory.

Northern State Teachers College. Aberdeen. Founded 1901. State control. Coeducational, 1,600 students. Faculty, 87. Quarter system. Liberal arts and teacher preparatory. Advanced degrees conferred.

Sioux Falls College. Sioux Falls. Founded 1883. Baptist control. Coeducational, 500 students. Faculty, 27. Semester system. Liberal arts, terminal-occupational, teacher preparatory.

South Dakota School of Mines and Technology. Rapid City. Founded 1885. State control. Coeducational, 1,000 students. Faculty, 78. Quarter system. Technical only. Advanced degrees conferred.

South Dakota State College of Agriculture and Mechanic Arts. Brookings. Founded 1881. State control. Coeducational, 3,000 students. Faculty, 361. Quarter system. Liberal arts, agriculture, engineering, home economics, nursing, pharmacy, science and applied arts. Advanced degrees conferred.

Southern State Teachers College. Springfield. Founded 1897. State control. Coeducational, 800 students. Faculty, 43. Quarter system. Professional and teacher preparatory.

State University of South Dakota. Vermillion. Founded 1882. State control. Coeducational, 2,500 students. Faculty, 214. Semester system. Liberal arts, business administration, education, fine arts, law, medicine, nursing. Advanced degrees conferred.

TENNESSEE

Austin Peay State College. Clarksville. Founded 1927. State control. Coeducational, 1,600 students. Faculty, 72. Quarter system. Liberal arts and teacher preparatory. Advanced degrees conferred.

Belmont College. Nashville. Founded 1951. Baptist control. Coeducational, 550 students. Faculty, 36. Quarter system. Liberal arts and teacher preparatory.

Bethel College. McKenzie. Founded 1842. Cumberland Presbyterian control. Coeducational, 550 students. Faculty, 31. Quarter system.

Carson-Newman College. Jefferson City. Founded 1851. Southern Baptist control. Coeducational, 1,300 students. Faculty, 68. Semester system. Liberal arts and teacher preparatory.

Christian Brothers College. Memphis. Founded 1871. Roman Catholic control. Men only, 750. Faculty, 45. Semester system. Liberal arts and terminal-occupational.

David Lipscomb College. Nashville. Founded 1891. Church of Christ control. Coeducational, 1,300 students. Faculty, 58. Quarter system. Liberal arts, terminal-occupational, teacher preparatory.

East Tennessee State College. Johnson City; branch at Kingsport. Founded 1911. State control. Coeducational, 4,500 students. Faculty, 196. Quarter system. Liberal arts, business and economics, education. Advanced degrees conferred.

Fisk University. Nashville. Founded 1866. Private control. Coeducational, 850 students. Faculty, 59. Semester system. Liberal arts and teacher preparatory. Advanced degrees conferred.

George Peabody College for Teachers. Nashville. Founded 1875. Private control. Coeducational, 1,600 students. Faculty, 106. Quarter system. Liberal arts, terminal-occupational, teacher preparatory. Advanced degrees conferred.

Knoxville College. Knoxville. Founded 1875.

United Presbyterian control. Coeducational, 650 students. Faculty, 48. Semester system. Liberal arts and teacher preparatory.

Lambuth College. Jackson. Founded 1843. Methodist control. Coeducational, 600 students. Faculty, 34. Semester system. Liberal arts and teacher preparatory.

Lane College. Jackson. Founded 1882. Christian Methodist Episcopal control. Coeducational, 500 students. Faculty, 23. Quarter system. Liberal arts and teacher preparatory.

Le Moyne College. Memphis. Founded 1870. American Missionary control. Coeducational, 550 students. Faculty, 29. Semester system. Liberal arts and teacher preparatory.

Lincoln Memorial University. Harrogate. Founded 1897. Private control. Coeducational, 400 students. Faculty, 31. Quarter system. Liberal arts and teacher preparatory.

Maryville College. Maryville. Founded 1819. Presbyterian control. Coeducational, 800 students. Faculty, 67. Semester system. Liberal arts and teacher preparatory.

Meharry Medical College. Nashville. Founded 1876. Private control. Coeducational, 400 students. Faculty, 196. Quarter system. Professional only. Advanced degrees conferred.

Memphis College of Music. (*See* Southwestern at Memphis.)

Memphis State University. Memphis. Founded 1912. State control. Coeducational, 5,400 students. Faculty, 216. Semester system. Liberal arts, business administration, education. Advanced degrees conferred.

Middle Tennessee State College. Murfreesboro. Founded 1909. State control. Coeducational, 3,000 students. Faculty, 116. Quarter system. Liberal arts and teacher preparatory. Advanced degrees conferred.

Milligan College. Milligan College. Founded 1881. Private control. Coeducational, 550 students. Faculty, 25. Semester system. Liberal arts and teacher preparatory.

Southern Missionary College. Collegedale. Founded 1916. Seventh Day Adventist control. Coeducational, 600 students. Faculty, 59. Semester system. Liberal arts, terminal-occupational, teacher preparatory.

Southwestern at Memphis. Memphis. Founded 1848. Presbyterian control. Coeducational, 750 students. Faculty, 79. Semester system. Liberal arts and Memphis College of Music.

Tennessee Agricultural and Industrial State University. Nashville. Founded 1912. State control. Coeducational, 3,400 students. Faculty, 232. Quarter system. Liberal arts, agriculture and home economics, education, engineering. Advanced degrees conferred.

Tennessee Polytechnic Institute. Cookeville. Founded 1916. State control. Coeducational, 2,800 students. Faculty, 168. Quarter system. Liberal arts, agriculture and home economics, business administration, education, engineering. Advanced degrees conferred.

Tennessee Wesleyan College. Athens. Founded 1857. Methodist control. Coeducational, 600 students. Faculty, 43. Quarter system. Liberal arts, terminal-occupational, teacher preparatory.

Union University. Jackson. Founded 1825. Southern Baptist control. Coeducational, 750 students. Faculty, 42. Semester system. Liberal arts and teacher preparatory.

University of Chattanooga. Chattanooga. Founded 1886. Private control. Coeducational, 2,300 students. Faculty, 133. Semester system. Liberal arts, applied arts, conservatory. Advanced degrees conferred.

University of the South. Sewanee. Founded 1857. Protestant Episcopal control. Men only, 700. Faculty, 57. Semester system. Liberal arts and theology. Advanced degrees conferred.

University of Tennessee. Knoxville; campuses also at Memphis, Martin, Nashville. Founded 1794. State control. Coeducational, 17,200 students. Faculty, 841. Quarter system. At Knoxville and Martin, liberal arts, agriculture, business administration, education, engineering, home economics, law; at Memphis, biological science, dentistry, medicine, nursing, pharmacy; at Nashville, social work. Advanced degrees conferred.

Vanderbilt University. Nashville. Founded 1872. Private control. Coeducational, 3,700 students. Faculty, 644. Semester system. Liberal arts, divinity, engineering, law, medicine, nursing. Advanced degrees conferred.

TEXAS

Abilene Christian College. Abilene. Founded 1906. Private control. Coeducational, 2,500 students. Faculty, 143. Semester system. Liberal arts and teacher preparatory. Advanced degrees conferred.

Agricultural and Mechanical College of Texas. College Station. (*See* Texas Agricultural and Mechanical College System.)

Austin College. Sherman. Founded 1849. Southern Presbyterian control. Coeducational, 900 students. Faculty, 63. Semester system. Liberal arts and teacher preparatory. Advanced degrees conferred.

Baylor University. Waco; medical branch at Houston, dental branch at Dallas. Founded 1845. Southern Baptist control. Coeducational, 6,000 students. Faculty, 308. Semester system. Liberal arts, business, dentistry, education, law, medicine, music, nursing. Advanced degrees conferred.

Bishop College. Dallas. Founded 1880. Baptist control. Coeducational, 550 students. Faculty, 42. Semester system. Liberal arts and teacher preparatory.

Brite College of the Bible. Fort Worth. (*See* Texas Christian University.)

East Texas Baptist College. Marshall. Founded 1917. Southern Baptist control. Coeducational, 500 students. Faculty, 37. Semester system. Liberal arts and teacher preparatory.

East Texas State College. Commerce. Founded 1889. State control. Coeducational, 3,400 students. Faculty, 152. Semester system. Liberal arts and teacher preparatory. Advanced degrees conferred.

Hardin-Simmons University. Abilene. Founded 1891. Southern Baptist control. Coeducational, 1,600 students. Faculty, 77. Semester system. Liberal arts and music. Advanced degrees conferred.

Howard Payne College. Brownwood. Founded 1889. Southern Baptist control. Coeducational, 1,100 students. Faculty, 60. Semester system. Liberal arts and teacher preparatory. Advanced degrees conferred.

Huston-Tillotson College. Austin. Founded 1952. Private control. Coeducational, 450 students. Faculty, 33. Semester system. Liberal arts and teacher preparatory.

Incarnate Word College. San Antonio. Founded 1900. Roman Catholic control. Women only, 1,100. Faculty, 77. Semester system. Liberal arts and teacher preparatory. Advanced degrees conferred.

Lamar State College of Technology. Beaumont. Founded 1923. District control. Coeducational, 4,900 students. Faculty, 184. Semester system. Liberal arts, business, education, engineering, fine and applied arts.

Mary Hardin-Baylor College. Belton. Founded 1845. Southern Baptist control. Women only, 650. Faculty, 53. Semester system. Liberal arts and teacher preparatory.

McMurry College. Abilene. Founded 1923. Methodist control. Coeducational, 1,200 students. Faculty, 47. Semester system. Advanced degrees conferred.

Midwestern University. Wichita Falls. Founded 1922. Municipal control. Coeducational, 1,700 students. Faculty, 101. Semester system. Liberal arts, business, education, fine arts. Advanced degrees conferred.

North Texas State University. Denton. Founded 1890. State control. Coeducational, 8,000 students. Faculty, 389. Semester system. Liberal arts, business, education, home economics, music. Advanced degrees conferred.

Our Lady of the Lake College. San Antonio. Founded 1912. Roman Catholic control. Women only, 800. Faculty, 75. Semester system. Liberal arts, education, social service. Advanced degrees.

Pan American College. Edinburg. Founded 1927. County control. Coeducational, 2,000 students. Faculty, 85. Semester system. Liberal arts, terminal-occupational, teacher preparatory.

Prairie View Agricultural and Mechanical College. Prairie View. (*See* Texas Agricultural and Mechanical College System.)

Rice University. Houston. Founded 1912. Private control. Coeducational, 2,000 students. Faculty, 159. Semester system. Liberal arts, engineering. Advanced degrees conferred.

St. Edward's University. Austin. Founded 1885. Roman Catholic control. Men only, 550. Faculty, 28. Semester system. Liberal arts and teacher preparatory.

St. Mary's University of San Antonio. San Antonio. Founded 1852. Roman Catholic control. Coeducational, 2,100 students. Faculty, 118. Semester system. Liberal arts, business administration, law. Advanced degrees conferred.

Sam Houston State Teachers College. Huntsville. Founded 1879. State control. Coeducational, 4,500 students. Faculty, 158. Semester system. Liberal arts and teacher preparatory. Advanced degrees conferred.

Southern Methodist University. Dallas. Founded 1915. Methodist control. Coeducational, 7,500 students. Faculty, 368. Semester system. Liberal arts, business, engineering, law, music, theology. Advanced degrees conferred.

Southwest Texas State College. San Marcos. Founded 1903. State control. Coeducational, 2,700 students. Faculty, 123. Semester system. Liberal arts, terminal-occupational, teacher preparatory. Advanced degrees conferred.

Southwestern Baptist Theological Seminary. Fort Worth. Founded 1908. Southern Baptist control. Coeducational, 1,900 students. Faculty, 61. Semester system. Church music, religious education, theology. Advanced degrees conferred.

Southwestern University. Georgetown. Founded 1840. Methodist control. Coeducational, 650 students. Faculty, 52. Semester system. Liberal arts and fine arts.

Stephen F. Austin State College. Nacogdoches. Founded 1917. State control. Coeducational, 2,100 students. Faculty, 100. Semester system. Liberal arts and teacher preparatory. Advanced degrees conferred.

Sul Ross State College. Alpine. Founded 1920. State control. Coeducational, 1,200 students. Faculty, 58. Semester system. Liberal arts and teacher preparatory. Advanced degrees conferred.

Texas Agricultural and Mechanical College System. College Station. The 4 schools in this system are under state control; 2 of the schools do not offer accredited 4-year courses and are not included here. *Agricultural and Mechanical College of Texas.* College Station. Founded 1876. Men only, 7,200. Faculty, 556. Semester system. Liberal arts, agriculture, engineering, veterinary medicine. Advanced degrees conferred. *Prairie View Agricultural and Mechanical College.* Prairie View. Founded 1876. Coeducational, 2,600 students. Faculty, 152. Semester system. Liberal arts, terminal-occupational, teacher preparatory. Advanced degrees conferred.

Texas Christian University. Fort Worth. Founded 1873. Christian Churches control. Coeducational, 6,300 students. Faculty, 427. Semester system. Liberal arts, Brite College of the Bible, business, education, fine arts, nursing. Advanced degrees conferred.

Texas College of Arts and Industries. Kingsville. Founded 1925. State control. Coeducational, 3,400 students. Faculty, 140. Semester system. Liberal arts, agriculture, business administration, engineering, teacher education. Advanced degrees conferred.

Texas Lutheran College. Seguin. Founded 1891. Lutheran control. Coeducational, 650 students. Faculty, 40. Semester system. Liberal arts, terminal-occupational, teacher preparatory.

Texas Southern University. Houston. Founded 1947. State control. Coeducational, 3,300 students. Faculty, 172. Semester system. Liberal arts, business, law, pharmacy, vocational-industrial education. Advanced degrees conferred.

Texas Technological College. Lubbock. Founded 1923. State control. Coeducational, 9,200 students. Faculty, 417. Semester system. Liberal arts, agriculture, business administration, engineering, home economics. Advanced degrees conferred.

Texas Wesleyan College. Fort Worth. Founded 1891. Methodist control. Coeducational, 1,200 students. Faculty, 53. Semester system. Liberal arts and teacher preparatory. Advanced degrees conferred.

Texas Western College. El Paso. Founded 1913. State control. Coeducational, 4,100 students. Faculty, 193. Semester system. Liberal arts and engineering. Advanced degrees conferred. Part of the University of Texas.

Texas Woman's University. Denton. Founded 1901. State control. Women only, 2,600. Faculty, 165. Semester system. Liberal arts; education; fine arts; health, physical education and recreation; household arts and sciences; library science; nursing; occupational therapy. Advanced degrees conferred.

Trinity University. San Antonio. Founded 1869. Presbyterian control. Coeducational, 1,700 students. Faculty, 110. Semester system. Liberal arts and teacher preparatory. Advanced degrees conferred.

University of Houston. Houston. Founded 1934. Private control. Coeducational, 11,400 students. Faculty, 565. Semester system. Liberal arts, architecture, business administration, education, engineering, law, optometry, pharmacy, technology. Advanced degrees conferred.

University of St. Thomas. Houston. Founded 1947. Roman Catholic control. Coeducational, 550 students. Faculty, 40. Semester system. Liberal arts and teacher preparatory.

University of Texas. Austin; medical branches at Dallas, Galveston, Houston; branch, Texas Western College, El Paso (see entry). Founded 1883. State control. Coeducational, 21,700 students. Faculty, 1,207. Semester system. Liberal arts, architecture, business administration, dentistry, education, engineering, fine arts, law, library school, medicine, nursing, pharmacy, social work. Advanced degrees conferred.

Wayland Baptist College. Plainview. Founded 1908. Southern Baptist control. Coeducational, 550 students. Faculty, 38. Semester system. Liberal arts and teacher preparatory.

West Texas State College. Canyon. Founded 1910. State control. Coeducational, 3,300 students. Faculty, 164. Semester system. Liberal arts, terminal-occupational, teacher preparatory. Advanced degrees conferred.

Wiley College. Marshall. Founded 1873. Methodist control. Coeducational, 500. Faculty, 32. Semester system. Liberal arts and teacher prep.

UTAH

Brigham Young University. Provo. Founded 1875. Latter Day Saints control. Coeducational, 11,600 students. Faculty, 502. Semester system. Liberal arts, biological and agricultural sciences, commerce, education, family living, fine arts, nursing, physical and engineering sciences, physical education. Advanced degrees conferred.

Carbon College. Price. (*See* University of Utah.)

College of Southern Utah. Cedar City. (*See* Utah State University.)

Snow College. Ephraim. (*See* Utah State University.)

University of Utah. Salt Lake City; branch, Carbon College, Price. Founded 1850. State control. Coeducational, 11,600 students. Faculty, 556. Quarter system. Liberal arts, business, education, engineering, fine arts, law, medicine, mines and mineral industries, nursing, pharmacy, social work. Advanced degrees conferred.

Utah State University. Logan; branches—Snow College at Ephraim; College of Southern Utah, Cedar City. Founded 1888. State control. Coeducational, 7,300 students. Faculty, 258. Quarter system. Liberal arts; agriculture; business and social sciences; education; engineering; forestry, range and wildlife; family life. Advanced degrees conferred.

Westminster College. Salt Lake City. Founded 1875. Interdenominational control. Coeducational, 500 students. Faculty, 47. Semester system. Liberal arts and teacher preparatory.

VERMONT

Bennington College. Bennington. Founded 1932. Private control. Women only, 350. Faculty, 52. Semester system.

Middlebury College. Middlebury. Founded 1800. Private control. Coeducational, 1,300 students. Faculty, 94. Semester system. Liberal arts and teacher preparatory. Advanced degrees conferred.

Norwich University. Northfield. Founded 1819. Private control. Men only, 1,000. Faculty, 60. Semester system.

St. Michael's College. Winooski. Founded 1904. Roman Catholic control. Men only, 900. Faculty, 53. Semester system. Liberal arts and teacher preparatory. Advanced degrees conferred.

University of Vermont and State Agricultural College. Burlington. Founded 1791. State control. Coeducational, 3,400 students. Faculty, 412. Semester system. Liberal arts, agriculture, education and nursing, medicine, technology. Advanced degrees conferred.

VIRGINIA

Bridgewater College. Bridgewater. Founded 1880. Church of the Brethren control. Coeducational, 600 students. Faculty, 41. Semester system. Liberal arts and teacher preparatory.

Clinch Valley College of the University of Virginia. Wise. (*See* University of Virginia.)

College of William and Mary. Williamsburg; divisions at Norfolk and Richmond. Founded 1693. State control. Coeducational, 12,100 students. Faculty, 157. Semester system. Liberal arts, education, law. Richmond Professional Institute, social work. Advanced degrees conferred.

Eastern Mennonite College. Harrisonburg. Founded 1917. Mennonite control. Coeducational, 550 students. Faculty, 38. Semester system. Liberal arts and teacher preparatory.

Emory and Henry College. Emory. Founded 1836. Methodist control. Coeducational, 750 students. Faculty, 41. Semester system. Liberal arts, terminal-occupational, teacher preparatory.

George Mason College of the University of Virginia. Fairfax. (*See* University of Virginia.)

Hampton Institute. Hampton. Founded 1868. Private control. Coeducational, 1,400 students. Faculty, 123. Semester system. Liberal arts, terminal-occupational, teacher preparatory. Advanced degrees conferred.

Hollins College. Hollins College. Founded 1842. Private control. Women only, 700. Faculty, 61. Semester system. Advanced degrees conferred.

Longwood College. Farmville. Founded 1884. State control. Women only, 1,100. Faculty, 64. Semester system. Liberal arts and teacher preparatory. Advanced degrees conferred.

Lynchburg College. Lynchburg. Founded 1903. Disciples of Christ control. Coeducational, 800 students. Faculty, 50. Semester system. Liberal arts and teacher preparatory.

Madison College. Harrisonburg. Founded 1909. State control. Primarily women, limited enrollment for men; 1,500 students. Faculty, 104. Semester system. Liberal arts and teacher preparatory. Advanced degrees conferred.

Mary Washington College of the University of Virginia. Fredericksburg. Founded 1908. State control. Women only, 1,700. Faculty, 113. Semester system. Advanced degrees conferred.

Medical College of Virginia. Richmond. Founded 1838. State control. Coeducational, 1,200 students. Faculty, 630. Quarter system. Professional only. Advanced degrees conferred.

Radford College. Radford. (*See* Virginia Polytechnic Institute.)

Randolph-Macon College. Ashland. Founded 1830. Methodist control. Men only, 600. Faculty, 43. Semester system.

Randolph-Macon Woman's College. Lynchburg. Founded 1891. Methodist control. Women only, 700. Faculty, 79. Semester system. Liberal arts and teacher preparatory.

Richmond College. (*See* University of Richmond.)

Richmond Professional Institute. (*See* College of William and Mary.)

Roanoke College. Salem. Founded 1842. Lutheran control. Coeducational, 700 students. Faculty, 41. Semester system. Liberal arts and teacher preparatory.

Sweet Briar College. Sweet Briar. Founded 1901. Private control. Women only, 550. Faculty, 68. Semester system.

University of Richmond. Richmond. Founded 1830. Baptist control. Coordinated, Richmond College (men), Westhampton College (women); 3,800 students. Faculty, 185. Semester system. Liberal arts, business administration, law. Advanced degrees conferred.

University of Virginia. Charlottesville; branches—Clinch Valley College, Wise, and George Mason College, Fairfax; Mary Washington College, Fredericksburg (see entry). Founded 1819. State control. Men, women, 10,900. Faculty, 470. Semester system. Liberal arts, architecture, business administration; at Clinch Valley College, commerce, education, engineering; at George Mason College, law, medicine, and nursing. Advanced degrees conferred.

Virginia Military Institute. Lexington. Founded 1839. State control. Men only, 1,000. Faculty, 95. Semester system.

Virginia Polytechnic Institute. Blacksburg; branch at Danville, women's division at Radford. Founded 1872. State control. Coeducational, 6,900

students. Faculty, 401. Quarter system. Agriculture, applied science and business administration, engineering and architecture; at Radford College, liberal arts and teacher preparatory. Advanced degrees conferred.

Virginia State College. Petersburg. Founded 1882. State control. Coeducational, 3,400 students. Faculty, 157. Semester system. Liberal arts, terminal-occupational, teacher preparatory. Advanced degrees conferred.

Virginia Union University. Richmond. Founded 1865. Baptist control. Coeducational, 1,000 students. Faculty, 63. Semester system. Liberal arts and religion.

Washington and Lee University. Lexington. Founded 1749. Private control. Men only, 1,200. Faculty, 81. Semester system. Liberal arts, commerce, law.

Westhampton College. Richmond. (*See* University of Richmond.)

WASHINGTON, D.C.

(*See* District of Columbia)

WASHINGTON

Central Washington State College. Ellensburg. Founded 1891. State control. Coeducational, 2,300 students. Faculty, 136. Quarter system. Primarily teacher preparatory. Advanced degrees.

Eastern Washington State College. Cheney. Founded 1890. State control. Coeducational, 3,200 students. Faculty, 121. Quarter system. Liberal arts and teacher preparatory. Advanced degrees conferred.

Gonzaga University. Spokane. Founded 1887. Roman Catholic control. Coeducational, 1,900 students. Faculty, 145. Semester system. Liberal arts, economics and business administration, education, engineering, law, nursing. Advanced degrees conferred.

Pacific Lutheran University. Tacoma. Founded 1894. Lutheran control. Coeducational, 1,700 students. Faculty, 87. Semester system. Liberal arts, business administration, education, fine and applied arts, nursing. Advanced degrees conferred.

Seattle Pacific College. Seattle. Founded 1891. Free Methodist control. Coeducational, 1,200 students. Faculty, 55. Quarter system. Liberal arts, terminal-occupational, teacher preparatory. Advanced degrees conferred.

Seattle University. Seattle. Founded 1891. Roman Catholic control. Coeducational, 3,200 students. Faculty, 199. Quarter system. Liberal arts, commerce, education, engineering, nursing. Advanced degrees conferred.

University of Puget Sound. Tacoma. Founded 1888. Methodist control. Coeducational, 2,800 students. Faculty, 102. Semester system. Liberal arts, business administration and economics, education, music, occupational therapy. Advanced degrees conferred.

University of Washington. Seattle. Founded 1861. State control. Coeducational, 22,500 students. Faculty, 998. Quarter system. Liberal arts, architecture and urban planning, business administration, dentistry, education, engineering, fisheries, forestry, law, medicine, nursing, pharmacy, social work. Advanced degrees conferred.

Walla Walla College. College Place. Founded 1892. Seventh Day Adventist control. Coeducational, 1,300 students. Faculty, 81. Quarter system. Liberal arts, nursing, theology. Advanced degrees conferred.

Washington State University. Pullman. Founded 1890. State control. Coeducational, 7,300 students. Faculty, 461. Semester system. Liberal arts, agriculture, economics and business, education, engineering and mineral technology, home economics, pharmacy, physical education, veterinary medicine. Advanced degrees conferred.

Western Washington State College. Bellingham. Founded 1899. State control. Coeducational, 3,200 students. Faculty, 130. Quarter system. Liberal arts and teacher preparatory. Advanced degrees conferred.

Whitman College. Walla Walla. Founded 1859. Private control. Coeducational, 900 students. Faculty, 58. Semester system. Liberal arts and teacher preparatory.

Whitworth College. Spokane. Founded 1890. Private control. Coeducational, 1,400 students. Faculty, 57. Semester system. Liberal arts and teacher preparatory. Advanced degrees conferred.

WEST VIRGINIA

Alderson-Broaddus College. Philippi. Founded 1871. Baptist control. Coeducational, 500 students. Faculty, 33. Semester system. Liberal arts, terminal-occupational, teacher preparatory.

Bethany College. Bethany. Founded 1840. Private control. Coeducational, 700 students. Faculty, 51. Semester system.

Bluefield State College. Bluefield. Founded 1895. State control. Coeducational, 600 students. Faculty, 39. Semester system. Liberal arts, terminal-occupational, teacher preparatory.

Concord College. Athens. Founded 1875. State control. Coeducational, 1,600 students. Faculty, 76. Semester system. Liberal arts and teacher prep.

Davis and Elkins College. Elkins. Founded 1904. Presbyterian control. Coeducational, 550 students. Faculty, 38. Semester system. Liberal arts terminal-occupational, teacher preparatory.

Fairmont State College. Fairmont. Founded 1867. State control. Coeducational, 1,400 students. Faculty, 61. Semester system. Liberal arts and teacher preparatory.

Glenville State College. Glenville. Founded 1873. State control. Coeducational, 800 students. Faculty, 37. Semester system. Liberal arts, terminal-occupational, teacher preparatory.

Marshall University. Huntington. Founded 1837. State control. Coeducational, 4,400 students. Faculty, 231. Semester system. Liberal arts, applied science, teachers college. Advanced degrees conferred.

Morris Harvey College. Charleston. Founded 1888. Private control. Coeducational, 2,500 students. Faculty, 76. Semester system. Liberal arts and teacher preparatory.

Shepherd College. Shepherdstown. Founded 1871. State control. Coeducational, 950 students. Faculty, 38. Semester system. Liberal arts, terminal-occupational, teacher preparatory.

West Liberty State College. West Liberty. Founded 1837. State control. Coeducational, 1,500 students. Faculty, 56. Semester system. Liberal arts, terminal-occupational, teacher preparatory.

West Virginia Institute of Technology. Montgomery. Founded 1895. State control. Coeducational, 950 students. Faculty, 58. Semester system. Technical only.

West Virginia State College. Institute. Founded 1891. State control. Coeducational, 2,000 students. Faculty, 99. Semester system. Liberal arts, terminal-occupational, teacher preparatory.

West Virginia University. Morgantown; branch at Charleston. Founded 1867. State control. Coeducational, 7,800 students. Faculty, 484. Semester system. Liberal arts; agriculture, forestry, home economics; commerce; dentistry; education; engineering; journalism; law; medical sciences; mines; music; nursing; pharmacy; physical education; Kanawha Graduate Center of Science and Engineering, Charleston. Advanced degrees conferred.

West Virginia Wesleyan College. Buckhannon. Founded 1890. Methodist control. Coeducational, 1,200 students. Faculty, 56. Semester system. Liberal arts and teacher preparatory.

WISCONSIN

Alverno College. Milwaukee. Founded 1890. Roman Catholic control. Women only, 1,100. Faculty, 80. Semester system. Liberal arts, music, nursing.

Beloit College. Beloit. Founded 1847. Private control. Coeducational, 1,000 students. Faculty, 99. Semester system. Liberal arts and teacher preparatory. Advanced degrees conferred.

Carroll College. Waukesha. Founded 1846. Presbyterian control. Coeducational, 850 students. Faculty, 59. Semester system. Liberal arts and teacher preparatory.

Edgewood College of the Sacred Heart. Madison. Roman Catholic control. Women only, 500. Faculty, 28. Semester system. Liberal arts and teacher preparatory.

Institute of Paper Chemistry, The. Appleton. (*See* Lawrence College.)

Lawrence College. Appleton. Founded 1847. Private control. Coeducational, 1,100 students. Faculty, 85. Semester system. Liberal arts; The Institute of Paper Chemistry. Advanced degrees conferred.

Marquette University. Milwaukee. Founded 1881. Roman Catholic control. Coeducational, 10,000 students. Faculty, 989. Semester system. Liberal arts, business, dentistry, engineering, journalism, law, medicine, nursing, speech. Advanced degrees conferred.

Milwaukee School of Engineering. Milwaukee. Founded 1903. Private control. Men only, 1,100. Faculty, 99. Quarter system. Technical and terminal-occupational.

Milwaukee-Downer College. Milwaukee. Founded 1851. Private control. Women only, 200. Faculty, 46. Semester system. Liberal arts and teacher preparatory.

Mount Mary College. Milwaukee. Founded 1913. Roman Catholic control. Women only, 1,100. Faculty, 62. Semester system. Liberal arts and teacher preparatory.

Ripon College. Ripon. Founded 1851. Private control. Coeducational, 650 students. Faculty, 65. Semester system. Liberal arts and teacher prep.

St. Norbert College. West De Pere. Founded 1898. Roman Catholic control. Coeducational, 1,100 students. Faculty, 72. Semester system. Liberal arts, terminal-occupational, teacher preparatory.

Stout State College. Menomonie. Founded 1903. State control. Coeducational, 1,500 students. Faculty, 76. Quarter system. Primarily teacher preparatory. Advanced degrees conferred.

University of Wisconsin. Madison; campus also at Milwaukee; Freshman-Sophomore Centers at Green Bay, Kenosha, Manitowoc, Marinette, Fox Valley (Menasha), Racine, Sheboygan, Wausau. Founded 1849. State control. Coeducational, 30,000 students. Faculty, 2,167. Semester system. Liberal arts, agriculture, commerce, education, engineering, law, medicine, pharmacy; at Milwaukee, liberal arts, education. Advanced degrees conferred.

Wisconsin State College. Eau Claire. Founded 1916. State control. Coeducational, 2,200 students. Faculty, 108. Semester system. Liberal arts and teacher preparatory.

Wisconsin State College. La Crosse. Founded 1909. State control. Coeducational, 2,000 students. Faculty, 112. Semester system. Liberal arts and teacher preparatory. Advanced degrees conferred.

Wisconsin State College. Oshkosh. Founded 1871. State control. Coeducational, 3,200 students. Faculty, 102. Semester system. Liberal arts and teacher preparatory.

Wisconsin State College. River Falls. Founded 1874. State control. Coeducational, 1,700 students. Faculty, 80. Quarter system. Liberal arts and teacher preparatory.

Wisconsin State College. Stevens Point. Founded 1894. State control. Coeducational, 2,600 students. Faculty, 90. Semester system. Liberal arts and teacher preparatory.

Wisconsin State College. Superior. Founded 1896. State control. Coeducational, 1,500 students. Faculty, 74. Semester system. Liberal arts and teacher preparatory. Advanced degrees conferred.

Wisconsin State College. Whitewater. Founded 1868. State control. Coeducational, 2,500 students. Faculty, 101. Semester system. Liberal arts and teacher preparatory.

Wisconsin State College and Institute of Technology. Platteville. Founded 1866. State control. Coeducational, 2,100 students. Faculty, 109. Semester system. Liberal arts and teacher preparatory.

WYOMING

University of Wyoming. Laramie. Founded 1887. State control. Coeducational, 4,500 students. Faculty, 300. Semester system. Liberal arts, agriculture, commerce and industry, education, engineering, law, nursing, pharmacy. Advanced degrees conferred.

GLOSSARY OF FOREIGN GEOGRAPHICAL TERMS

Arab	Arabic	Jap	Japanese
Bantu	Bantu	Kor	Korean
Bur	Burmese	Mal	Malayan
Celt	Celtic	Mong	Mongolian
Chn	Chinese	Nor	Norwegian
Czech	Czech	Per	Persian
Dan	Danish	Pol	Polish
Du	Dutch	Port	Portuguese
Fin	Finnish	Rom	Romanian
Fr	French	Rus	Russian
Ger	German	Siam	Siamese
Hung	Hungarian	So. Slav	Southern Slavonic
Ice	Icelandic		
India	India	Sp	Spanish
Indian	American Indian	Swe	Swedish
It	Italian	Tib	Tibetan
		Tur	Turkish

a, Dan., Nor., Swe ... river
aan, Du ... at, on
abad, India, Per ... dwelling, town
abu, abou, Arab ... father
alf, elf, Swe ... river
alp, Ger ... mountain
alt, Ger ... old
alta,-o, It., Port., Sp ... high
altopiano, It ... plateau
amarillo, Sp ... yellow
arroyo, Sp. . brook, dry bed of stream
as, Dan, Nor., Swe ... hill, ridge
austral, Sp ... southern
baai, Du ... bay
bab, Arab ... gate, strait
bach, Ger ... brook, stream
backe, Swe ... hill
bad, Ger ... bath
bahía, Port., Sp ... bay, gulf
baia, It ... bay, gulf
baie, Fr ... bay, gulf
bajo, Sp ... low, lower
bakke, Dan., Nor ... hill
balkan, Tur ... mountain range
bana, Jap ... cape
batang, Mal ... river
belyy, belaya, Rus ... white
ben, Celt ... mountain, summit
bender, bandar, Arab., India ... market town, port
beni, bani, Arab ... sons of, tribe of
berg, Du., Ger., Nor., Swe., mountain, hill
bir, Arab ... well
birket, Arab ... pool
bjeli,-a, -o, So. Slav ... white
bjerg, bjaerg, Dan., Nor ... mountain
blanco, Sp ... white
blau, Ger ... blue
bleu, Fr ... blue
bogaz, boghaz, Tur ... strait
bois, Fr ... forest, wood
boloto, Rus ... marsh
bolshoy, bolshoye, Rus ... great
boreal, Sp ... northern
borg, Dan., Nor., Swe ... castle
borgo, It ... town
bosch, Du ... forest, wood
bouche, Fr ... river, mouth
bourg, Fr ... town, borough
bro, Dan., Nor., Swe ... bridge
brücke, brücken, Ger. . bridge, bridges
brun, Fr ... brown
bucht, Ger ... bay, bight
bugt, Dan., Nor., Swe ... bay, gulf
bujuk, buyuk, Tur ... great
burg, Du., Ger ... castle, town
burun, burnu, Tur ... cape
by, Dan., Nor., Swe ... town, village
cabeza, Sp ... summit
cabo, Port., Sp ... cape
cairn, carn, Celt ... rocky headland
campo, It., Port., Sp ... field
campos, Port. (Brazil) ... plains
cañon, Sp ... canyon
cap, Fr ... cape
capo, It ... cape
casa, It., Port., Sp ... house
castello, It., Port ... castle, fort
castillo, Sp ... castle, fort
catingas, Port. (Brazil) ... open brushlands
cayo, Sp ... rock, shoal, islet
central, Fr ... middle
cerro, Sp ... hill
chai, ciai, Tur ... river
champ, Fr ... field
chateau, Fr ... castle
cherniy, chernyaya, Rus ... black
chin, Chn ... market town
chott, shat, Arab ... salt river or lake
cidade, Port ... city
città, It ... town, city
ciudad, Sp ... town, city
col, Fr ... pass
colina, Sp ... hill
colorado, Sp ... red
cordillera, Sp ... mountain chain
costa, It., Port., Sp ... coast
côte, Fr ... coast
crkva, So. Slav ... church
crni, So. Slav ... black
cuchilla, Sp ... mountain range
daal, dal, Du ... valley
dagh, dag, Tur ... mountain
dake, take, Jap ... peak, ridge
dal, Dan., Du., Nor., Swe ... valley
dar, Arab ... land, country
darya, daria, Per ... river, sea
dasht, Per ... plain, desert
deccan, India ... south
deir, Arab ... convent
denis, -z, Tur ... sea, lake
désert, Fr ... desert
deserto, It ... desert
desierto, Sp ... desert
détroit, Fr ... strait
djebel, jebel, Arab ... mountain
dorf, Ger ... village
dorp, Du ... village
drift, Du., Ger ... current
duinen, Du ... dunes
dun, Celt ... fortified hill
dyk, Du ... dam, dyke
dzong, Tib ... fort, administrative capital
eau, Fr ... water
ecuador, Sp ... equator
eiland, Du ... island
elf, älf, Swe ... river
elv, Dan., Nor ... river
erg, Arab ... dune, region of dunes
eski, Tur ... old
est, Fr ... east
estado, Sp ... state
este, It., Port., Sp ... east
estrecho, Sp ... strait
étang, Fr ... pond, lake
état, Fr ... state
étroit, Fr ... narrow
feld, Ger ... field, plain
fels, Ger ... rock
festung, Ger ... fort
fiume, It ... river
fjäll, Swe ... mountain
fjärd, Swe ... bay, inlet
fjeld, Nor ... mountain, hill
fjord, Dan., Nor ... fiord, inlet
fjördur, Ice ... fiord, inlet
fleuve, Fr ... river
flod, Dan., Swe., ... river
fluss, Ger ... river
foce, It ... river mouth
fontein, Du ... a spring
fors, Swe ... waterfall, torrent
forst, Ger ... forest
fos, Dan., Nor ... waterfall
fuente, Sp ... spring, fountain
fuerte, Sp ... fort
furt, Ger ... ford
gamla, Swe ... old
gamle, Dan., Nor ... old
gat, Dan., Nor ... passage, channel
gebel, Arab ... mountain
gebergte, Du ... mountain range
gebiet, Ger ... district, territory
gebirge, Ger ... range, mountains
ghat, India ... mountain pass, river passage
gobi, Mong ... desert
gol, golu, Tur ... lake
golf, Du., Ger ... gulf, bay
golfe, Fr ... gulf, bay
golfo, It., Port., Sp ... gulf, bay
gong, India ... village
gora, Pol., Rus., So. Slav ... mountain
gornji, -a, -o, So. Slav ... upper
gorny, Pol ... upper
gorod, grad, Rus., So. Slav ... town
grand, grande, Fr ... large, great
grande, It., Port., Sp ... large, great
grod, gorod, Pol., Rus ... town
grön, Dan ... green
groot, Du ... great
gross, Ger ... great
guba, Rus ... bay, gulf
gunto, Jap ... archipelago
haf, Swe ... sea
hafen, Ger ... port, harbor
haff, Ger ... gulf, inland sea
hai, Chn ... sea, lake
hamn, Swe ... harbor
hamun, Per ... swampy lake, plain
haus, hausen, Ger ... house, houses
haut, Fr ... high, summit, upper
hav, Dan., Nor ... sea, ocean
havn, Dan., Nor ... harbor, port
havre, Fr ... harbor, port
haz, -a, Hung ... house, dwelling of
heim, Ger ... hamlet
hem, Swe ... hamlet
higashi, Jap ... east
hinterland, Ger ... back country
hissar, hisar, Tur ... castle, fort
ho, Chn ... river
hoch, Ger ... high
hoek, Du ... cape
hof, Ger ... court, farm house
höfn, Ice ... harbor
hoku, Jap ... north
holm, Dan., Nor., Swe ... island
hora, Czech ... mountain
horn, Ger ... peak
hoved, Dan., Nor ... cape, headland
hsien, Chn ... district, district capital
hügel, Ger ... hill
huk, Dan., Nor., Swe ... point
hus, Dan., Nor., Swe ... house
hwang, Chn ... yellow
ile, Fr ... island
ilha, Port ... island
indre, Dan., Nor ... inner
indsö, Dan., Nor ... lake
inférieur, Fr ... lower
insel, Ger ... island
insjö, Swe ... lake
irmak, Tur ... river
isla, Sp ... island
isola, It ... island
istmo, It., Sp ... isthmus
jebel, djebel, Arab ... mountain
jima, shima, Jap ... island
jökel, Nor ... glacier
joki, Fin ... river
jökull, Ice ... ice-covered mountain
kaap, Du ... cape
kafr, kefr, Arab ... village
kaikyo, Jap ... strait
kala, kalat, Arab., Per. ... castle, fortress, village
kale, Tur ... castle, fort
kang, Chn ... village
kap, Ger ... cape
kara, Tur ... black
kaupunki, Fin ... town, city
kebir, Arab ... great
kefr, kafr, Arab ... village
ken, Jap ... prefecture
kend, kand, Per ... village
khrebet, Rus ... mountain range
ki, Jap ... tree, forest
kil, cill, Celt ... church, cell
kio, kyo, Jap ... town, capital
kis, Hung ... little, small
klein, Du., Ger ... small
köbstad, Dan ... city
kol, Mong ... lake
kong, Chn ... river
kopf, Ger ... head, summit, peak
köping, Swe ... market, borough
kraal, Du ... native village
krasniy, krasnaya, Rus. ... beautiful, fair, red
kuchuk, Tur ... small
kuh, koh, Per ... mountain
kul, Tur ... lake
kum, qum, Tur ... desert
kuppe, Ger ... summit
küste, Ger ... coast
kyzyl, kizil, Tur ... red
laag, Du ... low
lac, Fr ... lake
lago, It., Sp ... lake
lâgoa, Port ... lagoon
laguna, It., Port., Sp ... lagoon, lake
lahti, Fin ... bay, gulf
län, Swe ... county
landsby, Dan., Nor ... village
lilla, Swe ... small
lille, Dan., Nor ... small
liman, Tur ... bay, port
llanos, Sp ... prairies, plains
loch, Celt ... lake, bay (Scotland)
lough, Celt ... lake, bay (Ireland)
maha, India ... great
malyy, malaya, Rus ... small
mar, Port., Sp ... sea
mare, It., Rom ... sea
mare, Rom ... great
mark, Ger ... boundary, limit
mato, Port ... jungle, copse
medio, Sp ... middle
meer, Du., Ger ... lake, sea
mer, Fr ... sea
mesa, Sp ... flat-topped mountain
midden, Du ... middle
mina, Port., Sp ... mine
mittel, Ger ... middle
mont, Fr ... mount, mountain
montagna, It ... mountain
montagne, Fr ... mountain
montaña, Sp ... mountain
monte, It., Port., Sp. mount, mountain
more, Rus., So. Slav ... sea
morro, Port., Sp ... hill
moyen, Fr ... middle
mühle, Ger ... mill
mund, munde, Ger ... river mouth
mündung, Ger ... river mouth
mura, Jap ... village
nada, Jap ... sea
nadi, India ... river, creek
naes, näs, Dan., Nor., Swe ... cape
nagar, nagon, India ... town, city
nagy, Hung ... large, great
naka, Jap ... middle
neder, Du ... low
nedre, Nor ... lower
negro, It., Port., Sp ... black
nejd, Arab ... highland
neu, Ger ... new
nez, Fr ... point, cape
nieder, Ger ... low, lower
nieuw, Du ... new
nizhniy, nizhnyaya, Rus ... lower
noir, Fr ... black
nong, Siam ... marsh, pond, lake
noord, Du ... north
nor, Tib ... lake
nord, Dan., Fr., Ger., It., Nor ... north
norr, norra, Swe ... north
norte, Port., Sp ... north
nos, Rus ... cape
nouvelle, Fr ... new
novi, -a, -o, So. Slav ... new
novo, Port ... new
novy, -e, -a, Czech ... new
novyy, novaya, Rus ... new
nowy, Pol ... new
nuevo, Sp ... new
nuovo, It ... new
ny, Dan., Swe ... new
nyasa, Bantu ... lake
o, Jap ... great, large
ö, Dan., Nor., Swe ... island
ober, Ger ... upper
occidental, Sp ... western
odde, Dan., Nor ... point, cape
oedjoeng, Mal ... cape
oeste, Port., Sp ... west
oost, Du ... east
op, Du ... on
oriental, Sp ... eastern
oro, Sp ... gold
ost, Ger., Swe ... east
öst, öster, östre, Dan., Nor., Swe. ... east, eastern
ostrog, Rus ... fort
ostrov, Rus ... island
ouest, Fr ... west
ozero, Rus ... lake
paa, Fin ... mountain
padang, Mal ... plain, field
pampas, Sp. (Argentina) .. grassy plains
para, Indian (Brazil) ... river
pas, Fr ... channel, strait, pass
paso, Sp ... mountain pass
passo, It., Port ... mountain pass
patam, India ... city, town
pequeño, Sp ... small
peresheyek, Rus ... isthmus
petit, petite, Fr ... small, little
pic, Fr ... mountain peak
piccolo, It ... small
pico, Port., Sp ... mountain peak
piedra, Sp ... stone, rock
pik, Rus ... peak
planalto, Port ... plateau
plata, Sp ... silver
playa, Sp ... shore, beach
po, Chn ... lake
pointe, Fr ... point
polder, Du., Ger ... reclaimed marsh
polje, So. Slav ... field
poluostrov, Rus ... peninsula
pont, Fr ... bridge
ponta, Port ... point, headland
ponte, It., Port ... bridge
pore, pur, India ... city, town
porto, It., Port ... port, harbor
prado, Sp ... field, meadow
presqu'ile, Fr ... peninsula
proliv, Rus ... strait
pu, Chn ... commercial village
pueblo, Sp ... town, village
puerto, Sp ... port, harbor
punkt, Ger ... point
punt, Du ... point
punta, It., Sp ... point
pur, pura, India ... city, town
puy, Fr ... peak
rann, India ... wasteland
ras, Arab ... cape, summit
reka, Rus., So. Slav ... river
retto, Jap ... archipelago
ria, Sp ... river mouth
ribeira, -ão, Port ... stream, river
rio, It., Port ... river
río, Sp ... river
rivière, Fr ... river
roca, Sp ... rock
rouge, Fr ... red
rud, Per ... river
saari, Fin ... island
sable, Fr ... sand
sahra, Arab ... desert
sal, Sp ... salt
samar, Mong ... path, route
san, Chn., Jap., Kor ... mountain, hill
san, santa, santo, It., Port., Sp ... saint
são, Port ... saint
sat, satu, Rom ... village
schloss, Ger ... castle, fort
sebkha, Arab ... salt marsh
see, Ger ... lake, sea
sehir, shehr, Tur ... town
selvas, Port. (Brazil) ... tropical rain forests
serra, It., Port ... pass, mountain ridge
serranía, Sp ... mountain ridge
seto, Jap ... strait, channel
severnaya, Rus ... north
shahr, shehr, Per ... town
shan, Chn ... range, mountain, hill
shat, chott, Arab ... salt river or lake
shima, sima, Jap ... island
shimo, Jap ... lower
shiu, Chn., Jap ... province
shoto, Jap ... archipelago
si, Chn ... west, western
sierra, Sp ... mountain range
sint, Du ... saint
sjö, Nor., Swe ... lake, sea
sö, Dan., Nor ... lake, sea
söder, Swe ... south
soengai, sungei, Mal ... river
sopka, Rus ... extinct volcano
source, Fr ... spring
spitze, Ger ... summit, point
sredniy, srednyaya, Rus ... middle
staat, Ger ... state
stad, Dan., Du., Nor., Swe. . city, town
stadt, Ger ... city, town
stari, -a, -o, So. Slav ... old
stary, Czech., Pol ... old
staryy, staraya, Rus ... old
stato, It ... state
sten, Dan., Nor., Swe ... stone
step, Rus ... treeless plain, steppe
stor, Dan., Nor., Swe ... great, large
straat, Du ... strait
strand, Dan., Du., Ger., Nor., Swe. ... shore, beach
stretto, It ... strait
strom, Ger ... stream
ström, Dan., Nor., Swe ... river
stroom, Du ... stream, river
su, suyu, Tur ... water, river
sud, Fr., Sp ... south
süd, Ger ... south
sul, Port ... sound
sund, Dan., Nor., Swe ... sound
supérieure, Fr ... upper
sur, Fr ... on
sur, Sp ... south
syd, Dan., Nor., Swe ... south
tafelland, Du., Ger. plateau, tableland
tagh, Mong., Tur ... mountain
tai, Jap ... large, great
take, dake, Jap ... peak, ridge
tandjoeng, tanjong, Mal ... cape
tao, -u, Chn ... island
targ, targu, Rom ... market, town
tash, Per., Tur ... rock, stone
tau, Tur ... mountain range
tell, tel, Arab ... hill
terra, It ... land
terre, Fr ... earth, land
thal, Ger ... valley
tierra, Sp ... earth, land
torp, Swe ... village, cottage
torre, It., Port., Sp ... tower
tsi, Chn ... village, borough
tsu, Jap ... port
tundra, Rus ... marshy arctic plains
tung, Chn ... east, eastern
turn, turnu, Rom ... tower
tuz, Tur ... salt
udd, udde, Swe ... cape
ufer, Ger ... beach, shore, river bank
uj, Hung ... new
ulan, Mong ... red
umi, Jap ... sea, gulf
unter, Ger ... lower
ura, Jap ... bay, shore, creek
ust, Rus ... river mouth
vall, Swe ... coast
valle, Port., Sp ... valley
vallée, Fr ... valley
var, Hung ... fortress
varos, Hung., So. Slav ... town
vecchio, It ... old
veld, Du ... open plain, field
velho, Port ... old
velikiy, Rus., So. Slav ... great
verde, It., Port., Sp ... green
verkhniy, verkhnyaya, Rus. ... upper, higher
vert, Fr ... green
ves, Czech ... village
vest, Dan., Nor., Swe ... west
viejo, Sp ... old
vieux, Fr ... old
vik, Swe ... bay
villa, Port., Sp ... small town
villar, Sp ... village, hamlet
ville, Fr ... town, city
vishni, visni, Rus ... high
vostok, Rus ... east
wadi, wad, Arab. . intermittent stream
wald, Ger ... forest, woodland
wan, Chn., Jap ... bay, gulf
weiler, Ger ... hamlet, village
weiss, Ger ... white
westersch, Du ... western
wüste, Ger ... desert
yama, Jap ... mountain
yeni, Tur ... new
yokara, Tur ... upper
yug, Rus ... south
zaki, saki, Jap ... cape
zaliv, Rus ... bay, gulf
zapad, Rus ... west
zee, Du ... sea
zemlya, Rus ... land
zuid, Du ... south

GLOSSARY OF MAP TERMINOLOGY

A

Altitude. The height of an object or elevation above a given level.

Analemma. A graphic scale, usually drawn in the form of a figure 8, showing the overhead sun's latitude for every day in the year as well as the difference between sun time and mean time.

Antarctic Circle. The geographic parallel of 66° 33′ S., enclosing the area within which the sun is continuously above the horizon on December 22, and below the horizon on June 21.

Antipodes. Two places on the surface of the globe diametrically opposite to each other, i.e., North Pole and South Pole; England and the Antipodes Is.

Antoeci. Two places on the same meridian and in corresponding latitudes north and south of the Equator.

Aphelion. That point in the earth's orbit which is most distant from the sun, 94,560,000 miles.

Archipelago. A group of islands more or less adjacent to each other and arranged in groups covering portions of the sea.

Arctic Circle. The geographic parallel of 66° 33′ N., enclosing the area within which the sun is continuously above the horizon on June 21, and below the horizon on December 22.

Atlas. A bound collection of maps. First used in this sense by Mercator in the 16th century.

Atmosphere. The ocean of air surrounding the earth.

Atoll. A coral island in the form of a ring, more or less continuous, around an interior lagoon.

Autumnal Equinox. The time when the overhead sun crosses the Equator on its apparent migration from north to south, or about September 23, and the length of day and night is approximately the same in all latitudes.

Axis. The straight line passing through the center of the earth, about which the earth rotates.

Azimuth. A great circle direction, or the angle measured clockwise between any meridian and an intersecting great circle.

Azimuthal Projection. A map projection on which the directions of all lines radiating from a central point or pole are the same as the directions of the corresponding lines on the sphere. When centered on one of the poles, sometimes called a "polar projection."

B

Bank. An elevation of the ocean bottom above which the water is relatively shallow but sufficient for navigation, yet without any island rising from it to the surface.

Bar. A ridge of sand or other substance extending across the mouth of a river or harbor, and which may obstruct navigation.

Barrier Ice. The edge of a great glacier which enters the sea but remains attached to the land.

Basin. The area drained by a river and its tributaries.

Bathymetric Chart. A topographic map of the bed of the ocean showing ocean depth contours.

Bay. A penetration of the sea into the coast, and it is usually very much wider in the middle than at the entrance.

Bayou. A sluggish watercourse, usually the outlet of a lake or of a river through its delta.

Beach. The area between high and low water forming the margin of sea or lake.

Bight. A comparatively slight indentation in the coastline between distant headlands.

Bluff. A cliff or headland with an almost perpendicular face.

Butte. A conspicuous, isolated hill or mountain.

C

Calms, Belt of. A zone on either side of the Trade Winds where calms of long duration prevail.

Canal. An artificial watercourse. Sometimes applied to a natural waterway which has the appearance of an artificial canal.

Canyon. A deep gorge or ravine through which a river flows.

Cape. A point of land projecting into a body of water.

Cartography. The art or science of making maps.

Celestial Globe. A sphere on whose surface is drawn a map or representation of the heavens.

Central Meridian. The vertical meridian of a map projection around which the map is centered.

Chart. A map for use in marine or aeronautical navigation.

Civil Time. Solar time in a day that is considered as beginning at midnight. It may be counted in two series of 12 hours each (A.M. and P.M.) or in a single series of 24 hours beginning at midnight. Civil Time may be either true solar time or mean solar time.

Cliff. Land projecting nearly vertically from the water or from the surrounding land.

Climate. The aggregate weather conditions of a given region over a long period of time.

Co-Latitude. The difference between the latitude of a place and 90°, or its distance in degrees from one of the poles of the earth.

Conformal Projection. A map projection on which all small or elementary figures upon the surface of the earth retain true shape, the meridians and parallels being at right angles to one another.

Conic Projection. A map projection which can be imagined as drawn on the surface of a cone. The meridians appear as straight lines along which the parallels, as concentric circles, may be spaced in such a way as to give some desired quality, such as conformality or equal area.

Continent. One of the main continuous bodies of land on the earth's surface. The number of continents considered to exist varies with usage from five to seven, i.e., America (North and South); Eurasia (Europe and Asia); Africa; Antarctica; and Australia.

Continental Divide. The height of land which separates the streams flowing into one ocean from those flowing into another.

Continental Shelf. The zone of the continental margin extending from the shore line to the depth, usually about 100 fathoms or 200 meters, where there is a marked or rather steep descent toward the ocean depth.

Contour Line. A line drawn on a map to indicate points of the same height or depth.

Coordinates, Geographical. The intersecting lines of latitude and longitude which determine the geographical position of any given place.

Coulee. A dry ravine or gulch originally formed by running water.

Crater. The bowl-shaped cavity at the summit or on the side of a volcano.

Cultural Feature. Any man-made feature of the earth's surface shown on a map.

Cyclone. Technically, an atmospheric movement in which the wind blows spirally around and in toward a center. Popularly used as a synonym for tornado.

Cylindrical Projection. A map projection produced by projecting the geographic meridians and parallels onto a cylinder which is tangent to the surface of a sphere, and then developing the cylinder into a plane.

D

Day. A measure of time based upon one complete rotation of the earth. A solar day is measured from a transit of the sun across a given meridian to its next successive transit across the same meridian. A civil day is a solar day beginning at midnight.

Degree. A unit of measurement equal to 1/360 of a circle. A degree of latitude on the earth's surface is roughly equivalent to 69 statute miles. A degree of longitude varies in length but is always equivalent to about 4 minutes of time.

Delta. The tract of land formed by the deposit of silt at the mouth of a river.

Desert. A region of considerable extent which is almost destitute of vegetation, chiefly because of insufficient moisture.

Doldrums. The equatorial belt of calms and variable winds.

Downs. Certain hilly districts in southern England underlain by chalk and hence unforested.

Downstream. The direction in which a stream is flowing.

E

Eastern Hemisphere. Usually considered in cartography to be half of the earth extending from pole to pole between 20° W. and 160° E., including Old World continents of Eurasia, Africa, and Australia.

Ecliptic. A great circle sometimes shown on a map or globe and representing the apparent annual path of the sun across the surface of the earth.

Elevation. The vertical distance of a point above or below a reference surface, usually mean sea level.

Equal Area Projection. A map projection on which a constant ratio of areas is preserved; that is, any given part of the map bears the same relation to the area on the sphere which it represents, as the whole map bears to the entire area represented.

Equator. The great circle around the earth equidistant from the poles.

Equatorial Current. The westward drift of surface water on either side of the Equator in the Trade Wind belts.

Estuary. The coastal section of a river which is to a greater or lesser extent invaded by the sea and subject to tidal influence.

F

Fathom. A unit of measurement used for soundings, equal to 1.83 meters or 6 feet.

Ferrell's Law. Moving bodies on the earth's surface (such as air and water) are deflected to the right in the Northern Hemisphere and to the left in the Southern Hemisphere, as a result of the earth's rotation.

Firth. A long arm of the sea, partially landlocked.

Fjord. A long, narrow arm of the sea between high lands.

G

Geodesy. The investigation of scientific questions connected with the shape and dimensions of the earth.

Geographical Mile. See Nautical Mile.

Geographic Center. That point on which any area would balance if it were a plate of uniform thickness. The geographic center of the conterminous United States is at latitude 39° 50', longitude 98° 35', in the eastern part of Smith County, Kansas.

Geography. The scientific description and explanation of the earth's regions.

Glacier. A field or body of land-formed ice moving slowly down a mountainside or valley.

Global Geography. Those aspects of the science of geography which are directly related to the spherical shape of the earth.

Globe. A spherical map of the earth or heavens.

Globe Gore. A section of a map so shaped that it may be mounted on a sphere without appreciable stretching or shrinking.

Gnomonic Projection. A perspective map projection on a plane tangent to the surface of a sphere, having the point of projection at the center of the sphere. It is the only map projection on which all great circles represented are straight lines. Chiefly used for navigational charts.

Great Circle. The line of intersection of the surface of a sphere and any plane which passes through the center of the sphere.

Great Circle Direction. The great circle direction of point A from point B is the angle, measured clockwise from true North, formed by the meridian of point B and the great circle passing through points A and B. See Azimuth.

Great Circle Distance. The distance between any two points, measured either in degrees or miles, along the great circle connecting them.

Greenwich Civil Time. Mean solar time for the Greenwich Meridian, counted from midnight.

Greenwich Meridian. See Prime Meridian.

Gulf. A relatively large portion of the sea partly enclosed by land.

Gulf Stream. The warm current which flows out of the Gulf of Mexico through the Straits of Florida and northward through the Atlantic Ocean until it merges with the West Wind Drift.

H

Hachures. Lines used in shading elevations on a map to outline them and to indicate slope.

Hemisphere. Any half of the earth's surface. See Northern, Southern, Eastern, Western, Land, and Water Hemisphere.

Hill. A natural elevation of the earth's surface above surroundings but lower than a mountain.

Horizon. The line at which the earth and sky appear to meet.

Horizon Ring. A graduated ring fitted to a globe in such manner that its plane contains the center of the globe, and it can be adjusted into the plane of the horizon for a given point.

Horse Latitudes. Zones of high atmospheric pressure with calms and variable breezes, which border the polar edges of the Trade Wind areas.

Hurricane. A violent and destructive storm of the cyclonic type, originating in the Tropics.

Hydrography. The science of measuring and studying oceans, seas, rivers, and other waters, with their marginal land areas, especially for the purpose of aiding navigation. As a map feature, the pattern of rivers, lakes, seas, and oceans shown on a map.

Hypsometric Map. A map colored to show variations in elevation above sea level.

I

Iceberg. A mass of land ice which has broken away from its parent formation on the coast, and either floats in the sea or is stranded on a shoal.

Ice Cap. An ice sheet of vast extent covering the topographic features of a continental land mass.

Inch. A unit of lineal measurement; 1/12 of a foot, or 0.0254 of a meter.

Inclination of the Earth. The tilt of the earth's axis in relation to the plane of the earth's orbit. The angle of inclination is $23\frac{1}{2}°$ from vertical.

Insolation. The amount of solar radiation or heat received at a given place over a given period of time.

International Date Line. The line extending from pole to pole along the 180th meridian, with local variations, where each new calendar day begins with the passing of the midnight hour. Travelers crossing the line going west must advance their calendar one day, while those going east must retard the calendar one day.

Interrupted Map Projection. A projection in which the pattern of meridians and parallels is interrupted or broken so that certain areas may be centered upon different central meridians. Goode's Interrupted Homolosine Projection is probably the best known example.

Island. A body of land surrounded by water and smaller than a continent.

Isobar. A line on a map connecting places on the earth at which the barometric pressure is the same at a given time or on the average for a given period of time.

Isogonic Lines. Lines drawn so as to connect points on the earth's surface where the magnetic variation is the same.

Isohyet. A line drawn on a map to connect points having equal rainfall during a given period.

Isopleth. Any line on a map connecting points of equal density or value of distribution. A generic term for such lines as isotherms, isobars, isohyets, etc.

Isotherm. A line on a map connecting places on the earth at which the mean temperature is the same.

Isthmus. A narrow strip of land with water on both sides connecting two larger bodies of land.

K

Kilometer. A unit of length; 1,000 meters; 3,280.84 feet; approximately ⅝ of a mile.

L

Lake. A large sheet of water surrounded by land on all sides.

Land Forms. The shapes into which the earth's surface is sculptured by natural forces.

Land Hemisphere. That half of the earth, centered near Nantes, France, which includes the greatest possible land area.

Latitude. The angular distance in degrees of a point on the earth north or south of the Equator.

Lithosphere. The solid mass of the earth.

Longitude. The angular distance (degrees) of a place east or west of the Prime Meridian.

M

Magnetic Declination. From any given place, the angle of magnetic North from true North.

Magnetic Poles. The two locations representing the poles of unlike magnetism belonging to the earth as a magnetized body. The **North Magnetic Pole** is currently located at approximately 73°N. Lat. and 100°W. Long. on Prince of Wales Island. The **South Magnetic Pole** is currently located at approximately 71°S. Lat. and 149°E. Long. in Antarctica.

Map. A graphic representation, on a plane, of certain selected features of a part or the whole of the earth's surface.

Map Grid. The framework of parallels and meridians by means of which map features are located.

Map Projection. A network of lines representing parallels of latitude and meridians of longitude, derived by geometrical construction or mathematical analysis.

Map Scale. The relationship which exists between a distance on a map and the corresponding distance on the earth. It may be expressed as an equivalence, one inch equals 16 statute miles; as a fraction or ratio, 1:1,000,000; or as a bar graph subdivided to show the distance which each of its parts represents on the earth.

Mean Solar Time. Also called Mean Time. Time measured by the daily motion of a fictitious body called the "mean sun." Since the apparent sun travels in the ecliptic with a variable motion, it cannot be used to measure time, and the "mean sun," supposedly moving uniformly in the Celestial Equator, is used.

Mercator Projection. A conformal projection on which the meridians and parallels are shown as parallel straight lines at right angles to one another, the divisions of latitude being expanded north and south of the Equator in the same proportion as the divisions of longitude have been lengthened by projection. On this projection a line of constant bearing, or rhumb line, is represented by a straight line.

Meridian. A great circle on the earth's surface which passes through the terrestrial poles.

Meridian Ring. A graduated ring fitted to a globe in such manner that its plane contains the poles of the globe, and it can be adjusted into the plane of any given meridian of the globe.

Mesa. A flat-topped mountain or hill, usually bounded on at least one side by a steep cliff.

Meter. A unit of length equivalent in the United States to exactly 39.37 inches.

Mile. A unit of distance. See Nautical Mile and Statute Mile.

Monsoon. A periodic, seasonal movement of air from land to water and vice versa.

Moraine. A mound or ridge of unstratified rock material deposited by a glacier.

Mountain. A natural elevation of the earth's surface rising to a great height.

N

Nadir. The point in the heavens diametrically opposite to the Zenith, or the point directly under the observer.

Nautical Mile. A unit commonly used for measuring

distances at sea; the length of a minute of latitude; 1,853 meters or 6,080 feet. Also called Geographical Mile.

Noon. The instant when the sun's rays fall vertically on any given meridian.

North. The direction along any meridian toward the North Pole.

Northern Hemisphere. That half of the earth north of the Equator.

North Pole. The end or pole of the earth's axis pointing toward the star Polaris.

O

Ocean. A vast expanse of salt water bordered by the continents. The oceans are usually considered to be four in number: Pacific, Atlantic, Indian, and Arctic.

Ocean Basin. A large submarine cavity of more or less round or oval form.

Ocean Current. A specific portion of any ocean moving in a definite direction. It may also be called a stream or drift.

Ocean Deep. Smaller areas, within the great ocean basins, whose depths exceed 16,500 feet.

Oceanography. The science of the oceans, their forms, physical features, and phenomena.

Ocean Trench. A long, narrow, oceanic depression with relatively steep sides.

Orbit of the Earth. The curve which the earth describes in the heavens as it revolves around the sun.

P

Parallels. Small circles on the earth's surface, or lines on a map, perpendicular to the axis of the earth and marking latitude north or south of the Equator.

Peninsula. A piece of land nearly surrounded by water.

Perihelion. That point in the orbit of the earth when the earth is nearest to the sun, 91,450,000 miles.

Perioeci. A point in the same latitude but with a difference in longitude of 180°.

Physical Feature. Usually considered to be any natural feature of the earth's surface shown on a map.

Physical Map. A map in which natural regions and physical features are emphasized by the use of different colors.

Plane of the Ecliptic. The plane of the earth's orbit.

Planet. A celestial body revolving around the sun in a nearly circular orbit, such as the earth.

Plateau. An elevated area of comparatively flat or level land.

Polar Ice Pack. The entire area of thick and closely packed polar ice, more than one year old.

Polar Projection. One in which the meridians appear as straight lines radiating from the pole and the parallels of latitude as concentric circles with the pole as center.

Political Map. A map in which political divisions and boundaries are emphasized through the use of color.

Pond. A sheet of shallow water, generally without outlet, located in the interior of the land.

Prime Meridian. The meridian on the earth's surface from which longitude is measured, generally the meridian of Greenwich, England, on modern maps.

Projection. Any method of delineating on a plane surface the whole or a part of the surface of the earth, including parallels of latitude and meridians of longitude. See Azimuthal, Conformal, Conic, Cylindrical, Equal Area, Gnomonic, Mercator, and Polar Projections.

R

Reef. A rocky or coral elevation in the ocean bottom, which may be uncovered at times.

Representative Fraction. The fraction with unity or one as the numerator, denoting the scale of a map. Abbreviated as "R.F."

Revolution. The movement of the earth around the sun. One complete revolution of the earth requires one year, or 365 days, 5 hours, 48 minutes, 46 seconds.

Rhumb Line. A line which makes equal angles with all the meridians it crosses.

Right Bank. The right bank of a river is the one on the right hand when facing downstream; that of a channel, the same, when facing in the direction of the ebb tide.

Roaring Forties. A term used by sailors to describe the stormy regions between 40° and 50° from the Equator in both the northern and southern oceans.

Rotation. The movement of the earth around its axis. One complete rotation determines the length of one day.

S

Savanna. Originally, an extensive treeless plain, but now more frequently used to mean a tropical landscape of scattered trees and extensive grasslands.

Sea. A mass of salt water more or less confined by portions of the continent or by chains of islands, and forming a basin distinct from the great masses of water.

Sea Level. The level of the surface of the sea considered at any moment at a given place.

Small Circle. Any circle on a sphere smaller than a great circle. Thus, all parallels on a globe or map except the Equator.

Solar System. The sun and all celestial bodies revolving around it, together with their satellites.

Solar Time. Also called true solar time. Time measured by the apparent daily motion of the sun.

Solstice. The time at which the overhead sun is at its greatest distance from the Equator. In the Northern Hemisphere the summer solstice occurs about June 21, and the winter solstice about December 22.

South. The direction along any meridian toward the South Pole.

Southern Hemisphere. That half of the earth south of the Equator.

South Pole. The opposite end of the earth's axis from the North Pole.

Standard Parallel. A parallel of latitude which is used as a control line in the computation of a map projection, and is therefore true to scale.

Statute Mile. A unit of distance generally used in measurements on land, and equal to 5,280 feet.

Steppe. The grassy plains of European and Asiatic Russia.

Strait. A relatively narrow waterway between two larger bodies of water.

T

Temperate Zones. The two belts or zones of the earth lying between the Tropics and the Polar Circles.

Time. The measurable aspect of duration based upon the happening of periodic events, such as: the rotation of the earth (day), the revolution of the moon around the earth (month), and the revolution of the earth around the sun (year). See Civil Time, Greenwich Civil Time, Mean Solar Time, Solar Time.

Time Zone. A belt or zone, extending from north to south across a country, which is given a designated time by law. The United States has four standard time zones, namely: Eastern, Central, Mountain, and Pacific.

Topography. The features of the actual surface of the earth, considered collectively as to form. A single feature, such as a mountain or valley, is called a topographic feature.

Torrid Zone. A term formerly used to describe the belt or zone of the earth's surface bounded by the Tropic of Cancer and the Tropic of Capricorn. Better geographical form today is "Tropical Zone."

Trade Winds. The regular easterly winds which prevail over the oceans on either side of the Equator to about 30° north and south latitudes.

Transverse Projection. A map projection which is turned 90° from its usual orientation, and consequently is centered upon some other great circle than a meridian.

Triangulation. The measurement and calculation of a system of triangles connecting stations covering a particular region, for the purpose of fixing and plotting the positions of same on a chart or map.

Tropic. A line on a map or globe, usually broken or dotted, marking the limit reached by the overhead or vertical sun in its apparent annual migration. The northern line is called the **Tropic of Cancer**, and the southern line the **Tropic of Capricorn**. Both are about $23\frac{1}{2}°$ from the Equator.

Trough. A long and broad depression in the ocean bottom, with gently sloping sides.

Tundra. The marshy, treeless plains of northern Asia and northern North America.

Twilight. The periods of partial daylight after sunset and before sunrise, when light from the sun is reflected from the atmosphere overhead.

Typhoon. A violent and destructive storm similar to a hurricane, that occurs in the western Pacific Ocean.

V

Vernal Equinox. The date when the overhead sun crosses the Equator on its apparent migration from south to north, or about March 21, and the length of day and night is approximately the same in all latitudes.

Volcano. A more or less conical hill or mountain from which, when active, steam, gasses, ashes, or molten rocks are ejected.

W

Water Hemisphere. That half of the earth centered near New Zealand which includes the greatest possible water area.

Westerlies. The prevailing winds of the middle latitudes, that is, between 30° and 60° in north and south latitudes.

Western Hemisphere. Usually considered in cartography to be that half of the earth extending from pole to pole between 160° E. and 20° W., thus including the Americas and Greenland.

West Wind Drift. A general term applied to the eastward movement of oceanic water under the influence of the westerly winds.

Y

Year. An interval of time based upon the revolution of the earth in its orbit around the sun. It is equal to 365.24220 mean solar days.

Z

Zenith. The point in the celestial sphere directly over a given point on the earth.

236

Explanation of the Index and Abbreviations

This universal index includes in a single alphabetical list all important names that appear on the reference maps. For ease in index usage and interpretation, two kinds of type are used to distinguish political names from those of physical features and points of interest. All political names (cities, towns, counties, districts, states, provinces, countries, etc.) are set in roman type. Physical features, and points of interest (rivers, mountains, lakes, bays, hills, straits, islands, national parks, etc.) are set in *italic* type. The supplementary descriptive information with each index entry varies with its political or physical classification.

POLITICAL NAMES

The more important names and political divisions shown on the maps are listed in the index. Each place name is followed by its location; the population figure, when available; the map index key; and the page number of the map.

County and State locations are given for all places in the United States. Province and country locations are given for all places in Canada. All other place name entries show only country locations.

The index reference key, always a letter and figure combination, and the map page are the last items in each entry. Because some places are shown on both a main map and an inset map, more than one index key may be given for a single map page. Reference also may be made to more than a single map. In each case, however, the index key *letter and figure* precede the map *page number* to which reference is made. A lower case key letter indicates reference to an inset map which has been keyed separately.

All major and minor political divisions are followed by both a descriptive term (co., dist., reg., prov., dept., state, etc.), indicating political status, and by the country in which they are located. U.S. counties are listed with State locations; all others are given with country references.

POPULATION FIGURES

The populations in the index are based upon the latest available complete census figures and official estimates for each country. In some cases these populations may differ from those for the same places appearing in the tables elsewhere in the atlas. In these tables, all the populations refer to a single date, to facilitate comparisons. In some instances this date may be more recent than the last complete census used for the populations in the index.

For some larger cities a second population figure is given accompanied by a star (★). The second figure indicates the population of the city's entire metropolitan area including suburbs, as: Chicago, 3,550,404 (★6,517,600). A triangular symbol (▲) denotes a population figure for an *entire* township, district, or other minor civil division.

PHYSICAL NAMES AND POINTS OF INTEREST

The more important physical names and points of interest that are shown on the maps are listed in the index. Each entry is followed by a descriptive term (*bay, hill, range, riv., tombs, nat. park, mtn., isl.*, etc.), to indicate its nature.

Country locations are given for each name, except for features entirely within States of the United States or provinces of Canada, in which case these divisions are also given.

Some names are included in the index that were omitted from the maps because of scale size or lack of space. These entries may be identified by an asterisk (*) and reference is given to the approximate location of the place.

A long name may appear on the map in a shortened form, with the full name given in the index. The part of the name not on the map then appears in brackets, thus: Laval [des Rapides].

In the index, when more than one name with the same spelling is shown, including both political and physical names, the order of precedence is as follows: *first*, place names, *second*, political divisions, and *third*, physical features.

ABBREVIATIONS

admin	administered
Afg	Afghanistan
Afr	Africa
Ala	Alabama
Alb	Albania
Alg	Algeria
Alsk	Alaska
Alta	Alberta
Am	American
And	Andorra
Ang	Angola
Ant	Antarctica
Arc	Arctic
arch	archipelago
Arg	Argentina
Ariz	Arizona
Ark	Arkansas
Atl. O	Atlantic Ocean
Aus	Austria
Austl	Australia, Australian
auton	autonomous
Ba. Is	Bahama Islands
Barb	Barbados
Bas	Basutoland
B.C	British Columbia
Bech	Bechuanaland
Bel	Belgium, Belgian
Bhu	Bhutan
Bis. Arch	Bismarck Archipelago
Bol	Bolivia
Br	British
Braz	Brazil
Br. Gu	British Guiana
Br. Hond	British Honduras
Bul	Bulgaria
Bur	Burma
Calif	California
Cam	Cameroon
Camb	Cambodia
Can	Canada
Can. Is	Canary Islands
Cen. Afr. Rep	Central African Republic
Cen. Am	Central America
Cey	Ceylon
C.H	Court House
chan	channel
co	county
Col	Colombia
Colo	Colorado
Con. B. Congo; Capital: Brazzaville
Con. L. Congo, The; Capital: Léopoldville
Conn	Connecticut
cont	continent
C.R	Costa Rica
C.V. Is	Cape Verde Islands
Cyp	Cyprus
C.Z	Canal Zone
Czech	Czechoslovakia
Dah	Dahomey
Dan	Danish
D.C	District of Columbia
Del	Delaware
Den	Denmark
dept	department
dep	dependency, dependencies
des	desert
dist	district
div	division
Dom. Rep	Dominican Republic
Ec	Ecuador
Eg	Egypt
Eng	England
est	estuary
Eth	Ethiopia
Eur	Europe
Falk. Is	Falkland Islands
Fed	Federation
Fin	Finland
Fla	Florida
Fr	France, French
Fr. Gu	French Guiana
Fr. Som	French Somaliland
Ga	Georgia
Gam	Gambia
Ger	Germany
Gib	Gibraltar
Grc	Greece
Grnld	Greenland
Guad	Guadeloupe
Guat	Guatemala
Hai	Haiti
Haw	Hawaii
hbr	harbor
Hond	Honduras
Hung	Hungary
I	island
I.C	Ivory Coast
Ice	Iceland
Ill	Illinois
incl	includes, including
Ind	Indiana
Indon	Indonesia
Indian res	Indian reservation
I. of Man	Isle of Man
Ire	Ireland
is	islands
isl	island
Isr	Israel
isth	isthmus
It	Italy
Jam	Jamaica
Jap	Japan
Kans	Kansas
Ken	Kenya
Kor	Korea
Kuw	Kuwait
Ky	Kentucky
La	Louisiana
Leb	Lebanon
Le. Is	Leeward Islands
Lib	Liberia
Liech	Liechtenstein
Lux	Luxembourg
Mala	Malaysia
Malag	Malagasy Republic
Man	Manitoba
Mart	Martinique
Mass	Massachusetts
Maur	Mauritania
Md	Maryland
Medit	Mediterranean
Mex	Mexico
Mich	Michigan
Minn	Minnesota
Miss	Mississippi
Mo	Missouri
Mong	Mongolia
Mont	Montana
Mor	Morocco
Moz	Mozambique
mtn	mount, mountain
mts	mountains
mun	municipality
Mus. & Om	Muscat & Oman
N.A	North America
nat. mon	national monument
nat. park	national park
N.B	New Brunswick
N.C	North Carolina
N. Cal	New Caledonia
N. Dak	North Dakota
Nebr	Nebraska
Nep	Nepal
Neth	Netherlands
Neth. W. I	Netherlands West Indies
Nev	Nevada
Newf	Newfoundland
New Hebr	New Hebrides
N. Gui	New Guinea Territory
N.H	New Hampshire
Nic	Nicaragua
Nig	Nigeria
N. Ire	Northern Ireland
N.J	New Jersey
N. Mex	New Mexico
Nor	Norway, Norwegian
N. Rh	Northern Rhodesia
N.S	Nova Scotia
N.W. Ter	Northwest Territory
N.Y	New York
Nya	Nyasaland
N.Z	New Zealand
occ	occupied area
Okla	Oklahoma
Ont	Ontario
Oreg	Oregon
Pa	Pennsylvania
Pac. O	Pacific Ocean
Pak	Pakistan
Pan	Panama
Pap	Papua
Par	Paraguay
par	parish
P.E.I	Prince Edward Island
pen	peninsula
Peru	Peru
Phil	Philippines
plat	plateau
Pol	Poland
pol. dist	political district
pop	population
Port	Portugal, Portuguese
Port. Gui	Portuguese Guinea
Port. Timor	Portuguese Timor
poss	possession
P.R	Puerto Rico
pref	prefecture
prot	protectorate
prov	province
pt	point
Que	Quebec
reg	region, regions
rep	republic
res	reservoir
R.I	Rhode Island
riv	river
Rom	Romania
S.A	South America
S. Afr	South Africa
Sal	El Salvador
Sam	Samoa
Sask	Saskatchewan
Sau. Ar	Saudi Arabia
S.C	South Carolina
Scot	Scotland
S. Dak	South Dakota
Sen	Senegal
S.L	Sierra Leone
Sol. Is	Solomon Islands
Som	Somali Republic
Sov. Un	Soviet Union
Sp	Spain
S. Rh	Southern Rhodesia
St., Ste	Saint, Sainte
Sud	Sudan
Sur	Surinam
S.W. Afr	South-West Africa
Swaz	Swaziland
Swe	Sweden
Switz	Switzerland
Syr	Syria
Tan	Tanganyika
Tenn	Tennessee
ter	territory
Tex	Texas
Thai	Thailand
Trin	Trinidad & Tobago
Tr. Coast	Trucial Coast
trust	trusteeship
Tun	Tunisia
Tur	Turkey
U.A.R	United Arab Republic
Ug	Uganda
U.K	United Kingdom
Ur	Uruguay
U.S	United States
Va	Virginia
val	valley
Ven	Venezuela
Viet	Vietnam
Vir. Is	Virgin Islands
vol	volcano
Vt	Vermont
Wash	Washington
W.I	West Indies
Win. Is	Windward Islands
Wis	Wisconsin
W. Sam	Western Samoa
W. Va	West Virginia
Wyo	Wyoming
Yugo	Yugoslavia
Zan	Zanzibar

Index to Political-Physical Maps

A

Aachen, Ger., 169,800 (★430,000) D6 15, C3 16
Aalen, Ger., 31,800 E5 17
Aalsmeer, Neth., 5,400 B4 15
Aalst, Bel., 45,092 D4 15
Aalten, Neth., 7,200 C6 15
Äänekoski, Fin., 7,200 F11 25
Aaqraba, Syr. A8 32
Aarau, Switz., 17,045 (★34,815) B5 19
Aare, riv., Switz B5 19
Aargau, canton, Switz., 360,940 B5 19
Aaronsburg, Centre, Pa., 420 E7 114
Aarschot, Bel., 12,123 D4 15
Aba, Con. L., 3,719 A5 48
Aba, Nig., 57,787 E6 45
Abacaxis, riv., Braz C3 59
Ābādān, Iran, 226,103 F4 41
Abādeh, Iran, 7,448 F6 41
Abadla, Alg C4 44
Abaeté, Braz., 7,988 E1 57
Abaetetuba, Braz., 11,196 C5 59
Abaí, Par E4 55
Abajo, Bernalillo, N. Mex. (part of Albuquerque) G5 107
Abajo, peak, Utah F6 119
Abakan, Sov. Un., 62,000 D12 28
Abala, Con. B F3 46
Abancay, Peru, 6,828 D3 58
Abanda, Chambers, Ala., 135 B4 78
Abanilla, Sp., 8,220 C5 20
Abano Terme, It., 2,436 D7 18
Abar el Brins, Eg., U.A.R. D2 32
Abarqū, Iran, 16,000 F6 41
'Abasān, Gaza Area, 2,000 C6 32
Abashiri, Jap., 27,800 D12 37
Abau, Pap m12 50
Abba, Irwin, Ga E3 87
'Abbasiya, Sud., 2,846 C3 47
Abbaye, pt., Mich B2 98
Abbeville, Henry, Ala., 2,524 D4 78
Abbeville, Fr., 22,005 B4 14
Abbeville, Wilcox, Ga., 872 E3 87
Abbeville, Vermilion, La., 10,414 E3 95
Abbeville, Lafayette, Miss., 275 A4 100
Abbeville, Abbeville, S.C., 5,436 C5 115
Abbeville, co., S.C., 21,417 C3 115
Abbey, Sask., Can., 336 G1 70
Abbey, head, Scot F5 13
Abbeydorney, Ire., 164 E2 11
Abbeyfeale, Ire., 1,272 E2 11
Abbeyleix, Ire., 1,085 E4 11
Abbiategrasso, It., 21,652 B2 21
Abbot, butte, Oreg E4 113
Abbotsford, B.C., Can., 888 A10 68
Abbotsford, Que., Can., 500 D5 73
Abbotsford, Clark and Marathon, Wis., 1,171 D3 124
Abbott, Scott, Ark., 150 B1 81
Abbottsburg, Bladen, N.C., 157 C5 109
Abbottstown, Adams, Pa., 561 G8 114
Abbot Village, Piscataquis, Maine, 100 (404▲) C3 96
Abbyville, Reno, Kans., 118 B3, E5 93
Abdulino, Sov. Un., 27,500 E19 9
Abe, lake, Eth C5 47
Abéché, Chad, 9,877 C4 46
Abee, Alta., Can., 53 B4 69
Abejorral, Col., 5,129 B2 60
Abell, St. Marys, Md., 10 D4 85
Abengourou, I.C., 14,000 E4 45
Åbenrå, Den., 14,219 C3 24
Åbenrå, co., Den., 48,676 D3 24
Åbenrå, fjord, Den C3 24
Abeokuta, Nig., 84,451 E5 45
Aberayron, Wales, 1,220 B3 12
Aberchirder, Scot., 775 C6 13
Abercorn, Que., Can., 378 D5 73
Abercorn, N. Rh., 2,300 C5 48
Abercrombie, Richland, N. Dak., 244 C9 110
Aberdare, Wales, 39,044 C4 12
Aberdaron, Wales, 1,275 B3 12
Aberdeen, Sask., Can., 284 E2 70
Aberdeen, Bingham, Idaho, 1,484 G6 89
Aberdeen, Butler, Ky., 84 C3 94
Aberdeen, Harford, Md., 9,679 A5 85
Aberdeen, Monroe, Miss., 6,450 B5 100
Aberdeen, Moore, N.C., 1,531 B4 109
Aberdeen, Brown, Ohio, 774 D4 111
Aberdeen, Scot., 185,379 C6 13
Aberdeen, S. Afr., 4,647 D3 49
Aberdeen, Brown, S. Dak., 23,073 B7 116
Aberdeen, Grays Harbor, Wash., 18,741 C2 122
Aberdeen, co., Scot., 321,757 C5 13
Aberdeen, lake, N.W. Ter., Can D13 66
Aberdeen, mtn., B.C., Can D1 69
Aberdovey, Wales, 1,262 B3 12
Aberfeldy, Scot., 1,469 D5 13
Aberfoyle, Scot., 1,133 D4 13
Abergavenny, Wales, 9,625 C4 12
Abernant, Tuscaloosa, Ala., 250 B2 78
Abernathy, Hale and Lubbock, Tex., 2,491 C2 118
Abernethy, Sask., Can., 310 G4 70
Abert, lake, Oreg E6 113
Abert, rim, Oreg E6 113
Abertillery, Wales, 25,160 C4 12
Aberystwyth, Wales, 10,418 B3 12
Abesville, Stone, Mo., 50 E4 101
Abeytas, Socorro, N. Mex C3 107
Abhā, Sau. Ar B5 47
Abidjan, I.C., 180,000 E4 45
Abie, Butler, Nebr., 117 C9 103
Abilene, Alta., Can., 25 B5 69
Abilene, Dickinson, Kans., 6,746 D6 93
Abilene, Taylor, Tex., 90,368 C3 118
Abingdon, Eng., 14,283 C6 12
Abingdon, Knox, Ill., 3,469 C3 90
Abingdon, Jefferson, Iowa, 55 C5 92
Abingdon, Harford, Md., 400 B5 85
Abingdon, Washington, Va., 4,758 B3 121
Abington, Windham, Conn., 130 B8 84
Abington, Plymouth, Mass., 4,500 (10,607▲) B6, E3 97
Abington, Montgomery, Pa., 8,000 A11 114
Abiquiu, flood control res., N. Mex A3 107
Abisko, Swe C8 25
Abita Springs, St. Tammany, La., 655 B7, D5 95
Abitibi, co., Que., Can., 108,313 E10 72
Abitibi, lake, Ont., Que., Can E9 72, B2 73
Abitibi, riv., Ont., Can E9 72
Ablon, Fr., 5,086 h10 14
Abluilluk, fiord, Que., Can f8 75
Åbo, see Turku, Fin.
Abo, Torrance, N. Mex B5 107
Aboisso, I.C., 3,500 E4 45
Abomey, Dah., 18,900 E5 45
Abong Abeng, mtn., Indon J2 38
Abony, Hung., 12,633 B5 22
Aboriginal, reserve, Austl E4 50
Abou Deïa, Chad, 825 C3 46
Abou Kémal, Syr., 8,200 C8 43, E13 31
Abound, Sask., Can., 25 G3 70
Aboyne, Scot., 1,651 C6 13
Abra, prov., Phil., 116,700 *B6 35
Abraham, Millard, Utah D3 119
Abraham Lincoln, nat. historical park, Ky C4 94
Abrams, Oconto, Wis., 230 D5 124
Abrantes, Port., 3,507 C1 20
Abrego, Col., 2,250 B3 60
Abri, Sud A3 47
Abruzzi and Molise, reg., It C4 21
Abruzzi e Molise, pol. dist., It., 1,564,318 *C4 21
Absaraka, Cass, N. Dak., 40 C8 110
Absaroka, range, Wyo A3 125
Absaroka, ridge, Wyo C2 125
Absarokee, Stillwater, Mont., 600 E7 102
Absecon, Atlantic, N.J., 4,320 E3 106
Absecon, inlet, N.J E4 106
Abu Ali, isl., Sau. Ar H4 41
Abu Dis, Jordan, 2,000 h12 32
Abū Ḩadrīyah, Sau. Ar H4 41
Abu Hamed, Sud., 1,450 B3 47
Abū Ḩammād al Maḩaţţah, Eg., U.A.R., 1,000 D3 32
Abū Ḩummuş, Eg., U.A.R., 2,000 C2 32
Abuja, Nig E6 45
Abu Kershola, Sud., 4,154 C3 47
Abū Kabīr, Eg., U.A.R., 3,000 D3 32
Abūksāh, Eg., U.A.R E2 32
Abumombazi, Con. L A3 48
Abu Musa, isl., Sau. Ar I7 41
Abunã, riv., Bol., Braz C4 58
Abū Qīr, Eg., U.A.R., 2,000 C2 32
Abū Qīr, bay, Eg., U.A.R C2 32
Abū Qurqāş, Eg., U.A.R., 5,000 I8 31
Abu Shagara, cape, Sud A4 47
Abū Sunbul, ruins, Eg., U.A.R E6 43
Abu Tabari (Well), Sud B2 47
Abu 'Uruq (Well), Sud B3 47
Abū Zanīmah, Eg., U.A.R D6 43
Aby, Swe., 2,795 u34 25
Abyei, Sud D2 47
Abyssinia, see Ethiopia, Afr.
Academia, Knox, Ohio, 900 B5 111
Academy, Charles Mix, S. Dak., 25 D6 116
Acadia, par., La., 49,931 D3 95
Acadia, nat. park, Maine D4 96
Acadia Valley, Alta., Can., 239 D5 69
Acajutla, Sal., 3,659 D3 62
Acala, Hudspeth, Tex., 30 F1 118
Acámbaro, Mex., 26,011 C4, m13 63
Acaponeta, Mex., 8,453 C3 63
Acapulco de Juárez, Mex., 48,846 D5 63
Acaraí, mts., Braz B3 59
Acaraú, Braz., 3,042 B2 57
Acaray, riv., Par E4 55
Acari, Braz., 1,867 h5 57
Acarí, Peru, 561 E3 58
Acarigua, Ven., 31,737 B4 60
Acatlán, Mex., 7,086 D5, n14 63
Acayucan, Mex., 12,854 D6 63
Accident, Garrett, Md., 237 D1 85
Accomac, Accomack, Va., 414 D7 121
Accomack, co., Va., 30,635 D7 121
Accord, Plymouth, Mass., 150 D3 97
Accord, Ulster, N.Y., 400 D6 108
Accord, pond, Mass D3 97
Accoville, Logan, W. Va., 800 D3, D5 123
Accra, Ghana, 388,231 (491,060▲) E4 45
Accrington, Eng., 40,987 A5 12
Ace, Polk, Tex., 10 D5 118
Aceguá, Ur E2 56
Acequia, Minidoka, Idaho, 107 G5 89
Achacachi, Bol., 3,621 C2 55
Achaea (Akhaia), prov., Grc., 239,206 *C3 23
Achao, Chile, 707 C2 54
Achar, Ur E1 56
Acharacle, Scot D3 13
Achen, lake, Aus B7 18
Acheng, China, 5,000 D3 37
Achenkirch, Aus B7 18
Achern, Ger., 6,100 E3 17
Achisay, Sov. Un., 10,000 G22 9
Achicourt, Fr., 5,190 D2 15
Achill, head, Ire D1 11
Achill, isl., Ire D1 11
Achille, Bryan, Okla., 294 D5 112
Achill Sound, Ire., 277 D2 11
Achinsk, Sov. Un., 57,000 D12 28
Achiras, Arg A4 54
Achourat (Well), Mali B4 45
Acipayam, Tur., 3,900 D7 23
Acireale, It., 26,000 F5 21
Ackerly, Dawson, Tex., 550 C2 118
Ackerman, Choctaw, Miss., 1,382 B4 100
Ackia Battleground, nat. mon., Miss A5 100
Ackley, Hardin and Franklin, Iowa, 1,731 B4 92
Acklins, isl., Ba. Is D7 64
Acmar, St. Clair, Ala., 250 B3 78
Acme, Alta., Can., 328 D4 69
Acme, Concordia, La., 50 C4 95
Acme, Grand Traverse, Mich., 120 D5 98
Acme, Kanawha, W. Va., 500 D6 123
Acme, Sheridan, Wyo., 125 A6 125
Acmetonia, Allegheny, Pa., 1,500 *A6 114
Acobamba, Peru, 1,912 D3 58
Acoma, Lincoln, Nev., 4 F7 104
Acomayo, Peru, 2,120 D3 58
Aconcagua, prov., Chile, 139,878 A2 54
Aconcagua, peak, Arg A3 54
Acopiara, Braz., 3,953 C3 57
Acorn, Monroe, Tenn. D9 117
Acosta, Somerset, Pa., 500 F3 114
Acoyapa, Nic., 1,148 E5 62
Acquarossa, Switz D6 19
Acqui, It., 12,200 B2 21
Acre, Isr., 25,128 B7 32
Acre, state, Braz., 160,208 C4 58
Acre, bay, Isr B7 32
Acre, riv., Braz D4 58
Acree, Dougherty, Ga., 125 E3 87
Acres Homes, Harris, Tex., 5,000 *E5 118
Acton, Los Angeles, Calif., 500 E4 82
Acton, Ont., Can., 3,578 D4 72
Acton, Eng., 65,274 k11 10
Acton, Marion, Ind., 650 E6, I8 91
Acton, York, Maine, 65 (501▲) E2 96
Acton, Middlesex, Mass., 400 (7,238▲) B5, C1 97
Acton, Yellowstone, Mont., 15 E8 102
Acton Vale, Que., Can., 3,957 D5 73
Actopan, Mex., 7,581 m14 63
Açu, Braz., 8,158 C3 57
Açuruá, plat., Braz D2 57
Acushnet, Bristol, Mass., 3,000 (5,755▲) C6 97
Acworth, Cobb, Ga., 2,359 B2 87
Acworth, Sullivan, N.H., 75 (371▲) D2 105
Ada, Ghana, 2,327 E5 45
Ada, Ottawa, Kans., 155 C6 93
Ada, Norman, Minn., 2,064 C2 99
Ada, Hardin, Ohio, 3,918 B4 111
Ada, Okinawa 52
Ada, Pontotoc, Okla., 14,347 C5 112
Ada, Mercer, W. Va., 250 D3 123
Ada, Yugo., 11,534 C5 22
Ada, co., Idaho, 93,460 F2 89
Ada, cape, Sov. Un G19 9
Adair, Sask., Can G4 70
Adair, McDonough, Ill., 250 C3 90
Adair, Adair and Guthrie, Iowa, 742 C3 92
Adair, Mayes, Okla., 434 A6 112
Adair, co., Iowa, 10,893 C3 92
Adair, co., Ky., 14,699 C4 94
Adair, co., Mo., 20,105 A5 101
Adair, co., Okla., 13,112 B7 112
Adairsville, Barton, Ga., 1,026 B2 87
Adairville, Logan, Ky., 848 D2 94
Adak, isl., Alsk E5 79
Adam, is., Md D5 85
Adama, see Hadama, Eth.
Adamana, nat. mon., Ariz C6 80
Adamantina, Braz., 18,164 C2 56
Adamello, peak, It E5 16
Adamitullo, Eth D4 47
Adams, Decatur, Ind., 350 F6 91
Adams, Lawrence, Ky., 350 B7 94
Adams, Berkshire, Mass., 12,391 A1 97
Adams, Mower, Minn., 806 G7 99
Adams, Hinds, Miss., 20 C3 100
Adams, Gage, Nebr., 387 D9 103
Adams, Jefferson, N.Y., 1,914 B4 108
Adams, Walsh, N. Dak., 360 A7 110
Adams, Texas, Okla., 165 D3 112
Adams, Umatilla, Oreg., 192 B8 113
Adams, Robertson, Tenn., 500 A5 117
Adams, Adams, Wis., 1,301 E4 124
Adams, co., Colo., 120,296 B6 83
Adams, co., Idaho, 2,978 E2 89
Adams, co., Ill., 68,467 D2 90
Adams, co., Ind., 24,643 C8 91
Adams, co., Iowa, 7,468 C3 92
Adams, co., Miss., 37,730 D2 100
Adams, co., Nebr., 28,944 D7 103
Adams, co., N. Dak., 4,449 C3 110
Adams, co., Ohio, 19,982 D4 111
Adams, co., Pa., 51,906 G7 114
Adams, co., Wash., 9,929 B7 122
Adams, co., Wis., 7,566 D4 124
Adams, lake, B.C., Can D8 68
Adams, mtn., Mass A2 97
Adams, mtn., N.H B4 105
Adams, mtn., Vt C3 120
Adams, mtn., Wash C4 122
Adams, pt., Mich C7 98
Adam's Bridge, shoals, India G6 39
Adamsburg, Union, S.C., 150 B4 115
Adams Center, Jefferson, N.Y., 900 B5 108
Adamson, Pittsburg, Okla., 160 C6 112
Adams Run, Charleston, S.C., 250 F1 115
Adamston, Ocean, N.J., 800 C4 106
Adamstown, Ire., 157 E5 11
Adamstown, Frederick, Md., 310 B3 85
Adamstown, Lancaster, Pa., 1,190 F9 114
Adamsville, Jefferson, Ala., 2,095 E4 78
Adamsville, Que., Can., 390 D5 73
Adamsville, Fulton, Ga. (part of Atlanta) B5 87
Adamsville, Crawford, Pa., 250 C1 114
Adamsville, Newport, R.I., 250 C12 84
Adamsville, McNairy, Tenn., 1,046 B3 117
Adamsville, Lampasas, Tex., 100 D3 118
Adamsville, Beaver, Utah E3 119
Adana, Tur., 230,000 D10 31
Adanac, Sask., Can., 36 E1 70
Adapazari (Adabazar), Tur., 80,200 B8 31
Adarama, Sud., 981 B3 47
Adare, Ire., 590 E3 11
Adare, cape, Ant B30 5
Adavale, Aust., 82 B5 51
Adaza, Greene, Iowa, 25 B3 92
Adda, riv., It B2 21
Aḑ Ḑab'ah, Eg., U.A.R C5 43
Ad Dammān, Sau. Ar *H4 41
Addicks, Harris, Tex., 75 F4 118
Addielee, Adair, Okla B7 112
Addieville, Washington, Ill., 231 E4 90
Ad Dilinjāt, Eg., U.A.R., 1,000 D2 32
Addington, Jefferson, Okla., 144 C4 112
Addis, West Baton Rouge, La., 590 D4 95
Addis Ababa, Eth., 500,000 D4 47
Addison, Winston, Ala., 343 A2 78
Addison, Du Page, Ill., 9,046 F2 90
Addison, Washington, Maine, 350 (744▲) D5 96
Addison, Lenawee, Mich., 575 G6 98
Addison, Steuben, N.Y., 2,185 C3 108
Addison, Addison, Vt., 60 (645▲) C2 120
Addison, co., Vt., 20,076 C2 120
Ad Dīwān, Eg., U.A.R., 1,126 E6 43
Ad Diwaniya, Iraq, 27,839 E2 41
Addy, Stevens, Wash., 245 A8 122
Addyston, Hamilton, Ohio, 1,376 D2 111
Adel, Cook, Ga., 4,321 E3 87
Adel, Dallas, Iowa, 2,060 C3 92
Adel, Lake, Oreg., 100 E7 113
Adelaide, Austl., 23,051 (★630,000) F6 50, G2 51
Adelaide, isl., Ant C6 5
Adelanto, San Bernardino, Calif., 950 E5 82
Adelboden, Switz., 2,881 D4 19
Adelie, coast, Ant C27 5
Adelina, Calvert, Md., 100 D4 85
Adeline, St. Mary, La E4 95
Adelino, Valencia, N. Mex C3 107
Adell, Sheboygan, Wis., 398 E6 124
Adelphi, Polk, Iowa, 40 A8 92
Adelphi, Prince Georges, Md., 8,000 *C4 85
Adelphi, Ross, Ohio, 441 C5 111
Adelphia, Monmouth, N.J., 300 C4 106
Adelup, pt., Guam 52
Aden, Aden, 99,285 C6 47
Adena, Jefferson, Ohio, 1,317 B7 111
Adenau, Ger., 2,900 C1 17
Aden [Prot.], Br. dep., Asia, 758,000 H7 33, C6 47
Adgateville, Jasper, Ga., 100 C3 87
Adger, Jefferson, Ala., 500 B2, E4 78
Adige, riv., It B3 21
Adigrāt, Eth C4 47
Adilabad, India, 20,970 H7 40
Adin, Modoc, Calif., 175 B3 82
Adindān, Eg., U.A.R E6 43
Adirondack, mts., N.Y A6 108
Adi Ugri, Eth., 5,000 C4 47
Adiyaman, Tur., 17,000 C12 31
Adjud, Rom., 6,119 B8 22
Adjuntas, P.R., 5,318 C4 65
Adjuntas, mun., P.R., 19,658 C3 65
Adlavik, isl., Newf., Can g10 75
Admaston, Ont., Can., 200 B8 72
Admiral, Sask., Can., 139 H1 70
Admirals Beach, Newf., Can., 122 E5 75
Admiralty, isl., Alsk m22 79
Admiralty, is., Bis. Arch G8 7
Admire, Lyon, Kans., 149 D7 93
Adobe, creek, Colo C7 83
Adobe Creek, res., Colo C7 83
Adobes, Presidio, Tex F2 118
Adok, Sud D3 47
Adola, Eth D4 47
Adolfo Alsina, Arg., 5,836 B4 54
Adolphus, Allen, Ky., 200 D3 94
Adona, Perry, Ark., 154 B3 81
Adony, Hung., 3,641 B4 22
Adorf, Ger., 8,316 C7 17
Adour, riv., Fr F3 14
Adra, Sp., 7,923 D4 20
Adrano, It., 31,532 F5 21
Adrar, Alg., 1,865 (35,588▲) D4 44
Adrar, sand dunes, Maur B2 45
Adrar des Iforas, reg., Maur B5 45
Adrar Nahalet, Alg E5 44
Adré, Chad, 1,521 C4 46
Adri, Libya, 935 D2 43
Adria, It., 12,100 B4 21
Adrian, Emanuel and Johnson, Ga., 568 D4 87
Adrian, Lenawee, Mich., 20,347 G6 98
Adrian, Nobles, Minn., 1,215 G3 99
Adrian, Bates, Mo., 1,082 C3 101
Adrian, LaMoure, N. Dak., 64 C7 110
Adrian, Malheur, Oreg., 300 D9 113
Adrian, Oldham, Tex., 258 B1 118
Adrian, Upshur, W. Va., 475 C4 123
Adrianople, see Edirne, Tur.
Adriatic, sea, Eur E8 18
Adrigole, Ire., 65 F2 11
Aduwā, Eth., 5,000 C4 47
Advance, Baxter, Ark A3 81
Advance, Boone, Ind., 463 D4 91
Advance, Stoddard, Mo., 692 D8 101
Advocate Harbour, N.S., Can., 311 D5 74
Ady, Potter, Tex., 100 B2 118
Adzopé, I.C., 3,400 E4 45
Aebelö, isl., Den C4 24
Aegean, sea, Grc C5 23
Aegeri, lake, Switz B6 19
Aerö, isl., Den D4 24
Aerösköbing, Den., 1,273 D4 24
Aesch, Switz., 3,981 B4 19
Aetna, Alta., Can., 61 E4 69
Aetna, Barber, Kans., 50 E5 93
Aetna, Hickman, Tenn., 50 B4 117
Aetolia and Acarnania (Aitolia kai Akarnania), prov., Grc., 237,738 *C3 23
'Afar Tall, Iraq D14 31
Affinity, Raleigh, W. Va., 150 D3 123
Affoltern am Albis, Switz., 4,904 B5 19
Affton, St. Louis, Mo., 18,000 *B8 101
Afghanistan, country, Asia, 13,000,000 F9 33, 41
Afgoi, Som., 3,000 (14,400▲) E6 47
Aflou, Alg., 3,370 (38,360▲) C5 44
Afmadu, Som., 1,700 E5 47
Afognak, Alsk., 190 D9 79
Afognak, isl., Alsk D9 79
Afono, bay, Am. Sam 52
Afonso Cláudio, Braz., 2,823 F2 57
Africa, cont., 233,718,700 F22 3, 42
Afton, Union, Iowa, 773 C3 92
Afton, Cheboygan, Mich., 100 C6 98
Afton, Washington, Minn., 158 E8 99
Afton, Chenango, N.Y., 956 C5 108
Afton, Ottawa, Okla., 1,111 A7 112
Afton, Greene, Tenn., 125 C11 117
Afton, Dickens, Tex., 60 C2 118
Afton, Lincoln, Wyo., 1,337 C2 125
Aftout, sand dunes, Alg D4 44
Afula, Isr., 13,816 B7 32
Afyonkarhisar, Tur., 38,400 C8 31
Agadem, Niger C7 45
Agadès, Niger, 4,700 C6 45
Agadir, Mor., 16,695 C3 44
Agalak (Well), Niger C6 45
Agalta, mts., Hond C5 62
Agamiaure, mts., Braz B4 59
Agana, Guam, 1,642 (★17,000) 52
Agana, bay, Guam 52
Agar, India F5 40
Agar, Sully, S. Dak., 139 C5 116
Agartala, India, 54,878 D9 39
Agassiz, B.C., Can., 478 A11, E7 68
Agassiz, cape, Ant C6 5
Agat, bay, Guam 52
Agata, isl., Fiji Is 52
Agate, Elbert, Colo., 150 B7 83
Agate, Sioux, Nebr., 10 B2 103
Agate, Rolette, N. Dak., 15 A6 110
Agate, pass, Colo C4 83
Agate Beach, Lincoln, Oreg., 800 C3 113
Agattu, isl., Alsk E2 79
Agawam, Hampden, Mass., 5,000 (15,718▲) B2 97
Agawam, Teton, Mont., 30 B4 102
Agayman, Sov. Un H10 27
Agbor, Nig., 3,937 E6 45
Agboville, I.C., 13,000 E4 45
Agde, Fr., 7,696 F5 14
Agedábia, see Ajdábiyah, Libya.
Agematsu, Jap., 4,600 n16 37
Agen, Fr., 32,800 E4 14
Agency, Wapello, Iowa, 702 D5 92
Agency, Buchanan, Mo., 240 B3 101
Agenda, Republic, Kans., 124 C6 93
Agersö, isl., Den C5 24
Ag fayan, bay, Guam 52
Āghā Jārī, Iran F4 41
Aghéila, see Al Uqaylah, Libya.
Aghleam, Ire C1 11
A Ghlo, mtn., Scot D5 13
Agincourt, Ont., Can., 1,738 E7 72
Agingan, pt., Saipan 52
Agira, It., 14,079 F5 21
Agnes, lake, Ont., Can B7 99
Agno, Phil., 2,556 n12 35
Agnone, It., 9,888 D5 21
Agnos, Fulton, Ark., 75 A4 81
Agogna, riv., It D5 18
Agordat, Eth., 3,000 B4 47
Agordo, It., 3,603 C8 18
Agra, India, 462,020 (★508,680) C6 38, D6 40
Agra, Phillips, Kans., 277 C4 93
Agra, Lincoln, Okla., 265 B5 112
Agrado, Col., 2,546 C2 60
Agri, riv., It D6 21
Agria, pt., Phil o13 35
Agricola, Coffey, Kans., 15 D8 93
Agricola, George, Miss., 125 E5 100
Agrigento, It., 40,500 (47,919▲) F4 21
Agrinion, Grc., 24,763 C3 23
Agropoli, It., 7,470 D5 21
Agryz, Sov. Un., 19,100 D19 9
Agua Caliente, Maricopa, Ariz., 5 E2 80
Aguacate, P.R B2 65
Aguacate, Cuba., 3,324 k12 64
Aguada, P.R., 3,759 B2 65
Aguada, mun., P.R., 23,234 B2 65
Aguadas, Col., 8,064 B2 60
Aguadilla, P.R., 15,943 B2 65
Aguadilla, mun., P.R., 47,864 B2 65
Aguadilla, bay, P.R B2 65
Aguadulce, Pan., 6,010 F7 62
Agua Dulce, mts., Ariz E2 80
Agua Fria, Santa Fe, N. Mex., 150 F6 107
Agua Fria, riv., Ariz D3, G1 80
Aguaí, Braz., 7,047 m8 56
Aguán, riv., Hond C4 62
Agua Nueva, Jim Hogg, Tex., 50 F3 118
Aguapeí, riv., Braz C2 56
Aguapey, riv., Arg E4 55
Agua Preta, Braz., 3,241 k6 57
Agua Prieta, Mex., 15,275 A3 63
Aguaray, Arg D3 55
Aguas Buenas, P.R., 2,470 B6 65
Aguas Buenas, mun., P.R., 17,034 B6 65
Aguascalientes, Mex., 126,222 C4, m12 63
Aguascalientes, state, Mex., 236,574 C4, k12 63
Agudos, Braz., 6,564 C3 56
Agueda, Port., 4,911 B1 20
Agueda, riv., Sp B2 20
Agueloc, Mali C5 45
Aguelt Nemadi (Well), Maur C2 45
Aguéraktem (Well), Mali B3 45
Aguila, Maricopa, Ariz., 250 D2 80
Aguila, pt., P.R D2 65
Aguilar, Las Animas, Colo., 777 D6 83
Aguilar, Sp., 13,843 D3 20
Aguilas, Sp., 11,634 D5 20
Aguililla, Mex., 2,743 D4, n12 63
Agujita, Mex., 5,463 B4 63
Agulhas, cape, S. Afr D3 49
Agusan, prov., Phil., 272,000 *D7 35
Ahaggar, mts., Alg E6 44
Ahar, Iran, 18,886 B3 41
Ahascragh, Ire., 234 D3 11
Ahaus, Ger., 9,700 A2 17
Ahiri, India, 5,000 H8 40
Ahlen, Ger., 40,500 B2 17
Ahlhorn, Ger., 884 F2 24
Ahmadabad, India, 1,149,918 (★1,300,000) D5 39, F4 40
Ahmadnagar (Ahmednagar), India, 119,020 E5 39
Ahmadpur, Pak., 20,423 C5 39
Ahmar, mts., Eth D5 47
Ahmar, sand dunes, Alg., Mali D4 44
Ahmic, lake, Ont., Can B5 72
Aho, Jap., 3,210 o15 37
Ahoskie, Hertford, N.C., 4,583 A7 109
Ahousat, B.C., Can., 15 E4 68
Ahrdorf, Ger., 170 C1 17
Ahrensbök, Ger., 7,300 D4 24
Ahrweiler, Ger., 8,700 C2 17
Ahsahka, Clearwater, Idaho, 150 C2 89
Ahtanum, Yakima, Wash., 350 C5 122
Ahtanum, creek, Wash C5 122
Ahuacatlán, Mex., 5,004 m11 63
Ahuachapán, Sal., 13,298 D3 62
Ahualulco de Mercado, Mex., 8,036 m12 63
Ahukini, Kauai, Haw., 150 B2 88
Åhus, Swe., 4,400 C8 24
Ahvāz (Nāsiri), Iran, 120,098 F4 41
Ahvenanmaa (Åland), dept., Fin., 22,787 *G8 25
Ahvenanmaa, is., Fin G8 25
Ahwahnee, Madera, Calif., 450 D4 82
Aibonito, P.R., 5,477 C5 65
Aibonito, mun., P.R., 18,360 C5 65
Aichach, Ger., 6,500 E6 17
Aichi, pref., Jap., 4,206,313 *I8 37
Aidin, see Aydin, Tur.
Aiea, Honolulu, Haw., 11,826 B4, g10 88
Aigaleos, mts., Grc g11 23
Aigen, Aus., 1,941 D6 16
Aigle, Switz., 4,381 D2 19
Aiguá, Ur., 2,879 E2 56
Aigues-Mortes, Fr., 3,746 F6 14
Aiguilles, Fr., 250 E2 18
Aigun, China, 25,000 A10 34
Aihsien, China C8 38
Aija, Peru, 1,427 C2 58
Aikawa, Jap., 7,800 G9 37
Aiken, Aiken, S.C., 11,243 D4 115
Aiken, co., S.C., 81,038 D4 115
Aikens, lake, Man., Can D4 71
Aiken South, Aiken, S.C., 2,980 *D4 115
Aiken West, Aiken, S.C., 2,602 *D4 115
Aikin, Cecil, Md., 50 A5 85
Ailey, Montgomery, Ga., 469 D4 87
Ailsa Craig, Ont., Can., 533 D3 72
Aima, lake, Braz D5 60
Aimorés, Braz., 11,448 E2 57
Aimwell, Marengo, Ala., 50 C2 78
Ain, dept., Fr., 327,146 C1 18
Ain, riv., Fr C1 18
Ainabo, Som E6 47
Aïn Beïda, Alg., 15,130 (18,866▲) B6 44
Ain Ben Tili, Maur A3 45
Ain el Bagha, Eg., U.A.R E5 32
Ain Oadeis, Eg., U.A.R D6 32
Ainos, mtn., Grc C3 23
Aïn-Oussera, Alg G8 30
Aïn-Sefra, Alg., 4,637 (20,501▲) C4 44
Ain-Sefra, reg., Alg D4 44
Ainslie, lake, N.S., Can C8 74
Ain Sudr, Eg., U.A.R E5 32
Ainsworth, Washington, Iowa, 371 C6 92
Ainsworth, Brown, Nebr., 1,982 B6 103
Aïn Témouchent, Alg., 25,187 B4 44
Aïoun el Atrous, Maur C3 45
Aipe, Col., 2,221 C2 60
Aiquile, Bol., 3,465 C2 55
Air (Azbine), reg., Niger C6 45
Airabu, isl., Indon K7 38
Airai, Palau Is 52
Airaines, Fr., 1,673 E9 12
Aird, isl., Ont., Can B8 98
Aird Bremish, isl., Scot B1 13

Airdrie, Alta., Can., 524....D3 69
Airdrie, Scot., 33,620.......E5 13
Aire, Fr., 3,544..........F3 14
Aire, riv., Eng..........A6 12
Aire, riv., Fr...........E5 15
Aire-sur-la-Lys, Fr., 5,528...D2 15
Aireys, Dorchester, Md., 50..C6 85
Air Line, Hart, Ga., 200.....B3 87
Airolo, Switz., 2,023........C6 19
Airukiijl, isl., Bikini........ 52
Aisch, riv., Ger..........D5 17
Aisén, prov., Chile, 37,085...D2 54
Aisne, dept., Fr., 512,920....E3 15
Aisne, riv., Fr..........C5 14
Aissa, mtn., Alg..........C4 44
Aitape, N. Gui...........h11 50
Aitkin, Aitkin, Minn., 1,829 .D5 99
Aitkin, co., Minn., 12,162...D5 99
Aitolia kai Akarnania (Aetolia and Acarnania), prov., Grc., 237,738.........*C3 23
Aitolikon, Grc., 5,959.......C3 23
Aiud, Rom., 11,886..........B6 22
Aiún, Sp. Sahara, 1,369.....D2 44
Aiuruoca, Braz., 1,591.......g5 56
Aiwa, isl., Fiji Is......... 52
Aix [-en-Provence], Fr., 67,943................F6 14
Aix-la-Chapelle, see Aachen, Ger.
Aix-les-Bains, Fr., 18,132....D1 18
Aiyina, Grc., 6,217.........D4 23
Aiyina (Aegina), isl., Grc.....D4 23
Aiyion, Grc., 17,762........C4 23
Aiyon, Palau Is........... 52
Ajaccio, Fr., 41,006.........D2 21
Ajaccio, gulf, Fr...........D2 21
Ajana, Austl., 80...........E1 50
Ajanta, India, 3,560.........G6 40
Ajax, Ont., Can., 7,755.....D5 72
Ajax, mtn., Mont..........E3 102
Ajayan, bay, Guam.......... 52
Ajdábiyah (Agedábia), Libya, 16,386..............C4 43
Ajigasawa, Jap., 8,100.......F10 37
'Ajlūn, Jordan, 2,518.......B7 32
Ajlune, Lewis, Wash., 100...C3 122
Ajmer, India, 231,240.........C5 38, D5 40
Ajmer, state, India, 693,372 .*D5 40
Ajo, Pima, Ariz., 7,049.......E3 80
Ajoe, is., W. Irian..........E8 35
Ajuana, riv., Braz..........D4 60
Akaishi-Sammyaku, mts., Jap..n17 37
Akalkot, India, 21,278.......I6 40
Akaroa, N.Z., 632.........O14 51
Akasha, Sud..............A3 47
Akashi, Jap., 129,780........I7 37
Akaska, Walworth, S. Dak., 90....................B5 116
Ak-Bulak, Sov. Un., 10,000 .E20 9
Akcaabat, Tur., 4,414.......B12 31
Akcadag, Tur., 3,540........C11 31
Akdag-madeni, Tur., 4,000.C10 31
Akchar, sand dunes, Maur.....B2 45
Akechi, Jap., 4,800..........n16 37
Akela, Luna, N. Mex., 18.....E2 107
Akeley, Hubbard, Minn., 434.................C4 99
Aken, Ger., 12,700..........B7 17
Akers, Tangipahoa, La., 150.................B7, D5 95
Akershus, co., Nor., 234,500.*H4 25
Akesum, Eth.............C4 47
Aketi, Con. L., 12,100.......A3 48
Akhaia (Achaea), prov., Grc., 239,206..............C3 23
Akhaltsikhe, Sov. Un., 10,245...............B14 31
Akharnai, Grc., 11,290......g11 23
Akhḍar, mts., Libya..........C4 43
Akheloos, riv., Grc..........C3 23
Akhiok, Alsk., 84...........D9 79
Akhisar, Tur., 40,000.......C6 31
Akhmīm, Eg., U.A.R., 38,000 D6 43
Akhtopol, Bul., 1,052.......D8 22
Akhtyrka, Sov. Un., 55,200.F10 27
Aki, Jap., 13,100...........J6 37
Akiachak, Alsk., 229........C7 79
Akiak, Alsk., 187............C7 79
Akimiski, isl., N.W. Ter., Can.F16 67
Akirkeby, Den., 1,461........A3 26
Akita, Jap., 166,000 (203,661▲)..............G10 37
Akita, pref., Jap., 1,335,580..............*G10 37
Akjoujt, Maur., 2,200.......C2 45
Akkeshi, Jap., 16,100.......E12 37
Aklan, prov., Phil., 226,500 .*D7 35
Aklavik, N.W. Ter., Can., 599.................C5 66
Akobo, Sud.................D3 47
Akobo, riv., Eth...........D3 47
Akola, India, 115,760........D6 39
Akonolinga, Cam...........E2 46
Akpatok, isl., N.W. Ter., Can.D19 67
Akra, India................D8 39
Akra, Pembina, N. Dak., 26..A8 110
Akranes, Ice., 3,913........n21 25
Akritas, cape, Grc..........D3 23
Akritis, isl., Grc...........D6 23
Akron, Hale, Ala, 604........C2 78
Akron, Washington, Colo., 1,890...............A7 83
Akron, Fulton, Ind., 958.....B5 91
Akron, Plymouth, Iowa, 1,351.................B1 92
Akron, Tuscola, Mich., 503..E7 98
Akron, Erie, N.Y., 2,841....B2 108
Akron, Summit, Ohio, 290,351 (★573,800).......A6 111
Akron, Lancaster, Pa., 2,167 .F9 114
Aksaray, Tur., 20,000.......C10 31
Aksehir, Tur., 20,600........C8 31
Aksehir, lake, Tur...........C8 31
Akseki, Tur., 2,700..........D8 31
Aksenovo-Zilovskove, Sov. Un., 10,000.................D14 28
Akshimrau, Sov. Un.........E4 29
Aksu, China...............E11 28
Aksuat, Sov. Un., 3,000.....E21 9
Aktogay, Sov. Un..........D9 29
Aktumsyk, Sov. Un.........D5 29
Aktyubinsk, Sov. Un., 107,000.................C5 29
Akure, Nig., 38,853.........E6 45
Akureyri, Ice., 8,957.......n23 25
Akuseki, isl., Jap...........L4 37
Akutan, Alsk., 107..........E6 79
Akyab, Bur., 42,329.........D9 39
Alabama, state, U.S., 3,266,740..........D10 77, 78
Alabama, riv., Ala...........D2 78
Alabama Port, Mobile, Ala., 200................E1 78
Alabaster, Shelby, Ala., 1,623..................B3 78
Alabaster, Iosco, Mich., 125..D7 98
Alabat, isl., Phil...........o14 35
Al Abyār, Libya, 150........F3 31
Alachua, Alachua, Fla., 1,974..................C4 86
Alachua, co., Fla., 74,074....C4 86
Aladdin, Crook, Wyo., 10...A8 125
Alagna, It., 516............D3 18
Alagoa Grande, Braz., 12,115..............C3, h6 57
Alagôas, state, Braz., 1,271,062...........C3, k5 57
Alagoinhas, Braz., 38,246....D3 57
Alagón, Sp., 5,270..........B5 20
Alagón, riv., Sp...........B2 20
Al-Ahmadi, Kuw., 7,205.....G4 41
Al 'Ajamīyīn, Eg., U.A.R., 2,000..................E2 32
Alajuela, C.R., 19,900.......E5 62
Alajuela, prov., C.R., 232,000..............*E5 62
Alakanuk, Alsk., 278........C7 79
Alakol, lake, Sov. Un........D10 29
Alalān, Iran, 10,000........C4 41
Alamance, Alamance, N.C., 450....................B4 109
Alamance, co., N.C., 85,674 ..B4 103
Alameda, Alameda, Calif., 63,855................B5 82
Alameda, Sask., Can., 312...H4 70
Alameda, Bernalillo, N. Mex., 5,000.........B3, G5 107
Alameda, co., Calif., 908,209 .D3 82
Al 'Āmerīyah, Eg., U.A.R., 10,000..................G7 31
Alaminos, Phil., 3,014.......n12 35
Alamito, creek, Tex.........F2 118
Alamo, Contra Costa, Calif., 1,791................*B6 82
Alamo, Wheeler, Ga., 822....D4 87
Alamo, Montgomery, Ind., 144.....................E3 91
Alamo, Mex., 6,375.........m15 63
Alamo, Lincoln, Nev., 250...F6 104
Alamo, Williams, N. Dak., 182...................A2 110
Alamo, Crockett, Tenn., 1,665 .B2 117
Alamo, Hidalgo, Tex., 4,121 .F3 118
Alamogordo, Otero, N. Mex., 21,723...................E4 107
Alamogordo, res., N. Mex.....C5 107
Alamo Heights, Bexar, Tex., 7,552............B4, E3 118
Alamo Hueco, mts., N. Mex....F1 107
Alamos, Mex., 2,872.........B3 63
Alamosa, Alamosa, Colo., 6,205..................D5 83
Alamosa, co., Colo., 10,000..D5 83
Alamosa, creek, Colo.........D4 83
Alamosa, riv., N. Mex........D2 107
Åland (Ahvenanmaa), dept., Fin., 22,787..............*G8 25
Alanreed, Gray, Tex. 200....B2 118
Alanson, Emmet, Mich., 290.C6 98
Alantika, mts., Cam..........D2 46
Alanya, Tur., 10,100.........D9 31
Alaotra, lake, Malag.........g9 49
Alapah, mtn., Alsk...........B9 79
Alapaha, Berrien, Ga., 631...E3 87
Alapaha, riv., Ga...........E3 87
Alapayevsk, Sov. Un., 41,100.................B6 29, D21 9
Al 'Aqabah, Jordan, 2,835...E7 32
Alaqua, Walton, Fla., 200...G3 86
Al 'Arīsh, Eg., U.A.R., 10,791..................C6 43
Alarka, Swain, N.C., 300....D2 109
Alasehir, Tur., 13,900.......C7 31
Ala Shan, mts., China........D6 34
Al Ashkharah, Mus. & Om..D2 39
Alaska, state, U.S., 226,167..........E4 76, 79
Alaska, gulf, Alsk..........D10 79
Alaska, pen., Alsk...........D8 79
Alaska, range, Alsk..........C9 79
Alassio, It., 8,544..........C2 21
Alatyr, Sov. Un., 34,700.....C3 29
Alau, isl., Haw.............C6 88
Alausí, Ec., 4,812...........B2 58
Alava, prov., Sp., 138,934..*A4 20
Alava, cape, Wash...........A1 122
Alaverdi, Sov. Un., 10,000..B15 31
Al 'Aynā, Jordan 1,000.....D7 32
Al 'Ayzarīyah (Bethany), Jordan, 2,000...........h12 32
Al 'Azair, Iraq..............F3 41
Alba, It., 12,000...........B2 21
Alba, Antrim, Mich., 200....D6 98
Alba, Jasper, Mo., 336......D3 101
Alba, Bradford, Pa., 192.....C8 114
Alba, Wood, Tex., 742.......C5 118
Albacete, Sp., 59,600 (74,417▲)..................C5 20
Albacete, prov., Sp., 370,976...................*C5 20
Alba de Tormes, Sp., 3,666...B3 20
Ålbaek, Den., 877...........A4 24
Ålbaek, bay, Den............A4 24
Alba Iulia, Rom., 14,776....B6 22
Al Bājhūr, Eg., U.A.R., 1,000...................D3 32
Albalate [del Arzobispo], Sp., 3,844...............B5 20
Al Ballaḥ, Eg., U.A.R., 1,000...................D4 32
Albanel, lake, Que., Can......F18 67
Albania, country, Eur., 1,625,378...........G13 8, B3 23
Albano, lake, It.............h9 21
Albano Laziale, It., 11,700..............D4, h9 21
Albany, Austl., 10,526.......F2 50
Albany, P.E.I., Can., 226....C6 74
Albany, Dougherty, Ga., 55,890..................E2 87
Albany, Whiteside, Ill., 637..B3 90
Albany, Delaware, Ind., 2,132..................D7 91
Albany, Clinton, Ky., 1,887 .D4 94
Albany, Livingston, La., 557.A6 95
Albany, Stearns, Minn., 1,375....................E4 99
Albany, Gentry, Mo., 1,662..A3 101
Albany, Albany, N.Y., 129,726 (★592,400).........C7 108
Albany, Athens, Ohio, 629...C5 111
Albany, Bryan, Okla., 150...D5 112
Albany, Linn, Oreg., 12,926............C1, C3 113
Albany, Shackelford, Tex., 2,174..................C3 118
Albany, Orleans, Vt., 169 (560▲)...............B4 120
Albany, Green, Wis., 892.....F4 124
Albany, Albany, Wyo., 100..D6 125
Albany, co., N.Y., 272,926...C7 108
Albany, co., Wyo., 21,290....D7 125
Albany, riv., Ont., Can.......D9 72
Albany South, Dougherty, Ga., 1,200................*E2 87
Al Barkāt, Libya..........E2 43
Al Batānūn, Eg., U.A.R., 1,000.................D2 32
Al Bawīṭī, Eg., U.A.R......D5 43
Albay, prov., Phil., 516,000................*C6 35
Al Bayḍā, Libya...........C4 43
Albee, Grant, S. Dak., 42...B9 116
Albemarle, Stanly, N.C., 12,261..................B3 109
Albemarle, co., Va., 30,969..D4 121
Albemarle, sound, N.C........A7 109
Albenga, It., 8,700..........B2 21
Alberche, riv., Sp...........B3 20
Alberene, Albemarle, Va., 300....................D4 121
Ålberga, Swe., 267.........u34 25
Albergaria-a-Velha, Port.....B1 20
Alberhill, Riverside, Calif., 200.....................F3 82
Alberni, B.C., Can., 4,616...E5 68
Albers, Clinton, Ill., 566.....E4 90
Albersdorf, Ger., 2,800......D3 24
Albert, N.B., Can., 245......D5 74
Albert, Fr., 10,247..........B5 14
Albert, Barton, Kans., 221...D4 93
Albert, Harding, N. Mex., 5..B6 107
Albert, Caddo, Okla., 135...B3 112
Albert, Tucker, W. Va., 45...B5 123
Albert, co., N.B., Can., 12,485..................D5 74
Albert, lake, Ug............A5 48
Alberta, Wilcox, Ala., 225...C2 78
Alberta, Stevens, Minn., 149.E2 99
Alberta, Brunswick, Va., 430.E5 121
Alberta, prov., Can., 1,331,944.............F9 68, 69
Alberta, mtn., Alta., Can......C2 69
Albert Canyon, B.C., Can., 30.....................D9 68
Albert City, Buena Vista, Iowa, 721..................B3 92
Albert Lea, Freeborn, Minn., 17,108..................G5 99
Albert Markham, mtn., Ant...A28 5
Albert Nile, riv., Ug..........A5 48
Alberton, P.E.I., Can., 820..C5 74
Alberton, Mineral, Mont., 356....................C2 102
Albertson, Nassau, N.Y., 9,700...................*G2 84
Albertville, Marshall, Ala., 8,250..................A3 78
Albertville, Con. L., 29,500..C4 48
Albertville, Sask., Can., 100..D3 70
Albertville, Fr., 12,159.......E7 14
Albertville, Wright, Minn., 279.....................E5 99
Albeuve, Switz., 601........C3 19
Albi, Fr., 38,709............F5 14
Albia, Monroe, Iowa, 4,582..C5 92
Albin, Tallahatchie, Miss., 40.B3 100
Albin, Laramie, Wyo., 172...D8 125
Albina, Sur................A4 59
Albion, Mendocino, Calif., 300.....................C2 82
Albion, Cassia, Idaho, 415...G5 89
Albion, Edwards, Ill., 2,025..E5 90
Albion, Noble, Ind., 1,325...B7 91
Albion, Marshall, Iowa, 588..B5 92
Albion, Kennebec, Maine, 100 (974▲)...............D3 96
Albion, Calhoun, Mich., 12,749...................F6 98
Albion, Carter, Mont., 3....E12 102
Albion, Boone, Nebr., 1,982..C7 103
Albion, Orleans, N.Y., 5,182..................B2 108
Albion, Pushmataha, Okla., 161.....................C6 112
Albion, Erie, Pa., 1,630.....C1 114
Albion, Providence, R.I., 400....................B11 84
Albion, Whitman, Wash., 291.....................C8 122
Albion, Dane, Wis., 150.....F4 124
Al Bīrah, Jordan...........h11 32
Alborán, isl., Sp...........E4 20
Ålborg, Den., 85,800 (★119,063)..............A3 24
Ålborg, co., Den., 232,885...B3 24
Ålborg, bay, Den............B4 24
Alborn, St. Louis, Minn., 50.D6 99
Albox, Sp., 9,908...........D4 20
Albreda, B.C., Can., 52......C8 68
Albright, Preston, W. Va., 304...................*B5 123
Albrightsville, Carbon, Pa., 90.....................D10 114
Al Bu'ayrāt, Libya.C3 43, I14 30
Albufeira, Port., 3,160.......D1 20
Albuñol, Sp., 7,989.........D4 20
Albuquerque, Bernalillo, N. Mex., 201,189 (★266,300)............B3, G5 107
Alburg, Grand Isle, Vt., 426 (1,123▲)..............B2 120
Alburnett, Linn, Iowa, 341...B6 92
Alburquerque, Sp., 9,756.....C2 20
Alburtis, Lehigh, Pa., 1,086 .E10 114
Albury, Austl., 22,983........H6 51
Al Busaiya, Iraq............F3 41
Al Buşayli, Eg., U.A.R., 1,000...................C2 32
Alca, Peru, 713.............E3 58
Alcabideche, Port., 9,485.....f9 20
Alcácer do Sal, Port., 4,040...C1 20
Alcalá de Chisvert, Sp., 4,394.B6 20
Alcalá [de Guadaira], Sp., 25,000 (31,004▲)............D3 20
Alcalá de Henares, Sp., 25,123...............B4, p18 20
Alcalá de los Gazules, Sp., 9,693....................D3 20
Alcalá la Real, Sp., 9,979....D4 20
Alcalde, Rio Arriba, N. Mex., 650.................A4 107
Alcamo, It., 43,097..........F4 21
Alcanadre, riv., Sp...........B5 20
Alcanar, Sp., 6,944..........B6 20
Alcañiz, Sp., 10,035.........B5 20
Alcântara, Braz., 1,300.......B2 57
Alcantarilla, Sp., 15,748.....D5 20
Alcaraz, Sp., 5,864..........C4 20
Alcaudete, Sp., 18,442.......D3 20
Alcázar de San Juan, Sp., 24,963...................C4 20
Alcazarquivir, Mor., 34,035..B3 44
Alcester, Union, S.Dak., 479.D9 116
Alcira, Sp., 26,669..........C5 64
Alco, Stone, Ark., 25........B3 81
Alco, Vernon, La., 40.......C2 95
Alcoa, Blount, Tenn., 6,395..............D10, E11 117
Alcobendas, Sp., 4,778.......o17 20
Alcochete, Port., 6,494.......f10 20
Alcolu, Clarendon, S.C., 275.D7 115
Alcomdale, Alta., Can., 100..C4 69
Alcona, co., Mich., 6,352....D7 98
Alcora, Sp., 3,782...........B5 20
Alcorcón, Sp., 3,356.........p17 20
Alcorisa, Sp., 3,497.........B5 20
Alcorn, Jackson, Ky., 5.....C5 94
Alcorn, Claiborne, Miss., 1,100...................D2 100
Alcorn, co., Miss., 25,282....A5 100
Alcova, Natrona, Wyo., 75...C6 125
Alcovy, mtn., Ga............C3 87
Alcoy, Sp., 51,096...........C5 20
Alda, Hall, Nebr., 229.......D7 103
Aldama, Mex., 5,194........B3 63
Aldan, Delaware, Pa., 4,324................*G11 114
Aldan, Sov. Un., 10,000.....D15 28
Aldan, plat., Sov. Un........D15 28
Aldan, riv., Sov. Un.........C16 28
Aldeburgh, Eng., 2,972.......B9 12
Aldecoa, Cuba.............h11 64
Alden, Jefferson, Ala. (part of Graysville)..............E4 78
Alden, McHenry, Ill., 200...A5 90
Alden, Hardin, Iowa, 838....B4 92
Alden, Rice, Kans., 239......D5 87
Alden, Antrim, Mich., 190...D5 98
Alden, Freeborn, Minn., 215 .G5 99
Alden, Erie, N.Y., 2,042.....C2 108
Alden, Luzerne, Pa., 1,000...D9 114
Alden Bridge, Bossier, La., 200......................B2 95
Alder, Madison, Mont., 150..E4 102
Alder, brook, Vt............B4 120
Alder, mtn., Scot...........D4 13
Alder, peak, Mont...........E3 102
Alderney, isl., Guernsey......F5 10
Aldershot, Eng., 31,260 (★150,000)................C7 12
Alderson, Alta., Can., 100...D5 69
Alderson, Pittsburg, Okla., 207......................C6 112
Alderson, Greenbrier and Monroe, W. Va., 1,225....D4 123
Aldersyde, Alta., Can., 78....D4 69
Alderwood Manor, Snohomish, Wash., 4,000..............B3 122
Aldie, Loudoun, Va., 100....................A4, C5 121
Aldin, pt., Fiji Is.......... 52
Aldine, Salem, N.J..........D2 106
Aldora, Lamar, Ga., 535.....C2 87
Aldrich, Shelby, Ala., 800...B3 78
Aldrich, Wadena, Minn., 90......................D4 99
Aldrich, Polk, Mo., 181......D4 101
Aldridge, Walker, Ala., 100..B2 78
Ale, riv., Scot.............E6 13
Aledo, Mercer, Ill., 3,080....B3 90
Aledo, Parker, Tex., 450....B5 118
Aleg, Maur., 1,000..........C2 45
Alegre, Braz., 7,487.........C4 56
Alegrete, Braz., 33,735.......D1 56
Alegria, Phil., 2,624........*C6 35
Aleknagik, Alsk., 231........D8 79
Aleksander, wadi, Isr........f10 32
Aleksandriya, Sov. Un., 43,300...................G9 27
Aleksandrov, Sov. Un., 30,600....................C12 27
Aleksandrov-Gay, Sov. Un...C3 29
Aleksandrovskoye, Sov. Un., 12,000..................A9 29
Aleksandrovsk [-Sakhalinskiy], Sov. Un., 39,400..........D17 28
Aleksandrow, Pol., 7,577.....B5 26
Aleksandrow, Pol., 6,926.....C5 26
Aleksin, Sov. Un., 9,300....D11 27
Aleksinac, Yugo., 8,741......D5 22
Alelai, pt., W. Sam.......... 52
Aleman, Sierra, N. Mex.......E3 107
Alemania, Arg.............E2 55
Alençon, Fr., 25,584.........C3 14
Alenquer, Braz., 7,027.......C4 59
Alenquer, Port., 2,498.......C1 20
Alentejo, reg., Port., 685,300.C2 20
Alepokhori, Grc...........g10 23
Aleppo (Halab), Syr., 425,500...................D11 31
Aleria, Fr., 907.............C2 21
Alert, Decatur, Ind., 80.....F6 95
Alert, Franklin, N.C.........A5 109
Alert Bay, B.C., Can., 825...D4 68
Alès, Fr., 41,360...........E6 14
Alessandria, It., 92,760......B2 21
Ålestrup, Den., 1,763........B3 24
Ålesund, Nor., 19,200 (★30,500)................F2 25
Alet, Fr., 695..............F5 14
Aleutian, is., Alsk..........E3 79
Aleutian, range, Alsk.........D9 79
Alex, Grady, Okla., 545......C4 112
Alexander, Pulaski and Saline, Ark., 177..............C3, D5 81
Alexander, Man., Can., 269..E1 71
Alexander, Burke, Ga., 150..C5 87
Alexander, Morgan, Ill., 300.D3 90
Alexander, Franklin, Iowa, 294.....................B4 92
Alexander, Rush, Kans., 153.D4 93
Alexander, Genesee, N.Y., 335.....................C2 108
Alexander, Buncombe, N.C., 75......................D3 109
Alexander, McKenzie, N. Dak., 269.....................B2 110
Alexander, Upshur, W. Va., 25......................C4 123
Alexander, co., Ill., 16,061....F4 90
Alexander, co., N.C., 15,625..B2 109
Alexander, arch., Alsk.......D12 79
Alexander, lake, Minn........D4 99
Alexander Bay, Newf., Can., 20.....................D4 75
Alexander Bay, S. Afr., 2,066.................C2 49
Alexander I, isl., Ant.........B5 5
Alexander City, Tallapoosa, Ala., 13,140..............C4 78
Alexander Humboldt, mts., Ant.B14 5
Alexander Mills, Rutherford, N.C., 947...........B3, D4 109
Alexandra, N.Z., 2,296.....P12 51
Alexandra, isl., Sov. Un......A10 4
Alexandretta, gulf, Tur.......D10 31
Alexandria, Calhoun, Ala., 200.....................B4 78
Alexandria, B.C., Can., 205..C6 68
Alexandria, Ont., Can., 2,487....................B10 72
Alexandria (Al Iskandarīyah), Eg., U.A.R., 1,277,819 (★1,325,000)...............C5 43
Alexandria, Madison, Ind., 5,582....................D6 91
Alexandria, Campbell, Ky., 1,318...................A7, B5 94
Alexandria, Rapides, La., 40,279..................C3 95
Alexandria, Douglas, Minn., 6,713.....................E3 99
Alexandria, Clark, Mo., 452..A6 101
Alexandria, Thayer, Nebr., 257......................D8 103
Alexandria, Grafton, N.H., 75 (370▲)...............C3 105
Alexandria, Licking, Ohio, 452.....................B5 111
Alexandria, Huntingdon, Pa., 381......................E5 114
Alexandria, Rom., 19,294....D7 22
Alexandria, Hanson, S. Dak., 614.....................D8 116
Alexandria, DeKalb, Tenn., 599.....................C7 117
Alexandria (Independent City), Va., 91,023..........B5, C5 121
Alexandria Bay, Jefferson, N.Y., 1,583...........A5, B1 108
Alexandria Southwest, Rapides, La., 2,782............*C3 95
A exandrina, lake, Austl......G2 51
Alexandroupolis, Grc., 18,712....................B5 23
Alexis, Warren and Mercer, Ill., 878....................B3 90
Alexis, riv., Newf., Can......B3 79
Alexis Creek, B.C., Can., 134.C6 68
Alexo, Alta., Can..........C3 69
Aley, riv., Sov. Un..........E25 9
Aleysk, Sov. Un., 10,000...C10 29
Alfalfa, co., Okla., 8,445....A3 112
Al Fandaqūmīyah, Jordan, 3,000...............B7, f11 32
Al Farāfirah, Eg., U.A.R., 749.....................D5 43
Alfaro, Sp., 8,570..........A5 20
Alfatar, Bul., 3,346.........D8 22
Al Fayyūm (El Faiyūm), Eg., U.A.R., 84,800............E2 32
Alfeld, Ger., 13,100.........B4 17
Alfenas, Braz., 16,051....C3, k9 56
Alfios (Alpheus), riv., Grc.....D3 23
Alfonsine, It., 4,606.........E8 18
Alford, Eng., 2,134..........A8 12
Alford, Jackson, Fla., 380.B1, G3 86
Alford, Pike, Ind., 25........H3 91
Alford, Berkshire, Mass., 30 (256▲)................B1 97
Alford, Scot., 1,248.........C6 13
Alfordsville, Daviess, Ind., 121......................C4 91
Alfortville, Fr., 32,332......g10 14
Alfred, Ont., Can., 1,257...B10 72
Alfred, York, Maine, 300 (1,201▲)...............E2 96
Alfred, Allegany, N.Y., 2,807....................C3 108
Alfred, La Moure, N. Dak., 150......................C6 110
Alfred Station, Allegany, N.Y., 200..................C3 108
Al Fuqaha, Libya..........D3 43
Algā, Eth.................D5 47
Algarrobal, Chile..........E1 55
Algarrobo, Chile...........E1 55
Algarrobo del Aguilla, Arg..B3 54
Algarve, prov., Port., 328,231...............*D2 20
Algarve, reg., Port., 314,800 .D1 20
Algeciras, Sp., 66,317........D3 20
Algemesí, Sp., 19,057.......C5 20
Alger, see Algiers, Alg.
Alger, Arenac, Mich., 60....D6 98
Alger, Hardin, Ohio, 1,068...B4 111
Alger, co., Mich., 9,250.....B4 98
Algeria, country, Afr., 10,003,000............D5 42, D4 44
Algete, Sp., 1,150..........o18 20
Al Ghayatah, Eg., U.A.R., 1,000....................D2 32
Alghero, It., 23,000........D2 21
Al Ghurdaqah, Eg., U.A.R., 2,727....................D6 43
Algiers (Alger), Alg., 361,285 (★587,570)..........F8 30, B5 44
Algoa, bay, S. Afr...........D4 49
Algodones, Sandoval, N. Mex., 300..........B3, G5 107
Algoma, Pontotoc, Miss., 190.A4 100
Algoma, Kewaunee, Wis., 3,855....................D6 124
Algoma, co., Ont., Can., 82,059...................A2 72
Algoma Mills, Ont., Can., 373......................A2 72
Algona, Kossuth, Iowa, 5,702.....................A3 92
Algona, King, Wash., 1,311...............B3, D2 122
Algonac, St. Clair, Mich., 3,190.....................F8 98
Algonquin, Ont., Can., 120..C9 72
Algonquin, McHenry, Ill., 2,692.................A5, E2 90
Algonquin, park, Ont., Can.....B6 72
Algonquin Park, Ont., Can., 110.....................B6 72
Algood, Putnam, Tenn., 886.C8 117
Algorta, Ur...............E1 56
Al Ḥāfir (Oasis), Sau. Ar...H13 31
Alhama, Sp., 9,849..........D4 20
Alhama, Sp., 6,442..........D5 20
Alhambra, Los Angeles, Calif., 54,807..............F2 82
Alhambra, Madison, Ill., 537.E4 90
Alhambra, Jefferson, Mont., 40......................D5 102
Al Ḥāmūl, Eg., U.A.R., 20......................D3 32
Alhaurín el Grande, Sp., 11,525..................D3 20
Al Ḥawṣah (Oasis), Jordan.H11 31
Al Ḥayy, Eg., U.A.R., 1,000.E3 32
Al Ḥayyānīyah (Oasis), Sau. Ar.................H14 31
Alhos Vedros, Port., 2,509....f9 20
Al Ḥudaydah, Yemen, 40,000...................C5 47
Al Hufūf (Hofuf), Sau. Ar., 90,000...................I4 41
Al Ḥumaymah, Jordan, 1,000.....................E7 32
Al Ḥuṣn, Jordan, 1,000.......B7 32
Aliakmon, riv., Grc..........B3 23
Alibag, India, 9,909........H4 40
Alicahue, Chile...........A2 54
Alicante, Sp., 121,527.......C5 20
Alicante, prov., Sp., 711,942..................*C5 20
Alicante, gulf, Sp...........C5 20
Alice, Ont., Can., 97........B7 72
Alice, Cass, N. Dak., 124....C8 110
Alice, Jim Wells, Tex., 20,861....................F3 118
Alice, lake, Minn...........C7 99
Alicel, Union, Oreg., 35.....B9 113
Alice Southwest, Jim Wells, Tex., 1,813.............*F3 118
Alice Springs, Austl., 4,648...D5 50
Aliceville, Pickens, Ala., 3,194...................B1 78
Aliceville, Coffey, Kans., 100.D8 93
Alicia, Lawrence, Ark., 236...B4 81
Alicudi, isl., It.............E5 21
Alida, Sask., Can., 241......H5 70
Alief, Harris, Tex., 300......F4 118
Aligarh, India, 185,020......C6 39
Ali Gharbi, Iraq, 3,377......E3 41
Aligūdarz, Iran, 8,459.......E4 41
Alindao, Cen. Afr. Rep.....E4 46
Aline, Chandler, Ga., 100...D4 87
Aline, Alfalfa, Okla., 314.....A3 112
Alingsås, Swe., 17,500.......I5 25
Alipur Duar, India, 28,927.D12 40
Aliquippa, Beaver, Pa., 26,369.................A5, E1 114
Alisal, Monterey, Calif., 16,473......................C6 82
Al Iskandarīyah, see Alexandria, Eg., U.A.R.
Al Isma 'īlīyah, Eg., U.A.R., 82,100...................G9 31, D4 32
Alistrati, Grc., 4,951........B5 23
Aliwal North, S. Afr., 10,706..D4 49
Alix, Franklin, Ark., 350.....B2 81
Alix, Alta., Can., 631.......C4 69
Al Jaghbūb, Libya, 196.....D4 43
Al Jāniyah, Jordan, 1,000...h11 32
Al Jawf (Jauf), Sau. Ar., 10,000....................D7 43
Al Jawsh, Libya, 2,680......C2 43
Aljezur, Port., 5,286.........D1 20
Al Jīzah (El Giza) (Gizeh), Eg., U.A.R., 98,000........C6 43
Aljustrel, Port., 5,844.......D1 20
Alkabo, Divide, N. Dak., 70..A2 110
Al Kadhimain, Iraq, 126,443.................F15 31
Alkali, creek, Wyo...........D4 125
Alkali, lake, Nev............B2 104
Alkali, lake, Oreg...........E6 113
Alkaline, lake, N. Dak........C6 110
Al Karak, Jordan, 5,539.....C7 32
Al Kawm aṭ Ṭawīl, Eg., U.A.R., 1,000.....................C3 32
Al Khābūrah, Mus. & Om..D2 39
Al Khalīl (Hebron), Jordan, 35,983....................C7 32
Al Khalūf, Mus. & Om.....D2 39
Al Khārijah, Eg., U.A.R., 11,155...................D6 43
Al Khasab, Mus. & Om.....H8 41
Al Khaṭaṭibah, Eg., U.A.R., 1,000....................D2 32
Al Khums, Libya, 62,272....C2 43
Al Khurmah, Sau. Ar.........A5 47
Alkionidhes, gulf, Grc.......g10 23
Alkionidhes, is., Grc.........g9 23
Alkmaar, Neth., 44,100 (★71,000)....................B4 15
Al Kubr, isl., Kuw...........G4 41
Al Kūbrī, Eg., U.A.R., 1,000.....................D4 32
Al Kufrah, oasis, Libya, Eg., U.A.R.....................E4 43
Al Kuh, cape, Iran...........I8 41
Al Kuntillah, Eg., U.A.R., 1,000.....................E6 32
Allada, Dah., 4,700..........E5 45
Al Lādhiqīyah (Latakia), Syr., 67,600.............E10 31
Allagash, Aroostook, Maine, 500 (557▲)...............A3 96
Allagash, lake, Maine........B3 96
Allagash, riv., Maine.........B3 96
Allahabad, India, 411,955 (★430,730)..........C7 39, E8 40
Allakaket, Alsk., 115........B9 79
Allamakee, co., Iowa, 15,982 .A6 92
Allamoore, Hudspeth, Tex., 20......................F2 118
Allamuchy, Warren, N.J., 150.....................B3 106
Allamuchy, mtn., N.J..........B3 106
Allan, Sask., Can., 417......F2 70
Allanmyo, Bur., 15,580.....E10 39
Allāqui, wadi, Eg., U.A.R......E6 43
Allardt, Fentress, Tenn., 650......................C9 117
Allariz, Sp., 9,403..........A2 20
Allaykha, Sov. Un., 800....B36 4
Alleene, Little River, Ark., 120......................D1 81
Allegan, Allegan, Mich., 4,827....................F5 98
Allegan, co., Mich., 57,729...F4 98
Allegany, Cattaraugus, N.Y., 2,064....................C2 108
Allegany, Coos, Oreg., 40...D2 113
Allegany, co., Md., 84,169....A1 85
Allegany, co., N.Y., 43,978...C2 108
Allegany, Indian res., N.Y......C2 108
Alleghany, Sierra, Calif., 200.....................C3 82
Alleghany, Alleghany, Va., 150......................D2 121
Alleghany, co., N.C., 7,734...A2 109
Alleghany, co., Va., 12,128...D3 121
Allegheny, co., Pa., 1,628,587................E2 114
Allegheny, mts., U.S..........C11 71
Allegheny, plat., Pa., W. Va..............E1 114, C3 123
Allegheny, riv., N.Y., Pa.......E2 114
Allegheny Heights, mtn., Md....D1 85
Allegros, peak, N. Mex........C1 107
Alleman, Polk, Iowa, 100.....A7 92
Allemands, St. Charles, La., 1,167.................C6, E5 95
Allen, Clarke, Ala., 50......D2 78
Allen, Lyons, Kans., 205....D7 93
Allen, Floyd, Ky., 400......C7 94
Allen, Wicomico, Md., 175...D6 85
Allen, Dixon, Nebr., 350....B9 103
Allen, Hughes and Pontotoc, Okla., 1,005..............C5 112
Allen, Cumberland, Pa., 300 .F7 114
Allen, Bennett, S. Dak., 100..D4 116
Allen, co., Ind., 232,196.....B7 91
Allen, co., Kans., 16,369.....E8 93
Allen, co., Ky., 12,269.......D3 94
Allen, co., Ohio, 103,691....B3 111
Allen, par., La., 19,867......D3 95
Allen, lake, Ire.............C3 11
Allen, mtn., N.Z............Q11 51
Allendale, Wabash, Ill., 465..E6 90
Allendale, Worth, Mo., 136..A3 101
Allendale, Bergen, N.J., 4,092....................A4 106
Allendale, Allendale, S.C., 3,114.....................F5 115
Allendale, co., S.C., 11,362...E5 115
Allende, Mex., 9,938........B4 63

Allenford, Ont., Can., 209...C3 72
Allenhurst, Brevard, Fla., 200...D6 86
Allenhurst, Liberty, Ga., 200..E5 87
Allenhurst, Monmouth, N.J., 795...C4 106
Allenhurst, Matagorda, Tex., 40...G4 118
Allen Park, Wayne, Mich., 37,494...F7 98
Allenport, Washington, Pa., 981...*F2 114
Allenspark, Boulder, Colo., 40...A5 83
Allenstein, see Olsztyn, Pol.
Allenstown, Merrimack, N.H., (1,789▲)...D4 105
Allensville, Todd, Ky., 286...D2 94
Allensville, Vinton, Ohio, 100.C5 106
Allensville, Mifflin, Pa., 300..E6 114
Allenton, Wilcox, Ala., 100..B2 78
Allenton, St. Louis, Mo., 350.B7 101
Allenton, Washington, Wis., 350...E5 124
Allentown, Wilkinson, Ga., 225...D3 87
Allentown, Monmouth, N.J., 1,393...C3 106
Allentown, Allegany, N.Y., 400...C2 108
Allentown, Lehigh, Pa., 108,347 (★299,700)...E11 114
Allentown, King, Wash., 600...D1 122
Allenville, Moultrie, Ill., 191.D5 90
Allenville, Mackinac, Mich., 35...C6 98
Allenwood, Monmouth, N.J., 350...C4 106
Allenwood, Union, Pa., 300..D8 114
Alleppey, India, 138,834...G6 39
Aller (Cabañaquinta), Sp., 828...A3 20
Aller, riv., Ger...A6 17
Allerton, Champaign and Vermilion, Ill., 282...D6 90
Allerton, Wayne, Iowa, 692..D4 92
Allerton, pt., Mass...B6 97
Alley Spring, Shannon, Mo., 30...D6 101
Allgood, Blount, Ala., 147...B3 78
Alliance, Alta., Can., 291...C5 69
Alliance, Box Butte, Nebr., 7,845...B3 103
Alliance, Pamlico, N.C., 200..B7 109
Alliance, Stark, Ohio, 28,362 B6 111
Allier, dept., Fr., 380,221...*D5 14
Allier, riv., Fr...D5 14
Al Lifiyah (Oasis) Sau. Ar..G14 31
Alligator, Bolivar, Miss., 227 A3 100
Alligator, lake, Maine...D4 96
Alligator, lake, N.C....B7 109
Alligator, riv., N.C....B7 109
Allihies, Ire., 77...F1 11
Allimaso, creek, N. Mex...C5 107
Allingåbro, Den., 1,312...B4 24
Allinge, Den., 2,114...C8 24
Allison, La Plata, Colo., 125..D3 83
Allison, Butler, Iowa, 952...B5 92
Allison, McKinley, N. Mex., 25...B1 107
Allison, Fayette, Pa., 1,285...G2 114
Allisona, Williamson, Tenn., 100...B5 117
Allison Harbour, B.C., Can D4 68
Allisonia, Pulaski, Va., 160...E2 121
Allison Park, Allegheny, Pa., 5,000...A6 114
Alliston, Ont., Can., 2,426...C5 72
Al Līth, Sau. Ar., 10,000...A5 47
Alloa, Scot., 13,895...D5 13
Allons, Overton, Tenn., 270..C8 117
Allouez, Keweenaw, Mich., 175...A2 98
Allouez, Brown, Wis., 9,557.*A5 124
Alloway, Salem, N.J., 800...D2 106
Alloway, creek, N.J...D2 106
Allred, Overton, Tenn., 100..C8 117
All-Sabieh, Fr. Som...C5 47
Allsboro, Colbert, Ala., 50...A1 78
Allschwil, Switz., 12,875...A4 19
Al Luḩayyah, Yemen, 5,000 .B5 47
Allumette, lake, Ont., Can...B7 72
Allyns Point, New London, Conn., 75...D8 84
Alma, Crawford, Ark., 1,370..B1 85
Alma, N.B., Can., 476...D5 74
Alma, Ont., Can., 167...D4 72
Alma, Que., Can., 13,309...A6 73
Alma, Park, Colo., 107...B4 83
Alma, Bacon, Ga., 3,515...E4 87
Alma, Marion, Ill., 358...E5 90
Alma, Wabaunsee, Kans., 838.C7 93
Alma, Gratiot, Mich., 8,978..E6 98
Alma, Lafayette, Mo., 390...B4 101
Alma, Liberty, Mont., 35...B6 102
Alma, Harlan, Nebr., 1,342...D6 103
Alma, Robeson, N.C., 60...C4 109
Alma, Stephens, Okla., 120...C4 112
Alma, Tyler, W. Va., 250...B4 123
Alma, Buffalo, Wis., 1,008...D2 124
Alma, hill, N.Y...C2 108
Al Ma'ādī, Eg., U.A.R., 4,000...E3 32
Alma-Ata, Sov. Un., 508,000...E9 29
Alma Center, Jackson, Wis., 464...D3 124
Almada, Port., 11,995...C1, f9 20
Almadén, Sp., 13,443...C3 20
Al Madīnah (Medina), Sau. Ar., 80,000...E7 43
Al-Mafraq, Jordan, 1,000...F11 31
Almagro, Sp., 9,681...C4 20
Al Maḩallah al Kubrá, Eg., U.A.R., 132,300...D3 32
Al Maḩmūdīyah, Eg., U.A.R., 1,000...C2 32
Al Maḩsamah, Eg., U.A.R., 1,000...D4 32
Al Maḩtab, Sau. Ar...H11 31
Al Makīlī, Libya...B3 31
Almanor, lake, Calif...B3 82
Almansa, Sp., 15,391...C5 20
Al Manşūrah, Eg., U.A.R., 119,000...C3 32, C6 43
Al Manzilah, Eg., U.A.R., 2,000...C3 32
Almanzora, riv., Sp...D4 20
Almargem [do Bispo], Port., 4,867...f9 20
Al Marj (Barce), Libya, 12,000...C4 43, F3 31
Almartha, Ozark, Mo., 225..E5 101
Al Maţarīyah, Eg., U.A.R., 2,000...C4 32
Almaville, Rutherford, Tenn., 60...B5 117
Almazán, Sp., 3,958...B4 20
Al Mazār, Jordan, 1,000...C7 32
Almeirim, Braz., 2,082...C4 56
Almeirim, Port., 7,104...C1 25
Almelo, Neth., 51,300...B6 15
Almelund, Chisago, Minn., 150...E6 99
Almena, Norton, Kans., 555..C4 93
Almena, Barron, Wis., 398...C1 124
Almendares, Cuba...h11 64
Almendralejo, Sp., 21,884...C2 20
Almeria, Loup, Nebr., 18...C6 103
Almería, Sp., 86,808...D4 20
Almería, prov., Sp., 360,777.*D4 20
Almería, gulf, Sp...D4 20
Almería, riv., Sp...D4 20
Älmhult, Swe., 5,300...B8 24
Al Minyā, Eg., U.A.R., 93,300...D6 43
Almira, Lincoln, Wash., 414.B7 122
Almirante, Pan., 3,521...F6 62
Almirante Brown, Arg...g7 54
Almiros, Grc., 7,034...C4 23
Al Mishab, cape, Sau. Ar...G4 41
Almo, Cassia, Idaho 100...G5 89
Almo, Calloway, Ky., 150...B3 94
Almodôvar, Port., 4,390...D1 20
Almodóvar, Sp., 14,633...C3 20
Almogía, Sp., 8,341...D3 20
Almon, Newton, Ga., 300...C3 87
Almonaster, Sp., 4,770...D2 20
Almond, Randolph, Ala., 50 .B4 78
Almond, Allegany and Steuben, N.Y., 696...C3 108
Almond, Swain, N.C., 100...D2 109
Almond, Portage, Wis., 391..D4 124
Almond, riv., Scot...D5 13
Almonesson, Gloucester, N.J., 1,500...*D2 106
Almont, Gunnison, Colo., 11 .C4 83
Almont, Lapeer, Mich., 1,279...F7 98
Almont, Morton, N. Dak., 190...C4 110
Almonte, Marin, Calif., 600.*B5 82
Almonte, Ont., Can., 2,960...B8 72
Almonte, Sp., 11,538...D2 20
Almonte, riv., Sp...C3 20
Almora, India, 16,004...C6 39
Almoradí, Sp., 3,998...C5 20
Al Mughaiyir, Jordan, 1,000.g12 32
Al Mukallā, Aden, 25,000...C6 47
Al Mukhā (Mocha) (Mokha), Yemen, 5,000...C5 47
Almuñécar, Sp., 6,235...D4 20
Al Muwayliḩ, Sau. Ar...I10 31
Almy, Uinta, Wyo., 25...D2 125
Almyra, Arkansas, Ark., 240.C4 81
Aln, riv., Eng...E7 13
Alness, Scot., 1,019...C4 13
Alnwick, Eng., 7,482...C6 10
Aloha, Washington, Oreg., 4,000...B2 113
Along, bay, Viet...B7 38
Alonsa, Man., Can., 133...D2 71
Alor, isl., Indon...G6 35
Álora, Sp., 5,960...D3 20
Alor Star, Mala., 52,915...I4 38
Alorton (Fireworks), St. Clair, Ill., 3,282...*E3 90
Alosno, Sp., 5,814...D2 20
Alpachiri, Arg...B4 54
Alpaugh, Tulare, Calif., 600 .E4 82
Alpbach, Aus., 1,576...E5 16
Alpena, Boone, Ark., 283...A2 81
Alpena, Alpena, Mich., 14,682...C7 98
Alpena, Jerauld, S. Dak., 407 .C7 116
Alpena, Randolph, W. Va., 50...C5 123
Alpena, co., Mich., 28,556...C7 98
Alpenrose, Switz...C8 19
Alpes-Maritimes, dept., Fr., 618,265...*F7 14
Alpha, Henry, Ill., 637...B3 90
Alpha, Fayette, Iowa, 119...B5 92
Alpha, Clinton, Ky., 100...D4 94
Alpha, Iron, Mich., 317...B2 98
Alpha, Jackson, Minn., 207..G4 99
Alpha, Warren, N.J., 2,406..B2 106
Alpha, Greene, Ohio, 250...C3 111
Alpharetta, Fulton, Ga., 1,349...B2 87
Alphen aan den Rijn, Neth., 24,900...B4 15
Alpheus, see Alfios, riv., Grc.
Alpiarça, Port., 6,680...C1 20
Alpine, Talladega, Ala., 100..B3 78
Alpine, Apache, Ariz., 300..D6 80
Alpine, Clark, Ark., 75...C2 81
Alpine, San Diego, Calif., 1,044...*F5 82
Alpine, Bonneville, Idaho, 50.F7 89
Alpine, Bergen, N.J., 921...D6 106
Alpine, Benton, Oreg., 115..C3 113
Alpine, Overton, Tenn., 65..C8 117
Alpine, Brewster, Tex., 4,740...F2 118
Alpine, co., Calif., 397...C4 82
Alpoca, Wyoming, W. Va., 400...D3 123
Alps, mts., Eur...F10 8
Al Qaddāḩīyah, Libya...I14 30
Al Qāhirah, see Cairo, Eg., U.A.R.
Al Qanāyāt, Eg., U.A.R., 12,385...D3 32
Al Qanţarah, Eg., U.A.R., 2,000...D4 32
Al Qaryah ash Sharqīyah, Libya...C2 43
Al Qaşabāt, Libya, 3,190...C2 43
Al Qaşr, Eg., U.A.R...D5 43
Al Qaţīf, Sau. Ar., 5,000...H5 41
Al Qaţrānah, Jordan, 4,000.G11 31
Al Qaţrūn, Libya, 1,674...E2 43
Al Qulaiyaba (Oasis), Sau. Ar...H14 31
Al Qunaytirah, Syr., 17,100.F10 31
Al Qurna, Iraq, 3,156...F3 41
Al Quşaymah, Eg., U.A.R., 1,000...D6 32
Al Quşayr, Eg., U.A.R...D6 43
Alright, isl., Que., Can...B8 74
Alrö, Den...C4 24
Als, Den., 685...B4 24
Alsace, former prov., Fr., 1,317,000...C7 14
Alsask, Sask., Can., 230...F1 70
Alsasua, Sp., 5,927...A4 20
Alsdorf, Ger., 31,000...D6 15
Alsea, Benton, Oreg., 200...C3 113
Alsek, riv., Alsk., Yukon, Can.D12 79
Alsen, Cavalier, N. Dak., 228 .A7 110
Alsenbrück-Langmeil, Ger., 740...D2 17
Alsey, Scott, Ill., 248...D3 90
Alsfeld, Ger., 9,900...C4 17
Alsike, Swe., 271...t35 25
Alsip, Cook, Ill., 3,770...*F3 90
Alstead, Cheshire, N.H., 325 (843▲)...D2 105
Alstead, riv., Sask., Can...B2 70
Alstead Center, Cheshire, N.H., 40...D2 105
Alston, Eng., 1,724...F6 13
Alston, Montgomery, Ga., 154...D4 87
Alsuma, Tulsa, Okla., 500..A6 112
Alta, Buena Vista, Iowa, 1,393...B2 92
Alta, Teton, Wyo...B2 125
Altadena, Los Angeles, Calif., 40,568...*F2 82
Altaelv, riv., Nor...C10 25
Alta Gracia, Arg., 11,570...A4 54
Altagracia, Ven., 14,000...A3 60
Altagracia de Orituco, Ven., 13,860...B4 60
Alta Hill, Nevada, Calif., 1,078...*C3 82
Al Tahoe, El Dorado, Calif., 1,500...C4 82
Altai, mts., Asia...B3 34
Alta Loma, San Bernardino, Calif., 850...*F5 82
Alta Loma, Galveston, Tex., 1,020...G5 118
Altamaha, riv., Ga...E4 87
Altamaha, sound, Ga...E5 87
Altamahaw, Alamance, N.C., 625...A4 109
Altamira, Braz., 2,939...C4 59
Altamira, Chile...E2 55
Altamirano, Arg...g7 54
Altamont, Alameda, Calif., 20...B6 82
Altamont, Man., Can., 123..E2 71
Altamont, Effingham, Ill., 1,656...D5 90
Altamont, Labetta, Kans., 672...E8 93
Altamont, Daviess, Mo., 190.B3 101
Altamont, Albany, N.Y., 1,365...C6 108
Altamont, Klamath, Oreg., 10,811...E5 113
Altamont, Deuel, S. Dak., 77...C9 116
Altamont, Grundy, Tenn., 552...D8 117
Altamont, Duchesne, Utah, 102...C5 119
Altamonte Springs, Seminole, Fla., 1,212...D5 86
Altamura, It., 43,735...D6 21
Altamura, isl., Mex...C3 63
Altapass, Mitchell, N.C., 150.C4 109
Altar, Mex., 1,116...A2 63
Altario, Alta., Can., 72...D5 69
Alta Vista, Chickasaw, Iowa, 276...A5 92
Alta Vista, Wabaunsee, Kans., 400...D7 93
Altavista, Jim Hogg, Tex...F3 118
Altavista, Campbell, Va., 3,299...D3 121
Altdorf, Ger., 5,800...D6 17
Altdorf, Switz., 7,477...C6 19
Altefähr, Ger., 11,445...D7 24
Alte Mellum, isl., Ger...E2 24
Altenberg, Ger., 1,796...C8 17
Altenbruch, Ger., 4,000...E2 24
Altenburg, Ger., 46,800...C7 17
Altenburg, Perry, Mo., 260..D8 101
Altenkirchen, Ger., 4,500...C2 17
Altenkirchen, Ger., 1,478...D7 24
Altenmarkt, Ger., 2,800...A8 18
Altentreptow, Ger., 8,604..E7 24
Alter do Chão, Port., 4,633..C2 20
Altha, Calhoun, Fla., 413...B1 86
Altheim, Aus., 4,271...A9 18
Altheimer, Jefferson, Ark., 979...C4 81
Altinho, Braz., 3,825...C3, k6 57
Altkirch, Fr., 4,246...B3 18
Altman, Screven, Ga., 100...D5 87
Altman, Oswego, N.Y., 277..B5 108
Altmühl, riv., Ger...E6 17
Altnaharra, Scot...B4 13
Alto, Banks and Habersham, Ga., 275...B3 87
Alto, Richland, La., 100...F5 95
Alto, Kent, Mich., 250...F5 98
Alto, Lincoln, N. Mex., 8...D4 107
Alto, Franklin, Tenn., 85...B6 117
Alto, Cherokee, Tex., 869...D5 118
Alto Alentejo, prov., Port., 400,374...*C2 20
Alto Araguaia, Braz., 2,077..B2 56
Alto Cedro, Cuba, 679...E6 64
Alto Cuchumatanes, mts., Guat..C2 62
Alto Longá, Braz., 784...C2 57
Alto Molocuè, Moz...A6 49
Altomünster, Ger., 1,900...A7 18
Alton, Jefferson, Ala., 350.B3, E5 78
Alton, Ont., Can., 438...D4 72
Alton, Eng., 9,158...C7 12
Alton, Madison, Ill., 43,047..E3 90
Alton, Sioux, Iowa, 1,048...B1 92
Alton, Osborne, Kans., 299..C5 93
Alton, St. Tammany, La., 60 .B8 95
Alton, Oregon, Mo., 677...E6 101
Alton, Belknap, N.H., 300 (1,241▲)...D4 105
Alton, Franklin, Ohio, 180...C2 111
Alton, Washington, R.I., 300...D10 84
Alton, Kane, Utah, 116...F3 119
Alton, Upshur, W. Va., 70..C4 123
Altona, Man., Can., 2,026...E3 71
Altona, Ger. (part of Hamburg)...E3 24
Altona, Knox, Ill., 505...B3 90
Altona, DeKalb, Ind., 313...B7 91
Altona, Mecosta, Mich., 150.E5 98
Altona, Clinton, N.Y., 400...A3 108
Altonah, Duchesne, Utah, 10.C5 119
Alton Bay, Belknap, N.H., 300 (1,241▲)...D4 105
Alton North, Madison, Ill., 1,505...*E3 90
Altoona, Etowah, Ala., 744..A3 78
Altoona, Lake, Fla., 200...D5 86
Altoona, Polk, Iowa, 1,458...A7, C4 92
Altoona, Wilson, Kans., 490..E8 93
Altoona, Blair, Pa., 69,407 (★104,500)...E5 114
Altoona, Eau Claire, Wis., 2,114...D2 124
Alto Paraná, dept., Par., 9,531...E5 55
Alto Pass, Union, Ill., 323...F4 90
Alto Trombetas, riv., Braz...B3 59
Altötting, Ger., 9,200...A8 18
Altrincham, Eng., 41,104...A5 12
Altro, Breathitt, Ky., 300...C6 94
Altstätten, Switz., 8,751...B8 19
Altun Kopru, Iraq, 3,744...E15 31
Altura, Winona, Minn., 320..F7 99
Alturas, Modoc, Calif., 2,819.B3 82
Altus, Franklin, Ark., 392...B2 81
Altus, Jackson, Okla., 21,225...C2 112
Altus, res. and dam, Okla...C2 112
Altyn Tagh, mts., China...D2 34
Al 'Ujaylāt, Libya...H13 32
Alula, Som., 1,300...C7 47
Alum, creek, Ohio...C2 111
Alum Bank (Pleasantville), Bedford, Pa., 300...F4 114
Alum Bridge, Lewis, W. Va., 125...B4 123
Alum Creek, Kanawha, W. Va., 500...C5 123
Aluminé, Arg...B2 54
Alum Rock, Santa Clara, Calif., 18,942...*D3 82
Alunite, Clark, Nev., 10...H7 104
Alupka, Sov. Un., 21,200...I10 27
Al Uqaylah (Aghéila), Libya .C3 43
Al 'Uqayr, Sau. Ar...I5 41
Al Uqşur (Luxor), Eg., U.A.R., 30,100...D6 43
Al Uthaylah, Libya...C4 43
Alva, Lee, Fla., 300...F5 86
Alva, Harlan, Ky., 700...D6 94
Alva, Woods, Okla., 6,258...A3 112
Alva, Crook, Wyo., 60...A8 125
Alvarado, Mex., 12,424...D5 63
Alvarado, Marshall, Minn., 282...B2 99
Alvarado, Johnson, Tex., 1,907...B5, C4 118
Alvaton, Meriwether, Ga., 24...C2 87
Älvdalen, Swe., 1,771...G6 25
Alvear, Arg., 3,544...E4 55
Alvena, Sask., Can., 220...E2 70
Alverca [do Ribatejo], Port., 4,665...f9 20
Alverton, Westmoreland, Pa., 300...F2 114
Alvesta, Swe., 6,600...B8 24
Alvin, Vermilion, Ill., 281...C6 90
Alvin, Berkeley, S.C., 100...E8 115
Alvin, Brazoria, Tex., 5,643...E5, F5 118
Alvinston, Ont., Can., 652...E3 72
Alviso, Santa Clara, Calif., 1,174...*D2 82
Alvo, Cass, Nebr., 159..D9, E2 103
Alvord, Lyon, Iowa, 238...A1 92
Alvord, Wise, Tex., 694...C4 118
Alvord, lake, Oreg...E8 113
Alvordton, Williams, Ohio, 388...A3 111
Älvsborg, co., Swe., 376,500 .A6 24
Älvsjö, Swe...t36 25
Alvwood, Itasca, Minn., 20..C4 99
Al Wajh, Sau. Ar...D7 43
Alwar, India, 72,707...C6 39
Al Wāsiţah, Eg., U.A.R., 4,000...E3 32
Aly, Yell, Ark., 35...C2 81
Alyaty-Pristan, Sov. Un., 500.F3 29
Alyth, Scot., 1,862...D5 13
Alytus, Sov. Un., 9,084...A8 26
Alz, riv., Ger...D6 16
Alzada, Carter, Mont., 60..E12 102
Alzey, Ger., 11,900...D3 17
Ama, St. Charles, La., 600...C7 95
Amadeus, lake, Austl...D5 50
Amadi, Sud...D3 47
Amado, Santa Cruz, Ariz., 30...F4 80
Amador, co., Calif., 9,990...C3 82
Amagansett, Suffolk, N.Y., 1,095...D4 108
Amagasaki, Jap., 405,955...o14 37
Amager, isl., Den...C6 24
Amagon, Jackson, Ark., 234..B4 81
Amahai, Indon...F7 35
Amakusa, sea, Jap...J5 37
Amål, Swe., 9,200...H5 25
Amalfi, It., 4,700...D5 21
Amalga, Cache, Utah, 198...B4 119
Amalia, Taos, N. Mex., 10..A4 107
Amalias, Grc., 15,468...D3 23
Amalner, India, 46,963...G5 40
Amambay, dept., Par., 18,160...D4 55
Amamio (Amami), isl., Ryukyu Is...F10 34
Amana, Iowa, Iowa, 465...C6 92
Amaná, lake, Braz...D5 60
Amanda, Fairfield, Ohio, 732...C5 111
Amangeldy, Sov. Un...C7 29
Amantea, It., 5,822...E6 21
Amantes, pt., Guam... 52
Amapá, Braz., 1,591...B4 59
Amapá, ter., Braz., 68,889...B4 59
Amara, Iraq, 44,064...F3 41
Amaraji, Braz., 2,271...k6 57
Amarante, Braz., 3,199...C2 57
Amaranth, Man., Can., 294..D2 71
Amargosa, Braz., 6,059...D3 57
Amargosa, range, Calif...D5 82
Amargosa, riv., Calif...E5 82
Amarillo, Potter and Randall, Tex., 137,969 (★142,500)..B2 118
Amaro, mtn., It...C5 21
Amarousion, Grc., 20,135...g11 23
Amasa, Iron, Mich., 500...B2 98
Amasra, Tur., 1,379...B9 31
Amasya (Amasia), Tur., 28,200...B10 31
Amawalk, res., N.Y...D2 84
Amazon, Sask., Can., 75... F3 70
Amazon, see Solimões, riv., Braz., Peru
Amazon, see Amazonas, riv., S.A.
Amazonas, dept., Peru, 100,527...B2 58
Amazonas, comisaría, Col., 8,120...D3 60
Amazonas, state, Braz., 721,215...B3 58
Amazonas, ter., Ven., 11,757 .C4 60
Amazonas (Amazon), riv., S.A...D5 47, C3 59
Amazonia, Andrew, Mo., 326 .B3 101
Ambala, India, 76,204 (★181,747)...B6 40
Ambalavao, Malag., 4,000...h9 49
Ambam, Cam...E2 46
Ambanja, Malag...f9 49
Ambarchik, Sov. Un., 800...C19 28
Ambato, Ec., 33,908...B2 58
Ambato-Boeni, Malag...g9 49
Ambatosorata, Malag...g9 49
Amber, Grady, Okla., 300..B4 112
Amberg, Ger., 42,500...D6 17
Amberg, Marinette, Wis., 220...C6 124
Ambérieu [-en-Bugey], Fr., 7,748...E6 14
Amberley, Hamilton, Ohio, 2,951...*D2 111
Amberson, Franklin, Pa., 25..F6 114
Ambia, Benton, Ind., 351...D2 91
Ambikapur, India, 15,240...F9 40
Ambilobe, Malag...f9 49
Amble, Eng., 4,889...E7 13
Ambler, Montgomery, Pa., 6,765...A11, F11 114
Ambo, Eth...D4 47
Ambo, chan., Kwajalein... 52
Ambo, Peru, 1,243...D2 58
Amboear, W. Irian...F8 35
Amboina, Indon., 55,263...F7 35
Amboina, isl., Indon...F7 35
Amboise, Fr., 7,953...D4 14
Ambositra, Malag., 4,636...h9 49
Ambovombe, Malag., 2,250..k9 49
Amboy, San Bernardino, Calif., 125...E6 82
Amboy, Turner, Ga., 75...E3 87
Amboy, Lee, Ill., 2,067...B4 90
Amboy, Miami, Ind., 446...C6 91
Amboy, Blue Earth, Minn., 629...G4 99
Amboy, Clark, Wash., 150...D3 122
Ambridge, Beaver, Pa., 13,865...A5, E1 114
Ambridge Heights, Beaver, Pa., 5,106...*A5 114
Ambriz, Ang., 2,196...C1 48
Ambrizete, Ang., 1,147...C1 48
Ambrose, Coffee, Ga., 244...E3 87
Ambrose, Divide, N. Dak., 220...A2 110
Amchitka, isl., Alsk...E3 79
Amchitka, pass, Alsk...E4 79
Am Dam, Chad, 1,299...C4 46
Amden, Switz., 1,270...B7 19
Amderma, Sov. Un., 10,000..C9 28
Ameagle, Raleigh, W. Va., 500...D3, D6 123
Ameca, Mex., 17,396..C4, m11 63
Amecameca, Mex., 12,271..n14 63
Amechtil, sand dunes, Maur...C2 45
Ameland, isl., Neth...A5 15
Amelia, St. Mary, La., 950...C5, E4 95
Amelia, Holt, Nebr., 50...B7 103
Amelia, Clermont, Ohio, 913...C3 111
Amelia City, Nassau, Fla., 100.B6 86
Amelia, co., Va., 7,815...D4 121
Amelia, isl., Fla...B6 86
Amelia Court House, Amelia, Va., 800...D5 121
Amelinghausen, Ger., 920...E4 24
Amendolara, It., 3,297...E6 21
Amenia, Dutchess, N.Y., 117...D7 108
Amenia, Cass, N. Dak., 220..B8 110
American, highland, Ant...B21 5
American, riv., Calif...C3 82
American Bottom, valley, Oreg..C1 113
American Canyon, Napa, Calif., 800...*C2 82
American Falls, Power, Idaho, 2,123...G6 89
American Falls, dam, Idaho...G6 89
American Falls, res., Idaho...F5 89
American Fork, Utah, Utah, 6,373...C4 119
American Samoa, U.S. dep., Oceania, 20,051... 52
Americus, Sumter, Ga., 13,472...D2 87
Americus, Lyon, Kans., 300..D7 93
Amersfoort, Neth., 70,600 (★99,000)...B5 15
Amery, Man., Can., 35...A4 71
Amery, Polk, Wis., 1,769...C1 124
Ames, Story, Iowa, 27,003...B4 92
Ames, Cloud, Kans., 80...C6 93
Ames, Dodge, Nebr., 65...C9, D2 103
Ames, Major, Okla., 211...A3 112
Amesbury, Essex, Mass., 10,787...A6 97
Amfissa, Grc., 5,546...C4 23
Amga, Sov. Un., 800...C16 28
Amgu, Sov. Un...D8 37
Amgun, riv., Sov. Un...D16 28
Amhar, plat., Eth...D4 47
Amherst, N.S., Can., 10,788 .D5 74
Amherst, Phillips, Colo., 106 .A8 83
Amherst, Brevard, Fla., 1,500...*D6 86
Amherst, Hancock, Maine, 140 (168▲)...D4 96
Amherst, Hampshire, Mass., 10,306 (13,718▲)...B2 97
Amherst, Buffalo, Nebr., 220 .D6 103
Amherst, Hillsboro, N.H., 500 (2,051▲)...E3 105
Amherst, Erie, N.Y., 48,000.*C2 108
Amherst, Lorain, Ohio, 6,750.A5 111
Amherst, Marshall, S. Dak., 75...B8 116
Amherst, Lamb, Tex., 883...B1 118
Amherst, Amherst, Va., 1,200...D3 121
Amherst Portage, Wis., 596..D4 124
Amherst, co., Va., 22,953...D3 121
Amherst, isl., Que., Can...B7 74
Amherstburg, Ont., Can., 4,099...E1 72
Amherstdale, Logan, W. Va., 900...D3, D5 123
Amiata, mtn., It...C3 21
Amicalola, falls, Ga...B2 87
Amidon, Slope, N. Dak., 82..C2 110
Amiens, Fr., 105,433...C5 14
Aminuis, S.W. Afr...B2 49
'Āmir, cape, Libya...C4 43
Amirante, is., Indian O...G24 3
Amiret, Lyon, Minn., 75...F3 99
Amisk, Alta., Can., 127...C5 69
Amisk, lake, Sask., Can...C4 70
Amistad, Union, N. Mex., 35 .B6 107
Amite, Tangipahoa, La., 3,316...D5 95
Amite, co., Miss., 15,573...D3 100
Amite, riv., La...D5 95
Amity, Clark, Ark., 543...C2 81
Amity, De Kalb, Mo., 111...B3 101
Amity, Yamhill, Oreg., 620...B1, B3 113
Amity, Washington, Pa., 155.F1 114
Amity, hills, Oreg...B1 113
Amityville, Suffolk, N.Y., 8,318...E3, E7 108
Amizmiz, Mor., 4,036...I2 30
Amkyokyung, China...C11 40
Amlekhganj, Nep...D10 40
Amlwch, Wales, 2,910...A3 12
'Ammān, Jordan, 108,304..G10 31
Ammanford, Wales, 6,264...C3 12
Ammeloe, Ger., 7,200...A1 17
Ammendorf, Ger. (part of Halle)...B7 17
Ammer, lake, Ger...B7 18
Ammon, Bonneville, Idaho, 1,882...F7 89
Ammonoosuc, riv., N.H...B3 105
Amne Machin, mts., China...E4 34
Amo, Hendricks, Ind., 437...E4 91
Amol, Iran, 14,166...C6 41
Amonate, Tazewell, Va., 875...B3 121
Amoret, Bates, Mo., 261...C3 101
Amorgos, isl., Grc...D5 23
Amorita, Alfalfa, Okla., 74..A3 112
Amory, Monroe, Miss., 6,474 .B5 100
Amos, Que., Can., 6,080...E9 72
Amoy (Hsiamen), China, 224,300...G8 34
Ampanihy, Malag...h8 49
Amparo, Braz., 14,382..C3, m8 56
Amper, riv., Ger...E6 17
Ampezzo, It., 2,413...C8 18
Amposta, Sp., 8,444...B6 20
Amqui, Que., Can., 3,247..*B3 73
'Amrān, Yemen, 20,000...B5 47
Amraoti, India, 137,875 (★161,715)...G6 40
Amreli, India, 34,699...G3 40
Amriswil, Switz., 6,752...A7 19
Amritsar, India, 376,295 (★398,047)...B5 40
Amroha, India, 68,965...C7 40
Amrum, isl., Ger...D2 24
Amsteg, Switz...C6 19
Amsterdam, Decatur, Ga., 400...F2 87
Amsterdam, Bates, Mo., 118.C3 101
Amsterdam, Gallatin, Mont., 55...E5 102
Amsterdam, Neth., 866,300 (★1,680,000)...B4 15
Amsterdam, Montgomery, N.Y., 28,772...C6 108
Amsterdam, Jefferson, Ohio, 931...B7 111
Amsterdam, isl., Indian O...I2 2
Amstetten, Aus., 12,086...D7 16
Amston, Tolland, Conn., 300.C7 84
Am Timan, Chad, 1,859...C4 46
Amu Darya, riv., Sov. Un...H21 9
Amukta, pass, Alsk...E5 79
Amundsen, glacier, Ant...A32 5
Amundsen, gulf, N.W. Ter., Can...B8 66
Amundsen, sea, Ant...B1 5
Amur, riv., Sov. Un...E16 28
Amvrakia, gulf, Grc...C3 23
Amwaco, Kootenai, Idaho...E8 122
'Ana, Iraq, 5,860...E13 31
Anabar, riv., Sov. Un...B14 28
Anabta, Jordan, 3,000...f11 32
Anacoco, Vernon, La., 300..C2 95
Anaconda, Deer Lodge, Mont., 12,054...D4 102
Anaconda, Valencia, N. Mex., 350...B2 107
Anaconda, range, Mont...E3 102
Anacortes, Skagit, Wash., 8,414...A3 122
Anacostia, riv., Md...C1 85
Anadarko, Caddo, Okla., 6,299...B3 112
Anadia, Braz., 2,592...C3, k5 57
Anadyr, Sov. Un., 5,000...C20 28
Anadyr, gulf, Sov. Un...C21 28
Anadyr, range, Sov. Un...C20 28
Anadyr, riv., Sov. Un...C20 28
Anae, isl., Guam... 52
Anafi (Anaphe), isl., Grc...D5 23
Anagance, N.B., Can., 129..D4 74
Anaheim, Orange, Calif., 104,184...F3, F5 82
Anahim, lake, B.C., Can...C5 68
Anahola, Kauai, Haw., 326..A2 88
Anahuac, Chambers, Tex., 1,985...E5, F5 118
Anajás, Braz., 182...C5 59
Anakapalle, India, 46,402...I9 40
Anakie, Austl...A6 51
Anaklia, Sov. Un...A13 31
Analalava, Malag...f9 49
Ana Maria, gulf, Cuba...E4 64
Anambas, is., Indon...E3 35
Anamoose, McHenry, N. Dak., 503...B5 110
Anamosa, Jones, Iowa, 4,616.B6 92
Anamur, Tur., 6,600...D9 31
Anamur, cape, Tur...D9 31
Anandale, Rapides, La., 2,827...*C3 95
Anandpur, India...F10 40
Anantapur, India, 52,280...F6 39
Anantnag, India, 21,087...B6 39
Ananyev, Sov. Un., 5,000...H7 27
Anao, pt., Guam... 52
Anapa, Sov. Un., 32,900...I11 27
Anápolis, Braz., 48,847...E1 57
Anapú, riv., Braz...C4 59
Anār, Iran, 15,000...F7 41
Anārak, Iran, 1,270...E6 41
Anardara, Afr., 10,000...E10 41
Añasco, P.R., 2,068...B2 65
Añasco, mun., P.R., 17,200..B2 65
Anastasia, isl., Fla...C5 86
Anata, Jordan, 1,000...h12 32
Anatahan (Anatajan) isl., Mariana Is...E8 7
Anatone, Asotin, Wash., 90..C8 122
Añatuya, Arg., 9,310...E3 55
Anauá, riv., Braz...C5 60
Anawalt, McDowell, W. Va., 1,062...D3 123
Ancash, dept., Peru, 521,661.C2 58
Ancell, Scott, Mo. (part of Scott City)...D8 101
Ancenis, Fr., 5,095...D3 14
Anchi, China, 6,000...F8 34
Anchieta, Braz., 1,179...F2 57
Ancho, Lincoln, N. Mex., 10.D4 107
Anchor, Brazoria, Tex., 40..G5 118
Anchor, Hot Springs, Wyo...B4 125
Anchorage, Alsk., 44,237 (★80,000)...C10, g17 79
Anchorage, Jefferson, Ky., 1,170...A4 94
Anchor Bay Gardens, Macomb, Mich., 1,830...*A8 98
Anchor Point, Alsk., 171...h16, D9 79
Ancienne Lorette, Que., Can., 3,961...C6, C8 73
Anclote, keys, Fla...D4 86
Anco, Knott, Ky., 350...C6 94

Ancon, C.Z., 1,151........m11 62
Ancón, Peru, 1,097........D2 58
Ancona, It., 100,485......C4 21
Ancud, Chile, 6,410.......C2 54
Ancud, gulf, Chile........C2 54
Anda, Phil., 1,612.........*D6 35
Andacollo, Arg.............B2 54
Andahuaylas, Peru, 2,309...D3 58
Andale, Sedgwick, Kans.,
432..............B4, E6 93
Andalgalá, Arg., 5,016.....E2 55
Andalsnes, Nor., 1,943.....F2 25
Andalusia, Covington, Ala.,
10,263...........D3 78
Andalusia, Rock Island, Ill.,
560..............B3 90
Andalusia, reg., Sp.,
5,893,396..........D3 20
Andaman, is., India..........F9 39
Andaman, sea, Indian O.......F10 39
Andaman and Nicobar
Is., ter., 63,548..........*F9 39
Andamarca, Bol............C2 55
Andamarca, Peru, 1,576.....D3 58
Andapa, Malag..............f9 49
Andavaka, cape, Malag........k9 49
Andebu, Nor...............p28 25
Andeer, Switz., 988........C7 19
Andenne, Bel., 7,829.......D5 15
Anderlecht, Bel., 94,677...D4 15
Andermatt, Switz., 1,523...C6 19
Andernach, Ger., 20,800....C2 17
Anderslöv, Swe., 881........C7 24
Anderson, Lauderdale, Ala.,
450.................A2 78
Anderson, Arg..............g6 54
Anderson, Shasta, Calif.,
4,492...............B2 82
Anderson, Madison, Ind.,
49,061...............D6 91
Anderson, McDonald, Mo.,
992..................E3 101
Anderson, Anderson, S.C.,
41,316...............B2 115
Anderson, Franklin, Tenn.,
40...................B6 117
Anderson, Grimes, Tex., 500.D5 118
Anderson, co., Kans., 9,035.D8 93
Anderson, co., Ky., 8,618...C4 94
Anderson, co., S.C., 98,478..B2 115
Anderson, co., Tenn., 60,032.C9 117
Anderson, co., Tex., 28,162..D5 118
Anderson, riv., N.W. Ter., Can..C7 66
Anderson, riv., Ind.............H4 91
Anderson Dam, Elmore,
Idaho, 35............F3 89
Anderson East Side, Madison,
Ind., 3,778..........*D6 91
Anderson Ranch, res., Idaho....F3 89
Andersonville, Sumter, Ga.,
263..................D2 87
Andersonville, Franklin, Ind.,
250..................F7 91
Andes, Col., 6,905.........B2 60
Andes, Richland, Mont., 5..C12 102
Andes, Delaware, N.Y.,
399..................C6 108
Andes, lake, S. Dak............D7 116
Andes, mts., S.A.
D3 47, B2 48, C2 49, C2 58
Andhra Pradesh, state,
India, 35,983,447.........E6 39
Andidanob, mtn., Sud.........A4 47
Andikithira, isl., Grc..........E4 23
Andilamena, Malag..........g9 49
Andimeshk, Iran., 7,324.....E4 41
Anding, Yazoo, Miss., 30....C3 100
Andizhan, Sov. Un.,
141,000...............E8 29
Andkhui, Afg., 18,438.......C12 41
Andoas, Peru, 189..........B2 58
Andong, Kor., 45,700........H4 37
Andorra, And., 600.........A6 20
Andorra, country, Eur.,
5,664...........G9 8, A6 20
Andover, N.B., Can., 848....C2 74
Andover, Tolland, Conn.,
200 (1,771▲)..........C7 84
Andover, Eng., 16,974......C6 12
Andover, Henry, Ill., 295....B3 90
Andover, Butler, Kans., 186..B6 93
Andover, Essex, Mass.,
10,000 (17,134▲)......A5 97
Andover, Merrimack, N.H.,
350 (955▲)............D3 105
Andover, Sussex, N.J., 734...B3 106
Andover, Allegheny, N.Y.,
1,247.................C3 108
Andover, Ashtabula, Ohio,
1,116.................A7 111
Andover, Day, S. Dak., 224..B8 116
Andover, Windsor, Vt.
(215▲)................E3 120
Andöy, isl., Nor...............C6 25
Andrade, Hawaii, Haw., 130.D6 88
Andraitx, Sp., 5,077........C7 20
Andreanof, is., Alsk..........E4 79
Andreas, cape, Cyp...........E10 31
Andrelândia, Braz., 4,617....g5 56
Andrew, Alta., Can., 601....C4 69
Andrew, Jackson, Iowa, 349.B7 92
Andrew, co., Mo., 11,062....B3 101
Andrew, isl., N.S., Can.........D9 74
Andrew Jackson, mtn., Ant......B6 5
Andrews, Huntington, Ind.,
1,132.................C6 91
Andrews, Dorchester, Md.,
1,404.................D5 85
Andrews, Cherokee, N.C.,
1,404.................D2 109
Andrews, Harney, Oreg., 5..E8 113
Andrews, Georgetown and
Williamsburg, S.C., 2,995..E8 115
Andrews, Andrews, Tex.,
11,135................C1 118
Andrews, co., Tex., 13,450..C1 118
Andreyevka, Sov. Un.,
10,000................G11 27
Andria, It., 70,831.........D6 21
Androka, Malag..............h8 49
Andros, Grc., 2,238.........D5 23
Andros, isl., Ba. Is...........C5 64
Andros, isl., Grc..............D5 23
Androscoggin, co., Maine,
86,312................D2 96
Androscoggin, lake, Maine.....D2 96
Androscoggin, riv., Maine,
N.H.........B5 90, A4 105
Andrychow, Pol., 5,055......h10 26
Andsfjord, fiord, Nor..........C7 25
Andújar, Sp., 26,100........C3 20
Anéfis, Mali...............C5 45
Anegada, bay, Arg.............C4 54
Anegada, isl., Vir. Is..........m14 64
Anegada, passage, W.I.........m15 64
Anegam, Pima, Ariz., 100....E3 80
Añelo, Arg.................B3 54
Anerley, Sask., Can., 25.....F2 70
Aneroid, Sask., Can., 279...H2 70
Aneta, Nelson, N. Dak., 451..B8 110
Aneto, mtn., Sp...............A6 20
Angadanan, Phil., 2,510.....n13 35
Angamos, pt., Chile...........D1 55
Angangchi, China, 12,717....B9 34
Angara, riv., Sov. Un.........D13 28
Angarsk, Sov. Un., 154,000..A5 34
Angaur, isl., Palau Is...........52
Angel, fall, Ven...............B5 60
Angela, Rosebud, Mont., 5.D10 102
Ángel de la Guarda, is., Mex..B2 63
Angeles, Phil., 7,118.......o13 35
Angeles, P.R...............B3 65
Angeles, pt., Wash...........A2 122
Angelholm, Swe., 12,600.....I5 25
Angelica, Allegany, N.Y.,
898..................C2 108
Angelina, co., Tex., 39,814..D5 118
Angels Camp, Calaveras,
Calif., 1,121..........C3 82
Angelus, Sheridan, Kans., 70.C3 93
Angelus, Chesterfield, S.C.,
65....................B7 115
Ångermanälven, riv., Swe......F7 25
Angermünde, Ger., 11,700...B7 16
Angers, Que., Can., 575.....D2 73
Angers, Fr., 115,252........D3 14
Angicos, Braz., 1,551........C3 57
Angie, Washington, La., 254.D6 95
Angier, Hartnett, N.C., 1,249.B5 109
Angikuni, lake, N.W. Ter.,
Can..................D13 66
Angkor, ruins, Camb..........F5 38
Anglem, mtn., N.Z............Q11 51
Anglesey, co., Wales, 51,700.A3 12
Anglesey, is., Wales..........D4 10
Anglet, Fr., 16,150..........F3 14
Angleton, Brazoria, Tex.,
7,312.............E5, G5 118
Anglia, Sask., Can., 100.....F1 69
Angliers, Que., Can., 488..G17 67
Angling, riv., Man., Can.......A5 71
Anglo-Egyptian Sudan, see Sudan,
country, Afr.
Angmagssalik, Grnld........C18 4
Ango, Con. L...............A4 48
Angoche, isl., Moz...........A6 49
Angohrān, Iran, 5,000.......H8 41
Angol, Chile, 14,292........B2 54
Angola, Steuben, Ind., 4,746.A8 91
Angola, Labette, Kans., 50..E8 93
Angola, Erie, N.Y., 2,499....C1 108
Angola, Port. dep., Afr.,
4,832,677...........D2 48, H7 42
Angola, swamp, N.C..........C6 109
Angoon, Alsk., 395...D13, m22 79
Angora, Morrill, Nebr., 80..C2 103
Angostura, Dona Ana,
N. Mex...............E2 107
Angostura, res., S. Dak........D2 116
Angoulême, Fr., 48,190
(★80,000)..............E4 14
Angoumois, former prov., Fr.,
292,000...............*E4 14
Angra do Heroismo, Az. Is.,
10,296................g9 44
Angra dos Reis, Braz.,
10,634............C4, h5 56
Angren, Sov. Un., 63,000....E7 29
Anguilla, Sharkey, Miss., 580.B3 100
Anguilla, isl., St. Kitts-Nevis-
Anguilla.................m15 64
Anguillara Sabazia, It.,
2,774...................g8 21
Anguille, cape, Newf., Can.....E2 75
Anguille, mts., Newf., Can.....E2 75
Angumu, Con. L.............B4 48
Anguran, Iran..............H8 41
Angus, Ont., Can., 1,180.....C5 72
Angus, Boone, Iowa, 100....C3 92
Angus, Polk, Minn., 80......B2 99
Angus, co., Scot., 278,370...D5 13
Angusville, Man., Can.,
208....................D1 71
Angwin, Napa, Calif., 1,400..A5 82
Anhembi, Braz., 505.........m7 56
Anholt, isl., Den.............B5 24
Anholt By, Den.............B5 24
Anhsi, China...............C4 34
Anhua, China...............J4 36
Anhwei, prov., China,
33,560,000.............E8 34
Aniak, Alsk., 308...........C8 79
Aniene, riv., It...............h9 21
Animas, Hidalgo, N. Mex.,
40.....................F1 107
Animas, mts., N. Mex........F1 107
Animas, peak, N. Mex........F1 107
Animas, river, Colo...........D3 83
Anina, Rom., 11,837.........C5 22
Anita, Cass, Iowa, 1,233.....C3 92
Anita, Jefferson, Pa., 550....E4 114
Aniva (Rutaka), Sov. Un.,
10,000.................C11 37
Aniva, bay, Sov. Un...........C11 37
Aniva, cape, Sov. Un..........C11 37
Aniwa, Shawano, Wis., 247..C4 118
Anizy-le-Château, Fr., 1,526.E3 15
Anjar, India...............F3 40
Anjean, Greenbrier, W. Va.,
180....................C4 123
Anjo, Jap., 33,000..........o16 37
Anjou, former prov., Fr.,
631,000...............D3 14
Anjouan, isl., Comoro Is......f8 49
Anju, Kor., 21,861..........G2 37
Anjum, Neth., 1,002.........A6 15
Ankang, China..............H3 36
Anka, Nig..................D6 45
Ankara, Tur., 646,200........C9 31
Ankaratra, mts., Malag.......g9 49
Ankavandra, Malag..........g9 49
Ankazoabo, Malag., 18,000...h8 49
Ankeny, Polk, Iowa,
2,964..............A7, C4 92
An Khe, Viet...............F8 38
Anking (Hwaining),
China, 105,300...........I7 36
Anklam, Ger., 19,400........B6 16
Ankobar, Eth...............D4 47
Ankona, St. Lucie, Fla., 89..E6 86
Ankoro, Con. L.............C4 48
Ankuang, China, 5,000.......D1 36
Anlu, China................I5 36
Anlung, China, 9,000........F6 34
Anmoore, Harrison,
W. Va., 1,050.........B4, B7 123
Ann, cape, Ant...............C18 5
Ann, cape, Mass..............A6 97
Anna, Union, Ill., 4,280......F4 90
Anna, Shelby, Ohio, 701.....B3 111
Anna, Sov. Un., 5,000.......F13 27
Annabella, Sevier, Utah,
177...................E3 119
Annaberg-Buchholz, Ger.,
29,000.................C8 17
Annaburg, Ger., 5,184.......B8 17
Annada, Pike, Mo., 105......B7 101
An Nafud, des., Sau. Ar.......H13 31
An Najaf, Iraq, 74,089......I17 9
An Nakhl, Eg., U.A.R.,
3,000.................E5 32
Annalee Heights, Fairfax, Va.,
2,000................*B5 121
Annam, reg., Viet.,
8,000,000..............D7 38
Anna Maria, Manatee, Fla.,
690....................F1 86
Anna Maria, key, Fla..........F1 86
Annan, Scot., 5,572..........F5 13
Annan, riv., Scot..............E5 13
Annandale, Wright, Minn.,
984....................E4 99
Annandale, Hunterdon,
N.J., 500..............B3 106
Annandale, Fairfax, Va.,
5,000..................B5 121
Annapolis, Crawford, Ill.,
130....................D6 90
Annapolis, Parke, Ind., 100..E3 91
Annapolis, Anne Arundel,
Md., 23,385.............C5 85
Annapolis, Iron, Mo., 334...D7 101
Annapolis, Kitsap, Wash.,
600....................D1 122
Annapolis, co., N.S., Can.,
22,649.................E4 74
Annapolis, riv., N.S., Can.....E4 74
Annapolis Junction, Howard,
Md., 525...............B4 85
Annapolis Naval Academy, Md..C5 85
Annapolis Royal, N.S., Can.,
800....................E4 74
Annapurna 1, peak, Nep.......C9 40
Ann Arbor, Washtenaw,
Mich., 67,340.........A6, F7 98
An Nāri, mtn., Libya..........E4 43
An Nasiriya, Iraq, 25,515....F3 41
Annawan, Henry, Ill., 701....B4 90
An Nawfalīyah, Libya........C3 43
An Nazlah, Gaza Area,
4,000..................C6 32
Anne Arundel, co., Md.,
206,634................B4 85
Annecy, Fr., 43,255
(★57,000)..............E7 14
Annemasse, Fr., 13,814......D7 14
Annestown, Ire..............E4 11
Annieopsquotch, mts., Newf.,
Can...................D3 75
Anniston, Calhoun, Ala.,
33,657.................B4 78
Anniston, Mississippi, Mo.,
307....................E8 101
Annobón, isl., Afr.............F1 46
Annonay, Fr., 18,434........E6 14
Annotto Bay, Jam., 3,559....F5 64
Annsjon, lake, Swe...........u33 25
Annursnack, hill, Mass.......C1 97
Annville, Jackson, Ky., 400..C6 94
Annville, Lebanon, Pa.,
4,264..................F8 114
Anoka, Anoka, Minn.,
10,562.............E5, E7 99
Anoka, Boyd, Nebr., 32......B7 103
Anoka, co., Minn., 85,916....E5 99
Anona, Pinellas, Fla., 500....E1 86
Ano Nuevo, pt., Calif.........C5 82
Ano Theologos, Grc., 2,320..B5 23
Anou Mellène (Well), Mali...C5 45
Ano Viannos, Grc., 1,961.....E5 23
Anoyia, Grc., 3,072..........E5 23
Anpei, China, 2,000..........D3 36
Anping, China..............E6 36
Ansbach, Ger., 32,900.......D5 17
Anse au Loup, Newf.,
Can., 343..............C3 75
Anse d'Hainault, Hai., 1,000.F6 64
Anselm, Ransom, N. Dak.,
20.....................C8 110
Anselmo, Custer, Nebr., 269.C6 103
Anserma, Col., 7,767........B2 60
Anshan, China,
805,000...........C9 32, D10 36
Anshun, China, 41,000.......F6 34
Ansley, Pike, Ala., 90........D3 78
Ansley, Jackson, La., 400....B3 95
Ansley, Hancock, Miss., 60...E4 100
Ansley, Custer, Nebr., 714...C6 103
Anson, Somerset, Maine,
900 (2,252▲)...........D3 96
Anson, Jones, Tex., 2,890....C3 118
Anson, co., N.C., 24,962.....B3 109
Ansŏng, Kor., 15,800........H2 37
Ansongo, Mali, 400..........C5 45
Ansonia, New Haven, Conn.,
19,819.................D4 84
Ansonia, Darke, Ohio, 1,002.B3 111
Ansonville, Anson, N.C., 558.B3 109
Ansted, Fayette, W. Va.,
1,511.............C3, D7 123
Anstruther, Scot., 3,400......D6 13
Anta, China, 5,000...........C2 37
Anta, Peru, 1,542............D3 58
Antabamba, Peru, 2,127......D3 58
Antakya (Antioch), Tur.,
45,800.................D11 31
Antalaha, Malag., 10,300....f10 49
Antalāt, Libya..............C4 43
Antalya (Adalia), Tur.,
51,000.................D8 31
Antalya, gulf, Tur............D8 31
Antarctica, cont.................5
Ante, Brunswick, Va., 20......E5 121
Antelope, Sask., Can., 100...G1 70
Antelope, Marion, Kans., 50..D7 93
Antelope, Sheridan, Mont.,
150...................B12 102
Antelope, Wasco, Oreg., 46..C6 113
Antelope, co., Nebr., 10,176..B7 103
Antelope, creek, Wyo..........B7 125
Antelope, hills, Wyo..........C4 125
Antelope, isl., Utah...........B3 119
Antelope, lake, Sask., Can.....G1 70
Antelope, range, Nev..........D7 104
Antelope, res., Oreg...........E9 113
Antelope Mine, S. Rh........B4 49
Antequera, Sp., 29,000.......D3 20
Antes Fort, Lycoming, Pa.,
380....................D7 114
Anthon, Woodbury, Iowa,
681....................B2 92
Anthony, Hempstead, Ark.
120....................D2 81
Anthony, Marion, Fla., 500..C4 86
Anthony, Harper, Kans.,
2,744..................E5 93
Anthony, Dona Ana,
N. Mex., 900...........E3 107
Anthony, Kent, R.I., 2,800.C10 84
Anthony, El Paso, Tex.,
1,082.................*E3 107
Anthonys, creek, W. Va........D4 123
Anthony Wayne Village, Allen,
Ind., 3,000............*B7 91
Anti-Atlas, mts., Mor..........C3 44
Antibes, Fr., 35,439.........F7 14
Anticosti, isl., Que., Can......C3 73
Antietam, creek, Md...........A2 85
Antietam, nat. battlefield site, Md.B2 85
Antigo, Langlade, Wis., 9,691.C4 124
Antigonish, N.S., Can.,
4,344..................D7 74
Antigonish, co., N.S., Can.,
14,360.................D8 74
Antigua, Br. dep., N.A.,
54,354.................n16 64
Antigua, isl., N.A.............n16 64
Antigua Guatemala, Guat.,
10,996.................C2 62
Antilla, Cuba, 5,786.........E6 64
Antimony, Garfield, Utah,
161....................E4 119
Antioch, Contra Costa,
Calif., 17,305..........B6 82
Antioch, Lake, Ill.,
2,268..............A5, D2 90
Antioch, Clinton, Ind., 75....D4 91
Antioch, Sheridan, Nebr., 30.B3 103
Antioch, Davidson,
Tenn., 250.........A5, E9 117
Antioch, see Antakya, Tur.
Antionia, Jefferson, Mo., 150.C7 101
Antioquia, Col., 3,998........B2 60
Antioquia, dept., Col.,
1,747,580..............B2 60
Antique, prov., Phil.,
239,200...............*C6 35
Antiquity, Meigs, Ohio, 150.D6 111
Antler, Sask., Can., 149......H5 70
Antler, Bottineau, N. Dak.,
210....................A4 110
Antler, riv., Man., Can........E1 71
Antler, riv., Sask., Can.......H5 69
Antlers, Pushmataha, Okla.,
2,085..................C6 112
Antofagasta, Chile, 88,000...D1 55
Antofagasta, prov., Chile,
214,090................D1 55
Antofalla, vol., Arg............E2 55
Antoine, Pike, Ark., 163......C2 81
Anton, Washington,
Colo., 50..............B7 83
Anton, Hockley, Tex., 1,068.C1 118
Anton Chico, Guadalupe,
N. Mex., 400...........B4 107
Antongil, bay, Malag..........g9 49
Antonina, Braz., 8,520........D3 56
Antonino, Ellis, Kans., 50....D4 93
Antonio de Biedma, Arg.....D3 54
Antônio Dias, Braz., 1,338...E2 57
Antônio Enes, Moz., 11,628..A6 49
Antonito, Conejos, Colo.,
1,045..................D5 83
Antony, Fr., 46,483.........g10 14
Antratsit, Sov. Un., 5,000...q22 27
Antrim, Antrim, Mich., 200..D5 98
Antrim, Hillsboro, N.H., 850
(1,121▲)...............D3 105
Antrim, N. Ire., 1,448........C5 11
Antrim, Tioga, Pa., 375......C7 114
Antrim, co., Mich., 10,373...C5 98
Antrim, co., N. Ire.,
273,923................C5 11
Antrim, mts., N. Ire...........C5 11
Antsalova, Malag., 1,000.....g8 49
Antsirabe, Malag., 11,332....g9 49
Antsirane, see Diégo-Suarez,
Malag.
Antung, China,
360,000...........C9 32, F2 37
Antwerp, see Antwerpen, Bel.
Antwerp, Jefferson, N.Y.,
881...............A5, B1 108
Antwerp, Paulding, Ohio,
1,465..................A3 111
Antwerpen (Antwerp) (Anvers),
Bel., 253,295 (★1,000,000).C4 15
Antwerpen, prov., Bel.,
1,443,355..............C4 15
Anua, Am. Sam...............52
An Uaimh, Ire., 3,998.......D5 11
Anuradhapura, Cey., 18,528.G7 39
Anvers, see Antwerpen, Bel.
Anvers, isl., Ant..............C6 5
Anvik, Alsk., 120............C7 79
Anyang, China, 124,900......D7 34
Anyi, China................E7 34
Anyksciai, Sov. Un..........A8 26
Anza, Riverside, Calif., 220..F5 82
Anzá, Col., 610.............B2 60
Anzac, Valencia, N. Mex....B2 107
Anzhero-Sudzhensk,
Sov. Un., 119,000........B11 29
Anzin, Fr., 16,275...........D3 15
Anzio, It., 10,000.......D4, k9 21
Anzoátegui, state, Ven.,
382,002................B5 60
Aojidong, Kor., 39,616.......E5 37
Aomoen, isl., Bikini.............52
Aomon, isl., Eniwetok............52
Aomori, Jap., 159,000,
(202,211▲).............F10 37
Aomori, pref., Jap.,
1,426,606.............*F10 37
Aosta, It., 30,127............B1 21
Aoûk, riv., Cen. Afr. Rep......D3 46
Aoulef, Alg.................D5 44
Apa, riv., Par................D4 57
Apache, Caddo, Okla., 1,455.C3 112
Apache, co., Ariz., 30,438....B6 80
Apache, mts., Tex.............F2 118
Apache, peak, Ariz............F5 80
Apache Creek, Catron,
N. Mex., 100...........D1 107
Apache Junction, Pinal,
Ariz., 1,000...........G3 80
Apalachee, Morgan, Ga.,
158....................C3 87
Apalachee, bay, Fla...........B2 86
Apalachicola, Franklin, Fla.,
3,099..................C2 86
Apalachicola, bay, Fla........C2 86
Apalachicola, riv., Fla........B1 86
Apalachin, Tioga, N.Y., 350.C4 108
Apalona, Perry, Ind.........H4 91
Apan, Mex., 8,589...........n14 63
Apaporis, riv., Col............C3 60
Aparri, Phil., 14,000
(33,500▲)..............B6 35
Apatin, Yugo., 17,203........C4 22
Apatity, Sov. Un............D15 25
Apatzingán, Mex., 19,340....n12 63
Apeldoorn, Neth., 104,900....B5 15
Apennine, tunnel, It...........B3 21
Apennines, mts., It............C4 21
Apex, Wake, N.C., 1,368......B5 109
Apgar, Flathead, Mont., 50..B3 102
Api, peak, Nep...............C8 40
Apia, W. Sam., 16,000............52
Apia, hbr., W. Sam...............52
Apiacás, mts., Braz............E3 59
Apiaí, Braz., 2,728...........C3 56
Apiranthos, Grc., 2,438......D5 23
Apishapa, riv., Colo...........D6 83
Apison, Hamilton, Tenn.,
375...................E10 117
Apizaco, Mex., 15,622........n14 63
Aplao, Peru, 840.............E3 58
Aplin, Perry, Ark., 100.......C3 81
Aplington, Butler, Iowa, 840.B5 92
Apo, vol., Phil...............D6 35
Apodi, Braz., 2,512..........C3 57
Apohaqui, N.B., Can., 343..D4 74
Apolda, Ger., 29,300.........B6 17
Apolima, isl., W. Sam............52
Apolima, strait, W. Sam..........52
Apollo, Armstrong, Pa.,
2,694..................E2 114
Apolo, Bol., 1,043...........B2 55
Apopka, Orange, Fla.,
3,578..................D5 86
Aporé, riv., Braz..............B2 56
Apostle, is., Wis..............A3 124
Apóstoles, Arg., 3,385........E4 55
Appalachia, Wise, Va., 2,456.B2 121
Appalachian, mts., U.S........C11 77
Appam, Williams, N. Dak.,
60.....................A2 110
Appanoose, co., Iowa,
16,015.................D5 92
Appenzell, Switz., 5,001......B7 19
Appenzell, canton, Switz.,
61,863.................B7 19
Apperson, Osage, Okla., 15..A5 112
Appingedam, Neth., 7,800....A6 15
Apple, creek, N. Dak..........C5 110
Apple, riv., Ill...............F3 124
Apple, riv., Wis..............C1 124
Appleby, Eng., 1,751.........F6 13
Appleby, Nacogdoches, Tex.,
150....................D5 118
Apple Creek, Wayne, Ohio,
722....................B6 111
Applecross, Scot., 735........C3 13
Applegate, Jackson, Oreg.,
350....................E3 113
Applegate, Sanilac, Mich.,
244....................E8 98
Applegate, butte, Oreg.........E5 113
Applegate, riv., Oreg..........E3 113
Apple Grove, Louisa, Va.,
25.....................D5 121
Apple Grove, Mason,
W. Va., 600............C2 123
Apple Hill, Ont., Can., 384..B10 72
Apple River, N.S., Can., 78..D5 74
Apple River, Jo Daviess, Ill.,
477....................A3 90
Apple Springs, Trinity, Tex.,
200....................D5 118
Appleton, Pope, Ark., 150....B3 81
Appleton, Knox, Maine, 160
(672▲).................D3 96
Appleton, Swift, Minn.,
2,172..................E2 99
Appleton, Allendale, S.C.,
198....................E5 115
Appleton, Lawrence, Tenn.,
60.....................B4 117
Appleton, Klickitat, Wash.,
10.....................D4 122
Appleton, Outagamie, Wis.,
48,411 (★123,200)....A5, D5 124
Appleton City, St. Clair,
Mo., 1,075.............C3 101
Apple Valley, San Bernardino,
Calif., 950............*E5 82
Appleyard, Chelan, Wash.,
950....................B5 122
Appling, Columbia, Ga., 200.C4 87
Appling, co., Ga., 13,246....E4 87
Appomattox, Appomattox,
Va., 1,184.............D4 121
Appomattox, co., Va., 9,148.D4 121
Appomattox, riv., Va..........D5 121
Approuague, Fr., Gu., 856...B4 59
Apra, hbr., Guam.................52
Aprelsk, Sov. Un., 800.......D14 28
Aprilia, It...............D4, h9 21
Apsheron, pen., Sov. Un.......G19 9
Apsley, Ont., Can., 295......C6 72
Apt, Fr., 5,521..............F6 14
Aptos, Santa Cruz, Calif.,
950....................C6 82
Apuaí, riv., Braz..............D5 60
Apulia, reg., It.............D6 21
Apulyont, lake, Tur...........B7 23
Apure, state, Ven., 117,577..B3 60
Apure, riv., Ven..............B4 60
Apurímac, dept., Peru,
324,338................D3 58
Apurímac, riv., Peru..........D3 58
Aqaba, gulf, Afr., Asia........H10 31
'Aqiq, Sud..................B4 47
Aqir, Isr., 996...............h10 32
Aq Kupruk, Afg., 5,000......C13 41
Aqraba, Jordan, 3,000..B7, g12 32
Aquanish, Que., Can., 383...h9 75
Aquarius, mts., Ariz..........C2 80
Aquasco, Prince Georges,
Md., 400...............C4 85
Aquashicola, creek, Pa........B2 106
Aquebogue, Suffolk, N.Y.,
900....................F6 84
Aquidauana, Braz., 11,997...C1 56
Aquidneck, isl., R.I...........C11 84
Aquileia, It., 1,400..........D9 18
Aquiles Serdán, Mex., 3,927.B3 63
Aquin, Hai.................F7 64
Aquone, Macon, N.C., 150..D2 109
Arab, Marshall, Ala., 2,989..A3 78
Arabela, Lincoln, N. Mex.,
20.....................D4 107
Arabi, Crisp, Ga., 303.......E3 87
Arabi, St. Bernard, La.,
5,000..................C7 95
Arabia, see Saudi Arabia, Asia
Arabian, des., Eg., U.A.R......E3 32
Arabian, sea, Asia............H9 33
Arabs, gulf, Eg., U.A.R........G7 31
Araby, Yuma, Ariz., 5........E1 80
Aracaju, Braz., 112,516......D3 57
Aracati, Braz., 11,016........B3 57
Araçatuba, Braz., 53,563.....C2 56
Aracena, Sp., 7,737..........D2 20
Aracruz, Braz., 413..........E2 57
Araçuaí, Braz., 6,763........E2 57
Arad, Rom., 106,460.........B5 22
Araduey, riv., Sp.............B3 20
Arafura, sea, Indon...........G8 35
Arago, Coos, Oreg., 140......D2 113
Arago, cape, Oreg.............D2 113
Aragon, Polk, Ga., 1,023.....B1 87
Aragon, Catron, N. Mex., 63.D1 107
Aragon, reg., Sp., 1,105,498..B5 20
Aragón, riv., Sp..............A5 20
Aragua, state, Ven., 313,274.A4 60
Araguaia, riv., Braz..........D6 53
Aragua de Barcelona, Ven.,
6,830..................B5 60
Araguao, riv. mouth, Ven......B5 60
Araguari, Braz., 35,520......E1 57
Araguari, riv., Braz...........B4 59
Arai, Jap., 9,600............o16 37
Arāk (Sultanabad), Iran,
58,998.................D4 41
Arakabesan, isl., Palau Is........52
Arakan, range, Bur............E9 39
Arakhthos, riv., Grc...........C3 23
Araks, riv., Asia..............B3 41
Aral, sea, Sov. Un.............D5 29
Aral Karkum, des., Sov. Un.....D6 29
Aralsk, Sov. Un., 18,600.....D6 29
Aramac, Austl., 488.........D8 50
Aran, isl., Ire...............C3 11
Aran, is., Ire................D2 11
Aranda de Duero, Sp., 13,454.B4 20
Arandas, Mex., 17,110......m12 63
Aranjuez, Sp., 27,251........B4 20
Aransas, co., Tex., 7,006.....E4 118
Aransas, bay, Tex.............E4 118
Aransas Pass, San Patricio
and Aransas, Tex., 6,956..F4 118
Aranyaprathet, Thai., 11,601.F5 38
Araouane, Mali..............C4 45
Arapaho, Custer, Okla., 351..B3 112
Arapaho, peak, Colo...........A5 83
Arapahoe, Cheyenne, Colo.,
125....................C8 83
Arapahoe, Furnas, Nebr.,
1,084..................D6 103
Arapahoe, Pamlico, N.C.,
274....................B7 109
Arapahoe, Fremont, Wyo.,
30.....................C4 125
Arapahoe, co., Colo., 113,426.B6 83
Arapey, Ur..................E1 56
Arapey Grande, riv., Ur........E1 56
Arapkir, Tur., 6,900.........C12 31
Arar, wadi, Sau. Ar...........G13 31
Araranguá, Braz., 7,775......D3 56
Araraquara, Braz.,
58,076..............C3, k7 56
Araras, Braz., 23,898.....C3, m8 56
Araras, mts., Braz.............B2 56
Ararat, Choctaw, Ala., 50....D1 78
Ararat, Austl., 7,934.........H4 51
Ararat, Susquehanna, Pa.,
35....................C10 114
Ararat, Patrick, Va., 300.....E2 121
Ararat, mtn., Tur.............C15 31
Arareh, W. Irian.............F9 35
Araripe, mts., Braz...........C2 57
Araruna, Braz., 2,261.........h6 57
Aratane (Well), Maur.........C3 45
Arauca, Col., 2,028...........B3 60
Arauca, intendencia, Col.,
14,080.................B3 60
Arauca, riv., Ven..............B4 60
Arauco, Chile, 2,707.........B2 54
Arauco, prov., Chile, 89,211.B2 54
Aravaipa, Graham, Ariz......E5 80
Aravalli, range, India.........C5 39
Araxá, Braz., 24,041.........E1 57
Arayat, Phil., 791............o13 35
Arba, Randolph, Ind., 100..D8 91
Arba Jahan, Ken.............A6 48
Arbela, Scotland, Mo., 70....A5 101
Arboga, Swe., 11,100........t33 25
Arbois, Fr., 3,960............D6 14
Arboles, Archuleta, Colo.,
100....................D3 83
Arbon, Power, Idaho, 10.....G6 89
Arbon, Switz., 11,608
(★13,748)..............A7 19
Arbor, Middlesex, N.J.,
2,000.................*B4 106
Arborea, It., 2,681...........E2 21
Arborfield, Sask., Can., 579..D4 70
Arborg, Man., Can., 811.....D3 71
Arbor Terrace, St. Louis, Mo.,
1,225.................*C7 101
Arbor Vitae, Vilas, Wis.,
150....................C4 124
Arbroath, Scot., 19,533......D6 13
Arbuckle, Colusa, Calif.,
950....................C2 82
Arbuckle, lake, Fla............E5 86
Arbuckle, mts., Okla..........C4 112
Arbutus, Baltimore, Md.
(part of Halethorpe).....C2 85
Arcachon, Fr., 14,862
(★33,500)..............E3 14
Arcade, Jackson, Ga., 108....B3 87
Arcade, Wyoming, N.Y.,
1,930..................C2 108
Arcadia, Los Angeles, Calif.,
41,005.................F2 82
Arcadia, N.S., Can., 400....F3 74
Arcadia, De Soto, Fla., 5,889.E5 86
Arcadia, Hamilton, Ind.,
1,271..................D5 91
Arcadia, Carroll, Iowa, 437..B2 92
Arcadia, Crawford, Kans.,
507....................E9 93
Arcadia, Bienville, La., 2,547.B3 95
Arcadia, Manistee, Mich.,
600....................D4 98
Arcadia, Iron, Mo., 489......D7 101
Arcadia, Valley, Nebr., 446..C6 103
Arcadia, Hancock, Ohio, 610.A4 111
Arcadia, Oklahoma, Okla.,
400....................B4 112
Arcadia, Indiana, Pa., 500...E4 114
Arcadia, Washington, R.I.,
75....................C10 84
Arcadia, Spartanburg, S.C.,
2,458..................B4 115
Arcadia, Duchesne, Utah, 10.C5 119
Arcadia, Trempealeau, Wis.,
2,084..................D2 124
Arcadia (Arkadhia), prov.,
Grc., 135,042.........*D4 23
Arcanum, Darke, Ohio,
1,678..................C3 111
Arcata, Humboldt, Calif.,
5,235..................B1 82
Arc Dome, mtn., Nev..........E4 104
Arcelia, Mex., 8,526.........n13 63
Arch, Roosevelt, N. Mex., 20.C6 107
Archambault, lake, Que., Can..C3 73
Archangel, see Arkhangelsk,
Sov. Un.
Archibald, Lackawanna, Pa.,
5,471..................A9 114
Archbold, Fulton, Ohio,
2,348..................A3 111
Archdale, Randolph, N.C.,
1,520..................B4 109
Archer, Alachua, Fla., 707...C4 86
Archer, Madison, Idaho, 50..F7 89
Archer, O'Brien, Iowa, 209...A2 92
Archer, Sheridan, Mont., 5.B12 102
Archer, Merrick, Nebr., 80...C7 103
Archer, co., Tex., 6,110......C3 118
Archer City, Archer, Tex.,
1,974..................C3 118
Archers Post, Ken...........A6 48
Archerwill, Sask., Can., 340..E4 70
Arches, nat. mon., Utah.......E6 119
Archidona, Sp., 7,962........D3 20
Archie, Cass, Mo., 348.......C3 101
Archive, Sask., Can., 15.....G3 70

Archuleta, San Juan, N. Mex	A2	107
Archuleta, co., Colo., 2,629	D3	83
Arcila, Mor., 10,839	G3	30
Arcis-sur-Aube, Fr., 2,856	C6	14
Arco, Glynn, Ga., 5,417	E5	87
Arco, Butte, Idaho, 1,562	F5	89
Arco, Lincoln, Minn., 140	F2	99
Arcola, Sask., Can., 560	H4	70
Arcola, Douglas, Ill., 2,273	D5	90
Arcola, Allen, Ind., 275	B7	91
Arcola, Washington, Miss., 366	B3	100
Arcola, Dade, Mo., 105	D4	101
Arcola, Loudoun, Va., 40	A4	121
Arcos de la Frontera, Sp., 11,585	D3	20
Arcoverde, Braz., 18,008	C3, k5	57
Arctic, Alsk., 110	B10	79
Arctic, ocean	B33	4
Arctic Bay, N.W. Ter., Can	B15	66
Arcueil, Fr., 20,224	g10	14
Arda, riv., Bul	E7	22
Ardabīl, Iran, 65,742	B4	41
Ardagh, Ire., 122	E2	11
Ardahan, Tur., 7,200	B14	31
Ardakān, Iran, 6,026	F6	41
Ardara, Ire., 547	C3	11
Ardath, Sask., Can., 52	F2	70
Ardatov, Sov. Un., 5,000	D16	27
Ardbeg, Ont., Can., 67	B4	72
Ardea, It., 438	h9	21
Ardèche, dept., Fr., 248,516	*E6	14
Ardee, Ire., 2,710	D5	11
Arden, Little River, Ark., 25	D1	81
Arden, Sacramento, Calif., 25,000	*C3	82
Arden, Man., Can., 188	D2	17
Arden, Ont., Can., 227	C8	72
Arden (Ardentown), New Castle, Del., 1,500	A7	85
Arden, Den., 1,365	B3	24
Arden, Clark, Nev., 50	H6	104
Arden, Buncombe, N.C., 800	D3	109
Arden, Irion, Tex	D2	118
Ardena, Monmouth, N.J	C4	106
Arden Hills, Ramsey, Minn., 3,930	*E7	99
Ardennes, dept., Fr., 300,247	E4	15
Ardennes, mts., Bel	D5	15
Ardenno, It., 2,789	C5	18
Ardenvoir, Chelan, Wash., 200	B5	122
Arderin, mtn., Ire	D4	11
Ardestān, Iran, 5,669	E6	41
Ardglass, N. Ire., 735	C6	11
Ardgroom, Ire., 89	F2	11
Ardila, riv., Port	C2	20
Ardill, Sask., Can., 31	H3	70
Ardino, Bul., 1,469	E7	28
Ardley, Alta., Can., 100	C4	69
Ardmore, Limestone, Ala., 439	A3	78
Ardmore, Alta., Can., 172	B5	69
Ardmore, Ire., 290	F4	11
Ardmore, Prince Georges, Md., 750	C4	85
Ardmore, Carter, Okla., 20,184	C4	112
Ardmore, Delaware and Montgomery, Pa., 15,000	B11	114
Ardmore, Fall River, S. Dak., 73	D2	116
Ardmore, Giles, Tenn., 195	B5	117
Ardnaree, Ire	C2	11
Ardoch, Walsh, N. Dak., 106	A8	110
Ardpatrick, pt., Scot	E3	13
Ardres, Fr., 1,142	D9	12
Ardrossan, Austl., 558	G1	51
Ardrossan, Alta., Can., 84	C4	69
Ardrossan, Scot., 9,574	E4	13
Ardsley, Westchester, N.Y., 3,991	*D7	108
Ardvasar, Scot	C3	13
Arebeb (Well), Mali	B5	45
Arecibo, P.R., 28,826	B4	65
Arecibo, mun., P.R., 69,879	B4	65
Aredale, Butler, Iowa, 153	B4	92
Areia, Braz., 5,934	h6	57
Areia Branca, Braz., 8,904	B3	57
Arelee, Sask., Can., 125	E2	70
Arena, Sask., Can	H1	70
Arena, Burleigh, N. Dak., 6	B5	110
Arena, Iowa, Wis., 309	E4	124
Arena, pt., Calif	C2	82
Arena, pt., Mex	C3	63
Arenac, co., Mich., 9,860	D7	98
Arenal, P.R	D5	65
Arenas de San Pedro, Sp., 6,001	B3	20
Arenas Valley, Grant, N. Mex., 150	E1	107
Arendal, Nor., 11,300 (★20,000)	H3	25
Arendonk, Bel., 8,862	C5	15
Arendsee, Ger., 3,240	F5	24
Arendtsville, Adams, Pa., 588	G7	114
Arenillas, Ec	B1	58
Arenzville, Cass., Ill., 417	D3	90
Areopolis, Grc., 1,217	D4	23
Arequipa, Peru, 93,389	E3	58
Arequipa, dept., Peru., 307,943	E3	58
Arezzo, It., 39,000 (74,992▲)	C3	21
Arga, riv., Sp	A5	20
Argalasti, Grc., 3,021	C4	23
Arganda, Sp., 5,166	p18	20
Argelès-Gazost, Fr., 3,414	F3	14
Argenta, Macon, Ill., 860	D5	90
Argenta, It., 4,600	B3	21
Argenta, Beaverhead, Mont., 10	E4	102
Argentan, Fr., 12,757	C4	14
Argentat, Fr., 2,645	E5	14
Argenteuil, Fr., 82,321	C5, g10	14
Argenteuil, co., Que., Can., 31,830	D3	73
Argentia, Newf., Can., 493	E4	75
Argentina, country, S.A., 20,005,691	B4 53, C3 54, E2	55
Argentino, lake, Arg	E2	54
Argenton [-sur-Creuse], Fr., 6,344	D4	14
Argesul, riv., Rom	C7	22
Arghandab, res., Afg	E12	41
Argo, Jefferson, Ala., 150	B3	78
Argo, Cook, Ill. (part of Summit)	F2	90
Argo, Sud., 2,389	B3	47
Agolis, prov., Grc., 90,145	*D4	23
Argolis, gulf, Grc	D4	23
Argonia, Sumner, Kans., 553	E6	93
Argonne, Miner, S. Dak., 5	C8	116
Argonne, Forest, Wis., 150	C5	124
Argonne, forest, Fr	E4	15
Argonne, plat., Fr	C6	14
Argos, Grc., 16,712	D4	23
Argos, Marshall, Ind., 1,339	B5	91
Argos Orestikon, Grc., 4,292	B3	23
Argostolion, Grc., 8,711	C3	23
Argun, riv., Sov. Un	D14	28
Argungu, Nig	D5	45
Argusville, Cass, N. Dak., 118	B9	110
Argyle, N.S., Can., 56	F4	74
Argyle, Walton, Fla., 125	G3	86
Argyle, Clinch, Ga., 225	E4	87
Argyle, Lee, Iowa, 100	D6	92
Argyle, Sanilac, Mich., 150	E8	98
Argyle, Marshall, Minn., 789	B2	99
Argyle, Osage, Mo., 99	C5	101
Argyle, Washington, N.Y., 355	B7	108
Argyle, Lafayette, Wis., 786	F4	124
Argyle Downs, Austl., 153	C4	50
Argyll, co., Scot., 59,345	D3	13
Århus, Den., 119,568 (★177,234)	B4	24
Århus, co., Den., 210,409	B4	24
Århus, bay, Den	B4	24
Ariail, Pickens, S.C., 950	B2	115
Ariake, bay, Jap	K5	37
Ariano Irpino, It., 11,300	D5	21
Ariano nel Polesine, It., 2,505	E8	18
Ariari, riv., Col	C3	60
Arica, Chile, 37,000	C2	55
Arichat, N.S., Can., 667	D9	74
Ariège, dept., Fr., 137,192	*F4	14
Ariège, riv., Fr	F4	14
Ariel, Lafourche, La	C6	95
Ariel, Cowlitz, Wash., 75	D3	122
Ariesul, riv., Rom	B6	22
Arīḩā (Jericho), Jordan, 41,593	C7, h12	32
Arikaree, riv., Colo	B8	83
Arimo, Bannock, Idaho, 303	G6	89
Aringay, Phil., 2,028	n13	35
Arinos, riv., Braz	E3	59
Ario de Rosales, Mex., 9,196	D4, n13	63
Arion, Crawford, Iowa, 201	C2	92
Aripeko, Pasco, Fla., 250	D4	86
Aripine, Navajo, Ariz., 5	C5	80
Aripuanã, riv., Braz	D2	59
Arisaig, Scot., 1,002	D3	13
Arisaig, sound, Scot	D2	13
Arispe, Union, Iowa, 125	D3	92
Arista, Mercer, W. Va., 300	D3	123
Aristazabal, isl., B.C., Can	C3	68
Ariton, Dale, Ala., 687	D4	78
Arivaca, Pima, Ariz., 40	F4	80
Arivonimamo, Malag	g9	49
Arizaro, salt flat, Arg	D2	55
Arizola, Pinal, Ariz., 100	E4	80
Arizona, Arg	B3	54
Arizona, state, U.S., 1,302,161	D5	76, 80
Arizpe, Mex., 1,403	A2	63
Arjay, Bell, Ky., 500	D6	94
Arjeplog, Swe., 1,355	D7	25
Arjona, Col., 12,361	A2	60
Arjona, Sp., 8,154	D3	20
Arkabutla, Tate, Miss., 175	A3	100
Arkabutla, dam, Miss	A3	100
Arkabutla, res., Miss	A4	100
Arkadelphia, Cullman, Ala., 250	B3	78
Arkadelphia, Clark, Ark., 8,069	C2	81
Arkadhia (Arcadia), prov., Grc., 135,042	*D4	23
Arkaig, lake, Scot	D3	13
Arkansas, state, U.S., 1,786,272	C9	77, 81
Arkansas, co., Ark., 23,355	C4	81
Arkansas, riv., U.S	C9	77
Arkansas City, Desha, Ark., 783	D4	81
Arkansas City, Cowley, Kans., 14,262	D6	93
Arkansas Post, Arkansas, Ark., 25	C4	81
Arkansaw, Pepin, Wis., 400	D1	124
Arkhangelsk (Archangel), Sov. Un., 271,000	E19 25, C7	28
Arkhangelskoye, Sov. Un., 10,000	F13	27
Arkhara, Sov. Un	B5	37
Arkinda, Little River, Ark., 80	D1	81
Arklow, Ire., 5,390	E5	11
Arkoma, LeFlore, Okla., 1,862	B7	112
Arkona, Ont., Can., 447	D3	72
Arkonam, India, 30,658	F6	39
Arkport, Steuben, N.Y., 837	C3	108
Arkville, Delaware, N.Y., 500	C6	108
Arkwright, Bibb, Ga., 90	D3	87
Arkwright, Kent, R.I., 1,500	C10	84
Arkwright, Spartanburg, S.C., 1,656	*B4	115
Arlanza, riv., Sp	A4	20
Arlanza Village, Riverside, Calif., 4,000	*F5	82
Arlanzón, riv., Sp	A4	20
Arlberg, tunnel, Aus	E5	16
Arlee, Lake, Mont., 10	C2	102
Arles, Fr., 29,362	F6	14
Arley, Winston, Ala., 450	A2	78
Arline, Blount, Tenn	E11	117
Arlington, Maricopa, Ariz., 25	D3	80
Arlington, Kiowa, Colo., 5	C7	83
Arlington, Duval, Fla., 19,500	B5, B6	86
Arlington, Calhoun and Early, Ga., 1,462	E2	87
Arlington, Bureau, Ill., 254	B4	90
Arlington, Rush, Ind., 470	E6	91
Arlington, Fayette, Iowa, 614	B6	92
Arlington, Reno, Kans., 466	B3, E5	93
Arlington, Carlisle, Ky., 584	B2	94
Arlington, Middlesex, Mass., 49,953	C2	97
Arlington, Sibley, Minn., 1,601	F4	99
Arlington, Washington, Nebr., 740	C9, D2	103
Arlington, Dutchess, N.Y., 8,317	*D7	108
Arlington, Yadkin, N.C., 590	A3	109
Arlington, Hancock, Ohio, 955	B4	111
Arlington, Gilliam, Oreg., 643	B6	113
Arlington, Spartanburg, S.C., 500	B3	115
Arlington, Kingsbury, S. Dak., 996	C8	116
Arlington, Shelby, Tenn., 620	B2	117
Arlington, Tarrant, Tex., 44,775	B5	118
Arlington, Bennington, Vt., 1,111 (1,605▲)	E2	120
Arlington, Arlington, Va., 163,401	B5, C5	121
Arlington, Snohomish, Wash., 2,025	A3	122
Arlington, co., Va., 163,401	B5	121
Arlington Beach, Brookings, S. Dak., 20	C8	116
Arlington Heights, Cook, Ill., 33,424	A5, E2	90
Arlington Heights, Hamilton, Ohio, 1,355	*C3	111
Arlon, Bel., 13,272	E5	15
Arltunga, Austl., 25	D5	50
Arly, riv., Fr	D2	18
Arm, Lawrence, Miss., 75	D3	100
Arma, Crawford, Kans., 1,296	E9	93
Armada, Alta., Can., 75	D4	69
Armada, Macomb, Mich., 1,111	F8	98
Armagh, Que., Can., 914	C7	73
Armagh, N. Ire., 9,982	C5	11
Armagh, Indiana, Pa., 192	F3	114
Armagh, co., N. Ire., 117,580	C5	11
Armathwaite, Fentress, Tenn., 150	C9	117
Armavir, Sov. Un., 120,000	G17 9, I13	27
Armenia, Col., 57,098	C2	60
Armenia (S.S.R.) rep., Sov. Un., 1,893,000	G17	9
Armentières, Fr., 25,248	B5	14
Armeria, riv., Mex	n12	63
Armero, Col., 10,258	C3	60
Armidale, Austl., 12,877	E8	51
Armijo, Bernalillo, N. Mex., 2,500	G5	107
Armington, Tazewell, Ill., 327	C4	90
Armington, Cascade, Mont., 168	C6	102
Arminto, Natrona, Wyo., 15	B5	125
Armit, lake, Can	E5 64, C1	71
Armley, Sask., Can., 75	D3	70
Armona, Kings, Calif., 1,302	*D4	82
Armonk, Westchester, N.Y., 2,000	*D6	108
Armorel, Mississippi, Ark., 150	B6	81
Armour, Douglas, S. Dak., 875	D7	116
Armstead, Blount, Ala., 50	B3	78
Armstead, Beaverhead, Mont., 250	F4	102
Armstrong, B.C., Can., 1,288	D8	68
Armstrong, St. Johns, Fla., 250	C5	86
Armstrong, Vermilion, Ill., 200	C6	90
Armstrong, Emmet, Iowa, 958	A3	92
Armstrong, Howard, Mo., 387	B5	101
Armstrong, Kenedy, Tex., 45	F4	118
Armstrong, co., Pa., 79,524	E3	114
Armstrong, co., Tex., 1,966	B2	118
Armstrong, creek. W. Va	D6	123
Armstrong Creek, Forest, Wis., 250	C5	124
Armuchee, Floyd, Ga., 200	B1	87
Armyansk, Sov. Un., 5,800	H9	25
Arnaud, Man., Can., 88	E3	71
Arnaudville, St. Landry and St. Martin, La., 1,184	D4	95
Arneburg, Ger., 3,073	F5	24
Arnedo, Sp., 7,958	A4	20
Arnegard, McKenzie, N. Dak., 228	B2	110
Arneiroz, Braz., 447	C2	57
Arnett, Ellis, Okla., 547	A2	112
Arnett, Raleigh, W. Va., 300	D3, D6	123
Arney, riv., N. Ire	C4	11
Arnhem, Neth., 124,800 (★200,000)	C5	15
Arnhem, cape, Austl	B6	50
Arnhem Land, reg., Austl	B5	50
Arnissa, Grc., 2,915	B3	23
Arno, riv., It	C3	21
Arnold, Calaveras, Calif., 375	C3	82
Arnold, Ness, Kans., 120	D3	93
Arnold. Marquette, Mich., 45	B3	98
Arnold, St. Louis, Minn., 700	D6	99
Arnold, Jefferson, Mo., 550	C7	101
Arnold, Custer, Nebr., 844	C5	103
Arnold, Westmoreland, Pa., 9,437	A7	114
Arnold Mills, Providence, R.I., 250	B11	84
Arnoldsburg, Calhoun, W. Va., 170	C3	123
Arnolds Park, Dickinson, Iowa, 953	A2	92
Arnoldsville, Oglethorpe, Ga., 115	C3	87
Arnot, Tioga, Pa., 275	C7	114
Arnouville [-lès-Gonesse], Fr., 10,227	g10	14
Arnöy, isl., Nor	B9	25
Arnprior, Ont., Can., 5,474	B8	72
Arnsberg, Ger., 21,300	B3	17
Arnstadt, Ger., 26,400	C5	17
Arnstein, Ont., Can., 102	B5	72
Arnswalde, see Choszczno, Pol.		
Arntz, Navajo, Ariz., 10	C5	80
Årö, isl., Den	C3	24
Aroa, Ven., 3,930	A4	60
Aroab, S.W. Afr., 819	C2	49
Arock, Malheur, Oreg., 25	E9	113
Aroda, Madison, Va., 80	C4	121
Arolsen, Ger., 5,600	B4	17
Aroma, Sud., 3,451	B4	47
Aroma, Hamilton, Ind., 35	D6	91
Aroma Park, Kankakee, Ill., 744	B6	90
Aromas, Monterey, Calif., 700	C6	82
Aroostook, co., Maine, 106,064	B4	96
Aroostook, riv., Maine	B4	96
Aroostook Junction, N.B., Can., 683	C2	74
Aroroy, Phil., 2,445	*C6	35
Aros, Nor., 550	p28	25
Arosa, Switz., 2,600	C8	19
Aroya, Cheyenne, Colo., 24	C8	83
Arp, Smith, Tex., 812	C5	118
Arpajon, Fr., 5,935	F2	15
Arpin, Wood, Wis., 350	D3	124
Arques, Fr., 7,224	D2	15
Arques-la-Bataille, Fr., 2,586	E9	12
'Arrābah, Jordan, 4,000	B7	32
Arrah, India, 76,766	C7	39
Arraias, Braz., 1,446	D1	57
Arraiján, Pan., 2,192	m11	62
Arraiolos, Port., 2,413	C2	20
Ar Ramādī, Iraq, 67,395	F14	31
Ar Ramthā, Jordan, 1,000	B7	32
Arran, Sask., Can., 156	F5	70
Arran, Wakulla, Fla., 40	B2	80
Arran, isl., Scot	E3	13
Arrandale, B.C., Can	B3	68
Ar Raqqah, Syr., 11,411	E12	31
Arras, Fr., 41,761 (★60,000)	B5	14
Ar Rawḑah, Eg.. U.A.R., 1,000	H8	31
Arrecife, Can. Is., 12,886	D2	44
Arrecifes, Arg., 7,635	g6	54
Arrecifes, riv., Arg	f7	54
Arrey, Sierra, N. Mex., 10	E2	107
Arriaga, Mex., 11,601	m13	63
Arriba, Lincoln, Colo., 296	B7	83
Arriikan, isl., Bikini		52
Arrington, Atchinson, Kans., 100	A7, C8	93
Arrington, Nelson, Va., 250	D4	121
Arriola, Montezuma, Colo	D2	83
Ar Riyāḑ, see Riyadh, Sau. Ar.		
Arrochar, Scot., 1,367	D4	13
Arroio Grande, Braz., 5,623	E2	56
Arroll, Texas, Mo., 50	D6	101
Arrone, riv., It	h8	21
Arrow, lake, Ont., Can	A7	99
Arrow, lake, Ire	C3	11
Arrow, riv., Ont., Can	A7	99
Arrow, riv., Mont	C6	102
Arrowhead, B.C., Can., 117	D9	68
Arrow Rock, Saline, Mo., 245	B5	101
Arrowsic, Sagadahoc, Maine, 35 (177▲)	E3	96
Arrowsmith, McLean, Ill., 319	C5	90
Arrowwood, Alta., Can., 195	D4	69
Arroyo, P.R., 3,741	D6	65
Arroyo, mun., P.R., 13,315	D6	65
Arroyo Arenas, Cuba, 2,123	k11	64
Arroyo de la Luz, Sp., 9,781	C2	20
Arroyo Grande, San Luis Obispo, Calif., 3,291	E3	82
Arroyo Grande, Ur	f8	51
Arroyo Hondo, Taos, N. Mex., 541	A4	107
Arroyo Naranjo, Cuba, 2,755	k12	64
Arroyoseco, Taos, N. Mex., 50	A4	107
Års, Den., 3,206	B3	24
Arsenal, Jefferson, Ark., 1,200	*C3	81
Arsenyev, Sov. Un., 10,000	D6	37
Arta, Grc., 16,899	C3	23
Arta, prov., Grc., 82,630	*C3	23
Artaiyan (Oasis), Sau. Ar	H13	31
Artas, Campbell, S. Dak., 87	B6	116
Artashat, Sov. Un., 3,000	B2	41
Artem, Sov. Un., 60,000	E16	28
Artemisa, Cuba, 17,461	D2	64
Artemovsk, Sov. Un., 10,000	E27	9
Artemovsk, Sov. Un., 64,000	G12, q21	27
Artemovskiy, Sov. Un., 31,900	B6	29
Artemus, Knox, Ky., 950	D6	94
Artena, It., 7,707	h9	21
Arter, mtn., Wyo	C4	125
Artern, Ger., 7,601	B6	17
Artesia, Bech	B4	49
Artesia, Los Angeles, Calif., 9,993	F2	82
Artesia, Moffat, Colo., 318	A1	83
Artesia, Lowndes, Miss., 469	B5	100
Artesia, Eddy, N. Mex., 12,000	E5	107
Artesian, Sanborn, S. Dak., 330	C8	116
Artesia Wells, LaSalle, Tex., 75	E3	118
Arth, Switz., 6,321	B6	19
Arthabaska, Que., Can., 2,977	C6	73
Arthabaska, co., Que., Can., 45,301	C5	73
Arthur, Ont., Can., 1,124	D4	72
Arthur, Douglas and Moultrie, Ill., 2,120	D5	90
Arthur, Pike, Ind., 150	H3	91
Arthur, Ida, Iowa, 265	B2	92
Arthur, Arthur, Nebr., 165	C4	103
Arthur, Cass, N. Dak., 325	B8	110
Arthur, co., Nebr., 680	C4	103
Arthur, kill, N.J	E4	106
Arthur, lake, La	D3	95
Arthur's Town, Ba. Is., 450	C6	64
Artigas, Ur., 18,900	E1	56
Artigas, dept., Ur., 44,200	*E1	56
Artland, Sask., Can., 70	E1	70
Artois, former prov., Fr., 1,010,000	B5	14
Artois, hills, Fr	D2	15
Artvin, Tur., 7,900	B13	31
Aru, Con. L	A5	48
Aru, is., Indon	G8	35
Aru, pt., Ponape		52
Arua, Ug., 4,645	A5	48
Aruanã, Braz., 2,405	A2	56
Aruba, isl., Neth. W.I	A4	60
Arucas, Can. Is., 9,597	m14	20
Arun, riv., Eng	D7	12
Arundel, Que., Can., 409	D3	73
Arundel, Eng., 2,614	D7	12
Arundel Village, Anne Arundel, Md., 1,600	*B4	85
Arusha, Tan., 10,038	B6	48
Arusl, reg., Eth., 1,000,000	D4	47
Aruwimi, riv., Con. L	A4	48
Arvada, Jefferson, Colo., 19,242	B5	83
Arvada, Sheridan, Wyo., 100	A6	125
Arvagh, Ire., 512	D4	11
Arve, riv., Fr	D2	18
Arvida, Que., Can., 14,460	A6	73
Arvidsjaur, Swe., 2,900	E8	25
Arvigo, Switz., 102	D7	19
Arvika, Swe., 15,800	H5	25
Arvilla, Grand Forks, N. Dak., 100	B8	110
Arvin, Kern, Calif., 5,440	E4	82
Arvonia, Buckingham, Va., 700	D4	121
Arys, Sov. Un	E7 29, C22	9
Arzamas, Sov. Un., 39,000	B2	29
Arzew, Alg., 4,976	G6	30
Arzgir, Sov. Un	I15	27
Arzúa, Sp., 1,368	A1	20
As, Czech., 9,600	C2	26
Ås, Swe	A6	24
Asaa, Den., 1,265	A4	24
Asab, S.W. Afr	C2	49
Asahigawa, Jap., 188,309	E11	37
Asalā, Eth., 1,000	D4	47
Asan, Guam, 543		52
Asan, pt., Guam		52
Asansol, India, 103,405 (★250,000)	D8	39
Asaph, Tioga, Pa., 20	C7	114
Asbach, Ger., 920	C2	17
Asberrys, Tazewell, Va	B3	121
Asbest, Sov. Un., 65,000	B6	29
Asbestos, Que., Can., 11,083	D6	73
Asbury, Jasper, Mo., 186	D3	101
Asbury, Gloucester, N.J	D2	106
Asbury, Hunterdon, N.J., 300	B2	106
Asbury, Knox, Tenn., 300	E11	117
Asbury Park, Monmouth, N.J., 17,366	C4	106
Ascensión, Mex., 1,591	A3	63
Ascension, par., La., 27,927	D5	95
Ascension, isl., Atl. O	G9	6
Ascension Isl., Br. dep., Atl. O., 210	G4	42
Aschaffenburg, Ger., 54,100	D4	17
Aschersleben, Ger., 34,400	B6	17
Ascoli, Piceno, It., 29,500 (50,114▲)	C4	21
Ascona, Switz., 3,053	D6	19
Ascope, Peru, 3,727	C2	58
Ascot Corner, Que., Can., 494	D6	73
Ascutney, Windsor, Vt., 250	E4	120
Ascutney, mtn., Vt	E4	120
Åsele, Swe., 1,863 (5,737▲)	E7	25
Asenovgrad, Bul., 20,920	E7	22
Ash, Douglas, Oreg	D3	113
Ash, cave, Ohio	C5	111
Asha, Sov. Un., 32,000	B5	29
Ashanti, reg., Ghana	E4	45
Asharoken, Suffolk, N.Y., 253	F3	84
Ashaway, Washington, R.I., 1,298	D9	84
Ashbourne, Eng., 5,656	A6	12
Ashburn, Turner, Ga., 3,291	E3	87
Ashburn, Pike, Mo., 124	B6	101
Ashburn, Loudoun, Va., 200	B5	121
Ashburnham, Worcester, Mass., 900 (2,758▲)	A4	97
Ashburton, Eng., 2,715	D4	12
Ashburton, N.Z., 11,602	O13	51
Ashburton, riv., Austl	D2	50
Ashby, Bibb, Ala., 65	B3	78
Ashby, Middlesex, Mass., 500 (1,883▲)	A4	97
Ashby, Grant, Minn., 426	D3	99
Ashby, Grant, Nebr., 175	B4	103
Ashbyburg, Hopkins, Ky., 150	C2	94
Ashcroft, B.C., Can., 868	D7	68
Ashdod, Isr	C6, h9	32
Ashdot Ya'aqov, Isr., 1,393	B7	32
Ashdown, Little River, Ark., 2,725	D1	81
Ashe, co., N.C., 19,768	A2	109
Asheboro, Randolph, N.C., 9,449	B4	109
Asheboro South, Randolph, N.C., 1,515	*B4	109
Asheboro West, Randolph, N.C., 1,228	*B4	109
Ashepoo, riv., S.C	F6	115
Asher, Yuma Ariz, 5	E1	80
Asher, Pottawatomie, Okla., 343	C5	112
Ashern, Man., Can., 374	D2	71
Asherton, Dimmit, Tex., 1,890	E3	118
Asherville, Mitchell, Kans., 105	C6	93
Asheville, Buncombe, N.C., 60,182 (★105,000)	D3	109
Ashfield, Franklin, Mass., 350 (1,131▲)	A2	97
Ash Flat, Sharp, Ark., 192	A4	81
Ashford, Houston, Ala., 1,511	D4	78
Ashford, Eng., 27,962	C8	12
Ashford, Windham, Conn., 50 (1,315▲)	B8	84
Ashford, McDowell, N.C., 260	B2, C4	109
Ashford, Boone, W. Va., 250	C6	123
Ash Fork, Yavapai, Ariz., 750	B3	80
Ash Grove, Greene, Mo., 886	D4	101
Ashikaga, Jap., 110,972	H9, m18	37
Ashington, Eng., 27,294	E7	13
Ashio, Jap., 16,608	H9	37
Ashkhabad, Sov. Un., 187,000	H20	9
Ashkum, Iroquois, Ill., 601	C6	90
Ash Lake, St. Louis, Minn., 35	B6	99
Ashland, Clay, Ala., 1,610	B4	78
Ashland, Cass, Ill., 1,064	D3	90
Ashland, Clark, Kans., 1,312	E4	93
Ashland, Boyd, Ky., 31,283	B7	94
Ashland, Natchitoches, La., 300	B2	95
Ashland, Aroostook, Maine, 900 (1,980▲)	B4	96
Ashland, Middlesex, Mass., 7,779	D1	97
Ashland, Benton, Miss., 209	A4	100
Ashland, Boone, Mo., 495	C5	101
Ashland, Rosebud, Mont., 120	E10	102
Ashland, Saunders, Nebr., 1,989	C9, E2	103
Ashland, Grafton, N.H., 1,237 (1,473▲)	C3	105
Ashland, Ashland, Ohio, 17,419	B5	111
Ashland, Jackson, Oreg., 9,119	E4	113
Ashland, Schylkill and Columbia, Pa., 5,237	E9	114
Ashland, Hanover, Va., 2,773	D5	121
Ashland, Ashland, Wis., 10,132	B3	124
Ashland, co., Ohio, 38,771	B5	111
Ashland, co., Wis., 17,375	B3	124
Ashland, mtn., Oreg	E4	113
Ashland, res., Mass	D1	97
Ashland City, Cheatham, Tenn., 1,400	A4	117
Ashley, Washington, Ill., 662	E4	90
Ashley, DeKalb and Steuben, Ind., 721	A7	91
Ashley, Gratiot, Mich., 448	E6	98
Ashley, Pike, Mo., 350	B6	101
Ashley, McIntosh, N. Dak., 1,419	C6	110
Ashley, Delaware, Ohio, 907	B5	111
Ashley, Luzerne, Pa., 4,258	B8	114
Ashley, co., Ark., 24,220	D4	81
Ashley, creek, Utah	C6	119
Ashley, riv., S.C	F7	115
Ashley Falls, Berkshire, Mass., 300	B1	97
Ashmont, Alta., Can., 152	B5	69
Ashmore, Coles, Ill., 447	D5	90
Ashmūn, Eg., U.A.R., 26,100	D2	32
Ashnola, riv., Can., U.S	A5	122
Ashokan, Ulster, N.Y., 340	D6	108
Ashokan, res., N.Y	D6	108
Ashport, Lauderdale, Tenn., 150	B2	117
Ash Shabicha, Iraq	G14	31
Ash Shallāl, Eg., U.A.R	E6	43
Ash Shāriqah, Tr. Coast, 4,000	C2	39
Ash Shaţţ, Eg., U.A.R., 2,000	E4	32
Ash Shawāshinah, Eg., U.A.R., 1,000	E2	32
Ash Shawbak, Jordan	D7	32
Ash Shuwarif, Libya	D2	43
Ashtabula, Ashtabula, Ohio, 24,559	A7	111
Ashtabula, co., Ohio, 93,067	A7	111
Ashtabula Lake, res., N. Dak	B7	110
Ashton, Ont., Can., 94	B8	72
Ashton, Fremont, Idaho, 1,242	E7	89
Ashton, Lee, Ill., 1,024	B4	90
Ashton, Osceola, Iowa, 615	A2	92
Ashton, Montgomery, Md., 157	B3	84
Ashton, Osceola, Mich., 125	E5	98
Ashton, Sherman, Nebr., 320	C7	103
Ashton, Providence, R.I., 2,000	B11	84
Ashton, Spink, S. Dak., 182	B7	116
Ashuanipi, lake, Newf., Can	h8	75
Ashuelot, Cheshire, N.H., 300	E2	105
Ashuelot, riv., N.H	E2	105
Ashville, St. Clair, Ala., 973	B3	78
Ashville, Man., Can., 42	D1	71
Ashville, Pickaway, Ohio, 1,639	C5	111
Ashwaubenon, Brown, Wis., 2,657	*A5	124
Ashwood, Jefferson, Oreg., 15	C6	113
Ashwood, Maury, Tenn., 100	B4	117
Asia, cont., 1,493,600,000	C3	2, 33
Asiago, It., 7,228	D7	18
Asifabad, India, 6,190	H7	40
Asiga, pt., Guam		52
Asiga, pt. Tinian		52
Asinara, gulf, It	D2	21
Asinara, isl., It	D2	21
Asino, Sov. Un	B11	29
Asir, reg., Sau. Ar	B5	47
'Asirah ash Shamālīyah, Jordan, 3,000	g12	32
Asker, Nor., 398	p28	25
Askersund, Swe., 2,578	H6	25
Askew, Panola, Miss., 100	A3	100
Askim, Nor., 4,958	p29	25
Askim, Swe., 2,883	A5	24
Askö, isl., Swe	u35	25
Askov, Pine, Minn., 331	D6	99
Asmantay Matay, lake, Sov. Un.	F20	9
Asmar, Afg., 5,000	D15	41
Asmara, Eth., 120,000	B4	47
Asnaes, Den., 1,120	C5	24
Asnebumskit, hill, Mass	B4	97
Asnen, lake, Swe	B8	24
Asnières [-sur-Seine], Fr., 81,768	g10 14, F2	15
Asomante, P.R.	B4	65
Asopos, riv., Grc	g11	23
Asosā, Eth	C3	47
Asotin, Asotin, Wash., 745	C8	122
Asotin, co., Wash., 12,909	C8	122
Asotin, creek, Wash	C8	122
Aspe, Sp., 10,279	C5	20
Aspen, Pitkin, Colo., 1,101	B4	83
Aspen Hill, Giles, Tenn., 100	B5	117
Aspen, mts., Wyo	D3	125
Asperg, Ger., 8,600	E4	17
Aspermont, Stonewall, Tex., 1,286	C2	118
Aspers, Adams, Pa., 280	G7	114
Aspinwall, Crawford, Iowa, 95	C2	92
Aspinwall, Allegheny, Pa., 3,727	A6	114
Aspiring, mtn., N.Z	p12	51
Aspres [-sur-Buëch], Fr., 552	E1	18
Aspropirgos, Grc., 4,880	g11	23
Aspy, bay, N.S., Can	C9	74
Asquith, Sask., Can., 324	E2	70
Assab, Eth	C5	47
Assabet, riv., Mass	D1	97
Aş Şaff, Eg., U.A.R., 4,000	E3	32
Aş Şāfiyah, Jordan, 1,000	C7	32
Assaikwatamo, riv., Man., Can	A3	71
Aş Şaliḩīyah, Eg., U.A.R., 1,000	D3	32
As Sallūm, Eg., U.A.R., 1,011	C5	43
As Salman, Iraq	G15	31
As Salţ, Jordan, 15,478	F10	31
As Salwá, Sau. Ar	I5	41
Assam, state, India, 11,872,772	C9	39
Assaria, Saline, Kans., 322	D6	93
Assateague, isl., Va	D7	121
Assawompset, pond, Mass	C6	97
Assen, Neth., 29,500	B6	15
Assens, Den., 4,937	B4	24
Assens, co., Den., 58,005	C3	24
As Sinbillāwayn, Eg., U.A.R., 23,831	D3	32
Assiniboia, Sask., Can., 2,491	H3	70
Assiniboine, mtn., B.C., Can	F9 66, D10	68
Assiniboine, riv., Man., Sask., Can	E4	70
Assinie, I.C	E4	45
Assinika, riv., Man., Can	C4	71
Assinippi, Plymouth, Mass., 300	D4	97
Assis, Braz., 30,207	C2	56
Assisi, It., 5,100	C4	21
Assonet, Bristol, Mass., 550	C5	97
Aş Şūfīyah, Eg., U.A.R., 1,000	D3	32
As Sulaymānīyāh, Iraq., 35,352	E15	31
As Sulayyil, Sau. Ar	A6	47
As Sulţān, Libya	C3	43

Assumption, Christian, Ill., 1,439 D4 90
Assumption, par., La., 17,991 E4 95
As Suwaydā', Syr., 18,200 F11 31
As Suways, see Suez, Eg., U.A.R.
Astakos, Grc., 2,992 C3 23
Āstārā, Iran, 8,425 B4 41
Asten, Neth., 5,100 C5 15
Asterābād, see Gorgan, Iran
Asti, It., 45,000 (61,044▲) B2 21
Astipalaia, isl., Grc D6 23
Astola, isl., Pak I11 41
Aston, Delaware, Pa., 1,200 *G11 114
Aston, cape, N.W. Ter., Can B19 67
Aston Junction, Que., Can., 396 C5 73
Astor, Lake, Fla., 150 C5 86
Astorga, Sp., 10,101 A2 20
Astoria, Fulton, Ill., 1,206 C3 90
Astoria, Clatsop, Oreg., 11,239 A3 113
Astoria, Deuel, S. Dak., 176 C9 116
Astor Park, Lake, Fla., 85 C5 86
Astorville, Ont., Can., 200 A5 72
Astrakhan, Sov. Un., 313,000 D3 29
Astura, riv., It h9 21
Asturias, reg., Sp., 989,344 A2 20
Asuisui, cape, W. Sam 52
Asuke, Jap., 5,700 n16 37
Asunción, Par., 205,605 E4 55
Asuncion, isl., Mariana Is E8 7
Asunción Mita, Guat., 4,014 C3 62
Asunden, lake, Swe A7 24
Aswān, Eg., U.A.R., 28,400 E6 43
Aswān High, dam, Eg., U.A.R. E6 43
Asyūţ, Eg., U.A.R., 106,001 D6 43
Atacama, prov., Chile, 114,277 E1 55
Atacama, des., Chile D2 55
Atacama, salt flat, Chile D2 55
Atakpamé, Togo, 6,005 E5 45
Atalaia, Braz., 1,844 k6 57
Atalandi, Grc., 4,272 C4 23
Atalissa, Muscatine, Iowa, 212 C6 92
Atami, Jap., 52,163 n18 37
Atantano, Guam 52
Atar, Maur., 4,200 B2 45
Atārūt, Jordan, 1,000 h11 32
Atascadero, San Luis Obispo, Calif., 5,983 E3 82
Atascosa, co., Tex., 18,828 E3 118
Atasuskiy, Sov. Un., 2,800 D8 29
Atatyn Hiid, Mong A3 36
Ataúro, isl., Port. Timor G7 35
Atbara, Sud., 36,298 B3 47
Atbara, riv., Sud B4 47
Atbasar, Sov. Un., 22,300 C7 29
Atchafalaya, St. Martin, La., 10 D4 95
Atchafalaya, bay, La E4 95
Atchafalaya, riv., La D4 95
Atchison, Atchison, Kans., 12,529 A8, C8 93
Atchison, co., Kans., 20,898 C8 93
Atchison, co., Mo., 9,213 A2 101
Atchugau, mtn., Saipan 52
Atco, Bartow, Ga. (part of Cartersville) B2 87
Atco, Camden, N.J., 2,400 D3 106
Atenango, riv., Mex n14 63
Ath, Bel., 10,965 D3 15
Athabasca, Alta., Can., 1,487 B4 69
Athabasca, lake, Alta. and Sask., Can A1 70
Athabasca, riv., Alta., Can E10 66, B4 69
Athalia, Lawrence, Ohio, 341 D5 111
Athalmer, B.C., Can., 304 D9 68
Athapapuskow, lake, Man., Can B1 71
Atha Road, Ont., Can E7 72
Athboy, Ire., 680 D4 11
Athel, Wicomico, Md D6 85
Athelstan, Que., Can., 167 D3 73
Athelstan, Taylor, Iowa, 72 D3 92
Athena, Umatilla, Oreg., 950 B8 113
Athenry, Ire., 1,266 D3 11
Athens, Limestone, Ala., 9,330 A3 78
Athens, Howard, Ark., 67 C2 81
Athens, Ont., Can., 935 C9 72
Athens, Clarke, Ga., 31,355 C3 87
Athens (Athinai), Grc., 627,564 (★1,852,709) C4, h11 23
Athens, Menard, Ill., 1,035 D4 90
Athens, Fulton, Ind., 100 B5 91
Athens, Fayette, Ky., 225 C5 94
Athens, Claiborne, La., 406 B2 95
Athens, Somerset, Maine, 225 (602▲) D3 96
Athens, Calhoun, Mich., 966 F5 98
Athens, Greene, N.Y., 1,754 C7 108
Athens, Athens, Ohio, 16,470 C5 111
Athens, Bradford, Pa., 4,515 C8 114
Athens, McMinn, Tenn., 12,103 D9 117
Athens, Henderson, Tex., 7,086 C5 118
Athens, Mercer, W. Va., 1,086 D3 123
Athens, Marathon, Wis., 770 C3 124
Athens, co., Ohio, 46,998 C5 111
Athensville, Greene, Ill., 40 D3 90
Atherley, Ont., Can., 348 C5 72
Atherton, San Mateo, Calif., 7,717 *B5 82
Athertonville, Larue, Ky., 150 C4 94
Athi, riv., Ken B6 48
Athis-Mons, Fr., 24,004 h10 14
Athleague, Ire., 132 D3 11
Athlone, Ire., 9,624 D4 11
Athok, Bur., 4,770 E10 39
Athol, Kootenai, Idaho, 214 B2 89
Athol, Smith, Kans., 140 C5 93
Athol, Worcester, Mass., 11,637 A3 97
Athol, Spink, S. Dak., 100 B7 116
Athos, mtn., Grc B5 23
Ath Thamad, Eg., U.A.R., 3,000 D6 43
Athy, Ire., 3,842 D5 11
Atibaia, Braz., 8,957 m8 56
Atico, Peru, 373 E3 58
Atikameg, lake, Man., Can B1 71
Atikokan, Ont., Can., 6,674 E8 72
Atikonak, lake, Newf., Can h8 75
Atikonipi, lake, Que., Can C2 75
Atimonan, Phil., 4,027 p13 35
Atiquizaya, Sal., 6,338 D3 62
Atizapán de Zaragoza, Mex., 1,840 g9 63
Atka, Alsk., 119 E5 79
Atka, isl., Alsk E5 79
Atkarsk, Sov. Un., 39,800 F15 27
Atkins, Pope, Ark., 1,391 B3 81
Atkins, Benton, Iowa, 527 B6 92
Atkins, Bossier, La B2 95
Atkins, Smyth, Va., 400 B3, E1 121
Atkinson, Brantley, Ga., 50 E5 87
Atkinson, Henry, Ill., 944 B3 90
Atkinson, Holt, Nebr., 1,324 B7 103
Atkinson, Pender, N.C., 302 C5 109
Atkinson, co., Ga., 6,188 E4 87
Atkinson, lake, Man., Can A4 71
Atkinson Depot (Atkinson), Rockingham, N.H., 700 (1,017▲) E4 105
Atlanta, Columbia, Ark., 50 D2 81
Atlanta, Fulton and DeKalb, Ga., 487,455 (★1,011,100) B5, C2 87
Atlanta, Elmore, Idaho, 50 F3 89
Atlanta, Logan, Ill., 1,568 C4 90
Atlanta, Hamilton, Ind., 602 D5 91
Atlanta, Cowley, Kans., 267 E7 93
Atlanta, Winn, La., 300 C3 95
Atlanta, Montmorency, Mich., 450 C6 98
Atlanta, Macon, Mo., 386 B5 101
Atlanta, Phelps, Nebr., 107 D6 103
Atlanta, Steuben, N.Y., 500 C3 108
Atlanta, Pickaway, Ohio, 160 C4 111
Atlanta, Cass, Tex., 4,076 C5 118
Atlantic, Cass, Iowa, 6,890 C2 92
Atlantic, Carteret, N.C., 850 C7 109
Atlantic, co., N.J., 160,880 E3 106
Atlantic, ocean 6
Atlantic, peak, Wyo C3 125
Atlantic Beach, Duval, Fla., 3,125 B6 86
Atlantic Beach, Nassau, N.Y., 1,500 *G2 84
Atlantic City, Atlantic, N.J., 59,544 (★129,800) E4 106
Atlantic City, Fremont, Wyo., 25 C4 125
Atlantic Highlands, Monmouth, N.J., 4,119 C4 106
Atlantic Mine, Houghton, Mich., 400 A2 98
Atlántico, dept., Col., 514,490 A2 60
Atlas, Northumberland, Pa., 1,574 *E9 114
Atlas, mts., Afr C5 42
Atlee, Alta., Can., 50 D5 69
Atlin, B.C., Can., 150 D2 68
Atlin, lake, B.C., Can E6 66
Atlit, Isr., 1,300 B6 32
Atlixco, Mex., 30,433 n14 63
Atmore, Escambia, Ala., 8,173 D2 78
Atna, mts., B.C., Can B4 68
Atna, peak, B.C., Can C3 68
Atocha, Bol D2 55
Atoka, Eddy, N. Mex., 30 E5 107
Atoka, Atoka, Okla., 2,877 C5 112
Atoka, Tipton, Tenn., 357 B2 117
Atoka, co., Okla., 10,352 C5 112
Atoka, res., Okla C5 112
Atomic City, Bingham, Idaho, 143 F6 89
Atotonilco el Alto, Mex., 14,190 m12 63
Atoyac, Mex., 5,324 m12 63
Atoyac, riv., Mex n14 63
Atrato, riv., Col B2 60
Atrek, riv., Iran C8 41
Atrisco, Bernalillo, N. Mex., 3,000 *B5 107
Atsion, Burlington, N.J D3 106
Atsugi, Jap., 20,800 n18 37
Atsukeshi, bay, Jap E12 37
Atsuma, Jap E10 37
Atsumi, bay, Jap o16 37
At Tabbīn, Eg., U.A.R., 1,000 E3 32
Attachie, B.C., Can A7 68
Aţ Ţafīlah, Jordan, 8,588 D7 32
Aţ Ţā'if, Sau. Ar., 25,000 A5 47
Attala, co., Miss., 21,335 B4 100
Attalla, Etowah, Ala., 8,257 A3 78
Aţ Ţallāb, Libya E4 43
At Tamīmī, Libya C4 43
Attapulgus, Decatur, Ga., 567 F2 87
Aţţarah, Jordan, 1,000 g11 32
Attavyros, mtn., Grc D6 23
Attawapiskat, Ont., Can E9 72
Attawapiskat, riv., Ont., Can D9 72
Attawaugan, Windham, Conn., 500 B9 84
Aţ Tayyibah, Jordan, 2,000 C7, h12 32
Attean, pond, Maine C2 96
Attica, Fountain, Ind., 4,341 D3 91
Attica, Harper, Kans., 845 E5 93
Attica, Lapeer, Mich., 250 E7 98
Attica, Wyoming, N.Y., 2,758 C2 108
Attica, Seneca, Ohio, 965 A5 111
Attica, reg., Grc g11 23
Attica (Attiki), prov., Grc., 2,057,974 *C4 23
Attigny, Fr., 1,525 E4 15
Attikamagen, lake, Newf., Can g8 75
Attiki (Attica), prov., Grc., 2,057,974 *C4 23
Aţ Ţīnah, Eg., U.A.R., 1,000 C4 32
Attleboro, Bristol, Mass., 27,118 C5 97
Attleborough, Eng., 2,741 B9 12
Attopeu, Laos E7 38
Attow, mtn., Scot B4 10, C3 13
Attu, isl., Alsk E2 79
Aţ Ţūr, Eg., U.A.R D6 43
Attymon, Ire D3 11
Atuel, riv., Arg B3 54
Atuu, Am. Sam., 186 52
Atvidaberg, Swe., 7,800 H 25
Atwater, Merced, Calif., 7,318 D3 82
Atwater, Sask., Can., 95 G4 70
Atwater, Kandiyohi, Minn., 899 E4 99
Atwood, Ont., Can., 418 D3 72
Atwood, Logan, Colo., 55 A7 83
Atwood, Piatt and Douglas, Ill., 1,258 D5 90
Atwood, Kosciusko, Ind., 250 B6 91
Atwood, Rawlins, Kans., 1,906 C2 93
Atwood, Hughes, Okla., 200 C5 112
Atwood, Carroll, Tenn., 461 B3 117
Atwood, res., Ohio B6 111
Atwood Heights, Cook, Ill., 1,000 *F3 90
Atzcapotzalco (Azcapotzalco), Mex., 49,617 h9, n14 63
Atzmon, mtn., Isr B7 32
Aua, Am. Sam., 505 52
Auasc, Eth D5 47
Auau, chan., Haw C5 88
Auaz, mts., S.W. Afr B2 49
Aubagne, Fr., 12,612 F6 14
Aube, dept., Fr., 255,099 F4 15
Aube, riv., Fr F4 15
Aubenas, Fr., 5,754 E6 14
Auberry, Fresno, Calif., 400 D4 82
Aubervilliers, Fr., 70,632 g10 14
Aubière, Fr., 6,820 E5 14
Aubigny-sur-Nère, Fr., 3,244 D5 14
Aubin, Fr., 7,821 E5 14
Aubrey, Lee, Ark., 400 C5 81
Aubrey, Que., Can., 52 D4 73
Auburn, Lee, Ala., 16,261 C4 78
Auburn, Placer, Calif., 5,586 C3 82
Auburn, Ont., Can., 216 D3 72
Auburn, Barrow, Ga., 374 B3 87
Auburn, Sangamon, Ill., 2,209 D4 90
Auburn, DeKalb, Ind., 6,350 B7 91
Auburn, Sac, Iowa, 367 B3 92
Auburn, Shawnee, Kans., 200 D8 93
Auburn, Logan, Ky., 1,013 D3 94
Auburn, Androscoggin, Maine, 24,449 D2, D5 96
Auburn, Worcester, Mass., 14,047 B4 97
Auburn, Bay, Mich., 1,497 E6 98
Auburn, Lincoln, Miss., 250 D3 100
Auburn, Nemaha, Nebr., 3,229 D10 103
Auburn, Rockingham, N.H., 150 (1,292▲) D4 105
Auburn, Salem, N.J., 200 D2 106
Auburn, Cayuga, N.Y., 35,249 C4 108
Auburn, Walsh, N. Dak., 60 A8 116
Auburn, Schuylkill, Pa., 936 E9 114
Auburn, King, Wash., 11,933 B3, D2 122
Auburn, Ritchie, W. Va., 139 B4 123
Auburn, Lincoln, Wyo., 100 C2 125
Auburn Center, Schuylkill, Pa., 30 C9 114
Auburndale, Polk, Fla., 5,595 D5 86
Auburndale, Wood, Wis., 396 D3 124
Auburn Heights, Oakland, Mich., 2,500 F7 98
Auburntown, Cannon, Tenn., 256 B5 117
Aubusson, Fr., 5,669 E5 14
Auch, Fr., 18,918 F4 14
Auchel, Fr., 14,412 D2 15
Auchterarder, Scot., 2,426 D5 13
Aucilla, Jefferson, Fla., 240 B3 86
Aucilla, riv., Fla B3 86
Auckland, N.Z., 143,583 (★448,365) L15 51
Auckland, isl., Pac. O D29 5
Aude, dept., Fr., 269,782 *F5 14
Aude, riv., Fr F5 14
Audègle, Som., 2,000 (7,800▲) E5 47
Audet, Que., Can., 110 D7 73
Audierne, Fr., 3,782 D1 14
Audincourt, Fr., 12,433 D1 14
Audrain, co., Mo., 26,079 B6 101
Audubon, Audubon, Iowa, 2,928 C3 92
Audubon, Becker, Minn., 245 D3 99
Audubon, Camden, N.J., 10,440 D2 106
Audubon, co., Iowa, 10,919 C3 92
Audubon Park, Jefferson, Ky., 1,867 *A4 94
Audubon Park, Camden, N.J., 1,713 *C3 106
Audun-le-Roman, Fr., 2,848 E5 15
Aue, Ger., 31,200 C7 17
Auerbach, Ger., 19,400 C7 17
Au Fer, pt., La E4 95
Augathella, Austl., 624 B6 51
Augerville, New Haven, Conn. (part of Hamden and North Haven) D5 84
Aughnacloy, N. Ire., 805 C5 11
Aughrim, Ire., 528 E5 11
Auglaize, co., Ohio, 36,147 B3 111
Auglaize, riv., Ohio A3 111
Au Gres, Arenac, Mich., 584 D7 98
Augsburg, Ger., 208,700 (★280,000) E5 17
Augusta, Austl., 142 F2 50
Augusta, Woodruff, Ark., 2,272 B4 81
Augusta, Richmond, Ga., 70,626 (★160,600) C5 87
Augusta, Hancock, Ill., 915 C3 90
Augusta, Marion, Ind., 200 H3, H7 91
Augusta, It., 27,950 F5 21
Augusta, Butler, Kans., 6,434 B6, E7 93
Augusta, Bracken, Ky., 1,458 B6 94
Augusta, Kennebec, Maine, 21,680 D3 96
Augusta, Kalamazoo, Mich., 972 F5 98
Augusta, St. Charles, Mo., 206 C7 101
Augusta, Lewis and Clark, Mont., 450 C4 102
Augusta, Sussex, N.J., 70 A3 106
Augusta, Carroll, Ohio, 250 B6 111
Augusta, Eau Claire, Wis., 1,338 D2 124
Augusta, Hampshire, W. Va., 250 B6 123
Augusta, co., Va., 37,363 C3 121
Augusta Springs, Augusta, Va., 300 C3 121
Augustenborg, Den., 1,926 D3 24
Augustine, Socorro, N. Mex C2 107
Augustow, Pol., 12,700 B7 26
Auja, see Yarkon, riv., Isr.
Aul, mtn., Switz C7 19
Aulander, Bertie, N.C., 1,083 A6 109
Aulendorf, Ger., 5,000 B5 18
Aulnay-sous-Bois, Fr., 47,507 g11 14
Aulne, riv., Fr C1 14
Aulneau, pen., Ont., Can E4 71
Aulnoye [-Aymeries], Fr., 9,405 (★18,000) D3 15
Ault, Weld, Colo., 799 A6 83
Aultsville, Ont., Can., 301 C9 72
Aumale, Alg., 4,028 F8 30
Aumale, Fr., 1,757 E9 12
Aumsville, Marion, Oreg., 300 C2 113
Auneuil, Fr., 969 E2 15
Auning, Den., 1,314 B4 24
Aunis, former prov., Fr., 167,000 *D3 14
Auno, Nig D7 45
Aunuu, isl., Am. Sam 52
Aur, isl., Mala K6 38
Aura, Baraga, Mich., 40 B2 98
Aura, Gloucester, N.J D2 106
Auraiya, India, 17,463 D7 40
Aurangabad, India, 87,579 (★97,701) E6 39, G5 40
Auray, Fr., 8,118 D2 14
Aurelia, Cherokee, Iowa, 904 B2 92
Aurelian Springs, Halifax, N.C., 75 A6 109
Aurich, Ger., 13,000 B3 16
Aurillac, Fr., 24,563 E5 14
Aurora, Br. Gu., 290 A3 59
Aurora, Ont., Can., 8,791 D5 72
Aurora, Arapahoe and Adams, Colo., 48,548 B6 83
Aurora, Kane, Ill., 63,715 B5, F1 90
Aurora, Dearborn, Ind., 4,756 F8 91
Aurora, Buchanan, Iowa, 223 B6 92
Aurora, Cloud, Kans., 150 C6 93
Aurora, Hancock, Maine, 70 (75▲) D4 96
Aurora, St. Louis, Minn., 2,799 C6 99
Aurora, Lawrence, Mo., 4,683 E4 101
Aurora, Hamilton, Nebr., 2,576 D7 103
Aurora, Cayuga, N.Y., 834 C4 108
Aurora, Beaufort, N.C., 449 B7 109
Aurora, Portage, Ohio, 4,049 A6 111
Aurora, Marion, Oreg., 247 B2, B4 113
Aurora, Brookings, S. Dak., 232 C9 116
Aurora, Sevier, Utah, 465 E4 119
Aurora, Preston, W. Va., 400 B5 123
Aurora, Florence, Wis C5 124
Aurora, co., S. Dak., 4,749 D7 116
Aurora Center, Brookings, S. Dak., 232 D7 116
Aurskog, Nor p29 25
Aus, S. W. Afr., 687 C2 49
Au Sable, pt., Mich B4 98
Au Sable, pt., Mich D7 98
Au Sable, riv., Mich D6 98
Ausable, riv., N.Y B3 108
Au Sable Forks, Essex and Clinton, N.Y., 2,026 B3 108
Aussa, riv., It D9 18
Ausser-Rhoden, sub canton, Switz., 48,920 B7 19
Aust-Agder, co., Nor., 76,900 *H3 25
Austell, Cobb, Ga., 1,867 B4 87
Austin, Lonoke, Ark., 210 C4 81
Austin, Man., Can., 384 E2 71
Austin, Scott, Ind., 3,838 G6 91
Austin, Barren, Ky., 150 D3 94
Austin, Mower, Minn., 27,908 G6 99
Austin, Lewis and Clark, Mont., 20 D4 102
Austin, Lander, Nev., 500 D4 104
Austin, Grant, Oreg., 35 C8 113
Austin, Potter, Pa., 721 C5 114
Austin, Travis, Tex., 186,545 (★210,000) D4 118
Austin, Sevier. Utah, 100 E3 119
Austin, co., Tex., 13,777 E4 118
Austin, lake, Austl E2 50
Austinburg, Ashtabula, Ohio, 300 A7 111
Austin Lake, Kalamazoo, Mich., 3,520 *F5 98
Austintown, Mahoning, Ohio, 5,000 *A7 111
Austinville, Butler, Iowa, 130 B5 92
Austinville, Wythe, Va., 750 E2 121
Australia, country, Oceania, 10,508,189 H6 2, 50
Australian Alps, mts., Austl G8 50
Australian Capital Territory, Austl., 58,828 G7 51
Austria, country, Eur., 7,073,807 F12 8, E6 16
Austwell, Refugio, Tex., 287 E4 118
Autauga, co., Ala., 18,739 C3 78
Autaugaville, Autauga, Ala., 440 C3 78
Authie, riv., Fr D2 15
Autlán de Navarro, Mex., 17,069 D4, n11 63
Au Train, Alger, Mich., 100 B4 98
Autreyville, Colquitt, Ga., 50 E3 87
Autun, Fr., 15,305 D6 14
Auvergne, Jackson, Ark., 100 B4 81
Auvergne, mts., Fr E5 14
Auvergne, former prov., Fr., 838,000 E5 14
Auvers [-sur-Oise], Fr., 3,772 E2 15
Aux Barques, pt., Mich D8 98
Aux Chene, riv., La C7 95
Auxerre, Fr., 31,178 D5 14
Auxier, Floyd, Ky., 800 C7 94
Auxi-le-Château, Fr., 3,135 D10 12
Auxonne, Fr., 4,084 D6 14
Aux Pins, pt., Ont., Can E3 72
Auxvasse, Callaway, Mo., 534 B6 101
Ava, Jackson, Ill., 665 F4 90
Ava, Douglas, Mo., 1,581 E5 101
Ava, Noble, Ohio, 220 C6 111
Avakubi, Con. L A4 48
Avallon, Fr., 5,976 D5 14
Avalon, Los Angeles, Calif., 1,536 F4 82
Avalon, Stephens, Ga., 194 B3 87
Avalon, Livingston, Mo., 65 B4 101
Avalon, Cape May, N.J., 695 E3 106
Avalon, Allegheny, Pa., 6,859 A5 114
Avalon, pen., Newf., Can E5 75
Avalon, res., N. Mex E5 107
Avanos, Tur., 5,700 C10 31
Avant, Osage, Okla., 381 A5 112
Avard, Woods, Okla., 56 A3 112
Avaré, Braz., 20,334 C3 56
Āvāz, Iran E10 41
Aveiro, Port., 13,423 B1 20
Avella, Washington, Pa., 1,310 F1 114
Avellaneda, Arg., 329,626 A5, g7 54
Avellino, It., 33,500 (41,825▲) D5 21
Avenal, Kings, Calif., 3,147 E3 82
Avenches, Switz., 1,776 C3 19
Avenel, Middlesex, N.J., 9,000 E4 106
Avening, Ont., Can., 51 C4 72
Avenwedde, Ger., 8,500 B3 17
Avera, Jefferson, Ga., 197 C4 87
Averill, Clay, Minn., 54 D2 99
Averill Park, Rensselaer, N.Y., 900 C7 108
Aversa, It., 40,336 D5 21
Avery, Shoshone, Idaho, 450 B3 89
Avery, Monroe, Iowa, 200 C5 92
Avery, Lincoln, Okla., 30 B5 112
Avery, Red River, Tex., 343 C5 118
Avery, co., N.C., 12,009 A2 109
Avery Island, Iberia, La., 650 E4 95
Aves, is., Ven A4 60
Avesnes [-sur-Helpe], Fr., 6,151 D3 15
Avesta, Swe., 10,900 (★18,500) G7 25
Aveyron, dept., Fr., 290,442 *E5 14
Avezzano, It., 25,000 C4 21
Aviá Teraí, Arg E3 55
Aviemore, Scot C5 13
Avigliano, It., 4,554 D5 21
Avignon, Fr., 72,717 F6 14
Avila, San Luis Obispo, Calif., 300 E3 82
Ávila, Sp., 26,807 B3 20
Ávila, prov., Sp., 238,372 *B3 20
Avilés, Sp., 48,503 A3 20
Avilla, Noble, Ind., 919 B7 91
Avilla, Jasper, Mo., 135 D3 101
Avinger, Cass, Tex., 730 C5 118
Aviron Bay, Newf., Can E3 75
Avis, Clinton, Pa., 1,262 D7 114
Avisio, riv., It C7 18
Avize, Fr., 1,888 F4 15
Avlon, Grc., 2,231 g11 23
Avlum, Den., 1,253 B2 24
Avoca, Lawrence, Ala., 400 A2 78
Avoca, Benton, Ark., 90 A1 81
Avoca, Lawrence, Ind., 400 G4 91
Avoca, Pottawattamie, Iowa, 1,540 C2 92
Avoca, Assumption, La C5 95
Avoca, St. Clair, Mich., 300 E8 98
Avoca, Murray, Minn., 226 G3 99
Avoca, Cass, Nebr., 218 E3 103
Avoca, Steuben, N.Y., 1,086 C3 108
Avoca, Luzerne, Pa., 3,562 B9 114
Avoca, Jones, Tex., 200 C3 118
Avola, It., 27,453 F5 21
Avola, B.C., Can., 138 D8 68
Avon, Ont., Can., 97 E4 72
Avon, Eagle, Colo., 5 B4 83
Avon, Hartford, Conn., 5,273 B5 84
Avon, Fulton, Ill., 996 C3 90
Avon, Polk, Iowa, 25 A7 92
Avon, Franklin, Maine, (436▲) D2 96
Avon, Norfolk, Mass., 4,301 B5, E3 97
Avon, Stearns, Minn., 443 E4 99
Avon, Washington, Miss., 150 B2 100
Avon, Powell, Mont., 175 D4 102
Avon, Livingston, N.Y., 2,772 C3 108
Avon, Dare, N.C., 300 B8 109
Avon, Lorain, Ohio, 6,002 A5 111
Avon, Lebanon, Pa., 1,212 *F9 114
Avon, Bon Homme, S. Dak., 637 E7 116
Avon, riv., Eng E6 10, D6 12
Avon, riv., Eng D6 10, B5 12
Avon, riv., Eng C5 12
Avon, riv., Scot C5 13
Avon by the Sea, Monmouth, N.J., 1,707 C4 106
Avondale, Maricopa, Ariz., 6,151 G1 80
Avondale, Pueblo, Colo., 400 C6 83
Avondale, Prince Georges, Md., 2,000 *C4 85
Avondale, Clay, Mo., 663 E2 101
Avondale, Chester, Pa., 1,016 G10 114
Avondale Estates, DeKalb, Ga., 1,646 B5 87
Avondale, Washington, R.I., 200 D9 84
Avon Lake, Lorain, Ohio, 9,403 A5 111
Avonlea, Sask., Can., 354 G3 70
Avonmore, Westmoreland, Pa., 1,351 E3 114
Avon Park, Highlands, Fla., 6,073 E5 86
Avoyelles, par., La., 37,606 C3 95
Avranches, Fr., 8,854 C3 14
Awa, isl., Jap G9 37
Awaji, isl., Jap I7 37
Awalt, Franklin, Tenn., 40 B5 117
Awarta, Jordan, 2,000 B7 32
Awāsh, riv., Eth C5 47
Awashi, Okinawa 52
Awbāri, Libya D2 43
Awe, lake, Scot D3 13
Aweil, Sud., 2,438 D2 47
Awjilah, Libya D4 43
Axel, Neth., 5,200 C3 15
Axel Heiberg, isl., N.W. Ter., Can B24 4
Axial, Moffat, Colo., 5 A3 83
Axial, basin, Colo A2 83
Axim, Ghana, 4,647 F4 45
Axios (Vardar), riv., Grc B4 23
Axis, Mobile, Ala., 300 E1 78
Ax-les-Thermes, Fr., 1,628 F4 14
Axson, Atkinson, Ga., 320 E4 87
Axtell, Marshall, Kans., 493 C7 93
Axtell, Kearney, Nebr., 477 D6 103
Axtell, Sanpete, Utah, 10 D4 119
Ay, Fr., 6,806 C6 14
Ay, riv., Sov. Un B5 29
Ayabaca, Peru, 2,481 B2 58
Ayabe, Jap., 21,500 n14 37
Ayacucho, Arg., 9,220 B5 54
Ayacucho, Peru, 21,531 D3 58
Ayacucho, dept., Peru, 469,396 D3 58
Ayaguz, Sov. Un., 52,400 D10 29
Ayakhta, Sov. Un D28 9
Ayamonte, Sp., 8,908 D2 20
Ayan, Sov. Un., 9,300 D16 28
Ayapel, Col., 2,436 B2 60
Ayapel, mts., Col B3 60
Ayas, Tur., 4,300 B9 31
Ayaviri, Peru, 6,586 D3 58
Aycliffe, Eng., 594 F7 13
Aydar, riv., Sov. Un G13 27
Ayden, Pitt., N.C., 3,108 B6 109
Aydin (Aidin), Tur., 35,700 D6 31
Aydlett, Currituck, N.C., 100 A8 109
Ayer, Middlesex, Mass., 14,927 A4, C1 97
Ayers, Washington, Maine D5 96
Ayer's Cliff, Que., Can., 747 D5 73
Ayía, Grc., 2,823 C4 23
Ayía Paraskeví, Grc., 12,122 C6 23
Ayiassos, Grc., 5,692 C6 23
Ayion Oros (Mount Athos), prov., Grc., 2,687 *B5 23
Ayios Dhimitrios, Grc., 21,365 g11 23
Ayios Nikolaos, Grc., 3,319 E5 23
Aylen, lake, Ont., Can B7 72
Aylesbury, Sask., Can., 162 G3 70
Aylesbury, Eng., 27,891 C7 12
Aylesford, N.S., Can., 964 D5 74
Aylett, King William, Va., 60 D5 121
Aylmer, Pierce, N. Dak., 10 B5 110
Aylmer, lake, N.W. Ter., Can D11 66
Aylmer, mtn., Atla., Can D3 69
Aylmer East, Que., Can., 6,286 D2 73
Aylmer West, Ont., Can., 4,201 E4 72
Aylsham, Sask., Can., 251 D4 70
Aynor, Horry, S.C., 635 D9 115
Ayn Sīdī Muḥammad (Oasis), Libya H3 31
'Ayn Yabrūd, Jordan, 1,000 h11 32
Ayon, isl., Sov. Un C19 28
Ayora, Sp., 6,634 C5 20
Ayr, Adams, Nebr., 111 D7 103
Ayr, Cass, N. Dak., 81 B8 110
Ayr, Scot., 45,297 E4 13
Ayr, co., Scot., 342,855 E4 13
Ayr, riv., Scot E4 13
Ayrshire, Pike, Ind., 100 H3 91
Ayrshire, Palo Alto, Iowa, 298 A3 92
Aysén, see Aisén, prov., Chile
Ayshā, Eth C5 47
Ayton, Ont., Can., 375 C4 72
Aytos, Bul., 9,972 D8 22
Ayutla, Guat., 1,653 C1 62
Ayutla [de los Libres], Mex., 2,688 D5 63
Ayutthaya, Thai., 33,187 E4 38
Ayvacik, Tur., 1,900 C6 31
Ayvalik, Tur., 16,100 C6 31
Azalea Park, Orange, Fla., 6,500 *D5 86
Azalia, Bartholomew, Ind., 100 F6 91
Azama, Okinawa 52
Azamgarh, India, 32,391 D9 40
Azángaro, Peru, 2,619 D3 58
Azaouak Assakari, wadi, Mali C5 45
Azaouad, sand dunes, Mali C4 45
Azare, Nig D7 45
Azemmour, Mor., 12,449 C3 44
Azerbaidzhan, (S.S.R.), rep., Sov. Un., 3,973,000 G18 9
Azilal, Mor., 1,340 I3 30
Azilek (Well), Niger C6 45
Aziscoos, lake, Maine C1 96
Azle, Tarrant, Tex., 2,969 B5 118
Azogues, Ec., 6,579 B2 58
Azor, Isr., 3,300 g10 32
Azores, is., Atl. O h8 44
Azores Is., reg., Atl. O., 327,500 h8 44
Azov, Sov. Un., 10,000 H12 27
Azov, sea, Sov. Un I11 27
Azrou, Mor., 14,143 C3 44
Aztec, Yuma, Ariz., 30 E2 80
Aztec, San Juan, N. Mex., 4,137 A2 107
Aztec Ruins, nat. mon., N. Mex A1 107
Azua, Dom. Rep., 12,350 F8 64
Azuaga, Sp., 17,518 C3 20
Azuay, prov., Ec., 277,772 B2 58
Azuero, pen., Pan G7 62
Azul, Arg., 28,609 B5 54
Azul, range, Peru C2 58
Azurdui, Bol., 1,234 C3 55
Azusa, Los Angeles, Calif., 20,497 F3 82
Az Z̧āhiriyah, Jordan, 4,000 C6 32
'Az Z̧ahrān (Dhahran), 75,000 H5 41
Az Zaqazīg, Eg., U.A.R., 81,813 C6 43
Az Zargā', Jordan, 4,000 B8 32
'Azzās, cape, Libya C5 43
Az Zāwiyah, Jordan, 1,000 g11 32
Az Zawiyah, Libya, 8,428 C2 43
Azzel Matti, lake, Alg D5 44
Az Zubair, Iraq, 23,582 F3 41
Azzun, Jordan, 2,000 g11 32

B

Baagö, isl., Den C3 24
Baal, Ger., 1,850 C6 15
Baalbek, Leb., 9,623 E11 31
Baar, Switz., 9,114 B6 19
Baarle-Hertog, Bel., 2,044 C4 15
Baba, cape, Tur C6 23
Babadag, Rom., 5,549 C9 22
Babaeski, Tur., 11,700 B6 23
Babahoyo, Ec., 9,045 B2 58
Babana, Nig D5 45
Babanusa, Sud C2 47
Babati, Tan B6 48
Babayevo, Sov. Un., 10,000 B11 27
Babb, Glacier, Mont., 50 B3 102
Babb, creek, Pa C7 114
Babbie, Covington, Ala., 60 D3 78
Babbit, Mineral, Nev., 2,159 E3 104
Babbitt, St. Louis, Minn., 2,587 C7 99

Babcock, Wood, Wis., 125 D3 124
Bab el Mandeb, strait, Afr C5 47
Babelthuap, isl., Palau Is 52
Babenhausen, Ger., 3,800 A6 18
Babi, is., Indon K2 38
Babine, lake, B.C., Can B4 68
Babine, mts., B.C., Can B4 68
Babine, riv., B.C., Can B4 68
Babo, W. Irian F8 35
Bābol, Iran, 36,194 C6 41
Baboquivari, mts., Ariz F4 80
Baboua, Cen. Afr. Rep D2 46
Babson Park, Polk, Fla., 950 E5 86
Babuna, mts., Yugo E5 22
Babushkin, Sov. Un., 20,000 A6 34
Babuyan, chan., Phil *B6 35
Babuyan, is., Phil B6 35
Babyak, Bul., 6,211 E6 22
Babylon, Suffolk, N.Y., 11,062 E3 108
Baca, McKinley, N. Mex B1 107
Baca, co., Colo., 6,310 D8 83
Bacabal, Braz., 4,857 B2 57
Bacalar, Mex., 744 D7 63
Bacarra, Phil., 6,566 *B6 35
Bacău, Rom., 54,138 B8 22
Bacaville, Valencia, N. Mex. (part of Belen) C3 107
Baccalieu, isl., Newf., Can D5 75
Baccarat, Fr., 6,067 C7 14
Baccaro, pt., N.S., Can F4 74
Baceras, Mex., 1,016 A3 63
Bach, Huron, Mich., 100 E7 98
Bach, Switz C7 19
Bacharach, Ger., 2,000 C2 17
Bachau, India, 4,868 F3 40
Back, Scot B2 13
Back, riv., N.W. Ter., Can C12 66
Backa Palanka, Yugo., 16,487 C4 22
Bac Kan, Viet., 10,000 A6 38
Backa Topola, Yugo., 15,057 B4 22
Backbay, Princess Anne, Va., 500 E6 121
Backbone, mtn., Md D1 85
Back Creek, Frederick, Va B4 121
Backnang, Ger., 23,700 E4 17
Backoo, Pembina, N. Dak., 35 A8 110
Backus, Cass, Minn., 317 D4 99
Backway, bay, Newf., Can A2 75
Bac Lieu, Viet., 10,000 H6 38
Bacliff, Galveston, Tex., 1,707 *E5 118
Bac Ninh, Viet., 25,000 B7 38
Bacnotan, Phil., 1,069 n13 35
Bacobi, Navajo, Ariz., 100 B5 80
Bacolod, Phil., 52,000 (119,200▲) C6 35
Bacon, co., Ga., 8,359 E4 87
Bacone, Muskogee, Okla., 250 B6 112
Bacons Castle, Surry, Va B6, D6 121
Baconton, Mitchell, Ga., 564 E2 87
Bacoor, Phil, 6,276 o13 35
Bacova, Bath, Va., 175 C3 121
Bac Quang, Viet A6 38
Bacqueville-en Caux, Fr., 1,022 E8 12
Bacsalmas, Hung., 7,344 B4 22
Bacs-Kiskun, co., Hung., 587,028 *B4 22
Bacuit, Phil., 1,392 *C5 35
Bácum, Mex., 1,509 B3 63
Bad, riv., S. Dak C5 116
Bad, riv., Wis B3 124
Bad Aibling, Ger., 7,400 B8 18
Badajoz, Sp., 80,000 (96,317▲) C2 20
Badajoz, prov., Sp., 834,370 *C2 20
Badalona, Sp., 92,257 B7 20
Bad Axe, Huron, Mich., 2,998 E7 98
Bad Berneck, Ger., 3,300 C6 17
Bad Blankenburg, Ger., 6,601 C6 17
Bad Bramstedt, Ger., 6,200 E3 24
Baddeck, N.S., Can., 825 C9 74
Bad Doberan, Ger., 12,600 A5 16
Baden, Aus., 22,484 D8 16
Baden, Ont., Can., 977 D4 72
Baden, Prince Georges, Md., 12 C4 85
Baden, Beaver, Pa., 6,109 A5, E1 114
Baden, Switz., 13,949 (★43,545) B5 19
Baden, reg., Ger D4 16
Baden-Baden, Ger., 40,000 E3 17
Baden Baden, Bond, Ill. (part of Pierron) E4 90
Baden-Powell, mtn., Calif E5 82
Baden-Württemberg, state, Ger., 7,759,200 D4 17
Bad Freienwalde, Ger., 12,200 B6 16
Bad Friedrichshall, Ger., 8,600 D4 17
Badgastein, Aus., 5,742 B9 18
Badger, Man., Can E4 71
Badger, Newf., Can., 1,036 D3 75
Badger, Webster, Iowa, 340 B3 92
Badger, Roseau, Minn., 338 B2 99
Badger, Kingsbury, S. Dak., 117 C8 116
Badger, creek, Colo B7 83
Badger, mts., Wash B5 122
Badger Basin, Park, Wyo A3 125
Bad Godesberg, Ger., 65,100 C2 17
Bad Hersfeld, Ger., 23,000 C4 16
Bad Homburg, Ger., 37,340 C3 17
Bad Hönningen, see Hönningen, Ger.
Badi, Iraq E13 31
Badia Polesine, It., 3,825 D7 18
Badin, Stanly, N.C., 1,905 B3 109
Badin, Pak E2 40
Badin, lake, N.C B3 109
Bad Ischl, Aus., 12,703 E6 18
Bad Kissingen, Ger., 12,900 C5 17
Bad Kreuznach, Ger., 35,100 D2 17
Badlands, reg., N. Dak C2 110
Badlands, reg., S. Dak D3 116
Badlands, nat. mon., S. Dak D3 116
Bad Lauterberg, Ger., 10,100 B5 17
Bad Liebenwerda, Ger., 6,472 B8 17
Bad Lippspringe, Ger., 8,800 B3 17
Bad Mergentheim, Ger., 11,600 D4 17
Bad Muskau, Ger., 5,122 B9 17
Bad Nauheim, Ger., 13,400 C3 17
Bad Neustadt, Ger., 8,600 C5 17
Bad Oeynhausen, Ger., 14,100 A3 17
Bad Oldesloe, Ger., 16,000 B5 16
Badon, Viet D7 38
Bad Orb, Ger., 7,300 C4 17
Badoumé, Mali D2 45
Bad Pyrmont, Ger., 14,400 B4 17
Badra, Iraq, 3,638 E2 41
Bad Ragaz, Switz., 2,699 B7 19
Bad Reichenhall, Ger., 13,100 E6 16
Bad River, Indian res., Wis B3 124
Bad Salzuflen, Ger., 16,600 A3 17
Bad Salzungen, Ger., 10,000 C5 17
Bad Schandau, Ger., 5,095 C9 17
Bad Segeberg, Ger., 11,700 E4 24
Bad Sülze, Ger., 3,222 D6 24
Bad Tennstedt, Ger., 4,351 B5 17
Bad Tölz, Ger., 11,700 E5 16
Baduein, Eth D5 47
Badulla, Cey., 17,323 G7 39
Badwater, Natrona, Wyo B5 125
Badwater, riv., Wyo B5 125
Bad Wildungen, Ger., 11,200 B4 17
Bad Wörishofen, Ger., 7,500 A6 18
Bad Zwischenahn, Ger., 16,900 A8 15
Baena, Sp., 17,761 D3 20
Baependi, Braz., 5,109 C4 56
Baeza, Ec B2 58
Baeza, Sp., 15,176 D4 20
Bafang, Cam D2 46
Bafata, Port. Gui D2 45
Bafia, Cam E2 46
Baffin, bay, Can B18 67
Baffin, isl., N.W., Ter., Can C18 67
Baffins, bay, Tex F4 118
Bafoulabé, Mali, 2,700 D2 45
Bāfq, Iran, 6,000 F7 41
Bafra, Tur., 20,800 B10 31
Bafra, cape, Tur B11 31
Bāft, Iran, 8,693 G8 41
Bafwasende, Con. L A4 48
Baga, S.L E2 45
Bagabag, Phil., 2,477 n13 35
Bagaces, C.R., 4,079 E5 62
Bagamoyo, Tan., 3,861 C6 48
Bagana, Nig E6 46
Bagan Siapiapi, Indon., 15,321 E2 35, K4 38
Bagata, Con. L B2 48
Bagby, Mariposa, Calif., 20 D3 82
Bagdad, Yavapai, Ariz., 1,462 C2 80
Bagdad, Santa Rosa, Fla., 900 G2 86
Bagdad, Shelby, Ky., 500 B4 94
Bagé, Braz., 47,930 E2 56
Bagēmder, reg., Eth., 1,800,000 C4 47
Bagenkop, Den., 705 D4 24
Bagerhat, Pak., 7,431 F12 40
Baggs, Carbon, Wyo., 199 D5 125
Baghdad, Iraq, 355,958 (*650,000) E2 41
Bagheria, It., 34,201 E4 21
Baghlan, Afg., 24,410 C14 41
Baghnam Faoileann, riv., Scot C2 13
Bagley, Guthrie, Iowa, 406 C3 92
Bagley, Clearwater, Minn., 1,385 C3 99
Bagley, Grant, Wis., 275 F2 124
Bagleys Mills, Lunenburg, Va., 10 E4 121
Bagnacavallo, It., 4,693 E7 18
Bagnara Calabra, It., 10,000 E5 21
Bagnell, Miller, Mo., 62 C5 101
Bagnell, dam, Mo C5 101
Bagnères-de-Bigorre, Fr., 8,996 (10,314▲) F4 14
Bagnères-de-Luchon, Fr., 3,888 F4 14
Bagnolet, Fr., 31,576 g10 14
Bagnols [-sur-Cèze], Fr., 12,905 E6 14
Bagoe, riv., Mali D3 45
Bagolino, It., 5,549 D6 18
Bagot, Man., Can., 55 E2 71
Bagot, co., Que., Can., 21,390 D5 73
Bagotville, Que., Can., 5,629 A7 73
Bagrash Köl, lake, China C2 34
Bagrationovsk (Preussisch Eylau), Sov. Un., 5,000 A6 26
Baguezane, mtn., Niger C6 45
Baguio, Phil., 50,300 B6, n13 35
Bahado (Well), Som D6 47
Bahama, Durham, N.C., 200 A5 109
Bahama Islands, Br. dep., N.A., 131,000 C6 64
Bahawalnagar, Pak., 18,373 C4 40
Bahawalpur, Pak., 84,000 C5 39
Bahawalpur, prov., Pak., 1,823,125 *C5 39
Bāherdār-Giyorgis, Eth C4 47
Bahia, see Salvador, Braz.
Bahia, state, Braz., 5,990,605 D2 57
Bahía, is., Hond B4 62
Bahía Blanca, Arg., 136,000 B4 54
Bahía de Caráquez, Ec., 7,993 B1 58
Bahía Negra, Par., 590 D4 55
Bahraich, India, 56,033 D8 40
Bahrain (Bahrein), country, Asia, 109,650 H5 41
Bahrāmābād, Iran, 14,867 F8 41
Bahr el Abyad (White Nile), riv., Sud C3 47
Bahr el Arab, riv., Sud D2 47
Bahr el Azraq (Blue Nile), riv., Sud C3 47
Bahr el Ghazal, reg., Sud., 991,022 D2 47
Bahr el Ghazal, riv., Sud D2 47
Bahr el Je'bel (Mountain Nile), riv., Sud D3 47
Bahrgan, cape, Iran G4 41
Bahriyah, oasis, Eg., U.A.R D5 43
Bāhū Kalāt, Iran I10 41
Baia dos Tigres, Ang E1 48
Baia-Mare, Rom., 35,920 B6 22
Baião, Braz., 2,265 C5 59
Baïbokoum, Chad, 3,130 D3 46
Baidarik, riv., Mong B4 34
Baie Comeau, Que., Can., 7,956 *G19 67
Baie de Wasai, Chippewa, Mich., 30 B6 98
Baie d'Urfe, Que., Can., 3,549 *D8 73
Baie Johan Beetz, Que., Can., 237 h9 77
Baiersbronn, Ger., 9,200 E3 17
Baie Ste. Catherine, Que., Can., 256 A8 73
Baie St. Paul, Que., Can., 4,674 B7 73
Baie Verte, N.B., Can., 133 C5 74
Baie Verte, Newf., Can., 250 D3 75
Baile Átha Cliath, see Dublin, Ire.
Bailén, Sp., 11,245 C4 20
Băilești, Rom., 15,932 C6 22
Bailey, Park, Colo., 100 B5 83
Bailey, Lauderdale, Miss., 150 *C5 100
Bailey, Nash, N.C., 795 B5 109
Bailey, co., Tex., 9,090 B1 118
Bailey, brook, Maine B2 96
Bailey, isl., S.C G1 115
Bailey Island, Cumberland, Maine, 250 E5 96
Baileys Harbor, Door, Wis., 300 C6 124
Baileyton, Cullman, Ala., 100 A3 78
Baileyton, Greene, Tenn., 206 C11 117
Baileyville, Ogle, Ill., 200 A4 90
Baileyville, Nemaha, Kans., 200 C7 93
Bailieborough, Ire., 1,136 D5 11
Bainbridge, Decatur, Ga., 12,714 F2 87
Bainbridge, Putnam, Ind., 603 E4 91
Bainbridge, Chenango, N.Y., 1,712 C5 108
Bainbridge, Ross., Ohio, 1,001 C4 111
Bainbridge, is., Wash C1 122
Bain-de-Bretagne, Fr., 1,734 D3 14
Bains, West Feliciana, La., 85 D4 95
Bainville, Roosevelt, Mont., 285 B12 102
Baird, Sunflower, Miss., 175 B3 100
Baird, Callahan, Tex., 1,633 C3 118
Baird, Douglas, Wash B6 122
Baird, inlet, Alsk C7 79
Baird, mts., Alsk B7 79
Bairdford, Allegheny, Pa., 950 A6 114
Bairiki, isl., Tarawa 52
Bairnsdale, Austl., 7,428 H6 51
Bairoil, Sweetwater, Wyo., 300 C5 125
Baïse, riv., Fr F4 14
Bait, mts., B.C., Can B4 68
Baiting Hollow, Suffolk, N.Y. F6 84
Baixa Verde, Braz., 3,495 g6 57
Baixo Alentejo, prov., Port., 380,236 *C2 20
Baixo Longa, Ang E2 48
Baja, Hung., 30,355 B4 22
Baja California, state, Mex., 520,913 A1 63
Baja California Sur, ter., Mex., 83,433 B2 63
Bajadero, P.R B4 65
Bajmok, Yugo., 11,716 C4 22
Bakala, Cen. Afr. Rep D4 46
Bakar, Yugo., 2,026 C2 22
Bakel, Sen., 2,300 D2 45
Baker, San Bernardino, Calif., 200 E5 82
Baker, Okaloosa, Fla., 800 G2 86
Baker, Lemhi, Idaho, 200 D5 89
Baker, East Baton Rouge, La., 4,823 *D4 95
Baker, Clay, Minn., 60 D2 99
Baker, Fallon, Mont., 2,365 D12 102
Baker, White Pine, Nev., 30 E7 104
Baker, Benson, N. Dak., 45 A6 110
Baker, Texas, Okla., 70 D3 112
Baker, Baker, Oreg., 9,986 C9 113
Baker, co., Fla., 7,363 B4 86
Baker, co., Ga., 4,543 E2 87
Baker, co., Oreg., 17,295 C9 113
Baker, brook, Maine B3 96
Baker, butte, Ariz C4 80
Baker, isl., Pac. O F11 7
Baker, lake, N.W. Ter., Can D13 66
Baker, lake, Maine B3 96
Baker, mtn., Maine C3 96
Baker, mtn., Wash A4 122
Baker, riv., Wash A4 122
Baker, valley, Oreg C9 113
Bakerhill, Barbour, Ala., 125 D4 78
Baker Lake, N.W. Ter., Can., 386 D13 66
Bakers, Union, N.C., 80 B3 109
Bakers, Davidson, Tenn., 100 E9 117
Bakers, bayou, Ark D6 81
Bakers, isl., Mass C4 97
Bakers, riv., N.H C3 105
Bakersburg, Texas, Okla D3 112
Bakers Crossroads, White, Tenn C8 117
Bakersfield, Kern, Calif., 56,848 (*158,000) E4 82
Bakersfield, Ozark, Mo., 177 E5 101
Bakersfield, Pecos, Tex., 125 D1 118
Bakersfield, Franklin, Vt., 225 (664▲) B3 120
Bakers Mill, Hamilton, Fla., 30 B4 86
Bakerstown, Allegheny, Pa., 700 A6 114
Bakersville, Litchfield, Conn., 150 B4 84
Bakersville, Mitchell, N.C., 393 C4 109
Bakerton, see Elmora, Pa.
Bakerton, Jefferson, W. Va., 300 B7 123
Bakhchisaray, Sov. Un., 17,200 I9 27
Bakhmut, riv., Sov. Un q21 27
Bakir, riv., Tur C6 23
Bāko, Eth D4 47
Bako, I. C E3 45
Bakony Forest, mts., Hung B3 22
Bakouma, Cen. Afr. Rep D4 46
Bakoy, riv., Mali D3 45
Baku, Con. L A5 48
Baku, Sov. Un., 671,000 (*1,150,000) E3 29
Bakundi, Nig E7 45
Bakungan, Indon K2 38
Bakwanga, Con. L., 40,500 C3 48
Bala, Ont., Can., 452 C5 72
Bala, Riley, Kans., 50 C6 93
Bala, Tur., 3,100 C9 31
Bala, Wales, 1,603 B4 12
Bala, mts., Bol B2 55
Balabac, isl., Phil D5 35
Balabac, strait, Phil D5 35
Bala-Cynwyd, Montgomery, Pa., 8,000 *B11 114
Balad, Som E6 47
Balaghat, India, 18,990 G8 40
Balaguer, Sp., 8,342 B6 20
Bal'ah, Jordan, 4,000 f11 32
Balaka, Nya D6 48
Balakhna, Sov. Un., 31,800 D17 9
Balakhta, Sov. Un D27 9
Balaklava, Austl., 1,301 G2 51
Balakleya, Sov. Un., 5,000 G11 27
Balallan, Scot B2 13
Bala Murghab, Afg., 5,000 D11 41
Balancán, Mex., 1,980 D6 63
Balanda, Sov. Un., 10,000 F15 27
Balanga, Phil., 5,061 C6, o13 35
Balangiga, Phil., 5,264 *C7 35
Balaoan, Phil., 2,376 n13 35
Balashikha, Sov. Un., 60,000 D11 27
Balashov, Sov. Un., 67,000 C2 29
Balasore, India, 33,931 G11 40
Balassagyarmat, Hung., 12,457 A4 22
Balāṭ, Eg., U.A.R D5 43
Balaton, Lyon, Minn., 723 F3 99
Balaton, lake, Hung B3 22
Balayan, Phil., 6,033 p13 35
Balayan, bay, Phil p13 35
Balboa, C.Z., 3,139 m11 62
Balboa, dist., C.Z., 30,623 *m11 62
Balboa Heights, C.Z., 118 m11 62
Balbriggan, Ire., 2,943 D5 11
Balcarce, Arg., 15,210 B5 54
Balcarres, Sask., Can., 710 G4 70
Balch, Jackson, Ark., 100 B4 81
Balchen, glacier, Ant C34 5
Balchik, Bul., 6,011 D9 22
Balch Springs, Dallas, Tex., 6,821 *A6 118
Balclutha, N.Z., 3,935 Q12 51
Balcones Heights, Bexar, Tex., 950 *E3 118
Bald, hill, Conn B8 84
Bald, hill, R.I C10 84
Bald, mtn., Calif C3 82
Bald, mtn., Colo A5 83
Bald, mtn., Conn A3 84
Bald, mtn., Maine C2 96
Bald, mtn., N.J A4 106
Bald, mtn., Oreg C3 113
Bald, mtn., Oreg C9 113
Bald, mtn., S. Dak C7 116
Bald, mtn., Vt B5 120
Bald, mtn., Wyo A5 125
Bald, mts., N.C C3 109
Bald, mts., Tenn C11 117
Bald Eagle, Ramsey, Minn., 1,200 E7 99
Bald Eagle, lake, Minn C7 99
Bald Eagle, lake, Minn E7 99
Baldegger, lake, Switz B5 19
Baldhill, dam, N. Dak B8 110
Bald Knob, White, Ark., 2,096 B4 81
Bald Knob, mtn., Oreg E2 113
Bald Knob, mtn., Va D3 121
Bald Knob, mtn., W. Va C5 123
Bald Knoll, mtn., Wyo C2 125
Baldur, Man., Can., 370 E2 71
Baldwin, Duval, Fla., 1,272 B5 86
Baldwin, Habersham and Banks, Ga., 698 B3 87
Baldwin, Randolph, Ill., 336 E4 90
Baldwin, Jackson, Iowa, 228 B7 92
Baldwin, St. Mary, La., 1,548 E4 95
Baldwin, Baltimore, Md., 100 B5 85
Baldwin, Lake, Mich., 835 E5 98
Baldwin, Nassau, N.Y., 30,204 G2 84
Baldwin, Ashe, N.C., 50 A2 109
Baldwin, Burleigh, N. Dak., 60 B5 110
Baldwin, Allegheny, Pa., 24,489 *F1 114
Baldwin, St. Croix, Wis., 1,184 C1 124
Baldwin, co., Ala., 49,088 E2 78
Baldwin, co., Ga., 34,064 C3 87
Baldwin City, Douglas, Kans., 1,877 D8 93
Baldwin Heights, Gibson, Ind., 200 H2 91
Baldwin Park, Los Angeles, Calif., 33,951 *F3 82
Baldwinsville, Onondaga, N.Y., 5,985 B4 108
Baldwinton, Sask., Can., 63 E1 67
Baldwinville, Worcester, Mass., 1,631 A3 97
Baldwyn, Lee and Prentiss, Miss., 2,023 A5 100
Baldy, mtn., Calif E5 82
Baldy, mtn., B.C., Can D7 68
Baldy, mtn., Man., Can D1 71
Baldy, mtn., Colo C3 83
Baldy, mtn., Wyo C3 125
Baldy, peak, Ariz D6 80
Baldy, peak, N. Mex A4 107
Baleares (Balearic Islands), prov., Sp., 443,327 *C7 20
Balearic, is., Sp C6 20
Balembangan, isl., N. Bor *D5 35
Baler, Phil., 2,769 o13 35
Baleshare isl., Scot C1 13
Balesin, isl., Phil o14 35
Baleville, Sussex, N.J., 75 A3 106
Balfour, Henderson, N.C., 1,106 D3 109
Balfour, McHenry, N. Dak., 159 B5 110
Balgonie, Sask., Can., 430 G3 70
Bal Harbour, Dade, Fla., 727 *F3 86
Bali, Cam., 18,277 *E6 45
Bali, India, 9,855 E4 40
Bali, isl., Indon G5 35
Balikesir (Balikisri), Tur., 61,000 C6 23
Balikpapan, Indon., 88,534 F5 35
Balingen, Ger., 11,600 A4 18
Balintang, chan., Phil B6 35
Baliuag, Phil., 3,535 o13 35
Balje, Ger., 1,600 B4 16
Baljennie, Sask., Can E2 70
Balkan, Bell, Ky., 70 D6 94
Balkan, mts., Bul D7 22
Balkh, Afg., 12,466 C13 41
Balkhash, Sov. Un., 61,000 D8 29
Balkhash, lake, Sov. Un D9 29
Balki, Sov. Un., 25,000 H10 27
Balko, Beaver, Okla., 100 D3 112
Ball, mtn., Alta., Can D2 69
Ball, mtn., Conn A4 84
Balla, Ire., 324 D2 11
Ballachulish, Scot., 2,960 D3 13
Ballagh, Ire D3 11
Ballaghaderreen, Ire., 1,308 D3 11
Ballajá, P.R C2 65
Ballantine, Yellowstone, Mont., 250 E8 102
Ballantrae, Scot., 886 E4 13
Ballarat, Austl., 41,028 (★54,771) H4 51
Ballard, co., Ky., 8,291 A2 94
Ballard Vale, Essex, Mass., 1,000 A5, C2 97
Ballater, Scot., 1,132 C5 13
Ball Club, Itasca, Minn., 100 C5 99
Ballclub, lake, Minn C5 99
Ballé, Mali C3 45
Ballenas, bay, Mex B2 63
Ballenstedt, Ger., 11,000 B6 17
Balleny, is., Ant C29 5
Ballerup, Den., 9,392 C6 24
Ball Ground, Cherokee, Ga., 707 B2 87
Ballia, India, 38,216 E10 40
Balliguda, India G9 40
Ballina, Austl., 4,129 D9 51
Ballina, Ire., 6,027 C2 11
Ballina, Ire., 273 E3 11
Ballinakill, Ire., 315 E4 11
Ballinalack, Ire., 55 D4 11
Ballinamallard, N. Ire., 352 C4 11
Ballinascarthy, Ire F3 11
Ballinasloe, Ire., 5,711 D3 11
Ballindine, Ire., 222 D3 11
Ballineen, Ire., 270 F3 11
Ballingarry, Ire., 360 E3 11
Ballingeary, Ire., 180 F2 11
Ballinger, Runnels, Tex., 5,043 D3 118
Ballinluig, Scot D5 13
Ballinrobe, Ire., 1,165 D2 11
Ballinskelligs, bay, Ire F1 11
Ballintra, Ire., 250 C3 11
Ball Mountain, flood control res., Vt E3 120
Ballouville, Windham, Conn., 125 B9 84
Ballston, Polk, Oreg., 200 B3 113
Ballston Lake, Saratoga, N.Y., 700 C7 108
Ballston Spa, Saratoga, N.Y., 4,991 B7 108
Ballville, Sandusky, Ohio, 1,424 *A4 111
Ballwin, St. Louis, Mo., 5,710 B7 101
Bally, India, 101,159 *F12 40
Bally, Berks, Pa., 1,033 F10 114
Ballybofey, Ire., 1,030 C4 11
Ballybunion, Ire., 1,163 E2 11
Ballycanew, Ire., 168 E5 11
Ballycastle, Ire., 191 C2 11
Ballycastle, N. Ire., 2,643 B5 11
Ballyconneely, Ire D1 11
Ballyconnell, Ire., 542 C4 11
Ballycroy, Ire C2 11
Ballyduff, Ire., 379 E2 11
Ballyduff, Ire., 99 E3 11
Ballyferriter, Ire E1 11
Ballygar, Ire., 315 D3 11
Ballygawley, N. Ire., 427 C4 11
Ballyglunin, Ire D3 11
Ballygorman, Ire B4 11
Ballyhaunis, Ire., 1,174 D3 11
Ballyheige, Ire., 417 E2 11
Ballyheige, bay, Ire E2 11
Ballyhoura, mts., Ire E3 11
Ballyjamesduff, Ire., 581 D4 11
Ballykelly, N. Ire., 367 B4 11
Ballylongford, Ire., 594 E2 11
Ballymahon, Ire., 830 D4 11
Ballymena, N. Ire., 14,740 C5 11
Ballymoe, Ire., 117 D3 11
Ballymoney, N. Ire., 3,409 B5 11
Ballymote, Ire., 965 C3 11
Ballymurray, Ire D3 11
Ballynahinch, N. Ire., 2,038 C6 11
Ballyneety, Ire E3 11
Ballynoe, Ire., 102 E3 11
Ballysadare, Ire., 143 C3 11
Ballyshannon, Ire., 2,322 C3 11
Ballyteige, bay, Ire E5 11
Ballytore, Ire., 269 D5 11
Ballyvaughan, Ire., 152 D2 11
Ballyvourney, Ire., 321 F2 11
Balm, Hillsborough, Fla., 250 E4 86
Balmat, St. Lawrence, N.Y., 100 A5, B1 108
Balmazujvaros, Hung., 16,312 B5 22
Balmhorn, mtn., Switz D4 19
Balmoral, Man., Can., 103 D3 71
Balmoral Castle, Scot C5 13
Balmorhea, Reeves, Tex., 604 F2 118
Balmville, Orange, N.Y., 1,538 *D6 108
Baloda Bazar, India, 7,108 G9 40
Balonne, riv., Austl C7 51
Balotra, India, 12,110 E4 40
Balovale, N. Rh., 1,110 D3 48
Balrampur, India, 31,776 D9 40
Balranald, Austl., 1,331 G4 51
Balș, Rom., 6,956 C7 22
Balsam, Jackson, N.C., 300 D3 109
Balsam, lake, Wis C1 124
Balsam, lake, Ont., Can C6 72
Balsam Lake, Polk, Wis., 541 C1 124
Balsas, Braz., 4,690 C1 57
Balsas, riv., Braz C1 57
Balsas, riv., Braz D1 57
Balsas, riv., Mex D4 63
Balsthal, Switz., 5,735 B4 19
Balta, Pierce, N. Dak., 165 A5 110
Balta, Sov. Un., 47,400 H7 27
Baltic, New London, Conn., 1,366 C8 84
Baltic, Tuscarawas, Ohio, 537 B6 111
Baltic, Minnehaha, S. Dak., 278 D9 116
Baltic, sea, Eur I8 25
Baltīm, Eg., U.A.R., 2,000 C3 32
Baltimore, Ont., Can., 220 C6 72
Baltimore, Ire., 188 F2 11
Baltimore (Independent City), Md., 939,024 (★1,636,500) B4, C2 85
Baltimore, Fairfield, Ohio, 2,116 C3 111
Baltimore, co., Md., 492,428 B4 85
Baltinglass, Ire., 806 E5 11
Baltiysk (Pillau), Sov. Un., 30,000 A5 26
Baltra, isl., Ec g5 58
Baltrum, Ger., 770 A7 15
Baltrum, isl., Ger A7 15
Baluchistan, reg., Pak C4 39
Balya, Tur., 1,700 C6 23
Balzac, Alta., Can D3 69
Balzar, Ec., 3,015 B2 58
Bam, Iran, 13,938 G9 41
Bam, lake, China B3 39
Bama, Nig D7 45
Bamako, Mali, 68,200 D3 45
Bamba, Mali C4 45
Bambang, Phil., 4,225 n13 35
Bambari, Cen. Afr. Rep., 19,700 D4 46
Bamberg, Ger., 74,100 D5 17
Bamberg, Bamberg, S.C., 3,081 E5 109
Bamberg, co., S.C., 16,274 E9 115
Bambesa, Con. L A4 48
Bambuí, Braz., 8,148 F1 57
Bamburgh, Eng., 438 E7 13
Bambuto, mts., Nig., Cam E6 45
Bamenda [Mankou], Cam., 14,259 E7 45
Bampūr, Iran, 25,000 H10 41
Bampur, riv., Iran H9 41
Bams, butte, S. Dak B2 116
Baña Buey, Cuba k11 64
Banagher, Ire., 1,050 D4 11
Banalia, Con. L A4 48
Banamba, Mali D3 45
Banana, riv., Fla D6 86
Bananal, Braz., 2,189 h5 56
Bananal, isl., Braz E4 59
Bananeiras, Braz., 3,060 C3, h6 57
Banaras (Benares), India, 471,258 (*489,864) C7 39, E9 40
Banās, cape, Eg., U.A.R E6 43
Banas, riv., India F3 39
Banat, prov., Rom., 948,596 *C5 22
Banat, reg., Rom., Yugo C5 22
Ban Bangsaphan Yai, Thai., 3,362 G3 38
Banbridge, N. Ire., 6,115 C5 11
Banbury, Eng., 20,996 B6 12
Banchory, Scot., 1,918 C6 13
Bancker, Vermilion, La E3 95
Banco, Col., 9,636 B3 60
Bancroft, Ont., Can., 1,669 B7 72
Bancroft, Early, Ga., 100 E2 87
Bancroft, Caribou, Idaho, 416 G7 89
Bancroft, Kossuth, Iowa, 1,000 A3 92
Bancroft, Beauregard, La., 100 D2 95
Bancroft, Aroostook, Maine, 50 (94▲) C4 96
Bancroft, Hampshire, Mass., 40 B1 97
Bancroft, Shiawassee, Mich., 636 F6 98
Bancroft, Cuming, Nebr., 496 B9 103
Bancroft, Kingsbury, S. Dak., 86 C8 116
Bancroft, Portage, Wis., 250 D4 124
Banda, India, 37,744 E8 40
Banda, is., Indon F7 35
Banda, sea, Indon G7 35
Bandama, riv., I.C E3 45
Bandana, Ballard, Ky., 400 A2 94
Bandar, Afg., 5,000 D12 41
Bandar 'Abbās, Iran, 14,278 H8 41
Bandar Bahru, Mala., 1,187 J4 38
Bandar-e Chīrū, Iran H6 41
Bandar-e Deylam, Iran, 3,130 F5 41
Bandar-e Rīg, Iran, 2,250 G5 41
Bandar-e Shāh, Iran, 8,284 C7 41
Bandar-e Shāhpūr, Iran F4 41
Bandar Maharani (Muar), Mala., 39,046 K5 38
Bande, Sp., 6,275 A2 20
Banded, peak, Colo D4 83
Bandeira, peak, Braz F2 57
Bandelier, nat. mon., N. Mex B3 107
Bandera, Arg E3 55
Bandera, Bandera, Tex., 950 E3 118
Bandera, co., Tex., 3,892 E3 118
Banderas, bay, Mex m11 63
Bandholm, Den., 712 D5 24
Bandiagara, Mali, 3,800 D4 45
Bandikui, India, 10,638 D6 40
Bandirma, Tur., 28,900 B6 23
Bandjermasin, Indon., 212,683 F4 35
Bandoeng, see Bandung, Indon.
Bandon, Ire., 2,308 F3 11
Bandon, Coos, Oreg., 1,653 D2 113
Bandon, riv., Ire F3 11
Bandung (Bandoeng), Indon., 966,359 (*1,025,000) G3 35
Bandy, Tazewell, Va., 800 B3 121
Banes, Cuba, 20.257 E6 64
Banff, Alta., Can., 3,429 D3 69
Banff, Scot., 3,329 C6 13
Banff, co., Scot., 46,400 C5 13
Banff, nat. park, Alta., Can D3 69
Banfora, Upper Volta, 4,000 D4 45
Bangalore, India, 905,134 (*1,225,000) F6 39
Bangassou, Cen. Afr. Rep E4 46
Banggai, Indon F6 35
Banghāzī, see Bengasi, Libya
Bangka, isl., Indon F3 35
Bangkalan, Indon., 12,359 G4 35
Bangkok (Krung Thep), Thai., 1,394,513 (*2,075,000) F4 38
Bangor, Blount, Ala., 75 B3 78
Bangor, Sask., Can., 109 G4 70
Bangor, Penobscot, Maine, 38.912 D4 96
Bangor, Van Buren, Mich., 2,109 F4 98
Bangor, Franklin, N.Y., 250 A2 108
Bangor, N. Ire., 23,865 C5 11
Bangor, Northampton, Pa., 5,766 E11 114
Bangor, Wales, 13,977 A3 12
Bangor, La Crosse, Wis., 928 E3 124
Bangor Erris, Ire., 125 C2 11
Bangs, Brown, Tex., 967 D3 118
Bangs, mtn., Ariz A2 80
Bangued, Phil., 5,663 B6 35
Banguey (Banggi), isl., Mala D5 35
Bangui, Cen. Afr. Rep., 80,000 E3 46
Bangweulu, lake, N. Rh D5 48
Banhā, Eg., U.A.R., 42,600 D3 32
Ban Hat Yai, see Hadyai, Thai.
Ban Houei Sai, Laos B4 38
Baní, Dom. Rep., 14,472 F8 64
Bani, Phil., 2,565 n12 35
Bani, riv., Mali D3 45
Baniara, Pap k12 50
Baniloudi, Niger C5 45
Banī Na 'īn, Jordan, 3,000 C7 32
Banister, riv., Va E4 121
Bani Suheila, Gaza Area, 3,000 C6 32
Banī Suwayf, Eg., U.A.R., 65,600 D6 43
Bāniyās, Syr., 8,500 E10 31
Baniyas, Syr A7 32
Banjak, is., Indon K2 38
Banja Luka, Yugo., 50,463 C3 22
Banjuwangi, Indon., 25,185 G4 35
Bankasse, Mali D4 45
Bankfoot, Scot., 2,310 D5 13
Bankhead, Walker, Ala., 50 B2 78
Bankhead, lake, Ala B2 78
Banks, Pike, Ala., 201 D4 78

Banks, Bradley, Ark., 233....D3 81
Banks, Boise, Idaho, 50......E2 89
Banks, Tunica, Miss., 300...A3 100
Banks, Washington, Oreg., 347..........A1 113
Banks, co., Ga., 6,497.......B3 87
Banks, bay, Ec..........g5 58
Banks, isl., Austl..........B7 50
Banks, isl., B.C., Can........C2 68
Banks, isl., N.W. Ter., Can..B8 66
Banks, lake, Ga..........F3 87
Banks, pen., N.Z..........O14 51
Banksian, riv., Man., Can.....C4 71
Bankston, Fayette, Ala., 120..B2 78
Bankura, India, 62,833.....F11 40
Ban Me Thuot, Viet., 5,000...F8 38
Bann, riv., N. Ire..........B5 11
Banner, Calhoun, Miss., 75..A4 100
Banner, Sheridan, Wyo., 10..A6 125
Banner, co., Nebr., 1,269....C2 103
Banner Elk, Avery, N.C., 564..........A2, C4 109
Banner Hill, Unicoi, Tenn., 1,100..........*C11 117
Bannerman, Man., Can., 100.E2 71
Banner Town, Surry, N.C., 1,096..........*A3 109
Banning, Riverside, Calif., 10,250..........F5 82
Banning, Carroll, Ga., 150...C2 87
Banningville, Con. L., 4,753..B2 48
Bannock, co., Idaho, 49,342..G6 89
Bannock, pass, Idaho, Mont..........E5 89, F3 102
Bannock, peak, Idaho.........G6 89
Bannockburn, Ont., Can., 186..........C7 72
Bannu, Pak., 20,509 (★27,516)..........B5 39
Bañolas, Sp., 8,075..........A7 20
Baños, Ec., 2,768..........B2 58
Banqu, Con. L..........C3 48
Bansha, Ire., 244..........E3 11
Banska Bystrica, Czech., 22,600..........D5 26
Banska Stiavnica, Czech., 9,500..........D5 26
Bansko, Bul., 6,161..........E6 22
Banswara, India, 19,566.....F5 40
Bantam, Litchfield, Conn., 833..........C4 84
Bantam, lake, Conn..........C4 84
Bantam, riv., Conn..........B4 84
Banteer, Ire., 139..........E3 11
Bantry, Ire., 2,234..........F2 11
Bantry, McHenry, N. Dak., 93..........A5 110
Bantry, bay, Ire..........F2 11
Banyo, Cam..........D2 46
Banzyville, Con. L..........A3 48
Baoulé, riv., Mali..........D3 45
Baoulé, riv., Mali..........D3 45
Bapaume, Fr., 3,275..........D2 15
Bapchule, Pinal, Ariz., 100..........D4, H2 80
Baptist, Tangipahoa, La......A6 95
Baptiste, Ont., Can., 57.....B7 72
Baptistown, Hunterdon, N.J., 350..........B2 106
Baquba, Iraq, 13,203.......F15 31
Baquedano, Chile..........D2 55
Bar, Sov. Un., 22,100........G6 27
Bar, Yugo., 2,163..........D4 22
Bara, Sud., 4,885..........C3 47
Barabinsk, Sov. Un., 38,900..B9 29
Baraboo, Sauk, Wis., 7,660..E4 124
Baraboo, riv., Wis..........E3 124
Baracaldo, Sp., 77,802......A4 20
Baracoa, Cuba, 11,459.......E6 64
Barada, Richardson, Nebr., 58..........D10 103
Baradères, Haiti..........F7 64
Baradero, Arg., 10,194......E7 54
Baraga, Baraga, Mich., 991..B2 98
Baraga, co., Mich., 7,151....B2 98
Barahona, Dom. Rep., 20,398..........F8 64
Barajas de Madrid, Sp., 2,184 (pop. incl. in Madrid)....p17 20
Barak Khel, Afg., 5,000....E13 41
Barakot, India, 5,000.......G10 40
Baramula, India, 19,854.....B5 39
Baran, India, 22,764..........E6 40
Baranagar, India, 107,837.*F12 40
Baranof, Alsk., 13..........m22 79
Baranof, isl., Alsk..........m22 79
Baranovichi, Sov. Un., 63,000..........E5 27
Baranya, co., Hung., 285,316..........*B4 22
Barataria, Jefferson, La., 900..........C7, E5 95
Barataria, bay, La..........E6 95
Barataria, bayou, La..........C7 95
Baraya, Col., 1,736..........C2 60
Barbacena, Braz., 41,931..........C4, g6 56
Barbacoas, Col., 3,349......C2 60
Barbado, riv., Braz..........C3 55
Barbados, Br. dep., N.A., 232,085..........p17 64
Barbalha, Braz., 6,967.......C3 57
Barbastro, Sp., 10,227.......A6 20
Barbate, Sp., 10,720..........D3 20
Barbeau, Chippewa, Mich., 20..........B6 98
Barbee, lake, Ind..........B6 91
Barber, Ada, Idaho, 15.......F2 89
Barber, Golden Valley, Mont., 35..........D7 102
Barber, Cherokee, Okla......B7 112
Barber, co., Kans., 8,713.....E5 93
Barberton, Summit, Ohio, 33,805..........A6 111
Barberton, S. Afr.,11,016....E6 49
Barbertown, Hunterdon, N.J..........C2 106
Barberville, Volusia, Fla., 250..........C5 86
Barbezieux, Fr., 3,058.......E4 14
Barbour, co., Ala., 24,700...D2 78
Barbour, co., W. Va., 15,474.B4 123
Barboursville, Orange, Va., 150..........C4 121
Barboursville, Cabell, W. Va., 2,331..........C2 123
Barbourville, Knox, Ky., 3,211..........D6 94
Barbuda, isl., Barbados.....n16 64
Barby, Ger., 7,788..........B6 17
Barca de Alva, Port..........B2 20
Barcaldine, Austl., 1,705.....A5 51
Barcarena, Port., 2,552......F9 20
Barcarrota, Sp., 7,898.......C2 20
Barce, see Al Marj, Libya
Barcellona [Pozzo di Gotto], It., 20,500 (32,138▲)..........E5 21
Barcelona, Sp., 1,557,863 (★2,020,000)..........B7 20
Barcelona, Ven., 40,773.....A5 60
Barcelona, prov., Sp., 2,877,966..........*B7 20
Barceloneta, P.R., 762.......B4 65
Barceloneta, mun., P.R., 19,334..........B4 65
Barcelonnette, Fr., 2,432....E7 14
Barcelos, Braz., 1,094.......D5 60
Barcelos, Port., 7,875..........B1 20
Barclay, Osage, Kans., 50...D8 93
Barclay, Queen Annes, Md., 142..........B6 85
Barco, Currituck, N.C., 250.A8 109
Barcoo, riv., Austl..........B4 51
Bardejov, Czech., 6,572......D6 26
Bardera, Som., 1,500 (4,900▲)..........E5 47
Bardi, It., 8,159..........E5 18
Bardiyah, Libya..........C5 43
Bardley, Ripley, Mo., 50....E6 101
Bardolph, McDonough, Ill., 266..........C3 90
Bardonecchia, It., 2,273.....D2 18
Bardsey, isl., Wales..........B3 12
Bardstown, Nelson, Ky., 4,798..........C4 94
Bardstown Junction, Bullitt, Ky., 50..........C4 94
Bardswell, is., B.C., Can.....C3 68
Bardwell, Carlisle, Ky., 1,067.A2 94
Bare Hill, pond, Mass..........C7 97
Bareilly, India, 254,409 (★272,828)..........C7 40
Barenburg, Ger., 990..........F2 24
Barentin, Fr., 7,962..........C4 14
Barents, isl., Nor..........B12 4
Barents, sea, Sov. Un..........B6 28
Barentu, Eth..........B4 47
Baresville, York, Pa., 1,700.*G8 114
Barfleur, pt., Fr..........C3 14
Bargaintown, Atlantic, N.J..E3 106
Bargal, Som., 2,200..........C7 47
Bargarh, India, 15,375......G9 40
Barge, canal, N.Y..........B2 108
Bargersville, Johnson, Ind., 586..........E5 91
Barguzin, Sov. Un., 5,600..D13 28
Bar Harbour, Hancock, Maine, 2,444 (3,807▲)..........D4 96
Bari, It., 312,023..........D6 21
Baria, Viet., 5,000..........G7 38
Barika, Alg., 2,945 (71,235▲).B6 44
Barinas, P.R..........C3 65
Barinas, Ven., 25,707..........B3 60
Barinas, state, Ven., 139,271.B3 60
Baring, Washington, Maine, 130..........C5 96
Baring, Knox, Mo., 213......A5 101
Baripada, India, 20,301.....G11 40
Barirí, Braz., 8,403.......C3, m7 56
Bâris, Eg., U.A.R..........E6 43
Barisal, Pak., 89,694..........F13 40
Barisan, mts., Indon..........F2 35
Barium Springs, Iredell, N.C., 300..........B3 109
Bark, lake, Ont., Can..........B7 72
Bark, pt., Wis..........B2 124
Barker, Niagara, N.Y., 528...B2 108
Barkerville, B.C., Can., 62...C7 68
Barkeryd, Swe..........A8 24
Barkeyville, Venango, Pa., 200..........D2 114
Barkhamsted, Litchfield, Conn., 60 (1,370▲)..........B5 84
Barkhamsted, res., Conn..........B5 84
Barking, Eng., 72,282......k13 10
Barkly East, S. Afr., 3,648...D4 49
Barkol (Chensi), China, 10,000..........C3 34
Bark River, Delta, Mich., 200..........C3 98
Barksdale, Edwards, Tex., 150..........E2 118
Barlad, Rom., 32,043..........B8 22
Barladul, riv., Rom..........B8 22
Bar-le-Duc, Fr., 18,346......C6 42
Barlee, lake, Austl..........E2 50
Barletta, It., 68,035..........D6 21
Barling, Sebastian, Ark., 1,518..........B1 81
Barlow, Ballard, Ky., 731....A2 94
Barlow, Foster, N. Dak., 50..B6 110
Barlow Bend, Clarke, Ala., 200..........D2 78
Barmer, India, 27,600..........E3 40
Bar Mills, York, Maine, 400.E2 96
Barmouth, Wales, 2,348......B3 12
Barna, Ire., 143..........D2 11
Barnaby River, N.B., Can., 204..........C4 74
Barnard, Lincoln, Kans., 205.C5 93
Barnard, Nodaway, Mo., 237.A3 101
Barnard, Brown, S. Dak., 82.B7 116
Barnard, Windsor, Vt., 75 (435▲)..........D3 120
Barnard Castle, Eng., 4,969..F6 13
Barnardsville, Buncombe, N.C., 199..........D3 109
Barnasht, Eg., U.A.R..........E3 32
Barnaul, Sov. Un., 338,000..C10 29
Barnesdale, Ont., Can., 85...B5 72
Barnegat, Ocean, N.J., 287..D4 106
Barnegat, bay, N.J..........D4 106
Barnegat, inlet, N.J..........D4 106
Barnegat City, Ocean, N.J..D4 106
Barnes, Washington, Kans., 247..........C7 93
Barnes, Leake, Miss., 50.....C4 100
Barnes, Douglas, Oreg., 5,076..........*D3 113
Barnes, Warren, Pa., 220....C3 114
Barnes, co., N. Dak., 16,719..B7 110
Barnes, sound, Fla..........G6 86
Barnesboro, Cambria, Pa., 3,035..........E4 114
Barnes City, Mahaska and Poweshiek, Iowa, 273.....C5 92
Barnes Corners, Lewis, N.Y., 100..........B5 108
Barnesmore, Ire..........C4 11
Barneston, Gage, Nebr., 177.D9 103
Barnesville, Lamar, Ga., 4,919..........C2 87
Barnesville, Montgomery, Md., 145..........B3 85
Barnesville, Clay, Minn., 1,632..........D2 99
Barnesville, Belmont, Ohio, 4,425..........C6 111
Barnet, Caledonia, Vt., 250 (1,445▲)..........C4 120
Barnett, Morgan, Mo., 200..C5 101
Barnette, Clarke, Miss., 40..D5 100
Barneveld, Neth., 7,400......B5 15
Barneveld, Iowa, Wis., 420..E4 124
Barneville-sur-Mer, Fr., 656..C3 14
Barney, Brooks, Ga., 165....E3 87
Barney, Richland, N. Dak., 115..........C8 110
Barnhart, Jefferson, Mo., 400..........B8, C7 101
Barnhart, Irion, Tex., 250...D5 118
Barnsboro, Gloucester, N.J., 600..........D2 106
Barnsdall, Osage, Okla., 1,663..........A5 112
Barnsley, Eng., 74,650 (★190,000)..........A6 12
Barnstable, Barnstable, Mass., 800 (13,465▲)..........C7 97
Barnstable, co., Mass., 70,286..........C7 97
Barnstaple, Eng., 15,907.....C3 12
Barnstaple, bay, Eng..........C3 12
Barnstead, Belknap, N.H., 200 (850▲)..........D4 105
Barnston, Que., Can., 123...D6 73
Barnum, Webster, Iowa, 154 B3 92
Barnum, Carlton, Minn., 417..........D6 99
Barnwell, Alta., Can., 190...E4 69
Barnwell, Barnwell, S.C., 4,568..........E3 115
Barnwell, co., S.C., 17,659...E5 115
Baro, Nig., 217..........E6 45
Baro, riv., Eth..........D3 47
Baroda, India, 295,144 (★298,398)..........F4 40
Baron, Adair, Okla., 100....B7 112
Barons, Alta., Can., 345.....E4 69
Barotseland Protectorate, prov., N. Rh..........E3 48
Barquisimeto, Ven., 196,557..A4 60
Barr, Fr., 4,207..........C7 14
Barr, Tate, Miss., 275.......A4 100
Barr, Lauderdale, Tenn.....B2 117
Barra, Braz., 7,237..........D2 57
Barra, Braz., 1,539..........h5 57
Barra, isl., Scot..........D1 13
Barraba, Austl., 1,469.......E8 51
Barrack, pt., Ire..........F2 11
Barrackpore, India, 63,778..F12 40
Barrackville, Marion, W. Va., 950..........A7, B4 123
Barra do Corda, Braz., 3,723.C1 57
Barra do Piraí, Braz., 29,398..........C4, h6 56
Barra Mansa, Braz., 47,398..........C4, h5 56
Barranca, Peru, 192..........B2 58
Barrancabermeja, Col., 25,046..........B3 60
Barrancas, Ven., 4,034.......B5 60
Barranquilla, Col., 356,920...A3 60
Barranquitas, P.R., 4,684....C5 65
Barranquitas, mun., P.R., 18,978..........C5 65
Barraza, Chile..........A2 54
Barre, Que., Can..........C7 73
Barre, Worcester, Mass., 1,065 (3,479▲)..........B3 97
Barre, Washington, Vt., 10,387..........C4 120
Barre, lake, La..........E5 95
Barre des Écrins, mtn., Fr..........E7 14
Barreiras, Braz., 7,175.......D1 57
Barreirinhas, Braz., 2,184....B2 57
Barreiro, Braz., 870..........h5 56
Barreiro, Port, 22,190....C1, f9 20
Barreiros, Braz., 10,402..C3, k6 57
Barrellville, Allegany, Md., 300..........D2 85
Barren, co., Ky., 28,303.....D4 94
Barren, isl., Alsk..........h15 79
Barren, is., Malag..........g8 49
Barren, isl., Md..........D5 85
Barren, riv., Ky..........D3 94
Barren Plain, Robertson, Tenn., 100..........A5 117
Barrens, plat., Tenn..........B6 117
Barre Plains, Worcester, Mass., 300..........B3 97
Barretos, Braz., 39,950......F1 57
Barrett, Grant, Minn., 345...E3 99
Barrett, Grafton, N.H..........B3 105
Barrett, Harris, Tex., 1,200.*F5 118
Barrett, Boone, W. Va., 800..........D3, D6 123
Barretts, Lowndes, Ga., 100..F3 87
Barretts, hill, Mass..........C1 97
Barrhead, Alta., Can., 2,286..B3 69
Barrhead, Scot., 14,442......E4 13
Barrhill, Scot..........E4 13
Barrie, Ont., Can., 21,169...C5 72
Barrie, isl., Ont., Can..........B2 72
Barriere, B.C., Can., 472....D7 68
Barrigada, Guam, 1,729.......... 52
Barrigada, hill, Guam.......... 52
Barrington, N.S., Can., 261..F4 74
Barrington, Lake and Cook, Ill., 5,434..........A5, E2 90
Barrington, Strafford, N.H., 70 (1,036▲)..........D4 105
Barrington, Camden, N.J., 7,943..........*D2 106
Barrington, Bristol, R.I., 9,800 (13,826▲)..........C11 78
Barrington, lake, Man., Can....A1 71
Barrington, mtn., Austl..........F8 51
Barrington Hills, Cook and Lake, Ill., 1,726..........*A5 90
Barrington Passage, N.S., Can., 381..........F4 74
Barrio Azul, Cuba..........k11 64
Barr Lake, Adams, Colo.....B6 83
Barron, Barron, Wis., 2,338. C2 124
Barron, co., Wis., 34,270....C2 124
Barronett, Barron, Wis., 100.C2 124
Barrow, Alsk., 1,314..........A8 79
Barrow, Arg..........B4 54
Barrow, co., Ga., 14,485.....B3 87
Barrow, isl., Austl..........D2 50
Barrow, pt., Alsk..........A8 79
Barrow, riv., Ire..........E5 11
Barrow, strait, N.W. Ter., Can..........B14 5
Barrow Creek, Austl., 40....D5 50
Barrow-in-Furness, Eng., 64,824..........C5 10
Barrows, Man., Can., 123...C1 71
Barruelo de Santullán, Sp., 7,770..........A3 20
Barry, Pike, Ill., 1,422......D2 90
Barry, Wales, 42,039..........C4 12
Barry, co., Mich., 31,738....F5 98
Barry, co., Mo., 18,921......E4 101
Barrys Bay, Ont., Can., 1,366.B7 72
Barryton, Mecosta, Mich., 418..........E5 98
Barryville, Sullivan, N.Y., 400..........D6 108
Barsi, India, 50,389..........H5 40
Barsinghausen, Ger., 11,500..A4 17
Barsö, isl., Den..........C3 24
Barstow, San Bernardino, Calif., 11,644..........E5 82
Barstow, Ward, Tex., 707..........D1, F2 118
Bar-sur-Aube, Fr., 4,801.....C6 14
Bar [-sur-Seine], Fr., 2,559...C6 14
Bartelso, Clinton, Ill., 370...E4 90
Barth, Escambia, Fla., 200..G2 86
Barth, Ger., 12,200..........A6 16
Bartholomew, co., Ind., 48,198..........F6 91
Bartholomew, bayou, Ark.....C4 81
Bartica, Br. Gu., 2,352.......A3 59
Bartin, Tur., 11,700..........B9 31
Bartlesville, Washington, Okla., 27,893..........A6 112
Bartlett, Cook, Ill., 1,540....E2 90
Bartlett, Labette, Kans., 137..........E8 93
Bartlett, Wheeler, Nebr., 125.C7 103
Bartlett, Carroll, N.H., 500 (1,013▲)..........D4 105
Bartlett, Ramsey, N. Dak., 39..........A7 110
Bartlett, Washington, Ohio, 240..........C6 111
Bartlett, Wallowa, Oreg.....B9 113
Bartlett, Shelby, Tenn., 508..........B2, E8 117
Bartlett, Williamson and Bell, Tex., 1,540..........D4 118
Bartlett, res., Ariz..........D4, F3 80
Bartletts Ferry, dam, Ala., Ga..........C4 72, D1 84
Bartletts Harbour, Newf., Can., 185..........C3 75
Bartley, Red Willow, Nebr., 309..........D5 103
Bartley, Morris, N.J., 90.....B3 106
Bartolomeu Días, Moz..........B6 49
Barton, Colbert, Ala., 300...A2 78
Barton, Phillips, Ark., 250...C5 81
Barton, Ascension, La., 200..B5 95
Barton, Allegany, Md., 731..D2 85
Barton, Pierce, N. Dak., 80..A5 110
Barton, Belmont, Ohio, 975..........B7 111
Barton, Orleans, Vt., 1,169 (3,066▲)..........B4 120
Barton, Washington, Wis., 1,569..........E5 124
Barton, co., Kans., 32,368...D5 93
Barton, co., Mo., 11,113.....D3 101
Barton, riv., Vt..........B4 120
Barton-on-Humber, Eng., 6,584..........A7 12
Bartonsville, Windham, Vt., 70..........E3 120
Bartonville, Peoria, Ill., 7,253..........C4 90
Bartonwoods, Dekalb, Ga., 3,000..........*C2 87
Bartoszyce, Pol., 3,449......A6 26
Bartow, Polk, Fla., 12,849...E5 86
Bartow, Jefferson, Ga., 366..D4 87
Bartow, co., Ga., 28,267.....B2 87
Baruth, Ger., 2,232..........A8 17
Barvas, Scot..........B2 13
Barvenkovo, Sov. Un., 10,000..........G11 27
Barwani, India, 17,446.......F5 40
Barwice, Pol., 1,819..........F8 24
Barwick, Brooks and Thomas, Ga., 400..........F3 87
Barwick, Breathitt, Ky., 300.C6 94
Barwon, riv., Austl..........D7 51
Barybino, Sov. Un..........n17 27
Barycz, riv., Pol..........C4 26
Bärz Shovār, Iran..........F5 41
Basāki, Iran, 10,000..........F5 41
Basalt, Eagle, Colo., 213.....B3 83
Basalt, Bingham, Idaho, 275.F6 89
Basankusu, Con. L., 1,784...A2 48
Basantpur, India, 5,000.....D11 40
Basargechar, Sov. Un..........B15 31
Basbulak, Sov. Un..........E6 29
Basco, Hancock, Ill., 199.....C2 90
Bascom, Jackson, Fla., 250...B1 86
Bascom, Seneca, Ohio, 450...A4 111
Basehor, Leavenworth, Kans., 315..........B8 93
Basekpio, Con. L..........A3 48
Basel, Switz., 206,746 (★460,000)..........A4 19
Basel, canton, Switz., 373,870..........A4 19
Basel [Land], sub canton, Switz., 148,282..........A4 19
Basel-Stadt, sub canton, Switz., 225,588..........A4 19
Bashagird, range, Iran..........H9 41
Bashaw, Alta., Can., 614.....C4 69
Bashi, Clarke, Ala., 200......D2 78
Basiad, bay, Phil..........o14 35
Basic, Clarke, Miss., 70......C5 100
Basil, Fairfield, Ohio (part of Baltimore)..........C5 111
Basilan, isl., Phil..........D6 35
Basildon, see Billericay, Eng.
Basile, Evangeline, La., 1,932..........D3 95
Basilicata, pol. dists., It., 644,297..........*D5 21
Basilicata, reg., It..........D5 21
Basilio, Braz., 305..........E2 56
Basim, India..........G6 40
Basin, Jefferson, Mont., 300..D4 102
Basin, Big Horn, Wyo., 1,319..........A4 125
Basin, lake, Sask., Can..........E3 70
Basinger, Okeechobee, Fla., 150..........E5 86
Basingstoke, Eng., 25,940....C6 12
Basin Harbor, Addison, Vt..C2 120
Basin Mills, Penobscot, Maine (part of Orono).....D4 96
Baska, Yugo., 1,016..........C2 22
Baskahegan, lake, Maine..........C5 96
Baskett, Henderson, Ky., 300..........C2 94
Baskin, Pinellas, Fla., 500....E1 86
Baskin, Franklin, La., 238....C2 95
Basking Ridge, Somerset, N.J., 2,438..........B3 106
Basoko, Con. L..........A3 48
Basongo, Con. L..........B3 48
Basque Provinces, reg., Sp., 1,371,654..........A4 20
Basra, Iraq, 164,623..........F3 41
Bas-Rhin, dept., Fr., 770,150..........F7 15, A3 18
Bass, isl., Ohio..........A4 111
Bass, lake, Ind..........B4 91
Bass, strait, Austl..........I5 51
Bassac, Laos, 5,000..........E6 38
Bassano, Alta., Can., 815....D4 69
Bassano del Grappa, It., 23,800 (30,497▲)..........B3 21
Bassein, Bur., 77,905..........E9 39
Basses-Alpes, dept., Fr., 91,843..........E2 18
Basses-Pyrénées, dept., Fr., 466,038..........*F3 14
Basse-Terre, Guad. 9,124...o16 64
Basseterre, St. Kitts-Nevis-Anguilla, 15,742..........n15 64
Bassett, Los Angeles, Calif., 2,000..........*F3 82
Bassett, Chickasaw, Iowa, 130..........A5 92
Bassett, Allen, Kans., 67.....E8 93
Bassett, Rock, Nebr., 1,023...B6 103
Bassett, Henry, Va., 3,148....E3 121
Bassett, creek, Ala..........D2 78
Bassfield, Jefferson Davis, Miss., 295..........D4 100
Bassikounou, Maur..........C3 45
Bass River, N.S., Can., 447..D6 74
Bassum, Ger., 6,550..........F2 24
Basswood, Man., Can., 54...D1 71
Basswood, lake, Ont., Can.....B7 99
Båstad, Swe., 2,200..........*B6 24
Bastak, Iran, 7,500..........H7 41
Bastelica, Fr., 2,188..........D2 21
Basti, India, 38,403..........D9 40
Bastia, Fr., 50,117..........C2 21
Bastian, Bland, Va., 600.....D1 121
Bastogne, Bel., 5,927..........D5 15
Bastrop, Morehouse, La., 15,193..........B4 95
Bastrop, Bastrop, Tex., 3,001..........D4 118
Bastrop, co., Tex., 16,925....D4 118
Bastuträsk, Swe., 820..........E9 25
Basutoland, Br. dep., Afr., 685,000..........C4 49, I8 41
Basyūn, Eg., U.A.R., 1,000..D2 32
Bat, cave, Mo..........D5 101
Bata, Rio Muni, 842..........E1 46
Bataan, Phil., 2,186..........p13 35
Bataan, prov., Phil., 145,900..........*C6 35
Bataan, pen., Phil..........o13 35
Batabanó, Cuba, 3,177......D2 64
Batabanó, gulf, Cuba..........D2 64
Batac, Phil., 6,218..........*B6 35
Batala, India, 51,300..........B5 40
Batam, isl., Indon..........L5 38
Batan, is., Phil..........A6 35
Batanes, prov., Phil., 10,300..........*A6 35
Batangafo, Cen. Afr. Rep....D3 46
Batangas, Phil., 15,000 (82,800▲)..........C6, p13 35
Batangas, prov., Phil., 682,900..........*C6 35
Bataques, Mex., 21..........F6 82
Bataszek, Hung., 6,235.......B4 22
Batatais, Braz., 15,266.......F1 57
Batavia, Arg..........A3 54
Batavia, Boone, Ark., 40.....A2 75
Batavia, Kane, Ill., 7,496..........B5, F1 90
Batavia, see Djakarta, Indon.
Batavia, Jefferson, Iowa, 533.D5 92
Batavia, Genesee, N.Y., 18,210..........C2 108
Batavia, Clermont, Ohio, 1,729..........C3 111
Bataysk, Sov. Un., 70,000..H12 27
Batchelor, bay, N.C..........B7 109
Batchtown, Calhoun, Ill., 248..........D3 90
Bateman, Sask., Can., 106...H2 70
Bates, Scott, Ark., 106..........C1 81
Bates, Grant, Oreg., 200.....C8 113
Bates, co., Mo., 15,905......C3 101
Batesburg, Lexington and Saluda, S.C., 3,806..........D4 115
Bates City, Lafayette, Mo., 110..........B3 101
Batesland, Shannon, S. Dak., 95..........D3 116
Batesville, Barbour, Ala., 75..C4 78
Batesville, Independence, Ark., 6,207..........B4 81
Batesville, Ripley, Ind., 3,349..........F7 91
Batesville, Panola, Miss., 3,284..........A4 100
Batesville, Noble, Ohio, 160..C6 111
Batesville, Zavala, Tex., 500..E3 118
Bath, N.B., Can., 767..........C2 74
Bath, Ont., Can., 637..........C8 72
Bath, Eng., 80,856..........C5 12
Bath, Mason, Ill., 398..........C3 90
Bath, Franklin, Ind., 100.....E8 87
Bath, Sagadahoc, Maine, 10,717..........E3, E6 96
Bath, Clinton, Mich., 500.....F6 98
Bath, Grafton, N.H., 160 (604▲)..........B3 105
Bath, Steuben, N.Y., 6,166...C3 108
Bath, Beaufort, N.C., 346....B7 109
Bath, Northampton, Pa., 1,736..........E11 114
Bath, Aiken, S.C., 1,419.....D4 115
Bath, Brown, S. Dak., 80....B7 116
Bath, co., Ky., 9,114..........B6 94
Bath, co., Va., 5,335..........C3 121
Batha de Lairi, riv., Chad.....C3 46
Bathgate, Pembina, N. Dak., 175..........A8 110
Bathgate, Scot., 12,686......E8 13
Bathsheba, Barbados, 850...p17 64
Bath Springs, Decatur, Tenn., 50..........B3 117
Bathurst, Austl., 16,939.....F7 51
Bathurst, N.B., Can., 5,494..B4 74
Bathurst, Gam., 21,022......D1 45
Bathurst, cape, N.W. Ter., Can..........B7 66
Bathurst, inlet, N.W. Ter., Can..........C11 66
Bathurst, isl., Austl..........B4 50
Bathurst, isl., N.W. Ter., Can.A12 66
Bathurst Inlet, N.W. Ter., Can..........C11 66
Batie, Upper Volta..........E4 45
Batiscan, Que., Can., 231....C5 73
Batiscan, riv., Que., Can..........B5 73
Batkanu, S.L..........E2 45
Batna, Alg., 14,732 (26,413▲).B6 44
Batoche, Sask., Can., 165....E2 70
Baton Rouge, La., 152,419 (★248,700)..........B5, D4 95
Batouri, Cam..........E2 46
Batovi, riv., Braz..........E4 59
Batroun, Leb..........E10 31
Batson, Hardin, Tex., 650...D5 118
Batsto, riv., N.J..........D3 106
Battambang, Camb., 16,000..F5 38
Battery Park, Isle of Wight, Va., 240..........B6 121
Batticaloa, Cey., 17,662.....G7 39
Battiest, McCurtain, Okla., 50..........C7 112
Battir, Isr., 200..........k11 32
Battle, creek, Sask., Can..........H1 70
Battle, creek, Mont..........B7 102
Battle, mtn., Wyo..........D5 125
Battle, riv., Alta., Can..........C5 69
Battle, riv., Sask., Can..........E1 70
Battle, riv., Minn..........C4 99
Battleboro, Nash and Edgecombe, N.C., 364..........A6 109
Battle Creek, Sask., Can., 20..........H1 70
Battle Creek, Routt, Colo....A3 83
Battle Creek, Ida., Iowa, 786..........B2 92
Battle Creek, Calhoun, Mich., 48,774 (★107,300)..........F5 98
Battle Creek, Madison, Nebr., 587..........C8 103
Battleford, Sask., Can., 1,627..........E1 70
Battle Ground, Tippecanoe, Ind., 804..........C4 91
Battle Ground, Clark, Wash., 888..........D3 122
Battle Harbour, Newf., Can., 103..........B4, h11 75
Battle Lake, Otter Tail, Minn., 733..........D3 99
Battlement, mesa, Colo..........B2 83
Battle Mountain, Lander, Nev., 950..........C5 104
Battles, Wayne, Miss., 100...D5 100
Battleview, Burke, N. Dak., 55..........A3 110
Battonya, Hung., 9,216......C5 22
Battrum, Sask., Can., 35....G1 70
Batu, is., Indon..........F1 35
Batu, mtn., Eth..........D4 47
Batuc, Mex., 1,267..........B2 63
Batumi, Sov. Un., 87,000...B13 31
Batu Pahat, Mala., 39,294...L5 38
Baturadja, Indon., 2,955....F2 35
Baturino, Sov. Un..........B11 29
Baturité, Braz., 7,198.......B3 57
Bat Yam, Isr., 31,338..........g9 32
Bauan, Phil., 2,975..........p13 35
Bauang, Phil., 3,188......B6, n13 35
Baubau, Indon., 2,493.......G6 35
Bauchi, Nig., 1,344..........D6 45
Baudette, Lake of the Woods, Minn., 1,597..........B4 99
Baudouinville, Con. L.......C4 48
Bauer, rock, Pa..........B1 106
Baugé, Fr., 3,363..........D4 14
Bauld, cape, Newf., Can..........C4 75
Bauma, Switz., 3,214..........B6 19
Baume-les-Dames, Fr., 4,038.B2 18
Baunei, It., 4,033..........D2 21
Baures, Bol., 592..........B3 55
Baurette, Que., Can..........E9 73
Bauru, Braz., 85,237..........C3 56
Bautzen, Ger., 41,900.......B9 17
Bauxite, Saline, Ark., 950..........C3, E5 81
Bavaria, Saline, Kans., 76...D6 93
Bavaria, reg., Ger..........D5 16
Bavarian Alps, mts., Aus., Ger..B7 18
Bavay, Fr., 2,273..........D4 15
Båven, lake, Swe..........t34 25
Bavispe, Mex., 923..........A3 63
Bawcomville, Ouachita, La., 1,500..........*B3 95
Bawean, isl., Indon..........G4 35
Bawku, Ghana..........D4 45
Bawlf, Alta., Can., 203.......C4 69
Baxley, Appling, Ga., 4,268..E4 87
Baxter, Drew, Ark..........D4 81
Baxter, Jasper, Iowa, 681....C4 92
Baxter, Crow Wing, Minn., 1,037..........D4 99
Baxter, Putnam, Tenn., 853..C8 117
Baxter, Marion, W. Va., 600..........A7, B4 123
Baxter, co., Ark., 9,943.....A3 81
Baxter Estates, Nassau, N.Y., 932..........*D2 108
Baxter Springs, Cherokee, Kans., 4,498..........E9 93
Baxterville, Rio Grande, Colo., 50..........D4 83
Baxterville, Lamar, Miss., 175..........D4 100
Bay, Craighead, Ark., 627....B5 81
Bay, Gasconade, Mo., 54.....C6 101
Bay, co., Fla., 67,131..........G3 86
Bay, co., Mich., 107,042......E6 98
Bay, pt., S.C..........G7 115
Bay (Laguna de), lake, Phil...o13 35
Baya Dzur-Gunen, Mong., 5,000..........B5 34
Bay al Kabir, riv., Libya......C2 43
Bayambang, Phil., 3,945.....o13 35
Bayamo, Cuba., 20,178.......E5 64
Bayamón, P.R., 15,109.......B6 65
Bayamón, mun., P.R., 72,221.B6 65
Bayamón, riv., P.R..........B6 65
Bayan-Aul, Sov. Un., 2,600..........C9 29
Bayaney, P.R..........B3 65
Bayanga, Cen. Afr. Rep.....E3 43
Bayanovka, Sov. Un.........A5 29
Bayard, Duval, Fla., 400.B5, C6 86
Bayard, Guthrie, Iowa, 597..C3 92
Bayard, Morrill, Nebr., 1,519.C2 103
Bayard, Grant, N. Mex., 2,327..........E1 107
Bayard, Grant, W. Va., 484..B5 123
Baybay, Phil., 9,500 (51,400▲)..........*C6 35
Bayble, Scot..........B2 13
Bayboro, Pamlico, N.C., 545.B7 109
Bayboro, Horry, S.C., 25....D9 115
Bay Bulls, Newf., Can., 697..E5 75
Bayburt, Tur., 12,000.......B13 31
Bay Center, Pacific, Wash., 600..........C2 122
Bay City, Bay, Mich., 53,604.E7 98
Bay City, Tillamook, Oreg., 996..........B3 113
Bay City, Matagorda, Tex., 11,656..........E5, G4 118
Bay City, Grays Harbor, Wash..........C1 122
Bay City, Pierce, Wis., 327..D1 124
Baydaratskaya, bay, Sov. Un....B8 4
Bay de Verde, Newf., Can., 906..........D5 75
Bayern, state, Ger., 9,513,900..........D6 17, A7 18
Bayeux, Fr., 9,678..........C3 14
Bayfield, Ont., Can., 395....D3 72
Bayfield, La Plata, Colo., 322.D3 83
Bayfield, Bayfield, Wis., 969..........B3 124
Bayfield, co., Wis., 11,910...B2 124
Bay Harbour Islands, Dade, Fla., 3,249..........*F3 86
Bay Head, Ocean, N.J., 824.C4 106
Bayindir, Tur., 11,000.......C6 23
Baykal, lake, Sov. Un.......D13 28

Baykal, mts., Sov. Un....D13 28
Baykonur, Sov. Un....D7 29
Bay l'Argent, Newf., Can., 418....E4 75
Baylis, Pike, Ill., 284....D3 90
Bay Mills, Indian res., Mich....B6 98
Baylor, Valley, Mont., 66....B10 102
Baylor, co., Tex., 5,893....C3 118
Baymak, Sov. Un....C5 29
Bay Minette, Baldwin, Ala., 5,197....E2 78
Bay of Islands, bay, Newf., Can....k25 67
Bay of Whales, bay, Ant....B32 5
Bayombong, Phil., 6,929....n13 35
Bayonne, Fr., 36,941....F3 14
Bayonne, Hudson, N.J., 74,215....B4, E5 106
Bayou Barbary, Livingston, La....B6 95
Bayou Cane, Terrebonne, La., 3,173....*E5 95
Bayou Current, St. Landry, La. 75....D4 95
Bayou George, Bay, Fla., 100. G3 86
Bayou Goula, Iberville, La., 750....B4, D4 95
Bayou LaBatre, Mobile, Ala., 2,572....E1 78
Bayou Meto, Arkansas, Ark., 20....C4 81
Bayóvar, Peru....C1 58
Bay Park, Nassau, N.Y., 1,500....*G2 84
Bayport, N.S., Can., 141....E5 74
Bay Port, Huron, Mich., 400.E7 98
Bayport, Washington, Minn., 3,205....E6, E8 99
Bayport, Suffolk, N.Y., 3,000.D3 108
Bayram-Ali, Sov. Un., 25,000....C11 41
Bayramic, Tur., 4,100....C6 23
Bayreuth, Ger., 61,800....D6 17
Bay Ridge, Anne Arundel, Md., 750....C5 85
Bayrischzell, Ger., 1,700....B8 18
Bay Roberts, Newf., Can., 1,306....E5 75
Bays, lake, Ont., Can....B5 72
Bays, mtn., Tenn....C10 117
Bay St. Louis, Hancock, Miss., 5,073....E2, E4 100
Bayshore, San Mateo, Calif., 1,500....*D2 82
Bayshore, Charlevoix, Mich., 160....C5 98
Bay Shore, Suffolk, N.Y., 20,000....E3, E7 108
Bayshore Gardens, Manatee, Fla., 2,297....*E4 86
Bayside, Hancock, Maine, 70.D4 96
Bayside, Refugio, Tex., 300....E4 118
Bayside, Princess Anne, Va., 6,000....*B7 121
Bayside, Milwaukee, Wis., 3,181....*E6 124
Bay Springs, Jasper, Miss., 1,544....D4 100
Baysun, Sov. Un....B13 41
Bayt Dajan, Jordan, 5,000....g12 32
Baytillū, Jordan, 1,000....h11 32
Bayt Jālā, Jordan, 5,000.C7, k11 32
Bayt Laḥm (Bethlehem), Jordan, 19,155....C7, k11 32
Baytown, Harris, Tex., 28,159....E5, F6 118
Bay Trail, Sask., Can....E3 70
Bayview, Jefferson, Ala., 1,081....E4 78
Bayview, Humboldt, Calif., 1,800....*B1 82
Bayview, Kootenai, Idaho, 250....B2 89
Bayview, Essex, Mass. (part of Gloucester)....A6 97
Bay View, Emmet, Mich....C6 98
Bay Village, Cross, Ark., 150.B5 81
Bay Village, Cuyahoga, Ohio, 14,489....B1 111
Bayville, Ocean, N.J., 500....D4 106
Bayville, Nassau, N.Y., 3,962.F2 84
Baza, Sp., 14,880....D4 20
Bazaar, Chase, Kans., 100....D7 93
Bazaruto, isl., Moz....B6 49
Bazemore, Fayette, Ala., 100 .B2 78
Bazile Mills, Knox, Nebr., 45.B8 103
Bazine, Ness, Kans., 429....D4 93
Bazman Kuh, mtn., Iran....G10 41
Baztán, Sp., 1,534....A5 20
Beach, Ware, Ga., 53....E4 87
Beach (Dunes Park), Lake, Ill., 1,800....*A6 90
Beach, Golden Valley, N. Dak., 1,460....C1 110
Beach, Chesterfield, Va., 70....C7 121
Beach, pond, Conn., R.I....C9 84
Beachburg, Ont., Can., 500....B8 72
Beach City, Stark, Ohio, 1,151....B6 111
Beach Haven, Ocean, N.J., 1,041....D4 106
Beach Haven, Luzerne, Pa., 500....D9 114
Beach Haven, inlet, N.J....D4 106
Beach Haven Crest, Ocean, N.J. (part of Long Beach)....D4 106
Beach Haven Terrace, Ocean, N.J. (part of Long Beach)....D4 106
Beachlake, Wayne, Pa., 250.C11 114
Beachport, Austl., 293....G7 50
Beachville, Ont., Can., 849....D4 85
Beachville, St. Marys, Md., 350....D5 85
Beachwood, Ocean, N.J., 2,765....D4 106
Beachwood, Cuyahoga, Ohio, 6,089....*A6 111
Beachy, head, Eng....D8 12
Beacon, Mahaska, Iowa, 718.C5 92
Beacon, Marquette, Mich. (part of Champion)....B2 98
Beacon, Dutchess, N.Y., 13,922....D7 108
Beacon Falls, New Haven, Conn., 2,886....D4 84
Beacon Hill, Gulf, Fla., 150....C1, H3 86
Beacon Hill, Cowlitz, Wash., 1,019....*C3 122
Beaconsfield, Que., Can., 10,064....D8 73
Beaconsfield, Ringgold, Iowa, 71....D3 92
Beadle, Sask., Can., 50....F1 70
Beadle, co., S. Dak., 21,682....C7 116
Beadling, Allegheny, Pa., 1,500....*F1 114
Beagle, Miami, Kans., 150....D9 93
Bealanana, Malag....f9 49
Beal City, Isabella, Mich., 150....E6 98
Beale, cape, B.C., Can....E5 68
Beallsville, Monroe, Ohio, 491....C6 111
Beals, Henderson, Ky., 100....C2 94
Beals, Washington, Maine, 400 (640▲)....D5 96
Bealwood, Muscogee, Ga. (part of Columbus)....D2 87
Beamsville, Ont., Can., 2,198.D5 72
Bean City, Palm Beach, Fla., 150....F6 86
Bean Hill, New London, Conn....C8 84
Bean Lake, Platte, Mo., 245....B3 101
Bean Station, Grainger, Tenn., 100....C10 117
Bear, Bew Castle, Del., 65....A6 85
Bear, Adams, Idaho, 5....D2 89
Bear, cave, Mich....G4 98
Bear, cave, Mo....E5 101
Bear, creek, Ala....A1 78
Bear, creek, Colo....D8 83
Bear, creek, Kans....E2 93
Bear, creek, Miss....A5 100
Bear, creek, Oreg....E4 113
Bear, creek, Wyo....D8 125
Bear, inlet, N.C....C6 109
Bear, isl., Man., Can....B3 71
Bear, isl., Ire....F2 11
Bear, isl., Wis....A3 124
Bear, lake, Alta., Can....B1 69
Bear, lake, B.C., Can....A4 68
Bear. lake, Man., Can....B4 71
Bear, lake, Idaho, Utah .G7 83, A4 119
Bear, lake, Wis....C2 124
Bear, mtn., Ark....C5 81
Bear, mtn., Conn....A2 84
Bear, mtn., Ky....C5 90
Bear, mtn., Maine....D2 96
Bear, mtn., Oreg....D4 113
Bear, mtn., Vt....E2 120
Bear, mtn., Wyo....D6 125
*Bear, riv., Utah, Wyo.*A4 113, D2 125
Bear, swamp, Mass....E3 97
Bear Creek, Marion, Ala., 243....A2 78
Bearcreek, Carbon, Mont., 61....E7 102
Bear Creek, Dewey, S. Dak., 40....C4 116
Bear Creek, Outagamie, Wis., 455....D5 124
Bear Creek, flood control res., Pa....D10 114
Bearden, Ouachita, Ark., 1,268....D3 81
Bearden, Okfuskee, Okla., 150....B5 112
Bearden, Knox, Tenn., 3,600....D9, E11 117
Beardmore, glacier, Ant....A29 5
Beards Fork, Fayette, W. Va., 800....D7 123
Beardsley, Rawlins, Kans., 50....C2 93
Beardsley, Big Stone, Minn., 410....E2 99
Beardstown, Cass, Ill., 6,294....C3 90
Beardstown, Perry, Tenn., 50.B4 117
Bearfort, mtn., N.J....A4 106
Bear Lake, Manistee, Mich., 323....D4 98
Bear Lake, Warren, Pa., 260....C3 114
Bear Lake, co., Idaho, 7,148.G7 89
Bear Lodge, mts., Wyo....A8 125
Bearmouth, Granite, Mont., 10....D3 102
Béarn, former prov., Fr., 275,000....F3 14
Bear Paw, mtn., Mont....B7 102
Bear Pond, mts., Md....A2 85
Bear River, N.S., Can., 830....E4 74
Bear River, bay, Utah....B3 119
Bear River, divide, Wyo....D2 125
Bear River City, Box Elder, Utah, 447....B3 119
Bear Spring, Stewart, Tenn., 50....A4 117
Beartooth, pass, Wyo....A3 125
Beartooth, range, Mont....E7 102
Bear Town (McComb South), Pike, Miss., 1,865....D3 100
Beas de Segura, Sp., 9,251....C4 20
Beason, Logan, Ill., 250....C4 90
Beata, cape, Dom. Rep....G8 64
Beaties, butte, Oreg....E7 113
Beaton, B.C., Can., 50....D9 68
Beaton, riv., B.C., Can....A7 68
Beatrice, Monroe, Ala., 506....D2 78
Beatrice, Gage, Nebr., 12,132....D9 103
Beatrice, S. Rh....A5 49
Beattie, Marshall, Kans., 314.C7 93
Beatty, Sask., Can., 143....E3 70
Beatty, Carroll, Miss., 50....B4 100
Beatty, Nye, Nev., 450....G5 104
Beatty, Klamath, Oreg., 200....E5 113
Beatty Knob, hill, Ohio....C6 111
Beattyville, Lee, Ky., 1,048....C6 94
Beatyestown, Warren, N.J., 100....B3 106
Beau, lake, Maine....A3 96
Beaucaire, Fr., 8,243....F6 14
Beauce, co., Que., Can., 26,660....C7 73
Beauceville Est, Que., Can., 1,190....C7 73
Beauceville Ouest, Que., Can., 1,645....*C7 73
Beaucourt, Fr., 4,570....D7 14
Beaudry, Garland, Ark....C3 81
Beaufort, Mala., 2,000....D5 35
Beaufort, Franklin, Mo., 125....C6 101
Beaufort, Carteret, N.C., 2,922....C7 109
Beaufort, Beaufort, S.C., 6,298....G6 115
Beaufort, co., N.C., 36,014....B6 109
Beaufort, co., S.C., 44,187....G6 115
Beaufort, sea, N.A....B7 61
Beaufort West, S. Afr., 16,323....D3 49
Beaugency, Fr., 3,493....D4 14
Beauharnois, Que., Can., 8,704....D4, D8 73
Beauharnois, co., Que., Can., 49,667....D3 73
Beaulieu, Mahnomen, Minn., 40....C3 99
Beauly, Scot....C4 13
Beauly, firth, Scot....C4 13
Beauly, riv., Scot....C4 13
Beaumaris, Wales, 1,960....A3 12
Beaumaris, bay, Wales....A4 12
Beaumont, Bel., 1,725....D4 15
Beaumont, Riverside, Calif., 4,432....F4 82
Beaumont, Newf., Can., 340.D4 75
Beaumont, Que., Can., 500....C6 73
Beaumont, Fr., 2,702....F4 14
Beaumont, Butler, Kans., 150.E7 93
Beaumont, Perry, Miss., 926.D5 100
Beaumont, Jefferson, Tex., 119,175 (★266,600)....D5 118
Beaumont Place, Harris, Tex., 1,500....*E5 118
Beaumont [-sur-Oise], Fr., 6,787....E2 15
Beaune, Fr., 15,367....D6 14
Beauport, Que., Can., 9,192 .C9 73
Beauport Est., Que., Can., 1,417....*C9 73
Beaupre, Que., Can., 2,587....B7 73
Beauraing, Bel., 2,383....D4 15
Beauregard, Copiah, Miss., 193....D9 100
Beauregard, par., La., 19,191....D2 95
Beaurepaire, Que., Can., 2,400....D8 73
Beaurivage, Que., Can., 526 .C6 73
Beausejour, Man., Can., 1,770....D3 71
Beauty, Martin, Ky., 300....C7 88
Beauvais, Fr., 33,995....C4 14
Beauval, Sask., Can., 504....B2 70
Beauval, Fr., 2,173....D2 15
Beauvallon, Alta., Can., 71....C5 69
Beauvoir, Harrison, Miss., 50 .E2 100
Beaver, Alsk., 101....B10 79
Beaver, Carroll, Ark., 24....A2 81
Beaver, Barton, Kans., 125....D5 93
Beaver, Pike, Ohio, 341....C5 111
Beaver, Beaver, Okla., 2,087....A1, D4 112
Beaver, Tillamook, Oreg., 200....B3 113
Beaver, Beaver, Pa., 6,160....E1 114
Beaver, Beaver, Utah, 1,548....E3 119
Beaver (Glen Hedrick), Raleigh, W. Va., 1,230....D3, D7 123
Beaver, Marinette, Wis., 75....C5 124
Beaver, co., Okla., 6,965....A1 106
Beaver, co., Pa., 206,948....E1 114
Beaver, co., Utah, 4,331....E2 119
Beaver, brook, N.H....E4 105
Beaver, creek, Sask., Can....F2 70
Beaver, creek, Iowa....A7 92
Beaver, creek, Kans....C2 93
Beaver, creek, Kans....C5 93
Beaver, creek, Ky....C7 94
Beaver, creek, Md....A2 85
Beaver, creek, Mo....E5 101
Beaver, creek, Mont....B9 102
Beaver, creek, Mont....C12 102
Beaver, creek, Nebr....D5 103
Beaver, creek, N. Dak....C5 110
Beaver, creek, Okla....C3 112
Beaver, creek, Tenn....E11 117
Beaver, creek, Wyo....B8 125
Beaver, creek, Wyo....C8 125
Beaver, isl., Mich....C5 98
Beaver, lake, Nebr....C3 103
Beaver, lake, Wis....E5 124
Beaver, riv., N.Y....B5 108
Beaver, riv., Utah....E2 119
Beaver Bank, N.S., Can., 870.E6 74
Beaver Bay, Lake, Minn., 287....C7 99
Beaver City, Furnas, Nebr., 818....D6 103
Beaver Creek, Washington, Md., 150....A2 85
Beaver Creek, Rock, Minn., 250....G2 99
Beaver Creek, mtn., Ala....B3 78
Beaver Crossing, Seward, Nebr., 439....D8 103
Beaverdale, Cambria, Pa., 1,000....F4 114
Beaver Dam, Mohave, Ariz., 25....A2 80
Beaver Dam, Kosciusko, Ind., 100....B5 91
Beaver Dam, Ohio, Ky., 1,648....C3 94
Beaverdam, Allen, Ohio, 514.B4 111
Beaverdam, Hanover, Va., 50....D5 121
Beaver Dam, Dodge, Wis., 13,118....E5 124
Beaver Dams, Schuyler, N.Y., 280....C4 108
Beaverdell, B.C., Can., 332....E8 68
Beaver Falls, Lewis, N.Y., 640....B5 108
Beaver Falls, Beaver, Pa., 16,240....E1 114
Beaverfoot, riv., B.C., Can....D2 69
Beaverhead, co., Mont., 7,194....E3 102
Beaverhead, mts., Mont....E3 102
Beaverhead, riv., Mont....E4 102
Beaverhill, Overton, Tenn....C8 117
Beaverhill, lake, Alta., Can....C4 69
Beaverhill, lake, Man., Can....B4 71
Beaverlick, Boone, Ky., 175....A7 94
Beaverlodge, Alta., Can., 897 .B1 69
Beaver Meadows, Carbon, Pa., 1,392....E10 114
Beaver Point, Larimer, Colo., 50....A5 83
Beaver Run, res., Pa....E2 114
Beaver Springs, Snyder, Pa., 750....E7 114
Beavertail, pt., R.I....D11 84
Beaverton, Lamar, Ala., 162 .B1 78
Beaverton, Ont., Can., 1,099 .C5 72
Beaverton, Gladwin, Mich., 926....E6 98
Beaverton, Washington, Oreg., 5,937....B2, B3 113
Beavertown, Snyder, Pa., 738 .E7 114
Beaverville, Iroquois, Ill., 430....C6 90
Beawar, India, 53,931....D5 40
Beazley, Arg....A3 54
Bebedouro, Braz., 18,249....C3 56
Bebee, Pontotoc, Okla....C5 112
Bebra, Ger., 7,500....C4 17
Becancour, Que., Can., 320 .C5 73
Becancour, riv., Que., Can....C5 73
Beccles, Eng., 7,330....B9 12
Bečej, Yugo., 24,853....C5 22
Becerreá, Sp., 7,776....A2 20
Becharof, lake, Alsk....D8 79
Bechuanaland, Br. dep., Afr., 337,000....B3 49, I8 42
Bechyne, Czech., 2,251....D9 17
Beckemeyer, Clinton, Ill., 1,056....E4 90
Beckenham, Eng., 77,265....m12 10, C7 12
Becker, Sherburne, Minn., 279....E5 99
Becker, Monroe, Miss., 141 . B5 100
Becker, co., Minn., 23,959....D3 99
Becket, Berkshire, Mass., 350 (770▲)....B1 97
Beckham, co., Okla., 17,782 .B2 112
Beckley, Raleigh, W. Va., 18,642....D3, D7 123
Beckleys, Hartford, Conn....C6 84
Beckum, Ger., 20,600....B3 17
Beckville, Panola, Tex., 632 .C5 118
Beckwith, Lincoln, Wyo....D2 125
Beckwith, creek, La....D2 95
Bedale, Eng., 1,115....F7 13
Bédarieux, Fr., 7,263....F5 14
Bedburg, Ger., 9,100....C1 17
Beddington, Washington, Maine, (14▲)....D4 96
Bede, Man., Can....E1 71
Beder, Den., 306....B4 24
Bederkesa, Ger., 2,900....E2 24
Bedford, Que., Can., 2,855....D5 73
Bedford, N.S., Can., 2,021....E6 74
Bedford, Eng., 63,317....B7 12
Bedford, Lawrence, Ind., 13,024....G5 91
Bedford, Taylor, Iowa 1,807 .D6 92
Bedford, Trimble, Ky., 717....B4 94
Bedford, Middlesex, Mass., 10,969....B5, C2 97
Bedford, Calhoun, Mich., 150....F5 98
Bedford, Livingston, Mo., 15 .B4 101
Bedford, Hillsboro, N.H., 175 (3,636▲)....E3 105
Bedford, Westchester, N.Y., 893....D3, D7 108
Bedford, Cuyahoga, Ohio, 15,223....A6, B2 111
Bedford, Bedford, Pa., 3,696 . F4 114
Bedford, Tarrant, Tex., 2,706....*C4 118
Bedford, Bedford, Va., 5,921 .D3 121
Bedford, Lincoln, Wyo., 75....C2 125
Bedford, co., Eng., 380,704....B7 12
Bedford, co., Pa., 42,451....G4 114
Bedford, co., Tenn., 23,150....B5 117
Bedford, co., Va., 31,028....D3 121
Bedford Heights, Cuyahoga, Ohio, 5,275....*B2 111
Bedford Hills, Westchester, N.Y., 3,000....D3 108
Bedford Springs, Middlesex, Mass. (part of Bedford)....C2 97
Bedias, Grimes, Tex., 500....D5 118
Bédja, Tun., 34,645....B6 44
Bédja, prov., Tun., 248,525 .*B6 44
Bedlington, Eng., 29,373....C6 10
Bedminster, Somerset, N.J., 300....B3 106
Bedourie, Austl....B2 51
Bedrock, Montrose, Colo., 5....C2 83
Bedzin, Pol., 39,000....g10 26
Bee, Seward, Nebr., 149....C8 103
Bee, co., Tex., 23,755....E4 118
Beebe, White, Ark., 1,697....B4 81
Beebe, Que., Can., 1,363....D5 73
Beebe Plain, Orleans, Vt., 140....A4 120
Beebe River, Grafton, N.H., 250....C3 105
Bee Branch, Van Buren, Ark., 63....B3 81
Beech, fork, Ky....C4 94
Beech Bluff, Madison, Tenn., 150....B3 117
Beechbottom, Brooke, W. Va., 506....A4, B2 123
Beech Creek, Muhlenberg, Ky., 788....C2 94
Beech Creek, Clinton, Pa., 634....D6 114
Beecher, Will, Ill., 1,367....B6 90
Beecher City, Effingham, Ill., 452....D5 90
Beecher Falls, Essex, Vt., 350.A5 120
Beechey, head, B.C., Can....B9 68
Beech Grove, Greene, Ark., 60....A5 81
Beech Grove, Marion, Ind., 10,973....E5, H8 91
Beech Grove, McLean, Ky., 159....C2 94
Beechgrove, Coffee, Tenn., 150....B5 117
Beech Island, Aiken, S.C., 900....E4 115
Beechwood, Norfolk, Mass., 100....D4 97
Beechwood Village, Jefferson, Ky., 1,903....*A4 94
Beechy, Sask., Can., 402....G2 70
Beecroft, head, Austl....G8 51
Beedeville, Jackson, Ark., 150 .B4 81
Beef, isl., Vir. Is....f16 65
Beek, Neth., 6,000....D5 15
Beekman, Morehouse, La....B4 95
Beeler, Ness, Kans., 100....D4 93
Beelitz, Ger., 5,895....A7 17
Beemer, Cuming, Nebr., 667 .C9 103
Bee Ridge, Sarasota, Fla., 2,043....E4, F2 86
Beersheba, Isr., 43,158....C6 32
Beersheba Springs, Grundy, Tenn., 577....D8 117
Beerston, Delaware, N.Y....C5 108
Beersville, N.B., Can., 123....C5 74
Beer Tuvya, Isr., 574....k9 32
Beeskow, Ger., 7,571....A9 17
Beesleys Point, Cape May, N.J., 200....E3 106
Beeton, Ont., Can., 675....C5 72
Beetzendorf, Ger., 2,386....F5 24
Beeville, Bee, Tex., 13,811....E4 118
Befale, Con. L....A3 48
Beg, lake, N. Ire....C5 11
Bega, Austl., 3,858....H7 51
Beggs, Okmulgee, Okla., 1,114....B5 112
Bègles, Fr., 24,388....E3 14
Begovat, Sov. Un., 30,000 . G22 9
Béhague, pt., Fr. Gu....B4 59
Behbehān, Iran, 22,610....F5 41
Behm, canal, Alsk....n24 79
Beilngries, Ger., 3,300....D6 17
Beilul, Eth....C5 47
Beira, Moz., 25,000....A5 49
Beira, reg., Port., 2,039,800 . B2 20
Beira Alta, prov., Port., 703,231....*B2 20
Beira Baixa, prov., Port., 361,191....*C2 20
Beira Litoral, prov., Port., 985,135....*B1 20
Beirne, Clark, Ark., 300....D2 81
Beirut (Beyrouth), Leb., 400,000....F10 31
Beiseker, Alta., Can., 360....D4 69
Beitbridge, S. Rh., 395....B5 49
Beit Guvrin, Isr....C6 32
Beith, Scot., 6,908....E4 13
Beit Lid, Isr....f10 32
Beit-Shan, Isr., 9,572....B7 32
Beiuş, Rom., 6,467....B6 22
Beja, Port., 14,058....C2 20
Béjar, Sp., 13,522....B3 20
Bejestan, Iran....D9 41
Bejou, Mahnomen, Minn., 164....C3 99
Bejuco, Pan., 988....F8 62
Bekdash, Sov. Un....E4 29
Békés, Hung., 17,661 (21,699▲)....B5 22
Békés, co., Hung., 468,303....*B5 22
Bekescsaba, Hung., 40,965 (49,488▲)....B5 22
Bekily, Malag....h9 49
Bela, India, 21,397....E8 40
Bela, Pak., 3,063....C4 39
Bela Crkva, Yugo., 10,749....C5 22
Belaga, Mala., 258....E4 39
Bel Air, Harford, Md., 4,300 .A5 85
Belalcázar, Sp., 8,793....C3 20
Bel Alton, Charles, Md., 175 .D4 85
Belanger, riv., Man., Can....C3 71
Belanger, riv., Sask., Can....A2 70
Belas, Port., 8,514....f9 20
Bela Vista, Braz., 8,878....C1 56
Bela Vista, Moz....C5 49
Belawan, Indon....E1, m11 35
Belaya, riv., Sov. Un....B4 29
Belaya Glina, Sov. Un., 10,000....H13 27
Belaya Tserkov, Sov. Un., 76,000....G8 27
Belcamp, Harford, Md., 225 .B5 85
Belcher, Pike, Ky., 591....C7 94
Belcher, Caddo, La., 400....B2 95
Belcher, is., N.W. Ter., Can . F17 67
Belchertown, Hampshire, Mass., 900 (5,186▲)....B3 97
Belchirag, Afg., 5,000....D12 41
Belcourt, Rolette, N. Dak., 200....A6 110
Belden, Lee, Miss., 250....A5 100
Belden, Cedar, Nebr., 157....B8 103
Belden, Mountrail, N. Dak., 26....A3 110
Belding, Ionia, Mich., 4,887....E5 98
Belecke, Ger., 4,500....B3 17
Belém (Pará), Braz., 359,988 (★405,000)....B1 57
Belém, Port., 24,637....f9 20
Belén, Arg., 4,342....E2 55
Belen, Quitman, Miss., 90....A3 100
Belen, Valencia, N. Mex., 5,031....C3 107
Belén, Par., 6,647....D4 55
Belet Uin, Som., 7,800....E6 47
Belev, Sov. Un., 10,000....E11 27
Belfair, Mason, Wash., 400....B3 122
Belfast, P.E.I., Can., 145....C7 74
Belfast, Waldo, Maine, 6,140 .D3 96
Belfast, Allegany, N.Y., 790....C2 108
Belfast, N. Ire., 416,094 (★575,000)....C6 11
Belfast, Highland, Ohio ,100 .C4 111
Belfast, Marshall, Tenn., 200 .B5 117
Belfast, bay, N. Ire....C6 11
Belfield, Stark, N. Dak., 1,064....C2 110
Belfodiyo, Eth....C4 47
Belford, Eng., 891....E7 13
Belford, Monmouth, N.J., 3,500....C4 106
Belfort, Fr., 48,070 (★62,000)....D7 14
Belfort, Lewis, N.Y., 58....B5 108
Belfort, dept., Fr., 109,371....B2 18
Belfry, Pike, Ky., 950....C7 94
Belfry, Carbon, Mont., 250 . E7 102
Belgaum, India, 127,885 (★146,790)....E5 39
Belgica, mts., Ant....B16 5
Belgium, Vermilion, Ill., 494....C6 90
Belgium, Ozaukee, Wis., 643 .E6 124
Belgium, country, Eur., 9,189,741....E9 8, D4 15
Belgorod, Sov. Un., 81,000 . F11 27
Belgorod-Dnestrovskiy, Sov. Un., 38,600....H8 27
Belgrade, Kennebec, Maine, 250 (1,102▲)....D3 96
Belgrade, Sterns, Minn., 666....E4 99
Belgrade, Washington, Mo., 187....D7 101
Belgrade, Gallatin, Mont., 1,057....E5 102
Belgrade, Nance, Nebr., 224 .C7 103
Belgrade (Beograd), Yugo., 587,899....C5 22
Belgrade Lakes, Kennebec, Maine, 300....D3 96
Belgreen, Franklin, Ala., 200 .A2 78
Belhaven, Beaufort, N.C., 2,386....B7 109
Belice, riv., It....F4 21
Beli Lom, riv., Bul....D8 22
Belington, Barbour, W. Va., 1,528....B5 123
Belingwe, S. Rh., 1,100....B4 49
Belitung, isl., Indon....F3 35
Belize, Br. Hond., 32,813....B3 62
Belize, riv., Br. Hond., Guat....B3 56
Belk, Fayette, Ala., 150....B2 78
Belkino, Sov. Un....B11 37
Belknap, Johnson, Ill., 203....F5 90
Belknap, Davis, Iowa, 85....D5 92
Belknap, Sanders, Mont., 5....C1 102
Belknap, co., N.H., 28,912....C3 105
Belknap, crater, Oreg....C5 113
Belknap, mts., N.H.....C4 105
Belkofski, Alsk., 57....D7 79
Bell, Los Angeles, Calif., 19,450....F2 73
Bell, Gilchrist, Fla., 134....C4 86
Bell, Spokane, Wash....D8 122
Bell, co., Ky., 35,336....D6 94
Bell, co., Tex., 94,097....D4 118
Bell, isl., Newf., Can....C4 75
Bell, isl., Newf., Can....E5 75
Bella Bella, B.C., Can., 54....C3 68
Bellac, Fr., 4,022....D4 14
Bella Coola, B.C., Can., 345 .C4 68
Bella Coola, riv., B.C., Can....C4 68
Bell Acres, Allegheny, Pa., 1,283....*E1 114
Bellagio, It., 1,364....D5 18
Bellahy, Ire., 727....D3 11
Bellaire, Smith, Kans., 50....C5 93
Bellaire, Antrim, Mich., 689 .D5 98
Bellaire, Belmont, Ohio, 11,502....C7 111
Bellaire, Harris, Tex., 19,872....F5 118
Bellamy, Sumter, Ala., 750....C1 78
Bellary, India, 85,673....E6 39
Bella Unión, Ur., 2,519....E1 56
Bella Vista, Arg., 8,352....E2 55
Bella Vista, Arg., 7,922....E4 55
Bella Vista, Benton, Ark....A1 81
Bella Vista, Par., 5,762....D4 55
Bellbrook, Greene, Ohio, 941 .C3 111
Bell Buckle, Bedford, Tenn., 318....B5 117
Bellburn, Greenbrier, W. Va., 200....C4 123
Bell City, Calcasieu, La., 250.D3 95
Bell City, Stoddard, Mo., 409.D8 101
Belle, Maries, Mo., 1,016....C6 101
Belle, Kanawha, W. Va., 2,559....C3, C6 123
Belle, bay, Newf., Can....E4 75
Belle, isl., Newf., Can....C4 75
Belle, isl., Fr....D2 14
Belle, riv., La....C5 95
Belleair, Pinellas, Fla., 2,456 .E1 87
Belle Center, Logan, Ohio, 949....B4 111
Belle Chasse, Plaquemines, La., 700....C7, E5 95
Bellechasse, co., Que., Can., 26,054....C7 73
Belledune, N.B., Can., 163....B4 74
Belleek, N. Ire., 162....C3 11
Bellefont, Ford, Kans., 25....E4 93
Bellefontaine, Webster, Miss., 144....B4 100
Bellefontaine, Logan, Ohio, 11,424....B4 111
Bellefontaine Neighbors, St. Louis, Mo., 13,650....*A8 101
Bellefonte, Boone, Ark., 100....A2 81
Bellefonte, New Castle, Del., 1,536....A7 85
Bellefonte, Centre, Pa., 6,088....E6 114
Belle Fourche, Butte, S. Dak., 4,087....C2 116
Belle Fourche, res., S. Dak....C2 116
Belle Fourche, riv., S. Dak., Wyo....C2 110, A7 125
Bellegarde-sur-Valserine, Fr., 6,588....C1 18
Belle Glade, Palm Beach, Fla., 11,273....F6 86
Belle Haven, Accomack, Va., 371....D7 121
Belle Helene, Ascension, La....B5 95
Belle Isle, Orange, Fla., 2,344....D5 86
Belle Isle, strait, Newf., Can....C3 75
Belleisle Creek, N.B., Can., 75....D4 74
Bellemead, Prince Georges, Md., 1,400....*C4 85
Belle Mead, Somerset, N.J., 150....C3 106
Belle Meade, Davidson, Tenn., 3,082....A5, E9 117
Belle Mina, Limestone, Ala., 250....A3 78
Belleoram, Newf., Can., 570....E4 75
Belleplain, Cape May, N.J., 400....E3 106
Belle Plaine, Sask., Can., 117.G3 70
Belle Plaine, Benton, Iowa, 2,923....C5 92
Belle Plaine, Sumner, Kans., 1,579....E6 93
Belle Plaine, Scott, Minn., 1,931....F5 99
Bellerive, Que., Can., 1,000....E8 73
Belle Rive, Jefferson, Ill., 303.E5 90
Belle River, Ont., Can., 1,814....*E2 72
Bellerive Station, Que., Can., 96....C3 73
Bellerose, Assumption, La., 300....B5, D4 95
Bellerose, Nassau, N.Y., 1,083....*E3 108
Belleterre, Que., Can., 638....*E9 72
Belle Valley, Noble, Ohio, 438....C6 111
Belle Vernon, Fayette, Pa., 1,784....F2 114
Belleview, Marion, Fla., 864....C4 86
Belleview, Davidson, Tenn., 100....A5, E9 117
Belle View, Fairfax, Va., 3,500....*B5 121
Belleville, Yell, Ark., 273....B2 81
Belleville, N.S., Can., 276....F4 74
Belleville, Ont., Can., 30,655 C7 72
Belleville, St. Clair, Ill., 37,264....E4 90
Belleville, Republic, Kans., 2,940....C6 93
Belleville, Wayne, Mich., 1,921....A7 98
Belleville, Essex, N.J., 35,005....B4, D5 106
Belleville, Jefferson, N.Y., 275....B4 108
Belleville, Mifflin, Pa., 1,539....E6 114
Belleville, Washington, R.I. (part of North Kingstown)....C11 84
Belleville, Wood, W. Va., 50 .B3 123
Belleville, Dane and Green, Wis., 844....F4 124
Belleville North, Wayne, Mich., 1,128....*A7 98
Belleville [-sur-Saône], Fr., 2,976....D6 14
Bellevue, Alta., Can., 1,323....E3 69
Bellevue, Blaine, Idaho, 384....F4 89
Bellevue, Peoria, Ill., 1,561....*C4 90
Bellevue, Jackson, Iowa, 2,181....B7 92
Bellevue, Campbell, Ky., 9,336....A7 94
Bellevue, Talbot, Md., 267....C5 85
Bellevue, Eaton, Mich., 1,277 .F5 98
Bellevue, Sarpy, Nebr., 8,831....C9, E3 103
Bellevue, Huron and Sandusky, Ohio, 8,286....A5 111
Bellevue, Allegheny, Pa., 11,412....B5, F1 114
Bellevue, Clay, Tex., 309....C3 118
Bellevue, King, Wash., 12,809....D2 122
Belley, Fr., 4,609....E6 14
Bellflower, Los Angeles, Calif., 45,909....*F2 82
Bellflower, McLean, Ill., 389 .C5 90
Bellflower, Montgomery, Mo., 245....B6 101

BELL-BEVE 11A

Bell Gardens, Los Angeles, Calif., 26,467 ... *F2 82
Bellingham, Eng., 1,242 ... E6 13
Bellingham, Norfolk, Mass., 700 (6,774▲) ... B5, E1 97
Bellingham, Lac qui Parle, Minn., 327 ... E2 99
Bellingham, Whatcom, Wash., 34,688 ... A3 122
Bellingshausen, sea, Ant ... C3 5
Bellinzona, Switz., 13,435 (★17,716) ... D7 19
Bell Irving, riv., B.C., Can ... A3 68
Bellis, Atla., Can., 98 ... B4 69
Bellivela, Lib ... E3 45
Bellmawr, Camden, N.J., 11,853 ... D2 106
Bellmead, McLennan, Tex., 5,127 ... *D4 118
Bellmont, Wabash, Ill., 320 ... E6 90
Bellmore, Parke, Ind., 75 ... E3 91
Bellmore, Nassau, N.Y., 12,784 ... G2 84
Bello, Col., 28,398 ... B2 60
Bello, Cuba ... h11 64
Bellows Falls, Windham, Vt., 3,831 ... E4 120
Belloy, Alta., Can., 75 ... B1 69
Bellport, Suffolk, N.Y., 2,461 ... F5 84
Bells, Crockett, Tenn., 1,232 ... B2 117
Bells, creek, W. Va ... C6 123
Bellsburg, Dickson, Tenn., 75 ... A4 117
Bells Corners, Ont., Can., 1,900 ... A9 72
Bellton, Hall and Banks, Ga. (part of Lula) ... B3 87
Belltown, Sussex, Del., 300 ... C7 85
Belltown, Monroe, Tenn ... D9 117
Belluno, It., 22,500 (31,403▲) ... A4 21
Bell Ville, Arg., 15,796 ... A4 54
Bellville, Evans, Ga., 300 ... D5 87
Bellville, Richland, Ohio, 1,621 ... B5 111
Bellville, Austin, Tex., 2,218 ... E4 118
Bellvue, Larimer, Colo., 100 ... A5 83
Bellwood, Geneva, Ala., 273 ... D4 78
Bellwood, Cook, Ill., 20,729 ... F2 90
Bellwood, Butler, Nebr., 361 ... C8 103
Bellwood, Blair, Pa., 2,330 ... E5 114
Belly, riv., Alta., Can ... E4 69
Belmar, Monmouth, N.J., 5,190 ... C4 106
Bélmez, Sp., 8,068 ... C3 20
Belmond, Wright, Iowa, 2,506 ... B4 92
Belmont, San Mateo, Calif., 15,996 ... B5 82
Belmont, Man., Can., 378 ... E2 71
Belmont, N.S., Can., 307 ... D6 74
Belmont, Ont., Can., 649 ... E3 72
Belmont, Pinellas, Fla., 2,000 ... *E4 86
Belmont, Kingman, Kans., 200 ... E6 93
Belmont, Pointe Coupee, La ... C2 95
Belmont, Middlesex, Mass., 28,715 ... C2 97
Belmont, Tishomingo, Miss., 901 ... A5 100
Belmont, Golden Valley, Mont., 30 ... D7 102
Belmont, Dawes, Nebr ... B2 103
Belmont, Belknap, N.H., 600 (1,953▲) ... D4 105
Belmont, Allegany, N.Y., 1,146 ... C2 108
Belmont, Gaston, N.C., 5,007 ... B2 109
Belmont, Belmont, Ohio, 563 ... B6 111
Belmont, Coffee, Tenn ... B5 117
Belmont, Gonzales, Tex., 50 ... B4 118
Belmont, Rutland, Vt., 65 ... E3 120
Belmont, Whitman, Wash., 50 ... B8 122
Belmont, Pleasants, W. Va., 454 ... B3 123
Belmont, Lafayette, Wis., 616 ... F3 124
Belmont, co., Ohio, 83,864 ... C6 111
Belmonte, Braz., 7,897 ... E3 57
Belmore, Putnam, Ohio, 232 ... A4 111
Belmullet, Ire., 724 ... C2 11
Bel-Nor, St. Louis, Mo., 2,388 ... A8 101
Belo, Malag., 3,278 ... g8 49
Beloeil Village, Que., Can., 6,283 ... *D4 73
Belo Horizonte, Braz., 642,912 (★775,000) ... E2 57
Beloit, Lyon, Iowa, 110 ... A1 92
Beloit, Mitchell, Kans., 3,837 ... C5 93
Beloit, Mahoning, Ohio, 877 ... B7 111
Beloit, Rock, Wis., 35,199 ... F4 124
Beloit North (Perrygo Place), Rock, Wis., 4,475 ... *F4 124
Beloit West, Rock, Wis., 2,162 ... *F4 124
Belokany, Sov. Un., 10,000 ... B16 31
Belomorsk, Sov. Un., 17,400 ... E16 25
Belopolye, Sov. Un., 10,000 ... F10 27
Beloretsk, Sov. Un., 61,000 ... C5 29
Belot, Cuba ... h12 64
Belovo, Sov. Un., 115,000 ... E26 9
Beloye, lake, Sov. Un ... A11 27
Belozersk, Sov. Un., 16,800 ... A11 27
Belp, Switz., 4,922 ... C3 19
Belper, Eng., 15,563 ... A6 12
Belpre, Edwards, Kans., 211 ... E4 93
Belpre, Washington, Ohio, 5,418 ... C6 111
Bel-Ridge, St. Louis, Mo., 4,878 ... *A8 101
Belspring, Pulaski, Va., 400 ... D2 121
Belt, Cascade, Mont., 757 ... C6 102
Belt, creek, Mont ... C6 102
Belted, range, Nev ... F5 104
Belterra, Braz., 5,347 ... C4 59
Belton, Cass, Mo., 4,897 ... C3 101
Belton, Flathead, Mont ... B3 102
Belton, Anderson, S.C., 5,106 ... B3 115
Belton, Bell, Tex., 8,163 ... D4 118
Belton, res., Tex ... D7 118
Beltra, Ire ... C3 11
Beltrami, Polk, Minn., 186 ... C2 99
Beltrami, co., Minn., 23,425 ... B3 99
Beltsville, Prince Georges, Md., 3,500 ... B4 85
Beltsy, Sov. Un., 67,000 ... H6 27
Belturbet, Ire., 1,093 ... C4 11
Belukha, mtn., Sov. Un ... E11 28
Beluran, Mala., 50 ... D5 35
Belva, Woodward, Okla., 35 ... A2 112
Belvedere, Marin, Calif., 2,148 ... *B5 82
Belvedere, Aiken, S.C., 500 ... D4 115
Belvedere, Fairfax, Va., 1,100 ... *B5 121
Belvedere Marittimo, It., 9,069 ... E5 21
Belvidere, New Castle, Del., 1,000 ... A6 85
Belvidere, Boone, Ill., 11,223 ... A5 90
Belvidere, Kiowa, Kans., 125 ... E4 93
Belvidere, Thayer, Nebr., 185 ... D8 103
Belvidere, Warren, N.J., 2,636 ... B2 106
Belvidere, Perquimans, N.C., 75 ... A7 109
Belvidere, Jackson, S. Dak., 232 ... D4 116
Belvidere, Franklin, Tenn., 125 ... B5 117
Belvidere, mtn., Vt ... B3 120
Belvidere Center, Lamoille, Vt., 75 (155▲) ... B3 120
Belview, Redwood, Minn., 400 ... F3 99
Belvue, Pottawatomie, Kans., 179 ... C7 93
Belwood, Ont., Can., 117 ... D4 72
Belyando, riv., Austl ... D8 50
Belyy, Sov. Un., 17,200 ... D9 27
Belyy, isl., Sov. Un ... B9 28
Belyy Bom, Sov. Un ... E26 9
Belzig, Ger., 7,597 ... A7 17
Belzoni, Humphreys, Miss., 4,142 ... B3 100
Bembezar, riv., Sp ... D3 20
Bement, Piatt, Ill., 1,558 ... D5 90
Bemidji, Beltrami, Minn., 9,958 ... C4 99
Bemidji, lake, Minn ... C4 99
Bemis, Deuel, S. Dak., 50 ... C9 116
Bemis, Madison, Tenn., 3,127 ... B3 117
Bemis, Randolph, W. Va., 85 ... C5 123
Bemiss, Lowndes, Ga., 100 ... F3 87
Bemus Point, Chautauqua, N.Y., 443 ... C1 108
Bena, Cass, Minn., 286 ... C4 99
Benabarre, Sp., 1,231 ... A6 20
Bena Dibele, Con. L ... B3 48
Benadir, reg Som ... E5 47
Benalla, Austl., 8,259 ... H5 51
Benalto, Alta., Can., 147 ... C3 69
Benanee, Austl., 155 ... G4 51
Benares (Banaras), India, 471,258 (★489,864) ... C7 39, E9 40
Benatky nad Jizerou, Czech., 4,266 ... n18 26
Benavente, Sp., 11,080 ... A3 20
Benavides, Duval, Tex., 2,459 ... F3 118
Ben Avon, Allegheny, Pa., 2,553 ... *E1 114
Benbane, head, Ire ... B5 11
Benbecula, isl., Scot ... C1 13
Ben Bolt, Jim Wells, Tex., 200 ... F3 118
Benbrook, Tarrant, Tex., 3,254 ... *C4 118
Benbush, Tucker, W. Va., 100 ... B5 123
Ben Cat, Viet ... G7 38
Benchland, Judith Basin, Mont., 85 ... C7 102
Bend, B.C., Can ... C7 68
Bend, Deschutes, Oreg., 11,936 ... C5 113
Bend, San Saba, Tex., 25 ... D3 118
Bendale, Richland, S.C., 1,544 ... C5 115
Ben Davis, Marion, Ind., 900 ... H7 91
Ben Davis, pt., N.J ... E2 106
Bendeleben, mtn., Alsk ... B7 79
Bender, Sask., Can., 25 ... G4 70
Bender Beila, Som., 1,900 ... D7 47
Bendersville, Adams, Pa., 484 ... G7 114
Bendery, Sov. Un., 38,000 ... H7 27
Bendigo, Austl., 30,204 (★40,335) ... H5 51
Bendorf, Ger., 14,000 ... C2 17
Benedict, Wilson, Kans., 128 ... E8 93
Benedict, Charles, Md., 460 ... C4 85
Benedict, York, Nebr., 170 ... C8 103
Benedict, McLean, N. Dak., 129 ... B4 110
Benedict, Lee, Va., 30 ... B2 121
Benedict, mtn., Newf ... G10 75
Benedicta, Aroostook, Maine, 50 (200▲) ... C4 96
Benenitra, Malag ... h9 49
Beneraird, mtn., Scot ... E4 13
Benesov, Czech., 9,000 ... D3, o18 26
Bénestroff, Fr., 557 ... F6 16
Benevento, It., 40,000 (55,381▲) ... D5 21
Benevolence, Randolph, Ga., 123 ... E2 87
Benewah, co., Idaho, 6,036 ... B2 89
Benezett, Elk, Pa., 200 ... D5 114
Benfeld, Fr., 3,449 ... A3 18
Bengal, Latimer, Okla., 40 ... C6 112
Bengal, reg., India-Pak ... D8 39
Bengal, bay, India ... F8 39
Ben Gardane, Tun., 2,100 ... H12 28, C7 44
Bengasi (Banghāzī), Libya, 69,718 ... C4 43
Bengkalis, Indon., 3,291 ... E2 35
Bengkulu, In., 16,800 ... F2 35
Ben Goi, bay, Viet ... G8 38
Bengough, Sask., Can., 613 ... H3 70
Benguela, Ang., 15,399 ... D1 48
Benguela, dist., Ang., 487,871 ... D1 48
Benguerir, Mor., 4,325 ... H3 30
Benham, Harlan, Ky., 1,874 ... D7 94
Ben Hill, Fulton, Ga. (part of Atlanta) ... B5 87
Ben Hill, co., Ga., 13,633 ... E3 87
Beni, Con. L ... A4 48
Beni, Nig ... D7 45
Beni, dept., Bol., 119,770 ... B3 55
Beni, riv., Bol ... B2 55
Beni Abbes, Alg., 1,427 (12,418▲) ... C4 44
Benicarló, Sp., 10,627 ... B6 20
Benicia, Solano, Calif., 6,070 ... B5 82
Benicito, riv., Bol ... B2 55
Benin, Nig., 53,753 ... E6 45
Benin, bight, Afr ... E5 45
Beni Ounif, Alg., 877 ... C4 44
Benisa, Sp., 6,036 ... C6 20
Beni, Saf, Alg., 10,934 (21,098▲) ... B4 44
Benito, Man., Can., 427 ... D1 71
Benjamin, Knox, Tex., 338 ... C3 118
Benjamin, Utah, Utah, 100 ... C4 119
Benjamin, Constant, Braz., 3,224 ... B3 58
Benjes, Shelby, Tenn ... E8 117
Benkelman, Dundy, Nebr., 1,400 ... D4 103
Benkovac, Yugo., 1,367 ... C2 22
Benld, Macoupin, Ill., 1,848 ... D4 90
Ben Lomond, Sevier, Ark., 157 ... D1 81
Ben Lomond, Santa Cruz, Calif., 1,814 ... C5 82
Benmore, head, Ire ... B5 11
Bennane, head, Scot ... E3 13
Benndale, George, Miss., 450 ... E5 100
Bennet, Lancaster, Nebr., 381 ... D9, F2 103
Bennett, B.C., Can ... E6 66
Bennett, Adams, Colo., 287 ... B6 83
Bennett, Cedar, Iowa, 374 ... C7 92
Bennett, Lea., N. Mex., 150 ... E6 101
Bennett, Chatham, N.C., 222 ... B4 109
Bennett, Douglas, Wis., 50 ... B2 124
Bennett, co., S. Dak., 3,053 ... D4 116
Bennett, creek, Md ... B3 85
Bennett, isl., Sov. Un ... B17 28
Bennett, lake, Man., Can ... C3 71
Bennette, butte, Oreg ... E2 113
Bennetts, Miami, Ind., 130 ... C5 91
Bennettsbridge, Ire., 325 ... E4 11
Bennettsville, Marlboro, S.C., 6,963 ... B8 115
Bennetsville Southwest, Marlboro, S.C., 1,022 ... *B8 115
Benningen, Switz., 7,864 ... A4 19
Bennington, Bear Lake, Idaho, 100 ... G7 89
Bennington, Ottawa, Kans., 535 ... C6 93
Bennington, Douglas, Nebr., 341 ... D3 103
Bennington, Hillsboro, N.H., 400 (591▲) ... D3 105
Bennington, Bryan, Okla., 226 ... D5 112
Bennington, Bennington, Vt., 8,023 (13,002▲) ... F2 120
Bennington, co., Vt., 25,088 ... E2 120
Bennis Church, Isle of Wight, Va ... B6 121
Bennt Jbail, Leb., 4,000 ... A7 32
Benoit, Bolivar, Miss., 453 ... B2 100
Benoit, Bayfield, Wis., 30 ... B2 124
Benoni, S. Afr., 122,502 ... *C4 49
Benoud, Alg ... C5 44
Bénoué, riv., Cam ... D2 46
Benque Viejo, Br., Hond., 1,561 ... B3 62
Bens, pt., N.Y ... E8 84
Bensané, Guinea ... D2 45
Bensberg, Ger., 30,000 ... C2 17
Bensenville, Du Page, Ill., 11,057 ... B6, E2 90
Bensheim, Ger., 24,100 ... D3 17
Benson, Cochise, Ariz., 2,494 ... F5 80
Benson, Sask., Can., 137 ... H4 70
Benson, Woodford, Ill., 427 ... C4 90
Benson, De Soto, La., 100 ... C2 95
Benson, Harford, Md., 325 ... A5 85
Benson, Swift, Minn., 3,678 ... E3 99
Benson, Johnston, N.C., 2,355 ... B5 109
Benson, Somerset, Pa., 350 ... *F4 114
Benson, Rutland, Vt., 60 (549▲) ... D2 120
Benson, co., N. Dak., 9,435 ... B6 110
Benson Mines, St. Lawrence, N.Y., 100 ... A5 108
Bens Run, Tyler, W. Va., 200 ... B3 123
Bent, Otero, N. Mex., 10 ... D4 107
Bent, co., Colo., 7,419 ... D7 83
Bentheim, Ger., 6,700 ... A2 17
Bentiu, Sud ... D2 47
Bentley, Alta., Can., 588 ... C3 69
Bentley, Pottawattamie, Iowa, 50 ... C2 92
Bentley, Sedgwick, Kans., 204 ... B5, E6 93
Bentley, Grant, La., 200 ... C3 95
Bentley, Bay, Mich., 100 ... E6 98
Bentley, Hettinger, N. Dak., 70 ... C3 110
Bentley Creek, Bradford, Pa., 120 ... C8 114
Bentley Springs, Baltimore, Md., 120 ... A4 85
Bentleyville, Washington, Pa., 3,160 ... F1 114
Benton, Saline, Ark., 10,399 ... C3 81
Benton, Mono, Calif., 100 ... D4 82
Benton, N.B., Can., 184 ... C2 74
Benton, Franklin, Ill., 7,023 ... E5 90
Benton, Elkhart, Ind., 250 ... A6 91
Benton, Ringgold, Iowa, 84 ... D3 92
Benton, Butler, Kans., 452 ... B6, E6 93
Benton, Marshall, Ky., 3,074 ... B3 94
Benton, Bossier, La., 1,336 ... B2 95
Benton, Kennebec, Maine, 100 (1,521▲) ... D3 96
Benton, Yazoo, Miss., 100 ... C3 100
Benton, Scott, Mo., 554 ... D8 101
Benton, Grafton, N.H., 35 (172▲) ... B3 105
Benton, Columbia, Pa., 981 ... D9 114
Benton, Polk, Tenn., 638 ... D9 117
Benton, Atascosa, Tex ... B3 118
Benton, Lafayette, Wis., 837 ... F3 124
Benton, co., Ark., 36,272 ... A1 81
Benton, co., Ind., 11,912 ... C3 91
Benton, co., Iowa, 23,422 ... B5 92
Benton, co., Minn., 17,287 ... E4 99
Benton, co., Miss., 7,723 ... A4 100
Benton, co., Mo., 8,737 ... C4 101
Benton, co., Oreg., 39,165 ... C3 113
Benton, co., Tenn., 10,662 ... A3 117
Benton, co., Wash., 62,070 ... C6 122
Benton City, Audrain, Mo., 155 ... B6 101
Benton City, Benton, Wash., 1,210 ... C6 122
Benton Harbor, Berrien, Mich., 19,136 ... F4 98
Benton Heights, Berrien, Mich., 6,112 ... *F4 98
Bentonia, Yazoo, Miss., 511 ... C3 100
Benton Ridge, Hancock, Ohio, 325 ... A4 111
Benton Station, Alta., Can., 100 ... D5 69
Bentonville, Benton, Ark., 4,015 ... A1 81
Bentonville, Adams, Ohio, 250 ... D4 111
Bentonville, Warren, Va., 350 ... C4 121
Bentree, Nicholas, W. Va ... C7 123
Benue, riv., Nig ... E6 45
Benwee, head, Ire ... C2 11
Benwood, Marshall, W. Va., 2,850 ... A4, B2 123
Benzie, co., Mich., 7,834 ... D4 98
Benzonia, Benzie, Mich., 407 ... D4 98
Beo, Indon ... E7 35
Beograd, see Belgrade, Yugo.
Béoumi, I.C., 3,200 ... E3 45
Beowawe, Eureka, Nev., 200 ... C5 104
Beppu, Jap., 107,734 ... J5 37
Berach, riv., India ... E5 40
Berat, Alb., 15,700 ... B2 23
Berat, pref., Alb., 176,000 ... *B2 23
Berber, Sud., 10,977 ... B3 47
Berbera, Som., 20,000 ... C6 47
Berbérati, Cen. Afr. Rep ... E3 46
Berberia, isl., P.R ... D5 65
Berbice, riv., Br. Gu ... A3 59
Berceto, It., 6,860 ... E6 18
Bercher, Switz., 407 ... C2 19
Berchogur, Sov. Un ... D5 29
Berchtesgaden, Ger., 4,800 ... E6 16
Berck-sur-Mer, Fr., 12,877 ... B4 14
Berclair, Leflore, Miss., 75 ... B3 100
Berclair, Shelby, Tenn. (part of Memphis) ... E8 117
Berclair, Goliad, Tex., 280 ... E4 118
Berdichev, Sov. Un., 56,000 ... G7 27
Berdsk, Sov. Un., 30,000 ... E25 9
Berea, Madison, Ky., 4,302 ... C5 94
Berea, Box Butte, Nebr., 23 ... B3 103
Berea, Barnes, N. Dak., 10 ... C7 110
Berea, Cuyahoga, Ohio, 16,592 ... A6 111
Berebere, Indon ... E7 35
Beregovo, Sov. Un., 5,000 ... G4 27
Bereku, Tan ... B6 48
Berenice, Butte, Idaho ... F6 89
Berenice, ruins, Eg., U.A.R ... E7 43
Berens, isl., Man., Can ... C3 71
Berens, riv., Man., Can ... C3 71
Berens River, Man., Can., 129 ... C3, D5 71
Beresford, Man., Can., 50 ... E1 71
Beresford, Lincoln and Union, S. Dak., 1,794 ... D9 116
Berettyoujfalu, Hung., 11,377 ... B5 22
Berezhany, Sov. Un., 5,000 ... G5 27
Berezina, riv., Sov. Un ... E7 27
Berezna, Sov. Un., 10,000 ... F8 27
Berezniki, Sov. Un., 117,000 ... B5 29
Berezovo, Sov. Un ... C21 9
Berezovskiy, Sov. Un., 28,300 ... D21 9
Berg, Nor ... q27 25
Berga, Sp., 9,822 ... A6 20
Berga, Swe ... B8 24
Bergama, Tur., 21,800 ... C6 23
Bergamasque Alps, mts., It ... C5 18
Bergamo, It., 114,907 (★265,000) ... B2 21
Bergedorf, Ger. (part of Hamburg) ... B5 16
Bergen, see Mons, Bel.
Bergen, Ger., 10,500 ... A6 16
Bergen [bei Celle] (Bergen-Belsen), Ger., 5,300 ... F3 24
Bergen, Genesee, N.Y., 964 ... B3 108
Bergen, McHenry, N. Dak., 52 ... A5 110
Bergen, Nor., 115,900 (★205,000) ... G1 25
Bergen, co., N.J., 780,255 ... B4 106
Bergen aan Zee, Neth., 221 ... B4 15
Bergenfield, Bergen, N.J., 27,203 ... B5, D5 106
Bergen op Zoom, Neth., 35,500 ... C4 15
Berger, Franklin, Mo., 187 ... C6 101
Bergerac, Fr., 21,236 ... E4 14
Bergerville, Que., Can., 455 ... C9 73
Bergheim, Kendall, Tex., 25 ... A3 118
Bergholz, Jefferson, Ohio, 955 ... B7 111
Bergisch Gladbach, Ger., 41,900 ... C2 17
Bergland, Ont., Can ... B4 99
Bergland, Ontonagon, Mich., 600 ... A5 98
Bergman, Boone, Ark., 100 ... A2 81
Bergoo, Webster, W. Va., 460 ... C4 123
Bergton, Rockingham, Va., 100 ... C4 121
Berguent, Mor., 2,607 ... C4 44
Bergün, Switz., 551 ... C8 19
Bergzabern, Ger., 5,300 ... D3 17
Berhampore, India, 76,931 ... E12 40
Berhampur, India, 62,317 ... H10 40
Bering, sea, Alsk ... D4 79
Bering, strait, Alsk ... C5 79
Berino, Dona Ana, N. Mex., 10 ... E3 107
Berislav, Sov. Un., 10,000 ... H9 27
Berja, Sp., 6,425 ... D4 20
Berkeley, Alameda, Calif., 111,268 ... B5, D2 82
Berkeley, Ont., Can., 116 ... C4 72
Berkeley, Cook, Ill., 5,792 ... *F2 90
Berkeley, St. Louis, Mo., 18,676 ... A8 101
Berkeley, Providence, R.I., 1,000 ... B11 84
Berkeley, Berkeley, W. Va., 200 ... B7 123
Berkeley, co., S.C., 38,196 ... E8 115
Berkeley, co., W. Va., 33,791 ... B6 123
Berkeley, cape, Ec ... g5 58
Berkeley Heights, Union, N.J., 8,721 ... B4 106
Berkeley Springs (Bath), Morgan, W. Va., 1,138 ... B6 123
Berkey, Lucas, Ohio, 257 ... A1, A4 111
Berkley, Boone, Iowa, 58 ... C3 92
Berkley, Harford, Md., 48 ... A5 85
Berkley, Bristol, Mass., 130 (1,609▲) ... C5 97
Berkley, Oakland, Mich., 23,275 ... A7, F7 98
Berkovitsa, Bul., 6,870 ... D6 22
Berks, co., Pa., 275,414 ... F9 114
Berkshire, Prince Georges, Md., 1,200 ... *C4 85
Berkshire, Berkshire, Mass., 300 ... A1 97
Berkshire, Tioga, N.Y., 450 ... C4 108
Berkshire, Franklin, Vt., 35 (965▲) ... B3 120
Berkshire, co., Eng., 503,357 ... C6 12
Berkshire, co., Mass., 142,135 ... B1 97
Berkshire, hills, Mass ... B1 97
Berland, riv., Alta., Can ... C1 69
Berleburg, Ger., 6,400 ... B3 17
Berlengas, is., Port ... C1 20
Berlin (P.O.), Hartford, Conn., 3,500 (11,250▲) ... C6 84
Berlin, Colquitt, Ga., 419 ... E3 87
Berlin, Ger., 3,343,200 (★3,900,000) ... B6 16
Berlin, East, Ger., 1,055,300 ... *B6 16
Berlin, West, Ger., 2,197,600 (★4,000,000) ... *B6 16
Berlin, Worcester, Md., 2,046 ... D7 85
Berlin, Worcester, Mass., 450 (1,742▲) ... B4 97
Berlin, Coos, N.H., 17,821 ... B4 105
Berlin, Camden, N.J., 3,578 ... D3 106
Berlin, Rensselaer, N.Y., 800 ... C7 108
Berlin, La Moure, N. Dak., 78 ... C7 110
Berlin, Holmes, Ohio, 300 ... B6 111
Berlin, Roger Mills, Okla., 50 ... B2 112
Berlin, Somerset, Pa., 1,600 ... G4 114
Berlin, Washington, Vt., 100 (1,306▲) ... C3 120
Berlin, Green Lake and Waushara, Wis., 4,838 ... E5 124
Berlin, mtn., Mass ... A1 97
Berlin, mtn., Nev ... E4 104
Berlin, res., Ohio ... A6 111
Berlin Heights, Erie, Ohio, 721 ... A5 111
Bermejo, riv., Arg., Par ... E4 55
Bermeo, Sp., 13,781 ... A4 20
Bermuda, Br. dep., N.A., 53,326 ... D5 6, E14 77
Bermuda Hundred, Chesterfield, Va., 30 ... C7 121
Bern, Nemaha, Kans., 206 ... C8 93
Bern, Switz., 163,172 (★238,000) ... C3 19
Bern, canton, Switz., 889,523 ... C4 19
Bernalillo, Sandoval, N. Mex., 2,574 ... B3, G5 107
Bernalillo, co., N. Mex., 262,199 ... C3 107
Bernard, Sask., Can ... G2 70
Bernard, Dubuque, Iowa, 173 ... B7 92
Bernard, Hancock, Maine, 200 ... D4 96
Bernard, isl., Truk ... 52
Bernard, lake, Ont., Can ... B5 72
Bernardo, Socorro, N. Mex ... C3 107
Bernardston, Franklin, Mass., 600 (1,370▲) ... A2 97
Bernardsville, Somerset, N.J., 5,515 ... B3 106
Bernasconi, Arg., 2,094 ... B4 54
Bernau, Ger., 13,600 ... B6 16
Bernau, Ger., 1,300 ... B8 18
Bernay, Fr., 7,418 ... C4 14
Bernburg, Ger., 44,500 ... B6 17
Berndorf, Aus., 8,992 ... E8 16
Berne, Adams, Ind., 2,644 ... C8 91
Berne, Albany, N.Y., 330 ... C6 108
Berneray, isl., Scot ... C1 13
Bernese Alps, mts., Switz ... D3 19
Bernice, Union, La., 1,641 ... B3 95
Bernice, Delaware, Okla., 100 ... A7 112
Bernie, Stoddard, Mo., 1,578 ... E8 101
Bernina, pass, Switz ... D9 19
Bernina, peak, Switz ... D8 19
Bernstadt, Laurel, Ky., 425 ... C5 94
Bernville, Berks, Pa., 884 ... F9 114
Beroroha, Malag ... h9 49
Beroun, Czech., 15,600 ... D3, o17 26
Beroun, Pine, Minn., 100 ... C6 99
Berounka, riv., Czech ... o17 26
Berre, Fr., 8,677 ... F6 14
Berrechid, Mor., 13,780 ... C3 44
Berrien, co., Ga., 12,038 ... E3 87
Berrien, co., Mich., 149,865 ... G4 98
Berrien Springs, Berrien, Mich., 1,953 ... G4 98
Berry, Fayette, Ala., 645 ... B2 78
Berry, Harrison, Ky., 279 ... B5 94
Berry, former prov., Fr., 464,000 ... D4 14
Berrydale, Santa Rosa, Fla., 25 ... G2 86
Berryessa Lake, res., Calif ... C2 82
Berry Hill, Davidson, Tenn., 1,551 ... A5 117
Berryman, Crawford, Mo., 150 ... D6 101
Berry Mills, N.B., Can., 318 ... C5 74
Berry Mills, Franklin, Maine, 75 ... D2 96
Berryton, Chattooga, Ga., 300 ... B1 87
Berryville, Carroll, Ark., 1,999 ... A2 81
Berryville, Clarke, Va., 1,645 ... B5 121
Bersenbrück, Ger., 3,600 ... B7 15
Bershad, Sov. Un., 10,000 ... G7 27
Bertha, Todd, Minn., 562 ... D3 99
Berthier, co., Que., Can., 27,325 ... C3 73
Berthierville, Que., Can., 3,708 ... C4 73
Berthold, Ward, N. Dak., 431 ... A4 110
Berthoud, Larimer, Colo., 1,014 ... A5 83
Berthoud, pass, Colo ... B5 83
Bertie, Bertie, N.C., 303 ... B7 109
Bertie, co., N.C., 24,350 ... A6 109
Bertraghboy, bay, Ire ... D2 11
Bertram, Burnet, Tex., 850 ... D3 118
Bertrand, Berrien, Mich., 3,500 ... G4 98
Bertrand, Mississippi, Mo., 465 ... E8 101
Bertrand, Phelps, Nebr., 691 ... D6 103
Bertrandville, Plaquemines, La., 100 ... C7, E5 95
Bertrix, Bel., 4,466 ... E5 15
Berwick, N.S., Can., 1,282 ... D5 74
Berwick, Warren, Ill., 160 ... C3 90
Berwick, Polk, Iowa, 150 ... A7 92
Berwick, St. Mary, La., 3,880 ... C5, E4 95
Berwick, York, Maine, 1,557 (2,738▲) ... E2 96
Berwick, McHenry, N. Dak., 56 ... A5 110
Berwick, Columbia, Pa., 13,353 ... D9 114
Berwick, co., Scot., 22,441 ... E6 13
Berwick-on-Tweed, Eng., 12,166 ... C6 10
Berwind, McDowell, W. Va., 950 ... D3 123
Berwyn, Alta., Can., 347 ... A2 69
Berwyn, Cook, Ill., 54,224 ... F2 90
Berwyn, Custer, Nebr., 104 ... C6 103
Berwyn, Chester, Pa., 5,000 ... A10 114
Berwyn Heights, Prince Georges, Md., 2,376 ... B4 85
Besa, Indon ... E7 35
Besançon, Fr., 95,642 ... D7 14, B2 18
Beskids, mts., Czech., Pol ... D5 26
Besnard, lake, Sask., Can ... B3 70
Besalampy, Malag ... g8 49
Besni, Tur., 11,200 ... D11 31
Bessèges, Fr., 5,770 ... E6 14
Bessemer, Jefferson, Ala., 33,054 ... B3, E4 78
Bessemer, Gogebic, Mich., 3,304 ... A5 98
Bessemer, Lawrence, Pa., 1,491 ... E1 114
Bessemer City, Gaston, N.C., 4,017 ... B2 109
Bessie, Jackson, N.C ... D3 109
Bessie, Washita, Okla., 226 ... B3 112
Bessmay, Jasper, Tex., 250 ... D6 118
Best, Reagan, Tex., 30 ... D2 118
Bestobe, Sov. Un ... C8 29
Bestwater, Benton, Ark ... A1 81
Betafo, Malag ... g9 49
Betanzos, Sp., 7,561 ... A1 20
Bétaré Oya, Cam ... D2 46
Bethalto, Madison, Ill., 3,235 ... E3 90
Bethanie, S.W. Afr., 1,053 ... C2 49
Bethany, Ont., Can., 279 ... C6 72
Bethany, New Haven, Conn., 2,384 ... D5 84
Bethany, Moultrie, Ill., 1,118 ... D5 90
Bethany, see Al 'Ayzariyah, Jordan
Bethany, Caddo, La., 160 ... B1 95
Bethany, Harrison, Mo., 2,771 ... A3 101
Bethany, Butler, Ohio, 165 ... C2 111
Bethany, Oklahoma, Okla., 12,342 ... B4 112
Bethany, Brooke, W. Va., 992 ... B2 123
Bethany Beach, Sussex, Del., 170 ... C7 85
Bethayres-Huntingdon Valley, Montgomery, Pa., 2,500 ... *F11 114
Bethel, Alsk., 1,258 ... C7 79
Bethel, Fairfield, Conn., 8,200 ... D3 84
Bethel, Sussex, Del., 236 ... C6 85
Bethel, Bath, Ky., 200 ... B6 94
Bethel, Oxford, Maine, 1,117 (2,408▲) ... D2 96
Bethel, Anoka, Minn., 302 ... E5 99
Bethel, Shelby, Mo., 152 ... B5 101
Bethel, Pitt, N.C., 1,578 ... B6 109
Bethel, Clermont, Ohio, 2,019 ... D3 111
Bethel, McCurtain, Okla., 150 ... C7 112
Bethel, Lane, Oreg., 1,500 ... *C3 113
Bethel (Bethel Park), Allegheny, Pa., 23,650 ... B5 114
Bethel, Berks, Pa., 450 ... F9 114
Bethel, Windsor, Vt., 100 (1,356▲) ... D3 120
Bethel Springs, McNairy, Tenn., 533 ... B3 117
Bethera, Berkeley, S.C., 165 ... E8 115
Bethesda, Independence, Ark., 115 ... B4 81
Bethesda, Montgomery, Md., 56,527 ... C1, C3 85
Bethesda, Belmont, Ohio, 1,178 ... B6 111
Bethesda, Wales, 4,151 ... A3 12
Bethevan, Grant, Ind ... C6 91
Bethlehem, Litchfield, Conn., 500 (1,486▲) ... C4 84
Bethlehem, Barrow, Ga., 297 ... C3 87
Bethlehem, Clark, Ind., 150 ... G7 91
Bethlehem, see Bayt Laḥm, Jordan
Bethlehem, Caroline, Md., 85 ... C6 85
Bethlehem, Grafton, N.H., 450 (898▲) ... B3 105
Bethlehem, Lehigh and Northampton, Pa., 75,408 ... E11 114
Bethlehem, S. Afr., 24,125 ... C4 49
Bethlehem, Ohio, W. Va., 2,308 ... B2 123
Bethpage, Nassau, N.Y., 30,000 ... G3 84
Bethpage, Sumner, Tenn., 400 ... A5 117
Bethulie, S. Afr., 3,488 ... D4 49
Bethune, Sask., Can., 339 ... G3 70
Bethune, Kit Carson, Colo., 70 ... B8 83
Béthune, Fr., 23,445 (★45,000) ... D2 15
Bethune, Kershaw, S.C., 579 ... C7 115
Betioky, Malag., 755 ... h8 49
Betoota, Austl ... B3 51
Betpak-Dala, Sov. Un ... B8 29
Betpak-Dala (Golodnaya Steppe), steppe, Sov. Un ... D7 29
Betroka, Malag., 2,524 ... h9 49
Betsiamites, Que., Can ... A2 73
Betsiboka, riv., Malag ... g9 49
Betsie, pt., Mich ... D4 98
Betsy Layne, Floyd, Ky., 800 ... C7 94
Bette, peak, Libya ... E3 43
Bette, range, Nev ... C3 104
Bettendorf, Scott, Iowa, 11,534 ... C7, D7 92
Betteravia, Santa Barbara, Calif., 335 ... E3 82
Betterton, Kent, Md., 328 ... B5 85
Bettiah, India, 39,990 ... D10 40
Bettles (Bettles Field), Alsk., 77 ... B9 79
Bettsville, Seneca, Ohio, 776 ... A4 111
Bettyhill, Scot ... B4 13
Betul, India, 19,860 ... G6 40
Betwa, riv., India ... E7 40
Betzdorf, Ger., 10,100 ... C2 17
Beuel, Ger., 31,800 ... C2 17
Beulah, Lee, Ala., 200 ... C4 78
Beulah, Man., Can., 80 ... D1 71
Beulah, Pueblo, Colo., 500 ... C6 83
Beulah, Benzie, Mich., 436 ... D4 98
Beulah, Bolivar, Miss., 421 ... B3 100
Beulah, Mercer, N. Dak., 1,318 ... B4 110
Beulah, Malheur, Oreg ... D8 113
Beulah, Crook, Wyo., 45 ... A8 125
Beulah, lake, Miss ... B2 100
Beulaville, Duplin, N.C., 1,062 ... C6 109
Bevans, Sussex, N.J., 75 ... A3 106
Beverley, Eng., 16,024 ... D6 10
Beverley Station, Sask., Can., 92 ... G2 70
Beverly, Alta., Can., 9,041 ... C4 69
Beverly, Elbert, Ga., 132 ... B4 87
Beverly, Lincoln, Kans., 199 ... C6 93
Beverly, Bell, Ky., 30 ... D6 94
Beverly, Essex, Mass., 36,108 ... A6, C3 97
Beverly, Burlington, N.J., 3,400 ... C3 106
Beverly, Washington, Ohio, 1,194 ... C6 111
Beverly, Knox, Tenn., 700 ... E11 117

Beverly, Randolph, W. Va., 441 C5 123
Beverly Farms, Essex, Mass. (part of Beverly) C4 97
Beverly Gardens, Montgomery, Ohio, 2,200 *C3 111
Beverly Hills, Los Angeles, Calif., 30,817 F2 82
Beverly Hills, Oakland, Mich., 8,633 *F7 98
Beverly Park, Snohomish, Wash., 950 B3 122
Beverly Shores, Porter, Ind., 773 A4 91
Beverstedt, Ger., 1,900 E2 24
Beverwijk, Neth., 36,600 B4 15
Bevier, Macon, Mo., 781 B5 101
Bewdley, Ont., Can., 300 C6 72
Bex, Switz., 4,661 D3 19
Bexar, Marion, Ala., 200 A1 78
Bexar, co., Tex., 687,151 E3 118
Bexhill-on-Sea, Eng., 28,926 D8 12
Bexley, Eng., 89,629 C8 12
Bexley, George, Miss., 10 E5 100
Bexley, Franklin, Ohio, 14,319 C2 111
Beya, Sov. Un., 82,331 E27 9
Beyer, Indiana, Pa., 200 E3 114
Beykoz, Tur., 45,800 (part of Istanbul) B7 23
Beyla, Guinea, 6,100 E3 45
Beypazari, Tur., 8,900 B8 31
Beyram, Iran H6 41
Beyrouth, see Beirut, Leb.
Beysehir, Tur., 5,900 D8 31
Beysehir, lake, Tur D8 31
Bezau, Aus., 1,484 B5 18
Bezerros, Braz., 7,737 C3, k6 57
Bezhetsk, Sov. Un., 25,000 C11 27
Bezhitsa, Sov. Un., 85,000 E10 27
Béziers, Fr., 73,538 F5 14
Bezmer, Bul., 2,452 D8 22
Bezons, Fr., 22,061 g9 14
Bezwada, see Vijayavada, India
Bhadra, India, 10,000 C5 40
Bhadrachalam, India I8 40
Bhadrakh, India, 25,285 G11 40
Bhag, Pak C4 39
Bhagalpur, India, 143,850 C8 39, E11 40
Bhakkar, Pak., 12,397 B5 39
Bhamo, Bur., 9,817 D10 39
Bhandara, India, 27,710 G7 40
Bharatpur, India, 49,776 D6 40
Bharatpur, India F8 40
Bhatinda, India, 52,253 B5 40
Bhatpara, India, 147,630 D8 39, F12 40
Bhavnagar, India, 171,039 (★176,473) D5 39, G4 40
Bhawani-Patna, India, 14,300 H9 40
Bheigeir, mtn., Scot E2 13
Bhilsa, India, 10,000 F6 40
Bhima, riv., India E6 39
Bhind, India, 28,208 D7 40
Bhiwani, India, 58,194 C6 40
Bhola, Pak., 6,198 F13 40
Bhopal, India, 185,374 (★222,948) D6 39, F6 40
Bhopal, state, India, 836,474 *F6 40
Bhopalpatnam, India H8 40
Bhor, India, 8,627 H4 40
Bhreac, mtn., Scot D3 13
Bhubuneswar, India, 38,211 D8 39
Bhuj, India, 40,180 F2 40
Bhusaval, India, 73,994 (★79,121) G5 40
Bhutan, country, Asia, 300,000 G12 33, C9 39
Biābānak, Iran, 6,000 E7 41
Biaboye, Con. L A4 48
Bia fra, bight, Afr E1 46
Biak, isl., W. Irian F9 35
Biala, Eg., U.A.R., 10,005 C3 32
Biala, riv., Pol h10 26
Biala Podlaska, Pol., 15,300 B7 26
Biala Przemsza, riv., Pol g11 26
Bialogard, Pol., 15,200 A4 26
Bialowieza, Pol., 1,000 B7 26
Bialystok, Pol., 95,000 B7 26
Bianco, It., 5,057 E6 21
Biang, Indon E7 35
Biarritz, Fr., 25,231 F3 14
Biasca, Switz., 3,349 D6 19
Bibā, Eg., U.A.R., 22,100 D6 43
Bibai, Jap., 62,500 (87,345▲) E10 37
Bibb, co., Ala., 14,357 C2 78
Bibb, co., Ga., 141,249 D3 87
Bibb City, Muscogee, Ga., 1,213 D2 87
Biberach, Ger., 21,500 D4 16
Biberist, Switz., 7,188 B4 19
Bibi, New Gui k12 50
Bible Grove, Clay, Ill., 250 E5 90
Bibo, Valencia, N. Mex B2 107
Bic, Que., Can., 1,177 A9 73
Bic, isl., Que., Can A9 73
Bicester, Eng., 5,513 C6 12
Biche, lake, Alta., Can B4 69
Bichl, Ger., 1,300 B7 18
Bickerdike, Alta., Can., 210 C2 69
Bickerton, cape, Ant C26 5
Bickett, Knob, mtn., W. Va D4 123
Bickle Knob, mtn., W. Va C5 123
Bickleton, Klickitat, Wash., 100 D5 122
Bickmore, Clay, W. Va., 130 C7 123
Bicknell, Knox, Ind., 3,878 G3 91
Bicknell, Wayne, Utah, 366 E4 119
Bicske, Hung., 8,148 B4 22
Bida, Nig., 19,346 E6 45
Bidar, India, 32,420 I6 40
Biddeford, York, Maine, 19,255 E2, E4 96
Biddiyā, Jordan, 2,000 g11 32
Biddle, Powder River, Mont., 25 E11 102
Bidean Nam Bian, mtn., Scot D3 13
Bideford, Eng., 10,265 C3 12
Bideford, bay, Eng C3 12
BidonCinq, Alg., 2 E5 44
Bidwell, Gallia, Ohio, 500 D5 111
Bieber, Lassen, Calif., 300 B3 82
Biebrza, riv., Pol B7 26
Bie Cuando Cubango, dist., Ang., 567,253 D2 48
Biedenkopf, Ger., 6,700 C3 17
Biel, Switz., 59,216 (★70,150) B3 19
Biel, Switz, 102 D5 19
Biel (Bienne), lake, Switz B3 19
Bielawa, Pol., 28,000 C4 26
Bield, Man., Can., 35 D1 71
Bielefeld, Ger., 174,600 (★290,000) A3 17
Biella, It., 50,209 B2 21
Bielsko-Biala, Pol., 76,000 D5, h10 26
Bielsk Podlaski, Pol., 6,203 B7 26
Bienfait, Sask., Can., 842 H4 70
Bien Hoa, Viet., 5,000 G7 38
Bienne, see Biel, Switz.
Bienville, Bienville, La., 305 B3 95
Bienville, par., La., 16,726 B2 95
Bienville, lake, Que., Can B2 73
Bierawka, riv Pol g9 26
Biere, Switz., 1,166 C1 19
Bierun Nowy, Pol g10 26
Bierun Stary, Pol., 3,702 g10 26
Biese, riv., Ger F5 24
Bietigheim, Ger., 16,600 E4 17
Bievre, riv., Fr g9 14
Bièvres, Fr., 2,712 g9 14
Biferno, riv., It D5 21
Big, bayou, Ark D4 81
Big, creek, Ark C5 81
Big, creek, B.C., Can D6 68
Big, creek, Ind G7 91
Big, creek, Ind H2 91
Big, creek, Kans D4 93
Big, creek, La B4 95
Big, creek, Mo A3 101
Big, creek, Mo C3 101
Big, creek, Tenn E8 117
Big, isl., N.W. Ter., Can D18 67
Big, lake, Maine C5 96
Big, riv., Sask., Can D2 70
Big, riv., Mo C7 101
Biga, Tur., 10,800 B6 23
Big Arm, Lake, Mont., 60 C2 102
Big Bald, mtn., N.B., Can B3 74
Big Bald, mtn., Ga B2 87
Big Bald, mtn., Tenn D11 117
Big Bay, Marquette, Mich., 250 B3 98
Big Bay de Noc, bay, Mich C4 98
Big Bear City, San Bernardino, Calif., 800 E5 82
Big Bear Lake, San Bernardino, Calif., 1,562 *E5 82
Big Beaver, Sask., Can., 144 H3 70
Big Beaver, Beaver, Pa., 2,381 *E1 114
Bigbee, Monroe, Miss., 75 A5 100
Bigbee Valley, Noxubee, Miss., 70 B5 100
Big Belt, mts., Mont D5 102
Big Bend, Shasta, Calif., 230 B3 82
Big Bend, McLean, N. Dak., 39 B4 110
Big Bend, Waukesha, Wis., 797 F1 124
Big Bend, nat. park, Tex E1 118
Big Birch, lake, Minn E4 99
Big Black, riv., Miss C3 100
Big Blue, riv., Ind E6 91
Big Blue, riv., Kans C7 93
Big Blue, riv., Nebr D9 103
Bigbone, Boone, Ky., 50 A6 94
Big Bow, Stanton, Kans., 100 E2 93
Big Bushkill, creek, Pa A2 106
Big Bull, mtn., Tenn C11 117
Big Cabin, Craig, Okla., 228 A6 112
Big Cabin, creek, Okla A6 112
Big Cane, St. Landry, La., 100 D4 95
Big Canyon, Murray, Okla., 175 C4 112
Big Canyon, creek, Tex D1 118
Big Clifty, Grayson, Ky., 550 C3 94
Big Corney, creek, Ark D3 81
Big Cow, creek, Kans D5 93
Big Creek, Fresno, Calif., 400 D4 82
Big Creek, B.C., Can., 149 D6 68
Big Creek, Valley, Idaho, 34 D3 89
Big Creek, Clay, Ky., 250 C6 94
Bigcreek, Sullivan, Tenn C11 117
Big Creek, Logan, W. Va., 500 D2 123
Big Cypress, swamp, Fla F5 86
Big Delta, Alsk., 20 C10 79
Big Dry, creek, Mont C10 102
Big Duke, dam, N.C B3 109
Big Eau Pleine, res., Wis D3 124
Big Eddy, Wasco, Oreg B5 113
Bigej, isl., Kwajalein 52
Big Elk, creek, Md A6 85
Bigelow, Perry, Ark., 231 B3 81
Bigelow, Nobles, Minn., 256 G3 99
Bigelow, Holt, Mo., 100 A2 101
Bigelow, mtn., Maine C2 96
Big Falls, Koochiching, Minn., 526 B5 99
Bigflat, Baxter, Ark., 217 A3 81
Big Flats, Chemung, N.Y., 900 C4 108
Bigfoot, Frio, Tex., 100 E3 118
Bigfork, Itasca, Minn., 464 C5 99
Bigfork, Flathead, Mont., 500 B2 102
Big Fork, riv., Minn B5 99
Biggar, Sask., Can., 2,702 E1 70
Biggar, Scot., 1,403 E5 13
Biggerann, isl., Kwajalein 52
Biggers, Randolph, Ark., 274 A5 81
Biggersville, Alcorn, Miss., 135 A5 100
Biggleswade, Eng., 8,047 B7 12
Biggs, Butte, Calif., 831 C3 82
Biggs, Sherman, Oreg., 10 B6 113
Biggsville, Henderson, Ill., 345 C3 90
Big Gully, creek, Sask., Can D1 70
Big Hickory, mtn., Pa C11 114
Big Hickory, pass, Fla F5 86
Big Hole, pass, Idaho, Mont D5 89
Big Hole, peak, Mont C1 102
Big Hole, riv., Mont E4 102
Big Hole Battlefield, nat. mon., Mont E3 102
Bighorn, Treasure, Mont D9 102
Big Horn, Sheridan, Wyo., 100 A5 125
Big Horn, co., Mont., 10,007 E9 102
Big Horn, co., Wyo., 11,898 A4 125
Big Horn, basin, Wyo A4 125
Big Horn, mts., Ariz D2 80
Big Horn, mts., Wyo A5 125
Big Horn, riv., Mont., Wyo A4 125
Big Hurricane, cavern, Ark A3 81
Bigi, isl., Kwajalein 52
Big Indian, riv., Mich B4 98
Big Island, Rapides, La C3 95
Big Island, Bedford, Va., 500 D3 121
Big Kandiyohi, lake, Minn F4 99
Big Knob, mtn., Va B2 121
Big Lake, Sherburne, Minn., 610 E5 99
Big Lake, Reagan, Tex., 2,668 D2 118
Bigler, Clearfield, Pa., 500 E5 114
Biglerville, Adams, Pa., 923 G7 114
Big Lick, Cumberland, Tenn D8 117
Big Lost, riv., Idaho F5 89
Big Marco, bay, Fla G5 86
Big Marine, lake, Minn E7 99
Big Maumelle Lake, res., Ark C3 81
Big Monon, creek, Ind C4 91
Big Moose, Herkimer, N.Y., 125 B6 108
Big Muddy, creek, Mont B12 102
Big Muddy, lake, Sask., Can H3 70
Big Muddy, riv., Ill F4 90
Bignona, Sen., 2,450 D1 45
Bigonville, Lux., 491 E5 15
Big Payette, lake, Idaho E2 89
Big Pine, Inyo, Calif., 556 D4 82
Big Pine, Madison, N.C D3 109
Big Pine, hill, Pa B9 114
Big Pine, mtn., Calif E4 82
Big Piney, Pulaski, Mo., 55 D5 101
Big Piney, Sublette, Wyo., 663 C2 125
Big Pipe, creek, Md A3 85
Bigpoint, Jackson, Miss., 100 E5 100
Bigpool, Washington, Md., 50 A1 85
Big Prairie, Holmes, Ohio, 200 B5 110
Big Rapids, Mecosta, Mich., 8,686 E5 98
Big River, Sask., Can., 896 D2 70
Big Rock, Kane, Ill., 300 B5 90
Big Rock, Stewart, Tenn., 220 A4 117
Big Rock, mtn., Ark D6 81
Big Run, Jefferson, Pa., 857 E4 114
Big Sable, pt., Mich D4 98
Big Sand, creek, Mont B6 102
Big Sandy, Chouteau, Mont., 954 B6 102
Big Sandy, Benton, Tenn., 492 A3 117
Big Sandy, Upshur, Tex., 848 C5 118
Big Sandy, Sublette, Wyo C3 125
Big Sandy, creek, Colo C8 83
Big Sandy, lake, Sask., Can C3 70
Big Sandy, res., Wyo C3 125
Big Sandy, riv., Ariz C2 80
Big Sandy, riv., Ky., W. Va B7 94
Big Sandy, riv., Tenn B3 117
Big Satilla, creek, Ga E4 87
Big Savage, mtn., Md D2 85
Bigsby, isl., Ont., Can A4 99
Big Shiney, mtn., Pa B9 114
Big Sioux, riv., Iowa, S. Dak B1 92, E9 116
Big Slough, creek, Kans B5 93
Big Smoky, valley, Nev E4 104
Big Snow, mtn., Mont D7 102
Big Soldier, creek, Kans C8 93
Big South, butte, Idaho F5 89
Big Spring, Breckinridge, Ky., 150 C3 94
Big Spring, Washington, Md., 110 A2 85
Big Spring, Meigs, Tenn., 50 D9 117
Big Spring, Howard, Tex., 31,230 C2 118
Big Spring, valley, Ala A3 78
Big Springs, Douglas, Kans., 50 B7 93
Big Springs, Deuel, Nebr., 506 C3 103
Big Springs, Calhoun, W. Va., 50 C3 123
Big Spruce Knob, mtn., W. Va C4 123
Bigstick, lake, Sask., Can G1 70
Big Stone, co., Minn., 8,954 E2 99
Bigstone, lake, Man., Can C4 71
Bigstone, lake, Sask., Can C4 70
Big Stone, lake, Minn E2 99
Big Stone, lake, S. Dak B9 116
Bigstone, riv., Man., Can B4 71
Big Stone City, Grant, S. Dak., 718 B9 116
Big Stone Gap, Wise, Va., 4,688 B2 121
Big Sunflower, riv., Miss B3 100
Big Sur, Monterey, Calif., 125 D3 82
Big Thompson, riv., Colo A5 83
Big Timber, Sweet Grass, Mont., 1,660 E7 102
Big Top, mtn., Tenn B5 117
Big Trout, lake, Ont., Can D8 72
Biguaçu, Braz., 2,184 D3 56
Big Valley, Alta., Can., 461 C4 69
Big Walnut, creek, Ohio C2 111
Big Wells, Dimmit, Tex., 801 E3 118
Bigwood, Ont., Can., 80 A4 72
Big Wood, riv., Idaho F4 89
Bihać, Yugo., 15,552 C2 22
Bihar, India, 78,581 E10 40
Bihar, state, India, 46,455,610 D8 39
Bihar (Hajdu-Bihar), co., Hung *B5 22
Bihor, mtn., Rom B6 22
Biiziri, isl., Eniwetok 52
Bijagós, is., Port Gui D1 45
Bijapur, India, 78,854 I5 40
Bījār, Iran, 12,928 D3 41
Bijeljina, Yugo., 15,682 C4 22
Bijelo Polje, Yugo., 5,807 D3 22
Bijnor, India, 33,821 C7 40
Bijou, El Dorado, Calif., 2,000 C4 82
Bijou, creek, Colo B6 83
Bijou Hills, Brule, S. Dak., 20 D6 116
Bikampur, India, 1,000 D4 40
Bikaner, India, 150,634 C5 39, D4 40
Bikeman, isl., Tarawa 52
Bikin, Sov. Un., 10,000 C7 37
Bikin, riv., Sov. Un C7 37
Bikini, isl., Bikini 52
Bikini, isl., Marshall Is 52
Bikita, S. Rh B5 49
Bilaspur, India, 86,706 F9 40
Bilauktaung, range, Thai F3 38
Bilbao, Sp., 297,942 (★565,000) A4 20
Bilbays, Eg., U.A.R., 29,200 D3 32
Bíldudalur, Ice., 377 n21 25
Bileća, Yugo., 2,489 D4 22
Bilecik, Tur., 7,500 B7 31
Bilgoraj, Pol., 4,745 C7 26
Bili, Con. L A4 48
Bilin, Bur., 5,265 E10 39
Bilina, Czech., 10,400 C8 17
Bill, Converse, Wyo., 10 B7 125
Billerica, Middlesex, Mass., 1,600 (17,864▲) A5, C2 97
Billericay (Basildon), Eng., 88,459 C8 12
Billet, Lawrence, Ill., 70 E6 90
Billings, Christian, Mo., 602 D4 101
Billings, Yellowstone, Mont., 52,851 E8 102
Billings, Noble, Okla., 510 A4 112
Billings, co., N. Dak., 1,513 B2 110
Billings Bench, Yellowstone, Mont., 1,500 *E8 102
Billings Heights, Yellowstone, Mont., 2,500 *E8 102
Billingsley, Autauga, Ala., 179 C3 78
Bill Williams, mtn., Ariz B3 80
Bilma, Niger, 1,100 C7 45
Biloela, Austl., 2,048 B8 51
Biloxi, Harrison, Miss., 44,053 E2, E5 100
Biloxi, bay, Miss E3 100
Biloxi, riv., Miss E4 100
Bilqas Qism Awwal, Eg., U.A.R., 43,200 C3 32
Biltine, Chad, 1,217 C4 46
Biltmore Forest, Buncombe, N.C., 1,004 D3 109
Bilzen, Bel., 6,426 D5 15
Bim, Boone, W. Va., 300 D6 123
Bimberi, peak, Austl G7 51
Bimini, is., Ba. Is E12 77
Binalbagan, Phil., 6,979 *C6 35
Binalonan, Phil., 3,010 n13 35
Binan, Phil., 2,726 o13 35
Bindle, Austl C7 51
Bindloss, Alta., Can., 72 D5 69
Bindura, S. Rh., 2,500 A5 49
Binfield, Blount, Tenn., 100 D9 117
Binford, Griggs, N. Dak., 261 B7 110
Binga, Con. L A3 48
Bingamon, creek, W. Va A6 123
Bingen, Hempstead, Ark., 150 D2 81
Bingen, Ger., 20,200 D2 17
Bingen, Klickitat, Wash., 636 D4 122
Binger, Caddo, Okla., 603 B3 112
Bingham, Fayette, Ill., 122 D4 90
Bingham, Somerset, Maine, 1,180 (1,308▲) C3 96
Bingham, Sheridan, Nebr., 100 B3 103
Bingham, Socorro, N. Mex., 10 D3 80
Bingham, Dillon, S.C., 95 C8 115
Bingham, co., Idaho, 28,218 F6 89
Bingham Canyon, Salt Lake, Utah, 1,516 C3 119
Bingham Lake, Cottonwood, Minn., 254 G3 99
Binghamton, Broome, N.Y., 75,941 (★212,600) C4 108
Binghamville, Franklin, Vt B3 120
Bingley, Eng., 22,308 G6 13
Bingöl (Çapakçur), Tur., 8,500 C13 31
Binh Dinh, Viet F8 38
Binnaway, Austl., 761 E7 51
Binscarth, Man., Can., 456 D1 71
Bintan, is., Indon L6 38
Bintang, mtn., Mala J4 38
Bintuhan, Indon F2 38
Bintulu, Mala., 5,307 E4 35
Binyamina, Isr., 2,550 B6 32
Bío-Bío, prov., Chile, 167,286 B2 54
Bío Bío, riv., Chile B2 54
Bippus, Huntington, Ind., 275 C6 91
Bir, India, 33,066 H5 40
Bira, Sov. Un B6 37
Bi'r Abu Hadd (Well), Eg., U.A.R E6 43
Bir Abu Reida, Eg., U.A.R D4 32
Birakan, Sov. Un., 5,000 B5 37
Bi'r al 'Abd, Eg., U.A.R., 1,000 C5 32
Bi'r al 'Ajramiyah (Oasis), Eg., U.A.R D5 32
Bi'r al 'Aṭash (Well), Libya D4 43
Bi'r al Ghaylānīyah (Well), Libya D2 43
Bi'r al Ghuzayl (Well), Libya D2 43
Bi'r al Hadīrah (Oasis), Eg., U.A.R D6 32
Bi'r al Haysi, Eg., U.A.R E6 32
Bi'r al Jidy (Oasis), Eg., U.A.R D5 32
Bi'r al Jilbānah (Oasis), Eg., U.A.R D4 32
Bi'r al Jufayr (Oasis), Eg., U.A.R D4 32
Bi'r al Ḥakayyim, Libya G4 31
Bi'r al Lahtān (Oasis), Eg., U.A.R D5 32
Bi'r al Mazār, Eg., U.A.R., 1,000 C5 32
Bi'r al 'Uẓaym, Libya G4 31
Bi'r an Nuṣf (Well), Eg., U.A.R C5 43
Birao, Cen. Afr. Rep C4 46
Bi'r ar Rummānah, Eg., U.A.R., 1,000 C4 32
Bi'r ash Shaqqah, Libya G5 31
Bi'r Bayzah (Well), Eg., U.A.R D6 43
Bi'r Bū Jarrārah (Well), Libya D2 43
Birch, Cherokee, N.C D2 109
Birch, isl., B.C., Can D8 68
Birch, isl., Man., Can C2 71
Birch, lake, Sask., Can D2 70
Birch, lake, Minn C6 99
Birch, riv., W. Va C4 123
Birch Cliff, Ont., Can., 960 D5, E7 72
Birch Hills, Sask., Can., 534 E6 70
Birch Island, B.C., Can., 180 D8 68
Birch River, Man., Can., 799 C1 71
Birch Run, Saginaw, Mich., 844 E2 98
Birch Tree, Shannon, Mo., 420 E6 101
Birchwood, Hamilton, Tenn., 250 D8 117
Birchwood, Washburn, Wis., 433 C2 124
Bird, creek, Okla A6 112
Bird, isl., N.C D5 109
Bird, isl., S.C D11 115
Bird City, Cheyenne, Kans., 678 C2 93
Bird Island, Renville, Minn., 1,384 F4 99
Birds, Lawrence, Ill., 235 E6 90
Birdsboro, Berks, Pa., 3,025 F10 114
Birds Creek, Ont., Can., 77 B7 72
Birdseye, Dubois, Ind., 366 H4 91
Birdseye, Utah, Utah, 25 D4 119
Birdsville, Austl., 45 B2 51
Birdsville, Livingston, Ky., 92 A3 94
Birdtail, creek, Man., Can D1 71
Birdtown, Swain, N.C., 300 D3 109
Birdum, Austl., 41 C5 50
Birdwood, creek, Nebr C4 103
Birecik, Tur., 13,100 D11 31
Bir el Ksaib (Well), Mali B3 45
Bir el Nuss, Eg., U.A.R G6 31
Bi'r Fajr (Oasis), Sau. Ar H11 31
Birganj, Nepal, 10,037 D10 40
Bi'r Ghētta (Oasis), Eg., U.A.R E5 32
Bi'r Ḥasanah (Oasis), Eg., U.A.R D5 32
Bi'r Hooker (Oasis), Eg., U.A.R D2 32
Birimbāl, Eg., U.A.R., 1,000 C2 32
Bīrjand, Iran, 23,488 E9 41
Bi'r Jumayl (Oasis), Eg., U.A.R D5 32
Birkat as Sab', Eg., U.A.R., 1,000 D3 32
Birkat Qaran, lake, Eg., U.A.R E2 32
Birkenfeld, Ger., 5,100 D2 17
Birkenhead, Eng., 141,683 A4 12
Birket Idku, lake, Eg., U.A.R C2 32
Bi'r Māliḥah (Oasis), Eg., U.A.R E5 32
Birmingham, Jefferson, Ala., 340,887 (★624,000) B3, E4 78
Birmingham, Sask., Can., 50 G4 70
Birmingham, Eng., 1,105,651 (★2,570,000) B6 12
Birmingham, Van Buren, Iowa, 441 D6 92
Birmingham, Jackson, Kans., 20 A6 93
Birmingham, Oakland, Mich., 25,525 A7, F7 98
Birmingham, Clay, Mo., 201 E2 101
Birmingham, Burlington, N.J., 25 D3 106
Bi'r Misāḥah (Well), Eg., U.A.R E5
Bi'r Murākh, (Oasis), Eg., U.A.R E6 32
Bi'r Murr (Well), Eg., U.A.R E6 43
Birnamwood, Shawano, Wis., 568 D4 124
Birney, Rosebud, Mont., 27 E10 102
Birnie, Man., Can., 88 D2 71
Birnin Kebbi, Nig., 12,270 D5 45
Birni-Nkoni, Niger, 6,000 D6 45
Birobidzhan, Sov. Un., 41,000 E16 28
Biron, Wood, Wis., 726 D4 124
Bir Ounane (Well), Mali B4 45
Bi'r Rawḍ Sālim (Oasis), Eg., U.A.R D5 32
Birr, Ire., 3,221 D4 11
Birs, riv., Switz B3 19
Birsay, Sask., Can., 177 F2 70
Bir Sejri (Oasis), Syr F12 31
Birsk, Sov. Un., 26,200 D20 9
Bir Taba (Oasis), Eg., U.A.R. E6 32
Bir Tarfawi (Well), Sud A3 47
Birtle, Man., Can., 846 D1 71
Bi'r Umm Ḥuṣayrah (Oasis), Eg., U.A.R D6 32
Birvulevo, Sov. Un., 33,400 n17 27
Bir Zreigat (Well), Maur B3 45
Bisbee, Cochise, Ariz., 9,914 F6 80
Bisbee, Towner, N. Dak., 388 A6 110
Biscay, bay, Fr., Sp E2 14
Biscayne, bay, Fla G6 86
Biscayne, key, Fla F3 86
Biscayne Park, Dade, Fla., 2,911 F3 86
Bisceglie, It., 41,451 D6 21
Bischheim, Fr., 12,355 F7 15
Bischofshofen, Aus., 8,287 B9 18
Bischofswerda, Ger., 11,300 B9 17
Bischwiller, Fr., 8,198 F7 15
Biscoe, Montgomery, N.C., 1,053 B4 109
Biscoe, King and Queen, Va., 75 D5 121
Biscoe, is., Ant C6 5
Biscoe, mtn., Ant C18 5
Bisevo, isl., Yugo D2 22
Bishnupur, India, 30,958 F11 40
Bishop, Inyo, Calif., 2,875 D4 82
Bishop, Oconee, Ga., 214 C3 87
Bishop, Worcester, Md., 5 D7 85
Bishop, Nueces, Tex., 3,722 F4 118
Bishop Auckland, Eng., 35,276 F7 13
Bishopric, Sask., Can., 81 H3 70
Bishops, Windham, Conn C9 84
Bishop's Castle, Eng., 1,229 B4 12
Bishop's Falls, Newf., Can., 4,099 D4 75
Bishops Head, Dorchester, Md., 327 D5 85
Bishops Mills, Ont., Can., 67 C9 72
Bishop's Stortford, Eng., 18,308 C8 12
Bishipton, Que., Can., 345 D6 73
Bishopville, Worcester, Md., 100 D7 85
Bishopville, Lee., S.C., 3,586 C7 115
Biskra, Alg., 52,511 C6 44
Bislig, Phil., 1,086 *D7 35
Bismarck, Hot Spring, Ark., 250 D5 81
Bismarck, Vermilion, Ill., 400 C6 90
Bismarck, St. Francois, Mo., 1,237 D7 101
Bismarck, Burleigh, N. Dak., 27,670 C5 110
Bismarck, arch., New Gui G8 7, h12 50
Bismarck, cape, Grnld B16 4
Bismarck, range, N. Gui k11 50
Bismark, Ger., 3,412 F5 24
Bison, Rush, Kans., 291 D4 93
Bison, Garfield, Okla., 200 A4 112
Bison, Perkins, S. Dak., 457 B3 116
Bison, lake, Alta., Can A2 69
Bison, peak, Colo B5 83
Bissamcuttack, India, 5,000 H9 40
Bissau, Port. Gui., 18,309 D1 45
Bissett, Man., Can., 857 D4 71
Bistineau, lake, La B2 95
Bistrita, Rom., 20,292 B7 22
Bistrita, riv., Rom B7 22
Bitam, Gabon E2 46
Bitburg, Ger., 7,300 E6 15
Bitche, Fr., 4,277 E7 15
Bitely, Newaygo, Mich., 200 E5 98
Bititu, isl., Tarawa 52
Bitjoli, Indon E7 35
Bitlis, Tur., 16,600 C14 31
Bitola, Yugo., 49,101 E5 22
Bitonto, It., 33,300 D6 21
Bitter, creek, Wyo D4 125
Bitter, lake, Sask., Can G1 70
Bitter, lake, S. Dak B8 116
Bitter Creek, Sweetwater, Wyo., 70 D4 125
Bitterfeld, Ger., 31,000 (★125,000) B7 17
Bitterfontein, S. Afr D2 49
Bittern, lake, Sask., Can D3 70
Bittern Lake, Alta., Can., 76 C4 69
Bitterroot, range, U.S B4 77
Bitterroot, riv., Mont D2 102
Bittinger, Garrett, Md., 20 D1 85
Bityug, riv., Sov. Un F13 27
Biu, Nig D7 45
Bivalve, Wicomico, Md., 230 D6 85
Bivalve, Cumberland, N.J., 600 E2 106
Biwa, lake, Jap n15 37
Biwabik, St. Louis, Minn., 1,836 C6 99
Bixby, Tulsa, Okla., 1,711 B6 112
Bixby, Steele, Minn., 100 G5 99
Bixby, Iron, Mo., 50 D6 101
Bixby, Davie, N.C., 100 B3 109
Biya, riv., Sov. Un E26 9
Biysk, Sov. Un., 162,000 C11 29
Bizerte, Tun., 44,721 B6 44
Bizerte, prov., Tun., 258,544 *B6 44
Bizuta, Mong., 10,000 B8 34
Bjelovar, Yugo., 15,637 C3 22
Bjerringbro, Den., 3,582 B3 24
Björlanda, Swe A5 24
Bjornefjord, fjord, Nor G1 25
Blabon, Steele, N. Dak., 40 B8 110
Black, Geneva, Ala., 133 D4 78
Black, Parmer, Tex., 125 B1 118
Black, bayou, La E5 95
Black, butte, Mont F5 102
Black, butte, Wyo A5 125
Black, canyon, Colo C3 83
Black, creek, Miss E4 100
Black, creek, S.C B7 115
Black, creek, Vt B3 120
Black, fork, Ohio B5 111
Black, head, Newf., Can E5 75
Black, hills, S. Dak C2 116
Black, hills, Wyo A9 125
Black, isl., Man., Can D3 71
Black, lake, Sask., Can E11 66
Black, lake, Mich C6 98
Black, lake, N.Y B1 108
Black, mesa, Ariz A5 80
Black, mesa, Okla D1 112
Black, mtn., Ariz E4 80
Black, mtn., Colo A5 83
Black, mtn., Ky D7 94
Black, mtn., N.C D3 109
Black, mtn., Wyo A5 125
Black, mtn., Wyo C7 125
Black, mts., Ariz B1 80
Black, mts., N.B., Can B2 74
Black, mts., Wales C4 12
Black, peak, Ariz C1 80
Black, pond, Maine B3 96
Black, range, N. Mex C2 107
Black, riv., Ariz D5 80
Black, riv., Ark B4 81
Black, riv., Man., Can D4 71
Black, riv., Que., Can A7 72
Black, riv., La C4 95
Black, riv., Maine B3 96
Black, riv., Mich C6 98
Black, riv., Mich E8 98
Black, riv., Mo E7 101
Black, riv., N.J B3 106
Black, riv, N.Y B5 108
Black, riv., N.C C5 109
Black, riv., S.C D8 115
Black, riv., S.C E2 115
Black, riv., Vt B4 120
Black, riv., Vt E3 120
Black, riv., Viet B5 38
Black, riv., Wis D3 124
Black, sea, Eur G15 8
Blackall, Austl., 2,217 B5 51
Black Bear, bay, Newf., Can B4 75
Black Bear, creek, Okla A4 112
Black Bear Bay, Newf., Can., 8 B4 75
Blackbeard, isl., Ga E5 87
Black Bear Island, lake, Sask., Can B3 70
Blackburn, Eng., 106,114 (★230,000) A5 12
Blackburn, Saline, Mo., 310 B4 101
Blackburn, Pawnee, Okla., 129 A5 112
Blackburn, mtn., Alsk C11 79
Black Canyon, Yavapai, Ariz., 150 C3 80
Black Canyon of the Gunnison, nat. mon., Colo C3 83
Black Cape, Que., Can., 131 B3 73
Black Creek, Wilson, N.C., 310 B6 109
Black Creek, Outagamie, Wis., 707 A5, D5 124
Black Diamond, Jefferson, Ala., 250 E4 78
Black Diamond, Alta., Can., 1,043 D3 69
Black Diamond, King, Wash., 1,026 B3, D2 122
Black Dome, mtn., B.C., Can C4 68
Blackdown, hills, Eng D4 12
Blackduck, Beltrami, Minn., 765 C4 99
Black Eagle, Cascade, Mont., 2,000 C5 102
Black Earth, Dane, Wis., 784 E4 124
Blackey, Buchanan, Va., 200 B3 121
Blackfalds, Alta., Can., 477 C4 69
Blackfeet, Indian res., Mont B4 102
Blackfoot, Alta., Can., 91 C5 69
Blackfoot, Bingham, Idaho, 7,378 F6 89
Blackfoot, Glacier, Mont., 80 B4 102
Blackfoot, res., Idaho G7 89
Blackfoot, riv., Mont D3 102
Blackfoot, mts., Idaho F7 89
Blackford, co., Ind., 14,792 D7 91
Black Forest, El Paso, Colo., 1,000 C6 83
Black Forest, mts., Ger E3 17
Black Fork, Scott, Ark., 30 C1 81
Blackfork, Lawrence, Ohio, 300 D5 111
Black Hawk, Ont., Can., 75 B5 99
Black Hawk, Gilpin, Colo., 171 B5 83
Black Hawk, Carroll, Miss., 100 B4 100
Black Hawk, Meade, S. Dak., 200 C2 116
Black Hawk, co., Iowa, 122,482 B5 92
Blackhead, bay, Newf., Can D5 75
Blackie, Alta., Can., 184 D4 69
Blackjack, St. Louis, Mo., 215 A8 101
Black Jack, mtn., Ga A5 87
Black Lake, Que., Can., 4,180 C6 73
Black Lake, bayou, La B2 95
Blacklick, Franklin, Ohio, 200 C2 111
Black Lick, Indiana, Pa., 700 F3 114

Blackman, Okaloosa, Fla., 50 G2 86
Blackmore, mtn., Mont E6 102
Black Mountain, Buncombe, N.C., 1,313 D4 109
Black Oak, Craighead, Ark., 220 B5 81
Black Oak, Lake, Ind., 15,000 *A3 91
Black Oak, ridge, Tenn D9 117
Black Pine, peak, Idaho G5 89
Black Point, Marin, Calif., 650 B5 82
Blackpool, Eng., 152,133 (★255,000) A4 12
Black River, Jam., 2,616 F5 64
Blackriver, Alcona, Mich., 90 D7 98
Black River, Jefferson, N.Y., 1,237 A5 108
Black River Falls, Jackson, Wis., 3,195 D3 124
Black Rock, Lawrence, Ark., 554 A4 81
Black Rock, McKinley, N. Mex., 35 B1 107
Black Rock, Millard, Utah, 15 E3 119
Black Rock, des., Nev B3 104
Black Rock, des., Utah D3 119
Black Rock, mts., Nev B3 104
Blacks, fork, Utah B6 119
Blacks, fork, Wyo D3 125
Blacksburg, Cherokee, S.C., 2,174 A4 115
Blacksburg, Montgomery, Va., 7,070 D2 121
Blacks Harbour, N.B., Can., 1,297 D3 74
Blackshear, Pierce, Ga., 2,482 E4 87
Blacksod, bay, Ire C1 11
Black Springs, Montgomery, Ark., 75 C2 81
Black Springs, Washoe, Nev., 100 D2 104
Black Squirrel, creek, Colo C6 83
Blackstairs, mtn., Ire E5 11
Blackstairs, mts., Ire E5 11
Blackstock, Ont., Can., 265 C6 72
Blackstock, Chester and Fairfield, S.C., 175 B5 115
Blackstone, Worcester, Mass., 2,000 (5,130▲) B4 97
Blackstone, Caldwell, N.C A2 109
Blackstone, Nottoway, Va., 3,659 D5 121
Blackstone, riv., Alta., Can C2 69
Blacksville, Monongalia, W. Va., 211 B4 123
Blacktail, Flathead, Mont., 10 B3 102
Black Thunder, creek, Wyo B8 125
Blackton, Monroe, Ark., 50 C4 81
Blackville, N.B., Can., 484 C4 74
Blackville, Barnwell, S.C., 1,901 E5 115
Black Volta, riv., Ghana E4 45
Black Walnut, pt., Md C5 85
Black Warrior, riv., Ala C2 78
Blackwater, Ire., 216 E5 11
Blackwater, Cooper, Mo., 284 C5 101
Blackwater, Lee, Va., 50 B2 121
Blackwater, res., N.H D3 105
Blackwater, res., Scot D3 13
Blackwater, riv., Eng C8 12
Blackwater, riv., Fla G2 86
Blackwater, riv., Ire E3 11
Blackwater, riv., Md D5 85
Blackwater, riv., N.H D3 105
Blackwater, riv., Va E6 121
Blackwell, Conway, Ark., 750 B3 81
Blackwell, St. Francis, Mo., 100 C7 101
Blackwell, Kay, Okla., 9,588 A4 112
Blackwell, Tioga, Pa., 75 C7 114
Blackwell, Nolan, Tex., 314 C2 118
Blackwell, Forest, Wis., 100 C5 124
Blackwood, Camden, N.J., 3,000 *D2 106
Bladen, Webster, Nebr., 322 D7 103
Bladen, co., N.C., 28,881 C5 109
Bladenboro, Bladen, N.C., 774 C5 109
Bladensburg, Prince Georges, Md., 3,103 C2 85
Blades, Sussex, Del., 729 C6 85
Bladon Springs, Choctaw, Ala., 150 D1 78
Bladworth, Sask., Can., 190 F2 70
Blagodarnoye, Sov. Un., 29,500 D2 29
Blagoevgrad (Gorna-Dzhumaya), Bul., 14,066 D6 22
Blagoveshchensk, Sov. Un., 99,000 D15 28
Blagoveshchensk, Sov. Un., 10,000 D20 9
Blain, Fr., 2,009 D3 14
Blain, Perry, Pa., 336 F7 114
Blaine, Pottawatomie, Kans., 78 C7 93
Blaine, Aroostook, Maine, 375 (945▲) B5 96
Blaine, Anoka, Minn., 7,570 E7 99
Blaine, Sunflower, Miss., 125 B3 100
Blaine, Belmont, Ohio, 750 B1 111
Blaine, Whatcom, Wash., 1,735 A3 122
Blaine, co., Idaho, 4,598 F4 89
Blaine, co., Mont., 8,091 B7 102
Blaine, co., Nebr., 1,016 C6 103
Blaine, co., Okla., 12,077 B3 112
Blaine, creek, Ky B7 94
Blaine Lake, Sask., Can., 641 E2 70
Blair, Doniphan, Kans., 75 C8 93
Blair, Washington, Nebr., 4,931 C9 103
Blair, Grafton, N.H., 60 C3 105
Blair, Jackson, Okla., 893 C2 112
Blair, Fairfield, S.C., 75 C5 115
Blair, Logan, W. Va., 500 D5 123
Blair, Trempealeau, Wis., 909 D2 124
Blair, co., Pa., 137,270 E5 114
Blair-Atholl, Scot., 1,868 D5 13
Blairgowrie [& Rattray], Scot., 5,168 D5 13
Blair Mills, Anderson, S.C., 75 B2 115
Blairmore, Alta., Can., 1,980 E3 69
Blairsburg, Hamilton, Iowa, 287 B4 92
Blairsden, Plumas, Calif., 90 C3 82
Blairstown, Benton, Iowa, 583 C5 92
Blairstown, Henry, Mo., 177 C4 101
Blairstown, Warren, N.J., 550 B3 106
Blairsville, Union, Ga., 437 B3 87
Blairsville, Indiana, Pa., 4,930 F3 114
Blairton, Berkeley, W. Va., 200 B7 123
Blaisdell, Mountrail, N. Dak., 62 A3 110
Blaj, Rom., 8,731 B7 22
Blakeley, Scott, Minn., 100 F5 99
Blakeley, Kanawha, W. Va., 165 C6 123
Blakely, Garland, Ark., 250 C2 81
Blakely, Early, Ga., 3,580 E2 87
Blakeman, Rawlins, Kans., 12 C2 93
Blakes, pt., Mich A2 98
Blakesburg, Wapello, Iowa, 401 D5 92
Blakeslee, Williams, Ohio, 156 A3 111
Blakeslee, Monroe, Pa., 225 D10 114
Blalock, Gilliam, Oreg., 20 B6 113
Blâmont, Fr., 1,409 F6 15
Blanc, cape, Maur B1 45
Blanc, cape, Tun F11 30
Blanc, mtn., Fr E7 14
Blanca, Costilla, Colo., 233 D5 83
Blanca, bay, Arg B4 54
Blanca, cape, Oreg E2 113
Blanca, peak, Colo D5 83
Blanca, pt., Mex B2 63
Blanca, range, Peru C2 58
Blanchard, Bonner, Idaho, 100 A2 89
Blanchard, Page, Iowa, 174 D2 92
Blanchard, Caddo, La., 500 B2 95
Blanchard, Piscataquis, Maine, 10 (57▲) C3 96
Blanchard, Isabella, Mich., 275 E5 98
Blanchard, San Miguel, N. Mex B4 107
Blanchard, Trail, N. Dak., 40 B8 110
Blanchard, McClain, Okla., 1,377 B4 112
Blanchard, Centre, Pa., 600 D6 114
Blanchard, Skagit, Wash., 200 A3 122
Blanchard, riv., Ohio A3 111
Blanchardville, Lafayette, Wis., 632 F4 124
Blanche, Ont., Can E8 72
Blanche, Que., Can., 95 D2 73
Blanche, Lincoln, Tenn., 200 B5 117
Blanche, lake, Austl D2 51
Blanchester, Clinton, Ohio, 2,944 C4 111
Blanco, San Juan, N. Mex., 20 A2 107
Blanco, Blanco, Tex., 789 D3 118
Blanco, co., Tex., 3,657 D3 118
Blanco, cape, C.R F5 62
Blanco, cape, Oreg E2 113
Blanco, creek, N. Mex C6 107
Blanco, riv., Arg E2 55
Blanco, riv., Bol B3 55
Blanco, riv., Mex n15 63
Blanco, riv., P.R C3 65
Blanc Sablon, Que., Can., 252 C3, h10 75
Bland, Gasconade, Mo., 654 C6 101
Bland, Bland, Va., 500 D1 121
Bland, co., Va., 5,982 D1 121
Blandburg, Cambria, Pa., 900 E5 114
Blandford [Forum], Eng., 3,558 D5 12
Blandford, Hampden, Mass., 600 (636▲) B2 97
Blanding, San Juan, Utah, 1,805 F6 119
Blandinsville, McDonough, Ill., 853 C3 90
Blandville, Ballard, Ky., 133 A2 94
Blandy, Baldwin, Ga C3 87
Blaney, Kershaw, S.C., 329 C6 115
Blanford, Vermilion, Ind., 800 E2 91
Blangy-sur-Bresle, Fr., 2,925 E9 12
Blankenberge, Bel., 10,199 C3 15
Blankenburg, Ger , 19,500 B5 17
Blankenfelde, Ger., 6,667 A8 17
Blankenheim Ger., 1,100 C1 17
Blanket, Brown, Tex., 320 D3 118
Blantyre, Nya., 62,400 A6 49
Blarney, Ire., 995 F3 11
Blasdell, Erie., N.Y., 3,909 C2 108
Blatna, Czech., 3,209 D8 17
Blato, Yugo., 5,140 D3 22
Blaubeuren, Ger., 7,800 E4 17
Blauvelt, Rockland, N.Y., 3,000 *C5 108
Blawnox, Allegheny, Pa., 2,085 A6 114
Blaye, Fr., 4,291 E3 14
Blazon, Lincoln, Wyo D2 125
Blazowa, Pol., 4,002 D7 26
Bleckede, Ger., 4,000 E4 24
Bleckley, co., Ga., 9,642 D3 87
Bled-Grad, Yugo., 4,120 B1 22
Bledow, Pol g10 26
Bledsoe, Cochran, Tex., 100 C1 118
Bledsoe, co., Tenn., 7,811 D8 117
Bleecker, Lee, Ala., 175 C4 78
Bleicherode, Ger., 7,923 B5 17
Blekinge, co., Swe., 145,000 B8 24
Blencoe, Monona, Iowa, 286 C1 92
Blende, Pueblo, Colo., 600 C6 83
Blendecques, Fr., 3,943 D10 12
Blenheim, Ont., Can., 2,844 E2 72
Blenheim, Marlboro, S.C., 185 B8 115
Blenheim, N.Z., 11,957 N14 51
Blessing, Matagorda, Tex., 700 E4 118
Blessington, Ire., 491 D5 11
Blevins, Hempstead, Ark., 198 D2 81
Blewett, Uvalde, Tex., 300 E3 118
Blida, Alg., 67,913 B5 44
Blija, Neth., 784 A5 15
Blind, pass, Fla F4 86
Blind, riv., La B6 95
Blind River, Ont., Can., 3,633 A2 72
Bliss, Gooding, Idaho, 91 G4 89
Bliss, Wyoming, N.Y., 400 C2 108
Blissfield, Lenawee, Mich., 2,653 G7 98
Blita, Togo E5 45
Blitzen, Harney, Oreg E7 113
Blocher, Scott, Ind., 250 G6 91
Blocker, Pittsburg, Okla., 85 B6 112
Block, isl., R.I E10 84
Block Island, Newport, R.I., 400 (486▲) E10 84
Block Island, sound, R.I E9 84
Blockton, Taylor, Iowa, 343 D3 92
Blodgett, Scott, Mo., 203 D8 101
Blodgett Landing, Merrimack, N.H., 30 D2 105
Bloedel, B.C., Can D5 68
Bloemfontein, S. Afr., 112,606 (★145,273) C4 49
Blois, Fr., 33,838 D4 14
Blokhus, Den A3 24
Blomberg, Ger., 7,000 B4 17
Blönduós, Ice., 599 n22 25
Blonie, Pol., 6,416 m13 26
Blood, mtn., Ga B3 87
Bloodroot, mtn., Vt D3 120
Bloodsworth, isl., Md D5 85
Bloodvein, riv., Man., Can D3 71
Bloody Foreland, pt., Ire B3 11
Bloom, Ford, Kans., 100 E4 93
Bloomdale, Wood, Ohio, 669 A4 111
Bloomer, Chippewa, Wis., 2,834 C2 124
Bloomfield, Sonoma, Calif., 80 B4 82
Bloomfield, Ont., Can., 769 D7 72
Bloomfield, Hartford, Conn., 5,000 (13,613▲) B6 84
Bloomfield, Bibb, Ga., 4,381 *D3 87
Bloomfield, Greene, Ind., 2,224 F4 91
Bloomfield, Davis, Iowa, 2,771 D5 92
Bloomfield, Nelson, Ky., 916 C4 94
Bloomfield, Stoddard, Mo., 1,330 E8 101
Bloomfield, Dawson, Mont., 50 C12 102
Bloomfield, Knox, Nebr., 1,349 B8 103
Bloomfield, Essex, N.J., 51,867 D4 106
Bloomfield, San Juan, N. Mex., 1,292 A2 107
Bloomfield, Essex, Vt., 120 (212▲) B5 120
Bloomfield Hills, Oakland, Mich., 2,378 A7 98
Bloomfield Station, N.B., Can., 68 D4 74
Bloomfield Station, P.E.I., Can., 77 C5 74
Bloomfield Village, Oakland, Mich., 3,500 *F7 98
Bloomingburg, Fayette, Ohio, 719 C4 111
Bloomingdale, Chatham, Ga., 1,000 D5 87
Bloomingdale, Du Page, Ill., 1,262 E2 90
Bloomingdale, Parke, Ind., 455 E3 91
Bloomingdale, Van Buren, Mich., 471 F5 98
Bloomingdale, Passaic, N.J., 5,293 A4 106
Bloomingdale, Essex, N.Y., 490 B3 108
Blooming Grove, Pike, Pa., 250 D11 114
Blooming Grove, Navarro, Tex., 725 C4 118
Blooming Grove, Dane, Wis., 8,500 *E4 124
Blooming Prairie, Steele, Minn., 1,778 G5 99
Bloomington, San Bernardino, Calif., 3,500 *F3 82
Bloomington, Bear Lake, Idaho, 254 G7 89
Bloomington, McLean, Ill., 36,271 C5 90
Bloomington, Monroe, Ind., 31,357 F4 91
Bloomington, Hennepin, Minn., 50,498 E6 99
Bloomington, Garrett, Md., 338 D1 85
Bloomington, Franklin, Nebr., 176 D6 103
Bloomington, Victoria, Tex., 1,756 E4 118
Bloomington, Grant, Wis., 735 F3 124
Bloomsbûrg, Columbia, Pa., 10,655 E9 114
Bloomsbury, Hunterdon, N.J., 838 B2 106
Bloomsdale, Ste. Genevieve, Mo., 400 C7 101
Bloomville, Delaware, N.Y., 250 C6 108
Bloomville, Seneca, Ohio, 836 A4 111
Blossburg, Jefferson, Ala., 500 E4 78
Blossburg, Tioga, Pa., 1,956 C7 114
Blosseville, coast, Grnld C17 4
Blossom, Lamar, Tex., 545 C5 118
Blossom Hill, Lancaster, Pa., 1,000 *F9 114
Blount, co., Ala., 25,449 B3 78
Blount, co., Tenn., 57,525 D10 117
Blount Springs, Blount, Ala., 50 B3 78
Blounts Creek, Beaufort, N.C., 50 B7 109
Blountstown, Calhoun, Fla., 2,375 B1 86
Blountsville, Blount, Ala., 672 A3 78
Blountville, Sullivan, Tenn., 600 C11 117
Blovice, Czech., 2,388 D8 17
Blowing Rock, Watauga and Caldwell, N.C., 711 A2 109
Bloxom, Accomack, Va., 349 D7 121
Blucher, Sask., Can., 45 F2 70
Bludenz, Aus., 11,127 E4 16
Blue, Greenlee, Ariz., 10 D6 80
Blue, bayou, La E5 95
Blue, creek, Nebr C3 103
Blue, creek, W. Va C6 123
Blue, hill, Kans C4 93
Blue, hills, Kans C5 93
Blue, mesa, Colo C3 83
Blue, mound, Kans C4 93
Blue, mtn., Ark C1 81
Blue, mtn., N.B., Can B3 74
Blue, mtn., Newf., Can j26 67
Blue, mtn., Maine D2 96
Blue, mtn., Mont C12 102
Blue, mtn., N.H A4 105
Blue, mtn., N.Y B6 108
Blue, mtn., Pa B1 106, F6 114
Blue, mts., Austl F8 50
Blue, mts., Oreg C7 113
Blue, mts., Tex D3 118
Blue, mts., Wash C8 122
Blue, pt., N.Y G4 84
Blue, riv., Ind H5 91
Blue, riv., Mo E2 101
Blue, riv., Okla D5 112
Blue Ash, Hamilton, Ohio, 8,341 D2 111
Blue Bell, Montgomery, Pa., 1,000 *F11 114
Blueberry, creek, B.C., Can D2 69
Blueberry, riv., B.C., Can A7 68
Blue Buck, pt., La E2 95
Blue Creek, Box Elder, Utah B3 119
Bluecreek, Stevens, Wash., 45 A8 122
Blue Creek, Kanawha, W. Va., 310 C3, C6 123
Blue Diamond, Clark, Nev., 300 G6 104
Blue Earth, Faribault. Minn., 4,200 G4 93
Blue Earth, co., Minn., 44,385 G4 99
Blue Earth, riv., Iowa, Minn A3 92
Blue Eye, Stone, Mo., 75 E4 101
Bluefield, Tazewell, Va., 4,235 B3 121
Bluefield, Mercer, W. Va., 19,256 D3 123
Bluefields, Nic., 11,900 D6 62
Blue Grass, Scott, Iowa, 568 C7 92
Bluegrass, Knox, Tenn E10 117
Blue Grass, Highlands, Va., 75 C3 121
Blue Hill, Hancock, Maine, 500 (1,270▲) D4 96
Blue Hill, Webster, Nebr., 723 D7 103
Blue Hill, range, Mass D3 97
Blue Hill, range, N.H D4 105
Blue Hill Falls, Hancock, Maine, 100 D4 96
Blue Hills, Hartford, Conn., 4,000 *B6 84
Blue Hills, res., Mass D3 97
Blue Hills of Couteau, hills, Newf., Can E3 75
Bluehole, Clay, Ky., 200 C6 94
Blue Island, Cook, Ill., 19,618 B6, F3 90
Bluejacket, Craig, Okla., 245 A6 112
Bluejoint, lake, Oreg E7 113
Blue Lake, Humboldt, Calif., 1,234 B2 82
Blue Mound, Macon, Ill., 1,038 D4 90
Blue Mound, Linn, Kans., 319 D8 93
Blue Mound, Tarrant, Tex., 1,253 *B5 118
Blue Mountain, Calhoun, Ala., 446 B4 78
Blue Mountain, Logan, Ark., 94 B2 81
Blue Mountain, Moffat, Colo., 30 A2 83
Blue Mountain, Tippah, Miss., 741 A4 100
Blue Mountain, dam and res., Ark B2 81
Blue Mountain Lake, Hamilton, N.Y., 250 B6 108
Blue Mud, bay, Austl B6 50
Blue Nile, reg., Sud., 2,069,646 C3 47
Blue Pennant, Boone, W. Va., 300 D3, D6 123
Blue Point, Suffolk, N.Y., 2,300 G4 84
Blue Rapids, Marshall, Kans., 1,426 C7 93
Blue Ridge, Alta., Can., 233 B3 69
Blue Ridge, Fannin, Ga., 1,406 B2 87
Blue Ridge, Shelby, Ind., 150 E6 91
Blue Ridge, Botetourt, Va., 900 D3 121
Blue Ridge, dam, Ga B2 87
Blue Ridge, lake, Ga B2 87
Blue Ridge, mts., U.S C11 77
Blue Ridge Summit, Franklin, Pa., 800 G7 114
Blue River, B.C., Can., 390 C8 68
Blue River, Grant, Wis., 356 E3 124
Bluesky, Alta., Can., 108 A1 69
Blue Springs, Barbour, Ala., 94 D3 78
Blue Springs, Jackson, Mo., 2,555 E2 101
Blue Springs, Union, Miss., 99 A5 100
Blue Springs, Gage, Nebr., 509 D6 103
Blue Springs, Socorro, N. Mex C3 107
Blue Stack, mts., Ire C3 11
Bluestem, Lincoln, Wash B7 122
Bluestone, res., Va., W. Va D2 121, D4 123
Bluestone, riv., W. Va D3 123
Bluevale, Ont., Can., 148 D3 72
Bluewater, Valencia, N. Mex., 950 B2 107
Bluff, Fayette, Ill., 160 E4 90
Bluff, N.Z., 3,042 Q12 51
Bluff, San Juan, Utah, 100 F6 119
Bluff, creek, Kans E4 93
Bluff, creek, Kans E6 93
Bluff, creek, Okla A4 112
Bluff, mtn., N.C A2 109
Bluff, mtn., Vt B5 120
Bluff City, Nevada, Ark., 140 D2 81
Bluff City, Harper, Kans., 152 E6 93
Bluff City, Henderson, Ky., 250 C2 94
Bluff City, Sullivan, Tenn., 948 C11 117
Bluff Creek, res., Okla B4 112
Bluff Dale, Erath, Tex., 400 C3 118
Bluff Park, Jefferson, Ala., 3,000 *B3 78
Bluffs, Scott, Ill., 779 D3 90
Bluffton, Yell, Ark., 50 C2 81
Bluffton, Wells, Ind., 6,238 C7 91
Bluffton, Otter Tail, Minn., 211 D3 99
Bluffton, Allen, Ohio, 2,591 B4 111
Bluffton, Beaufort, S.C., 356 G6 115
Bluford, Jefferson, Ill., 388 E5 90
Bluit, Roosevelt, N. Mex D6 107
Blumenau, Braz., 46,591 D3 56
Blumengard Colony, Faulk, S. Dak., 70 B6 116
Blumenhof, Sask., Can., 79 G2 70
Blunt, Hughes, S. Dak., 532 C6 116
Bly, Riverside, Calif., 1,554 *F3 82
Bly, Klamath, Oreg., 600 E5 113
Bly, ridge, Oreg E5 113
Blying, sound, Alsk h17 79
Blyn, Clallam, Wash., 50 A3 122
Blyth, Ont., Can., 724 D3 72
Blyth, Eng., 35,933 C6 10
Blyth, riv., Eng E7 13
Blythe, Riverside, Calif., 6,023 F6 82
Blythe, Richmond and Burke, Ga., 172 C4 87
Blythedale, Harrison, Mo., 179 A4 101
Blytheville, Mississippi, Ark., 25,883 B6 81
Blythewood, Richland, S.C., 300 C6 115
Bnei Braq, Isr., 47,080 g10 32
Bo, S.L E2 45
Boac, Phil., 3,262 C6 35
Boaco, Nic., 3,078 D5 62
Boakview, Ont., Can., 165 B4 72
Boalsburg, Centre, Pa., 800 E6 114
Board Camp, Polk, Ark., 70 C1 81
Boardman, Mahoning, Ohio, 20,000 *A7 111
Boardman, Morrow, Oreg., 153 B7 113
Boardmans Bridge, Litchfield, Conn., 165 C3 84
Boatland, Fentress, Tenn C8 117
Boa Vista, Braz., 10,592 C5 60
Boaz, Marshall, Ala., 4,654 A3 78
Boaz, Richland, Wis., 117 E3 124
Bobbili, India, 25,592 H9 40
Bobbio, It., 6,970 E5 18
Bobcaygeon, Ont., Can., 1,210 C6 72
Bobigny, Fr., 37,010 g10 14
Böblingen, Ger., 25,400 E4 17
Bobo, Coahoma, Miss., 150 A3 100
Bobo-Dioulasso, Upper Volta, 45,000 D4 45
Bobov Dol, Bul., 2,627 D6 22
Bobr, riv., Pol C3 26
Bobrawa, riv., Ger B10 17
Bobrinets, Sov. Un., 10,000 G9 27
Bobrka, Sov. Un., 10,000 G5 27
Bobruysk, Sov. Un., 104,000 E7 27
Bobtown, Greene, Pa., 1,167 G2 114
Bobures, Ven., 9,397 B3 60
Boca Chica, is., Fla H5 86
Boca Chica Central, P.R D4 65
Boca Ciega, bay, Fla E1 86
Bôca do Acre, Braz., 2,994 C4 58
Boca Grande, Lee, Fla., 400 F4 86
Bocaiúva, Braz., 5,952 E2 57
Bocaranga, Cen. Afr. Rep D3 46
Boca Raton, Palm Beach, Fla., 6,961 F6 86
Bocas del Toro, Pan., 2,459 F6 62
Bocay, Nic C5 62
Boccea (Buxus), It h8 21
Bochnia, Pol., 11,000 D6 26
Bocholt, Ger., 45,700 B1 17
Bochov, Czech., 1,084 C8 17
Bochum, Ger., 361,400 B2 17
Bock, Mille Lacs, Minn., 91 E5 99
Bockhorn, Ger., 7,100 E2 24
Bockum-Hövel, Ger., 24,300 B2 17
Boda, Cen. Afr. Rep E3 46
Bodafors, Swe., 2,500 A8 24
Bodaybo, Sov. Un., 14,700 D14 28
Bodcau, creek, Ark D2 81
Bodcaw, Nevada, Ark., 100 D2 81
Boddam, Scot C7 13
Bode, Humboldt, Iowa, 430 B3 92
Bode, riv., Ger B6 17
Bodega, head, Calif C2 82
Bodele, reg., Afr B3 42
Boden, Mercer, Ill B3 90
Boden, Swe., 13,700 E9 25
Bodenham, Giles, Tenn B4 117
Bodensee (Constance), lake, Ger., Switz A7 19
Boderg, lake, Ire D3 11
Bodhia, mtn., Eg., U.A.R E5 32
Bodie, isl., N.C B8 109
Bodine, mtn., B.C., Can B5 68
Bodines, Lycoming, Pa., 90 D8 114
Bodkin, pt., Md B5 85
Bodmin, Eng., 6,209 D3 12
Bodmin, moor, Eng D3 12
Bodo, Alta., Can., 54 C5 69
Bodö, Nor., 12,700 D6 25
Bodrog, riv., Hung A5 22
Bodrum (Halicarnassus), Tur., 5,000 D6 31
Bodva, riv., Hung A5 22
Boelus (Howard City), Howard, Nebr., 181 C7 103
Boende, Con. L B3 48
Boeotia (Voiotia), prov., Grc., 114,256 *C4 23
Boerne, Kendall, Tex., 2,169 A3, E3 118
Boeuf, bayou, La C3 95
Boeuf, lake, La E5 95
Boeuf, riv., Ark., La D4 81, B4 95
Boffa, Guinea, 1,000 D2 45
Bogale, Bur., 23,211 E10 39
Bogallua, isl., Eniwetok 52
Bogalusa, Washington, La., 21,423 D6 95
Bogan, isl., Eniwetok 52
Bogan, riv., Austl E6 51
Bogandé, Upper Volta D4 45
Bogard, Carroll, Mo., 277 B4 101
Bogart, Clark and Oconee, Ga., 403 C3 87
Bogata, Red River, Tex., 1,112 C5 118
Bog Brook res., N.Y D2 84
Bogatynia, Pol., 2,851 C9 17
Bogbonga, Con. L A2 48
Bogen, Ger., 3,200 E7 17
Bogenfels, S.W. Afr C2 49
Bogense, Den., 2,968 C4 24
Boger City, Lincoln, N.C., 2,200 *B2 109
Boggeragh, mts., Ire E3 11
Boggerik, isl., Kwajalein 52
Boggs, cape, Ant B6 5
Boghari, Alg., 10,166 (11,518▲) B5 44
Boghé, Maur., 1,200 C2 45
Bogno, Switz., 145 D7 19
Bognor Regis, Eng., 28,144 D7 33
Bogo, Phil., 6,371 *C6 35
Bogodukhov, Sov. Un., 10,000 F10 27
Bogon, isl., Eniwetok 52
Bogor (Buitenzorg), Indon., 146,907 G3 35
Bogoroditsk, Sov. Un., 10,000 E12 27
Bogorodsk, Sov Un., 36,200 C14 27
Bogotá, Col., 954,120 (★1,075,000) C3 60
Bogota, Bergen, N.J., 7,965 D5 106
Bogota, Dyer, Tenn., 250 A2 117
Bogotol, Sov. Un., 25,000 D26 9
Bogra, Pak., 25,303 E12 40
Boguchany, Sov. Un., 10,000 D28 9
Bogue, Graham, Kans., 234 C4 93
Bogue Chitto, Lincoln, Miss., 400 D3 100
Bogue Chitto, riv., La., Miss D5 95, D3 100
Bogue Phalia, riv., Miss B3 100
Boguslav, Sov. Un., 10,000 G8 27
Bohain-en-Vermandois, Fr., 6,726 E3 15
Boharm, Sask., Can., 100 G3 70
Bohemia, Suffolk, N.Y., 2,000 *G4 84
Bohemia (Čechy), reg., Czech., 6,035,500 D3 26
Bohemian Forest, mts., Ger D7 17
Bohemian-Moravian, highlands, Czech D3 26
Böhme, riv., Ger F3 24
Bohmte, Ger., 4,000 A3 17
Bohol, prov., Phil., 602,500 *D6 35
Bohol, isl., Phil D6 35
Bohotleh, Som D6 47
Boice, Wayne, Miss., 30 D5 100
Boiestown, N.B., Can., 343 C3 74
Boiling Springs, Cleveland, N.C., 1,311 B2 109
Boiling Springs, Cumberland, Pa., 1,182 F7 114
Bois, lake, N.W. Ter., Can C8 66
Bois Blanc, isl., Mich C6 98
Boischatel, Que., Can., 1,576 C6, C9 73
Bois-Colombes, Fr., 29,938 h10 14
Boisdale, N.S., Can., 133 C9 74
Boisdale, inlet, Scot C1 13
Bois D'Arc, Greene, Mo., 152 D4 101
Bois des Filion, Que., Can., 2,499 D8 73
Bois de Sioux, riv., Minn E2 99
Boise, Ada, Idaho, 34,481 F2 89
Boise, co., Idaho, 1,646 F3 89
Boise City, Cimarron, Okla., 1,978 D2 112
Boise, riv., Idaho F2, F3 89
Boissevain, Man., Can., 1,303 E1 71
Boissevain, Tazewell, Va., 600 B3 121
Boisvert, Que., Can., 75 B9 71
Boizenburg, Ger., 11,600 E4 24
Bojador, cape, Sp. Sahara D1 44
Bojnūrd, Iran, 15,293 C8 41
Bokanda, I.C E4 45
Bokchito, Bryan, Okla., 620 C5 112
Bokeelia, Lee, Fla., 150 F4 86
Boké, Guinea, 5,400 D2 45
Boko, Con. L C2 48
Bokobyaadaa, isl., Bikini 52
Bokoro, Chad, 1,739 C3 46
Bokororyuru, isl., Bikini 52
Bokoshe, Le Flore, Okla., 431 B7 112
Boksburg, S. Afr., 71,029 *C4 49
Bokungu, Con. L B3 48
Bol, Chad, 1,073 C2 46
Bolafa, Con. L A3 48
Boland, riv., Ont., Can B8 98
Bolanos, mtn., Guam 52
Bolaños, riv., Mex m12 63
Bolar, Bath, Va., 45 C3 121
Bolbec, Fr., 12,212 C4 14
Bolckow, Andrew, Mo., 232 A3 101
Bolding, Union, Ark., 75 D3 81
Bole, Ghana, 1,813 E4 45
Boleko, Con. L B2 48
Boles, Scott, Ark., 120 C1 81
Boleslawiec, Pol., 23,000 C3 26
Boley, Okfuskee, Okla., 573 B5 112
Bolgrad, Sov. Un., 10,000 I7 27
Boli, see Bolu, Tur.
Bolia, Con. L B2 48
Boligee, Greene, Ala., 134 C1 78
Bolinao, Phil., 2,041 n12 35
Bolinao, cape, Phil n12 35
Boling, Wharton, Tex., 950 E5, G4 118
Bolinger, Choctaw, Ala., 200 D1 78
Bolívar, Arg., 14,010 B4 54
Bolívar, Col., 6,121 B2 60
Bolivar, Polk, Mo., 3,512 D4 101
Bolivar, Allegany, N.Y., 1,405 C2 108
Bolivar, Tuscarawas, Ohio, 932 B6 111
Bolivar, Westmoreland, Pa., 716 F3 114
Bolívar, Peru C2 58
Bolivar, Hardeman, Tenn., 3,338 B3 117
Bolivar, Jefferson, W. Va., 754 B7 123
Bolivar, co., Miss., 54,464 B3 100
Bolívar, dept., Col., 737,890 B2 60
Bolívar, prov., Ec., 123,205 B2 58
Bolívar, state, Ven., 213,543 B4 60
Bolivar, lake, Miss B2 100
Bolivar (La Columna), peak, Ven B3 60
Bolivia, Brunswick, N.C., 201 C5 109
Bolivia, country, S.A., 3,311,000 E4 53, C2 55
Bolkhov, Sov. Un., 10,000 E11 27
Bollebygd, Swe., 708 A6 24
Bolling, Butler, Ala., 200 D3 78
Bollingen, Switz., 14,914 C4 19
Bollinger, co., Mo., 9,167 D7 101
Bollnäs, Swe., 10,679 (16,699▲) G7 25
Bollullos, Sp., 10,947 D2 20
Bolmen, lake, Swe B7 24
Bolnisi, Sov. Un., 6,065 B15 31
Bolobo, Con. L B2 48
Bologna, It., 444,872 E7 18, B3 21
Bologoye, Sov. Un., 22,000 C10 27
Bolomba, Con. L A2 48
Bolon, lake, Sov. Un B8 37
Bolotnoye, Sov. Un., 26,500 B10 29
Bolovens, plat., Laos E7 38
Bolsena, lake, It C3 21
Bolshaya Boktybay, mtn., Sov. Un D5 29
Bolshaya Irgiz, riv., Sov. Un C3 29
Bolshaya Lepetikha, Sov. Un., 5,000 H9 27
Bolshaya Uzen, riv., Sov. Un D3 29

Bolshaya Viska, Sov. Un., 10,000 ... G8 27
Bolshaya Yugan, riv., Sov. Un B8 29
Bolshevik, isl., Sov. Un ... B5 4
Bolshoya Hamenka, riv., Sov. Un ... q22 27
Bolshoy Tokmak, Sov. Un., 10,000 ... H10 27
Bolsward, Neth., 8,300 ... A5 15
Bolton, Ont., Can., 2,104 ... D5 72
Bolton, Tolland, Conn., 500 (2,933▲) ... B7 84
Bolton, Eng., 160,887 ... A5 12
Bolton, Worcester, Mass., 150 (1,264▲) ... B4, C1 97
Bolton, Hinds, Miss., 797 ... C3 100
Bolton, Columbus, N.C., 617 . C5 109
Bolton, Chittenden, Vt., 40 (237▲) ... C3 120
Bolton, lake, Man., Can ... B4 71
Bolton, riv., Man., Can ... B4 71
Bolton Landing, Warren, N.Y., 900 ... B7 108
Boltonville, Orange, Vt., 105 . C4 120
Bolu, Tur., 13,700 ... B8 31
Bolus, head, Ire ... F1 11
Bolvadin, Tur., 16,100 ... C8 31
Bolzano, It., 88,799 ... A3 21
Boma, Con. L., 31,500 ... C1 48
Bomarton, Baylor, Tex., 100 . C3 118
Bomba, Libya ... F4 31
Bombala, Austl., 1,389 ... H7 51
Bombay, India, 4,152,056 (*4,300,000) ... E5 39, H4 40
Bombay, Franklin, N.Y., 400 . A2 108
Bombay Hook, isl., Del ... B7 85
Bombetoka, bay, Malag ... g9 49
Bomboma, Con. L ... A2 48
Bom Conselho, Braz., 6,840 ... C3, k5 57
Bom Despacho, Braz., 13,568 . E1 57
Bom Jardim, Braz., 1,894 ... g5 56
Bom Jardim, Braz., 2,500 ... h6 57
Bom Jesus, Braz., 1,431 ... C2 57
Bom Jesus da Lapa, Braz., 6,107 ... D2 57
Bomokandi, riv., Con. L ... A4 48
Bomongo, Con. L ... A2 48
Bomoseen, Rutland, Vt., 85 . D2 120
Bomoseen, lake, Vt ... D2 120
Bomu, riv., Cen. Afr. Rep., Con. L ... A3 48
Bon, cape, Tun ... B7 44
Bonabéri, Cam., 13,268 ... *E1 46
Bon Accord, Alta., Can., 175 . C4 69
Bonaigarh, India ... G10 40
Bon Air, Talladega, Ala., 297 . B3 78
Bon Air, Delaware, Pa., 1,000 ... *G11 114
Bon Air, White, Tenn., 200 . D8 117
Bon Air, Chesterfield, Va., 1,500 ... C7 121
Bonaire, Houston, Ga., 200 . D3 87
Bonaire, isl., Neth. W.I ... A4 60
Bonanza, Sebastian, Ark., 247 ... B1 81
Bonanza, Saguache, Col., 19 . C4 83
Bonanza, Klamath, Oreg., 297 ... E5 113
Bonanza, peak, Wash ... A5 122
Bonaparte, Van Buren, Iowa, 574 ... D6 92
Bonaparte, mtn., Wash ... A6 122
Bonaparte, riv., B.C., Can ... D7 68
Bon Aqua, Hickman, Tenn., 150 ... B4 117
Boñar, Sp., 3,682 ... A3 20
Bonar Bridge, Scot ... C4 13
Bonarlaw, Ont., Can., 82 ... C7 72
Bonaventure, Que., Can., 804 ... *B3 73
Bonaventure, co., Que., Can., 42,962 ... A3 74
Bonavista, Newf., Can., 4,186 ... D5 75
Bonavista, bay, Newf., Can ... D5 75
Bonavista, cape, Newf., Can ... D5 75
Bonavista North, dist., Newf., Can ... D4 75
Bonavista South, dist., Newf., Can ... D4 75
Bond, Eagle, Colo., 150 ... B4 83
Bond, Jackson, Ky., 600 ... C5 94
Bond, Stone, Miss., 250 ... E4 100
Bond, co., Ill., 14,060 ... E4 90
Bondeno, It., 6,000 ... E7 18
Bondo, Con. L ... A3 48
Bondoukou, I.C., 5,100 ... E4 45
Bondsville, Hampden, Mass., 950 ... B3 97
Bonduel, Shawano, Wis., 876 . D5 124
Bondurant, Polk, Iowa, 389 ... A7, C4 92
Bondurant, Sublette, Wyo., 35 ... B2 125
Bondville, Bennington, Vt., 50 ... E3 120
Bondy, Fr., 38,039 ... g10 14
Bône, Alg., 114,068 . F10 30, B6 44
Bone, gulf, Indon ... F6 35
Bone, lake, Wis ... C1 124
Bone Cave, Van Buren, Tenn., 50 ... D8 117
Bone Gap, Edwards, Ill., 245 ... E5 90
Bonesteel, Gregory, S. Dak., 452 ... D7 116
Bonetrail, Williams, N. Dak., 7 ... A2 110
Bonfield, Ont., Can., 714 ... A5 72
Bonfield, Custer, Mont., 3 . D11 102
Bonfouca, St. Tammany, La., 60 ... B8 95
Bongā, Eth ... D4 47
Bongandanga, Con. L ... A3 48
Bongor, Chad ... C3 46
Bong Son, Viet ... E8 38
Bonham, Fannin, Tex., 7,357 . C4 118
Bon Homme, co., S. Dak., 9,229 ... D8 116
Bon Homme Colony, Bon Homme, S. Dak., 100 . E8 116
Bonidee, bayou, La ... B4 95
Bonifacio, Fr., 1,895 ... D2 21
Bonifacio, strait, It ... D2 21
Bonifay, Holmes, Fla., 2,222 . G3 86
Bonilla, Beadle, S. Dak., 75 . C7 116
Bonin (Ogasawara) Islands, U.S. occ., Asia, 148 ... E8 7
Bonita, Graham, Ariz., 10 ... E6 80
Bonita, San Diego, Calif., 2,000 ... *E2 82
Bonita, Morehouse, La., 574 . B4 95
Bonita, Lauderdale, Miss., 500 ... C5 100
Bonita Springs, Lee, Fla., 356 ... F5 86
Bonito, Braz., 5,427 ... k6 57
Bonlee, Chatham, N.C., 300 . B4 109
Bonn, Ger., 143,900 (*280,000) ... C2 17
Bonne, bay, Newf., Can ... D2 75
Bonneau, Berkeley, S.C., 402 . E8 121
Bonne Bay, Newf., Can ... D3 75
Bonner, Missoula, Mont., 400 ... D3 102
Bonner, co., Idaho, 15,587 . A2 89
Bonnerdale, Hot Spring, Ark., 40 ... C2 81
Bonners Ferry, Boundary, Idaho, 1,921 ... A2 89
Bonner Springs, Wyandotte, Kans., 3,171 ... B8, C9 93
Bönnerup Strand, Den ... B4 24
Bonnétable, Fr., 2,435 ... C4 14
Bonnet, lake, Man., Can ... D4 71
Bonnet Carre, spillway and floodway, La ... B7 95
Bonne Terre, St. Francois, Mo., 3,219 ... D7 101
Bonneville, Fr., 2,913 ... C2 18
Bonneville, Mulnomah, Oreg., 150 ... B5 113
Bonneville, Fremont, Wyo., 50 ... B4 125
Bonneville, co., Idaho, 46,906 ... F7 89
Bonneville, dam, Oreg., Wash ... B4 113, D4 122
Bonneville, peak, Idaho ... G6 89
Bonney Lake, Pierce, Wash., 645 ... B3 122
Bonnie, Jefferson, Ill., 215 ... E5 90
Bonnie Doone, Cumberland, N.C., 4,481 ... *B5 109
Bonnieville, Hart, Ky., 376 . C4 94
Bonnots Mill, Osage, Mo., 210 ... C6 101
Bonny, Nig., 8,690 ... F6 45
Bonny, res., Colo ... B8 83
Bonny Blue, Lee, Va., 504 ... B2 121
Bonnyville, Alta., Can., 1,736 . B5 69
Bono, Craighead, Ark., 339 . B5 81
Bono, Lawrence, Ind., 75 ... G5 91
Bono, Lucas, Ohio, 450 ... A4 111
Bonorva, It., 6,669 ... D2 21
Bonsal, Wake, N.C., 100 ... B5 109
Bon Secour, Baldwin, Ala., 600 ... E2 78
Bonsecours, Que., Can., 224 . D5 73
Bonsucesso, Braz ... m7 56
Bontang, Indon ... E5 35
Bonthain, Indon, 6,711 ... G5 35
Bonthe, S.L., 4,404 ... E2 45
Bontoc, Phil., 4,471 ... *B6 35
Bon Wier, Newton, Tex., 350 . D6 118
Book, Catahoula, La., 20 ... C4 95
Book, cliffs, Colo ... B1 83
Booker, Lipscomb and Ochiltree, Tex., 817 ... A2 118
Booker T. Washington, nat. mon., Va ... D13 121
Booligal, Austl., 26 ... F5 51
Boolyglass, Ire ... E4 11
Boom, Pickett, Tenn., 50 ... C8 117
Boomer, Fayette, W. Va., 1,657 ... C3, D6 123
Boon, Wexford, Mich., 150 ... D5 98
Boone, Pueblo, Colo., 548 ... C6 83
Boone, Boone, Iowa, 12,468 . B4 92
Boone, Boone, Nebr., 60 ... C8 103
Boone, Watauga, N.C., 3,686 . A2 109
Boone, Norfolk, Va., 30 ... B6 121
Boone, co., Ark., 16,116 ... A2 81
Boone, co., Ill., 20,326 ... A5 90
Boone, co., Ind., 27,543 ... D4 91
Boone, co., Iowa, 28,037 ... B9 92
Boone, co., Ky., 21,940 ... B5 94
Boone, co., Mo., 55,202 ... B5 101
Boone, co., Nebr., 9,134 ... C7 103
Boone, co., W. Va., 28,764 . C3 123
Boone, riv., Iowa ... B4 92
Boone Grove, Porter, Ind., 175 ... B3 91
Boones Mill, Franklin, Va., 371 ... D3 121
Booneville, Logan, Ark., 2,690 ... B2 75
Booneville, Owsley, Ky., 143 ... C6 94
Booneville, Prentiss, Miss., 3,480 ... A5 100
Boons, pond, Mass ... C1 97
Boonsboro, Washington, Md., 1,211 ... A2 85
Boon Terrace, Washington, Pa., 1,100 ... *F1 114
Boonton, Morris, N.J., 7,981 ... B4 106
Boonville, Mendocino, Calif., 950 ... C2 82
Boonville, Warrick, Ind., 4,801 ... H3 91
Boonville, Cooper, Mo., 7,090 ... C5 101
Boonville, Oneida, N.Y., 2,403 ... B5 108
Boonville, Yadkin, N.C., 539 ... A3 109
Boot, Eng ... F5 13
Booth, Autauga, Ala., 250 ... C3 78
Booth, hill, Conn ... D4 84
Boothbay, Lincoln, Maine, 600 (1,617▲) ... E3 96
Boothbay Harbor, Lincoln, Maine, 1,850 (2,252▲) ... E3 96
Boothby, cape, Ant ... C18 5
Boothia, gulf, N.W. Ter., Can ... B15 66
Boothia, pen., N.W. Ter., Can . B14 66
Booths, creek, W. Va ... A7 123
Boothspoint, Dyer, Tenn., 30 ... A2 117
Boothville, Plaquemines, La., 550 ... E6 95
Boothwyn, Delaware, Pa., 5,000 ... *G11 114
Bootle, Eng., 82,829 ... A4 12
Booué, Gabon ... E2 46
Boporo, Lib ... E2 45
Boppard, Ger., 8,600 ... C2 17
Boque, inlet, N.C ... C6 109
Boquerón, P.R ... C2 65
Boqueron, dept., Par., 28,082 . D3 55
Boqueron, pass, Peru ... C2 58
Boquerón, riv., Pan ... k11 62
Boquete, Pan., 2,611 ... F6 62
Bor, Sov. Un., 31,700 ... D17 9
Bor, Sud., 1,632 ... D3 47
Bor, Yugo., 18,612 ... C6 22
Borah, peak, Idaho ... E5 89
Borama, Som ... C5 47
Borama, dist., Som ... C5 47
Borås, Swe., 67,700 . A6 24, I5 25
Borāzjān, Iran, 8,543 ... G5 46
Borba, Braz., 1,304 ... C3 59
Borborema, plat., Braz ... h5 57
Borculo, Neth., 2,596 ... B6 15
Bordeaux, Fr., 249,688 (*480,000) ... E3 14
Bordeaux, Davidson, Tenn., 500 ... E9 117
Bordeaux, Thurston, Wash . C2 122
Borden, Sask., Can., 208 ... E2 70
Borden (New Providence), Clark, Ind., 327 ... H6 91
Borden, co., Tex., 1,076 ... C2 118
Borden, isl., Can ... B26 4
Borden Springs, Cleburne, Ala., 500 ... B4 78
Bordentown, Burlington, N.J., 4,974 ... C3 106
Border, Lincoln, Wyo ... C2 125
Borderland, Mingo, W. Va., 300 ... D2 123
Bordertown, Austl., 1,546 ... H3 51
Bordighera, It., 8,621 ... C1 21
Bordj Amquid, Alg ... D6 44
Bordj Flamand, Alg ... D5 44
Bordj Fouchet, Alg ... D3 44
Bordj-Ménaïel, Alg., 27,920 . *B5 44
Bordj Ouallen, Alg ... E5 44
Bordj Viollette, Alg ... D4 44
Bordley, Union, Ky., 150 ... C2 94
Bordulac, Foster, N. Dak., 200 ... B7 110
Bordzon, Mong ... C1 36
Boré, Mali ... C4 45
Boren, lake, Swe ... u33 25
Borensberg, Swe., 1,230 ... u33 25
Borgå, Fin., 11,800 ... G11 25
Börger, Ger., 1,700 ... B7 15
Borger, Neth., 1,411 ... B6 15
Borger, Hutchinson, Tex., 20,911 ... B2 118
Borgholm, Swe., 2,500 ... I7 25
Borghorst, Ger., 15,500 ... A2 17
Borgne, lake, La ... D6 95
Borgne, riv., Switz ... D3 19
Borgomanero, It., 7,900 ... B2 21
Borgo Piave, It ... k9 21
Borgo Val di Taro, It., 11,809 ... B2 21
Borgo Valsugana, It., 5,141 . C7 18
Borikhane, Laos ... C5 38
Borinquen, P.R ... B2 65
Borislav, Sov. Un., 47,500 ... G4 27
Borisoglebsk, Sov. Un., 56,000 ... C2 29
Borisov, Sov. Un., 63,000 ... D7 27
Borispol, Sov. Un., 25,000 ... F8 27
Borja, Sp., 4,381 ... B5 20
Borjas Blancas, Sp., 5,082 ... B6 20
Borken, Ger., 12,300 ... B1 17
Borken, Ger., 4,300 ... B4 17
Børkop, Den., 1,051 ... C3 24
Borkou, reg., Chad ... B3 46
Borkum, Ger., 5,800 ... A6 15
Borkum, isl., Ger ... B3 16
Borlänge, Swe., 26,700 ... G6 25
Bormes, Fr., 772 ... F7 14
Bormida, riv., It ... E4 18
Bormio, It., 3,293 ... C6 18
Borna, Ger., 17,800 ... B7 17
Borne, Neth., 9,800 ... B6 15
Borneo, British, see Brunei, North Borneo, and Sarawak, reg., Mala.
Borneo, North, see North Borneo, reg., Mala.
Borneo (Kalimantan), isl., Asia . E4 35
Bornheim, Ger., 12,000 ... C2 17
Bornholm, co., Den., 48,632 ... C8 24
Bornholm, isl., Den ... C8 24
Bor Nor, lake, China ... B8 34
Bornos, Sp., 8,697 ... D3 20
Boromlya, Sov. Un., 10,000 . F10 27
Boromo, Upper Volta ... D4 45
Boron, Kern, Calif., 950 ... E5 82
Borongan, Phil., 2,965 ... *C7 35
Borovan, Bul., 5,905 ... D6 22
Borovichi, Sov. Un., 47,100 . B9 27
Borovsk, Sov. Un., 31,200 . D20 9
Borovskoye, Sov. Un., 5,000 . q21 27
Borrby, Swe., 1,139 ... C8 24
Borre, Den ... D6 24
Borrego Springs, San Diego, Calif ... F5 82
Borris, Ire., 413 ... E5 11
Borrisokane, Ire., 750 ... E3 11
Borrisoleigh, Ire., 465 ... E4 11
Borroloola, Austl ... C6 50
Borsod-Abauj-Zemplen, co., Hung., 583,001 ... *A5 22
Bort-les-Orgues, Fr., 5,115 ... E5 14
Boruca, C.R., 682 ... F6 62
Borūjerd, Iran, ..49,186 ... E4 41
Borup, Norman, Minn., 145 . C2 99
Borzhomi, Sov. Un., 8,218 . B14 25
Borzna, Sov. Un., 10,000 ... F9 27
Borzonasca, It., 3,985 ... E5 18
Bosa, It., 8,169 ... D2 21
Bosanska Dubica, Yugo., 6,253 ... C3 22
Bosanska Gradiska, Yugo., 6,373 ... C3 22
Bosanska Kostajnica, Yugo., 2,037 ... C3 22
Bosanski Novi, Yugo., 7,082 . C3 22
Bosanski Petrovac, Yugo., 3,374 ... C3 22
Bosanski Samac, Yugo., 3,607 ... C4 22
Bosaso, Som., 5,200 ... C6 47
Boscawen, Merrimack, N.H., 350 (2,181▲) ... D3 105
Bosco, Ouachita, La., 100 ... B3 95
Boscobel, Grant, Wis., 2,608 . E3 124
Bosdagan, Tur., 3,842 ... D7 23
Boshrūyeh, Iran ... E8 41
Boskovice, Czech., 6,396 ... D4 26
Bosler, Albany, Wyo., 75 ... D7 125
Bosna, riv., Yugo ... C3 22
Bosnek, W. Irian, 4,500 ... F9 35
Bosnia, reg., Yugo ... C3 22
Bosnia-Hercegovina, rep., Yugo., 3,274,886 ... *C3 22
Bosobolo, Con. L ... A2 48
Bosporus, strait, Tur ... B7 31
Bosque, Valencia, N. Mex., 10 ... C3 107
Bosque, co., Tex., 10,809 ... D4 118
Bossangoa, Cen. Afr. Rep ... D3 46
Bossembélé, Cen. Afr. Rep ... D3 46
Bossier, Bossier, La., 32,776 . B2 95
Bossier, par., La., 57,622 ... B2 95
Bosso, Niger ... D7 45
Boston, Eng., 24,903 ... B7 12
Boston, Thomas, Ga., 1,357 . F3 87
Boston, Wayne, Ind., 240 ... E8 91
Boston, Suffolk, Mass., 697,197 (*2,913,500) . B5, D3 97
Boston, Summit, Ohio, 400 ... B2 111
Boston, Allegheny, Pa., 2,300 ... *F2 114
Boston, bay, Mass ... B6 97
Boston, mts., Ark ... B2 81
Boston, mts., Okla ... B7 112
Boston Bar, B.C., Can., 629 . E7 68
Boston Heights, Summit, Ohio, 831 ... B2 111
Bostonnais, riv., Que., Can ... B5 73
Bostwick, Putnam, Fla., 400 . C5 86
Bostwick, Morgan, Ga., 272 . C3 87
Bostwick, Nuckolls, Nebr., 50 . D7 103
Boswarlos, Newf., Can., 313 . D2 75
Boswell, Izard, Ark., 10 ... A3 81
Boswell, B.C., Can., 100 ... E9 68
Boswell, Benton, Ind., 957 ... C3 91
Boswell, Choctaw, Okla., 753 . C6 112
Boswell, Somerset, Pa., 1,508 . F3 114
Bosworth, Carroll, Mo., 465 . B4 101
Botetourt, co., Va., 16,715 ... D3 121
Botev, peak, Bul ... D7 22
Botevgrad, Bul., 5,925 ... D6 22
Botha, Alta., Can., 112 ... C4 69
Bothell, King, Wash., 2,237 . B3 122
Bothnia, gulf, Eur ... F9 25
Bothwell, Ont., Can., 819 ... E3 70
Botiala, Som ... C6 47
Botijas, P.R ... C5 65
Botkinburg, Van Buren, Ark., 175 ... B3 81
Botkins, Shelby, Ohio, 854 ... B3 111
Botkyrka, Swe., 4,348 ... t35 25
Botolan, Phil., 1,963 ... o13 35
Botoşani, Rom., 29,569 ... B8 22
Botsford, Fairfield, Conn., 150 ... D3 84
Bottineau, Bottineau, N. Dak., 2,613 ... A5 110
Bottineau, co., N. Dak., 11,315 ... A4 110
Bottrop, Ger., 111,500 ... B1 17
Botucatu, Braz., 33,878 . C3, m7 56
Botwood, Newf., Can., 2,800 . D4 75
Bouaflé, I.C., 1,700 ... E3 45
Bouaké, I.C., 45,000 ... E3 45
Bouar, Cen. Afr. Rep., 20,700 ... D3 46
Bou Arfa, Mor., 8,775 ... C4 44
Boucau, Fr., 5,400 ... F3 14
Boucher, lake, Que., Can ... C2 77
Boucherville, Que., Can., 7,403 ... D4, D9 73
Bouches-du-Rhône, dept., Fr., 1,248,355 ... *F6 14
Bouchette, Que., Can., 464 . C2 73
Boudenib, Mor ... C4 44
Bou Djébena, Mali ... C4 45
Boudreaux, lake, La ... E5 95
Boudry, Switz., 3,086 ... C2 19
Boufarik, Alg., 23,024 ... *B5 44
Bougainville, isl., Solomon Is ... G9 7
Bougaroun, cape, Alg ... F10 30
Bougie, Alg., 43,934 ... B6 44
Bougival, Fr., 7,296 ... g9 14
Bougouni, Mali, 3,100 ... D3 45
Bouillon, Bel., 3,017 ... E5 15
Boulder, Austl., 5,773 ... F3 51
Boulder, Boulder, Colo., 37,718 ... A5 83
Boulder, Jefferson, Mont., 1,394 ... D4 102
Boulder, Garfield, Utah, 108 . F4 119
Boulder, Sublette, Wyo., 30 . C3 125
Boulder, co., Colo., 74,254 ... A5 83
Boulder City, Clark, Nev., 4,059 ... H7 104
Boulder Creek, Santa Cruz, Calif., 1,306 ... C5 82
Boulevard, San Diego, Calif., 25 ... F5 82
Boulevard Heights, Prince Georges, Md., 750 ... C2 85
Boulloum (Well), Niger ... C7 45
Boulogne-Billancourt, Fr., 106,641 ... C5, g9 14
Boulogne-sur-Mer, Fr., 49,281 (*98,000) . D9 12, B4 14
Boulsa, Upper Volta ... D4 45
Boumalne, Mor ... C3 44
Bouna, I.C., 2,300 ... E4 45
Boundary, Alsk., 10 ... C11 79
Boundary, co., Idaho, 5,809 . A2 89
Boundary, peak, Nev ... F3 104
Boundary, plat., Sask., Can ... H1 70
Bound Brook, Somerset, N.J., 10,263 ... B3 106
Boundiali, I.C., 2,800 ... E3 45
Bountiful, Davis, Utah, 17,039 ... C4 119
Bounty, Sask., Can., 87 ... F2 70
Bounty, is., Pac. O ... J11 7
Bourbeuse, riv., Mo ... C6 101
Bourbon, Marshall, Ind., 1,522 ... B5 91
Bourbon, Crawford, Mo., 779 ... C6 101
Bourbon, co., Kans., 16,090 . E9 93
Bourbon, co., Ky., 18,178 ... B5 94
Bourbonnais, Kankakee, Ill., 3,336 ... B6 90
Bourbonnais, former prov., Fr., 318,000 ... D5 14
Bourbonne-les-Bains, Fr., 2,617 ... A1 18
Bourem, Mali, 1,700 ... C4 45
Bourg, Terrebonne, La., 900 . E5 95
Bourg-Bruche, Fr., 444 ... A3 18
Bourg-de-Péage, Fr., 7,804 ... E6 14
Bourg [-en-Bresse], Fr., 32,596 ... D6 14
Bourges, Fr., 60,632 ... D5 14
Bourget, Ont., Can., 769 ... B9 72
Bourget, lake, Fr ... D1 18
Bourg-la-Reine, Fr., 17,694 . g10 14
Bourgoin, Fr., 9,240 ... E6 14
Bourg-St.-Andéol, Fr., 3,152 . E6 14
Bourg-St. Maurice, Fr., 2,666 ... D2 18
Bourjeïmat (Well), Maur ... C1 45
Bourke, Austl., 3,001 ... E5 51
Bour-khaya, bay, Sov. Un ... B2 4
Bourlamaque, Que., Can., 3,344 ... *E9 72
Bourne, Eng., 5,339 ... B7 12
Bourne, Barnstable, Mass., 750 (14,001▲) ... C6 97
Bournemouth, Eng., 153,965 (*275,000) ... D6 12
Bourneville, Ross, Ohio, 275 . C4 111
Bou Saada, Alg., 11,661 ... B5 44
Bouse, Yuma, Ariz., 150 ... D2 80
Bousso, Chad, 2,573 ... C3 46
Boutaiha, Syr ... B7 32
Boutilimit, Maur ... C2 45
Boutte, St. Charles, La., 155 . C6 95
Bouvet, isl., Atl. O ... D13 5
Bovey, Itasca, Minn., 1,086 . C5 99
Bovill, Latah, Idaho, 357 ... C2 89
Bovina, Lincoln, Colo., 25 ... B7 85
Bovina, Warren, Miss., 25 ... C3 100
Bovina, Parmer, Tex., 1,029 . B1 181
Bovina Center, Delaware, N.Y., 275 ... C6 108
Bow, Merrimack, N.H., 300 (1,340▲) ... D3 105
Bow, lake, N.H ... D4 105
Bow, riv., Alta., Can ... D4 69
Bowbells, Burke, N. Dak., 687 ... A3 110
Bowden, Alta., Can., 437 ... D3 69
Bowden, Creek, Okla., 35 ... A5 112
Bowdens, Duplin, N.C., 300 . B5 109
Bowdle, Edmunds, S. Dak., 678 ... B6 116
Bowdoin, Sagadahoc, Maine, 75 (668▲) ... D6 96
Bowdoinham, Sagadahoc, Maine, 375 (1,131▲) ... D3, D6 96
Bowdon, Carroll, Ga., 1,548 . C1 87
Bowdon, Wells, N. Dak., 259 ... B6 110
Bowen, Austl., 5,160 ... C8 50
Bowen, Hancock, Ill., 559 ... C2 90
Bowen, cape, N.W. Ter., Can . B18 67
Bowens Mill, Ben Hill, Ga., 25 ... E3 87
Bowers, Kent, Del., 324 ... B7 85
Bowers, coulee, Wash ... B7 122
Bowers Hill, Norfolk, Va., 800 ... B6 121
Bowerston, Harrison, Ohio, 463 ... B6 111
Bowersville, Hart, Ga., 293 . B3 87
Bowersville, Greene, Ohio, 327 ... C4 111
Bowesmont, Pembina, N. Dak., 175 ... A8 110
Bowie, Cochise, Ariz., 500 ... E6 80
Bowie, Delta, Colo., 100 ... C3 83
Bowie, Prince Georges, Md., 1,072 ... B4 85
Bowie, Montague, Tex., 4,566 ... C4 118
Bowie, co., Tex., 59,971 ... C5 118
Bowie, creek, Miss ... D4 100
Bow Island, Alta., Can., 1,122 ... E5 69
Bowlegs, Seminole, Okla., 200 ... B5 112
Bowler, Shawano, Wis., 274 ... D5 124
Bowling Green, Hardee, Fla., 1,171 ... E5 86
Bowling Green, Clay, Ind., 229 ... F3 91
Bowling Green, Warren, Ky., 28,338 ... D3 94
Bowling Green, Pike, Mo., 2,650 ... B6 101
Bowling Green, Wood, Ohio, 13,574 ... A2, A4 111
Bowling Green, York, S.C., 700 ... A5 115
Bowling Green, Caroline, Va., 528 ... C5 121
Bowlus, Morrison, Minn., 263 ... E4 99
Bowman, Elbert, Ga., 654 ... B3 87
Bowman, Bowman, N. Dak., 1,730 ... C2 110
Bowman, Orangeburg, S.C., 1,106 ... E6 115
Bowman, co., N. Dak., 4,154 ... C2 110
Bowman, creek, Pa ... A7 114
Bowman, mtn., B.C., Can ... D7 68
Bowmanville, Ont., Can., 7,397 ... D6 72
Bowmont, Canyon, Idaho, 25 ... F2 89
Bowmore, Scot ... E2 13
Bowness, Alta., Can., 9,184 . D3 69
Bowokan, is., Indon ... F6 35
Bowring, Osage, Okla., 100 . A5 112
Bowringpet, India, 11,360 ... F6 39
Bowser, lake, B.C., Can ... A2 68
Bowsman, Man., Can., 519 . C1 71
Bowstring, lake, Minn ... C5 99
Bowstring, riv., Minn ... C5 99
Bow Valley, Cedar, Nebr., 75 . B8 103
Box, creek, Wyo ... B7 125
Boxboro, Middlesex, Mass., 100 (744▲) ... A5, C1 97
Box Butte, co., Nebr., 11,688 ... B2 103
Box Butte, creek, Nebr ... B3 103
Box Butte Table, plat., Nebr ... B2 103
Box Elder, Hill, Mont., 230 . B6 102
Box Elder, Pennington, S. Dak., 56 ... C2 116
Box Elder, co., Utah, 25,061 . B2 119
Boxelder, creek, Colo ... A6 83
Boxelder, creek, Mont ... C8 102
Box Elder, creek, Mont ... E12 102
Boxford, Essex, Mass., 300 (2,010▲) ... A6 97
Boxholm, Boone, Iowa, 250 . B3 92
Boxmeer, Neth., 5,600 ... C5 15
Boxtel, Neth., 11,000 ... C5 15
Boyacá, dept., Col., 824,700 . B3 60
Boyce, Rapides, La., 1,094 ... C3 95
Boyce, Ellis, Tex., 60 ... B5 118
Boyce, Clarke, Va., 384 ... B4 121
Boyceville, Dunn, Wis., 660 . C1 124
Boyd, Taylor, Fla., 250 ... B3 86
Boyd, Lac qui Parle, Minn., 419 ... F3 99
Boyd, Carbon, Mont., 30 ... E7 102
Boyd, Wasco, Oreg., 30 ... B5 113
Boyd, Wise, Tex., 581 ... A5, C4 118
Boyd, Chippawa, Wis., 622 . D2 124
Boyd, co., Ky., 52,163 ... B7 94
Boyd, co., Nebr., 4,513 ... B7 103
Boyd, lake, Maine ... C4 96
Boydell, Ashley, Ark., 85 ... D4 81
Boyden, Sioux, Iowa, 562 ... A2 92
Boyds, Montgomery, Md., 85 ... B3 85
Boyds Cove, Newf., Can., 368 ... D4 75
Boyds Creek, Sevier, Tenn . D10 117
Boydsville, Clay, Ark., 25 ... A5 81
Boydton, Mecklenburg, Va., 449 ... E4 121
Boyer, riv., Iowa ... C2 92
Boyera, Con. L ... C2 48
Boyer Knob, mtn., Md ... D3 85
Boyero, Lincoln, Colo., 66 ... C7 83
Boyers, Butler, Pa., 550 ... D2 114
Boyertown, Berks, Pa., 4,067 . F10 114
Boyes, Carter, Mont., 15 ... E11 102
Boyes Hot Springs, Sonoma, Calif., 2,462 ... B5 82
Boykin, Miller, Ga., 601 ... E2 87
Boykins, Southampton, Va., 710 ... E5 121
Boyle, Alta., Can., 346 ... B4 69
Boyle, Ire., 1,739 ... D3 11
Boyle, Bolivar, Miss., 848 ... B3 100
Boyle, co., Ky., 21,257 ... C5 94
Boylston, Montgomery, Ala., 1,300 ... C3 78
Boylston, N.S., Can., 121 ... D8 74
Boylston Center (Boylston), Worcester, Mass., 400 (2,367▲) ... B4 97
Boyne, riv., Ire ... D5 11
Boyne City, Charlevoix, Mich., 2,797 ... C5 98
Boyne Falls, Charlevoix, Mich., 260 ... C6 98
Boynton, Muskogee, Okla., 604 ... B6 112
Boynton, Somerset, Pa., 300 . G3 114
Boynton Beach, Palm Beach, Fla., 10,467 ... F6 86
Boy River, Cass, Minn., 51 . C4 99
Boysen, res., Wyo ... B4 125
Boys Ranch, Oldham, Tex., 350 ... B1 118
Boys Town, Douglas, Nebr., 997 ... D3 103
Boz, cape, Tur ... B7 23
Bozcaada, Tur., 1,800 ... C6 23
Bozcaada (Tenedos), isl., Tur . C6 23
Bozeman, Gallatin, Mont., 13,361 ... E5 102
Bozeman, pass, Mont ... E6 102
Bozman, Talbot, Md., 150 ... C5 85
Bozoum, Cen. Afr. Rep ... D3 46
Bozovici, Rom., 3,431 ... C6 22
Bozrah, New London, Conn., (1,590▲) ... *C8 84
Bozuyuk, Tur., 9,200 ... C8 31
Bra., It., 13,100 ... B1 21
Braband, Den., 5,139 ... B4 24
Brabant, isl., Ant ... C6 5
Brabant, lake, Sask., Can ... A4 70
Brabant, prov., Bel., 1,992,139 ... D4 15
Brac, isl., Yugo ... D3 22
Bracadale, Scot., 969 ... C2 13
Bracadale, bay, Scot ... C2 13
Bracciano, It., 6,783 ... g8 21
Bracciano, lake, It ... C4 21
Bracebridge, Ont., Can., 2,927 ... B5 72
Braceville, Grundy, Ill., 558 . B5 90
Brach, Libya, 3,874 ... D2 43
Bracken, Sask., Can., 132 ... H1 70
Bracken, co., Ky., 7,422 ... B5 94
Brackenridge, Allegheny, Pa., 5,697 ... A7 114
Brackettville, Kinney, Tex., 1,662 ... E2 118
Brackley, Eng., 3,202 ... B6 12
Brackwede, Ger., 26,000 ... B3 17
Brad, Rom., 9,963 ... B6 22
Bradano, riv., It ... D6 21
Bradbury Heights, Prince Georges, Md., 1,100 ... *C3 85
Braddock, Camden, N.J., 300 ... D3 106
Braddock, Emmons, N. Dak., 141 ... C5 110
Braddock, Allegheny, Pa., 12,337 ... B6 114
Braddock, pt., S.C ... G6 115
Braddock Heights, Frederick, Md., 600 ... B2 85
Braddock Hills, Allegheny, Pa., 2,414 ... *B5 114
Braddyville, Page, Iowa, 176 ... D2 92
Braden, Fayette, Tenn., 500 . B2 117
Bradenton, Manatee, Fla., 19,380 ... E4, F2 86
Bradenton Beach, Manatee, Fla., 1,124 ... *E4 86
Bradenville, Westmoreland, Pa., 1,000 ... F3 114
Bradford, White, Ark., 779 ... B4 81
Bradford, Ont., Can., 2,342 . C5 72
Bradford, Eng., 295,768 ... A6 12
Bradford, Stark, Ill., 857 ... B4 90
Bradford, Franklin, Iowa, 200 ... B4 92
Bradford, Penobscot, Maine, 150 (670▲) ... C4 96
Bradford, Merrimack, N.H., 300 (508▲) ... D6 105
Bradford, Darke and Miami, Ohio, 2,148 ... B3 111
Bradford, McKean, Pa., 15,061 ... C4 114
Bradford, Washington, R.I., 950 ... D10 84
Bradford, Gibson, Tenn., 763 ... A3 117
Bradford, Orange, Vt., 760 (1,619▲) ... D4 120
Bradford, co., Fla., 12,446 ... C4 88
Bradford, co., Pa., 54,925 ... C8 114
Bradford, mtn., Conn ... B3 84
Bradford Center, Penobscot, Maine, 175 ... C4 96
Bradfordsville, Marion, Ky., 387 ... C4 94
Bradfordwoods, Allegheny, Pa., 866 ... A5 114
Bradgate, Humboldt, Iowa, 166 ... B3 92
Bradley, Escambia, Ala., 100 . D3 78
Bradley, Lafayette, Ark., 712 ... D2 81
Bradley, Monterey, Calif., 60 . E3 82
Bradley, Polk, Fla., 1,035 ... E5 86
Bradley, Kankakee, Ill., 8,082 ... B6 90
Bradley, Penobscot, Maine, 500 (951▲) ... D4 96
Bradley, Jefferson, Ohio, 500 ... B7 111
Bradley, Grady, Okla., 294 ... C4 112
Bradley, Greenwood, S.C., 135 ... C3 115
Bradley, Clark, S. Dak., 188 . B8 116
Bradley, co., Ark., 14,029 ... D3 81
Bradley, co., Tenn., 38,324 ... D9 117
Bradley Beach, Monmouth, N.J., 4,204 ... C4 106
Bradley Gardens, Somerset, N.J., 1,800 ... *B3 106
Bradleyton, Crenshaw, Ala., 100 ... D3 78
Bradner, Wood, Ohio, 994 ... A4 111
Bradore, bay, Newf., Can ... C3 75
Bradore, hills, Newf., Can ... C3 75
Bradshaw, Vigo, Ind., 150 ... E2 91
Bradshaw, York, Nebr., 306 . D8 103
Bradshaw, Taylor, Tex., 75 ... C3 118
Bradshaw, McDowell, W. Va., 950 ... D3 123
Bradshaw, mts., Ariz ... C3 80
Bradstreet, Hampshire, Mass., 100 ... B2 97
Bradwardine, Man., Can., 75 ... D1 71
Bradwell, Sask., Can., 115 ... F2 70
Brady, Pondera, Mont., 180 . B5 102

Brady, Lincoln, Nebr., 273...C5 103
Brady, Cherokee, N.C...D2 109
Brady, McCulloch, Tex., 5,338...C3 118
Brady, Lincoln, W. Va...C2 123
Brady, mts., Tex...D3 118
Brady's Hot Springs, Churchill, Nev., 5...D3 104
Bradyville, Cannon, Tenn., 75...B5 117
Braedstrup, hill, Den...C3 24
Braemar, Scot., 1,291...C5 13
Braeside, Ont., Can., 528...B8 72
Braga, Port., 41,000...B1 20
Bragado, Arg., 16,104...B4, g6 54
Bragança, Braz., 12,848...B1 57
Bragança, Port., 8,245...B2 20
Bragança Paulista, Braz., 27,328...C3, m8 56
Braggadocio, Pemiscot, Mo., 450...E8 101
Bragg City, Pemiscot, Mo., 274...E8 101
Braggs, Muskogee, Okla., 279...B6 112
Braham, Isanti, Minn., 728...E5 99
Brahmani, riv., India...G10 40
Brahmaputra, riv., Asia...C9 39
Brahme, riv., Ger...D3 24
Braidwood, Will, Ill., 1,944...B5 90
Brail, Switz...C9 19
Brăila, Rom., 102,500...C8 22
Brainard, Butler, Nebr., 300...C9 103
Braine-le-Comte, Bel., 10,779.D4 15
Brainerd, Butler, Kans., 65...B6 93
Brainerd, Crow Wing, Minn., 12,898...D4 99
Braintree, Norfolk, Mass., 31,069...B5, D3 97
Braintree, Orange, Vt., 100 (536▲)...D3 120
Braintree, mtn., Vt...D3 120
Braintree Highlands, Norfolk, Mass. (part of Braintree)...D3 97
Braithwaite, Plaquemines, La., 375...C7 95
Brake, Ger., 15,900...E2 24
Brakel, Ger., 6,400...B4 17
Brakpan, S. Afr., 77,777...C4 49
Bralorne, B.C., Can., 670...D6 68
Braman, Kay, Okla., 336...A4 112
Bramhapuri, India...G7 40
Bramminge, Den., 2,900...C2 24
Brampton, Ont., Can., 18,467...D9, E6 72
Brampton, Eng., 3,130...F6 13
Bramtpon, Sargent, N. Dak., 70...D8 110
Bramche, Ger., 10,200...B7 15
Bramwell, Mercer, W. Va., 1,195...D3 123
Branaman, Pinal, Ariz., 4...D5 80
Branch, Franklin, Ark., 258...B2 81
Branch, Newf., Can., 556...E5 75
Branch, Acadia, La., 100...D3 95
Branch, Scott, Miss., 30...C4 100
Branch, Manitowoc, Wis., 190...B6 124
Branch, co., Mich., 34,903...G5 98
Branch, pond, Maine...D4 96
Branch, riv., Wis...A6 124
Branchland, Lincoln, W. Va., 500...C2 123
Branchport, Yates, N.Y., 350.C3 108
Branchville, St. Clair, Ala., 50...B3 78
Branchville, Fairfield, Conn., 200...D3 84
Branchville, Sussex, N.J., 963.A3 106
Branchville, Orangeburg, S.C., 1,182...E6 115
Branchville, Southampton, Va., 158...E5 121
Branco, riv., Braz...C5 60
Branco, riv., Braz...E2 59
Brandberg, mtn., S.W. Afr...B1 49
Brande, Den., 4,151...C3 24
Brandenburg, Ger., 86,700...A7 17
Brandenburg, Meade, Ky., 1,542...C3 94
Brandenburg, reg., Ger...B6 16
Brandenburg, former state, Ger., 2,527,492...C8 17
Brandon, Man., Can., 28,166...E2 71
Brandon, Kiowa, Colo., 75...C8 83
Brandon, Hillsborough, Fla., 1,665...E4 86
Brandon, Buchanan, Iowa, 322...B6 92
Brandon, Douglas, Minn., 353...E3 99
Brandon, Rankin, Miss., 2,139...C4 100
Brandon, Perkins, Nebr., 35...D4 103
Brandon, Greenville, S.C., 1,000...B3 115
Brandon, Minnehaha, S. Dak., 200...D9 116
Brandon, Rutland, Vt., 1,675 (3,329▲)...D2 120
Brandon, Prince George, Va...D6 121
Brandon, Fond du Lac, Wis., 758...E5 124
Brandon, head, Ire...E1 11
Brandon, hill, Ire...E1 11
Brandsville, Howell, Mo., 128...E6 101
Brandt, Miami, Ohio, 450...C3 111
Brandt, Susquehanna, Pa., 60...C10 114
Brandt, Deuel, S. Dak., 148...C9 116
Brandvlei, S. Afr., 1,417...D3 49
Brandys nad Labem, Czech., 13,300...n18 26
Brandy Station, Culpeper, Va., 200...C5 121
Brandywine, Prince Georges, Md., 80...C4 85
Brandywine, Pendleton, W. Va., 200...C5 123
Brandywine, creek, Del...A6 85
Branford, New Haven, Conn., 2,371 (16,610▲)...D5 84
Branford, Suwannee, Fla., 663...C4 86
Branford Point, New Haven, Conn., 100...D5 84
Braniewo, Pol., 1,373...A5 26
Bransfield, strait, Ant...C7 5
Bransk, Pol., 2,542...B7 26
Branson, Las Animas, Colo., 124...D7 83
Branson, Taney, Mo., 1,887...E4 101
Brant, Alta., Can., 76...D4 69
Brant, co., Ont., Can., 83,839...D4 72
Brantford, Ont., Can., 55,201 (*56,741)...D4 72
Brantford, Eddy, N. Dak., 55...B7 110
Brant Lake, Warren, N.Y., 800...B7 108
Brantley, Crenshaw, Ala., 1,014...D3 78
Brantley, co., Ga., 5,891...E4 87
Brant Rock, Plymouth, Mass., 325...B6 97
Brantwood, Price, Wis., 40...C3 124
Bras d'Apic, Que., Can., 56...C7 73
Bras d'Or, lake, N.S., Can...D9 74
Brasfield, Prairie, Ark., 200...C4 81
Brashear, Adiar, Mo., 309...A5 101
Brasiléia, Braz., 1,852...D4 58
Brasília, Braz., 150,000...B3 56, E1 57
Brasília, Braz., 3,182...E2 57
Brașov, Rom., 123,882...C7 22
Brass, Nig...F6 45
Brass, is., Vir. Is...f15 65
Brasstown, Clay, N.C., 200...D2 109
Brasstown Bald, mtn., Ga...B3 87
Brassua, lake, Maine...C3 96
Bratenahl, Cuyahoga, Ohio, 1,332...B2 111
Bratislava, Czech., 242,100...D4 26
Bratsk, Sov. Un., 63,000...D29 9
Bratslav, Sov. Un., 10,000...G7 27
Bratt, Escambia, Fla., 150...G1 86
Brattleboro, Windham, Vt., 11,734...F3 120
Braunau [am Inn], Aus., 14,449...D6 16
Braunfels, Ger., 3,700...C3 17
Braunschweig (Brunswick), Ger., 246,200...A5 17
Brava, Som., 3,000 (7,000▲)...E5 47
Brave, Greene, Pa., 400...G1 114
Bråviken, lake, Swe...u34 25
Bravo, riv., Chile...D2 54
Brawley, Imperial, Calif., 13,752...F6 82
Brawsley, peaks, Calif...C4 82
Braxton, Simpson, Miss., 191...C4 100
Braxton, co., W. Va., 15,152...C4 123
Bray, Ire., 11,688...D5 11
Bray, head, Ire...D5 11
Bray, head, Ire...F1 11
Braymer, Caldwell, Mo., 874...B4 101
Brayton, Audubon, Iowa, 225...C3 92
Brayton, Bledsoe, Tenn...D8 117
Brazeau, riv., Alta., Can...C2 69
Brazeau, mtn., Alta., Can...C2 69
Brazil, Clay, Ind., 8,853...E3 91
Brazil, Appanoose, Iowa, 300...D5 92
Brazil, Gibson, Tenn., 200...B2 117
Brazil, country, S.A., 70,967,185.D5 53, C3 56, C2 57
Brazil Lake, N.S., Can., 87...F4 74
Brazilton, Crawford, Kans., 80...E9 93
Brazoria, Brazoria, Tex., 1,291...E5, G4 118
Brazoria, co., Tex., 76,204...E5 118
Brazos, Rio Arriba, N. Mex...A3 107
Brazos, co., Tex., 44,895...D4 118
Brazos, peak, N. Mex...A3 107
Brazos, riv., Tex...D4 118
Brazzaville, Con. B., 99,000...F3 46
Brčko, Yugo., 17,834...C4 22
Brda, riv., Pol...B4 26
Brea, Orange, Calif., 8,487...F3 82
Brea, pt., P.R...D3 65
Breaden, lake, Austl...E4 51
Bread Loaf, mtn., Vt...D3 120
Breakenridge, mtn., B.C., Can...E7 68
Breaker, pt., Am. Sam...52
Breakeyville, Que., Can., 460...C9 73
Breakneck, hill, Md...D2 85
Breathitt, co., Ky., 15,490...C6 94
Breaux Bridge, St. Martin, La., 3,303...D4 95
Brechin, Scot., 7,114...D6 13
Brechin, Ont., Can., 264...C5 72
Breckenridge, Summit, Colo., 393...B4 83
Breckenridge, Gratiot, Mich., 1,131...E6 98
Breckenridge, Wilkin, Minn., 4,335...D2 99
Breckenridge, Caldwell, Mo., 605...B4 101
Breckenridge, Stephens, Tex., 6,273...C3 118
Breckenridge Hills, St. Louis, Mo., 6,299...*A8 101
Breckenridge Station, Que., Can...A9 72
Breckinridge, Garfield, Okla., 42...A4 112
Breckinridge, co., Ky., 14,734...C3 94
Breckinridge, mtn., Calif...E4 84
Brecknock, co., Wales, 55,544...C4 12
Brecksville, Cuyahoga, Ohio, 5,435...A6, B2 111
Breclav, Czech., 11,800...D4 26
Brecon, Wales, 6,466...C4 12
Brecon Beacons, mts., Wales...C4 12
Breda, Carroll, Iowa, 543...B3 92
Breda, Neth., 108,700...C4 15
Bredaryd, Swe., 723...B7 24
Bredasdorp, S. Afr., 4,686...D3 49
Bredenbury, Sask., Can., 484...G4 70
Bredstedt, Ger., 4,100...D2 24
Bredy, Sov. Un., 3,800...E21 9
Bree, Bel., 7,031...C5 15
Breedenton, Meigs, Tenn...D9 117
Breese, Clinton, Ill., 2,461...E4 90
Bregalnica, riv., Yugo...E6 22
Bregenz, Aus., 21,428...E4 16
Bregovo, Bul., 5,271...C6 22
Breien, Morton, N. Dak., 14...C5 110
Breil [sur Roya], Fr., 1,994...F7 14
Breisach, Ger., 4,900...A3 18
Breitenbush, Marion, Oreg...C5 113
Brejo, Braz., 3,084...B2 57
Bremen, Cullman, Ala., 65...B2 78
Bremen, Haralson, Ga., 3,132...C1 87
Bremen, Ger., 564,500...B4 16, E2 24
Bremen, Marshall, Ind., 3,062...B5 91
Bremen, Muhlenberg, Ky., 328...C2 94
Bremen, Wells, N. Dak., 87...B6 110
Bremen, Fairfield, Ohio, 1,417...C5 111
Bremen, state, Ger., 706,400...E2 24
Bremer, co., Iowa, 21,108...B5 92
Bremerhaven, Ger., 141,800...B4 16, E2 24
Bremerton, Kitsap, Wash., 28,922...B3, D1 122
Bremerton East (Enetai), Kitsap, Wash., 2,539...*B3 122
Bremervörde, Ger., 9,300...E3 24
Bremgarten, Switz., 4,155...*B5 19
Bremo Bluff, Fluvanna, Va., 100...D4 121
Bremond, Robertson, Tex., 803...D4 118
Brendon, hills, Eng...C4 12
Brenham, Washington, Tex., 7,740...D4 118
Brenish, Scot...B1 13
Brenner, pass, Aus., It.E5 16, C7 18
Brent, Bibb, Ala., 1,879...C2 78
Brent, Ont., Can., 56...A6 72
Brent, Escambia, Fla., 7,000...G2 86
Brenta, riv., It...D7 18
Brentford, Spink, S. Dak., 96...B7 116
Brentford [& Chiswick], Eng., 54,832...m11 10
Brenton, Wyoming, W. Va., 500...*D3 123
Brenton, pt., R.I...D11 84
Brentwood, Contra Costa, Calif., 2,186...B6 82
Brentwood, Prince Georges, Md., 3,693...C1 85
Brentwood, St. Louis, Mo., 12,250...C7 101
Brentwood, Rockingham, N.H., 75 (1,072▲)...E4 105
Brentwood, Suffolk, N.Y., 15,387...F4 84
Brentwood, Allegheny, Pa., 13,706...B6 114
Brentwood, Williamson, Tenn., 350...A5, E9 117
Brescia, It., 172,744.D6 18, B3 21
Breskens, Neth., 3,700...C3 15
Breslau, see Wroclaw, Pol.
Bressanone, It., 9,100...A3 21
Bressler, Dauphin, Pa., 1,000...*F8 114
Bressuire, Fr., 6,528...D3 14
Brest, Fr., 136,104...C1 14
Brest, Sov. Un., 78,000...E4 27
Breteuil, Fr., 2,748...E2 15
Brétigny [-sur-Orge], Fr., 6,830...F2 15
Breton, Alta., Can., 428...C3 69
Breton, isl., La...E6 95
Breton, sound, La...E6 95
Breton, strait, Fr...D3 14
Bretton Woods, Ocean, N.J., 1,292...C4 106
Bretton Woods, Eaton, Mich., 1,500...*F6 98
Bretton Woods, Coos, N.H., 15...B4 105
Breuil, It., 142...D3 18
Brevard, Transylvania, N.C., 4,857...D3 109
Brevard, co., Fla., 111,435...D6 86
Breves, Braz., 2,051...C4 59
Brevik, Nor., 2,400...p27 25
Brevnov, Czech. (part of Prague)...n17 26
Brevoort, lake, Mich...B6 98
Brewarrina, Austl., 1,225...D6 51
Brewer, Penobscot, Maine, 9,009...D4 96
Brewer, Perry, Mo., 200...D8 101
Brewers, Hond...C5 62
Brewster, Thomas, Kans., 317...C2 93
Brewster, Barnstable, Mass., 500 (1,236▲)...C7 97
Brewster, Nobles, Minn., 500.G3 99
Brewster, Blaine, Nebr., 44...C6 103
Brewster, Putnam, N.Y., 1,714...D3, D7 108
Brewster, Stark, Ohio, 2,025...B6 111
Brewster, Okanogan, Wash., 940...A6 122
Brewster, co., Tex., 6,434...E1 118
Brewster, isl., Mass...D3 97
Brewton, Escambia, Ala., 6,309...D2 78
Brewton, Laurens, Ga., 100...D4 87
Brežice, Yugo., 2,625...C2 22
Breznice, Czech., 2,385...D8 17
Breznik, Bul., 197...D6 22
Brezno [nad Hronom], Czech., 9,900...D5 26
Bria, Cen. Afr. Rep...D4 46
Brian Boru, peak, B.C., Can...B4 68
Briançon, Fr., 7,570...E7 14
Brian Head, mtn., Utah...F3 119
Briar Bluff, Henry, Ill...D7 92
Briarcliff, Delaware, Pa., 9,000...*G11 114
Briarcliff Manor, Westchester, N.Y., 5,105...D7 108
Briare, Fr., 4,114...D5 14
Brice, Franklin, Ohio, 180...C2 111
Bricelyn, Faribault, Minn., 542...G5 99
Briceville, Anderson, Tenn., 1,217...C9 117
Brickaville, Malag...g9 49
Brickeys, Lee, Ark., 62...C5 81
Bricktown, Ocean, N.J., 800.C4 106
Brickyard, Russell, Ala., 100...C4 78
Bridal Veil, Multnomah, Oreg., 60...B4 113
Bridal Veil, falls, Utah...C4 119
Bride, riv., Ire...E3 11
Brideport, Saginaw, Mich., 1,326...E7 98
Bridesville, B.C., Can., 112...E8 68
Bridge, Cassia, Idaho, 10...G5 89
Bridge, Coos, Oreg., 200...D3 113
Bridge, riv., B.C., Can...D6 68
Bridgeboro, Worth, Ga., 200...E3 87
Bridgeboro, Burlington, N.J., 950...C3 106
Bridge City, Jefferson, La., 3,000...*C7 95
Bridge City, Orange, Tex., 4,677...*D6 118
Bridgedale, Jefferson, La., 15,000...*C7 95
Bridgeford, Sask., Can., 66...G2 70
Bridgehampton, Suffolk, N.Y., 900...D4 108
Bridge Lake, B.C., Can...D7 68
Bridgeland, Duchesne, Utah, 10...C5 119
Bridgend, Wales, 15,156...C4 12
Bridgeport, Jackson, Ala., 2,906...A4 78
Bridgeport, Mono, Calif., 300...C4 82
Bridgeport, Ont., Can., 1,672...*D4 72
Bridgeport, Fairfield, Conn., 156,748 (*322,800)...E4 84
Bridgeport, Lawrence, Ill., 2,260...E6 90
Bridgeport, Marion, Ind., 700...H7 91
Bridgeport, Saginaw, Mich., 1,326...E7 98
Bridgeport, Morrill, Nebr., 1,645...C2 103
Bridgeport, Gloucester, N.J., 500...D2 106
Bridgeport, Belmont, Ohio, 3,824...B7 111
Bridgeport, Caddo, Okla., 139...B3 112
Bridgeport, Baker, Oreg., 10.C9 113
Bridgeport, Montgomery, Pa., 5,306...A10, F11 114
Bridgeport, Wise, Tex., 3,218.C4 118
Bridgeport, Douglas, Wash., 876...B6 122
Bridgeport, Harrison, W. Va., 4,199...B4, B7 123
Bridger, Carbon, Mont., 824.E8 102
Bridger, Ziebach, S. Dak...C4 116
Bridger, basin, Wyo...D3 125
Bridger, mts., Wyo...B4 125
Bridger, peak, Wyo...D5 125
Bridger, range, Mont...E6 102
Bridgeton, Parke, Ind., 350...E3 91
Bridgeton, St. Louis, Mo., 7,820...A8, C7 101
Bridgeton, Cumberland, N.J., 20,966...E2 106
Bridgeton, Craven, N.C., 638...B6 109
Bridgetown, Barbados, 11,289 (*110,000)...p17 64
Bridgetown, N.S., Can., 1,043...E4 74
Bridgetown, Hamilton, Ohio, 5,000...*D2 111
Bridgeview, Cook, Ill., 7,334...*F3 90
Bridgeville, N.S., Can., 162...D7 74
Bridgeville, Sussex, Del., 1,469...C6 85
Bridgeville, Allegheny, Pa., 7,112...B5 114
Bridgewater, Austl., 267...o15 50
Bridgewater, N.S., Can., 4,497...E5 74
Bridgewater, Litchfield, Conn., 250 (898▲)...C3 84
Bridgewater, Adair, Iowa, 225...C3 92
Bridgewater, Aroostook, Maine, 700 (999▲)...B5 96
Bridgewater, Plymouth, Mass., 4,296 (10,276▲)...C6 97
Bridgewater, Grafton, N.H., 40 (293▲)...C3 105
Bridgewater, McCook, S. Dak., 694...D8 116
Bridgewater, Windsor, Vt., 175 (776▲)...D3 120
Bridgewater, Rockingham, Va., 1,815...C4 121
Bridgewater Corners, Windsor, Vt., 175 (776▲)...D3 120
Bridgman, Berrien, Mich., 1,454...G4 98
Bridgnorth, Eng., 7,552...B5 12
Bridgton, Cumberland, Maine, 1,715 (2,707▲)...D2 96
Bridgwater, Eng., 25,582...C5 12
Bridgwater, bay, Eng...C4 12
Bridlington, Eng., 26,007...C6 10
Bridport, Eng., 6,517...D5 12
Bridport, Addison, Vt., 100 (653▲)...D2 120
Brielle, Monmouth, N.J., 2,619...C4 106
Brienz, Switz., 2,864...C5 19
Brienz, lake, Switz...C5 19
Brier, creek, Ga...C5 87
Briercrest, Sask., Can., 175...G3 70
Brierville, Alta., Can...B5 69
Brierfield, Bibb, Ala., 90...B3 78
Brier Hill, St. Lawrence, N.Y., 400...B1 108
Brig, Switz., 4,647...D4 19
Brigantine, Atlantic, N.J., 4,201...E4 106
Brigantine, beach, N.J...E4 106
Brig Bay, Newf., Can., 157...C3 75
Brigden, Ont., Can., 620...E2 72
Brigg, Eng., 4,906...A7 12
Briggs, Burnet, Tex., 150...D4 118
Briggsdale, Weld, Colo., 120.A6 83
Briggsville, Yell, Ark., 60...C2 81
Briggsville, Berkshire, Mass., 300...A1 97
Briggsville, Marquette, Wis., 200...E4 124
Brigham City, Box Elder, Utah, 11,728...B3 119
Brig, Harbour, isl., Newf., Can...A3 75
Brighouse, B.C., Can., 146...A9 68
Bright, Ont., Can., 307...D4 72
Bright, Niobrara, Wyo...B8 125
Bright, lake, Ont., Can...B7 98
Brighton, Jefferson, Ala., 2,884...B3, E4 78
Brighton, Ont., Can., 2,403...C7 72
Brighton, Adams, Colo., 7,055...B6 83
Brighton, Eng., 162,757 (*400,000)...D7 12
Brighton, Highlands, Fla., 50.E5 86
Brighton, Macoupin and Jersey, Ill., 1,248...D3 90
Brighton, Washington, Iowa, 724...C6 92
Brighton, Somerset, Maine, (62▲)...C3 96
Brighton, Livingston, Mich., 2,282...F7 98
Brighton, Monroe, N.Y., 27,849...*B3 108
Brighton, Tillamook, Oreg., 110...B3 113
Brighton, Tipton, Tenn., 652...B2 117
Brighton, Kenosha, Wis., 35...F1 124
Brightsand, lake, Sask., Can...D1 70
Brightshade, Clay, Ky., 200...C6 94
Brightwaters, Suffolk, N.Y., 3,193...E3 108
Brigie, hill, Scot...F3 13
Brignac, Ascension, La...B6 95
Brignoles, Fr., 5,347...F7 14
Brigus, Newf., Can., 704...E5 75
Brihuega, Sp., 2,229...B4 20
Brijnagar, India, 14,643...E6 40
Brilhante, riv., Braz...C2 56
Brilliant, Marion, Ala., 749...A2 78
Brilliant, B.C., Can., 590...E9 68
Brilliant, Jefferson, Ohio, 2,174...B7 111
Brillion, Calumet, Wis., 1,783...A6, D5 124
Brilon, Ger., 11,900...B3 17
Brimfield, Peoria, Ill., 656...C4 90
Brimhall, McKinley, N. Mex., 200...B1 107
Brimley, Chippewa, Mich., 400...B6 98
Brimson, Grundy, Mo., 107...A4 101
Brimson, St. Louis, Minn., 50...C7 99
Brindisi, It., 70,657...D6 21
Bringhurst, Carroll, Ind., 275...C4 91
Brinje, Yugo., 997...C2 22
Brinkhaven (Gann), Knox, Ohio, 191...B5 111
Brinkley, Monroe, Ark., 4,636...C4 81
Brinkman, Greer, Okla., 14...B2 112
Brinkworth, Austl., 218...F2 51
Brinnon, Jefferson, Wash., 100...B3 122
Brinsmade, Benson, N. Dak., 110...A6 110
Brinson, Decatur, Ga., 246...F2 87
Brion, isl., Que., Can...B8 74
Brione-Verzasca, Switz., 337.D6 19
Brioude, Fr., 6,184...E5 14
Brisbane, Austl., 593,668 (*625,000)...C9 51
Brisbane, San Mateo, Calif., 5,000...*B5 82
Briscoe, Wheeler, Tex., 100...B2 118
Briscoe, co., Tex., 3,577...B2 118
Brissago, Switz., 1,845...D6 19
Bristol, N.B., Can., 643...C2 74
Bristol, Prowers, Colo., 250...C8 83
Bristol, Hartford, Conn., 45,499...C5 84
Bristol, Eng., 436,440 (*595,000)...C5 12
Bristol, Liberty, Fla., 614...B2 86
Bristol, Kendall, Ill. (part of Yorkville)...B5 90
Bristol, Elkhart, Ind., 991...A6 91
Bristol, St. Landry, La., 65...D3 95
Bristol, Anne Arundel, Md., 16...C4 85
Bristol, Grafton, N.H., 1,054 (1,470▲)...C3 105
Bristol, Bucks, Pa., 12,364...F12 114
Bristol, Bristol, R.I., 14,570...C11 84
Bristol, Day, S. Dak., 562...B8 116
Bristol, Sullivan, Tenn., 17,582...C11 117
Bristol, Addison, Vt., 1,421 (2,159▲)...C2 120
Bristol (Independent City), Va., 17,144...B2 121
Bristol, Harrison, W. Va., 300...B6 123
Bristol, Kenosha, Wis., 350...F1, F5 124
Bristol, co., Mass., 398,488...C5 97
Bristol, co., R.I., 37,146...C11 84
Bristol, bay, Alsk...D7 79
Bristol, chan., Eng...C4 12
Bristol, pond, Vt...C2 120
Bristol Ferry, Newport, R.I., 175...C11 84
Bristol Silver Mines, Lincoln, Nev., 10...E7 104
Bristol Terrace No. 2, Bucks, Pa., 1,300...*F12 114
Bristolville, Trumbull, Ohio, 400...A7 111
Bristow, Perry, Ind., 100...H4 91
Bristow, Butler, Iowa, 268...B5 92
Bristow, Boyd, Nebr., 153...B7 103
Bristow, Creek, Okla., 4,795...B5 112
Bristow, Prince William, Va., 50...B4 121
Britannia Bay, Ont., Can., 270...A9 72
Britannia Beach, B.C., Can., 771...E6 68
British, mts., Alsk...B12 79
British Borneo, see Brunei, North Borneo, and Sarawak, reg., Mala.
British Columbia, prov., Can., 1,629,082...E7 66, 68
British Guiana, dep., S.A., 524,000...C5 53, C3 59
British Honduras, dep., N.A., 90,381...H12 61, B3 62
British North Borneo, see North Borneo, reg., Mala.
Britstown, S. Afr., 2,834...D3 49
Britt, Ont., Can., 621...B4 72
Britt, Hancock, Iowa, 2,042...A4 92
Britt, St. Louis, Minn., 5...C6 99
Brittany (Bretagne), former prov., Fr., 3,072,000...C2 14
Britton, Lenawee, Mich., 622...G7 98
Britton, Oklahoma, Okla. (part of Oklahoma City)...B4 112
Britton, Marshall, S. Dak., 1,442...B8 116
Brive [-la-Gaillarde], Fr., 40,175...E4 14
Briviesca, Sp., 3,779...A4 20
Brixham, Eng., 10,679...D4 12
Brno, Czech., 314,400...D4 26
Broach, India, 73,639...G4 40
Broad, bay, Scot...B2 13
Broad, riv., Ga...B4 87
Broad, riv., S.C...C5 115
Broad, run, Va...B4 121
Broadacres, Sask., Can., 87...E1 70
Broadacres, Marion, Oreg., 65...B2 113
Broadalbin, Fulton, N.Y., 1,438...B6 108
Broadback, riv., Que., Can...F17 67
Broadbent, Coos, Oreg., 300.D2 113
Broad Brook, Hartford, Conn., 1,389...B6 84
Broad Creek, Princess Anne, Va...B7 121
Broaddus, San Augustine, Tex., 200...D5 118
Broadford, Scot...C3 13
Broadford, Smyth, Va., 600...B3 121
Broad Haven, bay, Ire...C2 11
Broadhead, creek, Pa...A2 106
Broadkill, riv., Del...C7 85
Broadkill Beach, Sussex, Del., 40...C7 85
Broadland, Beadle, S. Dak., 33...C7 116
Broadlands, Que., Can., 215...B2 73
Broadlands, Champaign, Ill., 44...D5 90
Broad Law, mtn., Scot...E5 13
Broadmead, Polk and Yamhill, Oreg., 15...B1 113
Broadmoor, El Paso, Colo., 2,000...C6 83
Broad Top, Huntingdon, Pa., 334...F5 114
Broadus, Powder River, Mont., 628...E11 102
Broadview, Sask., Can., 1,008...G4 70
Broadview, Cook, Ill., 8,588.*F2 90
Broadview, Monroe, Ind., 1,865...*F4 91
Broadview, Yellowstone, Mont., 160...D8 102
Broadview, Curry, N. Mex., 70...C6 107
Broadview Heights, Cuyahoga, Ohio, 6,209...B2 111
Broadwater, Morrill, Nebr., 235...C3 103
Broadwater, co., Mont., 2,804...D5 102
Broadway, Warren, N.J., 250.B2 106
Broadway, Lee, N.C., 466...B4 109
Broadway, Union, Ohio, 300.B4 111
Broadway, Rockingham, Va., 646...C4 121
Broager, Den., 1,601...D3 24
Brochet, Man., Can., 158...E12 66
Brock, Sask., Can., 222...F2 70
Brock, Nemaha, Nebr., 213.D10 103
Brockdell, Bledsoe, Tenn...D7 117
Brocken, mtn., Ger...B5 17
Brocket, Alta., Can., 75...E4 69
Brocket, Ramsey, N. Dak., 153...A7 110
Brockport, Monroe, N.Y., 5,256...B3 108
Brockport, Elk., Pa., 450...D4 114
Brocksburg, Keya Paha, Nebr., 15...B6 103
Brockton, Plymouth, Mass., 72,813...B5 97
Brockton, Roosevelt, Mont., 367...B12 102
Brockton, res., Mass...E3 97
Brockville, Ont., Can., 17,744...C9 72
Brockway, McCone, Mont., 185...C11 102
Brockway, Jefferson, Pa., 2,563...D4 114
Brockwell, Izard, Ark., 30...A4 81
Brocton, Edward, Ill., 380...D6 90
Brocton, Chautauqua, N.Y., 1,416...C1 108
Brod, Yugo., 28,729...C4 22
Broderick, Yolo., Calif., 5,000...A6 82
Broderick, Sask., Can., 141...F2 70
Brodeur, pen., N.W., Ter., Can.B15 66
Brodhead, Rockcastle, Ky., 762...C5 94
Brodhead, Green, Wis., 2,444...F4 124
Brodheadsville, Monroe, Pa., 500...E11 114
Brodick, Scot...E3 13
Brodnax, Brunswick and Mecklenburg, Va., 561...E4 121
Brodnica, Pol., 12,600...B5 26
Brody, Sov. Un., 10,000...F5 27
Brogan, Malheur Oreg., 100.C9 113
Brokaw, Marathon, Wis., 319...C4 124
Broken Arrow, Tulsa, Okla., 5,928...A6 112
Broken Bow, Custer, Nebr., 3,482...C6 103
Broken Bow, McCurtain, Okla., 2,087...C7 112
Brokenburg, Spotsylvania, Va., 100...C5 121
Broken Hill, Austl., 31,267...E3 51
Broken Hill, N. Rh., 22,000 (*35,300)...C4 48
Brome, Ger., 1,800...F4 24
Brome, co., Que., Can., 13,691...D5 73
Brome, lake, Que., Can...D5 73
Bromhead, Sask., Can., 98...H4 70
Bromide, Coal and Johnston, Okla., 264...C5 112
Bromley, Eng., 68,169...m13 10
Bromptonville, Que., Can., 2,726...D6 73
Bromsgrove, Eng., 34,474...B5 12
Bromyard, Eng., 1,681...B5 12
Bronaugh, Vernon, Mo., 173...D3 101
Bronco, Yoakum, Tex., 30...C1 118
Brönderslev, Den., 9,454...A3 24
Brong-Ahafo, reg., Ghana...E4 45
Bronnitsy, Sov. Un., 1,000...n18 27
Brönnöysund, Nor., 1,900...E5 25
Bronson, Levy, Fla., 707...C4 86
Bronson, Woodbury, Iowa, 225...B1 92
Bronson, Bourbon, Kans., 354...E8 93
Bronson, Branch, Mich., 2,267...G5 98
Bronson, Sabine, Tex., 500...D5 118
Bronte, Coke, Tex., 999...D2 118
Bronwood, Terrell, Ga., 400...E2 87
Bronx, borough and co., N.Y., 1,424,815 (part of New York City)...D2 108
Bronx, riv., N.Y...D6 108
Bronxville, Westchester, N.Y., 6,744...D2 108
Brook, Newton, Ind., 845...C3 91
Brookdale, Pierce, Wash. (part of Parkland)...B3 122
Brooke, Stafford, Va., 100...C5 121
Brooke, co., W. Va., 28,940...A4 123
Brooker, Bradford, Fla., 292...C4 86
Brookeville, Montgomery, Md., 140...B3 85
Brookfield, N.S., Can., 653...D6 74
Brookfield, Fairfield, Conn., 500 (3,405▲)...D3 84
Brookfield, Tift, Ga., 500...E3 87
Brookfield, Cook, Ill., 20,429.F2 90
Brookfield, Worcester, Mass., 950 (1,751▲)...B3 97
Brookfield, Linn, Mo., 5,694.B4 101
Brookfield, Carroll, N.H., 40 (145▲)...C4 105
Brookfield, Madison, N.Y., 425...C5 108

Brookfield, Orange, Vt., 35 (597▲) ... C3 120
Brookfield, Waukesha, Wis., 19,812 ... E1 124
Brookfield Center, Fairfield, Conn., 400 ... D3 84
Brookfield Mines, N.S., Can., 53 ... E5 74
Brookford, Catawba, N.C., 596 ... B2 109
Brookhaven, DeKalb, Ga., 8,000 ... *A5 87
Brookhaven, Lincoln, Miss., 9,885 ... D3 100
Brookhaven, Suffolk, N.Y., 900 ... F5 84
Brookhaven, Delaware, Pa., 5,280 ... *G11 114
Brooking, Sask., Can., 110 ... H3 70
Brookings, Curry, Oreg., 2,637 ... E2 113
Brookings, Brookings, S. Dak., 10,558 ... C9 116
Brookings, co., S. Dak., 20,046 ... C9 116
Brookland, Craighead, Ark., 301 ... B5 81
Brookland, Potter, Pa., 40 ... C6 114
Brooklands, Man., Can., 4,369 ... *E3 71
Brooklands, Oakland, Mich., 1,800 ... *F7 98
Brooklandville, Baltimore, Md., 100 ... C2 85
Brooklawn, Camden, N.J., 2,504 ... D2 106
Brooklet, Bulloch, Ga., 557 ... D5 87
Brooklin, Ont., Can., 1,531 ... D6 72
Brooklin, Hancock, Maine, 400 (525▲) ... D4 96
Brookline, Norfolk, Mass., 54,044 ... B5, D2 97
Brookline, Jackson, Mich., 1,600 ... *F6 98
Brookline, Hillsboro, N.H., 300 (795▲) ... E3 105
Brookline, Windham, Vt., 89 (127▲) ... E3 120
Brooklyn, Conecuh, Ala., 150 ... D3 78
Brooklyn, N.S., Can., 755 ... E5 74
Brooklyn, Windham, Conn., 900 (3,312▲) ... B9 84
Brooklyn, Stewart, Ga., 30 ... D2 87
Brooklyn (Lovejoy), St. Clair, Ill., 1,922 ... B8 101
Brooklyn, Schuyler, Ill., 140 ... C3 90
Brooklyn, Morgan, Ind., 866 ... E5 91
Brooklyn, Poweshiek, Iowa, 1,415 ... C5 92
Brooklyn, Jackson, Mich., 986 ... F6 98
Brooklyn, Forrest, Miss., 500 ... D4 100
Brooklyn, Cuyahoga, Ohio, 10,733 ... B2 111
Brooklyn, Susquehanna, Pa., 200 ... C10 114
Brooklyn, Pacific, Wash., 25 ... C2 122
Brooklyn, Dane and Green, Wis., 590 ... F4 124
Brooklyn, borough (Kings co.), N.Y., 2,627,319 (part of New York City) ... E1 108
Brooklyn Center, Hennepin, Minn., 24,356 ... E6 99
Brooklyn Park, Anne Arundel, Md., 1,800 ... *B4 85
Brooklyn Park, Hennepin, Minn., 10,197 ... *E6 99
Brookmere, B.C., Can., 100 ... E7 68
Brookneal, Campbell, Va., 1,070 ... D4 121
Brook Park, Pine, Minn., 108 ... E5 99
Brook Park, Cuyahoga, Ohio, 12,856 ... B2 111
Brookport, Massac, Ill., 1,154 ... F5 90
Brooks, Covington, Ala., 50 ... D3 78
Brooks, Alta., Can., 2,827 ... D5 69
Brooks, Adams, Iowa, 185 ... D3 92
Brooks, Bullitt, Ky., 150 ... A4 94
Brooks, Waldo, Maine, 550 (758▲) ... D3 96
Brooks, Red Lake, Minn., 148 ... C3 99
Brooks, Marion, Oreg., 250 ... B2 113
Brooks, Adams, Wis., 130 ... E4 124
Brooks, co., Ga., 15,292 ... F3 89
Brooks, co., Tex., 8,609 ... F3 118
Brooks, pen., B.C., Can. ... D4 68
Brooks, range, Alsk. ... B8 79
Brooksburg, Jefferson, Ind., 129 ... G7 91
Brooksby, Sask., Can., 74 ... E3 70
Brookshire, Waller, Tex., 1,339 ... E5, F4 118
Brookside, Jefferson, Ala., 999 ... E4 78
Brookside, Fremont, Colo., 163 ... C5 83
Brookside, New Castle, Del., 5,000 ... A6 85
Brookside, Morris, N.J., 300 ... B3 106
Brookston, White, Ind., 1,202 ... C4 91
Brookston, St. Louis, Minn., 144 ... D6 99
Brookston, Forest, Pa., 70 ... C4 114
Brooksville, Hernando, Fla., 3,301 ... D4 86
Brooksville, Bracken, Ky., 601 ... B5 94
Brooksville, Pottawatomie, Okla., 200 ... B5 112
Brooksville, Noxubee, Miss., 857 ... B5 100
Brookton, Washington, Maine, 160 ... C5 96
Brooktondale, Tompkins, N.Y., 300 ... C4 108
Brookvale, Clear Creek, Colo., 100 ... B5 83
Brookview, Duval, Fla., 1,500 ... *B5 86
Brookview, Dorchester, Md., 83 ... C6 85
Brookville, Ogle, Ill., 90 ... A4 90
Brookville, Franklin, Ind., 2,596 ... F7 91
Brookville, Saline, Kans., 246 ... D6 93
Brookville, Norfolk, Mass., 3,000 ... E3 97
Brookville, Ocean, N.J. ... D4 106
Brookville, Nassau, N.Y., 1,468 ... *F2 84
Brookville, Montgomery, Ohio, 3,184 ... C3 111
Brookville, Jefferson, Pa., 4,620 ... D3 114
Brookwood, Tuscaloosa, Ala., 350 ... B2 78
Broom, inlet, Scot. ... C3 13
Broomall, Delaware, Pa., 19,772 ... *B10 114
Broome, Austl., 1,222 ... C3 50
Broome, Sterling, Tex. ... D2 118
Broome, co., N.Y., 212,661 ... C5 108
Broomes Island, Calvert, Md., 450 ... D4 85
Broomfield, Boulder and Jefferson, Colo., 4,535 ... B5 83
Broomtown, val., Ala. ... A4 78
Brooten, Stearns, Minn., 661 ... E3 99
Brora, Scot. ... B5 13
Brora, riv., Scot. ... B4 13
Brösarp, Swe., 202 ... C8 24
Broseley, Butler, Mo., 200 ... E7 101
Brosna, riv., Ire. ... D4 11
Brotas de Macaubas, Braz., 1,163 ... D2 57
Brothers, Deschutes, Oreg., 10 ... D6 113
Brothers, is., Thai ... I3 38
Brotherton, Putnam, Tenn., 165 ... C8 117
Brothertown, Calumet, Wis., 65 ... B5 124
Brou, Fr., 2,526 ... C4 14
Brough, Eng., 631 ... F6 13
Brough, head, Scot. ... A5 13
Brough Ness, cape, Scot. ... B6 13
Broughton, Hamilton, Ill., 235 ... F5 90
Broughton, Clay, Kans., 90 ... C6 93
Broughton, Allegheny, Pa., 3,500 ... *B6 114
Broughton Station, Que., Can., 85 ... C6 73
Broussard, Lafayette, La., 1,600 ... D4 95
Brovst, Den., 1,640 ... A3 24
Broward, co., Fla., 333,946 ... F6 86
Browerville, Todd, Minn., 744 ... D4 99
Brown, co., Ill., 6,210 ... D3 90
Brown, co., Ind., 7,024 ... F5 91
Brown, co., Kans., 13,229 ... C8 93
Brown, co., Minn., 27,676 ... F4 99
Brown, co., Nebr., 4,436 ... B6 103
Brown, co., Ohio, 25,178 ... D4 111
Brown, co., S. Dak., 34,106 ... B7 116
Brown, co., Tex., 24,728 ... D3 118
Brown, co., Wis., 125,082 ... D6 124
Brown, cape, N.W. Ter., Can. ... C16 67
Brown or Roan, cliffs, Utah ... D6 119
Brown, pt., Wash. ... C1 122
Brown, riv., Vt. ... B2 120
Brown City, Sanilac, Mich., 993 ... E8 98
Brown Deer, Milwaukee, Wis., 11,280 ... E2 124
Brownell, Ness, Kans., 118 ... D4 93
Brownfield, Pope, Ill., 80 ... F5 90
Brownfield, Terry, Tex., 10,286 ... C1 118
Browning, Sask., Can., 50 ... H4 70
Browning, Schuyler, Ill., 300 ... C3 90
Browning, Linn and Sullivan, Mo., 412 ... A4 101
Browning, Glacier, Mont., 2,011 ... B3 102
Browning, entrance, B.C., Can. ... C2 68
Brownington, Henry, Mo., 130 ... C4 101
Brownington, Orleans, Vt., 50 (599▲) ... B4 120
Brownlee, Sask., Can., 153 ... G2 70
Brownlee, Cherry, Nebr., 36 ... B5 103
Brownlee, Baker, Oreg., 800 ... C10 113
Brownlee, dam, Idaho ... D2 89
Brown Lee Park, Calhoun, Mich., 3,307 ... *F5 98
Browns, Edwards, Ill., 251 ... E6 90
Browns, inlet, N.C. ... C6 109
Browns, val., Ala. ... B3 78
Brownsboro, Oldham, Ky., 50 ... A4 94
Brownsboro, Jackson, Oreg. ... E4 113
Brownsburg, Que., Can., 3,617 ... D3 73
Brownsburg, Hendricks, Ind., 4,478 ... E5 91
Brownsburg, Rockbridge, Va., 300 ... D3 121
Brownsdale, Mower, Minn., 622 ... G6 99
Brownsmead, Clatsop, Oreg., 140 ... A3 113
Browns Mills, Burlington, N.J., 792 ... D3 106
Brownstown, Sevier, Ark. ... D1 81
Brownstown, Fayette, Ill., 659 ... D5 90
Brownstown, Jackson, Ind., 2,158 ... G5 91
Brownstown, Cambria, Pa., 1,379 ... *F4 114
Brownstown, Lancaster, Pa., 600 ... F9 114
Brownstown, Yakima, Wash., 65 ... C5 122
Brownstown, head, Ire. ... E4 11
Browns Valley, Montgomery, Ind., 75 ... E4 91
Browns Valley, Traverse, Minn., 1,033 ... E2 99
Brownsville, Union, Ind., 250 ... E8 91
Brownsville, Edmonson, Ky., 473 ... C3 94
Brownsville, Ouachita, La., 3,000 ... *B3 95
Brownsville, Houston, Minn., 382 ... G7 99
Brownsville, Linn, Oreg., 875 ... C4 113
Brownsville, Fayette, Pa., 6,055 ... F2 114
Brownsville, Haywood, Tenn., 5,424 ... B2 117
Brownsville, Cameron, Tex., 48,040 (★141,000) ... F4 118
Brownsville, Windsor, Vt., 70 ... E4 120
Brownton, McLeod, Minn., 698 ... F4 99
Brownton, Barbour, W. Va., 500 ... B4 123
Browntown, Green, Wis., 263 ... F4 124
Brownvale, Alta., Can., 237 ... A1 69
Brownville, Jefferson, Ala., 534 ... E4 78
Brownville, Tuscaloosa, Ala., 220 ... B2 78
Brownville, Piscataquis, Maine, 900 ... C3 96
Brownville, Nemaha, Nebr., 243 ... D10 103
Brownville, Jefferson, N.Y., 1,082 ... A5 108
Brownville Junction, Piscataquis, Maine, 900 ... C3 96
Brownwood, Stoddard, Mo., 150 ... D8 101
Brownwood, Brown, Tex., 16,974 ... D3 118
Brownwood, Orange, Tex., 1,286 ... *D6 118
Brownwood, lake, Tex. ... D3 118
Broxton, Coffee, Ga., 907 ... E4 87
Broye, riv., Switz. ... C2 19
Brozas, Sp., 5,634 ... C2 20
Bruay-en-Artois, Fr., 30,902 (★110,000) ... B5 14, D2 15
Bruce, Alta., Can., 171 ... C4 69
Bruce, Walton, Fla., 225 ... G3 86
Bruce, Calhoun, Miss., 1,698 ... B4 100
Bruce, Brookings, S. Dak., 272 ... C9 116
Bruce, Rusk., Wis., 815 ... C2 124
Bruce, co., Ont., Can., 43,036 ... C3 72
Bruce, mtn., Austl. ... D2 50
Bruce Crossing, Ontonagon, Mich., 130 ... A6 98
Brucefield, Ont., Can., 89 ... D3 72
Bruceton, Carroll, Tenn., 1,158 ... A3 117
Bruceton Mills, Preston, W. Va., 209 ... B5 123
Brucetown, Frederick, Va., 200 ... B4 121
Bruceville, Knox, Ind., 623 ... G3 91
Bruchsal, Ger., 22,600 ... D3 17
Bruck, Aus., 16,087 ... E7 16
Bruck, Aus., 6,791 ... E8 16
Brück, Ger., 2,583 ... A7 17
Brückenau, Ger., 5,500 ... C4 17
Bruck-Fusch (Bruck an der Grossglocknerstrasse), Aus., 3,258 ... B8 18
Bruederheim, Alta., Can., 289 ... C4 69
Bruels, pt., Calif. ... E2 82
Bruff, Ire., 545 ... E3 11
Brugg, Switz., 6,683 ... B5 19
Brugge (Bruges), Bel., 52,220 (★98,000) ... C3 15
Brühl, Ger., 35,300 ... C1 17
Bruin, Butler, Pa., 706 ... D2 114
Bruins, Crittenden, Ark. ... C5 81
Brule, Keith, Nebr., 370 ... C4 103
Brule, Douglas, Wis., 100 ... B2 124
Brule, co., S. Dak., 6,319 ... D6 116
Brule, lake, Minn. ... A7 99
Brule, riv., Mich. ... C2 98
Brule, riv., Wis. ... B2 124
Brule Lake Station, Ont., Can., 50 ... B6 72
Brumath, Fr., 6,801 ... C7 14
Brumley, Miller, Mo., 74 ... C5 101
Brundidge, Pike, Ala., 2,523 ... D4 78
Bruneau, Owyhee, Idaho, 200 ... G3 89
Bruneau, riv., Idaho ... G3 89
Bruneau, riv., Nev. ... B6 104
Brunei, Mala., 9,702 ... E4 35
Brunei, reg., Mala., 40,657 ... I14 33, E4 35
Brunete, Sp., 897 ... p16 20
Brunette, isl., Newf., Can. ... E4 75
Brunette Downs, Austl. ... C6 50
Bruni, Webb, Tex., 200 ... F3 118
Brunico, It., 6,637 ... C3 20
Brünig-Hasliberg, Switz. ... C5 19
Bruning, Thayer, Nebr., 289 ... D8 103
Brunkild, Man., Can., 120 ... E3 71
Brunnen, Switz. ... C6 19
Bruno, Sask., Can., 750 ... E3 70
Bruno, Pine, Minn., 116 ... D6 99
Bruno, Butler, Nebr., 155 ... C9 103
Bruno, Atoka, Okla., 30 ... C5 112
Brunsbüttelkoog, Ger., 8,600 ... E3 24
Brunson, Hampton, S.C., 603 ... F5 115
Brunsville, Plymouth, Iowa, 128 ... B1 92
Brunswick, Glynn, Ga., 21,703 ... E5 87
Brunswick, see Braunschweig, Ger.
Brunswick, Lake, Ind., 60 ... B2 91
Brunswick, Cumberland, Maine, 12,500 (15,797▲) ... E3, E5 96
Brunswick, Frederick, Md., 3,555 ... B2 85
Brunswick, Chariton, Mo., 1,493 ... B4 101
Brunswick, Antelope, Nebr., 254 ... B8 103
Brunswick, Columbus, N.C., 169 ... C5 109
Brunswick, Medina, Ohio, 11,725 ... A6 111
Brunswick, Brunswick, Va. ... E5 121
Brunswick, co., N.C., 20,278 ... C5 109
Brunswick, co., Va., 17,779 ... E5 121
Brunswick, pen., Chile ... h11 54
Bruree, Ire., 308 ... E3 11
Brusett, Garfield, Mont., 12 ... C9 102
Brush, Morgan, Colo., 3,621 ... A7 83
Brush, mtn., Va. ... D1 121
Brushy, mts., N.C. ... A2 109
Brusly, East Baton Rouge, La., 544 ... D4 95
Brusque, Braz., 16,127 ... D3 56
Brussels (Bruxelles), Bel., 170,489 (★1,930,000) ... D4 15
Brussels, Ont., Can., 844 ... D3 72
Brussels, Calhoun, Ill., 201 ... E3 90
Brussels, Door, Wis., 200 ... D6 124
Bruton, Eng., 1,614 ... E5 12
Bruyères-en-Vosges, Fr., 3,639 ... A2 18
Bryan, Williams, Ohio, 7,361 ... A3 111
Bryan, Brazos, Tex., 27,542 ... D4 118
Bryan, co., Ga., 6,226 ... D5 87
Bryan, co., Okla., 24,252 ... D5 112
Bryansk, Sov. Un., 231,000 ... E10 27
Bryans Road, Charles, Md., 50 ... C3 85
Bryant, Saline, Ark., 737 ... C3, D5 81
Bryant, Sask., Can., 45 ... H4 70
Bryant, Palm Beach, Fla., 400 ... F6 86
Bryant, Fulton, Ill., 346 ... C3 90
Bryant, Jay, Ind., 316 ... C8 91
Bryant, Clackamas, Oreg. (part of Oswego) ... B2 113
Bryant, Hamlin, S. Dak., 522 ... C8 116
Bryant, Langlade, Wis., 100 ... C4 124
Bryant, creek, Mo. ... D5 101
Bryant, mtn., Mass. ... B2 97
Bryant Pond, Oxford, Maine, 600 ... D2 96
Bryantown, Charles, Md., 5 ... C4 85
Bryantsville, Garrard, Ky., 73 ... C5 94
Bryantville, Plymouth. Mass., 350 ... B6 97
Bryce Canyon, Garfield, Utah, 5 ... F3 119
Bryce Canyon, nat. park, Utah ... F3 119
Bryceland, Bienville, La., 89 ... B3 95
Bryceville, Nassau, Fla., 75 ... B5 86
Bryn Athyn, Montgomery, Pa., 1,057 ... A12 114
Brynica, riv., Pol. ... g9 26
Bryn Mawr, Delaware and Montgomery, Pa., 9,000 ... A11 114
Bryn Mawr, King, Wash., 10,000 ... D2 122
Bryson, Jack, Tex., 545 ... C3 118
Bryson City, Swain, N.C., 1,084 ... D2 109
Bryte, Yolo, Calif., 3,000 ... *A6 82
Brzeg, Pol., 24,000 ... C4 26
Brzesko, Pol., 2,684 ... D6 26
Brzezinka, Pol., 4,770 ... g10 26
Brzeziny, Pol., 6,008 ... C5 26
Brzozow, Pol., 3,725 ... D7 26
B-Say-Tah, Sask., Can., 36 ... G4 70
Buayan, Phil., 4,434 ... *D7 35
Bubiyan, isl., Kuw. ... G4 41
Bucaramanga, Col., 153,790 ... B3 60
Buccaneer, arch., Austl. ... C3 50
Buchanan, Sask., Can., 462 ... F4 70
Buchanan, Haralson, Ga., 753 ... C1 87
Buchanan, Lib. ... E2 45
Buchanan, Berrien, Mich., 5,341 ... G4 98
Buchanan, Westchester, N.Y., 2,019 ... *D7 108
Buchanan, Stutsman, N. Dak., 76 ... B7 110
Buchanan, Henry, Tenn., 50 ... A3 117
Buchanan, Botetourt, Va., 1,349 ... D3 121
Buchanan, co., Iowa., 22,293 ... B6 92
Buchanan, co., Mo., 90,581 ... B3 101
Buchanan, co., Va., 36,724 ... B3 121
Buchan Ness, cape, Scot. ... C7 13
Buchans, Newf., Can., 2,413 ... D3 75
Buchardo, Arg. ... A4 54
Bucharest (Bucureşti), Rom., 1,177,661 ... C8 22
Buchen, Ger., 4,700 ... D4 17
Buchholz, Ger., 8,600 ... E3 24
Buchloe, Ger., 5,700 ... D5 16
Buchon, pt., Calif. ... E3 82
Buchs, Switz., 6,345 ... B7 19
Buchtel, Athens, Ohio, 499 ... C5 111
Buck, creek, Ind. ... H8 91
Buck, creek, Ky. ... C5 94
Buck, isl., Vir. Is. ... h17 65
Buck, mtn., Wash. ... A6 122
Buckatunna, Wayne, Miss., 50 ... D5 100
Buck Creek, Tippecanoe, Ind., 270 ... D4 91
Bückeburg, Ger., 11,700 ... A4 17
Buckeye, Maricopa, Ariz., 2,286 ... D3, G1 80
Buckeye, Hardin, Iowa, 190 ... B4 92
Buckeye, Lea, N. Mex., 225 ... E6 107
Buckeye, Pocahontas, W. Va., 350 ... C4 123
Buckeye Lake, Licking, Ohio, 2,129 ... C5 111
Buckeystown, Frederick, Md., 400 ... B3 85
Buckfield, Oxford, Maine, 400 (982▲) ... D2 96
Buckhannon, Upshur, W. Va., 6,389 ... C4 123
Buckhaven & Methil, Scot., 21,104 ... D5 13
Buckhead, Morgan, Ga., 169 ... C3 87
Buck Hill Falls, Monroe, Pa., 160 ... D11 114
Buckhorn, Maricopa, Ariz., 400 ... G3, D4 80
Buckhorn, Grant, N. Mex., 87 ... D1 107
Buckhorn, Weston, Wyo. ... A8 125
Buckhorn, flood control res., Ky. ... C6 94
Buckhorn Knob, mtn., W. Va. ... D4 123
Buckhurst Hill, Eng. (part of Chigwell) ... k13 10
Buckie, Scot., 7,666 ... C6 13
Buckingham, Que., Can., 7,421 ... D2 73
Buckingham, Weld, Colo., 50 ... A7 83
Buckingham, Buckingham, Va., 218 ... D4 121
Buckingham, co., Eng., 486,183 ... C7 12
Buckingham, co., Va., 10,877 ... D4 121
Buckland (Elephant Point), Alsk., 87 ... B7 79
Buckland, Que., Can., 314 ... C7 73
Buckland, Hartford, Conn. (part of Manchester) ... B6 84
Buckland, Franklin, Mass., 150 (1,664▲) ... A2 97
Buckland, Auglaize, Ohio, 300 ... B3 111
Buckley, Iroquois, Ill., 690 ... C5 90
Buckley, Wexford, Mich., 247 ... D5 98
Buckley, Pierce, Wash., 3,538 ... B3 122
Bucklin, Ford, Kans., 752 ... E4 93
Bucklin, Linn, Mo., 639 ... B5 111
Buckman, Morrison, Minn., 166 ... E4 99
Bucknell Manor, Fairfax, Va., 2,000 ... *B5 121
Buckner, Lafayette, Ark., 289 ... D2 81
Buckner, Franklin, Ill., 610 ... F4 90
Buckner, Oldham, Ky., 200 ... A4 94
Buckner, Jackson, Mo., 1,198 ... E3 101
Buckner, creek, Kans. ... D4 93
Bucknum, Natrona, Wyo. ... B6 125
Buckroe Beach, Va. (part of Hampton) ... B7 121
Bucks, Mobile, Ala., 100 ... E1 78
Bucks, co., Pa., 308,567 ... F11 114
Bucks Harbor, Washington, Maine, 150 ... D5 96
Buckskin, Gibson, Ind., 280 ... H3 91
Buckskin, mts., Ariz. ... C2 80
Bucksport, Hancock, Maine, 2,327 (3,466▲) ... D4 96
Bucksport, Horry, S.C., 600 ... D9 115
Bucks Store, McKinley, N. Mex. ... B2 107
Bucktail, Arthur, Nebr., 10 ... C4 103
Buckville, Garland, Ark., 75 ... C2 81
Bucoda, Thurston, Wash., 390 ... C3 122
Bucovina, see Bukovina, prov., Rom.
Bucovina, reg., Rom., Sov. Un. ... B7 22
Buco Zau, Ang. ... B1 48
Bucouche, N.B., Can., 1,537 ... C5 74
Bucyrus, Miami, Kans., 125 ... D9 93
Bucyrus, Adams, N. Dak., 60 ... C3 110
Bucyrus, Crawford, Ohio, 12,276 ... B5 111
Bud, Wyoming, W. Va., 400 ... D3 123
Buda, Bureau, Ill., 732 ... B4 90
Buda, Hays, Tex., 451 ... D4 118
Buda, Port. Gui. ... D1 45
Budapest, Hung., 1,807,299 (★2,090,000) ... B4 22
Budaun, India, 58,770 ... C7 40
Budd, coast, Ant. ... C24 5
Budd, lake, N.J. ... B3 106
Budd Lake, Morris, N.J., 1,520 ... B3 106
Buddon Ness, cape, Scot. ... D6 13
Buddtown, Burlington, N.J., 100 ... D3 106
Bude, Eng., 5,095 ... D3 12
Bude, Franklin, Miss., 1,185 ... D3 100
Bude, bay, Eng. ... D3 12
Budennovsk, Sov. Un., 10,000 ... G17 9
Büdingen, Ger., 6,300 ... C4 17
Budjala, Con. L. ... A2 48
Budnany, Czech., 690 ... o17 26
Budrio, It., 16,364 ... B3 21
Budyne, Czech., 1,220 ... n17 26
Buechel, Jefferson, Ky., 8,000 ... A4, B4 94
Buena, Atlantic, N.J., 3,243 ... D3 106
Buena, Yakima, Wash., 670 ... C5 122
Buena, riv., Chile ... C2 54
Buena Park, Orange, Calif., 46,401 ... *F3 82
Buenaventura, Col., 35,087 ... C2 60
Buenaventura, Mex., 2,613 ... B3 63
Buena Vista, Monroe, Ala., 50 ... D2 78
Buena Vista, Bol., 435 ... C3 55
Buena Vista, Chaffee, Colo., 1,806 ... C4 83
Buena Vista, Marion, Ga., 1,574 ... D2 87
Buena Vista, Chickasaw, Miss., 150 ... B5 100
Buena Vista, Mora, N. Mex., 5 ... B4 107
Buena Vista, Scioto, Ohio, 250 ... D4 111
Buena Vista, Carroll, Tenn., 200 ... B3 117
Buena Vista (Independent City), Va., 6,300 ... D3 121
Buena Vista, co., Iowa, 21,189 ... B2 92
Buenos Aires, Arg., 2,966,816 (★6,975,000) ... A5, g7 54
Buenos Aires, C.R., 3,089 ... F6 62
Buenos Aires, prov., Arg., 6,734,548 ... B4, B5 54
Buenos Aires, lake, Arg. ... D2 54
Bueyeros, Harding, N. Mex., 12 ... B6 107
Buffalo, Chambers, Ala., 25 ... C4 78
Buffalo, Alta., Can., 100 ... D5 69
Buffalo, Sangamon, Ill., 356 ... D4 90
Buffalo, White, Ind., 250 ... C4 91
Buffalo, Scott, Iowa, 1,088 ... C7, D7 92
Buffalo, Wilson, Kans., 422 ... E8 93
Buffalo, Larue, Ky., 201 ... C4 94
Buffalo, Wright, Minn., 2,322 ... E5 99
Buffalo, Dallas, Mo., 1,477 ... D4 101
Buffalo, Fergus, Mont., 60 ... D7 102
Buffalo, Erie, N.Y., 532,759 (★1,330,000) ... C2 108
Buffalo, Cass, N. Dak., 234 ... C8 110
Buffalo, Guernsey, Ohio, 800 ... C6 111
Buffalo, Harper, Okla., 1,618 ... A2 112
Buffalo, Union, S.C., 1,209 ... B4 115
Buffalo, Harding, S. Dak., 652 ... B2 116
Buffalo, Humphreys, Tenn. ... B4 117
Buffalo, Leon, Tex., 1,108 ... D4 118
Buffalo, Putnam, W. Va., 396 ... C3 123
Buffalo, Buffalo, Wis., 484 ... D2 124
Buffalo, Johnson, Wyo., 2,907 ... A6 125
Buffalo, co., Nebr., 26,236 ... D6 103
Buffalo, co., S. Dak., 1,547 ... C6 116
Buffalo, co., Wis., 14,202 ... D2 124
Buffalo, creek, Kans. ... C5 93
Buffalo, creek, W. Va. ... A6 123
Buffalo, creek, W. Va. ... B2 123
Buffalo, creek, Wyo. ... B7 125
Buffalo, lake, Alta., Can. ... C4 69
Buffalo, lake, N.W. Ter., Can. ... D9 66
Buffalo, lake, Wis. ... E4 124
Buffalo, riv., Ark. ... B3 81
Buffalo, riv., Minn. ... D2 99
Buffalo, riv., Tenn. ... B4 117
Buffalo, riv., Wis. ... D2 124
Buffalo Bill, res., Wyo. ... A3 125
Buffalo Center, Winnebago, Iowa, 1,140 ... A4 92
Buffalo Cove, Caldwell, N.C. ... A2 109
Buffalo Creek, Jefferson, Colo., 75 ... B5 83
Buffalo Fork, creek, Wyo. ... B2 125
Buffalo Gap, Sask., Can., 60 ... H3 70
Buffalo Gap, Custer, S. Dak., 194 ... D2 116
Buffalo Grove, Cook, Ill., 1,497 ... *E2 90
Buffalo Lake, Renville, Minn., 707 ... F4 99
Buffalo Mills, Bedford, Pa., 100 ... G4 114
Buffalo Ridge, Patrick, Va. ... E2 121
Buffalo Valley, Putnam, Tenn., 175 ... C8 117
Buford, Rio Blanco, Colo., 10 ... B3 83
Buford, Gwinnett, Ga., 4,168 ... B2 87
Buford, Williams, N. Dak., 65 ... A2 110
Buford, Highland, Ohio, 350 ... C4 111
Buford, Albany, Wyo., 20 ... D7 125
Bufords, Giles, Tenn. ... B4 117
Bug, riv., Pol., Sov. Un. ... B6 26, F5 27
Buga, Col., 32,016 ... C2 60
Bugach, Sov. Un., 10,000 ... G5 27
Buganda, reg., Ug. ... A5 48
Bugene, Tan. ... B5 48
Bugojno, Yugo., 5,411 ... C3 22
Bugulma, Sov. Un., 62,000 ... E8 29
Buguruslan, Sov. Un., 35,700 ... E19 9, C4 29
Buhl, Tuscaloosa, Ala., 200 ... B2 78
Bühl, Fr., 2,931 ... D7 14
Bühl, Ger., 9,100 ... E3 17
Buhl, Twin Falls, Idaho, 3,059 ... G4 89
Buhl, St. Louis, Minn., 1,526 ... C6 99
Buhler, Reno, Kans., 888 ... A4, D6 93
Bühlertal, Ger., 7,600 ... E3 17
Buhuşi, Rom., 12,382 ... B8 22
Buick, Elbert, Colo., 20 ... B7 83
Buie, inlet, Scot. ... D2 13
Buies Creek, Hartnett, N.C., 435 ... B5 109
Builth Wells, Wales, 1,602 ... B4 12
Buitenpost, Neth., 1,515 ... A6 15
Buitenzorg, see Bogor, Indon.
Bujalance, Sp., 11,475 ... D3 20
Bukama, Con. L. ... C4 48
Bukavu, Con. L., 33,500 ... B4 48
Bukene, Tan. ... B5 48
Bukhara, Sov. Un., 79,000 ... H21 8
Bukidnon, prov., Phil., 195,600 ... *D6 35
Bukoba, Tan., 5,297 ... B5 48
Bukovina (Bucovina), prov., Rom., 300,751 ... *B7 22
Bukuru, Nig., 8,450 ... E6 45
Bula, Indon. ... F8 35
Bulacan, prov., Phil., 557,700 ... *C6 35
Bülach, Switz., 8,188 ... A6 19
Buladean, Mitchell, N.C., 200 ... C4 109
Buladeen, Carter, Tenn., 100 ... C12 117
Bulan, Perry, Ky., 500 ... C6 94
Bulan, Phil., 14,279 ... *C6 35
Bulawayo, S. Rh., 160,000 (★195,500) ... B4 49
Bulfontein, S. Afr. ... C4 49
Bulgaria, country, Eur., 7,022,206 ... G14 8, D7 22
Bulgroo, Austl. ... B4 51
Bulhar, Som. ... C5 47
Bulkley, mts., B.C., Can. ... B4 68
Bulkley, riv., B.C., Can. ... B4 68
Bull, Wayne, W. Va., 150 ... D2 123
Bull, bay, S.C. ... F9 115
Bull, creek, S. Dak. ... B2 116
Bull, isl., S.C. ... F4, F8 115
Bull, mtn., Mont. ... D4 102
Bullaque, riv., Sp. ... C3 20
Bullange, Bel., 2,216 ... D6 15
Bullard, Twiggs, Ga., 120 ... D3 87
Bullas, Sp., 8,936 ... C5 20
Bullay, Ger., 930 ... C2 17
Bulle, Switz., 5,983 ... C3 19
Bullfrog, creek, Utah ... F5 119
Bullhead, Corson, S. Dak., 200 ... B4 116
Bullhead, mtn., N.Y. ... B6 108
Bullhead City, Mohave, Ariz., 250 ... B1 80
Bullion, Ascension, La. ... B5 95
Bullitt, co., Ky., 15,726 ... C4 94
Bullittsville, Boone, Ky., 100 ... A6 94
Bullmoose, mtn., B.C., Can. ... B7 68
Bulloch, co., Ga., 24,263 ... D5 87
Bullock, Granville, N.C., 150 ... A5 109
Bullock, co., Ala., 13,462 ... C4 78
Bulloo, riv., Austl. ... C5 51
Bull River, B.C., Can., 125 ... E10 68
Bull Ruffin, mtn., N.C. ... A2 109
Bull Run, mts., Nev. ... B5 104
Bull Run, mts., Va. ... C5 121
Bull Run, ridge, Tenn. ... E11 117
Bull Run, riv., Va. ... B4 121
Bullrun Rock, mtn., Oreg. ... C8 113
Bulls Gap, Hawkins, Tenn., 682 ... C10 117
Bull Shoals, Marion, Ark., 268 ... *A3 81
Bull Shoals, res., Ark., Mo. ... A3 81
Bulnes, Chile, 5,147 ... B2 54
Bulo Burti, Som., 3,300 ... E6 47
Buloir, is., Alsk. ... E3 79
Bulsar, India, 25,440 ... G4 40
Bulu, Con. L. ... A4 48
Buluan, Phil., 3,296 ... D6 35
Bulun, Sov. Un., 800 ... B15 28
Bulungu, Con. L. ... B2 48
Bulun Tukhoi (Puluntohai), China, 10,000 ... B2 34
Bulwark, Alta., Can., 50 ... C5 69
Bulyea, Sask., Can., 193 ... G3 70
Bumba, Con. L., 3,531 ... A3 48
Bumble Bee, Yavapai, Ariz., 10 ... C3 80
Bumping, riv., Wash. ... C4 122
Bumpus Mills, Stewart, Tenn., 200 ... A4 117
Buna, Pap. ... k12 50
Buna, Jasper, Tex., 950 ... D6 118
Bunbury, Austl., 13,186 ... F2 50
Bunceton, Cooper, Mo., 468 ... C5 101
Bunch, Adair, Okla., 80 ... B7 112
Bunche Park, Dade, Fla., 5,000 ... *F3 86
Buncombe, Johnson, Ill., 200 ... F5 90
Buncombe, co., N.C., 130,074 ... D3 109
Buncrana, Ire., 2,960 ... B4 11
Bundaberg, Austl., 22,799 ... B9 51
Bünde, Ger., 10,700 ... A3 17
Bundi, India, 26,478 ... E5 40
Bundick, creek, La. ... D2 95
Bundoran, Ire., 1,326 ... C3 11
Bunessan, Scot. ... D2 13
Bungay, Eng., 3,581 ... B9 12
Bungo, strait, Jap. ... J6 37
Bunia, Con. L., 5,000 ... A5 48
Bunker, Reynolds, Mo., 250 ... D6 101
Bunker Hill, Macoupin, Ill., 1,524 ... D4 90
Bunker Hill, Miami, Ind., 1,049 ... C5 91
Bunker Hill, Russell, Kans., 200 ... D5 93
Bunker Hill, Coos, Oreg., 1,655 ... *D2 113
Bunker Hill, Giles, Tenn., 75 ... B5 117
Bunker Hill, Harris, Tex., 2,216 ... *E5 118
Bunker Hill, Berkeley, W. Va., 400 ... B6 123
Bunker Hill, peak, Nev. ... D4 104
Bunkerville, Clark, Nev., 210 ... G7 104
Bunkeya, Con. L. ... D4 48

Bunkie, Avoyelles, La., 5,188....D3 95
Bunn, Dallas, Ark., 25....C3 81
Bunn, Franklin, N.C., 332....B5 109
Bunnahowen, Ire....C2 11
Bunnell, Flagler, Fla., 1,860....C5 86
Bunnlevel, Harnett, N.C., 187....B5 109
Bunny Run, Oakland, Mich., 1,058....*F7 98
Bū Nujaym, Libya....C3 43
Buo Ha, Viet., 10,000....A6 38
Bupto, China....B10 40
Bura, Ken....B7 48
Bur A'caba, Som., 2,500 (10,600▲)....E5 47
Buram, Sud....C2 47
Buran, Som....C6 47
Burao, Som., 10,000....E6 47
Buras, Plaquemines, La., 4,000....E6 95
Buraydah, Sau. Ar., 20,000....*G7 33
Burayk, Libya, 683....D2 43
Burbank, Los Angeles, Calif., 90,155....F2 82
Burbank, Santa Clara, Calif., 5,000....*B6 82
Burbank, Osage, Okla., 238....A5 112
Burbank, Clay, S. Dak., 112....E9 116
Burchard, Pawnee, Nebr., 132....D9 103
Burdekin, riv., Austl....C8 50
Burden, Cowley, Kans., 580....E7 93
Burdett, Alta., Can., 229....E5 69
Burdett, Pawnee, Kans., 250....D4 93
Burdett, Schuyler, N.Y., 420....C4 108
Burdette, Mississippi, Ark., 115....B6 81
Burdick, Morris, Kans., 100....D7 93
Burditt, lake, Ont., Can....B5 99
Burdock, Fall River, S. Dak., 5....D2 116
Burdur, Tur., 25,400....D8 23
Burdwan, India, 108,224....F11 40
Bureau, Bureau, Ill., 401....B4 90
Bureau, co., Ill., 37,594....B4 90
Bureinsky, range, Sov. Un....B6 37
Büren, Ger., 5,900....B3 17
Büren, Switz., 2,432....B3 19
Bureya, Sov. Un., 5,000....B4 37
Bureya, riv., Sov. Un....A5 37
Burford, Ont., Can., 1,074....D4 72
Burford, lake, N. Mex....A3 107
Burg, Ger., 29,300....B4 17
Burg al'Arab, Eg., U.A.R....G7 31
Burg [auf Fehmarn], Ger., 6,049....D5 24
Burgas, Bul., 43,684....D8 22
Burgas, co., Bul., 629,593....*D8 22
Burgas, gulf, Bul....D8 22
Bur Gavo, Som....F5 47
Burgaw, Pender, N.C., 1,750....C6 109
Burgdorf, Ger., 12,400....F3 24
Burgdorf, Idaho, Idaho....D3 89
Burgdorf, Switz., 13,936 (*16,966)....B4 19
Burgenland, state, Aus., 271,001....*E8 16
Burgeo, Newf., Can., 1,138....E3 75
Burgeo and La Poile, dist., Newf., Can....E2 75
Burgesdorp, S. Afr., 6,184....D4 49
Burgess, Horry, S.C., 100....D11 115
Burgess Hill, Eng., 13,990....D7 12
Burgettstown, Washington, Pa., 2,383....F1 114
Burghead, Scot., 1,346....C5 13
Burghausen, Ger., 13,200....A8 18
Burgin, Mercer, Ky., 879....C5 94
Burglengenfeld, Ger., 8,100....D7 17
Burgoon, Sandusky, Ohio, 243....A4 111
Burgos, Phil., 1,420....n12 35
Burgos, Sp., 82,177....A4 20
Burgos, prov., Sp., 380,791....*A4 20
Burgreuland, Bel., 1,985....D6 15
Burgstädt, Ger., 17,300....C7 17
Burgsteinfurt, Ger., 12,200....A2 17
Burgsvik, Swe., 378....I8 25
Burgundy (Bourgogne), former prov., Fr., 1,264,000....D6 14
Burhanpur, India, 82,090....G6 40
Burhave, Ger., 2,400....E2 24
Buri, Braz., 1,844....m7 56
Burien, King, Wash., 10,000....*D1 122
Burin, Newf., Can., 1,116....E4 75
Burin, Jordan, 2,000....g12 32
Burin, dist., Newf., Can....E4 75
Burin, pen., Newf., Can....E4 75
Buriram, Thai., 12,352....E5 38
Burk, chan., B.C., Can....C4 68
Burkburnett, Wichita, Tex., 7,621....B3 118
Burke, Shoshone, Idaho, 300....B3 89
Burke, Franklin, N.Y., 273....A3 108
Burke, Gregory, S. Dak., 811....D6 116
Burke, Fairfax, Va., 150....B4 121
Burke, co., Ga., 20,596....C4 87
Burke, co., N.C., 52,701....B2 109
Burke, co., N. Dak., 5,886....A3 110
Burke City, St. Louis, Mo., 1,500....A8 101
Burkes Garden, Tazewell, Va., 100....B3, D1 121
Burkesville, Cumberland, Ky., 1,688....D4 94
Burket, Kosciusko, Ind., 259....B6 91
Burketon Station, Ont., Can., 152....C6 72
Burketown, Austl., 59....C6 50
Burkett, Coleman, Tex., 150....C3 118
Burkettsville, Mercer and Darke, Ohio, 290....B3 111
Burkeville, Newton, Tex., 300....D6 118
Burkeville, Nottoway, Va., 705....C4 121
Burkley, Carlisle, Ky., 150....B2 94
Burks Falls, Ont., Can., 926....B5 72
Burkville, Lowndes, Ala., 100....C3 78
Burkville, Franklin, Mass. (part of Conway)....A2 97
Burleigh, Cape May, N.J., 100....E3 106
Burleigh, co., N. Dak., 34,016....C5 110
Burleson, Johnson, Tex., 2,345....B5 118
Burleson, co., Tex., 11,177....D4 118
Burley, Cassia, Idaho, 7,508....G5 89
Burley, Kitsap, Wash., 250....D1 122
Burlingame, San Mateo, Calif., 24,036....B5 82
Burlingame, Osage, Kans., 1,151....D8 93
Burlington, Newf., Can., 263....D3 75
Burlington, Ont., Can., 47,008....D5 72
Burlington, Kit Carson, Colo., 2,090....B8 83
Burlington, Hartford, Conn., 700 (2,790▲)....B5 84
Burlington, Kane, Ill., 360....A5 90
Burlington, Carroll, Ind., 500....D5 91
Burlington, Des Moines, Iowa, 32,430....D6 92
Burlington, Coffey, Kans., 2,113....D8 93
Burlington, Boone, Ky., 350....A5, A6 94
Burlington, Penobscot, Maine, 150 (353▲)....C4 96
Burlington, Middlesex, Mass., 12,852....C2 97
Burlington, Burlington, N.J., 12,687....C3 106
Burlington, Alamance, N.C., 33,199....A4 109
Burlington, Ward, N. Dak., 262....A4 110
Burlington, Alfalfa, Okla., 174....A3 112
Burlington, Bradford, Pa., 115....C8 114
Burlington, Chittenden, Vt., 35,531....C2 120
Burlington, Skagit, Wash., 2,968....A3 122
Burlington, Mineral, W. Va., 200....B6 123
Burlington, Racine, Wis., 5,856....F1, F5 124
Burlington, Big Horn, Wyo., 100....A4 125
Burlington, co., N.J., 224,499....D3 106
Burlington Beach, Ont., Can., 3,314....*D5 72
Burlington Junction, Nodaway, Mo., 650....A2 101
Burli-Tyube, Sov. Un....D9 29
Burlyu-Tobe, Sov. Un., 10,000....F24 9
Burma, country, Asia, 19,856,000....G12 33, D10 39
Burmis, Alta., Can., 67....E3 69
Burnaby, isl., B.C., Can....C2 68
Burnet, Burnet, Tex., 2,214....D3 118
Burnet, co., Tex., 9,265....D3 118
Burnett, Vigo, Ind., 270....E3 91
Burnett, co., Wis., 9,214....C1 124
Burnett, riv., Austl....B8 51
Burnettsville, White, Ind., 452....C4 91
Burney, Shasta, Calif., 1,294....B3 82
Burney, Decatur, Ind., 250....F6 91
Burneyville, Love, Okla., 30....D4 112
Burnham, Cook, Ill., 2,478....*F3 90
Burnham, Waldo, Maine, 100 (755▲)....D3 96
Burnham, Howell, Mo., 50....E6 101
Burnham, San Juan, N. Mex....A1 107
Burnham, Mifflin, Pa., 2,755....E6 114
Burnham-on-Sea, Eng., 9,850....C5 12
Burnie, Austl., 14,201....o15 50
Burning Springs, Clay, Ky., 350....C6 94
Burning Springs, Wirt, W. Va., 25....C3 123
Burnley, Eng., 80,588 (*165,000)....A5 12
Burns, Austl., 42....F3 51
Burns, Eagle, Colo., 30....B4 83
Burns, Marion, Kans., 314....A6, D7 93
Burns, Smith, Miss., 100....C4 100
Burns, Harney, Oreg., 3,523....D7 113
Burns, Dickson, Tenn., 386....A4 117
Burns, Laramie, Wyo., 225....D8 125
Burns City, Martin, Ind., 180....G4 91
Burns Flat, Washita, Okla., 2,280....B2 112
Burnside, Hartford, Conn. (part of East Hartford)....B6 84
Burnside, Hancock, Ill., 125....C2 90
Burnside, Pulaski, Ky., 575....D5 94
Burnside, Ascension, La., 20....B5 95
Burnside, Neshoba, Miss., 100....C4 100
Burnside, Clearfield, Pa., 307....E4 114
Burns Lake, B.C., Can., 1,041....B5 68
Burnstad, Logan, N. Dak., 50....C6 110
Burnsville, Dallas, Ala., 250....C3 78
Burnsville, N.B., Can., 209....B4 74
Burnsville, Tishomingo, Miss., 416....A5 100
Burnsville, Yancy, N.C., 1,388....C4 109
Burnsville, Braxton, W. Va., 728....C4 123
Burnt, riv., Alta., Can....B1 69
Burnt, riv., Oreg....C9 113
Burnt Cabins, Fulton, Pa., 150....F6 114
Burnt Corn, Monroe, Ala., 300....D2 78
Burnt Fort, Camden, Ga., 50....F5 87
Burntisland, Scot., 6,036....D5 13
Burnt Prairie (Liberty), White, Ill., 118....E5 90
Burnt River, Ont., Can., 121....C6 72
Burntroot, lake, Ont., Can....B6 72
Burntside, lake, Minn....C6 99
Burntwood, lake, Man., Can....B1 71
Burntwood, riv., Man., Can....B2 71
Burnwell, Walker, Ala., 500....E4 78
Burnwell, Kanawha, W. Va., 210....D6 123
Burqā, Jordan, 3,000....B7, f11 32
Burr, Yellow Medicine, Minn., 50....F2 99
Burr, Otoe, Nebr., 81....D9 103
Burra, Austl., 1,382....F2 51
Burray, isl., Scot....B6 13
Burr Hill, Orange, Va., 106....C5 121
Burriana, Sp., 15,154....C5 20
Burro, mts., Mex....E1 116
Burroak, Winneshiek, Iowa, 200....A6 92
Burr Oak, Jewell, Kans., 473....C5 93
Burr Oak, St. Joseph, Mich., 867....G5 98
Burr Oak Lake, res., Ohio....C5
Burroughs, Chatham, Ga., 30....E5 87
Burrow, head, Scot....F4 13
Burrows, Carroll, Ind., 200....C4 91
Burrton, Harvey, Kans., 774....A4, D6 93
Burruyacú, Arg., 3,034....E3 55
Burrville, Litchfield, Conn. (part of Torrington)....B4 84
Burrville, Morgan, Tenn., 230....C9 117
Burrwood, Plaquemines, La., 400....F6 95
Burry Port, Wales, 5,671....C3 12
Bursa, Tur., 153,600....B7 23
Būr Sa'īd, see Port Said, Eg., U.A.R.
Burstall, Sask., Can., 266....G1 70
Burt, Kossuth, Iowa, 620....A3 92
Burt, Hettinger, N. Dak., 75....C3 110
Burt, Cannon, Tenn., 25....B5 117
Burt, co., Nebr., 10,192....C9 103
Burt, lake, Mich....C6 98
Būr Tawfīq, Eg., U.A.R., 1,000....E4 32
Burton, Tulare, Calif., 4,635....*D4 82
Burton, Adams, Ill., 50....D2 90
Burton, Keya Paha, Nebr., 17....B6 103
Burton, Geauga, Ohio, 1,085....A6 111
Burton, Washington, Tex., 400....D4 118
Burton, King, Wash., 400....D1 122
Burton, Wetzel, W. Va., 250....B4 123
Burton, lake, Ga....B3 87
Burton-on-Trent, Eng., 50,766....B6 12
Burton Port, Ire., 224....C3 11
Burträsk, Swe., 1,350....E9 25
Burtrum, Todd, Minn., 160....E4 99
Burtts Corner, N.B., Can., 389....C3 74
Burtus, Eg., U.A.R....D3 32
Burtville, Potter, Pa., 50....C5 114
Buru, isl., Indon....F7 35
Burullus, lake, Eg., U.A.R....C5 32
Burūn, cape, Eg., U.A.R....C5 32
Burundi, country, Afr., 2,300,000....B4 48
Burwash, Ont., Can., 616....A4 72
Burwell, Garfield, Nebr., 1,425....C6 103
Bury, Que., Can., 440....D6 73
Bury, Eng., 59,984....A5 12
Burynshik, Sov. Un....D4 29
Bury St. Edmunds, Eng., 21,144....B8 12
Busby, Alta., Can., 85....C4 69
Busby, Big Horn, Mont., 100....E10 102
Busch, Pike, Mo., 50....B6 101
Büschfeld, Ger., 1,200....D1 17
Bush, Laurel, Ky., 20....C5 94
Bush, Simpson, Miss., 50....D4 100
Bush, riv., B.C., Can....D2 69
Bush, riv., Md....B5 97
Bush, riv., N. Ire....B5 11
Bush City, Anderson, Kans., 50....D8 93
Būshehr, Iran, 27,317....G5 41
Bushenyi, Ug....B5 48
Bushimoie, riv., Con. L....C3 48
Bushkill, Pike, Pa., 500....D11 114
Bushkill, creek, Pa....B2 106
Bushland, Potter, Tex., 200....B1 118
Bushmills, N. Ire., 935....B5 11
Bushnell, Sumter, Fla., 644....D4 86
Bushnell, McDonough, Ill., 3,710....C3 90
Bushnell, Kimball, Nebr., 266....C2 103
Bushnell, Brookings, S. Dak., 92....C9 116
Bushong, Lyon, Kans., 51....D7 93
Bushton, Rice, Kans., 499....D5 93
Businga, Con. L....A3 48
Busk, Sov. Un., 13,900....G5 27
Buskerud, co., Nor., 168,500....*H4 25
Busko, Pol., 5,975....C6 26
Buşrá ash Shām, Syr....F11 31
Bussa, Nig....D5 45
Busselton, Austl., 3,495....F2 50
Busseto, It., 9,959....E6 18
Bussey, Marion, Iowa, 557....C5 92
Bussum, Neth., 39,800....B5 15
Bustamante, Arg....D3 54
Bustamante, Zapata, Tex., 100....F3 118
Busto Arsizio, It., 64,367 (*320,000)....B2 21
Büsum, Ger., 4,200....D2 24
Buta, Con. L., 11,200....A3 48
Buta Ranquil, Arg....B3 54
Bute, co., Scot., 15,129....E3 13
Bute, inlet, B.C., Can....D5 68
Bute, isl., Scot....E3 13
Butedale, B.C., Can., 15....C3 68
Buthier, riv., It....D3 18
Butkhak, Afg., 10,000....D14 41
Butler, Choctaw, Ala., 1,765....C1 78
Butler, Taylor, Ga., 1,346....D2 87
Butler, Montgomery, Ill., 249....D4 90
Butler, DeKalb, Ind., 2,176....B8 91
Butler, Pendleton, Ky., 450....B5, B7 94
Butler, Bates, Mo., 3,791....C3 101
Butler, Morris, N.J., 5,414....B4 106
Butler, Richland, Ohio, 976....B5 111
Butler, Custer, Okla., 351....B2 112
Butler, Butler, Pa., 20,975....E2 114
Butler, Day, S. Dak., 62....B8 116
Butler, Johnson, Tenn., 608....C12 117
Butler, Waukesha, Wis., 2,274....E1 124
Butler, co., Ala., 24,560....D3 78
Butler, co., Iowa, 17,467....B5 92
Butler, co., Kans., 38,395....E7 93
Butler, co., Ky., 9,586....C3 94
Butler, co., Mo., 34,656....E7 101
Butler, co., Nebr., 10,312....C8 103
Butler, co., Ohio, 199,076....C1 111
Butler, co., Pa., 114,639....E2 114
Butler, lake, Fla....E2 86
Butler's Bridge, Ire., 157....C4 11
Butlers Landing, Clay, Tenn....C8 117
Butlerville, Jennings, Ind., 240....F6 91
Butman, Gladwin, Mich....D6 98
Butner, Granville, N.C., 1,000....*A5 109
Butovo, Sov. Un....n17 27
Buttahatchie, riv., Ala....B1 78
Buttahatchie, riv., Miss....B5 100
Butte, Silver Bow, Mont., 27,877....D4 102
Butte, Boyd, Nebr., 526....B7 103
Butte, McLean, N. Dak., 257....B5 110
Butte, co., Calif., 82,030....C3 82
Butte, co., Idaho, 3,498....F5 89
Butte, co., S. Dak., 8,592....C2 116
Butte, mts., Nev....D6 104
Butte City, Glenn, Calif., 5....C3 82
Butte City, Butte, Idaho, 104....F5 89
Butte Falls, Jackson, Oreg., 384....E4 113
Butterfield, Hot Spring, Ark., 200....D6 81
Butterfield, Watonwan, Minn., 601....G4 99
Butterfield, Barry, Mo., 125....E4 101
Butternut, Ashland, Wis., 499....B3 124
Butternut, lake, Wis....C5 124
Butternut Ridge, N.B., Can., 190....D4 74
Butters, Bladen, N.C., 225....C5 109
Butterworth, Mala., 42,504....J4 38
Butterworth, S. Afr., 2,354....D4 49
Buttevant, Ire., 981....E3 11
Butt of Lewis, cape, Scot....B2 13
Buttonwillow, Kern, Calif., 950....E4 82
Butts, co., Ga., 8,976....C3 87
Buttzville, Warren, N.J., 300....B2 106
Butuan, Phil., 15,000 (82,800▲)....D7 35
Butung, isl., Indon....F6 35
Buturlinovka, Sov. Un., 39,800....C2 29
Butzbach, Ger., 9,900....C3 17
Bützow, Ger., 10,900....B5 16
Buxtehude, Ger., 15,700....B4 16
Buxton, Br. Gu., 5,164....A3 59
Buxton, Eng., 19,236....A6 12
Buxton, Dare, N.C., 500....B8 109
Buxton, Traill, N. Dak., 321....B8 110
Buxton, Washington, Oreg., 100....A1 113
Buxton, mtn., B.C., Can....D3 68
Buy, Sov. Un., 25,500....B2 29
Buyck, St. Louis, Minn., 20....B6 99
Buynaksk, Sov. Un., 27,300....E3 29
Buyukliman, see Vakfıkebir, Tur.
Büyük Menderes (Maeander), riv., Tur....D7 23
Buzachi, pen., Sov. Un....G19 9
Buzancy, Fr., 461....E4 15
Buzău, Rom., 47,595....C8 22
Buzau, riv., Rom....C8 22
Buzaymah, Libya....E4 43
Buzet, Yugo., 444....C1 22
Buzuluk, Sov. Un., 57,000....C4 29
Buzzards, bay, Mass....C6 97
Buzzards Bay, Barnstable, Mass., 2,170....C6 91
Byala, Bul., 1,388....D7 22
Byala Slatina, Bul., 9,357....D6 22
Byam Martin, isl., N.W. Ter., Can....A12 66
Byars, McClain, Okla., 256....C4 112
Bybee, Cocke, Tenn., 50....C10 117
Bydgoszcz, Pol., 231,000....B5 26
Byelorussia (S.S.R.) (White Russia), rep., Sov. Un., 8,226,000....E5 27
Byemoor, Alta., Can., 129....D4 69
Byer, Jackson, Ohio, 100....C5 111
Byers, Arapahoe, Colo., 500....B6 83
Byers, Pratt, Kans., 52....E5 93
Byesville, Guernsey, Ohio, 2,447....C6 111
Byfield, Essex, Mass., 600....A6 97
Bygland, Nor....H2 25
Byhalia, Marshall, Miss., 674....A4 100
Byington, Knox, Tenn., 700....E11 117
Bykovo, Sov. Un., 10,000....G15 27
Bykovo, Sov. Un., 1,000....n18 27
Bylas, Graham, Ariz., 700....D5 80
Bylot, isl., N.W. Ter., Can....B17 67
Byng Inlet, Ont., Can., 625....B4 72
Bynum, Teton, Mont., 85....C4 102
Bynum, Chatham, N.C., 400....B4 109
Bynumville, Chariton, Mo., 78....B5 101
Byram, Hinds, Miss., 350....C3 100
Byram, Hunterdon, N.J....C2 106
Byram, riv., Conn....A5 106
Byrdstown, Pickett, Tenn., 613....C8 117
Byrnedale, Elk, Pa., 500....D4 114
Byrock, Austl., 139....E6 51
Byromville, Dooly, Ga., 349....D3 87
Byron, Contra Costa, Calif., 500....B6 82
Byron, Peach, Ga., 1,138....D3 87
Byron, Ogle, Ill., 1,578....A4 90
Byron, Oxford, Maine, 60 (108▲)....D2 96
Byron, Olmsted, Minn., 660....F6 99
Byron, Thayer, Nebr., 147....D8 103
Byron, Alfalfa, Okla., 82....A3 112
Byron, Big Horn, Wyo., 417....A4 125
Byron, cape, Austl....D9 51
Byrum, Den., 395....A5 24
Bysice, Czech., 1,170....n18 26
Bystra, riv., Czech....o18 26
Bystrzyca, Pol., 9,564....C4 26
Bytom, Pol., 182,000....C5, g9 26
Bytosh, Sov. Un., 10,000....E10 27
Bytow, Pol., 3,810....A4 23

C

Ca, riv., Viet....C6 38
Caacupé, Par., 15,834....E4 55
Caaguazú, dept., Par., 2,610....E4 55
Caamano, sound, B.C., Can....C3 68
Caapucú, Par., 6,608....E4 55
Caazapá, Par., 16,588....E4 55
Caazapá, dept., Par., 3,003....E4 55
Cabaceiras, Braz., 581....h5 57
Cabalete, isl., Phil....o13 35
Caballo, pt., P.R....g12 65
Caballo, res., N. Mex....E2 107
Cabana, Peru, 2,560....C2 58
Cabanaconda, Peru, 2,278....E3 58
Cabanatuan, Phil., 20,000 (70,400▲)....G6 37, o13 35
Cabano, Que., Can., 2,695....B9 73
Cabarrus, co., N. Car., 68,137....B3 109
Cabarruyan, isl., Phil....n12 35
Cabbage, swamp, Fla....C6 86
Cabedelo, Braz., 10,738....C4, h6 57
Cabell, co., W. Va., 108,202....C2 123
Cabery, Ford and Kankakee, Ill., 293....C5 90
Cabeza del Buey, Sp., 11,737....C3 20
Cabezas, Bol., 298....C3 55
Cabimas, Ven., 93,347....A3 60
Cabin, creek, Okla....A6 112
Cabin, creek, W. Va....D6 123
Cabincreek, Kanawha, W. Va., 800....C3, C6 123
Cabinda, Ang., 1,554....C1 48
Cabinda, dist., Ang., 58,453....C1 48
Cabinet, Bonner, Idaho, 10....A2 89
Cabinet, mts., Mont....B1 102
Cabin John, Montgomery, Md., 2,000....C1, C3 85
Cable, Bayfield, Wis., 262....B2 124
Cabo, Braz., 6,029....k6 57
Cabo Blanco, Arg....D3 54
Cabo Delgado, prov., Moz....D6 48
Cabo Frio, Braz., 13,117....C4 56
Cabo Gracias a Dios, Nic., 590....C6 62
Cabonga, res., Que., Can....G17 67
Cabool, Texas, Mo., 1,284....D5 101
Caboolture, Austl., 1,926....C9 51
Cabo Raso, Arg....C3 54
Caborca, Mex., 9,285....A2 63
Cabo Rojo, P.R., 3,086....C2 65
Cabo Rojo, mun., P.R., 24,868....C2 65
Cabot, Lonoke, Ark., 1,321....C3 81
Cabot, Butler, Pa., 400....E2 114
Cabot, Washington, Vt., 244 (763▲)....C4 120
Cabot, head, Ont., Can....B3 72
Cabot, mtn., N.H....B6 120
Cabot, strait, Newf., Can....E2 75
Cabra, Sp., 15,995....D3 20
Cabras, isl., Guam....52
Cabras, isl., P.R....C8 65
Cabrera, isl., Sp....C7 20
Cabri, Sask., Can., 711....G1 70
Cabri, lake, Sask., Can....F1 70
Cabriel, riv., Sp....C5 20
Cabrillo, nat. mon., Calif....E2 82
Cabrobó, Braz., 2,889....C3 57
Cabullón, pt., P.R....D4 65
Čacak, Yugo., 27,356....D5 22
Caçapava, Braz., 7,987....C3 56
Caçapava do Sul, Braz., 6,712....E2 56
Cacapon, mtn., W. Va....B6 123
Cacequi, Braz., 8,458....D2 56
Cáceres, Braz., 8,246....C4 55
Cáceres, Col., 305....B2 60
Cáceres, Sp., 48,005....C2 20
Cáceres, prov., Sp., 544,407....C2 20
Cachan, Fr., 23,282....g10 14
Cache, Alexander, Ill., 60....F4 90
Cache, Comanche, Okla., 1,003....C3 112
Cache, co., Utah, 35,788....B4 119
Cache, creek, Calif....C2 82
Cache, creek, Okla....C3 112
Cache, peak, Idaho....G5 89
Cache, riv., Ark....B4 81
Cache, riv., Ill....F4 90
Cache Bay, Ont., Can., 810....A5 72
Cache la Poudre, riv., Colo....A5 83
Cachi, Arg....E2 55
Cachi, mts., Arg....D2 55
Cachimbo, mts., Braz....D4 59
Cachinal, Chile....D2 55
Cachoeira, Braz., 11,415....D3 57
Cachoeira do Sul, Braz., 38,661....E2 56
Cachoeiro do Itapemirim, Braz., 39,470....C4 56
Cacique, Pan., 102....h11 62
Cacola, Ang....D2 48
Caconda, Ang., 4,851....D2 48
Cacouna, Que., Can., 834....B8 73
Cactus, Maricopa, Ariz., 300....D4, G2 80
Cactus, flat, Nev....F5 104
Cactus, Moore, Tex., 900....*A2 118
Cactus, Webb, Tex....F3 118
Cactus, peak, Nev....F5 104
Cactus, range, Nev....F5 104
Cactus Lake, Sask., Can., 10....E1 70
Cadboro Bay, B.C., Can....B9 68
Caddo, Bryan, Okla., 814....C5 112
Caddo, Stephens, Tex., 120....C3 118
Caddo, co., Okla., 28,621....B3 112
Caddo, par., La., 223,859....B2 95
Caddo, creek, Okla....C4 112
Caddo, lake, La., Tex....B1 95
Caddo, mts., Ark....C2 81
Caddo, riv., Ark....C2 81
Caddoa, Bent, Colo., 25....C8 83
Caddo Gap, Montgomery, Ark., 75....C2 81
Cade, St. Martin, La., 500....D4 95
Cadena, mtn., P.R....B2 65
Cadena, pt., P.R....B2 65
Cades, Williamsburg, S. C., 125....D8 115
Cades, Gibson, Tenn., 15....B3 117
Cadig, mtn., Phil....o14 35
Cadillac, Que., Can., 1,077....*E9 72
Cadillac, Sask., Can., 245....H2 70
Cadillac, Fr., 2,324....E3 14
Cadillac, Wexford, Mich., 10,112....D5 98
Cadiz, San Bernardino, Calif., 50....E6 79
Cadiz, Trigg, Ky., 1,980....D2 94
Cadiz, Harrison, Ohio, 3,259....B6 111
Cádiz, Sp., 117,871....D2 20
Cádiz, prov., Sp., 818,847....*D2 20
Cádiz, gulf, Sp....D2 20
Cadogan, Alta., Can., 109....C5 69
Cadogan, Armstrong, Pa., 562....E2 114
Cadomin, Alta., Can., 106....C2 69
Cadott, Chippewa, Wis., 881....D2 124
Cadotte, riv., Alta., Can....A2 69
Cadron, creek, Ark....B3 81
Cadwell, Laurens, Ga., 360....D3 87
Cadyville, Clinton, N.Y., 500....A3 108
Caen, Fr., 91,336 (*120,000)....C3 14
Caernarvon, Wales, 8,998....A3 12
Caernarvon, co., Wales, 121,194....A3 12
Caernarvon, bay, Wales....A3 12
Caerphilly, Wales, 36,008....C4 12
Caetité, Braz., 4,823....D2 57
Cafayate, Arg....E2 55
Cagayan, Phil., 23,000 (68,000▲)....D6 35
Cagayan, prov., Phil., 445,700....*B6 35
Cagayan, is., Phil....D6 35
Cagayan, riv., Phil....B6 35
Cagayan Sulu, isl., Phil....D5 35
Cagiallo, Switz., 319....D6 19
Cagle, Sequatchie, Tenn....D8 117
Cagli, It., 3,235....C4 21
Cagliari, It., 183,784....E2 21
Cagliari, gulf, It....E2 21
Cagnes-sur-Mer, Fr., 15,392....F7 14
Caguán, riv., Col....C3 60
Caguas, P.R., 32,015....C6 65
Caguas, mun., P.R., 65,098....C6 65
Caha, mts., Ire....F2 11
Cahaba, riv., Ala....C2 78
Cahaba Heights, Jefferson, Ala., 2,000....*B3 78
Cahaba Valley, creek, Ala....E5 78
Caherdaniel, Ire., 83....F1 11
Cahir, Ire., 1,662....E4 11
Cahirciveen, Ire., 1,659....F1 11
Cahokia, St. Clair, Ill., 15,829....B8 101
Cahokia, creek, Ill....A8 101
Cahone, Dolores, Colo., 50....D2 83
Cahore, pt., Ire....E5 11
Cahors, Fr., 17,046....E4 14
Cahuinari, riv., Col....D3 60
Caí, Braz., 3,361....E2 56
Caiapó, mts., Braz....B2 56
Caiapônia, Braz., 2,476....B2 56
Caibarién, Cuba, 22,657....D4 64
Caicara, Ven., 3,082....B4 60
Caicedonia, Col., 10,681....C2 60
Caicó, Braz., 15,826....C3 57
Caicos, is., Turks & Caicos Is....E7 64
Caicos, passage, Ba. Is....E7 64
Caigo o no Caigo, pt., P.R....e10 65
Cailloma, Peru, 923....E3 58
Caillou, bay, La....E5 95
Caillou, lake, La....E5 95
Caiman, pt., Phil....o12 35
Caimanera, Cuba, 4,035....F6 64
Caimito, riv., Pan....m11 62
Cain City, Gillespie, Tex., 25....D3 118
Cains, riv., N.B., Can....C4 74
Cains Store, Pulaski, Ky., 250....C5 94
Cainsville, Harrison, Mo., 495....A4 101
Cainsville, Wilson, Tenn., 40....B5 117
Caintown, Ont., Can., 92....C9 72
Caird, coast, Ant....B10 5
Cairnbrook, Somerset, Pa., 1,100....F4 114
Cairngorm, mts., Scot....C5 13
Cairns, Austl., 25,204....C8 50
Cairnsmore, mtn., Scot....E4 13
Cairn Table, mtn., Scot....E5 13
Cairo (Al Qāhirah), Eg., U.A.R., 2,877,195 (*3,300,000)....D6 43, D3 32
Cairo, Grady, Ga., 7,427....F2 87
Cairo, Alexander, Ill., 9,348....F4 90
Cairo, Pratt, Kans., 50....E5 93
Cairo, Randolph, Mo., 210....B5 101
Cairo, Hall, Nebr., 503....C7 103
Cairo, Greene, N.Y., 650....C7 108
Cairo, Allen, Ohio, 566....B3 111
Cairo, Ritchie, W. Va., 418....B3 123
Caithness, co., Scot., 27,345....B5 13
Caiundo, Ang....E2 48
Cajabamba, Peru, 3,196....C2 58
Cajacay, Peru, 1,094....D2 58
Cajamarca, Peru, 18,324....D2 58
Cajamarca, dept., Peru, 651,068....C2 58
Cajatambo, Peru, 2,561....C2 58
Cajàzeiras, Braz., 15,884....C3 57
Čajniče, Yugo., 1,051....D4 22
Čakovec, Yugo., 9,647....B3 22
Cakovice, Czech., 3,902....n18 26
Calabar, Nig., 46,705....F6 45
Calabogie, Ont., Can., 300....B8 72
Calabogie, lake, Ont., Can....B8 72
Calabozo, Ven., 15,286....B4 60
Calabria, pol. dist., It., 2,045,047....*E5 21
Calabria, reg., It....E6 21
Calafat, Rom., 8,069....D6 22
Calahorra, Sp., 14,462....A4 20
Calais, Fr., 70,372....B4 14
Calais, Washington, Maine, 4,223....C5 96
Calais, Washington, Vt., 150 (684▲)....C4 120
Calalaste, mts., Arg....E2 55
Calama, Chile, 12,955....D2 55
Calamar, Col., 5,393....A3 60
Calamba, Phil., 7,600....o13 35
Calamba, Phil., 19,097....*C6 35
Calamian Group, is., Phil....C6 35
Calamocha, Sp., 2,545....B5 20
Calamus, Clinton, Iowa, 435....C7 92
Calamus, riv., Nebr....B6 103
Calanda, Sp., 2,928....B5 20
Calañas, Sp., 4,059....D2 20
Calanscio, des., Libya....D4 43
Calanscio, sand sea, Libya....D4 43
Calansho, sand sea, Libya....H5 31
Calapan, Phil., 6,113....C6 35
Calapooya, mts., Oreg....D4 113
Călărași, Rom., 25,555....C8 22
Calarcá, Col., 15,707....C2 60
Calasparra, Sp., 8,155....C5 20
Calataga, Phil., 1,852....*C6 35
Calatayud, Sp., 15,792....B5 20
Calau, Ger., 5,372....B8 17
Calauag, Phil., 4,260....p14 35
Calauag, bay, Phil....o14 35
Calaveras, Wilson, Tex....B4 118
Calaveras, co., Calif., 10,289....C3 82
Calayan, Phil., 1,584....*B6 35
Calbayog, Phil., 8,000 (77,800▲)....*C6 35
Calbe, Ger., 16,600....B6 17
Calbuco, Chile, 2,049....C2 54
Calca, Peru, 3,037....D3 58
Calcasieu, par., La., 145,475....D2 95
Calcasieu, lake, La....E2 95
Calcasieu, pass, La....E2 95
Calcasieu, riv., La....D2 95
Calceta, Ec., 3,680....B1 58
Calcha, Bol....D2 55
Calchaquí, riv., Arg....E2 55
Calcutta, India, 2,927,289 (*6,350,000)....D8 39, F12 40
Calcutta, Columbiana, Ohio, 2,221....B7 111
Caldaro, It., 4,523....C7 18
Caldas, dept., Col., 1,212,450....B2 60
Caldas da Rainha, Port., 10,039....C1 20
Calder, Sask., Can., 232....F5 70
Calder, Shoshone, Idaho, 75....B2 89
Calder, lake, Scot....B5 13
Caldera, Chile, 1,525....E1 55
Calderwood, Blount, Tenn., 70....D10 117
Caldew, riv., Eng....F5 13
Caldron Falls, res., Wis....C5 124
Caldwell, Canyon, Idaho, 12,230....F2 89
Caldwell, Sumner, Kans., 1,788....E6 93
Caldwell, Essex, N.J., 6,942....B4 106
Caldwell, Noble, Ohio, 1,999....C6 111

Caldwell, Burleson, Tex., 2,204 D4 118
Caldwell, Greenbrier, W. Va., 200 D4 123
Caldwell, co., Ky., 13,073 C2 94
Caldwell, co., Mo., 8,830 B3 101
Caldwell, co., N.C., 49,552 B2 109
Caldwell, co., Tex., 17,222 E4 118
Caldwell, par., La., 9,004 B3 95
Cale, Nevada, Ark., 85 D2 81
Caledon, N. Ire., 350 C5 11
Caledonia, N.S., Can., 404 E4 74
Caledonia, Ont., Can., 2,198 D5 72
Caledonia, Boone, Ill., 225 A5 90
Caledonia, Kent, Mich., 739 F5 98
Caledonia, Huston, Minn., 2,563 G7 99
Caledonia, Lowndes, Miss., 289 B5 100
Caledonia, Washington, Mo., 119 D7 101
Caledonia, Livingston, N.Y., 1917 C3 108
Caledonia, Traill, N. Dak., 150 B9 110
Caledonia, Marion, Ohio, 673 B5 111
Caledonia, Elk, Pa., 300 D5 114
Caledonia, Goochland, Va D4 121
Caledonia, Racine, Wis., 50 F2 124
Caledonia, co., Vt., 22,786 C4 120
Caledonian, canal, Scot C4 13
Calella, Sp., 7,947 B7 20
Calenzana, Fr., 1,384 C2 21
Calera, Shelby, Ala., 1,928 B3 78
Calera, Chile, 13,047 A2 54
Calera, Bryan, Okla., 692 D5 112
Caleta Buena, Chile C2 55
Caleta Olivia, Arg D3 54
Calexico, Imperial, Calif., 7,992 F6 82
Calf of Man, isl., I. of Man C4 10
Calgary, Alta., Can., 249,641 (*279,062) D3 69
Calha, pt., Port F9 20
Calhan, El Paso, Colo., 347 B6 83
Calheta, Port., 5,376 h11 20
Calhoun, Lowndes, Ala., 325 C3 78
Calhoun, N.B., Can., 70 C5 74
Calhoun, Gordon, Ga., 3,587 B2 87
Calhoun, Richland, Ill., 188 E5 90
Calhoun, McLean, Ky., 817 C2 94
Calhoun, Ouachita, La., 600 B3 95
Calhoun, Henry, Mo., 374 C4 101
Calhoun, McMinn, Tenn., 350 D9 117
Calhoun, co., Ala., 95,878 B4 78
Calhoun, co., Ark., 5,991 D3 81
Calhoun, co., Fla., 7,422 B1 86
Calhoun, co., Ga., 7,341 E2 87
Calhoun, co., Ill., 5,933 D3 90
Calhoun, co., Iowa, 15,923 B3 92
Calhoun, co., Mich., 138,858 F5 98
Calhoun, co., Miss., 15,941 B4 100
Calhoun, co., S.C., 12,256 D6 115
Calhoun, co., Tex., 16,592 E4 118
Calhoun, co., W. Va., 7,948 C3 123
Calhoun City, Calhoun, Miss., 1,714 B4 100
Calhoun Falls, Abbeville, S.C., 2,525 C3 115
Cali, Col., 429,170 C2 60
Calico Rock, Izard, Ark., 773 A3 81
Calicut, see Kozhikode, India.
Caliente, Kern, Calif., 100 E4 82
Caliente, Lincoln, Nev., 792 F7 104
Califon, Hunterdon, N.J., 777 B3 106
California, Harrison, Iowa C2 92
California, Campbell, Ky., 163 A7 94
California, Montieau, Mo., 2,788 C5 101
California, Washington, Pa., 5,978 F2 114
California, state, U.S., 15,717,204 C3 76, 82
California, gulf, Mex B2 63
Caliman, mts., Rom B7 22
Calingasta, Arg A3 54
Calio, Cavalier, N. Dak., 101 A7 110
Calion, Union, Ark., 544 D3 81
Calipatria, Imperial, Calif., 2,548 F6 82
Calistoga, Napa, Calif., 1,514 A5, C2 82
Call, Newton, Tex D6 118
Callabonna, lake, Austl D3 51
Callabonna, riv., Austl D3 51
Callafo, Eth D5 47
Callahan, mtn., Nev D5 104
Callahan, Siskiyou, Calif., 50 B2 82
Callahan, Nassau, Fla., 782 B5, B6 86
Callahan, Webb, Tex., 25 F3 118
Callahan, co., Tex., 7,929 C3 118
Callan, Ire., 1,346 E4 11
Callan, Menard, Tex D3 118
Callander, Ont., Can., 1,236 A5 72
Callander, Scot., 1,654 D4 13
Callands, Pittsylvania, Va., 20 E3 121
Callao, Macon, Mo., 329 B5 101
Callao, Peru, 84,232 D2 58
Callao, Juab, Utah, 5 D2 119
Callao, Northumberland, Va., 150 D6 121
Callao, dept., Peru, 476 *D2 58
Callaway, St. Marys, Md., 25 D4 85
Callaway, Becker, Minn., 235 D3 99
Callaway, Custer, Nebr., 603 C6 103
Callaway, Franklin, Va., 130 D2 121
Callaway, co., Mo., 23,858 C6 101
Callender, Webster, Iowa, 358 B3 92
Callensburg, Clarion, Pa., 280 D2 114
Callery, Butler, Pa., 419 E1 114
Callicoon, Sullivan, N.Y., 750 D5 108
Callicoon Center, Sullivan, N.Y., 200 D6 108
Calliham, McMullen, Tex., 200 E3 118
Calling, lake, Alta., Can B4 69
Calling Lake, Alta., Can., 211 B4 69
Calloway, co., Ky., 20,972 B3 94
Calmar, Alta., Can., 700 C4 69
Calmar, Winneshiek, Iowa, 954 A6 92
Calne, Eng., 6,559 C5 12
Caloocan, Phil., 58,208 o13 35
Caloosahatchee, riv., Fla F5 86
Calpella, Mendocino, Calif., 500 C2 82
Calpet, Sublette, Wyo., 65 C2 125
Caltagirone, It., 36,000 F5 21
Caltanissetta, It., 48,500 (63,027*) F5 21
Caltra, Ire D3 11
Caluire [-et-Cuire], Fr., 25,754 E6 14
Calulo, Ang., 2,763 D1 48
Calumet, Walker, Ala., 75 B2 78
Calumet, Que., Can., 889 D3 73
Calumet, O'Brien, Iowa, 225 B2 92
Calumet, Houghton, Mich., 1,139 A2 98
Calumet, Itasca, Minn., 799 C5 99
Calumet, Canadian, Okla., 354 B3 112
Calumet, Westmoreland, Pa., 1,241 *F3 114
Calumet, co., Wis., 22,268 D5 124
Calumet, lake, Ill F3 90
Calumet City, Cook, Ill., 25,000 B6, F3 90
Calumet Park, Cook, Ill., 8,448 *F3 90
Calumet Sag, chan., Ill F2 90
Calumetville, Fond du Lac, Wis., 100 B5 124
Calva, Graham, Ariz D5 80
Calvados, dept., Fr., 480,686 *C3 14
Calvario, Cuba, 848 k12 64
Calvary, Grady, Ga., 400 F2 87
Calvary, Fond du Lac, Wis., 100 B5 124
Calvert, Washington, Ala., 450 D1 78
Calvert, Robertson, Tex., 2,073 D4 118
Calvert, co., Md., 15,826 C4 79
Calvert, isl., B.C., Can D3 68
Calvert City, Marshall, Ky., 1,505 A3 94
Calverton, Fauquier, Va., 220 C5 121
Calverton Park, St. Louis, Mo., 1,714 A8 101
Calvi, Fr., 2,523 C2 21
Calvillo, Mex., 5,593 m12 63
Calvin, Winn, La., 232 C3 95
Calvin, Cavalier, N. Dak., 104 A7 110
Calvin, Hughes, Okla., 331 C5 112
Calvin, Lee, Va., 25 B2 121
Calvinia, S. Afr., 5,189 D2 49
Calvörde, Ger., 3,129 A6 17
Calw, Ger., 9,700 E3 17
Calwa, Fresno, Calif., 8,000 *D4 82
Calypso, Duplin, N.C., 633 B5 109
Calzada de Calatrava, Sp., 9,140 C4 20
Cam, riv., Eng B8 12
Camabatela, Ang., 3,383 C2 48
Camacho, Mex., 1,692 C4 63
Camagüey, Cuba, 110,388 E5 64
Camagüey, prov., Cuba, 618,256 E5 64
Camak, Warren, Ga., 285 C4 87
Camamu, Braz., 3,052 D3 57
Camaná, Peru, 2,253 E3 58
Camanche, Clinton, Iowa, 2,225 C7 92
Camano, isl., Wash A3 122
Camaquã, Braz., 9,732 E2 56
Camaquã, riv., Braz E2 56
Camargo, Bol., 1,609 D2 55
Camargo, Douglas, Ill., 276 D5 90
Camargo, see Ciudad Camargo, Mex.
Camargo, Dewey, Okla., 254 A2 112
Camargo, Lincoln, Tenn B5 117
Camarillo, Ventura, Calif., 2,359 *E4 82
Camarines Norte, prov., Phil., 188,600 *C6 35
Camarines Sur, prov., Phil., 826,600 *C6 35
Camarón, cape, Hond C5 62
Camarones, Arg C3 54
Camarones, bay, Arg C3 54
Camas, Jefferson, Idaho, 5 E6 89
Camas, Sanders, Mont., 80 C2 102
Camas, Clark, Wash., 5,666 D3 122
Camas, co., Idaho, 917 F4 89
Camas Valley, Douglas, Oreg., 130 D3 113
Camataquí, Bol., 539 D2 55
Camau, pt., Viet H6 38
Cambalache Central, P.R B4 65
Cambay, India, 51,291 D5 40
Cambay, gulf, India G4 40
Cambodia, country, Asia, 3,800,000 F6 38
Camborne [-Redruth], Eng., 36,090 D2 12
Cambra, Luzerne, Pa., 120 D9 114
Cambrai, Fr., 32,897 B5 14
Cambray, Luna, N. Mex E2 107
Cambria, San Luis Obispo, Calif., 700 E3 82
Cambria, Wayne, Iowa, 103 D4 92
Cambria, Hillsdale, Mich., 250 G6 98
Cambria, Columbia, Wis., 589 E4 124
Cambria, co., Pa., 203,283 E4 114
Cambrian, mts., Wales B4 12
Cambrian Park, Santa Clara, Calif., 4,000 *D3 82
Cambridge, Eng., 95,358 B8 12
Cambridge, Washington, Idaho, 473 E2 89
Cambridge, Henry, Ill., 1,665 B3 90
Cambridge, Story, Iowa, 587 A7, C4 92
Cambridge, Cowley, Kans., 140 E7 93
Cambridge, Somerset, Maine, 150 (354*) C3 96
Cambridge, Dorchester, Md., 12,239 C5 85
Cambridge, Middlesex, Mass., 107,716 B5, D3 97
Cambridge, Isanti, Minn., 2,728 E5 99
Cambridge, Furnas, Nebr., 1,090 D5 103
Cambridge, Burlington, N. J., 1,900 *C3 106
Cambridge, Washington, N.Y., 1,748 B7 108
Cambridge, Guernsey, Ohio, 14,562 B6 111
Cambridge, Lamoille, Vt., 217 (1,295*) B3 120
Cambridge, Dane, Wis., 605 F4 124
Cambridge, co., Eng., 189,913 B8 12
Cambridge, res., Mass C2 97
Cambridge Bay, N.W. Ter., Can., 531 C12 66
Cambridge City, Wayne, Ind., 2,569 E7 91
Cambridge Junction, Lamoille, Vt., 50 B3 120
Cambridgeport, Windham, Vt., 100 E3 120
Cambridge Springs, Crawford, Pa., 2,031 C1 114
Cambuí, Braz., 3,556 m8 56
Camby, Marion, Ind., 400 E5, I7 91
Camden, Wilcox, Ala., 1,121 D2 78
Camden, Ouachita, Ark., 15,823 D3 81
Camden, Kent, Del., 1,125 B6 85
Camden, Schuyler, Ill., 116 C3 90
Camden, Carroll, Ind., 601 C4 91
Camden, Knox, Maine, 3,523 (3,988*) D3 96
Camden, Hillsdale, Mich., 434 G6 98
Camden, Madison, Miss., 250 C4 100
Camden, Ray, Mo., 310 B3 101
Camden, Camden, N.J., 117,159 D2 106
Camden, Oneida, N.Y., 2,694 B5 108
Camden, Camden, N.C., 300 A7 109
Camden, Preble, Ohio, 1,308 C3 111
Camden, Kershaw, S.C., 6,842 C6 115
Camden, Benton, Tenn., 2,774 A3 117
Camden, Polk, Tex., 1,131 D5 118
Camden, Lewis, W. Va., 100 B4 123
Camden, co., Ga., 9,975 F5 87
Camden, co., Mo., 9,116 C5 101
Camden, co., N.J., 392,035 D3 106
Camden, co., N.C., 5,598 A7 109
Camden on Gauley, Webster, W. Va., 301 C4 123
Camdenton, Camden, Mo., 1,405 D5 101
Camel Back, mtn., Ariz G2 80
Camelback, mtn., Pa A2 106
Camels Hump, mtn., Vt C3 120
Cameo, Mesa, Colo., 10 B2 83
Camerino, It., 4,200 C4 21
Cameron, Coconino, Ariz., 50 B4 80
Cameron, Warren, Ill., 250 C3 90
Cameron, Cameron, La., 950 E2 95
Cameron, Clinton and De Kalb, Mo., 3,674 B3 101
Cameron, Madison, Mont., 10 E5 102
Cameron, Quay, N. Mex., 5 C6 107
Cameron, Moore, N.C., 298 B4 109
Cameron, LeFlore, Okla., 211 B7 112
Cameron, Cameron, Pa., 80 D5 114
Cameron, Calhoun, S.C., 607 D6 115
Cameron, Milam, Tex., 5,640 D4 118
Cameron, Marshall, W. Va., 1,652 B4, C2 123
Cameron, Barron, Wis., 982 C2 124
Cameron, co., Pa., 7,586 D5 114
Cameron, co., Tex., 151,098 F4 118
Cameron, par., La., 6,909 E2 95
Cameron, hills, N.W. Ter., Can D9 66
Cameroon, country, Afr., 4,129,000 F7 42, D2 46
Cameroon, mtn., Cam F6 45
Cametá, Braz., 5,695 C5 59
Cameta, Sharkey, Miss B3 100
Camiling, Phil., 7,765 o13 35
Camilla, Mitchell, Ga., 4,753 E2 87
Camillus, Onondaga, N.Y., 1,416 *B4 108
Caminha, Port., 2,190 B1 20
Camino, El Dorado, Calif., 700 C3 82
Camlachie, Ont., Can., 206 D2 72
Cammack (Cammack Village), Pulaski, Ark., 1,355 C3, D5 81
Cammal, Lycoming, Pa., 100 D7 114
Cammin, see Kamien Pomorski, Pol.
Camocim, Braz., 10,788 B2 57
Camooweal, Austl., 192 C6 50
Camp, co., Tex., 7,849 C5 118
Campagna di Roma, reg., It h9 21
Campaign, Warren, Tenn., 500 D8 117
Camp Allison, Sutton, Tex D2 118
Campana, Arg., 14,452 g7 54
Campana, isl., Chile D1 54
Campanario, Sp., 9,660 C3 20
Campania, pol. dist., It, 4,760,759 *D5 21
Campania, reg., It D5 21
Campania, isl., B.C., Can C3 68
Campanquiz, mts., Peru B2 58
Campbell, Clarke, Ala D2 78
Campbell, Searcy, Ark B3 81
Campbell, Santa Clara, Calif., 11,863 C5 82
Campbell, Osceola, Fla., 285 D5 86
Campbell, Wilkin, Minn., 365 D2 99
Campbell, Dunklin, Mo., 1,964 E7 101
Campbell, Franklin, Nebr., 424 D7 103
Campbell, Steuben, N.Y., 650 C3 108
Campbell, Mahoning, Ohio, 13,406 A7 111
Campbell, Albemarle, Va., 40 C4 121
Campbell, co., Ky., 86,803 B5 94
Campbell, co., S. Dak., 3,531 B5 116
Campbell, co., Tenn., 27,936 C9 117
Campbell, co., Va., 32,958 D3 121
Campbell, co., Wyo., 5,861 A7 125
Campbell, cape, N.Z N15 51
Campbell, creek, W. Va C6 123
Campbell, hill, Ohio B4 111
Campbell, isl., B.C., Can C3 68
Campbell, isl., Pac. O D29 5
Campbell, lake, Oreg E7 113
Campbell, mtn., Yukon, Can D5 66
Campbellford, Ont., Can., 3,478 C7 72
Campbell Hill, Jackson, Ill., 263 F4 90
Campbell Island, B.C., Can., 150 C3 68
Campbellpur, Pak., 10,135 (*17,689) B5 39
Campbell River, B.C., Can., 3,737 D4 68
Campbells Bay, Que., Can., 1,024 B8 72
Campbellsburg, Washington, Ind., 612 G5 91
Campbellsburg, Henry, Ky., 348 B4 94
Campbellsport, Fond du Lac, Wis., 1,472 E5 124
Campbell Station, Jackson, Ark., 181 *B4 81
Campbellsville, Taylor, Ky., 6,966 C4 94
Campbellsville, Giles, Tenn., 150 B4 117
Campbellton, N.B., Can., 9,873 A3 74
Campbellton, Newf., Can., 636 D4 75
Campbellton, Jackson, Fla., 309 B1, G3 86
Campbelltown, Lebanon, Pa., 1,061 *F9 114
Campbeltown, Scot., 6,525 E3 13
Campcreek, Greene, Tenn., 130 C11 117
Camp Crook, Harding, S. Dak., 90 B2 116
Camp Douglas, Juneau, Wis., 489 E3 124
Campeche, Mex., 44,426 D6 63
Campeche, state, Mex., 164,256 D6 63
Campeche, gulf, Mex C6 63
Campechuela, Cuba, 2,782 E5 64
Camperdown, Austl., 3,446 I4 51
Campgrove, Marshall, Ill., 300 B4 90
Camp Hill, Tallapoosa, Ala., 1,270 C4 78
Camp Hill, Cumberland, Pa., 8,559 F8 114
Campiglia Marittima, It., 2,716 C3 21
Câmpina, Rom., 18,680 C7 22
Campina Grande, Braz., 116,226 C3, h6 57
Campinas, Braz., 179,797 C3, m8 56
Campion, Larimer, Colo., 350 A5 83
Campione d'Italia, It D4 18
Camp Lister, B.C., Can., 163 E9 68
Camp Namanu, Clackamas, Oreg B4 113
Campo, San Diego, Calif., 300 F5 82
Campo, Cam E1 46
Campo, Baca, Colo., 235 D8 83
Campoalegre, Col., 5,997 C2 60
Campobasso, It., 26,500 (34,011*) D5 21
Campobello, Spartanburg, S.C., 420 A3 115
Campobello, isl., N.B., Can E3 74
Campo Belo, Braz., 15,742 F1 57
Campo de Criptana, Sp., 14,608 C4 20
Campo Gallo, Arg E3 55
Campo Grande, Braz., 64,477 C2 56
Campo Maior, Braz., 13,939 B2 57
Campo Maior, Port., 8,086 C2 20
Campo Real, Sp., 1,921 p18 20
Campos, Braz., 90,601 C4 56
Camposampiero, It., 6,870 D7 18
Campos Novos, Braz., 3,589 D2 56
Campos Sales, Braz., 3,390 C2 57
Campo Tencia, mtn., Switz D6 19
Campo Tures, It., 3,016 C7 18
Camp Point, Adams, Ill., 1,092 C2 90
Camp Ruby, Polk, Tex., 110 D5 118
Camp San Saba, McCulloch, Tex., 75 D3 118
Camp Sealth, King, Wash D1 122
Camp Sherman, Jefferson, Oreg., 25 C5 113
Camp Springs, Prince Georges, Md., 50 C2 85
Campti, Natchitoches, La., 1,045 C2 95
Campton, Okaloosa, Fla., 100 G2 86
Campton, Walton, Ga., 200 C3 87
Campton, Wolfe, Ky., 484 C6 94
Campton, Grafton, N.H., 350 (1,058*) C3 105
Campton Heights, Humboldt, Calif., 1,200 *B1 82
Camptown, Bradford, Pa., 200 C9 114
Câmpulung, Rom., 18,880 C7 22
Câmpulung Moldovenesc, Rom., 13,627 B7 22
Campus Livingston, Ill., 165 B5 90
Camp Verde, Yavapai, Ariz., 600 C4 80
Campville, Alachua, Fla., 150 C4 86
Camp Wood, Yavapai, Ariz., 40 C3 80
Camp Wood, Real, Tex., 879 E2 118
Camrose, Alta., Can., 6,939 C4 69
Camuy, P.R., 2,341 B3 65
Camuy, mun., P.R., 19,739 B3 65
Camuy, riv., P.R B3 65
Canaan, Searcy, Ark B3 81
Canaan, Litchfield, Conn., 1,146 A3 84
Canaan, Jefferson, Ind., 100 G7 91
Canaan, Benton, Miss., 150 A4 100
Canaan, Grafton, N.H., 450 (1,507*) C2 105
Canaan, Columbia, N.Y., 500 C7 108
Canaan, Essex, Vt., 275 (1,094*) A5 120
Canaan, riv., N.B., Can C4 74
Canaan Center, Grafton, N.H., 140 C2 105
Canaan Station, N.B., Can., 88 C4 74
Canaan Street, Grafton, N.H., 150 C2 105
Canada, Pike, Ky., 400 C7 94
Canada, country, N.A., 18,238,247 D10 61, 66
Canada, bay, Newf., Can C3 75
Cañada de Gómez, Arg., 12,354 A4 54
Canada de los Alamos, Santa Fe, N. Mex., 30 F6 107
Canada Falls Deadwater, lake, Maine C2 96
Canadensis, Monroe, Pa., 800 D11 114
Canadian, Pittsburg, Okla., 255 B6 112
Canadian, Hemphill, Tex., 2,239 B2 118
Canadian, co., Okla., 24,727 B4 112
Canadian, riv., U.S C7 76
Cañadón León, Arg D2 54
Canadys, Colleton, S.C E6 115
Caña Gorda, is., P.R D3 65
Canairiktok, riv., Newf., Can g9 75
Canajoharie, Montgomery, N.Y., 2,681 C6 108
Canakkale, Tur., 19,500 B6 23
Canal [Township], Venango, Pa. (685*) D2 114
Canal, riv., Ven C4 60
Canale, It., 4,863 E3 18
Canal Fulton, Stark, Ohio, 1,555 B6 111
Canalou, New Madrid, Mo., 447 E8 101
Canal Point, Palm Beach, Fla., 900 F6 86
Canal Winchester, Franklin, Ohio, 1,976 C2, C5 111
Canal Zone, U.S. dep., N.A., 42,122 I13 61, k11 62
Canandaigua, Ontario, N.Y., 9,370 C3 108
Canandaigua, lake, N.Y C3 108
Cananea, Mex., 19,895 A2 63
Cananéia, Braz., 1,948 D3 56
Cañar, Ec., 4,230 B2 58
Cañar, prov., Ec., 107,701 B2 58
Canary Islands, Sp. overseas prov., Afr., 944,448 D1 44
Canary, is., Afr m14 20
Cañas, C.R., 5,929 E5 62
Canaseraga, Allegany, N.Y., 730 C3 108
Canastota, Madison, N.Y., 4,896 B5 108
Canaveral, cape, Fla D6 86
Canavieiras, Braz., 10,264 E3 57
Cañazas, Pan., 1,105 F7 62
Canazei, It., 610 C7 18
Canbelego, Austl., 115 E6 51
Canberra, Austl., 56,449 G8 45, G7 51
Canby, Modoc, Calif., 500 B3 82
Canby, Yellow Medicine, Minn., 2,146 F2 99
Canby, Clackamas, Oreg., 2,168 B2, B4 113
Cancale, Fr., 3,394 C3 14
Canche, riv., Fr D1 14
Cancienne, Assumption, La., 25 C5 95
Cándala, Som., 1,800 (2,800*) C6 47
Candelaria, Cuba, 3,461 D2 64
Candelaria, Phil., 950 o12 35
Candelaria, Presidio, Tex., 45 F2 118
Candeleda, Sp., 6,983 B3 20
Candelero, pt., P.R C7 65
Candia, see Iraklion, Grc.
Candia, Rockingham, N.H., 200 (1,490*) D4 105
Candiac Station, Sask., Can., 55 G4 70
Candia Village, Rockingham, N.H., 55 D4 105
Candle, Alsk., 73 B7 79
Candle, lake, Sask., Can D3 70
Candler, Marion, Fla., 150 C5 86
Candler, co., Ga., 6,672 D4 87
Candlewood, hill, Conn D6 84
Candlewood, lake, Conn C3 84
Candlewood, mtn., Conn C3 84
Cando, Sask., Can., 247 E1 70
Cando, Towner, N. Dak., 1,566 A6 110
Candor, Tioga, N.Y., 956 C4 108
Candor, Montgomery, N.C., 593 B4 109
Cane, creek, Tenn B5 117
Cane, creek, Utah E6 119
Cane, riv., La C2 95
Canea, see Khania, Grc.
Cane Beds, Mohave, Ariz A3 80
Caneel, bay, Vir. Is f15 65
Canela, Braz., 7,058 D2 56
Canelones, Ur., 8,041 E1 56
Canelones, dept., Ur., 182,100 *E1 56
Cañete, Chile, 3,137 B2 54
Cañete, Peru, 4,794 D2 58
Cane Valley, Adair, Ky., 200 C4 94
Caney, Montgomery, Kans., 2,682 E8 93
Caney, Morgan, Ky., 400 C6 94
Caney, Atoka, Okla., 128 C5 112
Caney, fork, Tenn C8 117
Caney, ridge, Ky C3 94
Caney, riv., Okla A5 112
Caneyville, Grayson, Ky., 278 C3 94
Canfield, Lafayette, Ark., 200 D2 81
Canfield, Mahoning, Ohio, 3,252 A7 111
Cangallo, Peru, 902 D3 58
Cangamba, Ang., 269 D2 48
Cangas, Sp., 4,008 A1 20
Cangas de Narcea, Sp., 2,528 A2 20
Cangas de Onís, Sp., 9,936 A3 20
Canguaretama, Braz., 4,261 C3, h6 57
Cangussu, Braz., 3,257 E2 56
Canhotinho, Braz., 2,730 k5 57
Caniçado, Moz B5 49
Canicanian, Phil., 1,258 o13 35
Canicatti, It., 31,800 F4 21
Caniles, Sp., 8,040 D4 20
Canim, lake, B.C., Can D7 68
Canim Lake, B.C., Can., 106 D7 68
Canindé, Braz., 5,854 B3 57
Canistear, res., N.J A4 106
Canisteo, Steuben, N.Y., 2,731 C3 108
Canisteo, riv., N.Y C3 108
Canistota, McCook, S.Dak., 627 D8 116
Canjilon, Rio Arriba N. Mex., 100 A3 107
Çankiri (Changra), Tur., 20,000 B9 31
Canmer, Hart, Ky., 500 C4 94
Canmore, Alta., Can., 1,736 D3 69
Canna, isl., Scot D2 13
Canna, sound, Scot D2 13
Cannelburg, Daviess, Ind., 124 G4 91
Cannel City, Morgan, Ky., 300 C6 94
Cannelton, Perry, Ind., 1,829 I4 91
Cannelton, Fayette, W. Va., 400 C6 123
Cannes, Fr., 58,079 F7 14
Cannes, bayou, La D3 95
Canneto sull'Oglio, It., 4,963 D6 18
Cannich, Scot C4 13
Canning, N.S., Can., 850 D5 74
Canning, Hughes, S. Dak., 50 C5 116
Cannington, Ont., Can., 1,024 C5 72
Cannock, Eng., 42,186 B5 12
Cannon, Sussex, Del., 150 C6 85
Cannon, co., Tenn., 8,537 B5 117
Cannon, bay, Fla G5 86
Cannon, riv., Minn F6 99
Cannon Ball, Sioux, N. Dak., 300 C5 110
Cannonball, riv., N. Dak C4 110
Cannon Beach, Clatsop, Oreg., 495 B3 113
Cannondale, Fairfield, Conn., 300 E3 84
Cannon Falls, Goodhue, Minn., 2,055 F6 99
Cannon Mines, Washington, Mo., 40 C7 101
Cannonsburg, Boyd, Ky., 300 B7 94
Cannonsburg, Jefferson, Miss., 25 D2 100
Cannonsville, Del., N.Y., 400 C5 108
Cannonville, Garfield, Utah, 153 F3 119
Cano, Cuba, 1,258 k11 64
Canobie Lake, Rockingham, N.H., 250 E4 105
Canoe, Escambia, Ala., 500 D2 78
Canoe, B.C., Can., 504 D8 68
Canoe, riv., B.C., Can C8 68
Canoe Lake, Ont., Can., 25 B6 72
Canoinhas, Braz., 9,252 D2 56
Canon, Franklin and Hart, Ga., 626 B3 87
Canonchet, Washington, R.I., 100 D10 84
Canon City, Fremont, Colo., 8,973 C5 83
Canones, Rio Arriba, N. Mex., 90 A3 107
Canonsburg, Washington, Pa., 11,877 F1 114
Canoochee, riv., Ga D5 87
Canora, Sask., Can., 2,117 F4 70
Canosa, It., 34,015 D6 21
Canova, Miner, S. Dak., 247 D8 116
Canóvanas Central, P.R B7 65
Canso, N.S., Can., 1,151 D9 74
Canso, cape, N.S., Can G20 67
Cantabrian, mts., Sp A3 20
Cantal, dept., Fr., 172,977 *E5 14
Cantanhede, Port., 3,133 B1 20
Cantaura, Ven., 14,096 B5 60
Canterbury, Austl B3 51
Canterbury, Windham, Conn., 175 (1,857*) C9 84
Canterbury, Eng., 30,376 C9 12
Canterbury, Merrimack, N.H., 100 (674*) D3 105
Canterbury, bight, N.Z P13 51
Canterbury Station, N.B., Can., 595 D2 74
Can Tho, Viet., 27,000 G6 38
Cantilan, Phil., 4,141 *D7 35
Cantley, Que., Can., 359 D2 73
Canton, China, 1,840,000 G7 34
Canton, Hartford, Conn., 400 (4,783*) B5 84
Canton, Cherokee, Ga., 2,411 B2 87
Canton, Fulton, Ill., 13,588 C3 90
Canton, McPherson, Kans., 784 D6 93
Canton, Trigg, Ky., 350 D2 94
Canton, Oxford, Maine, 400 (728*) D2 96
Canton, Norfolk, Mass., 12,771 B5, D2 97
Canton, Fillmore, Minn., 467 G7 99
Canton, Madison, Miss., 9,707 C3 100
Canton, Lewis, Mo., 2,562 A6 101
Canton, Salem, N.J., 500 E2 106
Canton, St. Lawrence, N.Y., 5,046 B2 108
Canton, Haywood, N.C., 5,068 D3 109
Canton, Stark, Ohio, 113,631 (*281,900) B6 111
Canton, Blaine, Okla., 887 A3 112
Canton, Bradford, Pa., 2,102 C8 114
Canton, Lincoln, S. Dak., 2,511 D9 116
Canton, Van Zandt, Tex., 1,114 C5 118
Canton, Barron, Wis., 100 C2 124
Canton, res., Okla A3 112
Canton & Enderbury, Br. and U.S. dep., Oceania, 320 *G11 7
Canton Center, Hartford, Conn., 175 B5 84
Cantonment, Escambia, Fla., 2,499 G2 86
Canton Point, Oxford, Maine D2 96
Cantril, Van Buren, Iowa, 299 D5 92
Cantuar, Sask., Can., 50 G1 70
Cantwell, Choctaw, Ala D1 78
Cantwell, Alsk., 20 C10 79
Cantwell, St. Francois, Mo., 200 D7 101
Cañuelas, Arg., 5,614 B5, g7 54
Canutama, Braz., 977 D2 59
Canute, Washita, Okla., 370 B2 112
Canutillo, El Paso, Tex., 1,377 E1 118
Canwood, Sask., Can., 311 D2 70
Cany, creek, Tex G4 118
Cany-Barville, Fr., 1,045 E8 12
Canyon, Sandoval, N. Mex B3, F5 107
Canyon, Randall, Tex., 5,864 B2 118
Canyon, Yellowstone National Park, Wyo A2 125
Canyon, co., Idaho, 57,662 F2 89
Canyoncito, Santa Fe, N. Mex F6 107
Canyon City, Grant, Ore., 654 C8 113

Canyon Creek, Lewis and Clark, Mont., 10...D4 102
Canyon De Chelly, nat. mon., Ariz...A6 80
Canyon Ferry, res., Mont...D5 102
Canyonville, Douglas, Oreg., 1,089...E3 113
Cao Bang, Viet., 25,000...A7 38
Caorle, It., 3,207...D8 18
Capa, P.R...B2 65
Capa, Jones, S. Dak., 39...C5 116
Capac, St. Clair, Mich., 1,235...F8 98
Cap-à-l'Aigle, Que., Can., 659...B7 73
Capalonga, Phil., 1,344...o14 35
Capanaparo, riv., Ven...B4 60
Capanema, Braz., 9,678...B1 57
Capão Bonito, Braz., 6,829...C3, m7 56
Caparica, Port., 8,575...f9 20
Capasin, Sask., Can...D2 70
Capatárida, Ven., 1,277...A3 60
Cap Bon, prov., Tun., 240,350...*B7 44
Capbreton, Fr., 3,688...F3 14
Cap Chat, Que., Can., 2,035...G19 67
Cap de la Madeleine, Que., Can., 26,925...C5 73
Capdenac-Gare, Fr., 115...E5 14
Capdevila, Cuba...k11 64
Cape, isl., S.C...E9 115
Cape Breton, co., N.S., Can., 131,507...D10 74
Cape Breton, isl., N.S., Can...G20 67
Cape Breton Highlands, nat. park, N.S., Can...G20 67, C9 74
Cape Broyle, Newf., Can., 630...E5 75
Cape Canaveral, Fla., 3,500...D6 86
Cape Charles, Newf., Can...B4 75
Cape Charles, Northampton, Va., 2,041...D6 121
Cape Coast, Ghana, 56,730...E4 45
Cape Cod, bay, Mass...C7 97
Cape Cod, canal, Mass...C6 97
Cape Elizabeth (The Cape), Cumberland, Maine, 5,505...E2, E5 96
Cape Fear, riv., N.C...C5 109
Cape Girardeau, Cape Girardeau, Mo., 24,947...D8 101
Cape Girardeau, co., Mo., 42,020...D8 101
Cape Hatteras, nat. seashore recreational area, N.C...B8 109
Cape Henry, Princess Anne, Va...B7, E6 121
Cape Horn, Skamania, Wash...D3 122
Cape La Hune, Newf., Can., 84...E3 75
Capelinha, Braz., 3,262...E3 57
Capella, is., Vir. Is...f15 65
Capels, McDowell, W. Va., 800...D3 123
Cape May, Cape May, N.J., 4,477...F3 106
Cape May, co., N.J., 48,555...E3 106
Cape May Court House, Cape May, N.J., 1,749...E3 106
Cape May Point, Cape May, N.J., 263...F3 106
Cape Neddick, York, Maine, 600...E2 96
Cape of Good Hope, prov., S. Afr., 5,355,368...D3 49
Cape Porpoise, York, Maine, 400...E2 96
Cape Race, Newf., Can., 25...E5 75
Cape Ray, Newf., Can., 207...E2 75
Capers, Pueblo, Colo., 25...D6 83
Capers, inlet, S.C...F8 115
Capers, is., S.C...F8 115
Capers, is., S.C...G6 115
Capesterre, Guad., 3,725...n16 74
Capeto, Chile...D2 55
Cape Tormentine, N.B., Can., 345...C6 74
Cape Town, S. Afr., 508,341 (*807,211)...D2 49
Cape Verde Is., Port. dep., Afr., 147,328...E3 42
Cape Verde, is., Atl. O...E3 42
Capeville, Northampton, Va., 200...A7 121
Cape Vincent, Jefferson, N.Y., 770...A4 108
Cape York, pen., Austl...B7 50
Cap-Haïtien, Hai., 24,957...F7 64
Capilano, B.C., Can., 800...A9 68
Capilla del Monte, Arg., 2,522...A4 54
Capilla del Señor, Arg., 3,421...g7 54
Capinota, Bol., 1,734...C2 55
Capistrano Beach, Orange, Calif., 2,026...*F5 82
Capital Federal, fed. dist., Arg., 2,966,816...*g7 54
Capitan, Lincoln, N. Mex., 552...D4 107
Capitán Bado, Par., 4,104...D4 55
Capitol, Carter, Mont., 5...E12 102
Capitol, peak, Nev...B4 104
Capitola, Santa Cruz, Calif., 2,021...*C6 82
Capitola, Leon, Fla., 30...B2 86
Capitol Heights, Prince Georges, Md., 3,138...C2, C4 85
Capitol Reef, nat. mon., Utah...E4 119
Capivari, Braz., 10,961...m8 56
Capiz, prov., Phil., 314,800...*C6 35
Capleville, Shelby, Tenn., 350...E8 117
Capon Bridge, Hampshire, W. Va., 198...B6 123
Capon Springs, Hampshire, W. Va., 200...B6 123
Cappoquin, Ire., 806...E4 11
Capps, Henry, Ala...D4 78
Cappstown, Miller, Ark...D2 81
Capraia, isl., It...C2 21
Caprara, pt., It...C2 21
Capreol, Ont., Can., 3,003...E9 72
Caprera, isl., It...D2 21
Capri, isl., It...D5 21
Capricorn, is., Austl...J9 51
Caprock, Lea, N. Mex., 5...D6 107
Capron, Boone, Ill., 656...A5 90
Capron, Woods, Okla., 102...A3 112
Capron, Southampton, Va., 327...E5 121
Cap Rouge, Que., Can., 365...C9 73
Cap St. Ignace, Que., Can., 1,247...B7 73
Cap Sante, Que., Can., 546...C6 73
Capshaw, Limestone, Ala., 250...A3 78
Captain Cook, Hawaii, Haw., 1,687...D6 88
Captina, creek, Ohio...B1 111
Captiva, Lee, Fla., 200...F4 86
Captiva, isl., Fla...F4 86
Capua, It., 13,200...D5 21
Capulin, Conejos, Colo., 20...D4 83
Capulin, Rio Arriba, N. Mex., 90...A3 107
Capulin, Union, N. Mex., 200...A6 107
Capulin Mountain, nat. mon., N. Mex...A6 107
Caputa, Pennington, S. Dak., 40...D3 116
Caquetá, pol. div., Col., 63,970...C3 60
Caquetá, riv., Col...D3 60
Carabaña, Sp., 1,947...p18 20
Carabanchel Bajo, Sp. (part of Madrid)...p17 20
Carabobo, state, Ven., 381,636...A4 60
Caracal, Rom., 19,082...C7 22
Caracas, Ven., 786,710 (*1,550,000)...A4 60
Carajás, mts., Braz...D4 59
Carangola, Braz., 11,896...F2 57
Caransebes, Rom., 15,195...C6 22
Carapari, Bol., 351...D3 55
Carapeguá, Par., 23,675...E4 55
Caraquet, N.B., Can., 1,214...B5 74
Caraquet, bay, N.B., Can...B5 74
Caratinga, Braz., 22,275...E2 57
Caratunk, Somerset, Maine, (90*)...C3 96
Caraúbas, Braz., 3,066...C3 57
Caraúbas, Braz., 285...h5 57
Caravaca, Sp., 10,678...C5 20
Caravelas, Braz., 3,096...E3 57
Caravelí, Peru, 1,177...E3 58
Caraway, Craighead, Ark., 821...B5 81
Caraz, Peru, 3,065...C2 58
Caràzinho, Braz., 18,162...D2 56
Carballino, Sp., 9,639...A1 20
Caraballo, Sp., 3,110...A1 20
Carberry, Man., Can., 1,113...E2 71
Carbó, Mex., 2,213...B2 63
Carbon, Alta., Can., 371...D4 69
Carbon, Clay, Ind., 409...E3 91
Carbon, Adams, Iowa, 162...C3 92
Carbon, Kanawha, W. Va., 400...D6 123
Carbon, co., Mont., 8,317...E7 102
Carbon, co., Pa., 52,889...E10 114
Carbon, co., Utah, 21,135...D5 119
Carbon, co., Wyo., 14,937...E5 125
Carbonado, Pierce, Wash., 424...B3 122
Carbonara, cape, It...E2 21
Carbon Cliff, Rock Island, Ill., 1,268...*B3 90
Carbondale, Alta., Can., 52...C4 69
Carbondale, Garfield, Colo., 612...B3 83
Carbondale, Jackson, Ill., 14,670...F4 90
Carbondale, Osage Kans., 664...D8 93
Carbondale, Athens, Ohio, 350...C5 111
Carbondale, Lackawanna, Pa., 13,595...C10 114
Carbonear, Newf., Can., 4 234...E5 75
Carbonear-Bay de Verde, dist., Newf., Can...E5 75
Carbon Hill, Walker, Ala., 1,944...B2 78
Carbonia, It., 34,400...E2 21
Carbury, Bottineau, N. Dak., 40...A5 110
Carcagente, Sp., 17,937...C5 20
Carcans, lagoon, Fr...E3 14
Carcar, Phil., 5,344...*C6 35
Carcassonne, Fr., 40,897...F5 14
Carchi, prov., Ec., 84,738...A2 58
Carcross, Yukon, Can., 175...D6 66
Cardak, Tur., 144...D7 23
Cardale, Man., Can., 91...D1 71
Cardale, Fayette, Pa., 900...*G2 114
Cárdenas, Cuba, 43,750...D3 64
Cárdenas, Mex., 12,344...k14 63
Cárdenas, Mex., 3,010...D6 63
Cardenas, De Baca, N. Mex., 25...C5 107
Carderview, Johnson, Tenn...C12 117
Cardiel, lake, Arg...D2 54
Cardiff, Jefferson, Ala., 202...E4 78
Cardiff, Garfield, Colo., 120...B3 83
Cardiff, Harford, Md., 450...A5 85
Cardiff, Atlantic, N.J...E3 106
Cardiff, Wales, 256,270 (*600,000)...C4 12
Cardiff-by-the-Sea, San Diego, Calif., 1,500...*F5 82
Cardigan, P.E.I., Can., 193...C7 74
Cardigan, Wales, 3,780...B3 12
Cardigan, co., Wales, 53,564...B3 12
Cardigan, bay, P.E.I., Can...C7 74
Cardigan, bay, Wales...B3 12
Cardinal, Man., Can., 50...E2 71
Cardinal, Ont., Can., 1,944...C9 72
Cardinal, lake, Alta., Can...A1 69
Cardington, Morrow, Ohio, 1,613...B3 111
Cardona, Ur...E1 56
Cardross, Sask., Can., 15...H3 70
Cardston, Alta., Can., 2,801...E4 69
Cardville, Penobscot, Maine, 130...C4 96
Cardwell, Dunklin, Mo., 816...E7 101
Cardwell, Jefferson, Mont., 40...E5 102
Cardwell, mtn., Tenn...D8 117
Carega, mtn., It...D7 18
Carei, Rom., 16,780...B6 22
Carencro, Lafayette, La., 1,519...D3 95
Carenero, Ven...A4 60
Carentan, Fr., 5,256...C3 14
Caretta, McDowell, W. Va., 1,092...D3 123
Carey, Blaine, Idaho, 200...F5 89
Carey, Wyandot, Ohio, 3,722...B4 111
Carey, Childress, Tex., 100...B2 118
Careyhurst, Converse, Wyo...C7 125
Careysburg, Lib...E2 45
Careywood, Bonner, Idaho, 25...A2 89
Carhaix, Fr., 4,032...C2 14
Carhuaz, Peru, 2,359...C2 58
Cariamanga, Ec., 3,376...B2 58
Cariati, It., 5,204...E6 21
Caribana, pt., Col...B2 60
Caribbean, sea, N.A...H13 61
Caribe Central, P.R...D6 65
Caribes, is., P.R...D6 65
Cariboo, dist., B.C., Can...C6 68
Cariboo, mts., B.C., Can...F8 66
Caribou, Aroostook, Maine, 8,305 (12,464*)...B5 96
Caribou, co., Idaho, 5,976...G7 89
Caribou, isl., N.S., Can...D7 74
Caribou, lake, Maine...C3 96
Caribou, mts., Alta., Can...E9 66
Caribou, mtn., Idaho...F7 89
Caribou, mtn., Maine...C2 96
Caribou, range, Idaho...F7 89
Carichic, Mex., 965...B3 63
Carievale, Sask., Can., 264...H5 70
Carignan, Fr., 3,403...C6 15
Carinhanha, Braz., 2,163...D2 57
Carini, It., 16,723...E4 21
Carinish, Scot...C1 13
Carinthia, reg., Aus...E6 16
Caripe, Ven., 3,651...A5 60
Caripito, Ven., 19,003...A5 60
Cariris Velhos, mts., Braz...h5 57
Cariús, Braz., 1,523...C3 57
Carl Blackwell, lake, Okla...A4 112
Carle Place, Nassau, N.Y., 5,100...*G2 84
Carleton, Monroe, Mich., 1,379...F7 98
Carleton, Thayer, Nebr., 207...D8 103
Carleton, co., Ont., Can., 352,931...B9 72
Carleton, co., N.B., Can., 23,507...C2 74
Carleton, mtn., N.B., Can...B3 74
Carleton Place, Ont., Can., 4,796...B8 72
Carlile, Crook, Wyo., 5...A8 125
Carlin, Elko, Nev., 1,023...C5 104
Carlin Bay, Kootenai, Idaho...B8 122
Carlingford, Ire., 471...C5 11
Carlingford, bay, Ire., N. Ire...C5 11
Carlinville, Macoupin, Ill., 5,440...D4 90
Carlisle, Lonoke, Ark., 1,514...C4 81
Carlisle, Eng., 71,112...C5 10
Carlisle, Sullivan, Ind., 755...G3 91
Carlisle, Warren, Iowa, 1,317...A7, C4 92
Carlisle, Nicholas. Ky., 1,601...B5 94
Carlisle, Warren, Ohio, 671...C3 111
Carlisle, Cumberland, Pa., 16,623...F7 114
Carlisle, Union, S.C., 390...B5 115
Carlisle, Stewart, Tenn., 100...A4 117
Carlisle, Trinity, Tex., 35...D5 118
Carlisle, co., Ky., 5,608...B2 94
Carl Junction, Jasper, Mo., 1,220...D3 101
Carlock, McLean, Ill., 318...C4 90
Carloforte, It., 7,275...E2 21
Carlos, Allegany, Md., 135...D2 85
Carlos, Douglas, Minn., 262...E3 99
Carlos Casares, Arg., 7,558...B4 54
Carlos Chagas, Braz., 6,383...E2 57
Carlos Tejedor, Arg., 2,897...B4 54
Carlow, Ire., 7,708...E5 11
Carlow, co., Ire., 33,342...E5 11
Carloway, Scot...B2 13
Carlowville, Dallas, Ala., 250...C2 78
Carlsbad, San Diego, Calif., 9,253...F5 82
Carlsbad, Eddy, N. Mex., 25,541...E5 107
Carlsbad, Tom Green, Tex., 350...D2 118
Carlsbad Caverns, nat. park, N. Mex...E5 107
Carlsbad Springs, Ont., Can., 256...A10 72
Carlsborg, Clallam, Wash., 250...A2 122
Carlstadt, Bergen, N.J., 6,042...D5 106
Carlton, Clarke, Ala., 200...D2 78
Carlton, Sask., Can., 67...E2 70
Carlton, Eng., 38,790...B6 12
Carlton, Madison, Ga., 321...B3 87
Carlton, Dickinson, Kans., 78...D6 93
Carlton, Carlton, Minn., 862...D6 99
Carlton, Missoula, Mont., 84...D2 102
Carlton, Yamhill, Oreg., 959...B1, B3 113
Carlton, Hamilton, Tex., 220...D3 118
Carlton, co., Minn., 27,932...D6 99
Carluke, Scot., 11,415...C5 13
Carlyle, Sask., Can., 982...H4 70
Carlyle, Clinton, Ill., 2,903...E4 90
Carlyle, Wibaux, Mont., 18...D12 102
Carmacks, Yukon, Can., 218...D5 66
Carmagnola, It., 4,800...B1 21
Carman, Man., Can., 1,930...E2 71
Carman, Schenectady, N.Y., 1,500...*G6 108
Carman, hill, Pa...C6 114
Carmangay, Alta., Can., 297...D4 69
Carmans, riv., N.Y...F5 84
Carmanville, Newf., Can., 855...D4 75
Carmarthen, Wales, 13,249...C3 12
Carmarthen, co., Wales, 167,736...C3 12
Carmarthen, bay, Wales...C3 12
Carmaux, Fr., 14,565...E5 14
Carmel, Monterey, Calif., 4,580...C6, D3 82
Carmel, Hamilton, Ind., 1,442...E5 91
Carmel, Penobscot, Maine, 650 (1,206*)...D3 96
Carmel, Cumberland, N.J., 200...E2 106
Carmel, Putnam, N.Y., 2,735...D7 108
Carmel, mtn., Isr...B7 32
Carmel, pt., Calif...C5 82
Carmelo, Ur., 11,800...E1 56
Carmel Station, Sask., Can., 106...E3 70
Carmel Valley, Monterey, Calif., 1,143...*D3 82
Carmel Woods, Monterey, Calif., 1,043...*D3 82
Carmen, Santa Cruz, Ariz., 100...F4 80
Carmen, Lemhi, Idaho, 10...D5 89
Carmen, Alfalfa, Okla., 533...A3 112
Carmen, isl., Mex...B2 63
Carmen Alto, Chile...D2 55
Carmen de Areco, Arg., 4,411...A5, g7 54
Carmen del Paraná, Par., 10,699...E4 55
Carmen de Patagones, Arg., 5,243...C4 54
Carmen-Sylva, Rom., 3,286...C9 22
Carmi, White, Ill., 6,152...E5 90
Carmi, lake, Vt...B3 120
Carmichael, Sacramento, Calif., 20,455...A6 82
Carmichael, Sask., Can., 53...G1 70
Carmichaels, Greene, Pa., 788...G2 114
Carmine, Fayette, Tex., 500...E4 118
Carmona, Sp., 28,216...D3 20
Carmyllie, Scot., 665...D6 13
Carnarvon, Austl., 1,809...D1 50
Carnarvon, Sac, Iowa, 100...B2 92
Carnarvon, S. Afr., 3,762...D3 49
Carnation, King, Wash., 490...B4 122
Carnaxide, Port., 20,136...f9 20
Carncastle, N. Ire...C6 11
Carndonagh, Ire., 1,016...B4 11
Carnduff, Sask., Can., 957...H4 70
Carne, Luna, N. Mex., 12...E2 107
Carnegie, Randolph, Ga., 113...E2 86
Carnegie, Caddo, Okla., 1,500...B3 112
Carnegie, Allegheny, Pa., 11,887...B5, F1 114
Carnegie, lake, N.J...C3 106
Carneiro, Ellsworth, Kans., 56...D5 93
Carnes, Forrest, Miss., 30...E4 100
Carnesville, Franklin, Ga., 481...B3 87
Carnew, Ire., 551...E5 11
Carney, Menominee, Mich., 250...C3 98
Carney, Lincoln, Okla., 227...B4 112
Carneys Point, Salem, N.J., 3,500...D2 106
Carnforth, Eng., 4,113...F6 13
Carnic Alps, mts., Aus...E6 16
Car Nicobar, isl., India...G9 39
Carnlough, N. Ire., 585...C6 11
Carn Mairg, mtn., Scot...C4 13
Carnmor, mtn., Scot...C5 13
Carnot, Cen. Afr. Rep...D3 46
Carnoustie, Scot., 5,511...D6 13
Carnsore, pt., Ire...E5 11
Carnwath, Scot., 6,160...E5 13
Caro, Tuscola, Mich., 3,534...E7 98
Carol City, Dade, Fla., 21,749...E3 86
Caroleen, Rutherford, N.C., 1,168...B2 109
Carolina, Braz., 8,137...C1 57
Carolina, P.R., 3,075...B7 65
Carolina, mun., P.R., 40,923...B7 65
Carolina, Washington, R.I., 450...D10 84
Carolina, S. Afr., 4,336...C5 49
Carolina, Marion, W. Va., 700...A7 123
Carolina Beach, New Hanover, N.C., 1,192...C6 109
Caroline, Shawano, Wis., 350...D5 124
Caroline, co., Md., 19,462...C6 85
Caroline, co., Va., 12,725...B5 121
Caroline, atoll, Pac. O...G13 7
Caroline, is., Pac. O...F8 7
Caron, Sask., Can., 105...G3 70
Carona, Cherokee, Kans., 175...E9 93
Caron Brook, N.B., Can., 149...A4 96
Caroni, riv., Ven...B5 60
Carora, Ven., 22,251...A3 60
Carouge, Switz., 12,760...D1 19
Carp, Ont., Can., 498...A8, B8 72
Carp, Lincoln, Nev., 25...F7 104
Carp, lake, B.C., Can...B6 68
Carpathians, mts., Czech., Pol., Rom...D6, B7 22
Carpentaria, gulf, Austl...B6 50
Carpenter, Madison, Ill., 125...A9 101
Carpenter, Mitchell, Iowa, 177...A4 92
Carpenter, Copiah, Miss., 15...C3 100
Carpenter, Roger Mills, Okla...B2 112
Carpenter, Clark, S. Dak., 75...C8 116
Carpenter, Laramie, Wyo., 100...D8 125
Carpenter, dam, Ark...D6 81
Carpentersville, Kane, Ill., 17,424...A5 90
Carpentersville, Warren, N.J., 100...B2 106
Carpenterville, Curry, Oreg., 25...E2 107
Carpentras, Fr., 14,169...E6 14
Carpi, It., 21,000 (45,208*)...B3 21
Carpinteria, Santa Barbara, Calif., 4,998...E4 82
Carpio, Ward, N. Dak., 199...A4 110
Carr, Weld, Colo., 90...A6 83
Carr, lake, Ind...G4 91
Carrabasset, Franklin, Maine, 45...C2 96
Carrabelle, Franklin, Fla., 1,146...C2 86
Carradale, Scot...E3 13
Carrantuohill, mtn., Ire...E2 11
Carranza, cape, Chile...B2 54
Carrara, It., 64,901...B3 21
Carrboro, Orange, N.C., 1,997...B4 109
Carr Bridge, Scot...C5 13
Carrickfergus, N. Ire., 10,211...C6 11
Carrickmacross, Ire., 1,940...D5 11
Carrick-on-Shannon, Ire., 1,497...D4 11
Carrick-on-Suir, Ire., 4,672...E4 11
Carrie, mtn., Wash...B2 122
Carrier, Que., Can...C9 73
Carrier, Garfield, Okla., 150...A3 112
Carriere, Pearl River, Miss., 700...E4 100
Carriers Mills, Saline, Ill., 2,006...F5 90
Carrigahorig, Ire...D3 11
Carrigain, mtn., N.H...B4 105
Carrigaline, Ire., 688...F3 11
Carrigan, head, Ire...C3 11
Carrington, Foster, N. Dak., 2,438...B6 110
Carrión, riv., Sp...A3 20
Carrión de los Condes, Sp., 3,414...A3 20
Carrizal Bajo, Chile...E1 55
Carrizo, Navajo, Ariz...C5 80
Carrizo, creek, Ariz., N. Mex...C6 80
Carrizo, mts., Ariz., N. Mex...A6 80
Carrizo, peak, N. Mex...D4 107
Carrizo Arroya, riv., Tex...B1 118
Carrizo Springs, Dimmit, Tex., 5,699...E3 118
Carrizozo, Lincoln, N. Mex., 1,546...D4 107
Carroll, Man., Can., 78...E1 71
Carroll, Carroll, Iowa, 7,682...B3 92
Carroll, Penobscot, Maine, 25 (147*)...C4 96
Carroll, Wayne, Nebr., 220...B8 103
Carroll, Coos, N.H., 115 (295*)...C4 105
Carroll, Fairfield, Ohio, 444...C5 111
Carroll, co., Ark., 11,284...A2 81
Carroll, co., Ga., 36,451...C1 87
Carroll, co., Ill., 19,507...A4 90
Carroll, co., Ind., 16,934...C4 91
Carroll, co., Iowa, 23,431...B3 92
Carroll, co., Ky., 7,978...B4 94
Carroll, co., Md., 52,785...A3 85
Carroll, co., Miss., 11,177...B4 100
Carroll, co., Mo., 13,847...B4 101
Carroll, co., N.H., 15,829...C4 105
Carroll, co., Ohio, 20,857...B6 111
Carroll, co., Tenn., 23,476...B3 117
Carroll, co., Va., 23,178...E2 121
Carrolls, Cowlitz, Wash., 150...C3 122
Carrollton, Pickens, Ala., 894...B1 78
Carrollton, Carroll, Ga., 10,973...C1 87
Carrollton, Greene, Ill., 2,558...D3 90
Carrollton, Carroll, Ky., 3,218...B4 94
Carrollton (P.O.), Carroll, Md., 75...A4 85
Carrollton, Prince Georges, Md., 3,385...*C4 85
Carrollton, Saginaw, Mich., 6,718...E7 98
Carrollton, Carroll, Miss., 343...B4 100
Carrollton, Carroll, Mo., 4,554...B4 101
Carrollton, Carroll, Ohio, 2,786...B6 111
Carrollton, Dallas, Tex., 4,242...B5 118
Carrolltown, Cambria, Pa., 1,525...E4 114
Carrollville, Milwaukee, Wis. (part of Oak Creek)...F2, F6 124
Carron, inlet, Scot...B4 10
Carron, riv., Scot...C3 13
Carrot, riv., Man., Sask., Can...D4 70, C1 71
Carrot River, Sask., Can., 930...D4 70
Carrowkeel, Ire., 109...B4 11
Carrowmore, lake, Ire...C2 11
Carrsville, Ky., 166...A3 94
Carruthers, Sask., Can., 75...E1 70
Carrville, Tallapoosa, Ala., 1,081...C4 78
Carrying Place, Ont., Can., 197...C7 72
Carsamba, Tur., 14,800...B11 31
Carseland, Alta., Can., 117...D4 69
Carshalton, Eng., 61,000...m12 10
Carson, Washington, Ala., 20...D2 78
Carson (N. Wilmington), Los Angeles, Calif., 38,059...*F2 82
Carson, Pottawattamie, Iowa, 583...C2 92
Carson, Jefferson Davis, Miss., 250...D4 100
Carson, Grant, N. Dak., 501...C4 110
Carson, Baker, Oreg., 30...C9 113
Carson, Dinwiddie, Va., 160...D5 121
Carson, Skamania, Wash., 250...D4 122
Carson, co., Tex., 7,781...B2 118
Carson, riv., Nev...D3 104
Carson, sink, Nev...D3 104
Carson City, Montcalm, Mich., 1,201...E5 98
Carson City, Ormsby, Nev., 5,163...D2 104
Carsonville, Sanilac, Mich., 502...E8 98
Carsonville, St. Louis, Mo., 4,500...*A8 101
Carstairs, Alta., Can., 665...D3 69
Carstensz, mtn., W. Irian...F9 35
Carswell, McDowell, W. Va., 500...*D3 123
Cartagena, Chile, 2,384...A2 54
Cartagena, Col., 152,380...A2 60
Cartagena, Sp., 100,000...D5 20
Cartagena, lake, P.R...C2 65
Cartago, Col., 31,051...C2 60
Cartago, C.R., 18,900...F6 62
Cartago, prov., C.R., 154,200...*F6 62
Carta Valley, Edwards, Tex., 15...E2 118
Cartaxo, Port., 5,920...C1 20
Cartaya, Sp., 9,002...D2 20
Carter, Carter, Ky., 122...B6 94
Carter, Chouteau, Mont., 85...C6 102
Carter, Beckham, Okla., 364...B2 112
Carter, Tripp, S. Dak., 18...D5 116
Carter, Carter, Tenn., 200...C11 117
Carter, Forest, Wis., 100...C5 124
Carter, Uinta, Wyo., 85...D2 125
Carter, co., Ky., 20,817...B6 94
Carter, co., Mo., 3,973...E7 101
Carter, co., Mont., 2,493...E12 102
Carter, co., Okla., 39,044...C4 112
Carter, co., Tenn., 41,578...C11 117
Carter, caves and natural bridge, Ky...B6 94
Carteret (Roosevelt), Middlesex, N.J., 20,502...B4, E4 106
Carteret, co., N.C., 30,940...C7 109
Carter Lake, Pottawattamie, Iowa, 2,287...C2 92
Carters, Murray, Ga., 100...B2 87
Cartersburg, Hendricks, Ind., 300...E5 91
Carters Creek, Maury, Tenn...B5 117
Cartersville, Bartow, Ga., 8,668...B2 87
Cartersville, Rosebud, Mont...D10 102
Cartersville, Florence. S.C., 96...C7 115
Cartersville, Cumberland, Va., 85...D4 121
Cartertown, Allen, Ky., 100...D3 94
Carterville, Williamson, Ill., 2,643...F4 90
Carterville, Jasper, Mo., 1,443...D3 101
Carthage, Dallas, Ark., 528...C3 81
Carthage, Hancock, Ill., 3,325...C2 90
Carthage, Rush, Ind., 1,043...E6 91
Carthage, Leake, Miss., 2,442...C4 100
Carthage, Jasper, Mo., 11,264...D3 101
Carthage, Socorro, N. Mex...D3 107
Carthage, Jefferson, N.Y., 4,216...B5 108
Carthage, Moore, N.C., 1,190...B4 109
Carthage, Miner, S. Dak., 368...C8 116
Carthage, Smith, Tenn., 2,021...C8 117
Carthage, Panola, Tex., 5,262...C5 118
Carthage, Tun., 4,873...*F12 30
Carthage, ruins, Tun...F12 30
Cartwright, Man., Can., 482...E2 71
Cartwright, Newf., Can., 359...B3, h10 75
Cartwright, McKenzie, N. Dak., 45...B2 110
Caruaru, Braz., 64,471...C3, k6 57
Carúpano, Ven., 38,210...A5 60
Caruthersville, Pemiscot, Mo., 8,643...E8 101
Carvellachs, isl., Scot...D2 13
Carver, Plymouth, Mass., 300 (1,949*)...C6 97
Carver, Carver, Minn., 467...F5 99
Carver, co., Minn., 21,358...F5 99
Carville, Iberville, La., 950...B5 95
Carvin, Fr., 16,139...D3 15
Cary, Bleckley, Ga., 40...D3 87
Cary, McHenry, Ill., 3,204...A5, E2 90
Cary, Sharkey, Miss., 428...C3 100
Cary, Wake, N.C., 3,356...B5 109
Caryville, Washington, Fla., 730...G3 86
Caryville, Norfolk, Mass., 300...E1 97
Caryville, Campbell, Tenn., 950...C9 117
Casa, Perry, Ark., 184...B2 81
Casa Blanca, Cuba, 3,433...h12 64
Casa Blanca, Valencia, N. Mex., 135...B2 107
Casablanca, Mor., 965,277 (*1,025,000)...C3 44, H3 30
Casa Branca, Braz., 8,980...C3, k8 56
Casa de Oro, San Diego, Calif., 1,500...*E2 82
Casa de Piedra, P.R...B3 65
Casa Grande, Pinal, Ariz., 8,311...E4 80
Casa Grande, nat. mon., Ariz...E4 80
Casale Monferrato, It., 33,200 (40,827*)...B2 21
Casalmaggiore, It., 5,656...B3 21
Casanova, Fauquier, Va., 100...C5 121
Casape, It., 1,253...h9 21
Casa Piedra, Presidio, Tex., 10...F2 118
Casar, Cleveland, N.C., 310...B2 109
Casarano, It., 14,744...D7 21
Casas Adobes, Pima, Ariz., 5,000...*E5 80
Casas Grandes, Mex., 1,126...D5 107
Casas Grandes, riv., Mex...D4 107
Cascade, B.C., Can., 52...E8 68
Cascade, El Paso, Colo., 300...C6 83
Cascade, Valley, Idaho, 923...E2 89
Cascade, Dubuque and Jones, Iowa, 1,601...B6 92
Cascade, Wayne, Mo., 73...D7 101
Cascade, Cascade, Mont., 604...C5 102
Cascade, Coos, N.H., 500...B4 105
Cascade, Pittsylvania, Va., 500...E3 121
Cascade, Preston, W. Va., 165...B5 123
Cascade, Sheboygan, Wis., 449...E5 124
Cascade, co., Mont., 73,418...C5 102
Cascade, range, B.C., Can...E7 68
Cascade, range, U.S...B3 76
Cascade, tunnel, Wash...B5 122
Cascade Locks, Hood River, Oreg., 660...B5 113
Cascades Point, Que., Can...D8 73
Cascade Summit, Klamath, Oreg., 25...D4 113
Cascadia, Linn, Oreg., 100...C4 113
Cascais, Port., 7,887...C1, f9 20
Cascavel, Braz., 3,803...B3 57
Cascilla, Tallahatchie, Miss., 109...B3 100
Casco, Cumberland, Maine, 300 (947*)...E2 96
Casco, Kewaunee, Wis., 460...D6 124
Casco, bay, Maine...E3 96
Caselton, Lincoln, Nev., 15...F7 104
Caserta, It., 38,000 (50,381*)...D5 21
Caseville, Huron, Mich., 659...E7 98
Casey, Clark, Ill., 2,890...D6 90
Casey, Guthrie and Adair, Iowa, 589...C3 92
Casey, co., Ky., 14,327...C5 94
Casey, key, Fla...E4 86
Casey, mtn., Idaho...A2 89
Casey, pass, Fla...E4 86
Caseyville, St. Clair, Ill., 2,455...*E3 90
Cash, Craighead, Ark., 141...B5 81
Cash Corner, Cumberland, Maine (part of South Portland)...E5 96
Cashel, Ire...D2 11
Cashel, Ire., 2,679...E4 11
Cashiers, Jackson, N.C., 342...D3 109
Cashion, Maricopa, Ariz., 1,000...D3 80
Cashion, Kingfisher, Okla., 221...B4 112
Cashmere, Chelan, Wash., 1,891...B5 122
Cashton, Monroe, Wis., 828...E3 124
Cashtown, Adams, Pa., 270...G7 114
Casiguran, Phil., 260...n14 35
Casilda, Arg., 11,023...A4 54
Casilda, Cuba, 1,986...E4 64
Casino, Austl., 8,091...D9 51
Casiquiare, riv., Ven...C4 60
Casky, Christian, Ky., 75...D2 94
Caslav, Czech., 10,100...D3 26
Casma, Hartnett, N.C...B5 109
Casma, Peru, 2,676...C2 58
Casmalia, Santa Barbara, Calif., 200...E3 82
Caspar, Mendocino, Calif., 250...C2 82
Caspe, Sp., 9,033...B5 20

Casper, Natrona, Wyo., 38,930....C6 125
Caspian, Iron, Mich., 1,493....B2 98
Caspian, depression, Sov. Un....F19 9
Caspian, pond, Vt.....B4 120
Caspian, sea, Iran, Sov. Un....G19 9
Caspiana, Caddo, La, 250...B2 95
Cass, Sullivan, Ind., 350....F3 91
Cass, Pocahontas, W. Va., 327....C5 123
Cass, co., Ill., 14,539....D3 90
Cass, co., Ind., 40,931....C5 91
Cass, co., Iowa, 17,919....C3 92
Cass, co., Mich., 36,932....G4 98
Cass, co., Minn., 16,720....D4 99
Cass, co., Mo., 29,702....C3 101
Cass, co., Nebr., 17,821....D9 103
Cass, co., N. Dak., 66,947...C8 110
Cass, co., Tex., 23,496....C5 118
Cass, lake, Minn.....C4 99
Cass, riv., Mich.....E7 98
Cassá, Sp., 4,760....B7 20
Cassa, Platte, Wyo., 5....C8 125
Cassadaga, Chautauqua, N.Y., 820....C1 108
Cassandra, see Kassandra gulf, Grc.
Cass City, Tuscola, Mich., 1,945....E7 98
Cassel, Fr., 1,823....D2 15
Cassel, see Kassel, Ger.
Casselberry, Seminole, Fla., 2,463....*D5 86
Casselman, Ont., Can., 1,277....B9 72
Casselman, Somerset, Pa., 103....G3 114
Casselman, riv., Md.....D1 85
Casselton, Cass, N. Dak., 1,394....C8 110
Cassia, Lake, Fla., 300....D5 86
Cassia, co., Idaho, 16,121....G5 89
Cassidy, B.C., Can., 136....A9 68
Cassie, Wayne, W. Va....D2 123
Cassils, Alta., Can., 25....D4 69
Cassino, It., 8,000 (21,105▲)...B4 21
Cassiporé, riv., Braz.....B4 59
Cass Lake, Cass, Minn., 1,586....C4 99
Cassoday, Butler, Kans., 180....D7 93
Cassopolis, Cass, Mich., 2,027....G4 98
Casstown, Miami, Ohio, 366....B3 111
Cassville, Bartow, Ga., 200...B2 87
Cassville, Barry, Mo., 1,451..E4 101
Cassville, Ocean, N.J., 600...C4 106
Cassville, Grant, Wis., 1,290..F3 124
Castalia, Winneshiek, Iowa, 216....A6 92
Castalia, Nash, N.C., 267....A5 109
Castalia, Erie, Ohio, 954....A5 111
Castalian Springs, Sumner, Tenn., 150....A5 117
Castana, Monona, Iowa, 230....B2 92
Castanea, Clinton, Pa., 1,218....D7 114
Castanheira de Pêra, Port., 5,830....B1 20
Castaños, Mex., 2,607....B4 63
Castel di Guido, It. (part of Rome)....h8 21
Castelfranco Emilia, It., 5,260....E7 18
Castelfranco Veneto, It., 7,000....D7 18
Castel Gandolfo, It., 2,600...h9 21
Castel Giuliano, It....g8 21
Casteljaloux, Fr., 5,031....E3 14
Castella, Shasta, Calif., 300..B2 82
Castellamonte, It., 8,207....D3 18
Castelli, Arg., 3,263....B5 54
Castelli, Arg....E3 55
Castellón (Castellón de la Plana), prov., Sp., 339,229....*C5 20
Castellón de la Plana, Sp., 62,493....C5 20
Castel Madama, It., 4,509...h9 21
Castelnaudary, Fr., 8,071...F5 14
Castelo, Braz., 5,729....F7 57
Castelo Branco, Port., 13,056.C2 20
Castelo de Vide, Port., 3,379.C2 20
Castel San Pietro dell'Emilia, It., 4,080....E7 18
Castelsarrasin, Fr., 6,774....E4 14
Castelvetrano, It., 31,282...F4 21
Casterton, Austl., 2,442....H3 51
Castile, Wyoming, N.Y., 1,146....C2 108
Castilla, Peru, 8,892....C1 58
Castillo de la Punta, fort, Cuba..h12 64
Castillo del Morro, fort, Cuba....h12 64
Castillo de San Marcos, nat. mon., Fla.....C5 86
Castillos, Ur., 4,204....E2 56
Castine, Hancock, Maine, 550 (824▲)....D4 96
Castle, Okfuskee, Okla., 149..B5 112
Castle, butte, Idaho....C3 89
Castle, hbr., Bermuda....E14 77
Castle, mtn., Yukon, Can.....D6 66
Castle, mts., Mont.....D6 102
Castle, rock, Oreg.....C8 113
Castlebar, Ire., 5,482....D2 11
Castlebay, Scot....D1 13
Castlebellingham, Ire., 656..D5 11
Castleberry, Conecuh, Ala., 669....D2 78
Castleblayney, Ire., 2,127....C5 11
Castle Butte, Navajo, Ariz...B5 80
Castle Cliff, Washington, Utah, 4....F2 119
Castlecomer, Ire., 1,129....E4 11
Castle Dale, Emery, Utah, 617....D4 119
Castlederg, N. Ire., 890....C4 11
Castledome, mts., Ariz.....D1 80
Castle Douglas, Scot., 3,253..F5 13
Castlefinn, Ire., 565....C4 11
Castleford, Twin Falls, Idaho, 274....G4 89
Castlegar, B.C., Can., 2,253..E9 68
Castle Gate, Carbon, Utah, 321....D5 119
Castlegregory, Ire., 235....E1 11
Castle Hayne, New Hanover, N.C., 500....C6 109
Castle Hill, Black Hawk, Iowa, 932....B5 92
Castle Hills, Bexar, Tex., 2,622....*E3 118
Castle Hot Springs, Yavapai, Ariz....D3 80
Castleisland, Ire., 1,718....E2 11
Castlemaine, Austl., 7,217...H5 51
Castlemaine, Ire., 171....E2 11
Castlemartyr, Ire., 284....F3 11
Castle Park, San Diego, Calif., 1,800....*E2 82
Castlepollard, Ire., 778....D4 11
Castlereagh, Ire., 1,568....D3 11
Castle Rock, Douglas, Colo., 1,152....B6 83
Castle Rock, Dakota, Minn., 100....F5 99
Castle Rock, Butte, S. Dak., 15....B2 116
Castle Rock, Summit, Utah, 15....B4 119
Castle Rock, Cowlitz, Wash., 1,424....C3 122
Castle Rock, butte, S. Dak.....B2 116
Castle Rock, mtn., Va.....D4 121
Castle Rock, res., Wis.....E4 124
Castle Shannon, Allegheny, Pa., 11,836....F1 114
Castleton, Ont., Can., 267...C7 72
Castleton, Stark, Ill., 175....B4 90
Castleton, Marion, Ind., 267.H8 91
Castleton, Reno, Kans., 50...B4 93
Castleton, Rutland, Vt., 375 (1,902▲)....D2 120
Castleton Corners, Rutland, Vt., 140....D2 120
Castleton-on-Hudson, Rensselaer, N.Y., 1,752....C7 108
Castletown, Ire., 182....D4 11
Castletown, I. of Man, 1,549..F4 13
Castletown, Scot....B5 13
Castletownroche, Ire., 381...E3 11
Castlewellan, N. Ire., 1,241..C6 11
Castlewood, Hamlin, S. Dak., 500....C8 116
Castlewood, Russell, Va., 500.B2 121
Castolon, Brewster, Tex., 10..G2 118
Castor, Alta., Can., 1,025....C5 69
Castor, Bienville, La., 142...B2 95
Castor, creek, La.....B3 95
Castor, riv., Mo.....D7 101
Castorland, Lewis, N.Y., 321....B5 108
Castres, Fr., 36,978....F5 14
Castries, St. Lucia, 17,505...o16 64
Castro, Braz., 9,249....C3 56
Castro, Chile, 6,283....C2 54
Castro, Sp., 11,842....D3 20
Castro, co., Tex., 8,923....B1 118
Castro Alves, Braz., 7,388...D3 57
Castro Daire, Port., 4,547...B2 20
Castro Marim, Port., 4,613...D2 20
Castropol, Sp., 7,368....A2 20
Castrop-Rauxel, Ger., 87,900....B2 17
Castro Urdiales, Sp., 6,222...A4 20
Castro Valley, Alameda, Calif., 37,120....*B5 82
Castro Verde, Port., 2,794...D1 20
Castrovillari, It., 12,500....E6 21
Castroville (Del Monte Junction), Monterey, Calif., 2,838....C6, D3 82
Castroville, Medina, Tex., 1,508....E3 118
Castrovirreyna, Peru, 872...D2 58
Castuera, Sp., 10,166....C3 20
Caswell, co., N.C., 19,912....A4 109
Cat, isl., Ba. Is.....C6 64
Cat, isl., Miss.....E4 100
Cat, isl., S.C.....E9 115
Catacamas, Hond., 2,039....C5 62
Catacaos, Peru, 8,526....C1 58
Catacocha, Ec., 2,754....B2 58
Catacombs, tombs, It.....h9 21
Cataguases, Braz., 21,476....C4, g6 56
Catahoula, par., La., 11,421..C4 95
Catahoula, lake, La.....C3 95
Cataiñgan, Phil., 7,462....*C6 35
Catalão, Braz., 11,471....E1 57
Catalca, Tur., 4,800....B7 25
Catalina, Newf., Can., 800..D5 75
Catalina, pt., Guam....52
Catalonia, reg., Sp., 3,925,779....B6 20
Catamarca, Arg., 31,067....E2 55
Catamarca, prov., Arg., 172,407....E2 55
Catanauan, Phil., 3,273....p14 35
Catanduanes, prov., Phil., 156,800....*C6 35
Catanduanes, isl., Phil.....C6 35
Catanduva, Braz., 37,307....C3 56
Catania, It., 363,928....F5 21
Catania, gulf, It.....F5 21
Cataño, P.R., 8,276....B6 65
Cataño, mun., P.R., 25,208..B6 65
Catanzaro, It., 74,037....E6 21
Cataouatche, lake, La.....C7 95
Cataract, Owen, Ind., 40....F4 91
Cataract, Monroe, Wis., 200....D3 124
Cataract, lake, Ind.....E4 91
Catarina, Dimmit, Tex., 160.E3 118
Catarman, Phil., 11,293....*C6 35
Catarroja, Sp., 11,680....C5 20
Catasauqua, Lehigh, Pa., 5,062....E11 114
Catatula, Harris, Ga., 350....D2 87
Cataumet, Barnstable, Mass., 300....C6 97
Catawba, Catawba, N.C., 504....B2 109
Catawba, Clark, Ohio, 355..C4 111
Catawba, York, S.C., 225....B6 115
Catawba, Roanoke, Va., 25..C2 121
Catawba, Marion, W. Va., 400....A7 123
Catawba, Price, Wis., 230....C3 124
Catawba, co., N.C., 73,191..B2 109
Catawba, isl., Ohio....A4 111
Catawba, res., S.C.....A5 115
Catawba, riv., N.C.....B3 109
Catawissa, Columbia, Pa., 1,824....E9 114
Catbalogan, Phil., 10,757....C6 35
Cat Creek, Petroleum, Mont., 200....C8 102
Cateechee, Pickens, S.C., 600....B2 115
Cateel, Phil., 2,333....*D7 35
Cater, Sask., Can....D1 70
Cates, Fountain, Ind., 150...E3 91
Catesby, Ellis, Okla., 5....A2 112
Catete, Ang., 716....C1 48
Cathance, Sagadahoc, Maine....D5 96
Cathance, lake, Maine....D5 96
Catharine, Ellis, Kans., 225..D4 93
Cathay, Mariposa, Calif., 110....D3 82
Cathay, Wells, N. Dak., 110..B6 110
Cathedral, bluffs, Colo.....B2 83
Cathedral, cave, Mo.....C6 101
Cathedral, mtn., Tex.....F2 118
Cathedral, peak, Calif.....D3 82
Cathedral City, Riverside, Calif., 1,855....F5 82
Catherine, Wilcox, Ala., 150.C2 78
Catherine, lake, Ark.....C3 81
Catherine, mtn., Utah....E3 119
Cathlamet, Wahkiakum, Wash., 615....C2 122
Cathro, Alpena, Mich., 150..C7 98
Cat Law, mtn., Scot.....D5 13
Catlettsburg, Boyd, Ky., 3,874....B7 94
Catlin, Vermilion, Ill., 1,263.C6 90
Catlin, Parke, Ind., 150....E3 91
Cato, Faulkner, Ark., 50....C3 81
Cato, Cayuga, N.Y., 476....B4 108
Catoche, cape, Mex.....C7 63
Catoctin, creek, Md.....B2 85
Catoctin, mtn., Md.....B2 85
Catoctin Furnace, Frederick, Md., 150....A3 85
Catonsville, Baltimore, Md., 37,372....B4, C2 85
Catoosa, Rogers, Okla., 638..A6 112
Catoosa, Mortan, Tenn....C9 117
Catoosa, co., Ga., 21,101....B1 87
Catriló, Arg....B4 54
Catrimani, riv., Braz.....C5 60
Catron, New Madrid, Mo., 177....E8 101
Catron, co., N. Mex., 2,773..D1 107
Catskill, Greene, N.Y., 5,825.C7 108
Catskill, mts., N.Y.....C6 108
Catt, mtn., B.C., Can.....B3 68
Cattaraugus, Cattaraugus, N.Y., 1,258....C2 108
Cattaraugus, co., N.Y., 80,187....C2 108
Cattaraugus, creek, N.Y.....C2 108
Cattaraugus, Indian res., N.Y....C2 108
Cattolica, It., 12,969....C4 21
Catuá, riv., Braz.....C2 59
Catumbela, riv., Ang.....D1 48
Cauayan, Phil., 3,809....*D6 35
Cauca, dept., Col., 481,980..C2 60
Cauca, riv., Col.....B2 60
Cauca, val., Col.....B2 60
Caucasia, Col., 897....B2 60
Caucasus, mts., Sov. Un.....G17 9
Caucomgomac, lake, Maine....B3 96
Caudebec-les-Elbeuf, Fr., 9,270....C4 14
Caudéran, Fr., 28,715....E3 14
Caudete, Sp., 7,544....C5 20
Caudry, Fr., 13,207....B5 14
Caughnawaga, Que., Can., 2,200....D4, D9 73
Caungula, Ang....C2 48
Cauquenes, Chile, 14,849....B2 54
Caura, riv., Ven.....B5 60
Causapscal, Que., Can., 3,463....*B3 73
Causey, Roosevelt, N. Mex., 5....D6 107
Caussade, Fr., 3,163....E4 14
Cauthron, Scott, Ark., 60....C1 81
Cautín, prov., Chile, 393,041.B2 54
Caution, cape, B.C., Can.....D4 68
Cauvery, riv., India....F6 39
Caux, Fr. Gu., 206....B4 59
Cavado, riv., Port.....B1 20
Cavaillon, Fr., 12,067....F6 14
Cavalier, Pembina, N. Dak., 1,423....A8 110
Cavalier, co., N. Dak., 10,064....A7 110
Cavalla, riv., I.C.....E3 45
Cavallo, mtn., It.....C8 18
Cavan, Ire., 3,208....D4 11
Cavan, co., Ire., 56,594....D4 11
Cavanal, mtn., Okla.....B7 112
Cavarzere, It., 5,600....B4 21
Cave, It., 5,856....h9 21
Cave City, Sharp, Ark., 540..B4 81
Cave City, Barren, Ky., 1,418....C4 94
Cavecreek, Maricopa, Ariz., 450....D4, F2 80
Cave Creek, Roane, Tenn....D9 117
Cave Creek, res., Ariz.....F2 80
Cave in Rock, Hardin, Ill., 495....F5 90
Cave Junction, Josephine, Oreg., 248....E3 113
Cavell, Sask., Can., 25....E1 70
Cavendish, Alta., Can., 75....D5 69
Cavendish, Clearwater, Idaho, 25....C2 89
Cavendish, Windsor, Vt., 250 (1,223▲)....E3 120
Cave of the Winds, cave, Vt....B3 120
Cave Spring, Floyd, Ga., 1,153....B1 87
Cave Spring Onyx, caverns, Mo...E6 101
Cave Springs, Benton, Ark., 281....A1 81
Cave Springs, cave, Ga.....B1 87
Cavetown, Washington, Md., 282....A2 85
Cavite, Phil., 54,900....o13 35
Cavite, prov., Phil., 379,900....*C6 35
Cavour, Beadle, S. Dak., 140.C7 116
Cavour, Forest, Wis., 130....C5 124
Cawdor, Scot., 823....C5 13
Cawker City, Mitchell, Kans., 686....C5 93
Cawnpore, see Kanpur, India.
Cawood, Harlan, Ky., 800...D6 94
Caxias, Braz., 19,092....D2 57
Caxias do Sul., Braz., 60,607.D2 56
Caxito, Ang., 8,690....C1 48
Cayambe, Ec., 7,364....A2 58
Cayce, Fulton, Ky., 125....B2 94
Cayce, Lexington, S.C., 8,517....D5 115
Cayenne, Fr. Gu., 10,961....B4 59
Cayeux-sur-Mer, Fr., 1,994..D9 12
Cayey, P.R., 19,738....C6 65
Cayey, mun., P.R., 38,061...C6 65
Cayey Central, P.R....C6 65
Cayley, Alta., Can., 146....D4 69
Cayman Brac, isl., Cayman Is...F4 64
Cayman Islands, Br. dep., N.A., 7,616....F3 64
Cayo, Br. Hond., 1,812....B3 62
Cayo, isl., Cuba....D5 64
Cayo Largo, isl., Cuba....E3 64
Cayuga, Ont., Can., 897....E5 72
Cayuga, Vermillion, Ind., 904....E3 91
Cayuga, Hinds, Miss....C3 100
Cayuga, Sargent, N. Dak., 195....C8 110
Cayuga, Ashland, Wis....B3 124
Cayuga, co., N.Y., 73,942...C4 108
Cayuga, lake, N.Y.....C4 108
Cayuga Heights, Tompkins, N.Y., 2,788....*C4 108
Cayuse, Umatilla, Oreg., 80..B8 113
Cazalla de la Sierra, Sp., 9,284....D3 20
Cazaux, lagoon, Fr.....E3 14
Cazenovia, Madison, N.Y., 2,584....C5 108
Cazenovia, Richland, Wis., 351....E3 124
Cazombo, Ang., 2,212....D3 48
Cazorla, Sp., 8,699....D4 20
Cea, riv., Sp.....A3 20
Ceanannas, Ire., 2,193....D5 11
Ceará, state, Braz., 3,337,856.C3 57
Ceará Mirim, Braz., 8,290.C3, g6 57
Ceará Mirim, riv., Braz.....g6 57
Cebolla, Rio Arriba, N. Mex., 100....A3 107
Cebollar, Arg....E2 55
Cebollati, riv., Ur.....E2 56
Cebreros, Sp., 4,019....B3 20
Cebu, Phil., 259,200....C6 35
Cebu, prov., Phil., 1,340,400....*C6 35
Cebu, isl., Phil.....C6 35
Cecil, Montgomery, Ala., 150....C3 78
Cecil, Cook, Ga., 279....E3 87
Cecil, Gloucester, N.J....D3 106
Cecil, Paulding, Ohio, 288...A3 111
Cecil, Morrow, Oreg., 5....B7 113
Cecil, Washington, Pa., 900..F1 114
Cecil, Shawano, Wis., 357...D5 124
Cecil, co., Md., 48,408....A6 85
Cecile, Caddo, La....B2 95
Cecilia, Hardin, Ky., 500....C4 94
Cecilton, Cecil, Md., 596....B6 85
Cecina, It., 6,741....C3 21
Cedar, Smith, Kans., 73....C5 93
Cedar, Washington, Maine..C5 96
Cedar, Leelanau, Mich., 150.D5 98
Cedar, Anoka, Minn., 57....E5 99
Cedar, co., Iowa, 17,791....C6 92
Cedar, co., Mo., 9,185....D4 91
Cedar, co., Nebr., 13,368....B8 103
Cedar, bay, N.C.....B7 109
Cedar, creek, Colo.....A7 83
Cedar, creek, Ind.....B7 91
Cedar, creek, Iowa....D6 92
Cedar, creek, Mo.....C5 101
Cedar, creek, N.J.....D4 106
Cedar, creek, N. Dak.....C3 110
Cedar, creek, Ohio....A2 111
Cedar, isl., N.C.....C7 109
Cedar, isl., S.C.....E9 115
Cedar, isl., Va.....D7 121
Cedar, keys, Fla.....C3 86
Cedar, lake, Man., Can.....C1 71
Cedar, lake, Ont., Can.....A6 72
Cedar, mtn., Calif.....B3 82
Cedar, mtn., Wyo.....A5 125
Cedar, mts., Nev.....E4 104
Cedar, mts., Oreg.....D9 113
Cedar, pt., Fla.....D4 86
Cedar, pt., Md.....D5 85
Cedar, pt., Ohio....A5 111
Cedar, riv., Iowa, Minn.....C6 92
Cedar, riv., Nebr.....C7 103
Cedar, riv., Wash.....B4 122
Cedar, riv., Wis.....C2 124
Cedar, swamp, Mass.....C3 97
Cedar, val., Wash.....C4 122
Cedar Bluff, Cherokee, Ala., 687....A4 78
Cedar Bluff, Caldwell, Ky., 75....C2 94
Cedarbluff, Clay, Miss., 100..B5 100
Cedar Bluff, Tazewell, Va., 995....*A3 121
Cedar Bluff, res., Kans.....D4 93
Cedar Bluffs, Decatur, Kans., 60....C3 93
Cedar Bluffs, Saunders, Nebr., 585....C9, D2 103
Cedar Breaks, nat. mon., Utah..F3 119
Cedar Brook, Camden, N.J., 600....D3 106
Cedarburg, Ozaukee, Wis., 5,191....E1, E6 124
Cedar Canyon, petrified forest, N. Dak.....C2 110
Cedar City, Callaway, Mo., 600....C5 101
Cedar City, Iron, Utah, 7,543....F2 119
Cedar Creek, Gila, Ariz., 75.D5 80
Cedarcreek, Taney, Mo., 200.E5 101
Cedar Creek, Cass, Nebr., 110....C9, E3 103
Cedar Creek, Greene, Tenn..C11 117
Cedar Creek, Bastrop, Tex., 35....D4 118
Cedar Crest, Bernalillo, N. Mex., 150....G5 107
Cedaredge, Delta, Colo., 549 C3 83
Cedar Falls, Black Hawk, Iowa, 21,195....B5 92
Cedar Falls, Randolph, N.C., 500....B4 109
Cedar Gap, Wright, Mo., 50....D5 101
Cedar Glades, Garland, Ark....C2, C5 81
Cedar Grove, Bay, Fla., 676.*G3 86
Cedar Grove, Franklin, Ind., 232....F8 91
Cedar Grove, Essex, N.J., 14,603....*B4 106
Cedar Grove, Sante Fe, N. Mex....G6 107
Cedar Grove, Kanawha, W. Va., 1,569....C3, C6 123
Cedar Grove, Sheboygan, Wis., 1,175....E6 124
Cedar-Hammock, Manatee, Fla., 3,089....*E4 86
Cedar Heights, Prince Georges, Md., 1,900....*C4 85
Cedar Hill, Jefferson, Mo., 300....B7 101
Cedar Hill, San Juan, N. Mex., 10....A2 107
Cedar Hill, Robertson, Tenn., 530....A5 117
Cedar Hill, Dallas, Tex., 1,848....B5 118
Cedar Hills, Duval, Fla., 8,000....*B5 86
Cedar Hills, Washington, Oreg., 4,000....*B4 113
Cedarhurst, Nassau, N.Y., 6,954....E2 108
Cedar Island (Roe), Carteret, N.C., 225....C7 109
Cedar Key, Levy, Fla., 668..C3 86
Cedar Knob, mtn., Mo.....E5 101
Cedar Knolls, Morris, N.J., 3,300....B4 106
Cedar Lake, Morgan, Ala., 150....A3 78
Cedar Lake, Lake, Ind., 5,766....B3 91
Cedar Lake, Matagorda, Tex., 130....E5 118
Cedar Lodge, Davidson, N.C. (part of Fair Grove)....B3 109
Cedar Mills, Meeker, Minn., 96....F4 99
Cedar Point, LaSalle, Ill., 308....B4 90
Cedar Point, Chase, Kans., 87....D7 93
Cedar Rapids, Linn, Iowa, 92,035 (●119,600)....C6 92
Cedar Rapids, Boone, Nebr., 512....C7 103
Cedar Ridge, Coconino, Ariz....A4 80
Cedar River, Menominee, Mich., 50....C3 98
Cedar Run, Ocean, N.J., 350....D4 106
Cedar Run, Lycoming, Pa., 50....C7 114
Cedars, Que., Can., 385....D8 73
Cedar Springs, Ont., Can., 292....E2 72
Cedar Springs, Early, Ga., 100....E1 87
Cedar Springs, Kent, Mich., 1,768....E5 98
Cedar Terrace, Richland, S.C., 1,000....*D5 115
Cedartown, Polk, Ga., 9,340..B1 87
Cedarvale, B.C., Can., 79...B3 68
Cedar Vale, Chautauqua, Kans., 859....E7 93
Cedarvale, Torrance, N. Mex., 20....C4 107
Cedar Valley, Utah, Utah, 150....C3 119
Cedarville, Crawford, Ark., 52....B1 81
Cedarville, Modoc, Calif., 750....B3 82
Cedarville, Stephenson, Ill., 570....A4 90
Cedarville, Allen, Ind., 240..B7 91
Cedarville, Mackinac, Mich., 300....B6 98
Cedarville, Cumberland, N.J., 1,095....E2 106
Cedarville, Herkimer, N.Y., 125....C5 108
Cedarville, Greene, Ohio, 1,702....C4 111
Cedarwood, Pueblo, Colo., 20....D6 83
Cedarwood Park, Ocean, N.J., 1,052....*C4 106
Cedoux, Sask., Can., 74....H4 70
Cedros, Hond., 1,095....C4 62
Cedros, isl., Mex.....B1 63
Ceduna, Austl., 1,292....F5 50
Cedynia, Pol., 286....F8 24
Ceepeecee, B.C., Can., 85...E4 68
Cefalù, It., 10,700....E5 21
Cega, riv., Sp.....B3 20
Cegléd, Hung., 30,310 (37,943▲)....B4 22
Ceglie Messapico, It., 16,500....D6 21
Cehegín, Sp., 7,084....C5 20
Ceiba, P.R., 1,644....B8, f12 65
Ceiba, mun., P.R., 9,075 B8, f12 65
Cela, Ang....D2 48
Celakovice, Czech., 6,041...n18 26
Celanese Village, Floyd, Ga., 1,500....*B1 87
Celaya, Mex., 58,762....m13 63
Celebes (Sulawesi), isl., Indon...F6 35
Celebes, sea, Indon.....E6 35
Celendín, Peru, 4,045....C2 58
Celica, Ec., 1,627....B2 58
Celilo, Wasco, Oreg., 15....B6 113
Celina, Mercer, Ohio, 7,659..B3 111
Celina, Clay, Tenn., 1,228...C8 117
Celina, Collin, Tex., 1,204...C4 118
Celje, Yugo., 16,487....B2 22
Celle, Ger., 58,500....B5 16
Cellina, riv., It.....C8 18
Celoron, Chautauqua, N.Y., 1,507....C1 108
Celriver, York, S.C., 255....B6 115
Cement, Caddo, Okla., 959....C3 112
Cement City, Lenawee, Mich., 471....F6 98
Cementon, Lehigh, Pa., 1,800....E10 114
Cemiskezek, Tur., 1,200....C12 31
Cemmaes, head, Wales....B3 12
Cencia, Eth....D4 47
Cenis, mtn., Fr.....E7 14
Cenon, Fr., 13,821....E3 14
Centenary, Marion, S.C., 300....C9 115
Centennial, Albany, Wyo., 50....D6 125
Centennial, range, Idaho, Mont.....E7 83, F4 102
Centennial, val., Mont.....F4 102
Center, Saguache, Colo., 1,600....D4 83
Center, Jackson, Ga., 137....B3 87
Center, Howard, Ind., 250..D5 91
Center, Metcalfe, Ky., 115...C4 94
Center, Attala, Miss....C4 100
Center, Ralls, Mo., 484....B6 101
Center, Knox, Nebr., 147....B8 103
Center, Oliver, N. Dak., 476....B4 110
Center, Pontotoc, Okla., 150....C5 112
Center, McCook, S. Dak., 20....D8 116
Center, Shelby, Tex., 4,510..D5 118
Center Barnstead, Belknap, N.H., 150....D4 105
Center Belpre, Washington, Ohio, 100....C6 111
Centerbrook, Middlesex, Conn., 600....D7 84
Centerburg, Knox, Ohio, 963....B5 111
Center City, Chisago, Minn., 293....E6 99
Center City, Mills, Tex., 30..D3 118
Center Conway, Carroll, N.H., 200....C4 105
Center Cross, Essex, Va., 200....D6 121
Centereach, Suffolk, N.Y., 8,524....*E3 108
Center Effingham, Carroll, N.H., 50....C4 105
Centerfield, Sanpete, Utah, 475....D4 119
Center Harbor, Belknap, N.H., 200 (511▲)....C4 105
Center Hill, White, Ark., 306.B4 81
Center Hill, Sumter, Fla., 529....D5 86
Center Hill, Fulton, Ga. (part of Atlanta)....B5, C2 87
Center Hill, res., Tenn.....C8 117
Center Island, pt., N.Y.....F2 84
Center Junction, Jones, Iowa, 201....B6 92
Center Line, Macomb, Mich., 10,164....F7 98
Center Lovell, Oxford, Maine, 100....D2 96
Center Montville, Waldo, Maine (366▲)....D3 96
Center Moreland, Wyoming, Pa., 150....A8 114
Center Moriches, Suffolk, N.Y., 2,521....D4 108
Center Ossipee, Carroll, N.H., 500....C4 105
Center Point, Jefferson, Ala., 7,500....E5 78
Center Point, Howard, Ark., 136....C2 81
Centerpoint, Clay, Ind., 268..F3 91
Center Point, Linn, Iowa, 1,236....B6 92
Center Point, Turner and Yankton, S. Dak., 20....D8 116
Center Point, Kerr, Tex., 1,000....E3 118
Center Point, Doddridge, W. Va., 125....A6, B4 123
Centerport, Suffolk, N.Y., 3,628....*D3 108
Center Ridge, Conway, Ark., 100....B3 81
Center Rutland, Rutland, Vt., 225....D2 120
Center Sandwich, Carroll, N.H., 250....C4 105
Center Square, Montgomery, Pa., 900....A11 114
Center Stafford, Stafford, N.H., 40....D4 105
Center Star, Lauderdale, Ala., 200....A2 78
Centerton, Benton, Ark., 177.A1 81
Centerton, Morgan, Ind., 500....E5 91
Centerton, Salem, N.J., 100..D2 106
Centertown, Ohio, Ky., 327..C3 94
Centertown, Cole, Mo., 190..C5 101
Centertown, Warren, Tenn., 169....D8 117
Center Tuftonboro, Carroll, N.H., 100....C4 105
Center Valley, Outagamie, Wis., 25....A5 124
Centerview, Johnson, Mo., 208....C4 101
Centerville, Yavapai, Ariz., 25....C3 80
Centerville, Yell, Ark., 200..B2 81
Centerville, New Haven, Conn. (part of Hamden)..D5 84
Centerville, Gwinnett, Ga...B6 87
Centerville, St. Clair, Ill., 12,769....E3 90
Centerville, Wayne, Ind., 2,378....E8 91
Centerville, Appanoose, Iowa, 6,629....D5 92
Centerville, Linn, Kans., 250....D8 93
Centerville, St. Mary, La., 537....E4 95
Centerville, Washington, Maine, 40 (47▲)....D5 96
Centerville, Anoka, Minn., 338....E7 99
Centerville, Reynolds, Mo., 163....D7 101
Centerville, Silver Bow, Mont., 950....D4 102
Centerville, Hunterdon, N.J., 120....B3 106
Centerville, Franklin, N.C., 300....A5 109
Centerville, Montgomery, Ohio, 3,490....C3 111
Centerville, Crawford, Pa., 238....C2 114
Centerville, Washington, Pa., 5,088....F2 114
Centerville, Turner, S. Dak., 887....D9 116
Centerville, Hickman, Tenn., 1,678....B4 117
Centerville, Leon, Tex., 836.D5 118
Centerville, Davis, Utah, 2,361....C4 119
Centerville, Klickitat, Wash., 100....D5 122
Cento, It., 7,700....E7 18
Centrahoma, Coal, Okla., 148....C5 112
Central, Elmore, Ala., 40....C3 78
Central, Graham, Ariz., 110..E6 80
Central, Caribou, Idaho, 10..G7 89
Central, St. James, La....B5 95
Central, Grant, N. Mex., 1,075....E1 107
Central, Pickens, S.C., 1,473..B2 115
Central, dept., Par., 174,789....*E4 55
Central, prov., N. Rh....D4 48
Central, prov., Nya....D5 48
Central, reg., Ghana....E4 45
Central, reg., Ken....B6 48
Central, reg., Tan., 886,306..C6 48
Central, plat., Tan.....C5 48
Central, range, Dom. Rep.....F8 64
Central African Republic, country, Afr., 1,227,000....D4 46, F8 42
Central Aguirre, P.R., 1,689..D6 65
Central America, reg., N.A., 9,699,000....H12 61, 62
Central Avenue Park, Lucas, Ohio, 900....*A2 105
Central Barren, Harrison, Ind., 100....H5 91
Central Bridge, Schoharie, N.Y., 400....C6 108
Central Butte, Sask., Can., 459....G2 70
Central City, Gilpin, Colo., 250....B5 83
Central City, Marion, Ill., 1,422....E4 90
Central City, Linn, Iowa, 1,087....B6 92
Central City, Muhlenberg, Ky., 3,694....C2 94
Central City, Merrick, Nebr., 2,406....C8 103

Central City, Somerset, Pa., 1,604 F4 114
Central City, Lawrence, S. Dak., 247 C2 116
Central College, Franklin, Ohio, 120 B2 111
Central Falls, Randolph, N.C., 400 B4 109
Central Falls, Providence, R.I., 19,858 B11 84
Centralhatchee, Heard, Ga., 174 C1 87
Central Heights, Gila, Ariz., 2,486 D5 80
Central Heights, Cerro Gordo, Iowa, 900 A4 92
Centralia, Ont., Can., 179 D3 72
Centralia, Clinton and Marion, Ill., 13,904 E4 90
Centralia, Nemaha, Kans., 527 C7 93
Centralia, Boone, Mo., 3,200 B5 101
Centralia, Craig, Okla., 80 A6 112
Centralia, Columbia, Pa., 1,435 *E9 114
Centralia, Lewis, Wash., 8,586 C3 122
Centralia, Braxton, W. Va., 145 C4 123
Central Islip, Suffolk, N.Y., 10,000 D3, E7 108
Central Java, see Java, isl., Indon.
Central Lake, Antrim, Mich., 692 C5 98
Central Makran, range, Iran, Pak C3 39
Central Nyack, Rockland, N.Y., 1,400 *D7 108
Central Park, Vermilion, Ill., 2,676 *C6 90
Central Park, Grays Harbor, Wash., 1,622 *C2 122
Central Point, Jackson, Oreg., 2,289 E4 113
Central Provinces, see Madhya Pradesh, state, India
Central Siberian, uplands, Sov. Un C13 28
Central Square, Oswego, N.Y., 935 B4 108
Central Sumatra, see Sumatra, isl., Indon.
Central Valley, Shasta, Calif., 2,854 B2 82
Central Valley, Orange, N.Y., 950 D2, D6 108
Central Village, Windham, Conn., 800 C9 84
Central Village, Bristol, Mass., 200 C5 97
Centre, Cherokee, Ala., 2,392 A4 78
Centre, co., Pa., 78,580 E6 114
Centre, mtn., Idaho D3 89
Centre Hall, Centre, Pa., 1,109 E6 114
Centre Square, Gloucester, N.J D2 106
Centreville, Bibb, Ala., 1,981 C2 78
Centreville, St. Joseph, Mich., 971 G5 98
Centreville, N.B., Can., 352 C2 74
Centreville, N.S., Can., 304 E3 74
Centreville, Queen Annes, Md., 1,863 B5 85
Centreville, St. Joseph, Mich., 971 *G5 98
Centreville, Amite and Wilkinson, Miss., 1,737 D2 100
Centreville, Wilson, Tenn., 50 A5 117
Centreville, Fairfax, Va., 600 B4 121
Centro Comunal Nogueras, P.R C6 65
Centropolis, Franklin, Kans., 100 D8 93
Centro Puntas, P.R B1 65
Centuria, Polk, Wis., 551 C1 124
Century, Escambia, Fla., 2,046 G2 86
Century, Barbour, W. Va., 700 B4 123
Ceram, isl., Indon F7 35
Cercany, Czech., 1,261 o18 26
Cereal, Alta., Can., 195 D5 69
Cereales, Arg B4 54
Ceredo, Wayne, W. Va., 1,387 C2 123
Ceres, Stanislaus, Calif., 4,406 D3 82
Ceres, It., 562 D3 18
Ceres, Allegany, N.Y., 450 C2 108
Ceres, Bland, Va., 75 B3, D1 121
Ceresco, Saunders, Nebr., 429 C9, E2 103
Ceres Northwest, Stanislaus, Calif., 1,126 *D3 82
Céret, Fr., 4,548 F5 14
Cereté, Col., 6,161 B2 60
Cerignola, It., 49,287 D5 21
Cerillos, P.R C6 65
Cerknica, Yugo., 1,404 C2 22
Cernavodă, Rom., 8,802 C8 22
Cernay, Fr., 8,372 D7 14
Cerralvo, Mex., 3,050 B5 63
Cerralvo, isl., Mex C3 63
Cerrillos, Santa Fe, N. Mex., 150 B3, G6 107
Cerritos, Mex., 9,849 C4 63
Cerro Azul, Peru, 1,372 D2 58
Cerro Bolívar, Ven B5 60
Cerro Colorado, Ur E1 56
Cerro de Pasco, Peru, 22,688 D2 58
Cerrode de Punta, peak, P.R C4 65
Cerrogordo, Little River, Ark D1 81
Cerro Gordo, Piatt, Ill., 1,067 D5 90
Cerro Gordo, Columbus, N.C., 306 C5 109
Cerrogordo, McCurtain, Okla., 60 D7 112
Cerro Gordo, co., Iowa, 49,894 A4 102
Cerro Largo, dept., Ur., 67,500 *E2 56
Cerro Negro, Chile D2 55
Cerulean, Trigg, Ky., 206 D2 94
Cervantes, Phil., 682 B6 35
Cervantes, Sp., 7,011 A2 20
Cervera del Río Alhama, Sp., 7,101 A5 20
Cerveteri, It., 3,759 g8 21
Cervignano del Friuli, It., 7,729 D9 18
Cervione, Fr., 1,162 C2 21
Cesena, It., 38,000 (79,704*) B4 21
Cesenatico, It., 5,935 B4 21
Ceska Kamenice, Czech., 4,885 C9 17
Ceska Lipa, Czech., 14,000 C3 26
Ceska Trebova, Czech., 13,000 D4 26
Ceske Budejovice, Czech., 63,900 D3 26
Cesky Brod, Czech., 5,754 n18 26
Cesky Krumlov, Czech., 8,900 D3 26
Cesme, Tur., 3,700 C6 23
Cess, riv., Lib E3 45
Cessford, Alta., Can., 160 D5 69
Cessnock, Austl., 35,282 F8 51
Cetina, riv., Yugo D3 22
Cetinje, Yugo., 9,345 D4 22
Cetraro, It., 9,366 E5 21
Cetti, bay, Guam 52
Ceuta, Sp., poss., Afr., 73,182 B3 44, G4 30
Cévennes, mts., Fr E5 14
Cevio, Switz., 504 D6 19
Ceyhan, riv., Tur D11 31
Ceylon, Ont., Can., 95 C4 72
Ceylon, Martin, Minn., 554 G4 99
Ceylon, country, Asia, 8,929,000 I11 33, G7 39
Ceylon Station, Sask., Can., 288 H3 70
Cezma, riv., Yugo C3 22
Chabarovice, Czech., 3,388 C8 17
Chacabuco, Arg., 12,530 A4, g6 54
Chacahoula, Terregonne, La., 50 C5, E5 95
Chachapoyas, Peru, 6,480 C2 58
Chachoengsao, Thai., 19,849 F4 38
Chachran, Pak., 2,954 C3 40
Chachro, Pak E3 40
Chaco, prov., Arg., 535,443 E3 55
Chaco, riv., N. Mex A1 107
Chaco Canyon, San Juan, N. Mex A2 107
Chaco Canyon, nat. mon., N. Mex B2 107
Chacon, Mora, N. Mex., 10 A4 107
Chad, country, Afr., 2,675,000 B3 46, E7 42
Chad, lake, Chad C2 46
Chadbourn, Columbus, N.C., 2,323 C5 109
Chadds Ford, Delaware, Pa., 140 G10 114
Chadiza, N. Rh D5 48
Chadron, Dawes, Nebr., 5,079 B3 103
Chadwick, Carroll, Ill., 602 A4 90
Chadwick, Christian, Mo., 175 E4 101
Chadwicks, Oneida, N.Y., 1,000 *B5 108
Chaerhsen, China A9 36
Chaffee, Scott, Mo., 2,862 D8 101
Chaffee, Cass, N. Dak., 106 C8 110
Chaffee, co., Colo., 8,298 C4 83
Chaffins, Worcester, Mass., 3,500 *B4 97
Chagny, Fr., 4,065 D6 14
Chagodoshcha, riv., Sov. Un B10 27
Chagos, is., Indian O G1 2
Chagres, riv., Pan k12 62
Chagrin Falls, Cuyahoga, Ohio, 3,458 A6 111
Chagyl, Sov. Un E5 29
Chahal, Guat., 329 C3 62
Chahar Burjak, Afg., 5,000 F11 41
Chahba, Syr., 2,700 F11 31
Chāh Bahār, Iran, 5,189 I10 41
Chaibasa, India, 22,019 D8 39
Chainat, Thai., 4,431 E4 38
Chaires, Leon, Fla., 85 B2 86
Chakai, Pak A5 39
Chake Chake, Zan C6 48
Chakradharpur, India, 20,260 (*30,906) D8 39
Chakwal, Pak., 13,319 B5 39
Chala, Peru, 721 E3 58
Chalatenango, Sal., 5,183 C3 62
Chalchuapa, Sal., 13,680 D3 62
Chalcidice (Khalkidhiki), prov., Grc., 79,849 *B4 23
Chalco, Sarpy, Nebr D2 103
Chalender, Coconino, Ariz., 20 B3 80
Chalfant, Allegheny, Pa., 1,414 *F2 114
Chalfont, Bucks, Pa., 1,410 F11 114
Chalhuanca, Peru, 2,538 D3 58
Chaling, China K5 36
Chalk River, Ont., Can., 1,135 A7 72
Chalkville, Jefferson, Ala., 1,000 *E4 78
Chalkyitsik, Alsk., 57 B11 79
Challans, Fr., 2,915 D3 14
Challapata, Bol., 2,529 C2 55
Challis, Custer, Idaho, 732 E4 89
Chalmers, White, Ind., 548 C4 91
Chalmette, St. Bernard, La., 10,000 C7, E6 95
Chalmette, nat. hist. park and cem., La C7 95
Châlons-sur-Marne, Fr., 41,705 C6 14
Chalon-sur-Saône, Fr., 43,655 D6 14
Chalus, Iran C5 41
Chalybeate, Tippah, Miss., 199 A5 100
Chalybeate, Van Buren, Tenn D8 117
Chalybeate Springs, Harnett, N.C., 200 B5 109
Cham, Ger., 9,200 D7 17
Cham, Switz., 6,483 B5 19
Cham, pt., Okinawa 52
Chama, Costilla, Colo., 350 D5 83
Chama, Rio, Arriba, N. Mex., 900 A3 107
Chamaicó, Arg B4 54
Chamalières, Fr., 14,700 E5 14
Chaman, Pak., 7,161 B4 39
Chamba, Tan D6 48
Chamberino, Dona Ana, N. Mex., 10 E3 107
Chamberlain, Sask., Can., 162 B3 70
Chamberlain, Brule, S. Dak., 2,598 D6 116
Chamberlain, lake, Maine B3 96
Chamberlayne Heights, Henrico, Va., 1,000 *D5 121
Chamberlin, mtn., Alsk B10 79
Chambers, Apache, Ariz., 200 B6 80
Chambers, Holt, Nebr., 396 B7 103
Chambers, co., Ala., 37,828 C4 78
Chambers, co., Tex., 10,379 E5 118
Chambers, isl., Wis C6 124
Chambersburg, Franklin, Pa., 17,670 G6 114
Chambéry, Fr., 44,246 (*57,000) E6 14
Chambezi, riv., N. Rh D5 48
Chamblee, DeKalb, Ga., 6,635 A5 87
Chambly, Que., Can., 3,737 D4 73
Chambly, Fr., 5,331 E2 15
Chambly, co., Que., Can., 146,745 D4 73
Chambly Canton, Que., Can., 1,885 *D4 73
Chambord, Que., Can., 1,188 A5 73
Chamcook, N.B., Can., 327 D2 74
Chamdo (Changtu), China E4 34
Chamical, Arg., 2,702 F2 55
Chamois, Osage, Mo., 658 C6 101
Chamonix-Mont-Blanc, Fr., 3,164 E7 14, D2 18
Champ, Somerset, Md., 100 D6 85
Champagne, Yukon, Can., 56 D5 66
Champagne, former prov., Fr., 1,504,000 C5 14
Champagnole, Fr., 7,531 D6 14
Champaign, Champaign, Ill., 49,583 C5 90
Champaign, co, Ill., 132,436 C5 90
Champaign, co., Ohio, 29,714 B4 111
Champdore, lake, Que., Can g8 75
Champerico, Guat., 982 C2 62
Champéry, Switz., 810 D2 19
Champex, Switz D3 19
Champigneulles, Fr., 5,854 F6 15
Champigny-sur-Marne, Fr., 57,876 g11 14
Champion, Alta., Can., 419 D4 69
Champion, Marquette, Mich., 750 B3 98
Champion, Chase, Nebr., 140 D4 103
Champion (Champion Heights), Trumbull, Ohio, 2,500 *A7 111
Champlain, Que., Can., 750 C5 73
Champlain, Clinton, N.Y., 1,549 A3 108
Champlain, co., Que., Can., 111,953 B4 73
Champlain, lake, Que., Can E4 73
Champlain, lake, N.Y., Vt B2 120
Champlin, Hennepin, Minn., 1,271 E6 99
Champlin, Juab, Utah D3 119
Champlitte, Fr., 1,423 D6 14
Champneys West, Newf., Can., 266 D5 75
Champotón, Mex., 2,853 D6 63
Champua, India, 5,000 F10 40
Chana, Ogle, Ill., 200 B4 90
Chañaral, Chile, 2,980 E1 55
Chanārān, Iran, C9 41
Chañarcillo, Chile E1 55
Chanca, riv., Port D2 20
Chancay, Peru, 2,761 D2 58
Chance, Clarke, Ala., 400 D2 78
Chance, Somerset, Md., 275 D6 85
Chancellor, Geneva, Ala., 150 D4 78
Chancellor, Turner, S. Dak., 214 D8 116
Chanchelulla, mtn., Calif B2 82
Chanco, Chile, 1,931 B2 54
Chanda, India, 51,484 E6 39
Chandalar, riv., Alsk B10 79
Chandausi, India, 48,557 C7 40
Chandeleur, is., La E7 95
Chandeleur, sound, La E6 95
Chandigarh, India, 89,321 (*99,262) B6 39, B6 40
Chandler, Maricopa, Ariz., 9,531 D4, G2 80
Chandler, Que., Can., 3,406 B3 73
Chandler, Warrick, Ind., 1,784 H3 91
Chandler, Murray, Minn., 388 G3 99
Chandler, Clay, Mo E2 101
Chandler, Lincoln, Okla., 2,524 B5 112
Chandler, Henderson, Tex., 500 C5 118
Chandler Heights, Maricopa, Ariz., 400 G3 80
Chandlers, brook, N.H A4 105
Chandlers Valley, Warren, Pa C8 114
Chandlerville, Cass, Ill., 718 C3 90
Chandos, lake, Ont., Can C7 72
Chandpur, Pak., 32,048 D9 39
Chandrakona, India, 7,383 F11 40
Chandrapur, India E5 40
Chaneliak, Alsk., 25 C7 79
Chaneysville, Bedford, Pa., 50 G5 114
Chānf, Iran H10 41
Chang, riv., China C11 39
Changchih, China G7 36
Changchiu, China F7 36
Changchow, China, 296,500 E8 34
Changchun (Hsinking), China, 975,000 C10 34, E2 37
Change Islands, Newf., Can., 804 D4 75
Changewater, Warren, N.J., 100 B3 106
Changhsing, China, 12,000 I8 36
Changhua, China, 6,000 I8 36
Changhua, Taiwan, 52,340 *G9 34
Changjin, Kor., 5,000 F3 37
Changkiakow, see Kalgan, China
Changli, China E8 36
Changling, China, 5,000 D1 37
Chang-pai (Paektu-san), mts., China E3 37
Changpa Shan, peak, Kor F4 37
Changpei, China C7 36
Changping, China D7 36
Changsha, China, 703,000 F7 34, J5 36
Changsŏng, Kor., 17,661 F2 37
Changte, China, 94,800 F7 34
Changting, China, 27,000 F8 34
Changtu, see Chamdo, China
Changtze, China D6 34
Changwu, China, 8,000 C10 36
Changwu, China, 10,000 D6 36
Changyeh, China D5 34
Changyon, Kor., 18,072 G2 37
Chanhassen, Carver, Minn., 244 F5 99
Chankiang, China, 166,000 G7 34
Channahon, Will, Ill., 400 B5 90
Channel, is., Eur F5 10
Channel Islands, nat. mon., Calif E4 82
Channel Lake, Lake, Ill., 1,969 *D2 90
Channel-Port aux Basques, Newf., Can., 4,105 E2 75
Channelview, Harris, Tes., 7,500 F5 118
Channing, Dickinson, Mich., 600 B2 98
Channing, Hartley, Tex., 351 B1 118
Chantada, Sp., 2,262 A2 20
Chanthaburi, Thai., 10,649 F5 38
Chantilly, Fr., 8,197 C5 14
Chantilly, Fairfax, Va., 400 B4 121
Chantrey, inlet, N.W. Ter., Can C13 66
Chanute, Neosho, Kans., 10,849 E8 93
Chanute, Pickett, Tenn., 25 C8 117
Chany, lake, Sov. Un D24 9
Chanyu, China B10 36
Chao, lake, China I7 36
Chaoan (Chaochow), China, 101,300 G8 34
Chaoan, China, 23,000 G8 34
Chaochou, China, 5,000 D2 37
Chao Phraya (Menam), riv., Thai E4 38
Chaotung, China C2 37
Chaotung, China F5 34
Chaoyang, China, 66,000 *G8 34
Chaoyang, China, 16,000 C4 39
Chapada, mts., Braz B1 56
Chapadinha, Braz., 3,698 B2 57
Chapala, lake, Mex m12 63
Chapanoke, Perquimans, N.C., 50 A7 109
Chaparral, Col., 11,705 C2 60
Chapayevsk, Sov. Un., 85,000 C5 29
Chapel Hill, Sevier, Ark C1 81
Chapelhill, Allen, Ky., 300 D3 94
Chapel Hill, Orange, N.C., 12,573 B4 109
Chapel Hill, Marshall, Tenn., 630 B5 117
Chapelle, San Miguel, N. Mex B4 107
Chaperito, San Miguel, N. Mex B5 107
Chapin, Morgan, Ill., 477 D3 90
Chapin, Franklin, Iowa, 200 B4 92
Chapin, Lexington, S.C., 358 C5 115
Chapleau, Ont., Can., 3,350 E9 72
Chaplin, Sask., Can., 479 G2 70
Chaplin, Windham, Conn., 130 (1,230*) B8 84
Chaplin, Nelson, Ky., 350 C4 94
Chaplin, lake, Sask., Can G2 70
Chaplygin, Sov. Un., 10,000 E12 27
Chapman, Butler, Ala., 617 D3 78
Chapman, Dickinson, Kans., 1,095 D6 93
Chapman, Aroostook, Maine, 40 (376*) B4 96
Chapman, Merrick, Nebr., 303 C7 103
Chapman, cape, N.W. Ter., Can C15 66
Chapman, lake, Ind B6 91
Chapman Camp, B.C., Can., 649 E9 68
Chapman Ranch, Nueces, Tex., 200 F4 118
Chapmanville, Logan, W. Va., 1,241 D2 119
Chappaqua, Westchester, N.Y., 6,000 *D7 108
Chappaquiddick, isl., Mass D7 97
Chappell, Deuel, Nebr., 1,280 C3 103
Chappells, Newberry, S.C., 128 C4 115
Chapra, India, 75,580 (*88,264) C7 39, E10 40
Chaptico, creek, Md D4 85
Chaqui, Bol., 291 C2 55
Char (Well), Maur B2 45
Charadai, Arg E4 55
Charagua, Bol., 1,185 C3 55
Charalá, Col., 3,309 B3 60
Charaña, Bol., 794 C2 55
Charcas, Mex., 9,058 C4 63
Charco, Goliad, Tex., 50 E4 118
Charco Hondo, P.R B4 65
Charcot, isl., Ant C5 5
Chard, Eng., 5,778 D5 12
Chardon, Geauga, Ohio, 3,154 A6 111
Chardzhou, Sov. Un., 72,000 H21 9
Charente, dept., Fr., 327,658 *E4 14
Charente, riv., Fr E3 14
Charente-Maritime, dept., Fr., 470,897 *D3 14
Charenton, St. Mary, La., 650 E4 95
Charenton-le-Pont, Fr., 22,530 g10 14
Charette, Que., Can., 583 C5 73
Chargoggagoggmanchauggagogg-chaubunagungamaugg, lake, Mass *B4 97
Chari, riv., Chad C3 46
Charikar, Afg., 21,070 D14 41
Charing, Taylor, Ga., 100 D2 87
Chariton, Lucas, Iowa, 5,042 C4 92
Chariton, co., Mo., 12,720 B4 101
Chariton, riv., Iowa, Mo A5 101
Charity, Br. Gu., 838 A3 59
Charity, Dallas, Mo D4 101
Charkhlik, China, 5,000 D2 34
Charlack, St. Louis, Mo., 1,493 *A8 101
Charlemagne, Que., Can., 3,068 D4, D9 73
Charlemont, Franklin, Mass., 500 (897*) A2 97
Charleroi, Bel., 26,175 (*375,000) D4 15
Charleroi, Washington, Pa., 8,148 F2 114
Charles, co., Md., 32,572 C3 85
Charles, cape, Va D6 121
Charles, mound, Ill A3 90
Charles, riv., Mass B5 97
Charles A. Goodwin, dam, Conn A4 84
Charlesbourg, Que., Can., 14,308 C9 73
Charlesburg, Calumet, Wis., 100 B5 124
Charles City, Floyd, Iowa, 9,964 A5 92
Charles City, Charles City, Va., 20 D5 121
Charles City, co., Va., 5,492 D5 121
Charles Mill, res., Ohio B5 111
Charles Mix, co., S. Dak., 11,785 D7 116
Charleston, Franklin, Ark., 1,036 B1 81
Charleston, Coles, Ill., 10,505 D5 90
Charleston, Penobscot, Maine, 175 (750*) C3 96
Charleston, Tallahatchie, Miss., 2,528 A3 100
Charleston, Mississippi, Mo., 5,911 E8 101
Charleston, Coos, Oreg., 500 D2 113
Charleston, Charleston, S.C., 65,925 (*203,100) F3, F8 115
Charleston, Bradley, Tenn., 764 D9 117
Charleston, Wasatch, Utah, 223 C4 119
Charleston, Kanawha, W. Va., 85,796 (*213,900) C3, C6 123
Charleston, co., S.C., 216,382 F8 115
Charleston, harbor, S.C F8 115
Charleston, peak, Nev G6 104
Charleston Heights, Charleston, S.C., 25,000 F2 115
Charlestown, Clark, Ind., 5,726 H6 91
Charlestown, Ire., 727 D3 11
Charlestown, St. Kitts-Nevis-Anguilla, 1,852 n15 64
Charlestown, Cecil, Md., 711 A6 85
Charlestown, Sullivan, N.H., 1,173 (2,576*) D2 105
Charlestown, Washington, R.I., 300 (1,966*) D10 84
Charles Town, Jefferson, W. Va., 3,329 B7 123
Charlesville, Con. L C3 48
Charleville, Austl., 5,154 C6 51
Charleville, Fr., 24,668 (*56,000) C6 14
Charlevoix, Charlevoix, Mich., 2,751 C5 98
Charlevoix, co., Mich., 13,421 C5 98
Charlevoix, lake, Mich C5 98
Charlevoix East, co., Que., Can., 16,450 B7 73
Charlevoix West, co., Que., Can., 14,562 B7 73
Charlie, lake, B.C., Can A7 68
Charlieu, Fr., 4,911 D6 14
Charlo, Lake, Mont., 200 C2 102
Charlo Station, N.B., Can., 409 B3 74
Charlotte, Clinton, Iowa, 417 C7 92
Charlotte, Washington, Maine, (260*) C5 96
Charlotte, Eaton, Mich., 7,657 F6 98
Charlotte, Mecklenburg, N.C., 201,564 (*272,700) B3 109
Charlotte, Dickson, Tenn., 551 A4 117
Charlotte, Atascosa, Tex., 1,465 E3 118
Charlotte, Chittenden, Vt., 160 (1,271*) C2 120
Charlotte, co., N.B., Can., 23,285 D3 74
Charlotte, co., Fla., 12,594 F4 86
Charlotte, co., Va., 13,368 D4 121
Charlotte, harbor, Fla F4 86
Charlotte, lake, B.C., Can C5 68
Charlotte Amalie, Vir. Is., 12,880 m14 64, f15 65
Charlotte Court House, Charlotte, Va., 555 D4 121
Charlotte Hall, St. Marys, Md., 83 D4 85
Charlotte Harbor, Charlotte, Fla., 500 F4 86
Charlottenburg, Ger A8 17
Charlottenberg, Sur A4 59
Charlottesville, Hancock, Ind., 500 E6 91
Charlottesville (Independent City), Va., 29,427 C4 121
Charlottetown, P.E.I., Can., 18,318 C6 74
Charlotte Waters, Austl E5 50
Charlson, McKenzie, N. Dak., 15 A3 110
Charlton, co., Ga., 5,313 F4 87
Charlton City, Worcester, Mass., 750 B4 97
Charlton Depot, Worcester, Mass., 250 B4 97
Charmco, Greenbrier, W. Va., 700 *D4 123
Charmes, Fr., 5,177 C7 14
Charmey, Switz., 1,144 C3 19
Charny, Que., Can., 4,189 C6, C9 73
Charouïn, Alg D4 44
Charron, lake, Man., Can C4 71
Charskiy, Sov. Un D10 29
Charter Oak, Los Angeles, Calif., 4,000 *F3 82
Charter Oak, Crawford, Iowa, 665 B2 92
Charters Towers, Austl., 7,633 D8 50
Chartierville, Que., Can., 188 D6 73
Chartley, Bristol, Mass., 700 C5 97
Chartrand, Ont., Can., 210 A10 72
Chartres, Fr., 31,495 (*48,000) C4 14
Charysh, riv., Sov. Un C10 29
Chascomús, Arg., 9,105 B5 55
Chase, B.C., Can., 990 D8 68
Chase, Rice, Kans., 922 D5 93
Chase, Lake, Mich., 185 E5 98
Chase, co., Kans., 3,921 D7 93
Chase, co., Nebr., 4,317 D4 103
Chase, mtn., Maine B4 96
Chaseburg, Vernon, Wis., 242 E2 124
Chase City, Mecklenburg, Va., 3,207 E4 121
Chaseley, Wells, N. Dak., 72 B6 110
Chasicó, Arg C3 54
Chaska, Carver, Minn., 2,501 F5 99
Chasong, Kor., 5,000 F3 37
Chasov Yar, Sov. Un., 19,000 q20 27
Chassahowitzka, bay, Fla D4 86
Chasseral, mtn., Switz B2 19
Chasseron, mtn., Switz C2 19
Chastang, Mobile, Ala., 200 D1 78
Chataignier, Evangeline, La., 550 D3 95
Chatanika, Alsk., 14 B10 79
Chatawa, Pike, Miss., 100 D3 100
Chatcolet, Benewah, Idaho, 101 B2 89
Châteaubriant, Fr., 10,852 D3 14
Chateau d'Oex, Switz., 3,378 D3 19
Château-du-Loir, Fr., 4,707 D4 14
Châteaudun, Fr., 11,982 C4 14
Chateaugay, Franklin, N.Y., 1,097 A3 108
Château-Gontier, Fr., 7,065 D3 14
Chateauguay, Que., Can., 7,570 D4, D8 73
Chateauguay, co., Que., Can., 34,042 D4 73
Chateauguay Heights, Que., Can., 1,231 D8 73
Chateauguay Station, Que., Can., 3,265 *D8 73
Châteauneuf [-sur-Loire], Fr., 4,350 D5 14
Château-Renault, Fr., 4,238 D4 14
Châteauroux, Fr., 45,063 D4 14
Château-Salins, Fr., 2,174 F6 15
Château-Thierry, Fr., 10,006 C5 14
Châtellerault, Fr., 27,079 D4 14
Châtel-St.-Denis, Switz., 2,666 C2 19
Chatfield, Crittenden, Ark., 100 B5 81
Chatfield, Man., Can., 158 D3 71
Chatfield, Fillmore and Olmstead, Minn., 1,841 G6 99
Chatfield, Crawford, Ohio, 263 B5 111
Chatham, Alsk., 5 m22, D12 79
Chatham, N.B., Can., 7,109 B4 74
Chatham, Ont., Can., 29,826 E2 72
Chatham, Eng., 48,989 (*190,000) C8 12
Chatham, Sangamon, Ill., 1,069 D4 90
Chatham, Jackson, La., 758 B3 95
Chatham, Barnstable, Mass., 1,479 (3,273*) C8 97
Chatham, Alger, Mich., 175 B4 98
Chatham, Washington, Miss., 50 B2 100
Chatham, Carroll, N.H., 30 (150*) B4 105
Chatham, Morris, N.J., 9,517 B4 106
Chatham, Columbia, N.Y., 2,426 C7 108
Chatham, Medina, Ohio, 200 A5 111
Chatham, Pittsylvania, Va., 1,822 E3 121
Chatham, co., Ga., 188,299 E5 87
Chatham, co., N.C., 26,785 B4 109
Chatham, is., Pac. O I11 7
Chatham, sound, B.C., Can B2 68
Chatham, strait, Alsk m22 79
Chatillon, It., 2,335 D3 18
Châtillon-sur-Seine, Fr., 5,518 D6 14
Chatkal, range, Sov. Un E8 29
Chatom, Washington, Ala., 993 D1 78
Chatrapur, India, 7,835 H10 40
Chatsworth, Ont., Can., 419 C4 72
Chatsworth, Murray, Ga., 1,184 B2 87
Chatsworth, Livingston, Ill., 1,330 C5 90
Chatsworth, Sioux, Iowa, 84 B1 92
Chatsworth, Burlington, N.J., 295 D3 106
Chattahoochee, Gadsden, Fla., 9,699 B2 86
Chattahoochee, Fulton, Ga. (part of Atlanta) B5, C2 87
Chattahoochee, co., Ga., 13,011 D2 87
Chattahoochee, riv., U.S D10 77
Chattanooga, Mercer, Ohio, 150 B3 111
Chattanooga, Comanche, Okla., 356 C3 112
Chattanooga, Hamilton, Tenn., 130,009 (*286,700) D8, E10 117
Chattaroy, Spokane, Wash., 150 B8 122
Chattaroy, Mingo, W. Va., 950 D2 123
Chatteris, Eng., 5,490 B8 12
Chattooga, co., Ga., 19,954 B1 87
Chattooga, ridge, S.C B1 115
Chattooga, riv., S.C B1 115
Chatuge, lake, N.C D2 109
Chatwood, Chester, Pa., 3,621 *G10 114
Chaudiere, riv., Que., Can C7 73
Chau Doc, Viet., 5,000 G6 38
Chaudrant, bayou, La B3 95
Chauk, Bur., 24,466 D9 39
Chaumont, Fr., 21,717 C6 14
Chaumont, Jefferson, N.Y., 523 A4 108
Chauncey, Dodge, Ga., 330 D3 87
Chauncey, Athens, Ohio, 996 C5 111
Chauncey, Logan, W. Va., 300 D3 123
Chauncy, pond, Mass D1 97
Chauny, Fr., 12,626 C5 14
Chautauqua, Chautauqua, Kans., 205 E7 93
Chautauqua, Chautauqua, N.Y., 383 C1 108
Chautauqua, co., Kans., 5,956 E7 93
Chautauqua, co., N.Y., 145,377 C1 108
Chautauqua, lake, N.Y C1 108
Chauvigny, Fr., 4,024 D4 14
Chauvin, Alta., Can., 395 C5 69
Chauvin, Terrebonne, La., E5 95
Chavanga, Sov. Un D17 25
Chaves, Braz., 428 C5 59
Chaves, Port., 11,286 B2 20
Chaves, co., N. Mex., 57,649 D5 107
Chavies, Perry, Ky., 250 C6 94
Chavinda, Mex., 5,418 m12 63
Chavuma, N. Rh D3 48
Chazelles [-sur-Lyon], Fr., 5,076 E6 14
Chazy, Clinton, N.Y., 750 A3 108
Cheadle, Alta., Can., 64 D4 69
Cheaha, mtn., Ala B4 78
Cheapside, Northampton, Va., 150 A7 121
Cheat, mtn., W. Va C5 123

Cheat, riv., W. Va. B5 123
Cheatham, co., Tenn., 9,428 A4 117
Cheb, Czech., 20,600 C2 26
Chebacco, lake, Mass C4 97
Chebanse, Iroquois and Kankakee, Ill., 995 C6 90
Chebeague Island, Cumberland, Maine, 300 E2, E5 96
Cheboksary, Sov. Un., 123,000 B3 29
Cheboygan, Cheboygan, Mich., 5,859 C6 98
Cheboygan, co., Mich., 14,550 C6 98
Chech, sand dunes, Alg., Mali E4 44
Checheng, China G6 36
Chechon, Kor., 24,600 H4 37
Checkerberry, Chittenden, Vt., 80 B2 120
Checotah, McIntosh, Okla., 2,614 B6 112
Chedabucto, bay, N.S., Can D8 74
Cheddar, Eng., 2,600 C5 12
Cheddar, Anderson, S.C., 500 B3 115
Cheduba, isl., Bur E9 39
Cheektowaga (Cheektowaga-Northwest), Erie, N.Y., 52,362 C2 108
Cheektowaga-Southwest, Erie, N.Y., 12,766 *C2 108
Cheepie, Austl., 60 C5 51
Cheesman, lake, Colo B5 83
Chefoo, China, 116,000 F9 36
Chefuncte, riv., La D5 95
Chehalem, mts., Oreg B1 113
Chehalis, Lewis, Wash., 5,199 C3 122
Chehalis, riv., Wash C2 122
Cheikh Meskine, Syr. B8 32
Cheikh Saad, Syr. B7 32
Cheju, Kor., 34,300 (68,100▲) J3 37
Cheju, isl., Kor J3 37
Chekhov, Sov. Un., 10,000 C10 37
Chekiang, prov., China, 25,280,000 F9 34
Chekunda, Sov. Un A6 37
Chela, mts., Ang E1 48
Chelan, Sask., Can., 115 E4 70
Chelan, Chelan, Wash., 2,402 B6 122
Chelan, co., Wash., 40,744 B5 122
Chelan, lake, Wash A5 122
Chelan, range, Wash B5 122
Chelan Falls, Chelan, Wash., 150 B6 122
Cheleiros, Port., 1,253 f9 20
Cheleken, Sov. Un B6 41
Cheleken, isl., Sov. Un B6 41
Chelforó, Arg B3 54
Cheli, China G5 34
Chelia, mtn., Alg B6 44
Chéliff, riv., Alg B5 44
Chelkar, Sov. Un., 19,300 D5 29
Chelkar, lake, Sov. Un E19 9
Chelkar Tengiz, lake, Sov. Un F21 9
Chellala, Alg., 2,899 G8 30
Chelles, Fr., 28,382 F2 15
Chełm [Lubelski], Pol., 27,000 C7 26
Chelmno, Pol., 14,400 B5 26
Chelmsford, Ont., Can., 2,c59 *E9 72
Chelmsford, Eng., 49,810 C8 12
Chelmsford, Middlesex, Mass., 3,500 (15,130▲) A5, C2 97
Chelmza, Pol., 12,200 B5 26
Chelsea, Que., Can., 327 D2 73
Chelsea, Tama, Iowa, 453 C5 92
Chelsea, Kennebec, Maine, 125 (1,893▲) D3 96
Chelsea, Suffolk, Mass., 33,749 B5, D3 97
Chelsea, Washtenaw, Mich., 3,355 F6 98
Chelsea, Rogers, Okla., 1,541 A6 112
Chelsea, Faulk, S. Dak., 53 B7 116
Chelsea, Orange, Vt., 500 (957▲) D4 120
Chelsea, Taylor, Wis., 110 C3 124
Cheltenham, Eng., 71,968 C5 12
Cheltenham, Prince Georges, Md., 500 C4 85
Cheltenham, Montgomery, Pa., 7,000 *A12 114
Chelva, Sp., 4,400 C5 20
Chelyabinsk, Sov. Un., 733,000 (*910,000) B6 29
Chelyan, Kanawha, W. Va., 500 C3, C6 117
Chelyuskin, cape, Sov. Un B13 28
Chemainus, B.C., Can., 1,518 B9, E6 70
Chemal, Sov. Un E26 9
Chemawa, Marion, Oreg C1, C4 113
Chemillé, Fr., 3,109 D3 14
Chemnitz, see Karl-Marx-Stadt, Ger.
Chemquassabamticook, lake, Maine B3 96
Chemult, Klamath, Oreg., 150 D5 113
Chemung, McHenry, Ill., 150 A5 90
Chemung, co., N.Y., 98,706 C4 108
Chen, mtn., Sov. Un C17 28
Chenab, riv., Pak B5 39
Chenachane (Oasis), Alg D4 44
Chenan, China H3 36
Chenango, Brazoria, Tex G5 118
Chenango, co., N.Y., 43,243 C5 108
Chenango, riv., N.Y. C5 108
Chenango Bridge, Broome, N.Y., 2,000 *C5 108
Chenango Forks, Broome, N.Y., 400 C5 108
Chenchi, China, 2,000 K4 34
Chene, bayou, La C5 95
Chenega, Alsk., 20 C10, g17 79
Chenequa, Waukesha, Wis., 445 E1 124
Chéneville, Que., Can., 746 D2 73
Cheney, Sedgwick, Kans., 1,101 B4, E6 93
Cheney, Spokane, Wash., 3,173 B8, D7 122
Cheneyville, Rapides, La., 1,037 C3 95
Cheng, China J9 36
Chengan, China, 14,000 J2 36
Chengane, riv., Moz B5 49
Chengchan Tow, pt., China D9 34
Chengchow, China, 766,000 G5 36
Chenghai, China, 33,000 G8 34
Chenghsien, see Chengchow, China
Chenghwa, see Sharasume, China
Chengkou, China, 13,000 I3 36
Chengku, China, 10,000 H2 36
Chengteh (Jehol), China, 92,900 C8 34, D7 36
Chengtu, China, 1,107,000 E5 34
Chengyangkuan, China H6 36
Chenhsien, China F7 34
Chenkang, China G4 34
Chennan, China, 5,000 F5 34
Chenoa, McLean, Ill., 1,523 C5 90
Chenpa, China, 3,000 H2 36
Chenping, China I3 36
Chensi, see Barkol, China
Chentung, China, 5,000 B10 36
Chenyuan, China, 11,000 K3 36
Cheoah, Graham, N.C. D2 109
Cheo Reo, Viet F8 38
Chepachet, Providence, R.I., 800 B10 84
Chepén, Peru, 8,214 C2 58
Chepes, Arg., 2,131 A3 54
Chepo, Pan., 1,664 F8 62
Chepstow, Wales, 6,041 C5 12
Cheptsa, riv., Sov. Un D19 9
Chequamegon, bay, Wis B3 124
Cher, dept., Fr., 293,514 *D5 14
Cher, riv., Fr D4 14
Cheran, India, 5,000 E13 40
Cheraw, Otero, Colo., 173 C7 83
Cheraw, Marion, Miss., 50 D4 100
Cheraw, Chesterfield, S.C., 5,171 B8 115
Cherbourg, Fr., 37,486 (*70,000) C3 14
Cherchel, Alg., 7,805 B5 44
Cherchen, China, 5,000 A8 39
Cherdyn, Sov. Un A5 29
Cheremkhovo, Sov. Un., 122,000 D13 28
Cheren, Eth., 2,700 B4 47
Cherepanovo, Sov. Un., 23,100 C10 29
Cherepovets, Sov. Un., 113,000 B1 29
Cherhill, Alta., Can., 72 C3 59
Cheriton, Northampton, Va., 761 D7 121
Cherkassy, Sov. Un., 95,000 G9 27
Cherkessk, Sov. Un., 45,000 G17 9
Cherlakskiy, Sov. Un C8 29
Chernigov, Sov. Un., 101,000 F8 27
Chernigovka, Sov. Un., 3,300 H11 27
Chernikovsk, Sov. Un., 206,000 *E20 9
Chernobay, Sov. Un., 25,000 G9 27
Chernobyl, Sov. Un., 10,000 F8 27
Chernofski, Alsk., 5 E6 79
Chernogor, Sov. Un., 30,000 E27 9
Chernomorskoye, Sov. Un., 3,800 I9 27
Chernovtsy, Sov. Un., 147,000 A7 22, G5 27
Chernoye, Sov. Un r32 25
Chernyakhovsk (Insterburg), Sov. Un., 33,000 D3 27
Cherokee, Colbert, Ala., 1,349 A2 78
Cherokee, Cherokee, Iowa, 7,724 B2 92
Cherokee, Crawford, Kans., 797 E9 93
Cherokee, Swain, N.C., 500 D3 109
Cherokee, Alfalfa, Okla., 2,410 A3 112
Cherokee, San Saba, Tex., 150 D3 118
Cherokee, co., Ala., 16,303 A4 78
Cherokee, co., Ga., 23,001 B2 87
Cherokee, co., Iowa, 18,598 B2 92
Cherokee, co., Kans., 22,279 E9 93
Cherokee, co., N.C., 16,335 D2 109
Cherokee, co., Okla., 17,762 B6 112
Cherokee, co., S.C., 35,205 A4 115
Cherokee, co., Tex., 33,120 D5 118
Cherokee, Indian res., Tenn C10 117
Cherokee Falls, Cherokee, S.C., 350 A5 115
Cherokee, lake, Okla A7 112
Cherokee, lake, Tenn C10 117
Cherokee Ranch, Berks, Pa., 1,200 *F10 114
Cherokee Sound, Ba. Is., 320 B5 64
Cherquenco, Chile, 1,677 B2 54
Cherrapunji, India C9 39
Cherry, Hartford, Conn B5 84
Cherry, Bureau, Ill., 501 B4 90
Cherry, Lauderdale, Tenn., 75 B2 117
Cherry, co., Nebr., 8,218 B4 103
Cherry, creek, Colo B6 83
Cherry, creek, S. Dak C3 116
Cherry, pt., Va D6 121
Cherry Creek, White Pine, Nev., 50 D7 104
Cherry Creek, Chautauqua, N.Y., 649 C1 108
Cherry Creek, Ziebach, S. Dak., 150 C4 116
Cherryfield, Washington, Maine, 750 (780▲) D5 96
Cherry Fork, Adams, Ohio, 185 D4 111
Cherrygrove, Fillmore, Minn., 60 G6 99
Cherry Grove, Hamilton, Ohio, 2,000 *C3 111
Cherry Grove, Washington, Oreg., 300 B1 113
Cherry Hill, Polk, Ark., 150 C1 81
Cherry Hill, Cecil, Md., 150 A6 85
Cherry Hill, pt., N.Y. E8 84
Cherry Hills Village, Arapahoe, Colo., 1,931 *B6 83
Cherry Run, Morgan, W. Va., 100 B6 123
Cherry Tree, Indiana, Pa., 469 E4 114
Cherryvale, Montgomery, Kans., 2,783 E8 93
Cherry Valley, Cross, Ark., 455 B5 81
Cherry Valley, Ont., Can., 272 D7 72
Cherry Valley, Winnebago, Ill., 875 A5 90
Cherry Valley, Worcester, Mass., 1,500 *B4 97
Cherry Valley, Otsego, N.Y., 668 C6 108
Cherryville, Gaston, N.C., 3,607 B2 109
Cherskiy, mts., Sov. Un C17 28
Chesaning, Saginaw, Mich., 2,770 E6 98
Chesapeake, Lawrence, Ohio, 1,396 D5 111
Chesapeake (Independent City), Va., 73,647 E6, B7 121
Chesapeake, Kanawha, W. Va., 2,699 C3 123
Chesapeake, bay, U.S. C12 77
Chesapeake Beach, Calvert, Md., 731 C4 85
Chesapeake City, Cecil, Md., 1,104 A6 85
Chesapeake and Delaware, canal, Del A6 85
Chesaw, Okanogan, Wash., 50 A6 122
Chesham, Cheshire, N.H., 150 E2 105
Cheshire, New Haven, Conn., 4,072 (13,383▲) D5 84
Cheshire, Berkshire, Mass., 1,078 (2,472▲) A1 97
Cheshire, Gallia, Ohio, 369 D5 111
Cheshire, co., Eng., 1,367,860 A5 12
Cheshire, co., N.H., 43,342 E2 105
Cheshire, res., Mass A1 97
Cheshskaya, bay, Sov. Un B18 9
Chesilhurst, Camden, N.J., 384 D3 106
Chesley, Ont., Can., 1,697 C3 72
Chesnaye, Man., Can C5 71
Chesnee, Spartanburg, S.C., 1,045 A4 115
Chesney, Union, Tenn C10 117
Chesnokovka, Sov. Un., 20,000 E25 9
Chester, Crawford, Ark., 99 B1 81
Chester, Plumas, Calif., 1,553 B3 82
Chester, N.S., Can., 990 E5 74
Chester, Middlesex, Conn., 1,414 (2,520▲) D7 84
Chester, Eng., 59,283 A5 12
Chester, Nassau, Fla., 100 B5, B6 86
Chester, Dodge, Ga., 377 D3 87
Chester, Freemont, Idaho, 100 F7 89
Chester, Randolph, Ill., 4,460 F4 90
Chester, Wayne, Ind., 130 E8 91
Chester, Howard, Iowa, 211 A5 92
Chester, Queen Annes, Md., 900 C5 85
Chester, Hampden, Mass., 950 (1,155▲) B2 97
Chester, Liberty, Mont., 1,158 B6 102
Chester, Thayer, Nebr., 480 D8 103
Chester, Rockingham, N.H., 400 (1,053▲) E4 105
Chester, Morris, N.J., 1,074 B3 106
Chester, Orange, N.Y., 1,492 D2, D6 108
Chester, Meigs, Ohio, 200 C6 111
Chester, Major, Okla., 150 A3 112
Chester, Delaware, Pa., 63,658 B10, G11 114
Chester, Chester, S.C., 6,906 B5 115
Chester, Lake, S. Dak., 200 D9 116
Chester, Windsor, Vt., 923 (2,318▲) E3 120
Chester, Chesterfield, Va., 2,000 C7, D5 121
Chester, Spokane, Wash D8 122
Chester, Hancock, W. Va., 3,787 A2, A4 123
Chester, co., Pa., 210,608 G10 114
Chester, co., S.C., 30,888 B5 115
Chester, co., Tenn., 9,569 B3 117
Chester, creek, Pa D1 106
Chester, riv., Md B5 85
Chester Basin, N.S., Can., 490 E5 74
Chester Depot, Windsor, Vt., 350 E3 120
Chesterfield, New London, Conn., 100 D8 84
Chesterfield, Eng., 67,833 (*127,000) A6 12
Chesterfield, Macoupin, Ill., 280 D3 90
Chesterfield, Madison, Ind., 2,588 D6 91
Chesterfield, Hampshire, Mass., 180 (556▲) B2 97
Chesterfield, Cheshire, N.H., 90 (1,405▲) E2 105
Chesterfield, Burlington, N.J., 150 C3 106
Chesterfield, Chesterfield, S.C., 1,532 B7 115
Chesterfield, Henderson, Tenn., 200 B3 117
Chesterfield, Chesterfield, Va., 135 C7, D5 121
Chesterfield, co., S.C., 33,717 B7 115
Chesterfield, co., Va., 71,197 D5 121
Chesterfield, inlet, N.W. Ter., Can D14 66
Chesterfield, is., Pac. O H9 7
Chesterfield Inlet, N.W. Ter., Can., 146 D14 66
Chesterhill, Morgan, Ohio, 876 C6 111
Chesterton, Porter, Ind., 4,335 A3 91
Chestertown, Kent, Md., 3,602 B5 85
Chestertown, Warren, N.Y., 200 B7 108
Chesterville, Ont., Can., 1,248 B9 72
Chesterville, Pontotoc, Miss A5 100
Chesterville, Morrow, Ohio, 275 B5 111
Chestnut, Logan, Ill., 300 C4 90
Chestnut Hill, New London, Conn., 75 C7 84
Chestnut Mound, Smith, Tenn., 125 C8 117
Chestoa, Unicoi, Tenn., 150 C11 117
Chesuncook, Piscataquis, Maine, 10 B3 96
Chesuncook, lake, Maine C3 96
Cheswick, Allegheny, Pa., 2,734 A6 114
Cheswold, Kent, Del., 281 B6 85
Chetac, lake, Wis C2 124
Chetco, riv., Oreg E2 113
Chetek, Barron, Wis., 1,729 C2 124
Chetek, lake, Wis C2 124
Cheticamp, N.S., Can., 1,223 C8 74
Chetopa, Labette, Kans., 1,538 E8 93
Chetumal, bay, Mex D7 63
Chevalon, creek, Ariz C5 80
Cheverie, N.S., Can., 195 D5 74
Cheverly, Prince Georges, Md., 5,223 *C1 85
Cheviot, Sask., Can., 23 E2 70
Cheviot, N.Z., 440 O14 51
Cheviot, Hamilton, Ohio, 10,701 C3, D2 111
Cheviot, hills, Scot., Eng E6 13
Chevreuil, bayou, La C6 95
Chevreuil, pt., La E4 95
Chevy Chase, Montgomery, Md., 2,405 C1, C3 85
Chevy Chase Heights, Indiana, Pa., 1,160 *F3 114
Chevy Chase Lake, Montgomery, Md., 2,500 *C3 85
Chevy Chase Section Four, Montgomery, Md., 2,243 *C3 85
Chevy Chase View, Montgomery, Md., 1,000 *C3 85
Chewack, creek, Wash A5 122
Chewalla, McNairy, Tenn., 150 B3 117
Chewelah, Stevens, Wash., 1,525 A8 122
Chewey, Adair, Okla., 30 A7 112
Chew Road, Camden, N.J. D3 106
Chews, Camden, N.J., 1,500 *C4 106
Chewsville, Washington, Md., 250 A2 85
Cheyenne, Roger Mills, Okla., 930 B2 112
Cheyenne, Winkler, Tex., 15 D1 118
Cheyenne, Laramie, Wyo., 43,505 D8 125
Cheyenne, co., Colo., 2,789 C8 83
Cheyenne, co., Kans., 4,708 C2 93
Cheyenne, co., Nebr., 14,828 C2 103
Cheyenne, pass, Wyo D7 125
Cheyenne, riv., S. Dak., Wyo C4 116
Cheyenne Agency, Dewey, S. Dak C5 116
Cheyenne River, Indian res., S. Dak B5 116
Cheyenne Wells, Cheyenne, Colo., 1,020 C8 83
Chhatarpur, India, 22,146 E7 40
Chhindwara, India, 37,244 F7 40
Chi, China G6 36
Chia, China E4 36
Chiahsing, China, 78,300 I9 36
Chiai, Taiwan, 83,400 (165,200▲) *G9 34
Chialing, riv., China J2 36
Chiamboni (Dicks Head), cape, Ken B7 48
Chiamussu, see Kiamusze, China
Chian, China F3 37
Chian, China, 52,800 K6 36
Chianghua, China *F7 34
Chiang Kham, Thai C4 38
Chiang Khong, Thai., 3,320 B4 38
Chiangling, China, 15,000 I5 36
Chiang Mai, Thai., 66,019 C3 38
Chiang Rai, Thai., 13,494 C3 38
Chiangyin, China, 57,000 I9 36
Chiangyu, China, 15,000 E5 34
Chiaochia, China, 13,000 F5 34
Chiaoho, China, 5,000 E3 37
Chiaotso, China G5 36
Chiapa de Corzo, Mex., 6,972 D6 63
Chiapas, state, Mex., 1,215,475 D6 63
Chiari, It., 9,000 B2 21
Chiashan, China H8 36
Chiasso, Switz., 7,377 (*12,216) E7 19
Chiatura, Sov. Un., 20,000 A14 31
Chiautla de Tapia, Mex., 3,554 n14 63
Chiavari, It., 24,603 B2 21
Chiavenna, It., 6,211 C5 18
Chiawuli Tak, Pima, Ariz., 100 F4 80
Chiayu, China, 18,000 J5 60
Chiba, Jap., 241,615 I10, n19 37
Chiba, pref., Jap., 2,306,010 *I10 37
Chibana, Okinawa 52
Chibemba, Ang., 354 E1 48
Chibougamau, Que., Can., 4,765 G18 61, B2 73
Chibuto, Moz B5 49
Chicago, Cook, Ill., 3,550,404 (*6,517,600) B6, F3 90
Chicago Heights, Cook, Ill., 34,331 B6, F3 84
Chicago Ridge, Cook, Ill., 5,748 *F3 90
Chicago Sanitary and Ship, canal, Ill F2 90
Chicamacomico, creek, Md D6 85
Chicamuxen, Charles, Md C3 85
Chicapa, riv., Ang C3 48
Chichagof, hbr., Alsk E2 79
Chichagof, isl., Alsk k22 79
Chichaoua, Mor C3 44
Chichen-Itzá, ruins, Mex C7 63
Chichester, Eng., 20,118 D7 12
Chichester, Merrimack, N.H., 130 (821▲) D4 99
Chichiang, China, 20,000 J2 36
Chichibu, Jap., 43,700 n18 37
Ch'i-ch'i-ha-erh, see Tsitsihar, China
Chichun, China, 32,000 I6 36
Chickahominy, riv., Va D5 121
Chickaloon, Alsk., 43 C10, g17 79
Chickamauga, Walker, Ga., 1,824 B1 87
Chickamauga, dam, Tenn E10 117
Chickamauga, lake, Tenn D8 117
Chickasaw, Mobile, Ala., 10,002 E1 78
Chickasaw, co., Iowa, 15,034 A5 92
Chickasaw, co., Miss., 16,891 B5 100
Chickasawhay, riv., Miss D5 100
Chickasha, Grady, Okla., 14,866 B4 112
Chicken, Alsk., 23 C11 79
Chiclana, Sp., 21,524 D2 20
Chiclayo, Peru, 38,517 C2 58
Chico, riv., Arg C3 54
Chico, Butte, Calif., 14,757 C3 82
Chico, Park, Mont., 24 E6 102
Chico, Wise, Tex., 654 C4 118
Chico, Kitsap, Wash., 300 B3 122
Chicoa, Moz A5 49
Chicomo, Moz B5 49
Chicontepec, Mex., 2,859 m14 63
Chicopee, Hall, Ga., 900 B3 88
Chicopee, Crawford, Kans., 200 E9 93
Chicopee, Hampden, Mass., 61,553 B2 97
Chicora, Wayne, Miss., 100 D5 100
Chicora (Millerstown), Butler, Pa., 1,156 E2 114
Chicot, Chicot, Ark., 25 D4 81
Chicot, co., Ark., 18,990 D4 81
Chicot, isl., La E6 95
Chicot, lake, Ark D4 81
Chicot, pt., La E6 95
Chicota, Lamar, Tex., 370 C5 118
Chicoutimi, Que., Can., 31,657 (*105,009) A6, B2 73
Chicoutimi, co., Que., Can., 157,196 A6 73
Chicoutimi, riv., Que., Can A6 73
Chicoutimi Nord, Que., Can., 11,229 A6 73
Chico Vecino, Butte, Calif., 4,688 *C3 82
Chiderly, riv., Sov. Un E23 9
Chidester, Ouachita, Ark., 348 D2 81
Chidley, cape, Newf., Can f8 75
Chieerete, isl., Bikini 52
Chiefland, Levy, Fla., 1,459 C4 86
Chiefs, pt., Ont., Can C3 72
Chiehhsiu, China F4 36
Chiem, lake, Ger E6 16
Chien, China G3 36
Chienan, China D8 36
Chienchang, China F2 37
Chienchiang, China, 11,000 J3 36
Chienli, China, 20,000 J5 36
Chienning, China, 8,000 K7 36
Chienou, China, 12,000 K8 36
Chienping, China D8 36
Chienshan, China I7 36
Chienshih, China, 15,000 I3 36
Chienshui, China, 28,000 G5 34
Chientang, riv., China J8 36
Chiente, China, 12,000 J8 36
Chienyang, China, 12,000 F8 34
Chierhkalang, China C10 36
Chieri, It., 11,900 B1 21
Chiers, riv., Fr E5 15
Chiese, riv., It D6 18
Chieti, It., 36,000 (47,792▲) C5 21
Chigirin, Sov. Un., 10,000 G9 27
Chignahuapan, Mex., 3,873 n14 63
Chignecto, bay, N.S., Can D5 75
Chignik, Alsk D8 79
Chigubo, Moz B5 49
Chigwell, Eng., 61,001 k13 10
Chihchiang, China F6 34
Chihchiang, China, 19,000 I4 36
Chihfeng, China, 40,000 C8 36
Chihli (Pohai), gulf, China D8 34
Chihsi, China, 50,000 B11 34
Chihtan, China D6 34
Chihuahua, Mex., 149,437 B3 63
Chihuahua, state, Mex., 1,235,891 B3 63
Chiili, Sov Un E7 29
Chikan, China A9 34
Chikaskia, creek, Okla A4 112
Chikaskia, riv., Kans E6 93
Chiko, China B4 37
Chikura, Jap., 10,900 o18 37
Chikwolnepy, stream, N.H. A4 105
Chilapa, Mex., 7,105 D5, o14 63
Chilca, Peru, 1,341 D2 58
Chilcat, pass, B.C., Can E5 66
Chilcotin, riv., B.C., Can C6 68
Childersburg, Talladega, Ala., 4,884 B3 78
Childress, Childress, Tex., 6,399 B2 118
Childress, co., Tex., 8,421 B2 118
Childs, lake, Fla E5 86
Chile, country, S.A., 7,339,546 G3 53, B2 54, E2 55
Chilecito, Arg., 6,121 E2 55
Chilete, Peru, 476 C2 58
Chilhowee, Johnson, Mo., 339 C4 101
Chilhowee, Blount, Tenn D9 117
Chilhowee, mtn., Tenn E12 117
Chilhowie, Smyth, Va., 1,169 B3 121
Chili, Miami, Ind., 145 C5 91
Chili, Clark, Wis., 225 D3 124
Chilibre, Pan., 1,220 k11 62
Chilili, Bernalillo, N. Mex., 100 C3 107
Chilin, see Kirin, China
Chilka, lake, India H10 40
Chilko, lake, B.C., Can D6 68
Chilko, riv., B.C., Can D6 68
Chillán, Chile, 65,000 B2 54
Chillicothe, Peoria, Ill., 3,054 C4 90
Chillicothe, Wapello, Iowa, 148 C5 92
Chillicothe, Livingston, Mo., 9,236 B4 101
Chillicothe, Ross, Ohio, 24,957 C5 111
Chillicothe, Hardeman, Tex., 1,161 B3 118
Chilliwack, B.C., Can., 8,259 B11, E7 68
Chilliwack, lake, B.C., Can A4 122
Chilliwack, riv., B.C., Can A4 122
Chillum, Prince Georges, Md., 10,000 *C3 85
Chilmark, Dukes, Mass., 150 (238▲) D6 97
Chilocco, Kay, Okla., 80 A4 112
Chiloé, prov., Chile, 98,662 C2 54
Chiloé, isl., Chile C2 54
Chiloquin, Klamath, Oreg., 945 E5 113
Chilpancingo [de los Bravos], Mex., 17,942 D5, o14 63
Chilson, Essex, N.Y., 50 B7 108
Chilton, Casey, Ky., 65 C5 94
Chilton, Falls, Tex., 400 D4 118
Chilton, Calumet, Wis., 2,578 B5, D5 124
Chilton, co., Ala., 25,693 C3 78
Chilung (Keelung), Taiwan, 165,200 (212,500▲) F9 34
Chilwa, lake, Nya A6 49
Chimacum, Jefferson, Wash., 100 A3 122
Chimalhuacán, Mex., 2,433 h10 63
Chimaltenango, Guat., 6,138 C2 62
Chimán, Pan., 535 F8 62
Chimay, Bel., 3,180 D4 15
Chimayo, Rio Arriba, N. Mex., 700 A4 107
Chimbay, Sov. Un., 16,100 E5 29
Chimborazo, prov., Ec., 241,258 B2 58
Chimborazo, vol., Ec B2 58
Chimbote, Peru, 4,243 C2 58
Chimen, China J7 36
Chimkent, Sov. Un., 171,000 E7 29
Chimney Rock, Archuleta, Colo., 5 D3 83
Chimney Top, mtn., Tenn C11 117
Chimo, China F9 36
Chin, Alta., Can., 50 E4 69
Chin, China F5 36
Chin, cape, Ont., Can B3 72
Chin, riv., China G5 36
China, Mex., 2,496 B5 63
China (excl. Taiwan), country, Asia, 646,530,000 F12 33, E5 34
China, lake, Maine D3 96
China Grove, Pike, Ala., 50 C4 78
China Grove, Rowan, N.C., 1,500 B3 109
China Hat, mtn., Oreg D5 113
Chinaja, Guat B2 62
Chinan, see Tsinan, China
Chinandega, Nic., 19,800 D4 62
Chinati, peak, Tex F2 118
Chincha Alta, Peru, 15,053 D2 58
Chincheng, China G5 36
Chincheros, Peru, 1,330 D3 58
Chinchiang, see Chuanchow, China
Chinchilla, Austl., 3,072 C8 51
Chinchilla, Lackawanna, Pa., 1,100 *D10 114
Chinchilla, Sp., 7,616 C5 20
Chinchow (Chinhsien), China, 352,200 D9 36
Chincoteague, Accomack, Va., 2,131 D7 121
Chincoteague, bay, Md., Va D7 85
Chinde, Moz., 5,000 A6 49
Chindo, Kor., 4,100 I3 37
Chindwin, riv., Bur D9 39
Ching, China I8 36
Ching, China K3 36
Ching, riv., China G3 36
Chingcheng, China, 5,000 C3 37
Chingchiang, China, 77,000 J6 36
Chingchien, China, 2,000 F4 36
Chingchuan, China E2 36
Chingford, Eng., 45,777 k13 10, C8 12
Chinghai, China D5 34
Chinghai (Tsinghai), prov., China, 2,050,000 D4 34
Chinghsing, China, 5,000 A10 36
Chingliu, China, 10,000 F8 34
Chingmen, China, 12,000 I5 36
Chingning, China G1 36
Chingola, N. Rh., 12,000 C4 48
Chingpeng, China C7 36
Chingpien, China F3 36
Chingpu, China, 45,000 I9 36
Chingshan, China, 19,000 I5 36
Chingtai, China D5 36
Chingte, China I8 36
Chingtzukuan, China H4 36
Chinguar, Ang., 1,675 D2 48
Chinguetti, Maur B2 45
Chingyang, China, 5,000 D6 34
Chingyuan, China, 20,000 E2 37
Chingyun, China F7 36
Chinhae, Kor., 60,000 (67,400▲) I4 37
Chinhsi, China D9 36
Chinhsien, China E9 36
Chinhsien, see Chinchow, China
Chinhua, China, 46,200 J8 36
Chining, China, 19,000 C7 34
Chining, China, 86,200 G7 36
Chinipas, Mex., 474 B3 63
Chinit, riv., Camb F6 38
Chinju, Kor., 68,000 (86,900▲) I4 37
Chinkiang, China, 207,600 E8 34
Chinko, riv., Cen. Afr. Rep D4 46
Chinle, Apache, Ariz., 150 A6 80
Chinle, creek, Ariz A6 80
Chinle, val., Ariz A6 80
Chinnampo (Nampo), Kor G2 37
Chinniuchen, China I6 36
Chinnur, India, 9,645 H7 40
Chino, San Bernardino, Calif., 10,305 F3 82
Chinon, Fr., 4,872 D4 14
Chinook, Alta., Can., 114 D5 69
Chinook, Blaine, Mont., 2,326 B7 102
Chinook, Pacific, Wash., 350 C2 122
Chino Valley, Yavapai, Ariz., 300 B3 80
Chinquapin, Duplin, N.C., 800 C6 109
Chinsali, N. Rh., 169 D5 48
Chinsha, riv., China F4 34
Chinshanchen, China, 1,000 A10 34
Chinta, China C4 34
Chinteche, Nya D5 48
Chinú, Col., 4,987 B2 60
Chinwangtao, China, 186,800 D8 34
Chinyang, China G5 36
Chinyun, China, 5,000 J9 36
Chiôco, Moz A5 49
Chioggia, It., 47,151 B4 21
Chios (Khios), prov., Grc., 62,223 *C6 23
Chip, lake, Alta., Can C3 69
Chipewyan, riv., Alta., Can A4 69
Chipinga, S. Rh., 1,400 B5 49
Chipita Park, El Paso, Colo C6 83
Chip Lake, Alta., Can C3 69
Chipley, Washington, Fla., 3,159 G9 86
Chipman, N.B., Can., 1,490 C4 74
Chipmonk, Cattaraugus, N.Y. C2 108
Chippawa, Ont., Can., 3,256 D5 72
Chippawa Hill, Ont., Can., 230 C3 72
Chippenham, Eng., 17,525 C5 12
Chippewa, co., Mich., 32,655 B6 98
Chippewa, co., Minn., 16,320 E3 99
Chippewa, co., Wis., 45,096 C2 124
Chippewa, lake, Wis C2 124
Chippewa, riv., Minn E3 99
Chippewa, riv., Wis D2 124
Chippewa Falls, Chippewa, Wis., 11,708 D2 124
Chippewa Lake, Mecosta, Mich., 150 E5 98
Chiputneticook, lakes, Maine C5 96
Chiquimula, Guat., 8,840 C3 62
Chiquimulilla, Guat., 3,541 C2 62
Chiquinquirá, Col., 10,143 B3 60
Chiquita, lake, Arg A4 54
Chira, riv., Ec B1 58
Chiradzulu, Nya A6 49
Chiras, Afg., 5,060 D12 41
Chirchik, Sov. Un., 76,000 E7 29
Chireno, Nacogdoches, Tex., 300 D5 118
Chiricahua, nat. mon., Ariz F6 80
Chiricahua, peak, Ariz F6 80
Chirikof, isl., Alsk D8 79
Chiriquí, gulf, Pan G7 62
Chiriquí Grande, Pan., 98 F6 62
Chiri-San, peak, Kor I3 37
Chiromo, Nya A6 49
Chirpan, Bul., 13,231 D7 22

Chirripó Grande, mtn., C.R. F6 62
Chirundu, S. Rh. E4 48
Chisago, co., Minn., 13,419 E6 99
Chisago City, Chisago, Minn., 772 E6 99
Chisamba, N. Rh., 113 C4 48
Chisana, Alsk., 5 C11 79
Chiselville, Bennington, Vt. (part of East Arlington) E2 120
Chisholm, Montgomery, Ala., 1,500 *C3 78
Chisholm, Franklin, Maine, 1,193 D2 96
Chisholm, St. Louis, Minn., 7,144 C6 99
Chisholm Mills, Alta., Can., 197 B3 69
Chishui, China I6 36
Chisimaio (Kismayu), Som., 8,000 (9,800▲) F5 47
Chisineu-Cris, Rom., 6,124 B5 22
Chislehurst [& Sidcup], Eng., 86,907 m13 10
Chisos, mts., Tex E1 118
Chistochina, Alsk., 28 C11, f19 79
Chistopol, Sov. Un., 55,000 B4 29
Chistyakovo, Sov. Un., 90,000 q21 27
Chiswick, Eng. (part of Brentford & Chiswick) m11 10
Chita, Sov. Un., 182,000 D14 28
Chitai (Kuchêng), China, 25,000 C2 34
Chitaldroog, India, 33,336 F6 39
Chitek, Sask., Can., 83 D2 70
Chitek, lake, Man., Can C2 71
Chitembo, Ang D2 48
Chitimacha, Indian res., La E4 95
Chitina, Alsk., 31 C11, g19 79
Chitral, Pak A5 39
Chitré, Pan., 9,120 F7 62
Chittagong, Pak., 180,000 (★364,205) D9 39
Chittenango, Madison, N.Y., 3,180 *B5 108
Chittenden, co., Vt., 74,425 C2 120
Chittenden, res., Vt D3 120
Chittoor, India, 47,876 F6 39
Chiuchiu, Chile D2 55
Chiuchuan, China D4 34
Chiumbe, riv., Ang C3 48
Chiume, Ang E3 48
Chiupu, China I7 36
Chiusi, It., 8,498 C3 21
Chivasso, It., 8,800 B1 21
Chivato, mesa, N. Mex B2 107
Chivay, Peru, 2,027 E3 58
Chivilcoy, Arg., 23,386 A4, g6 54
Chivington, Kiowa, Colo., 55 C8 83
Chiyang, China, 15,000 F7 34
Chizhapka, riv., Sov. Un B9 29
Chloe, Valencia, N. Mex C3 107
Chloride, Mohave, Ariz., 135 B1 80
Chmielnik, Pol., 3,171 C6 26
Cho, China E7 36
Choapa, riv., Chile A2 54
Choapan, Mex., 710 D5 63
Chobe, riv., Bech., S.W. Afr A3 49
Choccolocco, Calhoun, Ala., 100 B4 78
Choccolocco, creek, Ala B4 78
Chochiwon, Kor., 16,700 H3 37
Chocó, dept., Col., 139,380 C2 60
Chocó, bay, Col C2 60
Chocó, range, Col B2 60
Chocolate, mts., Ariz D1 80
Chocolate Bayou, Brazoria, Tex., 50 G5 118
Chocontá, Col., 2,442 B3 60
Choconut [Township], Susquehanna, Pa., (325▲) C9 114
Chocorua, Carroll, N.H., 225 C4 105
Chocorua, mtn., N.H C4 105
Chocowinity, Beaufort, N.C., 580 B6 109
Choctaw, Choctaw, Ala., 400 C1 78
Choctaw, Oklahoma, Okla., 623 B4 112
Choctaw, co., Ala., 17,870 C1 78
Choctaw, co., Miss., 8,423 B4 100
Choctaw, co., Okla., 15,637 C6 112
Choctaw, Indian res., Miss C4 100
Choctaw Bluff, Clarke, Ala., 450 D2 78
Choctawhatchee, bay, Fla G2 86
Choctawhatchee, riv., Ala., Fla D4 78
Chodziez, Pol., 7,694 B4 26
Choele Choel, Arg B3 54
Choibalsan, Mong B7 34
Choiceland, Sask., Can., 415 D3 70
Choiseul, isl., Sol. Is G9 7
Choisy-le-Roi, Fr., 40,950 g10 14
Choix, Mex., 1,922 B3 63
Chojna, Pol., 1,484 B3 26
Chojnice, Pol., 20,000 B4 26
Chojnow, Pol., 5,467 C3 26
Chokai-San, peak, Jap G10 37
Choke, mts., Eth C4 47
Chokio, Stevens. Minn., 498 E2 99
Chokoloskee, Collier, Fla., 155 G8 86
Cholame, San Luis Obispo, Calif., 20 E3 82
Cholet, Fr., 36,565 D3 14
Cholo, Nya A6 49
Cholo, riv., China B7 34
Choluteca, Hond., 5,275 D4 62
Choluteca, riv., Hond D4 62
Choma, N. Rh., 3,900 E4 48
Chomedey, Que., Can., 30,445 D4, D8 73
Chomo Lhari, peak, Bhu D12 40
Chomutov, Czech., 33,200 C2 26
Chonan, Kor., 36,500 H3 37
Chon Buri, Thai., 33,086 F4 38
Chone, Ec., 8,030 B1 58
Chŏngjin, Kor., 184,301 F4 37
Chongju, Kor., 92,300 H3 37
Chongsong, Kor., 6,834 E4 37
Chonju, Kor., 152,000 (188,000▲) I3 37
Chonos, arch., Chile C2 54
Chonzie, mtn., Scot D5 13
Chopin, riv., Braz D2 57
Choptank, Caroline, Md., 150 C6 85
Choptank, riv., Md C6 85
Chorrera, Cuba k12 64
Chorrillos, Peru, 6,996 D2 58
Chortkov, Sov. Un., 10,000 I9 27
Chorum (Corum), Tur., 22,707 B10 31
Chorwon, Kor., 30,085 G3 37
Chorzow, Pol., 147,000 C5, g9 26
Chosan, Kor., 18,239 F2 37
Chosen, see Korea, Asia
Chosen, Palm Beach, Fla., 1,858 F6 86
Chōshi, Jap., 71,500 (91,470▲) I10, n19 37
Chosica, Peru, 4,160 D2 58
Chos Malal, Arg B2 54
Choszczno, Pol., 2,052 B3 26
Chota, Peru, 2,705 C2 58
Choteau, Teton, Mont., 1,966 C4 102
Chotetov, Czech., 675 n18 26
Chouchiakou, China, 85,500 H6 36
Choudrant, Lincoln, La., 465 B3 95
Choushan, is., China I10 36
Chouteau, Mayes, Okla., 958 A6 112
Chouteau, co., Mont., 7,348 C6 102
Choutsun, China F7 36
Chowan, co., N.C., 11,729 A7 109
Chowan, riv., N.C A7 109
Chowchilla, Madera, Calif., 4,525 D3 82
Choyren, Mong B6 34
Chrisman, Edgar, Ill., 1,221 D6 90
Chrisney, Spencer, Ind., 380 H3 91
Christchurch, Eng., 26,498 D6 12
Christchurch, N.Z., 151,671 (★220,510) O14 51
Christian, co., Ill., 37,207 D4 90
Christian, co., Ky., 56,904 D2 94
Christian, co., Mo., 12,359 E2 101
Christian, isl., Ont., Can C4 72
Christian, sound, Alsk m22 79
Christiana, New Castle, Del., 500 A6 85
Christiana, Lancaster, Pa., 1,069 G6 114
Christiana, S. Afr., 5,848 C4 49
Christiana, Rutherford, Tenn., 250 B5 117
Christianburg, Sanpete, Utah D4 119
Christiansburg, Champaign, Ohio, 788 B3 111
Christiansburg, Montgomery, Va., 3,653 D2 121
Christiansfeld, Den., 819 (★1,562) C3 24
Christiansted, Vir. Is., 5,137 n14 64
Christie, mtn., N.W. Ter., Can D7 66
Christina, Fergus, Mont., 20 C7 102
Christina, lake, Alta., Can B5 69
Christina, lake, B.C., Can A7 122
Christina, lake, Minn D3 99
Christina, riv., Alta., Can A5 69
Christine, Richland, N. Dak., 125 C9 110
Christine, Atascosa, Tex., 276 E3 118
Christmas, Gila, Ariz., 190 E5 80
Christmas, Orange, Fla., 250 D5 86
Christmas, isl., Indian O G4 2
Christmas, isl., Oceania F12 7
Christmas, lake, Oreg D6 113
Christmas Island, Austl. dep., Oceania, 3,099 F12 7
Christopher, Franklin, Ill., 2,854 F4 90
Christopher, King, Wash D2 122
Christoval, Tom Green, Tex., 500 D2 118
Chromo, Archuleta, Colo., 10 D4 83
Chronister, Cherokee, Okla B7 112
Chrudim, Czech., 15,500 D3 26
Chrysler, Monroe, Ala., 250 D2 78
Chrzanow, Pol., 21,000 g10 26
Chu, China E6 34
Chu, China H8 36
Chu, Sov. Un., 14,800 E8 29
Chu, riv., Sov. Un G23 9
Chualar, Monterey, Calif., 450 C6 82
Chüan, China, 12,000 F7 34
Chüanchow (Chinchiang), China, 107,700 G8 34
Chuangho, China E10 36
Chubbuck, Bannock, Idaho, 1,590 G6 89
Chubut, prov., Arg., 142,195 C3 54
Chucheng, China G8 36
Chuchi, China, 30,000 J9 36
Chuchou, China, 127,300 K5 36
Chu Chua, B.C., Can., 50 D7 68
Chuckatuck, Nansemond, Va., 250 B6 121
Chuckey, Greene, Tenn., 250 C11 117
Chudovo, Sov. Un., 10,000 G8 27
Chuehshan, China H6 36
Chüfou, China, 5,000 *D8 34
Chugach, is., Alsk h16 79
Chugach, mts., Alsk g17 79
Chugiak, Alsk., 1,500 g17, C10 79
Chuguchak, China, 5,000 B1 34
Chuguyev, Sov. Un., 10,000 G11 27
Chugwater, Platte, Wyo., 287 D8 125
Chugwater, creek, Wyo D8 125
Chuho, China, 5,000 D3 37
Chuhsien, China, 32,000 J8 36
Chühsien (Chu), China, 11,000 G8 36
Chui Chuischu, Pinal, Ariz., 50 E4 80
Chuius, mtn., B.C., Can B5 68
Chukai, Mala., 10,803 J5 38
Chukudu Kraal, Bech B3 49
Chukut Kuk, Pima, Ariz., 60 F3 80
Chula, Tift, Ga., 300 E3 87
Chula, Livingston, Mo., 285 B3 101
Chula, Amelia, Va., 125 D5 121
Chulahoma, Marshall, Miss., 30 A4 100
Chula Vista, San Diego, Calif., 42,034 E2, F5 82
Chulu, China F6 36
Chulucanas, Peru, 14,874 B1 58
Chulumani, Bol., 2,362 C2 55
Chulym, Sov. Un., 17,200 B10 29
Chulym, riv., Sov.Un. D27 9, D11 28
Chumar, India A7 40
Chumatien, China, 50,000 E7 32, H6 36
Chumbicha, Arg E2 55
Chumikan, Sov. Un., 1,100 D16 28
Chumphon, Thai., 11,411 G3 38
Chumysh, riv., Sov. Un C10 29
Chunchon, Kor., 83,000 H3 37
Chunchula, Mobile, Ala., 300 E1 78
Chungan, China, 6,000 K7 36
Chungchou, China, 17,000 E6 34
Chunghsiang, China I5 36
Chungju, Kor., 47,600 (68,600▲) H3 37
Chungking (Chungching), China, 1,800,000 (2,121,000▲) G13 33, J2 36
Chungtien, China, 3,000 F4 34
Chungwei, China, 11,000 D6 34
Chunky, Newton, Miss., 224 C5 100
Chunya, Tan., 1,664 C5 48
Chupaca, Peru, 4,482 D2 58
Chupadera, mesa, N. Mex D3 107
Chupadero, Santa Fe, N. Mex F6 107
Chuquibamba, Peru, 2,480 E3 58
Chuquicamata, Chile, 29,000 D2 55
Chuquisaca, dept., Bol., 282,980 C2 55
Chur, Switz., 24,825 C8 19
Church, mtn., B.C., Can B11 68
Churchbridge, Sask., Can., 347 G5 70
Church Creek, Dorchester, Md., 146 C5 85
Church Hill, Queen Annes, Md., 263 B6 85
Church Hill, Hawkins, Tenn., 769 C11 117
Churchill, Man., Can., 1,878 C5 71
Churchill, Cassia, Idaho G5 89
Church Hill, Jefferson, Miss., 50 D2 100
Churchill, Allegheny, Pa., 3,428 *F2 114
Churchill, co., Nev., 8,452 D3 104
Churchill, cape, Man., Can C5 71
Churchill, lake, Sask., Can B1 70
Churchill, lake, Maine B3 96
Churchill, mtn., B.C., Can E6 68
Churchill, riv., Man., Sask., Can E11 66
Churchland, Norfolk, Va., 3,000 B6 121
Church Point, Acadia, La., 3,606 D3 95
Church Point, N.S., Can., 326 E3 74
Church Rock, McKinley, N. Mex., 300 B1 107
Churchs Ferry, Ramsey, N. Dak., 161 A6 110
Church Stretton, Eng., 2,712 B5 12
Churchton, Anne Arundel, Md., 600 C4 85
Churchville, Harford, Md., 100 A5 85
Churchville, Monroe, N.Y., 1,003 B3 108
Churchville, Bucks, Pa., 1,500 *F11 114
Churchville, Augusta, Va., 400 C3 21
Churdan, Greene, Iowa, 586 B3 92
Churu, India, 41,727 C5 40
Churubusco, Whitley, Ind., 1,284 B7 91
Churubusco, Clinton, N.Y., 200 A3 08
Churuguara, Ven., 4,446 A4 60
Churumuco, Mex., 1,131 n13 63
Chushan, China, 12,000 E7 34
Chuska, mtns., N. Mex A1 07
Chusovaya, riv., Sov. Un D20 9
Chusovoy, Sov. Un., 63,000 B5 29
Chust, Sov. Un., 15,000 G23 9
Chute, Shipshaw, Que., Can., 165 A6 73
Chybie, Pol h9 26
Ciales, P.R., 3,275 B5 65
Ciales, mun., P.R., 18,106 B4 65
Ciano d'Enza, It., 5,398 E6 18
Cibolo, creek, Tex B4 18
Cicekdag, Tur., 2,200 C10 31
Cicero, Cook, Ill., 69,130 F3 90
Cicero, Hamilton, Ind., 1,284 D5 91
Cícero Dantas, Braz., 2,972 D3 57
Cide, Tur., 2,100 B9 31
Cidra, P.R., 3,191 C6 65
Cidra, mun., P.R., 21,891 C6 65
Ciechanow, Pol., 20,000 B6 26
Ciego de Avila, Cuba, 35,178 E4 64
Ciempozuelos, Sp., 9,042 B4 20
Ciénaga, Col., 24,358 A3 60
Cienega, Otero, N. Mex E4 107
Ciénaga de Oro, Col., 6,108 B2 60
Cienfuegos, Cuba, 57,991 D3 64
Cieszyn, Pol., 23,000 D5 26
Cieza, Sp., 20,000 C5 20
Ciezkowice, Pol., 3,704 g10 26
Cihanbeyli, Tur., 3,035 C9 31
Cihuatlán, Mex., 3,571 n11 63
Cima, San Bernardino, Calif., 35 E6 82
Cimarron, Montrose, Colo., 5 C3 83
Cimarron, Gray, Kans., 1,115 E3 93
Cimarron, Colfax, N. Mex., 997 A5 107
Cimarron, co., Okla., 4,496 D2 112
Cimarron, riv., U.S C8 76
Cincinnati, Appanoose, Iowa, 583 D5 92
Cincinnati, Hamilton, Ohio, 502,550 (★1,203,300) C3, D2 111
Cincinnatus, Cortland, N.Y., 600 C5 108
Cinco Bayou, Okaloosa, Fla., 643 *G2 86
Cine, Tur., 6,500 D7 23
Cinto, mtn., Fr C2 21
Cipolletti, Arg., 2,763 B3 54
Cipres, Hidalgo, Tex F3 118
Circle, Alsk., 41 B11 79
Circle, McCone, Mont., 1,117 C11 102
Circle, cliffs, Utah F4 119
Circle Pines, Anoka, Minn., 2,789 *E5 99
Circle Springs, Alsk., 20 B11 79
Circleville, Jackson, Kans., 151 C8 93
Circleville, Pickaway, Ohio, 11,059 C5 111
Circleville, Piute, Utah, 478 E3 119
Circleville, Pendleton, W. Va., 150 C5 123
Cirencester, Eng., 11,836 C6 12
Cirque, mtn., Newf., Can f9 75
Cirrik, Alb., 2,533 E4 22
Cisco, Piatt, Ill., 398 C5 90
Cisco, Magoffin, Ky., 185 C6 94
Cisco, Eastland, Tex., 4,499 C3 118
Cisco, Grand, Utah, 40 E6 119
Cisne, Wayne, Ill., 615 E5 90
Cisneros, Col., 5,489 B2 60
Ciso, Murray, Ga., 150 B2 87
Cispus, riv., Wash C4 122
Cissna Park, Iroquois, Ill., 803 C6 90
Cistern, Fayette, Tex., 50 E4 118
Cisterna di Latina, It., 8,000 h9 21
Cistierna, Sp., 4,659 A3 20
Citra, Marion, Fla., 750 C4 86
Citronelle, Mobile, Ala., 1,918 D1 78
Citrus, co., Fla., 9,268 D4 86
Citrus Heights, Sacramento, Calif., 9,000 *C3 82
Citrus Park, Hillsborough, Fla. E2 86
Cittadella, It., 5,002 B3 21
Città di Castello, It., 11,300 C4 21
City, isl., N.Y B5 106
City Mills, Norfolk, Mass., 150 E2 97
City Park, Christian, Ill., 1,133 *D4 90
City Point, Brevard, Fla., 450 D6 86
City Terrace, Los Angeles, Calif., 6,000 *F4 82
City View, Greenville, S.C., 2,475 B2 115
Ciucul, mts., Rom B7 22
Ciudad Altamirano, Mex., 5,960 n13 63
Ciudad Bolívar, Ven., 56,032 B5 60
Ciudad Bolivia, Ven., 1,719 B3 60
Ciudad Camargo, Mex., 11,940 B3 63
Ciudad Camargo, Mex., 18,850 F3 118
Ciudad Chetumal, Mex., 12,775 D7 63
Ciudad de las Casas, Mex., 23,355 D6 63
Ciudad del Carmen, Mex., 20,901 D6 63
Ciudad del Maíz, Mex., 4,169 C5 63
Ciudad de Valles, Mex., 23,620 C5, m14 63
Ciudad Dr. Hernández Alvarez, Mex., 8,553 m13 63
Ciudadela, Sp., 9,804 B7 20
Ciudad García [Salinas], Mex., 14,948 C4 63
Ciudad Guzman, Mex., 30,971 D4, n12 63
Ciudad Hidalgo, Mex., 17,060 n13 63
Ciudad Juárez, Mex., 261,683 A3 63
Ciudad Lerdo, Mex., 17,537 B4 63
Ciudad Madero, Mex., 53,526 C5 63
Ciudad Mante, Mex., 22,701 C5 63
Ciudad Melchor, Múzquiz, Mex., 12,659 B4 63
Ciudad Obregón, Mex., 68,010 B3 63
Ciudad Real, Sp., 37,081 C4 20
Ciudad Real, prov., Sp., 583,948 *C4 20
Ciudad Rodrigo, Sp., 9,919 B2 20
Ciudad Serdán, Mex., 9,584 n15 63
Ciudad Victoria, Mex., 50,727 C5 63
Cividale del Friuli, It., 7,074 A4 21
Civitanova Marche, It., 12,300 C4 21
Civitavecchia, It., 38,138 C3 21
Civray, Fr., 2,867 D4 14
Civril, Tur., 4,241 C7 23
Cizre, Tur., 6,700 D14 31
C. J. Strike, res., Idaho G3 89
Clachan, Scot C1 13
Clackamas, Clackamas, Oreg., 750 B2 113
Clackamas, co., Oreg., 113,038 B4 113
Clackmannan, co., Scot., 41,391 D5 13
Clacton-on-Sea, Eng., 27,543 C9 12
Claflin, Barton, Kans., 891 D5 93
Claiborne, Monroe, Ala., 50 D2 78
Claiborne, Talbot, Md., 130 C5 85
Claiborne, co., Miss., 10,845 D3 100
Claiborne, co., Tenn., 19,067 C10 117
Claiborne, par., La., 19,407 B2 95
Clair, N.B., Can., 770 B1 74
Clair, Sask., Can., 128 E3 70
Claire, lake, Alta., Can D1 69
Claire, riv., Viet B6 38
Claire City, Roberts, S. Dak., 86 B8 116
Clairemont, Kent, Tex., 100 C2 118
Clairfield, Claiborne, Tenn., 500 C10 117
Clair Haven, Macomb, Mich., 1,365 *F8 98
Clairmont, Alta., Can., 292 B1 69
Clairmont Springs, Clay, Ala., 30 B4 78
Clairton, Allegheny, Pa., 18,389 F2 114
Clallam, co., Wash., 30,022 B2 122
Clallam Bay, Clallam, Wash., 300 A1 122
Clam, lake, Wis C1 124
Clamart, Fr., 47,991 g10 14
Clamecy, Fr., 5,520 D5 14
Clan Alpine, mts., Nev D4 104
Clancey, Jefferson, Mont., 300 D5 102
Clandeboye, Ont., Can., 116 D3 72
Clandonald, Alta., Can., 211 C5 69
Clanton, Chilton, Ala., 5,683 C3 78
Clanwilliam, Man., Can., 182 D2 71
Clanwilliam, S. Afr., 2,214 D2 49
Clapham, Union, N. Mex A6 107
Clapperton, isl., Ont., Can A2 72
Clara, Ire., 2,477 D4 11
Clara, Wayne, Miss., 500 D5 100
Clara Barton, Middlesex, N.J., 3,500 *B4 106
Clara City, Chippewa, Minn., 1,358 F3 99
Clare, Austl., 1,622 F2 51
Clare, Hamilton, Ind., 55 D6 91
Clare, Webster, Iowa, 245 B3 92
Clare, Clare, Mich., 2,442 E6 98
Clare, co., Ire., 73,702 E2 11
Clare, co., Mich., 11,647 E6 98
Clare, isl., Ire D1 11
Clare, riv., Ire D3 11
Clarecastle, Ire., 703 E3 11
Claregalway, Ire D3 11
Claremont, Los Angeles, Calif., 12,633 F3 82
Claremont, Dodge, Minn., 466 F6 99
Claremont, Sullivan, N.H., 13,563 D2 105
Claremont, Catawba, N.C., 728 B2 109
Claremont, Brown, S. Dak., 247 B7 116
Claremont, Surry, Va., 377 D6 121
Claremont Junction, Sullivan, N.H. (part of Claremont) D2 105
Claremore, Rogers, Okla., 6,639 A6 112
Claremorris, Ire., 1,519 D2 11
Clarence, Cedar, Iowa, 859 C6 92
Clarence, Natchitoches, La., 286 C2 95
Clarence, Shelby, Mo., 1,103 B5 101
Clarence, Erie, N.Y., 1,456 *C2 108
Clarence, Centre, Pa., 500 D6 114
Clarence, isl., Atl. O C7 5
Clarence, strait, Austl B5 50
Clarence Town, Ba. Is., 344 D6 64
Clarenceville, Que., Can., 362 D4 73
Clarendon, Monroe, Ark., 2,293 C4 81
Clarendon, Warren, Pa., 825 C3 114
Clarendon, Donley, Tex., 2,172 B2 118
Clarendon, Rutland, Vt., 75 (1,091▲) D3 120
Clarendon, co., S.C., 29,490 D7 115
Clarendon, riv., Vt E2 120
Clarendon Hills, Du Page, Ill., 5,885 F2 90
Clarendon Springs, Rutland, Vt. D2 120
Clarendon Station, Ont., Can., 75 C8 72
Clarenville, Newf., Can., 1,195 D4 75
Claresholm, Alta., Can., 2,143 D4 69
Clareton, Weston, Wyo., 5 B8 125
Clareville, Bee, Tex E4 118
Claridge, Westmoreland, Pa., 1,100 F2 114
Clarinda, Page, Iowa, 5,901 D2 92
Clarington, Monroe, Ohio, 394 C7 111
Clarington, Forest, Pa., 95 D3 114
Clarion, Wright, Iowa, 3,232 B4 92
Clarion, Clarion, Pa., 4,958 D3 114
Clarion, co., Pa., 37,408 D3 114
Clarion, riv., Pa D3 114
Clarissa, Todd, Minn., 569 D4 96
Clarita, Coal, Okla., 125 C5 112
Clark, Routt, Colo., 5 A4 83
Clark, Randolph, Mo., 260 B5 101
Clark, Union, N.J., 12,195 *B4 106
Clark, Coshocton and Holmes, Ohio, 190 B6 111
Clark, Mercer, Pa., 312 *D1 114
Clark, Clark, S. Dak., 1,484 C8 116
Clark, Park, Wyo., 10 A3 125
Clark, co., Ark., 20,950 C2 81
Clark, co., Idaho, 915 E6 89
Clark, co., Ill., 16,546 D6 90
Clark, co., Ind., 62,795 H6 91
Clark, co., Kans., 3,396 E4 93
Clark, co., Ky., 21,075 C5 94
Clark, co., Mo., 8,725 A6 101
Clark, co., Nev., 127,016 G6 104
Clark, co., Ohio, 131,440 C4 111
Clark, co., S. Dak., 7,134 C8 116
Clark, co., Wash., 93,809 D3 122
Clark, co., Wis., 31,527 D3 124
Clark, lake, Alsk C9 79
Clark, mtn., Va C5 121
Clarkdale, Yavapai, Ariz., 1,095 C5 80
Clarkdale, Cobb, Ga., 700 A4 87
Clarke, co., Ala., 25,738 D2 78
Clarke, co., Ga., 45,363 C3 87
Clarke, co., Iowa, 8,222 D4 92
Clarke, co., Miss., 16,493 C5 100
Clarke, co., Va., 7,942 B4 121
Clarke, lake, Sask., Can C2 70
Clarke City, Que., Can., 816 F19 77
Clarkedale, Crittenden, Ark., 50 E8 117
Clarkes Harbour, N.S., Can., 945 F4 74
Clarkesville, Habersham, Ga., 1,352 B3 87
Clarkfield, Yellow Medicine, Minn., 1,100 F3 99
Clark Fork, Bonner, Idaho, 452 A2 89
Clark Fork, riv., Idaho, Mont B3 89
Clark Hill, res., Ga C4 88
Clarkia, Shoshone, Idaho, 200 C2 89
Clark Mills, Oneida, N.Y., 1,148 *B5 108
Clarkrange, Fentress, Tenn., 150 C8 117
Clarks, Caldwell, La., 940 B3 95
Clarks, Merrick, Nebr., 439 C8 103
Clarks, riv., Ky A2 94
Clarksboro, Gloucester, N.J., 500 D2 106
Clarksburg, Yolo, Calif., 350 A6 82
Clarksburg, Ont., Can., 404 C4 72
Clarksburg, Decatur, Ind., 300 F7 91
Clarksburg, Montgomery, Md., 900 B3 85
Clarksburg, Berkshire, Mass., 120 (1,741▲) A1 97
Clarksburg, Moniteau, Mo., 357 C5 101
Clarksburg, Monmouth, N.J., 300 C4 106
Clarksburg, Ross, Ohio, 438 C4 111
Clarksburg, Carroll, Tenn., 250 B3 117
Clarksburg, Harrison, W. Va., 28,112 B4, B6 123
Clarks Corners, Windham, Conn., 40 B8 84
Clarksdale, Coahoma, Miss., 21,105 A3 100
Clarksdale, De Kalb, Mo., 242 B3 101
Clarks Falls, New London, Conn., 100 D9 84
Clarks Fork of the Yellowstone, riv., Wyo., Mont F7 100
Clarks Green, Lackawanna, Pa., 1,256 *D10 114
Clarks Grove, Freeborn, Minn., 353 G5 99
Clarks Hill, Tippecanoe, Ind., 654 D4 91
Clarks Hill, McCormick, S.C., 25 D3 115
Clarkson, Ont., Can., 1,400 D5 72
Clarkson, Grayson, Ky., 645 C3 94
Clarkson, Webster, Miss., 100 B4 100
Clarkson, Colfax, Nebr., 797 C8 103
Clarks Point, Alsk., 138 D8 79
Clarks Summit, Lackawanna, Pa., 3,693 A9 114
Clarkston, De Kalb, Ga., 1,524 B5 87
Clarkston, Oakland, Mich., 769 F7 98
Clarkston, Gallatin, Mont., 5 D5 102
Clarkston, Cache, Utah, 490 B3 119
Clarkston, Asotin, Wash., 6,209 C8 122
Clarksville, Johnson, Ark., 3,919 B2 81
Clarksville, N.S., Can., 75 D6 74
Clarksville, Sussex, Del., 200 C7 85
Clarksville, Kootenai, Idaho, 44 B2 89
Clarksville, Clark, Ind., 8,088 H6 91
Clarksville, Butler, Iowa, 1,328 B5 92
Clarksville, Howard, Md., 180 B4 85
Clarksville, Ionia, Mich., 371 F5 98
Clarksville, Pike, Mo., 638 B7 101
Clarksville, Clinton, Ohio, 583 C4 111
Clarksville, Montgomery, Tenn., 22,021 A4 117
Clarksville, Red River, Tex., 3,851 C5 118
Clarksville, Mecklenburg, Va., 1,530 E4 121
Clarkton, Dunklin, Mo., 1,049 E8 101
Clarkton, Bladen, N.C., 662 C5 109
Clarno, Wheeler, Oreg C6 113
Claryville, Campbell, Ky., 20 A7 94
Claryville, Perry, Mo., 30 D8 101
Clashmore, Ire., 175 E4 11
Clatonia, Gage, Nebr., 203 D9 103
Clatskanie, Columbia, Oreg., 797 A3 113
Clatsop, co., Oreg., 27,380 A3 113
Claude, Armstrong, Tex., 895 B2 118
Claudy, N. Ire., 286 C4 11
Claunch, Socorro, N. Mex., 80 C3 107
Clausthal-Zellerfeld, Ger., 15,200 B5 17
Claveria, Phil., 5,046 *B6 35
Clavet, Sask., Can., 60 F2 70
Clawson, Oakland, Mich., 14,795 A7 98
Clawson, Emery, Utah, 130 D4 119
Claxton, Evans, Ga., 2,672 D5 87
Clay, Webster, Ky., 1,343 C2 94
Clay, Onondaga, N.Y., 500 B2 102
Clay, Burleson, Tex., 250 D4 118
Clay, Clay, W. Va., 486 C3, C7 123
Clay, co., Ala., 12,400 B4 78
Clay, co., Ark., 21,258 A5 81
Clay, co., Fla., 19,535 B5 86
Clay, co., Ga., 4,551 E2 87
Clay, co., Ill., 15,815 E5 90
Clay, co., Ind., 24,207 F3 91
Clay, co., Iowa, 18,504 A2 92
Clay, co., Kans., 10,675 C6 93
Clay, co., Ky., 20,748 C6 94
Clay, co., Minn., 39,080 D2 99
Clay, co., Miss., 18,933 B5 100
Clay, co., Mo., 87,474 B3 101
Clay, co., Nebr., 8,717 D7 103
Clay, co., N.C., 5,526 D2 109
Clay, co., S. Dak., 10,810 E8 116
Clay, co., Tenn., 7,289 C8 117
Clay, co., Tex., 8,351 C3 118
Clay, co., W. Va., 11,942 C3 123
Clay Center, Clay, Kans., 4,613 C6 93
Clay Center, Clay, Nebr., 792 D7 103
Clay Center, Ottawa, Ohio, 446 A2, A4 111
Clay City, Clay, Ill., 1,144 E5 90
Clay City, Clay, Ind., 950 F3 91
Clay City, Powell, Ky., 764 C6 94
Claycomo, Clay, Mo., 1,423 *B3 101
Clay Cross, Eng., 9,173 A6 12
Claydon, Sask., Can., 105 H1 70
Clayhatchee, Dale, Ala., 300 D4 78
Claymont, New Castle, Del., 10,000 A7 85
Clayoquot, sound, B.C., Can E4 68
Claypool, Gila, Ariz., 1,800 D5 80
Claypool, Kosciusko, Ind., 452 B6 91
Claysburg, Blair, Pa., 1,439 F5 114
Clay Springs, Navajo, Ariz., 200 C1 80
Clay Spur, Weston, Wyo., 15 A8 125
Claysville, Washington, Pa., 986 F1 114
Clayton, Barbour, Ala., 1,313 D4 78
Clayton, Ont., Can., 84 B6 72
Clayton, Kent, Del., 1,028 B6 85
Clayton, Rabun, Ga., 1,507 B3 87
Clayton, Custer, Idaho, 125 E3 89
Clayton, Adams, Ill., 774 C3 90
Clayton, Hendricks, Ind., 653 E4 91
Clayton, Clayton, Iowa, 130 B6 92
Clayton, Norton and Decatur, Kans., 161 C3 93
Clayton, Concordia, La., 882 C4 95
Clayton, St. Louis, Mo., 15,245 B8, C7 101
Clayton, Gloucester, N.J., 4,711 D2 106
Clayton, Union, N. Mex., 3,314 A6 107
Clayton, Jefferson, N.Y., 1,996 A4, B1 108
Clayton, Johnston, N.C., 3,302 B5 109
Clayton, Pushmataha, Okla., 615 C6 112
Clayton, Obion, Tenn., 45 A2 117
Clayton, Stevens, Wash., 240 B8 122
Clayton, Polk, Wis., 324 C1 124
Clayton, co., Ga., 46,365 C2 87
Clayton, co., Iowa, 21,962 B6 92
Clayton Lake, Aroostook, Maine B3 96
Clayville, Oneida, N.Y., 686 C5 108
Clayville, Providence, R.I., 150 B10 84
Clear, cape, Alsk h18 79
Clear, cape, Ire F2 11
Clear, creek, Ariz C4 80
Clear, creek, Tenn C9 117
Clear, creek, Wyo A6 125
Clear, fork, Ohio B5 111
Clear, fork, Tenn C9 117

Clear, fork, W. Va...D3 123
Clear, fork, W. Va...D6 123
Clear, isl., Ire...F2 11
Clear, lake, Calif...C2 82
Clear, lake, Man., Can...D2 71
Clear, lake, Ont., Can...B7 72
Clear, lake, Que., Can...B4 73
Clear, lake, Iowa...A4 92
Clear, riv., Alta., Can...A1 69
Clear, stream, N.H...A4 105
Clear Boggy, creek, Okla...C5 112
Clearbrook, Clearwater, Minn., 650...C3 99
Clearco, Greenbrier, W. Va., 35...C4 123
Clear Creek, Monroe, Ind., 250...F4 91
Clear Creek, Raleigh, W. Va., 400...D6 123
Clear Creek, co., Colo., 2,793...B5 83
Clearfield, Taylor and Ringgold, Iowa, 504...D3 92
Clearfield, Rowan, Ky., 550...B6 94
Clearfield, Clearfield, Pa., 9,270...D5 114
Clearfield, Davis, Utah, 8,833...B3 119
Clearfield, co., Pa., 81,534...D5 114
Clear Fork, Wyoming, W. Va., 210...D3 123
Clear Lake, Cerro Gordo, Iowa, 6,158...A4 92
Clear Lake, Sherburne, Minn., 316...E5 99
Clear Lake, Beaver, Okla., 112...A1, D4 122
Clear Lake, Deuel, S. Dak., 1,137...C9 116
Clearlake, Skagit, Wash., 600...A3 122
Clear Lake, Polk, Wis., 724...C1 124
Clear Lake, res., Calif...B3 82
Clear Lake Shores, Tex., 700...F5 118
Clearmont, Nodaway, Mo., 292...A2 101
Clearmont, Sheridan, Wyo., 154...A6 125
Clear Spring, Washington, Md., 488...A2 85
Clearview, Okfuskee, Okla., 500...B5 112
Clearville, Bedford, Pa., 140...G5 114
Clearwater, Man., Can., 121...E2 71
Clearwater, Pinellas, Fla., 34,653...E1, E4 86
Clearwater, Idaho, Idaho, 40...C3 89
Clearwater, Sedgwick, Kans., 1,073...E6 93
Clearwater, Wright, Minn., 274...E4 99
Clearwater, Antelope, Nebr., 418...B7 103
Clearwater, Aiken, S.C., 1,450...E4 115
Clearwater, Jefferson, Wash., 70...B1 122
Clearwater, co., Idaho, 8,548...C3 89
Clearwater, co., Minn., 8,864...C3 99
Clearwater, creek, Kans...B4 93
Clearwater, lake, B.C., Can...C7 68
Clearwater, lake, Que., Can...A2 73
Clearwater, mts., Idaho...C2 89
Clearwater, res., Mo...D7 101
Clearwater, riv., Alta., Can...A5 69
Clearwater, riv., Alta., Can...D3 69
Clearwater, riv., B.C., Can...C7 68
Clearwater, riv., Idaho...C2 89
Clearwater, riv., Minn...C3 99
Clearwater Lake, Oneida, Wis., 200...C4 124
Cleaton, Muhlenberg, Ky., 700...C2 94
Cleator Moor, Eng., 6,411...F5 13
Clebit, McCurtain, Okla., 250...C7 112
Cleburne, Riley, Kans., 150...C7 93
Cleburne, Johnson, Tex., 15,381...B5, C4 118
Cleburne, co., Ala., 10,911...B4 78
Cleburne, co., Ark., 9,059...B3 81
Cle Elum, Kittitas, Wash., 1,814...B5 122
Cle Elum, res., Wash...B4 122
Cle Elum, riv., Wash...B4 122
Cleethorpes, Eng., 32,705...A7 12
Cleeves, Sask., Can., 50...D1 70
Clegg, Live Oak, Tex...E3 118
Cleggan, Ire...D1 11
Cleghorn, Cherokee, Iowa, 228...B2 92
Cleghorn, Eau Claire, Wis., 80...D2 124
Clem, Carroll, Ga., 180...C1 87
Clementon, Camden, N.J., 3,766...D3 106
Clements, Chase, Kans., 50...D7 93
Clements, St. Marys, Md., 150...D4 85
Clements, Redwood, Minn., 269...F3 99
Clementsport, N.S., Can., 519...E4 74
Clementsvale, N.S., Can., 266...E4 74
Clemmons, Forsyth, N.C., 700...A3 109
Clemons, Marshall, Iowa, 198...B4 92
Clemons, Washington, N.Y., 85...B7 108
Clemscot, Carter, Okla., 250...C4 112
Clemson, Pickens, S.C., 1,587...*B2 115
Clemson College, Oconee, S.C., 3,500...*B2 115
Clendenin, Kanawha, W. Va., 1,510...C3, C6 123
Clendening, res., Ohio...B6 111
Cleo, Kimble, Tex., 30...D3 118
Cleona, Lebanon, Pa., 1,988...F9 114
Cleo Springs, Major, Okla., 236...A3 112
Clermont, Austl., 1,737...D8 51
Clermont, Que., Can., 3,114...B7 73
Clermont, Lake, Fla., 3,313...D5 86
Clermont, Fr., 7,509...E2 15
Clermont, Hall, Ga., 268...B3 87
Clermont, Marion, Ind., 1,058...E5, H7 91
Clermont, Fayette, Iowa, 570...A6 92
Clermont, Bullitt, Ky., 100...C4 94
Clermont, Cape May, N.J., 250...E3 106
Clermont, McKean, Pa., 250...C5 114
Clermont, co., Ohio, 80,530...C3 111
Clermont, hill, Calif...C3 82
Clermont-en-Argonne, Fr., 864...E5 15
Clermont-Ferrand, Fr., 127,684...E5 14
Clermont-l'Hérault, Fr., 5,538...F5 14
Clervaux, Lux., 1,459...D6 15
Cleve, see Kleve, Ger.
Clevedon, Eng., 9,700...C5 12
Cleveland, Blount, Ala., 500...A3 78
Cleveland, Conway, Ark., 79...B3 81
Cleveland, Charlotte, Fla., 200...F5 86
Cleveland, White, Ga., 657...B3 87
Cleveland, Kingman, Kans...B3 93
Cleveland, Le Sueur, Minn., 389...F5 99
Cleveland, Bolivar, Miss., 10,172...B3 100
Cleveland, Cass, Mo., 216...C3 101
Cleveland, Blaine, Mont., 10...B7 102
Cleveland, Mora, N. Mex., 10...A4 107
Cleveland, Oswego, N.Y., 732...B5 108
Cleveland, Rowan, N.C., 594...B3 109
Cleveland, Stutsman, N. Dak., 169...C6 110
Cleveland, Cuyahoga, Ohio, 876,050 (★2,090,800)...A6, B2 111
Cleveland, Pawnee, Okla., 2,519...A5 112
Cleveland, Greenville, S.C., 250...A2 115
Cleveland, Bradley, Tenn., 16,196...D9 117
Cleveland, Liberty, Tex., 5,838...D5 118
Cleveland, Emery, Utah, 261...D5 119
Cleveland, Russell, Va., 415...B3 121
Cleveland, Maitowoc, Wis., 687...B6 124
Cleveland, co., Ark., 6,944...D3 81
Cleveland, co., N.C., 66,048...B2 109
Cleveland, co., Okla., 47,600...B4 112
Cleveland, hills, Eng...F7 13
Cleveland Heights, Cuyahoga, Ohio, 61,813...A6, B2 111
Clevelândia, Braz., 1,671...D2 56
Clever, Christian, Mo., 283...D4 101
Cleves, Hamilton, Ohio, 2,076...*C3 111
Clew, bay, Ire...D2 11
Clewiston, Hendry, Fla., 3,114...F6 86
Clibreck, mtn., Scot...B4 13
Clichy [-la-Garenne], Fr., 56,316...C5, g10 14
Clifden, Ire., 986...D1 11
Cliff, Grant, N. Mex., 175...E1 107
Cliff Dwellings, ruins, Ariz...A5 80
Clifford, Ont., Can., 542...D4 72
Clifford, Bartholomew, Ind., 241...F6 91
Clifford, Lapeer, Mich., 389...E7 98
Clifford, Traill, N. Dak., 109...B8 110
Clifford, Susquehanna, Pa., 250...C10 114
Clifford, Amherst, Va., 135...D3 121
Cliffside, B.C., Can...B9 68
Cliffside, Rutherford, N.C., 1,275...B2, D4 109
Cliffside Park, Bergen, N.J., 17,642...B5, D5 106
Clifftop, Fayette, W. Va., 300...D7 123
Cliffwood, Monmouth, N.J., 3,000...*C4 106
Cliffwood Beach, Middlesex and Monmouth, N.J., 3,000...*C4 106
Clifton, Greenlee, Ariz., 4,191...D6 80
Clifton, Mesa, Colo., 300...B2 83
Clifton, Franklin, Idaho, 150...G7 89
Clifton, Iroquois, Ill., 1,018...C6 90
Clifton, Washington and Clay, Kans., 746...C6 93
Clifton, Passaic, N.J., 82,084...B4, D5 106
Clifton, Spartanburg, S.C., 1,249...B4 115
Clifton, Wayne, Tenn., 708...B4 117
Clifton, Bosque, Tex., 2,335...D4 118
Clifton, Fairfax, Va., 230...B4, C5 121
Clifton, Weston, Wyo...B8 125
Clifton City, Cooper, Mo., 125...C4 101
Clifton Forge (Independent City), Va., 5,268...D3 121
Clifton Heights, Delaware, Pa., 8,005...B11 114
Clifton Hill, Randolph, Mo., 207...B5 101
Clifton Springs, Ontario, N.Y., 1,953...C3 107
Cliftonville, Noxubee, Miss., 250...B5 100
Clifty, Decatur, Ind., 197...F6 91
Clifty White and Cumberland, Tenn., 75...D8 117
Climax, Sask., Can., 425...H1 70
Climax, Lake, Colo., 1,609...B4 83
Climax, Decatur, Ga., 329...F2 87
Climax, Greenwood, Kans., 81...E7 93
Climax, Kalamazoo, Mich., 587...F5 98
Climax, Polk, Minn., 310...C2 99
Climax Springs, Camden, Mo., 93...C4 101
Climbing Hill, Woodbury, Iowa, 100...B1 92
Clinch, co., Ga., 6,545...F4 87
Clinch, mtn., Va...B2 121
Clinch, mts., Tenn...C10 117
Clinch, riv., Tenn...D9 117
Clinch, riv., Va...B2 121
Clinchburg, Washington, Va., 450...B3 121
Clincho, Dickenson, Va., 975...B2 121
Clinchfield, Houston, Ga., 250...D3 87
Clinchfield, McDowell, N.C., 950...*D4 109
Clinchmore, Campbell, Tenn., 250...C9 117
Clinchport, Scott, Va., 302...B2 121
Cline, Uvalde, Tex., 50...E2 118
Cline, falls, Oreg...C5 113
Clines Corners, Torrance, N. Mex...C4 107
Clingmans Dome, mtn., Tenn...D10 117
Clint, El Paso, Tex., 800...F1 118
Clinton, Greene, Ala., 150...C1 78
Clinton, Van Buren, Ark., 744...B3 81
Clinton, B.C., Can., 1,011...D7 68
Clinton, Ont., Can., 3,491...D3 72
Clinton, Middlesex, Conn., 4,166...D6 84
Clinton, Jones, Ga...C3 87
Clinton, DeWitt, Ill., 7,355...C5 90
Clinton, Vermillion, Ind., 5,843...E3 91
Clinton, Clinton, Iowa, 33,589...C7 92
Clinton, Hickman, Ky., 1,647...B2 94
Clinton, East Feliciana, La., 1,568...D4 94
Clinton, Kennebec, Maine, 800 (1,729▲)...D3 96
Clinton, Prince Georges, Md., 1,578...C4 85
Clinton, Worcester, Mass., 12,848...B4 97
Clinton, Lenawee, Mich., 1,481...F7 98
Clinton, Big Stone, Minn., 565...E2 98
Clinton, Hinds, Miss., 3,438...C3 100
Clinton, Henry, Mo., 6,925...C4 101
Clinton, Missoula, Mont., 160...D3 102
Clinton, Sheridan, Nebr., 46...B3 103
Clinton, Hunterdon, N.J., 1,158...B3 106
Clinton, Oneida, N.Y., 1,855...B5 108
Clinton, Sampson, N.C., 7,461...C5 109
Clinton, Summit, Ohio, 924...*B6 111
Clinton, Custer, Okla., 9,617...B3 112
Clinton, Laurens, S.C., 7,937...C4 115
Clinton, Anderson, Tenn., 4,943...C9, E11 117
Clinton, Davis, Utah, 1,025...B3 119
Clinton, Rock, Wis., 1,274...F5 124
Clinton, co., Ill., 24,029...E4 90
Clinton, co., Ind., 30,765...D4 91
Clinton, co., Iowa, 55,060...C7 92
Clinton, co., Ky., 8,886...D4 94
Clinton, co., Mich., 37,969...F6 98
Clinton, co., Mo., 11,588...B3 101
Clinton, co., N.Y., 72,722...A3 108
Clinton, co., Ohio, 30,004...C4 111
Clinton, co., Pa., 37,619...D6 114
Clinton, res., N.J...A4 106
Clinton Colden, lake, N.W. Ter., Can...D11 66
Clintonville, New Haven, Conn. (part of North Haven)...D5 84
Clintonville, Bourbon, Ky., 200...B5 94
Clintonville, Venango, Pa., 311...D2 114
Clintonville, Greenbrier, W. Va., 25...D4 123
Clintonville, Waupaca, Wis., 4,778...D5 124
Clintwood, Dickenson, Va., 1,400...B2 121
Clio, Barbour, Ala., 929...D4 78
Clio, Wayne, Iowa, 120...D4 92
Clio, Livingston, La...B6 95
Clio, Genesee, Mich., 2,212...E7 98
Clio, Marlboro, S.C., 847...B8 115
Clio, Roane, W. Va., 300...C3 123
Clipper, Whatcom, Wash., 65...A3 122
Clipperton, isl., Pac. O...F15 7
Clisham, mtn., Scot...C1 13
Clitherall, Ottertail, Minn., 138...D3 99
Clive, Alta., Can., 251...C4 69
Cliza, Bol., 3,121...C2 55
Cloan, Sask., Can...E1 70
Cloe, Jefferson, Pa., 500...E4 114
Cloghan, Ire...C4 11
Cloghan, Ire., 303...D4 11
Clogher, Ire., 75...E1 11
Clogheen, Ire., 576...E3 11
Clogher, head, Ire...D5 11
Clonakilty, Ire., 2,417...F3 11
Clonakilty, bay, Ire...F3 11
Cloncurry, Austl., 2,438...D7 50
Clonegall, Ire., 158...E5 11
Clones, Ire., 2,107...C4 11
Clonmany, Ire., 238...B4 11
Clonmel, Ire., 10,640...E4 11
Clonmellon, Ire., 271...D4 11
Clontarf, Swift, Minn., 139...E3 99
Cloone, Ire., 106...D4 11
Clo-oose, B.C., Can...A1 122
Cloppenburg, Ger., 15,200...F2 24
Clopton, Dale, Ala., 125...D4 78
Cloquet, Carlton, Minn., 9,013...D6 99
Cloquet, riv., Minn...C6 99
Closter, Bergen, N.J., 7,767...B5, D5 106
Clothier, Logan, W. Va., 500...D5 123
Cloud, co., Kans., 14,407...C6 93
Cloud, mtn., Newf., Can...C3 75
Cloud, peak, Wyo...A5 125
Cloud Chief, Washita, Okla., 75...B3 112
Cloudcroft, Otero, N. Mex., 464...E4 107
Cloud Lake, Palm Beach, Fla., 148...*F6 86
Cloudy, bay, N.Z...N15 51
Clover, York, S.C., 3,500...A5 115
Clover, Halifax, Va., 261...C4 121
Clover Botton, Jackson, Ky., 190...C5 94
Cloverdale, Lauderdale, Ala., 800...A2 78
Cloverdale, Sonoma, Calif., 2,848...C2 82
Cloverdale, B.C., Can., 569...B10 68
Cloverdale, Du Page, Ill., 40...F2 90
Cloverdale, Putnam, Ind., 741...E4 91
Cloverdale, Tillamook, Oreg., 200...B3 113
Cloverdale, Botetourt, Va., 500...D3 121
Cloverleaf, Harris, Tex., 3,000...*E5 118
Clover Lick, Pocahontas, W. Va., 200...C5 123
Cloverport, Breckinridge, Ky., 1,334...C3 94
Cloverton, Pine, Minn., 150...D6 99
Clovis, Fresno, Calif., 7,704...D4 82
Clovis, Curry, N. Mex., 23,713...C7 107
Cloyne, Ire., 612...F3 11
Cluanie, lake, Scot...C3 13
Cluj, Rom., 154,752...B6 22
Clune, Indiana, Pa., 400...E3 114
Cluny, Alta., Can., 174...D4 69
Cluny, Fr., 4,032...D6 14
Clusone, It., 6,832...D5 18
Cluster Springs, Halifax, Va., 150...E4 121
Clute, Brazoria, Tex., 4,501...G5 116
Clutier, Tama, Iowa, 292...B5 92
Clyattville, Lowndes, Ga., 150...F3 87
Clyde, Alta., Can., 259...B4 69
Clyde, N.W. Ter., Can...B19 67
Clyde, Cloud, Kans., 1,025...C6 93
Clyde, Nodaway, Mo., 90...A3 101
Clyde, Wayne, N.Y., 2,693...B4 108
Clyde, Haywood, N.C., 680...D3 109
Clyde, Cavalier, N. Dak., 100...A7 110
Clyde, Sandusky, Ohio, 4,826...A5 111
Clyde, Callahan, Tex., 1,116...C3 118
Clyde, firth, Scot...E4 13
Clyde, riv., N.S., Can...E4 74
Clyde, riv., Scot...E4 13
Clydebank, Scot., 49,654...E4 13
Clyde Hill, King, Wash., 1,871...*D2 122
Clyde Park, Park, Mont., 253...E6 102
Clymer, Chautauqua, N.Y., 500...C1 108
Clymer, Indiana, Pa., 2,251...E3 114
Clyo, Effingham, Ga., 250...D5 87
Cnocmoy, mtn., Scot...E3 13
Cnossus, ruins, Grc...E5 23
Coachella, Riverside, Calif., 4,854...F4 82
Coachella, canal, Calif...F12 82
Coachford, Ire., 275...F3 11
Coahoma, Coahoma, Miss., 300...A3 100
Coahoma, Howard, Tex., 1,239...C2 118
Coahoma, co., Miss., 46,212...A3 100
Coahuila, state, Mex., 896,509...B4 63
Coal, co., Okla., 5,546...C5 112
Coal, creek, Ind...D3 91
Coal, creek, Okla...B6 112
Coal, fork, W. Va...D6 123
Coal, lake, W. Va...C3 123
Coal, riv., W. Va...C3 123
Coal Bluff, Vigo, Ind., 100...E3 91
Coal Branch Station, N.B., Can., 142...C4 74
Coalburn, Scot...E5 13
Coal City, Grundy, Ill., 2,852...B5 90
Coal City, Owen, Ind., 235...F3 91
Coalcoman, Mex., 7,371...D4, h12 63
Coal Creek, B.C., Can., 130...E10 68
Coalcreek, Fremont, Colo., 206...C5 83
Coaldale, Alta., Can., 2,592...E4 69
Coaldale, Fremont, Colo., 15...C5 83
Coaldale, Schuylkill, Pa., 3,949...E10 114
Coalfield, Morgan, Tenn., 250...C9 117
Coal Fork, Kanawha, W. Va., 1,000...C3, C6 123
Coalgate, Coal, Okla., 1,689...C5 112
Coalgood, Harlan, Ky. 800...D6 94
Coal Grove, Lawrence, Ohio, 2,961...D5 111
Coal Harbour, B.C., Can., 137...D4 68
Coal Hill, Johnson, Ark., 704...B2 81
Coalhurst, Alta., Can., 190...E4 69
Coaling, Tuscaloosa, Ala., 200...B2 78
Coalinga, Fresno, Calif., 5,965...D3 82
Coalmont, B.C., Can., 66...E7 68
Coalmont, Jackson, Colo., 8...A4 83
Coalmont, Clay, Ind., 500...F3 91
Coalmont, Grundy, Tenn., 458...D8 117
Coalport, Clearfield, Pa., 821...E4 114
Coalridge, Sheridan, Mont., 36...B12 102
Coalridge, Noble, Ohio, 70...C6 111
Coalspur, Alta., Can., 75...C2 69
Coalton, Montgomery, Ill., 352...D4 90
Coalton, Jackson, Ohio, 648...B4 111
Coalton, Okmulgee, Okla., 200...B6 112
Coaltown, Lawrence, Pa., 1,033...*D1 114
Coal Valley, Walker, Ala., 653...B2 78
Coal Valley, Rock Island, Ill., 435...D7 92
Coalville, Eng., 26,159...B6 12
Coalville, Webster, Iowa, 300...B3 92
Coalville, Summit, Utah, 907...C4 119
Coalwood, Powder River, Mont., 20...E11 102
Coalwood, McDowell, W. Va., 1,199...*D3 123
Coamo, P.R., 12,146...C5 65
Coamo, mun., P.R., 26,082...C5 65
Coari, Braz., 5,908...D5 60
Coast, reg., Ken...B6 48
Coast, mts., B.C., Can...C4 68
Coast, ranges, U.S...B3 76
Coastal, plain, N.C., Va...C5 108, D6 121
Coatbridge, Scot., 53,946...E4 13
Coatepec, Mex., 18,081...n15 63
Coatepeque, Guat., 6,281...C2 62
Coatesville, Hendricks, Ind., 497...E4 91
Coatesville, Chester, Pa., 12,971...G10 114
Coaticook, Que., Can., 6,906...D6 73
Coatopa, Sumter, Ala., 20...C1 78
Coats, Pratt, Kans., 152...E5 93
Coats, Harnett, N.C., 1,049...B5 109
Coats, isl., N.W., Ter., Can...D16 67
Coats Land, reg., Ant...B10 5
Coatsville, Schuyler, Mo., 100...A5 101
Coatzacoalcos (Puerto México), Mex., 36,989...D6 63
Cobalt, Ont., Can., 2,209...E9 72
Cobalt, Middlesex, Conn., 200...C6 84
Cobalt, Lemhi, Idaho, 250...E4 89
Cobán, Guat., 7,911...C2 72
Cobar, Austl., 2,178...E5 51
Cobb, Sumter, Ga., 90...E3 87
Cobb, Caldwell, Ky., 75...D2 94
Cobb, Iowa, Wis., 387...F3 124
Cobb, co., Ga., 114,174...C2 87
Cobb, isl., Md...D4 85
Cobb, riv., Minn...G5 99
Cobble Hill, B.C., Can., 97...B9 68
Cobble Mountain, res., Mass...B2 97
Cobbosseecontee, lake, Maine...D3 96
Cobbtown, Tattnall, Ga., 280...D4 87
Cobbville, Telfair, Ga...E4 87
Cobden, Ont., Can., 942...B8 72
Cobden, Union, Ill., 918...E4 90
Cobequid, mts., N.S., Can...D5 74
Cobh, Ire., 5,266...F3 11
Cobham, riv., Man., Can...C4 71
Cobija, Bol., 1,726...B2 55
Cobija, Chile...D1 55
Coble, Hickman, Tenn., 100...B4 117
Cobleskill, Schoharie, N.Y., 3,471...C6 108
Coboconk, Ont., Can., 506...C6 72
Cobourg, Ont., Can., 10,646...D6 72
Cobourg, pen., Austl...B5 50
Cobre, Elko, Nev., 25...B7 98
Cobun, creek, W. Va...A7 123
Coburg, Ger., 44,200...C5 17
Coburg, Montgomery, Iowa, 54...D2 92
Coburg, Blaine, Mont., 7...B8 102
Coburg, Lane, Oreg., 754...C3 113
Coburn, Centre, Pa., 280...E7 114
Coburn, Wetzel, W. Va., 125...A6, B4 123
Coburn, mtn., Maine...C2 96
Coburn Gore, Franklin, Maine, 80...C2 96
Coca, Ec...B2 58
Cocanada, see Kakinada, India
Cocentaina, Sp., 8,108...C5 20
Cochabamba, Bol., 80,795...C2 55
Cochabamba, dept., Bol., 490,475...C2 55
Coche, isl., Ven...A5 60
Cocheco, riv., N.H...D4 105
Cochem, Ger., 6,900...D7 15
Cochetopa, creek, Colo...C4 83
Cochetopa, hills, Colo...C4 83
Cochin, India, 35,076 (★300,000)...G6 39
Cochin China, reg., Viet., 6,500,000...G7 38
Cochise, Cochise, Ariz., 90...E6 80
Cochise, co., Ariz., 55,039...F5 80
Cochise Head, mtn., Ariz...E6 80
Cochite Pueblo, Indian res., N. Mex...F6 107
Cochiti, Sandoval, N. Mex...F5 107
Cochituate, Middlesex, Mass., 4,000...D1 97
Cochituate, lake, Mass...D1 97
Cochran, Bleckley, Ga., 4,714...D3 87
Cochran, co., Tex., 6,417...C1 118
Cochrane, Ont., Can., 4,521...E9 73
Cochrane, Alta., Can., 857...D3 69
Cochrane, Buffalo, Wis., 455...D2 124
Cochrane, co., Ont., Can., 95,666...E9 73
Cochrane, mtn., Arg...D2 54
Cochranton, Crawford, Pa., 1,139...C1 114
Cochranville, Chester, Pa., 350...G10 114
Cockburn, Austl., 101...F3 51
Cockburn, isl., Ont., Can...C7 98
Cocke, co., Tenn., 23,390...D10 117
Cockermouth, Eng., 5,823...F5 13
Cockeysville, Baltimore, Md., 2,582...B4 85
Cockrell Hill, Dallas, Tex., 3,104...B5 118
Cockroach, isl., Vir. Is...f14 65
Cockrum, De Soto, Miss., 150...A4 100
Coco, riv., Nic...C6 62
Cocoa, Brevard, Fla., 12,294...D6 86
Cocoa Beach, Brevard, Fla., 3,475...D6 86
Cocoa West, Brevard, Fla., 3,975...*D6 86
Cocobeach, Gabon...E2 46
Cocolalla, Bonner, Idaho, 20...A2 89
Coconino, co., Ariz., 41,857...B3 80
Coconino, plat., Ariz...B3 80
Cocos, isl., Guam...52
Cocos (Keeling) Is., Aust. dep., Oceania, 606...G3 2
Cocula, Mex., 10,148...C4, m12 63
Cocuy, Col., 2,973...B3 60
Cod, cape, Mass...B7 97
Cod, isl., Newf., Can...g9 75
Codajás, Braz., 1,505...D5 60
Codajás, lake, Braz...D5 60
Codell, Rooks, Kans., 100...C4 93
Coden, Mobile, Ala., 950...E1 78
Coderre, Sask., Can., 229...G2 70
Codesa, Alta., Can., 65...B1 69
Codette, Sask., Can., 186...D3 70
Codigoro, It., 6,631...E8 18
Codington, co., S. Dak., 20,220...C8 116
Codó, Braz., 11,089...B2 57
Codogno, It., 10,500...B2 21
Codpa, Chile...C2 55
Codroipo, It., 4,305...D8 15
Codroy, Newf., Can., 258...E2 78
Codroy Pond, Newf., Can., 51...D2 75
Codrul, mts., Rom...B6 22
Cody, Cherry, Nebr., 230...B4 103
Cody, Park, Wyo., 4,838...A3 125
Codys, N.B., Can., 55...D4 76
Coeburn, Wise, Va., 2,471...B2 121
Coe Hill, Ont., Can., 239...C7 72
Coen, Austl., 77...B7 50
Coesfeld, Ger., 20,300...B2 17
Coesse, Whitley, Ind., 225...B7 91
Coeur d'Alene, Kootenai, Idaho, 14,291...B2 89
Coeur d'Alene, lake, Idaho...B2 89
Coeur d'Alene, mtn., Idaho...D9 122
Coeur d'Alene, mts., Idaho...B2 89
Coevorden, Neth., 5,800...B6 15
Coffee, Bacon, Ga., 50...E4 87
Coffee, co., Ala., 30,583...D3 78
Coffee, co., Tenn., 28,603...B5 117
Coffee Creek, Fergus, Mont., 100...C6 102
Coffeen, Montgomery, Ill., 502...D4 90
Coffee Springs, Geneva, Ala., 205...D4 78
Coffeeville, Clarke, Ala., 250...D1 78
Coffeeville, Yalobusha, Miss., 813...B4 100
Coffey, Daviess, Mo., 190...A3 110
Coffey, co., Kans., 8,403...D8 93
Coffeyville, Montgomery, Kans., 17,382...E8 93
Coffin, isl., Que., Can...B8 74
Coff's Harbour, Austl., 7,185...E9 51
Cofield, Hertford, N.C., 350...A7 109
Cogdell, Clinch, Ga., 210...E4 87
Coggon, Linn, Iowa, 672...B6 92
Coghinas, riv., It...D2 21
Coglar, buttes, Oreg...E6 113
Cognac, Fr., 20,798...E3 14
Cogswell, Sargent, N. Dak., 305...C8 110
Cohagen, Garfield, Mont., 42...C10 102
Cohansey, creek, N.J...E2 106
Cohasset, Norfolk, Mass., 2,748 (5,840▲)...B6, D4 97
Cohasset, Itasca, Minn., 605...C5 99
Cohocton, Steuben, N.Y., 929...C3 108
Cohocton, riv., N.Y...C3 108
Cohoes, Albany, N.Y., 20,129...C7 108
Cohutta, Whitfield, Ga., 325...B2 87
Cohutta, mtn., Ga...B2 87
Coila, Carroll, Miss., 200...B4 100
Coimbatore, India, 286,305 (★410,000)...F6 39
Coimbra, Port., 45,500...B1 20
Coin, Page, Iowa, 346...D2 92
Coín, Sp., 11,828...D3 20
Coinjock, Currituck, N.C., 150...A8 109
Coipasa, lake, Bol...C2 55
Coipasa, salt flat, Bol...C2 55
Coire, riv., Scot...B4 13
Cojedes, state, Ven., 72,652...B4 60
Cojutepeque, Sal., 11,617...D3 62
Cokato, Wright, Minn., 1,356...E4 99
Coke, co., Tex., 3,589...D2 118
Cokeburg, Washington, Pa., 989...*F1 114
Cokedale, Las Animas, Colo., 219...D6 83
Coker, Tuscaloosa, Ala., 400...B2 78
Cokesbury, Hunterdon, N.J., 100...B3 106
Cokesbury, Greenwood, S.C...C3 115
Coketon, Brooke, W. Va., 140...B2 123
Coketon, Tucker, W. Va., 130...B5 123
Cokeville, Lincoln, Wyo., 545...C2 125
Colac, Austl., 9,257...I4 51
Colares, Port., 4,976...f9 20
Colatina, Braz., 26,757...E2 57
Colbeck, cape, Ant...B33 5
Colbert, Madison, Ga., 425...B3 87
Colbert, Bryan, Okla., 671...D5 112
Colbert, co., Ala., 46,506...A2 78
Colborne, Ont., Can., 1,336...C7 72
Colbún, Chile...B2 54
Colburn, Bonner, Idaho, 50...A2 89
Colburn, Tippecanoe, Ind., 280...C4 91
Colby, Thomas, Kans., 4,210...C2 93
Colby, Aroostook, Maine, 200...B4 96
Colby, Marathon and Clark, Wis., 1,085...D3 124
Colbyville, Washington, Vt., 120...C3 120
Colchagua, prov., Chile, 158,024...A2 54
Colchester, New London, Conn., 2,260 (4,648▲)...C7 84
Colchester, Eng., 65,072...C8 12
Colchester, McDonough, Ill., 1,495...C3 90
Colchester, Chittenden, Vt., 225 (4,718▲)...B2 120
Colchester, co., N.S., Can., 34,307...D6 74
Colcord, Delaware, Okla., 173...A7 112
Cold, lake, Alta., Sask., Can...B5 69, B1 70
Cold, riv., N.H...C4 105
Cold, riv., N.H...D2 105
Coldbrook, N.B., Can., 507...D4 74
Cold Fell, mtn., Eng...F6 13
Cold Hollow, mts., Vt...B3 120
Colditz, Ger., 6,971...B7 17
Cold Lake, Alta., Can., 1,307...B5 69
Cold River, Cheshire, N.H., 50...D2 105
Cold Spring, Campbell, Ky., 1,095...A7, B5 94
Cold Spring, Stearns, Minn., 1,760...E4 99
Cold Spring, Cape May, N.J., 350...F3 106
Cold Spring, Putnam, N.Y., 2,083...*D7 108
Coldspring, San Jacinto, Tex., 500...D5 118
Cold Spring, pond, Newf., Can...D3 75
Cold Spring Harbor, Cape May, N.J...F3 106
Cold Spring Harbor, Suffolk, N.Y., 1,705...F3 84
Cold Springs, Kiowa, Okla., 75...C3 112
Coldstream, Scot., 1,227...E6 13
Cold Stream, pond, Maine...C4 96
Coldwater, Ont., Can., 726...C5 72
Coldwater, Comanche, Kans., 1,164...E4 93
Coldwater, Branch, Mich., 8,880...G5 98
Coldwater, Tate, Miss., 1,264...A4 100
Coldwater, Mercer, Ohio, 2,766...B3 111
Coldwater, creek, Okla...D3 112
Coldwater, creek, Tex...A1 118
Coldwater, riv., Miss...A3 100
Cole, McClain, Okla., 100...B4 112
Cole, co., Mo., 40,761...C5 101
Colebrook, Litchfield, Conn., 125 (791▲)...B4 84
Colebrook, Coos, N.H., 1,550 (2,389▲)...B1 105
Colebrook River, Litchfield, Conn...B4 84
Cole Camp, Benton, Mo., 853...C4 101
Coleharbor, McLean, N. Dak., 210...B4 110
Cole Lake, Ont., Can...C8 72
Coleman, Alta., Can., 1,713...E3 69
Coleman, P.E.I., Can., 160...C5 74
Coleman, Sumter, Fla., 921...D4 86
Coleman, Randolph, Ga., 220...E2 87
Coleman, Hancock, Ga...C4 87
Coleman, Kent, Md., 275...B5 85
Coleman, Midland, Mich., 1,264...E6 98
Coleman, Johnston, Okla., 150...C5 112

Coleman, Coleman, Tex., 6,371....D3 118
Coleman, Marinette, Wis., 718....C5 124
Coleman, co., Tex., 12,458....D3 118
Coleman, riv., Austl....B7 50
Colemans Falls, Bedford, Va., 180....D3 121
Colerain, Bertie, N.C., 340....A7 109
Coleraine, Austl., 1,503....H3 51
Coleraine, N. Ire., 11,912....B5 11
Coleraine, Itasca, Minn., 1,346....C5 99
Coleraine Station, Que., Can., 1,043....D6 73
Coleridge, Alta., Can., 160....E5 69
Coleridge, Cedar, Nebr., 604....B8 103
Coleridge, Randolph, N.C., 500....B4 109
Coles, Amite, Miss., 150....D2 100
Coles, co., Ill., 42,860....D5 90
Colesberg, S. Afr., 4,855....D4 49
Colesburg, Camden, Ga., 75....F5 87
Colesburg, Delaware, Iowa, 365....B6 92
Colesburg, Hardin, Ky., 150....C4 94
Colestin, Jackson, Oreg....E4 113
Colesville, Sussex, N.J., 100....A3 106
Coleta, Whiteside, Ill., 197....B4 90
Coleville, Sask., Can., 503....F1 70
Coleville, Mono, Calif., 275....C4 82
Coleyville, Cottle, Tex., 100....B2 118
Colfax, Placer, Calif., 915....C3 82
Colfax, Sask., Can., 69....H3 70
Colfax, McLean, Ill., 894....C5 90
Colfax, Clinton, Ind., 725....D4 91
Colfax, Jasper, Iowa, 2,331....C4 92
Colfax, Grant, La., 1,934....C3 95
Colfax, Richland, N. Dak., 98....C9 110
Colfax, Whitman, Wash., 2,860....C8 122
Colfax, Dunn, Wis., 885....D2 124
Colfax, co., Nebr., 9,595....C8 103
Colfax, co., N. Mex., 13,806....A5 107
Colgan, Divide, N. Dak., 22....A2 110
Colgate, Sask., Can., 101....H4 70
Colgate, Steele, N. Dak., 60....B8 110
Colgate, Washington, Wis., 60....E1 124
Colico, It., 4,475....C5 18
Colijnsplaat, Neth., 1,686....C5 15
Colima, Mex., 44,860....D4, n12 63
Colima, state, Mex., 157,338....D4, n12 63
Colinas, Braz., 2,972....C2 57
Colinet, Newf., Can., 261....E5 75
Colington, Dare, N.C., 140....A8 109
Colinton, Alta., Can., 114....B4 69
Coll, isl., Scot....D2 13
Collbran, Mesa, Colo., 310....B3 83
College, Alsk., 1,755....C10 79
College Corner, Preble and Butler, Ohio, 439....C3 111
Collegedale, Hamilton, Tenn., 1,500....E10 117
College Gardens, Stanislaus, Calif., 4,132....*D3 82
College Grove, Williamson, Tenn., 200....B5 117
College Heights, Drew, Ark., 1,000....D4 81
College Heights, Darlington, S.C., 1,330....C7 115
College Park, St. Johns, Fla., 950....C5 86
College Park, Fulton and Clayton, Ga., 23,469....B5, C2 87
College Park, Prince Georges, Md., 18,482....C2, C4 85
College Place, Richland, S.C....C5 115
College Place, Walla Walla, Wash., 4,031....C7 122
College Springs, Page, Iowa, 290....D2 92
College Station, Brazos, Tex., 11,396....D4 118
College View, Arapahoe, Colo., 1,800....*B6 83
Collegeville, Saline, Ark., 30....D5 81
Collegeville, Jasper, Ind., 1,400....C3 91
Collegeville (P.O.), Stearns, Minn., 1,600....*E4 99
Collegeville, Montgomery, Pa., 2,254....F11 114
Colleton, co., S.C., 27,816....F6 115
Colley, Sullivan, Pa., 75....C9 114
Colleyville, Tarrant, Tex., 1,491....*B5 118
Collie, Austl., 7,547....F2 50
Collier, co., Fla., 15,753....F5 86
Collier, bay, Austl....C3 50
Colliers, Brooke, W. Va., 900....A2 123
Collierville, Shelby, Tenn., 2,020....B2 117
Collin, co., Tex., 41,247....C4 118
Collingdale, Delaware, Pa., 10,268....B11 114
Collingswood, Camden, N.J., 17,370....D2 106
Collingsworth, co., Tex., 6,276....B2 118
Collingwood, Ont., Can., 8,385....C4 72
Collins, Drew, Ark., 107....D4 81
Collins, Ont., Can., 80....E8 72
Collins, Tattnall, Ga., 565....D4 87
Collins, Whitley, Ind., 95....B7 91
Collins, Story, Iowa, 435....C4 92
Collins, Covington, Miss., 1,537....D4 100
Collins, St. Clair, Mo., 177....D4 101
Collins, Teton, Mont., 15....C5 102
Collins, Manitowoc, Wis., 190....B6 124
Collins Park, New Castle, Del., 2,500....*A6 85
Collins Park, Catron, N. Mex., 50....D1 107
Collinston, Morehouse, La., 497....B4 95
Collins View, Multnomah, Oreg., 1,500....*B4 113
Collinsville, DeKalb, Ala., 1,199....A4 78
Collinsville, Hartford, Conn., 1,682....B5 84
Collinsville, DeKalb, Ga., 120....B5 87
Collinsville, Madison and St. Clair, Ill., 14,217....E4 90
Collinsville, Middlesex, Mass. (part of Dracut)....A5 97
Collinsville, Lauderdale, Miss., 400....C5 100
Collinsville, Tulsa, Okla., 2,526....A6 112
Collinsville, Henry, Va., 3,586....*E6 121
Collinwood, Wayne, Tenn., 596....B4 117
Collipulli, Chile, 4,057....B2 54
Collis, Traverse, Minn., 30....E2 99
Collister, Ada, Idaho, 5,436....F2 89
Collo, Alg., 4,518 (6,960▲)....B6 44
Collooney, Ire., 553....C3 11
Collyer, Trego, Kans., 253....C3 93
Colman, Moody, S. Dak., 505....D9 116
Colmar, Fr., 47,305....C7 14, A3 18
Colmar, Bell, Ky., 500....D6 94
Colmar Manor, Prince Georges, Md., 1,772....C2 85
Colmenar de Oreja, Sp., 5,547....B4 20
Colmenar Viejo, Sp., 8,375....B4, o17 20
Colmesneil, Tyler, Tex., 500....D5 118
Colmor, Colfax, N. Mex., 35....A5 107
Colo, Story, Iowa, 574....B4 92
Cologne, see Köln, Ger.
Cologne, Carver, Minn., 454....F5 99
Cologne, Atlantic, N.J., 600....D3 106
Coloma, Berrien, Mich., 1,473....F4 98
Coloma, Waushara, Wis., 312....D4 124
Colomb-Béchar, Alg., 18,090 (43,250▲)....C4 44
Colombes, Fr., 76,918....g10 14
Colombey-les-Belles, Fr., 685....F5 15
Colombia, Col., 1,217....C3 60
Colombia, country, S.A., 12,939,140....C3 53, C3 60
Colombier, Switz., 2,652....C2 19
Colombo, Cey., 424,816 (★725,000)....G6 39
Colome, Tripp, S. Dak., 398....D6 116
Colón, Arg., 8,335....A5 54
Colón, Cuba, 15,755....D3 64
Colon, St. Joseph, Mich., 1,055....G5 98
Colon, Saunders, Nebr., 110....D2 103
Colon, Lee, N.C., 350....B4 109
Colón, Pan., 59,598....F8, k11 62
Colón, Archipielago de (Galápagos Is.), prov., Ec., 1,546....g5 58
Colon, lake, Man., Can....B4 71
Colona, Ouray, Colo., 75....C3 83
Colonia, Middlesex, N.J., 6,000....*B4 100
Colonia, Ur., 7,500....E1 56
Colonia, dept., Ur., 95,400....*E1 56
Colonia Dora, Arg., 2,183....E3 55
Colonia Gustavo A. Madero, Mex., 60,239....h9 63
Colonia Las Heras, Arg....D3 54
Colonial Beach, Westmoreland, Va., 1,769....C6 121
Colônia Leopoldina, Braz., 1,694....k6 57
Colonial Heights, Sullivan, Tenn., 2,312....*C11 118
Colonial Heights (Independent City), Va., 9,587....C7, D5 121
Colonial Manor, Gloucester, N.J., 1,300....*D2 106
Colonial Park, Dauphin, Pa., 2,500....*F8 114
Colonia Mennonita, Par., 2,247....D4 55
Colonia Paraíso, P.R....B8, f12 65
Colonia Providencia, P.R....D6 65
Colonias, Guadalupe, N. Mex....B5 107
Colonia Sarmiento, Arg., 3,648....D3 54
Colonia Suiza, Ur., 4,307....g8 54
Colonie, Albany, N.Y., 6,992....*C7 108
Colonna, It., 1,695....h9 21
Colonne, cape, It....E6 21
Colonsay, Sask., Can., 278....F3 70
Colonsay, isl., Scot....D2 13
Colony, Anderson, Kans., 419....D8 93
Colony, Washita, Okla., 200....B3 112
Colony, Crook, Wyo....A8 125
Colorado, C.R....E6 62
Colorado, state, U.S., 1,753,947....C6 76, 83
Colorado, co., Tex., 18,463....E4 118
Colorado, desert, Calif....F6 82
Colorado, nat. mon., Colo....B2 83
Colorado, plat., Ariz....A3 80
Colorado, riv., Arg....B3 54
Colorado, riv., Tex....D3 118
Colorado, riv., U.S....C5 76
Colorado City, Mitchell, Tex., 6,457....C2 118
Colorado River, Indian res., Ariz....D1 80
Colorados, arch., Cuba....D1 64
Colorado Springs, El Paso, Colo., 70,194 (★139,500)....C6 83
Colored Hill, Mercer, W. Va., 1,115....*D3 123
Coloso, Chile....D1 55
Coloso, P.R....B2 65
Coloso Central, P.R....B2 65
Colotepec, Mex., 455....D5 63
Colotlán, Mex., 6,281....C4, k12 63
Colquechaca, Bol., 1,070....C2 55
Colquitt, Miller, Ga., 1,556....E2 87
Colquitt, Claiborne, La., 40....B3 95
Colquitt, co., Ga., 34,048....E3 87
Colrain, Franklin, Mass., 180 (1,426▲)....A2 97
Colstrip, Rosebud, Mont., 200....E10 102
Colt, St. Francis, Ark., 394....B5 81
Coltauco, Chile....A2 54
Colton, San Bernardino, Calif., 18,666....F3 82
Colton, St. Lawrence, N.Y....B2 108
Colton, Clackamas, Oreg., 150....B4 113
Colton, Minnehaha, S. Dak., 593....D9 116
Colton, Utah, Utah, 12....D5 119
Colton, Whitman, Wash., 253....C8 122
Colton, hill, Vt....D4 120
Coltons Point, St. Marys, Md., 118....D4 85
Colts Neck, Monmouth, N.J., 400....C4 106
Colulli, Eth....C5 47
Columbia, Houston, Ala., 783....D4 78
Columbia, Tolland, Conn., 200 (2,163▲)....C7 84
Columbia, Monroe, Ill., 3,174....E3 90
Columbia, Adair, Ky., 2,255....C4 94
Columbia, Caldwell, La., 1,021....B3 95
Columbia, Marion, Miss., 7,117....D4 100
Columbia, Boone, Mo., 36,650....C5 101
Columbia, Warren, N.J., 400....B2 106
Columbia, Tyrrell, N.C., 1,099....B7 108
Columbia, Lancaster, Pa., 12,075....F9 114
Columbia, Richland, S.C., 97,433 (★213,400)....D5 115
Columbia, Brown, S. Dak., 272....B7 116
Columbia, Maury, Tenn., 17,624....B4 117
Columbia, Carbon, Utah, 300....D5 119
Columbia, Fluvanna, Va., 86....D4 121
Columbia, co., Ark., 26,400....D2 81
Columbia, co., Fla., 20,077....B4 86
Columbia, co., Ga., 13,423....C4 87
Columbia, co., N.Y., 47,322....C7 108
Columbia, co., Oreg., 22,379....B3 113
Columbia, co., Pa., 53,489....D9 114
Columbia, co., Wash., 4,569....C7 122
Columbia, co., Wis., 36,708....E4 124
Columbia, basin, Wash....C6 122
Columbia, cape, N.W. Ter., Can....A22 4
Columbia, dam, Ala., Ga....D4 78, D1 87
Columbia, lake, B.C., Can....D10 68
Columbia, mtn., Alta., Can....C2 69
Columbia, mts., Mex....B2 63
Columbia, res., Conn....C7 84
Columbia, riv., Can., U.S....G9 66, A3 76
Columbia Bridge, Coos, N.H....B1 105
Columbia City, Whitley, Ind., 4,803....B7 91
Columbia City, Columbia, Oreg., 423....B4 113
Columbia Cross Roads, Bradford, Pa., 200....C8 114
Columbia Falls, Washington, Maine, 300 (442▲)....D5 96
Columbia Falls, Flathead, Mont., 2,132....B2 102
Columbia Heights, Anoka, Minn., 17,533....E7 99
Columbia Heights, Cowlitz, Wash., 2,227....*C3 122
Columbiana, Shelby, Ala., 2,264....B3 78
Columbiana, Columbiana, Ohio, 4,164....B7 111
Columbiana, co., Ohio, 107,004....B7 111
Columbia Park (Pittsburg East), Contra Costa, Calif., 1,977....*B5 82
Columbia Park, Hamilton, Ohio....A6 111
Columbia Station, Lorain, Ohio, 1,000....*A5 111
Columbiaville, Lapeer, Mich., 878....E7 98
Columbine, Routt, Colo., 4....A4 83
Columbine, Natrona, Wyo., 20....B6 125
Columbus, Hempstead, Ark., 275....D2 81
Columbus, Muscogee, Ga., 116,779 (★201,500)....D2 87
Columbus, Bartholomew, Ind., 20,778....F6 91
Columbus, Cherokee, Kans., 3,395....E9 93
Columbus, Hickman, Ky., 357....B2 94
Columbus, Lowndes, Miss., 24,771....B5 100
Columbus, Stillwater, Mont., 1,281....E7 102
Columbus, Platte, Nebr., 12,476....C8 103
Columbus, Burlington, N.J., 600....C3 106
Columbus, Luna, N. Mex., 307....F2 107
Columbus, Polk, N.C., 725....D4 109
Columbus, Burke, N. Dak., 672....A3 110
Columbus, Franklin, Ohio, 471,316 (★715,400)....C2, C5 111
Columbus, Warren, Pa., 500....C2 114
Columbus, Colorado, Tex., 3,656....E4 118
Columbus, Columbus, Wis., 3,467....E4 124
Columbus, co., N.C., 48,973....C5 109
Columbus City, Louisa, Iowa, 327....C6 92
Columbus Grove, Putnam, Ohio, 2,104....B3 111
Columbus Junction, Louisa, Iowa, 1,016....C6 92
Columbus Manor, Cook, Ill., 2,500....*B6 90
Colusa, Colusa, Calif., 3,518....C2 82
Colusa, co., Calif., 12,075....C2 82
Colver, Cambria, Pa., 1,261....E4 114
Colville, Stevens, Wash., 3,806....A8 122
Colville, lake, N.W. Ter., Can....C7 66
Colville, lake, Wash....B7 122
Colville, riv., Alsk....B9 79
Colville, riv., Wash....B8 122
Colvin, mtn., Ala....B4 78
Colvos, pass, Wash....D1 122
Colwich, Sedgwick, Kans., 703....B5, E6 93
Colwyn, Delaware, Pa., 3,074....*G11 114
Colwyn Bay, Wales, 23,090....A4 12
Colyell, creek, La....B6 95
Comacchio, It., 10,200....B4 21
Comal, Comal, Tex., 25....B4 118
Comal, co., Tex., 19,844....E3 118
Comalapa, Guat., 7,404....C2 62
Comanche, Stephens, Okla., 2,082....C4 112
Comanche, Comanche, Tex., 3,415....D3 118
Comanche, co., Kans., 3,271....E4 93
Comanche, co., Okla., 90,803....C3 112
Comanche, co., Tex., 11,865....C3 118
Comarapa, Bol., 1,096....C3 55
Comayagua, Hond., 4,828....C4 62
Combahee, riv., S.C....F6 115
Combarbalá, Chile, 2,112....A2 54
Combeaufontaine, Fr., 412....B1 18
Comber, Ont., Can., 601....E2 72
Comber, N. Ire., 3,980....C6 11
Combermere, Ont., Can., 82....B7 72
Combined Locks, Outagamie, Wis., 1,421....A5 124
Combomune, Moz....B5 49
Combs, Madison, Ark., 200....B2 81
Combs, Perry, Ky., 900....C6 94
Comer, Barbour, Ala., 100....C4 78
Comer, Madison, Ga., 882....B3 87
Comeragh, mts., Ire....E4 11
Comerío, P.R., 5,232....C6 65
Comerío, mun., P.R., 18,583....C6 65
Comertown, Sheridan, Mont., 40....B12 102
Comfort, Jones, N.C., 400....B6 109
Comfort, Marion, Tenn., 35....D8 117
Comfort, Kendall, Tex., 950....E3 118
Comfort, Boone, W. Va., 250....C3, D6 123
Comfort, pt., La....E6 95
Comfrey, Brown, Minn., 616....F4 99
Comilla, Pak., 47,526....D9 39
Comines, Bel., 8,373....D2 15
Comins, Oscoda, Mich., 80....D6 98
Comiso, It., 25,904....F5 21
Comitán, Mex., 15,378....D6 63
Comite, riv., La....D4 95
Commack, Suffolk, N.Y., 9,613....*F3 84
Commentry, Fr., 9,711....D5 14
Commerce, Conecuh, Ala....D3 78
Commerce, Los Angeles, Calif., 9,555....*F2 82
Commerce, Jackson, Ga., 3,551....B3 87
Commerce, Oakland, Mich., 1,200....*A7 98
Commerce, Scott, Mo., 247....D8 101
Commerce, Ottawa, Okla., 2,378....A7 112
Commerce, Hunt, Tex., 5,789....C5 118
Commerce Town, Adams, Colo., 8,970....B6 83
Commercial Point, Pickaway, Ohio, 308....C4 111
Commercy, Fr., 7,043....C6 14
Commiskey, Jennings, Ind., 100....G6 91
Commissioners, lake, Que., Can....A5 73
Committee, bay, N.W. Ter., Can....C15 66
Commonwealth, bay, Ant....C27 5
Communism, peak, Sov. Un....H23 9
Como, Que., Can., 807....D8 63
Como, It., 81,983 (★125,000)....B2 21
Como, Panola, Miss., 789....A4 100
Como, Hertford, N.C., 150....A6 109
Como, Henry, Tenn., 160....A3 117
Como, lake, It....B2 21
Comobabi, Pima, Ariz., 75....E4 80
Comodoro Rivadavia, Arg., 25,651....D3 54
Comoé, riv., I.C....E3 45
Comoro, is., Afr....f9 49
Comoro Islands, Fr. dep., Afr., 183,133....f9 49
Comox, B.C., Can., 1,756....E5 68
Compass Lake, Jackson, Fla., 300....B1, G3 86
Compeer, Alta., Can., 58....D5 69
Competition, Laclede, Mo., 80....D5 101
Compiègne, Fr., 24,427....C5 14
Compo Beach, Fairfield, Conn. (part of Westport)....E3 84
Comptche, Mendocino, Calif., 175....C2 82
Compton, Los Angeles, Calif., 71,812....F2 82
Compton, Que., Can., 543....D6 73
Compton, co., Que., Can., 24,410....D6 73
Compton, Lee, Ill., 366....B4 90
Comstock, Kalamazoo, Mich., 3,000....F5 98
Comstock, Clay, Minn., 138....D2 99
Comstock, Custer, Nebr., 235....C6 103
Comstock, Washington, N.Y., 100....B7 108
Comstock, Val Verde, Tex., 300....E2 118
Comstock, Barron, Wis., 60....C1 124
Comstock Park, Kent, Mich., 2,500....*E5 98
Conakry, Guinea, 43,000 (★109,500)....E2 45
Conanicut, isl., R.I....C11 84
Conasauga, Polk, Tenn., 150....D9 117
Concarneau, Fr., 11,691....D2 14
Conceição do Norte, Braz., 245....D1 57
Concepción, Arg., 12,338....E2 55
Concepción, Bol., 1,056....C3 55
Concepción, Bol., 860....D3 55
Concepcion, Santa Barbara, Calif....E3 82
Concepción, Chile, 150,000 (★295,000)....B2 54
Concepción, Guat., 1,855....C2 62
Concepción, Pan., 6,532....F6 62
Concepción, Par., 32,556....D4 55
Concepcion, Phil., 3,014....o13 35
Concepcion, Duval, Tex., 150....F3 118
Concepción, dept., Par., 14,640....D4 55
Concepción, prov., Chile, 537,711....B2 54
Concepción del Oro, Mex., 8,379....C4 63
Concepción del Uruguay, Arg., 31,498....A5 54
Conception, Nodaway, Mo., 450....A3 101
Conception, bay, Newf., Can....E5 75
Conception, pt., Calif....E3 82
Conception Junction, Nodaway, Mo., 253....A3 101
Conchas, Braz., 2,519....m7 56
Conchas Dam, San Miguel, N. Mex., 40....B5 107
Conchas, res., N. Mex....B5 107
Conche, Newf., Can., 498....C4 75
Conchi, Chile....D2 55
Conchillas, Ur....g7 54
Concho, Apache, Ariz., 150....C6 80
Concho, Canadian, Okla., 500....B4 112
Concho, co., Tex., 3,672....D3 118
Conconully, Okanogan, Wash., 108....A6 122
Concord, Contra Costa, Calif., 36,208....D2 82
Concord, Sussex, Del., 400....C6 85
Concord, Gadsden, Fla., 80....B2 86
Concord, Pike, Ga., 333....C2 87
Concord, Morgan, Ill., 210....D3 90
Concord, Lewis, Ky., 83....B6 94
Concord, Middlesex, Mass., 3,188 (12,517▲)....B5, C2 97
Concord, Jackson, Mich., 990....F6 98
Concord, Dixon, Nebr., 150....B9 103
Concord, Merrimack, N.H., 28,991....D3 105
Concord, Cabarrus, N.C., 17,799....B3 109
Concord, Franklin, Pa., 175....F6 114
Concord, Knox, Tenn., 250....D9, E11 117
Concord, Essex, Vt., 389 (956▲)....C5 120
Concord, Campbell, Va., 400....D4 121
Concord, riv., Mass....A5 97
Concordia, Arg., 64,000....A5 54
Concordia, Col., 3,906....B2 60
Concordia, Cloud, Kans., 7,022....C6 93
Concordia, Lafayette, Mo., 1,471....C4 101
Concordia, par., La., 20,467....C4 95
Concord North, Cabarrus, N.C., 2,199....B3 109
Concrete, Cavalier, N. Dak., 125....A8 110
Concrete, Skagit, Wash., 840....A4 122
Conda, Caribou, Ida., 200....G7 89
Conde, Braz., 4,190....D3 57
Conde, Spink, S. Dak., 388....B7 116
Condé [-sur-l'Escaut], Fr., 14,066....D4 15
Condé [-sur-Marne], Fr., 528....E4 15
Condé-sur-Noireau, Fr., 6,231....C3 14
Condeúba, Braz., 1,628....D2 57
Condino, It., 1,297....D6 18
Condit, Delaware, Ohio, 35....B5 111
Condobolin, Austl., 3,147....F6 51
Condom, Fr., 4,473....F4 14
Condon, Missoula, Mont., 25....C3 102
Condon, Gilliam, Oreg., 1,149....B6 113
Condor, range, Peru....B2 58
Cone, Crosby, Tex., 120....C2 118
Conecuh, co., Ala., 17,762....D2 78
Conecuh, riv., Ala....D3 78
Conegliano, It., 11,300....B4 21
Conehatta, Newton, Miss., 50....C4 100
Conejos, Conejos, Colo., 175....D4 83
Conejos, co., Colo., 8,428....D4 83
Conejos, creek, Colo....D4 83
Conejos, peak, Colo....D4 83
Conemaugh, Cambria, Pa., 3,334....F4 114
Conemaugh River, flood control res., Pa....F3 114
Cones, Coos, N.H....B1 105
Conestee, Greenville, S.C., 750....B3 115
Conesville, Muscatine, Iowa, 248....C6 92
Conesville, Coshocton, Ohio, 451....B6 111
Conetoe, Edgecomb, N.C., 147....B6 109
Coney, isl., N.Y....E1 108
Conflans, Fr., 2,802....E5 15
Conflans-Ste. Honorine, Fr., 21,874....g9 14
Confluence, Leslie, Ky., 175....C6 94
Confluence, Somerset, Pa., 938....G3 114
Confolens, Fr., 2,462....D4 14
Confusion, bay, Newf., Can....C4 75
Confusion, range, Utah....D2 119
Cong, Ire., 178....D2 11
Congamond, lakes, Conn....A6 84
Congaree, riv., S.C....D6 115
Conger, Freeborn, Minn., 694....G5 99
Congers, Rockland, N.Y., 3,000....*D7 108
Congleton, Eng., 16,802....A5 12
Congo, Perry, Ohio, 100....C5 111
Congo, Hancock, W. Va., 50....A2 123
Congo (Brazzaville), country, Afr., 795,000....B2 48, G7 42
Congo (Léopoldville), country, Afr., 13,284,300....B3 48, G8 42
Congo, basin, Con. L....A3 48
Congo, riv., Afr....G7 42
Congress, Yavapai, Ariz., 150....C3 80
Congress, Sask., Can., 86....H2 70
Conical, peak, Mont....D6 102
Coniclere, Pan....k10 62
Conifer, St. Lawrence, N.Y., 200....A6, B2 108
Conil, Sp., 9,002....D2 20
Coniston, Ont., Can., 2,692....E9 72
Coniston Water, lake, Eng....F5 13
Conjeeveram, see Kancheepuram, India
Conklin, Alta., Can., 69....B5 69
Conklin, Ottawa, Mich., 270....E5 98
Conklin, pt., N.Y....G3 84
Conlen, Dallam, Tex., 35....A1 118
Conn, lake, Ire....C2 11
Connacht, prov., Ire., 419,465....D2 11
Conneaut, Ashtabula, Ohio, 10,557....A7 111
Conneaut, creek, Ohio....A7 111
Conneaut Lake, Crawford, Pa., 700....C1 114
Conneautville, Crawford, Pa., 1,100....C1 114
Connecticut, state, U.S., 2,535,234....B13 77, 84
Connecticut, lake, N.H....A2 105
Connecticut, riv., U.S....C6 84, B2 97, D2 105
Connell, Franklin, Wash., 906....C7 122
Connell, mtn., B.C., Can....A1 102
Connellsville, Fayette, Pa., 12,814....F2 114
Connelsville, Adair, Mo., 80....A5 101
Connemara, mts., Ire....D2 11
Conner, Aroostook, Maine, 180....B4 96
Conner, Ravalli, Mont., 5....E2 102
Connersville, Fayette, Ind., 17,698....E7 91
Connerville, Johnston, Okla., 200....C5 112
Conning Towers, Conn., 3,457....*D8 84
Connor, N.B., Can., 244....B1 74
Conococheague, creek, Md....A2 85
Conover, Catawba, N.C., 2,281....B2 109
Conowingo, Cecil, Md., 25....A5 85
Conowingo, dam, Md....A5 85
Conquest, Sask., Can., 295....F2 70
Conrad, Grundy, Iowa, 799....B5 92
Conrad, Pondera, Mont., 2,665....B5 102
Conrad, Potter, Pa....C6 114
Conran, New Madrid, Mo., 100....E8 101
Conrath, Rusk, Wis., 121....C2 124
Conroe, Montgomery, Tex., 9,192....D5 118
Conroy, Iowa, Iowa, 150....C6 92
Consecon, Ont., Can., 359....D7 72
Conselheiro Lafaiete, Braz., 29,208....F2 57
Consett, Eng., 38,927....F7 13
Conshohocken, Montgomery, Pa., 10,259....A11, F11 114
Consolación del Sur, Cuba, 5,392....D2 64
Consort, Alta., Can., 557....C5 69
Constable, Franklin, N.Y., 275....A2 108
Constableville, Lewis, N.Y., 439....B5 108
Constance, Boone, Ky., 230....*A7 94
Constance, mtn., Wash....B2 122
Constancia Central, P.R....B5 65
Constancia Central, P.R....D4 65
Constanța, Rom., 99,690....C9 22
Constantia, Oswego, N.Y., 800....B4 108
Constantina, Sp., 11,910....D3 20
Constantine, Alg., 148,725....B6 44, F10 30
Constantine, St. Joseph, Mich., 1,710....G5 98
Constantine, dept., Alg., 3,108,165....*F10 30
Constantinople, see Istanbul, Tur.
Constitución, Chile, 8,285....B2 54
Constitution, De Kalb, Ga., 900....*B5 88
Consuegra, Sp., 10,572....C4 20
Consul, Marengo, Ala....C2 78
Consul, Sask., Can., 172....H1 70
Consumnes, riv., Calif....B6 82
Consumo, P.R....C2 65
Contact, Elko, Nev., 25....B7 104
Contai, India, 5,000....G11 40
Contamana, Peru, 2,860....C3 58
Contendas, Braz....D2 57
Content, keys, Fla....H5 86
Continental, Pima, Ariz., 75....F5 80
Continental, Hidalgo, N.Mex....F1 107
Continental, Putnam, Ohio, 1,147....A3 111
Continental, divide, Can., U.S....E6 66, B6 76
Continental, res., Colo....D3 83
Continental Divide, McKinley, N. Mex., 100....B1 107
Contoocook, Merrimack, N.H., 900....D3 105
Contoocook, riv., N.H....D3 105
Contraalmirante Cordero, Arg....B3 54
Contra Costa, co., Calif., 409,030....D2 82
Contratación, Col., 3,303....B3 60
Contrecoeur, Que., Can., 2,007....D4 63
Contreras, Mex., 10,112....h9 53
Cnotreras, Socorro, N. Mex., 25....C3 107
Contrexéville, Fr., 2,155....A1 18
Contumaza, Peru, 1,911....C2 58
Contwoyto, lake, N.W. Ter., Can....C11 66
Convent, St. James, La., 400....B6, D5 95
Converse, Miami, Ind., 1,044....C6 91
Converse, Sabine, La., 291....C2 95
Converse, Spartanburg, S.C., 950....B4 115
Converse, co., Wyo., 6,366....B7 125
Converse, lake, Ala....E1 78
Convoy, Van Wert, Ohio, 976....A3 111
Conway, Faulkner, Ark., 12,500....B3 81
Conway, Orange, Fla., 1,000....*D5 86
Conway, Taylor, Iowa, 82....D3 92
Conway, McPherson, Kans., 75....D6 93
Conway, Franklin, Mass., 600 (875▲)....A2 97
Conway, Leake, Miss., 96....C4 100
Conway, Laclede, Mo., 500....D5 101
Conway, Carroll, N.H., 1,143 (4,298▲)....C4 105
Conway, Northampton, N.C., 662....A6 109
Conway, Walsh, N. Dak., 67....A8 110
Conway, Beaver, Pa., 1,926....E1 114
Conway, Horry, S.C., 8,563....D9 115
Conway, Carson, Tex., 40....B2 118
Conway, Wales, 11,392....A4 12
Conway, co., Ark., 15,430....B3 81
Conway, lake, N.H....C4 105
Conway Springs, Sumner, Kans., 1,057....E6 93
Conway Station, P.E.I., Can., 90....C6 74
Conyers, Rockdale, Ga., 2,881....B6, C2 87
Cooch-Behar, India, 41,922....D12 40
Cook, Austl., 205....F5 50
Cook, Lake, Ind., 250....B3 91
Cook, St. Louis, Minn., 527....C6 99
Cook, Johnson, Nebr., 313....D9 103
Cook, co., Ga., 11,822....E3 87
Cook, co., Ill., 5,129,725....B6 90
Cook, co., Minn., 3,377....A6 99
Cook, cape, B.C., Can....D4 68
Cook, inlet, Alsk....D9 79
Cook, mtn., W. Va....D6 123
Cook, pt., Md....C5 85

Cook, strait, N.Z. N15 51
Cooke, Park, Mont., 100 E7 102
Cooke, co., Tex., 22,560 C4 118
Cooke City, Park, Mont., 30 E7 102
Cookeville, Putnam, Tenn., 7,805 C8 117
Cook Islands, N.Z. dep., Oceania, 15,079 H12 7
Cooks, Schoolcraft, Mich., 60 C4 98
Cooks, peak, N. Mex E2 107
Cooksburg, Forest, Pa., 15 D3 114
Cooks Falls, Delaware, N.Y., 150 D5 108
Cooks Harbour, Newf., Can., 342 C4 75
Cookshire, Que., Can., 1,412 D6 73
Cooks Knob, mtn., Mo C4 101
Cookstown, Ont., Can., 1,025 C5 72
Cookstown, Burlington, N.J., 75 C3 106
Cookstown, N. Ire., 4,964 C5 11
Cooksville, Ont., Can., 1,750 D5, E6 72
Cooksville, McLean, Ill., 221 C5 90
Cooktown, Austl., 397 C8 50
Cookville, Titus, Tex., 300 C5 118
Coolaney, Ire., 124 C3 11
Cooleemee, Davie, N.C., 1,609 B3 109
Coolgardie, Austl., 963 F3 50
Coolidge, Pinal, Ariz., 4,990 E4 80
Coolidge, Thomas, Ga., 679 E3 87
Coolidge, Hamilton, Kans., 117 D2 93
Coolidge, Limestone, Tex., 913 D4 118
Coolin, Bonner, Idaho, 100 A2 89
Cool Ridge, Raleigh, W. Va., 400 D3 123
Coolspring, Jefferson, Pa., 150 D3 114
Cool Valley, St. Louis, Mo., 1,492 *C7 101
Coolville, Athens, Ohio, 443 C6 111
Cooma, Austl., 8,717 H7 51
Coonabarabran, Austl., 2,548 E7 51
Coonamble, Austl., 3,235 E7 51
Coon Rapids, Carroll, Iowa, 1,560 C3 92
Coon Rapids, Anoka, Minn., 14,931 E7 99
Coon Valley, Vernon, Wis., 536 E2 123
Cooper, Chilton, Ala., 200 C3 78
Cooper, Fairfield, Conn D3 84
Cooper, Green, Iowa, 125 C3 92
Cooper, Wayne, Ky., 250 D5 94
Cooper, Washington, Maine, 65 (106▲) D5 96
Cooper, Osage, Okla A5 112
Cooper, Delta, Tex., 2,213 C5 118
Cooper, co., Mo., 15,448 C5 101
Cooper, mtn., B.C., Can D2 69
Cooper, riv., S.C E8 115
Co-Operative, McCreary, Ky., 350 D5 94
Cooper Lake, Albany, Wyo., 20 D7 125
Coopersburg, Lehigh, Pa., 1,800 E10 114
Coopers Mills, Lincoln, Maine, 150 D3 96
Cooperstown, Otsego, N.Y., 2,553 C6 108
Cooperstown, Griggs, N. Dak., 1,424 B7 110
Cooperstown, Venango, Pa., 267 D2 114
Cooperstown, Manitowoc, Wis., 100 A6 124
Coopersville, Ottawa, Mich., 1,584 E5 98
Cooperton, Kiowa, Okla., 106 C3 112
Coopertown, Robertson, Tenn., 50 A5 117
Cooper Village, Gloucester, N.J., 1,500 *D2 106
Coorg, state, India, 229,405 *F6 39
Coos, co., N.H., 37,140 A4 105
Coos, co., Oreg., 54,955 D2 113
Coos, riv., Oreg D3 113
Coosa, co., Ala., 10,726 C3 78
Coosa, riv., Ala., Ga A4 78
Coosada, Elmore, Ala., 235 C3 78
Coosawattee, riv., Ga B2 87
Coosawhatchie, Jasper, S.C., 160 F6 115
Coosawhatchie, riv., S.C F5 115
Coos Bay, Coos, Oreg., 7,084 D2 113
Coos Junction, Coos, N.H. A3 105
Cootamundra, Austl., 5,938 G7 51
Cootehill, Ire., 1,296 C4 11
Cooter, Pemiscot, Mo., 477 E7 101
Copacabana, Arg E2 55
Copacabana, Bol., 2,019 C2 55
Copainalá, Mex., 1,932 D6 63
Copake, Columbia, N.Y., 600 C7 108
Copalis Beach, Grays Harbor, Wash., 350 B1 122
Copalis Crossing, Grays Harbor, Wash., 100 B1 122
Copán, Hond., 1,045 C3 62
Copan, Washington, Okla., 617 A6 112
Cope, Washington, Colo., 125 B8 83
Cope, Orangeburg, S.C., 227 E5 115
Copeland, Collier, Fla., 500 G5 86
Copeland, Boundary, Idaho, 25 A2 89
Copeland, Gray, Kans., 247 E3 93
Copemish, Manistee, Mich., 232 D5 98
Copen, Braxton, W. Va., 100 C4 123
Copenhagen (København), Den., 721,381 (★1,350,000) C6 24
Copenhagen, Lewis, N.Y., 673 B5 108
Copenhagen (København), co., Den., 1,269,366 C6 24
Copetonas, Arg B4 54
Copeville, Collin, Tex., 300 A6 118
Copiague, Suffolk, N.Y., 14,081 G3 84
Copiah, co., Miss., 27,051 D3 100
Copiapó, Chile, 28,000 E1 55
Copinsay, isl., Scot B6 41
Coplay, Lehigh, Pa., 3,701 *E11 114
Copley, Summit, Ohio, 4,000 *A6 111
Copparo, It., 5,800 B3 21
Copper, mts., Ariz E2 80
Copper, ridge, Tenn E11 117
Copper, riv., Alsk C11 79
Copperas Cove, Coryell, Tex., 4,567 D4 118
Copper Center, Alsk., 151 C10, g19 79
Copper Cliff, Ont., Can., 3,600 E9 72
Copperfield, Washoe, Nev., 250 D2 104
Copper Harbor, Keweenaw, Mich., 60 A3 98
Copper Hill, Hunterdon, N.J. C3 106
Copperhill, Polk, Tenn., 631 D9 117
Copper Hill, Floyd, Va., 40 D2 121
Coppermine, N.W. Ter., Can., 230 C9 66
Coppermine, riv., N.W. Ter., Can C10 66
Copper Mountain, B.C., Can., 1,039 E7 68
Coppet, Switz., 774 D1 19
Coquet, riv., Eng E7 13
Coquí, P.R., 2,088 D6 65
Coquilhatville, Con. L., 37,500 B2 48
Coquille, Coos, Oreg., 4,730 D2 113
Coquille, riv., Oreg E3 113
Coquimbana, Chile E1 55
Coquimbo, Chile, 31,000 E1 55
Coquimbo, prov., Chile, 306,384 A2 54
Cora, Iberville, La B5 95
Cora, Logan, W. Va., 500 *D2 123
Cora, Sublette, Wyo., 5 C3 125
Corabia, Rom., 11,502 D7 22
Coracora, Peru, 3,671 E3 58
Coral, Montcalm, Mich., 225 E5 98
Coral, Indiana, Pa., 600 F3 114
Coral, sea, Pac. O G9 7
Coral Gables, Dade, Fla., 34,793 F3, G6 86
Coral Rapids, Ont., Can., 60 E9 72
Coral Ridge, see Fairdale, Ky.
Coralville, Johnson, Iowa, 2,357 C6 92
Coram, Flathead, Mont., 400 B2 102
Coram, Suffolk, N.Y., 400 D4 108
Coraopolis, Allegheny, Pa., 9,643 A5, F1 114
Corato, It., 39,452 D6 21
Corazón, Ec., 1,051 B2 58
Corbeil, Fr., 26,804 C5 14
Corbeny, Fr., 540 E3 15
Corbett, Delaware, N.Y., 150 C5 108
Corbetton, Ont., Can., 89 C4 72
Corbie, Fr., 4,657 C5 14
Corbin, B.C., Can E10 68
Corbin, Sumner, Kans., 100 E6 93
Corbin, Whitley and Knox, Ky., 7,119 D5 94
Corbin, Livingston, La., 100 A6 95
Corbin, Jefferson, Mont., 50 D4 102
Corbin City, Atlantic, N.J., 271 E3 106
Corby, Eng., 625 B7 12
Corby, Eng., 36,322 B7 12
Corcoran, Kings, Calif., 4,976 D4 82
Corcoran, Hennepin, Minn., 1,237 *E5 99
Corcovado, gulf, Chile C2 54
Cord, Independence, Ark., 100 B4 81
Cordaville, Worcester, Mass., 250 D1 97
Cordele, Crisp, Ga., 10,609 E3 87
Cordelia, Solano, Calif., 200 B5 82
Cordell (New Cordell), Washita, Okla., 3,589 B3 112
Cordell Hull, glacier, Ant B35 5
Corder, Lafayette, Mo., 506 B4 101
Cordesville, Berkeley, S.C., 500 E8 115
Cordillera, dept., Par., 145,232 D4 55
Cordillera Central, range, Bol., Col C2 55, C2 60
Cordillera Occidental, range, Bol., Col C2 55, C2 60
Cordillera Occidental o de la Costa, range, Peru D3 58
Cordillera Oriental, range, Bol., Col C3 55, B3 60
Córdoba, Arg., 580,000 A4 54
Córdoba, Mex., 47,390 n15 63
Córdoba, Sp., 198,148 D3 20
Córdoba, prov., Arg., 1,759,997 A4 54
Córdoba, dept., Col., 357,000 B2 60
Córdoba, prov., Sp., 798,437 *D3 20
Cordon, Phil., 1,674 n13 35
Cordova, Walker, Ala., 3,184 B2 78
Cordova, Alsk., 1,128 C10, g19 79
Cordova, Rock Island, Ill., 502 B3 90
Cordova, Talbot, Md., 245 C6 85
Cordova, Teton, Mont., 2 C5 102
Cordova, Seward, Nebr., 152 D8 103
Cordova, Richmond, N.C., 950 C4 109
Córdova, Peru, 534 D2 58
Cordova, Orangeburg, S.C., 209 E6 115
Cordova, see Córdoba, prov. and city, Sp.
Cordova, Shelby, Tenn., 350 B2 117
Cordova, peak, Alsk g19 79
Cordova Mines, Ont., Can., 265 C6 72
Corea, Hancock, Maine, 160 D5 96
Coreaú, Braz., 2,118 B2 57
Corella, Sp., 5,591 A5 20
Corfu, Genesee, N.Y., 616 C2 108
Corfu, see Kerkira, isl., Grc.
Cori, It., 9,681 h9 21
Coria, Sp., 8,204 C2 20
Corigliano, Calabro, It., 16,300 E6 21
Corinna, Penobscot, Maine, 650 (1,895▲) D3 96
Corinne, Box Elder, Utah, 510 B3 119
Corinne, Wyoming, W. Va., 1,273 D3 123
Corinth, see Korinthos, Grc.
Corinth, Heard, Ga., 105 C2 87
Corinth, Grant, Ky., 238 B5 94
Corinth, Alcorn, Miss., 11,453 A5 100
Corinth, Saratoga, N.Y., 3,193 B7 108
Corinth, Williams, N. Dak., 38 A2 110
Corinth, Orange, Vt., 25 (775▲) C4 120
Corinth, Preston, W. Va., 175 B5 123
Corinth, bay, Grc h9 23
Corinth, canal, Grc h10 23
Corinth, gulf, Grc C4 23
Corinthia (Korinthia), prov., Grc., 112,505 *D4 23
Corinto, Braz., 12,247 E2 57
Corinto, Nic., 7,400 D4 62
Corisco, isl., Afr E1 46
Cork, Ire., 77,980 (★118,000) F3 11
Cork, co., Ire., 330,443 F3 11
Cork Station, N.B., Can., 84 D3 74
Corleone, It., 14,682 F4 21
Corlu, Tur., 22,000 B6 31
Cormack, mtn., Newf., Can D4 75
Cormeilles-en-Parisis, Fr., 11,486 g9 14
Cormoran, reef, Palau Is 52
Cormorant, Man., Can., 272 B1 71
Cormorant, lake, Man., Can B1 71
Corn, Washita, Okla., 317 B3 112
Corn, creek, Ariz B5 80
Corn, is., Cen. Am D6 62
Corncake, inlet, N.C D6 109
Cornelia, Habersham, Ga., 2,936 B3 87
Cornelius, Mecklenburg, N.C., 1,444 B2 109
Cornelius, Washington, Oreg., 1,146 B1 113
Cornell, Livingston, Ill., 524 C5 90
Cornell, Chippewa, Wis., 1,685 C2 124
Corner Brook, Newf., Can., 25,185 D3 75
Cornersville, Marshall, Tenn., 314 B5 117
Cornerville, Lincoln, Ark., 100 D4 81
Corney, creek, La B3 95
Cornfield, pt., Conn D7 84
Cornfield, Apache, Ariz., 200 B6 80
Corn Hill, N.B., Can., 93 D4 74
Cornimont, Fr., 2,592 B2 18
Corning, Clay, Ark., 2,565 A5 81
Corning, Tehama, Calif., 3,006 C2 82
Corning, Sask., Can., 107 H4 70
Corning, Adams, Iowa, 2,041 D3 92
Corning, Nemaha, Kans., 240 C7 93
Corning, Holt, Mo., 128 A2 101
Corning, Steuben, N.Y., 17,085 C3 108
Corning, Perry, Ohio, 1,065 C5 111
Corning, Lincoln, Wis C4 124
Cornish, Weld, Colo., 5 A5 83
Cornish, York, Maine, 600 (816▲) E2 96
Cornish, Jefferson, Okla., 127 C4 112
Cornish Flat, Sullivan, N.H., 90 D2 105
Cornishville, Mercer, Ky., 250 C5 94
Corno, mtn., It C4 21
Cornplanter, Indian res., Pa C4 114
Cornucopia, Baker, Oreg., 355 C9 113
Cornucopia, Bayfield, Wis., 200 B2 124
Cornudas, Hudspeth, Tex F1 118
Cornville, Yavapai, Ariz., 35 C4 80
Cornville, Somerset, Maine, 150 (585▲) D3 96
Cornwall, Ba. Is., 436 C5 64
Cornwall, Ont., Can., 43,639 B10 72
Cornwall, Litchfield, Conn., 150 (1,051▲) B3 84
Cornwall (P.O.), Orange, N.Y., 2,824 *D6 108
Cornwall, Lebanon, Pa., 1,934 F9 114
Cornwall, Rockbridge, Va., 100 D3 121
Cornwall, co., Eng., 341,746 D3 12
Cornwall Bridge, Litchfield, Conn., 200 B3 84
Cornwallis, Ritchie, W. Va., 100 B3 123
Cornwallis, isl., N.W. Ter., Can A13 66
Cornwall [-on-the-Hudson], Orange, N.Y., 2,785 D6 108
Cornwell, Highlands, Fla., 50 E5 86
Cornwells Heights, Bucks, Pa., 10,000 *A12 113
Coro, Ven., 44,757 A4 60
Coroatá, Braz., 7,720 B2 57
Corocoro, Bol., 4,431 C2 55
Coroico, Bol., 2,235 C2 55
Corolla, Currituck, N.C., 50 A8 109
Coromandel, N.Z., 709 L15 51
Coromandel, coast, India F7 39
Corona, Walker, Ala., 100 B2 78
Corona, Riverside, Calif., 13,336 F3, F5 82
Corona, Lincoln, N. Mex., 420 C4 107
Corona, Roberts, S. Dak., 150 B9 116
Coronaca, Greenwood, S.C., 150 C3 115
Coronach, Sask., Can., 395 H3 70
Coronado, San Diego, Calif., 18,039 E2, F5 82
Coronation, Alta., Can., 864 C5 69
Coronation, gulf, N.W. Ter., Can C10 66
Coronda, Arg., 4,656 A4 54
Coronel, Chile, 28,000 B2 48
Coronela, Cuba k11 64
Coronel Brandsen, Arg., 3,803 B5, g7 54
Coronel Dorrego, Arg., 7,245 B4 54
Coronel Oviedo, Par., 34,442 E4 55
Coronel Pringles, Arg., 12,844 B4 54
Coronel Pringles, Arg C4 54
Coronel Suárez, Arg., 11,133 B4 54
Coronie, Sur A3 59
Corowa, Austl., 2,593 G6 51
Corozal, Br. Hond., 2,190 A3 62
Corozal, Col., 7,240 B2 60
Corozal, P.R., 3,166 B5 65
Corozal, mun., P.R., 23,570 B5 65
Corpataux, Switz., 359 C3 19
Corpen, Arg D3 54
Corps, Fr., 608 E1 18
Corpus Christi, Nueces, Tex., 167,690 (★195,200) F4 118
Corral, Chile, 5,525 B2 54
Corral, Camas, Idaho, 20 F4 89
Corral de Almaguer, Sp., 8,261 C4 20
Corrales, Col., 1,226 B3 60
Corralillo, Cuba, 1,073 D3 64
Corral Viejo, P.R. C4 65
Corraun, Ire D2 11
Corravillers, Fr., 208 B2 18
Correctionville, Woodbury, Iowa, 912 B2 92
Corregidor, isl., Phil o13 35
Correll, Big Stone, Minn., 101 E2 99
Corrente, Braz., 2,214 D2 57
Correntes, Braz., 3,788 k5 57
Correntina, Braz., 2,636 D2 57
Correo, Valencia, N. Mex., 35 C2 107
Corrèze, dept., Fr., 237,926 *E4 14
Corrib, lake, Ire D2 11
Corrientes, Arg., 104,000 E4 55
Corrientes, prov., Arg., 543,226 E4 55
Corrientes, cape, Cuba E1 64
Corrientes, cape, Mex C3 63
Corrientes, riv., Arg E4 55
Corrientes, riv., Peru B2 58
Corrigan, Polk, Tex., 986 D5 118
Corriganville, Allegany, Md., 602 D2 85
Corrine, key, Fla G6 86
Corro, isl., Az. Is g8 44
Corrofin, Ire., 362 E3 11
Corry, Erie, Pa., 7,744 C2 114
Corryton, Knox, Tenn., 300 C10 117
Corse (Corsica), dept., Fr., 275,465 *C2 21
Corse, cape, Fr C2 21
Corsica, Jefferson, Pa., 431 D3 114
Corsica, P.R. B2 65
Corsica, Douglas, S. Dak., 479 D7 116
Corsica (Corse), dept., Fr., 224,266 C2 21
Corsicana, Navarro, Tex., 20,344 C4 118
Corson, Minnehaha, S. Dak., 75 D9 116
Corson, co., S. Dak., 5,798 B4 116
Corson, inlet, N.J E3 106
Cortada Central, P.R. D5 65
Cortaro, Pima, Ariz., 50 E4 80
Corte, Fr., 5,268 C2 21
Corte Alto, Chile C2 54
Cortegana, Sp., 7,179 D2 20
Cortelyou, Washington, Ala., 225 D1 78
Corte Madera, Marin, Calif., 5,962 *B5 82
Cortemilia, It., 3,459 E4 18
Cortes, Sp., 5,050 D3 20
Cortez, Montezuma, Colo., 6,764 D2 83
Cortez, Manatee, Fla., 900 F2 86
Cortez, mts., Nev C5 104
Cortina d'Ampezzo, It., 3,900 C8 18
Cortland, DeKalb., Ill., 461 B5 90
Cortland, Jackson, Ind., 170 G6 91
Cortland, Gage, Nebr., 285 D9 103
Cortland, Cortland, N.Y., 19,181 C4 108
Cortland, Trumbull, Ohio, 1,957 A7 111
Cortland, co., N.Y., 41,113 C4 108
Cortona, It., 4,000 C3 21
Coruche, Port., 2,925 C1 20
Çorum (Chorum), Tur., 34,600 B10 31
Corumbá, Braz., 36,744 B1 56
Corumba, riv., Braz E1 57
Corunna, Ont., Can., 1,942 E2 72
Corunna, DeKalb, Ind., 361 B7 91
Corunna, Shiawassee, Mich., 2,764 F6 98
Corvallis, Ravalli, Mont., 390 D2 102
Corvallis, Benton, Oreg., 20,669 C1, C3 113
Corwin, Harper, Kans., 40 E5 93
Corwin Springs, Park, Mont., 10 E6 102
Corwith, Hancock, Iowa, 448 B4 92
Cory, Clay, Ind., 200 F3 91
Corydon, Harrison, Ind., 2,701 H5 91
Corydon, Wayne, Iowa, 1,687 D4 92
Corydon, Henderson, Ky., 746 C2 94
Corydon, Warren, Pa., 200 C4 114
Coryell, co., Tex., 23,961 D4 118
Coryville, McKean, Pa., 250 C5 114
Cosalá, Mex., 1,694 C3 63
Cosby, Andrew, Mo., 119 B3 101
Cosby, Cocke, Tenn., 50 D10 117
Cos Cob, Fairfield, Conn. (part of Greenwich) E2 84
Coscomatepec, Mex., 5,649 n15 63
Cosenza, It., 78,611 E6 21
Cosgrave, Pershing, Nev., 5 C3 104
Coshocton, Coshocton, Ohio, 13,106 B6 111
Coshocton, co., Ohio, 32,224 B6 111
Cosmopolis, Grays Harbor, Wash., 1,312 C2 122
Cosmos, Meeker, Minn., 487 F4 99
Cosne [-sur-Loire], Fr., 8,802 D5 14
Cossatot, mts., Ark C2 81
Cossatot, riv., Ark C1 81
Cossonay, Switz., 1,284 C2 19
Costa Mesa, Orange, Calif., 37,550 F3 82
Costa Rica, country, N.A., 1,251,300 H12 61, E5 62
Costello, Potter, Pa., 100 C5 114
Costelloe, Ire D2 11
Costermanville, see Bukavu, Con. L.
Costigan, Penobscot, Maine, 70 C4 96
Costigan, mts., N.B. Can C2 74
Costilla, Taos, N. Mex., 300 A4 107
Costilla, co., Colo., 4,219 D5 83
Costilla, peak, N. Mex A4 107
Coswig, Ger., 13,700 B7 17
Coswig, Ger., 17,600 B8 17
Cotabato, Phil., 17,000 (37,300▲) D6 35
Cotabato, prov., Phil., 1,152,500 *D6 35
Cotagaita, Bol., 1,353 D2 55
Cotahuasi, Peru, 1,354 D2 58
Cotati, Sonoma, Calif., 1,852 B5 83
Coteau, Burke, N. Dak., 100 A3 110
Coteau, plat., Sask., Can G3 70
Coteau Landing, Que., Can., 544 D3 73
Coteau Station, Que., Can., 1,032 *D3 73
Coteaux, Haiti F6 64
Côte-d'Or, dept., Fr., 387,869 *D6 14
Côte d'Or, ridge, Fr D6 14
Cote St. Luc, Que., Can., 13,266 *D4 73
Cote St. Michel, Que., Can., 55,978 *D9 73
Côtes de Fer, Haiti F7 64
Côtes de L'ile de France, mts., Fr F3 15
Côtes de Meuse, mts., Fr E5 15
Côtes-du-Nord, dept., Fr., 501,923 *C2 14
Cotesfield, Howard, Nebr., 81 C7 103
Cotija, Mex., 8,006 n12 63
Cotingo, riv., Braz C5 60
Coto Laurel, P.R. C4 65
Cotonou, Dah., 21,140 E5 45
Cotopaxi, Fremont, Colo., 120 C5 83
Cotopaxi, prov., Ec., 179,965 B2 58
Cotopaxi, vol., Ec B2 58
Cotswold, hills, Eng C5 12
Cottage, Aroostook, Maine, 15 A4 96
Cottage City, Prince Georges, Md., 1,099 *C1 85
Cottage Grove, Washington, Minn., 160 E7 99
Cottage Grove, Lane, Oreg., 3,895 D3 113
Cottagegrove, Henry, Tenn., 130 A3 117
Cottage Grove, Dane, Wis., 413 *E4 124
Cottage Grove, dam, Oreg D3 113
Cottage Hills, Madison, Ill., 3,976 *E3 90
Cottage Home, Wilson, Tenn A5 117
Cottageville, Colleton, S.C., 520 F7 115
Cottageville, Jackson, W. Va., 400 C3 123
Cottam, Ont., Can., 520 E2 72
Cottbus, Ger., 67,700 B9 17
Cottel, isl., Newf., Can D5 75
Cotter, Baxter, Ark., 683 A3 81
Cotter, cape, Ant B30 5
Cottian Alps, mts., Fr E7 14
Cottingham, Eng., 42,388 A7 12
Cottle, co., Tex., 4,207 B2 118
Cottle Knob, mtn., W. Va C4 123
Cottleville, St. Charles, Mo., 200 A7 101
Cotton, Mitchell, Ga., 108 E2 87
Cotton, St. Louis, Minn., 70 C6 99
Cotton, co., Okla., 8,031 C3 112
Cotton Center, Hale, Tex., 150 B2 118
Cottondale, Tuscaloosa, Ala., 900 B2 78
Cottondale, Jackson, Fla., 849 B1, G3 86
Cotton Mills, Grayson, Tex., 950 *C4 118
Cotton Plant, Woodruff, Ark., 1,704 B4 81
Cottonport, Avoyelles, La., 1,581 D3 95
Cottonton, Russell, Ala., 120 C4 78
Cotton Town, Yell, Ark., 40 B2 81
Cottontown, Sumner, Tenn., 50 A5 117
Cotton Valley, Webster, La., 1,145 B2 95
Cottonwood, Houston, Ala., 953 D4 78
Cottonwood, Yavapai, Ariz., 1,879 C3 80
Cottonwood, Shasta, Calif., 700 B2 82
Cottonwood, Idaho, Idaho, 1,081 C2 89
Cottonwood, Lyon, Minn., 717 F3 99
Cottonwood, Coal, Okla., 100 C5 112
Cottonwood, Jackson, S. Dak., 38 D4 116
Cottonwood, Callahan, Tex., 100 C3 118
Cottonwood, co., Minn., 16,166 F3 99
Cottonwood, creek, Wyo B4 125
Cottonwood. creek, Wyo C2 125
Cottonwood, creek, Wyo D2 125
Cottonwood, riv., Kans D7 93
Cottonwood, riv., Minn F3 99
Cottonwood, wash, Utah F6 119
Cottonwood Falls, Chase, Kans., 971 D7 93
Cottonwood Heights, Salt Lake, Utah, 1,500 *C4 119
Cotuit, Barnstable, Mass., 700 C7 97
Cotulla, La Salle, Tex., 3,960 E3 118
Couchiching, lake, Ont., Can C5 72
Couchville, Davidson, Tenn E9 117
Couchwood, Webster, La., 200 B2 95
Coudekerque-Branche, Fr., 20,757 B5 14
Couderay, Sawyer, Wis., 113 C2 124
Coudersport, Potter, Pa., 2,889 C5 114
Coudres, isl., Que., Can B7 73
Couëron, Fr., 5,727 D3 14
Cougar, peak, Oreg E6 113
Cougar, rock, Oreg C4 113
Coul, pt., Scot E2 13
Coulee, Montrail, N. Dak., 100 A3 110
Coulee, creek, Wash D7 122
Coulee City, Grant, Wash., 654 B6 122
Coulee Dam, Douglas, Grant, and Okanogan, Wash., 1,344 B7 122
Coulommiers, Fr., 8,561 C5 14
Coulonge, riv., Que., Can A8 72
Coulter, Man., Can., 50 E1 71
Coulter, Franklin, Iowa, 315 B4 92
Coulterville, Mariposa, Calif., 180 D3 82
Coulterville, Randolph, Ill., 1,022 E4 90
Coulwood, Russell, Va., 125 B3 121
Counce, Hardin, Tenn., 600 B3 117
Council, Clinch, Ga., 60 F4 87
Council, Adams, Idaho, 827 E2 89
Council, Buchanan, Va., 50 B3 121
Council Bluffs, Pottawattamie, Iowa, 55,641 C2 92
Council Grove, Morris, Kans., 2,664 D7 93
Council Hill, Muskogee, Okla., 130 B6 112
Counselor, Sandoval, N. Mex. A2 107
Country Club Hills, Cook, Ill., 4,771 *B6 90
Country Club Hills, St. Louis, Mo., 1,763 *A8 101
Country Homes, Spokane, Wash., 1,600 *B8 122
Countyline, Carter and Stephens, Okla., 500 C4 112
Coupar Angus, Scot., 2,049 D5 13
Coupeville, Island, Wash., 740 A3 122
Coupon, Cambria, Pa., 250 E4 114
Courantyne, riv., Br. Gu A3 59
Courbevoie, Fr., 59,491 g10 14
Courcelles, Que., Can., 773 D7 73
Courland (Kurisches Haff), lagoon, Sov. Un A6 26
Courmayeur, It., 719 D2 18
Coursan, Fr., 3,212 F5 14
Courtenay, B.C., Can., 3,485 E5 68
Courtenay, Stutsman, N. Dak., 168 B7 110
Courtland, Lawrence, Ala., 495 A2 78
Courtland, Sacramento, Calif., 550 B6 82
Courtland, Ont., Can., 605 E4 72
Courtland, Republic, Kans., 384 C6 93
Courtland, Nicollet, Minn., 239 F4 99
Courtland, Panola, Miss., 242 A4 100
Courtland, Southampton, Va., 855 E5 121
Courtmacsherry, Ire., 205 F3 11
Courtney, Love, Okla., 40 D4 112
Courtrai, see Kortrijk, Bel.
Courtright, Ont., Can., 532 E2 72
Courtrock, Grant, Oreg C7 113
Courval, Sask., Can., 108 G2 70
Courville, Que., Can., 4,670 C6 73
Coushatta, Red River, La., 1,663 B2 95
Coutances, Fr., 7,806 C3 14
Coutras, Fr., 3,688 E3 14
Coutts, Alta., Can., 469 E5 69
Couvin, Bel., 3,840 D4 15
Covada, Ferry, Wash A7 122
Cove, Polk, Ark., 320 C1 81
Cove, Garrett, Md D1 85
Cove, Mille Lacs, Minn., 45 D5 99
Cove, Union, Oreg., 311 B9 113
Cove, Scot C3 13
Cove, Chambers, Tex., 25 F5 118
Cove, Cache, Utah B4 119
Cove, King, Wash., 150 D1 122
Cove, isl., Ont., Can B3 72
Cove, pt., Md D5 85
Cove City, Craven, N.C., 551 B6 109
Cove City, Orange, Tex., 1,749 *D6 118
Covedale, Hamilton, Ohio, 10,000 D2 111
Cove Fort, Millard, Utah E3 119
Covelo, Mendocino, Calif., 600 C2 82
Coventry, Tolland, Conn., 3,568 (6,356▲) B7 84
Coventry, Eng., 305,060 (★560,000) B6 12
Coventry, Kent, R.I., 4,500 (15,432▲) C10 84
Coventry, Orleans, Vt., 130 (458▲) B4 120
Coventry Center, Kent, R.I., 500 C10 84
Cove Orchard, Yamhill, Oreg B1, B3 113
Covered Bridge, N.H B2 105
Covert, Osborne, Kans., 50 C5 93
Covert, Van Buren, Mich., 500 F4 98
Covesville, Albemarle, Va., 150 D4 121
Covilhã, Port., 20,423 B2 20
Covin, Fayette, Ala., 100 B2 78
Covina, Los Angeles, Calif., 20,124 *F3 82
Covington, Newton, Ga., 8,167 C3 87
Covington, Fountain, Ind., 2,759 D3 91
Covington, Kenton, Ky., 60,376 A5, A7 94
Covington, St. Tammany, La., 6,754 B7, D5 95
Covington, Baraga, Mich., 200 B2 98
Covington, Miami, Ohio, 2,473 B3 111
Covington, Garfield, Okla., 687 A4 112
Covington, Tioga, Pa., 673 C7 114
Covington, Tipton, Tenn., 5,298 B2 117
Covington (Independent City), Va., 11,062 D3 121
Covington, King, Wash., 50 D2 122
Covington, co., Ala., 35,631 D3 78
Covington, co., Miss., 13,637 D4 100
Cow, creek, Wash C7 122
Cow, lake, Oreg D9 113
Cowal, lake, Austl F6 51
Cowan, Delaware, Ind., 250 D7 91
Cowan, Franklin, Tenn., 1,979 B5 117
Cowan, lake, Austl F3 50
Cowan, lake, Sask., Can C2 70
Cowan, riv., Sask., Can C2 70
Cowangie, Austl., 159 G3 51
Cowan Knob, mtn., Ark B2 81
Cowansville, Que., Can., 7,050 D5 73
Cowansville, Armstrong, Pa., 400 E2 114
Coward, Florence, S.C., 150 D8 115
Coward Springs, Austl E6 50
Cowarts, Houston, Ala., 200 D4 78
Cowcreek, Owsley, Ky C6 94
Cowden, Shelby, Ill., 575 D5 90
Cowdenbeath, Scot., 11,918 D5 13
Cowdrey, Jackson, Colo., 100 A4 83
Cowell, Newton, Ark., 35 B2 81
Cowen, Webster, W. Va., 475 D4 123
Cowen, mtn., Mont E6 102
Cowes, Eng., 16,974 D6 12

Coweta, Wagoner, Okla., 1,858 .... B6 112
Coweta, co., Ga., 28,893 .... C2 87
Cowgill, Caldwell, Mo., 259 .. B4 101
Cow Head, Newf., Can., 544 .... D3, k10 75
Cowichan, lake, B.C., Can E5 68
Cowichan Station, B.C., Can., 97 .... B9 68
Cowiche, Kakima, Wash., 175 .... C5 122
Cowles, Webster, Nebr., 55 .. D7 103
Cowles, San Miguel, N. Mex . F6 107
Cowley, Alta., Can., 127 .... E3 69
Cowley, Big Horn, Wyo., 459 . A3 125
Cowley, co., Kans., 37,861 .. E7 93
Cowlic, Pima, Ariz., 80 .... F4 80
Cowlington, LeFlore, Okla., 74 .... B7 112
Cowlitz, co., Wash., 57,801 . C3 122
Cowlitz, riv., Wash C3 122
Cowpasture, riv., Va C3 121
Cowpen, mtn., Ga B2 87
Cowpens, Spartanburg, S.C., 2,038 .... A4 115
Cowra, Austl., 6,289 .... F7 51
Cowskin, creek, Kans B5 93
Cow Springs, Coconino, Ariz., 3 .... A5 80
Coxburg, Benton, Tenn B3 117
Cox City, Grady, Okla., 150 . C4 112
Coxim, Braz., 1,371 .... B2 56
Coxipi, riv., Que., Can C2 75
Coxsackie, Greene, N.Y., 2,849 .... C7 108
Cox's Cove, Newf., Can., 630 .... D2 75
Coxs Mills, Gilmer, W. Va., 75 .... B4 123
Coy, Wilcox, Ala., 100 .... D2 78
Coy, Lonoke, Ark., 206 .... C4 81
Coya, Chile D2 55
Coyame, Mex., 790 .... B3 63
Coyanosa, draw, Tex D1 118
Coyle, Logan, Okla., 292 .... B4 112
Coyoacán, Mex., 46,031 . h9, n14 63
Coyote, Lincoln, N. Mex., 50 . D4 107
Coyote, Rio Arriba, N. Mex., 50 .... A3 107
Coyote, basin, Colo A2 83
Coytesville, Bergen, N.J. (part of Fort Lee) D5 106
Coyville, Wilson, Kans., 133 . E8 93
Cozad, Dawson, Nebr., 3,184 . D6 103
Cozahome, Searcy, Ark., 50 .. A3 81
Cozumel, Mex., 2,330 .... C7 63
Cozumel, isl., Mex C7 63
Crab, creek, Wash B7 122
Crab, creek, Wash C6 122
Crab Orchard, Lincoln, Ky., 808 .... C5 94
Crab Orchard, Johnson, Nebr., 103 .... D9 103
Crab Orchard, Cumberland, Tenn., 700 .... D9 117
Crab Orchard, Raleigh, W. Va., 1,953 .... D7 123
Crab Orchard, lake, Ill F5 90
Crab Orchard, mts., Tenn D9 117
Crabtree, Van Buren, Ark B3 81
Crabtree, Linn, Oreg., 250 .... C2, C4 113
Crabtree, Westmoreland, Pa., 950 .... F3 114
Crabtree Bald, mtn., N.C D3 109
Crabtree Mills, Que., Can., 1,313 .... D4 73
Cracking, riv., Sask., Can D4 70
Cracow, Austl., 417 .... B8 51
Cracow (Krakow), Pol., 479,000 .... C5 26
Cradle, mtn., Austl o15 50
Cradock, S. Afr., 19,476 .... D4 49
Crafton, Allegheny, Pa., 8,418 .... B5 114
Craftsbury, Orleans, Vt., 175 (674▲) B4 120
Craftsbury Common, Orleans, Vt., 90 .... B4 120
Cragford, Clay, Ala., 100 .... B4 78
Cragged, mtn., N.H C4 105
Cragmor, El Paso, Colo. (part of Ivywild) C6 83
Crags, mts., Idaho C3 89
Craig, Alsk., 273 .... D13, n23 79
Craig, Moffat, Colo., 3,984 . A3 83
Craig, Monroe, Fla., 50 .... H6 86
Craig, Plymouth, Iowa, 117 .. B1 92
Craig, Holt, Mo., 488 .... A2 101
Craig, Lewis and Clark, Mont., 50 .... C5 102
Craig, Burt, Nebr., 378 .... C9 103
Craig, co., Okla., 16,303 .... A6 112
Craig, co., Va., 3,356 .... D2 121
Craig, creek, Va D2 121
Craig Beach, Mahoning, Ohio, 1,139 .... *A7 111
Craigfield, Williamson, Tenn . B4 117
Craig Harbor, N. W. Ter., Can B23 4
Craighead, co., Ark., 47,253 . B5 81
Craighouse, Scot E3 13
Craighurst, Ont., Can., 56 ... C5 72
Craigmont, Lewis, Idaho, 703 .... C2 89
Craigmyle, Alta., Can., 107 . D4 69
Craigs Road Station, Que., Can., 75 .... D8 73
Craigsville, Augusta, Va., 978 .... C3 121
Craigsville, Nicholas, W. Va., 250 .... C4 123
Craigville, Wells, Ind., 190 . C7 91
Craigville, Koochiching, Minn., 35 .... C5 99
Craik, Sask., Can., 606 .... F3 70
Crail, Scot., 1,066 .... D6 13
Crailsheim, Ger., 14,400 .... D5 17
Craiova, Rom., 96,929 .... C6 22
Cramerton, Gaston, N.C., 3,123 .... B2 109
Crampel, Alg., 161 .... C4 44
Crampton, Henry, Ill C4 90
Cranberry, Venango, Pa., 275 .... D2 114
Cranberry, hill, Conn D6 84
Cranberry, lake, N.Y A6 108
Cranberry Isles, Hancock, Maine, 135 (181▲) D4 96
Cranberry Lake, B.C., Can., 1,350 .... E5 68
Cranberry Lake, St. Lawrence, N.Y., 150 .... A6, B2 108
Cranberry Portage, Man., Can., 838 .... B1 71
Cranbrook, B.C., Can., 5,549 .... E10 68
Cranbury, Fairfield, Conn. (part of Norwalk) E3 84
Cranbury, Middlesex, N.J., 1,038 .... C3 106
Cranbury Station, Middlesex, N.J., 50 .... C4 106
Crandall, Man., Can., 105 .... D1 71
Crandall, Murray, Ga., 208 .. B2 87
Crandall, Clarke, Miss., 75 .. D5 100
Crandall, Day, S. Dak., 30 .. B8 116
Crandall, Kaufman, Tex., 640 . B6 118
Crandon, Forest, Wis., 1,679 . C5 124
Crane, Stone, Mo., 954 .... E4 101
Crane, Richland, Mont., 85 . C12 102
Crane, Harney, Oreg., 75 ... D8 113
Crane, Crane, Tex., 3,796 .. D1 118
Crane, co., Tex., 4,699 .... D1 118
Crane, creek, Ohio A2 111
Crane, lake, Sask., Can G1 70
Crane, lake, Ill C3 90
Crane, lake, Minn B6 99
Crane, mtn., Oreg E6 113
Crane Creek, res., Idaho E2 89
Crane Hill, Cullman, Ala., 90 .... A2 78
Crane Lake, St. Louis, Minn., 100 .... B6 99
Crane Neck, pt., N.Y F4 84
Crane Prairie, res., Oreg D5 113
Cranesville, Erie, Pa., 575 ... C1 114
Crane Valley, Sask., Can., 113 .... H3 70
Cranfield, Adams, Miss., 100 . D2 100
Cranford, Union, N.J., 26,424 .... B4 106
Crannell, Fresno, Calif B1 82
Cransac, Fr., 4,765 .... E5 14
Cranston, Providence, R.I., 66,766 .... B11 84
Crapo, Dorchester, Md., 40 .. D5 85
Crary, Ramsey, N. Dak., 195 . A7 110
Craryville, Columbia, N.Y., 100 .... C7 108
Crasna, riv., Rom B6 22
Crater, lake, Oreg E4 113
Crater Lake, Klamath, Oreg., 50 .... E4 113
Crater Lake, nat. park, Oreg E4 113
Craters of the Moon, nat. mon., Idaho F5 89
Crateús, Braz., 14,572 .... C2 57
Crato, Braz., 27,649 .... C3 57
Craughwell, Ire., 159 .... D3 11
Craven, Sask., Can., 147 .... G3 70
Craven, co., N.C., 58,773 .... B6 109
Cravinhos, Braz., 6,294 .... k8 56
Crawford, Delta, Colo., 147 .. C3 83
Crawford, Oglethorpe, Ga., 541 .... C3 87
Crawford, Washington, Maine (83▲) C5 96
Crawford, Lowndes, Miss., 317 .... B5 100
Crawford, Dawes, Nebr., 1,588 .... B2 103
Crawford, Roger Mills, Okla., 50 .... B2 112
Crawford, co., Ark., 21,318 .. B1 81
Crawford, co., Ga., 5,816 .... D3 87
Crawford, co., Ill., 20,751 ... D6 90
Crawford, co., Ind., 8,379 ... H4 91
Crawford, co., Iowa, 18,569 . B2 92
Crawford, co., Kans., 37,032 . E9 93
Crawford, co., Mich., 4,971 .. D6 98
Crawford, co., Mo., 12,647 ... D6 101
Crawford, co., Ohio, 46,775 .. B5 111
Crawford, co., Pa., 77,956 .... C1 114
Crawford, co., Wis., 16,351 ... E3 124
Crawford, lake, Maine C5 96
Crawford House, Coos, N.H., 10 .... B4 105
Crawfordsville, Crittenden, Ark., 744 .... B5 81
Crawfordsville, Montgomery, Ind., 14,231 .... D4 91
Crawfordsville, Washington, Iowa, 317 .... C6 92
Crawfordsville, Linn, Oreg., 170 .... C4 113
Crawfordville, Wakulla, Fla., 650 .... B2 86
Crawfordville, Taliaferro, Ga., 786 .... C4 87
Crayne, Crittenden, Ky., 125 . A3 94
Crazy, mts., Mont D6 102
Crazy, peak, Mont D6 102
Crazy Woman, creek, Wyo A6 125
Creagerstown, Frederick, Md., 75 .... A3 85
Creal Springs, Williamson and Marion, Ill., 784 .... F5 90
Cream, hill, Conn B3 84
Creamridge, Monmouth, N.J., 50 .... C3 106
Crean, lake, Sask., Can C2 70
Crediton, Ont., Can., 429 ... D3 72
Crediton, Eng., 4,422 .... D4 12
Cree, lake, Sask., Can E11 66
Creede, Mineral, Colo., 350 .. D4 83
Creedmoor, Granville, N.C., 862 .... A5 109
Creek, co., Okla., 40,495 .... B5 112
Creekside, Indiana, Pa., 482 . E3 114
Creelman, Sask., Can., 196 .. H4 70
Creemore, Ont., Can., 850 .... C4 72
Creggan, N. Ire C4 11
Cregganbaun, Ire D2 11
Creighton, Ont., Can., 1,727 . E9 72
Creighton, Sask., Can., 1,729 .... *C5 70
Creighton, Cass, Mo., 228 ... C3 101
Creighton, Knox, Nebr., 1,388 .... B8 103
Creighton, Allegheny, Pa., 2,865 .... A6 114
Creil, Fr., 19,235 (*48,000) . C5 14
Crellin, Garrett, Md., 425 .. D1 85
Crema, It., 30,035 .... B2 21
Cremona, It., 73,902 .... B3 21
Crenshaw, Panola, Miss., 1,382 .... A3 100
Crenshaw, Jefferson, Pa., 350 . D4 114
Crenshaw, co., Ala., 14,909 . D3 78
Creola, Mobile, Ala., 500 .... E1 78
Creole, Cameron, La., 150 ... E2 95
Crépy-en-Valois, Fr., 7,379 .. E2 15
Cres, isl., Yugo C2 22
Cresaptown, Allegany, Md., 1,680 .... D2 85
Cresbard, Faulk, S. Dak., 229 .... B7 116
Crescent, McIntosh, Ga., 80 .... E5 87
Crescent, Pottawattamie, Iowa, 296 .... C2 92
Crescent, St. Louis, Mo., 300 . B7 101
Crescent, Albany, N.Y., 75 .. C7 107
Crescent, Logan, Okla., 1,264 .... B4 112
Crescent, Klamath, Oreg., 350 .... D5 113
Crescent, lake, Fla C5 86
Crescent, lake, Oreg D5 113
Crescent, lake, Wash A2 122
Crescent, range, N.H B4 105
Crescent Beach, New London, Conn. (part of Niantic) ... D8 84
Crescent Beach, Cumberland, Maine (part of Cape Elizabeth) E5 96
Crescent City, Del Norte, Calif., 2,958 .... B1 82
Crescent City, Putnam, Fla., 1,629 .... C5 86
Crescent City, Iroquois, Ill., 393 .... C6 90
Crescent City Northwest, Del Norte, Calif., 3,086 .... *B1 82
Crescent Junction, Grand, Utah, 12 .... E6 119
Crescent Lake, Klamath, Oreg., 175 .... D5 113
Crescent Springs, Kenton, Ky., 946 .... A7 94
Cresco, Howard, Iowa, 3,809 . A5 92
Cresco, Monroe, Pa., 500 ... D11 114
Crespo, see Villa Crespo, Arg.
Cresskill, Bergen, N.J., 7,290 . D5 106
Cresson, Cambria, Pa., 2,659 . F4 114
Cresson, Hood, Tex., 300 .... B5 118
Cressona, Schuylkill, Pa., 1,854 .... E9 114
Crest, Fr., 4,494 .... E6 14
Cresta, Switz D8 19
Crested Butte, Gunnison, Colo., 289 .... C4 83
Crest Hill, Will, Ill., 5,887 . *F2 90
Cresthill, Fauquier, Va., 50 . C5 121
Crestline, San Bernardino, Calif., 1,290 .... F3 82
Crestline, Lincoln, Nev., 15 . F7 104
Crestline, Crawford, Ohio, 5,521 .... B5 111
Crestmont, Montgomery, Pa., 2,000 .... *F11 114
Crestmore, San Bernardino and Riverside, Calif., 2,000 ... *E5 82
Crestmoor, Morris, N.J., 30 . B3 106
Creston, B.C., Can., 2,460 ... E9 68
Creston, Newf., Can., 837 ... E4 75
Creston, Ogle, Ill., 454 .... B5 90
Creston, Union, Iowa, 7,667 . C3 92
Creston, Flathead, Mont., 25 . B2 102
Creston, Platte, Nebr., 177 .. C8 103
Creston, Wayne, Ohio, 1,522 . B6 111
Creston, Malheur, Oreg D8 113
Creston, Calhoun, S.C., 250 . D6 115
Creston, Cumberland, Tenn., 25 .... C8 117
Creston, Lincoln, Wash., 317 . B7 122
Creston, Wirt, W. Va., 75 ... C3 123
Crestone, Saguache, Colo., 51 .... D5 83
Crestone, peak, Colo D5 83
Crestview, Okaloosa, Fla., 7,467 .... G2 86
Crestview, Lawrence, Tenn . B4 117
Crestwood, Cook, Ill., 3,918 . *B6 90
Crestwood, Oldham, Ky., 600 .... A4, B4 94
Crestwood, St. Louis, Mo., 11,106 .... *B8 101
Crestwood, Norfolk, Va., 2,200 .... *E6 121
Crestwynd, Sask., Can., 42 . G3 70
Creswell, Washington, N.C., 402 .... B7 109
Creswell, Lane, Oreg., 760 .. D3 113
Crete, Will, Ill., 3,463 .... B6 90
Crete, Saline, Nebr., 3,546 .. D9 103
Crete, Sargent, N. Dak., 200 . C8 110
Crete, isl., Grc E5 23
Crete, sea, Grc E5 23
Cretone, It., 374 .... g9 21
Creuse, dept., Fr., 163,515 .. *D4 14
Creuse, riv., Fr D4 14
Creutzwald, Fr., 13,649 .... C7 14
Creve Coeur, Tazewell, Ill., 6,684 .... C4 90
Creve Coeur, St. Louis, Mo., 5,122 .... *C7 101
Crevillente, Sp., 14,047 .... C5 20
Crewe, Eng., 53,394 .... A5 12
Crewe, Nottoway, Va., 2,012 . D4 121
Crewkerne, Eng., 4,215 .... E5 12
Criam More, mtn., Scot B5 13
Cricket, Wilkes, N.C., 950 .. A2 109
Cricket, mts., Utah E2 119
Cridersville, Auglaize, Ohio, 1,053 .... B3 111
Crieff, Scot., 5,773 .... D5 13
Criffel, mtn., Scot F5 13
Criglersville, Madison, Va., 45 .... C4 121
Crikvenica, Yugo., 3,637 .... C2 22
Crimea, see Krym, Sov. Un.
Crimmitschau, Ger., 31,300 . C7 17
Crinan, Scot D3 13
Cripple Creek, Teller, Colo., 614 .... C5 83
Cripple Creek, Wythe, Va., 300 .... E1 121
Crisana-Maramures, prov., Rom., 1,391,672 .... *B6 22
Crisfield, Somerset, Md., 3,540 .... E6 85
Crisman, Porter, Ind. (part of Portage) A3 91
Crisp, Ellis, Tex B6 118
Crisp, co., Ga., 17,768 .... E3 87
Criss Creek, B.C., Can D7 68
Crissolo, It., 701 .... E3 21
Cristal, mts., Gabon E2 46
Cristalina, Braz., 3,810 .... E1 57
Cristobal, C.Z., 817 .... k11 62
Cristobal, dist., C.Z., 11,499 .... *k11 62
Cristóbal Colón, mtn., Col A3 60
Crisul Alb, riv., Rom B5 22
Crittenden, Grant, Ky., 287 .... B5, B7 94
Crittenden, Nansemond, Va., 250 .... B6 121
Crittenden, co., Ark., 47,564 . B5 81
Crittenden, co., Ky., 8,648 .. A3 94
Crivitz, Ger., 5,879 .... E5 24
Crivitz, Marinette, Wis., 650 .... C6 124
Crna, riv., Yugo E5 22
Crna Gora (Montenegro), rep., Yugo., 419,625 .... *D4 22
Črnomelj, Yugo., 2,326 .... C2 22
Croatia, rep., Yugo., 4,148,122 .... *C2 22
Croatia, reg., Yugo C2 22
Croche, riv., Que., Can A5 73
Crocheron, Dorchester, Md., 125 .... D5 85
Crocker, Pulaski, Mo., 821 .. D5 101
Crocker, Clark, S. Dak., 77 .. B8 116
Crockett, Contra Costa, Calif., 4,500 .... B5 82
Crockett, Houston, Tex., 5,356 .... D5 118
Crockett, co., Tenn., 14,594 . B2 117
Crockett, co., Tex., 4,209 .... D2 118
Crockett Mills, Crockett, Tenn., 125 .... B2 117
Crocketville, Hampton, S.C., 75 .... F5 115
Crofton, Christian, Ky., 892 . F5 94
Crofton, Knox, Nebr., 604 .... B8 103
Croghan, Lewis, N.Y., 821 ... B2 108
Crohy, head, Ire C3 11
Croix, Fr., 20,081 .... D3 15
Croker, cape, Ont., Can C4 72
Croker, isl., Austl B5 50
Crolly, Ire B3 11
Cromarty, Scot., 605 .... C4 13
Cromer, Man., Can., 71 .... E1 71
Cromer, Eng., 4,895 .... B9 12
Cromona (Haymond), Letcher, Ky., 950 .... C7 94
Cromwell, Choctaw, Ala., 150 .... C1 78
Cromwell, Middlesex, Conn., 6,780 .... C6 84
Cromwell, Noble, Ind., 451 . B6 91
Cromwell, Union, Iowa, 138 . C3 92
Cromwell, Ohio, Ky., 200 .... C3 94
Cromwell, Carlton, Minn., 187 .... D6 99
Cromwell, N.Z., 942 .... P12 51
Cromwell, Seminole, Okla., 269 .... B5 112
Cromwell, Pierce, Wash D1 122
Crook, Logan, Colo., 209 .... A8 83
Crook, Eng., 25,218 .... F7 13
Crook, co., Oreg., 9,430 .... C6 113
Crook, co., Wyo., 4,691 .... A8 125
Crooked, creek, Ark A3 81
Crooked, creek, Kans B7 93
Crooked, creek, Kans E3 93
Crooked, creek, Pa C7 114
Crooked, isl., Ba. Is D6 64
Crooked, lake, Newf., Can D3 75
Crooked, lake, Fla E5 86
Crooked, lake, Minn B7 99
Crooked, riv., B.C., Can B6 68
Crooked, riv., Oreg C6 113
Crooked Creek, Alsk., 92 .... C8 79
Crooked Creek, Tioga, Pa., 90 .... C7 114
Crooked Creek, res., Pa E3 114
Crooked River, Sask., Can., 163 .... E4 70
Crooks, Minnehaha, S. Dak., 135 .... D9 116
Crookston, Polk, Minn., 8,546 .... C2 99
Crookston, Cherry, Nebr., 139 .... B5 103
Crookstown, Ire., 124 .... F3 11
Crooksville, Perry, Ohio, 2,958 .... C5 111
Crookwell, Austl., 2,340 .... G7 51
Croom, Ire., 720 .... E3 11
Croom Station, Prince Georges, Md C4 85
Cropper, Shelby, Ky., 250 ... B4 94
Cropsey, McLean, Ill., 170 ... C5 90
Crosby, Crow Wing, Minn., 2,629 .... D5 99
Crosby, Amite, and Wilkinson, Miss., 705 .... D2 100
Crosby, Divide, N. Dak., 1,759 .... A2 110
Crosby, McKean, Pa., 350 .... C5 114
Crosby, Harris, Tex., 1,200 .. F5 118
Crosby, co., Tex., 10,347 .... C2 118
Crosby, mtn., Wyo B3 125
Crosbyton, Crosby, Tex., 2,088 .... C2 118
Crosland, Colquitt, Ga., 95 ... E3 87
Cross, Berkeley, S.C., 100 .. E7 115
Cross, Grimes, Tex., 25 .... D2 118
Cross, co., Ark., 19,551 .... B5 81
Cross, cape, S.W. Afr B1 49
Cross, creek, W. Va B2 123
Cross, isl., Maine D5 96
Cross, lake, Man., Can B3 71
Cross, lake, Man., Can C2 71
Cross, lake, La B2 95
Cross, lake, Maine A4 96
Cross, mtn., Tenn C9 117
Cross, mts., Ark C1 81
Cross, riv., Nig E6 45
Cross, sound, Alsk k21 79
Cross Anchor, Spartanburg, S.C., 500 .... B4 115
Crossbost, Scot B2 13
Cross Canyon, Apache, Ariz .. B6 80
Cross City, Dixie, Fla., 1,857 . C3 86
Cross Creek, N.B., Can., 252 . C3 74
Cross Creek, Brooke, W. Va . *A4 123
Crossdoney, Ire D4 11
Crosses, Madison, Ark B2 81
Crossett, Ashley, Ark., 5,370 . D4 81
Cross Fell, mtn., Eng F6 13
Crossfell Edge, mts., Eng F6 13
Crossfield, Alta., Can 593 ... D3 69
Crossgar, N. Ire., 843 .... C6 11
Crosshaven, Ire., 858 .... F3 11
Cross Hill, Laurens, S.C., 441 .... C4 115
Cross Keys, Bibb, Ga., 1,000 .... *D3 87
Cross Keys, Gloucester, N.J., 140 .... D2 106
Crosslake, Crow Wing, Minn., 165 .... D4 99
Crossley, Ocean, N.J D4 106
Crossmaglen, N. Ire., 932 .... C5 11
Cross Mill, McDowell, N.C., 700 .... D4 109
Crossmolina, Ire., 777 .... C2 11
Cross Plains, Ripley, Ind., 160 .... G7 91
Cross Plains, Robertson, Tenn., 200 .... A5 117
Cross Plains, Callahan, Tex., 1,168 .... C3 118
Cross Plains, Dane, Wis., 1,066 .... E4 124
Cross River, res., N.Y D2 84
Cross Roads, San Bernardino, Calif., 150 .... E6 82
Crossroads, Lea, N. Mex., 15 . D6 107
Cross Roads Ohio, N.S., Can., 143 .... D7 86
Cross Timbers, Hickory, Mo., 186 .... D4 101
Crosstown, Perry, Mo., 83 ... D8 101
Cross Village, Emmet, Mich., 50 .... C5 98
Crossville, DeKalb, Ala., 579 .... A4 78
Crossville, White, Ill., 874 ... E5 90
Crossville, Cumberland, Tenn., 4,668 .... D8 117
Crosswicks, Burlington, N.J., 550 .... C3 106
Croswell, Sanilac, Mich., 1,817 .... E8 98
Crothersville, Jackson, Ind., 1,449 .... G6 91
Croton, Lee, Iowa, 60 .... D6 92
Croton, Newaygo, Mich E5 98
Croton (Hartford), Licking, Ohio, 397 .... B5 111
Crotone, It., 43,256 .... E6 21
Croton Falls, Westchester, N.Y., 900 .... D7 108
Croton Falls, res., N.Y D2 84
Croton-on-Hudson, Westchester, N.Y., 6,812 ... D3, D7 108
Crottendorf, Ger., 5,562 .... C7 17
Crouch, Boise, Idaho, 89 .... E3 89
Crouse, Lincoln, N.C., 901 .. B2 109
Crouseville, Aroostook, Maine, 230 .... B4 96
Crow, creek, Colo A6 83
Crow, creek, Wyo D8 125
Crow, Indian res., Mont E9 102
Crow, riv., Minn F5 99
Crow Agency, Big Horn, Mont., 600 .... E9 102
Crowder, Panola and Quitman, Miss., 528 .... A3 100
Crowder, Pittsburg, Okla., 254 .... B6 112
Crowell, Foard, Tex., 1,703 . C3 118
Crowheart, Fremont, Wyo., 5 .... B3 125
Crowley, Crowley, Colo., 265 .... C7 83
Crowley, Acadia, La., 15,617 . D3 95
Crowley, Tarrant, Tex., 583 . B5 118
Crowley, co., Colo., 3,978 .... C7 83
Crowley, lake, Calif D4 82
Crowleys, ridge, Mo E7 101
Crown City, Gallia, Ohio, 323 .... D5 111
Crown King, Yavapai, Ariz., 50 .... C3 80
Crown Point, Lake, Ind., 8,443 .... B3 91
Crown Point, Jefferson, La., 175 .... C7 95
Crownpoint, McKinley, N. Mex., 300 .... B1 107
Crown Point, Essex, N.Y., 500 .... B7 108
Crows Nest, B.C., Can., 59 . E10 68
Crows Nest, Marion, Ind., 122 .... H8 91
Crows Nest, mtn., S. Dak C2 116
Crowsnest, pass, Alta., Can E3 69
Crow Wing, co., Minn., 32,134 .... D4 99
Crow Wing, riv., Minn D4 99
Croydon, Austl., 86 .... C7 50
Croydon, Eng., 252,387 .... C7 12, E6, m12 10
Croydon, Sullivan, N.H., 100 (312▲) D2 105
Croydon, Bucks, Pa., 9,000 . F12 114
Croydon, Morgan, Utah, 91 . B4 119
Croydon, mtn., N.H D2 105
Croydon, peak, N.H D2 105
Croydon, riv., N.H D2 105
Croydon Flat, Sullivan, N.H., 130 .... D2 105
Crozet, Albemarle, Va., 900 .. C4 121
Crozet, is., Indian O J24 3
Crozier, Mohave, Ariz., 10 .. B2 80
Crozier, Goochland, Va., 100 . D2 121
Crozon, Fr., 2,130 .... C1 14
Cruachan, mtn., Scot D3 13
Cruagh, isl., Ire D1 11
Cruce Magueyes, P.R B4 65
Crucero, Peru, 226 .... D3 58
Cruces, Cuba, 10,704 .... D3 64
Crucible, Greene, Pa., 1,064 . G1 114
Cruden, Scot., 2,294 .... C7 13
Cruger, Holmes, Miss., 362 . B3 100
Crum, Wayne, W. Va., 500 .. D2 123
Crumlin, N. Ire., 822 .... C5 11
Crum Lynne, Delaware, Pa., 3,500 .... *B10 114
Crummock Water, lake, Eng F5 13
Crump, Bay, Mich., 50 .... E6 98
Crump, lake, Oreg E7 113
Crumpton, Queen Annes, Md., 300 .... B6 85
Crumstown, St. Joseph, Ind., 250 .... A5 91
Cruseilles, Fr., 865 .... C2 18
Crusheen, Ire., 97 .... E3 11
Cruso, Haywood, N.C., 175 .. D3 109
Crutchfield, Surry, N.C A3 109
Crutwell, Sask., Can., 115 .. D2 70
Cruz, cape, Cuba F5 64
Cruz Alta, Arg., 4,196 .... A4 54
Cruz Alta, Braz., 33,190 .... D2 56
Cruz Bay, Vir. Is., 200 .... f15 65
Cruz del Eje, Arg., 15,563 .. A4 54
Cruzeiro, Braz., 27,005 .... C4 56
Cruzeiro do Sul, Braz., 4,807 . C3 58
Cruz Grande, Chile E1 55
Crysler, Ont., Can., 498 .... B9 72
Crystal, Montcalm, Mich., 600 .... E6 98
Crystal, Hennepin, Minn., 24,283 .... E6 99
Crystal, Coos, N.H., 40 .... A4 105
Crystal, San Juan, N. Mex .. A1 107
Crystal, Pembina, N. Dak., 372 .... A8 110
Crystal, Klamath, Oreg E4 113
Crystal, bay, Fla D4 86
Crystal, caverns, Mo E4 101
Crystal, lake, Conn B7 84
Crystal, lake, Mich D4 98
Crystal, lake, N.H D4 105
Crystal, lake, Vt B4 120
Crystal, mtn., N.H B2 105
Crystal, pond, Conn B8 84
Crystal, riv., Colo B3 83
Crystal Bay, Washoe, Nev., 500 .... D2 104
Crystal Beach, Ont., Can., 1,886 .... E5 72
Crystal Beach, Pinellas, Fla., 1,000 .... D4 86
Crystal City, Man., Can., 542 .... E2 71
Crystal City, Jefferson, Mo., 3,678 .... B8, C7 101
Crystal City, Zavala, Tex., 9,101 .... E3 118
Crystal Falls, Iron, Mich., 2,203 .... B2 98
Crystal Hill, Halifax, Va., 150 .... E4 121
Crystal Lake, Tolland, Conn., 640 .... B7 84
Crystal Lake, Washington, Fla., 100 .... G3 86
Crystal Lake, McHenry, Ill., 8,314 .... A5, E1 90
Crystal Lake, Hancock, Iowa, 267 .... A4 92
Crystal Lake, cave, Iowa B7 92
Crystal Lakes, Clark, Ohio, 1,569 .... C3 111
Crystal River, Citrus, Fla., 1,423 .... D4 86
Crystal Springs, Garland, Ark., 100 .... C2, C5 81
Crystal Springs, Sask., Can., 113 .... E3 70
Crystal Springs, Pasco, Fla., 350 .... D4 86
Crystal Springs, Copiah, Miss., 4,496 .... D3 100
Crystal Springs, Kidder, N. Dak., 30 .... C6 110
Crystal Valley, Oceana, Mich., 100 .... E4 98
Csongrad, Hung., 16,801 (20,690▲) B4 22
Csongrad, co., Hung., 335,565 .... *B5 22
Csorna, Hung., 9,208 .... B3 22
Cuajimalpa, Mex., 3,504 .... h9 63
Cuando, riv., N. Rh., Ang E3 48
Cuangar, Ang., 136 .... E2 48
Cuango, riv., Ang C2 48
Cuanza, riv., Ang C2 48
Cuanza Norte, dist., Ang., 263,101 .... C1 48
Cuanza Sul, dist., Ang., 405,012 .... D1 48
Cua Rao, Viet C6 38
Cuarto, riv., Arg A4 54
Cuatro Calles, P.R. (part of Arrayo) D6 65
Cuauhtémoc, Mex., 14,639 ... B3 63
Cuautepec, Mex., 5,122 .... g9 63
Cuautla [Morelos], Mex., 11,847 .... n14 63
Cuba, Sumter, Ala., 390 .... C1 78
Cuba, Fulton, Ill., 1,380 .... C3 90
Cuba, Republic, Kans., 336 .. C6 93
Cuba, Crawford, Mo., 1,672 .... C6 101
Cuba, Sandoval, N. Mex., 800 .... A3 107
Cuba, Allegany, N.Y., 1,949 . C2 108
Cuba, Clinton, Ohio, 150 .... C4 111
Cuba, Port., 4,394 .... C2 20
Cuba, country, N.A., 5,829,029 .... G13 61, E4 64
Cuba City, Grant, Wis., 1,673 .... F3 124
Cubal, Ang D1 48
Cuba Landing, Humphreys, Tenn B4 117
Cubango, riv., Ang E2 48
Cubero, Valencia, N. Mex., 225 .... B2 107
Cubia, Ang E3 48
Cub Run, Hart, Ky., 250 .... C3 94
Cucamonga, San Bernardino, Calif., 2,500 .... F3 82
Cuchara, Huerfano, Colo., 20 .... D6 83
Cuchillo, Sierra, N. Mex., 200 . D2 107
Cuchivero, riv., Ven B4 60
Cuchora, riv., Colo D6 83
Cuckfield, Eng., 20,113 .... C7 12
Cúcuta, Col., 86,300 (116,700▲) B3 60
Cudahy, Los Angeles, Calif., 12,000 .... *F2 82
Cudahy, Milwaukee, Wis., 17,975 .... E2, F6 124
Cuddalore, India, 79,168 .... F6 39
Cuddapah, India, 49,027 .... F6 39
Cuddy, Allegheny, Pa., 1,400 .... *B5 114
Cudgewa, Austl., 192 .... H6 51
Cudjos, cave, Ky., Va D6 94
Cudworth, Sask., Can., 628 . E3 70
Cue, Austl., 511 .... E2 50
Cuéllar, Sp., 7,514 .... B3 20
Cuenca, Ec., 46,428 .... B2 58
Cuenca, Sp., 27,007 .... B4 20
Cuenca, prov., Sp., 315,433 . *B4 20
Cuenca, mts., Sp B4 20
Cuencamé [de Ceniceros], Mex., 2,321 .... C4 63
Cuernavaca, Mex., 35,824 .... D5, n14 63
Cuero, DeWitt, Tex., 7,338 . E4 118
Cuervo, Cuba k12 64
Cuervo, Guadalupe, N. Mex., 160 .... B5 107
Cuetzalan [del Progreso], Mex., 4,006 .... m15 63
Cuevas, Harrison, Miss., 500 . E2 100
Cuevas, Sp., 9,530 .... D5 20
Cuglieri, It., 4,435 .... D2 21
Cuiabá, Braz., 43,112 .... B1 56
Cuiabá, riv., Braz B1 56
Cuicas, Ven., 963 .... B3 60
Cuicatlán, Mex., 1,986 .... o15 63
Cuilapa, Guat., 2,746 .... C2 62
Cuilco, Guat., 519 .... C2 62
Cuilcagh, mtn., Ire C4 11
Cuillin, hills, Scot C2 13
Cuillin, sound, Scot C2 13
Cuipo, Pan., 99 .... k10 62
Cuito, riv., Ang E2 48
Cuito Cuanavale, Ang E2 48
Cuitzéo, lake, Mex n13 63
Cuivre, riv., Mo B6 101
Culberson, co., Tex., 2,794 .. F2 118
Culbertson, Roosevelt, Mont., 919 .... B12 102
Culbertson, Hitchcock, Nebr., 803 .... D5 103
Culcairn, Austl G6 51
Culdaff, bay, Ire B4 11
Culdesac, Nez Perce, Idaho, 209 .... C2 89
Culebra, P.R., 498 .... f13 65
Culebra, mun., P.R., 573 .... f13 65
Culebra, isl., P.R f13 65
Culebra, peak, Colo D5 83
Culebrinas, riv., P.R B2 65
Culebrita, isl., P.R f14 65
Culhuacán, Mex., 2,087 .... h9 63
Culiacán, Mex., 84,602 .... C3 63
Culion, Phil., 3,279 .... C6 35
Cúllar de Baza, Sp., 9,502 .. D4 20
Cullen, Webster, La., 2,194 . B2 95
Cullen, Frederick, Md., 550 .. A3 85
Cullen, Scot., 1,327 .... C6 13
Culleoka, Maury, Tenn., 300 . B4 117
Cullera, Sp., 14,103 .... C5 20
Cullion, N. Ire C4 11
Cullison, Pratt, Kans., 129 ... E5 93
Cullman, Cullman, Ala., 10,883 .... A3 78

Cullman, co., Ala., 45,572...A3 78
Culloden, Monroe, Ga., 260...D2 87
Culloden, Cabell, W. Va., 700...*C2 123
Culloden, pt., N.Y...E9 84
Cullom, Livingston, Ill., 555.C5 90
Cullowhee, Jackson, N.C., 1,500...D3 109
Culmore, Fairfax, Va., 1,700...*B5 121
Culoz, Fr., 2,317...D1 18
Culp Creek, Lane, Oreg., 100...D4 113
Culpeper, Van Buren, Ark...B3 81
Culpeper, Culpeper, Va., 2,412...C4 121
Culpeper, co., Va., 15,088..C5 121
Cultus, lake, B.C., Can...A4 121
Culuene, riv., Braz...A2 56
Culver, Marshall, Ind., 1,558...B5 91
Culver, Ottawa, Kans., 200..D6 93
Culver, St. Louis, Minn., 60..D6 99
Culver, Jefferson, Oreg., 301 C5 113
Culver City, Los Angeles, Calif., 32,163...F2 82
Culvers, lake, N.J...A3 106
Culverton, Hancock, Ga...C4 87
Culzean, bay, Scot...E3 13
Cumaná, Ven., 71,563...A5 60
Cumberland, B.C., Can., 1,303...E5 68
Cumberland, Ont., Can., 360...A10 72
Cumberland, Marion, Ind., 872...E6, H8 91
Cumberland, Cass, Iowa, 425...C3 92
Cumberland, Harlan, Ky., 4,271...D7 94
Cumberland, Cumberland, Maine, 600 (2,765▲)...E2, E5 96
Cumberland, Allegany, Md., 33,415...D2 85
Cumberland, Webster, Miss., 145...B4 100
Cumberland, Cumberland, N.J...E3 106
Cumberland, Cumberland, N.C., 500...B5 109
Cumberland, Guernsey, Ohio, 493...C6 111
Cumberland, Marshall, Okla., 250...C5 112
Cumberland, Providence, R.I., 7,500 (18,792▲)...B11 84
Cumberland, Cumberland, Va., 250...D4 121
Cumberland, King, Wash., 160...D2 122
Cumberland, Barron, Wis., 1,860...C1 124
Cumberland, co., N.S., Can., 37,767...D5 74
Cumberland, co., Eng., 294,162...*C5 10
Cumberland, co., Ill., 9,936..D5 90
Cumberland, co., Ky., 7,835.D4 94
Cumberland, co., Maine, 182,751...E2 96
Cumberland, co., N.J., 106,850...E2 106
Cumberland, co., N.C., 148,418...B5 109
Cumberland, co., Pa., 124,816...F7 114
Cumberland, co., Tenn., 19,135...D8 117
Cumberland, co., Va., 6,360.D4 121
Cumberland, isl., Ga...F5 87
Cumberland, lake, Sask., Can...C4 70
Cumberland, lake, Ky...D4 94
Cumberland, mtn., Tenn...C9 117
Cumberland, pen., N.W. Ter., Can...C20 67
Cumberland, plat., Ala., Ky., Tenn...A3 78, C6 94, D7 117
Cumberland, riv., U.S...C11 77
Cumberland, sound, N.W. Ter., Can...C19 67
Cumberland Center, Cumberland, Maine, 600 (2,765▲).E5 96
Cumberland, City, Stewart, Tenn., 314...A4 117
Cumberland Furnace, Dickson, Tenn., 250...A4 117
Cumberland Gap, Claiborne, Tenn., 291...C10 117
Cumbres, pass, Colo...D4 83
Cumbrian, mts., Eng...F5 13
Cuming, co., Nebr., 12,435..C9 103
Cumming, Forsyth, Ga., 1,561...B2 87
Cumming, Warren, Iowa, 148...A7 92
Cummings, Atchison, Kans., 50...A7 93
Cummings, Traill, N. Dak., 51...B8 110
Cummingsville, Van Buren, Tenn...D8 117
Cummington, Hampshire, Mass., 200 (550▲)...B2 97
Cumnock, Lee, N C., 200...B4 109
Cumnock, Scot., 5,403...E4 13
Cumpas, Mex., 2,314...A3 63
Cumra, Tur., 7,100...D9 31
Cunard, Fayette, W. Va., 450...D7 123
Cunco, Chile, 2,728...B2 54
Cuncumen, Chile...A2 54
Cundiff, Adair, Ky., 125...D4 94
Cundinamarca, dept., Col., 1,840,890...C3 60
Cundys Harbor, Cumberland, Maine, 125...E6 96
Cunene, riv., Ang., S.W. Afr...A1 49
Cuneo, It., 34,000 (46,065▲)..B1 21
Cuney, Cherokee, Tex., 220...C5 118
Cunnamulla, Austl., 2,234...D5 52
Cunningham, Kingman, Kans., 618...E5 93
Cunningham, Carlisle, Ky., 300...A2 94
Cunningham, Montgomery, Tenn., 40...A4 117
Cupar, Sask., Can., 578...G3 70
Cupar, Scot., 5,495...D6 13
Cupertino, Santa Clara, Calif., 3,664...*C5 82
Cuprum, Adams, Idaho, 20..D2 89
Curacá, Braz., 1,264...C3 57
Curaçao, isl., Neth. W.I...A4 60
Curacautín, Chile, 9,201...B2 54
Curacó, riv., Arg...B3 54
Curañilahue, Chile, 3,995...B2 54
Curaray, riv., Ec., Peru...B2, B3 58
Curdsville, Daviess, Ky., 175.C2 94
Curepto, Chile, 1,739...B2 54
Curiapo, Ven., 374...B5 60
Curicó, Chile, 35,000...A2 54
Curicó, prov., Chile, 107,160.A2 54
Curicuriari, riv., Braz...D4 60
Curimata, riv., Braz...h6 57
Curitiba, Braz., 344,560...D3 56
Curitibanos, Braz., 8,339...D2 56
Curlew, Palo Alto, Iowa, 134...B3 92
Curlew, Ferry, Wash., 100...A7 122
Curlew, creek, Wash...A7 122
Curlew, lake, Wash...A7 122
Curon Venosta, It., 2,970...C6 18
Currais Novos, Braz., 7,782...C3, h5 57
Curralinho, Braz., 361...C5 59
Curran, Ont., Can., 113...B10 72
Curran, Alcona, Mich., 50...D7 98
Currans, Ire...E2 11
Currant, creek, Colo...C5 83
Curreeny, Ire...E3 11
Current, riv., Ark., Mo...D6 101
Currie, Murray, Minn., 438..F3 99
Currie, Elko, Nev., 25...C7 104
Currituck, Currituck, N.C., 250...A7 109
Currituck, co., N.C., 6,601...A7 109
Curry, Alsk., 20...C10, f17 79
Curry, co., N. Mex., 32,691..C6 107
Curry, co., Oreg., 13,983...E2 113
Curryville, Gordon, Ga., 40..B1 87
Curryville, Pike, Mo., 287...B6 101
Curtea de Arges, Rom., 10,764...C7 22
Curtice, Lucas and Ottawa, Ohio, 500...A2, A4 111
Curtici, Rom., 8,050...B5 22
Curtin, Douglas, Oreg., 40...D3 113
Curtis, Clark, Ark., 350...D2 81
Curtis, Mackinac, Mich., 40...B5 98
Curtis, Frontier, Nebr., 868..D5 103
Curtis, Woodward, Okla., 25.A2 112
Curtis, isl., Austl...A8 51
Curtiss, Clark, Wis., 147...D3 124
Curtisville, Allegheny, Pa., 1,376...A6, E2 114
Curuá, riv., Braz...D4 59
Curuá do Sul, riv., Braz...C4 59
Curuçá, Braz., 3,871...B1 57
Curug, Yugo., 9,457...C5 22
Curuguaty, Par., 2,342...D4 55
Curupira, mts., Ven...C5 60
Curuzú Cuatiá, Arg., 15,440.E4 55
Curve, Lauderdale, Tenn., 40...B2 117
Curvelo, Braz., 21,772...E2 57
Curver, mtn., Switz...C7 19
Curwensville, Clearfield, Pa., 3,231...E4 114
Cusco, Peru, 53,242...D3 58
Cusco, dept., Peru, 649,643..D3 58
Cushendall, N. Ire., 618...B5 11
Cushendunn, N. Ire...B5 11
Cushing, Woodbury, Iowa, 261...B2 92
Cushing, Knox, Maine, 130 (479▲)...D3 96
Cushing, Howard, Nebr., 56.C7 103
Cushing, Payne, Okla., 8,619...B5 112
Cushing, Nacogdoches, Tex., 388...D5 118
Cushman, Independence, Ark., 241...B4 81
Cushman, Hampshire, Mass., 300...B2 97
Cushman, Golden Valley, Mont., 47...D7 102
Cushman, Lane, Oreg., 150..D2 113
Cushman, mtn., N.H...C3 105
Cushman, res., Wash...B2 122
Cusick, Pend Oreille, Wash., 299...A8 122
Cusihuiriachic, Mex., 380...B3 63
Cusset, Fr., 11,468...D5 14
Cusseta, Chambers, Ala., 180...C4 78
Cusseta, Chattahoochee, Ga., 768...D2 87
Custar, Wood, Ohio, 246...A4 111
Custer, Mason, Mich., 365...E4 98
Custer, Yellowstone, Mont., 275...D9 102
Custer, McLean, N. Dak., 20...B4 110
Custer, Custer, Okla., 448...B3 112
Custer, Custer, S. Dak., 2,105...D2 116
Custer, Whatcom, Wash., 400...A3 122
Custer, co., Colo., 1,305...C5 83
Custer, co., Idaho, 2,996...E4 89
Custer, co., Mont., 13,227..D11 102
Custer, co., Nebr., 16,517...C6 103
Custer, co., Okla., 21,040...B2 112
Custer, co., S. Dak., 4,906..D2 116
Custer, peak, S. Dak...C2 116
Custer Battlefield, nat. mon., Mont...E9 102
Custer City, McKean, Pa., 400...C4 114
Cut Bank, Glacier, Mont., 4,539...B4 102
Cutbank, riv., Alta., Can...B1 69
Cutchogue, Suffolk, N.Y., 950...D4 108
Cutervo, Peru, 3,481...C2 58
Cuthbert, Randolph, Ga., 4,300...E2 87
Cut Knife, Sask., Can., 487..E1 70
Cutler, Tulare, Calif., 2,191...*D4 82
Cutler, Ont., Can., 200...A2 72
Cutler, Carroll, Ind., 200...D4 91
Cutler, Washington, Maine, 200 (654▲)...D5 96
Cutler City, Lincoln, Oreg., 525...C3 113
Cutler Ridge, Dade, Fla., 7,005...*G6 86
Cut Off, Lafourche, La., 700.E5 95
Cutshin, Leslie, Ky., 450...C6 94
Cuttack, India, 146,308...D8 38, G10 40
Cutten, Humboldt, Calif., 1,572...*B1 82
Cutter, Sierra, N. Mex., 20..D2 107
Cut Throat, isl., Newf., Can...A3 82
Cuttingsville, Rutland, Vt., 100...E3 120
Cuttyhunk, isl., Mass...D6 97
Cutzamala, riv., Mex...n13 63
Cuvo, riv., Ang...D1 48
Cuxhaven, Ger., 44,100...B4 16
Cuyahoga, co., Ohio, 1,647,895...A6 111
Cuyahoga, riv., Ohio...A6 111
Cuyahoga Falls, Summit, Ohio, 47,922...H6 111
Cuyama, riv., Calif...E4 82
Cuyamaca, peak, Calif...F5 82
Cuyamungue, Santa Fe, N. Mex., 120...F6 107
Cuylerville, Livingston, N.Y., 350...C3 108
Cuyo, Phil., 2,326...*C6 35
Cuyo, is., Phil...C6 35
Cuyuna, Crow Wing, Minn., 86...D5 99
Cuyuni, riv., Br. Gu...A3 59
Cuzco, Dubois, Ind., 30...H4 91
Cwmamman, Wales, 4,272...C4 12
Cybinka, Pol., 1,025...A9 17
Cyclades (Kikladhes), prov., Grc., 99,959...*D5 23
Cyclades, is., Grc...D5 23
Cyclone, McKean, Pa., 450..C4 114
Cygnet, Wood, Ohio, 593...A4 111
Cylinder, Palo Alto, Iowa, 161...A3 92
Cynthiana, Posey, Ind., 663...H2 91
Cynthiana, Harrison, Ky., 5,641...B5 94
Cynthiana, Pike, Ohio, 150..C4 111
Cypres, lake, Que., Can...C3 73
Cypress, Hale, Ala., 125...C2 78
Cypress, Orange, Calif., 1,753...*F2 82
Cypress, Jackson, Fla., 260..B1 86
Cypress, Johnson, Ill., 264...F4 90
Cypress, Harris, Tex., 25...F4 118
Cypress, bayou, Ark...B4 81
Cypress, hills, Sask., Can...H1 70
Cypress, lake, Sask., Can...H1 70
Cypress, lake, Fla...D5 86
Cypress Hills, park, Sask., Can.H1 70
Cypress Hills, peak, Sask., Can..H1 70
Cypress Inn, Wayne, Tenn., 250...B4 117
Cypress River, Man., Can., 288...E2 71
Cyprus, country, Asia, 528,618...E9 31
Cyrenaica, prov., Libya, 291,328...D4 43
Cyrene, Decatur, Ga., 200...F2 87
Cyril, Caddo, Okla., 1,284...C3 112
Cyril, riv., Man., Can...B4 71
Cyrus, Pope, Minn., 362...E3 99
Czar, Alta., Can., 196...C5 69
Czarna, riv., Pol...k14 26
Czarna Przemsza, riv., Pol...g10 26
Czarnkow, Pol., 4,394...B4 26
Czechoslovakia, country, Eur., 13,741,700..F12 8, D4 26
Czersk, Pol., 7,092...B4 26
Częstochowa, Pol., 164,000..C5 26
Człuchow, Pol., 3,711...B4 26

D

Daaquam, Que., Can., 195..C7 73
Daarburuk, Som...D5 47
Dab, Pol., 9,096...B3 26
Dab, Pol., 13,666...g9 26
Dabā, Tr. Coast...I8 41
Dab'ah, Jordan...C8 32
Dabakala, I.C., 900...E4 45
Dabaro, Som...B6 44
Dabdab, Libya...D2 43
Dabeiba, Col., 2,832...B2 60
Dablice, Czech., 5,378...n18 26
Dabney, Vance, N.C., 25...A5 109
Dabneys, Louisa, Va., 15...D5 121
Dabola, Guinea, 3,700...D2 45
Dabra-Berhām, Eth...D4 47
Dabra-Mārk'os, Eth...C4 47
Dabra-Tābor, Eth...C4 47
Dabrowa, Pol., 4,520...C6 26
Dabrowa Gornicza, Pol., 42,000...C5, g10 26
Dabrowa Grodzienska, Pol., 3,026...B7 26
Dacca, Pak., 556,712 (*720,000)...D9 39, F13 40
Dachau, Ger., 29,100...D5 16
Dacoma, Woods, Okla., 219.A3 112
Dacono, Weld, Colo., 302...A6 83
Dacula, Gwinnett, Ga., 440..C3 87
Dacura, Nic...D5 62
Dacus, Montgomery, Tex...D5 118
Dacusville, Pickens, S.C., 175...B2 115
Dadanawa, Br. Gu...B3 59
Dadar, Eth...D5 47
Daday, Tur., 1,600...B9 31
Daddy, creek, Tenn...C9 117
Dade, co., Fla., 935,047...G6 86
Dade, co., Ga., 8,666...B1 87
Dade, co., Mo., 7,577...D4 101
Dade City, Pasco, Fla., 4,759...D4 86
Dadeville, Tallapoosa, Ala., 2,940...C4 78
Dadeville, Dade, Mo., 142...D4 101
Dadu, Pak., 13,716...D1 40
Daet, Phil., 20,000 (35,600▲)...*C6 35
Dafoe, Sask., Can., 79...F3 70
Dafoe, riv., Man., Can...B4 71
Dagahabur, Eth...D5 47
Dagana, Sen., 4,000...C1 45
Daggett, San Bernardino, Calif., 400...E5 82
Daggett, Menominee, Mich., 296...C3 98
Daggett, co., Utah, 1,164...C6 119
Dagmar, Sheridan, Mont...B12 102
Dagsboro, Sussex, Del., 477..C7 85
Dagupan, Phil., 17,000 (63,100▲)...n13 35
Daguscahonda, Elk, Pa., 150...D4 114
Dagus Mines, Elk, Pa., 450..D4 114
Dahan-i-Kashan, Afg., 5,000...D13 41
Dahan-i-Kusnak, Afg., 5,000...D11 41
Dahanu, India, 9,648...H4 40
Dahinda, Knox, Ill., 225...C3 90
Dahlen, Nelson, N. Dak., 100...A8 110
Dahlgren, Hamilton, Ill., 480...E5 90
Dahlgren, King George, Va., 475...C5 121
Dahlia, Guadalupe, N. Mex., 25...B4 107
Dahlonega, Lumpkin, Ga., 2,604...B3 87
Dahme, Ger., 6,391...B8 17
Dahomey, country, Afr., 2,000,000...F6 42, E5 45
Dahshūr, Eg., U.A.R...E3 32
Daigle, Aroostook, Maine, 150...A4 96
Daigleville, Terrebonne, La., 5,906...*E6 95
Dailey, Logan, Colo., 10...A8 83
Dailey, Randolph, W. Va., 800...C5 123
Daimiel, Sp., 19,625...C4 20
Daingean, Ire., 679...D4 11
Daingerfield, Morris, Tex., 3,133...C5 118
Dairen (Talien), China, 950,000 (1,508,000▲)...D9 34
Dairy, Klamath, Oreg., 50...E5 113
Dairyland, Douglas, Wis., 25...B1 124
Dairy Valley, Los Angeles, Calif., 3,508...*F2 82
Daisetta, Liberty, Tex., 1,500...D5 118
Daisy, Pike, Ark., 86...C2 81
Daisy, Evans, Ga., 229...D5 87
Daisy, Atoka, Okla., 25...C6 112
Daisy (Melville), Hamilton, Tenn., 1,508...D8, E10 117
Daisy, Stevens, Wash., 30...A7 122
Daisytown, Washington, Pa., 1,396...*F2 114
Dajabón, Dom. Rep., 3,230..F8 64
Dajarra, Austl., 182...D6 50
Dakar, Sen., 231,000...D1 45
Dakhilah, oasis, Eg., U.A.R..D5 43
Dakoro, Niger...D6 45
Dakota, Stephenson, Ill., 363...A4 90
Dakota, Winona, Minn., 339...G7 99
Dakota, co., Minn., 78,303...F5 99
Dakota, co., Nebr., 12,168...B9 103
Dakota City, Humboldt, Iowa, 706...B3 92
Dakota City, Dakota, Nebr., 928...B9 103
Dakwa, Con. L...A4 48
Dalaba, Guinea, 5,500...D2 45
Dalälven, riv., Swe...G6 25
Dalaman, Tur...D7 25
Dalaman, riv., Tur...D7 25
Dalark, Dallas, Ark., 123...C3 81
Dalarö, Swe., 523...t36 25
Dalat, Viet., 5,217...G8 38
Dalay Sayn Shanda, Mong..B4 36
Dalbandin, Pak...C3 39
Dalbeattie, Scot., 3,104...F5 13
Dalbo, Isanti, Minn., 60...E5 99
Dalby, Austl., 7,400...C8 51
Dalby, Swe., 1,461...C7 24
Dalcahue, Chile...C2 54
Dale, Hamilton, Ill., 140...F5 90
Dale, Spencer, Ind., 900...H4 91
Dale, Pottawatomie, Okla., 400...B4 112
Dale, Grant, Oreg., 5...C8 113
Dale, Cambria, Pa., 2,807..*F4 114
Dale, co., Ala., 31,066...D4 78
Dale Hollow, lake, Tenn...C8 117
Dalemead, Alta., Can., 50...D4 69
Daleview, Sheridan, Mont..B12 102
Daleville, Dale, Ala., 693...D4 78
Daleville, Delaware, Ind., 1,548...D6 91
Daleville, Lauderdale, Miss., 125...C5 100
Dalhart, Dallam and Hartley, Tex., 5,160...A1 118
Dalhousie, N.B., Can., 5,856.A3 74
Dalhousie, India, 2,739...A5 40
Dalhousie Junction, N.B., Can., 172...A3 74
Dalías, Sp., 3,540...D4 20
Daliburgh, Scot...C1 13
Daliyat el Karmil, Isr., 3,144.B7 32
Dalkeith, Gulf, Fla., 35...B1 86
Dalkeith, Scot., 8,864...E5 13
Dalkena, Pend Oreille, Wash., 30...A8 122
Dalkey, Ire...D5 11
Dall, isl., Alsk...n23 79
Dall, mtn., Alsk...f15 79
Dallam, co., Tex., 6,302...A1 118
Dallas, Blount, Ala., 150...B3 78
Dallas, Paulding, Ga., 2,065.C2 87
Dallas, Franklin, Maine, 50 (77▲)...D2 96
Dallas, Gaston, N.C., 3,270..B2 109
Dallas, Polk, Oreg., 5,072...C3 113
Dallas, Luzerne, Pa., 2,586...B8, D10 114
Dallas, Gregory, S. Dak., 212...D6 116
Dallas, Dallas, Tex., 679,684 (*1,022,300)..B5, C4 118
Dallas, Marshall, W. Va., 100...B2 123
Dallas, Barron, Wis., 401...C2 124
Dallas, co., Ala., 56,667...C2 78
Dallas, co., Ark., 10,522...D3 81
Dallas, co., Iowa, 24,123...C3 92
Dallas, co., Mo., 9,314...D4 101
Dallas, co., Tex., 951,527...C4 118
Dallas Center, Dallas, Iowa, 1,083...C4 92
Dallas City, Hancock and Henderson, Ill., 1,276...C2 90
Dallas Mine, Dallas, Iowa, 165...A7 92
Dallastown, York, Pa., 3,615.G8 114
Dalmacio Vélez, Arg...A4 54
Dalmally, Scot., 876...D4 13
Dalmatia, Northumberland, Pa., 600...E8 114
Dalmatia, reg., Yugo...D3 22
Dalmellington, Scot., 4,702...E4 13
Dalmeny, Sask., Can., 415...E2 70
Dalnyaya, Sov. Un., 10,000.D10 37
Daloa, I.C., 13,000...E3 45
Dalroy, Alta., Can., 50...D4 69
Dalry, Scot., 6,764...E4 13
Dalrymple, mtn., Austl...D8 50
Dalton, Randolph, Ark., 25..A4 81
Dalton, Whitfield, Ga., 17,868...B2 87
Dalton, Berkshire, Mass., 6,436...B1 97
Dalton, Otter Tail, Minn., 239...D3 99
Dalton, Chariton, Mo., 197..B5 101
Dalton, Cheyenne, Nebr., 503...C3 103
Dalton, Coos, N.H., 50 (567▲) B3 105
Dalton, Livingston, N.Y., 500...C3 108
Dalton, Wayne, Ohio, 1,067..B6 111
Dalton, Lackawanna, Pa., 1,229...C10 114
Dalton City, Moultrie, Ill., 386...D5 90
Daltonganj, India, 25,270...E8 40
Dalton Gardens, Kootenai, Idaho, 1,083...B2 89
Dalwhinnie, Scot...D4 13
Daly, riv., Austl...B5 50
Daly City, San Mateo, Calif., 44,791...B5 82
Daly Waters, Austl...C5 50
Dalzell, Sumter, S.C., 80...C7 115
Dam, Sur...B3 59
Dama, is., Viet...H6 38
Damanhūr, Eg., U.A.R., 99,900...C6 43
Damão, India, 10,000...G4 40
Damar, Rooks, Kans., 361..C4 93
Damar, isl., Indon...G7 35
Damar, is., Indon...G7 35
Damara, Cen. Afr. Rep...D3 46
Damariscotta, Lincoln, Maine, 800 (1,093▲)...D3 96
Damariscotta, lake, Maine...D3 96
Damas, see Damascus, Syr.
Damas, pass, Arg...A2 54
Damascus, Faulkner, Ark., 400...B3 81
Damascus, Early, Ga., 297...E2 87
Damascus, Gordon, Ga., 200.B2 87
Damascus, Montgomery, Md., 1,500...B3 85
Damascus, Wayne, Pa., 250.C11 114
Damascus (Dimashq), Syr., 530,000...F11 31
Damascus, Washington, Va., 1,485...B3 121
Damaturu, Nig., 2,379...D7 45
Damba, Ang., 1,367...C2 48
Dambidolo, Eth...D3 47
Dambovita, riv., Rom...C7 22
D'Ambre, cape, Malag...f9 49
Dame-Marie, cape, Hai...F6 64
Dam Gamad, Sud...C2 47
Dāmghān, Iran, 12,235...C7 41
Damietta (Dumyāţ), Eg., U.A.R., 63,100...C6 43
Damietta, riv. mouth, Eg., U.A.R...C3 32
Damietta Branch, riv., Eg., U.A.R...D3 32
Dāmiyā, Jordan, 1,000...g13 32
Dammartin-en-Goële, Fr., 1,678...E2 15
Damme, Ger., 9,300...B8 15
Dam No. 17, Ala...B2 78
Damodar, riv., India...F10 40
Damoh, India, 46,656...F7 40
Damon, Brazoria, Tex., 750..G4 118
Damongo, Ghana...E4 45
Dampier, strait, W. Irian...F8 35
Dan, riv., N.C., Va..A4 103, E4 121
Dana, Sask., Can., 68...E3 70
Dana, LaSalle, Ill., 240...C5 90
Dana, Vermillion, Ind., 811..E2 91
Dana, Greene, Iowa, 123...B3 92
Dana, Henderson, N.C., 200.D4 109
Danakil, depression, Eth...C5 47
Danané, I.C., 4,000...E3 45
Dana Point, Orange, Calif., 1,186...*F5 82
Danburg, Wilkes, Ga., 108..C4 87
Danbury, Fairfield, Conn., 22,928...D3 84
Danbury, Woodbury, Iowa, 510...B2 92
Danbury, Red Willow, Nebr., 185...D5 103
Danbury, Merrimack, N.H., 200 (435▲)...C3 105
Danbury, Stokes, N.C., 175..A3 109
Danbury, Brazoria, Tex., 600...G5 118
Danbury, Burnett, Wis., 300.C1 124
Danby, San Bernardino, Calif., 20...E6 82
Danby, Tompkins, N.Y...C4 108
Danby, Rutland, Vt., 250 (891▲)...E3 120
Danby Four Corners, Rutland, Vt., 40...E2 120
Dancy, Pickens, Ala., 100...B1 78
Dancy, Marathon, Wis., 100.D4 124
Dancyville, Haywood, Tenn., 50...B2 117
Dande, riv., Ang...C1 48
Dandenong, Austl., 24,914...H5 51
Dandridge, Jefferson, Tenn., 829...C10 117
Dandy, York, Va., 400...A6 121
Dane, Dane, Wis., 394...E4 124
Dane, co., Wis., 222,095...E4 124
Danevang, Wharton, Tex., 300...E4 118
Danforth, Ont., Can., 450...E7 72
Danforth, Iroquois, Ill., 394..C6 90
Danforth, Washington, Maine, 800 (821▲)...C5 96
Danforth, hills, Colo...A2 83
Dānglā Eth...C4 47
Dangrek, mts., Thai...E6 38
Dania, Broward, Fla., 7,065...E3, F6 86
Daniel, Sublette, Wyo., 110..C2 125
Daniel, mtn., Wash...B4 122
Daniels, Howard, Md., 750..B4 85
Daniels, co., Mont., 3,755...B11 102
Daniel's Harbour, Newf., Can., 403...C3 75
Daniels Landing, Perry, Tenn...B4 117
Danielson, Windham, Conn., 4,642...B9 84
Danielsville, Madison, Ga., 362...B3 87
Danilov, Sov. Un., 16,600...B2 29
Danilov Grad, Yugo., 1,373..D4 22
Danjo, isl., Jap...J4 37
Dankhar, India...A7 40
Danlí, Hond., 3,757...C4 62
Dannebrog, Howard, Nebr., 277...C7 103
Dannemora, Clinton, N.Y., 4,835...A3 108
Dannenberg, Ger., 3,500...E5 24
Danner, Malheur, Oreg...E9 113
Dannevirke, N.Z., 5,508...N16 51
Dannike, Swe., 218...A7 24
Danripple, Halifax, Va...E3 121
Dans, mtn., Md...D2 85
Dansalan, Phil., 4,882...*D6 35
Dansville, Livingston, N.Y., 5,460...C3 108
Dante, Som...C7 47
Dante, Charles Mix, S. Dak., 102...D7 116
Dante, Knox, Tenn., 600...E11 117
Dante, Russell, Va., 1,436...B2 121
Danube, Renville, Minn., 494...F3 99
Danube, riv., Eur...G14 8
Danube, riv. mouths, Eur...F14 8
Danubyu, Bur., 10,000...D1 38
Danvers, McLean, Ill., 783..C4 90
Danvers, Essex, Mass., 21,926...A6, C3 97
Danvers, Swift, Minn., 132..E3 99
Danvers, Fergus, Mont., 25..C7 102
Danville, Morgan, Ala., 140...A2 78
Danville, Yell, Ark., 955...B2 81
Danville, Contra Costa, Calif., 3,585...B6 82
Danville, Que., Can., 2,562..D5 73
Danville, Twiggs and Wilkinson, Ga., 459...D3 87
Danville, Vermilion, Ill., 41,856...C6 90
Danville, Hendricks, Ind., 3,287...E3 91
Danville, Des Moines, Iowa, 579...D6 92
Danville, Harper, Kans., 118.E6 93
Danville, Boyle, Ky., 9,010..C5 94
Danville, Rockingham, N.H., 175 (605▲)...E4 105
Danville, Knox, Ohio, 926...B5 111
Danville, Highland, Ohio, 50...C4 111
Danville, Montour, Pa., 6,889...E8 114
Danville, Houston, Tenn...A4 117
Danville, Caledonia, Vt., 300 (1,368▲)...C4 120
Danville (Independent City), Va., 46,577...E3 121
Danville, Ferry, Wash., 80...A7 122
Danville Boone, W. Va., 507...C3, D5 123
Danville East (Mechanicsville), Montour, Pa., 1,758...*E8 114
Danzig, see Gdańsk, Pol.
Danzig, gulf, Pol...A5 26
Daosa, India, 14,612...D6 40
Daphne, Baldwin, Ala., 1,527...E2 78
Dapp, Alta., Can., 71...B4 69
Dārāb, Iran, 7,403...G7 41
Darabani, Rom., 11,379...A8 22
Daraj, Libya...C2 43
Dār al Ḩamrā, Sau. Ar...I12 31
Darasun, Sov. Un., 18,000..D14 28
Darawah, Eg., U.A.R...D3 32
Darbhanga, India, 103,016...C8 39, D10 40
D'Abronne, bayou, La...B3 95
Darbun, Marion, & Walthall, Miss., 150...D3 100
Darby, Ravalli, Mont., 398..D2 102
Darby, Delaware, Pa., 14,059...B11, G11 114
Darby, creek, Ohio...C4 111
Darbyville, Pickaway, Ohio, 213...C4 111
Dar Chebika, Mor...D2 44
D'Arcy Station, Sask., Can., 65...F1 70
Dardanelle, Yell, Ark., 2,098.B2 81
Dardanelle, Tuolumne, Calif...C4 82
Dardanelles, strait, Tur...B6 23
Darden, Henderson, Tenn., 250...B3 117
Dardenne, creek, Mo...A7 111
Dardens, Martin, N.C., 175..B7 109
Dare, co., N.C., 5,936...B8 109
Darende, Tur., 6,900...C11 29
Dar es Salaam, Tan., 128,742...C6 48
Daretown, Salem, N.J., 75...D2 106
Darfo, It., 8,711...D6 18
Darfur, Watonwan, Minn., 191...F4 99
Darfur, reg., Sud., 1,328,765..C1 47
Dargai, Pak...B5 39
Dargan, Washington, Md., 150...B2 85
Darganata, Sov. Un...E6 29
Dargaville, N.Z., 3,733...K14 51
Darien, Fairfield, Conn., 18,437...E3 84
Darien, McIntosh, Ga., 1,569...E3 87
Darien, Walworth, Wis., 805...F5 124
Darien, mts., Pan...F9 62
Dariense, mts., Nic...D5 62
Darjeeling, India, 40,651...D12 40
Darke, co., Ohio, 45,612...B3 111
Darkesville, Berkeley, W. Va., 200...*B7 123
Darkharbor, Waldo, Maine, 100...D4 96
Darling, Quitman, Miss., 250...A3 100
Darling, lake, N. Dak...A6 110
Darling, range, Austl...F2 50
Darling, riv., Austl...E5 51
Darlingford, Man., Can., 189...E2 71
Darlington, Wilcox, Ala., 100...C3 78
Darlington, Eng., 84,162...C6 10
Darlington, Walton, Fla., 1,100...G3 86
Darlington, Butte, Idaho, 10...F5 89
Darlington, Montgomery, Ind., 668...D4 91
Darlington, Harford, Md., 250...A5 85
Darlington, Gentry, Mo., 169...A3 101
Darlington, Darlington, S.C., 6,710...C8 115
Darlington, Lafayette, Wis., 2,349...F3 124
Darlington, cape, Ant...B7 5
Darlington, co., S.C., 52,928.C8 115
Darlove, Washington, Miss., 50...B3 100
Darłowo, Pol., 5,262...A4 26
Dar Mazar, Iran...G8 41
Darmody, Sask., Can., 48..G2 70
Darmstadt, Ger., 136,400 (*215 000)...D3 17
Darnah (Derna), Libya, 15,891...C4 43
Darnell, West Carroll, La., 65...B4 95
Darnestown, Montgomery, Md., 150...B3 85
Darney, Fr., 1,810...A2 18

Darnley, cape, Ant........C19 5
Daro, Mala........E4 35
Daroca, Sp., 3,378........B5 20
Darr, Dawson, Nebr., 11....D6 103
Darrah, mtn., B.C., Can......A2 102
Darrington, Snohomish, Wash., 1,272........A4 122
Darrouzett, Lipscomb, Tex., 375........A2 118
Darrow, Ascension, La., 400..B5 95
Darsser, pt., Ger........D6 24
Dart, cape, Ant........B36 5
Dart, riv., Eng........D4 12
Dartford, Spokane, Wash., 30........D7 122
Dartmoor, moor, Eng........D4 12
Dartmouth, N.S., Can., 46,966........E6 74
Dartmouth, Eng., 5,757.....D4 12
Dartmouth, Bristol, Mass., 700 (14,607▲)........C5 97
Dartry, mts., Ire........C3 11
Daru, isl., Pap........k11 50
Daruvar, Yugo., 6,280......C3 22
Darvel, bay, Mala........E5 35
Darwin, Austl., 12,335......B5 51
Darwin, Inyo, Calif., 450....D5 82
Darwin, isl., Ec........f5 58
Daryacheh-i-Namakzar, salt lake, Iran........E10 41
Darya yi Namak, salt lake, Iran........D5 41
Dasē, Eth., 40,000........C4 47
Dash Point, Pierce, Wash., 300........D1 122
Dasht, riv., Pak........C3 39
Dasht-i-Daqq-i-Tundi, salt lake, Afg........E10 41
Dasht-i-Kavir, salt plain, Iran..D7 41
Dasht-i-Lut, plain, Iran......E8 41
Dasht-i-Margo, des., Afg......F11 41
Dashwood, Ont., Can., 433..D3 72
Dasol, bay, Phil........o12 35
Dassel, Meeker, Minn., 863..E4 99
Date, Yavapai, Ariz., 25.....C3 80
Dateland, Yuma, Ariz., 30...E2 80
Datia, India, 29,430........C7 40
Datil, Catron, N. Mex., 50...C2 107
Datto, Clay, Ark., 167.......A5 81
Datu, cape, Indon., Mala......K8 38
Datze, riv., Ger........E7 24
Dauchite, bayou, La........B2 95
Daudnagar, India, 13,320........E10 40
Daufuskie, isl., S.C.........G6 115
Daufuskie Island, Beaufort, S.C., 225........G6 115
Daugava (Dvina), riv., Sov. Un.C5 27
Daugavpils, Sov. Un., 70,000.D6 27
Daulatabad, Afg., 5,000.....E11 41
Daulat, Yar, Afg., 5,000....D12 41
Daule, Ec........A2 58
Daule, Ec., 4,697........B2 58
Daun, Ger., 3,400........C1 17
Dauphin, Man., Can., 7,374.D1 71
Dauphin, Dauphin, Pa., 638.F8 114
Dauphin, co., Pa., 220,255...F8 114
Dauphin, isl., Ala........E1 78
Dauphin, lake, Man., Can......D2 71
Dauphin, riv., Man., Can......D2 71
Dauphiné, former prov., Fr., 1,016,000........E7 16
Daus, Sequatchie, Tenn., 250........D8 117
Davant, Plaquemines, La., 415........C8, E6 95
Davao, Phil., 135,000 (231,800▲)........D7 35
Davao, prov., Phil., 903,200.*D7 35
Davao, gulf, Phil........D7 35
Dāvar Panāh, Iran, 10,000.H11 41
Daveluyville, Que., Can., 733........C5 73
Davenport, Santa Cruz, Calif., 500........C5, D2 82
Davenport, Polk, Fla., 1,209.D5 86
Davenport, Scott, Iowa, 88,981 (★260,300)....C7, D7 92
Davenport, Thayer, Nebr., 416........D8 103
Davenport, Delaware, N.Y., 260........C6 108
Davenport, Cass, N. Dak., 143........C8 110
Davenport, Lincoln, Okla., 813........B5 112
Davenport, Red River, Tex..C5 118
Davenport, Lincoln, Wash., 1,494........B7 122
Davenport Downs, Austl.....B3 51
Daventry, Eng., 5,846.......B6 12
Davey, Lancaster, Nebr., 121.E2 103
David, Floyd, Ky., 600......C7 94
David, Pan., 22,924........F6 62
David City, Butler, Nebr., 2,304........C8 103
David-Gorodok, Sov. Un., 10,000........E6 27
Davidson, Sask., Can., 928...F2 70
Davidson, Mecklenburg, N.C., 2,573........B3 109
Davidson, Tillman, Okla., 429........C2 112
Davidson, Fentress, Tenn., 200........C8 117
Davidson, co., N.C., 79,493..B3 109
Davidson, co., Tenn., 399,743........A5 117
Davidson, mts., Alsk........B11 79
Davie, Broward, Fla., 1,500........E3, F6 86
Davie, co., N.C., 16,728.....B3 109
Daviess, co., Ind., 26,636....G3 91
Daviess, co., Ky., 70,588.....C2 94
Daviess, co., Mo., 9,502......B3 101
Davin, Sask., Can., 60.......G3 70
Davis, Yolo, Calif., 8,910.A6, C3 82
Davis, Stephenson, Ill., 494..A4 90
Davis, Carteret, N.C., 500...C7 109
Davis, Murray, Okla., 2,203.C4 112
Davis, Turner, S. Dak., 124..D9 116
Davis, Atascosa, Tex........E3 118
Davis, Tucker, W. Va., 898..B5 123
Davis, co., Iowa, 9,199......D5 92
Davis, co., Utah, 64,760.....C3 119
Davis, bay, Ant........C26 5
Davis, creek, W. Va........C6 123
Davis, dam, Nev........H7 104
Davis, isl., Fla........E2 86
Davis, lake, Oreg........D5 113
Davis, mtn., Pa........G3 114
Davis, mts., Tex........F2 118
Davis, sea, Ant........C22 5
Davis, strait, Can........C21 67
Davisboro, Washington, Ga., 417........D4 87
Davis City, Decatur, Iowa, 347........D4 92
Davis Dam, Mohave, Ariz., 200........B1 80
Davis Inlet, Newf., Can., 98..g9 75
Davis Junction, Ogle, Ill., 250........A4 90
Davison, Genesee, Mich., 3,761........E2 98
Davison, co., S. Dak., 16,681.D7 116
Davis Station, Clarendon, S.C., 60........D7 115
Daviston, Tallapoosa, Ala., 129........C4 78
Davisville, Crawford, Mo., 100........D6 101
Davisville, Washington, R.I., 1,800........C11 84
Davle, Czech., 1,490.......o17 26
Davos, Switz., 9,588.......C8 19
Davy, McDowell, W. Va., 1,331........C3 123
Dāwā, riv., Eth........E5 47
Dawes, co., Nebr., 9,536.....B2 103
Dawley, Eng., 9,553........B5 12
Dawn, Livingston, Mo., 200..B4 101
Dawn, Deaf Smith, Tex., 60..B1 118
Dawros, head, Ire........C3 11
Dawson, Yukon, Can., 881...D5 66
Dawson, Terrell, Ga., 5,062..E2 87
Dawson, Dallas, Iowa, 257...C3 92
Dawson, Lac qui Parle, Minn., 1,766........F2 99
Dawson, Richardson, Nebr., 263........D10 103
Dawson, Colfax, N. Mex....A5 107
Dawson, Kidder, N. Dak., 206........C6 110
Dawson, Fayette, Pa., 707...F2 114
Dawson, Navarro, Tex., 911........D4 118
Dawson, co., Ga., 3,590......B2 87
Dawson, co., Mont., 12,314.C11 102
Dawson, co., Nebr., 19,405..D6 103
Dawson, co., Tex., 19,185.....C1 118
Dawson, bay, Man., Can......D5 70
Dawson, mtn., B.C., Can......D9 68
Dawson, range, Yukon, Can....D5 66
Dawson, riv., Austl........D8 50
Dawson Creek, B.C., Can., 10,946........B7 68
Dawson-Lambton, glacier, Ant..B10 5
Dawson Springs, Hopkins, Ky., 3,002........*C2 94
Dawsonville, Dawson, Ga., 307........B2 87
Dax, Fr., 17,051........F3 14
Day, Lafayette, Fla., 125....B3 86
Day, co., S. Dak., 10,516....B8 116
Day, mtn., Calif........B6 82
Dayang, Bunting, isl., Thai.....I3 38
Daykin, Jefferson, Nebr., 144.D8 103
Daylesford, Austl., 2,776....H5 51
Daylight, Warren, Tenn., 20........D8 117
Dayr Abū Sa'īd, Jordan, 1,000........B7 32
Dayr al Balaḩ, Gaza Area, 3,000........C6 32
Dayr az Zawr, Syr., 42,000..E13 31
Dayr Dibwān, Jordan, 3,000.h12 32
Dayr Istiyā, Jordan, 2,000...g11 32
Dayrūṭ, Eg., U.A.R., 5,000..D6 43
Daysland, Alta., Can., 539...C4 69
Daysville, Cumberland, Tenn., 100........D9 117
Dayton, Marengo, Ala., 99...C2 78
Dayton, Franklin, Idaho, 212........G7 89
Dayton, Tippecanoe, Ind., 700........D4 91
Dayton, Webster, Iowa, 820..B3 92
Dayton, Campbell, Ky., 9,050........A7 94
Dayton, Howard, Md., 200...B4 85
Dayton, Hennepin and Wright, Minn., 456........E2 99
Dayton, Lake, Mont., 57.....C2 102
Dayton, Lyon, Nev., 150.....D2 104
Dayton, Middlesex, N.J., 500.C3 106
Dayton, Eddy, N. Mex.......E5 107
Dayton, Cattaraugus, N.Y., 250........C2 108
Dayton, Montgomery, Ohio, 262,332 (★648,600)........C3 111
Dayton, Yamhill, Oreg., 673........B1, B3 113
Dayton, Armstrong, Pa., 769.E3 114
Dayton, Rhea, Tenn., 3,500..D7 117
Dayton, Liberty, Tex., 3,367........D5, F5 118
Dayton, Rockingham, Va., 930........C4 121
Dayton, Columbia, Wash., 2,913........C8 122
Dayton, Sheridan, Wyo., 333.A5 125
Daytona Beach, Volusia, Fla., 37,395........C6 86
Daytona Beach Shores (South Peninsula), Volusia, Fla., 3,741........*C6 86
Dayville, Windham, Conn., 900........B9 84
Dayville, Grant, Oreg., 234..C7 113
Dazey, Barnes, N. Dak., 226..B7 110
Dazgīr, Iran........C2 41
De Aar, S. Afr., 14,357.....D3 49
Dead, creek, Vt........C2 120
Dead, lake, Fla........B1 86
Dead, lake, Minn........D3 99
Dead, riv., Maine........C2 96
Dead, sea, Isr., Jordan......C7 32
Dead Diamond, riv., N.H......A2 105
Dead Indian, peak, Wyo......A3 125
Deadman, creek, Wash........D7 122
Deadmans, bay, Fla........C3 86
Deadmans, pt., Newf., Can.....D5 75
Dead River, Somerset, Maine........C2 96
Deadwood, Lawrence, S. Dak., 3,045........C2 116
Deadwood, res., Idaho........E3 89
Deaf Smith, co., Tex., 13,187.B1 118
Deakin, bay, Ant........C28 5
Deal, Eng., 24,791........C9 12
Deal, Monmouth, N.J., 1,889.C4 106
Deal, isl., Md........D6 85
Deale, Anne Arundel, Md., 900........C4 85
Deal Island, Somerset, Md., 600........D6 85
Dean, Scott, Tenn., 50.......C9 117
Dean, chan., B.C., Can......C4 68
Dean, riv., B.C., Can........C4 68
Deanburg, Chester, Tenn., 60.B3 117
Dean Dale, Jefferson, Ohio (part of Mingo Junction)........B2 123
Deaneyville, Hempstead, Ark........D2 81
Deán Funes, Arg., 13,840....A4 54
Deans, Middlesex, N.J., 300..C3 106
Deans Island, Mississippi, Ark........B5 81
Deanville, Lewis, W. Va., 350........B4 123
Deanwood, Ware, Ga., 2,758.E4 87
Dearborn, Wayne, Mich., 112,007........A7, F7 98
Dearborn, Platte, Mo., 444...B3 101
Dearborn, co., Ind., 28,674...F8 91
Dearborn Heights, Cook, Ill., 1,300........*B6 90
Dearg, mtn., Scot........C4 13
Dearing, McDuffie, Ga., 403.C4 87
Dearing, Montgomery, Kans., 249........E8 93
DeArmanville, Calhoun, Ala., 250........B4 78
Deary, Latah, Idaho, 349....C2 89
Dease, arm, N.W. Ter., Can...C8 66
Dease, strait, N.W. Ter., Can.C11 66
Dease Lake, B.C., Can.......E6 66
Death, val., Calif........D5 82
Death Valley, Inyo, Calif., 75........D5 82
Death Valley, nat. mon., Calif...D5 82
Deatsville, Elmore, Ala., 200.C3 78
Deauville, Fr., 5,051........C4 14
Deaver, Big Horn, Wyo., 121.A4 125
De Baca, co., N. Mex., 2,991.C5 107
Debaltsevo, Sov. Un., 33,800........q21 27
Debar, Yugo., 6,341........E5 22
DeBary, Volusia, Fla., 2,362........D5 86
Debden, Sask., Can., 402....D2 70
Debec, N.B., Can., 236.......C2 74
De Beque, Mesa, Colo., 172..B2 83
De Berry, Panola, Tex., 250..C5 118
Debica, Pol., 19,000........C6 26
Debin, Bul., 2,745........E7 22
Deblin, Pol., 1,000........C6 26
Deblois, Washington, Maine, 20 (26▲)........D4 96
Debno, Pol., 3,341........B3 26
Debo, lake, Mali........C4 45
De Borgia, Mineral, Mont., 100........C1 102
Deboullie, mtn., Maine........B4 96
Debovo, Bul., 2,028........D7 22
Debrecen, Hung., 129,679....B5 22
De Cade, lake, La........E5 95
Decamere, Eth........B4 47
Decatur, Morgan, Ala., 29,217........A3 78
Decatur, Benton, Ark., 415...A1 81
Decatur, De Kalb, Ga., 22,026........B5, C2 87
Decatur, Macon, Ill., 78,004 (★111,300)........D5 90
Decatur, Adams, Ind., 8,327........C8 91
Decatur, Decatur, Iowa, 203.D4 92
Decatur, Van Buren, Mich., 1,827........F5 98
Decatur, Newton, Miss., 1,340........C4 100
Decatur, Burt, Nebr., 786....B9 103
Decatur, Meigs, Tenn., 681..D9 117
Decatur, Wise, Tex., 3,563...C4 118
Decatur, co., Ga., 25,203.....F2 87
Decatur, co., Ind., 20,019....F6 91
Decatur, co., Iowa, 10,539...D4 92
Decatur, co., Kans., 5,778....C3 93
Decatur, co., Tenn., 8,324...B3 117
Decatur, lake, Ill........D5 90
Decaturville, Decatur, Tenn., 571........B3 117
Decazeville, Fr., 11,855......E5 14
Deccan, reg., India........E6 39
Deception, lake, Sask., Can.....A3 70
Deception, mtn., N.W. Ter., Can........C5 66
Decherd, Franklin, Tenn., 1,704........B5 117
Decimal, Man., Can........E4 71
Decin, Czech., 39,400.......C3 26
Decize, Fr., 6,594........D5 14
Decker, Man., Can., 67.......D1 71
Decker, Knox, Ind., 317.....G3 91
Decker, Big Horn, Mont., 15........E10 102
Deckers, Douglas, Colo., 50...B5 83
Deckerville, Sanilac, Mich., 798........E8 98
Declo, Cassia, Idaho, 237.....G5 89
De Cocksdorp, Neth., 347...A4 15
Decorah, Winneshiek, Iowa, 6,435........A6 92
Decota, Kanawha, W. Va., 500........D6 123
De Coursey, Kenton, Ky., 50........A7 94
Dededo, Guam, 2,247........ 52
Dedemsvaart, Neth., 5,800...B6 15
Dedham, Carroll, Iowa, 322..C3 92
Dedham, Norfolk, Mass., 23,869........B5, D2 97
Dedinovo, Sov. Un., 5,000...n19 27
Dédougou, Upper Volta, 2,900........D4 45
Dedza, Nya., 2,325........D6 48
Dee, Hood River, Oreg., 50..B5 113
Dee, riv., Eng........A5 12
Dee, riv., Scot........C6 13
Dee, riv., Wales........D5 10
Deedsville, Miami, Ind., 120.C5 91
Deel, riv., Ire........E3 11
Deemer, Neshoba, Miss., 100.C4 100
Deep, creek, Del........C6 85
Deep, creek, Mont........C4 102
Deep, creek, Utah........C2 119
Deep, creek, Utah........F2 119
Deep, creek, Wash........D7 122
Deep, entrance, Eniwetok........ 52
Deep, fork, Okla........B5 112
Deep, inlet, Newf., Can........g10 75
Deep Bay, Nya........D5 48
Deep Brook, N.S., Can., 462.E4 74
Deep Creek, Norfolk, Va., 350........B6, E6 121
Deepcreek, Spokane, Wash., 100........D7 122
Deep Creek, lake, Md........D1 85
Deepdale, Man., Can., 80....D1 71
Deep Gap, Watauga, N.C., 130........A2 109
Deep Gap, mtn., N.C........A2 109
Deephaven, Hennepin, Minn., 3,286........*F3 99
Deep Red, run, Okla........C3 112
Deep River, Middlesex, Conn., 2,116 (2,968▲)........D7 84
Deep River, Poweshiek, Iowa, 329........C5 92
Deep Run, Lenoir, N.C., 183.B6 109
Deepstep, Washington, Ga., 139........C4 87
Deepwater, Henry, Mo., 712.C4 101
Deepwater, Salem, N.J., 700........D1 106
Deep Water, Fayette, W. Va., 900........D6 123
Deepwater, pt., Del........B7 85
Deer, Newton, Ark., 150.....B2 81
Deer, creek, Ind........C5 91
Deer, creek, Md........A5 85
Deer, creek, Miss........B3 100
Deer, creek, Ohio........C4 111
Deer, isl., Maine........D4 96
Deer, isl., Mass........D3 97
Deer, isl., Miss........E2 100
Deer, lake, Newf., Can......D3 75
Deer, lake, Minn........C5 99
Deer, mtn., Maine........C2 96
Deer, peak, Colo........C5 83
Deer, pond, Newf., Can......D4 75
Deer Creek, Tazewell and Woodford, Ill., 583........C4 90
Deer Creek, Carroll, Ind., 200........C5 91
Deer Creek, Otter Trail, Minn., 312........D3 99
Deer Creek, Grant, Okla., 215........A4 112
Deerfield, Lake, Ill., 11,786........A6, E2 90
Deerfield, Kearny, Kans., 442........E2 93
Deerfield, Franklin, Mass., 500 (3,338▲)........A2 97
Deerfield, Lenawee, Mich., 866........G7 98
Deerfield, Vernon, Mo., 200..D3 101
Deerfield, Rockingham, N.H., 75 (714▲)........D4 105
Deerfield, Portage, Ohio, 700.A6 111
Deerfield, Dane, Wis., 795...E4 124
Deerfield, riv., Mass., Vt.......F3 120
Deerfield Beach, Broward, Fla., 9,573........F6 86
Deerfield Street, Cumberland, N.J., 400........D2 106
Deer Flat, res., Idaho........F2 89
Deer Grove, Whiteside, Ill., 86........B4 90
Deering, Alsk., 95........B7 79
Deering, Pemiscot, Mo., 122.E8 101
Deering, Hillsboro, N.H., 30 (345▲)........D3 105
Deering, McHenry, N. Dak., 117........A4 110
Deer Isle, Hancock, Maine, 600 (1,129▲)........D4 96
Deer Lake, Newf., Can., 3,481........D3 75
Deer Lodge, Powell, Mont., 4,681........D5 102
Deer Lodge, Morgan, Tenn., 250........C9 117
Deer Lodge, co., Mont., 18,640........D3 102
Deer Park, Washington, Ala., 200........D1 78
Deer Park, Osceola, Fla., 100.D6 86
Deer Park, Garrett, Md., 379.D1 85
Deer Park, Suffolk, N.Y., 11,725........F3 84
Deer Park, Hamilton, Ohio, 8,423........D2 111
Deer Park, Harris, Tex., 4,865........*E5 118
Deer Park, Spokane, Wash., 1,333........B8 122
Deer Park, St. Croix, Wis., 221........C1 124
Deer River, Itasca, Minn., 992........C5 99
Deer River, Lewis, N.Y., 170.B5 108
Deerton, Alger, Mich., 50....B3 98
Deer Trail, Arapahoe, Colo., 764........B6 83
Deerwood, Crow Wing, Minn., 527........D5 99
Deeson, Bolivar, Miss., 100..A3 100
Deeth, Elko, Nev., 75.......B6 104
Defensa Central, P.R. (part of Aquas)........C6 65
Defense Highway, Prince Georges, Md., 1,000......*C4 85
Deferiet, Jefferson, N.Y., 470.A5 108
Defiance, Shelby, Iowa, 386..C2 92
Defiance, St. Charles, Mo., 110........B7 101
Defiance, McKinley, N. Mex., 12........B1 107
Defiance, Defiance, Ohio, 14,553........A3 111
Defiance, co., Ohio, 31,508...A3 111
Defiance, mtn., Oreg........B5 113
Deford, Tuscola, Mich., 150...E7 98
De Forest, Dane, Wis., 1,223.E4 124
De Funiak Springs, Walton, Fla., 5,282........G3 86
Degeberga, Swe., 590.......C8 25
Degerfors, Swe., 9,500.......E4 25
Degerön, Swe........u33 25
Degersheim, Switz., 3,221.....B7 19
Deggendorf, Ger., 17,100.....E7 17
De Graff, Swift, Minn., 196..E3 99
Degraff, Logan, Ohio, 996....B4 111
Dehak, Iran, 5,000........G10 41
Dehart, Wilkes, N.C........A2 109
Deh Bid, Iran, 10,000.......F6 41
Dehibat, Tun., 1,579........C7 44
Deh-i-Haji, Afg........F12 41
Deh Pāin, Iran, 10,000......G8 41
Dehra Dun, India, 126,918 (★156,341)........B6 39, B7 40
Deh Titan, Afg., 5,000......E11 41
Deinze, Bel., 6,004........D3 15
Dej, Rom., 19,281........B6 22
Dejvice, Czech. (part of Prague)........n17 26
De Kalb, De Kalb, Ill., 18,486........B5 90
De Kalb, Kemper, Miss., 880........C5 100
De Kalb, Buchanan, Mo., 304........B3 101
De Kalb, St. Lawrence, N.Y., 40........B1 108
De Kalb, Bowie, Tex., 2,042........C5 118
De Kalb, co., Ala., 41,417....A4 78
De Kalb, co., Ga., 256,182....C2 87
De Kalb, co., Ill., 51,714.....B5 90
De Kalb, co., Ind., 28,271....B7 91
De Kalb, co., Mo., 7,226.....B3 101
De Kalb, co., Tenn., 10,174..D8 117
De Kalb Junction, St. Lawrence, N.Y., 260...B2 108
De Kays, Sussex, N.J........A4 106
Dekese, Con. L........B3 48
Dekoa, Cen. Afr. Rep........D3 46
Delacroix, St. Bernard, La., 650........C8 95
Delafield, Waukesha, Wis., 2,334........*E5 124
Delagoa, bay, Moz........C5 49
Delagua, Las Animas, Colo..D6 83
Del Aire, Los Angeles, Calif., 5,000........*F2 82
Delake, Lincoln, Oreg., 803..C3 113
De Lamere, Sargent, N. Dak., 170........C8 110
Delanco, Burlington, N.J., 4,011........C3 106
De Land, Volusia, Fla., 10,775........C5 86
De Land, Piatt, Ill., 422.....C5 90
Delaney, Madison, Ark., 75..B2 81
Delano, Kern, Calif., 11,913..E4 82
Delano, Wright, Minn., 1,612........E5 99
Delano, Polk, Tenn., 200....D9 117
Delano, peak, Utah........E3 119
Delano Mines, Elko, Nev., 5..B7 104
Delanson, Schenectady, N.Y., 398........C6 108
Delaplaine, Greene, Ark., 186........A5 81
Delaronde, lake, Sask., Can.....C2 70
Delavan, Tazewell, Ill., 1,377........C4 90
Delavan, Morris, Kans., 200........D7 93
Delavan, Faribault, Minn., 322........G5 99
Delavan, Walworth, Wis., 4,846........F5 124
Delavan Lake, Walworth, Wis., 1,884........*F5 124
Delaware, Logan, Ark., 300..B2 81
Delaware, Ont., Can., 466...E3 72
Delaware, Delaware, Iowa, 167........B6 92
Delaware, Warren, N.J., 275........B2 106
Delaware, Delaware, Ohio, 13,282........B4 111
Delaware, Nowata, Okla., 540........A6 112
Delaware, co., Ind., 110,938.D7 91
Delaware, co., Iowa, 18,483..B6 92
Delaware, co., N.Y., 43,540.C6 108
Delaware, co., Ohio, 36,107..B4 111
Delaware, co., Okla., 13,198.A7 112
Delaware, co., Pa., 553,154........G11 114
Delaware, state, U.S., 446,292........C12 77, 85
Delaware, bay, Del........C7 85
Delaware, mts., Tex........F2 118
Delaware, res., Ohio........B4 111
Delaware, riv., Del., N.J., N.Y., Pa........C3 106
Delaware, riv., Kans........A7 93
Delaware City, New Castle, Del., 1,658........A6 85
Delaware Water Gap, Monroe, Pa., 554.......E11 114
Delaware, water gap, Pa......B2 106
Delbarton, Mingo, W. Va., 1,122........D2 123
Delburne, Alta., Can., 450...C4 69
Delcambre, Vermilion and Iberia, La., 1,857........E4 95
Del City, Oklahoma, Okla., 12,934........B4 112
Delco, Columbus, N.C., 466..C5 118
Deleau, Man., Can., 200....E1 71
Delémont, Switz., 9,542.....B3 19
De Leon, Comanche, Tex., 2,022........C3 118
De Leon Springs, Volusia, Fla., 900........C5 86
De Léry, Que., Can., 1,957..D9 73
Delevan, Cattaraugus, N.Y., 777........C2 108
Delfore, Craighead, Ark.....B5 81
Delft, Cottonwood, Minn., 125........G3 99
Delft, Neth., 74,500........B4 15
Delfzijl, Neth., 16,600......A6 15
Delgado, cape, Moz........D7 48
Delger Tsogtuin Huryee, Mong........A2 36
Delgo, Sud........A3 47
Delhi, Merced, Calif., 1,175.*D3 82
Delhi, Ont., Can., 3,427.....E4 72
Delhi, India, 2,061,758 (★2,658,612)......C6 36, C6 40
Delhi, Delaware, Iowa, 464..B6 92
Delhi, Richland, La., 2,514..B4 95
Delhi, Redwood, Minn., 124.F3 99
Delhi, Delaware, N.Y., 2,307.C6 108
Delhi, Beckham, Okla., 50...B2 112
Delhi, ter., India, 2,658,612........*C6 40
Delhi Hills, Hamilton, Ohio, 5,000........*D2 111
Delia, Alta., Can., 287......D4 69
Delia, Jackson, Kans., 163...C8 93
Delight, Pike, Ark., 446.....C2 81
Delina, Marshall, Tenn......B5 117
Delisle, Que., Can., 1,302....A6 73
Delisle, Sask., Can., 508......F2 70
DeLisle, Harrison, Miss., 800........E2, E4 100
Delitzsch, Ger., 22,900......B7 17
Dell, Mississippi, Ark., 383...B5 81
Dell, Beaverhead, Mont., 30..F4 102
Delle, Fr., 5,189........D7 14
Delle, Tooele, Utah, 25.....C3 119
Dellenbaugh, mtn., Ariz........A2 80
Dell Rapids, Minnehaha, S. Dak., 1,863........D9 116
Dellrose, Lincoln, Tenn., 175.B5 117
Dellroy, Carroll, Ohio, 391...B6 111
Dells, gorge, Wis........E4 124
Dellwood, Jackson, Fla., 100.B1 86
Dellwood, Washington, Minn., 310........E6, E7 99
Dellwood, St. Louis, Mo., 4,720........*C7 101
Dellwood, Yamhill, Oreg.....B1 113
Dellys, Alg., 5,774 (21,591▲).B5 44
Delmar, Winston, Ala., 100...A2 78
Del Mar, San Diego, Calif., 3,124........E2 82
Delmar, Sussex, Del., 934....D6 85
Delmar, Clinton, Iowa, 556..C7 92
Delmar, Wicomico, Md., 1,291........D6 85
Delmar, Albany, N.Y., 7,000........*C7 108
Del Mar-Heights (Marro Beach), San Luis Obispo, Calif., 1,907........*E3 82
Delmas, Sask., Can., 234....E1 70
Delmenhorst, Ger., 57,300...B4 17
Delmita, Starr, Tex., 70......F3 118
Delmont, Cumberland, N.J., 300........E3 106
Delmont, Westmoreland, Pa., 1,313........*F2 114
Delmont, Douglas, S. Dak., 363........D7 116
Del Monte Heights, Monterey, Calif., 1,174........*C6 82
Del Monte Park, Monterey, Calif., 2,177........*D3 82
Del Norte, Rio Grande, Colo., 1,856........D4 83
Del Norte, co., Calif., 17,771........B2 82
Deloit, Crawford, Iowa, 222..B2 92
Delong, Fulton, Ind., 130....B5 91
Deloraine, Man., Can., 916..E1 71
Deloro, Ont., Can., 157......C7 72
Del Paso Heights, Sacramento, Calif., 11,495........*A6 82
Delphi, Carroll, Ind., 2,517..C4 91
Delphia, Musselshell, Mont., 10........D8 102
Delphos, Ottawa, Kans., 619........C6 93
Delphos, Allen, Ohio, 6,961..B3 111
Delray Beach, Palm Beach, Fla., 12,230........F6 86
Del Ray Oaks, Monterey, Calif., 1,831........*C6 82
Del Rio, Cocke, Tenn., 25..D10 117
Del Rio, Val Verde, Tex., 18,612........E2 118
Del Rosa, San Bernardino, Calif., 6,000........*E5 82
Delson Village, Que., Can., 2,075........D9 73
Delta, Clay, Ala., 150.......B4 78
Delta, Ont., Can., 417.......C8 72
Delta, Delta, Colo., 3,832....C2 83
Delta, Keokuk, Iowa, 514....C5 92
Delta, Madison, La., 111.....B4 95
Delta, Cape Girardeau, Mo., 416........D8 101
Delta, Fulton, Ohio, 2,376....A4 111
Delta, York, Pa., 822........G9 114
Delta, Millard, Utah, 1,576..D3 119
Delta, co., Colo., 15,602.....C3 83
Delta, co., Mich., 34,298.....C4 98
Delta, co., Tex., 5,860.......C5 118
Delta, res., N.Y........B5 108
Delta Amacuro, ter., Ven., 33,979........B5 60
Delta Beach, Man., Can., 61........D2 71
Delta City, Sharkey, Miss., 300........B3 100
Delta Farms, Lafourche, La..E5 95
Deltaville, Middlesex, Va., 800........D6 121
Delton, Barry, Mich., 250....F5 98
Delungra, Austl., 585.......D8 51
Delvin, Ire., 346........D4 11
Delvine, Alb., 3,207........C3 23
Delyatin, Sov. Un., 14,000..G5 27
Demaine, Sask., Can., 166...G2 70
Demarcation, bay, Alsk........B11 79
Demarest, Bergen, N.J., 4,231........D5 106
Demavend, mtn., Iran........D6 41
Demba, Con. L........C3 48
Demchok, China, 5,000......A7 40
Demidov, Sov. Un., 10,000..D8 27
Deming, Hamilton, Ind., 50........D5 91
Deming, Luna, N. Mex., 6,764........E2 107
Deming, Whatcom, Wash., 250........A3 122
Demirci, Tur., 8,700........C7 23
Demir Kapija, Yugo........E6 22
Demmin, Ger., 16,400.......B6 16
Demmitt, Alta., Can........B1 69
Demnat, Mor., 6,223........I3 30
Demopolis, Marengo, Ala., 7,377........C2 78
Demorest, Habersham, Ga., 1,029........B3 87
Demorestville, Ont., Can., 120........C7 72
DeMossville, Pendleton, Ky., 90........B7 94
Demotte, Jasper, Ind., 700...B3 91
Dempster, Hamlin, S. Dak., 95........C9 116
Demyanka, riv., Sov. Un........B8 29
Demyanovka, Sov. Un.......C7 29
Denain, Fr., 29,467 (★65,000)........D3 15
Denali, Alsk., 3........C10 79
Denau, Sov. Un., 15,000....B13 41
Denaud, Hendry, Fla., 120...F5 86
Denbigh, McHenry, N. Dak., 25........A5 104
Denbigh, Wales, 8,044.......A4 12
Denbigh, co., Wales, 173,843........A4 12
Den Burg, Neth., 2,888......A4 15
Dendermonde, Bel., 9,815....C4 15
Dendron, Sask., Can........G1 70
Dendron, Surry, Va., 403....D6 121
Denezhkin Kamen, mtn., Sov. Un........A5 29
Denham, Pulaski, Ind., 180..B4 91
Denham, Pine, Minn., 71.....D6 99
Denham Springs, Livingston, La., 5,991........A5, D5 95
Den Helder, Neth., 49,200...B4 15
Denhoff, Sheridan, N. Dak., 164........B5 110
Denholm, Sask., Can., 72....E1 70
Den Hoorn, Neth., 454.......A4 15
Denia, Sp., 7,876........C6 20
Deniau, Que., Can., 215......B8 73
Deniliquin, Austl., 5,574....G5 51
Denio, Humboldt, Nev., 30...B3 104
Denison, Crawford, Iowa, 4,930........B2 92
Denison, Jackson, Kans., 184........A7 93
Denison, Grayson, Tex., 22,748........C4 118
Denison, dam, Okla........D5 112
Denizli, Tur., 49,000........D7 23
Denman, Buffalo, Nebr., 25..D7 103
Denman, glacier, Ant........C22 5
Denmark, Jackson, Ark......B4 81
Denmark, N.S., Can., 73....D6 74
Denmark, Lee, Iowa, 235....D6 92
Denmark, Lincoln, Kans., 50........C5 93
Denmark, Oxford, Maine, 160 (376▲)........E2 96
Denmark, Curry, Oreg., 25..E2 113
Denmark, Bamberg, S.C., 3,221........E5 115
Denmark, Madison, Tenn., 58........B2 117
Denmark, Brown, Wis., 1,106........A6, D6 124
Denmark, country, Eur., 4,585,256........D10 8, 24
Denmark, strait, Arc. O........C17 4

Dennard, Van Buren, Ark., 100 B3 81
Denning, Franklin, Ark., 227 B2 81
Denning, San Augustine, Tex. D5 118
Dennis, Labette, Kans., 150 E8 93
Dennis, Barnstable, Mass., 550 (3,727▲) C7 97
Dennis, Tishomingo, Miss., 125 A5 100
Dennis Port, Barnstable, Mass., 1,271 C7 97
Dennison, Goodhue and Rice, Minn., 179 F5 99
Dennison, Tuscarawas, Ohio, 4,158 B6 111
Denniston, Halifax, Va., 50 E4 121
Dennisville, Cape May, N.J., 500 E3 106
Dennysville, Washington, Maine, 250 (303▲) D5 96
Denny Terrace, Richland, S.C., 3,000 C5 115
Den Oever, Neth., 1,497 B5 15
Denoya, Osage, Okla A5 112
Denpasar, Indon., 16,639 G5 35
Dent, Otter Tail, Minn., 176 D3 99
Dent, co., Mo., 10,445 D6 101
Dent du Midi, mtn., Switz D2 19
Denton, Jeff Davis, Ga., 255 E4 87
Denton, Doniphan, Kans., 161 C8 93
Denton, Caroline, Md., 1,938 C6 85
Denton, Wayne, Mich., 200 A7 98
Denton, Fergus, Mont., 410 C7 102
Denton, Lancaster, Nebr., 94 D9 103
Denton, Davidson, N.C., 852 B3 109
Denton, Cocke, Tenn., 100 D10 117
Denton, Denton, Tex., 26,844 C4 118
Denton, co., Tex., 47,432 C4 118
Denton, creek, Tex A5 118
D'Entrecasteaux, pt., Austl F2 50
D'Entrecasteaux, is., Pap k13 50
Dentsville, Richland, S.C., 1,500 C6 115
Denver, Denver, Colo., 493,887 (★858,300) B6 78
Denver, Hancock, Ill., 120 C2 90
Denver, Miami, Ind., 565 C5 91
Denver, Bremer, Iowa, 831 B5 92
Denver, Worth, Mo., 116 A3 101
Denver, Lincoln, N.C., 113 B2 109
Denver, Lancaster, Pa., 1,875 F9 114
Denver, Humphreys, Tenn., 100 A4 117
Denver, Marshall, W. Va., 45 C2 123
Denver, Preston, W. Va., 50 B5 123
Denver, co., Colo., 493,887 B6 83
Denver City, Yoakum, Tex., 4,302 C1 118
Denville, Morris, N.J., 10,632 B4 106
Denzil, Sask., Can., 328 E1 70
Deoghar, India, 30,813 E11 40
Deora, Baca, Colo., 5 D8 83
De Panne, Bel., 6,407 C2 15
Depauville, Jefferson, N.Y., 330 A4 108
Depauw, Harrison, Ind., 120 H5 91
De Pere, Brown, Wis., 10,045 A6, D5 124
Depew, Erie, N.Y., 13,580 C2 108
Depew, Creek, Okla., 686 B5 112
Depoe Bay, Lincoln, Oreg., 750 C3 113
Deport, Lamar and Red River, Tex., 639 C5 118
Deposit, Broome and Delaware, N.Y., 2,025 C5 108
Depot Harbour, Ont., Can., 480 B4 72
Depue, Bureau, Ill., 1,920 B4 90
Deputy, Jefferson, Ind., 300 G6 91
DeQueen, Sevier, Ark., 2,859 C1 81
Dequen, Que., Can., 267 A5 73
DeQuincy, Calcasieu, La., 3,928 D2 95
Der'a, Syr., 15,635 F11 31
Dera Ghazi Khan, Pak., 36,239 B5 39, B3 40
Dera Ismail Khan, Pak., 39,846 (★41,663) B3 40
Derbent, Sov. Un., 50,000 E3 29, G18 9
Derbetovka, Sov. Un., 10,000 I14 27
Derby, Austl., 326 C3 50
Derby, N.B., Can., 111 C4 74
Derby, Adams, Colo., 10,124 *B6 83
Derby, New Haven, Conn., 12,132 D4 84
Derby, Eng., 132,325 (★260,000) B6 12
Derby, Perry, Ind., 60 H4 91
Derby, Lucas, Iowa, 151 D4 92
Derby, Sedgwick, Kans., 6,458 E6 93
Derby, Piscataquis, Maine, 500 C4 96
Derby, Pearl River, Miss., 200 E4 100
Derby, Erie, N.Y., 3,500 C2 108
Derby, Pickaway, Ohio, 365 C4 111
Derby, Frio, Tex., 40 E3 118
Derby, Orleans, Vt., 433 (2,506▲) B4 120
Derby, Wise, Va., 800 B2 121
Derby, co., Eng., 877,548 A6 12
Derby Junction, N.B., Can., 146 C4 74
Derby Line, Orleans, Vt., 849 A4 120
Derecske, Hung., 9,970 B5 22
Dereno, DeBaca, N. Mex C6 107
Derg, lake, Ire E3 11
De Ridder, Beauregard, La., 7,188 D2 95
Derik, Tur., 3,586 D13 31
Derita, Mecklenburg, N.C., 1,500 B3 109
Derma, Calhoun, Miss., 578 B4 100
Dermott, Chicot, Ark., 3,665 D4 81
Dermott, Scurry, Tex., 40 C2 118
Derna, see Darnah, Libya
Derniere, isl., La E5 95
Derravaragh, lake, Ire D4 11
Derry, Natchitoches, La., 75 C3 95
Derry, Rockingham, N.H., 4,468 (6,987▲) E4 105
Derry, Sierra, N. Mex., 10 E2 107
Derry, Westmoreland, Pa., 3,426 F3 114
Derrybrien, Ire D3 11
Derudeb, Sud B4 47
DeRuyter, Madison, N.Y., 627 C5 108
Derventa, Yugo., 9,795 C3 22
Derwent, Alta., Can., 281 C5 69
Derwent, Guernsey, Ohio, 225 C6 111
Derwent, riv., Eng F5 13
Derwent, riv., Eng F6 13
Derwent, riv., Eng F8 13
Derwood, Montgomery, Md., 110 B3 85
Derzhavinskoye, Sov. Un E22 9
Desaguadero, riv., Arg A3 54
Desaguadero, riv., Bol C2 55
Des Allemands, bayou, La C6 95
Des Allemands, lake, La E5 95
Des Arc, Prairie, Ark., 1,482 C4 81
Des Arc, Iron, Mo., 275 D7 101
Des Arc, bayou, Ark B4 81
Des Arc, mtn., Mo D7 101
Desatoya, mts., Nev D4 104
Desatoya, peak, Nev D4 104
Desbiens, Que., Can., 1,970 A6 73
Desboro, Ont., Can., 160 C4 72
Descalabrado, P.R C5 65
Descalvado, Braz., 7,220 k8 56
Deschaillons, Que., Can. 415 C5 73
Deschaillons sur St. Laurent, Que., Can., 1,283 *C5 73
Deschambault, Que., Can., 1,056 C6 73
Deschenes, Que., Can., 2,090 *A9 72
Deschutes, co., Oreg., 23,100 D5 113
Deschutes, peak, Wash C3 122
Deschutes, riv., Oreg B6 113
Desembarcadero Mosquito, P.R., 4 g12 65
Desengaño, cape, Arg D3 54
Desenzano del Garda, It., 6,500 D6 18
Deseret, Millard, Utah, 310 D3 119
Deseronto, Ont., Can., 1,797 C7 72
Desert, game range, Nev G6 104
Desert, mtn., W. Va C6 123
Desert, peak, Utah B2 119
Desert, val., Nev B3 104
Deserta Grande, isl., Port h12 20
Desert Center, Riverside, Calif., 200 F6 82
Desert Hot Springs, Riverside, Calif., 1,472 *F5 82
Desha, Independence, Ark., 350 B4 81
Desha, co., Ark., 20,770 D4 81
Deshler, Thayer, Nebr., 956 D8 103
Deshler, Henry, Ohio, 1,824 A4 111
Deshu, Afg., 5,000 F11 41
Desio, It., 23,750 D5 18
Des Lacs, Ward, N. Dak., 185 A4 110
Des Lacs, riv., N. Dak A4 110
Desloge, St. Francois, Mo., 2,308 D7 101
Desmet, Benewah, Idaho, 100 B2 89
De Smet, Missoula, Mont., 25 D2 102
De Smet, Kingsbury, S. Dak., 1,324 C8 116
Desmochado, Par., 2,398 E4 55
Des Moines, Polk, Iowa, 208,982 (★261,900) A7, C4 92
Des Moines, Union, N. Mex., 207 A6 107
Des Moines, King, Wash., 1,987 B3, D1 122
Des Moines, co., Iowa, 44,605 D6 92
Des Moines, riv., U.S B9 77
Desna, riv., Sov. Un F8 27
Desolacion, isl., Chile h11 54
Desolation, lake, Newf., Can g9 75
De Soto, Sumter, Ga., 282 E2 87
De Soto, Jackson, Ill., 723 F4 90
De Soto, Dallas, Iowa, 273 C3 92
De Soto, Johnson, Kans., 1,271 B8, D9 93
De Soto, Clarke, Miss., 240 D5 100
De Soto, Jefferson, Mo., 5,804 C7 101
De Soto, Dallas, Tex., 1,969 *C4 118
De Soto, Crawford and Vernon, Wis., 357 E2 124
De Soto, co., Fla., 11,683 E5 86
De Soto, co., Miss., 23,891 A3 100
De Soto, par., La., 24,248 B2 95
De Soto City, Highlands, Fla., 245 E5 86
Despard, Harrison, W. Va., 1,763 B7 123
Desperes, St. Louis, Mo., 4,362 B8 101
Des Plaines, Cook, Ill., 41,209 A6, E2 90
Des Plaines, riv., Ill., Wis B5 90, F2 124
Dessau, Ger., 94,000 B7 17
Destin, Okaloosa, Fla., 900 G2 86
Destrehan, St. Charles, La., 330 C7, E5 95
Desvres, Fr., 5,518 D9 12
Detlor, Ont., Can., 65 B7 72
Detmold, Ger., 31,200 B3 17
Detour, Carroll, Md., 100 A3 85
De Tour, Chippewa, Mich., 669 C7 98
Detour, pt., Mich C4 98
Detroit, Lamar, Ala., 113 A1 78
Detroit, Dickinson, Kans., 100 D6 93
Detroit, Somerset, Maine, 250 (564▲) D3 96
Detroit, Wayne, Mich., 1,670,144 (★4,028,500) A8, F7 98
Detroit, Marion, Oreg., 206 C4 113
Detroit, Red River, Tex., 576 C5 118
Detroit, res., Oreg C4 113
Detroit, riv., Ont., Can., Mich F7 98
Detroit Beach, Monroe, Mich., 1,571 *G7 98
Detroit Lakes, Becker, Minn., 5,633 D3 99
Dett, S. Rh., 820 A4 49
Detva, Czech., 7,786 D5 26
Deuel, co., Nebr., 3,125 C3 103
Deuel, co., S. Dak., 6,782 C9 116
Deurne, Bel., 68,703 C4 15
Deux Frères, isl., Viet H7 38
Deux Rivieres, Ont., Can., 163 A6 72
Deux-Sèvres, dept., Fr., 321,118 *D3 14
Deva, Rom., 16,879 C6 22
De Valls Bluff, Prairie, Ark., 654 C4 81
Devarkonda, India, 8,311 I7 40
Devavanya, Hung., 10,828 B5 22
Deventer, Neth., 56,600 B6 15
Devereux, Hancock, Ga., 200 C3 87
Deveron, riv., Scot C6 13
De View, bayou, Ark B4 81
Devil River, peak, N.Z N14 51
Devils, isl., Fr. Gu A4 59
Devils, lake, N. Dak A6 110
Devils, riv., Tex E2 118
Devils Den, Kern, Calif., 150 E4 82
Devils Elbow, Pulaski, Mo., 250 D5 101
Devils Knob, mtn., Va D3 121
Devils Lake, Ramsey, N. Dak., 6,299 B7 110
Devils Lake, Indian res., N. Dak B6 110
Devils Postpile, nat. mon., Calif D4 82
Devils Slide, Morgan, Utah, 250 B4 119
Devils Tower, Crook, Wyo., 10 A8 125
Devils Tower, nat. mon., Wyo A8 125
Devils Track, lake, Minn B7 99
Devine, Medina, Tex., 2,522 E3 118
Devizes, Eng., 8,497 C6 12
Devol, Cotton, Okla., 117 C3 112
Devoll, riv., Alb B3 23
Devon, Alta., Can., 1,418 *C4 69
Devon, New Haven, Conn. (part of Milford) E4 84
Devon, Bourbon, Kans., 100 E9 93
Devon, Toole, Mont., 43 B5 102
Devon, Chester, Pa., 1,500 *G11 114
Devon, co., Eng., 822,906 D4 12
Devon, isl., N.W. Ter., Can A15 66
Devonport, Austl., 13,068 o15 50
Devonport, N.Z., 10,976 L15 51
Dewalt, Fort Bend, Tex., 100 F5 118
Dewar, Black Hawk, Iowa, 68 B5 92
Dewar, Okmulgee, Okla., 817 B6 112
Dewar Lake, Sask., Can., 75 F1 70
Dewberry, Alta., Can., 179 C5 69
Dewees, inlet, S.C F8 115
Dewees, isl., S.C F4 115
Deweese, Clay, Nebr., 100 D7 103
Dewey, Yavapai, Ariz., 50 C3 80
Dewey, Beaverhead, Mont., 35 E4 102
Dewey, Washington, Okla., 3,994 A6 112
Dewey, Custer, S. Dak., 55 D1 116
Dewey, co., Okla., 6,051 B2 112
Dewey, co., S. Dak., 5,257 B4 116
Dewey, res., Ky C7 94
Dewey Mills, Windsor, Vt., 100 D4 120
Deweyville, Newton, Tex., 750 D6 118
De Winton, Alta., Can., 51 D3 69
DeWitt, Arkansas, Ark., 3,019 C4 81
DeWitt, DeWitt, Ill., 245 C5 90
DeWitt, Clinton, Iowa, 3,224 C7 92
DeWitt, Clinton, Mich., 1,238 F6 98
DeWitt, Carroll, Mo., 174 B4 101
DeWitt, Saline, Nebr., 504 D9 103
DeWitt, Onondaga, N.Y., 3,500 *B4 108
Dewitt, Dinwiddie, Va., 100 D5 121
DeWitt, co., Ill., 17,253 C5 90
DeWitt, co., Tex., 20,683 E4 118
Dewsbury, Eng., 52,942 (★150,000) A6 12
Dewy Rose, Elbert, Ga., 150 B4 87
Dexter, Laurens, Ga., 359 D3 87
Dexter, Dallas, Iowa, 670 C3 92
Dexter, Cowley, Kans., 291 E7 93
Dexter, Calloway, Ky., 250 B3 94
Dexter, Penobscot, Maine, 2,720 (3,951▲) C3 96
Dexter, Washtenaw, Mich., 1,702 F7 98
Dexter, Mower, Minn., 313 G6 99
Dexter, Stoddard, Mo., 5,519 E8 101
Dexter, Chaves, N. Mex., 885 D5 107
Dexter, Jefferson, N.Y., 1,009 A4 108
Dexter, Lane, Oreg., 200 D4 113
Dexter, lake, Fla C5 86
Deyhuk, Iran, 10,000 E8 41
Deyyer, Iran, 5,000 H5 41
Dezfūl, Iran, 52,121 E4 41
Dezhnev, cape, Sov. Un C21 28
Dezh Shāhpūr, Iran D3 41
Dhaain, Sau. Ar I5 41
Dhahran, see Az Zahrān, Sau. Ar.
Dhamtari, India, 31,552 G8 40
Dhanbad, India, 46,756 (★125,000) F11 40
Dhankuta, Nep., 4,194 D11 40
D'Hanis, Medina, Tex., 850 E3 118
Dhar, India, 28,325 F5 40
Dharamjaygarh, India, 5,000 F9 40
Dharmapuri, India, 28,031 F6 39
Dharmsala, India, 10,255 A6 40
Dharwar, India, 77,163 E6 39
Dhāt al Ḩajj, Sau. Ar H11 31
Dhaulagiri, peak, Nep C9 40
Dhekelia, Grc., 388 g11 23
Dhenousa, isl., Grc D5 23
Dhībān, Jordan, 2,000 G10 31
Dhidhimotikhon, Grc., 8,111 B6 23
Dhílos (Delos), isl., Grc D5 23
Dhimitsana, Grc., 1,710 D4 23
Dhodekanisos, prov., Grc., 123,021 *D6 23
Dholpur, India, 27,412 D6 40
Dhond, India, 9,947 (★18,849) H5 40
Dhoraji, India, 48,951 G3 40
Dhritsa, Grc., 696 g11 23
Dhubri, India, 28,355 E12 40
Dhulia, India, 98,893 G5 40
Dia, isl., Grc E5 23
Diablerets, mtn., Switz D3 19
Diablo, Contra Costa, Calif., 2,096 *D2 82
Diablo, canyon, Ariz C4 80
Diablo, dam, Wash A4 122
Diablo, isl., P.R f12 65
Diablo, mtn., Calif B6 82
Diablo, mtn., Oreg E6 113
Diablo, range, Calif D3 82
Diablo Heights, C.Z., 1,647 k11 62
Diagonal, Ringgold, Iowa, 443 D3 92
Dial, Fannin, Ga., 25 B2 87
Diamante, Arg., 13,600 A4 54
Diamantina, Braz., 14,252 E2 57
Diamantina, riv., Austl D7 50
Diamond, Ont., Can A8 72
Diamond, Parke, Ind., 45 E3 91
Diamond, Plaquemines, La., 200 E6 95
Diamond, Newton, Mo., 453 E3 101
Diamond, Portage, Ohio, 280 A6 111
Diamond, Harney, Oreg., 10 D8 113
Diamond, Kanawha, W. Va., 600 C6 123
Diamond, Platte, Wyo., 398 D8 125
Diamond, cave, Ark B2 81
Diamond, head, Haw g10 88
Diamond, lake, Oreg D4 113
Diamond, mts., Nev D6 104
Diamond, peak, Oreg D4 113
Diamond, pt., Indon k11 35
Diamond Bluff, Pierce, Wis., 150 D1 124
Diamond City, Alta., Can., 78 E4 69
Diamond Harbour, India, 10,135 F12 40
Diamond Lake, Lake, Ill., 700 E2 90
Diamond Point, Warren, N.Y., 400 B7 108
Diamond Springs, Eldorado, Calif., 617 C3 82
Diamond Springs, Princess Anne, Va., 1,500 *E6 121
Diamondville, Lincoln, Wyo., 398 D6 125
Diana, Giles, Tenn., 100 B5 117
Diana, Webster, W. Va., 250 C4 123
Dianalund, Den., 2,145 C5 24
Dianópolis, Braz., 2,145 D1 57
Díapaga, Upper Volta D5 45
Diapitan, bay, Phil n14 35
Diarbekr (Diyarbakir), Tur., 80,600 D3 31
Dias Creek, Cape May, N.J. E3 106
Diaz, Jackson, Ark., 348 B4 81
Dibaya, Con. L C3 48
Diber (Dibra), pref., Alb., 96,000 *B3 23
D'Iberville, Harrison & Jackson, Miss., 3,005 E2, E5 100
Diboll, Angelina, Tex., 2,506 D5 118
Dibra (Diber), pref., Alb., 96,000 *B3 23
Dibrell, Warren, Tenn., 90 D8 117
Dibrugarh, India, 58,480 C9 39
Dickens, Clay, Iowa, 241 A2 92
Dickens, Lincoln, Nebr., 25 D5 103
Dickens, Dickens, Tex., 302 C2 118
Dickens, co., Tex., 4,963 C2 118
Dickenson, co., Va., 20,211 B2 121
Dickerson, Montgomery, Md., 246 B3 85
Dickey, Aroostook, Maine A4 96
Dickey, La Moure, N. Dak., 143 C4 110
Dickey, co., N. Dak., 8,147 C7 110
Dickeyville, Grant, Wis., 671 F3 124
Dickie, Hot Springs, Wyo., 15 B4 125
Dickinson, Clarke, Ala., 250 D2 78
Dickinson, Stark, N. Dak., 9,971 C3 110
Dickinson, Galveston, Tex., 4,715 *F5 118
Dickinson, co., Iowa, 12,574 A2 92
Dickinson, co., Kans., 21,572 D6 93
Dickinson, co., Mich., 23,917 B3 98
Dickinson, dam, N. Dak C2 110
Dickinson Center, Franklin, N.Y., 240 A2 108
Dickson, Dickson, Tenn., 5,028 A4 117
Dickson, co., Tenn., 18,839 A4 117
Dickson City, Lackawanna, Pa., 7,738 A9, D10 114
Dicle, riv., Tur D13 31
Didsbury, Alta., Can., 1,254 D7 69
Didwana, India, 13,547 D5 40
Die, Fr., 2,824 E6 14
Dieburg, Ger., 9,500 D3 17
Diégo-Suarez (Antsirane), Malag., 30,254 f9 49
Diégo-Suarez, prov., Malag f9 49
Diekirch, Lux., 4,397 E6 15
Diemel, riv., Ger B3 17
Diem Zubeir, Sud D2 47
Dien Bien Phu, Viet B5 38
Diepholz, Ger., 10,900 B4 16
Dieppe, N.B., Can., 4,032 C5 74
Dieppe, Fr., 29,967 C4 14
Dierks, Howard, Ark., 1,276 C1 81
Diessen, Ger., 5,000 B7 18
Diest, Bel., 9,816 D5 15
Dieterich, Effingham, Ill., 591 D5 90
Dietikon, Switz., 14,920 B5 19
Dietrich, Lincoln, Idaho, 118 G4 89
Dieuze, Fr., 3,431 F6 15
Diever, Neth., 857 B6 15
Diez, Ger., 9,600 C3 17
Dif, Som E5 47
Differdange, Lux., 17,637 E5 15
Difficult, Smith, Tenn., 150 C8 117
Difficulty, Carbon, Wyo C6 125
Digby, N.S., Can., 2,308 E4 74
Digby, co., N.S., Can., 20,216 E4 74
Dighton, Lane, Kans., 1,526 D3 93
Dighton, Bristol, Mass., 700 (3,769▲) C5 97
Dighton, Osceola, Mich., 75 D5 98
Digne, Fr., 10,299 E7 14
Digoel, riv., W. Irian G10 35
Digoin, Fr., 8,529 D5 14
Digos, Phil., 21,455 *D7 35
Dijon, Fr., 139,694 D6 14
Dike, Grundy, Iowa, 630 B5 92
Dikhil, Fr. Som., 500 C5 47
Dikili, Tur., 6,100 C6 23
Dikili, cape, Tur B8 23
Dikirnis, Eg., U.A.R., 5,000 C3 32
Diksmuide, Bel., 3,812 C2 15
Dikson, Sov. Un., 10,000 B10 28
Dikwa, Nig., 5,242 D7 45
Dilāram, Afg., 5,000 E11 41
Dili (Dilli), Port. Timor, 1,795 G7 35
Dilia, Guadalupe, N. Mex., 10 B4 107
Dilke, Sask., Can., 187 G3 70
Dilkon, Navajo, Ariz B5 80
Dillard, Rabun, Ga., 204 B3 87
Dillard, Carter, Okla., 125 C4 112
Dill City, Washita, Okla., 623 B2 112
Dille, Clay, W. Va., 250 C4 123
Dillen, Johnson, Ark B2 81
Dillenburg, Ger., 10,700 C3 17
Diller, Jefferson, Nebr., 286 D9 103
Dilley, Frio, Tex., 2,118 E3 118
Dilli (Dili), Port. Timor, 1,795 G7 35
Dilliner, Greene, Pa., 100 F2 114
Dilling, Sud., 5,295 C2 47
Dillingen [an der Donau], Ger., 11,200 E5 17
Dillingen, Ger., 17,700 D1 17
Dillingham, Alsk., 424 D8 79
Dillon, Summit, Colo., 814 B4 83
Dillon, Beaverhead, Mont., 3,690 E4 102
Dillon, Dillon, S.C., 6,173 C9 115
Dillon, co., S.C., 30,584 C9 115
Dillonvale, Jefferson, Ohio, 1,232 B7 111
Dillsboro, Dearborn, Ind., 745 F7 91
Dillsboro, Jackson, N.C., 140 D3 109
Dillsburg, York, Pa., 1,322 F7 114
Dillwyn, Buckingham, Va., 515 D4 121
Dilolo, Con. L D3 48
Dilworth, Clay, Minn., 2,102 D2 99
Dimashq, see Damascus, Syr.
Dimbelenge, Con. L C3 48
Dimbokro, I.C., 3,800 E4 45
Dimitrovgrad, Bul., 45,000 D7 22
Dimitrovgrad, Yugo., 3,669 D6 22
Dimitrovo (Pernik), Bul., 28,504 D6 22
Dimmit, co., Tex., 10,095 E3 118
Dimmitt, Castro, Tex., 2,935 B1 118
Dimock, Hutchinson, S. Dak., 150 D8 116
Dimondale, Eaton, Mich., 866 F6 98
Di Monte Santu, cape, It D2 21
Dinagat, isl., Phil C7 35
Dinajpur, Pak., 35,687 E12 40
Dinan, Fr., 12,847 C3 14
Dinant, Bel., 6,851 D4 15
Dinapore, India, 35,159 E10 40
Dinar, Tur., 9,400 C8 23
Dinard [-St. Enogat], Fr., 9,270 C2 14
Dinaric Alps, mts., Yugo D3 22
Dindigul, India, 92,947 F6 39
Dinero, Live Oak, Tex., 75 E4 118
Dines, Sweetwater, Wyo D3 125
Dingalan, bay, Phil o13 35
Dingelstädt, Ger., 5,602 B5 17
Dingess, Mingo, W. Va., 400 D2 123
Dingle, Bear Lake, Idaho, 100 G7 89
Dingle, Ire., 1,460 E1 11
Dingle, Phil., 1,328 *C6 35
Dingle, bay, Ire E1 11
Dingmans Ferry, Pike, Pa., 300 D12 114
Dingo, Austl., 102 A7 51
Dingolfing, Ger., 10,500 E7 17
Dinguiraye, Guinea, 2,900 D2 45
Dingwall, N.S., Can., 269 C9 74
Dingwall, Scot., 3,752 C4 13
Dinh Lap, Viet, 10,000 B7 38
Dinkelsbühl, Ger., 7,900 D5 17
Dinkey Creek, Fresno, Calif., 300 D4 82
Dinklage, Ger., 7,300 F2 24
Dinorwick, lake, Ont., Can E5 71
Dinosaur, nat. mon., Colo., Utah A1 83, C6 119
Dinsdale, Tama, Iowa, 80 B5 92
Dinslaken, Ger., 43,800 B1 17
Dinsmore, Sask., Can., 433 F2 70
Dinsmore, Duval, Fla., 2,000 B5 86
Dinuba, Tulare, Calif., 6,103 D4 82
Dinwiddie, Dinwiddie, Va., 200 D5 121
Dinwiddie, co., Va., 22,183 D5 121
Dioïla, Mali D3 45
Dioka, Mali D2 45
Diourbel, Sen., 20,600 D1 45
Diphu, pass, China, India C10 39
Diplo, Pak E2 40
Dipolog, Phil., 8,770 *D6 35
Dipper Harbour, N.B., Can., 150 D3 74
Dippoldiswalde, Ger., 5,937 C8 17
Dirēdawā, Eth., 30,000 D5 47
Diriamba, Nic., 13,100 E4 62
Dirico, Ang E3 48
Dirk Hartog, isl., Austl E1 50
Dirkou, Niger C7 45
Dirmil, Tur., 602 D7 23
Dirranbandi, Austl., 514 D7 51
Dirty Devil, riv., Utah E5 119
Disappointment, cape, Wash C1 122
Disappointment, lake, Austl D3 50
Disautel, Okanogan, Wash., 50 A6 122
Discovery, cape, N.W. Ter., Can A23 4
Discovery, inlet, Ant B31 5
Disentis (Mustèr), Switz., 2,376 C6 19
Dishman, Spokane, Wash., 5,000 D7 122
Dishnā, Eg., U.A.R., 16,336 D6 43
Disko, Fulton, Ind., 140 B6 91
Disko, isl., Grnld C20 4
Disley, Sask., Can., 62 G3 70
Dismal, swamp, N.C., Va A7 109, E6 121
Disney, Mayes, Okla., 224 A6 112
Disputanta, Prince George, Va., 350 D5 121
Disraeli, Que., Can., 3,067 D6 73
Diss, Eng., 3,682 B9 12
Distant, Armstrong, Pa., 600 E3 114
District Heights, Prince Georges, Md., 7,534 *C4 85
District of Columbia, U.S., 763,956 C12 77, C7 85
Distrito Federal, fed. dist., Braz., 141,742 *B3 56
Distrito Federal, fed. dist., Mex., 4,829,402 D5 63
Distrito Federal, fed. dist., Ven., 1,257,515 A4 60
Disūq, Eg., U.A.R., 36,600 C2 32
Dittlinger, Comal, Tex., 100 B4 118
Ditzum, Ger., 820 A7 15
Diu, India, 4,856 G3 40
Divènié, Con. B F2 46
Divernon, Sangamon, Ill., 997 D4 90
Dives-sur-Mer, Fr., 6,258 C4 14
Divide, Silver Bow, Mont., 5 E4 102
Divide, co., N. Dak., 5,566 A2 110
Divide, peak, Wyo D5 125
Dividend, Utah, Utah, 10 D3 119
Dividing, creek, Md D6 85
Dividing Creek, Cumberland, N.J., 600 E2 106
Divinópolis, Braz., 41,544 F2 57
Divion, Fr., 11,300 D2 15
Divisov, Czech., 904 o18 26
Divnoye, Sov. Un., 17,300 D2 29
Divo, I.C., 4,800 E3 45
Divrigi, Tur., 8,900 C12 31
Dix, Jefferson, Ill., 181 E5 90
Dix, Kimball, Nebr., 420 C2 103
Dix, dam, Ky C5 94
Dix, hills, N.Y F3 84
Dix, riv., Ky C5 94
Dixence, riv., Switz D3 19
Dixfield, Oxford, Maine, 1,298 (2,323▲) D2 96
Dixiana, Jefferson, Ala. 500 B3 78
Dixiana, Lexington, S.C., 150 D5 115
Dixie, Escambia, Ala., 100 D3 78
Dixie, Woodruff, Ark., 40 B4 81
Dixie, Ont., Can., 551 E6 72
Dixie, Brooks, Ga., 220 F3 87
Dixie, Idaho, Idaho, 25 D3 89
Dixie, Caddo, La., 250 B2 95
Dixie, Stephens, Okla C4 112
Dixie, Walla Walla, Wash., 250 C7 122
Dixie, Nicholas, W. Va., 650 C3, C7 123
Dixie, co., Fla., 4,479 C3 86
Dixie, butte, Oreg C8 113
Dixie Union, Ware, Ga., 100 E4 87
Dixmoor, Cook, Ill., 3,076 *F3 90
Dixon, Solano, Calif., 2,970 A6, C3 82
Dixon, Lee, Ill., 19,565 B4 90
Dixon, Scott, Iowa, 280 C7 92
Dixon, Webster, Ky., 541 C2 94
Dixon, Pulaski, Mo., 1,473 D5 101
Dixon, Sanders, Mont., 132 C2 102
Dixon, Dixon, Nebr., 139 B9 103
Dixon, Rio Arriba, N. Mex., 800 A4 107
Dixon, Onslow, N.C C6 109
Dixon, Van Wert, Ohio, 80 B3 111
Dixon, Gregory, S. Dak., 26 D6 116
Dixon, Carbon, Wyo., 108 D5 125
Dixon, co., Nebr., 8,106 B9 103
Dixon entrance, B.C., Can B2 68
Dixons Mills, Marengo, Ala., 50 C2 78
Dixon Springs, Pope, Ill., 40 F5 90
Dixonville, Escambia, Ala., 200 D2 78
Dixonville, Indiana, Pa., 900 E4 114
Dixville, Que., Can., 521 D6 73
Dixville, peak, N.H A4, B1 105
Dixville Notch, Coos, N.H., 10 B1 105
Diyarbakir, Tur., 80,600 D13 31
Dizy-le-Gros, Fr., 1,057 E4 15
Dizy-Magenta, Fr., 2,079 E3 15
Dja, riv., Cam E2 46
Djado, Niger B7 45
Djafou, Alg D5 44
Djailolo, Indon E7 35
Djakarta (Jakarta, Batavia), Indon., 2,922,000 G3 35
Djakovica, Yugo., 20,741 D5 22
Djakovo, Yugo., 12,069 C4 22
Djambala, Con. B F2 46
Djambi, Indon., 108,834 F2 35
Djanet (Ft. Charlet), Alg., 74 E6 44
Djaravica, peak, Yugo D5 22
Djedeida, Tun F11 30
Djelfa, Alg., 10,070 (110,681▲) C5 44
Djema, Cen. Afr. Rep D5 46
Djemadja, isl., Indon K6 38
Djenné, Mali, 5,000 D4 45
Djerba, isl., Tun C7 44, H12 30
Djerid, salt lake, Tun C6 44
Djibo, Upper Volta D4 45
Djibouti, Fr. Som., 31,500 C5 47
Djidjelli, Alg., 31,580 B6 44
Djiring, Viet G8 38
Djolu, Con. L A3 48
Djouf, basin, Maur B3 45
Djougou, Dah., 4,900 E5 45
Djugu, Con L A5 48
Djurdjevac, Yugo., 6,408 B3 22
Djurjura, mtn., Alg B5 44
Djursholm, Swe., 7,400 t36 25
D'Lo, Simpson, Miss., 428 D4 100
Dmitriyevka, Sov. Un., 10,000 H12, r21 27
Dmitriyev-Lgovskiy, Sov. Un., 10,000 E10 27
Dmitrov, Sov. Un., 10,000 C11 27
Dmitrovsk-Orlovskiy, Sov. Un., 10,000 E10 27
Dnepr, riv., Sov. Un H9 27
Dneprodzerzhinsk, Sov. Un., 203,000 G10 27
Dnepropetrovsk, Sov. Un., 707,000 (★795,000) G10 27
Dnestr, riv., Sov. Un H7 27
Dno, Sov. Un., 10,000 C7 27
Doaktown, N.B., Can., 595 C3 74
Doba, Chad, 7,375 D3 46
Dobbinton, Ont., Can., 110 C3 72
Dobbs Ferry, Westchester, N.Y., 9,260 D1 108
Dobbyn, Austl., 109 C6 50
Döbeln, Ger., 28,900 B8 17
Doblas, Arg B4 54
Dobo, Indon G8 35
Doboj, Yugo., 13,445 C4 22
Doboy, sound, Ga E5 87
Dobrejovice, Czech., 2,188 o18 26
Dobrich, see Tolbukhin, Bul.
Dobris, Czech., 4,130 o17 26
Dobrogea, prov., Rom., 503,217 *C9 22
Dobrovice, Czech., 2,137 n18 26
Dobrovolsk, Sov. Un A7 26
Dobruja, reg., Bul., Rom C9 22
Dobsina, Czech., 4,215 D6 26
Dobson, Surry, N.C., 684 A3 109
Doce, riv., Braz E2 57
Docena, Jefferson, Ala., 1,400 *E4 78
Dock Junction, Glynn, Ga., 3,920 *E5 87
Dockton, King, Wash., 400 D1 122
Doctor Arroyo, Mex., 3,055 C4 63
Doctors, lake, Fla C6 86
Doctors Inlet, Clay, Fla., 600 C6 86
Doctortown, Wayne, Ga., 200 E5 87
Doddridge, Miller, Ark., 500 D2 81
Doddridge, co., W. Va., 6,970 B4 123

Dodds, Alta., Can. C4 69
Doddsville, Sunflower, Miss., 190 D3 100
Dodecanese, prov., Grc., 123,021 D6 23
Dodecanese, is., Grc D6 23
Dodge, Worcester, Mass., 400 B4 97
Dodge, Dodge, Nebr., 649 C9 103
Dodge, Dunn, N. Dak., 226 B3 110
Dodge, Trempealeau, Wis., 130 D2 124
Dodge, co., Ga., 16,483 D3 87
Dodge, co., Minn., 13,259 G6 99
Dodge, co., Nebr., 32,471 C9 103
Dodge, co., Wis., 63,170 E5 124
Dodge Center, Dodge, Minn., 1,441 F6 99
Dodge City, Ford, Kans., 13,520 E3 93
Dodgeville, Bristol, Mass. (part of Attleboro) C5 97
Dodgeville, Ashtabula, Ohio, 150 A7 111
Dodgeville, Iowa, Wis., 2,911 F3 124
Dodoma, Tan., 13,435 C6 48
Dodsland, Sask., Can., 365 F1 70
Dodson, Winn, La., 512 B3 95
Dodson, Phillips, Mont., 313 B8 102
Dodson, Collingsworth, Tex., 308 B2 118
Doe, riv., B.C., Can B7 68
Doebay, San Juan, Wash., 85 A3 122
Doe River, B.C., Can., 275 B7 68
Doerun, Colquitt, Ga., 1,037 E3 87
Doe Run, St. Francois, Mo., 600 D7 101
Doetinchem, Neth., 17,500 C6 15
Doeville, Johnson, Tenn C12 117
Dog, isl., Fla C2 86
Dog, lake, Man., Can D2 71
Dog, riv., Vt C3 120
Dog, rocks, Vir. Is f15 65
Dogie, Niobrara, Wyo B8 125
Dogiuma, Som E5 47
Dog Keys, pass, Miss E3 100
Dogondoutchi, Niger, 4,000 D5 45
Dog Pound, Alta., Can., 26 D3 69
Dogtooth, mts., B.C., Can D9 68
Dogubayazit, Tur., 6,800 C15 31
Doha, Qatar, 10,000 I5 41
Dohad, India, 35,483 F5 40
Dohat es Salwa, bay, Sau. Ar I5 41
Doheny, Que., Can., 76 B5 73
Doi Angka, mtn., Thai C3 38
Doire, riv., It D2 18
Dois Irmãos, mts., Braz C2 57
Dojran, lake, Yugo E6 22
Dokkum, Neth., 7,400 A6 15
Doksy, Czech., 3,061 C9 17
Doland, Spink, S. Dak., 481 C7 116
Dolavón, Arg C3 54
Dolbeau, Que., Can., 6,052 G18 67
Dolchburg, Maverick, Tex E2 118
Dol [-de-Bretagne], Fr., 4,130 C3 14
Dole, Fr., 24,525 D6 14
Dolega, Pan., 831 F6 62
Doleib Hill, Sud D3 47
Doles, Worth, Ga., 60 E3 87
Dolgelley, Wales, 2,267 B4 12
Dolgeville, Fulton and Herkimer, N.Y., 3,058 B6 108
Dolina, Sov. Un., 5,000 D8 26
Dolinsk (Ochiai), Sov. Un., 30,000 C11 37
Dolinskoye, Sov. Un B9 22
Dolisie, Con. B F2 46
Dollar Bay, Houghton, Mich., 500 A2 98
Dollard, Sask., Can., 165 H1 70
Dollart, bay, Neth A7 15
Dollarville, Luce, Mich., 100 B5 98
Dolliver, Emmet, Iowa, 122 A3 92
Dolo, Eth E5 47
Dolomite, Jefferson, Ala., 1,300 B3, E4 78
Dolomites, mts., It C7 18
Dolores, Arg., 14,438 B5 54
Dolores, Montezuma, Colo., 805 D2 83
Dolores, Guat., 512 B3 62
Dolores, Mex., 137 B3 63
Dolores, Webb, Tex F3 118
Dolores, Ur., 13,300 E1 56
Dolores, co., Colo., 2,196 D2 83
Dolores, riv., Colo., Utah C2 83, E7 119
Dolores Hidalgo, Mex., 11,733 m13 63
Dolphin and Union, straits, N.W. Ter., Can C9 66
Dolton, Cook, Ill., 18,746 F3 90
Dolton, Turner, S. Dak., 71 D8 116
Dolzhanskaya, Sov. Un., 5,000 q22 27
Dom, mtn., Switz D4 19
Domadare, Som E5 47
Domain, Man., Can., 50 E3 71
Domanovici, Yugo., 1,900 D3 22
Domazlice, Czech., 8,500 D2 26
Dombarovskiy, Sov. Un C5 29
Dombås, Nor., 522 F3 25
Dombasle [-sur-Meurthe], Fr., 9,367 F6 15
Dombe Grande, Ang D1 48
Dombey, Beaver, Okla D3 112
Dombóvár, Hung., 15,355 B4 22
Domburg, Neth., 1,400 C3 15
Dome, Yuma, Ariz., 50 E1 80
Dome, peak, N.W. Ter., Can D7 66
Dome Rock, mts., Ariz D1 80
Domeyko, Chile, 1,517 D2 55
Domeyko, range, Chile D2 55
Domfront, Fr., 2,653 C3 14
Domingo, Sandoval, N. Mex., 75 B3 107
Domingo, Indian res., N. Mex G6 107
Dominguez, Los Angeles, Calif., 6,000 *F2 82
Dominica, C.R F6 62
Dominica, Br. dep., N.A., 59,479 o16 64
Dominica, isl., N.A o16 64
Dominican Republic, country, N.A., 3,013,525 H13 61, F8 64
Dominion, N.S., Can., 2,999 C9 74
Dominion, cape, N.W. Ter., Can C18 67
Dominion, lake, Newf., Can h9 75
Dominion City, Man., Can., 534 E3 71
Domino, Cass, Tex C5 118
Domino Harbour, Newf., Can B4, h11 75
Dömitz, Ger., 4,585 B5 16
Dommel, riv., Neth C5 15
Domodedovo, Sov. Un., 17,600 n17 27
Domodossola, It., 16,728 C4 18
Dom Pedrito, Braz., 15,429 E2 56
Dompierre [-sur-Authie], Fr., 324 D9 12
Domremy, Sask., Can., 234 E3 70
Domuyo, mtn., Arg B2 54
Domvraina, Grc., 1,993 g9 23
Don, pen., B.C., Can C3 68
Don, riv., Scot C6 13
Don, riv., Sov. Un C1 29
Dona Ana, Dona Ana, N. Mex., 100 E3 107
Dona Ana, co., N. Mex., 59,948 E2 107
Donaghadee, N. Ire., 3,226 C6 11
Donald, Ont., Can., 86 C6 72
Donald, Marion, Oreg., 201 B2 113
Donald, Taylor, Wis C3 124
Donald, lake, Ont., Can D4 71
Donalda, Alta., Can., 289 C4 69
Donalds, Abbeville, S.C., 416 C3 115
Donaldson, Hot Spring, Ark., 500 C3 81
Donaldson, Marshall, Ind., 120 B5 91
Donaldson, Kittson, Minn., 64 B2 99
Donaldson, Schuylkill, Pa., 637 E9 114
Donaldsonville, Ascension, La., 6,082 B5, D4 95
Donaldsonville, Seminole, Ga., 2,621 E2 87
Donard, Ire., 175 D5 11
Donat-Ems, Switz., 2,694 C7 19
Donau (Danube), riv., Ger E6 17
Donaueschingen, Ger., 10,700 B4 18
Donauwörth, Ger., 10,200 E5 17
Donavon, Sask., Can., 51 F2 70
Don Benito, Sp., 25,248 C3 20
Doncaster, Eng., 86,402 (★145,000) A6 12
Doncaster, Charles, Md., 250 D3 85
Dondo, Ang., 645 C1 48
Dondo, Moz A5 49
Donegal, Ire., 1,458 C3 11
Donegal, co., Ire., 113,842 C3 11
Donegal, bay, Ire C3 11
Donegal, mts., Ire C3 11
Donegal, pt., Ire E2 11
Donelson, Davidson, Tenn., 17,195 A5, E9 117
Doneraile, Darlington, S.C., 1,043 C8 115
Donets, riv., Sov. Un G13 27
Donetsk, Sov. Un., 749,000 (★1,600,000) H11, r28 27
Dongara, Austl., 269 E1 50
Donggala, Indon., 3,821 F5 35
Dong Hoi, Viet., 10,000 D7 38
Dongo, Con. L A2 48
Dongola, Sud., 3,350 B3 47
Dongola, Union, Ill., 757 F4 90
Dongou, Con. B E2 46
Donington, Eng., 1,917 B7 12
Doniphan, Doniphan, Kans., 150 C8 93
Doniphan, Ripley, Mo., 1,421 E7 101
Doniphan, Hall, Nebr., 390 D7 103
Doniphan, co., Kans., 9,574 C8 93
Donji Vakuf, Yugo., 3,756 C3 22
Donkey, creek, Wyo A7 125
Donkin, N.S., Can., 1,010 C10 74
Donley, co., Tex., 4,449 B2 118
Don Martin, lake, Mex B4 63
Donna, Hidalgo, Tex., 7,522 F3 118
Donnacona, Que., Can., 4,812 C6, C8 73
Donnellson, Montgomery and Bond, Ill., 292 D4 90
Donnellson, Lee, Iowa, 709 D6 92
Donnelly, Alta., Can., 289 B2 69
Donnelly, Valley, Idaho, 161 E2 89
Donnelly, Stevens, Minn., 358 E2 99
Donnels, Rutherford, Tenn B5 117
Donner, Terrebonne, La., 300 C5, E5 95
Donnybrook, Ward, N. Dak., 196 A4 110
Donora, Washington, Pa., 11,131 F2 114
Donovan, Johnson, Ga., 100 D4 87
Donovan, Iroquois, Ill., 320 C6 90
Donzère, Fr., 1,544 E6 14
Dooagh, Ire., 387 D1 11
Doogort, Ire D1 11
Doole, McCulloch, Tex., 55 D3 118
Dooley, Sheridan, Mont., 20 B12 102
Dooling, Dooly, Ga., 300 D3 87
Doolittle, Phelps, Mo., 449 D6 101
Dooly, co., Ga., 11,474 D3 87
Doon, Lyon, Iowa, 436 A1 92
Doon, lake, Scot E4 13
Doon, riv., Scot E4 13
Doonbeg, Ire., 224 E2 11
Door, co., Wis., 20,685 D6 124
Dora, Walker, Ala., 1,776 B2 78
Dora, Ozark, Mo., 100 E5 101
Dora, Roosevelt, N. Mex., 200 D6 107
Dora, Coos, Oreg., 100 D3 113
Dora Baltea, riv., It B1 21
Dorado, P.R., 2,120 B5 65
Dorado, mun., P.R., 13,460 B5 65
Doran, Wilkin, Minn., 136 D2 99
Doraville, DeKalb, Ga., 4,437 A5 87
Dorcas, Okaloosa, Fla., 100 G2 86
Dorcheat, creek, Ark D2 81
Dorchester, N.B., Can., 1,779 D5 74
Dorchester, Eng., 12,266 D5 12
Dorchester, Liberty, Ga., 50 E2 87
Dorchester, Allamakee, Iowa, 97 A6 92
Dorchester, Saline, Nebr., 460 D8 103
Dorchester, Grafton, N.H., 10 (91▲) C3 105
Dorchester, Cumberland, N.J., 250 E3 106
Dorchester, Dorchester, S.C., 400 E7 115
Dorchester, Clark, Wis., 504 D3 124
Dorchester, co., Que., Can., 34,711 C7 73
Dorchester, co., Md., 29,666 D5 85
Dorchester, co., S.C., 24,383 E7 115
Dorchester, cape, N.W. Ter., Can C17 67
Dorcyville, Iberville, La., 400 B5 95
Dordogne, dept., Fr., 375,455 *E4 14
Dordogne, riv., Fr E3 14
Dordrecht, Neth., 83,800 (★130,000) C4 15
Dordrecht, S. Afr., 4,019 D4 49
Dore, McKenzie, N. Dak., 15 B2 110
Dore, lake, Ont., Can B7 72
Dore, lake, Sask., Can C2 70
Doré, riv., Sask., Can C2 70
Dorena, Mississippi, Mo., 400 E8 101
Dorena, Lane, Oreg., 225 D4 113
Dorena, dam, Oreg D3 113
Dorenlee, Alta., Can., 50 C4 69
Dores, Scot., 607 C4 13
Dores do Indaiá, Braz., 10,354 E1 57
Dorfen, Ger., 4,200 A8 18
Dorgali, It., 7,189 D2 21
Dori, Upper Volta, 3,500 D4 45
Dorion-Vaudreuil, Que., Can., 4,996 *D8 73
Dormans, Fr., 1,530 E3 15
Dormont, Allegheny, Pa., 13,098 B5 114
Dornbirn, Aus., 28,075 B5 18
Dornie, Scot C3 13
Dornoch, Scot., 933 C4 13
Dornoch, firth, Scot C5 13
Dorogobuzh, Sov. Un., 17,900 D9 27
Dorohoi, Rom., 14,771 B8 22
Dorothy, Alta., Can., 25 D4 69
Dorothy, Red Lake, Minn., 60 C2 99
Dorothy, Atlantic, N.J., 450 E3 106
Dorothy, Raleigh, W. Va., 350 D6 123
Dorr, Allegan, Mich., 275 F5 98
Dorrance, Russell, Kans., 331 D5 93
Dorrigo, Austl., 1,027 E9 51
Dorris, Siskiyou, Calif., 973 B3 82
Dorset, Ont., Can., 220 B6 72
Dorset, Hubbard, Minn., 50 D4 99
Dorset, Ashtabula, Ohio, 350 A7 111
Dorset, Bennington, Vt., 300 (1,150▲) E2 120
Dorset, co., Eng., 309,176 D5 12
Dorset, mtn., Vt E2 120
Dorset, peak, Vt E2 120
Dorsey, Madison, Ill., 100 A8 101
Dorsey, Anne Arundel and Howard, Md., 500 B4 85
Dorsey, Itawamba, Miss., 130 A5 100
Dorsten, Ger., 36,300 B1 17
Dortmund, Ger., 641,500 B2 17
Dortmund Ems, canal, Ger B2 17
Dorton, Pike, Ky., 700 C7 94
Dortyol, Tur., 10,200 D11 31
Dorum, Ger., 2,900 E2 24
Doruma, Con. L A4 48
Dorval, Que., Can., 18,592 D8 73
Dorya, riv., Eth D5 47
Dos Bahias, cape, Arg C3 54
Dos Bocas, P.R B4 65
Dos Hermanas, Sp., 22,700 D3 20
Doshi, Afg., 10,000 D14 41
Dos Palos, Merced, Calif., 2,373 D3 82
Dosquet, Que., Can., 394 C6 73
Dos Rios, Mendocino, Calif., 18 C2 82
Dosse, riv., Ger E2 28
Dosso, Niger, 2,500 D5 45
Dothan, Houston, Ala., 31,440 D4 78
Dothan, Fayette, W. Va., 400 D7 123
Dott, Mercer, W. Va., 250 D3 123
Doty, Lewis, Wash., 260 C2 122
Doty, isl., Wis A5 124
Douai, Fr., 47,639 (★135,000) B5 14
Douala (Duala), Cam., 125,000 E1 46
Douarnenez, Fr., 19,887 C1 14
Double, mtn., Ala E5 78
Double Bayou, Chambers, Tex., 120 F5 118
Double Mer, lake, Newf., Can A2 75
Double Oak, mtn., Ala E5 78
Double Springs, Winston, Ala., 811 A2 78
Double Springs, Putnam, Tenn., 100 C8 117
Doubletop, peak, Wyo C2 125
Doubs, Frederick, Md., 200 B3 85
Doubs, dept., Fr., 384,881 B2 18
Doubs, riv., Fr., Switz B2 19, D7 14
Doubtful, sound, N.Z P11 51
Doucette, Tyler, Tex., 500 D5 118
Douds, Van Buren, Iowa, 250 D5 92
Doué [-la-Fontaine], Fr., 3,895 D3 14
Douentza, Mali, 2,250 D4 45
Dougherty, Cerro Gordo, Iowa, 398 B4 92
Dougherty, Murray, Okla., 294 C4 112
Dougherty, co., Ga., 75,680 E2 87
Douglas, Alsk., 1,042 D13, k22 79
Douglas, Cochise, Ariz., 11,925 F6 80
Douglas, Ont., Can., 365 B8 72
Douglas, Coffee, Ga., 8,736 E4 87
Douglas, I. of Man, 18,837 C4 10
Douglas, Worcester, Mass., 397 (2,559▲) B4 97
Douglas, Allegan, Mich., 602 F4 98
Douglas, Olmsted, Minn., 126 F6 99
Douglas, Otoe, Nebr., 197 D9 103
Douglas, Ward, N. Dak., 210 B4 110
Douglas, Garfield, Okla., 74 A4 112
Douglas, S. Afr., 3,974 C3 49
Douglas, Converse, Wyo., 2,822 C7 125
Douglas, co., Colo., 4,816 B6 83
Douglas, co., Ga., 16,741 C2 87
Douglas, co., Ill., 19,243 D5 90
Douglas, co., Kans., 43,720 D8 93
Douglas, co., Minn., 21,313 E3 99
Douglas, co., Mo., 9,653 E5 101
Douglas, co., Nebr., 343,490 C9 103
Douglas, co., Nev., 3,481 E2 104
Douglas, co., Oreg., 68,458 D3 113
Douglas, co., S. Dak., 5,113 D7 116
Douglas, co., Wash., 14,890 B6 122
Douglas, co., Wis., 45,008 B2 124
Douglas, chan., B.C., Can C3 68
Douglas, creek, Colo B2 83
Douglas, lake, Mich C6 98
Douglas, lake, Tenn D10 117
Douglas, pt., Ont., Can C3 72
Douglas Lake, B.C., Can., 166 D7 68
Douglass, Butler, Kans., 1,058 E7 93
Douglas Station, Man., Can., 289 E2 71
Douglastown, N.B., Can., 615 C4 74
Douglasville, Baldwin, Ala E2 78
Douglasville, Douglas, Ga., 4,462 C2 87
Doullens, Fr., 6,321 B5 14
Doulus, head, Ire F1 11
Doumé, Cam E2 46
Douna, Mali D3 45
Dour, Bel., 10,785 D3 15
Dourada, mts., Braz D1 57
Dourado, Braz., 10,757 m7 56
Dourdan, Fr., 4,293 F2 15
Douro (Duero), riv., Port B1 20
Douro Litoral, prov., Port., 1,240,149 *B1 20
Dousman, Waukesha, Wis., 410 E5 124
Douthat, Ottawa, Okla., 250 A7 112
Douz, Tun., 4,993 C6 44
Dove Creek, Dolores, Colo., 986 D2 83
Dover, Pope, Ark., 525 B2 81
Dover, Kent, Del., 7,250 B6 85
Dover, Eng., 35,248 C9 12
Dover, Hillsborough, Fla., 800 D4 86
Dover, Screven, Ga., 150 D5 87
Dover, Bonner, Idaho, 250 A2 89
Dover, Shawnee, Kans., 150 D8 93
Dover, Mason, Ky., 718 B6 94
Dover, Norfolk, Mass., 1,400 (2,846▲) D2 97
Dover, Olmsted, Minn., 312 G6 99
Dover, Lafayette, Mo., 172 B4 101
Dover, Stafford, N.H., 19,131 D5 105
Dover, Morris, N.J., 13,034 B3 106
Dover, Craven, N.C., 651 B6 109
Dover, Tuscarawas, Ohio, 11,300 B6 111
Dover, Kingfisher, Okla., 350 B4 112
Dover, York, Pa., 975 F8 114
Dover, Stewart, Tenn., 736 A4 117
Dover, riv., Alta., Can A4 69
Dover, strait, Eng E7 10
Doverel, Terrell, Ga., 25 E2 87
Dover-Foxcroft, Piscataquis, Maine, 2,481 (4,173▲) C3 96
Doverhill, Martin, Ind., 100 G4 91
Dover Plains, Dutchess, N.Y., 950 D7 108
Dover South Mills, Piscataquis, Maine C3 96
Dovesville, Darlington, S.C., 100 C8 115
Dovray, Murray, Minn., 113 F3 99
Dovrefjell, mts., Nor F3 25
Dow, Pittsburg, Okla., 300 C6 112
Dow, Jersey, Ill., 150 D3 90
Dow, lake, Bech B3 49
Dowa, Nya., 1,085 D5 48
Dowagiac, Cass, Mich., 7,208 G4 98
Dow City, Crawford, Iowa, 531 C2 87
Dowdell Knob, mtn., Ga D2 87
Dowell, Jackson, Ill., 453 F4 90
Dowelltown, De Kalb, Tenn., 279 C8 117
Dowling, Alta., Can D4 69
Dowling, lake, Alta., Can D4 69
Dowling Park, Suwannee, Fla., 150 B3 86
Down, co., N. Ire., 267,013 C5 11
Downer, Clay, Minn., 100 D2 99
Downer, Gloucester, N.J D2 106
Downers Grove, Du Page, Ill., 21,154 F2 90
Downey, Los Angeles, Calif., 82,505 *F2 82
Downey, Bannock, Idaho, 726 G6 89
Downham Market, Eng., 2,650 B8 12
Downhill, N. Ire B5 11
Downieville, Sierra, Calif., 400 C3 82
Downing, Schuyler, Mo., 463 A5 101
Downing, Dunn, Wis., 241 C1 124
Downingtown, Chester, Pa., 5,598 F10 114
Downpatrick, N. Ire., 4,219 C6 11
Downpatrick, head, Ire C2 11
Downs, Macon, Ala., 30 C4 78
Downs, McLean, Ill., 654 C5 90
Downs, Osborne, Kans., 1,206 C5 93
Downs, mtn., Wyo B3 125
Downsville, Union, La., 150 B3 95
Downsville, Delaware, N.Y., 400 C6 108
Downsville, Dunn, Wis., 275 D2 124
Downton, mtn., B.C., Can C5 68
Dows, Wright and Franklin, Iowa, 882 B4 92
Doyle, Lassen, Calif., 150 B3 82
Doyle, Livingston, La. (part of Livingston) A6 95
Doyle, White, Tenn., 500 D8 117
Doyle, creek, Kans A6 93
Doyles, Newf., Can., 210 E2 75
Doylestown, Wayne, Ohio, 1,873 B6 111
Doylestown, Bucks, Pa., 5,917 F11 114
Doyleville, Gunnison, Colo., 35 C4 83
Doyline, Webster, La., 1,061 B2 95
Doyon, Ramsey, N. Dak., 90 A7 110
Dozier, Crenshaw, Ala., 335 D3 78
Dra, plat., Alg D3 44
Dra, wadi, Mor D3 44
Drabenderhöhe, Ger., 8,738 C2 17
Drachten, Neth., 9,249 A6 15
Dracut, Middlesex, Mass., 10,000 (13,674▲) A5 97
Draganeşti, Rom., 3,965 C7 22
Draganovo, Bul., 5,465 D7 22
Drăgăsani, Rom., 9,963 C7 22
Dragerton, Carbon, Utah, 2,959 D5 119
Dragoon, Cochise, Ariz., 100 E5 80
Dragoon, creek, Kans D8 93
Dragör, Den., 6,986 C6 24
Draguignan, Fr., 11,814 F7 14
Drain, Douglas, Oreg., 1,052 D3 113
Drake, Yavapai, Ariz., 20 C3 80
Drake, Sask., Can., 215 F3 70
Drake, Larimer, Colo., 40 A5 83
Drake, McHenry, N. Dak., 752 B5 110
Drake, creek, Ky D3 94
Drake Passage, strait, Ant J16 3
Drakensberg, mts., S. Afr J8 42
Drakes, bay, Calif B5 82
Drakesboro, Muhlenberg, Ky., 832 C2 94
Drakes Branch, Charlotte, Va., 759 E4 121
Drakesville, Davis, Iowa, 197 D5 92
Draketown, Haralson, Ga., 100 C1 87
Drama, Grc., 32,195 B5 23
Drama, prov., Grc., 121,006 *B5 23
Drama, riv., Pol g9 26
Drammen, Nor., 31,300 (★57,500) H4, p28 25
Dramselv, riv., Nor p27 25
Drance, riv., Fr C2 18
Drancy, Fr., 65,890 g10 14
Dranesville, Fairfax, Va A4 121
Draper, Rockingham, N.C., 3,382 A3 109
Draper, Jones, S. Dak., 215 D5 116
Draper, Salt Lake, Utah, 1,000 C4 119
Draperstown, N. Ire., 592 C5 11
Drasco, Cleburne, Ark., 75 B4 81
Drau, riv., Aus E6 16
Drava, riv., Yugo B2 22
Draveil, Fr., 18,124 F2 15
Dravograd, Yugo., 2,131 B2 22
Dravosburg, Allegheny, Pa., 3,458 F2 114
Drawenburg, see Drawsko, Pol.
Drawsko, Pol., 3,504 B3 26
Drayton, Ont., Can., 646 D4 72
Drayton, Pembina, N. Dak., 940 A8 110
Drayton, Spartanburg, S.C., 1,128 B4 115
Drayton Plains, Oakland, Mich., 6,000 F7 98
Drayton Valley, Alta., Can., 3,854 C3 69
Drebkau, Ger., 2,518 B9 17
Drelsdorf, Ger., 910 D3 24
Drenthe, prov., Neth., 314,400 B6 15
Dresbach, Winona, Minn., 350 G7 99
Dresden, Ont., Can., 2,346 E2 72
Dresden, Ger., 491,700 (★700,000) B8 17
Dresden, Decatur, Kans., 134 C3 93
Dresden, Sagadahoc, Maine D3 96
Dresden, Cavalier, N. Dak., 65 A7 110
Dresden, Muskingum, Ohio, 1,338 B5 111
Dresden, Weakley, Tenn., 1,510 A3 117
Dresden Village, Macomb, Mich., 5,500 *F7 98
Dresser, Polk, Wis., 498 C1 124
Dresslerville, Douglas, Nev., 150 E2 104
Dreux, Fr., 21,588 C4 14
Drew, Penobscot, Maine, (43▲) C4 96
Drew, Sunflower, Miss., 2,143 B3 100
Drew, Douglas, Oreg., 25 E4 113
Drew, co., Ark., 15,213 D4 81
Drewrys Bluff, Chesterfield, Va., 250 C7, D5 121
Drewryville, Southampton, Va., 200 E5 121
Drews, res., Oreg E6 113
Drewsey, Harney, Oreg., 39 D8 113
Drewsville, Cheshire, N.H., 100 D2 105
Drexel, Cass, Mo., 651 C3 101
Drexel, Mineral, Mont., 14 C1 102
Drexel, Burke, N.C., 1,146 B2 109
Drexel, Montgomery, Ohio, 2,500 C3 111
Drexel Gardens, Marion, Ind., 1,000 *E5 91
Drexel Hill, Delaware, Pa., 39,000 *B11 114
Drezna, Sov. Un n18 27
Driffield, Eng., 6,890 F8 13
Drift, Floyd, Ky., 800 C7 94
Drifton, Luzerne, Pa., 900 D10 114
Driftpile, Alta., Can., 62 B3 69
Driftwood, Alfalfa, Okla., 32 A3 112
Driftwood, Cameron, Pa., 203 D5 114
Driftwood, creek, Kans C3 93
Driftwood, creek, Nebr B3 93
Driggs, Teton, Idaho, 824 F7 89
Drimoleague, Ire., 369 F2 11
Drin, gulf, Alb B2 23
Drina, riv., Yugo C4 22
Drinkwater, Sask., Can., 138 G3 70
Dripping Springs, Hays, Tex., 150 D3 118
Driscoll, Burleigh, N. Dak., 220 C5 110
Driscoll, Nueces, Tex., 669 F4 118
Driskill, mtn., La B3 95
Drissa, Sov. Un., 10,000 D6 27
Driver, Mississippi, Ark., 150 B5 81
Driver, Nansemond, Va., 160 B6 121
Drøbak, Nor., 2,700 H4, p28 25
Drogheda, Ire., 17,085 D5 11
Drogobych, Sov. Un., 42,000 G4 27
Droitwich, Eng., 7,975 B5 12
Dromahair, Ire., 229 C3 11
Drôme, dept., Fr., 304,227 *E6 14
Drôme, riv., Fr E6 14
Dromore, N. Ire., 2,503 C5 11
Dromore West, Ire., 99 C3 11
Dronero, It., 6,776 E3 18
Dronfield, Eng., 7,628 A6 12
Dronninglund, Den., 1,647 A4 24
Dropmore, Man., Can., 50 D1 71
Druid, Sask., Can., 52 F1 70
Druid Hills, DeKalb, Ga., 2,000 *C2 87
Drulingen, Fr., 875 F7 15
Drum, isl., S.C F3 115
Drumahoe, N. Ire C4 11
Drumbeg, Scot B3 13
Drumbo, Ont., Can., 416 D4 72
Drumcliffe, Ire C3 11
Drumheller, Alta., Can., 2,931 D4 69
Drumlish, Ire., 343 D4 11
Drummond, Fremont, Idaho, 31 E7 89
Drummond, Granite, Mont., 577 D3 102
Drummond, Garfield, Okla., 281 A3 112
Drummond, Bayfield, Wis., 450 B2 124
Drummond, co., Que., Can., 58,220 D5 73
Drummond, isl., Mich B7 98
Drummond, lake, Va E6 121
Drummond Island, Chippewa, Mich., 150 B7 98
Drummondville, Que., Can., 27,909 (★39,307) D5 73
Drummondville Ouest, Que., Can., 2,057 *D5 73
Drummore, Scot F4 13
Drumod, Ire., 103 D4 11
Drumquin, N. Ire., 307 C4 11
Drumright, Creek, Okla., 4,190 B5 112
Drumshanbo, Ire., 565 C3 11
Druzhkovka, Sov. Un., 39,800 q20 27
Drweca, riv., Pol B5 26
Dry, creek, Kans B6 93
Dry, fork, Mo D3 101
Dry, fork, Mo D6 101
Dry, fork, W. Va B5 123
Dry, fork, W. Va D3 123
Dry, hill, Mass A3 97
Dryad, Lewis, Wash., 100 C2 122
Dryberry, lake, Ont., Can E5 71
Dry Branch, Bibb, Ga., 200 D3 87
Drybranch, Kanawha, W. Va., 800 C6 123
Dry Creek, Beauregard, La., 50 D2 95
Dryden, Ont., Can., 5,728 E8 72
Dryden, Franklin, Maine, 625 D2 96
Dryden, Lapeer, Mich., 531 F7 98
Dryden, Tompkins, N.Y., 1,263 C4 108
Dryden, Josephine, Oreg E3 113
Dryden, Terrell, Tex., 160 D1 118
Dryden, Chelan, Wash., 300 B5 122
Dryfork, Carroll, Ark A2 81
Dry Fork, Pittsylvania, Va., 250 E3 121
Dryfork, Randolph, W. Va., 25 C5 123
Dryhead, Carbon, Mont., 30 E8 102
Drymen, Scot., 1,221 D4 13
Dry Mills, Cumberland, Maine, 500 E5 96
Dry Prong, Grant, La., 360 C3 95
Dry Ridge, Grant, Ky., 802 B5 94
Dry Run, Franklin, Pa., 250 F6 114
Dry Tortugas, is., Fla H4 86
Dsalatu, Mong B2 36
Dsamdo, China, 5,000 B11 40
Dschang, Cam D2 46
Duala (Douala), Cam., 125,000 E1 46
Duane, Franklin, N.Y., 5 A3 108
Duaringa, Austl., 218 A7 51
Duart, pt., Scot D3 13
Duarte, Los Angeles, Calif., 13,962 *F3 82
Dubach, Lincoln, La., 1,013 B3 95
Dubawnt, lake, N.W. Ter., Can D12 66
Du Bay, res., Wis D4 124
Dubayy, Tr. Coast, 40,000 I7 41
Dubberly, Webster, La., 249 B2 95
Dubbo, Austl., 14,121 F7 51
Dubbs, Tunica, Miss., 100 A3 100
Düben, Ger., 6,599 B7 17
Dübendorf, Switz., 11,784 B6 19
Dublin, Montgomery, Ala., 100 C3 78
Dublin, Ont., Can., 301 D3 72
Dublin, Laurens, Ga., 13,814 D4 87
Dublin, Wayne, Ind., 1,021 E7 91
Dublin (Baile Átha Cliath), Ire., 537,448 (★690,000) D5 11
Dublin, Coahoma, Miss., 200 A3 100
Dublin, Cheshire, N.H., 225 (684▲) E2 105
Dublin, Bladen, N.C., 366 C5 109
Dublin, Franklin, Ohio, 552 C2 111
Dublin, Bucks, Pa., 517 F11 114
Dublin, Erath, Tex., 2,443 C3 118
Dublin, Pulaski, Va., 1,427 D2 121
Dublin, co., Ire., 718,332 D5 11
Dublin, bay, Ire D5 11
Dublin Gulch, Silver Bow, Mont., 2,450 *D4 102
Dublin Shore, N.S., Can., 116 E5 74
Dublon, isl., Truk 52
Dubois, Clark, Idaho, 447 E6 89
Dubois, Washington, Ill., 229 E4 90
Dubois, Dubois, Ind., 510 H4 91
Du Bois, Pawnee, Nebr., 218 D9 103
Du Bois, Clearfield, Pa., 10,667 D4 114
Dubois, Fremont, Wyo., 574 B3 125
Dubois, co., Ind., 27,463 H4 91
Duboistown, Lycoming, Pa., 1,358 D7 114
Du Bose Park, Kershaw, S.C., 900 *C6 115
Dubossary, Sov. Un., 25,000 H7 27
Dubovka, Sov. Un., 12,300 D2 29
Dubrovka, Sov. Un s31 25
Dubrovnik, Yugo., 22,961 D4 22
Dubsdread, Orange, Fla., 1,400 *D5 86
Dubuc, Sask., Can., 183 G4 70
Dubulu, Con. L A3 48
Dubuque, Dubuque, Iowa, 56,606 B7 92
Dubuque, co., Iowa, 80,048 B7 92
Du Chein, bayou, Ky B2 94
Duchesne, Duchesne, Utah, 770 C5 119
Duchesne, co., Utah, 7,179 C5 119
Duchess, Austl., 73 D6 50
Duchess, Alta., Can., 218 D5 69
Duchov, Czech., 8,229 C2 26
Ducie, isl., Pac. O H14 7
Duck, Dare, N.C., 60 A8 109
Duck, creek, Del B7 85
Duck, creek, Nev D7 104
Duck, creek, Ohio C6 111
Duck, creek, Wis A5 124
Duck, lake, Man., Can B2 71
Duck, lake, Maine C4 96
Duck, mtn., Man., Can D1 71
Duck, riv., Tenn B4 117
Duck Creek, Brown, Wis. (part of Howard) D5 124
Duck Hill, Montgomery, Miss., 674 B4 100

Duck Lake, Sask., Can., 668 .... E2 70
Duck Mountain, park, Sask., Can F12 66, F5 70
Duck Pond, pt., N.Y. E6 84
Duck River, Hickman, Tenn., 100 .... B4 117
Ducktown, Polk, Tenn., 741 .... D9 117
Duckwater, Nye, Nev., 20 .... E6 104
Duckwater, peak, Nev E6 104
Ducor, Tulare, Calif., 150 .... E4 82
Dudelange, Lux., 14,617 .... E6 15
Duderstadt, Ger., 10,700 .... B5 17
Dudhi, India, 5,000 .... E9 40
Dudinka, Sov. Un., 17,000 .... C11 28
Dudley, Eng., 61,748 .... B5 12
Dudley, Laurens, Ga., 360 .... D3 87
Dudley, Worcester, Mass., 200 (6,510▲) B4 97
Dudley, Stoddard, Mo., 287 .... E7 101
Dudley, Wayne, N.C., 158 .... B5 109
Dudley, Huntingdon, Pa., 295 .... F5 114
Dudweiler, Ger., 28,900 .... D2 17
Duékoué, I.C., 3,700 .... E3 45
Duenweg, Jasper, Mo., 529 .... D3 101
Duero (Douro), riv., Sp B3 20
Due West, Abbeville, S.C., 1,166 .... C3 115
Duff, Sask., Can., 84 .... G4 70
Duff, Dubois, Ind., 80 .... H3 91
Duff, Rock, Nebr B6 103
Duff, Campbell, Tenn., 200 .... C9 117
Duff, reef, Fiji Is 52
Duffee, Newton, Miss., 50 .... C5 100
Duffer, peak, Nev B3 104
Dufferin, co., Ont., Can., 16,095 .... C4 72
Duffield, Alta., Can., 66 .... C3 69
Dufftown, Scot., 1,555 .... C5 13
Duffy, Monroe, Ohio, 270 .... C7 111
Dufrost, Man., Can., 96 .... E3 71
Dufur, Wasco, Oreg., 488 .... B2 113
Dugdemona, bayou, La B3 95
Dugdown, mtn., Ga C1 87
Dugger, Sullivan, Ind., 1,062 .... F3 91
Dug Hill, ridge, Md A4 85
Dugi Otok, isl., Yugo C2 22
Dugway, Tooele, Utah C3 119
Dugway, range, Utah C2 119
Duhamel, Que., Can., 187 .... D2 73
Duich, inlet, Scot C3 13
Duisburg, Ger., 503,000 .... B1 17, C3 16
Duitama, Col., 7,723 .... B3 60
Duiwelskloof, S. Afr., 2,809 .... B5 49
Duke, Calhoun, Ala., 150 .... B4 78
Duke, Phelps, Mo., 18 .... D5 101
Duke (East Duke), Jackson, Okla., 333 .... C2 112
Duke, isl., Alsk n24 79
Duke, Center, McKean, Pa., 800 .... C5 114
Dukedom, Weakley, Tenn., 125 .... A2 117
Duke Ernst, bay, Ant B9 5
Dukes, co., Mass., 5,829 .... D6 97
Duk Fadiat, Sud D3 47
Duki, Pak B4 39
Dukla, pass, Czech., Pol D6 26
Dulawan, Phil., 16,376 .... *D6 35
Dulce, Rio Arriba, N. Mex., 500 .... A3 107
Dulce, riv., Arg E3 55
Dülken, Ger., 20,900 .... C6 15
Dull Center, Converse, Wyo B8 125
Dülmen, Ger., 16,700 .... B2 17
Dulowa, Pol g11 26
Duluth, Gwinnett, Ga., 1,483 .... A5, B2 87
Duluth, St. Louis, Minn., 106,884 (*165,200) D6 99
Dūmā, Jordan, 2,000 .... g12 32
Dumaguete, Phil., 13,000 (35,300▲) D6 35
Dumaran, isl., Phil C5 35
Dumas, Desha, Ark., 3,540 .... D4 81
Dumas, Webster, Ga., 170 .... D2 87
Dumas, Tippah, Miss., 200 .... A5 100
Dumas, Moore, Tex., 8,477 .... B2 118
Dumba, Con. L B3 48
Dumbarton, Scot., 26,335 .... E4 13
Dumbarton (Dunbarton), co., Scot., 184,546 .... D4 13
Dumboa, Nig D7 45
Dumfries, Scot., 27,275 .... E5 13
Dumfries, Prince William, Va., 1,368 .... C5 121
Dumfries, co., Scot., 88,423 .... E5 13
Dumiat, see Damietta, Eg., U.A.R.
Dumka, India, 18,720 .... E11 40
Dummer, Sask., Can., 125 .... H3 70
Dummer, Coos, N.H., (202▲) A4 105
Dümmer, lake, Ger F2 24
Dumoine, riv., Que., Can A7 72
Dumont, Butler, Iowa, 719 .... B5 92
Dumont, Traverse, Minn., 226 .... E2 99
Dumont, Bergen, N.J., 18,882 .... B5, D5 106
Dumont, King, Tex., 150 .... C2 118
Dumyāṭ, see Damietta, Eg., U.A.R.
Duna (Danube), riv., Czech., Hung B4 22
Dunaff, head, Ire B4 11
Dunafoldvar, Hung., 11,251 .... B4 22
Dunany, pt., Ire D5 11
Dunapataj, Hung., 4,846 .... B4 22
Dunarea (Danube), riv., Rom C8 22
Dunaujvaros, Hung., 31,048 .... B4 22
Dunav (Danube), riv., Yugo C4 22
Dunavant, Jefferson, Kans B7 93
Dunay, Sov. Un E6 37
Dunayevsty, Sov. Un., 10,000 .... G6 27
Dunbar, Butler, Ky., 100 .... C3 94
Dunbar, Otoe, Nebr., 232 .... D9, F3 103
Dunbar, Fayette, Pa., 1,536 .... G2 114
Dunbar, Scot., 4,003 .... E6 13
Dunbar, Marlboro, S.C., 150 .... B8 115
Dunbar, Kanawha, W. Va., 11,006 .... C3, C6 123
Dunbar, Marinette, Wis., 75 .... C5 124
Dunbarton, Merrimack, N.H., 85 (632▲) D3 105
Dunbeath, Scot B5 13
Dunblane, Sask., Can., 429 .... F2 70
Dunblane, Scot., 2,922 .... D5 13
Dunboyne, Ire., 521 .... D5 11
Dunbridge, Wood, Ohio, 300 .... A2 111
Duncan, Greenlee, Ariz., 862 .... E6 80
Duncan, B.C., Can., 3,726 .... B9, E6 68
Duncan, Mercer, Ky., 50 .... C5 94
Duncan, Bolivar, Miss., 465 .... A3 100
Duncan, Platte, Nebr., 294 .... C8 103
Duncan, Stephens, Okla., 20,009 .... C4 112
Duncan, Umatilla, Oreg B8 113
Duncan, Spartanburg, S.C., 1,186 .... B3 115
Duncan, Spokane, Wash D7 122
Duncan, Fremont, Wyo B3 125
Duncan, lake, B.C., Can D2 69
Duncan, riv., B.C., Can D9 68
Duncan Falls, Muskingum, Ohio, 750 .... C6 111
Duncannon, Perry, Pa., 1,800 .... F7 114
Duncansby, head, Scot B6 13
Duncansville, Blair, Pa., 1,396 .... F5 114
Duncanville, Dallas, Tex., 3,774 .... B5 118
Duncombe, Webster, Iowa, 355 .... B4 92
Duncormick, Ire., 90 .... E5 11
Dundalk, Ont., Can., 852 .... C4 72
Dundalk, Ire., 19,790 .... C5 11
Dundalk, Baltimore, Md., 82,428 .... B4, C3 85
Dundalk, bay, Ire D5 11
Dundas, Ont., Can., 12,912 .... D5 72
Dundas, Richland, Ill., 200 .... E5 90
Dundas, Rice, Minn., 488 .... F5 99
Dundas, Vinton, Ohio, 450 .... C5 111
Dundas, Lunenburg, Va., 200 .... E4 121
Dundas, Calumet, Wis., 60 .... A5 124
Dundas, co., Ont., Can., 17,162 .... B9 72
Dundas, isl., B.C., Can B2 68
Dundas, lake, Austl F3 50
Dundas, strait, Austl B5 50
Dundas, Harbour, N.W. Ter., Can B16 66
Dundee, Polk, Fla., 1,554 .... D5 86
Dundee (West Dundee), Kane, Ill., 2,530 .... A5, E2 90
Dundee, Delaware, Iowa, 185 .... B6 92
Dundee, Ohio, Ky., 150 .... C3 94
Dundee, Monroe, Mich., 2,377 .... G7 98
Dundee, Nobles, Minn., 148 .... G3 99
Dundee, Tunica, Miss., 200 .... A3 100
Dundee, Yates, N.Y., 1,468 .... C4 108
Dundee, Yamhill, Oreg., 318 .... B1 113
Dundee, Scot., 182,959 .... D5 13
Dundee, S. Afr., 10,943 .... C5 49
Dundee, Archer, Tex., 80 .... C3 118
Dundee, Fond du Lac, Wis., 100 .... E5 124
Dundonald, chan., Bermuda E13 77
Dundrum, Ire., 733 .... D5 11
Dundrum, Ire., 86 .... E3 11
Dundrum, bay, N. Ire C6 11
Dundurn, Sask., Can., 411 .... F2 70
Dundy, co., Nebr., 3,570 .... D4 103
Dunean, Greenville, S.C., 3,950 .... *B3 115
Dunedin, Pinellas, Fla., 8,444 .... D4, E1 86
Dunedin, N.Z., 73,245 (*105,003) P13 51
Dunellen, Middlesex, N.J., 6,840 .... B4 106
Dunfanaghy, Ire., 324 .... B4 11
Dunfermline, Sask., Can., 30 .... E2 70
Dunfermline, Fulton, Ill., 284 .... C3 90
Dunfermline, Scot., 47,159 .... D5 13
Dungannon, Ont., Can., 166 .... D3 72
Dungannon, N. Ire., 6,494 .... C5 11
Dungannon, Scott, Va., 444 .... B2 121
Dungarpur, India, 12,755 .... F4 40
Dungarvan, Ire., 5,188 .... E4 11
Dungarvan, harbor, Ire E4 11
Dungarvan, riv., N.B., Can C3 74
Dungeness, Clallam, Wash., 75 .... A2 122
Dungeness, cape, Eng D8 12
Dungeness, riv., Wash B2 122
Dungiven, N. Ire., 799 .... C5 11
Dungloe, Ire., 793 .... C3 11
Dungu, Con. L A4 48
Dunham, Que., Can., 434 .... D5 73
Dunkard, creek, W. Va B4 123
Dunkeld, Scot., 833 .... D5 13
Dunken, Chaves, N. Mex E4 107
Dunkerque, Fr., 27,616 (*120,000) B5 14
Dunkerrin, Ire E4 11
Dunkerton, Black Hawk, Iowa, 507 .... B5 92
Dunkineely, Ire., 261 .... C3 11
Dunkirk, Sask., Can., 50 .... G3 70
Dunkirk, Jay and Blackford, Ind., 3,117 .... D7 91
Dunkirk, see Dunkerque, Fr.
Dunkirk, Toole, Mont., 14 .... B5 102
Dunkirk, Chautauqua, N.Y., 18,205 .... C1 108
Dunkirk, Hardin, Ohio, 1,006 .... B5 111
Dunklin, co., Mo., 39,139 .... E7 101
Dunkwa, Ghana, 6,827 .... E4 45
Dún Laoghaire, Ire., 47,792 .... D5 11
Dunlap, Peoria, Ill., 564 .... C4 90
Dunlap, Elkhart, Ind., 1,935 .... A6 91
Dunlap, Harrison, Iowa, 1,254 .... C2 92
Dunlap, Morris, Kans., 134 .... D7 93
Dunlap, Dawes, Nebr B3 103
Dunlap, De Baca, N. Mex., 5 .... C5 107
Dunlap, Sequatchie, Tenn., 1,026 .... D8 117
Dunlap, lake, Ont., Can B8 98
Dunlavin, Ire., 416 .... D5 11
Dunleer, Ire., 529 .... D5 11
Dunlo, Cambria, Pa., 950 .... F4 114
Dunloup, creek, W. Va D7 123
Dunmanway, Ire., 1,411 .... F2 11
Dunmor, Muhlenberg, Ky., 158 .... C2 94
Dunmore, Ire., 500 .... D3 11
Dunmore, Lackawanna, Pa., 18,917 .... A9, D10 114
Dunmore, lake, Vt D2 120
Dunmore East, Ire., 547 .... E5 11
Dunn, Harnett, N.C., 7,566 .... B5 109
Dunn, Scurry, Tex., 85 .... C2 118
Dunn, co., N. Dak., 6,350 .... B3 110
Dunn, co., Wis., 26,156 .... D2 124
Dunn Center, Dunn, N. Dak., 250 .... B3 110
Dunnamanagh, N. Ire., 352 .... C4 11
Dunnegan, Polk, Mo., 150 .... D4 101
Dunnell, Martin, Minn., 260 .... G4 99
Dunnellon, Marion, Fla., 1,079 .... C4 86
Dünnern, riv., Switz B4 19
Dunnet, bay, Scot B5 13
Dunnet, head, Scot B5 13
Dunnfield, Warren, N.J. B2 106
Dunning, Blaine, Nebr., 210 .... C5 103
Dunn Loring, Fairfax, Va., 1,500 .... *C5 121
Dunnsville, Essex, Va., 50 .... D6 121
Dunnville, Ont., Can., 5,181 .... E5 72
Dunnville, Casey, Ky., 240 .... C5 94
Du Noir, Fremont, Wyo B3 125
Dunoon, Scot., 9,211 .... E4 13
Dunowen, Ire F3 11
Dunphy, Eureka, Nev., 15 .... C5 104
Dunrea, Man., Can., 196 .... E2 71
Dunrobin, Ont., Can., 125 .... A8, B8 72
Duns, Scot., 1,838 .... E6 13
Dun Seilcheig, lake, Scot C4 13
Dunseith, Rollette, N. Dak., 1,017 .... A5 110
Dunsmuir, Siskiyou, Calif., 2,873 .... B2 82
Dunstable, Eng., 25,618 .... C7 12
Dunstable, Middlesex, Mass., 300 (824▲) A5 97
Dun [-sur-Auron], Fr., 4,008 .... D5 14
Dun [-sur-Meuse], Fr., 717 .... E5 15
Dunton, Dolores, Colo D2 83
Duntroon, Ont., Can., 67 .... C4 72
Dunvegan, Scot C2 13
Dunvegan, bay, Scot C2 13
Dunvegan, head, Scot C1 13
Dunville, Newf., Can., 1,121 .... E5 75
Dunwoody, DeKalb, Ga., 300 .... A5 87
Duong Dong, Viet G5 38
Duoro, Guadalupe, N. Mex C4 107
Du Page, co., Ill., 313,459 .... B5 90
Du Page, riv., Ill F2 90
Duparquet, Que., Can., 978 .... *E9 72
Dupax, Phil., 2,365 .... n13 35
Duplin, co., N.C., 40,270 .... C6 109
Dupnitsa, see Stanke Dimitrov, Bul.
Dupo, St. Clair, Ill., 2,937 .... E3 90
Dupont, Adams, Colo., 650 .... B6 83
Du Pont, Clinch, Ga., 210 .... F4 87
Dupont, Jefferson, Ind., 375 .... G6 91
Dupont, Putnam, Ohio, 239 .... A3 111
Dupont, Luzerne, Pa., 3,669 .... B9 114
Dupont, Charleston, S.C., (part of Saint Andrews) F2 115
DuPont, Pierce, Wash., 354 .... B3 122
Dupont, bayou, La C7 95
Dupontonia, Davidson, Tenn., 1,896 .... *A5 117
Dupree, Ziebach, S. Dak., 548 .... B4 116
Dupuyer, Pondera, Mont., 115 .... B4 102
Duque de Braganca, Ang., 2,037 .... C2 48
Duquesne, Allegheny, Pa., 15,019 .... B6, F2 114
Du Quoin, Perry, Ill., 6,558 .... E4 90
Dura, Jordan, 10,000 .... C7 32
Durán, Ec B2 58
Duran, Torrence, N. Mex., 10 .... C4 107
Durance, riv., Fr F6 14
Durand, Meriwether, Ga., 195 .... D2 87
Durand, Winnebago, Ill., 797 .... A4 90
Durand, Shiawassee, Mich., 3,312 .... F6 98
Durand, Pepin, Wis., 2,039 .... D2 124
Durango, La Plata, Colo., 10,530 .... D3 83
Durango, Mex., 97,520 .... C4 63
Durango, state, Mex., 754,220 .... C4 63
Durant, Cedar, Iowa 1,266 .... C7 92
Durant, Holmes, Miss., 2,617 .... B2 100
Durant, Bryan, Okla., 10,467 .... D5 112
Durants Neck, Perquimans, N.C., 150 .... A7 109
Duratón, riv., Sp B4 20
Durazno, Ur., 18,900 .... E1 56
Durazno, dept., Ur., 49,500 .... *E1 56
Durban, Man., Can., 118 .... D1 71
Durban, S. Afr., 560,010 (*659,934) C5 49
Durbin, St. Johns, Fla., 100 .... C6 86
Durbin, Cass, N. Dak., 35 .... C8 110
Durbin, Pocahontas, W. Va., 431 .... C5 123
Durbin, creek, Fla C6 86
Durbodden, Switz C8 19
Düren, Ger., 49,100 .... C3 16, D6 15
Durfee, hill, R.I B10 84
Durgapur, India, 10,000 .... E13 40
Durge Nuur, lake, Mong B3 34
Durham, Butte, Calif., 700 .... C3 82
Durham, Ont., Can., 2,180 .... C4 72
Durham, Middlesex, Conn., 700 (3,096▲) D6 84
Durham, Eng., 20,484 .... C6 10
Durham, Marion, Kans., 183 .... D6 93
Durham Androscoggin, Maine, 75 (1,086▲) D5 96
Durham, Lewis, Mo., 100 .... B6 101
Durham, Stafford, N.H., 4,688 (5,504▲) D5 105
Durham, Durham, N.C., 78,302 (*106,200) B5 109
Durham, Roger Mills, Okla., 75 .... B2 112
Durham, co., Ont., Can., 39,916 .... C6 72
Durham, co., Eng., 1,517,039 .... *C6 10
Durham, co., N.C., 111,995 .... A5 109
Durham, creek, Pa B2 106
Durham Bridge, N.B., Can., 193 .... C3 74
Durham Center, Middlesex, Conn., 250 .... D6 84
Durham Downs, Austl., 1,000 .... C3 51
Durham Hill, Waukesha, Wis., 130 .... F1 124
Durhamville, Lauderdale, Tenn., 35 .... B2 117
Durkee, Baker, Oreg., 200 .... C9 113
Durmersheim, Ger., 6,400 .... E3 17
Durness, Scot., 413 .... B4 13
Durness Kyle, bay, Scot B4 13
Durnford, pt., Sp. Sahara E1 44
Durrell, Newf., Can., 273 .... D4 75
Durres, Alb., 32,300 .... B2 23
Durres, pref., Alb., 92,000 .... *B2 23
Durrow, Ire., 439 .... E4 11
Durrus, Ire., 108 .... F2 11
Dursey, isl., Ire F1 11
Dursunbey, Tur., 5,900 .... C7 23
Duru, Con. L A4 48
D'Urville, cape, W. Irian F9 35
D'Urville, isl., N.Z N14 51
Duryea, Luzerne, Pa., 5,626 .... B8, D10 114
Dushanbe, Sov. Un., 248,000 .... H22 9
Dushaymi, isl., Eg., U.A.R C2 32
Dushore, Sullivan, Pa., 731 .... C9 114
Dusky, sound, N.Z P11 51
Dusniky, Czech., 1,880 .... n17 26
Duson, Lafayette, La., 1,033 .... D3 95
Düsseldorf, Ger., 702,600 (*1,020,000) B1 17
Dustin, Hughes, Okla., 457 .... B5 112
Dusty, Socorro, N. Mex., 41 .... D2 107
Dutch, creek, Ark C2 81
Dutchess, co., N.Y., 176,008 .... D7 108
Dutch Harbor, Alsk., 3 .... E6 79
Dutchman's Cap, isl., Vir. Is f14 65
Dutch Town, Ascension, La B5 95
Dutchtown, Cape Girardeau, Mo., 140 .... D8 101
Duthie, Shoshone, Idaho, 75 .... B3 89
Dutton, Jackson, Ala., 200 .... A4 78
Dutton, Ont., Can., 784 .... E3 72
Dutton, Teton, Mont., 504 .... C5 102
Dutton, mtn., Utah E3 119
Dutzow, Warren, Mo., 100 .... C7 101
Duval, Sask., Can., 186 .... F3 70
Duval, co., Fla., 455,411 .... B5 86
Duval, co., Texas, 13,398 .... F3 118
Duvergé, Dom. Rep., 6,701 .... F8 64
Duvno, Yugo., 1,610 .... D3 22
Duxbury, Plymouth, Mass., 1,069 (4,727▲) B6 97
Duxbury, Washington, Vt., 150 (546▲) C3 120
Duyuk, Jordan, 1,000 .... h12 32
Duzce, Tur., 18,200 .... B8 31
Dvina (Daugava), riv., Sov. Un D7 27
Dvina, Northern, riv., Sov. Un C17 9
Dvinskaya, bay, Sov. Un E18 25
Dvůr Králove [nad Labem], Czech., 15,000 .... C3 26
Dwarka, India, 14,314 .... F2 40
Dwarka, pt., India *F2 40
Dwight, Livingston, Ill., 3,086 .... B5 90
Dwight, Morris, Kans., 281 .... D7 93
Dwight, Hampshire, Mass B3 97
Dwight, Butler, Nebr., 209 .... C8 103
Dwight, Richland, N. Dak., 101 .... C9 110
Dwyer, Grant, N. Mex E2 107
Dwyer, Platte, Wyo., 20 .... C8 125
Dyakovo, Sov. Un r22 27
Dyce, Scot., 1,699 .... C6 13
Dycusburg, Crittenden, Ky., 99 .... A3 94
Dyer, Crawford, Ark., 450 .... B1 81
Dyer, Lake, Ind., 3,993 .... B2 91
Dyer, Esmeralda, Nev., 20 .... F3 104
Dyer, Gibson, Tenn., 1,909 .... A3 117
Dyer, co., Tenn., 29,537 .... A2 117
Dyer, bay, Ont., Can B3 72
Dyers Bay, Ont., Can., 85 .... B3 72
Dyersburg, Dyer, Tenn., 12,499 .... A2 117
Dyersville, Dubuque and Delaware, Iowa, 2,818 .... B6 92
Dyess, Mississippi, Ark., 185 .... B5 81
Dyje, riv., Czech D3 26
Dyment, Ont., Can., 78 .... E5 71
Dyrlow, see Darlowo, Pol.
Dysart, Sask., Can., 296 .... G3 70
Dysart, Tama, Iowa, 1,197 .... B5 92
Dzamiin Ude, Mong B4 36
Dzaoudzi, Comoro Is., 4,000 .... f9 49
Dzerzhinsk, Sov. Un., 176,000 .... D17 9, C14 27
Dzerzhinsk, Sov. Un E6 27
Dzerzhinsk, Sov. Un., 26,200 .... q20 27
Dzhabhan, riv., Mong B3 34
Dzhalal-Abad, Sov. Un., 24,900 .... E8 29
Dzhambul, Sov. Un., 131,000 .... E8 29
Dzhankoy, Sov. Un., 10,000 .... I10 27
Dzhebel, Sov. Un B7 41
Dzhetygara, Sov. Un., 18,000 .... C6 29
Dzhezkazgan, Sov. Un., 29,000 .... D7 29
Dzhizak, Sov. Un., 15,000 .... G22 9
Dzhugdzhur, mts., Sov. Un D16 28
Dzhulfa, Sov. Un C15 31
Dzhusaly, Sov. Un D6 29
Działdowo, Pol., 5,139 .... B6 26
Działoszyce, Pol., 2,306 .... C6 26
Dziedzice, Pol., 2,921 .... h10 26
Dzierzlow, see Darlowo, Pol.
Dzierzoniow, Pol., 27,000 .... C4 26
Dzilam González, Mex., 1,930 .... C7 63
Dzioua, Alg C6 44
Dzitbalché, Mex., 3,617 .... C6 63
Dzungaria, reg., China B2 34

E

Eads, Kiowa, Colo., 929 .... C8 83
Eads, Shelby, Tenn., 250 .... B2 117
Eagan, Claiborne, Tenn., 500 .... C10 117
Eagar, Apache, Ariz., 873 .... C6 80
Eagarville, Macoupin, Ill., 149 .... *D4 90
Eagle, Alsk., 92 .... C11 79
Eagle, Eagle, Colo., 546 .... B4 83
Eagle, Ada, Idaho, 500 .... F2 89
Eagle, Cass, Nebr., 302 .... D9, E2 103
Eagle, Wyoming, N.Y., 100 .... C2 108
Eagle, Fayette, W. Va., 200 .... C3, D6 123
Eagle, Waukesha, Wis., 620 .... F5 124
Eagle, Park, Wyo A4 125
Eagle, co., Colo., 4,677 .... B4 83
Eagle, cave, Wis E3 124
Eagle, cliff, Mont C1 102
Eagle, creek, Ind H7 91
Eagle, creek, Ky B5 94
Eagle, key, Fla G6 86
Eagle, lake, Calif B3 82
Eagle, lake, Maine A4 96
Eagle, lake, Maine B3 96
Eagle, lake, Ont., Can E5 71
Eagle, lake, Wis C4 124
Eagle, mtn., Tex F1 118
Eagle, peak, Calif B3 82
Eagle, riv., Newf., Can B2 75
Eagle Bend, Todd, Minn., 611 .... D3 99
Eagle Bridge, Rensselaer, N.Y., 250 .... C7 108
Eagle Butte, Dewey, S. Dak., 495 .... C4 116
Eagle Cap, mtn., Oreg B9 113
Eagle City, Blaine, Okla., 70 .... B3 112
Eagle Cliff, mtn., Idaho B3 89
Eagle Creek, Benton, Tenn., 35 .... B4 117
Eagle Grove, Wright, Iowa, 4,381 .... B4 92
Eaglehill, creek, Sask., Can F2 70
Eagle Lake, Polk, Fla., 1,364 .... E5 86
Eagle Lake, Aroostook, Maine, 900 (1,138▲) A4 96
Eagle Lake, Blue Earth, Minn., 506 .... F3 99
Eagle Lake, Colorado, Tex., 3,565 .... E4 118
Eagle Mills, Ouachita, Ark., 200 .... D3 81
Eagle Mountain, lake, Tex B5 118
Eagle Nest, Colfax, N. Mex., 300 .... A4 107
Eagle Nest, res., N. Mex A4 107
Eagle Pass, Maverick, Tex., 12,094 .... E2 118
Eagle Point, Jackson, Oreg., 752 .... E4 113
Eagle River, Keweenaw, Mich A2 98
Eagle River, Vilas, Wis., 1,367 .... C4 124
Eagle Rock, Botetourt, Va., 450 .... D3 121
Eaglesham, Alta., Can., 223 .... B2 69
Eagles Mere, Sullivan, Pa., 138 .... D8 114
Eagle Springs, Moore, N.C., 200 .... B4 109
Eagle Tail, mts., Ariz D2 80
Eagleton, Polk, Ark C1 81
Eagleton Village, Blount, Tenn., 5,068 .... *D10 117
Eagletown, McCurtain, Okla., 400 .... C7 112
Eagleville, Tolland, Conn., 200 .... B7 84
Eagleville, Harrison, Mo., 341 .... A3 101
Eagleville, Montgomery, Pa., 500 .... A10 114
Eagleville, Rutherford, Tenn., 363 .... B5 117
Eakly, Caddo, Okla., 217 .... B3 112
Ealing, Eng., 183,151 .... k11 10, C7 12
Eardley, Que., Can., 45 .... A8 72
Eardley, lake, Man., Can C3 71
Earl, isl., Newf., Can B3 75
Earle, Crittenden, Ark., 2,391 .... B5 81
Earl Grey, Sask., Can., 258 .... G3 70
Earlham, Madison, Iowa, 788 .... C3 92
Earlimart, Tulare, Calif., 2,897 .... E4 82
Earling, Shelby, Iowa, 431 .... C2 92
Earling, Logan, W. Va., 600 .... D5 123
Earlington, Hopkins, Ky., 2,786 .... C2 94
Earl Park, Benton, Ind., 551 .... C3 91
Earlsboro, Pottawatomie, Okla., 257 .... B5 112
Earlton, Ont., Can., 665 .... E9 72
Earlton, Neosho, Kans., 104 .... E8 93
Earlville, LaSalle, Ill., 1,420 .... B5 90
Earlville, Delaware, Iowa, 668 .... B6 92
Earlville, Chenango and Madison, N.Y., 1,004 .... C5 108
Early, Sac, Iowa., 824 .... B2 92
Early, co., Ga., 13,151 .... E8 87
Early Branch, Hampton, S.C., 200 .... F6 115
Earn, lake, Scot D4 13
Earn, riv., Scot D5 13
Earth, Lamb, Tex., 1,104 .... B1 118
Easingwold, Eng., 2,591 .... F7 13
Easky, Ire., 317 .... C3 11
Easley, Pickens, S.C., 8,283 .... B2 115
East, div., Ice., 9,794 .... *n15 25
East, bay, Tex F5 118
East, brook, N.H B4 105
East, butte, Mont B5 102
East, cape, Fla G5 86
East, cape, N.Z L17 51
East, cape, P.R e10 65
East, chan., Man., Can B3 71
East, isl., Que., Can B8 74
East, mtn., Mass A1 97
East, pass, Fla C2 87
East, pt., P.E.I., Can C8 74
East, pt., Que., Can B8 75
East, pt., Mass B6 97
East, pt., N.J E2 106
East, pt., N.Y D9 84
East, riv., Ont., Can B5 72
East, riv., N.Y E6 106
East, riv., Wis A6 124
East, riv., Wyo C3 125
Eastaboga, Talladega, Ala., 125 .... B3 78
Eastabutchie, Jones, Miss., 105 .... D4 100
East Acworth, Sullivan, N.H D2 105
East Alamosa, Alamosa, Colo., 800 .... D5 83
East Alburgh, Grand Isle, Vt., 75 .... B2 120
East Alliance, Mahoning, Ohio, 1,275 .... *B6 111
East Alstead, Cheshire, N.H., 75 .... D2 105
East Alton, Madison, Ill., 7,630 .... E3 90
East Andover, Oxford, Maine, 150 .... D2 96
East Andover, Merrimack, N.H., 250 .... D3 105
East Angus, Que., Can., 4,756 .... D6 73
East Arcadia, Bladen, N.C. C5 109
East Arlington, Bennington, Vt., 500 .... E2 120
East Ashtabula, Ashtabula, Ohio, 4,179 .... A7 111
East Aurora, Erie, N.Y., 6,791 .... C2 108
East Baldwin, Cumberland, Maine, 150 .... E2 96
East Bangor, Northampton, Pa., 970 .... E11 114
East Bank, Kanawha, W. Va., 1,023 .... C6 123
East Barnard, Windsor, Vt., 45 .... D3 120
East Barnet, Eng., 40,599 .... k12 10
East Barnet, Caledonia, Vt., 100 .... C4 120
East Barre, Washington, Vt., 550 .... C4 120
East Barre, riv., Vt C4 120
East Barrington, Strafford, N.H., 100 .... B4 105
East Baton Rouge, par., La., 230,058 .... D4 95
East Bend, Yadkin, N.C., 446 .... A3 109
East Bengal, reg., Pak E12 40
East Berkshire, Franklin, Vt., 200 .... B3 120
East Berlin, Hartford, Conn., 900 .... C6 84
East Berlin, Adams, Pa., 1,037 .... G8 114
East Bernard, Wharton, Tex., 900 .... E4 118
East Bernstadt, Laurel, Ky., 594 .... C5 94
East Bethel, Anoka, Minn., 1,408 .... *E5 99
East Bethel, Windsor, Vt., 50 .... D3 120
East Blue Hill, Hancock, Maine, 200 .... D4 96
East Bonne Terre, St. Francois, Mo., 150 .... D7 101
East Boothbay, Lincoln, Maine, 500 .... E3 96
Eastborough, Sedgwick, Kans., 1,001 .... B5 93
Eastbourne, Eng., 60,897 .... D8 12
East Brady, Clarion, Pa., 1,282 .... E2 114
East Braintree, Man., Can., 50 .... E4 71
East Braintree, Norfolk, Mass. (part of Braintree) D3 97
East Braintree, Orange, Vt., 75 .... D3 120
East Branch, Delaware, N.Y., 300 .... D5 108
East Branch, res., N.Y D2 84
East Branch Clarion River, res., Pa C4 114
East Brewster, Barnstable, Mass., 400 .... C7 97
East Brewton, Escambia, Ala., 2,511 .... D2 78
East Bridgewater, Plymouth, Mass., 3,000 (6,139▲) B6 97
Eastbrook, Hancock, Maine, 100 (167▲) D4 96
East Brookfield, Worcester, Mass., 1,150 (1,533▲) *B3 97
East Brookfield, Orange, Vt., 40 .... C3 120
East Brooklyn, Windham, Conn., 1,213 .... *B9 84
East Broughton, Que., Can., 1,136 .... C6 73
East Broughton Station, Que., Can., 1,136 .... C6 73
East Brownfield, Oxford, Maine, 200 .... E2 96
East Brunswick, Middlesex, N.J., 12,000 .... *B4 106
East Burke, Caledonia, Vt., 110 .... B5 120
East Butler, Butler, Pa., 1,007 .... E2 114
East Calais, Washington, Vt., 110 .... C4 120
East Canaan, Litchfield, Conn., 570 .... A3 84
East Candia, Rockingham, N.H., 200 .... D4 105
East Canon, Fremont, Colo., 1,101 .... C5 83
East Canterbury, Merrimack, N.H D3 105
East Canton, Stark, Ohio, 1,521 .... *B6 111
East Carondelet, St. Clair, Ill., 463 .... B8 101
East Carroll, par., La., 14,433 .... B4 95
East Carver, Plymouth, Mass., 200 .... C6 97
East Charleston, Orleans, Vt., 60 .... B5 120
East Chelmsford, Middlesex, Mass., 1,500 .... *A5 97
Eastchester, Westchester, N.Y., 12,000 .... *E7 108
East Chicago, Lake, Ind., 57,669 .... A3 91
East Chicago Heights, Cook, Ill., 4,295 .... *F3 90
East China, sea, China F9 34
East Chop, pt., Mass D6 97
East Cleveland, Cuyahoga, Ohio, 37,991 .... B2 111
East Clifton, Que., Can., 670 .... D6 73
East Columbia, Brazoria, Tex., 125 .... G4 118
East Columbus, Bartholomew, Ind., 1,912 .... F6 91
East Concord, Essex, Vt., 250 .... C5 120
East Conemaugh (Conemaugh), Cambria, Pa., 3,334 .... F4 114
East Conway, Carroll, N.H., 100 .... B4 105
East Corinth, Penobscot, Maine, 250 .... D3 96
East Corinth, Orange, Vt., 200 .... C4 120
East Cote Blanche, bay, La E4 95
East Coulee, Alta., Can., 683 .... D4 69
East Craftsbury, Orleans, Vt., 45 .... B4 120
East Dennis, Barnstable, Mass., 250 .... C7 97
East Dereham, Eng., 7,197 .... B8 12
East Derry, Rockingham, N.H., 200 .... E4 105
East Des Moines, riv., Iowa A3 92
East Detroit, Macomb, Mich., 45,756 .... A8 98

East Dixmont, Penobscot, Maine, 100 ... D3 96
East Dorset, Bennington, Vt., 140 ... E2 120
East Douglas, Worcester, Mass., 1,695 ... B4 97
East Dover, Piscataquis, Maine, 70 ... C3 96
East Dublin, Laurens, Ga., 1,677 ... *D4 87
East Dubuque, Jo Daviess, Ill., 2,082 ... A3 90
East Dummerston, Windham, Vt., 85 ... F3 120
East Dundee, Kane, Ill., 2,221 ... *A5 90
East Eddington, Penobscot, Maine, 300 ... D4 96
East Ellijay, Gilmer, Ga., 501 . B2 87
East Ely, White Pine, Nev., 1,796 ... D7 104
Eastend, Sask., Can., 767 ... H1 70
East End, Vir. Is ... f16 65
East Enterprise, Switzerland, Ind., 300 ... G8 91
Easter, isl., see Rapa Nui, isl., Pac. O.
Eastern, prov., N. Rh ... D5 48
Eastern, reg., Ghana ... E4 45
Eastern, reg., Nig., 7,218,000 . E6 45
Eastern, reg., Tan., 1,084,484 ... C6 48
Eastern, reg., Ug ... A5 48
Eastern, bay, Md ... C5 85
Eastern, isl., Newf., Can ... C4 75
Eastern, isl., Midway Is ... 52
Eastern, pt., Mass ... C4 97
Eastern Ghats, range, India ... E7 39
Eastern Neck, isl., Md ... B5 85
Eastern Plain, pt., N.Y. ... E8 84
Eastern Valley, Jefferson, Ala., 1,000 ... *E4 78
East Fairfield, Franklin, Vt., 150 ... B3 120
East Fairview, McKenzie, N. Dak., 200 ... B1 110
East Falmouth, Barnstable, Mass., 1,655 ... C9 97
East Farmington Heights, Hartford, Conn., 1,800 ... *B6 84
East Farms, Spokane, Wash . D8 122
East Faxon, Lycoming, Pa., 3,641 ... *D8 114
East Fayetteville, Cumberland, N.C., 2,797 ... *B5 109
East Feliciana, par., La., 20,198 ... D4 95
East Flanders, prov., Bel., 1,272,005 ... C3 15
East Flat Rock, Henderson, N.C., 1,700 ... D3 109
East Flevoland Polder, reg., Neth ... B5 15
East Florenceville, N.B., Can., 443 ... C2 74
Eastford, Windham, Conn., 350 (746▲) ... B8 84
East Franklin, Franklin, Vt., 35 ... B3 120
East Freetown, Bristol, Mass., 300 ... C9 97
East Frisian, is., Ger ... A7 15
East Fultonham, Muskingum, Ohio, 500 ... C5 111
East Gaffney, Cherokee, S.C., 4,779 ... A4 115
East Galesburg, Knox, Ill., 660 ... C3 90
East Gary, Lake, Ind., 9,309 . A3 91
East Gastonia, Gaston, N.C., 3,326 ... *B2 109
East Gate, Churchill, Nev., 10 ... D4 104
Eastgate, King, Wash., 3,000 ... *B3 122
East Georgia, Franklin, Vt., 40 ... B2 120
East Glacier Park, Glacier, Mont., 350 ... B3 102
East Glastonbury, Hartford, Conn., 500 ... C6 84
East Glenville, Schenectady, N.Y., 5,000 ... *C7 108
East Grafton, Grafton, N.H., 100 ... C3 105
East Granby, Hartford, Conn., 200 (2,434▲) ... B6 84
East Grand Forks, Polk, Minn., 6,998 ... C2 99
East Granville, Addison, Vt., 75 ... C3 120
East Greenbush, Rensselaer, N.Y., 1,325 ... *C7 108
East Greenville, Montgomery, Pa., 1,931 ... F10 114
East Greenwich, Kent, R.I., 6,100 ... C11 84
East Grinstead, Eng., 15,421 ... C7 12
East Haddam, Middlesex, Conn., 500 (3,637▲) ... D7 84
East Ham, Eng., 105,359 ... k13 10
Eastham, Barnstable, Mass., 300 (1,200▲) ... C8 97
East Hampden, Penobscot, Maine, 1,500 ... *D4 96
East Hampton, Middlesex, Conn., 3,000 (5,403▲) ... C6 84
Easthampton, Hampshire, Mass., 12,326 ... B2 97
East Hampton, Suffolk, N.Y., 1,772 ... D4 108
East Hardwick, Caledonia, Vt., 195 ... B4 120
East Harpswell, Cumberland, Maine, 50 ... E5 96
East Hartford, Hartford, Conn., 43,977 ... B6 84
East Hartland, Hartford, Conn., 330 ... B5 84
East Haven, New Haven, Conn., 21,388 ... D5 84
East Haven, Essex, Vt., 30 (164▲) ... B5 120
East Haverhill, Grafton, N.H., 75 ... B3 105
East Hazelcrest, Cook, Ill., 1,457 ... *F3 90
East Hebron, Oxford, Maine, D2 96
East Hebron, Grafton, N.H., 60 ... C3 105
East Helena, Lewis and Clark, Mont., 1,490 ... D5 102
East Herkimer, Herkimer, N.Y., 1,068 ... B6 108
East Hickory, Forest, Pa., 200 . C3 114
East Highgate, Franklin, Vt., 110 ... B3 120
East Hills, Nassau, N.Y., 7,184 ... *E3 108
East Holden, Penobscot, Maine, 200 ... D4 96
East Holderness, Grafton, N.H., 85 ... C3 105
East Holliston, Middlesex, Mass., 121 ... D1 97
East Islip, Suffolk, N.Y., 7,000 ... G4 84
East Jamaica, Windham, Vt., 50 ... E3 120
East Jamestown, Fentress, Tenn ... C9 117
East Java, see Java, isl., Indon:
East Jordan, N.S., Can., 97 . F4 74
East Jordan, Charlevoix, Mich., 1,919 ... C5 98
East Juliette, Jones, Ga., 201 . C3 87
East Keansburg, Monmouth, N.J., 3,000 ... *C4 106
East Kelowna, B.C., Can., 403 ... E8 68
East Kent, Litchfield, Conn . C3 84
East Kildonan, Man., Can., 27,305 ... E3 71
East Killingly, Windham, Conn., 560 ... B9 84
East Kingsford, Dickinson, Mich., 1,063 ... *C2 98
East Kingston, Rockingham, N.H., 300 (574▲) ... E4 105
East Knox, Waldo, Maine . D3 96
Eastlake, Manistee, Mich., 436 ... D4 98
East Lake, Dare, N.C., 90 ... B8 109
Eastlake, Lake, Ohio, 12,467 . A6 111
Eastland, Eastland, Tex., 3,292 ... C3 118
Eastland, co., Tex., 19,526 . C3 118
East Lansdowne, Delaware, Pa., 3,224 ... *G11 114
East Lansing, Ingham, Mich., 30,198 ... F6 98
East Laport, Jackson, N.C., 50 ... D3 109
East Laurinburg, Scotland, N.C., 695 ... C4 109
Eastlawn, Washtenaw, Mich., 2,500 ... *F7 98
East Lebanon, York, Maine, 40 ... E2 96
East Lee, Berkshire, Mass., 200 ... B1 97
Eastleigh, Eng., 36,577 ... D6 12
East Lempster, Sullivan, N.H., 150 ... D2 105
East Lexington, Middlesex, Mass. (part of Lexington) . C2 97
East Liberty, Logan, Ohio, 385 ... B4 111
East Litchfield, Litchfield, Conn., 75 ... B4 84
East Livermore, Androscoggin, Maine, 150 ... D2 96
East Liverpool, Columbiana, Ohio, 22,306 ... B7 111
East London, S. Afr., 113,746 (*115,677) ... D4 49
East Longmeadow, Hampden, Mass., 10,294 ... B2 97
East Los Angeles, Los Angeles, Calif., 98,000 ... *E4 82
East Lothian, co., Scot., 52,653 ... E6 13
East Lowell, Penobscot, Maine ... C4 96
East Lumberton, Robeson, N.C. (part of Lumberton) . C5 109
East Lyme, New London, Conn., 850 (6,782▲) ... D8 84
East Lynn, Wayne, W. Va., 250 ... C2 123
East Lynne, Cass, Mo., 243 . C3 101
East Machias, Washington, Maine, 700 (1,198▲) ... D5 96
East McKeesport, Allegheny, Pa., 3,470 ... F2 114
East Madison, Somerset, Maine, 250 ... D3 96
Eastmain, Que., Can., 212 . F17 67
Eastman, Que., Can., 637 ... D5 73
Eastman, Dodge, Ga., 5,118 . D3 87
Eastman, Crawford, Wis., 348 ... E2 124
East Mansfield, Bristol, Mass., 200 ... B5 97
East Marion, Suffolk, N.Y., 270 ... E7 84
East Marion, McDowell, N.C., 2,442 ... *D2 109
East Massapequa, Nassau, N.Y., 6,000 ... *G3 84
East Matunuck, Washington, R.I., 270 ... D10 84
East Meadow, Nassau, N.Y., 46,036 ... *E7 108
East Meredith, Delaware, N.Y., 150 ... C6 108
East Middlebury, Addison, Vt., 320 ... D2 120
East Middletown, Orange, N.Y., 1,752 ... *D6 108
East Milford, Hillsboro, N.H. (part of Milford) ... E3 105
East Millcreek, Salt Lake, Utah, 6,000 ... C4 119
East Millinocket, Penobscot, Maine, 2,392 ... C4 96
East Millstone, Somerset, N.J., 700 ... B3 106
East Milton, Norfolk, Mass. (part of Milton) ... D9 97
East Mobridge, Walworth, S. Dak. (part of Mobridge) ... B5 116
East Moline, Rock Island, Ill., 16,732 ... B3 90
East Monkton, Addison, Vt . C2 120
East Monroe, Highland, Ohio, 100 ... C4 111
East Montpelier, Washington, Vt., 200 (1,200▲) ... C4 120
East Moriches, Suffolk, N.Y., 1,210 ... F5 84
East Morris, Litchfield, Conn., 125 ... C4 84
East Naples, Collier, Fla., 1,500 ... F5 86
East Newark, Hudson, N.J., 1,872 ... B4 106
East New Market, Dorchester, Md., 225 ... C6 85
East Newnan, Coweta, Ga., 500 ... C2 87
East Nishnabotna, riv., Iowa . C2 92
East Northport, Suffolk, N.Y., 8,381 ... F3 84
East Norton, Bristol, Mass., 200 ... C5 97
East Norwalk, Fairfield, Conn. (part of Norwalk) ... D3 84
East Norwich, Nassau, N.Y., 2,500 ... *F2 84
East Olympia, Thurston, Wash., 500 ... C3 122
East Omaha, Douglas, Nebr., 684 ... *D3 103
Easton, Fresno, Calif., 2,500 ... *D4 82
Easton, Fairfield, Conn., 600 (3,407▲) ... D3 84
Easton, Mason, Ill., 361 ... C4 90
Easton, Leavenworth, Kans., 320 ... B8 93
Easton, Aroostook, Maine, 550 (1,389▲) ... B5 96
Easton, Talbot, Md., 6,337 . C5 85
Easton, Bristol, Mass., 100 (9,078▲) ... B5 97
Easton, Faribault, Minn., 411 ... G5 99
Easton, Buchanan, Mo., 198 . B3 101
Easton, Grafton, N.H., 50 (740▲) ... B3 105
Easton, Washington, N.Y., 200 ... B7 108
Easton, Northampton, Pa., 31,955 ... E11 114
Easton, Kittitas, Wash., 250 . B4 122
Easton, res., Conn ... E3 84
East Orange, Essex, N.J., 77,259 ... B4 106
East Orange, Orange, Vt., 60 . C4 120
East Orland, Hancock, Maine, 140 ... D4 96
East Orleans, Barnstable, Mass., 300 ... C8 97
East Orrington, Penobscot, Maine, 600 ... D4 96
East Otis, Berkshire, Mass., 100 ... B1 97
East Otto, Cattaraugus, N.Y., 300 ... C2 108
Eastover, Richland, S.C., 713 ... D6 115
East Pakistan, prov., Pak., 50,835,721 ... *D9 39
East Palatka, Putnam, Fla., 1,133 ... C5 86
East Palestine, Columbiana, Ohio, 5,232 ... B7 111
East Palo Alto, San Mateo, Calif., 12,000 ... *B5 82
East Parsonfield, York, Maine, 100 ... E2 96
East Patchogue, Suffolk, N.Y., 5,500 ... F5 84
East Paterson, Bergen, N.J., 19,344 ... B4 106
East Peacham, Caledonia, Vt., 40 ... C3 120
East Pea Ridge, Cabell, W. Va., 1,500 ... *C2 123
East Peoria, Tazewell, Ill., 12,310 ... C4 90
East Pepperell, Middlesex, Mass., 1,200 ... A4 97
East Peru, Oxford, Maine, 75 ... D2 96
East Petersburg, Lancaster, Pa., 2,053 ... F9 114
East Pine, B.C., Can., 98 ... B7 68
East Pines, Prince Georges, Md., 1,800 ... *C4 85
East Pittsburgh, Allegheny, Pa., 4,122 ... F2 114
Eastpoint, Franklin, Fla., 700 . C2 86
East Point, Fulton, Ga., 35,633 ... B5, C2 87
East Point, Red River, La., 100 ... B2 95
East Poland, Androscoggin, Maine, 100 ... D2 96
East Poplar, Sask., Can ... H3 70
Eastport, Newf., Can., 438 . D5 75
Eastport, Boundary, Idaho, 100 ... A2 89
Eastport, Washington, Maine, 2,537 ... D6 96
Eastport, Suffolk, N.Y., 950 . D4 108
East Portal, Gilpin, Colo., 30 ... B5 83
East Port Chester, Fairfield, Conn. (part of Greenwich) ... F2 84
East Porterville, Tulare, Calif., 2,000 ... *D4 82
East Poultney, Rutland, Vt., 300 ... D2 120
East Prairie, Mississippi, Mo., 3,449 ... E8 101
East Princeton, Worcester, Mass., 150 ... B4 97
East Prospect, York, Pa., 623 . G8 114
East Providence, Providence, R.I., 41,955 ... B11 84
East Punjab, state, India, 16,000,000 ... B6 40
East Putney, Windham, Vt., 40 ... F4 120
East Quincy, Plumas, Calif., 1,020 ... *C3 82
East Quogue, Suffolk, N.Y., 1,000 ... F6 84
East Rainelle, Greenbrier, W. Va., 1,244 ... D4 123
East Randolph, Cattaraugus, N.Y., 594 ... C2 108
East Randolph, Orange, Vt., 130 ... D3 120
East Retford, Eng., 17,788 ... A7 12
East Richmond, Contra Costa, Calif., 4,000 ... *B5 82
East Ridge, Hamilton, Tenn., 19,570 ... E10 117
East Riding, Yorkshire, see York, co., Eng.
East Rindge, Cheshire, N.H., 225 ... E3 105
East River, New Haven, Conn., 200 ... D6 84
East River, mtn., Va ... D1 121
East River Sheet Harbour, N.S., Can., 104 ... E7 74
East Rochester, Strafford, N.H. (part of Rochester) ... D5 105
East Rochester, Monroe, N.Y., 8,152 ... B3 108
East Rockaway, Nassau, N.Y., 10,721 ... G2 84
East Rockingham, Richmond, N.C., 3,211 ... C4 109
East Rockwood, Wayne, Mich., 1,000 ... *F7 98
East Roxbury, Washington, Vt ... C3 120
East Rutherford, Bergen, N.J., 7,769 ... D5 106
East Ryegate, Caledonia, Vt., 170 ... C4 120
East St. Johnsbury, Caledonia, Vt., 150 ... C5 120
East St. Louis, St. Clair, Ill., 81,712 ... E3 90
East Salt, creek, Colo ... B2 83
East Sandwich, Barnstable, Mass., 250 ... C7 97
East Saugus, Essex, Mass. (part of Saugus) ... C3 97
East Sebago, Cumberland, Maine, 75 ... E2 96
East Selkirk, Man., Can., 401 ... D3 71
East Setauket, Suffolk, N.Y., 1,127 ... F4 84
East Siberian, sea, Sov. Un ... B18 28
Eastside, Jackson, Miss., 4,318 ... E3 100
Eastside, Coos, Oreg., 1,380 . D2 113
Eastside Galesburg, Knox, Ill., 1,147 ... *C3 90
East Sioux Falls, Minnehaha, S. Dak., 45 ... D9 116
East Smithfield, Bradford, Pa., 350 ... C8 114
East Somerset, Pulaski, Ky., 3,645 ... C5 94
Eastsound, San Juan, Wash., 150 ... A3 122
East Sparta, Stark, Ohio, 961 . B6 111
East Spencer, Rowan, N.C., 2,171 ... B3 109
East Spokane, Spokane, Wash., 6,000 ... B8, D7 122
East Springfield, Otsego, N.Y., 150 ... C6 108
East Springfield, Erie, Pa., 511 ... C1 114
East Stanwood, Snohomish, Wash. (part of Stanwood) . A3 122
East Stoneham, Oxford, Maine, 160 ... D2 96
East Stroudsburg, Monroe, Pa., 7,674 ... D11 114
East Sudbury, Middlesex, Mass., 50 ... D1 97
East Sullivan, Hancock, Maine, 250 ... D4 96
East Sullivan, Cheshire, N.H., 75 ... E2 105
East Sumner, Oxford, Maine, 100 ... D2 96
East Swamp, creek, Pa ... C2 106
East Swanzey, Cheshire, N.H., 250 ... E2 105
East Syracuse, Onondaga, N.Y., 4,708 ... B4 108
East Taunton, Bristol, Mass. (part of Taunton) ... C5 97
East Tavaputs, plat., Utah ... D6 119
East Tawas, Iosco, Mich., 2,462 ... D7 98
East Templeton, Worcester, Mass., 900 ... A3 97
East Thermopolis, Hot Springs, Wyo., 281 ... B4 125
East Thetford, Orange, Vt., 100 ... D4 120
East Thomaston, Upson, Ga., 2,237 ... D2 87
East Tirol, reg., Aus ... C8 18
East Tohopekaliga, lake, Fla ... D5 86
East Towanda, Bradford, Pa., 200 ... C9 114
East Troy, Walworth, Wis., 1,455 ... F5 114
East Tulare, Tulare, Calif., 1,342 ... *D4 82
East Uniontown, Fayette, Pa., 2,424 ... *G2 114
East Vandergrift, Westmoreland, Pa., 1,388 ... *E2 114
East Vaughn, Guadalupe, N. Mex. (part of Vaughn) . C4 107
Eastview, Ont., Can., 24,555 . A9 72
Eastville, Northampton, Va., 261 ... D7 121
East Wakefield, Carroll, N.H., 50 ... C4 105
East Walker, riv., Nev ... E2 104
East Wallingford, Rutland, Vt., 150 ... E3 120
East Walpole, Norfolk, Mass., 2,000 ... D2 97
East Wareham, Plymouth, Mass., 950 ... C6 97
East Washington, Sullivan, N.H., 100 ... D2 105
East Washington, Washington, Pa., 2,483 ... F1 114
East Waterboro, York, Maine, 180 ... E2 96
East Waterford, Oxford, Maine, 75 ... D2 96
East Waterford, Juniata, Pa., 200 ... F6 114
East Wellington, B.C., Can., 128 ... A8 68
East Wenatchee, Douglas, Wash., 383 ... B5 122
East Wenatchee Beach, Douglas, Wash., 2,327 ... *B5 122
East Weymouth, Norfolk, Mass. (part of Weymouth) . D3 97
East Whittier, Los Angeles, Calif., 19,884 ... *F4 82
East Willington, Tolland, Conn., 150 ... B8 84
East Williston, Nassau, N.Y., 2,940 ... G2 84
East Wilmington, New Hanover, N.C., 5,520 ... *C6 109
East Wilton, Franklin, Maine, 250 ... D2 96
East Windsor Hill, Hartford, Conn., 400 ... B6 84
East Wolfeboro, Carroll, N.H., 340 ... C4 105
Eastwood, Jefferson, Ky., 250 . A4 94
Eastwood, Kalamazoo, Mich., 6,000 ... *E5 98
East Woodstock, Windham, Conn., 150 ... B9 84
East Worcester, Otsego, N.Y., 275 ... C6 108
East York, York, Pa., 1,800 . *G8 114
Eaton, Weld, Colo., 1,267 ... A6 83
Eaton, Crawford, Ill., 100 ... D6 90
Eaton, Delaware, Ind., 1,529 . D7 91
Eaton, Washington, Maine, 75 ... C5 96
Eaton, Madison, N.Y., 350 ... C5 108
Eaton, Preble, Ohio, 5,034 ... C3 111
Eaton, Gibson, Tenn., 125 ... B2 117
Eaton, co., Mich., 49,684 ... F6 98
Eaton Center, Carroll, N.H., 75 (151▲) ... C4 105
Eatonia, Sask., Can., 609 ... F1 70
Eaton Rapids, Eaton, Mich., 4,052 ... F6 98
Eatons, neck, N.Y ... F3 84
Eatons, pt. N.Y ... F3 84
Eatonton, Putnam, Ga., 3,612 ... C3 87
Eatontown, Monmouth, N.J., 10,334 ... C4 106
Eatonville, Pierce, Wash., 896 ... C3 122
Eau Claire, Berrien, Mich., 562 ... G4 98
Eau Claire, Chippewa and Eau Claire, Wis., 37,987 ... D2 124
Eau Claire, co., Wis., 58,300 . D2 124
Eau Claire, riv., Wis ... D3 124
Eau Galle, Dunn, Wis., 235 . D2 124
Eau Gallie, Brevard, Fla., 12,300 ... D6 86
Eau Pleine, riv., Wis ... D3 124
Eauze, Fr., 1,823 ... F4 14
Ebadon, isl., Kwajalein ... 52
Ebb, Madison, Fla., 40 ... B3 86
Ebbw Vale, Wales, 28,631 ... C4 12
Ebeleben, Ger., 2,535 ... B5 17
Ebeltoft, Den., 2,227 ... B4 24
Ebenezer, Sask., Can., 137 ... F4 70
Ebenezer, Holmes, Miss., 75 . C3 100
Ebenezer, Knox, Tenn ... E11 117
Eben Junction, Alger, Mich., 75 ... B4 98
Ebensburg, Cambria, Pa., 4,111 ... F4 114
Ebermannstadt, Ger., 2,600 . D6 17
Ebersbach, Ger., 11,100 ... B9 17
Ebersbach, Ger., 9,200 ... E4 17
Eberswalde, Ger., 32,100 ... B6 16
Ebingen, Ger., 21,100 ... D4 16
Eboli, It., 20,500 (25,634▲) ... D5 21
Ebolowa, Cam., 1,200 ... E2 46
Ebony, Crittenden, Ark ... E8 81
Ebro, Washington, Fla., 125 ... G3 86
Ebro, Clearwater, Minn., 75 ... C3 99
Ebro, riv., Sp ... B6 20
Eburne, B.C., Can., 1,000 . A9 68
Eccles, Raleigh, W. Va., 1,145 ... D3, D6 123
Eceabat (Maydos), Tur., 2,800 ... B6 23
Echague, Phil., 2,568 ... n13 35
Echallens, Switz., 1,428 ... C2 19
Echeconnee, creek, Ga ... D3 87
Echigawa, Jap., 3,400 ... n15 37
Echo, Dale, Ala., 100 ... D4 78
Echo, Rapides, La., 350 ... C3 95
Echo, Yellow Medicine, Minn., 459 ... F3 99
Echo, Umatilla, Oreg., 456 . B7 113
Echo, Summit, Utah, 130 ... C4 119
Echo, lake, Ont., Can ... B7 98
Echo, lake, Maine ... D2 96
Echo, pond, Vt ... B5 120
Echo, riv., Ont., Can ... B7 98
Echoing, lake, Ont., Can ... B5 71
Echoing, riv., Man., Can ... B6 71
Echola, Tuscaloosa, Ala., 100 ... B2 78
Echols, co., Ga., 1,876 ... F4 87
Echt, Neth., 4,700 ... C5 15
Echt, Scot., 1,027 ... C6 13
Echternach, Lux., 3,389 ... E6 15
Echuca, Austl., 6,443 ... H5 51
Écija, Sp., 36,200 ... D3 20
Eckelson, Barnes, N. Dak., 90 ... C7 110
Eckerman, Chippewa, Mich., 200 ... B6 98
Eckernförde, Ger., 19,500 ... A4 16
Eckernförder Bucht, bay, Ger . D4 24
Eckert (Orchard City), Delta, Colo., 1,021 ... C3 83
Eckerty, Crawford, Ind., 150 ... H4 91
Eckhart, Allegany, Md., 900 ... D3 85
Eckley, Yuma, Colo., 207 ... A8 83
Eckley, Luzerne, Pa., 340 ... E10 114
Eckman, Bottineau, N. Dak., 5 ... A4 110
Eckman, McDowell, W. Va., 1,125 ... *D3 123
Eckville, Alta., Can., 580 ... C3 69
Eclectic, Elmore, Ala., 926 ... C3 78
Eclipse, Nansemond, Va., 290 ... B6, E6 121
Econfina, Bay, Fla., 130 ... G3 86
Econfina, riv., Fla ... B3 86
Economy, N.S., Can., 142 ... D6 74
Economy, Wayne, Ind., 280 . E7 91
Economy, Beaver, Pa., 5,925 ... *E1 114
Ecorces, riv., Que., Can ... A6 73
Ecorse, Wayne, Mich., 17,328 ... A7 98
Ecru, Pontotoc, Miss., 442 ... A4 100
Ector, co., Tex., 90,995 ... D1 118
Ecuador, country, S.A., 3,796,000 ... D3 53, 58
Ecum Secum, N.S., Can., 243 ... E7 74
Ecum Secum Bridge, N.S., Can., 222 ... E7 74
Eda, Swe., 6,344 ... H5 25
Edam, Sask., Can., 277 ... D1 70
Edam, Neth ... B5 15
Eday, isl., Scot ... A6 13
Edberg, Alta., Can., 179 ... C4 69
Edcouch, Hidalgo, Tex., 2,814 ... F4 118
Edd, Eth ... C5 47
Ed Da'ein, Sud ... C2 47
Ed Damer, Sud., 5,458 ... B3 47
Ed Debba (Well), Sud ... B3 47
Eddiceton, Franklin, Miss., 300 ... D3 100
Eddington, Penobscot, Maine, 130 (958▲) ... D4 96
Eddington, Bucks, Pa. (part of Cornwells Heights) ... A12 114
Ed Dueim, Sud., 12,319 ... C3 47
Eddy, Sanders, Mont., 26 ... C1 102
Eddy, co., N. Mex., 50,783 ... E5 107
Eddy, co., N. Dak., 4,936 ... B7 110
Eddy, mtn., Calif ... B2 82
Eddystone, Delaware, Pa., 3,006 ... B10 114
Eddystone Rocks, reef, English Chan ... E4 10
Eddyville, Pope, Ill., 125 ... F5 90
Eddyville, Wapello and Mahaska, Iowa, 1,014 ... C5 92
Eddyville, Lyon, Ky., 1,858 . A3 94
Eddyville, Dawson, Nebr., 119 ... C6 103
Ed Dzong, China, 1,000 ... A13 40
Ede, Neth., 21,200 (57,900▲) . B5 15
Ede, Nig., 44,808 ... *E5 45
Edea, Cam ... E2 46
Eden, Graham, Ariz., 65 ... E6 80
Eden, Austl., 1,246 ... H7 51
Eden, Man., Can., 146 ... D2 71
Eden, Effingham, Ga., 400 ... D5 87
Eden, Jerome, Idaho, 426 ... G4 89
Eden, Hancock, Maine, 140 ... D4 96
Eden, Somerset, Md., 105 ... D6 85
Eden, Yazoo, Miss., 218 ... C3 100
Eden, Cascade, Mont., 6 ... C5 102
Eden, Erie, N.Y., 2,366 ... C2 108
Eden, Marshall, S. Dak., 136 ... B8 116
Eden, Concho, Tex., 1,486 ... D3 118
Eden, Lamoille, Vt., 65 (430▲) ... B3 120
Eden, Fond du Lac, Wis., 312 ... E5 124
Eden, Sweetwater, Wyo., 220 ... C3 125
Eden, riv., Eng ... C5 10
Eden, riv., Scot ... D5 13
Edenberry, Ire., 2,996 ... D4 13
Edenborn, Fayette, Pa., 800 ... G2 114
Eden Mills, Lamoille, Vt., 60 ... B3 120
Edenton, Chowan, N.C., 4,458 ... A7 109
Eden Valley, Meeker and Stearns, Minn., 793 ... E4 99
Edenwold, Sask., Can., 165 . G3 70
Edenwold, Davidson, Tenn., 500 ... A5, E9 117
Eder, riv., Ger ... B4 17
Edesville, Kent, Md., 210 ... B5 85
Edgar, Carbon, Mont., 178 . E8 102
Edgar, Clay, Nebr., 730 ... D8 103
Edgar, Marathon, Wis., 803 . D4 120
Edgar, co., Ill., 22,550 ... D6 90
Edgard, St. John the Baptist, La., 750 ... B6, D5 95
Edgar Springs, Phelps, Mo., 150 ... D6 101
Edgartown, Dukes, Mass., 1,181 (1,474▲) ... D6 97
Edge, isl., Nor ... B12 4
Edgecombe, co., N.C., 54,226 ... B6 109
Edgecumbe, cape, Alsk ... m22 79
Edgefield, Edgefield, S.C., 2,876 ... D4 115
Edgefield, co., S.C., 15,735 . D4 115
Edgeley, Sask., Can., 67 ... G3 70
Edgeley, La Moure, N. Dak., 992 ... C7 110
Edgely, Bucks, Pa., 950 ... *F12 114
Edgemere, Bonner, Idaho, 10 ... A2 89
Edgemere, Baltimore, Md., 2,200 ... *B5 85
Edgemont, Cleburne, Ark., 40 ... B3 81
Edgemont, Riverside, Calif., 1,628 ... *F5 82
Edgemont, Jefferson, Colo., 1,500 ... *B5 83
Edgemont, Fall River, S. Dak., 1,772 ... D2 116
Edgemont, chan., Fla ... F1 86
Edgemont, key, Fla ... F1 86
Edgemoor, Montgomery, Md. (part of Bethesda) ... C1 85
Edgemoor, Chester, S.C., 275 ... B5 115
Edgemoor, Anderson, Tenn., 50 ... E11 117
Edgerly, Calcasieu, La., 350 ... D2 95
Edgerton, Alta., Can., 295 ... C5 69
Edgerton, Johnson, Kans., 414 ... D8 93
Edgerton, Pipestone, Minn., 1,019 ... G2 99
Edgerton, Platte, Mo., 449 ... B3 101
Edgerton, Williams, Ohio, 1,566 ... A3 111
Edgerton, Rock, Wis., 4,000 . F4 124
Edgerton, Natrona, Wyo., 512 ... B6 125
Edgewater, Jefferson, Ala., 1,200 ... E7 78
Edgewater, Jefferson, Colo., 4,314 ... B5 83
Edgewater, Volusia, Fla., 2,051 ... D6 86
Edgewater, Anne Arundel, Md., 3,000 ... C4 85
Edgewater, Bergen, N.J., 4,113 ... D5 106
Edgewater Gulf Beach, Bay, Fla., 70 ... *G3 86
Edgewater Park, Harrison, Miss., 750 ... E2 100
Edgewater Park, Burlington, N.J., 2,866 ... C3 106
Edgewood, B.C., Can., 325 . E8 68
Edgewood, Effingham, Ill., 515 ... E5 90
Edgewood, Madison, Ind., 2,119 ... *D6 91
Edgewood, Clayton and Delaware, Iowa, 767 ... B6 92
Edgewood, Kenton, Ky., 1,100 ... A7 94
Edgewood, Harford, Md., 2,240 ... B5 85
Edgewood, Santa Fe, N. Mex., 25 ... B3, G6 107
Edgewood, Allegheny, Pa., 5,124 ... F2 114
Edgewood, Northumberland, Pa., 3,399 ... *E8 114
Edgewood, Pierce, Wash., 100 ... D2 122
Edgeworth, Allegheny, Pa., 2,030 ... A5 114
Edgigen, isl., Kwajalein ... 52
Edgin, Briscoe, Tex ... B2 118
Edgmoor, Chester, S.C. ... B5 115
Edgware, Eng. (part of Hendon) ... k11 10
Edhessa, Grc., 15,534 ... B4 23
Edina, Hennepin, Minn., 28,501 ... E6 99
Edina, Knox, Mo., 1,457 ... A5 101
Edinboro, Erie, Pa., 1,703 ... C1 114
Edinburg, Christian, Ill., 1,003 ... D4 90
Edinburg, Johnson, Ind., 3,664 ... F6 91
Edinburg, Leake, Miss., 200 ... C4 100
Edinburg, Saratoga, N.Y., 200 ... B6 108

Edinburg, Walsh, N. Dak., 330 ... A8 110
Edinburg, Hidalgo, Tex., 18,706 ... F3 118
Edinburg, Shenandoah, Va., 517 ... C4 121
Edinburgh, Scot., 468,378 (★600,000) ... E5 13
Edirne, Tur., 31,900 ... B6 23
Edison, Calhoun, Ga., 1,232 ... E2 87
Edison, Furnas, Nebr., 249 ... D6 103
Edison, Middlesex, N.J., 11,000 ... *B4 106
Edison, Morrow, Ohio, 559 ... B5 111
Edison, Skagit, Wash., 150 ... A3 122
Edisto, isl., S.C. ... F7 115
Edisto, riv., S.C. ... E6 115
Edisto Island, Charleston, S.C., 240 ... F7, G1 115
Edith, Woods, Okla ... A2 112
Edithburgh, Austl., 477 ... G1 51
Edith Ronne Land, reg., Ant ... B6 5
Edmeston, Otsego, N.Y., 600 ... C5 108
Edmond, Norton, Kans., 91 ... C4 93
Edmond, Oklahoma, Okla., 8,577 ... B4 112
Edmonds, Snohomish, Wash., 8,016 ... B3 122
Edmondson, Crittenden, Ark., 288 ... B5 81
Edmonson, co., Ky., 8,085 ... C3 94
Edmonston, Prince Georges, Md., 1,197 ... *C1 85
Edmonton, Alta., Can., 281,027 (★337,568) ... C4 69
Edmonton, Eng., 92,062 ... k12 10
Edmonton, Metcalfe, Ky., 749 ... C4 94
Edmore, Montcalm, Mich., 1,234 ... E5 98
Edmore, Ramsey, N. Dak., 405 ... A7 110
Edmund, lake, Man., Can ... B5 71
Edmunds, Washington, Maine, 35 ... D5 96
Edmunds, Stutsman, N. Dak., 50 ... B7 110
Edmunds, co., S. Dak., 6,079 ... B6 116
Edmundson, St. Louis, Mo., 1,428 ... *C7 101
Edmundston, N.B., Can., 12,791 ... B1 74
Edna, Labette, Kans., 442 ... E8 93
Edna, Jackson, Tex., 5,038 ... E4 118
Edna, Monongalia, W. Va., 150 ... A7 123
Edon, Williams, Ohio, 757 ... A3 111
Edouard, lake, Que., Can ... B5 75
Edremit, Tur., 22,200 ... C6 23
Edremit, gulf, Tur ... C6 23
Edsel Ford, ranges, Ant ... B34 5
Edson, Alta., Can., 3,198 ... C2 69
Edson, Sherman, Kans., 64 ... C2 93
Eduardo Castex, Arg., 4,020 ... B4 54
Eduni, mtn., N.W. Ter., Can ... D7 66
Edwall, Lincoln, Wash., 165 ... B8 122
Edwand, Alta., Can., 125 ... B4 69
Edward, lake, Con. L. ... B4 48
Edward VII, pen., Ant ... B33 5
Edward VIII, bay, Ant ... C18 5
Edwards, Ont., Can., 61 ... B10 72
Edwards, Eagle, Colo., 10 ... B4 83
Edwards, Peoria, Ill., 175 ... C4 90
Edwards, Hinds, Miss., 1,206 ... C3 100
Edwards, St. Lawrence, N.Y., 658 ... A5, B2 108
Edwards, co., Ill., 7,940 ... E5 90
Edwards, co., Kans., 5,118 ... E4 93
Edwards, co., Tex., 2,317 ... E2 118
Edwards, butte, Oreg ... B3 113
Edwards, plat., Tex ... D2 118
Edwardsburg, Cass, Mich., 902 ... G4 98
Edwardsport, Knox, Ind., 533 ... G3 91
Edwardsville, Cleburne, Ala., 168 ... B4 78
Edwardsville, Madison, Ill., 9,996 ... E4 90
Edwardsville, Wyandotte, Kans., 513 ... B8 93
Edwardsville, Luzerne, Pa., 5,711 ... B8 114
Edwight, Raleigh, W. Va., 70 ... D6 123
Edwin, Henry, Ala., 100 ... D4 78
Edzell, Scot., 946 ... D6 13
Eek, Alsk., 200 ... C7 79
Eekloo, Bel., 18,510 ... C3 15
Eel, riv., Calif ... B2 82
Eel, riv., Ind ... C5 91
Eel, riv., Ind ... F3 91
Eferding, Aus., 3,151 ... D7 16
Effie, Avoyelles, La., 100 ... C3 95
Effie, Itasca, Minn., 195 ... C5 99
Effingham, Effingham, Ill., 8,172 ... D5 90
Effingham, Atchison, Kans., 564 ... C8, A7 93
Effingham, Carroll, N.H., 40 (329▲) ... C4 105
Effingham, Florence, S.C., 100 ... C8 115
Effingham, co., Ga., 10,144 ... D5 87
Effingham, co., Ill., 23,107 ... D5 90
Efland, Orange, N.C., 500 ... A4 109
Eforie, Rom., 428 ... C9 22
Ega, riv., Sp ... A4 20
Egadi, is., It ... F4 21
Egan, Fulton, Ga. (part of East Point) ... B5 87
Egan, Moody, S. Dak., 310 ... D9 116
Egan, range, Nev ... E7 104
Eganville, Ont., Can., 1,549 ... B7 72
Egbert, Laramie, Wyo ... D9 125
Egegik, Alsk., 150 ... D8 79
Egeland, Towner, N. Dak., 190 ... A6 110
Egeln, Ger., 7,650 ... B6 17
Eger, Hung., 35,375 ... B5 22
Egersund, Nor., 3,900 ... H2 25
Egg, lake, Que., Can ... F18 67
Egg, lake, Sask., Can ... B3 70
Eggbornsville, Culpeper, Va ... C4 121
Eggebek, Ger., 1,850 ... D3 24
Eggenfelden, Ger., 5,800 ... E7 17
Egg Harbor, Door, Wis., 150 ... C6 124
Egg Harbor City, Atlantic, N.J., 4,416 ... D3 106
Egg Island, pt., N.J. ... E2 106
Egilsstadir, Ice., 83 ... n25 25
Égletons, Fr., 3,201 ... E4 14
Eglisau, Switz., 1,911 ... A6 19
Egmond aan Zee, Neth., 4,200 ... B4 15
Egmont, bay, P.E.I., Can ... C5 74
Egmont, cape, N.Z. ... M14 51
Egmont, mtn., N.Z. ... M15 51
Egnar, San Miguel, Colo., 15 ... D2 83
Egremont, Alta., Can., 119 ... B4 69
Egremont, Eng., 6,213 ... F5 13
Egremont, Berkshire, Mass., 100 (895▲) ... B1 97
Egridir, Tur., 7,100 ... D8 31
Egridir, lake, Tur ... C8 31
Egtved, Den., 1,012 ... C3 24
Egypt, Craighead, Ark., 225 ... B5 81
Egypt, Effingham, Ga., 100 ... D5 87
Egypt, Plymouth, Mass., 800 ... B6, D4 97
Egypt, Chickasaw, Miss., 110 ... B5 100
Egypt, Lehigh, Pa., 1,500 ... E10 114
Egypt, see United Arab Republic, country, Afr.
Ehime, pref., Jap., 1,500,687 ... *J6 37
Eholt, B.C., Can., 60 ... E8 68
Ehrenberg, Yuma, Ariz., 250 ... D1 80
Ehrenburg, Ger., 160 ... F2 24
Ehrenfeld, Cambria, Pa., 566 ... F4 114
Ehrhardt, Bamberg, S.C., 482 ... E5 115
Éibar, Sp., 31,725 ... A4 20
Eibau, Ger., 5,394 ... C9 17
Eibenstock, Ger., 8,250 ... C7 17
Eichstätt, Ger., 10,600 ... E6 17
Eider, riv., Ger ... A4 16
Eidsberg, Nor., 233 ... H4, p29 25
Eidson, Hawkins, Tenn., 30 ... C10 117
Eidsvold, Austl., 487 ... B8 51
Eidsvoll, Nor ... G4 25
Eifel, mts., Ger ... C1 17
Eigg, isl., Scot ... D2 13
Eightmile, Morrow, Oreg ... B7 113
Eights, coast, Ant ... B3 5
Eighty Eight, Barren, Ky., 150 ... D4 94
Eighty Mile, beach, Austl ... C3 50
Eikeren, lake, Nor ... p27 25
Eil, lake, Scot ... D3 13
Eildon, res., Austl ... H6 51
Eilenburg, Ger., 19,400 ... B7 17
Eil Malk, isl., Palau Is ... 52
Einbeck, Ger., 18,600 ... B4 17
Eindhoven, Neth., 168,900 (★245,000) ... C5 15
Ein Gev, Isr., 494 ... B7 32
Ein Harod, Isr., 1,478 ... B7 32
Ein Kerem, Isr. (part of Jerusalem) ... h11 32
Ein Netafim, Isr ... E6 32
Ein Scilleme, mts., Libya ... E4 43
Einsiedeln, Switz., 8,792 ... B6 19
Eirunepé, Braz., 3,023 ... C4 58
Eisenach, Ger., 48,100 ... C5 17
Eisenberg, Ger., 13,800 ... C6 17
Eisenhower, Dwight D., lock, N.Y. ... A6 108
Eisenhüttenstadt, Ger., 24,400 ... A9 17
Eisfeld, Ger., 5,586 ... C5 17
Eisleben, Ger., 34,500 ... B6 17
Eita, isl., Tarawa ... 52
Eitorf, Ger., 12,800 ... C2 17
Eitzen, Houston, Minn., 181 ... G7 99
Eix, Fr., 163 ... E5 15
Ejby, Den., 1,309 ... C3 24
Ejea de los Caballeros, Sp., 8,438 ... A5 20
Ejeda, Malag ... h8 49
Ejstrup, Den., 913 ... C3 24
Ejutla de Crespo, Mex., 4,288 ... D5 63
Ekalaka, Carter, Mont., 738 ... E12 102
Ekenäs, Fin., 5,500 ... G10 25
Ekeren, Bel., 21,452 ... C4 15
Ekhinos, Grc., 4,005 ... B5 23
Ekimchan, Sov. Un., 700 ... D16 28
Eklutna, Alsk., 50 ... C10, g17 79
Ekoln, lake, Swe ... t35 25
Ekron, Meade, Ky., 205 ... C3 94
Eksjö, Swe., 10,100 ... I6 25
Ekumaketo, Con. L ... B3 48
Ekwok, Alsk., 105 ... D8 79
El-Abiod Sidi Cheikh, Alg ... H7 30
Elaine, Phillips, Ark., 898 ... C5 81
El Alia, Alg ... C6 44
El Alto, P.R ... B5 65
Elamanchili, India, 13,556 ... I9 40
El Amparo, Cuba ... k11 64
Elamsville, Patrick, Va ... E2 121
Elamton, Morgan, Ky., 200 ... C6 94
Elamville, Barbour, Ala., 140 ... D4 78
Eland, Shawano, Wis., 213 ... D4 124
El Arahal, Sp., 17,361 ... D3 20
El Aricha, Alg ... G6 30
Elasson, Grc., 5,658 ... C4 23
El Atate, mtn., Jordan ... D7 32
El 'Atrun (Oasis), Sud ... B2 47
Elath (Eilat), Isr ... E6 32
El Auja, Isr ... D6 32
Elazig, Tur., 60,400 ... C12 31
El Azucar, res., Mex ... F3 118
Elba, Coffee, Ala., 4,321 ... D3 78
Elba, Van Buren, Ark., 11 ... B3 81
Elba, Cassia, Idaho, 70 ... G5 89
Elba, Lapeer, Mich ... E7 98
Elba, Winona, Minn., 152 ... F6 99
Elba, Howard, Nebr., 184 ... C7 103
Elba, Genesee, N.Y., 739 ... B2 108
Elba, dam, Ala ... D3 78
Elba, isl., It ... C3 21
El Barco, Sp., 7,041 ... A2 20
Elbasan, Alb., 26,000 ... B3 23
Elbasan, pref., Alb., 115,000 ... *B3 23
El Baúl, Ven., 1,522 ... B4 60
Elbe, riv., Ger ... B8 17
Elberfeld, Warrick, Ind., 485 ... H3 91
Elberon, Tama, Iowa, 211 ... B5 92
Elberon, Monmouth, N.J. (part of Long Beach) ... C5 106
Elbert, Elbert, Colo., 250 ... B6 83
Elbert, Throckmorton, Tex., 50 ... C3 118
Elbert, McDowell, W. Va., 950 ... *D3 123
Elbert, co., Colo., 3,708 ... B6 83
Elbert, co., Ga., 17,835 ... B4 87
Elbert, mtn., Colo ... B4 83
Elberta, Baldwin, Ala., 384 ... E2 78
Elberta, Houston, Ga., 644 ... D3 87
Elberta, Benzie, Mich., 552 ... D4 98
Elberta, Utah, Utah, 50 ... D4 119
Elberton, Elbert, Ga., 7,107 ... B4 87
Elberton, Whitman, Wash., 66 ... C8 122
Elbeuf, Fr., 18,988 ... C4 14
Elbigenalp, Aus., 589 ... B6 18
Elbing, Butler, Kans., 105 ... A6 93
Elbing, see Elblag, Pol.
Elbistan, Tur., 10,300 ... C11 31
Elblag, Pol. 77,000 ... A5 26
El Bolsón, Arg ... C2 54
El Bonillo, Sp., 5,286 ... C4 20
El Bordo, Col., 1,475 ... C2 60
El Borouj, Mor., 3,955 ... H3 30
Elbow, Sask., Can., 401 ... F2 70
Elbow, lake, Man., Can ... B1 71
Elbow, riv., Alta., Can ... D3 69
Elbow Lake, Grant, Minn., 1,521 ... E3 99
Elbridge, Obion, Tenn., 65 ... A2 117
Elbrus, mtn., Sov. Un ... G17 9
El Bur, Som., 2,300 ... E6 47
Elburg, Neth., 3,600 ... B5 15
El Burgo de Osma, Sp., 3,842 ... B4 20
Elburn, Kane, Ill., 960 ... B5 90
Elburz, mts., Iran ... C5 41
El Cajon, San Diego, Calif., 37,618 ... E2, F5 82
El Callao, Ven., 3,276 ... B5 60
El Campo, Wharton, Tex., 7,700 ... E4 118
El Campo North, Wharton, Tex., 1,086 ... *E4 118
El Campo South, Wharton, Tex., 1,884 ... *E4 118
El Capitan, mtn., Mont ... D2 102
El Capitan, res., Calif ... E2 82
El Carmen, Bol ... C4 55
El Carmen, Col., 9,647 ... B2 60
El Carre, Eth ... D5 47
El Carrizo, Mex., 163 ... B3 63
El Centro, Imperial, Calif., 18,340 ... F6 82
El Cerrito, Contra Costa, Calif., 25,437 ... B5 82
El Cerro, Bol., 117 ... C3 55
El Cerro, P.R ... C2 65
El Chaparro, Ven., 913 ... B4 60
Elche, Sp., 45,500 (73,320▲) ... C5 20
Elcho, Langlade, Wis., 525 ... C4 124
Elco, Alexander, Ill., 150 ... F4 90
El Cobre, Cuba ... k12 64
El Cuervo, butte, N. Mex ... G6 107
Eld, isl., Fiji Is ... 52
Elda, Sp., 28,151 ... C5 20
El Dar el Beda, ruins, Eg., U.A.R ... D3 32
Elde, riv., Ger ... B5 16
Elderon, Marathon, Wis., 177 ... D4 124
Eldersley, Sask., Can., 58 ... E4 70
Elderton, Armstrong, Pa., 387 ... E3 114
El Diviso, Col ... C2 60
El Djem, Tun., 5,122 ... G12 30
Eldon, Wapello, Iowa, 1,386 ... D5 92
Eldon, Miller, Mo., 3,158 ... C5 101
Eldora, Hardin, Iowa, 3,225 ... B4 92
Eldora, Cape May, N.J ... E3 106
El Dorado, Union, Ark., 25,292 ... D3 81
Eldorado, Braz., 1,524 ... C3 56
Eldorado, Ont., Can., 80 ... C7 72
Eldorado, Saline, Ill., 3,573 ... F5 90
El Dorado, Butler, Kans., 12,523 ... B6, E7 93
Eldorado, Dorchester, Md., 80 ... C6 85
Eldorado, Montgomery, N.C., 150 ... B4 109
Eldorado, Preble, Ohio, 449 ... C3 111
Eldorado, Jackson, Okla., 708 ... C2 112
Eldorado, Schleicher, Tex., 1,815 ... D2 118
El Dorado, co., Calif., 29,390 ... C3 82
Eldorado, mtn., Wash ... A4 122
Eldorado Springs, Boulder, Colo., 150 ... B5 83
El Dorado Springs, Cedar, Mo., 2,864 ... D3 101
Eldoret, Ken., 8,193 ... A6 48
Eldred, St. Lucie, Fla., 100 ... E6 86
Eldred, Greene, Ill., 302 ... D3 90
Eldred, Polk, Minn., 30 ... C2 99
Eldred, McKean, Pa., 1,107 ... C5 114
Eldridge, Walker, Ala., 350 ... B2 78
Eldridge, Scott, Iowa, 583 ... C7, D7 92
Eldridge, Laclede, Mo., 100 ... D5 101
Eldridge, Stutsman, N. Dak., 67 ... C7 110
Eleanor, Putnam, W. Va., 700 ... C3 123
Electra, Wichita, Tex., 4,759 ... B3 118
Electra, lake, Colo ... D3 83
Electric, peak, Mont., Wyo ... E6 102, A2 125
Electric City, Grant, Wash., 404 ... B6 122
Electric Mills, Kemper, Miss., 100 ... C5 100
Eleele, Kauai, Haw., 950 ... B2 88
El Ejemplo Central, P.R ... C7 65
Elektrogorsk, Sov. Un., 18,900 ... n18 27
Elektrostal, Sov. Un., 102,000 ... n18 27
Elenora, mtn., Utah ... F2 119
Elephant, butte, N. Mex ... D2 107
Elephant, isl., Atl. O ... C7 5
Elephant, mtn., Maine ... D2 96
Elephant, range, Camb ... G5 38
Elephant Butte, res., N. Mex ... D2 107
El Escorial, Sp., 3,763 ... o16 20
El Espino, P.R ... B2 65
Eleuthera, isl., Ba. Is ... C5 64
Eleva, Trempealeau, Wis., 548 ... D2 124
Eleven Mile Canyon, res., Colo ... C5 83
Eleven Point, riv., Mo ... F6 101
Elevsís (Eleusis), Grc., 15,527 ... g11 23
Elevsís, gulf, Grc ... g11 23
El Faiyum (Al Fayyūm), Eg., U.A.R., 84,800 ... E2 32
El Faro, P.R ... C3 65
El Fasher, Sud., 26,161 ... C2 47
El Ferrol, Sp., 74,799 ... A1 20
Elfers, Pasco, Fla., 450 ... D4 86
Elfin Cove, Alsk., 20 ... k21 79
Elfrida, Cochise, Ariz., 200 ... F6 80
Elfros, Sask., Can., 289 ... F4 70
El Fud, Eth ... D5 47
El Fuerte, Mex., 5,183 ... B3 63
El Galhak, Sud ... C3 47
El Geteina, Sud ... C3 47
Elgin, Lauderdale, Ala., 150 ... A2 78
Elgin, Santa Cruz, Ariz., 25 ... F5 80
Elgin, Man., Can., 259 ... E1 71
Elgin, N.B., Can., 315 ... D4 74
Elgin, Cook and Kane, Ill., 49,447 ... A5, E2 90
Elgin, Fayette, Iowa, 644 ... B6 92
Elgin, Chautauqua, Kans., 148 ... E7 93
Elgin, Wabasha, Minn., 521 ... F6 99
Elgin, Antelope, Nebr., 881 ... C7 103
Elgin, Lincoln, Nev., 25 ... F7 104
Elgin, Grant, N. Dak., 944 ... C4 110
Elgin, Comanche, Okla., 540 ... C3 112
Elgin, Union, Oreg., 1,315 ... B7 113
Elgin, Erie, Pa., 218 ... C2 114
Elgin, Scot., 11,971 ... C5 13
Elgin, Lancaster, S.C., 300 ... B6 115
Elgin, Bastrop, Tex., 3,511 ... D4 118
Elgin, co., Ont., Can., 62,862 ... E3 72
Elgin Mills, Ont., Can., 781 ... E6 72
El Gîza (Al Jîzah) (Gizeh), Eg., U.A.R., 98,000 ... C6 43
Elgol, Scot ... C2 13
El Goléa, Alg., 7,452 (12,486▲) ... C5 44
Elgon, mtn., Ug ... A5 48
El Grullo, Mex., 9,028 ... n11 63
El Hamurre, Som ... D6 47
El Haseke, Syr., 18,900 ... D13 31
El Hasiheisa, Sud., 6,600 ... C3 47
El Hawata, Sud., 3,921 ... C3 47
El Huecu, Arg ... B2 54
El Huerfanito, peak, N. Mex ... A2 107
Eli, Cherry, Nebr., 15 ... B4 103
Eliasville, Young, Tex., 750 ... C3 118
Elida, Roosevelt, N. Mex., 534 ... E6 107
Elida, Allen, Ohio, 1,215 ... B3 111
Elikon (Helicon), mtn., Grc ... g9 23
Elila, riv., Con. L ... B4 48
Elim, Alsk., 145 ... C7 79
Elim, Cambria, Pa., 1,000 ... *F4 114
El Indio, Maverick, Tex., 120 ... E2 118
Eliot, York, Maine, 1,730 (3,133▲) ... E2 96
Elis (Ilia), prov., Grc., 188,861 ... *D3 23
Elisabethville, Con. L., 183,000 ... D4 48
Elista, Sov. Un., 29,000 ... D2 29
Elizabeth, Fulton, Ark., 20 ... A3 81
Elizabeth, Elbert, Colo., 326 ... B6 83
Elizabeth, Cobb, Ga., 1,620 ... A4, C2 87
Elizabeth, Jo Daviess, Ill., 729 ... A3 90
Elizabeth, Harrison, Ind., 214 ... H6 91
Elizabeth, Allen, La., 1,030 ... D3 95
Elizabeth, Otter Tail, Minn., 168 ... D2 99
Elizabeth, Washington, Miss., 300 ... B3 100
Elizabeth, Union, N.J., 107,698 ... B4, E4 106
Elizabeth, Alleghany, Pa., 2,597 ... *F2 114
Elizabeth, Wirt, W. Va., 727 ... B3 123
Elizabeth, bay, Ec ... g5 58
Elizabeth, cape, Maine ... E5 96
Elizabeth, cape, Wash ... B1 122
Elizabeth, is., Mass ... D6 97
Elizabeth, mtn., N.B., Can ... B3 74
Elizabeth City, Pasquotank, N.C., 14,062 ... A7 109
Elizabeth Lake Estates, Oakland, Mich., 2,000 ... *F7 98
Elizabethton, Carter, Tenn., 10,896 ... C11 117
Elizabethtown, Hardin, Ill., 524 ... F5 90
Elizabethtown, Bartholomew, Ind., 417 ... F6 91
Elizabethtown, Hardin, Ky., 9,641 ... C4 94
Elizabethtown, Essex, N.Y., 779 ... A7, B3 108
Elizabethtown, Bladen, N.C., 1,625 ... C5 109
Elizabethtown, Lancaster, Pa., 6,780 ... F8 114
Elizabethville, Dauphin, Pa., 1,455 ... E8 114
Elizondo, see Baztan, Sp.
El Jebelein, Sud ... C3 47
El Jeib, wadi, Jordan ... D7 32
Elk, Mendocino, Calif., 175 ... C2 82
Elk, Pol., 16,900 ... B7 26
Elk, Spokane, Wash., 75 ... A8 122
Elk, Teton, Wyo., 15 ... B2 125
Elk, co., Kans., 5,048 ... E7 93
Elk, co., Pa., 37,328 ... D4 114
Elk, creek, Okla ... B2 112
Elk, creek, S Dak ... C3 116
Elk, isl., Man., Can ... D3 71
Elk, mts., Colo ... B3 83
Elk, mtn., N. Mex ... D1 107
Elk, mtn., Okla ... C3 112
Elk, mtn., S Dak ... D1 116
Elk, mtn., Wyo ... D6 125
Elk, riv., B.C., Can ... D10 68
Elk, riv., Colo ... A4 83
Elk, riv., Kans ... E7 93
Elk, riv., Md ... A6 85
Elk, riv., Tenn ... B5 117
Elk, riv., W. Va ... C3 123
Elk, riv., Wis ... C3 124
Elkader, Clayton, Iowa, 1,526 ... B6 92
El Kamlin, Sud., 4,341 ... B3 47
Elkatawa, Breathitt, Ky., 200 ... C6 94
Elk Basin, Park, Wyo ... A4 125
Elk City, Idaho, Idaho, 300 ... D3 89
Elk City, Montgomery, Kans., 498 ... E8 93
Elk City, Douglas, Nebr., 48 ... D2 103
Elk City, Beckham, Okla., 8,196 ... B2 112
Elk Creek, Glenn, Calif., 300 ... C2 82
Elk Creek, Johnson, Nebr., 170 ... D9 103
Elk Creek, Grayson, Va., 100 ... E1 121
Elk Falls, Elk, Kans., 179 ... E7 93
Elk Garden, Mineral, W. Va., 329 ... B5 123
Elk Grove, Sacramento, Calif., 2,205 ... A6, C3 82
Elk Grove Village, Cook, Ill., 4,771 ... *A6 90
El Khandaq, Sud., 11,995 ... B3 47
Elkhart, Logan, Ill., 418 ... C4 90
Elkhart, Elkhart, Ind., 40,274 ... A6 91
Elkhart, Polk, Iowa, 260 ... A7 92
Elkhart, Morton, Kans., 1,780 ... E2 93
Elkhart, Anderson, Tex., 780 ... D5 118
Elkhart, co., Ind., 106,790 ... A6 91
Elkhart Lake, Sheboygan, Wis., 651 ... B5, E5 124
Elk Head, mts., Colo ... A3 83
Elkhorn, Man., Can., 666 ... E1 71
Elk Horn, Shelby, Iowa, 679 ... C2 92
Elkhorn, Douglas, Nebr., 749 ... D2 103
Elkhorn, McDowell, W. Va., 900 ... *D3 123
Elkhorn, Walworth, Wis., 3,586 ... F5 124
Elkhorn, peaks, Idaho ... F7 89
Elkhorn, ridge, Oreg ... C8 113
Elkhorn, riv., Nebr ... B7 103
Elkhorn City, Pike, Ky., 1,085 ... C7 94
Elkhovo, Bul., 6,749 ... D8 22
Elkin, Surry, N.C., 2,868 ... A3 109
Elkins, Washington, Ark., 250 ... B1 81
Elkins, Merrimack, N.H., 165 ... D3 105
Elkins, Chaves, N. Mex., 42 ... D5 107
Elkins, Randolph, W. Va., 8,307 ... C5 123
Elkins Park, Montgomery, Pa., 12,000 ... A11 114
Elkinsville, Brown, Ind ... F5 91
Elk Island, nat. park, Alta., Can ... C4 69
Elk Lake, Deschutes, Oreg ... D5 113
Elkland, Webster, Mo., 200 ... D4 101
Elkland, Tioga, Pa., 2,189 ... C7 114
Elk Mills, Cecil, Md., 500 ... A6 85
Elkmont, Limestone, Ala., 169 ... A3 78
Elkmont, Sevier, Tenn., 50 ... D10 117
Elk Mound, Dunn, Wis., 379 ... D2 124
Elk Mountain, Buncombe, N.C., 300 ... D3 109
Elk Mountain, Carbon, Wyo., 190 ... D6 125
Elko, B.C., Can., 158 ... E10 68
Elko, Houston, Ga., 165 ... D3 87
Elko, Elko, Nev., 6,298 ... C6 104
Elko, Barnwell, S.C., 194 ... E5 115
Elko, co., Nev., 12,011 ... B6 104
Elkol, Lincoln, Wyo., 25 ... D2 125
Elk Park, Avery, N.C., 460 ... A2, C4 109
Elk Point, Alta., Can., 692 ... C5 69
Elk Point, Union, S. Dak., 1,378 ... E9 116
Elk Ranch, Carroll, Ark., 35 ... A2 81
Elk Rapids, Antrim, Mich., 1,015 ... D5 98
Elkridge, Howard, Md., 2,000 ... B4, C2 85
Elkridge, Fayette, W. Va., 400 ... D6 123
Elk River, Clearwater, Idaho, 382 ... C2 89
Elk River, Sherburne, Minn., 1,763 ... E5 99
Elk Run Heights, Black Hawk, Iowa, 1,124 ... *B5 92
Elk Springs, Moffat, Colo., 6 ... A2 83
Elkton, St. Johns, Fla., 400 ... C5 86
Elkton, Todd, Ky., 1,448 ... D2 94
Elkton, Cecil, Md., 5,989 ... A6 85
Elkton, Huron, Mich., 1,014 ... E7 98
Elkton, Mower, Minn., 147 ... G6 99
Elkton, Douglas, Oreg., 146 ... D3 113
Elkton, Brookings, S. Dak., 621 ... C9 116
Elkton, Giles, Tenn., 199 ... B5 117
Elkton, Rockingham, Va., 1,506 ... C4 121
Elk Valley, Campbell, Tenn., 250 ... C9 117
Elkville, Jackson, Ill., 743 ... F4 90
Ellabell, Bryan, Ga., 175 ... D5 87
Ellamar, Alsk., 20 ... g18 79
Ellamore, Randolph, W. Va., 450 ... C4 123
Ellaville, Schley, Ga., 905 ... D2 87
Ellef Ringnes, isl., N.W. Ter., Can ... B25 4
Ellen, mtn., Utah ... E3 119
Ellen, mtn., Vt ... C3 120
Ellenboro, Rutherford, N.C., 492 ... B2 109
Ellenboro, Ritchie, W. Va., 340 ... B3 123
Ellenburg Center, Clinton, N.Y., 250 ... A3 108
Ellendale, Sussex, Del., 370 ... C7 85
Ellendale, Terrebonne, La., 300 ... C6, E5 95
Ellendale, Steele, Minn., 501 ... G5 99
Ellendale, Dickey, N. Dak., 1,800 ... C7 110
Ellendale, Shelby, Tenn., 1,000 ... *B1 117
Ellensburg, Kittitas, Wash., 8,625 ... C5 122
Ellenton, Manatee, Fla., 950 ... E4, F2 86
Ellenton, Colquitt, Ga., 385 ... E3 87
Ellenton, Lycoming, Pa., 40 ... C8 114
Ellenton, Aiken, S.C ... E4 115
Ellenville, Ulster, N.Y., 5,003 ... D6 108
Ellenwood, Clayton, Ga., 220 ... B5 87
Eller, isl., Kwajalein ... 52
Ellerbe, Richmond, N.C., 843 ... B4 109
Ellershouse, N.S., Can., 397 ... E6 74
Ellerslie, Harris, Ga., 350 ... D2 87
Ellerslie, Allegany, Md., 560 ... D2 85
Ellesmere, isl., N.W. Ter., Can ... A22 4
Ellersmere Port, Eng., 44,714 ... A5 12
Ellettsville, Monroe, Ind., 1,222 ... F4 91
Ellice, is., Gilbert & Ellice Is ... G10 7
Ellichpur, India, 36,538 ... G6 40
Ellicott City, Howard and Baltimore, Md., 2,000 ... B4 85
Ellicottville, Cattaraugus, N.Y., 1,150 ... C2 108
Ellijay, Gilmer, Ga., 1,320 ... B2 87
Ellijoy, Macon, N.C ... D3 109
Ellington, Tolland, Conn., 400 (5,580▲) ... B7 84
Ellington, Reynolds, Mo., 812 ... D7 101
Ellinwood, Barton, Kans., 2,729 ... D5 93
Elliot, Windham, Conn ... B8 84
Elliot Lake, Ont., Can., 9,950 ... A2 72
Elliott, Ouachita, Ark., 100 ... D3 81
Elliott, Ford, Ill., 343 ... C5 90
Elliott, Montgomery, Iowa, 459 ... C2 92
Elliott, Dorchester, Md., 150 ... D5 85
Elliott, Grenada, Miss., 250 ... B4 100
Elliott, Ransom, N. Dak., 62 ... C8 110
Elliott, Lee, S.C., 270 ... C7 115
Elliott, co., Ky., 6,330 ... B6 94
Elliott, bay, Wash ... D1 122
Elliott, key, Fla ... G6 86
Elliott, lake, Man., Can ... C4 71
Ellis, Baxter, Ark., 250 ... A3 81
Ellis, Lemhi, Idaho ... E4 89
Ellis, Ellis, Kans., 2,218 ... D4 93
Ellis, Gage, Nebr., 75 ... D9 103
Ellis, Minnehaha, S. Dak., 75 ... D9 116
Ellis, co., Kans., 21,270 ... D4 93
Ellis, co., Okla., 5,457 ... A2 112
Ellis, co., Tex., 43,395 ... C4 118
Ellis, pond, Maine ... D2 96
Ellis, riv., N.H ... B4 105
Ellisburg, Jefferson, N.Y., 328 ... B4 108
Ellisburg, Potter, Pa., 40 ... C6 114
Ellisdale, Monmouth, N.J., 50 ... C4 106
Ellisgrove, Randolph, Ill., 218 ... E4 90
Ellison Bay, Door, Wis., 150 ... C5 124
Elliston, Austl., 127 ... F5 50
Elliston, Newf., Can., 678 ... D5 75
Elliston, Powell, Mont., 200 ... D4 102
Elliston, Montgomery, Va., 600 ... D2 121
Ellisville, Jones, Miss., 4,592 ... D4 100
Ellisville, St. Louis, Mo., 2,732 ... B7 101
Ellon, Scot., 1,456 ... C6 13
Ellore, India ... E7 39, I8 40
Elloree, Orangeburg, S.C., 1,031 ... D6 115
Ellport, Lawrence, Pa., 1,458 ... E1 114
Ellrich, Ger., 5,302 ... B5 17
Ells, riv., Alta., Can ... A4 69
Ellscott, Alta., Can., 15 ... B4 69
Ellsinore, Carter, Mo., 311 ... E7 101
Ellston, Ringgold, Iowa, 116 ... D3 92
Ellsworth, McLean, Ill., 224 ... C5 90
Ellsworth, Hamilton, Iowa, 493 ... B4 92
Ellsworth, Ellsworth, Kans., 2,361 ... D5 93
Ellsworth, Hancock, Maine, 4,444 ... D4 96
Ellsworth, Antrim, Mich., 386 ... C5 98
Ellsworth, Nobles, Minn., 634 ... G3 99
Ellsworth, Sheridan, Nebr., 11 ... B3 103
Ellsworth, Grafton, N.H., (3▲) ... C3 105
Ellsworth, Washington, Pa., 1,456 ... F1 114
Ellsworth, Pierce, Wis., 1,701 ... D1 124
Ellsworth, co., Kans., 7,677 ... D5 93
Ellsworth, highland, Ant ... B2 5
Ellsworth, mtn., Conn ... B3 84
Ellwangen, Ger., 12,500 ... E5 17
Ellwood City, Beaver and Lawrence, Pa., 12,413 ... E1 114
Ellzey, Levy, Fla., 25 ... C4 86
Elm, Camden, N.J ... D3 106
Elm, creek, Minn ... G4 99
Elm, creek, Tex ... A5 118
Elma, Howard, Iowa, 706 ... A5 92
Elma, Erie, N.Y., 4,000 ... *C2 108
Elma, Grays Harbor, Wash., 1,811 ... B2 122
Elmali, Tur., 7,800 ... D7 23
El Manteco, Ven., 899 ... B5 60
Elm City, Wilson, N.C., 729 ... B6 109
Elm Creek, Man., Can., 337 ... E3 71
Elm Creek, Buffalo, Nebr., 778 ... D6 103
Elmcrest, Genesee, Mich., 1,000 ... *E7 98
Elmdale, Chase, Kans., 114 ... D7 93
El Melik, riv., Sud ... B2 47
El Memrhar, Maur ... C1 45
Elmendorf, Socorro, N. Mex ... D3 107
Elmendorf, Bexar, Tex., 500 ... B4 118
El Mene, Ven., 3,160 ... A3 60
Elmer, Macon, Mo., 266 ... B5 101
Elmer, Salem, N.J., 1,505 ... D2 106
Elmer, Jackson, Okla., 120 ... C2 112
Elmer City, Okanogan, Wash., 265 ... B7 122
El Mesellemiya, Sud., 3,131 ... C3 47
Elm Grove, Waukesha, Wis., 4,994 ... *E5 124
Elmhurst, DuPage, Ill., 40,329 ... B6, F2 90
Elmhurst, Lakawanna, Pa., 788 ... A9, D10 114
El Minao, P.R ... B6 65
Elmira, Solano, Calif., 200 ... B6 82
Elmira, P.E.I., Can., 149 ... C7 74
Elmira, Ont., Can., 3,337 ... D4 72
Elmira, Bonner, Idaho, 20 ... A2 89
Elmira, Otsego, Mich., 145 ... C6 98
Elmira, Ray, Mo., 123 ... B3 101
Elmira, Chemung, N.Y., 46,517 ... C4 108
Elmira, Lane, Oreg., 500 ... C3 113
Elmira, Braxton, W. Va., 74 ... C4 123
Elmira Heights, Chemung, N.Y., 5,157 ... C4 108
El Mirage, Maricopa, Ariz., 1,723 ... G1 80
Elmo, Dickinson, Kans., 50 ... D6 93
Elmo, Nodaway, Mo., 213 ... A2 101
Elmo, Lake, Mont., 100 ... C2 102
Elmo, Emery, Utah, 175 ... D5 119
Elmo, Carbon, Wyo., 91 ... D6 125
Elmodel, Baker, Ga., 100 ... E2 87
El Modeno, Orange, Calif., 1,000 ... *F5 82
Elmont, Shawnee, Kans., 75 ... B8 93
Elmont, Nassau, N.Y., 42,000 ... *G2 84
Elmont, Hanover, Va., 150 ... D5 121
El Monte, Contra Costa, Calif., 4,186 ... *B5 82
El Monte, Los Angeles, Calif., 13,163 ... F3 82
Elmora (Bakerton), Cambria, Pa., 1,057 ... E4 114
Elmore, Elmore, Ala., 200 ... C3 78
Elmore, Faribault, Minn., 1,078 ... G4 99
Elmore, Ottawa, Ohio, 1,302 ... A4 111
Elmore, co., Ala., 30,524 ... C3 78

Elmore, co., Idaho, 16,719 ... F3 89
Elmore City, Garvin, Okla., 982 ... C4 112
El Morro, Las Animas, Colo., 20 ... D6 83
El Morro, Valencia, N. Mex., 10 ... B1 107
El Morro, nat. mon., N. Mex ... C1 107
Elm Park, West Feliciana, La ... D4 95
El Mraïti (Well), Mali ... C4 45
El Mreïti (Well), Maur ... B3 45
El Mreyer (Well), Maur ... B3 45
Elms, York, Maine ... E2 96
Elmsdale, N.S., Can., 535 ... E6 74
Elmsford, Westchester, N.Y., 3,795 ... D2 108
Elmshorn, Ger., 35,000 ... B4 16
Elm Springs, Washington, Ark., 238 ... A1 81
Elm Springs, Meade, S. Dak., 20 ... C3 116
Elm Springs Colony, Hutchinson, S. Dak., 100 ... D8 116
Elmsville, N.B., Can., 94 ... D3 74
Elmvale, Ont., Can., 957 ... C5 72
Elmwood, Ont., Can., 357 ... C3 72
Elmwood, Peoria, Ill., 1,882 ... C4 90
Elmwood, Cass, Nebr., 481 ... D9, E2 103
Elmwood, Pierce, Wis., 776 ... D1 124
Elmwood Park, Cook, Ill., 23,866 ... F2 90
Elmwood Place, Hamilton, Ohio, 3,813 ... D2 111
Elna, Johnson and Morgan, Ky., 150 ... C7 94
Elne, Fr., 4,458 ... F5 14
Elnora, Alta., Can., 214 ... D4 69
Elnora, Daviess, Ind., 824 ... G3 91
El Obeid, Sud., 52,372 ... C3 47
El Ocho, P.R ... B6 65
El Odaiya, Sud., 11,913 ... C2 47
Eloise, Polk, Fla., 3,256 ... E5 86
Eloise, Wayne, Mich ... A7 98
Elon College, Alamance, N.C., 1,284 ... A4 109
Elora, Ont., Can., 1,486 ... D4 72
Elora, Lincoln, Tenn., 300 ... B5 117
El Oro, prov., Ec., 103,950 ... B2 58
El Oued, Alg., 13,001 (86,092▲) ... C6 44
El Ouig, Mali ... C5 45
Eloy, Pinal, Ariz., 4,899 ... E4 80
El Palmito, Mex., 640 ... B3 63
El Pao, Ven ... B5 60
El Paraíso, Hond., 1,617 ... D4 62
El Pardo, Sp., 3,902 (part of Madrid) ... o17 20
El Paso, White, Ark., 120 ... B3 81
El Paso, Woodford, Ill., 1,964 ... C4 90
El Paso, El Paso, Tex., 276,687 (★599,000) ... E1 118
El Paso, co., Colo., 143,742 ... C6 83
El Paso, co., Tex., 314,070 ... E1 118
El Paso Gap, Eddy, N. Mex ... E5 107
Elphin, Ire., 494 ... D3 11
Elphinstone, Man., Can., 386 ... D1 71
El Polvorín, P.R ... B5 65
El Portal, Mariposa, Calif., 200 ... D7 82
El Portal, Dade, Fla., 2,079 ... G6 86
El Porto Beach, Los Angeles, Calif., 1,200 ... *F2 82
El Prado, Taos, N. Mex., 100 ... A4 107
El Puerto de Santa María, Sp., 35,505 ... D2 20
El Qaryatein, Syr ... E11 31
Elqui, riv., Chile ... F1 55
Elrama, Washington, Pa., 800 ... F2 114
El Real, Pan., 1,071 ... F8 62
El Reno, Canadian, Okla., 11,015 ... B4 112
El Rio, Ventura, Calif., 6,966 ... E4 82
El Rito, Rio Arriba, N. Mex., 50 ... A3 107
El Roboré, Bol., 3,715 ... C4 55
Elrod, Tuscaloosa, Ala., 200 ... B2 78
Elrosa, Stearns, Minn., 205 ... E4 99
Elrose, Sask., Can., 585 ... F1 70
Elroy, Juneau, Wis., 1,505 ... E3 124
Elsa, Hidalgo, Tex., 3,847 ... F3 118
Elsah, Jersey, Ill., 507 ... A8 101
El Salto, Mex., 5,926 ... C3 63
El Salvador, country, N.A., 2,501,278 ... H12, 55, D3 62
El Samán, Ven., 1,154 ... B4 60
El Sauce, Nic., 1,781 ... D4 62
Elsberry, Lincoln, Mo., 1,491 ... B6 101
El Segundo, Los Angeles, Calif., 14,219 ... F2 82
Elsie, Clinton, Mich., 933 ... E6 98
Elsie, Perkins, Nebr., 198 ... D4 103
Elsie, Clatsop, Oreg., 25 ... B3 113
Elsinore, Riverside, Calif., 2,432 ... F3, F5 82
Elsinore, Sevier, Utah., 483 ... E3 119
Elsinore, lake, Calif ... F3 82
Elsmere, New Castle, Del., 7,319 ... A6 85
Elsmere, Kenton, Ky., 4,607 ... A7, B5 94
Elsmere, Cherry, Nebr., 10 ... B5 103
Elsmere, Albany, N.Y. (part of Delmar) ... C7 108
Elsmore, Allen, Kans., 128 ... E8 93
El Sobrante, Contra Costa, Calif., 15,000 ... *B5 82
El Sombrero, Ven., 5,592 ... B4 60
Elsterwerda, Ger., 9,749 ... B8 17
Elstow, Sask., Can., 98 ... F2 70
Eltham, Eng. (part of London) ... *m13 10
El Tigre, Ven., 41,550 ... B5 60
El Tigrito, Ven., 19,658 ... B5 60
El Tocuyo, Ven., 14,803 ... B4 60
Elton, Jefferson Davis, La., 1,595 ... D3 95
Elton, Langlade, Wis., 50 ... C5 124
Eltopia, Franklin, Wash., 70 ... C6 122
El Toro, Orange, Calif., 300 ... F3 82
El Transito, Chile ... E1 55
El Triunfo, Hond., 893 ... D4 62
El Triunfo, Mex., 520 ... C2 63
Eltville, Ger., 6,878 ... C3 17
El Uaoh, Som ... E5 47
Elva, Man., Can., 58 ... E1 71
Elva, Scott, Tenn., 40 ... C9 117
Elvas, Port., 10,821 ... C2 20
Elvaston, Hancock, Ill., 232 ... C2 90
El Verano, Sonoma, Calif., 1,236 ... *C2 82
Elverson, Chester, Pa., 472 ... F10 114
Elverum, Nor., 3,595 ... G4 25
El Viejo, Nic., 8,900 ... D4 62
El Vigía, P.R., (part of Arecibo) 520 ... B4 65
El Vigía, Ven., 8,515 ... B3 60
Elvins, Francois, Mo., 1,818 ... D7 101
El Vista, Peoria, Ill., 2,000 ... *C4 90
El Volcán, Chile ... A2 54
El Wak, Ken ... A7 48
Elwha, riv., Wash ... A2 122
Elwood, Will, Ill., 746 ... B5 90
Elwood, Madison, Ind., 11,793 ... D6 91
Elwood, Doniphan, Kans., 1,191 ... C9 93
Elwood, Gosper, Nebr., 581 ... D6 103
Elwood, Atlantic, N.J., 400 ... D3 106
Elwood, Suffolk, N.Y., 2,000 ... *F3 84
Elwood, Box Elder, Utah, 345 ... B3 119
Elwood Park, Manatee, Fla., 450 ... E4 86
Elwyn, Delaware, Pa., 1,500 ... *G10 114
Ely, Eng., 9,815 ... B8 12
Ely, Linn, Iowa, 226 ... C6 92
Ely, St. Louis, Minn., 5,438 ... C7 99
Ely, White Pine, Nev., 4,018 ... D7 104
Ely, Orange, Vt., 40 ... D4 120
Elyria, McPherson, Kans., 60 ... D6 93
Elyria, Valley, Nebr., 89 ... C6 103
Elyria, Lorain, Ohio, 43,782 ... A5 111
Elysburg, Northumberland, Pa., 1,100 ... E8 114
Elysian, Le Sueur, Minn., 382 ... F5 99
Elzach, Ger., 2,400 ... A4 18
Emanuel, co., Ga., 17,815 ... D4 87
Emba, Sov. Un., 2,900 ... D5 29
Emba, riv., Sov. Un ... F20 9
Embarcación, Arg., 3,303 ... D3 55
Embarrass, St. Louis, Minn., 210 ... C6 99
Embarrass, Waupaca, Wis., 306 ... D5 124
Embarrass, riv., Ill ... E6 90
Embarrass, riv., Wis ... D5 124
Embden, pond, Maine ... E3 96
Embden, Cass, N. Dak., 61 ... C8 110
Embetsu, Jap., 3,000 ... D10 37
Embira, riv., Braz ... C3 58
Emblem, Big Horn, Wyo., 5 ... A4 125
Embreeville, Washington, Tenn., 50 ... C11 117
Embreeville Junction, Washington, Tenn., 1,204 ... *C11 117
Embro, Ont., Can., 552 ... D4 72
Embrun, Ont., Can., 1,112 ... B9 72
Embrun, Fr., 3,850 ... E7 14
Embu, Ken ... B6 48
Embudo, Rio Arriba, N. Mex., 10 ... A4 107
Emden, Ger., 45,700 ... B3 16
Emden, Logan, Ill., 502 ... C4 90
Emden, Shelby, Mo., 200 ... B6 101
Emelle, Sumter, Ala., 318 ... C1 78
Emerado, Grand Forks, N. Dak., 328 ... B8 110
Emerald, Austl., 2,029 ... A7 51
Emerson, Columbia, Ark., 350 ... D2 81
Emerson, Man., Can., 932 ... E3 71
Emerson, Bartow, Ga., 666 ... B2 87
Emerson, Mills, Iowa, 521 ... C2 92
Emerson, Dakota, Nebr., 803 ... B9 103
Emerson, Bergen, N.J., 6,849 ... D5 106
Emery, Hanson, S. Dak., 502 ... D8 116
Emery, Emery, Utah, 326 ... E4 119
Emery, co., Utah, 5,546 ... E4 119
Emery Mills, York, Maine, 100 ... E2 96
Emeryville, Alameda, Calif., 2,686 ... *B5 82
Emet, Tur., 4,100 ... C7 23
Emida, Benewah, Idaho, 125 ... B2 89
Emigrant, Park, Mont., 28 ... E6 102
Emigrant Gap, Placer, Calif., 50 ... C3 82
Emi Koussi, vol., Chad ... B3 46
Emil, riv., China ... F25 9
Emilia-Romagna, pol. dist., It., 3,666,680 ... E6 18
Emilia, reg., It ... B3 21
Emiliano Zapata, Mex., 7,483 ... D6 63
Emilio Meyer, Braz., 944 ... E2 56
Emily, Crow Wing, Minn., 351 ... D5 99
Emily, lake, Minn ... E3 99
Emine, cape, Bul ... D9 22
Eminence, Morgan, Ind., 200 ... E4 91
Eminence, Henry, Ky., 1,958 ... B4 94
Eminence, Shannon, Mo., 516 ... D6 101
Emington, Livingston, Ill., 133 ... C5 90
Emirau, isl., Bis. Arch ... h13 50
Emirdag, Tur., 10,000 ... C8 31
Emlenton, Venango, Pa., 844 ... D2 114
Emlyn, Whitley, Ky., 600 ... D5 94
Emma, Lafayette and Saline, Mo., 202 ... C4 101
Emma, lake, Sask., Can ... D3 70
Emmaus, Lehigh, Pa., 10,262 ... E11 114
Emme, riv., Switz ... B4 19
Emmen, Neth., 7,500 ... B6 15
Emmen, Switz., 16,856 ... B5 19
Emmendingen, Ger., 13,200 ... A3 18
Emmerich, Ger., 16,800 ... C3 16
Emmet, Nevada, Ark., 474 ... D2 81
Emmet, Holt, Nebr., 66 ... B7 103
Emmet, McLean, N. Dak., 10 ... B4 110
Emmet, co., Iowa, 14,871 ... A3 92
Emmet, co., Mich., 15,904 ... C6 98
Emmett, Gem, Idaho, 3,769 ... F2 89
Emmett, Pottawatomie, Kans., 128 ... C7 93
Emmett, St. Clair, Mich., 283 ... F8 98
Emmetsburg, Palo Alto, Iowa, 3,887 ... A3 92
Emmitsburg, Frederick, Md., 1,369 ... A3 85
Emmons, Freeborn, Minn., 408 ... G5 99
Emmons, co., N. Dak., 8,462 ... C5 110
Emo, Ont., Can., 630 ... E6 72
Emory, Rains, Tex., 559 ... C5 118
Emory, Summit, Utah ... B4 119
Emory, peak, Tex ... G2 118
Emory Gap, Roane, Tenn., 500 ... D9 117
Emory Land, glacier, Ant ... B34 5
Emory University, De Kalb, Ga., 4,200 ... *B5 87
Empangeni, S. Afr., 6,572 ... C5 49
Empedrado, Arg., 3,715 ... E4 55
Empire, Walker, Ala., 400 ... B2 78
Empire, Stanislaus, Calif., 1,635 ... *D3 82
Empire, Clear Creek, Colo., 110 ... B5 83
Empire, Dodge, Ga., 125 ... D3 87
Empire, Plaquemines, La., 450 ... E6 95
Empire, Leelanau, Mich., 448 ... D4 98
Empire, Washoe, Nev., 650 ... C2 104
Empire, Jefferson, Ohio, 551 ... B7 111
Empire, Coos, Oreg., 3,781 ... D2 113
Empoli, It., 28,500 ... C3 21
Emporia, Lyon, Kans., 18,190 ... D7 93
Emporia, Greensville, Va., 5,535 ... E5 121
Emporium, Cameron, Pa., 3,397 ... C5 114
Empress, Alta, Can., 405 ... D5 69
Emptinne, Bel., 715 ... D5 15
Emrick, Wells, N. Dak., 20 ... B6 110
Ems, riv., Ger ... A2 17
Emsdale, Ont., Can., 149 ... B5 72
Emsdetten, Ger., 25,000 ... A2 17
Ems-Jade, canal, Ger ... A7 15
Ems-Weser, canal, Ger ... A3 17
Emsworth, Allegheny, Pa., 3,341 ... A5 114
Emyvale, Ire., 255 ... C5 11
Ena (Nakatsu), Jap., 14,700 ... n16 37
Enard, bay, Scot ... B3 13
Ena-San, peak, Jap ... n16 37
Encampment, Carbon, Wyo., 333 ... D6 125
Encanto, cape, Phil ... o13 35
Encarnación, Par., 39,804 ... E4 55
Encarnación de Díaz, Mex., 8,638 ... m12 63
Enchant, Alta., Can., 97 ... E4 69
Enchi, Ghana, 2,064 ... E4 45
Encinal, LaSalle, Tex., 800 ... E3 118
Encinitas, San Diego, Calif., 2,786 ... F5 82
Encino, Torrence, N. Mex., 346 ... C4 107
Encino, Brooks, Tex., 400 ... F3 118
Encinoso, Lincoln, N. Mex ... D4 107
Encontrados, Ven., 9,565 ... B3 60
Encounter, bay, Austl ... G2 51
Endako, B.C., Can., 109 ... B5 68
Endako, riv., B.C., Can ... B5 68
Endau, Mala., 2,675 ... K5 38
Ende, Indon., 7,226 ... G6 35
Endeavor, Forest, Pa., 200 ... C3 114
Endeavor, Marquette, Wis., 280 ... E4 124
Endeavour, Sask., Can., 215 ... E4 70
Endee, Quay, N. Mex ... B6 107
Endelave, Den ... C4 24
Endelave, isl., Den ... C4 24
Enderby, B.C., Can., 1,075 ... D8 68
Enderby Land, reg., Ant ... B18 5
Enderlin, Ransom, N. Dak., 1,596 ... C8 110
Enders, Chase, Nebr., 100 ... D4 103
Enders, res., Kans ... B2 93
Endiang, Alta., Can., 59 ... D4 69
Endicott, Jefferson, Nebr., 166 ... D8 103
Endicott, Broome, N.Y., 18,775 ... C4 108
Endicott, Franklin, Va., 300 ... E2 121
Endicott, Whitman, Wash., 369 ... C8 122
Endicott, mts., Alsk ... B9 79
Endless, lake, Maine ... C4 96
Endwell, Broome, N.Y., 12,000 ... *C4 108
Ene, riv., Peru ... D3 58
Enebakk, Nor ... p29 25
Enez, Tur., 566 ... B6 23
Enfield, N.S., Can., 258 ... E6 74
Enfield, Hartford, Conn., 3,000 (31,464▲) ... B6 84
Enfield, Eng., 109,524 ... C7 12
Enfield, White, Ill., 791 ... E5 90
Enfield, Grafton, N.H., 1,121 (1,867▲) ... C2 105
Enfield, Tompkins, N.Y., 100 ... *C4 108
Enfield, Halifax, N.C., 2,978 ... A6 109
Enfield Center, Grafton, N.H., 200 ... C2 105
Engadine, Mackinac, Mich., 240 ... B5 98
Engaru, Jap., 11,000 ... D11 37
Engebi, isl., Eniwetok ... 52
Engelberg, Switz., 2,646 ... C5 19
Engelhard, Hyde, N.C., 600 ... B8 109
Engels, Sov. Un., 102,000 ... C3 29
Enghien, Bel., 4,225 ... D4 15
Enghien, Fr., 12,504 ... g10 14
England, Lonoke, Ark., 2,861 ... C4, E6 81
England, reg., U.K., 43,430,972 ... D6 10, B6 12
Englee, Newf., Can., 677 ... C3, h10 75
Englefeld, Sask., Can., 187 ... E3 70
Englevale, Ransom, N. Dak., 85 ... C8 110
Englewood, B.C., Can., 121 ... D4 68
Englewood, Arapahoe, Colo., 33,398 ... B6 83
Englewood, Duval, Fla., 5,000 ... *B5 86
Englewood, Sarasota, Fla., 2,877 ... F4 86
Englewood, Lawrence, Ind., 1,232 ... *G5 91
Englewood, Clark, Kans., 243 ... E7 93
Englewood, Bergen, N.J., 26,057 ... D5 106
Englewood, Montgomery, Ohio, 1,515 ... C3 111
Englewood, Coos, Oreg., 1,382 ... *D2 113
Englewood, Lawrence, S. Dak., 10 ... C2 116
Englewood, McMinn, Tenn., 1,574 ... D9 117
English, Crawford, Ind., 698 ... H5 91
English, chan., Eng ... F4 10
English, riv., Ont., Can ... F14 66
English, riv., Iowa ... C5, C6 92
English Bazar, India, 45,900 ... E12 40
English Bay, Alsk., 78 ... h16, D9 79
English Center, Lycoming, Pa., 100 ... D7 114
English Creek, Atlantic, N.J., 400 ... E3 106
English Harbour West, Newf., Can., 371 ... E4 75
English Lake, Starke, Ind., 200 ... B4 91
Englishtown, Monmouth, N.J., 1,143 ... C4 106
Enguera, Sp., 5,663 ... C5 20
Enhaut, Dauphin, Pa., 2,000 ... *F8 114
Enid, Tallahatchie, Miss., 128 ... A4 100
Enid, Richland, Mont., 6 ... C12 102
Enid, Garfield, Okla., 38,859 ... A4 112
Enid, res., Miss ... A4 100
Enigma, Berrien, Ga., 525 ... E3 87
Eniirikku, isl., Bikini ... 52
Enilda, Alta., Can., 106 ... B2 69
Eniwetok, isl., Marshall Is ... 52
Enka, Buncombe, N.C., 1,400 ... D3 109
Enkeldoorn, S. Rh., 800 ... A5 49
Enkhuizen, Neth., 10,500 ... B5 15
Enköping, Swe., 13,200 ... H7, t35 25
Enna, It., 28,323 ... F5 21
En Nahud, Sud., 16,499 ... C2 47
En Naqura, Leb., 3,000 ... A7 32
En Nebk, Syr., 10,400 ... E11 31
Ennedi, plat., Chad ... B4 46
Ennell, lake, Ire ... D4 11
Enngonia, Austl., 78 ... D5 51
Ennigerloh, Ger., 8,900 ... B3 17
Enning, Meade, S. Dak., 17 ... C3 116
Ennis, Ire., 5,699 (★8,410) ... E3 11
Ennis, Madison, Mont., 525 ... E5 102
Ennis, Ellis, Tex., 9,347 ... B6, C4 118
Enniscorthy, Ire., 5,754 ... E5 11
Ennis Creek, Clallam, Wash ... A2 122
Enniskillen, N. Ire., 7,438 ... C4 11
Ennistymon, Ire., 1,145 ... E2 11
Enns, Aus., 8,919 ... D7 16
Enns, riv., Aus ... D7 16
Ennylabegan, isl., Kwajalein ... 52
Enoch, Iron, Utah, 465 ... F2 119
Enochs, Bailey, Tex., 220 ... C1 118
Enochsburg, Franklin, Ind., 80 ... F7 91
Enola, Faulkner, Ark., 100 ... B3 81
Enola, Madison, Nebr., 20 ... C8 103
Enola, Cumberland, Pa., 4,500 ... F8 114
Enon, Bullock, Ala., 125 ... C4 78
Enon, Clark, Ohio, 1,227 ... C4 111
Enoree, Spartanburg, S.C., 950 ... B4 115
Enoree, riv., S.C ... B3 115
Enos, Newton, Ind ... B3 91
Enosburg Falls, Franklin, Vt., 1,321 ... B3 120
Enrich, riv., Scot ... C4 13
Enriquillo, Dom. Rep., 3,485 ... G8 64
Enriquillo, lake, Dom. Rep ... E8 64
Enschede, Neth., 126,100 ... B6 15
Ensenada, Arg., 24,925 ... g8 54
Ensenada, Mex., 42,770 ... A1 63
Ensenada, Rio Arriba, N. Mex., 125 ... A3 107
Ensenada, P.R., 3,299 ... D3 65
Enshih, China, 16,000 ... E6 34
Ensign, Alta., Can., 51 ... D4 69
Ensign, Gray, Kans., 255 ... E3 93
Ensign, Delta, Mich., 50 ... C4 98
Ensisheim, Fr., 4,498 ... D7 14
Ensley, Escambia, Fla., 1,836 ... G2 86
Ensley, Shelby, Tenn ... E8 117
Entebbe, Ug., 10,941 ... A5 48
Enterprise, Coffee, Ala., 11,410 ... D4 78
Enterprise, Shasta, Calif., 4,946 ... *B2 82
Enterprise, Ont., Can., 222 ... C8 72
Enterprise, Polk, Iowa, 26 ... A7 92
Enterprise, Dickinson, Kans., 1,015 ... D6 93
Enterprise, Clarke, Miss., 532 ... C5 100
Enterprise, Wallowa, Oreg., 1,932 ... B9 113
Enterprise, Davis, Utah ... B4 119
Enterprise, Washington, Utah, 859 ... F2 119
Enterprise, Harrison, W. Va., 900 ... A7, B4 123
Entiat, Chelan, Wash., 357 ... B5 122
Entiat, mts., Wash ... B5 122
Entiat, riv., Wash ... B5 122
Entrance, Alta., Can., 87 ... C2 69
Entraygues, Fr., 917 ... E5 14
Entre Minho e Douro, reg., Port., 2,067,900 ... B1 20
Entre Rios, prov., Arg., 803,505 ... A5, f7 54
Entroncamento, Moz ... A5 49
Entry, isl., Que., Can ... B8 74
Entwistle, Alta., Can., 411 ... C3 69
Enugu, Nig., 62,764 ... E6 45
Enumclaw, King, Wash., 3,269 ... B4 122
Envermeu, Fr., 795 ... E9 12
Enville, Chester, Tenn., 250 ... B3 117
Enyu, chan., Bikini ... 52
Enyu, isl., Bikini ... 52
Enz, riv., Ger ... E3 17
Enza, riv., It ... E6 18
Enzan, Jap., 12,500 ... n17 37
Eola, Avoyelles, La., 250 ... D3 95
Eola, hills, Oreg ... C1 113
Eolia, Pike, Mo., 400 ... B6 101
Eoline, Bibb, Ala., 250 ... C2 78
Epe, Neth., 4,700 ... B5 15
Epéna, Con. B ... E3 46
Epernay, Fr., 21,882 ... C5 14
Epes, Sumter, Ala., 337 ... C1 78
Ephesus, ruins, Tur ... D6 23
Ephraim, Sanpete, Utah, 1,801 ... D6 119
Ephraim, Door, Wis., 221 ... C6 124
Ephrata, Lancaster, Pa., 7,688 ... F9 114
Ephrata, Grant, Wash., 6,548 ... B6 122
Épila, Sp., 5,072 ... B5 20
Epileptic Village, Henry, Ind ... E7 91
Épinal, Fr., 30,313 ... A2 18, C7 14
Épinay [-sur-Seine], Fr., 34,167 ... g10 14
Epirus, reg., Grc ... C3 23
Epoufette, Mackinac, Mich., 50 ... B5 98
Epperson, Monroe, Tenn., 50 ... D9 117
Epping, Rockingham, N.H., 980 (2,006▲) ... D4 105
Epping, Williams, N. Dak., 151 ... A2 110
Eppingen, Ger., 5,500 ... D3 17
Epps, West Carroll, La., 411 ... B4 95
Epsie, Powder River, Mont., 5 ... E11 102
Epsom, Daviess, Ind., 60 ... G3 91
Epsom, Merrimack, N.H., 35 (1,002▲) ... D4 105
Epsom [& Ewell], Eng., 71,177 ... C7 12
Epworth, Dubuque, Iowa, 698 ... B7 92
Equality, Coosa, Ala., 160 ... C3 78
Equality, Gallatin, Ill., 665 ... F5 90
Equator, prov., Con. L., 1,752,200 ... A3 48
Equatoria, reg., Sud., 903,503 ... D3 47
Equeurdreville, Fr., 9,624 ... C3 14
Equinunk, Wayne, Pa., 150 ... C11 114
Erath, Vermilion, La., 2,019 ... E3 95
Erath, co., Tex., 16,236 ... C3 118
Erba, mtn., Sud ... A4 47
Erbach, Ger., 400 ... D3 17
Erbacon, Webster, W. Va., 200 ... C4 123
Erbendorf, Ger., 3,200 ... D7 17
Ercis, Tur., 9,900 ... C14 31
Erciyas, mtn., Tur ... C10 31
Ercsi, Hung., 6,296 ... B4 22
Érd, Hung., 23,177 ... B4 22
Erdek, Tur., 6,700 ... B6 23
Erdenheim, Montgomery, Pa., 3,700 ... A11 114
Erdeni Dzuu, Mong ... B5 34
Erding, Ger., 11,300 ... D5 16
Erebus, mtn., Ant ... B29 5
Erechim, Braz., 24,941 ... D2 56
Eregli, Tur., 6,360 ... B8 31
Eregli, Tur., 32,100 ... D10 31
Erepecurú, riv., Braz ... C5 59
Eressós, Grc., 3,301 ... C5 23
Eretria, see Nea Psara, Grc.
Erft, riv., Ger ... D6 15
Erfurt, Ger., 186,400 ... C6 17
Ergene, riv., Tur ... B6 23
Erhard, Otter Tail, Minn., 150 ... D2 99
Eria, riv., Sp ... A2 20
Eriboll, bay, Scot ... B4 13
Erice, It., 18,021 ... E4 21
Ericht, riv., Scot ... D4 13
Erick, Beckham, Okla., 1,342 ... B2 112
Erickson, Man., Can., 531 ... D2 71
Ericsburg, Koochiching, Minn., 140 ... B5 99
Ericson, Wheeler, Nebr., 157 ... C7 103
Erie, Weld, Colo., 875 ... A5 83
Erie, Whiteside, Ill., 1,215 ... B3 90
Erie, Neosho, Kans., 1,309 ... E8 93
Erie, Monroe, Mich., 500 ... G7 98
Erie, Cass, N. Dak., 150 ... B8 110
Erie, Erie, Pa., 138,440 (★212,000) ... B1 114
Erie, co., N.Y., 1,064,688 ... C2 108
Erie, co., Ohio, 68,000 ... A5 111
Erie, co., Pa., 250,682 ... C1 114
Erie, lake, U.S. and Can ... B11 77
Erigavo, Som ... C6 47
Erigavo, dist., Som ... C6 47
Eriksdale, Man., Can., 242 ... D2 71
Erimanthos, mtn., Grc ... D3 23
Erimo, cape, Jap ... F11 37
Erin, Ont., Can., 1,005 ... D4 72
Erin, Houston, Tenn., 1,097 ... A4 117
Eriskay, isl., Scot ... C1 13
Erisort, bay, Scot ... B2 13
Erithrai, Grc., 3,495 ... C4, g10 23
Eritrea, prov., Eth., 1,100,000 ... E9 41, B4 47
Erkelenz, Ger., 11,700 ... C6 15
Erken, lake, Swe ... t36 25
Erlangen, Ger., 69,600 ... D6 17
Erlanger, Kenton, Ky., 7,072 ... A5, A7 94
Erlton, Camden, N.J., 1,000 ... *D2 106
Erma, Cape May, N.J., 200 ... F3 106
Ermelo, Neth., 6,100 ... B5 15
Ermelo, S. Afr., 16,894 ... C5 49
Ermenek, Tur., 7,500 ... D9 31
Ermont, Fr., 19,263 ... g10 14
Ernakulam, India, 117,253 ... *G6 40
Erne, lake, N. Ire ... C4 11
Ernée, Fr., 3,901 ... C3 14
Ernest, Indiana, Pa., 950 ... E3 114
Ernestina, Arg ... g7 54
Ernestville, Unicoi, Tenn., 40 ... C11 117
Ernfold, Sask., Can., 133 ... G2 70
Eromanga, Austl., 31 ... C4 51
Eros, Jackson, La., 176 ... B3 95
Er Rahad, Sud., 6,706 ... C3 47
Er Rama, Isr., 2,621 ... B6 32
Errigal, mtn., Ire ... B3 11
Erris, head, Ire ... C1 11
Errol, Coos, N.H., 130 (220▲) ... B2 105
Errol, isl., La ... E7 95
Errol Heights, Multnomah, Oreg., 10,000 ... *B4 113
Er Roseires, Sud., 3,927 ... C3 47
Erskine, Alta., Can., 208 ... C4 69
Erskine, Polk, Minn., 614 ... C3 99
Erstein, Fr., 6,165 ... C7 14
Erstfeld, Switz., 4,126 ... C6 19
Erving, Franklin, Mass., 400 (1,272▲) ... A3 97
Erwin, Harnett, N.C., 3,183 ... B5 109
Erwin, Kingsbury, S. Dak., 157 ... C8 116
Erwin, Unicoi, Tenn., 3,210 ... C11 117
Erwin, Preston, W. Va., 100 ... B5 123
Erwood, Sask., Can., 125 ... E4 70
Erzgebirge (Ore), mts., Czech., Ger ... C8 17
Erzincan (Erzinjan), Tur., 36,500 ... C12 31
Erzurum, Tur., 91,200 ... C13 31
Esan, cape, Jap ... F10 37
Esashi, Jap., 7,700 ... D11 37
Esashi, Jap., 9,700 ... F10 37
Esbjerg, Den., 55,171 ... C2 24
Esbon, Jewell, Kans., 237 ... C5 93
Escada, Braz., 13,761 ... k6 57
Escalante, Garfield, Utah, 702 ... F4 119
Escalante, des., Utah ... F2 119
Escalante, hills, Colo ... A2 83
Escalante, riv., Utah ... F4 119
Escalon, San Joaquin, Calif., 1,763 ... D3 82
Escalón, Mex., 1,270 ... B4 63
Escambia, co., Ala., 33,511 ... D2 78
Escambia, co., Fla., 173,829 ... G1 86
Escambia, riv., Ala., Fla ... D2 78
Escanaba, Delta, Mich., 15,391 ... C3 98
Escanaba, riv., Mich ... B3 98
Escatawpa, Jackson, Miss., 1,464 ... E3, E5 100
Escatawpa, riv., Ala., Miss ... E5 100
Eschede, Ger., 3,600 ... F4 24
Eschenbach, Ger., 3,200 ... D6 17
Escholzmatt, Switz., 3,257 ... C4 19
Esch-sur-Alzette, Lux., 27,954 (★68,000) ... E5 15
Eschwege, Ger., 24,100 ... B5 17
Eschweiler, Ger., 39,600 ... C3 16
Escobal, C.Z., 580 ... k11 62
Escobar, Arg., 3,693 ... g7 54
Escobas, Zapata, Tex., 250 ... F3 118
Escoheag, Kent, R.I., 35 ... C9 84
Escondida, pt., Mex ... D5 63
Escondido, San Diego, Calif., 16,377 ... F5 82
Escondido, Lincoln, N. Mex ... D4 107
Escondido, riv., Nic ... D6 62
Escoublac-La-Baule, Fr., 13,004 ... D2 14
Escoumains, riv., Que., Can ... A8 73
Escuinapa, Mex., 9,875 ... C3 63
Escuintla, Guat., 9,760 ... C2 62
Escuminac, Que., Can., 156 ... A3 74
Escuminac, pt., N.B., Can ... B5 74
Eséka, Cam ... E2 46
Esens, Ger., 4,500 ... A7 15
Eşfahān (Isfahan), Iran, 254,876 ... E5 41
Esgueva, riv., Sp ... B3 20
Eshimba, Con. L ... C3 48
Eshowe, S. Afr., 4,919 ... C5 49
Esh Sheikh Jarrah, Jordan ... m14 32
Esh Shobek, Jordan, 1,000 ... D7 32
Esk, riv., Eng ... F5 13
Esk, riv., Scot ... E6 13
Eskbank, Sask., Can., 102 ... G2 70
Eskdale, Kanawha, W. Va., 800 ... C3, D6 123
Eske, lake, Ire ... C3 11
Eskifoca, Tur ... C6 23
Eskilstuna, Swe., 59,900 ... t34, H7 25
Eskimo, lakes, N.W. Ter., Can ... C6 66
Eskimo Point, N.W. Ter., Can., 168 ... D14 66
Eskisehir, Tur., 153,200 ... C8 23
Esko, Carlton, Minn., 240 ... D6 99
Eskridge, Wabaunsee, Kans., 519 ... D7 93
Esla, riv., Sp ... B3 20
Eslarn, Ger., 2,800 ... D7 17
Eslava, riv., Mex ... h9 63
Eslöv, Swe., 9,700 ... C7 24
Esme, Tur., 4,900 ... C7 23
Esmeralda, Ven ... C4 60
Esmeralda, co., Nev., 619 ... F4 104
Esmeraldas, Ec., 14,046 ... A2 58
Esmeraldas, prov., Ec., 87,870 ... A2 58
Esmeraldas, riv., Ec ... A2 58
Esmond, DeKalb, Ill., 85 ... A5 90
Esmond, Benson, N. Dak., 420 ... A6 110
Esmond, Providence, R.I., 4,500 ... B10 84
Esmond, Kingbury, S. Dak., 19 ... C8 116
Esmont, Albemarle, Va., 100 ... D4 121
Esom Hill, Polk, Ga., 200 ... C1 87
Esopus, Ulster, N.Y., 175 ... *D7 108
Espada, pt., Col ... A3 60
Espanola, Ont., Can., 5,353 ... A3 72
Espanola, Flagler, Fla., 80 ... C5 86
Espanola, Rio Arriba, N. Mex., 1,976 ... B3 107
Espanola, Spokane, Wash., 25 ... D7 122
Española, isl., Ec ... g6 58
Esparto, Yolo, Calif., 300 ... C2 82
Esperance, Austl., 1,111 ... F3 50
Esperanza, Arg., 10,035 ... A4 54
Esperanza, Pontotoc, Miss., 250 ... A4 100
Esperanza, Hudspeth, Tex., 40 ... F1 118
Esperanza, P.R ... g13 65
Espichel, cape, Port ... C1 20
Espinal, Col., 9,389 ... C3 60
Espinhaço, mts., Braz ... E2 57
Espírito Santo, Braz., 31,027 ... F2 57
Espírito Santo, state, Braz., 1,188,665 ... C4 56
Espiritu Santo, isl., New Hebr ... H10 7
Espita, Mex., 5,089 ... C7 63
Esposende, Port., 1,760 ... B1 20
Espy, Columbia, Pa., 1,375 ... D9 114
Espyville Station, Crawford, Pa., 200 ... C1 114
Esquatzel Coulee, creek, Wash ... C6 122
Esquel, Arg., 5,584 ... C2 54
Esquimalt, B.C., Can., 10,500 ... B9 E6 68
Esquina, Arg., 5,878 ... F4 55
Es Sanamein, Syr ... A8 32
Esschen, Bel., 9,850 ... C4 15
Essen, Ger., 726,600 (★4,800,000) ... B2 17
Essequibo, riv., Br. Gu ... B3 59
Essex, San Bernardino, Calif., 52 ... E6 82
Essex, Ont., Can., 3,428 ... E2 72
Essex, Middlesex, Conn., 1,470 (4,057▲) ... D7 84
Essex, Kankakee, Ill., 328 ... B5 90
Essex, Page, Iowa, 767 ... D2 92
Essex, Baltimore, Md., 35,205 ... B5 85
Essex, Stoddard, Mo., 511 ... E8 101
Essex, Flathead, Mont., 100 ... B3 102
Essex, Essex, N.Y., 12 ... B3 108
Essex (Essex Center), Chittenden, Vt., 300 (7,090▲) ... B2 118
Essex, co., Ont., Can., 258,218 ... E2 72
Essex, co., Eng., 2,286,970 ... C8 12
Essex, co., Mass., 568,831 ... A5 97
Essex, co., N.J., 923,545 ... B4 106
Essex, co., N.Y., 35,300 ... B7 108
Essex, co., Vt., 6,083 ... B5 120
Essex, co., Va., 6,690 ... D6 121
Essex Fells, Essex, N.J., 2,174 ... *B4 106
Essex Junction, Chittenden, Vt., 5,340 ... C2 120
Essexvale, S. Rh ... B4 49
Essexville, Bay, Mich., 4,590 ... E7 98
Essington, Delaware, Pa., 3,300 ... *B11 114
Esslingen, Ger., 83,200 ... E4 17
Es Suki, Sud., 7,388 ... C3 47
Estacada, Clackamas, Oreg., 957 ... B4 113
Estación Superi, Arg ... E2 55
Estados, isl., Arg ... h13 54
Estaire, Ont., Can ... A4 72
Estância, Braz., 16,106 ... D3 57
Estancia, Torrance, N. Mex., 797 ... C3 107
Estarreja, Port., 2,450 ... B1 20
Estcourt, Que., Can., 737 ... B8 73

Estcourt, S. Afr., 8,959......C4 49
Este, It., 10,600..........B3 21
Estelí, Nic., 8,500.........D4 62
Estella, Craig, Okla........A6 112
Estella, Sp., 8,236.........A4 20
Estelle Manor, Atlantic, N.J., 496........E3 106
Estelline, Hamlin, S. Dak., 722.........C9 116
Estelline, Hall, Tex., 346....B2 118
Estepa, Sp., 9,476.........D3 20
Estepona, Sp., 10,935.......D3 20
Esterbrook, Albany, Wyo....C7 125
Esterhazy, Sask., Can., 1,114.G4 70
Esternay, Fr., 691.......F3 15
Esternberg, Aus., 2,203......E8 17
Estero, Lee, Fla., 300.......F5 86
Estero, bay, Calif..........E3 82
Estero, isl., Fla...........F5 86
Estes Park, Larimer, Colo., 1,175.........A5 83
Estevan, Sask., Can., 7,728..H4 70
Estevan, isl., B.C., Can.......C3 68
Estevan Point, B.C., Can....E4 68
Esther, Alta., Can.......D5 69
Esther, St. Francois, Mo., 1,033.........*D7 101
Estherville, Emmet, Iowa, 7,927........A3 92
Estherwood, Acadia, La., 639.........D3 95
Estill, Washington, Miss., 75.B3 100
Estill, Hampton, S.C., 1,865..F5 115
Estill, co., Ky., 12,466......C6 94
Estillfork, Jackson, Ala., 100.A3 78
Estill Springs, Franklin, Tenn., 734.........B5 117
Estlin, Sask., Can., 63.......G3 70
Esto, Holmes, Fla., 148.......G3 86
Eston, Sask., Can., 1,695.....F1 70
Eston, Eng., 37,160.........F7 13
Estonia (S.S.R.), rep., Sov. Un., 1,221,000......D5 28
Estrées-St-Denis, Fr., 1,658...E2 15
Estrêla, Braz., 5,795........D2 56
Estrêla, mts., Port.........B2 20
Estremadura, prov., Port., 1,592,858.........*C1 20
Estremadura, reg., Port., 2,626,400.........C1 20
Estremadura, reg., Sp., 1,378,777.........C2 20
Estremoz, Port., 7,057.......C2 20
Estrondo, mts., Braz..C1 57, D5 59
Estuary, Sask., Can., 62......G1 70
Esztergom, Hung., 23,065...B4 22
Etah, Grnld.........B22 4
Étain, Fr., 3,764.........E5 15
Etamamu, riv., Que., Can......C1 75
Étampes, Fr., 13,515........C5 14
Etang du Nord, Que., Can., 1,090.........B8 74
Étang Saumatre, lake, Haiti.....F7 64
Etaples, Fr., 8,628........B4 14
Etawa, India, 69,681........E7 40
Etawah, India, 59,986.......C6 39, D7 40
Eten, isl., Truk.........52
Ethan, Davison, S. Dak., 297.........D8 116
Ethel, Ont., Can., 134........D3 72
Ethel, Attala, Miss., 566.....B4 100
Ethel, Macon, Mo., 149.......B5 101
Ethel, Lewis, Wash., 90.......C3 122
Ethel, Logan, W. Va., 650...D5 123
Ethel, mtn., Colo..........A4 83
Ethelbert, Man., Can., 556..D1 71
Ethelsville, Pickens, Ala., 62.........B1 78
Ethelton, Sask., Can., 56.....E3 70
Ether, Montgomery, N.C., 150.........B4 109
Ethete, Fremont, Wyo.......B4 125
Ethiopia (Abyssinia), country, Afr., 21,600,000..D4 47, F9 42
Ethridge, Toole, Mont., 51....B5 102
Ethridge, Lawrence, Tenn., 550.........B4 117
Etive, inlet, Scot.........D3 13
Etlan, Madison, Va., 100.....C4 121
Etna, Whitley, Ind., 125.....B6 91
Etna, Penobscot, Maine, 175 (486▲).........D3 96
Etna, Grafton, N.H., 150.....C2 105
Etna, Tompkins, N.Y., 270...C4 108
Etna, Licking, Ohio, 365......C5 111
Etna, Allegheny, Pa., 5,519..A6 114
Etna, Box Elder, Utah.......B2 119
Etna, Lincoln, Wyo., 100....B2 125
Etna, vol., It...........F5 21
Etna Green, Kosciusko, Ind., 483.........B5 91
Etolin, isl., Alsk........m23 79
Etolin, strait, Alsk.......C6 79
Eton, Murray, Ga., 275.......B2 87
Etosha, lake, S.W. Afr......A2 49
Etowah, Mississippi, Ark., 100.........B5 81
Etowah, McMinn., Tenn., 3,223.........D9 117
Etowah, co., Ala., 96,980....A3 78
Etowah, riv., Ga..........B2 87
Etra, Mercer, N.J., 50........C3 106
Etretat, Fr., 1,565..........C4 14
Etsin Gol, riv., China.......C5 34
Etta, Union, Miss., 75.......A4 100
Et Taiyiba, Isr., 7,601...B7, f11 32
Ettelbruck, Lux., 4,452......E6 15
Etter, Moore, Tex., 50........A1 118
Etters (Goldsboro), York, Pa., 542.........F8 114
Ettington, Sask., Can., 23...H2 70
Et Tira, Isr., 4,154.....B6, g10 32
Ettlingen, Ger., 19,400......E3 17
Ettrick, Chesterfield, Va., 2,998.........C7, D5 121
Ettrick, Trempealeau, Wis., 479.........D2 124
Ettrick, forest, Scot.......E5 13
Etumba, Fiji Is.........52
Etzatlán, Mex., 8,752......m11 63
Etzikom, Alta., Can., 101....E5 69
Etzikom, coulee, Alta., Can.....E5 69
Eu, Fr., 7,029..........B4 14
Euboea (Evvoia), prov., Grc., 166,097.........*C4 23
Eucha, Delaware, Okla., 150.A7 112
Eucla, Austl.........F4 50
Euclid, Polk, Minn., 200.....C2 99
Euclid, Cuyahoga, Ohio, 62,998.........A6, B2 111
Euclid, Butler, Pa., 150......E2 114
Euclid Center, Berrien, Mich., 2,343.........*F4 98
Euclid Heights, Garland, Ark., 2,030.........*C2 81
Eudora, Chicot, Ark., 3,598..D4 81
Eudora, Douglas, Kans., 1,526.........B8, D8 93
Eudora, De Soto, Miss......A3 100
Eudunda, Austl., 735.......G2 51
Eufaula, Barbour, Ala., 8,357.D4 78
Eufaula, McIntosh, Okla., 2,382.........B6 112
Eufola, Iredell, N.C., 35.....B3 109
Eugene, Vermillion, Ind., 300.........E3 91
Eugene, Cole, Mo., 151.......C5 101
Eugene, Lane, Oreg., 50,977 (*122,200).........C3 113
Eugenia, pt., Mex.........B1 63
Euless, Tarrant, Tex., 2,062.*B5 118
Eulia, Macon, Tenn.........C7 117
Eulonia, McIntosh, Ga., 200.........E5 87
Eunice, St. Landry, La., 11,326.........D3 95
Eunice, Lea, N. Mex., 3,531..E6 107
Eupen, Bel., 14,445.........D6 15
Euphrates, riv., Asia....I18 9, F2 41
Eupora, Webster, Miss., 1,468.........B4 100
Eure, Gates, N.C., 200......A7 109
Enre, dept., Fr., 361,904....*C4 14
Eure, riv., Fr..........C4 14
Eure-et-Loir, dept., Fr., 277,546.........*C4 14
Eureka, Humboldt, Calif., 28,137.........B1 82
Eureka, Marion, Fla., 50.....C5 86
Eureka, Woodford, Ill., 2,538.........C4 90
Eureka, Greenwood, Kans., 4,055.........E7 93
Eureka, St. Louis, Mo., 1,134.........B7 101
Eureka, Lincoln, Mont., 1,229.........B1 102
Eureka, Eureka, Nev., 500....D6 104
Eureka, Wayne, N.C., 246....B6 109
Eureka (Chambersburg), Gallia, Ohio, 205.........D5 111
Eureka, Aiken, S.C., 75......D4 115
Eureka, McPherson, S. Dak., 1,555.........B6 116
Eureka, Juab, Utah, 771.....D3 116
Eureka, Pleasants, W. Va., 100.........B3 123
Eureka, Winnebago, Wis., 300.........D5 124
Eureka, co., Nev., 767.......D5 104
Eureka Central, P.R.........C2 65
Eureka East (Ryans Slough), Humboldt, Calif., 3,634..*B1 82
Eureka Springs, Carroll, Ark., 1,437.........A2 81
Euroa, Austl., 3,040.........H5 51
Europe, cont., 553,300,000.........C22 3, 8
Euros, riv., Tur..........B6 31
Euskirchen, Ger., 20,300.....C1 17
Eustis, Lake, Fla., 6,189.....D5 86
Eustis, Franklin, Maine, 100 (666▲).........C2 96
Eustis, Frontier, Nebr., 386..D5 103
Eutaw, Greene, Ala., 2,784...C2 78
Eutawville, Orangeburg, S.C., 468.........E7 115
Eutin, Ger., 16,900.........D4 24
Eutsuk, lake, B.C., Can......C4 68
Eva, Morgan, Ala., 150.......A3 78
Eva, Texas, Okla., 10........D2 112
Eva, Benton, Tenn., 200......A3 117
Evadale, Jasper, Texas, 900.*D5 118
Evale, Ang.........E2 48
Evan, Brown, Minn., 153.....F4 99
Evangeline, par., La., 31,639.D3 95
Evanger, Nor., 139.........G2 25
Evans, Weld, Colo., 1,453....A6 83
Evans, Columbia, Ga., 600...C4 87
Evans, Vernon, La., 100......D2 95
Evans, Jackson, W. Va., 200.C3 123
Evans, co., Ga., 6,952.......D5 87
Evans, mtn., Colo..........B5 83
Evans, mtn., N.Z..........O13 51
Evans, strait, N.W. Ter., Can.D16 67
Evansburgh, Alta., Can., 452.........C3 69
Evans City (Evansburg), Butler, Pa., 1,825.........E1 114
Evansdale, Black Hawk, Iowa, 5,738.........B5 92
Evans Landing, Harrison, Ind.........H6 91
Evans Mills, Jefferson, N.Y., 618.........A5 108
Evansport, Defiance, Ohio, 290.........A3 111
Evanston, Weld, Colo., 200..A6 83
Evanston, Cook, Ill., 79,283.........A6, E3 90
Evanston, Breathitt, Ky., 300.C6 94
Evanston, Uinta, Wyo., 4,901.........D2 125
Evansville, Washington, Ark., 25.........B1 81
Evansville, Randolph, Ill., 846.........E4 90
Evansville, Vanderburgh, Ind., 141,543 (*200,300).......I2 91
Evansville, Douglas, Minn., 411.........D3 99
Evansville, Tunica, Miss., 15..A3 100
Evansville, Rock, Wis., 2,858.F4 124
Evansville, Natrona, Wyo., 678.........C6 125
Evant, Coryell, Tex., 700....D3 118
Evanton, Scot.........C4 12
Evaro, Missoula, Mont., 40..C2 102
Evart, Osceola, Mich., 1,775.E5 98
Evarts, Harlan, Ky., 1,473...D6 94
Eveleth, St. Louis, Minn., 5,721.........C6 99
Evening Shade, Sharp, Ark., 232.........A4 81
Even Yehuda, Isr., 2,549....f10 32
Everard, lake, Austl........F5 50
Everard, mtn., B.C., Can......D5 68
Everard, ranges, Austl.......E5 50
Everest, Brown, Kans., 348...C8 93
Everest, peak, Nep.........D11 40
Everett, Ont., Can., 426......C5 72
Everett, Middlesex, Mass., 43,544.........C3 97
Everett, Bedford, Pa., 2,279..F5 114
Everett, Snohomish, Wash., 40,304.........B3 122
Everett, mtn., Mass........B1 97
Everett, Glynn, Ga., 250.....E5 87
Everettville, Monongalia, W. Va., 300.........A7 123
Evergem, Bel., 11,332.......C3 15
Everglades, Collier, Fla., 552.G5 86
Everglades, nat. park, Fla......G6 86
Everglades, swamp, Fla........G6 86
Evergreen, Conecuh, Ala., 3,703.........D3 78
Evergreen, Jefferson, Colo., 950.........B5 83
Evergreen, Avoyelles, La., 325.........D3 95
Evergreen, Itawamba, Miss., 100.........A5 100
Evergreen, Columbus, N.C., 300.........C5 109
Evergreen, Appomattox, Va., 150.........D4 121
Evergreen Park, Cook, Ill., 25,284.........F3 90
Everly, Clay, Iowa, 668......A2 92
Everman, Tarrant, Tex., 1,076.........*B5 118
Everson, Fayette, Pa., 1,304..F2 114
Everson, Whatcom, Wash., 431.........A3 122
Everton, Boone, Ark., 118....A3 81
Everton, Dade, Mo., 261.....D4 101
Evesham, Sask., Can., 96.....E1 70
Evesham, Eng., 12,608.......B6 12
Évian-les-Bains, Fr., 3,738...C2 18
Evinayong, Rio Muni, 870...E1 46
Evington, Campbell, Va., 200.........D3 121
Evolène, Switz., 1,786.......D3 19
Evora, Port., 25,678........C2 20
Évreux, Fr., 23,647.........C4 14
Evritania, prov., Grc., 39,716.........*C3 23
Évron, Fr., 2,648.........C3 14
Évros, prov., Grc., 157,760..*B5 23
Évros, riv., Grc..........B6 23
Evrotas, riv., Grc.........D4 23
Evstratios, isl., Grc........C5 23
Evvoia (Euboea), prov., Grc., 166,097.........*C4 23
Evvoia (Northern), gulf, Grc...C4 23
Evvoia (Southern), gulf, Grc...g11 23
Evvoia (Eugoea), isl., Grc......C4 23
Ewa, Honolulu, Haw., 3,257.........B3, g9 88
Ewa, beach, Haw..........g9 87
Ewab, is., Indon..........G8 35
Ewa Beach, Honolulu, Haw., 2,459.........g8 88
Ewan, Whitman, Wash., 70..B8 122
Ewe, bay, Scot..........C3 13
Ewell, Somerset, Md., 380...E5 85
Ewen, Ontonagon, Mich., 500.........A6 98
Ewing, Franklin, Ill., 250.....E5 90
Ewing, Jackson, Ind., 500...G5 91
Ewing, Fleming, Ky., 375.....B6 94
Ewing, Lewis, Mo., 324.......A5 101
Ewing, Holt, Nebr., 583......B7 103
Ewing, Lee, Va., 500........B1 121
Ewing Township, Mercer, N.J., 26,628.........*C3 106
Ewo, Con. B.........F2 46
Exaltación, Bol., 405.......B2 55
Excel, Monroe, Ala., 313.....D2 78
Excel, Alta., Can., 75.......D5 69
Excello, Macon, Mo., 100....B5 101
Excelsior, Hennepin, Minn., 2,020.........F5 99
Excelsior, Richland, Wis., 150.........E3 124
Excelsior, mtn., Calif........C4 82
Excelsior, mts., Nev........E3 104
Excelsior Springs, Clay and Ray, Mo., 6,473.......B3, D2 101
Exchange, Montour, Pa., 150.D8 114
Exchange, Braxton, W. Va., 100.........C4 123
Excocesa, bay, Dom. Rep......F9 64
Excursion Inlet, Alsk........k22 79
Exe, riv., Eng..........D4 12
Executive Committee, range, Ant..B1 5
Exeland, Sawyer, Wis., 214..C4 124
Exeter, Tulare, Calif., 4,264..D4 82
Exeter, Ont., Can., 3,047....D3 72
Exeter, Eng., 80,215........D4 12
Exeter, Penobscot, Maine, 130 (707▲).........D3 96
Exeter, Barry, Mo., 294.......E4 101
Exeter, Fillmore, Nebr., 745..D8 103
Exeter, Rockingham, N.H., 7,243.........E5 105
Exeter, Luzerne, Pa., 4,747.........B8, D10 114
Exeter, Washington, R.I., 80 (2,298▲).........C10 84
Exeter, riv., N.H..........E5 105
Exira, Audubon, Iowa, 1,111.........C3 92
Exline, Appanoose, Iowa, 223.........D5 92
Exmoor, moor, Eng........C4 12
Exmore, Northampton, Va., 1,566.........D7 121
Exmouth, Eng., 19,740......D4 12
Exmouth, gulf, Austl........D1 50
Expanse, Sask., Can., 55.....H3 70
Experiment, Spalding, Ga., 2,497.........C2 87
Exploits, bay, Newf., Can.....D4 75
Exploits, riv., Newf., Can.....D3 75
Export, Westmoreland, Pa., 1,518.........F2 114
Exshaw, Alta., Can., 678....D3 69
Extension, B.C., Can., 171...A8 68
Extinct Volcanoes and Lava Beds, Ariz.........C7 80
Exuma, sound, Ba. Is........C5 64
Eya, riv., Sov. Un..........H12 27
Eyak, Alsk., 20.........C10, g19 79
Eyasi, lake, Tan..........B5 48
Eye, pen., Scot..........B2 13
Eyebrow, Sask., Can., 285...G2 70
Eyehill, creek, Alta., Can.........C5 69, E1 70
Eyemouth, Scot., 2,160......E6 13
Eynort, inlet, Scot.........C1 13
Eyota, Olmsted, Minn., 558..G6 99
Eyre, Austl.........F4 50
Eyre, Sask., Can., 15........F1 70
Eyre, lake, Austl..........E6 50
Eyre, pen., Austl..........F6 50
Eyrecourt, Ire., 355.........D3 11
Ezeiza, Arg.........g7 56
Ezine, Tur., 7,500..........C6 23

F

Faaborg, Den., 5,135.......C4 24
Fabens, El Paso, Tex., 3,134.........F1 118
Faber, Nelson, Va., 80......D4 121
Faber, lake, N.W. Ter., Can...D9 66
Fabius, Onondaga, N.Y., 378.........C5 108
Fabius San Hilo, pt., Tinian.....52
Fabriano, It., 12,700.......C4 21
Fabrica, Phil., 8,397.......*C6 35
Fabyan, Windham, Conn., 170.........A9 84
Fabyan House, Coos, N.H., 25.........B4 105
Facatativá, Col., 13,479....C3 60
Faceville, Decatur, Ga., 100..F2 87
Fachi (Well), Niger.........C7 45
Fackler, Jackson, Ala., 300...A4 78
Facpi, pt., Guam.........52
Factory, Wayne, Tenn......B4 117
Factoryville, Wyoming, Pa., 991.........C10 114
Fada, Chad.........B4 46
Fada, Mali.........C4 45
Fada-N'Gourma, Upper Volta, 5,200.........D5 45
Faddeyev, isl., Sov. Un.......B17 28
Fadian, pt., Guam.........52
Faejö, isl., Den..........D5 24
Faemö, isl., Den..........D5 24
Faenza, It., 26,500.........B3 17
Faeroe Is., Dan. dep., Eur., 34,596.........C7 8
Fafan, riv., Eth..........D5 47
Fafe, Port., 5,855..........B1 20
Fagaalu, Am. Sam., 531.......52
Fagaloa, bay, W. Sam........52
Fagamafute, Am. Sam........52
Făgăras, Rom., 17,256.......C7 22
Fagatogo, Am. Sam., 1,344.....52
Fågelbro, Swe.........t36 25
Fågered, Swe.........A6 24
Fagerhult, Swe., 471........B7 24
Fagernes, Nor., 982........G3 25
Fagersta, Swe., 15,500......G6 25
Fagnano, lake, Arg.........h12 54
Faguibine, lake, Mali........C4 47
Fagus, Butler, Mo., 90.......E7 101
Fahan, Ire., 322.........B4 11
Fahan, inlet, Saipan........52
Fahraj, Iran, 2,245.........H10 41
Fā'id, Eg., U.A.R., 1,000.....D4 32
Faido, Switz., 1,441.........D6 19
Faifo, Viet., 5,000.........E8 38
Failaka, isl., Kuw..........G4 41
Fairacres, Dona Ana, N. Mex., 5.........E3 107
Fairbank, Cochise, Ariz., 50..F5 80
Fairbank, Buchanan and Fayette, Iowa, 650.........B5 92
Fairbank, Talbot, Md., 175...C5 85
Fairbanks, Alsk., 13,311....C10 79
Fairbanks, Alachua, Fla., 70..C4 86
Fairbanks, Franklin, Maine, 50.........D2 96
Fair Bluff, Columbus, N.C., 1,030.........C4 109
Fairborn, Greene, Ohio, 19,453.........C3 111
Fairburn, Fulton, Ga., 2,470.........B4, C2 87
Fairburn, Custer, S. Dak., 47.D2 116
Fairbury, Livingston, Ill., 2,937.........C5 90
Fairbury, Jefferson, Nebr., 5,572.........D8 103
Fairchance, Fayette, Pa., 2,120.........G2 114
Fairchild, Eau Claire, Wis., 594.........D3 124
Fairdale, Wyandotte, Kans., 2,100.........B8, C9 93
Fairdale, Jefferson, Ky., 6,000.........A4 94
Fairdale, Walsh, N. Dak., 126.........A7 110
Fairdale, Susquehanna, Pa., 400.........C10 114
Fairfax, Chambers, Ala., 3,107.........C4 78
Fairfax, Marin, Calif., 5,813.........B5, D2 82
Fairfax, Man., Can., 75......E1 71
Fairfax, New Castle, Del., 1,000.........*A6 85
Fairfax, Linn, Iowa, 528....C6 92
Fairfax, Renville, Minn., 1,489.........F4 99
Fairfax, Atchison, Mo., 736..A2 101
Fairfax, Hamilton, Ohio, 2,430.........*C3 111
Fairfax, Osage, Okla., 2,076..A5 112
Fairfax, Allendale, S.C., 1,814.........F5 115
Fairfax, Gregory, S. Dak., 223.........D7 116
Fairfax, Franklin, Vt., 350 (1,244▲).........B2 120
Fairfax, Fairfax, Va., 13,585.........B4, C5 121
Fairfax, co., Va., 262,482....C5 121
Fairfax Station, Fairfax, Va., 175.........B4 121
Fairfield, Jefferson, Ala., 15,816.........B3, E4 78
Fairfield, Solano, Calif., 14,968.........B5, C2 82
Fairfield, Fairfield, Conn., 46,183.........E3 84
Fairfield, Camas, Idaho, 474.........F4 89
Fairfield, Wayne, Ill., 6,362..E5 90
Fairfield, Franklin, Ind., 175.E8 91
Fairfield, Jefferson, Iowa, 8,054.........C6 92
Fairfield, Nelson, Ky., 290...C4 94
Fairfield, Somerset, Maine, 3,766 (5,829▲).........D3 96
Fairfield, Benton, Mo., 46...C4 101
Fairfield, Teton, Mont., 752..C5 102
Fairfield, Clay, Nebr., 495...D7 103
Fairfield, Essex, N.J., 3,310..*B4 106
Fairfield, Hyde, N.C., 250....B7 109
Fairfield, Billings, N. Dak., 6..B2 110
Fairfield, Butler, Ohio, 9,734.........C2 111
Fairfield, Lane, Oreg., 2,000.........*C3 113
Fairfield, Adams, Pa., 519...G7 114
Fairfield, Freestone, Tex., 1,781.........D4 118
Fairfield, Franklin, Vt., 100 (1,225▲).........B3 120
Fairfield, Spokane, Wash., 367.........B8 122
Fairfield, co., Conn., 653,589.D3 84
Fairfield, co., Ohio, 63,912...C5 111
Fairfield, co., S.C., 20,713....C5 115
Fairfield, pond, Vt..........B3 120
Fairfield Highland, Jefferson, Ala., 4,500.........*B3 78
Fairford, Washington, Ala., 40.........D1 78
Fairford, Eng., 2,439........C6 12
Fairgrove, Tuscola, Mich., 609.........E7 98
Fair Grove, Greene, Mo., 275.........D4 101
Fairhaven, Bristol, Mass., 14,339.........C6 97
Fair Haven, St. Clair, Mich., 225.........F8 98
Fairhaven, Stearns, Minn., 125.........E4 99
Fair Haven, Monmouth, N.J., 5,678.........C4 106
Fair Haven, Cayuga, N.Y., 764.........B4 108
Fair Grove, Davidson, N.C., 1,500.........*B3 111
Fair Haven, Rutland, Vt., 2,378.........D2 120
Fairholme, Sask., Can., 58...D1 70
Fairhope, Baldwin, Ala., 4,858.........E2 78
Fairhope, Fayette, Pa., 1,700.........*F2 114
Fairland, Shelby, Ind., 750..E6 91
Fairland, Ottawa, Okla., 646.A7 112
Fair Lawn, Bergen, N.J., 36,421.........D5 106
Fairlawn, Pulaski, Va., 1,325.........*D2 121
Fair Lawn Heights, Summit, Ohio, 2,000.........*A6 111
Fairlee, Kent, Md., 200.....B5 85
Fairlee, Orange, Vt., 400 (569▲).........D4 120
Fairless Hills, Bucks, Pa., 8,000.........*F12 114
Fairlight Station, Sask., Can., 193.........H5 70
Fairmont, Will, Ill., 2,000..*F2 90
Fairmont, Martin, Minn., 9,745.........G4 99
Fairmont, Fillmore, Nebr., 829.........D8 103
Fairmont, Robeson, N.C., 2,286.........C4 109
Fairmont, Garfield, Okla., 115.........A4 112
Fairmont, Spartanburg, S.C., 300.........B3 115
Fairmont, Snohomish, Wash., 1,227.........*B3 122
Fairmont, Marion, W. Va., 27,477.........A7, B4 123
Fairmont, City, St. Clair, Ill., 2,688.........B8 90
Fairmont Heights, Prince Georges, Md., 2,308.......C2 85
Fairmount, Gordon, Ga., 619.........B2 87
Fairmount, Vermilion, Ill., 725.........C6 90
Fairmount, Grant, Ind., 3,080.........D6 91
Fairmount, Leavenworth, Kans., 20.........B8 93
Fairmount, Somerset, Md., 800.........D6 85
Fairmount, Onondaga, N.Y., 3,000.........*B4 108
Fairmount, Richland, N. Dak., 503.........C9 110
Fairmount Station, Sask., Can., 192.........F1 70
Fairoaks, Cross, Ark., 150....B4 81
Fair Oaks, Sacramento, Calif., 9,000.........A6 82
Fair Oaks, San Luis Obispo, Calif., 1,622.........*E3 76
Fair Oaks, Cobb, Ga., 7,969..A5 87
Fair Oaks, Jasper, Ind., 200..B3 91
Fairoaks, Allegheny, Pa., 1,239.........A5 114
Fair Plain, Berrien, Mich., 7,998.........*F4 98
Fairplay, Park, Colo., 404...B5 83
Fair Play, Washington, Md., 420.........A2 85
Fair Play, Polk, Mo., 335....D4 101
Fair Play, Oconee, S.C., 240.B2 115
Fairpoint, Belmont, Ohio, 600.........B7 111
Fairport, Muscatine, Iowa, 150.........C7 92
Fairport, Delta, Mich., 150..C4 98
Fairport, De Kalb, Mo., 88...B3 101
Fairport, Monroe, N.Y., 5,507.........B3 108
Fair Port, Northumberland, Va., 650.........D6 121
Fairport Harbor, Lake, Ohio, 4,267.........A6 111
Fairton, Cumberland, N.J., 900.........E2 106
Fairvalley, Woods, Okla.....A2 112
Fairview, Atla., Can., 1,506..A1 69
Fairview, Walker, Ga., 2,000.........*B1 87
Fairview, Fulton, Ill., 544...C3 90
Fairview, St. Clair, Ill., 850.........*E3 90
Fairview, Switzerland, Ind., 35.........G7 91
Fairview, Brown, Kans., 272.........C8 93
Fairview, Christian & Todd, Ky., 275.........D2 94
Fairview, Concordia, La......C4 95
Fairview, Oscoda, Mich., 250.........D6 98
Fairview, Newton, Mo., 249..E3 101
Fairview, Richland, Mont., 1,006.........C12 102
Fairview, Bergen, N.J., 9,399.........D5 106
Fairview, Burlington, N.J., 150.........D3 106
Fairview, Monmouth, N.J., 4,500.........*C4 106
Fairview, Dutchess, N.Y., 8,626.........*D7 108
Fairview, Cuyahoga, Ohio, 14,624.........B2 111
Fairview, Belmont and Guernsey, Ohio, 166.........B6 111
Fairview, Major, Okla., 2,213.........A3 112
Fairview, Multnomah, Oreg., 578.........B4 113
Fairview, Erie, Pa., 1,399....B1 114
Fairview, Northampton, Pa., 1,146.........*E11 114
Fairview, Northumberland, Pa., 2,100.........*E8 114
Fairview, Lincoln, S. Dak., 101.........D9 116
Fairview, Williamson, Tenn., 1,017.........B4 117
Fairview, Sanpete, Utah, 655.........D4 119
Fairview, Yakima, Wash., 2,758.........*C5 122
Fairview, Marion, W. Va., 653.........A7, B4 123
Fairview, Lincoln, Wyo., 100.........C2 125
Fairview, peak, Nev........D3 104
Fairview Park, Vermillion, Ind., 1,039.........E3 91
Fairview Park, Cuyahoga, Ohio, 14,624.........*A6 111
Fairview Shores, Orange, Fla., 900.........*D5 86
Fairvilla, Orange, Fla., 1,000.........*D5 86
Fairway, Johnson, Kans., 5,398.........D9 93
Fairweather, mtn., Alsk.......D12 79
Fairy Glen, Sask., Can., 53...D3 70
Fairyland, Walker, Ga., 1,000.........*B1 87
Faison, Duplin, N.C., 666....B5 109
Faith, Rowan, N.C., 494.....B3 109
Faith, Meade, S. Dak., 591...B3 116
Faithorn, Menominee, Mich., 35.........C3 98
Faix, Pickett, Tenn.........C8 117
Faizabad, Afg., 25,770......C15 41
Faizabad, India, 76,582.........C7 38, D9 40
Fajardo, P.R., 12,409....B8, f12 65
Fajardo, mun., P.R., 18,321.........B8, f12 65
Fajardo Central, P.R.....B8, f12 65
Fakenham, Eng., 2,933.......B8 12
Fakfak, W. Irian, 1,800......F8 35
Fakse, Den., 2,002..........C6 24
Fakse, bay, Den..........C6 24
Fakse Ladeplads, Den., 1,579.........C6 24
Faku, China.........C10 36
Fal, Vernon, La.........D2 95
Fal, riv., Eng..........D3 12
Falabequets, isl., Truk.......52
Falaise, Fr., 6,325..........C3 14
Falalu, isl., Truk..........52
Falas, isl., Truk..........52
Fălciu, Rom., 5,124........B9 22
Falcon, Nevada, Ark., 25....D2 81
Falcon, Quitman, Miss., 200.........A3 100
Falcon, El Paso, Colo., 75...C6 83
Falcon, Cumberland, N.C., 235.........B5 109
Falcon, Zapata, Tex., 150....F3 118
Falcón, state, Ven., 340,450..A3 60
Falconer, Chautauqua, N.Y., 3,343.........C1 108
Falcon Heights, Ramsey, Minn., 5,927.........E7 99
Falconwood, Erie, N.Y., 1,000.........*B2 108
Falealili, hbr., W. Sam........52
Falefa, hbr., W. Sam.........52
Falémé, riv., Mali, Sen.......D2 45
Faleshty, Sov. Un., 5,000.....B8 22
Falfurrias, Brooks, Tex., 6,515.........F3 118
Falher, Alta., Can., 741......B2 69
Falkenberg, Ger., 7,831.......B8 17
Falkenberg, Swe., 11,000.....B6 24
Falkenburg Station, Ont., Can., 50.........B5 72
Falkenstein, Ger., 14,900.....C7 17
Falkirk, McLean, N. Dak., 15.........B4 110
Falkirk, Scot., 38,043.......D5 13
Falkland, B.C., Can., 408...D8 68
Falkland Islands, Br. dep., S.A., 2,439.........I4 53
Falkner, Tippah, Miss., 200..A5 100
Falköping, Swe., 14,400.....H5 25
Falkville, Morgan, Ala., 682..A3 78
Fall, Ger.........B7 18
Fall, riv., Kans..........E7 93
Fall Branch, Washington, Tenn., 950.........C11 117
Fallbrook, San Diego, Calif., 4,814.........F5 82
Fall City, King, Wash., 560.........B4, D2 122
Fall Creek, Eau Claire, Wis., 710.........D2 124
Fallentimber, Cambria, Pa., 200.........E5 114
Falling, creek, Va..........C7 121
Falling, creek, W. Va........C6 123
Falling Creek, Chesterfield, Va.........C7 121
Falling Water, Hamilton, Tenn., 500.........E10 117
Falling Waters, Berkeley, W. Va., 200.........B6 123
Fallis, Lincoln, Okla., 42....B4 112
Fållnäs, Swe.........u35 25
Fallon, Prairie, Mont., 300.........D11 102
Fallon, Churchill, Nev., 2,734.........D3 104
Fallon, co., Mont., 3,997....D12 102
Fall River, Greenwood, Kans., 226.........E7 93
Fall River, Bristol, Mass., 99,942 (*139,200).......C5 97
Fallriver, Lawrence, Tenn., 40.........B4 117
Fall River, Columbia, Wis., 584.........E4 124
Fall River, co., S. Dak., 10,688.........D2 116
Fall River Mills, Shasta, Calif., 500.........B3 82
Fall River, res., Kans........E7 93
Fall Rock, Clay, Ky., 500....C6 94
Falls, Wyoming, Pa., 500....A8 114
Falls, co., Tex., 21,263......D4 118
Falls, riv., Wyo..........A2 125
Fallsburg, Lawrence, Ky., 200.........B7 94
Falls Church (Independent City), Va., 10,192.........B5 121
Falls City, Richardson, Nebr., 5,598.........D10 103
Falls City, Polk, Oreg., 653...C3 113
Falls Creek, Clearfield and Jefferson, Pa., 1,344.....D4 114
Fallsington, Bucks, Pa., 1,000.........*F12 114
Falls of Rough, Grayson, Ky., 40.........C3 94
Fallston, Harford, Md., 100..A5 85
Fallston, Cleveland, N.C., 500.........B2 109
Falls Village, Litchfield, Conn., 500.........B3 84
Falmouth, N.S., Can., 831....D5 74
Falmouth, Eng., 15,427......D2 12
Falmouth, Suwannee, Fla., 25.........B3 86
Falmouth, Jam., 4,126.......F5 64

Falmouth, Pendleton, Ky., 2,568 B5 94
Falmouth, Cumberland, Maine, 5,976 E5 96
Falmouth, Barnstable, Mass., 3,308 (13,037▲) C6 97
Falmouth, Missaukee, Mich., 250 D5 98
Falmouth, Stafford, Va., 1,478 C5 121
Falmouth, bay, Eng D2 12
Falmouth Foreside, Cumberland, Maine (part of Falmouth) E5 96
Falmouth Heights, Barnstable, Mass., 500 C6 97
Falo, isl., Truk 52
Faloma, Multnomah, Oreg A2 113
Falsa Chipana, pt., Chile D1 55
False, bay, S. Afr D2 49
False, cape, Fla D6 86
False, cape, Va E7 121
Falster, isl., Den D5 24
Falsterbo, Swe., 392 C6 24
Fălticeni, Rom., 13,305 B8 22
Falun, Saline, Kans., 100 D6 93
Falun, Swe., 18,700 G6 25
Famagusta, Cyp., 26,763 E9 31
Famatina, mts., Arg E2 55
Fame, McIntosh, Okla., 100 B6 112
Family, Glacier, Mont., 5 B4 102
Family, lake, Man., Can C4 71
Famous Ice Beds, Vt E3 120
Fancher, Orleans, N.Y., 130 B2 108
Fancy Farm, Graves, Ky., 375 B2 94
Fangcheng, China H5 36
Fanghsien, China, 12,000 E2 34
Fangshen, China C8 36
Fannettsburg, Franklin, Pa., 400 F6 114
Fannin, Rankin, Miss., 40 C4 100
Fannin, Goliad, Texas, 130 E4 118
Fannin, co., Ga., 13,620 B2 87
Fannin, co., Tex., 23,880 C4 118
Fanning, isl., Pac. O F12 7
Fannūj, Iran H9 41
Fanny Bay, B.C., Can., 132 E5 68
Fannystelle, Man., Can., 153 E3 71
Fano, It., 23,000 (41,033▲) C4 21
Fanö, bay, Den C2 24
Fanö, isl., Den C2 24
Fanwood, Union, N.J., 7,963 *B4 106
Fao, Iraq, 2,916 G4 41
Fâqûs, Eg., U.A.R., 2,000 D3 32
Faradje, Con. L A4 48
Farafangana, Malag., 7,300 h9 49
Farâfirah, oasis, Eg., U.A.R D5 43
Farah, Afg., 15,258 E11 41
Farallon, isl., Calif g4 82
Farallón, pt., Mex n11 63
Farallones, gulf, Calif B4 82
Faramana, Upper Volta D4 45
Faranah, Guinea, 2,250 D2 45
Faranuf, Valley, Mont C9 102
Farber, Audrain, Mo., 451 B6 101
Fareham, Eng., 58,277 D6 12
Farewell, cape, Grnld C18 3, C19 4
Farewell, cape, N.Z N14 51
Fargo, Monroe, Ark., 100 C4 81
Fargo, Clinch, Ga., 900 F4 87
Fargo, Cass, N. Dak., 46,662 C9 110
Fargo, Ellis, Okla., 291 A2 112
Far Hills, Somerset, N.J., 702 B3 106
Farî, Mali D2 45
Faria, riv., Jordan g12 32
Faribault, Rice, Minn., 16,926 F5 99
Faribault, co., Minn., 23,685 G5 99
Faridpur, Pak., 25,556 F12 40
Farilhoes, is., Port C1 20
Farina, Austl., 55 E2 51
Farina, Fayette, Ill., 692 E5 90
Farisita, Huerfano, Colo., 5 D5 83
Fâriskûr, Eg., U.A.R., 2,000 C3 32
Farley, Dubuque, Iowa, 920 B6 92
Farley, Franklin, Mass., 150 A3 97
Farley, Platte, Mo., 120 B3 101
Farley, Colfax, N. Mex., 60 A5 107
Farm, pond, Mass D2 97
Farmer, Defiance, Ohio, 120 A3 111
Farmer, Hanson, S. Dak., 95 D8 116
Farmer City, De Witt, Ill., 1,838 C5 90
Farmers, Rowan, Ky., 236 B6 94
Farmersburg, Sullivan, Ind., 1,027 F3 91
Farmersburg, Clayton, Iowa, 250 B6 92
Farmers Exchange, Hickman, Tenn., 15 B4 117
Farmersville, Lowndes, Ala., 80 C3 78
Farmersville, Tulare, Calif., 3,101 *D4 82
Farmersville, Montgomery, Ill., 495 D4 90
Farmersville, Collins, Tex., 2,021 C4 118
Farmerville, Union, La., 2,727 B3 95
Farmingdale, Monmouth, N.J., 959 C4 106
Farmingdale, Nassau, N.Y., 6,128 G3 84
Farmingdale, Pennington, S. Dak., 30 D3 116
Farmington, Washington, Ark., 216 A1 81
Farmington, Hartford, Conn., 2,500 (10,813▲) C5 84
Farmington, Kent, Del., 142 C6 85
Farmington, Fulton, Ill., 2,831 C3 90
Farmington, Van Buren, Iowa, 902 D6 92
Farmington, Franklin, Maine, 2,749 (5,001▲) D2 96
Farmington, Oakland, Mich., 6,881 A7 98
Farmington, Dakota, Minn., 2,300 F5 99
Farmington, St. Francois, Mo., 5,618 D7 101
Farmington, Teton, Mont., 12 C4 102
Farmington, Stafford, N.H., 2,241 (3,287▲) D4 105
Farmington, San Juan, N. Mex., 23,786 A1 107
Farmington, Davie, N.C., 300 A3 109
Farmington, Fayette, Pa., 200 G2 114
Farmington, Marshall, Tenn., 100 B5 117
Farmington, Davis, Utah, 1,951 C4 119
Farmington, Whitman, Wash., 176 B8 122
Farmington, Marion, W. Va., 709 A7 123
Farmington, riv., Conn B5 84
Farmingville, Suffolk, N.Y., 2,134 *D4 108
Farmland, Randolph, Ind., 1,102 D7 91
Farmville, Pitt, N.C., 3,997 B6 109
Farmville, Prince Edward, Va., 4,293 D4 121
Farnam, Dawson, Nebr., 258 D5 103
Farnams, Berkshire, Mass., 250 A1 97
Farnborough, Eng., 31,437 C7 12
Farne, is., Eng C6 10
Farner, Polk, Tenn., 150 D9 117
Farnham, Que., Can., 6,354 D5 73
Farnham, Erie, N.Y., 422 C1 108
Farnham, mtn., B.C., Can D9 68
Farnhamville, Calhoun, Iowa, 409 B3 92
Farnharm, Richmond, Va., 300 D6 121
Farnhurst, New Castle, Del., 60 A6 85
Farnigen, Switz C6 19
Farnumsville, Worcester, Mass. (part of South Grafton) B4 97
Faro, Braz., 1,434 C5 59
Faro, Port., 17,631 D2 20
Farragut, Fremont, Iowa, 495 D2 92
Farran's Point, Ont., Can., 150 *C9 72
Farrar, Polk, Iowa, 50 A4 92
Farrar, riv., Scot C4 13
Farrāshband, Iran G6 41
Farrell, Coahoma, Miss., 150 A3 100
Farrell, Mercer, Pa., 13,793 D1 114
Farrellton, Quebec, Can., 368 D2 73
Farrerdale, Sask., Can F3 70
Farrington, Chatham, N.C B4 109
Farris, Atoka, Okla., 25 C6 112
Farris, lake, Nor p27 25
Farrukhabad, India, 87,793 (*94,591) C6 39, D7 40
Farry, Woods, Okla A2 112
Farsala, Grc., 5,771 C4 23
Farsi, Afg., 10,000 E11 41
Farsö, Den., 1,581 B3 24
Farson, Wapello, Iowa, 121 C5 92
Farson, Sweetwater, Wyo., 20 C3 125
Farsund, Nor., 2,200 H2 25
Fårup, Den., 506 B3 24
Farwell, Clare, Mich., 737 E6 98
Farwell, Pope, Minn., 106 E3 99
Farwell (Fardale), Howard, Nebr., 137 C7 103
Farwell, Parmer, Tex., 1,009 B1 118
Fasā, Iran, 9,907 G6 41
Fasano, It., 16,100 D6 21
Fassett, Que., Can., 434 D3 73
Fastov, Sov. Un., 10,000 F7 27
Fatehpur, India, 28,232 C7 39
Fatsa, Tur., 6,800 B11 31
Fatuosofia, cape, W. Sam 52
Fatuvalu, W. Sam., 117 52
Faucett, Buchanan, Mo., 150 B3 101
Faulk, co., S. Dak., 4,397 B6 116
Faulkner, co., Ark., 24,303 B3 81
Faulkton, Faulk, S. Dak., 1,051 B6 116
Faunsdale, Marengo, Ala., 124 C2 78
Fauquembergues, Fr., 842 D10 12
Fauquier, B.C., Can., 114 E8 68
Fauquier, co., Va., 24,066 C5 121
Faussepoint, lake, La D4 95
Faust, Franklin, N.Y., 880 A6, B2 108
Favara, It., 27,909 F4 21
Faverges, Fr., 2,329 D1 18
Faversham, Eng., 12,983 C8 12
Favorita, Flagler, Fla., 70 C5 86
Faw, hbr., Fiji Is 52
Fawn, lake, Ont., Can F15 67
Fawnie, range, B.C., Can C5 68
Faxaflói, bay, Ice n21 25
Faxon, Comanche, Okla., 137 C3 112
Faxon, Lycoming, Pa., 1,841 *D8 114
Faxon, Benton, Tenn., 100 A3 117
Fay, Dewey, Okla., 150 B3 112
Fayd, Sau. Ar I14 31
Fayal, isl., Az. Is g8 44
Fayette, Fayette, Ala., 4,227 B2 78
Fayette, Boone, Ind., 80 E5 91
Fayette, Fayette, Iowa, 1,597 B6 92
Fayette, Jefferson, Miss., 1,626 D2 100
Fayette, Howard, Mo., 3,294 B5 101
Fayette, Fulton, Ohio, 1,090 A3 111
Fayette, Sanpete, Utah, 161 D4 119
Fayette, co., Ala., 16,148 B2 78
Fayette, co., Ga., 8,199 C2 87
Fayette, co., Ill., 21,946 E5 90
Fayette, co., Ind., 24,454 E7 91
Fayette, co., Iowa, 28,581 B6 92
Fayette, co., Ky., 131,906 C5 94
Fayette, co., Ohio, 24,775 C4 111
Fayette, co., Pa., 169,340 G2 114
Fayette, co., Tenn., 24,557 B2 117
Fayette, co., Tex., 20,384 E4 118
Fayette, co., W. Va., 61,731 C3 123
Fayette City, Fayette, Pa., 1,159 *F2 114
Fayette Corners, Fayette, Tenn B2 117
Fayetteville, Talladega, Ala., 200 B3 78
Fayetteville, Washington, Ark., 22,887 A1 81
Fayetteville, Fayette, Ga., 1,389 C2 87
Fayetteville, St. Clair, Ill., 294 B9 101
Fayetteville, Lawrence, Ind., 150 G4 91
Fayetteville, Onondaga, N.Y., 4,311 B5 108
Fayetteville, Cumberland, N.C., 47,106 (*132,000) B5 109
Fayetteville, Brown, Ohio, 389 C4 111
Fayetteville, Franklin, Pa., 950 G6 114
Fayetteville, Lincoln, Tenn., 6,804 B5 117
Fayetteville, Fayette, W. Va., 1,848 C3, D7 123
Fayetteville North, Cumberland, N.C., 3,071 *B5 109
Fays-Billot, Fr., 1,409 B1 18
Fayville, Worcester, Mass., 600 D1 97
Faywood, Grant, N. Mex., 10 E1 107
Fear, cape, N.C D6 109
Fearn, Scot., 1,317 C5 13
Feasterville, Bucks, Pa., 6,000 A12 114
Feather, riv., Calif C3 82
Feather Falls, Butte, Calif., 900 C3 82
Featherston, Pittsburg, Okla., 50 B6 112
Fécamp, Fr., 19,491 C4 14
Feche, Eth D4 47
Fédala, Mor., 35,010 C3 44
Federal, Ont., Can A9 72
Federal, Laramie, Wyo., 10 D7 125
Federal Dam, Cass, Minn., 185 C4 99
Federal Point, Putnam, Fla., 275 C5 86
Federalsburg, Caroline, Md., 2,060 C6 85
Federal Way, King, Wash., 7,000 *B3 122
Fedora, Miner, S. Dak., 150 C8 116
Fedscreek, Pike, Ky., 500 C7 94
Feeding Hills, Hampden, Mass., 5,000 B2 97
Fefan, isl., Truk 52
Fehmarn, isl., Ger D5 24
Feijó, Braz., 1,628 C3 58
Feilding, N.Z., 8,172 N15 51
Feira, N. Rh E4 48
Feira de Santana, Braz., 61,612 D3 57
Fejer, co., Hung., 300,119 *B4 22
Felanitx, Sp., 11,759 C7 20
Felch, Dickinson, Mich., 150 B3 98
Felda, Hendry, Fla., 200 F5 86
Feldbach, Aus., 3,687 E7 16
Feldberg, Ger., 2,862 E7 24
Feldberg, mtn., Ger B3 18
Feldkirch, Aus., 17,343 E4 16
Feliciano, P.R B2 65
Felicity, Clermont, Ohio, 878 D3 111
Felipe Carrillo Puerto, Mex., 595 D7 63
Felix, Chaves, N. Mex D5 107
Felix, cape, N.W. Ter., Can C13 66
Felixstowe, Eng., 17,254 C9 12
Felixville, East Feliciana, La., 35 D5 95
Fellbach, Ger., 26,000 E4 17
Fellingsbro, Swe., 5,348 t33 25
Fellows, Kern, Calif., 700 E4 82
Fellowship, Burlington, N.J., 200 D3 106
Fellsmere, Indian River, Fla., 732 E6 86
Felsenthal, Union, Ark., 250 D3 81
Felt, Teton, Idaho, 20 F7 89
Felt, Cimarron, Okla., 70 D1 112
Felton, Santa Cruz, Calif., 1,380 D2 82
Felton, Kent, Del., 422 D6 85
Felton, Haralson, Ga., 160 C1 87
Felton, Clay, Minn., 201 C2 99
Felton, York, Pa., 430 G8 114
Feltre, It., 8,500 A3 21
Femund, lake, Nor F4 25
Fence, Florence, Wis., 250 C5 124
Fence, lake, Wis C4 124
Fence, riv., Mich B2 98
Fence Lake, Valencia, N. Mex., 55 C1 107
Fender, Tift, Ga., 150 E3 87
Fenelon Falls, Ont., Can., 1,359 C6 72
Fénérive, Malag g9 49
Fengchen, China, 35,000 C7 36
Fengcheng, China, 21,000 J6 36
Fengchieh, China, 25,000 I3 36
Fenghsiang, China, 10,000 G2 36
Fenghsien, China, 12,000 E6 34
Fengning, China D7 36
Fengshan, China, 1,000 C4 37
Fengshan, For., 5,000 *G9 34
Fengtu, China, 14,000 J2 36
Fenick, isl., S.C G1 115
Fenn, Idaho, Idaho, 25 D2 89
Fennimore, Grant, Wis., 1,747 F3 124
Fennville, Allegan, Mich., 705 F4 98
Fenton, Sask., Can., 20 E3 70
Fenton, Kossuth, Iowa, 440 A3 90
Fenton, Jefferson Davis, La., 429 D3 95
Fenton, Genesee, Mich., 6,142 F7 98
Fenton, St. Louis, Mo., 1,059 B8 101
Fentress, Caldwell, Tex., 350 A4 118
Fentress, Norfolk, Va., 350 E6 121
Fentress, co., Tenn., 13,288 C9 117
Fenwick, Ont., Can., 685 D5 72
Fenwick, Nicholas, W. Va., 505 C4 123
Fenwood, Sask., Can., 157 G4 70
Fenyang, China, 30,000 D7 34, F4 36
Feodosiya, Sov. Un., 43,600 I10 27
Ferbane, Ire., 896 D4 11
Ferdig, Toole, Mont., 150 B5 102
Ferdinand, Idaho, Idaho, 176 C2 89
Ferdinand, Dubois, Ind., 1,427 H4 91
Ferdinandshof, Ger., 2,490 E7 24
Ferdows, Iran E9 41
Fère-Champenoise, Fr., 2,146 F4 15
Ferentino, It., 16,836 D4 21
Fergana, Sov. Un., 91,000 E8 29
Fergana, mts., Sov. Un E8 29
Fergus, Ont., Can., 3,831 D4 72
Fergus, Fergus, Mont., 6 C7 102
Fergus, co., Mont., 14,018 C7 102
Fergus Falls, Otter Tail, Minn., 13,733 D2 99
Ferguson, B.C., Can., 35 D9 68
Ferguson, Mashall, Iowa, 186 C5 92
Ferguson, Pulaski, Ky., 468 C5 94
Ferguson, St. Louis, Mo., 22,149 A8, C7 101
Ferguson, Wayne, W.Va., 100 C2 123
Ferguson Creek, Pike, Ky., 300 *C7 94
Fériana, Tun., 4,192 G11 30
Ferintosh, Alta., Can., 174 C4 69
Fermanagh, co., N. Ire., 51,613 C4 11
Ferme Neuve, Que., Can., 1,971 C2 73
Fermeuse, Newf., Can., 311 E5 75
Fermo, It., 11,400 C4 21
Fermoselle, Sp., 3,885 B2 20
Fermoy, Ire., 3,241 E3 11
Fern, Clarion, Pa., 50 D2 114
Fern, creek, Ky A4 94
Fernandina Beach, Nassau, Fla., 7,276 B5, B6 86
Fernandina (Narborough), isl., Ec g5 58
Fernando de Noronha, ter., Braz., 1,389 B4 57
Fernando de Noronha, isl., Braz B4 57
Fernando Póo, Sp. dep., Afr., 45,330 F6 42, E1 46
Fernando Póo, isl., Afr E1 46
Fernán-Núñez, Sp., 12,225 D3 20
Fernão Veloso, bay, Moz D7 48
Fernbank, Lamar, Ala., 80 B1 78
Fern Creek, Jefferson, Ky., 1,500 A5 94
Fern Crest Village, Broward, Fla., 93 *F6 86
Ferndale, Pulaski, Ark., 60 D5 81
Ferndale, Humboldt, Calif., 1,371 B1 82
Ferndale, Bell, Ky., 200 D6 94
Ferndale, Anne Arundel, Md., 2,500 B4 85
Ferndale, Oakland, Mich., 31,347 A7 98
Ferndale, Cambria, Pa., 2,717 F4 114
Ferndale, Northumberland, Pa., 1,900 *D8 114
Ferndale, Schuylkill, Pa., 40 E9 114
Ferndale, Whatcom, Wash., 1,442 A3 122
Ferney, Brown, S. Dak., 100 B7 116
Fernie, B.C., Can., 2,661 E10 68
Fernley, Lyon, Nev., 500 D2 104
Fern Ridge, dam, Oreg C3 113
Ferns, Ire., 557 E5 11
Fernwood, Benewah, Idaho, 250 B2 89
Fernwood, Pike, Miss., 600 D3 100
Fernwood, Oswego, N.Y., 2,108 B4 108
Ferozepore, India, 47,060 (*97,932) B5 40
Ferrara, It., 99,000 (152,654▲) E7 18, B3 21
Ferreira do Alentejo, Port., 5,205 C1 20
Ferreira do Zêzere, Port., 2,503 C1 20
Ferrellsburg, Lincoln, W. Va., 150 C2 123
Ferrelo, cape, Oreg E2 113
Ferreñafe, Peru, 8,812 C2 58
Ferriday, Concordia, La., 4,563 C4 95
Ferrier, Alta., Can C3 69
Ferrière, It., 6,433 E5 18
Ferris, Hancock, Ill., 208 C2 90
Ferris, Ellis, Tex., 1,807 C4, B6 118
Ferris, mts., Wyo C5 125
Ferrisburg, Addison, Vt., 170 (1,426▲) C2 120
Ferron, Emery, Utah, 386 D4 119
Ferros, Braz., 2,456 E2 56
Ferrum, Franklin, Va., 400 E2 121
Ferry, Oceana, Mich., 100 E4 98
Ferry, co., Wash., 3,889 A7 122
Ferry, pt., N.J E4 106
Ferryland, Newf., Can., 713 E5 75
Ferryland, dist., Newf., Can E5 75
Ferrysburg, Ottawa, Mich., 2,590 *E4 98
Ferryville, Tun., 29,353 F11 30
Ferryville, Crawford, Wis., 194 E2 124
Fertile, Worth, Iowa, 386 A4 92
Fertile, Polk, Minn., 968 C2 99
Fertilia, It D2 21
Fès (Fez), Mor., 216,133 C3 44, G4 30
Feshi, Con. L C2 48
Fessenden, Wells, N. Dak., 920 B6 110
Festina, Winneshiek, Iowa, 150 A6 92
Festus, Jefferson, Mo., 7,021 B8, C7 101
Fetești, Rom., 15,383 C8 22
Fethard, Ire., 962 E4 11
Fethiye, Tur., 7,700 D7 23
Fetisovo, Sov. Un E4 29
Fettercairn, Scot., 1,323 D6 13
Fetters Hot Springs, Sonoma, Calif., 750 *B5 82
Feucht, Ger., 7,300 D6 17
Feuchtwangen, Ger., 4,500 D5 17
Feudal, Sask., Can F2 70
Feurs, Fr., 6,252 E6 14
Feversham, Ont., Can., 151 C4 72
Fez (Fès), Mor., 830,999 C3 44, G4 30
Fezzan (Fazzān), prov., Libya, 54,438 D2 43
Ffestiniog, Wales, 6,923 D5 12
Fiambala, Arg E2 55
Fianarantsoa, Malag., 16,943 h9 49
Fianarantsoa, prov., Malag h9 49
Fianga, Chad, 1,952 D3 46
Fibre, Chippewa, Mich., 20 B6 98
Ficarolo, It., 4,821 E6 18
Fich, Eth D5 47
Fichot, is., Newf., Can C4 75
Fichtel Gebirge, mts., Ger C6 17
Ficklin, Wilkes, Ga., 100 C4 87
Ficksburg, S. Afr., 7,778 C4 49
Fidalgo, isl., Wash A3 122
Fidelis, Santa Rosa, Fla., 150 G2 86
Fidenza, It., 9,800 B3 21
Field, B.C., Can., 399 D9 68
Field, Curry, N. Mex., 25 C6 107
Field, Socorro, N. Mex C2 107
Field, lake, La E5 95
Fieldale, Henry, Va., 1,499 E3 121
Fielding, Sask., Can., 82 E2 70
Fielding, Box Elder, Utah, 270 B3 119
Fieldon, Jersey, Ill., 239 D3 90
Fields, Harney, Oreg., 10 E8 113
Fieldsboro, Burlington, N.J., 583 C3 106
Fier, Alb., 11,000 B2 23
Fierro, Grant, N. Mex., 250 E1 107
Fiesch, Switz., 567 D5 19
Fiesole, It., 5,200 C3 21
Fife, Pierce, Wash., 1,463 *B3 122
Fife, co., Scot., 320,541 E5 13
Fife, lake, Sask., Can H3 70
Fife Lake, Sask., Can., 144 H3 70
Fife Lake, Grand Traverse, Mich., 218 D5 98
Fife Ness, cape, Scot D6 13
Fifield, Price, Wis., 300 C3 124
Fifteen Mile, creek, Wyo A4 125
Fifteen Mile Falls, res., N.H. and Vt B3 105
5th. Cataract (Nile River), Sud B3 47
Fifty Lakes, Crow Wing, Minn., 143 D4 99
Figeac, Fr., 6,933 E4 14
Figtree, S. Rh B4 49
Figueira da Foz, Port., 10,486 B1 20
Figueras, Sp., 17,548 A7 20
Figuig, Mor., 12,108 C4 44
Figure Five, Crawford, Ark., 50 B1 81
Fiji, Br. dep., Oceania, 345,737 H11 7, 52
Filbert, McDowell, W. Va., 950 D3 123
Filchner, cape, Ant C22 5
Filchner, ice shelf, Ant B8 5
File, lake, Man., Can B1 71
File, riv., Man., Can B1 71
Filer, Twin Falls, Idaho, 1,249 G4 89
Filer City, Manistee, Mich., 500 D4 98
Filey, Eng., 4,705 F8 13
Fili, Grc., 1,748 g11 23
Filiatra, Grc., 9,209 D3 23
Filicudi, isl., It E5 21
Filingue, Niger D5 45
Filisur, Switz., 318 C8 19
Filley, Gage, Nebr., 149 D9 103
Fillmore, Ventura, Calif., 4,725 E4 82
Fillmore, Sask., Can., 340 H4 70
Fillmore, Montgomery, Ill., 360 D4 90
Fillmore, Putnam, Ind., 550 E4 91
Fillmore, Andrew, Mo., 254 A3 101
Fillmore, Dona Ana, N. Mex E3 107
Fillmore, Allegany, N.Y., 522 C2 108
Fillmore, Benson, N. Dak., 125 A6 110
Fillmore, Johnston, Okla., 100 C5 112
Fillmore, Millard, Utah, 1,602 E3 119
Fillmore, co., Minn., 23,768 G6 99
Fillmore, co., Nebr., 9,425 D8 103
Fimi, riv., Con. L B2 48
Finale Emilia, It., 6,157 E7 18
Fina Susu, mtn., Saipan 52
Finca Marini, P.R C2 65
Fincastle, Botetourt, Va., 403 D3 121
Finch, Ont., Can., 386 B9 72
Finchley, Eng., 69,311 k12 10
Finderne, Somerset, N.J., 1,100 *B3 106
Findhorn, Scot C5 13
Findhorn, riv., Scot C5 13
Findlater, Sask., Can., 128 G3 70
Findlay, Shelby, Ill., 759 D5 90
Findlay, Hancock, Ohio, 30,344 A4 111
Findley, mtn., B.C., Can D9 68
Fine, St. Lawrence, N.Y., 250 A5, B2 108
Finegayan, Guam 52
Finesville, Warren, N.J., 275 B2 106
Fingal, Barnes, N. Dak., 190 C8 110
Finger, McNairy, Tenn., 150 B3 117
Fingerville, Spartanburg, S.C., 350 A3 115
Fingoe, Moz A5 49
Finike, Tur., 2,900 D8 23
Finistère, dept., Fr., 727,847 *C2 14
Finisterre, cape, Sp A1 20
Finjasjon, lake, Swe B7 24
Finke, riv., Austl E5 50
Finksburg, Carroll, Md., 385 B4 85
Finland, Lake, Minn., 200 C7 99
Finland, country, Eur., 4,448,500 C14 8, E12 25
Finland, gulf, Fin., Sov. Un H11 25
Finlay, riv., B.C., Can A5 68
Finlay Forks, B.C., Can B6, D2 68
Finlayson, Pine, Minn., 213 D6 99
Finley, Steele, N. Dak., 808 B8 110
Finley, Pushmataha, Okla., 250 C6 112
Finley, Dyer, Tenn., 600 A2 117
Finleyson, Pulaski, Ga., 82 D3 87
Finmoore, B.C., Can., 135 C6 68
Finnegan, Alta., Can D4 69
Finney, co., Kans., 16,093 D3 93
Finneytown, Hamilton, Ohio, 5,000 *C3 111
Finnmark, co., Nor., 72,000 *B13 25
Finschhafen, N. Gui k12 50
Finspång, Swe., 15,200 u33 25
Finsteraarhorn, mtn., Switz C5 19
Finsterwalde, Ger., 20,700 B8 17
Fintona, N. Ire., 1,266 C4 11
Fintown, Ire C3 11
Finvoy, N. Ire., 1,026 B5 11
Finzel, Garrett, Md D2 85
Fionn, bay, Scot C3 13
Fiq, Syr., 1,494 F10 31
Fir, riv., Sask., Can D4 70
Firecrest, Pierce, Wash., 3,565 B3 122
Fire, isl., N.Y G4 84
Fire, isl., N.Y G5 84
Firebaugh, Fresno, Calif., 2,627 D3 82
Fire Island, beach, N.Y G4 84
Fire Island, inlet, N.Y G3 84
Firenze (Florence), It., 436,516 (*535,000) C3 21
Firenzuola, It., 743 B3 21
Firesteel, Dewey, S. Dak., 150 B4 116
Firestone, Weld, Colo., 276 A6 83
Firmat, Arg., 4,051 A4 54
Firminy, Fr., 26,065 E6 14
Fir Mountain, Sask., Can., 74 H2 70
Firozabad, India, 98,611 D7 40
Firsovo, Sov. Un C11 37
1st. Cataract (Nile River), Eg., U.A.R E6 43
Firth, Bingham, Idaho, 322 F6 89
Firth, Lancaster, Nebr., 277 D9 103
Firūzābād, Iran, 23,382 G6 41
Firūzkūh, Iran, 5,874 D6 41
Fish, creek, W. Va C2 123
Fish, mtn., Oreg D4 131
Fish, riv., Maine A4 96
Fish, riv., S.W. Afr C2 49
Fish Cove, pt., Newf., Can A3 75
Fish Creek, Door, Wis., 180 C5 124
Fisheating, creek, Fla F5 86
Fisher, Poinsett, Ark., 303 B5 81
Fisher, Champaign, Ill., 1,155 C5 90
Fisher, Polk, Minn., 326 C2 99
Fisher, Clarion, Pa., 150 D3 114
Fisher, Hardy, W. Va., 20 B5 123
Fisher, co., Tex., 7,865 C2 118
Fisher, bay, Man., Can D3 81
Fisher, peak, Va E2 121
Fisher, riv., Man., Can D3 71
Fisher, strait, N.W. Ter., Can D16 67
Fisher Branch, Man., Can., 369 D3 81
Fisherman, isl., Va B7 121
Fishers, Hamilton, Ind., 344 E5, G8 91
Fishers, isl., N.Y D5 108
Fishers, lake, N.S., Can E4 74
Fishers, peak, Colo D6 83
Fishers Island, sound, Conn., N.Y D8 84
Fishersville, Augusta, Va., 700 C4 121
Fishertown, Bedford, Pa., 250 F4 114
Fisherville, Ont., Can., 234 E5 73
Fisherville, Worcester, Mass. (part of South Grafton) B4 97
Fishguard, Wales, 4,898 C3 12
Fish Haven, Bear Lake, Idaho, 130 G7 89
Fishing, bay, Md D5 85
Fishing, creek, N.C A6 109
Fishing, creek, S.C B5 115
Fishing, creek, W. Va B4 123
Fishing, lake, Man., Can C4 71
Fishing Brook, mtn., N.Y B6 108
Fishing Creek, Dorchester, Md., 544 D5 81
Fishing Creek, Cape May. N.J., 300 E3 106
Fishing Creek, res., S.C B6 115
Fishing Ship Harbour, Newf., Can B4, h11 75
Fishkill, Dutchess, N.Y., 1,033 D7 108
Fishkill, creek, N.Y C1 84
Fish River, lake, Maine B4 96
Fish Springs, range, Utah D2 119
Fishtail, Stillwater, Mont., 50 E7 102
Fishtrap, Lincoln, Wash B8 122
Fishville, Grant, La., 150 C3 95
Fisk, Butler, Mo., 498 E7 101
Fiskburg, Kenton, Ky., 40 B5, A7 94
Fiskdale, Worcester, Mass., 900 B3 97
Fiske, Sask., Can., 89 F1 70
Fiskeville, Providence, R.I., 300 C10 84
Fismes, Fr., 3,222 E3 15
Fitch Bay, Que., Can., 189 D5 73
Fitchburg, Worcester, Mass., 43,021 A4 97
Fitchville, New London, Conn., 500 C8 84
Fithian, Vermilion, Ill., 495 C6 90
Fitler, Issaquena, Miss., 150 C2 100
Fittstown, Pontotoc, Okla., 200 B5 112
Fitzgerald, Ben Hill, Ga., 8,781 E3 87
Fitzhugh, Pontotoc, Okla., 150 C5 112
Fitzhugh, sound, B.C., Can D4 68
Fitzpatrick, Que., Can., 59 B5 73
Fitz Roy, Arg D3 54
Fitzroy, Austl., 88 C4 50
Fitz Roy, mtn., Arg D2 54
Fitzroy, riv., Austl C3 50
Fitzroy, riv., Austl D8 50
Fitzroy Harbor, Ont., Can., 253 B8 72
Fitzwilliam, Cheshire, N.H., 300 (966▲) E2 105
Fitzwilliam, isl., Ont., Can B3 72
Fitzwilliam Depot, Cheshire, N.H., 200 E2 105
Fiume, see Rijeka, Yugo.
Fiumicino, It., 1,121 h8 21
Five Islands, N.S., Can., 194 D5 74
Fivemile, creek, Wyo B4 125
Fivemiletown, N. Ire., 777 C4 11
Five Points, Chambers, Ala., 285 B4 78
Five Points, Los Angeles, Calif., 15,000 *F3 82
Five Points, Marion, Ind., 200 H8 91
Five Points, Wayne, Mich. (part of Redford Heights) A7 98
Five Points, Bernalillo, N. Mex., 2,500 *B3 107
Five Points, Lawrence, Tenn., 115 B4 117
Fizi, Con. L B4 48
Fjärås, Swe A6 24
Fjärdhundra, Swe., 320 t34 25
Fjerritslev, Den., 1,925 A3 24
Flagg, Castro, Tex B1 118
Flagler, Kit Carson, Colo., 693 B7 83
Flagler, co., Fla., 4,566 C5 86
Flagler Beach, Flagler, Fla., 970 C5 86
Flag Pond, Unicoi, Tenn., 75 C11 117
Flagstaff, Coconino, Ariz., 18,214 B4 80
Flagstaff, lake, Oreg E7 113
Flagtown, Somerset, N.J., 250 B3 106
Flamand, lake, Que., Can B4 73
Flambeau, res., Wis B3 124
Flambeau, riv., Wis C3 124
Flamborough, head, Eng C6 10
Flaming Gorge, dam, Utah B6 119
Flaming Gorge, res., Utah, Wyo B6 119, D3 125
Flanagan, Livingston, Ill., 841 C5 90
Flanders, Ont., Can., 57 E5 71
Flanders, Litchfield, Conn., 85 B3 84
Flanders, Morris, N.J., 500 B3 106
Flanders, Suffolk, N.Y., 1,248 F6 84

Flanders, see East Flanders and West Flanders, provs., Bel.
Flanders (Flandre), former prov., Fr., 2,099,000 ..... B5 14
Flanders, bay, N.Y. F6 84
Flandreau, Moody, S. Dak., 2,129 ..... C9 116
Flanigan, Washoe, Nev., 5 ..... C2 104
Flannan, is., Scot A3 10
Flasher, Morton, N. Dak., 515 ..... C4 110
Flat, Alsk., 14 ..... C8 79
Flat, Wolfe, Ky., 74 ..... C6 94
Flat, Phelps, Mo., 75 ..... D6 101
Flat, brook, N.J A3 106
Flat, isl., Newf., Can C3 75
Flat, is., Newf., Can B4 75
Flat, lake, La C5 95
Flat, riv., Mich E5 98
Flat Bay, Newf., Can., 388 ..... D2 75
Flatbrookville, Sussex, N.J., 15 ..... A3 106
Flatbush, Alta., Can., 91 ..... B3 69
Flat Creek, Walker, Ala., 800 ..... B2, E4 78
Flatcreek, Bedford, Tenn., 100 ..... B5 117
Flat Gap, Wise, Va B2 121
Flathead, co., Mont., 32,965 ..... B2 102
Flathead, lake, Mont C2 102
Flathead, mts., Mont B2 102
Flathead, range, Mont B3 102
Flathead, riv., B.C., Can., Mont E10 68, B2 102
Flathead, val., Mont C2 102
Flatlands, N.B., Can., 282 ..... B3 74
Flat Lick, Knox, Ky., 500 ..... D6 94
Flatonia, Fayette, Tex., 1,009 ..... E4 118
Flat River, St. Francois, Mo., 4,515 ..... D7 101
Flat River, res., R.I C10 84
Flat Rock, Jackson, Ala., 600 ..... A4 78
Flat Rock, Crawford, Ill., 497 ..... E6 90
Flat Rock, Shelby, Ind., 250 ..... F6 91
Flat Rock, Wayne, Mich., 4,696 ..... F7 98
Flat Rock, Henderson, N.C., 1,808 ..... D3 109
Flat Rock, Seneca, Ohio, 400 ..... A5 111
Flatrock, creek, Ind F6 91
Flatrock, lake, Man., Can B1 71
Flats, McPherson, Nebr., 4 ..... C4 103
Flats, Macon, N.C D2 109
Flattery, cape, Wash A1 122
Flattop, Platte, Wyo C8 125
Flat Tops, plat., Colo B3 83
Flatts, Bermuda E14 77
Flat Willow, creek, Mont D8 102
Flatwood, Wilcox, Ala., 300 ..... C2 78
Flatwoods, Greenup, Ky., 3,741 ..... B7 94
Flatwoods, Rapides, La., 150 ..... C3 95
Flat Woods, Perry, Tenn., 150 ..... B4 117
Flat Woods, Braxton, W. Va., 248 ..... C4 123
Flawil, Switz., 7,256 ..... B7 19
Flaxcombe, Sask., Can., 123 ..... F1 70
Flaxton, Burke, N. Dak., 375 ..... A3 110
Flaxville, Daniels, Mont., 262 ..... B11 102
Fleet, Alta., Can., 79 ..... C5 69
Fleet, Eng., 13,672 ..... C7 12
Fleeton, Northumberland, Va., 200 ..... D6 121
Fleetwing Estates, Bucks, Pa., 1,100 ..... *F12 114
Fleetwood, Eng., 27,760 ..... G5 13
Fleetwood, Ashe, N.C., 70 ..... A2 109
Fleetwood, Berks, Pa., 2,647 ..... F10 114
Fleischmanns, Delaware, N.Y., 450 ..... C6 108
Flekkefjord, Nor., 3,200 ..... H2 25
Fleming, Sask., Can., 187 ..... G5 70
Fleming, Logan, Colo., 384 ..... A8 83
Fleming, Liberty, Ga., 150 ..... E5 87
Fleming (Unionville), Centre, Pa., 371 ..... E6 114
Fleming, co., Ky., 10,890 ..... B6 94
Flemingsburg, Fleming, Ky., 2,067 ..... B6 94
Flemington, Liberty, Ga., 149 ..... E5 87
Flemington, Polk, Mo., 142 ..... D4 101
Flemington, Hunterdon, N.J., 3,232 ..... B3 106
Flemington, Clinton, Pa., 1,608 ..... D7 114
Flemington, Taylor, W. Va., 478 ..... B7 123
Flen, Swe., 5,700 ..... t34 25
Flensburg, Ger., 98,500 ..... A4 16, D3 24
Flensburg, Morrison, Minn., 280 ..... E4 99
Flensburg, fjord, Den D3 24
Flers, Fr., 14,634 ..... C3 14
Flesherton, Ont., Can., 515 ..... C4 72
Fletcher, Ont., Can., 109 ..... E2 72
Fletcher, Henderson, N.C., 800 ..... D3 109
Fletcher, Miami, Ohio, 569 ..... B3 111
Fletcher, Comanche, Okla., 884 ..... C3 112
Fletcher, Franklin, Vt., 125 (399▲) B3 120
Fletcher, mtn., Vt B3 120
Fletcher Park, Albany, Wyo C7 125
Fleur de Lys, Newf., Can., 457 ..... C3, h10 75
Fleurier, Switz., 3,814 ..... C2 19
Flieden, Ger., 3,400 ..... C4 17
Flims, Switz., 1,444 ..... C7 19
Flinders, is., Austl F5 50
Flinders, isl., Austl n15 50
Flinders, range, Austl F6 50
Flinders, riv., Austl C7 50
Flin Flon, Man., Can., 11,104 ..... B1 71
Flint, Morgan, Ala., 432 ..... A3 78
Flint, Steuben, Ind., 60 ..... A7 91
Flint, Genesee, Mich., 196,940 (★379,900) E7 98
Flint, Wales, 13,690 ..... A4 12
Flint, co., Wales, 149,888 ..... A4 12
Flint, isl., Pac. O G12 7
Flint, riv., Ga D2 87
Flint, riv., Mich E5 98
Flint, run, W. Va A6 123
Flint Creek, range, Mont D4 102
Flinthill, St. Charles, Mo., 200 ..... C7 101
Flint Hill, Rappahannock, Va., 200 ..... C4 121
Flintridge, Los Angeles, Calif., 5,000 ..... *F2 82
Flintoft, Sask., Can., 56 ..... H2 70
Flinton, Ont., Can., 182 ..... C7 72
Flintstone, Allegany, Md. D2 85
Flintville, Lincoln, Tenn., 175 ..... B5 117
Flipper, pt., Wake Isl 52
Flippin, Marion, Ark., 433 ..... A3 81
Flippin, Monroe, Ky., 150 ..... D4 94
Flixecourt, Fr., 3,285 ..... D2 15
Flöha, Ger., 6,876 ..... C8 17
Flom, Norman, Minn., 75 ..... C2 99
Flomaton, Escambia, Ala., 1,454 ..... D2 78
Flomot, Motley, Tex., 475 ..... B2 118
Floodwood, Dickinson, Mich B2 98
Floodwood, St. Louis, Minn., 677 ..... D6 99
Flora, Clay, Ill., 5,331 ..... E5 90
Flora, Carroll, Ind., 1,742 ..... C4 91
Flora, Madison, Miss., 743 ..... C3 100
Flora, Benson N. Dak., 38 ..... B6 110
Flora, Wallowa, Oreg., 75 ..... B9 113
Florahome, Putnam, Fla., 400 ..... C5 86
Florala, Covington, Ala., 3,011 ..... D3 78
Floral City, Citrus, Fla., 900 ..... D4 86
Floral Park, Silver Bow, Mont., 4,079 ..... *E4 102
Floral Park, Nassau, N.Y., 17,499 ..... D2 108
Flora Vista, San Juan, N. Mex., 700 ..... A1 107
Flordell Hills, St. Louis, Mo., 1,119 ..... *A8 101
Florence, Lauderdale, Ala., 31,649 ..... A2 78
Florence, Pinal, Ariz., 2,143 ..... E4 80
Florence, Los Angeles, Calif., 38,164 ..... *E4 82
Florence, Fremont, Colo., 2,821 ..... C5 83
Florence, Switzerland, Ind., 150 ..... G8 91
Florence, see Firenze, It.
Florence, Marion, Kans., 853 ..... D7 93
Florence, Boone, Ky., 5,837 ..... A5, A7 94
Florence, Hampshire, Mass. (part of Northampton) B2 97
Florence, Rankin, Miss., 360 ..... C3 100
Florence, Morgan, Mo., 65 ..... C5 101
Florence, Ravalli, Mont., 150 ..... D2 102
Florence, Burlington, N.J., 4,215 ..... C3 106
Florence, Lane, Oreg., 1,642 ..... *D2 113
Florence, Florence, S.C., 24,722 ..... C8 115
Florence, Codington, S. Dak., 216 ..... B8 116
Florence, Rutherford, Tenn B5 117
Florence, Williamson, Tex., 610 ..... D4 118
Florence, Rutland, Vt., 80 ..... D2 120
Florence, Snohomish, Wash., 45 ..... A3 122
Florence, Florence, Wis., 700 ..... C5 124
Florence, co., S.C., 84,438 ..... C8 115
Florence, co., Wis., 3,437 ..... C5 124
Florence Junction, Pinal, Ariz., 25 ..... D4 80
Florenceville, N.B., Can., 229 ..... C2 74
Florencia, Col., 8,119 ..... C2 60
Florenville, Bel., 2,378 ..... E5 15
Florenville, St. Tammany, La B8 95
Flores, Guat., 1,596 ..... B3 62
Flores, dept., Ur., 22,900 ..... *E1 56
Flores, isl., Az. Is g8 44
Flores, isl., Indon G6 35
Flores, sea, Indon G6 35
Flores Island, B.C., Can E4 68
Floresta, Braz., 2,377 ..... C3 57
Floresville, Wilson, Tex., 2,126 ..... E3, B4 118
Florey, Andrews, Tex., 50 ..... C1 118
Florham Park, Morris, N.J., 7,222 ..... B4 106
Floriano, Braz., 16,063 ..... C2 57
Florianópolis, Braz., 74,323 ..... D3 56
Florida, Cuba, 21,159 ..... E4 64
Florida, Berkshire, Mass., 80 (569▲) A1 97
Florida, Monroe, Mo., 55 ..... B6 101
Florida, Luna, N. Mex E2 107
Florida, Socorro, N. Mex., 50 ..... C3 107
Florida, Orange, N.Y., 1,550 ..... D2, D6 108
Florida, Henry, Ohio, 290 ..... A3 111
Florida [Adentro], P.R., 2,955 ..... B4 65
Florida, Ur., 15,500 ..... E1 56
Florida, dept., Ur., 61,200 ..... *E1 56
Florida, state, U.S., 4,951,560 ..... E11 77, 86
Florida, bay, Fla H6 86
Florida, cape, Fla G6 86
Florida, keys, Fla H6 86
Florida, mts., N. Mex E2 107
Florida, straits, Fla H6 86
Florida City, Dade, Fla., 4,114 ..... F3, G6 86
Florien, Sabine, La., 496 ..... C2 95
Florin, Lancaster, Pa., 1,518 ..... F8 114
Florina, Grc., 11,933 ..... B3 23
Florina, prov., Grc., 67,356 ..... *B3 23
Floris, Davis, Iowa, 187 ..... D5 92
Floris, Fairfax, Va., 75 ..... B4 121
Florissant, Teller, Colo., 40 ..... C5 83
Florissant, St. Louis, Mo., 38,166 ..... A8 101
Florosa, Okaloosa, Fla., 150 ..... G2 86
Flossmoor, Cook, Ill., 4,624 ..... F3 90
Flour Bluff, Nueces, Tex., 3,500 ..... *F4 118
Flournoy, Tehama, Calif., 50 ..... C2 82
Flourtown, Montgomery, Pa., 4,000 ..... A11 114
Flovilla, Butts, Ga., 284 ..... C3 87
Flower, riv., Vt E2 120
Floweree, Chouteau, Mont., 20 ..... C5 102
Flower Hill, Nassau, N.Y., 4,594 ..... E2 108
Flowers Cove, Newf., Can., 312 ..... C3, h10 75
Flower Station, Ont., Can., 74 ..... B8 72
Flower Village, Kern, Calif., 5,000 ..... *E4 82
Flowery Branch, Hall, Ga., 741 ..... B3 87
Flowood, Rankin, Miss., 486 ..... C3 100
Floyd, Floyd, Iowa, 401 ..... A5 92
Floyd, Roosevelt, N. Mex., 200 ..... C6 107
Floyd, Floyd, Va., 487 ..... E2 121
Floyd, co., Ga., 69,130 ..... B1 87
Floyd, co., Ind., 51,397 ..... H6 91
Floyd, co., Iowa, 21,102 ..... A5 92
Floyd, co., Ky., 41,642 ..... C7 94
Floyd, co., Tex., 12,369 ..... B2 118
Floyd, co., Va., 10,462 ..... E2 121
Floyd, mtn., Ariz B3 80
Floyd, riv., Iowa B1 92
Floydada, Floyd, Tex., 3,769 ..... C2 118
Floyd Dale, Dillon, S.C., 125 ..... C9 115
Floyds, canyon, Nev C4 104
Floyds, fork, Ky B4 94
Floydsburg, Oldham, Ky., 75 ..... A4 94
Floyds Knobs, Floyd, Ind., 300 ..... H6 91
Fluchthorn, mtn., Ger., Switz C9 19
Fluessenmeer, lake, Neth B5 15
Flumendosa, riv., It E2 21
Flums, Switz., 4,462 ..... B7 19
Flushing, Genesee, Mich., 3,761 ..... E7 98
Flushing, Belmont, Ohio, 1,189 ..... B6 111
Fluvanna, Scurry, Tex., 300 ..... C2 118
Fluvanna, co., Va., 7,227 ..... D4 121
Fly Creek, Otsego, N.Y., 300 ..... C6 108
Flying Fish, cape, Ant B3 5
Flying H, Chaves, N. Mex., 30 ..... D4 107
Flynns Lick, Jackson, Tenn., 50 ..... C8 117
Foam Lake, Sask., Can., 933 ..... F4 70
Foard, co., Tex., 3,125 ..... B3 118
Foard City, Foard, Tex C3 118
Foča, Yugo., 6,762 ..... D4 22
Fochabers, Scot C5 13
Fochimi Hoyoudine, well, Chad B3 46
Focşani, Rom., 28,244 ..... C8 22
Fogauso, cape, Am. Sam 52
Foggia, It., 118,608 ..... D5 21
Foggo, Nig D6 45
Fogliano, It k9 21
Fogo, Newf., Can., 1,184 ..... D4 75
Fogo, dist., Newf., Can D4 75
Fogo, cape, Newf., Can D4 75
Fogo, isl., C.V. Is *E4 42
Fogo, isl., Newf., Can k27 67
Fohnsdorf, Aus., 11,517 ..... E7 16
Föhr, isl., Ger A4 16
Foix, Fr., 7,164 ..... F4 14
Foix, former prov., Fr., 81,000 ..... *F4 14
Fokis (Phocis), prov., Grc., 47,842 ..... *C4 23
Folcroft, Delaware, Pa., 7,013 ..... *G11 114
Foley, Baldwin, Ala., 2,889 ..... E2 78
Foley, Taylor, Fla., 200 ..... B3 86
Foley, Benton, Minn., 1,112 ..... E5 99
Foley, Lincoln, Mo., 183 ..... B7 101
Foligno, It., 31,500 ..... C4 21
Folkestone, Eng., 44,129 ..... C9 12
Folkston, Charlton, Ga., 1,810 ..... F4 87
Folkstone, Onslow, N.C., 75 ..... C6 109
Follansbee, Brooke, W. Va., 4,052 ..... A4, B2 123
Follett, Lipscomb, Tex., 466 ..... A2 118
Follonica, It., 7,147 ..... C3 21
Folly, hill, Mass C3 97
Folly, isl., S.C F8 115
Folly Beach, Charleston, S.C., 1,137 ..... F18, G3 115
Follyfarm, Malheur, Oreg D8 113
Folly Lake, N.S., Can., 68 ..... D6 74
Folsom, Sacramento, Calif., 3,925 ..... A6, C3 82
Folsom, St. Tammany, La., 225 ..... D5 95
Folsom, Atlantic, N.J., 482 ..... D3 106
Folsom, Union, N. Mex., 142 ..... A6 107
Folsom, Delaware, Pa., 5,000 ..... *B11 114
Folsom, Wetzel, W. Va., 450 ..... A6, B4 123
Folsomville, Warrick, Ind., 130 ..... H3 91
Fomento, Cuba, 6,038 ..... D4 64
Fonda, Pocahontas, Iowa, 1,026 ..... B3 92
Fonda, Montgomery, N.Y., 1,004 ..... C6 108
Fond du Lac, Sask., Can., 253 ..... E11 66
Fond du Lac, Fond du Lac, Wis., 32,719 ..... B5, E5 124
Fond du Lac, co., Wis., 75,085 ..... E5 124
Fond du Lac, Indian res., Minn D6 99
Fonde, Bell, Ky., 200 ..... D6 94
Fondi, It., 13,700 ..... D4 21
Fonsagrada, Sp., 950 ..... A2 20
Fonseca, gulf, Cen. Am D7 62
Fontainebleau, Fr., 20,583 ..... C5 14
Fontana, San Bernardino, Calif., 14,659 ..... F3 82
Fontana, Miami, Kans., 138 ..... D9 93
Fontana, Walworth, Wis., 1,326 ..... F5 124
Fontana Dam, Graham, N.C., 250 ..... D2 109
Fontanelle, Adair, Iowa, 729 ..... C3 92
Fontanelle, Washington, Nebr., 50 ..... C9 103
Fontanet, Vigo, Ind., 200 ..... E3 91
Fonte Boa, Braz., 1,154 ..... D4 60
Fontenay-le-Comte, Fr., 10,109 ..... D3 14
Fontenay [-sous-Bois], Fr., 37,484 ..... g10 14
Fontenelle, Lincoln, Wyo C2 125
Fontenelle, mtn., Wyo C2 125
Fonthill, Ont., Can., 1,872 ..... *D5 72
Foochow (Minhow), China, 616,000 ..... F8 34
Foosland, Champaign, Ill., 150 ..... C5 90
Foothill, Spokane, Wash D8 122
Foothills, Alta., Can., 224 ..... C2 69
Footville, Rock, Wis., 675 ..... F4 124
Foping, China, 500 ..... H2 36
Forada, Douglas, Minn., 98 ..... E3 99
Foraker, Hardin, Ohio, 160 ..... B4 111
Foraker, Osage, Okla., 74 ..... A5 112
Forbach, Fr., 21,704 (★58,000) C7 14
Forbes, Austl., 6,826 ..... F7 51
Forbes, St. Louis, Minn., 25 ..... C6 99
Forbes, Holt, Mo., 100 ..... B2 101
Forbes, Mitchell, N.C., 100 ..... C4 109
Forbes, Dickey, N. Dak., 138 ..... D7 110
Forbes, mtn., Alta., Can D2 69
Forbing Park, Yavapai, Ariz., 300 ..... C3 80
Forcados, Nig., 3,001 ..... E6 45
Forcalquier, Fr., 2,050 ..... F6 14
Force, Elk, Pa., 400 ..... D4 114
Forchheim, Ger., 20,900 ..... D6 17
Ford, Kootenai, Idaho E8 89
Ford, Ford, Kans., 252 ..... E4 93
Ford, Clark, Ky., 250 ..... C5 94
Ford, Scot D3 13
Ford, co., Ill., 16,606 ..... C5 90
Ford, co., Kans., 20,938 ..... E4 93
Ford, riv., Mich C3 98
Ford City, Kern, Calif., 3,926 ..... E4 82
Ford City, Armstrong, Pa., 5,440 ..... E2 114
Fordland, Webster, Mo., 338 ..... D5 101
Fordlândia, Braz C3 59
Fords, Middlesex, N.J., 10,000 ..... *B4 106
Fords Prairie, Lewis, Wash., 1,404 ..... *C3 122
Fordsville, Ohio, Ky., 524 ..... C3 94
Fordville, Walsh, N. Dak., 367 ..... A8 110
Fordwich, Ont., Can., 267 ..... D3 72
Fordwick, Augusta, Va., 150 ..... C3 121
Fordyce, Dallas, Ark., 3,890 ..... D3 81
Fordyce, Cedar, Nebr., 143 ..... B8 103
Forécariah, Guinea, 5,300 ..... E2 45
Foreman (New Rocky Comfort), Little River, Ark., 1,001 ..... D1 81
Foremost, Alta., Can., 561 ..... E5 69
Forest, Bel., 51,503 ..... D4 15
Forest, Ont., Can., 2,188 ..... D2 72
Forest, Livingston, Ill., 1,040 ..... C5 90
Forest, Clinton, Ind., 400 ..... D5 91
Forest, Scott, Miss., 3,917 ..... C4 100
Forest, Hardin, Ohio, 1,314 ..... B4 111
Forest, Cherokee, Tex., 100 ..... D5 118
Forest, Bedford, Va., 250 ..... D3 121
Forest, co., Pa., 4,485 ..... C3 114
Forest, co., Wis., 7,542 ..... C5 124
Forest Acres, Richland, S.C., 3,842 ..... *D6 115
Forestburg, Alta., Can., 677 ..... C4 69
Forestburg, Sanborn, S. Dak., 150 ..... C8 116
Forest Center, Lake, Minn., 150 ..... C7 99
Forest City, Mason, Ill., 249 ..... C4 90
Forest City, Winnebago, Iowa, 2,930 ..... A4 92
Forest City, Washington, Maine, 25 ..... C5 96
Forest City, Holt, Mo., 435 ..... B2 101
Forest City, Rutherford, N.C., 6,556 ..... B2, D4 109
Forest City, Susquehanna, Pa., 2,651 ..... C11 114
Forestdale, Clarke, Ala D2 78
Forestdale, Providence, R.I., 500 ..... B10 84
Forest Dale, Rutland, Vt., 450 ..... D2 120
Forester, Scott, Ark C2 81
Forest Glen, Montgomery, Md., 215 ..... B3 85
Forest Green, Chariton, Mo., 100 ..... B5 101
Forestgrove, Fergus, Mont., 20 ..... D7 102
Forest Grove, Washington, Oreg., 5,628 ..... B1, B3 113
Forest Heights, Prince Georges, Md., 3,524 ..... *C1 85
Forest Hill, Ont., Can., 20,489 ..... E6 72
Forest Hill, Rapides, La., 302 ..... C3 95
Forest Hill, Harford, Md., 200 ..... A5 85
Forest Hills, Davidson, Tenn., 2,101 ..... *A5 117
Forest Hill, Tarrant, Tex., 3,221 ..... *C4 118
Forest Hills, Allegheny, Pa., 8,796 ..... F2 114
Forest Home, Butler, Ala., 200 ..... D3 78
Forest Homes, Madison, Ill., 2,025 ..... *E3 90
Forest Junction, Calumet, Wis., 200 ..... A6 124
Forest Knolls, Marin, Calif., 800 ..... B5 82
Forest Lake, Alger, Mich., 60 ..... B4 98
Forest Lake, Washington, Minn., 2,347 ..... E6 99
Forest Lawn, Alta., Can., 12,263 ..... D4 69
Foreston, N.B., Can., 40 ..... C2 74
Foreston, Mille Lacs, Minn., 266 ..... E5 99
Foreston, Clarendon, S.C., 210 ..... D7 115
Forest Park, Clayton, Ga., 14,201 ..... B5, C2 87
Forest Park, Cook, Ill., 14,452 ..... F2 90
Forestport, Oneida, N.Y., 250 ..... B2 108
Forest River, Walsh, N. Dak., 191 ..... A8 110
Forest Station, Washington, Maine, 35 ..... C5 96
Forest View, Cook, Ill., 1,042 ..... *F3 90
Forest View, Greenville, S.C., 1,000 ..... *B3 115
Forestville, Prince Georges, Md., 1,500 ..... *C4 85
Forestville, Chautauqua, N.Y., 905 ..... C1 108
Forestville, Butler, Pa., 300 ..... D1 114
Forestville, Door, Wis., 300 ..... D6 124
Forfar, Scot., 10,252 ..... D6 13
Forgan, Sask., Can., 93 ..... F2 70
Forgan, Beaver, Okla., 532 ..... A1, D4 112
Forges-les-Eaux, Fr., 2,914 ..... C4 14
Forget, Sask., Can., 220 ..... H4 70
Forge Village, Middlesex, Mass., 1,191 ..... A5, C1 97
Fork, Davie, N.C., 85 ..... B3 109
Fork, Dillon, S.C., 168 ..... C9 115
Fork, creek, W. Va D5 123
Forked Deer, Haywood, Tenn., 100 ..... B2 117
Forked Deer, riv., Tenn B2 117
Forked River, Ocean, N.J., 800 ..... D4 106
Forkland, Greene, Ala., 125 ..... C2 78
Fork Mountain, Anderson, Tenn., 150 ..... C9 117
Fork Ridge, Claiborne, Tenn., 200 ..... C10 117
Fork River, Man., Can., 174 ..... D1 71
Forks, Clallam, Wash., 1,156 ..... B1 122
Fork Shoals, Greenville, S.C., 200 ..... B3 115
Forks of Elkhorn, Franklin, Ky., 172 ..... B5 94
Forksville, Sullivan, Pa., 131 ..... D8 114
Fork Union, Fluvanna, Va., 200 ..... D4 121
Forkville, Scott, Miss., 100 ..... *C4 100
Forli, It., 59,500 (91,945▲) B4 21
Forman, Sargent, N. Dak., 530 ..... C8 110
Formazza, It., 570 ..... C4 18
Formby, Eng., 11,730 ..... A4 12
Formello, It., 2,067 ..... g8 21
Formentera, isl., Sp C6 20
Formiga, Braz., 18,763 ..... F1 57
Formosa, Arg., 16,506 ..... E4 55
Formosa, Van Buren, Ark., 190 ..... B3 81
Formosa, Braz., 9,449 ..... E1 57
Formosa, Ont., Can., 370 ..... C3 72
Formosa, prov., Arg., 178,458 ..... E4 55
Formosa, see Taiwan, country, Asia.
Formosa, bay, Ken B7 48
Formosa, strait, China G8 34
Formoso, Jewell, Kans., 192 ..... C6 93
Forney, Kaufman, Tex., 1,544 ..... B6, C4 118
Fornovo di Taro, It., 6,920 ..... E6 18
Forres, Scot., 4,780 ..... C5 13
Forrest, Austl., 20 ..... F4 50
Forrest, Livingston, Ill., 1,220 ..... C5 90
Forrest, Quay, N. Mex., 140 ..... C6 107
Forrest, co., Miss., 52,722 ..... D4 100
Forrest City, St. Francis, Ark., 10,544 ..... B5 81
Forreston, Ogle, Ill., 1,153 ..... A4 90
Forrest Station, Man., Can., 75 ..... E2 71
Forsan, Howard, Tex., 300 ..... C2 118
Forsayth, Austl., 100 ..... C7 50
Forserum, Swe., 1,786 ..... A8 24
Förslövsholm, Swe., 747 ..... B6 24
Forst, Ger., 28,700 ..... B9 17
Forsyth, Monroe, Ga., 3,697 ..... C3 87
Forsyth, Macon, Ill., 424 ..... D5 90
Forsyth, Marquette, Mich., 200 ..... B3 98
Forsyth, Taney, Mo., 489 ..... E4 101
Forsyth, Rosebud, Mont., 1,879 ..... D10 102
Forsyth, co., Ga., 12,170 ..... B2 87
Forsyth, co., N.C., 189,428 ..... A3 109
Fort Adams, Wilkinson, Miss., 100 ..... D2 100
Fort Albany, Ont., Can E9 72
Fortaleza, Braz., 354,942 (★525,000) B3 57
Fort Ann, Washington, N.Y., 453 ..... B7 108
Fort Apache, Navajo, Ariz., 60 ..... D6 80
Fort Apache, Indian res., Ariz C5 80
Fort-Archambault, Chad, 22,228 ..... D3 46
Fort Ashby, Mineral, W. Va., 900 ..... B6 123
Fort Assiniboine, Alta., Can., 216 ..... B3 69
Fort Atkinson, Winneshiek, Iowa, 353 ..... A6 92
Fort Atkinson, Jefferson, Wis., 7,908 ..... F5 124
Fort Augustus, Scot C4 13
Fort Banya, Ken A6 48
Fort Barnwell, Craven, N.C., 300 ..... B6 109
Fort Basinger, Highlands, Fla., 140 ..... E5 86
Fort Bayard, Grant, N. Mex E1 107
Fort Beaufort, S. Afr., 9,748 ..... D4 49
Fort Belfontaine, St. Louis, Mo., 40 ..... A8 101
Fort Belknap, Blaine, Mont., 200 ..... B8 102
Fort Belknap Agency, Blaine, Mont., 195 ..... B8 102
Fort Belknap, Indian res., Mont B8 102
Fort Bend, co., Tex., 40,527 ..... E5 118
Fort Benning, Chattahoochee, Ga D2 87
Fort Benton, Chouteau, Mont., 1,887 ..... C6 102
Fort Berthold, Indian res., N. Dak B3 110
Fort Bidwell, Modoc, Calif., 300 ..... B3 82
Fort Bidwell, Indian res., Calif B3 82
Fort Blackmore, Scott, Va., 250 ..... B2 121
Fort Bragg, Mendocino, Calif., 4,433 ..... C2 82
Fort Branch, Gibson, Ind., 1,983 ..... H2 91
Fort Bridger, Uinta, Wyo., 150 ..... D2 125
Fort Calhoun, Washington, Nebr., 458 ..... C9 103
Fort Chadbourne, Coke, Tex D2 118
Fort Charlet, see Djanet, Alg.
Fort Chimo, Que., Can., 480 ..... A3 73
Fort Chipewyan, Alta., Can., 717 ..... E10 66
Fort Clark, Oliver, N. Dak., 30 ..... B4 110
Fort Cobb, Caddo, Okla., 687 ..... B3 112
Fort Cobb, res., Okla B3 112
Fort Collins, Larimer, Colo., 25,027 ..... A5 83
Fort Collins West, Larimer, Colo., 1,569 ..... *A5 83
Fort Coulonge, Que., Can., 1,823 ..... B8 72
Fort Covington, Franklin, N.Y., 976 ..... A1 108
Fort Crampel, Cen. Afr. Rep D3 46
Fort Crook, Sarpy, Nebr., 75 ..... E3 103
Fort-Dauphin, Malag., 7,253 ..... h9 49
Fort Davis, Macon, Ala., 350 ..... C4 78
Fort Davis, Jeff Davis, Tex., 900 ..... F2 118
Fort Defiance, Apache, Ariz., 750 ..... B6 80
Fort-de-France, Mart., 60,648 ..... o16 64
Fort Deposit, Lowndes, Ala., 1,466 ..... D3 78
Fort Dick, Del Norte, Calif., 150 ..... B1 82
Fort Dix, Burlington, N.J C3 106
Fort Dodge, Webster, Iowa, 28,399 ..... B3 92
Fort Dodge, Ford, Kans E4 93
Fort Donelson, nat. military park and cemetery, Tenn A4 117
Fort Duchesne, Uintah, Utah, 200 ..... C6 119
Forteau, Newf., Can., 232 ..... C3 75
Fort Edward, Washington, N.Y., 3,737 ..... B7 108
Fort Erie, Ont., Can., 9,027 ..... E6 72
Fortescue, Holt, Mo., 78 ..... A2 101
Fortescue, Cumberland, N.J., 200 ..... E2 106
Fortescue, riv., Austl D2 50
Fort Fairfield, Aroostook, Maine, 3,082 (5,876▲) B5 96
Fort Fitzgerald, Alta., Can., 149 ..... E10 66
Fort Flatters, Alg D6 44
Fort Foureau, Cam., 1,004 ..... C2 46
Fort Frances, Ont., Can., 9,481 ..... E6 72
Fort Fraser, B.C., Can., 256 ..... B5 68
Fort Gaines, Clay, Ga., 1,320 ..... E1 87
Fort Garland, Costilla, Colo., 500 ..... D5 83
Fort Garry, Man., Can., 50 ..... E3 71
Fort Gay, Wayne, W. Va., 739 ..... C2 123
Fort George, B.C., Can., 300 ..... C6 68
Fort George, Que., Can., 1,074 ..... B2 73
Fort George, Duval Fla., 250 ..... B6 86
Fort George, riv., Que., Can B2 73
Fort Gibson, Muskogee, Okla., 1,407 ..... B6 112
Fort Good Hope, N.W. Ter., Can., 292 ..... C7 66
Fort-Gouraud, Maur B2 45
Fort Grahame, B.C., Can A5 68
Fort Green, Hardee, Fla., 50 ..... E5 86
Fort Griffin, Shackelford, Tex C3 118
Forth, firth, Scot D6 13
Forth, riv., Scot D4 13
Fort Hall, Bingham, Idaho, 700 ..... F6 89
Fort Hall, Ken B6 48
Fort Hall, Indian res., Idaho F6 89
Fort Hancock, Hudspeth, Tex., 600 ..... F1 118
Fort Harrison, Lewis and Clark, Mont., 350 ..... D4 102
Fort Henry, Stewart, Tenn., 50 ..... A3 117, 85
Fort Hill, Nya C5 48
Fort Hill, Somerset, Pa., 150 ..... G3 114
Fort Howard, Baltimore, Md., 375 ..... B5 85
Fort Huachuca, Cochise, Ariz F5 80
Fortierville, Que., Can., 558 ..... C5 73
Fortín Uno, Arg B3 54
Fortine, Lincoln, Mont., 150 ..... B2 102
Fort Jameson, N. Rh., 3,500 ..... D5 48
Fort Jefferson, nat. mon., Fla H4 86
Fort Jennings, Putnam, Ohio, 436 ..... B3 111
Fort Johnston, Nya., 950 ..... D6 48
Fort Jones, Siskiyou, Calif., 483 ..... B2 82
Fort Kent, Aroostook, Maine, 2,787 (4,761▲) A4 96
Fort Kent Village, Aroostook, Maine (part of Fort Kent) A4 96
Fort Keogh, Custer, Mont., 265 ..... D11 102
Fort Klamath, Klamath, Oreg., 400 ..... E5 113
Fort Knox, Hardin, Ky C4 94
Fort Lallemand, Alg C6 44
Fort-Lamy, Chad, 23,470 ..... C3 46
Fort Landing, Tyrrell, N.C B7 109
Fort Langley, B.C., Can., 962 ..... A10 68
Fort Laperrine (Tamanrasset), Alg., 1,714 (10,089▲) E6 44
Fort Laramie, Goshen, Wyo., 233 ..... C8 125
Fort Laramie, nat. mon., Wyo C8 125
Fort Lauderdale, Broward, Fla., 83,648 ..... E3, F6 86
Fort Lawn, Chester, S.C., 192 ..... B6 115
Fort Leavenworth, Leavenworth, Kans B8, C9 93
Fort Lee, Bergen, N.J., 21,815 ..... D5 106
Fort Liard, N.W., Ter., Can., 154 ..... D8 66
Fort-Liberté, Hai., 1,900 ..... F8 64
Fort Lincoln, Burleigh, N. Dak., 150 ..... C5 110
Fort Littleton, Fulton, Pa., 100 ..... F6 114
Fort Loramie, Shelby, Ohio, 687 ..... B3 111
Fort Loudon, Franklin, Pa., 500 ..... G6 114
Fort Loudon, lake, Tenn D9 117
Fort Lupton, Weld., Colo., 2,194 ..... A6 83
Fort Lyon, Bent, Colo., 260 ..... C7 83
Fort McDermitt, Indian res., Nev B4 104
Fort McDowell, Maricopa, Ariz., 150 ..... F2 80
Fort McDowell, Indian res., Ariz D4, F3 80
Fort McHenry, nat. mon. and historical shrine, Md C2 85
Fort McIntosh, Webb, Tex. (part of Laredo) F3 118
Fort McKavett, Menard, Tex., 150 ..... D2 118
Fort McKenzie, Que., Can., 181 ..... A2 73
Fort McKinley, Montgomery, Ohio, 1,000 ..... *C3 111
Fort Macleod, Alta., Can., 2,490 ..... E4 69
Fort MacMahon, Alg D5 44
Fort McPherson, N.W. Ter., Can., 509 ..... C6 66
Fort Madison, Lee, Iowa, 15,247 ..... D6 92
Fort Manning, Nya D5 48
Fort Matanzas, nat. mon., Fla C5 86

Fort Meade, Polk, Fla., 4,014 . E5 86
Fort Meade, Meade, S. Dak., 250 . C2 116
Fort Meadow, res., Mass . D1 97
Fort Mill, York, S.C., 3,315 . A6 115
Fort Miller, Washington, N.Y., 170 . B7 108
Fort Missoula, Missoula, Mont . D2 102
Fort Mitchell, Russell, Ala., 25 . C4 78
Fort Mitchell, Lunenburg, Va., 150 . E4 121
Fort Mohave, Indian res., Ariz . C1 80
Fort Morgan, Baldwin, Ala., 100 . E1 78
Fort Morgan, Morgan, Colo., 7,379 . A7 83
Fort Motte, Calhoun, S.C., 386 . D6 115
Fort Munro, Pak . C2 40
Fort Myers, Lee, Fla., 22,523 . F5 86
Fort Myers Beach, Lee, Fla., 2,463 . F5 86
Fort Nelson, B.C., Can., 1,607 . D2 66
Fort Norman, N.W. Ter., Can., 189 . D7 66
Fort Odgen, DeSoto, Fla., 150 . E5 86
Fort Oglethorpe, Catoosa and Walker, Ga., 2,251 . B1 87
Fort Payne, DeKalb, Ala., 7,029 . A4 78
Fort Peck, Valley, Mont., 950 . B10 102
Fort Peck, dam, Mont . B10 102
Fort Peck, res., Mont . C10 102
Fort Pierce, St. Lucie, Fla., 25,256 . E6 86
Fort Pierce, inlet, Fla . E6 86
Fort Pierre, Stanley, S. Dak., 2,649 . C5 116
Fort Pierre Bordes, see Tin Zaouaten, Alg.
Fort Plain, Montgomery, N.Y., 2,809 . C6 108
Fort Polignac, Alg . D6 44
Fort Portal, Ug., 8,317 . A5 48
Fort Providence, N.W. Ter., Can., 402 . D9 66
Fort Pulaski, nat. mon., Ga . D6 87, G6 115
Fort Qu'Appelle, Sask., Can., 1,521 . G4 70
Fort Randall, dam., S. Dak . D7 116
Fort Randall, res., S. Dak . D6 116
Fort Ransom, Ransom, N. Dak., 200 . C8 110
Fort Recovery, Mercer, Ohio, 1,336 . B3 111
Fort Reno, Canadian, Okla., 50 . B3 112
Fort Resolution, N.W. Ter., Can., 485 . D10 66
Fort Rice, Morton, N. Dak., 15 . C5 110
For Riley, Geary, Kans . C7 93
Fort Ripley, Crow Wing, Minn., 55 . D4 99
Fort Ritner, Lawrence, Ind., 120 . G5 91
Fort Robinson, Dawes, Nebr., 40 . B2 103
Fort Robinson, Sullivan, Tenn., 2,000 . *C5 117
Fort Rock, Lake, Oreg., 15 . D5 113
Fortrose, N.Z., 136 . Q12 51
Fortrose, Scot., 902 . C4 13
Fort Rosebery, N. Rh., 2,600 . D4 48
Fort Rousset, Con. B . F3 46
Fort St. James, B.C., Can., 1,081 . B5 68
Fort St. John, B.C., Can., 3,619 . A5 68
Fort Sandeman, Pak., 6,001 . B4 39
Fort Saskatchewan, Alta., Can, 2,960 . C4 69
Fort Scott, Bourbon, Kans., 9,410 . E9 93
Fort Selkirk, Yukon, Can . D5 66
Fort Severn, Ont., Can . D8 72
Fort Shaw, Cascade, Mont., 85 . C5 102
Fort Shawnee, Allen, Ohio, 4,000 . B3 111
Fort Sheridan, Lake, Ill . A6, E2 90
Fort Shevchenko, Sov. Un., 18,800 . E4 29
Fort-Sibut, Cen. Afr. Rep . D3 46
Fort Smith, Sebastian, Ark., 63,309 . B1 81
Fort Smith, Alta., Can . D1 69
Fort Smith, N.W. Ter., Can., 1,591 . D10 66
Fort Smith, lake, Ark . B1 81
Fort Spring, Fayette, Ky., 60 . B5 94
Fort Stanton, Lincoln, N. Mex . D4 107
Fort Steele, B.C., Can., 125 . E10 68
Fort Steilacoom, Pierce, Wash . D1 122
Fort Stockton, Pecos, Tex., 6,373 . D1 118
Fort Sumner, De Baca, N. Mex., 1,809 . C5 107
Fort Sumter, S.C. . F3 115
Fort Supply, Woodward, Okla., 394 . A2 112
Fort Supply, res. and dam, Okla . A2 112
Fort Thomas, Graham, Ariz., 200 . D6 80
Fort Thomas, Campbell, Ky., 14,896 . A7 94
Fort Thompson, Buffalo, S. Dak., 150 . C6 116
Fort Totten, Benson, N. Dak., 200 . A7 110
Fort Towson, Choctaw, Okla., 474 . C6 112
Fort Trinquet, Maur . A2 45
Fort Trumbull Beach, New Haven, Conn. (part of Milford) . E4 84
Fortuna, Humboldt, Calif., 3,523 . B1 82
Fortuna, Moniteau, Mo., 155 . C5 101
Fortuna, Divide, N. Dak., 185 . A2 110
Fortuna Ledge, Alsk., 71 . C7 79
Fortune, Newf., Can., 1,194 . F4 75
Fortune, bay, Newf., Can . E4 75
Fortune Bay and Hermitage, dist., Newf., Can . E4 75
Fortune Harbour, Newf., Can., 156 . D4 75
Fort Valley, Peach, Ga., 8,310 . D3 87
Fort Vermilion, Alta., Can., 768 . D1 69
Fort Victoria, S. Rh., 12,300 . B5 49
Fortville, Hancock, Ind., 2,209 . E6 91
Fort Walton Beach, Okaloosa, Fla., 12,147 . G2 86
Fort Washakie, Fremont, Wyo., 130 . B4 125
Fort Washington, Prince Georges, Md . C3 85
Fort Washington Montgomery, Pa., 2,500 . A11 114
Fort Wayne, Allen, Ind., 161,776 (★215,400) . B7 91
Fort White, Columbia, Fla., 425 . C4 86
Fort William, Ont., Can., 45,214 (★93,251) . E8 72
Fort William, Scot., 2,715 . D3 13
Fort William, mtn., Austl . B8 51
Fort Wingate, McKinley, N. Mex., 150 . B1 107
Fort Worth, Tarrant, Tex., 356,268 (★505,100) . B5, C4 118
Fort Wright, Kenton, Ky., 2,184 . A7 94
Fort Yates, Sioux, N. Dak., 900 . C5 110
Forty Fort, Luzerne, Pa., 6,431 . D10 114
Fort Yukon, Alsk., 701 . B10 79
Forward, Sask., Can., 23 . H3 70
Foshan, China, 5,000 . B11 34
Foshee, Escambia, Ala . D2 78
Foss, Washita, Okla., 289 . B2 112
Foss, dam, Okla . B2 112
Fossano, It., 11,000 . B1 21
Fosser, Nor., 113 . p29 25
Fossil, Wheeler, Oreg., 672 . B6 113
Fossil, Lincoln, Wyo . D2 125
Fossil, lake, Oreg . D6 113
Fossil, ridge, Colo . C4 83
Fossombrone, It., 4,659 . C4 21
Fosston, Sask., Can., 117 . E4 70
Fosston, Polk, Minn., 1,704 . C3 99
Foster, Que., Can., 453 . D5 73
Foster, Bracken, Ky., 114 . B5 94
Foster, Bates, Mo., 153 . C3 101
Foster, Pierce, Nebr., 60 . B8 103
Foster, Garvin, Okla., 25 . C4 112
Foster, Linn, Oreg., 250 . C4 113
Foster, Providence, R.I., 80 (2,097▲) . B10 84
Foster, Eau Claire, Wis., 80 . D2 124
Foster, co., N. Dak., 5,361 . B6 110
Foster, riv., Sask., Can . A3 70
Foster Brook, McKean, Pa., 950 . C4 114
Fosterburg, Madison, Ill., 100 . A8 90
Foster City, Dickinson, Mich., 200 . C3 98
Fosters, Tuscaloosa, Ala., 100 . B2 78
Fosters, pond, Mass . C2 97
Fosters Falls, Wythe, Va., 200 . E2 121
Foster Village, Honolulu, Haw., 2,300 . g10 88
Fosterville, Rutherford, Tenn., 150 . B5 117
Fostoria, Lowndes, Ala., 200 . C3 78
Fostoria, Clay, Iowa, 167 . A2 92
Fostoria, Pottawatomie, Kans., 90 . C7 93
Fostoria, Tuscola, Mich., 300 . E7 98
Fostoria, Seneca and Hancock, Ohio, 15,732 . A4 111
Fostoria, Montgomery, Tex., 666 . D5 118
Fougamou, Gabon . F2 46
Fougères, Fr., 24,279 . C3 14
Fouke, Miller, Ark., 394 . D2 81
Foula, isl., Scot . g9 10
Fouliang, China, 92,000 . J7 36
Fouling, China, 32,000 . F6 34
Foulness, isl., Eng . C8 12
Foulwind, cape, N.Z. . N13 51
Foumban, Cam., 18,000 . D2 46
Foum el Hassane, Mor . D3 44
Founing, China, 85,000 . E8 34
Fount, Knox, Ky., 25 . D6 94
Fountain, Monroe, Ala., 150 . D2 78
Fountain, El Paso, Colo., 1,602 . C6 83
Fountain, Bay, Fla., 195 . B1, G3 86
Fountain, Mason, Mich., 194 . D4 98
Fountain, Fillmore, Minn., 297 . G6 99
Fountain, co., Ind., 18,706 . D3 91
Fountain, creek, Colo . C6 83
Fountain City, Wayne, Ind., 833 . E8 91
Fountain City, Knox, Tenn., 10,365 . C10, E11 117
Fountain City, Buffalo, Wis., 934 . D2 124
Fountain Green, Sanpete, Utah, 544 . D4 119
Fountain Head, Washington, Md., 950 . *A2 85
Fountain Head, Sumner, Tenn., 200 . A5 117
Fountain Hill, Ashley, Ark., 230 . D4 81
Fountain Hill, Lehigh, Pa., 5,428 . E11 114
Fountain Inn, Greenville, S.C., 2,385 . B3 115
Fountain Place, Easton Baton Rouge, La., 5,000 . *D4 95
Fountain Run, Monroe, Ky., 300 . D4 94
Fountains, Pitt, N.C., 496 . B6 109
Fountaintown, Shelby, Ind., 300 . E6 91
Fountain Valley, Orange, Calif., 2,068 . *F5 82
Fountain Valley School, El Paso, Colo., 150 . C6 83
Fouping, China . E6 36
Four Buttes, Daniels, Mont., 45 . B11 102
Fourchambault, Fr., 6,240 . D5 14
Fourche, riv., Ark . D5 81
Fourche LaFave, riv., Ark . C2 81
Fourche Maline, creek, Okla . C6 112
Fourchu, N.S., Can., 174 . D9 74
Four Corners, Marion, Oreg., 4,743 . *C4 113
Four Corners, Weston, Wyo., 5 . A8 125
Four Holes, Orangeburg, S.C., 300 . E7 115
Four Lakes, Spokane, Wash., 250 . B8, D7 122
Fourmies, Fr., 14,508 . B6 14
Four Mile, Bell, Ky., 500 . D6 94
Four Mile, creek, Iowa . A7 92
Four Mountains, is., Alsk . E6 79
Fournier, Ont., Can., 227 . B10 72
Fournier, Aroostook, Maine, 100 . A4 96
Four Oakes, Johnston, N.C., 1,010 . B5 109
Four Points, Dougherty, Ga., 1,500 . E2 87
4th. Cataract (Nile River), Sud . B3 47
Fourup, isl., Truk . 52
Fouta Djallon, mts., Guinea . D2 45
Fouyang, China, 65,000 . H6 36
Foveaux, strait, N.Z. . O12 51
Fowey, Eng., 2,237 . D3 12
Fowler, Conecuh, Ala . D2 78
Fowler, Fresno, Calif., 1,892 . D4 82
Fowler, Otero, Colo., 1,240 . C6 83
Fowler, Adams, Ill., 160 . C2 90
Fowler, Benton, Ind., 2,491 . C3 91
Fowler, Meade, Kans., 717 . E3 93
Fowler, Clinton, Mich., 854 . E6 98
Fowler, Pondera, Mont., 37 . B5 102
Fowlerton, Grant, Ind., 297 . D6 91
Fowlerton, LaSalle, Tex., 270 . E3 118
Fowlerville, Livingston, Mich., 1,674 . F6 98
Fowliang (Kingtehchen), China . F8 34
Fowlkes, Dyer, Tenn., 250 . B2 117
Fowlstown, Decatur, Ga., 400 . F2 87
Fox, Stone, Ark., 100 . B3 81
Fox, Menominee, Mich . C3 98
Fox, Carbon, Mont., 10 . E7 102
Fox, Carter, Okla., 400 . C4 112
Fox, Grant, Oreg., 10 . C7 113
Fox, cape, B.C., Can . B2 68
Fox, isl., Wash . D1 122
Fox, is., Alsk . E5 79
Fox, lake, Ill . A5 90
Fox, riv., Man., Can . B1 71
Fox, riv., Ill . B5 90
Fox, riv., Iowa . D6 92
Fox, riv., Mich . B5 98
Fox, riv., Mo . A6 101
Fox, riv., Wis . D5 124
Foxboro, Ont., Can., 494 . C7 72
Foxboro, Norfolk, Mass., 5,000 (10,136) . B5 97
Foxboro, Douglas, Wis., 150 . B1 124
Foxburg, Clarion, Pa., 383 . D2 114
Fox Chapel, Allegheny, Pa., 3,302 . *F2 114
Foxdale, I. of Man . F4 13
Foxe, basin, N.W. Ter., Can . C17 67
Foxe, chan., N.W. Ter., Can . D17 67
Foxe, pen., N.W. Ter., Can . D17 67
Fox Farm, Laramie, Wyo., 1,371 . *D8 124
Foxford, Sask., Can., 50 . D3 70
Foxford, Ire., 876 . D2 11
Fox Harbour, Newf., Can., 232 . B4 75
Fox Hill (part of Hampton), Va . B7, D6 121
Foxholm, Ward, N. Dak., 200 . A4 110
Fox Island, Pierce, Wash., 150 . D1 122
Fox Lake, Lake, Ill., 3,700 . A5, E2 90
Fox Lake, Dodge, Wis., 1,181 . E5 124
Foxon, New Haven, Conn. (part of East Haven) . D5 84
Foxpark, Albany, Wyo., 150 . D6 125
Fox Point, Miwaukee, Wis., 7,315 . E2, E6 124
Fox River Grove, McHenry, Ill., 1,866 . E2 90
Fox River Heights, Kane, Ill., 700 . *E2 90
Foxton, N.Z., 2,628 . N15 51
Fox Valley, Sask., Can., 479 . G1 70
Foxville, Orange, Vt., 100 . C4 120
Foxwarren, Man., Can., 272 . D1 71
Foxworth, Marion, Miss., 950 . D4 100
Foyil, Rogers, Okla., 127 . A6 112
Foyle, bay, N. Ire . B4 11
Foyle, riv., N. Ire . C4 11
Foynes, Ire., 686 . E2 11
Foz do Iguaçu, Braz., 7,407 . D2 56
Frackville, Schuylkill, Pa., 5,654 . E9 114
Fraga, Sp., 8,691 . B6 20
Fraile Muerto, Ur., 1,876 . E2 56
Frametown, Braxton, W. Va., 500 . C4 123
Framingham, Middlesex, Mass., 44,526 . B5, D1 97
França, Braz., 646 . D2 57
Franca, Braz., 47,244 . F1 57
Francavilla Fontana, It., 27,000 (30,300▲) . D6 21
France, country, Eur., 46,520,271 . F9 8, 14
Frances, Crittenden, Ky., 200 . A3 94
Frances, Pacific, Wash., 100 . C2 122
Francestown, Hillsboro, N.H., 180 (495▲) . E3 105
Francesville, Pulaski, Ind., 1,002 . C4 91
Franceville, Gabon . F2 46
Franche-Comté, former prov., Fr., 757,000 . D6 14
Francis, Sask., Can., 150 . G4 70
Francis, Gallatin, Mont., 10 . D5 102
Francis, Pontotoc, Okla., 286 . C5 112
Francis, Summit, Utah, 252 . C4 119
Francis, Harrison, W. Va., 20 . A7 123
Francis, lake, N.H . A1 105
Francisco, Gibson, Ind., 565 . H3 91
Francis Creek, Manitowoc, Wis., 400 . A6 124
Francistown, Bech., 10,000 . B4 49
Francois, Newf., Can., 341 . E3 75
Francois, lake, B.C., Can . C5 68
Franconia, Grafton, N.H., 300 (419▲) . B3 105
Franconia, Fairfax, Va., 3,000 . *C5 121
Franconville, Fr., 11,185 . g9 14
Franeker, Neth., 9,400 . A5 15
Frank, Pocahontas, W. Va., 200 . C5 123
Frankenberg, Ger., 8,100 . B3 17
Frankenberg, Ger., 15,400 . C8 17
Frankenmuth, Saginaw, Mich., 1,728 . E7 98
Frankenthal, Ger., 33,900 . D3 17
Frankewing, Giles, Tenn., 100 . B5 117
Frankford, Ont., Can., 1,142 . C7 72
Frankford, Sussex, Del., 558 . C7 85
Frankford, Pike, Mo., 474 . B6 101
Frankford, Greenbrier, W. Va. 125 . D4 123
Frankfort, Will, Ill., 1,135 . F2 90
Frankfort, Clinton, Ind., 15,302 . D4 91
Frankfort, Marshall, Kans., 1,106 . C7 93
Frankfort, Franklin, Ky., 18,365 . B5 94
Frankfort, Waldo, Maine, 300 (692▲) . D4 96
Frankfort, Benzie, Mich., 1,690 . D4 98
Frankfort, Herkimer, N.Y., 3,872 . B5 108
Frankfort, Ross, Ohio, 871 . C4 111
Frankfort, Spink, S. Dak., 240 . C7 116
Frankfort, S. Afr., 4,787 . C4 49
Frankfurt [am Main], Ger., 683,100 (★1,385,000) . C4 16, C3 17
Frankfurt [an der Oder], Ger., 57,000 . A9 17
Frankische Saale, riv., Ger . C4 17
Franklin, Macon, Ala . C4 78
Franklin, Monroe, Ala., 100 . D2 78
Franklin, Izard, Ark., 75 . A4 80
Franklin, Man., Can., 73 . D2 71
Franklin, New London, Conn., 50 (974▲) . C8 84
Franklin, Heard, Ga., 603 . C1 87
Franklin, Ada, Idaho, 7,222 . *F2 89
Franklin, Franklin, Idaho, 446 . G7 89
Franklin, Morgan, Ill., 500 . D3 90
Franklin, Johnson, Ind., 9,453 . F5 91
Franklin, Lee, Iowa, 174 . D6 92
Franklin, Crawford, Kans., 620 . E9 93
Franklin, Simpson, Ky., 5,319 . D3 94
Franklin, St. Mary, La., 8,673 . E4 95
Franklin, Hancock, Maine, 300 (627▲) . D4 96
Franklin, Allegany, Md . D1 85
Franklin, Norfolk, Mass., 6,391 (10,530▲) . B5 97
Franklin, Oakland, Mich., 2,262 . *F7 98
Franklin, Renville, Minn., 548 . F4 99
Franklin, Howard, Mo., 355 . B5 101
Franklin, Franklin, Nebr., 1,194 . D7 103
Franklin, Merrimack, N.H., 6,742 . D3 105
Franklin, Sussex, N.J., 3,624 . A3 106
Franklin, Delaware, N.Y., 525 . C5 108
Franklin, Macon, N.C., 2,173 . D3 109
Franklin, Warren, Ohio, 7,917 . C3 111
Franklin, Cambria, Pa., 1,352 . *F4 114
Franklin, Venango, Pa., 9,586 . D2 114
Franklin, Williamson, Tenn., 6,977 . B5 117
Franklin, Robertson, Tex., 1,065 . D4 118
Franklin, Franklin, Vt., 185 (796▲) . B3 120
Franklin (Independent City), Va., 7,264 . E6 121
Franklin, Pendleton, W. Va., 758 . C5 123
Franklin, Milwaukee, Wis., 10,006 . *F1 124
Franklin, co., Ala., 21,988 . A4 78
Franklin, co., Ark., 10,213 . B2 81
Franklin, co., Fla., 6,576 . C2 86
Franklin, co., Ga., 13,274 . B3 87
Franklin, co., Idaho, 8,457 . G7 89
Franklin, co., Ill., 39,281 . E5 90
Franklin, co., Ind., 17,015 . F7 91
Franklin, co., Iowa, 15,472 . B4 92
Franklin, co., Kans., 19,548 . D8 93
Franklin, co., Ky., 29,421 . B5 94
Franklin, co., Maine, 20,069 . C2 96
Franklin, co., Mass., 54,864 . A2 97
Franklin, co., Miss., 9,286 . D3 100
Franklin, co., Mo., 44,566 . C6 101
Franklin, co., Nebr., 5,449 . D7 103
Franklin, co., N.Y., 44,742 . A6 108
Franklin, co., N.C., 28,755 . A5 109
Franklin, co., Ohio, 682,962 . B4 111
Franklin, co., Pa., 88,172 . G6 114
Franklin, co., Tenn., 25,528 . B5 117
Franklin, co., Tex., 5,101 . C5 118
Franklin, co., Vt., 29,474 . B2 120
Franklin, co., Va., 25,925 . E3 121
Franklin, co., Wash., 23,342 . C7 122
Franklin, dist., N.W. Ter., Can., 3,424 . B12 66
Franklin, par., La., 26,088 . B4 95
Franklin, isl., Ont., Can . B4 72
Franklin, lake, N.W. Ter., Can . C13 66
Franklin, lake, Nev . C6 104
Franklin, mts., N.W. Ter., Can . C7 66
Franklin, pt., Alsk . A8 79
Franklin, straits, N.W. Ter., Can . B13 66
Franklin D. Roosevelt, lake, Wash . A7 122
Franklin Grove, Lee, Ill., 773 . B4 90
Franklin, Hartsell, Brown and Norcot Mills (West Concord), Cabarrus, N.C. 5,510 . B3 109
Franklin Lakes, Bergen, N.J., 3,316 . *B4 106
Franklin Mine, Houghton, Mich., 150 . A2 98
Franklin Park, Cook, Ill., 18,322 . E2 90
Franklin Park, Somerset, N.J., 750 . C3 106
Franklin Park, Fairfax, Va., 1,300 . *C5 121
Franklin Square, Nassau, N.Y., 32,483 . G2 84
Franklinton, Washington, La., 3,141 . D5 95
Franklinton, Franklin, N.C., 1,513 . A5 109
Franklinville, Gloucester, N.J., 900 . D2 106
Franklinville, Cattaraugus, N.Y., 2,124 . C2 108
Franklinville, Randolph, N.C., 686 . B4 109
Franks, peak, Wyo . B3 125
Frankston, Anderson, Tex., 953 . C5 118
Franksville, Racine, Wis., 400 . F2 124
Frankton, Madison, Ind., 1,445 . D6 91
Franktown, Douglas, Colo., 50 . B6 83
Frankville, Washington, Ala., 500 . D1 78
Frankville, Garrett, Md . D1 85
Frankville, N.S., Can., 245 . D8 74
Frannie, Park, Wyo., 171 . A4 125
Franz Josef Land, reg., Sov. Un . B8 28
Frascati, It., 11,500 . D4, h9 21
Fraser, Grand, Colo., 253 . B5 83
Fraser, Boone, Iowa, 134 . B4 92
Fraser, Macomb, Mich., 7,027 . A8 98
Fraser (Great Sandy), isl., Austl . B9 51
Fraser, lake, B.C., Can . B5 68
Fraser, mtn., B.C., Can . C8 68
Fraser, reach, B.C., Can . C3 68
Fraser, riv., B.C., Can . C6 68
Fraser, riv., Newf., Can . g9 75
Fraserburg, S. Afr., 2,359 . D3 49
Fraserburgh, Scot., 10,462 . C6 13
Fraserdale, Ont., Can., 215 . E9 72
Fraserwood, Man., Can., 78 . D3 71
Frasne, Fr., 1,367 . C2 18
Frauenfeld, Switz., 14,702 . A6 19
Fray Bentos, Ur., 12,900 . E1 56
Frazee, Becker, Minn., 1,083 . D3 99
Frazer, Valley, Mont., 400 . B10 102
Frazeysburg, Muskingum, Ohio, 842 . B5 111
Frazier, Chaves, N. Mex . D5 107
Frazier, mtn., Calif . E4 82
Frazier Park, Kern, Calif., 250 . E4 82
Frederic, Crawford, Mich., 400 . D6 98
Frederic, Polk, Wis., 857 . C1 124
Frederica, Kent, Del., 863 . B7 85
Fredericia, Den., 29,870 . C3 24
Frederick, Weld, Colo., 595 . A6 83
Frederick, Schuyler, Ill., 175 . C3 90
Frederick, Frederick, Md., 21,744 . B3 85
Frederick, Tillman, Okla., 5,879 . C2 112
Frederick, Brown, S. Dak., 381 . B7 116
Frederick, co., Md., 71,930 . B3 85
Frederick, co., Va., 21,941 . B4 121
Frederick, sound, Alsk . m23 79
Frederick Junction, Frederick, Md., 11 . B3 85
Fredericksburg, Washington, Ind., 207 . H5 91
Fredericksburg, Chickasaw, Iowa, 797 . B5 92
Fredericksburg, Wayne, Ohio, 565 . B6 111
Fredericksburg, Lebanon, Pa., 700 . F9 114
Fredericksburg, Gillespie, Tex., 4,629 . D3 118
Fredericksburg (Independent City), Va., 13,639 . C5 121
Fredericks Hall, Louisa, Va., 60 . D5 121
Fredericktown, Cecil, Md., 130 . B6 85
Fredericktown, Madison, Mo., 3,848 . D7 101
Fredericktown, Knox, Ohio, 1,531 . B5 111
Fredericktown, Washington, Pa., 1,270 . F1 114
Fredericton, N.B., Can., 19,683 . D3 74
Fredericton Junction, N.B., Can., 641 . D3 74
Frederika, Bremer, Iowa, 249 . B5 92
Frederik Hendrik, isl., W. Irian . G9 35
Frederiksborg, co., Den., 162,889 . C6 24
Frederikshavn, Den., 22,522 . A4 24
Frederikssund, Den., 5,722 . C6 24
Frederiksted, Vir. Is., 277 . n14 60, k17 65
Frederiksvaerk, Den., 4,435 (★6,155) . C6 24
Fredonia, Chambers, Ala., 200 . C4 78
Fredonia, Coconino, Ariz., 643 . A3 80
Fredonia (Biscoe), Prairie, Ark., 350 . C4 81
Fredonia, Louisa, Iowa, 147 . C6 92
Fredonia, Wilson, Kans., 3,233 . E8 93
Fredonia, Caldwell, Ky., 427 . C1 94
Fredonia, Chautauqua, N.Y., 8,477 . C1 108
Fredonia, Logan, N. Dak., 141 . C6 110
Fredonia, Mercer, Pa., 657 . D1 114
Fredonia, Mason, Tex., 60 . D3 118
Fredonia, Ozaukee, Wis., 710 . E6 124
Fredonyer, peak, Calif . B3 82
Fredrikstad, Nor., 14,549 (★40,600) . H4, p28 25
Freeborn, Freeborn, Minn., 314 . G5 99
Freeborn, co., Minn., 37,891 . G5 99
Freeburg, St. Clair, Ill., 1,908 . E4 90
Freeburg, Osage, Mo., 399 . C6 101
Freeburg, Houston, Minn., 50 . G7 99
Freeburg, Snyder, Pa., 575 . E8 114
Freedom, Santa Cruz, Calif., 4,206 . C6 82
Freedom, Owen, Ind., 190 . F4 91
Freedom, Frontier, Nebr . D5 103
Freedom, Carroll, N.H., 250 (363▲) . C4 105
Freedom, Woods, Okla., 268 . A2 112
Freedom, Beaver, Pa., 2,895 . E1 114
Freedom, Outagamie, Wis., 300 . A5 124
Freedom, Lincoln, Wyo., 25 . C2 125
Freehold, Monmouth, N.J., 9,140 . C4 106
Freeland, Saginaw, Mich., 900 . E6 98
Freeland, Brunswick, N.C., 200 . C5 109
Freeland, Luzerne, Pa., 5,068 . D10 114
Freeland, Henry, Tenn., 50 . A3 117
Freeland, Natrona, Wyo . C6 125
Freeland, Park, Benton, Ind., 173 . C3 91
Freelandville, Knox, Ind., 720 . G3 91
Freels, cape, Newf., Can . D5 75
Freelton, Ont., Can., 293 . D4 72
Freeman, Cass, Mo., 391 . C3 101
Freeman, Hutchinson, S. Dak., 1,140 . D8 116
Freeman, Spokane, Wash., 55 . D8 122
Freeman, lake, Ind . C4 91
Freemansburg, Northampton, Pa., 1,652 . E11 114
Freemanville, Escambia, Ala., 150 . D2 78
Freemason, is., La . E7 95
Freemen, riv., Alta., Can . B3 69
Freemount, Ire., 124 . E3 11
Freeport, N.S., Can., 363 . E3 75
Freeport, Walton, Fla., 800 . G3 86
Freeport, Stephenson, Ill., 26,628 . A4 90
Freeport, Winneshiek, Iowa, 100 . A6 92
Freeport, Harper, Kans., 31 . E6 93
Freeport, Cumberland, Maine, 1801 (4,055▲) . E2, E5 96
Freeport, Barry, Mich., 495 . F5 98
Freeport, Stearns, Minn., 615 . E4 99
Freeport, Nassau, N.Y., 34,419 . G2 84
Freeport, Harrison, Ohio, 503 . B6 111
Freeport, Armstrong, Pa., 2,439 . E2 114
Freeport, Brazoria, Tex., 11,619 . E5, G5 118
Freer, Duval, Tex., 2,724 . F3 118
Free Soil, Mason, Mich., 209 . D4 98
Freestone, co., Tex., 12,525 . D4 118
Freetown, P.E.I., Can., 84 . C6 75
Freetown, S.L., 100,000 . E2 45
Freetown, Jackson, Ind., 450 . G5 91
Freetown, Suffolk, N.Y., 1,365 . *D4 108
Free Union, Albemarle, Va., 60 . C4 121
Freeville, Tompkins, N.Y., 471 . C4 108
Freewater, see Milton-Freewater, Oreg.
Freezeout, mts., Wyo . C6 125
Fregenal de la Sierra, Sp., 10,498 . C2 20
Fregene, It . h8 21
Freiberg, Ger., 47,400 . C8 17
Freiberger Mulde, riv., Ger . B8 17
Freiburg [an der Elbe], Ger., 2,600 . E3 24
Freiburg [im Breisgau], Ger., 145,000 . D3 16, D3 18
Freirina, Chile, 1,504 . E1 55
Freising, Ger., 27,600 . E6 17
Freistadt, Aus., 5,375 . D7 16
Freital, Ger., 37,600 . B8 17
Fréjus, Fr., 13,243 . F7 14
Frelighsburg, Que., Can., 361 . D5 73
Fremantle, Austl., 24,343 . F2 50
Fremont, Alameda, Calif., 43,790 . B5, D3 82
Fremont, Steuben, Ind., 937 . A8 91
Fremont, Mahaska, Iowa, 461 . C5 92
Fremont, Newaygo, Mich., 3,384 . E5 98
Fremont, Carter, Mo., 131 . E6 101
Fremont, Dodge, Nebr., 19,698 . C9, D2 103
Fremont, Rockingham, N.H., 300 (783▲) . E4 105
Fremont, Wayne, N.C., 1,609 . B6 109
Fremont, Sandusky, Ohio, 17,573 . A4 111
Fremont, Wayne, Utah, 125 . E4 117
Fremont, Waupaca, Wis., 575 . D5 124
Fremont, co., Colo., 20,196 . C5 83
Fremont, co., Idaho, 8,679 . E7 89
Fremont, co., Iowa, 10,282 . D2 92
Fremont, co., Wyo., 26,186 . B4 125
Fremont, isl., Utah . B3 119
Fremont, lake, Wyo . C3 125
Fremont, peak, Wyo . B3 125
Fremont, riv., Utah . E4 119
French, Colfax, N. Mex., 15 . A5 107
French, Carbon, Wyo . D6 125
French, creek, Pa . C1 114
French, pt., Newf., Can . C4 75
French, prairie, Oreg . B1 113
French, riv., Ont., Can . A4 72
French Broad, riv., N.C . D3 109
French Broad, riv., Tenn . D10 117
Frenchburg, Menifee, Ky., 296 . C6 92
French Camp, San Joaquin, Calif., 1,000 . B6 82
French Camp, Choctaw, Miss., 123 . B4 100
French Creek, Upshur, W. Va., 165 . C4 123
French Equatorial Africa, see Central African Rep., Chad, Congo, and Gabon, countries, Afr.
French Frigate, shoal, Haw . m14 88
Frenchglen, Harney, Oreg., 20 . E8 113
French Guiana, dep., S.A., 28,506 . C5 53, B4 59
French Gulch, Shasta, Calif., 300 . B2 82
French Lick, Orange, Ind., 1,954 . G4 91
Frenchman, Churchill, Nev., 2 . D3 104
Frenchman, bay, Maine . D4 96
Frenchman, creek, Nebr . D4 103
Frenchman, hills, Wash . C6 122
Frenchman, riv., Sask., Can . H1 70
Frenchman Butte, Sask., Can., 111 . D1 70
Frenchman Knob, peak, Ky . C4 94
Frenchmans Cove, Newf., Can., 162 . D2 75
Frenchmans Island, Newf., Can . B4, h11 75
French Oceania, see French Polynesia, dep., Oceania
French Park, Ire., 155 . D3 11
French Polynesia, dep., Oceania, 65,000 . H13 7
French Polynesia, see Gambier, Makatea, Marquesas, Rapa, Rurutu and Rimatara, Society, Tubuai and Raivavae Islands and Tuamotu Archipelago
French River, St. Louis, Minn., 250 . D7 99
French Settlement, Livingston, La., 350 . B6, D5 95
French Somaliland, dep., Afr., 69,000 . C5 47, E10 42

Frenchton, Upshur, W. Va., 150 ..... C4 123
Frenchtown, Missoula, Mont., 100 ..... C2 102
Frenchtown, Hunterdon, N.J., 1,340 ..... B2 106
French Village, N.S., Can., 101 ..... E6 74
Frenchville, Aroostook, Maine, 875 (1,421▲) A4 96
Frenchville, Clearfield, Pa D5 114
French West Africa, see Dahomey, Guinea, Ivory Coast, Mali, Mauritania, Niger, Senegal, and Upper Volta, countries, Afr.
Frenda, Alg., 8,579 ..... G7 30
Frenier, St. John the Baptist, La., 25 ..... B6 95
Fresco, riv., Braz D4 59
Freshfield, cape, Ant C28 5
Freshfield, mtn., B.C., Can D9 68
Freshford, Ire., 656 ..... E4 11
Freshwater, Newf., Can., 1,048 ..... *E5 75
Freshwater, Eng., 3,423 ..... D6 12
Fresko, I.C E3 45
Fresnes-en-Woëvre, Fr., 540 ..... E5 15
Fresnillo, Mex., 36,919 ..... C4 63
Fresno, Fresno, Calif., 133,929 (★228,000) D4 82
Fresno, Coshocton, Ohio, 200 ..... B6 111
Fresno, co., Calif., 365,945 ..... D3 82
Fresno, dam, Mont B7 102
Freudenstadt, Ger., 14,200 ..... E3 17
Frèvent, Fr., 4,386 ..... B5 14
Frewsburg, Chautauqua, N.Y., 1,623 ..... C1 108
Freyburg, Ger., 5,856 ..... B6 17
Freycinet's, pen., Austl o15 50
Freyenstein, Ger., 2,119 ..... E6 24
Freyung, Ger., 4,900 ..... E8 17
Friant, Fresno, Calif., 300 ..... D4 82
Friars Point, Coahoma, Miss., 1,029 ..... A3 100
Frias, Arg., 7,941 ..... E2 55
Fribourg, Switz, 32,583 ..... C3 19
Fribourg, canton, Switz., 159,194 ..... C2 19
Friday Harbor, San Juan, Wash., 706 ..... A2 122
Fridley, Anoka, Minn., 15,173 ..... E7 99
Fridtjof Nansen, peak, Ant A32 5
Fried, Stutsman, N. Dak., 20 .C7 110
Friedberg, Ger., 17,300 ..... C3 17
Friedberger Mulde, riv., Ger B8 17
Friedens, Somerset, Pa., 900 . F4 114
Friedland, Ger., 8,500 ..... B6 16
Friedrichrode, Ger., 7,254 ..... C5 17
Friedrichshafen, Ger., 37,100 . E4 16
Friedrichskoog, Ger., 2,900 ..... D2 24
Friedrichstadt, Ger., 3,000 ..... D3 24
Friedrichsthal, Ger., 15,000 ..... D2 17
Friend, Finney, Kans., 50 ..... D3 93
Friend, Saline, Nebr., 1,069 . D8 103
Friend, Wasco, Oreg., 15 ..... B5 113
Friendship, Hot Springs, Ark., 162 ..... C3 81
Friendship, Ripley, Ind., 125 . G7 91
Friendship, Knox, Maine, 350 (806▲) E3 96
Friendship, Anne Arundel, Md., 87 ..... C4 85
Friendship, Allegany, N.Y., 1,231 ..... C2 108
Friendship, Scioto, Ohio, 550 . D4 111
Friendship, Jackson, Okla., 100 ..... C2 112
Friendship, Crockett, Tenn., 399 ..... B2 117
Friendship, Adams, Wis., 560 . E4 124
Friendsville, Garrett, Md., 580 ..... D1 85
Friendsville, Blount, Tenn., 606 ..... D9, E11 117
Friendswood, Hendricks, Ind., 120 ..... I7 91
Friern Barnet, Eng., 28,807 . k12 10
Frierson, De Soto, La., 200 ..... B2 95
Fries, Grayson, Va., 1,039 ..... E2 121
Friesach, Aus., 3,388 ..... E7 16
Friesack, Ger., 3,551 ..... F6 24
Friesland, Columbia, Wis., 308 ..... E4 124
Friesland, prov., Neth., 479,900 ..... A5 15
Fries Mills, Gloucester, N.J., 50 ..... D2 106
Friesoythe, Ger., 5,300 ..... A7 15
Frijoles, C.Z. k11 62
Frio, co., Tex., 10,112 ..... E3 118
Frio, cape, S.W. Afr A1 49
Frio, riv., Tex E3 118
Friona, Parmer, Tex., 2,048 . B1 118
Fripps, inlet, S.C G7 115
Fripps, isl., S.C G7 115
Frisches Haff, see Vistula, lagoon, Sov. Un.
Frisco, Summit, Colo., 316 ..... B4 83
Frisco, Carton, N. Mex D1 107
Frisco, Dare, N.C., 150 ..... B8 109
Frisco, Pontotoc, Okla C5 112
Frisco, Beaver, Pa., 900 ..... *E1 114
Frisco, Collin, Tex., 1,184 . *C4 118
Frisco, mtn., Utah E2 119
Frisco City, Monroe, Ala., 1,177 ..... D2 78
Fristoe, Benton, Mo., 100 ..... C4 101
Fritch, Hutchison, Tex., 1,617 ..... B2 118
Fritzlar, Ger., 8,500 ..... B4 17
Friuli-Venezia Giulia, pol. dist., It., 1,204,298 ..... C8 18
Frobisher, Sask., Can., 335 . H4 70
Frobisher, bay, N.W. Ter., Can D19 67
Frobisher, lake, Sask., Can A1 70
Frobisher Bay, N.W. Ter., Can., 512 ..... C21 4, D9 67
Frog, mtn., Tenn D9 117
Frogmore, Beaufort, S.C., 50 . G6 115
Frohna, Perry, Mo., 216 ..... D8 101
Froid, Roosevelt, Mont., 418 ..... B12 102
Frolovo, Sov., Un., 25,600 ..... D2 29
Fromberg, Carbon, Mont., 367 ..... E8 102
Frome, Eng., 11,440 ..... C5 12
Frome, riv., Eng D5 12
Frompton, inlet, S.C G2 115
Front, range, Colo B5 83
Frontenac, Crawford, Kans., 1,713 ..... E9 93
Frontenac, Goodhue, Minn., 168 ..... F6 99
Frontenac, St. Louis, Mo., 3,089 ..... *B8 101
Frontenac, co., Ont., Can., 87,534 ..... C8 72
Frontenac, co., Que., Can., 30,600 ..... D7 73
Fronteras, Mex., 697 ..... G5 80
Frontier, Sask., Can., 274 ..... H1 70
Frontier, co., Nebr., 4,311 ..... D5 103
Frontignan, Fr., 5,341 ..... F5 14
Front Royal, Warren, Va., 7,949 ..... C4 121
Frosinone, It., 25,000 (31,155▲) D4 21
Frosses, Ire., 105 ..... C3 11
Frost, Faribault, Minn., 381 . G5 99
Frost, Navarro, Tex., 508 ..... C4 118
Frostburg, Allegany, Md., 6,722 ..... D2 85
Frostproof, Polk, Fla., 2,664 . E5 86
Frouard, Fr., 6,916 ..... C7 14
Froude, Sask., Can., 56 ..... H4 70
Froya, isl., Nor F3 25
Fruges, Fr., 2,499 ..... D10 12
Fruit, Madison, Ill., 25 ..... A9 90
Fruita, Mesa, Colo., 1,830 ..... B2 83
Fruitdale, Washington, Ala., 350 ..... D1 78
Fruitdale, Butte, S. Dak., 79 . C2 116
Fruitdale, Dallas, Tex., 1,418 ..... *C4 118
Fruithurst, Cleburne, Ala., 255 ..... B4 78
Fruitland, Payette, Idaho, 804 ..... F2 89
Fruitland, Rock Island, Ill., 1,000 ..... *B3 90
Fruitland, Muscatine, Iowa, 175 ..... C6 92
Fruitland, Wicomico, Md., 1,147 ..... D6 85
Fruitland, Cape Girardeau, Mo., 100 ..... D8 101
Fruitland, San Juan, N. Mex., 150 ..... A1 107
Fruitland, Gibson, Tenn., 150 ..... B3 117
Fruitland, Duchesne, Utah, 10 ..... C5 119
Fruitland Park, Lake, Fla., 774 ..... D5 86
Fruitland Park, Forrest, Miss., 55 ..... E4 100
Fruitport, Muskegon and Ottawa, Mich., 1,037 ..... E4 98
Fruitridge, Sacramento, Calif., 5,480 ..... *A6 82
Fruitvale, B.C., Can., 1,032 . E9 68
Fruitvale, Mesa, Colo., 300 ..... B2 83
Fruitvale, Adams, Idaho, 100 . E2 89
Fruitville, Sarasota, Fla., 2,131 ..... F2 86
Frunze, Sov. Un., 252,000 ..... E8 29
Frutal, Braz, 8,252 ..... E1 57
Frutigen, Switz., 5,565 ..... C4 19
Fry, Fannin, Ga., 75 ..... B2 87
Fryazevo, Sov. Un n18 27
Fryburg, Billings, N. Dak., 50 ..... C2 110
Fryburg, Clarion, Pa., 250 ..... D3 114
Frydek-Mistek, Czech., 27,900 ..... D5 26
Frydlant, Czech., 4,308 ..... C3 26
Frye, Oxford, Maine, 100 ..... D2 96
Fryeburg, Bienville, La., 100 . B2 95
Fryeburg, Oxford, Maine, 975 (1,874▲) D2 96
Fryele, Swe A8 24
Frys, Sask., Can., 75 ..... H5 70
Frystat, Czech., (pop. incl. in Karvinna) h9 26
Fthiotis (Phthiotis), prov., Grc., 160,035 ..... *C4 23
Fu, China F3 36
Fuan, China, 11,000 ..... K8 36
Fuchin, China, 40,000 ..... B11 34
Fuching, China, 25,000 ..... F8 34
Fuchou, China G6 34
Fuchow, China, 10,000 ..... F8 34
Fuencarral, Sp., 7,078 (part of Madrid) p17 20
Fuente Álamo, Sp., 9,270 ..... D5 20
Fuente de Cantos, Sp., 8,941 ..... C2 20
Fuente el Saz, Sp., 852 ..... o17 20
Fuenteovejuna, Sp., 5,914 ..... C3 20
Fuentesaúco, Sp., 3,036 ..... B3 20
Fuentes [de Andalucía], Sp., 11,640 ..... D3 20
Fuerte, riv., Mex B3 63
Fuerte Olimpo, Par D4 55
Fuerteventura, isl. Can. Is m14 20
Fuglebjerg, Den., 967 ..... C5 24
Fuhsien, China E9 36
Fuji, vol., Jap n17, I9 37
Fujimi, Jap., 4,500 ..... n17 37
Fujinomiya, Jap., 43,400 (76,645▲) n17 37
Fujiyoshida, Jap., 42,607 ..... n17 37
Fūkah, Eg., U.A.R G6 31
Fukien, prov., China, 14,650,000 ..... F8 34
Fukou, China G6 36
Fukuchiyama, Jap., 35,200 ..... I7, n14 37
Fukue, Jap., 8,700 ..... o16 37
Fukui, Jap., 149,823 ..... H8 37
Fukui, pref., Jap., 752,696 . *H8 37
Fukuoka, Jap., 647,122 (★790,000) J5 37
Fukuoka, Jap., 9,500 ..... F10 37
Fukuoka, pref., Jap., 4,006,679 ..... *J5 37
Fukushima, Jap., 103,000 (138,961▲) H10 37
Fukushima, Jap., 6,900 . I8, n16 37
Fukushima, pref., Jap., 2,051,137 ..... *H10 37
Fukuyama, Jap., 107,000 (140,603▲) I6 37
Fukuyama, see Matsumae, Jap.
Fulang, isl., Fiji Is 52
Fulanga, passage, Fiji Is 52
Fulda, Ger., 45,100 ..... C4 17
Fulda, Spencer, Ind., 150 ..... H4 91
Fulda, Murray, Minn., 1,202 . G3 99
Fulda, riv., Ger C4 17
Fullarton, Ont., Can., 59 ..... D3 72
Fullerton, Orange, Calif., 56,180 ..... F3 82
Fullerton, Greenup, Ky., 1,082 ..... B7 94
Fullerton, Baltimore, Md., 3,000 ..... B4, C3 85
Fullerton, Nance, Nebr., 1,475 ..... C8 103
Fullerton, Dickey, N. Dak., 181 ..... C7 110
Fullerton, Lehigh, Pa., 5,500 ..... *E11 114
Fulshear, Fort Bend, Tex., 300 ..... F4 118
Fulton, Clarke, Ala., 688 ..... D2 78
Fulton, Hempstead, Ark., 309 ..... D2 81
Fulton, Sonoma, Calif., 250 . A5 82
Fulton, Whiteside, Ill., 3,387 . B3 90
Fulton, Fulton, Ind., 410 ..... C5 91
Fulton, Bourbon, Kans., 207 . D9 93
Fulton, Fulton, Ky., 3,265 ..... B2 94
Fulton, Howard, Md., 140 ..... B4 85
Fulton, Kalamazoo, Mich., 225 ..... F5 98
Fulton, Keweenaw, Mich., 375 ..... A2 98
Fulton, Itawamba, Miss., 1,706 ..... A5 100
Fulton, Callaway, Mo., 11,131 ..... C6 101
Fulton, Oswego, N.Y., 14,261 ..... B4 108
Fulton, Morrow, Ohio, 292 . B5 111
Fulton, Hanson, S. Dak., 135 . D8 116
Fulton, Lauderdale, Tenn., 85 ..... B2 117
Fulton, Aransas, Tex., 900 . *E4 118
Fulton, co., Ark., 6,657 ..... A4 81
Fulton, co., Ga., 556,326 ..... C2 87
Fulton, co., Ill., 41,954 ..... C3 90
Fulton, co., Ind., 16,957 ..... B5 91
Fulton, co., Ky., 11,256 ..... B2 94
Fulton, co., N.Y., 51,304 ..... B6 108
Fulton, co., Ohio, 29,301 ..... A3 111
Fulton, co., Pa., 10,597 ..... G5 114
Fultondale, Jefferson, Ala., 2,820 ..... E4 78
Fultonham, Schoharie, N.Y., 100 ..... C6 108
Fulton Heights, Pueblo, Colo., 1,000 ..... *C6 83
Fultz, Carter, Ky., 150 ..... B6 94
Fumay, Fr., 6,185 ..... C6 14
Funabashi, Jap., 135,038 ..... n19 37
Funaka, see Oga, Jap.
Funazu, see Kamioka, Jap.
Funchal, Madeira Is., 43,300 ..... h12 20, C1 44
Fundación, Col., 6,620 ..... A3 60
Fundão, Port., 3,777 ..... B2 20
Fundy, bay, Can E3 74
Fundy, nat. park, N.B., Can D4 74
Funhalouro, Moz B5 49
Funing, bay, China K9 36
Funk, Phelps, Nebr., 141 ..... D6 103
Funk, isl., Newf., Can D5 75
Funkley, Beltrami, Minn., 28 . C4 99
Funkstown, Washington, Md., 968 ..... A2 85
Funter, Alsk., 10 ..... K22, D13 79
Funtua, Nig D6 45
Fuquay Springs, Wake, N.C., 3,389 ..... B5 109
Fur, isl., Den B3 24
Furancungo, Moz D5 48
Furano, Jap., 11,500 ..... E11 37
Fürg, Iran G7 41
Furka, pass, Switz C5 19
Furley, Sedgwick, Kans B5 93
Furlow, Lonoke, Ark., 30 ..... D6 81
Furman, Hampton, S.C., 244 ..... F5 115
Furmanov, Sov. Un., 37,300 ..... C13 27
Furmanovka, Sov. Un E8 29
Furman University, Greenville, S.C., 1,500 ..... *B3 115
Furnace, Worcester, Mass., 40 ..... B3 97
Furnace, brook, Vt D3 120
Furnas, co., Nebr., 7,711 ..... D6 103
Furnas, res., Braz C3, k6 56
Furneaux, is., Austl o15 50
Furnes, see Veurne, Bel.
Furness, Sask., Can., 78 ..... D1 70
Furqlus, Syr E11 31
Fürstenau, Ger., 4,800 ..... B7 15
Fürstenfeld, Aus., 6,415 ..... E8 16
Fürstenfeldbruck, Ger., 17,600 ..... A7 18
Fürstenwalde, Ger., 31,600 ..... A9 16
Fürth, Ger., 97,800 ..... D5 17
Furth im Wald, Ger., 8,200 . D7 17
Furu, mtn., Iwo 52
Furukawa, Jap., 6,200 . H8, m16 37
Fusagasugá, Col., 8,345 ..... C3 60
Fuschl, Aus., 684 ..... B9 18
Fuse, Jap., 212,754 ..... I7, o14 37
Fushih, see Yenen, China
Fushun, China, 985,000 ..... C9 38, D10 36
Fusilier, Sask., Can., 250 ..... F1 70
Fusin, China, 188,600 ..... C9 34
Füssen, Ger., 10,700 ..... B6 18
Fusung, China, 5,000 ..... E3 36
Futamata, Jap., 11,442 ..... o16 37
Futatsune, reef, Iwo 52
Futing, China, 13,000 ..... K9 36
Fuwah, Eg., U.A.R., 22,900 . C2 32
Fuyang, China, 15,000 ..... I8 36
Fuyu, China, 64,969 ..... C2 37
Fuyu, China, 45,000 . B9 34, D2 37
Fuyuan, China, 5,000 ..... B7 37
Fyn, isl., Den C4 24
Fyne, inlet, Scot D3 13

G

Gaastra, Iron, Mich., 582 ..... B2 98
Gabarouse, N.S., Can., 252 . D9 74
Gabbs, Nye, Nev., 770 ..... E4 104
Gabela, Ang., 4,996 ..... D1 48
Gaberones, Bech., 10,000 ..... B4 49
Gabès, Tun., 24,420 ..... C7 44
Gabès, prov., Tun., 242,684 . *C7 44
Gabès, gulf, Tun C7 44, G12 28
Gabilan, range, Calif C6 82
Gabin, Pol., 3,108 ..... B5 26
Gabon (Gabun), country, Afr., 440,000 ..... G7 42, F2 46
Gabriel, mtn., Ire F2 11
Gabriels, Franklin, N.Y., 450 . B3 108
Gabriola, B.C., Can., 35 ..... A9 68
Gabriola, isl., B.C., Can A9 68
Gabrovo, Bul., 21,268 ..... D7 22
Gabun (Gabon), country, Afr., 440,000 ..... G7 42, F2 46
Gach Sārān, Iran F5 41
Gackle, Logan, N. Dak., 523 . C6 110
Gacko, Yugo., 1,227 ..... D4 22
Gadag, India, 76,614 ..... E6 39
Gadmen, Switz., 510 ..... C5 19
Gadsby, Alta., Can., 98 ..... C4 69
Gadsden, Etowah, Ala., 58,088 ..... A3 78
Gadsden, Yuma, Ariz., 200 . E1 80
Gadsden, Richland, S.C., 150 ..... D6 115
Gadsden, Crockett, Tenn., 222 ..... B3 117
Gadsden, co., Fla., 41,989 ..... B2 86
Gadyach, Sov. Un., 10,000 . F9 27
Găeşti, Rom., 7,179 ..... C7 22
Gaeta, It., 20,569 ..... D4 21
Gaeta, gulf, It D4 21
Gaffney, Cherokee, S.C., 10,435 ..... A4 115
Gafsa, Tun., 24,345 ..... B11 30
Gafsa, prov., Tun., 255,685 . *C6 44
Gagan, isl., Kwajalein 52
Gage, Musselshell, Mont., 10 . D8 102
Gage, Luna, N. Mex., 10 ..... E1 107
Gage, Ellis, Okla., 482 ..... A2 112
Gage, co., Nebr., 26,818 ..... D9 103
Gage, cape, P.E.I., Can C5 74
Gages Lake, Lake, Ill., 3,395 ..... *A5 90
Gagetown, N.B., Can., 572 . D3 74
Gagetown, Tuscola, Mich., 376 ..... E7 98
Gaggenau, Ger., 12,500 ..... E3 17
Gagliano del Capo, It., 3,869 ..... E7 21
Gagnoa, I.C., 15,000 ..... E3 45
Gagny, Fr., 29,004 ..... g11 14
Gahanna, Franklin, Ohio, 2,717 ..... C2 111
Gaibandha, Pak., 14,310 ..... E12 40
Gail, Borden, Tex., 100 ..... C2 118
Gail, riv., Aus C9 18
Gaildorf, Ger., 4,600 ..... D4 17
Gaillac, Fr., 6,205 ..... F4 14
Gaillard, lake, Conn D5 84
Gaillimh, see Galway, Ire.
Gaines, Genesee, Mich., 387 . E6 98
Gaines, Tioga, Pa., 150 ..... C6 114
Gaines, co., Tex., 12,267 ..... C1 118
Gaines, creek, Okla B6 112
Gainesboro, Jackson, Tenn., 1,021 ..... C8 117
Gainesville, Sumter, Ala., 214 ..... C1 78
Gainesville, Alachua, Fla., 29,701 ..... C4 86
Gainesville, Hall, Ga., 16,523 . B3 87
Gainesville, Hancock, Miss., 100 ..... E4 100
Gainesville, Ozark, Mo., 266 . E5 101
Gainesville, Wyoming, N.Y., 369 ..... C2 108
Gainesville, Cooke, Tex., 13,083 ..... C4 118
Gainesville, Prince William, Va., 150 ..... B4 121
Gainesville Cotton Mills, Hall, Ga., 2,207 ..... B3 87
Gainesville East, Alachua, Fla., 2,393 ..... *C4 86
Gainesville North, Alachua, Fla., 4,290 ..... *C4 86
Gainesville West, Alachua, Fla., 2,725 ..... *C4 86
Gainsborough, Sask., Can., 411 ..... H5 70
Gainsborough, Eng., 17,276 . A7 12
Gairdner, lake, Austl F6 50
Gairloch, Scot., 1,991 ..... C3 13
Gairloch, bay, Scot C3 13
Gaither, Carroll, Md., 150 ..... B4 85
Gaithersburg, Montgomery, Md., 3,847 ..... B3 85
Gakhutt, Richland, N. Dak C9 110
Gakona, Alsk., 33 ..... C10, f19 79
Gala, riv., Scot E6 13
Galacz, see Galati, Rom.
Galahad, Alta., Can., 231 ..... C5 69
Galala el Bahariya, mts., Eg., U.A.R E4 32
Galap, Palau Is 52
Galapagar, Sp., 2,453 ..... o17 20
Galápagos Islands, see Colón, Archipiélago de, prov., Ec.
Galashiels, Scot., 12,374 ..... E6 13
Galata, Toole, Mont., 43 ..... B5 102
Galatea, Kiowa, Colo C8 83
Galaţi (Galatz), Rom., 95,646 ..... C8 22
Galatia, Saline, Ill., 830 ..... F5 90
Galatia, Barton, Kans., 73 ..... D5 93
Galatina, It., 18,200 ..... D7 21
Galatz (Galaţi), Rom., 95,646 ..... C8 22
Galax (Independent City), Va., 5,254 ..... E2 121
Galaxidhion, Grc., 2,240 ..... C4 23
Galbraith, Natchitoches, La., 125 ..... C3 95
Galchutt, Richland, N. Dak., 50 ..... D9 110
Gáldar, Can. Is., 6,165 (13,875▲) m14 20
Galdhöpiggen, mtn., Nor G3 25
Galeana, Mex., 744 ..... A3 63
Galen, Deer Lodge, Mont., 100 ..... D4 102
Galen, Macon, Tenn., 80 ..... C8 117
Galena, Alsk., 261 ..... C8 79
Galena, Jo Daviess, Ill., 4,410 ..... A3 90
Galena, Cherokee, Kans., 3,827 ..... E9 93
Galena, Kent, Md., 299 ..... B6 85
Galena, Stone, Mo., 389 ..... E4 101
Galena, Delaware, Ohio, 411 . B5 111
Galena Park, Harris, Tex., 10,852 ..... F5 118
Galeota, pt., Trin B5 60
Galera, pt., Chile B2 54
Galera, pt., Trin A5 60
Galera, riv., It h8 21
Gales, peak, Oreg B1 113
Galesburg, Knox, Ill., 37,243 . C3 90
Galesburg, Neosho, Kans., 128 ..... E8 93
Galesburg, Kalamazoo, Mich., 1,410 ..... F5 98
Galesburg, Traill, N. Dak., 166 ..... B8 110
Gales Creek, Washington, Oreg., 200 ..... B3 113
Gales Ferry, New London, Conn., 600 ..... D8 84
Galestown, Dorchester, Md., 151 ..... C6 85
Galesville, Anne Arundel, Md., 625 ..... C4 85
Galesville, Trempealeau, Wis., 1,199 ..... D2 124
Galeton, Weld, Colo., 200 ..... A6 83
Galeton, Potter, Pa., 1,646 ..... C6 114
Galetta, Ont., Can., 142 ..... B8 72
Galeville, Onondaga, N.Y., 1,500 ..... *B4 108
Galiano, isl., B.C., Can E6 68
Galich, Sov. Un., 14,700 ..... B2 29
Galicia, reg., Pol., Sov. Un D6 26
Galicia, reg., Sp., 2,602,962 . A1 20
Galien, Berrien, Mich., 750 . G4 98
Galilee, Washington, R.I., 300 ..... D10 84
Galilee, lake, Austl D8 50
Galilee, reg., Isr B7 32
Galilee, Sea of, see Tiberias, lake, Isr.
Galion, Crawford, Ohio, 12,650 ..... B5 111
Galisteo, Santa Fe, N. Mex., 125 ..... B3, G6 107
Galisteo, riv., N. Mex G6 107
Galiuro, mts., Ariz E5 80
Galivants Ferry, Horry, S.C., 25 ..... C9 115
Galkayu, Som., 8,400 ..... E6 47
Gallabat, Sud C4 47
Gallaher, Curry, N. Mex., 30 . C6 107
Gallan, head, Scot B1 13
Gallant, Etowah, Ala., 725 ..... B3 78
Gallarate, It., 35,477 ..... B2 21
Gällared, Swe A8 24
Gallatin, Daviess, Mo., 1,658 . B4 101
Gallatin, Sumner, Tenn., 7,901 ..... A5 117
Gallatin, co., Ill., 7,638 ..... F5 90
Gallatin, co., Ky., 3,867 ..... B5 94
Gallatin, co., Mont., 26,045 . E5 102
Gallatin, range, Mont., Wyo E5 102
Gallatin, riv., Mont E5 102
Gallatin Gateway, Gallatin, Mont., 160 ..... E5 102
Gallaway, Fayette, Tenn., 100 ..... B2 117
Galle, Cey., 55,825 ..... G7 39
Gallego, riv., Sp A5 20
Gallegos, riv., Arg E2 54
Gallet, lake, Que., Can C2 75
Galley, head, Ire F3 11
Gallia, co., Ohio, 26,120 ..... D5 111
Galliate, It., 12,549 ..... D4 18
Gallicano nel Lazio, It., 2,024 ..... h9 21
Gallina, Rio Arriba, N. Mex., 10 ..... A3 107
Gallinas, Lincoln, N. Mex., 15 ..... C5 107
Gallinas, mts., N. Mex C2 107
Gallinas, pt., Col A3 60
Gallion, Hale, Ala., 150 ..... C2 78
Gallion, Morehouse, La., 75 . B4 95
Gallipoli, It., 16,196 ..... D6 21
Gallipoli, see Gelibolu, Tur.
Gallipoli, pen., Tur B6 23
Gallipolis, Gallia, Ohio, 8,775 ..... D5 111
Gallipolis Ferry, Mason, W. Va., 400 ..... C2 123
Gallitzin, Cambria, Pa., 2,783 ..... F4 114
Gällivare, Swe., 25,363 ..... D9 25
Gallman, Copiah, Miss., 100 . D3 100
Gallneukirchen, Aus., 2,742 . E9 17
Gallo, mts., N. Mex C1 107
Gallo, riv., Sp B5 20
Galloo, isl., N.Y B4 108
Galloway, Greene, Mo., 200 . D4 101
Galloway, Franklin, Ohio, 250 ..... C2 111
Galloway, Barbour, W. Va., 300 ..... B4 123
Galloway, Marathon, Wis., 100 ..... D4 124
Galloway Mull, head, Scot F4 13
Gallstad, Swe., 362 ..... A7 24
Gallup, McKinley, N. Mex., 14,089 ..... B1 107
Galt, Sacramento, Calif., 1,868 ..... B6, C3 82
Galt, Ont., Can., 27,830 ..... D4 72
Galt, Wright, Iowa, 75 ..... B4 92
Galt, Grundy, Mo., 373 ..... A4 101
Galten, lake, Swe t34 25
Galtür, Aus., 523 ..... C6 18
Galty, mts., Ire E3 11
Galva, Henry, Ill., 3,060 ..... B3 90
Galva, Ida, Iowa, 469 ..... B2 92
Galva, McPherson, Kans., 442 ..... D6 93
Galvarino, Chile, 1,209 ..... B2 54
Galveston, Cass, Ind., 1,111 . C5 91
Galveston, Galveston, Tex., 67,175 (★138,700) E5, G5 118
Galveston, co., Tex., 140,364 . E5 118
Galveston, bay, Tsx E5 118
Galveston, isl., Tex E5 118
Gálvez, Arg., 7,891 ..... A5 54
Galvin, Lewis, Wash., 200 ..... C2 122
Galway, Ire., 22,028 ..... D2 11
Galway, co., Ire., 149,887 . D2, 3 11
Galway, bay, Ire D2 11
Gamaches, Fr., 3,194 ..... E9 12
Gamagori, Jap., 55,926 ..... o16 37
Gamaliel, Baxter, Ark., 150 . A3 81
Gamaliel, Monroe, Ky., 868 . D4 94
Gamarra, Col., 2,576 ..... B3 60
Gamba, Con. L B3 48
Gambaga, Ghana, 1,952 ..... D4 45
Gambēlā, Eth D3 47
Gambell, Alsk., 358 ..... C5 79
Gamber, Carroll, Md., 180 ..... B4 85
Gambia, Br., dep., Afr., 292,000 ..... D1 45, E4 42
Gambia, riv., Sen D2 45
Gambier, Knox, Ohio, 1,148 . B5 111
Gambier, is., Fr. Polynesia *H14 7
Gambo, Newf., Can., 480 ..... D4 72
Gamboma, Con. B F3 46
Gambrills, Anne Arundel, Md., 600 ..... B4 85
Game Lodge, Custer, S. Dak., 85 ..... D2 116
Gameleira, Braz., 5,078 . C3, k6 57
Gamerco, McKinley, N. Mex., 300 ..... B1 107
Gamina, Chile C2 55
Gammertingen, Ger., 2,000 . A5 18
Gammon, riv., Man., Can D4 71
Gamo-Gofa, reg., Eth., 900,000 ..... D4 47
Ganado, Jackson, Tex., 1,626 . E4 118
Ganālē, riv., Eth D5 47
Gananoque, Ont., Can., 5,096 ..... C8 72
Gand, see Gent, Bel.
Gandajika, Con. L C3 48
Gandamak, Afg., 10,000 ..... D15 41
Gandeeville, Roane, W. Va., 400 ..... C3 123
Gander, Newf., Can., 5,725 . D4 75
Gander, lake, Newf., Can D4 75
Gander, riv., Newf., Can D4 75
Gander Bay, Newf., Can., 350 ..... D4 75
Gandía, Sp., 15,812 ..... C5 20
Gandy, Logan, Nebr., 41 ..... C5 103
Gandy, Millard, Utah D1 119
Ganedidalem, Indon F7 35
Gang, canal, India C4 40
Gangapur, India, 22,591 ..... D6 40
Gangaw, Bur., 3,800 ..... D9 39
Ganges, B.C., Can., 400 ..... B9 68
Ganges, Fr., 4,262 ..... F5 14
Ganges, Richland, Ohio, 100 ..... B5 111
Ganges, riv., India, Pak D8 39
Ganges, riv. mouths, Pak G12 40
Gangtok, Sikkim, 2,744 ..... D12 40
Ganjur, China B8 34
Ganju-San, peak, Jap G10 37
Gannat, Fr., 5,376 ..... D5 14
Gannet, is., Newf., Can B3 75
Gannett, Blaine, Idaho, 20 ..... F4 89
Gannett, hill, N.Y C3 108
Gannett, peak, Wyo B3 125
Gannon, Allegany, Md., 450 ..... D2 85
Gannvalley, Buffalo, S. Dak., 100 ..... C7 116
Gano, Butler, Ohio, 150 ..... C2 111
Gano, Payne, Okla A5 112
Gans, Sequoyah, Okla., 234 . B7 112
Gansevoort, Saratoga, N.Y., 285 ..... B7 108
Ganta, Lib E3 45
Gantt, Covington, Ala., 375 . D3 78
Gantt, Greenville, S.C., 900 . B2 115
Gantt, dam, Ala D3 78
Gantts Quarry, Talladega, Ala., 238 ..... B3 78
Gao, Mali, 6,500 ..... C5 45
Gaoua, Upper Volta., 4,300 . D4 45
Gaoual, Guinea, 4,600 ..... D2 45
Gap, Fr., 20,478 ..... E7 14
Gap, Lancaster, Pa., 800 ..... G9 114
Gapan, Phil., 7,842 ..... o13 35
Gapland, Washington, Md., 200 ..... B2 85
Gap Mills, Monroe, W. Va., 95 ..... D4 123
Gara, lake, Ire D3 11
Garachiné, Pan., 1,326 ..... F8 62
Garad, Som D6 47
Garakayo, isl., Palau Is 52
Garanhuns, Braz., 34,050 . C3, k5 57
Garapan, Saipan, 2,977 52
Garapan, anchorage, Saipan 52
Garashiyoo, Palau Is 52
Garautha, India E7 40
Garber, Clayton, Iowa, 148 . B6 92
Garber, Garfield, Okla., 905 ..... A4 112
Garberville, Humboldt, Calif., 900 ..... B2 82
Garça, Braz., 18,155 ..... C3 57
Garching, Ger., 4,900 ..... A8 16
Garcia, Costilla, Colo., 160 . D5 83
Garciasville, Starr, Tex., 900 ..... F3 118
Gard, dept., Fr., 435,482 ..... *F6 14
Gard, riv., Fr F6 14
Garda, lake, It B3 21
Gardanne, Fr., 4,164 ..... F6 14
Gardar, Pembina, N. Dak., 92 ..... A8 110
Gardelegen, Ger., 12,400 ..... B5 16
Garden, Bartholomew, Ind., 250 ..... F6 91
Garden, Delta, Mich., 380 ..... C4 98
Garden, co., Nebr., 3,472 ..... C3 103
Garden, isl., Mich C5 98
Garden, riv., Ont., Can B6 98
Gardena, Los Angeles, Calif., 35,943 ..... *F2 82
Gardena, Boise, Idaho, 25 ..... F2 89
Gardena, Tazewell, Ill., 500 . C4 90
Gardena, Bottineau, N. Dak., 113 ..... A5 110
Garden City, Cullman, Ala., 536 ..... A3 78
Garden City, Duval, Fla., 800 ..... B6 86
Garden City (Chatham City), Chatham, Ga., 5,451 ..... *D5 87
Garden City, Ada, Idaho, 1,681 ..... *F2 89
Garden City, Finney, Kans., 11,811 ..... E3 93
Garden City, St. Mary, La., 300 ..... E4 95
Garden City, Wayne, Mich., 38,017 ..... A7 98
Garden City, Blue Earth, Minn., 300 ..... F4 99
Garden City, Franklin, Miss., 40 ..... D2 100
Garden City, Cass, Mo., 600 . C3 101
Garden City, Nassau, N.Y., 23,948 ..... G2 84
Garden City, Delaware, Pa., 2,000 ..... *G11 114
Garden City, Clark, S. Dak., 226 ..... C8 116
Garden City, Glasscock, Tex., 325 ..... D2 118
Garden City, Rich, Utah, 168 ..... B4 119
Garden City Park, Nassau, N.Y., 15,364 ..... G2 84
Gardendale, Jefferson, Ala., 4,712 ..... B3, E4 78
Gardendale, La Salle, Tex., 40 ..... E3 118
Garden Grove, Orange, Calif., 84,238 ..... *F3 82
Garden Grove, Decatur, Iowa, 335 ..... D4 92
Garden Home, Washington, Oreg., 2,000 ..... *B2 113
Garden Lakes, Floyd, Ga., 1,300 ..... *B1 87
Garden Plain, Sedgwick, Kans., 560 ..... B4, E6 93
Garden Prairie, Boone, Ill., 300 ..... A5 90
Gardenton, Man., Can., 104 . E6 71
Garden Valley, Boise, Idaho, 10 ..... E3 89
Garden View, Lycoming, Pa., 2,418 ..... *D7 114
Gardez, Afg., 17,540 ..... E14 41
Gardi, Wayne, Ga., 250 ..... E5 87
Gardiner, Kennebec, Maine, 6,897 ..... D3 96
Gardiner, Park, Mont., 400 . E6 102
Gardiner, Douglas, Oreg., 550 ..... D2 113
Gardiners, bay, N.Y E8 84
Gardiners, isl., N.Y D5 108
Gardiners, pt., N.Y E8 84
Gardner, Huerfano, Colo., 200 ..... D5 83

Gardner, Hardee, Fla., 80 ... E5 86
Gardner, Grundy, Ill., 1,041 ... B5 90
Gardner, Johnson, Kans., 1,619 ... D9 93
Gardner, Worcester, Mass., 19,038 ... A4 97
Gardner, Cass, N. Dak., 107 ... B9 110
Gardner, Weakley, Tenn., 40 ... A3 117
Gardner, canal, B.C., Can ... C3 68
Gardner, lake, Conn ... C8 84
Gardner, lake, Maine ... D5 96
Gardner, mtn., N.H ... B2 105
Gardner Pinnacles, isl., Haw ... k14 88
Gardnersville, Pendleton, Ky ... B7 94
Gardnerville, Douglas, Nev., 800 ... E2 104
Gardo, Som ... D6 47
Gardone Riviera, It., 1,259 ... D6 18
Gardone Val Trompia, It., 8,330 ... D6 18
Gårdstånga, Swe ... C7 24
Gardulā, Eth ... D4 47
Garesnica, Yugo., 2,332 ... C3 22
Garfield, Benton, Ark., 48 ... A2 81
Garfield, Chaffee, Colo., 20 ... C4 83
Garfield, Emanuel, Ga., 225 ... D4 87
Garfield, Pawnee, Kans., 278 ... D4 93
Garfield, Douglas, Minn., 240 ... E3 99
Garfield, Bergen, N.J., 29,253 ... D5 106
Garfield, Dona Ana, N. Mex., 100 ... E2 107
Garfield, Whitman, Wash., 607 ... B8 122
Garfield, co., Colo., 12,017 ... B2 83
Garfield, co., Mont., 1,981 ... C9 102
Garfield, co., Nebr., 2,699 ... C6 103
Garfield, co., Okla., 52,975 ... A4 112
Garfield, co., Utah, 3,577 ... F4 119
Garfield, co., Wash., 2,976 ... C8 122
Garfield, mtn., Mont ... F4 102
Garfield Heights, Cuyahoga, Ohio, 38,455 ... B2 111
Gargalianoi, Grc., 7,658 ... D3 23
Gargnano, It., 4,092 ... D6 18
Gargzdai, Sov. Un ... A6 26
Garhchiroli, India ... G8 40
Garibaldi, Braz., 3,635 ... D2 56
Garibaldi, Tillamook, Oreg., 1,163 ... B3 113
Garibaldi, mtn., B.C., Can ... E6 68
Garibaldi, park, B.C., Can ... E6 68
Garies, S. Afr., 1,103 ... D2 49
Garissa, Ken ... B6 48
Garita, P.R ... C2 65
Garland, Miller, Ark., 377 ... D2 81
Garland, Bourbon, Kans., 250 ... E9 93
Garland, Penobscot, Maine, 200 (568▲) ... C3 96
Garland, Anne Arundel, Md., 1,200 ... *B4 85
Garland, Custer, Mont., 4 ... D11 102
Garland, Seward, Nebr., 198 ... D9 103
Garland, Sampson, N.C., 642 ... C5 109
Garland, Warren, Pa., 450 ... C3 114
Garland, Tipton, Tenn., 168 ... B2 117
Garland, Dallas, Tex., 38,501 ... B6 118
Garland, Box Elder, Utah, 1,119 ... B3 119
Garland, Park, Wyo., 50 ... A4 125
Garland, co., Ark., 46,697 ... C2 81
Garm, Sov. Un ... H23 9
Garmisch-Partenkirchen, Ger., 25,000 ... E5 16
Garmouth, Scot ... C5 13
Garnavillo, Clayton, Iowa, 662 ... B6 92
Garneill, Fergus, Mont., 35 ... D7 102
Garner, White, Ark., 120 ... B4 81
Garner, Hancock, Iowa, 1,990 ... A4 92
Garner, Wake, N.C., 3,451 ... B5 109
Garnet, Granite, Mont., 3 ... D3 102
Garnet, range, Mont ... D3 102
Garnett, Anderson, Kans., 3,034 ... D8 93
Garnett, Hampton, S.C., 100 ... F5 115
Garnish, Newf., Can., 500 ... E4 75
Garo, Park, Colo ... B5 83
Garonne, riv., Fr ... E3 14
Garou, lake, Mali ... C4 45
Garoua, Cam ... D2 46
Garrard, co., Ky., 9,747 ... C5 94
Garretson, Minnehaha, S. Dak., 850 ... D9 116
Garrett, Douglas, Ill., 249 ... D5 90
Garrett, DeKalb, Ind., 4,364 ... B7 91
Garrett, Floyd, Ky., 800 ... C7 94
Garrett, Somerset, Pa., 617 ... G3 114
Garrett, Albany, Wyo., 5 ... C7 125
Garrett, co., Md., 20,420 ... D1 85
Garrett Park, Montgomery, Md., 965 ... B3 85
Garrett Park Estates, Montgomery, Md., 2,000 ... *B3 85
Garrettsville, Portage, Ohio, 1,662 ... A6 111
Garrick, Sask., Can., 450 ... D3 70
Garrison, Benton, Iowa, 421 ... B5 92
Garrison, Lewis, Ky., 350 ... B6 94
Garrison, Baltimore, Md., 500 ... B4 85
Garrison, Crow Wing, Minn., 118 ... D5 99
Garrison, Powell, Mont., 60 ... D4 102
Garrison, Putnam, N.Y., 900 ... D3, D7 108
Garrison, McLean, N. Dak., 1,794 ... B4 110
Garrison, Nacogdoches, Tex., 951 ... D5 118
Garrison, Millard, Utah, 100 ... E1 119
Garrison, dam, N. Dak ... B4 110
Garrison, res., N. Dak ... A2, B3 110
Garrisonville, Stafford, Va., 200 ... C5 121
Garristown, Ire., 121 ... D5 11
Garrochales, P.R ... B4 65
Garrote, Pan., 166 ... h11 62
Garrovillas, Sp., 5,764 ... C2 20
Garruk, Pak ... C4 39
Garry, lake, N.W., Ter., Can ... C12 66
Garry, riv., Scot ... D5 13
Garryowen, Big Horn, Mont., 4 ... E9 102
Garske, Ramsey, N. Dak., 50 ... A7 110
Garstang, Eng., 1,439 ... G6 13
Garth, Jackson, Ala., 100 ... A3 78
Garthby Station, Que., Can., 505 ... D6 73
Gartok, China, 5,000 ... B7 39
Gartz, Ger., 3,672 ... E8 24
Garvagh, N. Ire., 550 ... C5 11
Garve, Scot ... C4 13
Garvin, Lyon, Minn., 205 ... F3 99
Garvin, McCurtain, Okla., 138 ... D7 112
Garvin, co., Okla., 28,290 ... C4 112
Garwin, Tama, Iowa, 546 ... B5 92
Garwolin, Pol., 5,315 ... C6 26
Garwood, Union, N.J., 5,426 ... B4 106
Garwood, Colorado, Tex., 580 ... E4 118
Gary, Lake, Ind., 178,320 ... A3 91
Gary, Norman, Minn., 262 ... C2 99
Gary, Hidalgo, N. Mex ... E1 107
Gary, Deuel, S. Dak., 471 ... C9 116
Gary, Panola, Tex., 200 ... C5 118
Gary, McDowell, W. Va., 1,393 ... D3 123
Garysburg, Northampton, N.C., 181 ... A6 109
Garyville, St. John the Baptist, La., 2,389 ... B6, D5 95
Garza, co., Tex., 6,611 ... C2 118
Garza Little Elm, res., Tex ... A5 118
Garzón, Col., 5,750 ... C2 60
Gas, Allen, Kans., 342 ... E8 93
Gas City, Grant, Ind., 4,469 ... D6 91
Gasconade, Gasconade, Mo., 333 ... C6 101
Gasconade, co., Mo., 12,195 ... C6 101
Gasconade, riv., Mo ... C6 101
Gascons, Que., Can., 207 ... B4 73
Gascony (Gascogne), former prov., Fr., 996,000 ... E3 14
Gascoyne, Bowman, N. Dak., 50 ... C2 110
Gashaka, Cam., 1,088 ... E7 45
Gashua, Nig ... D7 45
Gasmata, New Britain I ... k13 50
Gasparilla, pass, Fla ... F4 86
Gaspé, Que., Can., 2,603 ... B3 73
Gaspé, pen., Que., Can ... B3 73
Gaspé East, co., Que., Can., 41,333 ... *B3 73
Gaspé West, co., Que., Can., 20,529 ... *B3 73
Gasport, Niagara, N.Y., 900 ... B2 108
Gasque, Baldwin, Ala., 100 ... E2 78
Gassaway, Cannon, Tenn., 70 ... B5 117
Gassaway, Braxton, W. Va., 1,223 ... C4 123
Gassetts, Windsor, Vt., 50 ... E3 120
Gassol, Nig ... E7 45
Gassville, Baxter, Ark., 233 ... A3 81
Gaston, Delaware, Ind., 801 ... D6 91
Gaston, Northampton, N.C., 1,214 ... A6 109
Gaston, Washington, Oreg., 320 ... B1 113
Gaston, Lexington, S.C., 175 ... D5 115
Gaston, co., N.C., 127,074 ... B2 109
Gastonia, Gaston, N.C., 37,276 ... B2 109
Gastre, Arg ... C3 54
Gata, cape, Sp ... E4 20
Gata, mts., Sp ... B2 20
Gatchel, Perry, Ind ... H4 91
Gatchina, Sov. Un., 47,900 ... B8 27, s31 25
Gate, Beaver, Okla., 130 ... A1, D4 112
Gate City, Scott, Va., 2,142 ... B2 121
Gatehouse of Fleet, Scot. 820 ... F4 13
Gates, Custer, Nebr., 14 ... C6 103
Gates, Monroe, N.Y., 10,000 ... *B3 108
Gates, Gates, N.C., 350 ... A7 109
Gates, Lauderdale, Tenn., 291 ... B2 117
Gates, co., N.C., 9,254 ... A7 109
Gateshead, Eng., 103,232 ... C6 10
Gates Mills, Cuyahoga, Ohio, 1,588 ... *A6 111
Gatesville, Gates, N.C., 460 ... A7 109
Gatesville, Coryell, Tex., 4,626 ... D4 118
Gateway, Mesa, Colo., 110 ... C2 83
Gateway, Jefferson, Oreg ... C5 113
Gatineau, Que., Can., 13,022 ... D2 73
Gatineau, co., Que., Can., 44,308 ... C2 73
Gatineau, riv., Que., Can ... D2 73
Gatliff, Whitely, Ky., 225 ... D5 94
Gatlinburg, Sevier, Tenn., 1,764 ... D10 117
Gato, Archuleta, Colo., 20 ... D3 83
Gatooma, S. Rh., 13,500 ... A4 49
Gattman, Monroe, Miss., 145 ... B5 100
Gatun, C.Z., 692 ... k11 62
Gatun, lake, C.Z ... k11 62
Gatzke, Marshall, Minn., 45 ... B3 99
Gauhati, India, 100,707 ... C9 39
Gauja, riv., Sov. Un ... C5 27
Gauko-Otavi, S.W. Afr ... A1 49
Gaula, riv., Nor ... F4 25
Gauley, riv., W. Va ... C4 123
Gauley Bridge, Fayette, W. Va., 950 ... C3, D7 123
Gauley Mills, Webster, W. Va., 180 ... C4 123
Gause, Milam, Tex., 500 ... D4 118
Gauss, pen., Grnld ... B17 4
Gautier, Jackson, Miss., 800 ... E3 100
Gavāter, Iran ... I10 41
Gavdhos, isl., Grc ... E5 23
Gave de Pau, riv., Fr ... F3 14
Gave d'Oloron, riv., Fr ... F3 14
Gaviota, Santa Barbara, Calif., 50 ... E3 82
Gavins Point, dam, Nebr ... B8 103
Gävle, Swe., 55,900 ... G7 25
Gävleborg, co., Swe., 294,300 ... *G7 25
Gävlebukten, bay, Swe ... G7 25
Gavrilov Posad, Sov. Un ... C13 27
Gavrilovka, Sov. Un., 10,000 ... G11 27
Gawler, Austl., 5,639 ... G2 51
Gawler, ranges, Austl ... F6 50
Gay, Meriwether, Ga., 194 ... C2 87
Gay, Keweenaw, Mich., 240 ... A2 98
Gay, head, Mass ... D6 97
Gaya, Niger, 3,100 ... D5 45
Gaya, India, 151,105 ... D7 39, E10 40
Gaya, Nig., 12,996 ... F6 45
Gay Head, Dukes, Mass., 80 (103▲) ... D6 97
Gaylesville, Cherokee, Ala., 144 ... A4 78
Gaylord, Smith, Kans., 239 ... C5 93
Gaylord, Otsego, Mich., 2,568 ... C6 98
Gaylord, Sibley, Minn., 1,631 ... F4 99
Gaylord, Coos, Oreg., 25 ... E2 113
Gaylordsville, Litchfield, Conn., 500 ... C3 84
Gayndah, Austl., 1,805 ... B8 51
Gays, Moultrie, Ill., 263 ... D5 90
Gaysin, Sov. Un., 10,000 ... G7 27
Gays, Mills, Crawford, Wis., 634 ... E3 118
Gaysville, Windsor, Vt., 100 ... D3 120
Gayville, Yankton, S. Dak., 261 ... E8 116
Gaza, O'Brien, Iowa, 100 ... A2 92
Gaza, see Ghazzah, Gaza Area.
Gaza, prov., Moz ... B5 49
Gaza Area, Eg. occ., Asia, 377,000 ... C6 32
Gazak, Iran ... H9 41
Gaziantep, Tur., 125,500 ... D11 31
Gdańsk (Danzig), Pol., 286,000 (★500,000) ... A5 26
Gdov, Sov. Un, 7,900 ... B6 27
Gdynia, Pol., 148,000 ... A5 26
Gearhart, Clatsop, Oreg., 725 ... A3 113
Gearhart, mtn., Oreg ... E6 113
Geary, N.B., Can., 938 ... D3 74
Geary, Canadian and Blaine, Okla., 1,416 ... B3 112
Geary, co., Kans., 28,779 ... D7 93
Geauga, co., Ohio, 47,573 ... A6 111
Gebeit Mines, Sud ... A4 47
Gebo, Hot Springs, Wyo., 80 ... B4 125
Gebze, Tur., 7,900 ... B7 23
Ged, Calcasieu, La., 30 ... D2 95
Gedaref, Sud., 17,537 ... C4 47
Geddes, Charles Mix, S. Dak., 380 ... D7 116
Gedera, Isr., 4,000 ... h10 32
Gedern, Ger., 3,000 ... C4 17
Gedinne, Bel., 940 ... E4 15
Gediz, Tur., 7,300 ... C7 23
Gediz (Hermus), riv., Tur ... C6 23
Gedser, Den., 1,262 ... D5 24
Geel, Bel., 27,007 ... C5 15
Geelong, Austl., 18,022 (★104,000) ... I5 51
Geelvink, bay, W. Irian ... F9 35
Geeraardsbergen, Bel., 9,582 ... D3 15
Gees Bend (Boykin), Wilcox, Ala., 100 ... C2 78
Geetbets, Bel., 3,187 ... D5 15
Geff (Jeffersonville), Wayne, Ill., 330 ... E5 90
Geidam, Nig., 11,032 ... D7 45
Geiger, Sumter, Ala., 104 ... C1 78
Geisenfeld, Ger., 2,800 ... E6 17
Geislingen, Ger., 25,800 ... E4 17
Geismar, Ascension, La., 100 ... B5 95
Geist, res., Ind ... E6 91
Geistown, Cambria, Pa., 3,186 ... F4 114
Geita, Tan, 365 ... B5 48
Gela, It., 54,774 ... F5 21
Gelatt, Susquehanna, Pa., 60 ... C10 114
Gelderland, prov., Neth., 1,287,800 ... B6 15
Geldermalsen, Neth., 1,752 ... C5 15
Geldern, Ger., 10,200 ... C3 16
Geldrop, Neth., 19,400 ... C5 15
Gelert, Ont., Can., 200 ... C6 72
Gelfingen, Switz., 493 ... B5 19
Gelib, Som., 3,000 (10,000▲) ... E5 47
Gelibolu (Gallipoli), Tur., 13,000 ... B6 23
Gelligaer, Wales, 34,572 ... C4 12
Gellinam, isl., Kwajalein ... 52
Gelnhausen, Ger., 7,800 ... C4 17
Gelsenkirchen, Ger., 382,700 ... B2 17
Gelting, Ger., 1,600 ... D3 24
Gem, Alta., Can., 25 ... D4 69
Gem, Shoshone, Idaho 350 ... B3 89
Gem, Thomas, Kans., 116 ... C3 93
Gem, Braxton, W. Va., 30 ... C4 123
Gem, co., Idaho, 9,127 ... E2 89
Gemas, Mala., 4,841 ... k5 38
Gembloux, Bel., 5,875 ... D4 15
Gemena, Con. L ... A2 48
Gemert, Neth., 7,300 ... C5 15
Gemlik, Tur., 12,700 ... B7 23
Gemmell, Koochiching, Minn., 60 ... C4 99
Gemona, del Friuli, It., 6,772 ... C9 18
Gemünden, Ger., 4,200 ... C4 17
Gem Village, La Plata, Colo., 95 ... D3 83
Genarp, Swe., 808 ... C7 24
Gene Autry, Carter, Okla., 110 ... C4 112
Geneina, Sud., 11,817 ... C1 47
General Acha, Arg., 4,709 ... B4 54
General Alvarado, Arg ... B5 54
General Alvear, Arg., 2,548 ... B4 54
General Alvear, Arg ... A3 54
General Belgrano, Arg., 3,789 ... B5 54
General Bravo, Mex., 1,225 ... G3 118
General Conesa, Arg ... C4 54
General La Madrid, Arg., 3,572 ... B4 54
General Lavalle, Arg ... B5 54
General Madariaga, Arg., 7,073 ... B5 54
General Paz, Arg ... E4 55
General Pico, Arg., 11,121 ... B4 54
General Pinedo, Arg., 2,198 ... E3 55
General Roca, Arg., 7,449 ... B3 54
General San Martín (San Martín), Arg., 269,514 ... g7 54
General Sarmiento, Arg ... g7 54
General Toshevo, Bul., 2,102 ... D9 22
General Viamonte, Arg., 5,324 ... B4 54
General Villagas, Arg., 4,738 ... A4 54
Genesee, Latah, Idaho, 535 ... C2 89
Genesee, Genesee, Mich., 700 ... E7 98
Genesee, Potter, Pa., 400 ... C6 114
Genesee, co., Mich., 374,313 ... E7 98
Genesee, co., N.Y., 53,994 ... B2 108
Genesee, riv., N.Y ... C3 108
Genesee Depot, Waukesha, Wis., 160 ... E1 124
Geneseo, Henry, Ill., 5,169 ... B3 90
Geneseo, Rice, Kans., 558 ... D5 93
Geneseo, Livingston, N.Y., 3,284 ... C3 108
Geneseo, Sargent, N. Dak., 106 ... C8 112
Geneva, Geneva, Ala., 3,840 ... D4 78
Geneva, Talbot, Ga., 261 ... D2 87
Geneva, Bearlake, Idaho, 551 ... G7 89
Geneva, Kane, Ill., 7,646 ... B5, F1 90
Geneva, Adams, Ind., 1,053 ... C8 91
Geneva, Franklin, Iowa, 219 ... B4 92
Geneva, Henderson, Ky., 150 ... C2 94
Geneva, Fillmore, Nebr., 2,352 ... D8 103
Geneva, Ontario, N.Y., 17,286 ... C4 108
Geneva, Ashtabula, Ohio, 5,677 ... A7 111
Geneva, Crawford, Pa., 250 ... C1 114
Geneva, co., Ala., 22,310 ... D4 78
Geneva, see Léman, lake, Switz.
Geneva, lake, Wis ... F5 124
Genève (Geneva), Switz., 176,183 (★270,000) ... D1 19
Genève, canton, Switz., 259,234 ... D1 19
Genevia, Pulaski, Ark., 6,000 ... C3, D6 81
Genichesk, Sov. Un., 25,000 ... H10 27
Genil, riv., Sp ... D3 20
Génissiat, Fr., 322 ... C1 18
Genk, Bel., 47,416 ... D5 15
Gennep, Neth., 5,400 ... C5 15
Gennevilliers, Fr., 42,595 ... g10 14
Geno, Harris, Tex ... E5, F5 118
Genoa, Miller, Ark., 90 ... D2 81
Genoa, Lincoln, Colo., 185 ... B7 83
Genoa, DeKalb, Ill., 2,330 ... A5 90
Genoa, see Genova, It.
Genoa, Nance, Nebr., 1,009 ... C8 103
Genoa, Douglas, Nev., 115 ... E2 104
Genoa, Cayuga, N.Y., 160 ... C4 108
Genoa, Ottawa, Ohio, 1,957 ... A2, A4 111
Genoa, Stark, Ohio, 1,000 ... *B6 111
Genoa, Harris, Tex., 200 ... E5, F5 118
Genoa, Vernon, Wis., 325 ... E2 124
Genoa, gulf, It ... B2 21
Genoa City, Walworth, Wis., 1,005 ... F1, F5 124
Genola, Morrison, Minn., 108 ... E4 99
Genolier, Switz., 286 ... D1 19
Genou, Chouteau, Mont., 46 ... B5 102
Genova (Genoa), It., 784,194 (★825,000) ... E4 18, B2 21
Genovesa, isl., Ec ... f6 58
Gent (Gand) (Ghent), Bel., 157,811 (★330,000) ... C3 15
Gentbrugge, Bel., 22,222 ... C3 15
Genthin, Ger., 15,200 ... A7 17
Gentian, Muscogee, Ga., 800 ... D2 87
Gentilly, Que., Can., 677 ... C5 73
Gentilly, Fr., 19,211 ... g10 14
Gentilly, Polk, Minn., 100 ... C2 99
Gentry, Benton, Ark., 686 ... A1 81
Gentry, Gentry, Mo., 98 ... A3 101
Gentry, co., Mo., 8,793 ... A3 101
Gentryville, Spencer, Ind., 297 ... H3 91
Gentryville, Gentry, Mo., 80 ... A3 101
Genzano di Roma, It., 12,727 ... h9 21
Geographe, bay, Austl ... F2 50
Geographe, chan., Austl ... D1 50
Geographical Society, isl., Grnld ... B17 4
Geographic Center of North America, N. Dak ... A6 110
Geographic Center of United States, S. Dak ... B7 76
Geokchay, Sov. Un., 10,000 ... G18 9
George, Lyon, Iowa, 1,200 ... A2 92
George, Northampton, N.C., 200 ... A6 109
George, S. Afr., 14,505 ... D3 49
George, co., Miss., 11,098 ... E5 100
George, Grant, Wash., 50 ... C6 122
George, bay, N.S., Can ... D8 74
George, hill, Md ... D1 85
George, isl., Newf., Can ... A3 75
George, lake, Austl ... G7 51
George, lake, N.S., Can ... E4 74
George, lake, Ont., Can ... B6 98
George, lake, Fla ... C5 86
George, lake, N.Y ... B7 108
George, riv., Que., Can ... A3 73
George V, coast, Ant ... C27 5
George B. Stevenson, flood control res., Pa ... D5 114
George River, Que., Can., 144 ... f8 75
Georges, Mills, Sullivan, N.H., 100 ... B1 105
Georgesville, Franklin, Ohio, 200 ... C1 111
Georgetown, White, Ark., 200 ... B4 81
George Town, Ba., 445 ... D6 64
Georgetown, Br. Gu., 92,000 ... A3 59
Georgetown, Ont., Can., 10,298 ... D5 72
Georgetown, P.E.I., Can., 744 ... C7 74
Georgetown, Cayman Is ... F3 64
Georgetown, Clear Creek, Colo., 307 ... B5 83
Georgetown, Fairfield, Conn., 1,100 ... D3 84
Georgetown, Sussex, Del., 1,765 ... C7 85
Georgetown, Putnam, Fla., 500 ... C5 86
Georgetown, Gam ... D2 45
Georgetown, Quitman, Ga., 554 ... E1 87
Georgetown, Bear Lake, Idaho, 551 ... G7 89
Georgetown, Vermilion, Ill., 3,544 ... D6 90
Georgetown, Floyd, Ind., 643 ... H6 91
Georgetown, Scott, Ky., 6,986 ... B5 94
Georgetown, Grant, La., 321 ... C3 95
Georgetown, Kent, Md., 50 ... B6 85
Georgetown, Essex, Mass., 2,005 (3,755▲) ... A6 97
Georgetown, Clay, Minn., 178 ... C2 99
Georgetown, Copiah, Miss., 285 ... D3 100
Georgetown, Madison, N.Y., 240 ... C5 108
Georgetown, Brown, Ohio, 2,674 ... D4 111
Georgetown, Beaver, Pa., 246 ... A2 123
Georgetown, Georgetown, S.C., 12,261 ... E9 115
Georgetown, Hamilton, Tenn., 75 ... D9 117
Georgetown, Williamson, Tex., 5,218 ... D4 118
Georgetown, co., S.C., 34,798 ... D9 115
Georgeville, Que., Can., 320 ... D5 73
George Washington Birthplace, nat. mon., Va ... C6 121
George West, Live Oak, Tex., 1,878 ... E3 118
Georgia (Georgian S.S.R.), rep., Sov. Un., 4,200,000 ... G17 9
Georgia, state, U.S., 3,943,116 ... D11 77, 87
Georgia, mtn., Vt ... B2 20
Georgia, straits, B.C., Can ... E5 68
Georgia Center, Franklin, Vt., 100 (1,079▲) ... B2 120
Georgian S.S.R. (Georgia), rep., Sov. Un., 4,200,000 ... G17 9
Georgian, bay, Ont., Can ... B3 72
Georgiana, Butler, Ala., 2,093 ... D3 78
Georgian Bay Island, nat. park, Ont., Can ... C5 72
Georgia Southern, Bulloch, Ga., 1,400 ... D5 87
Georgina, riv., Austl ... D6 50
Georgiyevsk, Sov. Un., 10,000 ... G17 9
Georgsheil, Ger., 99 ... A7 15
Gera, Ger., 101,400 ... C7 17
Gerald, Sask., Can., 131 ... G5 70
Gerald, Franklin, Mo., 474 ... C6 101
Geraldine, De Kalb, Ala., 340 ... A4 78
Geraldine, Chouteau, Mont., 364 ... C6 102
Geraldton, Austl., 10,894 ... E1 50
Geraldton, Ont., Can., 3,375 ... E8 72
Gerard, Somerset, Maine ... C2 96
Gérardmer, Fr., 8,970 ... A2 18
Gerber, Tehama, Calif., 700 ... B2 82
Gerbstedt, Ger., 6,326 ... B6 17
Gercüs, Tur., 2,300 ... D13 31
Gerdau, riv., Ger ... F4 24
Gerdine, mtn., Alsk ... g15 79
Gerede, Tur., 5,400 ... B9 31
Gérgal, Sp., 3,934 ... D4 20
Gerin, Que., Can., 136 ... C4 73
Gering, Scotts Bluff, Nebr., 4,585 ... C2 103
Gerlach, Washoe, Nev., 170 ... C2 104
Gerlachovka, mtn., Czech ... D6 26
Germania, Atlantic, N.J ... D3 106
Germania, Potter, Pa., 150 ... C6 114
Germansen, lake, B.C., Can ... B5 68
Germantown, N.B., Can., 75 ... D5 74
Germantown, Fairfield, Conn., 2,893 ... *D3 84
Germantown, Clinton, Ill., 983 ... E4 90
Germantown, Bracken and Mason, Ky., 251 ... B6 94
Germantown, Montgomery, Md., 125 ... B3 85
Germantown, Montgomery, Ohio, 3,399 ... C3 111
Germantown, Shelby, Tenn., 1,104 ... B2 117
Germantown, Washington, Wis., 622 ... E1, E5 124
German Valley, Stephenson, Ill., 224 ... A4 90
Germany, reg., Eur ... E11 8, 16
Germany, East, country, Eur., 17,079,300 ... B6 16
Germany, West, country, Eur., 56,173,300 ... C4 16
Germersheim, Ger., 7,500 ... D3 17
Germfask, Schoolcraft, Mich., 125 ... B5 98
Germiston, S. Afr., 148,102 ... C4 49
Gernrode, Ger., 6,033 ... B6 17
Gernsheim, Ger., 7,100 ... D3 17
Gero, Jap, 16,163 ... n16 37
Gerolzhofen, Ger., 5,500 ... D5 17
Gerona, Phil., 2,471 ... o13 35
Gerona, Sp., 32,784 ... B7 20
Gerona, prov., Sp., 351,369 ... *B7 20
Geronimo, Comanche, Okla., 199 ... C3 112
Geronimo, Guadalupe, Tex., 110 ... B4 118
Gerrard, B.C., Can., 50 ... D9 68
Gerrardstown, Berkeley, W. Va., 240 ... B6 123
Gerrish, Merrimack, N.H., 30 ... D3 105
Gers, dept., Fr., 182,264 ... *F4 14
Gersfeld, Ger., 2,100 ... C4 17
Gersthofen, Ger., 10,800 ... E5 17
Gerty, Hughes, Okla., 135 ... C5 112
Gervais, Marion, Oreg., 438 ... B2, B4 113
Géryville, Alg., 7,614 (62,408▲) ... C5 44
Gerze, Tur., 4,272 ... B10 31
Geseke, Ger., 11,400 ... B3 17
Gesher Haziv, Isr ... A7 32
Gessie, Vermillion, Ind., 130 ... D3 91
Getafe, Sp., 21,895 ... B4, p17 20
Gethsemani, Que., Can., 147 ... h9 75
Gettysburg, Darke, Ohio, 443 ... B3 111
Gettysburg, Adams, Pa., 7,960 ... G7 114
Gettysburg, Potter, S. Dak., 1,950 ... C6 122
Getúlio Vargas, Braz., 5,705 ... D2 56
Getz, ice shelf, Ant ... B35 5
Geuda Springs, Sumner and Cowley, Kans., 223 ... E6 93
Gevelsberg, Ger., 31,700 ... B2 17
Gex, Fr., 1,295 ... C2 18
Geyser, Judith Basin, Mont., 175 ... C6 102
Geyserville, Sonoma, Calif., 225 ... C2 82
Geyve, Tur., 3,700 ... B8 23
Gézenti, Chad, 106 ... A3 46
Ghadir, as Sufi (Oasis), Iraq ... F13 31
Ghagar, res., India ... E9 40
Ghaggar, riv., India ... C5 40
Ghana, country, Afr., 6,690,730 ... F5 42, E4 45
Ghanzi, Bech ... B3 49
Ghardaïa, Alg., 14,046 (48,080▲) ... C5 44
Ghardaïa, reg., Alg ... C5 44
Gharyān, Libya, 2,796 ... C2 43
Ghāt, Libya, 1,508 ... E2 43
Ghats, mts., India ... E5, E7 39
Ghazal, riv., Chad ... C3 46
Ghazipur, India, 37,147 ... E9 40
Ghazni, Afg., 27,084 ... E14 41
Ghazzah (Gaza), Gaza Area, 37,820 ... C6 32
Gheen, St. Louis, Minn., 50 ... C6 99
Gheens, Lafourche, La., 100 ... C6, E5 95
Ghent, see Gent, Bel.
Ghent, Carroll, Ky., 342 ... B4 94
Ghent, Lyon, Minn., 326 ... F3 99
Ghent, Columbia, N.Y., 350 ... C7 108
Gheorgheni, Rom., 11,969 ... B7 22
Gherla, Rom., 7,617 ... B6 22
Ghimir, see Ginir, Eth.
Ghio, Richmond, N.C ... C4 109
Ghītah, Eg., U.A.R., 1,000 ... D3 32
Ghizao, Afg., 5,000 ... E13 41
Gholson, Noxubee, Miss., 50 ... C5 100
Ghor, riv., Afg ... E11 41
Ghudāmis, Libya ... C1 43
Ghurian, Afg., 10,000 ... D10 41
Giahel, riv., Som ... C6 47
Giamda (Taichao), China ... E3 34
Giannutri, isl., It ... C3 21
Giaveno, It., 9,692 ... D3 18
Gibara, Cuba, 8,045 ... E5 64
Gibbon, Sibley, Minn., 896 ... F4 99
Gibbon, Buffalo, Nebr., 1,083 ... D7 103
Gibbon, Umatilla, Oreg., 80 ... B8 113
Gibbonsville, Lemhi, Idaho, 125 ... D5 89
Gibbs, Adair, Mo., 158 ... A5 101
Gibbs, Obion, Tenn., 100 ... A3 117
Gibbs City, Iron, Mich ... B2 98
Gibbstown, Gloucester, N.J., 4,065 ... D2 106
Gibeon, S.W. Afr., 485 ... C2 49
Gibraleón, Sp., 8,865 ... D2 20
Gibraltar, Gib., 23,232 (★85,000) ... D3 20
Gibraltar, Br. dep., Eur., 23,232 ... *D3 20
Gibraltar, Wayne, Mich., 2,196 ... *F7 98
Gibraltar, bay, Sp ... D3 20
Gibraltar, pt., Eng ... A8 12
Gibraltar, strait, Afr., Eur ... G4 30
Gibsland, Bienville, La., 1,150 ... B2 95
Gibson, Glascock, Ga., 479 ... C4 87
Gibson, Terrebonne, La., 280 ... C5, E5 95
Gibson, Monroe, Miss., 50 ... B5 100
Gibson, Dunklin, Mo., 100 ... E8 101
Gibson, Scotland, N.C., 501 ... C4 109
Gibson, Gibson, Tenn., 297 ... B3 117
Gibson, co., Ind., 29,949 ... H2 91
Gibson, co., Tenn., 44,699 ... A3 117
Gibson, des., Austl ... D3 50
Gibson, isl., Md ... B5 85
Gibsonburg, Sandusky, Ohio, 2,540 ... A2, A4 111
Gibson City, Ford, Ill., 3,453 ... C5 90
Gibsonia, Allegheny, Pa., 1,150 ... A6 114
Gibsons, B.C., Can., 1,091 ... E6 68
Gibsonton, Hillsborough, Fla., 1,673 ... E2 86
Gibsonville, Guilford and Alamance, N.C., 1,784 ... A4 109
Gidam, India ... H8 40
Giddings, Lee, Tex., 2,821 ... D4 118
Gideon, New Madrid, Mo., 1,411 ... E8 101
Gien, Fr., 8,812 ... D5 14
Giessen, Ger., 66,300 (★100,000) ... C3 17
Giffard, Que., Can., 10,129 ... C9 73
Gifford, Hot Spring, Ark., 150 ... D6 81
Gifford, Indian River, Fla., 3,509 ... E6 86
Gifford, Nez Perce, Idaho, 50 ... C2 89
Gifford, Champaign, Ill., 609 ... C5 90
Gifford, McKean, Pa., 430 ... C4 114
Gifhorn, Ger., 17,700 ... F4 24
Gift, Tipton, Tenn., 40 ... B2 117
Gifu, Jap., 304,492 ... I8, n15 37
Gifu, pref., Jap., 1,638,399 ... *I8 37
Giganta, mts., Mex ... B2 63
Gigante, Col., 2,607 ... C2 60
Gigha, isl., Scot ... E3 13
Gigha, sound, Scot ... E3 13
Gig Harbor, Pierce, Wash., 1,094 ... B3, D1 122
Giglio, isl., It ... C3 21
Giguela, riv., Sp ... C4 20
Gihon, riv., Vt ... B3 120
Gijón, Sp., 124,714 ... A3 20
Gil, isl., B.C., Can ... C3 68
Gila, Grant, N. Mex., 10 ... E1 108
Gila, co., Ariz., 25,745 ... D5 80
Gila, mts., Ariz ... D6 80
Gila, riv., Ariz., N. Mex ... E2 80, D1 107
Gila, riv., Eth ... D3 47
Gila Bend, Maricopa, Ariz., 2,132 ... E3 80
Gila Bend, Indian res., Ariz ... D3 80
Gila Bend, mts., Ariz ... D2 80
Gila Center, Yuma, Ariz., 5 ... E1 80
Gila Cliff Dwellings, nat. mon., N. Mex ... D1 107
Gila River, Indian res., Ariz ... D4 80
Gilbert, Maricopa, Ariz., 1,833 ... D4, G2 80
Gilbert, Searcy, Ark., 52 ... B3 81
Gilbert, Story, Iowa, 318 ... B4 92
Gilbert, Franklin, La., 472 ... B4 95
Gilbert, St. Louis, Minn., 2,591 ... C6 99
Gilbert, Lexington, S.C., 171 ... D5 115
Gilbert, Mingo, W. Va., 874 ... D3 123
Gilbert, is., Gilbert & Ellice Is ... G 7 10
Gilbert, riv., Austl ... C7 50
Gilbert, riv., Newf., Can ... B3 75
Gilbert & Ellice Islands, Br. dep., Oceania, 36,000 ... G10 7
Gilbert Plains, Man., Can., 849 ... D1 71
Gilberton, Schuylkill, Pa., 1,712 ... *E9 114
Gilbertown, Choctaw, Ala., 270 ... D1 78
Gilbertville, Black Hawk, Iowa, 533 ... B5 92
Gilbertsville, Marshall, Ky., 231 ... A3 94
Gilbertsville, Otsego, N.Y., 522 ... C5 108
Gilbertsville, Montgomery, Pa., 750 ... F10 114
Gilbertville, Worcester, Mass., 1,202 ... B3 97
Gilboa, Cheshire, N.H., 60 ... E2 105
Gilboa, Putnam, Ohio, 207 ... A4 111
Gilby, Grand Forks, N. Dak., 281 ... A8 110
Gilchrist, Klamath, Oreg., 500 ... D5 113
Gilchrist, co., Fla., 2,868 ... C4 86
Gilcrest, Weld, Colo., 357 ... A6 83
Gildford, Hill, Mont., 300 ... B6 102
Gilead, Tolland, Conn., 75 ... C7 84
Gilead, Oxford, Maine, 75 (136▲) ... D2 96
Gilead, Thayer, Nebr., 79 ... D8 103
Giles, Donley, Tex., 25 ... B2 118
Giles, co., Tenn., 22,410 ... B4 117
Giles, co., Va., 17,219 ... C2 121

Gilford, Balknap, N.H., 165 (2,043▲) C4 105
Gilford, N. Ire C5 11
Gilford, isl., B.C., Can D4 68
Gilford Park, Ocean, N.J., 1,560 *D4 106
Gilgandra, Austl., 2,245 E7 51
Gilgit, Pak., 4,671 A5 39
Gilgunnia, Austl., 20 F6 51
Gill, Weld, Colo., 150 A6 83
Gill, Franklin, Mass., 100 (1,203▲) A2 97
Gill, lake, Ire C3 11
Gillam, Man., Can., 332 A4 71
Gilleleje, Den., 2,219 B6 24
Gillen, lake, Austl E3 50
Gillespie, Macoupin, Ill., 3,569 D4 90
Gillespie, co., Tex., 10,048 D3 118
Gillespie, dam, Ariz D3 80
Gillett, Arkansas, Ark., 674 C4 81
Gillett, Manatee, Fla., 200 F2 86
Gillett, Bradford, Pa., 200 C8 114
Gillett, Karnes, Tex., 100 B4 118
Gillett, Oconto, Wis., 1,374 D5 124
Gillette, Campbell, Wyo., 3,580 A7 125
Gillett Grove, Clay, Iowa, 185 A2 92
Gillham, Sevier, Ark., 177 C1 81
Gilliam, Caddo, La., 300 B2 95
Gilliam, Saline, Mo., 249 B4 101
Gilliam, co., Oreg., 3,069 B6 113
Gilles Point, N.S., Can., 100 C9 74
Gillingham, Eng., 72,611 C8 12
Gillisburg, Amite, Miss., 75 D3 100
Gillises Mills, Hardin, Tenn B3 117
Gilluly, Utah, Utah, 12 D4 119
Gilly, Bel., 23,858 D4 15
Gilman, Eagle, Colo., 356 B4 83
Gilman, New London, Conn., 150 C8 84
Gilman, Iroquois, Ill., 1,704 C5 90
Gilman, Marshall, Iowa, 491 C5 92
Gilman, Benton, Minn., 146 E5 99
Gilman, Lewis and Clark, Mont., 15 C4 102
Gilman, Essex, Vt., 600 C5 120
Gilman, Taylor, Wis., 379 C3 124
Gilman City, Harrison, Mo., 379 A4 101
Gilmanton, Belknap, N.H., 185 (736▲) D4 105
Gilmanton, Buffalo, Wis., 200 D2 124
Gilmanton Iron Works, Belknap, N.H., 150 D4 105
Gilmer, Upshur, Tex., 4,312 C5 118
Gilmer, Gilmer, W. Va., 150 C4 123
Gilmer, co., Ga., 8,922 B2 87
Gilmer, co., W. Va., 8,050 C4 123
Gilmore, Crittenden, Ark., 438 B5 81
Gilmore, Lemhi, Idaho, 5 E5 89
Gilmore City, Humboldt, Iowa, 688 B3 92
Gilmour, Ont., Can., 69 C7 72
Gilpin, Casly, Ky., 50 C5 94
Gilpin, co., Colo., 685 B5 83
Gilpin, Dickens, Tex C2 118
Gilroy, Santa Clara, Calif., 7,348 C6, D3 82
Gilroy, Sask., Can., 91 G2 70
Gilson, Knox, Ill., 200 C3 90
Gilsum, Cheshire, N.H., 175 (528▲) D2 105
Giltner, Hamilton, Nebr., 293 D7 103
Gima, Okinawa 52
Gimli, Man., Can., 1,841 D3 71
Gingoog, Phil., 9,331 *D6 35
Ginir, Eth D5 47
Ginosa, It., 17,800 D6 21
Ginzo, Sp., 9,130 A2 20
Gioia del Colle, It., 24,000 D6 21
Gioiosa Ionica, It., 5,002 E6 21
Gi-Paraná, riv., Braz E2 59
Gipsera, Switz C3 19
Girard, Burke, Ga., 248 C5 87
Girard, Macoupin, Ill., 1,734 D4 90
Girard, Crawford, Kans., 2,350 E9 93
Girard, Richland, La., 100 B4 95
Girard, Trumbull, Ohio, 12,997 A7 111
Girard, Erie, Pa., 2,451 C1 114
Girard, Kent, Tex., 100 C2 118
Girardot, Col., 35,665 C3 60
Girardville, Schuylkill, Pa., 2,958 E9 114
Girdletree, Worcester, Md., 300 D7 85
Giresun (Kerasund), Tur., 19,900 B12 31
Giri, riv., Con. L A2 48
Giridih, India, 36,881 E11 40
Girishk, Afg., 5,000 F12 41
Girna, riv., India G5 40
Giromagny, Fr., 3,181 D7 14
Girón, Ec., 1,623 B2 58
Gironde, dept., Fr., 935,448 *E3 14
Gironde, riv., Fr E3 14
Giroux, Man., Can., 51 E3 71
Girouxville, Alta., Can., 318 B2 69
Girvan, Scot., 6,159 E4 13
Girvin, Sask., Can., 143 F3 70
Girvin, Pecos, Tex., 35 D1 118
Gisborne, N.Z., 21,769 (*25,065) M17 51
Gisburn, lake, Newf., Can E4 75
Giscome, B.C., Can., 646 B6 68
Gislaveds, Swe., 6,900 A7 24
Gisors, Fr., 6,398 C4 14
Gitano, Jones, Miss., 40 D4 100
Giuba (Juba), riv., Som E5 47
Giulianova, It., 9,100 C4 21
Giumbo, Som F5 47
Giurgiu, Rom.,3 2,613 D7 22
Givataim, Isr., 30,822 g10 32
Give, Den., 1,800 C3 24
Givet, Fr., 7,444 B6 14
Givhans, Dorchester, S.C., 200 E1, E7 115
Givors, Fr., 17,066 (*27,000) E6 14
Givry, isl., Turk 52
Gizeh, see Al Jizah, Eg., U.A.R.
Gizhduvan, Sov. Un., 15,000 A12 41
Gizhiga, Sov. Un C19 28
Gizycho, Pol., 12,400 A6 26
Gjerrild, Den., 300 B4 24
Gjinokaster, Alb., 13,100 B3 23
Gjinokaster, pref., Alb., 168,000 *B3 23
Gjoa Haven, N.W. Ter., Can., 98 C13 66
Gjövik, Nor., 7,900 G4 25
Glace Bay, N.S., Can., 24,186 (*43,800) C10 74
Glacier, B.C., Can., 67 D9 68
Glacier, Whatcom, Wash., 50 A4 122
Glacier, co., Mont., 11,565 B3 102
Glacier, bay, Alsk k21 79
Glacier, nat. park, B.C., Can D9 68
Glacier, nat. park, Mont B3 102
Glacier Bay, nat. mon., Alsk D12 79
Glacier, peak, Wash A4 122
Gladbeck, Ger., 84,200 B1 17
Gladbrook, Tama, Iowa, 949 B5 92
Glade, Phillips, Kans., 133 C4 93
Glade, Catahoula, La C4 95
Glade, creek, Wash C6 122
Glade, creek, W. Va D7 123
Glade Park, Mesa, Colo., 5 B2 83
Glades, swamp, N.J E2 106
Glades, co., Fla., 2,950 F5 86
Glade Spring, Washington, Va., 1,407 B3 121
Glade Valley, Alleghany, N.C., 100 A2 109
Gladeville, Wilson, Tenn., 120 A5 117
Gladewater, Gregg and Upshur, Tex., 5,742 C5 118
Gladiola, Lea, N. Mex D6 107
Gladmar, Sask., Can., 107 H3 70
Gladsakse, Den., 64,693 *C6 24
Gladstone, Austl., 7,181 A8 51
Gladstone, Austl., 1,063 F2 51
Gladstone, Man., Can., 994 D2 71
Gladstone, Henderson, Ill., 356 C3 90
Gladstone, Delta, Mich., 5,267 C3 98
Gladstone, Clay, Mo., 14,502 *B3 101
Gladstone, Jefferson, Nebr., 55 D8 103
Gladstone (Peapack-Gladstone), Somerset, N.J., 1,804 B3 106
Gladstone, Union, N. Mex., 6 A6 107
Gladstone, Stark, N. Dak., 185 C3 110
Gladstone, Clackamas, Oreg., 3,854 B2, B4 113
Gladstone, Nelson, Va., 150 D4 121
Glad Valley, Ziebach, S. Dak., 30 B4 116
Gladwin, Gladwin, Mich., 2,226 E6 98
Gladwin, co., Mich., 10,769 D6 98
Gladwyne, Montgomery, Pa., 2,500 *A11 114
Glady, Randolph, W. Va., 100 C5 123
Gladys, Campbell, Va., 180 D3 121
Glåma, riv., Nor G5, p29 25
Glamis, Sask., Can., 61 F2 70
Glamoč, Yugo., 986 C3 22
Glamorgan, co., Wales, 1,227,828 C4 12
Glan, Phil., 856 *D7 35
Glan, lake, Swe u33 25
Glan, riv., Ger D2 17
Glancy, Copiah, Miss., 30 D3 100
Gland, Switz., 1,545 D1 19
Glandorf, Putnam, Ohio, 747 A3 111
Glanshammar, Swe., 234 t33 25
Glarnisch, mtn., Switz C6 19
Glarus, Switz., 5,852 B7 19
Glarus, canton, Switz., 40,148 C7 19
Glarus Alps, mts., Switz C6 19
Glasco, Cloud, Kans., 812 C6 93
Glasco, Ulster, N.Y., 950 C7 108
Glascock, co., Ga., 2,672 C4 87
Glasford, Peoria, Ill., 1,012 C4 90
Glasgo, New London, Conn., 210 C9 84
Glasgow, Barren Ky., 10,069 C4 94
Glasgow, Howard, Mo., 1,200 B5 101
Glasgow, Valley, Mont., 6,398 B10 112
Glasgow, Scot., 1,054,913 (*1,880,000) E4 13
Glasgow, Rockbridge, Va., 1,091 D3 121
Glasgow, Kanawha, W. Va., 914 C6 123
Glaslyn, Sask., Can., 269 D1 70
Glasnevin, Sask., Can., 35 H3 70
Glass, butte, Oreg D6 113
Glass, Obion, Tenn., 50 A2 117
Glass, mts., Tex D1 118
Glassboro, Gloucester, N.J., 10,253 D2 106
Glasscock, co., Tex., 1,118 D2 118
Glasson, Ire., 95 D4 11
Glassport, Allegheny, Pa., 8,418 F2 114
Glasston, Pembina, N. Dak., 60 A8 110
Glastenbury, mtn., Vt F12 120
Glastonbury, Hartford, Conn., 3,400 (14,497▲) C6 84
Glastonbury, Eng., 5,796 C5 12
Glauchau, Ger., 33,700 C7 17
Glazier, Hemphill, Tex., 55 A2 118
Glazier, lake, Maine A3 96
Glazov, Sov. Un., 61,000 B4 29
Glazypeau, mtn., Ark C5 81
Gleason, Weakley, Tenn., 900 A3 117
Gleason, Lincoln, Wis., 200 C4 124
Gleasondale, Middlesex, Mass. (part of Hudson) C1 91
Gleichen, Alta., Can., 426 D4 68
Glen, Beaverhead, Mont., 35 E4 102
Glen, Sioux, Nebr., 50 B2 103
Glen, Carroll, N.H., 190 B4 105
Glen, Clay, W. Va., 45 C3, C6 123
Glen, canyon, Utah G4 119
Glen, lake, Mich D4 98
Glenada, Lane, Oreg., 75 D2 113
Glenade, Ire C3 11
Glenade, Alcoe, Roane, Tenn D9 117
Glen Allan, Washington, Miss., 900 B2 100
Glen Allen, Fayette, Ala., 131 B2 78
Glenallen, Henrico, Va., 500 D5 121
Glen Almond, Que., Can., 219 D2 73
Glen Alpine, Burke, N.C., 734 B2, D4 109
Glenalum, Mingo, W. Va., 250 D3 123
Glenamoy, Ire C2 11
Glénans, is., Fr D1 14
Glen Arbor, Leelanau, Mich., 500 D4 98
Glenarden, Prince Georges, Md., 1,336 *C4 85
Glenarm, N. Ire., 591 C6 11
Glenavon, Sask., Can., 377 G3 70
Glen Avon Heights, Riverside, Calif., 3,416 *F5 82
Glenbain, Sask., Can., 60 H2 70
Glenbarr, Scot E3 13
Glenbeigh, Ire., 150 E2 11
Glenbeulah, Sheboygan, Wis., 428 B6 124
Glenboro, Man., Can., 797 E2 71
Glenbrook, Douglas, Nev., 20 D2 104
Glenburn, Renville, N. Dak., 363 A4 110
Glen Burnie, Anne Arundel, Md., 15,000 B4 85
Glenbush, Sask., Can., 56 D2 70
Glen Campbell, Indiana, Pa., 400 E4 114
Glen Canyon, dam, Ariz A4 80
Glencar, Ire E2 11
Glen Carbon, Madison, Ill., 1,241 E4 90
Glencliff, Grafton, N.H., 70 C3 105
Glencliff, Davidson, Tenn., 3,000 *A5 117
Glencoe, Etowah, Ala., 2,592 B4 78
Glencoe, Ont., Can., 1,156 E3 72
Glencoe, Cook, Ill., 10,472 E3 90
Glencoe, Gallatin, Ky., 500 B5 94
Glencoe, Baltimore, Md., 200 A4 85
Glencoe, McLeod, Minn., 3,216 F4 99
Glencoe, St. Louis, Mo., 300 A7 101
Glencoe, Lincoln, N. Mex., 10 D4 107
Glencoe, Belmont, Ohio, 380 B1 123
Glencoe, Payne, Okla., 284 A5 112
Glencoe, val., Scot D4 13
Glencolumbkille, Ire., 95 C3 11
Glen Cove, Nassau, N.Y., 23,817 D2 108
Glencross, Dewey, S. Dak., 45 B5 116
Glencullen, Multnomah, Oreg., 1,000 B2 113
Glendale, Maricopa, Ariz., 15,696 D3, G2 80
Glendale, Lincoln, Ark., 40 D4 81
Glendale, Los Angeles, Calif., 119,442 F2 82
Glendale, Walton, Fla., 100 G3 86
Glendale, Franklin, Idaho, 30 G7 89
Glendale, Hardin, Ky., 300 C4 94
Glendale, St. Louis, Mo., 7,048 *C7 101
Glendale, Hamilton, Ohio, 2,823 C3, D2 111
Glendale, Douglas, Oreg., 748 E3 113
Glendale, Providence, R.I., 500 B10 84
Glendale, Spartanburg, S.C., 600 B4 115
Glendale, Davidson, Tenn., 9,000 *A5 117
Glendale, Maury, Tenn., 50 B5 117
Glendale, Kane, Utah, 223 E3 119
Glen Dale, Marshall, W. Va., 1,905 B4 123
Glendale, Milwaukee, Wis., 10,432 *E6 124
Glendale Colony, Spink, S. Dak., 100 C7 116
Glen Dean, Breckinridge, Ky., 150 C3 94
Glendevey, Larimer, Colo., 30 A5 83
Glendive, Dawson, Mont., 7,058 C12 102
Glendo, Platte, Wyo., 292 C8 125
Glendo, res., Wyo C8 125
Glendon, Alta., Can., 315 B5 69
Glendon, Lincoln, Maine, 50 D3 96
Glendora, Los Angeles, Calif., 20,752 F3 82
Glendora, Tallahatchie, Miss., 147 B3 100
Glendora, Camden, N.J., 5,000 *D2 106
Glendowan, Ire C4 11
Glen Easton, Marshall, W. Va., 85 B4, C2 123
Glen Echo, Montgomery, Md., 310 C1 85
Glen Echo Heights, Montgomery, Md., 700 *C3 85
Glen Elder, Mitchell, Kans., 444 C5 93
Glenelg, Howard, Md., 200 B4 85
Glenelg, Scot., 1,486 C3 13
Glenella, Man., Can., 219 D2 71
Glen Ellen, Sonoma, Calif., 700 B5 82
Glen Ellis, falls, N.H B4 105
Glen Ellyn, DuPage, Ill., 15,972 F2 90
Glen Ewen, Sask., Can., 289 H4 70
Glengarriff, Ire., 392 F2 11
Glen Ferris, Fayette, W. Va., 300 D7 123
Glenfield, Lewis, N.Y., 270 B5 108
Glenfield, Foster, N. Dak., 129 B7 110
Glenfinnan, Scot D3 13
Glenflora, Wharton, Tex., 350 E4 118
Glen Gardner, Hunterdon, N.J., 787 B3 106
Glengarriff, Ire., 392 F2 11
Glengarry, co., Ont., Can., 19,217 B10 72
Glenham, Walworth, S. Dak., 171 B5 116
Glen Haven, Larimer, Colo., 25 A5 83
Glen Haven, De Kalb, Ga., 2,000 *C2 87
Glen Head, Nassau, N.Y., 4,900 *F2 84
Glen Innes, Austl., 5,769 D8 51
Glenisla, Scot., 440 D5 13
Glen Jean, Fayette, W. Va., 900 D3, D7 123
Glen Kerr, Sask., Can., 15 G2 70
Glenluce, Scot., 1,919 F4 13
Glen Lyon, Luzerne, Pa., 4,173 D9 114
Glen Mary, Scott, Tenn., 175 C9 117
Glenmont, Holmes, Ohio, 283 B5 111
Glen Moore, Lancaster, Pa., 1,700 *G10 114
Glenmora, Rapides, La., 1,447 D3 95
Glenmore, Van Wert, Ohio, 100 B3 111
Glenmorgan, Austl., 127 C7 51
Glenn, Allegan, Mich., 200 F4 98
Glenn, co., Calif., 17,245 C2 82
Glenn Allen, Alsk., 169 F19 79
Glennamaddy, Ire., 234 D3 11
Glenn Dale, Prince Georges, Md., 205 C4 85
Glennie, Alcona, Mich., 90 D7 98
Glennonville, Dunklin, Mo., 100 E7 101
Glenns Ferry, Elmore, Idaho, 1,374 G3 89
Glennville, Tattnall, Ga., 2,791 E5 87
Glen Oak Acres, Cook, Ill., 1,500 *A6 90
Glen Oaks, Yavapai, Ariz., 20 C3 80
Glenolden, Delaware, Pa., 7,249 B11 114
Glenoma, Lewis, Wash., 50 C3 122
Glenpool, Tulsa, Okla., 353 B5 112
Glen Raven, Alamance, N.C., 2,418 A4 109
Glen Riddle, Delaware, Pa., 800 B10 114
Glen Ridge, Palm Beach, Fla., 226 *F6 86
Glenridge, Marion, Ill E4 90
Glen Ridge, Essex, N.J., 8,322 B4 106
Glenrio, Quay, N. Mex., 40 B6 107
Glenrio, Deaf Smith, Tex., 25 B1 118
Glen Robertson, Ont., Can., 409 B10 72
Glen Rock, Bergen, N.J., 12,896 B4, D5 106
Glen Rock, York, Pa., 1,546 G8 114
Glen Rock, Princess Anne, Va. (part of Norfolk) B7 121
Glenrock, Converse, Wyo., 1,584 C7 125
Glen Rose, Somervell, Tex., 1,422 C4 118
Glen St. Mary, Baker, Fla., 329 B4 86
Glens Falls, Warren, N.Y., 18,580 B7 108
Glenshaw, Allegheny, Pa., 24,939 *B6 114
Glenside, Sask., Can., 143 F2 70
Glenside, Montgomery, Pa., 22,500 *A11 114
Glentana, Valley, Mont., 70 B10 102
Glenties, Ire., 828 C3 11
Glentworth, Sask., Can., 150 H2 70
Glen Ullin, Morton, N. Dak., 1,210 C4 110
Glenview, Cook, Ill., 18,132 E2 90
Glenview Countryside, Cook, Ill., 2,000 *E2 90
Glenvil, Clay, Nebr., 323 D7 103
Glenville, Russell, Ala., 100 C4 78
Glenville, Nevada, Ark D2 81
Glenville, Fairfield, Conn. (part of Greenwich) E2 84
Glenville, Ire., 146 E3 11
Glenville, Freeborn, Minn., 643 G5 99
Glenville, Jackson, N.C., 250 D3 109
Glenville, Gilmer, W. Va., 1,828 C4 123
Glen White, Raliegh, W. Va., 800 D3 123
Glenwillard, Allegheny, Pa., 1,100 *A5 114
Glen Wilton, Botetourt, Va., 300 D3 121
Glenwood, Grenshaw, Ala., 416 D3 78
Glenwood, Pike, Ark., 840 C2 81
Glenwood, Newf., Can., 1,130 D4 75
Glenwood, Volusia, Fla., 200 C5 86
Glenwood, Wheeler, Ga., 682 D4 87
Glenwood, Fayette, Ind., 382 E7 91
Glenwood, Mills, Iowa, 4,783 C2 92
Glenwood, Aroostook, Maine, (30▲) C4 96
Glenwood, Howard, Md., 25 B3 85
Glenwood, Cass, Mich., 115 F4 98
Glenwood, Pope, Minn., 2,631 E3 99
Glenwood, Schuyler, Mo., 242 A5 101
Glenwood, Sussex, N.J., 200 A4 106
Glenwood, Catron, N. Mex., 150 D1 107
Glenwood, Washington, Oreg., 500 A1, B3 113
Glenwood, Sevier, Utah, 277 E4 119
Glenwood, Klickitat, Wash., 300 C4 122
Glenwood, Mason, W. Va., 250 C2 123
Glenwood City, St. Croix, Wis., 835 C1 124
Glenwood Landing, Nassau, N.Y., 3,400 *F2 84
Glenwood Park, Orange, N.Y., 1,317 *D6 108
Glenwood Springs, Garfield, Colo., 3,637 B3 83
Glenwoodville, Alta., Can., 274 E4 69
Gletsch, Switz C5 19
Glezen, Pike, Ind., 180 H3 91
Glidden, Sask., Can., 145 F1 70
Glidden, Carroll, Iowa, 993 B3 92
Glidden, Ashland, Wis., 700 B3 124
Glide, Douglas, Oreg., 200 D3 113
Glien Albyn, mtn., Scot D3 13
Glifadha, Grc., 12,361 h11 23
Glin, Ire., 763 E2 11
Gliwice, Pol., 135,000 C5, g9 26
Globe, Gila, Ariz., 6,217 D5 80
Globe, Carter, Ky., 300 B6 94
Globino, Sov. Un., 10,000 G9 27
Głogów, Pol., 1,681 C3 26
Glorenza, It., 792 A3 21
Glória, Braz., 1,062 C3 57
Gloria, Plaquemines, La C7 95
Glória de Goitá, Braz., 1,959 k6 57
Glorieta, Santa Fe, N. Mex., 100 B4, F6 107
Glorieuses, is., Afr f9 49
Glossop, Eng., 17,490 A6 11
Gloster, Gwinnett, Ga., 60 A5 87
Gloster, De Soto, La., 250 B2 95
Gloster, Amite, Miss., 1,369 D2 100
Glostrup, Den., 21,845 C6 24
Gloucester, Ont., Can B9 72
Gloucester, Eng., 69,687 (*100,000) C5 12
Gloucester, Essex, Mass., 25,789 A6, C4 97
Gloucester, Gloucester, Va., 500 D6 121
Gloucester, co., N.B., Can., 66,343 B4 74
Gloucester, co., Eng., 1,000,493 C5 12
Gloucester, co., N.J., 134,840 D2 106
Gloucester, co., Va., 11,919 D6 121
Gloucester City, Camden, N.J., 15,511 D2 106
Glouster, Athens, Ohio, 2,255 C5 111
Glover, Iron. Mo., 400 D7 101
Glover, Dickey, N. Dak., 75 C7 110
Glover, Orleans, Vt., 230 (683▲) B4 120
Glover, isl., Newf., Can D3 75
Glovergap, Marion, W. Va., 125 A6 123
Gloversville, Fulton, N.Y., 21,741 B6 108
Glovertown, Newf., Can., 604 D4 75
Gloverville, Aiken, S.C., 1,551 E4 115
Glubczyce, Pol., 5,020 C4 26
Glubokoye, Sov. Un., 25,000 D6 27
Glucholazy, Pol., 7,658 C4 26
Gluck, Anderson, S.C. (part of Anderson) C2 115
Glücksburg, Ger., 5,800 D3 24
Glückstadt, Ger., 12,200 E3 24
Gluek, Chippewa, Minn., 60 E3 99
Glukhov, Sov. Un., 10,000 F9 27
Glussk, Sov. Un., 10,000 E7 27
Glyndon, Baltimore, Md., 915 B4 85
Glyndon, Clay, Minn., 489 D2 99
Glyngöre, Den., 930 B2 24
Glynn, co., Ga., 41,954 E5 87
Gmünd, see Schwäbisch Gmünd, Ger.
Gmunden, Aus., 12,518 E6 16
Gnadenhutten, Tuscarawas, Ohio, 1,257 B6 111
Gnesta, Swe., 3,200 t35 25
Gniezno, Pol., 44,000 B4 26
Gnjilane, Yugo., 12,508 D5 22
Gnoien, Ger., 5,368 E6 24
Goa, Damão, and Diu, ter., India, 684,941 E5 39, H10 33
Goalpara, India, 13,692 D13 40
Goat, isl., Am. Sam 52
Goat, mtn., Mont C3 102
Goat, riv., B.C., Can C7 68
Goat Fell, mtn., Scot E3 13
Goat River, B.C., Can C7 68
Goat Rock, dam, Ala C4 78
Goat Rock, dam, Ga D1 87
Goba, Eth D5 47
Gobabis, S.W. Afr., 1,997 B2 49
Goback, mtn., N.H A3, B1 105
Gobernador Udaondo, Arg g7 54
Gobey, Morgan, Tenn., 95 C9 117
Gobi, des., China, Mong C5 34
Gobles, Van Buren, Mich., 816 F5 98
Goch, Ger., 15,200 C3 16
Godar-i-Shah, Afg., 5,000 G10 41
Godavari, riv., India E6 39
Godchaux, Lafourche, La C6 95
Godda, India, 7,500 E11 40
Goddard, Sedgwick, Kans., 533 B5, E6 93
Goderich, Ont., Can., 6,411 D3 72
Godfrey, Ont., Can., 75 C8 72
Godfrey, Morgan, Ga., 181 C3 87
Godfrey, Madison, Ill., 1,231 E3 90
Godhavn, Grnld., 319 C20 4
Godhra, India, 52,167 F5 40
Godley, Johnson, Tex., 401 B5 118
Gods, lake, Man., Can B4 71
Gods, riv., Man., Can A5 71
Gods Lake, Man., Can., 80 B4 71
Godthaab, Grnld., 970 C20 4
Godwin, Cumberland, N.C., 149 B5 109
Godwin Austen, peak, India A6 39
Godwinsville, Dodge, Ga., 150 D3 87
Goehner, Seward, Nebr., 106 D8 103
Goes, Neth., 15,200 C3 15
Goessel, Marion, Kans., 327 D6 93
Goff, Nemaha, Kans., 259 C8 93
Goffs, San Bernardino, Calif., 25 E6 82
Goffs Falls, Hillsboro, N.H. (part of Manchester) E4 105
Goffstown, Hillsboro, N.H., 1,052 (7,230▲) D3 105
Gogebic, co., Mich., 24,370 A5 98
Gogebic, lake, Mich A5 98
Gohfeld, Ger., 14,400 A3 17
Göhren, Ger., 2,624 D7 24
Goiana, Braz., 19,026 C4, h6 57
Goiandira, Braz., 3,169 E1 57
Goianinha, Braz., 1,427 h6 57
Goiás, Braz., 7,121 B2 56
Goiás, state, Braz., 1,954,862 D1 57
Goito, It., 9,630 B3 21
Gojām, reg., Eth., 1,600,000 C4 47
Gojo, Jap., 18,800 o14 37
Göksun, Tur., 3,700 C11 31
Gokwe, S. Rh A4 49
Gol, Nor G3 25
Gola, isl., Ire B3 11
Golam, head, Ire D2 11
Golchikha, Sov. Un., 1,300 B11 28
Golconda, Pope, Ill., 864 F5 90
Golconda, Humboldt, Nev., 225 C4 104
Gold, Potter, Pa., 40 C6 114
Gold, range, B.C., Can D1 69
Gold Acres, Lander, Nev., 60 C5 104
Goldap, Pol., 632 A7 26
Goldbach, Ger., 6,600 C4 17
Gold Bar, Snohomish, Wash., 315 B4 122
Gold Beach, Curry, Oreg., 1,765 E2 113
Goldberg, Ger., 5,507 E6 24
Goldboro, N.S., Can., 142 D8 74
Gold Bridge, B.C., Can., 153 D6 68
Goldbutte, Toole, Mont B5 102
Gold Coast, see Ghana, country, Afr.
Goldcreek, Powell, Mont., 60 D4 102
Golddust, Lauderdale, Tenn., 25 B2 117
Golden, Crenshaw, Ala D3 78
Golden, B.C., Can., 1,776 D9 68
Golden, Jefferson, Colo., 7,118 B5 83
Golden, Idaho, Idaho, 50 D3 89
Golden, Adams, Ill., 491 C2 90
Golden, Ire., 153 E4 11
Golden, Tishomingo, Miss., 121 A5 100
Golden, Barry, Mo., 50 E4 101
Golden, Santa Fe, N. Mex., 30 B3, G6 107
Golden, Okanogan, Wash A6 122
Golden Acres, Harris, Tex., 2,500 *F5 118
Golden Beach, Dade, Fla., 413 E3 86
Golden City, Barton, Mo., 714 D3 101
Goldendale, Klickitat, Wash., 2,536 D5 122
Goldengate, Wayne, Ill., 156 E5 90
Golden Gate, entrance, Calif B5 82
Golden Glades, Dade, Fla., 3,000 *F3 86
Golden Hill, Dorchester, Md., 300 D5 85
Golden Hill, Olmstead, Minn., 2,190 *F6 99
Golden Hinde, mtn., B.C., Can E5 68
Golden Lake, Ont., Can., 254 B7 72
Golden Meadow, Lafourche, La., 3,097 E5 95
Golden Prairie, Sask., Can., 206 G1 70
Golden Spike, nat. historical site, Utah B3 119
Golden Valley, Hennepin, Minn., 14,559 E3 99
Goldenvalley, Mercer, N. Dak., 286 B3 110
Golden Valley, co., Mont., 1,203 D7 102
Golden Valley, co., N. Dak., 3,100 B2 110
Goldenville, N.S., Can., 68 D7 74
Goldfield, Teller, Colo., 75 C5 83
Goldfield, Wright, Iowa, 682 B4 92
Goldfield, Esmeralda, Nev., 300 F4 104
Goldfield, mts., Ariz G3 80
Gold Hill, Lee, Ala., 195 C4 78
Gold Hill, Rowan, N.C., 249 B3 109
Gold Hill, Jackson, Oreg., 608 E3 113
Gold Hill, Tooele, Utah C2 119
Goldonna, Natchitoches, La., 292 B3 95
Gold Point, Esmeralda, Nev., 40 F4 104
Goldpoint, Hamilton, Tenn E10 117
Goldsand, lake, Man., Can A1 71
Goldsboro, Caroline, Md., 204 B6 85
Goldsboro, Wayne, N.C., 28,873 B6 109
Goldsmith, Tipton, Ind., 200 D5 91
Goldsmith, Ector, Tex., 670 D1 118
Goldston, Chatham, N.C., 374 B4 109
Goldstone, mtn., Idaho D5 89
Goldstream, riv., B.C., Can D1 69
Goldthwaite, Mills, Tex., 1,383 D3 118
Goleen, Ire., 68 F2 11
Golela, S. Afr C5 49
Goleniow, Pol., 1,713 B3 26
Goleta, Santa Barbara, Calif., 4,000 E4 82
Golf, Cook, Ill., 409 *E3 90
Golf Manor, Hamilton, Ohio, 4,648 *D2 111
Golfview, Palm Beach, Fla., 131 *F6 86
Golfview Heights, Palm Beach, Fla., 1,500 *F6 86
Goliad, Goliad, Tex., 1,782 E4 118
Goliad, co., Tex., 5,429 E4 118
Golling, Aus., 2,845 B9 18
Golovin, Alsk., 22 C7 79
Golovnino, Kur. Is., Sov. Un E12 37
Golpāyegān, Iran, 20,844 E5 41
Golspie, Scot., 1,323 C5 13
Goltry, Alfalfa, Okla., 313 A3 112
Golts, Kent, Md., 100 B6 85
Golva, Golden Valley, N. Dak., 162 C2 110
Golyamo Konare, Bul., 7,153 D7 22
Goma, Con. L B4 48
Gombari, Con. L A4 48
Gombe, Nig., 18,483 D7 45
Gomel, Sov. Un., 184,000 E8 27
Gomera, isl., Can. Is D1 44
Gomez, Martin, Fla., 250 E6 86
Gómez Palacio, Mex., 60,765 B4 63
Gommern, Ger., 6,227 A6 17
Gonaïves, Hai., 13,534 E7 64
Gonbad-e Kāvūs, Iran, 9,637 C7 41
Gonda, India, 43,496 C7 39
Gondar, Eth., 25,000 C4 47
Gondia, India, 56,320 G8 40
Gondrecourt-le-Château, Fr., 1,088 F5 15
Gönen, Tur., 10,800 B6 23
Gonesse, Fr., 8,517 g10 14
Gongola, riv., Nig D7 45
Gonic, Strafford, N.H. (part of Rochester) D5 105
Gonvick, Clearwater, Minn., 363 C3 99
Gonzales, Monterey, Calif., 2,138 D3 82
Gonzales, Ascension, La., 3,252 B5, D5 95
Gonzales, Gonzales, Tex., 5,829 E4 118
Gonzales, co., Tex., 17,845 E4 118
Gonzalez, Escambia, Fla., 300 G2 86
González, Mex., 1,913 C5 63
González Chaves, Arg., 4,718 B4 54
Goochland, Goochland, Va., 200 D5 121
Goochland, co., Va., 9,206 D5 121
Goodbee, St. Tammany, La B7 95
Goode, Bedford, Va., 250 D3 121
Goodell, Hancock, Iowa, 231 B4 92
Goodenough, cape, Ant C25 5
Gooderham, Ont., Can., 275 C6 72
Goodeve, Sask., Can., 212 F4 70
Goodfield, Woodford, Ill., 286 C4 90

Good Hart, Emmet, Mich., 50 ....C5 98
Good Hope, Walton, Ga., 165 ....C3 87
Good Hope, McDonough, Ill., 394 ....C3 90
Good Hope, St. Charles, La. (part of Norco) C7 95
Good Hope, Leake, Miss., 125 ....C4 100
Good Hope, Fayette, Ohio, 300 ....C4 111
Good Hope, cape, S. Afr D2 49
Good Hope, mtn., B.C., Can D5 68
Goodhue, Goodhue, Minn., 566 ....F6 99
Goodhue, co., Minn., 33,035 ..F6 99
Gooding, Gooding, Idaho, 2,750 ....G4 89
Gooding, co., Idaho, 9,544 ...F4 89
Goodland, Collier, Fla., 100 ..G5 86
Goodland, Newton, Ind., 1,202 ....C3 91
Goodland, Sherman, Kans., 4,459 ....C2 93
Goodland, Choctaw, Okla., 50 ....D6 112
Goodlands, Man., Can., 138 ..E1 71
Goodlettsville, Davidson, Tenn., 3,163 ....A5, E9 117
Goodman, Holmes, Miss., 932 ....C4 100
Goodman, McDonald, Mo., 540 ....E3 101
Goodman, Marinette, Wis., 550 ....C5 124
Goodnews Bay, Alsk., 154 ..D7 79
Goodnight, Armstrong, Tex., 125 ....B2 118
Good Pasture, Pueblo, Colo ..C6 83
Good Pine, La Salle, La., 600 ....C3 95
Goodrich, Morgan, Colo., 10 .A6 83
Goodrich, Linn, Kans., 35 ..D9 93
Goodrich, Sheridan, N. Dak., 392 ....B5 110
Goodrich, Polk, Tex., 800 ..D5 118
Goodridge, Pennington, Minn., 134 ....B3 99
Goodspeeds, Middlesex, Conn D7 84
Good Spirit Lake, park, Sask., Can F4 70
Goodspring, Giles, Tenn., 25 ....B4 117
Goodsprings, Walker, Ala., 900 ....B2 78
Goodsprings, Clark, Nev H6 104
Good Thunder, Blue Earth, Minn., 468 ....F4 99
Goodview, Winona, Minn., 1,348 ....F7 99
Good Water, Coosa, Ala., 2,023 ....B3 78
Goodwater, Sask., Can., 87 .H4 70
Goodwater, McCurtain, Okla., 100 ....D7 112
Goodwell, Texas, Okla., 771 .D7 112
Good Will Farm, Somerset, Maine, 100 ....D3 96
Goodwin, Yavapai, Ariz., 20 .C3 80
Goodwin, Deuel, S. Dak., 113 ....C3 116
Goodyear, Maricopa, Ariz., 1,654 ....G1 80
Goodyear, Windham, Conn ..B9 84
Goole, Eng., 18,875 ....A7 12
Goondiwindi, Austl., 3,274 ..D8 51
Goor, Neth., 7,600 ....B6 15
Goose, bay, Newf., Can B1 75
Goose, creek, Idaho G5 89
Goose, creek, Va D3 121
Goose, creek, Wyo A6 125
Goose, isl., B.C., Can D3 68
Goose, lake, Calif B3 82
Goose, lake, Man., Can B1 71
Goose, lake, Sask., Can F2 70
Goose, lake, Oreg E6 113
Goose, pond, N.H C2 105
Goose, river, Newf., Can h9 75
Goose Bay, Newf., Can., 3,040 ....B1, h9 75
Gooseberry, creek, Wyo A4 125
Goose Creek, Berkeley, S.C., 830 ....E2, F7 115
Goose Creek, res., S.C E2 115
Goose Egg, Natrona, Wyo ..C6 125
Gooselake, Clinton, Iowa, 191 ....C7 92
Goosport, Calcasieu, La., 16,778 ....*D2 95
Gopalganj, India, 14,090 ..D10 40
Goplo, lake, Pol B5 26
Göppingen, Ger., 48,900 ....E4 17
Gora, Pol., 3,526 ....C4 26
Gora Kalwaria, Pol., 3,687 ..C6 26
Gorakhpur, India, 180,255 ....C7 39, D9 40
Gorday, Worth, Ga E3 87
Gordes, Tur., 5,300 ....C7 23
Gordo, Pickens, Ala., 1,714 ..B2 78
Gordon, Houston, Ala., 222 ..D4 78
Gordon, Wilkinson, Ga., 1,793 ....D3 87
Gordon, Butler, Kans., 100 ..B6 93
Gordon, Claiborne, La B3 95
Gordon, Sheridan, Nebr., 2,223 ....B3 103
Gordon, Schuylkill, Pa., 888 ....*E9 114
Gordon, Palo Pinto, Tex., 349 ....C3 118
Gordon, Douglas, Wis., 350 ..B2 124
Gordon, co., Ga., 19,228 ....B2 87
Gordon, lake, Alta., Can A5 69
Gordonhorne, peak, B.C., Can ..D8 68
Gordonsburg, Lewis, Tenn., 315 ....B4 117
Gordonsville, Freeborn, Minn., 100 ....G5 99
Gordonsville, Smith, Tenn., 249 ....C8 117
Gordonsville, Orange, Va., 1,109 ....C4 121
Gordonville, Cape Girardeau, Mo., 92 ....D8 101
Gore, N.S., Can., 144 ....D6 74
Gore, Eth., 10,000 ....D4 47
Gore, N.Z., 7,270 ....Q12 51
Gore, Hocking, Ohio, 200 ..C5 111
Gore, Sequoyah, Okla., 334 ..B6 112
Gore, Frederick, Va., 200 ....B4 121
Gore, canyon, Colo B4 83
Gore, mtn., Vt B5 120
Gore, pt., Alsk h16 79
Gore, range, Colo B4 83
Gore Bay, Ont., Can., 716 ..B2 72
Goree, Knox, Tex., 543 ....C3 118
Gore Springs, Grenada, Miss., 100 ....B4 100
Goreville, Johnson, Ill., 625 ..F5 90
Gorey, Ire., 2,671 ....E5 11
Gorgān (Asterābād), Iran, 28,380 ....C7 41
Gorgas, Walker, Ala., 950 ..B2 78
Gorge High, dam, Wash A4 122
Gorgona, isl., Col C2 60
Gorham, Jackson, Ill., 378 ..F4 90
Gorham, Russell, Kans., 429 .D4 93
Gorham, Cumberland, Maine, 2,322 (5,767▲) E2, E4 96
Gorham, Coos, N.H., 1,945 (3,039▲) B4 105
Gorham, Ontario, N.Y., 500 .C3 108
Gori, Sov. Un., 33,000 ....E2 29
Gorin, Scotland, Mo., 410 ..A5 101
Gorinchem, Neth., 21,600 ...C5 15
Goris, Sov. Un., 4,660 ....C16 31
Gorizia, It., 42,187 ....B4 21
Gorki, Sov. Un., 10,000 ....D8 27
Gorkiy (Gorky), Sov. Un., 1,003,000 (*1,340,000) B2 29
Gorkiy, res., Sov. Un B2 29
Görlev, Den., 1,379 ....C5 24
Gorlice, Pol., 6,100 ....D6 26
Görlitz, Ger., 89,000 ....B9 17
Gorlovka, Sov. Un., 307,000 ....G12, q21 27
Gorm, lake, Scot E2 13
Gorman, Garrett, Md., 83 ..D1 85
Gorman, Humphreys, Tenn., 50 ....A4 117
Gorman, Eastland, Tex., 1,142 ....C3 118
Gormania, Grant, W. Va., 150 ....B5 123
Gormley, Ont., Can., 108 ....E6 74
Gorna Dzhumaya, see Blagoevgrad, Bul.
Gorna-Oryakhovitsa, Bul., 10,303 ....D7 22
Gornji Milanovac, Yugo., 4,493 ....C5 22
Gorno-Altaysk, Sov. Un., 29,000 ....E26 9
Gornozavodsk, Sov. Un., 20,000 ....C10 37
Gorodenka, Sov. Un., 10,000 .G5 27
Gorodets, Sov. Un., 30,000 ..D17 9
Gorodishche, Sov. Un., 10,000 ....q22 27
Gorodnya, Sov. Un., 10,000 ..F8 27
Gorodok, Sov. Un A5 34
Gorodok, Sov. Un., 10,000 ..G4 27
Gorodok, Sov. Un., 10,000 ..D7 27
Goroke, Austl., 379 ....H3 51
Gorong, is., Indon F8 35
Gorontalo, Indon., 71,232 ....E6 35
Gorrie, Ont., Can., 319 ....D3 72
Gorskoye, Sov. Un., 10,000 ..q21 27
Gort, Ire., 1,044 ....D3 11
Gortahork, Ire B3 11
Gorumna, isl., Ire D2 11
Goryn, riv., Sov. Un F6 27
Gorzow [Wielkopolski], Pol., 56,000 ....B3 26
Gosainthan, peak, Nep C10 40
Gosford, Austl., 7,317 ....F8 51
Goshen, Pike, Ala., 260 ....D3 78
Goshen, Tulare, Calif., 1,061 ....*D4 82
Goshen, N.S., Can., 199 ....D8 74
Goshen, Litchfield, Conn., 400 (1,288▲) B4 84
Goshen, Elkhart, Ind., 13,718 ....A6 91
Goshen, Oldham, Ky., 50 ..A4 94
Goshen, Hampshire, Mass., 250 (385▲) B2 97
Goshen, Sullivan, N.H., 125 (351▲) D2 105
Goshen, Cape May, N.J., 500 .E3 106
Goshen, Orange, N.Y., 3,906 ....D2, D6 108
Goshen, Lane, Oreg., 300 ....D4 113
Goshen, Utah, Utah, 426 ....D4 119
Goshen, Rockbridge, Va., 99 ....D3 121
Goshen, co., Wyo., 11,941 ..C8 125
Goshen Hole, sink, Wyo D8 125
Goshen Springs, Rankin, Miss., 50 ....C4 100
Goshute, Indian res., Utah D2 119
Goshute, mts., Nev C7 104
Goslar, Ger., 41,400 ....B5 17
Gosper, co., Nebr., 2,489 ....D6 103
Gospić, Yugo., 6,857 ....C2 22
Gosport, Clarke, Ala., 25 ....D2 78
Gosport, Eng., 62,436 ....E6 12
Gosport, Owen, Ind., 646 ....F4 91
Goss, De Soto, La B2 95
Goss, Marion, Miss., 30 ....D4 100
Gossau, Switz., 9,731 ....B7 19
Gossburg, Coffee, Tenn., 100 .B5 117
Gossville, Merrimack, N.H., 155 ....D4 105
Gostivar, Yugo., 12,776 ....E5 22
Gostyn, Pol., 8,021 ....C4 26
Gostynia, riv., Pol g9 26
Gostynin, Pol., 7,357 ....B5 26
Gotebo, Kiowa, Okla., 538 ..B3 112
Göteborg, Swe., 408,300 (*505,000) A5 24, I4 25
Göteborg och Bohus, co., Swe., 633,000 ....A5 24
Gotel, mts., Cam E7 45
Gotemba, Jap., 28,700 ....n17 37
Gotha, Ger., 56,200 ....C5 17
Gotham, Richland, Wis., 250 ....E3 124
Gothenburg, Dawson, Nebr., 3,050 ....D5 103
Gothic, mesas, Ariz., N. Mex A6 80, A1 107
Gotland, co., Swe., 53,700 ..*I8 25
Gotland, isl., Swe I8 25
Goto, is., Jap J4 37
Gotse Delchev, Bul., 11,061 ..E6 22
Gott, peak, B.C., Can D6 68
Göttingen, Ger., 80,900 (*110,000) B4 17
Gottwaldov (Zlin), Czech., 54,200 ....D4 26
Gouda, Neth., 43,400 ....A6 15
Goudeau, Avoyelles, La., 130 .D3 95
Goudiri, Sen D2 45
Goudswaard, Neth., 730 ....C4 15
Gough, Burke, Ga., 300 ....C4 87
Gough, isl., Atl., O I9 6
Gough, lake, Alta., Can C4 69
Gouin, res., Que., Can G17 67
Goulais, riv., Ont., Can p18 72
Goulburn, Austl., 20,544 ....G7 51
Gould, Lincoln, Ark., 1,210 ..D4 81
Gould, Que., Can., 100 ....D6 73
Gould, Jackson, Colo., 60 ....A4 83
Gould, Harmon, Okla., 241 ..C2 112
Gouldbusk, Coleman, Tex., 75 ....D3 118
Gould City, Mackinac, Mich., 100 ....B5 98
Goulding, Escambia, Fla., 900 ....*G1 86
Goulds, Dade, Fla., 5,121 ....F3 86
Gouldsboro, Hancock, Maine, 280 (1,100▲) D4 96
Gouldsboro, Wayne, Pa., 450 ....D11 114
Goulimine, Mor., 8,015 ....D2 44
Goundam, Mali, 6,600 ....C4 45
Gouré, Niger, 1,000 ....D7 45
Gourma-Rarous, Mali C4 45
Gournay-en-Bray, Fr., 3,195 .C4 14
Gouro, Chad B3 46
Gouverneur, Sask., Can., 70 .H2 70
Gouverneur, St. Lawrence, N.Y., 4,946 ....B1 108
Govan, Sask., Can., 380 ....F3 70
Govan, Bamberg, S.C., 138 ..E5 115
Gove, Gove, Kans., 228 ....D3 93
Gove, co., Kans., 4,107 ....D3 93
Govenlock, Sask., Can., 176 .H1 70
Governador, Rio Arriba, N. Mex., 10 ....A2 107
Governador Valadares, Braz., 70,494 ....E2 57
Government Camp, Clackamas, Oreg., 40 ....B5 113
Governor Generoso, Phil., 1,659 ....*D7 35
Governor's Harbour, Ba. Is., 516 ....C5 64
Gowan, St. Louis, Minn., 15 ....D6 99
Gowan, riv., Man., Can B4 71
Gowanda, Cattaraugus and Erie, N.Y., 3,352 ....C2 108
Gowen, Montcalm, Mich., 200 ....E5 98
Gowen, Latimer, Okla., 350 .C6 112
Gower, Clinton, Mo., 406 ....B3 101
Gowk, Iran, 6,285 ....G8 41
Gowna, lake, Ire D4 11
Gowrie, Webster, Iowa, 1,127 ....B3 92
Goya, Arg., 20,804 ....E4 55
Goz Béïda, Chad, 1,250 ....C4 46
Göz Regeb, Sud B4 47
Graaff-Reinet, S. Afr., 16,703 ....D3 49
Graal-Müritz, Ger., 3,875 ..D6 24
Graasten, Den., 2,308 ....D3 24
Grabill, Allen, Ind., 495 ....B8 91
Grabow, Ger., 8,708 ....E5 24
Grabow, bay, Ger D6 24
Gračac, Yugo., 2,831 ....C2 22
Gračanica, Yugo., 7,668 ....C4 22
Grace, Caribou, Idaho, 725 ..G7 89
Grace, Issaquena, Miss., 300 ....C3 100
Grace City, Foster, N. Dak., 100 ....B7 110
Graceham, Frederick, Md., 165 ....A3 85
Gracemont, Caddo, Okla., 306 ....B3 112
Grace Park, Delaware, Pa., 2,000 ....*B10 114
Graceton, Lake of the Woods, Minn., 45 ....B4 99
Graceville, Jackson, Fla., 2,307 ....G3 86
Graceville, Big Stone, Minn., 823 ....E2 99
Gracewood, Richmond, Ga., 500 ....C4 87
Gracey, Christian, Ky., 234 .D2 94
Gracias, Hond., 1,521 ....C3 62
Gracias a Dios, prov., Hond .C5 62
Graciosa, isl., Az. Is g8 44
Gradačac, Yugo., 5,897 ....C4 22
Gradaús, mts., Braz D4 59
Gradefes, Sp., 4,597 ....A3 20
Gradizhsk, Sov. Un., 10,000 .G9 27
Grado, It., 9,666 ....D9 18
Grado, Sp., 3,671 ....A2 20
Grady, Lincoln, Ark., 622 ..C4 81
Grady, Curry, N. Mex., 100 .C6 107
Grady, Jefferson, Okla., 25 ..C4 112
Grady, co., Ga., 18,015 ....F2 87
Grady, co., Okla., 29,590 ....C4 112
Gradyville, Adair, Ky., 150 ..C4 94
Gradyville, Delaware, Pa., 80 ....B10 114
Graehl, Alsk., 1,200 ....*C10 79
Graested, Den., 1,078 ....B6 24
Graettinger, Palo Alto, Iowa, 879 ....A3 92
Grafenau, Ger., 2,460 ....E8 17
Graford, Palo Pinto, Tex., 448 ....C3 118
Grafton, Austl., 15,526 ....D9 51
Grafton, Ont., Can., 348 ....D6 72
Grafton, Jersey, Ill., 1,084 ..E3 90
Grafton, Worth, Iowa, 273 ..A4 92
Grafton, Worcester, Mass., 2,200 (10,627▲) B4 97
Grafton, Fillmore, Nebr., 171 ....D8 103
Grafton, Grafton, N.H., 75 (348▲) C3 105
Grafton, Rensselaer, N.Y., 300 ....C7 108
Grafton, Walsh, N. Dak., 5,885 ....A8 110
Grafton, Lorain, Ohio, 1,683 .A5 111
Grafton, Windham, Vt., 150 (426▲) E3 120
Grafton, York, Va., 200 ....B6 121
Grafton, Taylor, W. Va., 5,791 ....B4, B7 123
Grafton, Ozaukee, Wis., 3,748 ....E2, E6 124
Grafton, co., N.H., 48,857 ..C3 105
Grafton Center, Grafton, N.H., 40 ....C3 105
Graham, Randolph, Ala., 50 ....B4 78
Graham, Bradford, Fla., 120 .C4 86
Graham, Appling, Ga., 130 ..E4 87
Graham, Muhlenberg, Ky., 550 ....C2 94
Graham, Nodaway, Mo., 215 ....A2 101
Graham, Alamance, N.C., 7,723 ....A4 109
Graham, Carter, Okla., 350 ..C4 112
Graham, Hickman, Tenn., 35 ....B4 117
Graham, Young, Tex., 8,505 ....C3 118
Graham, Pierce, Wash., 75 ..B3 122
Graham, co., Ariz., 14,045 ..E5 80
Graham, co., Kans., 5,586 ..C4 93
Graham, co., N.C., 6,432 ....D2 109
Graham, creek, Ind G6 91
Graham, isl., B.C., Can C1 68
Graham, lake, Maine D4 96
Graham, mtn., Ariz E6 80
Graham, reach, B.C., Can C3 68
Graham, riv., B.C., Can A6 68
Graham Bell, isl., Sov. Un A8 4
Grahamstown, S. Afr., 32,611 ....D4 49
Grahamsville, Sullivan, N.Y., 300 ....D6 108
Grahamville, Jasper, S.C., 200 ....G6 115
Grahn, Carter, Ky., 400 ....B6 94
Graian Alps, mts., Fr., It D3 18
Graiba, Tun G12 26
Grainfield, Gove, Kans., 389 .C3 93
Grainger, Alta., Can., 25 ....D4 69
Grainger, co., Tenn., 12,506 ....C10 117
Grainland, Sask., Can F2 70
Grainola, Osage, Okla., 67 ..A5 112
Grainton, Perkins, Nebr., 35 ....D4 103
Grain Valley, Jackson, Mo., 552 ....B3 101
Grajaú, Braz., 2,539 ....C1 57
Grajaú, riv., Braz B1 57
Grajewo, Pol., 6,171 ....B7 26
Gram, Den., 1,801 ....C3 24
Gram, isl., Thai F4 38
Grama, Dona Ana, N. Mex ..E2 107
Gramada, Bul., 4,662 ....D6 22
Gramalote, Col., 2,776 ....B3 60
Gramastetten, Aus., 2,443 ....E9 17
Grambling, Lincoln, La., 3,144 ....B3 95
Gramercy, St. James, La., 2,094 ....B6 95
Gramling, Spartanburg, S.C., 200 ....A3 115
Grammer, Bartholomew, Ind., 200 ....F6 91
Grammichele, It., 14,486 ....F5 21
Grampian, Clearfield, Pa., 529 ....E4 114
Grampian, mts., Scot D4 13
Gramsdale, Scot C1 13
Gramsh, Alb., 650 ....B3 23
Gramzow, Ger., 2,152 ....E8 24
Granada, Prowers, Colo., 593 ....C8 83
Granada, Martin, Minn., 418 ....G4 99
Granada, Nic., 34,400 ....E5 62
Granada, Sp., 157,178 ....D4 20
Granada, prov., Sp., 769,408 ....*D4 20
Granard, Ire., 1,044 ....D4 11
Granbury, Hood, Tex., 2,227 .C4 118
Granby, Que., Can., 31,463 .D5 73
Granby, Grand, Colo., 503 ..A5 83
Granby, Hartford, Conn., 700 (4,968▲) B5 84
Granby, Hampshire, Mass., 1,700 (4,221▲) B2 97
Granby, Newton, Mo., 1,808 ....E3 101
Granby, Essex, Vt., 15 (56▲) .B5 120
Gran Canaria, isl., Can. Is D1 44
Gran Chaco, plain, Arg., Par ..F4 55
Grand, co., Colo., 3,537 ....A4 83
Grand, co., Utah, 6,345 ....E6 119
Grand, bayou, La B5 95
Grand, bayou, La C6 95
Grand, canal, China G7 36
Grand, canal, Ire D4 11
Grand, canyon, Ariz A3 80
Grand, caverns, Tenn D9 117
Grand, falls, Newf., Can h8 75
Grand, falls, Maine C5 96
Grand, isl., La D6 95
Grand, isl., La E6 95
Grand, isl., Mich B4 98
Grand, isl., Vt B2 120
Grand, lake, N.B., Can D2 74
Grand, lake, N.B., Can D3 74
Grand, lake, Newf., Can D3 75
Grand, lake, La E3 95
Grand, lake, La E4 95
Grand, lake, La E6 95
Grand, lake, Maine C5 96
Grand, lake, Mich C7 98
Grand, lake, Ohio B3 111
Grand, mesa, Colo C2 83
Grand, riv., Newf., Can E2 75
Grand, riv., Ont., Can D4 72
Grand, riv., Mich E4 98
Grand, riv., Mo B4 101
Grand, riv., Ohio A7 111
Grand, riv., S. Dak B4 116
Grand Atlas, mts., Mor C3 44
Grand Bahama, isl., Ba. Is B4 64
Grand Bank, Newf., Can., 2,430 ....E4 75
Grand Banks, shoals, Newf., Can C17 3
Grand-Bassam, I.C., 13,000 ..E4 45
Grand Bay, Mobile, Ala., 600 ....E1 78
Grand Bay, N.B., Can., 725 .D3 74
Grand Bayou, Red River, La B2 95
Grand Beach, Man., Can., 74 ....D3 71
Grand Blanc, Genesee, Mich., 1,565 ....F7 98
Grand Bostonnais, lake, Que., Can B5 73
Grand-Bourg, Guad., 13,833 .o16 64
Grand Bruit, Newf., Can., 132 ....E2 75
Grand Cane, De Soto, La., 322 ....B2 95
Grand Canyon, Coconino, Ariz., 900 ....A3 80
Grand Canyon, nat. mon. and park, Ariz A3 80
Grand Cayman, isl., Cayman Is ..F3 64
Grand Centre, Alta., Can., 1,493 ....B5 69
Grand Cess, Lib F3 45
Grand Chenier, Cameron, La., 200 ....E3 95
Grand Combin, mtn., Switz E3 19
Grand Coteau, St. Landry, La., 1,165 ....D3 95
Grand Coulee, Grant, Wash., 1,058 ....B6 122
Grand Coulee, canyon, Wash ..B6 122
Grand Coulee, dam, Wash B6 122
Grande, Union, N. Mex A6 107
Grande, bay, Arg H3 54
Grande, hills, Ur E1 54
Grande, isl., Braz h5 56
Grande, mts., Mex G2 118
Grande, riv., Arg B3 54
Grande, riv., Bol C3 55
Grande, riv., Braz B2 56
Grande, riv., Braz D2 57
Grande, riv., Chile D2 55
Grande, riv., Nic D5 62
Grande, riv., Pan k11 62
Grande, riv., P.R B7 65
Grande, riv. mouth, Ven B5 60
Grande, Anse, N.B., Can., 522 ....B4 74
Grande Baie, Que., Can. (part of Port Alfred) A7 73
Grande Catwick, is., Viet G8 38
Grande de Añasco, riv., P.R ..B3 65
Grande de Arecibo, riv., P.R ..B4 65
Grande de Loíza, riv., P.R B6 65
Grande de Manatí, riv., P.R ..B4 65
Grande de Santiago, riv., Mex .m11 63
Grande Digue, N.B., Can., 183 ....C5 74
Grande Greve, Que., Can., 155 ....A4 73
Grande Ligne, Que., Can., 76 ....D4 73
Grande Miquelon, isl., Miquelon Isl E3 75
Grande Prairie, Alta., Can., 6,302 ....B1 69
Grand Erg Occidental, sand dunes, Alg C5 44
Grand Erg Oriental, sand dunes, Alg D6 44
Grande Riviere, Que., Can., 1,176 ....*B3 73
Grande Ronde, riv., Oreg., Wash B9 113
Grand-Etang, N.S., Can., 335 ....C8 74
Grandes-Bergeronnes, Que., Can., 779 ....A8 73
Grandes Piles, Que., Can., 670 ....C5 73
Grand Falls, N.B., Can., 3,983 ....B2 74
Grand Falls, Newf., Can., 6,064 ....D4 75
Grand Falls, Koochiching, Minn., 50 ....B5 99
Grandfalls, Ward, Tex., 1,012 ....D1 118
Grand Falls, dist., Newf., Can., 6,606 ....D3 75
Grand Falls, lake, Maine C5 96
Grandfather, mtn., N.C A2 109
Grandfield, Tillman, Okla., 1,606 ....C3 112
Grand Forks, B.C., Can., 2,347 ....E8 68
Grand Forks, Grand Forks, N. Dak., 34,451 ....B8 110
Grand Forks, co., N. Dak., 48,677 ....B8 110
Grand-Fougeray, Fr., 885 ..D3 14
Grand Gorge, Delaware, N.Y., 600 ....C6 108
Grand Gulf, Claiborne, Miss., 50 ....C2 100
Grand Harbour, N.B., Can., 439 ....E3 74
Grand Haven, Ottawa, Mich., 11,066 ....E4 98
Grand Hogback, mtn., Colo B3 83
Grandin, Putnam, Fla., 150 ..C5 86
Grandin, Carter, Mo., 259 ..E7 101
Grandin, Cass, N. Dak., 147 .B9 110
Grand Island, Hall, Nebr., 25,742 ....D7 103
Grand Island, Erie, N.Y., 1,700 ....*B2 108
Grand Isle, Jefferson, La., 2,074 ....E6 95
Grand Isle, Aroostook, Maine, 500 (978▲) A4 96
Grand Isle, Grand Isle, Vt., 100 (624▲) B2 120
Grand Isle, co., Vt., 2,927 ..B2 120
Grand Junction, Mesa, Colo., 18,694 ....B2 83
Grand Junction, Kootenai, Idaho D8 122
Grand Junction, Greene, Iowa, 949 ....B3 92
Grand Junction, Van Buren, Mich., 300 ....F4 98
Grand Junction, Hardeman, Tenn., 446 ....B2 117
Grand-Lahou, I.C., 2,700 ....E3 45
Grand Lake, Grand, Colo., 170 ....A5 83
Grand Lake, Cameron, La., 150 ....D2 95
Grand Lake Stream, Washington, Maine, 270 (219▲) .C5 96
Grand Ledge, Eaton, Mich., 5,165 ....F6 98
Grand-Lieu, lake, Fr D3 14
Grand Manan, chan., Can., U.S E3 74
Grand Manan, isl., N.B., Can ..E3 74
Grand Marais, Alger, Mich., 600 ....B5 98
Grand Marais, Cook, Minn., 1,301 ....B7 99
Grand Marsh, Adams, Wis., 130 ....E4 124
Grand Meadow, Mower, Minn., 837 ....G6 99
Grand'Mere, Que., Can., 15,806 ....C5 73
Grand Mound, Clinton, Iowa, 565 ....C7 92
Grândola, Port., 3,936 ....C1 20
Grand Pass, Saline, Mo., 120 ....B4 101
Grand Portage, Cook, Minn., 100 ....A8 99
Grand Portage, Indian res., Minn B8 99
Grand Prairie, Dallas, Tex., 30,386 ....B5 118
Grand Pre, N.S., Can., 270 ..D5 74
Grand Rapids, Man., Can., 986 ....C2 71
Grand Rapids, Kent, Mich., 177,313 (*367,300) F5 98
Grand Rapids, Itasca, Minn., 7,265 ....C5 99
Grand Rapids, La Moure, N. Dak., 150 ....C7 110
Grand Rapids, Wood, Ohio, 670 ....A1, A4 111
Grand Ridge, Jackson, Fla., 415 ....B1 86
Grand Ridge, LaSalle, Ill., 659 ....B5 90
Grand River, N.S., Can., 121 ....D9 74
Grand River, Decatur, Iowa, 284 ....D4 92
Grand River, valley, Colo., Utah D6 119
Grand Rivers, Livingston, Ky., 378 ....A3 94
Grand St. Bernard, pass, Switz., It D3 18
Grand Saline, Van Zandt, Tex., 2,006 ....C5 118
Grand Seboeis, lake, Maine B4 96
Grandson, Switz., 2,091 ....C2 19
Grand Terrace, San Bernardino, Calif., 1,450 ....*F5 82
Grand Terre, isl., La E6 95
Grand Teton, mtn., Wyo B2 125
Grand Teton, nat. park, Wyo ..B2 125
Grand Tower, Jackson, Ill., 847 ....F4 90
Grand Traverse, co., Mich., 33,490 ....D5 98
Grand Traverse, bay, Mich C5 98
Grand Turk, isl., W.I. Fed f14 65
Grand Valley, Ont., Can., 634 ....D4 72
Grand Valley, Garfield, Colo., 245 ....B2 83
Grand Valley, Warren, Pa., 175 ....C2 114
Grand View, Man., Can., 1,057 ....D1 71
Grand View, Owyhee, Idaho, 200 ....G2 89
Grandview, Edgar, Ill., 120 ..D6 90
Grandview, Sangamon, Ill., 2,214 ....*D4 90
Grand View, Spencer, Ind., 599 ....I4 91
Grandview, Louisa, Iowa, 300 ....C6 92
Grandview, Jackson, Mo., 6,027 ....C3, E2 101
Grandview, Cherokee, N.C., 30 ....D2 109
Grandview, Washington, Ohio, 100 ....C6 111
Grandview, Jefferson, Oreg ..C5 113
Grandview, Rhea, Tenn., 300 ....D9 117
Grandview, Johnson, Tex., 961 ....*C4 118
Grandview, Yakima, Wash., 3,366 ....C6 122
Grandview, Bayfield, Wis., 150 ....B2 124
Grandview Heights, Franklin, Ohio, 8,270 ....C2 111
Grandview Heights, Lancaster, Pa., 3,800 ....*F9 114
Grandville, Kent, Mich., 7,975 ....F5 98
Grandvilliers, Fr., 2,118 ....E1 15
Grand Wash, cliffs, Ariz A2 80
Grandy, Isanti, Minn., 175 ..E5 99
Grandy, Currituck, N.C., 400 ....A8 109
Grandy, isl., Newf., Can B4 75
Grangärde, Swe., 340 ....G6 25
Grange, Sharp, Ark., 35 ....B4 81
Grange, Eng., 3,117 ....F6 13
Grangeburg, Houston, Ala., 100 ....D4 78
Grangemont, Clearwater, Idaho C2 89
Grangemouth, Scot., 18,860 ....D5 13
Granger, St. Joseph, Ind., 125 ....A5 91
Granger, Dallas, Iowa, 468 ....A7, C4 92
Granger, Scotland, Mo., 146 ....A6 101
Granger, Williamson, Tex., 1,339 ....D4 118
Granger, Salt Lake, Utah, 1,300 ....*C3 119
Granger, Yakima, Wash., 1,424 ....C5 122
Granger, Sweetwater, Wyo., 159 ....D3 125
Granges, Fr., 1,985 ....A2 18
Grangeville, Idaho, Idaho, 3,642 ....D2 89
Grangeville, St. Helena, La., 40 ....D5 95
Grangeville, York, Pa., 1,100 ....*G8 114
Granite, Chaffee, Colo., 40 ..B4 83
Granite, Baltimore, Md., 600 ....B4 85
Granite, Greer, Okla., 952 ..C2 112
Granite, co., Mont., 3,014 ....D3 102
Granite, mtn., Ark D6 81
Granite, mts., Ariz E2 80
Granite, pass, Wyo A5 125
Granite, peak, Mont D4 102
Granite, peak, Mont E7 102
Granite, peak, Nev C2 104
Granite, peak, Wyo C4 125
Granite, pt., Newf., Can C3 75
Granite, range, Nev C2 104
Granite Canon, Laramie, Wyo., 30 ....D7 125
Granite City, Madison, Ill., 40,073 ....E3 90
Granite Falls, Yellow Medicine and Chippewa, Minn., 2,728 ....F3 99
Granite Falls, Caldwell, N.C., 2,644 ....B2 109
Granite Falls, Snohomish, Wash., 599 ....A4 122
Granite Quarry, Rowan, N.C., 1,059 ....B3 109
Graniteville, Middlesex, Mass., 850 ....C1 97
Graniteville, Iron, Mo., 300 ..D7 101
Graniteville, Aiken, S.C., 3,000 ....D4 115
Graniteville, Washington, Vt., 900 ....C4 120
Granja, Braz., 5,074 ....B2 57
Granja de Torrehermosa, Sp., 6,314 ....C3 20
Grannis, Polk, Ark., 185 ....C1 81
Grano, Renville, N. Dak., 14 ....A4 110
Granollers, Sp., 20,194 ....B7 20
Gran Quivira, Torrance, N. Mex., 20 ....C3 107
Gran Quivira, nat. mon., N. Mex C3 107
Gran Sabana, plat., Ven B5 60
Gransee, Ger., 6,092 ....E7 24
Grant, Marshall, Ala., 274 ..A3 78
Grant, Park, Colo., 100 ....B5 83
Grant, Brevard, Fla., 300 ....E6 86
Grant, Montgomery, Iowa, 180 ....C3 92
Grant, Boone, Ky., 175 ....A6 94
Grant, Allen, La., 150 ....D3 95

Grant, Newaygo, Mich., 732 .... E5 98
Grant, Beaverhead, Mont., 5 .... E3 102
Grant, Perkins, Nebr., 1,166 .... D4 103
Grant, Choctaw, Okla., 286 .... D6 112
Grant, co., Ark., 8,294 .... C3 81
Grant, co., Ind., 75,741 .... C6 91
Grant, co., Kans., 5,269 .... E2 93
Grant, co., Ky., 9,489 .... B5 94
Grant, co., Minn., 8,840 .... E2 99
Grant, co., Nebr., 1,009 .... C4 103
Grant, co., N. Mex., 18,700 .... E1 107
Grant, co., N. Dak., 6,248 .... C4 110
Grant, co., Okla., 8,140 .... A4 112
Grant, co., Oreg., 7,726 .... C7 113
Grant, co., S. Dak., 9,913 .... B8 116
Grant, co., Wash., 46,477 .... B6 122
Grant, co., W. Va., 8,304 .... B5 123
Grant, co., Wis., 44,419 .... F3 124
Grant, par., La., 13,330 .... C3 95
Grant, mtn., Nev D4 104
Grant, mtn., Nev E3 104
Grant, range, Nev. E6 104
Grant City, Worth, Mo., 1,061 .... A3 101
Grantham, Eng., 25,030 .... B7 12
Grantham, Sullivan, N.H., 170 (332▲) D2 105
Grantland, coast, N.W. Ter., Can A22 4
Granton, Ont., Can., 303 .... D3 72
Granton, Clark, Wis., 278 .... D3 124
Grant Orchards, Grant, Wash B6 122
Grantown-on-Spey, Scot., 1,581 .... C5 13
Grant Park, Kankakee, Ill., 757 .... B6 90
Grant, Valencia, N. Mex., 10,274 .... B2 107
Grantsboro, Pamlico, N.C., 150 .... B7 109
Grantsburg, Crawford, Ind., 85 .... H5 91
Grantsburg, Burnett, Wis., 900 .... C1 124
Grantsdale, Ravalli, Mont., 150 .... D2 102
Grants Lick, Campbell, Ky., 50 .... A7 94
Grants Mills, Providence, R.I., 60 .... A11 84
Grants Pass, Josephine, Oreg., 10,118 .... E3 113
Grantsville, Garrett, Md., 446 .... D1 85
Grantsville, Tooele, Utah, 2,166 .... C3 119
Grantsville, Calhoun, W. Va., 866 .... C3 123
Grant Town, Marion, W. Va., 1,105 .... A7, B4 123
Grantville Coweta, Ga., 1,158 .... C2 87
Grantville, Jefferson, Kans., 120 .... B7 93
Grantville, Dauphin, Pa., 200 .... F8 114
Granum, Alta., Can., 290 .... E4 69
Granville, Greenlee, Ariz D6 80
Granville, Fr., 9,827 .... C3 14
Granville, Putnam, Ill., 1,048 .... B4 90
Granville, Sioux, Iowa, 381 .... B2 92
Granville, Hampden, Mass., 250 (847▲) B2 97
Granville, Washington, N.Y., 2,715 .... B7 108
Granville, McHenry, N. Dak., 400 .... A5 110
Granville, Licking, Ohio, 2,868 .... B5 111
Granville, Jackson, Tenn., 150 .... C8 117
Granville, Addison, Vt., 110 (215▲) D3 120
Granville, Milwaukee, Wis. (part of Brown Deer) E1 124
Granville, co., N.C., 33,110 .... A5 109
Granville, lake, Man., Can A1 71
Granville Centre, N.S., Can., 217 .... E4 74
Granville Ferry, N.S., Can., 411 .... E4 74
Grão Mogol, Braz., 1,121 .... E2 57
Grapeland, Houston, Tex., 1,113 .... D5 118
Grapeview, Mason, Wash., 200 .... B3 122
Grapeville, Westmoreland, Pa., 1,600 .... *F2 114
Grapevine, Hopkins, Ky., 500 .... C2 94
Grapevine, Tarrant, Tex., 2,821 .... B5, C4 118
Gras, lake, N.W. Ter., Can D10 66
Grasmere, Owyhee, Idaho, 5 .... G3 89
Grasmere, Hillsboro, N.H., 400 .... D3 105
Grasonville, Queen Annes, Md., 925 .... C5 85
Grass, isl., Fla C3 86
Grass, lake, Ill A5 90
Grass, riv., Man., Can B2 71
Grass, riv., N.Y B2 108
Grasscreek, Fulton, Ind., 125 .... C5 91
Grass Creek, Hot Springs, Wyo., 60 .... B4 125
Grasse, Fr., 26,258 .... F7 14
Grasselli, Jefferson, Ala., 500 .... B3, E4 78
Grassflat, Clearfield, Pa., 850 .... D5 114
Grassington, Eng., 1,151 .... F7 13
Grasslake, Lake, Ill., 862 .... D2 90
Grass Lake, Jackson, Mich., 1,037 .... F6 98
Grassrange, Fergus, Mont., 222 .... C8 102
Grasston, Kanabec, Minn., 146 .... E5 99
Grass Valley, Nevada, Calif., 4,876 .... C3 82
Grass Valley, Sherman, Oreg., 234 .... B6 113
Grassy, brook, Vt E3 120
Grassy, lake, La C5 95
Grassy, mtn., Oreg D2 113
Grassy Butte, McKenzie, N. Dak., 80 .... B2 110
Grassy Cove, Cumberland, Tenn D9 117
Grassy Creek, Ashe, N.C., 25 .... A2 109
Grassy Knob, mtn., Mo D8 101
Grassy Lake, Alta., Can., 259 .... E5 69
Grates Cove, Newf., Can., 382 .... D5 75
Gratiot, Licking and Muskingum, Ohio, 222 .... C5 111
Gratiot, Lafayette, Wis., 294 .... F3 124
Gratiot, co., Mich., 37,012 .... E6 98
Gratis, Preble, Ohio, 586 .... C3 111
Graton, Sonoma, Calif., 1,055 .... *C2 82
Gratz, Dauphin, Pa., 704 .... E8 114
Graubünden, canton, Switz., 147,458 .... C7 19
Graue Hörner, mtn., Switz C7 19
Graulhet, Fr., 7,996 .... F5 14
Gravatá, Braz., 15,550 .... C3, k6 57
Grave, creek, W. Va B2 123
Grave, peak, Idaho C4 89
Gravelbourg, Sask., Can., 1,499 .... H2 70
Gravelines, Fr., 7,720 .... D10 12
Gravelly, Yell, Ark., 300 .... C2 81
Gravelridge, Bradley, Ark D3 81
Gravenhurst, Ont., Can., 3,077 .... C5 72
Graves, Terrell, Ga., 150 .... E2 87
Graves, Georgetown, S.C., 300 .... E9 115
Graves, co., Ky., 30,021 .... B2 94
Graves, is., Mass D4 97
Gravesend, Eng., 51,388 .... C8 12
Gravesend, bay, N.Y E5 106
Gravesville, Calumet, Wis., 200 .... B6 124
Gravette, Benton, Ark., 855 .... A1 81
Gravina [in Puglia], It., 31,977 .... D6 21
Gravity, Taylor, Iowa, 275 .... D3 92
Gravois Mills, Morgan, Mo., 30 .... C5 101
Grawn, Grand Traverse, Mich., 150 .... D5 98
Gray, Sask., Can., 79 .... G3 70
Gray, Fr., 6,995 .... D6 14
Gray, Jones, Ga., 1,320 .... C3 87
Gray, Audubon, Iowa, 162 .... C3 92
Gray, Knox, Ky., 800 .... D5 94
Gray, Terrebonne, La., 400 .... C6 95
Gray, Cumberland, Maine, 400 (2,184▲) E2, E5 96
Gray, Beaver, Okla., 10 .... D3 112
Gray, Somerset, Pa., 500 .... f3 114
Gray, co., Kans., 4,380 .... E3 93
Gray, co., Tex., 31,535 .... B2 118
Gray Court, Laurens, S.C., 473 .... B3 115
Gray Horse, Osage, Okla., 200 .... A5 112
Grayland, Grays Harbor, Wash., 550 .... C1 122
Grayling, Crawford, Mich., 2,015 .... D6 98
Grayling, Gallatin, Mont., 10 .... F5 102
Grayridge, Stoddard, Mo., 300 .... E8 101
Grayrocks, Platte, Wyo C8 125
Grays, Jasper, S.C., 100 .... F5 115
Grays, harbor, Wash C1 122
Grays, lake, Idaho F7 89
Grays, peak, Colo B5 83
Grays Branch, Greenup, Ky., 100 .... B7 94
Grays Harbor, co., Wash., 54,465 .... B2 122
Grayslake, Lake, Ill., 3,762 .... A5, E2 90
Grayson, Winston, Ala., 500 .... A2 78
Grayson, Sask., Can., 323 .... G4 70
Grayson, Gwinnett, Ga., 282 .... A6, C3 87
Grayson, Carter, Ky., 1,692 .... B7 94
Grayson, Caldwell, La., 428 .... B3 95
Grayson, co., Ky., 15,834 .... C3 94
Grayson, co., Tex., 73,043 .... C4 118
Grayson, co., Va., 17,390 .... E1 121
Grays River, Wahkiakum, Wash., 100 .... C2 122
Gray Station, Cumberland, Maine E5 96
Gray Summit, Franklin, Mo., 200 .... B7 101
Graysville, Jefferson, Ala., 2,870 .... E4 78
Graysville, Monroe, Ohio, 127 .... C6 111
Graysville, Greene, Pa., 160 .... G1 114
Graysville, Rhea, Tenn., 838 .... D8 117
Grayville, White and Edwards, Ill., 2,280 .... E5 90
Graz, Aus., 237,080 .... E7 16
Gready, Harbour, Newf., Can B3, h10 75
Great, basin, U.S C4 76
Great, bay, N.H D5 105
Great, bay, N.J D5 106
Great, chan., India G9 39
Great, falls, Tenn D8 117
Great, isl., Mass C7 97
Great, isl., N.C B7 109
Great, lake, Austl o15 50
Great, pt., Mass D7 97
Great, pond, Maine D3 96
Great, pond, Mass D3 97
Great, pond, N.Y E9 84
Great, sand sea, Eg., U.A.R D5 43
Great, sound, Bermuda E13 77
Great, val., U.S., B3 106, G6 114, D9 117, D2 121
Great Abaco, isl., Ba. Is B5 64
Great Alföld, reg., Hung B5 22
Great Arber, mtn., Ger D8 17
Great Artesian, basin, Austl D7 50
Great Australian, bight, Austl F5 50
Great Averill, pond, Vt B5 120
Great Barre, mtn., Conn A3 84
Great Barrier, isl., N.Z L15 51
Great Barrier, reef, Austl C8 50
Great Barrington, Berkshire, Mass., 4,000 (6,624▲) B1 97
Great Basin, boundary, Nev B4, E7 104
Great Basin, boundary, Oreg D7, E5, E8 113
Great Basin, reg., Nev C3, E3 104
Great Bear, lake, N.W. Ter., Can C8 66
Great Bend, Barton, Kans., 16,670 .... D5 93
Great Bend, Richland, N. Dak., 164 .... C9 110
Great Bend, Susquehanna, Pa., 777 .... C10 114
Great Bitter, lake, Eg., U.A.R D4 32
Great Blasket, isl., Ire E1 11
Great Britain & Northern Ireland, see United Kingdom, country, Eur.
Great Burnt, lake, Newf., Can D3 75
Great Cacapon, Morgan, W. Va., 800 .... B6 123
Great Captain, isl., Conn F2 84
Great Deer, Sask., Can E2 70
Great Divide, basin, Wyo D4 125
Great Dividing, range, Austl C7 50, E8 51
Great Duck, isl., Ont., Can B2 72
Great East, pond, Maine E2 96
Great Egg, bay, N.J E3 106
Great Egg, inlet, N.J E3 106
Great Egg Harbor, riv., N.J D3 106
Greater Antilles, is., W.I B3 53
Greater Khingan, mts., Asia E15 33
Greater Leech Lake, Indian res., Minn C4 99
Great Exuma, isl., Ba. Is D6 64
Great Falls, Man., Can., 164 .... D3 71
Great Falls, Cascade, Mont., 55,244 .... C5 102
Great Falls, Chester, S.C., 3,030 .... B6 115
Great Guana, cay, Ba. Is C4 64
Great Hog, neck, N.Y E7 84
Great Inagua, isl., Ba. Is E7 64
Great Karroo, plat., S. Afr D3 49
Great Kills, inlet, N.Y E5 106
Great Lakes, Lake, Ill E2 90
Great Meadows, Warren, N.J., 250 .... B3 106
Great Mecatina, isl., Que., Can C2 77
Great Mercury, isl., N.Z L15 51
Great Miquelon, isl., N.A m26 67
Great Misery, isl., Mass C4 97
Great Natuna, isl., Indon E3 35
Great Neck, Nassau, N.Y., 10,171 .... D2 108
Great Neck (Great Neck Plaza), Nassau, N.Y., 4,948 .... *D2 108
Great Neck Estates, Nassau, N.Y., 3,262 .... *D2 108
Great Nicobar, isl., India G9 39
Great Paternoster, isl., Indon G5 35
Great Peconic, bay, N.Y F6 84
Great Pond, Hancock, Maine, 37 .... D4 96
Great Quittacus, pond, Mass C6 97
Great Rann of Kutch, salt flat, India F2 40
Great River, Suffolk, N.Y., 350 .... G4 84
Great St. Bernard, pass, Switz E3 19
Great Salt Lake, des., Utah C2 119
Great Salt, lake, Utah B3 119
Great Salt, pond, R.I E10 84
Great Salt Plains, res. and dam, Okla A3 112
Great Sand, hills, Sask., Can G1 70
Great Sand Dunes, nat. mon., Colo D5 83
Great Sandy, des., Austl D3 50
Great Sea, reef, Fiji Is 52
Great Seneca, creek, Md B3 85
Great Skellig, isl., Ire F1 11
Great Slave, lake, N.W. Ter., Can D10 66
Great Smoky, mts., N.C., Tenn D2 109, D10 117
Great Smoky Mountains, nat. park, N.C., Tenn D2 109, D10 117
Great South, bay, N.Y G4 84
Great Stone Face (Old Man of the Mountain), mtn., N.H B3 105
Great Thatch., isl., Vir. Is f16 65
Great Victoria, des., Austl E4 50
Great Village, N.S., Can., 348 .... D6 74
Great Wall, wall, China D5 34
Great Wass, isl., Maine D5 96
Great Whale, riv., Que., Can B2 73
Great Works, Somerset, Maine (part of Old Town) D4 96
Great Yarmouth, Eng., 52,860 .... B9 12
Great Zab, riv., Iraq D14 31
Gredos, mts., Sp B3 20
Greece, Monroe, N.Y., 25,000 .... *B3 108
Greece, country, Eur., 8,388,553 .... H13 8, C4 23
Greeley, Weld, Colo., 26,314 .... A6 83
Greeley, Delaware, Iowa, 369 .... B6 92
Greeley, Anderson, Kans., 415 .... D8 93
Greeley, Greeley, Nebr., 656 .... C7 103
Greeley, Pike, Pa., 250 .... D12 114
Greeley, co., Kans., 2,087 .... D2 93
Greeley, co., Nebr., 4,595 .... C7 103
Greeleyville, Williamsburg, S.C., 504 .... D8 115
Green, Clay, Kans., 190 .... C6 93
Green, Ontonagon, Mich A5 98
Green, co., Ky., 11,249 .... C4 94
Green, co., Wis., 25,851 .... F4 124
Green, bay, Mich., Wis C3 98
Green, lake, B.C., Can D7 68
Green, lake, Sask., Can C2 70
Green, lake, Maine D4 96
Green, lake, Minn E4 99
Green, lake, Wis E5 124
Green, mts., Vt F2 120
Green, mts., Wyo C5 125
Green, pond, N.J A4 106
Green, ridge, Colo A5 83
Green, riv., N.B., Can B1 74
Green, riv., Colo., Utah, Wyo E5 119, D3 125
Green, riv., Ill B4 90
Green, riv., Ky C2 94
Green, riv., Vt F3 120
Green, riv., Wash B2 122
Green, swamp, N.C C5 109
Green, val., Tex F2 118
Greenacres, Kern, Calif., 1,000 .... *E4 82
Green Acres, Nassau, N.Y., 2,500 .... *E2 108
Green Acres, Montgomery, Md., 1,000 .... *C3 85
Greenacres, Spokane, Wash., 2,074 .... B8, D8 122
Greenacres City, Palm Beach, Fla., 1,026 .... F6 86
Greenback, Loudon, Tenn., 285 .... D9 117
Greenbackville, Accomack, Va., 300 .... D7 121
Green Bank, Burlington, N.J., 100 .... D3 106
Green Bank, Pocahontas, W. Va., 200 .... C5 123
Green Bay, Prince Edward, Va., 100 .... D4 121
Green Bay, Brown, Wis., 62,888 .... A6, D6 124
Green Bay, dist, Newf., Can D3 75
Greenbelt, Prince Georges, Md., 7,479 .... B4 85
Greenbrae, Marin, Calif., 2,000 .... *B5 82
Greenbrier, Limestone, Ala., 125 .... A3 78
Greenbrier, Faulkner, Ark., 401 .... B3 81
Greenbrier, Robertson, Tenn., 1,238 .... A5 117
Greenbrier, co., W. Va., 34,446 .... D4 123
Greenbrier, riv., W. Va D4 123
Green Brook, Somerset, N.J., 3,622 .... *B4 106
Greenburr, Clinton, Pa., 100 .... D7 114
Greenbush, Plymouth, Mass., 400 .... D4 97
Greenbush, Alcona, Mich., 70 .... D7 98
Greenbush, Roseau, Minn., 706 .... B2 99
Greenbush, Sheboygan, Wis., 150 .... B6 124
Green Camp, Marion, Ohio, 492 .... B4 111
Greencastle, Putnam, Ind., 8,506 .... E4 91
Green Castle, Sullivan, Mo., 250 .... A5 101
Greencastle, Franklin, Pa., 2,988 .... G6 114
Green City, Sullivan, Mo., 628 .... A5 101
Green Court, Alta., Can., 59 .... B3 69
Green Cove Springs, Clay, Fla., 4,233 .... C5, C6 86
Greencreek, Idaho, Idaho, 50 .... C2 89
Green Creek, Cape May, N.J., 350 .... E3 106
Greendale, Dearborn, Ind., 2,861 .... F8 91
Greendale, Milwaukee, Wis., 6,843 .... F2, F5 124
Greendell, Sussex, N.J., 100 .... B3 106
Greene, Butler, Iowa, 1,427 .... B5 92
Greene, Androscoggin, Maine, 100 (1,226▲) D2 96
Greene, Chenango, N.Y., 2,051 .... C5 108
Greene, Kent, R.I., 150 .... C10 84
Greene, co., Ala., 13,600 .... C1 78
Greene, co., Ark., 25,198 .... A5 81
Greene, co., Ga., 11,193 .... C3 87
Greene, co., Ill., 17,460 .... D3 90
Greene, co., Ind., 26,327 .... F3 91
Greene, co., Iowa, 14,379 .... B3 92
Greene, co., Miss., 8,366 .... D5 100
Greene, co., Mo., 126,276 .... D4 101
Greene, co., N.Y., 31,372 .... C6 108
Greene, co., N.C., 16,741 .... B6 109
Greene, co., Ohio, 94,642 .... C4 111
Greene, co., Pa., 39,424 .... G1 114
Greene, co., Tenn., 42,163 .... C11 117
Greene, co., Va., 4,715 .... C4 121
Greeneville, Greene, Tenn., 11,759 .... C11 117
Greenfield, Monterey, Calif., 1,680 .... D3 82
Greenfield, Greene, Ill., 1,064 .... D3 90
Greenfield, Hancock, Ind., 9,049 .... E6 91
Greenfield, Adair, Iowa, 2,243 .... C3 92
Greenfield, Penobscot, Maine, 50 (100▲) C4 96
Greenfield, Franklin, Mass., 17,690 .... A2 97
Greenfield, Dade, Mo., 1,172 .... D4 101
Greenfield, Hillsboro, N.H., 200 (538▲) E3 105
Greenfield, Chaves, N. Mex., 100 .... D5 107
Greenfield, Highland, Ohio, 5,422 .... C4 111
Greenfield, Blaine, Okla., 128 .... B3 112
Greenfield, Weakley, Tenn., 1,779 .... A3 117
Greenfield, Nelson, Va., 40 .... D4 121
Greenfield, Milwaukee, Wis., 17,636 .... *F1 124
Greenfield Park, Que., Can., 7,807 .... D9 73
Greenfield Village, Salt Lake, Utah, 1,000 .... *D2 119
Green Forest, Carroll, Ark., 1,038 .... A2 81
Green Garden, Marquette, Mich B3 98
Green Grass, Dewey, S. Dak., 50 .... B4 116
Green Haven, Anne Arundel, Md., 1,302 .... *B2 85
Greenhill, Lauderdale, Ala., 200 .... A2 78
Greenhill, Warren, Ind., 200 .... D3 91
Green Hill, Washington, R.I., 100 .... D10 84
Greenhills, Hamilton, Ohio, 5,407 .... D2 111
Green Hills, Davidson, Tenn., 14,000 .... *A5 117
Greenhorn, mtn., Colo D6 83
Green Island, Jackson, Iowa, 97 .... B7 92
Green Island, Albany, N.Y., 3,533 .... *C7 108
Green Isle, Sibley, Minn., 331 .... F4 99
Green Lake, Sask., Can., 756 .... C2 70
Green Lake, Green Lake, Wis., 953 .... E5 124
Green Lake, co., Wis., 15,418 .... E4 124
Greenland, Washington, Ark., 127 .... B1 81
Greenland, Duval, Fla., 30 .... C6 86
Greenland, Ontonagon, Mich., 360 .... A6 98
Greenland, Rockingham, N.H., 375 (1,196▲) D5 105
Greenland, Dan. dep., N.A., 24,018 .... B19 4, B16 61
Greenland, sea, Arc. O B15 4
Greenlaw, Scot., 831 .... E6 13
Greenlawn, Suffolk, N.Y., 5,422 .... F3 84
Greenleaf, Washington, Kans., 562 .... C7 93
Greenleaf, Brown, Wis., 250 .... A6, D5 124
Greenlee, co., Ariz., 11,509 .... D6 80
Green Lowther, mtn., Scot E5 13
Greenmanorville, Hartford, Conn., 1,200 .... *B6 84
Green Meadows, Prince Georges, Md., 1,500 .... *C1 85
Greenmount, Carroll, Md., 150 .... A4 85
Green Mountain, Marshall, Iowa, 200 .... B5 92
Greenmountain, Yancey, N.C., 250 .... C4 109
Green Mountain, res., Colo B4 83
Greenock, Allegheny, Pa., 1,500 .... F2 114
Greenock, Scot., 74,578 (★108,000) E4 13
Greenough, Missoula, Mont., 10 .... D3 102
Green Pond, Bibb, Ala., 500 .... B2 78
Green Pond, Colleton, S.C., 285 .... F6 115
Green Pond, mtn., N.J B3 106
Greenport, Suffolk, N.Y., 2,608 .... D4 108
Green Ridge, Pettis, Mo., 375 .... C4 101
Green Ridge, Delaware, Pa., 3,500 .... *G11 114
Green River, Emery, Utah, 1,075 .... E5 119
Green River, Windham, Vt., 30 .... F3 120
Green River, Sweetwater, Wyo., 3,497 .... D3 125
Green River, res., Vt B3 120
Green Rock, Henry, Ill., 2,677 .... *B3 90
Greens, peak, Ariz C6 80
Greensboro, Hale, Ala., 3,081 .... C2 78
Greensboro, Gadsden, Fla., 709 .... B2 86
Greensboro, Greene, Ga., 2,773 .... C3 87
Greensboro, Henry, Ind., 232 .... E7 91
Greensboro, Caroline, Md., 1,160 .... C6 85
Greensboro, Guilford, N.C., 119,574 (★156,800) A4 109
Greensboro, Greene, Pa., 505 .... F2 114
Greensboro, Orleans, Vt., 115 (600▲) B4 120
Greensboro Bend, Orleans, Vt., 100 .... B4 120
Greensburg, Decatur, Ind., 7,492 .... F7 91
Greensburg, Kiowa, Kans., 1,988 .... E4 93
Greensburg, Green, Ky., 2,334 .... C4 94
Greensburg, St. Helena, La., 512 .... D5 95
Greensburg, Westmoreland, Pa., 17,383 .... F2 114
Greens Creek, Jackson, N.C. D3 109
Greens Fork, Wayne, Ind., 474 .... E7 91
Green Sea, Horry, S.C., 100 .... C10 115
Greenspond, Newf., Can., 784 .... D5 75
Green Spring, Hampshire, W. Va., 120 .... B6 123
Green Springs, Sandusky and Seneca, Ohio, 1,262 .... A4 111
Green Sulphur Springs, Summers, W. Va., 350 .... D4 123
Greensville, co., Va., 16,155 .... E5 121
Greentop, Schuyler, Mo., 311 .... A5 101
Greentown, Howard, Ind., 1,266 .... D6 91
Greentree, Allegheny, Pa., 5,226 .... *F1 114
Greenup, Cumberland, Ill., 1,477 .... D5 90
Greenup, Greenup, Ky., 1,240 .... B7 94
Greenup, co., Ky., 29,238 .... B7 94
Greenvale, Nassau, N.Y., 1,650 .... *F2 84
Green Valley, Tazewell, Ill., 552 .... C4 90
Green Valley, Lyon, Minn., 130 .... F3 99
Greenvalley, Shawano, Wis., 100 .... D5 124
Greenview, Menard, Ill., 796 .... C4 90
Greenview, Boone, W. Va., 222 .... D5 123
Greenville, Butler, Ala., 6,894 .... D3 78
Greenville, Plumas, Calif., 1,140 .... B3 82
Greenville, New London, Conn. (part of Norwich) C8 84
Greenville, Madison, Fla., 1,318 .... B3 86
Greenville, Meriwether, Ga., 726 .... C2 87
Greenville, Bond, Ill., 4,569 .... E4 90
Greenville, Floyd, Ind., 453 .... H6 91
Greenville, Clay, Iowa, 173 .... A2 92
Greenville, Muhlenberg, Ky., 3,198 .... C2 94
Greenville, Lib F3 45
Greenville, Piscataquis, Maine, 1,400 (2,025▲) C3 96
Greenville, Montcalm, Mich., 7,440 .... E5 98
Greenville, Washington, Miss., 41,502 .... B2 100
Greenville, Wayne, Mo., 282 .... D7 101
Greenville, Hillsboro, N.H., 1,251 (1,385▲) E3 105
Greenville, Greene, N.Y., 380 .... C6 108
Greenville, Westchester, N.Y., 2,800 .... *D2 108
Greenville, Pitt, N.C., 22,860 .... B6 109
Greenville, Darke, Ohio, 10,585 .... B3 111
Greenville, Mercer, Pa., 8,765 .... D1 114
Greenville, Providence, R.I., 3,000 .... B10 84
Greenville, Greenville, S.C., 66,188 (★164,500) B3 115
Greenville, Hunt, Tex., 19,087 .... C4 118
Greenville, Beaver, Utah, 10 .... E3 119
Greenville, Monroe, W. Va., 200 .... D4 123
Greenville, Outagamie, Wis., 200 .... A5 124
Greenville, chan., B.C., Can C3 68
Greenville, co., S.C., 209,776 .... B3 115
Greenville, creek, Ohio B3 111
Greenville Junction, Piscataquis, Maine, 500 .... C3 96
Greenwald, Stearns, Minn., 266 .... E4 99
Greenwater Lake, park, Sask., Can E4 70
Greenway, Clay, Ark., 179 .... A5 81
Green Way, Man., Can., 130 .... E2 71
Greenway, McPherson, S. Dak., 101 .... B6 116
Greenwich, Fairfield, Conn., 53,793 .... E2 84
Greenwich, Eng., 85,585 (part of London) C7 12, m12 10
Greenwich, Sedgwick, Kans., 55 .... B5 93
Greenwich, Cumberland, N.J., 450 .... E2 106
Greenwich, Washington, N.Y., 2,263 .... B7 108
Greenwich, Huron, Ohio, 1,371 .... A5 111
Greenwich, Piute, Utah, 5 .... E4 119
Greenwich, Prince William, Va., 100 .... B4 121
Greenwich, pt., Conn E2 84
Greenwich Hill, N.B., Can., 73 .... D3 74
Greenwood, Jefferson, Ala., 400 .... B3 78
Greenwood, Sebastian, Ark., 1,558 .... B1 81
Greenwood, B.C., Can., 932 .... E8 68
Greenwood, Sussex, Del., 768 .... C6 85
Greenwood, Jackson, Fla., 427 .... B1 86
Greenwood, Johnson, Ind., 7,169 .... E5, I8 91
Greenwood, McCreary, Ky., 200 .... D5 94
Greenwood, Caddo, La., 500 .... B2 95
Greenwood, Leflore, Miss., 20,436 .... B3 100
Greenwood, Jackson, Mo., 488 .... E2 101
Greenwood, Cass, Nebr., 403 .... D9, E2 103
Greenwood, Steuben, N.Y., 200 .... C3 108
Greenwood, Blair, Pa., 1,500 .... F5 114
Greenwood, Greenwood, S.C., 16,644 .... C3 115
Greenwood, Charles Mix, S. Dak., 120 .... E7 116
Greenwood, Wilson, Tenn A5 117
Greenwood, Clark, Wis., 1,041 .... D3 124
Greenwood, co., Kans., 11,253 .... E7 93
Greenwood, co., S.C., 44,346 .... C3 115
Greenwood, lake, Minn C7 99
Greenwood, lake, N.J., N.Y A4 106
Greenwood, lake, S.C C4 115
Greenwood Lake, Orange, N.Y., 1,236 .... *D6 108
Greenwood Mountain, Oxford, Maine, 200 .... D2 96
Greer, Clearwater, Idaho, 70 .... C2 89
Greer, Oregon, Mo., 40 .... E6 101
Greer, Greenville and Spartanburg, S.C., 8,967 .... B3 115
Greer, co., Okla., 8,877 .... C2 112
Greetsiel, Ger., 1,400 .... A7 15
Gregg, Man., Can., 30 .... D2 71
Gregg, co., Tex., 69,436 .... C5 118
Gregory, Livingston, Mich., 300 .... F6 98
Gregory, Currituck, N.C., 50 .... A7 109
Gregory, Gregory, S. Dak., 1,478 .... D6 116
Gregory, San Patricio, Tex., 1,970 .... *F4 118
Gregory, co., S. Dak., 7,399 .... D6 116
Gregory, lake, Austl E6 50
Gregory, range, Austl C7 50
Gregory, riv., Austl C6 50
Greifenhagen, see Gryfino, Pol.
Greifswald, Ger., 47,000 .... A6 16
Greifswalder, bay, Ger D7 24
Greig, Lewis, N.Y., 80 .... B5 108
Grein, Aus., 2,518 .... D7 16
Greinwich Terrace, Calcasieu, La., 2,000 .... *D2 95
Greiz, Ger., 39,100 .... C7 17
Grelton, Henry, Ohio, 150 .... A4 111
Gremyachinsk, Sov. Un., 30,000 .... D20 9
Grenaa, Den., 9,088 .... B4 24
Grenada, Siskiyou, Calif., 300 .... B2 82
Grenada, Grenada, Miss., 7,914 .... B4 100
Grenada, co., Miss., 18,409 .... B4 100
Grenada, Br. dep., N.A., 88,617 .... p16 64
Grenada, isl., N.A p16 64
Grenada, res., Miss B4 100
Grenada, Br. dep., N.A., 88,617 .... p16 64
Grenade [-sur-Garonne], Fr., 2,112 .... F4 14
Grenadines, is., Grenada-St. Vincent p16 64
Grenagh, Ire E3 11
Grenay, Fr., 8,730 .... D2 15
Grenchen, Switz., 18,000 .... B3 19
Grene, Renville, N. Dak., 15 .... A4 110
Grenfell, Sask., Can., 1,256 .... G4 70
Grenloch, Gloucester, N.J., 975 .... D2 106
Grenoble, Fr., 156,707 (★230,000) E6 14
Grenola, Elk, Kans., 349 .... E7 93
Grenora, Williams, N. Dak., 448 .... A2 110
Grenville, Que., Can., 1,330 .... D3 73
Grenville, Union, N. Mex., 55 .... A6 107
Grenville, Day, S. Dak., 151 .... B8 116
Grenville, co., Ont., Can., 22,864 .... C9 72
Grenville, cape, Austl B7 50
Grenville, pt., Wash B1 122
Gresham, Marion, S.C., 150 .... D9 115
Gresham, York, Nebr., 239 .... C8 103
Gresham, Multnomah, Oreg., 3,944 .... B4 113
Gresham, Shawano, Wis., 458 .... D5 124
Gressitt, King and Queen, Va., 300 .... D6 121
Gretna, Man., Can., 575 .... E3 71
Gretna, Gadsden, Fla., 647 .... B2 86
Gretna, Jefferson, La., 21,967 .... C7, E5 95
Gretna, Sarpy, Nebr., 745 .... C9, E2 103
Gretna, Scot., 3,037 .... F5 13
Gretna, Pittsylvania, Va., 900 .... E3 121
Greven, Ger., 23,000 .... A2 17
Grevena, Grc., 4,779 .... B3 23
Grevenbroich, Ger., 22,000 .... B1 17

Grevesmühlen, Ger., 11,100 . . E5 24
Grey, co., Ont., Can., 62,005 . C4 72
Grey, is., Newf., Can . . . C4 75
Grey, riv., N.Z. . . . O13 51
Greybull, Big Horn, Wyo., 2,286 . . . A4 125
Greybull, riv., Wyo . . . A4 125
Greycliff, Sweet Grass, Mont., 85 . . . E7 102
Grey Eagle, Todd, Minn., 372 . E4 99
Grey Islands Harbour, Newf., Can . . . C4 75
Greylock, mtn., Mass . . . A1 97
Greymouth, N.Z., 8,881 . . . O13 51
Greys, riv., Wyo . . . B2 125
Greystone, Moffat, Colo., 2 . . A2 83
Greystone, Litchfield, Conn., 100 . . . C4 84
Greystone, Vance, N.C., 60 . . A5 109
Greytown, N.Z., 1,580 . . . N15 51
Greytown, S. Afr., 7,737 . . . C5 49
Gribbell, isl., B.C., Can . . . C3 68
Gridley, Butte, Calif., 3,343 . . C3 82
Gridley, McLean, Ill., 889 . . . C5 90
Gridley, Coffey, Kans., 321 . . D8 93
Griegos, Bernalillo, N. Mex. . G5 107
Grier, Curry, N. Mex . . . C6 107
Griesheim, Ger., 13,700 . . . D3 17
Grieskirchen, Aus., 4,137 . . . D6 16
Griffin, Union, Ark., 20 . . . D3 81
Griffin, Sask., Can., 140 . . . H4 70
Griffin, Spalding, Ga., 21,735 . . . C2 87
Griffin, Posey, Ind., 212 . . . H2 91
Griffing Park, Jefferson, Tex., 2,267 . . . *E5 118
Griffins, Hartford, Conn . . . B6 84
Griffith, Austl., 7,700 . . . G6 51
Griffith, Lake, Ind., 9,483 . . . A3 91
Griffith, isl., Ont., Can . . . C4 72
Griffithsville, Lincoln, W. Va., 500 . . . C3 123
Griffithville, White, Ark., 172 . . . B4 81
Grifton, Pitt, N.C., 1,816 . . . B6 109
Griggs, Cimarron, Okla., 15 . . D2 112
Griggs, co., N. Dak., 5,023 . . B7 110
Griggsville, Pike, Ill., 1,240 . . D3 90
Grigoriopol, Sov. Un . . . B9 22
Grim, cape, Austl . . . o14 50
Grimari, Cen. Afr. Rep . . . D4 46
Grimes, Dale, Ala., 50 . . . D4 78
Grimes, Polk, Iowa, 697 . A7, C4 92
Grimes, Roger Mills, Okla., 15 . . . B2 112
Grimes, co., Tex., 12,709 . . . D5 118
Grimesland, Pitt, N.C., 362 . . B6 109
Grimma, Ger., 16,100 . . . B7 17
Grimmen, Ger., 11,000 . . . D7 24
Grimmialp, Switz . . . C4 19
Grimms Landing, Mason, W. Va., 300 . . . C3 123
Grimsby, Ont., Can., 5,148 . . D5 72
Grimsby, Eng., 96,665 (★135,000) . . . A7 12
Grimsey, isl., Ice . . . m23 25
Grimshaw, Alta., Can., 1,095 . . A2 69
Grimstad, Nor., 2,600 . . . H3 25
Grimsthorpe, Ont., Can., 110 . . . B2 72
Grindelwald, Switz., 3,244 . . . C5 19
Grindsted, Den., 5,289 . . . C2 24
Grindstone, Penobscot, Maine, 60 . . . C4 96
Grindstone, isl., Que., Can . . . B7 74
Grind Stone City, Huron, Mich . . . D8 98
Grindstone Island, Que., Can., 761 . . . B8 74
Grinem, isl., Eniwetok . . . 52
Grinnell, Poweshiek, Iowa, 7,367 . . . C5 92
Grinnell, Gove, Kans., 396 . . C3 93
Griquatown, S. Afr., 2,524 . . . C3 49
Grisdella, Garfield, Mont . . C10 102
Gris Nez, cape, Fr . . . D9 12
Griswold, Man., Can., 137 . . . E1 71
Griswold, Cass, Iowa, 1,207 . . C2 92
Griswold, Aroostook, Maine . . B4 96
Griswoldville, Hartford, Conn. (part of Wethersfield) . . . C6 84
Griswoldville, Franklin, Mass., 200 . . . A2 97
Grizzly, Jefferson, Oreg . . . C6 113
Grizzly, creek, Colo . . . A4 83
Grizzly, mtn., Oreg . . . C6 113
Grizzly, mtn., Oreg . . . E4 113
Grizzly Bear, mtn., N.W. Ter., Can . . . C8 66
Groais, isl., Newf., Can . . . C4 75
Grodekovo, Sov. Un., 5,000 . . D5 37
Grodkow, Pol., 2,953 . . . C4 26
Grodno, Sov. Un., 81,000 . . . E4 27
Grodzisk, Pol., 6,015 . . . B4 26
Grodzisk Mazowiecki, Pol., 17,800 . . . m13 26
Groenlo, Neth., 6,600 . . . B6 15
Groesbeck, Hamilton, Ohio, 9,000 . . . *C3 111
Groesbeck, Limestone, Tex., 2,498 . . . D4 118
Groix, isl., Fr . . . D2 14
Grojec, Pol., 6,841 . . . C6 26
Grombalia, Tun., 5,043 . . . F12 30
Gronau, Ger., 25,600 . . . A2 17
Gronau, Ger., 5,300 . . . A4 17
Groningen, Neth., 146,300 (★169,000) . . . A6 15
Groningen, prov., Neth., 477,700 . . . A6 15
Gronlid, Sask., Can., 152 . . . D3 70
Groom, Carson, Tex., 679 . . B2 118
Groom Creek, Yavapai, Ariz., 150 . . . C3 80
Groos, Delta, Mich . . . C3 98
Groote Eylandt, isl., Austl . . . B6 50
Grootfontein, S.W. Afr., 1,525 . . . A2 49
Groot Vloer, lake, S. Afr . . . C3 49
Groscap, Mackinac, Mich . . . C6 98
Groslay, Fr., 4,744 . . . g10 14
Gros Morne, Que., Can., 577 . A3 73
Gros Morne, mtn., Newf., Can . D3 75
Gros Pate, mtn., Newf., Can . . C3 75
Gross, Boyd, Nebr., 17 . . . B7 103
Grossbreitenbach, Ger., 4,361 . . . C6 17
Grosse, isl., Que., Can . . . B8 74
Grosse Ile, Wayne, Mich., 6,318 . . . *B7 98
Grossenhain, Ger., 19,500 . . . B8 17
Grosse Pointe, Wayne, Mich., 6,631 . . . A8 98
Grosse Pointe Farms, Wayne, Mich., 12,172 . . . A8 98
Grosse Pointe Park, Wayne, Mich., 15,457 . . . A8 98
Grosse Pointe Shores, Macomb and Wayne, Mich., 2,301 . *F8 98
Grosse Pointe Woods, Macomb and Wayne, Mich., 18,580 . F8 98
Grosser Plön, lake, Ger . . . D4 24
Grosser Priel, peak, Aus . . . E7 16
Grosse Tete, Iberville, La., 768 . . . D4 95
Grosseto, It., 36,000 (51,730▲) . C3 21
Grossevichi, Sov. Un . . . C9 37
Grossgerau, Ger., 12,200 . . . D3 17
Grossglockner, mtn., Aus . . . B8 18
Grossmont, San Diego, Calif., 1,000 . . . *F5 82
Grossotheim, Ger., 6,900 . . . D4 17
Grössraschen, Ger., 12,100 . . B9 17
Grossrörsdorf, Ger., 8,285 . . . B9 17
Grosvenor Dale, Windham, Conn., 530 . . . B9 84
Gros Ventre, range, Wyo . . . B2 125
Gros Ventre, riv., Wyo . . . B2 125
Groton, New London, Conn., 10,111 (29,937▲) . . . D8 84
Groton, Middlesex, Mass., 1,178 (3,904▲) . . . A4, C1 97
Groton, Grafton, N.H., 80 (99▲) . . . C3 105
Groton, Tompkins, N.Y., 2,123 . . . C4 108
Groton, Brown, S. Dak., 1,063 . . . B7 116
Groton, Caledonia, Vt., 387 (631▲) . . . C4 120
Grottaferrata, It., 5,377 . . . h9 21
Grottaglie, It., 23,223 . . . D6 21
Grottoes, Rockingham, Va., 969 . . . C4 121
Grouard, Alta., Can., 303 . . . B2 69
Grouse (Lost River), Custer, Idaho, 58 . . . F5 89
Grouse, creek, Kans . . . E7 93
Grouse, creek, Utah . . . B2 119
Grouse Creek, Box Elder, Utah, 49 . . . B2 119
Grouse Creek, mts., Utah . . . B2 119
Grovania, Houston, Ga., 186 . . . D3 88
Grove, Washington, Maine, 35 . . . C5 96
Grove, Delaware, Okla., 975 . A7 112
Grove, York, Va . . . A6 121
Grove, pt., Md . . . B5 85
Grove City, Meeker, Minn., 466 . . . E4 99
Grove City, Franklin, Ohio, 8,107 . . . C2, C3 111
Grove City, Mercer, Pa., 8,368 . . . D1 114
Grove Hill, Clarke, Ala., 1,834 . . . D2 78
Groveland, Tuolumne, Calif., 350 . . . D3 82
Groveland, Lake, Fla., 1,747 . D5 86
Groveland, Bryan, Ga., 100 . D5 87
Groveland, Essex, Mass., 1,600 (3,297▲) . . . A5 97
Groveland, Livingston, N.Y., 150 . . . C3 108
Groveport, Franklin, Ohio, 2,043 . . . C2, C5 111
Grover, Weld, Colo., 133 . . . A6 83
Grover, Cleveland, N.C., 538 . . . B2 109
Grover, Bradford, Pa., 160 . . C8 114
Grover, Dorchester, S.C., 300 . . . E6 115
Grover, Codington, S. Dak., 35 . . . C8 116
Grover, Lincoln, Wyo., 120 . C2 125
Grover City, San Luis Obispo, Calif., 5,210 . . . *E3 82
Grover Hill, Paulding, Ohio, 547 . . . A3 111
Grovertown, Starke, Ind., 175 . . . B4 91
Groves, Jefferson, Tex., 17,304 . . . E6 118
Grovespring, Wright, Mo., 92 . . . D5 101
Groveton, Coos, N.H., 2,004 . A3 105
Groveton, Allegheny, Pa., 1,300 . . . *B5 114
Groveton, Trinity, Tex., 1,148 . . . D5 118
Groveton, Fairfax, Va., 900 . B5 121
Grovetown, Columbia, Ga., 1,396 . . . C4 87
Grow, King, Tex . . . C2 118
Groznyy, Sov. Un., 270,000 . E3 29
Grubbs, Jackson, Ark., 360 . . B4 81
Gruber, lake, Ger . . . D4 24
Grudovo, Bul., 3,928 . . . D8 22
Grudziadz, Pol., 65,000 . . . B5 26
Grues, isl., Que., Can . . . B7 73
Gruetli, Grundy, Tenn., 400 . D8 117
Gruinard, bay, Scot . . . C3 13
Grulla, Starr, Tex., 1,436 . . . F3 118
Grünberg, Ger., 3,900 . . . C4 17
Grundy, Buchanan, Va., 2,287 . . . B3 121
Grundy, co., Ill., 22,350 . . . B5 90
Grundy, co., Iowa, 14,132 . . . B5 92
Grundy, co., Mo., 12,220 . . . A4 101
Grundy, co., Tenn., 11,512 . D8 117
Grundy Center, Grundy, Iowa, 2,403 . . . B5 92
Grünstadt, Ger., 7,800 . . . D3 17
Grunthal, Man., Can., 287 . . E3 71
Grüsch, Switz., 729 . . . C8 19
Gruver, Hansford, Tex., 1030 . A2 118
Gruyères, Switz., 1,349 . . . C3 19
Gruz, Yugo., 10,000 . . . D4 22
Gryazi, Sov. Un., 10,000 . . E12 27
Gryazovets, Sov. Un., 12,500 . B13 27
Gryfice, Pol., 10,100 . . . B3 26
Gryfino, Pol., 1,347 . . . B3 26
Grygla, Marshall, Minn., 192 . B3 99
Gstaad, Switz . . . D3 19
Guacanayabo, gulf, Cuba . . . E5 64
Gu Achi, Pima, Ariz., 250 . . . E3 80
Guadalajara, Mexico, 580,617 (★635,000) . . . C4, m12 63
Guadalajara, Sp., 21,230 . . . B4 20
Guadalajara, prov., Sp., 183,545 . . . *B4 20
Guadalcanal, Sp., 6,931 . . . C3 20
Guadalcanal, isl., Sol. Is . . . G9 7
Guadalhorce, riv., Sp . . . D3 20
Guadalimar, riv., Sp . . . C4 20
Guadalope, riv., Sp . . . B5 20
Guadalquivir, riv., Sp . . . D2 20
Guadalupe, Maricopa, Ariz., 1,200 . . . G2 80
Guadalupe, Santa Barbara, Calif., 2,614 . . . E3 82
Guadalupe, Conejos, Colo., 150 . . . D4 83
Guadalupe, Guadalupe, N. Mex . . . C4 107
Guadalupe, co., N. Mex., 5,610 . . . C5 107
Guadalupe, co., Tex., 29,017 . E4 118
Guadalupe, isl., Pac. O . . . E14 7
Guadalupe, mts., N. Mex . . . E4 107
Guadalupe, mts., Sp . . . C3 20
Guadalupe, peak, Tex . . . E2 118
Guadalupe, riv., Tex . . . B4 118
Guadalupe [Bravos], Mex., 1,214 . . . A3 63
Guadalupita, Mora, N. Mex., 470 . . . A4 107
Guadarrama, mts., Sp . . . B4 20
Guadarrama, riv., Sp . . . p17 20
Guadeloupe, Fr. dep., N.A., 251,000 . . . n16 64
Guadeloupe, passage, W.I . . . n16 64
Guadiana, riv., Port . . . D2 20
Guadiana Alto, riv., Sp . . . C4 20
Guadiana Menor, riv., Sp . . . D4 20
Guadiela, riv., Sp . . . B4 20
Guadix, Sp., 18,500 . . . D4 20
Guafo, isl., Chile . . . C2 54
Guagua, Phil., 5,914 . . . o13 35
Guainia, riv., Col . . . C4 60
Guairá, dept., Par., 100,446 . . E4 55
Guajará Mirim, Braz., 7,115 . B2 55
Guajataca, lake, P.R . . . B3 69
Guajataca, riv., P.R . . . B3 65
Guajira (La Guajira), pol. div., Col., 112,190 . . . A3 60
Gualaceo, Ec., 2,735 . . . B2 58
Gualala, Mendocino, Calif., 400 . . . C2 82
Gualán, Guat., 2,898 . . . C3 62
Gualaquiza, Ec., 259 . . . B2 58
Gualeguay, Arg., 23,517 . . . A5 54
Gualeguaychú, Arg., 37,109 . A5 54
Guam, U.S. dep., Oceania, 67,044 . . . 52
Guamá, riv., Braz . . . B1 57
Guamaní Central, P.R . . . D6 65
Guaminí, Arg., 2,273 . . . B4 54
Guamúchil, Mex., 5,865 . . . B3 63
Guanabacoa, Cuba, 32,490 . . D2 64
Guanabana, P.R . . . C2 65
Guanabara, state, Brazil, 3,307,163 . . . h6 56
Guanacaste, prov., C.R., 146,600 . . . *E5 62
Guanacaste, mts., C.R . . . E5 62
Guanahacabibes, gulf, Cuba . . . D1 64
Guanajay, Cuba, 12,908 . . . D2 64
Guanajibo, pt., P.R . . . C2 65
Guanajibo, riv., P.R . . . C2 65
Guanajuato, Mex., 28,135 . . . C4, m13 63
Guanajuato, state, Mex., 1,728,358 . . . C4 63
Guanambi, Braz., 5,268 . . . D2 57
Guanare, Ven., 16,935 . . . B4 60
Guanche, Pan., 30 . . . h11 62
Guandacol, Arg . . . E2 55
Guane, Cuba, 2,248 . . . D1 64
Guánica, P.R., 4,100 . . . D3 65
Guánica, mun., P.R., 13,767 . D3 65
Guánica, lake, P.R . . . C3 65
Guanillos del Norte, Chile . . . D1 55
Guaniquilla, pt., P.R . . . C2 65
Guano, isl., Vir. Is . . . f16 65
Guano, lake, Oreg . . . E7 113
Guanoco, Ven . . . A5 60
Guantánamo, Cuba, 64,671 . . E6 64
Guapí, Col., 1,882 . . . C2 60
Guaporé, riv., Bol . . . B3 55
Guaporé, riv., Braz . . . E2 59
Guaqui, Bol., 2,266 . . . C2 55
Guará, Braz., 5,652 . . . F1 57
Guara, mts., Sp . . . A5 20
Guarabira, Braz., 15,848 . C3, h6 57
Guaraguao, P.R . . . C4 65
Guaranda, Ec., 7,287 . . . B2 58
Guarapuava, Braz., 13,546 . . . D2 56
Guaraqueçaba, Braz., 741 . . . D3 56
Guaratinguetá, Braz., 38,293 . C3 56
Guarda, Port., 7,704 . . . B2 20
Guardafui, cape, Som . . . C7 47
Guardian, Webster, W. Va., 200 . . . C4 123
Guareña, Sp., 9,742 . . . C2 20
Guárico, riv., Ven . . . B4 60
Guárico, state, Ven., 244,966 . B4 60
Guarujá, Braz., 6,506 . . . n8 56
Guasave, Mex., 8,505 . . . B3 63
Guasdualito, Ven., 4,549 . . . B3 60
Guasipati, Ven., 3,382 . . . B5 60
Guastalla, It., 6,416 . . . E6 18
Guata, Hond., 497 . . . C4 62
Guatemala, Guat., 284,276 . . C2 62
Guatemala, country, N.A., 3,301,559 . . . H11 61, C2 62
Guateque, Col., 2,408 . . . C3 60
Guatimozin, Arg . . . A4 54
Guaviare, riv., Col . . . C3 60
Guaxupé, Braz., 14,168 . . C3, k8 56
Guayabal, Cuba, 569 . . . E5 64
Guayabal, lake, P.R . . . C4 65
Guayabero, riv., Col . . . C3 57
Guayama, P.R., 19,183 . . . D6 65
Guayama, mun., P.R., 33,678 . . . D6 65
Guayanés, riv., P.R . . . C7 65
Guayanilla, P.R., 3,067 . . . C3 65
Guayanilla, mun., P.R., 17,396 . . . C3 65
Guayaquil, gulf, Ec . . . B1 58
Guayaquil, Ec., 258,966 . . . B2 58
Guayas, prov., Ec., 663,516 . . B1 58
Guaymas, Mex., 34,845 . . . B2 63
Guaynabo, P.R., 3,343 . . . B6 65
Guaynabo, mun., P.R., 39,718 . . . B6 65
Guazacapán, Guat., 3,383 . . . C2 62
Gubakha, Sov. Un., 53,800 . . . B5 29
Gubat, Phil., 9,045 . . . *C6 35
Gubbio, It., 8,600 . . . C4 21
Gubin, Pol., 3,040 . . . C3 26
Gúdar, mts., Sp . . . B5 20
Gudauta, Sov. Un., 4,379 . A13 31
Gudenå, Den . . . B3 24
Gudenaa, riv., Den . . . B3 24
Guderup, Den . . . D3 24
Gudur, India, 25,618 . . . F6 39
Guebwiller, Fr., 10,568 . . . D7 14
Guecho, Sp., 22,951 . . . A4 20
Guékédou, Guinea, 1,400 . . . E2 45
Guelma, Alg., 17,225 (21,587▲) . . . B6 44
Guelph, Ont., Can., 39,838 (★41,767) . . . D4 72
Guelph, Dickey, N. Dak., 300 . C7 110
Güemes, Arg., 5,688 . . . D2 55
Guenette, Que., Can., 151 . . C2 73
Guercif, Mor., 5,579 . . . C4 44
Guéret, Fr., 11,384 . . . D4 14
Guerette, Aroostook, Maine, 160 . . . A4 96
Guerneville, Sonoma, Calif., 900 . . . A4 82
Guernica [y Luno], Sp., 3,381 . A4 20
Guernsey, Sask., Can., 79 . . . F3 70
Guernsey, Platte, Wyo., 800 . . C8 125
Guernsey, co., Ohio, 38,579 . . B6 111
Guernsey, Br. dep., Eur., 47,178 . . . F5 10
Guerra, Jim Hogg, Tex., 50 . . F3 118
Guerrara, Alg . . . C5 44
Guerrero, Mex., 1,786 . . . F3 118
Guerrero, state, Mex., 1,189,085 . . . D5 63
Gueugnon, Fr., 8,374 . . . D5 14
Gueydan, Vermilion, La., 2,156 . . . D3 95
Guffey, Park, Colo., 15 . . . C5 83
Guge, mtn., Eth . . . D4 47
Gugegwe, isl., Kwajalein . . . 52
Guiana, British, see British Guiana, dep., S.A.
Guiana, French, see French Guiana, dep., S.A.
Guiana, Netherlands, see Surinam, Neth. dep., S.A.
Guiana, highlands, Ven . . . C5 60
Guider, Cam . . . D2 46
Guide Rock, Webster, Nebr., 441 . . . D7 103
Guidonia, It., 22,205 . . . D4, h9 21
Guiglo, I.C., 2,700 . . . E3 45
Guild, Sullivan, N.H., 250 . . D2 105
Guild, Marion, Tenn., 400 . . D8 117
Guildford, Eng., 53,977 . . . C7 12
Guildhall, Essex, Vt., 100 (248▲) . . . B5 120
Guilford, New Haven, Conn., 2,420 (7,913▲) . . . D6 84
Guilford, Dearborn, Ind., 250 . . . F8 91
Guilford, Piscataquis, Maine, 1,372 (1,880▲) . . . C3 96
Guilford, Howard, Md., 175 . B4 85
Guilford, Nodaway, Mo., 125 . A3 101
Guilford, Chenango, N.Y., 450 . . . C5 108
Guilford, Guilford, N.C., 1,000 . . . *A4 109
Guilford, Windham, Vt., 100 (823▲) . . . F3 120
Guilford, co., N.C., 246,520 . . A4 109
Guilford College, Guilford, N.C., 1,700 . . . A4 109
Guillestre, Fr., 913 . . . E7 18
Guimarães, Braz., 1,512 . . . B2 57
Guimarães, Port., 18,294 . . . B1 20
Guin, Marion, Ala., 1,462 . . . B2 78
Guinea, Caroline, Va., 75 . . . C5 121
Guinea, country, Afr., 3,000,000 . . . E4 42, D2 45
Guinea, Portuguese, see Portuguese Guinea, dep., Afr.
Guinea, gulf, Afr . . . F6 42
Guinea Mills, Cumberland, Va . . . D4 121
Güines, Cuba, 29,226 . . . D2 64
Guines, Fr., 3,481 . . . D1 15
Guingamp, Fr., 8,912 . . . C2 14
Guion, Izard, Ark., 222 . . . B4 81
Guipúzcoa, prov., Sp., 478,337 . . . *A4 20
Guir, cape, Mor . . . C3 44
Güira de Melena, Cuba, 13,715 . . . D2 64
Güiria, Ven., 10,724 . . . A5 60
Guisborough, Eng., 12,079 . . F7 13
Guise, Fr., 6,284 . . . C5 14
Guists, creek, Ky . . . B4 94
Guitiriz, Sp., 1,006 . . . A2 20
Guiuan, Phil., 7,222 . . . *C7 35
Guján [-Mestras], Fr., 3,776 . E3 14
Gujarat, state, India, 20,633,350 . . . D5 39
Gujranwala, Pak., 196,154 . . . A5 40, B5 39
Gu Komelik, Pinal, Ariz., 30 . E4 80
Gulbarga, India, 97,069 . . . E6 39, I6 40
Guldborg, Den., 598 . . . D5 24
Gulf, Chatham, N.C., 180 . . . B4 109
Gulf, co., Fla., 9,937 . . . C1 86
Gulf creek, Okla . . . D2 112
Gulf Coastal, plain, Ark . . . C4 81
Gulf Crest, Mobile, Ala., 75 . . . D1 78
Gulf Hammock, Levy, Fla., 350 . . . C4 86
Gulfport, Pinellas, Fla., 9,730 . . . E4, F2 86
Gulfport, Henderson, Ill., 214 . . . C2 90
Gulfport, Harrison, Miss., 30,204 (★124,200) . . . E2, E4 100
Gulf Shores, Baldwin, Ala., 356 . . . E2 78
Gulf Stream, Palm Beach, Fla., 176 . . . *F6 86
Gulgong, Austl., 1,399 . . . F7 51
Gulkana, Alsk., 51 . . . C10, f19 79
Gull, isl., N.Y . . . E8 84
Gull, isl., N.C . . . B8 109
Gull, lake, Alta., Can . . . C4 69
Gull, lake, Minn . . . D4 99
Gullivan, bay, Fla . . . G5 86
Gulliver, Schoolcraft, Mich., 75 . . . C5 98
Gull Lake, Sask., Can., 1,038 . G1 70
Gullrock, lake, Ont., Can . . . E8 72
Gulluk, Tur., 826 . . . D6 23
Gully, Polk, Minn., 168 . . . C3 99
Gulu, Ug., 4,770 . . . A5 48
Gulyantsi, Bul., 4,741 . . . D7 22
Gulyay-Pole, Sov. Un., 10,000 . . . H11 27
Gumaca, Phil., 6,461 . . . p14 35
Gumba, Con. L . . . A3 48
Gumboro, Sussex, Del., 150 . D7 85
Gumefens, Switz., 330 . . . C3 19
Gumel, Nig., 10,406 . . . D6 45
Gumma, pref., Jap., 1,578,476 . . . *H9 37
Gummersbach, Ger., 32,000 . . B2 17
Gum Spring, mtn., Tenn . . . D8 117
Gümüşane (Gumush-Khaneh), Tur., 5,400 . . . B12 31
Guna, Eth . . . C4 47
Guna, India, 31,031 . . . E6 40
Guna, mtn., Eth . . . C4 47
Gunamitz, riv., N.B., Can . . . B2 74
Gundelfingen, Ger., 5,100 . . . E5 17
Gunflint, range, Minn . . . A7 99
Gungu, Con. L . . . C2 48
Gunisao, lake, Man., Can . . . C3 71
Gunisao, riv., Man., Can . . . C3 71
Gunlock, Washington, Utah, 90 . . . F2 119
Gunn, Smith, Miss., 50 . . . C4 100
Gunnedah, Austl., 6,546 . . . E8 51
Gunnison, Gunnison, Colo., 3,477 . . . C4 83
Gunnison, Bolivar, Miss., 448 . B2 100
Gunnison, Sanpete, Utah, 1,059 . . . D4 119
Gunnison, co., Colo., 5,477 . . C3 83
Gunnison, riv., Colo . . . C2 83
Gunnworth, Sask., Can . . . F1 70
Gunpowder, creek, Ky . . . A6 94
Gunpowder, riv., Md . . . B5 85
Gunpowder Falls, riv., Md . . . A4 85
Gunter, mtn., Ala . . . A3 78
Guntersville, Marshall, Ala., 6,592 . . . A3 78
Guntersville, dam, Ala . . . A3 78
Gunton, Man., Can., 88 . . . D3 71
Guntown, Lee, Miss., 269 . . . A5 100
Guntur, India, 187,122 . . . E7 39
Gunung Api, vol., Indon . . . G7 35
Gunungsitoli, Indon., 3,124 . . L2 38
Gunz, riv., Ger . . . A6 18
Günzburg, Ger., 11,800 . . . E5 17
Gunzenhausen, Ger., 9,300 . . D5 17
Gurabo, P.R., 3,957 . . . B7 65
Gurabo, mun., P.R., 16,603 . B7 65
Gurabo, riv., P.R . . . B6 65
Gura-Humorului, Rom., 7,216 . . . B7 22
Gurdaspur, India, 27,665 . . . A5 40
Gurdon, Clark, Ark., 2,166 . . D2 81
Gurgan, bay, Iran . . . C6 41
Gurgan, riv., Iran . . . C7 41
Gurguan, pt., Tinian . . . 52
Gurgueia, riv., Braz . . . C2 57
Gurk, riv., Aus . . . E7 16
Gurla Mandhata, peak, China . . . B8 40
Gurley, Madison, Ala., 850 . . . A3 78
Gurley, Cheyenne, Nebr., 329 . . . C3 103
Gurleyville, Tolland, Conn., 100 . . . B8 84
Gurnee, Lake, Ill., 1,831 . . . E2 90
Gurnet, pt., Mass . . . B6 97
Gurney, Iron, Wis., 100 . . . B3 124
Gurupá, Braz., 912 . . . C4 59
Gurupí, cape, Braz . . . B1 57
Gurupí, mts., Braz . . . B1 57
Gurupí, riv., Braz . . . B1 57
Gurvevsk, Sov. Un., 30,000 . E26 9
Guryev, Sov. Un., 86,000 . . . D4 29
Gusau, Nig., 40,202 . . . D6 45
Gusev (Gumbinnen), Sov. Un., 25,000 . . . A7 26
Gusher, Uintah, Utah, 65 . . . C6 119
Gusinje, Yugo., 2,757 . . . D4 22
Gus-Khrustalnyy, Sov. Un., 58,000 . . . D13 27
Gustavus, Alsk., 107 . . . k22, D12 79
Güsten, Ger., 8,160 . . . B6 17
Gustine, Merced, Calif., 2,300 . . . D3 82
Gustine, Comanche, Tex., 380 . . . D3 118
Güstrow, Ger., 38,000 . . . B6 16
Gutau, Aus., 2,292 . . . E9 17
Gütersloh, Ger., 52,300 . . . B3 17
Guthrie, Lawrence, Ind., 25 . G4 91
Guthrie, Todd, Ky., 1,211 . . . D2 94
Guthrie, Hubbard, Minn., 90 . . . C4 99
Guthrie, McHenry, N. Dak., 38 . . . A5 110
Guthrie, Logan, Okla., 9,502 . . . B4 112
Guthrie, King, Tex., 250 . . . C2 118
Guthrie, co., Iowa, 13,6 7 . . . C3 92
Guthrie Center, Guthrie, Iowa, 2,071 . . . C3 92
Guttenberg, Clayton, Iowa, 2,087 . . . B6 92
Guttenberg, Hudson, N.J., 5,118 . . . D5 106
Gu Vo, Pima, Ariz., 100 . . . E3 80
Guy, Faulkner, Ark., 300 . . . B3 81
Guy, Fort Bend, Tex., 100 . . G4 118
Guyandot, mtn., W. Va . . . D6 123
Guyandot, riv., W. Va . . . C2 123
Guyenne, former prov., Fr., 2,061,000 . . . E4 14
Guymon, Texas, Okla., 5,768 . . . D3 112
Guyot, mtn., Tenn . . . D10 117
Guyra, Austl., 1,628 . . . E8 51
Guys, McNairy, Tenn., 100 . B3 117
Guysborough, N.S., Can., 490 . . . D8 74
Guysborough, co., N.S., Can., 13,274 . . . D8 74
Guys Mills, Crawford, Pa., 350 . . . C2 114
Guyton, Effingham, Ga., 670 . . . D5 87
Gvardeysk, Sov. Un., 5,000 . . A6 26
Gwa, Bur., 5,000 . . . E9 39
Gwaai, S. Rh . . . A4 49
Gwabegar, Austl . . . E7 51
Gwadar (Gwadur), Pak . . . C3 39
Gwalior, India, 300,587 . . . C6 39, D7 40
Gwanda, S. Rh., 1,600 . . . B4 49
Gwane, Con. L . . . A4 48
Gweebarra, bay, Ire . . . C3 11
Gweesalia, Ire . . . C2 11
Gwelo, S. Rh., 35,100 . . . A4 49
Gwin, Holmes, Miss., 150 . . . B3 100
Gwinn, Marquette, Mich., 1,009 . . . B3 98
Gwinner, Sargent, N. Dak., 242 . . . C8 110
Gwinnett, co., Ga., 43,541 . . C2 87
Gwydyr, bay, Alsk . . . A10 79
Gwynn, Mathews, Va., 400 . D6 121
Gwynne, Alta., Can., 109 . . . C4 69
Gwynneville, Shelby, Ind., 300 . . . E6 91
Gwynns Falls, riv., Md . . . C2 85
Gyangtse, China, 10,000 . . . C8 39
Gydan, mts., Sov. Un . . . C18 28
Gympie, Austl., 11,094 . . . C9 51
Gyongyos, Hung., 28,668 . . . B4 22
Gyor, Hung., 70,812 . . . B3 22
Gyor-Sopron, co., Hung., 391,734 . . . *B3 22
Gypsum, Eagle, Colo., 358 . . B4 83
Gypsum, Saline, Kans., 593 . D6 93
Gypsum, Ottawa, Ohio, 400 . A5 111
Gypsumville, Man., Can., 235 . . . D2 71
Gypsy, Creek, Okla . . . B5 112
Gyula, Hung., 19,990 (24,609▲) . . . B5 22
Gzhatsk, Sov. Un., 16,000 . . D10 27

H

Haag, Ger., 2,300 . . . A8 18
Haakon, co., S. Dak., 3,303 . C4 116
Haaksbergen, Neth., 6,100 . . B6 15
Haamstede, Neth., 1,601 . . . C3 15
Haapamäki, Fin., 2,200 . . . F11 25
Haapsalu, Sov. Un., 10,000 . . B4 27
Ha Arava (Wadi el Araba), depression, Isr., Jordan . . . D7 32
Haarlem, Neth., 164,500 . . . B4 15
Haarlemmermeer, Neth., 4,300 (45,100▲) . . . B4 15
Hab, riv., Pak . . . I13 41
Habana (Havana), prov., Cuba, 1,538,803 . . . h11 64
Habermehl, peak, Ant . . . B13 5
Habersham, Habersham, Ga., 400 . . . B3 87
Habersham, co., Ga., 18,116 . B3 87
Haboro, Jap., 19,800 . . . D10 37
Hachiman, Jap., 11,700 . . . n15 37
Hachinohe, Jap., 131,000 (174,348▲) . . . F10 37
Hachioji, Jap., 158,443 . . I9, n18 37
Hachita, Grant, N. Mex., 200 . . . F1 107
Hacienda, Broward, Fla., 125 . . . *F6 86
Hack, mtn., Austl . . . E2 51
Hackamore, Modoc, Calif., 20 . . . B3 82
Hackberry, Mahave, Ariz., 75 . . . B2 80
Hackberry, Cameron, La., 800 . . . E2 95
Hackberry, Edwards, Tex . . . E2 118
Hackberry, creek, Kans . . . C3 93
Hackensack, Cass, Minn., 204 . . . D4 99
Hackensack, Bergen, N.J., 30,521 . . . B4, D5 106
Hackensack, riv., N.J . . . D5 106
Hacker Valley, Webster, W. Va., 200 . . . C4 115
Hacketstown, Ire., 509 . . . E5 11
Hackett, Sebastian, Ark., 328 . . . B1 81
Hackett, Alta., Can . . . C4 69
Hackettstown, Warren, N.J., 5,276 . . . B3 106
Hackleburg, Marion, Ala., 527 . . . A2 78
Haco, isl., Truk . . . 52
Hacoda, Geneva, Ala., 50 . . D3 78
Hadar, Pierce, Nebr., 100 . . B8 103
Hadarba, cape, Eg., U.A.R., Sud . . . A4 47
Haddam, Middlesex, Conn., 350 (3,466▲) . . . D6 84
Haddam, Washington, Kans., 311 . . . C6 93
Haddam Neck, Middlesex, Conn . . . D7 84
Haddington, Scot., 5,506 . . . E6 13
Haddock, Jones, Ga., 600 . . . C3 87
Haddonfield, Camden, N.J., 13,201 . . . D2 106
Haddon Heights, Camden, N.J., 9,260 . . . D2 106
Hadejia, Nig., 10,453 . . . D7 45
Hadera, Isr., 25,358 . . . B6 32
Haderslev, Den., 19,735 . . . C3 24
Haderslev, co., Den., 71,715 . C3 24
Hadhramaut, reg., Aden . . . C6 47
Hadim, Tur., 2,584 . . . D9 31
Haditha, Iraq, 5,434 . . . E14 31
Hadleigh, Eng., 3,460 . . . E8 12
Hadley, Hampshire, Mass., 1,000 (3,099▲) . . . B2 97
Hadley, Murray, Minn., 151 . . . F3 99
Hadley, Saratoga, N.Y., 500 . . . B7 108
Hadley, Mercer, Pa., 250 . . . D1 114
Hadley, bay, N.W. Ter., Can . . . B11 66
Hadley, lake, Maine . . . D5 96
Hadlock, Jefferson, Wash., 300 . . . A3 122
Hadlyme, New London, Conn., 375 . . . D7 84
Hadsten, Den., 2,525 . . . B4 24
Hadsund, Den., 3,424 . . . B4 24
Hadyai (Ban Hat Yai), Thai., 35,697 . . . I4 38
Haeju, Kor., 82,135 . . . G2 37
Haena, Kauai, Haw., 76 . . . A2 86
Hafford, Sask., Can., 511 . . . E2 70
Hafnarfjördur, Ice., 7,310 . . n22 25
Haft Gel, Iran . . . E4 41
Hafun, cape, Som . . . C7 47
Hagaman, Montgomery, N.Y., 1,292 . . . *C6 108
Hagan, Evans, Ga., 552 . . . D5 87
Hagar, mtn., Colo . . . B5 83
Hagari, riv., India . . . F6 39
Hagar Shores (Lake Michigan Beach), Berrien, Mich., 1,092 . . . *F4 98
Hagarville, Johnson, Ark., 150 . . . B2 81
Hagemeister, isl., Alsk . . . D7 79
Hagen, Sask., Can., 62 . . . E3 70
Hagen, Ger., 195,500 . . . B2 17
Hagenow, Ger., 10,300 . . . E5 24
Hagerman, Gooding, Idaho, 430 . . . G4 89
Hagerman, Chaves, N. Mex., 1,144 . . . D5 107
Hagermans Corners, Ont., Can., 218 . . . E7 72
Hagerstown, Wayne, Ind., 1,730 . . . E7 91
Hagerstown, Washington, Md., 36,660 . . . A2 85
Hagersville, Ont., Can., 2,075 . . . E4 72
Hagg, mtn., Ant . . . B5 5
Hagginwood, Sacramento, Calif., 11,469 . . . *A6 82
Hagi, Jap., 41,000 . . . I5 37
Ha Giang, Viet., 25,000 . . . A6 38
Hags, head, Ire . . . E2 11
Hague, Sask., Can., 413 . . . E2 70
Hague, Alachua, Fla., 75 . . . C4 86
Hague, The ('s Gravenhage), see The Hague, Neth.
Hague, Warren, N.Y., 200 . . . B7 108
Hague, Emmons, N. Dak., 197 . . . C3 110
Hague, cape, Fr . . . C3 14
Haguenau, Fr., 20,457 . . . C7 14
Hagues, peak, Colo . . . A5 83
Ha Ha, bay, Que., Can . . . C2 75
Hahira, Lowndes, Ga., 1,297 . . . F3 87
Hahnville, St. Charles, La., 1,297 . . . C6, E5 95
Haicheng, China . . . D10 36

Haichow, bay, China G8 36
Hai Duong, Viet., 25,000 B7 32
Haifa, Isr., 182,007 (★250,000) B6 32
Haig, lake, Alta., Can A3 69
Haig, mtn., Alta., Can E3 69
Haigerloch, Ger., 1,800 A4 18
Haigler, Dundy, Nebr., 268 D4 103
Haiku, Maui, Haw., 800 C5 88
Ḩā'il, Sau. Ar., 15,000 I14 31
Hailar (Hulun), China, 43,200 B8 34
Haile, Union, La., 160 B3 95
Hailesboro, St. Lawrence, N.Y., 250 B1 108
Hailey, Blaine, Idaho, 1,185 F4 89
Haileybury, Ont., Can., 2,638 E9 72
Haileyville, Pittsburg, Okla., 922 C6 112
Hailstone, Wasatch, Utah, 15 C4 119
Hailun, China, 47,684 B10 34
Hailung, China, 20,000 C10 34
Haimen, China, 95,000 I9 36
Haina, Hawaii, Haw., 670 C6 88
Hainan, isl., China C8 38
Hainan, strait, China B9 38
Hainaut, prov., Bel., 1,248,854 D3 15
Haines, Alsk., 392 D12, k22 79
Haines, Baker, Oreg., 331 C9 113
Hainesburg, Warren, N.J., 200 B2 106
Haines City, Polk, Fla., 9,135 D5 86
Haines Falls, Greene, N.Y., 600 C6 108
Hainesport, Burlington, N.J., 2,000 D3 106
Hainesville, Sussex, N.J., 150 A3 106
Hainichen, Ger., 11,200 C8 17
Haining, China, 5,000 I9 36
Haiphong, Viet., 143,000 B7 38
Hairy Hill., Alta., Can., 173 C5 69
Haiti, country, N.A., 3,097,220 H13 61, F7 64
Haiya, Sud B4 47
Haiyang, China F9 36
Haiyuan, China F1 36
Hajdu-Bihar, co., Hung., 393,332 *B5 22
Hajdúböszörmény, Hung., 26,918 (32,214▲) B5 22
Hajduhadhaz, Hung., 11,221 B5 22
Hajduki Weilkie, Pol g9 26
Hajdunanas, Hung., 14,003 (18,413▲) B5 22
Hajduszoboszlo, Hung., 16,709 (23,885▲) B5 22
Hajiabad Kavir, salt flats, Iran E9 41
Hajiki, cape, Jap G9 37
Hakalau, Hawaii, Haw., 800 D6 88
Hakari, Tur., 2,664 D14 31
Hakodate, Jap., 243,012 F10 37
Haku-San, peak, Jap m15 37
Halabja, Iraq D2 41
Halaib, Eg., U.A.R E7 43
Halal, mtn., Eg., U.A.R D5 32
Halaula, Hawaii, Haw., 600 C6 88
Halawa, Maui, Haw., 25 B5 88
Halawa, riv., Haw g10 88
Halawa Heights, Honolulu, Haw., 2,000 G10 88
Halberstadt, Ger., 45,500 B6 17
Halbrite, Sask., Can., 180 H4 70
Halbur, Carroll, Iowa, 214 B3 92
Halcyon Hot Springs, B.C., Can., 50 D9 68
Haldeman, Rowan, Ky., 170 B6 94
Halden, Nor., 10,000 H4 25
Haldimand, co., Ont., Can., 28,197 E5 72
Hale, Yuma, Colo., 5 B8 83
Hale, Iosco, Mich., 450 D7 98
Hale, Carroll, Mo., 504 B4 101
Hale, co., Ala., 19,537 C2 78
Hale, co., Tex., 36,798 B2 118
Haleakala, crater, Haw C5 88
Haleburg, Henry, Ala., 75 D4 78
Hale Center, Hale, Tex., 2,196 B2 118
Haledon, Passaic, N.J., 6,161 B4 106
Haleiwa, Honolulu, Haw., 2,504 B3, f9 88
Hales Bar, lake, Tenn D8 117
Hales Corners, Milwaukee, Wis., 5,549 F1 124
Halesite, Suffolk, N.Y., 2,857 *D3 108
Halesowen, Eng., 44,160 B5 12
Hales Point, Lauderdale, Tenn B2 117
Halesworth, Eng., 2,252 B9 12
Halethorpe, Baltimore, Md., 22,402 C2 85
Haley, Bowman, N. Dak., 112 D2 110
Haleyville, Winston, Ala., 3,740 A2 78
Haleyville, Cumberland, N.J., 150 E2 106
Half Hollow, hills, N.Y F3 84
Half Moon, Flathead, Mont., 150 B2 102
Half Moon Bay, San Mateo, Calif., 1,957 B5 82
Half-moon Bay (Oban), N.Z., 281 Q12 51
Halfway, Washington Md., 4,256 A2 85
Half Way, Polk, Mo., 150 D4 101
Halfway, Baker, Oreg., 505 C9 113
Halfway, Sublette, Wyo C2 125
Halfway, riv., B.C., Can A6 68
Ḩalḩūl, Jordan, 4,000 C7 32
Haliburton, Ont., Can., 853 B6 72
Haliburton, co., Ont., Can., 8,928 B6 72
Halibut, pt., Mass A6 97
Halicarnassus, see Bodrum, Tur.
Halifax, N.S., Can., 92,511 (★183,946) E6 74
Halifax, Eng., 96,073 (★168,000) A6 12
Halifax, Plymouth, Mass., 600 (1,599▲) C6 97
Halifax, Halifax, N.C., 370 A6 109
Halifax, Dauphin, Pa., 824 F8 114
Halifax, Windham, Vt., 100 (268▲) F3 120
Halifax, Halifax, Va., 792 E4 121
Halifax, co., N.S., Can., 225,723 E7 74
Halifax, co., N.C., 58,956 A6 109
Halifax, co., Va., 33,637 E4 121
Halifax, bay, Austl C8 50
Haliimaile, Maui, Haw., 600 C5 88
Halileh, cape, Iran G5 41
Halin, Som D6 47
Haliri, riv., Iran H9 41
Halkirk, Alta., Can., 172 C4 69
Hall, Aus., 10,016 E5 16
Hall, Morgan, Ind., 120 E4 91
Hall, Granite, Mont., 100 D3 102
Hall, Ontario, N.Y., 300 C3 108
Hall, co., Ga., 49,739 B3 87
Hall, co., Nebr., 35,757 D7 103
Hall, co., Tex., 7,322 B2 118
Hall, mtn., Wash A8 122
Hall, pen., N.W. Ter., Can D19 67
Halladale, riv., Scot B5 13
Hallam, Lancaster, Nebr., 264 D9 103
Hallam, peak, B.C., Can C8 68
Halland, co., Swe., 171,200 B6 24
Hallandale, Broward, Fla., 10,483 E3, G6 86
Hallands Väderö, isl., Swe B6 24
Halla San, peak, Kor J3 37
Hallboro, Man., Can., 90 D2 71
Halldale, Waldo, Maine D3 96
Halle, Bel., 19,339 D4 15
Halle, Ger., 276,200 (★425,000) B6 17
Halle, Ger., 7,500 A3 17
Halleck, Elko, Nev., 10 C6 104
Hallein, Aus., 13,329 E6 16
Hällestad, Swe., 380 u33 25
Hallett, Pawnee, Okla., 132 A5 112
Hallett, cape, Ant B30 5
Hallettsville, Lavaca, Tex., 2,808 E4 118
Halley, Desha, Ark., 213 D4 81
Halliday, Dunn., N. Dak., 509 B3 110
Hall Meadow Brook, res., Conn B4 84
Hallock, Kittson, Minn., 1,527 B2 99
Hallonquist, Sask., Can., 110 G2 70
Hallowell, Cherokee, Kans., 160 E9 93
Hallowell, Kennebeck, Maine, 3,169 D3 96
Halls, Lauderdale, Tenn., 1,890 B2 117
Halls, stream, N.H A1 105
Hallsberg, Swe., 6,000 t33 25
Hallsboro, Columbus, N.C., 250 C5 109
Halls Creek, Sumter, Ala C2 78
Halls Creek, Austl., 50 C4 50
Halls Crossroads, Knox, Tenn., 500 E11 117
Halls Harbour, N.S., Can., 98 D5 74
Hallson, Pembina, N. Dak., 4 A8 110
Halls Summit, Coffey, Kans., 40 D8 93
Hallstead, Susquehanna, Pa., 1,580 C10 114
Hall Summit, Red River, La., 170 B2 95
Hallsville, Boone, Mo., 363 B5 101
Hallsville, Ross, Ohio, 250 C5 111
Hallsville, Harrison, Tex., 684 C5 118
Hallton, Elk, Pa., 45 D4 114
Halltown, Lawrence, Mo., 86 D4 101
Hallville, New London, Conn., 100 D8 84
Hallwil, lake, Switz B5 19
Hallwood, Accomack, Va., 269 D7 121
Hallwood, Mason, W. Va C2 123
Halma, Kittson, Minn., 115 B2 99
Halmahera, isl., Indon E7 35
Halmstad, Swe., 39,700 B6 24
Hals, Den., 1,563 B4 24
Halsell, Choctaw, Ala., 200 C1 78
Halsey, Thomas, Nebr., 111 C5 103
Halsey, Sussex, N.J A3 106
Halsey, Linn, Oreg., 404 C3 113
Hälsingborg, Swe., 77,000 B6 24, I5 25
Halsö, isl., Swe A5 24
Halstad, Norman, Minn., 639 C2 99
Halstead, Eng., 6,465 C8 12
Halstead, Harvey, Kans., 1,598 A5, D6 93
Haltern, Ger., 14,700 B2 17
Haltia, mtn., Fin., Nor C9 25
Halton, co., Ont., Can., 106,967 D5 72
Halton City, Tarrant, Tex., 23,133 *C4 118
Haltwhistle, Eng., 3,745 F6 13
Halvorgate, Sask., Can G2 70
Ham, Fr., 4,204 E3 15
Hama, Okinawa 52
Hamada, Jap., 33,200 I6 37
Hamadān, Iran, 99,909 D4 46
Hamāh, Syr., 97,400 E11 31
Hamahika, isl., Okinawa 52
Hamakuapoko, Maui, Haw., 335 C5 88
Hamamatsu, Jap., 333,009 I8, o16 37
Hamar, Eddy, N. Dak., 84 B7 110
Hamar, Nor., 13,500 G4 25
Hamatombetsu, Jap., 4,700 D11 37
Hambantota, Cey., 4,345 G7 39
Hamber, park, B.C., Can C8 68
Hamberg, Wells, N. Dak., 64 B6 110
Hamblen, co., Tenn., 33,092 C10 117
Hamburg, Ashley, Ark., 2,904 D4 81
Hamburg, New London, Conn., 150 D7 84
Hamburg, Ger., 1,832,400 (★2,250,000) B5 16, E4 24
Hamburg, state, Ger., 1,832,400 E4 24
Hamburg, Calhoun, Ill., 264 D3 90
Hamburg, Fremont, Iowa, 1,647 D2 92
Hamburg, Franklin, Miss., 60 D2 100
Hamburg, Sussex, N.J., 1,532 A3 106
Hamburg, Erie, N.Y., 9,145 C2 108
Hamburg, Berks, Pa., 3,747 E10 114
Hamburg, Aiken, S.C., 150 E4 115
Hamburg, Hardin, Tenn B3 117
Hamburg, Marathon, Wis., 200 C4 124
Hamden, New Haven, Conn., 41,056 D5 84
Hamden, Delaware, N.Y., 300 C6 108
Hamden, Vinton, Ohio, 1,035 B5 111
Häme, dept., Fin., 519,015 *G11 25
Hämeenlinna, Fin., 28,300 G11 25
Hamel, Madison, Ill., 362 A9 101
Hamel (Medina), Hennepin, Minn., 1,472 *E5 99
Hameln, Ger., 50,400 A4 17
Hamer, Jefferson, Idaho, 144 F6 89
Hamer, Dillon, S.C., 170 C9 115
Hamersley, plat., Austl D2 50
Hamersville, Brown, Ohio, 224 D4 111
Hamhŭng, Kor., 112,184 G3 37
Hami (Kumul) (Qomul), China, 30,000 C3 34
Hamill, Tripp, S. Dak., 40 D6 116
Hamilton, Marion, Ala., 1,934 A2 78
Hamilton, Alsk., 35 C7 79
Hamilton, Austl., 9,498 H4 51
Hamilton, Bermuda, 2,878 (★15,371) E14 77
Hamilton, Glenn, Calif., 700 C2 82
Hamilton, Ont., Can., 273,991 (★395,189) D5 72
Hamilton, Moffat, Colo., 22 A3 83
Hamilton, Harris, Ga., 396 D2 87
Hamilton, Hancock, Ill., 2,228 C2 90
Hamilton, Steuben, Ind., 380 A8 91
Hamilton, Marion, Iowa, 197 C5 92
Hamilton, Greenwood, Kans., 400 E7 93
Hamilton, Boone, Ky A6 94
Hamilton, Essex, Mass., 350 (5,488▲) A6, C4 97
Hamilton, Allegan, Mich., 700 F4 98
Hamilton, Monroe, Miss., 115 B5 100
Hamilton, Caldwell, Mo., 1,701 B3 101
Hamilton, Ravalli, Mont., 2,475 D2 102
Hamilton, Madison, N.Y., 3,348 C5 108
Hamilton, N.Z., 42,212 (★50,505) L15 51
Hamilton, Martin, N.C., 565 B6 109
Hamilton, Pembina, N. Dak., 217 A8 110
Hamilton, Butler, Ohio, 72,354 (★103,200) C2, C3 111
Hamilton, Grant, Oreg., 20 C7 113
Hamilton, Washington, R.I. (part of North Kingstown) C11 84
Hamilton, Scot., 41,928 E4 13
Hamilton, Hamilton, Tex., 3,106 D3 118
Hamilton, Loudoun, Va., 403 B5 121
Hamilton, Skagit, Wash., 271 A4 122
Hamilton, co., Fla., 7,705 B3 86
Hamilton, co., Ill., 10,010 E5 90
Hamilton, co., Ind., 40,132 D5 91
Hamilton, co., Iowa, 20,032 B4 92
Hamilton, co., Kans., 3,144 E2 93
Hamilton, co., Nebr., 8,714 D7 103
Hamilton, co., N.Y., 4,267 B6 108
Hamilton, co., Ohio, 864,121 C3 111
Hamilton, co., Tenn., 237,905 D8 117
Hamilton, co., Tex., 8,488 D3 118
Hamilton, inlet, Newf., Can A3 75
Hamilton, lake, Ark C2 81
Hamilton, mtn., Alsk C8 79
Hamilton, mtn., Nev D6 104
Hamilton, mtn., N.Y B6 108
Hamilton, res., Mass B3 97
Hamilton, riv., Newf., Can h9 75
Hamilton, sound, Newf., Can D4 75
Hamilton Acres, Alsk., 960 *C10 79
Hamilton Dome, Hot Springs, Wyo., 150 B4 125
Hamilton Park, Lancaster, Pa., 3,500 *F9 114
Hamiltons Fort, Iron, Utah, 47 F2 119
Hamilton Square (Hamilton Township), Mercer, N.J., 65,035 C3 106
Hamina, Fin., 9,800 G12 25
Hamiota, Man., Can., 779 D1 71
Hamirpur, India, 10,000 C7 39
Hamirpur, India, 10,921 B6 40
Hamler, Henry, Ohio, 588 A3 111
Hamlet, Starke, Ind., 688 B4 91
Hamlet, Hayes, Nebr., 113 D4 103
Hamlet, Richmond, N.C., 4,460 C4 109
Hamlet, Williams, N. Dak., 20 A2 110
Hamletsburg, Pope, Ill., 107 F5 90
Hamlin, Audubon, Iowa, 150 C3 92
Hamlin, Brown, Kans., 99 C8 93
Hamlin, Wayne, Pa., 275 D11 114
Hamlin, Jones and Fisher, Tex., 3,791 C2 118
Hamlin, Lincoln, W. Va., 850 C2 123
Hamlin, co., S. Dak., 6,303 C8 116
Hamlin, lake, Mich D4 98
Hammamet, gulf, Tun F12 30
Hammam Lif, Tun., 22,060 *B7 44
Hammar, Iraq F3 41
Hammarby, Swe t35 25
Hammarsjön, lake, Swe C8 24
Hamme, Bel., 16,794 C4 15
Hammel, Den., 2,162 B3 24
Hammelburg, Ger., 6,000 C4 17
Hammer, Roberts, S. Dak., 65 B8 116
Hammerfest, Nor., 5,900 B10 25
Hammersley Fork, Clinton, Pa., 15 D6 114
Hammett, Elmore, Idaho, 75 G3 89
Hamm [in Westfalen], Ger., 68,800 (★150,000) B2 17
Hammon, Roger Mills, Okla., 656 B2 112
Hammonasset, pt., Conn E6 84
Hammonasset, riv., Conn D6 84
Hammond, Piatt, Ill., 471 D5 90
Hammond, Lake, Ind., 111,698 A2 91
Hammond, Tangipahoa, La., 10,563 A6, D5 95
Hammond, Wabasha, Minn., 205 F6 99
Hammond, Carter, Mont., 13 E12 102
Hammond, St. Lawrence, N.Y., 314 B1 108
Hammond, Clatsop, Oreg., 480 A3 113
Hammond, St. Croix, Wis., 645 D1 124
Hammond, bay, Mich C6 98
Hammond East, Tangipahoa, La., 1,462 *A6 95
Hammondsport, Steuben, N.Y., 1,176 C3 108
Hammondsville, Jefferson, Ohio, 475 A1 123
Hammonton, Atlantic, N.J., 9,854 D3 106
Hamneda, Swe., 168 B7 24
Ham Nord, Que., Can., 573 D6 73
Hamoyet, mtn., Eth B4 47
Hampden, Newf., Can., 682 D3 75
Hampden, Hampden, Mass., 400 (2,345▲) B3 97
Hampden, N.Z., 303 P13 51
Hampden, Ramsey, N. Dak., 159 A7 110
Hampden, co., Mass., 429,353 B2 97
Hampden, Highlands, Penobscot, Maine, 1,000 (4,583▲) D4 96
Hampshire, Kane, Ill., 1,309 A5 90
Hampshire, Maury, Tenn., 150 B11 117
Hampshire (Southampton), co., Eng., 1,336,084 C6 12
Hampshire, co., Mass., 103,229 B2 97
Hampshire, co., W. Va., 11,705 B6 123
Hampshire Road, Rockingham, N.H E4 105
Hampstead, N.B., Can., 116 D3 74
Hampstead, Que., Can., 4,551 *D8 73
Hampstead, Carroll, Md., 696 A4 85
Hampstead, Rockingham, N.H., 300 (1,261▲) E4 105
Hampstead, Pender, N.C., 350 C6 109
Hampton, Calhoun, Ark., 1,011 D3 81
Hampton, N.B., Can., 571 D4 74
Hampton, Windham, Conn., 300 (934▲) B8 84
Hampton, Bradford, Fla., 340 C4 86
Hampton, Henry, Ga., 1,253 C2 87
Hampton, Rock Island, Ill., 742 D7 92
Hampton, Franklin, Iowa, 4,501 B4 92
Hampton, Livingston, Ky., 100 A3 94
Hampton, Dakota, Minn., 305 F6 99
Hampton, Washington, Miss B2 100
Hampton, Hamilton, Nebr., 331 D8 103
Hampton, Rockingham, N.H., 3,281 (5,379▲) E5 105
Hampton, Hunterdon, N.J., 1,135 B3 106
Hampton, Washington, N.Y., 225 B7 108
Hampton, Deschutes, Oreg., 10 D6 113
Hampton, Hampton, S.C., 2,486 F5 115
Hampton, Carter, Tenn., 1,048 C11 117
Hampton (Independent City), Va., 89,258 B6, D6 121
Hampton, Uinta, Wyo D2 125
Hampton, co., S.C., 17,425 F5 115
Hampton, beach, N.Y F7 84
Hampton, butte, Oreg D6 113
Hampton Bays, Suffolk, N.Y., 2,000 D4 108
Hampton Beach, Rockingham, N.H., 700 E2 105
Hampton Falls, Rockingham, N.H., 400 (885▲) E5 105
Hampton Roads, harbor, Va B6 121
Hamrā, plat., Libya D2 43
Hamrarne, strait, Swe C8 24
Hams, bluff, Vir. Is h17 65
Hams, fork, Wyo D2 125
Hamsfork, Lincoln, Wyo C2 125
Ham Sud, Que., Can., 59 D6 73
Hamtramck, Wayne, Mich., 34,137 A8 98
Hamun-i-Mashkel, lake, Pak J21 9
Hamun-i-Murgho, lake, Pak H12 41
Hamyang, Kor., 10,300 I3 37
Han, riv., China E7 34
Hana, Maui, Haw., 547 C6 88
Hanahan, Berkeley, S.C., 4,000 F2 115
Hanalei, Kauai, Haw., 364 A2 88
Hanalei, bay, Haw A2 88
Hanamaulu, Kauai, Haw., 950 B2 88
Hanapepe, Kauai, Haw., 1,383 B2 88
Hanau, Ger., 47,200 C3 17
Hanceville, Cullman, Ala., 1,174 A3 77
Hanceville, B.C., Can., 54 D6 68
Hancock, Litchfield, Conn., 25 C4 84
Hancock, Pottawattamie, Iowa, 252 C2 92
Hancock, Hancock, Maine, 350 (806▲) D4 96
Hancock, Washington, Md., 2,004 A1 85
Hancock, Berkshire, Mass., 200 (455▲) A1 97
Hancock, Houghton, Mich., 5,022 A2 98
Hancock, Stevens, Minn., 942 E3 99
Hancock, Hillsboro, N.H., 300 (722▲) E3 105
Hancock, Delaware, N.Y., 1,830 D5 108
Hancock, Addison, Vt., 160 (323▲) D3 120
Hancock, Morgan, W. Va., 100 B6 123
Hancock, Waushara, Wis., 367 D4 124
Hancock, co., Ga., 9,979 C3 87
Hancock, co., Ill., 24,574 C2 90
Hancock, co., Ind., 26,665 E6 91
Hancock, co., Iowa, 14,604 A4 92
Hancock, co., Ky., 5,330 C3 94
Hancock, co., Maine, 32,293 D4 96
Hancock, co., Miss., 14,039 E4 100
Hancock, co., Ohio, 53,686 A4 111
Hancock, co., Tenn., 7,757 C10 117
Hancock, co., W. Va., 39,615 A4 123
Hancock, lake, Fla E5 86
Hancock, mtn., N.H B3 105
Hancock, pond, Maine E2 96
Hancocks Bridge, Salem, N.J., 300 D2 106
Hand, co., S. Dak., 6,712 C6 116
Handa, Jap., 71,380 o15 37
Handa, isl., Scot B3 13
Handel, Sask., Can., 100 E1 70
Handeni, Tan C6 48
Handley, Tarrant, Tex. (part of Fort Worth) B5, C4 118
Handley, Kanawha, W. Va., 900 C6 123
Handsboro, Harrison, Miss., 1,577 E2, E4 100
Handsworth, Sask., Can H4 70
Haney, B.C., Can., 1,538 A10, E6 68
Hanford, Kings, Calif., 10,133 D4 82
Hanford, Benton, Wash C6 122
Hanford Northwest, Kings, Calif., 1,364 *D4 82
Hangchow, China, 784,000 E9 34, I9 36
Hangchow, bay, China I9 36
Hangelsberg, Ger., 976 A8 17
Hanging Rock, Lawrence, Ohio, 352 D5 111
Hangingstone, riv., Alta., Can A5 69
Hangö, Fin., 8,200 H10 25
Hanita, Isr A7 32
Hankinson, Claiborne, Miss C3 100
Hankinson, Richland, N. Dak., 1,285 C9 110
Hanks, Williams, N. Dak., 78 A2 110
Hanksville, Wayne, Utah, 100 E5 119
Hanley, Sask., Can., 455 F2 70
Hanley Falls, Yellow Medicine, Minn., 334 F3 99
Hanley Hills, St. Louis, Mo., 3,308 *A8 101
Hanlontown, Worth, Iowa, 193 A4 92
Hann, mtn., Austl C4 50
Hanna, Alta., Can., 2,645 D5 69
Hanna, LaPorte, Ind., 500 B4 91
Hanna, McIntosh, Okla., 233 B6 112
Hanna, Duchesne, Utah, 160 C5 119
Hanna, Carbon, Wyo., 625 D6 125
Hanna City, Peoria, Ill., 1,056 C4 90
Hannaford, Griggs, N. Dak., 277 B7 110
Hannagan, Greenlee, Ariz D6 80
Hannah, Cavalier, N. Dak., 253 A7 110
Hannawa Falls, St. Lawrence, N.Y., 400 B2 108
Hannibal, Marion and Ralls, Mo., 20,028 B6 101
Hannibal, Oswego, N.Y., 611 B4 108
Hannibal, Monroe, Ohio, 525 C7 111
Hannon, Macon, Ala., 50 C4 77
Hannover (Hanover), Ger., 573,100 (★710,000) A4 17
Hannover, Oliver, N. Dak., 25 B4 110
Hannoversch Münden, see Münden, Ger.
Hanöbukten, bay, Swe J6 25
Hanoi, Viet., 237,500 B6 38
Hanover, Stone, Ark., 10 B3 81
Hanover, Ont., Can., 4,401 C3 72
Hanover, New London, Conn., 250 C8 84
Hanover (Hannover), Ger., 573,100 (★710,000) A4 17
Hanover, Jo Daviess, Ill., 1,396 A3 90
Hanover, Jefferson, Ind., 1,170 G7 91
Hanover, Washington, Kans., 773 C7 93
Hanover, Oxford, Maine, 170 (240▲) D2 96
Hanover, Plymouth, Mass., 600 (5,923▲) B6, E4 97
Hanover, Jackson, Mich., 449 F6 98
Hanover, Fergus, Mont., 125 C7 102
Hanover, Grafton, N.H., 5,649 (7,329▲) C2 105
Hanover (East Hanover), Morris, N.J., 4,379 *B4 106
Hanover, Grant, N. Mex., 400 C1 107
Hanover, Licking, Ohio, 267 B5 111
Hanover, York, Pa., 15,538 G8 114
Hanover, Hanover, Va., 250 C5 121
Hanover, Wyoming, W. Va., 100 D3 123
Hanover, co., Va., 27,550 C5 121
Hanover, reg., Ger B4 16
Hanover, isl., Chile E2 54
Hanover Green, Luzerne, Pa., 1,000 *D10 114
Hanoverton, Columbiana, Ohio, 442 B7 111
Hansboro, Towner, N. Dak., 143 A6 110
Hansell, Franklin, Iowa, 168 B4 92
Hansen, Twin Falls, Idaho, 427 G4 89
Hansford, co., Tex., 6,208 A2 118
Hanska, Brown, Minn., 491 F4 99
Hans Lollik, is., Vir. Is f15 65
Hanson, Madison, Fla., 40 B3 86
Hanson, Hopkins, Ky., 376 C2 94
Hanson, Plymouth, Mass., 800 (4,370▲) B6 97
Hanson, Sequoyah, Okla B7 112
Hanson, co., S. Dak., 4,584 D8 116
Hanson, lake, Sask., Can C4 70
Hansted, see Hanstholm Havn, Den.
Hansthom Havn (Hansted), Den A2 24
Hanston, Hodgeman, Kans., 279 D4 93
Hantachi, China A3 37
Hants, co., N.S., Can., 26,444 D6 74
Hant's Harbour, Newf., Can., 487 D5 75
Hantsport, N.S., Can., 1,381 D5 74
Haofeng, China, 10,000 J3 36
Haoli, China, 90,000 B10 34
Haparanda, Swe., 3,400 E11 25
Hapeville, Fulton, Ga., 10,082 B5, C2 87
Happy, Randall and Swisher, Tex., 624 B2 118
Happy Camp, Siskiyou, Calif., 500 B2 82
Happy Jack, Coconino, Ariz., 350 C4 80
Hapsu, Kor F4 37
Ḩaql, Sau. Ar H10 31
Haque, Alachua, Fla., 120 C4 86
Harahan, Jefferson, La., 9,275 C7 95
Harald, isl., Sov. Un B32 4
Haralson, co., Ga., 14,543 C1 87
Hārar, Eth., 25,000 D5 47
Harardera, Som., 500 E6 47
Hārargē, reg., Eth., 1,600,000 D5 47
Hara Usa, lake, Mong B3 34
Harazé, Chad, 742 D4 46
Harbeson, Sussex, Del., 142 C7 85
Harbin, China, 1,814,000 B10 34, D3 37
Harbinger, Currituck, N.C., 100 A8 109
Harboöre, Den., 1,028 B2 24
Harbor, Curry, Oreg., 40 E2 113
Harbor Beach, Huron, Mich., 2,282 E7 98
Harborcreek, Erie, Pa., 800 B2 114
Harbor Isle, Nassau, N.Y., 1,300 *E3 108
Harbor Springs, Emmet, Mich., 1,433 C5 98
Harborton, Accomack, Va., 350 D7 121
Harbor View, Lucas, Ohio, 273 A2 111
Harbour, strait, B.C., Can B9 68
Harbour Breton, Newf., Can., 989 E4 75
Harbour Buffet, Newf., Can., 285 E4 75
Harbour Deep, Newf., Can., 200 C3 75
Harbour Grace, Newf., Can., 2,545 E5 75
Harbour Grace, dist., Newf., Can E5 75
Harbour Main, Newf., Can., 469 E5 75
Harbour Main-Bell Island, dist., Newf., Can E5 75
Harbour Mille, Newf., Can., 345 E4 75
Harbourton, Mercer, N.J., 50 C3 106
Harbourville, N.S., Can., 161 D5 76
Harbuck, Polk, Tenn., 50 D9 117
Harby, Den., 1,225 C4 24
Harcourt, N.B., Can., 231 C4 74
Harcourt, Webster, Iowa, 268 B3 92
Harcuvar, Yuma, Ariz D2 80
Harcuvar, mts., Ariz D2 80
Harda, India, 22,279 F6 40
Hardangerfjord, fjord, Nor H1 25
Hardangerjökelen, mtn., Nor G2 25
Hardangervidda, mts., Nor G2 25
Hardaway, Macon, Ala., 100 C4 78
Hardburly, Perry, Ky., 650 C6 94
Hardee, Issaquena, Miss., 40 C3 100
Hardee, co., Fla., 12,370 E5 86
Hardeeville, Beaufort and Jasper, S.C., 700 G5 115
Hardeman, co., Tenn., 21,517 B2 117
Hardeman, co., Tex., 8,275 B3 118
Hardenberg, Neth., 1,957 B6 15
Harden City, Pontotoc, Okla., 150 C5 112
Harderwijk, Neth., 17,700 B5 15
Hardesty, Texas, Okla., 187 D3 112
Hardin, Calhoun, Ill., 1,040 D3 90
Hardin, Marshall, Ky., 458 B3 94
Hardin, Ray, Mo., 727 B4 101
Hardin, Big Horn, Mont., 2,789 E9 102
Hardin, co., Ill., 5,879 F5 90
Hardin, co., Iowa, 22,533 B4 92
Hardin, co., Ky., 67,789 C4 94
Hardin, co., Ohio, 29,633 B4 111
Hardin, co., Tenn., 17,397 B3 117
Hardin, co., Tex., 24,629 D5 118
Harding, Norfolk, Mass., 200 B5, D2 97
Harding, Harding, S. Dak., 15 B2 116
Harding, Randolph, W. Va., 175 C5 123
Harding, co., N. Mex., 1,874 B5 107
Harding, co., S. Dak., 2,371 B2 116
Harding, lake, Ala., Ga C4 78
Hardingville, Gloucester, N.J D2 106
Hardinsburg, Washington, Ind., 218 H5 91
Hardinsburg, Breckinridge, Ky., 1,377 C3 94
Hardisty, Alta., Can., 582 C5 69
Hardisty, lake, N.W. Ter., Can D9 66
Hardman, Morrow, Oreg., 30 B7 113
Hardoi, India, 36,725 D8 40
Hardtner, Barber, Kans., 372 E5 93
Hardwar, India, 58,513 (★59,960) C7 40
Hardwick, Baldwin, Ga., 3,500 C3 87
Hardwick, Worcester, Mass., 200 (2,340▲) B3 97
Hardwick, Rock, Minn., 328 G2 99
Hardwick, Caledonia, Vt., 1,521 (2,349▲) B4 120
Hardwicke, N.B., Can., 131 B5 74
Hardy, Alg., 1,002 B5 44
Hardy, Sharp, Ark., 555 A4 81
Hardy, Sask., Can., 76 H3 70
Hardy, Pike, Ky., 854 C7 94
Hardy, Grenada, Miss., 125 B4 100
Hardy, Cascade, Mont., 25 C5 102
Hardy, Nuckolls, Nebr., 285 D8 103
Hardy, Kay, Okla., 6 A5 111
Hardy, co., W. Va., 9,308 B6 123
Hare, bay, Newf., Can C4 75
Hare, hill, Newf., Can E4 75
Hare Bay, Newf., Can., 1,467 D4 75
Harelson, East Baton Rouge, La., 150 B5 95
Haren, Neth., 3,787 A6 11

Harfleur, Fr., 10,514 C4 14
Harford, Susquehanna, Pa., 230 C10 114
Harford, co., Md., 76,722 A5 85
Hargeisa, Som., 53,000 D6 47
Hargeisa, dist., Som D6 47
Harghitei, mts., Rom B7 22
Hargill, Hidalgo, Tex., 750 F3 118
Hargrave, Man., Can., 51 E1 71
Hargrave, lake, Man., Can B2 71
Hargrave, riv., Man., Can B2 71
Hari, riv., Afg D10 41
Harihar, India, 22,829 F6 39
Harkers Island, Carteret, N.C., 1,362 C7 109
Harkus, isl., Sau. Ar H4 41
Harlan, Allen, Ind., 500 B8 91
Harlan, Shelby, Iowa, 4,350 C2 92
Harlan, Smith, Kans., 125 C5 93
Harlan, Harlan, Ky., 4,177 D6 94
Harlan, co., Ky., 51,107 D6 94
Harlan, co., Nebr., 5,081 D6 103
Harlan, res., Nebr D6 103
Harlau, Rom., 4,172 B8 22
Harlech, Wales B3 12
Harlem Hendry, Fla., 1,256 *F6 86
Harlem, Columbia, Ga., 1,423 C4 87
Harlem, Blaine, Mont., 1,267 B8 102
Harlem, riv., N.Y D5 106
Harleton, Harrison, Tex., 150 C5 118
Harley Dome, Grand, Utah, 5 D6 119
Harleyville, Dorchester, S.C., 561 E7 115
Harlingen, Neth., 12,000 A5 15
Harlingen, Somerset, N.J., 130 C3 106
Harlingen, Cameron, Tex., 41,207 F4 118
Harlow, Eng., 53,475 C8 12
Harlow, Benson, N. Dak., 90 A6 110
Harlowton, Wheatland, Mont., 1,734 D7 102
Härlunda, Swe B8 24
Harman, Randolph, W. Va., 128 C5 123
Harmancik, Tur C7 31
Harmans, Anne Arundel, Md., 300 B4 103
Harmarville, Allegheny, Pa., 2,000 A6 114
Harmersville, Salem, N.J., 300 E2 106
Harmon, Lee, Ill., 214 B4 90
Harmon, Red River, La., 80 B2 95
Harmon, Ellis, Okla., 15 A2 112
Harmon, co., Okla., 5,852 C2 112
Harmon, creek, W. Va A2 123
Harmon, riv., Alta., Can A2 69
Harmonsburg, Crawford, Pa., 300 C1 114
Harmony, Clay, Ind., 700 E3 91
Harmony, Somerset, Maine, 350 (712▲) D3 96
Harmony, Fillmore, Minn., 1,214 G6 99
Harmony, Warren, N.J., 125 B2 106
Harmony, Iredell, N.C., 322 B3 109
Harmony, Butler, Pa., 1,142 E1 114
Harmony, Providence, R.I., 800 B10 104
Harmony, Halifax, Va., 40 E4 121
Harmonyville, Windham, Vt., 50 E3 120
Harms, Lincoln, Tenn., 65 B5 117
Harned, Breckinridge, Ky., 375 C4 94
Harnett, co., N.C., 48,236 B5 109
Harney, Carroll, Md., 200 B3 85
Harney, co., Oreg., 6,744 D7 113
Harney, lake, Fla D6 86
Harney, lake, Oreg D7 113
Harney, peak, S. Dak D2 116
Harney, valley, Oreg D8 113
Härnösand, Swe., 17,200 F8 25
Haro, Sp., 8,554 A4 20
Haro, cape, Mex B2 63
Harold, Santa Rosa, Fla., 200 G2 86
Harper, Keokuk, Iowa, 177 C5 92
Harper, Harper, Kans., 1,899 E5 93
Harper, Lib., 5,000 F3 45
Harper, Malheur, Oreg., 100 D9 113
Harper, Gillespie, Tex., 300 D3 118
Harper, Kitsap, Wash., 500 D1 122
Harper, co., Kans., 9,541 E5 93
Harper, co., Okla., 5,956 A2 112
Harper, lake, Que., Can B4 73
Harpers Ferry, Allamakee, Iowa, 211 A6 92
Harpers Ferry, Jefferson, W. Va., 572 B7 123
Harpersville, Shelby, Ala., 667 B3 78
Harperville, Scott, Miss., 250 C4 100
Harper Woods, Wayne, Mich., 19,995 *A8 98
Harpeth, riv., Tenn A5 117
Harpster, Idaho, Idaho D3 89
Harpster, Wyandot, Ohio, 302 B4 111
Harptree, Sask., Can H3 70
Harquahala, mts., Ariz D2 80
Harrah, Oklahoma, Okla., 934 B4 112
Harrah, Yakima, Wash., 284 C5 122
Harray, bay, Scot A5 13
Harrell, Calhoun, Ark., 267 D3 81
Harricanaw, riv., Que., Can B2 73
Harrietta, Wexford, Mich., 119 D5 98
Harriettsville, Noble, Ohio, 100 C6 111
Harriman, Klamath, Oreg E4 113
Harriman, Roane, Tenn., 5,931 D9 117
Harrington, Kent, Del., 2,495 C6 85
Harrington, Washington, Maine, 500 (717▲) D5 96
Harrington, Lincoln, Wash., 575 B7 122
Harrington, lake, Maine C3 96
Harrington, sound, Bermuda E14 77
Harrington Harbour, Que., Can., 150 F21 67
Harrington Park, Bergen, N.J., 3,581 D5 106
Harriott, lake, Sask., Can A4 70
Harris, Sask., Can., 305 F2 70
Harris, Osceola, Iowa, 258 A2 92
Harris, Anderson, Kans., 36 D8 93
Harris, Chisago, Minn., 552 E6 99
Harris, Sullivan, Mo., 171 A4 101
Harris, McCurtain, Okla., 110 D7 112
Harris, Scot D2 13
Harris, Obion, Tenn., 120 A3 117
Harris, co., Ga., 11,167 D2 87
Harris, co., Tex., 1,243,158 E5 118
Harris, hill, Mass A3 97
Harris, lake, Fla D5 86
Harris, sound, Scot C1 13
Harrisburg, Poinsett, Ark., 1,907 B5 81
Harrisburg, Saline, Ill., 9,171 F5 90
Harrisburg, Boone, Mo., 124 B5 101
Harrisburg, Banner, Nebr., 100 C2 103
Harrisburg, Pickaway and Franklin, Ohio, 359 C2 111
Harrisburg, Linn, Oreg., 939 C3 113
Harrisburg, Dauphin, Pa., 79,697 (*257,600) F8 114
Harrisburg, Lincoln, S. Dak., 313 G9 116
Harris Grove, York, Va., 100 A6 121
Harris Hill, Erie, N.Y., 3,944 *C2 108
Harrismith, S. Afr., 13,753 C4 49
Harrison, Boone, Ark., 6,580 A2 81
Harrison, Washington, Ga., 209 D4 87
Harrison, Kootenai, Idaho, 249 B2 89
Harrison, Cumberland, Maine, 550 (1,014▲) D2 96
Harrison, Clare, Mich., 1,072 D6 98
Harrison, Madison, Mont., 151 E5 102
Harrison, Sioux, Nebr., 448 B2 103
Harrison, Hudson, N.J., 11,743 D5 106
Harrison, Westchester, N.Y., 19,201 D2 108
Harrison, Hamilton, Ohio, 3,878 C3 111
Harrison [Township] (Natrona Heights), Allegheny, Pa., 15,710 A7, E2 114
Harrison, Douglas, S. Dak., 80 D7 116
Harrison, Hamilton, Tenn., 200 E10 117
Harrison, co., Ind., 19,207 H5 91
Harrison, co., Iowa, 17,600 C2 92
Harrison, co., Ky., 13,704 B5 94
Harrison, co., Miss., 119,489 E4 100
Harrison, co., Mo., 11,603 A3 101
Harrison, co., Ohio, 17,995 B6 111
Harrison, co., Tex., 45,594 C5 118
Harrison, co., W. Va., 77,856 B4 123
Harrison, bay, Alsk A9 79
Harrison, cape, Newf., Can g10 75
Harrison, lake, B.C., Can E7 68
Harrisonburg, Catahoula, La., 594 C4 95
Harrisonburg (Independent City), Va., 11,916 C4 121
Harrison Hot Springs, B.C., Can., 475 A11 68
Harrison Valley, Potter, Pa., 375 C6 114
Harrisonville, Monroe, Ill., 100 E3 90
Harrisonville, Baltimore, Md., 180 B4 85
Harrisonville, Cass, Mo., 3,510 C3 101
Harrisonville, Gloucester, N.J., 250 D2 106
Harrisonville, Fulton, Pa., 30 G5 114
Harriston, Ont., Can., 1,631 D4 72
Harriston, Jefferson, Miss., 300 D2 100
Harrisville, Alcona, Mich., 487 D7 98
Harrisville, Simpson, Miss., 165 D3 100
Harrisville, Cheshire, N.H., 250 (459▲) E2 105
Harrisville, Lewis, N.Y., 842 A5 108
Harrisville, Harrison, Ohio, 343 B1 123
Harrisville, Butler, Pa., 896 D1 114
Harrisville, Providence, R.I., 1,024 B10 84
Harrisville, Weber, Utah B3 119
Harrisville, Ritchie, W. Va., 1,428 B3 123
Harrisville, Marquette, Wis., 100 E4 124
Harrod, Allen, Ohio, 563 B4 111
Harrods, creek, Ky A4 94
Harrodsburg, Monroe, Ind., 500 F4 91
Harrodsburg, Mercer, Ky., 6,061 C5 94
Harrogate, Eng., 56,332 D6 10
Harrold, Hughes, S. Dak., 255 C6 116
Harrop, lake, Man., Can C4 71
Harrow, Ont., Can., 1,787 E2 72
Harrow, Eng., 208,963 C7 12
Harrowby, Man., Can., 72 D1 71
Harrowsmith, Ont., Can., 469 C8 72
Harrys, riv., Newf., Can D2 75
Harshaw, Santa Cruz, Ariz F5 80
Harşova, Rom., 4,761 C8 22
Harstad, Nor., 3,900 C7 25
Hart, Sask., Can., 45 H3 70
Hart, Oceana, Mich., 1,990 E4 98
Hart, Castro, Tex., 577 B1 118
Hart, co., Ga., 15,229 B4 87
Hart, co., Ky., 14,119 C4 94
Hart, isl., Md B5 85
Hart, isl., N.Y B5 106
Hart, lake, Fla D5 86
Hart, lake, Oreg E7 113
Hart, mtn., Man., Can C1 71
Hartell, Alta., Can., 350 D3 69
Hartfield, Middlesex, Va., 200 D6 121
Hartford, Geneva, Ala., 1,956 D4 78
Hartford, Sebastian, Ark., 531 B1 81
Hartford, Hartford, Conn., 162,178 (*763,700) B6 84
Hartford, Madison, Ill., 2,355 E3 90
Hartford, Warren, Iowa, 271 B7, C4 92
Hartford, Lyon, Kans., 337 D8 93
Hartford, Ohio, Ky., 1,618 C3 94
Hartford, Oxford, Maine, 50 (325▲) D2 96
Hartford, Van Buren, Mich., 2,305 F4 98
Hartford, Burlington, N.J., 300 D3 106
Hartford, Washington, N.Y., 150 B7 108
Hartford, Minnehaha, S. Dak., 688 D9 116
Hartford, Cocke, Tenn., 100 D10 117
Hartford, Windsor, Vt., 450 (6,355▲) D4 120
Hartford, Mason, W. Va., 376 C3 123
Hartford, Washington, Wis., 5,627 E1, E5 124
Hartford, co., Conn., 689,555 B5 84
Hartford City, Blackford, Ind., 8,053 D7 91
Hartha, Ger., 8,522 B7 17
Hartington, Cedar, Nebr., 1,648 B8 103
Hartland, N.B., Can., 1,025 C2 74
Hartland, Somerset, Maine, 1,016 (1,447▲) D3 96
Hartland, Livingston, Mich., 200 F7 98
Hartland, Freeborn, Minn., 330 G5 99
Hartland, Ward, N. Dak., 35 A4 110
Hartland, Windsor, Vt., 200 (1,592▲) D4 120
Hartland, Waukesha, Wis., 2,088 E1, E5 124
Hartland, pt., Eng C3 12
Hartland Four Corners, Windsor, Vt., 125 D4 120
Hartlepool, Eng., 17,674 C6 10
Hartley, O'Brien, Iowa, 1,738 A2 92
Hartley, S. Rh., 2,900 A5 49
Hartley, Hartley, Tex., 185 B1 118
Hartley, co., Tex., 2,171 B1 118
Hartline, Grant, Wash., 206 B6 122
Hartly, Kent, Del., 164 B6 85
Hartman, Johnson, Ark., 299 B2 81
Hartman, Prowers, Colo., 164 C8 83
Hartney, Man., Can., 592 E1 71
Harts, pond, N.H C2 105
Hartsburg, Logan, Ill., 300 C4 90
Hartsburg, Boone, Mo., 158 C5 101
Hartsdale, Westchester, N.Y., 9,000 *D2 108
Hartsel, Park, Colo., 30 B5 83
Hartselle, Morgan, Ala., 5,000 A3 78
Hartsfield, Colquitt, Ga., 75 E3 87
Hartshorn, Texas, Mo., 135 D6 101
Hartshorne, Pittsburg, Okla., 1,903 C6 112
Hartsville, Bartholomew, Ind., 399 F6 91
Hartsville, Berkshire, Mass., 100 B1 97
Hartsville, Darlington, S.C., 6,392 C7 115
Hartsville, Trousdale, Tenn., 1,712 A5 117
Hartuv, Isr h11 32
Hartville, Stark, Ohio, 1,353 B6 111
Hartville, Wright, Mo., 486 D5 101
Hartville, Platte, Wyo., 177 C8 125
Hartwell, Hart, Ga., 4,599 B4 87
Hartwell, dam, Ga., S.C B4 87, C2 115
Hartwell, res., Ga., S.C B3 87, B1 115
Hartwick, Poweshiek, Iowa, 126 C5 92
Hartwick, Otsego, N.Y., 600 C5 108
Harty Station, Ont., Can., 87 *E8 72
Harut, riv., Afg E10 41
Harvard, McHenry, Ill., 4,248 A5 90
Harvard, Wayne, Iowa, 150 D4 92
Harvard, Worcester, Mass., 350 (2,563▲) A4, C1 97
Harvard, Clay, Nebr., 1,261 D7 103
Harvard, mtn., Colo C4 83
Harvel, Montgomery and Christian, Ill., 285 D4 90
Harvest, Madison, Ala., 300 A3 78
Harvey, N.B., Can., 142 D3 74
Harvey, Cook, Ill., 29,071 B6, F3 90
Harvey, Marion, Iowa, 270 C5 92
Harvey, Jefferson, La., 10,000 C7, E5 95
Harvey, Marquette, Mich., 350 B3 98
Harvey, Wells, N. Dak., 2,365 B6 110
Harvey, co., Kans., 25,865 D6 93
Harvey, creek, Pa B7 114
Harvey, lake, Pa A7 114
Harvey, mtn., Mass B1 97
Harveysburg, Warren, Ohio, 514 C3 111
Harvey Station, N.B., Can., 323 D3 74
Harveyton, Perry, Ky., 300 C6 94
Harveyville, Wabaunsee, Kans., 204 D8 93
Harviell, Butler, Mo., 177 E7 101
Harwich, Eng., 13,569 C9 12
Harwich, Barnstable, Mass., 800 (3,447▲) C7 97
Harwich Port, Barnstable, Mass., 800 (3,447▲) C7 97
Harwick, Allegheny, Pa., 1,500 *A6 114
Harwinton, Litchfield, Conn., 500 (3,344▲) B4 84
Harwood, Ont., Can., 185 C6 72
Harwood, Anne Arundel, Md., 125 C4 85
Harwood, Vernon, Mo., 89 D3 101
Harwood, Cass, N. Dak., 100 C9 110
Harwood Heights, Cook, Ill., 5,688 *B6 90
Harz, mts., Ger B5 17
Harzgerode, Ger., 6,202 B6 17
Hasā, Jordan, 1,000 D7 32
Hasa, reg., Sau. Ar *H4 41
Hasan Kiādeh, Iran E5 41
Hasbrouck Heights, Bergen, N.J., 13,046 B4 106
Hasdo, riv., India F9 40
Hase, riv., Ger B3 16
Haselünne, Ger., 5,200 B3 16
Hashimoto, Jap., 16,700 o14 37
Hasi Zegdou, Alg D4 44
Haskell, Saline, Ark., 215 C3 81
Haskell, Passaic, N.J. (part of Wanaque) A4 106
Haskell, Musogee, Okla., 1,887 B6 112
Haskell, Haskell, Tex., 4,016 C3 118
Haskell, co., Kans., 2,990 E3 93
Haskell, co., Okla., 9,121 B6 112
Haskell, co., Tex., 11,174 C3 118
Haskins, Wood, Ohio, 521 A2, A4 111
Haskins, Benton, Oreg C3 113
Haslach, Aus., 2,565 E9 17
Haslam, Shelby, Tex., 100 D5 118
Hasle, Den., 1,487 C8 24
Haslemere, Eng., 12,528 C7 12
Haslet, Tarrant, Tex., 200 A5 118
Haslett, Ingham, Mich., 1,500 *F6 98
Haslev, Den., 6,155 C6 24
Hassan, India, 32,172 F6 39
Hassayampa, Maricopa, Ariz., 10 D3 80
Hassayampa, creek, Ariz D3 80
Hassayampa, riv., Ariz F1 80
Hassell, Martin, N.C., 147 B6 109
Hasselt, Bel., 36,618 D5 15
Hasselt, Neth., 2,800 B6 15
Hassfurt, Ger., 6,800 C5 17
Hassleben, Ger., 624 E7 24
Hässleholm, Swe., 13,500 I5 25
Hassloch, Ger., 15,400 D3 17
Hastings, Ont., Can., 897 C7 72
Hastings, Eng., 66,346 D8 12
Hastings, St. Johns, Fla., 617 C5 86
Hastings, Mills, Iowa, 260 C2 92
Hastings, Barry, Mich., 6,375 F5 98
Hastings, Dakota and Washington, Minn., 8,965 F6, F7 99
Hastings, Adams, Nebr., 21,412 D7 103
Hastings, N.Z., 23,383 (*32,490) M16 51
Hastings, Barnes, N. Dak., 106 C7 110
Hastings, Jefferson, Okla., 200 C3 112
Hastings, Cambria, Pa., 1,751 E4 114
Hastings, Brazoria, Tex., 350 F5 118
Hastings, Wetzel, W. Va., 175 A6 123
Hastings, co., Ont., Can., 93,377 C7 72
Hastings-on-Hudson, Westchester, N.Y., 8,979 D1 108
Hästveda, Swe., 1,109 B7 24
Hasty, Newton, Ark., 65 A2 81
Hasty, Bent, Colo., 180 C8 83
Haswell, Kiowa, Colo., 169 C7 83
Hatboro, Montgomery, Pa., 7,315 F11 114
Hatch, Dona Ana, N. Mex., 888 E2 107
Hatch, Garfield, Utah, 198 F3 119
Hatchechubbee, Russell, Ala., 250 C4 78
Hatchet, creek, Ala C3 78
Hatchie, riv., Tenn B2 117
Hatchineha, lake, Fla D5 86
Hatchville, Barnstable, Mass., 250 C6 97
Hat Creek, Shasta, Calif B3 82
Hat Creek, Niobrara, Wyo., 5 C8 121
Hateg, Rom., 3,853 C6 22
Hatfield, Polk, Ark., 337 C1 81
Hatfield, Sask., Can., 15 F3 70
Hatfield, Spencer, Ind., 500 I3 91
Hatfield, Hampshire, Mass., 1,330 (2,350▲) B2 97
Hatfield, Harrison, Mo., 75 A3 101
Hatfield, Montgomery, Pa., 1,941 F11 114
Hatfield, S. Rh., 8,500 A5 49
Hathaway, Rosebud, Mont., 27 D10 102
Hathaway, Lake, Tenn A2 117
Hathorn, Marion, Miss D4 100
Hathorne, Essex, Mass. (part of Danvers) C3 97
Hathras, India, 64,045 D7 40
Ha Tien, Viet., 5,000 G6 38
Hatillo, P.R., 2,582 B3 65
Hatillo, mun., P.R., 20,238 B3 65
Ha Tinh, Viet., 5,000 C6 38
Hatley, Crisp, Ga., 60 E3 87
Hatley, Marathon, Wis., 306 D4 124
Hato Mayor, Dom. Rep., 5,775 F9 64
Hato Rey, P.R. (part of San Juan) B6 65
Hato Viejo, P.R B5 65
Hatta, India, 9,117 E7 40
Hattem, Neth., 5,300 B6 15
Hatteras, Dare, N.C., 700 B8 109
Hatteras, cape, N.C B8 109
Hatteras, inlet, N.C B8 109
Hattfjelldal, Nor., 159 E5 25
Hattiesburg, Forrest, Miss., 34,989 D4 100
Hattieville, Conway, Ark., 120 B3 81
Hatton, Lawrence, Ala., 250 A2 78
Hatton, Polk, Ark., 100 C1 81
Hatton, Sask., Can., 57 G1 70
Hatton, Traill, N. Dak., 856 B8 110
Hatton, Millard, Utah E3 119
Hatton Fields, Monterey, Calif., 2,362 *D3 82
Hatvan, Hung., 19,952 B4 22
Hatzic, B.C., Can., 724 A10 68
Haubourdin, Fr., 12,610 D2 15
Haubstadt, Gibson, Ind., 1,029 H2 91
Haugan, Mineral, Mont., 100 C1 102
Haugen, Barron, Wis., 265 C2 124
Haugesund, Nor., 27,100 H1 25
Haughton, Bossier, La., 611 B2 95
Haultain, riv., Sask., Can A2 70
Hauppauge, Suffolk, N.Y., 1,500 *D3 108
Hauraki, gulf, N.Z L15 51
Hausach, Ger., 3,600 E4 16
Hauser, Coos, Oreg., 250 D2 113
Hauser, dam, Mont D5 102
Hausstock, mtn., Switz C7 19
Haut, isl., Maine D4 96
Haute-Garonne, dept., Fr., 594,633 *F4 14
Haute-Loire, dept., Fr., 211,036 *E5 14
Haute-Marne, dept., Fr., 208,446 B1 18
Hautes-Alpes, dept., Fr., 85,436 E2 18
Haute-Saône, dept., Fr., 208,440 B2 18
Haute-Savoie, dept., Fr., 329,230 D2 18
Hautes-Pyrénées, dept., Fr., 211,433 *F4 14
Haute-Vienne, dept., Fr., 332,514 *E4 14
Hautmont, Fr., 18,594 B6 14
Haut-Rhin, dept., Fr., 547,920 A3 18
Hauula, Honolulu, Haw., 950 B4, f10 88
Hauzenberg, Ger., 2,700 E8 17
Havana, Hale, Ala., 100 C2 78
Havana, Yell, Ark., 277 B2 81
Havana (La Habana), Cuba, 785,455 (*1,217,674) D2, h11 64
Havana, Gadsden, Fla., 2,090 B2 86
Havana, Mason, Ill., 4,363 C3 90
Havana, Montgomery, Kans., 162 E8 93
Havana, Sargent, N. Dak., 206 D8 110
Havana, bay, Cuba h12 64
Havasu, creek, Ariz A3 80
Havasu, lake, Ariz., Calif B1 80, E6 82
Havel, riv., Ger A7 17
Havelberg, Ger., 7,027 F6 24
Havelländischer Grosser Haupt, canal, Ger F6 24
Havelock, Ont., Can., 1,260 C7 72
Havelock, Pocahontas, Iowa, 289 B3 92
Havelock, Craven, N.C., 2,433 C7 109
Havelock, Hettinger, N. Dak., 50 C3 110
Haven, Reno, Kans., 982 B4 93
Haven, Sheboygan, Wis., 50 B6 124
Havensville, Pottawatomie, Kans., 166 C7 93
Haverford, Delaware and Montgomery, Pa., 5,000 *B11 114
Haverfordwest, Wales, 8,872 C3 12
Haverhill, Eng., 5,446 D8 12
Haverhill, Palm Beach, Fla., 442 *F6 86
Haverhill, Butler, Kans., 25 B6 93
Haverhill, Essex, Mass., 46,346 A5 97
Haverhill, Grafton, N.H., 300 (3,127▲) B2 105
Haversham, Washington, R.I., 80 D10 84
Haverstraw, Rockland, N.Y., 5,771 D2, D7 108
Havertown (Llanerch), Delaware, Pa., 35,000 B11 114
Haviland, Kiowa, Kans., 725 E4 93
Haviland, Paulding, Ohio, 235 A3 111
Havlickuv Brod, Czech., 14,900 D3 26
Havre, see Le Havre, Fr.
Havre, Hill, Mont., 10,740 B7 102
Havre Aubert, Que., Can., 494 B8 74
Havre Boucher, N.S., Can., 504 D8 74
Havre de Grace, Harford, Md., 8,510 A5 85
Havre St. Pierre, Que., Can., 2,407 h9 75
Havza, Tur., 4,539 B10 31
Haw, riv., N.C B4 109
Hawaii, co., Haw., 61,332 D6 88
Hawaii, state, U.S., 632,772 F7 76, 88
Hawaiian Gardens, Los Angeles, Calif., 15,000 *F4 82
Hawarden, Sask., Can., 268 F2 70
Hawarden, Sioux, Iowa, 2,544 A1 92
Hawdon, lake, Austl H2 51
Hawea, lake, N.Z P12 51
Hawera, N.Z., 7,542 M15 51
Hawes, Garland, Ark., 100 C5 81
Hawes, Eng., 1,196 F6 13
Hawesville, Hancock, Ky., 882 C3 94
Hawes Water, lake, Eng F6 13
Hawi, Hawaii, Haw., 800 C6 88
Hawick, Scot., 16,204 E6 13
Hawke, bay, N.Z M16 51
Hawke, isl., Newf., Can B4 75
Hawke, riv., Newf., Can B3 75
Hawke Harbour, Newf., Can., 15 B4, h11 75
Hawker, Austl., 368 E2 51
Hawkesbury, Ont., Can., 8,661 B10 72
Hawkesbury, isl., B.C., Can C3 68
Hawkestone, Ont., Can., 202 C5 72
Hawkeye, Fayette, Iowa, 516 B6 92
Hawkins, Bannock, Idaho G6 89
Hawkins, Rusk, Wis., 402 C3 124
Hawkins, co., Tenn., 30,468 C10 117
Hawkins, peak, Utah F2 119
Hawkinsville, Pulaski, Ga., 3,967 D3 87
Hawk Point, Lincoln, Mo., 270 C6 101
Hawk Run, Clearfield, Pa., 850 E3 114
Hawksbill, mtn., Va C4 121
Hawkshaw, N.B., Can., 67 D2 74
Hawley, Clay, Minn., 1,270 D2 99
Hawley, Wayne, Pa., 1,433 D11 114
Hawley, Jones, Tex., 200 C3 118
Hawleyville, Fairfield, Conn., 150 D3 84
Haworth, Bergen, N.J., 3,215 D5 106
Haworth, McCurtain, Okla., 351 D7 112
Haw River, Alamance, N.C., 1,410 A4 109
Hawsh 'Īsá, Eg. U.A.R., 1,000 D2 32
Hawthorn, Washington, Ala., 35 D1 78
Hawthorn, Clarion, Pa., 612 D3 114
Hawthorne, Los Angeles, Calif., 33,035 F2 82
Hawthorne, Alachua, Fla., 1,167 C4 86
Hawthorne, Mineral, Nev., 2,838 E3 104
Hawthorne, Passaic, N.J., 17,735 B4 106
Hawthorne, Westchester, N.Y., 4,000 D2 108
Haxby, Garfield, Mont., 5 C10 102
Haxtun, Phillips, Colo., 990 A8 83
Hay, Austl., 3,133 G5 51
Hay, Wales, 1,321 B4 12
Hay, Whitman, Wash., 75 C8 122
Hay, isl., Ont., Can C4 72
Hay, riv., Austl D6 50
Hay, riv., Alta., N.W. Ter., Can E9 66
Hayange, Fr., 11,009 C7 14
Hayden, Blount, Ala., 187 B3 78
Hayden, Gila, Ariz., 1,760 E5 80
Hayden, Routt, Colo., 764 A3 83
Hayden, Jennings, Ind., 275 G6 91
Hayden, Union, N. Mex., 15 B6 107
Hayden, lake, Idaho B2 89
Hayden, lake, Maine D3 96
Haydenburg, Jackson, Tenn., 50 C8 117
Hayden Junction, Pinal, Ariz., 55 E5 80
Hayden Lake, Kootenai, Idaho, 247 B2 89
Haydens, Hartford, Conn., 125 B6 84
Haydenville, Hampshire, Mass., 750 B2 97
Haydenville, Hocking, Ohio, 800 C5 111
Hayes, Calcasieu, La., 800 D3 95
Hayes, Stanley, S. Dak., 21 C4 116
Hayes, co., Nebr., 1,919 D4 103
Hayes, mtn., Alsk C10 79
Hayes, riv., Man., Can B5 71
Hayes Center, Hayes, Nebr., 283 D4 103
Hayesville, Clay, N.C., 428 D2 109
Hayesville, Ashland, Ohio, 435 B5 111
Hayesville, Marion, Oregon, 4,568 *C4 113
Hayfield, Man., Can., 90 E1 71
Hayfield, Hancock, Iowa, 150 A4 92
Hayfield, Dodge, Minn., 889 G6 99
Hayfield, Frederick, Va., 150 B4 121
Hayford, Spokane, Wash., 350 D7 122
Hayfork, Trinity, Calif., 400 B2 82
Hay Lakes, Alta., Can., 233 C4 69
Haylow, Echols, Ga., 100 F4 87
Haymana, Tur., 2,791 C9 31
Haymarket, Prince William, Va., 257 B4, C5 121
Haymock, lake, Maine B3 96
Haymond, Franklin, Ind., 120 F7 91
Haymond, Brewster, Tex D1 118
Haynau, see Chojnow, Pol.
Haynes, Lee, Ark., 200 C5 81
Haynes, Alta., Can., 65 C4 69
Haynes, Sampson, N.C B5 103
Haynes, Adams, N. Dak., 111 D3 110
Haynesville, Claiborne, La., 3,031 B2 95
Haynesville, Aroostook, Maine, 100 (187▲) C5 96
Haynesville, Lowndes, Ala., 900 C3 78
Hay River, N.W. Ter., Can., 1,338 D9 66
Hays, Ellis, Kans., 11,947 D4 93
Hays, Blaine, Mont., 400 B8 102
Hays, Wilkes, N.C., 200 A2 109
Hays, co., Tex., 19,934 D3 118
Haysi, Dickenson, Va., 485 B2 122
Hay Springs, Sheridan, Nebr., 823 B3 103
Haystack, mtn., Okla B2 112
Haystack, mtn., Vt F3 120
Haystack, mtn., Vt B3 120
Haystack, peak, Utah D2 119
Haysville, Dubois, Ind., 500 H4 91
Haysville, Sedgwick, Kans., 5,836 B5 93
Haysville, Macon, Tenn., 100 C7 117
Hayter, Alta., Can., 50 C5 69
Hayti, Pemiscot, Mo., 3,737 E8 101
Hayti, Hamlin, S. Dak., 425 C8 116
Hayton, Calumet, Wis., 95 B6 124
Hayward, Alameda, Calif., 72,700 B5 82
Hayward, Freedom, Minn., 258 G5 99
Hayward, Garfield, Okla., 40 A4 112
Hayward, Sawyer, Wis., 1,540 B2 124
Haywood, Man., Can., 146 E2 71
Haywood, Chatham, N.C., 713 B4 109
Haywood, co., N.C., 39,711 D3 109
Haywood, co., Tenn., 23,393 B2 117
Hazard, Perry, Ky., 5,958 C6 94
Hazard, Sherman, Nebr., 104 C6 103
Hazardville, Hartford, Conn., 4,000 B6 84
Hazaribagh, India, 40,958 E10 40
Hazaribagh, range, India F9 40
Hazebrouck, Fr., 17,446 B5 14
Hazel, Calloway, Ky., 342 B3 94
Hazel, Hamlin, S. Dak., 128 C8 116
Hazel Crest, Cook, Ill., 6,205 F3 90
Hazel Dell, Cumberland, Ill., 130 D5 90
Hazel Dell, Clark, Washington, 2,500 *D3 122

Hazel Green, Madison, Ala., 125 ....A3 78
Hazel Green, Wolfe, Ky., 259 ....C6 94
Hazel Green, Grant, Wis., 807 ....F3 124
Hazel Hurst, McKean, Pa., 500 ....C4 114
Hazelhurst, Oneida, Wis., 225 ....C4 124
Hazel Park, Oakland, Mich., 25,631 ....F7 98
Hazel Run, Yellow Medicine, Minn., 115 ....F3 99
Hazelton, B.C., Can., 410 ....B4 68
Hazelton, Jerome, Idaho, 433 ....G4 89
Hazelton, Barber, Kans., 246 ....E5 93
Hazelton, Emmons, N. Dak., 451 ....C5 111
Hazelton, peak, Wyo A5 125
Hazelwood, St. Louis, Mo., 6,045 ....*C7 101
Hazelwood, Haywood, N.C., 1,925 ....D3 109
Hazelwood, King, Wash D2 122
Hazen, bay, Alsk C6 79
Hazen, Prairie, Ark., 1,456 ....C4 81
Hazen, Churchill, Nev., 50 ....D2 104
Hazen, Mercer, N. Dak., 1,222 ....B4 110
Hazenmore, Sask., Can., 149 ....H2 70
Hazlehurst, Jeff Davis, Ga., 3,699 ....E4 87
Hazlehurst, Copiah, Miss., 3,400 ....D3 100
Hazle Patch, Laurel, Ky., 200 ....C5 94
Hazlet, Sask., Can., 202 ....G1 70
Hazlet, Monmouth, N.J., 9,000 ....*C4 106
Hazleton, Gibson, Ind., 507 ....H2 91
Hazleton, Buchanan, Iowa, 665 ....B6 92
Hazleton, Luzerne, Pa., 32,056 ....E10 114
Hazor, Isr., 423 ....h9 32
Hazy, Raleigh, W. Va. D6 123
Headford, Ire., 567 ....D2 11
Head Harbor, isl. Maine D5 96
Headland, Henry, Ala., 2,650 ....D4 78
Headlight, Clinch, Ga., 6 ....F4 87
Head of Island, Livingston, La., 150 ....B6 95
Headquarters, Clearwater, Idaho, 300 ....C3 89
Headrick, Jackson, Okla., 152 ....C2 112
Heads, cape, Oreg E2 113
Heafford Junction, Lincoln, Wis., 100 ....C4 124
Healdsburg, Sonoma, Calif., 4,816 ....A4, C2 82
Healdton, Carter, Okla., 2,898 ....C4 112
Healdville, Rutland, Vt., 60 ....E3 120
Healing Springs, Bath, Va., 200 ....D3 121
Healy, Lane, Kans., 228 ....D3 93
Healy Fork, Alsk., 20 ....C10 79
Heanza, isl., Okinawa 52
Heard, co., Ga., 5,333 ....C1 87
Heard, isl., Indian O. J1 2
Heardmont, Elbert, Ga., 175 ....B4 87
Hearne, Sask., Can., 53 ....G3 70
Hearne, Robertson, Tex., 5,072 ....D4 118
Hearst, Ont., Can., 2,373 ....E9 72
Hearst, isl., Ant C6 5
Heart, hill, Sask., Can H5 70
Heart, lake, Wyo A2 125
Heart, riv., Alta., Can B2 69
Heart, riv., N. Dak C4 110
Heart Butte, Pondera, Mont., 100 ....B4 102
Heart's Content, Newf., Can., 607 ....E5 75
Heartstone, mtn., Md A2 85
Heartwell, Kearney, Nebr., 113 ....D7 103
Heartwellville, Bennington, Vt., 50 ....F2 120
Heaters, Braxton, W. Va., 90 ....C4 123
Heath, Franklin, Mass., 100 (304▲) A2 97
Heath, Fergus, Mont., 10 ....C7 102
Heath, Licking, Ohio, 2,426 ....*B5 111
Heatherton, N.S., Can., 236 ....D8 74
Heath Springs, Lancaster, S.C., 832 ....B6 115
Heathsville, Northumberland, Va., 225 ....D6 121
Heaton, Avery, N.C., 250 ....A2, C4 109
Heaton, Wells, N. Dak., 100 ....B6 110
Heavener, LeFlore, Okla., 1,891 ....C7 112
Hebardsville, Ware, Ga., 2,758 ....E4 87
Hebbronville, Jim Hogg, Tex., 3,987 ....F3 118
Hebel, Austl., 81 ....D6 51
Heber, Navajo, Ariz., 400 ....C5 80
Heber, Wasatch, Utah, 2,936 ....C4 119
Heber Springs, Cleburne, Ark., 2,265 ....B3 81
Hebert, Caldwell, La., 70 ....B3 95
Hebertville, Que., Can., 1,604 ....A6 73
Hebertville Station, Que., Can., 1,257 ....*A6 73
Hebgen, res., Mont F5 102
Hebo, Tillamook, Oreg., 200 ....B3 113
Hebrides, sea, Scot C2 13
Hebron, Newf., Can., 189 ....f9 75
Hebron, N.S., Can., 449 ....F3 74
Hebron, Tolland, Conn., 200 (1,819▲) C7 84
Hebron, McHenry, Ill., 701 ....A5 90
Hebron, Porter, Ind., 1,401 ....B3 91
Hebron, see Al Khalil, Jordan
Hebron, Boone, Ky., 300 ....A7 94
Hebron, Wicomico, Md., 754 ....D6 85
Hebron, Thayer, Nebr., 1,920 ....D8 103
Hebron, Grafton, N.H., 30 (153▲) C3 105
Hebron, Morton, N. Dak., 1,340 ....C3 110
Hebron, Licking, Ohio, 1,260 ....C5 111
Hebron, Denton, Tex., 80 ....A5 118
Hebron, Dinwiddie, Va., 20 ....D5 121
Hebron, Pleasants, W. Va., 100 ....B3 123
Heby, Swe., 2,151 ....t34 25
Hecate, strait, B.C., Can F6 66
Hecelchakán, Mex., 3,399 ....C6 63
Hechingen, Ger., 9,600 ....D4 16
Hechtel, Bel., 4,016 ....C5 15
Hecker, Monroe, Ill., 313 ....E3 90
Hecklingen, Ger., 6,845 ....B6 17
Hecla, Hooker, Nebr., 4 ....B4 103
Hecla, Brown, S. Dak., 444 ....B7 116
Hecla, Man., Can., 66 ....D3 71
Hecla, isl., Man., Can D3 71
Hectanooga, N.S., Can., 180 ....E3 74
Hector, Pope, Ark., 200 ....B3 81
Hector, Renville, Minn., 1,297 ....F4 99
Hede, Swe., 385 ....F5 25
Hedemora, Swe., 6,000 ....G6 25
Hedensö, Swe t34 25
Hedensted, Den., 1,717 ....C3 24
He Devil, mtn., Idaho D2 89
Hedgesville, Wheatland, Mont., 40 ....D7 102
Hedgesville, Berkeley, W. Va., 342 ....B7 123
Hedley, B.C., Can., 425 ....E7 68
Hedley, Donley, Tex., 494 ....B2 118
Hedmark, co., Nor., 177,200 ....*G4 25
Hedo, Okinawa 52
Hedo, pt., Okinawa 52
Hedon, Eng., 2,238 ....A7 12
Hedrick, Warren, Ind., 60 ....D2 91
Hedrick, Keokuk, Iowa, 762 ....C5 92
Hedwig Village, Harris, Tex., 1,182 ....*E5 118
Heek, Ger., 3,500 ....A2 17
Heel, pt., Wake Isl 52
Heeney, Summit, Colo B4 83
Heerde, Neth., 2,232 ....B6 15
Heerenveen, Neth., 8,400 (25,900▲) B5 15
Heerlen, Neth., 72,400 (*240,000) D5 15
Heessen, Ger., 17,100 ....B2 17
Heflin, Cleburne, Ala., 2,400 ....B4 78
Heflin, Webster, La., 289 ....B2 95
Hegeler, Vermilion, Ill., 1,640 ....*C6 90
Hegins, Schuylkill, Pa., 800 ....E9 114
Heglar, Cassia, Idaho G5 89
Heiberger, Perry, Ala., 50 ....C2 78
Heide, Ger., 20,000 ....A4 16
Heidelberg, Ger., 125,300 ....D3 17
Heidelberg, Jasper, Miss., 1,049 ....D5 100
Heidelberg, Allegheny, Pa., 2,118 ....*F1 114
Heidelberg, S. Afr., 9,292 ....D3 49
Heiden, Switz., 3,158 ....B8 19
Heidenau, Ger., 19,400 ....C8 17
Heidenheim, Ger., 48,800 ....E5 17
Heil, Grant, N. Dak., 100 ....C4 110
Heilbron, S. Afr., 7,182 ....C4 49
Heilbronn, Ger., 89,100 (*125,000) D4 17
Heiligenblut, Aus., 1,195 ....B8 18
Heiligenhafen, Ger., 8,900 ....D4 24
Heiligenstadt, Ger., 12,500 ....B5 17
Heilsberg, see Lidzbark Warminski, Pol.
Heilungkiang, prov., China, 14,860,000 ....B10 34
Heilwood, Indiana, Pa., 600 ....E4 114
Heimbach, Ger., 2,000 ....D6 15
Heimdal, Wells, N. Dak., 130 ....B6 110
Heinola, Fin., 11,000 ....G12 25
Heinsberg, Ger., 4,700 ....C6 15
Heisburg, Alta., Can., 135 ....C5 69
Heiskell, Knox, Tenn., 175 ....E11 117
Heisler, Alta., Can., 214 ....C4 69
Heislerville, Cumberland, N.J., 600 ....E3 106
Heizer, Barton, Kans., 90 ....D5 93
Hejaz, pol. div., Sau. Ar., 2,000,000 ....D7 43
Hejls, Den., 292 ....C3 24
Hekla, peak, Ice n23 25
Hekura, isl., Jap H8 37
Hel, Pol., 1,400 ....A5 26
Helagsfjaelet, mtn., Swe F5 25
Helbra, Ger., 9,500 ....B6 17
Heldt, Goshen, Wyo D8 125
Helena, Shelby, Ala., 523 ....B3 78
Helena, Phillips, Ark., 11,500 ....C5 81
Helena, Telfair, Ga., 1,290 ....D4 87
Helena, Jackson, Miss E3 100
Helena, Andrew, Mo., 166 ....B3 101
Helena, Lewis and Clark, Mont., 20,227 ....D5 102
Helena, St. Lawrence, N.Y., 160 ....A2 108
Helena, Sandusky, Ohio, 281 ....A4 111
Helena, Alfalfa, Okla., 580 ....A3 112
Helena, Newberry, S.C., 497 ....C4 115
Helensburgh, Scot., 9,605 ....D4 13
Helenwood, Scott, Tenn., 300 ....C9 117
Helgasjön, lake, Swe B8 24
Helgean, riv., Swe B8 24
Helgeroa, Nor., 331 ....p28 25
Helgoland, bay, Ger C3 16
Helgoland, isl., Ger D1 24
Helix, Umatilla, Oreg., 148 ....B8 113
Hellam, York, Pa., 1,234 ....G8 114
Hellenthal, Ger., 4,900 ....D6 15
Hellertown, Northampton, Pa., 6,716 ....E11 114
Hellier, Pike, Ky., 104 ....C7 94
Hellín, Sp., 12,900 ....C5 20
Hellville, Malag., 9,321 ....f9 49
Helm, Fresno, Calif., 125 ....D3 82
Helm, Washington, Miss B3 100
Helmand, riv., Afg F12 41
Helmbrechts, Ger., 8,300 ....C6 17
Helmer, Latah, Idaho, 30 ....C2 89
Helmer, Steuben, Ind., 110 ....A7 91
Helmeringhausen, S.W. Afr C2 49
Helmetta, Middlesex, N.J., 779 ....C4 106
Helmond, Neth., 43,400 ....C5 15
Helmsburg, Brown, Ind., 180 ....F5 91
Helmsdale, Scot B5 13
Helmsdale, riv., Scot B5 13
Helmsley, Eng., 1,292 ....F7 13
Helmstedt, Ger., 29,600 ....A5 17
Helmville, Powell, Mont., 50 ....D4 102
Helnaes, isl., Den C3 24
Heloise, Dyer, Tenn A2 117
Helotes, Bexar, Tex., 300 ....B3 118
Helper, Carbon, Utah, 2,459 ....D5 119
Helsingfors, see Helsinki, Fin.
Helsingör, Den., 26,658 (*32,636) B6 24
Helsinki, Fin., 452,800 (*545,000) G11 25
Helston, Eng., 7,085 ....D1 12
Heltonville, Lawrence, Ind., 400 ....G5 91
Helvecia, Arg., 3,390 ....A4 54
Helvetia Mines, Clearfield, Pa., 160 ....D4 114
Helvick, head, Ire E4 11
Hemaruka, Alta., Can., 40 ....D5 69
Hematite, Jefferson, Mo., 204 ....C7 101
Hemau, Ger., 2,600 ....D6 17
Hemel Hempstead, Eng., 55,164 ....C7 12
Hemet, Riverside, Calif., 5,416 ....F4, F5 82
Hemet East, Riverside, Calif., 1,936 ....*F5 82
Hemford, N.S., Can., 248 ....E5 74
Hemingford, Box Butte, Nebr., 904 ....B2 103
Hemingway, Williamsburg, S.C., 951 ....D1 115
Hemlock, Howard, Ind D5 91
Hemlock, Saginaw, Mich., 900 ....E6 98
Hemlock, Livingston, N.Y., 350 ....C3 108
Hemlock, Ashe, N.C A2 109
Hemlock (Eureka), Chester, S.C., 1,423 ....B5 115
Hemlock, res., Conn E3 84
Hemmingford, Que., Can., 778 ....D4 73
Hemnesberget, Nor., 1,212 ....D5 25
Hemp, Fannin, Ga., 40 ....B2 87
Hemphill, Sabine, Tex., 913 ....D6 118
Hemphill, co., Tex., 3,185 ....B2 118
Hemple, Clinton, Mo., 100 ....B3 101
Hempstead, Nassau, N.Y., 34,641 ....D2 108
Hempstead, Waller, Tex., 1,505 ....D4 118
Hempstead, co., Ark., 19,661 ....D2 81
Hempstead, hbr., N.Y F2 84
Henares, riv., Sp B4 20
Henau, Switz., 7,828 ....B7 19
Hendaye, Fr., 7,204 ....F3 14
Hendek, Tur., 9,900 ....B8 31
Henderson, Pike, Ala., 60 ....D3 78
Henderson, Arg., 3,928 ....B4 54
Henderson, Adams, Colo., 280 ....B6 83
Henderson, Mills, Iowa, 191 ....C2 92
Henderson, Henderson, Ky., 16,892 ....C2 94
Henderson, Caroline, Md., 129 ....B6 85
Henderson, Shiawassee, Mich., 250 ....E6 98
Henderson, Sibley, Minn., 728 ....F5 99
Henderson, York, Nebr., 730 ....D8 103
Henderson, Clark, Nev., 12,525 ....H7 104
Henderson, Jefferson, N.Y., 257 ....B4 108
Henderson, Vance, N.C., 12,740 ....A5 109
Henderson, Chester, Tenn., 2,691 ....B3 117
Henderson, Rusk, Tex., 9,666 ....C5 118
Henderson, Mason, W. Va., 601 ....C2 123
Henderson, co., Ky., 33,519 ....C2 94
Henderson, co., Ill., 8,237 ....C3 90
Henderson, co., N.C., 36,163 ....D3 109
Henderson, co., Tenn., 16,115 ....B3 117
Henderson, co., Tex., 21,786 ....C5 118
Henderson, pt., Miss E2 100
Hendersonville, Henderson, N.C., 5,911 ....D3 109
Hendersonville, Colleton, S.C., 200 ....F6 115
Hendersonville, Sumner, Tenn., 950 ....A5, E9 117
Hendon, Sask., Can., 111 ....E4 70
Hendon, Eng., 151,500 ....k12 10, C7 12
Hendon, Bledsoe, Tenn D8 117
Hendricks, Lincoln, Minn., 797 ....F2 99
Hendricks, Wilkes, N.C A2 109
Hendricks, co., Ind., 40,896 ....E4 85
Hendrix, San Juan, N. Mex A2 107
Hendrum, Norman, Minn., 305 ....C2 99
Hendry, co., Fla., 8,119 ....F5 86
Henefer, Summit, Utah, 408 ....B4 119
Hengchun, For., 5,000 ....*G9 34
Hengelo, Neth., 62,000 ....B6 15
Henghsien, China, 18,000 ....G6 34
Hengshan, China F3 36
Hengshan, China K5 36
Hengyang, China, 235,000 ....K5 36
Hénin-Liétard, Fr., 25,527 ....D2 15
Henley Harbour, Newf., Can., 80 ....B4, h11 75
Henley-on-Thames, Eng., 9,131 ....C7 12
Henlopen, cape, Del C7 85
Hennebont, Fr., 7,623 ....D2 14
Hennef, Ger., 13,200 ....C2 17
Hennepin, Putnam, Ill., 391 ....B4 90
Hennepin, Garvin, Okla., 325 ....C4 112
Hennepin, co., Minn., 842,854 ....E5 99
Hennessey, Kingfisher, Okla., 1,228 ....A4 112
Henniker, Merrimack, N.H., 850 (1,636▲) D3 105
Henning, Vermilion, Ill., 271 ....C6 90
Henning, Otter Trail, Minn., 980 ....D3 99
Henning, Lauderdale, Tenn., 466 ....B2 117
Henribourg, Sask., Can., 73 ....D3 70
Henrico, co., Va., 117,339 ....D5 121
Henrietta, Ray, Mo., 497 ....B4 101
Henrietta, Monroe, N.Y., 280 ....B3 108
Henrietta, Rutherford, N.C., 950 ....B2, D4 109
Henrietta, Clay, Tex., 3,062 ....C3 118
Henrietta Maria, cape, Ont., Can D9 72
Henrietta Northwest, Monroe, N.Y., 6,403 ....B3 108
Henrieville, Garfield, Utah, 152 ....F4 119
Henrique de Carvalho, Ang C3 48
Henry, Marshall, Ill., 2,278 ....B4 90
Henry, Scotts Bluff, Nebr., 138 ....C1 103
Henry, Elko, Nev., 8 ....B7 104
Henry, Codington, S. Dak., 276 ....C8 116
Henry, Henry, Tenn., 178 ....A3 117
Henry, Franklin, Va., 125 ....E3 121
Henry, Grant, W. Va B5 123
Henry, co., Ala., 75,286 ....D4 78
Henry, co., Ga., 17,619 ....C2 87
Henry, co., Ill., 49,317 ....B3 90
Henry, co., Ind., 48,899 ....E7 91
Henry, co., Iowa, 18,187 ....C6 92
Henry, co., Ky., 10,987 ....B4 94
Henry, co., Mo., 19,226 ....C4 101
Henry, co., Ohio, 25,392 ....A3 111
Henry, co., Tenn., 22,275 ....A3 117
Henry, co., Va., 40,335 ....E3 121
Henry, mts., Mont B1 102
Henry, mts., Utah E2 119
Henryetta, Okmulgee, Okla., 7,987 ....B6 112
Henry Kater, cape, N.W. Ter., Can C19 67
Henrys, fork, Utah, Wyo B5 119
Henryville, Que., Can., 711 ....D4 73
Henryville, Clark, Ind., 400 ....G6 91
Henryville, Nicholas, Ky., 125 ....B5 94
Hensall, Ont., Can., 926 ....D3 72
Hensel (Canton), Pembina, N. Dak., 130 ....A8 110
Henshaw, Union, Ky., 200 ....C1 94
Hensler, Oliver, N. Dak., 35 ....B4 110
Hensley, Pulaski, Ark., 350 ....C3 81
Henson, creek, Md C1 85
Henteyn Nuruu, mts., Mong B6 34
Henzada, Bur., 61,972 ....E10 39
Hepburn, Sask., Can., 294 ....E2 70
Hepburn, Page, Iowa, 49 ....D2 92
Hepburn, Hardin, Ohio, 175 ....B4 111
Hephzibah, Richmond, Ga., 676 ....C4 87
Hepler, Crawford, Kans., 178 ....E9 93
Heppenheim, Ger., 13,900 ....D3 17
Heppner, Morrow, Oreg., 1,661 ....B7 113
Hepworth, Ont., Can., 358 ....C3 72
Herat, Afg., 75,642 ....D11 41
Hérault, dept., Fr., 516,658 ....*F5 14
Herbert, Sask., Can., 1,008 ....G2 70
Herbert, peak, N.Z O14 51
Herbertingen, Ger., 2,000 ....A5 18
Herbertsville, Ocean, N.J., 50 ....C4 106
Herbesthal (Lontzen), Bel., 2,353 ....D5 15
Herb Lake, Man., Can., 65 ....B2 71
Herblay, Fr., 10,220 ....g9 14
Herblet, lake, Man., Can B2 71
Herborn, Ger., 10,100 ....C3 17
Hercegnovi, Yugo., 2,536 ....D4 22
Hercegovina, reg., Yugo D4 22
Herchen, Ger., 4,800 ....C2 17
Herculaneum, Jefferson, Mo., 1,767 ....B8, C7 101
Herd, Osage, Okla., 100 ....A5 112
Herdecke, Ger., 17,300 ....B2 17
Heredia, C.R., 19,000 ....E5 62
Heredia, prov., C.R., 77,300 ....*E5 62
Hereford, Cochise, Ariz., 150 ....F5 80
Hereford, Weld, Colo., 50 ....A6 83
Hereford, Eng., 40,431 ....B5 12
Hereford, Baltimore, Md., 380 ....A4 85
Hereford, Baker, Oreg., 35 ....C8 113
Hereford, Deaf Smith, Tex., 7,652 ....B1 118
Hereford, co., Eng., 130,919 ....B5 12
Hereford, inlet, N.J E3 106
Hérémence, Switz., 1,868 ....D3 19
Herencia, Sp., 8,606 ....C4 20
Herendeen Bay, Alsk., 22 ....D7 79
Herentals, Bel., 17,451 ....C4 15
Herford, Ger., 55,700 (*88,000) A3 17
Herfölge, Den., 1,138 ....C6 24
Héricourt, Fr., 7,160 ....B2 18
Herington, Dickinson, Kans., 3,702 ....D7 93
Heriot, Lee., S.C., 105 ....C7 115
Herisau, Switz., 14,361 ....B7 19
Herkimer, Marshall, Kans., 110 ....C7 93
Herkimer, Herkimer, N.Y., 9,396 ....B6 108
Herkimer, co., N.Y., 66,370 ....B6 108
Herman, Baraga, Mich., 55 ....B2 98
Herman, Grant, Minn., 764 ....E2 99
Herman, Washington, Nebr., 335 ....C9 103
Hermanas, Luna, N. Mex., 6 ....F2 107
Hermann, Gasconade, Mo., 2,536 ....C6 101
Hermano, peak, Colo D2 83
Hermanos, is., Ven A5 60
Hermansville, Menominee, Mich., 750 ....C3 98
Hermantown, St. Louis, Minn., 1,000 ....D6 99
Hermanville, Claiborne, Miss., 200 ....D3 100
Hermel, Leb E11 31
Hermeskeil, Ger., 4,500 ....D1 17
Herminie, Westmoreland, Pa., 1,571 ....F2 114
Hermiston, Umatilla, Oreg., 4,402 ....B7 113
Hermit, is., N. Gui h12 50
Hermitage, Bradley, Ark., 379 ....D3 81
Hermitage, Newf., Can., 417 ....E4 75
Hermitage, Point Coupee, La., 75 ....D4 95
Hermitage, Hickory, Mo., 328 ....D4 101
Hermitage, Davidson, Tenn., 100 ....E9 117
Hermitage, Vir. Is f16 65
Hermitage, bay, Newf., Can E3 75
Hermitage Springs, Clay, Tenn., 150 ....C8 117
Hermleigh, Scurry, Tex., 650 ....C2 118
Hermon, St. Lawrence, N.Y., 612 ....B2 108
Hermosa, Custer, S. Dak., 126 ....D2 116
Hermosa Beach, Los Angeles, Calif., 16,115 ....F2 82
Hermosillo, Mex., 96,122 ....B2 63
Hernad, riv., Hung A5 22
Hernandarias, Par E5 55
Hernandex, Rio Arriba, N. Mex., 12 ....A3 107
Hernando, Citrus, Fla., 301 ....D4 86
Hernando, De Soto, Miss., 1,898 ....A4 100
Hernando, co., Fla., 11,205 ....D4 86
Herndon, Jenkins, Ga., 50 ....D4 87
Herndon, Guthrie, Iowa, 125 ....C3 92
Herndon, Rawlins, Kans., 339 ....C3 93
Herndon, Christian, Ky., 150 ....D2 94
Herndon, Northumberland, Pa., 622 ....E8 114
Herndon, Fairfax, Va., 1,960 ....A4, C5 121
Herndon, Wyoming, W. Va., 200 ....D3 123
Herne, Ger., 113,200 ....B2 17
Herning, Den., 24,790 ....B2 24
Herod, Terrell, Ga., 100 ....E2 87
Herod, pt., N.Y F5 84
Heron, Sanders, Mont., 50 ....C1 102
Heron Bay, Ont., Can., 167 ....E8 72
Heron Lake, Jackson, Minn., 852 ....G3 99
Herouxville, Que., Can., 591 ....C5 73
Herreid, Campbell, S. Dak., 767 ....B5 110
Herrenberg, Ger., 8,800 ....E3 17
Herrick, Shelby, Ill., 440 ....D5 90
Herrick, Gregory, S. Dak., 160 ....D6 116
Herrick Center, Susquehanna, Pa., 80 ....C10 114
Herrick, creek, B.C., Can B7 68
Herrick, mtn., Vt D2 120
Herrin, Williamson, Ill., 9,474 ....F4 90
Herring, Roger Mills, Okla B2 112
Herring, bay, Md C4 85
Herring, run, Md C2 85
Herrington, LaCross, Wis., 2,405 ....*E2 124
Herrington, lake, Ky C5 94
Hersbruck, Ger., 8,300 ....D6 17
Herschel, Sask., Can., 188 ....F1 70
Herscher, Kankakee, Ill., 658 ....B5 90
Hersey, Osceola, Mich., 246 ....E5 98
Hershey, Lincoln, Nebr., 504 ....C5 103
Hershey, Dauphin, Pa., 6,851 ....F8 114
Herstal, Bel., 29,606 ....D5 15
Herten, Ger., 51,800 ....*C7 15
Hertford, Eng., 15,734 ....C7 12
Hertford, Perquimans, N.C., 2,068 ....A7 109
Hertford, co., Eng., 832,088 ....C7 12
Hertford, co., N.C., 22,718 ....A6 109
Hervás, Sp., 4,352 ....B3 20
Hervey, bay, Austl B9 51
Hervey Junction, Que., Can., 342 ....C5 73
Herxheim, Ger., 6,700 ....D3 17
Herzberg, Ger., 11,100 ....B5 17
Herzberg, Ger., 6,635 ....B8 17
Herzberg, Ger., 1,275 ....F6 24
Herzliya, Isr., 26,809 ....B6, g10 32
Herzogenaurach, Ger., 9,900 ....D5 17
Herzogenbuchsee, Switz., 4,641 ....B4 19
Hesdin, Fr., 3,200 ....D10 12
Hesel, Ger., 2,100 ....A7 15
Heshikiya, Okinawa 52
Hespeler, Ont., Can., 4,519 ....D4 72
Hesper, Yellowstone, Mont., 21 ....E8 102
Hesper, Benson, N. Dak., 25 ....B6 110
Hesperia, San Bernardino, Calif., 950 ....E5 82
Hesperia, Newaygo and Oceana, Mich., 822 ....E4 98
Hesperus, La Plata, Colo., 47 ....D2 83
Hesperus, peak, Colo D2 83
Hesse, reg., Ger C4 16
Hessel, Mackinac, Mich., 240 ....B6 98
Hesselager, Den C4 24
Hesselö, isl., Den B5 24
Hessen, state, Ger., 4,814,400 ....C3 17
Hessmer, Avoyelles, La., 433 ....C3 95
Hesston, Harvey, Kans., 1,103 ....A5, D6 93
Hester, Granville, N.C A5 109
Hetland, Kingsbury, S. Dak., 107 ....C8 116
Hetona, Okinawa 52
Hettick, Macoupin, Ill., 253 ....D3 90
Hettinger, Adams, N. Dak., 1,769 ....C3 110
Hettinger, co., N. Dak., 6,317 ....C3 110
Hettstedt, Ger., 17,000 ....B6 17
Heuvelton, St. Lawrence, N.Y., 810 ....B1 108
Heves, Hung., 9,923 ....B5 22
Heves, co., Hung., 348,756 ....*B5 22
Heward, Sask., Can., 136 ....H4 70
Hewins, Chautauqua, Kans., 110 ....E7 93
Hewitt, Todd, Minn., 267 ....D3 99
Hewitt, Passaic, N.J., 200 ....A4 106
Hewitt, Wood, Wis., 150 ....D3 124
Hewlett, Nassau, N.Y., 7,500 ....*G2 84
Hexham, Eng., 9,897 ....F6 13
Hext, Menard, Tex., 40 ....D3 118
Heyburn, Minidoka, Idaho, 829 ....G5 89
Heyworth, McLean, Ill., 1,196 ....C5 90
Hialeah, Dade, Fla., 66,972 ....G6, F3 86
Hialeah Gardens, Dade, Fla., 172 ....*G6 86
Hiattville, Bourbon, Kans., 125 ....E9 93
Hiawassee, Towns, Ga., 455 ....B3 87
Hiawatha, Linn, Iowa, 1,336 ....*C6 92
Hiawatha, Brown, Kans., 3,391 ....C8 93
Hiawatha, Schoolcraft, Mich B4 98
Hiawatha, Carbon and Emery, Utah, 439 ....D4 119
Hibbard, Navajo, Ariz C5 80
Hibbard, Madison, Idaho F7 89
Hibbard, Marshall, Ind., 150 ....B5 91
Hibbing, St. Louis, Minn., 17,731 ....C6 99
Hibernia, Morris, N.J., 450 ....B4 106
Hickiwan, Pima, Ariz., 100 ....E3 80
Hickman, Kent, Del., 100 ....C6 85
Hickman, Fulton, Ky., 1,537 ....B2 94
Hickman, Lancaster, Nebr., 288 ....D9 103
Hickman, Catron, N. Mex C2 107
Hickman, co., Ky., 6,747 ....B2 94
Hickman, co., Tenn., 11,862 ....B4 117
Hickman's Harbour, Newf., Can., 419 ....D5 75
Hickok, Grant, Kans., 45 ....E2 93
Hickory, Graves, Ky., 170 ....B2 94
Hickory, Newton, Miss., 539 ....C4 100
Hickory, Catawba, N.C., 19,328 ....B2 109
Hickory, Murray, Okla., 112 ....C5 112
Hickory, co., Mo., 4,516 ....D4 101
Hickory Corners, Barry, Mich., 200 ....F5 98
Hickory East, Catawba, N.C., 3,274 ....*B2 109
Hickory Flat, Benton, Miss., 344 ....A4 100
Hickory Grove, York, S.C., 287 ....B5 115
Hickory Hills, Cook, Ill., 2,707 ....*F2 90
Hickory Plains, Prairie, Ark., 300 ....C4 81
Hickory Point, Montgomery, Tenn A4 117
Hickory Ridge, Cross, Ark., 364 ....B5 81
Hickory Valley, Hardeman, Tenn., 179 ....B2 117
Hickory Withe, Fayette, Tenn., 100 ....B2 117
Hickox, Brantley, Ga., 71 ....E4 87
Hickson, Ont., Can., 151 ....D4 72
Hickson, lake, Sask., Can A3 70
Hicksville, Nassau, N.Y., 50,405 ....D3, E7 108
Hicksville, Defiance, Ohio, 3,116 ....A3 111
Hico, Hamilton, Tex., 1,020 ....D3 118
Hicoria, Highlands, Fla E5 86
Hidalgo, Jasper, Ill., 126 ....D5 90
Hidalgo, Hidalgo, Tex., 1,078 ....*F3 118
Hidalgo, co., N. Mex., 4,961 ....F1 107
Hidalgo, co., Tex., 180,904 ....F3 118
Hidalgo, state, Mex., 983,161 ....C5, m14 63
Hidalgo del Parral, Mex., 41,461 ....B3 63
Hiddenite, Alexander, N.C., 500 ....B2 109
Hiddensee, isl., Ger D7 24
Hidden Timber, Todd, S. Dak., 15 ....D5 116
Hideaway Park, Grand, Colo., 200 ....B5 83
Hierro, isl., Can. Is n13 20
Higashi (Dōgo), isl., Jap H6 37
Higbee, Randolph, Mo., 646 ....B5 101
Higby, Roane, W. Va., 158 ....C3 123
Higden, Cleburne, Ark., 40 ....B3 81
Higdon, Jackson, Ala., 200 ....A4 78
Higganum, Middlesex, Conn., 900 ....D6 84
Higgins, Lipscomb, Tex., 711 ....A2 118
Higgins, pond, Md C6 85
Higgins Lake, Roscommon, Mich., 100 ....D6 98
Higginson, White, Ark., 183 ....B4 81
Higginsport, Brown, Ohio, 412 ....D4 111
Higginsville, Lafayette, Mo., 403 ....B4 101
High, Carroll, Ark A2 81
High, des, Oreg D6 113
High, hill, N.Y F3 84
High, isl., Mich C5 98
High, peak, Phil o13 35
High Bluff, Man., Can., 77 ....D2 71
High Bridge, Jessamine, Ky., 250 ....C5 94
High Bridge, Hillsboro, N.H., 125 ....E2 105
High Bridge, Hunterdon, N.J., 2,148 ....B3 106
Highcoal, Boone, W. Va., 15 ....D6 123
Highest Point in Ala B4 78
Highest Point in Alsk C9 79
Highest Point in Ariz B4 80
Highest Point in Ark B2 81
Highest Point in Calif D4 82
Highest Point in Can D6 66
Highest Point in Colo B4 83
Highest Point in Conn A3 84
Highest Point in Del A6 85
Highest Point in Fla G3 86
Highest Point in Ga B3 87
Highest Point in Haw D6 88
Highest Point in Idaho E5 89
Highest Point in Ill A3 90
Highest Point in Ind D8 91
Highest Point in Iowa A2 92
Highest Point in Kans D1 93
Highest Point in Ky D7 94
Highest Point in La B3 95
Highest Point in Maine C4 96
Highest Point in Md D1 85
Highest Point in Mass A1 97
Highest Point in Mich B2 98
Highest Point in Minn C9 99
Highest Point in Miss A5 100
Highest Point in Mo D7 101
Highest Point in Mont E7 102
Highest Point in Nebr C1 103
Highest Point in Nev F3 104
Highest Point in N.H B4 105
Highest Point in N.J A3 106
Highest Point in N. Mex A4 107
Highest Point in N.Y A7 108
Highest Point in N.A C9 79
Highest Point in N.C D4 109
Highest Point in N. Dak C2 110
Highest Point in Ohio B4 111
Highest Point in Okla D1 112
Highest Point in Oreg B5 113
Highest Point in Pa G3 114
Highest Point in R.I B9 84
Highest Point in S.C A2 115
Highest Point in S. Dak D2 116
Highest Point in Tenn D10 117
Highest Point in Tex E2 118
Highest Point in U.S C9 79
Highest Point in Utah C5 119
Highest Point in Vt B3 120
Highest Point in Va B3 121
Highest Point in Wash C4 122
Highest Point in W. Va C5 123
Highest Point in Wis D4 124

Highest Point in Wyo........B3 125
Highfalls, Moore, N.C., 200..B4 109
High Falls, res., Wis.........C5 124
Highfield, Washington, Md.,
  500....................A3 85
Highgate, Ont., Can., 374...E3 72
Highgate Center, Franklin,
  Vt., 300 (1,608▲)........B2 120
Highgate Falls, Franklin, Vt.,
  200....................B2 120
Highgate Springs, Franklin,
  Vt., 125................B2 120
Highgrove, Riverside, Calif.,
  2,000..................*F5 82
High Hill, Montgomery, Mo.,
  173....................C6 101
High Hill, lake, Man., Can...B4 71
High Hill, riv., Alta., Can....A5 69
High Island, Galveston, Tex.,
  800....................E5 118
High Knob, mtn., Md.........B3 85
High Knob, mtn., Va.........B4 121
High Knob, mtn., W. Va......C5 123
High Knob, mtn., W. Va......B6 123
Highland, San Bernardino,
  Calif., 7,000............F3 82
Highland, Clay, Fla., 100....B4 86
Highland, Madison, Ill.,
  4,943..................E4 90
Highland, Lake, Ind., 16,284.A3 91
Highland, Doniphan, Kans.,
  755....................C8 93
Highland, Oakland, Mich.,
  375....................A7 98
Highland, Ulster, N.Y.,
  2,931..................D7 108
Highland, Highland, Ohio,
  265....................C4 111
Highland, Iowa, Wis., 741...E3 124
Highland, co., Ohio, 29,716..C4 111
Highland, co., Va., 3,221....C3 121
Highland, lake, Maine.......E5 96
Highland, peak, Calif.......C4 82
Highland, pt., Fla..........G5 86
Highland Beach, Palm Beach,
  Fla., 65...............*F6 86
Highlandale, Leflore, Miss..B3 100
Highland City, Polk, Fla.,
  1,020..................*E5 86
Highland Creek, Ont., Can.,
  1,300..................D5 72
Highland Crest, Wyandotte,
  Kans., 4,000...........*B8 93
Highland Falls, Orange, N.Y.,
  4,469..................*D7 108
Highland Grove, Ont., Can.,
  84....................B6 72
Highland Heights, Campbell,
  Ky., 3,491.............*A5 94
Highland Heights, Cuyahoga,
  Ohio, 2,929............*A6 111
Highland Home, Crenshaw,
  Ala., 200..............D3 78
Highland Lake, Cumberland,
  Maine (part of Westbrook).E5 96
Highland Park, Hartford, Conn.
  (part of Manchester)....B7 84
Highland Park, Lake, Ill.,
  25,532................A6, E2 90
Highland Park, Wayne, Mich.,
  38,063.................A8 98
Highland Park, Middlesex,
  N.J., 11,049...........C4 106
Highland Park, Mifflin, Pa.,
  1,534.................*E6 114
Highland Park, Sullivan,
  Tenn. (part of Kingsport).C11 117
Highland Park, Dallas, Tex.,
  10,411................B5 118
Highland Park, Norfolk, Va.,
  2,500.................*E6 121
Highlands, Broward, Fla.,
  5,000.................*F6 86
Highlands, Coos, N.H.......B4 105
Highlands, Monmouth, N.J.,
  3,536..................C5 106
Highlands, Macon, N.C., 597.D3 109
Highlands, Harris, Tex.,
  4,336..................F5 118
Highlands, co., Fla., 21,338..E5 86
Highland Springs, Henrico,
  Va., 5,000............C7, D5 121
Highlandville, Winneshiek,
  Iowa, 57...............A6 92
Highmore, Hyde, S. Dak.,
  1,078.................C6 116
Highpine, York, Maine......E2 96
High Point, Garland, Ark....C6 81
Highpoint, Winston, Miss.,
  75....................B4 100
High Point, Guilford, N.C.,
  62,063 (*100,600)......B3 109
High Point, King, Wash., 100.D2 122
High Point, mtn., N.J.......A3 106
High Point, mtn., W. Va.....C4 123
High Prairie, Alta., Can.,
  1,756.................B2 69
High Ridge, Jefferson, Mo.,
  250...................B7 101
High River, Alta., Can.,
  2,276.................D4 69
Highrock, York, Pa.........G9 114
Highrock, lake, Man., Can...B1 101
High Rock, mtn., Md........D1 85
High Rock, res., N.C.......B3 109
High Shoals, Morgan and
  Oconee, Ga., 217.......C3 87
Highshoals, Gaston, N.C.,
  900...................B2 109
Highspire, Dauphin, Pa.,
  2,999.................F8 114
Highsplint, Harlan, Ky.,
  500...................D6 94
High Springs, Alachua, Fla.,
  2,329.................C4 86
Hightstown, Mercer, N.J.,
  4,317.................C3 106
High Veld, plain, S. Afr....C4 49
Highway City, Fresno, Calif.,
  1,381................*D4 82
Highway Village, Nucces, Tex.,
  1,927................*F4 118
Highwood, Lake, Ill.,
  4,499................A6, E2 90
Highwood, Gladwin, Mich...E6 98
Highwood, Choteau, Mont.,
  200...................C6 102
Highwood, mts., Mont......C6 102
Highwood, peak, Mont.....C6 102
High Wycombe, Eng.,
  50,301..............E6 10, C7 12
Higley, Maricopa, Ariz.,
  100...................D5 80
Higley, Dom. Rep., 10,084..F9 64
Hiiumaa, isl., Sov. Un.....B4 27
Hijo, Phil., 3,606........*D7 35
Hika, Manitowoc, Wis., 150..B6 124

Hikone, Jap., 41,400......n15 37
Hikurangi, mtn., N.Z......L17 51
Hilbert, Calumet, Wis.,
  736..................B5, D5 124
Hilda, Alta., Can., 194....D5 69
Hilda, Barnwell, S.C., 259..E5 115
Hildburghausen, Ger., 7,870.C5 17
Hilden, N.S., Can., 554....D6 74
Hilden, Ger., 37,600......B1 17
Hildesheim, Ger., 96,300...A4 17
Hildreth, Franklin, Nebr.,
  305..................D6 103
Hiles, Forest, Wis., 125...C5 124
Hilgard, Union, Oreg., 15..B8 113
Hilger, Fergus, Mont., 60..C7 102
Hilham, Overton, Tenn.,
  164..................C8 117
Hill, Merrimack, N.H.,
  190 (396▲)...........C3 105
Hill, co., Mont., 18,653...B6 102
Hill, co., Tex., 23,650....D4 118
Hill, lake, Ark..........D6 81
Hilla (Hillah), Iraq,
  46,441...........117 9, E2 41
Hillared, Swe., 323.......A7 24
Hillburn, Rockland, N.Y.,
  1,114...............*D6 108
Hill Center, Merrimack, N.H.C3 105
Hill City, Camas, Idaho, 30.F3 89
Hill City, Graham, Kans.,
  2,421................C4 93
Hill City, Aitkin, Minn., 429.D5 99
Hill City, Pennington, S. Dak.,
  419..................D2 116
Hillcrest, Warren, N.J.,
  1,922...............*B2 106
Hillcrest, Broome, N.Y.,
  1,500...............*C5 108
Hillcrest, Rockland, N.Y.,
  1,800...............*D1 108
Hillcrest Center, Kern, Calif.,
  15,000..............*E4 82
Hillcrest Heights, Prince Georges,
  Md., 15,295.........*C3 85
Hillcrest Mines, Alta., Can.,
  618..................E3 69
Hillegom, Neth., 12,400....B4 15
Hiller, Fayette, Pa., 1,746.*F2 114
Hilleröd, Den., 11,605
  (*18,147)............C6 24
Hillesheim, Ger., 1,443....C1 17
Hillgrove, Los Angeles, Calif.,
  3,500...............*F3 82
Hillhead, Marshall. S. Dak.,
  20...................B8 116
Hilliard, Alta., Can., 85...C4 69
Hilliard, Nassau, Fla., 1,075.B5 86
Hilliards, Franklin, Ohio,
  5,633...............C2 111
Hilliards, Butler, Pa., 200..D2 114
Hilliard, Ont., Can., 77....D7 72
Hillsburg, Clinton, Ind.,
  245..................D5 91
Hillman, Montmorency,
  Mich., 445...........C7 98
Hillman, Morrison, Minn.,
  80...................D5 99
Hillmond, Sask., Can......D1 70
Hillrose, Morgan, Colo., 157.A7 83
Hills, Johnson, Iowa, 310...C6 92
Hills, Rock, Minn., 516....G2 99
Hills, hill, Newf., Can....D3 75
Hills, lake, Newf., Can....D3 75
Hillsboro, Lawrence, Ala.,
  218..................A2 78
Hillsboro, Jasper, Ga., 150..C3 87
Hillsboro, Montgomery, Ill.,
  4,232................D4 90
Hillsboro, Fountain, Ind.,
  517..................D3 91
Hillsboro, Henry, Iowa,
  218..................D6 92
Hillsboro, Marion, Kans.,
  2,441................D7 93
Hillsboro, Caroline, Md.,
  201..................C6 85
Hillsboro, Scott, Miss., 100.C4 100
Hillsboro, Jefferson, Mo.,
  457.................B7, C7 101
Hillsboro, Carbon, Mont.,
  25...................E8 102
Hillsboro, Hillsboro, N.H.,
  1,645 (2,310▲).......D3 105
Hillsboro, Sierra, N. Mex.,
  250..................E2 107
Hillsboro, Orange, N.C.,
  1,349................A4 109
Hillsboro, Traill, N. Dak.,
  1,278................B8 110
Hillsboro, Highland, Ohio,
  5,474................C2 111
Hillsboro, Washington, Oreg.,
  8,232..............B1, B3 113
Hillsboro, Coffee, Tenn.,
  200..................B6 117
Hillsboro, Hill, Tex., 7,402.C4 118
Hillsboro, Pocahontas,
  W. Va., 210..........C4 123
Hillsboro, Vernon, Wis.,
  1,366................E3 124
Hillsboro, co., N.H.,
  178,161..............E3 105
Hillsboro, canal, Fla......F6 86
Hillsboro, riv., Fla.......D4 86
Hillsboro Beach, Broward,
  Fla., 437...........*F6 86
Hillsboro Lower Village,
  Hillsboro, N.H., 100..D3 105
Hillsboro Upper Village,
  Hillsboro, N.H., 75....D3 105
Hillsborough, San Mateo, Calif.,
  7,554..............*B5 82
Hillsborough, N.B., Can.,
  679..................D5 74
Hillsborough, co., Fla.,
  397,788..............D4 86
Hillsborough, bay, P.E.I., Can.C6 74
Hillsborough, bay, Fla.....E2 86
Hillsburgh, Ont., Can., 440.D4 72
Hills Creek, res., Oreg....D4 113
Hillsdale, Ont., Can., 255..C5 72
Hillsdale, Rock Island, Ill.,
  490..................B3 90
Hillsdale, Vermillion, Ind.,
  250..................E3 91
Hillsdale, Miami, Kans.,
  142..................D9 93
Hillsdale, Hillsdale, Mich.,
  7,629................G6 98
Hillsdale, St. Louis, Mo.,
  2,788...............*A8 101
Hillsdale, Pearl River, Miss.E4 100
Hillsdale, Bergen, N.J.,
  8,734................B5 106
Hillsdale, Columbia, N.Y.,
  400..................C7 108
Hillsdale, Garfield, Okla.,
  60...................A4 112
Hillsdale, Macon, Tenn.,
  25...................C7 117

Hillsdale, Barron, Wis., 125.C2 124
Hillsdale, Laramie, Wyo.,
  100..................D8 125
Hillsdale, co., Mich.,
  34,742...............G6 98
Hillsgrove, Sullivan, Pa.,
  200..................D8 114
Hillside, Fremont, Colo., 5..C5 83
Hillside, Cook, Ill., 7,794.*F2 90
Hillside, Union, N.J.,
  22,304...............E4 106
Hillside Manor, Nassau, N.Y.,
  8,000...............*G2 84
Hillston, Austl..........F5 51
Hillsview, McPherson,
  S. Dak., 44..........B6 116
Hillsville, Lawrence, Pa.,
  950..................D1 114
Hillsville, Carroll, Va., 905.E2 121
Hilltonia, Screven, Ga., 353.D5 87
Hilltop, Cochise, Ariz.....F6 80
Hilltop, Camden, N.J.,
  1,000...............*D2 106
Hilltop, Fayette, W. Va.,
  765..................D7 123
Hilltown, N. Ire., 209.....C5 11
Hillview, Greene, Ill., 305..D3 90
Hillville, Haywood, Tenn.,
  30...................B2 117
Hillwood, Coosa, Ala......C3 78
Hillwood, Fairfax, Va.,
  1,400...............*B5 121
Hilly, Lincoln, La........B3 95
Hilo, Hawaii, Haw.,
  25,966..............D6, n16 88
Hilo, bay, Haw...........D6 88
Hilpoltstein, Ger., 3,900..D6 17
Hilpsford, pt., Eng.......F5 13
Hilton, Early, Ga., 90.....E1 87
Hilton, Monroe, N.Y.,
  1,334................B3 108
Hiltonhead, Beaufort, S.C.,
  300..................G6 115
Hilton Head, isl., S.C.....G6 115
Hiltons, Scott, Va., 250...B2 121
Hilts, Siskiyou, Calif., 500.B2 82
Hilversum, Neth., 102,000..B5 15
Hima, Clay, Ky., 500......C6 94
Himachal Pradesh, ter., India,
  1,351,144...........*B6 40
Himalaya, mts., Asia.....G11 33
Himeji, Jap., 328,689.....I7 37
Himes, Big Horn, Wyo.....A4 125
Himrod, Yates, N.Y., 200..C4 108
Hims, Hai., 4,511........F7 64
Hinchinbrook, isl., Alsk..g18 79
Hinchinbrook, isl., Austl..C8 50
Hinchcliffe, Sask., Can....E4 70
Hinckley, Eng., 41,573....B6 12
Hinckley, DeKalb, Ill., 940.B5 90
Hinckley, Somerset, Maine,
  175..................D3 96
Hinckley, Pine, Minn., 851.D6 99
Hinckley, Medina, Ohio, 300.A6 111
Hinckley, Millard, Utah, 397.D3 119
Hinckley, res., N.Y.......B5 108
Hindarabi, isl., Iran.....H6 41
Hindenburg, see Zabrze, Pol.
Hindian, Iran, 2,000.....F4 41
Hindman, Knott, Ky., 793..C7 94
Hindmarsh, lake, Austl....G3 51
Hinds, co., Miss., 187,045..C3 100
Hinds, hill, Newf., Can....D3 75
Hinds, lake, Newf., Can....D3 75
Hindsboro, Douglas, Ill., 376.D5 90
Hindsville, Madison, Ark.,
  150..................A2 81
Hindubagh, Pak...........B4 39
Hindu Kush, mts., Afg....D14 41
Hines, Dixie, Fla., 30.....C3 86
Hines, Beltrami, Minn., 75.C4 99
Hines, Harney, Oreg., 1,207.D7 113
Hines, riv., Alta., Can....A1 69
Hinesburg, Chittenden, Vt.,
  200 (1,180▲).........C2 120
Hinesburg, pond, Vt......C2 120
Hines Creek, Alta., Can.,
  398..................A1 69
Hinesville, Liberty, Ga.,
  3,174................E5 87
Hingham, Plymouth, Mass.,
  10,500 (15,378▲).....B6, D3 97
Hingham, Hill, Mont., 254..B6 102
Hingham, bay, Mass.......D3 97
Hingham Center, Plymouth,
  Mass..................D3 97
Hingoli, India, 23,407....H6 40
Hinis, Tur., 2,511.......C13 31
Hinkle, Alcorn, Miss., 75..A5 100
Hinkley, San Bernardino,
  Calif., 75...........E5 82
Hinnom, val., Isr., Jordan.m14 32
Hinnöy, isl., Nor........C6 25
Hinojosa [del Duque], Sp.,
  14,767...............C4 20
Hinsdale, Du Page and Cook,
  Ill., 12,859.........F2 90
Hinsdale, Berkshire, Mass.,
  990 (1,414▲).........B1 97
Hinsdale, Valley, Mont., 350.B9 102
Hinsdale, Cheshire, N.H.,
  1,235 (2,187▲).......E2 105
Hinsdale, Cattaraugus, N.Y.,
  200..................C2 108
Hinsdale, co., Colo., 208..D3 83
Hinterrhein, Switz., 86....C7 19
Hinton, Alta., Can., 3,529..C2 69
Hinton, Plymouth, Iowa, 403.B1 92
Hinton, Caddo, Okla., 907..B3 112
Hinton, Summers, W. Va.,
  5,197................D4 123
Hintonville, Perry, Miss.,
  150..................D5 100
Hinze, Winston, Miss......B4 100
Hinzir, cape, Tur........D10 31
Hirado, isl., Jap........J4 37
Hiraiwa, cape, Iwo.......E2 96
Hiram, Paulding, Ga., 358..C2 87
Hiram, Oxford, Maine, 110
  (699▲)...............E2 96
Hiram, Portage, Ohio, 1,011.A6 111
Hirara, Ryukyu Is., 28,504.*G10 37
Hiratsuka, Jap.,108,279.I9, n18 37
Hiraya, Jap.............n14 37
Hire, Cherry, Nebr.......B4 103
Hirosaki, Jap., 80,000....F10 37
Hiroshima, Jap., 431,336
  (*560,000)...........I6 37
Hiroshima, pref., Jap.....*I6 37
Hirson, Fr., 11,715......C5 14
Hirtshals, Den., 4,177....A3 24
Hisaronu, Tur., 6,586....B9 31
Hiseville, Barren, Ky., 196.C4 94
Hissar, India, 60,222....C6 39
Hit, Iraq, 4,830.........F14 31
Hita, Jap., 37,200.......*J5 37
Hitachi, Jap., 161,226...H10 37
Hitchcock, Sask., Can., 66.H4 70

Hitchcock, Blaine, Okla.,
  134..................B3 112
Hitchcock, Beadle, S. Dak.,
  193..................C7 116
Hitchcock, Galveston, Tex.,
  5,216................*E5 118
Hitchcock, co., Nebr., 4,829.D4 103
Hitchin, Eng., 24,243.....C7 12
Hitchins, Carter, Ky., 600..B7 94
Hitchita, McIntosh, Okla.,
  120..................B6 112
Hitchland, Hansford, Tex.,
  25...................A2 118
Hiteman, Monroe, Iowa,
  200..................C5 92
Hitoyoshi, Jap., 27,200...J5 37
Hitra, isl., Nor.........F3 25
Hitterdal, Clay, Minn., 235.D2 99
Hitzacker, Ger., 3,800....E5 24
Hivonnait, Arg..........E4 55
Hiwassee, Pulaski, Va., 400.E2 121
Hiwassee, lake, N.C......D2 109
Hiwassee, riv., Tenn......D9 117
Hixson, Hamilton, Tenn.,
  1,500..............D8, E10 117
Hixton, Jackson, Wis., 310..D2 124
Hjallerup, Den., 1,241....A4 24
Hjälmaren (Hjälmar), lake,
  Swe..................H7 25
Hjälmseryd, Swe.........A8 24
Hjo, Swe., 4,600.........H6 25
Hjörring, Den., 15,038....A3 24
Hjörring, co., Den., 173,233.A3 24
Hlohovec, Czech., 12,700..D4 26
Hlomsak, Thai, 7,906.....D4 38
Ho, China..............F4 36
Hoa Binh, Viet, 25,000...B6 38
Hoagland, Allen, Ind., 500..C8 91
Hoagland, Logan, Nebr....C5 103
Hoback, Teton, Wyo......B2 125
Hoback, riv., Wyo........B2 125
Hoban Heights, Wyoming,
  Pa..................A8, D10 114
Hobart, Austl., 54,021
  (*116,000)...........o15 50
Hobart, Lake, Ind., 18,680..A3 91
Hobart, Delaware, N.Y., 585.C6 108
Hobart, Kiowa, Okla.,
  5,132................C3 112
Hobbema, Alta., Can., 61..C4 69
Hobbieville, Greene, Ind.,
  150..................G4 91
Hobbs, Tipton, Ind., 250...D6 91
Hobbs, Lea, N. Mex.,
  26,275...............E6 107
Hobbs Island, Madison, Ala.A3 78
Hobe Sound, Martin, Fla.,
  900..................E6 86
Hobgood, Halifax, N.C., 630.A6 109
Hoboken, Bel., 30,557....C4 15
Hoboken, Brantley, Ga.,
  552..................E4 87
Hoboken, Hudson, N.J.,
  48,441...............D5 106
Hobro, Den., 8,208.......B3 24
Hobsogol Dalay, lake, Mong.A5 34
Hobson, Navajo, Ariz.....C5 80
Hobson, Judith Basin, Mont.,
  207..................D7 102
Hobson, White Pine, Nev...C6 104
Hobson, Nansemond, Va.,
  250..................B6 121
Hobson, lake, B.C., Can...C7 68
Hobson City, Calhoun, Ala.,
  770..................B4 78
Hobucken, Pamlico, N.C.,
  500..................B7 109
Hochatown, McCurtain,
  Okla., 100...........C7 112
Hochfeld, S.W. Afr.......B2 49
Hochfelden, Fr., 2,755....F7 15
Hochien, China..........E7 36
Hochih, China..........G6 34
Hochst, Ger............C3 17
Höchstadt, Ger., 3,400...D5 17
Hochuan, China, 50,000..I2 36
Hockenheim, Ger., 13,200..D3 17
Hockessin, New Castle, Del.,
  305..................A6 85
Hocking, co., Ohio, 20,168.C5 111
Hocking, riv., Ohio......C5 111
Hockingport, Athens, Ohio,
  150..................C5 111
Hockinson, Clark, Wash., 50.D3 122
Hockley, Harris, Tex., 200..E5 118
Hockley, co., Tex., 22,340..C1 118
Hodgdon, Aroostook, Maine,
  225 (926▲)...........B5 96
Hodge, Jackson, La., 878...B3 95
Hodgeman, co., Kans., 3,115.D4 93
Hodgenville, Larue, Ky.,
  1,985................C4 94
Hodges, Franklin, Ala., 194.A2 78
Hodges, Dawson, Mont.,
  50...................D12 102
Hodges, Greenwood, S.C.,
  200..................C3 115
Hodges, hill, Newf., Can...D4 75
Hodgeville, Sask., Can., 388.G2 70
Hodgkins, Cook, Ill., 1,126.*F2 90
Hodgson, Man., Can., 222..D3 71
Hodmezovasarhely, Hung.,
  40,047 (53,505▲).....B5 22
Hodna, lake, Alg.........G9 30
Hodonin, Czech., 18,000..D4 26
Hoea, Hawaii, Haw., 170...C6 88
Hoehne, Las Animas, Colo.,
  290..................D6 83
Hoek Nederburgh, pt., Indon.F6 35
Hoek van Holland, Neth.,
  2,245 (part of Rotterdam).C4 15
Hoensbroek, Neth., 21,700..D5 15
Hoeryong, Kor., 24,330....E4 37
Hoey, Sask., Can., 131....E3 70
Hof, Nor...............p28 25
Hof, Ger., 57,100.......C6 17
Hofei, China, 304,000....I7 36
Hoffman, Grant, Minn., 605.E3 99
Hoffman, Richmond, N.C.,
  344..................B4 109
Hoffman, Okmulgee, Okla.,
  248..................B6 112
Hoffman Estates, Cook, Ill.,
  8,296...............*E2 90
Hofgeismar, Ger., 8,200...B4 17
Hofn, Ice., 525.........n25 25
Hofsjökull, glacier, Ice...n23 25
Hofu, Jap., 96,821......I5 37
Hog, fish, Fla..........C3 86
Hog, isl., Alsk.........C5 79
Hog, isl., Mich.........C5 98
Hog, isl., N.C.........B7 109
Hog, isl., Va..........D7 121
Hog, neck, N.Y.........E7 84
Hogan, Duval, Fla.......B5 86
Högänäs, Swe., 7,500....B6 24
Hogansburg, Franklin, N.Y.,
  150..................A2 108

Hogansville, Troup, Ga.,
  3,658................C2 87
Hog Back, mtn., Mont.....F4 102
Hogback, mtn., Nebr......C2 103
Hogback, mtn., Vt........C2 120
Hog Creek, pt., N.Y......E8 84
Hogeland, Blaine, Mont.,
  125..................B8 102
Hogem, pass, B.C., Can....B5 68
Hogglesville, Hale, Ala...C2 78
Hogsett, Mason, W. Va., 40.C2 123
Hoh, head, Wash.........B1 122
Hoh, riv., Wash.........B1 122
Hohenlinden, Webster, Miss.,
  50...................B4 100
Hohen Solms, Ascension,
  La., 200.............B5 95
Hohenwald, Lewis, Tenn.,
  2,194................B4 117
Hohes Venn, plat., Bel...D6 15
Hohe Tauern, mts., Aus...E6 16
Ho-Ho-Kus, Bergen, N.J.,
  3,988................A4 106
Hohsien, China, 15,000...G7 34
Hohwachter, bay, Ger.....D4 24
Hoihow, China, 135,300..B9 38
Hoihow, China..........B9 38
Hoima, Ug..............A5 48
Hoi Xuan, Viet.........B6 38
Höjer, Den., 1,400......D2 24
Hokah, Houston, Minn.,
  685..................G7 99
Hoke, co., N.C., 16,356..B4 109
Hokendauqua, Lehigh, Pa.,
  1,400...............*E11 114
Hokes, Bluff, Etowah, Ala.,
  1,619................A4 78
Hoking, China..........F5 34
Hokitika, N.Z., 3,007...O13 51
Hokkaido, isl., Jap......E11 37
Hokkaido, ter., Jap......*E10 37
Hokoda, Jap., 7,100
  (28,657▲)...........m19 37
Holabird, Hyde, S. Dak., 35.C6 116
Holap, isl., Truk........52
Holbaek, Den., 15,475....C5 24
Holbaek, co., Den., 127,127.C5 24
Holbeach, Eng., 4,805....B8 12
Holbrook, Navajo, Ariz.,
  3,438................C5 80
Holbrook, Oneida, Idaho,
  10...................G6 89
Holbrook, Norfolk, Mass.,
  6,000 (10,104▲).....B5, D3 97
Holbrook, Furnas, Nebr.,
  354..................D5 103
Holbrook, Suffolk, N.Y.,
  3,441...............*F4 84
Holcomb, Finney, Kans., 270.E3 93
Holcomb, Grenada, Miss.,
  97...................B4 100
Holcomb, Dunklin, Mo., 436.E7 101
Holcombe, Chippewa, Wis.,
  275..................C2 124
Holcut, Tishomingo, Miss..A5 100
Holden, Alta., Can., 556...C4 69
Holden, Livingston, La., 75.A6 95
Holden, Worcester, Mass.,
  2,000 (10,117▲)......B4 97
Holden, Johnson, Mo., 1,951.C4 101
Holden, Millard, Utah, 388..D3 119
Holden, Chelan, Wash.....A5 122
Holden, Logan, W. Va.,
  1,900................D2 123
Holdenville, Hughes, Okla.,
  5,712................B5 112
Holderness, Grafton, N.H.,
  300 (749▲)..........C3 105
Holdfast, Sask., Can., 366..G3 70
Holdingford, Stearns, Minn.,
  526..................E4 99
Holdman, Umatilla, Oreg..B8 113
Holdrege, Phelps, Nebr.,
  5,226................D6 103
Holeb, Somerset, Maine,
  25...................C2 96
Hole in the Mountain, peak, Nev.C6 104
Hölen, Nor., 334........p28 25
Holgate, Henry, Ohio, 1,374.A3 111
Holguin, Cuba, 57,573....C5 64
Holikachuk, Alsk., 122....C8 79
Hollabrunn, Aus., 5,832..D8 16
Holladay, Benton, Tenn., 175.B3 117
Holladay, Salt Lake, Utah,
  28,000...............C4 119
Holladay, Spotsylvania, Va.C5 115
Holland, Dubois, Ind., 661..H3 91
Holland, Grundy, Iowa, 264.B5 92
Holland, Allen, Ky., 150...D3 94
Holland, Hampden, Mass.,
  150 (561▲)...........B3 97
Holland, Ottawa, Mich.,
  24,777...............F4 98
Holland, Pipestone, Minn.,
  264..................G2 99
Holland, Pemiscot, Mo., 403.E8 101
Holland, Erie, N.Y., 950...C2 108
Holland, Lucas, Ohio,
  924................A1, A4 111
Holland, Josephine, Oreg.,
  25...................E3 113
Holland, Bell, Tex., 653...D4 118
Holland, Nansemond, Va.,
  338..................E6 121
Holland, isl., Md........D5 85
Holland, pt., Md........C4 85
Holland, straits, Md.....D5 85
Hollandale, Washington,
  Miss., 2,646.........B3 100
Hollandale, Iowa, Wis., 275.F4 124
Holland Center, Ont., Can.,
  72...................C4 72
Hollandia, see Kotabaru,
  W. Irian
Holland Patent, Oneida,
  N.Y., 538............B5 108
Hollansburg, Darke, Ohio,
  311..................C3 111
Hollenberg, Washington,
  Kans., 55............C7 93
Holley, Orleans, N.Y., 1,788.B2 108
Holley, Linn, Oreg., 100...C3 113
Holliday, Johnson, Kans., 80.B8 93
Holliday, Monroe, Mo., 181..B5 101
Holliday, Archer, Tex., 1,139.C3 118
Hollidaysburg, Blair, Pa.,
  6,475................F5 114

Hollins, Roanoke, Va., 1,000.D3 121
Hollis, Alsk., 292.......n23 79
Hollis, Perry, Ark., 15....C2 81
Hollis, Cloud, Kans., 75...C6 93
Hollis, Hillsboro, N.H.,
  175 (1,720▲).........E3 105
Hollis, Harmon, Okla.,
  3,006................C2 112
Hollis Center, York, Maine,
  75 (1,195▲)..........E2 96
Hollister, San Benito, Calif.,
  6,071..............C6, D3 82
Hollister, Twin Falls, Idaho,
  60...................G4 89
Hollister, Taney, Mo., 600..E4 101
Hollister, Halifax, N.C., 450.A6 109
Hollister, Tillman, Okla.,
  166..................C3 112
Hollister, Langdale, Wis.,
  200..................C5 124
Hollisterville, Wayne, Pa.,
  30...................D11 114
Holliston, Middlesex, Mass.,
  2,447 (6,222▲).......D1 97
Holloway, Swift, Minn., 242.E3 99
Holloway Terrace, New Castle,
  Del., 1,500.........*A6 85
Hollow Rock, Carroll, Tenn.,
  568..................A3 117
Hollsopple (P.O.), Somerset,
  Pa., 900............F4 114
Hollum, Neth., 878......A5 15
Holly, Prowers, Colo., 1,108.C8 83
Holly, Oakland, Mich., 3,269.F7 98
Holly Bluff, Yazoo, Miss., 250.C3 100
Holly Grove, Monroe, Ark.,
  672..................C4 81
Holly Hill, Volusia, Fla.,
  4,182................C5 86
Holly Hill, Orangeburg, S.C.,
  1,235................E7 115
Hollyoak, New Castle, Del.,
  1,000................A7 85
Holly Pond, Cullman, Ala.,
  193..................A3 78
Holly Ridge, Richland, La.,
  300..................B4 95
Holly Ridge, Onslow, N.C.,
  731..................C6 109
Holly Shelter, swamp, N.C..C6 109
Holly Springs, Dallas, Ark.,
  100..................D3 81
Holly Springs, Cherokee, Ga.,
  475..................B2 87
Holly Springs, Marshall,
  Miss., 5,621.........A4 100
Holly Springs, Wake, N.C.,
  558..................B5 109
Hollyville, Sussex, Del., 40.C7 85
Hollywood, Jackson, Ala.,
  246..................A4 78
Hollywood, Graham, Ariz.,
  125..................E6 80
Hollywood, Clark, Ark., 25..C2 81
Hollywood, Los Angeles, Calif.
  (part of Los Angeles)..F2 82
Hollywood, Broward, Fla.,
  35,237.............E3, F6 86
Hollywood, Habersham, Ga.,
  150..................B3 87
Hollywood, Calcasieu, La.,
  1,750...............*D2 95
Hollywood, St. Marys, Md.,
  260..................D4 85
Hollywood, Tunica, Miss.,
  147..................A3 100
Hollywood, Lincoln, N. Mex.,
  100..................D4 107
Hollywood, Montgomery, Pa.,
  2,000...............*F12 114
Hollywood, Charleston,
  S.C., 334............F1 115
Hollywood Heights, St. Clair,
  Ill., 1,000.........*E3 90
Hollywood Ridge Farms,
  Broward, Fla., 108...*F6 86
Holman, Mora, N. Mex., 250.A4 107
Holmdel Gardens, Reno, Kans.,
  1,436...............*D6 93
Holmen, LaCrosse, Wis., 635.E2 124
Holmes, Delaware, Pa.,
  3,000...............*B11 114
Holmes, Albany, Wyo., 15..D6 125
Holmes, co., Fla., 10,844..G3 86
Holmes, co., Miss., 27,096..B3 100
Holmes, co., Ohio, 21,591..B4 111
Holmes, mtn., Wyo........A2 125
Holmes Beach, Mantee, Fla.,
  1,143...............*E4 86
Holmeson, Monmouth, N.J...C4 106
Holmes Park, Jackson, Mo. (part
  of Kansas City)......E2 101
Holmes Run Acres, Fairfax, Va.,
  1,000...............*B5 121
Holmes Run Park, Fairfax, Va.,
  1,000...............*B5 121
Holmestrand, Nor., 2,100..p28 25
Holmesville, N.B., Can., 212.C2 74
Holmesville, Holmes, Ohio,
  422..................B6 111
Holmfield, Man., Can.....E2 71
Holmquist, Day, S. Dak., 35.B8 116
Holmsbu, Nor...........p28 25
Holmsund, Swe., 5,200....F9 25
Holon, Isr., 48,102......B6, g10 32
Holopaw, Osceola, Fla., 250.D5 86
Holsljunga, Swe.........A7 24
Holstebro, Den., 18,563...B2 24
Holsted, Den., 1,081.....C2 24
Holstein, Ont., Can., 185..C4 72
Holstein, Ida, Iowa, 1,413..B2 92
Holstein, Adams, Nebr., 205.D7 103
Holstein, Warren, Mo., 150..C6 101
Holston, riv., Tenn......C11 117
Holston, riv., Va........A2 121
Holston High Knob, mtn., Tenn.C11 117
Holsworthy, Eng., 1,619...D3 12
Holt, Tuscaloosa, Ala., 2,800.B2 78
Holt, Okaloosa, Fla., 600...G2 86
Holt, Ingham, Mich., 4,818..F6 98
Holt, Marshall, Minn., 114..B2 99
Holt, Clay and Clinton, Mo.,
  281..................B3 101
Holt, co., Mo., 7,885.....A2 101
Holt, co., Nebr., 13,722...B7 103
Holtland, Marshall, Tenn.,
  100..................B5 117
Holton, Ripley, Ind., 500..F7 91
Holton, Jackson, Kans.,
  3,028..............A6, C8 93
Holton, Tangipahoa, La.,
  250..................D5 95
Holton, Muskegon, Mich.,
  250..................E4 98
Holtville, Imperial, Calif.,
  3,080................F6 82
Holtwood, Lancaster, Pa.,
  250..................G9 114

Holualoa, Hawaii, Haw., 475 D6 88
Holy, isl., Eng C6 10
Holy, isl., Scot E3 13
Holy, isl., Wales A3 12
Holy Cross, Alsk., 256 C8 79
Holy Cross, Dubuque, Iowa, 157 B7 92
Holycross, Ire E4 11
Holy Cross, mtn., Colo B4 83
Holy Cross, nat. mon., Colo B4 83
Holyhead, Wales, 10,408 A3 12
Holyoke, Phillips, Colo., 1,555 A8 83
Holyoke, Hampden, Mass., 52,689 B2 97
Holyoke, Carlton, Minn., 70 D6 99
Holyoke, range, Mass B2 97
Holyrood, Ellsworth, Kans., 737 D5 93
Holy Trinity, Russell, Ala., 300 C4 78
Holzkirchen, Ger., 4,400 B7 18
Holzminden, Ger., 22,500 B4 17
Homalin, Bur., 10,000 D9 39
Homathko, riv., B.C., Can D5 68
Homberg, Ger., 35,100 B1 17
Homberg, Ger., 6,100 B4 17
Hombori, Mali C4 45
Homburg, Ger., 29,700 D2 17
Home, Marshall, Kans., 188 C7 93
Home, Baker, Oreg C9 113
Home, bay, N.W. Ter., Can C19 67
Homeacre, Butler, Pa., 3,508 *E2 114
Home Corner, Grant, Ind., 2,636 *C6 91
Homécourt, Fr., 10,141 C6 14
Homedale, Owyhee, Idaho, 1,381 F2 89
Homedale, Franklin, Ohio, 670 *C2 111
Home Gardens, Riverside, Calif., 1,541 *F5 82
Homelake, Rio Grande, Colo D4 83
Homeland, Polk, Fla., 400 E5 86
Homeland, Charlton, Ga., 508 F4 87
Home Place, Hamilton, Ind., 600 E5, H8 91
Homer, Alsk., 1,247 D9, L16 79
Homer, Banks, Ga., 612 B3 87
Homer, Champaign, Ill., 1,276 C6 90
Homer, Claiborne, La., 4,665 B2 95
Homer, Calhoun, Mich., 1,629 F6 98
Homer, Dakota, Nebr., 370 B9 103
Homer, Cortland, N.Y., 3,622 C4 108
Homer, Licking, Ohio, 250 B5 111
Homer City, Indiana, Pa., 2,471 E3 114
Homerville, Clinch, Ga., 2,634 E4 87
Homestake, Jefferson, Mont., 30 E4 102
Homestead, Dade, Fla., 9,152 F3, G6 86
Homestead, Iowa, Iowa, 150 C6 92
Homestead, Sheridan, Mont., 100 B12 102
Homestead, Blaine, Okla., 75 A3 112
Homestead, Baker, Oreg., 50 B10 112
Homestead, Allegheny, Pa., 7,502 B6 114
Homestead, nat. mon., Nebr D9 103
Homestead Valley, Marin, Calif., 1,000 *B5 82
Hometown, Cook, Ill., 7,479 *F3 90
Homewood, Jefferson, Ala., 20,289 E4 78
Homewood, Placer, Calif., 100 C3 82
Homewood, Cook, Ill., 13,371 B6, F3 90
Homewood, Scott, Miss., 100 C4 100
Homewood, Horry, S.C., 50 D9 115
Homeworth, Columbiana, Ohio, 600 B6 111
Hominy, Osage, Okla., 2,866 A5 112
Hominy, creek, Okla A5 112
Homochitto, riv., Miss D2 100
Homosassa, Citrus, Fla., 700 D4 86
Homosassa, pt., Fla D4 86
Homs, Syr., 137,200 E11 31
Hon, Scott, Ark., 100 C1 81
Honaker, Russell, Va., 851 B3 121
Honan, prov., China, 48,670,000 E7 34
Honaunau, Hawaii, Haw., 150 D6 88
Hon Chuoi, isl., Viet H6 38
Honda, Col., 16,051 B3 60
Honda, bay, Phil D5 35
Hondo, Lincoln, N. Mex., 100 D4 107
Hondo, Medina, Tex., 4,992 E3 118
Hondo, riv., Br. Hond A3 62
Hondo, riv., Mex h9 63
Honduras, country, N.A., 1,883,480 H12 61, C4 62
Honduras, cape, Hond B4 62
Honduras, gulf, Br. Hond B3 62
Honea Path, Anderson, S.C., 3,453 C3 115
Hönefoss, Nor., 4,400 G4 25
Honeoye Falls, Monroe, N.Y., 2,143 C3 108
Honesdale, Wayne, Pa., 5,569 C11 114
Honey, lake, Calif B3 82
Honey Brook, Chester, Pa., 1,023 F10 114
Honey Creek, Pottawattamie, Iowa, 45 C2 92
Honey Creek, Walworth, Wis., 140 F1 124
Honeycutt, Mitchell, N.C. C4 109
Honeyford, Grand Fork, N. Dak., 40 A8 110
Honey Grove, Juniata, Pa., 75 F6 114
Honey Grove, Fannin, Tex., 2,071 C5 118
Honeyville, Box Elder, Utah, 646 B3 119
Honfleur, Que., Can., 140 A6 73
Honfleur, Fr., 8,970 C4 14
Høng, Den., 1,950 C5 24
Honga, riv., Md D5 85
Hon Gay, Viet., 25,000 B7 38
Honger, mt., Vt C3 120
Hong Kong, Br. dep., Asia, 3,133,131 G14 33, G7 34
Hongwon, Kor., 25,663 F3 37
Honichin Hural, Mong C3 36
Honiton, Eng., 4,724 D4 12
Hon Me, isl., Viet C6 38
Honnef, Ger., 15,500 C2 17
Hönningen, Ger., 5,600 C2 17
Hönö, isl., Swe A5 24
Honobia, Le Flore, Okla., 30 C7 112
Honohina, Hawaii, Haw., 150 D6 88
Honokaa, Hawaii, Haw., 1,247 C6 88
Honokohau, Maui, Haw B5 88
Honolulu, Honolulu, Haw., 294,179 (★486,400) B4, g10 88
Honolulu, co., Haw., 500,409 B3 88
Honomu, Hawaii, Haw., 800 D6 88
Honor, Benzie, Mich., 278 D4 98
Honoraville, Crenshaw, Ala., 150 D3 78
Hon Quan, Viet G7 38
Honshū, isl., Jap I8 37
Honto, see Nevelsk, Sov. Un.
Hood, Sacramento, Calif., 225 B6 82
Hood, co., Tex., 5,443 C4 118
Hood, canal, Wash B2 122
Hood, mtn., Calif A5 82
Hood, mtn., Oreg B5 113
Hood, pt., Austl F2 50
Hoodoo, Coffee, Tenn., 50 B5 117
Hood River, Hood River, Oreg., 3,657 B5 113
Hood River, co., Oreg., 13,395 B5 113
Hoodsport, Mason, Wash., 580 B2 122
Hoogeveen, Neth., 11,500 B6 15
Hoogezand, Neth., 3,600 A6 15
Hooghly, riv., India F12 40
Hook, head, Ire E5 75
Hookena, Hawaii, Haw., 15 D6 88
Hooker, Pulaski, Mo., 40 D5 101
Hooker, Texas, Okla., 1,684 D3 112
Hooker, Turner, S. Dak., 20 D8 116
Hooker, co., Nebr., 1,130 C4 103
Hooker, mtn., B.C., Alta., Can C8 68, C1 69
Hookerton, Greene, N.C., 358 B6 109
Hooks, Bowie, Tex., 2,048 *C5 118
Hooksett, Merrimack, N.H., 900 (3,713▲) D4 105
Hoolehua, Maui, Haw., 973 B4 88
Hoonah, Alsk., 686 D12, k22 79
Hoopa, Humboldt, Calif., 500 B2 82
Hoopa Valley, Indian res., Calif B2 82
Hooper, Alamosa, Colo., 58 D5 83
Hooper, Dodge, Nebr., 832 C9 103
Hooper, Weber, Utah, 300 B3 119
Hooper, Whitman, Wash., 55 C7 122
Hooper, creek, Nebr E2 103
Hooper, isl., Md D5 85
Hooper, strait, Md D5 85
Hooper Bay, Alsk., 460 C6 79
Hoopersville, Dorchester, Md., 230 D5 85
Hoopeston, Vermilion, Ill., 6,606 C6 90
Hooping Harbour, Newf., Can., 166 C3 75
Hoople, Walsh, N. Dak., 334 A8 110
Hooppole, Henry, Ill., 227 B4 90
Hoorn, Neth., 16,100 B5 15
Hoosac, range, Mass., Vt A1 97
Hoosac, tunnel, Mass A2 97
Hoosac Tunnel, Berkshire, Mass., 40 A2 97
Hoosic, riv., N.Y., Vt A1 97, C7 108
Hoosick Falls, Rensselaer, N.Y., 4,023 C7 108
Hoosier, Sask., Can., 40 F1 72
Hoover, Butte, S. Dak., 5 B2 116
Hoover, dam, Ariz., Nev A1 80, H7 104
Hoover, res., Ohio B5 111
Hooverson Heights, Brooke, W. Va., 1,800 B2 123
Hooversville, Somerset, Pa., 1,120 F4 114
Hop, riv., Conn C7 84
Hopa, Tur., 4,900 B13 31
Hopatcong, Sussex, N.J., 3,391 B3 106
Hopatcong, lake, N.J B3 106
Hop Bottom, Susquehanna, Pa., 381 C10 114
Hope, Alsk., 44 C10, g17 79
Hope, Yuma, Ariz., 2 D2 80
Hope, Hempstead, Ark., 8,399 D2 81
Hope, B.C., Can., 2,751 E7 68
Hope, Bonner, Idaho, 96 A2 89
Hope, Bartholomew, Ind., 1,489 F6 91
Hope, Dickinson, Kans., 463 D6 93
Hope, Midland, Mich., 80 E6 98
Hope, Steel, Minn., 102 G5 99
Hope, Warren, N.J., 300 B3 106
Hope, Eddy, N. Mex., 108 E5 107
Hope, Steele, N. Dak., 390 B8 110
Hope, Providence, R.I., 500 C10 84
Hope, isl., B.C., Can D4 68
Hope, isl., Nor B12 4
Hope, lake, Scot B4 13
Hope, mtn., Scot B4 13
Hope, pt., Alsk B6 79
Hopedale, Newf., Can., 218 g9 75
Hopedale, Tazewell, Ill., 737 C4 90
Hopedale, St. Bernard, La., 720 E6 95
Hopedale, Worcester, Mass., 3,987 B4, E1 97
Hopedale, Harrison, Ohio, 930 B7 111
Hopeh (Hopei), prov., China, 42,500,000 D8 34
Hope Hull, Montgomery, Ala., 50 C3 78
Hopelawn, Middlesex, N.J., 2,000 *B4 106
Hopelchén, Mex., 2,037 D7 63
Hopeman, Scot C5 13
Hope Mills, Cumberland, N.C., 1,109 C5 109
Hopemont, Preston, W. Va B5 123
Hope Ranch, Santa Barbara, Calif., 1,000 *E4 82
Hopes Advance, cape, Que., Can A2 73
Hopeton, Woods, Okla., 70 A3 112
Hopetoun, Austl., 72 F3 50
Hopetoun, Austl., 779 G4 51
Hopetown, S. Afr., 2,631 C3 49
Hope Valley, Washington, R.I., 900 C10 84
Hopeville, Ont., Can., 59 C4 72
Hopeville, Clarke, Iowa, 75 D4 92
Hopewell, Cleburne, Ala B4 78
Hopewell, N.S., Can., 404 D7 74
Hopewell Somerset, Md D6 85
Hopewell, Copiah, Miss., 100 D3 100
Hopewell, Mercer, N.J., 1,928 C3 106
Hopewell, Muskingum, Ohio, 200 C5 111
Hopewell, Bedford, Pa., 301 F5 114
Hopewell (Independent City), Va., 17,895 C7, D5 121
Hopewell, Marion, W. Va., 1,230 *B4 123
Hopewell Cape, N.B., Can., 141 D5 74
Hopewell Junction, Dutchess, N.Y., 500 D7 108
Hopi, buttes, Ariz B5 80
Hopi, Indian res., Ariz A5 80
Hopkins, Fairfield, Conn E3 84
Hopkins, Ware, Ga F4 87
Hopkins, Allegan, Mich., 556 F5 98
Hopkins, Hennepin, Minn., 11,370 E6 99
Hopkins, Nodaway, Mo., 710 A3 101
Hopkins, Richland, S.C., 125 D6 115
Hopkins, co., Ky., 38,458 C2 94
Hopkins, co., Tex., 18,594 C5 118
Hopkinsville, Christian, Ky., 19,465 D2 94
Hopkinton, Delaware, Iowa, 768 B6 92
Hopkinton, Middlesex, Mass., 2,754 (4,932▲) B4, D1 97
Hopkinton, Merrimack, N.H., 500 (2,225▲) D4 105
Hopkinton, St. Lawrence, N.Y., 177 A2 108
Hopkinton, Washington, R.I., 120 (4,174▲) D9 84
Hopkinton-Everett, flood control res., N.H D3 105
Hopland, Mendocino, Calif., 600 C2 82
Hoppo, China, 67,000 G6 34
Hopson, Carter, Tenn., 150 C11 117
Hopwood, Fayette, Pa., 1,615 G2 114
Hopuiam, Grays Harbor, Wash., 10,762 C2 122
Horace, Greeley, Kans., 195 D2 93
Horace, Cass, N. Dak., 178 C9 110
Hor al Hammar, lake, Iraq F3 41
Horatio, Sevier, Ark., 722 D1 81
Horatio, Sumter, S.C., 500 C6 115
Horazdovice, Czech., 2,747 D8 17
Horb [am Neckar], Ger., 4,300 E3 17
Horburg, Alta., Can., 20 C3 69
Hörby, Den., 462 A4 24
Hörby, Swe., 3,600 C7 24
Horconcitos, Pan., 1,079 F6 62
Hordaland, co., Nor., 225,300 *G1 25
Hordio, Som C7 47
Hordville, Hamilton, Nebr., 128 C8 103
Horgen, Switz., 13,482 B6 19
Horicon, Dodge, Wis., 2,996 E5 124
Horizon, Sask., Can., 54 H3 70
Hormigueros, P.R., 1,647 C2 65
Hormi-Gueros, mun., P.R., 7,153 C2 65
Hormuz, isl., Iran H8 41
Hormuz, strait, Iran H8 41
Horn, Yuma, Ariz., 100 E2 80
Horn, Aus., 4,705 D7 16
Horn, cape, Chile k12 54
Horn, isl., Miss E5 100
Horn, lake, Miss A3 100
Horn, mtn., Ala B3 78
Horn, mts., N.W. Ter., Can D9 66
Horn pt., Ice m21 25
Horn, pond, Mass C2 97
Hornavan, lake, Swe D7 25
Hornbeak, Obion, Tenn., 307 A2 117
Hornbeck, Vernon, La., 374 C2 95
Hornburg, Ger., 3,400 A5 17
Horncastle, Eng., 3,768 A7 12
Hornell, Steuben, N.Y., 13,907 C3 108
Hornerstown, Monmouth, N.J., 145 C3 106
Hornersville, Dunklin, Mo., 752 E7 101
Hornick, Woodbury, Iowa, 275 B1 92
Horni Litvinov, Czech., 22,800 C8 17
Hornings Mills, Ont., Can., 211 C4 72
Horni Pocernice, Czech., 7,579 n18 26
Horn Island, pass, Miss E3 100
Horn Lake, DeSoto, Miss., 50 E8 117
Hornsby, Hardeman, Tenn., 228 B3 117
Hornsbyville, York, Va., 525 A6 121
Hornsea, Eng., 5,949 D6 10
Hornsey, Eng., 97,885 k12 10
Hornslet, Den., 1,637 B4 24
Hörnum, Ger., 1,000 A4 16
Horovice, Czech., 4,533 D8 17
Horqueta, Par., 17,389 D4 55
Horred, Swe., 504 A6 24
Horrel Hill, Richland, S.C., 125 D6 115
Horry, co., S.C., 68,247 D10 115
Hor Sanniyah, lake, Iraq E3 41
Horse, creek, Colo C7 83
Horse, creek, Mo D4 101
Horse, creek, Wyo D8 125
Horse, peak, N. Mex C1 107
Horse, riv., Alta., Can A5 69
Horseback Knob, hill, Ohio C4 111
Horse Branch, Ohio, Ky., 300 C3 94
Horse Cave, Hart, Ky., 1,780 C4 94
Horse Creek, Laramie, Wyo., 100 D7 125
Horsefly, B.C., Can., 54 C7 68
Horsefly, lake, B.C., Can C7 68
Horse Head, lake, N. Dak B6 110
Horseheads, Chemung, N.Y., 7,207 C4 108
Horse Heaven, Jefferson, Oreg C6 113
Horse Heaven, hills, Wash C6 122
Horse Islands, Newf., Can., 156 C4 75
Horseneck Beach, Bristol, Mass C5 97
Horsens, Den., 37,261 C3 24
Horseshoe Beach, Dixie, Fla C3 86
Horseshoe, cove, Fla C3 86
Horseshoe, lake, Man., Can C4 71
Horseshoe, lake, Ill A8 101
Horseshoe, pt., Fla C3 86
Horseshoe, res., Ariz C4 80
Horse Shoe Bend, Boise, Idaho, 480 F2 89
Horse Springs, Catron, N. Mex., 100 D1 107
Horsham, Austl., 9,241 H4 51
Horsham, Sask., Can G1 70
Horsham, Eng., 21,155 C7 12
Horsovsky Tyn, Czech., 2,393 D7 17
Horta, Az. Is., 8,564 g8 44
Horten, Nor., 13,500 H4, p28 25
Hortense, Brantley, Ga., 380 E5 87
Horton, Marshall, Ala., 50 A3 78
Horton, Bremer, Iowa, 75 B5 92
Horton, Brown, Kans., 2,361 C8 93
Horton, Vernon, Mo., 98 D3 101
Horton, Custer, Mont., 29 D10 102
Horton, pt., N.Y E7 84
Hortonia, lake, Vt D2 120
Hortonville, Outagamie, Wis., 1,366 D5 124
Hörve, Den., 906 C5 24
Hörviken, Swe., 563 B8 24
Hoschton, Jackson, Ga., 370 B3 87
Hosford, Liberty, Fla., 800 B2 86
Hoshab, Pak I11 41
Hoshan, China I7 36
Hoshiarpur, India, 50,739 B5 40
Hoskins, Wayne, Nebr., 179 B8 103
Hoskins, Benton, Oreg., 100 G3 113
Hoskins, Brazoria, Tex G5 118
Hoskins Junction, Brazoria, Tex G5 118
Hosmer, B.C., Can., 104 E10 68
Hosmer, Edmunds, S. Dak., 433 B6 116
Hospers, Sioux, Iowa, 600 A2 92
Hospital, Ire., 572 E3 11
Hospitalet, Sp., 122,813 B7 20
Hosston, Caddo, La., 400 B2 95
Hosta, butte, N. Mex B1 107
Hoste, isl., Chile k12 54
Hotchkiss, Delta, Colo., 626 C3 83
Hot Creek, range, Nev E5 104
Hötensleben, Ger., 5,570 A6 17
Hotevilla, Navajo, Ariz., 700 B5 80
Hotien, see Khotan, China
Hoting, Swe., 945 E7 25
Hot Lake, Union, Oreg B9 113
Hotomice, Czech., 1,609 o17 26
Hotse, China G6 36
Hot Spring, co., Ark., 21,893 C2 81
Hot Springs [National Park], Garland, Ark., 29,212 C2, C6 81
Hot Springs, Sanders, Mont., 585 C2 102
Hot Springs, see Truth or Consequences, N. Mex.
Hot Springs, Madison, N.C., 723 C3 109
Hot Springs, Fall River, S. Dak., 4,943 D2 116
Hot Springs, Brewster, Tex E1 118
Hot Springs, Bath, Va., 200 D3 121
Hot Springs, co., Wyo., 6,365 B4 125
Hot Springs, nat. park, Ark C2 81
Hot Springs, peak, Calif B3 82
Hot Springs, peak, Nev B4 104
Hot Sulphur Springs, Grand, Colo., 237 A4 83
Hottah, lake, N.W. Ter., Can C9 66
Hotte, mts., Haiti F7 64
Hot Wells, Hudspeth, Tex F2 118
Hou, riv., Laos B5 38
Houck, Apache, Ariz., 30 B6 80
Houffalize, Bel., 1,302 D5 15
Houghton, Oxford, Maine D2 96
Houghton, Houghton, Mich., 3,393 A2 98
Houghton, Allegany, N.Y., 1,200 C2 108
Houghton, Brown, S. Dak., 90 B7 116
Houghton, King, Wash., 2,426 D2 122
Houghton, co., Mich., 35,654 B2 98
Houghton, lake, Mich D6 98
Houghton Lake, Roscommon, Mich., 150 D6 98
Houghton Lake Heights, Roscommon, Mich., 1,195 D6 98
Houilles, Fr., 26,370 g9 14
Houlka, Chickasaw, Miss., 547 A4 100
Houlton, Aroostook, Maine, 5,976 (8,289▲) B5 96
Houlton, Columbia, Oreg B4 113
Houltonville, St. Tammany, La., 200 B7 95
Houma, Terrebonne, La., 22,561 C6, E5 95
Houmt-Souk, Tun H12 30
Houndé, Upper Volta, 1,200 D4 45
Hourn, inlet, Scot C3 13
Housatonic, Berkshire, Mass., 1,370 B1 97
Housatonic, riv., Conn., Mass D3 84
House, Quay, N. Mex., 125 C6 107
House, range, Utah D2 119
House, riv., Alta., Can A4 69
House Harbour, Que., Can., 400 B8 74
House Rock, Mohave, Ariz A3 80
House Springs, Jefferson, Mo., 375 B7 101
Houston, Perry, Ark., 206 B3 81
Houston, B.C., Can., 699 B4 68
Houston, Kent, Del., 421 C6 85
Houston, Suwannee, Fla., 50 B4 86
Houston, Jackson, Ind., 40 F5 91
Houston, Houston, Minn., 1,082 G7 99
Houston, Chickasaw, Miss., 2,577 B5 100
Houston, Texas, Mo., 1,660 D6 101
Houston, Shelby, Ohio, 95 B3 111
Houston, Washington, Pa., 1,865 F1 114
Houston, Harris, Tex., 938,219 (★1,251,700) E5, F5 118
Houston, co., Ala., 50,718 D4 78
Houston, co., Ga., 39,154 D3 87
Houston, co., Minn., 16,588 G7 99
Houston, co., Tenn., 4,794 A4 117
Houston, co., Tex., 19,376 D5 118
Houstonia, Pettis, Mo., 261 C4 101
Houtzdale, Clearfield, Pa., 1,239 E5 114
Hove, Eng., 72,843 D7 12
Hövelhof, Ger., 7,900 B3 17
Hoven, Potter, S. Dak., 568 B6 116
Hovenweep, nat. mon., Colo., Utah F6 119
Hoveyzeh, Iran F4 41
Hovland, Cook, Minn., 150 B7 99
Howar, riv., Sud B2 47
Howard, Fayette, Ala., 100 B2 78
Howard, Fremont, Colo., 43 C5 83
Howard, Dade, Fla F3 86
Howard, Taylor, Ga., 200 D2 87
Howard, Elk, Kans., 1,017 E7 93
Howard, Holmes, Miss., 100 B3 100
Howard, Steuben, N.Y., 140 C3 108
Howard, Knox, Ohio, 400 B5 111
Howard, Centre, Pa., 770 D6 114
Howard, Providence, R.I C11 84
Howard, Miner, S. Dak., 1,208 C8 116
Howard, Brown, Wis., 3,485 *D5 124
Howard, co., Ark., 10,878 C2 81
Howard, co., Ind., 69,509 C5 91
Howard, co., Iowa, 12,734 A5 92
Howard, co., Md., 36,152 B4 85
Howard, co., Mo., 10,859 B5 101
Howard, co., Nebr., 6,541 C7 103
Howard, co., Tex., 40,139 C2 118
Howard City, Montcalm, Mich., 1,004 E5 98
Howard Hanson, flood control res., Wash B3 102
Howard Lake, Wright, Minn., 1,007 E4 99
Howard Prairie, res., Oreg E4 113
Howards Grove, Sheboygan, Wis., 350 B6, E6 124
Howe, Butte, Idaho, 25 F6 89
Howe, Lagrange, Ind., 550 A7 91
Howe, LeFlore, Okla., 390 C7 112
Howe, cape, Austl H7 51
Howe, sound, B.C., Can E6 68
Howe Brook, Aroostook, Maine, 25 B4 96
Howell, Woodruff, Ark., 200 B4 81
Howell, Echols, Ga., 141 F3 87
Howell, Livingston, Mich., 4,861 F7 98
Howell, Lincoln, Tenn., 125 B5 117
Howell, Box Elder, Utah, 188 B3 119
Howell, co., Mo., 22,027 E6 101
Howells, Colfax, Nebr., 694 C8 103
Howes, Meade, S. Dak., 2 C2 116
Howes Mill, Dent, Mo D6 101
Howesville, Preston, W. Va., 100 B5 123
Howick, Que., Can., 647 D4 73
Howison, Harrison, Miss E4 100
Howland, Penobscot, Maine, 1,362 C4 96
Howland, isl., Pac. O F11 7
Howley, Newf., Can., 452 D3 75
Howley, mtn., Newf., Can D2 75
Howrah, India, 512,598 D8 39, F12 40
Howser, B.C., Can., 45 D9 68
Howson, peak, B.C., Can B4 68
Howth, Ire., 4,832 D5 11
Hoxie, Lawrence, Ark., 1,886 A5 81
Hoxie, Sheridan, Kans., 1,289 C3 93
Höxter, Ger., 15,200 B4 17
Hoy, isl., Scot B5 13
Hoya, Ger., 4,300 B4 16
Hoyang, China G4 36
Hoyerswerda, Ger., 24,500 B9 17
Hoylake, Eng., 32,268 A4 12
Hoyleton, Washington, Ill., 475 E4 90
Hoyo, mtn., Sp o17 20
Hoyt, Morgan, Colo., 5 B6 83
Hoyt, Jackson, Kans., 283 C8 93
Hoyt, Dawson, Mont., 4 D12 102
Hoyt, Haskell, Okla., 320 B6 112
Hoyt Lakes, St. Louis, Minn., 3,186 *C6 99
Hoyt Station, N.B., Can., 327 D3 74
Hoytsville, Summit, Utah, 250 C4 119
Hoytville, Wood, Ohio, 334 A4 111
Hoytville, Tioga, Pa C7 114
Hradec Kralove, Czech., 55,100 C3 26
Hranice, Czech., 10,800 D4 26
Hrinova, Czech., 6,831 D5 26
Hron, riv., Czech D5 26
Hrubieszow, Pol., 11,300 C7 26
Hsi, China F4 36
Hsiachiang, China, 5,000 K6 36
Hsiaching, China, 4,000 F6 36
Hsiamen, see Amoy, China
Hsian, China, 5,000 E2 37
Hsian, see Sian, China
Hsiang, riv., China J5 36
Hsianghsiang, China, 3,000 K5 36
Hsiangyang, China, 18,000 H5 36
Hsiapu, China, 13,000 K9 36
Hsichang (Sichang), China, 50,000 F5 34
Hsienning, China, 4,000 J6 36
Hsienyang, China G3 36
Hsifeng, China, 33,886 C11 36
Hsilung, China G6 34
Hsinchiang, China G4 36
Hsinchu, Taiwan, 97,600 (137,300▲) G9 34
Hsinfeng, China, 16,000 B3 37
Hsingan, mtn., China D7 36
Hsingcheng, China D9 36
Hsingshanchen, China C5 37
Hsingtai, China, 70,000 F6 36
Hsinhua, China K6 36
Hsining, see Sining, China
Hsinkao, mtn., Taiwan G9 34
Hsinking, see Changchun, China
Hsinmin, China, 35,000 C9 34, C10 36
Hsinning, China F7 36
Hsinpin, China F2 37
Hsintai, China G7 36
Hsintsai, China H6 36
Hsinyang, China, 50,000 H6 36
Hsinyeh, China H5 36
Hsipaw, Bur., 5,000 A2 38
Hsiushui, China, 10,000 J6 36
Hsuancheng, China I8 36
Hsuanhua, China, 114,100 C8 34, D6 36
Hsuchang, China, 58,000 G5 36
Hsui, China H8 36
Hsunho, China B4 37
Hsupu, China K4 36
Hsuwen, China, 15,000 B9 38
Huacho, Peru, 16,039 D2 58
Huachuca City, Cochise, Ariz., 1,330 F5 80
Huahua, riv., Nic C6 62
Huai, riv., China E7 34
Huaian, China, 35,000 H8 36
Huailai, China D6 36
Huaite, China, 1,000 E2 37
Huaiyang, China H6 36
Huaiyin, China, 77,000 H8 36
Huaiyuan, China H7 36
Huajuapan de León, Mex., 8,415 D5, o15 63
Hualalai, mtn., Haw D6 88
Hualgayoc, Peru, 1,173 C2 58
Hualien, Taiwan, 29,455 G9 34
Huallaga, riv., Peru C2 58
Huallanca, Peru C2 58
Hualpai, Indian res., Ariz B2 80
Hualpai, mts., Airz C2 80
Hualpai, peak, Ariz B2 80
Huamachuco, Peru, 2,324 C2 58
Huamantla, Mex., 9,811 n15 63
Huambo, dist., Ang., 598,545 D2 48
Huancabamba, Peru, 2,443 C2 58
Huancané, Peru, 2,236 E4 58
Huancavelica, Peru, 9,594 D2 58
Huancavelica, dept., Peru, 305,619 D2 58
Huancayo, Peru, 33,790 D2 58
Huanchaca, Bol D2 55
Huangan, China, 31,000 I6 36
Huangchuan, China H6 36
Huangkang, China, 53,000 I6 36
Huangmei, China, 21,000 I6 36
Huangping, China, 4,000 K2 36
Huangyen, China, 21,000 J9 36
Huangyuan, China D5 34
Huanjen, China, 5,000 F2 37
Huanta, Peru, 4,439 D3 58
Huánuco, Peru, 15,180 C2 58
Huánuco, dept., Peru, 315,025 C2 58
Huanuni, Bol., 5,696 C2 55
Huara, Chile, 1,794 C2 55
Huaral, Peru, 5,012 D2 58
Huarás, Peru, 14,250 C2 58
Huari, Bol., 1,070 C2 55
Huariaca, Peru, 1,593 D2 58
Huarina, Bol., 1,151 C2 55
Huarmey, Peru, 1,333 D2 58
Huasco, Chile, 1,537 E1 55
Huatabampo, Mex., 10,030 B3 63
Huatien, China, 35,000 C10 34
Huatusco, Mex., 8,673 n15 63
Huauchinango, Mex., 12,053 m14 63
Huaunta, Nic D6 62
Huayllay, Peru, 593 D2 58
Huaytará, Peru, 718 D2 58
Hubbard, Sask., Can., 169 F4 70
Hubbard, Hardin, Iowa, 806 B4 92
Hubbard, Hubbard, Minn., 100 D3 99
Hubbard, Rockingham, N.H E4 105
Hubbard, Trumbull, Ohio, 7,137 A7 111
Hubbard, Marion, Oreg., 526 B2, B4 113
Hubbard, Hill, Tex., 1,628 D4 118
Hubbard, co., Minn., 9,962 C4 99
Hubbard, lake, Mich D7 98
Hubbard Creek, res., Tex C3 118
Hubbard Lake, Alpena, Mich., 150 D7 98
Hubbards, N.S., Can., 525 E5 74
Hubbardston, Worcester, Mass., 500 (1,217▲) B3 97
Hubbardston, Ionia and Clinton, Mich., 381 E5 98
Hubbardton, Rutland, Vt., 25 (238▲) D2 120
Hubbardton, riv., Vt D2 120
Hubbell, Houghton, Mich., 1,429 A2 98
Hubbell, Thayer, Nebr., 126 D8 103
Huben, Aus E6 18
Huberdeau, Que., Can., 605 D3 73
Huber Heights, Montgomery, Ohio, 5,000 C3 111
Hubli, India, 171,326 E6 39
Huckleberry, mtn., Oreg D4 113
Huckleberry, mts., Wash A8 122
Hud, Scurry, Tex C2 118
Huddersfield, Eng., 130,302 (★205,000) A6 12
Huddleston, Bedford, Va., 85 D3 121
Hude, Ger., 7,900 E2 24
Hudiksvall, Swe., 12,000 G7 25
Hudin (Well), Som D6 47
Hudson, Ont., Can., 65 E5 71
Hudson, Que., Can., 1,671 D3, D8 73
Hudson, Weld, Colo., 430 A6 83
Hudson, Pasco, Fla., 800 D4 86
Hudson, McLean, Ill., 493 C5 90
Hudson, Steuben, Ind., 428 A7 91
Hudson, Black Hawk, Iowa, 1,085 B5 92
Hudson, Stafford, Kans., 201 D5 93
Hudson, Middlesex, Mass., 9,666 B4, D1 97
Hudson, Lenawee, Mich., 2,546 G6 98
Hudson, Hillsboro, N.H., 3,561 (5,876▲) E4 105
Hudson, Columbia, N.Y., 11,075 C7 108
Hudson, Caldwell, N.C., 1,536 B2 109
Hudson, Summit, Ohio, 2,438 A6 111
Hudson, Luzerne, Pa., 900 *B8 114
Hudson, Lincoln, S. Dak., 455 D9 116
Hudson, St. Croix, Wis., 4,325 D1 124
Hudson, Fremont, Wyo., 369 C4 125
Hudson, co., N.J., 610,734 B4 106
Hudson, bay, Can D15 67
Hudson, mtn., Maine B3 96
Hudson, riv., N.J., N.Y A5, D5 106
Hudson, strait, N.W. Ter., Can D18 67
Hudson Bay, Sask., Can., 1,601 E4 70
Hudson Center, Hillsboro, N.H., 35 E4 105
Hudson Falls, Washington, N.Y., 7,752 B7 108
Hudson Heights, Que., Can., 1,540 D8 73
Hudson Hope, B.C., Can., 66 A7 68
Hudsonville, Ottawa, Mich., 2,649 F5 98
Hudsonville, Marshall, Miss., 40 A4 100

Hudspeth, co., Tex., 3,343 ... F1 118
Hudwin, lake, Man., Can. ... C4 71
Hue, Viet., 113,000 ... D7 38
Hueco, mts., Tex. ... E1 118
Huedin, Rom., 5,134 ... B6 22
Huehuetenango, Guat., 6,187 ... C2 62
Huejutla, Mex., 3,682 ... m14 63
Huelma, Sp., 7,739 ... D4 20
Huelva, Sp., 74,384 ... D2 20
Huelva, prov., Sp., 399,934 ... *D2 20
Huentelauquén, Chile ... A2 54
Huércal-Overa, Sp., 13,016 ... D5 20
Huerfano, co., Colo., 7,867 ... D5 83
Huerfano, riv., Colo. ... D6 83
Huerva, riv., Sp. ... B5 20
Huésca, Sp., 24,377 ... A5 20
Huesca, prov., Sp., 233,543 ... *A5 20
Huéscar, Sp., 5,499 ... D4 20
Huesco, mts., Mex. ... F1 118
Huetamo de Núñez, Mex., 6,195 ... D4, n13 63
Huete, Sp., 3,209 ... B4 20
Hueysville, Floyd, Ky., 100 ... C7 94
Hueytown, Jefferson, Ala., 5,997 ... E4 78
Huff, Independence, Ark., 70 ... B4 81
Huff, Morton, N. Dak., 55 ... C5 110
Huffakers, Washoe, Nev., 150 ... D2 104
Huffman, Mississippi, Ark., 200 ... B6 81
Huffton, Brown, S. Dak., 10 ... B7 116
Huger, Berkeley, S.C., 500 ... E3 115
Hugh Butler Lake, res., Nebr. ... D5 103
Hughenden, Austl., 2,329 ... D7 50
Hughes, Alsk., 69 ... B9 79
Hughes, St. Francis, Ark., 1,960 ... C5 81
Hughes, Austl. ... F4 50
Hughes, co., Okla., 15,144 ... B5 112
Hughes, co., S. Dak., 12,725 ... C5 116
Hughes, range, B.C., Can. ... E10 68
Hughes, riv., Man., Can. ... A1 71
Hughes, riv., W. Va. ... B3 123
Hughes Springs, Cass, Tex., 1,813 ... *C5 118
Hughestown, Luzerne, Pa., 1,615 ... *D10 114
Hughesville, Charles, Md., 160 ... C4 85
Hughesville, Pettis, Mo., 134 ... C4 101
Hughesville, Judith Basin, Mont., 15 ... C6 102
Hughesville, Lycoming, Pa., 2,218 ... D8 114
Hughson, Stanislaus, Calif., 1,898 ... *D3 82
Hughton, Sask., Can., 86 ... F2 70
Hugo, Lincoln, Colo., 811 ... B7 83
Hugo, Washington, Minn., 538 ... E7 99
Hugo, Choctaw, Okla., 6,287 ... C6 112
Hugo, Josephine, Oreg. ... E3 113
Hugoton, Stevens, Kans., 2,912 ... E2 93
Huguley (Southwest Lanett), Chambers, Ala., 2,189 ... *C4 78
Huhehot (Kweisui), China, 314,000 ... C7 34, D4 36
Huichang, China, 11,000 ... F8 34
Huichapan, Mex., 2,197 ... m14 63
Huichon, Kor., 14,619 ... F3 37
Huila, dist., Ang., 592,451 ... E1 48
Huila, dept., Col., 330,270 ... C2 60
Huili, China ... F5 34
Huimin, China ... E8 36
Huinan, China, 1,000 ... E3 37
Huitzuco, Mex., 6,267 ... n14 63
Huixtla, Mex., 12,344 ... D6 63
Huiyang, China, 35,000 ... G7 34
Huizen, Neth., 17,100 ... B5 15
Hukou, China, 5,000 ... J7 36
Hukuntsi, Bech, 1,423 ... B3 49
Hula (Huleh), lake, Isr. ... A7 32
Hulah, Osage, Okla., 50 ... A5 112
Hulah, res., Okla. ... A5 112
Hulan, China, 60,000 ... B10 34
Hulbert, Crittenden, Ark., 500 ... B5 81
Hulbert, Chippewa, Mich., 300 ... B5 98
Hulbert, Cherokee, Okla., 500 ... B6 112
Hulberton, Orleans, N.Y., 400 ... B2 108
Hulda, Isr., 332 ... C6, h10 32
Huleh (Hula), lake, Isr. ... A7 32
Hulett, Crook, Wyo., 335 ... A8 125
Hulin, China, 1,000 ... D6 37
Hull, Tuscaloosa, Ala. ... B2 78
Hull, Que., Can., 56,929 ... D2 73
Hull, Eng., 303,268 (*360,000) ... A7 12
Hull, Pike, Ill., 535 ... D2 90
Hull, Sioux, Iowa, 1,289 ... A1 92
Hull, Plymouth, Mass., 7,055 ... B6, D3 97
Hull, Emmons, N. Dak., 28 ... C5 110
Hull, co., Que., Can., 84.803 ... D2 73
Hull, riv., Eng. ... G8 13
Hulls Cove, Hancock, Maine, 350 ... D4 96
Hullt, Marion, Oreg. ... C2, C4 113
Hulmeville, Bucks, Pa., 968 ... *F12 114
Hulst, Neth., 5,200 ... C4 15
Hulun, see Hailar, China.
Hulun Nor, lake, China ... B8 34
Hulutao, China, 30,000 ... D9 36
Ḩulwān, Eg., U.A.R., 8,000 ... E3 32
Humacao, P.R., 8,005 ... C7 65
Humacao, mun., P.R., 33,381 ... C7, g11 65
Humahuaca, Arg., 2,094 ... D2 55
Humaitá, Braz., 1,192 ... D1 59
Humansdorp, S. Afr., 3,117 ... D3 49
Humansville, Polk, Mo., 745 ... D4 101
Humarock, Plymouth, Mass., 150 ... E4 97
Humbe, Ang. ... E1 48
Humber, dist., Newf., Can. ... D3 75
Humber, riv., Eng. ... A7 12
Humbird, Clark, Wis., 300 ... D3 124
Humble, Harris, Tex., 1,711 ... E5, F5 118
Humble City, Lea, N. Mex., 33 ... E6 107
Humboldt, Yavapai, Ariz., 450 ... C3 80
Humboldt, Sask., Can., 3,245 ... E3 70
Humboldt, Coles, Ill., 342 ... D5 90
Humboldt, Humboldt. Iowa, 4,031 ... B3 92
Humboldt, Allen, Kans., 2,285 ... E8 93
Humboldt, Marquette, Mich. ... B3 98
Humboldt, Kittson, Minn., 169 ... B1 99
Humboldt, Richardson, Nebr., 1,322 ... D10 103
Humboldt, Pershing, Nev., 20 ... C3 104
Humboldt, Minnehaha, S. Dak., 446 ... D8 116
Humboldt, Gibson, Tenn., 8,482 ... B3 117
Humboldt, co., Calif., 104,892 ... B2 82
Humboldt, co., Iowa, 13,156 ... B3 92
Humboldt, co., Nev., 5,708 ... B3 104
Humboldt, range, Nev. ... C3 104
Humboldt, riv., Nev. ... C3 104
Hume, res., Austl. ... G6 51
Hume, Edgar, Ill., 449 ... D6 90
Hume, Bates, Mo., 369 ... C3 101
Hume, Allegany, N.Y., 350 ... C2 108
Hume, Fauquier, Va., 130 ... C5 121
Humenné, Czech., 11,100 ... D6 26
Humeston, Wayne, Iowa, 663 ... D4 92
Humlum, Den., 548 ... B2 24
Hummelstown, Dauphin, Pa., 4,474 ... *F8 114
Hummels Wharf, Snyder, Pa., 700 ... E8 114
Humnoke, Lonoke, Ark., 319 ... C4 81
Humpata, Ang., 490 ... E1 48
Humphrey, Arkansas and Jefferson, Ark., 649 ... C4 81
Humphrey, Clark, Idaho, 25 ... E6 89
Humphrey, Platte, Nebr., 801 ... C8 103
Humphreys, Terrebonne, La. ... C5 95
Humphreys, Sullivan, Mo., 163 ... A4 101
Humphreys, Jackson, Okla., 50 ... C2 112
Humphreys, co., Miss., 19,093 ... B3 100
Humphreys, co., Tenn., 11,511 ... A4 117
Humphreys, mtn., Calif. ... D4 82
Humphreys, peak, Ariz. ... B4 80
Humpolec, Czech., 5,083 ... D3 26
Humptulips, Grays Harbor, Wash., 110 ... B2 122
Humuula, Hawaii, Haw. ... D6 88
Hūn, Libya ... D3 43
Hunan, prov., China, 36,220,000 ... F7 34
Hunchun, China, 25,000 ... C11 34
Hundested, Den., 3,473 ... C5 24
Hundred, Wetzel, W. Va., 475 ... B4 123
Hunedoara, Rom., 36,498 ... C6 22
Hünfeld, Ger., 6,200 ... C4 17
Hungary, country, Eur., 9,976,530 ... F12 8, B4 22
Hunger, mtn., Mass. ... B1 97
Hungerford, Austl., 80 ... D5 51
Hunghae, Kor. ... H4 37
Hungnam, Kor., 143,600 ... G3 37
Hungshui, riv., China ... G6 34
Huningue, Fr., 4,963 ... B3 18
Hunnewell, Sumner, Kans., 83 ... E6 93
Hunnewell, Shelby, Mo., 284 ... B6 101
Hunsrück, mts., Ger. ... D2 17
Hunstanton, Eng., 4,843 ... B8 12
Hunt, Apache, Ariz. ... C6 80
Hunt, Johnson, Ark., 50 ... B2 81
Hunt, co., Tex., 39,399 ... C4 118
Hunt, mtn., Yukon, Can. ... D7 66
Hunt, mtn., Wyo. ... A5 125
Hunte, riv., Ger. ... B4 16
Hunter, Montgomery, Ala., 1,500 ... *C3 78
Hunter, Woodruff, Ark., 202 ... B4 81
Hunter, Mitchell, Kans., 229 ... C5 93
Hunter, De Soto, La. ... C2 95
Hunter, Carter, Mo., 105 ... E7 101
Hunter, Greene, N.Y., 457 ... C6 108
Hunter, Cass, N. Dak., 446 ... B8 110
Hunter, Garfield, Okla., 203 ... A4 112
Hunter, Comal, Tex., 50 ... A4, E3 118
Hunter, is., Austl. ... o14 50
Hunter, isl., B.C., Can. ... D3 68
Hunter, mtn., N.Y. ... C6 108
Hunterdon, co., N.J., 54,107 ... B3 106
Hunters, Stevens, Wash., 220 ... A7 122
Hunters, hot springs, Oreg. ... E6 113
Hunters, range, B.C., Can. ... D1 69
Huntersfield, mtn., N.Y. ... C6 108
Hunters Creek, Harris, Tex., 2,478 ... *E5 118
Hunters River, P.E.I., Can., 236 ... C6 74
Huntersville, Mecklenburg, N.C., 1,004 ... B3 109
Huntersville, Madison, Tenn. ... B3 117
Huntersville, Pocahontas, W. Va., 80 ... C4 119
Huntertown, Allen, Ind., 400 ... B7 91
Hunting, creek, Md. ... C4 85
Hunting, isl., S.C. ... G7 115
Huntingburg, Dubois, Ind., 4,146 ... H4 91
Huntingdon, B.C., Can., 122 ... B10 68
Huntingdon, Que., Can., 3,134 ... D3 73
Huntingdon, Eng., 8,812 ... B7 12
Huntingdon, Huntingdon, Pa., 7,234 ... F5 114
Huntingdon, Carroll, Tenn., 2,119 ... A3 117
Huntingdon, co., Que., Can., 14,752 ... D3 73
Huntingdon, co., Eng., 79,879 ... B7 12
Huntingdon, co., Pa., 39,457 ... F5 114
Huntingdon, isl., Newf., Can. ... B3 75
Huntington, Sebastian, Ark., 560 ... B1 81
Huntington, Huntington, Ind., 16,185 ... C7 91
Huntington, Hampshire, Mass., 900 (1,392▲) ... B2 97
Huntington, Warren, N.J., 1,879 ... *B2 106
Huntington (P.O.), Suffolk, N.Y., 11,255 ... D3, E7 108
Huntington, Baker, Oreg., 689 ... C9 113
Huntington, Angelina, Tex., 1,009 ... D5 118
Huntington, Emery, Utah, 787 ... D5 119
Huntington, Chittenden, Vt., 118 (518▲) ... C3 120
Huntington, Cabell and Wayne, W. Va., 83,627 (*231,100) ... C2 123
Huntington, co., Ind., 33,814 ... C6 91
Huntington, bay, N.Y. ... F3 84
Huntington Bay, Suffolk, N.Y., 1,267 ... D3 108
Huntington Beach, Orange, Calif., 11,492 ... F3 82
Huntington Beach (East Neck), Suffolk, N.Y., 3,789 ... *E7 108
Huntington Center, Chittenden, Vt., 100 ... C3 120
Huntington Park, Los Angeles, Calif., 29,920 ... F2 82
Huntington Station, Suffolk, N.Y., 23,438 ... F3 84
Huntington Woods, Oakland, Mich., 8,746 ... A7 98
Huntingtown, Calvert, Md., 165 ... C4 85
Huntland, Franklin, Tenn., 500 ... B5 117
Huntley, McHenry, Ill., 1,143 ... A5 90
Huntley, Faribault, Minn., 136 ... G4 99
Huntley, Yellowstone, Mont., 250 ... E8 102
Huntley, Harlan, Nebr., 91 ... D6 103
Huntley, Goshen, Wyo., 75 ... D8 125
Huntly, Scot., 3,952 ... C6 13
Hunts Point, N.S., Can., 169 ... F5 74
Huntsville, Madison, Ala., 72,365 ... A3 78
Huntsville, Madison, Ark., 1,050 ... A2 81
Huntsville, Ont., Can., 3,189 ... B5 72
Huntsville, Madison, Ind., 300 ... E6 91
Huntsville, Butler, Ky., 150 ... C3 94
Huntsville, Randolph, Mo., 1,526 ... B5 101
Huntsville, Scott, Tenn., 500 ... C9 117
Huntsville, Walker, Tex., 11,999 ... D5 118
Huntsville, Weber, Utah, 552 ... B4 119
Huntsville, Columbia, Wash., 100 ... C7 122
Hunucmá, Mex., 6,604 ... C7 63
Huon, gulf, N. Gui. ... k12 50
Hupeh (Hupei), prov., China, 30,790,000 ... E7 34
Hurd, cape, Ont., Can. ... B3 72
Hurdland, Knox, Mo., 205 ... A5 101
Hurdsfield, Wells, N. Dak., 183 ... B6 110
Hurffville, Gloucester, N.J., 200 ... D2 106
Hurlbutt, Fairfield, Conn. (part of Wilton) ... E3 84
Hurley, Jackson, Miss., 400 ... E5 100
Hurley, Stone, Mo., 117 ... E4 101
Hurley, Grant, N. Mex., 1,851 ... E1 107
Hurley, Ulster, N.Y., 350 ... D6 108
Hurley, Turner, S. Dak., 450 ... D8 116
Hurley, Buchanan, Va., 400 ... B3 121
Hurley, Iron, Wis., 2,763 ... B3 124
Hurleyville, Sullivan, N.Y., 800 ... D6 108
Hurliness, Scot. ... B5 13
Hurlock, Dorchester, Md., 1,035 ... C6 85
Hurmagai, Pak. ... G12 40
Huron, Fresno, Calif., 1,269 ... *D3 82
Huron, Lawrence, Ind., 225 ... G4 91
Huron, Atchison, Kans., 119 ... C8 93
Huron, Erie, Ohio, 5,197 ... A5 111
Huron, Beadle, S. Dak., 14,180 ... C7 116
Huron, co., Ont., Can., 53,805 ... D3 72
Huron, co., Mich., 34,006 ... E7 98
Huron, co., Ohio, 47,326 ... A5 111
Huron, lake, Can., U.S. ... B11 77
Huron, mts., Mich. ... B3 98
Huron, riv., Mich. ... A6 98
Huron, riv., Ohio ... A5 111
Hurricane, Washington, Utah, 1,251 ... F2 119
Hurricane, Putnam, W. Va., 1,970 ... C2 123
Hurricane, cliffs, Ariz. ... A2 80
Hurricane, creek, Ark. ... C3 81
Hurricane, creek, Ga. ... E4 87
Hurricane Mills, Humphreys, Tenn., 15 ... B4 117
Hurst, Williamson, Ill., 863 ... F4 90
Hurst, Tarrant, Tex., 10,165 ... *C4 118
Hürth, Ger., 45,700 ... C1 17
Hurtsboro, Russell, Ala., 1,056 ... C4 78
Hurup, Den., 2,160 ... B2 24
Huşi, Rom., 18,055 ... B9 22
Huskerville, Lancaster, Nebr., 550 ... *D9 103
Huskvarna, Swe., 13,800 ... I6 25
Huson, Missoula, Mont., 5 ... C2 102
Hussar, Alta., Can., 213 ... D4 69
Husser, Tangipahoa, La., 35 ... D5 95
Hussein-Dey, Alg., 62,134 ... *B5 44
Hustburg, Humphreys, Tenn., 150 ... B4 117
Hustisford, Dodge, Wis., 708 ... E5 124
Hustle, Essex, Va., 65 ... C5 121
Hustontown, Fulton, Pa., 150 ... F5 114
Hustonville, Lincoln, Ky., 387 ... C5 94
Husum, Ger., 23,800 ... A4 16
Husum, Klickitat, Wash., 15 ... D4 122
Hutch, isl., S.C. ... G1 115
Hutch, mtn., Ariz. ... C4 80
Hutchins, Dallas, Tex., 1,100 ... B5 118
Hutchinson, Reno, Kans., 37,574 ... A4, D6 93
Hutchinson, McLeod, Minn., 6,207 ... F4 99
Hutchinson, co., S. Dak., 11,085 ... D8 116
Hutchinson, co., Tex., 34,419 ... B2 118
Hutchinsons, Isl., Fla. ... E6 86
Hutsonville, Crawford, Ill., 583 ... D6 90
Huttig, Union, Ark., 936 ... D3 81
Hutto, Williamson, Tex., 442 ... D4 118
Hutton, Vernon, La., 50 ... C2 95
Hutton, Garrett, Md., 175 ... D1 85
Huttonsville, Randolph, W. Va., 242 ... C5 123
Hutton Valley, Howell, Mo., 125 ... E6 101
Huttwil, Switz., 4,664 ... B4 19
Huu, Indon. ... G5 35
Ḩuwwārah, Jordan, 2,000 ... g12 32
Huxford, Escambia, Ala., 100 ... D2 78
Huxley, Alta., Can., 102 ... D4 69
Huxley, Story, Iowa, 486 ... C4 92
Huy, Bel., 13,447 ... D5 15
Huyett, Washington, Md. ... A2 85
Hvalpsund, Den., 821 ... B3 24
Hvidbjerg, Den., 978 ... B2 24
Hvide Sande, Den., 1,514 ... C2 24
Hvidovre, Den., 38,411 ... *C6 24
Hvitsten, Nor., 188 ... p28 25
Hvittingfoss, Nor., 576 ... p28 25
Hwainan, China, 286,900 ... H7 36
Hwaining, see Anking, China
Hwang Ho (Yellow), riv., China ... D8 34, F4 36
Hwangling, China ... D6 34
Hweitseh, China ... F5 34
Hyak, Kittitas, Wash., 30 ... B4 122
Hyampom, Trinity, Calif., 30 ... B2 82
Hyannis, Barnstable, Mass., 5,139 ... C7 97
Hyannis, Grant, Nebr., 373 ... B4 103
Hyannis Port, Barnstable, Mass., 800 ... C7 97
Hyas, Sask., Can., 246 ... F4 70
Hyatts, Delaware, Ohio, 100 ... B4 111
Hyattstown, Montgomery, Md., 120 ... B3 85
Hyattsville, Prince Georges, Md., 15,168 ... C1, C4 85
Hyattsville, Big Horn, Wyo., 50 ... A5 125
Hybart, Monroe, Ala., 150 ... D2 78
Hyba, Ont., Can., 20 ... B7 72
Hybla Valley, Fairfax, Va., 1,500 ... *C5 121
Hyco, riv., Va. ... E4 121
Hydaburg, Alsk., 251 ... D13, n23 79
Hyde, co., N.C., 5,765 ... B7 109
Hyde, co., S. Dak., 2,602 ... C6 116
Hyden, Austl., 200 ... F2 50
Hyden, Leslie, Ky., 348 ... C6 94
Hyde Park, Br. Gu. ... A3 59
Hyde Park, Dutchess, N.Y., 1,979 ... D7 108
Hyde Park, Berks, Pa., 2,500 ... *F10 114
Hyde Park, Cache, Utah, 696 ... B4 119
Hyde Park, Lamoille, Vt., 474 (1,219▲) ... B3 120
Hyder, Alsk., 32 ... D13, n24 79
Hyder, Yuma, Ariz. ... E2 80
Hyderabad, India, 931,082 (*1,300,000) ... E6 39, I7 40
Hyderabad, Pak., 420,000 (*434,537) ... C4 39, E2 40
Hydetown, Crawford, Pa., 679 ... C2 114
Hyde Villa, Berks, Pa., 1,300 ... *F10 114
Hydeville, Rutland, Vt., 300 ... D2 120
Hydraulic, B.C., Can. ... C7 68
Hydro, Caddo, Okla., 697 ... B3 112
Hye, Blanco, Tex., 25 ... D3 118
Hyères, Fr., 18,671 ... F7 14
Hyères, is., Fr. ... F7 14
Hyesanjin, Kor. ... F4 37
Hygiene, Boulder, Colo., 125 ... A5 83
Hylo, Alta., Can. ... B4 69
Hylton, Nolan, Tex. ... C2 118
Hyman, Florence, S.C., 40 ... D8 115
Hyman, Mitchell, Tex. ... C2 118
Hymel, St. James, La., 250 ... B6 95
Hymera, Sullivan, Ind., 1,015 ... F3 91
Hyndman, Bedford, Pa., 1,124 ... G4 114
Hyndman, peak, Idaho ... F4 89
Hyner, Clinton, Pa., 75 ... D6 114
Hyogo, pref., Jap., 3,906,487 ... *I7 37
Hyopchon, Kor., 7,000 ... I4 37
Hyrum, Cache, Utah, 1,728 ... B4 119
Hyrynsalmi, Fin., 1,300 ... E13 25
Hysham, Treasure, Mont., 494 ... D9 102
Hyson, Ocean, N.J. ... C4 106
Hythe, Alta., Can., 449 ... B1 69
Hythe, Eng., 10,026 ... C9 12

I

Iablès, dunes, Alg. ... D4 44
Iaeger, McDowell, W. Va., 930 ... D3 123
Ialomita, riv., Rom. ... C8 22
Iamonia, lake, Fla. ... B2 86
Iantha, Barton, Mo., 147 ... D3 101
Iaşi (Jassy), Rom., 94,075 ... B8 22
Iatan, Platte, Mo., 95 ... B3 101
Iates, pt., Guam ... 52
Iatt, lake, La. ... C3 95
Iba, Phil., 3,064 ... o12 35
Ibadan, Nig., 459,196 ... E5 45
Ibagué, Col., 65,500 ... C2 60
Ibanda, Ug. ... B5 48
Ibapah, Tooele, Utah, 25 ... C2 119
Ibar, riv., Yugo. ... D5 22
Ibaraki, pref., Jap., 2,047,024 ... *H10 37
Ibarra, Ec., 14,221 ... A2 58
Ibbenbüren, Ger., 15,700 ... A2 17
Ibera, lake, Arg. ... E4 55
Iberia, Miller, Mo., 694 ... C5 101
Iberia, Morrow, Ohio, 200 ... B5 111
Iberia, par., La., 51,657 ... E4 95
Iberville, Que., Can., 7,588 ... D4 73
Iberville, Iberville, La., 150 ... B5, D4 95
Iberville, co., Que., Can., 18,080 ... D4 73
Iberville, par., La., 29,939 ... D4 95
Ibi, Nig., 6,183 ... E6 45
Ibiá, Braz., 6,999 ... E1 57
Ibiapaba, mts., Braz. ... B2 57
Ibicuí, riv., Braz. ... D2 56
Ibicuy, Arg. ... A5, f7 54
Ibiraputa, riv., Braz. ... E1 56
Ibitinga, Braz., 8,881 ... C3 56
Ibiza, Sp., 11,259 ... C6 20
Ibiza, isl., Sp. ... C6 20
Ibo, Moz., 5,000 ... D7 48
Ibu, Okinawa ... 52
Iburg, Ger., 3,200 ... A3 17
Ica, Peru, 25,273 ... D2 58
Ica, dept., Peru, 158,783 ... D2 58
Içá, riv., Braz. ... D4 60
Icacal, Pan., 86 ... k10 62
Icacos, isl., P.R. ... B8 65
Içana, riv., Braz. ... C4 60
Ice, bay, Ant. ... C18 5
Ice, cave, Iowa ... A6 92
Ice, mtn., B.C., Can. ... B7 68
Ice, mtn., W. Va. ... B6 123
Ice, pond, Pa. ... A1 106
Iceberg, pt., N.W. Ter., Can. ... A23 4
Ice Harbor, dam, Wash. ... C7 122
Iceland, country, Eur., 180,058 ... C5 8, n23 25
Icemorlee, Union, N.C. ... B3 109
Ichang, China, 81,000 ... I4 36
Icheng, China, 15,000 ... I5 36
Ichi Banare, isl., Okinawa ... 52
Ichikawa, Jap., 157,301 ... n18 37
Ichinomiya, Jap., 182,984 ... n15 37
Ichinoseki, Jap., 25,700 ... G10 37
Ichnya, Sov. Un., 10,000 ... F9 27
Ichuan, China ... F4 36
Ichun, China, 60,000 ... B11 34
Ichun, China, 19,000 ... K6 36
Icicle, creek, Wash. ... B5 122
Ickesburg, Perry, Pa., 300 ... F7 114
Icó, Braz., 5,586 ... C3 57
Icy, cape, Alsk. ... A7 79
Icy, strait, Alsk. ... k22 79
Ida, Caddo, La., 300 ... A2 95
Ida, Monroe, Mich., 700 ... G7 98
Ida, co., Iowa, 10,269 ... B2 92
Ida, lake, Minn. ... E3 99
Ida, mtn., Grc. ... E5 29
Idabel, McCurtain, Okla., 4,967 ... D7 112
Ida Grove, Ida, Iowa, 2,265 ... B2 92
Idah, Nig., 7,334 ... E6 45
Idaho, co., Idaho, 13,542 ... D3 89
Idaho, state, U.S., 667,191 ... B5 76, 89
Idaho City, Boise, Idaho, 188 ... F3 89
Idaho Falls, Bonneville, Idaho, 33,161 ... F6 89
Idaho Springs, Clear Creek, Colo., 1,480 ... B5 83
Idalia, Yuma, Colo., 75 ... B8 83
Idalou, Lubbock, Tex., 1,274 ... C2 118
Idamay, Lee, Ky., 100 ... C6 94
Idana, Clay, Kans., 100 ... C6 93
Idanha, Marion, Oreg., 295 ... C4 113
Idanha-a-Nova, Port., 4,459 ... C2 20
Idar-Oberstein, Ger., 30,200 ... D2 17
Idaville, White, Ind., 600 ... C4 91
Idd Abu Sufyan (Well), Sud. ... B2 47
Iddan, Som. ... D6 47
'Iddel Ghanam, Sud. ... C1 47
Iddesleigh, Alta., Can., 50 ... D5 69
Ideal, Macon, Ga., 432 ... D2 87
Ideal, Tripp, S. Dak., 30 ... D6 116
Idehan, des., Libya ... D2 43
Idehan Marzūq, dunes, Libya ... E2 43
Idell, Hunterton, N.J. ... C2 106
Ider, DeKalb, Ala., 140 ... A4 78
Idetown, Luzerne, Pa., 200 ... B7 114
Idfinā, Eg., U.A.R., 1,000 ... C2 32
Idfū, Eg., U.A.R., 20,700 ... E6 43
Idhra, Grc., 2,786 ... D7 23
Ídhra (Hydra), isl., Grc. ... D4 23
Idi, Indon. ... E1, m11 35
Idiofa, Con. L. ... C2 48
Idkū, Eg., U.A.R., 1,000 ... C2 32
Idlewild, Gibson, Tenn., 100 ... A3 117
Idleyld Park, Douglas, Oreg., 100 ... D4 113
Idlib, Syr., 23,700 ... E11 31
Idnah, Jordan, 3,000 ... C6 32
Idrigill, pt., Scot. ... C2 13
Idrija, Yugo., 6,024 ... B2 22
Idrinskoye, Sov. Un. ... E27 9
Idritsa, Sov. Un., 7,000 ... C7 27
Idro, lake, It. ... D6 18
Ie, isl., Okinawa ... 52
Ieper, Bel., 18,121 ... D2 15
Ierapetra, Grc., 5,521 ... E5 23
Ierissos, Grc., 2,768 ... B4 23
Iesi, It., 21,900 ... C4 21
Ifakara, Tan. ... C6 48
Ife, Nig., 111,000 ... E5 45
Iférouane, Niger ... C6 45
Ifni (city), see Sidi Ifni, Ifni
Ifni, Sp. dep., Afr., 52,000 ... D2 44
Iganga, Ug. ... A5 48
Igaraçu, Braz., 2,116 ... h6 57
Igara Paraná, riv., Col. ... D3 60
Igarapava, Braz., 9,083 ... C3 56
Igarape Açu, Braz., 4,195 ... B1 57
Igarapé Miri, Braz., 2,591 ... C5 59
Igarka, Sov. Un., 33,800 ... C11 28
Igdir, Tur., 12,700 ... C15 31
Iglesias, It., 28,004 ... E2 21
Igloo, Fall River, S. Dak., 750 ... D2 116
Ignace, Ont., Can., 517 ... E7 72
Ignacio, LaPlata, Colo., 609 ... D3 83
Igneada, Tur., 713 ... B6 23
Igny, Fr., 4,931 ... h9 14
Igoumenitsa, Grc., 2,448 ... C3 23
Igra, Sov. Un. ... B4 29
Iguaçu, cataracts, Braz. ... D2 56
Iguaçu, riv., Braz. ... D2 56
Iguala, Mex., 28,814 ... D4, n14 63
Igualada, Sp., 19,866 ... B6 20
Igualdad Central, P.R. ... B2 65
Iguape, Braz., 5,465 ... C3 56
Iguatemi, riv., Braz. ... D2 56
Iguatu, Braz., 16,540 ... C3 57
Iguidi, sand dunes, Alg., Maur. ... D3 44
Igurin, isl., Eniwetok ... 52
Iha, Okinawa ... 52
Ihlen, Pipestone, Minn., 111 ... G2 99
Ihosy, Malag. ... L7 49
Ihsien, China ... D9 36
Ihsien, China ... E8 36
Ihsien, China ... G7 36
Ihsing, China, 93,000 ... I8 36
Iida, Jap., 44,300 ... I8, n16 37
Iide-San, peak, Jap. ... H9 37
Iijima, Jap., 3,200 ... n16 37
Iisalmi, Fin., 6,000 ... F12 25
Iizuka, Jap., 60,431 (*120,000) ... *J5 37
Ijamsville, Frederick, Md., 120 ... B3 85
Ijebu Ode, Nig., 27,558 ... E5 45
IJmuiden, Neth., 42,300 ... B4 15
IJssel, riv., Neth. ... B6 15
IJsselmeer, see Zuider Zee, Neth.
Ijuí, Braz., 19,671 ... D2 56
Ijuí, riv., Braz. ... D2 56
Ikaalinen, Fin., 700 ... G10 25
Ikaria, isl., Grc. ... D6 23
Ikast, Den., 5,797 ... B3 24
Ikeda, Jap., 59,688 ... o14 37
Ikela, Con. L. ... B3 48
Ikerre, Nig., 35,584 ... *E6 45
Ikhtiman, Bul., 6,516 ... D6 22
Iki, isl., Jap. ... J4 37
Ila, Madison, Ga., 216 ... B3 87
Ilagan, Phil., 7,436 ... B6 35
Ilan, Taiwan, 38,910 ... *G9 34
Ilan (Sanhsing), China, 40,000 ... B10 34
Ilanz, Switz., 1,843 ... C7 19
Ilawa, Pol., 2,220 ... B5 26
Ilbunga, Austl. ... E6 51
Ilchester, Howard, Md., 60 ... B4 85
Ilderton, Ont., Can., 289 ... D3 72
Ile a la Crosse, Sask., Can., 570 ... B2 70
Ile-a-la-Crosse, lake, Sask., Can. ... B2 70
Ile Bizard, Que., Can., 750 ... D8 73
Ile-de-France, former prov., Fr., 7,733,000 ... C5 14
Ile de France, hills, Fr. ... F3 15
Ile Perrot, Que., Can., 3,106 ... D8 73
Ilesha, Nig., 72,029 ... E5 45
Ilfeld, San Miguel, N. Mex., 5 ... B4 107
Ilford, Man., Can., 165 ... A4 71
Ilford, Eng., 178,210 ... k13 10, C8 12
Ilfracombe, Eng., 8,701 ... C3 12
Ilgachuz, range, B.C., Can. ... C5 68
Ilgin, Tur., 3,100 ... C8 31
Ílhavo, Port., 6,969 ... B1 20
Ilhéus, Braz., 45,712 ... D3 57
Ili, Sov. Un. ... E9 29
Ili, riv., Sov. Un. ... G24 9
Ilia (Elis), prov., Grc., 188,861 ... *D3 23
Iliamna, (Newhalen), Alsk., 110 ... D8 79
Iliamna, lake, Alsk. ... D8 79
Iliamna, vol., Alsk. ... g15 79
Iliff, Logan, Colo., 204 ... A7 83
Iligan, Phil., 7,000 (60,100▲) ... *D6 35
Iliki, lake, Grc. ... g10 23
Ilinskaya, Sov. Un., 10,000 ... I13 27
Iliodhromia, isl., Grc. ... C4 23
Ilion, Herkimer, N.Y., 10,199 ... B5 108
Ilkeston, Eng., 34,672 ... B6 12
Ilkley, Eng., 18,519 ... G7 13
Ill, riv., Aus. ... E4 16
Illampu, mtn., Bol. ... C2 55
Illana, bay, Phil. ... D6 35
Illapel, Chile, 8,266 ... A2 54
Ille-et-Vilaine, dept., Fr., 614,268 ... *C3 14
Iller, riv., Ger. ... D5 16
Ille [-sur-la-Têt], Fr., 3,957 ... F5 14
Illianna, lake, Alsk. ... D8 79
Illiers, Fr., 2,174 ... C4 14
Illinois, state, U.S., 10,081,158 ... B10 77, 90
Illinois, bayou, Ark. ... B2 81
Illinois, peak, Idaho, Mont. ... B3 89
Illinois, riv., Ill. ... B3 90
Illinois, riv., Ark., Okla. ... B6 112
Illinois, riv., Oreg. ... E3 113
Illinois City, Rock Island, Ill., 200 ... B3 90
Illiopolis, Sangamon, Ill., 995 ... D4 90
Illkirch-Graffenstaden, Fr., 9,607 ... F7 15
Illmo, Scott, Mo., 1,174 ... D8 101
Illo, Nig. ... D5 45
Illora, Sp., 13,458 ... D4 20
Ilmen, lake, Sov. Un. ... B8 27
Ilmenau, Ger., 17,700 ... C5 17
Ilmenau, riv., Ger. ... B5 24
Ilo, Peru, 1,043 ... E3 58
Ilobu, Nig., 38,322 ... *E5 45
Ilocos Norte, prov., Phil., 287,000 ... *B6 35
Ilocos Sur, prov., Phil., 338,600 ... *B6 35
Iloilo, Phil., 120,000 (151,000▲) ... C6 35
Iloilo, prov., Phil., 966,100 ... *C6 35
Ilorin, Nig., 40,994 ... E5 45
Ilovaysk, Sov. Un., 10,000 ... r21 27
Ilovlinskaya, Sov. Un. ... G14 27
Ilsenburg, Ger., 6,713 ... B5 17
Ilubabor, reg., Eth., 1,300,000 ... D3 47
Ilwaco, Pacific, Wash., 518 ... C1 122
Ilwaki, Indon. ... G7 35
Ilza, Pol., 3,813 ... C6 26
Imabari, Jap., 81,000 (100,082▲) ... I6 37
Imambaba, Sov. Un. ... C11 41
Im Amguel, Alg. ... E6 44
Iman, Sov. Un., 18,000 ... E16 28
Iman, riv., Sov. Un. ... D7 37
Imathia, prov., Grc., 114,515 ... *B4 23
Imazu, Jap., 4,300 ... n15 37
Imbābah, Eg., U.A.R., 50,100 ... *C6 43
Imbabura, prov., Ec., 157,356 ... A2 58
Imbler, Union, Oreg., 137 ... B9 113
Imboden, Lawrence, Ark., 400 ... A4 81
Ime, mt., Scot. ... B3 13
Imgyt, marsh, Sov. Un. ... B8 29
Imías, Cuba ... E6 64
Imienpo, China, 1,000 ... D4 36
Imilac, Chile ... D2 55
Imlay, Pershing, Nev., 100 ... C3 104
Imlay City, Lapeer, Mich., 1,968 ... E7 98
Imlaystown, Monmouth, N.J., 150 ... C3 106
Immeln, lake, Swe. ... B8 24
Immenstadt, Ger., 10,000 ... E5 16
Immokalee, Collier, Fla., 3,224 ... F5 86
Imnaha, Wallowa, Oreg., 30 ... B10 113
Imnaha, riv., Oreg. ... B10 113
Imogene, Fremont, Iowa, 264 ... D2 92
Imola, It., 27,300 (51,289▲) ... B3 21
Imotski, Yugo., 3,785 ... D3 22
Imperatriz, Braz., 9,004 ... C1 57
Imperia, It., 34,995 ... C1 21
Imperial, Imperial, Calif., 3,007 ... F6 82
Imperial, Sask., Can., 557 ... F3 70
Imperial, Jefferson, Mo., 250 ... B8, C7 101
Imperial, Chase, Nebr., 1,423 ... D4 103
Imperial, Allegheny, Pa., 2,000 ... B8 114
Imperial, Pecos, Tex., 750 ... D1 118
Imperial, co., Calif., 72,105 ... F6 82
Imperial, diversion dam, Ariz. ... E1 80
Imperial, valley, Calif. ... F5 82

Imperial Beach, San Diego, Calif., 17,773 ... *E2 82
Imperoyal, N.S., Can., 490 . E6 74
Imphal, India, 67,717 ... D9 39
Imroz, isl., Tur ... B5 23
Imst, Aus., 5,074 ... E5 16
Imus, Phil., 3,715 ... o13 35
'Imwās, Jordan, 2,000 ... h10 32
Ina, Jap., 18,300 ... n16 37
Ina, Jefferson, Ill., 332 ... E5 90
Ina, riv., Pol ... B3 26
In Ahmar (Well), Maur ... C3 45
Inajá, Braz., 1,079 ... C3 57
In Alay, Mali ... C4 45
Inanwatan, W. Irian ... F8 35
Iñapari, Peru, 131 ... D4 58
Inarajan, Guam, 761 ... 52
Inarajan, bay, Guam ... 52
Inari, Fin., 300 ... C12 25
Inari, lake, Fin ... C12 25
Inatori, Jap., 9,100 ... o18 37
Inavale, Webster, Nebr., 150 . D7 103
Inawashiro, lake, Jap ... H10 37
In Azaoua (Oasis), Niger ... B6 45
In Belbel, Alg ... D5 44
In Beriem (Well), Mali ... C4 45
Inca, Sp., 13,816 ... C7 20
Incastro, riv., It ... h9 21
Ince, cape, Tur ... A10 31
Incesu, Tur., 5,900 ... C10 31
Inch, Ire ... E2 11
Inchard, bay, Scot ... B3 13
Inchelium, Ferry, Wash., 100 ... A7 122
Inchon, Kor., 402,000 ... H3 37
Indaal, inlet, Scot ... E2 13
Indalsälven, riv., Swe ... F7 25
Indan, Phil ... o14 35
Indang, Phil., 2,747 ... o13 35
Indaw, Bur., 2,138 ... D10 39
Independence, Inyo, Calif., 950 ... D4 82
Independence, Warren, Ind., 170 ... D3 91
Independence, Buchanan, Iowa, 5,498 ... B6 92
Independence, Montgomery, Kans., 11,222 ... E8 93
Independence, Kenton, Ky., 309 ... A7, B5 94
Independence, Tangipahoa, La., 1,941 ... D5 95
Independence, Hennepin, Minn., 1,446 ... *F5 99
Independence, Tate, Miss., 159 ... A4 100
Independence, Jackson, Mo., 62,328 ... B3, E2 101
Independence, Cuyahoga, Ohio, 6,868 ... B2 111
Independence, Polk, Oreg., 1,930 ... C1, C3 113
Independence, Grayson, Va., 679 ... E1 121
Independence, Trempealeau, Wis., 954 ... D2 124
Independence, co., Ark., 20,048 ... B4 81
Independence, mts., Nev ... C5 104
Independence, riv., N.Y ... B5 108
Independence, rock, Wyo ... C5 125
Indepedence Hill, Lake, Ind., 1,824 ... *B3 91
Independencia, Bol., 1,742 ... C2 55
Inderagiri, riv., Indon ... F2 35
Inderborskiy, Sov. Un ... D4 29
Index, peak, Wyo ... A3 125
India, country, Asia, 439,072,893 ... D6 39, G10 33
India, Portuguese, see Goa, Damão, and Diu, ter., India
Indiahoma, Comanche, Okla., 378 ... C3 112
Indialantic, Brevard, Fla., 1,653 ... *D6 86
Indian, bay, Fla ... D4 86
Indian, cave, Tenn ... C10 117
Indian, creek, Ind ... H5 91
Indian, creek, Md ... C4 85
Indian, creek, Ohio ... C3 111
Indian, creek, S. Dak ... B2 116
Indian, creek, Tenn ... B3 117
Indian, creek, W. Va ... D4 123
Indian, isl., N.C ... B7 109
Indian, lake, Mich ... C4 98
Indian, lake, N.Y ... B6 108
Indian, lake, Ohio ... B4 111
Indian, mtn., Conn ... B3 84
Indian, ocean, World ... G2 2
Indian, peak, Utah ... E2 119
Indian, peak, Wyo ... A3 125
Indian, pond, Maine ... B3 96
Indian, pond, Maine ... C3 96
Indian, pond, Maine ... D3 96
Indian, riv., Ont., Can ... B7 72
Indian, riv., Del ... C7 85
Indian, riv., Fla ... D6 86
Indian, riv., N.Y ... A5 108
Indian, rock, Oreg ... C8 113
Indian, stream, N.H ... A1 105
Indiana, Indiana, Pa., 13,005 . E3 114
Indiana, co., Pa., 75,366 ... E3 114
Indiana, state, U.S., 4,662,498 ... B10 77, 91
Indian Agency, La Plata, Colo., 180 ... D3 83
Indianapolis, Marion, Ind., 476,258 (★806,900) ... E5, H8 91
Indianapolis, Custer, Okla . B3 112
Indian Bay, Man., Can ... E4 71
Indian Bayou, Vermilion, La ... D3 95
Indian Brook, N.S., Can., 66 ... C9 74
Indian Cove, Owyhee, Idaho ... G3 89
Indian Creek, Dade, Fla., 60 ... *G6 86
Indian Gap, Hamilton, Tex., 40 ... D3 118
Indian Grave, mtn., Ga ... C2 87
Indian Harbour, Newf., Can ... A3, g10 75
Indian Head, Sask., Can., 1,802 ... G4 70
Indian Head, Charles, Md., 780 ... C3 85
Indian Hill, Hamilton, Ohio, 4,526 ... D2 111
Indian Lake, Hamilton, N.Y., 600 ... B6 108
Indian Mills, Burlington, N.J . D3 106
Indian Mound, Stewart, Tenn., 325 ... A4 117
Indianola, Vermilion, Ill., 295 ... D6 90
Indianola, Warren, Iowa, 7,062 ... B7, C4 92
Indianola, Sunflower, Miss., 6,714 ... B3 100
Indianola, Red Willow, Nebr., 754 ... D5 103
Indianola, Pittsburg, Okla., 234 ... C6 112
Indianola, Allegheny, Pa., 1,000 ... *E2 114
Indian Pass, Gulf, Fla., 50 ... C1 86
Indian Prairie, canal, Fla ... E5 86
Indian River, Ont., Can., 130 ... C6 72
Indian River (village), Washington, Maine, 100 .. D5 96
Indian River, Cheboygan, Mich., 300 ... C6 98
Indian River, co., Fla., 25,309 ... E6 86
Indian Rocks Beach, Pinellas, Fla., 1,940 ... E1 86
Indian Springs, Butts, Ga., 250 ... C3 87
Indian Springs, Martin, Ind., 120 ... G4 91
Indian Springs, Clark, Nev., 450 ... G6 104
Indian Town, Martin, Fla., 1,411 ... E6 87
Indian Trail, Union, N.C., 364 ... B3 109
Indian Valley, Adams, Idaho, 30 ... E2 89
Indian Valley, Floyd, Va., 75 ... E2 121
Indian Wells, Navajo, Ariz., 5 ... B5 80
Indiera Alta, P.R ... C3 65
Indiga, Sov. Un., 800 ... B18 9
Indigirka, riv., Sov. Un ... C17 28
Indio, Riverside, Calif., 9,745 ... F5 82
Indio, riv., Pan ... k10 62
Indio, riv., Pan ... k12 62
Indochina, reg., Asia ... D6 38, H13 33
Indonesia, country, Asia, 95,189,000 ... F6 35, J15 33
Indore, India, 394,941 ... D6 39, F5 40
Indravati, riv., India ... H8 40
Indre, dept., Fr., 251,432 ... *D4 14
Indre-et-Loire, dept., Fr., 395,210 ... *D4 14
Indus, Koochiching, Minn., 10 ... B5 99
Indus, riv., Pak ... C4 39
Indus, riv., mouths, Pak ... D4 39
Industrial, York, S.C., 1,000 ... B6 115
Industrial City, Jefferson, Ala., 1,000 ... *E4 78
Industry, McDonough, Ill., 514 ... C3 90
Industry, Clay and Dickinson, Kans., 7 ... C6 93
Industry, Beaver, Pa., 2,338 . *E1 114
Ine, Jap., 2,500 ... n14 37
Inebolu, Tur., 5,900 ... B9 31
Inez, Martin, Ky., 566 ... C7 94
Inez, Victoria, Tex., 400 ... E4 118
Infanta, Phil., 957 ... o12 35
Infanta, Phil., 2,412 ... o13 35
Infantas, Col ... B3 60
Infantes, Sp., 9,909 ... C4 20
Infiesto, Sp., 1,650 ... A3 20
Ingá, Braz., 6,383 ... h6 57
In Gall, Niger ... C6 45
Ingalls, Bradley, Ark., 100 ... D3 81
Ingalls, Madison, Ind., 873 .. E6 91
Ingalls, Gray, Kans., 174 ... E3 93
Ingalls, Menominee, Mich., 200 ... C3 98
Ingalls Park, Will, Ill., 6,840 ... *B5 114
Ingallston, Menominee, Mich ... C3 98
Ingelheim, Ger., 15,800 ... D3 17
Ingende, Con. L ... B2 48
Ingeniero Jacobacci, Arg., 2,257 ... C3 54
Ingeniero Luiggi, Arg ... B4 54
Ingenika, riv., B.C., Can ... A5 68
Ingenio, P.R ... B6 65
Inger, Itasca, Minn., 135 ... C5 99
Ingersheim, Fr., 3,006 ... A3 18
Ingersoll, Ont., Can., 6,874 . D4 72
Ingersoll, Alfalfa, Okla., 30 . A3 112
Ingham, Austl., 4,790 ... C8 50
Ingham, Lincoln, Nebr ... D5 103
Ingham, co., Mich., 211,296 . F6 98
Ingleford, Spokane, Wash ... D7 122
Ingleside, Queen Annes, Md., 110 ... B6 85
Ingleside, Adams, Nebr., 400 ... D7 103
Ingleside, San Patricio, Tex., 3,022 ... F4 118
Inglewood, Austl., 1,058 ... D8 51
Inglewood, Los Angeles, Calif., 63,390 ... F2 82
Inglewood, Ont., Can., 419 . D5 72
Inglewood, N.Z., 1,901 ... M15 52
Inglewood, Davidson, Tenn., 26,527 ... *A5 117
Inglis, Man., Can., 295 ... D1 71
Ingold, Sampson, N.C., 100 . C5 109
Ingolf, Ont., Can., 90 ... E4 71
Ingolstadt, Ger., 53,400 ... E6 17
Ingomar, Union, Miss., 262 . A4 100
Ingomar, Rosebud, Mont., 80 ... D9 102
Ingomar, Allegheny, Pa., 1,500 ... A5 114
Ingonish, N.S., Can., 375 ... C9 74
Ingornachoix, bay, Newf., Can . C3 75
Ingraham, lake, Fla ... G5 86
Ingram, Randolph, Ark ... A4 81
Ingram, Allegheny, Pa., 4,730 ... B5 114
Ingram, Kerr, Tex., 950 ... *D3 118
Ingram, Rusk, Wis., 99 ... C3 124
Ingramport, N.S., Can., 107 . E5 74
Ingrid Christensen, coast, Ant . C20 5
In Guezzam (Oasis), Alg., 3 .. F5 44
Ingul, riv., Sov. Un ... H9 27
Ingulets, riv., Sov. Un ... H9 27
Inhambane, Moz ... B6 49
Inhambane, prov., Moz ... B5 49
Inhambane, bay, Moz ... B6 49
Inhambupe, Braz., 3,811 ... D3 57
Inhaminga, Moz ... A6 49
Inharrime, Moz., 10,000 ... B6 49
Inhuçu, Braz., 938 ... B2 57
Inhumas, Braz., 8,298 ... B3 56
Iniesta, Sp., 4,694 ... C5 20
Ining (Kuldja), China, 108,200 ... E10 29
Inírida, riv., Col ... C4 60
Inishark, isl., Ire ... D1 11
Inishbofin, isl., Ire ... D1 11
Inishcrone, Ire., 533 ... C2 11
Inisheer, isl., Ire ... D2 11
Inishkea, is., Ire ... C1 11
Inishmaan, isl., Ire ... D2 11
Inishmore, isl., Ire ... D2 11
Inishmurray, isl., Ire ... C3 11
Inishowen, head, Ire ... B5 11
Inishtioge, Ire., 292 ... E4 11
Inishtooskert, isl., Ire ... E1 11
Inishtrahull, isl., Ire ... B4 11
Inishturk, isl., Ire ... D1 11
Inishvickillane, isl., Ire ... E1 11
Injasuti, peak, Bas., S. Afr ... D4 49
Injune, Austl., 322 ... B7 52
Inkerman, N.B., Can., 468 ... B5 74
Inkerman, Luzerne, Pa., 1,000 ... *B8 114
Inkom, Bannock, Idaho, 528 . G6 89
Inkster, Wayne, Mich., 39,097 ... A7 98
Inkster, Grand Forks, N. Dak., 282 ... A8 110
Inland, sea, Jap ... I6 37
Inlet, Hamilton, N.Y., 300 .. B6 108
Inlet, pt., N.Y ... E7 84
Inman, Fayette, Ga., 175 ... C2 87
Inman, McPherson, Kans., 729 ... D6 93
Inman, Holt, Nebr., 192 ... B7 103
Inman, Spartanburg, S.C., 1,714 ... A3 115
Inman Mills, Spartanburg, S.C., 1,769 ... *A3 115
Inn, riv., Aus., Ger., Switz ... E8 17, D6 16
Innamincka, Austl., 25 ... E7 50
Inner, sound, Scot ... C3 13
Inner Hebrides, is., Scot ... *C2 13
Innerleithen, Scot., 2,299 ... E5 13
Inner Mongolia, prov., China, 9,200,000 ... C8 34
Inner-Rhoden, sub canton, Switz., 12,943 ... B7 19
Innerthaler, lake, Switz ... B6 19
Innisfail, Austl., 6,917 ... C8 50
Innisfail, Alta., Can., 2,270 . C4 69
Innisfree, Alta., Can., 291 ... C5 69
Innokentyevskiy, Sov. Un ... B10 37
Innsbruck, Aus., 100,695 ... E5 16
Inny, riv., Ire ... D4 11
Inola, Rogers, Okla., 584 ... A6 112
Inongo, Con. L., 2,061 ... B2 48
Inowroclaw, Pol., 47,000 ... B5 26
I-n-Rabir, Alg ... E5 44
Ins, Switz., 2,486 ... B3 19
In Salah, Alg., 330 (17,511▲) . D5 44
Insch, Scot., 1,421 ... C6 13
Insein, Bur., 27,030 ... E10 39
Insinger, Sask., Can., 129 ... F4 70
Inspiration, Gila, Ariz., 500 ... D5 80
Institute, Kanawha, W. Va., 2,500 ... C3 123
Instow, Sask., Can., 45 ... H1 70
Intake, Dawson, Mont., 14 .. C12 102
Intercession City, Osceola, Fla., 500 ... D5 86
Intercity, Snohomish, Wash., 1,475 ... *B3 122
Interior, Jackson, S. Dak., 179 ... D4 116
Interior, Giles, Va., 40 ... D2 121
Interlachen, Putnam, Fla., 349 ... C5 86
Interlaken, Monmouth, N.J., 1,168 ... *C4 106
Interlaken, Seneca, N.Y., 780 . C4 108
Interlaken, Switz., 4,738 ... C4 19
International Falls, Koochiching, Minn., 6,778 . B5 99
International Peace Garden, park, Man., Can ... E1 71
Intersection, mtn., B.C., Can ... C8 68
Intervale, Cumberland, Maine ... E2, E5 96
Intervale, Carroll, N.H., 200 . B4 105
Intiyaco, Arg ... E3 55
Intracoastal, waterway, La., Tex ... E3 95, E5 118
Inubo, cape, Jap ... I10 37
Inútil, bay, Chile ... h12 54
Inuvik, N.W. Ter., Can., 1,248 ... C6 67
Inver, bay, Ire ... C3 11
Inveraray, Scot., 501 ... D3 13
Inverbervie, Scot., 921 ... D6 13
Invercargill, N.Z., 35,605 (★41,088) ... Q12 51
Inverell, Austl., 8,208 ... D8 51
Invergarry, Scot ... C4 13
Invergordon, Scot., 1,640 ... C4 13
Inver Grove, Dakota, Minn., 713 ... E7 99
Invermay, Sask., Can., 395 .. F4 70
Invermere, B.C., Can., 744 .. D9 68
Inverness, Bullock, Ala., 100 . D4 78
Inverness, Marin, Calif., 450 ... B4, C2 82
Inverness, B.C., Can., 15 ... B2 68
Inverness, Que., Can., 296 ... C6 73
Inverness, N.S., Can., 2,109 . C8 74
Inverness, Citrus, Fla., 1,878 . D4 86
Inverness, Cook, Ill., 1,110 .. *E2 90
Inverness, Sunflower, Miss., 1,039 ... B3 100
Inverness, Hill, Mont., 175 .. B6 102
Inverness, Scot., 29,773 ... C4 13
Inverness, co., N.S., Can., 18,718 ... C8 74
Inverness, co., Scot., 83,425 . C3 13
Inwood, Ont., Can., 201 ... E3 72
Inwood, Man., Can., 183 ... D3 71
Inwood, Marshall, Ind., 165 .. B5 91
Inwood, Lyons, Iowa, 638 ... A1 92
Inwood, Nassau, N.Y., 10,362 ... E2 108
Inwood, Berkeley, W. Va., 425 ... B6 123
Inyanga, S. Rh ... A5 49
Inyankara, creek, Wyo ... A8 125
Inyankara, mtn., Wyo ... A8 125
Inyo, co., Calif., 11,684 ... D5 82
Inyokern, Kern, Calif., 450 .. E5 82
Inza, Sov. Un., 2,000 ... E16 27
Inzano, lake, B.C., Can ... B5 68
Ioannina, Grc., 34,997 ... C3 23
Ioannina, prov., Grc., 155,326 ... *C3 23
Ioka, Duchesne, Utah ... C5 119
Iokanga, Sov. Un ... D18 25
Iola, Clay, Ill., 155 ... E5 90
Iola, Allen, Kans., 6,885 ... E8 93
Iola, Waupaca, Wis., 831 ... D4 124
Iona, N.S., Can., 179 ... D9 74
Iona, Bonneville, Idaho, 702 .. F7 89
Iona, Murray, Minn., 328 ... G3 99
Iona, Gloucester, N.J., 200 ... D2 106
Iona, Lyman, S. Dak., 25 ... D6 116
Iona, isl., Scot ... D2 13
Ione, Amador, Calif., 1,118 .. C3 82
Ione, Weld, Colo., 100 ... A6 83
Ione, Nye, Nev., 10 ... E4 104
Ione, Union, N. Mex., 3 ... B6 107
Ione, Morrow, Oreg., 350 ... B7 113
Ione, Pend Oreille, Wash., 648 ... A8 122
Ionia, Chickasaw, Iowa, 265 . A5 92
Ionia, Jewell, Kans., 100 ... C5 93
Ionia, Ionia, Mich., 6,754 ... F5 98
Ionia, Benton, Mo., 114 ... G4 101
Ionia, co., Mich., 43,132 ... F5 98
Ionian, is., Grc ... C3 23
Ionian, sea, Grc ... C2 23
Ios, isl., Grc ... D5 23
Iosco, co., Mich., 16,505 ... D7 98
Iosegun, riv., Alta., Can ... B2 69
Iosepa, Tooele, Utah, 15 ... C3 119
Iota, Acadia, La., 1,245 ... D3 95
Iowa, Calcasieu, La., 1,857 .. D2 95
Iowa, co., Iowa, 16,396 ... C5 92
Iowa, co., Wis., 19,631 ... E3 124
Iowa, state, U.S., 2,757,537 ... B9 77, 92
Iowa, lake, Iowa ... A3 92
Iowa, riv., Iowa ... C5 92
Iowa City, Johnson, Iowa, 33,443 ... C6 92
Iowa Falls, Hardin, Iowa, 5,565 ... B4 92
Iowa Park, Wichita, Tex., 3,295 ... C3 118
Ipameri, Braz., 8,987 ... E1 57
Ipava, Fulton, Ill., 623 ... C3 90
Ipel, riv., Czech ... D5 26
Iphigenia, sound, Alsk ... n23 79
Ipiales, Col., 11,569 ... C2 60
Ipin, China, 177,500 ... F5 34
Ipirá, Braz., 3,807 ... D3 57
Ipoh, Mala., 125,770 ... J4 38
Ipoly, riv., Hung ... B4 22
Ippy, Cen. Afr. Rep ... D4 46
Ipsala, Tur., 6,000 ... B6 23
Ipswich, Austl., 48,679 ... C9 51
Ipswich, Essex, Mass., 5,400 (8,544▲) ... A6 97
Ipswich, Eng., 117,325 ... B9 12
Ipswich, Edmunds, S. Dak., 1,131 ... B6 116
Ipswich, riv., Mass ... A5 97
Ipu, Braz., 7,724 ... B2 57
Ipueiras, Braz., 3,173 ... B2 57
Iquique, Chile, 50,000 ... D2 55
Iquitos, Peru, 40,408 ... B3 58
Ira, Jasper, Iowa, 95 ... C4 92
Ira, Scurry, Tex., 250 ... C2 118
Ira, Rutland, Vt., 50 (220▲) . D2 120
Iraan, Pecos, Tex., 1,255 ... D2 118
Iracoubo, Fr. Gu., 1,109 ... A4 59
Iraklion, Grc., 63,458 ... E5 23
Iraklion, prov., Grc., 208,374 ... *E5 23
Irala, Par., 1,174 ... E5 55
Iran (Persia), country, Asia, 20,042,000 ... F8 33, 41
Iran, mts., Indon., Mala ... E4 35
Iran, plat., Iran ... E7 41
Irapa, Ven., 3,657 ... A5 60
Irapuato, Mex., 83,505 . C4, m13 63
Iraq, country, Asia, 5,047,635 ... F14 31, E2 41
Irasburg, Orleans, Vt., 200 (711▲) ... B4 120
Irbid, Jordan, 23,157 ... B7 32
Irbil, Iraq, 34,313 ... D15 31
Irbit, Sov. Un., 41,200 ... D21 9
Irebu, Con. L ... B2 48
Iredell, co., N.C., 62,526 ... B3 109
Ireland, Dubois, Ind., 340 ... H3 91
Ireland, Coryell, Tex., 60 ... D4 118
Ireland (Eire), country, Eur., 2,818,341 ... E7 8, D3 10
Ireland, isl., Bermuda ... E13 77
Ireland, pt., Bermuda ... E13 77
Irene, Clay, S. Dak., 399 ... D8 116
Irerrer, riv., Alg ... E6 44
Ireton, Sioux, Iowa, 510 ... B1 92
Irgiz, Sov. Un., 1,900 ... D6 29
Irgiz, riv., Sov. Un ... F21 9
Irharhar, riv., Alg ... D6 44
Iri, Kor., 53,400 (65,700▲) .. I3 37
Iringa, Tan., 9,587 ... C6 48
Iriomote, isl., Ryukyu Is ... G9 34
Irion, co., Tex., 1,183 ... D2 118
Iriri, riv., Braz ... C4 59
Irish, sea, Eur ... D4 10
Irkutsk, Sov. Un., 380,000 . D13 28
Irma, Alta., Can., 425 ... C5 69
Irmino, Sov. Un., 10,000 ... q21 27
Irmo, Lexington, S.C., 359 . C5 115
Iron, St. Louis, Minn., 187 .. C6 99
Iron, co., Mich., 17,184 ... B2 98
Iron, co., Mo., 8,041 ... D7 101
Iron, co., Utah, 10,795 ... F2 119
Iron, co., Wis., 7,830 ... B3 124
Iron, mtn., Ariz ... D4 80
Iron, mtn., Fla ... E5 86
Iron, mts., Ire ... C4 11
Iron, mts., Tenn., Va . C12 117, E1 121
Iron Belt, Iron, Wis., 550 ... B3 124
Iron City, Seminole, Ga., 298 ... F2 87
Iron City, Lawrence, Tenn., 700 ... B4 117
Irondale, Jefferson, Ala., 3,501 ... E5 78
Irondale, Ont., Can., 50 ... C6 72
Irondale, Washington, Mo., 335 ... D7 101
Irondale, Jefferson, Ohio, 705 ... B7 111
Irondequoit, Monroe, N.Y., 55,337 ... B3 108
Iron Gate, Alleghany, Va., 716 ... C3 121
Iron Gate, gorge, Rom., Yugo . C6 22
Iron Lightning, Ziebach, S. Dak., 60 ... B4 116
Iron Mountain, Dickinson, Mich., 9,299 ... C2 98
Iron Mountain, St. Francois, Mo., 300 ... D7 101
Iron Mountain, Mineral, Mont ... C2 102
Iron Mountain, Laramie, Wyo., 15 ... D7 125
Iron Ridge, Dodge, Wis., 419 ... E5 124
Iron River, Iron, Mich., 3,754 ... B2 98
Iron River, Bayfield, Wis., 900 ... B2 124
Irons, Lake, Mich., 30 ... D5 98
Irons, mtn., Ark ... B3 81
Ironsburg, Monroe, Tenn ... D9 117
Ironside, Que., Can., 359 ... A9 72
Ironside, Malheur, Oreg., 30 ... C9 113
Ironspot, Muskingum, Ohio, 100 ... C5 111
Iron Springs, Yavapai, Ariz., 5 ... C3 80
Ironton, Crow Wing, Minn., 724 ... D5 99
Ironton, Iron, Mo., 1,310 ... D7 101
Ironton, Lawrence, Ohio, 15,745 ... D5 111
Ironwood, Gogebic, Mich., 10,265 ... A5 98
Iroquois, Ont., Can., 1,136 .. C9 72
Iroquois, Iroquois, Ill., 231 .. C6 90
Iroquois, Beadle and Kingsbury, S. Dak., 385 ... C8 116
Iroquois, co., Ill., 33,562 ... C5 90
Iroquois, riv., Ind ... C3 91
Iroquois Falls, Ont., Can., 1,681 ... E9 72
Irosin, Phil., 7,394 ... *C6 35
Irrawaddy, riv., Bur ... D10 39
Irricana, Alta., Can., 167 ... D4 69
Irrigon, Morrow, Oreg., 232 ... B7 113
Irt, riv., Eng ... F5 13
Irthing, Eng ... F6 13
Irtysh, Sov. Un ... C8 29
Irtysh, riv., Sov. Un ... E24 9
Irumu, Con. L ... A4 48
Irún, Sp., 29,814 ... A5 20
Irvine, Alta., Can., 240 ... E5 69
Irvine, Marion, Fla., 200 ... C4 86
Irvine, Estill, Ky., 2,955 ... C6 94
Irvine, Warren, Pa., 300 ... C3 114
Irvine, Scot., 16,910 ... E4 13
Irvine, riv., Scot ... E4 13
Irvines Landing, B.C., Can., 83 ... E5 68
Irving, Montgomery, Ill., 570 ... D4 90
Irving, Lane, Oreg., 200 ... C3 113
Irving, Dallas, Tex., 45,985 .. B5 118
Irving College, Warren, Tenn ... D8 117
Irvington, Mobile, Ala., 350 . E1 78
Irvington, Washington, Ill., 387 ... E4 90
Irvington, Breckinridge, Ky., 1,190 ... D3 94
Irvington, Douglas, Nebr., 100 ... D3 103
Irvington, Essex, N.J., 59,379 ... E4 106
Irvington, Westchester, N.Y., 5,494 ... D1 108
Irvington, Lancaster, Va., 570 ... D6 121
Irvona, Clearfield, Pa., 781 . E4 114
Irwin, Bonneville, Idaho, 330 . F7 89
Irwin, Shelby, Iowa, 425 ... C2 92
Irwin, Cherry, Nebr ... B4 103
Irwin, Westmoreland, Pa., 4,270 ... F2 114
Irwin, Lancaster, S.C., 1,113 ... *B6 115
Irwin, co., Ga., 9,211 ... E3 87
Irwindale, Los Angeles, Calif., 1,518 ... *E4 82
Irwinton, Wilkinson, Ga., 673 ... D3 87
Irwinville, Irwin, Ga., 300 .. E3 87
Isa, Nig ... D6 45
Isa, Okinawa ... 52
Isaac, lake, B.C., Can ... C7 68
Isaacs Harbour, N.S., Can., 139 ... D8 74
Isabel, Barber, Kans., 181 ... E5 93
Isabel, Dewey, S. Dak., 488 . B4 116
Isabel, mtn., Wyo ... C2 125
Isabela (Basilan), Phil., 9,000 (156,046▲) ... D6 35
Isabela, P.R., 7,302 ... A2 65
Isabela, mun., P.R., 28,754 . B2 65
Isabela, prov., Phil., 442,800 ... *B6 35
Isabela (Albemarle), isl., Ec ... g5 58
Isabella, Man., Can., 52 ... D1 71
Isabella, Worth, Ga., 100 ... E3 87
Isabella, Delta, Mich ... C4 98
Isabella, Lake, Minn., 50 ... C7 99
Isabella, Major, Okla., 100 . A3 112
Isabella, Fayette, Pa., 900 .. G2 114
Isabella, Polk, Tenn., 415 ... D9 117
Isabella, co., Mich., 35,348 . E6 98
Isabella, lake, Minn ... C7 99
Isabella, mts., Nic ... D5 62
Isaccea, Rom., 5,203 ... C9 22
Isafjördur, Ice., 2,694 ... m21 25
Isahaya, Jap., 31,900 ... J5 37
Isaka, Con. L ... B2 48
Isanga, Con. L ... B3 48
Isangi, Con. L ... A3 48
Isanti, Isanti, Minn., 521 ... E5 99
Isanti, co., Minn., 13,530 ... E5 99
Isar, riv., Ger ... E7 17
Isarco, riv., It ... A3 21
'Isawiyah, Jordan ... h12, m14 32
Isbergues, Fr., 4,277 ... D2 15
Isbister, riv., Man., Can ... C4 71
Ischia, isl., It ... D4 21
Ischua, Cattaraugus, N.Y., 200 ... C2 108
Iscia Baidoa (Isha Baidoa), Som., 11,000 (13,200▲) ... E5 47
Ise, riv., Ger ... F4 24
Isel, riv., Aus ... E6 16
Iselin, Middlesex, N.J., 14,000 ... B4 106
Iselin, Indiana, Pa., 500 ... E3 114
Isen, riv., Ger ... A8 18
Isenthal, Switz., 556 ... C6 19
Iseo, lake, It ... B3 21
Isère, dept., Fr., 729,789 ... D1 18
Isère, riv., Fr ... E6 14
Iserlohn, Ger., 55,300 (★95,000) ... B2 17
Isernia, It., 8,000 ... D5 21
Iseyin, Nig., 49,690 ... E5 45
Isezaki, Jap., 54,500 (84,250▲) ... H9, m18 37
Isfahan, see Esfahān, Iran
Isha Baidoa (Iscia Baidoa), Som., 11,000 (13,200▲) ... E5 47
Ishan, China ... G6 38
Ishawooa, Park, Wyo ... A3 125
Ishawooa Cone, mtn., Wyo ... A3 125
Ishikawa, Okinawa ... 52
Ishikawa, pref., Jap., 973,418 ... *H8 37
Ishim, Sov. Un., 39,000 ... B7 29
Ishim, riv., Sov. Un ... D23 9
Ishimbay, Sov. Un., 53,500 . C5 29
Ishinomaki, Jap., 83,947 ... G10 37
Ishioka, Jap., 17,800 ... m19 37
Ishkashim, Afg., 5,000 ... C15 41
Ishpeming, Marquette, Mich., 8,857 ... B3 98
Isigny-sur-Mer, Fr., 2,391 ... C3 14
Isil-Kul, Sov. Un ... C8 29
Isiolo, Ken ... A6 48
Isisford, Austl., 294 ... B5 51
Iskenderun (Alexandretta), Tur., 63,700 ... D11 31
Isker, riv., Bul ... D6 22
Iskilip, Tur., 12,300 ... B10 31
Iskitim, Sov. Un., 30,000 ... E25 9
Isla Cabellos, Ur., 1,485 ... E1 56
Isla Cristina, Sp., 8,276 ... D2 20
Islamorada, Monroe, Fla., 700 ... H6 86
Island, McLean, Ky., 462 ... C2 94
Island, co., Wash., 19,638 ... A3 122
Island, beach, N.J ... D4 106
Island, dam, Ala ... B3 78
Island, lake, Man., Can ... C4 71
Island, pond, N.H ... E4 105
Island, pond, Vt ... B5 120
Island Brook, Que., Can., 420 ... D6 73
Island City, Union, Oreg., 158 ... B8 113
Island Falls, Aroostook, Maine, 800 (1,018▲) ... B4 96
Island Grove, Alachua, Fla., 175 ... C4 86
Island Heights, Ocean, N.J., 1,150 ... D4 106
Island Lake, Lake and McHenry, Ill., 1,639 ... *E2 90
Island Lake, riv., Man., Can .. B4 71
Island No. 40, isl., Tenn ... E8 117
Island Park, Fremont, Idaho, 53 ... E7 89
Island Park, Nassau, N.Y., 3,846 ... *E3 108
Island Park, Newport, R.I., 1,147 ... C12 84
Island Park, res., Idaho ... E7 89
Island Pond, Essex, Vt., 1,319 ... B5 120
Islay, Alta., Can., 107 ... C5 69
Islay, isl., Scot ... E2 13
Isle, Mille Lacs, Minn., 529 ... D5 99
Isle, riv., Fr ... E4 14
Isle au Haut, Knox, Maine, 60 (68▲) ... D4 96
Isle aux Morts, Newf., Can., 884 ... E2 75
Isle La Motte, Grand Isle, Vt., 125 (238▲) ... B2 120
Isle Maligne, Que., Can., 2,070 ... A6 73
Isle of Ely, co., Eng., 89,112 ... *D7 10
Isle of Hope, Chatham, Ga., 1,500 ... *D6 87
Isle of Man, Br. dep., Eur., 48,151 ... C7 11
Isle of Palms, Charleston, S.C., 1,186 ... F3 115
Isle of Wight, Isle of Wight, Va., 60 ... B6, E6 121
Isle of Wight, co., Eng., 95,479 ... D6 12
Isle of Wight, co., Va., 17,164 ... E6 121
Isle of Wight, isl., Eng ... E6 10
Isle Pierre, B.C., Can., 63 ... C6 68
Isle Royale, isl., Mich ... A2 98
Isle Royale, nat. park, Mich ... A2 98
Islesboro, Waldo, Maine, 150 (444▲) ... D4 96
Islesford, Hancock, Maine, 100 ... D4 96
Islet, isl., Eniwetok ... 52
Isleta, Bernalillo, N. Mex., 750 ... C3 107
Isleta, Indian res., N. Mex ... C3 107
Isleton, Sacramento, Calif., 1,039 ... B6, C3 82
Isle Verte, Que., Can., 1,517 ... A8 73
Islington, Ont., Can., 2,550 . E6 72
Islington, Norfolk, Mass., 4,500 ... D2 97
Islip, Suffolk, N.Y., 8,000 ... G4 84
Islip Terrace, Suffolk, N.Y., 3,000 ... *G4 84
Islitas, Webb, Tex ... F3 118
Ismā'īliah, canal, Eg., U.A.R . D3 32
Ismay, Custer, Mont., 59 .. D12 102
Isnā, Eg., U.A.R., 27,900 ... D6 43
Isoka, N. Rh ... D5 48
Isola, Humphreys, Miss., 532 ... B3 100
Isola Capo Rizzuto, It., 9,218 ... E6 21
Isola della Scala, It., 9,693 .. D7 18
Isole Pelagie, is., It ... B2 43
Isoline, Cumberland, Tenn .. C8 117
Isparta, Tur., 36,200 ... D8 31
Ispir, Tur., 2,000 ... B13 31
Israel, country, Asia, 2,170,082 ... F6 33, 32
Israel, riv., N.H ... B3 105
Issano Landing, Br. Gu., 353 ... A3 59
Issaquah, King, Wash., 1,870 ... B3, D2 122
Issaquena, co., Miss., 3,576 ... C2 100
Issia, I.C., 1,800 ... E3 45
Issoire, Fr., 10,454 ... E5 14
Issoudun, Fr., 13,900 ... D4 14
Issue, Charles, Md., 35 ... D4 85
Issyk-kul, lake, Sov. Un ... G24 9
Issy-[les-Moulineaux], Fr., 51,776 ... g10 14
Istachatta, Hernando, Fla., 150 ... D4 86
Istanbul (Constantinople), Tur., 1,459,500 (★1,775,000) ... B7 23, B7 31
Istiaia, Grc., 5,147 ... C4 23
Istokpoga, lake, Fla ... E5 86
Istonio, see Vasto, It.
Istranca, mts., Tur ... B6 23
Istres, Fr., 6,426 ... F6 14
Istrian, pen., Yugo ... C2 22
Itá, Par., 18,777 ... E4 55
Itabaiana, Braz., 11,847 .. C3, h6 57
Itabaiana, Braz., 11,050 ... D3 57
Itabaianinha, Braz., 2,907 .. D3 57
Itabapoana, riv., Braz ... C4 56
Itaberaba, Braz., 8,555 ... D2 57
Itaberaí, Braz., 4,632 ... B3 56
Itabira, Braz., 15,539 ... E2 57
Itaboraí, Braz., 4,930 ... C4, h6 56
Itabuna, Braz., 54,268 ... D3 57
Itacaiunas, riv., Braz ... D6 59
Itacoatiara, Braz., 8,818 ... C3 59
Itaetê, Braz., 1,168 ... D2 57
Itaguaçu, Braz., 1,892 ... E2 57
Itaituba, Braz., 1,187 ... C3 59
Itajaí, Braz., 38,889 ... D3 57
Itajubá, Braz., 31,262 ... C3 57
Itala, Som., 800 ... E6 47

Italia, Nassau, Fla. B6 86
Italy, Ellis, Tex., 1,183 C4 118
Italy, country, Eur., 50,623,569 G11 8, 21
Italy Cross, N.S., Can., 158 E5 74
Itamaraca, isl., Braz h6 57
Itami, Jap., 86,455 o14 37
Itanhaém, Braz., 5,376 n8 56
Itapecuru, riv., Braz B2 57
Itapecuru-Mirim, Braz., 3,385 B2 57
Itapemirim, Braz., 4,095 C4 56
Itaperuna, Braz., 18,095 C4 56
Itapetininga, Braz., 29,468 C3, m7 56
Itapeva, Braz., 13,510 C3 56
Itapi, riv., Braz C3 59
Itapicuru, riv., Braz D3 57
Itapipoca, Braz., 7,186 B3 57
Itápolis, Braz., 7,430 k7 56
Itaporanga, Braz., 5,328 C3 57
Itapúa, dept., Par., 111,424 E4 55
Itaqui, Braz., 13,223 D1 56
Itararé, Braz., 12,812 C3 56
Itararé, riv., Braz C3 56
Itaretama, Braz., 1,559 C3, g5 57
Itarsi, India, 33,611 F6 40
Itasca, DuPage, Ill., 3,564 E2 88
Itasca, Hill, Tex., 1,383 C4 118
Itasca, co., Minn., 38,006 C5 99
Itasca, lake, Minn C3 99
Itatiaya, peak, Braz h5 56
Itatiba, Braz., 12,336 m8 56
Itatinga, Braz., 1,628 m7 56
Itaúna, Braz., 22,319 F2 57
Itawamba, co., Miss., 15,080 A5 100
Itcha, mts., B.C., Can C5 68
Itea, Grc., 2,532 C3 23
Itecoaí, riv., Braz C3 58
Ithaca, Gratiot, Mich., 2,611 E6 98
Ithaca, Saunders, Nebr., 126 E2 103
Ithaca, Tompkins, N.Y., 28,799 C4 108
Ithaki, Grc., 2,632 C3 23
Itháki (Ithaca), isl., Grc C3 23
Itigi, Tan C5 48
Itimbiri, riv., Con. L. A3 48
Itirapina, Braz., 2,730 m8 56
Ito, Jap., 43,600 o18 37
Itoko, Con. L. B3 48
Itoman, Okinawa, 6,872 52
Itsmina, Col., 2,755 B2 60
Itta Bena, Leflore, Miss., 1,914 B3 100
Itu, Braz., 23,435 C3, m8 56
Itu, China F8 36
Ituango, Col., 2,673 B2 60
Ituiutaba, Braz., 29,724 B3 56
Itumbiara, Braz., 12,575 B3 56
Ituna, Sask., Can., 837 F4 70
Itung, China C10 34
Ituxi, riv., Braz C4 58
Ityāy al Bārūd, Eg., U.A.R., 1,000 D2 32
Itzehoe, Ger., 36,100 B4 16
Iuka, Marion, Ill., 378 E5 90
Iuka, Pratt, Kans., 225 E5 93
Iuka, Tishomingo, Miss., 2,010 A5 100
Iva, W. Sam., 623 52
Iva, Anderson, S.C., 1,357 C2 115
Ivaí, riv., Braz C2 56
Ivalo, Fin., 2,600 C12 25
Ivangrad, Yugo., 7,020 D4 22
Ivanhoe, Austl., 351 F5 51
Ivanhoe, Tulare, Calif., 1,616 *D4 82
Ivanhoe, Lincoln, Minn., 719 F2 99
Ivanhoe, Sampson, N.C., 40 C5 109
Ivanhoe, Wythe, Va., 800 E2 121
Ivanić Grad, Yugo., 1,111 C3 22
Ivanovichi, Sov. Un D7 37
Ivano-Frankovsk, Sov. Un., 72,000 G5 27
Ivanovka, Sov. Un., 5,000 p21 27
Ivanovka, Sov. Un., 1,000 A4 37
Ivanovo, Sov. Un., 352,000 B2 29
Ivanteyevka, Sov. Un., 10,000 n17 27
Ivato, Malag g9 49
Ivaton, Lincoln, W. Va C5 123
Ivaylovgrad, Bul., 2,419 E8 22
Ivdel, Sov. Un., 40,000 A6 29
Ivesdale, Champaign and Piatt, Ill., 360 D5 90
Ivigtut, Grnld., 141 C19 4
Ivinheima, riv., Braz C2 57
Ivins, Washington, Utah, 77 F2 119
Ivohibe, Malag., 642 h9 49
Ivor, Southampton, Va., 398 E6 121
Ivory Coast, country, Afr F5 42, E3 45
Ivoryton, Middlesex, Conn., 950 D7 84
Ivösjön, lake, Swe B8 24
Ivrea, It., 23,723 B1 21
Ivry, Que., Can., 142 C3 73
Ivry-sur-Seine, Fr., 53,406 g10 14
Ivy, Dallas, Ark., 150 C3 81
Ivy, Monroe, Tenn D9 117
Ivy, Albemarle, Va., 250 C4 121
Ivy, mtn., Conn B4 84
Ivyton, Magoffin, Ky., 500 C7 94
Ivywild, El Paso, Colo., 11,065 C6 83
Iwakuni, Jap., 100,346 I6 37
Iwamisawa, Jap., 37,800 E10 37
Iwamurada, Jap., 7,600 m17 37
Iwanai, Jap., 21,100 E10 37
Iwate, pref., Jap., 1,448,517 *G10 37
Iwaya, Jap., 13,700 o14 37
Iwo, Nig., 100,006 E5 45
Iwo (Iō), isl., Pac. O 52
Iwon, Kor F4 37
Ixiamas, Bol., 292 B2 55
Ixmiquilpan, Mex., 1,739 m14 63
Ixmiquilpan, lake, Mex m14 63
Ixtacalco, Mex., 10,896 h9 63
Ixtapalapa, Mex., 17,372 h9 63
Ixtlán de Juárez, Mex., 1,108 o15 63
Ixtlán del Rió, Mex., 8,282 C4, m11 63
Iyang, China, 80,000 J5 36
Izabal, lake, Guat C3 62
Izamal, Mex., 8,675 C7 63
Izard, co., Ark., 6,766 A4 81
Izberbash, Sov. Un., 25,000 G18 9
Izee, Grant, Oreg C7 113
Izegem, Bel., 17,095 D3 15
Izhevsk, Sov. Un., 312,000 B4 29
Izhma, riv., Sov. Un C19 9
Izmail, Sov. Un., 52,000 I7 27
Izmir (Smyrna), Tur., 370,900 C6 23, C6 31
Izmir, bay, Tur C6 23
Izmit (Kocaeli), Tur., 73,700 B7 23
İznik, Tur., 6,300 B7 23
Iznik, lake, Tur B7 23
Izra', Syr., 2,900 B8 32
Izu, isl., Pac. O D8 7
Izúcar de Matamoros, Mex., 16,175 n14 63
Izuhara, Jap., 15,300 I4 37
Izumo, Jap., 34,400 I6 37
Izyum, Sov. Un., 34,000 G11 27

J

Jabal Al'Uwaynat, mtn., Sud A2 47
Jabal Ash Sham, mtn., Mus. & Om D2 39
Jabalón, riv., Sp C4 20
Jabalpur (Jubbulpore), India, 295,375 (★367,014) D6 39, F7 40
Jablonec [nad Nisou], Czech., 27,300 C3 26
Jablonna, Czech., 219 o17 26
Jablonna, Pol., 2,000 k13 26
Jablunkov, pass, Czech D5 26
Jaboatão, Braz., 33,963 C4, k6 57
Jaboticabal, Braz., 20,231 C3, k7 56
Jabrin, isl., Iran H5 41
Jaca, Sp., 9,856 A5 20
Jacala [de Ledesma], Mex., 1,612 m14 63
Jácana, P.R C5 65
Jacaraci, Braz., 650 D2 57
Jacaré, riv., Braz D2 57
Jacareí, Braz., 28,131 m9 56
Jacarèzinho, Braz., 14,813 C3 56
Jachal, Arg., 4,278 F2 55
Jachymov, Czech., 6,806 C2 26
Jacinto, Dallas, Ark D3 81
Jacinto, Alcorn, Miss., 75 A5 100
Jacinto City, Harris, Tex., 9,547 F5 118
Jack, co., Tex., 7,418 C3 118
Jack, mtn., Mont D4 102
Jack, mtn., Va C3 121
Jack, mtn., Wash A5 122
Jackfish, lake, Sask., Can D1 70
Jackfork, mtn., Okla C6 112
Jackman, Somerset, Maine, 300 (984▲) C2 96
Jackman Station, Somerset, Maine, 500 C2 96
Jacksboro, Campbell, Tenn., 800 C9 117
Jacksboro, Jack, Tex., 3,816 C3 118
Jacks Creek, Chester, Tenn., 150 B3 117
Jackson, Clarke, Ala., 4,959 D2 78
Jackson, Amador, Calif., 1,852 C3 82
Jackson, Butts, Ga., 2,545 C3 87
Jackson, Breathitt, Ky., 1,852 C6 94
Jackson, Calcasieu, La., 1,824 D4 95
Jackson, Waldo, Maine, 65 (220▲) D3 96
Jackson, Jackson, Mich., 50,720 (★121,400) F6 98
Jackson, Jackson, Minn., 3,370 G4 99
Jackson, Hinds, Miss., 144,422 (★191,200) C3 100
Jackson, Cape Girardeau, Mo., 4,875 D8 101
Jackson, Beaverhead, Mont., 100 E3 102
Jackson, Dakota, Nebr., 224 B9 103
Jackson, Carroll, N.H., 200 (315▲) B4 105
Jackson, Camden, N.J D3 106
Jackson, Northampton, N.C., 765 A6 109
Jackson, Jackson, Ohio, 6,980 C5 111
Jackson, Providence, R.I., 140 C10 84
Jackson, Aiken, S.C., 1,746 E4 115
Jackson, Madison, Tenn., 34,376 B3 117
Jackson, Washington, Wis., 458 E1, E5 124
Jackson, Teton, Wyo., 1,437 B2 125
Jackson, co., Ala., 36,681 A3 78
Jackson, co., Ark., 22,843 B4 81
Jackson, co., Colo., 1,758 A4 83
Jackson, co., Fla., 36,208 B1 86
Jackson, co., Ga., 18,499 B3 87
Jackson, co., Ill., 42,151 F4 90
Jackson, co., Ind., 30,556 G5 91
Jackson, co., Iowa, 20,754 B7 92
Jackson, co., Kans., 10,309 C8 93
Jackson, co., Ky., 10,677 C5 94
Jackson, co., Mich., 131,994 F6 98
Jackson, co., Minn., 15,501 G3 99
Jackson, co., Miss., 55,522 E5 100
Jackson, co., Mo., 622,732 C3 101
Jackson, co., N.C., 17,780 D3 109
Jackson, co., Ohio, 29,372 C5 111
Jackson, co., Okla., 29,736 C2 112
Jackson, co., Oreg., 73,962 E4 113
Jackson, co., S. Dak., 1,985 D4 116
Jackson, co., Tenn., 9,233 C8 117
Jackson, co., Tex., 14,040 E4 118
Jackson, co., W. Va., 18,541 C3 123
Jackson, co., Wis., 15,151 D3 124
Jackson, par., La., 15,828 B3 95
Jackson, lake, Fla B2 86
Jackson, lake, Fla E5 86
Jackson, lake, Ga C3 87
Jackson, lake, Wyo B2 125
Jackson, mtn., Maine D2 96
Jackson, mts., Nev B3 104
Jackson, riv., Va D3 121
Jacksonboro, Colleton, S.C., 400 F1, F7 115
Jacksonburg, Wetzel, W. Va., 300 A6, B4 123
Jackson Center, Shelby, Ohio, 980 B3 111
Jackson Center, Mercer, Pa., 292 D1 114
Jackson Hill, Sullivan, Ind., 150 F3 91
Jackson Junction, Winneshiek, Iowa, 89 A5 92
Jackson Mills, Ocean, N.J C4 106
Jacksonport, Jackson, Ark., 271 B4 81
Jackson's Arm, Newf., Can., 422 D3 75
Jacksons Gap, Tallapoosa, Ala., 500 C4 78
Jacksonville, Calhoun, Ala., 5,678 B4 78
Jacksonville, Pulaski, Ark., 14,488 C3, D6 81
Jacksonville, Duval, Fla., 201,030 (★456,700) B5, B6 86
Jacksonville, Telfair, Ga., 236 E3 87
Jacksonville, Morgan, Ill., 21,690 D3 90
Jacksonville, Shelby, Iowa, 150 C2 92
Jacksonville, Washington, Maine, 250 D5 96
Jacksonville, Randolph, Mo., 153 B5 101
Jacksonville, Onslow, N.C., 13,491 C6 109
Jacksonville, Athens, Ohio, 580 C5 111
Jacksonville, Jackson, Oreg., 1,172 E4 113
Jacksonville, Cherokee, Tex., 9,590 D5 118
Jacksonville, Windham, Vt., 240 F3 120
Jacksonville Beach, Duval, Fla., 12,049 B5, B6 86
Jacmel, Hai., 8,545 F7 64
Jaco, isl., Indon G7 35
Jacob Lake, Coconino, Ariz., 125 A3 80
Jacobabad, Pak., 22,835 C2 40
Jacobina, Braz., 12,373 D2 57
Jacobson, Aitkin, Minn., 85 C5 99
Jacobstown, Burlington, N.J., 150 C3 106
Jacobsville, Houghton, Mich., 100 B2 98
Jacoby, Pointe Coupee, La., 290 D4 95
Jacomino, Cuba, 6,121 h12 64
Jacques Cartier, Que., Can., 40,807 D9 73
Jacques, Cartier, co., Que., Can., 1,747,696 D8 73
Jacques Cartier, mtn., Que., Can A3 73
Jacques Cartier, riv., Que., Can B6 73
Jacquet, riv., N.B., Can B3 74
Jacquet River, N.B., Can., 500 B4 74
Jacuí, Braz., 1,372 k8 56
Jacuí, riv., Braz D2 56
Jacuípe, riv., Braz D3 57
Jacuitiba, Braz m8 57
Jacumba, San Diego, Calif., 650 F5 82
Jacundá, riv., Braz C4 59
Jaddi, cape, Pak I11 41
Jade, Ger., 3,600 A8 15
Jade, bay, Ger E2 24
Jadhamiyah, Sau. Ar I14 31
Jadotville, Con. L., 74,500 D4 48
Jādū, Libya, 960 C2 43
Jaegerspris, Den., 1,254 C5 24
Jaeggevarre, mtn., Nor C8 25
Jaén, Peru, 510 C2 58
Jaén, Sp., 64,917 D4 20
Jaén, prov., Sp., 736,391 *D4 20
Jaffa, see Tel Aviv-Jaffa, Isr.
Jaffa, cape, Austl H2 51
Jaffna, Cey., 76,664 G7 39
Jaffrey, Cheshire, N.H., 1,648 E2 105
Jaffrey Center, Cheshire, N.H., 300 E2 105
Jafura Dahana, des., Sau. Ar I4 41
Jagadhri, India, 32,637 (★84,337) B6 40
Jagdalpur, India, 20,412 H9 40
Jagin, riv., Iran H9 41
Jagok, lake, China B12 40
Jagst, riv., Ger D4 17
Jaguarão, Braz., 12,336 E2 56
Jaguariaíva, Braz., 6,465 C3 56
Jagüey Grande, Cuba, 4,374 D3 64
Jahrom, Iran, 23,390 G6 41
Jahra, Kuw G3 41
Jaicós, Braz., 1,308 C2 57
Jaipur, India, 403,444 C6 39, D5 40
Jaisalmer, India, 8,362 D3 40
Jajce, Yugo., 6,839 C3 22
Jajpur, India, 13,802 G11 40
Jakarta, see Djakarta, Indon.
Jakes Corner, Gila, Ariz., 15 C4 80
Jakin, Early, Ga., 176 E2 87
Jakobstad (Pietersaari), Fin., 7,378 F10 25
Jal, Lea, N. Mex., 3,051 E6 107
Jalaigai, pt., Guam 52
Jalalabad, Afg., 14,756 D15 41
Jalapa, Guat., 6,610 C2 62
Jalapa Enríquez (Jalapa), Mex., 51,166 D5, n15 63
Jalaun, India, 14,101 D7 40
Jalca Grande Amazonas, Peru, 1,189 C2 58
Jaleswar, India, 10,000 D10 40
Jalgaon, India, 80,351 G5 40
Jalisco, Mex., 3,154 m11 63
Jalisco, state, Mex., 2,402,884 C4, m12 63
Jalna, India, 67,158 H5 40
Jalón, riv., Sp B5 20
Jalor, India, 12,882 E4 40
Jalpa, Mex., 6,204 C4, m12 63
Jalpaiguri, India, 48,738 D12 40
Jalpan, Mex., 1,009 C5, m14 63
Jalu, Libya, 80 D4 43
Jālū, oasis, Libya D4 43
Jamachim, riv., Braz D3 59
Jamaica, Guthrie, Iowa, 256 C3 92
Jamaica, Windham, Vt., 250 (496▲) E3 120
Jamaica, country, N.A., 1,613,148 F5 64
Jamaica, bay, N.Y E2 108
Jamaica, chan., W.I G6 64
Jamalpur, Pak., 27,078 E12 40
Jamdena, isl., Indon G8 35
James, Jones, Ga., 150 D3 87
James, bay, Ont., Can F16 67
James, isl., B.C., Can C3 68
James, isl., Md C5 85
James, isl., S.C F8 115
James, lake, Ind A7 91
James, lake, N.C B2 109
James, pt., Md C5 85
James, range, Austl D5 50
James, riv., Alta., Can D3 69
James, riv., Mo E4 101
James, riv., N. Dak C7 110
James, riv., S. Dak C7 116
James, riv., Va D5 121
Jamesburg, Middlesex, N.J., 2,853 C4 106
James City, Craven, N.C., 1,474 B6 109
James City, Elk, Pa., 450 C4 114
James City, co., Va., 11,539 D6 121
James Craik, Arg., 2,409 A4 54
James Creek (Marklesburg), Huntingdon, Pa., 197 E5 114
James Island, B.C., Can., 163 B9 68
Jameson, Daviess, Mo., 177 A3 101
Jamesport, Daviess, Mo., 622 B4 101
Jamesport, Suffolk, N.Y., 700 F6 84
James Ross, isl., Ant C7 5
James Ross, strait, N.W. Ter., Can C13 66
Jamestown, Cherokee, Ala., 150 A4 78
Jamestown, Austl., 1,304 F2 51
Jamestown, Tuolumne, Calif., 900 D3 82
Jamestown, Ont., Can., 4,040 E8 72
Jamestown, Boulder, Colo., 107 A5 83
Jamestown, Clinton, Ill., 80 E4 90
Jamestown, Boone, Ind., 827 E4 91
Jamestown, Ire., 162 D3 11
Jamestown, Cloud, Kans., 422 C6 93
Jamestown, Russell, Ky., 792 D4 94
Jamestown, Bienville, La., 140 B2 91
Jamestown, Montieau, Mo., 216 C5 101
Jamestown, Chautauqua, N.Y., 41,818 C1 108
Jamestown, Guilford, N.C., 1,247 B4 109
Jamestown, Stutsman, N. Dak., 15,163 C7 110
Jamestown, Greene, Ohio, 1,730 C4 111
Jamestown, Mercer, Pa., 897 D1 114
Jamestown, Newport, R.I., 2,267 D11 85
Jamestown, Berkeley, S.C., 184 E8 115
Jamestown, Fentress, Tenn., 1,727 C9 117
Jamestown, James City, Va., 5 A6 121
Jamestown, res., N. Dak B7 110
Jamesville, Onondaga, N.Y., 1,000 *C4 108
Jamesville, Martin, N.C., 538 B7 109
Jamesville, Northampton, Va., 300 D7 121
Jamieson, Gadsden, Fla., 100 B2 86
Jamieson, Maheur, Oreg., 10 C9 113
Jamiltepec, Mex., 2,029 D5 63
Jamison, Keya Paha, Nebr., 200 B6 103
Jammain, Jordan, 2,000 g11 32
Jammer Bugt, bay, Den A3 24
Jammu, India, 102,738 (★108,257) A5 40, B5 39
Jammu and Kashmir, disputed reg., India, Pak., 3,560,976 B6 39
Jamnagar, India, 139,692 (★148,572) D5 39, F3 40
Jampur, Pak., 13,235 C3 40
Jamsah, Eg., U.A.R. D6 43
Jamshedpur, India, 291,791 (★328,044) D8 39, F11 40
Jämshög, Swe., 1,024 B8 24
Jämtland, co., Swe., 137,600 *F6 25
Jamul, San Diego, Calif., 350 E2, F5 82
Jamuna, riv., Pak E12 40
Jamundá, riv., Braz C3 59
Jan, lake, Sask., Can C4 72
Jándula, riv., Sp C4 20
Jane, McDonald, Mo. 100 E3 101
Jane Lew, Lewis, W. Va., 426 B4 123
Janesville, Lassen, Calif., 600 B3 82
Janesville, Bremer and Black Hawk, Iowa, 648 B5 92
Janesville, Waseca, Minn., 1,426 F5 99
Janesville, Rock, Wis., 35,164 F4 124
Janesville, N.B., Can., 155 B4 74
Jangipur, India, 24,201 E12 40
Janīn, Jordan, 12,663 B7 32
Jan Mayen, isl., Atl. O B15 4
Janos, Mex., 628 A3 63
Janoshalma, Hung., 10,767 B4 22
Janow, Man., Can., 230 E4 71
Janow Lubelski, Pol., 3,793 C7 26
Jansen, Sask., Can., 275 F3 70
Jansen, Jefferson, Nebr., 204 D8 103
Januária, Braz., 9,741 E2 57
Jaoho (Tuanshantzu), China, 30,000 B11 34
Jaozari, Afg., 5,000 D13 41
Japan, country, Asia, 83,408,026 F16 33, 37
Japan, sea, Asia G7 37
Japaratuba, Braz., 2,557 D3 57
Japen, isl., W. Irian F9 35
Japtan, isl., Eniwetok 52
Japurá, riv., Braz D4 60
Japvo, peak, India C9 39
Jaqué, Pan., 1,195 G8 62
Jara, peak, China E5 34
Jarablus, Syr., 8,023 D11 31
Jaraguá, Braz., 3,813 B3 56
Jaraguá do Sul, Braz., 4,382 D3 56
Jaral [del Progreso], Mex., 7,040 m13 63
Jarales, Valencia, N. Mex., 200 C3 107
Jarama, riv., Sp B4 20
Jarash, Jordan, 2,614 B7 32
Jarbalo, Leavenworth, Kans., 100 B8 93
Jarbidge, Elko, Nev., 30 B6 104
Jardim de Angicos, Braz., 159 g6 57
Jardim do Seridó, Braz., 2,734 C3 57
Jardine, Park, Mont., 150 E6 102
Jardines de la Reina, is., Cuba E4 64
Jardinópolis, Braz., 6,965 k8 56
Jarí, riv., Braz B4 59
Järlåsa, Swe., 259 t35 25
Järna, Swe., 2,427 t35 25
Jarnac, Fr., 4,574 E3 14
Jarny, Fr., 9,248 E6 15
Jarocin, Pol., 19,100 C4 26
Jaroslaw, Pol., 26,000 C7 26
Jaroso, Costilla, Colo., 100 D5 83
Jarrahi, riv., Iran F4 41
Jarratt, Sussex, Va., 608 E5 121
Jarrell, Carroll, Tenn., 30 A3 117
Jarrettsville, Harford, Md., 300 A5 85
Jarrow, Alta., Can., 60 C5 69
Jarvie, Alta., Can., 147 B3 69
Jarvis, Ont., Can., 733 E4 72
Jarvis, isl., Pac. O G12 7
Jarvisburg, Currituck, N.C., 500 A8 109
Jarvisville, Harrison, W. Va., 90 *B4 123
Jasien, Pol., 1,180 B10 17
Jāsk, Iran, 4,938 I8 41
Jask, bay, Iran I8 41
Jaslo, Pol., 3,563 D6 26
Jasmin, Sask., Can., 72 F4 70
Jasmine, White, Ark B4 81
Jason, isl., Ant C6 5
Jasonville, Greene, Ind., 2,436 F3 91
Jasper, Walker, Ala., 10,799 B2 78
Jasper, Newton, Ark., 273 B2 81
Jasper, Alta., Can., 2,360 C1 69
Jasper, Ont., Can., 221 C9 72
Jasper, Hamilton, Fla., 2,103 B4 86
Jasper, Pickens, Ga., 1,036 B2 87
Jasper, Dubois, Ind., 7,910 H4 91
Jasper, Pipestone and Rock, Minn., 850 G2 99
Jasper, Jasper, Mo., 746 D3 101
Jasper, Steuben, N.Y., 500 (983▲) C3 108
Jasper, Pike, Ohio, 150 C4 111
Jasper, Marion, Tenn., 1,450 D8 117
Jasper, Jasper, Tex., 4,889 D6 118
Jasper, co., Ga., 6,135 C3 87
Jasper, co., Ill., 11,346 E5 90
Jasper, co., Ind., 18,842 B3 91
Jasper, co., Iowa, 35,282 C4 92
Jasper, co., Miss., 16,909 C4 100
Jasper, co., Mo., 78,863 D3 101
Jasper, co., S.C., 12,237 G5 115
Jasper, co., Tex., 22,100 D6 118
Jasper, nat. park, Alta., Can C1 69
Jasper Place, Alta., Can., 30,530 *C4 69
Jassy (Iasi), Rom., 94,075 B8 22
Jastarnia, Pol., 2,600 A5 26
Jastrzebia Gora, Pol A5 26
Jastrzebie, Pol., 665 h9 26
Jaszapati, Hung., 9,423 B5 22
Jaszbereny, Hung., 22,284 (30,211▲) B4 22
Jath, India, 7,005 I5 40
Jati, Pak E2 40
Jatibonico, Cuba, 3,486 E4 64
Játiva, Sp., 19,896 C5 20
Jat Poti, Afg F12 41
Jaú, Braz., 31,229 C3, m7 57
Jaú, riv., Braz D5 60
Jauaperi, riv., Braz C5 60
Jauca, P.R D5 65
Jauja, Peru, 8,276 D2 58
Jaumave, Mex., 1,883 C5 63
Jaunpur, India, 61,851 E9 40
Java, Walworth, S. Dak., 406 B6 116
Java, Pittsylvania, Va., 25 E3 121
Java, isl., Indon G4 35
Java, sea, Indon G4 35
Javaés, riv., Braz E5 59
Jávea, Sp., 6,136 C6 20
Jawhar, India, 4,732 H4 40
Jawor, Pol., 11,900 C4 26
Jaworzno, Pol., 53,000 g10 26
Jay, Santa Rosa, Fla., 672 G2 86
Jay, Franklin, Maine, 350 (3,247▲) D2 96
Jay, Essex, N.Y., 400 B2 108
Jay, Delaware, Okla., 1,120 A7 112
Jay, co., Ind., 22,572 D7 91
Jay, peak, Vt B3 120
Jayanga, Peru, 3,413 C2 58
Jay Em, Goshen, Wyo., 30 C8 125
Jayess, Lawrence, Miss., 120 D3 100
Jaynes, Pima, Ariz., 340 E4 80
Jayton, Kent, Tex., 649 C2 118
Jayuya, P.R., 2,344 C4 65
Jayuya, mun., P.R., 14,633 C4 65
Jean, Clark, Nev., 50 H6 104
Jeanerette, Iberia, La., 5,568 E4 95
Jeannette, Westmoreland, Pa., 16,565 F2 114
Jebba, Nig., 768 E5 45
Jebel Aulia, Sud B3 47
Jebel Akhdhar, mtn., Mus. & Om D2 39
Jebel ed Druz, mtn., Syr F11 31
Jebel Marra, mtn., Sud C1 47
Jebel Neba, mtn., Jordan C7 32
Jebel Ram, mtn., Jordan E7 32
Jedburg, Dorchester, S.C., 350 E1 115
Jedburgh, Scot., 3,647 E6 13
Jeddito, Navajo, Ariz., 243 B5 80
Jeddo, St. Clair, Mich., 125 E8 98
Jedrzejow, Pol., 12,000 C6 26
Jeetze, riv., Ger E5 24
Jeff, Perry, Ky., 500 C6 94
Jeff Davis, co., Ga., 8,914 E4 87
Jeff Davis, co., Tex., 1,582 F2 118
Jeffers, Cottonwood, Minn., 489 E5 99
Jeffers, Madison, Mont., 40 E5 102
Jefferson, Marengo, Ala., 200 C2 78
Jefferson, Jefferson, Ark., 350 C9 81
Jefferson, Ont., Can., 151 E6 72
Jefferson, Park, Colo., 30 B5 83
Jefferson, Jackson, Ga., 1,746 B3 87
Jefferson, Greene, Iowa, 4,570 B3 92
Jefferson, Jefferson, La., 35,000 *E5 95
Jefferson, Lincoln, Maine, 200 (1,048▲) D3 96
Jefferson, Frederick, Md., 250 B2 85
Jefferson, Worcester, Mass., 800 B4 97
Jefferson, Coos, N.H., 100 (600▲) B4 105
Jefferson, Gloucester, N.J., 150 D2 106
Jefferson, Schoharie, N.Y., 400 C6 108
Jefferson, Ashe, N.C., 814 A2 109
Jefferson, Ashtabula, Ohio, 2,116 A7 111
Jefferson, Grant, Okla., 119 A4 112
Jefferson, Marion, Oreg., 716 C1, C4 113
Jefferson, Allegheny, Pa., 8,280 *F2 114
Jefferson, Chesterfield, S.C., 493 B7 115
Jefferson, Union, S. Dak., 443 E9 116
Jefferson, Marion, Tex., 3,082 C5 118
Jefferson, Jefferson, Wis., 4,548 E5 124
Jefferson, co., Ala., 634,864 B3 78
Jefferson, co., Ark., 81,373 C3 81
Jefferson, co., Colo., 127,520 B5 83
Jefferson, co., Fla., 9,543 B3 86
Jefferson, co., Ga., 17,468 C4 87
Jefferson, co., Idaho, 11,672 F6 89
Jefferson, co., Ill., 32,315 E5 90
Jefferson, co., Ind., 24,061 G7 91
Jefferson, co., Iowa, 15,818 C6 92
Jefferson, co., Kans., 11,252 C8 93
Jefferson, co., Ky., 610,947 B4 94
Jefferson, co., Miss., 10,142 D2 100
Jefferson, co., Mo., 66,377 C7 101
Jefferson, co., Mont., 4,297 D4 102
Jefferson, co., Nebr., 11,620 D8 103
Jefferson, co., N.Y., 87,835 A5 108
Jefferson, co., Ohio, 99,201 B7 111
Jefferson, co., Okla., 8,192 C4 112
Jefferson, co., Oreg., 7,130 C5 113
Jefferson, co., Pa., 46,792 D3 114
Jefferson, co., Tenn., 21,493 C10 117
Jefferson, co., Tex., 245,659 E5 118
Jefferson, co., Wash., 9,639 B2 122
Jefferson, co., W. Va., 18,665 B7 123
Jefferson, co., Wis., 50,094 E5 124
Jefferson, par., La., 208,769 E5 95
Jefferson, mtn., Nev E5 104
Jefferson, mtn., Oreg C5 113
Jefferson City, Cole, Mo., 28,228 C5 101
Jefferson City, Jefferson, Mont., 50 D5 102
Jefferson City, Jefferson, Tenn., 4,550 C10 117
Jefferson Davis, co., Miss., 13,540 D4 100
Jefferson Davis, par., La., 29,825 D3 95
Jefferson Highland, Coos, N.H B4 105
Jefferson Island, Iberia, La., 200 E4 95
Jefferson Island, Madison, Mont., 35 E5 102
Jeffersontown, Jefferson, Ky., 3,431 A4, B4 94
Jefferson Village, Fairfax, Va., 2,000 *B5 121
Jeffersonville, Twiggs, Ga., 1,013 D3 87
Jeffersonville, Clark, Ind., 19,522 H6 91
Jeffersonville, Montgomery, Ky., 500 C6 94
Jeffersonville, Sullivan, N.Y., 434 D6 108
Jeffersonville, Fayette, Ohio, 897 C4 111
Jeffersonville, Lamoille, Vt., 346 B3 120
Jeffrey, Boone, W. Va., 57 D3, D5 123
Jeffreys, creek, S.C. C8 115
Jeffs, York, Va. (part of Poquoson) B6 121
Jehossee, isl., S.C G1 115
Jejui-Guazú, riv., Par D4 55
Jekabpils, Sov. Un., 10,000 C5 27
Jekyll, isl., Ga E5 87
Jelenia Gora, Pol., 50,000 C3 26
Jelgava, Sov. Un., 35,000 C4 27
Jellico, Campbell, Tenn., 2,210 C9 117
Jelling, Den., 1,228 C3 24
Jelm, Albany, Wyo., 5 D6 125
Jelm, mtn., Wyo D7 125
Jelöy, isl., Nor p28 25
Jembongan, isl., N. Bor *D5 35
Jemez, dam, N. Mex G5 107
Jemez, riv., N. Mex G5 107
Jemez Pueblo, Sandoval, N. Mex., 900 B3, F5 107
Jemez Pueblo, Indian res., N. Mex F5 107
Jemez Springs, Sandoval, N. Mex., 223 B3, F5 107
Jemison, Chilton, Ala., 977 C3 78
Jemseg, N.B., Can., 110 D3 74
Jemtland, Aroostook, Maine, 40 A4 96
Jena, Ger., 81,400 C6 17
Jena, La Salle, La., 2,098 C3 95
Jenaz, Switz., 1,143 C8 19
Jenbach, Aus., 5,479 B7 18
Jenera, Hancock, Ohio, 272 B4 111
Jenison, Ottawa, Mich., 2,000 *F4 98
Jenkins, Letcher, Ky., 3,202 C7 94
Jenkins, Crow Wing, Minn., 144 D4 99
Jenkins, co., Ga., 9,148 D5 87
Jenkinsburg, Butts, Ga., 233 C2 87
Jenkinsville, Fairfield, S.C., 500 C5 115
Jenkintown, Montgomery, Pa., 5,017 A11, F11 114
Jenks, Tulsa, Okla., 1,734 A6 112
Jenner, Alta., Can., 27 D5 69
Jenners, Somerset, Pa., 900 F3 114
Jennie, Chicot, Ark., 100 D4 81
Jennings, Hamilton, Fla., 516 B3 86
Jennings, Decatur, Kans., 292 C3 93
Jennings, Jefferson Davis, La., 11,887 D3 95
Jennings, Garrett, Md., 150 D1 85
Jennings, Missaukee, Mich., 75 D5 98
Jennings, St. Louis, Mo., 19,965 C7 101
Jennings, Pawnee, Okla., 306 A5 112
Jennings, co., Ind., 17,267 G6 91
Jennings Lodge, Clackamas, Oreg., 1,000 *B4 113
Jenny Lake, Teton, Wyo B2 125
Jenny Lind, Sebastian, Ark., 150 B1 82
Jensen, Uintah, Utah, 300 C6 119
Jesen Beach, Martin, Fla., 900 E6 86
Jenshou, China, 36,000 B11 39
Jeptha Knob, peak, Ky B4 94
Jequié, Braz., 40,158 D2 57
Jequitinhonha, riv., Braz E2 57
Jerauld, co., S. Dak., 4,048 C7 116
Jérémie, Hai., 11,138 F6 64
Jeremoabo, Braz., 3,117 D3 57

Jerez de la Frontera, Sp., 97,500 (130,900▲) D2 20
Jerez de los Caballeros, Sp., 12,446 ....C2 20
Jericho, Crittenden, Ark., 80 ....B5 81
Jericho, Austl., 302 ....A6 51
Jericho, see Arīḥa, Jordan
Jericho, Nassau, N.Y., 10,795 ....F2 84
Jericho, Juab, Utah, 4 ....D3 119
Jericho, Chittenden, Vt., 275 (1,425▲) B3 120
Jericho, Nansemond, Va., 2,300 ....*E6 121
Jericho, hill, Mass D1 97
Jericho Center, Chittenden, Vt., 120 ....C3 120
Jerico Springs, Cedar, Mo., 179 ....D4 101
Jermyn, Lackawanna, Pa., 2,568 ....C10 114
Jermyn, Jack, Tex., 150 ....C3 118
Jerome, Yavapai, Ariz., 243 ..C3 80
Jerome, Drew, Ark., 76 ....D4 81
Jerome, Collier, Fla., 100 ....G5 86
Jerome, Jerome, Idaho, 4,761 ....G4 89
Jerome, Sangamon, Ill., 1,666 ....*D4 90
Jerome, Appanoose, Iowa, 160 ....D4 92
Jerome, Hillsdale, Mich., 200 ....F6 98
Jerome, Phelps, Mo., 250 ....D6 101
Jerome, Somerset, Pa., 1,241 ..F4 114
Jerome, Shenandoah, Va., 75 ....C4 121
Jerome, co., Idaho, 11,712 ....G4 89
Jeromesville, Ashland, Ohio, 540 ....B5 111
Jerris, inlet, B.C., Can D6 68
Jersey, Walton, Ga., 170 ....C3 87
Jersey, co., Ill., 17,023 ....D3 90
Jersey, Br. dep., Eur., 63,300 .F5 10
Jersey City, Hudson, N.J., 276,101 ....B4, E5 106
Jersey Shore, Lycoming, Pa., 5,613 ....D7 114
Jerseyville, Ont., Can., 146 ..D4 72
Jerseyville, Jersey, Ill., 7,420 .D3 90
Jerumenha, Braz., 1,473 ....C2 57
Jerusalem, Conway, Ark., 150 ....B3 81
Jerusalem, Monroe, Ohio, 317 ....C6 111
Jerusalem, Isr., 166,301 (*240,000) C7, h11 32
Jerusalem, Jordan, 46,713 ....C7, h11 32
Jervis, bay, Austl G8 51
Jesenik, Czech., 5,873 ....C4 26
Jessamine, co., Ky., 13,625 ..C5 94
Jesselton, Mala., 21,719 ....D5 35
Jessen, Ger., 5,043 ....B7 17
Jessie, Griggs, N. Dak., 70 ..B7 110
Jessieville, Garland, Ark., 200 ....C2 81
Jessnitz, Ger., 10,300 ....B7 17
Jessore, Pak., 24,146 ....F12 40
Jessup (Winton), Lackawanna, Pa., 5,456 ....A9 114
Jessup, lake, Fla D5 86
Jessups, Howard, Md., 100 ..B4 85
Jesterville, Wicomico, Md., 200 ....D6 85
Jesup, Wayne, Ga., 7,304 ....E5 87
Jesup, Buchanan, Iowa, 1,488 .B5 92
Jesus, mtn., Kans E4 93
Jesús Carranza, Mex., 2,088 .D5 63
Jesus del Monte, Cuba h12 64
Jesús María, Arg., 6,284 ....A4 54
Jet, Alfalfa, Okla., 339 ....A3 112
Jetafe, Phil., 2,420 ....*C5 35
Jetersville, Amelia, Va., 175 ..D4 121
Jetmore, Hodgeman, Kans., 1,028 ....D4 93
Jetson, Butler, Ky., 200 ....C3 94
Jette, Bel., 34,927 ....D4 15
Jeumont, Fr., 9,660 ....B4 14
Jever, Ger., 9,400 ....B3 16
Jewell, Hamilton, Iowa, 1,113 ....B4 92
Jewell, Jewell, Kans., 582 ....C5 93
Jewell, Defiance, Ohio, 150 ..A3 111
Jewell, Clatsop, Oreg., 30 ....B3 113
Jewell, co., Kans., 7,217 ....C5 93
Jewell, cave, Tenn A4 117
Jewell Cave, nat. mon., S. Dak .D2 116
Jewell Ridge, Tazewell, Va., 500 ....B2 121
Jewett, Cumberland, Ill., 238 ....D5 90
Jewett, Harrison, Ohio, 925 ..B6 111
Jewett, Leon, Tex., 445 ....D4 118
Jewett, lake, Sask., Can A3 70
Jewett City, New London, Conn., 3,608 ....C9 84
Jeypore, India, 25,291 ....H9 40
Jeziorna, Pol., 1,000 ....m14 26
Jhal, Pak C1 40
Jhal Jhao, Pak C4 39
Jhalrapatan, India, 9,128 ....E6 40
Jhang Maghiana, Pak., 73,402 ....B4 40
Jhansi, India, 140,217 (*169,712) .C6 39, E7 40
Jhawani, Nep., 3,880 ....D10 40
Jhelum, Pak., 47,409 (*56,617) B5 39
Jhelum, riv., Pak B5 39
Jhunjhunu, India, 24,962 ....C5 40
Jiachan, China B8 40
Jibhalanta (Uliassutai), Mong., 25,000 ....B4 34
Jicarilla, Lincoln, N. Mex ..D4 107
Jicarilla, Indian res., N. Mex .A2 107
Jicin, Czech., 12,200 ....C3 26
Jifna, Jordan, 2,000 ....h11 32
Jiggalong, Austl D3 50
Jigger, Franklin, La., 125 ....B4 95
Jiggitai, lake, China A8 39
Jiggs, Elko, Nev., 10 ....C6 104
Jihchao, China, 25,000 ....G8 36
Jihlava, Czech., 34,700 ....D3 26
Jijigā, Eth D5 47
Jijja, riv., Rom B8 22
Jijona, Sp., 6,942 ....C5 20
Jilf Al Kabir, plat., Eg., U.A.R E5 43
Jiloca, riv., Sp B5 20
Jilove, Czech., 2,256 ....o17 26
Jimā, Eth., 10,000 ....D4 47
Jimā, reg., Eth., 1,200,000 ..D4 47
Jimbolia, Rom., 11,281 ....C5 22
Jimena, Sp., 10,123 ....D3 20
Jiménez, Mex., 14,922 ....B4 63
Jim Falls, Chippewa, Wis., 250 ....C2 124
Jim Hogg, co., Tex., 5,022 ..F3 118
Jim Thorpe (Mauch Chunk), Carbon, Pa., 5,945 ....E10 114
Jim Wells, co., Tex., 34,548 ..F3 118
Jināh, Eg., U.A.R D6 43
Jindrichuv Hradec, Czech., 10,300 ....D3 26
Jinja, Ug., 29,741 ....A5 48
Jinotega, Nic., 5,500 ....D5 62
Jinotepe, Nic., 16,400 ....E4 62
Jipijapa, Ec., 7,605 ....B1 58
Jiquilisco, Sal., 4,416 ....D3 62
Jiquilpan de Juárez, Mex., 12,013 ....n12 63
Jirgā, Eg., U.A.R., 37,300 ..D6 43
Jirgalanta (Kobdo), Mong., 5,000 ....B3 34
Jirkov, Czech., 4,247 ....C8 17
Jisr esh Shughur, Syr., 10,400 ....E11 31
Jiul, riv., Rom C6 22
Jiyānklīs, Eg. U.A.R., 1,000 .D2 22
Jizān, Sau. Ar B5 47
Joanna, Laurens, S.C., 1,831 .C4 115
João Pessoa, Braz., 135,820 (*190,000) C4, h6 57
Joaquin, Shelby, Tex., 528 ..D6 118
Joaquín V. González, Arg., 2,132 ....E4 55
Job, Randolph, W. Va., 184 ..C5 123
Jobos, P.R., 815 ....D6 65
Jobstown, Burlington, N.J., 250 ....C4 106
Jockvale, Ont., Can., 50 ....B9 72
Jocobus, York, Pa., 968 ....*G8 114
Jocotepec, Mex., 8,015 ....m12 63
Jódar, Sp., 14,424 ....D4 20
Jo Daviess, co., Ill., 21,821 ..A3 90
Jodhpur, India, 224,760 ....C5 39, D4 40
Jodie, Fayette, W. Va., 600 ..C7 123
Jodoigne, Bel., 4,262 ....D4 15
Joe Batts Arm, Newf., Can., 1,058 ....D4 75
Joelton, Davidson, Tenn., 150 ....E9 117
Joensuu, Fin., 28,300 ....F13 25
Joes, Yuma, Colo., 110 ....B8 83
Joes, brook, Vt C4 120
Joes, creek, W. Va D6 123
Joeuf, Fr., 12,588 ....C6 14
Joffre, mtn., B.C., Can D10 68
Joggins, N.S., Can., 909 ....D5 74
Jogipet, India, 10,681 ....I7 40
Jogjakarta (Jokjakarta), Indon., 308,530 ....G4 35
Johannesburg, Kern, Calif., 250 ....E5 82
Johannesburg, Otsego, Mich., 120 ....D6 98
Johannesburg, S. Afr., 594,290 (*2,075,000) F5 49
John, isl., Ont., Can B8 98
John, pass, Fla F5 86
John Day, Grant, Oreg., 1,520 ....C8 113
John Day, riv., Oreg B6 113
John Long, mts., Mont D3 102
John Martin, res., Colo C7 83
Johns, Jefferson, Ala., 338 ....E4 78
Johns, creek, Ky C7 94
Johns, isl., S.C F7 115
Johns, pass, Fla E1 86
Johnsburg, McHenry, Ill., 340 ....E2 90
Johnsburg, Warren, N.Y., 225 ....B7 108
Johnsburg, Fond du Lac, Wis., 100 ....B5 124
John Sevier, Knox, Tenn., 800 ....E11 117
Johns Island, Charleston, S.C., 500 ....F2, F7 115
Johnson, Cochise, Ariz., 90 ..E5 80
Johnson, Washington, Ark., 221 ....A1 81
Johnson, Gibson, Ind., 200 ..H2 91
Johnson, Stanton, Kans., 860 .E2 93
Johnson, Nemaha, Nebr., 304 ....D10 103
Johnson, Lamoille, Vt., 941 (1,478▲) B3 120
Johnson, Whitman, Wash., 30 ....C8 122
Johnson, co., Ark., 12,421 ....B2 81
Johnson, co., Ga., 8,048 ....D4 87
Johnson, co., Ill., 6,928 ....F5 90
Johnson, co., Ind., 43,704 ....F5 91
Johnson, co., Iowa, 53,663 ....C6 92
Johnson, co., Kans., 143,792 .D9 93
Johnson, co., Ky., 19,748 ....C7 94
Johnson, co., Mo., 28,981 ....C4 101
Johnson, co., Nebr., 6,281 ..D9 103
Johnson, co., Tenn., 10,765 ....C12 117
Johnson, co., Tex., 34,720 ....C4 118
Johnson, co., Wyo., 5,475 ....A6 125
Johnsonburg, Warren, N.J., 230 ....B3 106
Johnsonburg, Elk, Pa., 4,966 .D4 114
Johnson City, Broome, N.Y., 19,118 ....C5 108
Johnson City, Washington, Tenn., 31,187 ....C11 117
Johnson City, Blanco, Tex., 611 ....D3 118
Johnson City Southeast, Washington, Tenn., 2,435 ....*C11 117
Johnson Creek, Jefferson, Wis., 686 ....E5 124
Johnsondale, Tulare, Calif., 300 ....D4 82
Johnsons Grove, Crockett, Tenn., 50 ....B2 117
Johnsonville, Florence, S.C., 882 ....D9 115
Johnsonville, Humphreys, Tenn., 559 ....A5 117
Johnston, Polk, Iowa, 1,000 ..A7 92
Johnston, Providence, R.I., 17,160 ....B10 84
Johnston, Edgefield, S.C., 2,119 ....D4 115
Johnston, co., N.C., 62,936 ....B5 109
Johnston, co., Okla., 8,517 ..C5 112
Johnston, isl., Oceania E11 7, n14 88
Johnston, key, Fla H5 86
Johnston City, Williamson, Ill., 3,891 ....F5 90
Johnstone, strait, B.C., Can D4 68
Johnston Island, U.S. dep., Oceania, 156 ....E11 7, n14 88
Johnstons, pt., Scot E3 12
Johnstons Station, Pike and Lincoln, Miss., 121 ....D3 100
Johnstown, Weld, Colo., 976 .A6 83
Johnstown, Brown, Nebr., 81 ....B5 103
Johnstown, Fulton, N.Y., 10,390 ....B6 108
Johnstown, Licking, Ohio, 2,881 ....B5 111
Johnstown, Cambria, Pa., 53,949 (*125,500) F4 114
Johnsville, Bradley, Ark D3 81
Johnsville, Frederick, Md., 160 ....A3 85
Johnstown, Grand Forks, N. Dak., 58 ....A8 110
Johnstown, Red River, Tex., 200 ....C5 118
Johnville, Que., Can., 161 ..D6 73
Johore, state, Mala., 926,850 .L5 38
Johore Bahru, Mala., 74,909 .L5 38
Joice, Worth, Iowa, 231 ....A4 92
Joice, Lincoln, Maine D3 96
Joigny, Fr., 7,144 ....D5 14
Joiner, Mississippi, Ark., 748 .B5 81
Joinville, Braz., 44,255 ....D3 56
Joinville, Fr., 4,015 ....C6 14
Joinville, isl., Ant C7 5
Jojutla, Mex., 11,489 ....n14 63
Jokaj, hbr., Ponape 52
Jokaj, isl., Ponape 52
Jokjakarta (Jogjakarta), Indon., 308,530 ....G4 35
Jokkmokk, Swe., 2,508 ....D8 25
Joliet, Will, Ill., 66,780 (*132,100) B5, F2 90
Joliet, Carbon, Mont., 452 ..E8 102
Joliette, Que., Can., 18,088 ..C4 73
Joliette, Pembina, N. Dak., 40 ....A8 110
Joliette, co., Que., Can., 44,969 ....C3 73
Jolley, Calhoun, Iowa, 120 ..B3 92
Jolly, Clay, Tex., 100 ....C3 118
Jolo, Phil., 22,000 (33,400▲) .D6 35
Jolo, isl., Phil D6 35
Jomalig, isl., Phil o14 35
Jonas Ridge, Burke, N.C., 50 ....B2, C4 109
Jonava, Sov. Un., 10,000 ....A8 26
Jones, Morehouse, La., 400 ..B4 95
Jones, Anne Arundel, Md., 100 ....B4 85
Jones, Oklahoma, Okla., 794 .B4 112
Jones, Phil., 1,838 ....n13 35
Jones, Spotsylvania, Va C5 121
Jones, co., Ga., 8,468 ....C3 87
Jones, co., Iowa, 20,693 ....B6 92
Jones, co., Miss., 59,542 ....D4 100
Jones, co., N.C., 11,005 ....B6 109
Jones, co., S. Dak., 2,066 ....D5 116
Jones, co., Tex., 19,299 ....C3 118
Jones, beach, N.Y G2 84
Jones, cape, N.W. Ter., Can .F16 67
Jones, sound, N.W. Ter., Can .B23 4
Jonesboro, Craighead, Ark., 21,418 ....B5 81
Jonesboro, Clayton, Ga., 3,014 ....C2 87
Jonesboro, Union, Ill., 1,636 .F4 90
Jonesboro, Grant, Ind., 2,260 ....D6 91
Jonesboro, Jackson, La., 3,848 ....B3 95
Jonesboro, Washington, Maine, 400 (428▲) D5 96
Jonesboro, Lee, N.C B4 109
Jonesboro, Washington, Tenn., 1,148 ....C11 117
Jonesburg, Montgomery, Mo., 415 ....C6 101
Jones Cove, Sevier, Tenn ..D10 117
Jones Creek, Brazoria, Tex., 950 ....*E5 118
Jones Mill, Hot Spring, Ark., 950 ....C3 81
Jones Mill, Burlington, N.J ..D3 106
Jones Mill, Henry, Tenn., 35 ....A3 117
Jones Mills, Westmoreland, Pa., 250 ....F3 114
Jonesport, Washington, Maine, 800 (1,563▲) D5 96
Jonestown, Coahoma, Miss., 889 ....A3 100
Jonestown, Yazoo, Miss., 200 .A3 100
Jonestown, Lebanon, Pa., 813 ....F9 114
Jonesville, LaSalle, Ill., 160 ..B4 90
Jonesville, Bartholomew, Ind., 196 ....F6 91
Jonesville, Catahoula, La., 2,347 ....C4 95
Jonesville, Hillsdale, Mich., 1,896 ....G6 98
Jonesville, Yadkin, N.C., 1,895 ....A3 109
Jonesville, Union, S.C., 1,439 ....B4 113
Jonesville, Chittenden, Vt., 50 ....C3 120
Jonesville, Lee, Va., 711 ....B1 121
Jongkha Dzong, China, 5,000 ....C10 40
Jönköping, Swe., 51,000 (*75,000) I6 25
Jönköping, co., Swe., 286,300 ....A8 24
Jonquière, Que., Can., 28,588 ....A6 73
Jonuta, Mex., 1,482 ....D6 63
Jonzac, Fr., 2,804 ....E3 14
Joplin, Jasper and Newton, Mo., 38,958 ....D3 101
Joplin, Liberty, Mont., 300 ..B6 102
Joppa, Cullman, Ala., 160 ....A3 78
Joppa, Massac, Ill., 578 ....F5 90
Jordan, Washington, Ala., 150 ....D1 78
Jordan, Baxter, Ark., 45 ....A3 81
Jordan, Fulton, Ky., 30 ....B2 94
Jordan, Scott, Minn., 1,479 ..F5 99
Jordan, Garfield, Mont., 557 ....C10 102
Jordan, Quay, N. Mex C6 107
Jordan, Onondaga, N.Y., 1,390 ....B4 108
Jordan, country, Asia, 1,329,174 ....F6 33, G11 31
Jordan, creek, Pa B1 106
Jordan, dam, Ala C3 78
Jordan, lake, Ala C3 78
Jordan, lake, N.S., Can E4 74
Jordan, riv., Asia B7 32
Jordan, riv., Utah C4 119
Jordan Valley, Malheur, Oreg., 204 ....E9 113
Jorhat, India, 24,953 ....C9 39
Jornada del Muerto, des., N. Mex E3 107
Jörns, Swe., 1,432 ....E9 25
Jos, Nig., 31,582 (*38,527) ..E6 45
José Batlle y Ordóñez, Ur., 1,781 ....E1 56
José de San Martín, Arg C2 54
Joseph, Wallowa, Oreg., 788 .B9 113
Joseph, Sevier, Utah, 117 ....E3 119
Joseph, lake, Ont., Can B5 72
Joseph Bonaparte, gulf, Austl ..B4 50
Joseph City, Navajo, Ariz., 500 ....C5 80
Josephine, co., Oreg., 29,917 .E3 113
Josephville, St. Charles, Mo., 50 ....A7 101
Joshua, Johnson, Tex., 764 ..B5 118
Joshua Tree, nat. mon., Calif ..F6 82
Jostedalsbre, glacier, Nor G2 25
Jost Van Dyke, is., Vir. Is f16 65
Jotunheimen, mts., Nor G3 25
Joubert, Douglas, S. Dak., 15 ....D7 116
Jourdanton, Atascosa, Tex., 1,504 ....E3 118
Joussard, Alta., Can., 249 ....B3 69
Joux, lake, Switz C1 19
Jouy-en-Josas, Fr., 3,321 ....g9 14
Jovellanos, Cuba, 10,444 ....D3 64
Joy, Mercer, Ill., 503 ....B3 90
Joyce, Winn, La., 600 ....C3 95
Joyce, Webb, Tex F3 118
Joyuda, P.R C2 65
Juab, Juab, Utah D4 119
Juab, co., Utah, 4,597 ....D2 119
Juana Díaz, P.R., 4,618 ....C4 65
Juana Díaz, mun., P.R., 30,043 ....C5 65
Juan Aldama, Mex., 7,672 ....C4 63
Juan de Fuca, strait, Wash A1 122
Juan Díaz, Pan., 821 ....k12 62
Juan E. Barra, Arg B4 54
Juan Fernández, is., Pac. O I17 7
Juanita, Archuleta, Colo., 15 ....D3 83
Juanita, Foster, N. Dak., 100 .B7 110
Juanita, King, Wash., 1,500 .*B3 121
Juanita Central, P.R B6 65
Juanita Junction, Delta, Colo .C3 83
Juanjui, Peru, 2,118 ....C2 58
Juan Lacaze, Ur., 8,400 ....g8 54
Juárez, Arg., 7,602 ....B5 54
Juárez, Mex., 1,198 ....B4 63
Juaso, Ghana E4 45
Juàzeiro, Braz., 21,196 ....C2 57
Juàzeiro do Norte, Braz., 53,421 ....C3 57
Juba, Sud., 10,660 ....E3 47
Jubari, Jap., 99,530 ....E10 37
Jubbah, Sau. Ar D3 31
Jubbulpore, see Jabalpur, India
Jubilee, lake, Newf., Can D4 75
Júcar, riv., Sp C5 20
Júcaro, Cuba, 868 ....E4 64
Juchipilar, Mex., 3,190 ....m12 63
Juchipila, riv., Mex m12 63
Juchitán [de Zaragoza], Mex., 20,031 ....D5 63
Jucuapa, Sal., 4,923 ....D3 62
Jud, La Moure, N. Dak., 156 .C7 110
Juda, Green, Wis., 400 ....F4 124
Judah, Alta., Can., 20 ....A2 69
Judayyidat 'Ar'ar, Sau. Ar. .G13 31
Juddah, Sau. Ar., 160,000 ....A4 47
Judds Bridge, Litchfield, Conn., 40 ....C3 84
Jude, isl., Newf., Can E4 75
Judean, hills, Isr C7 32
Judenburg, Aus., 9,869 ....E7 16
Judique, N.S., Can., 225 ....D8 74
Judith, isl., N.C B7 109
Judith, mts., Mont C7 102
Judith, riv., Mont C7 102
Judith Basin, co., Mont., 3,085 ....C6 102
Judith Gap, Wheatland, Mont., 185 ....D7 102
Judith Neck, pt., R.I D11 84
Judkin, Ector, Tex D1 118
Judson, Parke, Ind., 80 ....E3 91
Judson, Blue Earth, Minn., 150 ....F4 99
Judson, Morton, N. Dak., 70 ....C4 110
Judson, Greenville, S.C., 2,000 ....*B3 115
Judsonia, White, Ark., 1,470 .B4 81
Judyville, Warren, Ind., 70 ..D3 91
Juelsminde, Den., 950 ....C4 24
Juian, China, 36,000 ....K9 36
Juigalpa, Nic., 5,600 ....D5 62
Juist, Ger., 2,300 ....A6 15
Juist, isl., Ger B3 16
Juiz de Fora, Braz., 124,979 ....C4, g6 56
Jujuy, Arg., 31,091 ....D2 55
Jujuy, prov., Arg., 239,783 ..D2 55
Jukao, China, 80,000 ....H9 36
Jukkasjärvi, Swe., 540 ....D9 25
Julesburg, Sedgwick, Colo., 1,840 ....A8 83
Juliaca, Peru, 7,002 ....E3 58
Juliaetta, Latah, Idaho, 368 ..C2 89
Julian, San Diego, Calif., 400 ....F5 82
Julian, Nemaha, Nebr., 131 .D10 103
Julian, Randolph and Guilford, N.C., 200 ....B4 109
Julian, Boone, W. Va., 300 ....C3, D5 123
Julian Alps, mts., Yugo C2 22
Julianehaab, Grnld., 936 ..C19 4
Jülich, Ger., 14,700 ....D6 15
Julietta, Marion, Ind H8 91
Juliette, Monroe, Ga., 550 ....C3 87
Julimes, Mex., 1,209 ....B3 63
Júlio de Castilhos, Braz., 6,438 ....D2 56
Julita, Swe., 266 ....t34 25
Juliustown, Burlington, N.J., 300 ....C3 106
Jullundur, India, 222,569 (*265,030) .B5 40, B6 39
Jumas, riv., Braz C2 59
Jumento, is., Ba. Is D6 64
Jumet, Bel., 28,713 ....D5 15
Jumilla, Sp., 16,199 ....C5 20
Jumna, riv., India E8 40
Jump, riv., Wis C2 124
Jumping Branch, Summers, W. Va., 100 ....D4 123
Jumullong Manglo, mtn., Guam 52
Junagadh, India, 74,298 ....D5 39, G3 40
Junan, China, 30,000 ....H6 36
Junas, riv., Braz D5 60
Juncos, P.R., 6,247 ....C7 65
Juncos, mun., P.R., 21,496 ..C7 65
Juncos Central, P.R C7 65
Junction, Gallatin, Ill., 238 ..F5 90
Junction, Paulding, Ohio, 40 ....A3 111
Junction, Kimble, Tex., 2,441 ....D2 118
Junction, Piute, Utah, 219 ....E3 119
Junction City, Union, Ark., 749 ....D3 81
Junction City, Talbot, Ga., 226 ....D2 87
Junction City, Geary, Kans., 17,700 ....C7 93
Junction City, Boyle, Ky., 1,047 ....C5 94
Junction City, Union, La., 639 ....B3 95
Junction City, Perry, Ohio, 763 ....C5 111
Junction City, Lane, Oreg., 1,614 ....C3 113
Junction City, Portage, Wis., 381 ....D4 124
Jundah, Austl., 122 ....B4 51
Jundiaí, Braz., 79,536 ....C3, m8 56
Juneau, Alsk., 6,797 ....D13, k22 79
Juneau, Dodge, Wis., 1,718 ..E5 124
Juneau, co., Wis., 17,490 ....E3 124
Junee, Austl., 3,980 ....G6 51
June Lake, Mono, Calif., 25 ..D4 82
Jungcheng, China F10 36
Jungchiang, China, 10,000 ....L3 36
Jungfrau, mtn., Switz C4 19
Junghsien, China, 15,000 ....F6 34
Junglinster, Lux., 1,802 ....E6 15
Jungyun, China G7 34
Juniata, Sask., Can., 25 ....E2 70
Juniata, Adams, Nebr., 422 ..D7 103
Juniata, co., Pa., 15,874 ....E7 114
Juniata, riv., Pa E7 114
Juniata Terrace, Mifflin, Pa., 1,130 ....E6 114
Junín, Arg., 36,149 ....A4 54
Junín, Peru, 3,058 ....D2 58
Junín, dept., Peru, 431,082 ..D2 58
Junín, lake, Peru D2 58
Junín de los Andes, Arg B2 54
Junior, Barbour, W. Va., 552 ....C5 123
Junior, lake, Maine C4 96
Juniper, N.B., Can., 691 ....C2 74
Juniper, Talbot, Ga., 100 ....D2 87
Juniper, flat, Oreg E7 113
Juniper, mtn., Oreg E7 113
Junipero Serra, peak, Calif D3 82
Junius, Lake, S. Dak., 50 ....C8 116
Juno, Henderson, Tenn B3 117
Juno, Val Verde, Tex., 30 ....D2 118
Juno Beach, Palm Beach, Fla., 249 ....*F6 86
Junor, Sask., Can., 10 ....D2 70
Juntura, Malheur, Oreg., 98 ....D8 113
Jupiter, Palm Beach, Fla., 1,058 ....F6 86
Jupiter, inlet, Fla F6 86
Jupiter, isl., Fla F6 86
Jupuiá, Braz., 2,573 ....C3, n8 56
Jur, riv., Sud D2 47
Jura, dept., Fr., 225,682 ....C1 18
Jura, isl., Scot D3 13
Jura, mts., Switz., Fr C1 19
Jura, sound, Scot C3 13
Jurbarkas, Sov. Un., 5,000 ..A7 26
Jurby, head, I. of Man F4 13
Jurf ad Darāwish, Jordan, 1,000 ....D7 32
Juriti, Braz., 1,868 ....C3 59
Juruá, riv., Braz C4 58
Juruena, riv., Braz D3 59
Jussey, Fr., 2,235 ....B1 18
Justice, Man., Can., 40 ....D2 71
Justice, Cook, Ill., 2,803 ....*F2 90
Justiceburg, Garza, Tex., 50 ....C2 118
Justin, Denton, Tex., 622 ....A5, C4 118
Justo Daract, Arg., 4,663 ....A3 54
Jutai, riv., Braz B4 58
Jüterbog, Ger., 13,500 ....B8 17
Jutiapa, Guat., 5,162 ....C4 62
Juticalpa, Hond., 3,372 ....C4 62
Jutland, pen., Den B3 24
Juwain, Afg., 5,000 ....F10 41
Jūymand, Iran D8 41
Južna Morava, riv., Yugo D5 22
Jyderup, Den., 2,305 ....C5 24
Jyekundo (Yushu), China E4 34
Jyväskylä, Fin., 39,600 ....F11 25

K

Kaaawa, Honolulu, Haw., 450 ....f10 88
Kaala, peak, Haw f9 88
Kaalaea, Honolulu, Haw., 300 ....g10 88
Kabaena, isl., Indon G6 35
Kabala, S.L E2 45
Kabale, Ug., 10,919 ....B5 48
Kabalo, Con. L C4 48
Kabambare, Con. L B4 48
Kabarnet, Ken A6 48
Kabba, Nig E6 45
Kabelstation, Sur B3 59
Kabenga, Con. L C4 48
Kabetogama, lake, Minn B5 99
Kabinda, Con. L C4 48
Kabir Kuh, mts., Iran E3 41
Kabompo, riv., N. Rh D3 48
Kabong, Mala., 1,957 ....E4 35
Kabongo, Con. L C4 48
Kaboudia, cape, Tun G12 20
Kabul, Afg., 206,208 ....D14 41
Kabul, riv., Afg D14 41
Kacak, riv., Czech n17 26
Kachek, China C9 36
Kachess, res., Wash B4 122
Kachuga, Sov. Un., 53,500 .D13 28
Kackley, Republic, Kans., 50 ....C6 93
Kadan, Czech., 5,062 ....C8 17
Kade, Ghana, 2,077 ....E4 45
Kadet, chan., Ger D6 24
Kadinhan, Tur., 4,338 ....C9 31
Kadiri, India, 24,307 ....F6 39
Kadiyevka, Sov. Un., 191,000 ....G12, q21 27
Kadoka, Jackson, S. Dak., 840 ....D4 116
Kadugli, Sud., 4,716 ....C2 47
Kaduna, Nig., 22,511 (*38,794) D6 45
Kaduna, riv., Nig D6 45
Kaea, cape, Haw C5 88
Kaédi, Maur., 8,500 ....C2 45
Kaélé, Cam., 619 ....C2 46
Kaeleku, Maui, Haw., 5 ....C5 88
Kaena, pt., Haw C4 88
Kaena, pt., Haw B3, f8 88
Kaeryong, Kor., 6,200 ....H4 37
Kaesong, Kor., 139,900 ....H3 37
Kāf, Sau. Ar., 10,000 ....C7 43
Kafanchan, Nig., 7,016 ....E6 45
Kaffrine, Sen D1 45
Kafia Kingi, Sud D1 47
Kafr ad Dawwār, Eg., U.A.R., 10,000 ....C2 32
Kafr 'Ammār, Eg. U.A.R E3 32
Kafr ash Shaykh, Eg., U.A.R., 3,000 ....C2 32
Kafr az Zayyāt, Eg., U.A.R., 25,100 ....D2 32
Kafr Dā'ūd, Eg., U.A.R D2 32
Kafr Malik, Jordan, 2,000 ..h12 32
Kafr Qaddum, Jordan, 2,000 .g11 32
Kafr Qasim, Isr g10 32
Kafr Ṣaqr, Eg., U.A.R., 1,000 ....D3 32
Kafr Zibad, Jordan, 2,000 ..g11 32
Kafue, N. Rh., 2,100 ....E4 48
Kafue, riv., N. Rh D4 48
Kagan, Sov. Un., 30,000 ....B12 41
Kaganovich, Sov. Un., 10,000 ....E12 27
Kagawa, pref., Jap., 918,867 .*I7 37
Kagawong, Ont., Can., 58 ..B2 72
Kagera, riv., Tan B5 48
Kagizman, Tur., 7,200 ....B14 31
Kagman, mtn., Saipan 52
Kagoshima, Jap., 296,003 ..K5 37
Kagoshima, pref., Jap., 1,963,104 ....*K5 37
Kagoshima, bay, Jap K5 37
Kagul, Sov. Un., 20,000 ....C9 22
Kaguyak, Alsk., 36 ....D9 79
Kahajan, riv., Indon F4 35
Kahakuloa, Maui, Haw., 35 ..B5 88
Kahaluu, Hawaii, Haw., 50 ..D6 88
Kahaluu, Honolulu, Haw., 1,125 ....g10 88
Kahama, Tan., 1,866 ....B5 48
Kahana, Honolulu, Haw., 375 ....B4, f10 88
Kahemba, Con. L C2 48
Kahira, Okinawa 52
Kahla, Ger., 9,342 ....C6 17
Kahnūj, Iran H8 41
Kahlotus, Franklin, Wash., 131 ....C7 122
Kahoka, Clark, Mo., 2,160 .*A6 101
Kahoolawe, isl., Haw C5 88
Kahua, Hawaii, Haw., 81 ....C6 88
Kahuku, Oahu, Haw., 1,238 ....B4, f10 88
Kahuku, pt., Haw B4 88
Kahului, Maui, Haw., 4,223 ....C5 88
Kaiama, Nig E5 45
Kaiapoi, N.Z., 3,110 ....O14 51
Kaiba, isl., Sov. Un C10 37
Kaibab, Mohave, Ariz., 40 ..A3 80
Kaibab, Indian res., Ariz A3 80
Kaibab, plat., Ariz A3 80
Kaibito, Coconino, Ariz., 65 .A3 80
Kaieteur, falls, Br. Gu A3 59
Kaifeng, China, 299,100 ....G6 36
Kaihsien, China, 21,000 ....E6 36
Kaihua, China, 6,000 ....J8 36
Kaikoura, N.Z., 1,328 ....O14 51
Kailas, range, China C11 40
Kailu, China C9 36
Kailua, Honolulu, Haw., 25,622 ....g11 88
Kailua-Kona, Hawaii, Haw., 600 ....D6 88
Kaimana, W. Irian, 800 ....F8 35
Kaimare, Pap k11 50
Kaimu, Hawaii, Haw., 16 ..D7 88
Kainaliu, Hawaii, Haw., 510 .D6 88
Kainalu, Maui, Haw., 60 ....B5 88
Kainan, Jap., 42,100 ....I7 37
Kainan, bay, Ant B33 5
Kaintira, India, 5,000 ....G10 40
Kaiparowits, plat., Utah F4 119
Kaiping, China, 40,000 ....D10 36
Kaipokok, bay, Newf., Can g10 75
Kairouan, Tun., 33,968 ....E7 44
Kairouan, prov., Tun., 205,039 ....*B7 44
Kaiser, Price, Wis C3 124
Kaiserslautern, Ger., 86,300 .D2 17
Kaiserstuhl, Switz., 398 ....A5 19
Kaiser Wilhelm (Kiel), canal, Ger D3 24
Kaisiadorys, Sov. Un A8 26
Kaitaia, N.Z., 2,706 ....K14 51
Kaitangata, N.Z., 1,249 ....Q12 51
Kaithal, India, 34,890 ....C6 40
Kaitung, China B10 36
Kaiwi, chan., Haw B4 88
Kaiyuan, China, 25,000 ....E2 37
Kaizuka, Jap., 61,067 ....o14 37
Kajaani, Fin., 14,700 ....E12 25
Kajakai, res., Afg E12 41
Kajan, riv., Indon E5 35
Kajiado, Ken B6 48
Kajikazawa, Jap., 4,500 ....n17 37
Kaka, Sud C3 47
Kakagi, lake, Ont., Can E4 71
Kakamas, S. Afr F3 49
Kakamega, Ken A5 48
Kake, Alsk., 455 ....D13, m23 79
Kakegawa, Jap., 27,600 ....O17 37
Kaketsa, mtn., B.C., Can E6 66
Kakhovka, Sov. Un., 30,000 .H9 27
Kakhovka, res., Sov. Un H9 27
Kakinada, India, 122,865 ....E7 39, I9 40
Kakisa, lake, N.W. Ter., Can ..D9 66
Kakogawa, Jap., 57,800 (89,539▲) *I7 37
Kakumaa, Ken A5 48
Kakwa, riv., Alta., Can B1 69
Kalabagh, Pak., 10,523 ....B5 40
Kalabakan, Mala E5 35
Kalabana, Mali D3 45
Kalābishah, Eg., U.A.R E6 43
Kalabo, N. Rh., 1,710 ....D3 48
Kalach, Sov. Un., 16,900 ....C2 29
Kalach, canal, Sov. Un G14 27
Kaladar, Ont., Can., 251 ....C7 72
Ka Lae (South Cape), cape, Haw .E6 88
Kalahari, des., Bech B3 49
Kalahari Gemsbok, nat. park, Bech C3 49
Kalaheo, Kauai, Haw., 1,185 .B2 88
Kalai-Khumb, Sov. Un B15 41
Kalai-Mor, Sov. Un D11 41
Kalak, Iran I9 41
Kalama, Cowlitz, Wash., 1,088 ....C3 122
Kalamai (Kalamata), Grc., 38,211 ....D4 23
Kalamazoo, Kalamazoo, Mich., 82,089 (*170,000) ..E5 98
Kalamazoo, co., Mich., 169,712 ....F5 98

Kalamazoo, riv., Mich...F5 98
Kalambo, falls, Rh. & Nya., Tan...C5 48
Kalampaka, Grc., 4,043...C3 23
Kalankala (Well), Niger...C7 45
Kalaotoa, isl., Indon...G6 35
Kalapana, Hawaii, Haw., 60...D7 88
Kalat, Pak., 2,009...C4 39
Kalāteh Minār, Iran...D10 41
Kalat-i-Ghilzai, Afg., 5,000...E13 41
Kalaupapa, Maui, Haw., 270...B5 88
Kalavrita, Grc., 2,189...C4 23
Kalecik, Tur., 4,100...B9 31
Kaleden, B.C., Can., 350...E8 68
Kalehe, Con. L...B4 48
Kalem, Scott, Miss., 40...C4 100
Kaleva, Manistee, Mich., 348...D4 98
Kalewa, Bur., 2,263...D9 39
Kalgalaksha, Sov. Un...E16 25
Kalgan (Changkiakow), China, 229,300...C7 34
Kalgary, Crosby, Tex...C2 118
Kalima, Con. L...B4 48
Kalgin, isl., Alsk...g16 79
Kalgoorlie, Austl., 9,696 (*21,773)...F3 50
Kāl Gūsheh, Iran...F9 41
Kalimnos, Grc., 10,211...D6 23
Kalimnos, isl., Grc...D6 23
Kalinin, Sov. Un., 279,000...B1 29
Kaliningrad (Königsberg), Sov. Un., 226,000.A6 26, D3 27
Kaliningrad, Sov. Un., 74,000...n17 27
Kaliningrad, reg., Sov. Un...A6 26
Kalinkovichi, Sov. Un., 10,000...E7 27
Kali Sindh, riv., India...E6 40
Kalispel, Indian res., Wash...A8 122
Kalispell, Flathead, Mont., 10,151...B2 102
Kalisz, Pol., 70,000...C5 26
Kaliua, Tan...C5 48
Kalixälven, riv., Swe...D10 25
Kalīyā, Jordan...h12 32
Kalkan, Tur., 474...D7 23
Kalkaska, Kalkaska, Mich., 1,321...D5 98
Kalkaska, co., Mich., 4,382...D5 98
Kalkfeld, S.W. Afr., 835...B2 49
Kalkrand, S.W. Afr...B2 49
Kållerstad, Swe...A7 24
Kallithea, Grc., 54,720...h11 23
Kallsjön, lake, Swe...F5 25
Kalmalo, Nig...D6 45
Kalmar, Swe., 31,400...I7 25
Kalmar, Swe...t35 25
Kalmar, co., Swe., 234,800...*I7 25
Kalmarsund, sound, Swe...I7 25
Kalmykovo, Sov. Un...D4 29
Kalo, Webster, Iowa, 150...B3 92
Kalocsa, Hung, 13,663...B4 22
Kalohi, chan., Haw...C5 88
Kalomo, N. Rh., 1,185...E4 48
Kalona, Washington, Iowa, 1,235...C6 92
Kalpi, India, 17,278...D7 40
Kalskag, Alsk., 147...C7 79
Kaltag, Alsk., 165...C8 79
Kaluaaha, Maui, Haw., 50...B5 88
Kaluga, Sov. Un., 145,000...D11 27
Kalundborg, Den., 9,763...C5 24
Kalush, Sov. Un., 10,000...G5 27
Kaluszyn, Pol., 2,554...B6 26
Kalvehave, Den., 454...C6 24
Kalvesta, Finney, Kans., 52...D3 93
Kalvsjön, lake, Swe...A7 24
Kalyan, India, 73,482 (*194,334)...H4 40
Kama, Con. L...B4 48
Kama, res., Sov. Un...D8 28
Kama, riv., Sov. Un...D19 9
Kama, rock, Iwo... 52
Kamadia (Well), Niger...B7 45
Kamaing, Bur., 608...C10 39
Kamaishi, Jap., 87,511...G10 37
Kamakura, Jap., 98,617...n18 37
Kamananui, riv., Haw...f9 88
Kamaniskeg, lake, Ont., Can...B7 72
Kamaran, isl., Asia...B5 47
Kamarhati. India, 125,457...*F12 40
Kamarod, Pak...C3 39
Kamas, Summit, Utah, 749...C4 119
Kamatsi, lake, Sask., Can...A4 70
Kamba, pt., Fiji Is... 52
Kambara, isl., Fiji Is... 52
Kambove, Con. L...D4 48
Kamchalka, pen., Sov. Un...D18 28
Kamela, Union, Oreg., 10...B8 113
Kamenets-Podolskiy, Sov. Un., 25,000...G6 27
Kamenjak, cape, Yugo...C1 22
Kamenka, Sov. Un., 3,700...G7 27
Kamen na Obi, Sov. Un., 37,000...C10 29
Kamensk-Shakhtinskiy, Sov. Un., 60,000...G13 27
Kamensk-Uralskiy, Sov. Un., 151,000...B6 29
Kamenz, Ger., 14,900...B9 17
Kameoka, Jap., 17,300...o14 37
Kamet, peak, India...B7 40
Kami, isl., Jap...I4 37
Kamiah, Lewis, Idaho, 1,245...C2 89
Kamiak, mtn., Wash...C8 122
Kamichli, Syr., 34,200...D13 31
Kamien Pomorski, Pol., 1,576...A3 26
Kamina, Con. L., 32,100...C4 48
Kamioka, Jap., 20,100...m16 37
Kamloops, B.C., Can., 10,076...D7 68
Kamooloa, Honolulu, Haw., 150...f9 76
Kamori-Yama, peak, Jap...I6 37
Kamouraska, Que., Can., 518...B8 73
Kamouraska, co., Que., Can., 27,138...B8 73
Kampa Dzong, China, 5,000...C12 40
Kampala, Ug., 46,735 (*123,332)...A5 48
Kampar, Mala., 24,602...J4 38
Kampar, riv., Indon...E2 35
Kampen, Neth., 27,200...B5 15
Kampeska, lake, S. Dak...C8 116
Kamphaeng Phet, Thai., 7,045...D3 38
Kampot, Camb., 5,000...G6 38
Kampsville, Calhoun, Ill., 453...D4 90
Kampungbaru, Indon...E6 35
Kamrar, Hamilton, Iowa, 268...B4 92
Kamsack, Sask., Can., 2,968...F5 70
Kamuchawie, lake, Sask., Can...A4 70
Kamuela, Hawaii, Haw., 950...C6 88
Kamuli, Ug...A5 48
Kamyshebakha, Sov. Un., 10,000...q21 27
Kamyshin, Sov. Un., 62,000...C3 29
Kamyshlov, Sov. Un., 25,700...B6 29
Kan, riv., China...K6 36
Kana, Okinawa... 52
Kanab, Kane, Utah, 1,645...F3 119
Kanab, creek, Ariz...A3 80
Kanabec, co., Minn., 9,007...E5 99
Kanaga, isl., Alsk...E4 79
Kanagawa, pref., Jap., 3,443,176...*I9 37
Kanalaksiorvik, fiord, Newf., Can...f9 75
Kananaskis, riv., Alta., Can...D3 69
Kanapou, bay, Haw...C5 88
Kanarraville, Iron, Utah, 236...F2 119
Kanash, Sov. Un., 29,000...B3 29
Kanaskat, King, Wash., 100...D2 122
Kanathea, isl., Fiji Is... 52
Kanauga, Gillia, Ohio, 350...D5 111
Kanauj, India, 24,646...D7 40
Kanawha, Hancock, Iowa, 735...B4 92
Kanawha, co., W. Va., 252,925...C3 123
Kanawha, lake, W. Va...C4 123
Kanawha, riv., W. Va...C3 123
Kanayama, Jap., 5,000...n16 37
Kanazawa, Jap., 298,972...H8 37
Kanchanaburi, Thai., 13,357...E3 38
Kancheepuram (Conjeeveram), India, 92,714...F7 39
Kanchenjunga, peak, Nep...D12 40
Kandagach, Sov. Un...D5 29
Kandahar, Afg., 77,186...F12 41
Kandahar, Sask., Can., 111...F3 70
Kandalaksha, Sov. Un., 37,500...D15 25
Kandalakshskaya, bay, Sov. Un...D15 25
Kandangan, Indon., 9,774...F5 35
Kandavu, isl., Fiji Is... 52
Kandel, Ger., 5,700...D3 17
Kandersteg, Switz., 937...D4 19
Kandi, Dah., 5,200...D5 45
Kandira, Tur., 4,900...B8 23
Kandiyohi, Kandiyohi, Minn., 312...E4 99
Kandiyohi, co., Minn., 29,987...E3 99
Kandla, India...F3 40
Kandreho, Malag., 180...g9 49
Kandy, Cey., 57,013...G7 39
Kane, Greene, Ill., 469...D3 90
Kane, McKean, Pa., 5,380...C4 114
Kane, Big Horn, Wyo., 20...A4 125
Kane, co., Ill., 208,246...B5 90
Kane, co., Utah, 2,667...F3 119
Kane, basin, Arc. O...B21 4
Kaneohe, Honolulu, Haw., 14,414...B4, g10 88
Kaneohe, bay, Haw...f10 88
Kanevskaya, Sov. Un., 10,000...H12 27
Kangal, Tur., 3,700...C11 31
Kangaroo, isl., Austl...G6 50
Kangean, is., Indon...G5 35
Kanggye, Kor., 30,013...F3 37
Kangha, bay, Kor...H3 37
Kangley, La Salle, Ill., 267...B5 90
Kangnung, Kor., 37,300 (58,700▲)...H4 37
Kango, Gabon...E2 46
Kangoku, rock, Iwo... 52
Kangpao, China...D6 36
Kangting, China, 30,000...E5 34
Kanhsien, China, 98,600...F7 34
Kani, Bur...A1 38
Kani, I.C...E3 45
Kaniama, Con. L...C3 48
Kaniapiskau, lake, Que., Can...F19 67
Kaniapiskau, riv., Que., Can...A2 73
Kanin, cape, Sov. Un...B17 9
Kanjiža, Yugo., 10,709...B5 22
Kankakee, Kankakee, Ill., 27,666...B6 90
Kankakee, co., Ill., 92,063...B6 90
Kankakee, riv., Ill., Ind...B5 90, B4 91
Kankan, Guinea, 24,600...D3 45
Kanker, India, 6,437...G8 40
Kånna, Swe...B7 24
Kannan, China, 5,000...C1 37
Kannapolis, Cabarrus, N.C., 34,647...B3 109
Kano, Nig., 93,016 (*130,173)...D6 45
Kanona, N. Rh...D5 48
Kanopolis, Ellsworth, Kans., 732...D5 93
Kanopolis, res., Kans...D5 93
Kanorado, Sherman, Kans., 245...C1 93
Kanosh, Millard, Utah, 499...E3 119
Kanovnice, riv., Czech...n18 26
Kanoya, Jap., 31,700...K5 37
Kanpur, India, 881,177 (*1,000,000)...C7 39, D8 40
Kanrach, Pak...C4 39
Kansas, Walker, Ala., 211...B2 78
Kansas, Edgar, Ill., 815...D6 90
Kansas, Seneca, Ohio, 375...A4 111
Kansas, Delaware, Okla., 300...A7 112
Kansas, state, U.S., 2,178,611...C8 76, 93
Kansas, riv., Kans...C7 93
Kansas City, Wyandotte, Kans., 121,901...B9, C9 93
Kansas City, Jackson, Mo., 475,539 (*1,025,900)...B3, E2 95
Kansasville, Racine, Wis., 100...F1 124
Kansk, Sov. Un., 88,000...D12 28
Kansong, Kor., 5,000...G4 37
Kansu, prov., China, 12,800,000...D5 34
Kantang, Thai., 5,177...I3 38
Kantunilkín, Mex., 872...C7 63
Kanturk, Ire., 1,985...E3 11
Kanye, Bech., 22,922...B4 49
Kanyü, China, 5,000...G8 36
Kaoan, China, 10,000...J6 36
Kaohe, Hawaii, Haw., 35...D6 88
Kaohsiung, (Takao), Taiwan, 312,000 (411,700▲)...G9 34
Kaolack, Sen., 46,600...D1 45
Kao Tao, is., Viet...B7 38
Kaouar, oases, Niger...C7 45
Kaoyao, China, 49,000...G7 34
Kaoyu, China, 65,000...H8 36
Kaoyu, lake, China...H8 36
Kapaa, Kauai, Haw., 3,439...A2 88
Kapal, Sov. Un., 10,000...F24 9
Kapandrition, Grc., 1,462...g11 23
Kapanga, Con. L...C3 48
Kapapa, isl., Haw...g10 88
Kapapala, Hawaii, Haw., 65...D6 88
Kapenguria, Ken...A6 48
Kapfenberg, Aus., 23,859...E7 16
Kapiri Mposhi, N. Rh., 184...C4 48
Kapit, Mala., 1,398...E4 35
Kaplan, Vermilion, La., 5,267...D3 95
Kaplice, Czech., 1,588...E9 17
Kapoeta, Sud...E3 47
Kapoho, Hawaii, Haw., 250...D7 88
Kaposvar, Hung., 43,458...B3 22
Kaposvar, creek, Sask...G4 70
Kapsan, Kor...F4 37
Kapterko (Well), Chad...B4 46
Kapuas, riv., Indon...F4 35
Kapulena, Hawaii, Haw., 100...C6 88
Kapuskasing, Ont., Can., 6,870...E9 72
Kapustin Yar, Sov. Un...D3 29
Kaputar, mtn., Austl...E8 51
Kapuvár, Hung., 10,902...B3 22
Kara, Sov. Un...B21 9
Kara, mtn., Tur...D9 31
Kara, sea, Sov. Un...B9 28
Karabash, Sov. Un., 24,300...D21 9
Kara-Bogaz, gulf, Sov. Un...G19 9
Kara-Bogaz-Gol, Sov. Un...E4 29
Kara-Bogaz-Gol, gulf, Sov. Un...E4 29
Karaburun, Tur., 782...C6 23
Karachev, Sov. Un., 10,000...E10 27
Karachi, Pak., 1,600,000 (*1,912,598)...D4 39
Karaga, Sov. Un., 10,000...D19 28
Karaganda, Sov. Un., 441,000...D8 29
Karagin, isl., Sov. Un...D19 28
Karaisali, Tur., 1,051...D10 31
Karakas, Sov. Un...D10 29
Karakelong, isl., Indon...E7 35
Karakoram, pass, India, Pak...A6 39
Karakoram, range, Asia...A6 39
Karakose, Tur., 19,800...C14 31
Karakul, Sov. Un...B11 41
Karakum, des., Sov. Un...E5 29
Karaman, Tur., 21,700...D9 31
Karamea, N.Z., 220...N14 51
Karamea, bight, N.Z...N13 51
Karamursel, Tur., 6,200...B7 23
Karand, Iran, 15,000...D3 41
Karapinar, Tur., 10,800...D9 31
Karas, mts., S.W. Afr...C2 49
Kara Shahr (Yenki), China, 10,000...C2 34
Karasu, Sov. Un...C6 29
Karasu, Tur., 2,906...B8 23
Karasuk, Sov. Un., 17,300...C9 29
Karatal, riv., Sov. Un...D9 29
Karatas, cape, Tur...D10 31
Kara Tau, range, Sov. Un...E7 29
Karatsu, Jap., 55,000 (77,825▲)...J4 37
Kara-Tyube, Sov. Un...D4 29
Karauli, India, 23,696...D6 40
Karaul Keldy, Sov. Un...D5 29
Karawanken, mts., Aus., Yugo...E7 16, B2 22
Karbalā', Iraq, 44,600...F15 31
Karcag, Hung., 20,803 (26,098▲)...B5 22
Karczew, Pol., 3,000...m14 26
Kardhitsa, Grc., 23,708...C3 23
Kardhitsa, prov., Grc., 152,543...*C3 23
Karelia (Karelo-Finnish S.S.R.), former rep., Sov. Un...*C6 28
Karema, Tan., 882...C5 48
Karesuando, Swe., 405...C10 25
Kargasok, Sov. Un., 900...B10 29
Kargat, Sov. Un., 10,000...D25 9
Kargil, India, 5,000...B6 39
Kargopol, Sov. Un., 5,000...A12 27
Karguiri, Niger...D7 45
Kariai, Grc., 305...B5 23
Kariba, S. Rh., 6,170...*A4 49
Karibib, S.W. Afr., 1,395...B2 49
Karikal, India, 22,252...F6 39
Karikari, cape, N.Z...K14 51
Karima, Sud., 5,989...B3 47
Karimata, arch., Indon...F3 35
Karimundjawa, is., Indon...G4 35
Karin, Som...C6 47
Karisimbi, vol., Con. L., Rwanda...B4 48
Karistos, Grc., 3,118...C5 23
Kārīz, Iran...D10 41
Karkar, isl., N. Gui...h12 50
Karkaralinsk, Sov. Un., 12,200...D9 29
Karkaraly, mts., Sov. Un...D9 29
Karkaraly, peak, Sov. Un...F24 9
Karkinitskiy, bay, Sov. Un...I9 27
Karkur, Isr., 3,200...B6 32
Karla, lake, Grc...C4 23
Karl-Marx-Stadt (Chemnitz), Ger., 286,100 (*400,000)...C7 17
Karlobag, Yugo., 403...C2 22
Karlovac, Yugo., 38,803...C2 22
Karlovasi, Grc., 5,024...D6 23
Karlovy Vary, Czech., 42,800...C2 26
Karlsbad, see Karlovy Vary, Czech.
Karlshamn, Swe., 11,700...I6 25
Karlskoga, Swe., 36,100...H6 25
Karlskrona, Swe., 33,200...I6 25
Karlsruhe, Ger., 241,900 (*335,000)...D3 17
Karlsruhe, McHenry, N. Dak., 221...A5 110
Karlstad, Kittson, Minn., 720...B2 99
Karlstad, Swe., 43,600...H5 25
Karlstadt, Ger., 6,000...D4 17
Karluk, Alsk., 129...D9 79
Karmutzen, mtn., B.C., Can...D4 68
Karnak, Pulaski, Ill., 667...F4 90
Karnes, co., Tex., 14,995...E4 118
Karnes City, Karnes, Tex., 2,693...E4 118
Karns City, Butler, Pa., 404...E2 114
Kärnten (Carinthia), state, Aus., 495,226...C9 18
Karoi, S. Rh...A4 49
Karonga, Nya...C5 48
Karora, Sud...B4 47
Karow, Ger., 1,043...E6 24
Karpathos, isl., Grc...E6 23
Karpenision, Grc., 3,693...C3 23
Karpinsk, Sov. Un., 50,000...D21 9
Kars, Tur., 32,000...B14 31
Karsakpay, Sov. Un., 12,000...D7 29
Karshi, Sov. Un., 19,000...H22 9
Karshinskiy, Sov. Un...D4 29
Karstädt, Ger., 1,872...E5 24
Kartal, Tur., 10,800...B7 23
Kartaly, Sov. Un., 33,400...C6 29
Karthaus, Clearfield, Pa., 500...D5 114
Kartuzy, Pol., 5,991...A5 26
Karup, Den., 1,137...B3 24
Karup Å, riv., Den...B3 24
Karval, Lincoln, Colo., 80...C7 83
Karvia, Fin., 350...F10 25
Karviná, Czech., 46,800...D5 26
Karwar, India, 12,134...F5 39
Kås, Den., 1,005...A3 24
Kas, Sud...C1 47
Kaş, Tur., 1,500...D7 31
Kasaan, Alsk., 20...n23 79
Kasai, prov., Con. L., 2,119,500...B3 48
Kasai, riv., Afr...G8 42
Kasaji, Con. L...D3 48
Kasalinsk, Sov. Un., 62,000...F21 9
Kasama, N. Rh., 3,700...D5 48
Kasanga, Tan., 5,369...C5 48
Kasaragod, India, 27,635...F6 39
Kasba, India, 13,051...E11 40
Kasba, lake, N.W. Ter., Can...D12 66
Kasba Tadla, Mor., 11,733...C3 44
Kasempa, N. Rh., 225...C4 48
Kasenga, Con. L...D4 48
Kasese, Ug...A5 48
Kāshān, Iran, 45,955...E5 41
Kashegelok, Alsk., 5...C8 79
Kashgar (Sufu), China, 91,000...H24 9
Kashima, Jap., 3,800...n19 37
Kashipur, India, 24,258...C7 40
Kashira, Sov. Un., 10,000...D12 27
Kashirka, riv., Sov. Un...n18 27
Kashiwazaki, Jap., 39,800...H9 37
Kāshmar, Iran, 12,052...D9 40
Kashmir, see Jammu and Kashmir, state, India
Kashmor, Pak...C2 40
Kasilof, Alsk., 89...C9, g16 79
Kasimov, Sov. Un., 33,500...C2 29
Kaskaskia, Randolph, Ill., 97...F4 90
Kaskaskia, riv., Ill...D5 90
Kaskö, Fin., 1,500...F9 25
Kas Kong, Camb...G5 38
Kaslo, B.C., Can., 646...E9 68
Kasongo, Con. L...B4 48
Kasongo-Lunda, Con. L...C2 48
Kasos, isl., Grc...E6 23
Kasota, LeSueur, Minn., 649...F5 99
Kassala, Sud., 35,621...B4 47
Kassala, reg., Sud., 941,039...B4 47
Kassandra, gulf, Grc...B4 23
Kassel, Ger., 207,500 (*270,000)...B4 17
Kasserine, Tun., 5,825...G11 30
Kassol, passage, Palau Is...52
Kasson, Dodge, Minn., 1,732...F6 99
Kastamonu (Kastamuni), Tur., 19,500...B9 31
Kastelli, Grc., 1,882...E4 23
Kastellorizo, isl., Grc...D7 23
Kastoria, Grc., 10,162...B3 23
Kastoria, prov., Grc., 47,487...*B3 23
Kastro, Tur...B5 23
Kastron, Grc., 3,493...C5 23
Kasugai, Jap., 49,200 (77,174▲)...*n15 37
Kasukabe, Jap., 18,500...n18 37
Kasumi, Jap., 12,100...I7 37
Kasumiga-Ura, bay, Jap...H10 37
Kasungu, Nya...D5 48
Kasur, Pak., 63,086...B5 39, B5 40
Katada, Jap., 6,900...n14 37
Katahdin, mtn., Maine...C4 96
Katako-Kombe, Con. L...B3 48
Katalla, Alsk., 12...C11 79
Katanga, prov., Con. L., 1,664,200...C3 48
Katanning, Austl., 3,360...F2 50
Katerini, Grc., 28,046...B4 23
Katha, Bur., 7,714...D10 39
Katherine, Austl., 314...B5 50
Kathleen, Polk, Fla., 650...D4 86
Kathryn, Alta., Can., 55...D4 69
Kathryn, Barnes, N. Dak., 142...C8 110
Kathua, India, 9,647...A5 40
Kathwood, Lexington, S.C., 2,000...*D4 115
Katia Kingi, Sud...D1 47
Katihar, India, 46,837...E11 40
Katimik, lake, Man., Can...C2 71
Katiola, I.C., 7,900...E3 45
Katire, Sud., 699...E3 47
Katmai, nat. mon., Alsk...D9 79
Katmai, vol., Alsk...D9 79
Katmandu, Nep., 106,579...D10 40
Katni (Murwara), India, 46,169...F8 40
Katokhi, Grc., 1,750...C3 23
Katombe, Con. L...C3 48
Katonah, Westchester, N.Y., 3,000...A5 106
Katoomba (Blue Mountains), Austl., 28,119...F8 51
Katouna, Grc., 3,176...C3 23
Katowice (Stalinograd), Pol., 269,000 (*1,875,000)...C5, g10 26
Katrinah, mtn., Eg., U.A.R...D6 43
Katrine, lake, Scot...D4 13
Katrineholm, Swe., 19,000...H7, u34 25
Katsina, Nig., 52,672...D6 45
Katsina Ala, riv., Nig...E6 45
Katsuyama, Jap., 19,300...m15 37
Katta-Kurgan, Sov. Un., 47,500...H22 9
Kattegat, chan., Eur...I4 25
Katun, riv., Sov. Un...E26 9
Katwijk aan Zee, Neth., 25,200 (30,000▲)...B4 15
Katy, Fort Bend, Harris and Waller, Tex., 1,569...F4 118
Kauai, co., Haw., 28,176...B2 82
Kauai, chan., Haw...B3 88
Kauai, isl., Haw...A2 88
Kaufbeuren, Ger., 34,700...E5 16
Kaufman, Kaufman, Tex., 3,087...C4 118
Kaufman, co., Tex., 29,931...C4 118
Kauhajoki, Fin., 1,700...F10 25
Kauhava, Fin., 2,100...F10 25
Kaukau Veld, plain, Bech., S. W. Afr...B3 49
Kaukauna, Outagamie, Wis., 10,096...A5, D5 124
Kaulakahi, chan., Haw...A2 88
Kauliranta, Fin., 700...D10 25
Kaumakani, Kauai, Haw., 950...B2 88
Kaumalapau, Maui, Haw., 100...C5 88
Kauna, pt., Haw...D6 88
Kaunakakai. Maui, Haw., 900...B4 88
Kaunas, Sov. Un., 232,000...A7 26, D4 27
Kaupo, Maui, Haw., 20...C5 88
Kaura Namoda, Nig., 19,146...D6 45
Kautokeino, Nor., 284...C10 25
Kavacha, Sov. Un...C19 28
Kavaje, Alb., 13,700...B2 23
Kavali, India, 20,544...F6 39
Kavália, Grc., 44,517...B5 23
Kavália, gulf, Grc...B5 23
Kavália, prov., Grc., 140,751...*B5 23
Kavanaugh, Boyd, Ky., 25...B7 94
Kavar, Iran...G6 41
Kavarna, Bul., 5,625...D9 22
Kavieng, Bis. Arch., 190...h13 50
Kävlinge, Swe., 3,500...C7 24
Kavrina, Czech., 33,905...D5 26
Kaw, Kay, Okla., 457...A5 111
Kawada, Okinawa... 52
Kawagama, lake, Ont., Can...B6 72
Kawagoe, Jap., 70,000 (107,523▲)...n18 37
Kawaguchi, Jap., 170,066...n18 37
Kawaihae, Hawaii, Haw., 100...C6 88
Kawainae, bay, Haw...C6 88
Kawaihoa, pt., Haw...B1 88
Kawaikini, peak, Haw...A2 88
Kawailoa, Honolulu, Haw., 300...f9 88
Kawailoa Beach, Honolulu, Haw., 400...f9 88
Kawambwa, N. Rh., 610...C4 48
Kawardha, India...F8 40
Kawasaki, Jap., 632,975...I9, n18 37
Kawayan, Phil., 2,084...*C6 35
Kawbar, Jordan, 2,000...h11 32
Kawela, Maui, Haw., 20...B5 88
Kawhia, hbr., N.Z...M15 51
Kawich, range, Nev...F5 104
Kawinaw, lake, Man., Can...C2 71
Kawkareik, Bur., 10,000...D3 38
Kawkawlin, Bay, Mich., 300...E7 98
Kawm, Ḩamādah, Eg. U.A.R., 1,000...D2 32
Kawm Umbū, Eg., U.A.R., 41,200...E6 43
Kawnipi, lake, Ont., Can...B7 99
Kay, co., Okla., 51,042...A4 112
Kaya, Upper Volta, 4,000...D4 45
Kayangel, is., Palau Is... 52
Kayenta, Navajo, Ariz., 200...A5 80
Kayes, Con. B...F2 46
Kayes, Mali, 19,600...D2 45
Kayford, Kanawha, W. Va., 400...C3, D6 123
Kayjay, Knox, Ky., 150...D6 94
Kaylor, Hutchinson, S. Dak., 165...D8 116
Kayseri, Tur., 102,809...C10 31
Kaysville, Davis, Utah, 3,608...B4 119
Kayville, Sask., Can., 152...H3 70
Kazachye, Sov. Un., 900...B16 28
Kazakevichevo, Sov. Un., 1,000...B7 37
Kazakh S.S.R., rep., Sov. Un., 10,387,000...F19 9
Kazakh, hills, Sov. Un...D7 29
Kazan, Sov. Un., 693,000...B3 29
Kazan, riv., N.W. Ter., Can...D12 66
Kazanlŭk, Bul., 19,386...D7 22
Kazan-retto (Volcano), is., Pac. O...E8 7
Kazbek, mtn., Sov. Un...E2 29
Kāzerūn, Iran, 30,641...G5 41
Kazhim, Sov. Un...A4 29
Kazi-Magomed, Sov. Un...E3 29
Kazimierz, Pol., 2,929...C6 26
Kazim Pasa (Saray), Tur., 3,486...C15 31
Kazincbarcika, Hung., 15,285...A5 22
Kazumba, Con. L...C3 48
Kazym, riv., Sov. Un...C22 9
Kbele, peak, N.W. Ter., Can...D6 66
Kdyne, Czech., 2,177...D8 17
Kea, Grc., 2,200...D5 23
Kea, isl., Grc...D5 23
Keady, N. Ire., 1,638...C5 11
Keahole, pt., Haw...D5 88
Kealaikahiki, chan., Haw...C5 88
Kealakekua, Hawaii, Haw., 325...D6 88
Kealakekua, bay, Haw...D6 88
Kealia, Hawaii, Haw., 100...D6 88
Kealia, Kauai, Haw., 655...A2 88
Keams Canyon, Navajo, Ariz...B5 80
Keanae, Maui, Haw., 54...C5 88
Keansburg, Monmouth, N.J., 6,854...C4 106
Keaoi, isl., Haw...D6 88
Kearney, Ont., Can., 365...B5 72
Kearney, Clay, Mo., 678...B3, D2 101
Kearney, Buffalo, Nebr., 14,210...D6 103
Kearney, Johnson, Wyo...A6 125
Kearney, co., Nebr., 6,580...D7 103
Kearneyville, Jefferson, W. Va., 600...B7 123
Kearns, Salt Lake, Utah, 17,172...C3 119
Kearny, Pinal, Ariz., 902...D4 80
Kearny, Hudson, N.J., 37,472...D5 106
Kearny, co., Kans., 3,108...D2 93
Kearsarge, Houghton, Mich., 400...A2 98
Kearsarge, Carroll, N.H., 150...B4 105
Kearsarge, mtn., N.H...D3 105
Keasbey, Middlesex, N.J., 1,500...*B4 106
Keatchie, De Soto, La., 345...B2 95
Keating, Baker, Oreg., 10...C9 113
Keating, Clinton, Pa., 35...D6 114
Keatley, Sask., Can...E2 70
Keats, Riley, Kans., 85...C7 93
Keauhou, Hawaii, Haw., 250...D6 88
Keawakapu, Maui, Haw., 100...C5 88
Keawekaheka, pt., Haw...D6 88
Kebbi, Nig...D5 45
Kebili, Tun...H11 30
Kebnekaise, mtn., Swe...D8 25
Kebock, head, Scot...B2 13
Kechi, Sedgwick, Kans., 245...B5, E6 93
Kecskemet, Hung., 45,904 (66,819▲)...B4 22
Kedah, state, Mala., 701,964...I4 38
Kedainiai, Sov. Un., 8,602...A7 26
Keddie, Plumas, Calif., 300...B3 82
Kedges, straits, Md...D5 85
Kedgwick, N.B., Can., 1,095...B2 74
Kedgwick, riv., N.B., Can...B2 74
Kedleston, Sask., Can., 65...G3 70
Kédougou, Sen., 1,200...D2 45
Kedron, Cleveland, Ark., 30...C3 81
Keedysville, Washington, Md., 433...B2 85
Keefers, B.C., Can., 100...D7 68
Keefeton, Muskogee, Okla., 80...B6 112
Keegan, Aroostook, Maine, 800...A5 96
Keego Harbor, Oakland, Mich., 2,761...A7 98
Keei, Hawaii, Haw., 100...D6 88
Keeler, Sask., Can., 82...G3 70
Keeler, Inyo, Calif., 200...D5 82
Keeline, Niobrara, Wyo., 30...C8 125
Keeling, Haywood, Tenn., 25...B2 117
Keeling, Pittsylvania, Va., 30...E3 121
Keels, Newf., Can., 185...D5 75
Keelung, see Chilung, Taiwan
Keene, Kern, Calif., 120...E4 82
Keene, Ont., Can., 324...C6 72
Keene, Jessamine, Ky., 500...C5 94
Keene, Kearney, Nebr., 25...D6 103
Keene, Cheshire, N.H., 17,562...E2 105
Keene, Essex, N.Y., 150...B3, A7 108
Keene, McKenzie, N. Dak., 100...B3 110
Keene, Coshocton, Ohio, 150...B6 111
Keene, Johnson, Tex., 1,532...B5 118
Keenesburg, Weld, Colo., 409...A6 83
Keene Valley, Essex, N.Y., 500...A3, B7 108
Keeney Knob, mtn., W. Va...D4 123
Keensburg, Wabash, Ill., 263...E6 90
Keeper, hill, Ire...E3 11
Keeseville, Essex and Clinton, N.Y., 2,213...B3 108
Keetly, Wasatch, Utah, 60...C4 119
Keetmanshoop, S.W. Afr., 4,410...C2 49
Keewatin, Ont., Can., 2,197...E4 71
Keewatin, Itasca, Minn., 1,651...C5 99
Keewatin, dist., N.W. Ter., Can., 2,301...D14 66
Keewatin, riv., Man., Can...A1 71
Keezletown, Rockingham, Va., 175...C4 121
Kafallinia, prov., Grc., 47,314...*C3 23
Kefallinia (Cephalonia), isl., Grc...C3 23
Keflavik, Ice., 4,852...o21 25
Kegaska, Que., Can., 93...h9 75
Kegonsa, lake, Wis...F4 124
Kegueur Terbi, mtn., Chad...43 46
Keheili, Sud...B3 47
Kehl, Ger., 13,100...E2 17
Kehsi Mansam, Bur...B2 38
Keighley, Eng., 55,852...D6 10
Keimoes, S. Afr., 2,995...C3 49
Keiser, Mississippi, Ark., 516...B5 81
Keiser (Marion Heights), Northumberland, Pa., 1,132...*E8 108
Keitele, lake, Fin...F11 25
Keith, Austl...H3 51
Keith, Scot., 4,208...C6 13
Keith, co., Nebr., 7,958...C4 103
Keithley Creek, B.C., Can., 15...C7 68
Keithsburg, Mercer, Ill., 963...B3 90
Keitum, Ger...D2 24
Keizer, Marion, Oregon, 5,288...*C4 113
Kejimkujik, lake, N.S., Can...E4 74
Kekaha, Kauai, Haw., 2,082...B2 88
Kelantan, state, Mala., 505,522...I5 38
Kelantan, riv., Mala...J5 38
Keldron, Corson, S. Dak., 23...B4 116
Kelfield, Sask., Can., 32...F1 70
Kelford, Bertie, N.C., 362...A6 109
Kelheim, Ger., 11,900...E6 17
Kelibia, Tun...F12 30
Kelif, Sov. Un...C13 41
Kelilberg, mtn., Czech...C8 17
Kelkit, riv., Tur...B11 31
Kell, Marion, Ill., 194...E5 90
Kellé, Con. B...F2 46
Kellenhusen, Ger., 1,100...D5 24
Keller, Tarrant, Tex., 827...B5 118
Keller, Accomack, Va., 263...D7 121
Keller, Ferry, Wash., 25...A7 122
Kellerman, Tuscaloosa, Ala., 500...B2 78
Kellerton, Ringgold, Iowa, 341...D3 92
Kellerville, Adams, Ill., 20...D3 90
Kellet, cape, N.W. Ter., Can...B7 66
Kellettville, Forest, Pa., 30...C3 114
Kelley, Story, Iowa, 239...C4 92
Kelleys, isl., Ohio...A5 111
Kelleys Island, Erie, Ohio, 171...A5 111
Kelliher, Sask., Can., 461...F4 70
Kelliher, Beltrami, Minn., 297...C4 99
Kellmünz, Ger., 1,100...A6 18
Kellnersville, Manitowoc, Wis., 350...A6, D6 124
Kelloe, Man., Can., 80...D1 71
Kellogg, Shoshone, Idaho, 5,061...B2 89
Kellogg, Jasper, Iowa, 623...C5 92
Kellogg, Wabasha, Minn., 446...F6 99
Kells, range, Scot...E4 13
Kelly, Christian, Ky., 175...D2 94
Kelly, Caldwell, La., 450...C3 95
Kelly, Socorro, N. Mex., 20...C2 107
Kelly, Teton, Wyo., 20...B2 125
Kelly Brook, mtn., Maine...A3 96
Kelly Lake, St. Louis, Minn., 900...C5 99
Kellyton, Coosa, Ala., 450...C3 78
Kellytown, Henry, Ga., 200...C2 87
Kellyville, Sullivan, N.H., 140...D2 105
Kellyville, Creek, Okla., 501...B5 106
Kelme, Sov. Un...A7 26
Kélo, Chad, 6,067...D3 46
Kelowna, B.C., Can., 13,188...E8 68
Kelsey, Alta., Can., 51...C4 69
Kelsey, lake, Man., Can...D5 70
Kelsey, mtn., N.H...A4 105
Kelsey Bay, B.C., Can., 307...D4 68

Kelso, San Bernardino, Calif., 100 ... E6 82
Kelso, Dearborn, Ind., 150 ... F8 91
Kelso, Scott, Mo., 258 ... D8 101
Kelso, Traill, N. Dak., 27 ... B8 110
Kelso, Scot., 3,964 ... E6 13
Kelso, Cowlitz, Wash., 8,379 ... C3 122
Kelso Station, Sask., Can., 110 ... H6 70
Keltie, cape, Ant ... C26 5
Keltonburg, De Kalb, Tenn., 60 ... D8 117
Keltys, Angelina, Tex., 1,056 ... *D5 118
Kelvington, Sask., Can., 885 ... E4 70
Kelwood, Man., Can., 323 ... D2 71
Kem, Sov. Un., 10,000 ... E16 25
Kem, riv., Sov. Un ... E15 25
Ké-Macina, Mali, 1,200 ... D3 45
Kemah, Galveston, Tex., 950 ... *F5 118
Kembé, Cen. Afr. Rep ... E4 46
Kemerovo, Sov. Un., 298,000 ... B11 29
Kemi, Fin., 28,800 ... E11 25
Kemijärvi, Fin., 5,000 ... D12 25
Kemijärvi, lake, Fin ... D12 25
Kemijoki, riv., Fin ... D11 25
Kemme, Libya ... E3 43
Kemmerer, Lincoln, Wyo., 2,028 ... D2 125
Kemnath, Ger., 3,000 ... D6 17
Kemnay, Man., Can., 81 ... E1 71
Kemp, Bryan, Okla., 153 ... D5 112
Kemp, Kaufman, Tex., 816 ... C4 118
Kemp, coast, Ant ... C18 5
Kemp, lake, Tex ... C3 118
Kempen, Ger., 12,900 ... C6 15
Kemper, Dillon, S.C., 40 ... C9 115
Kemper, co., Miss., 12,277 ... C5 100
Kempner, Lampasas, Tex., 250 ... D3 118
Kempsey, Austl., 8,016 ... E9 51
Kempsville, Princess, Anne, Va., 500 ... B7 121
Kempten, Ger., 43,100 ... E5 16
Kempton, Ford, Ill., 252 ... C5 90
Kempton, Tipton, Ind., 480 ... D5 91
Kempton, Grand Forks, N. Dak., 58 ... B8 110
Kemptville, Ont., Can., 1,959 ... B9 72
Kenadsa, Alg ... C4 44
Kenai, Alsk., 778 ... C9, g16 79
Kenai, mts., Alsk ... h16 79
Kenai, pen., Alsk ... h16 79
Kenamu, riv., Newf., Can ... B1 79
Kenansville, Osceola, Fla., 250 ... E6 86
Kenansville, Duplin, N.C., 724 ... C6 109
Kenaston, Sask., Can., 423 ... F2 70
Kenaston, Ward, N. Dak., 30 ... A3 110
Kenberma, Plymouth, Mass. (part of Hull) ... D3 97
Kenbridge, Lunenburg, Va., 1,188 ... E4 121
Kendal, Eng., 18,595 ... C5 10
Kendal Green, Middlesex, Mass ... D2 97
Kendall, Dade, Fla., 3,000 ... F3 86
Kendall, Hamilton, Kans., 250 ... E2 93
Kendall, Orleans, N.Y., 220 ... B2 108
Kendall, Monroe, Wis., 528 ... E3 124
Kendall, co., Ill., 17,540 ... B5 90
Kendall, co., Tex., 5,889 ... E3 118
Kendallville, Noble, Ind., 6,765 ... B7 91
Kendal Station, Sask., Can., 161 ... G4 70
Kendari, Indon ... F6 35
Kendrapara, India, 15,830 ... G11 40
Kendrick, Marion, Fla., 500 ... C4 86
Kendrick, Latah, Idaho, 443 ... C2 89
Kendrick, Lincoln, Okla., 155 ... B5 112
Kenduskeag, Penobscot, Maine, 300 (584▲) ... D4 96
Kenedy, Karnes, Tex., 4,301 ... E4 118
Kenedy, co., Tex., 884 ... F4 118
Kenefic, Bryan, Okla., 125 ... C5 112
Kenel, Corson, S. Dak., 75 ... B5 116
Kenema, S.L ... E2 45
Kenesaw, Adams, Nebr., 546 ... D7 103
Kenge, Con. L ... B2 48
Keng Kabao, Laos ... D6 38
Keng Tung, Bur., 5,508 ... D10 39
Kenhardt, S. Afr., 2,832 ... C3 49
Kenhorst, Berks, Pa., 2,815 ... *F10 114
Kéniéba, Mali ... D2 45
Kenilworth, Eng., 14,427 ... B6 12
Kenilworth, Cook, Ill., 2,959 ... E3 90
Kenilworth, Union, N.J., 8,379 ... *B4 106
Kenilworth, Carbon, Utah, 500 ... D5 119
Kenimekh, Sov. Un ... A12 41
Kenitra, Mor., 86,775 ... C3 44
Kenly, Johnston, N.C., 1,147 ... B5 109
Kenmare, Ire., 1,046 ... F2 11
Kenmare, Ward, N. Dak., 2,463 ... A3 110
Kenmare, bay, Ire ... F2 11
Kenmawr, Allegheny, Pa., 3,000 ... *F1 114
Kenmore, Erie, N.Y., 21,261 ... C2 108
Kenmore, King, Wash., 1,000 ... *B3 122
Kenna, Roosevelt, N. Mex., 80 ... D6 107
Kenna, Jackson, W. Va., 200 ... C3 123
Kenneday, peak, Wyo ... D6 125
Kennan, Price, Wis., 162 ... C3 124
Kennard, Henry, Ind., 466 ... E6 91
Kennard, Washington, Nebr., 331 ... C9, D2 103
Kennard, Houston, Tex., 400 ... D5 118
Kennebago, lake, Maine ... C2 96
Kennebec, Lyman, S. Dak., 372 ... D6 116
Kennebec, co., Maine, 83,881 ... D3 96
Kennebec, riv., Maine ... D3 96
Kennebunk, York, Maine, 2,804 (4,551▲) ... E2 96
Kennebunkport, York, Maine, 700 (1,851▲) ... E2 96
Kennedale, Tarrant, Tex., 1,521 ... B5 118
Kennedy, Lamar, Ala., 379 ... B2 78
Kennedy, Sask., Can., 274 ... H4 70
Kennedy, Kittson, Minn., 458 ... B2 99
Kennedy, Cherry, Nebr., 11 ... B5 103
Kennedy, Chautauqua, N.Y., 500 ... C1 108
Kennedy, lake, Sask., Can ... D4 70
Kennedyville, Kent, Md., 350 ... B6 85
Kenner, Jefferson, La., 17,037 ... C7, E5 95
Kennesaw, Cobb, Ga., 1,507 ... A4, B2 87
Kennesaw, mtn., Ga ... C2 87
Kennet, riv., Eng ... C6 12
Kennetcook, N.S., Can., 375 ... D6 74
Kenneth City, Pinellas, Fla., 2,114 ... *E4 86
Kennett, Dunklin, Mo., 9,098 ... E7 101
Kennett Square, Chester, Pa., 4,355 ... G10 114
Kennewick, Benton, Wash., 14,244 ... C6 122
Kenney, De Witt, Ill., 400 ... C4 90
Kennington Cove, N.S., Can., 97 ... D10 74
Kennisis, lake, Ont., Can ... B6 72
Kennydale, King, Wash., 3,500 ... D2 122
Keno, Klamath, Oreg., 175 ... E5 113
Kenogami, Que., Can., 11,816 ... A6 73
Kenogami, lake, Que., Can ... A6 73
Kenora, Ont., Can., 10,904 ... E7 72
Kenora, co., Ont., Can., 51,474 ... D8 72
Kenosha, Kenosha, Wis., 67,899 ... F2, F6 124
Kenosha, co., Wis., 100,615 ... F5 124
Kenova, Wayne, W. Va., 4,577 ... C2 123
Kensal, Stutsman, N. Dak., 334 ... B7 110
Kensett, White, Ark., 905 ... B4 81
Kensett, Worth, Iowa, 409 ... A4 92
Kensico, res., N.Y ... D2 108
Kensington, Contra Costa, Calif., 6,161 ... *B5 82
Kensington, P.E.I., Can., 884 ... C6 74
Kensington, Hartford, Conn., 4,500 ... C5 84
Kensington, Smith, Kans., 619 ... C4 93
Kensington, Montgomery, Md., 2,175 ... B3 85
Kensington, Douglas, Minn., 324 ... E3 99
Kensington, Rockingham, N.H., 50 (708▲) ... E5 105
Kensington, Nassau, N.Y., 1,166 ... *E3 108
Kensington, Columbiana, Ohio, 350 ... B7 111
Kensington Estates, Montgomery, Md., 1,600 ... *B3 85
Kensington Park, Sarasota, Fla., 2,969 ... *E4 86
Kensington Park, Chatham, Ga., 1,000 ... *D6 87
Kenspur, Ravalli, Mont., 74 ... D2 102
Kent, Elmore, Ala., 500 ... C4 78
Kent, Litchfield, Conn., 400 (1,686▲) ... C3 84
Kent, Jefferson, Ind., 65 ... G6 91
Kent, Union, Iowa, 94 ... D3 92
Kent, Wilkin, Minn., 134 ... D2 99
Kent, Portage, Ohio, 17,836 ... A6 111
Kent, Sherman, Oreg., 65 ... B6 113
Kent, Culberson, Tex., 50 ... F2 118
Kent, King, Wash., 9,017 ... B3, D2 122
Kent, co., N.B., Can., 26,667 ... C4 74
Kent, co., Ont., Can., 89,247 ... E2 72
Kent, co., Del., 65,651 ... B6 85
Kent, co., Eng., 1,701,083 ... C8 12
Kent, co., Md., 15,481 ... B5 85
Kent, co., Mich., 363,187 ... E5 98
Kent, co., R.I., 112,619 ... C10 84
Kent, co., Tex., 1,727 ... C2 118
Kent, isl., Del ... B7 85
Kent, isl., Md ... C5 85
Kent, pt., Md ... C5 85
Kent, riv., Eng ... F6 13
Kent Bridge, Ont., Can., 118 ... E2 72
Kent City, Kent, Mich., 617 ... E5 98
Kentfield, Marin, Calif., 5,000 ... *B5 82
Kent Furnace, Litchfield, Conn., 100 ... C3 84
Kent Junction, N.B., Can., 182 ... C4 74
Kentland, Newton, Ind., 1,783 ... C3 91
Kentland, Prince Georges, Md., 1,800 ... *C4 85
Kenton, Man., Can., 222 ... E1 71
Kenton, Kent, Del., 249 ... B6 85
Kenton, Kenton, Ky., 240 ... A7 94
Kenton, Houghton, Mich., 200 ... B2 98
Kenton, Hardin, Ohio, 8,747 ... B4 111
Kenton, Cimarron, Okla., 100 ... D1 112
Kenton, Obion and Gibson, Tenn., 1,095 ... A2 117
Kenton, co., Ky., 120,700 ... B5 94
Kents Store, Fluvanna, Va., 20 ... D4 121
Kentucky, state, U.S., 3,038,156 ... C11 77, 94
Kentucky, dam, Ky ... A3 94
Kentucky, lake, Ky., Tenn ... D1 94
Kentucky, ridge, Ky ... D6 94
Kentucky, riv., Ky ... B5 94
Kent Village, Prince Georges, Md., 2,500 ... *C4 86
Kentville, N.S., Can., 4,612 ... D5 74
Kentwood, Tangipahoa, La., 2,607 ... D5 95
Kenvil, Morris, N.J., 2,000 ... B3 106
Kenville, Man., Can., 144 ... D1 71
Kenvir, Harlan, Ky., 950 ... D6 94
Kenwood, Hamilton, Ohio, 2,000 ... *C3 111
Kenwood, Delaware, Okla., 75 ... A7 112
Kenya, Br. dep., Afr., 7,287,000 ... F9 42, A6 48
Kenya, mtn., Ken ... B6 48
Kenyon, Goodhue, Minn., 1,624 ... F6 99
Kenyon, Washington, R.I., 250 ... D10 84
Keo, Lonoke, Ark., 237 ... C4, E6 81
Keokea, Maui, Haw ... C5 88
Keokuk, Lee, Iowa, 16,316 ... D6 92
Keokuk, co., Iowa, 15,492 ... C5 92
Keokuk, lock and dam, Iowa ... D6 92
Keoma, Alta., Can., 25 ... D4 69
Keonjhargarh, India, 12,624 ... G10 40
Keosauqua, Van Buren, Iowa, 1,023 ... D6 92
Keota, Weld, Colo., 13 ... A6 83
Keota, Keokuk, Iowa, 1,096 ... C6 92
Keota, Haskell, Okla., 579 ... B7 114
Kep, Camb ... G6 38
Kepno, Pol., 7,810 ... C5 26
Keppel, Sask., Can., 95 ... E2 70
Kepples, Butler, Pa ... E2 114
Kepsut, Tur., 4,000 ... C6 23
Kerala, state, India, 16,903,715 ... F6 39
Kerang, Austl., 3,727 ... G4 51
Kerasund (Giresun), Tur., 19,900 ... B12 31
Kerby, Josephine, Oreg., 600 ... E3 113
Kerby, peak, Oreg ... E3 113
Kerch, Sov. Un., 104,000 ... I11 27
Kerch, strait, Sov. Un ... I11 27
Kerchemya, Sov. Un ... C19 9
Kerema, Pap ... k12 50
Keremeos, B.C., Can., 563 ... E8 68
Kerempe, cape, Tur ... A9 31
Kerens, Navarro, Tex., 1,123 ... C4 118
Kerguelen, isl., Indian O ... J1 2
Kerhonkson, Ulster, N.Y., 690 ... D6 108
Kericho, Ken ... B6 48
Kerintji, mtn., Indon ... F2 35
Kerkennah, is., Tun ... C7 44
Kerkhoven, Swift, Minn., 645 ... E3 99
Kerki, Sov. Un., 21,600 ... H22 9
Kerkira, Grc., 26,991 ... C2 23
Kerkira, prov., Grc., 101,770 ... *C2 23
Kerkira (Corfu), isl., Grc ... C2 23
Kerkrade, Neth., 49,400 ... D6 15
Kermadec, is., Pac. O ... I11 7
Kerman, Fresno, Calif., 1,970 ... *D4 82
Kermān, Iran, 62,157 ... F8 41
Kermānshāh, Iran, 125,439 ... D3 41
Kermit, Divide, N. Dak., 23 ... A2 110
Kermit, Winkler, Tex., 10,465 ... D1 118
Kermit, Mingo, W. Va., 743 ... D2 123
Kermode, mtn., B.C., Can ... C2 68
Kern, co., Calif., 291,984 ... E4 82
Kern, riv., Calif ... E4 82
Kernersville, Forsyth, N.C., 2,942 ... A3 109
Kernville, Kern, Calif., 600 ... E4 82
Kernville, Lincoln, Oreg., 15 ... C3 113
Kerpen, Ger., 7,200 ... C1 17
Kerr, Lonoke, Ark ... D6 81
Kerr, Gallia, Ohio, 30 ... D5 111
Kerr, co., Tex., 16,800 ... D3 118
Kerr, lake, Fla ... C5 86
Kerr, John H., res., N.C., Va ... A5 109, E4 121
Kerrera, isl., Scot ... D3 13
Kerrick, Pine, Minn., 110 ... D6 99
Kerrick, Dallam, Tex., 50 ... A1 118
Kerrobert, Sask., Can., 1,220 ... F1 70
Kerrs Creek, Rockbridge, Va ... D3 121
Kerrville, Kerr, Tex., 8,901 ... D3 118
Kerry, co., Ire., 116,458 ... E2 11
Kerrykeel, Ire ... B4 11
Kersey, Weld, Colo., 378 ... A6 83
Kersey, Elk, Pa., 600 ... D4 114
Kershaw, Chouteau, Mont., 3 ... C6 102
Kershaw, Kershaw and Lancaster, S.C., 1,567 ... B6 115
Kershaw, co., S.C., 33,585 ... C6 115
Kersley, B.C., Can., 182 ... C6 68
Kerteminde, Den., 4,024 ... C4 24
Kerulen, riv., Mong ... B7 34
Kerza, Alg ... D4 44
Kerzers, Switz., 2,228 ... C3 19
Kesagami, lake, Ont., Can ... D9 72
Kesan, Tur., 15,100 ... B6 23
Kesarya (Sdot Yam), Isr., 458 ... B6 32
Kesch, peak, Switz ... C8 19
Kesennuma, Jap., 37,200 ... G10 37
Kesh, N. Ire., 202 ... C4 11
Keshena, Shawano, Wis., 500 ... D5 124
Kesley, Butler, Iowa, 115 ... B5 92
Kestenga, Sov. Un ... E14 25
Kesteven, Lincolnshire, see Lincoln, co., Eng.
Keswick, Ont., Can., 699 ... C5 72
Keswick, Eng., 4,752 ... F5 13
Keswick, Keokuk, Iowa, 265 ... C5 92
Keswick, Albermarle, Va., 300 ... C4 121
Keszthely, Hung., 14,854 ... B3 22
Ket, riv., Sov. Un ... D26 9
Keta, Ghana, 11,380 ... E5 45
Ketapang, Indon., 4,385 ... F3 35
Ketchikan, Alsk., 6,483 ... D13, n24 79
Ketchum, Blaine, Idaho, 746 ... F4 89
Ketchum, Craig, Okla., 255 ... A6 111
Ketchum, mtn., Tex ... D2 118
Ketona, Jefferson, Ala., 900 ... E5 78
Ketrzyn Mazowiecki, Pol., 13,900 ... A6 26
Kettering, Eng., 38,631 ... B7 12
Kettering, Montgomery, Ohio, 54,462 ... *C3 111
Kettle, creek, Pa ... C6 114
Kettle, riv., B.C., Can., Wash ... E8 68
Kettle, riv., Minn ... D6 99
Kettle Falls, Stevens, Wash., 905 ... A7 122
Kettleman City, Kings, Calif., 400 ... E4 82
Kettle River, Carlton, Minn., 234 ... D6 99
Kettle River, range, Wash ... A7 122
Kettlewell, Eng., 304 ... F6 13
Kettwig, Ger., 17,100 ... B1 17
Kety, Pol., 6,581 ... D5, h10 26
Ketzin, Ger., 5,107 ... A7 17
Keuka, lake, N.Y ... C3 108
Keuterville, Idaho, Idaho, 30 ... C2 89
Kevil, Ballard, Ky., 231 ... A2 94
Kevin, Toole, Mont., 375 ... B5 102
Kew, Eng. (part of Richmond) ... m11 10
Kewanee, Henry, Ill., 16,324 ... B4 90
Kewanee, Lauderdale, Miss., 225 ... C5 100
Kewanee, New Madrid, Mo., 200 ... E8 101
Kewanna, Fulton, Ind., 683 ... B5 91
Kewaskum, Washington, Wis., 1,572 ... E5 124
Kewaunee, Kewaunee, Wis., 2,772 ... D6 124
Kewaunee, co., Wis., 18,282 ... D6 124
Keweenaw, co., Mich., 2,417 ... A2 98
Keweenaw, bay, Mich ... B2 98
Keweenaw, pt., Mich ... A3 98
Keweenaw Bay, Baraga, Mich., 75 ... B2 98
Key, lake, Ire ... C3 11
Keyapaha, Tripp, S. Dak., 19 ... E6 116
Keya Paha, co., Nebr., 1,672 ... B6 103
Keya Paha, riv., Nebr., S. Dak ... D6 116
Key Biscayne, Dade, Fla., 2,500 ... *G6 86
Keyes, Stanislaus, Calif., 1,546 ... *D4 82
Keyes, Man., Can., 65 ... D2 71
Keyes, Cimarron, Okla., 627 ... D2 112
Keyesport, Clinton and Bond, Ill., 412 ... E4 90
Keyhole, res., Wyo ... A8 125
Key Junction, Ont., Can., 55 ... B4 72
Key Largo, Monroe, Fla., 900 ... G6 86
Keymar, Carroll, Md., 150 ... A3 85
Keynsham, Eng., 15,144 ... C5 12
Keyport, Monmouth, N.J., 6,440 ... C4 106
Keysburgh, Logan, Ky., 100 ... D2 94
Keyser, Mineral, W. Va., 6,192 ... B6 123
Keystone, Wells, Ind., 260 ... C7 91
Keystone, Benton, Iowa, 522 ... C5 92
Keystone, Keith, Nebr., 50 ... C4 103
Keystone, White Pine, Nev ... D7 104
Keystone, Tulsa, Okla., 151 ... A5 114
Keystone, Pennington, S. Dak., 500 ... D2 116
Keystone, McDowell, W. Va., 1,457 ... D3 123
Keystone Heights, Clay, Fla., 655 ... C5 86
Keystown, Sask., Can., 85 ... G3 70
Keysville, Hillsborough, Fla., 500 ... E4 86
Keysville, Burke, Ga., 250 ... C4 87
Keysville, Charlotte, Va., 733 ... D4 121
Keytesville, Chariton, Mo., 644 ... B5 101
Key West, Monroe, Fla., 33,956 ... H5 86
Key West, Dubuque, Iowa, 85 ... B7 92
Kezar, lake, Maine ... D2 96
Kezar, pond, Maine ... D2 96
Kezar Falls, York, Maine, 600 ... E2 96
Kezhma, Sov. Un., 3,000 ... D13 28
Kezmarok, Czech., 7,372 ... D6 26
Kfar Ata, Isr., 14,245 ... B7 32
Kfar Blum, Isr ... A7 32
Kfar Monash, Isr., 232 ... f10 32
Kfar Saba, Isr., 18,635 ... g10 32
Kfar Vitkin, Isr., 1,094 ... B6 32
Kfar Yona, Isr., 3,500 ... f10 32
Khabab, Syr ... A8 32
Khabarovsk, Sov. Un., 349,000 ... E16 26, B7 37
Khachmas, Sov. Un ... E3 29
Khadar Khel, Afg., 5,000 ... E14 41
Khairpur, Pak., 18,186 ... D2 40
Khalafābād, Iran ... F4 41
Khalij Surt, see Sidra, gulf, Libya.
Khalkí (Chalke), isl., Grc ... D6 23
Khalkidhiki (Chalcidice), prov., Grc., 79,849 ... *B4 23
Khalkidhiki, pen., Grc ... B4 23
Khalkis, Grc., 24,745 ... C4, g11 23
Khambhaliya, India, 20,064 ... D5 39
Khamgaon, India, 44,432 ... G6 40
Khanabad, Afg., 18,042 ... C14 41
Khanaqin, Iraq, 10,090 ... D2 41
Khān az Zabīb, Jordan, 1,000 ... C8 32
Khanderi, is., India ... H4 40
Khandwa, India, 63,505 ... G6 40
Khanh An, Viet ... H6 38
Khania, Grc., 38,467 ... E4 23
Khania, prov., Grc., 131,061 ... *E4 23
Khania, gulf, Grc ... E4 23
Khanka, lake, China, Sov. Un ... D6 37
Khanpur, Pak., 13,484 ... C5 39
Khanty-Mansiysk, Sov. Un., 19,000 ... A7 29
Khan Yunis, Gaza Area, 10,000 ... C6 32
Kharagpur, India, 147,253 ... D8 38, F11 40
Kharānaq, Iran ... E7 41
Kharan Kalat, Pak., 2,589 ... G12 41
Kharg, isl., Iran ... G5 41
Khargone, India, 30,652 ... G5 40
Khariar, India, 7,873 ... G9 40
Kharijah, oasis, Eg., U.A.R ... D6 43
Kharkov, Sov. Un., 976,000 (★1,175,000) ... G11 27
Kharmanli, Bul., 9,240 ... E7 22
Kharovsk, Sov. Un., 5,000 ... B13 27
Kharr, wadi, Sau. Ar ... G14 31
Khartoum, Sud., 93,103 (★260,000) ... B3 47
Khartoum, reg., Sud., 504,923 ... B3 47
Khartoum North, Sud., 39,082 ... B3 47
Khartsyzsk, Sov. Un., 10,000 ... q21 27
Khasavyurt, Sov. Un., 30,000 ... G18 9
Khash, Afg., 5,000 ... F11 41
Khāsh, Iran, 9,291 ... G10 41
Khashuri, Sov. Un., 5,347 ... A14 31
Khasi, hills, India ... C9 39
Khaskovo, Bul., 27,394 ... E7 22
Khatanga, Sov. Un., 10,000 ... B13 28
Khatanga, riv., Sov. Un ... B13 28
Khemmarat, Thai., 8,194 ... D6 38
Khenchela, Alg., 11,051 (12,196▲) ... B6 44
Khénifra, Mor., 18,503 ... C3 44
Kherson, Sov. Un., 174,000 ... H9 31
Kheta, riv., Sov. Un ... B12 28
Khilchipur, India, 6,970 ... E6 40
Khilok, Sov. Un., 18,600 ... D14 28
Khimki, Sov. Un., 43,000 ... n17 27
Khios, Grc., 24,053 ... C6 23
Khios (Chios), prov., Grc., 62,223 ... *C6 23
Khíos (Chios), isl., Grc ... C6 23
Khisfin, Syr ... B7 32
Khiva, Sov. Un., 19,000 ... E6 29
Khlebarovo, Bul., 5,829 ... D8 22
Khmelnik, Sov. Un., 10,000 ... G6 27
Khmelnitskiy, Sov. Un., 62,000 ... G6 27
Khochniye, Syr ... A7 32
Khodzheyli, Sov. Un., 15,000 ... E5 29
Khokhropar, Pak ... E3 40
Kholm, Sov. Un., 10,000 ... C8 27
Kholmsk (Maoka), Sov. Un., 33,000 ... E17 28
Khonak, Afg., 5,000 ... D13 41
Khong, Laos, 10,000 ... E6 38
Khong, riv., Laos ... E7 38
Khong Sedone, Laos, 10,000 ... E6 38
Khon Kaen, Thai., 17,952 ... D5 38
Khoper, riv., Sov. Un ... C2 29
Khor, riv., Sov. Un ... C7 37
Khor al Kalba, Mus. & Om ... I8 41
Khor-Anghar, Fr. Som ... C5 47
Khora Sfakion, Grc., 377 ... E5 23
Khorinsk, Sov. Un ... A6 34
Khorog, Sov. Un., 9,000 ... H23 9
Khorol, Sov. Un., 10,000 ... G9 27
Khorramābād, Iran, 38,676 ... E4 41
Khorramshahr, Iran, 43,850 ... F4 41
Khotan (Hotien), China, 50,000 ... A6 39
Khotan, riv., China ... A7 39
Khotin, Sov. Un., 10,000 ... G6 27
Khouribga, Mor., 40,838 ... *C3 44
Khrisoupolis, Grc., 5,037 ... B5 23
Khrom-Tau, Sov. Un ... C5 29
Khu Khan, Thai ... E6 38
Khulna, Pak., 127,970 ... D9 39
Khunzakh, Sov. Un ... A16 31
Khurda, India, 12,497 ... G10 40
Khurja, India, 41,491 ... C6 40
Khushab, Pak., 20,476 ... A4 40
Khust, Sov. Un., 10,000 ... G4 27
Khuzdar, Pak ... C4 39
Khvāf, Iran ... D10 41
Khvor, Iran ... E7 41
Khvormūj, Iran, 2,500 ... G5 41
Khvoy, Iran, 34,491 ... B2 41
Khyber, pass, Afg., Pak ... D15 41
Kia, isl., Fiji Is ... 52
Kialing, riv., China ... E6 34
Kiamichi, Pushmataha, Okla., 75 ... C6 114
Kiamichi, mtn., Okla ... C7 114
Kiamichi, riv., Okla ... C6 114
Kiamika, Que., Can., 168 ... C2 73
Kiamika, riv., Que., Can ... C2 73
Kiamusze (Chiamussu), China, 146,000 ... B11 34
Kiana, Alsk., 253 ... B7 79
Kiangsi, prov., China, 18,610,000 ... F9 34
Kiangsu, prov., China, 42,630,000 ... E8 34
Kiani, lake, P.R ... g12 65
Kiantajärvi, lake, Fin ... E13 25
Kiaohsien, China, 75,000 ... F9 36
Kiask, lake, Man., Can ... A3 71
Kiawah, isl., S.C ... F7 115
Kibangou, Con. B ... F2 46
Kibau, Tan ... C6 48
Kiberege, Tan ... C6 48
Kiblah, Miller, Ark., 195 ... D2 81
Kibombo, Con. L ... B4 48
Kibondo, Tan ... B5 48
Kibwezi, Ken ... B6 48
Kičevo, Yugo., 10,273 ... E5 22
Kickapoo, Leavenworth, Kans., 30 ... A8 93
Kickapoo, Polk, Tex ... D5 118
Kickapoo, creek, Ill ... C4 90
Kickapoo, riv., Wis ... E3 124
Kicking Horse, pass, Alta., B.C., Can ... D2 69
Kidal, Mali, 750 ... C5 45
Kidder, Caldwell, Mo., 224 ... B3 101
Kidder, Marshall, S. Dak., 142 ... B8 116
Kidder, co., N. Dak., 5,386 ... B6 110
Kidderminster, Eng., 40,822 ... B5 12
Kidira, Sen ... D2 45
Kidnappers, cape, N.Z ... M16 51
Kidron, val., Jordan ... m14 32
Kidsgrove, Eng., 19,726 ... A5 12
Kidugalo, Tan ... C6 48
Kidwelly, Wales, 2,879 ... C3 12
Kief, McHenry, N. Dak., 97 ... B5 110
Kiefer, Creek, Okla., 489 ... B5 112
Kiel, Ger., 273,300 ... A5 16, D4 24
Kiel, Calumet and Manitowoc, Wis., 2,524 ... B5, E5 124
Kiel, bay, Ger ... A5 16
Kiel (Kaiser Wilhelm), canal, Ger ... D3 24
Kielce, Pol., 89,000 ... C6 26
Kiental, Switz ... C4 19
Kiesling, Spokane, Wash ... D7 122
Kiester, Faribault, Minn., 741 ... G5 99
Kiev (Kiyev), Sov. Un., 1,174,000 (★1,260,000) ... F8 27
Kifisia, Grc., 14,193 ... g11 23
Kifisos, riv., Grc ... g11 23
Kifissos (Cephisus), riv., Grc ... g9 23
Kifri, Iraq, 4,760 ... D2 41
Kifta, Maur., 1,300 ... C2 45
Kigali, Rwanda, 4,000 ... B5 48
Kigi, Tur., 1,072 ... C13 31
Kiglapait, mtn., Newf., Can ... g9 75
Kigoma, Tan., 4,244 ... B4 48
Kihei, Maui, Haw., 95 ... C5 88
Kiholo, bay, Haw ... D6 88
Kii, strait, Jap ... I7 37
Kikinda, Yugo., 33,906 ... C5 22
Kikladhes (Cyclades), prov., Grc., 99,959 ... *D5 23
Kikongo, Con. L ... B2 48
Kikori, Pap ... k11 50
Kikwit, Con. L., 11,000 ... C2 48
Kila, Flathead, Mont., 75 ... B2 102
Kilauea, Kauai, Haw., 800 ... A2 88
Kilauea, crater, Haw ... D6 88
Kilbaha, Ire ... E2 11
Kilbeggan, Ire., 799 ... D4 11
Kilbourne, Mason, Ill., 352 ... C3 90
Kilbourne, West Carroll, La., 227 ... B4 95
Kilbourne, Delaware, Ohio, 220 ... B5 111
Kilbrannan, sound, Scot ... E3 13
Kilburn, N.B., Can., 137 ... C2 74
Kilcar, Ire., 229 ... C3 11
Kilchberg, Switz., 6,784 ... B6 19
Kilchoan, Scot., 948 ... D2 13
Kilchreest, Ire ... D3 11
Kilchrenan, Scot., 349 ... D3 13
Kilchu, Kor., 30,026 ... F4 37
Kilcock, Ire., 739 ... D5 11
Kilcolgan, Ire ... D3 11
Kilconnell, Ire., 113 ... D3 11
Kilcoole, Ire., 549 ... D5 11
Kilcormac, Ire., 1,018 ... D4 11
Kilcullen, Ire., 637 ... D5 11
Kildare, Que., Can., 465 ... C4 73
Kildare, Ire., 2,551 ... D5 11
Kildare, Kay, Okla., 124 ... A4 110
Kildare, co., Ire., 64,420 ... D5 11
Kildare, cape, P.E.I., Can ... C6 74
Kildorrery, Ire., 228 ... E3 11
Kilembe, Con. L ... C2 48
Kilfenora, Ire., 135 ... E2 11
Kilfinnane, Ire., 565 ... E3 11
Kilgarvan, Ire., 183 ... F2 11
Kilgore, Clark, Idaho, 20 ... E7 89
Kilgore, Cherry, Nebr., 157 ... B5 103
Kilgore, Gregg, Tex., 10,092 ... C5 118
Kilifi, Ken ... B6 48
Kilimanjaro, mtn., Tan ... B6 48
Kilis (Killis), Tur., 33,300 ... D11 31
Kiliya, Sov. Un., 10,000 ... I7 27
Kilkee, Ire., 1,392 ... E2 11
Kilkeel, N. Ire., 2,490 ... C5 11
Kilkenny, Ire., 1,159 ... E4 11
Kilkenny, LeSueur, Minn., 221 ... F5 99
Kilkenny, co., Ire., 61,668 ... E4 11
Kilkerrin, Ire., 199 ... D3 11
Kilkieran, Ire., 153 ... D2 11
Kilkieran, bay, Ire ... D2 11
Kilkis, Grc., 10,963 ... B4 23
Kilkis, prov., Grc., 102,812 ... *B4 23
Kill, Ire., 186 ... D5 11
Killadoon, Ire ... D2 11
Killadysert, Ire., 295 ... E2 11
Killala, Ire., 337 ... C2 11
Killala, bay, Ire ... C2 11
Killaloe, Ire., 835 ... E3 11
Killaloe Station, Ont., Can., 932 ... B7 72
Killaly, Sask., Can., 212 ... G4 70
Killam, Alta., Can., 552 ... C5 69
Killarney, Man., Can., 1,729 ... E2 71
Killarney, Ire., 6,825 ... E2 11
Killarney, Raleigh, W. Va., 200 ... D3 123
Killashandra, Ire., 397 ... C4 11
Killavally, Ire ... D2 11
Killbuck, Holmes, Ohio, 865 ... B6 111
Killdeer, Sask., Can., 52 ... H2 70
Killdeer, Dunn, N. Dak., 765 ... B3 110
Killdeer, mts., N. Dak ... B3 110
Killduff, Jasper, Iowa, 150 ... C5 92
Killeen, Bell, Tex., 23,377 ... D4 118
Killen, Lauderdale, Ala., 620 ... A2 78
Killenaule, Ire., 531 ... E4 11
Killeter, N. Ire ... C4 11
Killian, Livingstone, La., 22 ... B6 95
Killian, Richland, S.C., 800 ... C6 115
Killiecrankie, pass, Scot ... D4 13
Killimor, Ire., 195 ... D3 11
Killin, Scot., 1,199 ... D4 13
Killinek, isl., Newf., Can ... F8 75
Killington, peak, Vt ... D3 120
Killingworth, Middlesex, Conn., 150 (1,098▲) ... D6 84
Killini, Grc., 744 ... D3 23
Killis (Kilis), Tur., 33,300 ... D11 31
Killona, St. Charles, La., 650 ... C6 95
Killorglin, Ire., 1,100 ... E2 11
Killucan, Ire., 314 ... D4 11
Kill Van Kull, chan., N.Y ... E5 106
Killybegs, Ire., 1,065 ... C3 11
Killyleagh, N. Ire., 1,876 ... C6 11
Kilmacthomas, Ire., 446 ... E4 11
Kilmaine, Ire., 94 ... D2 11
Kilmallie, Scot ... D3 13
Kilmallock, Ire., 1,159 ... E3 11
Kilmanagh, Huron, Mich., 50 ... E7 98
Kilmarnock, Scot., 47,509 ... E4 13
Kilmarnock, Lancaster and Northumberland, Va., 927 ... D6 121
Kilmatinde, Tan ... C5 48
Kilmelfort, Scot., 279 ... D3 13
Kilmichael, Montgomery, Miss., 532 ... B4 100
Kilmonivaig, Scot ... D3 13
Kiln, Hancock, Miss., 300 ... E4 100
Kilosa, Tan., 3,743 ... C6 48
Kilrea, N. Ire., 954 ... C5 11
Kilrush, Ire., 2,861 ... E2 11
Kilsyth, Scot., 9,831 ... E4 13
Kiltamagh, Ire., 980 ... D3 11
Kiltoom, Ire ... D3 11
Kilwa, Con. L ... C4 48
Kilwa Kivinje, Tan ... C6 48
Kilworthy, Ont., Can., 125 ... C5 72
Kim, Las Animas, Colo., 400 ... D7 83
Kimamba, Tan., 1,330 ... C6 48
Kimball, Stearns, Minn., 535 ... E4 99
Kimball, Kimball, Nebr., 4,384 ... C2 103
Kimball, Brule, S. Dak., 912 ... D7 116
Kimball, McDowell, W. Va., 1,175 ... *D3 123
Kimball, co., Nebr., 7,975 ... C2 103
Kimballton, Audubon, Iowa, 380 ... C2 92
Kimberley, B.C., Can., 6,013 ... E9 68
Kimberly, S. Afr., 75,376 (★79,031) ... C3 49
Kimberlin Heights, Knox, Tenn., 275 ... E12 117
Kimberly, Jefferson, Ala., 763 ... B3 78
Kimberly, Twin Falls, Idaho, 1,298 ... G4 89
Kimberly, Aitkin, Minn., 22 ... D5 99
Kimberly, White Pine, Nev ... D6 104
Kimberly, Grant, Oreg., 10 ... C7 113
Kimberly, Fayette, W. Va., 900 ... D6 123
Kimberly, Outagamie, Wis., 5,322 ... A5 124
Kimble, co., Tex., 3,943 ... D3 118
Kimbles, Pike, Pa., 30 ... D11 114
Kimbolton, Guernsey, Ohio, 218 ... B6 111
Kimbrough, Wilcox, Ala., 150 ... C2 78
Kimchaek (Sŏngjin), Kor., 67,778 ... F4 37
Kimi, Grc., 4,071 ... C5 23
Kimiwan, lake, Alta., Can ... B2 69
Kimmell, Noble, Ind., 350 ... B6 92
Kimmins, Lewis, Tenn., 50 ... B4 117
Kimmswick, Jefferson, Mo., 303 ... B8 101
Kimowin, riv., Sask., Can ... A6 69
Kimry, Sov. Un., 40,000 ... C11 27
Kimsquit, B.C., Can., 25 ... C4 68

Kin, Okinawa 52
Kinabalu, mtn., Mala D5 35
Kinard, Calhoun, Fla., 150 B1 86
Kinards, Laurens and Newberry, S.C., 234 C4 115
Kinbasket, lake, B.C., Can D8 68
Kinbrace, Scot B5 13
Kinbrae, Nobles, Minn., 55 G3 99
Kinburn, Ont., Can., 172 B8 72
Kincaid, Sask., Can., 310 H2 70
Kincaid, Christian, Ill., 1,544 D4 90
Kincaid, Anderson, Kans., 220 D8 93
Kincardine, Ont., Can., 2,841 C3 72
Kincardine, co., Scot., 25,556 D6 13
Kinchafonee, riv., Ga E2 87
Kincorth, Sask., Can., 59 H1 70
Kinde, Huron, Mich., 624 E7 98
Kinder, Allen, La., 2,299 D3 95
Kinderhook, Pike, Ill., 276 D2 90
Kinderhook, Columbia, N.Y., 1,078 C7 108
Kindersley, Sask., Can., 2,990 F1 70
Kindia, Guinea, 13,000 D2 45
Kindred, Cass, N. Dak., 580 C8 110
Kindu [-Port Empain], Con. L., 14,200 B4 48
Kineo, mtn., Maine C3 96
Kineshma, Sov. Un., 90,000 B2 29
King, Ont., Can., 898 E6 72
King, Stokes, N.C., 950 A3 109
King, Waupaca, Wis., 500 D4 124
King, co., Tex., 640 C2 118
King, co., Wash., 935,014 B4 117
King, isl., Austl G7, n14 50
King, isl., Bur F3 38
King, isl., B.C., Can C4 68
King, mtn., Oreg D8 113
King, sound, Austl C3 50
King and Queen C. H., King and Queen, Va., 65 D6 121
King and Queen, co., Va., 5,889 D6 121
Kingaroy, Austl., 4,914 C8 51
Kingarrow, Ire C3 11
King Christian IX Land, reg., Grnld C18 4
King Christian X Land, reg., Grnld B17 4
King City, Monterey, Calif., 2,937 D3 82
King City, Gentry, Mo., 1,009 A3 101
King Cove, Alsk., 290 E7 79
King Ferry, Cayuga, N.Y., 300 C4 108
Kingfield, Franklin, Maine, 700 (864▲) D2 96
Kingfisher, Kingfisher, Okla., 3,249 B4 112
Kingfisher, co., Okla., 10,635 B4 112
King Frederik VI Coast, reg., Grnld C19 4
King Frederik VIII Land, reg., Grnld B17 4
King George, King George, Va., 240 C5 121
King George, co., Va., 7,243 C5 121
King Goerge, isl., Ant C7 5
King George, mtn., B.C., Can D10 68
King George, sound, Austl F2 50
King George IV, lake, Newf., Can D3 75
King Hill, Elmore, Idaho, 200 F3 89
King Lear, peak, Nev B3 104
King Leopold, range, Austl C4 50
Kingman, Mohave, Ariz., 4,525 B1 80
Kingman, Alta., Can., 108 C4 69
Kingman, Fountain, Ind., 461 E3 91
Kingman, Kingman, Kans., 3,582 B3, E5 93
Kingman, Penobscot, Maine, 250 C4 96
Kingman, co., Kans., 9,958 E5 93
King of Prussia, Montgomery, Pa., 6,000 *F11 114
Kings, Ogle, Ill., 225 A4 90
Kings, Warren, Miss., 65 B3 100
Kings, co., Calif., 49,954 D4 82
Kings, co., N.B., Can., 25,908 D4 74
Kings, co., N.S., Can., 41,747 E5 74
Kings, co., P.E.I., Can., 17,893 C7 74
Kings, co., (Brooklyn borough) N.Y. (part of New York City), 2,627,319 E1 108
Kings, peak, Utah C5 119
Kings, ridge, Tex A3 118
Kings, riv., Ark A2 81
Kings, riv., Calif D4 82
Kings, riv., Ire E4 11
Kings, riv., Nev B3 104
Kingsbridge, Eng., 3,283 D4 12
Kingsburg, Fresno, Calif., 3,093 D4 82
Kingsbury, LaPorte, Ind., 281 A4 91
Kingsbury, Piscataquis, Maine, 5 (80▲) C3 96
Kingsbury, Guadalupe, Tex., 300 B4 118
Kingsbury, co., S. Dak., 9,227 C8 116
Kings Canyon, Jackson, Colo., 10 A4 83
Kings Canyon, nat. park, Calif D4 82
Kingsclear, N.B., Can., 200 D3 74
Kingscote, Austl., 534 G6 50
Kings Creek, Somerset, Md D6 85
Kings Creek, Cherokee, S.C., 200 A5 116
Kingsdale, Pine, Minn., 60 D6 99
Kingsdown, Ford, Kans., 100 E4 93
Kingsey Falls, Que., Can., 531 D5 73
Kingsford, Dickinson, Mich., 5,084 C2 98
Kingsford Heights, La Porte, Ind., 1,276 *B4 91
Kings Gardens, Reno, Kans., 400 A4 93
Kingsgate, B.C., Can., 71 E9 68
Kingshill, Vir. Is k17 65
Kingsland, Cleveland, Ark., 249 D3 81
Kingsland, Camden, Ga., 1,536 F5 87
Kingsland, Llano, Tex., 150 D3 118
Kingsley, Plymouth, Iowa, 1,044 B2 92
Kingsley, Grand Traverse, Mich., 586 D5 98
Kingsley, dam, Nebr C4 103
King's Lynn, Eng., 27,554 B8 12
Kingsmere, lake, Sask., Can C2 70
Kings Mill, Gray, Tex., 75 B2 118
Kings Mills, Warren, Ohio, 700 C2, C3 111
Kings Mountain, Lincoln, Ky., 500 C5 94
Kings Mountain, Cleveland, N.C., 8,008 B2 109
Kings Park, Suffolk, N.Y., 4,949 D3 108
Kings Point, Newf., Can., 546 D3 75
Kings Point, Nassau, N.Y., 5,410 F1 84
Kingsport, Sullivan, Tenn., 26,314 C11 117
Kingston, Madison, Ark., 150 A2 81
Kingston, Austl., 806 H2 51
Kingston, N.S., Can., 1,210 E5 74
Kingston, Ont., Can., 53,526 (★63,419) C8 72
Kingston, Bartow, Ga., 695 B2 87
Kingston, Shoshone, Idaho, 200 B2 89
Kingston, De Kalb, Ill., 406 A5 90
Kingston, Jam., 421,718 F5 64
Kingston, Madison, Ky., 200 C5 94
Kingston, Somerset, Md., 5 D6 85
Kingston, Plymouth, Mass., 2,000 (4,302▲) C6 97
Kingston, Tuscola, Mich., 456 E7 98
Kingston, Meeker, Minn., 125 E4 99
Kingston, Caldwell, Mo., 311 B3 101
Kingston, Rockingham, N.H., 600 (1,672▲) E4 105
Kingston, Somerset, N.J., 850 C3 106
Kingston, Sierra, N. Mex., 10 E2 107
Kingston, Ulster, N.Y., 29,260 D6 108
Kingston, Ross, Ohio, 1,066 C5 111
Kingston, Marshall, Okla., 639 D5 112
Kingston, Luzerne, Pa., 20,261 B8, D10 114
Kingston, Washington, R.I., 2,616 D10 84
Kingston, Roane, Tenn., 2,010 D9 117
Kingston, Piute, Utah, 143 E3 119
Kingston, Fayette, W. Va., 400 D3, D6 123
Kingston, Green Lake, Wis., 343 E4 124
Kingston-on-Thames, Eng., 36,450 m11 10
Kingston Springs, Cheatham, Tenn., 400 A4 117
Kingstown, St. Vincent, 4,296 (★16,141) p16 64
Kingstree, Williamsburg, S.C., 3,847 D8 115
Kings Valley, Benton, Oreg., 250 C3 113
Kingsville, Ont., Can., 3,041 E2 72
Kingsville, Johnson, Mo., 225 C4 101
Kingsville, Ashtabula, Ohio, 950 A7 111
Kingsville, Kleberg, Tex., 25,297 F4 118
Kingswood, Breckinridge, Ky., 248 C3 94
Kingtehchen, see Fowliang, China.
Kington, Eng., 1,861 B4 12
Kingurutik, lake, Newf., Can g9 75
Kingussie, Scot., 1,079 C4 13
Kingwa, Que., Can D17 67
King William, King William, Va., 40 D5 121
King William, co., Va., 7,563 D5 121
King William, isl., N.W. Ter., Can C13 66
King William's Town, S. Afr., 14,646 D4 49
Kingwood, Preston, W. Va., 2,530 B5 123
Kinistino, Sask., Can., 764 E3 70
Kinkala, Con. B F2 46
Kinkora, P.E.I., Can., 271 C6 74
Kinley, Sask., Can., 119 E2 70
Kinloch, St. Louis, Mo., 6,501 A8 101
Kinlochewe, Scot C3 13
Kinloch Hourn, Scot C3 13
Kinmount, Ont., Can., 256 C6 72
Kinmundy, Marion, Ill., 813 E5 90
Kinnaird, B.C., Can., 2,123 E9 68
Kinnairds, head, Scot C7 13
Kinnelon, Morris, N.J., 4,431 B4 106
Kinney, co., Tex., 2,452 E2 118
Kinomoto, Jap., 5,600 n15 37
Kinosaki, Jap., 4,400 I7 37
Kinpoku, peak, Jap G9 37
Kinrooi, Bel., 2,022 C5 15
Kinross, P.E.I., Can., 66 C7 74
Kinross, Scot., 2,365 D5 13
Kinross, co., Scot., 6,704 D5 13
Kinsale, Ire., 1,587 F3 11
Kinsale, Westmoreland, Va., 250 C6 121
Kinsella, Alta., Can., 91 C5 71
Kinsey, Houston, Ala., 283 D4 78
Kinsey, Custer, Mont., 5 D11 102
Kinsley, Edwards, Kans., 2,263 E4 93
Kinsman, Trumbull, Ohio, 900 A7 111
Kinston, Coffee, Ala., 470 D3 78
Kinston, Lenoir, N.C., 24,819 B6 109
Kinta, Haskell, Okla., 233 B6 109
Kintampo, Ghana, 2,829 E4 45
Kintinku, Tan C6 48
Kinton, St. Francis, Ark B5 81
Kintore, Scot., 749 C6 13
Kintyre, Emmons, N. Dak., 102 C6 110
Kinuso, Alta., Can., 323 B3 69
Kinvara, Ire., 338 D3 11
Kinwood, Harris, Tex., 2,500 *F5 118
Kinyangiri, Tan., 1,540 B5 48
Kinyeti, mtn., Sud E3 47
Kinzig, riv., Ger C4 17
Kinzig, riv., Ger E2 17
Kinzua, Wheeler, Oreg., 300 C6 113
Kinzua, Warren, Pa., 420 C4 114
Kioa, isl., Fiji Is 52
Kiona, Benton, Wash., 150 C6 122
Kiowa, Elbert, Colo.,195 B6 83
Kiowa, Barber, Kans., 1,674 E5 93
Kiowa, Pittsburg, Okla., 607 C6 112
Kiowa, co., Colo., 2,425 C8 83
Kiowa, co., Kans., 4,626 E4 93
Kiowa, co., Okla., 14,825 C3 112
Kiowa, creek, Colo B6 83
Kiowa, creek, Okla A1 112
Kipahigan, lake, Man., Can B1 71
Kipahigan, lake, Sask., Can B5 70
Kipahulu, Maui, Haw., 25 C5 88
Kiparissia, Grc., 5,027 D3 23
Kiparissia, gulf, Grc D3 23
Kipili, Tan C5 48
Kipini, Ken B7 48
Kipling, Delta, Mich., 400 C3 98
Kipling Station, Sask., Can., 773 G4 70
Kipnuk, Alsk., 221 C7 79
Kipp, Saline, Kans., 80 D6 93
Kippen, Toole, Mont., 40 B5 102
Kipton, Lorain, Ohio, 353 A5 111
Kiptopeke, Northampton, Va B7 121
Kipushi, Con. L., 15,100 D4 48
Kirby, Pike, Ark., 300 C2 81
Kirby, Big Horn, Mont., 2 E10 102
Kirby, Hot Springs, Wyo., 82 B4 125
Kirbyville, Jasper, Tex., 1,680 D6 118
Kirchberg, Switz., 3,304 B4 19
Kirchdorf, Ger., 783 D5 24
Kirchhain [im Bezirk Kassel], Ger., 5,500 C3 17
Kirchhain [Niederlausitz], Ger., 7,636 B8 17
Kirchheim [unter Teck], Ger., 25,000 E4 17
Kirchheimbolanden, Ger., 5,200 D3 17
Kirchhunden, Ger., 10,500 B3 17
Kirdāsah, Eg., U.A.R., 1,000 D3 32
Kirensk, Sov. Un., 12,500 D13 28
Kirghiz, steppe, Sov. Un E9 33
Kirghiz S.S.R., rep., Sov. Un., 2,225,000 G23 9
Kirgiz, range, Sov. Un E8 29
Kiri, Con. L B2 48
Kirin (Chilin), China, 568,000 C10 34
Kirin, prov., China, 12,550,000 C10 34
Kirinian, isl., Eniwetok 52
Kirk, Yuma, Colo., 75 B3 83
Kirk, Klamath, Oreg., 20 E5 113
Kirkagac, Tur., 11,300 C6 23
Kirkby Lonsdale, Eng., 1,240 F6 13
Kirkby Stephen, Eng., 1,718 F6 13
Kirkcaldy, Alta., Can., 60 D4 69
Kirkcaldy, Scot., 52,371 D5 13
Kirkcolm, Scot., 1,635 F3 13
Kirkcudbright, Scot., 2,448 F4 13
Kirkcudbright, co., Scot., 28,877 E4 13
Kirkcudbright, bay, Scot F4 13
Kirkella, Man., Can., 60 D1 71
Kirkella, Sask., Can., 80 G5 70
Kirkenes, Nor., 3,596 C14 25
Kirkersville, Licking, Ohio, 417 C5 111
Kirkfield, Ont., Can., 211 C6 72
Kirkfield Park, Man., Can., 800 E3 71
Kirkintilloch, Scot., 18,257 E4 13
Kirkland, Yavapai, Ariz., 20 C3 80
Kirkland, Atkinson, Ga., 150 E4 87
Kirkland, DeKalb, Ill., 928 A5 90
Kirkland, Oneida, N.Y., 100 *B5 108
Kirkland, Williamson, Tenn., 50 B5 117
Kirkland, Childress, Tex., 300 B2 118
Kirkland, King, Wash., 6,025 B3, D2 122
Kirkland Junction, Yavapai, Ariz., 10 C3 80
Kirkland Lake, Ont., Can., 15,366 E9 72
Kirklareli, Tur., 20,200 B6 23
Kirklin, Clinton, Ind., 767 D5 91
Kirkman, Shelby, Iowa, 92 C2 92
Kirkmansville, Todd, Ky., 150 D2 94
Kirkoswald, Eng., 528 F6 13
Kirkpatrick, lake, Alta., Can D5 69
Kirkpatrick, peak, Ant A28 5
Kirksey, Calloway, Ky., 180 B3 94
Kirksey, Greenwood, S.C., 60 C3 115
Kirksville, Monroe, Ind., 20 F4 91
Kirksville, Madison, Ky., 100 C5 94
Kirksville, Adair, Mo., 13,125 A5 101
Kirkton, Ont., Can., 182 D3 72
Kirkuk, Iraq, 120,593 D2 41
Kirkville, Wapello, Iowa, 203 C5 92
Kirkwall, Scot., 4,315 B6 13
Kirkwood, New Castle, Del., 430 A6 85
Kirkwood, Warren, Ill., 771 C3 90
Kirkwood, Prince Georges, Md., 2,500 *C3 85
Kirkwood, St. Louis, Mo., 29,421 B7 101
Kirkwood, Broome, N.Y., 250 C5 108
Kirkwood, Lancaster, Pa., 100 G9 114
Kirkwood, S. Afr., 5,055 D4 49
Kirn, Ger., 9,600 D2 17
Kiron, Crawford, Iowa, 271 B2 92
Kirov, Sov. Un., 269,000 B3 29
Kirova, bay, Sov. Un B4 41
Kirovabad, Sov. Un., 116,000 E3 29
Kirovabad, Sov. Un., 123,000 H22 9
Kirovakan, Sov. Un., 57,000 B15 31
Kirovgrad, Sov. Un., 21,500 B6 29
Kirovograd, Sov. Un., 134,000 G9 27
Kirovsk, Sov. Un., 52,800 D15 25
Kirriemuir, Alta., Can., 54 D5 69
Kirriemuir, Scot., 3,485 D5 13
Kirsanov, Sov. Un., 10,000 E14 27
Kirsehir (Kir-Shehr), Tur., 20,200 C10 31
Kirther, range, Pak C4 39
Kirtland, San Juan, N. Mex., 800 A1 107
Kirtland, Lake, Ohio, 1,500 *A6 111
Kirtley, Niobrara, Wyo C8 125
Kiruna, Swe., 19,348 D9 25
Kirundu, Con. L B4 48
Kirwin, Phillips, Kans., 356 C4 93
Kirwin, res., Kans C4 93
Kiryu, Jap., 123,010 H9, m18 37
Kisaki, Tan C6 48
Kisalaya, Nic C5 62
Kisanga, Con. L C4 48
Kisarazu, Jap., 29,200 n18 37
Kisbey, Sask., Can., 254 H4 70
Kiselevsk, Sov. Un., 141,000 C11 29
Kisengi, Con. L D3 48
Kisengwa, Con. L C4 48
Kisenyi, Rwanda, 2,800 B5 48
Kishanganj, India, 27,002 D11 40
Kishangarh, India, 25,244 C5 39
Kishi, Nig E5 45
Kishinev, Sov. Un., 236,000 B9 22, H7 27
Kishiwada, Jap., 120,265 o14 37
Kishon, riv., Isr B7 32
Kishorganj, Pak., 19,067 D9 39
Kishorn, inlet, Scot C3 13
Kisii, Ken B5 48
Kisiju, Tan C6 48
Kisiwa, creek, Kans A4 93
Kisiwani, Tan B6 48
Kiska, isl., Alsk E3 79
Kiskatinaw, riv., B.C., Can B7 68
Kiskitto, lake, Man., Can B2 71
Kiskittogisu, lake, Man., Can B2 71
Kiskoros, Hung., 9,972 B4 22
Kiskundorozsma, Hung., 8,705 B5 22
Kiskunfelegyhaza, Hung., 22,925 (33,187▲) B4 22
Kiskunhalas, Hung., 18,314 (26,461▲) B4 22
Kiskunmajsa, Hung., 7,611 B4 22
Kislovodsk, Sov. Un., 82,000 E2 29
Kismayu [Chisimaio], Som., 8,000 (9,800▲) F5 47
Kismet, Seward, Kans., 150 E3 93
Kiso-Sammyaku, mts., Jap n16 37
Kissee Mills, Taney, Mo., 38 E4 101
Kisseynew, lake, Man., Can B1 71
Kissidougou, Guinea, 5,000 E2 45
Kissimmee, Osceola, Fla., 6,845 D5 86
Kissimmee, lake, Fla E5 86
Kissimmee, riv., Fla E5 86
Kissimmee Park, Osceola, Fla., 100 D5 86
Kississing, Man., Can., 183 B1 71
Kississing, lake, Man., Can B1 71
Kississing, riv., Man., Can B1 71
Kistigan, lake, Man., Can B5 71
Kistler, Logan, W. Va., 1,084 D3, D5 123
Kistna, riv., India E6 39
Kistrand, Nor B11 25
Kisujszallas, Hung., 11,304 (13,790▲) B5 22
Kisumu, Ken., 10,899 B5 48
Kisvarda, Hung., 13,284 A6 22
Kita, Iwo 52
Kita, Mali, 5,100 D3 45
Kita Iwo, isl., Pac. O E8 7
Kitakyūshū, Jap., 986,401 (★1,335,000) J5 37
Kitale, Ken., 6,338 A5 48
Kitamaki, Jap m17 37
Kitami, Jap., 38,200 E11 37
Kitangari, Tan D6 48
Kitano, pt., Iwo 52
Kita-Shiretoko, cape, Sov. Un B12 37
Kit Carson, Cheyenne, Colo., 356 C8 83
Kit Carson, co., Colo., 6,957 B8 83
Kitchener, Ont., Can., 74,485 (★154,864) D4 72
Kitchener, mtn., Alta., Can C2 69
Kitchen's Creek, falls, Pa D9 114
Kite, Johnson, Ga., 424 D4 87
Kitega, Burundi, 5,000 B5 48
Kitgum, Ug A5 48
Kithira, Grc., 1,002 D4 23
Kíthira (Cythera), isl., Grc D4 23
Kithnos, isl., Grc D5 23
Kiti, pt., Ponape 52
Kitimat, B.C., Can., 8,217 B3 68
Kitsap, co., Wash., 84,176 B3 122
Kitscoty, Alta., Can., 326 C5 69
Kittanning, Armstrong, Pa., 6,793 E2 114
Kittatinny, mts., N.J. A2 106
Kittery, York, Maine, 8,051 (10,689▲) E2 96
Kittery Point, York, Maine, 1,259 E2 96
Kittilä, Fin., 1,700 D11 25
Kittitas, Kittitas, Wash., 536 C5 122
Kittitas, co., Wash., 20,467 B5 122
Kittitas, val., Wash B5 122
Kittrell, Vance, N.C., 121 A5 109
Kitts, Harlan, Ky., 950 D6 94
Kittson, co., Minn., 8,343 B2 99
Kitty Hawk, Dare, N.C., 300 A8 109
Kitty Hawk, bay, N.C A8 109
Kitui, Ken B6 48
Kitwe, N. Rh., 35,000 (★89,500) C4 48
Kityang, China, 54,000 G8 34
Kitzbühel, Aus., 7,744 E6 16
Kitzingen, Ger., 17,800 D5 17
Kitzmiller, Garrett, Md., 535 D1 85
Kiukiang, China, 64,600 F8 34
Kiungshan (Chiungshan), China C9 38
Kivalina, Alsk., 142 B7 79
Kivu, prov., Con. L., 2,205,000 B4 48
Kivu, lake, Con. L B4 48
Kiwalik, Alsk B7 79
Kiya, riv., Sov. Un D26 9
Kiyan, Okinawa 52
Kiyev (Kiev), Sov. Un., 1,174,000 (★1,260,000) F8 27
Kiyiu, lake, Sask., Can F1 70
Kizel, Sov. Un., 61,000 B5 29
Kizil, riv., Tur B10 29
Kizlyar, Sov. Un., 33,200 E3 29
Kizyl-Arvat, Sov. Un., 26,000 H20 9
Kjakan, Nor C10 25
Kjelvik, Nor B12 25
Kjellerup, Den B3 24
Kladno, Czech., 49,600 C3, n17 26
Klagenfurt, Aus., 69,218 E7 16
Klagetoh, Apache, Ariz., 200 B6 80
Klaipeda (Memel), Sov. Un., 100,000 D3 27
Klamath, Del Norte, Calif., 300 B2 82
Klamath, co., Oreg., 47,475 E5 113
Klamath, Indian res., Oreg E5 113
Klamath, mts., Oreg E2 113
Klamath, riv., Calif., Oreg B2 82
Klamath Agency, Klamath, Oreg., 200 E5 113
Klamath Falls, Klamath, Oreg., 16,949 E5 113
Klang, Mala., 75,649 K4 38
Klanxbüll, Ger., 520 D2 24
Klarälven, riv., Swe G5 25
Klatovy, Czech., 14,000 D2 26
Klawock, Alsk., 251 D13, n23 79
Kleberg, Dallas, Tex., 3,572 *B6 118
Kleberg, co., Tex., 30,052 F4 118
Klecany, Czech., 1,700 n17 26
Kleena Kleene, B.C., Can D5 68
Klein, Musselshell, Mont., 20 D8 102
Kleinburg, Ont., Can., 275 E6 72
Klemme, Hancock, Iowa, 615 A4 92
Klerksdorp, S. Afr., 43,726 C4 49
Kletnya, Sov. Un., 10,000 E9 27
Kletsk, Sov. Un., 10,000 E6 27
Kleve, Ger., 21,500 C6 15
Klickitat, Klickitat, Wash., 850 D4 122
Klickitat, co., Wash., 13,455 D5 122
Klickitat, creek, Wash D5 122
Klickitat, riv., Wash C4 122
Klimovsk, Sov. Un., 30,000 n17 27
Klin, Sov. Un., 57,000 C11 27
Klinaklini, riv., B.C., Can D5 68
Kline, La Plata, Colo., 35 D2 83
Kline, Barnwell, S.C., 213 E5 115
Klingenthal, Ger., 15,500 C7 17
Klintsy, Sov. Un., 40,000 E9 27
Klippan, Swe., 8,300 B7 24
Klitmöller, Den., 471 A2 24
Ključ, Yugo., 2,322 C3 22
Klobuck, Pol., 6,533 C5 26
Klock, Ont., Can., 75 A6 72
Klodnica, riv., Pol g9 26
Klodzko, Pol., 20,200 C4 26
Klondike, DeKalb, Ga B5 87
Klondike, Lake, Ill., 400 A5 90
Klondike, Delta, Tex., 20 C5 118
Klondike, Milwaukee, Wis F2 124
Klondike, region, Yukon, Can D5 66
Klostermansfeld, Ger., 5,957 B6 17
Klosterneuburg, Aus., 22,787 D8 16
Klosters, Switz., 3,181 C8 19
Kloten, Nelson, N. Dak., 120 B7 110
Kloten, Switz., 8,446 B6 19
Klötze, Ger., 6,255 F5 24
Klotzsche, Ger., 9,848 B8 17
Klotzville, Assumption, La., 300 B5 95
Kluane, lake, Yukon, Can D5 66
Kluang, Mala., 31,181 K5 38
Kluczbork, Pol., 11,800 C5 26
Klucze, Pol g11 26
Klukwan, Alsk., 112 k22 79
Klütz, Ger., 3,179 E5 24
Klyazma, riv., Sov. Un B2 29
Klyuchevskaya, vol., Sov. Un D18 28
Knap, pt., Scot E3 13
Knapp, Dunn, Wis., 374 D1 124
Knaresborough, Eng., 9,311 F7 13
Knebel, Den B4 24
Knee, lake, Man., Can B4 71
Knee, lake, Sask., Can B2 70
Knesebeck, Ger., 2,500 F4 24
Knezha, Bul., 13,133 D7 22
Knickerbocker, Tom Green, Tex., 150 D2 118
Knierim, Calhoun, Iowa, 153 B3 92
Knife, riv., N. Dak B3 110
Knife River, Lake, Minn., 350 D7 99
Knifley, Adair, Ky., 250 C4 94
Knight, Uinta, Wyo D2 125
Knight, inlet, B.C., Can D5 68
Knightdale, Wake, N.C., 622 B5 109
Knighton, Wales, 1,817 B4 12
Knights Landing, Yolo, Calif., 725 C3 82
Knightstown, Henry, Ind., 2,496 E6 91
Knight's Town, Ire., 337 F1 11
Knightsville, Clay, Ind., 722 E3 91
Knightsville, dam, Mass B2 97
Knik, Alsk., 15 C10, g17 79
Knin, Yugo., 5,112 C3 22
Knippa, Uvalde, Tex., 350 E3 118
Knislinge, Swe., 2,145 B8 24
Knittlefeld, Aus., 14,259 E7 16
Knob, creek, Ky A4 94
Knobel, Clay, Ark., 339 A5 81
Knob Lick, Metcalfe, Ky., 100 C4 94
Knob Lick, St. Francois, Mo., 150 D7 101
Knobly, mtn., W. Va B5 123
Knob Noster, Johnson, Mo., 2,292 C4 101
Knobs, Fallon, Mont E12 102
Knobs, mtn., Pa D5 114
Knobsville, Fulton, Pa., 150 F6 114
Knock, Ire E2 11
Knockadoon, head, Ire F4 11
Knockanefune, mtn., Ire E2 11
Knocklong, Ire E3 11
Knockmealdown, mts., Ire D3 10
Knokke, Bel., 13,649 C3 15
Knolls, Tooele, Utah, 10 C2 119
Knollwood, Greene, Ohio, 2,500 *C3 111
Knollwood Park, Jackson, Mich., 2,100 *F6 98
Knops, pond, Mass C1 97
Knott, Howard, Tex., 100 C2 118
Knott, co., Ky., 17,362 C6 94
Knotts Island, Currituck, N.C., 140 A8 109
Knottsville, Daviess, Ky., 350 C3 94
Knowles, Beaver, Okla., 62 A1, D4 112
Knowlton, Que., Can., 1,396 D5 73
Knowlton, Marathon, Wis., 75 D4 124
Knox, Knox, Ill C3 90
Knox, Starke, Ind., 3,458 B4 91
Knox, Benson, N. Dak., 122 A6 110
Knox, Clarion, Pa., 1,247 D2 114
Knox, co., Ill., 61,280 B3 90
Knox, co., Ind., 41,561 G3 91
Knox, co., Ky., 25,258 D6 94
Knox, co., Maine, 28,575 D4 96
Knox, co., Mo., 6,558 A5 101
Knox, co., Nebr., 13,300 B8 103
Knox, co., Ohio, 38,808 B5 111
Knox, co., Tenn., 250,523 C10 117
Knox, co., Tex., 7,857 C3 118
Knox, cape, B.C., Can B1 68
Knox, coast, Ant C23 5
Knox, creek, Ky., Va., W. Va D2 123
Knox City, Knox, Mo., 330 A5 101
Knox City, Knox, Tex., 1,805 C3 118
Knox Dale, Jefferson, Pa., 350 D3 114
Knoxville, Greene, Ala., 100 C2 78
Knoxville, Johnson, Ark., 300 B2 81
Knoxville, Crawford, Ga., 200 D3 87
Knoxville, Knox, Ill., 2,560 C3 90
Knoxville, Marion, Iowa, 7,817 C4 92
Knoxville, Frederick, Md., 400 B2 85
Knoxville, Franklin, Miss., 150 D2 100
Knoxville, Ray, Mo., 103 B3 101
Knoxville, Tioga, Pa., 694 C7 114
Knoxville, Knox, Tenn., 111,827 (★286,000) D10, E11 117
Knysna, S. Afr., 11,085 D3 49
Knyszyn, Pol., 2,780 B7 26
Kobdo, riv., Mong B3 34
Kobe, Jap., 1,113,977 I7, o14 37
Kobelyaki, Sov. Un., 10,000 G10 27
Köbenhavn (Copenhagen), co., Den., 1,269,336 C6 24
Koblenz, Ger., 99,200 (★165,000) C2 17
Kobona, Sov. Un r32 25
Kobrin, Sov. Un., 25,000 E5 27
Kobroor, isl., Indon G8 35
Kobuk, Alsk., 54 B8 79
Kobuk, riv., Alsk B8 79
Kobuleti, Sov. Un., 10,000 B13 31
Koca (Xanthus), riv., Tur D7 23
Kocaba, riv., Czech o17 26
Kočani, Yugo., 10,986 E6 22
Kočevje, Yugo., 5,760 C2 22
Koch, mtn., Mont E5 102
Kochel, Ger., 3,900 B7 18
Kochel, lake, Ger B7 18
Kocher, riv., Ger D4 17
Kochi, Jap., 196,288 J6 37
Kochi, pref., Jap., 854,595 *J6 37
Kochiu, China, 159,700 G5 34
Kock, Pol., 2,381 C7 26
Kodak, Perry, Ky., 500 C6 94
Kodiak, Alsk., 2,628 D9 79
Kodiak, isl., Alsk D9 79
Kodok, Sud D3 47
Koehler, Colfax, N. Mex A5 107
Koepang, Indon., 7,171 H6 35
Koes, S.W. Afr., 422 C2 49
Koetaradja, Indon., 10,724 k11 35
Koetatjane, Indon K2 38
Kofa, mts., Ariz D2 80
Koffiefontein, S. Afr., 2,985 C4 49
Koforidua, Ghana, 17,806 E4 45
Kofu, Jap., 160,963 I9, n17 37
Koga, Jap., 42,474 m18 37
Köge, Den., 12,294 C6 24
Köge, bay, Den C6 24
Kogilnik, riv., Sov. Un B9 22
Kohala, Hawaii, Haw., 950 C6 88
Kohat, Pak., 30,719 (★40,841) B5 39
Kohatk, Pinal, Ariz., 50 E3 80
Kohima, India, 7,246 C9 39
Kohler, Sheboygan, Wis., 1,524 B6, E6 124
Kohls Ranch, Gila, Ariz C4 80
Kohlu, Pak C4 39
Kohtla-Jarve, Sov. Un., 58,000 B6 27
Köinge, Swe A6 24
Koje, isl., Kor I4 37
Kokadjo, Piscataquis, Maine, 10 C3 96
Kokadjo, lake, Maine C3 96
Kokai, see Kanggye, Kor.
Kokand, Sov. Un., 113,000 E8 29
Kokanee, peak, B.C., Can E2 69
Kokanee Glacier, park, B.C., Can E9 68
Kokchetav, Sov. Un., 59,000 C7 29
Kokkola, Fin., 16,200 F10 25
Koko, Nig., 7,624 D5 45
Koho, head, Haw B4 88
Kokomo (Recen), Summit, Colo., 14 B4 83
Kokomo, Maui, Haw., 250 C5 88
Kokomo, Howard, Ind., 47,197 D5 91
Kokomo, Marion, Miss., 250 D3 100
Koko-Nor (Tsinghai), lake, China D4 34
Kokopo, Bis. Arch., 220 h13 50
Kokos, is., Indon K1 38
Kokosing, riv., Ohio B5 111
Kokrines, Alsk., 20 C9 79
Koksilah, B.C., Can., 382 B9 68
Kokstad, S. Afr., 7,984 D4 49
Kola, Sov. Un C15 25
Kola, pen., Sov. Un D17 25
Kolan, China E4 36
Kolar Gold Fields, India, 146,811 F6 39
Kolarovo, Bul., 31,169 D8 22
Kolarovo, Czech., 10,800 E4 26
Kolbäck, Swe., 1,871 t34 25
Kolbano, Indon H6 35
Kolbio, Som F5 47
Kolbuszowa, Pol., 2,124 C6 26
Kolchugino, Sov., Un., 40,000 C12 27
Kolczewo, Pol., 2,000 E8 24
Kolda, Sen., 4,300 D2 45
Koldewey, isl., Grnld B16 4
Kolding, Den., 35,101 C3 24
Kole, Con. L B3 48
Kolec, Czech., 974 n17 26
Koleno, Sov. Un., 10,000 F13 27
Kolezhma, Sov. Un., 500 E16 25
Kolgueyv, isl., Sov. Un B18 9
Kolhapur, India, 187,442 (★193,186) E5 39, I5 40
Kolliganek, Alsk., 100 D8 79
Kolimbine, riv., Maur C2 45
Kolin, Czech., 23,200 C3 26
Kolin, Rapides, La., 50 C3 95
Kolin, Judith Basin, Mont., 5 C7 102
Kölleda, Ger., 7,672 B6 17

Köln (Cologne), Ger., 809,200
(*1,470,000).......C1 17
Kolno, Pol., 3,295.......B6 26
Kolo, Pol., 10,600.......B5 26
Koloa, Kauai, Haw., 1,426..B2 88
Kolobrzeg, Pol., 2,816.....A3 26
Kolokani, Mali.......D3 45
Kolola Springs, Lowndes,
Miss., 40.......B5 100
Kolomna, Sov. Un.,
124,000.......B1 29
Kolomyya, Sov. Un., 45,000.G5 27
Kolonodale, Indon.......F6 35
Kolp, riv., Sov. Un.......B11 27
Kolpashevo, Sov Un.,
32,100.......B10 29
Kolpino, Sov. Un., 50,000..s31 25
Kolva, riv., Sov. Un.......A5 29
Kolwa, Pak.......C3 39
Kolwezi, Con. L., 48,000...D4 48
Kolyberovo, Sov. Un.,
10,000.......n18 27
Kolyma, riv., Sov. Un.....C18 28
Komádi, Hung., 7,362.....B5 22
Komadugu, riv., Nig.......D6 45
Komandorski Village, Alameda,
Calif., 1,006.......*B6 82
Komandorskiye, is., Sov. Un..D19 28
Komarno, Man., Can., 101..D3 71
Komarno, Czech., 24,000...C5 26
Komarno, Sov. Un., 5,000..D7 26
Komarno, marsh, Sov. Un...B10 27
Komarom, Hung., 9,862....A4 22
Komárom, co., Hung.,
270,508.......*B4 22
Komatipoort, S. Afr., 2,031..C5 49
Komatsu, Jap., 54,900
(89,085*).......H8 37
Komatsushima, Jap., 25,100..I7 37
Komelik, Pima, Ariz., 65...F4 80
Komlo, Hung., 24,850.....B4 22
Kommunarsk, Sov. Un.,
107,000.......G12, q21 27
Kominato, Jap., 9,700.....n19 37
Komodo, isl., Indon.......G5 35
Komono, Con. B.......F2 46
Komoran, isl., W. Irian....G9 35
Komotini (Komotine), Grc.,
32,906.......D4 23
Kompong Cham, Camb.,
25,000.......F6 38
Kompong Chhnang, Camb.,
25,000.......F6 38
Kompong Kleang, Camb.,
10,000.......F6 38
Kompong Som, bay, Camb...G5 38
Kompong Speu, Camb.,
5,000.......G6 38
Kompong Thom, Camb.,
25,000.......F6 38
Komrat, Sov. Un., 10,000..H7 27
Komsomolsk [-na Amure],
Sov. Un., 189,000.......D16 28
Komvo, Pima, Ariz., 50....F3 80
Konawa, Seminole, Okla.,
1,555.......C5 111
Kondoa, Tan., 2,816.......B6 48
Kondopoga, Sov. Un.,
5,000.......F16 25
Kong, reg., I.C.......E4 45
Kong, isl., Camb.......G5 38
Kongeå, riv., Den.......C2 24
Kongju, Kor., 23,400.....H3 37
Kongolo, Con. L., 2,842...C4 48
Kongsberg, McHenry,
N. Dak., 20.......B5 110
Kongsberg, Nor.,
9,700.......H3, p27 25
Kongsmark, Den.......C2 24
Kongsvinger, Nor., 2,300...G5 25
Königsbrück, Ger., 5,100...B8 17
Königshofen, Ger., 3,200...C5 17
Königslutter, Ger., 8,600...A5 17
Königstein, Ger., 5,139....C9 17
Königs Wusterhausen, Ger.,
6,614.......A8 17
Konin, Pol., 13,500.......B5 26
Konispol, Alb., 4,169.....C3 23
Konitsa, Grc., 2,313.......B3 23
Konjic, Yugo., 5,891.......B3 22
Könnern, Ger., 6,053......B6 17
Konolfingen, Switz., 3,964..C4 19
Konomoc, lake, Conn.......D8 84
Konosha, Sov. Un., 10,000..A13 29
Konotop, Sov. Un., 56,000..F9 27
Konskie, Pol., 7,386.......C6 26
Konstantinovka, Sov. Un.,
93,000.......q20 27
Konstanz, Ger., 52,700.....E4 16
Konta, India, 5,000.......I8 40
Kontagora, Nig., 5,665.....D6 45
Kontcha, Cam.......D2 46
Kontich, Bel., 10,923.....C4 15
Kontiomäki, Fin., 850.....E13 25
Kontum, Viet., 10,000.....E8 38
Konya, Tur., 122,700.....D9 31
Konyang, Kor., 2,000......I3 37
Konzhakovskiy Kamen, mtn.,
Sov. Un.......B5 29
Koochiching, co., Minn.,
18,190.......B4 99
Koolau, range, Haw......g10 88
Koontz Lake, Starke, Ind.,
900.......B5 91
Koosharem Sevier, Utah,
148.......E4 119
Kooskia, Idaho, Idaho, 801..C3 89
Kootenai, Bonner, Idaho,
180.......A2 89
Kootenai, co., Idaho, 29,556.B2 89
Kootenai, riv., Mont......B1 101
Kootenay, lake, B.C., Can...D9 68
Kootenay, dist., B.C., Can..D9 68
Kootenay, lake, B.C., Can...E9 68
Kootenay, nat. park, B.C., Can..D9 68
Kootenay, riv., B.C., Can...E10 68
Kopaski, Ice., 83.......m24 25
Kopervik, Nor., 1,800.....H1 25
Kopet, mts., Iran.......B7 41
Kopeysk, Sov. Un.,
168,000.......G9 29
Köping, Swe., 17,700.....H6, t33 25
Koppany, riv., Hung......B4 22
Kopparberg, co., Swe.,
286,600.......*G6 25
Koppel, Beaver, Pa., 1,389..E1 114
Koprivnica, Yugo., 1,329...B3 22
Kopychintsy, Sov. Un.,
10,000.......G5 27
K'orahē, Eth.......D5 47
Koraluk, riv., Newf., Can...g9 73
Koram, Eth.......C4 47
Korarou, lake, Mali.......C4 45
Korba, Tun.......F12 30
Korbach, Ger., 15,100.....B3 17
Korbu, mtn., Mala.......J4 38

Korçë, Alb., 34,400.......B3 23
Korçë, pref., Alb.,
175,000.......*B3 23
Korcula, isl., Yugo.......D3 22
Kordofan, reg., Sud.,
1,761,968.......C2 47
Korea, reg., Asia.......F15 33, 37
Korea, North, country, Asia,
10,029,500.......F4 37
Korea, South, country, Asia,
24,994,117.......H3 37
Korea, bay, China.......D9 34
Korea, strait, Kor.......I4 37
Korets, Sov. Un., 10,000...F6 27
Korhogo, I.C., 14,000.....E3 45
Kori, creek, Pak.......D4 39, F2 40
Korinthia (Corinthia), prov.,
Grc., 112,505.......*D4 23
Korinthos (Corinth),
Grc., 15,892.......D4, h9 23
Koriyama, Jap., 102,636...H10 37
Korkino, Sov. Un., 87,000..D21 9
Kormatiki, cape, Cyp......E9 31
Kormend, Hung., 7,581....B3 22
Kornat, isl., Yugo.......D2 22
Korner, Glacier, Mont., 5...B4 102
Korneuburg, Aus., 8,276...D8 16
Kornwestheim, Ger., 26,300.E4 17
Koro, isl., Fiji Is.......52
Koromo, Jap., 46,822.....n16 37
Koronis, lake, Minn.......E4 99
Koropi, Grc., 7,128.......h11 23
Koror, isl., Palau Is.......52
Korosten, Sov. Un., 33,800..F7 27
Korotoyak, Sov. Un.,
11,500.......F12 27
Korovin, vol., Alsk.......E5 79
Korsakov (Otomari), Sov.
Un., 43,800.......C11 37
Korsör, Den., 14,276.....C5 24
Körti, Sud.......B3 47
Kortrijk, Bel., 43,606
(*90,000).......D3 15
Korumburra, Austl., 3,237..I5 51
Korville, Harris, Texas, 75..F4 118
Koryak, mts., Sov. Un......C19 28
Kos, Grc., 8,844.......D6 23
Kos, isl., Grc.......D6 23
Koschagyl, Sov. Un.......D4 29
Koscian, Pol., 12,900.....B4 26
Koscierzyna, Pol., 7,820...A4 26
Kosciusko, Attala, Miss.,
6,800.......B4 100
Kosciusko, co., Ind., 40,373.B6 91
Kosciusko, mtn., Austl.....H7 51
Koshan, China, 25,000....B10 34
Koshiki, isl., Jap.......K4 37
Koshkonong, Oregon, Mo.,
478.......E6 101
Koshkonong, lake, Wis.....F5 124
Kosi, riv., India.......C8 39
Košice, Czech., 79,600....D6 26
Koskaecodde, lake, Newf., Can..D4 75
Koslan, Sov. Un., 1,600...C18 9
Kosong, Kor., 14,842.....G4 37
Kosove (Kosovo), pref.,
Alb., 53,000.......*A3 23
Kosovska Mitrovica, Yugo.,
26,620.......D5 22
Kospash, Sov. Un., 30,000..D20 9
Kosse, Limestone, Tex., 354..D4 118
Kösslarn, Ger., 900......A9 18
Kossol, reef, Palau Is.......52
Kossuth, Alcorn, Miss., 178..A5 100
Kossuth, co., Iowa, 25,314..A3 92
Kostelec, Czech., 2,704....n18 26
Kostelec nad Cernymi Lesy,
Czech., 3,341.......o18 26
Kosti, Sud., 22,688......C3 47
Kostroma, Sov. Un.,
184,000.......B2 29
Kostrzyn, Pol., 23,000....B3 26
Koszalin, Pol., 44,000....A4 26
Koszeg, Hung., 9,818.....B3 22
Kotabaru, Indon., 38,075..F5 35
Kotabaru (Hollandia) W. Irian,
16,300.......F10 35
Kota Bharu, Mala., 38,103..J5 38
Kotah, India, 120,345.....E5 40
Kota Kota, Nya.......D5 48
Kotatengah, Indon.......L4 38
Kotel, Bul., 3,671.......D8 22
Kotelnich, Sov. Un., 27,000.B3 29
Kotelnikovskiy, Sov. Un....D2 29
Köthen, Ger., 37,600......B7 17
Kotido, Ug.......A5 48
Kotka, Fin., 30,300......G12 25
Kotlas, Sov. Un., 55,000...C18 9
Kotlik, Alsk., 57.......C7 79
Kotonkoro, Nig.......D6 45
Kotor, Yugo., 4,833......D4 22
Kotor, bay, Yugo.......D4 22
Kotor Varos, Yugo., 2,898..C3 22
Kotovsk, Sov. Un., 30,000..E13 27
Kotovsk, Sov. Un., 10,000..H7 27
Kotovskoye, Sov. Un.......B9 22
Kotri, Pak., 15,154......E2 40
Kotung, China, 5,000.....B3 37
Kotzebue, Alsk., 1,290....B7 79
Kotzebue, sound, Alsk.....B7 79
Kötzting, Ger., 3,200.....D7 17
Kouango, Cen. Afr. Rep....D4 46
Kouchibouguacis, riv., N.B.,
Can.......C4 74
Koudougou, Upper Volta,
8,000.......D4 45
Koula-Moutou, Gabon.....F2 46
Koulikoro, Mali, 4,350....D3 45
Kounradskiy, Sov. Un......D9 28
Kountze, Hardin, Tex., 1,768.D5 118
Kouri, isl., Okinawa.......52
Kourou, Fr. Gu., 582.....A4 59
Kouroussa, Guinea, 6,500..D3 45
Koutiala, Mali, 4,100.....D3 45
Kouts, Porter, Ind., 1,007..B3 91
Kouvola, Fin., 18,200....G12 25
Ko Vayo, Pima, Ariz., 50...E4 80
Kovel, Sov. Un., 42,600...F5 27
Kovrov, Sov. Un., 103,000..B2 29
Kowa, Jap., 10,300.......o15 37
Kowloon, Hong Kong,
726,976.......B8 34
Kowon, Kor., 5,000......G3 37
Koyuk, Alsk., 20.......C7 79
Koyukuk, Alsk., 128......C8 79
Koyukuk, riv., Alsk.......B9 79
Kozan, Tur., 15,200......D10 31
Kozani, Grc., 21,537.....B3 23
Kozani, prov., Grc., 190,835.*B3 23
Kozelsk, Sov. Un., 12,600..D10 27
Kozhikode, India, 192,521
(*248,548).......F6 39
Kozhva, Sov. Un., 800.....B20 9

Kozienice, Pol., 4,099.....C6 26
Kozle, Pol., 8,277.......C5 26
Kozloduy, Bul., 7,422......D6 22
Kozu, isl., Jap.......I9 37
Kozuchow, Pol., 8,782.....C3 26
Kpandu, Ghana, 4,040.....E5 45
Kra, isth., Thai.......G3 38
Kraemer, La Fourche, La.,
450.......C6 95
Kragerö, Nor., 4,600......H3 25
Kragujevac, Yugo., 52,491..D5 22
Krakow, Ger., 3,553.......E6 24
Krakow (Cracow), Pol.,
479,000.......C5 26
Krakower, lake, Ger.......E6 24
Kraljevo, Yugo., 20,409...D5 22
Ksar es Souk, Mor., 6,554..C4 44
Ktipas, mtn., Grc.......g10 23
Kralupy [nad Vltavou], Czech.,
11,100.......n17 26
Kraluv Dvur, Czech., 3,390..o17 26
Kramatorsk, Sov. Un.,
123,000.......G11, q20 27
Kramer, Warren, Ind., 150..B3 91
Kramer, Bottineau, N. Dak.,
175.......A5 110
Kramer Junction, San Bernardino,
Calif., 70.......E5 82
Kranidhion, Grc., 4,381...D4 23
Kranj, Yugo., 21,354.....B2 22
Kranzburg, Codington,
S. Dak., 156.......C9 116
Krasburg, S.W. Afr.......C2 49
Kraslice, Czech., 6,294...C2 26
Krasnaya Sloboda, Sov. Un..B25 9
Krasnik, Lubelski, Pol., 9,158.C7 26
Krasnoarmeysk, Sov. Un.,
54,000.......F15 27
Krasnoarmeysk, Sov. Un.,
2,000.......G15 27
Krasnodar, Sov. Un.,
343,000.......I12 27
Krasnodon, Sov. Un.,
10,000.......q22 27
Krasnogorsk, Sov. Un.,
25,000.......B11 29
Krasnograd, Sov. Un.,
10,000.......G10 27
Krasnokamsk, Sov. Un.,
56,000.......B5 29
Krasnoselkup, Sov. Un.....C9 28
Krasnoselye, Sov. Un.,
10,000.......G9 27
Krasnoslobodsk, Sov. Un.,
5,000.......D14 27
Krasnoturinsk, Sov. Un.,
64,000.......D21 9
Krasnoufimsk, Sov. Un.,
31,300.......B5 29
Krasnouralsk, Sov. Un.,
36,400.......D21 9
Krasnovishersk, Sov. Un.,
28,300.......A5 29
Krasnovodsk, Sov. Un.,
38,000.......E4 29
Krasnoyarsk, Sov. Un.,
468,000.......D27 9, D12 28
Krasnoye Selo, Sov. Un....s31 25
Krasnoznamenskiy, Sov. Un..C7 29
Krasnystaw, Pol., 10,300...C7 26
Krasnyy Kholm, Sov. Un.,
11,700.......B11 29
Krasnyy Kut, Sov. Un.....C3 29
Krasnyy Liman, Sov. Un.,
10,000.......q20 27
Krasnyy Luch, Sov. Un.,
98,000.......q21 27
Krasnyy Sulin, Sov. Un.,
66,600.......H13 27
Krasnyy Yar, Sov. Un.,
10,000.......F15 27
Kratie, Camb., 25,000.....F7 38
Kratovo, Yugo., 2,401....D6 22
Krause, pt., Vir. Is.......k17 65
Krebs, Pittsburg, Okla.,
1,342.......C6 112
Krefeld, Ger., 213,100
(*360,000).......B1 17
Kremenchug, Sov. Un.,
95,000.......G9 27
Kremenchug, res., Sov. Un...G9 27
Kremenets, Sov. Un., 10,000.F5 27
Kremennaya, Sov. Un.,
5,000.......p21 27
Kremlin, Hill, Mont., 125..B6 102
Kremlin, Garfield, Okla., 128.A4 112
Kremmling, Grand, Colo.,
576.......A4 83
Krems, Aus., 21,046......D7 16
Kreole, Jackson, Miss.,
1,870.......E3, E5 100
Kresgeville, Monroe, Pa.,
300.......E10 114
Kress, Swisher, Tex., 438..B2 118
Kreuzlingen, Switz., 12,597..A7 19
Kribi, Cam., 3,055.......E1 46
Krichev, Sov. Un., 5,000...E8 27
Krider, Roosevelt, N. Mex.,
12.......C6 107
Kriens, Switz., 14,029.....B5 19
Krilon, cape, Sov. Un......D11 37
Krilon, pen., Sov. Un......C11 37
Krishnagar, India, 70,440..F12 40
Krishnagiri, India, 19,774..F6 39
Kristiansand, Nor., 27,900
(*40,000).......H3 25
Kristianstad, Swe., 26,100..I6 25
Kristianstad, co., Swe.,
256,600.......B7 24
Kristiansund, Nor., 17,200..F2 25
Kristinestad, Fin., 2,700...F9 25
Kriva Palanka, Yugo., 2,848.D6 22
Krivorozhye, Sov. Un.,
10,000.......q21 27
Krivoy Rog, Sov. Un.,
436,000.......H9 27
Križevci, Yugo., 6,498....B3 22
Krk, isl., Yugo.......B2 22
Krka, riv., Yugo.......D2 22
Krnov, Czech., 21,600....C4 26
Krogsered, Swe.......A6 24
Krokeai, Grc., 3,012......D4 23
Kroken, Swe., 10,000.....H6, t33 25
Krokstadelva, Nor., 2,490..p28 25
Krolevets, Sov. Un., 10,000.F9 27
Kromeriz, Czech., 20,600..D4 26
Kromy, Sov. Un., 10,000...G9 27
Kronach, Ger., 10,200.....C6 17
Kronoberg, co., Swe.,
159,100.......B8 24
Kronshtadt, Sov. Un., 59,000.B7 27
Kröpelin, Ger., 4,839.....D5 24
Kropp, Ger., 1,500.......D4 24
Krosno, Pol., 20,000.....D6 26
Krosno Odrzanskie, Pol.,
5,841.......B3 26
Krotoszyn, Pol., 16,300...C4 26
Krotz Springs, St. Landry,
La., 1,057.......D4 95
Krško, Yugo., 629.......B2 22
Kruger, nat. park, S. Afr....B5 49

Krugersdorp, S. Afr., 89,947.C4 49
Kruje, Alb., 5,107.......B2 23
Krumbach, Ger., 7,900....A6 16
Krumovgrad, Bul., 1,249...E7 22
Krumroy, Summit, Ohio,
1,400.......*A6 111
Krusenstern Rock, reef, Haw..m12 88
Kruševac, Yugo., 32,140...D5 22
Kruševo, Yugo., 4,099.....E6 22
Kruszwica, Pol., 4,822....B5 26
Krydor, Sask., Can., 184...E2 70
Krym (Crimea), pen., Sov. Un..I10 27
Krymskaya, Sov. Un., 10,000.I11 27
Krynica, Pol., 2,649......D6 26
Krynki, Pol., 5,290.......B7 26
Ksar es Souk, Mor., 6,554..C4 44
Ktipas, mtn., Grc.......g10 23
Kuala Dungun, Mala., 12,515.J5 38
Kuala Krau, Mala., 1,271...K5 38
Kualakurun, Indon.......F4 35
Kuala Lipis, Mala., 8,753...J5 38
Kuala Lumpur, Mala.,
316,230 (*400,000).......K4 38
Kualapuu, Mala., Haw., 607.B4 88
Kuala Trengganu, Mala.,
29,446.......J5 38
Kuan, China.......F6 36
Kuandang, Indon.......I2 36
Kuangan, China, 18,000...I2 36
Kuangchang, China, 10,000.K7 36
Kuanghua, China, 29,000..H4 36
Kuangnan, China.......G6 34
Kuangte, China.......I8 36
Kuangyuan, China, 31,000..H1 36
Kuantan, Mala., 23,034...K5 38
Kuanti, China.......E4 37
Kuantien, China, 5,000....F2 37
Kuanyün, China, 65,000...G8 36
Kuba, Sov. Un.......E3 29
Kuban, riv., Sov. Un......F16 9
Kuchen, China.......H4 36
Kucheng, China, 20,000...H4 36
Kuching, Mala., 50,579...E4 35
Kuchino, isl., Jap.......L4 37
Kuchino Erabu, isl., Jap....K5 37
Kudat, Mala., 3,660......D5 35
Kudymkar, Sov. Un.,
20,000.......B4 29
Kueichi, China, 6,000.....I7 36
Kueichih, China.......I7 36
Kueisui, see Huhehot, China
Kueite, China.......D5 34
Kuerhlo (Korla), China,
10,000.......C2 34
Kufstein, Aus., 11,215....E6 16
Kuge, Jap.......n14 37
Kuh, cape, Iran.......I8 41
Kuh-e-Bozgūsh, mts., Iran..D4 41
Kuh-e-Sahand, mtn., Iran...C3 41
Kuhestak, Iran.......D6 41
Kuh-i-Birg, mtn., Iran......H10 41
Kuh-i-Birak, mtn., Iran....D10 41
Kuh-i-Furgun, mtn., Iran....D6 41
Kuh-i-Garrah, mtn., Iran...F5 41
Kuh-i-Gireh, mtn., Iran....H8 41
Kuh-i-Gugird, mtn., Iran....D5 41
Kuh-i-Hormuz, mtn., Iran...H7 41
Kuh-i-Huzar, mtn., Iran....G8 41
Kuh-i-Istin, mtn., Iran.....F10 41
Kuh-i-Kharman, mtn., Iran..G6 41
Kuh-i-Kurkhud, mtn., Iran..D5 41
Kuh-i-Kuru, mtn., Iran.....D5 41
Kuh-i-Mazar, mtn., Afg.....E13 41
Kuh-i-Murghum, mtn., Iran..E8 41
Kuh-i-Naibandan, mtn., Iran..E8 41
Kuh-i-Nila, mtn., Iran......E5 41
Kuh-i-Rahmand, mtn., Iran..D4 41
Kuh-i-Ran, mtn., Iran.......H9 41
Kuh-i-Saguch, mtn., Iran...F8 41
Kuh-i-Surkh, mtn., Iran....D9 41
Kuh-i-Tafrish, mtn., Iran...D5 41
Kūhpāyeh, Iran.......D5 41
Kuhsan, Afg., 10,000.....D10 41
Kuji, Jap., 14,750.......F10 37
Kuju-San, peak, Jap.......J5 37
Kukaiau, Hawaii, Haw., 80..C6 88
Kukawa, Nig.......D7 45
Kukës, B.S., 3,896.......A3 23
Kuki, Jap., 11,300.......m18 37
Kukong, China, 73,000....G7 34
Kuku, pt., Wake Isl.......52
Kukuhaele, Hawaii, Haw.,
375.......C6 88
Kula, Bul., 5,566.......C6 22
Kula, Tur., 9,180.......C7 23
Kula, Yugo., 13,612......C4 22
Ku Lao Cham, isl., Viet....E8 38
Ku Lao Re, isl., Viet.......E8 38
Kulebaki, Sov. Un., 34,000.D14 27
Kulhakangri, peak, China...C13 40
Kuling, China, 5,000......J7 36
Kulkarni, Lyon, Ky., 635...A3 94
Kulm, La Moure, N. Dak.,
664.......C7 110
Kulmbach, Ger., 23,500...C6 17
Kulpmont, Northumberland,
Pa., 4,288.......E9 114
Kulu, isl., Alsk.......m22 79
Kulun, China.......B9 36
Kulunda, Sov. Un., 800....C9 28
Kulunda, lake, Sov. Un.....C9 28
Kulyab, Sov. Un., 15,000..H22 9
Kuma, riv., Sov. Un.......F18 9
Kumagaya, Jap., 60,100
(98,168*).......m18 37
Kumai, Indon.......F4 35
Kumamoto, Jap., 373,922..J5 37
Kumamoto, pref., Jap.,
1,856,192.......*J5 37
Kumanovo, Yugo., 30,734..D5 22
Kumasi, Ghana, 220,922...E4 45
Kumba, Cam., 11,672.....*F6 45
Kumbakonam, India,
92,581 (*96,746).......F6 39
Kumhwa, Kor., 1,400.....G3 37
Kumi, Ug.......A5 48
Kumihama, Jap., 4,800....n14 37
Kumkale, Tur., 404.......C6 23
Kumla, Swe., 10,000.....H6, t33 25
Kummerower, lake, Ger.....t33 24
Kumo, Sov. Un., 66,000..D20 9
Kumta, India, 5,000.......E3 39
Kumukahi, cape, Haw......D7 88
Kuna, Ada, Idaho, 516....F2 89
Kunar, Afg., 5,000.......D15 41
Kunar, riv., Pak.,
Afg.......A5 39, D15 41
Kunckle, Luzerne, Pa., 250..A8 114
Kundal, riv., India.......B2 40
Kunchuling, China, 25,000..E2 37
Kunghit, isl., B.C., Can....C2 68
Kungrad, Sov. Un.......F10 9
Kungsbacka, Swe., 5,000...A4 24
Kungsör, Swe., 4,500.....t34 25
Kungtu, Con. L.......C2 48
Kungur, Sov. Un., 66,000..D20 9
Kunia, Honolulu, Haw., 570.g9 88
Kunkletown, Monroe, Pa.,
400.......B2 106
Kunlun, mts., China.......D3 34

Kunming (Yunnanfu), China,
880,000.......F5 34, C11 39
Kunsan, Kor., 90,500.....I3 37
Kunszentmarton, Hung.,
8,783.......B5 22
Kuntu, China.......B9 36
Künzelsau, Ger., 7,800....D4 17
Kuolayarvi, Sov. Un.,
10,000.......D13 25
Kuop (Royalists), is., Truk...52
Kuopio, Fin., 45,000.....F12 25
Kuopio, dept., Fin.,
453,881.......*F12 25
Kuoyang, China.......H7 36
Kupa, riv., Yugo.......C2 22
Kupang, Indon., 7,171....H6 35
Kupino, Sov. Un., 20,600..C9 29
Kupyansk, Sov. Un., 32,400.G11 27
Kur, riv., Sov. Un.......B7 29
Kura, riv., Sov. Un.......G17 9
Kurakhovka, Sov. Un.,
5,000.......q20 27
Kurashiki, Jap., 90,000
(125,097*).......I6 37
Kuray, Sov. Un.......E26 9
Kurdikos-Naumiestis, Sov.
Un.......A7 26
Kurdzhali, Bul., 10,480...E7 22
Kure, Jap., 210,032......I6 37
Kure, isl., Haw.......k12 88
Kuressaare, Sov. Un., 10,000.B4 27
Kureyka, riv., Sov. Un.....C12 28
Kurgaldzhina, Sov. Un......C8 29
Kurgan, Sov. Un., 164,000..B7 29
Kurgannaya, Sov. Un.,
10,000.......G17 9
Kurgan-Tyube, Sov. Un.,
20,000.......H22 9
Kuri, India, 10,000.......D3 40
Kurigram, Pak., 8,063....E12 40
Kuril, is., Sov. Un.......E18 28
Kuril, strait, Sov. Un......D18 28
Kurinskaya, cape, Sov. Un..H18 9
Kurkhera, India.......G8 40
Kurmuk, Sud., 1,647.....C3 47
Kurnool, India, 100,815...E6 38
Kuro, isl., Jap.......K4 37
Kuroki, Sask., Can., 158...F4 70
Kurovskoye, Sov. Un.,
27,600.......n18 27
Kurow, N.Z., 512.......P13 51
Kursk, Sov. Un., 222,000..F11 27
Kuršumlija, Yugo., 3,392...D5 22
Kurtalan, Tur., 1,591.....D13 31
Kurthwood, Vernon, La., 70.C2 95
Kurtistown, Hawaii, Haw.,
1,025.......D6 88
Kurtz, Jackson, Ind., 200...G5 91
Kuruman, S. Afr., 6,386...C3 49
Kurume, Jap., 155,041....J5 37
Kurunegala, Cey., 13,510..G7 39
Kurusku, Eg., U.A.R., 483..E6 43
Kusadasi, Tur., 5,442.....D6 23
Kusatsu, Jap., 17,800.....n14 37
Kusel, Ger., 5,500.......D2 17
Kushchevskaya, Sov. Un.,
10,000.......H12 27
Kushevat, Sov. Un., 1,000..B22 9
Kushiro, Jap., 150,624...E12 37
Kushka, Sov. Un.......H21 9
Kushmurun, Sov. Un.......C6 29
Kushmurun, lake, Sov. Un..C6 29
Kuskokwim, bay, Alsk.....D7 79
Kuskokwim, mts., Alsk.....C8 79
Kuskokwim, riv., Alsk.....C8 79
Küsnacht, Switz., 11,984...B6 19
Küssnacht, Switz., 6,287...B5 19
Kustanay, Sov. Un., 98,000..C6 29
Küsten, canal, Ger.......A7 15
Küstrin, see Kostrzyn, Pol.
Kutahya (Kutaiah), Tur.,
39,900.......C7 23
Kutaisi, Sov. Un.,
137,000.......G17 9, A14 31
Kutaka, isl., Okinawa.......52
Kut al Hai, Iraq, 10,199...E3 46
Kut al Imara, Iraq, 16,237..E2 46
Kutaradja, Indon., 34,207..k11 35
Kutatjane, Indon.......K2 38
Kutch, gulf, India.......D3 39
Kute, Alb.......B2 23
Kutina, Yugo., 7,132.....C3 22
Kutlay, Sov. Un.......A7 26
Kutna Hora, Czech., 16,900.D3 26
Kutno, Pol., 26,000......B5 26
Kuttawa, Lyon, Ky., 635...A3 94
Kutu, Con. L.......B2 48
Kutubdia, isl., Pak.......G13 40
Kutum, Sud., 7,708......C1 47
Kuty, Sov. Un.......A7 26
Kutztown, Berks, Pa.,
3,312.......E10 114
Kuusamo, Fin., 3,200.....E13 25
Kuvandyk, Sov. Un., 21,500.E20 9
Kuwait, Kuw, 104,551
(*125,929).......G4 41
Kuwait, country, Asia,
206,177.......C7 33, G3 41
Kuwana, Jap., 69,391....I8, n15 37
Kuyang, Mong.......D4 36
Kuybyshev, Sov. Un.,
26,500.......B9 29
Kuybyshev, Sov. Un.,
863,000 (*975,000).......C4 29
Kuybyshev, res., Sov. Un...D8 28
Kuybyshevka-Vostochnaya,
Sov. Un., 50,000.......D15 28
Kuyuan, China.......G2 36
Kuznetsk, Sov. Un.,
62,000.......C4 29
Kuznetsk Alatau, mts., Sov. Un.B11 9
Kuznetsk Basin, reg., Sov. Un..C10 9
Kuznetsova, Sov. Un.......C10 9
Kvichak, bay, Alsk.......D8 79
Kvarner, gulf, Yugo.......C2 22
Kwajalein, atoll, Marshall Is...52
Kwajalong, isl., Kwajalein...52
Kwakoegron, Sur.......A3 59
Kwale, Ken.......B6 48
Kwale, Nig.......E6 45
Kwamouth, Con. L.......B2 48
Kwando, riv., Afr.......D4 48
Kwang Mtoro, Tan.......B6 48
Kwangchow, bay, China....B9 34
Kwangju, Kor., 240,000
(315,000*).......I3 37
Kwango, riv., Con. L.......C2 48
Kwangsi Chuang, auton. reg.,
China, 19,390,000.......F7 34
Kwangtung, prov., China,
37,960,000.......G8 34
Kwangyang, Kor.......I3 37
Kwania, lake, Ug.......A5 48
Kwatisore, W. Irian.......F8 35
Kweichow, prov., China,
16,890,000.......F6 34
Kweichu, see Kweiyang, China

Kweihsien, China.......G6 34
Kweilin, China, 145,100...F7 34
Kweisui, see Huhehot, China
Kweiyang, China, 504,000..F6 34
Kwenge, riv., Con. L......C2 48
Kwidzyn, Pol., 20,000....B5 26
Kwigillingok, Alsk., 344...D7 99
Kwiguk, Alsk., 358.......C7 79
Kwilu, riv., Con. L.......B2 48
Kwobsis, China.......D7 34
Kyabé, Chad, 2,918......D3 46
Kyabra Creek, riv., Austl...B4 51
Kyaiklat, Bur., 15,781....E10 39
Kyaikto, Bur., 10,000....D2 38
Kyakhta, Sov. Un., 21,700..D13 28
Kyangin, Bur., 6,073.....E10 39
Kyaring, lake, China......B8 39
Kyaring, lake, China......E4 34
Kyaukpadaung, Bur., 5,000.C1 38
Kyaukpyu, Bur., 7,335....E9 39
Kyaukse, Bur., 8,659.....B2 38
Kybartai, Sov. Un., 7,274..A7 26
Kyelang, India.......A6 40
Kyje, Czech., 5,836.....D9, n18 26
Kyle, Sask., Can., 535....G1 70
Kyle, Scot., 1,718.......C3 13
Kyle, Shannon, S. Dak., 100.D3 116
Kyle, Hays, Tex., 1,023...E4 118
Kyles Ford, Hancock, Tenn.,
75.......C10 117
Kylestrome, Scot.......B3 13
Kyll, riv., Ger.......D1 17
Kyllburg, Ger., 1,250.....D6 15
Kymes, mtn., Ark.......B1 81
Kymi, dept., Fin., 247,913..*G12 25
Kymulga, cove, Ala.......B3 78
Kynlyn, New Castle, Del.,
1,600.......*A6 85
Kynuna, Austl., 32.......D7 50
Kyoga, cape, Jap.......H5 37
Kyoga, lake, Ug.......A5 48
Kyöngju, Kor., 42,100
(76,000*).......I4 37
Kyöngsong, Kor., 25,925...F4 37
Kyoto, Jap., 1,284,818
(*1,490,000).......I7, n14 37
Kyoto, pref., Jap.,
1,993,403.......*I7 37
Kyrenia, Cyp., 3,680.....E9 31
Kyritz, Ger., 8,679.......B6 17
Kyrkjebö, Nor., 139......G1 25
Kyrock, Edmonson, Ky., 150.C3 94
Kyshtovka, Sov. Un.......D24 9
Kyshtym, Sov. Un., 31,100..B6 29
Kyuquot, B.C., Can., 184...D4 68
Kyuquot, sound, B.C., Can...D4 68
Kyuroku, isl., Jap.......F9 37
Kyūshū, isl., Jap.......J5 37
Kyustendil, Bul., 19,309...D6 22
Kyzyl, Sov. Un., 38,000...D12 28
Kyzyl-Kiya, Sov. Un.,
28,600.......E8 29
Kyzyl-Kum, des., Sov. Un...E6 29
Kzyl-Orda, Sov. Un.,
71,000.......E7 29

L

Laa, Aus., 4,925.......D8 16
Laaber, riv., Ger.......E7 17
Laage, Ger., 3,824.......E6 24
La Almunia de Doña Godina,
Sp., 4,337.......C3 20
Laasphe, Ger., 5,600.....C3 17
La Asunción, Ven., 5,541..A5 63
Labadie, Franklin, Mo., 250.C7 101
Labadieville, Assumption, La.,
650.......C5, E5 95
La Baie, Que., Can., 554...C5 73
La Baie Shawinigan, Que.,
Can., 1,085.......*C5 73
La Banda, Arg., 16,953...E3 55
La Barca, Mex., 16,330.C4, m12 64
La Barge, Lincoln, Wyo., 700.C2 125
La Barge, creek, Wyo......C2 125
Labé, Guinea, 11,800.....D2 45
Labe (Elbe), riv., Czech....C3 26
Labelle, Que., Can., 1,224..C3 63
Labelle, co., Que., Can.,
29,084.......*C2 63
La Belle, Hendry, Fla., 1,262.F5 86
La Belle, Lewis, Mo., 866...A6 101
Laber, riv., Ger.......D6 17
Labette, Labette, Kans.,
114.......E8 93
Labette, co., Kans., 26,805..E8 93
Labette, creek, Kans.......E8 93
Labin, Yugo., 6,152.......C2 22
La Bisbal, Sp., 5,275.....B7 20
Labo, Phil., 3,322.......o14 35
Labo, mtn., Phil.......p14 35
La Boca, C.Z., 48.......m11 62
La Boca, P.R.......B4 65
Laboe, Ger., 3,600.......D4 24
La Bolt, Grant, S. Dak., 125..B9 116
Labouheyre, Fr., 2,169....E3 14
Laboulaye, Arg., 9,032....A4 54
Labrador, reg., Newf., Can.,
10,814.......g8 73
Labrador, highland, Newf....E20 67
Lábrea, Braz., 2,080.....D2 59
La Broquerie, Man., Can.,
321.......E3 71
Labuan, isl., Mala.......D5 35
Labuha, Indon.......F7 35
Labuhanbilik, Indon.......K4 38
Labuk, bay, Mala.......D5 35
Laburnum Manor, Henrico,
Va., 2,500.......*D5 121
La Cabaña, Cuba.......h12 64
Lac a Beauce, Que., Can.,
95.......B5 63
Lacadena, Sask., Can., 142..G1 70
L'Acadie, Que., Can., 327...D4 73
Lacamp, Vernon, La., 50...C3 95
La Capelle, Fr., 2,019.....C5 14
La Carlota, Arg., 4,501....A4 54
La Carlota, Phil., 7,213...C6 35
La Carolina, Sp., 12,138...C4 20
Lac au Saumon, Que., Can.,
1,548.......*B3 73
Lac aux Brochets, Que.,
Can.......B5 73
Lac aux Sables, Que., Can.,
857.......C5 73
Lac Baker, N.B., Can., 338..B1 74
Lac Beauport, Que., Can.,
115.......C9 73

Lac Bouchette, Que., Can., 911....A5 73
Laccadive, Minicoy and Amindivi Is., ter., India, 24,108....*F5 39
Laccadive, is., India....F5 39
Lac Carre, Que., Can., 601..C3 73
Lac Chat, Que., Can., 95....B5 73
Lac Court Oreilles, lake, Wis..C2 124
Lac Court Oreilles, Indian res., Wis....C2 124
Lac du Bonnet, Man., Can., 569....D3 71
Lac Du Flambeau, Vilas, Wis., 700....C4 124
Lac Du Flambeau, Indian res., Wis....C3 124
Lac Edouard, Que., Can., 240....B5 73
La Ceiba, Ven., 197....B3 60
La Ceiba, Hond., 24,868....C4 62
La Ceja, Col., 5,075....B2 60
La Center, Ballard, Ky., 882....A2 94
Lac Etchemin, Que., Can., 1,485....C7 73
Lacey, Thurston, Wash., 6,630....B3 122
Lacey Park, Bucks, Pa., 4,000....*F11 114
Laceys Spring, Morgan, Ala., 500....A3 78
Lac Frontiere, Que., Can., 264....C7 73
La Chambre, Fr., 773....D2 18
La Charité, Fr., 5,747....D5 14
La Charqueada, Ur., 1,189....E2 56
La Châtre, Fr., 4,025....D5 14
La Chaux-de-Fonds, Switz., 38,906....B2 19
Lachen, Switz., 3,913....B6 19
Lachine, Que., Can., 38,630....D4, D9 73
Lachine, Alpena, Mich., 120....C7 98
Lachlan, riv., Austl....F6 51
La Chorrera, Pan., 8,626....F8, m11 62
Lachute, Que., Can., 7,560..D3 73
La Cienega, Santa Fe, N. Mex., 115....F6 107
La Ciotat, Fr., 18,827....F6 14
La Cisa, pass., It....B3 21
Lackawanna, Erie, N.Y., 29,564....C2 108
Lackawanna, co., Pa., 234,531....D10 114
Lackawaxen, Pike, Pa., 250....D12 114
Lac la Biche, Alta., Can., 1,314....B5 69
Lac la Croix, lake, Minn....B6 99
Lac la Plonge, lake, Sask., Can..B2 70
Lac la Ronge, lake, Sask., Can...B3 70
Laclede, Bonner, Idaho, 428.A2 89
La Clede, Fayette, Ill., 140...E5 90
Laclede, Linn, Mo., 428....B4 101
Laclede, co., Mo., 18,991....D4 101
Lac Masson, Que., Can., 640....C3 73
Lac Megantic, Que., Can., 7,015....D7 73
Lacolle, Que., Can., 1,187...D4 63
La Colorado, Mex., 561....B2 63
Lacomb, Linn, Oreg., 100....C2, C4 113
Lacombe, Alta., Can., 3,029..C4 69
Lacombe, St. Tammany, La., 650....B7, D6 95
Lacombe, bayou, La....B8 95
Lacon, Marshall, Ill., 2,175..B4 90
Lacona, Warren, Iowa, 396..C4 92
Lacona, Oswego, N.Y., 556..B4 108
La Conception Station, Que., Can., 128....C3 73
Laconia, Belknap, N.H., 15,288....C4 105
Laconia, Fayette, Tenn., 40..B2 117
Laconia (Lakonia), prov., Grc., 118,661....*D4 23
Laconia, gulf, Grc....D4 23
La Conner, Skagit, Wash., 638....A3 122
Lacoochee, Pasco, Fla., 1,523....D4 86
La Coruña, Sp., 177,502....A1 20
La Coruña, prov., Sp., 971,641....*A1 20
Lacota, Van Buren, Mich., 135....F4 98
La Courneuve, Fr., 25,792..g10 14
Lac qui Parle, co., Minn., 13,330....F2 99
Lac qui Parle, lake, Minn....E2 99
Lac qui Parle, riv., Minn....E2 99
Lacreek, lake, S. Dak....D4 116
La Crescent, Houston, Minn., 2,624....G7 99
La Crescenta, Los Angeles, Calif., 12,000....*F2 82
La Cresta Village, Kern, Calif., 10,000....*E4 82
La Crosse, Izard, Ark., 30..A4 81
La Crosse, Alachua, Fla., 165....C4 86
La Crosse, La Porte, Ind., 632....B4 91
La Crosse, Rush, Kans., 1,767....D4 93
La Crosse, Mecklenburg, Va., 726....E4 121
Lacrosse, Whitman, Wash., 463....C8 122
La Crosse, La Crosse, Wis., 47,575....E2 124
La Crosse, co., Wis., 72,465..E2 124
La Crosse, riv., Wis....E3 124
La Cruz, Col., 2,745....C2 60
La Cruz, Mex., 2,151....C3 63
Lac Saguay, Que., Can., 187....C2 73
Lac Sainte Marie, Que., Can., 232....D2 73
La Cueva, Mora, N. Mex., 10.B4 107
Lac Vert, Sask., Can., 122..E3 70
La Cygne, Linn, Kans., 810..D9 93
Lacy-Lakeview, McLennan, Tex., 2,272....*D4 118
Ladd, Bureau, Ill., 1,255....B4 90
Ladder, creek, Kans....D2 93
Laddonia, Audrain, Mo., 671....B6 101
Ladera Heights, Los Angeles, Calif., 1,500....*F2 82
Ladgasht, Pak....C3 39
Ladies, isl., S.C....G6 115
Ladispoli, It., 464....h8 21
Lādīz, Iran, 10,000....G10 41
Ladner, B.C., Can., 2,000....A10, E6 68
Ladner, Harding, S. Dak., 10.B2 116
Ladoga, Montgomery, Ind., 974....E4 91
Ladoga, lake, Sov. Un....A8 27
Ladonia, Fannin, Tex., 890..C5 118
Ladora, Iowa, Iowa, 307....C5 92
La Dorado, Col., 14,577....B3 60
La Due, Henry Mo., 175..C4 101
Ladue, St. Louis, Mo., 9,466....*C7 101
Ladybank, Scot., 1,207....D5 13
Ladybrand, S. Afr., 7,045....C4 49
Lady Lake, Lake, Fla., 335..D5 86
Lady Laurier, mtn., B.C., Can..A6 68
Lady Newnes, ice shelf, Ant...B30 5
Ladysmith, B.C., Can., 2,173....B9, E6 68
Ladysmith, S. Afr., 22,955..C4 49
Ladysmith, Rusk, Wis., 3,584....C2 124
Lae, N. Gui., 948....k12 50
Laerdal, Nor....G2 25
Laesö, isl., Den....A4 24
La Esperanza, Cuba....k12 64
La Esperanza, Cuba, 1,038..D2 64
La Esperanza, Hond., 1,327..C3 62
La Esperanza, P.R....B6 65
La Estrada, Sp., 2,540....A1 20
La Estrella, Bol....C3 55
La Farge, Vernon, Wis., 833..E3 124
La Fargeville, Jefferson, N.Y., 425....A5, B1 108
Lafayette, Chambers, Ala., 2,605....C4 78
Lafayette, Contra Costa, Calif., 7,114....*B5 28
Lafayette, Boulder, Colo., 2,612....B5 83
La Fayette, Walker, Ga., 5,588....B1 87
La Fayette, Stark, Ill., 269...B4 90
Lafayette, Tippecanoe, Ind., 42,330....D4 91
La Fayette, Christian, Ky., 196....D2 94
Lafayette, Lafayette, La., 40,400....D3 95
Lafayette, Nicollet, Minn., 516....F4 99
Lafayette, Sussex, N.J., 350..A3 106
Lafayette, Onondaga, N.Y., 290....C4 108
Lafayette, Madison, Ohio, 150....C4 111
Lafayette, Yamhill, Oreg., 553....B1, B3 113
LaFayette, Washington, R.I. (part of North Kingstown)....C11 84
Lafayette, Macon, Tenn., 1,590....C7 117
Lafayette, co., Ark., 11,030..D2 81
Lafayette, co., Fla., 2,889....C3 86
Lafayette, co., Miss., 21,355..A4 100
Lafayette, co., Mo., 25,274..B4 101
Lafayette, co., Wis., 18,142..F3 124
Lafayette, par., La., 84,656..D3 95
Lafayette, mtn., N.H....B3 105
Lafayette Central, P.R....D6 65
Lafayette Hill, Montgomery, Pa., 3,500....*A11 114
Lafayette Southwest, Lafayette, La., 6,682....*D3 95
Lafayette Springs, Lafayette, Miss., 151....A4 100
Lafe, Greene, Ark., 150....A5 81
La Fère, Fr., 3,161....E3 15
La Feria, Cameron, Tex., 3,047....F4 118
La Ferté-Bernard, Fr., 5,979..C4 14
La Ferté-Macé, Fr., 4,109...C3 14
La Ferté-sous-Jouarre, Fr., 4,077....C5 14
Lafferty, Izard, Ark....B4 81
Lafferty, Belmont, Ohio, 950....B6 111
Lafia, Nig....E6 45
Lafiagi, Nig....E6 45
Lafitte, Jefferson, La., 257..C7 95
Lafleche, Sask., Can., 749...H2 70
La Flèche, Fr., 9,439....D3 14
La Follette, Campbell, Tenn., 6,204....C9 117
La Fontaine, Wabash, Ind., 779....C6 91
Lafontaine, Wilson, Kans., 100....E8 93
Lafourche, Lafourche, La., 500....C6, E5 95
Lafourche, par., La., 55,381..E5 95
Lafourche, bayou, La....E5 95
La France, Anderson, S.C., 900....B2 115
Lafrenais, lake, Que., Can....A4 73
La Garita, Saguache, Colo., 10....D4 83
Lagarto, Pan., 306....k10 62
Lage, Ger., 12,900....B3 17
Lågen, riv., Nor....G3, p27 25
Lagg, Scot....E3 13
Laggan, bay, Scot....E2 13
Laggan, lake, Scot....D4 13
Laghouat, Alg., 11,058 (43,220★)....C5 44
La Gloria, Col., 1,277....B3 60
Lagny, Fr., 11,945....C5 14
Lago, mtn., Wash....A5 122
Lagoa, Port., 2,249....D1 20
Lago Argentino, Arg....E2 54
Lago Buenos Aires, Arg....D2 54
Lagoon, Bladen, N.C....C5 109
Lago Posadsa, Arg....D2 54
Lagos, Chile, 2,106....B2 54
Lagos, Nig., 364,000....E5 45
Lagos, fed. reg., Nig., 272,000....*E5 45
Lagos, Port., 7,143....D1 20
Lagos de Moreno, Mex., 23,298....C4, m13 63
Lago Viedma, Arg....D2 54
La Grand' Combe, Fr., 14,440 (★22,000)....E5 14
La Grande, Union, Oreg., 9,014....B8 113
La Grange, Lee, Ark., 300...C5 81
La Grange, Austl....C3 50
La Grange, Brevard, Fla., 300....D6 86
La Grange, Troup, Ga., 23,632....C1 87
La Grange, Cook, Ill., 16,326....B6, F2 90
Lagrange, Lagrange, Ind., 1,990....A7 91
La Grange, Oldham, Ky., 2,168....B4 94
La Grange, Penobscot, Maine, 300 (424▲)....C4 96
La Grange, Lewis, Mo., 1,347....A6 101
La Grange, Lenoir, N.C., 2,133....B6 109
Lagrange, Lorain, Ohio, 1,007....A5 111
Lagrange, Fayette, Tenn., 217....B2 117
La Grange, Fayette, Tex., 3,623....E4 118
Lagrange, Goshen, Wyo., 176.D8 125
Lagrange, co., Ind., 17,380..A7 91
La Grange Highlands, Cook, Ill., 5,000....*F2 90
La Grange Park, Cook, Ill., 13,793....F2 90
La Guadeloupe, Que., Can., 1,728....D7 63
La Guaira, Ven., 20,275....A4 60
La Guajira, see Guajira, pol. div., Col.
La Guardia, Sp., 9,311....B1 20
Laguna, Yuma, Ariz., 30....E1 80
Laguna, Braz., 17,451....D3 56
Laguna, Valencia, N. Mex., 650....B2 107
Laguna, prov., Phil., 472,200....*C6 35
Laguna, dam, Ariz....E1 80
Laguna, Indian res., N. Mex..C2 107
Laguna Beach, Orange, Calif., 9,288....F5 82
Laguna Madre, lagoon, Tex....F4 118
Laguna Shores, Nueces, Tex., 1,500....*F4 118
Lagunillas, Bol., 840....C3 55
Lagunillas, Ven., 67,869....A3 60
Lagunitas, Sandoval, N. Mex.A2 107
Lagrue, bayou, Ark....C4 81
La Habana, see Havana, Cuba
La Habra, Orange, Calif., 25,136....F3 82
LaHabra Heights, Los Angeles, Calif., 1,500....*F3 82
Lahad Datu, Mala., 600....L5 35
Lahaina, Maui, Haw., 3,423....C5 88
La Harpe, Hancock, Ill., 1,322....C3 90
La Harpe, Allen, Kans., 529..E8 93
Lahat, Indon....F2 35
La Have, N.S., Can., 248....E5 74
La Have, isl., N.S., Can....E5 74
Lahave, riv., N.S., Can....E5 74
La Have Isl., N.S., Can., 200.E5 74
La Haye-Descartes, Fr., 1,679....D4 14
Laheria Sarai, India....D10 40
Laḥij, Aden, 10,000....C5 47
Lahinch, Ire., 389....E2 11
Lahn, riv., Ger....C3 17
Laholm, Swe., 3,500....B7 24
Laholmsbukten, bay, Swe....B6 24
Lahoma, Garfield, Okla., 160.A3 112
Lahontan, res., Nev....D2 104
Lahore, Pak., 1,225,000 (★1,325,000)....B5 39, B5 40
Lahr, Ger., 22,600....D3 16
Lahri, Pak....C2 40
Lahti, Fin., 66,600....G11 25
La Huaca, Peru, 2,012....B1 58
La Huerta, Eddy, N. Mex. (part of Carlsbad)....E5 107
Laï, Chad, 4,787....D3 46
Lai Chau, Viet....A5 38
Laichow, bay, China....F8 36
Laide, Scot....C3 13
Laidon, lake, Scot....D4 13
Laie, Honolulu, Haw., 1,767....B4, f10 88
Laifeng, China, 14,000....J3 36
Laigle, Fr., 5,520....C4 14
Laihka, Bur....B2 38
Laingsburg, Shiawassee, Mich., 1,057....F6 98
Laingsburg, S. Afr., 2,815..D3 49
Lair, Harrison, Ky., 102....B5 94
Laird, Sask., Can., 278....E2 70
Laird, Yuma, Colo., 156....A8 83
Lairg, Scot., 962....B4 13
Laiwui, Indon....F7 35
Laiyang, China, 40,000....F9 36
Laiyuan, China....E6 36
La Jara, Conejos, Colo., 724..D5 83
La Jara, Sandoval, N. Mex., 10....A3 107
Lajas, P.R., 914....C2 65
Lajas, mun., P.R., 15,375....C2 65
Lajes, Braz., 35,112....E2 56
Lajolla, pt., Calif....E2 82
Lajord, Sask., Can., 135....G3 70
La Jose (Newburg), Clearfield, Pa., 150....E8 114
Lajosmizse, Hung., 4,735....B4 22
Lojaya, Socorro, N. Mex., 150....C3 107
La Joya, Peru....E3 58
La Junta, Otero, Colo., 8,026....D7 83
Lak, riv., Ken....A7 48
Lake, Ascension, La....B6 95
Lake, Clare, Mich., 90....E5 98
Lake, Scott, Miss., 297....C4 100
Lake, co., Calif., 13,786....C2 82
Lake, co., Colo., 7,101....B4 83
Lake, co., Fla., 57,383....D5 86
Lake, co., Ill., 293,656....A6 90
Lake, co., Ind., 513,269....B3 91
Lake, co., Mich., 5,338....E5 98
Lake, co., Minn., 13,702....C7 99
Lake, co., Mont., 13,104....C2 102
Lake, co., Ohio, 148,700....A6 111
Lake, co., Oreg., 7,158....E6 113
Lake, co., S. Dak., 11,764...C8 116
Lake, co., Tenn., 9,572....A2 117
Lake, reg., Tan., 3,330,206..B5 48
Lake, creek, Wash....B7 122
Lake, district, Eng....C5 10
Lake, fork, Colo....C3 83
Lake, mtn., Wyo....D6 125
Lake, range, Nev....C2 104
Lake, swamp, S.C....D8 115
Lake Alfred, Polk, Fla., 2,191.D5 86
Lake Alma, Sask., Can., 171.H3 70
Lake Andes, Charles Mix, S. Dak., 1,097....D7 116
Lake Annis, N.S., Can., 65..E4 74
Lake Ariel, Wayne, Pa., 400....D11 114
Lake Arrowhead, San Bernardino, Calif., 500.E5, F3 82
Lake Arthur, Jefferson Davis, La., 3,541....D3 95
Lake Arthur, Chaves, N. Mex., 387....E5 107
Lake Barcroft, Fairfax, Va., 1,800....*B5 121
Lake Benton, Lincoln, Minn., 905....F2 99
Lake Beulah, Walworth, Wis., 80....F1 124
Lake Bluff, Lake, Ill., 3,494....A6, E2 90
Lake Bonaparte, Lewis, N.Y., 30....A5 108
Lake Bronson, Kittson, Minn., 421....B2 99
Lake Brown, Austl., 170....F2 50
Lake Bruce, Pulaski, Ind., 100....B5 91
Lake Burien Heights, King, Wash., 2,000....*B3 122
Lake Butler, Union, Fla., 1,311....B4 86
Lake Carey, Wyoming, Pa., 30....C10 114
Lake Cargelligo, Austl., 1,118....F6 51
Lake Chance, creek, Utah....F4 119
Lake Charles, Calcasieu, La., 63,392 (★127,200)....D2 95
Lake City, Craighead, Ark., 850....B5 81
Lake City, Modoc, Calif., 80.B3 82
Lake City, Hinsdale, Colo., 106....C3 83
Lake City, Columbia, Fla., 9,465....B4 86
Lake City, Clayton, Ga., 1,042....*C2 87
Lake City, Calhoun, Iowa, 2,114....B3 92
Lake City, Barber, Kans., 165....E5 93
Lake City, Missaukee, Mich., 718....D5 98
Lake City, Wabasha, Minn., 3,494....F6 99
Lake City, Erie, Pa., 1,722..B1 114
Lake City, Florence, S.C., 6,059....D8 115
Lake City, Marshall, S. Dak., 81....B8 116
Lake City, Anderson, Tenn., 1,914....C9 117
Lake Clear Junction, Franklin, N.Y., 300....B2 108
Lake Como, Jasper, Miss., 150....D4 100
Lake Como, Monmouth, N.J. (part of Spring Lake Heights)....C4 106
Lake Como, Wayne, Pa., 100....C11 114
Lake Cormorant, De Soto, Miss., 300....A3 100
Lake Cowichan, B.C., Can., 2,149....B8 68
Lakecreek, Jackson, Oreg., 15....E5 113
Lake Crystal, Blue Earth, Minn., 1,652....F4 99
Lakedale, Cumberland, N.C. (part of Fayetteville)....B5 109
Lake Delton, Sauk, Wis., 714....E4 124
Lake Dick, Jefferson, Ark....C4 81
Lake Elmo, Washington, Minn., 550....E7 99
Lake Elmore, Lamoille, Vt., 75....B3 120
Lake End, Red River, La., 40....C2 95
Lake Entiat, res., Wash....B5 122
Lake Erie Beach, Erie, N.Y., 2,117....*C1 108
Lake Fenton, Genesee, Mich., 1,500....*F7 98
Lakefield, Ont., Can., 2,167..C6 72
Lakefield, Jackson, Minn., 1,789....G3 99
Lake Forest, Duval, Fla., 8,000....*B5 86
Lake Forest, Lake, Ill., 10,687....A6, E2 90
Lake Forest, Greenville, S.C., 1,500....*B3 115
Lake Forest Park, King, Wash., 1,500....*B3 122
Lake Fork Valley, Idaho, 10....E2 89
Lake Fork, Logan, Ill., 130..D4 90
Lake Fork, riv., Utah....C5 119
Lake Fort Gibson, res., Okla....A6 112
Lake Geneva, Walworth, Wis., 4,929....F5 124
Lake George, Park, Colo., 35....C5 83
Lake George, Clare, Mich., 250....E6 98
Lake George, Warren, N.Y., 1,026....B7 108
Lake Greeson, res., Ark....C2 81
Lake Hamilton, Polk, Fla., 930....*D5 86
Lake Harbor, Palm Beach, Fla., 500....F6 86
Lake Helen, Volusia, Fla., 1,096....D5 86
Lake Hiawatha, Morris, N.J., 4,000....B4 106
Lake Hills, King, Wash., 4,500....*D2 122
Lake Holloway, Polk, Fla., 3,172....*D5 86
Lake Hopatcong (Espanong) Morris, N.J., 1,107....*B3 106
Lake Hubert, Crow Wing, Minn., 150....D4 99
Lake Hughes, Los Angeles, Calif., 375....E4 82
Lake Huntington, Sullivan, N.Y., 200....D6 108
Lakehurst, Ocean, N.J., 2,780....C4 106
Lake in the Hills, McHenry, Ill., 2,046....*A5 90
Lake Itasca, Clearwater, Minn., 30....C3 99
Lake Jackson, Brazoria, Tex., 9,651....G5 118
Lake James, Steuben, Ind., 900....A7 91
Lake June, Dallas, Tex. (part of Dallas)....*C4 118
Lake Katrine, Ulster, N.Y., 1,149....*D6 108
Lakeland, Polk, Fla., 41,350.D5 86
Lakeland, Lanier, Ga., 2,236....E3 87
Lakeland, La Porte, Ind. (part of Michigan City)...A4 91
Lakeland, Rankin, Miss., 900....C3 100
Lakeland Village, Riverside, Calif., 3,593....*F5 82
Lake Lemon, res., Ind....F5 91
Lake Lenore, Sask., Can., 447....E3 70
Lake Lillian, Kandiyohi, Minn., 335....F4 99
Lake Linden, Houghton, Mich., 1,314....A2 98
Lake Lindsey, Hernando, Fla., 100....D4 86
Lake Lotawana, Jackson, Mo., 1,499....*C3 101
Lake Louise, Alta., Can., 54....D2 69
Lake Luzerne, Warren, N.Y., 900....B7 108
Lake McDonald, Flathead, Mont....B3 102
Lake Madge, Sask., Can....F5 70
Lake Manawa, Pottawattamie, Iowa, 900....C2 92
Lake Mary, Seminole, Fla., 900....D5 86
Lakemba, isl., Fiji Is....52
Lakemba, passage, Fiji Is....52
Lake Mead, nat. recreation area, Ariz., Nev....A1 80, H7 104
Lake Merwin, res., Wash....D3 122
Lake Mills, Winnebago, Iowa, 1,758....A4 92
Lake Mills, Jefferson, Wis., 2,951....E5 124
Lake Minchumina, Alsk., 34.C9 79
Lake Mohawk, Sussex, N.J., 4,647....*A3 106
Lakemont, Blair, Pa., 1,500..F5 114
Lakemore, Summit, Ohio, 2,765....A6 111
Lakenan, Shelby, Mo., 83....B6 101
Lake Nebagamon, Douglas, Wis., 346....B2 124
Lake Norden, Hamlin, S. Dak., 390....C8 116
Lake Norman, res., N.C....B3 109
Lake Odessa, Ionia, Mich., 1,806....F5 98
Lake of the Rivers, lake, Sask., Can....H3 70
Lake of the Woods, co., Minn., 4,304....B4 99
Lake of the Woods, lake, Ont., Can., Minn..A4 99, E7 72
Lake Orion, Oakland, Mich., 2,698....F7 98
Lake Oswego, Oreg., 8,906....B2, B4 113
Lake O'the Pines, res., Tex....C5 118
Lake Ouachita, res., Ark....C2 81
Lake Ozark, Miller, Mo., 500....C5 101
Lake Park, Palm Beach, Fla., 3,589....F6 86
Lake Park, Lowndes, Ga., 338....F3 87
Lake Park, Dickinson, Iowa, 952....A2 92
Lake Park, Becker, Minn., 730....D2 99
Lake Placid, Highlands, Fla., 1,007....E5 86
Lake Placid, Essex, N.Y., 2,998....A7, B3 108
Lake Pleasant, Hamilton, N.Y., 100....B6 108
Lake Point, Tooele, Utah....C3 119
Lakeport, Lake, Calif., 2,303....C2 82
Lake Preston, Kingsbury, S. Dak., 955....C8 116
Lake Providence, East Carroll, La., 5,781....B4 95
Lake River, Ont., Can....D9 72
Lake Ronkonkoma, Suffolk, N.Y., 4,841....F4 84
Lake St. John, Concordia, La....C4 95
Lake St. John East, co., Que., Can., 43,920....A6 73
Lake St. John West, co., Que., Can., 61,310....A5 73
Lake Shore, Duval, Fla., 11,000....B6 86
Lake Shore, Ramsey, Minn..E7 99
Lakeshore, Hancock, Miss., 460....E4 100
Lake Shore, Utah, Utah....C4 119
Lakeside, Navajo, Ariz., 700..C6 80
Lakeside, San Diego, Calif., 6,000....E2, F5 82
Lakeside, Buena Vista, Iowa, 306....B2 92
Lakeside (Pineville Junction), Rapides, La., 1,233....*C3 95
Lakeside, Roosevelt, Mont., 16....B12 102
Lakeside, Sheridan, Nebr., 75....B3 103
Lakeside, Passaic, N.J., 70..A4 106
Lakeside, Ottawa, Ohio, 950.A5 111
Lakeside, Coos, Oreg., 134..D2 113
Lakeside, Box Elder, Utah, 6.B3 119
Lakeside, Henrico, Va., 19,000....*C7 121
Lakeside, Chelan, Wash. (part of Chelan)....B5 122
Lakeside, mts., Utah....C3 119
Lakeside Park, Kenton, Ky., 2,214....A7 94
Lake Stevens, Snohomish, Wash., 1,538....A3 122
Lake Success, Nassau, N.Y., 2,954....D2 108
Lakesville, Dorchester, Md., 15....D5 85
Lake Tawakoni, res., Tex....C4 118
Lake Telemark, Morris, N.J., 1,000....*B3 106
Lake Tomahawk, Oneida, Wis., 250....C4 124
Laketon, Wabash, Ind., 500..C6 91
Laketown, Rich, Utah, 211..B4 119
Lake Toxaway, Transylvania, N.C., 700....D3 109
Lake Traverse, Ont., Can., 30....B6 72
Lake Tschida, dam, N. Dak..C4 110
Lake Valley, Sask., Can., 85..G2 70
Lake Valley, Sierra, N. Mex., 8....E2 107
Lakeview, Riverside, Calif., 300....F3 82
Lakeview, Catoosa, Ga., 1,000....*B1 87
Lakeview, Bonner, Idaho, 25.B2 89
Lake View, Sac, Iowa, 1,165..B2 92
Lakeview, Calhoun, Mich., 10,384....*F5 98
Lakeview, Montcalm, Mich., 1,126....E5 98
Lake View, De Soto, Miss., 50....E8 117
Lakeview, Beaverhead, Mont., 10....F5 102
Lake View, Erie, N.Y., 4,000....*C2 102
Lakeview, Nassau, N.Y., 9,000....*E7 108
Lakeview, Logan, Ohio, 1,008....B4 111
Lakeview, Lake, Oreg., 3,260....E6 113
Lake View, Dillon, S.C., 865.C9 125
Lakeview, Dallas, Tex., 1,200....*B6 118
Lakeview, Hall, Tex., 219...B2 118
Lakeview, Jefferson, Tex., 3,849....*E6 118
Lakeview, Utah, Utah....C4 119
Lake Villa, Lake, Ill., 903...E2 90
Lake Village, Chicot, Ark., 2,998....D4 81
Lake Village, Newton, Ind., 300....B3 91
Lakeville, Litchfield, Conn., 950....B3 84
Lakeville, St. Joseph, Ind., 757....A5 91
Lakeville, Plymouth, Mass., 50 (3,209▲)....C6 97
Lakeville, Dakota, Minn., 924....F5 99
Lakeville, Livingston, N.Y., 500....C3 108
Lakeville, Ashtabula, Ohio, 4,181....A7 111
Lake Waccamaw, Columbus, N.C., 780....C5 109
Lake Wales, Polk, Fla., 8,346.E5 86
Lake Williams, Kidder, N. Dak., 45....B6 110
Lake Wilson, Murray, Minn., 436....G3 99
Lake Winola, Wyoming, Pa., 80....A8 114
Lakewood, Los Angeles, Calif., 67,126....*F2 82
Lakewood, Jefferson, Colo., 36,209....*B5 83
Lakewood, Duval, Fla., 7,000....*B5 86
Lakewood, Walton, Fla., 123.G3 86
Lakewood, Sommerset, Maine, 130....D3 96
Lakewood, Ocean, N.J., 16,020....C4 106
Lakewood, Eddy, N. Mex., 46....E5 107
Lakewood, Chautauqua, N.Y., 3,933....C1 108
Lakewood, Cuyahoga, Ohio, 66,154....A6, B2 111
Lakewood, Harris, Tex., 1,882....*E5 118
Lakewood, Snohomish, Wash., 100....A3 122
Lakewood, Oconto, Wis., 125....C5 124
Lakewood Center, Pierce, Wash., 20,000....*E1 122
Lake Worth, Palm Beach, Fla., 20,758....F6 86
Lake Worth Village, Tarrant, Tex., 3,833....*B5 112
Lake Worth, inlet, Fla....F6 86
Lake Zurich, Lake, Ill., 3,458.E2 90
Lakhi, Afg., 5,000....F12 41
Lakhimpur, India, 25,055..D8 40
Lakhpat, India....F2 40
Lakhtinskiy, Sov. Un., 10,000....s31 25
Lakin, Kearney, Kans., 1,432.E2 93
Lakoleh, Ken....A7 48
Lakonia (Laconia), prov., Grc., 118,661....*D4 23
Lakota, Kossuth, Iowa, 459..A3 92
Lakota, Nelson, N. Dak., 1,066....A7 110
Lakota, Cache, Utah....B4 119
Laksefjord, fjord, Nor....B12 25
La Laguna, Can. Is., 18,000 (57,344▲)....m13 20
La Lande, De Baca, N. Mex..C5 107
Lālī, Iran....E4 41
La Libertad, Guat., 632....B2 62
La Libertad, Sal., 4,877....D3 62
La Libertad, dept., Peru, 462,026....C2 58
La Liendre, San Miguel, N. Mex....B5 107
La Ligua, Chile, 3,178....A2 54
Lalín, Sp., 1,353....A1 20
La Línea, Sp., 59,456....D3 20
La Lisa, Cuba....k11 64
Lalitpur, India, 20,792....E7 40
Lalo, pt., Tinian....52
La Loma, Stanislaus, Calif., 5,700....*D3 82
La Loma, Guadalupe, N. Mex., 200....B4 107
La Louvière, Bel., 23,107....D4 15
La Luz, Otero, N. Mex., 250.E4 107
La Machine, Fr., 6,164....D5 14
La Maddalena, It., 11,169..D2 21
La Madeleine, Fr., 23,381..D3 15
La Madera, Rio Arriba, N. Mex., 100....A3 107
Lama-kara, Togo....E5 45
La Malbaie, Que., Can., 2,580....B7 73
Lamaline, Newf., Can., 530.E4 75
Lamar, Johnson, Ark., 514..B2 81
Lamar, Prowers, Colo., 7,369.C8 83
Lamar, Benton, Miss., 68....A4 100
Lamar, Barton, Mo., 3,608..D3 101
Lamar, Chase, Nebr., 50....D4 103
Lamar, Hughes, Okla., 150..B5 112
Lamar, Clinton, Pa., 250....D6 114
Lamar, Darlington, S.C., 1,121....C7 115
Lamar, co., Ala., 14,271....B1 78
Lamar, co., Ga., 10,240....C2 87
Lamar, co., Miss., 13,675....D4 100
Lamar, co., Tex., 34,234....C5 118
Lamar, riv., Wyo....A2 125
Lamarche, Fr., 932....A1 18
La Marque, Galveston, Tex., 13,969....G5 118
Lamartine, Que., Can., 66..B7 73
Lamas, Peru, 5,625....C2 58
Lamasco, Lyon, Ky., 100....A3 94
Lamb, co., Tex., 21,896....B1 118
Lamballe, Fr., 5,069....C2 14
Lambaréné, Gabon....F2 46
Lambasa, Fiji Is., 1,595....52
Lambay, isl., Ire....D5 11
Lambayeque, Peru, 6,846....C2 58
Lambayeque, dept., Peru, 229,958....C2 58

Lambert, Cook, Ill., 50......F2 90
Lambert, Quitman, Miss., 1,181......A3 100
Lambert, Richland, Mont., 250......C12 102
Lambert, Alfalfa, Okla., 21..A3 112
Lambert Lake, Washington, Maine, 150 (172▲)......C5 96
Lamberton, Redwood, Minn., 1,141......F3 99
Lambertville, N.B., Can., 247......D3 74
Lambertville, Monroe, Mich., 1,168......*G7 98
Lambertville, Hunterdon, N.J., 4,269......C3 106
Lambeth, Ont., Can., 2,293..E3 72
Lambourn, Eng., 1,941......C6 12
Lambrecht, Ger., 5,100......D3 17
Lambro, riv., It......D5 18
Lambrook, Phillips, Ark., 150......C5 81
Lambs, ferry, Ala......B2 78
Lamb's, head, Ire......F1 11
Lambsburg, Carroll, Va., 250......E2 121
Lambton, Que., Can., 699..D6 73
Lambton, co., Ont., Can., 102,131......E2 72
Lambton, cape, N.W. Ter., Can..B8 66
Lamdessar Timur, Indon......G8 35
Lame Deer, Rosebud, Mont., 350......E10 102
Lamego, Port., 7,449......B2 20
Lameque, N.B., Can., 1,082..B5 74
La Mère et L'Enfant, mtn., Viet.F8 38
La Mesa, San Diego, Calif., 30,441......F5 82
La Mesa, Dona Ana, N. Mex., 25......E3 107
Lamesa, Dawson, Tex., 12,438......C2 118
Lameshur, Vir. Is......f16 65
Lamia, Grc., 21,509......C4 23
La Minerve, Que., Can., 344......C3 73
La Mirada, Los Angeles, Calif., 22,444......*F4 82
Lammermuir, hills, Scot......E5 13
Lamlash, Scot......E3 13
Lamock, see Nan Peng, is., China
Lamoil, isl., Truk......52
La Moille, Bureau, Ill., 655..B4 90
Lamoille, Elko, Nev., 100....C6 104
Lamoille, co., Vt., 11,027....B3 120
Lamoille, riv., Vt......B3 120
Lamoine, Shasta, Calif., 75......B2 82
Lamoine, Hancock, Maine (484▲)......D4 96
La Moine, riv., Ill......C3 90
Lamon, bay, Phil......C6, o13 35
Lamona, Lincoln, Wash., 20..B7 122
Lamoni, Decatur, Iowa, 2,173......D4 92
Lamont, Kern, Calif., 6,177..E4 82
Lamont, Alta., Can., 705....C4 69
Lamont, Jefferson, Fla., 250..B3 86
Lamont, Fremont, Idaho, 5..F7 89
Lamont, Buchanan, Iowa, 554......B6 92
Lamont, Greenwood, Kans., 80......D7 93
Lamont, Grant, Okla., 543..A4 112
Lamont, Whitman, Wash., 111......B8 122
Lamont, Carbon, Wyo., 100.C5 125
LaMonte, Pettis, Mo., 801...C4 97
La Motte, Jackson, Iowa, 323.B7 92
La Motte, isl., Vt......B2 120
La Moure, La Moure, N. Dak., 1,068......C7 110
La Moure, co., N. Dak., 8,705......C7 110
Lampa, Peru, 2,619......E3 58
Lampang, Thai., 35,460......C3 38
Lampasas, Lampasas, Tex., 5,061......D3 118
Lampasas, co., Tex., 9,418...D3 118
Lampedusa, isl., It......G13 30
Lampertheim, Ger., 19,200..D3 17
Lampeter, Wales, 1,853......B3 12
Lamphun, Thai., 9,862......C3 38
Lampman, Sask., Can., 637......H4 70
Lamu, Ken., 5,868......B7 48
La Mure, Fr., 6,111......E6 14
Lamy, Santa Fe, N. Mex., 195......B4, G6 107
L'Anacoco, bayou, La......C2 95
Lanagan, McDonald, Mo., 357......E3 101
Lanai, isl., Haw......C5 88
Lanai City, Maui, Haw., 2,056......C5 88
Lanak La, pass, China, India..B6 39
Lanao del Norte, prov., Phil., 280,500......*D6 35
Lanao del Sur, prov., Phil., 531,300......*D6 35
Lanark, Carroll, Ill., 1,473...A4 90
Lanark, Scot., 8,436......E5 13
Lanark, Raleigh, W. Va., 400......D3, D7 123
Lanark, co., Ont., Can., 40,313......B8 72
Lanark, co., Scot., 1,626,317.E4 13
Lancashire, co., Eng., 5,131,646......A5 12
Lancaster, Los Angeles, Calif., 26,012......E4 82
Lancaster, N.B., Can., 13,848......D3 74
Lancaster, Ont., Can., 584......B10 72
Lancaster, Eng., 48,887 (★95,000)......C5 10
Lancaster, Atchison, Kans., 196......C8 93
Lancaster, Garrard, Ky., 3,021......C5 94
Lancaster, Worcester, Mass., 750 (3,958▲)......B4 97
Lancaster, Kittson, Minn., 462......B2 99
Lancaster, Schuyler, Mo., 740......A5 101
Lancaster, Coos, N.H., 2,392 (3,138▲)......B3 105
Lancaster, Erie, N.Y., 12,254......C2 108
Lancaster, Fairfield, Ohio, 29,916......C5 111
Lancaster, Lancaster, Pa., 61,055 (★125,100)......F9 114
Lancaster, Lancaster, S.C., 7,999......B6 115
Lancaster, Smith, Tenn., 150......C8 117
Lancaster, Dallas, Tex., 7,501......B5 118
Lancaster, Lancaster, Va., 100......D6 121
Lancaster, Grant, Wis., 3,703......F3 124
Lancaster, co., Nebr., 155,272......D9 103
Lancaster, co., Pa., 278,359..G9 114
Lancaster, co., S.C., 39,352..B6 115
Lancaster, co., Va., 9,174...D6 121
Lancaster, sound, N.W. Ter., Can......B15 66
Lance, creek, Wyo......C8 125
Lance, plat., Newf., Can......E4 75
Lance Creek, Niobrara, Wyo., 500......B8 125
Lancer, Sask., Can., 207....G1 70
Lancer, Floyd, Ky., 500......C7 94
Lanchi, China, 28,000......J8 36
Lanchow, China, 699,000...D5 34
Lanciano, It., 13,300......C5 21
Lancing, Morgan, Tenn., 250......C9 117
Lancut, Pol., 9,106......C7 26
Landa, Bottineau, N. Dak., 110......A5 110
Landaff, Grafton, N.H., 100 (289▲)......B3 105
Lândana, Ang., 819......C1 48
Landau [an der Isar], Ger., 6,000......E7 17
Landau [in der Pfalz], Ger., 28,700......D3 17
Landeck, Aus., 6,514......E5 16
Landen, Bel., 4,970......D5 15
Lander, Fremont, Wyo., 4,182......C4 125
Lander, co., Nev., 1,566....C4 104
Landerneau, Fr., 11,834......C1 14
Landes, Grant, W. Va., 20..C5 123
Landes, dept., Fr., 260,495......*F3 14
Landes, heath, Fr......E3 14
Landess, Grant, Ind., 200...C6 91
Landing, Morris, N.J., 1,068......*B3 106
Landing, lake, Man., Can......B3 71
Landis, Sask., Can., 248......E1 70
Landis, Rowan, N.C., 1,763......B3 109
Landisburg, Perry, Pa., 285..F7 114
Landisville, Atlantic, N.J. (part of Buena)......D3 106
Landisville, Bucks, Pa., 1,690......*F11 114
Landivisiau, Fr., 4,121......C2 14
Lando, Chester, S.C., 732....B5 115
Land O'Lakes, Vilas, Wis., 500......B4 124
Landover Hills, Prince Georges, Md., 1,850......*C1 85
Landquart, Switz......C8 19
Landquart, riv., Switz......C8 19
Landrum, Spartanburg, S.C., 1,930......A3 115
Landsberg, Ger., 13,400....D5 16
Landsdown, Hunterdon, N.J., 30......B3 106
Land's End, cape, Eng......D2 12
Landshut, Ger., 49,500......E7 17
Landskrona, Swe., 28,900....C6 24
Landsman, creek, Colo......B8 83
Landsort, Swe......u35 25
Landusky, Phillips, Mont., 30......C8 102
Landwasser, riv., Switz......C8 19
Lane, DeWitt, Ill., 150......C5 90
Lane, Franklin, Kans., 282..D8 93
Lane, White Pines, Nev., 10.D7 104
Lane, Williamsburg, S.C., 658......D8 115
Lane, Jerauld, S. Dak., 99...C7 116
Lane, Dyer, Tenn., 65......A2 117
Lane, co., Kans., 3,060......D3 93
Lane, co., Oreg., 162,890....D4 113
Laneburg, Nevada, Ark., 125......D2 81
La Negra, Chile......D2 55
Lanesboro, Carroll, Iowa, 258......B3 92
Lanesboro, Berkshire, Mass., 400 (2,933▲)......A1 97
Lanesboro, Fillmore, Minn., 1,063......G7 99
Lanesboro, Susquehanna, Pa., 502......C10 114
Lanesville, Harrison, Ind., 346......H6 91
Lanett, Chambers, Ala., 7,674......C4 78
Lanfine, Alta., Can., 75......D5 69
Lanford, Laurens, S.C., 300..B4 115
Lang, Sask., Can., 258......H3 70
Langaa, Den., 2,119......B3 24
Langadhas, Grc., 7,590......B4 23
Langak, lake, China......B8 40
Langa Langa, Con. L......B2 48
Langbank, Sask., Can., 75...G4 70
Langchung, China, 21,000...I2 36
Langdale, Chambers, Ala., 2,528......C4 78
Langdale, dam, Ala., Ga......C4 78, D1 87
Langdon, Alta., Can., 98...D4 69
Langdon, Clay, Iowa, 60....A2 92
Langdon, Reno, Kans., 97...E5 93
Langdon, Atchison, Mo., 60..A2 101
Langdon, Sullivan, N.H., 80 (338▲)......D2 105
Langdon, Cavalier, N. Dak., 2,151......A7 110
Langeac, Fr., 4,826......E5 14
L'Ange Gardien, Que., Can., 1,543......C9 73
Langeland, isl., Den......D4 24
Langelands Belt, strait, Den....D4 24
Langell, valley, Oreg......E5 113
Langeloth, Washington, Pa., 1,112......F1 114
Langen, Ger., 20,000......D3 17
Langenau, Ger., 8,000......E5 17
Langenburg, Sask., Can., 757......G5 70
Langensalza, Ger., 16,200...B5 17
Langenselbold, Ger., 8,700..C4 17
Langenthal, Switz., 10,974...B4 19
Langenzenn, Ger., 4,800......D5 17
Langeoog, Ger., 2,200......A7 15
Langeoog, isl., Ger......B3 16
Langerak, riv., Den......A4 24
Langesund, Nor., 2,200.H3, p27 25
Langevin, Que., Can., 540...C7 73
Langford, Marshall, S. Dak., 397......B8 116
Langford Station, B.C., Can., 1,204......B9 68
Långhalsen, lake, Swe......u34 25
Langham, Sask., Can., 429..E2 70
Länghem, Swe., 494......A7 24
Langholm, Scot., 2,369......E5 13
Langhorne, Bucks, Pa., 1,924......F12 114
Langhorne Manor, Bucks, Pa., 1,001......*F12 114
Långhundra, Swe......t36 25
Langjokull, glacier, Ice......n22 25
Langkawi, isl., Mala......I3 38
Langkha Tuk, mtn., Thai......H3 38
Langlade, co., Wis., 19,916...C4 127
Langley, Pike, Ark., 25......C2 81
Langley, B.C., Can., 2,365..B10 68
Langley, Floyd, Ky., 700....C7 94
Langley, Mayes, Okla., 205..A6 112
Langley, Aiken. S.C., 1,216..D4 115
Langley, Island, Wash., 448..A3 122
Langley Park, Prince Georges, Md., 11,510......*C4 85
Langleyville, Christian, Ill., 300......D4 90
Langlois, Curry, Oreg., 300..E2 113
Langnau in Emmental, Switz., 9,201......C4 19
Langogne, Fr., 4,184......E5 14
Langon, Fr., 4,197......E3 14
Langöy, isl., Nor......C6 25
Langreo, Sp., 7,138......A3 20
Langres, Fr., 9,577......D6 14
Langres, plat., Fr......D6 14
Langruth, Man., Can., 236...D2 71
Langsa, Indon., 4,749..E1, m11 35
Lang Son, Viet., 25,000......B7 38
Langstaff, Ont., Can., 1,153..E6 72
Langston, Jackson, Ala., 100.A3 78
Langston, Logan, Okla., 136..B4 112
Langton, Ont., Can., 379....E4 72
Langtry, Val Verde, Tex., 100......E2 118
Languedoc, former prov., Fr., 2,413,000......F5 14
L'Anguille. riv., Ark......B5 81
Lanham, Prince Georges, Md., 2,500......C4 85
Lanham, Gage, Nebr., 50...D9 103
Lanhsi, China, 5,000......C3 37
Lanier, co., Ga., 5,097......E3 87
Lanigan, Sask., Can., 516....F3 70
Lanigan, creek, Sask., Can......F3 70
Lanín, vol., Arg......B2 54
Lankin, Walsh, N. Dak., 303.A8 110
Lanklaar, Bel., 2,232......C5 15
Lannion, Fr., 6,497......C2 14
Lannon, Waukesha, Wis., 1,084......E1 124
L'Annonciation, Que., Can., 1,042......C3 73
Lanoka Harbor, Ocean, N.J., 100......D4 106
Lanoraie, Que., Can., 1,060.D4 73
Lansdale, Montgomery, Pa., 12,612......F11 114
Lansdowne, Ont., Can., 522......C8, E8 72
Lansdowne, Baltimore, Md., 13,134......B4 85
Lansdowne, Delaware, Pa., 12,601......B11, G11 114
L'Anse, Baraga, Mich., 2,397.B2 98
L'Anse, Indian res., Mich......B2 98
L'Anse Saint Jean, Que., Can., 265......A7 73
Lansford, Bottineau, N. Dak., 382......A4 110
Lansford, Carbon, Pa., 5,958.E10 114
Lansing, Ont., Can., 4,000...E6 72
Lansing, Cook, Ill., 19,366......B6, F3 90
Lansing, Allamakee, Iowa, 1,325......A6 92
Lansing, Leavenworth, Kans., 1,264......B8 93
Lansing, Ingham, Mich., 107,807 (★209,100)......F6 98
Lansing, Ashe, N.C., 278....A2 109
Lansing, Belmont, Ohio, 1,200......B1 123
Lansing, Fayette, W. Va., 500......D7 123
Lanslebourg, Fr., 570......D2 18
Lanta, isl., Thai......I3 38
Lantana, Palm Beach, Fla., 5,021......F6 86
Lantana, Cumberland, Tenn D8 117
Lantern, hill, Conn......D9 84
Lantry, Dewey, S. Dak., 22..B4 116
Lantsang, China......G4 34
Lantz, Barbour, W. Va., 30..C4 123
Lanús, Arg., 381,561......*g7 54
Lanusei, It., 5,449......E2 21
Lanuvio, It., 3,486......h9 21
Lanzarote, isl., Can. Is......D2 44
Laoag, Phil., 26,000 (50,100▲)......B6 35
Laoighis, co., Ire., 45,069...D4 11
Lao Kay, Viet., 25,000......A6 38
Laon, Fr., 25,078......C5 14
Laona, Forest, Wis., 950......C5 124
La Oroya, Peru, 17,076......D2 58
Laos, country, Asia, 1,200,000......B3 38
Laotto, Noble, Ind., 300......B7 91
Lap, isl., Truk......52
Lapa, Braz., 7,167......D3 56
La Palma, Pinal, Ariz., 30...E4 80
La Palma, Pan., 1,885......F8 62
La Palma, Sp., 8,669......D2 20
La Pampa, prov., Arg., 158,489......B3 54
La Paragua, Ven., 686......B5 60
La Passe, Ont., Can., 113....B8 72
La Patrie, Que., Can., 519..D6 73
La Paz, Arg., 15,006......A5 54
La Paz., Arg......A3 54
La Paz, Bol., 321,073......C2 55
La Paz, Col......A3 60
La Paz, Hond., 3,681......C4 62
Lapaz, Marshall, Ind., 545...B5 91
La Paz, Mex., 23,324......C2 63
La Paz, Mex., 3,330......C4 63
La Paz, dept., Bol., 948,446..C2 55
La Pedrera, Col......D4 60
Lapeer, Lapeer, Mich., 6,160......E7 98
Lapeer, co., Mich., 41,926...E7 98
Lapel, Madison, Ind., 1,772..D6 91
La Petite Riviere St. Francois, Que., Can., 4,707......B7 73
Lapi (Lappi), dept., Fin., 152,607......*D11 25
La Pica, P.R......C3 65
La Pice, St. James, La......B5 95
La Piedad, Mex., 24,306....m12 63
Lapine, Montgomery, Ala., 400......D3 78
La Pine, Deschutes, Oreg., 450......D5 113
Laplace, St. John the Baptist, La., 3,541......B6 95
Lapland, reg., Eur......C12 25
La Plant, Dewey, S. Dak., 40.B5 116
La Plata, Arg., 295,000 (★410,000)......A5, g8 54
La Plata, Col., 2,416......C2 60
La Plata, Charles, Md., 1,214.C4 85
La Plata, Macon, Mo., 1,365......A5 101
Laplata, San Juan, N. Mex., 10......A1 107
La Plata, co., Colo., 19,225..D3 83
La Plata, mts., Colo......D2 83
La Plata, peak, Colo......B4 83
La Plata, riv., Arg......B5 54
La Platte, Sarpy, Nebr., 125..E3 103
La Platte, riv., Vt......C2 120
La Playa, P.R......B4 65
La Plonge, lake, Sask., Can......B2 70
La Pobla de Lillet, Sp., 2,732......A6 20
La Poile, bay, Newf., Can......E2 75
Lapoint, Uintah, Utah, 25...C6 119
La Pointe, Ashland, Wis., 300......B3 124
Lapointe Station, Que., Can., 170......B8 73
La Pola, Sp., 5,430......A3 20
La Porte, Plumas, Calif., 50......C3 82
Laporte, Sask., Can., 78......F1 70
Laporte, Larimer, Colo., 300......A5 83
La Porte, La Porte, Ind., 21,157......A4 95
Laporte, Hubbard, Minn., 155......C4 99
Laporte, Sullivan, Pa., 195..D9 114
La Porte, Harris, Tex., 4,512.F5 118
La Porte, co., Ind., 95,111...A4 91
La Porte City, Black Hawk, Iowa, 1,953......B5 92
Läppe, Swe......t33 25
Lappeenranta, Fin., 21,600.G13 25
Lappi (Lapi), dept., Fin., 152,607......*D11 25
Laprairie, Que., Can., 7,328......D4 73
La Prairie, Itasca, Minn., 231......C5 99
Laprairie, co., Que., Can., 31,157......D4 73
La Presa, San Diego, Calif., 1,300......*E2 82
Laprida, Arg., 3,261......B4 54
La Providence, Que., Can., 4,251......*D5 63
La Pryor, Zavala, Tex., 900......E3 118
Lâseki, Tur., 3,100......B6 31
Laptev, sea, Sov. Un......B15 28
La Puebla, Sp., 9,931......C7 20
La Puebla de Montalbán, Sp., 7,700......C3 20
La Puente, Los Angeles, Calif., 24,723......*F3 82
La Puente, Rio Arriba, N. Mex., 150......A3 107
La Purísima, Mex., 557......B2 63
Lapush, Clallam, Wash., 250......B1 122
Lapwai, Nez Perce, Idaho, 500......C2 89
Laqiya Arba'in (Well), Sud......B2 47
La Quiaca, Arg., 6,768......D2 55
L'Aquila, It., 36,000 (56,019▲)......C4 21
Lār, Iran, 13,010......H7 41
Lara, state, Ven., 489,140...A3 60
Larache, Mor., 30,763......B3 47, G3 28
Laragh, Ire......D5 11
Larak, isl., Iran......H8 41
Laramie, Albany, Wyo., 17,520......D7 125
Laramie, co., Wyo., 60,149..D8 125
Laramie, basin, Wyo......C6 125
Laramie, mts., Colo., Wyo......A5 83, C7 125
Laramie, peak, Wyo......C7 125
Laramie, riv., Wyo......D7 125
La Rangu, Sud., 2,971......E2 47
Laranjeiras, Braz., 4,296....D7 57
Laranjeiras do Sul, Braz., 3,802......D2 56
Larantuka, Indon......G6 35
Larch, riv., Que., Can......A2 73
Larche, pass, Fr., It......E2 18
Larchmont, Westchester, N.Y., 6,789......D2 108
Larchwood, Lyon, Iowa, 531.A1 92
Lardeau, B.C., Can., 200....D9 68
Lardo, Valley, Idaho, 75....E2 89
L'Ardoise, N.S., Can., 275...D9 74
Laredo, Grundy, Mo., 370....A4 101
Laredo, Hill, Mont., 10......B7 102
Laredo, Sp., 6,206......A4 20
Laredo, Webb, Tex., 60,678 (★152,500)......F3 118
Laredo, sound, B.C., Can......C3 68
La Réole, Fr., 3,264......E4 14
Lares, P.R., 4,216......B3 65
Lares, mun., P.R., 26,922....B3 65
Largeau, Chad......B3 46
Largentière, Fr., 1,284......E6 14
Largo, Pinellas, Fla., 5,302......E1, E4 86
Largo, canyon, N. Mex......A2 107
Largo, key, Fla......G6 86
Largs, Scot., 9,100......E4 13
Lariat, Rio Grande, Colo., 900......D4 83
Lariat, Parmer, Tex., 80.....B1 118
La Ricamarie, Fr., 11,902....E6 14
Larimer, Westmoreland, Pa., 1,500......*F2 114
Larimer, co., Colo., 53,343..A5 83
Larimore, Grand Forks, N. Dak., 1,714......B8 110
Larino, It., 8,678......D5 21
La Rioja, Arg., 23,809......E2 55
La Rioja, Cuba......E5 64
La Rioja, prov., Arg., 128,270......E2 55
Lárisa (Larissa), Grc., 55,391.C4 23
Larisa (Larissa), prov., Grc., 237,776......*C4 23
La Riviere, Man., Can., 249......E2 71
Larkana, Pak., 33,414......D2 40
Lark Harbour, Newf., Can., 335......D2 75
Larkinburg, Jackson, Kans., 51......A7 93
Larkinsville, Jackson, Ala., 250......A3 78
Larkspur, Marin, Calif., 5,710......B5 82
Larkspur, Douglas, Colo., 175......B6 83
Larksville, Luzerne, Pa., 4,390......B8 114
Larnaca, Cyp., 17,867......E9 31
Larne, N. Ire., 16,341......C6 11
Larned, Pawnee, Kans., 5,001......D4 93
La Robla, Sp., 2,801......A3 20
La Rochelle, Fr., 66,590....D3 14
La Roche-sur-Foron, Fr., 4,025......C2 18
La Roche-sur-Yon, Fr., 24,019......D3 14
La Roda, Sp., 12,190......C4 20
La Romana, Dom. Rep., 24,058......F9 64
LaRonge, Sask., Can., 692...B3 70
La Ronge, lake, Sask., Can......B3 70
Larose, Lafourche, La., 2,796.E5 95
Larrabee, Cherokee, Iowa, 167......B2 92
Larrabee, Washington, Maine, 125......D5 96
Larrys River, N.S., Can., 239......D8 74
Larsen, ice shelf, Ant......C6 5
Larsen Bay, Alsk., 72......D9 79
Larslan, Valley, Mont., 5...B10 102
Larsmont, Lake, Minn., 203.D7 99
Larson, Burke, N. Dak., 62..A3 110
Larto, Catahoula, La., 125..C4 95
La Rue, Marion, Ohio, 842..B4 111
Larue, co., Ky., 10,346......C4 94
Laruns, Fr., 1,268......F3 14
Larvik, Nor., 10,600 (★18,500)......H4, p28 25
Laryak, Sov. Un......C25 9
La Sal, San Juan, Utah, 200..E6 119
La Sal, mts., Utah......E6 119
La Salle, Man., Can., 128....E3 71
La Salle, Que., Can., 30,904......*D9 73
La Salle, Weld, Colo., 1,070..A6 83
La Salle, LaSalle, Ill., 11,897.B4 90
La Salle, Watonwan, Minn., 147......F4 99
La Salle, Flathead, Mont., 50.B2 102
La Salle, co., Ill., 110,800....B5 90
La Salle, co., Tex., 5,972....E3 118
La Salle, par., La., 13,011...C3 95
Las Animas, Bent, Colo., 3,402......C7 83
Las Animas, co., Colo., 19,983......D6 83
Las Anod, Som......D6 47
Las Anod, dist., Som......D6 47
Las Arenas, P.R......C2 65
La Sarre, Que., Can., 3,944..E9 72
Lasauses, Conejos, Colo., 200.D5 83
Lasberg, Aus., 2,190......E9 17
Las Cabras, Chile, 1,032....A2 54
Lascahobas, Hai., 2,191......E8 64
Lascano, Ur., 3,731......E2 56
Lascassas, Rutherford, Tenn., 100......B5 117
L'Ascension, Que., Can., 1,197......A6 73
La Scie, Newf., Can., 939....D4 75
Las Coloradas, Arg......B2 54
Las Conchas, Arg., 24,809....g7 54
Las Cordilleras, dept., Par., 150,716......*E4 55
Las Cruces, Dona Ana, N. Mex., 29,367......E3 107
La Serena, Chile, 40,000....E1 55
La Seyne [-sur-Mer], Fr., 33,570......F6 14
Las Flores, Arg., 8,955......B5 54
Las Flores, P.R......C5 65
Lashburn, Sask., Can., 475..D1 70
Lashio, Bur., 4,638......D10 39
Lashkar, see Gwalior, India
La Sierra, Riverside, Calif., 7,000......F3 82
Lasithi (Lasithion), prov., Grc., 73,880......*E5 23
Lask, Pol., 3,819......C5 26
Las Lajas, Arg......B2 54
Las Lomitas, Arg......D3 55
Las Mareas, P.R......D6 65
Las Marías, P.R., 511......B3 65
Las Marías, mun., P.R., 9,237......C3 65
Las Matas de Farfan, Dom. Rep., 3,585......E8 64
Las Nutrias, Socorro, N. Mex., 149......C3 107
La Solana, Sp., 14,948......C4 20
Las Palmas, Pan., 753......F7 62
Las Palmas, P.R......C2 65
Las Palmas, P.R......D6 65
Las Palmas [de Gran Canaria], Can. Is., 193,862......D1 44
Las Palmas, prov., Sp., 453,793......*m14 20
Las Palomas, Sierra, N. Mex., 14......D2 107
La Spezia, It., 121,923 (★185,000)......B2 21
Las Piedras, P.R., 3,147.....C7 65
Las Piedras, Ur., 15,400.....E1 56
Las Piedras, mun., P.R., 17,047......C7 65
Las Piñas, P.R......B7 65
Las Pipinas, Arg......B5 54
Las Plumas, Arg......C3 54
Las Rosas, Arg., 6,153......A4 54
Las Rozas de Madrid, Sp., 1,196......D6, p17 20
Lassen, co., Calif., 13,597...B3 82
Lassen, peak, Calif......B3 82
Lassen Volcanic, nat. park, Calif.B3 82
Lasso, mtn., Tinian......52
L'Assomption, Que., Can., 4,448......D4 73
L'Assomption, co., Que., Can., 39,440......D4 73
Last, mtn., Sask., Can......F3 70
Las Tablas, Rio Arriba, N. Mex., 100......A3 107
Las Tablas, Pan., 3,504......G7 62
Last Chance, creek, Utah......F4 119
Las Termas, Arg., 4,699......E5 55
Last Mountain, lake, Sask., Can.F3 70
Lastourville, Gabon......F2 46
Lastrup, Ger., 4,900......B7 15
Lastrup, Morrison, Minn., 138......D4 99
Las Varillas, Arg., 5,950......A4 54
Las Vegas, Clark, Nev., 64,405 (★119,300)......G6 104
Las Vegas (city), San Miguel, N. Mex., 7,790......B1 107
Las Vegas (town), San Miguel, N. Mex., 6,028......B4 107
Las Vegas, P.R......C2 65
Las Vigas, Mex., 4,857......n15 63
Las Villas, prov., Cuba, 1,030,162......D4 64
La Tabatiere, Que., Can., 289......C2 75
Latacunga, Ec., 10,340......B2 58
La Tagua, Col......D3 60
Latah, Spokane, Wash., 190..B8 122
Latah, co., Idaho, 21,170....C2 89
Latah, creek, Wash......B8 122
Laterrière, Que., Can., 651...A6 73
La Teste-de-Buch, Fr., 8,907.E3 14
Latexo, Houston, Tex., 500..D5 118
Latham, Baldwin, Ala., 25...D2 78
Latham, Logan, Ill., 389....D4 90
Latham, Butler, Kans., 203..E7 93
Latham, Albany, N.Y., 5,000......*C7 108
Latheron, Scot., 2,593......B5 13
Lathrop, San Joaquin, Calif., 1,123......B6 82
Lathrop, Delta, Mich., 25....B3 98
Lathrop, Clinton, Mo., 1,006.B3 101
Lathrop Wells, Nye, Nev., 30......G5 104
Lathrup Village, Oakland, Mich., 3,556......*A7 98
Latimer, Franklin, Iowa, 445......B4 92
Latimer, Morris, Kans., 40..D7 93
Latimer, co., Okla., 7,738...C6 112
Latina, It., 24,500 (49,331▲)......D4, k9 21
Latium, reg., It......C4 21
La Toma, Ec......B2 58
Laton, Fresno, Calif., 1,052..D4 82
Latour, Johnson, Mo., 68....C3 101
La Tour-du-Pin, Fr., 4,694..E6 14
La Tremblade, Fr., 3,514....E3 14
Latrobe, Westmoreland, Pa., 11,932......F3 114
Latrūn, Jordan, 1,000..C6, h10 32
Latta, Dillon, S.C., 1,901....C9 115
Lattimer Mines, Luzerne, Pa., 600......E10 114
Lattimore, Cleveland, N.C., 257......B2 109
Lattingtown, Nassau, N.Y., 1,461......*F2 84
Latty, Paulding, Ohio, 286..A3 111
Latuda, Carbon, Utah, 75...D5 119
La Tuque, Que., Can., 13,023......B5 73
Latvia (S.S.R.), rep., Sov. Un., 2,142,000......D5 28
Lau, Nig......E7 45
Lau, is., Fiji Is......52
Lauchhammer, Ger., 28,000.B8 17
Laud, Whitley, Ind., 150....B7 91
Lauda, Ger., 4,800......D4 17
Lauder, Man., Can., 54......E1 71
Lauder, Scot., 597......E6 13
Lauderdale, St. James, La...B5 95
Lauderdale, Ramsey, Minn., 1,676......E7 99
Lauderdale, Lauderdale, Miss., 500......C5 100
Lauderdale, co., Ala., 61,622......A2 78
Lauderdale, co., Miss., 67,119......C5 100
Lauderdale, co., Tenn., 21,844......B2 117
Lauderdale-by-the-Sea, Broward, Fla., 1,327......*F6 86
Lauenburg, Ger., 10,700......E4 24
Lauf, Ger., 12,900......D6 17
Laufen, Ger., 3,600......B8 18
Laufen, Switz., 3,955......B4 19
Lauffen, Ger., 8,500......D4 17
Laughery, creek, Ind......F7 91
Laughlin, peak, N. Mex......A5 107
Laughlintown, Westmoreland, Pa., 500......F3 114
Lauingen, Ger., 8,500......E5 17
Laulau, Saipan......52
Laulii, Am. Sam., 393......52
Launceston, Austl., 38,118 (★56,721)......o15 50
Launceston, Eng., 4,518....D3 12
La Unión, Arg......B3 54
La Unión, Chile, 9,830......C2 54
La Unión, Col., 2,796......C2 60
La Unión, Mex., 1,336......D4, o13 63
La Union, Dona Ana, N. Mex., 200......F3 107
La Unión, Peru, 1,672......C2 58
La Unión, Sal., 10,687......D4 62
La Unión, Sp., 11,687......D5 20
La Union, prov., Phil., 295,200......*B6 35
Laupahoehoe, Hawaii, Haw., 500......D6 89
Laupheim, Ger., 8,700......D4 16
Laura, Austl., 44......C7 50
Laura, Sask., Can., 68......F2 70
Laura, Miami, Ohio, 526......C3 111
Laurbjerg, Den., 815......B3 24
Laurel, Ont., Can., 74......D4 72
Laurel, Sussex, Del., 2,709...C6 85
Laurel, Sarasota, Fla., 1,000......E4 86
Laurel, Franklin, Ind., 848..E7 91
Laurel, Marshall, Iowa, 223......C5 92
Laurel, Prince Georges, Md., 8,503......B4 85
Laurel, Jones, Miss., 27,889..D4 100
Laurel, Yellowstone, Mont., 4,601......E8 102
Laurel, Cedar, Nebr., 922...B8 103
Laurel, Suffolk, N.Y., 150...F6 84
Laurel, Washington, Oreg....B1 113
Laurel, Henrico, Va., 500...D3 121
Laurel, Whatcom, Wash....A3 122
Laurel, Klickitat, Wash., 5...D4 122
Laurel, co., Ky., 24,901.....C5 94
Laurel, creek, W. Va......D6 123
Laurel, creek, W. Va......D7 123
Laurel, fork, W. Va......C5 123
Laurel, ridge, W. Va......B5 123
Laurel, riv., Del......C6 85
Laurel, riv., Ky......D5 94
Laurel Bloomery, Johnson, Tenn., 100......C12 117
Laureldale, Atlantic, N.J., 400......D3 106
Laureldale, Berks, Pa., 4,051......F10 114
Laureles, Ur......E1 56
Laurel Fork, Carroll, Va., 25......E2 121
Laurel Gardens, Alleghany, Pa., 1,500......*B6 114
Laurel Heights, Snohomish, Wash., 900......*B3 122
Laurel Hill, Okaloosa, Fla., 411......G2 86

Laurel Hill, Scotland, N.C., 900 C4 109
Laurel Run, Luzerne, Pa., 855 B8 114
Laurel Springs, Camden, N.J., 2,028 *D3 106
Laurelton, Union, Pa., 400 E7 114
Laurelville, Hocking, Ohio, 539 C5 111
Laurence Harbor, Middlesex, N.J., 3,000 C4 106
Laurencekirk, Scot., 1,389 D6 13
Laurens, Pocahontas, Iowa, 1,799 B3 92
Laurens, Laurens, S.C., 9,598 B3 115
Laurens, co., Ga., 32,313 D4 87
Laurens, co., S.C., 47,609 C4 115
Laurentides, Que., Can., 1,698 D4 73
Laurentides, park, Que., Can B6 73
Lauria Inferiore, It., 4,530 D5 21
Laurie, lake, Man., Can A1 71
Laurie, riv., Man., Can A1 71
Laurier, Man., Can., 262 D2 71
Laurier, mtn., B.C., Can A6 68
Laurierville, Que., Can., 872 C6 73
Laurin, Madison, Mont., 50 E2 102
Laurinburg, Scotland, N.C., 8,242 C4 109
Laurium, Houghton, Mich., 3,058 A2 98
Laurot, is., Indon F5 35
Laurus, lake, Ont., Can D4 71
Lausanne, Switz., 126,328 (★173,000) C2 19
Lauscha, Ger., 6,506 C6 17
Lausitzer, mts., Czech., Ger C9 17
Laut, isl., Indon F5 35
Lautaro, Chile, 9,255 B2 54
Lauterbach, Ger., 9,800 C4 17
Lauterbrunnen, Switz., 3,216 C4 19
Lauterecken, Ger., 2,700 D2 17
Lauthala, is., Fiji Is 52
Lauzon, Que., Can., 11,533 C6, C9 73
Lauzon, lake, Ont., Can B8 98
Lava, Socorro, N. Mex., 4 D3 107
Lava, flow, N. Mex B2 107
Lava Bed, Idaho F4, F5, F6 89
Lava Beds, Nev C3 104
Lava Beds, nat. mon., Calif B3 82
Lavaca, Choctaw, Ala., 100 C1 78
Lavaca, Sebastian, Ark., 502 B1 81
Lavaca, co., Tex., 20,174 E4 118
Lava Hot Springs, Bannock, Idaho, 593 G6 89
Laval, Fr., 39,283 C3 14
Laval, co., Que., Can., 124,741 D4 73
La Vale, Allegany, Md., 4,031 *D2 85
Laval [des Rapides], Que., Can., 19,227 D8 73
Lavalle Arg A3 54
La Valle, Sauk, Wis., 417 E3 124
Lavalleja, dept., Ur., 62,200 *E1 56
Lavallette, Ocean, N.J., 832 D4 106
Lavalley, Costilla, Colo., 50 D5 83
Lavaltrie, Que., Can., 1,034 D4 73
Lavant, riv., Aus E7 16
Lavant Station, Ont., Can., 59 B8 72
Lavaur, Fr., 4,137 F4 14
Lavaveix-les-Mines, Fr., 1,061 D5 14
Laveaga, peak, Calif C6 82
Laveen, Maricopa, Ariz., 150 G2 80
La Vega, Dom. Rep., 19,884 F8 64
La Vela, Ven., 3,967 A4 60
Lavelanet, Fr., 7,648 F4 14
Lavello, It., 13,745 D5 21
La Ventana, Sandoval, N. Mex B3 107
La Vergne, Rutherford, Tenn., 800 A5, E9 117
La Verkin, Washington, Utah, 365 F2 119
La Verne, Los Angeles, Calif., 6,516 F3 82
Laverne, Harper, Okla., 1,937 A2 112
La Vernia, Wilson, Tex., 600 B4, E3 118
Laverton, Austl., 153 E3 50
La Veta, Huerfano, Colo., 632 D5 83
Laviana, Sp., 12,455 A3 20
La Victoria, Ven., 23,126 A4 60
Lavielle, lake, Ont., Can B6 72
La Villa, Hidalgo, Tex., 1,261 *F4 118
Lavina, Golden Valley, Mont., 212 D8 102
Lavinia, Carroll, Tenn., 100 B3 117
La Vista, DeKalb, Ga., 3,000 *C2 87
Lavon, Collin, Tex., 130 A6 118
Lavon, flood control res., Tex A6 118
Lavonia, Franklin, Ga., 2,088 B3 87
Lavoy, Alta., Can., 131 C5 69
Lavras, Braz., 23,793 C3 56
Lavras da Mangabeira, Braz., 2,835 C3 57
Lavras do Sul, Braz., 3,302 E2 56
Lavrion, Grc., 6,842 D5 23
Lawagan, Bis. Arch h13 50
Lawai, Kauai, Haw., 145 B2 88
Lawen, Harney, Oreg., 15 D8 113
Lawers, mtn., Scot D4 13
Lawford, lake, Man., Can B3 71
Lawit, mtn., Mala J5 38
Lawler, Chickasaw, Iowa, 532 A5 92
Lawler, Aitkin, Minn., 75 D5 99
Lawn, Newf., Can., 716 E4 75
Lawn, Lebanon, Pa., 175 F8 114
Lawn, Taylor, Tex., 310 C3 118
Lawndale, Los Angeles, Calif., 21,740 *F2 82
Lawndale, Wilkin, Minn., 20 D2 99
Lawndale, Cleveland, N.C., 723 B2 109
Lawnside, Camden, N.J., 2,155 *D2 106
Lawnton, Dauphin, Pa., 3,500 *F8 114
Lawra, Ghana D4 45
Lawrence, Marion, Ind., 10,103 E5, H8 91
Lawrence, Douglas, Kans., 32,858 B7, D8 93
Lawrence, Essex, Mass., 70,933 (★196,500) A5 97
Lawrence, Van Buren, Mich., 773 F4 98
Lawrence, Newton, Miss., 250 C4 100
Lawrence, Nuckolls, Nebr., 338 D7 103
Lawrence, Nassau, N.Y., 5,907 G2 84
Lawrence, N.Z., 638 P12 51
Lawrence, Pontotoc, Okla., 100 C5 112
Lawrence, Washington, Pa., 1,048 F1 114
Lawrence, co., Ala., 24,501 A2 78
Lawrence, co., Ark., 17,267 A4 81
Lawrence, co., Ill., 18,540 E6 90
Lawrence, co., Ind., 36,564 G4 91
Lawrence, co., Ky., 12,134 B7 94
Lawrence, co., Miss., 10,215 D3 100
Lawrence, co., Mo., 23,260 D4 101
Lawrence, co., Ohio, 55,438 D5 111
Lawrence, co., Pa., 112,965 D1 114
Lawrence, co., S. Dak., 17,075 C2 116
Lawrence, co., Tenn., 28,049 B4 117
Lawrence, lake, Ont., Can A5 99
Lawrenceburg, Dearborn, Ind., 5,004 F8 91
Lawrenceburg, Anderson, Ky., 2,523 B5 94
Lawrenceburg, Lawrence, Tenn., 8,042 B4 117
Lawrence Park, Erie, Pa., 4,403 *B1 114
Lawrenceport, Lawrence, Ind., 200 G5 91
Lawrence Station, N.B., Can., 210 D2 74
Lawrencetown, N.S., Can., 509 E4 74
Lawrenceville, Que., Can., 430 D5 73
Lawrenceville, Gwinnett, Ga., 3,804 A6, C3 87
Lawrenceville, Lawrence, Ill., 5,492 E6 90
Lawrenceville, Mercer, N.J., 2,000 C3 106
Lawrenceville, St. Lawrence, N.Y., 150 A2 108
Lawrenceville, Tioga, Pa., 548 C7 114
Lawrenceville, Brunswick, Va., 1,941 E5 121
Laws, Inyo, Calif., 75 D4 82
Lawshe, Adams, Ohio, 100 D4 111
Lawson, Union, Ark., 200 D3 81
Lawson, Sask., Can., 72 G2 70
Lawson, Ray, Mo., 778 B3 101
Lawsonville, Stokes, N.C., 70 A3 109
Lawtell, St. Landry, La., 500 D3 95
Lawtey, Bradford, Fla., 623 B4 86
Lawton, Cuba h12 64
Lawton, Woodbury, Iowa, 324 B1 92
Lawton, Carter, Ky., 300 B6 94
Lawton, Van Buren, Mich., 1,402 F5 98
Lawton, Ramsey, N. Dak., 159 A7 110
Lawton, Comanche, Okla., 61,697 C3 112
Lawton, Fayette, W.Va., 500 D4, D7 123
Lawver, Campbell, Wyo B7 125
Lawyersville, Schoharie, N.Y., 100 C6 108
Laxford, bay, Scot B3 13
Lay, Moffat, Colo., 15 A3 83
Lay, cape, Viet D7 38
Lay, dam, Ala D3 78
Lay, lake, Ala C3 78
Layland, Fayette, W. Va., 400 D7 123
Layman, Montgomery, Va D2 121
Laysan, isl., Haw k13 88
Layton, Sussex, N.J., 200 A3 106
Layton, Davis, Utah, 9,027 B4 119
Laytonsville, Montgomery, Md., 196 B3 85
Lazdijai, Sov. Un A7 26
Lazear, Delta, Colo., 75 C3 83
Lazio, pol. dist., It., 3,958,957 *D4 21
Lazy Lake, Broward, Fla., 49 *F6 86
Lea, Lea, N. Mex., 50 E6 107
Lea, co., N. Mex., 53,429 D6 107
Leach, Delaware, Okla., 35 A7 112
Leachville, Mississippi, Ark., 1,507 B5 81
Leacross, Sask., Can., 75 D3 70
Lead, Lawrence, S. Dak., 6,211 C2 116
Leadbetter, pt., Wash C1 122
Leader, Sask., Can., 1,211 G1 70
Leader, riv., Scot E6 13
Lead Hill, Boone, Ark., 102 A3 81
Leadhills, Scot., 1,362 E5 13
Lead Mountain, ponds, Maine D5 96
Leadore, Lemhi, Idaho, 141 E5 89
Leadpoint, Stevens, Wash., 25 A8 122
Leadville, Lake, Colo., 4,008 B4 83
Leadwood, St. Francois, Mo., 1,343 D7 101
Leaf, Greene, Miss., 350 D5 100
Leaf, lake, Sask., Can D4 70
Leaf, riv., Que., Can A2 73
Leaf, riv., Miss D4 100
Leaf River, Ogle, Ill., 546 A4 90
League City, Galveston, Tex., 2,622 *F5 118
Leah, Columbia, Ga., 300 C4 87
Leake, co., Miss., 18,660 C4 100
Leaksville, Greene, Miss., 1,014 D5 100
Leakey, Real, Tex., 587 E3 118
Leaksville, Rockingham, N.C., 6,427 A4 109
Leal, Barnes, N. Dak., 70 A7 110
Lealman, Pinellas, Fla., 13,000 E2 86
Leamington, Ont., Can., 9,030 E2 72
Leamington, Eng., 43,236 B6 12
Leamington, Millard, Utah, 190 D3 119
Leane, lake, Ire E2 11
Leapwood, McNairy, Tenn., 145 B3 117
Learned, Hinds, Miss., 96 C3 100
Leary, Calhoun, Ga., 848 E2 87
Leasburg, Crawford, Mo., 176 C6 101
Leasburg, Caswell, N.C., 275 A4 109
Leaside, Ont., Can., 18,579 E6 72
Leask, Sask., Can., 499 E2 70
Leatherman, Macon, N.C. D3 109
Leatherwood, Perry, Ky., 1,283 C6 94
Leatherwood, creek, W. Va C7 123
Leavenworth, Crawford, Ind., 387 H5 91
Leavenworth, Leavenworth, Kans., 22,052 B8, C9 93
Leavenworth, Chelan, Wash., 1,480 B5 122
Leavenworth, co., Kans., 48,524 C8 93
Leavittsburg, Trumbull, Ohio, 3,300 A7 111
Leawood, Johnson, Kans., 7,466 D9 93
Leba, Pol., 3,021 A4 26
Lebam, Pacific, Wash., 400 C2 122
Lebanon, New London, Conn., 300 (2,434▲) C7 84
Lebanon, Kent, Del., 110 B6 85
Lebanon, Levy, Fla., 15 C4 86
Lebanon, St. Clair, Ill., 2,863 E4 90
Lebanon, Boone, Ind., 9,593 D5 91
Lebanon, Smith, Kans., 583 C5 93
Lebanon, Marion, Ky., 4,813 C4 94
Lebanon, Laclede, Mo., 8,220 D5 101
Lebanon, Red Willow, Nebr., 143 D5 103
Lebanon, Grafton, N.H., 9,299 C2 105
Lebanon, Hunterdon, N.J., 880 B3 106
Lebanon, Warren, Ohio, 5,993 C3 111
Lebanon, Marshall, Okla., 100 D5 112
Lebanon, Linn, Oreg., 5,858 C4, D2 113
Lebanon, Lebanon, Pa., 30,045 F9 114
Lebanon, Potter, S. Dak., 198 B6 116
Lebanon, Wilson, Tenn., 10,512 A5 117
Lebanon, Russell, Va., 2,085 B3 121
Lebanon, country, Asia, 1,303,940 E10 31, F6 33
Lebanon, co., Pa., 90,853 F9 114
Lebanon Junction, Bullitt, Ky., 1,527 C4 94
Lebec, Kern, Calif., 400 E4 82
Lebedin, Sov. Un., 10,000 F10 27
Lebedyan, Sov. Un., 10,000 E12 27
Le Blanc, Fr., 5,279 D4 14
Le Blanc-Mesnil, Fr., 35,708 g10 14
Lebo, Con. L A3 48
Lebo, Coffey, Kans., 498 D8 93
Lebo, Meagher, Mont., 40 D6 102
Lebork, Pol., 21,000 A4 26
Le Bourg-d'Oisans, Fr., 1,308 D2 18
Le Bourget, Fr., 10,077 g10 14
Le Bouscat, Fr., 21,404 E3 14
Le Brassus, Switz C1 19
Lebret, Sask., Can., 316 G4 70
Lebrija, Sp., 12,297 D2 20
Lebu, Chile, 3,827 B2 54
Lebyazhye, Sov. Un B7 29
Le Cateau, Fr., 9,055 B5 14
Lecce, It., 75,297 D7 21
Lecco, It., 48,230 B2 21
Le Center, Le Sueur, Minn., 1,597 F5 99
Lech, riv., Ger E5 17
Le Chambon-Feugerolles, Fr., 20,320 E6 14
Le Château, Fr., 1,278 E3 14
Lechbruck, Ger., 2,000 B6 18
Le Chesnay, Fr., 13,249 g9 14
Le Chesne, Fr., 1,078 E4 15
L'Echourie, Que., Can., 430 A4 73
Leck, Ger., 4,400 D2 24
Le Claire, Scott, Iowa, 1,546 C7, D7 92
Leclercville, Que., Can., 446 C5 73
Lecompte, Rapides, La., 1,485 C3 89
Lecompton, Douglas, Kans., 304 B7, C8 93
Leconi, Gabon F2 46
Le Conte, mtn., Tenn D10 117
Le Coteau, Fr., 6,571 D6 14
Le Creusot, Fr., 33,737 D6 14
Le Croisic, Fr., 4,017 D2 14
Le Crotoy, Fr., 2,726 D9 12
Lectoure, Fr., 2,435 F4 14
Leczyca, Pol., 6,755 B5 26
Ledbetter, Richmond, N.C., 500 C4 109
Ledbury, Eng., 3,632 B5 12
Ledcice, Czech., 709 n17 26
Ledeberg, Bel., 11,232 C3 15
Ledesma, Arg., 4,476 D3 55
Ledesma, Sp., 2,702 B3 20
Ledford, Saline, Ill., 100 F5 90
Ledger, Pondera, Mont., 10 B5 102
Ledgewood, Morris, N.J., 800 B3 106
Ledo, India, 5,000 C10 39
Ledoux, Mora, N. Mex., 100 B4 107
Leduc, Alta., Can., 2,356 C4 69
Leduck, isl., Vir. Is f16 65
Ledyard, New London, Conn., 250 (5,395▲) D8 84
Ledyard, Kossuth, Iowa, 289 A3 92
Lee, Madison, Fla., 243 B3 86
Lee, Lee and DeKalb, Ill., 228 B5 90
Lee, Penobscot, Maine, 300 (555▲) C4 96
Lee, Berkshire, Mass., 3,078 (5,271▲) B1 97
Lee, Elko, Nev., 15 C6 104
Lee, Strafford, N.H., 50 (931▲) D4 105
Lee, co., Ala., 49,754 C4 78
Lee, co., Ark., 21,001 C5 81
Lee, co., Fla., 54,539 F5 86
Lee, co., Ga., 6,204 E2 87
Lee, co., Ill., 38,749 B4 90
Lee, co., Iowa, 44,207 D6 92
Lee, co., Ky., 7,420 C6 94
Lee, co., Miss., 40,589 A5 101
Lee, co., N.C., 26,561 B4 109
Lee, co., S.C., 21,832 C7 115
Lee, co., Tex., 8,949 D4 118
Lee, co., Va., 25,824 B1 121
Lee, creek, Ark., Okla B1 81, B7 112
Lee, lake, Miss B2 100
Lee, riv., Ire F3 11
Lee Center, Lee, Ill., 275 B5 90
Leech, lake, Sask., Can F4 70
Leech, lake, Minn C4 99
Leechburg, Armstrong, Pa., 3,545 E2 114
Leechville, Beaufort, N.C., 25 B7 109
Leedey, Dewey, Okla., 451 B2 112
Leedon Estates, Delaware, Pa., 1,800 *B11 114
Leeds, Jefferson and St. Clair, Ala., 6,162 C3, E5 78
Leeds, Eng., 510,597 (★1,335,000) A6 12
Leeds, Hampshire, Mass. (part of Northampton) B2 97
Leeds, Benson, N. Dak., 797 A6 110
Leeds, Chester, S.C., 120 B5 115
Leeds, Washington, Utah, 109 F2 119
Leeds, co., Ont., Can., 46,889 C8 72
Leeds Point, Atlantic, N.J., 350 E4 106
Leeds Village, Que., Can., 475 C6 73
Leedy, Tishomingo, Miss., 40 A5 100
Leek, Eng., 19,173 A5 12
Leelanau, Leelanau, Mich D5 98
Leelanau, co., Mich., 9,321 D5 98
Leelanau, lake, Mich C5 98
Leenane, Ire., 123 D2 11
Lee Park, Luzerne, Pa., 3,500 *D10 114
Leeper, Wayne, Mo., 350 D7 101
Leeper, Clarion, Pa., 250 D3 114
Lee Pope, Crawford, Ga., 30 D3 87
Leer, Ger., 20,500 B3 16
Lees, Bledsoe, Tenn., 50 D8 117
Leesburg, Cherokee, Ala., 150 A4 78
Leesburg, Lake, Fla., 11,172 D5 86
Leesburg, Lee, Ga., 774 E2 87
Leesburg, Kosciusko, Ind., 427 B6 91
Leesburg, Rankin, Miss C4 100
Leesburg, Cumberland, N.J., 625 E3 106
Leesburg, Highland, Ohio, 932 C4 111
Leesburg, Loudoun, Va., 2,869 B5 121
Lees Ferry, Mohave, Ariz A4 80
Leesport, Berks, Pa., 1,138 *F9 114
Lee's Summit, Jackson, Mo., 8,267 C3, E2 101
Leesville, Lawrence, Ind., 40 G5 91
Leesville, Vernon, La., 4,689 C2 95
Leesville, Carroll, Ohio, 287 B6 111
Leesville, Lexington, S.C., 1,619 D4 115
Leesville, Gonzales, Tex., 150 B4, E4 118
Leetes Island, New Haven, Conn., 400 D6 84
Leeton, Johnson, Mo., 371 C4 101
Leetonia, Columbiana, Ohio, 2,543 B7 111
Leetsdale, Allegheny, Pa., 2,153 A5 114
Leeuwarden, Neth., 80,928 A5 15
Leeuwin, cape, Austl F2 50
Lee Valley, Hawkins, Tenn C10 117
Lee Vining, Mono, Calif., 350 D4 82
Leeward Islands, see Antigua, Montserrat, St. Kitts-Nevis-Anguilla, and Virgin Is. (Br.), Br. dep., N.A.
Lefebvre, Que., Can B8 63
Le Ferriere, It., 1,000 h9 21
Leflore, Le Flore, Okla., 250 C7 112
Leflore, co., Miss., 47,142 B3 100
Le Flore, co., Okla., 29,106 C7 112
Lefor, Stark, N. Dak., 175 C3 110
Lefors, Gray, Tex., 864 B2 118
Le François, Mart., 2,189 o16 64
Lefroy, Ont., Can., 366 C5 72
Lefroy, lake, Austl F3 50
Legal, Alta., Can., 524 C4 69
Leganés, Sp., 8,539 p17 20
Legaspi, Phil., 28,000 (60,800▲) C6 35
Legau, Ger., 2,500 E5 16
Leghorn, see Livorno, It.
Legion, Kerr, Tex., 1,691 *E3 118
Legler, Ocean, N.J., 200 C4 106
Legnago, It., 7,500 D7 18
Legnano, It., 42,460 B2 21
Legnica, Pol., 64,000 C4 26
Le Gore, Frederick, Md., 80 A3 85
Le Grand, Merced, Calif., 900 D3 82
Le Grand, Marshall, Iowa, 465 B5 92
Leh, India, 3,720 B6 39
Le Havre, Fr., 183,776 (★220,000) C3 20
Lehi, Utah, Utah, 4,377 C4 119
Lehigh, Webster, Iowa, 846 B3 92
Lehigh, Marion, Kans., 178 D6 93
Lehigh, Coal, Okla., 296 C5 112
Lehigh, co., Pa., 227,536 E10 114
Lehigh, riv., Pa D10 114
Lehighton, Carbon, Pa., 6,318 E10 114
Lehman, Luzerne, Pa., 200 B7 114
Lehman Caves, nat. mon., Nev E7 104
Lehman Hot Springs, Umatilla, Oreg B8 113
Lehr, Logan, N. Dak., 381 C6 110
Lehrte, Ger., 21,300 A4 17
Lehua, isl., Haw A1 88
Leiah, Pak., 14,914 B5 39
Leibnitz, Aus., 6,356 E7 16
Leicester, Eng., 273,298 (★420,000) D6 10, B6 12
Leicester, Worcester, Mass., 1,750 (8,177▲) B4 97
Leicester, Livingston, N.Y., 365 C3 108
Leicester, Addison, Vt., 50 (551▲) D2 120
Leicester, co., Eng., 682,196 B6 12
Leicester Junction, Addison, Vt., 100 D2 120
Leichhardt, riv., Austl C7 50
Leiden, Neth., 96,700 (★118,000) B4 15
Leigh, Colfax, Nebr., 502 C8 103
Leigh Creek, Austl., 1,020 E2 51
Leighlinbridge, Ire., 457 E5 11
Leighton, Colbert, Ala., 1,158 A2 78
Leinan, Sask., Can., 10 G2 70
Leine, riv., Ger A4 17
Leinster, prov., Ire., 1,332,149 D4 11
Leinster, mtn., Ire E5 11
Leipalingis, Sov. Un A7 26
Leipers Fork, Williamson, Tenn., 250 B5 117
Leipsic, Kent, Del., 281 B6 85
Leipsic, Putnam, Ohio, 1,802 A4 111
Leipsic, riv., Del B6 85
Leipzig, Sask., Can., 106 E1 70
Leipzig, Ger., 585,300 (★800,000) B7 17
Leiria, Port., 7,123 C1 20
Leisnig, Ger., 9,500 B7 17
Leiston, Eng., 4,119 B9 12
Leisure City, Dade, Fla., 3,001 F3 86
Leitches Creek, N.S., Can., 94 C9 74
Leitchfield, Grayson, Ky., 2,982 C3 94
Leiter, Sheridan, Wyo., 5 A6 125
Leitersburg, Washington, Md A2 85
Leiters Ford, Fulton, Ind., 250 B5 91
Leith, Grant, N. Dak., 100 C4 110
Leith, Fayette, Pa., 1,622 *G2 114
Leitha, riv., Aus E8 16
Leitrim, co., Ire., 33,470 C3 11
Leixlip, Ire., 915 D5 11
Lejunior, Harlan, Ky., 900 D6 94
Lek, riv., Neth C4 15
Leka, isl., Nor E4 25
Le Kef, Tun., 14,743 B6 44
Le Kef, prov. Tun., 265,502 *B6 44
Le Kreider, Alg C5 44
Leksands, Swe., 1,800 G6 25
Leksula, Indon F7 35
Lekumbi, pt., Fiji Is 52
Lela, Wheeler, Tex., 100 B2 118
Leland, La Salle, Ill., 642 B5 90
Leland, Winnebago, Iowa, 209 A4 92
Leland, Leelanau, Mich., 400 C5 98
Leland, Washington, Miss., 6,295 B3 100
Leland, Josephine, Oreg., 75 E3 113
Leland, Jefferson, Wash., 30 B3 122
Leland Grove, Sangamon, Ill., 1,731 *D4 90
Leleiwi, pt., Haw D6 88
Leleque, Arg C2 54
Lelia Lake, Donley, Tex., 150 B2 118
Leloaloa, Am. Sam., 249 52
Le Locle, Switz., 13,762 B2 25
Lelom, reef, Truk 52
Le Loup, Franklin, Kans., 110 D8 93
Lelystad, Neth., 300 B5 15
Léman (Geneva), lake, Switz D1 19
Le Mans, Fr., 132,181 D4 14
Le Marin, Mart., 4,912 o16 64
Le Mars, Plymouth, Iowa, 6,767 B1 92
Lemasters, Franklin, Pa., 280 G6 114
Lemay (Luxemburg), St. Louis, Mo., 15,000 *C7 95
Lembach, Fr., 1,602 F7 15
Lembak, Indon F2 35
Lemberg, Sask., Can., 468 G4 70
Leme, Braz., 11,785 m8 56
Lemelerveld, Neth., 1,210 B6 15
Lemery, Phil., 6,050 p13 35
Lemeta, Alsk., 1,227 *C10 79
Lemgo, Ger., 21,400 A3 17
Lemhi, Lemhi, Idaho, 4 E5 89
Lemhi, co., Idaho, 5,816 E4 89
Lemhi, range, Idaho E5 89
Lemhi, riv., Idaho E5 89
Lemieux, Que., Can., 129 C5 73
Lemitar, Socorro, N. Mex., 600 C3 107
Lemmer, Neth., 4,237 B5 15
Lemmon, Perkins, S. Dak., 2,412 B3 116
Lemmon, mtn., Ariz E5 80
Lemon, isl., S.C G6 125
Lemon Fair, riv., Vt D2 120
Lemon Grove, San Diego, Calif., 19,348 E2, E5 82
Lemons, Putnam, Mo., 140 A4 101
Lemont, Cook, Ill., 3,397 B5, F2 90
Lemont, Centre, Pa., 1,153 E6 114
Lemonweir, riv., Wis E3 124
Lemoore, Kings, Calif., 2,561 D4 82
Le Moule (Moule), Guad., 5,284 n16 64
Le Moyen, St. Landry, La., 200 D3 95
Le Moyne, Que., Can., 8,057 *D9 73
Lemoyne, Keith, Nebr., 60 C4 103
Lemoyne, Wood, Ohio, 200 A2 111
Lemoyne, Cumberland, Pa., 4,662 *F8 114
Lempster, Sullivan, N.H., 75 (272▲) D2 105
Lemsford, Sask., Can., 63 G1 70
Lemvig, Den., 5,783 B2 24
Lena, Stephenson, Ill., 1,552 A4 90
Lena, Rapides, La., 200 C3 95
Lena, Leake, Miss., 307 C4 100
Lena, Morrow, Oreg B7 113
Lena, Sp., 2,252 A3 20
Lena, Oconto, Wis., 506 D5 124
Lena, riv., Sov. Un C15 28
Lenapah, Nowata, Okla., 322 A6 112
Lenawee, co., Mich., 77,789 G6 98
Lençóis, Braz., 2,483 D2 57
Lend, Aus., 2,175 B9 18
Lendinara, It., 16,673 D7 18
Lenexa, Johnson, Kans., 2,487 B8, D9 93
Leney, Sask., Can., 35 E2 70
Lengby, Polk, Minn., 181 C3 99
Lengede, Ger., 3,600 A5 17
Lengeh, Iran, 9,617 H7 41
Lenger, Sov. Un., 40,000 E7 29
Lengerich, Ger., 21,000 A2 17
Lenhartsville, Berks, Pa., 209 E10 114
Leninabad, Sov. Un., 83,000 E7 29
Leninakan, Sov. Un., 113,000 E2 29
Leningrad, Sov. Un., 2,997,000 (★3,800,000) H14, s31 25, B8 27
Lenino, Sov. Un., 19,500 n17 27
Leninogorsk, Sov. Un., 68,000 C10 29
Leninsk, Sov. Un., 5,000 G15 27
Leninsk-Kuznetskiy, Sov. Un., 138,000 C11 29
Lenk, Switz., 1,900 D3 19
Lenkoran, Sov. Un., 30,000 H18 9
Lenne, riv., Ger B2 17
Lennep, Meagher, Mont., 25 D6 102
Lennox, Los Angeles, Calif., 31,224 *F2 82
Lennox, Lincoln, S. Dak., 1,353 D9 116
Lennox, isl., Chile k12 54
Lennox and Addington, co., Ont., Can., 23,717 B7 72
Lennoxville, Que., Can., 3,699 D6 73
Leno, It., 4,550 D6 18
Lenoir, Caldwell, N.C., 10,257 B2 109
Lenoir, co., N.C., 55,276 B6 109
Lenoir City, Loudon, Tenn., 4,979 D9 117
Lenora, Norton, Kans., 512 C4 93
Lenora, Dewey, Okla., 20 A2 112
Lenorah, Martin, Tex., 75 C2 118
Lenore, Man., Can., 98 E1 71
Lenore, Nez Perce, Idaho, 40 C2 89
Lenore, Fremont, Wyo B3 125
Lenore, lake, Sask., Can E3 70
Lenore, lake, Wash B6 122
Lenox, Cook, Ga., 802 E3 87
Lenox, Taylor, Iowa, 1,178 D3 92
Lenox, Berkshire, Mass., 1,713 (4,253▲) B1 97
Lenox, Dent, Mo., 225 D6 101
Lenox, Dyer, Tenn., 250 A2 117
Lenox, Princess Anne, Va., 1,520 *B7 121
Lenox Dale, Berkshire, Mass., 500 B1 97
Lenox Avenue (Center Mills), Chemung, N.Y., 1,500 *C4 108
Lens, Fr., 42,590 (★260,000) B5 14
Lentner, Shelby, Mo., 100 B5 101
Lenwood, San Bernardino, Calif., 2,407 *E5 82
Lenzburg, St. Clair, Ill., 420 B9 90
Lenzburg, Switz., 6,378 B5 19
Lenzen, Ger., 3,480 E5 24
Lenzerheide, Switz C8 19
Léo, Upper Volta, 2,100 D4 45
Leo, Allen, Ind., 600 B7 91
Leo, Carbon, Wyo C6 125
Leoben, Aus., 36,257 E7 16
Léogane, Hai., 3,608 F7 64
Leola, Grant, Ark., 321 C3 81
Leola, McPherson, S. Dak., 833 B7 116
Leoma, Lawrence, Tenn., 150 B4 117
Leominster, Eng., 6,403 B5 12
Leominster, Worcester, Mass., 27,929 A4 97
Leon, Decatur, Iowa, 2,004 D4 92
Leon, Butler, Kans., 541 E7 93
León (León de los Aldamas), Mex., 209,469 C4, m13 63
Leon, Cattaraugus, N.Y., 125 (738▲) C1 108
León, Nic., 52,900 D4 62
Leon, Love, Okla., 109 D4 112
León, Sp., 73,483 A3 20
Leon, Mason, W. Va., 236 C3 123
Leon, co., Fla., 74,225 B2 86
Leon, co., Tex., 9,951 D5 118
León, prov., Sp., 584,594 *A3 20
León, reg., Sp., 1,291,452 B3 20
Leona, Doniphan, Kans., 110 C8 93
Leona, Leon, Tex., 150 D5 118
Leonard, Ont., Can., 62 A10 72
Leonard, Oakland, Mich., 359 F7 98
Leonard, Clearwater, Minn., 70 C3 99
Leonard, Shelby, Mo., 142 B5 101
Leonard, Cass, N. Dak., 232 C8 110
Leonard, Fannin, Tex., 1,117 C4 118
Leonardo, Monmouth, N.J., 2,000 C4 106
Leonardtown, St. Marys, Md., 1,281 D4 85
Leonardville, N.B., Can., 256 E2 74
Leonardville, Riley, Kans., 378 C7 93
Leonardsville, Madison, N.Y., 500 C5 108
Leonberg, Ger., 20,300 E4 17
Leone, bay, Am. Sam 52
Leonfelden, Aus., 2,546 *E9 17
Leonforte, It., 17,927 F5 21
Leongatha, Austl., 2,753 I5 51
Leonia, Holmes, Fla., 225 G3 86
Leonia, Bergen, N.J., 8,384 B5, D5 106
Leonidhion, Grc., 3,356 D4 23
Leonora, Austl., 452 E3 50
Leonville, St. Landry, La., 526 D3 95
Leopold, Bollinger, Mo., 140 D8 101
Léopold II, lake, Con. L B2 48
Leopold and Astrid, coast, Ant C21 5
Leopoldina, Braz., 17,726 C4, g6 56
Leopoldsburg, Bel., 9,375 C5 15
Léopoldville, Con. L., 390,000 B2 48
Léopoldville, prov., Con. L., 3,134,500 B2 48
Leora, Stoddard, Mo., 75 D7 101
Leota, Clare, Mich D6 98
Leota, Nobles, Minn., 275 G3 99
Leoti, Witchita, Kans., 1,401 D2 93
Leoville, Sask., Can., 416 D2 70
Leoville, Decatur, Kans., 90 C3 93
Leovo, Sov. Un., 10,000 H7 27
Lepanto, Poinsett, Ark., 1,585 B5 81
Lepe, Sp., 10,038 D2 20
Lepel, Sov. Un., 10,000 D7 27
Le Petit-Quevilly, Fr., 21,098 C4 14
Lepihué, Chile C2 54
L'Épiphanie, Que., Can., 2,663 D4 73
Le Plessis-Belleville, Fr., 1,016 E2 15
Le Plessis-Robinson, Fr., 18,449 g10 14
LePont, Switz C1 19
Lepontine Alps, mts., Switz D6 19
Le Portel, Fr., 11,198 D9 12
Lepreau, N.B., Can., 94 D3 74
Lepreau, pt., N.B., Can D3 74
Lepsinsk, Sov. Un F25 9
Lepsy, Sov. Un D9 29
Le Puy [-en-Velay], Fr., 25,125 E5 14
Le Quesnoy, Fr., 4,570 D3 15

Le Raincy, Fr., 14,908 ... g11 14
Le Raysville, Bradford, Pa., 371 ... C9 114
Lercara Friddi, It., 11,872 ... F4 21
Léré, Chad, 3,617 ... D2 46
Lérida, Sp., 51,200 (63,850▲) ... B6 20
Lérida, prov., Sp., 333,765 ... *B6 20
Lerma, Sp., 2,605 ... A4 20
Lerma, riv., Mex ... m13 63
Lermoos, Aus., 768 ... B6 18
Lerna, Coles, Ill., 296 ... D5 90
Leros, isl., Grc ... D6 23
Leross, Sask., Can., 117 ... F4 70
Leroux, wash, Ariz ... B5 80
Leroy, Washington, Ala., 100 ... D2 78
Leroy, Sask., Can., 515 ... F3 70
Le Roy, McLean, Ill., 2,088 ... C5 90
Leroy, Lake, Ind., 350 ... B3 91
Le Roy, Decatur, Iowa, 70 ... D8 92
Le Roy, Coffey, Kans., 601 ... B8 93
Le Roy, Osceola, Mich., 267 ... D5 98
Le Roy, Mower, Minn., 971 ... G6 99
Leroy, Blaine, Mont., 5 ... B7 102
Le Roy, Genesee, N.Y., 4,662 ... C3 108
Leroy, Pembina, N. Dak., 100 ... A8 110
Leroy, Medina, Ohio, 504 ... A6 105
Le Roy, Bradford, Pa., 150 ... C8 114
Le Roy, Uinta, Wyo ... D2 125
Lerum, Swe., 6,504 ... A6 24
Lerwick, Scot., 5,906 ... g10 10
Lery, lake, La ... C8 95
Lesbos (Lesvos), prov., Grc., 140,251 ... *C6 23
Les Cayes, Hai., 11,835 ... F7 64
Les Diablerets, Switz ... D3 19
Les Eboulements, Que., Can., 619 ... B7 73
Les Échelles, Fr., 863 ... D1 18
Le Sentier, Switz ... C1 19
Le Sépey, Switz ... D3 19
Les Escoumins, Que., Can., 685 ... A8 73
Les Etroits, Que., Can., 291 ... B9 73
Les Fonds, Que., Can., 195 ... C8 73
Lesh, Alb., 1,609 ... B2 23
Leshara, Saunders, Nebr., 103 ... D2 103
Les Haudères, Switz ... D3 19
Lesina, lake, It ... D5 21
Lesko, Pol., 2,129 ... D7 26
Leskovac, Yugo., 33,941 ... D5 22
Leslie, Searcy, Ark., 506 ... B3 81
Leslie, Sumter, Ga., 494 ... E2 87
Leslie, Custer, Idaho, 20 ... F5 89
Leslie, Ingham, Mich., 1,807 ... F6 98
Leslie, Greenbrier, W. Va., 500 ... C4 123
Leslie, co., Ky., 10,941 ... C6 94
Leslie Station, Sask., Can., 113 ... F4 70
Leslieville, Alta., Can., 178 ... C3 69
Les Lilas, Fr., 17,721 ... g10 14
Lesneven, Fr., 4,581 ... C1 14
Lesogorsk (Nayoshi), Sov. Un ... B11 37
Lesozavodsk, Sov. Un., 18,800 ... D6 37
Lesparre [-Médoc], Fr., 2,372 ... E3 14
Les Ponts-de-Cé, Fr., 2,786 ... D3 14
Les Sables-d'Olonne, Fr., 18,401 ... D3 14
Lesser Antilles, is., W.I ... B4 55
Lesser Khingan, mts., China ... B10 34
Lesser Slave, lake, Alta., Can ... B3 69
Lessines, Bel., 9,242 ... D3 15
Lesslie, York, S.C., 250 ... B6 115
Lester, Limestone, Ala., 250 ... A2 78
Lester, Ouachita, Ark ... D3 81
Lester, Lyon, Iowa, 239 ... A1 92
Lester, Delaware, Pa., 2,000 ... *B11 114
Lester, Raleigh, W. Va., 626 ... D6 123
Lester Prairie, McLeod, Minn., 966 ... F4 99
Lesterville, Reynolds, Mo., 196 ... D7 101
Lesterville, Yankton, S. Dak., 173 ... D8 116
Lestock, Sask., Can., 412 ... F3 70
Le Sueur, Le Sueur, Minn., 3,310 ... F5 99
Le Sueur, co., Minn., 19,906 ... F5 99
Les Verrières, Switz., 1,084 ... C1 19
Lesvos (Lesbos), prov., Grc., 140,251 ... *C6 23
Lésvos (Lesbos), isl., Grc ... C6 23
Leszno, Pol., 29,000 ... C4 26
Le Tarf, Alg ... F2 21
Letart Falls, Meigs, Ohio, 300 ... D6 111
Letcher, Sanborn, S. Dak., 296 ... D7 116
Letcher, co., Ky., 30,102 ... C7 94
Le Teil, Fr., 8,236 ... E6 14
Letellier, Man., Can., 266 ... E3 71
Letha, Gem, Idaho, 100 ... F2 89
Lethbridge, Alta., Can., 35,454 ... E4 69
Lethbridge, Newf., Can., 532 ... D5 75
Le Thillot, Fr., 2,994 ... B2 18
Letiahau, riv., Bech ... B3 49
Leticia, Col., 1,898 ... D4 60
Letohatchee, Lowndes, Ala., 250 ... C3 78
Letona, White, Ark., 141 ... B4 81
Le Touquet-Paris-Plage, Fr., 4,064 ... D9 12
Letpadan, Bur., 15,896 ... D1 38
Le Tréport, Fr., 6,136 ... B4 14
Letterfrack, Ire., 236 ... D2 11
Letterkenny, Ire., 4,329 ... C4 11
Lettermullen, Ire ... D2 11
Letts, Decatur, Ind., 150 ... F6 91
Letts, Louisa, Iowa, 392 ... C6 92
Lettsworth, Pointe Coupee, La., 175 ... D4 95
Leucadia, San Diego, Calif., 5,665 ... *F5 82
Leucas (Levkas), prov., Grc., 28,980 ... *C3 23
Leuk, Switz., 2,546 ... D4 19
Leukerbad, Switz., 619 ... D4 19
Leupp, Coconino, Ariz., 25 ... B5 80
Leupp Corners, Coconino, Ariz., 5 ... B5 80
Leushinskiy Tuman, lake, Sov. Un ... B6 29
Leutkirch, Ger., 7,200 ... B6 18
Leuven, Bel., 32,524 (*85,000) ... D4 15
Leuze, Bel., 7,002 ... D3 15
Levack, Ont., Can., 3,178 ... *E9 72
Levadhia, Grc., 12,609 ... C4, g9 23
Levallois-Perret, Fr., 61,804 ... C4, g10 14
Levan, Juab, Utah, 421 ... D4 119
Levanger, Nor., 1,700 ... F4 25
Levanna, mtn., It ... B1 21
Levant, Thomas, Kans., 125 ... C2 93
Levant, Penobscot, Maine, 250 (765▲) ... D4 96
Level Green, Westmoreland, Pa., 1,500 ... *B7 114
Levelland, Hockley, Tex., 10,153 ... C1 118
Leven, Scot., 8,872 ... D5 13
Leveque, cape, Austl ... C3 50
Leverett, Franklin, Mass., 100 (914▲) ... B2 97
Leverett, Niobrara, Wyo ... B8 125
Levering, Emmet, Mich., 300 ... C6 98
Leverkusen, Ger., 94,600 ... B1 17
Levey, Polk, Iowa, 100 ... A7 92
Levi, W. Sam., 2,152 ... 52
Levi, Braxton, W. Va ... C4 123
Levice, Czech., 13,800 ... D5 26
Levie, Fr., 3,409 ... D2 21
Le Vigan, Fr., 3,110 ... E5 14
Levin, N.Z., 7,934 ... N15 51
Levis, Que., Can., 15,112 ... C6, C9 73
Levis, co., Que., Can., 51,842 ... C6 73
Levisa, fork, Ky ... C7 94
Levithos, isl., Grc ... D6 23
Levittown, Burlington, N.J., 11,861 ... *C3 106
Levittown, Nassau, N.Y., 65,276 ... G2 84
Levittown, Bucks, Pa., 58,000 ... *F12 114
Levkas, Grc., 6,577 ... C3 23
Levkas (Leucas), prov., Grc., 28,980 ... *C3 23
Levkás (Leucas), isl., Grc ... C3 23
Levktra, Grc., 1,202 ... g10 23
Levoca, Czech., 7,584 ... D6 26
Levskigrad, Bul., 8,862 ... D7 22
Levuka, Fiji Is., 1,944 ... 52
Levy, Mora, N. Mex., 5 ... A5 107
Levy, co., Fla., 10,364 ... C4 86
Levy, lake, Fla ... C4 86
Lewellen, Garden, Nebr., 411 ... C3 103
Lewes, Sussex, Del., 3,025 ... C7 85
Lewes, Eng., 13,637 ... D8 12
Lewes, riv., Yukon, Can ... D5 66
Lewis, Vigo, Ind., 600 ... F3 91
Lewis, Cass, Iowa, 501 ... C2 92
Lewis, Edwards, Kans., 486 ... E4 93
Lewis, Essex, N.Y., 250 ... A7, B3 108
Lewis, Polk, Wis., 250 ... C1 124
Lewis, co., Idaho, 4,423 ... C2 89
Lewis, co., Ky., 13,115 ... B6 94
Lewis, co., Mo., 10,984 ... A6 101
Lewis, co., N.Y., 23,249 ... B5 108
Lewis, co., Tenn., 6,269 ... B4 117
Lewis, co., Wash., 41,858 ... C3 122
Lewis, co., W. Va., 19,711 ... C4 123
Lewis, creek, Vt ... C2 120
Lewis, hill, Newf., Can ... D2 75
Lewis, lake, Wyo ... A2 125
Lewis, mtn., Nev ... C5 104
Lewis, range, Can., U.S ... B3 102, G10 66
Lewis, riv., Wash ... C3 122
Lewis, riv., Wyo ... A2 125
Lewis and Clark, co., Mont., 28,006 ... C4 102
Lewis and Clark, cavern, Mont ... E5 102
Lewis and Clark, res., Nebr ... B8 103
Lewisberry, York, Pa., 314 ... F8 114
Lewisburg, Jefferson, Ala ... E4 78
Lewisburg, Logan, Ky., 512 ... D3 94
Lewisburg, St. Landry, La., 150 ... D3 95
Lewisburg, Preble, Ohio, 1,415 ... C3 111
Lewisburg, Union, Pa., 5,523 ... E8 114
Lewisburg, Marshall, Tenn., 6,338 ... B5 117
Lewisburg, Greenbrier, W. Va., 2,259 ... D4 123
Lewis Center, Delaware, Ohio, 220 ... B3 111
Lewis-Clark, lake, S. Dak ... E8 116
Lewisdale, Prince George, Md., 5,000 ... *C3 85
Lewis Gardens, Henrico, Va., 1,380 ... *D5 121
Lewisport, Hancock, Ky., 610 ... C3 94
Lewisporte, Newf., Can., 2,076 ... D4 75
Lewis Run, McKean, Pa., 714 ... C4 114
Lewis Station, Henry, Mo ... C4 97
Lewiston, Nez Perce, Idaho, 12,691 ... C2 89
Lewiston, Androscoggin, Maine, 40,804 ... D2, D5 96
Lewiston, Montmorency, Mich., 400 ... D6 98
Lewiston, Winona, Minn., 890 ... G7 99
Lewiston, Niagara, N.Y., 3,320 ... B1 108
Lewiston, Bertie, N.C., 360 ... A8 109
Lewiston, Cache, Utah, 1,336 ... B4 119
Lewiston Orchards, Nez Perce, Idaho, 9,680 ... C2 89
Lewistown, Fulton, Ill., 2,603 ... C3 90
Lewistown, Frederick, Md., 350 ... A3 85
Lewistown, Lewis, Mo., 454 ... A6 101
Lewistown, Fergus, Mont., 7,408 ... C7 102
Lewistown, Mifflin, Pa., 12,640 ... E6 114
Lewisville, Lafayette, Ark., 1,373 ... D2 81
Lewisville, Jefferson, Idaho, 385 ... F6 89
Lewisville, Henry, Ind., 592 ... E7 91
Lewisville, Watonwan, Minn., 375 ... G4 99
Lewisville, Monroe, Ohio, 193 ... C6 111
Lewisville, Chester, Pa., 200 ... G10 114
Lewisville, Denton, Tex., 3,956 ... A5, C4 118
Lewis with Harris, isl., Scot ... B2 13
Lewvan, Sask., Can., 69 ... H3 70
Lexa, Phillips, Ark., 400 ... C5 81
Lexie, Walthall, Miss., 150 ... D3 100
Lexington, Lauderdale, Ala., 315 ... A2 78
Lexington, Stone, Ark., 60 ... B3 81
Lexington, Oglethorpe, Ga., 376 ... C3 87
Lexington, McLean, Ill., 1,244 ... C5 90
Lexington, Scott, Ind., 350 ... G6 91
Lexington, Fayette, Ky., 62,810 (*124,000) ... B5 94
Lexington, Somerset, Maine ... C2 96
Lexington, Middlesex, Mass., 27,691 ... B5, C2 97
Lexington, Sanilac, Mich., 722 ... E8 98
Lexington, Anoka, Minn., 1,457 ... *E5 99
Lexington, Holmes, Miss., 2,839 ... B3 100
Lexington, Lafayette, Mo., 4,845 ... B4 101
Lexington, Dawson, Nebr., 5,572 ... D6 103
Lexington, Greene, N.Y., 400 ... C6 108
Lexington, Davidson, N.C., 16,093 ... B3 109
Lexington, Richland, Ohio, 1,311 ... B5 111
Lexington, Cleveland, Okla., 1,216 ... B4 112
Lexington, Morrow, Oreg., 240 ... B7 113
Lexington, Lexington, S.C., 1,127 ... D5 115
Lexington, Henderson, Tenn., 3,943 ... B3 117
Lexington, Lee, Tex., 711 ... D4 118
Lexington, Rockbridge, Va., 7,537 ... D3 121
Lexington, Cowlitz, Wash., 110 ... C3 122
Lexington, co., S.C., 60,726 ... D5 115
Lexington Park, St. Marys, Md., 7,039 ... D5 85
Leyba, San Miguel, N. Mex., 125 ... B4 107
Leyburn, Eng., 1,281 ... F7 13
Leyden, Franklin, Mass., 40 (343▲) ... A2 97
Leyland, Alta., Can., 25 ... C2 69
Leyond, riv., Man., Can ... D3 71
Leysdown, Eng., 347 ... C8 12
Leysin, Switz., 4,241 ... D3 19
Leyte, prov., Phil., 1,177,400 ... *C6 35
Leyte, isl., Phil ... C6 35
Leyton, Eng., 93,857 ... k12 10
Lezajsk, Pol., 4,957 ... C7 26
Lézignan, Fr., 6,939 ... F5 14
Lezirias, reg., Port ... f10 20
Lgov, Sov. Un., 10,000 ... F10 27
Lhasa, China, 50,000 ... F3 34
Lhatse Dzong, China, 5,000 ... C11 40
L'Hay-les-Roses, Fr., 17,968 ... g10 14
Lhokseumawe, Indon ... J2 38
Lhoksukon, Indon ... J2 38
Li, China ... J4 36
Lianga, Phil., 3,173 ... *D7 35
Lianga, bay, Phil ... D7 35
Liancheng, China, 13,000 ... D5 36
Liang East, mtn., Mala ... K4 38
Liangshan, China, 18,000 ... I2 36
Liao, China ... F5 36
Liao, riv., China ... C10 36
Liaocheng, China ... F6 36
Liaoning, prov., China, 24,090,000 ... C9 34
Liaotung, bay, China ... D9 36
Liaotung, pen., China ... E10 36
Liaoyang, China, 135,000 ... D10 36
Liard, riv., B.C., N.W. Ter., Can ... E7 66
Liart, Fr., 668 ... E4 15
Liathac, mtn., Scot ... C3 13
Líbano, Col., 12,019 ... C3 60
Libau, Man., Can., 84 ... D3 71
Libby, Lincoln, Mont., 2,828 ... B1 102
Libcice, Czech., 3,121 ... n17 26
Libechov, Czech., 1,092 ... n17 26
Libenge, Con. L., 2,747 ... A2 48
Liberal, Seward, Kans., 13,813 ... E3 93
Liberal, Barton, Mo., 612 ... D3 101
Liberec, Czech., 65,300 ... C3 26
Liberia, C.R., 7,000 ... E5 62
Liberia, country, Afr., 1,250,000 ... E2 45, F4 42
Libertad, Ven., 1,171 ... B4 60
Liberty, Sask., Can., 157 ... F3 70
Liberty, Bear Lake, Idaho, 60 ... G7 89
Liberty, Adams, Ill., 325 ... D2 90
Liberty, Union, Ind., 1,745 ... E8 91
Liberty, Montgomery, Kans., 233 ... E8 93
Liberty, Casey, Ky., 1,578 ... C5 94
Liberty, Waldo, Maine, 150 (458▲) ... D3 96
Liberty, Amite, Miss., 642 ... D3 100
Liberty, Clay, Mo., 8,909 ... B3, E2 101
Liberty, Gage, Nebr., 174 ... D9 103
Liberty, Sullivan, N.Y., 4,704 ... D6 108
Liberty, Randolph, N.C., 1,438 ... B4 109
Liberty, Allegheny, Pa., 3,624 ... *F2 114
Liberty, Tioga, Pa., 269 ... C7 114
Liberty, Pickens, S.C., 2,657 ... B2 115
Liberty, De Kalb, Tenn., 293 ... C8 117
Liberty, Liberty, Tex., 6,127 ... D5, F5 118
Liberty, Kittitas, Wash., 30 ... B5 122
Liberty, co., Fla., 3,138 ... B2 86
Liberty, co., Ga., 14,487 ... E5 87
Liberty, co., Mont., 2,624 ... B5 102
Liberty, co., Tex., 31,595 ... D5 118
Liberty Acres, Los Angeles, Calif., 5,200 ... *F2 82
Liberty Center, Wells, Ind., 275 ... C7 91
Liberty Center, Warren, Iowa, 120 ... C4 92
Liberty Center, Henry, Ohio, 867 ... A3 111
Liberty Corner, Somerset, N.J., 800 ... B3 106
Liberty Grove, Cecil, Md., 55 ... A5 85
Liberty Hill, Kershaw, S.C., 350 ... C6 115
Liberty Hill, Grainger, Tenn., 50 ... C10 117
Liberty Lake, Spokane, Wash., 800 ... D8 122
Liberty Mills, Wabash, Ind., 300 ... B6 91
Libertytown, Frederick, Md., 650 ... B3 85
Libertyville, Lake, Ill., 8,560 ... A6, E2 90
Libertyville, Jefferson, Iowa, 368 ... D5 92
Libiron, isl., Eniwetok ... 52
Liblar, Ger., 6,900 ... C1 17
Libourne, Fr., 15,170 (19,834▲) ... E4 14
Libramont, Bel., 2,445 ... E5 15
Library, Allegheny, Pa., 3,000 ... *F1 114
Libreville, Gabon, 19,700 ... E1 46
Libusin, Czech., 3,584 ... n17 26
Libya, country, Afr., 1,530,000 ... D2 43, D7 42
Libyan, des., Libya ... D8 42
Libyan, plat., Libya, Eg., U.A.R ... C4 43
Licantén, Chile ... B2 54
Licata, It., 38,655 ... F4 21
Lice, Tur., 6,700 ... C13 31
Licenza, It., 1,588 ... g9 21
Lichfield, Eng., 14,077 ... B6 12
Lichiang, China, 9,000 ... F5 34
Liching, China ... F8 36
Lichtenburg, S. Afr., 14,035 ... C4 49
Lichtenfels, Ger., 11,300 ... C6 17
Lichtenstein, Ger., 13,200 ... C7 17
Lick, creek, Ind ... H8 91
Lick, creek, Ky ... A7 94
Lick, creek, Tenn ... C11 117
Licking, Texas, Mo., 954 ... D6 101
Licking, co., Ohio, 90,242 ... B5 111
Licking, creek, Md ... A1 85
Licking, riv., Ky ... B6 94
Licking, riv., Ohio ... B5, C5 111
Licosa, cape, It ... D5 21
Lida, Esmeralda, Nev., 15 ... F4 104
Lida, Sov. Un., 28,000 ... E5 27
Lida, lake, Minn ... D3 99
Liddel, riv., Eng ... E6 13
Lidderdale, Carroll, Iowa, 201 ... B3 92
Lidgerwood, Richland, N. Dak., 1,081 ... C8 110
Lidice, Czech ... C3, n17 26
Lidingö, Swe., 30,800 ... H8, t36 25
Lidköping, Swe., 16,900 ... H5 25
Lido di Roma, It., 13,730 (part of Rome) ... h8 21
Lidzbark, Warminski, Pol., 10,000 ... A6 26
Liebenthal, Sask., Can ... G1 70
Liebenthal, Rush, Kans., 191 ... D4 93
Lieberose, Ger., 2,495 ... B9 17
Liechtenstein, country, Eur., 16,628 ... F11 8, B5 18
Liège, Bel., 153,240 (*550,000) ... D5 15
Liège, prov., Bel., 1,003,526 ... D5 15
Liege, riv., Alta., Can ... A4 69
Lieksa, Fin., 4,400 ... F14 25
Lienen, Ger., 6,600 ... A2 17
Lienhua, China, 4,000 ... K5 36
Lienyün, China, 85,000 ... G8 36
Lienz, Aus., 11,132 ... E6 16
Liepāja, Sov. Un., 76,000 ... C3 27
Lier, Bel., 28,755 ... C4 15
Lierneux, Bel., 2,864 ... D5 15
Liestal, Switz., 10,262 ... B4 19
Liévin, Fr., 35,127 ... D2 13
Lievre, riv., Que., Can ... D2 73
Liffey, riv., Ire ... D5 11
Lifford, Ire., 864 ... C4 11
Light, Dimmit, Tex., 25 ... E3 118
Lighthouse, inlet, S.C ... G3 115
Lighthouse, pt., Fla ... C2 86
Lighthouse, pt., La ... E3 95
Lighthouse, pt., Mich ... C5 98
Lighthouse Point, Broward, Fla., 2,453 ... *F6 86
Lightning, creek, Wyo ... B8 125
Lightning Ridge, Austl., 286 ... D6 51
Lignite, Burke, N.D., 355 ... A3 110
Lignum, Culpeper, Va., 120 ... C5 121
Ligny-en-Barrois, Fr., 4,910 ... F5 15
Ligonha, riv., Moz ... A6 49
Ligonier, Noble, Ind., 2,595 ... B6 91
Ligonier, Westmoreland, Pa., 2,276 ... F3 114
Ligovo, Sov. Un., 10,000 ... s31 25
Liguria, pol. dist., It., 1,735,349 ... *B2 21
Liguria, reg., It ... B2 21
Ligurian, sea, It ... C2 21
Ligurian Apennine, mts., It ... E4 18
Lihua, China ... F5 34
Lihue, Kauai, Haw., 3,908 ... B2 88
Likely, Modoc, Calif., 100 ... B3 82
Likhoslavl, Sov. Un., 5,000 ... C10 27
Likino-Dulevo, Sov. Un., 10,000 ... n18 27
Lilbourn, New Madrid, Mo., 1,216 ... E8 101
Lilburn, Gwinnett, Ga., 753 ... A5 87
L'Île-Rousse, Fr., 1,867 ... C2 21
Lilesville, Anson, N.C., 635 ... C4 109
Liling, China ... K5 36
Lille, Fr., 193,096 (*865,000) ... B5 14, D3 15
Lille, Union, La., 85 ... B3 89
Lille, Aroostook, Maine, 125 ... A4 96
Lille Belt, strait, Den ... D4 24
Lillehammer, Nor., 6,000 ... G4 25
Lillers, Fr., 5,511 ... D2 15
Lillesand, Nor., 1,100 ... H3 25
Lilleström, Nor., 10,600 ... H4, p29 25
Lillian, Baldwin, Ala., 500 ... E2 78
Lillian, Johnson, Tex., 90 ... B5 118
Lillie, glacier tongue, Ant ... C29 5
Lillington, Harnett, N.C., 1,242 ... B5 109
Lillis, Marshall, Kans., 50 ... C7 93
Lillooet, B.C., Can., 1,304 ... D7 68
Lillooet, range, B.C., Can ... E7 68
Lillooet, riv., B.C., Can ... D6 68
Lilly, Dooly, Ga., 136 ... D3 87
Lilly, Cambria, Pa., 1,642 ... F4 114
Lilly, fork, W. Va ... C6 123
Lilly Grove, Mercer, W. Va., 1,255 ... *D3 123
Lilongwe, Nya., 6,660 ... D5 48
Liloy, Phil., 7,628 ... *D6 35
Lily, Laurel, Ky., 400 ... C5 94
Lily, Day, S. Dak., 119 ... B8 116
Lily, Langlade, Wis., 60 ... C5 124
Lim, riv., Yugo ... D4 22
Lima, Adams, Ill., 160 ... C2 90
Lima, Beaverhead, Mont., 397 ... F4 102
Lima, Livingston, N.Y., 1,366 ... C3 108
Lima, Allen, Ohio, 51,037 ... B3 111
Lima, Par., 2,122 ... D4 55
Lima, Peru, 767,054 (*950,000) ... D2 58
Lima, dept., Peru, 1,104,617 ... D2 58
Lima, res., Mont ... F4 102
Lima, riv., Port ... B1 20
Lima Duarte, Braz., 2,788 ... C4, g6 56
Limanowa, Pol., 3,500 ... D6 26
Lima Ringmo, lake, China ... A9 40
Limassol, Cyp., 36,536 ... E9 31
Limavady, N. Ire., 4,324 ... B5 11
Limay, riv., Arg ... B3 54
Limay Mahuida, Arg ... B3 54
Limbach, Ger., 26,400 ... C7 17
Limbé, Hai., 3,212 ... F7 64
Limbe, Nya. (part of Blantyre-Limbe) ... A6 49
Limburg, Ger., 15,600 ... C3 17
Limburg, prov., Bel., 574,606 ... D5 15
Limburg, prov., Neth., 894,300 ... D5 15
Lime, Pueblo, Colo., 50 ... C6 83
Lime, Baker, Oreg., 75 ... C9 113
Lime, creek, Iowa ... A4 92
Lime City, Wood, Ohio, 100 ... A2 111
Limedale, Putnam, Ind., 60 ... E4 91
Limeil-Brévannes, Fr., 8,377 ... n10 14
Limeira, Braz., 45,256 ... C3, m8 56
Limerick, Sask., Can., 210 ... H2 70
Limerick, Ire., 50,786 ... E3 11
Limerick, York, Maine, 450 (907▲) ... E2 96
Limerick, co., Ire., 133,339 ... E2 11
Lime Rock, Litchfield, Conn., 220 ... B3 84
Lime Spring, Howard, Iowa, 581 ... A5 92
Limestone, Newton, Ark., 100 ... B2 81
Limestone, Hardee, Fla., 100 ... E5 86
Limestone, Aroostook, Maine, 1,772 (13,102▲) ... B5 96
Limestone, Cattaraugus, N.Y., 539 ... C2 108
Limestone, Clarion, Pa., 100 ... D3 114
Limestone, Washington, Tenn., 500 ... C11 117
Limestone, co., Ala., 36,513 ... A2 78
Limestone, co., Tex., 20,413 ... D4 118
Limestone, bay, Man., Can ... C2 71
Limestone, pt., Man., Can ... C2 71
Limestone, riv., Man., Can ... A4 71
Limeton, Warren, Va., 100 ... C4 121
Lime Village, Alsk., 15 ... C8 79
Limfjorden, fjord, Den ... B3 24
Limington, York, Maine, 125 (839▲) ... E2 96
Limko, China ... C8 38
Limkong, China ... B9 38
Limmen, bight, Austl ... B6 50
Limni, Grc., 3,398 ... C4 23
Límnos (Lemnos), isl., Grc ... C5 23
Limoeiro, Braz., 21,252 ... C3, h6 57
Limoeiro do Norte, Braz., 5,705 ... C3 57
Limoges, Ont., Can., 396 ... B9 72
Limoges, Fr., 117,827 ... E4 14
Limon, Lincoln, Colo., 1,811 ... B7 83
Limón, C.R., 16,500 ... F6 62
Limón, Hond., 1,082 ... C5 62
Limón, prov., C.R., 62,800 ... *F6 62
Limón, riv., P.R ... B6 65
Limones, P.R ... C7 65
Limousin, former prov., Fr., 487,000 ... E4 14
Limousin, plat., Fr ... E4 14
Limoux, Fr., 9,603 ... F5 14
Limpopo, riv., Afr ... I9 42
Linah, Sau. Ar ... H14 31
Linard, mtn., Switz ... C9 19
Linares, Chile, 26,000 ... B2 54
Linares, Col., 732 ... C2 60
Linares, Mex., 13,318 ... C5 63
Linares, Spain, 50,200 (60,068▲) ... C4 20
Linares, prov., Chile, 170,278 ... B2 54
Linaro, cape, It ... C3 21
Linchiang, China, 1,000 ... F3 37
Linching, China, 60,000 ... E6 36
Lincoln, Talladega, Ala., 629 ... B3 78
Lincoln, Washington, Ark., 820 ... B1 81
Lincoln, Arg., 12,695 ... A4 54
Lincoln, Placer, Calif., 3,197 ... C3 82
Lincoln, Sussex, Del., 400 ... C7 85
Lincoln, Eng., 77,065 ... A7 12
Lincoln, Upson, Ga., 1,840 ... D2 87
Lincoln, Bonneville, Idaho, 300 ... F6 89
Lincoln, Logan, Ill., 16,890 ... C4 90
Lincoln, Cass, Ind., 200 ... C5 91
Lincoln, Tama, Iowa, 183 ... B5 92
Lincoln (Lincoln Center), Lincoln, Kans., 1,717 ... C5 93
Lincoln, Clay, Ky., 200 ... C6 94
Lincoln, Penobscot, Maine, 3,616 (4,541▲) ... C4 96
Lincoln, Middlesex, Mass., 1,700 (5,613▲) ... C2 97
Lincoln, Alcona, Mich., 441 ... D7 98
Lincoln, Morrison, Minn., 25 ... D4 99
Lincoln, Benton, Mo., 446 ... C4 101
Lincoln, Lewis and Clark, Mont., 550 ... D4 102
Lincoln, Lancaster, Nebr., 128,521 (*145,400) ... D9, E2 103
Lincoln, Grafton, N.H., 900 (1,228▲) ... B3 105
Lincoln, Lincoln, N. Mex., 150 ... D4 107
Lincoln, Allegheny, Pa., 1,686 ... *F2 114
Lincoln, Providence, R.I., 7,600 (13,551▲) ... B11 84
Lincoln, Lincoln, Tenn., 100 ... B5 117
Lincoln, Addison, Vt., 250 (481▲) ... C3 120
Lincoln, Loudoun, Va., 150 ... B5 121
Lincoln, Lincoln, Wash., 140 ... B7 122
Lincoln, co., Ark., 14,447 ... D4 81
Lincoln, co., Ont., Can., 126,674 ... D5 72
Lincoln, co., Colo., 5,310 ... C7 83
Lincoln, co., Eng., 743,383 ... A7 12
Lincoln, co., Ga., 5,906 ... C4 87
Lincoln, co., Idaho, 3,686 ... G4 89
Lincoln, co., Kans., 5,556 ... C5 93
Lincoln, co., Ky., 16,503 ... C5 94
Lincoln, co., Maine, 18,497 ... D3 96
Lincoln, co., Minn., 9,651 ... F2 99
Lincoln, co., Miss., 26,759 ... D3 100
Lincoln, co., Mo., 14,783 ... B6 104
Lincoln, co., Mont., 12,537 ... B1 102
Lincoln, co., Nebr., 28,491 ... D5 103
Lincoln, co., Nev., 2,431 ... F6 104
Lincoln, co., N. Mex., 7,744 ... D4 107
Lincoln, co., N.C., 28,814 ... B2 109
Lincoln, co., Okla., 18,783 ... B5 112
Lincoln, co., Oreg., 24,635 ... C3 113
Lincoln, co., S. Dak., 12,371 ... D9 116
Lincoln, co., Tenn., 23,829 ... B5 117
Lincoln, co., Wash., 10,919 ... B7 122
Lincoln, co., W. Va., 20,267 ... C2 123
Lincoln, co., Wis., 22,338 ... C4 124
Lincoln, co., Wyo., 9,018 ... C2 125
Lincoln, par., La., 28,535 ... B3 95
Lincoln, mtn., Colo ... B4 83
Lincoln, mtn., Mass ... B3 97
Lincoln, mtn., N.H ... B3 105
Lincoln, plat., Wash ... C4 122
Lincoln, rock, Wash ... B5 122
Lincoln, sea, Arc. O ... A20 4
Lincoln, tomb, Ill ... D4 90
Lincoln, wolds, Eng ... A7 12
Lincoln Acres, San Diego, Calif., 3,500 ... *E2 82
Lincoln Center, Penobscot, Maine, 200 ... C4 96
Lincoln City, Spencer, Ind., 150 ... H4 91
Lincoln Heights, Hamilton, Ohio, 7,798 ... D2 111
Lincoln Highway, Lincoln, Nebr., 950 ... *C5 103
Lincoln Park, Fremont, Colo., 2,085 ... *C5 83
Lincoln Park, Upson, Ga., 1,840 ... *D2 87
Lincoln Park, Wayne, Mich., 53,933 ... A7 98
Lincoln Park, Morris, N.J., 6,048 ... B4 106
Lincoln Park, Berks, Pa., 1,500 ... *F10 114
Lincoln Park, Delaware, Pa., 1,500 ... *G11 114
Lincoln Park, Elizabeth City, Va. (part of Hampton) ... B6 121
Lincolnton, Lincoln, Ga., 1,450 ... C4 87
Lincolnton, Lincoln, N.C., 5,699 ... B2 109
Lincoln Valley, Sheridan, N. Dak., 90 ... B5 110
Lincoln Village, San Joaquin, Calif., 6,000 ... *D3 82
Lincoln Village, Franklin, Ohio, 2,400 ... *C1 111
Lincolnville, Marion, Kans., 244 ... D7 93
Lincolnville, Waldo, Maine, 200 (867▲) ... D3 96
Lincolnville, Charleston, S.C., 420 ... E2 115
Lincolnwood, Cook, Ill., 11,744 ... E3 90
Lincroft, Monmouth, N.J., 4,000 ... C4 106
L'Incudine, mtn., Fr ... D2 21
Lind, Adams, Wash., 697 ... C7 122
Linda, Yuba, Calif., 6,129 ... *C3 82
Lindale, Floyd, Ga., 2,600 ... B1 87
Lindale, Smith, Tex., 1,285 ... C5 118
Lindau, Ger., 24,200 ... E4 16
Lindbergh, Alta., Can., 85 ... C5 69
Lindbergh, Laramie, Wyo ... B8 125
Linden, Marengo, Ala., 2,516 ... C2 78
Linden, Sumter, Fla., 175 ... D4 86
Linden, Montgomery, Ind., 619 ... D4 91
Linden, Dallas, Iowa, 258 ... C3 92
Linden, Genesee, Mich., 1,146 ... F7 98
Linden, Clay, Mo. (part of Gladstone) ... E2 101
Linden, Union, N.J., 39,931 ... E4 106
Linden, Cumberland, N.C., 157 ... B5 109
Linden, Lycoming, Pa., 250 ... D7 114
Linden, Perry, Tenn., 1,086 ... B4 117
Linden, Cass, Tex., 1,832 ... C5 118
Linden, Iowa, Wis., 418 ... F3 124
Lindenborg, riv., Den ... B4 24
Lindenhurst, Lake, Ill., 1,259 ... *E2 90
Lindenhurst, Suffolk, N.Y., 20,905 ... G3 84
Lindenwold, Camden, N.J., 7,335 ... D3 106
Linderöd, Swe., 446 ... C7 24
Lindesberg, Swe., 6,000 ... H6 25
Lindhos, Grc., 793 ... D7 23
Lindi, Tan., 10,315 ... C6 48
Lindi, riv., Con. L ... A4 48
Lindlar, Ger., 11,300 ... B2 17
Lindome, Swe., 377 ... A6 24
Lindon, Washington, Colo., 20 ... B7 83
Lindon, Utah, Utah, 1,150 ... C4 119
Lindow, Ger., 3,414 ... F6 24
Lindrith, Rio Arriba, N. Mex., 10 ... A2 107
Lindsay, Tulare, Calif., 5,397 ... D4 82
Lindsay, Ont., Can., 11,399 ... C6 72
Lindsay, Dawson, Mont., 60 ... C11 102
Lindsay, Platte, Nebr., 218 ... C8 103
Lindsay, Garvin, Okla., 4,258 ... C4 112
Lindsborg, McPherson, Kans., 2,609 ... D6 93
Lindsey, Lincolnshire, see Lincoln, co., Eng.
Lindsey, Sandusky, Ohio, 581 ... A4 111
Lindstrom, Chisago, Minn., 835 ... E6 99
Lindy, Knox, Nebr., 45 ... B8 103
Line, mtn., N. Mex ... E1 107
Lineboro, Carroll, Md., 240 ... A4 85
Linefork, Letcher, Ky., 65 ... C7 94
Linesville, Crawford, Pa., 1,255 ... C1 114
Lineville, Clay, Ala., 1,612 ... B4 78
Lineville, Wayne, Iowa, 452 ... D4 92
Linfen, China ... F4 36
Linganore, creek, Md ... B3 85
Lingayen, Phil., 8,000 (45,000▲) ... B6, n13 35
Lingayen, gulf, Phil ... n13 35
Lingen, Ger., 25,200 ... B3 16
Lingga, arch., Indon ... F2 35
Lingle, Goshen, Wyo., 437 ... C8 125
Linglestown, Dauphin, Pa., 800 ... F8 114
Lingling (Yungchow), China, 25,000 ... F7 34
Lingo, Roosevelt, N. Mex., 10 ... D6 107
Lingshan, China, 4,000 ... G6 34
Lingwu, China, 4,000 ... E2 36
Lingyuan, China, 20,000 ... D8 36
Linhai, China, 26,000 ... J9 36
Linhares, Braz., 5,751 ... E2 57
Linho, China, 3,000 ... D2 36
Linhsi, China ... C8 36

Lini, China, 80,000 . . . . G8 36
Linière, Que., Can., 1,269 . . C7 73
Linju, China G5 36
Linköping, Swe., 66,000 . . . . . . H6 25
Linlithgow, Scot., 4,327 . . . . . D5 13
Linn, Washington, Kans., 466 . . . . C6 93
Linn, Osage, Mo., 1,050 . . . . C6 101
Linn, Hidalgo, Tex., 150 . . . F3 118
Linn, co., Iowa, 136,899 . . . . B6 92
Linn, co., Kans., 8,274 . . . . . D9 93
Linn, co., Mo., 16,815 . . . . . . B4 101
Linn, co., Oreg., 58,867 . . . . C4 113
Linn, mtn., Calif B2 82
Linn Creek, Camden, Mo., 174 . . . . C5 101
Linneus, Aroostook, Maine, 250 (607▲) B5 96
Linneus, Linn, Mo., 471 . . . . B4 101
Linn Grove, Adams, Ind., 250 . . . . C7 91
Linn Grove, Buena Vista, Iowa, 330 . . . . B2 92
Linnhe, inlet, Scot D3 13
Linnsburg, Montgomery, Ind., 100 . . . . D4 91
Lino Lakes, Anoka, Minn., 2,329 . . . . *E5 99
Linquère, Sen., 1,300 . . . . . . C1 45
Lins, Braz., 32,204 . . . . C3 56
Lintan, China E5 34
Linth, riv., Switz B7 19
Linthal, Switz., 2,645 . . . . . . C7 19
Linthicum Heights, Anne Arundel, Md., 6,000 . . . . . B4 85
Lintien, China, 5,000 . . . . . . C2 36
Lintlaw, Sask., Can., 271 . . . E4 70
Linton, Hancock, Ga., 150 . . C4 87
Linton, Greene, Ind., 5,736 . . F3 91
Linton, Trigg, Ky., 90 . . . . . . D2 94
Linton, Emmons, N. Dak., 1,826 . . . . C5 110
Linton, Davidson, Tenn., 45 . . . . A4 117
Linton Junction, Que., Can., 90 . . . . B5 73
Lintung, China B8 36
Linville, Avery, N.C., 500 . . . . A2, C4 109
Linville, Rockingham, Va., 180 . . . . C4 121
Linwood, Pike, Ala., 60 . . . . D4 78
Linwood, Jefferson, Ark C4 81
Linwood, Ont., Can., 374 . . . D4 72
Linwood, Bartow, Ga., 75 . . . B2 87
Linwood, Walker, Ga., 760 . . B1 87
Linwood, Scott, Iowa, 300 . . . . C7, D7 92
Linwood, Leavenworth, Kans., 375 . . . . B8 93
Linwood, Worcester, Mass., 950 . . . . B4 97
Linwood, Bay, Mich., 400 . . . E7 98
Linwood, Butler, Nebr., 151 C9 103
Linwood, Atlantic, N.J., 3,847 . . . . E3 106
Linwood, Davidson, N.C., 150 . . . . B3 109
Linwood, Delaware, Pa., 4,460 . . . . *G11 114
Linworth, Franklin, Ohio, 500 . . . . C2 111
Linyü, China, 60,000 . . . . D8 36
Linz, Aus., 195,978 . . . . D7 16
Lions, St. John the Baptist, La., 450 . . . . B6 95
Lions, gulf, Fr F5 14
Lions Head, Ont., Can., 416 . C3 72
Lionville, Chester, Pa., 300 . F10 114
Lipa, Phil., 12,000 (69,000▲) C6, p13 35
Lipan, Hood, Tex., 309 . . . . . C4 118
Lipari, It., 3,731 . . . . E5 21
Lipari, isl., It E5 21
Lipetsk, Sov. Un., 183,000 . . C11 29
Lipik, Yugo., 1,562 . . . . C3 22
Liping, China, 5,000 . . . . . . F6 34
Lipkany, Sov. Un., 10,000 . . A8 22
Lipno, Pol., 8,389 . . . . B5 26
Lipova, Rom., 10,064 . . . . . . B5 22
Lippe, riv., Ger B3 17
Lippstadt, Ger., 37,500 . . . . . B3 17
Lipscomb, Jefferson, Ala., 2,811 . . . . B3, E4 78
Lipscomb, Lipscomb, Tex., 100 . . . . A2 118
Lipscomb, co., Tex., 3,406 . . A2 118
Lipsos, isl., Grc D6 23
Lipton, Sask., Can., 409 . . . . G4 70
Lipu, China, 15,000 . . . . G7 34
Lira, Ug A5 48
Lircay, Peru, 2,012 . . . . D3 58
Liri, riv., It D4 21
Liria, Sp., 9,723 . . . . C5 20
Lisafa, Con. L B2 48
Lisala, Con. L., 1,682 . . . . . . A3 48
Lisbon, Linn, Iowa, 1,227 . . . C6 92
Lisbon, Claiborne, La., 229 . . B3 95
Lisbon, Androscoggin, Maine, 1,542 (5,042▲) D2, D5 96
Lisbon, Howard, Md., 109 . . . B3 85
Lisbon, Grafton, N.H., 1,220 (1,788▲) B3 105
Lisbon, St. Lawrence, N.Y., 280 . . . . A1 108
Lisbon, Ransom, N. Dak., 2,093 . . . . C8 110
Lisbon, Columbiana, Ohio, 3,579 . . . . B7 111
Lisbon (Lisboa), Port., 802,200 (*1,300,000) . . . C1, f9 20
Lisbon Center, Androscoggin, Maine, 350 . . . . D5 96
Lisbon Falls, Androscoggin, Maine, 2,640 . . . . D5, E2 96
Lisburn, N. Ire., 17,691 . . . . C5 11
Lisburne, cape, Alsk B6 79
Liscannor, bay, Ire E2 11
Liscarney, Ire D2 11
Lisco, Garden, Nebr., 140 . . . C3 103
Liscomb, N.S., Can., 78 . . . . . D7 74
Liscomb, Marshall, Iowa, 295 . . . . B4 92
Lisdoonvarna, Ire., 625 . . . . . D2 11
Lishih, China F4 36
Lishu, China, 5,000 . . . . E2 37
Lishui, China, 17,000 . . . . . . F8 34
Lishukou, China, 1,000 . . . . . D5 37
Lisianski, isl., Haw k13 88
Lisichansk, Sov. Un., 10,000 . . . . q21 27
Lisieux, Sask., Can., 72 . . . . H3 70
Lisieux, Fr., 21,156 . . . . C4 14
Liskeard, Eng., 4,490 . . . . . . D3 12
Liski, Sov. Un., 10,000 . . . . F12 27
Lisle, Ont., Can., 219 . . . . . . C5 72
Lisle, Du Page, Ill., 4,219 . . . F2 90
L'Isle, Fr., 3,162 . . . . B2 18
L'Isle, Switz., 678 . . . . C1 19
L'Islet, Que., Can., 816 . . . . B7 73
L'Islet, co., Que., Can., 24,798 . . . . B7 73
Lisman, Choctaw, Ala., 909 . . C1 78
Lismore, Austl., 18,931 . . . . . D9 51
Lismore, N.S., Can., 92 . . . . . D7 74
Lismore, Ire., 810 . . . . E4 11
Lismore, Nobles, Minn., 306 . G3 99
Lismore, isl., Scot D3 13
Lisnaskea, N. Ire., 977 . . . . . C4 11
Lissycasey, Ire., 118 . . . . E2 11
Listafjord, fjord, Nor H2 25
Lister, peak, Ant B29 5
Listie, Somerset, Pa., 450 . . . F3 114
Listonburg, Somerset, Pa., 150 . . . . G3 114
Listowel, Ont., Can., 4,002 . . D4 72
Listowel, Ire., 2,859 . . . . E2 11
Litavka, riv., Czech o16 26
Litchfield, Lassen, Calif., 55 . . B3 82
Litchfield, Litchfield, Conn., 1,363 (6,264▲) B4 84
Litchfield, Montgomery, Ill., 7,330 . . . . D4 90
Litchfield, Hillsdale, Mich., 993 . . . . F6 98
Litchfield, Meeker, Minn., 5,078 . . . . E4 99
Litchfield, Sherman, Nebr., 264 . . . . C6 103
Litchfield, Hillsboro, N.H., 100 (721▲) E4 105
Litchfield, co., Conn., 119,856 . . . . B3 84
Litchfield Park, Maricopa, Ariz., 900 . . . . G1 80
Litchville, Barnes, N. Dak., 345 . . . . C7 110
Liten, Czech., 936 . . . . o17 26
Lithgow, Austl., 14,230 . . . . . F8 51
Lithia Springs, Douglas, Ga., 222 . . . . B4 87
Lithinon, cape, Grc E5 23
Lithonia, De Kalb, Ga., 1,667 . . . . B5, C2 87
Lithopolis, Fairfield, Ohio, 411 . . . . C2 111
Lithuania (S.S.R.), rep., Sov. Un., 2,804,000 . . . . D5 28
Lititz, Lancaster, Pa., 5,987 . F9 114
Litokhoron, Grc., 5,032 . . . . . B4 23
Litomerice, Czech., 16,900 . . C3 26
Litomysl, Czech., 6,384 . . . . D4 26
Litslena, Swe t35 25
Littau, Switz., 8,715 . . . . B5 19
Little, butte, Idaho F6 89
Little, lake, La E5 95
Little, riv., Ark B5 81
Little, riv., Ark., Okla . D1 81, C6 112
Little (or Gray), riv., Newf., Can E3 75
Little, riv., Conn C8 84
Little, riv., Ga C3 87
Little, riv., Ga C4 87
Little, riv., Ga E3 87
Little, riv., Ky D2 94
Little, riv., La C3 95
Little, riv., N.C B4 109
Little, riv., Okla B5 112
Little, riv., S.C C3 115
Little, riv., Tenn E11 117
Little, riv., Va E2 121
Little Acres, Gila, Ariz., 300 . D5 80
Little Alföld, reg., Hung B3 22
Little America, Ant B32 5
Little Andaman, isl., India F9 39
Little Antietam, creek, Md . . . A2 85
Little Arkansas, riv., Kans . . . A4 93
Little Bahamas, bank, Ba. Is . . . B4 64
Little Bay Islands, Newf., Can., 426 . . . . D4 75
Little Bear, Laramie, Wyo . . D8 125
Little Beaver, creek, Colo B8 83
Little Beaver, creek, Kans C2 93
Little Belt, mts., Mont D6 102
Little Bighorn, riv., Mont., Wyo E9 102
Little Birch, Braxton, W. Va., 100 . . . . C4 123
Little Bitter, lake, Eg., U.A.R . D4 32
Little Black, riv., Maine A3 96
Little Blue, riv., Ind H5 91
Little Blue, riv., Kans., Nebr C7 93, D8 103
Little Bow, riv., Alta., Can . . . D4 69
Little Britain, Ont., Can., 290 . . . . C6 72
Little Brook, N.S., Can., 249 . E3 84
Little Bullhead, Man., Can., 15 . . . . D3 71
Little Cadotte, riv., Alta., Can . . A2 69
Little Canada, Ramsey, Minn., 3,512 . . . . *E5 99
Little Carpathians, mts., Czech . D4 26
Little Catalina, Newf., Can., 752 . . . . D5 75
Little Cayman, isl., Cayman Is . F3 64
Little Cedar, Mitchell, Iowa, 80 . . . . A5 92
Little Cedar, riv., Iowa A5 92
Little Chief, Osage, Okla., 150 . . . . A5 112
Little Chute, Outagamie, Wis., 5,099 . . . . A5, D5 124
Little Coal, riv., W. Va D5 123
Little Colorado, riv., Ariz A4 80
Little Compton, Newport, R.I., 275 (1,702▲) C12 84
Little Creek, Kent, Del., 306 . B7 85
Little Creek, peak, Utah F3 119
Little Current, Ont., Can., 1,527 . . . . B3 72
Little Cypress, creek, Tex F4 118
Little Deer Isle, Hancock, Maine, 200 . . . . D4 96
Little Des Allemands, lake, La . C6 95
Little Diomede, isl., Alsk B6 79
Little Dry, creek, Mont C10 102
Little Eagle, Corson, S. Dak., 125 . . . . B5 116
Little Egg, harbor, N.J D4 106
Little Egg, inlet, N.J E4 106
Little Falls, Morrison, Minn., 7,551 . . . . E4 99
Little Falls, Passaic, N.J., 9,430 . . . . B4 106
Little Falls, Herkimer, N.Y., 8,935 . . . . B6 108
Little Ferry, Bergen, N.J., 6,175 . . . . h8 106
Littlefield, Mohave, Ariz., 50 . A2 80
Littlefield, Lamb, Tex., 7,236 . . . . C1 118
Little Fishing, creek, W. Va . . . A6 123
Littlefork, Koochiching, Minn., 805 . . . . B5 99
Little Fork, riv., Minn B5 99
Little Frog, mtn., Tenn D9 117
Little Grand, lake, Newf., Can . D3 75
Little Gunpowder Falls, riv., Md A4 85
Little Haddam, Middlesex, Conn., 35 . . . . D7 84
Littlehampton, Eng., 15,647 . D7 12
Little Harbour Deep, Newf., Can., 53 . . . . C3 75
Little Hocking, Washington, Ohio, 500 . . . . C6 111
Little Humboldt, riv., Nev B4 104
Little Inagua, isl., Ba. Is E7 64
Little Juniper, mtn., Oreg D7 113
Little Kanawha, riv., W. Va B3, C3 123
Little Laramie, riv., Wyo D7 125
Littlelot, Hickman, Tenn., 100 . . . . B4 117
Little Lynches, riv., S.C C7 115
Little Manatee, riv., Fla F2 86
Little Manitou, lake, Sask., Can F3 70
Little Marais, Lake, Minn., 25 . . . . C7 99
Little Mazarn, creek, Ark C2 81
Little Meadows, Susquehanna, Pa., 301 . . . . C9 114
Little Mecatina, isl., Que., Can . C2 75
Little Mecatina, riv., Que., Can . . h9 75
Little Medicine, Albany, Wyo C7 125
Little Miami, riv., Ohio C3 111
Little Minch, chan., Scot C2 13
Little Miquelon, isl., N.A m26 67
Little Missouri, Billings, N. Dak., 30 . . . . C2 110
Little Missouri, riv., Ark D2 81
Little Missouri, riv., U.S A7 76
Little Moose, mtn., N.Y B6 108
Little Mountain, Newberry, S.C., 238 . . . . C5 115
Little Muddy, creek, N. Dak . . . A2 110
Little Muddy, riv., Ill E4 90
Little Nemaha, riv., Nebr . D9, E3 103
Little Ocmulgee, riv., Ga D3 87
Little Orleans, Allegany, Md., 70 . . . . A1 85
Little Osage, riv., Kans., Mo E9 87, D3 101
Little Otter, creek, Vt C2 120
Little Owyhee, riv., Nev B5 104
Little Paternoster, isl., Indon . . . F5 35
Little Patuxent, riv., Md B4 85
Little Peconic, bay, N.Y E7 84
Little Pee Dee, riv., S.C C9 115
Little Pend Oreille, riv., Wash . . A8 122
Little Pigeon, riv., Ind I3 91
Little Pine, creek, Pa D7 114
Little Pipe, creek, Md A3 85
Littleport, Clayton, Iowa, 119 . . . . B6 92
Little Powder, riv., Mont., Wyo E11 96, A7 125
Little Prairie, B.C., Can., 208 . . . . B7 68
Little Rann of Kutch, salt flat, India F3 40
Little Red, riv., Ark B4 81
Little River, Baldwin, Ala . . . D2 78
Little River, N.S., Can., 182 . E3 74
Little River, Rice, Kans., 552 . . . . D5 93
Little River, Horry, S.C., 25 . . . . D11 115
Little River, co., Ark., 9,211 . D1 81
Little River, inlet, N.C D5 109
Little River, inlet, S.C D11 115
Little River, res., Vt C3 120
Little Rock, Pulaski, Ark., 107,813 (*238,500) . . . C3, D6 81
Little Rock, Lyon, Iowa, 564 . A2 92
Little Rock, Bourbon, Ky., 300 . . . . B5 94
Little Rock, Dillon, S.C., 130 . . . . C9 115
Littlerock, Thurston, Wash., 250 . . . . C2 122
Little Rock, riv., Iowa A2 92
Little Sable, pt., Mich E4 98
Little Sachigo, lake, Ont., Can . B5 71
Little St. Bernard, pass, Fr. and It D2 18
Little Salt, lake, Utah F3 119
Little Sandy, creek, Wyo D3 125
Little Sandy, riv., Ky B7 94
Little Satilla, riv., Ga E4 87
Little Sauk, Todd, Minn., 100 . . . . E4 99
Little Sebago, lake, Maine E4 96
Little Sevier, riv., Utah F3 119
Little Silver, Monmouth, N.J., 5,202 . . . . C4 106
Little Sioux, Harrison, Iowa, 295 . . . . C1 92
Little Sioux, riv., Iowa B2 92
Little Smoky, riv., Alta., Can . . B2 69
Little Smoky, val., Nev D5 104
Little Snake, riv., Colo., Wyo A2 83, D5 125
Little Spokane, riv., Wash B8 122
Littlestown, Adams, Pa., 2,756 . . . . G7 114
Little Suamico, Oconto, Wis., 100 . . . . D5 124
Little Sugar, riv., N.H D2 105
Little Tallapoosa, riv., Ala., Ga B4 78, C1 87
Little Tenmile, creek, W. Va . . . A6 123
Little Tennessee, riv., Tenn . . . D8 117
Littleton, Jefferson, Ala., 75 . E4 78
Littleton, Arapahoe, Colo., 13,670 . . . . B6 83
Littleton, Schuyler, Ill., 176 . C3 90
Littleton, Aroostook, Maine, 180 (982▲) B5 96
Littleton, Middlesex, Mass., 700 (5,109▲) C1 97
Littleton, Grafton, N.H., 3,355 (5,003▲) B3 105
Littleton, Halifax and Warren, N.C., 1,024 . . . . . A6 109
Littleton, Wetzel, W. Va., 339 . . . . B4, C2 123
Littleton Common, Middlesex, Mass., 2,277 . . . . C1 97
Little Tucson, Pima, Ariz., 100 . . . . F4 80
Little Turtle, lake, Ont., Can . . B6 99
Little Valley, Calif B3 170
Little Valley, Cattaraugus, N.Y., 1,244 . . . . C2 108
Little Vermilion, lake, Ont., Can D5 71
Little Wabash, riv., Ill E5 90
Little West, fork, Tenn A4 117
Little White, riv., Ont., Can . . . B7 98
Little White, riv., S. Dak D4 116
Little Wolf, riv., Wis D4 124
Little Wood, riv., Idaho F4 89
Little Woods, Orleans, La. (part of New Orleans) B7 95
Little York, P.E.I., Can., 185 . C6 74
Little York, Warren, Ill., 329 . B3 90
Little York, Washington, Ind., 180 . . . . G6 91
Little York, Hunterdon, N.J., 120 . . . . B2 106
Little Zab, riv., Iraq E15 31
Lituhi, Tan D5 48
Litz Manor, Sullivan, Tenn. (part of Kingsport) C11 117
Liuan, China I7 36
Liucheng, China G6 34
Liucura, Chile, 1,094 . . . . B2 54
Liuho, China, 5,000 . . . . E2 37
Liupa, China, 1,000 . . . . E6 36
Liu Panshan, mts., China G2 36
Liu Pen Shan, mts., China K2 36
Liuyang, China J5 36
Livelong, Sask., Can., 140 . . D1 70
Lively, Ont., Can., 3,211 . . . *E9 72
Livengood, Alsk., 20 . . . . . . B10 79
Livenza, riv., It D8 18
Live Oak, Sutter, Calif., 2,276 . . . . C3 82
Live Oak, Suwannee, Fla., 6,544 . . . . B4 86
Live Oak, co., Tex., 7,846 . . E3 118
Livermore, Alameda, Calif., 16,058 . . . . B6 82
Livermore, Larimer, Colo., 20 . . . . A5 83
Livermore, Humboldt, Iowa, 545 . . . . B3 92
Livermore, McLean, Ky., 1,506 . . . . C2 94
Livermore, Androscoggin, Maine, 200 (1,363▲) D2 96
Livermore, peak, Tex F2 118
Livermore Falls, Androscoggin, Maine, 2,882 (3,343▲) D2 96
Livermore Falls, Grafton, N.H., 75 . . . . C3 105
Liverpool, Macon, Ala C4 78
Liverpool, N.S., Can., 3,712 . E5 74
Liverpool, Eng., 747,490 (*1,655,000) A5 12
Liverpool, Onondaga, N.Y., 3,487 . . . . B4 108
Liverpool, Perry, Pa., 894 . . . E7 114
Liverpool, Brazoria, Tex., 90 . . . . G5 118
Liverpool, Jackson, W. Va., 75 . . . . C3 123
Liverpool, bay, N.W. Ter., Can . B7 66
Liverpool, bay, Eng A4 12
Livinda, riv., Gabon, Con. B . . . E2 46
Livingston, Sumter, Ala., 1,544 . . . . C1 78
Livingston, Merced, Calif., 2,188 . . . . D3 82
Livingston, Guat., 2,606 . . . . C3 62
Livingston, Madison, Ill., 964 . . . . E4 90
Livingston, Rockcastle, Ky., 419 . . . . C5 94
Livingston, Livingston, La., 1,183 . . . . A6, D5 95
Livingston, Park, Mont., 8,229 . . . . E6 102
Livingston, Essex, N.J., 23,124 . . . . *B4 106
Livingston, Columbia, N.Y., 200 . . . . C7 108
Livingston, Orangeburg, S.C., 208 . . . . D5 115
Livingston, Overton, Tenn., 2,817 . . . . C8 117
Livingston, Polk, Tex., 3,398 . D5 118
Livingston, Grant and Iowa, Wis., 488 . . . . F3 124
Livingston, co., Ill., 40,341 . . C5 90
Livingston, co., Ky., 7,029 . . . A3 94
Livingston, co., Mich., 38,233 . . . . F6 98
Livingston, co., Mo., 15,771 . . B4 101
Livingston, co., N.Y., 44,053 . C3 108
Livingston, par., La., 26,974 . D5 95
Livingstone, N. Rh., 28,600 . . E4 48
Livingstone, range, Alta., Can . D3 69
Livingstone Cove, N.S., Can., 54 . . . . D7 74
Livingstonia, Nya D5 48
Livingston Manor, Sullivan, N.Y., 2,080 . . . . D6 108
Livno, Yugo., 5,170 . . . . D3 22
Livny, Sov. Un., 10,000 . . . . E11 27
Livonia, Washington, Ind., 150 . . . . G5 91
Livonia, Pointe Coupee, La., 430 . . . . D4 95
Livonia, Wayne, Mich., 66,702 . . . . F7 98
Livonia, Putnam, Mo., 154 . . A5 101
Livonia, Livingston, N.Y., 946 . . . . C3 108
Livorno (Leghorn), It., 161,077 . . . . C3 21
Livramento, Braz., 37,666 . . . E1 56
Livry-Gargan, Fr., 29,679 . . . g11 14
Liwale, Tan., 2,898 . . . . C6 48
Liyang, China, 21,000 . . . . . . I8 36
Lizard, creek, Iowa B3 92
Lizard, mts., B.C., Can E3 69
Lizard, pt., Eng E2 12
Lizard Head, pass, Colo D3 83
Lizard Head, peak, Wyo C3 125
Lizella, Bibb, Ga., 450 . . . . . D3 87
Lizemores, Clay, W. Va., 50 . . . . C3, C7 123
Lizotte, Que., Can A5 73
Lizton, Hendricks, Ind., 366 . E4 91
Ljubljana, Yugo., 133,386 . . . B2 22
Ljubuski, Yugo., 2,181 . . . . . D3 22
Ljungby, Swe., 9,300 . . . . I5 25
Ljusdal, Swe., 4,400 . . . . G7 25
Ljusterö, Swe., 1,132 . . . . . . t36 25
Llandilo, Wales, 1,906 . . . . . C4 12
Llandovery, Wales, 1,898 . . . . C4 12
Llandrindod Wells, Wales, 3,248 . . . . B4 12
Llandudno, Wales, 17,852 . . . A4 12
Llanelly, Wales, 29,994 . . . . . C3 12
Llanerch, see Havertown, Pa.
Llanes, Sp., 20,421 . . . . A3 20
Llanfyllin, Wales, 1,251 . . . . . B4 12
Llanfairfechan, Wales, 3,209 . . A3 12
Llangollen, Wales, 3,050 . . . . B4 12
Llangynog, Wales, 394 . . . . . . B4 12
Llanidloes, Wales, 2,375 . . . . B4 12
Llano, Llano, Tex., 2,656 . . . D3 118
Llano, Taos, N. Mex., 200 . . A4 107
Llano, co., Tex., 5,240 . . . . . D3 118
Llano Estacado, plain, Tex . . . C1 118
Llano Estacado, plat., N. Mex., Tex C6 107
Llanos, plains, Col., Ven C4 53
Llanquihue, lake, Chile C2 54
Llanquihue, prov., Chile, 165,956 . . . . C2 54
Llanrwst, Wales, 2,571 . . . . . A4 12
Llata, Peru, 1,741 . . . . C2 58
Llaves, Rio Arriba, N. Mex., 70 . . . . A3 107
Llerena, Sp., 8,217 . . . . C2 20
Llerena, pt., C.R F6 62
Lleyn, pen., Wales B3 12
Llico, Chile A2 54
Llivia, Sp., 637 . . . . F4 14
Llobregat, riv., Sp B6 20
Llorana, mtn., Pan h11 62
Lloyd, Jefferson, Fla., 300 . . B2 86
Lloyd, Greenup, Ky., 600 . . . B7 94
Lloyd, Blaine, Mont., 5 . . . . . B7 102
Lloyd, neck, N.Y F3 84
Lloyd, pt., N.Y F3 84
Lloyd Harbor, Suffolk, N.Y., 2,521 . . . . *F3 84
Lloydminster, Alta. and Sask., Can., 5,667 . . . . D1 70
Lloyd Place, Nansemond, Va., 2,282 . . . . *E6 121
Lluchmayor, Sp., 10,664 . . . C7 20
Llullaillaco, vol., Chile D2 55
Llyswen, Blair, Pa. (part of Altoona) F5 114
Loa, riv., Chile D2 55
Loa, Wayne, Utah, 359 . . . . . . E4 119
Loachapoka, Lee, Ala., 200 . . C4 78
Loami, Sangamon, Ill., 450 . . D4 90
Loan, lake, Maine B3 96
Loange, riv., Con. L C3 48
Löbau, Ger., 16,800 . . . . B9 17
Lobaye, riv., Cen. Afr. Rep E3 46
Lobelville, Perry, Tenn., 449 . . . . B4 117
Lobenstein, Ger., 4,194 . . . . . C6 17
Lobería, Arg., 7,916 . . . . B5 54
Lobito, Ang., 31,630 . . . . D1 48
Lobnya, Sov. Un., 1,000 . . . . m17 27
Lobo, Phil., 843 . . . . p13 35
Lobo, Culberson, Tex., 30 . . . F2 118
Lobos, Arg., 8,372 . . . . B5, g7 54
Lobos, isl., P.R f12 65
Lobos de Tierra, isl., Peru C1 58
Lobotsi, Bech C4 49
Lobster, lake, Maine C3 96
Lobstick, lake, Newf., Can g8 75
Loburg, Ger., 4,051 . . . . A7 17
Locarno, Switz., 10,155 (*17,420) D6 19
Locate, Custer, Mont., 25 . . D11 102
Lochaber, N.S., Can., 68 . . . . D7 74
Lochaline, Scot D3 13
Lochbroom, Scot B4 10
Lochcarron, Scot., 822 . . . . . C3 13
Lochdale, B.C., Can., 900 . . A10 68
Lochdonhead, Scot D3 13
Lochearn, Baltimore, Md., 25,000 . . . . *B4 85
Lochearnhead, Scot D4 13
Lochem, Neth., 6,600 . . . . . . B6 15
Loches, Fr., 5,902 . . . . D4 14
Lochgilphead, Scot., 1,208 . . D3 13
Lochiel, Santa Cruz, Ariz., 40 . . . . F5 80
Loching, China, 10,000 . . . . . . J9 36
Lochinver, Scot B3 13
Loch Lynn Heights, Garrett, Md., 476 . . . . D1 85
Lochmaben, Scot., 1,279 . . . . E5 13
Loch Maree, lake, Scot C3 13
Lochmere, Belknap, N.H., 225 . . . . D5 105
Lochnagar, mtn., Scot D5 13
Lochranza, Scot E3 13
Loch Raven, Baltimore, Md., 23,278 . . . . *B4 85
Loch Raven, res., Md B4 85
Lochsa, riv., Idaho C3 89
Loch Torridon, lake, Scot C3 13
Lochuan, China G3 36
Lochy, lake, Scot D4 13
Lockbourne, Franklin, Ohio, 460 . . . . C2 111
Locke, Elkhart, Ind., 80 . . . . B5 91
Locke, Cayuga, N.Y., 550 . . . C4 108
Locke, Shelby, Tenn., 50 . . . . B1, E8 117
Locke, Pend Oreille, Wash . . A8 122
Lockeford, San Joaquin, Calif., 900 . . . . B6 82
Locke Mills, Oxford, Maine, 160 . . . . D2 96
Lockeport, N.S., Can., 1,231 . F4 74
Lockerbie, Scot., 2,826 . . . . . E5 13
Lockesburg, Sevier, Ark., 511 . . . . D1 81
Lockhart, Covington, Ala., 799 . . . . D3 78
Lockhart, Orange, Fla., 1,500 . . . . *D5 86
Lockhart, Norman, Minn., 70 . . . . C2 99
Lockhart, Union, S.C., 128 . . B5 115
Lockhart, Caldwell, Tex., 6,084 . . . . A4, E5 118
Lock Haven, Clinton, Pa., 11,748 . . . . D7 114
Lockland, Hamilton, Ohio, 5,292 . . . . A7 94
Lockney, Floyd, Tex., 2,141 . B2 118
Lockport, Man., Can., 405 . . D3 71
Lockport, Will, Ill., 7,560 . . . . B5, F2 90
Lockport, Henry, Ky., 82 . . . . B4 94
Lockport, Lafourche, La., 2,221 . . . . C6, E5 95
Lockport, Niagara, N.Y., 26,443 . . . . B2 108
Lockport Station, Lafourche, La., 2,221 . . . . C6 95
Lockridge, Jefferson, Iowa, 206 . . . . C6 92
Lock Springs, Daviess, Mo., 117 . . . . B4 101
Lockwood, Sask., Can., 109 . . F3 70
Lockwood, Dade, Mo., 835 . . D4 101
Lockwood, Nicholas, W. Va., 912 . . . . C3, C7 123
Locminé, Fr., 1,898 . . . . D2 14
Loc Ninh, Viet G7 38
Loco, Stephens, Okla., 268 . . C4 112
Loco Hills, Eddy, N. Mex., 125 . . . . E6 107
Locumba, Peru, 634 . . . . E3 58
Locust, Monmouth, N.J., 700 . . . . C4 106
Locust, creek, Mo B4 101
Locust, fork, Ala B3 78
Locust, pt., Md B5 85
Locust Bayou, Calhoun, Ark., 150 . . . . D3 81
Locust Grove, Henry, Ga., 369 . . . . C2 87
Locust Grove, Washington, Md., 150 . . . . B2 85
Locust Grove, Mayes, Okla., 828 . . . . A6 112
Locust Hill, Ont., Can., 57 . . E7 72
Locust Valley, Nassau, N.Y., 3,700 . . . . F2 84
Lod (Lydda), Isr., 19,012 . . . . C6, h10 32
Loda, Iroquois, Ill., 585 . . . . C5 90
Löderup, Swe., 598 . . . . C8 24
Lodève, Fr., 6,869 . . . . F5 14
Lodeynoye Pole, Sov. Un., 10,000 . . . . A9 27
Lodge, Colleton, S.C., 181 . . . E6 115
Lodge, creek, Can., Mont B7 102, H1 70
Lodge Grass, Big Horn, Mont., 687 . . . . E9 102
Lodgepole, Cheyenne, Nebr., 492 . . . . C3 103
Lodgepole, Perkins, S. Dak., 35 . . . . B3 116
Lodgepole, creek, Nebr., Wyo C3 103, D8 125
Lodhran, Pak., 4,890 . . . . C3 40
Lodi, San Joaquin, Calif., 22,229 . . . . B6, C3 82
Lodi, Con. L B3 48
Lodi, It., 38,158 . . . . D5 18
Lodi, Bergen, N.J., 23,502 . . D5 106
Lodi, Seneca, N.Y., 396 . . . . C4 108
Lodi, Medina, Ohio, 2,213 . . A5 111
Lodi, Washington, Va., 35 . . . B3 121
Lodi, Columbia, Wis., 1,620 . . E4 124
Lodja, Con. L B3 48
Lodore, canyon, Colo A2 83
Lods, Fr., 342 . . . . B2 18
Lodwar, Ken A6 48
Łódź, Pol., 708,000 (*850,000) C5 26
Loeches, Sp., 1,719 . . . . p18 20
Loei, Thai., 7,700 . . . . D4 38
Loelli (Well), Sud D3 47
Loeriesfontein, S. Afr., 2,192 D2 49
Lofer, Aus., 1,519 . . . . B8 18
Lofgreen, Tooele, Utah, 20 . . C3 119
Lofoten, is., Nor C5 25
Lofton, Heard, Ga., 200 . . . . C1 87
Loftus, Eng., 8,111 . . . . F8 13
Logan (Hanaford), Franklin, Ill., 289 . . . . F5 90
Logan, Harrison, Iowa, 1,605 . . . . C2 92
Logan, Phillips, Kans., 846 . . C4 93
Logan, Gallatin, Mont., 200 . E5 102
Logan, Quay, N. Mex., 400 . . B6 107
Logan, Ward, N. Dak., 35 . . . A4 110
Logan, Hocking, Ohio, 6,417 . C5 111
Logan, Beaver, Okla., 10 . . . . A1, D4 112
Logan, Cache, Utah, 18,731 . B4 119
Logan, Logan, W. Va., 4,185 . . . . D3, D5 123
Logan, co., Ark., 15,957 . . . . B2 81
Logan, co., Colo., 20,302 . . . A7 83
Logan, co., Ill., 33,656 . . . . . C4 90
Logan, co., Kans., 4,036 . . . . D2 93
Logan, co., Ky., 20,896 . . . . . D3 94
Logan, co., Nebr., 1,108 . . . . C5 103
Logan, co., N. Dak., 5,369 . . C6 110
Logan, co., Ohio, 34,803 . . . . B4 111
Logan, co., Okla., 18,662 . . . B4 112
Logan, co., W. Va., 61,570 . . D3 123
Logan, creek, Nebr B9 103
Logan, mtn., Ariz A8 80
Logan, mtn., Yukon, Can D4 66
Logan, mtn., Wash A5 122
Logan, pass, Mont B3 102
Logan, peak, Ala A3 78
Logandale, Clark, Nev., 360 . G7 104
Logansport, Cass, Ind., 21,106 . . . . C5 91
Logansport, Butler, Ky., 125 . . . . C3 94
Logansport, De Soto, La., 1,371 . . . . C2 95
Loganton, Clinton, Pa., 385 . . . . D7 114
Loganville, Gwinnett and Walton, Ga., 926 . . . . C3 87
Loganville, York, Pa., 742 . . . G8 114
Loganville, Sauk, Wis., 220 . . E3 124
Logdell, Grant, Oreg C7 113
Loge, riv., Ang C1 48
Log Lane Village, Morgan, Colo., 310 . . . . A7 83
Logone, riv., Chad D3 46
Logroño, Sp., 61,292 . . . . A4 20
Logroño, prov., Sp., 229,852 . . . . *A4 20
Logrosán, Sp., 6,595 . . . . C3 20
Løgstør, Den., 3,435 . . . . B3 24
Logtown, Hancock, Miss., 250 . . . . E4 100
Løgumkloster, Den., 1,907 . . C2 24
Lohals, Den., 640 . . . . C4 24
Lohardaga, India, 13,203 . . . F10 40
Lohman, Blaine, Mont., 63 . . B7 102
Lohr, Ger., 11,100 . . . . D4 17
Lohrville, Calhoun, Iowa, 653 . . . . B3 92
Loi Mai, mtn., Bur B2 38
Loire, dept., Fr., 696,348 . . *E6 14
Loire, riv., Fr D3 14
Loire-Atlantique, dept., Fr., 803,372 . . . . *D3 14
Loiret, dept., Fr., 389,854 . . *D4 14
Loir-et-Cher, dept., Fr., 250,741 . . . . *D4 14
Lois, Moore, Tenn B5 117
Loíza, P.R., 3,097 . . . . B7 65
Loíza, mun., P.R., 28,131 . . B7 65
Loíza Aldea, P.R., 2,330 . . . . B7 65
Loja, Ec., 18,200 . . . . B2 58
Loja, Sp., 12,439 . . . . D3 20
Loja, prov., Ec., 252,737 . . . . B2 58
Lojar, cape, Indon F5 35
Lokandu, Con. L B4 48
Lokeren, Bel., 25,819 . . . . . . C3 15
Lokhvitsa, Sov. Un., 10,000 . F9 27
Lokichar, Ken A6 48
Lokichoggio, Ken A5 48
Lokitaung, Ken A6 48
Løkken, Den., 1,506 . . . . A3 24
Lokoja, Nig., 12,606 . . . . E6 45
Lokoro, riv., Con. L B3 48
Lokossa, Togo E5 45
Lokwei, China C9 38
Lol, riv., Sud D2 47
Lola, Livingston, Ky., 400 . . . A3 94
Lolland, isl., Den D5 24
Lollar, Ger., 4,500 . . . . C3 17
Lollie (Minter), Laurens, Ga., 147 . . . . D4 87
Lolo, Missoula, Mont., 100 . . D2 102
Lolo, pass, Idaho C4 89
Lolo Hot Springs, Missoula, Mont., 20 . . . . D2 102
Lom, Bul., 15,182 . . . . D6 22
Loma, Mesa, Colo., 100 . . . . B2 83

Loma, Chouteau, Mont., 110 C6 102
Loma, Butler, Nebr., 60 C9 103
Loma, Cavalier, N. Dak., 20 A7 110
Loma, mts., Guinea, S.L. E2 45
Loma Linda, San Bernardino, Calif., 6,000 *F3 82
Lomami, riv., Con. L. B3 48
Loman, Koochiching, Minn., 60 B5 99
Loma Prieta, mtn., Calif. C6 82
Lomas, Peru, 500 E3 58
Lomas de Zamora, Arg., 275,219 A5, g7 54
Lomax, Chilton, Ala., 200 C3 78
Lomax, Henderson, Ill., 535 C2 90
Lombard, Du Page, Ill., 25,296 F2 90
Lombard, Broadwater, Mont., 30 D5 102
Lombardia, pol. dist., It., 7,406,152 *B2 21
Lombardy, reg., It. B2 21
Lombez, Fr., 588 F4 14
Lomblen, isl., Indon. G6 35
Lombok, isl., Indon. G5 35
Lombok strait, Indon. G5 35
Lomé, Togo, 69,448 E5 45
Lomela, Con. L. B3 48
Lomela, riv., Con. L. B3 48
Lometa, Lampasas, Tex., 817 D3 118
Lomira, Dodge, Wis., 807 E5 124
Lomita, Los Angeles, Calif., 14,983 *F2 82
Lommel, Bel., 17,923 C5 15
Lomond, Alta., Can., 244 D4 69
Lomond, Newf., Can., 77 D3 75
Lomond, lake, Scot. D4 13
Lomonosov, Sov. Un., 30,000 s30 25
Lomonosovskaya, Sov. Un. C7 29
Lompoc, Santa Barbara, Calif., 14,415 E3 82
Lomza, Pol., 20,000 B7 26
Lon, Lincoln, N. Mex. C4 107
Lonaconing, Allegany, Md., 2,077 D2 85
Loncoche, Chile, 5,061 B2 54
Londesborough, Ont., Can., 113 D3 72
London, Pope, Ark., 282 B2 81
London, Ont., Can., 169,569 (*181,283) E3 72
London, Eng., 3,195,114 (Greater London, 8,346,137) (*10,830,000) E6, k12 10, D7 12
London, Laurel, Ky., 4,035 C5 94
London, Madison, Ohio, 6,379 C4 111
London, Kimble, Tex., 200 D3 118
London, Kanawha, W. Va., 500 *C5 123
London, co., Eng., 3,195,114 *E6 10
Londonbridge, Princess Anne, Va., 1,061 *E6 121
Londonderry, N.S., Can., 170 D6 74
Londonderry, N. Ire., 53,744 C4 11
Londonderry, Rockingham, N. H., 500 (2,457▲) E4 105
Londonderry, Ross, Ohio, 225 C5 111
Londonderry, Windham, Vt., 200 (898▲) E3 120
Londonderry, co., N. Ire., 111,565 C4 11
Londonderry, cape, Austl. B4 50
Londonderry, isl., Chile k11 54
London Mills, Fulton and Knox, Ill., 617 C3 90
Londrina, Braz., 74,110 C2 56
Lone, plat., Colo. D2 83
Lone Elm, Anderson, Kans., 69 D8 93
Lone Grove, Carter, Okla., 500 C4 112
Lonejack, Jackson, Mo., 180 C3 101
Lonely, Humphreys, Tenn., 150 A4 117
Lonely, isl., Ont., Can. B3 72
Lone Mountain, Claiborne, Tenn., 200 C10 117
Loneoak, McCracken, Ky., 2,104 A2 94
Lone Oak, Hunt, Tex., 495 C5 118
Lone Pine, Inyo, Calif., 1,310 D4 82
Lonepine, Evangeline, La. D3 95
Lonepine, Sanders, Mont., 16 C2 102
Lone Rock, Sask., Can., 191 D1 70
Lone Rock, Kossuth, Iowa, 185 A3 92
Lonerock, Gilliam, Oreg., 31 B7 113
Lone Rock, Richland, Wis., 563 E3 124
Lone Star, Calhoun, S.C., 350 D6 115
Lone Star, Morris, Tex., 1,513 *C5 118
Lone Tree, Johnson, Iowa, 717 C6 92
Lonetree, Uinta, Wyo., 5 D2 125
Lone Tree, creek, Colo. A6 83
Lone Wolf, Kiowa, Okla., 617 C2 112
Long, co., Ga., 3,874 E5 87
Long, bay, N.C. D5 109
Long, beach, N.Y. E7 84
Long, beach, N.Y. G2 84
Long, creek, Sask., Can. H3 70
Long, isl., Ba. Is. D6 64
Long, isl., N.S., Can. E3 64
Long, isl., Fla. H6 86
Long, isl., Maine D4 96
Long, isl., Mass. D3 97
Long, isl., N. Gui. k12 50
Long, isl., N.Y. D3 102, F3 81
Long, key, Fla. E4 86
Long, key, Fla. H6 86
Long, lake, N.B., Can. B3 74
Long, lake, La. E5 95
Long, lake, Maine A4 96
Long, lake, Maine D2 96
Long, lake, Mich. C7 98
Long, lake, Mich. D5 98
Long, lake, Minn. D4 99
Long, lake, N.Y. B6 108
Long, lake, N. Dak. C6 110
Long, lake, Wash. D1 122
Long, lake, Wis. C2 124
Long, mtn., N.H. A4 105
Long, pond, Fla. C4 86
Long, pond, Maine C2 96
Long, pond, Maine C3 96
Long, pond, Mass. C6 97
Long, strait, Sov. Un. B20 28
Longa, riv., Ang. D1 48
Longá, riv., Braz. B2 57
Longarone, It., 4,727 C8 18
Long Barn, Tuolumne, Calif., 200 C3 82
Long Beach, Los Angeles, Calif., 344,168 F2, F4 82
Longbeach, Manatee, Fla., 1,000 F2 86
Long Beach, La Porte, Ind., 2,007 A4 91
Long Beach, Pope, Minn., 236 E3 99
Long Beach, Harrison, Miss., 4,770 E2, E4 100
Long Beach, Nassau, N.Y., 26,473 E2, E3 108
Long Beach, Pacific, Wash., 665 C1 122
Long Beach Resort, Bay, Fla., 66 *G3 86
Longboat, inlet, Fla. F1 86
Longboat, key, Fla. E4 86
Longboat Key, Manatee and Sarasota, Fla., 1,000 *E4 86
Long Bottom, Meigs, Ohio, 135 C6 111
Long Branch, Ont., Can., 11,039 E6 72
Long Branch, Monmouth, N.J., 26,228 C5 106
Long Branch, Panola, Tex., 20 C5 118
Longbranch, Pierce, Wash., 150 B3 122
Long Bridge, Warren, N.J. B3 106
Long Cliff, Cass, Ind. C5 91
Long Corner, Howard, Md., 150 B3 85
Long Creek, Grant, Oreg., 295 C7 113
Longcreek, Oconee, S.C., 200 B1 115
Longdale, Blaine, Okla., 218 A3 112
Long Eaton, Eng., 30,464 B6 12
Long Eddy, Sullivan, N.Y., 300 D5 108
Longeten, riv., Switz. B4 19
Longford, Ire., 3,558 D4 11
Longford, Clay, Kans., 146 C6 93
Longford, co., Ire., 30,643 D4 11
Longford Mills, Ont., Can., 400 C5 72
Longfork, Dickenson, Va. B2 121
Long Grove, Scott, Iowa, 182 C7 92
Long Harbour, Newf., Can., 356 E5 75
Long Hill, Fairfield, Conn. (part of Trumbull) D4 84
Longhorn, Bexar, Tex. B4 118
Longhurst, Person, N.C., 1,546 A5 109
Longiram, Indon., 5,000 F5 35
Long Island, Jackson, Ala., 500 A4 78
Long Island, Phillips, Kans., 229 C4 93
Long Island, Cumberland, Maine (part of Portland) E5 96
Longisland, Catawba, N.C., 225 B3 109
Long Island, Campbell, Va., 75 D3 121
Long Island, sound, Conn., N.Y. E5 84, D3 108
Long Lake, Lake, Ill., 3,502 *E2 90
Long Lake, Iosco, Mich., 125 D7 98
Long Lake, Hennepin, Minn., 996 *F5 99
Long Lake, Hamilton, N.Y., 700 B6 108
Longlake, McPherson, S. Dak., 109 B6 116
Long Lake, Florence, Wis., 175 C5 124
Long Lane, Dallas, Mo., 110 D5 101
Longleaf, Rapides, La., 600 C3 95
Long Leaf Park (South Wilmington), New Hanover, N.C., 2,238 *C3 109
Longlegged, lake, Ont., Can. D4 71
Longmeadow, Hampden, Mass., 10,565 B2 97
Longmont, Boulder, Colo., 11,489 A5 83
Long Park, Montrose, Colo., 25 C2 83
Long Pine, Brown, Nebr., 487 B6 103
Long Plain, Bristol, Mass., 130 C6 97
Long Point, N.S., Can., 179 B8 74
Long Point, Livingston, Ill., 307 B5 90
Long Pond, Newf., Can., 1,244 E5 75
Long Pond, Somerset, Maine, 70 C2 96
Longport, Atlantic, N.J., 1,077 E3 106
Long Prairie, Todd, Minn., 2,414 E4 99
Long Range, mts., Newf., Can. D3, E2 75
Long Rapids, Alpena, Mich., 90 C7 98
Longreach, Austl., 3,806 D7 51
Long Ridge, Fairfield, Conn. (part of Stamford) E3 84
Longridge, Eng., 4,677 G6 13
Long Run, Doddridge, W. Va., 100 B4, B6 123
Longs, creek, Ark. A2 81
Longs, peak, Colo. A5 83
Long Savannah, Hamilton, Tenn. E10 117
Longstreet, DeSoto, La., 283 B2 95
Longton, Elk, Kans., 401 E7 93
Longtown, Eng., 2,577 E6 13
Longtown, Perry, Mo., 113 D8 101
Longueau, Fr., 5,316 E3 14
Longueuil, Que., Can., 24,131 D4, D9 73
Longuyon, Fr., 8,266 C6 14
Longvale, Mendocino, Calif., 130 C2 82
Long Valley, Coconino, Ariz. C4 80
Long Valley, Morris, N.J., 1,220 B3 106
Longvalley, Washabaugh, S. Dak., 10 D4 116
Long Valley Junction, Kane, Utah F3 119
Longview, Alta., Can., 246 D3 69
Longview, Champaign, Ill., 270 D5 90
Long View, Hardin, Ky., 500 C4 94
Longview, Oktibbeha, Miss., 227 B5 100
Longview, Catawba, N.C., 2,997 B2 109
Longview, Gregg, Tex., 40,050 C5 118
Longview, Cowlitz, Wash., 23,349 C3 122
Longville, Beauregard, La., 140 D2 95
Longville, Cass, Minn., 159 C4 99
Longwood, Seminole, Fla., 1,689 *D5 86
Longwood, Pettis, Mo., 35 C4 101
Longwood, Brunswick, N.C., 250 C5 109
Longwoods, Talbot, Md., 70 C5 85
Longwy, Fr., 21,939 (*75,000) C6 14
Long Xuyen, Viet., 5,000 G6 38
Loning, China G4 36
Lonkin, Bur., 5,000 C10 39
Lonoke, Lonoke, Ark., 2,359 C4, D6 81
Lonoke, co., Ark., 24,551 C4 81
Lonsdal, Nor. D6 25
Lonsdale, Garland, Ark., 95 C3, C6 81
Lonsdale, Ont., Can., 165 C7 72
Lonsdale, Rice, Minn., 541 F5 99
Lons-le-Saunier, Fr., 15,924 D6 14
Looe, Eng., 3,878 D3 12
Loogootee, Martin, Ind., 2,858 G4 91
Lookeba, Caddo, Okla., 158 B3 112
Lookout, Modoc, Calif. B3 82
Lookout, Pike, Ky., 900 C7 94
Lookout, Woods, Okla., 5 A2 112
Lookout, Fayette, W. Va., 600 C4, D7 123
Lookout, Albany, Wyo., 20 D7 125
Lookout, cape, N.C. C7 109
Lookout, mtn., Oreg. C6 113
Lookout, mtn., Oreg. C8 113
Lookout, mtn., Oreg. C9 113
Lookout, mtn., Tenn. E10 117
Lookout, mtn., Wash. D3 122
Lookout, pass, Mont. C1 102
Lookout, pt., Mich. D7 98
Lookout, ridge, Alsk. B8 79
Lookout Mountain, Hamilton, Tenn., 1,817 E10 117
Lookout Mountain, ridge, Ala., Ga., Tenn. A4 78
Lookout Point, res., Oreg. D4 113
Looma, Alta., Can., 15 C4 69
Loomis, Sask., Can., 75 H1 70
Loomis, Phelps, Nebr., 299 D6 103
Loomis, Davison, S. Dak., 75 D7 116
Loomis, Okanogan, Wash., 190 A6 122
Loon, creek, Sask., Can. G1 70
Loon, lake, Alta., Can. A3 69
Loon, riv., Alta., Can. A3 69
Loon, riv., Man., Can. A1 71
Loon Bay, Newf., Can., 144 D4 75
Loonhaunt, lake, Ont., Can. B5 99
Loon Lake, Stevens, Wash., 50 A8 122
Loon Lake, mts., N.Y. B3 108
Loop, Gaines, Tex., 100 C1 118
Loop, creek, W. Va. D6 123
Loop, head, Ire. E2 11
Loos, Fr., 18,367 B5 14
Loosahatchie, riv., Tenn. B2 117
Loose Creek, Osage, Mo., 200 C6 101
Looxahoma, Tate, Miss., 80 A4 100
Lopatka, cape, Sov. Un. D18 28
Lopei, China, 5,000 C5 37
Lopeno, Zapata, Tex., 10 F3 118
Lopez, Sullivan, Pa., 500 F9 114
Lopez, Phil., 3,644 p14 35
Lopez, cape, Gabon F1 46
Lopez, isl., Wash. A3 122
Loping, China, 18,000 J7 36
Lop Nor, lake, China D3 34
Lopori, riv., Con. L. A3 48
Lopphavet, sea, Nor. B9 25
Lora, Sp., 11,864 D3 20
Lora, riv., Afg. B3 39
Lorado, Logan, W. Va., 700 D3, D6 123
Lorain, Lorain, Ohio, 68,932 A5 111
Lorain, co., Ohio, 217,500 A5 111
Lorain, Cambria, Pa., 1,324 *F4 114
Loraine, Adams, Ill., 303 C2 90
Loraine, Renville, N. Dak., 54 A4 110
Loraine, Mitchell, Tex., 837 C2 118
Loralai, Pak., 4,437 B2 40
Loramie, res., Ohio B3 111
Loranger, Tangipahoa, La., 60 D5 95
Lorca, Sp., 21,000 (58,641▲) D5 20
Lord Howe, is., Pac. O. I9 7
Lordsburg, Hidalgo, N. Mex., 3,436 E1 107
Loreauville, Iberia, La., 655 D4 95
Loreburn, Sask., Can., 302 F2 70
Lore City, Guernsey, Ohio, 458 C6 111
Loree Estates, Santa Clara, Calif., 925 *C5 82
Lorena, Braz., 26,068 C3 56
Lorenz, Upshur, W. Va., 200 B4 123
Lorenzo, Jefferson, Idaho, 200 F7 89
Lorenzo, Cheyenne, Nebr., 100 C2 103
Lorenzo, Crosby, Tex., 1,188 C2 118
Loreto, Arg. E3 55
Loreto, Braz., 783 C1 57
Loreto, Col. D3 60
Loreto, It., 7,908 C4 21
Loreto, Mex., 1,409 B2 63
Loreto, Par., 11,321 D4 55
Loreto, dept., Peru, 350,306 C3 58
Loretta, Rush, Kans., 60 D4 93
Loretta, Sawyer, Wis., 110 C3 124
Lorette, Man., Can., 400 E3 71
Loretteville, Que., Can., 6,522 C6, C8 73
Loretto, Duval, Fla., 200 C6 86
Loretto, Marion, Ky., 500 C4 94
Loretto, Dickinson, Mich., 250 C3 98
Loretto, Boone, Nebr., 100 C7 103
Loretto, Cambria, Pa., 1,338 E4 114
Loretto, Lawrence, Tenn., 929 B4 117
Lorica, Col., 8,420 B2 60
Lorida, Highlands, Fla., 300 E5 86
Lorient, Fr., 60,566 D2 14
L'Orignal, Ont., Can., 1,189 B10 72
Lorimor, Union, Iowa, 460 C3 92
Loring, Phillips, Mont., 50 B9 102
Loris, Horry, S.C., 1,702 C10 115
Lorlie, Sask., Can., 56 G4 70
Lorman, Jefferson, Miss., 200 D2 100
Lorne, firth, Scot. D3 13
Lorne Park, Ont., Can., 555 E6 72
Lorneville, Ont., Can., 51 C5 72
Lörrach, Ger., 30,500 E3 16
Lorraine, Ellsworth, Kans., 157 D5 93
Lorraine, Jefferson, N.Y., 150 B5 108
Lorraine, plat., Fr. F6 15
Lorraine, former prov., Fr., 1,956,000 C6 14
Lorsch, Ger., 8,900 D3 17
Lorton, Otoe, Nebr., 58 D9 103
Lorup, Ger., 1,800 B7 15
Los Alamitos, Orange, Calif., 4,312 *F2 82
Los Alamos, Santa Barbara, Calif., 500 E3 82
Los Alamos, Los Alamos, N. Mex., 13,037 B3 107
Los Alamos, San Miguel, N. Mex. B4 107
Los Alamos, co., N. Mex., 13,037 B3 107
Los Altos, Santa Clara, Calif., 19,696 C5 82
Los Altos Hills, Santa Clara, Calif., 3,412 *C5 82
Los Amates, Guat., 628 C3 62
Los Amigos Privados, Cuba k11 64
Los Andes, Chile, 19,162 A2 54
Los Angeles, Los Angeles, Calif., 2,479,015 (*6,565,000) E4, F2 82
Los Angeles, Chile, 33,000 B2 54
Los Angeles, co., Calif., 6,038,771 E4, F2 82
Los Angeles, aqueduct, Calif. E4 82
Losantville (Bronson), Randolph, Ind., 251 D7 91
Los Banos, Merced, Calif., 6,090 D3 82
Los Barrios, Sp., 3,583 D3 20
Los Blancos, Arg. D3 55
Los Caños Central, P.R. B4 65
Los Cerrillos, Arg. A3 54
Los Chavez, Valencia, N. Mex. C3 107
Los Ebanos, Hidalgo, Tex., 750 F3 118
Los Fresnos, Cameron, Tex., 1,289 F4 118
Los Gatos, Santa Clara, Calif., 9,036 C5, D3 82
Loshan, China, 32,000 F5 34
Loshan, China, 5,000 H6 36
Los Herreras, Mex., 1,421 G3 63
Losinj, isl., Yugo. C2 22
Los Llanos, P.R. C5 65
Los Lunas, Valencia, N. Mex., 1,186 C3 107
Los Mochis, Mex., 37,682 B3 63
Los Molinos, Tehama, Calif., 900 B2 82
Los Muertos, P.R. B4 65
Los Nietos, Los Angeles, Calif., 9,000 *F2 82
Losombo, Con. L. A2 48
Los Palacios, Cuba, 4,008 D2 64
Los Palacios, Sp., 12,524 D3 20
Los Pinos, Cuba k11 64
Los Pinos, Rio Arriba, N. Mex., 130 A3 107
Los Pinos, riv., Colo. D3 83
Los Pozos, Chile E1 55
Los Rábanos, P.R. C3 65
Los Reyes, Mex., 9,777 D4, n12 63
Los Reyes [la Paz], Mex., 2,171 h10 63
Los Ríos, prov., Ec., 174,335 B2 58
Los Santos, Pan., 3,165 G7 62
Los Santos, Sp., 9,565 C2 20
Los Sarmientos, Arg. E2 55
Los Sauces, Chile, 2,158 B2 54
Lossiemouth, Scot., 5,855 C5 13
Lössnitz, Ger., 7,786 C7 17
Lost, creek, Wyo. C4 125
Lost, riv., Ind. G4 91
Lost, riv., Minn. C3 99
Lost, riv., Wash. A5 122
Lost, riv., W. Va. B6 123
Lostant, La Salle, Ill., 460 B4 90
Los Taques, Ven., 8,505 A3 60
Lost Cabin, Fremont, Wyo., 47 B5 125
Lost City, Hardy, W. Va., 75 C6 123
Lost Creek, Harrison, W. Va., 678 B4 123
Los Teques, Ven., 34,874 A4 60
Lost Hills, Kern, Calif., 200 E4 82
Lost Horse, plat., Wash. C4 122
Lostine, Wallowa, Oreg., 240 B9 113
Lost Nation, Clinton, Iowa, 567 C7 92
Lost River, Hardy, W. Va., 30 C6 123
Lost River, cave, Ky. D3 94
Lost River, range, Idaho E5 89
Lost River Glacial, caverns, N.H. B3 105
Lost Springs, Marion, Kans., 139 D7 93
Lost Springs, Converse, Wyo., 5 C7 125
Lost Trail, pass, Mont. E3 102
Lostwood, Mountrail, N. Dak., 34 A3 110
Los Vilos, Chile, 1,305 A2 54
Lot, Ponape 52
Lot, dept., Fr., 149,929 *E4 14
Lot, hbr., Ponape 52
Lot, riv., Fr. E4 14
Lota, Chile, 45,000 B2 54
Lotawana, lake, Mo. E2 101
Lotbinière, Que., Can., 561 C6 73
Lotbiniere, co., Que., Can., 30,234 C6 73
Lot-et-Garonne, dept., Fr., 275,028 *E4 14
Lothair, Perry, Ky., 1,082 C6 94
Lothair, Liberty, Mont., 50 B5 102
Lothrop, Alta., Can. B1 69
Loting, China, 10,000 A9 38
Loto, Con. L. B3 48
Lotofaga, W. Sam., 418 52
Lötschberg, tunnel, Switz. D4 19
Lott, Falls, Tex., 924 D4 118
Lotung, Taiwan G9 34
Lotus Point, Lake, Ill., 1,000 *A5 90
Louann, Ouachita, Ark., 261 D3 81
Loudima, Con. B. F2 46
Loudon, Merrimack, N.H., 350 (1,194▲) D4 105
Loudon, Loudon, Tenn., 3,812 D9 117
Loudon, co., Tenn., 23,757 D9 117
Loudonville, Albany, N.Y., 5,500 *C7 108
Loudonville, Ashland, Ohio, 2,611 B5 111
Loudoun, co., Va., 24,549 B5 121
Loudun, Fr., 5,587 D4 14
Louetta, Harris, Tex. F5 118
Louga, Sen., 13,200 C1 45
Loughborough, Eng., 38,621 B6 12
Lougheed, Alta., Can., 217 C5 69
Loughman, Polk, Fla., 250 D5 86
Loughrea, Ire., 2,784 D3 11
Loughros More, bay, Ire. C3 11
Loughton, Eng. (part of Chigwell) k13 10
Louin, Jasper, Miss., 389 C4 100
Louisa, Lawrence, Ky., 2,071 B7 94
Louisa, Louisa, Va., 576 C5 121
Louisa, co., Iowa, 10,290 C6 92
Louisa, co., Va., 12,959 C5 121
Louisa, lake, Fla. D5 86
Louisbourg, N.S., Can., 1,417 D10 74
Louisburg, Miami, Kans., 862 D9 93
Louisburg, Lac qui Parle, Minn., 91 E2 99
Louisburg, Dallas, Mo., 176 D4 101
Louisburg, Franklin, N.C., 2,862 A5 109
Louisburgh, Ire., 346 D2 11
Louisdale, N.S., Can., 793 D8 74
Louise, Troup, Ga., 120 C2 87
Louise, Humphreys, Miss., 481 B3 100
Louise, Wharton, Tex., 900 E4 118
Louise, Brooke, W. Va., 75 B2 123
Louise, isl., B.C., Can. C2 68
Louise, lake, Alsk. f18 79
Louis Gentil, Mor., 4,835 C3 44
Louisiade, arch., Sol. Is. G9 7
Louisiana, Pike, Mo., 4,286 B6 101
Louisiana, state, U.S., 3,257,022 D9 77, 95
Louisiana, pt., La. E2 95
Louis Trichardt, S. Afr., 9,703 B4 49
Louisville, Barbour, Ala., 890 D4 78
Louisville, Que., Can., 4,138 C5 67
Louisville, Boulder, Colo., 2,073 B5 83
Louisville, Jefferson, Ga., 2,413 C4 87
Louisville, Clay, Ill., 906 E5 90
Louisville, Pottawatomie, Kans., 204 C7 93
Louisville, Jefferson, Ky., 390,639 (*735,800) A4, B4 94
Louisville, Winston, Miss., 5,066 B4 100
Louisville, Cass, Nebr., 1,194 D9, E3 103
Louisville, St. Lawrence, N.Y., 150 A2 108
Louisville, Stark, Ohio, 5,116 B6 111
Louisville, Blount, Tenn., 200 D9, E11 117
Loukhi, Sov. Un. D15 25
Loulé, Port., 6,479 D1 20
Louny, Czech., 12,300 C2 26
Loup, co., Nebr., 1,097 C6 103
Loup, riv., Que., Can. B8 73
Loup, riv., Que., Can. C4 73
Loup, riv., Nebr. C8 103
Loup City, Sherman, Nebr., 1,415 C7 103
Lourdes, Newf., Can., 975 D2 75
Lourdes, Que., Can. D5 73
Lourdes, Que., Can., 174 C6 73
Lourdes, Fr., 16,023 F4 14
Lourenço Marques, Moz., 99,000 C5 49
Lourenço Marques, prov., Moz. C5 49
Loures, Port., 6,089 f9 20
Lousã, Port., 8,922 B1 20
Lousana, Alta., Can., 74 C4 69
Louth, Eng., 11,556 A8 12
Louth, Ire., 207 D5 11
Louth, co., Ire., 67,378 D5 11
Loutra Aidhipsou, Grc., 5,028 C4 23
Loutrákion (Loutrákion-Perakhóra), Grc., 6,168 h9 23
L'Outre, bayou, La. B3 95
Louvain, see Leuven, Bel.
Louvale, Stewart, Ga., 300 D2 87
Louviers, Douglas, Colo., 500 B6 83
Louviers, Fr., 13,160 C4 14
Louzi, Con. L. B1 48
Lovat, riv., Sov. Un. C8 27
Love, Sask., Can., 155 D3 70
Love, co., Okla., 5,862 D4 112
Love, pt., Md. B5 85
Lovech, Bul., 11,730 D7 22
Lovejoy, St. Clair, Ill., 1,922 *E3 90
Lovelaceville, Ballard, Ky., 200 A2 94
Lovelady, Houston, Tex., 466 D5 118
Loveland, Larimer, Colo., 9,734 A5 83
Loveland, Pottawattamie, Iowa, 100 C2 92
Loveland, Hamilton, Warren and Clermont, Ohio, 5,008 C3, D2 111
Loveland, Tillman, Okla., 90 C3 112
Loveland, pass, Colo. B5 83
Loveland Park, Warren, Ohio, 1,000 C2, C3 111
Lovell, Oxford, Maine, 65 (588▲) D2 96
Lovell, Logan, Okla., 27 A4 112
Lovell, Big Horn, Wyo., 2,451 A4 125
Lovelock, Pershing, Nev., 1,948 C3 104
Lovely, Martin, Ky., 600 C7 94
Loverna, Sask., Can., 145 F1 70
Loves Park, Winnebago, Ill., 9,086 A4 90
Lovett, Laurens, Ga., 61 D4 87
Lovett, Jennings, Ind., 40 G6 91
Lovettsville, Loudoun, Va., 217 B5 121
Lovewell, pond, Maine D2 96
Lovick, Jefferson, Ala., 250 E5 78
Lovilia, Monroe, Iowa, 630 C5 92
Loving, Eddy, N. Mex., 1,646 E5 107
Loving, co., Tex., 226 D1 118
Lovingston, Nelson, Va., 375 D4 121
Lovington, Moultrie, Ill., 1,200 D5 90
Lovington, Polk, Iowa, 800 A7 92
Lovington, Lea, N. Mex., 9,660 E6 107
Lovisa, Fin., 6,600 G12 25
Lovosice, Czech., 4,962 C9 17
Low, Que., Can., 422 D2 73
Low, cape, N.W. Ter., Can. D15 67
Low, pen., B.C., Can. E5 68
Lowa, Con. L. B4 48
Lowa, riv., Con. L. B4 48
Lowden, Cedar, Iowa, 641 C7 92
Lowe Farm, Man., Can., 310 E3 71
Lowell, Benton, Ark., 277 A1 81
Lowell, Lake, Ind., 2,270 B3 91
Lowell, Middlesex, Mass., 92,107 (*147,400) A5 97
Lowell, Kent, Mich., 2,545 F5 98
Lowell, Kearney, Nebr., 84 D7 103
Lowell, Gaston, N.C., 2,784 B2 109
Lowell, Washington, Ohio, 783 C6 111
Lowell, Lane, Oreg., 503 D4 113
Lowell, Orleans, Vt., 135 (617▲) B4 120
Lowell, Snohomish, Wash., 1,086 *B3 122
Lowell, Summers, W. Va., 100 D4 123
Lowell, isl., Mass. C4 97
Lowell, mts., Vt. B4 120
Lowellville, Mahoning, Ohio, 2,055 A7 111
Lower, bay, N.Y. E1 108
Lower, lake, Nev. B2 104
Lower Arrow, lake, B.C., Can. E8 68
Lower Bank, Burlington, N.J., 110 D3 106
Lower Brule, Lyman, S. Dak., 150 C6 116
Lower Burrell, Westmoreland, Pa., 11,952 *E2 114
Lower Cabot, Washington, Vt., 150 C4 120
Lower East Pubnico, N.S., Can., 114 F4 74
Lower Gilmanton, Belknap, N.H. D4 105
Lower Giuba, dist., Som. E5 47
Lower Hutt, N.Z., 53,044 *N15 51
Lower Island Cove, Newf., Can., 494 D5 75
Lower Lake, Lake, Calif., 550 C2 82
Lower Marlboro, Calvert, Md., 75 C4 85
Lower Matecumbe, key, Fla. H6 86
Lower Paia, Maui, Haw., 950 C5 88
Lower Peach Tree, Wilcox, Ala., 250 D2 78
Lower Red, lake, Minn. C3 99
Lower Salem, Washington, Ohio, 143 C6 111
Lower Saxony (Niedersachsen), state, Ger., 6,641,400 A4 17
Lower Southampton, N.B., Can., 72 C2 74
Lower Village, Lamoille, Vt., 75 C3 120
Lower West Pubnico, N.S., Can., 618 F4 74
Lowes, Graves, Ky., 200 A2 88
Lowestoft, Eng., 45,687 B9 12
Loweth, Meagher, Mont., 20 D6 102
Lowgap, Newton, Ark., 10 A2 81
Lowgap, Surry, N.C., 250 A3 109
Lowicz, Pol., 14,400 B5 26
Lowland, Pamlico, N.C., 500 B7 109
Lowman, Boise, Idaho, 100 E3 89
Lowmansville, Lawrence, Ky., 300 C7 94
Low Moor, Clinton, Iowa, 343 C7 92
Lowmoor, Alleghany, Va., 900 D3 121
Lowndes, Wayne, Mo., 150 D7 101
Lowndes, co., Ala., 15,417 C3 78
Lowndes, co., Ga., 49,270 F3 87
Lowndes, co., Miss., 46,639 B5 100
Lowndesboro, Lowndes, Ala., 250 C3 78
Lowndesville, Abbeville, S.C., 274 C2 115
Lowry, Pope, Minn., 294 E3 99
Lowry, Walworth, S. Dak., 44 B6 116
Lowry City, St. Clair, Mo., 437 C4 101
Lowrys, Chester, S.C., 298 B5 115
Lowryville, Hardin, Tenn. B3 117
Low Tatra, mts., Czech. D5 26
Lowville, Lewis, N.Y., 3,616 B5 108
Loxley, Baldwin, Ala., 831 E2 78
Loxton, Austl., 2,321 G3 51
Loyal, Kingfisher, Okla., 87 B3 112
Loyal, Clark, Wis., 1,146 D3 124
Loyalhanna, Westmoreland, Pa., 1,000 *F3 114
Loyalist, Alta., Can., 86 D5 69
Loyall, Harlan, Ky., 1,260 D6 94
Loyalsock, creek, Pa. D8 114
Loyalton, Sierra, Calif., 936 C3 82
Loyalton, Edmunds, S. Dak., 34 B6 116
Loyalty, is., Pac. O. H10 7
Loyang, China, 171,200 G5 36
Loysburg, Bedford, Pa., 200 F5 114
Loysville, Perry, Pa., 500 F7 114
Lozeau, Mineral, Mont., 18 C2 102
Lozère, dept., Fr., 81,868 *E5 14
Loznica, Yugo., 10,611 C4 22
Lozva, riv., Sov. Un. C21 9
Lozovatka, Sov. Un., 10,000 G9 27
Lozovaya, Sov. Un., 10,000 G11 27
Lozoya, canal, Sp. o17 20
Lualaba (Congo), riv., Con. L. B4 48
Lualualei, Honolulu, Haw., 2,100 g9 88
Luama, riv., Con. L. B4 48
Luampa, N. Rh. E3 48
Luan, riv., China C8 34
Luana, Clayton, Iowa, 276 A6 92
Luanda, Ang., 189,590 C1 48
Luanda, dist., Ang., 347,173 C1 48
Luang Prabang, Laos, 25,000 C5 38
Luanguinga, riv., Ang., N. Rh. D3 48
Luanshya, N. Rh., 18,000 (*56,900) C4 48
Luarca, Sp., 4,233 A2 20
Luashi, Con. L. D3 48
Lubaczow, Pol., 4,986 C7 26

Luban, Pol., 11,100...C3 26
Lubang, Phil., 3,149...p13 35
Lubang, is., Phil...C6 35
Lubao, Phil., 1,810...o13 35
Lubartow, Pol., 5,542...C7 26
Lubawa, Pol., 4,679...B5 26
Lübbecke, Ger., 10,500...A3 17
Lübben, Ger., 9,433...B8 17
Lübbenau, Ger., 12,000...B8 17
Lubbock, Lubbock, Tex., 128,691 (★144,300)...C2 118
Lubbock, co., Tex., 156,271...C2 118
Lubec, Washington, Maine, 1,289 (2,684▲)...D6 96
Lübeck, Ger., 234,600...B5 16, E4 24
Lübeck, bay, Ger...B5 16
Lubefu, Con. L...B3 48
Lubero, Con. L...B4 48
Lubicon, lake, Alta., Can...A3 69
Lubilash, riv., Con. L...C3 48
Lubin, Pol., 1,769...C4 26
Lublin, Pol., 181,000...C7 26
Lublin, Taylor, Wis., 160...C3 124
Lubliniec, Pol., 12,400...C5 26
Lubny, Sov. Un., 10,000...F9 27
Lubrín, Sp., 5,417...D4 20
Lubsko, Pol., 10,700...C3 26
Lübtheen, Ger., 5,707...E5 24
Lubudi, Con. L., 7,900...C4 48
Lubudi, riv., Con. L...D3 48
Lubue, Con. L...B2 48
Lubutu, Con. L...B4 48
Lucama, Wilson, N.C., 498...B5 109
Lucan, Ont., Can., 986...D3 72
Lucan, Ire., 1,657...D5 11
Lucan, Redwood, Minn., 216...F3 99
Lucania, see Basilicata, reg., It.
Lucania, mtn., Yukon, Can...D4 66
Lucas, Lucas, Iowa, 357...C4 92
Luca , Russell, Kans., 559...C5 93
Lucas, Barren, Ky., 150...D3 94
Lucas, Missaukee, Mich., 100...D5 98
Lucas, Richland, Ohio, 719...B5 111
Lucas, Gregory, S. Dak., 24...D6 116
Lucas, co., Iowa, 10,923...C4 92
Lucas, co., Ohio, 456,931...A4 111
Lucas, chan., Ont., Can...B3 72
Lucasville, Scioto, Ohio, 1,277...D5 111
Lucca, It., 88,428...C3 21
Lucca, Barnes, N. Dak., 37...C8 110
Luce, co., Mich., 7,827...B5 98
Luce, bay, Scot...F4 13
Lucea, Jam., 2,798...F4 64
Lucedale, George, Miss., 1,977...E5 100
Lucena, Phil., 28,000 (49,300▲)...C6, p13 35
Lucena, Sp., 19,400...D3 20
Lucena del Cid, Sp., 3,585...B5 20
Lučenec, Czech., 16,100...D5 26
Lucera, It., 24,500 (28,409▲)...D5 21
Lucerne, Lake, Calif., 402...C4 82
Lucerne, Weld, Colo., 75...A6 83
Lucerne, Cass, Ind., 215...C5 91
Lucerne, Putnam, Mo., 157...A4 101
Lucerne, see Luzern, Switz.
Lucerne, Hot Springs, Wyo., 25...B4 125
Lucerne, lake, Switz...B5 19
Lucernemines, Indiana, Pa., 1,524...E3 114
Lucerne Valley, San Bernardino, Calif., 900...E5 82
Lucero, Cuba...h12 64
Luceville, Que., Can., 1,419...A9 73
Luchi, China, 10,000...J4 36
Lüchow, Ger., 5,900...F5 24
Luchuan, China, 5,000...G7 34
Lucien, Franklin, Miss., 150...D3 100
Lucien, Noble, Okla., 100...A4 112
Lucile, Miller, Ga., 60...E2 87
Lucile, Idaho, Idaho, 20...D2 89
Lucin, Box Elder, Utah, 75...B2 119
Lucinda, Clarion, Pa., 300...D3 114
Lucipara, is., Indon...G7 35
Lucira, Ang...D1 48
Luck, Madison, N.C., 100...D3 109
Luck, Polk, Wis., 853...C1 124
Luckau, Ger., 6,145...B8 17
Luckenwalde, Ger., 28,600...A8 17
Luckey, Wood, Ohio, 946...A2, A4 111
Lucknow, Ont., Can., 1,031...D3 72
Lucknow, India, 595,440 (★675,000)...C7 39, D8 40
Lucknow, Dauphin, Pa., 900...*F8 114
Lucky Lake, Sask., Can., 426...G2 70
Lucky Peak, res., Idaho...F2 89
Lucky Strike, Alta., Can., 25...E5 69
Luçon, Fr., 7,599...D3 14
Lucy, St. John the Baptist, La., 725...B6 95
Lucy, Shelby, Tenn., 130...B2, E8 117
Luda Kamchiya, riv., Bul...D8 22
Ludden, Dickey, N. Dak., 59...C7 110
Ludell, Rawlins, Kans., 105...C3 93
Lüdenscheid, Ger., 58,200...B2 17
Lüderitz, S.W. Afr., 2,836...C2 49
Lüderitz, bay, S.W. Afr...C1 49
Ludgate, Ont., Can...B4 72
Ludhiana, India, 244,032...B6 39, B5 40
Lüdinghausen, Ger., 9,500...B2 17
Ludington, Mason, Mich., 9,421...E4 98
Ludlam, Dade, Fla., 20,000...*G6 86
Ludlow, San Bernardino, Calif., 250...E5 82
Ludlow, N.B., Can., 240...C3 74
Ludlow, Las Animas, Colo., 20...D6 83
Ludlow, Eng., 6,774...B5 12
Ludlow, Champaign, Ill., 460...C5 90
Ludlow, Kenton, Ky., 6,233...A7 94
Ludlow, Aroostook, Maine, 25 (274▲)...B4 96
Ludlow, Hampden, Mass., 8,000 (13,805▲)...B3 97
Ludlow, Scott, Miss., 40...C4 100
Ludlow, Livingston, Mo., 235...B4 101
Ludlow, McKean, Pa., 800...C4 114
Ludlow, Harding, S. Dak., 10...B2 116
Ludlow, Windsor, Vt., 1,658 (2,386▲)...E3 120
Ludlow, mtn., Vt...E3 120
Ludlow Falls, Miami, Ohio, 273...C3 111
Ludowici, Long, Ga., 1,578...E5 87
Ludvika, Swe., 12,300 (★18,000)...G6 25
Ludwigs, canal, Ger...D6 17
Ludwigsburg, Ger., 73,500...E4 17
Ludwigshafen, Ger., 165,800...D3 17
Ludwigslust, Ger., 12,000...B5 16
Luebo, Con. L...C3 48
Lueders, Jones, Tex., 654...C3 118
Luepa, Ven...B5 60
Lufkin, Angelina, Tex., 17,641...D5 118
Lufuna, Con. L...C2 48
Luga, Sov. Un., 43,000...B7 27
Luga, riv., Sov. Un...B7 27
Lugagnano Val d'Arda, It., 6,240...E5 18
Lugan, riv., Sov. Un...q22 27
Luganchik, riv., Sov. Un...q22 27
Lugano, Switz., 19,758 (★33,139)...D6 19
Lugano, lake, Switz...E6 19
Lugansk, Sov. Un., 300,000...G12, q22 27
Lugau, Ger., 10,600...C7 17
Lugela, riv., Moz...A6 49
Lugenda, riv., Moz...D6 48
Lugert, Kiowa, Okla., 20...C2 112
Lugh Ferrandi, Som., 2,800...E5 47
Lugnaquilla, mtn., Ire...E5 11
Lugo, It., 12,800...B3 21
Lugo, Sp., 40,900 (58,264▲)...A2 20
Lugo, prov., Sp., 479,530...*A2 20
Lugoff, Kershaw, S.C., 200...C6 115
Lugoj, Rom., 30,258...C5 22
Lugovoy, Sov. Un...E8 29
Luho, China...E5 34
Luhsien, China, 289,000...F6 34
Luichow, China, 158,800...G6 34
Luichow, pen., China...G6 34
Luing, isl., Scot...D3 13
Luís Correia, Braz., 1,523...B2 57
Luís Gomes, Braz., 1,480...C3 57
Luis Lopez, Socorro, N. Mex., 25...D3 107
Luitpold, coast, Ant...B9 5
Luitpold, mtn., N. Gui...k12 50
Luiza, Con. L...C3 48
Luján, Arg., 19,001...g7 54
Luján [de Cuyol], Arg., 3,542...A3 54
Lukachukai, Apache, Ariz., 20...A6 80
Luke, Allegany, Md., 587...D1 85
Lukenie, riv., Con. L...B2 48
Lukeville, Pima, Ariz., 30...F3 80
Lukolela, Con. L...B2 48
Lukovit, Bul., 7,755...D7 22
Lukow, Pol., 8,513...C7 26
Lukoyanov, Sov. Un., 5,000...D15 27
Lukuga, riv., Con. L...C4 48
Lukulu, N. Rh...D3 48
Lula, Hall, Ga., 557...B3 87
Lula, Coahoma, Miss., 484...A3 100
Lula, Pontotoc, Okla., 55...C5 112
Lula, Chester, Tenn...B3 117
Luleå, Swe., 31,600...E10 25
Luleälven, riv., Swe...D9 25
Luling, St. Charles, La., 2,122...C7 95
Luling, Caldwell, Tex., 4,412...B4, E4 118
Lulu, Columbia, Fla., 152...B4 86
Lulua, riv., Con. L...C3 48
Luluabourg, Con. L., 59,500...C3 48
Lulung, China...E8 36
Lum, Lapeer, Mich., 265...E7 98
Lumba, Con. L...A2 48
Lumber, Darlington, S.C...C8 115
Lumber, riv., N.C...C5 109
Lumber Bridge, Robeson, N.C., 100...C4 109
Lumber City, Telfair, Ga., 1,360...E4 87
Lumberport, Harrison, W. Va., 1,031...A6, B4 123
Lumberton, Lamar, Miss., 2,108...D4 100
Lumberton, Burlington, N.J., 600...D3 106
Lumberton, Rio Arriba, N. Mex., 250...A3 107
Lumberton, Robeson, N.C., 15,305...C5 109
Lumbo, Moz...D7 48
Lumby, B.C., Can., 842...D8 68
Luminao, reef, Guam...52
Lummi, Indian res., Wash...A3 122
Lumpkin, Stewart, Ga., 1,348...D2 87
Lumpkin, co., Ga., 7,241...B2 87
Lumsden, Newf., Can., 269...D5 75
Lumsden, Sask., Can., 685...G3 70
Lumsden, N.Z., 666...P12 51
Lumsden, Scot...C6 13
Lumut, mtn., Indon...F6 35
Lumut, pt., Indon...F3 35
Luna, Catron, N. Mex., 220...D1 107
Luna, Lincoln, N. Mex., 15...D4 107
Luna, co., N. Mex., 9,839...E2 107
Lunan, bay, Scot...D6 13
Luna Pier (Lakewood), Monroe, Mich., 1,815...G7 98
Lund, Caribou, Idaho, 10...G7 89
Lund, White Pine, Nev., 375...E6 104
Lund, Swe., 41,000...J5 25
Lund, Iron, Utah, 50...E2 119
Lunda, dist., Ang., 247,449...C2 48
Lundale, Logan, W. Va., 500...D3, D5 123
Lundar, Man., Can., 713...D2 71
Lundazi, N. Rh...D5 48
Lundbreck, Alta., Can., 112...E3 69
Lundby, Den., 549...C5 24
Lundby, Swe...t34 25
Lundell, Phillips, Ark., 150...C5 81
Lunderskov, Den., 1,181...C3 24
Lundu, Mala., 768...E3 35
Lundy, isl., Eng...C3 12
Lune, riv., Eng...F6 13
Lüneburg, Ger., 59,600...B5 16
Lunel, Fr., 7,758...F6 14
Lünen, Ger., 72,200...B2 17
Lunenburg, N.S., Can., 3,056...E5 74
Lunenburg, Worcester, Mass., 2,000 (6,334▲)...A4 97
Lunenburg, Essex, Vt., 250 (1,237▲)...C5 120
Lunenburg, Lunenburg, Va., 40...E5 121
Lunenburg, co., N.S., Can., 34,998...E5 74
Lunenburg, co., Va., 12,523...E4 121
Lunéville, Fr., 21,618...C7 14
Lung, China...G2 36
Lungchen, China, 1,000...B3 37
Lungchi, China, 81,200...G8 34
Lungchuan, China, 12,000...J8 36
Lungchuchai, China...H4 36
Lunghsi, China...E5 34
Lungkou, China...F9 36
Lungping, China...F6 36
Lungtsin, China...G6 34
Lungue-Bungo, riv., Ang...D3 48
Luni, riv., India...E3 40
Lunino, Sov. Un., 10,000...E15 27
Luning, Mineral, Nev., 55...E3 104
Lunyama, Con. L...C4 48
Lup, Thai., 6,940...D5 38
Lupani, S. Rh...A4 49
Lupei, China, 1,000...B9 36
Lupeni, Rom., 21,188...C6 22
Lupin, see Manchouli, China.
Lupton, Apache, Ariz., 250...B6 80
Lupton, Ogemaw, Mich., 100...D6 98
Lupton City, Hamilton, Tenn., 250...E10 117
Lupus, Moniteau, Mo., 75...C5 101
Luque, Par., 22,469...E4 55
Luquillo, P.R., 2,107...B8, f12 65
Luquillo, mun., P.R., 8,582...B8, f12 65
Luquillo, mts., P.R...B7 65
Luray, Russell, Kans., 328...C5 93
Luray, Clark, Mo., 154...A6 101
Luray, Hampton, S.C., 102...F5 115
Luray, Henderson, Tenn., 100...B3 117
Luray, Page, Va., 3,014...C4 121
Lure, Fr., 6,408...D7 14
Lurgan, N. Ire., 17,873...C5 11
Luribay, Bol., 392...C2 55
Lurín, Peru, 2,141...D2 58
Lúrio, Moz...D7 48
Lurio, riv., Moz...D6 48
Lurton, Newton, Ark., 100...B2 81
Lusaka, N. Rh., 75,000 (★85,800)...E4 48
Lusambo, Con. L., 9,800...B3 48
Lusby, Calvert, Md., 75...D5 85
Luscar, Alta., Can., 301...C2 69
Luseland, Sask., Can., 655...E1 70
Lushan, China...H5 36
Lushnjë, Alb., 8,585...B2 23
Lushoto, Tan., 1,270...B6 48
Lushton, York, Nebr., 45...D8 103
Lushun, see Port Arthur, China
Lusignan, Fr., 1,376...D4 14
Lusk, Throckmorton, Tex...C3 118
Lusk, Niobrara, Wyo., 1,890...C8 125
Lustre, Valley, Mont., 5...B11 102
Lutcher, St. James, La., 3,274...B6, D5 95
Lutesville, Bollinger, Mo., 658...D8 101
Luther, Lake, Mich., 325...D5 98
Luther, Carbon, Mont., 20...E7 102
Luther, Oklahoma, Okla., 517...B4 112
Luther, Hancock, Tenn...C10 117
Luthersville, Meriwether, Ga., 282...C2 87
Lutherville, Baltimore, Md., 12,265...B4 85
Luting, China...F5 34
Lütjenburg, Ger., 4,400...D4 24
Luton, Eng., 131,505 (★170,000)...C7 12
Luton, Woodbury, Iowa, 130...B1 92
Lutsen, Cook, Minn., 30...B6 99
Lutsk, Sov. Un., 63,000...F5 27
Luttrell, Union, Tenn., 600...C10 117
Lutts, Wayne, Tenn., 350...B4 117
Lutugino, Sov. Un., 5,000...q22 27
Lutz, Hillsborough, Fla., 700...D4 86
Lützow-Holm, bay, Ant...C16 5
Luverne, Crenshaw, Ala., 2,238...D3 78
Lu Verne, Kossuth and Humboldt, Iowa, 468...B3 92
Luverne, Rock, Minn., 4,249...G2 99
Luverne, Steele, N. Dak., 109...B8 110
Luvua, riv., Con. L...C4 48
Luwingu, N. Rh., 540...D5 48
Luwuk, Indon...F6 35
Luxapalila, creek, Ala...B2 78
Luxembourg, Lux., 71,653...E6 15
Luxembourg, country, Eur., 314,889...F10 8, E6 15
Luxembourg, prov., Bel., 216,848...E5 15
Luxemburg, Dubuque, Iowa, 159...B6 92
Luxemburg, Kewaunee, Wis., 730...D6 124
Luxeuil-les-Bains, Fr., 8,161...D7 14
Luxomni, Gwinnett, Ga., 180...A5 87
Luxor, see Al Uqşur, Eg., U.A.R.
Luxora, Mississippi, Ark., 1,236...B6 81
Luyanó, Cuba...h12 64
Luz, Braz., 5,633...E1 51
Luza, riv., Sov. Un...A3 29
Luzern, Switz., 67,433 (★123,500)...B5 19
Luzern, canton, Switz., 253,446...B5 19
Luzerne, Oscoda, Mich., 75...D6 98
Luzerne, Luzerne, Pa., 5,118...B8 114
Luzerne, co., Pa., 346,972...D9 114
Lužnice, riv., Czech...D3 26
Luzon, isl., Phil...B6, O13 35
Lvov, Sov. Un., 436,000...D8 24, G5 27
Lwanhsien, China...E8 34
Lwow, see Lvov, Sov. Un.
Lwowek Slaskie, Pol., 3,364...C3 26
Lyall, mtn., B.C., Can...D10 68
Lyall, mtn., N.Z...P11 51
Lyallpur, Pak., 425,248...B5 38, B4 40
Lyaskovets, Bul., 5,560...D7 22
Lybster, Scot...B5 13
Lychen, Ger., 3,649...E7 24
Lycksele, Swe., 5,800...E8 25
Lycoming, Oswego, N.Y., 150...B4 108
Lycoming, co., Pa., 109,367...D7 114
Lycoming, creek, Pa...D7 114
Lyden, Rio Arriba, N. Mex., 100...A3 107
Lydenburg, S. Afr., 7,393...C5 49
Lydia, Washington, Md...A2 85
Lydia, Darlington, S.C., 200...C7 115
Lydia Mills, Laurens, S.C., 1,177...*C4 115
Lydiatt, Man., Can., 50...E3 71
Lydick, St. Joseph, Ind., 1,217...A5 91
Lyell, isl., B.C., Can...C2 68
Lyell, mtn., Alta., B.C., Can...D2 69
Lyerly, Chattooga, Ga., 409...B1 87
Lyford, Parke, Ind., 400...E3 91
Lyford, Willacy, Tex., 1,554...F4 118
Lygnern, lake, Swe...A6 24
Lykens, Dauphin, Pa., 2,527...E8 114
Lykesland, Richland, S.C., 35...D6 115
Lyle, Mower, Minn., 607...G6 99
Lyle, Klickitat, Wash., 400...D4 122
Lyles, Hickman, Tenn., 500...B4 117
Lyleton, Man., Can., 123...E1 71
Lyman, Harrison, Miss., 225...E2, E4 100
Lyman, Scotts Bluff, Nebr., 626...C1 103
Lyman, Grafton, N.H., 25 (201▲)...B3 105
Lyman, Osage, Okla., 30...A5 112
Lyman, Spartanburg, S.C., 1,261...B3 115
Lyman, Lyman, S. Dak., 18...D6 116
Lyman, Wayne, Utah, 255...E4 119
Lyman, Skagit, Wash., 400...A3 122
Lyman, Uinta, Wyo., 425...D2 125
Lyman, co., S. Dak., 4,428...D6 116
Lyman, res., Ariz...C6 80
Lyme, Grafton, N.H., 250 (1,026▲)...C2 105
Lyme, bay, Eng...D4 12
Lyme Center, Grafton, N.H., 75...C2 105
Lymington, Eng., 28,642...D6 12
Lyna, riv., Pol., Sov. Un...B6 26
Lynas, pt., Wales...A3 12
Lynbrook, Nassau, N.Y., 19,881...G2 84
Lynch, Harlan, Ky., 3,810...D7 94
Lynch, Kent, Md., 150...B5 85
Lynch, Boyd, Nebr., 409...B7 103
Lynchburg, Highland, Ohio, 1,022...C4 111
Lynchburg, Lee, S.C., 544...C7 115
Lynchburg, Moore, Tenn., 396...B5 117
Lynchburg (Independent City), Va., 54,790...D3 121
Lynches, riv., S.C...D8 115
Lynch Station, Campbell, Va., 400...D3 121
Lynd, Lyon, Minn., 259...F3 99
Lyndeboro, Hillsboro, N.H., 25 (594▲)...E3 105
Lynden, Ont., Can., 532...D4 73
Lynden, Whatcom, Wash., 2,542...A3 122
Lyndes, neck, Conn...D7 84
Lyndhurst, Ont., Can., 192...C8 73
Lyndhurst, Bergen, N.J., 21,867...D5 106
Lyndhurst, Cuyahoga, Ohio, 16,805...B2 111
Lyndon, Whiteside, Ill., 677...B4 90
Lyndon, Osage, Kans., 953...D8 93
Lyndon, Jefferson, Ky., 5,000...A4 94
Lyndon, Caledonia, Vt., 250 (3,425▲)...B4 120
Lyndon Center, Caledonia, Vt., 274...B4 120
Lyndon Station, Juneau, Wis., 335...E4 124
Lyndonville, Orleans, N.Y., 755...B2 108
Lyndonville, Caledonia, Vt., 1,477...B4 120
Lyndora, Butler, Pa., 5,700...E2 114
Lyngby, Den., 63,712...C6 24
Lynhurst, Marion, Ind., 183...H7 91
Lynn, Winston, Ala., 531...A2 78
Lynn, Lawrence, Ark., 200...A4 81
Lynn, Randolph, Ind., 1,260...D8 91
Lynn, Essex, Mass., 94,478...B6, C3 97
Lynn, Susquehanna, Pa., 75...C10 114
Lynn, Box Elder, Utah...B2 119
Lynn, co., Tex., 10,914...C2 119
Lynn, canal, Alsk...k22 79
Lynn, lake, W. Va...B5 123
Lynn Creek, B.C., Can., 854...A10 68
Lynndyl, Millard, Utah, 145...D3 119
Lynne, Marion, Fla., 200...C5 86
Lynnfield, Essex, Mass., 8,398...C3 97
Lynn Garden, Sullivan, Tenn., 5,261...*C11 117
Lynn Grove, Calloway, Ky., 200...B3 94
Lynn Haven, Bay, Fla., 3,078...G3 86
Lynnhaven, Princess Anne, Va., 350...B7, E6 121
Lynnhaven Roads, hbr., Va...B7 121
Lynn Lake, Man., Can., 1,881...A1, C4 71
Lynnport, Lehigh, Pa., 100...E10 114
Lynnview, Jefferson, Ky., 1,711...*B4 94
Lynnville, Warrick, Ind., 409...H3 91
Lynnville, Jasper, Iowa, 411...C5 92
Lynnville, Graves, Ky., 130...B2 94
Lynnville, Giles, Tenn., 362...B5 117
Lynton, Eng., 1,198...C4 12
Lynnwood, Fayette, Pa., 2,230...*F2 114
Lynnwood, Luzerne, Pa., 1,200...*D10 114
Lynnwood, Snohomish, Wash., 7,207...*B3 122
Lynwood, Los Angeles, Calif., 31,614...F2 82
Lynxville, Crawford, Wis., 183...E2 124
Lyö, isl., Den...C4 24
Lyon (Lyons), Fr., 528,535 (★920,000)...E6 14
Lyon, Coahoma, Miss., 393...A3 100
Lyon, co., Iowa, 14,468...A1 92
Lyon, co., Kans., 26,928...D7 93
Lyon, co., Ky., 5,924...C1 94
Lyon, co., Minn., 22,655...F3 99
Lyon, co., Nev., 6,143...D2 104
Lyon, riv., Scot...D4 13
Lyon Mountain, Clinton, N.Y., 950...A3 108
Lyonnais, former prov., Fr., 1,621,000...E6 14
Lyons, Boulder, Colo., 706...A5 83
Lyons (Lyon), Fr., 471,270 (★755,000)...E6 14
Lyons, Toombs, Ga., 3,219...D4 87
Lyons, Cook, Ill., 9,936...F3 90
Lyons, Greene, Ind., 651...G3 91
Lyons, Rice, Kans., 4,592...D5 93
Lyons, Ionia, Mich., 687...F6 98
Lyons, Burt, Nebr., 972...C9 103
Lyons, Somerset, N.J...B3 106
Lyons, Wayne, N.Y., 4,673...B4 108
Lyons, Fulton, Ohio, 590...A3 111
Lyons, Linn, Oreg., 463...C4 113
Lyons, Minnehaha, S. Dak., 95...D9 116
Lyons, Burleson, Tex., 500...D4 118
Lyons, Walworth, Wis., 400...F1 124
Lyons, creek, Kans...D7 93
Lyons Falls, Lewis, N.Y., 887...B5 108
Lyracrompane, Ire...E2 11
Lys, riv., Bel., Fr...D2, D3 15
Lysa, Czech., 6,500...n18 26
Lysaya Gora, Sov. Un., 10,000...G8 27
Lysekil, Swe., 7,800...H4 25
Lysite, Fremont, Wyo., 70...B5 125
Lysogory, mts., Pol...C6 26
Lyss, Switz., 5,616...B3 19
Lyster Station, Que., Can., 912...C6 73
Lysva, Sov. Un., 76,000...B5 29
Lytham, Eng., 36,222...A5 12
Lytle, Atascosa, Tex., 798...E3 118
Lytton, B.C., Can., 422...D7 68
Lytton, Sac and Calhoun, Iowa, 376...B3 92
Lyubar, Sov. Un., 10,000...G6 27
Lyubertsy, Sov. Un., 100,000...n17 27

M

Ma'ād, Jordan, 1,000...B7 32
Maalaea, Maui, Haw., 200...C5 88
Maalaea, bay, Haw...C5 88
Ma'ale Aqrabim (Scorpion Pass), pass, Isr...D7 32
Maam Cross, Ire...D2 11
Ma'ān, Jordan, 4,509...G10 31
Maas, Ire...C3 11
Maas, riv., Neth...C5 15
Maaseik, Bel., 8,068...C5 15
Maastricht, Neth., 91,200 (★117,000)...D5 15
Mab, Allen, La...D3 95
Mabana, Island, Wash., 75...A3 122
Mabank, Kaufman, Tex., 944...C4 118
Mabber, cape, Som...D6 47
Mabe, Scott, Va...B2 121
Mabel, Fillmore, Minn., 815...G7 99
Mabel, lake, B.C., Can...D8 68
Mabelvale, Pulaski, Ark., 550...D5 81
Maben, Jefferson, Ala., 25...E4 78
Maben, Oktibbeha and Webster, Miss., 696...B4 100
Maberly, Ont., Can., 91...C8 72
Mabie, Randolph, W. Va., 400...C5 123
Mablethorpe, Eng., 5,389...A8 12
Mableton, Cobb, Ga., 7,127...A4 87
Mabote, Moz...B5 49
Mabou, N.S., Can., 293...C8 74
Mabrous (Well), Niger...B7 45
Mabscott, Raleigh, W. Va., 1,591...D3, D7 123
Mabton, Yakima, Wash., 958...C5 122
Macachín, Arg...B4 54
McAdam, N.B., Can., 2,472...D2 74
McAdams, Attala, Miss., 200...B4 100
McAdenville, Gaston, N.C., 748...*B2 109
McAdoo, Schuylkill, Pa., 3,560...E9 114
McAdoo, Dickens, Tex., 100...C2 118
Macaé, Braz., 19,830...C4 56
McAfee, De Kalb, Ga., 3,000...*C2 87
McAfee, Leake, Miss., 100...C4 100
McAfee, Sussex, N.J., 250...A3 106
Macaíba, Braz., 7,472...g6 57
Macalelon, Phil., 2,018...p14 35
McAlester, Pittsburg, Okla., 17,419...C6 112
McAlester, lake, Okla...B6 112
McAlister, Quay, N. Mex., 10...C6 107
McAlisterville, Juniata, Pa., 600...E7 114
McAllen, Hidalgo, Tex., 32,728...F3 118
McAllister, Madison, Mont., 20...E5 102
McAlmont, Pulaski, Ark...D6 81
McAlpin, Suwannee, Fla., 50...B4 86
MacAlpine, lake, N.W. Ter., Can...C12 66
Macamic, Que., Can., 1,388...*E9 72
McAndrews, Pike, Ky., 600...C7 94
Macao, Macao, 207,000...G7 34
Macao, Port. dep., Asia, 169,299...G7 34
Macapá, Braz., 27,585...B4 59
Macará, Ec., 2,702...B2 58
McArthur, Vinton, Ohio, 1,529...C5 111
MacArthur (Ormoc), Phil., 7,660...*C6 35
Macas, Ec., 1,079...B2 58
Macassar (Makasar), Indon., 367,882...G5 35
Macau, Braz., 11,876...C3 57
McAuley, Man., Can., 199...D1 71
McBain, Missaukee, Mich., 551...D5 98
McBaine, Boone, Mo., 40...C5 101
McBean, Richmond, Ga., 100...C5 87
McBee, Chesterfield, S.C., 512...C7 115
Macbeth, Berkeley, S.C., 80...E8 115
McBride, B.C., Can., 590...C7 68
McBride, Perry, Mo., 90...D8 101
McCabe, Roosevelt, Mont., 45...B12 102
McCain, Hoke, N.C...B4 109
McCall, Valley, Idaho, 1,423...E2 89
McCall, Ascension, La., 130...B5, D4 95
McCalla, Jefferson, Ala., 800...E4 78
McCall Creek, Franklin, Miss., 169...D3 100
McCallsburg, Story, Iowa, 272...B4 92
McCamey, Upton, Tex., 3,375...D1 118
McCammon, Bannock, Idaho, 557...G6 89
Maccan, N.S., Can., 300...D5 74
McCanna, Grand Forks, N. Dak., 80...A8 110
Maccarese, It., 4,000...h8 21
McCarthy, Alsk., 12...C11 79
McCarthy, mtn., Mont...E4 102
McCartys, Valencia, N. Mex., 25...B2 107
McCaskill, Hempstead, Ark., 62...D2 81
McCauley, isl., B.C., Can...C2 68
McCauley, Fisher, Tex., 125...C2 118
McCaysville, Fannin, Ga., 1,871...B2 87
McChesneytown, Westmoreland, Pa., 1,140...*F3 114
McClain, co., Okla., 12,740...B4 112
McClave, Bent, Colo., 90...C8 83
McCleary, Grays Harbor, Wash., 1,115...B2 122
McClelland, Woodruff, Ark., 65...B4 81
McClelland, Pottawattamie, Iowa, 150...C2 92
McClellanville, Charleston, S.C., 354...F9 115
Macclenny, Baker, Fla., 2,671...B4 86
Macclesfield, Eng., 37,578...A5 12
Macclesfield, Edgecombe, N.C., 473...B6 109
McClintock, chan., N.W. Ter., Can...B12 66
McClintock, mtn., Ant...B28 5
McCloud, Suskiyou, Calif., 2,140...B2 82
Maccluer, gulf, W. Irian...F8 35
McClure, Alexander, Ill., 400...F4 90
McClure, Henry, Ohio, 651...A4 111
McClure, Snyder, Pa., 1,001...E7 114
McClure, Dickenson, Va., 500...B2 121
McClusky, Sheridan, N. Dak., 751...B5 110
McColl, Marlboro, S.C., 2,479...B8 115
McComas, Mercer, W. Va., 950...D3 123
McComb, Pike, Miss., 12,020...D3 100
McComb, Hancock, Ohio, 1,176...A4 111
McConaughy, lake, Nebr...C4 103
McCondy, Chickasaw, Miss., 125...B5 100
McCone, co., Mont., 3,321...C11 102
McConnell, Stephenson, Ill., 225...A4 90
McConnell, Obion, Tenn., 50...A3 117
McConnells, York, S.C., 266...B5 115
McConnellsburg, Fulton, Pa., 1,245...G6 114
McConnellstown, Huntington, Pa., 325...F5 114
McConnelsville, Morgan, Ohio, 2,257...C6 111
McCook, Red Willow, Nebr., 8,301...D5 103
McCook, co., S. Dak., 8,268...D8 116
McCook Lake, Union, S. Dak., 500...E9 116
McCool, Attala, Miss., 211...B4 100
McCoole, Allegany, Md., 368...D2 85
McCool Junction, York, Nebr., 246...D8 103
McCord, Sask., Can., 135...H2 70
McCordsville, Hancock, Ind., 350...E6 91
McCorkle, Lincoln, W. Va., 300...C3, C5 123
McCormick, McCormick, S.C., 1,998...D3 115
McCormick, co., S.C., 8,629...D3 115
McCoy, Eagle, Colo., 25...B4 83
McCoysville, Juniata, Pa., 100...F6 114
McCracken, Rush, Kans., 406...D4 93
McCracken, co., Ky., 57,306...A2 94
McCreary, Man., Can., 579...D2 71
McCreary, co., Ky., 12,463...D5 94
McCredie, Callaway, Mo., 90...C6 101
McCrory, Woodruff, Ark., 1,053...B4 81
McCulloch, co., Tex., 8,815...D3 118
McCullom, Lee, Ala., 100...C4 78
McCullom Lake, McHenry, Ill., 759...*A5 90
McCullough, Escambia, Ala., 200...D2 78
McCullough, mtn., Nev...H6 104
McCullum, Walker, Ala., 125...B2 78
McCune, Crawford, Kans., 433...E8 93
McCurtain, Haskell, Okla., 528...B7 112
McCurtain, co., Okla., 25,851...C7 112
McCutchenville, Wyandot and Seneca, Ohio, 300...B4 111
McDade, Bastrop, Tex., 500...D4 118
McDaniel, Talbot, Md., 50...C5 85
McDavid, Escambia, Fla., 500...G2 86
McDermitt, Humboldt, Nev., 100...B4 104
McDermott, Scioto, Ohio, 900...D4 111
Macdhui, mtn., Scot...C5 13
Macdona, Bexar, Tex., 300...B3 118
McDonald, Rawlins, Kans., 323...C2 93
McDonald, Neshoba, Miss., 75...C4 100
McDonald, Lea, N. Mex., 5...D6 107
McDonald, Trumbull, Ohio, 2,727...A7 111
McDonald, Washington and Allegheny, Pa., 3,141...B5 114
McDonald, Bradley, Tenn., 125...D9, E11 117
MacDonald, Fayette, W. Va., 300...D3, D7 123
McDonald, co., Mo., 11,798...E3 101
McDonald, creek, Mont...C8 102
McDonald, isl., Indian O...J1 2
Macdonald, lake, Austl...D4 50
Macdonald, peak, Mont...C3 102
MacDonald, range, B.C., Can...E10 68

Macdonaldton, Somerset, Pa., 200 . . . . G4 114
MacDonnell, ranges, Austl. D5 50
McDonough, Henry, Ga., 2,224 . . . . C2 87
McDonough, Chenango, N.Y., 150 . . . . C5 108
McDonough, co., Ill., 28,928 . C3 90
McDougal, Clay, Ark., 200 . . A5 81
McDougal, mtn., Wyo C2 125
McDowell, Floyd, Ky., 500 . . C7 94
McDowell, Highland, Va., 127 . . . . C3 121
McDowell, co., N.C., 26,742 . D4 109
McDowell, co., W. Va., 71,359 . . . . D3 123
McDowell, mtn., Ariz F2 80
Macduff, Scot., 3,479 . . . . C6 13
McDuffie, co., Ga., 12,627 . . C4 87
Macedon, Wayne, N.Y., 645 . B3 108
Macedonia, Pottawattamie, Iowa, 290 . . . . C2 92
Macedonia, Summit, Ohio, 400 . . . . A6, B2 111
Macedonia, reg., Yugo., Grc. B4 23
Macedonia, rep., Yugo., 1,404,883 . . . . *D5 22
Maceió, Braz., 153,305 . . . C3, k6 57
McElhattan, Clinton, Pa., 480 . . . . D7 114
McElroy, creek, W. Va A6 123
Macenta, Guinea, 5,000 . . . . E3 45
Maceo, Daviess, Ky., 200 . . . C2 94
Macerata, It., 24,000 (38,338▲) C4 21
Maces, bay, N.B., Can D3 74
McEwen, Baker, Oreg C8 113
McEwen, Humphreys, Tenn., 979 . . . . A4 117
McFadden, Jackson, Ark B4 81
McFadden, Carbon, Wyo., 150 . . . . D6 125
McFaddin, Victoria, Tex., 300 . . . . E4 118
McFall, Gentry, Mo., 206 . . . A3 101
McFarlan, Anson, N.C., 161 . C3 109
Macfarlan, Ritchie, W. Va., 150 . . . . B3 123
McFarland, Kern, Calif., 3,686 . . . . E4 82
McFarland, Wabaunsee, Kans., 256 . . . . C7 93
McFarland, Marquette, Mich., 80 . . . . B3 98
McFarland, Dane, Wis., 1,272 . . . . E4 124
McGaffey, McKinley, N. Mex B1 107
McGaha, Adair, Ky., 200 . . C4 94
McGaheysville, Rockingham, Va., 250 . . . . C4 121
McGee, Sask., Can., 50 . . . . F1 70
McGehee, Desha, Ark., 4,448 . . . . D4 81
McGill, White Pine, Nev., 2,195 . . . . D7 104
MacGillicuddy's Reeks, mts., Ire. F2 11
McGillivray, lake, Que., Can . . . A7 72
McGillivray, range, B.C., Can E10 68
McGillivray Falls, B.C., Can D6 68
McGiverin, lake, Ont., Can B8 98
McGivney Junction, N.B., Can., 235 . . . . C3 74
McGrann, Armstrong, Pa., 800 . . . . *E2 114
McGrath, Alsk., 241 . . . . C8 79
McGrath, Aitkin, Minn., 96 . . . . D5 99
McGraw, Cortland, N.Y., 1,276 . . . . C4 108
McGraws, Wyoming, W. Va., 300 . . . . D3 123
Mac Gregor, Man., Can., 642 . E2 71
McGregor, Clayton, Iowa, 1,040 . . . . A6 92
McGregor, Aitkin, Minn., 283 . . . . D5 99
McGregor, Williams, N. Dak., 125 . . . . A3 110
McGregor, McLennan, Tex., 4,642 . . . . D4 118
McGregor, lake, Alta., Can . . . D4 69
McGregor, riv., B.C., Can . . . B7 68
McGrew, Scotts Bluff, Nebr., 90 . . . . C2 103
McGuffey, Hardin, Ohio, 647 . . . . B4 111
McGuire, mtn., Idaho D4 89
McGuires, Kootenai, Idaho . . B2 89
Mach, Pak., 3,211 . . . . C1 40
Machachi, Ec., 2,582 . . . . B2 58
Machado, Braz., 8,373 . . . C3, k9 56
Machakos, Ken B6 48
Machala, Ec., 7,491 . . . . B2 58
Machanao, mtn., Guam 52
Machar, lake, Ont., Can B5 72
Macheke, S. Rh A5 49
Macheng, China, 30,000 . . . . I6 36
McHenry, McHenry, Ill., 3,336 . . . . A5, E2 90
McHenry, Ohio, Ky., 446 . . . C3 94
McHenry, Garrett, Md., 60 . . D1 85
McHenry, Stone, Miss., 600 . . E4 100
McHenry, Foster, N. Dak., 155 . . . . B7 110
McHenry, co., Ill., 84,210 . . . A5 90
McHenry, co., N. Dak., 11,099 . . . . A4 110
Machens, St. Charles, Mo., 101 . . . . A8 101
Machete Central, P.R D6 65
Machias, Washington, Maine, 1,523 (2,614▲) D5 96
Machias, Cattaraugus, N.Y., 400 . . . . C2 108
Machias, bay, Maine D5 96
Machias, lakes, Maine C4 96
Machias, riv., Maine B4 96
Machias, riv., Maine D5 96
Machiasport, Washington, Maine, 280 (980▲) D5 96
Machico, Port., 4,734 . . . . h12 20
Machiques, Ven., 13,685 . . . . A3 60
Machrihanish, Scot E3 13
McHue, Independence, Ark., 25 . . . . B4 81
Machu Picchu, Peru D3 58
Machynlleth, Wales, 1,903 . . B4 12
Macia, Moz C5 49
Măcin, Rom., 6,533 . . . . C9 22
McIndoe Falls, Caledonia, Vt., 155 . . . . C4 120
McIntire, Mitchell, Iowa, 270 . . . . A5 92
McIntosh, Washington, Ala., 500 . . . . D1 78
McIntosh, Ont., Can., 110 . . . E5 71
McIntosh, Marion, Fla., 258 . . . . C4 86
McIntosh, Liberty, Ga., 150 . . E5 87
McIntosh, Polk, Minn., 785 . . C3 99
McIntosh, Torrance, N. Mex., 10 . . . . C3 107
McIntosh, Corson, S. Dak., 568 . . . . B4 116
McIntosh, co., Ga., 6,364 . . . E5 87
McIntosh, co., N. Dak., 6,702 . . . . C6 110
McIntosh, co., Okla., 12,371 . B6 112
McIntosh, lake, Sask., Can B3 70
McIntosh, run, Md D4 85
McIntyre, Wilkinson, Ga., 316 . . . . D3 87
McIntyre, Indiana, Pa., 425 . . E3 114
McIntyre, creek, Ohio B1 123
Mack, Mesa, Colo., 150 . . . . B2 83
Mack, Hamilton, Ohio, 3,500 . . . . D2 111
McKague, Sask., Can., 132 . . . E4 70
Mackay, Austl., 16,809 . . . . D8 50
MacKay, Alta., Can E10 66
Mackay, Custer, Idaho, 652 . . F5 89
Mackay, lake, Austl D4 50
MacKay, lake, N.W. Ter., Can D10 66
MacKay, riv., Alta., Can A4 69
Mackayville, Que., Can., 9,958 . . . . *D9 83
McKean, Erie, Pa., 442 . . . . C1 114
McKean, co., Pa., 54,517 . . . C4 114
McKee, Jackson, Ky., 234 . . . C6 94
McKee City, Atlantic, N.J., 300 . . . . E3 106
McKeesport, Allegheny, Pa., 45,489 . . . . B6, F2 114
McKees Rocks, Allegheny, Pa., 13,185 . . . . B5, F1 114
Mackenna, Arg A4 54
McKenney, Dinwiddie, Va., 519 . . . . E5 121
McKenzie, Butler, Ala., 558 . . D3 78
McKenzie, Burleigh, N. Dak., 100 . . . . C5 110
McKenzie, Carroll and Weakley, Tenn., 3,780 . . . A3 117
McKenzie, co., N. Dak., 7,296 . . . . B2 110
Mackenzie, dist., N.W. Ter., Can., 10,279 . . . . D10 66
MacKenzie, bay, Ant C20 5
Mackenzie, bay, N.W. Ter., Can C5 66
Mackenzie, mts., N.W. Ter., Can D6 66
McKenzie, pass, Oreg C4 113
Mackenzie, riv., N.W. Ter., Can C6 66
McKenzie Bridge, Lane, Oreg., 315 . . . . C4 113
McKerrow, Ont., Can., 221 . . A3 72
Mackey, Gibson, Ind., 92 . . . *H3 91
Mackeys, Washington, N.C., 200 . . . . B7 109
Mackey's Station, Ont., Can., 217 . . . . A7 72
Mackeyville, Clinton, Pa., 250 . . . . D7 114
Mackinac, co., Mich., 10,853 . B5 98
Mackinac, isl., Mich C6 98
Mackinac, straits, Mich C6 98
Mackinac Island, Mackinac, Mich., 942 . . . . C6 98
Mackinaw, Tazewell, Ill., 1,163 . . . . C4 90
Mackinaw, riv., Ill C4 90
Mackinaw City, Cheboygan and Emmet, Mich., 934 . . C6 98
McKinley, Marengo, Ala., 100 . . . . C2 78
McKinley, Hancock, Maine, 300 . . . . D4 96
McKinley, St. Louis, Minn., 408 . . . . C6 99
McKinley, Converse, Wyo., 35 . . . . C7 125
McKinley, co., N. Mex., 37,209 . . . . B1 107
McKinley, mtn., Alsk C9 79
McKinley Heights, Trumbull, Ohio, 1,500 . . . . *A7 111
McKinley Park, Alsk., 59 . . C10 79
McKinleyville, Humboldt, Calif., 950 . . . . *B1 82
McKinney, Lincoln, Ky., 300 . . . . C5 94
McKinney, Collin, Tex., 13,763 . . . . C4 118
McKinney, lake, Kans E2 93
McKinnon, Sweetwater, Wyo., 10 . . . . E3 125
Mackinnon Road, Ken B6 48
McKittrick, Kern, Calif., 135 . E4 82
McKittrick, Montgomery, Mo., 97 . . . . C6 101
McKittrick Summit, mtn., Calif . E4 82
Macklin, Sask., Can., 690 . . . E1 70
McKnight, Allegheny, Pa., 15,000 . . . . A5 114
McKnight, lake, Man., Can . . . A1 71
McKownville, Albany, N.Y., 2,000 . . . . *C7 108
Macksburg, Madison, Iowa, 174 . . . . C3 92
Macksburg, Washington, Ohio, 314 . . . . C6 111
Macks Creek, Camden, Mo., 123 . . . . D5 101
Macksville, Stafford, Kans., 546 . . . . E5 93
Macksville, Washington, Ky., 400 . . . . C4 94
Mackville, Outagamie, Wis., 50 . . . . A5 124
McLain, Greene, Miss., 600 . D5 100
McLaughlin, Alta., Can., 52 . C5 69
McLaughlin, Corson, S. Dak., 983 . . . . B5 116
McLaughlin, mtn., Oreg E4 113
McLaughlin, riv., Man., Can . . . C3 71
McLaurin, Forrest, Miss., 350 . . . . D4 100
Maclean, Austl., 1,804 . . . . D9 51
McLean, Sask., Can., 202 . . . G3 70
McLean, McLean, Ill., 758 . . C4 90
McLean, Pierce, Nebr., 73 . . B8 103
McLean, Gray, Tex., 1,330 . . B2 118
McLean, Fairfax, Va., 2,000 . A5 121
McLean, co., Ill., 83,877 . . . C5 90
McLean, co., Ky., 9,355 . . . . C2 94
McLean, co., N. Dak., 14,030 . . . . B4 110
McLean, lake, Sask., Can A1 70
McLean, mtn., Maine A4 96
McLeansboro, Hamilton, Ill., 2,951 . . . . E5 90
Maclear, S. Afr., 3,542 . . . . D4 49
McLemoresville, Carroll, Tenn., 285 . . . . B3 117
McLennan, Alta., Can., 1,078 . . . . B2 69
McLennan, co., Tex., 150,091 . . . . D4 118
McLeod, Sweet Grass, Mont., 15 . . . . E6 82
McLeod, Ransom, N. Dak., 300 . . . . C8 110
McLeod, co., Minn., 24,401 . . F4 99
McLeod, lake, B.C., Can B6 68
McLeod, peak, Mont C3 102
McLeod, riv., Alta., Can C2 69
MacLeod Lake, Br. Col., Can. E2 68
Macleods, isl., Scot C2 13
McLoud, Pottawatomie, Okla., 837 . . . . B4 112
McLoughlin, mtn., Oreg E4 113
McLouth, Jefferson, Kans., 494 . . . . B7, C8 93
McMahon, Sask., Can., 62 . . . G2 72
McMasterville, Que., Can., 2,075 . . . . *D4 73
McMechen, Marshall, W. Va., 2,999 . . . . B2, B4 123
McMichael, creek, Pa B2 106
McMillan, Luce, Mich., 300 . . B5 98
McMillan, Knox, Tenn E12 117
McMillan, lake, N. Mex E5 107
McMillan Manor, Ventura, Calif., 1,193 . . . . *E4 82
McMinn, co., Tenn., 33,662 . . D9 117
McMinnville, Yamhill, Oreg., 7,656 . . . . B1, B3 113
McMinnville, Warren, Tenn., 9,013 . . . . D8 117
McMorran, Sask., Can F1 70
McMullen, co., Tex., 1,116 . . E3 118
McMunn, Man., Can., 20 . . . . E4 71
McMurdo, Ant B29 5
McMurdo, sound, Ant B29 5
McMurray, Alta., Can., 1,186 . . . . A5 69
McMurray, Skagit, Wash., 75 . . . . A3 122
McNab, Hempstead, Ark., 142 . . . . D2 81
McNair, Jefferson, Miss., 50 . D2 100
McNair, Harris, Tex., 1,880 . *F5 118
McNairy, McNairy, Tenn., 100 . . . . B3 117
McNairy, co., Tenn., 18,085 . . B3 117
McNary, Apache, Ariz., 1,608 . . . . C6 80
McNary, Rapides, La., 250 . . D3 95
McNary, Umatilla, Oreg., 350 . . . . B7 113
McNary, Hudspeth, Tex., 50 . F1 124
McNeal, Cochise, Ariz., 50 . . F6 80
McNeal, Calhoun, Fla., 150 . . B1 86
McNeil, Columbia, Ark., 746 . D2 81
McNeil, Travis, Tex., 200 . . . D4 124
McNeil, isl., Wash D1 122
McNeill, Pearl River, Miss., 450 . . . . E4 100
McNeill, mtn., B.C., Can . . . B2 68
McNutt, isl., N.S., Can F4 74
MacNutt, Sask., Can., 211 . . . F5 70
Maco, Bol B2 55
Macocola, Ang C2 48
Macomb, McDonough, Ill., 12,135 . . . . C3 90
Macomb, Pottawatomie, Okla., 76 . . . . B4 112
Macomb, co., Mich., 405,804 . F8 98
Mâcon, Fr., 25,714 . . . . D6 14
Macon, Bibb, Ga., 69,764 (★170,700) D3 87
Macon, Macon, Ill., 1,229 . . D5 90
Macon, Noxubee, Miss., 2,432 . . . . B5 100
Macon, Macon, Mo., 4,547 . . B5 101
Macon, Franklin, Nebr., 25 . . D7 103
Macon, Warren, N.C., 187 . . . A5 109
Macon, Fayette, Tenn., 50 . . B2 117
Macon, co., Ala., 26,717 . . . . C4 78
Macon, co., Ga., 13,170 . . . . D2 87
Macon, co., Ill., 118,257 . . . D5 90
Macon, co., Mo., 16,473 . . . . B5 101
Macon, co., N.C., 14,935 . . . D2 109
Macon, co., Tenn., 12,197 . . C7 117
Macon, bayou, La B4 95
Macopin, Passaic, N.J A4 106
Macoun, Sask., Can., 193 . . . H4 70
Macoun, lake, Sask., Can A4 70
Macoupin, co., Ill., 43,524 . . D4 90
Macouria, Fr. Gu., 597 . . . . B4 59
Macovane, Moz B6 49
McPhail, riv., Man., Can . . . C3 71
McPherson, McPherson, Kans., 9,996 . . . . D6 93
McPherson, co., Kans., 24,285 . . . . D6 93
McPherson, co., Nebr., 735 . . C4 103
McPherson, co., S. Dak., 5,821 . . . . B6 116
McQuady, Breckinridge, Ky., 100 . . . . C3 94
Macquarie, isl., Pac. O J7 2
Macquarie, riv., Austl E6 51
McQueeney, Guadalupe, Tex., 900 . . . . B4 118
McRae, White, Ark., 428 . . . . B4 81
McRae, Telfair, Ga., 2,738 . . D4 87
McRoberts, Letcher, Ky., 1,363 . . . . C7 94
Mac-Robertson, coast, Ant C19 5
Macroom, Ire., 2,169 . . . . F3 11
Macrorie, Sask., Can., 182 . . . F2 70
McSherrystown, Adams, Pa., 2,839 . . . . G7 114
McTaggart, Sask., Can., 93 . . H3 70
McTavish, Man., Can., 120 . . E3 71
MacTier, Ont., Can., 851 . . . . B5 72
Macuelizo, Hond., 879 . . . . C3 62
Macungie, Lehigh, Pa., 1,266 . . . . E10 114
McVeigh, Pike, Ky., 800 . . . . C7 94
McVeytown, Mifflin, Pa., 488 . . . . F6 114
McVille, Nelson, N. Dak., 551 . . . . B7 110
Macwahoc, Aroostook, Maine, 100 (165▲) C4 96
McWilliams, Wilcox, Ala., 225 . . . . D2 78
Macy, Miami, Ind., 328 . . . . C5 91
Macy, Thurston, Nebr., 203 . . B2 103
Mad, creek, Ga B5 87
Mad, lake, Idaho D8 122
Mad, riv., Calif B2 82
Mad, riv., N.H C3 105
Mad, riv., Ohio C2 111
Mad, riv., Vt C3 120
Ma'dabā, Jordan, 8,545 . . . . C7 32
Madaba, Tan C6 48
Madagascar, isl., Afr g9 49
Madame, isl., N.S., Can D9 74
Madang, N. Gui., 379 . . . . k12 50
Madaoua, Niger, 2,100 . . . . D6 45
Madaripur, Pak., 21,693 . . . . F13 40
Madawaska, Ont., Can., 429 . B7 72
Madawaska, Aroostook, Maine, 4,035 (5,507▲) A4 96
Madawaska, co., N.B., Can., 38,983 . . . . B1 74
Madawaska, lake, Maine B4 96
Madawaska, riv., Ont., Can . . . B7 72
Madawaska, riv., Que., Can . . . B9 73
Madaya, Bur A2 38
Madbury, Strafford, N.H., 35 (556▲) D5 105
Madden, Leake, Miss., 130 . . C4 100
Madden, Natrona, Wyo B5 125
Madden, lake, Pan k11 62
Madden Dam, C.Z k11 62
Maddock, Benson, N. Dak., 740 . . . . B6 110
Maddox, St. Marys, Md., 5 . . D4 85
Madeira, Hamilton, Ohio, 6,744 . . . . D2 111
Madeira, isl., Port h11 20, C1 44
Madeira, riv., Braz D2 59
Madeira Is., reg., Atl. O., 268,900 . . . . h11 20, C1 44
Madelia, Watonwan, Minn., 2,190 . . . . F4 99
Madeline, Lassen, Calif., 60 . . B3 82
Madeline, isl., Wis B3 124
Madera, Madera, Calif., 14,430 . . . . D3 82
Madera, Mex., 7,327 . . . . B3 63
Madera, Clearfield, Pa., 900 . . . . E4 114
Madera, co., Calif., 40,468 . . D4 82
Madhupur, India, 19,519 . . . E11 40
Madhya Pradesh, state, India, 32,372,408 . . . . D6 39
Madill, Marshall, Okla., 3,084 . . . . C5 112
Madimba, Con. L C2 48
Madingou, Con. B F2 46
Madison, Madison, Ala., 1,435 . . . . A3 78
Madison, Montgomery, Ala., 900 . . . . C3 78
Madison, St. Francis, Ark., 750 . . . . B5 81
Madison, Sask., Can., 101 . . . F1 70
Madison, New Haven, Conn., 1,416 (4,567▲) D6 84
Madison, Madison, Fla., 3,239 . . . . B3 86
Madison, Morgan, Ga., 2,680 . . . . C3 87
Madison, Madison, Ill., 6,861 . . . . E3 90
Madison, Jefferson, Ind., 10,488 . . . . G7 91
Madison, Greenwood, Kans., 1,105 . . . . D7 93
Madison, Somerset, Maine, 2,761 (3,935▲) D3 96
Madison, Dorchester, Md., 200 . . . . C5 85
Madison, Lac qui Parle, Minn., 2,380 . . . . E2 99
Madison, Madison, Miss., 703 . . . . C3 100
Madison, Monroe, Mo., 528 . . B5 101
Madison, Madison, Nebr., 1,513 . . . . C8 103
Madison, Carroll, N.H., 130 (429▲) C4 105
Madison, Morris, N.J., 15,122 . . . . B4 106
Madison, Madison, N.Y., 327 . . . . C5 108
Madison, Rockingham, N.C., 1,912 . . . . A4 109
Madison, Lake, Ohio, 1,347 . . A6 111
Madison, Oconee, S.C., 300 . B1 115
Madison, Lake, S. Dak., 5,420 . . . . C8 116
Madison, Davidson, Tenn., 13,583 . . . . A5, E9 117
Madison, Madison, Va., 301 . C4 121
Madison, Boone, W. Va., 2,215 . . . . C3, D5 123
Madison, Dane, Wis., 126,706 (★179,200) E4 124
Madison, co., Ala., 117,348 . . A3 78
Madison, co., Ark., 9,068 . . . B2 81
Madison, co., Fla., 14,154 . . . B3 86
Madison, co., Ga., 11,246 . . . . B3 87
Madison, co., Idaho, 9,417 . . F7 89
Madison, co., Ill., 224,689 . . E4 90
Madison, co., Ind., 125,819 . . D6 91
Madison, co., Iowa, 12,295 . . C3 92
Madison, co., Ky., 33,482 . . . C5 94
Madison, co., Miss., 32,904 . . C4 100
Madison, co., Mo., 9,366 . . . D7 101
Madison, co., Mont., 5,211 . . E4 102
Madison, co., Nebr., 25,145 . . C8 103
Madison, co., N.Y., 54,635 . . C5 108
Madison, co., N.C., 17,217 . . C3 109
Madison, co., Ohio, 26,454 . . C4 111
Madison, co., Tenn., 60,655 . . B3 117
Madison, co., Tex., 6,749 . . . D5 118
Madison, co., Va., 8,187 . . . . C4 121
Madison, par., La., 16,444 . . B4 95
Madison, range, Mont E5 102
Madison, riv., Mont E5 102
Madison, riv., Wyo A2 125
Madisonburg, Centre, Pa., 150 . . . . E6 114
Madison College, Davidson, Tenn., 700 . . . . E9 117
Madison Heights, Oakland, Mich., 33,343 . . . . *F7 98
Madison Heights, Amherst, Va., 3,000 . . . . D3 121
Madison Lake, Blue Earth, Minn., 477 . . . . F5 99
Madisonville, Hopkins, Ky., 13,110 . . . . C2 94
Madisonville, St. Tammany, La., 860 . . . . B7, D5 95
Madisonville, Monroe, Tenn., 1,812 . . . . D9 117
Madisonville, Madison, Tex., 2,324 . . . . D5 118
Madjene, Indon F5 35
Madoc, Ont., Can., 1,347 . . . C7 72
Madoc, Daniels, Mont., 23 . B11 102
Madon, riv., Fr A2 18
Madona, Sov. Un., 5,000 . . . C6 27
Madonna di Campiglio, It., 245 . . . . C6 18
Madras, India, 1,729,141 (★2,000,000) F7 39
Madras, Jefferson, Oreg., 1,515 . . . . C5 113
Madras, state, India, 33,686,953 . . . . F6 39
Madre, mts., Guat C2 62
Madre, mts., Mex B6 63
Madre, mts., Phil B6 35
Madre de Dios, dept., Peru, 26,116 . . . . D3 58
Madre de Dios, isl., Chile E1 54
Madre de Dios, riv., Bol., Peru . B2 55
Madre del Sur, mts., Mex o13 63
Madre Occidental, mts., Mex . . . C3 63
Madre Oriental, mts., Mex C5 63
Madrid, Houston, Ala., 245 . . D4 78
Madrid, Boone, Iowa, 2,286 . . . . A7, C4 92
Madrid, Perkins, Nebr., 271 . . D4 103
Madrid, Santa Fe, N. Mex., 600 . . . . B3, G6 107
Madrid, St. Lawrence, N.Y., 800 . . . . A2 108
Madrid, Sp., 2,259,931 (★2,360,000) B4, p17 20
Madrid, prov., Sp., 2,606,254 . . . . *B4, p17 20
Madridejos, Phil., 3,041 . . . . *C6 35
Madridejos, Sp., 9,795 . . . . C4 20
Madrūsah, Libya E2 43
Madugula, India, 7,688 . . . . I9 40
Madura, India, 361,781 . . . . G6 39
Madura, isl., Indon G4 35
Mae, Grant, Wash B6 122
Maebashi, Jap., 135,000 (181,937▲) H9, m18 37
Mae Hong Son, Thai., 3,315 . C3 38
Maelamun, mtn., Thai E3 38
Maella, Sp., 2,638 . . . . B6 20
Maengsan, Kor G3 37
Maes, San Miguel, N. Mex . . B5 101
Maeser, Uintah, Utah, 929 . . C6 119
Maesteg, Wales, 21,652 . . . . C4 12
Maestra, mts., Cuba E5 64
Maevatanana, Malag., 6,660 . g9 49
Mafeking, Man., Can., 385 . . C1 71
Mafeking, S. Afr., 8,279 . . . . C4 49
Maffra, Austl., 3,404 . . . . H6 51
Mafra, Braz., 12,981 . . . . D3 56
Mafra, Port., 3,096 . . . . C1 20
Maga, mtn., Tinian 52
Magadan, Sov. Un., 65,000 . D18 28
Magadi, Ken B6 48
Magadi, lake, Ken B6 48
Magaguadavic, lake, N.B., Can . D2 74
Magalia, Butte, Calif., 125 . . C3 82
Magallanes, see Punta Arenas, Chile
Magallanes, prov., Chile, 73,037 . . . . E2, h11 54
Magallanes, strait, Chile h11 54
Maganga, Con. L A4 48
Magangué, Col., 17,114 . . . . B3 60
Maganjada Costa, Moz A6 49
Magaria, Niger D6 45
Magat, riv., Phil n13 35
Magazine, Logan, Ark., 463 . . B2 81
Magazine, mtn., Ark B2 81
Magdalena, Arg., 4,114 . . B5, g8 54
Magdalena, Bol., 1,724 . . . . B3 55
Magdalena, Mex., 9,413 . . . . A2 63
Magdalena, Socorro, N. Mex., 1,211 . . . . C2 107
Magdalena, dept., Col., 450,920 . . . . A3 60
Magdalena, isl., Chile C2 54
Magdalena, isl., Mex C2 63
Magdalena, lake, Mex m11 63
Magdalena, plain, Mex C2 63
Magdalena, riv., Col B3 60
Magdalena, riv., Mex h9 63
Magdalen Islands, co., Que., Can., 12,479 . . . . B8 74
Magdalo, Phil o13 35
Magdeburg, Ger., 262,400 (★360,000) A6 17
Magdiel, Isr., 5,400 . . . . g10 32
Magé, Braz., 10,712 . . . . h6 56
Magee, Simpson, Miss., 2,039 . . . . D4 100
Magelang, Indon., 91,636 . . G4 39
Mageney, Ire E5 11
Magenta, It., 18,417 . . . . B2 21
Maggia, riv., Switz D6 19
Maggie, Craig, Va., 25 . . . . D2 121
Maggie, creek, Nev C5 104
Maggiore, lake, It. and Switz . . C4 18
Maghāghah, Eg., U.A.R., 23,400 . . . . D6 43
Maghera, N. Ire., 1,613 . . . . C5 11
Magherafelt, N. Ire., 2,460 . . C5 11
Maghery, Ire C3 11
Magic, res., Idaho F4 89
Magic City, Wheeler, Tex . . B2 118
Magicienne, bay, Saipan 52
Magilligan, pt., Ire B5 11
Maglaj, Yugo., 4,579 . . . . C4 22
Maglic, Yugo., 2,182 . . . . D5 22
Maglie, It., 13,028 . . . . D7 21
Magna, Salt Lake, Utah, 6,442 . . . . C3 119
Magness, Independence, Ark., 140 . . . . B4 81
Magnet, Hot Spring, Ark., 200 . . . . C3, D6 81
Magnet, Man., Can., 45 . . . D2 71
Magnet, Cedar, Nebr., 116 . . B8 103
Magnetawan, Ont., Can., 205 . . . . B5 72
Magnetawan, riv., Ont., Can . . B4 72
Magnetic Springs, Union, Ohio, 344 . . . . B4 111
Magnisia, prov., Grc., 162,285 . . . . *C4 23
Magnitogorsk, Sov. Un., 328,000 . . . . C5 29
Magnolia, Columbia, Ark., 10,651 . . . . D3 81
Magnolia, Kent, Del., 310 . . B7 85
Magnolia, Putnam, Ill., 245 . B4 90
Magnolia, Harrison, Iowa, 215 . . . . C2 92
Magnolia, Larue, Ky., 300 . . C4 94
Magnolia, Harford, Md., 450 . . . . B5 85
Magnolia, Rock, Minn., 280 . G2 99
Magnolia, Pike, Miss., 2,083 . D3 100
Magnolia, Camden, N.J., 4,199 . . . . D2 106
Magnolia, Duplin, N.C., 629 . C5 109
Magnolia, Carroll and Stark, Ohio, 1,596 . . . . B6 111
Magnolia, Montgomery, Tex., 800 . . . . D5 118
Magnolia Springs, Baldwin, Ala., 350 . . . . E2 78
Magoari, cape, Braz B1 57
Magoffin, co., Ky., 11,156 . . C6 94
Magog, Que., Can., 13,139 . . D5 73
Magothy, riv., Md B4 85
Magou, Niger D5 45
Magoula, Grc., 458 . . . . g11 23
Magpie, Que., Can., 359 . . . h8 75
Magpie, riv., Que., Can h8 75
Magrath, Alta., Can., 1,338 . E4 69
Magruder, mtn., Nev F4 104
Magude, Moz B5 49
Magwe, Bur., 13,270 . . . . D9 39
Mahābād, Iran, 12,858 . . . . C2 41
Mahaffey, Clearfield, Pa., 582 . . . . E4 114
Mahagi, Con. L A5 48
Mahaicony, Br. Gu., 1,179 . . A3 59
Mahakam, riv., Indon F5 35
Mahalapye, Bech., 2,453 . . . . B4 49
Mahameru, mtn., Indon G4 35
Mahanadi, riv., India G10 40
Mahanayim, Isr., 188 . . . . m14 32
Mahanoro, Malag., 3,436 . . . . g9 49
Mahanoy City, Schuylkill, Pa., 8,536 . . . . E9 114
Maharashtra, state, India, 39,553,718 . . . . D5 39
Maha Sarakham, Thai., 15,709 . . . . D5 38
Mahaska, Washington, Kans., 160 . . . . C6 93
Mahaska, co., Iowa, 23,602 . C5 92
Mahbubnagar, India, 35,588 . I6 40
Mahdia, Tun., 10,880 . . . . B7 44
Mahe, India, 7,951 . . . . F6 39
Mahenge, Tan., 1,180 . . . . C6 48
Mahgara, isl., Eg., U.A.R C2 32
Mahi, riv., India F4 40
Mahia, pen., N.Z M16 51
Mahiganj, Pak E12 40
Mahitahi, N.Z., 19 . . . . O12 51
Mahnomen, Mahnomen, Minn., 1,462 . . . . C3 99
Mahnomen, co., Minn., 6,341 . . . . C3 99
Mahogany, peak, Nev B2 104
Mahomet, Champaign, Ill., 1,367 . . . . C5 90
Mahón, Sp., 16,619 . . . . C8 20
Mahone Bay, N.S., Can., 1,103 . . . . E5 74
Mahoning, co., Ohio, 300,480 . . . . B7 111
Mahoning, riv., Ohio A7 111
Mahoosuc, range, Maine, N.H D1 96, B4 105
Mahopac, Putnam, N.Y., 1,337 . . . . D3, D7 108
Mahopac, lake, N.Y D1 84
Mahto, Corson, S. Dak., 3 . . B5 116
Mahtomedi, Washington, Minn., 2,127 . . . . E7 99
Mahtowa, Carlton, Minn., 100 . . . . D6 99
Mahukona, Hawaii, Haw., 100 . . . . C6 88
Mahunda, Moz B6 48
Mahuta, Tan D6 48
Mahuva, India, 32,732 . . . . G3 40
Mahwah, Bergen, N.J., 3,200 . A4 106
Maîche, Fr., 3,361 . . . . B2 18
Maicurú, riv., Braz C4 59
Maida, Cavalier, N. Dak., 30 . A7 110
Maidani, cape, Iran I9 41
Maiden, Catawba, N.C., 2,039 . . . . B2 109
Maidenhead, Eng., 35,374 . . C7 12
Maiden Rock, Pierce, Wis., 189 . . . . D1 124
Maidens, isl., Ire C6 11
Maidstone, Sask., Can., 577 . . D1 70
Maidstone, Eng., 59,761 . . . . C8 12
Maidstone, lake, Vt B5 120
Maidsville, Monongalia, W. Va., 200 . . . . B5 123
Maiduguri, Nig., 54,646 . . . . D7 45
Maihar, India E8 40
Maikala, range, India F8 40
Maiko, riv., Con. L B4 48
Maikoor, isl., Indon G8 35
Maili, pt., Haw g9 88
Maillard, Que., Can B7 73
Maillor, Alg., 25 . . . . F9 30
Maimana, Afg., 25,698 . . . . D12 41
Main, pass, La E6 95
Main, riv., Ger C3, D4 17
Main, riv., Ire C5 11
Main-a-Dieu, N.S., Can., 382 . . . . C10 74
Mainburg, Ger., 5,400 . . . . E6 17
Main Centre, Sask., Can., 66 . . . . G2 70
Maine, Broome, N.Y., 600 . . C4 108
Maine, former prov., Fr., 682,000 . . . . C3 14
Maine, state, U.S., 969,265 . . . . A14 77, 96
Maine-et-Loire, dept., Fr., 556,272 . . . . *D3 14
Mainesburg, Tioga, Pa., 175 . C7 114
Maïné-Soroa, Niger D7 45
Mainland, see Pomona, isl., Scot.
Mainland, isl., Scot g10 10
Mainland, riv., Man., Can . . . C4 71
Maintirano, Malag., 2,594 . . g8 49
Main Topsail, mtn., Newf., Can. D3 75
Mainz, Ger., 134,400 . . . . D3 17
Maio, isl., C.V. Is *E3 42
Maipó, vol., Arg., Chile A2 54
Maipú, Arg., 5,469 . . . . B5 54
Maiquetía, Ven., 73,015 . . . . A4 60
Maira, riv., It E3 18
Maire, strait, Arg h12 54
Mairhofen, Aus., 2,351 . . . . B7 18
Mairiporã, Braz., 9,368 . . . . m8 56
Mairum, Pak G11 46
Maisi, cape, Cuba E6 64
Maison-Carrée, Alg., 55,144 . *B5 44
Maisons-Alfort, Fr., 51,186 . g10 14
Maisons-Laffitte, Fr., 19,132 . g9 14
Mait, Som C6 47
Maitland, Austl., 27,351 . . . . F8 51
Maitland, N.S., Can., 233 . . D6 74
Maitland, Ont., Can., 335 . . C9 72
Maitland (Lake Maitland), Orange, Fla., 3,570 . . . . *D5 86
Maitland, Holt, Mo., 427 . . . A2 101
Maize, Sedgwick, Kans., 623 . B5 93
Maizuru, Jap., 73,200 (99,615▲) I7, n14 37
Majagual, Col., 1,516 . . . . B3 60
Majé, see Magé, Braz.
Majestic, Pike, Ky., 503 . . . C7 94
Mājī, Eth D4 47
Majijo, peak, Ponape 52
Major, Sask., Can., 179 . . . . F1 70
Major, co., Okla., 7,808 . . . . A3 112
Majorsville, Marshall, W. Va., 40 . . . . B2 123
Majra, Afg., 5,000 . . . . D12 46
Majunga, Malag., 38,042 . . . . g9 49
Majunga, prov., Malag g9 49
Makah, Indian res., Wash A1 122
Makaha, Honolulu, Haw., 2,720 . . . . G9 88
Makalado, Con. L B4 48
Mak'alē, Eth C4 47

Makanalua, pen., Haw B2 88
Makanda, Jackson, Ill., 164 F4 90
Makapala, Hawaii, Haw., 100 C6 88
Makapuu, pt., Haw B4 88
Makarikari Pan, lake, Bech B4 49
Makarov (Shirutoru), Sov. Un., 5,000 B11 37
Makarska, Yugo., 3,637 D3 22
Makasar (Macassar), Indon., 367,882 G5 35
Makassar (Macassar), strait, Indon F5 35
Makat, Sov. Un., 3,500 D4 29
Makatea, isl., Fr. Polynesia *H13 7
Makawao, Maui, Haw., 950 C5 88
Makaweli, Kauai, Haw., 600 B2 88
Mak'dalā, Eth C4 47
Makelyville, Hyde, N.C. B7 109
Makena, Maui, Haw., 30 C5 88
Makeni, S.L. E2 45
Makenzen, see Orlyak, Bul.
Makeyevka, Sov. Un., 381,000 G11, 20 27
Makhachkala, Sov. Un., 129,000 E3 29
Makharadze, Sov. Un., 10,000 B14 31
Makhfar al Quwayrah, Jordan, 1,000 E7 32
Makhlata, Bul., 6,537 D7 22
Makin, isl., Gilbert Is F10 7
Makinak, Man., Can., 68 D2 71
Makinsk, Sov. Un C8 29
Makkah, see Mecca, Sau. Ar.
Makkinga, Neth., 443 B6 15
Makkovik, Newf., Can., 168 g10 75
Makkum, Neth., 2,086 A5 15
Mako, Hung., 29,935 B5 22
Makokou, Gabon E2 46
Makongai, isl., Fiji Is 52
Makoti, Ward, N. Dak., 214 B4 110
Makoua, Con. B F3 46
Makow, Pol., 2,642 B6 26
Makran, range, Pak H12 41
Makri, India, 5,000 H8 40
Maktar, Tun G11 30
Makteir, sand dunes, Maur B3 45
Mākū, Iran, 10,687 B2 41
Makumbi, Con. L C3 48
Makurazaki, Jap., 15,600 K5 37
Makurdi, Nig., 16,713 E6 45
Makushin, vol., Alsk E6 79
Mal, bay, Ire E2 11
Malabang, Phil., 2,377 *D6 35
Malabar, Brevard, Fla., 640 E6 86
Malabar, coast, India F5 39
Mal Abrigo, Ur E1 56
Malacca, Mala., 69,848 K5 38
Malacca, state, Mala., 291,211 K5 38
Malacca, strait, Asia K4 38
Malad City, Oneida, Idaho, 2,274 G6 89
Malafede, riv., It h8 21
Málaga, Col., 6,022 B3 60
Malaga, Gloucester, N.J., 200 D2 106
Malaga, Eddy, N. Mex., 50 E5 107
Málaga, Sp., 301,048 D3 20
Málaga, prov., Sp., 775,167 *D3 20
Málaga, bay, Sp D3 20
Malagash, N.S., Can., 69 D6 74
Malagasy Republic, country, Afr., 4,918,000 h9 49, I10 42
Malagón, Sp., 9,833 C4 20
Malaimbandy, Malag h9 49
Malaita, is., Solomon Is G9 7
Malakal, Sud., 9,680 D3 47
Malakoff, Fr., 33,603 g10 14
Malakoff, Henderson, Tex., 1,657 C4 118
Malalbergo, It., 6,351 E7 18
Malang, Indon., 332,023 G4 35
Malange, Ang., 12,815 C2 48
Malange, dist., Ang., 451,719 C2 48
Malanzán, Arg A3 54
Malar (Mälaren), lake, Swe t34 25
Mälaraspö, Swe t35 25
Mälaren (Malar), lake, Swe t34 25
Malargüe, Arg B3 54
Malartic, Que., Can., 6,998 E9 72
Malaspina, glacier, Alsk D11 79
Malaspina, Arg C3 54
Malatya, Tur., 84,200 C12 31
Malay, pen., Asia I13 33
Malaya, reg., Mala., 6,278,758 J4 38
Malaya Uzen, riv., Sov. Un F18 9
Malaya Vishera, Sov. Un., 25,000 B9 27
Malaybalay, Phil., 2,267 D7 35
Malāyer, Iran, 32,357 D4 41
Malaysia, country, Asia, 8,676,635 E2, 4 35, J5 38
Malazgirt, Tur., 5,100 C14 31
Malbaie, riv., Que., Can B7 73
Malbon, Austl., 73 D7 50
Malbork, Pol., 25,000 A5 26
Malchin, Ger., 6,825 B6 16
Malchow, Ger., 8,049 E6 24
Malcolm, Austl E3 50
Malcolm, Lancaster, Nebr., 116 D9 103
Malcolm, Poweshiek, Iowa, 416 C5 92
Malden, Middlesex, Mass., 57,676 C3 97
Malden, Dunklin, Mo., 5,007 E8 101
Malden, Whitman, Wash., 292 B8 122
Malden, Kanawha, W. Va., 1,000 C6 123
Malden, isl., Pac. O G12 7
Maldive Islands, country, Asia, 82,068 I10 33
Maldon, Eng., 10,507 C8 12
Maldonado, Ur., 12,100 E2 56
Maldonado, dept., Ur., 52,600 *E2 56
Malè, It., 2,841 C6 18
Malea, cape, Grc D4 23
Malegaon, India, 121,408 G5 40
Malema (Entre Rios), Moz D6 48
Malemba-Nkulu, Con. L C4 48
Maler-Lotla, India, 32,575 B5 40
Malesherbes, Fr., 2,029 C5 14
Malesus, Madison, Tenn., 350 B3 117
Malha (Well), Sud B2 47
Malheur, Malheur, Oreg C9 113
Malheur, co., Oreg., 22,764 D9 113
Malheur, lake, Oreg D8 113
Malheur, riv., Oreg D9 113
Mali, Guinea D2 45
Mali, country, Afr., 4,100,000 C4 45, E5 42
Mali, isl., Fiji Is 52
Malibu, Los Angeles, Calif., 2,000 *E4 82
Maligne, lake, Alta., Can C2 69
Malin, Ire., 164 B4 11
Malin, Klamath, Oreg., 568 E5 113
Malin, Sov. Un., 10,000 F7 27
Malin, head, Ire B4 11
Malinau, Indon E5 35
Malin Beg, Ire C3 11
Malindi, Ken., 3,292 B7 48
Malinec, Czech., 6,551 D5 26
Malinmore, head, Ire C3 11
Malino, Sov. Un n18 27
Malinta, Henry, Ohio, 339 A3 111
Malita, Phil., 5,125 *D7 35
Maliwun, Bur., 1,000 G3 38
Maljamar, Lea, N. Mex., 350 E6 107
Malkangiri, India, 5,000 H8 40
Malkara, Tur., 9,300 B6 23
Malko Turnovo, Bul., 3,489 E8 22
Mallard, Palo Alto, Iowa, 431 B3 92
Mallawi, Eg., U.A.R., 43,200 D6 43
Malleco, prov., Chile, 174,185 B2 54
Mallersdorf, Ger., 2,200 E7 17
Malles Venosta, It., 4,171 C6 18
Mallet Creek, Medina, Ohio, 150 A6 111
Malletts, bay, Vt B2 120
Mallorca, is., Sp C7 20
Mallory, Logan, W. Va., 1,133 *D3 123
Mallorytown, Ont., Can., 335 C9 72
Mallow, Ire., 5,545 E3 11
Malmbäck, Swe., 1,145 A4 24
Malmberget, Swe., 9,057 D9 25
Malmédy, Bel., 6,355 D6 15
Malmesbury, S. Afr., 8,206 D2 49
Malmköping, Swe., 1,861 t34 25
Malmo, Aitkin, Minn., 40 D5 99
Malmo, Saunders, Nebr., 135 C9, D1 103
Malmö, Swe., 233,400 C7 24, J5 25
Malmöhus, co., Swe., 632,200 C7 24
Maloarkhangelsk, Sov. Un., 10,000 E11 27
Maloja, Switz D8 19
Maloja, pass, Switz D8 19
Malolos, Phil., 3,500 (49,300▲) o13 35
Malone, Ont., Can., 50 C7 72
Malone, Jackson, Fla., 661 B1 86
Malone, Franklin, N.Y., 8,737 A2 108
Malone, Grays Harbor, Wash., 250 C2 122
Malone, Fond de Lac, Wis., 50 B5 124
Maloneyville, Knox, Tenn., 250 E11 117
Malonton, Man., Can., 50 D3 71
Malott, Okanogan, Wash., 350 A6 122
Maloy, Ringgold, Iowa, 68 D3 92
Maloyaroslavets, Sov. Un., 25,000 D11 27
Malpeque, bay, P.E.I., Can C6 74
Malpura, India, 10,622 D5 40
Malsch, Ger., 7,800 E3 17
Malshaya Uzen, riv., Sov. Un D3 29
Malta, Cassia, Idaho, 250 G5 89
Malta, De Kalb, Ill., 782 B5 90
Malta, Phillips, Mont., 2,239 B9 102
Malta, Morgan, Ohio, 983 C6 111
Malta, Br. dep., Eur., 319,620 G14 30
Malta Bend, Saline, Mo., 338 B4 101
Maltahohe, S.W., Afr., 1,044 B2 49
Malton, Ont., Can., 2,148 E6 72
Malton, Eng., 4,430 C6 10
Maltrata, Mex., 4,311 n15 63
Malung, Swe., 5,500 G5 25
Malvern, Geneva, Ala., 213 D4 78
Malvern, Hot Springs, Ark., 9,566 C3, D6 81
Malvern, Eng., 24,373 B5 12
Malvern, Mills, Iowa, 1,193 C2 92
Malvern, Carroll, Ohio, 1,320 B6 111
Malvern, Chester, Pa., 2,268 A10, F10 114
Malverne, Nassau, N.Y., 9,968 *G2 84
Mamala, bay, Haw g10 88
Mamanutha, is., Fiji Is 52
Mamanguape, Braz., 8,512 C3, h6 57
Mamantel, Mex., 89 D6 63
Mamaroneck, Westchester, N.Y., 17,673 D2 108
Mamba, Jap., 1,700 m17 37
Mambasa, Con. L A4 48
Mamberamo, riv., W. Irian F9 35
Mamers, Fr., 4,869 C4 14
Mamers, Harnett, N.C., 150 B5 109
Mameyes, P.R B7 65
Mamfe, Cam., 5,107 E6 45
Mamie, Currituck, N.C., 100 A8 109
Mamiña, Chile D2 65
Mammern, Switz., 424 A6 19
Mammoth, Pinal, Ariz., 1,913 E5 80
Mammoth, Juab, Utah, 100 D3 119
Mammoth, Kanawha, W. Va., 800 C3, C6 123
Mammoth Cave, nat. park, Ky C3 94
Mammoth Lakes, Mono, Calif., 300 D4 82
Mammoth Spring, Fulton, Ark., 825 A4 81
Mamolo, isl., Fiji Is 52
Mamoré, riv., Bol., Braz B2 55
Mamou, Guinea, 5,100 D2 45
Mamou, Evangeline, La., 2,928 D3 95
Mampawah, Indon E3 35
Mampong, Ghana, 3,948 E4 45
Mamry, lake, Pol A6 26
Mam Soul, mtn., Scot C3 13
Mamudju, Indon F5 35
Man, I.C., 17,000 E3 45
Man, Logan, W. Va., 1,486 D3, D5 123
Man, Isle of, see Isle of Man, Br. dep., Eur.
Man, riv., Sask., Can D4 70
Mana, Fr. Gu., 1,443 A4 59
Mana, Kauai, Haw., 225 A2 88
Manabí, prov., Ec., 477,503 B1 58
Manabo, Malag h8 49
Manacapuru, Braz., 2,584 D5 60
Manacor, Sp., 19,224 C7 20
Manado, Indon., 127,614 E6 35
Managua, Nic., 226,300 D4 62
Managua, lake, Nic D4 62
Manahawkin, Ocean, N.J., 400 D4 106
Manakara, Malag., 6,200 h9 49
Manakin, Goochland, Va., 330 C5 121
Manalapan, Monmouth, N.J., 50 C4 106
Manam, isl., N. Gui h12 50
Manama, Bahrain, 39,648 G8 33
Mañana, Pan F8 62
Manana, isl., Haw g11 88
Mananara, Malag., 2,852 g9 49
Mananara, riv., Malag h9 49
Mananjary, Malag., 11,269 h9 49
Manantenina, Malag., 750 h9 49
Manantico, creek, N.J E3 106
Manapire, riv., Ven B4 60
Manas (Suilai), China, 10,000 C2 34
Manasarowar, lake, China B9 40
Manasquan, Monmouth, N.J., 4,022 C4 106
Manasquan, riv., N.J C4 106
Manass, riv., China C2 34
Manassa, Conejos, Colo., 831 D5 83
Manassas, Tattnall, Ga., 154 D4 87
Manassas, Prince William, Va., 3,555 B4, C5 121
Manassas Park, Prince William, Va., 5,342 *C5 121
Manatee, co., Fla., 69,168 E4 86
Manatee, riv., Fla E4 86
Manatí, P.R., 9,682 B5 65
Manatí, mun., P.R., 29,354 B5 65
Manaus, Braz., 154,040 C2 59
Manavgat, Tur., 3,200 D8 31
Manawa, Waupaca, Wis., 1,037 D5 124
Manawan, lake, Sask., Can B4 70
Mancelona, Antrim, Mich., 1,141 D5 98
Mancha, reg., Sp C4 20
Manchac, bayou, La B5 95
Mancha Real, Sp., 7,855 D4 20
Manchaug, Worcester, Mass., 900 B4 97
Manche, dept., Fr., 446,878 *C3 14
Manchester, Hartford, Conn., 42,102 B6 84
Manchester, Eng., 661,041 (*2,835,000) D6 10, A5 12
Manchester, Meriwether and Talbot, Ga., 4,115 D2 87
Manchester, Scott, Ill., 282 D3 90
Manchester, Delaware, Iowa, 4,402 B6 92
Manchester, Dickinson, Kans., 153 C6 93
Manchester, Clay, Ky., 1,868 C6 94
Manchester, Kennebec, Maine, 300 (1,068▲) D3 96
Manchester, Carroll, Md., 1,108 A4 85
Manchester, Essex, Mass., 3,932 A6, C4 97
Manchester, Washtenaw, Mich., 1,568 F6 98
Manchester, St Louis, Mo., 2,021 B7 101
Manchester, Hillsboro, N.H., 88,282 (*111,900) E4 105
Manchester, Ontario, N.Y., 1,344 C3 108
Manchester, Adams, Ohio, 2,172 D4 111
Manchester, Grant, Okla., 162 A3 112
Manchester, York, Pa., 1,454 F8 114
Manchester, Kingsbury, S. Dak., 70 C8 116
Manchester, Coffee, Tenn., 3,930 B5 117
Manchester, Bennington, Vt., 403 (2,470▲) E2 120
Manchester Center, Bennington, Vt., 600 E2 120
Manchester Depot, Bennington, Vt., 800 E2 120
Manchouli (Lupin), China, 25,000 B8 34
Manchuria, reg., China, 41,700,000 B10 34
Mancos, Montezuma, Colo., 832 D2 83
Mancos, riv., Colo D2 83
Mandabe, Malag h8 49
Mandaguari, Braz., 8,210 C2 56
Mandal, Afg., 10,000 E10 41
Mandal, Nor., 5,200 H2 25
Mandalay, Bur., 190,000 B2 38, D10 39
Mandalaya, gulf, Tur D6 23
Mandali, Iraq, 9,722 E2 41
Mandan, Morton, N. Dak., 10,525 C5 110
Mandar, gulf, Indon F5 35
Mandara, mts., Cam C2 46
Mandarah, Eg., U.A.R., 2,000 C2 32
Mandaree, McKenzie, N. Dak., 135 B3 110
Mandarin, Duval, Fla., 900 B5, C6 86
Mandera, Ken A7 48
Manderfield, Bel., 1,344 D6 15
Manderfield, Beaver, Utah, 70 E3 119
Manderson, Shannon, S. Dak., 55 D3 116
Manderson, Big Horn, Wyo., 167 A5 125
Mandeville, Miller, Ark., 350 D2 81
Mandeville, Que., Can., 380 C4 73
Mandeville, St. Tammany, La., 1,740 B7, D5 95
Mandimba, Moz D6 48
Mandla, India, 19,416 E6 40
Mandra, Grc., 3,594 C4, g11 23
Mandritsara, Malag g9 49
Mandsaur, India, 41,876 E5 40
Manduria, It., 22,500 D6 21
Mandvi, India, 29,305 F2 40
Manell, chan., Guam 52
Manes, Wright, Mo., 65 D5 101
Manfalūt, Eg., U.A.R., 23,900 D6 43
Manfred, Wells, N. Dak., 79 B6 110
Manfredonia, It., 34,000 D5 21
Manfredonia, gulf, It D6 21
Manga, Braz., 2,000 D2 57
Mangalia, Rom., 4,792 D9 22
Mangalore, India, 142,669 (*170,253) F5 39
Mangaratiba, Braz., 2,741 h5 56
Mangas, mtn., N. Mex D1 107
Mangatarem, Phil., 2,392 o13 35
Mañgero, Phil., 254 *C6 35
Mangerton, mtn., Ire F2 11
Mangham, Richland, La., 521 B4 95
Mangla, Pak B5 39
Mango, isl., Fiji Is 52
Mangoky, riv., Malag h8 49
Mangole, isl., Indon F7 35
Mangonia Park, Palm Beach, Fla., 594 *F6 86
Mangotsfield, Eng., 24,092 C5 12
Mangrove, pt., Fla F2 86
Mangrove, swamp, Fla G6 86
Mangualde, Port., 3,093 B2 20
Manguéni, plat., Niger B7 45
Mangum, Greer, Okla., 3,950 C2 112
Mangyshlak, pen., Sov. Un E4 29
Manhasset, Nassau, N.Y., 8,300 F2 84
Manhattan, Will, Ill., 1,117 *B6 90
Manhattan, Putnam, Ind., 50 E4 91
Manhattan, Riley, Kans., 22,993 C7 93
Manhattan, Gallatin, Mont., 889 E5 102
Manhattan, Nye, Nev., 25 E4 104
Manhattan, borough (New York co.), N.Y., 1,698,281 (part of New York City) *D6 106
Manhattan, isl., N.Y D6 106
Manhattan Beach, Los Angeles, Calif., 34,513 F2 82
Manheim, Lancaster, Pa., 4,790 F9 114
Manheim, Preston, W. Va., 150 B5 123
Manhica, Moz C5 49
Manhuaçu, Braz., 10,546 F2 57
Mani, P.R C2 65
Maniagassa, isl., Saipan 52
Maniago, It., 7,518 C8 18
Manica e Sofala, prov., Moz A5 49
Manicoré, Braz., 2,268 D2 59
Manicouagan, riv., Que., Can F19 67
Manigotagan, Man., Can., 213 D3, D5 71
Manigotagan, lake, Man., Can D4 71
Manigotagan, riv., Man., Can D3 71
Manihiki, is., Pac. O G12 7
Manika E Sofala, prov., Moz A5 49
Manila, Navajo, Ariz., 40 C5 80
Manila, Mississippi, Ark., 1,753 B5 81
Manila, Phil., 1,145,700 (*2,500,000) C6, o13 35
Manila, Daggett, Utah, 329 C6 119
Manila, bay, Phil C6, o13 35
Manilla, Austl., 1,914 E8 51
Manilla, Rush, Ind., 400 E6 91
Manilla, Crawford, Iowa, 980 C2 92
Manipur, ter., India, 780,037 *D9 39
Manisa, Tur., 59,200 C6 23
Manistee, Manistee, Mich., 8,324 D4 98
Manistee, co., Mich., 19,042 D4 98
Manistee, riv., Mich D5 98
Manistique, Schoolcraft, Mich., 4,875 C4 98
Manistique, lake, Mich B5 98
Manistique, riv., Mich B4 98
Manito, Mason, Ill., 1,093 C4 90
Manito, lake, Sask., Can E1 70
Manitoba, prov., Can., 921,686 E13 66, 71
Manitoba, lake, Man., Can D2 71
Manitou, Man., Can., 863 E2 71
Manitou, Tillman, Okla., 269 C3 112
Manitou, isl., Mich A3 98
Manitou, lakes, Ont., Can E5 71
Manitou, riv., Ont., Can E5 71
Manitou Beach, Sask., Can., 159 F3 70
Manitou Beach, Lenawee, Mich., 1,544 *G6 98
Manitougan, riv., Que., Can B3 73
Manitoulin, co., Ont., Can., 11,176 B2 72
Manitoulin, isl., Ont., Can B2 72
Manitou Springs, El Paso, Colo., 3,626 C6 83
Manitowaning, Ont., Can., 400 B3 72
Manitowoc, Manitowoc, Wis., 32,275 B6, D6 124
Manitowoc, co., Wis., 75,215 D6 124
Manitowoc, riv., Wis B6 124
Manitowoc Rapids, Manitowoc, Wis., 400 B6 124
Maniwaki, Que., Can., 6,349 C2 73
Maniya (Oasis), Iraq G14 31
Manizales, Col., 147,210 B2 60
Manja, Malag., 2,253 h8 49
Manjra, riv., India E6 39
Mankato, Jewell, Kans., 1,231 C5 93
Mankato, Blue Earth, Minn., 23,797 F5 99
Mankono, I.C., 4,200 E3 45
Mankota, Sask., Can., 477 H2 70
Mankoya, N. Rh., 2,500 D3 48
Manley, Cass, Nebr., 113 E3 103
Manley, Hot Springs, Alsk., 72 B9 79
Manlius, Bureau, Ill., 374 B4 90
Manlius, Onondaga, N.Y., 1,997 C5 108
Manlléu, Sp., 9,410 A7 20
Manly, Worth, Iowa, 1,425 A4 92
Manly, Moore, N.C., 239 B4 109
Manlyville, Henry, Tenn., 40 A3 117
Manmad, India, 31,551 G5 40
Mann, ranges, Austl E5 50
Mannar, Cey., 5,190 G6 39
Mannar, gulf, India G6 39
Mannboro, Amelia, Va., 25 D5 121
Mannetto, hills, N.Y F3 84
Mannford, Creek, Okla., 358 A5 112
Mannheim, Ger., 313,900 (*1,125,000) D3 17
Manning, Dallas, Ark., 300 C3 81
Manning, Carroll, Iowa, 1,676 C2 92
Manning, Dunn, N. Dak., 50 B3 110
Manning, Washington, Oreg., 40 A1 113
Manning, Clarendon, S.C., 3,917 D7 115
Mannington, Christian, Ky., 500 C2 94
Mannington, Marion, W. Va., 2,996 A7, B4 123
Manns, creek, W. Va D7 123
Manns Choice, Bedford, Pa., 351 G4 114
Manns Harbor, Dare, N.C., 300 B8 109
Mannsville, Jefferson, N.Y., 446 B5 108
Mannsville, Johnston, Okla., 297 C5 112
Mannu, riv., It E2 21
Mannville, Alta., Can., 632 C5 69
Manö, isl., Den C2 24
Manokin, Somerset, Md., 80 D6 85
Manokin, riv., Md D6 85
Manokotak, Alsk., 149 D8 79
Manokwari, W. Irian F8 35
Manombo, Malag h8 49
Manomet, Plymouth, Mass., 800 C6 97
Manomet, pt., Mass C6 97
Manono, Con. L., 28,465 C4 48
Manono, isl., W. Sam 52
Manor, Sask., Can., 357 H4 70
Manor, Ware, Ga., 500 E4 87
Manor, Westmoreland, Pa., 1,136 *F2 114
Manor, Travis, Tex., 766 D4 118
Manorhamilton, Ire., 920 C3 11
Manorhaven, Nassau, N.Y., 3,566 *E3 108
Manorville, Suffolk, N.Y., 950 D4 108
Manorville, Armstrong, Pa., 557 E2 114
Manosque, Fr., 7,454 F6 14
Manotick, Ont., Can., 434 B9 72
Manpojin, Kor., 1,000 F3 37
Manresa, Sp., 52,216 B6 20
Mansalay, Phil., 1,294 *C6 35
Månsarp, Swe., 378 A8 24
Manseau, Que., Can., 837 C5 73
Mansel, isl., N.W. Ter., Can D16 67
Mansfield, Scott and Sebastian, Ark., 881 B1 81
Mansfield, Austl., 1,944 H6 51
Mansfield, Tolland, Conn., 100 (14,638▲) B7 84
Mansfield, Eng., 53,222 (*185,000) A6 12
Mansfield, Newton, Ga., 394 C3 87
Mansfield, Piatt, Ill., 743 C5 90
Mansfield, De Soto, La., 5,839 B2 95
Mansfield, Bristol, Mass., 5,500 (7,773▲) B5 97
Mansfield, Wright, Mo., 949 D5 101
Mansfield, Richland, Ohio, 47,325 B5 111
Mansfield, Tioga, Pa., 2,678 C7 114
Mansfield, Spink, S. Dak., 250 B7 116
Mansfield, Henry, Tenn., 110 A3 117
Mansfield, Tarrant, Tex., 1,375 B5 118
Mansfield, Douglas, Wash., 335 B6 122
Mansfield, flood control res., Ind E3 91
Mansfield, mtn., Vt B3 120
Mansfield Center, Tolland, Conn., 600 B8 84
Mansfield Depot, Tolland, Conn., 130 B7 84
Mansfield Southeast, Richland, Ohio, 2,961 *B5 111
Manson, Calhoun, Iowa, 1,789 B3 92
Manson, Chelan, Wash., 350 B5 122
Manson, creek, B.C., Can B5 68
Manson, riv., B.C., Can B6 68
Mansonville, Que., Can., 481 D5 73
Manso ou das Mortes, riv., Braz A2 56
Mansura, Avoyelles, La., 1,579 C3 95
Mant, hbr., Ponape 52
Mant, is., Ponape 52
Manta, Ec., 19,021 B1 58
Mantachie, Itawamba, Miss., 246 A5 100
Mantador, Richland, N. Dak., 98 C9 110
Mantagao, riv., Man., Can D3 71
Mantario, Sask., Can., 108 F1 70
Mantaro, riv., Peru D3 58
Mantas, Niger C5 45
Manteca, San Joaquin, Calif., 8,242 D3, B6 82
Mantee, Webster, Miss., 166 B4 100
Manteno, Kankakee, Ill., 2,225 B6 90
Manteo, Dare, N.C., 587 B8 109
Manteo, Buckingham, Va., 25 D4 121
Manter, Stanton, Kans., 183 E2 93
Mantes [-la-Jolie], Fr., 18,902 (*35,000) C4 14
Manthani, India, 7,779 H7 40
Manti, Sanpete, Utah, 1,739 D4 119
Mantilla, Cuba, 892 k12 64
Mantiqueira, mts., Braz C4 56
Mantoloking, Ocean, N.J., 160 C4 106
Manton, Wexford, Mich., 1,050 D5 98
Mantorville, Dodge, Minn., 498 F6 99
Mantova, It., 62,411 D6 18, B3 21
Mantua, Gloucester, N.J., 1,500 D2 106
Mantua, Portage, Ohio, 1,194 A6 111
Mantua, Box Elder, Utah, 275 B4 119
Manú, Peru D3 58
Manuan, lake, Que., Can B2 73
Manucan, Phil., 5,965 *D6 35
Manuel Alves, riv., Braz C1 57
Manuel Benavides, Mex., 798 B4 63
Manuelito, McKinley, N. Mex., 300 B1 107
Manui, isl., Indon F6 35
Manukau, hbr., N.Z L15 51
Manulla, riv., Ire D2 11
Manumuskin, Cumberland, N.J E3 106
Manumuskin, creek, N.J E3 106
Manvel, Grand Forks, N. Dak., 313 A8 110
Manvel, Brazoria, Tex., 50 F5 118
Manville, Somerset, N.J., 10,995 B3 106
Manville, Providence, R.I., 3,600 B11 84
Manville, Niobrara, Wyo., 124 C8 125
Many, Sabine, La., 3,164 C2 95
Manyara, lake, Tan B6 48
Manyas, lake, Tur B6 23
Manyberries, Alta., Can., 103 E5 69
Manych, canal, Sov. Un H14 27
Manych, lake, Sov. Un H14 27
Manych, riv., Sov. Un F17 9
Many Farms, Apache, Ariz A6 80
Manyoni, Tan., 1,388 C5 48
Manzala, lake, Eg., U.A.R C3 32
Manzanares, Sp., 17,847 C4 20
Manzanares, canal, Sp p17 20
Manzanares, riv., Sp o17 20
Manzanillo, Cuba, 42,252 E5 64
Manzanillo, Mex., 16,591 D4, n11 63
Manzanillo, P.R D4 65
Manzanita, Tillamook, Oreg., 363 B3 113
Manzanita Lake, Shasta, Calif., 5 B3 82
Manzano, Torrance, N. Mex., 100 C3 107
Manzanola, Otero, Colo., 562 C7 83
Manziana, It., 2,229 g8 21
Manzil, Jordan, 1,000 C8 32
Mao, Chad, 2,349 C3 46
Maoming, China, 14,000 G7 34
Mapai, Moz B5 49
Mapastepec, Mex., 3,580 D6 63
Mapaville, Jefferson, Mo., 300 C7 101
Mapes, Nelson, N. Dak., 50 A7 110
Mapia, is., W. Irian E8 35
Mapiri, Bol., 289 C2 55
Maple, Ont., Can., 1,552 E6 72
Maple, Bailey, Tex., 50 C1 118
Maple, peak, Ariz D6 80
Maple, riv., Iowa B2 92
Maple, riv., N. Dak C8 110
Maplebay, Polk, Minn., 75 C2 99
Maple Bluff, Dane, Wis., 1,565 *E4 124
Maple Creek, Sask., Can., 2,291 H1 70
Maple Grove, Hennepin, Minn., 2,213 *E5 99
Maple Heights, Cuyahoga, Ohio, 31,667 B2 111
Maple Hill, Wabaunsee, Kans., 244 C7 93
Maple Hill, Pender, N.C., 200 C6 109
Maple Lake, Wright, Minn., 1,018 E5 99
Maple Lane, St. Joseph, Ind., 2,500 *A5 91
Maple Mount, Daviess, Ky., 500 C2 94
Maple Rapids, Clinton, Mich., 683 E6 98
Maples, Allen, Ind., 125 B8 91
Maple Shade, Burlington, N.J., 12,947 D2 106
Maplesville, Clinton, Ala., 679 C3 78
Mapleton, Monona, Iowa, 1,686 B2 92
Mapleton, Bourbon, Kans., 127 D9 93
Mapleton, Aroostook, Maine, 900 (1,514▲) B4 96
Mapleton, Blue Earth, Minn., 1,107 G5 99
Mapleton, Cass, N. Dak., 180 C8 110
Mapleton, Lane, Oreg., 700 C3 113
Mapleton, Utah, Utah, 1,516 C4 119
Mapleton Depot, Huntingdon, Pa., 666 E6 114
Maple Valley, King, Wash., 800 D2 122
Mapleville, Providence, R.I., 800 B10 84
Maplewood, Calcasieu, La., 2,432 D2 95
Maplewood, York, Maine, 60 E2 96
Maplewood, Ramsey, Minn., 18,519 E5 99
Maplewood, St. Louis, Mo., 12,552 B8 101
Maplewood, Essex, N.J., 23,977 B4 106
Maplewood, Albany, N.Y., 1,000 *C7 108
Maplewood, Multnomah, Oreg., 2,000 *B4 113
Maplewood Park, St. Clair, Ill. (part of Cahokia) B8 101
Maplewood Park, Delaware, Pa., 1,800 *G11 114
Mapuera, riv., Braz C3 53
Ma'qalā', Sau. Ar H3 41
Maqnāh, Sau. Ar H10 31
Maquela do Zombo, Ang., 1,103 C2 48
Maquinchao, Arg C3 54
Maquoketa, Jackson, Iowa, 5,909 B7 92
Maquoketa, riv., Iowa B7 92
Maquon, Knox, Ill., 386 C3 90
Mar, mts., Braz D3 56
Marabá, Braz., 8,533 D5 59
Maracá, isl., Braz B4 59
Maracai, lake, Braz D4 60
Maracaibo, Ven., 432,902 A3 60
Maracaibo, lake, Ven B3 60
Maracaju, Braz., 1,848 C1 56
Maracay, Ven., 134,123 A4 60
Maracayo, P.R B3 65
Marādah, Libya, 1,299 C3 43
Maradi, Niger, 10,100 D6 45
Marāgheh, Iran, 36,551 C3 41
Maragogi, Braz., 1,031 k6 57
Maragogipe, Braz., 12,575 D3 57
Marais des Cygnes (Osage), riv., Kans D8 93
Marajó, isl., Braz C5 60
Maralal, Ken A6 48
Maramec, Pawnee, Okla., 169 A5 112
Marana, Pima, Ariz., 500 E4 80
Marand, Iran, 13,945 B2 41
Marandellas, S. Rh., 6,700 A5 49
Maranguape, Braz., 8,715 B3 57
Maranhão, state, Braz., 2,492,139 B1 57
Marano, lagoon, It D9 18

Maranoa, riv., Austl.....C7 51
Marañón, riv., Peru.....B2 58
Maras, Fr., 2,456.....D3 14
Maraş (Marash), Tur.,
54,600.....D11 31
Mărăşeşti, Rom., 5,604.....C8 22
Marash (Maras), Tur.,
54,600.....D11 31
Marathon, Monroe, Fla.,
950.....H5 86
Marathon, Grc., 2,515.....g11 23
Marathon, Buena Vista,
Iowa, 516.....B3 92
Marathon, Cortland, N.Y.,
1,079.....C4 108
Marathon, Brewster, Tex.,
800.....D1,F2 118
Marathon, Marathon, Wis.,
1,022.....D4 124
Maratua, isl., Indon.....E6 35
Maravtío, Mex., 5,182.....n13 63
Marãwah, Libya, 1,430.....F3 31
Marbach, Ger., 9,500.....E4 11
Marbella, Sp., 8,982.....D3 20
Marble, Madison, Ark., 150.....A2 81
Marble, Gunnison, Colo., 5.....B3 83
Marble, Itasca, Minn., 879.....C5 99
Marble, Cherokee, N.C., 356.....D2 109
Marble, Que., Can., 495.....A3 74
Maria, isl., Austl.....o15 50
Maria, riv., Arg.....B2 55
Maria Chiquita, Pan., 229.....k11 62
Mariager, Den., 1,500.....B3 24
Mariager, fjord, Den.....B4 24
María Grande, Arg., 3,400.....A5 54
Maria Grande, Pan., 4.....K11 62
María la Baja, Col., 4,182.....A2 60
Maria Luisa, Cuba.....k12 64
Marian, lake, Fla.....E5 86
Mariana, Cuba,
219,278.....k11 64
Marianna, Lee, Ark., 5,134.....C5 81
Marianna, Jackson, Fla.,
7,152.....B1 86
Marianna, Washington, Pa.,
1,088.....*F3 114
Marianske Lazne, Czech.,
12,600.....D2 26
Mariapolis, Man., Can., 258.....E2 71
Maria Stein, Mercer, Ohio,
230.....B3 111
Marias, pass, Mont.....B3 102
Marias, riv., Mont.....B5 102
Mariato, pt., Pan.....C7 62
Maria Van Diemen, cape, N.Z..K14 51
Mariaville, Hancock, Maine,
(144♦).....D4 96
Maribel, Pamlico, N.C., 250.....B7 109
Maribel, Manitowoc, Wis.,
250.....A6 124
Maribo, Den., 5,255.....D5 24
Maribo, co., Den., 133,870.....D5 24
Maribor, Yugo., 82,387.....B2 22
Maricá, Braz., 2,200.....h6 56
Maricaban, isl., Phil.....p13 35
Maricao, P.R., 1,475.....C3 65
Maricao, mun., P.R., 6,990.....C3 65
Maricopa, Pinal, Ariz., 400.....D3 80
Maricopa, Kern, Calif., 648.....E4 82
Maricopa, co., Ariz., 663,510.....D3 80
Maricopa, Indian res., Ariz.....E3 80
Maridi, Sud., 839.....E2 47
Marié, riv., Braz.....B4 60
Marie Byrd Land, reg., Ant.B36 4
Mariedam, Swe., 232.....u33 25
Mariefred, Swe., 1,673.....t35 25
Marie-Galante, isl., Guad.....o16 64
Mariehamn, Fin., 6,700.....G8 25
Mariembourg, Bel., 1,684.....D4 15
Mariemont, Hamilton, Ohio,
4,120.....C3 111
Marienberg, Ger., 8,281.....C8 17
Mariental, S.W. Afr., 1,803.....B2 49
Marienville, Forest, Pa., 800.....D3 114
Marienwerder, see Kwidzyn
Maries, co., Mo., 7,282.....C6 101
Mariestad, Swe., 11,900.....H5 25
Marietta, Duval, Fla.....B6 86
Marietta, Cobb, Ga.,
25,565.....A5 87
Marietta, Shelby, Ind., 300.....F6 91
Marietta, Lac qui Parle,
Minn., 327.....E2 99
Marietta, Prentiss, Miss., 160.....A5 100
Marietta, Washington, Ohio,
16,847.....C6 111
Marietta, Love, Okla., 1,933.....D4 112
Marietta, Lancaster, Pa.,
2,385.....F8 114
Marietta, Greenville, S.C.,
900.....A2 115
Marietta, Whatcom, Wash.,
300.....A2 122
Marietta East, Cobb, Ga.,
4,535.....*C2 87
Marieville, Que., Can.,
3,809.....D3 74
Mariinsk, Sov. Un., 36,000.....B11 29
Mariinskiy, Sov. Un., 10,000.E23 9
Marijampole, Sov. Un.,
130.....C3 29
Marília, Braz., 51,789.....C3 56
Marín, Sp., 7,261.....A1 20
Marin, co., Calif., 146,820.....C2 82
Marina, Monterey, Calif.,
3,310.....C6 82
Marina de Ravenna, It.....E8 18
Marin City, Marin, Calif.,
2,500.....*B5 82
Marinduque, prov., Phil.,
114,900.....*C6 35
Marine, Madison, Ill., 813.....E4 90
Marine City, St. Clair,
Mich., 4,404.....F8 98
Marine on St. Croix, Wash-
ington, Minn., 454.....E6 99
Marinette, Marinette, Wis.,
13,329.....C5 124
Marinette, co., Wis., 34,660.....C5 124
Maringa, riv., Con. L.....A3 48
Maringouin, Iberville, La.,
1,168.....D4 95
Marinha Grande, Port., 4,698.C1 20
Marino, It., 9,700.....h9 21
Marinovka, Sov. Un.....r21 27
Marion, Perry, Ala., 3,807.....C2 78
Marion, Crittenden, Ark.,
881.....B5 81
Marion, Hartford, Conn.,
500.....C5 84
Marion, Williamson, Ill.,
11,274.....F5 90
Marion, Grant, Ind., 37,854.....C6 91
Marion, Linn, Iowa, 10,882.....B6 92
Marion, Marion, Kans.,
2,169.....D6 93
Marion, Crittenden, Ky.,
2,468.....A3 94

Marion, Union, La., 685.....B3 95
Marion, Plymouth, Mass.,
1,160 (2,881♦).....C6 97
Marion, Osceola, Mich., 898.D5 98
Marion, Flathead, Mont., 40.B2 102
Marion, Red Willow, Nebr.,
172.....D5 103
Marion, Wayne, N.Y., 890.....B3 108
Marion, McDowell, N.C.,
3,345.....B2,D4 109
Marion, LaMoure, N. Dak.,
309.....C7 110
Marion, Marion, Ohio,
37,079.....B3 111
Marion, Marion, Oreg., 250.C2 113
Marion, Franklin, Pa., 650.....G6 114
Marion, Marion, S.C., 7,174.C9 115
Marion, Turner, S Dak.,
843.....D8 116
Marion, Smyth, Va., 8,385.....B3 121
Marion, Waupaca, Wis.,
1,200.....D5 124
Marion, co., Ala., 21,837.....A2 78
Marion, co., Ark., 6,041.....A3 81
Marion, co., Fla., 51,616.....C4 86
Marion, co., Ga., 5,477.....D2 87
Marion, co., Ill., 39,349.....E5 90
Marion, co., Ind., 697,567.....E5 91
Marion, co., Iowa, 25,886.....C4 92
Marion, co., Kans., 15,143.....D6 93
Marion, co., Ky., 16,887.....C4 94
Marion, co., Miss., 23,293.....D4 100
Marion, co., Mo., 29,522.....B6 101
Marion, co., Ohio, 60,221.....B4 111
Marion, co., Oreg., 120,888.....C4 113
Marion, co., S.C., 32,014.....C9 115
Marion, co., Tenn., 21,036.....D8 117
Marion, co., Tex., 8,049.....C5 118
Marion, co., W. Va., 63,717..B4 123
Marion, lake, S.C.....E7 115
Marion Center, Indiana, Pa.,
407.....E3 114
Marion Junction, Dallas, Ala.,
350.....B2 78
Marion Station, Somerset,
Md., 200.....D6 85
Marionville, Lawrence, Mo.,
1,251.....D4 101
Mariposa, Mariposa, Calif.,
550.....D4 82
Mariposa, co., Calif., 5,064..D3 82
Mariposas, Chile.....B2 54
Marismas, marshes, Sp.....D2 20
Marissa, St. Clair, Ill., 1,722.E4 90
Mariscal Estigarribia, Par.....D3 55
Maris Town, Madison, Miss.,
500.....C3 100
Mariveles, Phil., 1,828.....o13 35
Mark, Putnam, Ill., 445.....B4 90
Markakol, lake, Sov. Un.....E5 29
Markdale, Ont., Can., 1,090..C4 72
Marked Tree, Poinsett, Ark.,
3,216.....B5 101
Markerwaard Polder, reg.,
Neth.....B5 15
Markesan, Green Lake, Wis.,
1,060.....E5 124
Market Drayton, Eng.,
5,853.....B5 12
Market Harborough, Eng.,
11,556.....D7 12
Markethill, N. Ire., 813.....C5 11
Market Rasen, Eng., 2,257.....A7 12
Market Weighton, Eng.,
2,080.....G8 13
Markham, Ont., Can.,
4,294.....C10, D5 72
Markham, Cook, Ill., 11,704.*B3 90
Markham, Matagorda, Tex.,
800.....E4 118
Markham, Fauquier, Va.,
100.....C5 121
Markham, Grays Harbor,
Wash.....C2 122
Markham, mtn., Ant.....A28 5
Markhana, Afg., 5,000.....D13 41
Marki, Pol.....k14 26
Markinch, Sask., Can., 117.....G3 70
Markkleeberg, Ger., 20,500.....B7 17
Markland, Switzerland, Ind.,
75.....G8 91
Markle, Huntington and
Wells, Ind., 789.....C7 91
Markleeville, Alpine, Calif.,
50.....C4 82
Markleville, Madison, Ind.,
402.....E6 91
Markleysburg, Fayette, Pa.,
345.....G3 114
Markneukirchen, Ger., 8,903.C7 17
Markopoulon, Grc., 5,094.....h11 23
Markovka, Sov. Un., 5,000.....G12 27
Markovo, Sov. Un.....C20 29
Markranstädt, Ger., 10,300.....B7 17
Marks, Quitman, Miss.,
2,572.....A3 100
Marksboro, Warren, N.J.,
130.....B3 106
Markstay, Ont., Can., 306.....E9 72
Marksville, Avoyelles, La.,
4,257.....C3 95
Marktheidenfeld, Ger., 4,800.D4 17
Marktredwitz, Ger., 15,500.....D7 17
Markville, Pine, Minn., 100.....D6 99
Marl, Ger., 71,500.....*C3 16
Marland, Noble, Okla., 191.....A4 112
Marlbank, Ont., Can., 237.....C7 72
Marlboro, Alta., Can., 289.....C2 69
Marlboro, Hartford, Conn.,
700 (1,961♦).....C7 84
Marlboro, Middlesex,
Mass., 18,819.....B4, D1 97
Marlboro, Cheshire, N.H.,
1,097 (1,612♦).....E2 105
Marlboro, Monmouth, N.J.,
300.....C4 106
Marlboro, Ulster, N.Y.,
1,733.....D6 108
Marlboro, Windham, Vt.,
25 (347♦).....F3 120
Marlboro, co., S.C., 28,529.....B8 115
Marlborough, Br. Gu.....A3 59
Marlborough, prov., 4,843.....C6 42
Marle, Fr., 2,912.....E5 15
Marlene Village, Washington,
Oreg., 1,100.....*B3 113
Marlette, Sanilac, Mich.,
1,640.....E7 98
Marley, Anne Arundel, Md.,
1,500.....B4 85
Marlin, Falls, Tex., 6,918.....D4 118
Marlin (Krupp), Grant, Wash.,
99.....B7 122
Marlinton, Pocahontas,
W. Va., 1,586.....C4 123

Marlow, Baldwin, Ala., 100..E2 78
Marlow, Effingham, Ga.,
100.....D5 87
Marlow, Ger., 2,855.....D6 24
Marlow, Cheshire, N.H.,
500.....C5 100
Marlow, Stephens, Okla.,
4,027.....C4 112
Marlowe, Berkeley, W. Va.,
700.....B7 123
Marlton, Burlington, N.J.,
600.....D3 106
Marly-le-Roi, Fr., 10,193.....g9 14
Marmaduke, Greene, Ark.,
657.....A5 81
Marmande, Fr., 10,199.....E4 14
Marmara, isl., Tur.....B6 23
Marmaris, Tur., 3,400.....D7 23
Marmarth, Slope, N. Dak.,
319.....C2 110
Marmelos, riv., Braz.....D2 59
Mar Menor, lagoon, Sp.....D5 20
Marmet, Kanawha, W. Va.,
2,500.....C3, C6 123
Marmolada, mtn., It.....C7 18
Marmora, Ont., Can., 1,381.C7 72
Marmora, Cape May, N.J.,
100.....E3 106
Marmora, peak, It.....D2 21
Marmora, sea, Tur.....B6 23
Marnay, Fr., 869.....B1 18
Marne, Ger., 5,000.....B5 17
Marne, Cass, Iowa, 205.....C2 92
Marne, Ottawa, Mich., 450.....E5 98
Marne, canal, Fr.....F4 15
Marne, dept., Fr., 442,195.....F4 15
Marne, riv., Fr.....E5 15
Marne, au Rhin, canal, Fr.....F6 15
Mar Negro, P.R.....D5 65
Maro, reef, Haw.....k13 87
Maroa, Macon, Ill., 1,235.....C5 90
Maroantsetra, Malag., 4,412..g9 49
Maromokotro, mtn., Malag.....f9 49
Maroni, riv., Sur.....B4 60
Maroua, Cam.....B6 47
Marovoay, Malag., 7,217.....g9 49
Marpi, mtn., Saipan.....52
Marpi, pt., Saipan.....52
Marpo, pt., Tinian.....52
Marquam, Clackamas,
Oreg.....B2, B4 113
Marquand, Madison, Mo.,
392.....D7 101
Marquesas, is., Fr. Polynesia.G13 7
Marquesas, keys, Fla.....H4 86
Marquês de Valença, Braz.,
18,935.....C4, h6 56
Marquette, Man., Can., 87.....D3 51
Marquette, Clayton, Iowa,
572.....A6 92
Marquette, McPherson, Kans.,
607.....D6 93
Marquette, Marquette, Mich.,
19,824.....B3 98
Marquette, Hamilton, Nebr.,
210.....D8 103
Marquette, co., Mich.,
56,154.....B3 98
Marquette, co., Wis, 8,516..E4 124
Marquette Heights, Tazewell,
Ill., 2,517.....*C4 90
Marquez, Valencia, N. Mex.,
35.....B2 107
Marquez, Leon, Tex., 194.....D4 118
Marquis, Sask., Can., 170.....G3 70
Marquise, Fr., 4,802.....D9 12
Marrakech, Mor.,
243,134.....C3 44, I3 20
Marree, Austl., 190.....E6 50
Marrero, Jefferson, La.,
19,000.....C7 95
Marromeu, Moz., 5,000.....A6 49
Marroqui, pt., Sp.....D3 20
Marrowbone, Cumberland,
Ky., 200.....D4 94
Mars, Butler, Pa., 1,522.....E1 114
Mars, hill, Maine.....B5 96
Mars, riv., Que., Can.....A7 73
Marsabit, Ken.....A6 48
Marsala, It., 39,000.....F4 21
Marså Süsah, Libya.C4 43, F4 31
Marsden, Sask., Can., 208.....E1 70
Marseille, Franca, It., 3,394.....F5 14
Marseille (Marseilles), Fr.,
778,071 (*870,000).....F6 14
Marsh, dam, Ala.....C4 78
Martin, lake, Ala.....C4 78
Martina [Franca], It., 25,500.D6 21
Martindale, Que., Can., B5 73
Martindale, Caldwell, Tex.,
500.....A4 118
Martinez, Contra Costa,
Calif., 9,604.....B5, C2 82
Martínez, Cuba.....D2 64
Martinez, Columbia and
Richmond, Ga., 2,000.....C4 87
Martinique, Fr. dep., N.A.,
258,000.....o16 64
Martinique, passage, W.I.....o16 64
Martins, pond, Mass.....C3 97
Martinsburg, Washington,
Ind., 100.....H5 91
Martinsburg, Keokuk, Iowa,
172.....C5 92
Martinsburg, Audrain, Mo.,
330.....B6 101
Martinsburg, Dixon, Nebr.,
68.....B9 103
Martinsburg, Lewis, N.Y.,
250.....B5 108
Martinsburg, Knox, Ohio,
228.....B5 111
Martinsburg, Blair, Pa.,
1,772.....F5 114
Martinsburg, Berkeley,
W. Va., 15,179.....B6 123
Martins Creek, Warren, N.J.,
300.....*B2 106
Martins Creek, Northampton,
Pa., 900.....B2 106
Martinsdale, Meagher, Mont.,
150.....D6 102
Martins Ferry, Belmont,
Ohio, 11,919.....C7 111
Martins Mills, Wayne, Tenn.B4 117
Martins Springs, Marion,
Tenn.....D8 117
Martinsville, Clark, Ill.,
1,351.....D6 90
Martinsville, Morgan, Ind.,
7,525.....F5 91
Martinsville, Copiah, Miss.,
75.....D3 100
Martinsville, Harrison,
Mo.....A3 101
Martinsville, Somerset, N.J.,
J,700.....B3 106
Martinsville, Clinton, Ohio,
488.....C4 111

Marshalltown, New Castle, Del.,
1,800.....A6 85
Marshallton, Northumberland,
Pa., 2,316.....*E8 114
Marshalltown, Marshall,
Iowa, 22,521.....B5 92
Marshallville, Macon, Ga.,
1,308.....D3 87
Marshallville, Wayne, Ohio,
611.....B6 111
Marshes Siding, McCreary,
Ky., 500.....D5 94
Marshfield, Warren, Ind.,
100.....D3 91
Marshfield, Plymouth, Mass.,
1,500 (6,748♦).....B6, E4 97
Marshfield, Webster, Mo.,
2,221.....D5 101
Marshfield, Washington, Vt.,
313 (891♦).....C4 120
Marshfield, Wood, Wis.,
14,153.....D3 124
Marshfield Hills, Plymouth,
Mass., 500.....B6, E4 97
Marsh Harbour, Ba. Is., 289..B5 64
Mars Hill, Marion, Ind.,
1,000.....E5, H7 91
Mars Hill, Aroostook, Maine,
1,458 (2,062♦).....B5 96
Mars Hill, Madison, N.C.,
1,574.....D3 109
Marshville, Union, N.C.,
1,360.....B3 109
Marshyhope, creek, Del., Md.C6 85
Marsing, Owyhee, Idaho,
555.....F2 89
Marske-by-the Sea, Eng. (part
of Saltburn and Marske-by-
the-Sea).....F8 13
Marsland, Dawes, Nebr., 39..B2 103
Marstal, Den., 1,986
(*2,900).....D4 24
Marsteller, Cambria, Pa., 450.E4 114
Marston, New Madrid, Mo.,
631.....E8 101
Marston, Richmond, N.C.,
75.....C4 109
Marstons Mills, Barnstable,
Mass., 600.....C7 97
Mart, McLennan, Tex.,
2,197.....D4 118
Martaban, Bur., 5,639.....E10 39
Martaban, gulf, Bur.....E10 39
Martapura, Indon.....F4 35
Martel, Marion, Fla., 100.....C4 86
Martel, Fr., 882.....E4 14
Martelange, Bel., 1,583.....E5 15
Martell, Pierce, Wis., 150.....D1 124
Martelle, Jones, Iowa, 247.....B6 92
Martensdale, Warren, Iowa,
316.....C4 92
Martha, Jackson, Okla., 243..C2 112
Martha, Wilson, Tenn., 35.....A5 117
Marthasville, Warren, Mo.,
339.....C6 101
Martha's Vineyard, isl., Mass..D6 97
Marthaville, Natchitoches,
La., 181.....C2 95
Martí, Cuba, 652.....C5 64
Martigny-Ville, Switz.,
5,239.....D3 19
Martigues, Fr., 9,852.....F6 14
Martin, Stephens, Ga., 209.....B3 87
Martin, Floyd, Ky., 992.....C7 96
Martin, Allegan, Mich., 483.F5 98
Martin, Sherman, N. Dak.,
146.....B5 110
Martin, Ottawa, Ohio, 220.....A2 111
Martin, Allendale, S.C., 80.....E5 115
Martin, Bennett, S. Dak.,
1,184.....D4 116
Martin, Weakley, Tenn.,
4,750.....A3 117
Martin, co., Fla., 16,932.....E6 86
Martin, co., Ind., 10,602.....G4 91
Martin, co., Ky., 10,201.....C7 94
Martin, co., Minn., 26,986.....G4 99
Martin, co., N.C., 27,139.....B6 109
Martin, co., Tex., 5,068.....C2 118
Martin, dam, Ala.....C4 78
Martin, lake, Ala.....C4 78
Martin, bay, Newf., Can.....A4 75
Marseille-en-Beauvaisis, Fr.,
698.....E1 15
Marseilles (Marseille), Fr.,
778,071 (*870,000).....F6 14
Marseilles, La Salle, Ill.,
4,347.....B5 90
Marsh, Dawson, Mont., 75..D12 102
Marsh, creek, Mich.....C4 94
Marsh, fork, W. Va.....C3 123
Marsh, isl., La.....E4 95
Marsh, lake, Minn.....E2 99
Marshall, Searcy, Ark.,
1,095.....B3 81
Marshall, Sask., Can., 161.....D1 70
Marshall, Clark, Ill., 3,270.....D6 90
Marshall, Parke, Ind., 360.....E3 91
Marshall, Lib.....E2 45
Marshall, Calhoun, Mich.,
6,736.....F6 98
Marshall, Lyon, Minn.,
6,681.....F3 99
Marshall, Saline, Mo., 9,572.B4 101
Marshall, Madison, N.C.,
926.....D3 109
Marshall, Harrison, Tex.,
23.846.....C5 118
Marshall, Fauquier, Va., 500.C5 121
Marshall, Spokane, Wash.,
100.....D7 122
Marshall, Dane, Wis., 736.....E4 124
Marshall, Albany, Wyo.....C7 125
Marshall, co., Ala., 48,018.....A3 78
Marshall, co., Ill., 13,334.....B4 90
Marshall, co., Ind., 32,443.....B5 91
Marshall, co., Iowa, 37,984..C4 92
Marshall, co., Kans., 15,598.C7 93
Marshall, co., Ky., 16,736.....A3 94
Marshall, co., Minn., 14,262.B2 99
Marshall, co., Miss., 24,503..A4 100
Marshall, co., Okla., 7,263.....C5 112
Marshall, co., S. Dak., 6,663.B8 116
Marshall, co., Tenn., 16,859..B5 117
Marshall, co., W. Va., 38,041.B4 123
Marshall, is., Pac. O.....F10 7
Marshall, riv., N.T.....53
Marshallberg, Carteret, N.C.,
600.....C7 109
Marshall Hall, Charles, Md.,
20.....C3 85
Marshall Northeast, Harrison,
Tex., 1,192.....*C5 118
Marshalls Creek, Monroe, Pa.,
250.....D11 114

Masisea, Peru, 1,742........C3 58
Masisi, Con. L........B4 48
Masjed Soleymān, Iran., 44,651........F4 41
Mask, lake, Ire........D2 11
Maskell, Dixon, Nebr., 54........B9 103
Maskinongé, Que., Can., 893........C4 73
Maskinongé, co., Que., Can., 21,274........C4 73
Masnières, Fr., 2,399........D3 15
Mason, Effingham, Ill., 332........E5 90
Mason, Ingham, Mich., 4,522........F6 98
Mason, Lyon, Nev., 350........E2 104
Mason, Hillsboro, N.H., 50 (349▲)........F4 105
Mason, Warren, Ohio, 4,727........C2, C3 111
Mason, Okfuskee, Okla., 50........B5 112
Mason, Tipton, Tenn., 407........B2 117
Mason, Mason, Tex., 1,910........D3 118
Mason, Mason, W. Va., 1,005........B2 123
Mason, Bayfield, Wis., 100........B2 124
Mason, Sublette, Wyo........C2 125
Mason, co., Ill., 15,193........C4 90
Mason, co., Ky., 18,454........B6 94
Mason, co., Mich., 21,929........D4 98
Mason, co., Tex., 3,780........D3 118
Mason, co., Wash., 16,251........B2 122
Mason, co., W. Va., 24,459........C3 123
Mason, bayou, Ark........D4 81
Mason City, Mason, Ill., 2,160........C4 90
Mason City, Cerro Gordo, Iowa, 30,642........A4 92
Mason City, Custer, Nebr., 277........C6 103
Masonhall, Obion, Tenn., 200........A2 117
Masons, Coos, N.H........A3 105
Masontown, Fayette, Pa., 4,730........F2 114
Masontown, Preston, W. Va., 841........B5 123
Masonville, Desha, Ark., 50........D4 81
Masonville, Larimer, Colo., 200........A5 83
Masonville, Delaware, N.Y., 250........C5 108
Mass, Ontonagon, Mich., 500........A6 98
Massa, It., 56,988 (*129,000)........B3 21
Massaade, Syr........A7 32
Massabesic, lake, N.H........E4 105
Massabielle, Que., Can., 485........B6 73
Massac, co., Ill., 14,341........F5 90
Massachusetts, state, U.S., 5,148,578........B13 77, 97
Massachusetts, bay, Mass........B6 97
Massacre, bay, Am. Sam........52
Massacre, lake, Nev........B2 104
Massafra, It., 20,005........D6 21
Massakory, Chad, 1,033........C3 46
Massa Marittima, It., 5,700........C3 21
Massangena, Moz........B5 49
Massanutten, mtn., Va........C4 121
Massapê, Braz., 4,760........B2 57
Massapeag, New London, Conn., 100........D8 84
Massapequa, Nassau, N.Y., 30,000........G3 84
Massapequa Park, Nassau, N.Y., 19,904........*G3 108
Massapoag, pond, Mass........E2 97
Massaua, Eth., 25,000........B4 47
Massena, Cass, Iowa, 456........C3 92
Massena, St. Lawrence, N.Y., 15,478........A2 108
Masset, B.C., Can., 547........C1 68
Masset, inlet, B.C., Can........C1 68
Massey, Ont., Can., 1,324........A2 72
Massey, Kent, Md., 75........B6 85
Massif Central, mts., Fr........E5 14
Massif de Tarazit, plat., Niger........B6 45
Massillon, Stark, Ohio, 31,236........B6 111
Massinga, Moz........B6 49
Massive, mtn., Colo........B4 83
Masson, Que., Can., 1,933........D2 73
Massy, Fr., 19,137........h10 14
Masten, Kent, Del., 30........C6 85
Mastergeehy, Ire........F1 11
Masters, Weld, Colo., 3........A6 83
Masterton, N.Z., 15,128........N15 51
Mastic, Suffolk, N.Y., 1,600........*F5 108
Mastic Beach, Suffolk, N.Y., 3,035........*G5 84
Mastuj, Pak........k17 41
Mastung, Pak., 2,792........C1 40
Mastūrah, Sau. Ar........A4 47
Masuda, Jap., 27,400........I5 37
Masulipatnam, India, 101,417........E7 39
Masuria, reg., Pol........B6 26
Masury, Trumbull, Ohio, 5,900........A7 111
Masuya, Okinawa........52
Mata, Con. L........C3 48
Mata Armilla, Arg........D2 54
Matabeleland, prov., S. Rh........A4 49
Mataboor, W. Irian........F9 35
Matachewan, Ont., Can., 923........E9 72
Mata de São João, Braz., 8,117........D3 57
Matadi, Con. L., 59,000........C1 48
Matador, Sask., Can., 76........G2 70
Matador, Motley, Tex., 1,217........B2 118
Matafotufotu, cape, W. Sam........52
Matagalpa, Nic., 15,900........D5 62
Matagorda, Matagorda, Tex., 500........E5 118
Matagorda, co., Tex., 25,744........E5 118
Matagorda, bay, Tex........E4 118
Matagorda, isl., Tex........E4 118
Matagorda, pen., Tex........E5 118
Matak, isl., Indon........K7 38
Matam, Sen., 2,600........C2 45
Matamā, Eth........C4 47
Matamoras, Pike, Pa., 2,087........D12 114
Matamoros, Mex., 93,334........B5 63
Matamoros [de la Laguna], Mex., 13,572........B4 63
Ma'tan, Libya........D2 43
Ma'tan as Sarra (Well), Libya........E4 43
Ma'ţan Bishārah (Well), Libya........E4 43
Matane, Que., Can., 9,190........G19 67
Matane, co., Que., Can., 35,078........*B3 73
Ma'ţan Rāshidah, Libya........H3 31
Matanuska, riv., Alsk........g18 79
Matanuska, val., Alsk........g17 79
Matanzas, Cuba, 63,916........D3 64
Matanzas, prov., Cuba, 395,780........D3 64
Matanzas, inlet, Fla........C5 86
Matapan, cape, Grc........D4 23
Matapedia, co., Que., Can., 35,586........*B3 73
Matara, Cey., 27,735........G7 39
Mataram, Indon........G5 35
Matarani, Peru........E3 58
Matarka, Mor........H5 30
Mataró, Sp., 41,128........B7 20
Matatiele, S. Afr., 3,237........D4 49
Matautu, W. Sam., 392........52
Matawan, Monmouth, N.J., 5,097........C4 106
Matehuala, Mex., 19,738........C4 63
Mateko, Con. L........B2 48
Matera, It., 38,562........D6 21
Mateszalka, Hung., 11,249........B6 22
Mateur, Tun., 12,714........F11 30
Matewan, Mingo, W. Va., 896........D2 123
Matfield Green, Chase, Kans., 95........D7 93
Mathelo, Pak........D2 40
Mather, Man., Can., 125........E2 71
Mather, Greene, Pa., 1,033........G1 114
Mather, Juneau, Wis., 65........D3 124
Mather, peak, Wyo........A5 125
Matherville, Mercer, Ill., 612........B3 90
Matherville, Wayne, Miss., 100........D5 100
Matheson, Elbert, Colo., 150........B7 83
Matheson Island, Man., Can., 176........D3 71
Mathews, Montgomery, Ala., 10........C3 78
Mathews, Lafourche, La., 200........C6, E5 95
Mathews, Mathews, Va., 500........D5 121
Mathews, co., Va., 7,121........D6 121
Mathews, lake, Calif........F3 82
Mathias, Hardy, W. Va., 300........C6 123
Mathias Point, King George, Va........C5 121
Mathis, San Patricio, Tex., 6,075........E4 118
Mathiston, Webster and Choctaw, Miss., 597........B4 100
Mathura, India, 116,959 (*125,258)........C6 39, D6 40
Mati, Phil., 2,899........*D7 35
Matiakouali, Upper Volta........D5 45
Matinenda, lake, Ont., Can........B8 98
Matinicus, Knox, Maine, 75 (100▲)........E4 96
Matinicus, isl., Maine........E4 96
Matjan, is., Indon........G6 35
Matlock, Eng., 18,486........A6 12
Matlock, Sioux, Iowa, 103........A2 92
Matoaca, Chesterfield, Va., 2,000........C7 121
Matoaka, Mercer, W. Va., 613........D3 123
Mato Grosso, Braz., 520........B4 55
Mato Grosso, state, Braz., 910,262........B1 56
Matopiana, riv., Pol........f9 26
Matozinhos, Port., 33,400........B1 20
Maţraḩ, Mus. & Om., 3,500........D2 39
Matrosovo, Sov. Un........B11 37
Maţrūḩ, Eg., U.A.R., 3,047........G6 31
Matsqui, B.C., Can., 275........B10 68
Matsudo, Jap., 57,900 (86,372▲)........*n18 37
Matsue, Jap., 78,000 (106,476▲)........I6 37
Matsuida, Jap., 9,300........m17 37
Matsumae, Jap., 19,534........F10 37
Matsumoto, Jap., 108,000 (148,710▲)........H9, m16 37
Matsuyama, Jap., 238,604........J6 37
Matsuzaka, Jap., 57,600 (98,441▲)........I8, o15 37
Matsuzaki, Jap., 5,708........o17 37
Mattamiscontis, lake, Maine........C4 96
Mattamuskeet, lake, N.C........B7 109
Mattancheri, India, 83,896........G6 39
Mattapoisett, Plymouth, Mass., 1,640 (3,117▲)........C6 97
Mattaponi, riv., Va........D5 121
Mattawa, Ont., Can., 3,314........A6, B7 72
Mattawamkeag, Penobscot, Maine, 750 (945▲)........C4 96
Mattawamkeag, lake, Maine........C4 96
Mattawamkeag, riv., Maine........C4 96
Mattawana, Mifflin, Pa., 250........F6 114
Mattawin, riv., Que., Can........C4 73
Mattawoman, creek, Md........C3 85
Matterhorn, mtn., Switz........E4 19
Mattervisp, riv., Switz........E4 19
Matteson, Cook, Ill., 3,225........F3 90
Matthews, Jefferson, Ga., 106........C4 87
Matthews, Grant, Ind., 627........D7 91
Matthews, New Madrid, Mo., 450........E8 101
Matthews, Mecklenburg, N.C., 609........B3 109
Matthews, mtn., Mo........D7 101
Matthew Town, Ba. Is., 994........E7 64
Mattituck, Suffolk, N.Y., 1,274........D4 108
Mattoax, Amelia, Va........D5 121
Mattoon, Coles, Ill., 19,088........D5 90
Mattoon, Shawano, Wis., 435........D4 124
Mattson, Coahoma, Miss., 250........A3 100
Mattydale, Onondaga, N.Y., 9,000........*B4 108
Matucana, Peru, 1,746........D2 60
Matuku, isl., Fiji Is........52
Matun, Afg., 5,000........E14 41
Matundu, Con. L........A3 48
Matunuck, Washington, R.I., 70........D10 84
Maturín, Ven., 53,445........B5 60
Mau, Fiji Is........52
Mau, India, 48,785........E9 40
Mauá, Moz........D6 48
Mauban, Phil., 4,051........o13 35
Maubeuge, Fr., 27,214........B5 14
Maubin, Bur., 22,130........D1 38
Mauch Chunk, see Jim Thorpe, Pa.
Mauchline, Scot., 4,160........E4 13
Mauckport, Harrison, Ind., 107........H5 91
Maud, Butler, Ohio, 600........C2, C3 111
Maud, Pottawatomie and Seminole, Okla., 1,137........B5 112
Maud, Scot........C6 13
Maud, Bowie, Tex., 951........C5 118
Maudlow, Gallatin, Mont., 30........D5 102
Maués, Braz., 4,161........C3 59
Maues Guaçu, riv., Braz........C3 59
Mauganj, India, 1,000........E8 40
Maugansville, Washington, Md., 625........A2 85
Maughold, head, I. of Man........F4 13
Maui (incl. Kalawao), co., Haw., 42,855........B5 82
Maui, isl., Haw........C5 88
Mauk, Taylor, Ga., 100........D2 87
Mauldin, Greenville, S.C., 1,462........*B3 115
Maule, prov., Chile, 79,304........B2 54
Mauléon-Soule, Fr., 3,705........F3 14
Maumakeogh, mtn., Ire........C2 11
Maumee, Lucas, Ohio, 12,063........A2, A4 111
Maumee, bay, Ohio........A2 111
Maumee, riv., Ind., Ohio........B8 91, A4 111
Maun, Bech., 500........A3 49
Maunabo, P.R., 1,027........C7 65
Maunabo, mun., P.R., 10,785........C7 65
Mauna Kea, vol., Haw........D6 88
Maunaloa, Maui, Haw., 950........B4 88
Mauna Loa, vol., Haw........D6 88
Maunalua, bay, Haw........g11 88
Maungdaw, Bur., 3,846........D9 39
Maunie, White, Ill., 363........E5 90
Maupin, Wasco, Oreg., 381........B5 113
Maurepas, Livingston, La., 50........B6 95
Maurepas, lake, La........D5 95
Maurertown, Shenandoah, Va., 225........C4 121
Maurice, Sioux, Iowa, 237........B1 92
Maurice, Vermilion, La., 411........D3 95
Maurice, Lawrence, S. Dak., 25........C2 116
Maurice, riv., N.J........E2 106
Maurice River, Cumberland, N.J., 40........E2 106
Mauricetown, Cumberland, N.J., 250........E3 106
Mauriceville, Orange, Tex., 750........D2 95
Maurier, pt., Que., Can........C2 75
Maurine, Meade, S. Dak., 10........B3 116
Mauritania, country, Afr., 725,000........C2 45, D4 42
Mauritius, Br. dep., Afr., 516,556........H24 3
Maury, Greene, N.C., 285........B6 109
Maury, co., Tenn., 41,699........B4 117
Maury, isl., Wash........D1 122
Maury City, Crockett, Tenn., 624........B2 117
Mauston, Juneau, Wis., 3,531........E3 124
Mauterndorf, Aus., 1,615........E6 16
Maverick, Apache, Ariz., 50........D6 80
Maverick, co., Tex., 14,508........E2 118
Mavinga, Ang........E3 48
Mawer, Sask., Can., 72........G2 70
Mawkmai, Bur........B2 38
Mawlaik, Bur., 3,042........D9 39
Max, Dundy, Nebr., 150........D4 103
Max, McLean, N. Dak., 410........B4 110
Maxbass, Bottineau, N. Dak., 218........A4 110
Maxcanú, Mex., 5,127........C6 63
Maxeys, Oglethorpe, Ga., 149........C3 87
Maxie, Forrest, Miss., 60........E4 100
Maxim, Monmouth, N.J........C4 106
Maxinkuckee, lake, Ind........B5 91
Max Meadows, Wythe, Va., 900........E2 121
Maxstone, Sask., Can........H2 70
Maxton, Robeson, N.C., 1,755........C4 109
Maxville, Ont., Can., 804........B10 72
Maxville, Duval, Fla., 250........B4 86
Maxville, Granite, Mont., 40........D3 102
Maxwell, Colusa, Calif., 700........C2 82
Maxwell, Hancock, Ind., 280........E6 91
Maxwell, Story, Iowa, 773........A8, C4 92
Maxwell, Lincoln, Nebr., 324........C5 103
Maxwell, Colfax, N. Mex., 392........A5 107
Maxwell, Franklin, Tenn., 50........B5 117
Maxwell Colony, Hutchinson, S. Dak., 200........D8 116
May, Lemhi, Idaho, 60........E5 89
May, Harper, Okla., 114........A2 112
May, Brown, Tex., 400........D3 118
May, cape, N.J........C13 77
May, isl., Scot........D6 13
Maya, mts., Br. Hond........B3 62
Mayaguana, isl., Ba. Is........D7 64
Mayaguana, passage, Ba. Is........D7 64
Mayagüez, P.R., 50,147........C2 65
Mayagüez, mun., P.R., 83,850........C2 65
Mayagüez, bay, P.R........C2 65
Mayama, Con. B........C2 48
Mayarí, Cuba, 4,519........E6 64
Maybank, Forrest, Miss., 60........D4 100
Maybee, Monroe, Mich., 459........F7 98
Maybell, Moffat, Colo., 100........A2 83
Mayberry, Carroll, Md., 100........A3 85
Maybeury, McDowell, W. Va., 900........*D3 123
Maybole, Scot., 4,677........E4 13
Maybrook, Orange, N.Y., 1,348........D6 108
Maydee, Lauderdale, Tenn........B2 117
Maydi, Yemen........B5 47
Mayen, Ger., 17,300........C2 17
Mayenne, Fr., 10,270........C3 14
Mayenne, dept., Fr., 250,030........*C3 14
Mayenne, riv., Fr........D3 14
Mayer, Yavapai, Ariz., 300........C3 80
Mayersville, Issaquena, Miss., 450........C2 100
Mayerthorpe, Alta., Can., 663........C3 69
Mayes, co., Okla., 20,073........A6 112
Mayesville, Sumter, S.C., 5,000........D7 115
Mayetta, Jackson, Kans., 218........B6 93
Mayetta, Ocean, N.J., 150........D4 106
Mayfair, Sask., Can., 95........E2 70
Mayfair, Greenville, S.C., 5,000........*B3 115
Mayfield, Sumner, Kans., 119........E6 93
Mayfield, Graves, Ky., 10,762........B2 94
Mayfield, Fulton, N.Y., 818........B6 108
Mayfield, Butler, Ohio, 2,747........*C3 111
Mayfield, Cuyahoga, Ohio, 1,977........*A6 111
Mayfield, Lackawanna, Pa., 1,996........C10 114
Mayfield, Sanpete, Utah, 329........D4 119
Mayfield, Lewis, Wash., 45........C3 122
Mayfield, res., Wash........C3 122
Mayfield Heights, Cuyahoga, Ohio, 13,478........A6 111
Mayflower, Faulkner, Ark., 355........C3 81
Mayflower, Newton, Tex., 100........D6 118
Mayhew, Lowndes, Miss., 300........B5 100
Mayhill, Otero, N. Mex., 100........E4 107
Maykain, Sov. Un........C9 29
Mayking, Letcher, Ky., 750........C7 94
Maykop, Sov. Un., 88,000........G17 9
Mayland, Cumberland, Tenn., 450........C8 117
Maymont, Sask., Can., 239........E2 70
Maymyo, Bur., 22,287........D10 39
Maynard, Randolph, Ark., 201........A5 81
Maynard, Fayette, Iowa, 515........B6 92
Maynard, Allen, Ky., 50........D3 94
Maynard, Middlesex, Mass., 7,695........B5, C1 97
Maynard, Chippewa, Minn., 429........F3 99
Maynard, Belmont, Ohio, 600........B1 123
Maynardville, Union, Tenn., 620........C10 117
Mayne, B.C., Can., 139........B9 68
Maynooth, Ont., Can., 303........B7 72
Maynooth, Ire., 1,753........D5 11
Mayo, Lafayette, Fla., 687........B3 86
Mayo, Anne Arundel, Md., 500........C4 85
Mayo, Spartanburg, S.C., 500........A4 115
Mayo, co., Ire., 123,330........D2 11
Mayo, mts., Ire........C2 11
Mayo, riv., Arg........D2 54
Mayodan, Rockingham, N.C., 2,366........A4 109
Mayo Landing, Yukon, Can., 342........D5 66
Mayon, vol., Phil........C6 35
Mayotte, isl., Comoro Is........f9 49
Mayoumba, Gabon........F2 46
May Pen, Jam., 14,214........G5 64
Mayport, Duval, Fla., 450........B5, B6 86
Mays, Rush, Ind., 200........E7 91
Mays Landing, Atlantic, N.J., 1,404........E3 106
Mays Lick, Mason, Ky., 400........B6 94
Maysville, Madison, Ala., 400........A3 78
Maysville, Benton, Ark., 200........A1 81
Maysville, Banks, and Jackson, Ga., 553........B3 87
Maysville, Daviess, Ind........G3 91
Maysville, Scott, Iowa, 126........D7 92
Maysville, Mason, Ky., 8,484........B6 94
Maysville, DeKalb, Mo., 942........B3 101
Maysville, Jones, N.C., 892........C6 109
Maysville, Garvin, Okla., 1,530........C4 112
Maysville, Grant, W. Va., 120........B5 123
Maytown, Jefferson, Ala., 297........*B2 78
May Valley, Prowers, Colo........C8 83
Mayview, Lafayette, Mo., 270........B4 101
Mayville, Tuscola, Mich., 896........E7 98
Mayville, Cape May, N.J., 315........E3 106
Mayville, Chautauqua, N.Y., 1,619........C1 108
Mayville, Traill, N. Dak., 2,168........B8 110
Mayville, Gilliam, Oreg., 50........B6 113
Mayville, Dodge, Wis., 3,607........E5 124
Maywood, Los Angeles, Calif., 14,588........F2 82
Maywood, Cook, Ill., 27,330........F2 90
Maywood, Marion, Ind., 400........H7 91
Maywood, Lewis, Mo., 158........B6 101
Maywood, Frontier, Nebr., 337........D5 103
Maywood, Bergen, N.J., 11,460........D5 106
Maywood, Albany, N.Y., 1,500........*C7 108
Mayyat Yarqah (Oasis), Eg., U.A.R........E5 32
Maza, Arg........B4 54
Maza, Towner, N. Dak., 31........A6 110
Mazabuka, N. Rh., 4,400........E4 48
Mazagan, Mor., 40,302........C3 44
Mazagão, Braz., 919........C4 59
Mazamet, Fr., 14,863........F5 14
Mazán, Arg........E2 55
Mazapil, Mex., 1,742........C4 63
Mazara del Vallo, It., 36,827........F4 21
Mazāri 'an Nūbāni, Jordan, 2,000........g11 32
Mazar-i-Sharif, Afg., 41,960........C13 41
Mazarn, creek, Ark........C2 81
Mazarredo, Arg........D3 54
Mazarrón, Sp., 11,569........D5 20
Mazaruni, riv., Br. Gu........A2 59
Mazatenango, Guat., 11,067........C2 62
Mazatlán, Mex., 74,934........C3 63
Mazatzal, mts., Ariz........C4, F3 80
Mazatzal, peak, Ariz........C4 80
Mažeikiai, Sov. Un., 8,900........C4 27
Mazenod, Sask., Can., 121........H2 70
Mazeppa, Alta., Can., 35........D4 69
Mazeppa, Wabasha, Minn., 444........F6 99
Mazie, Mayes, Okla., 100........A6 112
Mazgirt, Tur., 1,500........C12 31
Mazomanie, Dane, Wis., 1,069........E4 124
Mazon, Grundy, Ill., 683........B5 90
Mazzarino, It., 17,789........F5 21
Mbabane, Swaz., 3,428........C5 49
Mbaïki, Cen. Afr. Rep........E3 46
Mbale, Ug., 13,569........A5 48
Mbalmayo, Cam........E2 46
Mbamba Bay, Tan........D5 48
Mbarara, Ug........B5 48
M'bari, riv., Cen. Afr. Rep........D4 46
Mbatiki, isl., Fiji Is........52
Mbenga, isl., Fiji Is........52
Mbenga, pass, Fiji Is........52
Mbeya, Tan., 6,932........C5 48
M'Bour, Sen., 8,800........D1 45
Mbout, Maur........C2 45
Mbulia, isl., Fiji Is........52
Mburucuyá, Arg., 2,555........E4 55
Mbya, bay, Fiji Is........52
Mchinja, Tan........C6 48
Mdandu, Tan........C5 48
Meacham, Sask., Can., 245........E3 70
Meacham, Umatilla, Oreg., 120........B8 113
Meacham Park, St. Louis, Mo., 1,800........*C7 101
Mead, Weld, Colo., 192........A6 83
Mead, Saunders, Nebr., 428........C9, D2 103
Mead, Bryan, Okla., 250........D5 112
Mead, Spokane, Wash., 800........B8, D7 122
Mead, lake, Nev........G7 104
Meade, Meade, Kans., 2,019........E3 93
Meade, co., Kans., 5,505........E3 93
Meade, co., Ky., 18,938........C3 94
Meade, co., S. Dak., 12,044........C3 116
Meade, peak, Idaho........G7 89
Meade River, Alsk., 5........A8 79
Meaderville, Silver Bow, Mont., 1,345........*D4 102
Meador, Allen, Ky., 40........D3 94
Meadow, Sarpy, Nebr........E3 103
Meadow, Perkins, S. Dak., 35........B3 116
Meadow, Terry, Tex., 484........C1 118
Meadow, Millard, Utah, 244........E3 119
Meadow, creek, W. Va........D7 123
Meadow, mtn., Md........D1 85
Meadow, riv., W. Va........C4 123
Meadow, val., Nev........F7 104
Meadow Bridge, Fayette, W. Va., 426........D4, D7 123
Meadowbrook, Allen, Ind., 1,500........*B7 91
Meadowbrook, Montgomery, Pa., 1,500........*F11 114
Meadowbrook, Harrison, W. Va., 500........A7 123
Meadow Creek, Boundary, Idaho, 15........A2 89
Meadow Creek, Summers, W. Va., 325........D4, D7 123
Meadowdale, Platte, Wyo........C8 125
Meadow Grove, Madison, Nebr., 430........B8 103
Meadowlands, St. Louis, Minn., 176........C6 99
Meadow Lands, Washington, Pa., 1,967........F1 114
Meadow Park, Palm Beach, Fla., 3,500........F6 86
Meadows, Adams, Idaho, 250........E2 89
Meadows, Coos, N.H., 40........B4 105
Meadow Valley, wash, Nev........F7 104
Meadowview, Washington, Va., 750........B3 121
Meadville, Franklin, Miss., 611........D3 100
Meadville, Linn, Mo., 447........B4 101
Meadville, Keya Paha, Nebr., 29........B6 103
Meadville, Crawford, Pa., 16,671........C1 114
Meaford, Ont., Can., 3,834........C4 72
Meagers Grant, N.S., Can., 127........E6 74
Meagher, co., Mont., 2,616........D6 102
Mealy, mts., Newf., Can........B2 75
Meandarra, Austl., 274........C7 51
Means, Menifee, Ky., 160........C6 94
Mearim, riv., Braz........C1 57
Mears, Oceana, Mich., 250........E4 98
Mea Shearim, Isr. (part of Jerusalem)........m14 32
Meath, co., Ire., 65,122........D5 11
Meathas Truim, Ire., 624........D4 11
Meath Park Station, Sask., Can., 184........D3 70
Meaux, Fr., 22,251........C5 14
Mebane, Alamance and Orange, N.C., 2,364........A4 109
Mecca, Riverside, Calif., 300........F5 82
Mecca, Parke, Ind., 500........E3 91
Mecca (Makkah), Sau. Ar., 200,000........A4 47
Mechanic Falls, Androscoggin, Maine, 2,195........D2 96
Mechanicsburg, Boone, Ind., 120........D5 91
Mechanicsburg, Champaign, Ohio, 1,810........B4 111
Mechanicsburg, Cumberland, Pa., 8,123........F7 114
Mechanicsville, Cedar, Iowa, 1,010........C6 92
Mechanicsville, St. Marys, Md., 175........D4 85
Mechanicsville, Hanover, Va., 5,000........D5 121
Mechanicville, Saratoga, N.Y., 6,831........C7 108
Mechant, lake, La........E4 95
Mechelen, Bel., 64,772........C4 15
Mecheria, Alg., 5,290 (39,347▲)........C4 44
Meck, isl., Kwajalein........52
Mecklenburg, co., N.C., 272,111........B3 109
Mecklenburg, co., Va., 31,428........E4 121
Mecklenburg, reg., Ger........B6 16
Mecklenburg, state, Sov. Zone, Ger., 2,139,640........E5 24
Mecklenburg, bay, Ger........D5 24
Meckling, Clay, S. Dak., 93........E8 116
Meconta, Moz........D6 48
Mecosta, Mecosta, Mich., 303........E5 98
Mecosta, co., Mich., 21,051........E5 98
Mecox, bay, N.Y........F4 84
Mecúfi, Moz........D7 48
Medak, India, 15,891........H7 40
Medan, Indon., 466,370........E1, m11 33, K3 38
Médanos, Arg., 2,229........B4, f7 54
Medanosa, pt., Arg........D3 54
Medart, Wakulla, Fla., 100........B2 86
Medaryville, Pulaski, Ind., 758........B4 91
Meddybemps, lake, Maine........C5 96
Médéa, Alg., 7,638 (26,350)........B5 44
Medeglia, Switz., 304........D6 19
Medellín, Col., 485,250 (*700,000)........B2 60
Medemblik, Neth., 5,100........B5 15
Médenine, Tun., 5,350........C7 44
Médenine, prov., Tun., 173,001........*C7 44
Méderdra, Maur........C1 45
Medfield, Norfolk, Mass., 2,424 (6,021▲)........D2 97
Medford, Middlesex, Mass., 64,971........B5, C3 97
Medford, Steele, Minn., 567........F5 99
Medford, Burlington, N.J., 1,480........D3 106
Medford, Grant, Okla., 1,223........A4 112
Medford, Jackson, Oreg., 24,425........E4 113
Medford, Taylor, Wis., 3,260........C3 124
Medford Lakes, Burlington, N.J., 2,876........D3 106
Medford Station, Suffolk, N.Y., 950........F5 84
Medgidia, Rom., 17,943........C9 22
Media, Henderson, Ill., 165........C3 90
Media, Delaware, Pa., 5,803........B10, G11 114
Mediapolis, Des Moines, Iowa, 1,040........C6 92
Medias, Rom., 32,503........B7 22
Medical Lake, Spokane, Wash., 4,765........B8, D7 122
Medicina, It., 3,881........E7 18
Medicine, creek, Mo........A4 101
Medicine, creek, Nebr........D5 103
Medicine Bow, Carbon, Wyo., 392........D6 125
Medicine Bow, mts., Colo. and Wyo........A4 77, D6 125
Medicine Bow, peak, Wyo........D6 125
Medicine Bow, riv., Wyo........D6 125
Medicine Hat, Alta., Can., 24,484........D5 69
Medicine Lake, Sheridan, Mont., 452........B12 102
Medicine Lodge, Barber, Kans., 3,072........E5 93
Medicine Lodge, riv., Kans........E5 93
Medicine Park, Comanche, Okla., 800........C3 112
Medill, Clark, Mo., 140........A6 101
Medimont, Kootenai, Idaho, 50........B2 89
Medina, Orleans, N.Y., 6,681........B2 108
Medina, Stutsman, N. Dak., 545........C6 110
Medina, Medina, Ohio, 8,235........A6 111
Medina, see Al Madīnah, Sau. Ar.
Medina, Gibson, Tenn., 722........B3 117
Medina, Bandera, Tex., 350........F3 118
Medina, King, Wash., 2,285........D2 122
Medina, co., Ohio, 65,315........A6 111
Medina, co., Tex., 18,904........E3 118
Medina del Campo, Sp., 14,327........B3 20
Medina de Ríoseco, Sp., 5,011........B3 20
Medinah, Du Page, Ill., 1,500........*B5 90
Medina Sidonia, Sp., 8,704........D3 20
Mediterranean, sea........D22 3, E10 28, E6 31
Medium, lake, Iowa........A3 92
Medjerda, riv., Tun. and Alg........F11, G10 30
Medjez-el-Bab, Tun., 3,340........F11 30
Medley, Dade, Fla., 112........*G6 86
Mednogorsk, Sov. Un., 32,400........C5 29
Medomak, Lincoln, Maine, 125........D3 96
Medon, Madison, Tenn., 97........B3 117
Medora, Man., Can., 90........E1 71
Medora, Macoupin, Ill., 447........D3 90
Medora, Jackson, Ind., 716........G5 91
Medora, Reno, Kans., 75........A4, D6 93
Medora, Billings, N. Dak., 133........C2 110
Medstead, Sask., Can., 199........D1 70
Meductic, N.B., Can., 232........D2 74
Meduxnekeag, riv., Maine........B5 96
Medveditsa, riv., Sov. Un........C2 29
Medvedovskaya, Sov. Un., 10,000........F16 9
Medvezhi, is., Sov. Un........B19 28
Medvezhyegorsk, Sov. Un., 19,200........F16 25
Medway, Penobscot, Maine, 150 (1,266▲)........C4 96
Medway, Norfolk, Mass., 1,602 (5,168▲)........B5, E1 97
Medzhibozh, Sov. Un., 10,000........G6 27
Meehan, Lauderdale, Miss., 200........C5 100
Meekatharra, Austl., 524........E2 50
Meeker, Rio Blanco, Colo., 1,655........A3 83
Meeker, Lincoln, Okla., 664........B5 112
Meeker, co., Minn., 18,887........E4 99
Meelpaeg, lake, Newf., Can........D3 75
Meeme, Manitowoc, Wis........B6 124
Meerane, Ger., 24,500........C7 17
Meerle, Bel., 2,666........C4 15
Meerut, India, 200,470 (*283,997)........C6 38, C6 40
Meeteetse, Park, Wyo., 514........A4 125
Meeting, lake, Sask., Can........D2 70
Meeting Creek, Alta., Can., 71........C4 69
Mega, Eth........E4 47
Megali, canal, Grc........g10 23
Megalopolis, Grc., 2,882........D4 23
Megantic, co., Que., Can., 57,400........C6 73
Megantic, lake, Que., Can........D7 73
Megantic, mtn., Que., Can........D6 73
Megara, Grc., 15,450........C4, g10 23
Megargel, Monroe, Ala., 115........D2 78
Megargel, Archer, Tex., 417........C3 118

Meggett, Charleston, S.C., 188 ... F1, F7 115
Mehan, Payne, Okla ... A5 112
Mehar, Pak ... C4 39
Meherpur, Pak., 7,174 ... F12 40
Meherrin, Lunenburg, Va., 300 ... D4 121
Meherrin, riv., Va ... E4 121
Mehkar, India, 11,872 ... G6 40
Mehoopany, Wyoming, Pa., 250 ... C9 114
Mehsana, India, 32,577 ... F4 40
Mehun-sur-Yèvre, Fr., 4,735 . D5 14
Meiganga, Cam ... D2 46
Meigs, Thomas and Mitchell, Ga., 1,236 ... F2 87
Meigs, co., Ohio, 22,159 ... C5 111
Meigs, co., Tenn., 5,160 ... D9 117
Meihsien, China, 85,000 ... G8 34
Meiktila, Bur., 25,180 ... D10 39
Meilap, Ponape ... 52
Meilen, Switz., 8,203 ... *B6 19
Meilleur, lake, Que., Can ... A4 73
Meiners Oaks, Ventura, Calif., 3,513 ... *E4 82
Meiningen, Ger., 23,700 ... C5 17
Meiringen, Switz., 3,749 ... C5 19
Meissen, Ger., 47,900 ... B8 17
Meitan, China, 4,000 ... K2 36
Meijicana, mtn., Arg ... E2 55
Mejillones, Chile, 1,056 ... D1 55
Mékambo, Gabon ... E2 46
Mekerrhane, lake, Alg ... D5 44
Mekhtar, Pak ... B2 40
Mekinac, lake, Que., Can ... B5 73
Mekinock, Grand Forks, N. Dak., 100 ... A8 110
Meknès, Mor., 175,943 ... C3 44, H4 30
Mekong, riv., Asia ... E6 38, B10 39
Mekoryuk, Alsk., 242 ... C6 79
Melambes, Grc., 1,414 ... E5 23
Melaval, Sask., Can., 113 ... H2 70
Melba, Canyon, Idaho, 197 . F2 89
Melber, McCracken, Ky., 219 ... A2 94
Melbern, Williams, Ohio, 145 ... A3 111
Melbeta, Scotts Bluff, Nebr., 118 ... C2 103
Melbourne, Izard, Ark., 571 ... A4 81
Melbourne, Austl., 76,834 (★1,930,000) ... H5 51
Melbourne, Ont., Can., 333 . E3 72
Melbourne, Brevard, Fla., 11,982 ... D6 86
Melbourne, Marshall, Iowa, 517 ... C4 92
Melbourne, Campbell, Ky., 299 ... A7 94
Melbourne, Harrison, Mo., 70 ... A4 101
Melbourne, Grays Harbor, Wash., 50 ... C2 122
Melbourne Beach, Brevard, Fla., 1,004 ... D6 86
Melcher, Marion, Iowa, 867 ... C4 92
Meldorf, Ger., 8,000 ... D3 24
Meldrim, Effingham, Ga., 220 ... D5 87
Meldrum Bay, Ont., Can., 74 ... B1 72
Meleb, Man., Can., 85 ... D3 71
Melekeiok, Palau ... 52
Melekess, Sov. Un., 54,000 . E18 9
Melenki, Sov. Un., 24,200 . D13 27
Melfi, Chad, 1,497 ... C3 46
Melfi, It., 18,208 ... D5 21
Melfort, Sask., Can., 4,039 ... E3 70
Melilla, Sp. poss., Afr., 79,056 ... B4 44
Melipilla, Chile, 11,525 ... A2 54
Melita, Man., Can., 1,038 ... E1 71
Melitopol, Sov. Un., 102,000 ... H10 27
Melito di Porto Salvo, It., 8,880 ... F5 21
Mella, riv., It ... D6 18
Melle, Fr., 3,762 ... D3 14
Melle, Ger., 9,300 ... A3 17
Mellen, Ashland, Wis., 1,182 ... B3 124
Mellette, Spink, S. Dak., 208 ... B7 116
Mellette, co., S. Dak., 2,664 . D4 116
Mellit, Sud ... C2 47
Mellott, Fountain, Ind., 312 . D3 91
Mellow Valley, Clay, Ala., 120 ... B4 78
Mellrichstadt, Ger., 4,000 ... C5 17
Mellville, Newport, R.I. ... C11 84
Mellwood, Phillips, Ark., 300 ... C5 81
Melmore, Seneca, Ohio, 300 . A4 111
Melnik, Czech., 13,100 ... n18 26
Melo, Ur., 27,100 ... E2 56
Melocheville, Que., Can., 1,666 ... D8 73
Melouprey, Camb., 10,000 ... F6 38
Melrose, N.B., Can., 178 ... C5 74
Melrose, N.S., Can., 56 ... D7 74
Melrose, Hartford, Conn., 100 ... B6 84
Melrose, Alachua, Fla., 180 . C4 86
Melrose, Monroe, Iowa, 214 ... D4 92
Melrose, Natchitoches, La., 200 ... C3 95
Melrose, Middlesex, Mass., 29,619 ... B5, C3 97
Melrose, Stearns, Minn., 2,135 ... E4 99
Melrose, Silver Bow, Mont., 150 ... E4 102
Melrose, Curry, N. Mex., 698 ... C6 107
Melrose, Paulding, Ohio, 260 ... A3 111
Melrose, Douglas, Oreg ... D3 113
Melrose, Jackson, Wis., 516 . D2 124
Melrose Park, Broward, Fla., 4,000 ... *F6 86
Melrose Park, Cook, Ill., 22,291 ... F2 90
Melrose Park, Cayuga, N.Y., 2,085 ... *C4 108
Melrose Park, Montgomery, Pa., 6,000 ... *G11 114
Mels, Switz., 5,254 ... B7 19
Melsetter, S. Rh., 400 ... A5 49
Melstone, Musselshell, Mont., 266 ... D9 102
Melstrand, Alger, Mich., 100 ... B4 98
Melsungen, Ger., 8,200 ... B4 17
Melton Mowbray, Eng., 15,913 ... B7 12
Melun, Fr., 26,873 ... C5 14
Melvaig, Scot ... C3 13
Melvern, Osage, Kans., 376 . D8 93
Melvich, Scot ... B5 13
Melville, Sask., Can., 5,191 . G4 70
Melville, St. Landry, La., 1,939 ... D4 95
Melville, Sweet Grass, Mont., 20 ... D7 102
Melville, Foster, N. Dak., 40 ... B6 110
Melville, bay, Grnld ... B21 4
Melville, cape, Austl ... B7 50
Melville, isl., Austl ... B5 50
Melville, isl., N.W. Ter., Can ... A10 66
Melville, lake, Newf., Can ... B2 75
Melville, pen., N.W. Ter., Can ... C16 66
Melvin, Choctaw, Ala., 300 ... D1 78
Melvin, Ford, Ill., 559 ... C5 90
Melvin, Osceola, Iowa, 364 . A2 92
Melvin, Floyd, Ky., 750 ... C7 94
Melvin, McCulloch, Tex., 401 ... D3 118
Melvin, lake, Ire ... C3 11
Melvindale, Wayne, Mich., 13,089 ... A7 98
Melvine, Bledsoe, Tenn., 60 ... D8 117
Melvin Village, Carroll, N.H., 270 ... C4 105
Melykut, Hung., 6,514 ... B4 22
Memba, Moz ... D7 48
Memel, see Klaipeda, Sov. Un.
Memmingen, Ger., 29,800 ... E5 16
Memorial, Clay, Tenn ... C8 117
Memphis, Manatee, Fla., 2,647 ... F2 86
Memphis, Clark, Ind., 200 . H6 91
Memphis, Macomb and St. Clair, Mich., 996 ... F8 98
Memphis, Scotland, Mo., 2,106 ... A5 101
Memphis, Saunders, Nebr., 77 ... E2 103
Memphis, Shelby, Tenn., 497,524 (★628,100) ... B1, E8 117
Memphis, Hall, Tex., 3,332 . B2 118
Memphis, ruins, Eg., U.A.R ... D6 43
Memphremagog, lake, Que., Can., Vt ... D5 73
Memramcook, N.B., Can., 402 ... C5 74
Mena, Polk, Ark., 4,388 ... C1 81
Menahga, Wadena, Minn., 799 ... D3 99
Menai, strait, Wales ... A3 12
Ménaka, Mali, 400 ... C5 45
Menam, see Chao Phraya, riv., Thai.
Menan, Jefferson, Idaho, 496 ... F7 89
Menands, Albany, N.Y., 2,314 ... C7 108
Menard, Menard, Tex., 1,914 ... D3 118
Menard, co., Ill., 9,248 ... C4 90
Menard, co., Tex., 2,964 ... D3 118
Menasha, Winnebago, Wis., 14,647 ... A5, D5 124
Menche, Guat ... B2 62
Mendawi, riv., Indon ... F4 35
Mende, Fr., 8,337 ... E5 14
Menden, Ger., 27,500 ... B2 17
Mendenhall, Simpson, Miss., 1,946 ... D4 100
Menderes (Scamander), riv., Tur ... C6 23
Mendes, Tattnall, Ga., 150 . F5 87
Méndez, Mex., 207 ... B5 63
Mendham, Sask., Can., 231 . G1 70
Mendham, Morris, N.J., 2,371 ... B3 106
Mendip, hills, Eng ... C5 12
Mendjalutung, Indon ... E5 35
Mendocino, Mendocino, Calif., 900 ... C2 82
Mendocino, co., Calif., 51,059 ... C2 82
Mendon, Adams, Ill., 784 ... C2 90
Mendon, Worcester, Mass., 900 (2,068▲) ... E1 97
Mendon, St. Joseph, Mich., 867 ... F5 98
Mendon, Chariton, Mo., 287 . B4 101
Mendon, Mercer, Ohio, 663 . B3 111
Mendon, Cache, Utah, 345 ... B4 119
Mendon, Rutland, Vt., 35 (461▲) ... D3 120
Mendota, Fresno, Calif., 3,086 ... D3 82
Mendota, La Salle, Ill., 6,154 ... B4 90
Mendota, Hemphill, Tex ... B2 118
Mendota, lake, Wis ... E4 124
Mendota Heights, Dakota, Minn., 5,028 ... *F5 99
Mendoza, Arg., 109,149 (★280,000) ... A3 54
Mendoza, prov., Arg., 825,535 ... A3 54
Mendoza, Pan., 212 ... k11 62
Mendrisio, Switz., 5,109 ... E6 19
Menemen, Tur., 15,100 ... C6 23
Menen, Bel ... C2 15
Menfi, It., 12,492 ... F4 21
Mengcheng, China, 5,000 ... H7 36
Menggala, Indon ... F2 35
Menglien, China ... G4 34
Mengtzu, China, 9,000 ... G5 34
Menifee, Conway, Ark., 300 . B3 81
Menifee, co., Ky., 4,276 ... C6 94
Menihek, lakes, Newf., Can ... g8 75
Menindee, Austl., 373 ... F4 51
Menindee, lake, Austl ... F4 51
Meningie, Austl., 376 ... G2 51
Menlo, Chattooga, Ga., 466 ... B1 87
Menlo, Guthrie, Iowa, 421 . C3 92
Menlo, Thomas, Kans., 99 . C3 93
Menlo, Pacific, Wash., 100 . C2 122
Menlo Park, San Mateo, Calif., 26,957 ... B5 82
Menlo Park, Middlesex, N.J., 400 ... B4 106
Menlo Park Terrace, Middlesex, N.J., 2,500 ... *B4 106
Menno, Hutchinson, S. Dak., 837 ... D8 116
Meno, Major, Okla., 118 ... A3 112
Menoken, Burleigh, N. Dak., 60 ... C5 110
Menola, Hertford, N.C., 35 . A6 109
Menominee, Menominee, Mich., 11,289 ... C3 98
Menominee, co., Mich., 24,685 ... C3 98
Menominee, co., Wis., 2,600 . C5 124
Menominee, Indian res., Wis ... C5 124
Menominee, riv., Mich., Wis ... C3 92, C6 124
Menomonee, riv., Wis ... E1 124
Menomonee Falls, Waukesha, Wis., 18,276 ... E1, E5 124
Menomonie, Dunn, Wis., 8,624 ... D2 124
Menorca, isl., Sp ... B7 20
Mentana, It., 5,489 ... g9 21
Mentawai, is., Indon ... F1 35
Mentmore, McKinley, N. Mex., 90 ... B1 107
Menton, Fr., 19,904 ... F7 14
Mentone, DeKalb, Ala., 250 . A4 78
Mentone, San Bernardino, Calif., 2,000 ... *F3 82
Mentone, Kosciusko, Ind., 813 ... B5 91
Mentone, Loving, Tex., 50 ... D1, F2 118
Mentor, Campbell, Ky., 350 ... A7 94
Mentor, Polk, Minn., 281 ... C2 99
Mentor, Lake, Ohio, 4,354 . A6 111
Mentor, Blount, Tenn., 350 . E11 117
Mentor Headlands, Lake, Ohio, 1,500 ... *A6 111
Mentor-on-the-Lake, Lake, Ohio, 3,290 ... A6 111
Menzel-Bourguiba, Tun., 34,842 ... *B6 44
Menzies, Austl., 217 ... E3 50
Menziken, Switz., 4,060 ... B5 19
Meoqui, Mex., 10,298 ... B3 63
Meota, Sask., Can., 281 ... D1 70
Meppel, Neth., 17,700 ... B6 15
Meppen, Ger., 14,900 ... B3 16
Mequellarë, Alb., 650 ... B3 23
Mequon, Ozaukee, Wis., 8,543 ... E2 124
Merabello, gulf, Grc ... E5 23
Meramec, caverns, Mo ... C6 101
Meramec, riv., Mo ... C7 101
Merano, It., 30,614 ... A3 21
Merasheen, Newf., Can., 291 . E4 75
Merasheen, isl., Newf., Can ... E4 75
Merauke, W. Irian, 3,600 ... G10 35
Meraux, St. Bernard, La., 500 ... C7 95
Merca, Som., 15,000 (59,000▲) ... E5 47
Mercara, India, 14,453 ... F6 39
Merced, Merced, Calif., 20,068 ... D3 82
Merced, co., Calif., 90,446 . D3 82
Merced, riv., Calif ... D3 82
Mercedes, Arg., 25,912 ... A3 54
Mercedes, Arg., 16,932 ... A5, g7 54
Mercedes, Arg., 14,813 ... E4 55
Mercedes, Hidalgo, Tex., 10,943 ... F3 118
Mercedes, Ur., 25,400 ... E1 56
Mercedita Cantral, P.R ... C4 65
Mercer, Somerset, Maine, 140 (272▲) ... D3 96
Mercer, Mercer, Mo., 368 ... A4 101
Mercer, McLean, N. Dak., 154 ... B5 110
Mercer, Mercer, Ohio, 90 ... B3 111
Mercer, Mercer, Pa., 2,800 . D1 114
Mercer, Madison, Tenn., 400 ... B2 117
Mercer, Iron, Wis., 700 ... B3 124
Mercer, co., Ill., 17,149 ... B3 90
Mercer, co., Ky., 14,596 ... C5 94
Mercer, co., Mo., 5,750 ... A4 101
Mercer, co., N.J., 266,392 . C3 106
Mercer, co., N. Dak., 6,805 . B4 110
Mercer, co., Ohio, 32,559 ... B3 111
Mercer, co., Pa., 127,519 ... D1 114
Mercer, co., W. Va., 68,206 . D3 123
Mercer, isl., Wash ... D2 122
Mercer Island, King, Wash., 1,700 ... B3, D2 122
Mercersburg, Franklin, Pa., 1,759 ... G6 114
Mercês, Braz., 2,897 ... g6 56
Merchant, Brunswick, Va ... E5 121
Merchantville, Camden, N.J., 4,075 ... *D2 106
Mercier, Brown, Kans., 50 ... C8 93
Mercoal, Alta., Can., 972 ... C2 69
Mercury, Nye, Nev., 300 ... G5 104
Mercury, McCulloch, Tex., 100 ... D3 118
Mercury, bay, N.Z ... L15 51
Mercy, cape, N.W. Ter., Can . C20 67
Merdjayoun, Leb ... F1 31
Meredith, Pitkin, Colo., 30 . B4 83
Meredith, Belknap, N.H., 950 (2,434▲) ... C3 105
Meredith, King, Wash ... D2 122
Meredith, lake, Colo ... C7 83
Meredith Center, Belknap, N.H., 100 ... C3 105
Meredosia, Morgan, Ill., 1,034 ... D3 90
Merefa, Sov. Un., 10,000 ... G11 31
Meregh, Som ... E6 47
Mergui, Bur., 33,697 ... F10 39
Mergui, arch., Bur ... F10 39
Merid, Sask., Can., 45 ... F1 70
Mérida, Mex., 170,513 ... C7 63
Mérida, Sp., 34,297 ... C2 20
Mérida, Ven., 40,404 ... B3 60
Mérida, state, Ven., 270,668 . B3 60
Meriden, New Haven, Conn., 51,850 ... C5 84
Meriden, Cherokee, Iowa, 192 ... B2 90
Meriden, Jefferson, Kans., 402 ... B7, C8 93
Meriden, Steele, Minn., 115 . F5 99
Meriden, Sullivan, N.H., 175 ... C2 105
Meriden, Laramie, Wyo., 5 . D8 125
Meridian, McIntosh, Ga., 150 ... E5 87
Meridian, Ada, Idaho, 2,081 ... F2 89
Meridian, Lauderdale, Miss., 49,374 ... C5 100
Meridian, Cayuga, N.Y., 379 ... B4 108
Meridian, Logan, Okla., 160 . B4 112
Meridian, Butler, Pa., 1,649 . *E2 114
Meridian, Bosque, Tex., 993 ... D4 118
Meridian Hills, Marion, Ind., 1,807 ... *E5 91
Meridiano, Arg ... B4 54
Meridianville, Madison, Ala., 250 ... A3 79
Merigold, Bolivar, Miss., 602 ... B3 100
Merigomish, N.S., Can., 203 ... D7 74
Merkarvia, Fin., 2,100 ... G9 25
Merino, Logan, Colo., 268 ... A7 83
Merino Village, Worcester, Mass., 3,099 ... B4 97
Merioneth, co., Wales, 39,007 ... B4 11
Merion Park, Montgomery, Pa., 1,000 ... *F11 114
Merion Station, Montgomery, Pa., 5,500 ... *B11 114
Merir, salt lake, Alg ... C6 44
Meriso, port, Guam ... 52
Meriwether, co., Ga., 19,756 . C2 87
Meriwether Lewis, nat. mon., Tenn ... B4 117
Merkel, Taylor, Tex., 2,312 . C2 118
Merkine, Sov. Un ... A8 26
Merkis, riv., Sov. Un ... D5 27
Merksem, Bel., 36,098 ... C4 15
Merksplas, Bel., 4,912 ... C4 15
Merlebach, Fr., 8,715 ... E7 15
Merlin, Ont., Can., 595 ... E2 72
Merlin, Josephine, Oreg., 300 ... E3 113
Merlo, Arg., 8,385 ... g7 54
Mermentau, Acadia, La., 716 ... D3 95
Mermentau, riv., La ... E2 95
Mern, Den., 516 ... C6 24
Merna, Custer, Nebr., 349 ... C6 103
Merna, Sublette, Wyo ... C2 125
Merom, Sullivan, Ind., 352 . F2 91
Merowe, Sud., 1,620 ... B3 47
Merriam, Johnson, Kans., 5,084 ... B8 93
Merrick, Nassau, N.Y., 18,789 ... G2 84
Merrick, co., Nebr., 8,363 ... C7 103
Merrick, mtn., Scot ... E4 13
Merrickville, Ont., Can., 947 ... C9 72
Merricourt, Dickey, N. Dak., 66 ... C7 110
Merrifield, Crow Wings, Minn., 250 ... D4 99
Merrifield, Fairfax, Va., 1 000 ... B5 121
Merrill, Plymouth, Iowa, 645 . B1 92
Merrill, Saginaw, Mich., 963 ... E6 98
Merrill, George, Miss., 150 . E5 100
Merrill, Klamath, Oreg., 804 ... E5 113
Merrill, Lincoln, Wis., 9,451 . C4 124
Merrillan, Jackson, Wis., 591 ... D3 124
Merrillville, Lake, Ind., 3,120 . B3 91
Merrimac, Taylor, Ky., 250 . C4 94
Merrimac, Essex, Mass., 1,800 (3,261▲) ... A5 97
Merrimac, Sauk, Wis., 297 ... E4 124
Merrimack, Hillsboro, N.H., 500 (2,989▲) ... E4 105
Merrimack, co., N.H., 67,785 ... D3 101
Merrimack, riv., Mass., N.H ... A5 91, D3 105
Merrimacport, Essex, Mass., 275 ... A6 97
Merriman, Cherry, Nebr., 285 ... B4 103
Merrimon, Carteret, N.C., 200 ... C7 109
Merrionette Park, Cook, Ill., 2,354 ... *F3 90
Merritt, B.C., Can., 3,039 ... D7 68
Merritt, Pamlico, N.C., 100 . B7 109
Merritt Island, Brevard, Fla., 3,554 ... D6 86
Merritt-Peck Colonies, Fresno, Calif., 1,299 ... *D4 82
Merriwa, Austl., 1,073 ... F8 51
Merriweather, Ontonagon, Mich., 90 ... A5 98
Mer Rouge, Morehouse, La., 853 ... B4 95
Merrow, Tolland, Conn., 75 . B7 84
Merry Hill, Bertie, N.C., 75 ... A7 109
Merrymeeting, lake, N.H ... D4 105
Merry Oaks, Chatham, N.C., 77 ... B4 109
Merryville, Beauregard, La., 1,232 ... D2 95
Mersa Fatma, Eth ... B5 47
Merseburg, Ger., 47,200 ... B6 17
Mers-el-Kebir, Alg., 3,977 ... C6 30
Mersey, riv., Eng ... A5 12
Mershon, Pierce, Ga., 300 ... E4 87
Mersin, Tur., 68,600 ... D10 31
Mersing, Mala., 7,228 ... K5 38
Merthyr Tydfil, Wales, 59,008 ... C4 12
Merti, Ken ... A6 48
Mértola, Port., 6,439 ... D2 20
Merton, Waukesha, Wis., 407 ... E1 124
Mertz, glacier, Ant ... C27 5
Mertzon, Irion, Tex., 584 ... D2 118
Méru, Fr., 5,435 ... C5 14
Meru, Ken ... A6 48
Meru, mtn., Tan ... B6 48
Merundung, isl., Indon ... K8 38
Mervelier, Switz., 548 ... B4 19
Merville, Fr., 4,936 ... D2 15
Mervin, Sask., Can., 193 ... D1 70
Merwin, Bates, Mo., 76 ... C3 101
Méry [-sur-Seine], Fr., 1,135 ... F3 15
Merzifon, Tur., 22,200 ... B10 31
Merzig, Ger., 12,100 ... E6 15
Mesa, Maricopa, Ariz., 33,772 ... D4, G2 80
Mesa, Mesa, Colo ... B2 83
Mesa, Adams, Idaho, 30 ... E2 89
Mesa, Chaves, N. Mex ... D5 108
Mesa, co., Colo., 50,715 ... B2 83
Mesa, mts., Colo ... D3 83
Mesa, peak, Colo ... D4 83
Mesa De Maya, plat., Colo ... D7 83
Mesa Verde, nat. park, Colo ... D2 83
Mesa Verde, plat., Colo ... D2 83
Mesabi, iron range, Minn ... C5 99
Mesagne, It., 23,000 ... D6 21
Mescal, Pima, Ariz ... F5 80
Mescalero, Otero, N. Mex., 370 ... D4 107
Mescalero Apache, Indian res., N. Mex ... D4 107
Meschede, Ger., 12,600 ... B3 17
Mesegon, isl., Truk ... 52
Mesena, Warren, Ga., 200 ... C4 87
Meservey, Cerro Gordo, Iowa, 331 ... B4 92
Meshcherskoye, Sov. Un ... n17 27
Meshchovsk, Sov. Un., 6,800 ... D10 27
Meshed, see Mashhad, Iran
Meshkovskiy, Sov. Un., 5,000 ... G13 27
Meshomasic, mtn., Conn ... C6 84
Meshoppen, Wyoming, Pa., 470 ... C9 114
Meshra'er Req, Sud ... D2 47
Mesic, Pamlico, N.C., 350 ... B7 109
Mesick, Wexford, Mich., 304 . D5 98
Mesilinka, riv., B.C., Can ... A5 68
Mesilla, Dona Ana, N. Mex., 1,264 ... E3 107
Mesilla Park, Dona Ana, N. Mex., 2,400 ... E3 107
Mesita, Costilla, Colo., 50 ... D5 83
Meskéné, Syr ... D12 31
Mesocco, Switz., 1,324 ... D7 19
Mesologion, Grc., 11,266 ... C3 23
Mesquite, Clark, Nev., 600 . G7 104
Mesquite, Dona Ana, N. Mex., 210 ... E3 107
Mesquite, Dallas, Tex., 27,526 ... B6 118
Messalonskee, lake, Maine ... D3 96
Messen, Switz., 682 ... B3 19
Messenia (Messinia), prov., Grc., 211,970 ... *D4 23
Messex, Washington, Colo., 20 ... A7 83
Messina, It., 254,715 ... E5 21
Messina, S. Afr., 10,225 ... B5 49
Messina, strait, It ... E5 21
Messini, Grc., 7,725 ... D3 23
Messini, gulf, Grc ... D4 23
Messinia (Messenia), prov., Grc., 211,970 ... *D4 23
Messkirch, Ger., 3,800 ... B5 18
Mesta, riv., Bul ... E6 22
Mestre, It., 73,435 ... B4 21
Mesudiye, Tur., 2,500 ... B11 31
Meta, Osage, Mo., 360 ... C5 101
Meta, dept., Col., 74,780 ... C3 60
Meta, riv., Ven ... B4 60
Metabetchouan, Que., Can., 1,962 ... A6 73
Metabetchouan, riv., Que., Can . A5 73
Metairie, Jefferson, La., 65,000 ... *C7 95
Metaline, Pend Oreille, Wash., 299 ... A8 122
Metaline Falls, Pend Oreille, Wash., 469 ... A8 122
Metalton, Carroll, Ark., 25 . A2 81
Metamora, Woodford, Ill., 1,808 ... C4 90
Metamora, Franklin, Ind., 400 ... F7 91
Metamora, Lapeer, Mich., 452 ... F9 98
Metamora, Fulton, Ohio, 598 ... A4 111
Metán, Arg., 6,915 ... E3 55
Metangula, Moz ... D5 48
Metapán, Sal., 4,073 ... C3 62
Metaponto, It., 445 ... D6 21
Metarica, Moz ... D6 48
Metarma, cape, Eg., U.A.R ... E4 32
Metascouac, lake, Que., Can ... B6 73
Metasville, Wilkes, Ga., 73 . C4 87
Metcalf, Thomas, Ga., 241 ... F3 87
Metcalf, Edgar, Ill., 278 ... D6 90
Metcalfe, Ont., Can., 385 ... B9 72
Metcalfe, Washington, Miss., 500 ... B2 100
Metcalfe, co., Ky., 8,367 ... D4 94
Metchosin, B.C., Can., 370 ... B9 68
Metedeconk, riv., N.J ... C4 106
Meteghan, N.S., Can., 925 ... E3 74
Meteghan River, N.S., Can., 423 ... E3 74
Meteghan Station, N.S., Can., 327 ... E4 74
Meteor, crater, Ariz ... C4 80
Method, Wake, N.C. (part of Raleigh) ... B5 109
Methoni, Grc., 2,087 ... D3 23
Methow, Okanogan, Wash., 75 ... A5 121
Methuen, Essex, Mass., 28,114 ... A5 97
Methven, Scot., 1,671 ... D5 13
Metiskow, Alta., Can., 99 ... C5 69
Metković, Yugo., 4,488 ... D3 22
Metlakatla, B.C., Can., 20 ... B2 68
Metlakatla, Alsk., 798 . D13, n14 79
Meto, Lonoke, Ark ... D6 81
Meto, bayou, Ark ... C4 81
Metolius, Jefferson, Oreg., 270 ... C5 113
Metolius, bench, Oreg ... C5 113
Metonga, lake, Wis ... C4 124
Metropolis, Massac, Ill., 7,339 ... F4 90
Metropolitan, Dickinson, Mich., 100 ... C3 98
Metsovon, Grc., 2,773 ... C3 23
Mettewee, riv., Vt ... E2 120
Metten, Ger., 2,800 ... E7 17
Meeter, Candler, Ga., 2,362 ... D4 87
Mettet, Bel., 3,160 ... D4 15
Mettingen, Ger., 9,100 ... A2 17
Mettmann, Ger., 24,600 ... B1 17
Metuchen, Middlesex, N.J., 14,041 ... B4 106
Metulla, Isr., 178 ... A7 32
Metz, Fr., 102,771 (★150,000) ... C7 14
Metz, Steuben, Ind., 200 ... A8 91
Metz, Persque Isle, Mich., 60 ... C7 98
Metz, Vernon, Mo., 137 ... C3 101
Metzeral, Fr., 1,104 ... A3 18
Metzger, Washington, Oreg. (part of Tigard) ... B2 113
Meudon, Fr., 34,878 ... g9 14
Meulaboh, Indon., 2,575 ... J2 38
Meung [-sur-Loire], Fr., 521 ... D4 14
Meureudoe, Indon ... J2 38
Meurthe, riv., Fr ... F6 15
Meurthe-et-Moselle, dept., Fr., 678,078 ... F6 15
Meuse, dept., Fr., 215,985 . F51 15
Meuse, hills, Fr ... C6 14
Meuse, riv., Bel., Fr ... C6 14, D5 15, A1 18
Meuselwitz, Ger., 10,500 ... B7 17
Mew, isl., Ire ... C6 11
Mexcala, riv., Mex ... n12 63
Mexhoma, Cimarron, Okla . D1 112
Mexia, Monroe, Ala., 100 ... D2 78
Mexia, Limestone, Tex., 6,121 ... D4 118
Mexicali, Mex., 172,554 ... A1 63
Mexican Springs, McKinley, N. Mex., 30 ... B1 107
Mexico, Miami, Ind., 600 ... C5 91
Mexico, Crittenden, Ky., 200 ... A3 94
Mexico, Oxford, Maine, 5,043 ... D2 96
Mexico, Audrain, Mo., 12,889 ... B6 101
Mexico, Oswego, N.Y., 1,465 ... B4 108
Mexico, Juniata, Pa., 400 ... E7 114
Mexico, country, N.A., 34,625,903 ... G10 61, 63
México, state, Mex., 1,883,291 ... D5, n14 63
Mexico, gulf, N.A ... G12 61
Mexico City, Mex., 2,697,994 (★4,800,000) ... D5, h9, n14 63
Meyādīn, Syr., 7,200 ... E13 31
Meyers, Garland, Ark., 75 ... C5 81
Meyers Chuck, Alsk., 27 ... n23 79
Meyersdale, Somerset, Pa., 2,901 ... G3 114
Meyronne, Sask., Can., 161 ... H2 70
Meyssac, Fr., 531 ... E4 14
Mèze, Fr., 4,546 ... F5 14
Mezen, Sov. Un., 7,300 ... B17 9
Mézenc, mtn., Fr ... E6 14
Mezenskaya, bay, Sov. Un ... B17 9
Meziadin, lake, B.C., Can ... A3 68
Mézières, Fr., 11,799 ... C6 14
Mézières, Switz., 504 ... C2 19
Mezokovesd, Hung., 18,640 . B5 22
Mezotur, Hung., 18,337 (23,632▲) ... B5 22
Mezotur, Hung., 16,000 ... B5 22
Mezquital, Mex., 832 ... C4 63
Mezzana, It., 1,055 ... C6 18
Mezzolombardo, It., 4,873 ... C7 18
Mga, Sov. Un ... s32 25
Mga, riv., Sov. Un ... s32 25
Mglin, Sov. Un., 10,000 ... E9 27
Mhor, lake, Scot ... C4 13
Mhow, India, 48,032 ... F5 40
Miahuatlán, Mex., 7,420 ... D5 63
Miajadas, Sp., 8,632 ... C3 20
Miajlar, India, 5,000 ... D3 40
Miami, Gila, Ariz., 3,350 ... D5 80
Miami, Man., Can., 349 ... E2 71
Miami, Dade, Fla., 291,688 (★1,212,000) . F3, G6 86
Miami, Miami, Ind., 300 ... C5 91
Miami, Saline, Mo., 156 ... B4 101
Miami, Colfax, N. Mex., 150 ... A5 107
Miami, Ottawa, Okla., 12,869 ... A7 113
Miami, S. Rh ... A4 49
Miami, Roberts, Tex., 656 ... B2 118
Miami, Kanawha, W. Va., 500 ... D6 123
Miami, co., Ind., 38,000 ... C5 91
Miami, co., Kans., 19,884 ... D9 93
Miami, co., Ohio, 72,901 ... B3 111
Miami, canal, Fla ... F6 86
Miami, riv., Ohio ... C3 111
Miami Beach, Dade, Fla., 63,145 ... F3, G6 86
Miamisburg, Montgomery, Ohio, 9,893 ... C3 111
Miami Shores, Dade, Fla., 8,865 ... F3, G6 86
Miami Shores, Montgomery, Ohio, 1,200 ... *C3 111
Miami Springs, Dade, Fla., 11,229 ... F3, G6 86
Miamitown, Hamilton, Ohio, 500 ... D2 111
Mīāndasht, Iran ... C8 41
Miandrivazo, Malag., 1,505 . g9 49
Mīāneh, Iran, 14,758 ... C3 41
Mianus, Fairfield, Conn. (part of Greenwich) ... E2 84
Mianus, riv., Conn ... A5 106
Mianwali, Pak., 23,341 ... A3 40
Miaoli, For., 5,000 ... *G9 34
Miarinarivo, Malag ... g9 49
Miass, Sov. Un., 107,000 ... C6 29
Miass, riv., Sov. Un ... B6 29
Miasteczko Slaskie, Pol., 2,212 ... f9 26
Miastko, Pol., 3,417 ... A4 26
Mica, Spokane, Wash., 75 ... B8, D8 122
Mica, peak, Idaho ... D8 122
Mica, riv., Rom ... B7 22
Micanopy, Alachua, Fla., 658 ... C4 86
Micay, Col ... C2 60
Micco, Brevard, Fla., 200 ... E6 86
Miccosukee, Leon, Fla., 120 . B2 86
Miccosukee, lake, Fla ... B2 86
Michael, lake, Newf., Can ... A2 75
Michalovce, Czech., 16,300 . D6 26
Michalow, Pol., 1,000 ... k13 26
Michaud, pt., N.S., Can ... D9 74
Michaudville, Que., Can., 205 ... D4 73
Michel, B.C., Can., 417 ... E10 68
Michel, lake, Newf., Can ... C3 75
Michelson, mtn., Alsk ... B11 79
Michendorf, Ger., 3,055 ... A8 17
Michiana Shores, La Porte, Ind., 229 ... *A4 91
Michichi, Alta., Can., 52 ... D4 69
Michie, McNairy, Tenn., 200 ... B3 117
Michigamme, Marquette, Mich ... B2 98
Michigamme, lake, Mich ... B2 98
Michigan, Nelson, N. Dak., 551 ... A7 110
Michigan, state, U.S., 7,823,194 ... B11 77, 98
Michigan, creek, Colo ... A4 83
Michigan, isl., Wis ... B3 124
Michigan, lake, U.S ... B10 77
Michigan, prairie, Wash ... C7 122
Michigan Center, Jackson, Mich., 4,611 ... F6 98
Michigan City, La Porte, Ind., 36,653 ... A4 91
Michigan City, Benton, Miss., 50 ... A4 100
Michigantown, Clinton, Ind., 513 ... D5 91
Michikamau, lake, Newf., Can . g8 75
Michoacán, state, Mex., 1,862,568 ... D4, n13 63
Michurin, Bul., 1,896 ... D8 22
Michurinsk, Sov. Un., 84,000 ... C2 29
Mickleyville, Marion, Ind., 950 ... H7 91
Micoud, St. Lucia, 1,350 ... p16 64
Micro, Johnston, N.C., 350 . B5 109
Midai, isl., Indon ... K7 38
Midale, Sask., Can., 645 ... H4 70
Midas, Bonner, Idaho, 50 ... A2 89

Middelburg, Neth., 22,600...C2 15
Middelfart, Den., 8,801 (★11,701)...C3 24
Middelharnis, Neth., 5,100..C4 15
Middelkerke, Bel., 4,610...C2 15
Middle, riv., B.C., Can...B5 68
Middle, riv., Iowa...C3 92
Middle, riv., Minn...B2 99
Middle Alkali, lake, Nev...B2 104
Middle Amana, Iowa, Iowa, 250...C6 92
Middle Andaman, isl., India...F9 39
Middleboro, Plymouth, Mass., 6,003 (11,065▲)...C6 97
Middlebourne, Tyler, W. Va., 711...B4 123
Middlebranch, Holt, Nebr...B7 103
Middlebranch, Stark, Ohio, 600...B6 111
Middlebro, Man., Can., 147..E4 71
Middle Brook, Newf., Can., 744...D4 75
Middlebrook, Augusta, Va., 140...C3 121
Middleburg, Clay, Fla., 750..B5 86
Middleburg, Casey, Ky., 150.C5 94
Middleburg, Schoharie, N.Y., 1,317...C6 108
Middleburg, Vance, N.C., 170...A5 109
Middleburg, Logan, Ohio, 300...B4 111
Middleburg, Snyder, Pa., 1,366...E7 114
Middleburg, S. Afr., 8,711...D4 49
Middleburg, Hardeman, Tenn., 25...B2 117
Middleburg, Loudoun, Va., 761...C5 121
Middleburgh Heights, Cuyahoga, Ohio, 7,282...B2 111
Middlebury, New Haven, Conn., 2,000 (4,785▲)...C4 84
Middlebury, Elkhart, Ind., 917...A6 91
Middlebury, Addison, Vt., 3,688 (5,305▲)...C2 120
Middlebury, riv., Vt...D2 120
Middle Bushkill, creek, Pa...A2 106
Middlefield, Middlesex, Conn., 400 (3,255▲)...C6 84
Middlefield, Hampshire, Mass., 100 (315▲)...B1 97
Middlefield, Geauga, Ohio, 1,467...A6 111
Middleford, Sussex, Del., 50..C6 85
Middle Granville, Washington, N.Y., 869...B7 108
Middle Ground, isl., Midway Is... 52
Middle Haddam, Middlesex, Conn., 500...C6 84
Middle Island, Suffolk, N.Y., 950...F5 84
Middle Island, creek, W. Va...B4 123
Middle Lake, Sask., Can., 238...E3 70
Middle Loup, riv., Nebr...C6 103
Middle Musquodoboit, N.S., Can., 558...D6 74
Middle Nodaway, riv., Iowa...C3 92
Middle Park, basin, Colo...A4 83
Middle Patuxent, riv., Md...B4 85
Middle Point, Van Wert, Ohio, 571...B3 111
Middleport, Niagara, N.Y., 1,882...B2 108
Middleport, Meigs, Ohio, 3,373...C5 111
Middle Raccoon, riv., Iowa...C3 92
Middle Ridge, mts., Newf., Can.D4 75
Middle River, Baltimore, Md., 10,825...*B5 85
Middle River, Marshall, Minn., 414...B2 99
Middlesboro, Bell, Ky., 12,607...D6 94
Middlesbrough, Eng., 157,308 (★530,000)...C6 10
Middlesex, Br. Hond., 207...B3 62
Middlesex, Middlesex, N.J., 10,520...B4 106
Middlesex, Yates, N.Y., 300 (765▲)...C3 108
Middlesex, Nash, N.C., 588..B5 109
Middlesex, Washington, Vt., 80 (770▲)...C3 120
Middlesex, co., Ont., Can., 221,422...D3 72
Middlesex, co., Conn., 88,865...D6 84
Middlesex, co., Eng., 2,230,093...C7 11
Middlesex, co., Mass., 1,238,742...A5 97
Middlesex, co., N.J., 433,856...C4 106
Middlesex, co., Va., 6,319...D6 121
Middlesex Falls, res., Mass...C3 97
Middle Stewiacke, N.S., Can., 217...D6 74
Middleton, N.S., Can., 1,921...E4 74
Middleton, Elbert, Ga., 106..B4 87
Middleton, Canyon, Idaho, 541...F2 89
Middleton, Essex, Mass., 2,200 (3,718▲)...A5, C3 97
Middleton, Gratiot, Mich., 550...E6 98
Middleton, Strafford, N.H., 100 (349▲)...D4 105
Middleton, Hardeman, Tenn., 461...B3 117
Middleton, Dane, Wis., 4,410...E4 124
Middleton, isl., Alsk...D10 79
Middletown, Lake, Calif., 450...C2 82
Middletown, Middlesex, Conn., 33,250...C6 84
Middletown, New Castle, Del., 2,191...B6 85
Middletown, Logan and Menard, Ill., 543...C4 90
Middletown, Henry, Ind., 2,033...D6 91
Middletown, Des Moines, Iowa, 245...D6 92
Middletown, Jefferson, Ky., 2,764...A4 94
Middletown, Frederick, Md., 1,036...B2 85
Middletown, Montgomery, Mo., 199...B6 101
Middletown, Monmouth, N.J., 3,500...C4 106
Middletown, Orange, N.Y., 23,475...D6 108
Middletown, Hyde, N.C., 200...B7 109
Middletown, Butler, Ohio, 42,115...C3 111
Middletown, Dauphin, Pa., 11,182...F8 114
Middletown, Newport, R.I., 12,675...C11 84
Middletown, Frederick, Va., 378...B4 121
Middletown Heights, Delaware, Pa., 1,000...*G11 114
Middletown Springs, Rutland, Vt., 275 (381▲)...E2 120
Middle Valley, Morris, N.J., 250...B3 106
Middleville, Ont., Can., 84...B8 70
Middleville, Barry, Mich., 1,196...F6 98
Middleville, Sussex, N.J., 75...A3 106
Middleville, Herkimer, N.Y., 648...B6 108
Middle Water, Hartley, Tex., 20...B1 118
Midelt, Mor., 6,504...H4 30
Midfield, Jefferson, Ala., 3,556...*E4 78
Midgic Station, N.B., Can., 284...F5 74
Midhurst, Ont., Can., 340...C5 72
Midkiff, Lincoln, W. Va., 300...C2 123
Midland, Sebastian, Ark., 261...B1 75
Midland, Riverside, Calif., 500...F6 82
Midland, Ont., Can., 8,656..C5 72
Midland, Greene, Ind., 475...F3 91
Midland, Acadia, La., 500...D3 95
Midland, Allegany, Md., 737...D2 85
Midland, Midland, Mich., 27,779...E6 98
Midland, Cabarrus, N.C., 750...B3 109
Midland, Clinton, Ohio, 367...C4 111
Midland, Beaver, Pa., 6,425..E1 114
Midland, Washington, Pa., 1,317...*F1 108
Midland, Haakon, S. Dak., 401...C4 116
Midland, Midland, Tex., 62,625...D1 118
Midland, Fauquier, Va., 100...C5 121
Midland, Clark, Wash., 900.*D3 122
Midland, Pierce, Wash., 4,000...*B3 122
Midland, co., Mich., 51,450..E6 98
Midland, co., Tex., 67,717...D1 118
Midland, basin, Colo...A2 83
Midland City, Dale, Ala., 854...D4 78
Midland Park, Sedgwick, Kans., 3,000...B5 93
Midland Park, Bergen, N.J., 7,543...B4 106
Midland Park, Charleston, S.C., 800...F2 115
Midlandvale, Alta., Can., 449...D4 69
Midleton, Ire., 2,772...F3 11
Midlothian, Cook, Ill., 8,749.F3 90
Midlothian, Allegany, Md., 525...D2 85
Midlothian, Ellis, Tex., 1,521...B5, C4 118
Midlothian, Chesterfield, Va., 400...C6, D5 121
Midlothian, co.. Scot., 580,332...E5 13
Midnapore, Alta., Can., 399..D3 69
Midnapore, India, 59,532...F11 40
Midnight, Humphreys, Miss., 150...B3 100
Midongy du Sud, Malag...h9 49
Midstate Mill (Amerotron Mill), Robeson, N.C., 1,090...*C4 109
Midvale, Washington, Idaho, 211...E2 89
Midvale, Tuscarawas, Ohio, 683...B6 111
Midvale, Salt Lake, Utah, 5,802...C4 119
Midville, Burke, Ga., 676...D4 87
Midway, Bullock, Ala., 594...C4 78
Midway, B.C., Can., 391...E8 68
Midway, Gadsden, Fla., 200..B2 86
Midway, Woodford, Ky., 1,044...B5 94
Midway, Multnomah, Oreg., 19,000...*B4 113
Midway, Adams, Pa., 1,568..G7 114
Midway, Greene, Tenn., 225...C11 117
Midway, Wasatch, Utah, 713...C4 119
Midway, King, Wash., 1,000.D2 122
Midway, range, B.C., Can...E8 68
Midway City, Orange, Calif., 2,500...*F3 82
Midway Islands, U.S. dep., Oceania, 2,356... 52
Midway Park, Onslow, N.C., 4,164...*C6 109
Midway Village, Oklahoma, Okla., 2,292...*B4 110
Midwest, Natrona, Wyo., 900...B6 125
Midwest City, Oklahoma, Okla., 36,058...B4 112
Midyat, Tur., 9,600...D13 31
Midye, Tur., 1,318...B7 31
Mie, pref., Jap., 1,485,054..*I8 37
Miechow, Pol., 6,878...C6 26
Miechowice, Pol., 13,831...g9 26
Miedzychod, Pol., 4,632...B3 26
Miedzyrzec, Pol., 8,696...C7 26
Miedzyrzecz, Pol., 4,385...B3 26
Miedzyrzecz, Pol...g10 26
Miedzyzdroje, Pol., 1,949...A3 26
Mielec, Pol., 22,000...C6 26
Mien, lake, Swe...B8 24
Mienia, riv., Pol...m14 26
Mienning, China, 11,000...F5 34
Mienyang, China, 28,000...I5 36
Mier, Mex., 1,018...F2 118
Mieres, Sp., 17,100...A3 20
Miesbach, Ger., 5,200...B7 18
Mīēso, Eth...D5 47
Mifflin, Juniata, Pa., 745...E7 114
Mifflin, Chester, Tenn., 50...B3 117
Mifflin, Kenedy, Tex...F4 118
Mifflin, co., Pa., 44,348...E6 114
Mifflinburg, Union, Pa., 2,476...E7 114
Mifflintown, Juniata, Pa., 887...E7 114
Mifflinville, Columbia, Pa., 1,027...D7 114
Migdal, Isr., 295...B7 32
Migdal Ashqelon, Isr., 14,400...C6 32
Migennes, Fr., 6,352...D5 14
Migiurtinia, dist., Som...D6 47
Migliarino, It., 3,062...E7 18
Mignon, Talladega, Ala., 2,271...B3 78
Miguel Alves, Braz., 1,537...B2 57
Miguel Auza, Mex., 7,140...C4 63
Mihai-Viteazu, Rom., 2,598.C9 22
Mihara, Jap., 71,000 (80,395▲)...*I6 37
Mijares, riv., Sp...B5 20
Mijdaḥah, Aden Prot...C6 47
Mikado, Sask., Can., 115...F4 70
Mikado, Alcona, Mich., 125..D7 92
Mikana, Barron, Wis., 120...C2 118
Mikhaylov, Sov. Un., 23,600...D12 27
Mikhaylovgrad, Bul., 8,067..D6 22
Mikhaylovka, Sov. Un., 31,000...C2 29
Mikhaylovka, Sov. Un., 10,000...H10 27
Mikhaylovskiy, Sov. Un...C9 29
Mikindani, Tan., 4,807...D7 48
Mikkalo, Gilliam, Oreg., 15..B6 113
Mikkeli, Fin., 19,800...G12 25
Mikkeli, dept., Fin., 235,403...*G12 25
Mikolow, Pol., 16,100...g9 26
Mikonos, isl., Grc...D5 23
Mikope, Con. L...C3 48
Mikulczyce, Pol...g9 26
Mikulov, Czech., 5,220...D4 26
Mikura, isl., Jap...J9 37
Mila, Northumberland, Va., 25...D6 121
Milaca, Mille Lacs, Minn., 1,821...E5 99
Milagro, Arg...A3 54
Milam, Hardy, W. Va., 65..C5 123
Milam, co., Texas, 22,263...D4 118
Milan, Que., Can., 129...D7 73
Milan, Telfair and Dodge, Ga., 786...D3 87
Milan, Rock Island, Ill., 3,065...B3 90
Milan, Ripley, Ind., 1,174...F7 91
Milan, Sumner, Kans., 144..E7 93
Milan, Monroe and Washtenaw, Mich., 3,616...F7 98
Milan, Chippewa, Minn., 482...E3 99
Milan, Sullivan, Mo., 1,670..A4 101
Milan, Coos, N.H., 100 (661▲)...A4 105
Milan, Valencia, N. Mex., 2,658...*B2 107
Milan, Erie, Ohio, 1,309...A5 111
Milan, Bradford, Pa., 180...C8 114
Milan, Gibson, Tenn., 5,208...B3 117
Milan, Spokane, Wash., 70..B8 122
Milan, Marathon, Wis., 150..D3 124
Milano, It., 1,582,534 (★2,600,000)...D5 18, B2 21
Milano, Milam, Tex., 600...D4 118
Milas, Tur., 11,700...D5 23
Milazzo, It., 12,900...E5 21
Milbank, Grant, S. Dak., 3,500...B9 116
Milbanke, sound, B.C., Can...C3 68
Milbridge, Washington, Maine, 675 (1,101▲)...D5 96
Milburn, Carlisle, Ky., 400..B2 94
Milburn, Custer, Nebr., 16..C6 103
Milburn, Johnston, Okla., 228...C5 112
Milburn, Fayette, W. Va., 200...D6 123
Milden, Sask., Can., 388...F2 70
Mildenhall, Eng., 6,742...B8 12
Mildmay, Ont., Can., 847...C3 72
Mildred, Sask., Can., 58...D2 70
Mildred, Allen, Kans., 60...D8 93
Mildred, Prairie, Mont., 50...D12 102
Mildred, Sullivan, Pa., 800...D9 114
Mildura, Austl., 12,279...G4 51
Mileai, Grc., 1,983...C4 23
Miles, Austl., 1,457...C8 51
Miles, Jackson, Iowa, 376...B7 92
Miles, Runnels, Tex., 626...D2 118
Milesburg, Centre, Pa., 1,158.E6 114
Miles City, Custer, Mont., 9,665...D11 102
Milestone, Sask., Can., 465..H3 70
Milesville, Haakon, S. Dak., 20...C4 116
Miletus, ruins, Tur...D6 23
Milevsko, Czech., 3,182...D9 17
Miley, Hampton, S.C., 450...F5 115
Milfay, Creek, Okla., 130...B5 111
Milford, Lassen, Calif., 5...B3 82
Milford, New Haven, Conn., 41,662...E4 84
Milford, Sussex and Kent, Del., 5,795...C7 85
Milford, Iroquois, Ill., 1,699..C6 90
Milford, Kosciusko, Ind., 1,167...B6 91
Milford, Dickinson, Iowa, 1,476...A2 92
Milford, Geary, Kans., 318..C7 93
Milford, Bracken, Ky., 100..B5 94
Milford, Penobscot, Maine, 800 (1,572▲)...D4 96
Milford, Worcester, Mass., 15,749...B4, E1 97
Milford, Oakland, Mich., 4,323...A6, F7 98
Milford, Seward, Nebr., 1,462...D8 103
Milford, Hillsboro, N.H., 3,916 (4,863▲)...E3 105
Milford, Hunterdon, N.J., 1,114...B2 106
Milford, Otsego, N.Y., 548...C6 108
Milford, Hamilton and Clermont, Ohio, 4,131...C3 111
Milford, Pike, Pa., 1,198...D12 114
Milford, Beaver, Utah, 1,471.E3 119
Milford, Caroline, Va., 250..C5 121
Milford Center, Union, Ohio, 794...B4 111
Milford Haven, Wales, 12,802...C2 12
Milford Mills, Baltimore, Md., 5,000...*B4 85
Milford Station, N.S., Can., 528...D6 74
Milhurst, Monmouth, N.J., 100...C4 106
Miliana, Alg., 5,983 (15,666▲)...B5 44
Milicz, Pol., 2,929...C4 26
Miling, Austl., 230...F2 50
Milk, riv., Alta., Can...E5 69
Milk, riv., Mont...B8 102
Milk River, Alta., Can., 801...E4 69
Mil Kuh, mtn., Afg...E10 41
Mill, brook, Vt...B5 120
Mill, creek, Ind...F4 91
Mill, creek, Kans...C6 93
Mill, creek, Kans...D7 93
Mill, creek, N.J...D4 106
Mill, creek, Ohio...B4 111
Mill, creek, Tenn...E9 117
Mill, creek, W. Va...C3 91
Mill, riv., Mass...D1 97
Milladore, Wood, Wis., 239..D4 124
Millard, Pike, Ky., 10...C7 94
Millard, Adair, Mo., 250...A5 101
Millard, Douglas, Nebr., 1,014...D3 103
Millard, co., Utah, 7,866...D2 119
Millarton, Stutsman, N. Dak., 40...C7 110
Millau, Fr., 21,229...E5 14
Millbank, Ont., Can., 235...D4 72
Millboro, Tripp, S. Dak., 34.D6 116
Millboro, Bath, Va., 300...D3 121
Millbrae, San Mateo, Calif., 15,873...B5 82
Millbrook, Elmore, Ala., 900...C3 78
Millbrook, Ont., Can., 891..C6 72
Millbrook, Mecosta, Mich., 100...E5 98
Millbrook, Morris, N.J., 900...*B3 106
Millbrook, Dutchess, N.Y., 1,717...D7 108
Millbrook, Wake, N.C., 75...B5 109
Millburn, Lake, Ill., 125...D2 90
Millburn, Essex, N.J., 18,799...B4 106
Millbury, Worcester, Mass., 6,000 (9,623▲)...B4 97
Millbury, Wood, Ohio, 730...A2, A4 111
Mill City, Pershing, Nev., 10.C3 104
Mill City, Marion, Oreg., 1,289...C4 113
Mill City, Wyoming, Pa., 200...A8 114
Millcreek, Union, Ill., 102...F4 90
Mill Creek, La Porte, Ind., 150...A4 91
Millcreek, Madison, Mo., 100...D7 101
Mill Creek, Deer Lodge, Mont., 200...D4 102
Mill Creek, Johnston, Okla., 287...C5 111
Mill Creek, Huntingdon, Pa., 400...E6 114
Mill Creek, Randolph, W. Va., 817...C5 123
Mill Creek, falls, Oreg...E4 113
Milldale, Hartford, Conn., 950...C5 84
Milledgeville, Baldwin, Ga., 11,117...C3 87
Milledgeville, Carroll, Ill., 1,208...B4 90
Milledgeville, McNairy, Tenn., 300...B3 111
Mille lac, Indian res., Minn...D5 99
Mille Lacs, co., Minn., 14,560...E5 99
Mille Lacs, lake, Minn...D5 99
Millen, Jenkins, Ga., 3,633..D5 87
Miller, Lyon, Kans., 75...D8 93
Miller, De Soto, Miss., 150..A4 100
Miller, Buffalo, Nebr., 137...D6 103
Miller, Lawrence, Ohio, 180.D5 111
Miller, Sherman, Oreg...B6 113
Miller, Hand, S. Dak., 2,081.C7 116
Miller, co., Ark., 31,686...D2 81
Miller, co., Ga., 6,908...E2 87
Miller, co., Mo., 13,800...C5 101
Miller, flat, Oreg...C5 113
Miller, isl., Md...B5 86
Miller, peak, Alsk...C11 79
Miller, peak, Ariz...F5 80
Miller Dale Colony, Hand, S. Dak., 50...C6 116
Miller Heights, Northampton, Pa., 1,500...*E11 114
Mille-Roches, Ont., Can., 900...B10 72
Millerovo, Sov. Un., 32,400..D2 29
Miller Place, Suffolk, N.Y., 900...F5 84
Miller Run, riv., Vt...B4 120
Millers, Carroll, Md., 160...A4 85
Millers, riv., Mass...A3 97
Millersburg, Elkhart, Ind., 489...A6 91
Millersburg, Iowa, Iowa, 186.C5 92
Millersburg, Bourbon, Ky., 913...B5 94
Millersburg, Presque Isle, Mich., 280...C6 98
Millersburg, Holmes, Ohio, 3,101...B6 111
Millersburg, Linn, Oreg...C1 113
Millersburg, Dauphin, Pa., 2,984...E8 114
Millers Falls, Franklin, Mass., 1,199...A3 97
Millers Ferry, Wilcox, Ala., 250...C2 78
Millersport, Fairfield, Ohio, 752...C5 111
Millerstown, Perry, Pa., 675..E7 114
Millersview, Concho, Tex., 200...D3 118
Millersville, Marion, Ind., 65...H8 91
Millersville, Lancaster, Pa., 3,883...F9 114
Millerton, N.B., Can., 232...C4 74
Millerton, Wayne, Iowa, 90..D4 92
Millerton, Dutchess, N.Y., 1,027...D7 108
Millerton, McCurtain, Okla., 150...D6 112
Millerton, Tioga, Pa., 400...C8 114
Millertown, Newf., Can., 365.D3 75
Millertown Junction, Newf., Can., 186...D3 75
Millerville, Clay, Ala., 100...B4 78
Millerville, Worcester, Mass., 1,200...*B4 97
Millerville, Douglas, Minn., 119...D3 99
Milles Iles, riv., Que., Can...D8 73
Millet, Alta., Can., 403...C4 69
Millett, Allendale, S.C., 50...E4 115
Millett, La Salle, Tex., 150...E3 118
Mill Fork, Utah, Utah, 7...D4 119
Millgrove, Blackford, Ind., 130...D7 91
Mill Grove, Mercer, Mo., 139...A4 101
Mill Hall, Clinton, Pa., 1,891...D7 114
Millhaven, Screven, Ga., 60..D5 87
Millheim, Centre, Pa., 780...E7 114
Millhousen, Decatur, Ind., 212...F7 91
Millicent, Austl., 3,401...H3 51
Millicent, Alta., Can., 109...D5 69
Milligan, Okaloosa, Fla., 800.G2 86
Milligan, Fillmore, Nebr., 323...D8 103
Milligan, mtn., B.C., Can...B5 68
Milligan College, Carter, Tenn., 200...C11 117
Milliken, Weld, Colo., 630...A6 83
Millikin, East Carroll, La., 100...B4 95
Millington, Kent and Queen Annes, Md., 408...B6 85
Millington, Tuscola, Mich., 1,159...E7 98
Millington, Morris, N.J., 1,182...*B3 106
Millington, Coos, Oreg., 300.D2 113
Millington, Shelby, Tenn., 6,059...B2 117
Millinocket, Penobscot, Maine, 7,453...C4 96
Millinocket, lake, Maine...B4 96
Millinocket, lake, Maine...C4 96
Mill Iron, Carter, Mont., 5..E12 102
Millis, Norfolk, Mass., 2,588 (4,374▲)...B5, D2 97
Millom, Eng., 7,116...F5 13
Mill Plain, Fairfield, Conn., 170...D2 84
Mill Point, Pocahontas, W. Va., 50...C4 123
Millport, Lamar, Ala., 943...B1 78
Millport, Potter, Pa., 125...C5 114
Millport, Scot., 1,592...E4 13
Millrift, Pike, Pa., 110...D12 114
Mill River, Berkshire, Mass., 250...B1 97
Mill Run, Fayette, Pa., 300..G3 114
Millry, Washington, Ala., 645.D1 78
Mills, Keya Paha, Nebr., 37..B6 103
Mills, Harding, N. Mex., 136.A5 107
Mills, Potter, Pa., 150...C6 114
Mills, Juab, Utah...D3 119
Mills, Natrona, Wyo., 1,477..C6 125
Mills, co., Iowa, 13,050...C2 92
Mills, co., Tex., 4,467...D3 118
Mills, lake, N.W. Ter., Can...D9 66
Millsboro, Sussex, Del., 536..C6 85
Millsboro, Washington, Pa., 1,179...G1 114
Mill Shoals, White, Ill., 322..E5 90
Millside, New Castle, Del., 1,000...*A6 85
Mill Spring, Wayne, Mo., 226...D7 101
Millstadt, St. Clair, Ill., 1,830...E3 90
Millston, Jackson, Wis., 200..D3 124
Millstone, New London, Conn., 250...D8 84
Millstone, Somerset, N.J., 409...B3 106
Millstone, riv., N.J...C3 106
Mill Stream, Austl...D2 50
Millstreet, Ire., 1,283...E2 11
Milltown, Chambers, Ala., 45...B4 78
Milltown, N.B., Can., 1,892.D2 74
Milltown, Crawford and Harrison, Ind., 793...H5 91
Milltown, Adair, Ky., 150...C4 94
Milltown, Washington, Maine (part of Calais)...C5 96
Mill Town, Madison, Miss., 300...*C3 100
Milltown, Missoula, Mont., 650...D3 102
Milltown, Middlesex, N.J., 5,435...C4 102
Milltown, Hutchinson, S. Dak., 52...D8 116
Milltown, Polk, Wis., 608...C1 124
Millvale, Allegheny, Pa., 6,624...B6 114
Mill Valley, Marin, Calif., 10,411...D2 82
Mill Village, N.S., Can., 249.E5 74
Mill Village, Erie, Pa., 336..C2 114
Millville, N.B., Can., 390...C2 74
Millville, Sussex, Del., 231...C7 85
Millville, Woodford, Ky...B5 94
Millville, Worcester, Mass., 1,567...B4 97
Millville, Cumberland, N.J., 19,096...E2 106
Millville, Butler, Ohio, 676..C2 111
Millville, Columbia, Pa., 952.D9 114
Millville, Jefferson, W. Va., 350...B7 123
Millwood, Clarke, Va., 400..B4 121
Millwood, Spokane, Wash., 1,776...D8 122
Millwood, Jackson, W. Va., 75...C3 123
Milmay, Atlantic, N.J., 450..E3 106
Milmont Park, Delaware, Pa., 2,000...*B11 114
Milner, B.C., Can., 324...A10 68
Milner, Routt, Colo., 100...A3 83
Milner, Lamar, Ga., 305...C2 87
Milner, dam, Idaho...G5 89
Milner Ridge, Man., Can., 63...D3 71
Milnesand, Roosevelt, N. Mex., 10...D6 107
Milnor, Sargent, N. Dak., 658...C8 110
Milo, Alta., Can., 167...D4 69
Milo, Warren, Iowa, 468...C4 92
Milo, Piscataquis, Maine, 1,802 (2,756▲)...C4 96
Milo, Vernon, Mo., 108...D3 101
Milo, Bledsoe, Tenn...D9 117
Milolii, Hawaii, Haw., 95...D6 88
Milos, isl., Grc...D5 23
Milparinka, Austl., 83...D3 51
Milpitas, Santa Clara, Calif., 6,572...B6 82
Milroy, Rush, Ind., 690...F7 91
Milroy, Redwood, Minn., 268...F3 99
Milroy, Mifflin, Pa., 1,666...E6 114
Milstead, Rockdale, Ga., 1,047...B6, C3 87
Miltenberg, Ger., 8,100...D4 17
Milton, N.S., Can., 1,121...E5 74
Milton, Litchfield, Conn., 75.B3 84
Milton, Sussex, Del., 1,617..C7 85
Milton, Santa Rosa, Fla., 4,108...G2 86
Milton, Madison, Ill. (part of Alton)...A8 101
Milton, Pike, Ill., 309...D3 90
Milton, Wayne, Ind., 700...E7 91
Milton, Van Buren, Iowa, 609...D5 92
Milton, Sumner, Kans., 100..E6 93
Milton, Trimble, Ky., 365...B4 94
Milton, Lafayette, La., 150..D3 95
Milton, Norfolk, Mass., 26,375...B5, D3 97
Milton, Strafford, N.H., 650 (1,418▲)...D5 105
Milton, Morris, N.J., 400...A3 106
Milton, Ulster, N.Y., 800...D7 108
Milton, N.Z., 1,922...Q12 51
Milton, Caswell, N.C., 235...A4 109
Milton, Cavalier, N. Dak., 264...A7 110
Milton, Northumberland, Pa., 7,972...D8 114
Milton, Morgan, Utah...B4 119
Milton, Chittenden, Vt., 817 (2,022▲)...B2 120
Milton, Pierce, Wash., 2,218.D1 122
Milton, Cabell, W. Va., 1,714...C2 123
Milton, Rock, Wis., 1,671...F5 124
Milton, res., Colo...A6 83
Milton, res., Ohio...A7 111
Miltona, Douglas, Minn., 163...D3 99
Miltona, lake, Minn...D3 99
Miltonboro, Chittenden, Vt..B2 120
Milton-Freewater, Umatilla, Oreg., 4,110...B8 113
Milton Junction, Rock, Wis., 1,433...F5 124
Milton Mills, Strafford, N.H., 275...D5 105
Miltonvale, Cloud, Kans., 814...C6 93
Milton West, Ont., Can., 5,629...D5 72
Miltown Malbay, Ire., 700...E2 11
Milverton, Ont., Can., 1,111...D4 72
Milwaukee, Northampton, N.C., 311...A6 109
Milwaukee, Milwaukee, Wis., 741,324 (★1,240,700)..E2, E6 124
Milwaukee, co., Wis., 1,036,041...E5 124
Milwaukee, riv., Wis...E2 124
Milwaukie, Clackamas, Oreg., 9,099...B2, B4 113
Mimbres, Grant, N. Mex., 75.E2 107
Mimbres, Luna, N. Mex...F2 107
Mimbres, mts., N. Mex...D2 107
Mimico, Ont., Can., 18,212...D5, D9 72
Mimizan, Fr., 855...E3 14
Mimon, Czech., 4,605...C9 17
Mimongo, Gabon...F2 46
Mims, Brevard, Fla., 1,307..D6 86
Min, riv., China...F8 34
Mina, Mineral, Nev., 300...E3 104
Mina, Edmunds, S. Dak., 50..B7 116
Mināb, Iran...H8 41
Minago, riv., Man., Can...B2 71
Minam, Wallowa, Oreg., 75..B9 113
Minamata, Jap., 30,400...J5 37
Minami, Iwo... 52
Minami-Iwo, isl., Pac. O...E8 7
Minas, Cuba, 3,305...E5 74
Minas, Ur., 24,400...E1 56
Minas, basin, N.S., Can...D5 74
Minas, chan., N.S., Can...D5 74
Minas de Oro, Hond., 1,407..C4 62
Minas de Ríotinto, Sp., 9,060...D2 20
Minas Gerais, state, Braz., 9,798,880...B4, g3, k4 56
Minatare, Scotts Bluff, Nebr., 894...C2 103
Minatitlán, Mex., 34,980...D6 63
Minato, see Nakaminto, Jap.
Minato, Jap., 7,100...n18 37
Minbu, Bur., 9,096...D9 39
Minburn, Alta., Can., 164...C5 69
Minburn, Dallas, Iowa, 357..C3 92
Minch, chan., Scot...B3 13
Minchin, China...D5 34
Mincio, riv., It...D6 18
Minco, Grady, Okla., 1,021..B4 112
Mindanao, isl., Phil...D7 35
Mindanao, sea, Phil...D6 35
Mindel, riv., Ger...A6 18
Mindelheim, Ger., 8,500...A6 18
Minden, Ont., Can., 658...C6 72
Minden, Ger., 48,700 (★87,000)...A3 17
Minden, Pottawattamie, Iowa, 355...C2 92
Minden, Webster, La., 12,785...B2 95
Minden, Kearney, Nebr., 2,383...D7 103
Minden, Douglas, Nev., 400..E2 104
Minden, Fayette, W. Va., 1,114...D3, D7 123
Minden City, Sanilac, Mich., 369...E8 98
Mindenmines, Barton, Mo., 356...D3 101
Mindoro, La Crosse, Wis., 200...D2 124
Mindoro, isl., Phil...C6 35
Mindoro, strait, Phil...C6 35
Mine Centre, Ont., Can., 88..E5 71
Minechoag, mtn., Mass...B3 97
Minehead, Eng., 7,674...C4 12
Mine Hill, Morris, N.J., 3,362...*B3 106
Mineiros, Braz., 5,105...B2 56
Mine La Motte, Madison, Mo., 100...D7 101
Mineola, Mills, Iowa, 150...C2 92
Mineola, Clark, Kans., 679..E3 93
Mineola, Nassau, N.Y., 20,519...D2, D3, E7 108
Mineola, Wood, Tex., 3,810.C5 118
Miner, Scott, Mo., 548...E8 101
Miner, Park, Mont., 5...E6 102
Miner, co., S. Dak., 5,398...D8 116
Mineral, Bureau, Ill., 330...B4 90
Mineral, Louisa, Va., 366...C5 121
Mineral, Lewis, Wash., 400..C3 122
Mineral, co., Colo., 424...D4 83
Mineral, co., Mont., 3,037...C1 102
Mineral, co., Nev., 6,329...E3 104
Mineral, co., W. Va., 22,354...B6 123
Mineral, mtn., Ariz...D2 80

Mineral, mts., Utah E3 119
Mineral City, Tuscarawas, Ohio, 917 B6 111
Mineral del Oro, Mex., 4,283 n13 63
Mineral Hills, Iron, Mich., 311 B2 98
Mineral Park, Bradley, Tenn., 25 E10 117
Mineral Point, Washington, Mo., 332 D7 101
Mineral Point, Iowa, Wis., 2,385 F3 124
Mineral Ridge, Mahoning and Trumbull, Ohio, 2,000 *A7 111
Mineral Springs, Howard, Ark., 616 D2 81
Mineral Wells, De Soto, Miss., 210 A4 100
Mineral Wells, Palo Pinto, Tex., 11,053 C3 118
Minersville, Meigs, Ohio, 350 C6 111
Minersville, Schuylkill, Pa., 6,606 E9 114
Minersville, Beaver, Utah, 580 E3 119
Minerva, Terrebonne, La. C6 95
Minerva, Essex, N.Y., 100 B7 108
Minerva, Carroll and Stark, Ohio, 3,833 B6 111
Minerva Park, Franklin, Ohio, 1,169 *C5 111
Minervino Murge, It., 18,427 D6 21
Minetto, Oswego, N.Y., 800 B4 108
Mineville, Essex, N.Y., 1,181 A7 108
Mingan, Que., Can., 50 G20 67
Mingan, passage, Que., Can. h8 75
Mingchiang, China A7 38
Mingechaur, Sov. Un., 30,000 G18 9
Mingechaur, res., Sov. Un. E3 29
Mingenew, Austl., 317 E2 50
Mingo, Jasper, Iowa, 260 C4 92
Mingo, Thomas, Kans., 10 C3 93
Mingo, co., W. Va., 39,742 D2 123
Mingo, creek, S.C. D9 115
Mingo Junction, Jefferson, Ohio, 4,987 B7 111
Mingoyo, Tan. D6 48
Mingshui, China, 5,000 C2 37
Mingshui, China C4 34
Minho, prov., Port., 825,788 *B1 20
Minho, riv., Port. B1 20
Minhow, see Foochow, China
Minicoy, isl., India G5 39
Minidoka, Minidoka, Idaho, 154 G5 89
Minidoka, co., Idaho, 14,394 G5 89
Minidoka, dam, Idaho G5 89
Minier, Tazewell, Ill., 847 C4 90
Miniota, Man., Can., 248 D1 71
Minipi, lake, Newf., Can. h9 75
Minisink, isl., N.J. A3 106
Minitonas, Man., Can., 606 C1 71
Mink Creek, Franklin, Idaho, 109 G7 89
Minna, Nig., 12,810 E6 45
Minna, bluff, Ant. B29 5
Minneapolis, Ottawa, Kans., 2,024 C6 93
Minneapolis, Hennepin, Minn., 482,872 (★1,491,000) E7, F5 99
Minneapolis, Avery, N.C., 200 C4 103
Minnedosa, Man., Can., 2,211 D2 71
Minnedosa, riv., Man., Can. D1 71
Minnehaha, Clark, Wash., 2,000 *D3 122
Minnehaha, co., S. Dak., 86,575 D6 116
Minneiska, Wabasha, Minn., 110 F7 99
Minneota, Lyon, Minn., 1,297 F3 99
Mineral, Tehama, Calif., 125 B3 82
Minnesota, state, U.S., 3,413,864 A9 77, 99
Minnesota, riv., Minn. F3 99
Minnesota Lake, Faribault, Minn., 697 G5 99
Minnetonka, Hennepin, Minn., 25,037 *F5 99
Minnetonka, lake, Minn. F5 99
Minnetrista, Hennepin, Minn., 2,076 *E5 99
Minnewanka, lake, Alta., Can. D3 69
Minnewaska, lake, Minn. E3 99
Minnewaukan, Benson, N. Dak., 420 A6 110
Minne Maud, creek, Utah D5 119
Mino, Jap., 18,700 n15 37
Miño, riv., Sp. A1 20
Minoa, Onondaga, N.Y., 1,838 *B4 108
Minocqua, Oneida, Wis., 700 C4 124
Minokamo [Ota], Jap., 31,144 n15 37
Minong, Washburn, Wis., 348 B2 124
Minonk, Woodford, Ill., 2,001 C4 90
Minooka, Grundy, Ill., 539 B5 90
Minor Hill, Giles, Tenn., 400 B11 117
Minot, Plymouth, Mass., 200 D4 97
Minot, Ward, N. Dak., 33,477 A4 110
Minotola, Atlantic, N.J. (part of Buena) D3 106
Minquadale, New Castle, Del., 1,200 *A6 85
Minsen, Ger., 2,100 A7 15
Minsk, Sov. Un., 570,000 E6 27
Minsk, Mazowiecki, Pol., 20,000 B6, m15 26
Minster, Auglaize, Ohio, 2,193 *B3 111
Minstra, mts., Sp. B4 20
Minster, Dallas, Ala., 50 C2 78
Minter City, Leflore, Miss., 250 B3 100
Minto, Alsk., 161 C10 79
Minto, Man., Can., 171 E1 71
Minto, N.B., Can., 1,319 C3 74
Minto, Walsh, N. Dak., 642 A8 110
Minto, lake, Que., Can. A2 73
Minto, mtn., N.W. Ter., Can. D16 67
Minto, pass, Oreg. C5 113
Minton, Sask., Can., 208 H3 70
Mintons Corner, Brevard, Fla., 2,400 *D6 86
Minturn, Lawrence, Ark., 61 B4 81
Minturn, Eagle, Colo., 662 B4 83
Minturn, Hancock, Maine, 120 D4 96
Minturno, It., 3,125 D4 21
Minûf, Eg., U.A.R., 36,900 D3 32
Minusinsk, Sov. Un., 44,600 E27 9
Minvoul, Gabon E2 46
Minya al Qamḩ, Eg., U.A.R., 1,000 D3 32
Minya Konka, peak, China F5 34
Mio, Oscoda, Mich., 500 D6 98
Miola, Clarion, Pa., 15 D3 114
Miquelon, cape, Miquelon Isl. E3 75
Miquelon, isl., St. Pierre and Miquelon E3 75
Mira, Caddo, La., 75 B2 95
Mira, Port., 2,258 B2 20
Mira, riv., Port. D1 20
Miracle Hot Springs, Kern, Calif., 40 E4 80
Mirador, Braz., 818 C2 57
Miraflores, Col., 2,456 C3 60
Miraflores, Peru, 16,146 E3 58
Miraflores, locks, C.Z. m11 62
Miragoâne, Hai., 2,499 F7 64
Mira Gut, N.S., Can., 99 C9 74
Miraj, India, 53,345 I5 40
Miraleste, Los Angeles, Calif., 1,800 *E4 82
Mira Loma, Riverside, Calif., 3,982 F3 82
Miramar, San Diego, Calif., 75 E2 82
Miramar, Cuba h11 64
Miramar, Broward, Fla., 5,485 E3 86
Miramichi, bay, N.B., Can. B4 74
Miramichi, riv., N.B., Can. B4 74
Miranda, Braz., 2,075 C1 56
Miranda, Col., 4,082 C2 60
Miranda, Faulk, S. Dak., 65 F7 116
Miranda, state, Ven., 492,349 A4 60
Miranda, riv., Braz. C1 56
Miranda de Ebro, Sp., 27,881 A4 20
Miranda do Douro, Port., 1,331 B2 20
Mirande, Fr., 3,024 F4 14
Mirandela, Port., 3,418 B2 20
Mirando City, Webb, Tex., 600 F3 118
Mirandola, It., 8,000 C3 21
Mirassol, Braz., 13,674 C3 56
Mirebalais, Hai., 5,200 F7 64
Mirecourt, Fr., 8,572 C7 14
Mirepoix, Fr., 2,477 F4 14
Mirgorod, Sov. Un., 10,000 G9 27
Miri, Mala., 13,350 E4 35
Mirialguda, India, 10,024 I7 40
Miriam Vale, Austl., 297 B8 51
Mirond, lake, Sask., Can. B4 70
Mirow, Ger., 3,801 E6 15
Mirpur-Khas, Pak., 40,420 E2 40
Mirror, Alta., Can., 577 C4 69
Mirror Lake, Carroll, N.H., 60 C4 105
Mirzapur, India, 100,097 C7 39, E9 40
Misakubo, Jap., 5,000 n16 37
Misamis Occidental, prov., Phil., 251,100 *D6 35
Misamis Oriental, prov., Phil., 387,800 *D6 35
Misantla, Mex., 9,078 D5, n15 63
Misburg, Ger., 14,400 A4 17
Miscou, isl., N.B., Can. B5 74
Miscouche, P.E.I., Can., 676 D6 74
Misenheimer, Stanly, N.C., 850 B3 108
Mishan, China, 5,000 D6 37
Mishawaka, St. Joseph, Ind., 33,361 A5 91
Mishicot, Manitowoc, Wis., 762 A7, D6 124
Mishima, Jap., 62,966 I9, n17 37
Misiones, prov., Arg., 331,000 E4 55
Misiones, dept., Par., 43,449 E4 55
Miskitos, is., Nic. C5 62
Miskolc, Hung., 143,364 A5 22
Misoöl, isl., W. Irian F7 35
Mispillion, riv., Del. C7 85
Misquah, hills, Minn. A7 99
Misquamicut, Washington, R.I., 250 D9 84
Miṣr al Jadīdah, Eg., U.A.R. D3 32
Misrātah, Libya, 59,902 C3 43
Misratah, cape, Libya H14 30
Missaukee, co., Mich., 6,784 D5 98
Missaukee, lake, Mich. D5 98
Missinaibi, riv., Can. G16 67
Mission, Johnson, Kans., 4,626 *B9 93
Mission, Todd, S. Dak., 611 D5 116
Mission, Hidalgo, Tex., 14,081 F3 118
Mission, Indian res., Calif. F5 82
Mission, range, Mont. C3 102
Mission City, B.C., Can., 3,251 A10, E6 68
Mission Hill, Yankton, S. Dak., 165 E8 116
Mission Hills, Johnson, Kans., 3,621 D9 93
Missisquoi, Franklin, Vt., 100 B3 120
Missisquoi, co., Que., Can., 29,526 D5 73
Missisquoi, bay, Vt. A2 120
Missisquoi, riv., Vt. B3 120
Mississagi, riv., Ont., Can. B7 98
Mississagi, strait, Ont., Can. C7 98
Mississinewa, riv., Ind. C6 91
Mississippi, co., Ark., 70,174 B5 81
Mississippi, co., Mo., 20,695 E8 101
Mississippi, state, U.S., 2,178,141 D10 77, 100
Mississippi, delta, La. E6 95
Mississippi, riv., U.S. D9 77
Mississippi, sound, Miss. E5 100
Mississippi City, Harrison, Miss., 4,169 E2, E4 100
Missoula, Missoula, Mont., 27,090 D2 102
Missoula, co., Mont., 44,663 D2 102
Missour, Mor. C4 44
Missouri, state, U.S., 4,319,813 C9 77, 101
Missouri, buttes, Wyo. A8 125
Missouri, caverns, Mo. C6 101
Missouri, riv., U.S. B8 77
Missouri City, Clay, Mo., 404 E2 101
Missouri City, Fort Bend, Tex., 604 F5 118
Missouri Valley, Harrison, Iowa, 3,567 C2 92
Mistaken, pt., Newf., Can. E5 75
Mistassini, Que., Can., 3,461 *G18 67
Mistassini, lake, Que., Can. B2 73
Mistastin, lake, Newf., Can. g9 75
Mistelbach [an der Zaya], Aus., 5,434 D8 16
Misti, vol., Peru E3 58
Miston, Dyer, Tenn., 100 A2 117
Mistretta, It., 9,979 F5 21
Mita, pt., Mex. C3 63
Mitake, Jap., 8,200 I8, n16 37
Mitcham, Eng., 63,653 m12 10
Mitchell, Bullock, Ala., 25 C4 78
Mitchell, Austl., 1,822 C6 52
Mitchell, Ont., Can., 2,247 D3 72
Mitchell, Glascock, Ga., 184 C4 87
Mitchell, Madison, Ill., 150 A8 101
Mitchell, Lawrence, Ind., 3,552 G5 91
Mitchell, Mitchell, Iowa, 237 A5 92
Mitchell, Sabine, La., 300 C2 95
Mitchell, Scotts Bluff, Nebr., 1,920 C2 103
Mitchell, Wheeler, Oreg., 236 C6 113
Mitchell, Davison, S. Dak., 12,555 D7 116
Mitchell, co., Ga., 19,652 E2 87
Mitchell, co., Iowa, 14,043 A5 92
Mitchell, co., Kans., 8,866 C5 93
Mitchell, co., N.C., 13,906 C4 109
Mitchell, co., Tex., 11,255 C2 118
Mitchell, dam, Ala. C3 78
Mitchell, isl., La. E6 95
Mitchell, lake, Ala. C3 78
Mitchell, lake, Mich. D5 98
Mitchell, mtn., N.C. D4 109
Mitchell, riv., Austl. C5 52
Mitchellsberg, Boyle, Ky., 500 C5 94
Mitchellville, Polk, Iowa, 957 A8, C4 92
Mitchellville, Sumner, Tenn., 184 A5 117
Mitchelstown, Ire., 2,655 E3 11
Mīt Fāris, Eg., U.A.R., 1,000 C3 32
Mīt Ghamr, Eg., U.A.R., 34,400 D3 32
Mitilini (Mytilene), Grc., 25,758 C6 23
Mitishto, riv., Man., Can. B2 71
Mitla, pass, Eg., U.A.R. D4 32
Mito, Jap., 110,000 (139,389▲) H10, m19 37
Mitre, mtn., N.Z. N15 51
Mitsinjo, Malag. g9 49
Mittelland, canal, Ger. B5 16
Mittenwald, Ger., 8,500 B7 18
Mittersill, Aus., 3,502 B8 18
Mitterteich, Ger., 6,500 D7 18
Mittl Isar, canal, Ger. A7 18
Mittweida, Ger., 20,900 C7 17
Mitú, Col., 211 C3 60
Mitūbis, Eg., U.A.R., 1,000 C2 32
Mitumba, mts., Con. L. C4 48
Mitwaba, Con. L. C4 48
Mitzic, Gabon E2 46
Miura (Misaki), Jap., 19,334 n18 37
Mixcoac, Mex. (part of Mexico City) h9 63
Mixquiahuala, Mex., 7,184 m14 63
Mixteco, riv., Mex. o14 63
Miyagi, pref., Jap., 1,743,195 *G10 37
Miyake, isl., Jap. I9 37
Miyako, Jap., 37,400 G10 37
Miyakonojo, Jap., 54,600 (92,230▲) K5 37
Miyan Kaleh, pen., Iran C6 41
Miyazaki, Jap., 115,000 (158,328▲) K5 37
Miyazaki, pref., Jap., 1,134,590 *K5 37
Miyazu, Jap., 21,800 n14 37
Mizdah, Libya C2 43
Mize, Smith, Miss., 371 D4 100
Mizen, head, Ire. F2 11
Mizil, Rom., 7,460 C8 22
Mizpah, Koochiching, Minn., 140 C4 99
Mizpah, Custer, Mont. D11 102
Mizpah, Atlantic, N.J., 350 E3 106
Mizpah, creek, Mont. E11 102
Mizque, Bol., 870 C2 55
Mjölby, Swe., 10,700 H6 25
Mjösa, lake, Nor. G4 25
Mkalama, Tan. B5 48
M. Kemalpasa, see Mustafa Kemalpasa, Tur.
Mkushi, N. Rh. C4 48
Mlada Boleslav, Czech., 25,700 C3, n18 26
Mlanje, Nya. A6 49
Mlawa, Pol., 14,100 B6 26
Mljet, isl., Yugo. D3 22
Mnichovo Hradiste, Czech., 3,733 C9 17
Mo, Nor., 8,300 D6 25
Moa, isl., Indon. G7 35
Moab, Grand, Utah, 4,682 E6 119
Moala, isl., Fiji Is. 52
Moamba, Moz. C5 49
Moanda, Gabon F2 46
Moapa, Clark, Nev., 20 G7 104
Moapa River, Indian res., Nev. G7 104
Moar, lake, Man., Can. C4 71
Moark, Clay, Ark., 130 A5 81
Moate, Ire., 1,261 D4 11
Moaula, Hawaii, Haw., 65 D6 88
Mobara, Jap., 19,300 n19 37
Mobaye, Cen. Afr. Rep. E4 46
Mobeetie, Wheeler, Tex., 450 B2 118
Moberly, Randolph, Mo., 13,170 B5 101
Moberly, lake, B.C., Can. B7 68
Mobile, Mobile, Ala., 202,779 (★304,000) E1 78
Mobile, Maricopa, Ariz., 35 D3 80
Mobile, Newf., Can., 80 E5 75
Mobile, co., Ala., 314,301 E1 78
Mobile, bay, Ala. E1 78
Mobile, riv., Ala. E1 78
Mobley, Stewart, Tenn. A3 117
Mobridge, Walworth, S. Dak., 4,391 B5 116
Mobula, Con. L. A4 48
Moca, Dom. Rep., 13,829 F8 64
Moca, P.R., 1,938 B2 65
Moca, mun., P.R., 21,990 B2 65
Mocajuba, Braz., 1,352 C5 59
Moçambique, Moz., 9,222 A7 49
Mocambique, prov., Moz. D6 48
Moçâmedes, Ang., 7,185 E1 48
Moçâmedes, dist., Ang., 43,393 E1 48
Mocanaqua, Luzerne, Pa., 1,104 D9 114
Moccasin, Mohave, Ariz., 65 A3 80
Moccasin, Judith Basin, Mont., 150 C7 100
Mocha (Mokha) (Al Mukhā), Yemen, 5,000 C5 47
Mocha, isl., Chile B2 54
Mochudi, Bech., 11,767 B4 49
Mocimboa da Praia, Moz. D7 48
Möchlen, lake, Swe. B8 24
Mocksville, Davie, N.C., 2,379 B3 109
Moclips, Grays Harbor, Wash., 500 B1 122
Mocoa, Col., 1,698 C2 60
Mococa, Braz., 14,306 C3, k8 56
Mocomoco, Bol., 977 C2 55
Mocorito, Mex., 2,472 B3 63
Moctezuma, Mex., 2,151 B3 63
Moctezuma, riv., Mex. m14 63
Mocuba, Moz., 1,000 A6 49
Modale, Harrison, Iowa, 276 C2 92
Modane, Fr., 4,735 D2 18
Mode, Shelby, Ill., 125 D5 90
Model, Las Animas, Colo., 25 D6 83
Model, Stewart, Tenn., 100 A4 117
Model, res., Colo. D6 83
Modena, It., 139,183 E6 19, B3 21
Modena, Mercer, Mo., 66 A4 101
Modena, Ulster, N.Y., 450 D6 108
Modena, Iron, Utah, 62 F2 119
Modena, Buffalo, Wis., 125 D2 124
Modeste, Ascension, La., 250 B5 95
Modesto, Stanislaus, Calif., 36,585 D3 82
Modesto, Macoupin, Ill., 228 D4 90
Modica, It., 30,000 F5 21
Modlin, Pol. B6, k13 26
Mödling, Aus., 17,274 D8 16
Modoc, Emanuel, Ga., 33 D4 87
Modoc, Randolph, Ind., 238 D7 91
Modoc, Scott, Kans., 73 D2 93
Modoc, co., Calif., 8,308 B3 82
Modoc Point, Klamath, Oreg., 75 E5 113
Modrany, Czech., 10,100 n17 26
Moecherville, Kane, Ill., 1,200 *B5 90
Moen, isl., Truk 52
Moengo, Sur. A4 59
Moenkopi, Coconino, Ariz., 500 A4 80
Moenkopi, wash, Ariz. A5 80
Moerbeke, Bel., 5,164 C3 15
Moers, Ger., 46,700 B1 17
Moesa, riv., Switz. D7 19
Moesala, isl., Indon. L3 38
Moesi, riv., Indon. F2 35
Moeskroen, see Mouscron, Bel.
Moffat, Saguache, Colo., 104 C5 83
Moffat, co., Colo., 7,061 A2 83
Moffat, Scot., 1,917 E5 13
Moffat, railroad tunnel, Colo. B5 83
Moffett, Sequoyah, Okla., 357 B7 112
Moffit, Burleigh, N. Dak., 97 C5 110
Moga, Con. L. B4 48
Mogadiscio, Som., 77,000 E6 47
Mogador, Mor., 26,392 C3 44
Mogadore, Portage and Summit, Ohio, 3,851 A6 111
Mogaung, Bur., 2,940 C10 39
Mogfog, Guam 52
Mogi das Cruzes, Braz., 63,748 C3, m8 56
Mogielnica, Pol., 4,667 C6 26
Mogi Guaçu, riv., Braz. k8 56
Mogilev, Sov. Un., 134,000 E8 27
Mogilev-Podolskiy, Sov. Un., 46,300 G6 27
Mogilno, Pol., 5,193 B4 26
Mogi-Mirim, Braz., 18,345 m8 56
Mogincual, Moz. A7 49
Mogocha, Sov. Un., 18,000 D14 28
Mogochin, Sov. Un., 3,500 B10 29
Mogok, Bur., 8,369 D10 39
Mogollon, Carton, N. Mex., 25 D1 107
Mogollon, plat., Ariz. C4 80
Mogote, Conejos, Colo., 30 D4 83
Mogotes, pt., Arg. B5 54
Mogpog, Phil., 1,331 p13 35
Moguer, Sp., 7,222 D2 20
Mohacs, Hung., 15,860 (18,045▲) C4 22
Mohaleshoek, Bas. D4 49
Mohall, Renville, N. Dak., 956 A4 110
Mohammed, cape, Eg., U.A.R. I10 31
Mohave, co., Ariz., 7,736 B1 80
Mohave, lake, Ariz. B1 80
Mohave, mts., Ariz. C1 80
Mohawk, Yuma, Ariz., 20 E2 80
Mohawk, Keweenaw, Mich., 600 A2 98
Mohawk, Herkimer, N.Y., 3,533 C5 108
Mohawk, Lane, Oreg., 100 C4 113
Mohawk, lake, N.J. A3 106
Mohawk, mtn., Conn. B3 84
Mohawk, riv., N.H. B1 105
Mohawk, riv., N.Y. C6 108
Mohegan, New London, Conn., 700 D8 84
Mohegan Lake, Westchester, N.Y., 1,500 *D7 108
Mohican, riv., Ohio B5 111
Mohill, Ire., 905 D4 11
Mohler, Lewis, Idaho, 20 C2 89
Mohler, Lincoln, Wash., 30 B7 122
Mohnton, Berks, Pa., 2,223 F10 114
Mohon, Fr., 9,043 C6 14
Mohoro, Tan., 1,160 C6 48
Mohrland, Emery, Utah D4 119
Mohulu, Con. L. B4 48
Moiese, Lake, Mont., 5 C2 102
Moineşti, Rom., 12,934 B8 22
Mointy, Sov. Un. D8 29
Moio, Eth. D4 47
Moira, Franklin, N.Y., 50 A2 108
Moisie, Que., Can., 511 h8 75
Moisie, riv., Que., Can. h8 75
Moissac, Fr., 4,770 E4 14
Moïssala, Chad, 3,203 D3 46
Moita, Port., 3,797 f10 20
Mojave, Kern, Calif., 1,845 E4 82
Mojave, desert, Calif. E5 82
Mojave, riv., Calif. E5 82
Mokameh, India, 35,743 E10 40
Mokami, hill, Newf., Can. g9 75
Mokane, Callaway, Mo., 419 C6 101
Mokapu, pt., Haw. g11 88
Mokelumne, riv., Calif. C3 82
Mokelumne Hill, Calaveras, Calif., 425 C3 82
Mokena, Will, Ill., 1,332 F2 90
Mokepa, Con. L. A4 48
Mokha (Mocha) (Al Mukhā), Yemen, 5,000 C5 47
Mokhotlong, Bas. C4 49
Moknine, Tun., 17,699 *B7 44
Mokolo, Cam. C2 46
Mokolo, Con. L. B2 48
Mokpo, Kor., 129,700 I3 37
Moksha, riv., Sov. Un. C2 29
Mokuaweoweo, crater, Haw. D6 88
Mokuleia, Honolulu, Haw., 200 B3, f9 88
Moku Manu, isl., Haw. g11 88
Mol, Bel., 24,794 C5 15
Mol, Yugo., 8,079 C5 22
Mola di Bari, It., 22,852 D6 21
Molalla, Clackamas, Oreg., 1,501 B4 113
Molaoi, Grc., 3,018 D4 23
Molasses, pond, Maine D4 96
Mold, Wales, 6,857 A4 12
Moldavia (Moldova), prov., Rom., 2,598,258 *B8 22
Moldavia, reg., Rom. B8 22
Moldavia (S.S.R.), rep., Sov. Un., 3,040,000 E5 28
Molde, Nor., 8,100 F2 25
Moldova (Moldavia), prov., Rom., 2,598,258 *B8 22
Moldova, riv., Rom. B8 22
Moledet, Isr., 315 B7 32
Molena, Pike, Ga., 279 D2 87
Molengraaff, mts., Indon. F5 35
Molepolole, Bech., 14,805 B5 49
Môle St. Nicolas, Hai., 1,700 F7 64
Molfetta, It., 61,684 D6 21
Molina, Chile, 6,123 B2 54
Molina, Mesa, Colo., 10 B2 83
Molina de Aragón, Sp., 3,181 B5 20
Molina de Segura, Sp., 8,578 C5 20
Moline, Rock Island, Ill., 42,705 B3 90
Moline, Elk, Kans., 698 E7 93
Moline, Allegan, Mich., 550 F5 98
Moline, Wood, Ohio, 350 A2 111
Moline Acres, St. Louis, Mo., 3,132 *C7 101
Molinella, It., 2,950 E7 18
Molino, Escambia, Fla., 800 G2 86
Molinos, Arg. E2 55
Moliro, Con. L. C5 48
Moliterno, It., 5,657 D5 21
Möll, riv., Aus. C9 18
Mölle, Swe., 480 B6 24
Mollendo, Peru, 14,893 E3 58
Mollusk, Lancaster, Va., 325 D6 121
Mölndal, Swe., 26,500 I5 25
Molochansk, Sov. Un., 10,000 H10 27
Molodechno, Sov. Un., 26,000 D6 27
Molokai, isl., Haw. B5 88
Molokini, isl., Haw. C5 88
Molopo, riv., Bech., S. Afr. C3 49
Molotovsk, Sov. Un., 30,000 E18 25
Molotovskoye, Sov. Un., 10,000 I13 27
Moloundou, Cam. E3 46
Mols, isl., Den. B4 24
Molsheim, Fr., 4,955 F7 15
Molson, Man., Can., 51 D3 71
Molson, Okanogan, Wash., 80 A6 122
Molson, lake, Man., Can. B3 71
Molson, riv., Man., Can. B3 71
Molt, Stillwater, Mont., 5 E8 102
Molteno, S. Afr., 4,377 D4 49
Molucca, is., Indon. F7 35
Molucca, passage, Indon. E7 35
Molucca, sea, Indon. E7 35
Molus, Harlan, Ky., 500 D6 94
Moma, Moz. A6 49
Mombango, Con. L. A3 48
Mombasa, Ken., 189,800 B6 48
Mombetsu, Jap., 28,200 D11 37
Momboyo, riv., Con. L. B3 48
Momchilgrad, Bul., 3,150 E7 22
Momence, Kankakee, Ill., 2,949 B6 90
Momi, Fiji Is., 691 52
Momostenango, Guat., 5,002 C2 62
Mompano, Con. L. A3 48
Mompog, pass, Phil. p14 37
Mompós, Col., 9,192 B3 60
Mön, isl., Den. D6 24
Mona, Richland, Mont., 5 B12 102
Mona, Juab, Utah, 347 D4 119
Mona (Granville), Monongalia, W. Va., 806 A7 123
Mona, isl., P.R. e10 65
Mona, passage, W.I. F9 64
Monaca, Beaver, Pa., 8,394 E1 114
Monaco, Monaco, 22,297 (★37,500) F7 14
Monaco, country, Eur., 22,297 F7 14
Monadhliath, mts., Scot. C4 13
Monadnock, mtn., N.H. E2 105
Monagas, state, Ven., 246,217 B5 60
Monaghan, Ire., 4,013 C5 11
Monaghan, Greenville, S.C., 1,200 B2 115
Monaghan, co., Ire., 47,088 C4 11
Monahans, Ward, Tex., 8,567 D1 118
Monamolin, Ire. E5 11
Monango, Dickey, N. Dak., 133 C7 110
Mona ó Carreta, pt., C.R. F6 62
Monarch, Alta., Can., 109 E4 69
Monarch, Cascade, Mont., 20 C6 102
Monarch (Monarch Mills), Union, S.C., 1,990 B4 115
Monarch, mtn., B.C., Can. D5 68
Monarda, Aroostook, Maine, 50 C4 96
Monashee, mts., B.C., Can. D8 68
Monastir, Tun. B7 44
Monastyrshchina, Sov. Un., 5,300 D8 27
Mona Vatu, mtn., Fiji Is. 52
Monaville, Logan, W. Va., 825 *D5 123
Monção, Braz., 1,132 B1 57
Moncayo, mtn., Sp. B5 20
Monchegorsk, Sov. Un., 30,000 D15 25
Mönchengladbach, Ger., 152,200 (★330,000) C3 16, C6 15
Monchique, Port., 2,169 D1 20
Monchique, mts., Port. D1 20
Moncks Corner, Berkeley, S.C., 2,030 E7 115
Monclo, Logan, W. Va., 60 D3, D5 123
Monclova, Mex., 43,333 B4 63
Monclova, Lucas, Ohio, 125 A1 111
Moncton, N.B., Can., 43,840 (★55,768) C5 74
Moncure, Chatham, N.C., 400 B4 109
Mond, lake, Ger. B9 18
Mondamin, Harrison, Iowa, 436 C1 92
Mondego, cape, Port. B1 20
Mondego, riv., Port. B1 20
Mondoñedo, Sp., 9,153 A2 20
Mondorf-les-Bains, Lux., 1,757 E6 15
Mondovi, It., 8,800 (★13,000) B1 21
Mondovi, Buffalo, Wis., 2,320 D2 124
Monee, Will, Ill., 646 B6 90
Monegaw Springs, St. Clair, Mo., 53 C4 101
Monemvasia, Grc., 638 D4 23
Monero, Rio Arriba, N. Mex., 300 A3 107
Monessen, Westmoreland, Pa., 18,424 F2 114
Moneta, Bedford, Va., 170 D3 121
Moneta, Fremont, Wyo., 10 B6 125
Monett, Barry and Lawrence, Mo., 5,359 E4 100
Monetta, Aiken and Saluda, S.C., 242 D4 115
Monette, Craighead, Ark., 981 B5 81
Money, Leflore, Miss., 100 B3 100
Moneymore, N. Ire., 807 C5 11
Monfalcone, Italy, 26,818 B4 21
Monforte de Lemos, Sp., 13,502 A2 20
Monfort Heights, Hamilton, Ohio, 6,000 *D2 111
Monga, Con. L. A3 48
Mongala, riv., Con. L. A3 48
Mongalla, Sud. D3 47
Monges, is., Ven. A3 60
Mong Hpayak, Bur. B3 38
Mong Hsat, Bur. B3 38
Monghyr, India, 89,768 (★146,807) C8 38, E11 40
Mong Mit, Bur., 5,000 D10 39
Mong Nai, Bur. B3 38
Mongo, Chad, 2,721 C3 46
Mongo, Lagrange, Ind., 225 A7 91
Mongolia, Ont., Can. E7 72
Mongolia, country, Asia, 900,000 E13 31, B5 34
Mongolia, plat., Mong. B6 34
Mong Pan, Bur. B3 38
Mongu, N. Rh., 3,000 E3 48
Monhegan, isl., Maine E3 96
Moniaive, Scot., 1,219 E5 13
Monico, Oneida, Wis., 150 C4 120
Monida, Beaverhead, Mont., 23 F4 102
Monida, pass, Idaho F4 102
Monie, Somerset, Md., 70 D6 85
Moniquirá, Col., 3,230 B3 60
Moniteau, co., Mo., 10,500 C5 101
Monito, isl., P.R. e10 65
Monitor, Alta., Can., 86 D5 69
Monitor, Marion, Oreg., 100 B2 113
Monitor, Chelan, Wash., 400 B5 122
Monitor, range, Nev. E5 104
Monitor, val., Nev. E5 104
Monivea, Ire., 222 D3 11
Monkey River, Br. Hond., 407 B3 62
Monkira, Austl. B3 51
Monkman, pass, B.C., Can. B7 68
Monkoto, Con. L. B3 48
Monkton, Ont., Can., 392 D3 72
Monkton, Addison, Vt., 130 (551▲) C2 120
Monktonridge, Addison, Vt., 75 C2 120
MonLouis, Mobile, Ala., 300 E1 78
Monmouth, Warren, Ill., 10,372 C3 90
Monmouth, Jackson, Iowa, 271 B7 92
Monmouth, Kennebec, Maine, 500 (1,884▲) D2 96
Monmouth, Polk, Oreg., 2,229 C1, C3 113
Monmouth, Wales, 5,432 C5 12
Monmouth, co., Wales, 443,689 C5 12
Monmouth, co., N. J., 334,401 C4 106
Monmouth Beach, Monmouth, N.J., 1,363 C5 106
Monmouth Junction, Middlesex, N.J., 700 C3 106
Monnikendam, Neth., 3,800 B5 15
Monnomouth, mtn., B.C., Can. D6 68
Monnow, riv., Eng. C5 12
Mono, co., Calif., 2,213 D4 82
Mono, lake, Calif. D4 82
Mono, pt., Nic. E6 62
Monocacy, riv., Md. B3 85
Mono Lake, Mono, Calif., 150 C4 82
Monolith, Kern, Calif., 450 E4 82
Monomonac, lake, Mass. F4 105
Monomoy, isl., Mass. C7 97
Monomoy, pt., Mass. C7 97
Monon, White, Ind., 1,417 C4 91
Monona, Clayton, Iowa, 1,346 A6 92
Monona, Dane, Wis., 8,178 E4 118
Monona, co., Iowa, 13,916 B1 92
Monona, lake, Wis. E4 124
Monongah, Marion, W. Va., 1,321 A7, B4 123
Monongahela, Washington, Pa., 8,388 F2 114
Monongahela, riv., Pa., W. Va. B5 123
Monongalia, co., W. Va., 55,617 B4 123
Monopoli, It., 24,000 D6 21
Monor, Hung., 14,830 B4 22
Monóvar, Sp., 9,933 C5 20
Monoville, Smith, Tenn., 65 C8 117
Monowi, Boyd, Nebr., 40 B7 103
Monowice, Pol. g10 26
Monreale, It., 18,000 E4 21
Monroe, Monroe, Ark., 150 C4 81
Monroe, Fairfield, Conn., 1,000 (6,402▲) D4 84
Monroe, Walton, Ga., 6,826 C3 87
Monroe, Adams, Ind., 499 C8 91
Monroe, Tippecanoe, Ind., 30 D4 91
Monroe, Jasper, Iowa, 1,366 C4 92
Monroe, Ouachita, La., 52,219 B3 95
Monroe, Waldo, Maine, 180 (497▲) D3 96

Monroe, Monroe, Mich., 22,968...G7 98
Monroe, Platte, Nebr., 261...C8 103
Monroe, Grafton, N.H., 185 (421▲)...B2 105
Monroe, Sussex, N.J., 110...A3 106
Monroe, Orange, N.Y., 3,323...D2, D6 108
Monroe, Union, N.C., 10,882...C3 109
Monroe, Butler, Ohio, 1,475...*C3 111
Monroe, Le Flore, Okla., 135...C7 112
Monroe, Benton, Oreg., 374..C3 113
Monroe, Turner, S. Dak., 156...D8 116
Monroe, Overton, Tenn., 69.G8 117
Monroe, Sevier, Utah, 965...E3 119
Monroe, Amherst, Va., 800...D3 121
Monroe, Snohomish, Wash., 1,901...B4 122
Monroe, Greene, Wis., 8,050.F4 124
Monroe, co., Ala., 22,372....D2 78
Monroe, co., Ark., 17,327....C4 81
Monroe, co., Fla., 47,921....H5 86
Monroe, co., Ga., 10,495....D3 87
Monroe, co., Ill., 15,507.....E4 90
Monroe, co., Ind., 59,225....F4 91
Monroe, co., Iowa, 10,463...C5 92
Monroe, co., Ky., 11,799....D4 94
Monroe, co., Mich., 101,120.G7 98
Monroe, co., Miss., 33,953...B5 100
Monroe, co., Mo., 10,688....B5 101
Monroe, co., N.Y., 586,387...B3 108
Monroe, co., Ohio, 15,268...C6 111
Monroe, co., Pa., 39,567...D11 114
Monroe, co., Tenn., 23,316..D9 117
Monroe, co., W. Va., 11,584.D4 123
Monroe, co., Wis., 31,241...E3 124
Monroe Bridge, Franklin, Mass., 125...A2 97
Monroe Center, Ogle, Ill., 300...A5 90
Monroe City, Knox, Ind., 505...G3 91
Monroe City, Monroe and Marion, Mo., 2,337...B6 101
Monroeton, Bradford, Pa., 502...C9 114
Monroeville, Monroe, Ala., 3,632...D2 78
Monroeville, Allen, Ind., 1,294...C8 91
Monroeville, Salem, N.J., 200...D2 106
Monroeville, Huron, Ohio, 1,371...A5 111
Monroeville, Allegheny, Pa., 22,446...*B6 114
Monrovia, Los Angeles, Calif., 27,079...F3 82
Monrovia, Morgan, Ind., 450...E5 91
Monrovia, Atchison, Kans., 20...A7 93
Monrovia, Lib., 41,391...E2 45
Mons, Bel., 26,973...D3 15
Monsanto, St. Clair, Ill., 324.B8 101
Monschau, Ger., 2,500...D6 15
Monselice, It., 6,610...D7 18
Monserrate Central, P.R....B4 65
Monsey, Rockland, N.Y., 3,000...*D6 108
Möns Klint, isl., Den...D6 24
Monson, Piscataquis, Maine, 700 (852▲)...C3 96
Monson, Hampden, Mass., 2,413 (6,712▲)...B3 97
Monson Junction, Piscataquis, Maine...C3 96
Montabaur, Ger., 6,200...C2 17
Montagnana, It., 5,237...D7 18
Montagne Tremblante, park, Que., Can...C3 73
Montague, Siskiyou, Calif., 782...B2 82
Montague, P.E.I., Can., 1,126...C7 74
Montague, Franklin, Mass., 700 (7,836▲)...A2 97
Montague, Muskegon, Mich., 2,366...E4 98
Montague, Chouteau, Mont., 16...C6 102
Montague, Montague, Tex., 400...C4 118
Montague, co., Tex., 14,893.C4 118
Montague, isl., Alsk...D10 79
Montague, isl., Mex...A2 63
Montague City, Franklin, Mass., 600...A2 97
Montalbán, Sp., 2,200...B5 20
Montalegre, Port., 1,799...B2 20
Mont Alto, Franklin, Pa., 1,039...G6 114
Montalto, mtn., It...E5 20
Montalvo, Ventura, Calif., 2,028...*E4 82
Montana, Alsk., 39...f17 76
Montana, Johnson, Ark., 25..B2 81
Montana, state, U.S., 674,767...A6 76, 102
Montana-Vermala, Switz., 1,543...D3 19
Montánchez, Sp., 4,190...C2 20
Montandon, Northumberland, Pa., 350...E8 114
Montargis, Fr., 15,996...D5 14
Montataire, Fr., 9,630...E2 15
Montauban, Que., Can., 623.C5 73
Montauban, Fr., 30,242...E4 14
Montauban-les-Mines, Que., Can., 318...C5 73
Montauk, Dent, Mo., 60...D6 101
Montauk, Suffolk, N.Y., 900.D5 108
Montauk, pt., N.Y...E9 84
Monta Vista, Santa Clara, Calif., 2,000...*B5 82
Montbard, Fr., 6,386...D6 14
Montbéliard, Fr., 21,699...D7 14
Mont Belvieu, Chambers, Tex., 950...E5, F5 118
Montblanch, Sp., 4,598...B6 20
Montbrison, Fr., 9,051...E5 14
Montcalm, co., Que., Can., 18,766...C3 73
Montcalm, co., Mich., 35,795...E5 98
Montcalm, peak, Fr...F4 14
Mont Carmel, Que., Can., 895...B8 73
Montceau-les-Mines, Fr., 29,364 (*51,000)...D6 14
Mont Cenis, pass, Fr., It...D2 18
Mont Cenis, tunnel, Fr., It...D2 18
Montcerf, Que., Can., 216...C1 73
Montchanin [-les-Mines], Fr., 6,405...D6 14
Montclair (Monte Vista), San Bernardino, Calif., 13,546..F3 82
Montclair, Essex, N.J., 43,129...B4 106
Mont Clare, Montgomery, Pa., 1,124...A10 114
Montcoal, Raleigh, W. Va., 275...D3, D6 123
Montcornet, Fr., 1,439...E4 15
Mont-de-Marsan, Fr., 20,203...F3 14
Montdidier, Fr., 5,430...C5 14
Monteagle, Grundy and Marion, Tenn., 700...D8 117
Monteagudo, Arg...D2 56
Monte Alegre, Braz...C2 56
Monte Alegre, Braz., 3,911...C4 59
Monte Azul, Braz., 4,860...E2 57
Montebello, Los Angeles, Calif., 32,097...F2 82
Montebello, Que., Can., 1,486...D3 73
Montebello, P.R...B4 65
Montebello, Nelson, Va., 50..D3 121
Monte Bello, is., Austl...D2 50
Montebelluna, It., 5,348...D8 18
Monte Carlo, Monaco, 9,516 (part of Monaco)...*F7 14
Monte Carmelo, Braz., 10,016...E1 57
Monte Caseros, Arg., 11,409..A5 54
Montecatini Terme, It., 17,787...C3 21
Montecelio, It., 5,000...g9 21
Montecito, Santa Barbara, Calif., 5,000...*E4 82
Monte Coman, Arg...A3 54
Montecristi, Dom. Rep., 4,600...F8 64
Montecristi, Ec., 1,872...B1 58
Monte Cristo, Bol...B3 55
Montecristo, isl., It...C3 21
Montefrío, Sp., 5,137...D4 20
Monte Gargano, pen., It...D5 20
Montego Bay, Jam., 23,471..F5 64
Montegut, Terrebonne, La., 588...E5 95
Monteiro, Braz., 6,028...C3 57
Monteith, mtn., B.C., Can...B6 68
Montelavar, Port., 5,373...F9 20
Montélimar, Fr., 15,914...E6 14
Montell, Uvalde, Tex., 50...E2 118
Montellano, Sp., 9,334...D3 20
Montello, Elko, Nev., 150...B7 104
Montello, Marquette, Wis., 1,021...E4 124
Montemorelos, Mex., 11,525.B5 63
Montemor-o-Novo, Port., 5,047...C1 20
Monte Ne, Benton, Ark., 200.A1 81
Montenegro, Braz., 14,491...D2 56
Montenegro, reg., Yugo...D4 22
Montenegro (Crna Gora), rep., Yugo., 471,433...*D4 22
Monte Patria, Chile...A2 54
Monte Plata, Dom. Rep., 2,202...F9 64
Monte Porzio Catone, It., 3,203...h9 21
Montepuez, Moz...D6 48
Montepulciano, It., 17,552...C3 21
Monte Quemado, Arg., 2,512.E3 55
Montereau, Fr., 14,121...C5 14
Monterey, Butler, Ala., 25...D3 78
Monterey, Monterey, Calif., 22,618...C6, D3 82
Monterey, Pulaski, Ind., 278.B4 91
Monterey, Owen, Ky., 211...B5 94
Monterey, Berkshire, Mass., 100 (480▲)...B1 97
Monterey, Hamilton, Ohio, 1,500...*D2 111
Monterey, Putnam, Tenn., 2,069...C8 117
Monterey, Highland, Va., 270...C3 121
Monterey, co., Calif., 198,351...D3 82
Monterey, bay, Calif...D3 82
Monterey Park, Los Angeles, Calif., 37,821...F2 82
Montería, Col., 27,300...B2 60
Monteros, Arg., 7,745...E2 55
Monterotondo, It., 9,340...g9 21
Monterrey, Mex., 596,993 (*665,000)...B4 63
Montesano, Grays Harbor, Wash., 2,486...C2 122
Monte Sano, mts., Ala...A3 78
Monte Sant'Angelo, It., 21,601...D5 21
Monte Santo, Braz., 1,607...D3 57
Montes Claros, Braz., 40,545.E2 57
Monte Serena, Santa Clara, Calif., 1,506...*D3 82
Montevallo, Shelby, Ala., 2,755...B3 78
Montevarchi, It., 9,100...C3 21
Montevideo, Chippewa, Minn., 5,693...F3 99
Montevideo, Ur., 954,700 (*1,050,000)...E1 56
Montevideo, dept., Ur., 954,700...*E1 56
Monteview, Jefferson, Idaho, 10...F6 89
Monte Vista, Rio Grande, Colo., 3,385...D4 83
Monte Vista, Webster, Miss., 50...B4 100
Monte Vista, Pierce, Wash., 1,500...*B3 122
Montevue, Frederick, Md., 100...B3 85
Montezuma, Macon, Ga., 3,744...D2 87
Montezuma, Parke, Ind., 1,231...E3 91
Montezuma, Poweshiek, Iowa, 1,416...C5 92
Montezuma, Gray, Kans., 543...E3 93
Montezuma, Mercer, Ohio, 287...B3 111
Montezuma, Chester, Tenn., 100...B3 117
Montezuma, co., Colo., 14,024...D2 83
Montezuma, creek, Utah...F6 119
Montezuma Castle, nat. mon., Ariz...C4 80
Montfaucon, Switz., 524...B3 19
Montfort, Que., Can., 76...D3 73
Montfort, Grant, Wis., 538..F3 124
Montfort-sur-Meu, Fr., 2,031...C3 14
Montgomery, Montgomery, Ala., 134,393 (*155,200)..C3 78
Montgomery, Alta., Can., 5,077...D3 69
Montgomery, Chatham, Ga., 350...E5 87
Montgomery, Kane, Ill., 2,122...F1 90
Montgomery, Daviess, Ind., 446...G3 91
Montgomery, Grant, La., 866...C3 95
Montgomery, Hampden, Mass., 80 (333▲)...B2 97
Montgomery, Hillsdale, Mich., 362...G6 98
Montgomery, Le Sueur, Minn., 2,118...F5 99
Montgomery, Orange, N.Y., 1,312...D6 108
Montgomery, Hamilton, Ohio, 3,075...*C3 111
Montgomery, Pak., 75,000...B4 40
Montgomery, Lycoming, Pa., 2,150...D8 114
Montgomery, Franklin, Vt., 250 (876▲)...B3 120
Montgomery, Wales, 970...B4 12
Montgomery, Fayette and Kanawha, W. Va., 3,000.C3, C6 123
Montgomery, co., Ala., 169,210...C3 78
Montgomery, co., Ark., 5,370...C2 81
Montgomery, co., Ga., 6,284.D4 87
Montgomery, co., Ill., 31,244.D4 90
Montgomery, co., Ind., 32,089...D4 91
Montgomery, co., Iowa, 14,467...C2 92
Montgomery, co., Kans., 45,007...E8 93
Montgomery, co., Ky., 13,461...B6 94
Montgomery, co., Md., 340,928...B3 85
Montgomery, co., Miss., 13,320...B4 100
Montgomery, co., Mo., 11,097...C6 101
Montgomery, co., N.Y., 57,240...C6 108
Montgomery, co., N.C., 18,408...B3 109
Montgomery, co., Ohio, 527,080...C3 111
Montgomery, co., Pa., 516,682...F10 114
Montgomery, co., Tenn., 55,645...A4 117
Montgomery, co., Tex., 26,839...D5 118
Montgomery, co., Va., 32,923...D2 121
Montgomery, co., Wales, 44,228...B4 12
Montgomery, peak, Calif...D4 82
Montgomery Center, Franklin, Vt., 375...B3 120
Montgomery City, Montgomery, Mo., 1,918...C4 101
Montgomery Creek, Shasta, Calif., 150...B3 82
Monthermé, Fr., 2,995...E4 15
Monthey, Switz., 6,834...D2 19
Monthois, Fr., 444...E4 15
Monticello, Drew, Ark., 4,412...D4 81
Monticello, Jefferson, Fla., 2,490...B3 86
Monticello, Jasper, Ga., 1,931...C3 87
Monticello, Piatt, Ill., 3,219..C5 90
Monticello, White, Ind., 4,035...C4 91
Monticello, Jones, Iowa, 3,190...B6 92
Monticello, Wayne, Ky., 2,940...D5 94
Monticello, Aroostook, Maine, 625 (1,109▲)...B5 96
Monticello, Wright, Minn., 1,477...E5 99
Monticello, Lawrence, Miss., 1,432...D3 100
Monticello, Lewis, Mo., 159..A6 101
Monticello, Sierra, N. Mex., 150...D2 107
Monticello, Sullivan, N.Y., 5,222...D5 108
Monticello, San Juan, Utah, 1,845...F6 119
Monticello, Green, Wis., 789.F4 124
Mont Ida, Anderson, Kans., 50...D8 93
Montier, Shannon, Mo., 61..E6 101
Montier-en-Der, Fr., 1,808...F4 15
Montiers-sur-Saulx, Fr., 514..F5 15
Montigny-Lencoup, Fr., 468...F3 15
Montigny-lès-Metz, Fr., 22,388...E6 15
Montijo, Port., 13,306...C1, f10 20
Montijo, Sp., 14,961...C2 20
Montilla, Sp., 19,755...D3 20
Montivilliers, Fr., 8,427...C3 14
Mont Joli, Que., Can., 6,178...G19 67
Mont Laurier, Que., Can., 5,859...C2 73
Montluçon, Fr., 55,184...D5 14
Montmagny, Que., Can., 6,850...C7 73
Montmagny, co., Que., Can., 26,450...C7 73
Montmartre, Sask., Can., 482...G3 70
Montmédy, Fr., 2,061...E5 15
Montmélian, Fr., 1,583...D2 18
Montmirail, Fr., 3,010...F3 15
Montmorenci, Tippecanoe, Ind., 100...D4 91
Montmorenci, Aiken, S.C., 150...D4 115
Montmorency, Que., Can., 5,985...C6, C9 73
Montmorency, Fr., 16,369..g10 14
Montmorency, co., Mich., 4,424...C6 98
Montmorency No. 1, co., Que., Can., 20,734...B6 73
Montmorency No. 2, co., Que., Can., 4,974...*C7 73
Montmorency, riv., Que., Can...B6 73
Montmorillon, Fr., 4,766...D4 14
Montney, B.C., Can...A7 68
Monto, Austl., 1,795...B8 51
Montone, riv., It...B3 21
Montoro, Sp., 14,950...C3 20
Montour, Gem, Idaho, 75...F2 89
Montour, Tama, Iowa, 452...C5 92
Montour, co., Pa., 16,730...D8 114
Montour Falls, Schuyler, N.Y., 1,533...C4 108
Montoursville, Lycoming, Pa., 5,211...D8 114
Montoya, Quay, N. Mex., 70...B5 107
Montpelier, Bear Lake, Idaho, 3,146...G7 89
Montpelier, Blackford, Ind., 1,954...C7 91
Montpelier, Muscatine, Iowa, 105...C7 92
Montpelier, St. Helena, La., 197...D5 95
Montpelier, Clay, Miss., 230..B4 100
Montpelier, Stutsman, N. Dak., 97...C7 110
Montpelier, Williams, Ohio, 4,131...A3 111
Montpelier, Washington, Vt., 8,782...C3 120
Montpelier Station, Orange, Va., 150...C4 121
Montpellier, Que., Can., 318.D2 73
Montpellier, Fr., 118,864....F5 14
Montreal, Que., Can., 1,191,062 (*2,109,509)...D4, D9 73
Montreal, Iron, Wis., 1,361..B3 124
Montreal, lake, Sask., Can...C3 70
Montreal, riv., Sask., Can...C3 70
Montreal East, Que., Can., 5,884...*D9 73
Montreal Lake, Sask., Can., 69...C3 73
Montreal North, Que., Can., 48,433...D9 73
Montreal South, Que., Can., 5,319...*D9 73
Montreal West, Que., Can., 6,446...*D9 73
Montreat, Buncombe, N.C., 500...D4 109
Montreuil, Fr., 76,252...g10 14, F2 15
Montreuil-sur-Mer, Fr., 3,253...B4 14
Montreux, Switz., 12,222 (*18,478)...D2 19
Montricher, Switz., 543...C1 19
Mont Rolland, Que., Can., 1,457...D3 73
Montrose, Baldwin, Ala., 500...E2 78
Montrose, Ashley, Ark., 399.D4 81
Montrose, Los Angeles, Calif., 6,000...*E4 82
Montrose, Montrose, Colo., 5,044...C3 83
Montrose, Laurens, Ga., 236.D3 87
Montrose, Effingham, Ill., 320...D5 90
Montrose, Lee, Iowa, 632...D6 92
Montrose, Jewell, Kans., 105.C5 93
Montrose, Natchitoches, La..C2 95
Montrose, Genesee, Mich., 1,466...E7 98
Montrose, Jasper, Miss., 169..C4 100
Montrose, Henry, Mo., 526..C4 101
Montrose, Westchester, N.Y., 1,800...*D7 108
Montrose, Berks, Pa., 1,100...*F10 114
Montrose, Susquehanna, Pa., 2,363...C10 114
Montrose, Scot., 10,702...D6 13
Montrose, McCook, S. Dak., 430...D8 116
Montrose, co., Colo., 18,286..C2 83
Montrose Hill, Allegheny, Pa., 2,000...*E2 114
Montross, Westmoreland, Va., 394...C6 121
Mont Royal, Que., Can., 21,182...*D9 73
Montrouge, Fr., 45,260...g10 14
Mont-St. Martin, Fr., 7,016..C6 14
Mont-St. Michel, Fr., 268...C3 14
Montserrat, Johnson, Mo., 150...C4 101
Montserrat, Br. dep., N.A., 12,157...n15 64
Montserrat, isl., N.A...n15 64
Montserrat, peak, Sp...B6 20
Montserrat Island, ter., W.I. Fed., 12,157...n15 64
Mont Tremblant, Que., Can., 303...C3 73
Montvale, Bergen, N.J., 3,699...A4, C5 106
Montvale, Bedford, Va., 500...D3 121
Mont Vernon, Hillsboro, N.H., 150 (585▲)...E3 105
Montville, New London, Conn., 1,060 (7,759▲)...D8 84
Montville, Berkshire, Mass., 40...B1 97
Montville, Morris, N.J., 4,000...B4 106
Montzen, Bel., 2,499...D5 15
Monument, El Paso, Colo., 204...B6 83
Monument, Logan, Kans., 150...C2 93
Monument, Lea, N. Mex., 400...E6 107
Monument, Grant, Oreg., 214...C7 113
Monument, Centre, Pa., 200.D6 114
Monument, peak, Colo...B3 83
Monument, peak, Idaho...D2 89
Monument, peak, Oreg...C4 113
Monumental, buttes, Idaho...B3 89
Monument Beach, Barnstable, Mass., 600...C6 97
Monument Valley, ruins, Ariz...A5 80
Monywa, Bur., 26,172...D10 39
Monza, It., 84,445..D5 18, B2 21
Monze, N. Rh., 1,800...E4 48
Monzón, Peru, 514...C2 58
Monzón, Sp., 9,020...B6 20
Moodus, Middlesex, Conn., 1,103...D7 84
Moodus, lake, Conn...C7 84
Moody, York, Maine, 200...E2 96
Moody, Howell, Mo., 61...E6 101
Moody, McLennan, Tex., 1,074...D4 118
Moody, co., S. Dak., 8,810..C9 116
Moodyville, Pickett, Tenn...C8 117
Mooers, Clinton, N.Y., 543..A3 108
Mooers Forks, Clinton, N.Y., 300...A3 108
Moon, lake, Miss...A3 100
Moonachie, Bergen, N.J., 3,052...*B4 106
Moon Crest, Allegheny, Pa., 1,500...*E1 114
Moon Run, Allegheny, Pa., 650...B5 114
Moonshine, hill, Mass...A2 97
Moonta, Austl., 1,151...F6 50
Moora, Austl., 1,145...F2 50
Moorcroft, Crook, Wyo., 826.A8 125
Moore, Butte, Idaho, 358....F5 89
Moore, Fergus, Mont., 216..D7 102
Moore, Cleveland, Ohio, 1,783...B4 112
Moore, Spartanburg, S.C., 150...B4 115
Moore, Frio, Tex., 600...E3 118
Moore, Emery, Utah, 25....D4 119
Moore, Tucker, W. Va., 40..B3 123
Moore, co., N.C., 36,733....B4 109
Moore, co., Tenn., 3,454....B5 117
Moore, co., Tex., 14,773.....B2 118
Moore, res., N.H., Vt.B3 105, C5 102
Moorefield, Ont., Can., 253..D4 72
Moorefield, Nicholas, Ky., 200...B6 94
Moorefield, Frontier, Nebr., 55...D7 103
Moorefield, Hardy, W. Va., 1,434...B6 123
Moorefield, riv., W. Va...C5 123
Moore Haven, Glades, Fla., 790...F5 86
Mooreland, Henry, Ind., 477.E7 91
Mooreland, Woodward, Okla., 871...A2 112
Mooreland Heights, Knox, Tenn., 900...*D10 117
Moorepark, Man., Can., 150.D2 71
Mooresburg, Hawkins, Tenn., 200...C10 117
Moores Corner, Franklin, Mass., 100...B3 97
Moores Hill, Dearborn, Ind., 476...F7 91
Moores Mills, N.B., Can., 116...D2 74
Moorestown, Missaukee, Mich., 65...D5 98
Moorestown, Burlington, N.J., 12,497...D3 106
Mooresville, Morgan, Ind., 3,856...E5 91
Mooresville, Livingston, Mo., 117...B4 101
Mooresville, Iredell, N.C., 6,918...B2 109
Mooresville, Marshall, Tenn., 150...B5 117
Mooreton, Richland, N. Dak., 164...C9 110
Mooreville, Lee, Miss., 200..A5 100
Moorewood, Custer, Okla., 35...B2 112
Moorhead, Monona, Iowa, 313...C2 92
Moorhead, Clay, Minn., 22,934...D2 99
Moorhead, Sunflower, Miss., 1,754...B3 100
Moorhead, Powder River, Mont., 5...E11 102
Mooring, Lake, Tenn., 100..A2 117
Mooringsport, Caddo, La., 864...B2 95
Moor Lake Station, Ont., Can., 80...A7 72
Moorland, Webster, Iowa, 281...B3 92
Moorman, Muhlenberg, Ky., 250...C2 94
Moorpark, Ventura, Calif., 2,902...E4 82
Moors, The, moors, Scot...F4 13
Moosburg, Ger., 10,200...E6 17
Moose, Teton, Wyo., 15...B2 125
Moose, creek, Wyo...D2 125
Moose, hill, Mass...B5 97
Moose, isl., Man., Can...D3 71
Moose, lake, B.C., Can...C1 69
Moose, lake, Man., Can...B1 71
Moose, mtn., N.H...C2 105
Moose, mtn., N.Y...B6 108
Moose, pond, Maine...D3 96
Moose, riv., Ont., Can...F16 67
Moose, riv., Maine...C2 96
Moose, riv., N.H...B4 105
Moose, riv., N.Y...B5 108
Moose, riv., Vt...B5 120
Moose Creek, Ont., Can., 429...B10 72
Moose Creek, buttes, Idaho...C3 89
Moosehead, Piscataquis, Maine, 15...C3 96
Moosehead, lake, Maine...C3 96
Mooseheart, Kane, Ill., 1,100...B5, F1 90
Moosehorn, Man., Can., 208.D2 71
Moosehorn, Franklin, Maine.C2 96
Moose Jaw, Sask., Can., 33,206...C1, G3 70
Moosejaw, creek, Sask., Can...G3 70
Moose Lake, Man., Can., 283...C1 71
Moose Lake, Carlton, Minn., 1,514...D6 99
Moose Lake, res., Wis...B3 124
Mooseleuk, stream, Maine...B4 96
Mooselookmeguntic, lake, Maine...D2 96
Moose Mountain, creek, Sask., Can...H4 70
Moose Mountain, park, Sask., Can...G12 66
Moose Pass, Alsk., 136.C10, g17 79
Moose River, Somerset, Maine, 180 (205▲)...C2 96
Moosic, Lackawanna, Pa., 4,243...B9 114
Moosic, mts., Pa...A9 114
Moosilauke, mtn., N.H...B3 105
Moosomin, Sask., Can., 1,781...G5 70
Moosonee, Ont., Can., 975...E9 72
Moosup, Windham, Conn., 2,760...C9 84
Moosup Valley, Providence, R.I., 100...C9 84
Mopang, lakes, Maine...D5 96
Mopeia, Moz...A6 49
Mopti, Mali, 12,700...D4 45
Moquegua, Peru, 4,582...E3 58
Moquegua, dept., Peru, 41,072...E3 58
Mor, Hung., 11,482...B4 22
Mor, isl., Truk...52
Mora, Cam...C2 46
Mora, Atkinson, Ga., 150...E4 87
Mora, Kanabec, Minn., 2,329...E5 99
Mora, Mora, N. Mex., 400..B4 107
Mora, P.R...B2 65
Mora, Sp., 10,657...C4 20
Mora, Swe., 6,032 (12,893▲).G6 25
Mora, co., N. Mex., 6,028...A5 107
Mora, riv., N. Mex...B5 107
Morača, riv., Yugo...D4 22
Morada, San Joaquin, Calif., 2,156...*D3 82
Moradabad, India, 180,000 (*191,828)...C6 39, C7 40
Morada Nova, Braz., 2,475..C3 57
Mora de Ebro, Sp., 3,340...B6 20
Morafenobé, Malag...g8 49
Morag, Pol., 2,746...B5 26
Moraine, Montgomery, Ohio, 2,262...*C3 111
Morales, Guat., 3,587...C3 62
Morales, Mex., 3,338...k13 63
Morales, Jackson, Tex., 40...E4 118
Moramanga, Malag., 3,750..g9 49
Moran, Clinton, Ind., 60...D4 91
Moran, Allen, Kans., 549...E8 93
Moran, Mackinac, Mich., 180...B6 98
Moran, Shackelford, Tex., 392...C3 118
Moran, Teton, Wyo., 10...B2 125
Morann, Clearfield, Pa., 450.E5 114
Morant Bay, Jam., 5,053...G5 64
Morar, lake, Scot...D3 13
Morat, lake, Switz...C3 18
Morata de Tajuña, Sp., 3,801...p18 20
Moratalla, Sp., 5,879...C5 20
Morattico, Lancaster, Va., 250...D6 121
Morava, riv., Czech...D4 26
Morava, riv., Yugo...C5 22
Moravia (Morava), reg., Czech., 3,549,700...D4 26
Moravia, Appanoose, Iowa, 621...D5 92
Moravia, Cayuga, N.Y., 1,575...C4 108
Morawhanna, Br. Gu., 305..A3 59
Moray, co., Scot., 49,156...C5 13
Moraya, Bol...D2 55
Moray, firth, Scot...C5 13
Morbach, Ger., 2,400...D2 17
Morbihan, dept., Fr., 530,833...*D2 14
Morcenx, Fr., 2,370...E3 14
Morco, Anderson, Tenn., 150...C9 117
Morden, Man., Can., 2,793..E2 71
Morden, N.S., Can., 126...D5 74
More, mtn., Scot...C1 13
More, mtn., Scot...D2 13
More, mtn., Scot...D4 13
More Assynt, mtn., Scot...B4 13
Moreau, riv., S. Dak...B3 116
Moreauville, Avoyelles, La., 815...C4 95
Morecambe, Eng., 40,950...F6 13
Morecambe, bay, Eng...C5 10
Moree, Austl., 6,795...D7 51
Morehead, Rowan, Ky., 4,170...B6 94
Morehead City, Carteret, N.C., 5,583...C7 109
Morehouse, New Madrid, Mo., 1,417...E8 101
Morehouse, par., La., 33,709.B4 95
Mörel, Switz., 456...D5 19
Moreland, Pope, Ark., 55...B3 81
Moreland, Coweta, Ga., 329.C2 87
Moreland, Bingham, Idaho, 250...F6 89
Moreland, Lincoln, Ky., 300.C5 94
Moreland, Lycoming, Pa...D8 114
Moreland Hills, Cuyahoga, Ohio, 2,188...*A6 111
Morelia, Mex., 100,258.D4, n13 63
Morell, P.E.I., Can., 387...C7 74
Morelos, Mex., 5,037...B5 63
Morelos, state, Mex., 381,346...D5, n14 63
Morelos, dam, Ariz...E1 80
Morena, mts., Sp...C3 20
Morenci, Greenlee, Ariz., 2,431...D6 80
Morenci, Lenawee, Mich., 2,053...G6 98
Moreno, pt., Fla...G2 86
Möre og Romsdal, co., Nor., 213,400...*F2 25
Moresby, isl., B.C., Can...C2 68
Moreton, bay, Austl...C9 51
Moreton, isl., Austl...C9 51
Moreton-in-Marsh, Eng., 1,798...C6 12
Moretown, Washington, Vt., 150 (788▲)...C3 120
Moreuil, Fr., 3,609...E2 15
Morewood, Ont., Can., 189..B9 72
Morez, Fr., 5,777...D7 14
Morgan, Austl., 434...G2 51
Morgan, Calhoun, Ga., 293..E2 87
Morgan, Pendleton, Ky., 50..B5 94
Morgan, Redwood, Minn., 975...F4 99
Morgan, Phillips, Mont., 4...B9 102
Morgan, Morrow, Oreg...B7 113
Morgan, Bosque, Texas, 381.C4 118
Morgan, Morgan, Utah, 1,299...B4 119
Morgan, Orleans, Vt., 45 (260▲)...B4 120
Morgan, co., Ala., 60,454...A3 78
Morgan, co., Colo., 21,192...A7 83
Morgan, co., Ga., 10,280...C3 87
Morgan, co., Ill., 36,571...D3 90
Morgan, co., Ind., 33,875...F5 91
Morgan, co., Ky., 11,056...C6 94
Morgan, co., Mo., 9,476...C5 101
Morgan, co., Ohio, 12,747...C6 111
Morgan, co., Tenn., 14,304..C9 117
Morgan, co., Utah, 2,837...B4 119
Morgan, co., W. Va., 8,376..B6 123
Morgan, isl., S.C...G6 115
Morgan, pt., Conn...E5 84
Morgana, Edgefield, S.C., 122...D3 115
Morgan Center, Orleans, Vt., 80...B5 120
Morgan City, St. Mary, La., 13,540...C5, E4 95
Morgan City, Leflore, Miss., 250...B3 100
Morganfield, Union, Ky., 3,741...C2 94
Morgan Hill, Santa Clara, Calif., 3,151...C6, D3 82
Morganton, Van Buren, Ark., 75...B3 81
Morganton, Fannin, Ga., 211.B2 87
Morganton, Burke, N.C., 9,186...B2 109

Morgantown, Morgan, Ind., 971....F5 91
Morgantown, Butler, Ky., 1,318....C3 94
Morgantown, Marion, Miss., 310....D4 100
Morgantown, Berks, Pa., 550....F10 114
Morgantown, Monongalia, W. Va., 22,487....A7, B5 123
Morganville, Dade, Ga., 150....B1 87
Morganville, Clay, Kans., 226....C6 93
Morganville, Monmouth, N.J., 400....C4 106
Morganza, Pointe Coupee, La., 937....D4 95
Morges, Switz., 8,420....C1 19
Morgex, It., 837....D3 18
Morghāb, Iran....F6 41
Morhange, Fr., 4,786....F6 15
Mori, Jap., 12,500....E10 37
Moriah, Essex, N.Y., 250....A7 108
Moriah, mtn., Nev....D7 104
Moriah, mtn., N.H....B4 105
Moriarty, Torrance, N. Mex., 720....C3 107
Morice, lake, B.C., Can....B4 68
Morice, riv., B.C., Can....B4 68
Moriches, bay, N.Y....F5 84
Moriguchi, Jap., 102,295....*o14 37
Morin Heights, Que., Can., 133....D3 73
Morinville, Alta., Can., 935....C4 69
Morioka, Jap., 157,441....G10 37
Morisset Station, Que., Can., 139....C7 73
Morjärv, Swe., 655....D10 25
Morkill, riv., B.C., Can....C7 68
Mörkö, Swe....t35 25
Morlaix, Fr., 18,866....C2 14
Morland, Graham, Kans., 317....C3 93
Morley, Alta., Can., 100....D3 69
Morley, Mecosta, Mich., 445....E5 98
Morley, Scott, Mo., 472....D8 101
Morley, St. Lawrence, N.Y., 300....A2 108
Morman, range, Nev....G7 104
Mormon Lake, Coconino, Ariz....C4 80
Morningside, Prince Georges, Md., 1,708....*C4 85
Morningside, Hennepin, Minn., 1,981....*F5 99
Morningside, Beadle, S. Dak., 150....C7 116
Morning Sun, Louisa, Iowa, 875....C6 92
Mornington, isl., Chile....D1 54
Morning View, Kenton, Ky., 150....A7 94
Moro, Lee, Ark., 340....C5 81
Moro, Madison, Ill., 250....A8 101
Moro, Sherman, Oreg., 327....B6 113
Moro, creek, Ark....D3 81
Moro, gulf, Phil....D6 35
Morobay, Bradley, Ark., 40....D3 81
Morobe, N. Gui....k12 50
Morocco, Newton, Ind., 1,341....C3 91
Morocco, country, Afr., 11,626,470....C5 42, C3 44
Morococha, Peru, 1,522....D2 58
Morogoro, Tan., 14,507....C6 48
Moroleón, Mex., 17,955....m13 63
Morombe, Malag....h8 49
Morón, Cuba, 18,629....D4 64
Morona, riv., Peru....B2 58
Morondava, Malag., 5,300....h8 49
Morón de la Frontera, Sp., 25,800 (35,248▲)....D3 20
Moroni, Sanpete, Utah, 879....D4 119
Morong, Phil., 2,115....o13 35
Morotai, isl., Indon....E7 35
Moroto, Ug....A5 48
Morovis, P.R., 2,428....B5 65
Morovis, mun., P.R., 18,094....B5 65
Morozovsk, Sov. Un., 28,000....D2 29
Morpeth, Ont., Can., 211....E3 73
Morpeth, Eng., 12,430....E7 13
Morphou, Cyp., 6,097....E9 31
Morral, Marion, Ohio, 493....B4 111
Morrill, Brown, Kans., 299....C8 93
Morrill, Scotts Bluff, Nebr., 884....C2 103
Morrill, co., Nebr., 7,057....C2 103
Morrilton, Conway, Ark., 5,997....B3 81
Morrin, Alta., Can., 316....D4 69
Morrinhos, Braz., 9,879....E1 57
Morrinsville, N.Z., 4,111....L15 51
Morris, Jefferson, Ala., 638....B3 78
Morris, Man., Can., 1,370....E3 71
Morris, Litchfield, Conn., 150 (1,190▲)....C4 84
Morris, Quitman, Ga., 150....E2 87
Morris, Grundy, Ill., 7,935....B5 90
Morris, Ripley, Ind., 400....F7 91
Morris, Stevens, Minn., 4,199....E3 99
Morris, Otsego, N.Y., 677....C5 108
Morris, Okmulgee, Okla., 982....B6 112
Morris, co., Kans., 7,392....D7 93
Morris, co., N.J., 261,620....B3 106
Morris, co., Tex., 12,576....C5 118
Morris, isl., S.C....F8 115
Morris, mtn., N.Y....A6 108
Morrisburg, Ont., Can., 1,820....C9 72
Morris Chapel, Hardin, Tenn., 100....B3 117
Morrisdale, Clearfield, Pa., 800....E5 114
Morris Jesup, cape, Grnld....A18 4
Morrison, Jefferson, Colo., 426....B5 83
Morrison, Whiteside, Ill., 4,159....B4 90
Morrison, Gasconade, Mo., 232....C6 101
Morrison, Noble, Okla., 256....A6 112
Morrison, Warren, Tenn., 294....D8 117
Morrison, Brown, Wis., 125....A6 124
Morrison, co., Minn., 26,641....D4 99
Morrison, hill, Conn....A5 84
Morrison City, Sullivan, Tenn., 2,426....C11 117
Morrisonville, Christian, Ill., 1,129....D4 90
Morris Plains, Morris, N.J., 4,703....B4 106
Morris Run, Tioga, Pa., 400....C7 114
Morriston, Levy, Fla., 100....C4 86
Morristown, Maricopa, Ariz., 200....D3, F1 80
Morristown, Shelby, Ind., 709....E6 91
Morristown, Rice, Minn., 616....F5 99
Morristown, Morris, N.J., 17,712....B4 106
Morristown, St. Lawrence, N.Y., 541....B1 108
Morristown, Corson, S. Dak., 219....B4 116
Morristown, Hamblen, Tenn., 21,267....C10 117
Morristown, Lamoille, Vt., 65 (3,347▲)....B3 120
Morristown, nat. historical park, N.J....B3 106
Morrisville, Polk, Mo., 228....D4 101
Morrisville, Madison, N.Y., 1,304....C5 108
Morrisville, Bucks, Pa., 7,790....F12 114
Morrisville, Lamoille, Vt., 2,047....B3 120
Morrisville, Fauquier, Va., 100....C5 121
Morrito, Nic., 387....E5 62
Morro, Ec....B1 58
Morro Bay, San Luis Obispo, Calif., 3,692....E3 82
Morro do Chapéu, Braz., 2,039....D2 57
Morropón, Peru, 3,909....C2 58
Morrosquillo, gulf, Col....B2 59
Morrow, Clayton, Ga., 580....B5, C2 87
Morrow, St. Landry, La., 400....D3 95
Morrow, Warren, Ohio, 1,477....C3 111
Morrow, co., Ohio, 19,405....B5 111
Morrow, co., Oreg., 4,871....B7 113
Morrowville, Washington, Kans., 195....C6 93
Morse, Sask., Can., 458....G2 70
Morse, Acadia, La., 682....D3 95
Morse, Hansford, Tex., 150....A2 118
Morse, Ashland, Wis., 45....B3 124
Morse, res., Ind....D5 91
Morse Bluff, Saunders, Nebr., 119....C9 103
Morse Mill, Jefferson, Mo., 62....B7 101
Morses, creek, N.J....E4 106
Morshansk, Sov. Un., 48,000....C2 29
Mortagne [-au-Perche], Fr., 3,909....C4 14
Mortara, It., 10,500....B2 21
Morteau, Fr., 5,395....D7 14
Morteros, Arg., 5,593....A4 54
Mortlach, Sask., Can., 329....G2 70
Morton, Tazewell, Ill., 5,325....C4 90
Morton, Renville, Minn., 624....F4 99
Morton, Scott, Miss., 2,260....C4 100
Morton, Delaware, Pa., 2,207....*G11 114
Morton, Cochran, Tex., 2,731....C1 118
Morton, Lewis, Wash., 1,183....C3 122
Morton, Freemont, Wyo., 15....B4 125
Morton, co., Kans., 3,354....E2 93
Morton, co., N. Dak., 20,992....C4 110
Morton Grove, Cook, Ill., 20,533....E3 90
Mortons Gap, Hopkins, Ky., 1,308....C2 94
Morup, Swe....B6 24
Moruya, Austl., 1,181....G8 51
Morvan, mts., Fr....D6 14
Morven, Brooks, Ga., 476....F3 87
Morven, Anson, N.C., 518....C3 109
Morven, mtn., Scot....B5 13
Morvern, Scot., 457....D3 13
Morvi, India, 50,192....F3 40
Moryakovskiy Zaton, Sov. Un....B10 29
Morye, Sov. Un....r32 25
Mosalsk, Sov. Un., 5,700....D10 27
Mosbach, Ger., 11,300....D4 17
Mosby, Clay, Mo., 293....D2 101
Mosby, Garfield, Mont., 5....D9 102
Mosca, Alamosa, Colo., 150....D5 83
Moscarello, riv., It....k9 21
Moscos, is., Bur....E2 38
Moscow, Jefferson, Ark., 100....C4 81
Moscow, Latah, Idaho, 11,183....C2 89
Moscow, Rush, Ind., 180....F6 91
Moscow, Muscatine, Iowa, 208....C6 92
Moscow, Stevens, Kans., 211....E2 93
Moscow, Clermont, Ohio, 438....D3 111
Moscow, Lackawanna, Pa., 1,212....B9 114
Moscow (Moskva), Sov. Un., 6,208,000 (★8,075,000)....D11, n17 27
Moscow, Fayette, Tenn., 368....B2 117
Moscow, Lamoille, Vt., 100....C3 120
Moscow Mills, Lincoln, Mo., 360....C7 101
Mosel, riv., Ger....C2 17
Moseley, Powhatan, Va., 100....D2 121
Moselle, Jones, Miss., 500....D4 100
Moselle, dept., Fr., 919,412....E6 15
Moselle, riv., Fr....C7 14
Mosers River, N.S., Can., 382....E7 74
Moses, Union, N. Mex....A6 107
Moses, lake, Wash....B6 122
Moses Coulee, canyon, Wash....B6 122
Moses Lake, Grant, Wash., 11,299....B6 122
Mosgiel, N.Z., 6,456....P13 51
Moshelm, Greene, Tenn., 300....C11 117
Mosher, Mellette, S. Dak., 25....D5 116
Mosherville, N.S., Can., 87....D5 74
Moshi, Tan., 13,726....B6 48
Moshupa, Bech....B4 49
Mosier, Wasco, Oreg., 252....B5 113
Mosinee, Marathon, Wis., 2,067....D4 124
Mosjöen, Nor., 4,600....E5 25
Moskee, Crook, Wyo., 15....A8 125
Moskva, see Moscow, Sov. Un.
Moskva, riv., Sov. Un....n17 27
Mosonmagyarovar, Hung., 21,199....B3 22
Mosquera, Col., 318....C2 60
Mosquero, Harding, N. Mex., 310....B6 107
Mosquito, Newf., Can., 67....E5 75
Mosquito, creek, Iowa....D5 92
Mosquito, lagoon, Fla....D6 86
Mosquito Coast, reg., Hond., Nic....D6 62
Mosquito Creek, res., Ohio....A7 111
Mosquitos, gulf, Pan....F7 62
Moss, Jasper, Miss., 125....D4 100
Moss, Nor., 20,600....H4, p28 25
Moss, Clay, Tenn., 200....C8 117
Moss, mtn., Ark....C3 81
Mossaka, Con. B....F3 46
Mossbank, Sask., Can., 568....H3 70
Moss Bluff, Liberty, Tex....F5 118
Mossel Bay, S. Afr., 12,178....D3 49
Mossendjo, Con. B....F2 46
Moss Glen, falls and chasm, Vt....C3 120
Moss Landing, Monterey, Calif., 400....C6 82
Mossleigh, Alta., Can., 4,646....D4 69
Mossoro, Braz., 38,833....C3 56
Moss Point, Jackson, Miss., 6,631....E3, E5 100
Moss Vale, Austl., 3,040....G8 51
Mossville, Newton, Ark., 40....B2 81
Mossville, Calcasieu, La., 1,500....*D2 95
Mossy, riv., Man., Can....D2 71
Mossy, river, Sask., Can....C4 70
Mossy Head, Walton, Fla., 150....G2 86
Mossyrock, Lewis, Wash., 344....C3 122
Most, Czech., 44,500....C2 24
Mostaganem, Alg., 60,186....B5 44
Mostar, Yugo., 35,242....D3 22
Móstoles, Sp., 2,886....p17 20
Mosul, Iraq, 179,646....H17 9, D14 31
Motagua, riv., Guat....C3 62
Motala, Swe., 27,100....H6 25
Motatán, Ven., 4,358....B3 60
Mothe, isl., Fiji Is.... 52
Motherwell [& Wishaw], Scot., 72,799....E5 13
Motihari, India, 32,620....D10 40
Motilla del Palancar, Sp., 4,398....C5 20
Motley, Morrison, Minn., 430....D4 99
Motley, co., Tex., 2,870....B2 118
Moto, see Oshima, Jap.
Moto, mtn., Iwo.... 52
Motil, Sp., 19,185....D4 20
Motrul, riv., Rom....C6 22
Motsa, Isr....h11 32
Mott, Hettinger, N. Dak., 1,463....C3 110
Motta di Livenza, It., 9,255....D8 19
Motueka, N.Z., 3,310....N14 51
Motuhora, N.Z., 91....M16 51
Motul [de Felipe Carrillo Puerto], Mex., 9,965....C7 63
Motupe, Peru, 4,396....C2 58
Moturiki, isl., Fiji Is.... 52
Mou, Den., 677....B4 24
Mouchoir, passage, Ba. Is....E8 64
Moudhros (Mudros), Grc., 1,720....C5 23
Moudjéria, Maur....C2 45
Moudon, Switz., 2,806....C2 19
Moule, see Le Moule, Guad.
Moulins, Fr., 23,909....D5 14
Moulki, Grc., 1,447....g10 23
Moulmein, Bur., 102,777....D2 38, E10 39
Moulouya, riv., Mor....C4 44
Moulton, Lawrence, Ala., 1,716....A2 78
Moulton, Appanoose, Iowa, 773....D5 92
Moulton, valley, Ala....A2 78
Moultonboro, Carroll, N.H., 150 (840▲)....C4 105
Moultonville, Carroll, N.H., 215....C4 105
Moultrie, Colquitt, Ga., 15,764....E3 87
Moultrie, co., Ill., 13,635....D5 90
Moultrie, lake, S.C....E7 115
Mound, Hennepin, Minn., 5,440....F5 99
Mound Bayou, Bolivar, Miss., 1,354....B3 100
Mound City, Pulaski, Ill., 1,669....F4 90
Mound City, Linn, Kans., 661....D9 93
Mound City, Holt, Mo., 1,249....A2 101
Mound City, Campbell, S. Dak., 144....B5 116
Mound City Group, nat. mon., Ohio....C4 111
Moundou, Chad, 21,866....D3 46
Moundridge, McPherson, Kans., 1,214....D6 93
Mounds, Crittenden, Ark....E8 81
Mounds, Pulaski, Ill., 1,835....F4 90
Mounds, Creek, Okla., 674....B5 112
Mounds View, Ramsey, Minn., 6,416....*F5 99
Moundsville, Marshall, W. Va., 15,163....B2, B3 123
Mound Valley, Labette, Kans., 481....E8 93
Moundville, Hale, Ala., 922....C2 78
Moundville, Vernon, Mo., 136....D3 101
Mounier, mtn., Fr....E7 14
Mountain, Pembina, N. Dak., 218....A8 110
Mountain, Ritchie, W. Va., 200....B4 123
Mountain, Oconto, Wis., 400....C5 124
Mountain, prov., Phil., 436,200....*B6 35
Mountain, lake, Sask., Can....B3 70
Mountainair, Torrance, N. Mex., 1,605....C3 107
Mountain Ash, Whitley, Ky., 275....D5 94
Mountain Brook, Jefferson, Ala., 12,680....E5 78
Mountainburg, Crawford, Ark., 402....B1 81
Mountain City, Rabun, Ga., 550....B3 87
Mountain City, Elko, Nev., 75....B6 104
Mountain City, Johnson, Tenn., 1,379....C12 117
Mountain Creek, Chilton, Ala., 300....C3 78
Mountain Dale, Sullivan, N.Y., 900....D6 108
Mountain Dale, Unicoi, Tenn....C11 117
Mountain Fork, riv., Okla....C7 112
Mountain Grove, Ont., Can., 135....C8 72
Mountain Grove, Wright, Mo., 3,176....D5 101
Mountain Grove, Bath, Va., 25....C3 121
Mountain Home, Baxter, Ark., 2,105....A3 81
Mountain Home, Elmore, Idaho, 10,075....F3 89
Mountainhome, Monroe, Pa., 500....D11 114
Mountain Home, Kerr, Tex., 25....D3 118
Mountain Home or Needle, mts., Utah....E2 119
Mountain Iron, St. Louis, Minn., 1,808....C6 99
Mountain Lake, Cottonwood, Minn., 1,943....G4 99
Mountain Lake Park, Garrett, Md., 975....D1 85
Mountain Lakes, Morris, N.J., 4,037....B4 106
Mountain Nile, see Bahrel Je'bel, riv., Sud.
Mountain Park, Alta., Can., 415....C2 69
Mountain Park, Otero, N. Mex., (part of High Rolls-Mountain Park)....E4 107
Mountain Park, Kiowa, Okla., 403....C3 112
Mountain Pine, Garland, Ark., 1,279....C2, C5 81
Mountain Province, see Mountain, prov., Phil.
Mountainside, Union, N.J., 6,325....B4 106
Mountain Valley, Garland, Ark., 150....C2 81
Mountain View, Stone, Ark., 983....B3 81
Mountain View, Santa Clara, Calif., 30,889....B5 82
Mountain View, Alta., Can., 84....E4 69
Mountain View, Clayton, Ga., 1,500....*C2 87
Mountainview, Hawaii, Haw., 747....D6 88
Mountain View, Ada, Idaho, 4,898....*F2 89
Mountain View, Howell, Mo., 936....D6 101
Mountain View, Kiowa, Okla., 864....B3 112
Mountain View, Asotin, Wash....C8 122
Mountain View, Natrona, Wyo., 1,721....*C6 125
Mountain View, Uinta, Wyo., 400....D2 125
Mountain Village, Alsk., 300....C7 79
Mount Airy, Habersham, Ga., 417....B3 87
Mount Airy, Carroll and Frederick, Md., 1,352....B3 85
Mount Airy, Hunterdon, N.J., 60....C3 106
Mount Airy, Surry, N.C., 7,055....A3 109
Mount Airy, Sequatchie, Tenn., 25....D8 117
Mount Airy, Pittsylvania, Va., 45....E3 121
Mount Albert, Ont., Can., 561....C5 72
Mount Andrew, Barbour, Ala., 50....D4 78
Mount Angel, Marion, Oreg., 1,428....B2, B4 113
Mount Arlington, Morris, N.J., 1,246....B3 106
Mount Athos (Ayion Oros), prov., Grc., 2,687....*B5 23
Mount Auburn, Christian, Ill., 502....D4 90
Mount Auburn, Benton, Iowa, 186....B5 92
Mount Ayr, Newton, Ind., 186....C3 91
Mount Ayr, Ringgold, Iowa, 1,738....D3 92
Mount Baldy, San Bernardino, Calif., 225....F3 82
Mount Barker, Austl., 1,872....G2 51
Mount Bellew, Ire., 306....D3 11
Mount Berry, Floyd, Ga., 1,000....B1 87
Mount Bethel, Somerset, N.J., 150....B3 106
Mount Blanchard, Hancock, Ohio, 432....B4 111
Mount Blanc, tunnel, Fr., It....D3 18
Mount Brydges, Ont., Can., 1,016....E3 72
Mount Calvary, Fond du Lac, Wis., 650....B5, E5 124
Mount Carmel, Montgomery, Ala., 140....C3 78
Mount Carmel, Newf., Can., 673....E5 75
Mount Carmel, Santa Rosa, Fla., 150....G2 86
Mount Carmel, Wabash, Ill., 8,594....E6 90
Mount Carmel, Franklin, Ind., 142....F8 91
Mount Carmel, Cavalier, N. Dak., 50....A7 110
Mount Carmel, Clermont, Ohio, 400....D2 111
Mount Carmel, Northumberland, Pa., 10,760....E9 114
Mount Carmel, McCormick, S.C., 109....C3 115
Mount Carmel, Kane, Utah, 110....F3 119
Mount Carmel Junction, Kane, Utah, 9....F3 119
Mount Carroll, Carroll, Ill., 2,056....A4 90
Mount Clare, Harrison, W. Va., 900....B4 123
Mount Clemens, Macomb, Mich., 21,016....A8, F8 98
Mount Clinton, Rockingham, Va., 170....C4 121
Mount Cory, Hancock, Ohio, 301....B4 111
Mount Croghan, Chesterfield, S.C., 145....B7 115
Mount Darwin, S. Rh., 485....A5 49
Mount Desert, isl., Maine....D4 96
Mount Dora, Lake, Fla., 3,756....D5 86
Mount Dora, Union, N. Mex., 55....A6 107
Mount Eaton, Wayne, Ohio, 265....B6 111
Mount Eden, Spencer, Ky., 350....B4 94
Mount Edgecumbe, Alsk., 1,884....*m22 79
Mount Elgin, Ont., Can., 225....E4 72
Mount Enterprise, Rusk, Tex., 400....D5 118
Mount Ephraim, Camden, N.J., 5,447....*D2 106
Mount Erie, Wayne, Ill., 134....E5 90
Mount Etna, Huntington, Ind., 192....C6 91
Mount Etna, Adams, Iowa, 100....C3 92
Mount Forest, Ont., Can., 2,623....D4 72
Mountforest, Bay, Mich., 35....E6 98
Mount Gambier, Austl., 15,388....H3 51
Mount Gay, Logan, W. Va., 3,386....*D3 123
Mount Gilead, Montgomery, N.C., 1,229....B4 109
Mount Gilead, Morrow, Ohio, 2,788....B5 111
Mount Hamilton, Santa Clara, Calif., 75....C6 82
Mount Healthy, Hamilton, Ohio, 6,553....D2 111
Mount Hebron, Greene, Ala., 40....C1 78
Mount Hebron, Siskiyou, Calif., 150....B3 82
Mount Hermon, Washington, La., 90....D5 95
Mount Hermon, Franklin, Mass., 600....A3 97
Mount Heron, Buchanan, Va., 100....B3 121
Mount Holly, Union, Ark., 300....D3 81
Mount Holly, Burlington, N.J., 13,271....C4 106
Mount Holly, Gaston, N.C., 4,037....B2 109
Mount Holly, Berkeley, S.C., 150....E2, E7 115
Mount Holly, Rutland, Vt., 70 (517▲)....E3 120
Mount Holly Springs, Cumberland, Pa., 1,840....F7 114
Mount Hood, Hood River, Oreg., 50....B5 113
Mount Hope, Lawrence, Ala., 100....A2 78
Mount Hope, Austl., 129....F5 51
Mount Hope, Ont., Can., 807....D5 72
Mount Hope, Tolland, Conn., 60....B8 84
Mount Hope, Sedgwick, Kans., 539....B4, E6 93
Mount Hope, Morris, N.J., 500....B3 106
Mount Hope, Spokane, Wash....D8 122
Mount Hope, Fayette, W. Va., 2,000....D3, D7 123
Mount Hope, Grant, Wis., 218....F3 124
Mount Hope, riv., Conn....B8 84
Mount Horeb, Dane, Wis., 1,991....F4 124
Mount Houston, Harris, Tex., 2,500....*F5 118
Mount Ida, Montgomery, Ark., 805....C2 81
Mount Idaho, Idaho, Idaho, 90....D2 89
Mount Isa, Austl., 13,358....D6 50
Mount Jackson, Shenandoah, Va., 722....C4 121
Mount Jewett, McKean, Pa., 1,226....C4 114
Mount Joy, Scott, Iowa, 35....D7 92
Mount Joy, Lancaster, Pa., 3,292....F9 114
Mount Judea, Newton, Ark., 60....B2 81
Mount Juliet, Wilson, Tenn., 750....A5 117
Mount Kisco, Westchester, N.Y., 6,805....D3, D7 108
Mountlake Terrace, Snohomish, Wash., 9,122....B3 122
Mount Laurel, Burlington, N.J., 200....D3 106
Mount Lebanon, Allegheny, Pa., 35,361....F1 114
Mount Lookout, Nicholas, W. Va., 250....D7 123
Mount McGregor, Saratoga, N.Y....B7 108
Mount McKinley, nat. park, Alsk....C9 79
Mount Magnet, Austl., 631....E2 50
Mount Meigs, Montgomery, Ala., 85....C3 78
Mountmellick, Ire., 2,436....D4 11
Mount Misery, pt., N.Y....F4 84
Mount Montgomery, Mineral, Nev., 10....F3 104
Mount Morgan, Austl., 4,000....A8 51
Mount Moriah, Harrison, Mo., 225....A4 101
Mount Morris, Ogle, Ill., 3,075....A4 90
Mount Morris, Genesee, Mich., 3,484....E7 98
Mount Morris, Livingston, N.Y., 3,250....C3 108
Mount Morris, Greene, Pa., 700....G1 114
Mount Olive, Jefferson, Ala., 1,800....B3, E4 78
Mount Olive, Izard, Ark., 50....B3 81
Mount Olive, Macoupin, Ill., 2,295....D4 90
Mount Olive, Covington, Miss., 841....D4 100
Mount Olive, Wayne, N.C., 4,673....B5 109
Mount Olive, Knox, Tenn., 500....E11 117
Mount Oliver, Allegheny, Pa., 5,980....F1 114
Mount Olivet, Robertson, Ky., 386....B5 94
Mount Orab, Brown, Ohio, 1,058....C4 111
Mount Pearl Park-Glendale, Newf., Can., 1,979....*E5 75
Mount Penn, Berks, Pa., 3,574....*F10 114
Mount Perry, Marion, Ind....I8 91
Mount Pisgah, Augusta, Va., 100....C4 121
Mount Pleasant, Izard, Ark., 250....B4 81
Mount Pleasant, New Castle, Del., 65....A6 85
Mount Pleasant, Gadsden, Fla., 100....B2 86
Mount Pleasant, Henry, Iowa, 7,339....D6 92
Mount Pleasant, Frederick, Md., 100....B3 85
Mount Pleasant, Isabella, Mich., 14,875....E6 98
Mount Pleasant, Marshall, Miss., 150....A4 100
Mount Pleasant, Hunterdon, N.J., 145....B2 106
Mount Pleasant, Cabarrus, N.C., 1,041....B3 109
Mount Pleasant, Jefferson, Ohio, 656....B1 123
Mount Pleasant, Westmoreland, Pa., 6,107....F2 114
Mount Pleasant, Charleston, S.C., 5,116....E3, F8 115
Mount Pleasant, Maury, Tenn., 2,921....B4 117
Mount Pleasant, Titus, Tex., 8,027....C5 118
Mount Pleasant, Sanpete, Utah, 1,572....D4 119
Mount Pocono, Monroe, Pa., 935....D11 114
Mount Prospect, Cook, Ill., 22,945....A6, E2 90
Mount Pulaski, Logan, Ill., 1,689....C4 90
Mountrail, co., N. Dak., 10,077....A3 110
Mount Rainier, nat. park, Wash....C4 122
Mount Rainier, Prince Georges, Md., 9,855....C1 85
Mount Revelstoke, nat. park, B.C., Can....D8 68
Mount Robson, park, B.C., Can....C8 68
Mount Royal, Gloucester, N.J., 800....D2 106
Mount Rushmore, nat. memorial, S. Dak....D2 106
Mounts, bay, Eng....D2 12
Mount Savage, Allegany, Md., 1,639....D2 85
Mount Shasta, Siskiyou, Calif., 2,272....B2 82
Mount Sherman, Newton, Ark....A2 81
Mount Sidney, Augusta, Va., 500....C4 121
Mount Solon, Augusta, Va., 140....C3 121
Mount Sterling, Choctaw, Ala., 125....C1 78
Mount Sterling, Brown, Ill., 2,262....D3 90
Mount Sterling, Montgomery, Ky., 5,310....B6 94
Mount Sterling, Haywood, N.C....D3 109
Mount Sterling, Madison, Ohio, 1,338....C4 111
Mount Stewart, P.E.I., Can., 433....C7 74
Mount Storm, Grant, W. Va., 160....B5 123
Mount Summit, Henry, Ind., 424....D7 91
Mount Tabor, Rutland, Vt., 45 (165▲)....E3 120
Mount Tom, Hampshire, Mass. (part of Easthampton)....B2 97
Mount Trumbull, Mohave, Ariz....E5 80
Mount Uniacke, N.S., Can., 531....E6 74
Mount Union, Henry, Iowa, 176....C6 92
Mount Union, Huntingdon, Pa., 4,091....E6 114
Mount Upton, Chenango, N.Y., 400....C5 108
Mount Vernon, Mobile, Ala., 553....D1 78
Mount Vernon, Faulkner, Ark., 200....B3 81
Mount Vernon, Montgomery, Ga., 1,166....D4 87
Mount Vernon, Jefferson, Ill., 15,566....E5 90
Mount Vernon, Posey, Ind., 5,970....I2 91
Mount Vernon, Linn, Iowa, 2,593....C6 92
Mount Vernon, Rockcastle, Ky., 1,177....C5 94
Mount Vernon, Kennebec, Maine, 225 (596▲)....D3 96
Mount Vernon, Somerset, Md., 250....D6 85
Mount Vernon, Lawrence, Mo., 2,381....D4 101
Mount Vernon, Erie, N.Y., 4,000....*C2 108
Mount Vernon, Westchester, N.Y., 76,010....D2 108
Mount Vernon, Knox, Ohio, 13,284....B5 111
Mount Vernon, Lucas, Ohio, 1,000....*A2 111
Mount Vernon, Grant, Oreg., 502....C7 113
Mount Vernon, Davison, S. Dak., 379....D7 116
Mount Vernon, Monroe, Tenn., 600....D9 117
Mount Vernon, Franklin, Tex., 1,338....C5 118
Mount Vernon, Fairfax, Va....B5, C5 121
Mount Vernon, Skagit, Wash., 7,921....A3 122
Mount Victory, Hardin, Ohio, 598....B4 111
Mountville, Troup, Ga., 139....C2 87
Mountville, Lancaster, Pa., 1,411....*F9 114

Mountville, Laurens, S.C., 139...C4 115
Mount Viso, mtn., It...E3 18
Mount Washington, Bullitt, Ky., 1,173...A4, B4 94
Mount Washington, Coos, N.H., 7...B4 105
Mount Wolf, York, Pa., 1,514 .F8 114
Mount Zion, Carroll, Ga., 211...C1 87
Mount Zion, Macon, Ill., 925...*D5 90
Moura, Braz., 185...D5 60
Moura, Port., 9,509...C2 20
Mourdi, depression, Chad...B4 46
Mourmelon-le-Grand, Fr., 3,518...E4 15
Mourne, mts., N. Ire...C5 13
Mouscron, Bel., 36,554...D3 15
Mousie, Knott, Ky., 500...C7 94
Mouth of Keswick, N.B., Can., 289...C3 74
Moutiers [-Tarentaise], Fr., 3,788...E7 14
Moutier, Switz., 7,472...B3 19
Mouton, isl., N.S., Can...F5 74
Mouy, Fr., 3,781...E2 15
Movico, Mobile, Ala. (part of Mount Vernon)...D1 78
Moville, Woodbury, Iowa, 1,156...B1 92
Moville, Ire., 1,097...B4 11
Mowbullan, mtn., Austl...C8 51
Moweaqua, Shelby, Ill., 1,614...D4 90
Mower, co., Minn., 48,498...G6 99
Mowich, Klamath, Oreg...D5 113
Mowrystown, Highland, Ohio, 416...C4 111
Moxahala, Perry, Ohio, 300..C5 111
Moxee City, Yakima, Wash., 499...C5 122
Moxico, dist., Ang., 267,069.D3 48
Moxie, mtn., Maine...C3 96
Moxie, pond, Maine...C3 96
Moxley, Jefferson, Ga., 50...D4 87
Moy, Scot., 528...C6 13
Moy, riv., Ire...C2 11
Moyālē, Eth...E4 47
Moyale, Ken...A6 48
Moyamba, S.L...E2 45
Moycullen, Ire., 127...D2 11
Moyers, Pushmataha, Okla., 150...C6 112
Moyeuvre-Grande, Fr., 15,146...E6 15
Moyie, B.C., Can., 137...E10 68
Moyie, range, B.C., Can...E9 68
Moyie Springs, Boundary, Idaho, 196...A2 89
Moylan, Delaware, Pa., 1,000...*G10 114
Moynalty, Ire., 128...D5 11
Moyobamba, Peru, 8,850...C2 58
Moyock, Currituck, N.C., 350...A7 109
Moyuta, Guat., 1,478...C2 62
Mozambique, Port. dep., Afr., 6,234,000...I9 42, B5 49
Mozambique, chan., Afr...g9 49
Mozart, Sask., Can., 64...F3 70
Mozdok, Sov. Un., 32,000...E2 29
Mozhaysk, Sov. Un., 10,000.D11 27
Mozhga, Sov. Un., 10,000...D19 9
Mozyr, Sov. Un., 25,000...E7 27
Mpanda, Tan...C5 48
Mpika, N. Rh., 117...D5 48
Mporokoso, N. Rh...C5 48
Mpouia, Con. B...F3 46
Mpulungu, N. Rh...C5 48
Mpwapwa, Tan., 1,612...C6 48
Mragowo, Pol., 3,254...B6 26
M'Raier, Alg., 6,935...C6 44
Mrewa, S. Rh., 790...A5 49
Msec, Czech., 1,007...n16 26
Msene, Czech., 886...n17 26
M'Sila, Alg., 8,645 (71,627▲).B5 44
Msta, riv., Sov. Un...B9 27
Mstislavl, Sov. Un., 10,000...D8 27
Mtakuja, Tan...C5 48
Mtoko, S. Rh., 975...A5 49
Mtorashanga, S. Rh., 2,270..A5 49
Mtsensk, Sov. Un., 10,000..E11 27
Mtubatuba, S. Afr...C5 49
Mtwara, Tan., 10,459...D7 48
Muang Fang, Thai., 4,563...C3 38
Muang Hot, Thai...C3 38
Muang Nan, Thai., 13,624..C4 38
Muari, cape, Pak...I13 41
Mubi, Nig...D7 45
Mucajaí, riv., Braz...C5 60
Mücheln, Ger., 13,200...B6 17
Muchkap, Sov. Un., 10,000..F14 27
Much Wenlock, Eng., 14,929...B5 12
Muck, isl., Scot...D2 13
Muckalee, creek, Ga...D2 87
Muckleshoot, Indian res., Wash.D2 122
Mucuburi, Moz...D6 48
Mucuri, Braz., 603...E3 57
Mucurí, riv., Braz...E2 57
Mucusso, Ang...E3 48
Mud, creek, Ala...E4 78
Mud, creek, Ga...B5 87
Mud, creek, Iowa...D3 92
Mud, creek, Okla...C4 112
Mud, lake, Maine...A5 96
Mud, lake, Minn...C5 99
Mud, lake, Nev...B2 104
Mud, lake, Nev...C2 104
Mud, riv., Minn...B3 99
Mud, riv., W. Va...C2 123
Mudanya, Tur., 5,900...B7 23
Mudawwarah, Jordan, 1,000...H10 31
Mud Butte, Meade, S. Dak., 7...C3 116
Muddo Gashi, Ken...A6 48
Muddy, Saline, Ill., 95...F5 90
Muddy, creek, Colo...A4 83
Muddy, creek, Kans...B7 93
Muddy, creek, Ky...D2 94
Muddy, creek, Wyo...B4 125
Muddy, creek, Wyo...C6 125
Muddy, creek, Wyo...D2 125
Muddy, creek, Wyo...E3 125
Muddy, fork, Ind...A3 91
Muddy, lake, Sask., Can...E1 70
Muddy, mts., Nev...G7 104
Muddy, peak, Nev...G7 104
Muddy, riv., Utah...E4 119
Muddy Boggy, creek, Okla...C6 112
Muddy Creek, mtn., Ark...C2 81
Müden, Ger., 1,600...F4 24
Mudgee, Austl., 5,313...F7 51
Mudjatik, riv., Sask., Can...A2 70
Mud Lake, Jefferson, Idaho, 187...F6 89
Mud Lick, creek, Ky...A6 94
Mudon, Bur., 20,123...E10 39
Mudros, see Moudhros, Grc.
Mudugh, dist., Som...D6 47
Mueda, Mos...D6 48
Muenster, Sask., Can., 182..E3 70
Muenster, Cooke, Tex., 1,190...C4 118
Muertos, isl., P.R...D4 65
Muff, Ire., 154...B4 11
Mufulira, N. Rh., 18,000 (*67,900)...C4 48
Mugía, Sp., 7,061...A1 20
Mugla, Tur., 14,000...D7 23
Muglad, Sud., 3,735...C2 47
Mugodzhary, mts., Sov. Un...D5 29
Mugulo, Pap...k11 51
Mugwump, lake, Oreg...E7 113
Muhammadābād, Iran...C9 41
Muhammad Qol, Sud...A4 47
Muheza, Tan...C6 48
Muhinga, Burundi...B5 48
Mühlacker, Ger., 12,100...E3 17
Mühldorf, Ger., 10,800...D6 16
Muhlenberg, co., Ky., 27,791.C2 94
Muhlenburg Park, Berks, Pa., 1,000...*F10 114
Mühlhausen, Ger., 47,100...B5 17
Mühlig-Hofmann, mts., Ant..B14 5
Muhutwe, Tan...B5 48
Mui, isl., Eniwetok... 52
Muir, Ionia, Mich., 610...E6 98
Muirkirk, Prince Georges, Md., 100...B4 85
Muirkirk, Scot., 3,721...E4 13
Mui Ron, cape, Viet...C7 38
Muir Woods, nat. mon., Calif..B5 82
Mukachevo, Sov. Un., 33,000...G4 27
Mukah, Mala., 4,701...E4 35
Mukden (Shenyang), China, 2,411,000...C9 34, D10 36
Mukeru, Palau... 52
Mukilteo, Snohomish, Wash., 1,128...B3 122
Mukutawa, riv., Man., Can...C3 71
Mukwonago, Waukesha, Wis., 1,877...F1, F5 124
Mula, Sp., 14,312...C5 20
Mulas, pt., P.R...g13 65
Mulat, isl., Yugo...C2 22
Mulberry, Crawford, Ark., 934...B1 81
Mulberry, Butte, Calif., 2,643...*C3 82
Mulberry, Polk, Fla., 2,922..E5 86
Mulberry, Clinton, Ind., 1,062...D4 91
Mulberry, Crawford, Kans., 642...E9 93
Mulberry, Clermont, Ohio, 349...C3 111
Mulberry, Lincoln, Tenn., 200...B5 117
Mulberry, fork, Ala...B3 78
Mulberry, mtn., Ark...B3 81
Mulberry, riv., Ark...B2 81
Mulberry Gap, pass., N.C...A2 109
Mulberry Grove, Bond, Ill., 745...E4 90
Mulchén, Chile, 7,324...B2 54
Mulde, riv., Ger...B7 17
Muldoon, Fayette, Tex., 600.E4 118
Muldraugh, Meade, Ky., 1,743...C4 94
Muldrow, Sequoyah, Okla., 1,137...B7 112
Mule, creek, Kans...E4 93
Mule Creek, Grant, N. Mex., 10...D1 107
Mule Creek, Niobrara, Wyo., 30...B8 125
Mulegé, Mex., 945...B2 63
Mu-leng, China, 5,000...D5 37
Muleng, riv., China...D6 37
Muleshoe, Bailey, Tex., 3,871...B1 118
Mulga, Jefferson, Ala., 482...B3, E4 78
Mulga Mine, Jefferson, Ala., 950...*B2 78
Mulgrave, N.S., Can., 1,145.D8 74
Mulgrave, isl., Austl...B7 51
Mulhacén, mtn., Sp...D4 20
Mulhall, Logan, Okla., 253..A4 112
Mülheim [an der Ruhr], Ger., 187,000...B1 17
Mulhouse, Fr., 108,995 (*160,000)...D7 14, B3 18
Muliama, Bis. Arch...h13 50
Mulino, Clackamas, Oreg., 250...B2, B4 113
Mulitapuili, cape, W. Sam... 52
Mull, head, Scot...B6 13
Mull, isl., Scot...D3 13
Mull, sound, Scot...D2 13
Mullaghareirk, mts., Ire...E2 11
Mullaghcleevan, mtn., Ire...D5 11
Mullan, Shoshone, Idaho, 1,477...B3 89
Mullan, pass, Mont...D4 102
Mullen, Hooker, Nebr., 811..B4 103
Mullens, Wyoming, W. Va., 3,544...D3 123
Müller, mts., Indon...E4 35
Mullet, key, Fla...F1 86
Mullet, lake, Mich...C6 98
Mullet, pen., Ire...C1 11
Mullet, riv., Wis...B6 124
Mullewa, Austl., 627...E2 50
Müllheim, Ger., 17,000...E2 16
Mullica, riv., N.J...D3 106
Mullica Hill, Gloucester, N.J., 750...D2 106
Mulliken, Eaton, Mich., 484.F6 98
Mullin, Mills, Tex., 219...D3 118
Mullinavat, Ire., 339...E4 11
Mullingar, Ire., 5,894...D4 11
Mullins, Marion, S.C., 6,229...C9 115
Mullinville, Kiowa, Kans., 385...E4 93
Müllrose, Ger., 2,917...A9 17
Mulobezi, N. Rh...E4 48
Mulrany, Ire., 127...D2 11
Mulroy, bay, Ire...B4 11
Mulseryd, Swe...A7 24
Multan, Pak., 340,000 (*358,201)...B5 39, B3 40
Multnomah, co., Oreg., 522,813...B4 113
Mulvane, Sumner and Sedgwick, Kans., 2,981...E6 93
Mulvihill, Man., Can., 175..D2 71
Mumbwa, N. Rh., 700...C4 48
Mummy, range, Colo...A5 83
Mumper, Garden, Nebr...C3 103
Mun, riv., Thai...D5 38
Muna, Mex., 3,966...C7 63
Munascong, lake, Mich...B6 98
Münchberg, Ger., 10,800...C6 17
München (Munich), Ger., 1,084,500 (*1,390,000)...D5 16, A7 18
München-Gladbach, see Mönchengladbach, Ger.
Münchenstein, Switz., 10,345.A4 19
Muncie, Delaware, Ind., 68,603 (*100,500)...D7 91
Muncie, Wyandotte, Kans., 1,000...*C9 93
Muncy, Lycoming, Pa., 2,830...D8 114
Muncy Valley, Sullivan, Pa., 160...D8 114
Mundare, Alta., Can., 603...C4 69
Munday, Knox, Tex., 1,978..C3 118
Mundelein, Lake, Ill., 10,526...A5, E2 90
Mundell, Carroll, Ark...A2 81
Münden, Ger., 20,200...B4 17
Munden, Republic, Kans., 177...C6 93
Munden, Princess Anne, Va., 35...E6 121
Munford, Talladega, Ala., 549...B4 78
Munford, Tipton, Tenn., 1,014...B2 117
Munfordville, Hart, Ky., 1,157...C4 94
Mugallala, creek, Austl...D6 51
Mungana, Austl., 99...C7 50
Mungari, Moz...A5 49
Mungbere, Con. L...A4 48
Munger, Bay, Mich., 200...E7 98
Mungindi, Austl., 915...D7 51
Munhall, Allegheny, Pa., 17,312...B6 114
Munich, see München, Ger.
Munich, Cavalier, N. Dak., 213...A7 110
Munising, Alger, Mich., 4,228...B4 98
Munith, Jackson, Mich., 250...F6 98
Munjor, Ellis, Kans., 150...D4 93
Munnsville, Madison, N.Y., 391...C5 108
Munoz, Phil., 6,272...o13 35
Munoz Gamero, pen., Chile...h11 54
Munsan, Kor., 5,000...H3 37
Munsey Park, Nassau, N.Y., 2,847...E3 108
Münsingen, Ger., 3,500...E4 17
Münsingen, Switz., 6,051...C4 19
Munson, Alta., Can., 82...D4 69
Munson, Santa Rosa, Fla., 50...G2 86
Munson, Clearfield, Pa., 450...E5 114
Munsonville, Cheshire, N.H., 100...D2 105
Munster, Fr., 4,965...C7 14
Munster, Ger., 11,500...B5 16
Munster, Lake, Ind., 10,313..A2 91
Munster, prov., Ire., 849,203...E2 11
Münstereifel, Ger., 4,200...C1 17
Münster [in Westfalen], Ger., 182,700...B2 17
Munsungan, lake, Maine...B3 96
Muntenia, prov., Rom., 4,991,289...*C8 22
Muntenia, reg., Rom...C7 22
Muntok, Indon., 6,929...F3 35
Muong Hou Nua, Laos, 10,000...A4 38
Muong Hou Tai, Laos...A4 38
Muong Hun Xieng Hung, Laos...B5 38
Muong Lane, Laos...C5 38
Muong May, Laos...E7 38
Muong Phalane, Laos...D6 38
Muong Sing, Laos, 10,000..B4 38
Muong Soui, Laos...C5 38
Muong Sung, Laos...B5 38
Muonio, Fin., 1,100...D10 25
Muonio, riv., Fin...D10 25
Muotathal, Switz., 2,592...C6 19
Muqdadiyah, Iraq, 4,203...E2 41
Mur, riv., Aus...E7 16
Mura, riv., Yugo...B3 22
Murakami, Jap., 23,200...G9 37
Murano, It...D8 18
Murashi, Sov. Un., 12,900...B3 29
Murat, Fr., 2,438...E5 14
Murat, mtn., Tur...C7 23
Murat, riv., Tur...C13 31
Murau, Aus., 2,755...E7 16
Muravera, It., 4,417...E2 21
Muravyevo, Sov. Un...C11 37
Murcheh Khvort, Iran...E5 41
Murchison, N.Z., 580...N14 51
Murchison, falls, Ug...A5 48
Murchison, riv., Austl...E2 50
Murcia, Sp., 126,000 (249,738▲)...D5 20
Murcia, prov., Sp., 800,463..*D5 20
Murcia, reg., Sp., 1,171,439..C5 20
Murcki, Pol., 2,800...g10 26
Murdab, bay, Iran...C4 41
Murderkill, riv., Del...B7 85
Murdo, Jones, S. Dak., 783...D5 116
Murdock, Charlotte, Fla., 100...E4 86
Murdock, Kingman, Kans., 100...B4, E6 93
Murdock, Swift, Minn., 381..E3 99
Murdock, Cass, Nebr., 247...E2 103
Muresul, riv., Rom...B7 22
Muret, Fr., 3,323...F4 14
Murfreesboro, Pike, Ark., 1,096...C2 81
Murfreesboro, Hertford, N.C., 2,643...A6 109
Murfreesboro, Rutherford, Tenn., 18,991...B5 117
Murghab, riv., Afg...D12 41
Murgha Kibzai, Pak....B4 39, B2 40
Murgon, Austl., 2,099...C8 51
Muri, Nig...E7 45
Muri, Switz., 7,855...C4 19
Muriaé, Braz., 22,571...C4 56
Murici, Braz., 5,868...k6 57
Muriel, lake, Alta., Can...B5 69
Murit, mtn., Tur...C7 23
Müritz, lake, Ger...B6 16
Murmansk, Sov. Un., 237,000...C15 25
Murnau, Ger., 6,200...B7 18
Murom, Sov. Un., 81,000...B2 29
Muroran, Jap., 145,679...E10 37
Muros, Sp., 2,420...A1 20
Muroto, cape, Jap...J7 37
Murphy, Owyhee, Idaho, 50...F2 89
Murphy, Jefferson, Mo., 200...B7 101
Murphy, Cherokee, N.C., 2,235...D2 109
Murphy, Mayes, Okla...A6 112
Murphy, Josephine, Oreg., 150...E3 113
Murphy, isl., S.C...E9 115
Murphy, lake, B.C., Can...C7 68
Murphysboro, Jackson, Ill., 8,673...F4 90
Murphytown, Wood, W. Va., 200...B3 123
Murray, Shoshone, Idaho, 100...B3 89
Murray, Clarke, Iowa, 613..C4 92
Murray, Calloway, Ky., 9,303...B3 94
Murray, Cass, Nebr., 279...D10, E3 103
Murray, Salt Lake, Utah, 16,806...C4 119
Murray, co., Ga., 10,447...B2 87
Murray, co., Minn., 14,743..F3 99
Murray, co., Okla., 10,622...C4 112
Murray, head, P.E.I., Can...C7 74
Murray, lake, Okla...C4 112
Murray, lake, S.C...C5 115
Murray, riv., Austl...G2 51
Murray, riv., B.C., Can...B7 68
Murray Bridge, Austl., 5,404...G2 51
Murray City, Hocking, Ohio, 717...C5 111
Murry Harbour, P.E.I., Can., 407...D7 74
Murray River, P.E.I., Can., 504...D7 74
Murraysville, Jackson, W. Va., 25...D3 123
Murrayville, B.C., Can., 431...B10 68
Murrayville, Hall, Ga., 300..B3 87
Murrayville, Morgan, Ill., 420...D3 90
Murrysville, Westmoreland, Pa., 1,200...*F2 114
Murrells Inlet, Georgetown, S.C., 800...D9 109
Murrells, inlet, S.C...D10 115
Mürren, Switz...C4 19
Murrhardt, Ger., 9,300...E4 17
Murrieta, Riverside, Calif., 500...F5 82
Murrumbidgee, riv., Austl...G5 51
Murrupula, Moz...A6 49
Murrysville, Westmoreland, Pa., 1,200...B7 114
Murska Sobota, Yugo., 6,562...B3 22
Murtaugh, Twin Falls, Idaho, 214...G4 89
Murten, Switz., 3,330...C3 19
Murtle, lake, B.C., Can...C8 68
Murua, is., Pap...k13 51
Murud, India, 9,744...H4 40
Murwara, see Katni, India
Murwillumbah, Austl., 7,153...D9 51
Mürz, riv., Aus...E7 16
Murzynowo, Pol., 1,000...F8 24
Mürzzuschlag, Aus., 11,586..E7 16
Mus, Tur., 12,000...C13 31
Musala, isl., Indon...L3 38
Musala, peak, Bul...D6 22
Musan, Kor., 20,717...E4 37
Musangoi, Con. L...C3 48
Musa Qal'a, Afg., 5,000...E12 41
Musashino, Jap., 120,337..*n18 37
Musāsh as Sirr, Eg., U.A.R..D5 32
Muscat, Mus. & Om., 5,000.D2 39
Muscat & Oman, country, Asia, 550,000...G8 33, D2 39
Muscatatuck, riv., Ind...G5 91
Muscatine, Muscatine, Iowa, 20,997...C6 92
Muscatine, co., Iowa, 33,840.C6 92
Muscle Shoals, Colbert, Ala., 4,084...A2 78
Musclow, mtn., B.C., Can...C4 68
Muscoda, Grant, Wis., 927..E3 124
Muscogee, co., Ga., 158,623.D2 87
Musconetcong, mtn., N.J...B2 106
Musconetcong, riv., N.J...B3 106
Muscotah, Atchison, Kans., 228...A7, C8 93
Muscoy, San Bernardino, Calif., 9,000...*E5 82
Muse, LeFlore, Okla., 100...C7 113
Muse, Washington, Pa., 1,386...F1 114
Museville, Pittsylvania, Va..E3 121
Musgrave, ranges, Austl...E5 51
Musgrave Harbour, Newf., Can., 422...D5 75
Mushāsh Abū Khawf, Eg., U.A.R...E6 32
Mushie, Con. L...B2 48
Mushketovo, Sov. Un., 10,000...r20 27
Musi, riv., Indon...F2 35
Music, mtn., Ariz...B2 80
Muskeget, chan., Mass...D7 97
Muskeget, isl., Mass...D7 97
Muskego, Waukesha, Wis., 2,000...F1 124
Muskego, lake, Wis...F1 124
Muskegon, Muskegon, Mich., 46,485 (*167,400)...E4 98
Muskegon, co., Mich., 149,943...E4 98
Muskegon, lake, Mich...E4 98
Muskegon, riv., Mich...D4 98
Muskegon Heights, Mich., 19,552...E4 98
Muskingum, co., Ohio, 79,159...B6 111
Muskingum, riv., Ohio...C6 111
Muskö, Swe...u36 25
Muskogee, Muskogee, Okla., 38,059...B6 112
Muskogee, co., Okla., 61,866...B6 112
Muskoka, co., Ont., Can., 26,705...B5 72
Muskoka, lake, Ont., Can...C5 72
Muskrat, creek, Wyo...B5 125
Muskwa, riv., Alta., Can...A3 69
Musmār, Sud...B4 47
Musoma, Tan., 7,207...B5 48
Musquacook, lakes, Maine...B3 96
Musquaro, lake, Que., Can...h9 75
Musquash, N.B., Can., 88...D3 74
Musquash, mtn., Maine...C5 96
Musquodoboit Harbour, N.S., Can., 654...E7 74
Musselburgh, Scot., 17,273..E5 13
Musselshell, Musselshell, Mont., 350...D8 102
Musselshell, co., Mont., 4,888...D8 102
Musselshell, riv., Mont...D9 102
Mussende, Ang...D2 48
Mussidan, Fr., 3,024...E4 14
Mussoorie, India, 7,133...B7 40
Mustafa Kemalpasa, Tur., 209,000...B7 31
Mustang, Canadian, Okla., 198...B4 112
Mustang, draw, Tex...C1 118
Musters, lake, Arg...D3 54
Mustinka, riv., Minn...E2 99
Mustvee, Sov. Un., 7,000...B6 27
Musumusu, W. Sam., 881... 52
Muswellbrook, Austl., 5,718..F8 51
Mut, Eg., U.A.R., 2,529...D5 43
Mut, Tur., 4,600...D9 31
Mutá, pt., Braz...D3 57
Mutan, riv., China...D4 37
Mutankiang, China, 151,400...C10 34, D4 37
Mutarara, Moz...A6 49
Mutayyin, Yemen...B5 47
Mutcho, pt., Saipan... 52
Mutena, Con. L...C3 48
Mutok, hbr., Ponape... 52
Mutombo Mukulu, Con. L..C3 48
Mutsu, bay, Jap...F10 37
Mutton, mts., Oreg...C5 113
Mutton Bay, Que., Can., 236...C2, h10 75
Muttontown, Nassau, N.Y., 1,265...*F2 84
Muttra, see Mathura, India
Mutual, Woodward, Okla., 84...A2 112
Mutzig, Fr., 3,087...F7 15
Muxima, Ang., 143...C1 48
Muynak, Sov. Un...E5 29
Muyunkum, des., Sov. Un...E7 29
Muzaffargarh, Pak., 11,271..B3 40
Muzaffarnagar, India, 87,622...*C6 40
Muzaffarpur, India, 109,048...C8 39, D10 40
Muzambinho, Braz., 6,731..k8 56
Muzhi, Sov. Un...B21 9
Muzon, cape, Alsk...n23 79
Muztagh, mtn., China...F11 28
Muztagh Ata, mtn., China...H24 9
Mwambo, Tan...D7 48
Mwanza, Tan., 19,877...B5 48
Mwaya, Tan., 2,270...C5 48
Mweenish, isl., Ire...D2 11
Mweka, Con. L...B3 48
Mwene-Ditu, Con. L...C3 48
Mwenga, Con. L...B4 48
Mweru, lake, Con. L., N. Rh...C4 48
Mwimba, Con. L...C3 48
Mwinilunga, N. Rh., 140...D3 48
Myakka, riv., Fla...E4 86
Myakka City, Manatee, Fla., 100...E4 86
Myaungmya, Bur., 20,770...D1 38
Mycenae, ruins, Grc...D4 23
Myers, Treasure, Mont., 20..D9 102
Myers, Charleston, S.C., 1,000...F3 115
Myerstown, Lebanon, Pa., 3,268...F9 114
Myersville, Frederick, Md., 355...A2 85
Myingyan, Bur., 36,536...D10 39
Myitkyina, Bur., 12,833...C10 39
Myitnge, riv., Bur...B2 38
Myjava, Czech., 9,935...D4 26
Mylo, Rolette, N. Dak., 103..A6 110
Mymensingh, Pak., 45,315...D9 39, E13 40
Myn-Aral, Sov. Un...D8 29
Mynard, Cass, Nebr., 40...E3 103
Myrick, Jones, Miss., 120...D5 100
Myrnam, Alta., Can., 441...C5 69
Myrtle, Man., Can., 125...E3 71
Myrtle, Ont., Can., 55...C6 72
Myrtle, Freeborn, Minn., 89...G5 99
Myrtle, Union, Miss., 313...A4 100
Myrtle, Oregon, Mo., 75...E6 101
Myrtle, Nansemond, Va., 25...B6 121
Myrtle Beach, New Haven, Conn. (part of Milford)...E4 84
Myrtle Beach, Horry, S.C., 7,834...D10 115
Myrtle Creek, Douglas, Oreg., 2,231...D3 113
Myrtle Grove, Escambia, Fla., 3,000...G2 86
Myrtle Point, Coos, Oreg., 2,886...D2 113
Myrtlewood, Marengo, Ala., 403...C2 78
Mysen, Nor., 2,525...p29 25
Myslenice, Pol., 6,520...D5 26
Mysliborz, Pol., 3,887...B3 26
Myslowice, Pol., 40,000...g10 26
Mysore, India, 253,865...F6 39
Mysore, state, India, 23,586,772...F6 39
Mystic, New London, Conn., 2,536...D9 84
Mystic, Irwin, Ga., 274...E3 87
Mystic, Appanoose, Iowa, 761...D5 91
Mystic, Pennington, S. Dak., 13...C2 116
Mystic, cavern, Ark...A2 81
Mystic, lakes, Mass...C2 97
My Tho, Viet., 5,000...G7 38
Mytilene, see Mitilini, Grc.
Mytischi, Sov. Un., 104,000...n17 27
Myton, Duchesne, Utah, 329...C5 119
Mze, riv., Czech...D8 17
Mzimba, Nya., 1,335...D5 48
Mzuzu, Nya...D5 48

N

Naab, riv., Ger...D7 17
Naalehu, Hawaii, Haw., 950...D6 88
Naan, Isr., 981...h10 32
Naantali, Fin., 2,700...G10 25
Naas, Ire., 4,023...D5 11
Naas Harbour, B.C., Can., 100...B3 68
Nabberu, lake, Austl...E3 50
Nabburg, Ger., 4,000...D7 17
Nabesna, Alsk., 12...C11 79
Nabeul, Tun., 14,047...B7 44
Naboonspruit, S. Afr., 3,036...B4 49
Nabnasset, Middlesex, Mass., 1,381...*A5 97
Nabq, Eg., U.A.R...H10 31
Nābulus, Jordan, 42,499.B7, g12 30
Nacala, Moz...D7 48
Nacaome, Hond., 2,630...D4 62
Nacfa, Eth...B4 47
Na Cham, Viet...A7 38
Naches, Yakima, Wash., 680...C5 122
Naches, riv., Wash...C4 122
Nachingwea, Tan., 1,693...D6 48
Náchod, Czech., 17,800...C4 26
Nacimiento, Chile, 2,815...B2 54
Nacimiento, res., Calif...E3 82
Nacmine, Alta., Can., 285...D4 69
Naco, Cochise, Ariz., 500...F6 80
Nacogdoches, Nacogdoches, Tex., 12,674...D5 118
Nacogdoches, co., Tex., 28,046...D5 118
Nacozari, Mex., 3,562...A3 63
Nada, Iron, Utah...E3 119
Nadarzyn, Pol., 2,000...m13 26
Nadeau, Menominee, Mich., 220...C3 98
Naden, harbour, B.C., Can...B1 68
Nadiad, India, 78,952...F4 40
Nadina River, B.C., Can...C4 68
Nadir, Vir. Is...f15 65
Nadlac, Rom., 12,284...B5 22
Nadvornaya, Sov. Un., 10,000...G5 27
Nady, Arkansas, Ark...D4 81
Nadym, riv., Sov. Un...B23 9
Naesong, Kor., 5,000...H4 37
Naestved, Den., 19,617 (*26,856)...C5 24
Naf, Cassia, Idaho, 10...G5 89
Nafada, Nig...D7 45
Nafishah, Eg., U.A.R., 1,000.D4 32
Nafutan, mtn., Saipan... 52
Nafutan, pt., Saipan... 52
Naga, Phil., 8,000 (55,723▲)...C6 35
Naga, hills, India, Bur...C10 39
Nagahama, Jap., 33,700...*n15 37
Nagaland, state, India, 369,200...*C9 39
Naga-Naga, Phil., 5,310...*D6 35
Nagano, Jap., 160,522...H9 37
Nagano, pref., Jap., 1,981,433...*H9 37
Nagaoka, Jap., 99,000 (148,254▲)...H9 37
Nagarote, Nic., 6,500...D4 62
Nagar-Parkar, Pak...E3 40
Nagasaki, Jap., 344,153...J4 37
Nagasaki, pref., Jap., 1,760,421...*J4 37
Nagaur, India, 24,296...D4 40
Nagcarlan, Phil., 5,031...o13 35
Nagchhu Dzong, China...E3 34
Nageezi, San Juan, N. Mex., 3...A2 107
Nagēlē, Eth...D4 47
Nagercoil, India, 106,207...G6 39
Naggar, India...A6 40
Nagles, mts., Ire...E3 11
Nago, Okinawa, 13,820... 52
Nago, bay, Okinawa... 52
Nagog, pond, Mass...C1 97
Nagold, Ger., 8,900...E3 17
Nagoya, Jap., 1,591,935 (*1,945,000)...I8, n15 37
Nagpur, India, 643,659 (*700,000)...D6 39, G7 40
Nags Head, Dare, N.C., 200...B8 109
Naguabo, P.R., 3,396...C8, g12 65
Naguabo, mun., P.R., 17,195...C7, g11 65
Naguilian, Phil., 2,188...n13 35
Nagykanizsa, Hung., 34,222..B3 22
Nagykoros, Hung., 19,908 (25,861▲)...B4 22
Naha, Okinawa, 63,630... 52
Nahalat Yehuda, Isr., 5,000...h10 32
Nahant, Essex, Mass., 3,960..C3 97
Nahariya, Isr., 14,537...A7 32
Nahe, riv., Ger...D2 17
Nahiku, Maui, Haw., 15...C5 88
Nahma, Delta, Mich., 300...C4 98
Nahmakanta, lake, Maine...C3 96
Nahr al Khābūr, riv., Syr...E13 31
Nahr el Harir, riv., Syr...B7 32
Nahuel Huapí, Arg...C2 54
Nahuel Huapí, lake, Arg...C2 54
Nahuel Niyeu, Arg...C3 54
Nahuelquir, Arg...C3 54
Nahunta, Brantley, Ga., 952.E5 87
Naiau, isl., Fiji Is... 52
Naic, Phil., 6,580...o13 35
Naicam, Sask., Can., 672...E3 70
Naila, Ger., 6,500...C6 17
Nain, Newf., Can., 465...g9 75
Nā'īn, Iran, 6,790...E6 41
Naingoro, pass, Fiji Is... 52
Naini-Tal, India, 14,995...C7 40
Nairai, isl., Fiji Is... 52
Nairn, Scot., 4,899...C5 13
Nairn, co., Scot., 8,421...C5 13
Nairn, riv., Scot...C4 13
Nairobi, Ken., 297,000...B6 48
Nairobi, prov. dist., Ken...*B6 48
Naivasha, Ken...B6 48
Najibabad, India, 34,310...C7 40
Najin, Kor., 34,338...E5 37
Nakadomari, Okinawa... 52
Nakadori, isl., Jap...J4 37
Nakagusuku, bay, Okinawa... 52
Nakalau, Fiji Is., 139... 52
Nak'amet, Eth., 15,000...D4 47
Nakaminato, Jap., 34,665...m19 37
Nakamura, Jap., 15,600...J6 37
Nakano, isl., Jap...L4 37
Nakatsu, Jap., 40,900...J5 37
Nakatsu, see Ena, Jap.
Nakaushi, Okinawa... 52
Nakchak, China...B9 40
Nakhichevan na Arakse, Sov. Un., 27,000...F3 29
Nakhodka, Sov. Un., 71,000..E6 37
Nakhon Pathom, Thai., 28,648...F4 38
Nakhon Ratchasima, Thai., 43,698...E5 38
Nakhon Sawan, Thai., 33,678...E4 38
Nakhon Si Thammarat, Thai., 26,056...H3 38

Nakhrachi, Sov. Un........B7 29
Nakina, Columbus, N.C.,
100......................C5 109
Nakiri, Jap., 7,000......o15 37
Naklo, Pol., 12,000........A6 26
Naknek, Alsk..............D8 79
Naknek, lake, Alsk........D8 79
Nakorokoro, pt., Fiji Is...52
Nakskov, Den., 16,639....D5 24
Nakuru, Ken., 17,625......B6 48
Nakusp, B.C., Can., 992...D9 68
Nal, riv., Pak...........H12 41
Nalchik, Sov. Un., 98,000..E2 29
Nallen, Fayette, W. Va.,
185...................C4, D7 123
Nalón, riv., Sp...........A2 20
Naloto, Fiji Is., 222......52
Nālūt, Libya, 4,850.......C2 43
Nam (Tengri), lake, China..E3 34
Namai, bay, Palau Is......52
Namaka, Alta., Can., 68...D5 69
Namakan, lake, Minn......B6 99
Namangan, Sov. Un.,
134,000................E8 29
Namanock, isl., N.J......A3 106
Namanyere, Tan...........C5 48
Namapa, Moz.............D6 48
Namasgali, Ug...........A5 48
Namatanai, Bis. Arch....h13 50
Nambe, Santa Fe, N. Mex.,
350.................B4, F6 107
Nambour, Austl., 5,336....C9 51
Namcha Barwa, mtn., China..F4 34
Nam Dinh, Viet., 25,000...B7 38
Nämdö, isl., Swe..........t36 25
Namekagon, lake, Wis.....B2 124
Namekagon, riv., Wis.....B2 124
Nametil, Moz.............A6 49
Nametu, lake, Man., Sask.,
Can................C5 64, B1 71
Namib, des., S.W. Afr.....B1 49
Namiquipa, Mex., 489.....B3 63
Namlea, Indon............F7 35
Namoi, riv., Austl........E7 51
Namoli, Fiji Is., 501......52
Nampa, Alta., Can., 271...A2 69
Nampa, Canyon, Idaho,
18,897.................F2 89
Nampala, Mali...........C3 45
Nampo, see Chinnampo, Kor.
Nampula, Moz., 2,561.....A6 49
Namsen, riv., Nor........E4 25
Namsos, Nor., 5,300......E4 25
Namu, isl., Bikini........52
Namua, isl., W. Sam......52
Namur, Bel., 32,511......D4 15
Namur, Que., Can., 224...D3 73
Namur, prov., Bel., 369,432..D4 15
Namur, isl., Kwajalein....52
Namuruputh, Sud.........E4 47
Namutoni, S. W. Afr......A2 49
Namwala, N. Rh., 500.....B4 48
Namwon, Kor., 25,900.....I3 37
Namyslow, Pol., 4,095....C4 26
Nan, riv., Thai...........C4 38
Nana, riv., Cen. Afr. Rep...D3 46
Nanafalia, Marengo, Ala.,
250.....................C2 78
Nanaimo, B.C., Can.,
14,135..............A8, E6 68
Nanaimo, dist., B.C., Can...E4 68
Nanakuli, Honolulu, Haw.,
2,745...............B3, g9 88
Nanam, Kor., 21,258......F4 37
Nanao, Jap., 29,700......H8 37
Nanatsu, isl., Jap........H8 37
Nanawan, riv., Man., Can...C3 71
Nance, co., Nebr., 5,635...C7 103
Nanchang, China, 19,000...J4 36
Nancheng, China, 508,000..J6 36
Nancheng, China, 59,000...H2 36
Nancheng, China, 18,000...K7 36
Nanchuan, China, 7,000...J2 36
Nanchung, China, 164,700..J2 36
Nancy, Fr., 128,677
(*220,000).........C7 14, F6 15
Nancy, Pulaski, Ky., 300...C5 94
Nancy, creek, Ga..........A5 87
Nanda Devi, peak, India...B7 40
Nander, India, 81,087.....H6 40
Nandurbar, India, 41,055..G5 40
Nanduri, Fiji Is., 126.....52
Nanfeng, China, 14,000...K7 36
Nanga-Eboko, Cam.......E2 46
Nanga Parbat, peak, India..A5 39
Nangis, Fr., 3,761........F3 15
Nang Rong, Thai..........C4 38
Nanhai (Fatshan), China,
122,500................*G7 36
Nani, Afg., 5,000........E14 41
Nanjemoy, Charles, Md.,
50.....................D3 85
Nankapenparam, reef, Ponape..52
Nankin, Ashland, Ohio, 400..B5 111
Nanking, China,
1,419,000.........E8 34, H8 36
Nankou, pass, China......D7 36
Nankung, China..........F6 36
Nannine, Austl., 153......E2 50
Nanning (Yunning), China,
264,000................G6 34
Nanoose Bay, B.C., Can.,
56.....................A8 68
Nan Peng (Lamock), is., China..C4 37
Nanping, China, 29,000...K8 36
Nansemond, Nansemond, Va.,
120...................D11 121
Nansemond, co., Va.,
31,366................E6 121
Nansemond, riv., Va......B6 121
Nan Shan, mts., China....D4 34
Nansio, Tan.............B5 48
Nanson, Rolette, N. Dak., 25..A6 110
Nantel, Que., Can., 53....C3 73
Nanterre, Fr., 83,416....g9 14
Nantes, Fr., 240,028
(*335,000)............D3 14
Nanteuil-le-Haudouin, Fr.,
1,716..................E2 15
Nanticoke, Ont., Can., 121..E4 72
Nanticoke, Wicomico, Md.,
450....................D6 85
Nanticoke, Luzerne, Pa.,
15,601............B7, D10 114
Nanticoke, riv., Del., Md....D6 85
Nantien, China...........J9 36
Nanton, Alta., Can., 1,054..D6 69
Nantua, Fr., 3,261........F6 14
Nantucket, Nantucket, Mass.,
2,804 (3,559*).........D7 97
Nantucket, co., Mass., 3,559..D7 97
Nantucket, isl., Mass......D7 97
Nantucket, sound, Mass....C7 97
Nantung, China, 260,400...H9 36
Nantuxent, pt., N.J.......E2 106
Nantwich, Eng., 10,454...A5 12

Nasa Manned, spacecraft center,
Tex..................F5 118
Nasarawa, Nig...........E6 45
Nasca, Peru, 2,175.......D3 58
Naschitti, San Juan,
N. Mex., 50..........A1 107
Nase, Jap...............F10 34
Naseby, N.Z., 154.......P13 51
Nash, Walsh, N. Dak., 33..A8 110
Nash, Grant, Okla., 230...A3 112
Nash, Bowie, Tex., 1,124..C5 118
Nash, co., N.C., 61,002...A6 109
Nash, pt., Wales.........C4 12
Nash, stream, N.H.......A4 105
Nashawena, isl., Mass.....D6 97
Nash Creek, N.B., Can., 250..B3 74
Nashmead, Mendocino,
Calif., 40.............C2 82
Nashoba, Pushmataha, Okla.,
100...................C6 112
Nashoba, hill, Mass......C1 97
Nashua, Chickasaw, Iowa,
1,737.................B5 92
Nashua, Wilkin, Minn.,
146...................D2 99
Nashua, Clay, Mo., 300...D2 101
Nashua, Valley, Mont., 796..B10 102
Nashua, Hillsboro, N.H.,
39,096...............E4 105
Nashua, riv., Mass.,
N.H............B4 97, E3 105
Nashville, Howard, Ark.,
3,579.................D2 81
Nashville, Berrien, Ga.,
4,070.................A3 87
Nashville, Washington, Ill.,
2,606.................E4 90
Nashville, Brown, Ind., 489..F5 91
Nashville, Kingman, Kans.,
137...................E5 93
Nashville, Barry, Mich.,
1,525.................F5 98
Nashville, Nash., N.C., 1,423..B6 109
Nashville, Holmes, Ohio,
234...................B5 111
Nashville, Davidson, Tenn.,
170,874 (*411,500)..A5, E9 117
Nashwaak, riv., N.B., Can...C3 74
Nashwauk, Itasca, Minn.,
1,712.................C5 99
Nass, riv., B.C., Can......B3 68
Nassau, Ba. Is., 6,000
(*46,125).............C5 64
Nassau, Lac qui Parle, Minn..E2 99
Nassau, Rensselaer, N.Y.,
1,248.................C7 107
Nassau, co., Fla., 17,189...B5 86
Nassau, co., N.Y., 1,300,171..E7 108
Nassau, gulf, Chile.......k12 54
Nassau, riv., Fla.........B5 86
Nassau, sound, Fla.......B5 86
Nassauville, Nassau, Fla.,
50..................B5, B6 86
Nassawadox, Northampton,
Va., 650..............D7 121
Nässjö, Swe., 18,400.....I6 25
Nasugbu, Phil., 5,251....o13 35
Nasukoin, mtn., Mont....B2 102
Natá, Pan., 2,319........F7 62
Natagaima, Col., 4,107...C2 60
Natal, Braz., 154,276....C3, g6 57
Natal, B.C., Can., 829....E10 68
Natal, prov., S. Afr.,
2,979,920.............C5 49
Natalbany, Tangipahoa, La.,
350...................D5 95
Natalia, Medina, Tex., 1,154..E3 118
Natanes, plat., Ariz......D5 80
Natanya, Isr., 40,907....B6, f10 32
Natashquan, Que., Can.,
100..................F20 68
Natashquan, riv., Que., Can...h9 75
Natawahunan, lake, Man., Can..B3 71
Natchaug, riv., Conn.....B8 84
Natchez, Adams, Miss.,
23,791................D2 100
Natchitoches, Natchitoches,
La., 13,924...........C2 95
Natchitoches, par., La.,
35,653................C2 95
Natewa, bay, Fiji Is......52
Nathalie, Halifax, Va., 125..E4 121
Nathilau, pt., Fiji Is......52
Nathrop, Chaffee, Colo., 25..C4 83
Natick, Middlesex, Mass.,
28,831...........B5, D2 97
Natimuk, Austl., 595.....H3 51
Nazas, riv., Mex.........B4 63
Nation, Pierce, Wash., 60...C3 122
National, Monongalia, W. Va.,
60.................A7, B4 123
National City, San Diego,
Calif., 32,771.........E2 82
National City, Iosco, Mich.,
90....................D7 98
National Garden, Volusia,
Fla., 100.............C5 86
National Park, Gloucester, N.J.,
3,380.................*D2 106
Natitingou, Dah., 1,900...D5 45
Natividade, Braz., 1,243...D1 57
Natividad, Mex., 1,872....h9 63
Natoena, isl., Indon......E3 35
Natoma, Osborne, Kans.,
775...................C4 93
Natron, lake, Tan........B6 48
Natrona, co., Wyo., 49,623..C6 125
Natrona Heights, see Harrison
[Township], Pa.
Nattarö, isl., Swe........u36 25
Natuna, mtn., Bur........C2 38
Natuna, isl., Indon......E3 35
Natural, bridge, Va......D3 121
Natural Bridge, Winston,
Ala., 100..............A2 78
Natural Bridge, Jefferson,
N.Y., 500............A5 108
Natural Bridge, Rockbridge,
Va., 150.............D3 121
Natural Bridge, Utah.....D3 119
Natural Bridges, nat. mon.,
Utah..................F6 119
Naturaliste, cape, Austl...F2 50
Natural Steps, Pulaski, Ark.,
60...................D5 81

Naturita, Montrose, Colo.;
979...................*C2 83
Naubinway, Mackinac,
Mich., 100............B5 98
Naucalpan de Juárez, Mex.,
10,372................h9 63
Nauders, Aus., 1,150.....C5 16
Naudville, Que., Can.,
4,475.................*A6 73
Nauen, Ger., 12,300......B6 16
Naugatuck, New Haven,
Conn., 19,511.........A8 110
Naugatuck, riv., Conn....C4 84
Naughton, Ont., Can., 954..A3 72
Naumburg [an der Saale], Ger.,
37,400................B6 17
Naumburg [in Hessen], Ger.,
2,300.................B4 17
Naunhof, Ger., 5,868.....B7 17
Naupe, Peru...........C2 58
Nauru, Austl., N.Z. and Br.
trust, Oceania, 3,269...G10 7
Naushon, isl., Mass......D6 97
Nautla, Mex., 1,437...C5, m15 63
Nauvoo, Walker, Ala., 318..B2 78
Nauvoo, Hancock, Ill., 1,039..C2 90
Nauwigewauk, N.B., Can.,
83....................D4 74
Nava, lake, Sp..........A3 20
Nava del Rey, Sp., 3,860..A3 20
Navahermosa, Sp., 4,761..C3 20
Navajo, Apache, Ariz., 15..B6 80
Navajo, Daniels, Mont., 5..B11 102
Navajo, co., Ariz., 37,994..B5 80
Navajo, Indian res., Ariz...A4 80
Navajo, mtn., Ariz.......A5 80
Navajo, mtn., Utah......F5 119
Navajo, nat. mon., Ariz....A5 80
Navajo, res., Colo.,
N. Mex...........D3 83, A2 107
Navajo, riv., Colo.......D4 83
Naval Base, Charleston,
S.C.................F3 115
Navalcarnero, Sp., 4,681..B3 20
Navalmoral de la Mata, Sp.,
9,073.................C3 20
Navan, Ont., Can., 190..A10, B9 72
Navarin, cape, Sov. Un....C20 28
Navarino, Shawano, Wis.,
D5 124
Navarino, isl., Chile.....k12 54
Navarra (Navarre,), prov., Sp.,
402,042..............*A5 20
Navarre, Stark, Ohio, 1,698..B6 111
Navarre, (Navarra), prov.,
Sp., 402,042..........*A5 20
Navarre, reg., Sp., 402,042..A5 20
Navarre, Arg., 2,567...A5, g7 54
Navarro, co., Tex., 34,423..C4 118
Navas de Tolosa, Sp., 1,134..C4 20
Navasota, Grimes, Tex.,
4,937.................D4 118
Navassa, Brunswick, N.C.,
500..................C5 109
Navassa, isl., W.I........F6 64
Nave, isl., Scot.........E2 13
Navia, riv., Sp.........A2 20
Navidad, Chile..........A2 54
Naviti, isl., Fiji Is......52
Navojoa, Mex., 30,762...B3 63
Navola, Fiji Is., 215......52
Navolato, Mex., 7,119...C3 63
Navpaktos, Grc., 6,550...C3 23
Navplion, Grc., 8,456....D4 23
Navrongo, Ghana, 1,170..D4 45
Navsari, India, 51,300....G5 40
Navy Yard City, Kitsap, Wash.,
3,341................*B3 122
Nawabshah, Pak., 34,205..D2 40
Nawada, India, 17,468...E10 40
Nawngleng, riv., Sp......*D6 35
Naxcivan, Sov. Un., 25,000..D7 9
Naxos, Grc., 2,546.......D5 23
Naxos, isl., Grc.........D5 23
Nay, Fr., 3,444..........F3 14
Nayarit, state, Mex.,
391,970...........C4, m11 63
Nāy Band, Iran..........E8 41
Nāy Band, cape, Iran....E8 41
Naylor, Lowndes, Ga., 272..F3 87
Naylor, Ripley, Mo., 499...E7 101
Nayoro, Jap., 23,400....D11 37
Naytahwaush, Mahnomen,
Minn., 300...........C3 99
Nazaka, Okinawa........52
Nazaré, Braz., 14,644....D3 57
Nazaré, Port., 9,241.....C1 20
Nazaré da Mata, Braz.,
9,246.................h6 57
Nazareth, Isr., 25,066....B7 32
Nazareth, Northampton, Pa.,
6,209................E11 114
Nazareth, Castro, Tex., 100..B1 118
Nazas, riv., Mex.........B4 63
Naze, cape, Nor........H2 25
Naze, cape, Jap.........C9 17
Naze, headland, Eng.....H2 12
Näzerābād, Iran.........H11 41
Nāzik, Iran.............B2 41
Nazilli, Tur., 36,000.....D7 23, D7 9
Nazimovo, Sov. Un......D27 9
Naziya, Sov. Un., 10,000..s32 25
Nazko, riv., B.C., Can....C6 68
Nazreth (Ādāmā), Eth....D4 47
Nazyvayevsk, Sov. Un....B8 29
Ncheu, Nya............B5 48
Ndala, Tan.............B5 48
Ndélé, Cen. Afr. Rep....D4 46
N'Dende, Gabon........F2 46
Ndjolé, Gabon.........F2 46
Ndola, N. Rh., 84,500...C4 48
Ndravuni, isl., Fiji Is....52
Neagh, lake, N. Ire......C5 11
Neah Bay, Clallam, Wash.,
900..................A1 122
Neal, Greenwood, Kans., 122..E7 93
Nea Palatia, Grc., 999...g11 23
Neapolis, Grc., 17,586...D4 23
Neapolis, Grc., 3,908....C5 23
Nea Psara (Eretria), Grc.,
1,731................g11 23
Near, is., Alsk.........E2 79
Neath, Wales, 30,884....C4 12
Neavitt, Talbot, Md., 312..C5 85
Neba, Jap., 900.........n16 37
Nebbish, Beltrami, Minn., 30..C4 99
Nebit-Dag, Sov. Un., 40,000..B7 41
Nebo, Pike, Ill., 441......D3 90
Nebo, Hopkins, Ky., 338...C2 94
Nebo, LaSalle, La., 150...C5 95
Nebo, McDowell, N.C.,
235..............B2, D4 109

Nebo, mtn., Utah.........D4 119
Nebraska, Jennings, Ind., 120..F7 96
Nebraska, state, U.S.,
1,411,330.........B8 76, 103
Nebraska City, Otoe, Nebr.,
7,252.............D10, F3 103
Nebuzely, Czech., 465....n18 26
Necedah, Juneau, Wis., 691..D3 124
Nechako, mts., B.C., Can...B3 68
Nechako, riv., B.C., Can...C5 68
Neche, Pembina, N. Dak.,
545...................A8 110
Nechí, Col.............B3 60
Nechi, riv., Col.........B3 60
Neckar, riv., Ger........D4 17
Neck City, Jasper, Mo., 110..D3 101
Necker, isl., Haw.......m15 88
Necoc*lea, Arg., 17,808...B5 54
Nederburgh, cape, Indon...F6 35
Nederland, Boulder, Colo.,
272...................B5 83
Nederland, Jefferson, Tex.,
12,036................E6 118
Neder Rijn, riv., Neth....C5 15
Nedrow, Onondaga, N.Y.,
2,000................*C4 108
Nee, res., Colo.........C8 83
Needham, Norfolk, Mass.,
25,793................D2 97
Needham Heights, Norfolk, Mass.
(part of Needham)....D2 97
Needle, riv., Wyo.......A5 125
Needle, mtn., Colo......D3 83
Needle or Mountain Home, mts.,
Utah..................E3 119
Needles, San Bernardino,
Calif., 4,590.........E6 82
Needmore, Echols, Ga., 80..F4 87
Needmore, Lawrence, Ind.,
150...................G4 91
Needmore, Fulton, Pa., 145..G5 114
Needville, Fort Bend, Tex.,
861................E5, F4 118
Neel, gap, Ga..........B3 87
Neeley, Power, Idaho, 40..G6 89
Neelin, Man., Can., 72...E2 71
Neely, Greene, Miss., 100..D5 100
Neelyville, Butler, Mo., 385..E7 101
Neosho, riv., Kans.,
Okla..............B6 93, B6 112
Neenah, Winnebago, Wis.,
18,057...........A5, D5 124
Neepawa, Man., Can., 3,197..D2 71
Neerpelt, Bel., 7,600....C5 15
Neeses, Orangeburg, S.C.,
347..................D5 115
Neffs, Belmont, Ohio, 950...B7 111
Neffsville, Lancaster, Pa., 900..F9 114
Neftegorsk, Sov. Un.,
10,000...............G16 9
Negapatinnam, India, 57,854..F6 39
Negaunee, Marquette, Mich.,
D3 98
Negev, reg., Isr........D6 32
Negley, Columbiana, Ohio,
500..................B7 111
Negly, Red River, Tex.,
75...................C5 118
Negoi, peak, Rom.......C6 22
Negotin, Yugo., 8,639...C6 22
Negra, Torrance, N. Mex...C4 107
Negra, range, Peru......C2 58
Negrais, cape, Bur......E9 30
Negreira, Sp., 7,725....A1 20
Negrine, Alg...........G10 30
Negri, Sembilan, state, Mala.,
364,524..............K4 38
Negrita, C.R...........E6 62
Negro, mtn., Md........D1 85
Negro, riv., Arg........B3 54
Negro, riv., Bol........B1 56
Negro, riv., Braz.......D1 56
Negro, riv., Braz.......D2 56
Negro, riv., S.A.......C4 55
Negro, riv., Ur.........E1 56
Negros, isl., Phil.......D6 35
Negros Occidental, prov.,
Phil., 1,332,000....*C6 35
Negros Oriental, prov.,
Phil., 598,800......*D6 35
Negru Voda, Rom., 3,154..D9 22
Neguac, N.B., Can., 559...B5 74
Nehalem, Tillamook, Oreg.,
233..................B4 113
Nehalem, riv., Oreg.....B4 113
Nehbandán, Iran.........F10 41
Neheim-Hüsten, Ger.,
33,900................B4 17
Neichiang, China, 190,200..F5 34
Neiden, Nor...........C13 25
Neidpath, Sask., Can., 80...G2 70
Neihart, Cascade, Mont.,
D6 82
Neihsiang, China, 18,000..H5 36
Neihuang, China........G6 36
Neilburg, Sask., Can., 327..E1 70
Neill, pt., Wash........D1 122
Neillsville, Clark, Wis., 2,728..D3 124
Neisse, riv., Ger., Pol....B9 17
Neiva, Col., 41,100......C2 60
Nejd, pol. div., Sau. Ar.,
5,000,000............*D10 41
Nejdek, Czech., 5,748...C7 17
Nekoma, Rush, Kans., 100..D4 93
Nekoma, Cavalier, N. Dak.,
143..................A7 110
Nekoosa, Wood, Wis., 2,515..D4 124
Neksö, Den., 3,220.....A3 26
Nelagoney, Osage, Okla.,
80...................A5 112
Nelbi, Sov. Un.........C2 45
Nelidovo, Sov. Un., 25,000..C9 27
Neligh, Antelope, Nebr.,
1,776................B7 103
Nell, isl., Kwajalein.....52
Nellore, India, 106,776...F6 39
Nellysford, Nelson, Va., 140..D4 121
Nelma, Sov. Un.........C8 9
Nelson, Yavapai, Ariz., 20..B3 80
Nelson, B.C., Can., 7,074...E9 68
Nelson, Pickens and Cherokee,
Ga., 658.............B2 87
Nelson, Lee, Ill., 283.....B4 90
Nelson, Saline, Mo., 126..B4 101
Nelson, Nuckolls, Nebr., 695..D7 103
Nelson, Clark, Nev., 50...H7 104
Nelson, Cheshire, N.H.,
75 (222*)...........D4 105
Nelson, N.Z., 23,971
(*25,321)...........N14 51
Nelson, Tioga, Pa., 400...C7 114
Nelson, Buffalo, Wis., 250..D1 124
Nelson, co., S.Y., 22,168..C4 94
Nelson, co., N. Dak., 7,034..B7 110
Nelson, co., Va., 12,752..D4 121

Nelson, lake, Man., Can....B1 71
Nelson, range, B.C., Can...E9 68
Nelson, riv., Man., Can.....A4, B3 71
Nelson Forks, B.C., Can....D2 68
Nelsonville, Athens, Ohio,
4,834................C5 111
Nelspruit, S. Afr., 15,344...C5 49
Néma, Maur, 3,000......C3 45
Nemacolin, Greene, Pa.,
1,404................G2 114
Nemadji, riv., Minn.,
Wis.............D6 99, B1 124
Nemaha, Sac, Iowa, 151...B2 92
Nemaha, Nemaha, Nebr.,
232..................D10 103
Nemaha, co., Kans., 12,897..C7 93
Nemaha, co., Nebr., 9,099..D10 103
Nemaha, co., Nebr.......D10 103
Nemakssiai, Sov. Un......A7 26
Neman, Sov. Un., 20,000...A7 27
Nemea, Grc., Sov. Un.....E5 27
Nemeiben, lake, Sask., Can...B3 70
Nemi, lake, It..........h9 21
Nemo, Lawrence, S. Dak.,
100..................C2 116
Nemours, Alg., 6,148
(13,245*)............B4 44
Nemours, Fr., 6,345.....C5 14
Nemunas, riv., Sov. Un....A6 26
Nemuro, Jap., 42,740...E12 37
Nemuro, strait, Jap.....E12 37
Nenagh, Ire., 4,317.....E3 11
Nenana, Alsk., 286.....C10 79
Nene, riv., Eng........B8 12
Nenzel, Cherry, Nebr., 43..B4 103
Neodesha, Wilson, Kans.,
3,594.................E8 93
Neoga, Cumberland, Ill.,
1,145.................D5 90
Neola, Pottawattamie, Iowa,
870...................C2 92
Neola, Duchesne, Utah, 600..C5 119
Neopit, Shawano, Wis.,
1,359................D5 124
Neosho, Newton, Mo., 7,452..E3 101
Neosho, co., Kans., 19,455..E8 93
Neosho, co., Kans.......E8 93
Neosho, riv., Kans.,
Okla.............93, B6 112
Neosho Falls, Woodson, Kans.,
222...................D8 93
Neosho Rapids, Lyons, Kans.,
D8 93
Nepal, country, Asia,
8,700,000...........G11 33, D10 40
Nepaug, res., Conn......B5 84
Nepewassi, lake, Ont., Can...A4 72
Nephi, Juab, Utah, 2,566...D4 119
Nepomuk, Czech., 1,046...D8 17
Neponset, Bureau, Ill., 495...B6 90
Neponset, riv., Mass.....D2 97
Nepton, Fleming, Ky., 125...B6 94
Neptune, Monmouth, N.J.,
16,000...............C4 106
Neptune, Cheatham, Tenn...A4 117
Neptune Beach, Duval, Fla.,
2,868...............B5, B6 86
Néra, Fr., 3,917........E4 14
Nérac, Fr., 3,917.......E4 14
Nerane, Sp.............A7 32
Nerchinsk, Sov. Un., 28,700..D14 28
Nerekhta, Sov. Un., 19,900..C13 27
Neretva, riv., Yugo......D3 22
Neris, riv., Sov. Un......A8 26
Nero, Plaquemines, La.,
210..................C8, E6 95
Nerpichye, Sov. Un., 10,000..B12 37
Nerstrand, Rice, Minn., 217..F5 99
Nerva, Sp., 12,686.....D2 20
Nesbit, DeSoto, Miss., 130..A3 100
Nesbitt, Man., Can., 86...C2 71
Nesco, Atlantic, N.J., 300...D3 106
Nesconset, Suffolk, N.Y.,
1,964................F4 84
Nescopeck, Luzerne, Pa.,
1,934................D9 114
Nesebur, Bul., 2,289....D8 22
Neshaminy, creek, Pa....C2 106
Neshanic, riv., N.J......C3 106
Neshanic Station, Somerset,
N.J., 400...........B3 106
Nesher, Isr., 1,700......52
Neshkoro, Marquette, Wis.,
368..................E4 124
Neshoba, Neshoba, Miss.,
250..................C4 100
Neshoba, co., Miss., 20,927..C4 100
Neskaupstadur, Ice., 1,482..n26 25
Neskowin, Tillamook, Oreg.,
50...................B3 113
Nesle, Fr., 2,417........E2 15
Nesmith, Williamsburg, S.C.,
D8 115
Nesodden, Nor.........p28 25
Nes op Ameland, Neth., 782..A5 15
Nespelem, Okanogan, Wash.,
358..................A7 122
Nesquehoning, Carbon, Pa.,
2,714...............E10 114
Ness, lake, Scot........C4 13
Ness City, Ness, Kans.,
1,653................D4 93
Ness, co., Kans., 5,470...D3 93
Nesselwang, Ger., 2,900..B6 18
Nesslau, Switz., 2,002...B7 19
Nessmersiel, Ger., 530...A1 15
Nesterov, Sov. Un......C4 26
Nesterov, Sov. Un......C7 26
Nestoria, Baraga, Mich., 20..B2 98
Nestorion, Grc., 3,197...B3 23
Nestorville, Barbour, W. Va.,
100..................B5 123
Nestos, riv., Grc.......B5 23
Nesvizh, Sov. Un., 10,000...E6 27
Netarts, Tillamook, Oreg.,
600..................B3 113
Netawaka, Jackson, Kans.,
225..................C8 93
Netcong, Morris, N.J., 2,765..B3 106
Netherhill, Sask., Can., 111..F1 70
Netherlands, country, Eur.,
11,556,000..........E10 8, B5 15
Netherlands Antilles (Nether-
lands West Indies), dep.,
N.A., 187,000..*B4 53, *n15 64
Netherlands Guiana, see Surinam,
Neth. dep., S.A.
Netherlands Indies, see
Indonesia, country, Asia
Netherlands New Guinea, see
West Irian, Indon. dep.,
Asia
Netherlands West Indies (Neth-
erlands Antilles), dep., N.A.,
187,000.........*B4 53, *n15 64
Nethy Bridge, Scot......C5 13

Netolice, Czech., 2,066......D9 17
Netrakona, Pak., 12,924....E13 40
Nett, lake, Minn............B5 99
Nettie, Nicholas, W. Va., 600..........C4 123
Nettilling, lake, N.W. Ter., Can.........C18 67
Nett Lake, St. Louis, Minn., 250..........B5 99
Nettleton, Craighead, Ark. (part of Jonesboro).......B5 81
Nettleton, Lee and Monroe, Miss., 1,389.........A5 100
Nettleton, Caldwell, Mo., 63.B4 101
Nettuno, It., 14,100........k9 21
Netvorice, Czech., 619......o18 26
Neubeckum, Ger., 8,900.....B3 17
Neubert, Knox, Tenn., 600..........D10, E11 117
Neubrandenburg, Ger., 33,400........B6 16
Neuburg, Ger., 16,500.....E6 17
Neuchâtel, Switz., 33,430 (★45,930)........B2 19
Neuchâtel, canton, Switz., 147,633.........C2 19
Neuchâtel, lake, Switz.......C2 19
Neudorf, Sask., Can., 404...G4 70
Neue Elde, riv., Ger........E5 24
Neuenegg, Switz., 2,921.....C3 19
Neuenkirchen, Ger., 1,500...E3 24
Neuenrade, Ger., 5,600.....B2 17
Neuerburg, Ger., 1,550.....D6 15
Neufchâteau, Bel., 2,696.....E5 15
Neufchâteau, Fr., 4,527......C6 14
Neufchâtel-en-Bray, Fr., 5,590.........C4 14
Neufchâtel [-sur-Aisne], Fr., 434.........E4 15
Neufelden, Aus., 1,034.....D6 16
Neugersdorf, Ger., 12,300...C9 17
Neuhaldensleben, Ger., 22,000.........B5 16
Neuhardt, Crittenden, Ark...E8 81
Neuharlingersiel, Ger., 180..A7 15
Neuhaus, Ger., 10,500......B3 17
Neuhaus, Ger., 4,608.......C6 17
Neuhaus, Ger., 2,774.......E4 24
Neuilly [-sur-Marne], Fr., 15,082.........g11 14
Neuilly [-sur-Seine], Fr., 72,773.........g10 14
Neu-Isenburg, Ger., 25,400..C3 17
Neukirchen am Grossvenediger, Aus., 1,932.......B8 18
Neumarkt, Ger., 3,811.....A8 15
Neumarkt [in der Oberpfalz], Ger., 15,800.......D6 17
Neumarkt-Sankt Viet, Ger., 3,500.........A8 18
Neumarkt, see Sroda Slaska, Pol.
Neumünster, Ger., 75,100.........A4 16, D3 24
Neunburg vorm Wald, Ger., 3,500.........D7 17
Neunkirch, Switz., 1,208....A6 19
Neunkirchen, Aus., 10,027...E8 16
Neunkirchen, Ger., 44,000...D2 17
Neuquén, Arg., 7,498.......B3 54
Neuquén, prov., Arg., 111,008........B3 54
Neuquén, riv., Arg.........B3 54
Neurara, Chile.............D2 55
Neuruppin, Ger., 22,000....B6 16
Neuse, riv., N.C...........B5 109
Neusiedler, lake, Aus.......E8 16
Neuss, Ger., 92,900........B1 17
Neustadt, Ont., Can., 493...C3 72
Neustadt am Rübenberge, Ger., 9,100.........F3 24
Neustadt [an der Aisch], Ger., 8,900........D5 17
Neustadt [an der Dosse], Ger., 2,152.........B6 16
Neustadt [an der Orla], Ger., 10,600.........C6 17
Neustadt an der Waldnaab, Ger., 5,400........D7 17
Neustadt [an der Weinstrasse], Ger., 31,600.......D3 17
Neustadt [bei Coburg], Ger., 12,600.........C6 17
Neustadt-Glewe, Ger., 6,069.E5 24
Neustadt [im Schwarzwald], Ger., 6,900.........B4 18
Neustadt [in Holstein], Ger., 14,500.........D4 24
Neustadt [in Sachsen], Ger., 6,094..........B9 17
Neustadt [Kreis Marburg], Ger., 5,000.........C4 17
Neustrelitz, Ger., 28,000...B6 16
Neu-Ulm, Ger., 24,300.....D5 16
Neuville, Que., Can., 802..........C6, C8 73
Neuville-lès-Dieppe, Fr., 6,723..........E9 12
Neuwerk, isl., Ger.........E2 24
Neuwied, Ger., 26,400.....C2 17
Neva, Johnson, Tenn., 75...C12 117
Neva, riv., Sov. Un.........s31 25
Nevada, Story, Iowa, 4,227...B4 92
Nevada, Vernon, Mo., 8,416.D3 101
Nevada, Wyandot, Ohio, 919..........B4 111
Nevada, co., Ark., 10,700....D2 81
Nevada, co., Calif., 20,911...C3 82
Nevada, state, U.S., 285,278...........C4 76, 104
Nevada, mts., Sp...........D4 20
Nevada City, Nevada, Calif., 2,353..........C3 82
Nevada Mills, Steuben, Ind., 15..........A7 91
Nevado, mtn., Arg..........B3 54
Nevado del Huilo, mtn., Col....C2 60
Nevel, Sov. Un., 10,000.....C7 27
Nevelsk (Honto), Sov. Un., 10,980..........C10 37
Nevers, Fr., 39,085........D5 14
Neversink, res., N.Y........D8 108
Neversink, riv., N.Y........D6 108
Neve Sha'anan, Isr. (part of Jerusalem)........m14 32
Nevesinje, Yugo., 2,272.....D4 22
Neville, Sask., Can., 208....H2 70
Néville, Fr., 690...........E8 12
Neville Island, Allegheny, Pa., 2,400..........*B5 114
Nevils, Bulloch, Ga., 110....D5 87
Nevin, Wales, 2,248........B3 12
Nevinnomyssk, Sov. Un., 27,400.........G17 9
Nevinville, Adams, Iowa, 61..C3 92
Nevis, Hubbard, Minn., 344.D4 99
Nevis, bay, Scot............C3 13
Nevis, isl., St. Kitts-Nevis-Anguilla........n15 64
Nevis, mtn., Scot..........D3 13
Nevrokop, see Gotse Delchev, Bul.
Nevsehir, Tur., 18,800.....C10 25
New, inlet, Fla..........C2 86
New, inlet, N.C..........D6 109
New, riv., Ariz..........F1 80
New, riv., N.Z...........P12 51
New, riv., N.C...........C6 109
New, riv., Va., W. Va.........C3 123, D2 121
New Agat, Guam, 1,340.......52
Newala, Tan., 3,000........D6 48
New Albany, N.S., Can., 147.E5 74
New Albany, Floyd, Ind., 37,812..........H6 91
New Albany, Wilson, Kans., 104..........E8 93
New Albany, Union, Miss., 5,151..........A4 100
New Albany, Franklin, Ohio, 307..........C2 111
New Albany, Bradford, Pa., 359..........C9 114
New Albin, Allamakee, Iowa, 643.........A6 92
Newald, Forest, Wis., 175...C5 124
New Alexandria, Fairfax, Va., 1,500..........*C5 121
New Almelo, Norton, Kans., 100..........C3 93
New Alsace, Dearborn, Ind., 200..........F7 91
New Amsterdam, Br. Gu., 12,304..........A3 59
Newark, Independence, Ark., 728..........B4 81
Newark, Alameda, Calif., 9,884..........B5 82
Newark, New Castle, Del., 11,404.........A6 85
Newark, Eng., 24,610.......A7 12
Newark, Kendall, Ill., 489...B5 90
Newark, Greene, Ind., 100...F4 91
Newark, Worcester, Md., 175..........D7 85
Newark, Knox, Mo., 116.....B6 101
Newark, Essex, N.J., 405,220..........B4, E4 106
Newark, Wayne, N.Y., 12,868..........B3 108
Newark, Licking, Ohio, 41,790..........B5 111
Newark, Marshall, S. Dak., 39..........B8 116
Newark, Wise, Tex., 392.....A5 118
Newark, bay, N.J...........E5 106
Newark Valley, Tioga, N.Y., 1,234..........C4 108
New Athens, St. Clair, Ill., 1,923..........E4 90
New Athens, Harrison, Ohio, 472..........B6 111
New Auburn, Sibley, Minn., 299..........F4 99
New Auburn, Chippewa, Wis., 383..........C2 124
New Augusta, Marion, Ind., 225..........H3, H7 91
New Augusta, Perry, Miss., 275..........D4 100
Newaygo, Newaygo, Mich., 1,447..........E5 98
Newaygo, co., Mich., 24,160..E5 98
New Baden, Clinton and St. Clair, Ill., 1,464........E4 90
New Baltimore, Macomb and St. Clair, Mich., 3,159.....F8 98
New Baltimore, Greene, N.Y., 500..........C6 108
New Baltimore, Hamilton, Ohio, 260..........D2 111
New Baltimore, Somerset, Pa., 263..........G4 114
New Beaver, Lawrence, Pa., 1,338..........*E1 114
New Bedford, Bureau, Ill., 166..........B4 90
New Bedford, Bristol, Mass., 102,477 (★146,400)......C6 97
Newberg, Yamhill, Oreg., 4,204..........B1, B4 113
New Berlin, Duval, Fla., 400.B6 86
New Berlin, Sangamon, Ill., 627..........D4 90
New Berlin, Chenango, N.Y., 1,262..........C5 108
New Berlin, Union, Pa., 654..E8 114
New Berlin, Guadalupe, Tex., 70..........B4 118
New Berlin, Waukesha, Wis., 15,788..........E1 124
New Berlinville, Berks, Pa., 1,151..........*F10 114
Newbern, Hale, Ala., 316....C2 78
Newbern, Jersey, Ill., 35.....D3 90
Newbern, Bartholomew, Ind., 40..........F6 91
New Bern, Craven, N.C., 15,717..........B6 109
Newbern, Dyer, Tenn., 1,695.A2 117
Newberry, San Bernardino, Calif., 200..........E5 82
Newberry, Alachua, Fla., 1,105..........C4 86
Newberry, Greene, Ind., 256..........G3 91
Newberry, Luce, Mich., 2,612..........B5 98
Newberry, Newberry, S.C., 8,208..........C4 115
Newberry, co., S.C., 29,416..C4 115
New Bethlehem (Deanville), Clarion, Pa., 1,599......D3 114
New Blaine, Logan, Ark., 200..........B2 81
New Bloomfield, Callaway, Mo., 359..........C5 101
New Bloomfield (Bloomfield), Perry, Pa., 987.........F7 114
New Bloomington, Marion, Ohio, 368..........B4 111
Newborn, Newton, Ga., 283..C3 87
Newboro, Ont., Can., 301....C8 72
New Boston, Mercer, Ill., 726.B3 90
New Boston, Lee, Iowa, 55...D6 92
New Boston, Berkshire, Mass., 100..........B1 97
New Boston, Wayne, Mich., 900..........A7 98
New Boston, Hillsboro, N.H., 300 (925▲)..........E3 105
New Boston, Scioto, Ohio, 3,984..........D5 111
New Boston, Bowie, Tex., 2,773..........C5 118
New Braintree, Worcester, Mass., 400 (509▲)......B3 97
New Braunfels, Comal, Tex., 15,631..........A4, E3 118
New Bremen, Lewis, N.Y., 150..........B5 108
New Bremen, Auglaize, Ohio, 1,972..........B2 111
Newbridge, Ire..........D3 11
New Bridge, Baker, Oreg., 75..........C9 113
New Brigden, Alta., Can., 75..........D5 69
New Brighton, Ramsey, Minn., 6,448..........*E5 99
New Brighton, Beaver, Pa., 8,397..........E1 114
New Britain, Hartford, Conn., 82,201..........C5 84
New Britain, Bucks, Pa., 1,109..........*F11 114
New Britain, isl., Bis. Arch.....G9 7
New Brockton, Coffee, Ala., 1,093..........D4 78
Newbrook, Alta., Can., 202..B4 69
New Brunswick, Middlesex, N.J., 40,139..........C4 106
New Brunswick, prov., Can., 597,936..........G19 67, 74
New Brunswick Heights, Middlesex, N.J., 1,000...*C4 106
New Buffalo, Berrien, Mich., 2,128..........G4 98
Newburg, Warrick, Ind., 1,450..........I13 85
Newburg, Jasper, Iowa, 100..C5 92
Newburg, Charles, Md., 400.D4 85
Newburg, Phelps, Mo., 884..D6 101
Newburg, Bottineau, N. Dak., 158..........A5 110
Newburg, Cumberland, Pa., 283..........F7 114
Newburg, Preston, W. Va., 494..........B5 123
Newburg Center, Penobscot, Maine, 80..........D4 96
Newburgh, Ont., Can., 569..C8 72
Newburgh, Penobscot, Maine, 75 (636▲)..........D3 96
Newburgh, Orange, N.Y., 30,979 (★104,000)......D7 108
Newburgh, Scot., 2,079.....C6 13
Newburgh Heights, Cuyahoga, Ohio, 3,512..........B2 111
Newburgh Junction, Orange, N.Y..........A4 106
New Burlington, Clinton and Greene, Ohio, 400......C4 105
New Burnside, Johnson, Ill., 227..........F5 90
Newbury, Ont., Can., 328....E3 72
Newbury, Eng., 20,386......C6 12
Newbury, Merrimack, N.H., 100 (342▲)..........D2 105
Newbury, Orange, Vt., 391 (1,452▲)..........C4 120
Newburyport, Essex, Mass., 14,004..........A6 97
New Caledonia, Fr. dep., Oceania, 63,000........H9 7
New Cambria, Saline, Kans., 187..........D6 93
New Cambria, Macon, Mo., 270..........B5 101
New Canaan, Fairfield, Conn., 13,466..........E3 84
New Canton, Pike, Ill., 499..D2 90
New Canton, Buckingham, Va., 350..........D4 121
New Carlisle, Que., Can., 1,333..........A4 74
New Carlisle, St. Joseph, Ind., 1,376..........A4 91
New Carlisle, Clark, Ohio, 4,107..........C3 111
New Cassel, Nassau, N.Y., 7,000..........*G2 84
New Castile, reg., Sp., 4,210,817..........C4 20
New Castle, Jefferson, Ala., 950..........B3, E5 78
Newcastle, Austl., 142,580 (★235,000)..........F8 51
Newcastle, Placer, Calif., 900..........C3 82
Newcastle, Ont., Can., 1,272.D6 72
Newcastle, N.B., Can., 5,236..........B4 74
New Castle, Garfield, Colo., 447..........B3 83
New Castle, New Castle, Del., 4,469..........A6 85
New Castle, Henry, Ind., 20,349..........E7 91
Newcastle, Ire., 2,527......E2 11
Newcastle, Ire..........E4 11
New Castle, Henry, Ky., 699.B4 94
Newcastle, Lincoln, Maine, 430 (1,101▲)..........D3 96
Newcastle, Dixon, Nebr., 357.B9 103
New Castle, Rockingham, N.H., 823..........D5 105
Newcastle, N. Ire., 3,722....C6 11
Newcastle, McClain, Okla., 80..........B4 112
New Castle, Lawrence, Pa., 44,790..........D1 114
Newcastle, S. Afr., 17,418....C4 49
Newcastle, Young, Tex., 617.C3 118
Newcastle, Iron, Utah, 100...F2 119
New Castle, Craig, Va., 200..D2 121
Newcastle, Weston, Wyo., 4,345..........B8 125
New Castle, co., Del., 307,446..........A6 85
Newcastle Bridge, N.B., Can., 483..........C3 74
Newcastle Emlyn, Wales, 648.B3 12
Newcastle Mine, Alta., Can., 900..........D4 69
New Castle Northwest, Lawrence, Pa., 2,007.....*D1 114
Newcastle-on-Tyne, Eng., 269,389 (★1,145,000)......C6 10
Newcastle-under-Lyme, Eng., 76,433..........A5 12
Newcastle Waters, Austl., 82..C5 50
Newcestown, Ire..........F3 11
New Chicago, Lake, Ind., 2,312..........A3 91
New City, Rockland, N.Y., 4,000..........D3, D7 108
New Columbia, Union, Pa., 550..........D8 114
Newcomb, San Juan, N. Mex., 5..........A1 107
Newcomb, Essex, N.Y., 575..........B6 108
Newcomb, Campbell, Tenn., 288..........C9 117
Newcomerstown, Tuscarawas, Ohio, 4,273..........B6 111
New Concord, Calloway, Ky., 75..........B3 94
New Concord, Muskingum, Ohio, 2,127..........C6 111
New Croton, res., N.Y.........D2 84
New Cumberland, Cumberland, Pa., 9,257..........F8 114
New Cumberland, Hancock, W. Va., 2,076.........A2, A4 123
New Cumnock, Scot., 3,871...E4 13
Newdale, Man., Can., 299...D1 71
Newdale, Fremont, Idaho, 272..........F7 89
New Dayton, Alta., Can., 102..........E4 69
Newdegate, Austl., 75.......F2 50
New Delhi, India, 261,545..........C6 39, C6 40
New Denmark, N.B., Can., 79..........C2 74
New Denver, B.C., Can., 564.E9 68
New Diggings, Lafayette, Wis., 85..........F3 124
New Durham, Strafford, N.H., 200 (474▲)..........D4 105
New Eagle, Washington, Pa., 2,670..........*F2 114
New Echota Marker, nat. mon., Ga..........B2 87
New Edinburg, Cleveland, Ark., 300..........D3 81
New Effingham, Roberts, S. Dak., 280..........B9 116
New Egypt, Ocean, N.J., 1,737..........C3 106
Newell, Randolph, Ala., 100..B4 78
Newell, Union, Ark., 165.....D3 81
Newell, Buena Vista, Iowa, 893..........B2 92
Newell, Butte, S. Dak., 797...C2 116
Newell, Hancock, W. Va., 1,842..........A2, A4 123
New Ellenton, Aiken, S.C., 2,309..........*E4 115
Newellton, Tensas, La., 1,453.B4 95
New Empire, Ormsby, Nev., 400..........D2 104
New England, Hettinger, N. Dak., 1,095..........C3 110
New Enterprise, Bedford, Pa., 250..........F5 114
New Era, Oceana, Mich., 403..........E4 98
New Fairfield, Fairfield, Conn., 200 (3,355▲)........D3 84
Newfane, Niagara, N.Y., 1,423..........B2 108
Newfane, Windham, Vt., 146 (714▲)..........E3 120
Newfield, York, Maine, 150 (319▲)..........E2 96
Newfield, Gloucester, N.J., 1,299..........D2 106
Newfields, Rockingham, N.H., 400 (737▲)..........D5 105
New Florence, Montgomery, Mo., 616..........C6 101
New Florence, Westmoreland, Pa., 958..........F3 114
Newfolden, Marshall, Minn., 370..........B2 99
Newfound, gap, Tenn........D10 117
Newfound, lake, N.H........C3 105
Newfoundland, Morris, N.J., 450..........A4 106
Newfoundland, Wayne, Pa., 400..........D11 114
Newfoundland, prov., Can., 457,853..........G21 67, 75
Newfoundland, mts., Utah.....B2 119
New Franken, Brown, Wis., 150..........D6 124
New Franklin, Howard, Mo., 1,096..........B5 101
New Freedom, York, Pa., 1,395..........G8 114
New Galilee, Beaver, Pa., 593..........E1 114
New Galloway, Scot., 327.....E4 13
Newgate, B.C., Can., 56.....E10 68
New Germany, N.S., Can., 623..........E5 74
New Glarus, Green, Wis., 1,468..........F4 124
New Glasgow, N.S., Can., 9,782..........D7 74
New Gloucester, Cumberland, Maine, 200 (3,047▲)...D5, E2 96
New Goshen, Vigo, Ind., 500.E3 91
New Grenada, Fulton, Pa., 75..........F5 114
New Gretna, Burlington, N.J., 800..........D4 106
New Guinea, isl., Pac. O..........G8 7, F9 35
New Guinea, Territory of, Austl. trust., Oceania, 1,409,000..........h12 50
Newgulf, Wharton, Tex., 1,419..........*E5 118
Newhall, Los Angeles, Calif., 4,705..........E4, F2 82
Newhall, Benton, Iowa, 495..C6 92
Newhall, Cumberland, Maine, 250..........E4 96
Newhall, McDowell, W. Va., 500..........D3 123
New Hamburg, Ont., Can., 2,181..........D4 72
New Hampshire, state, U.S., 606,921..........B13 77, 105
New Hampton, Chickasaw, Iowa, 3,456..........A5 92
New Hampton, Harrison, Mo., 289..........A3 101
New Hampton, Belknap, N.H., 250 (862▲)..........C3 105
New Hanover, Monroe, Ill., 60..........B8 101
New Hanover, co., N.C., 71,742..........C6 109
New Harbour, Newf., Can., 806..........E5 75
New Harmony, Posey, Ind., 1,121..........H2 91
New Harmony, Washington, Utah, 105..........F2 119
New Hartford, Litchfield, Conn., 1,034 (3,033▲).....B5 84
New Hartford, Butler, Iowa, 649..........B5 92
New Hartford, Oneida, N.Y., 2,468..........*B5 108
New Haven, New Haven, Conn., 152,048 (★320,800).D5 84
New Haven, Gallatin, Ill., 642..........F2 90
New Haven, Allen, Ind., 3,396..........B7 91
New Haven, Mitchell, Iowa, 125..........A5 92
New Haven, Nelson, Ky., 1,009..........C4 94
New Haven, Macomb, Mich., 1,198..........F8 98
New Haven, Franklin, Mo., 1,223..........C6 101
New Haven, Oswego, N.Y., 250..........B4 108
New Haven, Huron, Ohio, 300..........A5 111
New Haven, Addison, Vt., 150 (922▲)..........C2 120
New Haven, Mason, W. Va., 1,314..........C3 123
New Haven, Crook, Wyo., 15.A8 125
New Haven, co., Conn., 660,315..........D4 84
New Haven, hbr., Conn........D5 84
New Haven, riv., Vt.........C2 120
New Hazelton, B.C., Can., 150..........B4 68
New Hebrides, Fr. and Br. dep., Oceania, 49,000........H10 7
New Hebrides, is., Oceania....H10 7
Newhebron, Lawrence, Miss., 271..........D4 100
Newhill, Wake, N.C., 200....B5 109
New Holland, Hall, Ga., 1,000..........B3 87
New Holland, Logan, Ill., 314..........C4 90
New Holland, Fayette and Pickaway, Ohio, 798.......C4 111
New Holland, Lancaster, Pa., 3,425..........F9 114
New Holland, Douglas, S. Dak., 110..........D7 116
New Holstein, Calumet, Wis., 2,401..........B6, E5 124
New Hope, Madison, Ala., 953..........A3 78
Newhope, Pike, Ark., 70.....C2 81
New Hope, Nelson, Ky., 250.C4 94
New Hope, Hennepin, Minn., 3,552..........*E5 99
New Hope, Bucks, Pa., 958..........F12 114
New Hope, Florence, S.C....C8 115
New Hope, mtn., Ala........E4 78
New Hradec, Dunn, N. Dak., 38..........B3 110
New Hudson, Oakland, Mich., 450..........A6 98
New Hyde Park, Nassau, N.Y., 10,808..........C2 84
New Iberia, Iberia, La., 29,062..........D4 95
Newington, Hartford, Conn., 17,664..........C6 84
Newington, Screven, Ga., 399..........D5 87
Newington, Rockingham, N.H., 125 (1,045▲)..........D5 105
Newington, Fairfax, Va., 130.B5 121
New Inn, Ire..........D3 11
New Inn, Ire., 164..........E4 11
New Ipswich, Hillsboro, N.H., 300 (1,455▲)..........E3 105
New Jersey, state, U.S., 6,066,782..........B13 77, 106
New Johnsonville, Humphreys, Tenn., 559..........A4 117
New Kensington, Westmoreland, Pa., 23,485..........A7, E2 114
New Kent, New Kent, Va., 25..........D6 121
New Kent, co., Va., 4,504....D6 121
Newkirk, Guadalupe, N. Mex., 150..........B5 107
New Kirk, Kay, Okla., 2,092.A4 113
New Knoxville, Auglaize, Ohio, 792..........B3 111
Newland, Avery, N.C., 564..........A2, C4 109
Newland, Richmond, Va., 90..........C6 121
New Lebanon, Sullivan, Ind., 130..........F3 91
New Lebanon, Columbia, N.Y., 350..........C7 108
New Lebanon, Montgomery, Ohio, 1,459..........*C3 111
New Leipzig, Grant, N. Dak., 390..........C4 110
New Lenox, Will, Ill., 1,750..........B6, F2 90
New Lexington, Tuscaloosa, Ala., 40..........B2 78
New Lexington, Perry, Ohio, 4,514..........C5 111
New Liberty, Owen, Ky., 250.B5 94
New Lima, Seminole, Okla., 87..........B5 112
New Limerick, Aroostook, Maine, 200 (394▲)........E5 96
Newlin, Beauregard, La.....D2 95
New Lisbon, Henry, Ind., 300..........E7 91
New Lisbon, Burlington, N.J., 200..........D3 106
New Lisbon, Juneau, Wis., 1,337..........E3 124
New Liskeard, Ont., Can., 4,896..........E9 72
Newllano, Vernon, La., 264..C2 95
Newlon, Upshur, W. Va., 25..........C4 123
New London, New London, Conn., 34,182 (★104,600).D8 84
New London, Howard, Ind., 240..........D5 91
New London, Henry, Iowa, 1,694..........D6 92
New London, Kandiyohi, Minn., 721..........E4 99
New London, Ralls, Mo., 875..........B6 101
New London, Merrimack, N.H., 1,007 (1,738▲).......D3 105
New London, Stanly, N.C., 223..........B3 109
New London, Huron, Ohio, 2,392..........A5 111
New London, Rusk, Tex., 950..........*C5 118
New London, Waupaca and Outagamie, Wis., 5,288...D5 124
New London, co., Conn., 185,745..........C8 84
New Lothrop, Shiawassee, Mich., 510..........E6 98
New Lowell, Ont., Can., 203..........C5 72
New Lyme, Ashtabula, Ohio, 30..........A7 111
New Madison, Darke, Ohio, 910..........C3 111
New Madrid, New Madrid, Mo., 2,867..........E8 101
New Madrid, co., Mo., 31,350..........E8 101
New Malden, Eng., 46,587...m11 10
Newman, Stanislaus, Calif., 2,148..........D3 82
Newman, Douglas, Ill., 1,097..........D6 90
Newman, Jefferson, Kans., 75..........B7 93
Newman, Otero, N. Mex., 25..........E3 107
Newman, lake, Wash.........B8 122
Newman, Grove, Madison, Nebr., 880..........C8 103
Newmans, lake, Fla..........C4 86
Newmanstown, Lebanon, Pa., 1,200..........F9 114
New Marion, Ripley, Ind., 150..........F7 91
New Market, Madison, Ala., 400..........A3 78
Newmarket, Ont., Can., 8,932..........C5 72
Newmarket, Eng., 11,207....B8 12
New Market, Montgomery, Ind., 578..........E4 91
New Market, Taylor, Iowa, 506..........D3 92
New Market, Ire., 791......E2 11
New Market, Frederick, Md., 358..........B3 85
New Market, Scott, Minn., 211..........F5 99
Newmarket, Rockingham, N.H., 2,745 (3,153▲).....D5 105
New Market, Middlesex, N.J., 3,500..........*B4 106
New Market, Jefferson, Tenn., 750..........C10 117
New Market, Shenandoah, Va., 783..........C4 121
Newmarket-on-Fergus, Ire., 807..........E3 11
New Marlboro, Berkshire, Mass., 75 (1,083▲)........B1 97
New Marshfield, Athens, Ohio, 350..........C5 110
New Martinsville, Wetzel, W. Va., 5,607..........B5 123
New Matamoras, Washington, Ohio, 925..........C6 111
New Meadows, Adams, Idaho, 647..........E2 89
New Melle, St. Charles, Mo., 300..........C7 101
New Mexico, state, U.S., 951,023..........D6 76, 107
New Miami, Butler, Ohio, 2,360..........C3 111
New Middleton, Smith, Tenn., 100..........C7 117
New Middletown, Harrison, Ind., 132..........H5 91
New Milford, Litchfield, Conn., 3,023 (8,318▲).....C3 84
New Milford, Bergen, N.J., 18,810..........D5 106
New Milford, Portage, Ohio, 500..........A6 111
New Milford, Susquehanna, Pa., 1,129..........C10 114
New Mills, N.B., Can., 230...B3 74
New Milton, Doddridge, W. Va., 100..........B4 119
New Minden, Washington, Ill., 166..........E4 90
New Munich, Stearns, Minn., 296..........E4 99
New Munster, Kenosha, Wis., 250..........F1 124
Newnan, Coweta, Ga., 12,169..........C2 87
Newnata, Stone, Ark., 60.....B3 81
Newnham, Eng., 1,217.......C5 12
New Norfolk, Austl., 5,445...o15 50
New Norway, Alta., Can., 263..........C4 69
New Offenburg, Ste. Genevieve, Mo., 200.....D7 101
New Orleans, Orleans, La., 627,525 (★885,200)....C7, E5 95
New Osgoode, Sask., Can., 60..........E4 70
New Oxford, Adams, Pa., 1,407..........G7 114
New Palestine, Hancock, Ind., 725..........E6 91
New Paltz, Ulster, N.Y., 3,041..........D6 108
New Paris, Elkhart, Ind., 900.B6 91
New Paris, Preble, Ohio, 1,679..........C3 111
New Paris, Bedford, Pa., 232.F4 114
New Perlican, Newf., Can., 427..........E5 74
New Philadelphia, Washington, Ind., 35..........G5 91
New Philadelphia, Tuscarawas, Ohio, 14,241..........B6 111
New Philadelphia (Silver Creek), Schuylkill, Pa., 1,702.....E9 114
New Pine Creek, Modoc, Calif., 100..........B3 82
New Pine Creek, Lake, Oreg., 150..........E6 113
New Plymouth, Payette, Idaho, 940..........F2 89
New Plymouth, N.Z., 29,368 (★32,387)..........M15 51
New Plymouth, Vinton, Ohio, 130..........C5 111
New Point, Decatur, Ind., 319..........F7 91
New Point, Holt, Mo., 85....A2 101
New Point Comfort, pt., Va...D6 121
Newport, Jackson, Ark., 7,007..........B4 81
Newport, N.S., Can., 419....E6 74
Newport, New Castle, Del., 1,239..........A6 85
Newport, Eng., 4,370.......B5 12
Newport, Eng., 105,285.....C5 12
Newport, Eng., 19,482......D6 12
Newport, Wakulla, Fla., 150.B2 86
Newport, Vermillion, Ind., 627..........E3 91
Newport, Ire., 459.........D2 11
Newport, Ire., 581..........E3 11
Newport, Campbell, Ky., 30,070..........A7, B5 94
Newport, Penobscot, Maine, 1,589 (2,322▲)..........D3 96
Newport, Monroe, Mich., 650..........G7 98
Newport, Washington, Minn., 2,349..........E7 99
Newport, Rock, Nebr., 162...B6 103

Newport, Sullivan, N.H., 3,222 (5,458▲) D2 105
Newport, Cumberland, N.J., 980 E2 106
Newport, Herkimer, N.Y., 827 B5 108
Newport, Carteret, N.C., 861 C7 108
Newport, Washington, Ohio, 480 C6 111
Newport, Lincoln, Oreg., 5,344 C3 113
Newport, Perry, Pa., 1,861 F7 114
Newport, Newport, R.I., 47,049 D11 84
Newport, Cocke, Tenn., 6,448 D10 117
New Port, York, S.C., 50 B5 115
Newport, Clay, Tex., 100 C3 118
Newport, Orleans, Vt., 5,019 B4 120
Newport, Giles, Va., 100 D2 121
Newport, Wales, 108,107 C5 12
Newport, Pend Oreille, Wash., 1,513 A8 122
Newport, co., R.I., 81,891 C11 84
Newport Beach, Orange, Calif., 26,564 F3, F4 82
Newport Center, Orleans, Vt., 288 B4 120
New Portland, Somerset, Maine, 175 (620▲) D2 96
Newport News (Independent City), Va., 113,662 (*219,200) B6, E6 121
New Port Richey, Pasco, Fla., 3,520 D4 86
Newport Station, N.S., Can., 446 E5 74
Newportville, Bucks, Pa., 700 *F12 114
New Prague, Le Sueur and Scott, Minn., 2,533 F5 99
New Preston, Litchfield, Conn., 900 C3 84
New Providence, Hardin, Iowa, 206 B4 92
New Providence, Union, N.J., 10,243 *B4 106
New Providence, Pan., 191 k11 62
New Providence, Lancaster, Pa., 200 G9 114
New Providence, Montgomery, Tenn., 4,451 A4 117
New Providence, isl., Ba. Is. C5 64
Newquay, Eng., 11,877 D2 12
New Quay, Wales, 951 B3 12
New Raymer, Weld, Colo., 91 A7 83
New Richland, Waseca, Minn., 1,046 G5 99
New Richmond, Que., Can., 411 A4 74
New Richmond, Montgomery, Ind., 394 D4 91
New Richmond, Clermont, Ohio, 2,834 D3 111
New Richmond, St. Croix, Wis., 3,316 C1 124
New Riegel, Seneca, Ohio, 349 A4 111
New River, Maricopa, Ariz., 75 D3 80
New River, Bradford, Fla., 20 C4 86
New River, gorge, W. Va. D7 123
New River, inlet, Fla E3 86
New River, inlet, N.C. C6 109
New River, Scott, Tenn., 300 C9 117
New Roads, Pointe Coupee, La., 3,965 D4 95
New Rochelle, Westchester, N.Y., 76,812 D2, D3 108
New Rockford, Eddy, N. Dak., 2,177 B6 110
New Ross, N.S., Can., 242 E5 74
New Ross, Montgomery, Ind., 332 E4 91
New Ross, Ire., 4,494 E3 11
Newry, Oxford, Maine, 125 (260▲) D2 96
Newry, N. Ire., 12,450 C5 11
Newry, Oconee, S.C., 762 B2 115
New Salem, Pike, Ill., 172 D3 90
New Salem, Rush, Ind., 250 E7 91
New Salem, Cowley, Kans., 60 E7 93
New Salem, Franklin, Mass., 250 (397▲) A3 97
New Salem, Morton, N. Dak., 986 C4 110
New Salem, Fayette, Pa., 860 *G2 114
New Salisbury, Harrison, Ind., 200 H5 91
New Sarpy, St. Charles, La., 1,259 C7 95
New Sharon, Mahaska, Iowa, 1,063 C5 92
New Sharon, Franklin, Maine, 250 (712▲) D2 96
New Shrewsbury, Monmouth, N.J., 7,313 C4 106
New Siberian, see Novosibirskiye, is., Sov. Un.
Newsite, Tallapoosa, Ala., 100 B4 78
New Site, Prentiss, Miss., 7 A5 100
New Smyrna Beach, Volusia, Fla., 8,781 C6 86
Newsom, Davidson, Tenn. E9 117
Newsoms, Southampton, Va., 423 E5 121
New South Wales, state, Austl., 3,917,016 F8 50
New Straitsville, Perry, Ohio, 1,019 C5 111
New Summerfield, Castro, Tex., 525 B1 118
New Sweden, Aroostook, Maine, 250 (713▲) B4 96
New Tazewell, Claiborne, Tenn., 768 C10 117
Newton, Dale, Ala., 958 D4 78
Newton, Baker, Ga., 529 E2 87
Newton, Jasper, Ill., 2,901 E5 90
Newton, Jasper, Iowa, 15,381 C4 92
Newton, Harvey, Kans., 14,877 A5, D6 93
Newton, Middlesex, Mass., 92,384 B5, D2 97
Newton, Newton, Miss., 3,178 C4 100
Newton, Rockingham, N.H., 175 (1,419▲) E4 105
Newton, Sussex, N.J., 6,563 A3 106
Newton, Catawba, N.C., 6,658 B2 109
Newton, Newton, Tex., 1,233 D6 118
Newton, Cache, Utah, 480 B4 119
Newton, Roane, W. Va., 100 C3 123
Newton, Manitowoc, Wis., 60 B6 124
Newton, co., Ark., 5,963 B4 81
Newton, co., Ga., 20,999 C3 87
Newton, co., Ind., 11,502 B3 91
Newton, co., Miss., 19,517 C4 100
Newton, co., Mo., 30,093 E3 101
Newton, co., Tex., 10,372 D6 118
Newton Abbot, Eng., 18,066 D4 12
Newton Brook, Ont., Can., 1,600 E6 72
Newton Falls, St. Lawrence, N.Y., 900 A6, B2 108
Newton Falls, Trumbull, Ohio, 5,038 A7 111
Newton Grove, Yellowstone, Mont., 2 D9 102
Newton Grove, Sampson, N.C., 477 B5 109
Newton Hamilton, Mifflin, Pa., 338 F6 114
Newton Junction, Rockingham, N.H., 225 E4 105
Newton Station, B.C., Can., 550 A10 68
Newton Stewart, Scot., 1,980 F4 13
Newtonville, Ont., Can., 331 D6 72
Newtonville, Spencer, Ind., 125 H4 91
Newtonville, Atlantic, N.J., 350 D3 106
Newtonville, Albany, N.Y., 1,000 *C7 108
New Toronto, Ont., Can., 13,384 D5, E6 72
Newtown, Newf., Can., 590 D5 75
Newtown, Fairfield, Conn., 1,261 (11,373▲) D3 84
Newtown, Fountain, Ind., 321 D3 91
Newtown, Sullivan, Mo., 265 A4 101
New Town, Mountrail, N. Dak., 1,586 B3 110
Newtown, Hamilton, Ohio, 1,750 C3 111
Newtown, Bucks, Pa., 2,323 F12 114
Newtown, Luzerne, Pa., 2,400 *B8 114
Newtown, King and Queen, Va., 65 D5 121
Newtown [& Llanllwchaiarn], Wales, 5,512 B4 12
Newtownards, N. Ire., 13,090 C6 11
Newtown Crommelin, N. Ire C5 11
Newtownhamilton, N. Ire., 589 C5 11
Newtownmountkennedy, Ire., 935 D5 11
Newtown Square, Delaware, Pa., 9,270 D2 106
Newtownstewart, N. Ire., 1,265 C4 11
New Trenton, Franklin, Ind., 150 F8 91
New Ulm, Brown, Minn., 11,114 F4 99
New Ulm, Austin, Tex., 500 E4 116
New Underwood, Pennington, S. Dak., 462 C3 116
New Vienna, Dubuque, Iowa, 265 B6 92
New Vienna, Clinton, Ohio, 858 C2 111
New Village, Warren, N.J., 350 B2 106
Newville, Henry, Ala., 546 D4 78
Newville, N.S., Can., 125 D5 74
Newville, Cumberland, Pa., 1,656 F7 114
Newville, Braxton, W. Va., 75 C4 123
New Vineyard, Franklin, Maine, 250 (357▲) D2 96
New Virginia, Warren, Iowa, 381 C4 92
New Washington, Clark, Ind., 700 G6 91
New Washington, Crawford, Ohio, 1,162 B5 111
New Waterford, N.S., Can., 10,592 C9 74
New Waterford, Columbiana, Ohio, 711 B7 111
New Waverly, Cass, Ind., 200 C5 91
New Waverly, Walker, Tex., 426 D5 118
New Westminster, B.C., Can., 33,654 A10, E6 68
New Whiteland, Johnson, Ind., 3,488 *E5 91
New Wilmington, Lawrence, Pa., 2,203 D1 113
New Windsor, Mercer, Ill., 658 B3 90
New Windsor, Carroll, Md., 738 A3 85
New Windsor, Orange, N.Y., 4,041 *D6 108
New Woodstock, Madison, N.Y., 375 C5 108
New World, isl., Newf., Can D4 75
New Year, lake, Nev B2 104
New York, Bronx, Kings, New York, Queens, Richmond, N.Y., 7,781,984 (*15,404,300) D1, D3, E7 108
New York, co. (Manhattan borough), N.Y., 1,698,281 (part of New York City) D1 108
New York, state, U.S., 16,782,304 B12 76, 108
New York, peak, Calif E6 82
New York Mills, Otter Tail, Minn., 828 D3 99
New York Mills, Oneida, N.Y., 3,788 *B5 108
New Zealand, country, Oceania, 2,414,984 I10 7, 51
New Zealand Dependencies, see Cook Islands, Niue, and Tokelau
New Zion, Clarendon, S.C., 200 D7 115
Ney, Defiance, Ohio, 338 A3 111
Ney, lake, Ont., Can B5 71
Neya, Sov. Un B2 29
Neye, isl., Guam G7 52
Neyriz, Iran, 19,439 G7 41
Neyshābūr, Iran, 25,820 C9 41
Nezhin, Sov. Un., 59,000 F8 27
Nezperce, Lewis, Idaho, 667 C2 89
Nez Perce, co., Idaho, 27,066 C2 89
Nez Perce, pass, Idaho, Mont D4 89
Nezpique, bayou, La D3 95
Ngabang, Indon E3 35
Ngabe, Con. B F3 46
Ngala, Nig D7 45
Ngaloa, bay, Fiji Is 52
Ngaloa, hbr., Fiji Is 52
Ngamea, isl., Fiji Is 52
N'Gaoundéré, Cam D2 46
Ngara, Tan B5 48
Ngardmau, Palau 52
Ngaruangl, reef, Palau Is 52
Ngaruawahia, N.Z., 3,273 L15 51
Ngatapa, N.Z., 284 M16 51
Ngau, isl., Fiji Is 52
Ngele Levu, isl., Fiji Is 52
Ngemelis, is., Palau Is 52
Ngesebus, isl., Palau Is 52
Ngidinga, Con. L C2 48
Ngong, Ken B6 48
Ngoring Nor, lake, China E4 34
Ngounié, riv., Gabon F2 46
Nguigmi, Niger, 2,400 D7 45
Nguru, Nig., 23,084 D7 45
Nha Trang, Viet., 5,000 F8 38
Nhill, Austl., 2,233 H3 51
Niafounké, Mali, 4,300 C4 45
Niagara, Grand Forks, N. Dak., 157 B8 110
Niagara, Marinette, Wis., 2,098 C5 124
Niagara, co., N.Y., 242,269 B2 108
Niagara, cave, Minn G6 99
Niagara, riv., Ont., Can D5 72
Niagara Falls, Ont., Can., 53,365 (*56,621) D5 72
Niagara Falls, Niagara, N.Y., 102,394 B1 108
Niagara-on-the-Lake, Ont., Can., 2,712 D5 72
Niagara University, Niagara, N.Y., 1,500 *B1 108
Niamey, Niger, 30,000 D5 45
Niangara, Con. L., 3,301 A4 48
Niangua, Webster, Mo., 287 D5 101
Niangua, riv., Mo D5 101
Niantic, New London, Conn., 2,788 D8 84
Niantic, Macon, Ill., 629 D4 90
Niarada, Sanders, Mont., 5 C2 102
Niassa, prov., Moz D6 48
Nibe, Den., 2,494 B3 24
Nibbe, Yellowstone, Mont., 20 E8 102
Nicaragua, country, N.A., 1,551,500 H12 61, D5 62
Nicaragua, lake, Nic E5 62
Nicastro, It., 21,500 E6 21
Nicatous, lake, Maine C4 96
Nice, Fr., 292,958 F7 14
Nice, former prov., Fr., 339,000 *F7 14
Niceville, Okaloosa, Fla., 4,517 G2 86
Nichicun, lake, Que., Can B2 73
Nicholas, co., Ky., 6,677 B6 94
Nicholas, co., W. Va., 25,414 C4 123
Nicholas, chan., Cuba D3 64
Nicholasville, Jessamine, Ky., 4,275 C5 94
Nicholls, Coffee, Ga., 930 E4 87
Nicholls, pt., N.Y G4 84
Nichols, Fairfield, Conn. (part of Trumbull) E4 84
Nichols, Muscatine, Iowa, 329 C6 92
Nichols, Greene, Mo., 100 D4 101
Nichols, Tioga, N.Y., 663 C4 108
Nichols, Marion, S.C., 617 C9 115
Nichols Hills, Oklahoma, Okla., 4,897 B4 112
Nicholson, Pearl River, Miss., 500 E4 100
Nicholson, Wyoming, Pa., 942 C10 114
Nicholson, riv., Austl C6 50
Nicholville, St. Lawrence, N.Y., 400 A2 108
Nickelsville, Scott, Va., 291 B2 121
Nickerie, riv., Sur A3 59
Nickerson, Reno, Kans., 1,091 A3, D5 93
Nickerson, Dodge, Nebr., 168 C9 103
Nickerson, hill, Conn D7 84
Nickleville, Venango, Pa., 50 D2 114
Nicobar, is., India G9 39
Nicodemus, Graham, Kans., 300 C4 93
Nicola, B.C., Can., 159 D7 68
Nicola, riv., B.C., Can D7 68
Nicolet, Que., Can., 4,441 C5 73
Nicolet, co., Que., Can., 30,827 C5 73
Nicolet, lake, Mich B6 98
Nicolet, riv., Que., Can C5 73
Nicollet, Nicollet, Minn., 493 F4 99
Nicollet, co., Minn., 23,196 F4 99
Nicolls' Town, Ba. Is., 441 C4 63
Nicoma Park, Oklahoma, Okla., 1,263 B4 112
Nicosia, Cyp., 48,864 (*81,744) E9 31
Nicosia, It., 17,600 F5 21
Nicotera, It., 9,143 E5 21
Nicoya, C.R., 10,461 E5 62
Nicoya, gulf, C.R F5 62
Nicoya, pen., C.R F5 62
Nictaux Falls, N.S., Can., 373 E5 74
Nicuadala, Moz A6 49
Nida, riv., Pol C6 26
Nidan, rock, Iwo 52
Nidd, riv., Eng F7 13
Nidda, Ger., 4,400 C4 17
Nidda, riv., Ger C4 17
Nidwalden, sub canton, Switz., 22,188 C5 19
Nidzica, Pol., 2,852 B6 26
Niebüll, Ger., 6,300 D2 24
Niederbronn-les-Bains, Fr., 4,074 F7 15
Niedermarsberg, Ger., 9,000 B3 17
Niederösterreich (Lower Austria), state, Aus., 1,374,012 *D7 16
Niedersachsen (Lower Saxony), state, Ger., 6,641,400 A4 17
Nielsville, Polk, Minn., 183 C2 99
Niemen, see Nemunas, riv., Sov. Un.
Niemodlin, Pol., 2,580 C4 26
Nienburg, Ger., 22,100 B4 16
Niers, riv., Ger C6 15
Nierstein, Ger., 5,500 D3 17
Niesky, Ger., 7,436 B9 17
Nieszawa, Pol., 2,403 B5 26
Niete, mtn., Lib E3 45
Nietleben, Ger., 5,208 B6 17
Nieuw Amsterdam, Sur A3 59
Nieuweroord, Neth., 753 B6 15
Nieuw Nickerie, Sur., 3,472 A3 59
Nieuwpoort, Bel., 6,899 C2 15
Nieuw Singkel, see Singkil Baru, Indon.
Nièvre, dept., Fr., 245,921 *D5 14
Nigadoo, N.B., Can., 361 B4 74
Nigde, Tur., 18,000 D10 29
Nigei, isl., B.C., Can D4 68
Nigel, S. Afr., 34,008 *C4 49
Niger, country, Afr., 2,850,000 E6 42, C6 45
Niger, riv., Afr E6 42
Niger, riv. mouths, Nig F6 45
Nigeria, country, Afr., 34,228,000 E6 45, F6 42
Nigg, Scot., 573 C5 13
Nighthawk, Okanogan, Wash., 15 A6 122
Nightingale, isl., Viet B7 38
Nigrita, Grc., 8,399 B4 23
Nihing, riv., Pak H11 41
Nihoa, isl., Haw m15 88
Nii, isl., Japan I9 37
Niigata, Jap., 314,528 H9 37
Niigata, pref., Jap., 2,442,037 *H9 37
Niihama, Jap., 125,688 J6 37
Niihau, isl., Haw B1 88
Niijimahon, Jap., 4,827 o18 37
Niimi, Jap., 15,100 I6 37
Níjar, Sp., 2,052 D4 20
Nijkerk, Neth., 8,200 B5 15
Nijmegen, Neth., 131,600 (*155,000) C5 15
Nikep, Allegany, Md., 215 D2 85
Nikitinka, Sov. Un D9 27
Nikitovka, Sov. Un., 10,000 q21 27
Nikki, Dah E5 45
Nikko, Jap., 33,348 H9 37
Nikolayev, Sov. Un., 242,000 H9 27
Nikolayevsk, Sov. Un., 40,000 D17 26
Nikolayevskiy, Sov. Un., 30,000 F15 27
Nikolsk, Sov. Un B3 29
Nikolski, Alsk., 92 E6 79
Nikonovskoye, Sov. Un n18 27
Nikopol, Bul., 5,409 D7 22
Nikopol, Sov. Un., 92,000 H10 27
Niksar, Tur., 10,600 B11 31
Nikshahr, Iran H10 41
Niksic, Yugo., 20,165 D4 22
Niland, Imperial, Calif., 800 F6 82
Nile, Allegany, N.Y., 100 C2 108
Nile, riv., Sud A3 47
Nile, riv., Afr D9 43
Niles, Cook, Ill., 25,073 E2 90
Niles, Ottawa, Kans., 105 D6 93
Niles, Berrien, Mich., 13,842 G4 98
Niles, Trumbull, Ohio, 19,545 A7 111
Nilvange, Fr., 9,337 C6 15
Nilwood, Macoupin, Ill., 274 D4 90
Nimaj, India, 10,000 E5 40
Nimba, mts., Guinea, Lib E3 45
Nimes, Fr., 99,802 F6 14
Nimmons, Clay, Ark., 154 A5 81
Nimmonsburg, Broome, N.Y *C5 108
Nimpkish, riv., B.C., Can D4 68
Nimrod, Perry, Ark C2 81
Nimrod, res., Ark C2 81
Nimule, Sud E3 47
Ninaview, Bent, Colo., 5 D7 83
Nine Mile, creek, Kans B8 93
Nine Mile, hill, Tenn E9 117
Nine Mile, pt., Mich C6 98
Nine Mile Falls, Spokane, Wash., 80 D7 122
Ninette, Man., Can., 673 E2 71
Ninety Mile, beach, Austl H7 51
Ninety Six, Greenwood, S.C., 1,435 C3 115
Nineveh, Johnson, Ind., 300 F5 91
Nineveh, Greene, Pa., 100 G1 114
Nineveh, ruins, Iraq D14 31
Ninga, Man., Can., 129 E2 71
Ningan, China, 40,000 C10 34, D4 37
Ningchiang, China H2 36
Ningchin, China, 15,000 F6 36
Ningching, China F4 34
Ningerh, China D11 34
Ninghai, China, 13,000 J9 36
Ninghsien, China, 5,000 D6 34
Ninghua, China, 21,000 K7 36
Ningming, China A7 38
Ningpo (Ninghsien), China, 237,500 F9 34
Ningsia, see Yinchwan, China
Ningsia Hui, auton. reg., China, 1,822,000 D6 34
Ningte, China, 30,000 K8 36
Ningtu, China, 32,000 K6 36
Ningwu, China E5 36
Ninh Binh, Viet., 25,000 B6 38
Ninilchik, Alsk., 169 C9, g16 79
Ninnekah, Grady, Okla., 300 C4 112
Ninnescah, riv., Kans E5 93
Ninnis, glacier, Ant C27 5
Ninole, Hawaii, Haw., 112 D6 88
Ninove, Bel., 11,831 D4 15
Nioaque, Braz., 2,578 C1 56
Niobe, Ward, N. Dak., 67 A3 110
Niobrara, Knox, Nebr., 736 B7 103
Niobrara, co., Wyo., 3,750 B8 125
Niobrara, creek, Wyo C8 125
Niobrara, riv., Nebr B7 103
Nioki, Con. L C2 48
Nioro, Mali, 6,100 C3 45
Nioro, Sen D1 45
Niort, Fr., 37,512 D3 14
Niota, Hancock, Ill., 250 C2 90
Niota, McMinn, Tenn., 679 D9 117
Niotaze, Chautauqua, Kans., 124 E8 93
Nipawin, Sask., Can., 3,836 D3 70
Nipigon, lake, Ont., Can E8 72
Nipishish, lake, Newf., Can g9 75
Nipisiguit, bay, N.B., Can B4 74
Nipisiguit, riv., N.B., Can B3 74
Nipissing, co., Ont., Can., 70,568 A5 72
Nipissing, lake, Ont., Can A5 72
Nipissing Junction, Ont., Can., 323 A5 72
Nipomo, San Luis Obispo, Calif., 550 E3 82
Nippers Harbour, Newf., Can., 236 D4 75
Nipton, San Bernardino, Calif., 40 E6 82
Niquelândia, Braz., 1,262 A3 56
Niquero, Cuba, 5,437 E5 64
Nirasaki, Jap., 11,500 n17 37
Nirmal, India, 19,896 H7 40
Niš (Nish), Yugo., 81,073 D5 22
Nisa, Port., 5,617 C2 20
Nisava, riv., Yugo D6 22
Nish (Niš), Yugo., 62,100 D5 22
Nishi, Iwo 52
Nishino, isl., Pac. O E8 7
Nishinomiya, Jap., 262,608 *o14 37
Nishinotoro, see Dalnyaya, Sov. Un.
Nishnabotna, riv., Iowa D2 92
Nisiros, isl., Grc D6 23
Nisko, Pol., 6,590 C7 26
Nisland, Butte, S. Dak., 211 C2 116
Nisqually, riv., Wash C3 122
Nissan, riv., Swe B7 24
Nissum, fjord, Den B2 24
Nisswa, Crow Wing, Minn., 742 D4 99
Nistowiak, lake, Sask., Can B3 70
Nisula, Houghton, Mich., 45 B2 98
Niterói, Braz., 228,826 C4, h6 56
Nith, riv., Scot E5 13
Nitra, Czech., 34,200 D5 26
Nitra, riv., Czech D5 26
Nitro, Kanawha and Putnam, W. Va., 6,894 C3 123
Nitta Yuma, Sharkey, Miss., 150 B3 100
Nittenau, Ger., 3,300 D7 17
Niue, N.Z. dep., Oceania, 4,553 *H11 7
Niut, range, B.C., Can D5 68
Nivelles, Bel., 14,345 D4 15
Nivernais, former prov., Fr., 236,000 D5 14
Nivernais, hills, Fr D5 14
Niverville, Man., Can., 474 E3 71
Nivskiy, Sov. Un D15 25
Niwka, Pol., 2,500 g10 26
Niwot, Boulder, Colo., 150 A5 83
Nixa, Christian, Mo., 944 D4 101
Nixburg, Coosa, Ala., 30 C3 78
Nixon, Washoe, Nev., 200 D2 104
Nixon, Middlesex, N.J., 14,000 *B4 106
Nixon, Gonzales, Tex., 1,751 E4, B4 118
Nizamabad, India, 79,093 H7 40
Nizamsagar, lake, India H6 40
Nizhmozero, Sov. Un E17 25
Nizhne-Chirskaya, Sov. Un., 2,000 G14 27
Nizhne-Kolymsk, Sov. Un C19 28
Nizhneudinsk, Sov. Un., 35,900 D12 28
Nizhneye, Sov. Un., 10,000 q21 27
Nizhniye Narykary, Sov. Un C21 9
Nizhniy-Lomov, Sov. Un., 5,000 E14 27
Nizhniy-Tagil, Sov. Un., 355,000 B5 29
Nizhnyaya Tunguska, riv., Sov. Un C12 28
Nizhnyaya Tura, Sov. Un., 20,400 D21 9
Nizmennyy, cape, Sov. Un E7 37
Nizza Monferrato, It., 9,372 E4 18
Njombe, Tan., 7,560 C5 48
Njurunda, Swe., 1,540 F7 25
Nkai, S. Rh A4 49
Nkangsamba, Cam E1 46
Nkata Bay, Nya D5 48
Noah, Coffee, Tenn B5 117
Noakhali, Pak., 16,677 F13 40
Noank, New London, Conn., 1,116 D9 84
Noasca, It., 963 D3 17
Noatak, Alsk., 275 B7 79
Noatak, riv., Alsk B7 79
Nobal, Lincoln, N. Mex., 20 D4 107
Nobel, Ont., Can., 693 B4 72
Nobeoka, Jap., 100,000 (122,527▲) J5 37
Noble, Walker, Ga., 200 B1 87
Noble, Richland, Ill., 761 E5 90
Noble, Sabine, La., 206 C2 95
Noble, Cleveland, Okla., 995 B4 112
Noble, co., Ind., 28,162 B7 91
Noble, co., Ohio, 10,982 C6 111
Noble, co., Okla., 10,376 A4 111
Nobleboro, Lincoln, Maine, 75 (679▲) D3 96
Nobleford, Alta., Can., 309 E4 69
Noble Lake, Jefferson, Ark., 100 C4 81
Nobles, co., Minn., 23,365 G3 99
Noblesville, Hamilton, Ind., 7,664 D5 91
Nobleton, Hernando, Fla., 150 D4 86
Noboribetsu, Jap E10 37
Nobscot, hill, Mass D1 97
Nocatee, De Soto, Fla., 750 E5 86
Nochixtlán, Mex., 2,571 o15 63
Nocona, Montague, Tex., 3,127 C4 118
Noda, see Chekhov, Sov. Un.
Nodaway, Adams, Iowa, 204 D3 92
Nodaway, co., Mo., 22,215 A3 101
Nodaway, riv., Iowa, Mo D2 92, A2 101
Node, Niobrara, Wyo., 15 C8 125
Nodoa, China C8 38
Noel, McDonald, Mo., 736 E3 101
Noelville, Ont., Can., 393 A4 72
Noemfoor, isl., W. Irian F8 35
Nogal, riv., Som D6 47
Nogales, Santa Cruz, Ariz., 7,286 F5 80
Nogales, Mex., 37,655 A2 63
Nogales, Mex., 11,179 n15 63
Nogara, It., 7,239 D7 18
Nogata, Jap., 62,179 J5 37
Nogent-en-Bassigny, Fr., 4,208 C6 14
Nogent-le-Rotrou, Fr., 9,428 C4 14
Nogent [-sur-Marne], Fr., 24,501 g10 14
Nogent [-sur-Seine], Fr., 3,777 C5 14
Noginsk, Sov. Un., 97,000 n18 27
Nogoa, riv., Austl B6 51
Nogoyá, Arg., 12,051 A5 54
Nograd, co., Hung., 236,393 *A4 22
Nogueira, Sp., 7,791 A2 20
Nohar, India, 13,728 C5 40
Noheji, Jap., 9,900 F10 37
Nohly, Richland, Mont., 9 C12 102
Noho, China, 5,000 B2 37
Noire, riv., Upper Volta D4 45
Noirmoutier, isl., Fr D2 14
Noisy-le-Sec, Fr., 31,187 g10 14
Nojima, cape, Jap I9 37
Nokkeushi, see Kitami, Jap.
Nok Kundi, Pak G11 41
Nokomis, Sask., Can., 560 F3 70
Nokomis, Sarasota, Fla., 2,253 E4 86
Nokomis, Montgomery, Ill., 2,476 D4 90
Nokomis, lake, Wis C4 124
Nokrek, peak, India E13 40
Nola, Scott, Ark., 25 C2 81
Nola, Cen. Afr. Rep E3 46
Nola, It., 16,400 D5 21
Nola, Lawrence, Miss., 125 D3 100
Nolan, Nolan, Tex., 100 C2 118
Nolan, Mingo, W. Va., 787 D2 123
Nolan, co., Tex., 18,963 C2 118
Nolensville, Williamson, Tenn., 400 B5 117
Nolin, flood control res., Ky C3 94
Nolin, riv., Ky C3 94
Nolinsk, Sov. Un., 9,600 B3 29
Noma, Holmes, Fla., 344 G3 86
Noma, cape, Jap K5 37
No Mans Land, isl., Mass D6 97
Nome, Alsk., 2,316 C6 79
Nome, Barnes, N. Dak., 145 C8 110
Nomeny, Fr., 964 F6 15
Nominingue, Que., Can., 744 C2 73
Nomme, Sov. Un., 18,026 *B5 27
Nonacho, lake, N.W. Ter., Can D11 66
Nonconnah, Shelby, Tenn E8 117
Nonconnah, creek, Tenn E8 117
Nondalton, Alsk., 205 C8 79
Nonesuch, riv., Maine E4 96
Nong Khai, Thai., 16,925 D5 38
Nongoma, S. Afr., 1,650 C5 49
Nonni, riv., China B9 34, B2 37
Nonoava, Mex., 1,582 B3 63
Noodle, Jones, Tex., 60 C3 118
Nooksack, Whatcom, Wash., 318 A3 122
Nooksack, riv., Wash A3 122
Noon, hill, Mass D2 97
Noonan, Divide, N. Dak., 625 A2 110
Noord Pagai, isl., Indon F2 35
Noordoostelijke Polder (Northeastern Polder), reg., Neth., 15,938 B5 15
Noordwijk-Binnen, Neth., 8,300 B4 15
Noorvik, Alsk., 384 B7 79
Nooseneck, Kent, R.I., 40 C10 84
Nootka, isl., B.C., Can E4 68
Nootka, sound, B.C., Can E4 68
Nopal, De Witt, Tex., 15 E4 118
No Point, pt., Md D5 85
Noquebay, lake, Wis C6 124
Nóqui, Ang., 406 C1 48
Nora, Sask., Can., 72 E4 70
Nora, Jo Daviess, Ill., 229 A4 90
Nora, Marion, Ind., 200 H8 91
Nora, Nuckolls, Nebr., 60 D8 103
Nora, Swe., 4,100 H6 25
Noranda, Que., Can., 11,477 G17 67
Nora Springs, Floyd, Iowa, 1,275 A5 92
Nor-Bayazet, Sov. Un., 8,986 B15 29
Norbeck, Faulk, S. Dak., 10 B6 116
Norborne, Carroll, Mo., 965 B4 101
Norcatur, Decatur, Kans., 302 C3 93
Norco, Riverside, Calif., 4,964 *F3 82
Norco, St. Charles, La., 4,682 B6, D5 95
Norcross, Gwinnett, Ga., 1,605 A5, C2 87
Norcross, Grant, Minn., 153 E2 99
Nord, canal, Fr D3 15
Nord, dept., Fr., 2,293,112 D3 14
Nord, mts., Hai F7 64
Nördborg, Den., 2,563 C3 24
Nordby, Den., 1,975 C2 24
Nordby, Den., 313 C4 24
Nordegg, Alta., Can., 1,000 C2 69
Nordegg, riv., Alta., Can C3 69
Norden, Ger., 16,200 B3 16
Norden, Keya Paha, Nebr., 15 B5 103
Nordenham, Ger., 26,900 B4 16
Norderney, Ger., 7,300 A7 15
Norderney, isl., Ger A7 15
Nordfjord, fjord, Nor G1 25
Nordhausen, Ger., 39,900 B5 17
Nordheim, Winnebago, Wis., 900 *D5 124
Nordhorn, Ger., 39,400 B3 16
Nordland, Jefferson, Wash., 100 A3 122
Nordland, co., Nor., 225,394 *D6 25
Nördlingen, Ger., 14,400 E5 17
Nordmaling, Swe., 1,150 F8 25
Nordman, Bonner, Idaho, 25 A2 89
Nordreisa, Nor C9 25
Nordrhein-Westfalen (North Rhine-Westphalia), state, Ger., 15,901,700 B2 17
Nordstrand, isl., Ger A4 16
Nord-Tröndelag, co., Nor., 112,185 *E4 25
Nordvik, Sov. Un., 2,500 B14 28
Nore, riv., Ire E4 11
Norene, Wilson, Tenn., 70 A5 117
Norfield, Lincoln, Miss., 50 D3 100
Norfolk, Litchfield, Conn., 850 (1,827▲) B4 84
Norfolk, Norfolk, Mass., 350 (3,471▲) E2 97
Norfolk, Madison, Nebr., 13,640 B8 103
Norfolk, St. Lawrence, N.Y., 1,353 A2 108
Norfolk (Independent City), Va., 304,869 (*574,900) B7, E6 121
Norfolk, co., Ont., Can., 50,475 E4 72
Norfolk, co., Eng., 561,980 B9 12
Norfolk, co., Mass., 510,256 B5 97
Norfolk, is., Pac. O H10 7
Norfolk Highlands, Norfolk, Va., 1,000 *E6 121
Norfolk Island, Austl. dep., Oceania H10 7

Norfork, Baxter, Ark., 283...A3 81
Norfork, dam, Ark...A3 81
Norfork, lake, Ark...A3 81
Norge, Grady, Okla., 60...C4 112
Norias, Kenedy, Tex., 25...F4 118
Norilsk, Sov. Un., 109,000..C11 28
Norland, Ont., Can., 257...C6 72
Norland, Dickenson, Va...B2 121
Norlina, Warren, N.C., 927..A5 109
Norma, Salem, N.J., 700...E2 106
Norma, Renville, N. Dak., 84.A4 110
Norma, Scott, Tenn., 250...C9 117
Normal, Madison, Ala., 1,500...A3 78
Normal, McLean, Ill., 13,357...C5 90
Norman, Montgomery, Ark., 482...C2 81
Norman, Jackson, Ind., 130..G5 91
Norman, Kearney, Nebr., 57.D7 103
Norman, Richmond, N.C., 220...B4 109
Norman, Cleveland, Okla., 33,412...B4 112
Norman, co., Minn., 11,253..C2 99
Norman, isl., Vir. Is...F16 65
Norman, riv., Austl...C7 50
Normanby, riv., Austl...B7 50
Normandin, Que., Can., 1,838...*G18 67
Normandy, Duval, Fla., 1,900...*B5 86
Normandy, former prov., Fr., 2,407,000...C4 14
Normandy, St. Louis, Mo., 4,452...A8 101
Normandy, Bedford, Tenn., 119...B5 117
Normandy, hills, Fr...C3 14
Normandy Beach, Ocean, N.J., 300...C4 106
Normandy Park, King, Wash., 3,224...*D2 122
Normangee, Madison and Leon, Tex., 718...D4 118
Normanna, Bee, Tex., 130..E4 118
Norman Park, Colquitt, Ga., 891...E3 87
Normanton, Austl., 234...C7 50
Normantown, Toombs, Ga., 49...D4 87
Norman Wells, N.W. Ter., Can., 297...C7 66
Nornalup, Austl...F2 50
Noroton Heights, Fairfield, Conn. (part of Darien)...E3 84
Norphlet, Union, Ark., 457..D3 81
Norquay, Sask., Can., 498..F4 70
Ñorquincó, Arg...C2 54
Norrahammar, Swe., 5,600..A8 24
Norra Värmdö, Swe...t36 25
Norrbotten, co., Swe., 262,600...*E10 25
Nörre Åby, Den., 1,844...C3 24
Nörre Alslev, Den., 1,062...D5 24
Nörresundby, Den., 10,456..A3 24
Norridge, Cook, Ill., 14,087.*E3 90
Norridgewock, Somerset, Maine, 850 (1,634▲)...D3 96
Norrie, Marathon, Wis., 65..D4 124
Norris, Fulton, Ill., 307...C3 90
Norris, Madison, Mont., 185.E5 102
Norris, Pickens, S.C., 594...B2 115
Norris, Mellette, S. Dak., 100...D4 116
Norris, Anderson, Tenn., 1,389...C9 117
Norris, dam, Tenn...C9 117
Norris, lake, Tenn...C10 117
Norris Arm, Newf., Can., 1,226...D4 75
Norris City, White, Ill., 1,243.F5 90
Norris Point, Newf., Can., 711...D3 75
Norristown, Montgomery, Pa., 38,925...A10, F11 114
Norrisville, Harford, Md., 75...A4 85
Norrköping, Swe., 91,400...H7, u34 25
Norrtälje, Swe., 8,900...H8, t36 25
Norseman, Austl., 2,104...F3 50
Norte, chan., Braz...B4 59
Norte, range, Braz...E3 59
Norte de Santander, dept., Col., 403,420...B3 60
North, Orangeburg, S.C., 1,047...D5 115
North, Mathews, Va., 150...D6 121
North, div., Ice., 28,388...*n23 25
North, brook, Vt...B3 120
North, cape, N.Z...K14 51
North, cape, Nor...A12 4, B11 25
North, cape, P.R...e10 65
North, chan., Ont., Can...A2 72
North, chan., N. Ire...C4 10
North, creek, Ga...B5 87
North, fork, Wyo...A3 125
North, head, Newf., Can...D2 75
North, inlet, S.C...E9 115
North, isl., N.Z...M14 51
North, isl., S.C...E9 115
North, is., La...E7 95
North, mtn., Okla...C3 112
North, mtn., Pa...D9 114
North, mtn., Va...C4 121
North, park, Pa...A5 114
North, pass, La...E6 95
North, plains, N. Mex...C1 107
North, pt., Md...B5 85
North, pt., Md...C5 85
North, pt., Mich...C7 98
North, pond, Maine...D3 96
North, pond, Mass...D1 97
North, riv., Ala...B2 78
North, riv., Newf., Can...B3 75
North, riv., Fla...C6 86
North, riv., Mass...E4 97
North, riv., Vt...F3 120
North, riv., Va...D3 121
North, sea, Eur...D9 8
North, sound, Ire...D2 11
North, sound, Scot...A6 13
North Abington, Plymouth, Mass., 4,900...B6, E3 97
North Acton, Middlesex, Mass., 100...C1 97
North Adams, Berkshire, Mass., 19,905...A1 97
North Adams, Hillsdale, Mich., 494...G6 98
North Agawam, Hampden, Mass., 2,000...*B2 97
North Albany, Benton, Oreg.C1 113
Northallerton, Eng., 6,720...F7 13
Northam, Austl., 7,200...F2 50
Northam, Eng., 6,572...C3 12
North America, cont., 229,700,000...C14 3, 61
North Amherst, Hampshire, Mass., 1,009...B2 97
North Amity, Aroostook, Maine, (206▲)...C5 96
North Amityville, Suffolk, N.Y., 6,000...*G3 84
Northampton, Austl., 626...E1 50
Northampton, Eng., 105,361...D6 10, B7 12
Northampton, Hampshire, Mass., 30,058...B2 97
Northampton, Northampton, Pa., 8,866...E11 114
Northampton, co., Eng., 398,132...B7 12
Northampton, co., N.C., 26,811...A6 109
Northampton, co., Pa., 201,412...E11 114
Northampton, co., Va., 16,966...D7 121
North Andaman, isl., India...F9 39
North Andover, Essex, Mass., 10,908...A5 97
North Anna, riv., Va...C5 121
North Anson, Somerset, Maine, 700...D3 96
North Apollo, Armstrong, Pa., 1,741...E2 114
North Arlington, Bergen, N.J., 17,477...D5 106
North Asheboro (Balfours), Randolph, N.C., 3,805...*B4 109
North Ashford, Windham, Conn...B8 84
North Atlanta, DeKalb, Ga., 12,661...C2 87
North Atlantic, ocean...D18 3
North Attleboro, Bristol, Mass., 14,777...C5 97
North Augusta, Ont., Can., 247...C9 72
North Augusta, Aiken, S.C., 10,348...D4 114
North Aurora, Kane, Ill., 2,088...F1 90
North Babylon, Suffolk, N.Y., 12,000...*G3 84
North Baltimore, Wood, Ohio, 3,011...A4 111
North Bancroft, Aroostook, Maine...C5 96
North Bangor, Franklin, N.Y., 570...A2 108
North Battleford, Sask., Can., 11,230...E1 70
North Battle Mountain, Lander, Nev., 10...C5 104
North Bay, Ont., Can., 23,781...A5, E9 72
North Bay, Dade, Fla., 2,006...*G6 86
North Bay, riv., Newf., Can...E4 75
North Beach, Calvert, Md., 606...C4 85
North Belle Vernon, Westmoreland, Pa., 3,148...*F2 114
North Bellingham, Norfolk, Mass., 250...E1 97
North Bellmore, Nassau, N.Y., 19,639...*G2 84
North Bellport, Suffolk, N.Y., 2,000...*G5 84
North Belmont, Gaston, N.C., 3,000...B2 109
North Bend, B.C., Can., 340.E7 68
North Bend, Dodge, Nebr., 1,174...C9 103
North Bend, Hamilton, Ohio, 622...D1 111
North Bend, Coos, Oreg., 7,512...D2 113
North Bend, Clinton, Pa., 800...D6 114
North Bend, King, Wash., 945...B4 122
North Bend, Jackson, Wis., 100...D2 124
North Bennington, Bennington, Vt., 1,437...F2 120
North Bergen, Hudson, N.J., 42,387...D5 106
North Berwick, York, Maine, 1,295 (1,844▲)...E2 96
North Berwick, Scot., 4,161..D6 13
North Billerica, Middlesex, Mass., 3,000...A5, C2 97
North Bloomfield, Trumbull, Ohio, 350...A7 111
North Bonneville, Skamania, Wash., 494...D4 122
North Borneo, reg., Mala., 334,141...I14 33, D5 35
Northboro, Page, Iowa, 135..D2 92
Northboro, Worcester, Mass., 2,516 (6,687▲)...B4 97
North Brabant, prov., Neth., 1,512,800...C5 15
North Braddock, Allegheny, Pa., 13,204...F2 114
North Bradford, Penobscot, Maine, 100...C4 96
North Bradley, Midland, Mich., 220...E6 98
North Branch, Allegany, Md., 250...D2 85
North Branch, Lapeer, Mich., 901...E7 98
North Branch, Chisago, Minn., 949...E6 99
North Branch, Somerset, N.J., 250...B2 106
North Branford, New Haven, Conn., 450 (6,771▲)...D5 84
North Brewer, Penobscot, Maine (part of Brewer)...D4 96
Northbridge, Worcester, Mass., 2,128 (10,800▲)...B4 97
North Bridgton, Cumberland, Maine, 300...D2 96
North Brook, Ont., Can., 369...C7 72
Northbrook, Cook, Ill., 11,635...E2 90
North Brookfield, Worcester, Mass., 2,615 (3,616▲)...B3 97
North Brooksville, Hancock, Maine, 100...D4 96
North Brother, mtn., Maine...C4 96
North Brunswick, Middlesex, N.J., 10,099...*C4 106
North Buena Vista, Clayton, Iowa, 50...B7 92
North Calais, Washington, Vt., 35...C4 120
North Caldwell, Essex, N.J., 4,163...B4 106
North Canadian, riv., Okla...A1 112
North Canton, Hartford, Conn., 250...B5 84
North Canton, Cherokee, Ga., 1,996...B2 87
North Canton, Stark, Ohio, 7,727...B6 111
North Cape, Racine, Wis., 200...F1 124
North Carolina, state, U.S., 4,556,155...C12 77, 109
North Carrollton, Carroll, Miss., 521...B4 100
North Carver, Plymouth, Mass., 400...C6 97
North Catasauqua, Northampton, Pa., 2,805...*E11 114
North Cedar, Black Hawk, Iowa, 1,500...*B5 92
North Charleroi, Washington, Pa., 2,259...*F2 114
North Charleston, Charleston, S.C., 22,339...F2, F8 115
North Charlestown, Sullivan, N.H., 75...D2 105
North Chelmsford, Middlesex, Mass., 3,500...A5 97
North Chicago, Lake, Ill., 22,938...A6, E2 90
North Chili, Monroe, N.Y., 2,000...*B3 108
North Chillicothe, Peoria, Ill., 2,259...C4 90
North City, King, Wash., 2,000...*B3 122
North Clarendon, Rutland, Vt. 200...D13 120
North Cohasset, Norfolk, Mass., 150...D4 97
North Cohocton, Steuben, N.Y., 250...C3 108
North College Hill, Hamilton, Ohio, 12,035...D2 111
North Collins, Erie, N.Y., 1,574...C2 108
North Conway, Carroll, N.H., 1,104...B4 105
North Corbin, Laurel, Ky., 950...*D5 94
North Cornville, Somerset, Maine...D3 96
North Coroin, Laurel, Ky., 950...D5 94
North Coventry, Tolland, Conn., 100...B7 84
Northcraft, Fulton, Pa...G5 114
North Creek, Warren, N.Y., 900...B7 108
Northcrest, Del Norte, Calif., 1,945...*B1 82
North Dakota, state, U.S., 632,446...A7 76, 110
North Danville, Caledonia, Vt., 80...C4 120
North Dartmouth, Bristol, Mass., 4,000...C5 97
North Decatur, De Kalb, Ga., 10,000...*C2 87
North Derby, Orleans, Vt., 81...A4 120
North Dighton, Bristol, Mass., 1,167...C5 97
North Dixmont, Penobscot, Maine, 100...D3 96
North Dorset, Bennington, Vt., 25...E2 120
North Downs, hills, Eng...C8 12
North Druid Hills, De Kalb, Ga., 4,000...*C2 87
North East, Cecil, Md., 1,628...A6 85
North East, Erie, Pa., 4,217..B2 114
Northeast, cape, Alsk...C6 79
North East, isl., Nor...A12 4
North East, is., Truk...52
Northeast, pass, La...E6 95
Northeast, pond, N.H...D5 105
Northeast, riv., Md...A6 85
North East Carry, Piscataquis, Maine, 9...C3 96
Northeastern Polder (Noordoostelijke Polder), prov., Neth., 28,800...B5 15
Northeast Foreland, pt., Grnld..A16 4
North East Frontier Agency, ter., India, 336,558...*C10 39
Northeast Harbor, Hancock, Maine, 750...D4 96
Northeast New Guinea, reg., N. Gui., 875,000...k12 50
North Eastham, Barnstable, Mass., 200...C8 97
North Easton, Bristol, Mass., 4,000...B5 97
Northeast Providence, chan., Ba. Is...C5 64
North Edmonton, Alta., Can., 1,200...C4 69
North Egremont, Berkshire, Mass., 150...B1 97
Northeim, Ger., 19,300...B5 17
North Emporia, Greensville, Va. (part of Emporia)...E5 121
North End, Grafton, N.H...C3 105
North English, Iowa, Iowa, 1,004...C5 92
North Enid, Garfield, Okla., 286...A4 112
Northern, prov., Nya...D5 48
Northern, prov., N. Rh...D5 48
Northern, reg., Ghana...E4 45
Northern, reg., Ken...A6 48
Northern, reg., Nig., 16,840..D6 45
Northern, reg., Sud., 873,059.B2 47
Northern, reg., Tan., 771,426.B6 48
Northern, reg., Ug...A5 48
Northern, head, N.B., Can...E3 74
Northern Bight, Newf., Can., 40...D4 75
Northern Dvina, riv., Sov. Un..C17 9
Northern Indian, lake, Man., Can...A3 71
Northern Ireland, reg., U.K., 1,425,462...C3 10
Northern Light, lake, Ont., Can.A7 99
Northern Rhodesia, Br. dep., Afr., 2,480,000...D4 48
Northern Sporades, is., Grc...C5 23
Northern Territory, ter., Austl., 27,095...B5 50
North Esk, riv., Scot...D6 13
North Fabius, riv., Mo...A6 101
North Fairfield, Huron, Ohio, 547...A5 111
North Falmouth, Barstable, Mass., 300...C6 97
North Fayette, Kennebec, Maine, 220...D2 96
North Ferrisburg, Addison, Vt., 200...C2 120
Northfield, B.C., Can., 610..A8 68
Northfield, Litchfield, Conn., 350...C4 84
Northfield, Cook, Ill., 4,005.*E3 90
Northfield, Washington, Maine, 50 (79▲)...D5 96
Northfield, Franklin, Mass., 1,179 (2,320▲)...A3 97
Northfield, Rice, Minn., 8,707...F5 99
Northfield, Merrimack, N.H., 1,243 (1,784▲)...D3 105
Northfield, Atlantic, N.J., 5,849...E3 106
Northfield, Summit, Ohio, 1,055...B2 111
Northfield (P.O.), Summit, Ohio, 2,427...*B2 111
Northfield, Motley, Tex., 60..B2 118
Northfield, Washington, Vt., 2,159 (4,511▲)...C3 120
Northfield, Jackson, Wis., 50...D2 124
Northfield, mts., Vt...C3 120
Northfield Center, Washington, Vt., 100...C3 120
Northfield Falls, Washington, Vt., 325...C3 120
North Fond du Lac, Fond du Lac, Wis., 2,549...B5, E5 124
Northford, New Haven, Conn., 300...D5 84
North Fork, Madera, Calif., 200...D4 82
North Fork, Lemhi, Idaho, 30.D5 89
North Fork Republican, riv., Colo.A8 83
North Foster, Providence, R.I., 80...B10 84
North Fox, isl., Mich...C5 98
North Franklin, New London, Conn., 60...C8 84
North Freedom, Sauk, Wis., 579...E4 124
North Frisian, is., Ger...A4 16
North Fryeburg, Oxford, Maine, 150...D2 96
North Galiano, B.C., Can., 83...B9 68
North Gamboa (Gamboa), C.Z., 3,489...k11 62
Northgate, Sask., Can., 70...H4 70
Northgate, Burke, N. Dak., 65...A3 110
North Glen Ellyn, Du Page, Ill., 1,500...*F2 90
North Gower, Ont., Can., 306...B9 72
North Grafton, Worcester, Mass., 2,600...B4 97
North Granby, Hartford, Conn., 200...B5 84
North Gray, Cumberland, Maine...E5 96
North Great River, Suffolk, N.Y., 1,500...*G4 84
North Greenville, Washington, Miss., 2,516...*B2 100
North Grosvenor Dale, Windham, Conn., 1,874...B9 84
North Groton, Grafton, N.H.C3 105
North Guilford, New Haven, Conn...D6 84
North Guilford, Piscataquis, Maine, 30...C3 96
North Gulfport, Harrison, Miss., 3,323...*E4 100
North Haledon, Passaic, N.J., 6,026...*B4 106
North Hampton, Rockingham, N.H., 678 (1,910▲)...E5 105
North Hampton, Clark, Ohio, 495...C4 111
North Hanover, Plymouth, Mass., 450...E3 97
North Harlowe, Craven, N.C...C7 109
North Hartland, Windsor, Vt., 150...D4 120
North Hartland, res., Vt...D4 120
North Hartsville, Darlington, S.C., 1,899...*C7 115
North Hatfield, Hampshire, Mass., 85...B2 97
North Hatley, Que., Can., 719...D6 73
North Haven, New Haven, Conn., 15,935...D5 84
North Haven, Knox, Maine, 330 (384▲)...D4 102
North Haverhill, Grafton, N.H., 300...B3 105
North Head, N.B., Can., 609.E3 74
North Henderson, Mercer, Ill., 210...B3 90
North Henderson, Vance, N.C., 1,995...*A5 109
North Hero, Grand Isle, Vt., 50 (328▲)...B2 120
North Highlands, Sacramento, Calif., 21,271...*A6 82
North Holland (Noordholland), prov., Neth., 2,073,100...B4 15
North Holston, Smyth, Va., 200...B3 121
North Horn, lake, Tenn...E8 117
North Hornell, Steuben, N.Y., 917...*C3 108
North Horr, Ken...A6 48
North Hudson, Essex, N.Y., 65...A7 108
North Hudson, St. Croix, Wis., 1,019...*D1 124
North Hyde Park, Lamoille, Vt., 230...B3 120
North Industry, Stark, Ohio, 1,800...B6 111
North Irwin, Westmoreland, Pa., 1,143...*F2 114
North Isleboro, Waldo, Maine, 50...D4 96
North Jackson, Mahoning, Ohio, 700...A7 111
North Java, Wyoming, N.Y., 250...C2 108
North Jay, Franklin, Maine, 500...D2 96
North Judson, Starke, Ind., 1,942...B4 91
North Kamloops, B.C., Can., 6,456...D7 68
North Kansas City, Clay, Mo., 5,657...E2 101
North Kennebunkport (Arundel), York, Maine (907▲)...E2 96
North Kingstown, Washington, R.I., 10,000 (18,977▲)...C11 84
North Kingsville, Ashtabula, Ohio, 1,854...A7 111
North La Junta, Otero, Colo., 950...C7 83
Northlake, Cook, Ill., 12,318...*F2 90
North Lake, Marquette, Mich., 400...B3 98
North Lake, Waukesha, Wis., 300...E1 124
Northland, Marquette, Mich., 45...B3 98
North Laramie, riv., Wyo...C7 125
North Larchmont, Westchester, N.Y., 9,000...*D2 108
North Las Vegas, Clark, Nev., 18,422...G6 104
North Lawrence, St. Lawrence, N.Y., 400...A2 108
North Leominster, Worcester, Mass. (part of Leominster).A4 97
North Lewisburg, Champaign, Ohio, 879...B4 111
North Liberty, St. Joseph, Ind., 1,241...A5 91
North Liberty, Johnson, Iowa, 334...C6 92
North Lima, Allen, Ohio, 600...B3 111
North Lima, Mahoning, Ohio, 600...B7 111
North Lindenhurst, Suffolk, N.Y., 10,000...G3 84
North Little Rock, Pulaski, Ark., 58,032...C3, D6 81
North Loup, Valley, Nebr., 453...C7 103
North Loup, riv., Nebr...B5 103
North Lubec, Washington, Maine, 250...D5 96
North Madison, New Haven, Conn., 40...D6 84
North Madison, Jefferson, Ind. (part of Madison)...G7 91
North Magnetic Pole, Can...B12 66
North Mam, peak, Colo...B3 83
North Manchester, Wabash, Ind., 4,377...C6 91
North Manitou, isl., Mich...C5 98
North Mankato, Nicollet, Minn., 5,927...F4 99
North Massapequa, Nassau, N.Y., 12,300...*G3 84
North Merrick, Nassau, N.Y., 12,976...*G2 84
North Miami, Dade, Fla., 28,708...F3, G6 86
North Miami, Ottawa, Okla., 472...A7 112
North Miami Beach, Dade, Fla., 21,405...E3 86
North Middleboro, Plymouth, Mass., 400...C6 97
North Middletown, Bourbon, Ky., 291...B5 94
North Monmouth, Kennebec, Maine, 300...D2 96
North Montpelier, Washington, Vt., 140...C4 120
Northmoor, Platte, Mo., 696...E2 101
North Muskegon, Muskegon, Mich., 3,855...E4 98
North Newcastle, Lincoln, Maine...D3 96
North New Hyde Park, Nassau, N.Y., 9,930...*E7 108
North Newport, Sullivan, N.H...D2 105
North New Portland, Somerset, Maine, 300...D2 96
North New River, canal, Fla..F6 86
North Newry, Oxford, Maine...D2 96
North Newton (Bethel College), Harvey, Kans., 890...A5 93
North Newtown, Fairfield, Conn...D3 84
North Norway, Oxford, Maine, 100...D2 96
North Norwich, Chenango, N.Y., 235...C5 108
North Ogden, Weber, Utah, 2,621...B4 119
North Ogden, peak, Utah...C2 119
North Olmsted, Cuyahoga, Ohio, 16,290...B1 111
Northome, Koochiching, Minn., 291...C4 99
North Orange, Franklin, Mass., 65...A3 97
North Oxford, Worcester, Mass., 1,466...B4 97
North Pacific, ocean...E9 2
North Palm Beach, Palm Beach, Fla., 2,684...*F6 86
North Park, Winnebago, Ill., 1,500...*A4 90
North Park, basin, Colo...A4 83
North Parsonfield, York, Maine, 100...E2 96
North Patchogue, Suffolk, N.Y., 6,000...*D4 108
North Pekin, Tazewell, Ill., 2,025...*C4 90
North Pelham, Westchester, N.Y., 5,326...*D1 108
North Penobscot, Hancock, Maine, 125...D4 96
North Pitcher, Chenango, N.Y., 75...C5 108
North Plain, Middlesex, Conn., 40...D7 84
North Plainfield, Somerset, N.J., 16,993...B4 106
North Plains, Washington, Oreg., 500...A1, B3 113
North Platte, Lincoln, Nebr., 17,184...C5 103
North Platte, riv., U.S...B7 76
North Pleasanton, Atascosa, Tex., 1,018...*E3 118
North Plymouth, Plymouth, Mass., 100...C6 97
North Pole, Arc. O...4
North Pole, mtn., Idaho...D3 89
North Pomfret, Windsor, Vt., 50...D4 120
Northport, Tuscaloosa, Ala., 5,245...B2 78
Northport, Waldo, Maine, 100 (648▲)...D4 96
Northport, Leelanau, Mich., 530...C5 98
Northport, Morrill, Nebr., 110...C2 103
Northport, Suffolk, N.Y., 5,972...D3 108
Northport, Stevens, Wash., 482...A8 122
North Powder, Union, Oreg., 399...B9 113
North Pownal, Cumberland, Maine, 55...E5 96
North Pownal, Bennington, Vt. 275...F2 120
North Prairie, Waukesha, Wis., 489...F5 124
North Providence, Providence, R.I., 18,220...B11 84
North Range Corner, N.S., Can., 61...E4 74
North Reading, Middlesex, Mass., 8,331...C3 97
North Rhine-Westphalia (Nordrhein-Westfalen), state, Ger., 15,852,500...B2 17
North Richland Hills, Tarrant, Tex., 8,662...*B5 118
Northridge, Montgomery, Ohio, 1,500...*C3 111
North Ridgeville, Lorain, Ohio, 8,057...A5, B1 111
North Riding, Yorkshire, see York, co., Eng.
North Rim, Mohave, Ariz...A3 80
North River Bridge, N.S., Can., 93...C9 74
North Riverside, Cook, Ill., 8,401...*F3 90
North Robinson, Crawford, Ohio, 289...B5 111
North Rochester, Strafford, N.H. (part of Rochester)..D5 105
North Rockville Centre, Nassau, N.Y., 1,500...*G2 84
North Ronaldsay, isl., Scot...A5 10
North Rose, Wayne, N.Y., 350...B4 108
North Royalton, Cuyahoga, Ohio, 9,290...B2 111
North Rustico, P.E.I., Can., 780...C6 74
Norths, coast, Ant...C25 5
Norths, highland, Ant...C25 5
North Sacramento, Sacramento, Calif., 12,922.A6 82
North St. Paul, Ramsey, Minn., 8,520...E7 99
North Salem, Hendricks, Ind., 626...E4 91
North Salem, Rockingham, N.H., 400...E4 105
North Salt Lake, Davis, Utah, 1,655...C4 119
North Sandwich, Carroll, N.H., 75...C4 105
North Santee, riv., S.C...E9 115
North Saskatchewan, riv., Alta., Sask., Can...F10 66
North Scarboro, Cumberland, Maine, 75...E5 96
North Scituate, Plymouth, Mass., 3,421...D4 97
North Scituate, Providence, R.I., 500...B10 84
North Seaford, Nassau, N.Y., 3,000...*G3 84
North Searsmont, Waldo, Maine, 50...D3 90
North Sherburne, Rutland, Vt...D3 120
North Shore, St. Tammany, La...B8 95
North Shreveport, Caddo, La., 7,701...*B2 95
North Shrewsbury, Rutland, Vt., 40...D3 120
Northside, Granville, N.C., 100...A5 109
North Sioux City, Union, S. Dak., 736...E9 116
North Slidell, St. Tammany, La. (part of Slidell)...B8 95
North Springfield, Erie, Pa., 250...C1 114
North Springfield, Windsor, Vt., 600...E3 120
North Springfield, Fairfax, Va., 5,000...*B5 121
North Springfield, flood control res., Vt...E3 120
Northstar, Gratiot, Mich., 200...E6 98
North Star, Darke, Ohio, 169...B3 111
North Sterling, Windham, Conn...B9 84
North Stonington, New London, Conn., 800 (1,982▲)...D9 84
North Stradbroke, isl., Austl...C9 51
North Stratford, Coos, N.H., 600...A3, B1 105
North Street, St. Clair, Mich., 50...E8 98
North Sudbury, Middlesex, Mass...C1 97
North Sumatra, see Sumatra, isl., Indon.
North Sunderland, Eng., 1,580...E7 13
North Sutton, Merrimack, N.H., 160...D3 105
North Swansea, Bristol, Mass., 250...C5 97
North Sydney, N.S., Can., 8,657...C9 74
North Syracuse, Onondaga, N.Y., 7,412...B4 108
North Tarrytown, Westchester, N.Y., 8,818...D2, D3 108
North Terre Haute, Vigo, Ind., 1,100...E3 91
North Thetford, Orange, Vt., 100...D4 120
North Thompson, riv., B.C., Can...D8 68
North Tolsta, Scot...B2 13
North Tonawanda, Niagara, N.Y., 34,757...B2 108
North Troy, Orleans, Vt., 961...B4 120
North Truchas, peak, N. Mex..B4 107
North Truro, Barnstable, Mass., 450...B7 97
North Tunica, Tunica, Miss., 1,025...*A3 100
North Turner, mtn., Maine...C4 96
North Twin, lake, Newf., Can...D3 75
North Twin, lake, Wis...B4 124
North Tyne, riv., Eng...E6 13
North Uist, isl., Scot...C1 13
Northumberland, Coos, N.H., 100 (2,586▲)...A3 105
Northumberland, Northumberland, Pa., 4,156...E8 114
Northumberland, co., N.B., Can., 47,223...B3 66

Northumberland, co., Ont., Can., 41,892 ... C7 72
Northumberland, co., Eng., 818,988 ... *C6 10
Northumberland, co., Pa., 104,138 ... D8 114
Northumberland, co., Va., 10,185 ... D6 121
Northumberland, is., Austl ... D9 50
Northumberland, strait, Can ... C5 74
North Umpqua, riv., Oreg ... D4 113
North Uvalde, Uvalde, Tex. (part of Uvalde) ... E3 118
North Uxbridge, Worcester, Mass., 1,882 ... B4 97
Northvale, Bergen, N.J., 2,892 ... A5, C5 106
North Valley Stream, Nassau, N.Y., 5,000 ... *E2 108
North Vancouver, B.C., Can., 23,656 ... A10, E6 68
North Vandergrift, Armstrong, Pa., 1,827 ... *E2 114
North Vassalboro, Kennebec, Maine, 778 ... D3 96
North Vernon, Jennings, Ind., 4,307 ... F6 91
Northview, Webster, Mo., 200 ... D4 101
Northville, Litchfield, Conn., 155 ... C3 84
Northville, Wayne and Oakville, Mich., 3,967 ... A7 98
Northville, Fulton, N.Y., 1,156 ... B6 108
Northville, Spink, S. Dak., 153 ... B7 116
North Virginia Beach, Princess Anne, Va., 2,587 ... *E7 121
North Wabiskaw, lake, Alta., Can ... A3 69
North Waldoboro, Lincoln, Maine, 262 ... D3 96
North Wales, Montgomery, Pa., 3,673 ... F11 114
North Walpole, Cheshire, N.H., 950 ... D2 105
North Walsham, Eng., 5,010 ... B9 12
North Warren, Knox, Maine, 75 ... D3 96
North Warren, Warren, Pa., 1,458 ... C3 114
North Washington, Chickasaw, Iowa, 156 ... A5 92
North Waterford, Oxford, Maine, 300 ... D2 96
North Weare, Hillsboro, N.H., 250 ... D3 105
North Webster, Kosciusko, Ind., 494 ... B6 91
North West, cape, Austl ... D1 50
North West, harbor, N.Y ... E7 84
Northwest, highlands, Scot ... D3 13
North Westchester, New London, Conn., 100 ... C7 84
North Western, prov., N. Rh ... D3 48
North-West Frontier, reg., Pak ... A3 40
North Westminster (Gageville), Windham, Vt., 368 ... E4 120
Northwest Park Apartments, Montgomery, Md., 3,000 ... *B3 85
Northwest Polder, reg., Neth ... B5 15
North Westport, Bristol, Mass., 3,000 ... C5 97
Northwest Providence, chan., W.I ... B4 64
North West River, Newf., Can., 753 ... B1, h9 75
Northwest St. Augustin, riv., Que., Can ... C2 75
Northwest Territories, ter., Can., 22,998 ... C13 66
North Weymouth, Norfolk, Mass. (part of Weymouth) ... D3 97
North Whittier Heights, Los Angeles, Calif., 12,200 ... *E4 82
Northwich, Eng., 19,374 ... A5 12
North Wilbraham, Hampden, Mass., 2,000 ... B3 97
North Wildwood, Cape May, N.J., 3,598 ... F3 106
North Wilkesboro, Wilkes, N.C., 4,197 ... A2 109
North Williston, Chittenden, Vt., 65 ... C2 120
North Wilmington, Middlesex, Mass., 900 ... C2 97
North Wilton, Fairfield, Conn., 50 ... E3 84
North Windham, Windham, Conn., 250 ... B8 84
North Windham, Cumberland, Maine, 900 ... E2, E4 96
North Windham, Windham, Vt ... E3 120
North Winterport, Waldo, Maine ... D4 96
North Woburn Junction, Middlesex, Mass. (part of Woburn) ... C2 97
Northwood, Worth, Iowa, 1,768 ... A4 92
Northwood, Kalamazoo, Mich., 4,000 ... *E5 98
Northwood, Rockingham, N.H., 350 (1,034▲) ... D4 105
Northwood, Grand Forks, N. Dak., 1,195 ... B8 110
North Woodbury, Litchfield, Conn., 750 ... C4 84
Northwood Center, Rockingham, N.H., 120 ... D4 105
Northwood Narrows, Rockingham, N.H., 200 ... D4 105
Northwood Ridge, Rockingham, N.H., 125 ... D4 105
Northwoods, DeKalb, Ga., 1,000 ... *A5 87
Northwoods, St. Louis, Mo., 4,701 ... *C7 101
North Woodstock, Windham, Conn., 350 ... B9 84
North Woodstock, Grafton, N.H., 600 ... B3 105
North Yarmouth, Cumberland, Maine, 150 (1,140) ... E5 96
North York, York, Pa., 2,290 ... G8 114
North York, moors, Eng ... C6 10
North Zurich, Madison, Tex., 400 ... D4 118
Norton, N.B., Can., 846 ... D4 74
Norton, Eng., 4,773 ... F8 13
Norton, Norton, Kans., 3,345 ... C4 93
Norton, Bristol, Mass., 1,501 (6,818▲) ... C5 97
Norton, Jackson, N.C. ... D3 109
Norton, S. Rh., 1,800 ... A5 49
Norton, Runnels, Tex., 90 ... D2 118
Norton, Essex, Vt., 100 (241▲) ... A5 120
Norton (Independent City), Va., 5,013 ... B2 121
Norton, Randolph, W. Va., 400 ... C5 123
Norton, co., Kans., 8,035 ... C4 93
Norton, bay, Alsk ... C8 79
Norton, pond, Vt ... B5 120
Norton, res., Mass ... B12 84
Norton, sound, Alsk ... C6 79
Norton Center, Summit, Ohio, 2,000 ... *A6 111
Nortons Corner, Maricopa, Ariz., 100 ... G2, D4 80
Nortonville, Jefferson, Kans., 595 ... A7, C8 93
Nortonville, Hopkins, Ky., 775 ... C2 94
Nortonville, La Moure, N. Dak., 105 ... C7 110
Nortorf, Ger., 5,900 ... D3 24
Norvegia, cape, Ant ... B11 5
Norvell, Crittenden, Ark., 362 ... B5 81
Norvello, Mecklenburg, Va., 75 ... E4 121
Norvelt, Westmoreland, Pa., 1,211 ... *F3 114
Norwalk, Los Angeles, Calif., 88,739 ... *F2 82
Norwalk, Fairfield, Conn., 67,775 ... E3 84
Norwalk, Warren, Iowa, 1,328 ... A7, C4 92
Norwalk, Huron, Ohio, 12,900 ... A5 111
Norwalk, Monroe, Wis., 484 ... E3 124
Norwalk, is., Conn ... E3 84
Norwalk, riv., Conn ... E3 84
Norway, Benton, Iowa, 516 ... C6 92
Norway, Republic, Kans., 100 ... C6 93
Norway, Oxford, Maine, 2,654 (3,733▲) ... D2 96
Norway, Dickinson, Mich., 3,171 ... C3 98
Norway, Coos, Oreg., 175 ... D2 113
Norway, Orangeburg, S.C., 525 ... E5 115
Norway, country, Eur., 3,596,300 ... C11 8, E5 25
Norway, isl., Viet ... B7 38
Norway, lake, Minn ... E3 99
Norway House, Man., Can., 543 ... C3 71
Norwayne, Wayne, Mich., 6,000 ... *A7 98
Norwegian, sea, Eur ... n26 25
Norwell, Plymouth, Mass., 800 (5,207▲) ... D4 97
Norwich, Ont., Can., 1,703 ... E4 72
Norwich, New London, Conn., 38,506 ... C8 84
Norwich, Eng., 119,904 (★165,000) ... B9 12
Norwich, Kingman, Kans., 430 ... E6 93
Norwich, Chenango, N.Y., 9,175 ... C5 108
Norwich, McHenry, N. Dak., 75 ... A5 110
Norwich, Windsor, Vt., 500 (1,790▲) ... D4 120
Norwichtown, New London, Conn. (part of Norwich) ... C8 84
Norwood, Ont., Can., 1,060 ... C7 72
Norwood, San Miguel, Colo., 443 ... C2 83
Norwood, Warren, Ga., 294 ... C4 87
Norwood, Polk, Iowa, 1,500 ... A7 92
Norwood, East Feliciana, La., 427 ... D4 95
Norwood, Norfolk, Mass., 24,898 ... B5, D2 97
Norwood, Carver, Minn., 945 ... F5 99
Norwood, Wright, Mo., 263 ... D5 101
Norwood, Bergen, N.J., 2,852 ... C5 106
Norwood, St. Lawrence, N.Y., 2,200 ... A2 108
Norwood, Stanly, N.C., 1,844 ... B3 109
Norwood, Hamilton, Ohio, 34,580 ... D2 111
Norwood Delaware, Pa., 6,729 ... G11 114
Norwood, Knox, Tenn., 5,000 ... *D10 117
Norwood, lake, N.C ... B3 108
Nosbonsing, lake, Ont., Can ... A5 72
Noshiro, Jap., 36,500 ... F10 37
Nösjön, lake, Swe ... A6 24
Noss, head, Scot ... B6 13
Nossa Senhora das Dores, Braz., 4,740 ... D7 57
Nossen, Ger., 8,505 ... B8 17
Nosy-bé, isl., Malag ... f9 49
Notasulga, Macon, Ala., 884 ... C4 78
Notch, mtn., Mass ... A3 97
Notch, peak, Utah ... D2 119
Notch Hill, B.C., Can., 104 ... D8 68
Notchland, Carroll, N.H. ... B4 105
Notec, riv., Pol ... B3 26
Notikewin, Alta., Can., 88 ... A2 69
Notikewin, riv., Alta., Can ... A2 69
Noto, It., 21,000 ... F5 21
Noto, gulf, It ... F5 21
Noto, isl., Jap ... H8 37
Notodden, Nor., 7,600 ... H3 25
Notre Dame, N.B., Can., 311 ... C5 74
Notre Dame, St. Joseph, Ind., 6,500 ... *A5 91
Notre Dame, bay, Newf., Can ... D4 75
Notre Dame, mts., Que., Can ... D6 73
Notre Dame [de la Salette], Que., Can., 444 ... D2 73
Notre Dame de Lourdes, Man., Can., 511 ... E2 71
Notre Dame [de Rimouski], Que., Can., 390 ... A9 73
Notre Dame [des Bois], Que., Can., 243 ... D6 73
Notre Dame du Lac, Que., Can., 1,680 ... B9 73
Notre Dame du Laus, Que., Can., 565 ... C2 73
Notre Dame Junction, Newf., Can., 40 ... D4 75
Nott, St. Tammany, La ... B7 95
Nottawa, Ont., Can., 272 ... C4 72
Nottawasaga, bay, Ont., Can ... C4 72
Nottaway, riv., Que., Can ... F17 67
Nottely, res., Ga ... B2 87
Nottingham, Eng., 311,645 (★615,000) ... B6 12
Nottingham Rockingham, N.H., 100 (623▲) ... D4 105
Nottingham, Bucks, Pa., 2,500 ... *A12 114
Nottingham, co., Eng., 902,966 ... A7 12
Nottingham, isl., N.W. Ter., Can ... D17 67
Nottoway, Nottoway, Va., 100 ... D4 121
Nottoway, co., Va., 15,141 ... D4 121
Nottoway, riv., Va ... E5 121
Notukeu, creek, Sask., Can ... H2 70
Notus, Canyon, Idaho, 324 ... F2 83
Nouakchott, Maur., 6,000 ... C1 45
Nouméa, N. Cal., 13,000 ... H10 7
Nouna, Upper Volta ... D4 45
Nounan, Bear Lake, Idaho, 10 ... G7 89
Noupoort, S. Afr., 6,322 ... D3 49
Noutonice, Czech., 172 ... n17 26
Nouvelle-Anvers, Con. L ... A2 48
Nouzonville, Fr., 6,954 ... C6 14
Nova, Ashland, Ohio, 271 ... A5 111
Nova Chaves, Ang., 293 ... D3 48
Nova Cruz, Braz., 6,780 ... C3, h6 57
Nova Freixo, Moz ... D6 48
Nova Friburgo, Braz., 49,901 ... C4, h6 56
Nova Gaia, Ang., 1,427 ... D2 48
Nova Granada, Braz., 5,134 ... C3 56
Nova Iguaçu, Braz., 134,708 ... h6 56
Nova Lima, Braz., 21,135 ... E2 57
Nova Lisboa, Ang., 37,381 ... D2 48
Nova Lusitania, Moz ... A5 49
Nova Mambone, Moz ... B6 49
Novar, Ont., Can., 344 ... B5 72
Novara, It., 87,704 ... D4 18, B2 21
Nova Russas, Braz., 4,666 ... B2 57
Nova Scotia, prov., Can., 737,007 ... H20 67, 74
Nova Sofala, Moz ... B5 49
Nova Soure, Braz., 1,760 ... D3 57
Novato, Marin, Calif., 17,881 ... C2 82
Nova Varos, Yugo., 3,186 ... D4 22
Nova Venécia, Braz., 4,567 ... E2 57
Novaya Astrakhan, Sov. Un., 5,000 ... p21 27
Novaya Kazanka, Sov. Un., 10,000 ... F18 9
Novaya Ladoga, Sov. Un., 25,000 ... A9 27
Novaya Lyalya, Sov. Un., 17,700 ... B6 29
Novaya Odessa, Sov. Un ... H8 27
Novaya Sibir, isl., Sov. Un ... B17 28
Novaya Zemlya, is., Sov. Un ... B8 28
Nova Zagora, Bul., 11,031 ... D8 22
Novelda, Sp., 8,887 ... C5 20
Novelty, Knox, Mo., 176 ... A5 101
Nové Mesto [nad Vahom], Czech., 12,400 ... D5 26
Nove Straseci, Czech., 2,764 ... C8 17
Nové Zámky, Czech., 22,000 ... E5 26
Novgorod, Sov. Un., 68,000 ... B8 27
Novgorod-Severskiy, Sov. Un., 10,000 ... E9 27
Novi, Oakland, Mich., 6,390 ... A7 98
Novice, Coleman, Tex., 227 ... D3 118
Novigrad, Yugo., 531 ... C2 22
Novi Ligure, It., 26,972 ... B2 21
Novinger, Adair, Mo., 621 ... A5 101
Novi Pazar, Bul., 5,461 ... D8 22
Novi Pazar, Yugo., 20,712 ... D5 22
Novi Sad, Yugo., 102,385 ... C4 22
Noviye Senzhary, Sov. Un., 5,000 ... G10 27
Novo-Annenskiy, Sov. Un., 18,900 ... C2 29
Novoaydar, Sov. Un., 10,000 ... G12, q21 27
Novocherkassk, Sov. Un., 96,000 ... H13 27
Novoekonomicheskoye, Sov. Un., 10,000 ... q20 27
Novograd-Volynskiy, Sov. Un., 38,000 ... F6 27
Novogrudok, Sov. Un., 25,000 ... E5 27
Novo-Kazalinsk, Sov. Un ... D6 29
Novokurovka, Sov. Un ... B7 37
Novokuznetsk, Sov. Un., 405,000 ... E26 9, D11 28
Novo Mesto, Yugo., 6,844 ... C2 22
Novomirgorod, Sov. Un., 10,000 ... G8 27
Novomoskovsk, Sov. Un., 112,000 ... C1 29
Novomoskovsk, Sov. Un., 37,500 ... G10 27
Novonazyvayevka, Sov. Un., 10,000 ... D23 9
Novopokrovskaya, Sov. Un., 10,000 ... F17 9
Novo Redondo, Ang., 1,016 ... D1 48
Novorossiysk, Sov. Un., 101,000 ... I11 27
Novo-Selo, Bul., 4,307 ... C6 22
Novoshakhtinsk, Sov. Un., 108,000 ... H12 27
Novosibirsk, Sov. Un., 963,000 (★1,025,000) ... B10 29
Novosibirskiye (New Siberian), is., Sov. Un ... B17 28
Novosil, Sov. Un., 10,000 ... E11 27
Novoukrainka, Sov. Un., 10,000 ... G8 27
Novouzensk, Sov. Un., 32,600 ... C3 29
Novoyeniseyskaya, Sov. Un ... B11 29
Novozybkov, Sov. Un., 35,000 ... E8 27
Nový Bohumín, Czech., 11,000 ... D5 26
Novy Bydzov, Czech., 6,120 ... C3 26
Novy Jicin, Czech., 16,600 ... D5 26
Novyy Dombass, Sov. Un., 10,000 ... q21 27
Novyy Oskol, Sov. Un., 10,000 ... F11 27
Novyy Port, Sov. Un ... C10 28
Novyy Vasyugan, Sov. Un ... B9 29
Nowagrod, Pol ... B10 17
Nowa Sol, 25,000 ... C3 26
Nowata, Nowata, Okla., 4,163 ... A6 112
Nowata, co., Okla., 10,848 ... A6 112
Nowe Warpno, Pol., 2,154 ... B3 26
Nowlin, Haakon, S. Dak., 40 ... C4 116
Nowogard, Pol., 2,446 ... B3 26
No Wood, Washakie, Wyo ... B5 125
Nowood, creek, Wyo ... A5, B5 125
Nowra, Austl., 6,221 ... G8 51
Nowshera, Pak., 23,122 ... B5 39
Nowy Dwor, Pol., 5,046 ... k13 26
Nowy Sacz, Pol., 34,000 ... D6 26
Nowy Targ, Pol., 12,600 ... D6 26
Nowy Tomysl, Pol., 2,700 ... B4 26
Noxapater, Winston, Miss., 549 ... C4 100
Noxen, Wyoming, Pa., 800 ... A7, D9 114
Noxon, Sanders, Mont., 150 ... C1 102
Noxon, res., Mont ... C1 102
Noxontown, pond, Del ... B6 85
Noxubee, co., Miss., 16,826 ... B5 100
Noxubee, riv., Miss ... B5 100
Noxville, Kimble, Tex ... D3 118
Noya, Sp., 4,236 ... A1 20
Noyack, bay, N.Y ... E7 84
Noyes, Kittson, Minn., 127 ... B1 99
Noyon, Fr., 9,317 ... C5 14
Nsawam, Ghana, 8,731 (★11,273) ... E4 45
Nsukka, Nig ... E6 45
Nuanetsi, S. Rh ... B5 49
Nubanusit, lake, N.H ... E2 105
Nuberg, Hart, Ga., 100 ... B4 87
Nubian, des., Sud ... A3 47
Ñuble, prov., Chile, 284,516 ... B2 54
Nuckolls, co., Nebr., 8,217 ... D7 103
Nucla, Montrose, Colo., 906 ... C2 83
Nudos Ojos del Salado, mtn., Arg ... E2 55
Nueces, co., Tex., 221,573 ... F4 118
Nueces, riv., Tex ... E3 118
Nueltin, lake, N.W. Ter., Can ... D13 66
Nueva Casas Grandes, Mex., 11,735 ... A3 63
Nueva Ecija, prov., Phil., 608,700 ... *B6 35
Nueva Esparta, state, Ven., 89,492 ... A5 60
Nueva Esparta, is., Ven ... A5 60
Nueva Gerona, Cuba, 2,935 ... E2 64
Nueva Imperial, Chile, 6,450 ... B2 54
Nueva Lubecka, Arg ... C2 54
Nueva Palmira, Ur., 5,100 ... E1 56
Nueva Rosita, Mex., 32,294 ... B4 63
Nueva San Salvador, Sal., 26,911 ... D3 62
Nueva Vizcaya, prov., Phil., 138,000 ... *B6 35
Nueve de Julio, Arg., 13,678 ... B4 54
Nuevitas, Cuba, 12,390 ... E5 64
Nuevo, gulf, Arg ... C4 54
Nuevo Laredo, Mex., 93,787 ... B5 63
Nuevo Leon, state, Mex., 1,063,399 ... B4 63
Nuevo Mundo, mtn., Bol ... D2 55
Nugget, Lincoln, Wyo ... D2 125
Nuits [-St Georges], Fr., 3,970 ... D6 14
Nukheila (Merga) (Well), Sud ... B2 47
Nukualofa, Tonga, 7,500 ... H11 7
Nukus, Sov. Un., 42,000 ... E5 29
Nulato, Alsk., 283 ... C8 79
Nules, Sp., 8,460 ... C5 20
Nulki, hills, B.C., Can ... C5 68
Nullarbor, Austl ... F5 50
Nullarbor, plain, Austl ... F4 50
Numa, Appanoose, Iowa, 202 ... D5 92
Numan, Nig., 1,209 ... E7 45
Numazu, Jap., 142,609 (★230,000) ... I9, n7 37
Nu Mine, Armstrong, Pa., 500 ... E3 114
Nunda, Livingston, N.Y., 1,224 ... C3 108
Nunda, Lake, S. Dak., 106 ... C8 116
Nuneaton, Eng., 56,598 ... B6 12
Nunez, Emanuel, Ga., 79 ... D4 87
Nungan, China, 35,000 ... C10 34
Nunica, Ottawa, Mich., 275 ... E4 98
Nunivak, isl., Alsk ... C6 79
Nunkiang, China, 5,000 ... B10 34
Nunn, Weld, Colo., 228 ... A6 83
Nunnelly, Hickman, Tenn., 200 ... B4 117
Nuns, isl., Que., Can ... D9 73
Nunukan Timur, isl., Indon ... E5 35
Nuoro, It., 23,033 ... D2 21
Nuquí, Col., 576 ... B2 60
Nuremberg, see Nürnberg, Ger.
Nuremberg, Luzerne and Schuylkill, Pa., 900 ... E9 114
Nuri, Mex., 808 ... B3 60
Nurmes, Fin., 2,200 ... F13 25
Nürnberg, Ger., 454,500 (★655,000) ... D6 17
Nurrari, lakes, Austl ... E5 50
Nursery, Victoria, Tex., 350 ... E4 118
Nürtingen, Ger., 20,500 ... E4 17
Nusaybin, Tur., 5,000 ... D13 31
Nushagak, Alsk., 7 ... D8 79
Nushki, Pak., 2,142 ... C4 39
Nusle, Czech., (part of Prague) ... n17 26
Nutak, Newf., Can., 122 ... g9 75
Nuthe, riv., Ger ... A8 17
Nuthegan, riv., Vt ... B5 120
Nutley, Essex, N.J., 29,513 ... B4, D5 106
Nut Mountain, Sask., Can., 164 ... E4 70
Nutrioso, Apache, Ariz., 200 ... D6 80
Nutt, Luna, N. Mex., 5 ... E2 107
Nutter Fort, Harrison, W. Va., 2,440 ... B4 123
Nutting Lake, Middlesex, Mass., 1,600 ... C2 97
Nuuanu Pali, pass, Haw ... g10 88
Nuutele, isl., W. Sam ... 52
Nuutoi, cape, W. Sam ... 52
Nuwara Eliya, Cey., 11,983 ... G7 39
Nuwaybi'al Muzayyinah, Eg., U.A.R ... H10 31
Nuweiba, Eg., U.A.R ... H10 31
Nuzvid, India, 18,974 ... E8 40
Nvalat, Isr ... C6, n10 32
Nyac, Alsk., 54 ... C8 79
Nyack, Flathead, Mont, 15 ... B3 102
Nyack, Rockland, N.Y., 6,062 ... D1, D3 108
Nyala, Sud., 12,278 ... C1 47
Nyamandhlovu, S. Rh., 10,000 ... A4 49
Nyamtumbo, Tan ... D6 48
Nyandoma, Sov. Un., 17,800 ... A12 27
Nyanza, Rwanda, 1,300 ... B4 48
Nyanza, reg., Ken ... A5 48
Nyasa, lake, Afr ... H9 42
Nyasaland, Br. dep., Afr., 2,890,000 ... D5 48
Nyborg, Den., 11,667 ... C4 24
Nybro, Swe., 8,700 ... I6 25
Nyda, Sov. Un ... B23 9
Nye, Stillwater, Mont., 200 ... E7 102
Nye, co., Nev., 4,374 ... E5 104
Nyeri, Ken ... B6 48
Nygami, lake, Bech ... B3 49
Nyimba, N. Rh ... D5 48
Nyiregyhaza, Hung., 39,662 (56,875▲) ... B5 22
Nykøbing [på] Falster, Den., 17,850 ... D5 24
Nykøbing [på Mors], Den., 9,326 ... B2 24
Nykøbing [på Sjaelland], Den., 4,803 ... C5 24
Nyköping, Swe., 24,300 ... H7, u34 25
Nyland Acres, Ventura, Calif., 1,619 ... *E4 82
Nylstroom, S. Afr., 6,662 ... B4 49
Nymagee, Austl., 197 ... F6 51
Nymburk, Czech., 12,600 ... C3, n19 26
Nymindegab, Den ... C2 24
Nynäshamn, Swe., 9,400 ... H7, u35 25
Nyngan, Austl., 2,414 ... E6 51
Nyon, Switz., 7,643 ... D1 19
Nyong, riv., Cam ... E2 46
Nyonga, Tan ... C5 48
Nyota, Blount, Ala ... B3 78
Nyrany, Czech., 4,073 ... D2 26
Nyrsko, Czech., 2,559 ... D8 17
Nysa, Pol., 23,000 ... C4 26
Nysa, riv., Pol ... C4 26
Nyssa, Malheur, Oreg., 2,611 ... D10 113
Nysted, Den., 1,328 ... D5 24
Nyunzu, Con. L ... C4 48
Nzega, Tan ... B5 48
Nzérékoré, Guinea, 10,800 ... E3 45

O

Oacoma, Lyman, S. Dak., 312 ... D6 116
Oahe, res., S. Dak ... C5 116
Oahu, isl., Haw ... B4 88
Oak, Nuckolls, Nebr., 125 ... D8 103
Oak, hill, Mass ... B4 97
Oak, isl., Wis ... B3 124
Oak, lake, Man., Can ... E1 71
Oak, mtn., Ala ... B3, F5 78
Oak, mtn., Ga ... D2 87
Oakalla, Burnet, Tex., 85 ... D4 118
Oak Bay, N.B., Can., 287 ... D2 74
Oak Bluffs, Dukes, Mass., 1,419 ... D6 97
Oakboro, Stanly, N.C., 581 ... B3 109
Oakbrook Terrace, DuPage, Ill., 1,121 ... *F2 90
Oakburn, Man., Can., 357 ... D1 71
Oak City, Martin, N.C., 574 ... B6 109
Oak City, Millard, Utah, 312 ... D3 119
Oak Creek, Routt, Colo., 666 ... A4 83
Oak Creek, Milwaukee, Wis., 9,372 ... *F6 124
Oak Creek, canyon, Ariz ... C4 80
Oakdale, Stanislaus, Calif., 4,980 ... D3 82
Oakdale, New London, Conn., 150 ... D8 84
Oakdale, Washington, Ill., 175 ... E4 90
Oakdale, Allen, La., 6,618 ... D3 95
Oakdale, Worcester, Mass., 325 ... B4 97
Oakdale, Antelope, Nebr., 397 ... B8 103
Oakdale, Broome, N.Y., 1,500 ... *C5 108
Oakdale, Suffolk, N.Y., 950 ... G4 84
Oakdale, Allegheny, Pa., 1,695 ... B5 114
Oakdale, Morgan, Tenn., 470 ... D9 117
Oakengates, Eng., 12,158 ... B5 12
Oakes, Dickey, N. Dak., 1,650 ... C7 110
Oakesdale, Whitman, Wash., 474 ... B8 122
Oakey, Austl., 1,871 ... C8 51
Oakfield, Worth, Ga., 141 ... E3 87
Oakfield, Aroostook, Maine, 560 (848▲) ... B4 96
Oakfield, Genesee, N.Y., 2,070 ... B2 108
Oakfield, Madison, Tenn., 175 ... B3 117
Oakfield, Fond du Lac, Wis., 772 ... E5 124
Oakford, Menard, Ill., 301 ... C4 90
Oakford, Howard, Ind., 250 ... D5 91
Oakford, Bucks, Pa., 2,000 ... A12, F12 114
Oak Forest, Cook, Ill., 5,850 ... F3 90
Oakgrove, Carroll, Ark., 151 ... A2 81
Oak Grove, De Kalb, Ga., 3,500 ... *C2 87
Oak Grove, Christian, Ky., 100 ... D2 94
Oak Grove, West Carroll, La., 1,797 ... B4 95
Oak Grove, Livingston, Mich., 125 ... F7 98
Oak Grove, Jackson, Mo., 1,100 ... *B3 101
Oak Grove, Clackamas, Oreg., 4,000 ... B2, B4 113
Oak Grove, Dillon, S.C ... C8 115
Oak Grove, Westmoreland, Va., 70 ... C6 121
Oakham, Eng., 4,571 ... B7 12
Oakham, Worcester, Mass., 150 (524▲) ... B3 97
Oakharbor, Ottawa, Ohio, 2,903 ... A4 111
Oak Harbor, Island, Wash., 3,942 ... A3 122
Oak Hill, Volusia, Fla., 758 ... D6 86
Oak Hill, Clay, Kans., 69 ... C6 93
Oak Hill, Manistee, Mich., 750 ... D4 98
Oak Hill, Crawford, Mo., 20 ... C6 101
Oak Hill, Jackson, Ohio, 1,748 ... D5 111
Oak Hill, Davidson, Tenn., 4,490 ... *A5 117
Oak Hill, Fayette, W. Va., 4,711 ... D3, D7 123
Oakhurst, Monmouth, N.J., 4,374 ... C4 106
Oakhurst, Tulsa and Creek, Okla., 3,000 ... *A5 112
Oak Lake, Man., Can., 430 ... E1 71
Oakland, Lauderdale, Ala., 75 ... A2 78
Oakland, Marion, Ark., 15 ... A3 81
Oakland, Alameda, Calif., 367,548 ... B5, D2 82
Oakland, Ont., Can., 264 ... D4 72
Oakland, Coles, Ill., 939 ... D5 90
Oakland, Pottawattamie, Iowa, 1,340 ... C2 92
Oakland, Warren, Ky., 148 ... C3 94
Oakland, Kennebec, Maine, 1,880 (3,075▲) ... D3 96
Oakland, Garrett, Md., 1,977 ... D1 85
Oakland, Yalobusha, Miss., 488 ... A4 100
Oakland, St. Louis, Mo., 1,552 ... *C7 101
Oakland, Burt, Nebr., 1,429 ... C9 103
Oakland, Bergen, N.J., 9,446 ... A4 106
Oakland, Transylvania, N.C., 60 ... D3 109
Oakland, Marshall, Okla., 288 ... C5 112
Oakland, Douglas, Oreg., 856 ... D3 113
Oakland, Lawrence, Pa., 2,303 ... *D1 114
Oakland, Susquehanna, Pa., 889 ... C10 114
Oakland, Providence, R.I., 450 ... B10 84
Oakland, Fayette, Tenn., 306 ... B2 117
Oakland, co., Mich., 690,259 ... F7 98
Oakland City, Gibson, Ind., 3,016 ... H3 91
Oaklandon, Marion, Ind., 500 ... H8 91
Oakland Park, Broward, Fla., 5,331 ... E3, F6 86
Oakland Park, Franklin, Ohio, 10,000 ... *C5 111
Oaklawn, Cook, Ill., 31,476 ... F3 90
Oaklawn, Sedgwick, Kans., 5,000 ... B5 93
Oaklawn, St. Tammany, La., 150 ... B8 95
Oakley, Contra Costa, Calif., 950 ... *B6 82
Oakley, Cassia, Idaho, 613 ... G5 89
Oakley, Logan, Kans., 2,190 ... C3 93
Oakley, Saginaw, Mich., 417 ... E6 98
Oakley, Hinds, Miss., 300 ... C3 100
Oakley, Buncombe, N.C. (part of Asheville) ... D3 109
Oakley, Berkeley, S.C., 200 ... E7 115
Oakley, Overton, Tenn ... C8 117
Oakley, Summit, Utah, 247 ... C4 119
Oakley, Lincoln, Wyo ... D2 125
Oakley Park, Oakland, Mich., 1,100 ... *A7 98
Oaklyn, Camden, N.J., 4,778 ... D2 106
Oakman, Walker, Ala., 849 ... B2 78
Oakman, Gordon, Ga., 156 ... B2 87
Oak Mills, Atchison, Kans., 38 ... A8 93
Oakmont, Allegheny, Pa., 7,504 ... A6, E2 114
Oakner, Man., Can., 115 ... D1 71
Oak Orchard, Sussex, Del., 300 ... C7 85
Oak Park, Emanuel, Ga., 302 ... D4 87
Oak Park, Cook, Ill., 61,093 ... F2 90
Oak Park, Oakland, Mich., 36,632 ... F7 98
Oak Point, Man., Can., 238 ... D2 71
Oak Ridge, Morehouse, La., 287 ... B4 95
Oak Ridge, Cape Girardeau, Mo., 175 ... D8 101
Oak Ridge, Guilford, N.C., 500 ... A4 109
Oakridge, Lane, Oreg., 1,973 ... D4 113
Oak Ridge, Armstrong, Pa., 400 ... E3 114
Oak Ridge, Anderson and Roane, Tenn., 27,169 ... C9 117
Oak River, Man., Can., 243 ... D1 71
Oaks, Delaware, Okla., 75 ... A7 112
Oaks, Montgomery, Pa., 950 ... A10 114
Oakton, Hickman, Ky., 225 ... B2 94
Oakton, Fairfax, Va., 350 ... B4 121
Oaktown, Knox, Ind., 798 ... G3 91
Oak Vale, Lawrence and Jefferson Davis, Miss., 99 ... D4 100
Oakvale, Mercer, W. Va., 267 ... D4 123
Oak Valley, Elk, Kans., 50 ... E8 93
Oak View, Ventura, Calif., 2,448 ... *E4 82
Oak View, Montgomery, Md., 1,000 ... *C3 85
Oakville, Man., Can., 377 ... E3 71
Oakville, Ont., Can., 10,366 ... D5 72
Oakville, Litchfield, Conn., 6,000 ... C4 84
Oakville, Louisa, Iowa, 346 ... C6 92
Oakville, Delaware, Ind., 250 ... D7 91
Oakville, St. Louis, Mo., 1,300 ... g13 101
Oakville, Shelby, Tenn., 6,000 ... B2, E8 117
Oakville, Appomattox, Va ... D4 121
Oakville, Grays Harbor, Wash., 377 ... C2 123
Oakway, Oconee, S.C., 300 ... B1 115
Oakwood, Hall, Ga., 218 ... B3 87
Oakwood, Vermilion, Ill., 861 ... C6 90
Oakwood, Walsh, N. Dak., 60 ... A8 110
Oakwood, Marion, Mo. (part of Hannibal) ... B6 101
Oakwood (Oakwood Village), Cuyahoga, Ohio, 3,283 ... *B2 111
Oakwood (Far Hills), Montgomery, Ohio, 10,493 ... C3 111
Oakwood, Paulding, Ohio, 686 ... A3 111
Oakwood, Dewey, Okla., 122 ... B3 112
Oakwood, Lawrence, Pa., 2,267 ... *D1 114
Oakwood, Leon, Tex., 716 ... D5 118

Otis, San Juan, N. Mex., 600 A2 107
Otis, Eddy, N. Mex., 100 E5 107
Otis, res., Mass B1 97
Otisco, Clark, Ind., 250 G6 91
Otisco, Waseca, Minn., 100 G5 99
Otisco, Onondaga, N.Y., 150 C4 108
Otis Orchards, Spokane, Wash., 750 D8 122
Otisville, Genesee, Mich., 701 E7 98
Otjiwarongo, S.W. Afr., 2,383 B2 49
Otley, Marion, Iowa, 177 C4 93
Oto, Woodbury, Iowa, 221 B2 93
Otočac, Yugo., 3,517 C2 22
Otoe, Otoe, Nebr., 225 D9, E3 103
Otoe, co., Nebr., 16,503 D9 103
Otradnoye, Sov. Un s31 25
Otranto, It., 4,309 D7 21
Otsego, Allegan, Mich., 4,142 F5 98
Otsego, co., Mich., 7,545 C6 98
Otsego, co., N.Y., 51,942 C5 108
Otsego, lake, N.Y C6 108
Otsego Lake, Otsego, Mich D6 98
Otsu, Jap., 38,402 *o14 37
Otsu, Jap., 113,547 o14 37
Otta, isl., Truk 52
Ottarp, Swe C6 24
Ottauquechee, riv., Vt D3 120
Ottawa, Ont., Can., 268,206 (*429,750) A9, B9 72
Ottawa, La Salle, Ill., 19,408 B5 90
Ottawa, Franklin, Kans., 10,673 D8 93
Ottawa, Putnam, Ohio, 3,245 A3 111
Ottawa, Boone, W. Va., 200 D5 123
Ottawa, co., Kans., 6,779 C6 93
Ottawa, co., Mich., 98,719 F4 98
Ottawa, co., Ohio, 35,323 A4 111
Ottawa, co., Okla., 28,301 A7 112
Ottawa, is., N.W. Ter., Can E16 67
Ottawa, riv., Ont., Que., Can G17 67
Ottawa, riv., Ohio A2 111
Ottawa Hills, Lucas, Ohio, 3,870 A2, A4 111
Ottawa Lake, Monroe, Mich., 250 G7 98
Otter, Powder River, Mont., 50 E10 102
Otter, brook, N.H E2 105
Otter, creek, Utah E4 119
Otter, creek, Vt C2 120
Otter, lake, Sask., Can B3 70
Otter, riv., Alta., Can A2 69
Otter, riv., Eng D4 12
Otter, riv., Va D3 121
Otterean, riv., Nor H2 25
Otterbein, Benton, Ind., 788 D3 91
Otterberg, Ger., 4,000 D2 17
Otter Brook, flood control res., N.H E2 105
Otterburn, Eng., 624 E6 13
Otterburne, Man., Can., 258 E3 71
Otter Creek, Levy, Fla., 400 C4 86
Otter Creek, Hancock, Maine, 300 D4 96
Otter Lake, Ont., Can B5 72
Otter Lake, Lapeer, and Genesee, Mich., 562 E7 98
Otter Lake, Oneida, N.Y., 85 B5 108
Otterndorf, Ger., 6,600 E2 24
Otter River, Worcester, Mass., 498 A3 97
Ottersburg, Ger., 3,200 E3 24
Ottertail, Otter Tail, Minn., 164 D3 99
Otter Tail, co., Minn., 48,960 D3 99
Otter Tail, lake, Minn D3 99
Otter Tail, riv., Minn D2 99
Otterup, Den., 1,687 C4 24
Otterville, Ont., Can., 725 E4 72
Otterville, Buchanan, Iowa, 25 B6 92
Otterville, Cooper, Mo., 416 C4 101
Otthon, Sask., Can., 38 F4 70
Ottignies, Bel., 5,103 D4 15
Öttingen, Ger., 3,800 E5 17
Otto, Big Horn, Wyo., 50 A4 125
Ottokee, Fulton, Ohio, 65 A3 111
Ottone, It., 3,183 E5 18
Ottosen, Humboldt, Iowa, 92 B3 92
Ottoville, Putnam, Ohio, 793 B3 111
Ottsville, Bucks, Pa., 60 F11 114
Ottumwa, Wapello, Iowa, 33,871 C5 92
Ottumwa, Coffey, Kans., 49 D8 93
Ottumwa, Haakon, S. Dak., 5 C4 116
Ottway, Greene, Tenn C11 117
Otukamamoan, lake, Ont., Can A6 99
Oturkpo, Nig., 1,367 E6 45
Otuzco, Peru, 3,534 C2 58
Otway, Carteret, N.C., 350 C7 109
Otway, cape, Austl I4 51
Otway, sound, Chile h11 54
Otwell, Craighead, Ark., 90 B5 81
Otwell, Pike, Ind., 550 H3 91
Otwock, Pol., 36,000 m14 26
Ötztal Alps, mts., Aus., It A3 21
Ouachita, co., Ark., 31,641 D3 81
Ouachita, par., La., 101,663 B3 95
Ouachita, mts., Okla C6 112
Ouachita, riv., Ark D3 81
Ouachita, riv., La B3 95
Ouaddaï, reg., Chad C4 46
Ouadi Rimé, riv., Chad C3 46
Ouagadougou, Upper Volta, 63,000 D4 45
Ouahigouya, Upper Volta, 10,000 D4 45
Ouaka, riv., Cen. Afr. Rep D4 46
Ouakoro, Mali D3 45
Oualata, Maur C3 45
Oallam, Niger D5 45
Ouanda Djallé, Cen Afr. Rep D6 46
Ouarane, sand dunes, Maur B3 45
Ouargla, Alg., 6,456 (27,360*) C6 44
Ouchina (Well), Mali C5 45
Ouddorp, Neth., 2,680 C3 15
Oudenaarde, Bel., 6,923 D3 15
Oude-Pekela, Neth., 5,900 A7 15
Oudtshoorn, S. Afr., 22,229 D3 49
Oued Zem, Mor., 18,640 H3 30
Oueiba (Well), Chad B4 46
Ouelle, I.C E4 45
Ouelle, riv., Que., Can B8 73
Ouellette, Aroostook, Maine A4 96
Ouesso, Con. B E3 46
Ouezzane, Mor., 26,203 C3 44
Ougarta, Alg D4 44
Oughter, lake, Ire C4 11
Oughterard, Ire., 618 D2 11
Ouidah, Dah., 14,000 E5 45
Oujda, Mor., 128,645 G6 28, C4 44
Oulchy-le-Château, Fr., 651 E3 15
Oullins, Fr., 24,356 E6 14
Oulu, Fin., 58,300 E11 25
Oulu, Bayfield, Wis B2 124
Oulu, dept., Fin., 338,243 *E11 25
Ouluj̈arvi, lake, Fin E12 25
Oum Chalouba, Chad B4 46
Oum el Asel (Oasis), Mali B4 45
Oum Hadjer, Chad, 1,208 C3 46
Ounasjoki, riv., Fin D11 25
Ounianga Kébir, Chad B4 46
Ouray, Ouray, Colo., 785 C3 83
Ouray, Uintah, Utah, 60 C6 119
Ouray, co., Colo., 1,601 C3 83
Ouray, peak, Colo C4 83
Ouri, Chad A3 46
Ouricurí, Braz., 3,159 C2 57
Ourinhos, Braz., 25,717 C3 56
Ourique, Port., 2,180 D1 20
Ouro Fino, Braz., 8,044 C3, m8 56
Ouro Prêto, Braz., 14,722 F2 57
Ourthe, riv., Bel D5 57
Ourukaen, isl., Bikini 52
Ouse, riv., Eng D6 10
Outagamie, co., Wis., 101,794 D5 124
Outat el Hadj, Mor H5 30
Outer, isl., Que., Can C2 75
Outer, isl., Wis A3 124
Outer Hebrides, is., Scot C1 13
Outer Santa Barbara, chan., Calif F4 82
Outes, Sp., 869 A1 20
Outing, Cass, Minn., 70 D5 99
Outjo, S.W. Afr., 1,398 B2 49
Outlook, Sask., Can., 1,340 F2 70
Outlook, Sheridan, Mont., 226 B12 102
Outlook, Yakima, Wash., 325 C5 122
Outlying Islands (U.S.), see Baker, Canton, Enderbury, Howland, Jarvis, Johnston and Sand, Midway, Palmyra, Wake Islands and Kingman Reef.
Outreau, Fr., 13,548 B4 14
Outremont, Que., Can., 30,753 D9 73
Ouville-la-Rivière, Fr., 440 E8 12
Ouyen, Austl., 1,695 G4 51
Ouzinkie, Alsk., 214 D9 79
Ovada, It., 9,594 E4 18, B2 21
Oval, Lycoming, Pa., 100 D7 114
Oval peak, Wash A5 122
Ovalau, isl., Fiji Is 52
Ovalle, Chile, 26,000 A2 54
Ovando, Powell, Mont., 100 C3 102
Ovar, Port., 7,298 B1 20
Overath, Ger., 11,500 C2 17
Overbrook, Osage, Kans., 509 D8 93
Overbrook, Love, Okla., 50 C4 112
Overflowing, riv., Man., Sask., Can D5 70, C1 71
Overgaard, Navajo, Ariz., 150 C5 80
Overijsche, Bel., 11,290 D4 15
Overijssel, prov., Neth., 783,400 B6 15
Overisel, Allegan, Mich., 150 F4 98
Overland, St. Louis, Mo., 22,763 *C7 101
Overland Park, Johnson, Kans., 21,110 B9 93
Overlea, Baltimore, Md., 10,795 B4, C3 85
Overly, Bottineau, N. Dak., 65 A5 110
Overstreet, Gulf, Fla., 100 B1, H3 86
Overton, Dawson, Nebr., 523 D6 103
Overton, Clark, Nev., 700 G7 104
Overton, Rusk, Tex., 1,950 C5 118
Overton, co., Tenn., 14,661 C8 117
Övertorneå, Swe., 1,536 D10 25
Ovett, Jones, Miss., 290 D4 100
Ovid, Sedgwick, Colo., 571 A8 83
Ovid, Bear Lake, Idaho, 200 G7 89
Ovid, Madison, Ind., 95 D6 91
Ovid, Clinton, Mich., 1,505 E6 98
Ovid, Seneca, N.Y., 789 C4 108
Oviedo, Seminole, Fla., 1,926 D5 86
Oviedo, Sp., 127,058 A3 20
Oviedo, prov., Sp., 989,344 *A3 20
Ovitt, Loup, Nebr B6 103
Ovruch, Sov. Un., 10,000 F7 27
Owaneco, Christian, Ill., 290 D4 90
Owanka, Pennington, S. Dak., 60 C3 116
Owasco, lake, N.Y C4 108
Owasso, Tulsa, Okla., 2,032 A6 112
Owatonna, Steele, Minn., 13,409 F5 99
Owego, Tioga, N.Y., 5,417 C4 108
Owen, Clark, Wis., 1,098 D3 124
Owen, co., Ind., 11,400 F4 91
Owen, co., Ky., 8,237 B5 94
Owen, lake, Wis B2 124
Owen, mtn., N.Z N14 52
Owen, sound, Ont., Can C4 72
Owendale, Huron, Mich., 298 E7 98
Owens, Cumberland, N.C., 5,207 *B5 109
Owens, creek, Md A3 85
Owens, lake, Calif D4 82
Owens, riv., Calif D4 82
Owensboro, Daviess, Ky., 42,471 C3 94
Owensboro East, Daviess, Ky., 2,244 *C3 94
Owensboro West, Daviess, Ky., 1,366 *C3 94
Owensburg, Greene, Ind., 400 G4 91
Owens Cross Roads, Madison, Ala., 600 A3 78
Owen Sound, Ont., Can., 17,421 C4 72
Owen Stanley, range, Pap k12 50
Owensville, Saline, Ark., 100 C3, C6 81
Owensville, Gibson, Ind., 1,121 H2 91
Owensville, Gasconade, Mo., 2,379 C6 101
Owensville, Clermont, Ohio, 609 C3 111
Owenton, Owen, Ky., 1,376 B5 94
Owerri, Nig., 2,069 E6 45
Owey, isl., Ire B3 11
Owikeno, lake, B.C., Can D4 68
Owings, Calvert, Md., 200 C4 85
Owings Mills, Baltimore, Md., 3,810 *B4 85
Owingsville, Bath, Ky., 1,040 B6 94
O-Wi-Yu-Kuts, mtn., Colo A2 83
Owl, creek, Wyo B4 125
Owl Creek, mts., Wyo B4 125
Owl River, Alta., Can B5 69
Owls Head, Knox, Maine, 450 (994*) D3 96
Owosso, Shiawassee, Mich., 17,006 E6 98
Owsley, co., Ky., 5,369 C6 94
Owyhee, Elko, Nev., 500 B5 104
Owyhee, co., Idaho, 6,375 G2 89
Owyhee, dam Oreg D9 113
Owyhee, res., Oreg D9 113
Owyhee, riv., Idaho, Nev., Oreg E9 113
Öxabäck, Swe A6 24
Oxbow, Sask., Can., 1,359 H4 70
Oxbow, Aroostook, Maine, 35 (139*) B4 96
Oxbow, Oakland, Mich., 1,700 *F7 98
Oxbox, dam, Idaho E2 89
Oxelösund, Swe., 10,000 u35 25
Oxford, Calhoun, Ala., 3,603 B4 78
Oxford, Izard, Ark., 191 A3 81
Oxford, N.S., Can., 1,471 D6 74
Oxford, New Haven, Conn., 400 (3,292*) *D4 84
Oxford, Eng., 106,124 (*175,000) C6 12
Oxford, Sumter, Fla., 287 D4 86
Oxford, Newton, Ga., 1,047 C3 87
Oxford, Franklin, Idaho, 83 G7 89
Oxford, Benton, Ind., 1,108 C3 91
Oxford, Johnson, Iowa, 633 C6 92
Oxford, Sumner, Kans., 989 E6 93
Oxford, De Soto, La., 350 C2 95
Oxford, Oxford, Maine, 550 (1,658*) D2 96
Oxford, Talbot, Md., 852 C5 85
Oxford, Worcester, Mass., 6,985 B4 97
Oxford, Oakland, Mich., 2,357 F7 98
Oxford, Lafayette, Miss., 5,283 A4 100
Oxford, Furnas and Harlan, Nebr., 1,090 D6 103
Oxford, Warren, N.J., 950 B3 106
Oxford, Chenango, N.Y., 1,871 C5 108
Oxford, Granville, N.C., 6,978 A5 109
Oxford, Butler, Ohio, 7,828 C3 111
Oxford, Chester, Pa., 3,376 G10 114
Oxford, Marquette, Wis., 548 E4 124
Oxford, co., Ont., Can., 70,499 D4 72
Oxford, co., Eng., 309,458 C6 12
Oxford, co., Maine, 44,345 D2 96
Oxford, lake, Man., Can B4 71
Oxford Junction, N.S., Can., 170 D6 74
Oxford Junction, Jones, Iowa, 725 C7 92
Oxley, Searcy, Ark., 75 B3 81
Oxly, Ripley, Mo., 100 E7 101
Oxnard, Ventura, Calif., 40,265 E4 82
Oxnard Beach, Ventura, Calif., 2 000 *E4 82
Oxon Hill, Prince Georges, Md., 900 C1 85
Oxus (Amu Darya), riv., Afg., Sov. Un C14 41
Oyat, riv., Sov. Un A10 27
Oyem, Gabon E2 46
Oyen, Alta., Can., 780 D5 69
Oyens, Plymouth, Iowa, 114 B1 92
Öyeren, lake, Nor p29 25
Oykell, riv., Scot C4 13
Oykell Bridge, Scot C4 13
Oymyakon, Sov. Un C17 28
Oyo, Nig., 72,133 E5 45
Oyonnax, Fr., 14,830 D6 14
Oyrot-Tura, see Gorno-Altaysk, Sov. Un.
Oyster, bay, N.Y F3 84
Oyster, keys, Fla G6 86
Oyster Bay, Nassau, N.Y., 6,500 D3, E7 108
Oyster Bay Cove, Nassau, N.Y., 988 *F2 84
Ozamiz, Phil., 13,000 (46,700*) D6 35
Ozan, Hempstead, Ark., 95 D2 81
Ozark, Dale, Ala., 9,534 D4 78
Ozark, Franklin, Ark., 1,965 B2 81
Ozark, Mackinac, Mich., 10 B6 98
Ozark, Christian, Mo., 1,536 D4 101
Ozark, co., Mo., 6,744 E5 101
Ozark, escarpment, Ark., Mo E7 101
Ozark, plat., Okla A7 106, A2 81
Ozarks, lake, Mo C5 101
Ozarowice, Pol g10 26
Ozaukee, co., Wis., 38,441 E6 124
Ozawkie, Jefferson, Kans., 265 B7 93
Ozd, Hung., 34,155 A5 22
Ozero Imandra, lake, Sov. Un D15 25
Ozette, lake, Wash A1 122
Ozieri, It., 11,884 D2 21
Ozona, Pinellas, Fla., 700 E1 86
Ozona, Pearl River, Miss., 300 E4 100
Ozona, Crockett, Tex., 3,361 D2 118
Ozorkow, Pol., 14,800 C5 26
Ozu, Jap., 17,100 J6 37
Ozuluama, Mex., 1,836 C5, m15 63

P

Pa, China E7 36
Paan, China E4 34
Paar, riv., Ger E6 17
Paarl, S. Afr., 41,540 D2 49
Paauhau, Hawaii, Haw., 500 C6 88
Paauilo, Hawaii, Haw., 1,059 C6 88
Pabbay, isl., Scot C1 13
Pabianice, Pol., 56,000 C5 26
Pablo, Lake, Mont., 150 C2 102
Pabna, Pak., 32,240 E12 40
Pacaás Novos, mts., Braz E2 59
Pacajá, riv., Braz C4 59
Pacajá Grande, riv., Braz C4 59
Pacaraima, mts., Braz C5 60
Pacasmayo, Peru, 6,615 C2 58
Pace, Santa Rosa, Fla., 900 G2 86
Pachai, China, 3,000 K2 36
Pachaug, New London, Conn., 150 C9 84
Pachaug, lake, Conn C9 84
Pacheco, Contra Costa, Calif., 1,518 *C2 82
Pachmarhi, India F7 40
Pachuca [de Soto], Mex., 64,564 C5, m14 63
Pachung, China, 16,000 I2 36
Pachuta, Clarke, Miss., 271 C5 100
Pacific, B.C., Can., 74 B3 68
Pacific, Franklin and St. Louis, Mo., 2,795 B7, C7 101
Pacific, King, Wash., 1,577 D2 122
Pacific, co., Wash., 14,674 C2 122
Pacific, creek, Wyo C3 125
Pacific, mtn., Calif F2 82
Pacific, ocean F6, H7 2, 7
Pacifica, San Mateo, Calif., 20,995 *B5 82
Pacific City, Tillamook, Oreg., 350 B3 113
Pacific Grove, Monterey, Calif., 12,121 C6, D3 82
Pacific Islands Trust Territory, U.S. trust., Oceania, 55,000 *F10 7
Pacific Junction, Mills, Iowa, 560 C2 92
Packard, Piscataquis, Maine C4 96
Packer, Windham, Conn., 40 C9 84
Pack Monadnock, mtn., N.H E3 105
Packwash, lake, Ont., Can D5 71
Packwaukee, Marquette, Wis., 300 E4 124
Packwood, Jefferson, Iowa, 215 C5 92
Pacolet, Spartanburg, S.C., 1,252 B4 115
Pacolet, riv., S.C A4 115
Pacolet Mills, Spartanburg, S.C., 1,476 B4 115
Pacora, Pan., 1,334 k12 62
Pacov, Czech., 2,816 D9 17
Pacquet, Newf., Can., 328 C4 75
Pactola, res., S. Dak C2 116
Pactolus, Pitt, N.C., 211 B6 109
Paczkow, Pol., 6,955 C4 26
Padagi, Iran F10 41
Padang, Indon., 143,615 F2 35
Padangpandjang, Indon., 25,509 F1 35
Padangsidempuan, Indon., 5,709 m11, E1 35
Padauiri, riv., Braz C5 60
Padding, lake, Sask., Can E2 70
Paddockwood, Sask., Can., 219 D3 70
Paddy Knob, mtn., W. Va C5 123
Paden, Tishomingo, Miss., 134 A5 100
Paden, Okfuskee, Okla., 417 B5 112
Paden City, Tyler and Wetzel, W. Va., 3,137 B4 123
Paderborn, Ger., 54,000 B3 17
Pad Idan, Pak B2 40
Padilla, Bol., 2,462 C3 55
Padilla, creek, N. Mex D5 107
Padova (Padua), It., 197,680 D7 18, B3 21
Padre, isl., Tex F4 118
Padroni, Logan, Colo., 110 A7 83
Padstow, Eng., 2,676 D3 12
Padua (Padova), It., 197,680 D7 18, B3 21
Paducah, McCracken, Ky., 34,479 A2 94
Paducah, Cottle, Tex., 2,392 B2 118
Padwa, India, 5,000 H9 40
Paepaealeia, cape, W. Sam 52
Paeroa, N.Z., 2,894 L15 51
Paestum, ruins, It D5 21
Pag, isl., Yugo C2 22
Pagadian, Phil., 9,839 *D6 35
Pagai Selatan, isl., Indon F2 35
Pagai Utara, isl., Indon F2 35
Pagan, isl., Mariana Is E8 7, F10 34
Pagat, pt., Guam 52
Pagatan, Indon F4 35
Pagato, riv., Sask., Can A4 70
Page, Coconino, Ariz., 2,960 A4 80
Page, Holt, Nebr., 230 B7 103
Page, Cass, N. Dak., 432 B8 110
Page, Le Flore, Okla., 35 C7 112
Page, Fayette, W. Va., 800 C3, D6 123
Page, co., Iowa, 21,023 D2 92
Page, co., Va., 15,572 C4 121
Page City, Logan, Kans., 100 C2 93
Pagedale, St. Louis, Mo., 5,106 *A8 101
Pageland, Chesterfield, S.C., 2,020 B7 115
Page Manor, Montgomery, Ohio, 5,000 C3 111
Pages, Essex, Vt C5 120
Pageton, McDowell, W. Va., 400 *D3 123
Paglia, mtn., Switz D7 19
Pagny [-sur-Moselle], Fr., 3,301 F6 15
Pago, bay, Guam 52
Pago, pt., Guam 52
Pagoda, peak, Colo A3 83
Pago Pago, Am. Sam., 1,251 52
Pago Pago, hbr., Am. Sam 52
Pagosa Junction, Archuleta, Colo., 50 D3 83
Pagosa Springs, Archuleta, Colo., 1,374 D3 83
Paguate, Valencia, N. Mex., 500 B2 107
Pahala, Hawaii, Haw., 1,602 D6 88
Pahang, state, Mala., 313,058 J5 38
Pahang, riv., Mala K5 38
Paharpur, Pak A3 40
Pahlavi, Iran, 37,511 C4 41
Pahoa, Hawaii, Haw., 1,046 D7 88
Pahokee, Palm Beach, Fla., 4,709 F6 86
Pahra, Afg., 5,000 D10 41
Pahranagat, range, Nev F6 104
Pahrump, Nye, Nev., 35 G5 104
Pahute, mesa, Nev F5 104
Pahute, peak, Nev B2 104
Paia, Maui, Haw., 3,195 C5 88
Paichuan, China, 5,000 C3 37
Paignton, Eng., 30,289 D4 12
Paiho, China H4 36
Päijänne, lake, Fin G11 25
Pailingmiao, China D4 36
Pailo, Bledsoe, Tenn., 15 D8 117
Paincourt, Ont., Can., 289 E2 72
Paincourtville, Assumption, La., 500 B5 95
Paine Oeste, mtn., Chile E2 54
Painesville, Lake, Ohio, 16,116 A6 111
Paint, Somerset, Pa., 1,275 *F4 114
Paint, creek, Ohio C4 111
Paint, creek, W. Va D6 123
Paint, lake, Man., Can B3 71
Paint, mtn., W. Va D6 123
Paint, riv., Mich B2 98
Paint Bank, Craig, Va., 100 D2 121
Painted Desert, Ariz A4 80
Painted Post, Steuben, N.Y., 2,570 C3 108
Painter, Accomack, Va., 349 D7 121
Painter, Park, Wyo A3 125
Paint Lick, Garrard, Ky., 200 C5 94
Paint Rock, Jackson, Ala., 264 A3 78
Paint Rock, Concho, Tex., 400 D3 118
Paintsville, Johnson, Ky., 4,025 C7 94
Paisley, Ont., Can., 759 C3 72
Paisley, Lake, Fla., 350 D5 86
Paisley, Lake, Oreg., 219 E6 113
Paisley, Scot., 95,753 E4 13
Paita, Peru, 7,177 C1 58
Paja, Pan., 570 m11 62
Pajala, Swe., 1,407 D10 25
Pajaro, Monterey, Calif., 1,273 *C6 82
Pajaros, isl, Mariana Is E8 7
Pajero, riv., Calif C6 82
Pajon, pt., Guam 52
Pakanbaru, Indon 69,147 E2 35
Pakaraima, mts., Br. Gu A2 59
Pakenham, Ont., Can., 350 B8 72
Pakhoi, China, 50,000 G6 34
Pakistan, country, Asia, 93,831,982 G9, G11 33 C4, D9 39, D2, D12 40
Pakli, China C8 38
Pakokku, Bur., 30,943 D10 39
Pakowki, lake, Alta., Can E5 69
Pakra, riv., Yugo C3 22
Paks, Hung., 10,452 B4 22
Pakse, Laos, 25,000 E6 38
Pala, Chad, 7,451 D3 46
Palacois, Matagorda, Tex., 3,676 E4 118
Palafrugell, Sp., 7,905 B7 20
Palagruža, is., Yugo D3 22
Palaiokhora, Grc., 2,130 E4 23
Palaion Faliron, Grc., 22,157 h11 23
Palaiseau, Fr., 16,326 h9 14
Palamas, Grc., 5,370 C4 23
Palamós, Sp., 7,639 B7 20
Palana, Sov. Un., 1,000 D18 28
Palanga, Sov. Un., 4,600 D3 27
Palanpur, India, 29,139 E4 40
Palapag, Phil., 7,862 C7 35
Palapye, Bech., 1,042 B4 49
Palas de Rey, Sp., 769 A2 20
Palatine, Cook, Ill., 15,189 A5, E2 90
Palatine, Salem, N.J D2 106
Palatka, Putnam, Fla., 11,028 C5 86
Palau (Pelew), is., Pac. O C6 52
Palauig, Phil., 1,986 o12 35
Palauli, W. Sam., 603 52
Palauli, bay, W. Sam 52
Palawan, prov., Phil., 162,900 *C5 35
Palawan, isl., Phil C5 35
Palazzolo Acreide, It., 11,024 F5 21
Palco, Rooks, Kans., 575 C4 93
Palembang, Indon., 458,661 F2 35
Palena, riv., Chile C2 54
Palencia, Sp., 48,216 A3 20
Palencia, prov., Sp., 231,977 *A3 20
Palenque, ruins, Mex B1 62
Palenville, Greene, N.Y., 350 C6 108
Palermo, It., 587,985 E4 21
Palermo, Cape May, N.J., 100 E3 106
Palermo, Mountrail, N. Dak., 188 A3 110
Palestine, St. Francis, Ark., 532 C5 81
Palestine, Crawford, Ill., 1,564 D6 90
Palestine, Darke, Ohio, 257 B3 111
Palestine, Anderson, Tex., 13,974 D5 118
Palestine, Wirt, W. Va., 100 B3 123
Palestine (Gaza Strip), see Gaza Area, Eg. occ., Asia
Palestine, reg., Asia C7 32
Palestine Potash Company, Jordan h13 32
Palestrina, It., 9,154 h9 21
Paletwa, Bur., 5,000 D9 39
Palezgir Chauki, Pak B2 40
Palghat, India, 77,620 F6 39
Palgrave, Ont., Can., 234 D5 72
Palhoça, Braz., 1,456 D3 56
Pali, India, 33,303 E4 40
Palidoro (Baebiana), It., 2,169 h8 21
Palikea, mtn., Haw g9 88
Palikir, pass, Ponape 52
Palikir, pt., Ponape 52
Palimé, Togo E5 45
Palisade, Mesa, Colo., 860 B2 83
Palisade, Aitkin, Minn., 180 D5 99
Palisade, Hitchcock, Nebr., 544 D4 103
Palisade, Eureka, Nev., 25 C5 104
Palisades, Bonneville, Idaho, 75 F7 89
Palisades, Douglas, Wash., 15 B6 122
Palisades, cliffs, N.J D5 106
Palisades, Park, Bergen, N.J., 11,943 D5 106
Palliseul, Bel., 1,667 E5 15
Palitana, India, 24,581 G3 40
Palk, strait, Cey., India G6 39
Pallaskenry, Ire., 315 E3 11
Pallasovka, Sov. Un C3 29
Palling, B.C., Can., 65 B5 68
Palliser, cape, N.Z N15 51
Pall Mall, Fentriss, Tenn., 125 C9 117
Palls, King William, Va D5 121
Palma, Moz D7 48
Palma, Torrance, N. Mex C4 107
Palma, isl., Can. Is m13 20
Palma, riv., Braz D1 57
Palma del Río, Sp., 11,324 D3 20
Palma [de Mallorca], Sp., 159,084 C7 20
Palmares, Braz., 17,327 C3, k6 57
Palmas, Braz., 5,540 D2 56
Palmas, cape, I.C F3 45
Palmas, Atlas, P.R B4 65
Palmas de Monte Alto, Braz., 1,279 D2 57
Palma Soriano, Cuba, 25,241 E6 64
Palm Bay, Brevard, Fla., 2,808 D6 86
Palm Beach, Palm Beach, Fla., 6,055 F6 86
Palm Beach, co., Fla., 228,106 F6 86
Palm Beach Shores, Palm Beach, Fla., 885 *F6 86
Palm City, San Diego, Calif. (part of San Diego) E2 82
Palmdale, Los Angeles, Calif., 11,522 E4 82
Palmdale, Glades, Fla., 150 F5 86
Palm Desert, Riverside, Calif., 1,295 *F5 82
Palmeira, Braz., 5,916 D3 56
Palmeira dos Indios, Braz., 15,642 C3, k5 57
Palmeir in Has, pt., Ang C1 48
Palmer, Alsk., 1,181 C10, g17 79
Palmer, Sask., Can., 104 H2 70
Palmer, Christian, Ill., 265 D4 90
Palmer, Pocahontas, Iowa, 271 B3 92
Palmer, Washington, Kans., 169 C6 93
Palmer, Hampden, Mass., 3,888 B3 97
Palmer, Marquette, Mich., 850 B3 98
Palmer, Merrick, Nebr., 418 C7 103
Palmer, Grundy, Tenn., 1,069 D8 117
Palmer, Ellis, Tex., 613 B6 118
Palmer, King, Wash., 25 D2 122
Palmer, pen., Ant B6 5
Palmer Heights, Northampton, Pa., 2,597 *E11 114
Palmer Lake, El Paso, Colo., 542 B6 83
Palmer Park, Prince Georges, Md., 4,000 *C4 85
Palmerston, Ont., Can., 1,554 D4 72
Palmerston, N.Z., 868 P13 51
Palmerston North, N.Z., 41,014 (*43,185) N15 51
Palmersville, Weakley, Tenn., 150 A3 117
Palmerton, Carbon, Pa., 5,942 E10 114
Palmerville, Austl C7 50
Palmetto, Manatee, Fla., 5,556 E4, F2 86
Palmetto, Fulton and Coweta, Ga., 1,466 C2 87
Palmetto, St. Landry, La., 430 D4 95
Palm Harbor, Pinellas, Fla., 1,000 E2 86
Palmi, It., 15,400 E5 21
Palmira, Col., 54,293 C2 60
Palmira, Cuba, 5,865 D3 64
Palmira, Ec B2 58
Palms, Sanilac, Mich., 50 E8 98
Palms, is., S.C F8 115
Palm Springs, Riverside, Calif., 13,468 F5 82
Palm Springs, Palm Beach, Fla., 2,503 F6 86
Palm Valley, St. Johns, Fla., 300 C6 86
Palmyra, Macoupin, Ill., 811 D4 90
Palmyra, Harrison, Ind., 470 H5 91
Palmyra, Warren, Iowa, 50 B7 92
Palmyra, Marion, Mo., 2,933 B6 101
Palmyra, Otoe, Nebr., 377 D9, E2 103
Palmyra, Burlington, N.J., 7,036 C2 106
Palmyra, Wayne, N.Y., 3,476 B3 108
Palmyra, Halifax, N.C., 50 A6 109
Palmyra, Lebanon, Pa., 6,999 F8 114
Palmyra, Syr., 8,500 E12 31
Palmyra, Montgomery, Tenn., 100 A4 117
Palmyra, Fluvanna, Va., 350 D4 121
Palmyra, Jefferson, Wis., 1,000 F5 124
Palmyra, isl., Pac. O F12 7
Palmyras, pt., India G11 40
Palo, Linn, Iowa, 387 B6 92
Palo, Ionia, Mich., 250 E4 98
Palo (Alsium), It., 363 h8 21
Palo Alto, Santa Clara, Calif., 52,287 B5, D2 82
Palo Alto, Schuylkill, Pa., 1,445 *E9 114
Palo Alto., co., Iowa, 14,736 A3 92
Palo Blanco, P.R B4 65
Paloduro, Armstrong, Tex., 30 B2 118
Palo Duro, canyon, Tex B2 118
Paloich, Sud C3 47
Palomar, mtn., Calif F5 82
Palomas, mtn., Ariz D2 80
Palombara Sabina, It., 6,215 g9 21
Palominos, isl., P.R f12 65
Palo Pinto, Palo Pinto, Tex., 525 C3 118
Palo Pinto, co., Tex., 20,516 C3 118
Palopo, Indon., 4,208 F6 35
Palos, cape, Sp D5 20
Palo Seco, P.R., 709 B6 65
Palos Heights, Cook, Ill., 3,775 *F3 90
Palos Hills, Cook, Ill., 3,766 *F2 90
Palos Park, Cook, Ill., 2,169 F2 90
Palos Verdes Estates, Los Angeles, Calif., 9,564 *F2 83
Palourde, lake, La E4 95

Palouse, Whitman, Wash., 926 C8 122
Palouse, riv., Wash C7 122
Palo Verde, Maricopa, Ariz., 200 D3, G1 80
Palo Verde, Imperial, Calif., 150 F6 82
Palpa, Peru, 2,171 D3 58
Pålsboda, Swe., 1,140 t33 25
Palua, Ven B5 60
Paluan, Phil., 1,835 *C6 35
Palung, China D4 34
Pama, Upper Volta D5 45
Pambrun, Sask., Can., 117 H2 70
Pamekasan, Indon., 24,912 G4 35
Pamiers, Fr., 13,297 F4 14
Pamir, mts., China, Sov. Un H23 9
Pamlico, co., N.C., 9,850 B7 109
Pamlico, riv., N.C B7 109
Pamlico, sound, N.C B7 109
Pampa, Gray, Tex., 24,664 B2 118
Pampa Grande, Bol., 727 C3 55
Pampanga, prov., Phil., 615,800 *B6 35
Pampanga, riv., Phil o13 35
Pampa Peñon, Chile D2 55
Pampas, Peru, 1,622 D3 58
Pampas, reg., Arg G4 55
Pampilhosa do Botão, Port., 2,779 B1 20
Pamplico, Florence, S.C., 988 D8 115
Pamplin, Appomattox and Prince Edward, Va., 312 D4 121
Pamplona, Col., 16,396 B3 60
Pamplona, Sp., 97,880 A5 20
Pamunkey, riv., Va D5 121
Pana, Ont., Can., 85 B10 72
Pana, Christian, Ill., 6,432 D4 90
Panaca, Lincoln, Nev., 500 F7 104
Panacea, Wakulla, Fla., 600 B2 86
Panache, lake, Ont., Can A3 72
Panagyurishte, Bul., 12,015 D7 22
Panaitan, isl., Indon G3 35
Panama, Montgomery, Ill., 487 D4 90
Panama, Shelby, Iowa, 257 C2 92
Panama, Lancaster, Nebr., 155 D9 103
Panama, Chautauqua, N.Y., 450 C1 108
Panama, LeFlore, Okla., 937 B7 112
Panamá, Pan., 273,440 F8, m11 62
Panama, country, N.A., 1,075,541 I12 61, F7 62
Panama, bay, Pan F8 62
Panama, canal, C.Z F8, k11 62
Panama, gulf, Pan F8 62
Panama Canal Zone, see Canal Zone, U.S. dep., N.A.
Panama City, Bay, Fla., 33,275 G3 86
Panama City Beach, Bay, Fla., 36 G3 86
Panamint, range, Calif D5 82
Panao, Peru, 954 C2 58
Panarea, isl., It E5 21
Panaro, riv., It B3 21
Panay, isl., Phil C6 35
Pancake, range, Nev E5 104
Pancevo, Yugo., 40,740 C5 22
Panco, Clay, Ky., 200 C6 94
Pancoastburg, Fayette, Ohio, 75 C4 111
Panda, Moz B5 49
Pandale, Val Verde, Tex., 10 D2 123
Pandan, Phil., 3,379 *C6 35
Pandharpur, India, 45,421 I5 40
Pandhurna, India G7 40
Pando, Ur., 6,248 E1 56
Pando, dept., Bol., 19,804 B2 55
Pandora, Putnam, Ohio, 782 B4 111
Pandora, Wilson, Tex., 250 B4, E4 123
Pandu, Con. L A2 48
Panelas, Braz., 1,717 k6 57
Panet, Que., Can., 571 C7 73
Panevezys, Sov. Un., 37,100 D5 27
Panfilov, Sov. Un., 27,300 E9 28
Pangala, Con. B F2 46
Pangasinan, prov., Phil., 1,124,800 *B6 35
Pangburn, White, Ark., 489 B4 81
Pangfou, see Pengpu, China
Pangi, Con. L B4 48
Pangim, Goa, 14,213 E5 39
Pañginay, Phil o13 35
Pangkalanbuun, Indon F4 35
Pangkalangresik, Indon F2 35
Pangkalpinang, Indon., 58,540 F3 35
Pangkong, lake, China B6 39
Pangman, Sask., Can., 260 H3 70
Pangnirtung, N.W. Ter., Can., 114 C19 67
Pangong, lake, India B6 39
Pangsau, pass, Bur C10 39
Panguitch, Garfield, Utah, 1,435 F3 119
Panhandle, Carson, Tex., 1,958 B2 123
Panipat, India, 67,026 C6 40
Paniqui, Phil., 4,789 o13 35
Panjang, isl., Viet H5 38
Panjao, Afg., 10,000 D13 41
Panjgur, Pak., 754 C3 39
Panjpai, Pak G13 41
Pankshin, Nig., 5,654 E6 45
Panmunjom, Kor H3 37
Panna, India, 16,737 E8 40
Panola, Sumter, Ala., 250 C1 78
Panola, co., Miss., 28,791 A3 100
Panola, co., Tex., 16,870 C5 123
Panora, Guthrie, Iowa, 1,019 C3 92
Panshan, China D9 36
Panshih, China, 5,000 E2 37
Pantar, isl., Indon G6 35
Pantego, Beaufort, N.C., 262 B7 109
Pantelleria, It., 2,294 F3 21
Pantelleria, isl., It F3 21
Panther, mtn., N.Y B6 108
Pantin, Fr., 46,292 g10 14
Panton, Addison, Vt., 35 (352▲) C2 120
Pantsyan, China C5 36
Pánuco, Mex., 8,688 C5, k14 63
Pánuco, riv., Mex k14 63
Panulcillo, Chile F1 55
Panzi, Con. L C2 48
Panzós, Guat., 573 C3 62
Paochang, China D6 36
Paocheng, China, 10,000 H2 36
Paochi, China, 130,100 G2 36
Paoching, China, 5,000 C6 37
Pão de Açúcar, Braz., 4,729 C3 57
Paokang, China, 10,000 I4 36
Paokuotu, China C9 36
Paola, It., 9,197 E6 21
Paola, Miami, Kans., 4,784 D9 93
Paoli, Phillips, Colo., 81 A8 83
Paoli, Orange, Ind., 2,754 G5 91
Paoli, Garvin, Okla., 358 C4 112
Paoli, Chester, Pa., 5,000 A10 114
Paonia, Delta, Colo., 1,083 C3 83
Paoshan, China, 12,000 F4 34
Paote, China E4 36
Paoti, China E7 36
Paoting (Tsingyuan), China, 265,000 E6 36
Paotow, China, 400,000 D4 36
Paoua, Cen. Afr. Rep D3 46
Paoying, China, 85,000 H8 36
Pap, mtn., Scot D3 13
Papa, Hawaii, Haw., 50 D6 88
Papa, Hung., 25,629 B3 22
Papaaloa, Hawaii, Haw., 500 D6 88
Papagayo, gulf, C.R E5 62
Papago, Indian res., Ariz E3 80
Papaikou, Hawaii, Haw., 1,427 D6 88
Papalaulelei, cape, W. Sam 52
Papantla, Mex., 18,579 C5, m15 63
Papatele, mtn., Am. Sam 52
Papaya, Phil., 7,301 o13 35
Papeete, Fr. Oceania, 12,428 H13 7
Papenburg, Ger., 15,000 B3 16
Paphos, Cyp., 7,283 E9 31
Papigi, see Parigi, Indon.
Papillion, Sarpy, Nebr., 2,235 C9, E3 103
Papillion, creek, Nebr E3 103
Papineau, co., Que., Can., 32,697 D2 73
Papineau, lake, Ont., Can B7 72
Papineauville, Que., Can., 1,300 D2 73
Paposo, Chile E1 55
Papua, Austl., dep., Oceania, 374,000 k12 50
Papua, gulf, N. Gui k11 50
Papudo, Chile A2 54
Papun, Bur., 1,881 E10 39
Papy, pt., Fla E2 86
Paquette, Que., Can., 93 D6 73
Paquetville, N.B., Can., 380 B4 74
Pará, state, Braz., 1,550,935 B1 57
Pará, riv., Braz B1 57
Parabel, Sov. Un B10 29
Paracale, Phil., 3,204 o14 35
Paracatu, Braz., 5,909 E1 57
Paracatu, riv., Braz E1 57
Paracel, is., China B4 35
Parachinar, Pak E15 46
Paraćin, Yugo., 15,627 D5 22
Paracurú, Braz., 2,484 B3 57
Parade, Dewey, S. Dak., 10 B4 116
Paradis, St. Charles, La., 800 C6 95
Paradise, Butte, Calif., 8,268 C3 82
Paradise, Stanislaus, Calif., 5,616 *D3 82
Paradise, Russell, Kans., 134 C5 93
Paradise, Sanders, Mont., 280 C2 102
Paradise, Wallowa, Oreg B9 113
Paradise, Lancaster, Pa., 750 F9 114
Paradise Hill, Sask., Can., 282 D1 70
Paradise Valley, Maricopa, Ariz., 1,000 G2 80
Paradise Valley, Alta., Can., 182 C5 69
Paradise Valley, Humboldt, Nev., 65 B4 104
Paradox, Montrose, Colo., 50 C2 83
Paradox, valley, Colo C2 83
Paragon, Morgan, Ind., 560 F4 91
Paragonah, Iron, Utah, 300 F3 119
Paragould, Greene, Ark., 9,947 A5 81
Paragua, riv., Ven B5 60
Paraguaçu, riv., Braz D3 57
Paraguaçu Paulista, Braz., 11,391 C2 56
Paraguai, riv., Braz B1 56
Paraguaná, pen., Ven A4 60
Paraguarí, Par., 12,405 E4 55
Paraguarí, dept., Par., 159,161 E4 55
Paraguay, country, S.A., 1,638,000 F5 53, D4 55
Paraguay, riv., Par D4 55
Paraíba, state, Braz., 2,018,023 C3, h6 57
Paraíba, riv., Braz C4 56
Paraíba do Sul, Braz., 7,675 h6 56
Paraíso, Mex., 2,804 D6 63
Paraisópolis, Braz., 6,582 C3 56
Parakhino-Poddubye, Sov. Un., 20,000 B9 27
Parakou, Dah., 5,700 E5 45
Paraloma, Sevier, Ark., 94 D1 81
Param, isl., Ponape 52
Paramaribo, Sur., 75,233 A3 59
Paramé, Fr., 8,811 C2 14
Paramirim, Braz., 1,776 D2 57
Paramithia, Grc., 2,956 C3 23
Paramonga, Peru D2 58
Paramount, Los Angeles, Calif., 27,249 *F2 82
Paramount, Washington, Md., 200 A2 85
Paramus, Bergen, N.J., 23,238 *B4 106
Paraná, Arg., 110,000 A4 54
Paranã, Braz., 970 D1 57
Paraná, state, Braz., 4,227,763 C2 56
Paraná, riv., Arg., Braz E4 55, C2 56
Paranã, riv., Braz D1 57
Paranaguá, Braz., 27,728 D3 56
Paranaíba, Braz., 3,853 B2 56
Paranaíba, riv., Braz B2 56
Paranam, Sur A3 59
Paranapanema, riv., Braz C2 56
Paranatinga, riv., Braz E3 59
Paraopeba, Braz., 3,678 E1 57
Parapetí, riv., Bol C3 55
Paratinga, Braz., 2,403 D2 57
Paray-le-Monial, Fr., 9,557 D6 14
Parbati, riv., India E6 40
Parbhani, India, 36,795 H6 40
Parchim, Ger., 18,900 B5 16
Parchment, Kalamazoo, Mich., 1,565 F5 98
Parcperdue, Iberia, La D4 95
Parczew, Pol., 6,173 C7 26
Pardeeville, Columbia, Wis., 1,331 E4 124
Pardess Hanna, Isr., 7,073 B6 32
Pardo, riv., Braz C2 56
Pardo, riv., Braz D3 57
Pardo, riv., Braz E2 57
Pardoe, Mercer, Pa., 120 D1 114
Pardoo, Austl D2 50
Pardubice, Czech., 52,700 C3 26
Pardum, Blaine, Nebr B5 103
Parecis, mts., Braz E2 59
Paredes, de Nava, Sp., 4,733 A3 20
Parent, Que., Can., 1,298 G18 67
Parepare, Indon., 62,683 F5 35
Parfúri, Moz B5 49
Parga, Grc., 1,722 C3 23
Pargolovo, Sov. Un., 5,000 r31 25
Parguera, P.R D2 65
Parhams, Catahoula, La., 35 C4 95
Paria, gulf, Ven A5 60
Paria, pen., Ven A5 60
Paria, riv., Utah F4 119
Pariaguán, Ven., 6,094 B5 60
Paricutin, vol., Mex n12 63
Parigi, Indon F6 35
Parika, Br. Gu., 577 A3 59
Parima, mts., Braz C5 60
Pariñas, pt., Peru B1 58
Parintins, Braz., 9,068 C3 59
Pario, riv., Ariz A4 80
Paris, Logan, Ark., 3,007 B2 81
Paris, Ont., Can., 5,820 D4 72
Paris, Fr., 2,790,091 (*7,750,000) C5, g10 14, F2 15
Paris, Bear Lake, Idaho, 746 G7 89
Paris, Edgar, Ill., 9,823 D6 90
Paris, Bourbon, Ky., 7,791 B5 94
Paris, Oxford, Maine, 200 (3,601▲) D2 96
Paris, Calvert, Md., 100 C4 85
Paris, Mecosta, Mich., 180 E5 98
Paris, Lafayette, Miss., 102 A4 100
Paris, Monroe, Mo., 1,393 B5 101
Paris, Greenville, S.C., 1,000 *B3 115
Paris, Henry, Tenn., 9,325 A3 117
Paris, Lamar, Tex., 20,977 C5 118
Paris, Fauquier, Va., 100 B5 121
Paris, Kenosha, Wis F1 124
Paris Crossing, Jennings, Ind., 135 G6 91
Paris, Oswego, N.Y., 125 B4 108
Parishville, St. Lawrence, N.Y., 400 A2 108
Parisville, Que., Can., 510 C6 63
Park, Grove, Kans., 218 C3 93
Park, co., Colo., 1,822 B5 83
Park, co., Mont., 13,168 E6 102
Park, co., Wyo., 16,874 A3 125
Park, plat., Colo D6 83
Park, range, Colo., Wyo B4 83
Parkbeg, Sask., Can., 80 G2 70
Park City, Lake, Ill., 1,408 *A6 90
Park City, Sedgwick, Kans., 2,687 B5 93
Park City, Barren, Ky., 497 C4 94
Park City, Stillwater, Mont., 350 E8 102
Park City, Summit, Utah, 1,366 C4 119
Park Court, Alta., Can C3 69
Parkdale, Ashley, Ark., 448 D4 81
Parkdale, P.E.I., Can., 1,735 *C6 74
Parkdale, Fremont, Colo., 35 C5 83
Parkdale, Hood River, Oreg., 400 B5 113
Parkdale, Douglas, Wis B2 124
Parke, co., Ind., 14,804 E3 91
Parker, Yuma, Ariz., 1,642 C1 80
Parker, Douglas, Colo., 100 B6 83
Parker, Bay, Fla., 2,669 G3 86
Parker, Fremont, Idaho, 284 F7 89
Parker, Linn, Kans., 181 D8 93
Parker, Polk, Oreg C1 113
Parker, Armstrong, Pa., 945 D2 114
Parker, Turner, S. Dak., 1,142 D8 116
Parker, Spotsylvania, Va., 100 C5 121
Parker, Yakima, Wash., 300 C5 122
Parker, co., Tex., 22,880 C4 118
Parker, dam, Ariz C1 80
Parker, peak, S. Dak D2 116
Parker City, Randolph, Ind., 1,181 D7 91
Parker Dam, Yuma, Ariz C1 80
Parker Dam, San Bernardino, Calif., 250 E6 82
Parker Head, Sagadahoc, Maine, 100 E6 96
Parkersburg, Richland, Ill., 253 E5 90
Parkersburg, Butler, Iowa, 1,468 B5 92
Parkersburg, Sampson, N.C., 65 C5 109
Parkersburg, Wood, W. Va., 44,797 B3 123
Parkers Prairie, Otter Tail, Minn., 884 D3 99
Parkerton, Converse, Wyo., 50 C7 125
Parkertown, Ocean, N.J., 400 D4 106
Parkerview, Sask., Can., 59 F4 70
Parkerville, Morris, Kans., 59 D7 93
Parkes, Austl., 8,223 F7 51
Parkesburg, Chester, Pa., 2,759 G10 114
Park Falls, Price, Wis., 2,919 C3 124
Parkfield, Monterey, Calif., 40 E3 82
Park Forest, Cook and Will, Ill., 29,993 F3 90
Park Grove, Valley, Mont., 100 B10 102
Park Hall, St. Marys, Md., 300 D5 85
Park Head, Ont., Can., 58 C3 72
Park Hill, Ont., Can., 1,169 D3 72
Park Hill, Cherokee, Okla., 150 B7 112
Park Hills, Kenton, Ky., 4,076 A7 94
Parkhurst, Aroostook, Maine (part of Presque Isle) B5 96
Parkin, Cross, Ark., 1,489 B5 81
Parkland, Alta., Can., 96 D4 69
Parkland, Bucks, Pa., 1,200 *F12 114
Parkland, Pierce, Wash., 15,000 E1 122
Parklawn, Fairfax, Va., 1,000 *C5 121
Parkman, Sask., Can., 70 H5 70
Parkman, Piscataquis, Maine, 65 (530▲) C3 96
Parkman, Geauga, Ohio, 300 A6 111
Parkman, Sheridan, Wyo., 25 A5 125
Park Place, Clackamas, Oreg., 100 B2 113
Park Place, Greenville, S.C., 1,600 *B3 115
Park Rapids, Hubbard, Minn., 3,047 D3 99
Park Ridge, Cook, Ill., 35,430 E2 90
Park Ridge, Bergen, N.J., 6,389 C5 106
Park Ridge Manor, Cook, Ill., 1,000 *E2 90
Park River, Walsh, N. Dak., 1,813 A8 110
Parkrose, Multnomah, Oreg., 24,000 *B4 113
Parks, St. Martin, La., 413 D4 95
Parks, Dundy, Nebr., 120 D4 103
Parks, Stephens, Tex C3 118
Parkside, Sask., Can., 149 D2 70
Parkside, Delaware, Pa., 2,426 *G11 114
Parksley, Accomack, Va., 850 D7 121
Parkston, Hutchinson, S. Dak., 1,514 D8 116
Parksville, B.C., Can., 1,183 E5 68
Parksville, Sullivan, N.Y., 900 D6 108
Parksville, McCormick, S.C., 164 D3 115
Parkton, Baltimore, Md., 180 A4 85
Parkton, Robeson, N.C., 906 C4 109
Park Valley, Box Elder, Utah, 25 B2 119
Park View, Rio Arriba, N. Mex., 300 A3 107
Parkview, Cuyahoga, Ohio, 2,018 *A6 111
Parkview, Allegheny, Pa., 1,000 *E2 114
Parkville, Baltimore, Md., 27,236 C3 85
Parkville, Platte, Mo., 1,229 B3, E2 101
Parkville, York, Pa., 1,300 *G8 114
Parkway Village, Jefferson, Ky., 949 *B4 94
Parkwood, Montgomery, Md., 1,400 *B3 85
Parla, Sp., 1,781 p17 20
Parlakimedi, India, 22,708 H10 40
Parlier, Fresno, Calif., 1,366 *D4 82
Parlin, Gunnison, Colo., 5 C4 83
Parlin, Middlesex, N.J. (part of Sayreville) C4 106
Parma, Canyon, Idaho, 1,295 F2 89
Parma, It., 141,203 E6 18, B3 21
Parma, Jackson, Mich., 770 F6 98
Parma, New Madrid, Mo., 1,060 E8 101
Parma, Cuyahoga, Ohio, 82,845 A6, B2 111
Parma, riv., It E6 18
Parmachenee, lake, Maine C2 96
Parma Heights, Cuyahoga, Ohio, 18,100 B2 111
Parmele, Martin, N.C., 323 B6 109
Parmelee, Todd, S. Dak., 140 D4 116
Parmer, co., Tex., 9,583 B1 118
Parnaguá, Braz., 508 D2 57
Parnaíba, Braz., 39,951 B2 57
Parnaíba, riv., Braz B2 57
Parnassos, mtn., Grc C4 23
Parnell, Iowa, Iowa, 200 C6 92
Parnell, Nodaway, Mo., 260 A3 101
Parnis, mtn., Grc g11 23
Pärnu, Sov. Un., 43,000 B5 27
Parole, Anne Arundel, Md. (part of Annapolis) C4 85
Paron, Saline, Ark., 100 C3 81
Paropamisus, range, Afg D12 41
Paros, Grc., 3,174 D5 23
Paros, isl., Grc D5 23
Parow, S. Afr., 39,185 *D2 49
Parowan, Iron, Utah, 1,486 F3 119
Parr, Jasper, Ind., 90 B3 91
Parral, Chile, 10,717 B2 54
Parramore, isl., Va P7 121
Parran, Calvert, Md., 200 C4 85
Parran, Churchill, Nev., 15 D3 104
Parras de la Fuente, Mex., 19,499 B4 63
Parrett, riv., Eng C5 12
Parrish, Walker, Ala., 1,608 B2 78
Parrish, Manatee, Fla., 800 E4, F2 86
Parrish, Langlade, Wis., 30 C4 124
Parrott, Terrell, Ga., 280 E2 87
Parrott, Pulaski, Va., 650 D2 121
Parrottsville, Cocke, Tenn., 91 C10 117
Parrs, ridge, Md A3 85
Parrsboro, N.S., Can., 1,834 D5 74
Parry, Sask., Can., 76 H3 70
Parry, cape, N.W. Ter., Can B8 66
Parry, isl., Ont., Can B4 72
Parry, isl., Eniwetok 52
Parry, is., N.W. Ter., Can B25 4
Parry, mtn., B.C., Can C3 68
Parry Sound, Ont., Can., 6,004 B4, F9 72
Parry Sound, co., Ont., Can., 29,632 B4 72
Parsberg, Ger., 2,800 D6 17
Parshall, Grand, Colo., 100 B4 83
Parshall, Mountrail, N. Dak., 1,216 B3 110
Parsippany, Morris, N.J., 3,500 *B4 106
Parsnip, peak, Nev E7 104
Parsnip, riv., B.C., Can B6 68
Parsons, Labette, Kans., 13,929 E8 93
Parsons, Decatur, Tenn., 1,859 B3 117
Parsons, Tucker, W. Va., 1,798 B5 123
Parsonsburg, Wicomico, Md., 500 D7 85
Parson's Pond, Newf., Can., 337 C3, h10 75
Partabgarh, India, 14,573 E5 40
Partabpur, India H8 40
Parthenay, Fr., 8,350 D3 14
Parthenen, Aus., 624 C6 18
Parthenon, Newton, Ark., 125 B2 81
Partinico, It., 26,119 E4 21
Partlow, Spotsylvania, Va., 30 C5 121
Partridge, Reno, Kans., 221 B4, E5 93
Partridgeberry, hills, Newf., Can D4 75
Partry, Ire D2 11
Parvan, Afg., 5,000 D11 41
Pasadena, Los Angeles, Calif., 116,407 E4, F2 82
Pasadena, Newf., Can., 288 D3 75
Pasadena, Anne Arundel, Md., 2,000 B2 85
Pasadena, Ocean, N.J D4 106
Pasadena, Harris, Tex., 58,737 F5 118
Pasadena Hills, St. Louis, Mo., 1,315 *C7 101
Pasadena Park, Spokane, Wash., 2,000 *B8 122
Pasaje, Ec., 4,864 B2 58
Pasamonte, Union, N. Mex A6 107
Pasay (Rizal), Phil., 132,200 C6, o13 35
Pasayton, riv., B.C., Can. and Wash A5 122
Pascagoula, Jackson, Miss., 17,155 E3, E5 100
Pascagoula, bay, Miss E3 100
Pascagoula, riv., Miss E5 100
Pascani [-Gară], Rom., 15,008 B8 22
Paschall, Warren, N.C A5 109
Pasco, Franklin, Wash., 14,522 C6 122
Pasco, co., Fla., 36,785 D4 86
Pasco, dept., Peru, 130,507 D2 58
Pascoag, Providence, R.I., 2,983 B10 84
Pascoag, res., R.I B9 84
Pascola, Pemiscot, Mo., 228 E8 101
Pascualitos, Mex., 237 F6 82
Pas-de-Calais, dept., Fr., 1,366,282 D2 15
Paseley, cape, Austl F3 50
Pasewalk, Ger., 12,400 B7 16
Pashkovo, Sov. Un., 1,000 B5 37
Pasig, Phil., 61,900 o13 35
Paslek, Pol., 3,278 A5 26
Pasni, Pak., 6,168 C3 39
Paso del Limay, Arg C2 54
Paso de los Indios, Arg C3 54
Paso de los Libres, Arg., 11,665 E4 55
Paso de los Toros, Ur., 8,989 E1 56
Paso Río Mayo, Arg D2 54
Paso Robles, (El Paso de Robles) San Luis Obispo, Calif., 6,677 E3 82
Pasqua, Sask., Can., 59 G3 70
Pasque, isl., Mass D6 97
Pasquia, hills, Sask., Can D4 70
Pasquia, riv., Sask., Man., Can D5 70, C1 73
Pasquo, Davidson, Tenn., 50 E9 117
Pasquotank, co., N.C., 25,630 A7 109
Pasrur, Pak., 9,403 A5 40
Passaconaway, mtn., N.H C4 105
Passadumkeag, Penobscot, Maine, 300 (355▲) C4 96
Passadumkeag, mtn., Maine C4 96
Passage East, Ire., 494 E5 11
Passage West, Ire., 2,561 F3 11
Passaic, Passaic, N.J., 53,963 B4, D5 106
Passaic, co., N.J., 406,618 A4 106
Passaic, riv., N.J B4 106
Passau, Ger., 31,800 E8 17
Pass Christian, Harrison, Miss., 3,881 E2, E4 100
Passekeag, N.B., Can., 180 D4 74
Passero, cape, It F5 21
Passo Fundo, Braz., 47,299 D2 56
Passos, Braz., 28,555 F1 57
Passugg-Araschgen, Switz C8 19
Passumpsic, Caledonia, Vt., 180 C4 120
Passumpsic, riv., Vt C4 120
Passwang, mtn., Switz B4 19
Pastaza, prov., Ec B2 58
Pastaza, riv., Ec., Peru B2 58
Pasto, Col., 59,500 C2 60
Pastora, peak, Ariz A6 80
Pastoria, Jefferson, Ark C3 81
Pasto Viejo Central, P.R C7 65
Pastura, Guadalupe, N. Mex., 30 C5 107
Pasuruan, Indon., 62,872 G4 35
Paswegin, Sask., Can., 110 F3 70
Patagonia, Santa Cruz, Ariz., 540 F5 80
Patan, India, 50,264 F4 40
Patan, Nep., 42,183 D10 40
Patapedia, riv., N.B., Can B2 74
Patapsco, Carroll, Md., 120 A4 85
Patapsco, res., Md B4 85
Patapsco, riv., Md B4 85
Pataskala, Licking, Ohio, 1,046 C5 111
Pataz, Peru, 214 C2 58
Patchewollock, Austl., 256 G4 51
Patchogue, Suffolk, N.Y., 8,838 D4 108
Patea, N.Z., 1,989 M15 51
Pateley Bridge, Eng., 1,671 F7 13
Pateras, mtn., Grc g10 23
Paternò, It., 42,935 F5 21
Pateros, Okanogan, Wash., 673 A6 122
Paterson, Passaic, N.J., 143,663 B4, D5 106
Paterson, Benton, Wash., 45 D6 122
Patesville, Hancock, Ky., 75 C3 94
Pathankot, India, 46,330 A5 40
Pathfinder, res., Wyo C6 125
Pathfork, Harlan, Ky., 500 D6 94
Pathlow, Sask., Can., 108 E3 70
Pati, pt., Guam 52
Patiala, India, 125,234 B6 39, B6 40
Patiala and East Punjab States Union, state, India, 3,493,685 *B6 40
Patiali, India, 4,616 D7 40
Patillas, P.R., 1,888 C6 65
Patillas, mun., P.R., 17,106 C6 65
Patmos, Hempstead, Ark., 120 D2 81
Patmos, isl., Grc D6 23
Patna, India, 363,700 (*450,000) C8 39, E10 40
Patnanongan, isl., Phil o14 35
Patoka, Marion, Ill., 601 E4 90
Patoka, Gibson, Ind., 579 H2 91
Patoka, riv., Ind H2 91
Paton, Greene, Iowa, 370 B3 92
Patos, Braz., 27,275 C3 57
Patos de Minas, Braz., 31,471 E1 57
Patrai, Grc., 95,364 C3 23
Patrai, gulf, Grc C3 23
Patricia, Alta., Can., 71 D5 69
Patricia, Dawson, Tex., 60 C1 118
Patrick, Cherokee, N.C., 200 D2 109
Patrick, Chesterfield, S.C., 393 B7 115
Patrick, co., Va., 15,282 E2 121
Patrick, mtn., Maine D3 96
Patricksburg, Owen, Ind., 350 F4 91
Patrick Springs, Patrick, Va., 500 E2 121
Patriot, Switzerland, Ind., 277 G8 91
Patrocínio, Braz., 13,933 E1 57
Patroon, Shelby, Tex., 500 D6 118
Patsaliga, creek, Ala D3 78
Patsville, Elko, Nev., 5 B6 104
Pattagumpus, Penobscot, Maine C4 96
Pattani, Thai., 16,534 I4 38
Pattaquattie, hill, Mass B3 97
Patten, Penobscot, Maine, 1,099 C4 96
Pattenburg, Hunterdon, N.J., 250 B2 106
Patterson, Woodruff, Ark., 324 B4 81
Patterson, Stanislaus, Calif., 2,429 B6 82
Patterson, Pierce, Ga., 719 E4 87
Patterson, Lemhi, Idaho, 24 E5 89
Patterson, Greene, Ill., 130 D3 90
Patterson, Madison, Iowa, 157 C4 92
Patterson, Harvey, Kans B4 93
Patterson, St. Mary, La., 2,923 E4 95
Patterson, Wayne, Mo., 125 D7 101
Patterson, Putnam, N.Y., 800 D7 108
Patterson, Buchanan, Va., 400 B3 121
Patterson, creek, W. Va B5 123
Patterson, mtn., Calif C4 82
Patterson Creek, mtn., W. Va B6 123
Patterson Knob, mtn., Tenn E9 117
Patterson Gardens, Monroe, Mich., 1,747 *G7 98
Pattison, Claiborne, Miss., 150 D3 100
Patton, Bollinger, Mo., 108 D7 101
Patton, Cambria, Pa., 2,880 E4 114
Pattonsburg, Daviess, Mo., 753 A3 101
Patu, Braz., 2,367 C3 57
Patuakhali, Pak., 10,289 F13 40
Patuca, riv., Hond C5 62
Patung, China, 16,000 I4 36
Patuxent, Anne Arundel, Md., 60 B4 85
Patuxent, riv., Md B3 85
Patzau, Douglas, Wis., 50 B1 124
Pátzcuaro, Mex., 14,281 D4, n13 63
Pâtzcuaro, lake, Mex n13 63
Patzicía, Guat., 5,021 C2 62
Patzún, Guat., 5,103 C2 62
Pau, Fr., 59,937 (*78,000) F3 24
Paucarbamba, Peru, 1,738 D3 58
Paucartambo, riv., Peru D3 58
Paudalho, Braz., 6,889 C3, h6 57
Pau dos Ferros, Braz., 4,298 C3 57
Paugh Lake, Ont., Can B7 72
Pauillac, Fr., 2,353 E3 14
Pauk, Bur., 5,000 D9 39
Paul, Minidoka, Idaho, 701 G5 89
Paul, Midland, Tex C2 118
Paul, isl., Newf., Can g9 75
Paul, stream, Vt B5 120
Paulden, Yavapai, Ariz., 15 C3 80
Paulding, Ontonagon, Mich., 75 A6 98
Paulding, Jasper, Miss., 180 C4 100
Paulding, Paulding, Ohio, 2,936 A3 111
Paulding, co., Ga., 13,101 C2 87
Paulding, co., Ohio, 16,792 A3 111
Paulette, Noxubee, Miss., 50 B5 100
Paulina (Remy), St. James, La., 1,014 *D5 95
Paulina, Warren, N.J., 75 B3 106
Paulina, Crook, Oreg., 30 C7 113
Paulina, mts., Oreg D5 113
Paulina, peak, Oreg D5 113
Pauline, Adams, Nebr., 85 D7 103
Pauline, Spartanburg, S.C., 200 B4 115
Pauline, mtn., B.C., Alta., Can C8 68, C1 69
Paulins, hill, N.J A3 106
Paulis, Con. L., 9,700 A4 48
Paulistana, Braz., 1,105 C2 57
Paull, lake, Sask., Can A3 70
Paull, riv., Sask., Can B3 70
Paullina, O'Brien, Iowa, 1,329 B9 22
Paulo Afonso, falls, Braz C3 57
Pauloff Harbor, Alsk., 77 E7 79
Paulsboro, Gloucester, N.J., 8,121 D2 106
Paul Smiths, Franklin, N.Y., 350 B2 108
Paul Spur, Cochise, Ariz., 30 F6 80
Pauls Valley, Garvin, Okla., 6,856 C4 112
Paungde, Bur., 17,286 E10 39
Paupack, Pike, Pa., 250 D11 114
Pauwela, Maui, Haw., 300 C5 88
Pavia, It., 74,962 D5 18, B2 21
Pavilion, Genesee, N.Y., 538 C2 108
Pavilion, key, Fla G5 86
Pavillion, Fremont, Wyo., 190 B4 125
Pavlodar, Sov. Un., 107,000 C9 29
Pavlof, vol., Alsk D7 79
Pavlograd, Sov. Un., 43,000 G10 27
Pavlovo, Sov. Un., 53,000 D14 27
Pavlovsk, Sov. Un., 30,000 s31 25
Pavlovsk, Sov. Un., 30,000 F13 27

Pavlovskaya, Sov. Un., 10,000 ... H13 27
Pavlovskiy Posad, Sov. Un., 58,000 ... n18 27
Pavo, Thomas and Brooks, Ga., 817 ... F3 87
Pavonia, Richland, Ohio, 190 ... B5 111
Pawcatuck, New London, Conn., 6,000 ... D9 84
Paw Creek, Mecklenburg, N.C., 2,000 ... B3 109
Pawhuska, Osage, Okla., 5,414 ... A5 112
Pawlet, Rutland, Vt., 165 (1,112▲) ... E2 120
Pawleys Island, Georgetown, S.C., 500 ... E9 115
Pawling, Dutchess, N.Y., 1,734 ... D7 108
Pawnee, Sangamon, Ill., 1,517 ... D4 90
Pawnee, Pawnee, Okla., 2,303 ... A5 112
Pawnee, Bee, Tex., 175 ... E4 118
Pawnee, co., Kans., 10,254 ... D4 93
Pawnee, co., Nebr., 5,356 ... D9 103
Pawnee, co., Okla., 10,884 ... A5 112
Pawnee, creek, Colo ... A7 83
Pawnee, riv., Kans ... D4 93
Pawnee City, Pawnee, Nebr., 1,343 ... D9 103
Pawnee Rock, Barton, Kans., 380 ... D5 93
Pawpaw, Lee, Ill., 725 ... B5 90
Paw Paw, Van Buren, Mich., 2,970 ... F5 98
Paw Paw, Morgan, W. Va., 789 ... B6 123
Pawpaw, creek, W. Va ... A7 123
Paw Paw, riv., Mich ... F4 98
Paw Paw Lake, Berrien, Mich., 3,518 ... *F4 98
Pawtuckaway, pond, N.H ... D5 105
Pawtowicze, Pol ... h9 26
Pawtucket, Providence, R.I., 81,001 ... B11 84
Pax, Fayette, W. Va., 408 ... D6 123
Paxico, Wabaunsee, Kans., 276 ... C7 93
Paxoi, isl., Grc ... C3 23
Paxtang, Dauphin, Pa., 1,916 ... *F8 114
Paxton, Walton, Fla., 215 ... G3 86
Paxton, Ford, Ill., 4,370 ... C5 90
Paxton, Sullivan, Ind., 275 ... F3 91
Paxton, Worcester, Mass., 600 (2,399▲) ... B4 97
Paxton, Keith, Nebr., 566 ... C4 103
Paxtonville, Snyder, Pa., 350 ... E7 114
Paxville, Clarendon, S.C., 216 ... D7 115
Payen, China, 5,000 ... C3 37
Payenhala, China ... A10 36
Payerne, Switz., 6,024 ... C2 19
Payette, Payette, Idaho, 4,451 ... E2 89
Payette, co., Idaho, 12,363 ... E2 89
Payette, lake, Idaho ... E2 89
Payette, riv., Idaho ... F2 89
Payintala, see Tungliao, China
Payne, Bibb, Ga., 346 ... D3 81
Payne, Paulding, Ohio 1,287 ... A3 111
Payne, co., Okla., 44,231 ... A5 112
Payne, lake, Que., Can ... A2 73
Payne Bay, Que., Can., 83 ... D18 67
Paynes, Tallahatchie, Miss., 25 ... B3 100
Paynes Creek, Tehama, Calif., 50 ... B3 82
Paynesville, Ontonagon, Mich., 55 ... A6 98
Paynesville, Stearns, Minn., 1,754 ... E4 99
Paynesville, Pike, Mo., 150 ... B7 101
Payneville, Meade, Ky., 113 ... C3 94
Paynton, Sask., Can., 216 ... E1 70
Paysandú, Ur., 44,800 ... E1 56
Paysandú, dept., Ur., 78,400 ... *E1 56
Payson, Gila, Ariz., 800 ... C4 80
Payson, Adams, Ill., 502 ... D2 90
Payson, Lincoln, Okla., 30 ... B5 112
Payson, Utah, Utah, 4,237 ... C4 119
Paz, bay, Mex ... C2 63
Pazardzhik, Bul., 30,430 ... D7 22
Pazin, Yugo., 3,004 ... C1 22
Pea, riv., Ala ... D3 78
Peabody, Marion, Kans., 1,309 ... A6, D6 93
Peabody, Essex, Mass., 32,202 ... A6, C3 97
Peabody, riv., N.H ... B4 105
Peace, riv., Alta., B.C., Can ... E9 66
Peace, riv., Fla ... E5 86
Peace Dale, Washington, R.I., 2,000 ... D11 84
Peace River, Alta., Can., 2,543 ... A2 69
Peach, co., Ga., 13,846 ... D3 87
Peacham, Caledonia, Vt., 75 (433▲) ... C4 120
Peach Creek, Logan, W. Va., 700 ... D5 123
Peache, pt., Mass ... C4 97
Peachland, B.C., Can., 575 ... E8 68
Peachland, Anson, N.C., 563 ... C3 109
Peach Orchard, Clay, Ark., 348 ... A5 81
Peach Orchard Knob, peak, Ky ... C2 94
Peach Springs, Mohave, Ariz., 600 ... B2 80
Peacock, Lake, Mich., 20 ... D5 98
Peacock, peak, Ariz ... B2 80
Peacock, pt., Wake Isl ... 52
Peak, Newberry, S.C., 86 ... C5 115
Peaked, mtn., Maine ... B4 96
Peak Hill, Austl, 46 ... E2 50
Peale, isl., Wake Isl ... 52
Peale, mtn., Utah ... E6 119
Pea Patch, isl., Del ... D1 106
Pearce, Cochise, Ariz., 25 ... F6 81
Pearce, Alta., Can., 41 ... E4 69
Pearcy, Garland, Ark., 100 ... C2, D5 81
Pea Ridge, Benton, Ark., 380 ... A1 81
Pearisburg, Giles, Va., 2,268 ... D2 121
Pearl, Pike, Ill., 348 ... D3 90
Pearl, Rankin, Miss., 5,081 ... *C3 100
Pearl hbr., Haw ... B4 88
Pearl, riv., Miss ... C3 100
Pearland, Brazoria, Tex., 1,497 ... F5 118
Pearl and Hermes, reef, Haw ... k12 88
Pearl Beach, St. Clair, Mich., 1,224 ... *F8 98
Pearl City, Honolulu, Haw., 8,200 ... B4, g10 88
Pearl City, Stephenson, Ill., 488 ... A4 90
Pearlington, Hancock, Miss., 500 ... E4 100
Pearl River, St. Tammany, La., 964 ... D6 95
Pearl River, Rockland, N.Y., 9,000 ... D1 108
Pearl River, co., Miss., 22,411 ... E4 100
Pear Ridge, Jefferson, Tex., 3,470 ... E2 95
Pearsall, Frio, Tex., 4,957 ... E3 118
Pearsoll, peak, Oreg ... E3 113
Pearson, Cleburne, Ark., 40 ... B3 81
Pearson, Atkinson, Ga., 1,615 ... E4 87
Pearson, Langlade, Wis., 80 ... C4 124
Pearsonia, Osage, Okla ... A5 112
Peary Land, reg., Grnld ... A18 4
Pease, Mille Lacs, Minn., 191 ... E5 99
Peasleeville, Clinton, N.Y., 40 ... B3 108
Peavine, Cumberland, Tenn ... C9 117
Pebane, Moz., 5,000 ... A6 49
Pebble Beach, Monterey, Calif., 900 ... *D3 82
Pebworth, Owsley, Ky., 200 ... C6 94
Peć, Yugo., 28,297 ... D5 22
Peçanha, Braz., 3,602 ... E2 57
Pecan Island, Vermilion, La ... E3 95
Pecatonica, Winnebago, Ill., 1,659 ... A4 90
Pecatonica, riv., Wis ... F3 124
Pechenga, Sov. Un., 13,200 ... C14 25
Pechora, riv., Sov. Un ... C20 9
Pechorskaya, bay, Sov. Un ... B19 9
Peck, Nez Perce, Idaho, 186 ... C2 89
Peck, Sedgwick & Sumner, Kans., 120 ... E6 93
Peck, Sanilac, Mich., 548 ... E8 98
Peckerwood Lake, res., Ark ... C4 81
Peckham, Kay, Okla., 100 ... A4 112
Peckville (Blakely), Lackawanna, Pa., 6,374 ... A9 114
Peconic, Suffolk, N.Y., 400 ... E7 84
Peconic, riv., N.Y ... F6 84
Pecos, San Miguel, N. Mex., 584 ... B4, F6 107
Pecos, Reeves, Tex., 12,728 ... D1, F2 118
Pecos, co., Tex., 11,957 ... D1 118
Pecos, riv., N. Mex., Tex ... E5 107, D2 118
Pecs, Hung., 114,713 ... B4 22
Peculiar, Cass, Mo., 458 ... C3 101
Pedasí, Pan., 988 ... G7 62
Peddocks, isl., Mass ... D3 97
Pedernal, Torrance, N. Mex ... C4 107
Pedernal, peak, N. Mex ... C4 107
Pedlar Mills, Amherst, Va., 35 ... D3 121
Pedley, Riverside, Calif., 1,600 ... *F3 82
Pedra Azul, Braz., 8,238 ... F2 57
Pedraza, see Ciudad Bolivia, Ven.
Pedregal, Pan., 631 ... F6 62
Pedregal, Pan ... k12 62
Pedregal, Ven., 1,166 ... A3 60
Pedreiras, Braz., 10,189 ... B2 57
Pedricktown, Salem, N.J., 645 ... D2 106
Pedro, Pennington, S. Dak., 5 ... C3 116
Pedro Afonso, Braz., 3,175 ... C1 57
Pedro Avelino, Braz., 1,399 ... C3 57
Pedro de Valdivia, Chile, 10,989 ... D2 55
Pedro Juan Caballero, Par., 12,521 ... D4 55
Pedro Luro, Arg ... B4 54
Pedro Miguel, C.Z., 603 ... k11 62
Pedro Velho, Braz., 2,041 ... h6 57
Peebinga, Austl., 67 ... G3 51
Peebles, Sask., Can., 66 ... G4 70
Peebles, Adams, Ohio, 1,601 ... D4 111
Peebles, Scot., 5,545 ... E5 13
Peebles, co., Scot., 14,117 ... E5 13
Pee Dee, riv., N.C., S.C ... D9 115
Peekaboo, mtn., Maine ... C5 96
Peekskill, Westchester, N.Y., 18,737 ... D3, D7 108
Peel, N.B., Can., 139 ... C2 74
Peel, I. of Man, 2,487 ... F4 13
Peel, co., Ont., Can., 111,575 ... D5 72
Peel, riv., N.W. Ter., Can ... C5 66
Peel, sound, N.W. Ter., Can ... B13 66
Peeled Chestnut, White, Tenn ... D8 117
Peel Fell, mtn., Scot ... E6 13
Pe Ell, Lewis, Wash., 593 ... C2 122
Peely (Warrior Run), Luzerne, Pa., 833 ... *D10 108
Peene, riv., Ger ... B6 16
Peer, Bel., 5,838 ... C5 15
Peerless, Daniels, Mont., 110 ... B11 102
Peerless, lake, Alta., Can ... A3 69
Peers, Alta., Can., 128 ... C3 69
Peesane, Sask., Can., 65 ... E4 70
Peetz, Logan, Colo., 218 ... A7 83
Peever, Roberts, S. Dak., 208 ... B9 116
Pefferlaw, Ont., Can., 411 ... C5 72
Pegan, hill, Mass ... D2 97
Pegasus, bay, N.Z ... O14 51
Pegau, Ger., 6,754 ... B7 17
Peggs, Cherokee, Okla., 28 ... A6 112
Peggy, Atascosa, Tex., 10 ... E3 118
Pegnitz, Ger., 8,100 ... D6 17
Pegnitz, riv., Ger ... D6 17
Pego, Sp., 8,291 ... C5 20
Pegram, Cheatham, Tenn., 400 ... A4 117
Pegu, Bur., 47,378 ... E10 39
Pegu Yoma, mts., Bur ... C1 38
Pehan, China, 18,000 ... B10 34
Pehčevo, Yugo., 1,750 ... E6 22
Pehuajo, Arg., 13,537 ... B4 54
Pei, China ... G7 36
Pei, China, 69,000 ... G7 36
Peine, Ger., 29,800 ... A5 17
Peiping, see Peking, China
Peipus, lake, Sov. Un ... B6 27
Peixe, Braz., 822 ... D1 57
Pejepscot, Sagadahoc, Maine, 200 ... E2, E5 96
Pekalongan, Indon., 100,261 ... G3 35
Pekan, Mala., 2,070 ... K5 38
Pekia, China, 1,000 ... A8 40
Pekin, Tazewell, Ill., 28,146 ... C4 90
Pekin (New Pekin), Washington Ind., 661 ... G5 91
Pekin, Nelson, N. Dak., 180 ... B7 110
Peking (Peiping), China, 3,500,000 (6,230,000▲) ... D8 34, E7 36
Pelagie, is., It ... G13 30
Pelagos, isl., Grc ... C5 23
Pelahatchie, Rankin, Miss., 1,066 ... C4 100
Pelaihari, Indon ... F4 35
Pelat, mtn., Fr ... E7 14
Pelee, pt., Ont., Can ... F2 72
Peleliu, isl., Palau Is ... 52
Pelham, Shelby, Ala., 500 ... B3 78
Pelham, Mitchell, Ga., 4,609 ... E2 87
Pelham, Hampshire, Mass., 150 (805▲) ... B3 97
Pelham, Hillsboro, N.H., 150 (2,605▲) ... E4 105
Pelham, Westchester, N.Y., 1,964 ... *D1 108
Pelham, Caswell, N.C., 200 ... A4 109
Pelham, Greenville, S.C., 500 ... B3 115
Pelham, Grundy, Tenn., 300 ... D8 117
Pelham Manor, Westchester, N.Y., 6,114 ... D1 108
Pelhrimov, Czech., 7,200 ... D3 26
Pelican, Alsk., 135 ... k21 79
Pelican, De Soto, La., 250 ... C2 95
Pelican, bay, Man., Can ... C1 71
Pelican, butte, Oreg ... E4 113
Pelican, lake, Man., Can ... C1 71
Pelican, lake, Minn ... B6 99
Pelican, lake, Minn ... D3 99
Pelican, lake, Minn ... D3 99
Pelican, lake, Minn ... D4 99
Pelican, lake, Minn ... E5 99
Pelican, lake, Sask., Can ... B4 70
Pelican, lake, Wis ... C4 124
Pelican, mts., Alta., Can ... B4 69
Pelican Lake, Palm Beach, Fla., 500 ... F6 86
Pelican Lake, Oneida, Wis., 300 ... C4 124
Pelican Rapids, Man., Can., 213 ... C1 71
Pelican Rapids, Otter Tail, Minn., 1,693 ... D2 99
Pelion, Lexington, S.C., 233 ... D5 115
Pella, Marion, Iowa, 5,198 ... C5 92
Pella, prov., Grc., 133,224 ... *B4 23
Pell City, Saint Clair, Ala., 4,165 ... B3 78
Pellejas Central, P.R ... C4 65
Pellettown, Salem, N.J ... A3 106
Pellice, riv., It ... E3 18
Pell Lake, Walworth, Wis., 900 ... *F5 124
Pellston, Emmet, Mich., 429 ... C6 98
Pellton, Ont., Can ... A8 98
Pellville, Hancock, Ky., 119 ... C3 94
Pellworm, isl., Ger ... A4 16
Pelly, Sask., Can., 476 ... F5 70
Pelly, lake, N.W. Ter., Can ... C12 66
Pelly, mts., Yukon, Can ... D6 66
Pelly, riv., Yukon, Can ... D6 66
Pelly Crossing, Yukon, Can., 151 ... D5 66
Peloncillo, mts., Ariz., N. Mex ... E6 80, E1 107
Peloponnesos (Peloponnesus), reg., Grc ... D4 23
Pelotas, Braz., 121,280 ... E2 56
Pelotas, riv., Braz ... D2 56
Pelto, lake, La ... E5 95
Pelzer, Anderson, S.C., 106 ... B3 115
Pelzer North, Anderson, S.C., 1,400 ... *B3 115
Pemadumcook, lake, Maine ... C3 96
Pemanggil, isl., Mala ... K6 38
Pemaquid, Lincoln, Maine, 200 ... E3 96
Pematangsiantar, Indon., 112,687 ... K3 38
Pemba, N. Rh., 189 ... E4 48
Pemba, isl., Zan ... B6 48
Pemberton, B.C., Can., 181 ... D6 68
Pemberton, Burlington, N.J., 1,250 ... D3 106
Pemberton, Shelby, Ohio, 240 ... B3 111
Pemberville, Wood, Ohio, 1,237 ... A2, A4 111
Pembina, Pembina, N. Dak., 625 ... A8 110
Pembina, co., N. Dak., 12,946 ... A8 110
Pembina, mts., N. Dak ... A7 110
Pembina, riv., Alta., Can ... C3 69
Pembina, riv., Man., Can ... E2 71
Pembine, Marinette, Wis., 550 ... C6 124
Pembroke, Ont., Can., 16,791 ... B7, E10 72
Pembroke, Bryan, Ga., 1,450 ... D5 87
Pembroke, Christian, Ky., 517 ... D2 94
Pembroke, Washington, Maine, 500 (871▲) ... D5 96
Pembroke, Plymouth, Mass., 1,300 (4,919▲) ... B6 97
Pembroke, Merrimack, N.H., (3,514▲) ... D4 105
Pembroke, Genesee, N.Y., 100 ... C2 108
Pembroke, Robeson, N C., 1,372 ... C4 109
Pembroke, Giles, Va., 1,038 ... D2 121
Pembroke, Wales, 12,737 ... C3 12
Pembroke, co., Wales, 93,980 ... C3 12
Pembroke Pines, Brownard, Fla., 1,429 ... *F6 86
Pembuang, Indon ... F4 35
Pemigewasset, riv., N.H ... C3 105
Pemiscot, co., Mo., 38,095 ... E8 101
Pemuco, Chile, 1,703 ... B2 54
Penablanca, Sandoval, N. Mex., 150 ... B3, F6 107
Penafiel, Port., 4,361 ... B1 20
Peñafiel, Sp ... B3 20
Peñalara, mtn., Sp ... B4 20
Penalosa, Kingman, Kans., 84 ... E5 93
Penamacor, Port., 2,740 ... B2 20
Peña Negra, mts., Sp ... A2 20
Penang, Mala., 234,903 (★325,000) ... J4 38
Penang (Penang and Province Wellesley), state, Mala., 572,100 ... J4 38
Penang, isl., Mala ... J4 38
Penápolis, Braz., 14,400 ... C2 56
Peñaranda de Bracamonte, Sp., 5,943 ... B3 20
Pen Argyl, Northampton, Pa., 3,693 ... E11 114
Peña Roya, mtn., Sp ... B5 20
Peñarroya-Pueblonuevo, Sp., 24,152 ... C3 20
Peñas, cape, Sp ... A3 20
Penas, gulf, Chile ... D2 54
Penasco, Taos, N. Mex., 627 ... A4 107
Penasse, Lake of the Woods, Minn., 3 ... A4 99
Penawawa, Whitman, Wash., 30 ... C8 122
Penbrook, Dauphin, Pa., 3,671 ... F8 114
Pence, Warren, Ind., 100 ... D2 91
Penck, cape, Ant ... C21 5
Penck, trough, Ant ... B12 5
Pendelikon (Pentelicus), mtn., Grc ... g11 23
Pendleton, Anderson, S.C., 2,358 ... B2 115
Pendembu, S.L. ... E2 45
Pender, B.C., Can ... B9 68
Pender, Thurston, Nebr., 1,165 ... B9 103
Pender, co., N.C., 18,508 ... C6 109
Pender, isl., B.C., Can ... B9 68
Pendergrass, Jackson, Ga., 215 ... B3 87
Pendle Hill, mtn., Eng ... G6 13
Pendleton, Madison, Ind., 2,472 ... E6 91
Pendleton, Umatilla, Oreg., 14,434 ... B8 113
Pendleton, co., Ky., 9,968 ... B5 94
Pendleton, co., W. Va., 8,093 ... C5 123
Pend Oreille, co., Wash., 6,914 ... A8 122
Pend Oreille, lake, Idaho ... A2 89
Pend Oreille, riv., Wash ... A8 122
Pendroy, Teton, Mont., 60 ... B4 102
Penedo, Braz., 17,084 ... D3 57
Penetanguishene, Ont., Can., 5,340 ... C5 72
Penfield, Greene, Ga., 105 ... C3 87
Penfield, Champaign, Ill., 275 ... C6 90
Penfield, Monroe, N.Y., 3,500 ... *B3 108
Penfield, Clearfield, Pa., 700 ... D4 114
Penfield Junction, Lorain, Ohio, 2,300 ... *A5 111
Pengan, China, 13,000 ... I2 36
Penganga, riv., India ... H7 40
Penge, Con. L ... C3 48
Pengibu, isl., Indon ... L7 38
Penglai, China ... F9 36
Pengpu, China, 253,000 ... H7 36
Pengshui, China, 11,000 ... J3 36
Penguin, is., Newf., Can ... E3 75
Penhold, Alta., Can., 319 ... C4 69
Penhook, Franklin, Va., 45 ... E3 121
Peniche, Port., 10,057 ... C1 20
Penicuik, Scot., 5,824 ... E5 13
Peninsula, Washington, Minn., 80 ... E7 99
Peninsula, Summit, Ohio, 644 ... A6 111
Penistaja, Sandoval, N. Mex ... B2 107
Penitas, Hidalgo, Tex., 700 ... F3 118
Penitente, mts., Braz ... C1 57
Penitentiary, mtn., Ala ... A2 78
Penki, China, 449,000 ... C9 34
Penn, Sask., Can ... D2 70
Penn, Ramsey, N. Dak., 70 ... A6 110
Pennant, pt., N.S., Can ... E6 74
Pennant Station, Sask., Can., 289 ... G1 70
Pennask, mtn., B.C., Can ... E7 68
Penndel, Bucks, Pa., 2,158 ... *F12 114
Penne, It., 5,054 ... C4 21
Pennell, mtn., Utah ... F5 119
Pennellville, Oswego, N.Y., 200 ... B4 108
Penneshaw, Austl., 149 ... G1 51
Penney Farms, Clay, Fla., 545 ... C5 86
Pennfield, N.B., Can., 334 ... D3 74
Penn Grove, Sonoma, Calif., 530 ... B5 82
Penn Hills, Allegheny, Pa., 51,512 ... *B6 114
Penniac, N.B., Can., 58 ... C3 74
Pennine Alps, mts., It., Switz ... D4 19
Pennine Chain, mts., Eng ... C5 10
Pennington, Choctaw, Ala., 400 ... C1 78
Pennington, Mercer, N.J., 2,063 ... C3 106
Pennington, co., Minn., 12,468 ... B2 99
Pennington, co., S. Dak., 58,195 ... D2 116
Pennington, mtn., Maine ... B4 96
Pennington Gap, Lee, Va., 1,799 ... B2 121
Pennock, Kandiyohi, Minn., 257 ... E3 99
Pennsauken, Camden, N.J., 33,771 ... D2 106
Pennsboro, Ritchie, W. Va., 1,660 ... B4 123
Pennsburg, Montgomery, Pa., 1,698 ... F11 114
Penns Grove, Salem, N.J., 6,176 ... D2 106
Pennside, Berks, Pa., 3,000 ... *F10 114
Pennsuco, Dade, Fla., 117 ... F3 86
Pennsville, Salem, N.J., 7,000 ... D2 106
Pennsville, Morgan, Ohio, 160 ... C6 111
Pennsylvania, state, U.S., 11,319,366 ... B12 77, 114
Pennsylvania Furnace, Huntingdon, Pa., 50 ... E5 114
Penn Valley, Montgomery, Pa., 3,500 ... *F11 114
Pennville, Jay, Ind., 730 ... D7 91
Pennville, York, Pa., 900 ... *G8 114
Penn Wynne, Montgomery, Pa., 4,500 ... *G11 114
Penn Yan, Yates, N.Y., 5,770 ... C3 108
Penny Highland, mts., N.W. Ter., Can ... C19 67
Penny Hill, New Castle, Del., 1,000 ... *A6 85
Penobscot, Hancock, Maine, 150 (706▲) ... D4 96
Penobscot, co., Maine, 126,346 ... C4 96
Penobscot, bay, Maine ... D4 96
Penobscot, lake, Maine ... C2 96
Penobscot, riv., Maine ... C4 96
Penobsquis, N.B., Can., 259 ... D4 74
Penokee, Graham, Kans., 92 ... C4 93
Penola, Caroline, Va., 25 ... D5 121
Penong, Austl., 118 ... F5 50
Penonomé, Pan., 4,266 ... F7 62
Penrith, Eng., 10,931 ... F6 13
Penrose, Fremont, Colo., 200 ... C5 83
Pensacola, Escambia, Fla., 56,752 (★165,400) ... G2 86
Pensacola, Mayes, Okla., 55 ... A6 112
Pensacola, mts., Ant ... A9 5
Pensacola, dam, Okla ... A6 112
Pensaukee, Oconto, Wis., 250 ... D6 124
Pense, Sask., Can., 374 ... G3 70
Pentagon, mtn., Mont ... C3 102
Pentecost, Sunflower, Miss., 100 ... B3 100
Pentecoste, Braz., 5,620 ... B3 57
Penticton, B.C., Can., 13,859 ... E8 68
Pentland, firth, Scot ... B5 13
Pentland, hills, Scot ... E5 13
Penton, DeSota, Miss., 100 ... A3 100
Penton, Salem, N.J., 150 ... D2 106
Pentwater, Oceana, Mich., 1,030 ... E4 98
Peñuelas, P.R., 2,261 ... C4 65
Peñuelas, mun., P.R., 14,887 ... C4 65
Penwell, Ector, Tex., 350 ... D1 118
Pen-y-Ghent, mtn., Eng ... F6 13
Penza, Sov. Un., 277,000 ... C2 29
Penzance, Navajo, Ariz., 10 ... C5 80
Penzance, Sask., Can., 99 ... F3 70
Penzance, Eng., 19,433 ... D2 12
Penzberg, Ger., 10,300 ... E5 16
Penzhina, riv., Sov. Un ... C19 28
Penzhino, Sov. Un., 600 ... C19 28
Penzlin, Ger., 3,811 ... E7 24
Peoa, Summit, Utah, 203 ... C4 119
Peone, Spokane, Wash ... D8 122
Peoples, Jackson, Ky., 200 ... C5 94
Peoria, Maricopa, Ariz., 2,593 ... D3, G1 80
Peoria, Peoria, Ill., 103,162 (★265,000) ... C4 90
Peoria, Amite, Miss., 100 ... D3 100
Peoria, Union, Ohio, 200 ... B4 111
Peoria, co., Ill., 189,044 ... C4 90
Peoria Heights, Peoria, Ill., 7,064 ... C4 90
Peotone, Will, Ill., 1,788 ... B6 90
Pep, Roosevelt, N. Mex., 4 ... D6 107
Pepacton, res., N.Y ... C6 108
Pepeekeo, Hawaii, Haw., 750 ... D6 88
Pepin, Pepin, Wis., 825 ... D1 124
Pepin, co., Wis., 7,332 ... D2 124
Pepin, lake, Wis ... D1 124
Pepperell, Middlesex, Mass., 700 (4,336▲) ... A4 97
Pepper Pike, Cuyahoga, Ohio, 3,217 ... *B2 111
Pepperton, Butts, Ga., 523 ... C3 87
P.E.P.S.U., see Patiala and East Punjab States Union, state, India
Peqin, Alb., 3,069 ... B2 23
Pequabuck, Litchfield, Conn., 400 ... C5 84
Pequaming, Baraga, Mich ... B2 98
Pequannock, Morris, N.J., 4,600 ... B4 106
Pequení, riv., Pan ... k12 62
Pequest, riv., N.J ... B3 106
Pequirí, riv., Braz ... B1 56
Pequirí, riv., Braz ... C2 56
Pequop, Elko Nev., 10 ... B7 104
Pequot Lakes, Crow Wing, Minn., 461 ... D4 99
Perak, isl., Indon ... J2 38
Perak, riv., Mala ... J4 38
Perak, state, Mala., 1,221,446 ... J4 38
Perakhóra (Loutrákion-Perakhóra), Grc., 6,168 ... g9 23
Perales, riv., Sp ... p16 20
Perales de Tajuña, Sp., 1,924 ... p18 20
Peralta, Valencia, N. Mex., 300 ... C3 107
Perce Station, Que., Can ... *B3 73
Perchas, P.R ... B3 65
Percival, Fremont, Iowa, 300 ... D2 92
Percy, Fr., 661 ... C3 14
Percy, Randolph, Ill., 810 ... E4 90
Percy, Marion, Iowa, 30 ... C4 92
Percy, Washington, Miss., 100 ... B3 100
Percy, Coos, N.H., 40 ... A4 105
Perdido, Baldwin, Ala., 900 ... D2 78
Perdido, bay, Fla ... G1 86
Perdido, mtn., Sp ... A6 20
Perdido, riv., Fla ... G1 86
Perdue, Sask., Can., 436 ... E2 70
Pereira, Col., 76,262 ... C2 60
Pereira Barreto, Braz., 7,173 ... C2 56
Perekop, Sov. Un., 10,000 ... H9 27
Pere Marquette, riv., Mich ... E4 98
Perené, riv., Peru ... D3 58
Pereslavl-Zalesskiy, Sov. Un., 22,200 ... C12 27
Pereyaslav-Khmelnitskiy, Sov. Un., 10,000 ... F8 27
Perez, Phil., 1,262 ... o13 35
Pergamino, Arg., 32,382 ... A4 54
Pergine, Valsugana, It., 3,879 ... A3 21
Perham, Aroostook, Maine, 125 (512▲) ... B4 96
Perham, Otter Tail, Minn., 2,019 ... D3 99
Peribonca, Que., Can., 496 ... A5 73
Perico, Dallam, Tex., 30 ... A1 118
Périgueux, Fr., 38,529 ... E4 14
Perijá, mts., Ven ... B3 60
Perim, isl., Aden ... F6 43
Perkasie, Bucks, Pa., 4,650 ... F11 114
Perkins, Sacramento, Calif., 300 ... *A6 82
Perkins, Que., Can., 318 ... D2 73
Perkins, Jenkins, Ga., 250 ... D5 87
Perkins, Delta, Mich., 250 ... C3 98
Perkins, Payne, Okla., 769 ... B4 112
Perkins, co., Nebr., 4,189 ... D4 103
Perkins, co., S. Dak., 5,977 ... B3 116
Perkinston, Stone, Miss., 350 ... E4 100
Perkinstown, Taylor, Wis ... C3 124
Perkinsville, Windsor, Vt., 167 ... E3 120
Perkiomen, creek, Pa ... C2 106
Perla, Hot Spring, Ark., 279 ... C3, D6 81
Perlas, arch., Pan ... F8 62
Perlas, lagoon, Nic ... D6 62
Perleberg, Ger., 13,500 ... B5 16
Perley, Norman, Minn., 165 ... C2 99
Perlis, state, Mala., 90,885 ... I4 38
Perl-Mack, Adams, Colo., 3,000 ... *B6 83
Perm, Sov. Un., 678,000 ... D20 9
Perma, Sanders, Mont., 60 ... C2 102
Përmet, Alb., 2,302 ... B3 23
Pernambuco, state, Braz., 4,136,900 ... C3, k6 57
Pernell, Garvin, Okla., 150 ... C4 112
Pernik, see Dmitrovo, Bul.
Péronne, Fr., 5,081 ... C5 14
Perot, bayou, La ... C7 95
Perote, Bullock, Ala., 45 ... D4 78
Perpignan, Fr., 83,025 ... F5 14
Perquimans, co., N.C., 9,178 ... A7 109
Perrégaux, Alg., 27,367 ... *B5 44
Perrin, Jack, Tex., 300 ... C3 118
Perrin, Gloucester, Va., 300 ... D6 121
Perrine, Dade, Fla., 6,424 ... F3, G6 86
Perrineville, Monmouth, N.J., 250 ... C4 106
Perris, Riverside, Calif., 2,950 ... F3, F5 82
Perros-Guirec, Fr., 3,495 ... C2 14
Perrot, isl., Que., Can ... D8 73
Perry, Perry, Ark., 224 ... B3 81
Perry, Taylor, Fla., 7,979 ... B3 86
Perry, Houston, Ga., 6,032 ... D3 87
Perry, Pike, Ill., 442 ... D3 90
Perry, Dallas, Iowa, 6,442 ... C3 92
Perry, Jefferson, Kans., 495 ... B7, C8 93
Perry, Vermilion, La., 150 ... E3 95
Perry, Washington, Maine, 125 (564▲) ... D5 96
Perry, Shiawassee, Mich., 1,370 ... F6 98
Perry, Ralls, Mo., 802 ... B6 101
Perry, Wyoming, N.Y., 4,629 ... C3 108
Perry, Lake, Ohio, 885 ... A6 111
Perry, Noble, Okla., 5,210 ... A4 112
Perry, Aiken, S.C., 196 ... D5 115
Perry, Box Elder, Utah, 587 ... B3 119
Perry, co., Ala., 17,358 ... C2 78
Perry, co., Ark., 4,927 ... C3 81
Perry, co., Ill., 19,184 ... E4 90
Perry, co., Ind., 17,232 ... H4 91
Perry, co., Ky., 34,961 ... C6 94
Perry, co., Miss., 8,745 ... D4 100
Perry, co., Mo., 14,642 ... D8 101
Perry, co., Ohio, 27,864 ... C5 111
Perry, co., Pa., 26,582 ... F7 114
Perry, co., Tenn., 5,273 ... B4 117
Perry, peak, Mass ... B1 97
Perry, stream, N.H ... A1 105
Perrydale, Polk, Oreg., 125 ... B3, C1 113
Perryman, Harford, Md., 700 ... B5 85
Perryopolis, Fayette, Pa., 1,799 ... *F2 114
Perrysburg, Cattaraugus, N.Y., 434 ... C2 108
Perrysburg, Wood, Ohio, 5,519 ... A2, A4 111
Perry's Victory and International Peace Memorial, nat. mon., Ohio ... A5 111
Perrysville, Vermillion, Ind., 497 ... D3 91
Perrysville, Ashland, Ohio, 769 ... B5 111
Perryton, Ochiltree, Tex., 7,903 ... A2 118
Perryvale, Alta., Can., 14 ... B4 69
Perryville, Alsk., 93 ... D8 79
Perryville, Perry, Ark., 719 ... B3 81
Perryville, Boyle, Ky., 715 ... C5 94
Perryville, Ouachita, La., 100 ... B4 95
Perryville, Cecil, Md., 674 ... A5 85
Perryville, Perry, Mo., 5,117 ... D8 101
Perryville, Washington, R.I., 200 ... D10 84
Perryville, Decatur, Tenn., 250 ... B3 117
Persan, Fr., 5,196 ... E2 15
Persepolis, ruins, Iran ... G6 41
Pershing (East Germantown), Wayne, Ind., 367 ... E7 91
Pershing, Marion, Iowa, 275 ... C5 92
Pershing, Gasconade, Mo., 40 ... C6 101
Pershing, co., Nev., 3,999 ... C3 104
Persia, Harrison, Iowa, 322 ... C2 92
Persia, see Iran, country, Asia
Persia, Hawkins, Tenn., 150 ... C10 117
Persian, gulf, Asia ... J19 9
Persimmon Grove, Campbell, Ky., 100 ... A7 94
Persinger, Nicholas, W. Va., 25 ... C4 123
Person, co., N.C., 24,361 ... A4 109
Perstorps, Swe., 4,151 ... B7 24
Pertek, Tur., 3,100 ... C12 31
Perth, Austl., 94,508 (★450,000) ... F2 50
Perth, Ont., Can., 5,360 ... C8 72
Perth, N.B., Can., 909 ... C2 74
Perth, Sumner, Kans., 100 ... E6 93
Perth, Towner, N. Dak., 73 ... A6 110
Perth, Scot., 41,199 ... D5 13
Perth, co., Ont., Can., 57,452 ... D3 72
Perth, co., Scot., 127,018 ... D4 13
Perth Amboy, Middlesex, N.J., 38,007 ... B4, E4 106
Perthshire, Bolivar, Miss., 500 ... B3 100
Pertuis, Fr., 6,774 ... F6 14
Pertuis Breton, bay, Fr ... D3 14
Peru, La Salle, Ill., 10,460 ... B4 90
Peru, Miami, Ind., 14,453 ... C5 91
Peru, Madison, Iowa, 265 ... C4 92
Peru, Chautauqua, Kans., 340 ... E7 93
Peru, Berkshire, Mass., 50 (197▲) ... B1 97
Peru, Nemaha, Nebr., 1,151 ... D10 103
Peru, Clinton, N.Y., 900 ... B3 108
Peru, Bennington, Vt., 40 (194▲) ... E3 120
Peru, country, S.A., 8,132,793 ... E3 53, D3 58
Perugia, It., 71,500 (112,511▲) ... C4 21
Peruque, St. Charles, Mo., 80 ... A7 101
Péruwelz, Bel., 7,668 ... D3 15
Pervomaysk, Sov. Un., 30,000 ... G8 27
Pervomaysk, Sov. Un., 10,000 ... q21 27
Pervouralsk, Sov. Un., 101,000 ... B5 29
Pesaro, It., 45,000 (65,973▲) ... C4 21
Pescadero, San Mateo, Calif., 400 ... C5, D2 82
Pescara, It., 87,436 ... C5 21
Pescara, riv., It ... C4 21
Peschani, cape, Sov. Un ... G19 9
Peschanyy, cape, Sov. Un ... B10 37
Peshastin, Chelan, Wash., 600 ... B5 122
Peshawar, Pak., 160,000 (★218,691) ... B5 39
Peshkopi, Alb., 2,524 ... B3 23
Peshtera, Bul., 8,946 ... D7 22

Peshtigo, Marinette, Wis., 2,504 C6 124
Peshtigo, riv., Wis C5 124
Peski, Sov. Un., 5,000 C7 29
Peski, Sov. Un., 25,500 F14 27
Peski, Sov. Un. n18 27
Peski Muyun-Kum, des., Sov. Un. C23 9
Peso da Régua, Port., 5,623 B2 20
Pesotum, Champaign, Ill., 468 D5 90
Pesqueira, Braz., 19,778 C3, k5 57
Pest, co., Hung., 783,108 *B4 22
Petaca, Rio Arriba, N. Mex., 150 A4 107
Petah Tiqva, Isr., 37,500 B6, g10 32
Petal, Forrest, Miss., 4,007 D4 100
Petaluma, Sonoma, Calif., 14,035 B5, C2 82
Pétange, Lux., 11,623 E5 15
Petatlán, Mex., 3,630 D4 63
Petauke, N. Rh., 1,410 D5 48
Petawawa, Ont., Can., 4,509 B7 72
Petawawa, riv., Ont., Can A7 72
Petenwell, res., Wis D4 124
Peter, isl., Vir. Is F16 65
Peterborough, Austl., 3,430 F2 51
Peterborough, Ont., Can., 47,185 (*49,902) C6 72
Peterborough, Eng., 62,031 B7 12
Peterborough (Peterboro), Hillsboro, N.H., 1,931 E3 105
Peterborough, co., Ont., Can., 76,375 C6 72
Peter I, isl., Ant C3 5
Petergof, see Petrodvorets, Sov. Un.
Peterhead, Scot., 12,497 C7 13
Peterman, Monroe, Ala., 600 D2 78
Peter Pond, lake, Sask., Can B1 70
Peters, creek, W. Va C7 123
Peters, mtn., Va., W. Va D4 123
Petersburg, Alsk., 1,502 D13, m23 79
Petersburg, Menard, Ill., 2,359 C4 90
Petersburg, Pike, Ind., 2,939 H3 91
Petersburg, Boone, Ky., 390 A6 94
Petersburg, Monroe, Mich., 1,018 G7 98
Petersburg, Boone, Nebr., 400 C7 103
Petersburg, Cape May, N.J., 200 E3 106
Petersburg, Rensselaer, N.Y., 445 C7 108
Petersburg, Nelson, N. Dak., 272 A8 110
Petersburg, Mahoning, Ohio, 600 B7 111
Petersburg, Huntingdon, Pa., 552 E5 114
Petersburg, Lincoln and Marshall, Tenn., 423 B5 117
Petersburg, Hale, Tex., 1,400 C2 118
Petersburg (Independent City), Va., 36,750 (*100,300) C7, D5 121
Petersburg, Grant, W. Va., 2,079 B5 123
Petersdorf, Ger., 1,800 D5 24
Petersfield, Man., Can., 157 D3 71
Petersfield, Eng., 7,379 D7 12
Petersham, Worcester, Mass., 450 (890▲) B3 97
Peters Landing, Perry, Tenn., 25 B4 117
Peterson, Clay, Iowa, 565 B2 92
Peterson, Fillmore, Minn., 283 G7 99
Peterstown, Monroe, W. Va., 616 D4 123
Petersville, Lewis, Ky., 50 B6 94
Peter The Great, bay, Sov. Un E5 37
Petilia Policastro, It., 8,662 E6 21
Pétionville, Hai., 9,570 F7 64
Petit, lake, La C8 95
Petit Bois, isl., Miss E5 100
Petitcodiac, N.B., Can., 902 D4 74
Petitcodiac, riv., N.B., Can C5 74
Petite Amite, riv., La B6 95
Petite-de-Grat Bridge, N.S., Can., 295 D9 74
Petite Lieure, riv., Que., Can A5 73
Petite Miquelon, isl., Miquelon Isl E3 75
Petite Riviere Bridge, N.S., Can., 310 E5 74
Petit-Etang, N.S., Can., 489 C9 74
Petite Vallee, Que., Can., 237 A3 73
Petit-Goâve, Hai., 5,536 F7 64
Petitjean, Mor., 19,478 G4 30
Petit Jean, creek, Ark B2 81
Petit Jean, mtn., Ark B2 81
Petit Jean, mtn., Ark B3 81
Petit Rocher, N.B., Can., 590 B4 74
Peto, Mex., 6,995 C7 63
Petorca, Chile, 1,098 A2 54
Petoskey, Emmet, Mich., 6,138 C6 98
Petrey, Crenshaw, Ala., 165 D3 78
Petrich, Bul., 13,456 E6 22
Petrified, forest, Miss C3 100
Petrified, forest, N. Dak B2 110
Petrified, forest, S. Dak B2 116
Petrified Forest, nat. mon., Ariz A6 80
Petrified Wood, park, S. Dak B3 116
Petrikov, Sov. Un., 10,000 E7 27
Petrinja, Yugo., 7,345 C3 22
Petrodvorets, Sov. Un., 30,000 s30 25
Petrohué, Chile C2 54
Petrokrepost, Sov. Un., 30,000 B8 27, s32 25
Petrolândia, Braz., 2,669 C3 57
Petroleum, Wells, Ind., 300 C7 91
Petroleum, co., Mont., 894 C8 102
Petrolia, Ont., Can., 3,708 E2 72
Petrolia, Allen, Kans., 125 E8 93
Petrolia, Butler, Pa., 527 D2 114
Petrolia, Clay, Tex., 631 B3 118
Petrolina, Braz., 14,652 C2 57
Petropavlovsk, Sov. Un., 140,000 C7 29
Petropavlovskiy, Sov. Un. D3 29
Petropavlovsk [-Kamchatskiy], Sov. Un., 96,000 D18 28
Petrópolis, Braz., 93,849 C4, h6 56
Petros, Morgan, Tenn., 850 C9 117
Petroseni, Rom., 23,052 C6 22
Petrovgrad, see Zrenjanin, Yugo.
Petrovsk, Sov. Un., 10,000 E15 27
Petrovskoye, Sov. Un., 29,000 D2 29
Petrovsk-Zabaykalskiy, Sov. Un., 59,000 D13 28
Petrozavodsk, Sov. Un., 139,000 A10 27
Petterill, riv., Eng F6 13
Pettibone, Kidder, N. Dak., 205 B6 110
Pettigoe, Ire., 313 C4 11
Pettigrew, Madison, Ark., 150 B2 81
Pettis, co., Mo., 35,120 C4 101
Pettisville, Fulton, Ohio, 320 A3 111
Pettit, Washington, Miss B2 100
Pettus, Lonoke, Ark C4, D6 81
Pettus, Bee, Tex., 450 E4 118
Petty Harbour, Newf., Can., 908 E5 75
Petukhovo, Sov. Un., 12,300 B7 29
Petway, Cheatham, Tenn A4 117
Pevely, Jefferson, Mo., 416 B8 101
Pewamo, Ionia, Mich., 416 E6 98
Pewaukee, Waukesha, Wis., 2,484 E1, E5 124
Pewaukee, lake, Wis E1 124
Pewee Valley, Oldham, Ky., 881 A4, B4 94
Peyton, El Paso, Colo., 110 B6 83
Pézenas, Fr., 7,198 F5 14
Pfaffenhofen, Ger., 8,600 E6 17
Pfäffikon, Switz., 5,735 B6 19
Pfarrkirchen, Ger., 6,000 E7 17
Pfeifer, Ellis, Kans., 200 D4 93
Pforzheim, Ger., 82,500 (*120,000) E3 17
Pfunds, Aus., 1,794 C6 18
Pfungstadt, Ger., 13,100 D3 17
Phair, Aroostook, Maine (part of Presque Isle) B5 96
Phalodi, India, 15,722 C5 40
Phaltan, India, 19,003 E5 40
Phangan, isl., Thai H4 38
Phan Rang, Viet., 5,000 G8 38
Phan Thiet, Viet., 5,000 G8 38
Pharoah, Okfuskee, Okla., 250 B5 112
Pharr, Hidalgo, Tex., 14,106 F3 118
Phatthalung, Thai., 10,147 I4 38
Pheba, Clay, Miss., 351 B5 100
Phelps, Pike, Ky., 725 C7 94
Phelps, Ontario, N.Y., 1,887 C3 108
Phelps, Vilas, Wis., 500 B4 124
Phelps, co., Mo., 25,396 D6 101
Phelps, co., Nebr., 9,800 D6 102
Phelps, lake, N.C B7 109
Phelps City, Atchison, Mo., 81 A2 101
Phenix City, Russell, Ala., 27,630 C4 78
Phet Buri, Thai., 14,288 F3 38
Phetchabun, Thai., 5,733 D4 38
Phichit, Thai., 9,477 D4 38
Philadelphia, Neshoba, Miss., 5,017 C4 100
Philadelphia, Marion, Mo., 166 B6 101
Philadelphia, Jefferson, N.Y., 868 A5 108
Philadelphia, Philadelphia, Pa., 2,002,512 (*3,969,500) B11, G11 114
Philadelphia, Loudon, Tenn., 500 D9 117
Philadelphia, co., Pa., 2,002,512 G11 114
Philbrook, Todd, Minn., 100 D4 99
Phil Campbell, Franklin, Ala., 898 A2 78
Philip, Haakon, S. Dak., 1,114 C4 116
Philipp, Tallahatchie, Miss., 250 B3 100
Philippeville, Alg., 70,406 B6 44, F10 30
Philippeville, Bel., 1,559 D4 15
Philippi, Barbour, W. Va., 2,228 B4 123
Philippine, is., Phil B6 35
Philippine, sea, Phil A6 35
Philippines, country, Asia, 27,455,500 H15 33, B6 35
Philippolis, S. Afr., 2,078 D4 49
Philipsburg, Granite, Mont., 1,107 D3 102
Philipsburg, Centre, Pa., 3,872 E5 114
Philipsville, Ont., Can., 97 C8 72
Philleo, lake, Wash D7 122
Phillip, isl., Austl I5 51
Phillippy, Lake, Tenn., 100 A2 111
Phillips, Franklin, Maine, 600 (1,021▲) D2 96
Phillips, Hamilton, Nebr., 192 D7 103
Phillips, Coal, Okla., 91 C5 112
Phillips, Hutchinson, Tex., 3,605 B2 118
Phillips, Price, Wis., 1,524 C3 124
Phillips, co., Ark., 43,997 C5 81
Phillips, co., Colo., 4,440 A8 83
Phillips, co., Kans., 8,709 C4 93
Phillips, co., Mont., 60,027 B8 102
Phillips, brook, N.H A4, B1 105
Phillips, isl., S.C G6 115
Phillipsburg, Que., Can., 379 D4 73
Phillipsburg, Tift, Ga., 2,037 *E3 87
Phillipsburg, Phillips, Kans., 3,233 C4 93
Phillipsburg, Laclede, Mo., 142 D5 101
Phillipsburg, Warren, N.J., 18,502 B2 106
Phillipsdale, Providence, R.I. (part of East Providence) B11 84
Phillipston, Worcester, Mass., 100 (695▲) A3 97
Phillipsville, Haywood, N.C., 1,311 *D3 109
Philmont, Columbia, N.Y., 1,750 C7 108
Philo, Champaign, Ill., 740 D5 90
Philo (Taylorsville), Muskingum, Ohio, 913 C6 111
Philomath, Benton, Oreg., 1,359 C3 113
Philpott, res., Va E2 121
Philrich, Hutchinson, Tex., 2,067 *B2 118
Phippen, Sask., Can., 60 E1 70
Phippsburg, Routt, Colo., 150 A4 83
Phippsburg, Sagadahoc, Maine, 100 (1,121▲) E6 96
Phitsanulok, Thai., 32,659 D4 38
Phlox, Langlade, Wis., 150 C4 124
Phnom Penh, Camb., 420,000 G6 38
Phocis (Fokis), prov., Grc., 47,842 *C4 23
Phoebus, Va. (part of Hampton) B6 121
Phoenicia, Ulster, N.Y., 475 C6 108
Phoenix, Maricopa, Ariz., 439,170 (*619,600) D3, G2 80
Phoenix, Cook, Ill., 4,203 F3 90
Phoenix, Baltimore, Md., 200 A4 85
Phoenix, Keweenaw, Mich., 45 A2 98
Phoenix, Oswego, N.Y., 2,408 B4 108
Phoenix, Jackson, Oreg., 769 E4 113
Phoenix, Charlotte, Va., 259 D4 121
Phoenix, is., Pac. O G11 7
Phoenixville, Windham, Conn., 100 B8 84
Phoenixville, Chester, Pa., 13,797 A10, F10 114
Phong Saly, Laos, 10,000 B5 38
Phrae, Thai., 15,616 C4 38
Phthiotis (Fthiotis), prov., Grc., 160,035 *C4 23
Phuket, Thai., 27,021 I3 38
Phuket, isl., Thai I3 40
Phu Lai Leng, mtn., Laos C6 38
Phu Quoc, isl., Viet G6 38
Phutthaisong, Thai E5 38
Piaanu, pass, Truk 52
Piacenza, It., 88,541 B2 21, D5 18
Pialba, Austl., 3,544 B9 51
Pianosa, isl., It C3 21
Pianosa, isl., It C5 21
Piapot, Sask., Can., 246 H1 70
Piaseczno, Pol., 13,800 m14 26
Piatra-Neamt, Rom., 32,648 B8 22
Piatt, co., Ill., 14,960 D5 90
Piauí, state, Braz., 1,263,368 C2 57
Piauí, mts., Braz C2 57
Piauí, riv., Braz C2 57
Piave, riv., It D8 18
Piazza Armerina, It., 24,887 F5 21
Pibor, riv., Sud D3 47
Pibor Post, Sud D3 47
Pibroch, Alta., Can., 118 B4 69
Picabo, Blaine, Idaho, 75 F4 89
Picacho, Pinal, Ariz., 400 E4 80
Picacho, Lincoln, N. Mex., 75 D4 107
Picacho, peak, Calif F6 83
Picadome, Fayette, Ky., 900 B5 94
Picard, Que., Can B8 73
Picardy (Picardie), former prov., Fr., 1,051,000 C5 14
Picayune, Pearl River, Miss., 7,834 E4 100
Piccadilly, Newf., Can., 427 D2 75
Pic de Tio, mtn., Guinea E3 45
Pic du Midi d'Ossau, mtn., Fr F3 14
Piceance, creek, Colo B2 83
Pichanal, Arg D3 55
Picher, Ottawa, Okla., 2,553 A7 112
Pichilemu, Chile A2 54
Pichincha, prov., Ec., 424,655 B2 58
Pickard, pt., Mass C4 97
Pickardville, Alta., Can., 140 B4 69
Pickardville, Sheridan, N. Dak., 23 B5 110
Pickaway, co., Ohio, 35,855 C4 111
Pick City, Mercer, N. Dak., 101 B4 110
Pickens, Desha, Ark., 100 D4 81
Pickens, Holmes, Miss., 727 C4 100
Pickens, Pickens, S.C., 2,198 B2 115
Pickens, Randolph, W. Va., 300 C4 123
Pickens, co., Ala., 21,882 B1 78
Pickens, co., Ga., 8,903 B2 87
Pickens, co., S.C., 46,030 B2 115
Pickensville, Pickens, Ala., 150 B1 78
Pickerel, Ont., Can., 116 B4 72
Pickerel, Langlade, Wis., 60 C5 124
Pickerel, lake, Wis C4 124
Pickerel, riv., Ont., Can B4 72
Pickering, Ont., Can., 1,755 D5 72
Pickering, Eng., 4,193 F8 13
Pickering, Nodaway, Mo., 234 A3 101
Pickerington, Fairfield, Ohio, 634 C2, C5 111
Pickett, co., Tenn., 4,431 C8 118
Pickford, Chippewa and Mackinac, Mich., 650 B6 98
Pickleville, Rich, Utah, 94 B4 119
Pickrell, Gage, Nebr., 130 D9 103
Pickstown, Charles Mix, S. Dak., 600 E7 116
Pickwick, Winona, Minn., 150 G7 99
Pickwick Dam, Hardin, Tenn., 25 B3 117
Pickwick, lake, Ala., Miss., Tenn A1 78
Pico, isl., Az. Is g8 44
Pico Rivera, Los Angeles, Calif., 49,150 *F2 82
Picos, Braz., 8,176 C2 57
Pictograph Rocks, Ariz D1 80
Picton, Ont., Can., 4,862 D7 72
Picton, N.Z., 2,315 N15 51
Pictou, N.S., Can., 4,534 D7 74
Pictou, co., N.S., Can., 43,908 D7 74
Pictou, isl., N.S., Can D7 74
Pictou Landing, N.S., Can., 405 D7 74
Pic Tousidé, mtn., Chad A3 46
Picture, gorge, Oreg C7 113
Picture Butte, Alta., Can., 978 E4 69
Pictured, cave, Wis E3 124
Pictured Rocks, Ariz E4 80
Picture Rocks, Lycoming, Pa., 594 D8 114
Picuí, Braz., 2,140 h5 57
Picún-Leufú, Arg B3 54
Pidcock, Brooks, Ga., 250 F3 87
Pidurutalagala, peak, Cey G7 39
Piedade, Braz., 4,812 m8 56
Piedmont, Calhoun, Ala., 4,794 B4 78
Piedmont, Alameda, Calif., 11,117 B5 82
Piedmont, Que., Can., 272 D3 73
Piedmont, Greenwood, Kans., 250 E7 93
Piedmont, Wayne, Mo., 1,555 D7 101
Piedmont, Canadian, Okla., 146 B4 112
Piedmont, Anderson and Greenville, S.C., 2,108 B3 115
Piedmont, Meade, S. Dak., 200 C2 116
Piedmont, Mineral, W. Va., 2,307 B5 123
Piedmont, Uinta, Wyo., 10 D2 125
Piedmont, reg., It B1 21
Piedmont, res., Ohio B6 111
Piedmont, upland, U.S C2 106, C2 109, E3 121
Piedrabuena, Sp., 6,210 C3 20
Piedra Negra, pt., Mex D5 63
Piedras, pt., Arg B5 54
Piedras, riv., Pan k12 62
Piedras, riv., Peru D3 58
Piedras Blancas, pt., Calif E3 82
Piedras Negras, Guat B2 62
Piedras Negras, Mex., 42,649 B4 63
Piedra Sola, Ur E1 56
Piekietko, Pol., 1,000 k13 26
Pieksämäki, Fin., 10,600 F12 25
Piélagos, Sp., 8,040 A4 20
Pielisjarvi, lake, Fin F13 25
Piemonte, pol., dist., It., 3,914,250 *B1 21
Piendamó, Col., 1,615 C2 60
Pienkaun, China E4 36
Piennes, Fr., 4,244 E5 15
Pierce, Weld, Colo., 424 A6 83
Pierce, Polk, Fla., 800 E5 86
Pierce, Clearwater, Idaho, 522 C3 89
Pierce, Pierce, Nebr., 1,216 B8 103
Pierce, Tucker, W. Va., 50 B5 123
Pierce, co., Ga., 9,678 E4 87
Pierce, co., Nebr., 8,722 B8 103
Pierce, co., N. Dak., 7,394 A5 110
Pierce, co., Wash., 321,590 C3 122
Pierce, co., Wis., 22,503 D1 124
Pierce, lake, Man., Can B5 71
Pierce, lake, Fla E5 86
Pierce, pond, Maine C2 96
Pierce Bridge, Grafton, N.H. B3 105
Pierce City, Lawrence, Mo., 1,006 E3 101
Piercefield, St. Lawrence, N.Y., 250 A6, B2 108
Pierces, Cape May, N.J., 175 E3 106
Pierce Station, Obion, Tenn., 25 A3 117
Pierceton, Kosciusko, Ind., 1,186 B6 91
Pierceville, Finney, Kans., 175 E3 93
Piercy, Mendocino, Calif., 50 C2 82
Pieria, prov., Grc., 97,697 *B4 23
Piermont, Grafton, N.H., 125 (477▲) C2 105
Piermont, Rockland, N.Y., 1,906 C6 108
Piermont, mtn., N.H C2 105
Pierowall, Scot A5 13
Pierpont, Ashtabula, Ohio, 250 A7 111
Pierpont, Day, S. Dak., 258 B8 116
Pierre, Hughes, S. Dak., 10,088 C5 116
Pierre, bayou, Miss D3 100
Pierrefitte-sur-Aire, Fr., 256 F5 15
Pierrefitte [-sur-Seine], Fr., 14,770 g10 14
Pierreville, Que., Can., 1,559 C5 73
Pierron, Bond and Madison, Ill., 371 E4 100
Pierson, Man., Can., 229 E1 71
Pierson, Volusia, Fla., 716 C5 86
Pierson, Woodbury, Iowa, 425 B2 92
Pierz, Morrison, Minn., 816 E4 99
Piestany, Czech., 18,700 D4 26
Pietermaritzburg, S. Afr., 91,178 (*96,236) C5 49
Pietersaari (Jakobstad), Fin., 7,378 F10 25
Pietersburg, S. Afr., 28,071 B4 49
Pie Town, Catron, N. Mex., 120 C1 107
Pietrasanta, It., 6,600 C3 21
Piet Retief, S. Afr., 8,604 C5 49
Pieve di Cadore, It., 3,893 A4 21
Pigeon, Huron, Mich., 1,191 E7 98
Pigeon, bay, Man., Can C3 71
Pigeon, creek, Ala D3 78
Pigeon, creek, Ind H3 91
Pigeon, lake, Alta., Can C3 69
Pigeon, mtn., Ga B1 87
Pigeon, mtn., N.Y B6 108
Pigeon, pt., Calif D2 82
Pigeon, pt., U.S., Can A8 99
Pigeon, riv., Man., Can C3 71
Pigeon, riv., Ind A6 91
Pigeon, riv., U.S., Can A8 99
Pigeon, riv., Wis B6 124
Pigeon Cove, Essex, Mass., 1,064 A6 97
Pigeon Falls, Trempealeau, Wis., 207 D2 124
Pigeon Forge, Sevier, Tenn., 950 D10 117
Pigeon River, Cook, Minn., 35 A8 99
Pigg, riv., Va E3 121
Piggott, Clay, Ark., 2,776 A5 81
Pigüé, Arg., 5,869 B4 54
Pihrai, isl., Eniwetok 52
Pijijiapan, Mex., 3,307 D6 63
Pike, Pike, Ark., 50 C2 81
Pike, Grafton, N.H., 200 B2 105
Pike, Wyoming, N.Y., 286 345 C2 108
Pike, Yamhill, Oreg B1 113
Pike, co., Ala., 25,987 D4 78
Pike, co., Ark., 7,864 C2 81
Pike, co., Ga., 7,138 C2 87
Pike, co., Ill., 20,552 D3 90
Pike, co., Ind., 12,797 H3 91
Pike, co., Ky., 68,264 C7 94
Pike, co., Miss., 35,063 D3 100
Pike, co., Mo., 16,706 B6 101
Pike, co., Ohio, 19,380 C4 111
Pike, co., Pa., 9,158 D11 114
Pike, riv., Wis C5 124
Pike Rigg, mtn., Eng F6 13
Pike Road, Beaufort, N.C., 400 B7 109
Pikes, beach, N.Y F6 84
Pikes, peak, Colo C5 83
Pikes Rocks, mts., Pa C3 114
Pikesville, Baltimore, Md., 18,737 C2 85
Piketberg, S. Afr., 3,312 D2 49
Piketon, Pike, Ohio, 1,244 C4 111
Pikeview, El Paso, Colo., 200 C6 83
Pikeville, Pike, Ky., 4,754 C7 94
Pikeville, Wayne, N.C., 525 B6 109
Pikeville, Bledsoe, Tenn., 951 D8 117
Pikwitonei, Man., Can., 175 B3 71
Piła, Pol., 27,000 B4 26
Pilar, Braz., 7,201 C3, k6 57
Pilar, Taso, N. Mex A4 107
Pilar, Par., 10,062 E4 55
Pilar de Goiás, Braz., 232 A3 56
Pilar [do Sul], Braz., 1,789 m8 56
Pilatus, peak, Switz C5 19
Pilcomayo, riv., Par D3 55
Pilger, Sask., Can., 133 E3 70
Pilger, Stanton, Nebr., 491 B8 103
Pilgram Knob, Buchanan, Va., 100 B3 121
Pilgrim Gardens, Delaware, Pa., 5,000 *B11 114
Pilibhit, India, 57,527 C7 40
Pilica, riv., Pol C5 26
Pillager, Cass, Minn., 338 D4 99
Pillar, pt., Calif B5 82
Pillaro, Ec., 2,814 B2 58
Pillau, see Baltiysk, Sov. Un.
Pilley's Island, Newf., Can., 478 D4 75
Pillow (Uniontown), Dauphin, Pa., 348 E8 114
Pillowville, Weakley, Tenn A3 117
Pillsbury, Barnes, N. Dak., 76 B8 110
Pilos, Grc., 3,130 D3 23
Pilot, peak, Nev B7 104
Pilot, range, N.H A4 105
Pilot Butte, Sask., Can., 381 G3 70
Pilot Grove, Cooper, Mo., 680 C5 101
Pilot Knob, Iron, Mo., 524 D7 101
Pilot Knob, mtn., Ark B2 81
Pilot Knob, mtn., Ark C1 81
Pilot Knob, mtn., Idaho D3 89
Pilot Knob, mtn., Mo E4 101
Pilot Knob, mtn., Tenn C8 117
Pilot Knob, ridge, Kans B8 93
Pilot Mound, Man., Can., 802 E2 71
Pilot Mound, Boone, Iowa, 196 B3 92
Pilot Mountain, Surry, N.C., 1,310 A3 109
Pilot Point, Alsk., 61 D8 79
Pilot Point, Denton, Tex., 1,254 C4 118
Pilot Rock, Umatilla, Oreg., 1,695 B8 113
Pilot Station, Alsk., 219 C7 79
Pilottown, Plaquemines, La., 150 E6 95
Pilsen, Kewaunee, Wis., 30 A6 124
Pilvo, Sov. Un A11 37
Pima, Graham, Ariz., 806 E6 80
Pima, co., Ariz., 265,660 E3 80
Pimba, Austl., 150 F6 50
Pimmit Hills, Fairfax, Va., 1,000 *B5 121
Pimple, hill, Pa A2 106
Piña, Pan., 229 k10 62
Pinal, co., Ariz., 62,673 E4 80
Pinal, mts., Ariz D5 80
Pinarbasi, Tur., 5,800 C11 31
Pinar del Río, Cuba, 38,885 D2 64
Pinar del Río, prov., Cuba, 448,442 D2 64
Pinardville, Hillsboro, N.H., 1,500 *E4 105
Pinas, Arg A3 54
Pincher Creek, Alta., Can., 2,961 E4 69
Pincher Station, Alta., Can., 79 E4 69
Pinchi Lake, B.C., Can B5 68
Pinckard, Dale, Ala., 578 D4 78
Pinckney, Crittenden, Ark E8 81
Pinckney, Livingston, Mich., 732 F7 98
Pinckney, isl., S.C G6 115
Pinckneyville, Perry, Ill., 3,085 E4 90
Pinckneyville, Wilkinson, Miss., 50 D2 100
Pinconning, Bay, Mich., 1,329 E7 98
Pinczow, Pol., 3,701 C6 26
Pindall, Searcy, Ark., 100 A3 81
Pindamonhangaba, Braz., 19,144 C3 56
Pindare, riv., Braz B1 57
Pindus, mts., Grc C3 23
Pine, Gila, Ariz., 150 C4 80
Pine, Jefferson, Colo., 35 B5 83
Pine, Ripley, Mo., 150 E6 101
Pine, San Miguel, N. Mex., 100 B4 107
Pine, co., Minn., 17,004 D6 99
Pine, cape, Newf., Can E5 75
Pine, creek, Nev C5 104
Pine, creek, Pa D7 114
Pine, creek, Wash B8 122
Pine, hill, Conn C4 84
Pine, key, Fla F1 86
Pine, lake, Ind A4 91
Pine, lake, Minn D3 99
Pine, lake, Wis C4 124
Pine, mtn., Conn B5 84
Pine, mtn., Conn B3 84
Pine, mtn., Ga D2 87
Pine, mtn., Ky., Tenn D6 94, C9 117
Pine, mtn., Mass B3 97
Pine, mtn., Okla C6 112
Pine, mtn., Oreg D6 113
Pine, mtn., Tenn C9 117
Pine, pt., Fla C3 86
Pine, ridge, Nebr B2 103
Pine, ridge, Wyo C8 125
Pine, riv., Alta., Can B4 69
Pine, riv., B.C., Can B6 68
Pine, riv., Man., Can D1 71
Pine, riv., Mich D7 98
Pine, riv., N.H C4 105
Pine, riv., Wis C4 124
Pine Apple, Wilcox, Ala., 355 D3 78
Pine Bank, Greene, Pa., 20 G1 114
Pine Beach, Ocean, N.J., 985 *D4 106
Pine Bluff, Jefferson, Ark., 44,037 C3 81
Pinebluff, Moore, N.C., 509 B4 109
Pinebluff, lake, Sask., Can C4 70
Pine Bluffs, Laramie, Wyo., 1,121 D8 125
Pine Bluff Southeast, Jefferson, Ark., 2,679 *C4 81
Pinebur, Marion, Miss., 40 D4 100
Pine Bush, Orange, N.Y., 1,016 D6 108
Pine Castle, Orange, Fla., 5,000 D5 86
Pine City, Monroe, Ark., 125 C4 81
Pine City, Pine, Minn., 1,972 E6 99
Pine City, Whitman, Wash., 50 B8 122
Pine Creek, Austl., 91 B5 50
Pinecreek, Roseau, Minn., 34 B3 99
Pine Creek, Beaver, Utah E3 119
Pine Creek, gorge, Pa C7 114
Pinecrest, Tuolumne, Calif., 2,000 C4 82
Pinecroft, Spokane, Wash D8 122
Pineda, Brevard, Fla., 150 D6 86
Pinedale, Navajo, Ariz., 80 C5 80
Pinedale, Fresno, Calif., 3,202 *D4 82
Pinedale, Sublette, Wyo., 965 C3 125
Pine Falls, Man., Can., 1,244 D3 71
Pine Forest, mts., Nev B3 104
Pinega, Sov. Un., 3,600 C17 9
Pine Grove, Appling, Ga., 100 E4 87
Pine Grove, St. Helena, La., 150 D5 95
Pine Grove, Schuylkill, Pa., 2,267 E9 114
Pine Grove, Wetzel, W. Va., 760 A6, B4 123
Pine Grove, Brown, Wis., 25 A6 124
Pine Grove Mills, Centre, Pa., 900 E6 114
Pine Hall, Stokes, N.C., 400 A3 109
Pine Hill, Wilcox, Ala., 367 D2 78
Pine Hill, Austl A6 51
Pine Hill, Rockcastle, Ky., 600 C5 94
Pine Hill, Camden, N.J., 3,939 D3 106
Pine Hills, Orange, Fla., 7,000 *D5 86
Pinehurst, Dooly, Ga., 457 D3 87
Pinehurst, Shoshone, Idaho, 1,432 B2 89
Pinehurst, Middlesex, Mass., 1,991 C2 97
Pinehurst, Moore, N.C., 1,124 B4 109
Pinehurst, Orange, Tex., 1,703 *D6 118
Pinehurst, Snohomish, Wash., 3,000 B3 122
Pine Island, Goodhue, Minn., 1,308 F6 99
Pine Island, sound, Fla F4 86
Pine Knot, McCreary, Ky., 750 D5 94
Pine Lake, De Kalb, Ga., 738 A5 88
Pinelake, La Porte, Ind., 1,400 *A4 91
Pineland, Jasper, S.C., 82 F5 115
Pineland, Sabine, Tex., 1,236 D6 118
Pine Lawn, St. Louis, Mo., 5,943 C7 101
Pine Level, Montgomery, Ala., 150 C3 78
Pine Level, Johnston, N.C., 833 B5 109
Pinellas, co., Fla., 374,665 E4 86
Pinellas, pt., Fla F2 86
Pinellas Park, Pinellas, Fla., 10,848 E1, E4 86
Pine Lodge, Lincoln, N. Mex D4 107
Pine Meadow, Litchfield, Conn., 400 B5 84
Pine Mountain, Harris, Ga., 790 D2 87
Pineora, Effingham, Ga., 210 D5 87
Pine Orchard, New Haven, Conn., 350 D5 84
Pine Park, Grady, Ga., 82 F2 87
Pine Plains, Dutchess, N.Y., 665 D7 108
Pine Point, Cumberland, Maine, 800 E2, E5 96
Pine Prairie, Evangeline, La., 387 D3 95
Pine Ridge, Montgomery, Ark., 100 C2 81
Pine Ridge, Adams, Miss., 50 D2 100
Pine Ridge, Shannon, S. Dak., 1,256 D3 116
Pine Ridge, Indian res., S. Dak D3 116
Pine River, Man., Can., 429 D1 71
Pine River, Arenac, Mich D5 98
Pine River, Cass, Minn., 775 D4 99
Pinerolo, It., 24,500 B1 21
Piñeros, isl., P.R B8 65
Pines Lake, Passiac, N.J. (part of Wayne) B4 106
Pinetop, Navajo, Ariz., 400 C3 80
Pinetops, Edgecombe, N.C., 1,372 B6 109
Pinetown, Beaufort, N.C., 215 B7 109
Pinetta, Madison, Fla., 200 B3 86
Pine Valley, San Diego, Calif., 150 F5 82
Pine Valley, Hillsboro, N.H., 100 E3 105
Pine Valley, LeFlore, Okla C7 112
Pineview, Wilcox, Ga., 369 D3 87
Pine View, Harnett, N.C B4 109
Pine Village, Warren, Ind., 309 D3 91
Pineville, Bell, Ky., 3,181 D6 94
Pineville, Rapides, La., 8,636 C3 95
Pineville, Smith, Miss., 100 C4 100
Pineville, McDonald, Mo., 454 E3 101
Pineville, Mecklenburg, N.C., 1,514 B3 109
Pineville, Berkeley, S.C., 100 E7 115
Pineville, Wyoming, W. Va., 1,137 D3 123
Pinewald, Ocean, N.J., 400 D4 106
Pinewater, range, Nev G6 104
Pinewood, Beltrami, Minn., 75 C3 99
Pinewood, Sumter, S. C., 570 D7 115
Piney, Man., Can., 197 E4 71
Piney, buttes, Mont C10 102
Piney, creek, Mo D5 101
Piney, creek, W. Va D7 123
Piney, fork, W. Va A6 123
Piney, pt., Fla C3 86
Piney Flats, Sullivan, Tenn., 300 C11 117
Piney Fork, Jefferson, Ohio, 800 B7 111
Piney Point, Manatee, Fla., 30 F2 86
Piney Point, Harris, Tex., 1,790 *E5 118

Piney Swamp Knob, mtn., W. Va. B5 123
Piney View, Raleigh, W. Va., 800 D7 123
Piney Woods, Rankin, Miss. C3 100
Pingchiang, China J5 36
Pingchuan, China D8 36
Pinghsiang, China, 22,000 K5 36
Pingliang, China, 55,000 G2 36
Pinglo, China, 10,000 G7 38
Pinglo, China, 5,000 E2 36
Pingnan, China, 5,000 G7 34
Pingree, Bingham, Idaho, 100 F6 89
Pingree, Stutsman, N. Dak., 151 B7 110
Pingting, China F5 36
Pingtingshan, China, 50,000 H5 36
Pingtu, China, 10,000 F8 36
Pingtung, Taiwan, 81,020 (113,300▲) *G9 35
Pingwu, China, 3,000 E5 34
Pinhal, Braz., 14,260 C3, m8 56
Pinhal Novo, Port., 6,429 f10 20
Pinhead, buttes, Oreg C5 113
Pinheiro, Braz., 6,537 B1 57
Pinhel, Port., 3,312 B2 20
Pinhsien, China, 6,000 D3 37
Pini, isl., Indon E1 35
Pinios, riv., Grc C4 23
Pink, cliffs, Utah E3 119
Pinkham, Sask., Can., 91 F1 70
Pink Hill, Lenoir, N.C., 457 B6 109
Pinkney (South Gastonia), Gaston, N.C., 3,762 *B2 109
Pinkstaff, Lawrence, Ill., 300 E6 90
Pinnacle, Pulaski, Ark., 25 C3, D5 81
Pinnacle, Stokes, N.C., 400 A3 109
Pinnacle, butte, Wyo B3 125
Pinnacle, mtn., Mo B7 101
Pinnacle, peak, Wyo B2 125
Pinnacle, peaks, B.C., Can D1 69
Pinnacles, nat. mon., Calif D3 82
Pinnaroo, Austl., 621 G3 51
Pinneberg, Ger., 28,400 E3 24
Pinnebog, Huron, Mich., 110 E7 98
Pinó Hachado, pass, Arg B2 54
Pinola, Simpson, Miss., 116 D4 100
Pinole, Contra Costa, Calif., 6,064 B5 82
Pinon, Navajo, Ariz., 100 A5 80
Pinon, Otero, N. Mex., 40 E4 107
Pinopolis, Berkeley, S.C., 311 E7 115
Pinopolis, dam, S.C E8 115
Pinos, Cuba h12 64
Pinos, Mex., 3,327 C4 63
Pinos, isl., Cuba E2 64
Pinos, mtn., Calif E4 82
Pinos, pt., Calif D3 82
Pinos Altos, Grant, N. Mex., 150 E1 107
Pinoso, Sp., 5,114 C5 20
Pinos-Puente, Sp., 8,653 D4 20
Pinotpandian, Phil o13 35
Pinsk, Sov. Un., 39,000 E6 27
Pinsk, marshes, Sov. Un E6 27
Pinson, Jefferson, Ala., 1,121 E5 78
Pinson, Madison, Tenn., 240 B3 117
Pinsonfork, Pike, Ky., 800 *C7 94
Pinta, isl., Ec f5 58
Pintada, Guadalupe, N. Mex C4 107
Pintados, Chile D2 55
Pintendre, Que., Can., 159 C9 73
Pinto, Allegany, Md., 150 D2 85
Pinto, Sp., 5,360 p17 20
Pinto, butte, Sask., Can H2 70
Pinto, creek, Sask., Can H2 70
Pintura, Washington, Utah, 5 F2 119
Pinware, riv., Newf., Can C3 75
Pinyon, peak, Idaho E4 89
Pinzolo, It., 3,469 C6 18
Pinzón, isl., Ec g5 58
Pioche, Lincoln, Nev., 900 F7 104
Piombino, It., 36,102 C3 21
Pioneer, Humboldt, Iowa, 448 B3 92
Pioneer, West Carroll, La., 154 B4 95
Pioneer, Williams, Ohio, 855 A3 111
Pioneer, mts., Idaho F5 89
Pioneer, mts., Mont E3 102
Pioneer Mine, B.C., Can., 226 D6 68
Piopolis, Que., Can., 139 D7 73
Piotrkow [Trybunalski], Pol., 53,000 C5 26
Piove di Sacco, It., 5,349 D8 18
Pipe, Fond du Lac, Wis., 90 B5 124
Pipe, creek, Ind D6 91
Pipe, creek, Ohio B1 123
Piper, Bibb, Ala., 15 B2 78
Piper, Wyandotte, Kans., 240 B8 93
Piper, Fertus, Mont., 14 D7 102
Piper City, Ford, Ill., 807 C5 90
Piperi, isl., Grc C5 23
Pipers Gap, Carroll, Va., 25 E2 121
Piperville, Ont., Can A10 72
Pipe Spring, nat. mon., Ariz A3 80
Pipestem, Summers, W. Va., 25 D4 123
Pipe Stem, creek, N. Dak B6 110
Pipestone, Man., Can., 226 E1 71
Pipestone, Pipestone, Minn., 5,324 G2 99
Pipestone, co., Minn., 13,605 F2 99
Pipestone, creek, Man., Can E1 71
Pipestone, creek, Sask., Can G5 70
Pipestone, nat. mon., Minn G2 99
Pipestone, pass, Mont E4 102
Pipmuacan, lake, Que., Can B2 74
Piqua, Woodson, Kans., 258 E8 93
Piqua, Miami, Ohio, 19,219 B1 111
Piracanjuba, Braz., 3,869 E1 57
Piracicaba, Braz., 80,670 C3, m8 56
Piracicaba, riv., Braz m8 56
Piraçununga, Braz., 12,546 C3, k8 56
Piracuruca, Braz., 4,320 B2 57
Piraeus, see Piraievs, Grc.
Piraievs (Piraeus), Grc., 183,877 D4, h11 23
Pirajú, Braz., 10,658 C3 56
Pirajuí, Braz., 6,465 C3 56
Piramida, mtn., Sov. Un D12 28
Piran, Yugo., 5,464 B4 21
Piranê, Arg., 3,561 E4 55
Piranga, Braz., 2,169 F2 57
Piranhas, Braz., 1,021 C3 57
Pirapóra, Braz., 13,772 E2 57
Pireway, Columbus, N.C., 100 C5 109
Pirgos, Grc., 20,558 D3 23
Piriápolis, Ur., 4,400 E1 56
Pirin, mts., Bul E6 22
Piripiri, Braz., 9,635 B2 57
Pirmasens, Ger., 53,200 D2 17
Pirna, Ger., 41,100 C8 17
Pirot, Yugo., 18,585 D6 22
Pirovskoye, Sov. Un D27 9
Pirtleville, Cochise, Ariz., 850 F6 80
Piru, Indon F7 35
Piryatin, Sov. Un., 10,000 F9 27
Pis, isl., Truk 52
Pisa, It., 90,928 C3 21
Pisagua, Chile, 419 C1 55
Piscataqua, riv., N.H., Maine D5 105
Piscataquis, co., Maine, 17,379 C3 96
Piscataquis, riv., Maine C4 96
Piscataquog, riv., N.H D5 105
Piscataway, creek, Md C4 85
Pisciotta, It., 4,025 D5 21
Pisco, Peru, 17,222 D2 58
Piseco, lake, N.Y B6 108
Pisek, Czech., 19,500 D3 24
Pisek, Walsh, N. Dak., 176 A8 110
Pisgah, Jackson, Ala., 214 A4 78
Pisgah, Harrison, Iowa, 343 C2 92
Pisgah, Charles, Md., 500 C3 85
Pisgah, mtn., Wyo B8 125
Pisgah Forest, Transylvania, N.C., 700 D3 109
Pishin, Pak., 3,106 B1 40
Pishin Lora, riv., Pak C3 39
Pishukan, cape, Pak I11 41
Pisia, riv., Pol m12 26
Pisinimo, Pima, Ariz., 200 E3 80
Pismo Beach, San Luis Obispo, Calif., 1,762 E3 82
Pisogne, It., 6,848 D6 18
Pissis, mtn., Arg E2 55
Pistal River, Curry, Oreg., 50 E2 113
Pisticci, It., 13,300 D6 21
Pistoia, It., 58,500 (84,561▲) C3 21
Pistolet, bay, Newf., Can C4 75
Pisuerga, riv., Sp A3 20
Pisz, Pol., 1,028 B6 26
Pita, Guinea, 6,800 D2 45
Pitalito, Col., 3,616 C2 60
Pitangui, Braz., 7,421 E2 57
Pitcairn, Allegheny, Pa., 5,383 F2 114
Pitcairn, Br. dep., Oceania, 125 H14 7
Pitchfork, Park, Wyo A3 125
Piteå, Swe., 7,400 E9 25
Piteälven, riv., Swe E9 25
Pitești, Rom., 38,333 C7 22
Pithiviers, Fr., 7,318 C5 14
Piti, Guam, 777 52
Pitkas Point, Alsk., 28 C7 79
Pitkin, Gunnison, Colo., 94 C4 83
Pitkin, Vernon, La., 400 D3 95
Pitkin, co., Colo., 2,381 B4 83
Pitlochry, Scot., 2,501 D5 13
Pitman, Gloucester, N.J., 8,644 D2 106
Pito, Pan F9 62
Pitreville, St. Landry, La., 30 D3 95
Pitrufquén, Chile, 5,193 B2 54
Pitsburg, Darke, Ohio, 394 C3 111
Pitt, Lake of the Woods, Minn., 25 B4 99
Pitt, co., N.C., 69,942 B6 109
Pitt, isl., B.C., Can C3 68
Pittman Center, Sevier, Tenn., 45 D10 117
Pitts, Wilcox, Ga., 388 E3 87
Pittsboro, Hendricks, Ind., 826 E5 91
Pittsboro, Calhoun, Miss., 205 B4 100
Pittsboro, Chatham, N.C., 1,215 B4 109
Pittsburg, Contra Costa, Calif., 19,062 B5 82
Pittsburg, Williamson, Ill., 485 F5 90
Pittsburg, Carroll, Ind., 250 C4 91
Pittsburg, Crawford, Kans., 18,678 E9 93
Pittsburg, Laurel, Ky., 810 C5 94
Pittsburg, Coos, N.H., 200 (639▲) A1 105
Pittsburg, Pittsburg, Okla., 195 C6 112
Pittsburg, Columbia, Oreg., 50 B3 113
Pittsburg, Camp, Tex., 3,769 C5 118
Pittsburg, co., Okla., 34,360 C6 112
Pittsburgh, Allegheny, Pa., 604,332 (*1,957,700) B6, F1 114
Pittsburg Landing, Hardin, Tenn., 100 B3 117
Pittsfield, Pike, Ill., 4,089 D3 90
Pittsfield, Somerset, Maine, 3,232 D3 96
Pittsfield, Berkshire, Mass., 57,879 B1 97
Pittsfield, Washtenaw, Mich., 1,500 *F7 98
Pittsfield, Merrimack, N.H., 1,407 D4 105
Pittsfield, Warren, Pa., 500 C3 114
Pittsfield, Rutland, Vt., 100 (254▲) D3 120
Pittsford, Hillsdale, Mich., 450 G6 98
Pittsford, Monroe, N.Y., 1,749 B3 108
Pittsford, Rutland, Vt., 671 (2,225▲) D2 120
Pittsford Mills, Rutland, Vt., 200 D2 120
Pittston, Luzerne, Pa., 12,407 B8, D10 114
Pittstown, Hunterdon, N.J., 230 B3 106
Pittsview, Russell, Ala., 300 C4 78
Pittsville, Wicomico, Md., 488 D7 85
Pittsville, Johnson, Mo., 85 C4 101
Pittsville, Wood, Wis., 661 D3 124
Pittsylvania, co., Va., 58,296 E3 121
Pitzuwo, China E10 36
Piuka, Jap., 5,200 D11 37
Piura, Peru, 23,678 C1 58
Piura, dept., Peru, 431,487 C1 58
Piute, co., Utah, 1,436 E3 119
Piuthan, Nep., 1,350 C9 40
Piva, riv., Yugo D4 22
Pixley, Tulare, Calif., 1,327 E4 82
Piyang, China H5 36
Pizzo, It., 9,560 E6 21
Placentia, Orange, Calif., 5,861 *F3 82
Placentia, Newf., Can., 1,610 E5 75
Placentia, bay, Newf., Can E4 75
Placentia-St. Marys, dist., Newf., Can E5 75
Placentia West, dist., Newf., Can E4 75
Placer, Josephine, Oreg., 125 E3 113
Placer, co., Calif., 56,998 C3 82
Placerville, El Dorado, Calif., 4,439 C3 82
Placerville, San Miguel, Colo., 50 C2 83
Placerville, Boise, Idaho, 12 F3 89
Placetas, Cuba, 25,226 D4 64
Placid, lake, N.Y B3 108
Placida, Charlotte, Fla., 150 F4 86
Placita, Sierra, N. Mex., 55 D2 107
Placitas, Sandoval, N. Mex., 60 B3, G5 107
Plaffeien, Switz., 1,352 C3 19
Plage Laval, Que., Can., 3,818 *D8 73
Plain, Sauk, Wis., 677 E3 124
Plain City, Union and Madison, Ohio, 2,146 B4 111
Plain City, Weber, Utah, 1,152 B3 119
Plain Dealing, Bossier, La., 1,357 B2 95
Plainfield, Windham, Conn., 2,044 C9 84
Plainfield, Dodge, Ga., 84 D3 87
Plainfield, Will, Ill., 2,183 B5, F2 90
Plainfield, Hendricks, Ind., 5,460 E5 91
Plainfield, Bremer, Iowa, 445 B5 92
Plainfield, Hampshire, Mass., 100 (375▲) A2 97
Plainfield, Sullivan, N.H., 125 (1,071▲) C2 105
Plainfield, Union, N.J., 45,330 B4 106
Plainfield, Blount, Tenn., 2,127 *D10 117
Plainfield, Washington, Vt., 507 (966▲) C4 120
Plainfield, Waushara, Wis., 660 D4 124
Plainfield Heights, Kent, Mich., 1,000 *F5 98
Plains, Sumter, Ga., 572 D2 87
Plains (West Plains), Meade, Kans., 780 E3 93
Plains, Sanders, Mont., 769 C2 102
Plains, Luzerne, Pa., 8,500 B8 114
Plains, Yoakum, Tex., 1,195 C1 118
Plainsboro, Middlesex, N.J., 600 C3 106
Plainsville, Luzerne, Pa., 500 *B8 114
Plainview, Yell, Ark., 548 C2 81
Plainview, Wabasha, Minn., 1,833 F6 99
Plainview, Pierce, Nebr., 1,467 B8 103
Plainview, Nassau, N.Y., 27,710 *G3 84
Plainview, Hale, Tex., 18,735 B2 118
Plainville, Hartford, Conn., 13,149 C5 84
Plainville, Adams, Ill., 227 D2 90
Plainville, Daviess, Ind., 545 G3 91
Plainville, Rooks, Kans., 3,104 C4 93
Plainville, Norfolk, Mass., 3,810 B5 97
Plainwell, Allegan, Mich., 3,125 F5 98
Plaisance, Que., Can., 451 D2 73
Plaisance, Hai, 2,800 F7 64
Plaisance-du-Gers, Fr., 1,490 F4 14
Plaisted, Aroostook, Maine, 200 A4 96
Plaistow, Rockingham, N.H., 1,500 (2,915▲) E4 105
Plamondon, Alta., Can., 133 B4 69
Plana, Czech., 2,485 D7 17
Planada, Merced, Calif., 1,704 *D3 82
Planaltina, Braz., 1,385 E1 57
Plandome, Nassau, N.Y., 1,379 *B5 106
Plandome Heights, Nassau, N.Y., 1,025 *B5 106
Plankinton, Aurora, S. Dak., 644 D7 116
Plano (Doyle Colony), Tulare, Calif., 1,500 *D4 82
Plano, Kendall, Ill., 3,343 B5 90
Plano, Appanoose, Iowa, 87 D4 92
Plano, Collin, Tex., 3,695 A6, C4 118
Plantagenet, Ont., Can., 854 B10 72
Plantation, Broward, Fla., 4,772 *F6 86
Plant City, Hillsborough, Fla., 15,711 D4 86
Plantersville, Dallas, Ala., 650 C3 78
Plantersville, Lee, Miss., 572 A5 100
Plantersville, Gerogetown, S.C., 150 D9 115
Plantsite, Greenlee, Ariz., 1,552 D6 80
Plantsville, Hartford, Conn., 2,793 C5 84
Plaquemine, Iberville, La., 7,689 B5, D4 95
Plaquemines, par., La., 22,545 E6 95
Plaquemine Southwest, Iberville, La., 1,272 *D4 95
Plasencia, Sp., 21,297 B2 20
Plast, Sov. Un., 28,100 E21 9
Plaster Rock, N.B., Can., 1,267 C2 74
Plastun, Sov. Un D8 37
Plasy, Czech., 1,472 D8 17
Plata, mtn., Switz D8 19
Plata, riv., Arg B5 54
Plata Central, P.R B2 65
Platea, Erie, Pa., 357 C1 114
Plateau City, Mesa, Colo., 50 B3 83
Plate Cove, Newf., Can., 442 D5 75
Platina, Shasta, Calif., 40 B2 82
Platinum, Alsk., 43 D7 79
Platner, Washington, Colo., 40 A7 83
Plato, Sask., Can., 178 F1 70
Plato, Col., 8,039 B3 60
Plato, Texas, Mo., 140 D5 101
Platt, nat. park, Okla C5 112
Platte, Charles Mix, S. Dak., 1,167 D7 116
Platte, co., Mo., 23,350 B3 101
Platte, co., Nebr., 23,992 C8 103
Platte, co., Wyo., 7,195 C7 125
Platte, riv., Iowa, Mo D3 92, B3 101
Platte, riv., Minn E4 99
Platte, riv., Nebr D6 103
Platte Arkansas, divide, Colo B7 83
Platte Center, Platte, Nebr., 402 C8 103
Platte City, Platte, Mo., 1,188 B3, D1 101
Plattenville, Assumption, La., 300 B5 95
Platter, Bryan, Okla., 200 D5 112
Platteville, Weld, Colo., 582 A6 83
Platteville, Grant, Wis., 6,957 F3 124
Plattling, Ger., 8,400 E7 17
Plattsburg, Clinton, Mo., 1,663 B3 101
Plattsburgh, Clinton, N.Y., 20,172 A3 108
Plattsmouth, Cass, Nebr., 6,244 C10, E3 103
Plattsville, Ont., Can., 456 D4 72
Plattville, Kendall, Ill., 125 B5 90
Plaucheville, Avoyelles, La., 288 D3 95
Plauen, Ger., 78,900 C7 17
Plauer See, lake, Ger E6 24
Plave, riv., Ger A7 65
Playa de Fajardo (Puerto Real), P.R., (part of Fajardo), 2,143 B8, f12 17
Playa de Guayanés, P.R C7 65
Playa de Guayanilla, P.R., 1,635 C3 65
Playa de Humacao (Punta Santiago), P.R C8, g12 65
Playa de Marianao, Cuba h11 64
Playa de Naguabo, P.R. C8, g12 65
Playa de Ponce, P.R., (part of Ponce), 15,040 D4 65
Playa Grande, P.R B2 65
Playas, Hidalgo, N. Mex., 25 F1 107
Playa Salinas, P.R D5 65
Playa Sardinera, P.R B8, f12 65
Playgreen, lake, Man., Can B2 71
Plaza, Mountrail, N. Dak., 385 A4 110
Plazuela Central, P.R B4 65
Pleasant, bay, Que., Can B8 74
Pleasant, bay, Maine D5 96
Pleasant, lake, Ariz D3 80
Pleasant, lake, Maine C5 96
Pleasant, mtn., N.B., Can D3 74
Pleasant, mtn., Maine D4 96
Pleasant, pond, Maine B4 96
Pleasant, pond, Maine C3 96
Pleasant, pond, N.H D4 105
Pleasant, riv., Maine C3 96
Pleasant Bay, N.S., Can., 109 C9 74
Pleasant Beach, Kitsap, Wash. (part of Port Blakely) D1 122
Pleasant City, Guernsey, Ohio, 491 C6 111
Pleasantdale, Sask., Can., 151 E3 70
Pleasant Dale, Seward, Nebr., 190 D8 103
Pleasant Gap, Centre, Pa., 1,389 E6 114
Pleasant Garden, Guilford, N.C., 1,000 B4 108
Pleasant Grove, Jefferson, Ala., 3,097 E4 78
Pleasant Grove, Panola, Miss., 60 A3 100
Pleasant Grove, Utah, Utah, 4,772 C4 119
Pleasant Hill, Dallas, Ala., 150 C3 78
Pleasant Hill, Contra Costa, Calif., 23,844 *B5 82
Pleasant Hill, Pike, Ill., 950 D3 90
Pleasant Hill, Sabine, La., 907 C2 95
Pleasant Hill, De Soto, Miss., 150 A4 100
Pleasant Hill, Cass, Mo., 2,689 C3 101
Pleasant Hill, Curry, N. Mex C6 107
Pleasant Hill, Miami, Ohio, 1,060 B1 111
Pleasant Hill, Lebanon, Pa., 1,200 *F9 114
Pleasant Hill, Lancaster, S.C., 150 B6 115
Pleasant Hill, Nansemond, Va., 2,636 *E6 121
Pleasant Hills, Allegheny, Pa., 8,573 *B5, F1 114
Pleasant Hope, Polk, Mo., 216 D4 101
Pleasant Lake, Steuben, Ind., 600 A7 91
Pleasant Lake, Barnstable, Mass., 200 C7 97
Pleasant Lake, Benson, N. Dak., 15 A6 110
Pleasant Mills, Adams, Ind., 160 C8 91
Pleasant Mount, Wayne, Pa., 160 C11 114
Pleasanton, Alameda, Calif., 4,203 B6 82
Pleasanton, Decatur, Iowa, 103 D4 92
Pleasanton, Linn, Kans., 1,098 D9 93
Pleasanton, Buffalo, Nebr., 199 D6 103
Pleasanton, Catron, N. Mex D1 107
Pleasanton, Atascosa, Tex., 3,467 E3 118
Pleasant Plain, Jefferson, Iowa, 147 C6 92
Pleasant Plains, Independence, Ark., 112 B4 81
Pleasant Plains, Sangamon, Ill., 518 D4 90
Pleasant Prairie, Spokane, Wash D7 122
Pleasant Prairie, Kenosha, Wis., 400 F2 124
Pleasant Ridge, Oakland, Mich., 3,807 A7 98
Pleasant Ridge, Princess Anne, Va., 200 E6 121
Pleasants, co., W. Va., 7,124 B3 123
Pleasant Shade, Smith, Tenn., 125 C8 116
Pleasant Valley, Litchfield, Conn., 225 B5 84
Pleasant Valley, Scott, Iowa, 500 D7 92
Pleasant Valley, Carroll, Md., 150 A3 85
Pleasant Valley, Clay, Mo., 1,109 *B3 101
Pleasant Valley, Jasper, Mo., 25 E2 101
Pleasant Valley, Dutchess, N.Y., 700 D7 108
Pleasant Valley, Logan, Okla B4 112
Pleasant Valley, Baker, Oreg., 25 C9 113
Pleasant Valley, Fairfax, Va B4 121
Pleasant View, Jefferson, Colo., 1,500 *B5 83
Pleasant View, Montezuma, Colo., 98 D2 83
Pleasant View, Schuyler, Ill., 125 C3 90
Pleasant View, Shelby, Ind., 100 I8 91
Pleasant View, Whitley, Ky., 500 D5 94
Pleasant View, Cheatham, Tenn., 150 A4 117
Pleasant View, Weber, Utah, 927 B3 119
Pleasantville, Sullivan, Ind., 175 G3 91
Pleasantville, Marion, Iowa, 1,025 C4 92
Pleasantville, Atlantic, N.J., 15,172 E3 106
Pleasantville, Westchester, N.Y., 5,877 A5 106, D3 108
Pleasantville, Fairfield, Ohio, 741 C5 111
Pleasantville, Venango, Pa., 940 C2 114
Pleasure Beach, New London, Conn., 1,264 *E4 84
Pleasure Ridge Park, Jefferson, Ky., 10,612 A4 94
Pleasureville, Henry and Shelby, Ky., 466 B4 94
Pleasureville, York, Pa., 1,000 *G8 114
Plebo, Lib F3 45
Pledger, Matagorda, Tex., 300 G4 118
Pleiku, Viet., 10,000 F8 38
Plenita, Rom., 6,735 C6 22
Plenty, Sask., Can., 245 F1 70
Plentywood, Sheridan, Mont., 2,121 B12 102
Plesetsk, Sov. Un., 18,000 C17 9
Plessis, Jefferson, N.Y., 120 A5, B1 108
Plessisville, Que., Can., 6,570 C6 73
Pleszew, Pol., 8,760 C4 26
Pletcher, Chilton, Ala., 60 C3 78
Pletipi, lake, Que., Can B2 73
Plettenberg, Ger., 28,400 B2 17
Plettenberg, West Feliciana, La., 35 D4 95
Pleven, Bul., 38,997 D7 22
Pleven, co., Bul., 1,056,436 *D7 22
Plevna, Reno, Kans., 117 E5 93
Plevna, Knox, Mo., 110 B5 101
Plevna, Fallon, Mont., 263 D12 102
Pliny, range, N.H B4 105
Pljevlja, Yugo., 10,240 D4 22
Plochingen, Ger., 11,400 E4 17
Plock, Pol., 42,000 B5 26
Ploërmel, Fr., 3,840 D2 14
Ploești, Rom., 114,560 C8 22
Plomarion, Grc., 5,889 C6 23
Plomb du Cantal, mtn., Fr E5 14
Plombières-les-Bains, Fr., 1,297 B2 18
Plomosa, mts., Ariz D1 80
Plön, Ger., 10,900 D4 24
Plonsk, Pol., 7,578 B6 26
Plougastel [-Daoulas], Fr., 2,351 C1 14
Plovdiv, Bul., 125,440 D7 22
Plovdiv, co., Bul., 876,993 *D7 22
Plover, Pocahontas, Iowa, 182 B3 92
Plover, Portage, Wis., 400 D4 124
Plover, riv., Wis D4 124
Pluck, Ire C4 11
Plum, Bourbon, Ky., 50 B5 94
Plum, Allegheny, Pa., 10,241 *F2 114
Plum, bayou, Ark D6 81
Plum, creek, Nebr B5 103
Plum, creek, Nebr D5 103
Plum, isl., Mass A6 97
Plum, isl., N.Y E8 84
Pluma, Lawrence, S. Dak C2 116
Plumas, Man., Can., 344 D2 71
Plumas, co., Calif., 11,620 B3 82
Plum Bayou, Jefferson, Ark., 100 C3 81
Plum Branch, McCormick, S.C., 139 D3 115
Plumbridge, N. Ire C4 11
Plum City, Pierce, Wis., 384 D1 124
Plum Coulee, Man., Can., 510 E3 71
Plumerville, Conway, Ark., 586 B3 81
Plum Gut, strait, N.Y E8 84
Plummer, Benewah, Idaho, 344 B2 89
Plummer, Red Lake, Minn., 283 C2 99
Plumpoint, Calvert, Md., 550 C4 85
Plumsteadville, Bucks, Pa., 250 F11 114
Plumtree, S. Rh., 1,200 B4 49
Plumwood, Madison, Ohio, 700 B4 111
Plunkett, Sask., Can., 136 F3 70
Plunkettville, McCurtain, Okla., 50 C7 112
Plush, Lake, Oreg., 40 E7 113
Plymouth, Litchfield, Conn., 950 (8,981▲) C4 84
Plymouth, Eng., 204,279 (*250,000) D3 12
Plymouth, Orange, Fla., 250 D5 86
Plymouth, Hancock, Ill., 781 C3 90
Plymouth, Marshall, Ind., 7,558 B5 91
Plymouth, Cerro Gordo, Iowa, 422 A4 92
Plymouth, Penobscot, Maine, 170 (949▲) D3 96
Plymouth, Plymouth, Mass., 11,000 (14,445▲) C6 97
Plymouth, Wayne, Mich., 8,766 A7, F7 98
Plymouth, Hennepin, Minn., 9,576 *F5 99
Plymouth, Jefferson, Nebr., 372 D9 103
Plymouth, Grafton, N.H., 2,244 C3 105
Plymouth, Washington, N.C., 4,666 B7 109
Plymouth, Huron and Richland, Ohio, 1,822 B5 111
Plymouth, Luzerne, Pa., 10,401 B8, D10 114
Plymouth, Box Elder, Utah, 231 B3 119
Plymouth, Montserrat, 1,924 n15 64
Plymouth, Putnam, W. Va., 100 C3 123
Plymouth, Sheboygan, Wis., 5,128 B6, E6 124
Plymouth, co., Iowa, 23,906 B1 92
Plymouth, co., Mass., 248,449 C6 97
Plymouth, bay, Mass C6 97
Plymouth, rock, Mass C6 97
Plymouth Meeting, Montgomery, Pa., 4,000 *A11 114
Plymouth Union, Windsor, Vt., 75 D3 120
Plymouth Valley, Montgomery, Pa., 1,700 *F11 114
Plympton, N.S., Can., 291 E4 74
Plympton, Plymouth, Mass., 300 (821▲) C6 97
Plymptonville, Clearfield, Pa., 1,220 *D5 114
Plzen, Czech., 137,800 D2 26
Po, China H6 36
Pô, Upper Volta D4 45
Po, riv., It D7 18
Po, riv., Va C5 121
Poages Mill, Roanoke, Va D2 121
Poai, China G5 36
Poamoho, riv., Haw f9 88
Pobé, Dah., 3,900 E5 45
Pobedino, Sov. Un., 10,000 B11 37
Poblado Cerrillos, P.R C4 65
Poblado Cerro Gordo, P.R B5 65
Poblado Mediana Alita, P.R B7 65
Poblado Paso Seco, P.R C5 65
Poblado Sabalos, P.R C2 65
Poca, Putnam, W. Va., 607 C3 123
Pocahontas, Randolph, Ark., 3,665 A5 81
Pocahontas, Bond, Ill., 718 E4 90
Pocahontas, Pocahontas, Iowa, 2,011 B3 92
Pocahontas, Hardeman, Tenn., 50 B3 117
Pocahontas, Tazewell, Va., 1,313 B3 121
Pocahontas, co., Iowa, 14,234 B3 92
Pocahontas co., W. Va., 10,136 C4 123
Pocasset, Barnstable, Mass., 900 C6 97
Pocasset, Grady, Okla., 180 B4 112
Pocatalico, Kanawha, W. Va., 400 C3 123
Pocatalico, riv., W. Va C3 123
Pocatello, Bannock and Power, Idaho, 27,140 G6 89
Pochep, Sov. Un., 10,000 E9 27
Pochutla, Mex., 3,084 D5 63
Pocitos, Arg D2 55
Pocola, Le Flore, Okla B7 112
Pocomoke, riv., Md D7 85
Pocomoke, sound, Md E6 85
Pocomoke City, Worcester, Md., 3,329 D6 85
Poconé, Braz., 4,702 B1 56
Pocono, lake, Pa A1 106
Pocono, mts., Pa A2 106
Pocono Pines, Monroe, Pa., 400 D11 114
Poços de Caldas, Braz., 32,291 C3, k8 56
Pocotopaug, lake, Conn C7 84
Podborany, Czech., 2,853 C8 17
Podgorica, see Titograd, Yugo.
Podgornoye, Sov. Un D25 9
Podhorany, Czech., 149 n17 26
Podkamennaya Tunguska, Sov. Un C27 9
Podkamennaya Tunguska, riv., Sov. Un C12 28
Podmokly, Czech (part of Decin) C3 26
Podolsk, Sov. Un., 139,000 B1 29
Podor, Sen., 3,400 C2 45
Podporozhye, Sov. Un., 10,000 G16 25
Poe, Greenville, S.C., 1,000 *B3 115
Poel, isl., Ger E5 24
Pofadder, S. Afr., 2,030 C2 49
Poge, cape, Mass D7 97
Poggibonsi, It., 7,236 C3 21
Pöggstall, Aus., 921 D7 16
Pogradec, Alb., 5,643 B3 23
Pogranichnoye, Sov. Un A11 37
Pohang, Kor., 47,500 (59,600▲) H4 37
Pohopoco, creek, Pa B1 106
Poindexter, Louisa, Va C4 121
Poinsett, co., Ark., 30,834 B5 81
Poinsett, lake, Fla D6 86
Poinsett, lake, S. Dak C8 116
Poinsetta, Hamlin, S. Dak., 50 C8 116
Point, lake, N.W. Ter., Can C10 66
Point Arena, Mendocino, Calif., 596 C2 82
Point Baker, Alsk., 20 m23 79
Point Cedar, Hot Spring, Ark., 150 C2, D5 81
Point Clear, Baldwin, Ala., 600 E2 78
Point Comfort, Calhoun Tex., 1,453 *E4 118
Pointe a Gatineau, Que., Can., 8,854 *A9 72
Pointe a la Hache, Plaquemines, La., 600 E6 95
Pointe-à-Pitre, Guad., 26,160 n16 64
Pointe-au-Baril-Station, Ont., Can., 315 B4 72
Pointe au Pic, Que., Can., 1,333 B7 73
Pointe aux Trembles, Que., Can., 21,926 D9 73
Pointe Bleue, Que., Can., 290 A5 73
Pointe Claire, Que., Can., 22,709 D4, D8 73
Pointe Coupee, par., La., 21,841 D4 95
Pointe du Bois, Man., Can., 284 D4 71

Pointe du Chene, N.B., Can., 534 C5 74
Pointe du Lac, Que., Can., 934 C5 73
Point Edward, Ont., Can., 2,744 E2 72
Pointe Noire, Con. B., 31,199 F2 46
Pointe Verte, N.B., Can., 297 B4 74
Point Hope, Alsk., 324 B6 79
Point Judith, Washington, R.I. D11 84
Point Lay, Alsk., 20 B7 79
Point Leamington, Newf., Can., 901 D4 75
Point Lookout, St. Marys, Md., 102 D5 85
Point Lookout, Taney, Mo., 200 E4 101
Point Marion, Fayette, Pa., 1,853 F2 114
Point of Rocks, Frederick, Md., 326 B2 85
Point of Rocks, Sweetwater, Wyo., 55 D4 125
Point Pelee, nat. park, Ont., Can F2 72
Point Pleasant, Tensas, La. B4 95
Point Pleasant, Ocean, N.J., 10,182 C4 106
Point Pleasant, Bucks, Pa., 400 F11 114
Point Pleasant, Bland, Va. D1 121
Point Pleasant, Mason, W. Va., 5,785 C2 123
Point Pleasant Beach, Ocean, N.J., 3,873 C4 106
Point Reyes Station, Marin, Calif., 600 B5 82
Point Roberts, Whatcom, Wash., 180 A2 122
Point Sapin, N.B., Can., 638 C5 74
Point Washington, Walton, Fla., 90 G3 86
Poirino, It., 5,713 E3 18
Poison, creek, Wyo B5 125
Poissy, Fr., 28,499 C4 14
Poitiers, Fr., 62,178 D4 14
Poitou, former prov., Fr., 1,037,000 D3 14
Poix, Fr., 1,278 E9 12
Poix,Terron, Fr., 721 E4 15
Pojo, Bol., 1,047 C3 55
Pojoaque, Santa Fe, N. Mex., 250 B3, F6 107
Pok, Ponape 52
Pokaran, India, 5,284 D3 40
Pokataroo, Austl., 52 D7 51
Pokegama, lake, Minn C5 99
Pokegamma, lake, Wis C2 124
Pokhara, Nep., 5,000 C9 40
Poko, Con. L. A4 48
Pokotu, China B9 34
Pokrov, Sov. Un., 8,000 n19 27
Pokrovskoye, Sov. Un., 10,000 H12 27
Pola, Phil., 1,538 *C6 35
Polacca, Navajo, Ariz., 300 B5 80
Polān, Iran I10 41
Poland, Herkimer, N.Y., 564 B5 108
Poland, Mahoning, Ohio, 2,766 *A7 111
Poland, country, Eur., 29,731,200 E12 8, C5 26
Poland Spring, Androscoggin, Maine, 100 D2 96
Polar, Kent, Tex C2 118
Polar, Langlade, Wis., 110 C5 124
Polar, caves, N.H C3 105
Polaris, Beaverhead, Mont., 20 E3 102
Polavaram, India I8 40
Polbain, Scot B3 13
Polch, Ger., 3,500 C2 17
Polcura, Chile B2 54
Pole, mtn., Wyo D7 125
Polebridge, Flathead, Mont., 5 B2 102
Polecat, creek, Okla B5 112
Polesella, It., 4,871 E7 18
Polessk, Sov. Un A6 26
Polevskoy, Sov. Un., 43,400 C21 9
Polgar, Hung., 7,697 B5 22
Poli, Cam D2 46
Poli, China, 35,000 B11 34
Policastro, gulf, It E5 21
Police, Pol., 2,000 E8 24
Poligny, Fr., 3,869 D6 14
Polikhnitos, Grc., 6,071 C6 23
Polillo, Phil., 2,400 o13 35
Polillo, isl., Phil o13 35
Polillo, is., Phil C6, o14 35
Polillo, strait, Phil o13 35
Polis, Cyp., 1,727 E9 31
Polish, mtn., Md D3 85
Poliyiros, Grc., 3,382 B4 23
Polk, Polk, Nebr., 433 C8 103
Polk, Ashland, Ohio, 358 B5 111
Polk, Venango, Pa., 3,574 D2 114
Polk, Obion, Tenn., 300 A2 117
Polk, co., Ark., 11,981 C1 81
Polk, co., Fla., 195,139 E5 86
Polk, co., Ga., 28,015 C1 87
Polk, co., Iowa, 266,315 C4 92
Polk, co., Minn., 36,182 C2 99
Polk, co., Mo., 13,753 D4 101
Polk, co., Nebr., 7,210 C8 103
Polk, co., N.C., 11,395 D4 109
Polk, co., Oreg., 26,523 C3 113
Polk, co., Tenn., 12,160 D9 117
Polk, co., Tex., 13,861 D5 118
Polk, co., Wis., 24,968 C1 124
Polk City, Polk, Fla., 203 D5 86
Polk City, Polk, Iowa, 567 A7, C4 92
Polkton, Anson, N.C., 530 B3 109
Polkville, Smith, Miss., 50 C4 100
Pollard, Escambia, Ala., 210 D2 78
Pollard, Clay, Ark., 170 A5 81
Pollensa, Sp., 8,984 C7 20
Polley, Taylor, Wis C3 124
Pollock, Idaho, Idaho, 40 D2 89
Pollock, Grant, La., 366 C3 95
Pollock, Sullivan, Mo., 202 A4 101
Pollock, Cambell, S. Dak., 417 B5 116
Pollockville, Alta., Can., 42 D5 69
Pollocksville, Jones, N.C., 416 C6 109
Polo, Ogle, Ill., 2,551 B4 90
Polo, Caldwell, Mo., 469 B3 101
Polochic, riv., Guat C3 62
Pologi, Sov. Un., 10,000 H11 27
Polonnoye, Sov. Un., 10,000 F6 27
Polotsk, Sov. Un., 30,000 D7 27
Polson, Lake, Mont., 2,314 C2 102
Poltava, Sov. Un., 150,000 G10 27
Poltimore, Que., Can., 156 D2 73
Polunochnoye, Sov. Un A6 29
Polvadera, Socorro, N. Mex., 300 C3 107
Polwarth, Sask., Can D2 70
Polyanovgrad, Bul., 10,225 D8 22
Polyarnyy, Sov. Un., 30,000 C15 25
Pomabamba, Peru, 2,346 C2 58
Pomarão, Port D2 20
Pomaria, Newberry, S.C., 230 C4 115
Pomba, riv., Braz g6 56
Pombal, Braz., 8,120 C3 57
Pombal, Port., 2,541 C1 20
Pombetsu, Jap., 6,600 E11 37
Pomerania, reg., Pol B4 26
Pomeranian, bay, Pol D8 24
Pomerene, Cochise, Ariz., 150 E5 80
Pomeroy, Calhoun, Iowa, 816 B3 92
Pomeroy, N. Ire., 341 C5 12
Pomeroy, Meigs, Ohio, 3,345 C5 111
Pomeroy, Chester, Pa., 1,085 *G10 114
Pomeroy, Garfield, Wash., 1,677 C8 122
Pomezia, It., 4,488 h9 21
Pomfret, Windham, Conn., 600 (2,136▲) B9 84
Pomfret, Charles, Md., 300 C3 85
Pomfret, Windsor, Vt., 20 (600▲) D3 120
Pomme de Terre, flood control res., Mo D4 101
Pomme de Terre, riv., Minn E3 99
Pomme de Terre, riv., Mo D4 101
Pomoho Camp, Honolulu, Haw., 350 f9 88
Pomona, Los Angeles, Calif., 67,157 (★212,000) E5, F5 82
Pomona, Franklin, Kans., 489 D8 93
Pomona, Howell, Mo., 300 E6 101
Pomona, Atlantic, N.J., 400 E3 106
Pomona, Guilford, N.C. (part of Greensboro) A4 109
Pomona (Mainland), isl., Scot A5 10
Pomona Park, Putnam, Fla., 516 C5 86
Pomonkey, Charles, Md., 400 C3 85
Pomorie, Bul., 4,721 D8 22
Pompano Beach, Broward, Fla., 15,992 E3, F6 86
Pompeii, ruins, It D5 21
Pompeyevka, Sov. Un B5 37
Pompeys Pillar, Yellowstone, Mont., 100 E9 102
Pompton Lakes, Passaic, N.J., 9,445 A4 106
Pompton Plains, Morris, N.J., 5,800 B4 106
Pomquet, N.S., Can., 347 D8 74
Ponaganset, lake, R.I B10 84
Ponape, Ponape 52
Ponape, isl., Caroline Is 52
Ponask, lake, Ont., Can C5 71
Ponask, riv., Ont., Can B5 71
Ponca, Dixon, Nebr., 924 B9 103
Ponca, creek, Nebr B7 99
Ponca, creek, S. Dak D6 116
Ponca, Indian res., Nebr B7 103
Ponca City, Kay, Okla., 24,411 A4 112
Ponce, P.R., 114,286 C4 65
Ponce, mun., P.R., 145,586 C4 65
Ponce de Leon, Holmes, Fla., 400 G3 86
Ponce de Leon, bay, Fla G5 86
Ponce de Leon, inlet, Fla C6 86
Poncha, pass, Colo C4 83
Poncha Springs, Chaffee, Colo., 201 C4 83
Ponchatoula, Tangipahoa, La., 4,727 B6, D5 95
Pond, creek, Colo C7 83
Pond, creek, Ky A4 94
Pond, creek, Okla B3 112
Pond, fork, W. Va D3 123
Pond, lake, Newf., Can C3 75
Pond, riv., Ky C2 94
Pondcreek, Grant, Okla., 935 A4 112
Pondera, co., Mont., 7,653 B4 102
Ponderay, Bonner, Idaho, 231 A2 89
Ponderosa, Sandoval, N. Mex., 100 B3, F5 107
Pondfork, Ozark, Mo E5 101
Pondicherry, India, 40,421 (★110,000) F6 39
Pondicherry, ter., India, 369,079 *F6 39
Pond Inlet, N.W., Ter., Can., 53 B17 67
Pondosa, Siskiyou, Calif., 310 B3 82
Pondosa, Union, Oreg., 160 C9 113
Pond Point, New Haven, Conn. (part of Milford) E4 84
Ponds, isl., Newf., Can B4 75
Pone, isl., Md D5 85
Poneloya, Nic D4 62
Ponemah, Beltrami, Minn., 50 B4 99
Ponemah, Hillsboro, N.H., 30 E3 105
Poneto, Wells, Ind., 289 C7 91
Ponferrada, Sp., 37,053 A2 20
Pongo de Manseriche, gorge, Peru B2 58
Pongoma, Sov. Un E16 25
Ponhook, lake, N.S., Can E4 74
Ponkapog, pond, Mass D3 97
Ponoka, Alta., Can., 3,938 C4 69
Ponoy, Sov. Un D19 25
Ponoy, riv., Sov. Un D18 25
Ponpanoosuc, Windsor, Vt., 180 D4 120
Pons, Fr., 3,129 E3 14
Ponsford, Becker, Minn., 150 D3 99
Ponta Delgada, Az. Is., 22,300 h9 44
Ponta de Pedras, Braz., 1,907 C5 59
Ponta Grossa, Braz., 77,803 D2 56
Pontal, Braz., 2,521 k7 56
Pont-à-Mousson, Fr., 12,802 C7 14
Ponta Porã, Braz., 9,160 C1 56
Pontarlier, Fr., 15,382 D7 14
Pontchartrain, lake, La D5 95
Pont-de-Roide, Fr., 3,744 B2 18
Pontedera, It., 12,400 C3 21
Ponte de Sor, Port., 3,827 C1 20
Ponte di Legno, It., 2,292 C6 18
Pontefract, Eng., 27,114 A6 12
Ponteix, Sask., Can., 887 H2 70
Ponte Nova, Braz., 22,536 F2 57
Pontevedra, Sp., 21,400 A1 20
Pontevedra, prov., Sp., 680,229 *A1 20
Ponte Vedra Beach, St. Johns, Fla., 800 B5, B6 86
Pont-Faverger [-Moronvilliers], Fr., 1,221 E4 12
Ponthierville, Con. L. B4 48
Pontiac, Livingston, Ill., 8,435 C5 90
Pontiac, Oakland, Mich., 82,233 A7, F7 98
Pontiac, Richland, S.C., 50 C6 115
Pontiac, co., Que., Can., 19,947 A7 72
Pontianak, Indon., 146,547 F3 35
Pontine, is., It D4 21
Pontivy, Fr., 8,775 C2 14
Pont-l'Abbé, Fr., 5,541 D1 14
Pontoise, Fr., 15,853 C4 14
Pontoosuc, lake, Mass B1 97
Pontotoc, Pontotoc, Miss., 2,108 A4 100
Pontotoc, Johnston, Okla., 60 C5 112
Pontotoc, Mason, Tex., 100 D3 118
Pontotoc, co., Miss., 17,232 A4 100
Pontotoc, co., Okla., 28,089 C5 112
Pontotoc, ridge, Miss A5 100
Pontremoli, It., 4,839 B2 21
Pontresina, Switz., 1,067 D8 19
Pontrilas, Sask., Can., 87 D3 70
Pont Rouge, Que., Can., 2,988 C6, C8 73
Pont-Ste. Maxence, Fr., 7,261 E2 15
Pont-St. Esprit, Fr., 3,691 E6 14
Pont Viau, Que., Can., 16,077 *D8 73
Pontypool, Ont., Can., 207 C6 72
Pontypool, Wales, 39,879 C4 12
Pontypridd, Wales, 35,536 C4 12
Pony, Madison, Mont., 50 E5 102
Ponza, isl., It D4 21
Poobah, lake, Ont., Can B7 99
Poole, Eng., 88,088 D6 12
Poole, Webster, Ky., 300 C2 94
Poole, Buffalo, Nebr., 19 D7 103
Pooler, Chatham, Ga., 1,073 D5 87
Pooles, isl., Md B5 85
Poolesville, Montgomery, Md., 298 B3 85
Pooleville, Carter, Okla., 50 C4 112
Poolewe, Scot C3 13
Poolville, Parker, Tex., 500 C4 118
Poona, India, 597,562 (★800,000) E5 39, H4 40
Pooncarie, Austl., 66 F4 51
Poopó, Bol., 736 C2 55
Poopó, lake, Bol C2 55
Poor Knights, is., N.Z. K15 51
Popasnaya, Sov. Un., 10,000 q21 27
Popayán, Col., 37,400 C2 60
Pope, Marengo, Ala., 300 C2 78
Pope, Man., Can., 35 D1 71
Pope, Panola, Miss., 246 A4 100
Pope, Perry, Tenn B4 117
Pope, co., Ark., 21,177 B2 81
Pope, co., Ill., 4,061 F5 90
Pope, co., Minn., 11,914 E3 99
Popejoy, Franklin, Iowa, 190 B4 92
Poperinge, Bel., 12,350 D2 15
Popham, pt., Ont., Can B3 72
Popham Beach, Sagadahoc, Maine, 35 E3 96
Poplar, Tulare, Calif., 1,478 *D4 82
Poplar, Roosevelt, Mont., 1,565 B11 102
Poplar, Mitchell, N.C., 275 C4 109
Poplar, Douglas, Wis., 475 B2 124
Poplar, isl., Md C5 85
Poplar, mtn., Ky D4 94
Poplar, riv., Man., Can C3 71
Poplar, riv., Sask., Can H2 70
Poplar, riv., Mont B11 102
Poplar Bluff, Butler, Mo., 15,926 E7 101
Poplar Branch, Currituck, N.C., 290 A8 109
Poplar Creek, B.C., Can., 50 D9 68
Poplar Creek, Montgomery, Miss., 75 B4 100
Poplarfield, Man., Can., 142 D3 71
Poplar Grove, Phillips, Ark., 169 C5 81
Poplar Grove, Boone, Ill., 460 A5 90
Poplar Heights, Fairfax, Va., 1,000 *B5 121
Poplar Plains, Fleming, Ky., 200 B6 94
Poplar Point, Man., Can., 257 D3 71
Poplars, Calvert, Md., 150 C4 85
Poplarville, Pearl River, Miss., 2,136 E4 100
Popocatepetl, vol., Mex n14 63
Popokabaka, Con. L. C2 48
Popovka, Sov. Un., 10,000 H11 27
Popovo, Bul., 6,469 D8 22
Poppel, Bel., 2,084 C5 15
Popple, riv., Wis C5 124
Poquetanuck, New London, Conn., 200 D8 84
Poquonock, Hartford, Conn., 600 B6 84
Poquonock Bridge, New London, Conn., 3,000 D8 84
Poquoson, York, Va., 4,278 B6 121
Porbandar, India, 74,476 G2 40
Porcher, isl., B.C., Can C2 68
Porcuna, Sp., 10,516 D3 20
Porcupine, Shannon, S. Dak., 25 D3 116
Porcupine, cape, Newf., Can B3 75
Porcupine, mtn., Man., Sask., Can E5 70, C1 71
Porcupine, mts., Mich A5 98
Porcupine, riv., Alsk., Yukon, Can C4 66
Porcupine Plain, Sask., Can., 766 E4 70
Pordenone, It., 34,055 D8 18, B4 21
Poreč, Yugo., 2,992 C1 22
Pori, Fin., 54,000 G9 25
Porjus, Swe., 1,309 D8 25
Porkhov, Sov. Un., 10,000 C7 27
Porkkala, Fin G11 25
Porlamar, Ven., 20,807 A5 60
Pornic, Fr., 2,460 D2 14
Poroma, Bol., 171 C2 55
Poronaysk (Shikuka), Sov. Un., 33,000 B11 37
Porong, riv., Camb F6 38
Porpoise, bay, Ant C26 5
Porpoise, pt., Fla G5 86
Porrentruy, Switz., 7,095 B3 19
Porsangerfjord, fjord, Nor B12 25
Porsgrunn, Nor., 10,700 p27 25
Portachuelo, Bol., 2,456 C3 55
Portacloy, Ire C2 11
Port Adelaide, Austl., 38,923 G2 51
Portadown, N. Ire., 18,605 C5 11
Portaferry, N. Ire., 1,406 C6 11
Portage, Alsk., 71 g17 79
Portage, P.E.I., Can., 55 C5 74
Portage, Porter, Ind., 13,659 *A3 91
Portage, Aroostook, Maine, 540 (458▲) B4 96
Portage, Kalamazoo, Mich., 6,000 *E5 98
Portage, Cascade, Mont., 20 C5 102
Portage, Cambria, Pa., 3,933 F4 114
Portage, Box Elder, Utah, 189 B3 119
Portage, Columbia, Wis., 7,822 E4 124
Portage, co., Ohio, 91,798 A6 111
Portage, co., Wis., 36,964 D4 124
Portage, bay, Man., Can D2 71
Portage, head, Wash A1 122
Portage, isl., N.B., Can B5 74
Portage, lake, Maine B4 96
Portage, river, Ohio A4 111
Portage Des Sioux, St. Charles, Mo., 371 A8 101
Portage Lakes, Summit, Ohio, 10,000 *B6 111
Portage la Prairie, Man., Can., 12,388 D5, E2 71
Portageville, New Madrid, Mo., 2,505 E8 101
Portageville, Wyoming, N.Y., 450 C2 108
Portal, Cochise, Ariz., 50 F6 80
Portal, Bullock, Ga., 494 D5 87
Portal, Burke, N. Dak., 351 A3 110
Port Alberni, B.C., Can., 11,560 E5 68
Portalegre, Port., 10,510 C2 20
Portales, Roosevelt, N. Mex., 9,695 C6 107
Portales, Presidio, Tex G2 118
Port Alexander, Alsk., 18 D13, m22 79
Port Alfred, Que., Can., 9,066 A7 73
Port Alfred, S. Afr., 6,171 D4 49
Port Alice, B.C., Can., 1,065 D4 68
Port Allegany, McKean, Pa., 2,742 C5 114
Port Allen, West Baton Rouge, La., 5,026 B5, D4 95
Port Angeles, Clallam, Wash., 12,653 A2 122
Port Angeles East, Clallam, Wash., 1,283 *A2 122
Port Anson, Newf., Can., 407 D4 75
Port Antonio, Jam., 7,827 F5 64
Port Aransas, Nueces, Tex., 824 F4 118
Portarlington, Ire., 2,846 D4 11
Port Arthur, Ont., Can., 45,276 E8 72
Port Arthur (Lushun), China, 126,000 D9 34
Port Arthur, Jefferson, Tex., 66,676 E6 118
Portaskaig, Scot E2 13
Port Augusta, Austl., 9,711 F6 50
Port au Persil, Que., Can., 216 B8 73
Port au Port, Newf., Can., 482 D2 75
Port au Port, bay, Newf., Can D2 75
Port au Port, pen., Newf., Can D2 75
Port-au-Prince, Haiti, 134,117 F7 64
Port Austin, Huron, Mich., 706 D8 98
Port Barre, St. Landry, La., 1,876 D4 95
Port Bergé, Malag., 1,538 g9 49
Port Blair, Andaman Is., 14,075 F9 39
Port Blakely, Kitsap, Wash., 500 D1 122
Port Blandford, Newf., Can., 716 D4 75
Port Bolivar, Galveston, Tex., 600 E5, G5 118
Port Borden, P.E.I., Can., 696 C6 74
Port Burwell, Ont., Can., 777 E4 72
Port Byron, Rock Island, Ill., 1,153 B3 90
Port Byron, Cayuga, N.Y., 1,201 B4 108
Port Carbon, Schuylkill, Pa., 2,775 *E9 114
Port Carling, Ont., Can., 529 B5 72
Port Chalmers, N.Z., 3,120 P13 51
Port Charlotte, Charlotte, Fla., 3,197 *F4 86
Port Chester, Westchester, N.Y., 24,960 D2, D3, E7 108
Port Chicago, Contra Costa, Calif., 1,746 B5 82
Port Clements, B.C., Can., 81 C1 68
Port Clinton, Ottawa, Ohio, 6,870 A5 111
Port Clyde, N.S., Can., 102 F4 74
Port Clyde, Knox, Maine, 350 E3 96
Port Colborne, Ont., Can., 14,886 E5 72
Port Colden, Warren, N.J., 200 B3 106
Port Coquitlam, B.C., Can., 8,111 A10, E6 68
Port Crane, Broome, N.Y., 550 C5 108
Port Credit, Ont., Can., 7,203 D5, E6 72
Port Daniel, Que., Can., 276 A5 74
Port-de-Bouc, Fr., 12,510 F6 14
Port De Grave, dist., Newf., Can E5 75
Port-de-Paix, Hai., 6,309 E7 64
Port Deposit, Cecil, Md., 953 A5 85
Port Dickinson, Broome, N.Y., 2,295 C5 108
Port Dickson, Mala., 4,415 K4 38
Port Discovery, Jefferson, Wash A3 122
Port Douglas, Austl., 122 C8 50
Port Douglas, Essex, N.Y B3 108
Port Dover, Ont., Can., 3,064 E4 72
Port Eads, Plaquemines, La., 12 E6 95
Port Edwards, Wood, Wis., 1,849 D4 124
Portel, Braz., 1,821 C4 59
Port Elgin, N.B., Can., 661 C5 74
Port Elgin, Ont., Can., 1,632 C3 72
Port Elizabeth, Cumberland, N.J., 300 E3 106
Port Elizabeth, S. Afr., 249,211 (★274,180) D4 49
Port Ellne, Scot E2 13
Porteña, Arg A4 54
Porter, Jefferson, Ala., 130 B3, E4 78
Porter, Porter, Ind., 2,189 A3 91
Porter, Yellow Medicine, Minn., 261 F2 98
Porter, Wagoner, Okla., 492 B6 112
Porter, Montgomery, Tex., 900 D5 118
Porter, Grays Harbor, Wash., 200 C2 122
Porter, co., Ind., 60,279 B3 91
Porter, lake, Sask., Can A2 70
Porter Corners, Saratoga, N.Y., 200 B7 108
Porterdale, Newton, Ga., 2,365 C3 87
Port Erin, I. of Man, 1,265 F4 13
Portersville, Butler, Pa., 344 E1 114
Porterville, Tulare, Calif., 7,991 D4 82
Porterville, Kemper, Miss., 120 C5 100
Porterville West (Burton), Tulare, Calif., 4,635 *D4 82
Port Essington, B.C., Can., 66 B3 68
Port Étienne, Maur., 1,200 B1 45
Port Ewen, Ulster, N.Y., 2,622 D7 108
Port Fairy, Austl., 2,434 I4 51
Port Fouad (Port Fuad), (Būr Fu'ād), Eg., U.A.R., 1,000 C4 32
Port-Francqui, Con. L., 3,553 B3 48
Port Fuad (Port Fouad), (Būr Fu'ād), Eg., U.A.R., 1,000 C4 32
Port Gamble, Kitsap, Wash., 400 B3 122
Port Gamble, Indian res., Wash B3 122
Port-Gentil, Gabon, 16,600 F1 46
Port George, N.S., Can., 74 D4 74
Port Gibson, Claiborne, Miss., 2,861 D3 100
Port Glasgow, Scot., 22,551 E9 13
Portglenone, N. Ire., 609 C5 11
Port Graham, Alsk., 139 D9, h16 79
Port Greville, N.S., Can., 128 D5 74
Port Hammond, B.C., Can., 1,267 A10 68
Port Harcourt, Nig., 71,634 F6 45
Port Hardy, B.C., Can., 606 D4 68
Port Hastings, N.S., Can., 326 D8 74
Port Hawkesbury, N.S., Can., 1,346 D8 74
Porthcawl, Wales, 11,082 C4 12
Port Hedland, Austl., 328 D2 50
Port Heiden, Alsk., 74 D8 79
Port Henry, Essex, N.Y., 1,767 A7 108
Port Herald, Nya A6 49
Port Hill, P.E.I., Can., 150 C6 74
Port Homer, Jefferson, Ohio, 80 A2 123
Port Hood, N.S., Can., 511 C8 74
Port Hope, Ont., Can., 8,091 D6 72
Port Hope, Huron, Mich., 349 E8 98
Port Hope Simpson, Newf., Can., 402 B3, h10 75
Port Hueneme, Ventura, Calif., 11,067 E4 82
Port Huron, St. Clair, Mich., 36,084 (★124,000) F8 98
Portia, Lawrence, Ark., 333 A4 81
Portillo, Cuba F5 64
Portimão, Port., 12,066 D1 20
Portis, Osborne, Kans., 232 C5 93
Port Isabel, Cameron, Tex., 3,575 F4 118
Port Jackson, bay, Austl F8 51
Port Jefferson, Suffolk, N.Y., 2,336 D3 108
Port Jefferson, Shelby, Ohio, 438 B3 111
Port Jefferson Station, Suffolk, N.Y., 1,041 *D3 108
Port Jervis, Orange, N.Y., 9,268 D6 108
Port Kent, Essex, N.Y., 200 B3 108
Portland, Ashley, Ark., 566 D4 81
Portland, Austl., 6,014 I3 51
Portland, Ont., Can., 254 C8 72
Portland, Fremont, Colo., 73 C5 83
Portland, Middlesex, Conn., 7,496 C6 84
Portland, Eng., 11,542 D5 12
Portland, Walton, Fla., 200 G3 86
Portland, Jay, Ind., 6,999 D8 91
Portland, Cumberland, Maine, 72,566 (★142,700) E2, E5 96
Portland, Ionia, Mich., 3,330 F6 98
Portland, Callaway, Mo., 125 C6 101
Portland, Chautauqua, N.Y., 400 C1 108
Portland, Traill, N. Dak., 606 B8 110
Portland, Meigs, Ohio, 200 C6 111
Portland, Multnomah, Oreg., 372,676 (★731,200) B2, B4 113
Portland, Northampton, Pa., 589 E11 114
Portland, Sumner, Tenn., 2,424 A5 117
Portland, San Patricio, Tex., 2,538 F4 118
Portland, canal, Alsk n24 79
Portland, canal, B.C., Can B3 68
Portland, inlet, B.C., Can B2 68
Portland, promontory, Que., Can A2 73
Portland Bill, pt., Eng E5 12
Portland Mills, Elk, Pa., 135 D4 114
Portlandville, Otsego, N.Y., 270 C6 108
Portlaoighise, Ire., 3,133 D4 11
Port Lavaca, Calhoun, Tex., 8,864 E4 118
Portlaw, Ire., 1,113 E4 11
Port Leyden, Lewis, N.Y., 898 B5 108
Port Lincoln, Austl., 7,508 F6 50
Port Loko, S.L E2 45
Port Loring, Ont., Can., 261 B5 72
Port-Louis, Fr., 4,140 D2 14
Port Louis, Mauritius, 60,500 H24 3
Port Ludlow, Jefferson, Wash., 300 B3 122
Port Macquarie, Austl., 5,951 E9 51
Portmadoc, Wales, 3,419 B3 12
Portmagee, Ire., 137 F1 11
Portmahomack, Scot C5 13
Port Mahon, Kent, Del., 20 B7 85
Port Maitland, Ont., Can., 103 E5 72
Port Maitland, N.S., Can., 469 F3 74
Port Maria, Jam., 3,997 F5 64
Port Matilda, Centre, Pa., 697 E5 114
Port Mayaca, Martin, Fla., 75 F6 86
Port Medway, N.S., Can., 310 E5 74
Port Menier, Que., Can., 475 k8 75
Port Moller, Alsk., 10 D7 79
Port Monmouth, Monmouth, N.J., 4,000 C4 106
Port Moody, B.C., Can., 4,789 A10, E6 68
Port Moresby, Pap., 2,503 k12 50
Port Morien, N.S., Can., 517 C10 74
Port Morris, Morris, N.J., 1,789 B3 106
Port Mouton, N.S., Can., 262 F5 74
Port Murray, Warren, N.J., 300 B3 106
Port Musgrave, bay, Austl B7 50
Portnahaven, Scot E2 13
Port Neches, Jefferson, Tex., 8,696 E6 118
Port Nelson, Man., Can. A5, C5 71
Portneuf, Que., Can., 1,380 C6 73
Portneuf, co., Que., Can., 50,711 C5 73
Portneuf-sur-Mer, Que., Can., 289 A8 73
Port Nolloth, S. Afr., 2,592 C2 49
Portnoo, Ire., 173 C3 11
Port Norris, Cumberland, N.J., 1,789 E2 106
Pôrto (Oporto), Port., 303,400 (★750,000) B1 20
Pôrto Alegre, Braz., 617,629 (★850,000) E2 56
Pôrto Alexandre, Ang., 2,874 E1 48
Pôrto Amboim, Ang., 1,537 D1 48
Pôrto Amélia, Moz., 10,000 D7 48
Portobelo, Pan., 591 F8, h11 62
Pôrto Calvo, Braz., 3,876 C3, k6 57
Port O'Connor, Calhoun, Tex., 500 E4 118
Pôrto de Mós, Port, 4,402 C1 20
Port de Moz, Braz., 879 C4 59
Pôrto Esperanca, Braz., 486 B1 56
Pôrto Feliz, Braz. 11,786 C3, m8 56
Portoferraio, It., 6,000 C3 21
Portofino, It., 1,011 B2 21
Port of Ness, Scot B2 13
Port-of-Spain, Trin., 94,100 (★170,000) A5 60
Porto Garibaldi, It E8 18
Portogruaro, It., 7,100 B4 21
Pôrto Guaíra, Braz C2 56
Portola, Plumas, Calif., 1,874 C3 82
Portomaggiore, It., 4,300 B3 21
Pôrto Mendes, Braz C2 56
Pôrto Murtinho, Braz., 4,476 D4 55
Pôrto Nacional, Braz., 4,926 D1 57
Porto Novo, Dah., 30,800 E5 45
Port Orange, Volusia, Fla., 1,801 C6 86
Port Orchard, Kitsap, Wash., 2,778 B3, D1 122
Port Orford, Curry, Oreg., 1,171 E2 113
Porto Santo, isl., Madeira, Is C1 44
Portoscuso, It., 2,262 E2 21
Pôrto Seguro, Braz., 2,697 E3 57
Porto Torres, It., 9,166 D2 21
Pôrto União, Braz., 9,954 D2 56
Porto-Vecchio, Fr., 2,711 D2 21
Pôrto Velho, Braz., 19,387 D2 59
Portoviejo, Ec., 18,082 B1 58
Portpatrick, Scot., 1,063 F3 13
Port Penn, New Castle, Del., 271 A6 85
Port Perry, Ont., Can., 2,262 C6 72
Port Phillip, bay, Austl G7 50
Port Pirie, Austl., 14,003 F1 51
Port Radium, N.W., Ter., Can., 412 C9 66
Port Reading, Middlesex, N.J., 3,000 E4 106
Portree, Scot., 1,767 C2 13
Portreeve, Sask., Can., 103 G1 70
Port Renfrew, B.C., Can., 279 E5 68
Port Republic, Atlantic, N.J., 561 D4 106
Port Republic, Rockingham, Va., 500 C4 121
Port Rexton, Newf., Can., 438 D5 75
Port Richey, Pasco, Fla., 1,931 D4 86
Port Rowan, Ont., Can., 787 E4 72
Port Royal, Henry, Ky., 90 B4 94
Port Royal, Juniata, Pa., 805 E7 114
Port Royal, Beaufort, S.C., 686 G6 115
Port Royal, Montgomery, Tenn., 30 A4 117
Port Royal, Caroline, Va., 128 C5 121
Port Royal, bay, Bermuda E13 77

Port Royal, isl., S.C. G6 115
Port Royal, sound, S.C. G7 115
Portrush, N. Ire., 4,263 B5 12
Port Said (Būr Sa'īd), Eg., U.A.R., 212,973 C9 31, C4 32, C6 43
Port St. Joe, Gulf, Fla., 4,217 C1, H3 86
Port St. Johns, S. Afr., 1,024 D4 49
Port-St. Louis [-du-Rhône], Fr., 4,262 F6 14
Portsalon, Ire. B4 12
Port Sanilac, Sanilac, Mich., 361 E8 98
Port Saunders, Newf., Can., 504 C3, h10 75
Port Sewall, Martin, Fla., 200 E6 86
Port Shepstone, S. Afr., 4,238 D5 49
Port Simpson, B.C., Can., 600 B2 68
Portsmouth, Ont., Can. (part of Kingston) C8 72
Portsmouth, Eng., 215,198 (*415,000) D6 12
Portsmouth, Shelby, Iowa, 232 C2 92
Portsmouth, Rockingham, N.H., 26,900 D5 105
Portsmouth, Scioto, Ohio, 33,637 D5 111
Portsmouth, Newport, R.I., 3,000 (8,251▲) C12 84
Portsmouth, (Independent City), Va., 80,039 B6, E6 121
Portsmouth, Dominica, 1,725 o16 64
Portsoy, Scot., 1,690 C6 13
Port Stanley, Ont., Can., 1,460 E3 72
Port Stanley (Stanley), Falk. Is., 1,250 I5 53
Port Stephens, bay, Austl. F9 52
Port Sudan, Sud., 47,562 B4 47
Port Sulphur, Plaquemines, La., 2,868 E6 95
Port-sur-Saône, Fr., 1,725 B2 18
Port Talbot, Wales, 50,223 (*120,000) C4 12
Port Tampa, Hillsborough, Fla., 1,764 E2, E4 86
Port Tobacco, Charles, Md., 75 C3 85
Port Townsend, Jefferson, Wash., 5,074 A3 122
Portugal, country, Eur., 8,889,300 H7 8, C1 20
Portugal Cove South, Newf., Can., 304 E5 75
Portugalete, Sp., 22,584 A4 20
Portugalia, Ang. C3 48
Portuguesa, riv., Ven. B4 60
Portuguesa, state, Ven., 203,707 B4 60
Portuguese Guinea, dep., Afr., 559,000 E4 42, D1 45
Portuguese India, see Goa, Damão, and Diu, ter., India
Portuguese Timor, dep., Asia, 439,000 G7 35
Portumna, Ire., 836 D3 11
Port Union, Newf., Can., 645 D5 75
Port Union, Butler, Ohio, 75 C2 111
Port-Vendres, Fr., 4,504 F5 14
Portville, Cattaraugus, N.Y., 1,336 C2 108
Port Vincent, Livingston, La., 340 B6, D5 95
Port Vue, Allegheny, Pa., 6,635 *F2 114
Port Wakefield, Austl., 429 G2 52
Port Washington, Nassau, N.Y., 15,657 D2 108
Port Washington, Tuscarawas, Ohio, 526 B6 111
Port Washington, Ozaukee, Wis., 5,984 E6 124
Port Wentworth, Chatham, Ga., 3,705 D5 87
Port William, Clinton, Ohio, 360 C4 111
Port William, Scot. F4 13
Port Wing, Bayfield, Wis., 250 B2 124
Porum, Muskogee, Okla., 573 B6 112
Porvenir, Chile h11 54
Porvenir, Presidio, Tex. F2 118
Porz, Ger., 50,900 C2 17
Posadas, Arg., 37,588 E4 55
Posadas, Sp., 8,999 D3 20
Poschiavo, Switz., 3,743 D9 19
Posen, Cook, Ill., 4,517 F3 90
Posen, Presque Isle, Mich., 341 C7 98
Posen, see Poznan, Pol.
Posey, co., Ind., 19,214 H2 91
Poseyville, Posey, Ind., 997 H2 91
Poshan, China, 250,000 (800,000▲) F8 36
Poshekhonye-Volodarsk, Sov. Un. B12 27
Poso, Indon., 2,875 F6 35
Poso, lake, Indon. F6 35
Posse, Braz., 1,953 D1 57
Pössneck, Ger., 19,600 C6 17
Post, Crook, Oreg., 5 C6 113
Post, Garza, Tex., 4,663 C2 118
Postell, Cherokee, N.C. D2 109
Post Falls, Kootenai, Idaho, 1,983 B2 89
Postmasburg, S. Afr., 4,701 C3 49
Post Mills, Orange, Vt., 200 D4 120
Postojna, Yugo., 4,848 C2 22
Postville, Allamakee, Iowa, 1,554 A6 92
Poston, Florence, S.C., 250 D9 115
Potaro Landing, Br. Gu., 353 A3 60
Potato Creek, Washabaugh, S. Dak., 40 D4 116
Potchefstroom, S. Afr., 41,927 C4 49
Poteau, LeFlore, Okla., 4,428 B7 112
Poteau, mtn., Ark., Okla. C1 81
Poteau, riv., Okla. B7 112
Poteet, Atascosa, Tex., 2,811 E3 118
Potenza, It., 29,500 (43,545▲) D5 21
Potenza, riv., It. C4 21
Potgietersrus, S. Afr., 11,438 B4 49
Poth, Wilson, Tex., 1,119 E3 118
Potholes, res., Wash. B6 122
Poti, Sov. Un., 43,000 G17 9
Poti, riv., Braz. C2 57
Potiskum, Nig., 14,692 D7 45
Potlatch, Latah, Idaho, 880 C2 89
Poto, Peru, 247 D4 58
Potocho, China A11 40
Potomac, Vermilion, Ill., 661 C6 90
Potomac, Montgomery, Md., 150 B3 85
Potomac, Missoula, Mont., 40 D3 102
Potomac, riv., Md., Va., W. Va. C12 77, A1, D4 85
Potomac Park, Allegany, Md., 1,016 *D2 85
Potosí, Bol., 45,758 C2 55
Potosi, Washington, Mo., 2,805 D7 101
Potosi, Grant, Wis., 589 F3 124
Potosí, dept., Bol., 534,399 C2 55
Pototan, Phil., 4,812 *C6 35
Potrerillos, Chile E2 55
Potrerillos, Hond., 831 C4 62
Potro, mtn., Arg. E2 55
Potsdam, Ger., 114,500 A8 17
Potsdam, St. Lawrence, N.Y., 7,765 A2 108
Potsdam, Miami, Ohio, 282 C3 111
Pottawatomie, co., Kans., 11,957 C7 93
Pottawatomie, co., Okla., 41,486 B5 112
Pottawatomie, creek, Kans. D8 93
Pottawatomie Indian res., Kans. C8 93
Pottawattamie, co., Iowa, 83,102 C2 92
Pottawattamie Park, La Porte, Ind., 292 *A4 91
Potter, Polk, Ark., 120 C1 81
Potter, Atchison, Kans., 109 A8, C8 93
Potter, Cheyenne, Nebr., 554 C2 103
Potter, Calumet, Wis., 225 B6 124
Potter, co., Pa., 16,483 C6 114
Potter, co., S. Dak., 4,926 B6 116
Potter, co., Tex., 115,580 B2 118
Potter Hill, Washington, R.I., 200 D9 84
Potter Place, Merrimack, N.H., 50 D3 105
Pottersdale, Clearfield, Pa., 50 D5 114
Potters Mills, Centre, Pa., 130 E6 114
Pottersville, Howell, Mo., 45 E5 101
Pottersville, Somerset, N.J., 200 B3 106
Pottersville, Warren, N.Y., 500 B7 108
Pottersville, Newport, R.I., 75 C12 84
Potter Valley, Mendocino, Calif., 220 C2 82
Potterville, Taylor, Ga., 400 D2 87
Potterville, Eaton, Mich., 1,028 *F6 98
Potterville, Bradford, Pa., 65 C9 114
Potts, creek, Va., W. Va. D2 121
Potts, mtn., Va. D2 121
Potts Camp, Marshall, Miss., 429 A4 100
Pottstown, Montgomery, Pa., 26,144 F10 114
Pottsville, Pope, Ark., 250 B2 81
Pottsville, Schuylkill, Pa., 21,659 E9 114
Potwin, Butler, Kans., 635 B6, E7 93
Pouce Coupé, B.C., Can., 669 B7 68
Pouch Cove, Newf., Can., 1,324 E5 75
Poughkeepsie, Sharp, Ark., 250 A4 81
Poughkeepsie, Dutchess, N.Y., 38,330 (*124,700) D7 108
Poulan, Worth, Ga., 736 E3 87
Poulo Condore, is., Viet. H7 38
Poulsbo, Kitsap, Wash., 1,505 B3 122
Poultney, Rutland, Vt., 1,810 (3,009▲) D2 120
Poultney, riv., Vt. E2 120
Pound, Wise, Va., 1,135 B2 121
Pound, Marinette, Wis., 273 C5 124
Pound, gap, Va. A2 121
Pouso Alegre, Braz., 18,852 C3, m9 56
Pouxeux, Fr., 1,783 A2 18
Povenets, Sov. Un., 5,000 F16 25
Poverty, bay, N.Z. M17 51
Póvoa de Varzim, Port., 16,913 B1 20
Povorino, Sov. Un., 24,000 G14 29
Povorotnyy, cape, Sov. Un. E6 37
Povungnituk, Que., Can., 434 E17 67
Powassan, Ont., Can., 1,064 A5 72
Poway, San Diego, Calif., 1,921 F5 82
Powder, riv., Mont., Wyo. B6 125
Powder, riv., Oreg. C9 113
Powderhorn, Gunnison, Colo., 5 C3 83
Powder River, Natrona, Wyo., 50 B6 125
Powder River, co., Mont., 2,485 E11 102
Powder River, pass, Wyo. A5 125
Powder Springs, Cobb, Ga., 746 A4 87
Powder Springs, Grainger, Tenn., 200 C10 117
Powderville, Powder River, Mont., 5 E11 102
Powe, Stoddard, Mo., 72 E7 101
Powell, McDonald, Mo., 100 E3 101
Powell, Delaware, Ohio, 390 B4 111
Powell, Bradford, Pa., 250 C8 114
Powell, Knox, Tenn., 500 E11 117
Powell, Navarro, Tex., 250 C4 118
Powell, Park, Wyo., 4,740 A4 125
Powell, co., Ky., 6,674 C6 94
Powell, co., Mont., 7,002 D4 102
Powell, mtn., Colo. B4 83
Powell, mtn., N. Mex. B1 107
Powell, mtn., Tenn. C10 117
Powell, peak, Ariz. C1 80
Powell, riv., Tenn., Va. C10 117
Powell Butte, Crook, Oreg., 40 C5 113
Powell Creek, Austl. C5 50
Powell Park, basin, Colo. A2 83
Powell River, B.C., Can., 5,500 E5 68
Powellsville, Bertie, N.C., 259 A7 109
Powellton, Fayette, W. Va., 1,256 C3, D6 123
Powellville, Wicomico, Md., 500 D7 85
Powelton, Hancock, Ga., 25 C4 87
Powelton, Centre, Pa. (part of Sandy Ridge) E5 114
Power, Teton, Mont., 5 C5 102
Power, Brooke, W. Va., 175 A4, B2 123
Power, co., Idaho, 4,111 G6 89
Power, head, Ire. F3 11
Powers, Jay, Ind., 40 D7 91
Powers, Menominee, Mich., 415 C3 98
Powers, Coos, Oreg., 1,366 E2 113
Powers Lake, Burke, N. Dak., 633 A3 110
Powersville, Putnam, Mo., 189 A4 101
Powerview, Man., Can., 902 *D3 71
Poweshiek, co., Iowa, 19,300 C5 92
Powhatan, Jefferson, Ala., 350 B2, E4 78
Powhatan, Lawrence, Ark., 136 A4 81
Powhatan, Natchitoches, La., 300 C2 95
Powhatan, Powhatan, Va., 300 D5 121
Powhatan, co., Va., 6,747 D5 121
Powhatan Point, Belmont, Ohio, 2,147 C7 111
Powhattan, Brown, Kans., 128 C8 93
Pownal, Cumberland, Maine, 50 (778▲) E5 96
Pownal, Bennington, Vt., 325 (1,509▲) E2 120
Pownal Center, Bennington, Vt., 75 F2 120
Poyang, China, 42,000 J7 36
Poyang, lake, China J7 36
Poyarkovo, Sov. Un. B4 37
Poyen, Grant, Ark., 312 C3 81
Poygan, lake, Wis. D5 124
Poynette, Columbia, Wis., 1,090 E8 124
Poyntzpass, N. Ire., 282 C5 11
Poy Sippi, Waushara, Wis., 450 D4 124
Požarevac, Yugo., 24,293 C5 22
Poznan, Pol., 408,000 B4 26
Poznan, prov., Pol., 1,994,000 A9 17
Pozo Almonte, Chile D2 55
Pozoblanco, Sp., 16,020 C3 20
Pozo Redondo, mts., Ariz. E3 80
Pozuelo de Alarcón, Sp., 9,412 p17 20
Pozuzo, Peru, 132 D2 58
Pozzallo, It., 12,253 F5 21
Pozzuoli, It., 51,308 D5 21
Prachatice, Czech., 5,100 E9 17
Prachin Buri, Thai., 13,308 E4 38
Prachuap Khiri Khan, Thai., 6,799 G3 38
Praco, Jefferson, Ala., 900 E4 78
Practicos, pt., Cuba E5 64
Prada, Switz. C8 19
Prade Ranch, Real, Tex., 15 E3 118
Prades, Fr., 5,676 F5 14
Prado, basin, Calif. F3 82
Praestø, Den., 1,528 C6 24
Praestø, co., Den., 123,382 C6 24
Praga, Pol. (part of Warsaw) k14 26
Prague, Czech., 1,003,300 (*1,100,000) C3, n17 26
Prague, Saunders, Nebr., 372 C9 103
Prague, Lincoln, Okla., 1,545 B5 112
Praha, see Prague, Czech.
Praia, C.V. Is., 9,980 *E3 42
Prainha, Braz., 778 C4 59
Prairie, Wilcox, Ala., 50 C2 78
Prairie, Monroe, Miss., 112 B5 100
Prairie, co., Ark., 10,515 C4 81
Prairie, co., Mont., 2,318 D11 102
Prairie, bayou, Ark. D6 81
Prairie, riv., Minn. C5 99
Prairie, riv., Wis. C4 124
Prairie City, McDonough, Ill., 613 C3 90
Prairie City, Jasper, Iowa, 943 C4 92
Prairie City, Grant, Oreg., 801 C8 113
Prairie City, Perkins, S. Dak., 50 B3 116
Prairie, Creek, Vigo, Ind., 240 F2 91
Prairie Dog, creek, Kans., Nebr. C3 93, D6 103
Prairie Dog Town, fork, Okla. C2 112
Prairie du Chien, Crawford, Wis., 5,649 E2 124
Prairie du Rocher, Randolph, Ill., 679 E3 90
Prairie du Sac, Sauk, Wis., 1,676 E4 124
Prairie Farm, Barron, Wis., 350 C2 124
Prairie Grove, Washington, Ark., 1,056 B1 81
Prairie Hill, Chariton, Mo., 84 B5 101
Prairie Home, Cooper, Mo., 213 C5 101
Prairie Home, Lancaster, Nebr., 29 E2 103
Prairie Point, Noxubee, Miss., 75 B5 100
Prairie River, Sask., Can., 84 E4 70
Prairieton, Vigo, Ind., 250 F3 91
Prairie View, Logan, Ark., 165 B2 81
Prairie View, Phillips, Kans., 188 C4 93
Prairie View, Waller, Tex., 2,326 *D5 118
Prairie Village, Johnson, Kans., 25,356 *B9 93
Prairieville, Ascension, La., 150 B5 95
Pralognan-la-Vanoise, Fr., 249 D2 18
Pran Buri, Thai. F3 38
Praszka, Pol., 3,013 C5 26
Prata, Braz., 4,725 E1 57
Prather, Clark, Ind. A4 94
Prato, It., 111,285 (*143,000) C3 21
Prats-de-Mollo, Fr., 682 F5 14
Pratt, Man., Can. E2 71
Pratt, Pratt, Kans., 8,156 E5 93
Pratt, Kanawha, W. Va., 602 *C6 123
Pratt, co., Kans., 12,122 E5 93
Pratteln, Switz., 9,492 A4 19
Prattsburg, Steuben, N.Y., 690 C3 108
Prattsville, Grant, Ark., 150 C3 81
Prattsville, Greene, N.Y., 600 C6 108
Prattville, Autauga, Ala., 6,616 C3 78
Prattville, Hillsdale, Mich., 175 G6 98
Prattville, Tulsa, Okla., 2,530 *A5 112
Pratum, Marion, Oreg., 100 C2 113
Pravdinsk (Friedland), Sov. Un., 10,000 A6 26
Pravia, Sp., 1,804 A2 20
Prawle, pt., Eng. D4 12
Pray, Park, Mont., 15 E6 102
Preble, Adams, Ind., 170 C7 91
Preble, Brown, Wis., 12,245 A6 124
Preble, co., Ohio, 32,498 C3 111
Predazzo, It., 3,783 C7 18
Predivinsk, Sov. Un. D27 9
Preeceville, Sask., Can., 924 F4 70
Preetz, Ger., 12,800 D4 24
Pregel, riv., Sov. Un. A6 26
Pregnall, Dorchester, S.C., 100 E7 115
Prelate, Sask., Can., 618 G1 70
Premier, B.C., Can., 450 A3 68
Premont, Jim Wells, Tex., 3,049 F3 118
Prémontré, Fr., 1,872 E3 15
Prenter, Boone, W. Va., 350 C3, D6 123
Prentice, Price, Wis., 427 C3 124
Prentiss, Penobscot, Maine, 200 (227▲) C4 96
Prentiss, Jefferson Davis, Miss., 1,321 D4 100
Prentiss, Macon, N.C. D3 109
Prentiss, co., Miss., 17,949 A5 100
Prenzlau, Ger., 19,700 B6 16
Prerov, Czech., 30,500 D4 26
Prescott, Yavapai, Ariz., 12,861 C3 80
Prescott, Nevada, Ark., 3,533 D2 81
Prescott, Ont., Can., 5,366 C9 72
Prescott, Adams, Iowa, 331 C3 92
Prescott, Linn, Kans., 278 D9 93
Prescott, Ogemaw, Mich., 308 D7 98
Prescott, Columbia, Oreg., 129 A4 113
Prescott, Walla Walla, Wash., 269 C7 122
Prescott, Pierce, Wis., 1,536 D1 124
Prescott, co., Ont., Can., 27,226 B10 72
Presho, Lyman, S. Dak., 881 D5 116
Presidencia Roque Sáenz Peña, Arg., 23,100 E3 55
Presidente Epitácio, Braz., 10,425 C2 56
Presidente Hayes, dept., Par., 23,490 D4 55
Presidente Prudente, Braz., 54,055 C2 56
Presidential, range, N.H. B4 105
Presidents, isl., Tenn. E8 117
Presidio, Presidio, Tex., 1,062 G2 118
Presidio, co., Tex., 5,460 F2 118
Preslav, Bul., 4,114 D8 22
Presov, Czech., 35,100 D6 26
Presque Isle, Aroostook, Maine, 12,886 B4 96
Presque Isle, Presque Isle, Mich., 60 C7 98
Presque Isle, Vilas, Wis., 200 B4 124
Presque Isle, co., Mich., 13,117 C6 98
Pressath, Ger., 3,700 D6 17
Pressmens Home, Hawkins, Tenn., 400 C10 117
Prestatyn, Wales, 10,771 A4 12
Presteigne, Wales, 1,190 B4 12
Prestice, Czech., 4,199 D8 17
Preston, Ont., Can., 11,557 D4 72
Preston, New London, Conn., 200 (4,992▲) C9 84
Preston, Eng., 113,208 (*225,000) A5 12
Preston, Webster, Ga., 232 D2 87
Preston, Franklin, Idaho, 3,640 G7 89
Preston, Jackson, Iowa, 819 B7 92
Preston, Pratt, Kans., 278 E5 93
Preston, Bath, Ky., 200 B6 94
Preston, Caroline, Md., 469 C6 85
Preston, Fillmore, Minn., 1,491 G6 99
Preston, Kemper, Miss., 300 C5 100
Preston, Hickory, Mo., 117 D4 101
Preston, Okmulgee, Okla., 200 B6 112
Preston, King, Wash., 250 D2 122
Preston, co., W. Va., 27,233 B5 123
Preston, peak, Calif. B2 82
Preston Park, Wayne, Pa., 50 C11 114
Prestonburg, Floyd, Ky., 3,133 C7 94
Prestonville, Carroll, Ky., 211 B4 94
Prestrud, inlet, Ant. B33 5
Prestwick, Washington, Ala., 200 D2 78
Prestwick, Scot., 12,564 E4 13
Presumpscot, riv., Maine E4 96
Preti, mtn., It. C8 18
Preto, riv., Braz. h6 56
Prêto, riv., Braz. D2 57
Prêto, riv., Braz. C2 59
Pretoria, S. Afr., 303,684 (*422,590) C4 49
Pretty Boy, res., Md. A4 85
Pretty Prairie, Reno, Kans., 525 B4, E5 93
Preveza, Grc., 11,172 C3 23
Preveza, prov., Grc., 62,523 *C3 23
Prewitt, res., Colo. A7 82
Prey Veng, Camb., 5,000 G6 38
Pribilof, is., Alsk. D5 79
Priboj, Yugo., 5,501 D4 22
Příbram, Czech., 25,700 D3 26
Price, Que., Can., 3,049 *B3 73
Price, Queen Annes, Md., 126 B6 18
Price, Rusk, Tex., 800 C5 118
Price, Carbon, Utah, 6,802 D5 119
Price, co., Wis., 14,370 C3 124
Price, isl., B.C., Can. C3 14
Price, riv., Utah D5 119
Pricedale, Pike, Miss., 80 D3 100
Pricedale, Westmoreland, Pa., 1,300 *F2 114
Price Hill, Raleigh, W. Va., 200 D3, D7 123
Priceville, Ont., Can., 141 C4 72
Prichard, Mobile, Ala., 47,371 E1 78
Prichard, Wayne, W. Va., 350 C2 123
Prides Crossing, Essex, Mass. (part of Beverly) C4 97
Priddy, Mills, Tex., 300 D3 118
Priego, Sp., 13,801 D3 20
Prienai, Sov. Un. A7 26
Prieska, S. Afr., 6,464 C3 49
Priest, lake, Idaho A2 89
Priestly, mtn., B.C., Can. B3 68
Priestly, mtn., Maine B3 96
Priest Rapids, dam, Wash. C6 122
Priest Rapids, res., Wash. C6 122
Priest River, Bonner, Idaho, 1,749 A2 89
Prijedor, Yugo., 11,632 C3 22
Prijepolje, Yugo., 4,627 D4 22
Prikumsk, Sov. Un. E2 29
Prilep, Yugo., 37,486 E5 22
Priluki, Sov. Un., 40,000 F9 27
Prim, pt., P.E.I., Can. C6 74
Primate, Sask., Can., 129 E1 70
Primera, Cameron, Tex., 1,066 *F4 118
Primero, riv., Arg. A4 54
Primghar, O'Brien, Iowa, 1,131 A2 92
Primolano, It. D7 18
Primorsk, Sov. Un., 8,000 G13 25
Primorsk (Fischhausen), Sov. Un., 5,000 A6 26
Primorsko-Akhtarsk, Sov. Un., 10,000 H12 27
Primos, Delaware, Pa., 1,000 *B4 114
Primrose, Boone, Nebr., 117 C7 103
Primrose, Schuylkill, Pa., 950 *E9 114
Primrose, Providence, R.I., 75 B10 84
Primrose, lake, Sask., Can. B1 70
Prince, Sask., Can., 66 E1 70
Prince, Lincoln, Nev., 10 F7 104
Prince, co., P.E.I., Can., 40,894 C5 74
Prince, inlet, S.C. F4 115
Prince, lake, Va. B6 121
Prince Albert, Sask., Can., 24,168 D3 70
Prince Albert, nat. park, Sask., Can. F11 66
Prince Albert, sound, N.W. Ter., Can. B10 66
Prince Alfred, cape, N.W. Ter., Can. B8 66
Prince Charles, isl., N.W. Ter., Can. C17 67
Prince Charles Foreland, pt., Nor. B13 4
Prince Edward, co., Ont., Can., 21,108 C7 72
Prince Edward, co., Va., 14,121 D4 121
Prince Edward, is., Indian O. J23 3
Prince Edward Island, prov., Can., 104,629 G20 67, C6 74
Prince Edward Island, nat. park, P.E.I., Can. C6 74
Prince Frederick, Calvert, Md., 500 C4 85
Prince George, B.C., Can., 13,877 C6 70
Prince George, Prince George, Va., 80 C7, D5 121
Prince George, co., Va., 20,270 D5 121
Prince George, isl., Sov. Un. A10 4
Prince Georges, co., Md., 357,395 C4 85
Prince Harald, coast, Ant. C16 5
Prince of Wales, cape, Alsk. B6 79
Prince of Wales, isl., Alsk. n23 79
Prince of Wales, isl., Austl. B7 50
Prince of Wales, isl., N.W. Ter., Can. B13 66
Prince of Wales, strait, N.W. Ter., Can. B9 66
Prince Olav, coast, Ant. C17 5
Prince Patrick, isl., N.W. Ter., Can. B27 4
Prince Regent, inlet, N.W. Ter., Can. B14 66
Prince Rupert, B.C., Can., 11,987 B2 68
Prince Rupert, dist., B.C., Can. B4 68
Princesa Isabel, Braz., 3,184 C3 57
Princes Lake, Ont., Can., 104 B6 72
Princes Risborough, Eng., 3,829 C7 12
Princess Anne, Somerset, Md., 1,351 D6 85
Princess Anne, Princess Anne, Va., 250 B7, E6 121
Princess Astrid, coast, Ant. B14 5
Princess Charlotte, bay, Austl. B7 50
Princess Martha, coast, Ant. B11 5
Princess Ragnhild, coast, Ant. B15 5
Princess Royal, isl., B.C., Can. C3 68
Princeton, Jackson, Ala., 150 A3 78
Princeton, Dallas, Ark., 57 D3 81
Princeton, B.C., Can., 2,163 E7 68
Princeton, Newf., Can., 174 D5 75
Princeton, Ont., Can., 480 D4 72
Princeton, Colusa, Calif., 250 C2 82
Princeton, Dade, Fla., 1,719 F3, G6 86
Princeton, Latah, Idaho, 90 C2 89
Princeton, Gibson, Ind., 7,906 H2 91
Princeton, Scott, Iowa, 580 C7 92
Princeton, Franklin, Kans., 174 D8 93
Princeton, Caldwell, Ky., 5,618 C2 94
Princeton, Washington, Maine, 600 (829▲) C5 96
Princeton, Worcester, Mass., 400 (1,360▲) B4 97
Princeton, Marquette, Mich., 180 B3 98
Princeton, Mille Lacs, Minn., 2,353 E5 99
Princeton, Mercer, Mo., 1,443 A4 101
Princeton, Mercer, N.J., 11,890 C3 106
Princeton, Johnston, N.C., 948 B4 109
Princeton, Butler, Ohio, 70 C2 111
Princeton, Harney, Oreg., 10 D8 113
Princeton, Greenville and Laurens, S.C., 167 B3 115
Princeton, Mercer, W. Va., 8,393 D3 123
Princeton, Green Lake, Wis., 1,509 E4 124
Princeton, mtn., Colo. C4 83
Princeton Junction, Mercer, N.J., 450 C3 106
Princeville, Que., Can., 3,174 C6 73
Princeville, Peoria, Ill., 1,281 C4 90
Princeville, Edgecombe, N.C., 797 B6 109
Prince William, N.B., Can., 269 D2 74
Prince William, co., Va., 50,164 C5 121
Prince William, sound, Alsk. g18 79
Principe, chan., B.C., Can. C3 68
Príncipe, isl., Sao Tome & Principe E1 46
Prineville, Crook, Oreg., 3,263 C6 113
Prineville, res., Oreg. C6 113
Prineville Southeast, Crook, Oreg., 1,299 *C6 113
Pringle, Luzerne, Pa., 1,418 *D10 114
Pringle, Custer, S. Dak., 145 D2 116
Prinsburg, Kandiyohi, Minn., 462 F3 99
Prinzapolca, Nic., 7,200 D6 62
Prinzapolca, riv., Nic. D6 62
Prior Lake, Scott, Minn., 848 F3 99
Priozersk, Sov. Un., 32,800 G14 25
Pripyat (Pripet), riv., Sov. Un. E7 27
Prishib, Sov. Un. B4 41
Pristina, Yugo., 38,891 D5 22
Pritchard, isl., S.C. G7 115
Pritchardville, Beaufort, S.C., 50 G6 115
Pritchett, Baca, Colo., 247 D8 83
Pritchett, Upshur, Tex., 100 C2 118
Pritzerbe, Ger., 1,943 F6 24
Pritzwalk, Ger., 9,416 B6 16
Privas, Fr., 8,663 E6 14
Privert, lake, Que., Can. g8 75
Privolnoye, Sov. Un., 9,000 H9 27
Prizren, Yugo., 28,050 D5 22
Prizzi, It., 9,752 F4 21
Probolinggo, Indon., 68,378 G4 35
Procious, Clay, W. Va., 75 C7 119
Procter, B.C., Can., 213 E9 68
Proctor, Logan, Colo., 35 A8 83
Proctor, Lee, Ky., 150 C6 94
Proctor, St. Louis, Minn., 2,963 D6 99
Proctor, Lake, Mont., 10 C2 102
Proctor, Elko, Nev., 10 C7 104
Proctor, Swain, N.C. D2 109
Proctor, Adair, Okla., 65 B7 112
Proctor, Comanche, Tex., 130 D3 118
Proctor, Rutland, Vt., 1,978 (2,102▲) D2 120
Proctor, Wetzel, W. Va., 130 C1 123
Proctorsville, Windsor, Vt., 476 E3 120
Proctorville, Robeson, N.C., 188 C4 109
Proctorville, Lawrence, Ohio, 831 D5 111
Proença-a-Nova, Port., 6,340 C2 20
Progreso, Hond., 6,921 C4 62
Progreso, Mex., 13,659 C7 63
Progreso, Hidalgo, Tex., 900 *F4 118
Progress, Pike, Miss., 120 D3 100
Progress, Dauphin, Pa., 1,700 *F8 114
Prohner, bay, Ger. D7 24
Project City, Shasta, Calif., 950 B2 82
Prokhladnyy, Sov. Un., 28,000 E2 29
Prokopyevsk, Sov. Un., 292,000 C11 29
Prokuplje, Yugo., 13,690 D5 22
Prole, Warren, Iowa, 580 B7 92
Proletarsk, Sov. Un., 10,000 q21 27
Proletarskaya, Sov. Un., 10,000 H13 27
Prome, Bur., 36,997 E10 39
Promise, Wallowa, Oreg. B9 113
Promise City, Wayne, Iowa, 161 D4 92
Promontory, Box Elder, Utah B3 119
Promontory, pt., Utah B3 119
Promontory Point, Box Elder, Utah B3 119
Pronsfeld, Ger., 740 D6 15
Prophetstown, Whiteside, Ill., 1,802 B4 90
Propriá, Braz., 15,947 D3 57
Prorer, bay, Ger. D7 24
Prorva, Sov. Un. D4 29
Proskurov, see Khmelnitsky, Sov. Un.
Prosna, riv., Pol. C5 26
Prosnica, riv., Pol. A3 26
Prosotsani, Grc., 6,276 B4 23
Prospect, N.S., Can., 171 E6 74
Prospect, New Haven, Conn., 4,367 C5 84
Prospect, Jefferson, Ky., 100 A4 94
Prospect, Marion, Ohio, 1,607 B4 111
Prospect, Jackson, Oreg., 350 E4 113
Prospect, Butler, Pa., 903 E1 114
Prospect, Giles, Tenn., 200 B5 117
Prospect, Prince Edward, Va., 125 D4 121
Prospect, Waukesha, Wis. (part of New Berlin) E1 124
Prospect, hill, Mass. D6 97
Prospect, hill, Mass. D4 97
Prospect, hill, Mass. D2 97
Prospect, hill, Oreg. C1 113
Prospect, mtn., Oreg. C2 113
Prospect Harbor, Hancock, Maine, 350 D4 96
Prospect Heights, Cook, Ill., 2,500 *E2 90
Prospect Hill, Caswell, N.C., 65 A4 109
Prospect Park, Passaic, N.J., 5,201 B4 106

Prospect Park, Delaware, Pa., 6,596 B11 114
Prospect Plains, Middlesex, N.J., 100 C4 106
Prosper, Cass, N. Dak., 35 C9 110
Prosper, Windsor, Vt. D3 120
Prosperity, Newberry, S.C., 757 C4 115
Pros·er, Adams, Nebr., 70 D7 103
Prosser, Benton, Wash., 2,763 C6 122
Prostejov, Czech., 33,500 D4 26
Prostki, Pol B7 26
Protection, Comanche, Kans., 780 E4 93
Protem, Taney, Mo., 33 E5 101
Protivin, Howard, Iowa, 302 A5 92
Proton Station, Ont., Can., 53 C4 72
Prouts Neck, Cumberland, Maine E5 96
Provadiya, Bul., 8,730 D8 22
Provadiya, riv., Bul D8 22
Provencal, Natchitoches, La., 570 C2 95
Provence, former prov., Fr., 1,750,000 F7 14
Providence, Polk, Fla., 100 C4 86
Providence, Webster, Ky., 3,771 C2 94
Providence, Cecil, Md., 85 A6 85
Providence, Providence, R.I., 207,498 (★804,300) B11 84
Providence, Davidson, Tenn., 3,830 *A5 117
Providence, Cache, Utah, 1,189 B4 119
Providence, Adams, Wash C7 122
Providence, co., R.I., 568,778 B10 84
Providence, canyons, Ga D2 87
Providence Bay, Ont., Can., 156 B2 72
Providence Forge, New Kent, Va., 130 D5 121
Providencia, isl., Col D7 62
Provincetown, Barnstable, Mass., 3,389 B7 97
Proving Ground, Carroll, Ill., 185 A3 90
Provins, Fr., 10,310 C5 14
Provo, Sevier, Ark., 160 C1 81
Provo, Fall River, S. Dak., 160 D2 116
Provo, Utah, Utah, 36,047 (★101,000) C4 119
Provo, riv., Utah C4 119
Provolt, Jackson, Oreg E3 113
Provost, Alta., Can., 1,022 C5 69
Prowers, co., Colo., 13,296 D8 83
Proyecto St. Just., P.R B7 65
Prozor, Yugo., 5,506 D3 22
Pruden, Claiborne, Tenn., 60 C10 117
Prudence, isl., R.I C11 84
Prudence Island, Newport, R.I., 60 C11 84
Prudentópolis, Braz., 4,524 D2 56
Prudenville, Roscommon, Mich., 100 D6 98
Prud'homme, Sask., Can., 264 E3 70
Prudnik, Pol., 15,200 C4 26
Prüm, Ger., 4,000 C3 16
Pruszkow, Pol., 37,000 B6, m13 26
Prut, riv., Sov. Un H7 27
Pruzhany, Sov. Un., 10,000 E5 27
Prydz, bay, Ant C21 5
Pryor, Huerfano, Colo., 26 D6 83
Pryor, Big Horn, Mont., 5 E8 102
Pryor (Pryor Creek), Mayes, Okla., 6,476 A6 112
Pryor, mts., Mont E8 102
Pryorsburg, Graves, Ky., 250 B2 94
Przasnysz, Pol., 7,015 B6 26
Przedborz, Pol., 3,503 C5 26
Przemsza, riv., Pol g10 26
Przemysl, Pol., 46,000 D7 26
Przeworsk, Pol., 8,569 C7 26
Przewoz, Pol B9 17
Przhevalsk, Sov. Un., 30,000 G24 9
Psakhna, Grc., 4,309 C4 23
Psara, isl., Grc C5 23
Psary, Pol g10 26
Psel, riv., Sov. Un G9 27
Pskov, Sov. Un., 93,000 C7 27
Pskov, lake, Sov. Un B6 27
Pszczyna, Pol., 12,800 D5, h9 26
Pszczynka, riv., Pol h9 26
Ptarmigan, mtn., Wyo A3 125
Ptich, riv., Sov. Un E7 27
Ptolemais, Grc., 12,747 B3 23
Ptuj, Yugo., 7,365 B2 22
Pu, China G6 36
Puako, Hawaii, Haw D6 88
Puán, Arg., 3,191 B4 54
Puapua, W. Sam., 236 52
Pubnico, N.S., Can., 337 F4 74
Pucallpa, Peru, 2,368 C3 58
Pucheng, China, 20,000 K8 36
Puck, Pol., 3,946 A5 26
Puckaway, lake, Wis E4 124
Puckett, Rankin, Miss., 302 C4 100
Pudasjärvi, Fin E12 25
Pudozh, Sov., Un., 2,000 A11 27
Puebla, Mex., 287,952 D5, n14 63
Puebla, state, Mex., 1,957,380 D5, n14 63
Puebla de Don Fadrique, Sp., 7,142 D4 20
Puebla del Caramiñal, Sp., 8,337 A1 20
Pueblito de Ponce, P.R B3 65
Pueblo, Pueblo, Colo., 91,181 (★111,000) C6 83
Pueblo, co., Colo, 118,707 C6 83
Pueblo, mts., Oreg E8 113
Pueblo Colorado, wash, Ariz B6 80
Pueblo Hundido, Chile E2 55
Pueblo Nuevo, Pan., 1,447 m11 62
Pueblo Nuevo, P.R B3 65
Pueblo Nuevo, Ven., 2,837 A4 60
Puebloviejo, Ec., 1,206 B2 58
Puenteareas, Sp., 14,634 A1 20
Puente Ceso, Sp., 9,780 A1 20
Puentedeume, Sp., 7,755 A2 20
Puente-Genil, Sp., 25,000 D3 20
Puerco, riv., Ariz B6 80
Puertecito, Socorro, N. Mex C2 107
Puerto Acosta, Bol., 1,302 C2 55
Puerto Asién, Chile, 3,767 D2 54
Puerto Alegre, Bol B3 55
Puerto Alvaro Obregón, Mex., 8,320 D6 63
Puerto Armuelles, Pan., 10,712 F6 62
Puerto Arroyo Verde, Arg C3 54
Puerto Asís, Col C2 60
Puerto Ayacucho, Ven., 5,418 B4 60
Puerto Baquerizo, Ec g6 58
Puerto Barrios, Guat., 15,155 C3 62
Puerto Belgrano, Arg B4 54
Puerto Bermúdez, Peru D3 58
Puerto Berrío, Col., 8,947 B3 60
Puerto Bolívar, Ec., 2,000 B2 58
Puerto Cabello, Ven., 50,973 A4 60
Puerto Cabezas, Nic., 6,000 C6 62
Puerto Carreño, Col., 540 B4 60
Puerto Casado, Par., 6,708 D4 55
Puerto Chicama, Peru, 2,274 C2 58
Puerto Colombia, Col., 5,689 A3 60
Puerto Constanza, Arg f7 54
Puerto Cortés, Hond., 17,412 C4 62
Puerto Cumarebo, Ven., 7,951 A4 60
Puerto de Cabras, Can. Is., 1,459 (4,029▲) D2 44
Puerto de la Cruz, Can. Is., 5,855 m13 20
Puerto de la Paloma, Ur E2 56
Puerto del Son, Sp., 2,221 A1 20
Puerto de Luna, Guadalupe, N. Mex., 70 C5 107
Puerto Deseado, Arg., 3,392 D3 54
Puerto Eten, Peru, 2,576 C2 58
Puerto Guaraní, Par., 2,758 D4 55
Puerto Heath, Bol B2 55
Puerto Iguazú, Arg E5 55
Puerto Jobos, P.R., 699 D6 65
Puerto la Cruz, Ven., 54,694 A5 60
Puerto Leguízamo, Col., 1,433 D3 60
Puertollano, Sp., 53,136 C3 20
Puerto Madryn, Arg., 3,441 C3 54
Puerto Maldonado, Peru, 1,285 D4 58
Puerto Manatí, Cuba E5 64
Puerto Mineral, Arg E4 55
Puerto Montt, Chile, 41,000 C2 54
Puerto Morazán, Nic D4 62
Puerto Natales, Chile, 8,140 E2, h11 54
Puerto Ordoz, Ven B5 60
Puerto Padre, Cuba, 6,949 E5 64
Puerto Páez, Ven., 677 B4 60
Puerto Peñasco, Mex., 2,517 A2 63
Puerto Pinasco, Par., 7,576 D4 55
Puerto Piramides, Arg C4 54
Puerto Plata, Dom. Rep., 19,073 F8 64
Puerto Princesa, Phil., 3,326 D5 35
Puerto Real, P.R. C2 65
Puerto Real, Sp., 10,033 D2 20
Puerto Rico, U.S. dep., N.A., 2,349,544 65
Puerto Sastre, Par., 5,058 D4 55
Puerto Siles, Bol., 357 B2 55
Puerto Suarez, Bol., 1,159 C4 55
Puerto Sucre, Bol., 1,470 B2 55
Puerto Supe, Peru, 2,180 D2 58
Puerto Tejada, Col., 8,535 C2 60
Puerto Vallarta, Mex., 7,397 C3, m11 63
Puerto Varas, Chile, 5,797 C2 54
Puerto Victoria, Peru C3 58
Puerto Viejo, C.R F6 62
Puerto Villamizar, Col B3 60
Puerto Visser, Arg D3 54
Puerto Wilches, Col., 3,451 B3 60
Pugachev, Sov. Un., 33,600 C3 29
Pugal, India, 5,000 C4 40
Puget, sound, Wash B3 122
Puget Island, Wahkiakum, Wash C2 122
Puget-Théniers, Fr., 919 F7 14
Pughtown, Hancock, W. Va., 400 A2 123
Puglia, pol. dist., It., 3,421,217 *D6 21
Pugwash, N.S., Can., 815 D6 74
Puhi, Kauai, Haw, 650 B2 88
Puhsi, China, 5,000 B2 37
Puigcerdá, Sp., 4,276 A6 20
Pujehun, S.L E2 45
Pujilí, Ec., 2,149 B2 58
Pukalani, Maui, Haw., 600 C5 88
Pukchong, Kor., 30,709 F4 37
Pukë, Alb., 976 A2 23
Pukeamaru, mtn., N.Z L17 51
Pukeashun, mtn., B.C., Can D8 68
Pukekohe, N.Z., 5,798 L15 51
Pukou, China H8 36
Pukwana, Brule, S. Dak., 247 D6 116
Pula, It., 3,642 E2 21
Pula, Yugo., 36,838 C1 22
Pulacayo, Bol., 7,984 D2 55
Pulantien, China, 12,000 D9 34
Pulaski, Candler, Ga., 155 D5 87
Pulaski, Pulaski, Ill., 415 F4 90
Pulaski, Davis, Iowa, 299 D5 92
Pulaski, Scott, Miss., 200 C4 100
Pulaski, Oswego, N.Y., 2,256 B4 108
Pulaski, Giles, Tenn., 6,616 B4 117
Pulaski, Pulaski, Va., 10,469 D2 121
Pulaski, Brown, Wis., 1,540 D5 124
Pulaski, co., Ark., 242,980 C3 81
Pulaski, co., Ga., 8,204 D3 87
Pulaski, co., Ill., 10,490 F4 90
Pulaski, co., Ind., 12,837 B4 91
Pulaski, co., Ky., 34,403 C5 94
Pulaski, co., Mo., 46,567 D5 101
Pulaski, co., Va., 27,258 D2 121
Pulawy, Po., 11,900 C6 26
Pul-i-Khumri, Afg., 12,246 D14 41
Pull, pt., Vir. Is h17 65
Pullman, Allegan, Mich., 200 F4 98
Pullman, Whitman, Wash., 12,957 C8 122
Pullman, Ritchie, W. Va., 162 B4 123
Pulog, mtn., Phil n13 35
Pulozero, Sov. Un C15 25
Pulpit Rocks, cavern, N.H E3 105
Puluntai, China A9 39, D3 34
Puluntohai, see Bulun Tukhoi, China
Pultusk, Pol., 8,787 B6 26
Pumphrey, Anne Arundel, Md., 700 C2 85
Pumpkin, buttes, Wyo B7 125
Pumpkin, creek, Mont E11 102
Pumpkin, creek, Nebr C2 103
Pumpville, Val Verde, Tex., 40 E2 118
Puna, Bol., 852 C2 55
Puná, isl., Ec B1 58
Punakha (Punaka), Bhu D12 40
Punat, Yugo., 1,900 C2 22
Punata, Bol., 5,014 C2 55
Punchaw, B.C., Can C6 68
Pungo, Beaufort, N.C., 160 B7 109
Pungo, Princess Anne, Va., 300 B7 121
Pungsan, Kor., 17,453 F4 37
Pungsan, Kor., 4,300 H4 37
Punjab, reg., India, Pak B5 39
Punjab, state, India, 20,306,812 B6 39
Punnichy, Sask., Can., 408 F3 70
Puno, Peru, 18,852 E3 58
Puno, dept., Peru, 646,385 E3 58
Punta, Torrance, N. Mex C3 107
Punta, mtn., P.R C4 65
Punta Alta, Arg., 19,852 B4 54
Punta Arenas, Chile, 48,000 h11 54
Punta de Díaz, Chile E1 55
Punta de Piedras, Ven., 2,257 A5 60
Punta de Vacas, Arg A3 54
Punta Gorda, Br. Hond., 1,758 B3 62
Punta Gorda, Charlotte, Fla., 3,157 F4 86
Punta Moreno, Peru C2 60
Punta Rassa, Lee, Fla., 20 F5 86
Puntarenas, C.R., 19,500 F5 62
Puntarenas, prov., C.R., 148,100 *F5 62
Punta Fijo, Ven., 15,441 A3 60
Puntzi, lake, B.C., Can C5 68
Punxsutawney, Jefferson, Pa., 8,805 E4 114
Puposky, Beltrami, Minn., 40 C4 99
Puquio, Peru, 6,183 D3 58
Púquios, Chile E2 55
Pur, riv., Sov. Un B24 9
Puranpur, India, 11,280 C8 40
Purcell, Jasper, Mo., 265 D3 101
Purcell, McClain, Okla., 3,729 B4 112
Purcell, mts., Mont B1 102
Purcell, range, B.C., Can E9 68
Purcellville, Loudoun, Va., 1,419 B5 121
Purdham, hill, Ark D6 81
Purdin, Linn., Mo., 207 B4 101
Purdum, Blaine, Nebr., 25 B5 103
Purdy, Barry, Mo., 467 E4 101
Purdy, Greensville, Va., 120 E5 121
Purépero, Mex., 10,092 n12 63
Purgatoire, riv., Colo D7 83
Purgatory, peak, Colo D5 83
Puri, India, 60,815 H10 40
Puritan, Gogebic, Mich., 150 A5 98
Purmerend, Neth., 10,800 B4 15
Purna, riv., India H6 40
Purnea, India, 40,602 E11 40
Purple Springs, Alta., Can., 54 E5 69
Purpula, mtn., Sov. Un D14 28
Pursat, Camb., 25,000 F5 38
Pursglove, Monongalia, W. Va., 600 B4 123
Purulia, India, 48,134 F11 40
Purús, riv., Braz., Peru D3 56, D2 59
Purvis, Lamar, Miss., 1,614 D4 100
Purvis, Nanesmond, Va., 60 B6 121
Puryear, Henry, Tenn., 408 A3 117
Puryong, Kor E4 37
Pusan, Kor., 1,162,600 I4 37
Pushaw, lake, Maine D4 96
Pushkin, Sov. Un., 51,000 B8 27
Pushkino, Sov. Un., 10,000 C11, m17 27
Pushmataha, Choctaw, Ala., 200 C1 78
Pushmataha, co., Okla., 9,088 C6 112
Pushthrough, Newf., Can., 247 E3 75
Puspokladany, Hung., 15,735 B5 22
Pustelnik, Pol., 5,000 k14 26
Pustunich, Mex., 343 D6 63
Pustynnove, Sov. Un B8 29
Putah, creek, Calif A5 82
Putao, Bur., 10,000 C10 39
Puteaux, Fr., 39,640 g9 14
Putien, China, 14,000 F8 34
Putilovo, Sov. Un s32 25
Puting, cape, Indon F4 35
Putivl, Sov. Un., 10,000 F9 27
Putlitz, Ger., 2,763 E6 24
Putnam, Marengo, Ala., 175 C1 78
Putnam, Windham, Conn., 6,952 (8,412▲) B9 84
Putnam, Dewey, Okla., 88 B3 112
Putnam, Callahan, Tex., 203 C3 118
Putnam, co., Fla., 32,212 C5 86
Putnam, co., Ga., 7,798 C3 87
Putnam, co., Ill., 4,570 B4 90
Putnam, co., Ind., 24,927 E4 91
Putnam, co., Mo., 6,999 A4 101
Putnam, co., N.Y., 31,722 D7 108
Putnam, co., Ohio, 28,331 B3 111
Putnam, co., Tenn., 29,236 C8 117
Putnam, co., W. Va., 23,561 C3 123
Putnam Station, Washington, N.Y., 10 B7 108
Putnamville, Putnam, Ind., 140 E4 91
Putney, Dougherty, Ga., 300 E2 87
Putney, Brown, S. Dak., 15 B7 116
Putney, Windham, Vt., 250 (1,177▲) F3 120
Putney, Kanawha, W. Va., 40 C3, C6 123
Putre, Chile C2 55
Puttalam, Cey., 10,162 G6 39
Putte, Neth., 2,000 C4 15
Püttlingen, Ger., 14,200 D1 17
Putumayo, comisaría, Col., 35,000 C2 60
Putumayo, riv., Col D3 60
Puturge, Tur., 2,383 C12 31
Puukolii, Maui, Haw., 500 C5 88
Puu Konahuanui, mtn., Haw g10 88
Puulavesi, lake, Fin G12 25
Puunene, Maui, Haw., 3,054 C5 88
Puurs, Bel., 6,042 C4 15
Puu Waawaa, peak, Haw D6 88
Puxico, Stoddard, Mo., 743 E7 101
Puyallup, Pierce, Wash., 12,063 B3, E2 122
Puyallup, riv., Wash C3 122
Puyang, China G6 36
Puy-de-Dôme, dept., Fr., 508,928 *E5 14
Puy de Dome, mtn., Fr E5 14
Puy de Sancy, mtn., Fr E5 14
Puyehue, Chile C2 54
Puyo, Ec., 1,092 B2 58
Puzim, cape, Iran I10 41
Pweto, Con. L C4 48
Pwllheli, Wales, 3,642 B3 12
Pyapon, Bur., 19,174 E10 39
Pyasina, riv., Sov. Un B12 28
Pyatigorsk, Sov. Un., 73,000 E2 29
Pyatt, Marion, Ark., 144 A3 81
Pyatte, Avery, N.C A2, C4 109
Pyavzero, lake, Sov. Un D14 25
Pyhäjärvi, lake, Fin G10 25
Pyinbugyi, Bur F3 38
Pyinmana, Bur., 22,066 E10 39
Pyland, Chickashaw, Miss B4 100
Pyles, fork, W. Va A6 123
Pymatuning, creek, Ohio A7 111
Pymatuning, res., Pa C1 114
Pyongchang, Kor., 4,300 H4 37
Pyongtaek, Kor., 15,200 H3 37
Pyŏngyang, Kor., 653,100 G2 37
Pyote, Ward, Tex., 420 D1 118
Pyramid, lake, Nev C2 104
Pyramid, mtn., B.C., Can E7 66
Pyramid, mts., N. Mex E1 107
Pyramid, peak, Calif C3 82
Pyramid, peak, Wyo B2 125
Pyramid Lake, Indian res., Nev C2 104
Pyramids, ruins, Eg., U.A.R. D6 43
Pyrenees, mts., France, Sp A5 20, G8 8
Pyrénées-Orientales, dept., Fr., 251,231 *F5 14
Pyrites, St. Lawrence, N.Y., 150 B2 108
Pyrmont, Carroll, Ind., 70 D4 91
Pyrzyce, Pol., 4,179 B3 26
Pysht, Clallam, Wash., 50 A1 122
Pyskowice, Pol., 21,000 g9 26
Pytalovo, Sov. Un., 3,000 C6 27
Pyu, Bur., 10,443 E10 39

Q

Qabātiyah, Jordan, 3,000 B7 32
Qadīmah, Sau. Ar A4 47
Qais, isl., Iran H6 41
Qala Bist., Afg., 5,000 F12 41
Qala-i-Ghor (Taiwara), Afg., 5,000 E12 41
Qalamshāh, Eg., U.A.R H8 31
Qala Nau, Afg., 5,000 D11 41
Qala Salih, Iraq, 4,022 F3 41
Qala Shaharak, Afg., 5,000 D12 41
Qala Sharqat, Iraq E14 31
Qala Sikar, Iraq, 4,913 F3 41
Qal'at al Mu'azzam, Sau. Ar I11 31
Qaliya, cape, Kuw G4 41
Qalqīlah, Jordan, 8,000 B6, g10 32
Qalyūb, Eg., U.A.R., 35,800 D3 32
Qamīnis, Libya, 1,100 C4 43
Qana, Leb., 5,000 A7 32
Qārah, Eg., U.A.R H6 31
Qareh Dāgh, mtn., Iran B2 41
Qaru, isl., Arabia G4 41
Qārūm, lake, Eg., U.A.R. D5 43
Qaryat al 'Ulyā, Sau. Ar H3 41
Qaryat el Inab, Isr., 662 h11 32
Qaşr al Kharānah, Jordan G11 31
Qaşr Bāni Walīd, Libya, 2,520 C2 43
Qatar, country, Asia, 20,000 I5 41
Qatia, Eg., U.A.R., 1,000 D4 32
Qattah, Eg., U.A.R D2 32
Qattara (Qaţţārah), depression, Eg., U.A.R. D5 43
Qaţţarah, see Qattara, depression, Eg., U.A.R.
Qāyen, Iran, 10,000 E9 41
Qazvīn, Iran, 66,420 C4 41
Qeshm, Iran, 15,000 H8 41
Qeshm, isl., Iran H7 41
Qila Saifullah, Pak B2 40
Qina, Eg., U.A.R., 47,700 D6 43
Qiryat Anavim, Isr., 286 C7, h11 32
Qiryat Hayim, Isr., (part of Haifa) B7 32
Qiryat Ono, Isr., 8,137 g10 32
Qiumbele, Ang C2 48
Qizil Unzun, riv., Iran C4 41
Qom, Iran, 96,499 D5 41
Quabbin, res., Mass B3 97
Quaco, head, N.B., Can D4 74
Quaddick, res., Conn B9 84
Quakenbrück, Ger., 7,800 B3 16
Quaker City, Guernsey, Ohio, 583 C6 111
Quaker Hill, New London, Conn., 1,671 D8 84
Quakertown, Hunterdon, N.J., 150 B3 106
Quakertown, Bucks, Pa., 6,305 F11 114
Qualicum Beach, B.C., Can., 759 E5 68
Quality, Butler, Ky., 75 C3 94
Qualls, Cherokee, Okla B6 112
Quamba, Kanabec, Minn., 95 E5 99
Quanah, Hardeman, Tex., 4,564 B3 118
Quang Ngai, Viet., 5,000 E8 38
Quang Tri, Viet., 13,425 D7 38
Quannapowite, lake, Mass C3 97
Quantico, Wicomico, Md., 300 D6 85
Quantico, Prince William, Va., 1,015 C5 121
Quapaw, Ottawa, Okla., 850 A7 112
Qu'Appelle, Sask., Can., 565 G4 70
Qu'Appelle, riv., Sask., Can G4 70
Quaraí, Braz., 10,575 E1 56
Quarai, Torrance, N. Mex C3 107
Quarai, riv., Braz E1 56
Quari, riv., Braz C2 59
Quarry, Manitowoc, Wis., 35 B6 124
Quarryville, N.B., Can., 256 C4 74
Quarryville, Lancaster, Pa., 1,427 G9 114
Quartu Sant'Elena, It., 22,916 E2 21
Quartz, mtn., Oreg D4 113
Quartz, peak, Nev G6 104
Quartz Hill, Los Angeles, Calif., 3,325 *E4 82
Quartz Mountain, Lake, Oreg., 5 E6 113
Quartzsite, Yuma, Ariz., 350 D1 80
Quasqueton, Buchanan, Iowa, 373 B6 92
Quassapaug, pond, Conn C4 84
Quatsino, sound, B.C., Can D3 68
Quay, Quay, N. Mex., 10 C6 107
Quay, Pawnee and Payne, Okla., 51 A5 112
Quay, co., N. Mex., 12,279 C6 107
Qūchān, Iran, 21,250 C9 41
Quealy, Sweetwater, Wyo., 42 D3 125
Quebec, Quebec, Can., 171,979 (★357,568) C6, C9 73
Quebec, co., Que., Can., 331,307 B5 73
Quebec, prov., Can., 5,259,211 F19 67, 73
Quebec-Ouest, Que., Can., 8,733 *C9 73
Quebeck, White, Tenn., 250 D8 117
Quebradillas, P.R., 2,131 B3 65
Quebradillas, mun., P.R., 13,075 B3 65
Quechee, Windsor, Vt., 300 D4 120
Quecreek, Somerset, Pa., 800 F3 113
Quedlinburg, Ger., 30,700 B6 17
Queen, Eddy, N. Mex E5 107
Queen, cape, N.W. Ter., Can D17 67
Queen Anne, Talbot, Md., 283 C6 85
Queen Annes, co., Md., 16,569 B5 85
Queen Bess, mtn., B.C., Can D5 68
Queen Charlotte, B.C., Can., 283 C1 68
Queen Charlotte, is., B.C., Can C1 68
Queen Charlotte, mts., B.C., Can C1 68
Queen Charlotte, sound, B.C., Can F7 66
Queen Charlotte, strait, B.C., Can D4 68
Queen City, Schuyler, Mo., 599 A5 101
Queen City, Cass, Tex., 1,081 *C6 118
Queen Creek, Maricopa, Ariz., 550 G3 80
Queen Fabiola, mts., Ant B12 5
Queen Mary, coast, Ant C22 5
Queen Maud, gulf, N.W. Ter., Can C12 66
Queen Maud, range, Ant A30 5
Queen Maud Land, reg., Ant B13 5
Queens, borough and co., N.Y., 1,809,578 (part of New York City) D2 108
Queens, co., N.B., Can., 11,640 D4 74
Queens, co., N.S., Can., 13,155 E4 74
Queens, co., P.E.I., Can., 45,842 C6 74
Queens, chan., Austl B4 50
Queens, sound, B.C., Can D3 68
Queensborough, Ont., Can., 117 C7 72
Queen Shoals, Kanawha, W. Va., 65 C6 123
Queensland, Ben Hill, Ga., 60 E3 87
Queensland, state, Austl., 1,518,828 D7 50
Queensport, N.S., Can., 199 D8 74
Queenstown, Austl., 4,601 o15 50
Queenstown, Br., Gu., 1,067 A3 59
Queenstown, Alta., Can., 67 D4 69
Queenstown, N.B., Can., 138 D3 74
Queenstown, Queen Annes, Md., 355 C5 85
Queenstown, N.Z., 1,321 P12 51
Queenstown, S. Afr., 33,182 D4 49
Queen Victoria, sea, Arc. O A11 4
Queguay, Ur E1 56
Queguay Grande, riv., Ur E1 56
Queimadas, Braz., 3,553 D3 57
Queixadas, Braz., 745 E2 57
Quela, Ang C2 48
Quelimane, Moz., 8,000 A6 49
Quemado, Catron, N. Mex., 150 C1 107
Quemado, Maverick, Tex., 300 E2 118
Quemado de Güines, Cuba, 3,276 D3 64
Quemoy (Chinmen), China, 6,000 *G8 34
Quemú Quemú, Arg B4 54
Quenemo, Osage, Kans., 434 D8 93
Quentin, Franklin, Miss., 300 D3 100
Que Que, S. Rh., 10,000 (★22,800) A4 49
Quequén, Arg., 4,760 B5 54
Querétaro, Mex., 67,277 C4, m13 63
Querétaro, state, Mex., 353,154 C4, m13 63
Querfurt, Ger., 7,976 B6 17
Quesada, Sp., 7,609 D4 20
Quesnel, B.C., Can., 4,673 C6 68
Quesnel, lake, B.C., Can C7 68
Quesnel, riv., B.C., Can C6 68
Questa, Taos, N. Mex., 900 A4 107
Quetena, Bol., 183 D2 55
Quetta, Pak., 75,000 (★106,633) B4 39, B1 40
Quezaltenango, Guat., 27,672 C2 62
Quezaltepeque, Sal., 8,471 D3 62
Quezon, prov., Phil., 656,900 *C6 35
Quezon City, Phil., 397,400 C6, o13 35
Quiani, riv., Braz D5 60
Quibala, Ang., 263 D1 48
Quibell, Ont., Can., 61 E5 71
Quibdó, Col., 9,700 B2 60
Quiberon, Fr., 4,103 D2 14
Quiberon, pen., Fr D2 14
Quicksburg, Shenandoah, Va., 150 C4 121
Quidnessett, Washington, R.I., 1,000 C11 84
Quidnick, Kent, R.I., 2,000 C10 84
Quidnick, res., R.I C10 84
Quietus, Big Horn, Mont., 5 E10 102
Quiindy, Par., 2,150 E4 55
Quijotoa, Pima, Ariz., 50 E3 80
Quilá, Mex., 1,290 C3 63
Quilan, cape, Chile C2 54
Quilcene, Jefferson, Wash., 600 B3 122
Quilengues, Ang., 472 D1 48
Quill, lakes, Sask., Can F3 70
Quillaga, Chile D2 55
Quillan, Fr., 4,424 F5 14
Quill Lake, Sask., Can., 529 E3 70
Quillota, Chile, 27,000 A2 54
Quilmes, Arg., 310,000 g7 54
Quilon, India, 91,018 G6 39
Quilpie, Austl., 640 C5 51
Quilpué, Chile, 26,000 A2 64
Quilty, Ire., 200 E2 11
Quimby, Cherokee, Iowa, 369 B2 92
Quimby, Arrostook, Maine, 200 B4 96
Quimilí, Arg., 3,686 E3 51
Quimper, Fr., 45,989 C2 14
Quimperlé, Fr., 10,272 D2 14
Quinaby, Marion, Oreg., 50 B4, C1 113
Quinapoxet, Worcester, Mass B4 97
Quinault, Grays Harbor, Wash., 300 B2 122
Quinault, Indian res., Wash B1 122
Quinault, lake, Wash B2 122
Quinault, riv., Wash B1 122
Quincy, Plumas, Calif., 1,700 C3 82
Quincy, Gadsden, Fla., 8,874 B2 86
Quincy, Adams, Ill., 43,793 D2 90
Quincy, Owen, Ind., 150 F4 91
Quincy, Greenwood, Kans., 90 E7 93
Quincy, Lewis, Ky., 300 B6 94
Quincy, Norfolk, Mass., 87,409 B3, D3 97
Quincy, Branch, Mich., 1,602 G6 98
Quincy, Monroe, Miss., 125 B5 100
Quincy, Hickory, Mo., 90 C4 101
Quincy, Grafton, N.H., 85 C3 105
Quincy, Logan, Ohio, 668 B4 111
Quincy, Columbia, Oreg., 200 A3 113
Quincy, Franklin, Pa., 400 G6 114
Quincy, Grant, Wash., 3,269 B6 122
Quincy, bay, Mass D3 97
Quinebaug, Windham, Conn., 350 A9 84
Quinebaug, riv., Conn C9 84
Quines, Arg., 3,038 A3 54
Quinhagak, Alsk, 228 D7 79
Qui Nhon, Viet., 10,000 F8 38
Quinlan, Woodward, Okla., 75 A2 112
Quinlan, Hunt, Tex., 621 C4 118
Quinn, Pennington, S. Dak., 162 D3 116
Quinn, riv., Nev B4 104
Quinn Canon, mts., Nev F6 104
Quinnesec, Dickinson, Mich., 400 C3 98
Quinnipiac, riv., Conn D5 84
Quinnville, Providence, R.I., 400 B11 84
Quintana de la Serena, Sp., 7,861 C3 20
Quintanar, Sp., 9,483 C4 20
Quintana Roo, ter., Mex., 52,312 D7 63
Quinter, Gove, Kans., 776 C3 93
Quintero, Chile, 5,563 A2 54
Quinton, Sask., Can., 195 F3 70
Quinton, Pulaski, Ky., 200 D5 94
Quinton, Salem, N.J., 600 D2 106
Quinton, Pittsburg, Okla., 898 B6 112
Quinwood, Greenbrier, W. Va., 506 C4 123
Quipapá, Braz., 3,421 C3, k6 57
Quipungo, Ang D1 48
Quirauk, mtn., Md A2 85
Quirindi, Austl., 2,790 E8 51
Quiriquire, Ven., 7,520 B5 60
Quirke, lake, Ont., Can B8 98
Quiroga, Sp., 8,380 A2 20
Quirpon, isl., Newf., Can C4 75
Quissanga, Moz D4 48
Quitaque, Briscoe, Tex., 586 B2 118
Quita Sueño Bank, shoals, Caribbean Sea C7 62
Quitman, Cleburne, Ark., 305 B3 81
Quitman, Brooks, Ga., 5,071 F3 87
Quitman, Jackson, La., 185 B3 95
Quitman, Clarke, Miss., 2,030 C5 100
Quitman, Nodaway, Mo., 113 A2 101
Quitman, Wood, Tex., 1,237 C5 118
Quitman, co., Ga., 2,432 E1 87
Quitman, co., Miss., 21,019 A3 100
Quitman, mts., Tex F1 118
Quito, Ec., 209,932 B2 58
Quitsna, Bertie, N.C B6 109
Quixadá, Braz., 8,747 B3 57
Quixeramobim, Braz., 6,384 C3 57
Qulin, Butler, Mo., 587 E7 101
Qum, riv., Iran E5 41
Quoi, isl., Truk 52
Quonochontaug, Washington, R.I., 50 D10 84
Quonset Point, Washington, R.I C11 84
Quorn, Austl., 566 F2 51
Quoyness, Scot B5 13
Qūs, Eg., U.A.R., 23,600 D6 43
Qusrah, Jordan, 2,000 g12 32

R

Raab, riv., Aus E7 16
Raabs, Aus., 1,132 D7 16
Raahe, Fin., 4,900 E11 25
Raalte, Neth., 3,301 B6 15
Raanana, Isr., 10,656 g10 32
Raasay, isl., Scot C3 13
Rab, isl., Yugo C2 22
Raba, Indon, 6,781 G5 35
Rába (Raab), riv., Hung B3 22
Rabastens, Fr., 2,491 F4 14
Rabat, Mor., 227,445 (★310,000) G3 30, C3 44
Rabaul, Bis. Arch., 2,950 G9 7
Rabbit, creek, S. Dak B3 116
Rabbit Ear, pass, Colo A4 83
Rabbit Hash, Boone, Ky., 50 A6 94
Rabbit Lake, Sask., Can., 196 D2 70
Rabch, riv., Iran D9 41
Rabun, Baldwin, Ala., 300 D2 78
Rabun, co., Ga., 7,456 B3 87
Rabun, gap, Ga B3 87
Rabun Bald, mtn., Ga B3 87
Råby, Swe u34 25
Rača Kragujevačka, Yugo., 1,000 C5 22
Raccoon, creek, Ind E4 91
Raccoon, creek, Ohio D5 111
Raccoon, mtn., Ala B3 78
Raccoon, pt., La E5 95
Raccoon, riv., Iowa C3 92
Raccourci, isl., La D4 95

Race, cape, Newf., Can....E5 75
Race, pt., Mass....B7 97
Race, pt., N.Y.....D8 84
Race, strait, N.Y.....E8 84
Raceland, Greenup, Ky., 1,115....B7 94
Raceland, Lafourche, La., 3,666....C6, E5 95
Racepond, Charlton, Ga., 250....F4 87
Rachaya, Leb....F10 31
Rach, Gia, Viet., 24,000....G6 38
Raciborz, Pol., 32,000....C5 26
Racine, Mower, Minn., 180....G6 99
Racine, Newton, Mo., 150....E3 101
Racine, Meigs, Ohio, 499....D6 111
Racine, Boone, W. Va., 600....C3, D6 123
Racine, Racine, Wis., 89,144 (*113,500)....F2, F6 124
Racine, co., Wis., 141,781....F5 124
Racineves, Czech., 626....n17 26
Rackwick, Scot....B5 13
Raco, Chippewa, Mich., 100....B6 98
Rădăuti, Rom., 15,949....B7 22
Radcliff, Hardin, Ky., 3,384....C4 94
Radcliff, Vinton, Ohio, 200....C5 111
Radcliffe, Hardin, Iowa, 615....B4 92
Råde, Nor....p28 25
Radeberg, Ger., 16,300....B8 17
Radebeul, Ger., 40,300....B8 17
Radersburg, Broadwater, Mont., 100....D5 102
Radford (Independent City), Va., 9,371....D2 121
Radiant Valley, Prince Georges, Md., 1,500....*C4 85
Radisson, Sask., Can., 515....E2 70
Radisson, Sawyer, Wis., 179....C2 124
Radium, Stafford, Kans., 64....D5 93
Radium, Marshall, Minn., 40....B2 99
Radium Hot Springs, B.C., Can., 306....D9 68
Radium Springs, Dona Ana, N. Mex., 30....E3 107
Radley, Crawford, Kans., 235....E9 93
Radnice, Czech., 2,067....D8 17
Radnor, Delaware, Pa., 1,000....*A10 114
Radnor, co., Wales, 18,431....B4 12
Radnor, forest, Wales....D5 12
Radom, Pol., 130,000....C6 26
Radomir, Bul., 5,778....D6 22
Radomsko, Pol., 27,000....C5 26
Radomyshl, Sov. Un., 25,000....F7 27
Radoviš, Yugo., 6,195....E6 22
Radstadt, Aus., 3,311....E6 16
Radville, Sask., Can., 1,067....H3 70
Radway, Alta., Can., 183....B4 69
Radzionkow, Pol., 24,000....g9 26
Radzymin, Pol., 4,356....B6, k14 26
Radzyn., Pol., 4,694....C7 26
Rae, N.W. Ter., Can., 522....D9 66
Rae, strait, N.W. Ter., Can....C13 66
Rae Bareli, India, 29,940....D8 40
Raeford, Hoke, N.C., 3,058....C4 109
Raesfeld, Ger., 3,600....B1 17
Raeville, Boone, Nebr., 80....C7 103
Rafaela, Arg., 23,665....A4 54
Rafah, Gaza Area, 2,000....C6 32
Rafai, Cen. Afr. Rep....D4 46
Raft, riv., Idaho....G5 89
Raft River, mts., Utah....B2 119
Rag, mtn., Tenn....D10 117
Raga, Sud....D2 47
Ragan, Harlan, Nebr., 90....D6 103
Raga Tsangpo, riv., China....C11 40
Ragay, Phil., 1,956....p14 35
Ragay, gulf, Phil....p14 35
Ragged, isl., Maine....E4 96
Ragged, lake, Maine....C3 96
Ragged, pt., Md....C5 85
Ragged Top, mtn., Wyo....D7 125
Raggon, isl., S.C....G1 115
Ragland, St. Clair, Ala., 1,166....B3 78
Ragland, Quay, N. Mex....C6 107
Raglesville, Daviess, Ind., 45....G4 91
Ragley, Beauregard, La., 25....D2 95
Rago, Kingman, Kans., 60....E5 93
Ragsdale, Knox, Ind., 210....G3 91
Ragunda, Swe., 202....F7 25
Ragusa, It., 47,000 (57,311▲)....F5 21
Rahab el Berdi, Sud....C1 47
Rahden, Ger., 3,500....F2 24
Rahimyar, Khan, Pak., 14,919....C3 40
Rahway, Union, N.J., 27,699....B4, E4 106
Rahway, riv., N.J....E4 106
Raichur, India, 63,329....E6 39
Raiford, Union, Fla., 300....B4 86
Raigarh, India, 36,933....G9 40
Railroad, val., Nev....E6 104
Rainbow, Alsk., 20....g17 79
Rainbow, Hartford, Conn....B6 84
Rainbow, Cascade, Mont., 50....C5 102
Rainbow, falls, Tenn....D10 117
Rainbow, lake, Maine....C3 96
Rainbow, pt., Fla....D4 86
Rainbow, res., Wis....C4 124
Rainbow Bridge, nat. mon., Utah....F5 119
Rainbow City, Etowah, Ala., 1,625....*A3 78
Rainbow Springs, Macon, N.C....D2 109
Rainelle, Greenbrier, W. Va., 649....D4 123
Raines, Crisp, Ga., 35....E3 87
Raines, Shelby, Tenn., (part of Whitehaven)....B1, E8 117
Rainier, Columbia, Oreg., 1,152....A4 113
Rainier, Thurston, Wash., 245....C3 122
Rainier, mtn., Wash....C4 122
Rains, co., Tex., 2,993....C5 118
Rainsboro, Highland, Ohio, 190....C4 111
Rainsville, DeKalb, Ala., 398....A4 78
Rainy, lake, Ont., Can., Minn....B5 99
Rainy, mtn., Okla....C3 112
Rainy, riv., Ont., Can., Minn....B4 99
Rainy River, Ont., Can., 1,168....*E7 72
Rainy River, co., Ont., Can., 26,531....E7 72
Raipur, India, 139,792....D7 39, G8 40
Ra'is, Sau. Ar....E7 43
Raisin, Victoria, Tex., 50....E4 118
Raivavae, isl., Fr. Polynesia....*H13 7
Raja, cape, Indon....K2 38
Rajahmundry, India, 130,002....E7 39, I8 40
Rajapalaiyam, India, 71,203....*G6 40
Rajasthan, state, India, 20,155,602....C5 39
Rajkot, India, 193,498 (*194,145)....D5 39, F3 40
Rajmahal, India, 6,801....E11 40
Rajnandgaon, India, 44,678....G8 40
Rajpipla, India, 21,426....G4 40
Rakaia, riv., N.Z.....O13 51
Rake, Winnebago, Iowa, 328....A4 92
Rakhov, Sov. Un....A7 22
Rakkestad, Nor., 1,613....p29 25
Rakovnik, Czech., 12,000....C2 26
Rakvere, Sov. Un., 25,000....B6 27
Raleigh, Newf., Can., 307....C4 75
Raleigh, Levy, Fla., 150....C4 86
Raleigh, Saline, Ill., 225....F5 90
Raleigh, Rush, Ind., 120....E7 91
Raleigh, Smith, Miss., 614....C4 100
Raleigh, Wake, N.C., 93,931 (*130,200)....B5 109
Raleigh, Grant, N. Dak., 125....C4 110
Raleigh, Shelby, Tenn., 6,000....B2, E8 117
Raleigh, Raleigh, W. Va., 750....D7 123
Raleigh, co., W. Va., 77,826....D3 123
Raleigh, bay, N.C....C7 109
Raley, Alta., Can., 15....E4 69
Rallaovia (Wells), Maur....B2 45
Ralls, Crosby, Tex., 2,229....C2 117
Ralls, co., Mo., 8,078....B6 101
Rally Hill, Maury, Tenn....B5 117
Ralph, Tuscaloosa, Ala., 150....B2 78
Ralph, Dickinson, Mich., 40....B3 98
Ralph, Harding, S. Dak., 20....B2 116
Ralphton, Somerset, Pa., 150....F3 114
Ralston, Carroll and Greene, Iowa, 143....B3 92
Ralston, Douglas, Nebr., 2,977....D3 103
Ralston, Morris, N.J., 90....B3 106
Ralston, Pawnee, Okla., 411....A5 112
Ralston, Lycoming, Pa., 400....D8 114
Ralston, Weakley, Tenn., 50....A3 117
Ralston, Park, Wyo., 20....A4 125
Ralston, val., Nev....E4 104
Ram, riv., Alta., Can....C3 69
Ram, head, Vir. Is....f16 65
Rama, Sask., Can., 288....F4 70
Rama, Nic., 600....D5 62
Ramadan, see Turabah, Sau. Ar.
Ramah, Newf., Can....f9 75
Ramah, El Paso, Colo., 109....B6 83
Ramah, McKinley, N. Mex., 175....B1 107
Rām Allāh, Jordan, 17,145....C7, h11 32
Ramapo, mts., N.J....A4 106
Ramapo, riv., N.J....A4 106
Ramat Gan, Isr., 90,234....B6, g10 32
Ramat Hakovesh, Isr....g10 32
Ramat Hasharon, Isr., 2,650....g10 32
Ramban, India, 1,490....B6 39
Rambervillers, Fr., 7,042....C7 14
Rambi, isl., Fiji Is.... 52
Rambouillet, Fr., 11,382....C4 14
Ramea, Newf., Can., 970....E3 75
Ramea, is., Newf., Can....E3 75
Ramenskoye, Sov. Un., 10,000....D12, n18 27
Ramer, Montgomery, Ala., 500....C3 78
Ramer, McNairy, Tenn., 358....B3 117
Rameshk, Iran, 10,000....H9 41
Ramganga, riv., India....D7 40
Rāmhormoz, Iran, 17,267....F4 41
Ramhurst, Murray, Ga., 100....B2 87
Ramirez, Duval, Tex., 50....F3 118
Ramiriquí, Col., 881....B3 60
Ramirito, Jim Hogg, Tex., 5....F3 118
Ramle, Isr., 22,444....C6, h10 32
Ramnäs, Swe., 1,614....t34 25
Râmnicu-Sărat, Rom., 19,095....C8 22
Râmnicu-Vâlcea, Rom., 18,984....C7 22
Ramon, Lincoln, N. Mex....C4 107
Ramona, San Diego, Calif., 2,449....F5 82
Ramona, Marion, Kans., 132....D6 93
Ramona, Washington, Okla., 546....A6 112
Ramona, Lake, S. Dak., 247....C8 116
Ramor, lake, Ire....D4 11
Rampart, Alsk., 49....B9 79
Rampart, range, Colo....B5 83
Rampur, India, 135,407....C6 39, C7 40
Rampur, India, 5,000....G9 40
Rampur-Baolia, Pak., 39,993....E12 40
Ramsay, Gogebic, Mich., 1,158....A5 98
Ramsay, Silver Bow, Mont., 100....D4 102
Ramsayville, Ont., Can., 61....A9 72
Ramseur, Randolph, N.C., 1,258....B4 109
Ramsey, Eng., 5,697....B7 12
Ramsey, Fayette, Ill., 815....D4 90
Ramsey, I. of Man, 3,764....F4 13
Ramsey, Bergen, N.J., 9,527....A4 106
Ramsey, co., Minn., 422,525....E5 99
Ramsey, co., N. Dak., 13,443....A7 110
Ramsgate, Eng., 36,906....C9 12
Ramshorn, mtn., Wyo....B3 125
Ramu, riv., N. Gui....k11 50
Ranaghat, India, 35,266....F12 40
Ranburne, Cleburne, Ala., 317....B4 78
Rancagua, Chile, 50,000....A2 50
Ranchcreek, Powder River, Mont., 3....E11 102
Rancherie, rock, Oreg....C6 113
Ranches of Taos, Taos, N. Mex., 1,668....A4 107
Ranchester, Sheridan, Wyo., 235....A5 125
Ranchi, India, 122,416 (*140,253)....D8 39, F10 40
Rancho Cordova, Sacramento, Calif., 7,429....*C3 82
Ranchvale, Curry, N. Mex....C6 107
Ranco, Chile....C2 54
Ranco, lake, Chile....C2 54
Rancocas, Burlington, N.J., 300....C3 106
Rancocas, creek, N.J....C3 106
Rand, Jackson, Colo., 12....A4 83
Rand, Kanawha, W. Va., 3,500....C6, C3 123
Randado, Jim Hogg, Tex., 300....F3 118
Randalia, Fayette, Iowa, 114....B6 92
Randall, Hamilton, Iowa, 201....B4 92
Randall, Jewell, Kans., 201....C5 93
Randall, Morrison, Minn., 516....D4 99
Randall, Burnett, Wis....C1 124
Randall, co., Tex., 33,913....B2 118
Randallstown, Baltimore, Md., 2,000....B4 85
Randers, Den., 42,238 (*54,780)....B4 24
Randers, co., Den., 170,802....B4 24
Randfontein, S. Afr., 41,499....*C4 49
Randle, Lewis, Wash., 100....C4 122
Randleman, Randolph, N.C., 2,232....B4 109
Randlett, Cotton, Okla., 356....C3 112
Randlett, Uintah, Utah, 10....C6 119
Randolph, Pinal, Ariz., 400....E4 80
Randolph, Fremont, Iowa, 257....D2 92
Randolph, Kennebec, Maine, 1,724....D3 96
Randolph, Norfolk, Mass., 18,900....B5, D3 97
Randolph, Pontotoc, Miss., 131....A4 100
Randolph, Cedar, Nebr., 1,063....B8 103
Randolph, Coos, N.H., 25 (140▲)....B4 105
Randolph, Cattaraugus, N.Y., 1,414....C2 108
Randolph, Portage, Ohio, 450....A6 111
Randolph, Tipton, Tenn....B2 117
Randolph, Rich, Utah, 537....B4 119
Randolph, Orange, Vt., 2,122 (3,414▲)....D3 120
Randolph, Columbia and Dodge, Wis., 1,507....E4 124
Randolph, co., Ala., 19,477....B4 78
Randolph, co., Ark., 12,520....A4 81
Randolph, co., Ga., 11,078....E2 87
Randolph, co., Ill., 29,988....E4 90
Randolph, co., Ind., 28,434....D7 91
Randolph, co., Mo., 22,014....B5 101
Randolph, co., N.C., 61,497....B4 109
Randolph, co., W. Va., 26,349....C5 123
Randolph Center, Orange, Vt., 140....D3 120
Randolph Field, Bexar, Tex....B4, E3 118
Randolph Hills, Montgomery, Md., 2,000....*B3 85
Random, isl., Newf., Can....D5 75
Random Lake, Sheboygan, Wis., 858....E6 124
Randow, riv., Ger....B7 16
Randsburg, Kern, Calif., 300....E5 82
Ranfurly, Alta., Can., 133....C5 69
Rangaunu, bay, N.Z.....K14 51
Range, Grant, Oreg....C8 113
Rangely, Rio Blanco, Colo., 1,464....A2 83
Rangeley, Franklin, Maine, 749 (1,087▲)....D2 96
Rangeley, lake, Maine....D2 96
Ranger, Gordon, Ga., 161....B2 87
Ranger, Cherokee, N.C....D2 109
Ranger, Eastland, Tex., 3,313....C3 118
Ranger, Lincoln, W.Va., 300....C2 123
Ranger, lake, N. Mex....D6 107
Rangiora, N.Z., 3,540....O14 51
Rangitata, riv., N.Z.....O13 51
Rangoon, Bur., 752,000....D2 38, E10 39
Rangpur, Pak., 31,759....E12 40
Rangsum (Tungpu), China....E4 34
Ranier, Koochiching, Minn., 262....B5 99
Raniganj, India, 30,113....D8 39
Rankin, Vermilion, Ill., 761....C6 90
Rankin, Allegheny, Pa., 5,164....F2 114
Rankin, Cocke, Tenn....C10 117
Rankin, Upton, Tex., 1,214....D2 118
Rankin, co., Miss., 34,322....C4 100
Ranlo, Gaston, N.C., 2,000....*B2 109
Rannes, Austl., 63....B8 51
Rannoch, lake, Scot....D4 13
Rann of Kutch, swamp, India....D4 39
Ranshaw, Northumberland, Pa., 1,078....*E9 114
Ransom, La Salle, Ill., 415....B5 90
Ransom, Ness, Kans., 387....D4 93
Ransom, Lackawanna, Pa., 150....A8 114
Ransom, co., N. Dak., 8,078....C8 110
Ransomville, Niagara, N.Y., 950....B2 108
Ranson, Jefferson, W. Va., 1,974....B7 123
Rantauparapat, Indon....K3 38
Rantem, Newf., Can., 15....E5 75
Rantoul, Champaign, Ill., 22,116....C5 90
Rantoul, Franklin, Kans., 157....D8 93
Rantowles, Charleston, S.C....F2 115
Ranum, Den., 1,153....B3 24
Raon [-l'Etape], Fr., 7,606....A2 18
Raoui, sand dunes, Alg....D4 44
Rapallo, It., 20,606....B2 21
Rapa, isl., Fr. Polynesia....*H13 7
Rapa Nui (Easter), isl., Pac. O....H15 7
Rapelje, Stillwater, Mont., 115....E7 102
Raphine, Rockbridge, Va., 300....D3 121
Raphoe, Ire., 818....C4 11
Rapid, riv., Minn....B4 99
Rapidan, Culpeper, Va., 220....C4 121
Rapid City, Man., Can., 467....D1 71
Rapid City, Kalkaska, Mich., 300....D5 98
Rapid City, Pennington, S. Dak., 42,399....C2 116
Rapides, Rapides, La....C3 95
Rapides, par., La., 111,351....C3 95
Rapid River, Delta, Mich., 550....C3 98
Rappahannock, co., Va., 5,368....C4 121
Rappahannock, riv., Va....C5 121
Rappahannock Academy, Caroline, Va., 5....C5 121
Rapperswil, Switz., 7,585....B6 19
Raquette, lake, N.Y.....B6 108
Raquette, riv., N.Y.....B2 108
Raquette Lake, Hamilton, N.Y., 150....B6 108
Rarden, Scioto, Ohio, 250....D4 111
Rardin, Coles, Ill., 130....D5 90
Raritan, Henderson, Ill., 182....C3 90
Raritan, Somerset, N.J., 6,137....B3 106
Raritan, bay, N.J.....C4 106
Raritan, riv., N.J.....C4 106
Rarous (Well), Niger....C6 45
Rasa, pt., Arg....C4 54
Ras al Bidiya, cape, Sau. Ar....H4 41
Ra's al Khaymah, Tr. Coast, 5,000....C2 39
Ra's al 'Ushsh, Eg., U.A.R....C4 32
Ra's an Naqb, Jordan, 1,000....E7 32
Rasar, Blount, Tenn....D10 117
Ras at Tannura, cape, Sau. Ar....H5 41
Rasbokil, Swe....t35 25
Ras Dashan, mtn., Eth....C4 47
Raseiniai, Sov. Un., 6,181....A7 26
Ras el Ain, Syr....D13 31
Ras el Hadd, cape, Om....D2 39
Ras el Milh, cape, Libya....F5 31
Ras el Tin, cape, Libya....F4 31
Ras en Nagura, cape, Leb....A6 32
Rashad, Sud., 1,683....C3 47
Rashid, see Rosetta, Eg., U.A.R.
Rashkov, Sov. Un....B9 22
Rasht, Iran, 109,491....C4 41
Raška, Yugo., 2,290....D5 22
Raska, riv., Yugo....D5 22
Ras Madraka, cape, Om....E6 39
Raso, cape, Arg....C3 54
Raso, cape, Braz....B5 59
Raspberry, peak, Ark....C1 81
Rasskazovo, Sov. Un., 43,500....E13 27
Rastatt, Ger., 24,100....E3 17
Rastede, Ger., 14,200....A8 15
Raszyn, Pol....m13 26
Rat, is., Alsk....E3 79
Rat, riv., Man., Can....A2 71
Rat, riv., Man., Can....E3 71
Ratangarh, India, 26,631....E5 40
Rat Buri, Thai., 19,092....F3 38
Ratcliff, Logan, Ark., 147....B2 81
Ratekau, Ger., 8,900....E4 24
Rathbun, Appanoose, Iowa, 203....D5 92
Rathdrum, Kootenai, Idaho, 710....B2 89
Rathdrum, Ire., 1,128....E5 11
Rathdrum, prairie, Idaho....D8 122
Rathenow, Ger., 28,600....B6 16
Rathfriland, N. Ire., 1,558....C5 11
Rathkeale, Ire., 1,459....E3 11
Rathlin, isl., N. Ire....B5 11
Rathlin, sound, N. Ire....B5 11
Rath Luire, Ire., 1,956....E3 11
Rathmelton, Ire., 808....B4 11
Rathmore, Ire., 417....E2 11
Rathmullen, Ire., 491....B4 11
Rathnew, Ire., 861....D5 11
Rathowen, Ire., 119....D4 11
Rathwell, Man., Can., 197....E2 71
Ratibor, see Raciborz, Pol.
Ratingen, Ger., 36,000....B1 17
Ratlam, India, 87,472....F5 40
Ratnagiri, India, 31,091....I4 40
Ratner, Sask., Can., 60....D3 70
Raton, Colfax, N. Mex., 8,146....A5 107
Raton, mesa, Colo....D6 83
Raton, pass, Colo....D6 83
Ratónes, is., P.R.....D5 61
Rattan, Pushmataha, Okla., 300....C6 112
Rattenberg, Aus., 745....E5 16
Rattlesnake, creek, Kans....E4 93
Rattlesnake, creek, Ohio....C4 111
Rattlesnake, creek, Wash....C6 122
Rattlesnake, flat, Wash....C7 122
Rattlesnake, hill, Conn....A7 84
Rattlesnake, hills, Wash....C6 122
Rattlesnake, mtn., Conn....C5 84
Rattlesnake, range, Wyo....C5 125
Rattling Brook, Newf., Can., 162....D3 75
Rattray, head, Scot....C7 13
Rättvik, Swe., 1,950....G6 25
Ratzeburg, Ger., 11,400....E4 24
Ratzeburger, lake, Ger....E4 24
Raub, Benton, Ind., 100....C3 91
Raub, McLean, N. Dak., 15....B3 110
Rauch, Arg., 5,274....B5 54
Rauch, Koochiching, Minn....C5 99
Raue, isl., Nor....p28 25
Raufarhofn, Ice., 483....m24 25
Rauland, Nor....H3 25
Rauma, Fin., 21,700....G9 25
Raus, Bedford, Tenn., 50....B5 117
Rausu, Dake, peak, Jap....D12 37
Rautalampi, Fin., 1,300....F12 25
Rauville, Codington, S. Dak....B8 116
Ravalli, Lake, Mont., 50....C2 102
Ravalli, co., Mont., 12,341....D2 102
Ravanna, Mercer, Mo., 127....A4 101
Rävar, Iran, 5,074....F8 41
Rava-Russkaya, Sov. Un., 10,000....F4 27
Raven, Tazewell, Va., 900....B3 121
Raven, headland, Ire....B1 12
Ravena, Albany, N.Y., 2,410....C7 108
Ravendale, Lassen, Calif., 40....B3 82
Ravenden, Lawrence, Ark., 231....A4 81
Ravenden Springs, Randolph, Ark., 126....A4 81
Ravenel, Charleston, S.C., 527....F1 115
Ravenglass, Eng., 417....F5 13
Ravenna, It., 54,000 (115,525▲)....E8 19, B4 21
Ravenna, Estill, Ky., 921....C6 94
Ravenna, Muskegon, Mich., 801....E5 98
Ravenna, Buffalo, Nebr., 1,417....C7 103
Ravenna, Portage, Ohio, 10,918....A6 111
Raven Park, basin, Colo....A2 83
Raven Rock, Hunterdon, N.J....C2 106
Ravensburg, Ger., 31,300....E4 16
Ravenscrag, Sask., Can., 59....H1 70
Ravenscroft, White, Tenn., 140....D8 117
Ravensdale, King, Wash., 250....D2 122
Ravensthorpe, Austl., 116....F3 51
Ravenswood, Marion, Ind., 618....H8 91
Ravenswood, Jackson, W. Va., 3,410....C3 123
Ravensworth, Ont., Can., 65....B5 72
Raventon, Lincoln, N. Mex....D4 107
Ravenwood, Nodaway, Mo., 282....A3 101
Ravi, riv., Pak....B4 40
Ravia, Johnston, Okla., 307....C5 112
Ravinia, Charles Mix, S. Dak., 164....D7 116
Rawalpindi, Pak., 250,000 (*340,175)....B5 39
Rawa Mazowiecka, Pol., 6,908....C6 26
Rawdon, Que., Can., 2,388....C4 73
Rawhide, creek, Who....C8 125
Rawhide, lake, Ont., Can....B8 98
Rawicz, Pol., 11,600....C4 26
Rawlings, Allegany, Md., 180....D2 85
Rawlings, Brunswick, Va., 50....E5 121
Rawlinna, Austl., 124....F4 51
Rawlins, Carbon, Wyo., 8,968....D5 125
Rawlins, co., Kans., 5,279....C2 93
Rawlins, hills, Wyo....C5 125
Rawson, Arg....C3 54
Rawson, Arg., 2,425....g6 54
Rawson, McKenzie, N. Dak., 28....B2 110
Rawson, Hancock, Ohio, 407....B4 111
Rawsonville, Windham, Vt., 30....E3 120
Ray, Pinal, Ariz., 1,468....D5 80
Ray, Steuben, Ind., 200....A8 91
Ray, Koochiching, Minn., 55....B5 99
Ray, Williams, N. Dak., 1,049....A2 110
Ray, co., Mo., 16,075....B3 101
Ray, cape, Newf., Can....E2 75
Raya, isl., Indon....J1 38
Rayagada, India, 14,537....H9 40
Ray-Aleksandrovka, Sov. Un....q20 27
Raybon, Brantley, Ga., 400....E5 87
Raychikhinsk, Sov. Un., 30,000....B4 37
Ray City, Berrien, Ga., 713....E3 87
Raygorodka, Sov. Un....q22 27
Rayland, Jefferson, Ohio, 694....B2 123
Rayle, Wilkes, Ga., 200....C4 87
Raymilton, Venango, Pa., 50....D2 114
Raymond, Madera, Calif., 300....D4 82
Raymond, Alta., Can., 2,362....E4 69
Raymond, Coweta, Ga., 300....C2 87
Raymond, Bear Lake, Idaho, 35....G7 89
Raymond, Montgomery, Ill., 871....D4 90
Raymond, Rice, Kans., 143....D5 93
Raymond, Cumberland, Maine 300 (732▲)....E4 96
Raymond, Kandiyohi, Minn., 608....E3 99
Raymond, Hinds, Miss., 1,381....C3 100
Raymond, Sheridan, Mont., 40....B12 102
Raymond, Lancaster, Nebr., 223....D9 103
Raymond, Rockingham, N.H., 800 (1,867▲)....D4 105
Raymond, Clark, S. Dak., 168....C8 116
Raymond, Pacific, Wash., 3,301....C2 122
Raymondville, Texas, Mo., 202....D6 101
Raymondville, St. Lawrence, N.Y., 200....A2 108
Raymondville, Willacy, Tex., 9,385....F4 118
Raymore, Sask., Can., 503....F3 70
Raymore, Cass, Mo., 268....C3 101
Rayne, Acadia, La., 8,634....D3 95
Raynesford, Judith Basin, Mont., 40....C6 102
Raynham, Bristol, Mass., 350 (4,150▲)....C5 97
Raynham Center, Bristol, Mass., 500....C5 97
Rayo, Socorro, N. Mex....C3 107
Rayón, Mex., 1,351....B2 63
Rayong, Thai., 10,776....F4 38
Raystown, branch, Pa....F5 114
Raytown, Taliaferro, Ga., 100....C4 87
Raytown, Jackson, Mo., 17,083....E2 101
Rayville, Richland, La., 4,052....B4 95
Rayville, Ray, Mo., 200....B3 101
Razan, Iran, 3,195....D4 41
Razgrad, Bul., 15,023....D8 22
Razlog, Bul., 6,857....E6 22
Razmak, Pak....B4 40
Razorback, mtn., B.C., Can....D5 68
Ré, isl., Fr....D3 14
Rea, Andrew, Mo., 90....A3 101
Rea, lake, Ire....D3 11
Reader, Ouachita, Ark., 86....D2 81
Reader, Wetzel, W. Va., 500....A6, B4 123
Reader, lake, Man., Can....C1 71
Readfield, Kennebec, Maine, 200 (1,029▲)....D3 96
Reading, Eng., 119,870 (*160,000)....C7 12
Reading, Lyon, Kans., 249....D8 93
Reading, Middlesex, Mass., 19,259....A5, C3 97
Reading, Hillsdale, Mich., 1,128....G6 98
Reading, Nobles, Minn., 160....G3 99
Reading, Hamilton, Ohio, 12,832....C3, D2 111
Reading, Berks, Pa., 98,177 (*192,500)....F10 114
Reading, Windsor, Vt., 165 (472▲)....E3 120
Reading Center, Schuyler, N.Y., 130....C4 108
Readland, Chicot, Ark., 100....D4 81
Readlyn, Sask., Can., 85....H3 70
Readlyn, Bremer, Iowa, 547....B5 92
Readsboro, Bennington, Vt., 577 (783▲)....F3 120
Readstown, Vernon, Wis., 469....E3 124
Readyville, Cannon, Tenn., 100....B5 117
Reagan, Henderson, Tenn., 150....B3 117
Reagan, co., Tex., 3,782....D2 118
Reagan Wells, Uvalde, Tex., 35....E3 118
Real, co., Tex., 2,079....E3 118
Real, range, Bol....C2 55
Realicó, Arg....B4 54
Realitos, Duval, Tex., 400....F3 118
Ream, Camb., 5,000....G5 38
Ream, McDowell, W. Va., 800....*D3 123
Reamstown, Lancaster, Pa., 950....F9 114
Reardan, Lincoln, Wash., 474....B8 122
Reasnor, Jasper, Iowa, 224....C4 92
Reata, Mex., 208....B4 63
Reaville, Hunterdon, N.J., 75....C3 106
Rebecca, Turner, Ga., 278....E3 87
Rebersburg, Centre, Pa., 500....E7 114
Rebiana (Oasis), Libya....E4 43
Rebiano, sand sea, Libya....E3 43
Reboly, Sov. Un....F14 25
Rebun, isl., Jap....D10 37
Recalde, Arg....B4 54
Recanati, It., 6,288....C4 21
Recherche, arch., Austl....F3 50
Rechitsa, Sov. Un., 33,800....E8 27
Recife, Braz., 788,569 (*1,100,000)....C4, k6 57
Recinto, Chile....B2 54
Recklinghausen, Ger., 130,600....B2 17
Recluse, Campbell, Wyo., 15....A7 125
Recogne, Bel., 746....E5 15
Reconnaissance, mtn., Guam.... 52
Reconquista, Arg., 12,729....E4 55
Recovery, Decatur, Ga....F2 87
Recreo, Arg., 5,463....E2 55
Rector, Clay, Ark., 1,757....A5 81
Red, bay, N. Ire....B5 11
Red, creek, Miss....E4 100
Red, isl., Newf., Can....E4 75
Red, lake, Ont., Can....D4 71
Red, mtn., Ala....B2 78
Red, mtn., Calif....B2 82
Red, mtn., Mont....C4 102
Red, peak, Colo....B4 83
Red, peak, Idaho....E3 89
Red, riv., Man., Can....E3 71
Red, riv., Ky....C6 94
Red, riv., Tenn....A4 117
Red, riv., U.S.....D9 77
Red, riv., Viet....B6 38
Red, riv., Wis....D2 124
Red, sea, Afr., Asia....D9 42
Redang, isl., Mala....J5 38
Redange, Lux., 1,693....E5 15
Red Bank, Monmouth, N.J., 12,482....C4 106
Red Bank, Lexington, S.C., 350....D5 115
Red Banks, Marshall, Miss., 250....A4 100
Red Bank-White Oak, Hamilton, Tenn., 10,777....*D8 117
Red Bay, Franklin, Ala., 1,954....A1 78
Red Bay, Newf., Can., 261....C3, h10 75
Redbay, Walton, Fla., 450....G3 86
Redberry, lake, Sask., Can....E2 70
Redbird, Holt, Nebr....B7 103
Redbird, Wagoner, Okla., 310....B6 112
Red Bird, creek, Ky....C6 94
Red Bluff, Tehama, Calif., 7,202....B2 82
Red Bluff, res., Tex....E2 118
Red Boiling Springs, Macon, Tenn., 597....C8 117
Red Bud, Randolph, Ill., 1,942....E4 90
Redby, Beltrami, Minn., 300....C4 99
Redcar, Eng., 31,460....F7 13
Red Cedar, lake, Wis....C2 124
Red Cedar, riv., Wis....C2 124
Redcliff, Alta., Can., 2,221....D5 69
Redcliff, Eagle, Colo., 586....B4 83
Red Cliff, Bayfield, Wis., 100....B3 124
Red Cliff, Indian res., Wis....B3 124
Red Cloud, Webster, Nebr., 1,525....D7 103
Red Cloud, peak, Colo....D3 83
Red Creek, Wayne, N.Y., 689....B4 108
Red Deer, Alta., Can., 19,612....C4 69
Red Deer, lake, Man., Can....C1 71
Red Deer, riv., Alta., Sask., Can....D5 69
Red Deer, riv., Man., Sask., Can....E4 70
Reddell, Evangeline, La., 500....D3 95
Reddick, Marion, Fla., 594....C4 86
Reddick, Kankakee and Livingston, Ill., 205....B5 90
Reddies River, Wilkes, N.C....A2 109
Redding, Jefferson, Ala....E4 78
Redding, Shasta, Calif., 12,773....B2 82
Redding, Fairfield, Conn., 200 (3,359▲)....D3 84
Redding, Ringgold, Iowa, 129....D3 92
Redding Ridge, Fairfield, Conn., 325....D3 84
Redditch, Eng., 34,077....B6 12
Rede, riv., Eng....E6 13
Redelm, Ziebach, S. Dak., 10....B4 116
Redenção, Braz., 2,631....B3 57
Redeye, riv., Minn....D3 99
Redeyef, Tun....G11 30
Red Feather Lakes, Larimer, Colo., 150....A5 83
Redfield, Jefferson, Ark., 242....C3 81
Redfield, Dallas, Iowa, 966....C3 92
Redfield, Bourbon, Kans., 133....E9 93
Redfield, Oswego, N.Y., 185....B5 108
Redfield, Spink, S. Dak., 2,952....C7 116
Redford, Clinton, N.Y., 350....A3 108
Redford, Presidio, Tex., 300....G2 118
Redford Heights, Wayne, Mich., 71,276....*F7 98
Redgranite, Waushara, Wis., 588....D4 124
Redgut, bay, Ont., Can....B6 98
Red Hill, Catron, N. Mex., 5....C1 107
Red Hill, Montgomery, Pa., 1,086....*F11 114
Red Hook, Dutchess, N.Y., 1,719....D7 108
Redhouse, Madison, Ky., 250....C5 94
Red House, Charlotte, Va., 50....D4 121
Red House, Putnam, W. Va., 350....C3 123
Redig, Harding, S. Dak., 5....B2 116
Red Indian, lake, Newf., Can....D3 75
Redington, Morrill, Nebr., 15....C2 103
Redington Beach, Pinellas, Fla., 1,368....*E4 86

Redington Shores, Pinellas, Fla., 917 ... *E4 86
Red Jacket, Mingo, W. Va., 950 ... D2 123
Red Key, Jay, Ind., 1,746 ... D7 91
Red Lake, Ont., Can., 2,051 . D5 71
Redlake, Beltrami, Minn., 400 ... C3 99
Reed Lake, co., Minn., 5,830 . C2 99
Red Lake, Indian res., Minn ... B3 99
Red Lake, riv., Minn ... C2 99
Red Lake Falls, Red Lake, Minn., 1,520 ... C2 99
Redlands, San Bernardino, Calif., 26,829 ... E5, F3 82
Redlawn, Mecklenburg, Va., 40 ... E4 121
Red Level, Covington, Ala., 327 ... D3 78
Red Lick, Jefferson, Miss., 250 ... D3 100
Red Lion, Logan, Colo ... A8 83
Red Lion, Burlington, N.J. . D3 106
Red Lion, York, Pa., 5,594 . G8 114
Red Lodge, Carbon, Mont., 2,278 ... E7 102
Redmesa, La Plata, Colo., 100 ... D2 83
Redmon, Edgar, Ill., 175 ... D6 90
Redmond, Deschutes, Oreg., 3,340 ... C5 113
Redmond, Sevier, Utah 413 . E4 119
Redmond, King, Wash., 1,426 ... C2 122
Red Mountain, San Bernardino, Calif., 350 ... E5 82
Red Mountain, pass, Colo ... D3 83
Rednitz, riv., Ger ... D6 17
Red Oak, Fulton, Ga., 800 ... B5 87
Red Oak, Montgomery, Iowa, 6,421 ... D2 92
Red Oak, Nash., N.C., 250 . A6 109
Red Oak, Latimer, Okla., 453 ... C6 112
Red Oak, Ellis, Tex., 415 ... B5 118
Redoak, Charlotte, Va., 50 ... E4 121
Redon, Fr., 6,444 ... D3 14
Redonda, isl., B.C., Can ... D5 68
Redondela, Sp., 3,261 ... A1 20
Redondo, Port., 4,103 ... C2 20
Redondo, King, Wash., 600 . D1 122
Redondo, peak, N. Mex ... F5 107
Redondo Beach, Los Angeles, Calif., 46,986 ... F2 82
Redore, St. Louis, Minn., 170 ... C6 99
Redoubt, mtn., Alsk ... g15 79
Redowl, Meade, S. Dak., 10 . C3 116
Red Pass, B.C., Can., 70 ... C8 68
Red Pheasant, Sask., Can., 10 . E1 70
Red Rapids, N.B., Can., 54 . C2 74
Red River, co., Tex., 15,682 . C5 118
Red River, par., La., 9,978 ... B2 95
Red River, Hot Springs, Idaho ... D3 89
Red River of the North, riv., Minn., N. Dak ... B9 110
Red Rock, Pinal, Ariz., 30 ... E4 80
Redrock, Newton, Ark ... B2 81
Red Rock, Beaverhead, Mont., 12 ... F4 102
Redrock, Grant, N. Mex ... E1 107
Redrock, Noble, Okla., 262 . A4 112
Red Rock, Bastrop, Tex., 50 . E4 118
Red Rock, pass, Idaho, Mont ... E7 89
Red Rock, riv., Mont ... F4 102
Redruth, Eng., 9,600 (part of Camborne-Redruth) ... D2 12
Red Slate, mtn., Calif ... D4 82
Red Springs, Robeson, N.C., 2,767 ... C4 109
Redstone, B.C., Can., 25 ... C6 68
Redstone, Pitkin, Colo., 160 ... B3 171
Redstone, Sheridan, Mont., 150 ... B12 102
Redstone, Carroll, N.H., 150 . B4 105
Redstone Park, Madison, Ala., 1,000 ... *A2 78
Red Sucker, lake, Man., Can ... B5 71
Red Sucker, riv., Man., Can ... B5 71
Red Table, mtn., Colo ... B4 83
Red Tank, C.Z. ... k11 62
Redtop, Dallas, Mo., 128 ... D4 101
Redvale, Montrose, Colo., 10 ... C2 83
Redvers, Sask., Can., 642 ... H5 70
Redwater, Alta., Can., 1,135 . C4 69
Redwater, creek, Mont ... C11 102
Red Wharf, bay, Wales ... A3 12
Red Wing, Huerfano, Colo., 15 ... D5 83
Red Willow, Alta., Can., 95 . C4 69
Red Willow, co., Nebr., 12,940 ... D5 103
Red Willow, creek, Colo ... A8 83
Red Willow, creek, Nebr ... D5 103
Red Willow, riv., B.C., Can ... B7 68
Redwine, Morgan, Ky., 88 ... C6 94
Red Wing, Goodhue, Minn., 10,528 ... F6 99
Redwood, Warren, Miss., 25 . C3 100
Redwood, Jefferson, N.Y., 524 ... A5, B1 108
Redwood, co., Minn., 21,718 . F3 99
Redwood, riv., Minn ... F3 99
Redwood City, San Mateo, Calif., 46,290 ... B5, D2 82
Redwood Estates, Santa Clara, Calif., 930 ... *C6 82
Redwood Falls, Redwood, Minn., 4,285 ... F3 99
Redwood Valley, Mendocino, Calif., 200 ... C2 82
Ree, lake, Ire ... D4 11
Reed, Greer, Okla., 100 ... C2 112
Reed, Lane, Oreg ... C3 113
Reed, lake, Man., Can ... B1 71
Reed City, Osceola, Mich., 2,184 ... E5 98
Reeder, Adams, N. Dak., 321 . C3 110
Reedley, Fresno, Calif., 5,850 ... D4 82
Reedpoint, Stillwater, Mont., 130 ... E7 102
Reedsburg, Sauk, Wis., 4,371 ... E3 124
Reeds Ferry, Hillsboro, N.H., 300 ... E4 105
Reeds Lake, Kent, Mich., 10,924 ... F5 98
Reedsport, Douglas, Oreg., 2,998 ... D3 113
Reeds Spring, Stone, Mo., 327 ... E4 101
Reedsville, Meigs, Ohio, 300 . C6 111
Reedsville, Mifflin, Pa., 950 ... E6 114
Reedsville, Preston, W. Va., 398 ... B5 123
Reedsville, Manitowoc, Wis., 830 ... A6, D6 124
Reedville, Northumberland, Va., 400 ... D6 121
Reedy, Roane, W. Va., 352 . C3 123
Reedy, creek, W. Va ... C3 123
Reedy, lake, Fla ... E5 86
Reedy, riv., S.C ... C3 115
Reef, pt., N.Z ... K14 51
Reefton, N.Z., 1,787 ... O13 51
Ree Heights, Hand, S. Dak., 188 ... C6 116
Reelfoot, lake, Tenn ... A2 117
Reelsville, Putnam, Ind., 100 . E4 91
Reeman, Newaygo, Mich., 120 ... E4 98
Reengus, India, 5,549 ... D5 40
Reere, isl., Bikini ... 52
Rees, Franklin, Ohio, 650 ... C2 111
Reese, Greenwood, Kans., 400 ... E7 93
Reese, Tuscola, Mich., 711 . E7 98
Reese, Weber, Utah ... B3 119
Reese, riv., Nev ... D4 104
Reese Village, Lubbock, Tex., 1,433 ... *C2 118
Reeseville, Dodge, Wis., 491 . E5 124
Reesville, Clinton, Ohio, 250 . C4 111
Reeth, Eng., 588 ... F7 13
Reeves, co., Tex., 17,644 ... D1 118
Reeves, mtn., Austl ... F7 51
Reevesville, Johnson, Ill., 150 ... F5 90
Reevesville, Dorchester, S.C., 268 ... E6 115
Reform, Pickens, Ala., 1,241 . B1 78
Reform, Choctaw, Miss., 300 . B4 100
Refresco, Chile ... E2 55
Reftele, Swe., 1,027 ... A7 24
Refugio, Refugio, Tex., 4,944 ... E4 118
Refugio, co., Tex., 10,975 ... E4 118
Rega, riv., Pol ... B3 26
Regan, Burleigh, N. Dak., 104 ... B5 110
Regat, riv., Ger ... D5 17
Regen, Ger., 5,400 ... E8 17
Regen, riv., Ger ... D6 16
Regensburg, Ger., 125,000 . D7 17
Regent, Man., Can., 65 ... E1 71
Regent, Hettinger, N. Dak., 388 ... C3 110
Reger, Sullivan, Mo., 77 ... A4 101
Reggane, Alg ... D5 44
Reggio [di] Calabria, It., 124,000 (153,380▲) ... E5 21
Reggio nell'Emilia, It., 86,000 (116,445▲) . B3 21, E6 18
Reghin, Rom., 18,091 ... B7 22
Regina, Sask., Can., 112,141 . G3 70
Regina, Phillips, Mont., 13 ... C9 102
Regina, Sandoval, N. Mex., 30 ... A3 107
Regina Beach, Sask., Can., 319 ... G3 70
Region Occidental (Chaco), dept., Par., 54,277 ... *D3 55
Register, Bulloch, Ga., 300 . D5 87
Regla, Cuba, 26,755 ... h12 64
Regna, Swe ... u33 25
Rego, Orange, Ind., 20 ... H5 91
Reguengos de Monsaraz, Port., 4,873 ... C2 20
Rehau, Ger., 10,200 ... C7 17
Rehna, Ger., 3,519 ... E5 24
Rehoboth, Bristol, Mass., 250 (4,953▲) ... C5 97
Rehoboth, McKinley, N. Mex., 100 ... B1 107
Rehoboth, S.W Afr., 2,954 ... B3 49
Rehoboth, bay, Del ... C7 85
Rehoboth Beach, Sussex, Del., 1,507 ... C7 85
Rehovot, Isr., 28,740 ... C6, h10 32
Reï Bouba, Cam ... D2 46
Reichenbach, Ger., 29,600 ... C7 17
Reichenbach, Ger., 370 ... A4 18
Reids Grove, Dorchester, Md., 25 ... C6 85
Reidsville, Tattnall, Ga., 1,229 ... D4 87
Reidsville, Rockingham, N.C., 14,267 ... A4 109
Reidville, Spartanburg, S. C., 242 ... B3 115
Reigate, Eng., 53,710 ... C7 12
Reiley, East Feliciana, La ... D5 95
Reims, Fr., 133,914 . C6 14, E4 15
Reina Adelaida, arch., Chile ... E2 54
Reinach, Switz., 5,174 ... B5 19
Reinbeck, Grundy, Iowa, 1,621 ... B5 92
Reindeer, isl., Man., Can ... C3 71
Reindeer, lake, Sask., Can ... A4 70
Reindeer, riv., Sask., Can ... B4 70
Reinfeld, Ger., 5,700 ... E4 24
Reinosa, Sp., 10,044 ... A3 20
Reipetown, White Pine, Nev., 100 ... D6 104
Reisterstown, Baltimore, Md., 3,300 ... B4 85
Reit im Winkl, Ger., 2,300 ... B8 18
Reitz, S. Afr., 4,990 ... C4 49
Rekarne, Swe., 214 ... t34 25
Reliance, N.W. Ter., Can . D11 66
Reliance, Lyman, S. Dak., 201 ... D6 116
Reliance, Polk, Tenn., 400 . D9 117
Reliance, Sweetwater, Wyo., 300 ... D3 125
Relizane, Alg., 27,120 ... B5 45
Rémada, Tun., 1,866 ... C7 44
Remagen, Ger., 7,200 ... C2 17
Remanso, Braz., 5,125 ... C2 57
Rembert, Sumter, S.C., 400 . C6 115
Rembertow, Pol., 22,000 . B6, k14 26
Rembrandt, Buena Vista, Iowa, 265 ... B2 92
Remecó, Arg ... B4 54
Remedios, Cuba, 10,602 ... B4 64
Remedios, Pan., 1,125 ... F7 62
Remer, Cass, Minn., 492 ... C5 99
Remerton, Lowndes, Ga., 571 ... F3 87
Remington, Jasper, Ind., 1,207 ... C3 91
Remington, Fauquier, Va., 288 ... C5 121
Remiremont, Fr., 9,350 ... C7 14
Remlap, Blount, Ala., 115 ... B3 78
Remmel, dam, Ark ... D6 81
Remmel, mtn., Wash ... A5 121
Rems, riv., Ger ... E4 17
Remscheid, Ger., 126,900 ... B2 17
Remsen, Plymouth, Iowa, 1,338 ... B2 92
Remsen, Oneida, N.Y., 567 . B5 108
Remus, Mecosta, Mich., 600 . E5 98
Renaix, see Ronse, Bel.
Renault, Alg., 2,113 ... F7 30
Renault, Monroe, Ill., 200 ... E4 90
Rencona, San Miguel, N. Mex., 5 ... B4, G6 107
Rencontre East, Newf., Can., 293 ... E4 75
Rendsburg, Ger., 35,700 ... A4 16
Rendville, Perry, Ohio, 197 . C5 111
Renens, Switz., 10,698 ... C2 19
Renews, Newf., Can., 567 ... E5 75
Renfrew, Ont., Can., 8,935 . B8 72
Renfrew, Greenville, S.C., 200 ... B2 115
Renfrew, co., Ont., Can., 89,635 ... B7 72
Renfrew, co., Scot., 338,815 ... E4 13
Renfrow, Grant, Okla., 38 ... A4 112
Rengo, Chile, 9,115 ... A2 54
Reni, Sov. Un ... C9 22
Renick, Randolph, Mo., 190 . B5 101
Renick (Falling Springs), Greenbrier, W. Va., 265 . D4 123
Renk, Sud ... C3 47
Renmark, Austl ... G3 51
Renner, Minnehaha, S. Dak., 100 ... D9 116
Renner, Collin, Tex., 212 ... A5 118
Rennerod, Ger., 1,900 ... C3 17
Rennert, Robeson, N.C., 194 . C4 109
Rennes, Fr., 151,948 ... C3 14
Rennick, bay, Ant ... C29 5
Rennie, Man., Can., 135 ... E4 71
Reno, Bond, Ill., 100 ... E4 90
Reno, Leavenworth, Kans., 30 ... B8 93
Reno, Washoe, Nev., 51,470 . D2 104
Reno, Venango, Pa., 600 ... D2 114
Reno, co., Kans., 59,055 ... E5 93
Reno, lake, Minn ... E3 99
Reno, riv., It ... B3 21
Renohill, Johnson, Wyo ... B6 125
Renous, N.B., Can., 350 ... C4 74
Renous, riv., N.B., Can ... C3 74
Renovo, Clinton, Pa., 3,316 . D6 114
Renown, Sask., Can., 75 ... F3 70
Rensselaer, Jasper, Ind., 4,740 ... C3 91
Rensselaer, Rensselaer, N.Y., 10,506 ... C7 108
Rensselaer, co., N.Y., 142,585 ... C7 108
Rensselaer Falls, St. Lawrence, N.Y., 375 ... B1 108
Rensselaerville, Albany, N.Y., 340 ... C6 108
Rentiesville, McIntosh, Okla., 122 ... B6 112
Renton, King, Wash., 18,453 ... B3, D2 122
Rentz, Laurens, Ga., 307 ... D4 87
Renuss, riv., Switz ... C6 19
Renville, Renville, Minn., 1,373 ... F3 99
Renville, co., Minn., 23,249 . F3 99
Renville, co., N. Dak., 4,698 . A4 110
Renwick, Humboldt, Iowa, 477 ... B4 92
Repton, Conecuh, Ala., 314 . D2 78
Republic, Republic, Kans., 333 ... C6 93
Republic, Marquette, Mich., 950 ... B3 98
Republic, Greene, Mo., 1,519 . D4 101
Republic, Seneca, Ohio, 729 . A4 111
Republic, Fayette, Pa., 1,921 . G2 114
Republic, Ferry, Wash., 1,064 . A7 122
Republic, co., Kans., 9,768 ... C6 93
Republican, riv., U.S ... C7 76
Republican City, Harlan, Nebr., 189 ... D6 103
Repulse Bay, N.W. Ter., Can., 116 ... C15 66
Requa, Del Norte, Calif., 150 . B1 82
Requena, Sp., 8,228 ... C5 20
Rerik, Ger., 4,132 ... D5 24
Reşadiye, Tur., 2,400 ... D6 23
Rescue, Isle of Wight, Va., 325 ... B6, D6 121
Reserve, Sask., Can., 202 ... E4 70
Reserve, Brown, Kans., 138 . C8 93
Reserve, St. John the Baptist, La., 5,297 ... B6 95
Reserve, Sheridan, Mont., 250 ... B12 102
Reserve, Catron, N. Mex., 300 ... D1 107
Reserve, Sawyer, Wis., 140 ... C2 124
Reshef, Isr ... g10 32
Resina, It., 45,148 ... *D5 21
Resistencia, Arg., 80,000 ... E4 55
Reşita, Rom., 41,241 ... C5 22
Resko, Pol., 1,314 ... B3 26
Resolute, N.W., Ter., Can . B14 66
Resolution, isl., N.W. Ter., Can ... D20 67
Restigouche, co., N.B., Can., 40,973 ... B2 74
Restigouche, riv., N.B., Can ... B2 74
Reston, Man., Can., 529 ... E1 71
Reszel, Pol., 5,693 ... A6 26
Retalhuleu, Guat., 9,304 ... C2 62
Retamito, Arg ... A3 54
Rethel, Fr., 7,359 ... C6 14
Rethimni (Rethymnon), prov., Grc., 69,943 ... *E5 23
Rethimnon, Grc., 14,999 ... E5 23
Reti, Pak ... C2 40
Retie, Bel., 5,820 ... C5 15
Retlaw, Alta., Can., 75 ... D4 69
Retsil, Kitsap, Wash ... D1 122
Retsof, Livingston, N.Y., 275 . C3 108
Reubens, Lewis, Idaho, 113 . C2 89
Reunion, Fr. dep., Afr., 241,708 ... H24 3
Reus, Sp., 41,014 ... B6 20
Reutlingen, Ger., 67,400 (*110,000) ... E4 17
Reutte, Aus., 4,285 ... E5 18
Reva, Harding, S. Dak., 5 ... B2 116
Revda, Sov. Un., 57,000 ... B5 29
Revel, Fr., 4,288 ... F5 14
Revelo, McCreary, Ky., 500 . D5 94
Revelstoke, B.C., Can., 3,624 . D8 68
Reventazón, Peru ... C1 58
Revenue, Sask., Can., 101 ... E1 70
Revere, Suffolk, Mass., 40,080 ... C3 97
Revere, Redwood, Minn., 201 ... F3 99
Revere, Clark, Mo., 190 ... A6 101
Reverie, Tipton, Tenn., 150 . B2 117
Revigny, Fr., 3,287 ... F5 15
Revillagigedo, isl., Alsk ... n24 79
Revillagigedo, is., Mex ... D2 63
Revillo, Grant, S. Dak., 202 . B9 116
Revin, Fr., 11,244 ... E4 15
Revivim, Isr ... C6 32
Revloc, Cambria, Pa., 900 ... F4 114
Revnice, Czech., 3,033 ... o17 26
Rew, McKean, Pa., 400 ... C4 114
Rewa, Fiji Is ... 52
Rewa, India, 43,065 ... E8 40
Reward, Sask., Can., 94 ... E1 70
Rewari, India, 36,994 ... C6 40
Rewey, Iowa, Wis., 219 ... F3 124
Rex., Clayton, Ga., 120 ... B5 87
Rex, mtn., Ant ... B4 5
Rexburg, Madison, Idaho, 4,767 ... F7 89
Rexford, Thomas, Kans., 245 ... C3 93
Rexford, Lincoln, Mont., 300 . B1 102
Rexford, Saratoga, N.Y., 200 . C7 108
Rexford, Carter, Tenn ... C11 117
Rexroat, Carter, Okla., 60 ... C4 112
Rexton, N.B., Can., 668 ... C5 74
Rexton, Mackinac, Mich., 90 ... B5 98
Reyburn, Hot Spring, Ark ... D6 81
Reydell, Jefferson, Ark., 50 ... C4 81
Reydon, Roger Mills, Okla., 183 ... B2 112
Reyes, pt., Calif ... C2 82
Reykjavík, Ice., 73,388 (*90,000) ... n22 25
Reynaud, Sask., Can., 175 ... E3 70
Reyno, Randolph, Ark., 348 . A5 81
Reynolds, Taylor, Ga., 1,087 . D2 87
Reynolds, Mercer and Rock Island, Ill., 494 ... B3 90
Reynolds, White, Ind., 547 . C4 91
Reynolds, Reynolds, Mo., 50 . D6 101
Reynolds, Jefferson, Nebr., 131 ... D8 103
Reynolds, Grand Forks and Trail, N. Dak., 269 ... B8 110
Reynolds, co., Mo., 5,161 ... D6 101
Reynoldsburg, Franklin, Ohio, 7,793 ... C2, C5 111
Reynolds Corners, Lucas, Ohio, 7,000 ... A2 111
Reynolds Knob, mtn., W. Va ... C5 123
Reynoldsville, Jefferson, Pa., 3,158 ... D4 114
Reynosa, Mex., 74,113 ... B5 63
Rezāiyeh (Rizaiyeh, Urmia), Iran, 67,605 ... C2 41
Rezé, Fr., 28,276 ... D3 14
Rezekne, Sov. Un., 13,139 ... C6 27
Rezina, Sov. Un ... B9 22
Rhaetian Alps, mts., It., Switz . D7 19
Rhame, Bowman, N. Dak., 254 ... C2 110
Rhazale, Syr ... B8 32
Rhea, co., Tenn., 15,863 ... D9 117
Rheda, Ger., 13,500 ... B3 17
Rheden, Neth., 7,800 ... B6 15
Rhein, Sask., Can., 367 ... F4 70
Rhein (Rhine), riv., Ger ... C6 15
Rheine, Ger., 44,300 ... A2 17
Rheinfelden, Switz., 5,197 ... A4 19
Rheinhausen, Ger., 68,100 ... B1 17
Rheinland-Pfalz (Rhineland-Palatinate), state, Ger., 3,417,100 ... D2 17
Rheinsberg, Ger., 4,215 ... E6 24
Rheinwaldhorn, mtn., Switz ... C7 19
Rhenen, Neth., 6,900 ... C5 15
Rhethymnon (Rethimni), prov., Grc., 69,943 ... *E5 23
Rheydt, Ger., 94,000 ... C6 15
Rhin, canal, Ger ... F6 24
Rhinau, Fr., 1,681 ... A3 18
Rhine, Dodge, Ga., 485 ... E3 87
Rhine, riv., Eur ... C2 17
Rhinebeck, Dutchess, N.Y., 2,093 ... D7 108
Rhineland, Montgomery, Mo., 190 ... C6 101
Rhineland, reg., Ger ... C3 16
Rhinelander, Oneida, Wis., 8,790 ... C4 124
Rhineland-Palatinate (Rheinland-Pfalz), state, Ger., 3,417,100 ... D2 17
Rhinow, Ger., 2,451 ... F6 24
Rhode Island, state, U.S., 859,488 ... B13 77, 84
Rhodell, Raleigh, W. Va., 626 ... D3 123
Rhodes, Marshall, Iowa, 358 . C4 92
Rhodes, Gladwin, Mich., 75 . E6 98
Rhodes, see Rodhos, isl., Grc.
Rhodes, peak, Idaho ... C4 89
Rhodesia, see Northern Rhodesia, and Southern Rhodesia, Br. dep., Afr.
Rhodesia and Nyasaland, Fed. of, see Northern Rhodesia, Southern Rhodesia, and Nyasaland, Br. dep., Afr.
Rhodes Point, Somerset, Md., 97 ... E5 85
Rhodhiss, Caldwell and Burke, N.C., 837 ... B2 109
Rhodope (Rodhopi), prov., Grc., 109,201 ... *B5 23
Rhodope, mts., Bul ... E7 22
Rhome, Wise, Tex., 412 ... A5 118
Rhondda, Wales, 100,314 ... C4 12
Rhône, dept., Fr., 1,116,664 ... *E6 14
Rhône, riv., Fr., Switz ... D3 19
Rhyl, Wales, 21,825 ... A4 12
Riacho de Santana, Braz., 1,832 ... D2 57
Rialto, San Bernardino, Calif., 18,567 ... *F3 82
Rialto, Tipton, Tenn ... B2 117
Rianjo, Sp., 9,971 ... A1 20
Rib, mtn., Wis ... D4 124
Rib, riv., Wis ... C3 124
Ribadavia, Sp., 7,031 ... A1 20
Ribadeo, Sp., 9,567 ... A2 20
Ribadesella, Sp., 8,228 ... A3 20
Ribas, do Rio Pardo, Braz., 1,175 ... C2 56
Ribatejo, prov., Port., 464,874 ... *C1 20
Ribáuè, Moz ... D6 48
Ribble, riv., Eng ... G6 13
Ribe, Den., 7,809 ... C2 24
Ribe, co., Den., 178,501 ... C2 24
Ribeauvillé, Fr., 4,314 ... C7 14
Ribeira do Pombal, Braz., 4,254 ... D3 57
Ribeirão Bonito, Braz., 1,921 ... m7 56
Ribeirão Branco, Braz., 618 . n7 56
Ribeirão Prêto, Braz., 116,153 ... C3, k8 56
Ribera, It., 18,547 ... F4 21
Ribérac, Fr., 2,242 ... E4 14
Riberalta, Bol., 6,549 ... B2 54
Rib Falls, Marathon, Wis., 80 ... D4 124
Rib Lake, Taylor, Wis., 794 . C3 124
Ribnitz, Ger., 14,800 ... D6 24
Ribstone, Alta., Can., 68 ... C5 69
Ricany, Czech., 6,376 ... o18 26
Ricardo, De Baca, N. Mex ... C5 107
Riccione, It., 10,600 ... B4 21
Rice, San Bernardino, Calif., 50 ... E6 82
Rice, Benton, Minn., 387 ... E4 99
Rice, Prince Edward, Va., 300 ... D4 121
Rice, co., Kans., 13,909 ... D5 93
Rice, co., Minn., 38,988 ... F5 93
Rice, creek, Minn ... E7 99
Rice, lake, Ont., Can ... C6 72
Rice, lake, Minn ... C3 99
Rice, lake, Minn ... D5 99
Rice, mtn., N.H ... B2 105
Riceboro, Liberty, Ga., 259 . E5 87
Rice Lake, Barron, Wis., 7,303 ... C2 124
Rices Landing, Greene, Pa., 693 ... G1 114
Riceton, Sask., Can., 130 ... G3 70
Riceville, Mitchell and Howard, Iowa, 898 ... A5 92
Riceville, McMinn, Tenn., 500 ... D9 117
Riceville, Pittsylvania, Va ... E3 121
Rich, Coahoma, Miss., 100 . A3 100
Rich, Mor., 2,455 ... H4 30
Rich, co., Utah, 1,685 ... B4 119
Rich, mtn., Ark., Okla ... C1 81
Rich, mtn., Va ... B3 121
Rich, mtn., W. Va ... C5 123
Richard, Sask., Can., 91 ... E2 70
Richard City, Marion, Tenn., 224 ... D8 117
Richards, Vernon, Mo., 133 . D3 101
Richards, Grimes, Tex., 500 . D5 118
Richards, lake, Ariz ... C5 80
Richard's Harbour, Newf., Can., 114 ... E3 75
Richardson, Lawrence, Ky., 125 ... C7 94
Richardson, Dallas, Tex., 16,810 ... B6 118
Richardson, co., Nebr., 13,903 ... D10 103
Richardson, lakes, Maine ... D2 96
Richardson, mts., Alsk., Yukon, Can ... C5 66
Richardsons, Tipton, Tenn . B2 117
Richardson Station, Sask., Can., 43 ... G3 70
Richards Spur, Comanche, Okla., 100 ... C3 112
Richardton, Stark, N. Dak., 792 ... C3 110
Richburg, Allegany, N.Y., 493 ... C2 108
Richburg, Chester, S.C., 235 . B5 115
Rich Creek, Giles, Va., 748 ... D2 121
Richdale, Alta., Can., 60 ... D5 69
Richelieu, co., Que., Can., 38,565 ... D4 73
Richey, Dawson, Mont., 480 . C11 102
Richfield, Tehama, Calif., 250 ... C2 82
Richfield, Lincoln, Idaho, 329 ... F4 89
Richfield, Morton, Kans., 122 ... E2 93
Richfield, Hennepin, Minn., 42,523 ... E7 99
Richfield, Sarpy, Nebr., 48 . E3 103
Richfield, Stanly, N.C., 293 . B3 109
Richfield, Juniata, Pa., 400 . E7 114
Richfield, Sevier, Utah, 4,412 ... E3 119
Richfield, Washington, Wis., 250 ... E1 124
Richfield Springs, Otsego, N.Y., 1,630 ... C6 108
Richford, Tioga, N.Y., 150 ... C4 108
Richford, Franklin, Vt., 1,663 (2,316▲) ... B3 120
Rich Fountain, Osage, Mo., 150 ... C6 101
Rich Hill, Bates, Mo., 1,699 . C3 101
Richibucto, N.B., Can., 1,375 . C5 74
Rich Lake, Alta., Can ... B5 69
Richland, Pasco, Fla ... D4 86
Richland, Stewart, Ga., 1,472 ... D2 87
Richland, Spencer, Ind., 100 . I3 91
Richland, Keokuk, Iowa, 546 . C6 92
Richland, Pulaski, Mo., 1,662 . D5 101
Richland, Valley, Mont., 95 ... B10 102
Richland, Colfax, Nebr., 139 . C8 103
Richland, Altantic, N.J., 500 . E3 106
Richland, Oswego, N.Y., 300 . B4 108
Richland, Baker, Oreg., 228 . C9 113
Richland, Lebanon, Pa., 1,276 ... *F9 114
Richland, Union, S. Dak., 55 ... E9 116
Richland, Benton, Wash., 23,548 ... C6 122
Richland, co., Ill., 16,299 ... E5 90
Richland, co., Mont., 10,504 . C12 102
Richland, co., N. Dak., 18,824 ... C9 110
Richland, co., Ohio, 117,761 . B5 111
Richland, co., S.C., 200,102 . D6 115
Richland, co., Wis., 17,684 ... E3 124
Richland, par., La., 23,824 ... B4 95
Richland, creek, Ill ... B5 101
Richland, creek, Tenn ... B5 117
Richland Center, Richland, Wis., 4,746 ... E3 124
Richland Hills, Tarrant, Tex., 7,804 ... *B5 118
Richlands, Onslow, N.C., 1,079 ... C6 109
Richlands, Tazewell, Va., 4,963 ... B3 121
Richland Springs, San Saba, Tex., 331 ... D3 118
Richlandtown, Bucks, Pa., 741 ... F11 114
Richlea, Sask., Can., 84 ... F1 70
Richmond, Austl., 775 ... D7 50
Richmond, Contra Costa, Calif., 71,854 ... B5, D2 82
Richmond, Ont., Can., 1,215 ... B9 72
Richmond, P.E.I., Can., 167 . C5 74
Richmond, Que., Can., 4,072 ... D5 73
Richmond, Eng., 5,764 ... F7 13
Richmond, Eng., 41,002 ... m11 10
Richmond, McHenry, Ill., 855 ... A5, D1 90
Richmond, Wayne, Ind., 44,149 ... E8 91
Richmond, Washington, Iowa, 150 ... C6 92
Richmond, Franklin, Kans., 352 ... D8 93
Richmond, Madison, Ky., 12,168 ... C5 94
Richmond, Sagadahoc, Maine, 1,412 ... D3, D6 96
Richmond, Berkshire, Mass., 130 (890▲) ... B1 97
Richmond, Macomb, Mich., 2,667 ... F8 98
Richmond, Stearns, Minn., 751 ... E4 99
Richmond, Ray, Mo., 4,604 . B4 101
Richmond, Cheshire, N.H., 70 (295▲) ... E2 105
Richmond, N.Z., 3,482 ... N14 51
Richmond, Jefferson, Ohio, 728 ... A1 123
Richmond, Wheeler, Oreg . C7 113
Richmond, S. Afr., 2,410 ... C5 49
Richmond, S. Afr ... D3 49
Richmond, Bedford, Tenn., 25 ... B5 117
Richmond, Fort Bend, Tex., 3,668 ... E5, F4 118
Richmond, Cache, Utah, 977 ... B4 119
Richmond, Chittenden, Vt., 765 (1,303▲) ... C3 120
Richmond (Independent City), Va., 219,958 (*409,100) ... C7, D5 121
Richmond, co., Que., Can., 42,232 ... D5 73
Richmond, co., N.S., Can., 11,374 ... D9 74
Richmond, co., Ga., 135,601 . C4 87
Richmond (Staten Island), borough and co., N.Y., 221,991 (part of New York City) ... E1 108
Richmond, co., N.C., 39,202 ... B4 109
Richmond, co., Va., 6,375 ... D6 121
Richmond Beach, King, Wash., 2,000 ... B3 122
Richmond Dale, Ross, Ohio, 800 ... C5 111
Richmond Heights, Dade, Fla., 4,311 ... F3 86
Richmond, Heights, St. Louis, Mo., 15,622 ... B8 101
Richmond Heights, Cuyahoga, Ohio, 5,068 ... *B2 111
Richmond Heights, Henrico, Va., 100 ... C7 121
Richmond Highlands, King, Wash., 6,000 ... *B3 122
Richmond Hill, Alamance, N.C., 2,943 ... *A4 109
Richmond Hill, Ont., Can., 16,446 ... E6, D5 72
Richmondville, Schoharie, N.Y., 743 ... C6 108
Richmound, Sask., Can., 215 ... G1 70
Rich Mountain, Polk, Ark ... C1 81
Rich Square, Northampton, N.C., 1,134 ... A6 109
Richthofen, mtn., Colo ... A5 83
Richton, Perry, Miss., 1,089 . D5 100
Richton Park, Cook, Ill., 933 ... *B6 90
Richvalley, Wabash, Ind., 150 ... C6 91
Richview, Washington, Ill., 255 ... E4 90
Richville, Tuscola, Mich., 400 ... E7 98
Richville, Otter Tail, Minn., 91 ... D3 99
Richville, St. Lawrence, N.Y., 292 ... f9 108
Richwood, Dooly, Ga., 35 ... D3 87
Richwood, Boone, Ky., 100 . A7 94
Richwood, Union, Ohio, 2,137 ... B4 111
Richwood, Nicholas, W. Va., 4,110 ... C4 123
Richwoods, Washington, Mo., 250 ... C7 101
Ricketts, Crawford, Iowa, 133 ... B2 92
Rickman, Overton, Tenn., 400 ... C8 117
Rickreall, Polk, Oreg., 150 ... C1, C3 113
Rico, Dolores, Colo., 353 ... D2 83
Rico, mts., Colo ... D2 83
Ridder, see Leninogorsk, Sov. Un.
Riddle, Owyhee, Idaho, 20 . G2 89
Riddle, Camden, N.C., 150 . A7 109
Riddle, Douglas, Oreg., 992 ... E3 113
Riddlesburg, Bedford, Pa., 420 ... F5 114
Riderwood, Choctaw, Ala., 150 ... C1 78
Ridge, Henrico, Va., 20,000 . *D5 121
Ridge, mtn., Va ... B2 121
Ridgebury, Fairfield, Conn., 175 ... D2 84
Ridgecrest, Kern, Calif., 5,099 ... *E5 82
Ridgecrest, Buncombe, N.C., 300 ... D4 109
Ridgecrest, King, Wash., 3,000 ... *B3 122
Ridgedale, Sask., Can., 191 . D3 70
Ridgedale, Taney, Mo., 135 . E4 101
Ridgedale, Knox, Tenn., 1,000 ... *D10
Ridge Farm, Vermilion, Ill., 894 ... D6 90
Ridgefield, Fairfield, Conn., 2,954 ... D3 84
Ridgefield, Bergen, N.J., 10,788 ... D5 106
Ridgefield Park, Bergen, N.J., 12,701 ... B4, D5 106
Ridgeland, Madison, Miss., 875 ... C3 100
Ridgeland, Jasper, S.C., 1,192 ... G6 115
Ridgeland, Dunn., Wis., 288 . C2 124
Ridgeley, Mineral, W. Va., 1,229 ... B6 123
Ridgely, Caroline, Md., 886 . C6 85
Ridgely, Lake, Tenn., 1,464 . A2 117
Ridgeside, Hamilton, Tenn., 448 ... E10 117
Ridge Spring, Saluda, S.C., 649 ... D4 115
Ridgetop, Davidson and Robertson, Tenn., 372 ... A5 117
Ridgetown, Ont., Can., 2,603 ... E3 72
Ridgeview, Miami, Ind., 439 ... C5 91

Ridgeview, Dewey, S. Dak., 40 ..... B5 116
Ridgeview, Boone, W. Va., 425 ..... C3, D5 123
Ridgeville, Man., Can., 86 ... E3 71
Ridgeville, McIntosh, Ga., 200 ..... E5 87
Ridgeville, Randolph, Ind., 950 ..... D7 91
Ridgeville, Frederick, Md., 200 ..... B3 85
Ridgeville, Caswell, N.C. A4 109
Ridgeville, Dorchester, S.C., 611 ..... E2, E7 115
Ridgeville Corners, Henry, Ohio, 400 ..... A3 111
Ridgeway, Ont., Can., 1,871 ..... E5 72
Ridgeway, Winneshiek, Iowa, 267 ..... A6 92
Ridgeway, Lenawee, Mich., 180 ..... F7 98
Ridgeway, Harrison, Mo., 470 ..... A4 101
Ridgeway, Warren, N.C., 250 ..... A5 109
Ridgeway, Logan and Hardin, Ohio, 448 ..... B4 111
Ridgeway, Fairfield, S.C., 417 ..... C6 115
Ridgeway, Henry, Va., 524 ... E3 121
Ridgeway, Iowa, Wis., 455 ... F3 124
Ridgeway, branch, N.J. C4 106
Ridgewood, Will, Ill., 5,586 . *B5 90
Ridgewood, Bergen, N.J., 25,391 ..... B4 106
Ridgewood Heights, Montgomery, Ohio, 1,500 ..... *C3 111
Ridgway, Ouray, Colo., 254 . C3 83
Ridgway, Gallatin, Ill., 1,055 . F4 90
Ridgway, Carter, Mont., 5 .. E12 102
Ridgway, Elk, Pa., 6,387 .... D4 114
Riding Mountain, nat. park, Man., Can D1 71
Ridley Farms, Delaware, Pa., 1,500 ..... *G11 114
Ridley Park, Delaware, Pa., 7,387 ..... B11 114
Ridlonville, Oxford, Maine (part of Mexico) D2 96
Ridott, Stephenson, Ill., 221 . A4 90
Ridotta Capuzzo, Libya, 1,983 ..... C4 43
Ridpath, Sask., Can., 65 ..... F1 70
Ried, Aus., 9,471 ..... D6 16
Riedland, McCracken, Ky ... A2 94
Riegelsville, Warren, N.J., 300 ..... B2 106
Riegelsville, Bucks, Pa., 953 .. B2 106
Riehen, Switz., 18,077 ..... A4 19
Rienza, riv., It C7 18
Rienzi, Alcorn, Miss., 375 ... A5 100
Riesa, Ger., 36,800 ..... B8 17
Riesel, McLennan, Tex., 503 . D4 118
Rieth, Umatilla, Oreg., 300 .. B8 113
Rieti, It., 23,500 (35,441▲) ... C4 21
Riffe, Lewis, Wash., 250 ..... C3 122
Riffle, Braxton, W. Va., 10 .. C4 123
Rifle, Garfield, Colo., 2,135 .. B3 83
Rifle, riv., Mich D6 98
Rift Valley, reg., Ken A6 48
Riga, Sov. Un., 607,000 ..... C5 27
Riga, gulf, Sov. Un C4 27
Rigaud, Que., Can., 1,990 .. D3 73
Rigby, Jefferson, Idaho, 2,281 ..... F7 89
Riggins, Idaho, Idaho, 588 .. D2 89
Riggisberg, Switz., 1,949 ..... C3 19
Rigi, peak, Switz B6 19
Rigili, isl., Eniwetok 52
Rigo, Pap k12 50
Rigolet, Newf., Can., 108 ..... A2, g10 75
Riihimäki, Fin., 20,200 ..... G11 25
Rijeka, Yugo., 100,339 ..... C2 22
Rijssen, Neth., 14,300 ..... B6 15
Rikers, isl., N.Y. D6 106
Riley, Vigo, Ind., 248 ..... F3 91
Riley, Riley, Kans., 575 ..... C7 93
Riley, Marion, Ky., 150 ..... C4 94
Riley, Socorro, N. Mex C2 107
Riley, co., Kans., 41,914 ..... C7 93
Riley, mtn., N. Mex F3 107
Rillito, Pima, Ariz., 250 ..... E4 80
Rilly-la-Montagne, Fr., 1,106 ..... E4 15
Rimatara, isl., Fr. Polynesia .. *H12 7
Rimbey, Alta., Can., 1,266 ... C3 69
Rimbo, Swe., 1,682 ..... t36 25
Rimersburg, Clarion, Pa., 1,323 ..... D2 114
Rimini, It., 63,000 ..... B4 21
Rimini, Lewis and Clark, Mont., 75 ..... D4 102
Rimouski, Que, Can., 17,739 ..... A9 73
Rimouski, co., Que., Can., 65,295 ..... A9 73
Rimouski, riv., Que., Can A9 73
Rimouski Est, Que., Can., 1,581 ..... A9 73
Rimrock, Yavapai, Ariz., 20 . C4 80
Rimrock, lake, Wash C4 122
Rinard, Calhoun, Iowa, 99 .. B3 92
Rinchhen Ling, China F3 34
Rincon, Effingham, Ga., 1,057 ..... D5 87
Rincon, Dona Ana, N. Mex., 300 ..... E2 107
Rincón, P.R., 1,094 ..... B1 65
Rincón, mun., P.R., 8,706 .. B1 65
Rinconada, Arg D2 65
Rinconada, Taos, N. Mex ... A4 107
Rincón de Romos, Mex., 5,856 ..... k12 63
Rindge, Cheshire, N.H., 100 (941▲) E2 105
Rindjani, peak, Indon G5 35
Riner, Montgomery, Va., 125 ..... D2 121
Rineyville, Hardin, Ky., 350 ..... C3 94
Ringe, Den., 2,936 ..... C4 24
Ringelheim, Ger., 2,640 ..... A5 17
Ringgold, Catoosa, Ga., 1,311 ..... B1 87
Ringgold, Bienville, La., 953 ..... B2 95
Ringgold, Washington, Md., 75 ..... A2 86
Ringgold, McPherson, Nebr., 23 ..... C5 103
Ringgold, Montague, Tex., 350 ..... C4 118
Ringgold, Pittsylvania, Va., 150 ..... E3 121
Ringgold, co., Iowa, 7,910 .. D3 92
Ringgold, is., Fiji Is 52
Ringim, Nig D6 45
Ringkøbing, Den., 4,869 ..... B2 24
Ringkøbing, co., Den., 198,389 ..... B2 24
Ringkøbing, fjord, Den C2 24
Ringling, Meagher, Mont., 65 ..... D6 102
Ringling, Jefferson, Okla., 1,170 ..... C4 112
Ringoes, Hunterdon, N.J., 550 ..... C3 106
Ringold, McCurtain, Okla., 50 ..... C6 112
Ringos Mills, Fleming, Ky., 45 ..... B6 94
Ringsaker, Nor G4 25
Ringsjön, lake, Swe C7 24
Ringso, isl., Swe u35 25
Ringsted, Den., 9,694 ..... C5 24
Ringsted, Emmet, Iowa, 559 ..... A3 92
Ringvassöy, isl., Nor C8 25
Ringville, Ire., 313 ..... E4 11
Ringwood, Passaic, N.J., 4,182 ..... A4 106
Ringwood, Major, Okla., 232 ..... A3 112
Riñihue, Chile B2 54
Rinteln, Ger., 9,700 ..... A4 17
Rio, Martin, Fla., 600 ..... E6 86
Rio, Knox, Ill., 177 ..... B3 90
Rio, Hampshire, W. Va., 100 . B6 123
Rio, Columbia, Wis., 788 ..... E4 124
Rio Arriba, co., N. Mex., 24,193 ..... A2 107
Río Balsas, Mex., 814 .. D5, o14 63
Riobamba, Ec., 29,611 ..... B2 58
Rio Blanco, Rio Blanco, Colo., 5 ..... B3 83
Río Blanco, P.R C7, g11 65
Rio Blanco, co., Colo., 5,150 . B2 83
Rio Branco, Braz., 17,245 ..... C4 58
Río Branco, Ur., 2,697 ..... E2 56
Rio Branco, ter., Braz., 29,489 ..... C5 60
Rio Branco do Sul, Braz., 715 ..... D3 56
Rio Bravo del Norte, see Rio Grande, riv., U.S., Mex.
Río Bueno, Chile, 6,259 ..... C2 54
Río Caribe, Ven., 7,188 ..... A5 60
Rio Chama, riv., N. Mex A3 107
Río Chico, Ven., 2,584 ..... A4 60
Rio Claro, Braz., 48,548 . C3, m8 56
Río Colorado, Arg., 3,304 .. B4 54
Río Cuarto, Arg., 70,000 ..... A4 54
Rio de Janeiro, Braz., 3,307,163 (*4,700,000) . C4, h6 56
Rio de Janeiro, state, Braz., 3,402,728 ..... C4, h6 56
Río de Jesús, Pan., 1,086 ..... G7 62
Rio Dell, Humboldt, Calif., 3,222 ..... B1 82
Río de Oro, Col., 1,679 ..... B3 60
Río de Oro, Sp. overseas prov., Afr., 1,304 ..... E2 44
Río Gallegos, Arg., 5,880 ..... E3, h12 54
Río Grande, Arg h12 54
Rio Grande, Braz., 83,189 ... E2 56
Río Grande, Mex., 8,208 ..... C4 63
Rio Grande, Cape May, N.J., 950 ..... E3 106
Río Grande, Nic., 173 ..... D6 62
Rio Grande, Gallia, Ohio, 333 ..... D5 111
Río Grande, P.R., 2,763 ..... B7 65
Rio Grande, co., Colo., 11,160 ..... D4 83
Río Grande, mun., P.R., 17,233 ..... B7, f11 65
Rio Grande, res., Colo D3 83
Rio Grande (Rio Bravo del Norte), riv., U.S., Mex B4 63, E7 76
Rio Grande City, Starr, Tex., 5,835 ..... F3 118
Rio Grande do Norte, state, Braz., 1,157,258 ..... C3, g6 56
Rio Grande do Sul, state, Braz., 5,448,823 ..... E2 56
Ríohacha, Col., 5,953 ..... A3 60
Río Hato, Pan., 2,725 ..... F7 62
Río Hondo, Mex., 1,718 ..... h9 63
Rio Hondo, Cameron, Tex., 1,344 ..... F4 118
Rio Hondo, riv., N. Mex D5 107
Rioja, Peru, 3,694 ..... C2 58
Río Jueyes, P.R C5 75
Rio Linda, Sacramento, Calif., 6,000 ..... *A6 82
Riom, Fr., 14,418 ..... E5 14
Rio Maior, Port., 3,055 ..... C1 20
Río Martin, Mor., 3,725 ..... G4 30
Rio Mulato, Bol., 381 ..... C2 55
Rio Muni, Sp. dep., Afr., 169,670 ..... E2 46
Rion, Fairfield, S.C., 500 ..... C5 115
Río Negro, Braz., 10,225 ..... D3 56
Río Negro, dept., Ur., 192,595 ..... *E1 56
Río Negro, prov., Arg., 41,400 ..... C3 54
Rionero in Vulture, It., 14,378 ..... D5 21
Rio Pardo, Braz., 14,412 ..... D2 56
Rio Pardo de Minas, Braz., 1,169 ..... E2 57
Rio Penasco, riv., N. Mex E5 107
Río Piedras, P.R. (part of San Juan), 65,406 ..... B6 65
Rio Pomba, Braz., 6,083 ..... g6 56
Río Primero, Arg A4 54
Rio Pueblo, Taos, N. Mex ... A4 107
Rio Puerco, riv., N. Mex B3 107
Rio Salado, riv., N. Mex C2 107
Río Seco, Chile D1 55
Ríosucio (Caldas dept.), Col., 7,363 ..... B2 60
Ríosucio (Choco dept.), Col., 847 ..... B2 60
Río Tercero, Arg., 10,683 ... A4 54
Rio Verde, Braz., 11,268 ..... B2 56
Río Verde (Ríoverde), Mex., 14,294 ..... m13 63
Rio Vista, Solano, Calif., 2,616 ..... B3, C3 82
Ripley, Riverside, Calif., 5 ... F6 82
Ripley, Ont., Can., 464 ..... C3 72
Ripley, Eng., 17,601 ..... A6 12
Ripley, Brown, Ill., 167 ..... C3 90
Ripley, Somerset, Maine, 125 (317▲) D3 96
Ripley, Tippah, Miss., 2,668 . A5 100
Ripley, Chautauqua, N.Y., 1,247 ..... C1 108
Ripley, Brown, Ohio, 2,174 .. D4 111
Ripley, Payne, Okla., 263 ..... A5 112
Ripley, Lauderdale, Tenn., 3,782 ..... B2 117
Ripley, Jackson, W. Va., 2,756 ..... C3 123
Ripley, co., Ind., 20,641 ..... F7 91
Ripley, co., Mo., 9,096 ..... E7 101
Riplinger, Clark, Wis., 100 .. D3 124
Ripogenus, lake, Maine C3 96
Ripoll, Sp., 9,034 ..... A7 20
Ripon, San Joaquin, Calif., 1,894 ..... B6 82
Ripon, Que., Can., 576 ..... D2 73
Ripon, Eng., 10,490 ..... C6 10
Ripon, Fond du Lac, Wis., 6,163 ..... E5 124
Rippey, Greene, Iowa, 331 .. C3 92
Ripple, mtn., B.C., Can A8 122
Ripples, N.B., Can., 233 ..... C3 74
Rippowam, riv., Conn., N.Y. ... A5 106
Ripton, Addison, Vt., 70 (131▲) D2 120
Ririe, Bonneville and Jefferson, Idaho, 560 ..... F7 89
Risco, New Madrid, Mo., 502 . E8 101
Rishiri, isl., Jap D10 37
Rishon le-Zion, Isr., 27,998 ..... C6, h10 32
Rising City, Butler, Nebr., 308 ..... C8 103
Rising Star, Eastland, Tex., 997 ..... C3 118
Rising Sun, Ohio, Ind., 2,230 ..... G8 91
Risingsun, Polk, Iowa, 50 ... A7 92
Rising Sun., Cecil, Md., 824 . A5 86
Risingsun, Wood, Ohio, 815 . A4 111
Risle, riv., Fr C4 14
Rison, Cleveland, Ark., 889 . D3 81
Risör, Nor., 3,100 ..... H3 25
Ritchey, Newton, Mo., 128 .. E3 101
Ritchie, co., W. Va., 10,877 .. B3 123
Ritidian, pt., Guam 52
Ritter, Grant, Oreg., 15 ..... C7 113
Ritter, mtn., Calif D4 82
Rittman, Wayne, Ohio, 5,410 ..... B6 111
Ritzville, Adams, Wash., 2,173 ..... B7 122
Riva, It., 6,839 ..... B3 21
Rivadavia, Arg., 5,643 ..... A3 54
Rivadavia, Arg., 4,925 ..... B4 54
Rivadavia, Arg D3 55
Rivadavia, Chile E1 55
Rivanna, riv., Va D4 121
Rivas, Nic., 8,700 ..... E5 62
Rive-de-Gier, Fr., 16,565 ... E6 14
River, Huntington, Ind., 75 .. C6 91
Rivera, Arg., 2,569 ..... B4 54
Rivera, Ur., 36,700 ..... E1 56
Rivera, dept., Ur., 68,700 .. *E1 56
Riverbank, Stanislaus, Calif., 2,786 ..... D3 82
River Bourgeois, N.S., Can., 432 ..... D9 74
River Cess, Lib E3 45
Riverdale, Fresno, Calif., 1,012 ..... D4 82
Riverdale, Clayton, Ga., 1,045 ..... *C2 87
Riverdale, Cook, Ill., 12,008 . F3 90
Riverdale, Sumner, Kans., 60 ..... E6 93
Riverdale, Prince Georges, Md., 4,389 ..... C2, C4 85
Riverdale, Essex, Mass. (part of Gloucester) A6 97
Riverdale, Gratiot, Mich., 380 ..... E6 98
Riverdale, Buffalo, Nebr., 144 ..... D6 103
Riverdale, Morris, N.J., 2,596 ..... B4 106
Riverdale, McLean, N. Dak., 1,055 ..... B4 110
Riverdale, Multnomah, Oreg., 1,500 ..... *B4 113
Riverdale, Weber, Utah, 1,848 ..... *B4 119
Riverdale Heights, Prince Georges, Md., 1,800 ..... *C4 85
River Edge, Bergen, N.J., 13,264 ..... D5 106
River Falls, Covington, Ala., 401 ..... D3 78
River Falls, Pierce and St. Croix, Wis., 4,857 ..... D1 124
River Falls, dam, Ala D3 78
River Forest, Cook, Ill., 12,695 ..... F2 90
Rivergaro, It., 4,810 ..... E5 18
River Grove, Cook, Ill., 8,464 ..... E2 90
Riverhead, Suffolk, N.Y., 5,830 ..... D4 108
River Herbert, N.S., Can., 1,382 ..... D5 74
River Hills, Milwaukee, Wis., 1,257 ..... E2 124
Riverhurst, Sask., Can., 281 . G2 70
Riverina, reg., Austl G5 51
River John, N.S., Can., 397 .. D7 74
River Jordan, B.C., Can., 217 ..... A1 122
Riverland Terrace, Charleston, S.C., 2,400 ..... *F2 115
River Oaks, Tarrant, Tex., 8,444 ..... *C4 118
River of Ponds, Newf., Can., 228 ..... C3 75
River Pines, Middlesex, Mass., 800 ..... *C2 97
River Plaza, Monmouth, Pa., 4,500 ..... *C4 114
Riverport, N.S., Can., 369 ... E5 74
River Rouge, Wayne, Mich., 18,147 ..... A7, F7 98
Rivers, Man., Can., 1,574 ... D1 71
Rivers, inlet, B.C., Can D4 68
Riversdale, N.S., Can., 58 ... D7 74
Riverside, St. Clair, Ala., 159 . B3 78
Riverside, Riverside, Calif., 84,332 ..... F3, F5 82
Riverside, Ont., Can., 18,089 . E2 72
Riverside, Douglas, Colo., 180 ..... B6 83
Riverside, Fairfield, Conn. (part of Greenwich) E2 84
Riverside, Colquitt, Ga., 329 . E3 87
Riverside, Cook, Ill., 9,750 ... F2 90
Riverside, Washington, Iowa, 656 ..... C6 92
Riverside, Charles, Md., 100 . D3 85
Riverside, Platte, Mo., 1,315 . B3 101
Riverside, Burlington, N.J., 8,474 ..... C3 106
Riverside, Steuben, N.Y., 1,030 ..... *C3 108
Riverside, Malheur, Oreg., 5 ..... D8 113
Riverside, Northumberland, Pa., 1,580 ..... E8 114
Riverside, Greenville, S.C., 1,200 ..... *B3 100
Riverside, Box Elder, Utah, 150 ..... B3 119
Riverside, Okanogan, Wash., 201 ..... A6 122
Riverside, Carbon, Wyo., 87 ..... D6 125
Riverside, Co., Calif., 306,191 ..... F5 82
Riverside, res., Colo A6 83
Riverside Park, Burlington, N. J., 800 ..... *C3 106
River Sioux, Harrison, Iowa, 150 ..... C2 92
Riversville, Fairfield, Conn. (part of Greenwich) E2 84
Riverton, Colbert, Ala., 50 . A1 78
Riverton, Man., Can., 808 .. D3 71
Riverton, Litchfield, Conn., 240 ..... B4 84
Riverton, Sangamon, Ill., 1,591 ..... D4 90
Riverton, Fremont, Iowa, 399 ..... D2 92
Riverton, Cherokee, Kans., 250 ..... E9 93
Riverton, Wicomico, Md., 100 ..... C6 85
Riverton, Crow Wing, Minn., 121 ..... D4 99
Riverton, Franklin, Nebr., 303 ..... D7 103
Riverton, Coos, N.H., 200 ... B3 105
Riverton, Burlington, N.J., 3,324 ..... C2 106
Riverton, N.Z., 1,225 ..... Q12 51
Riverton, Coos, Oreg., 250 .. D2 113
Riverton, Salt Lake, Utah, 1,993 ..... C4 119
Riverton, Washington, Vt., 100 ..... C3 120
Riverton, Warren, Va., 250 .. C4 121
Riverton, Pendleton, W. Va., 150 ..... C5 123
Riverton, Fremont, Wyo., 6,845 ..... B4 125
Riverton Heights, King, Wash., 19,000 ..... *B3 122
River Vale, Bergen, N.J., 5,616 ..... *A5 106
River View, Chambers, Ala., 1,171 ..... C4 78
Riverview, Kern, Calif., 7,000 ..... *E4 82
Riverview, Duval, Fla., 4,000 ..... B6 86
Riverview, Hillsborough, Fla., 1,000 ..... E2 86
Riverview, Wayne, Mich., 7,237 ..... A7 98
Riverview, St. Louis, Mo., 3,706 ..... *C7 101
Riverview (Pasco West) Franklin, Wash., 2,894 ..... *C6 122
Riverview, dam, Ala., Ga C4 88
Riverville, Amherst, Va., 50 . D4 121
Rives, Dunklin, Mo., 134 ... E7 101
Rives, Obion, Tenn., 291 ..... A2 117
Rivesaltes, Fr., 5,910 ..... F5 14
Rives Junction, Jackson, Mich., 300 ..... F6 98
Rivesville, Marion, W. Va., 1,191 ..... A7, B4 123
Riviera, Kleberg, Tex., 600 . F4 118
Riviera Beach, Palm Beach, Fla., 13,046 ..... F6 86
Riviere-a-Claude, Que., Can., 253 ..... A3 73
Riviere a Pierre, Que., Can., 812 ..... C5 73
Riviere Bleue, Que., Can., 1,540 ..... B8 73
Riviere des Prairies, Que., Can., 10,054 ..... *D9 73
Riviere du Loup, Que., Can., 10,835 ..... B8 73
Riviere du Loup, co., Que., Can., 40,239 ..... B8 73
Riviere du Milieu, Que., Can B5 73
Riviere du Moulin, Que., Can., 4,386 ..... A6 73
Riviere la Madeleine, Que., Can., 289 ..... A3 73
Riviere Ouelle, Que., Can., 89 ..... B7 73
Riviere Raquette, Que., Can., 119 ..... C7 73
Riviere Trois Pistoles, Que., Can., 357 ..... A8 73
Riviere Verte, N.B., Can., 918 ..... B1 74
Riviere Verte, Que., Can., 715 ..... B8 73
Rivulet, Mineral, Mont., 30 .. C2 102
Rixford, McKean, Pa., 650 ... C4 114
Riyadh (Ar Riyāḍ) Sau. Ar., 150,000 ..... I3 41
Rizaiyeh (Rezāiyeh, Urmia), Iran, 67,605 ..... C2 41
Rizal, prov., Phil., 1,463,500 ..... *C6 35
Rize, Tur., 22,300 ..... B13 31
Rizokarpaso, Cyp., 3,667 .. E10 31
Rjukan, Nor., 5,677 ..... H3 25
Roachdale, Putnam, Ind., 927 ..... E4 91
Road Forks, Hidalgo, N. Mex E1 107
Roads, Jackson, Ohio, 130 ... C5 111
Roadstown, Cumberland, N.J., 130 ..... E2 106
Road Town, Vir. Is., 891 ..... m14 64, f16 65
Roan or Brown, cliffs, Colo., Utah D6 119
Roan, mtn., Tenn C11 117
Roan, plat., Colo B2 83
Roane, co., Tenn., 39,133 ... D9 117
Roane, co., W. Va., 15,720 . C3 123
Roan High Knob, peak, Tenn . C11 117
Roan Mountain, Carter, Tenn., 800 ..... C11 117
Roann, Wabash, Ind., 478 ... C6 91
Roanne, Fr., 51,723 ..... D5 14
Roanoke, Randolph, Ala., 5,288 ..... B4 78
Roanoke, Woodford, Ill., 1,821 ..... C4 90
Roanoke, Huntington, Ind., 935 ..... C7 91
Roanoke, Jefferson Davis, La., 600 ..... D3 95
Roanoke, Denton, Tex., 585 . A5 118
Roanoke (Independent City), Va., 97,110 (*160,400) D3 121
Roanoke, Lewis, W. Va., 100 ..... C4 123
Roanoke, co., Va., 61,693 ... D2 121
Roanoke, isl., N.C B8 109
Roanoke, pt., N.Y F6 84
Roanoke, riv., N.C., Va E4 121
Roanoke Rapids, Halifax, N.C., 13,320 ..... A6 109
Roanoke Rapids Lake, res., N.C A6 109
Roaring, fork, Colo B4 83
Roaring Branch, Lycoming, Pa., 250 ..... C8 114
Roaring Bulls, is., Mass D4 97
Roaring Creek, Columbia, Pa E9 114
Roaring Spring, Blair, Pa., 2,937 ..... F5 114
Roaring Springs, Motley, Tex., 398 ..... C2 118
Roaringwater, bay, Ire F2 11
Roark, Leslie, Ky., 500 ..... C6 94
Roatán, Hond., 1,094 ..... B4 62
Roba, Macon, Ala., 50 ..... C4 78
Robanna, Gloucester, N.J. ... D2 106
Robards, Henderson, Ky., 375 ..... C2 94
Robāṭ-e Khān, Iran, 10,000 . E8 41
Robb, Alta., Can., 271 ..... C2 69
Robbin, Kittson, Minn., 70 .. B1 99
Robbins, Cook, Ill., 7,511 ... F3 90
Robbins, Moore, N.C., 1,294 . B4 109
Robbins, Scott, Tenn., 550 . C9 117
Robbins, pt., Md B5 85
Robbinsdale, Hennepin, Minn., 16,381 ..... E5, E6 99
Robbinston, Washington, Maine, 250 (476▲) C5 96
Robbinsville, Mercer, N.J., 300 ..... C3 106
Robbinsville, Graham, N.C., 587 ..... D2 109
Robbs, Pope, Ill., 100 ..... F5 90
Robe, riv., Ire D2 11
Robeline, Natchitoches, La., 308 ..... C2 95
Roberdel, Richmond, N.C., 379 ..... C4 109
Robersonville, Martin, N.C., 1,684 ..... B6 109
Robert, Tangipahoa, La., 150 ..... A7, D5 95
Robert, cape, Ont., Can A2 72
Roberta, Crawford, Ga., 714 . D2 87
Robert English, coast, Ant B5 5
Robert Lee, Coke, Tex., 990 ..... D2 118
Roberts, Jefferson, Idaho, 422 . F6 89
Roberts, Ford, Ill., 504 ..... C5 90
Roberts, Newton, Miss., 25 .. C4 100
Roberts, Carbon, Mont., 240 . E7 102
Roberts, St. Croix, Wis., 308 . D1 124
Roberts, co., S. Dak., 13,190 . B8 116
Roberts, co., Tex., 1,075 ..... B2 118
Roberts, pt., Wash A2 122
Roberts Creek, mtn., Nev D5 104
Robertsdale, Baldwin, Ala., 1,474 ..... E2 78
Robertsdale, Huntingdon, Pa., 800 ..... F3 114
Robertson, S. Afr., 8,166 ... D3 49
Robertson, Uinta, Wyo., 15 . D2 125
Robertson, co., Ky., 2,443 ... B5 94
Robertson, co., Tenn., 27,335 . A5 117
Robertson, co., Tex., 16,157 . D4 118
Robertson, bay, Ant B30 5
Robertson, lake, Que., Can C2 75
Robertsonville, Que., Can., 1,156 ..... C6 73
Robertsport, Lib E2 45
Robertstown, White, Ga., 400 ..... B3 87
Robertsville, Litchfield, Conn., 85 ..... B4 84
Roberval, Que., Can., 7,739 . A5 73
Robeson, co., N.C., 89,102 ... C4 109
Robesonia, Berks, Pa., 1,579 . *F9 114
Robinette, Baker, Oreg C9 113
Robin Hood's Bay, Eng F8 13
Robins, Linn, Iowa, 426 ..... B6 92
Robins, Guernsey, Ohio, 250 . C6 111
Robins, isl., N.Y F7 84
Robinson, Crawford, Ill., 7,226 ..... D6 90
Robinson, Brown, Kans., 317 . C8 93
Robinson, Kidder, N. Dak., 155 ..... B6 110
Robinson, Indiana, Pa., 900 . F3 114
Robinson, McLennan, Tex., 2,111 ..... *D4 118
Robinson, fork, W. Va A6 123
Robinson, fork, W. Va C7 123
Robinson, mtn., Wash A5 122
Robinsons, Aroostook, Maine, 125 ..... B5 96
Robinson's, Newf., Can., 322 . D2 75
Robinsonville, Tunica, Miss., 115 ..... A3 100
Robinvale, Austl., 194 ..... G4 51
Robinwood, Jefferson, Ala., 1,000 ..... *B3 78
Roblin, Man., Can., 1,368 ... D1 71
Roblin, riv., Man., Can ... A5, D5 71
Robsart, Sask., Can., 110 ... H1 70
Robson, B.C., Can., 909 ..... E9 68
Robson, mtn., B.C., Can C8 68
Robstown, Nueces, Tex., 10,266 ..... F4 118
Roby, Tex., Mo., 100 ..... D5 101
Roby, Fisher, Tex., 913 ..... C2 118
Roca, Lancaster, Nebr., 123 ..... F2 103
Roca, cape, Port f9 20
Rocafuerte, Ec., 2,788 ..... B1 58
Rocafuerte, Peru B2 58
Rocanville, Sask., Can., 496 ..... G5 70
Rocas, is., Braz B4 57
Rocca Massima, It., 1,961 ... h9 21
Roccastrada, It., 3,109 ..... C3 21
Rocha, Ur., 18,200 ..... E2 56
Rocha, dept., Ur., 53,400 .. *E2 56
Rochdale, Eng., 85,785 ..... A5 12
Rochdale, Worcester, Mass., 1,058 ..... B4 97
Rochechouart, Fr., 1,936 ... E4 14
Rochefort, Bel., 4,003 ..... D5 15
Rochefort, Fr., 28,648 ..... E3 14
Roche Harbor, San Juan, Wash., 20 ..... A2 122
Rochelaise Central, P.R C2 65
Rochelle, Wilcox, Ga., 1,235 . E3 87
Rochelle, Ogle, Ill., 7,008 .. B4 90
Rochelle, Grant, La., 175 ... C3 95
Rochelle, McCulloch, Tex., 300 ..... D3 118
Rochelle Park, Bergen, N.J., 6,119 ..... D5 106
Roche Percee, Sask., Can., 177 ..... H4 70
Rocheport, Boone, Mo., 375 . C5 101
Rochester, Alta., Can., 83 .. B4 69
Rochester, Eng., 50,121 ..... C8 12
Rochester, Sangamon, Ill., 742 ..... D4 90
Rochester, Fulton, Ind., 4,883 ..... B5 91
Rochester, Cedar, Iowa, 40 .. C6 92
Rochester, Butler, Ky., 314 ... C3 94
Rochester, Plymouth, Mass., 300 (1,559▲) C6 97
Rochester, Oakland, Mich., 5,431 ..... F7 98
Rochester, Olmsted, Minn., 40,663 ..... F6 99
Rochester, Strafford, N.H., 15,927 ..... D5 105
Rochester, Monroe, N.Y., 318,611 (*594,500) B3 108
Rochester, Lorain, Ohio, 226 . A5 111
Rochester, Beaver, Pa., 5,952 . E1 114
Rochester, Haskell, Tex., 625 . C3 118
Rochester, Windsor, Vt., 350 (879▲) D3 120
Rochester, Thurston, Wash., 350 ..... C2 122
Rochester, Racine, Wis., 413 . F1 124
Rochfort Bridge, Alta, Can., 85 ..... C3 69
Rochfort Bridge, Ire., 365 ... D4 11
Rochikarai, isl., Bikini 52
Rochlitz, Ger., 7,872 ..... B7 17
Rociada, San Miguel, N. Mex., 50 ..... B4 107
Rock, Cowley, Kans., 139 ... E7 93
Rock, Delta, Mich., 475 ..... B3 98
Rock, co., Minn., 11,864 ..... G2 99
Rock, co., Nebr., 2,554 ..... B6 103
Rock, co., Wis., 113,913 ..... F4 124
Rock, creek, Ill B4 90
Rock, creek, Kans B4 93
Rock, creek, Md B3 85
Rock, creek, Nebr E1 103
Rock, creek, Nev C5 104
Rock, creek, Wash D5 122
Rock, creek, Wash E8 122
Rock, creek, Wyo D6 125
Rock, isl., Fla C3 86
Rock, isl., Wis C7 124
Rock, lake, Man., Can E2 71
Rock, lake, Wash B8 122
Rock, mtn., Ala B2 78
Rock, mtn., Va E2 121
Rock, riv., Ill., Wis . B4 90, E5 124
Rock riv., Iowa, Minn . A1 92, G2 99
Rock, riv., Wash B8 122
Rockawalkin, Wicomico, Md . D6 85
Rockaway, Morris, N.J., 5,413 ..... B3 106
Rockaway, Tillamook, Oreg., 771 ..... B3 113
Rockaway, beach, N.Y E5 106
Rockaway, inlet, N.Y E5 106
Rockaway Beach, Taney, Mo., 117 ..... E4 101
Rock Bay, B.C., Can., 60 ... D5 68
Rock Bluff, Liberty, Fla., 200 . B2 86
Rockbridge, Greene, Ill., 253 . D3 90
Rockbridge, Hocking, Ohio, 400 ..... C5 111
Rockbridge, co., Va., 24,039 . D3 121
Rockcastle, co., Ky., 12,334 . C5 94
Rockcastle, riv., Ky C5 94
Rockcliffe Park, Ont., Can., 2,084 ..... *A9 72
Rock Creek, B.C., Can., 222 . A6 122
Rock Creek, Jefferson, Kans., 100 ..... B7 93
Rock Creek, Ashtabula, Ohio, 673 ..... A7 111
Rock Creek, Gilliam, Oreg ... B6 113
Rock Creek, Pickett, Tenn ... C9 117
Rock Creek, butte, Oreg C8 113
Rock Creek Hills, Montgomery, Md., 1,500 ..... *B3 85
Rockdale, Will, Ill., 1,272 ..... B5, F2 90
Rockdale, Dubuque, Iowa, 60 . B7 92
Rockdale, Baltimore, Md., 1,500 ..... *B4 85
Rockdale, Maury, Tenn., 40 . B4 117
Rockdale, Milam, Tex., 4,481 ..... D4 118
Rockdale, co., Ga., 10,572 ... C3 87
Rockeagle, Goshen, Wyo ... D8 125
Rock Eagle, mound, Ga C3 87
Rockefeller, plat., Ant B36 5
Rock Elm, Pierce, Wis., 75 . D1 124
Rocker, Silver Bow, Mont., 110 ..... E4 102
Rockerville, Pennington, S. Dak., 20 ..... D2 116
Rockfall, Middlesex, Conn., 550 ..... C6 84
Rock Falls, Whiteside, Ill., 10,261 ..... B4 90
Rock Falls, Cerro Gordo, Iowa, 156 ..... A4 92
Rockfield, Carroll, Ind., 350 . C4 91
Rockfield, Warren, Ky., 150 . D3 94
Rockfield, Washington, Wis., 200 ..... E1 124
Rockford, Coosa, Ala., 328 .. C3 78
Rockford, Winnebago, Ill., 126,706 (*191,100) A4 90
Rockford, Floyd, Iowa, 941 .. A5 92
Rockford, Kent, Mich., 2,074 . E5 98
Rockford, Wright and Hennepin, Minn., 533 ..... E5 99
Rockford, Mercer, Ohio, 1,155 ..... B3 111
Rockford, Blount, Tenn., 900 ..... D10, E11 117
Rockford, Spokane, Wash., 369 ..... B8 122
Rockford Bay, Kootenai, Idaho, 25 ..... D8 122
Rockglen, Sask., Can., 492 .. H3 70
Rock Hall, Kent, Md., 1,073 . B5 85
Rockham, Faulk, S. Dak., 197 ..... C10 116
Rockhampton, Austl., 44,128 . A8 51
Rockhaven, Sask., Can., 83 .. E1 70
Rock Hill, St. Louis, Mo., 6,523 ..... *C7 101
Rock Hill, York, S.C., 29,404 ..... B6 115
Rockhill Furnace, Huntingdon, Pa., 566 ..... F6 114
Rockholds, Whitley, Ky., 500 ..... D5 94
Rockingham, Bacon, Ga., 40 ..... E4 87
Rockingham, Ray, Mo., 28 .. B4 101
Rockingham, Richmond, N.C., 5,512 ..... C4 109
Rockingham, Windham, Vt., 65 (5,704▲) E3 120
Rockingham, co., N.H., 99,029 ..... D4 105

Rockingham, co., N.C., 69,629 A4 109
Rockingham, co., Va., 40,485 C4 121
Rock Island, Que., Can., 1,608 D5 73
Rock Island, Rock Island, Ill., 51,863 B3 90
Rock Island, Douglas, Wash., 260 B5 122
Rock Island, co., Ill., 150,991 B3 90
Rocklake, Towner, N. Dak., 350 A6 110
Rockland, Ont., Can., 3,037 B9 72
Rockland, Power, Idaho, 258 G6 89
Rockland, Knox, Maine, 8,769 D3 96
Rockland, Plymouth, Mass., 13,119 B6, E3 97
Rockland, Ontonagon, Mich., 500 A6 98
Rockland, Sullivan, N.Y., 300 D6 108
Rockland, co., N.Y., 136,803 D6 108
Rockledge, Brevard, Fla., 3,481 D6 86
Rockledge, Laurens, Ga., 100 D4 87
Rockledge, Montgomery, Pa., 2,587 A11 114
Rocklin, Placer, Calif., 1,495 *C3 82
Rockmart, Polk, Ga., 3,938 C1 87
Rock Point, Apache, Ariz., 50 A6 80
Rock Point, Charles, Md., 400 D4 85
Rockport, Hot Spring, Ark., 162 D6 81
Rockport, Mendocino, Calif., 90 C2 82
Rockport, Pike, Ill., 300 D2 90
Rockport, Spencer, Ind., 2,474 I3 91
Rockport, Ohio, Ky., 396 C3 94
Rockport, Knox, Maine, 900 (1,893▲) D3 96
Rockport, Essex, Mass., 3,511 A6 97
Rockport, Copiah, Miss D3 100
Rock Port, Atchison, Mo., 1,310 A2 101
Rockport, Warren, N.J. B3 106
Rockport, Aransas, Tex., 2,989 E4 118
Rockport, Skagit, Wash., 185 A4 122
Rock Rapids, Lyon, Iowa, 2,780 A1 92
Rock Rift, Delaware, N.Y., 150 C5 108
Rock River, Albany, Wyo., 497 D7 125
Rock Run, Cherokee, Ala., 150 A4 78
Rocks, Harford, Md., 150 A5 85
Rock Springs, Rosebud, Mont., 5 D10 102
Rocksprings, Edwards, Tex., 1,182 D2 118
Rock Springs, Sauk., Wis., 463 E4 124
Rock Springs, Sweetwater, Wyo., 10,371 D3 125
Rockstone, Br. Gu A3 59
Rockton, Winnebago, Ill., 1,833 A4 90
Rockton, Clearfield, Pa., 200 D4 114
Rocktown, Hunterdon, N.J C3 106
Rockvale, Fremont, Colo., 413 C5 83
Rockvale, Rutherford, Tenn., 150 B5 117
Rock Valley, Sioux, Iowa, 1,693 A1 92
Rockville, Clarke, Ala D2 78
Rockville, Tolland, Conn., 9,478 B7 84
Rockville, Parke, Ind., 2,756 E3 91
Rockville, Montgomery, Md., 26,090 B3 85
Rockville, Stearns, Minn., 357 E4 99
Rockville, Bates, Mo., 355 C3 101
Rockville, Sherman, Nebr., 153 C7 103
Rockville, Malheur, Oreg D9 113
Rockville, Washington, R.I., 250 C9 84
Rockville, Washington, Utah, 142 F2 119
Rockville Centre, Nassau, N.Y., 26,355 E3 108
Rockwall, Rockwall, Tex., 2,166 B6, C4 118
Rockwall, co., Tex., 5,878 C4 118
Rockwell, Cerro Gordo, Iowa, 772 B4 92
Rockwell, Rowan, N.C., 948 B3 109
Rockwell City, Calhoun, Iowa, 2,313 B3 92
Rockwood, Franklin, Ala., 200 A2 78
Rockwood, Ont., Can., 863 D4 72
Rockwood, Randolph, Ill., 98 F4 90
Rockwood, Somerset, Maine, 200 C3 96
Rockwood, Wayne, Mich., 2,026 F7 98
Rockwood, Somerset, Pa., 1,101 G3 114
Rockwood, Roane, Tenn., 5,345 D9 117
Rockwood, Coleman, Tex., 80 D3 118
Rocky, Washita, Okla., 343 B2 112
Rocky, bay, Newf., Can B4 75
Rocky, lake, Man., Can B1 71
Rocky, lake, Maine D5 96
Rocky, mtn., Mont C4 102
Rocky, mts., N.A E9 61
Rocky, pt., N.Y E7 84
Rocky, riv., Alta., Can C2 69
Rocky, riv., N.C B3 109
Rocky, riv., Ohio B2 111
Rocky, riv., S.C C2 115
Rocky Bar, Elmore, Idaho, 5 F3 89
Rocky Boys, Indian res., Mont B7 102
Rocky Comfort, McDonald, Mo., 151 E3 101
Rocky Coulee, creek, Wash B6 122
Rockyford, Alta., Can., 288 D4 69
Rocky Ford, Otera, Colo., 4,929 C7 83
Rocky Ford, Screven, Ga., 241 D8 87
Rocky Fork Lake, res., Ohio C4 111
Rocky Gap, Bland, Va., 250 D1 121
Rockygrove, Venango, Pa., 3,168 D2 114
Rocky Hill, Hartford, Conn., 7,404 C6 84
Rocky Hill, Edmonson, Ky., 178 C3 94
Rocky Hill, Somerset, N.J., 528 C3 106
Rockyhock, Chowan, N.C., 100 A7 109
Rocky Knob, hill, Ohio C5 111
Rocky Mount, Edgecombe and Nash, N.C., 32,147 B6 109
Rocky Mount, Franklin, Va., 1,412 D3 121
Rocky Mountain, nat. park, Colo A5 83
Rocky Mountain House, Alta., Can., 2,360 C3 69
Rocky Point, Benewah, Idaho, 30 B2 89
Rocky Point, Suffolk, N.Y., 2,261 F5 84
Rocky Point, Pender, N.C., 416 C6 109
Rocky Point, Kitsap, Wash., 1,000 *B3 122
Rockypoint, Campbell, Wyo., 25 A7 125
Rocky Ripple, Marion, Ind., 967 H8 91
Rocky River, Cuyahoga, Ohio, 18,097 A6, B2 111
Rocky River, Van Buren, Tenn D8 117
Rocky Top, mtn., Oreg C4 113
Roda, Wise, Va., 300 B2 121
Rodach, Ger., 4,500 C5 17
Rodanthe, Dare, N.C., 95 B8 109
Rödby, Den., 3,551 D5 24
Rödby Havn, Den D5 24
Roddickton, Newf., Can., 1,185 C3, h10 75
Rödding, Den., 628 C3 24
Rödding, see Spöttrup, Den.
Roddy, Rhea, Tenn., 60 D9 117
Rödekro, Den., 1,621 C3 24
Rodeo, Contra Costa, Calif., 5,400 *B5 82
Rodeo, Hidalgo, N. Mex., 150 F1 107
Roder, riv., Ger B8 17
Roderfield, McDowell, W. Va., 1,020 *D3 123
Roderick, isl., B.C., Can C3 68
Rodessa, Caddo, La., 700 B1 95
Rodewisch, Ger., 12,800 C7 17
Rodez, Fr., 20,924 E5 14
Rodgers Forge, Baltimore, Md., 7,645 *B4 85
Rodhopi (Rhodope), prov., Grc., 109,201 *B5 23
Ródhos, Grc., 27,393 D7 23
Ródhos (Rhodes), isl., Grc D6 23
Roding, Ger., 4,000 D7 17
Roding, riv., Eng k13 10
Rodinga, Austl D5 50
Rodman, Palo Alto, Iowa, 144 A3 92
Rodman, Chester, S.C., 225 B5 115
Rodnei, mts., Rom B7 22
Rodney, Ont., Can., 1,041 E3 72
Rodney, Jefferson, Miss., 110 D2 100
Rodney, pond, Newf., Can D4 75
Rodney Village, Kent, Del., 1,200 B6 85
Rödovre, Den., 39,345 *C6 24
Roduco, Gates, N.C., 200 A7 109
Roe, Monroe, Ark., 100 C4 81
Roebling, Burlington, N.J., 3,272 C3 106
Roebourne, Austl., 136 D2 50
Roebuck, Spartanburg, S.C., 300 B4 115
Roebuck, bay, Austl C3 50
Roebuck Plaza, Jefferson, Ala., 1,000 *B3 78
Roeland Park, Johnson, Kans., 8,949 B9 93
Roer, riv., Ger D6 15
Roermond, Neth., 34,500 C5 15
Roeselare, Bel., 35,645 D3 15
Roessleville, Albany, N.Y., 19,000 *C7 108
Roes Welcome, sound, N.W. Ter., Can D15 66
Roetgen, Ger., 2,900 D6 15
Roff, Pontotoc, Okla., 638 C5 112
Rogachev, Sov. Un., 25,000 E8 27
Rogaland, co., Nor., 239,000 *H1 25
Roganville, Jasper, Tex., 240 D6 118
Rogatica, Yugo., 3,044 D4 22
Roger, lake, Conn D7 84
Roger Mills, co., Okla., 5,090 B2 112
Rogers, Benton, Ark., 7,600 A1 81
Rogers, Windham, Conn., 650 B9 84
Rogers, Hennepin, Minn., 378 E5 99
Rogers, Colfax, Nebr., 162 C9 103
Rogers, Roosevelt, N. Mex., 60 D6 107
Rogers, Barnes, N. Dak., 119 B7 110
Rogers, Columbiana, Ohio, 295 B7 111
Rogers, Bell, Tex., 936 D4 118
Rogers, co., Okla., 20,614 A6 112
Rogers, mtn., Va B3 121
Rogers, pass, Mont C4 102
Rogers City, Presque Isle, Mich., 4,722 C7 98
Rogers Heights, Prince Georges, Md., 2,000 *C4 85
Rogerson, Twin Falls, Idaho, 100 G4 89
Rogersville, Lauderdale, Ala., 766 A2 78
Rogersville, N.B., Can., 1,040 C4 74
Rogersville, Webster, Mo., 447 D4 101
Rogersville, Greene, Pa., 300 G1 114
Rogersville, Hawkins, Tenn., 3,121 C10 117
Roggen, Weld, Colo., 75 A6 83
Rogozno, Pol., 5,536 B4 26
Rogue, riv., Oreg E2 113
Rogue River, Jackson, Oreg., 520 E3 113
Rogue River, range, Oreg E3 113
Rohnerville, Humboldt, Calif., 1,000 *B1 82
Rohrersville, Washington, Md., 170 B2 85
Rohri, canal, Pak D2 40
Rohtak, India, 88,193 *C6 40
Rohwer, Desha, Ark., 86 D4 81
Roi Et, Thai, 12,119 D5 38
Roi, isl., Kwajalein 52
Roig Central, P.R C7 65
Roissy-en-France, Fr., 1,243 g11 14
Roj, isl., Ponape 52
Rojas, Arg., 6,608 A4 54
Rojo, cape, P.R D2 67, n13 64
Rokan, riv., Indon L4 38
Rokeby, Sask., Can., 35 F4 70
Rokel, riv., S.L E2 45
Rokitno, Sov. Un., 9,000 F6 27
Rokycany, Czech., 12,000 D2 26
Roland, Pulaski, Ark., 550 C3, D5 81
Roland, Man., Can., 374 E3 71
Roland, Story, Iowa, 748 B4 92
Roland, Sequoyah, Okla., 100 B7 112
Roland, lake, Md C2 85
Roland Terrace, Anne Arundel, Md., 2,000 *B4 85
Rolesville, Wake, N.C., 358 B5 109
Rolette, Rolette, N. Dak., 524 A6 110
Rolette, co., N. Dak., 10,641 A6 110
Rolfe, Pocahontas, Iowa, 819 B3 92
Rolfsö, isl., Nor B10 25
Roll, Yuma, Ariz., 100 E2 80
Roll, Blackford, Ind., 115 C7 91
Roll, Roger Mills, Okla., 25 B2 112
Rolla, B.C., Can., 69 B7 68
Rolla, Morton, Kans., 464 E2 93
Rolla, Phelps, Mo., 11,132 D6 101
Rolla, Rolette, N. Dak., 1,398 A6 110
Rolleville, Ba. Is., 604 D5 64
Rolling, fork, Ark C1 81
Rollingbay, Kitsap, Wash., 600 D1 122
Rolling Fork, Sharkey, Miss., 1,619 C3 100
Rolling Fork, riv., Ky C4 94
Rolling Hills, Los Angeles, Calif., 1,664 *F2 82
Rolling Hills, Sedgwick, Kans., 2,000 B5 93
Rolling Hills Estates, Los Angeles, Calif., 3,941 *F2 82
Rolling Meadows, Cook, Ill., 10,879 *E2 90
Rolling Prairie, LaPorte, Ind., 700 A4 91
Rollingstone, Winona, Minn., 392 F7 99
Rollingwood, Contra Costa, Calif., 2,200 *B5 82
Rollins, Lake, Mont., 107 C2 102
Rollinsford, Strafford, N.H., 150 (1,935▲) D5 105
Rollo, Switz., 2,677 D1 19
Roma, Austl., 5,571 C7 51
Roma, see Rome, It.
Roma, Starr, Tex., 1,496 F3 118
Romain, cape, S.C F9 115
Romaine, riv., Que., Can h9 75
Roman, Rom., 27,948 B8 22
Romania, country, Eur., 17,489,794 F13 8, B7 22
Roman Nose, mtn., Md D1 85
Roman Nose, mtn., Oreg D3 113
Romano, cape, Fla G5 86
Romano, isl., Cuba D5 64
Romanshorn, Switz., 7,755 A7 19
Romans [sur-Isère], Fr., 26,377 E6 14
Romanzof, cape, Alsk C6 79
Romanzof, mts., Alsk B11 79
Romayor, Liberty, Tex., 200 D5 118
Rombauer, Butler, Mo., 150 E7 101
Romblon, Phil., 5,000 (16,700▲) *C6 35
Romblon, prov., Phil., 132,000 *C6 35
Rome, Floyd, Ga., 32,226 B1 87
Rome, Jefferson, Ill., 181 E5 90
Rome, Peoria, Ill., 1,347 C4 90
Rome, Perry, Ind., 60 I4 91
Rome, Henry, Iowa, 117 D6 92
Rome (Roma), It., 2,188,160 (★2,340,000) D4, h8 21
Rome, Sunflower, Miss., 279 B3 100
Rome, Douglas, Mo., 20 E5 101
Rome, Oneida, N.Y., 51,646 B5 108
Rome, Franklin, Ohio, 1,500 *C5 111
Rome, Bradford, Pa., 274 C9 114
Rome City, Noble, Ind., 900 B7 91
Romema, Isr. (pop. Incl. in Jerusalem) m14 32
Romeo, Conejos, Colo., 339 D4 83
Romeo, Will, Ill., 3,574 *B5 90
Romeo, Macomb, Mich., 3,327 F7 98
Romero, Hartley, Tex., 20 B1 118
Romeroville, San Miguel, N. Mex., 30 B4 107
Romeville, St. James, La., 150 B6 95
Romford, Litchfield, Conn C3 84
Romilly [-sur-Seine], Fr., 15,753 C5 14
Rominger, Watauga, N.C., 100 A2 109
Romita, Mex., 10,377 m13 63
Romney, Tippecanoe, Ind., 350 D4 91
Romney, Hampshire, W. Va., 2,203 B6 123
Romny, Sov. Un., 10,000 F9 27
Römö, isl., Den C2 24
Romona, Owen, Ind F4 91
Romont, Switz., 2,892 C2 19
Romorantin, Fr., 11,777 D4 14
Rompin, riv., Mala K5 38
Romsdalsfjord, fjord, Nor F2 25
Romsey, Eng., 6,229 D6 12
Romulus, Wayne, Mich., 350 A7 98
Romulus, Seneca, N.Y., 250 C4 108
Romurikku, isl., Bikini 52
Ron, Viet D7 38
Rona, isl., Scot C2 13
Ronald, Kittitas, Wash., 250 B4 122
Ronan, Lake, Mont., 1,334 C2 102
Roncador, mts., Braz E4 59
Roncador Bank, shoals, Caribbean Sea D7 62
Ronceverte, Greenbrier, W. Va., 1,882 D4 123
Ronciglione, It., 7,079 C4 21
Ronda, Wilkes, N.C., 510 A3 109
Ronda, Sp., 16,100 D3 20
Rönde, Den., 1,384 B4 24
Rondo, Lee, Ark., 219 C5 81
Rondônia, prov., Braz., 70,783 E2 59
Rondout, Lake, Ill., 200 E2 90
Roneys Point, Ohio, W. Va., 175 B2 123
Rong, isl., Camb G5 38
Rongerik, atoll, Marshall Is F10 7
Ronkiti, Ponape 52
Ronkiti, hbr., Ponape 52
Ronkonkoma, Suffolk, N.Y., 5,666 F4 84
Ronkonkoma, lake, N.Y F4 84
Rönne, Den., 13,195 J6 25
Ronne, entrance, Ant B5 5
Rönne, riv., Swe B6 24
Rönneburg, Ger., 12,000 C3 17
Ronse, Bel., 25,106 D3 15
Ronuro, riv., Braz E7 59
Roodepoort-Maraisburg, U.S. Afr., 95,211 *C4 49
Roodeschool, Neth., 748 A6 15
Roodhouse, Greene, Ill., 2,352 D3 90
Roof, butte, Ariz A6 80
Rooks, co., Kans., 9,734 C4 93
Roopville, Carroll, Ga., 203 C1 87
Roosendaal, Neth., 38,800 C4 15
Roosevelt, Gila, Ariz., 80 C3, D4 80
Roosevelt, White, Ark B4 81
Roosevelt, Roseau, Minn., 145 B3 99
Roosevelt (Jersey Homesteads), Monmouth, N.J., 764 C4 106
Roosevelt, Nassau, N.Y., 12,883 G2 84
Roosevelt, Kiowa, Okla., 495 C2 112
Roosevelt, Kimble, Tex., 80 D3 118
Roosevelt, Duchesne, Utah, 1,812 C5 119
Roosevelt, Klickitat, Wash., 60 D5 122
Roosevelt, co., Mont., 11,731 B11 102
Roosevelt, co., N. Mex., 16,198 C6 107
Roosevelt, isl., Ant B32 5
Roosevelt, riv., Braz D2 59
Roosevelt Park, Muskegon, Mich., 2,578 E4 98
Root, mtn., B.C., Can D1 68
Root, riv., Minn G7 99
Root, riv., Wis F2 124
Ropczyce, Pol., 2,822 C6 26
Roper, Washington, N.C., 771 B7 109
Roper, riv., Austl B5 50
Ropesville, Hockley, Tex., 423 C1 118
Roque Bluffs, Washington, Maine, 60 (152▲) D5 96
Roquefort, Fr., 1,465 E3 14
Roques, is., Ven A4 60
Roques, pt., Cuba h11 64
Roquetas, Sp., 5,514 B6 20
Rora, head, Scot B5 13
Roraima, mtn., Ven B5 60
Rorke, lake, Man., Can B5 71
Rorketon, Man., Can., 273 D2 71
Röros, Nor., 2,643 F4 25
Rorschach, Switz., 12,759 (★22,358) B7 19
Rosa, St. Landry, La., 80 D3 95
Rosa, mtn., It., Switz D3 18
Rosa, pt., Mex B3 63
Rosaire, Que., Can., 544 C7 73
Rosales, Phil., 4,411 o13 35
Rosalia, Butler, Kans., 200 E7 93
Rosalia, Whitman, Wash., 585 B8 122
Rosalie, Thurston, Nebr., 182 B9 103
Rosalind, Alta., Can., 197 C4 69
Rosamond, Kern, Calif., 700 E4 82
Rosamond, Christian, Ill., 250 D4 90
Rosa Morada, Mex., 1,664 k11 63
Rosario, Arg., 595,000 A4 54
Rosário, Braz., 6,999 B2 57
Rosario, Mex., 11,608 C3 63
Rosario, Par., 6,058 D4 55
Rosario, Phil., 3,150 p13 35
Rosario, P.R C2 65
Rosario, Ur., 7,600 E1 56
Rosario, de la Frontera, Arg., 4,927 E3 55
Rosário do Sul, Braz., 15,786 E2 56
Rosário Oeste, Braz., 2,607 A1 56
Rosario Tala, Arg., 10,584 A5 54
Rosati, Phelps, Mo., 200 C6 101
Rosboro, Pike, Ark., 75 C2 81
Rosburg, Wahkiakum, Wash., 50 C2 122
Roscoe, Winnebago, Ill., 556 A4 90
Roscoe, Stearns, Minn., 168 E4 99
Roscoe, St. Clair, Mo., 125 D4 101
Roscoe, Carbon, Mont., 45 E7 102
Roscoe, Keith, Nebr., 90 C4 103
Roscoe, Sullivan, N.Y., 900 D6 108
Roscoe, Washington, Pa., 1,315 F2 114
Roscoe, Edmunds, S. Dak., 532 B6 116
Roscoe, Nolan, Tex., 1,490 C2 118
Roscommon, Ire., 1,600 D3 11
Roscommon, Roscommon, Mich., 867 D6 98
Roscommon, co., Ire., 59,217 D3 11
Roscommon, co., Mich., 7,200 D6 98
Roscrea, Ire., 3,372 E4 11
Rose, Rock, Nebr., 200 B6 103
Rose, Wayne, N.C B5 109
Rose, Mayes, Okla., 45 A6 112
Rose, peak, Ariz D6 80
Roseau, Roseau, Minn., 2,146 B3 99
Roseau, Dominica, 10,422 o16 64
Roseau, co., Minn., 12,154 B3 99
Roseau, riv., Man., Can E3 71
Roseau, riv., Minn B2 99
Roseberry, B.C., Can., 52 D9 68
Rose Blanche, Newf., Can., 626 E2 75
Roseboro, Sampson, N.C., 1,354 C5 109
Rose Bud, White, Ark., 120 B3 81
Rosebud, Alta., Can., 99 D4 69
Rosebud, Gasconade, Mo., 288 C6 101
Rosebud, Rosebud, Mont., 250 D10 102
Rosebud, Harding, N. Mex B6 107
Rosebud, Todd, S. Dak., 600 D5 116
Rosebud, Falls, Tex., 1,644 D4 118
Rosebud, co., Mont., 6,187 D10 102
Rosebud, creek, Mont E10 102
Rosebud, Indian res., S. Dak D5 116
Rosebud, riv., Alta., Can D4 69
Roseburg, Douglas, Oreg., 11,467 D3 113
Rosebush, Isabella, Mich., 400 E6 98
Rose City, Ogemaw, Mich., 435 D6 98
Rose Creek, Mower, Minn., 351 G6 99
Rosedale, B.C., Can., 654 A11 68
Rosedale, Manatee, Fla., 4,085 *E4 86
Rosedale, Parke, Ind., 726 E3 91
Rosedale, Bolivar, Miss., 2,339 B2 100
Rosedale, Anderson, Tenn., 150 C9 117
Rosedale, Pierce, Wash., 30 D1 122
Rosedale, Braxton, W. Va., 100 C4 123
Rosedale Abbey, Eng F8 13
Rosedale Station, Alta., Can., 1,200 D4 69
Roseglen, McLean, N. Dak., 40 B4 110
Rosehearty, Scot., 1,140 C6 11
Rose Hill, Jasper, Ill., 117 D5 90
Rose Hill, Mahaska, Iowa, 223 C5 92
Rose Hill, Butler, Kans., 273 E6 93
Rose Hill, Jasper, Miss., 100 C4 100
Rose Hill, Duplin, N.C., 1,292 C5 109
Rose Hill, Lee, Va., 600 B1 121
Roseisle, Man., Can., 66 E2 71
Roseland, Sonoma, Calif., 4,510 *C2 82
Roseland, Indian River, Fla., 145 E6 86
Roseland, St. Joseph, Ind., 971 A5 91
Roseland, Tangipahoa, La., 1,254 D5 95
Roseland, Adams, Nebr., 163 D7 103
Roseland, Essex, N.J., 2,804 *B4 106
Roseland, Richland, Ohio, 8,204 *B5 111
Roselawn, Newton, Ind., 150 B3 91
Roselle, DuPage, Ill., 3,581 E2 90
Roselle, Union, N.J., 21,032 E4 106
Roselle Park, Union, N.J., 12,546 E4 106
Rose Lynn, Alta., Can., 10 D5 69
Rosemark, Shelby, Tenn., 250 B2 117
Rosemary, Alta., Can., 210 D4 69
Rosemead, Los Angeles, Calif., 15,476 *F2 82
Rosemere, Que., Can., 6,158 D8 73
Rose Mills, Nelson, Va D4 121
Rosemont, Cook, Ill., 978 *A6 90
Rosemont, St. Clair, Ill., 4,000 *E3 90
Rosemont, Hunterdon, N.J., 90 C3 106
Rosemont, Delaware and Montgomery, Pa., 4,000 *A10 114
Rosemont, Taylor, W. Va., 500 B7 123
Rosemount, Dakota, Minn., 1,068 F5 99
Rosenberg, Fort Bend, Tex., 9,698 F4 118
Rosendael, Fr., 19,960 B5 14
Rosendale, Andrew, Mo., 234 A3 101
Rosendale, Ulster, N.Y., 1,033 *D6 108
Rosendale, Fond du Lac, Wis., 415 E5 124
Roseneath, Ont., Can., 78 C6 72
Rosenfeld, Man., Can., 316 E3 71
Rosenhayn, Cumberland, N.J., 600 E2 106
Rosenheim, Ger., 31,600 E6 16
Rosepine, Vernon, La., 414 D2 95
Rose Prairie, B.C., Can A7 68
Roseray, Sask., Can G1 70
Roseto, Northampton, Pa., 1,630 E11 114
Roseton, Orange, N.Y., 180 D6 108
Rosetown, Sask., Can., 2,450 F1 70
Rosetta, Johnson, Ark B2 81
Rosetta (Rashīd), Eg., U.A.R., 32,800 C6 43
Rosetta, Wilkinson, Miss., 129 D2 100
Rosetta, riv. mouth, Eg., U.A.R C2 43
Rosette, Box Elder, Utah B2 119
Rose Valley, Sask., Can., 627 E4 70
Rosevear, Alta., Can., 15 C2 69
Roseville, Placer, Calif., 13,421 C3 82
Roseville, Warren, Ill., 1,065 C3 90
Roseville, Macomb, Mich., 50,195 A8 98
Roseville, Ramsey, Minn., 23,997 E7 99
Roseville, Muskingum and Perry, Ohio, 1,749 C5 111
Roseville, Stafford, Va., 30 C5 121
Rosewood, Humboldt, Calif., 1,100 *B1 82
Rosewood, Champaign, Ohio, 275 B4 111
Rosewood Heights, Madison, Ill., 4,572 *E3 90
Rosharon, Brazoria, Tex., 280 G5 118
Rosheim, Fr., 3,004 F7 15
Rosh Ha'ayin, Isr., 7,500 g10 32
Rosholt, Roberts, S. Dak., 423 B9 116
Rosholt, Portage, Wis., 497 D4 124
Rosh Pina, Isr., 1,200 B7 32
Rosiclare, Hardin, Ill., 1,700 F5 90
Rosie, Independence, Ark., 150 B4 81
Rosier, Burke, Ga., 30 D4 87
Rosières-en-Santerre, Fr., 2,381 E2 15
Rosignol, Br. Gu., 1,204 A3 59
Rosillo, peak, Tex E1 118
Rosine, Ohio, Ky., 350 C3 94
Rosiorii-de-Vede, Rom., 17,320 C7 22
Rosita, Custer, Colo., 10 C5 83
Roskilde, Den., 31,928 C6 24
Roskilde, co., Den., 82,223 C6 24
Roslavl, Sov. Un., 36,000 E9 27
Roslin, Fentress, Tenn., 50 C9 117
Roslyn, Nassau, N.Y., 2,681 *E3 108
Roslyn, Montgomery, Pa., 8,500 *A11 114
Roslyn, Day, S. Dak., 256 B8 116
Roslyn, Kittitas, Wash., 1,283 B4 122
Roslyn Estates, Nassau, N.Y., 1,289 *F2 84
Roslyn Harbor, Nassau, N.Y., 925 *F2 84
Roslyn Heights, Nassau, N.Y., 4,000 F2 84
Rosman, Transylvania, N.C., 419 D3 109
Rosny-sous-Bois, Fr., 21,001 g10 14
Ross, Marin, Calif., 2,551 *B5 82
Ross, Winston, Miss., 50 B4 100
Ross, N.Z., 503 O13 51
Ross, Mountrail, N. Dak., 167 A3 110
Ross, Butler, Ohio, 800 C2, C3 111
Ross, co., Ohio, 61,215 C4 111
Ross, dam, Wash A4 122
Ross, ice shelf, Ant A31 5
Ross, isl., Ant B29 5
Ross, isl., Bur F3 38
Ross, isl., Man., Can B3 71
Ross, lake, Wash A4 122
Ross, mtn., N.Z N15 51
Ross, sea, Ant B31 5
Rossa, Switz., 155 D7 19
Ross and Cromarty, co., Scot., 57,607 C4 13
Rossano, It., 12,400 E6 21
Rossburg, Darke, Ohio, 295 B3 111
Rossburn, Man., Can., 591 D1 71
Rosscarbery, Ire., 380 F2 11
Rosseau, Ont., Can., 233 B5 72
Rosseau, Ontonagon, Mich A6 98
Rossendale, Man., Can., 150 E2 71
Rosser, Kaufman, Tex., 300 B6 118
Rosses, bay, Ire B3 11
Rosses Point, Ire., 319 C3 11
Rossford, Wood, Ohio, 4,406 A2, A4 111
Ross Fork, Fergus, Mont., 10 C7 102
Rossie, St. Lawrence, N.Y., 150 B1 108
Rossignol, lake, N.S., Can E4 74
Rossinière, Switz., 504 D3 19
Rossiter, Indiana, Pa., 950 E4 114
Rossland, B.C., Can., 4,354 E9 68
Rosslare, Ire., 529 E5 11
Rosslau, Ger., 16,000 B7 17
Rosslea, N. Ire., 203 C4 11
Rossmoor, Orange, Calif., 9,000 *F2 82
Rossmoyne, Hamilton, Ohio, 2,000 *D2 111
Rosso, Maur., 2,300 C1 45
Ross-on-Wye, Eng., 5,643 C5 12
Rossosh, Sov. Un., 10,000 F12 27
Rosston, Nevada, Ark., 250 D2 81
Rosston, Harper, Okla., 58 A2 112
Rossville, Walker, Ga., 4,665 B1 87
Rossville, Vermilion, Ill., 1,470 C6 90
Rossville, Clinton, Ind., 831 D4 85
Rossville, Allamakee, Iowa, 100 A6 92
Rossville, Shawnee, Kans., 797 C8 93
Rossville, Fayette, Tenn., 183 B2 117
Rossville, Atascosa, Tex E3 118
Rossway, N.S., Can., 192 E4 74
Rosthern, Sask., Can., 1,264 E2 70
Rostock, Ger., 161,800 A6 14, D6 24
Rostov, Sov. Un., 29,200 C12 27
Rostov [-na-Donu], Sov. Un., 645,000 (★735,000) H12 27
Rosul, pass, Rom C6 22
Roswell, El Paso, Colo. (part of Colorado Springs) C6 83
Roswell, Fulton, Ga., 2,983 A5, B2 87
Roswell, Canyon, Idaho, 100 F2 89
Roswell, Chaves, N. Mex., 39,593 D5 107
Roswell, Miner, S. Dak., 39 C8 116
Rota, Sp., 16,856 D2 20
Rotan, Fisher, Tex., 2,788 C2 118
Rotenburg, Ger., 14,500 B4 16
Rotenburg [an der Fulda], Ger., 7,700 B4 17
Roth, Bottineau, N. Dak., 25 A5 110
Roth, Ger., 10,300 D6 17
Rothaar, mts., Ger B3 17
Rothbury, Eng., 1,648 E6 13
Rothbury, Oceana, Mich., 200 E4 98
Röthenbach, Ger., 9,600 D6 17
Röthenbach im Emmental, Switz., 1,368 C4 19
Rothenburg [in der Lausitz], Ger., 2,587 B9 17
Rothenburg ob der Tauber, Ger., 11,100 D5 17
Rothenthurm, Switz., 1,159 B6 19
Rother, riv., Eng D8 12
Rotherham, Eng., 85,346 A6 12
Rothes, Scot., 1,105 B5 13
Rothesay, N.B., Can., 782 D4 74
Rothesay, Scot., 7,656 E3 13
Rothsay, Wilkin, Minn., 457 D2 99
Rothschild, Marathon, Wis., 2,550 D4 124
Rothsville, Lancaster, Pa., 900 F9 114
Rothville, Chariton, Mo., 138 B4 101
Roto, Austl., 127 F5 51
Rotondella, It., 5,643 D6 21
Rotorua, N.Z., 19,360 M16 51
Rott, riv., Ger E8 17
Rottenburg, Ger., 10,800 E3 17
Rottenmann, Aus., 4,139 E7 16
Rotterdam, Neth., 729,800 (★990,000) C4 15
Rotterdam, Schenectady, N.Y., 16,871 *C6 108
Rotterdam Junction, Schenectady, N.Y., 756 C6 108
Rottumeroog, isl., Neth A6 15
Rottweil, Ger., 17,900 D4 16
Rotuma, isl., Pac. O G10 7
Roubaix, Fr., 112,856 B5 14, D3 15
Roudnice nad Labem, Czech., 8,683 n17 26
Rouen, Fr., 120,857 (★325,000) C4 14
Rouge, riv., Que., Can D3 73
Rougemont, Fr., 878 B2 18
Rougemont, Durham, N.C., 400 A5 109
Rough, riv., Ky C3 94
Roughneck, peak, Idaho E3 89
Rougon, Pointe Coupee, La., 375 D4 95
Rouleau, Sask., Can., 436 G3 70
Roulette, Potter, Pa., 700 C5 108
Round, hill, Conn E2 84
Round, hill, Va B4 121

Round, isl., Miss. E3 100
Round, lake, Ont., Can. B7 72
Round, lake, Wis. B2 124
Round, mtn., Austl. E9 51
Round, mtn., Kans. D4 73
Round Bay, Anne Arundel, Md., 600 B4 85
Round Harbour, Newf., Can., 63 D4 75
Roundhead, Hardin, Ohio, 150 B4 111
Round Hill, Alta., Can., 160 C4 69
Round Hill, N.S., Can., 300 E4 74
Round House, mtn., Kans. D4 93
Round Island, passage, Fiji Is. 52
Round Knob, mtn., Tenn. C3 117
Round Lake, Lake, Ill., 997 E2 90
Round Lake, Nobles, Minn., 449 G3 99
Roundlake, Bolivar, Miss., 175 A3 100
Round Lake, Saratoga, N.Y., 900 C7 108
Round Lake Beach, Lake, Ill., 5,011 *E2 90
Round Lake Heights, Lake, Ill., 900 *E2 90
Round Lake Park, Lake, Ill., 2,565 *E2 90
Round Mountain, Franklin, Maine C2 96
Round Mountain, Nye, Nev., 300 E4 104
Round Mountain, Blanco, Tex., 75 D3 118
Round Oak, Jones, Ga., 200 C3 87
Round Pond, St. Francis, Ark., 50 B5 81
Round Pond, Lincoln, Maine, 300 E3 96
Round Rock, Apache, Ariz., 20 A6 80
Round Rock, Williamson, Tex., 1,878 D4 118
Rounds, B.C., Can. E3 68
Round Spring, caverns, Mo. D6 101
Roundstone, Ire., 250 D2 11
Roundup, Musselshell, Mont., 2,842 D8 102
Round Valley, Indian res., Calif. C2 82
Roura, Fr. Gu., 437 B4 59
Rousay, isl., Scot. A5 13
Rouses Point, Clinton, N.Y., 2,160 A3 108
Rouseville, Venango, Pa., 923 D2 114
Rousseau, Ontonagon, Mich. A6 98
Roussillon, former prov., Fr., 217,000 *F5 14
Routhierville, Que., Can., 219 B2 73
Routon, Henry, Tenn., 40 A3 117
Routt, co., Colo., 5,900 A3 83
Rouville, co., Que., Can., 25,979 D4 73
Rouyn, Que., Can., 18,716 E9 72
Rouzerville, Franklin, Pa., 900 G6 114
Rovaniemi, Fin., 21,500 D11 25
Rovato, It., 6,288 B2 21
Rovenki, Sov. Un., 10,000 G12, q22 27
Rover, Yell, Ark., 200 C2 81
Rover, Bedford, Tenn., 120 B5 117
Roveredo, Switz., 1,878 D7 19
Rovereto, It., 25,638 B3 21
Rovigno, see Rovinj, Yugo.
Rovigo, It., 33,000 (45,649▲) B3 21
Rovinj, Yugo., 7,156 C1 22
Rovno, Sov. Un., 68,000 F6 27
Rovnoye, Sov. Un., 10,000 F16 27
Rowan, Wright, Iowa, 273 B4 92
Rowan, co., Ky., 12,808 B6 94
Rowan, co., N.C., 82,817 B3 109
Rowan, lake, Ont., Can. E5 71
Rowan Mill, Rowan, N.C., 1,089 *B3 109
Rowayton, Fairfield, Conn. (part of Norwalk) E3 84
Rowe, Franklin, Mass., 130 (231▲) A2 97
Rowe, San Miguel, N. Mex., 350 B4, G6 107
Rowena, Runnels, Tex., 300 D2 118
Rowes Run, Fayette, Pa., 950 *F2 114
Rowesville, Orangeburg, S.C., 398 E6 109
Rowland, Lincoln, Ky., 200 C5 94
Rowland, Robeson, N.C., 1,408 C4 109
Rowland, Pike, Pa., 100 D11 114
Rowlesburg, Preston, W. Va., 970 B5 123
Rowlett, Dallas, Tex., 1,015 *C4 118
Rowletts, Hart, Ky., 275 C4 94
Rowley, Alta., Can., 65 D4 69
Rowley, Buchanan, Iowa, 234 B6 92
Rowley, Essex, Mass., 1,223 A6 97
Rox, Lincoln, Nev., 15 F7 104
Roxabell, Ross, Ohio, 175 C4 111
Roxana, Sussex, Del., 100 D7 85
Roxana, Madison, Ill., 2,090 A8 101
Roxas (Capiz), Phil., 18,000 (49,400▲) C6 35
Roxboro, Person, N.C. 5,147 A5 109
Roxburgh, N.Z., 771 P12 51
Roxburgh, co., Scot., 43,171 E6 13
Roxbury, Litchfield, Conn., 225 (912▲) C3 84
Roxbury, McPherson, Kans., 135 D6 93
Roxbury, Oxford, Maine, 250 (344▲) D2 96
Roxbury, Cheshire, N.H., 50 (137▲) E2 105
Roxbury, Delaware, N.Y., 500 C6 108
Roxbury, Washington, Vt., 225 (364▲) C3 120
Roxbury Falls, Litchfield, Conn. C3 84
Roxie, Franklin, Miss., 585 D2 100
Roxobel, Bertie, N.C., 452 A6 109
Roxton, Lamar, Tex., 950 C5 118
Roxton Falls, Que., Can., 972 D5 73
Roxton Pond, Que., Can., 770 D5 73
Roy, Flagler, Fla., 350 C5 86
Roy, Bienville, La., 250 B2 95
Roy, Fergus, Mont., 175 C8 102
Roy, Harding, N. Mex., 633 B5 107
Roy, Weber, Utah, 9,239 B3 119
Royal, Galland, Ark., 25 C2, C5 81
Royal, Sumter, Fla., 200 D4 86
Royal, Cary, Iowa, 475 A2 92
Royal, Antelope, Nebr., 93 B7 103
Royal, Beaufort, N.C., 200 B7 109
Royal, Carbon, Utah, 100 D5 119
Royal, Carbon, Utah, 100 D5 119
Royal, canal, Ire. D4 11
Royal, gorge, Colo. C5 83
Royal, riv., Maine E5 96
Royal Center, Cass, Ind., 966 C4 91
Royal Mills, Wake, N.C., 600 A5 109
Royal Oak, B.C., Can. B9 68
Royal Oak, Talbot, Md., 500 C5 85
Royal Oak, Oakland, Mich., 80,612 A7, E7 98
Royal Oak Township, Oakland, Mich., 8,147 *A7 98
Royalties, Alta., Can., 156 D3 69
Royalton, Franklin, Ill., 1,225 F4 90
Royalton, Boone, Ind., 60 H7 91
Royalton, Magoffin, Ky., 300 C6 94
Royalton, Morrison, Minn., 580 E4 99
Royalton, Dauphin, Pa., 1,128 *F8 114
Royalton, Windsor, Vt., 150 (1,388▲) D3 120
Royalton, Waupaca, Wis., 300 D5 124
Royalty, Ward, Tex., 200 D1 118
Royan, Fr., 16,521 E3 14
Roy Brown, Lander, Nev., 10 D4 104
Roye, Fr., 4,912 C5 14
Royersford, Montgomery, Pa., 3,969 F10 114
Roy Hill, Austl. D2 50
Roy Knob, mtn., Tenn. D9 117
Royse City, Collin and Rockwall, Tex., 1,274 C4 118
Royston, Franklin, Hart and Madison, Ga., 2,333 B3 87
Roysville, Lib. E2 45
Rozay-en-Brie, Fr., 1,483 F2 15
Rozel, Pawnee, Kans., 207 D4 93
Rozet, Campbell, Wyo., 15 A7 125
Roznava, Czech., 10,200 D6 26
Rtishchevo, Sov. Un., 32,000 C2 29
Ruac, isl., Truk 52
Ruaha, Tan. C6 48
Ruaha, riv., Tan. C6 48
Ruapehu, mtn., N.Z. M15 51
Rub al Khali, des., Sau. Ar. B6 47
Rubezhnoye, Sov. Un., 50,000 p21 27
Rubidoux (West Riverside), Riverside, Calif., 8,000 *F3 82
Rubinéia, Braz. C2 56
Rubonia, Manatee, Fla., 400 F2 86
Rubtsovsk, Sov. Un., 123,000 C10 29
Ruby, Alsk., 157 C8 79
Ruby, Rapides, La., 25 C3 95
Ruby, Chesterfield, S.C., 284 B7 115
Ruby, lake, Nev. C6 104
Ruby, mts., Nev. C6 104
Ruby, range, Colo. C3 83
Ruby, range, Mont. E4 102
Rubys Inn, Garfield, Utah, 50 F3 119
Ruby Valley, Elko, Nev., 20 C6 104
Ruby Valley, Indian res., Nev. C6 104
Ruchi, isl., Eniwetok 52
Rucker, Rutherford, Tenn., 50 B5 117
Ruda, riv., Pol. g9 26
Rudbar, Afg., 5,000 F11 41
Rudd, Floyd, Iowa, 436 A5 92
Ruddell, Sask., Can., 100 E2 70
Ruddles Mills, Bourbon, Ky., 75 B5 94
Ruddock, St. John the Baptist, La. B6 95
Rüdesheim, Ger., 7,200 D3 16
Rudha Hunish, isl., Scot. C2 13
Rudköbing, Den., 4,336 D4 24
Rudolf, lake, Ken. A6 48
Rudnichnyy, Sov. Un. B4 29
Rudolph, Wood, Ohio, 350 A4 111
Rudolph, Kenedy, Tex., 5 F4 118
Rudolph, isl., Sov. Un. A8 4
Rudolstadt, Ger., 27,700 C6 17
Rūd Sar, Iran, 10,000 C5 41
Rudyard, Chippewa, Mich., 600 B6 98
Rudyard, Hill, Mont., 650 B6 102
Rue, Fr., 1,787 D9 12
Rueil-Malmaison, Fr., 54,786 g9 14
Ruelle [-sur-Touvre], Fr., 5,855 E4 14
Rueun, Switz., 506 C7 19
Rufa'a, Sud., 9,137 C3 47
Rufe, McCurtain, Okla., 80 C6 112
Ruffec, Fr., 4,009 D4 14
Ruffin, Rockingham, N.C., 500 A4 109
Ruffin, Colleton, S.C., 250 E6 115
Rufiji, riv., Tan. C6 48
Rufina Central, P.R. C3 65
Rufino, Arg., 10,987 A4 54
Rufisque, Sen., 37,500 D1 45
Rufus, Sherman, Oreg., 150 B6 113
Rufus Woods Lake, res., Wash. A6 122
Rugby, Eng., 51,651 B6 12
Rugby, Pierce, N. Dak., 2,972 A6 110
Rugeley, Eng., 13,012 B6 12
Rügen isl., Ger. A6 16
Rugged, mtn., B.C., Can. D4 68
Rugozero, Sov. Un. E15 25
Ruhla, Ger., 9,226 C5 17
Ruhland, Ger., 4,533 B8 17
Ruhr, riv., Ger. B3 17
Rui, Afg., 5,000 D13 41
Ruidosa, Presidio, Tex., 150 F2 118
Ruidoso, Lincoln, N. Mex., 1,557 D4 107
Ruidoso Downs, Lincoln, N. Mex., 407 D4 107
Ruiz, Mex., 6,490 m11 63
Rujiyoru, isl., Eniwetok 52
Rukoji, isl., Bikini 52
Rukoji, pass, Bikini 52
Rukuruku, bay, Fiji Is. 52
Rukwa, lake, Tan. C5 48
Rule, Haskell, Tex., 1,347 C3 118
Ruleton, Sherman, Kans., 50 C2 93
Ruleville, Sunflower, Miss., 1,902 B3 100
Rulo, Richardson, Nebr., 412 D10 103
Rum, creek, W. Va. D5 123
Rum, isl., Ba. Is. D6 64
Rum, isl., Scot. C2 13
Rum, riv., Minn. E5 99
Rum, sound, Scot. D2 13
Ruma, Yugo, 19,570 C4 22
Rumaitha, Iraq, 4,468 F2 41
Rumbek, Sud., 2,944 D2 47
Rumbley, Somerset, Md., 105 D6 85
Rumburk, Czech., 6,759 C9 17
Rumely, Alger, Mich., 25 B3 98
Rumford, Oxford, Maine, 7,233 D2 96
Rumford, Fall River, S. Dak., 35 D2 116
Rumford Corner, Oxford, Maine, 60 D2 96
Rumilly, Fr., 3,940 D1 18
Rum Jungle, Austl. B5 50
Rummerfield, Bradford, Pa., 50 C9 114
Rumney, Grafton, N.H., 200 (820▲) C3 105
Rumney Depot, Grafton, N.H., 110 C3 105
Rumoi, Jap., 29,100 E10 37
Rump, mtn., Maine C1 96
Rumpi, Nya. D5 48
Rumsey, Alta., Can., 123 D4 69
Rumsey, McLean, Ky., 252 C2 94
Rumson, Monmouth, N.J., 6,405 C4 106
Runcorn, Eng., 26,035 A5 12
Runge, Karnes, Tex., 1,036 E4 118
Rungwa, Tan. C5 48
Runnells, Polk, Iowa, 322 A8, C4 92
Runnels, co., Tex., 15,016 D3 118
Runnelstown, Perry, Miss., 125 D4 100
Runnemede, Camden, N.J., 8,396 D2 106
Running, creek, Colo. B6 83
Runnymede, Sask., Can., 120 F5 70
Runtu, S.W. Afr. A2 49
Rupanco, Chile C2 54
Rupat, isl., Indon. L4 38
Rupert, Minidoka, Idaho, 4,153 G3 89
Rupert, Bennington, Vt., 150 (603▲) E2 120
Rupert, Greenbrier, W. Va., 921 D4 123
Rupert, riv., Que., Can. B2 73
Rupert House, Que., Can., 528 B2 73
Rupununi, riv., Br. Gu. B3 59
Rural Hall, Forsyth, N.C., 1,503 A3 109
Rural Retreat, Wythe, Va., 413 B3, E1 121
Rurrenabaque, Bol., 1,225 B2 55
Rurui, cape, Jap. D13 37
Rurutu, isl., Fr. Polynesia *H12 7
Rusagonis, N.B., Can., 262 D3 74
Rusapi, S. Rh., 4,400 A5 49
Ruschuk (Ruse), Bul., 53,420 D7 22
Ruschuk (Ruse), co., Bul., 527,708 *D8 22
Ruse (Ruschuk), Bul., 53,420 D7 22
Ruse (Ruschuk), co., Bul., 527,708 *D8 22
Rusera, India, 13,142 E11 40
Rush, Marion, Ark. A3 81
Rush, El Paso, Colo., 25 C6 83
Rush, Ire., 2,118 D5 11
Rush, Boyd, Ky., 300 B7 94
Rush, co., Ind., 20,393 E6 91
Rush, co., Kans., 6,160 D4 93
Rush, creek, Colo. C7 83
Rush, creek, Nebr. C3 103
Rush, creek, Ohio B4 111
Rush, creek, Okla. C4 112
Rush, lake, Sask., Can. G2 70
Rush, lake, Minn. D3 99
Rush, lake, Wis. E5 124
Rush, riv., Wis. D1 124
Rush, val., Utah C3 119
Rush Center, Rush, Kans., 278 D4 93
Rush City, Chisago, Minn., 1,108 E6 99
Rushden, Eng., 17,370 B7 12
Rushford, Fillmore, Minn., 1,335 G7 99
Rushford, Allegany, N.Y., 400 C2 108
Rush Hill, Audrain, Mo., 132 B6 101
Rush Lake, Sask., Can., 213 G2 70
Rushmere, Isle of Wight, Va., 125 B6 121
Rushmore, Nobles, Minn., 382 G3 99
Rush Springs, Grady, Okla., 1,303 C4 112
Rushsylvania, Logan, Ohio, 601 B4 111
Rushville, Schuyler, Ill., 2,819 C3 90
Rushville, Rush, Ind., 7,264 E7 91
Rushville, Buchanan, Mo., 253 B2 101
Rushville, Sheridan, Nebr., 1,228 B3 103
Rushville, Ontario and Yates, N.Y., 465 C3 108
Rushville, Susquehanna, Pa., 40 C9 114
Rusk, Cherokee, Tex., 4,900 D5 118
Rusk, co., Tex., 36,421 C5 118
Rusk, co., Wis., 14,794 C2 124
Ruskin, Hillsborough, Fla., 1,894 E4, F2 86
Ruskin, Nuckolls, Nebr., 203 D8 103
Ruskin, Dickson, Tenn., 50 A4 117
Ruso, McLean, N. Dak., 31 B5 110
Russas, Braz., 7,102 B3 57
Russell, White, Ark., 203 B4 81
Russell (Russell City), Alameda, Calif., 1,100 *B5 82
Russell, Ont., Can., 528 B9, B10 72
Russell, Man., Can., 1,263 D1 71
Russell, Clay, Fla., 300 C6 86
Russell, Lucas, Iowa, 577 D4 92
Russell, Russell, Kans., 6,113 D5 93
Russell, Greenup, Ky., 1,458 B7 94
Russell, Hampden, Mass., 600 (1,366▲) B2 97
Russell, Lyon, Minn., 449 F3 99
Russell, St. Lawrence, N.Y., 500 B2 108
Russell, N.Z., 441 K15 51
Russell, Bottineau, N. Dak., 25 A5 110
Russell, Greer, Okla., 100 C2 112
Russell, Warren, Pa., 800 C3 114
Russell, co., Ala., 46,351 C4 78
Russell, co., Ont., Can., 20,892 B9 68
Russell, co., Kans., 11,348 D5 93
Russell, co., Ky., 11,076 D4 94
Russell, co., Va., 26,290 B3 121
Russell, fork, Ky. C7 94
Russell, lake, Alta., Can. A3 69
Russell, lake, Man., Can. A1 71
Russell, mtn., Alsk. f15 80
Russell Gardens, Nassau, N.Y., 1,156 *E3 108
Russell Konda, India H10 40
Russell Springs, Logan, Kans., 93 D2 93
Russell Springs, Russell, Ky., 1,125 C4 94
Russellton, Allegheny, Pa., 1,613 *A6 114
Russellville, Franklin, Ala., 6,628 A2 78
Russellville, Pope, Ark., 8,921 B2 81
Russellville, Lawrence, Ill., 197 E6 90
Russellville, Putnam, Ind., 372 E4 91
Russellville, Logan, Ky., 5,861 D3 94
Russellville, Cole, Mo., 442 C5 101
Russellville, Brown, Ohio, 412 D4 111
Russellville, Berkeley, S.C., 100 E8 115
Russellville, Hamblen, Tenn., 750 C10 117
Rüsselsheim, Ger., 39,500 D3 17
Russia, Shelby, Ohio, 300 B3 111
Russia, see Soviet Union, country, Eur., Asia
Russian, riv., Calif. C2 82
Russian Soviet Federated Socialist Republic, rep., Sov. Un., 120,554,000 D17 9
Russiaville, Howard, Ind., 1,064 D5 91
Russum, Claiborne, Miss., 30 D2 100
Rustad, Clay, Minn., 25 D2 99
Rustburg, Campbell, Va., 350 D3 121
Rustenburg, S. Afr., 21,016 *C4 49
Rustico, P.E.I., Can., 74 C6 74
Ruston, Lincoln, La., 13,991 B3 95
Ruston, Pierce, Wash., 694 B3, D1 122
Ruswil, Switz., 4,657 B5 19
Ruszow, Pol. B10 17
Rutana, Burandi B5 48
Rutba, Iraq F13 31
Rutchenkovo, Sov. Un. r20 27
Rute, Sp., 10,077 D3 20
Ruteng, Indon. G6 35
Ruth, Huron, Mich., 210 E8 98
Ruth, Lincoln, Miss., 150 D3 100
Ruth, White Pine, Nev., 800 D7 104
Ruth, Rutherford, N.C., 529 D4 109
Ruth, lake, Minn. E6 99
Ruthenia, reg., Sov. Un. D7 26
Rutherford, Napa, Calif., 150 A5 82
Rutherford, Bergen, N.J., 20,473 B4, D5 106
Rutherford, Gibson, Tenn., 983 A3 117
Rutherford, co., N.C., 45,091 B2 109
Rutherford, co., Tenn., 52,368 B5 117
Rutherford, fork, Tenn. A3 117
Rutherford Heights, Dauphin, Pa., 1,700 *F8 114
Rutherfordton, Rutherford, N.C., 3,392 B2, D4 109
Rutherglen, Ont., Can., 50 A5 72
Rutherglen, Scot., 25,067 E4 13
Rutherton, Rio Arriba, N. Mex., 25 A3 107
Ruthilda, Sask., Can., 86 F1 70
Ruthin, Wales, 3,502 A4 12
Ruthton, Pipestone, Minn., 476 F2 99
Ruthven, Wilcox, Ala., 25 D2 78
Ruthven, Palo Alto, Iowa, 712 A3 92
Rüti, Switz, 8,282 B6 19
Rutland, B.C., Can., 1,495 E8 68
Rutland, La Salle, Ill., 509 C4 90
Rutland, Humboldt, Iowa, 221 B3 92
Rutland, Worcester, Mass., 1,774 B4 97
Rutland, Sargent, N. Dak., 308 C8 110
Rutland, Meigs, Ohio, 687 C5 111
Rutland (Roseville), Tioga, Pa., 162 C8 114
Rutland, Lake, S. Dak., 100 C9 116
Rutland, Rutland, Vt., 18,325 D3 120
Rutland, co., Eng., 23,956 B7 12
Rutland, co., Vt., 46,719 D2 120
Rutland Station, Sask., Can., 115 E1 70
Rutledge, Crenshaw, Ala., 276 D3 78
Rutledge, Morgan, Ga., 478 C3 87
Rutledge, Pine, Minn., 146 D6 99
Rutledge, Scotland, Mo., 158 A5 101
Rutledge, Grainger, Tenn., 793 C10 117
Rutshuru, Con. L. B4 48
Rutter, Ont., Can., 216 A4 72
Ruunitto, isl., Eniwetok 52
Ruvo [di Puglia], It., 23,746 D6 21
Ruvuma, riv., Moz., Tan. D6 48
Ruwandiz, Iraq, 3,320 C2 41
Ruweiha, ruins, Jordan D7 32
Ruwenzori, mts., Con. L., Ug. A4 48
Ruxton, Baltimore, Md., 2,100 C2 85
Ruza, Sov. Un., 7,000 D11 27
Ruzayevka, Sov. Un., 34,500 C3 29
Ruzomberok, Czech., 18,600 D5 26
Rwanda, country, Afr., 2,725,000 B4 48
Ry, Den., 2,004 B3 24
Ryan, Delaware, Iowa, 347 B6 92
Ryan, Jefferson, Okla., 978 C4 112
Ryan, peak, Idaho F4 89
Ryan, riv., Ala. A2 78
Ryan Park, Carbon, Wyo., 100 E6 125
Ryazan, Sov. Un., 240,000 C1 29
Ryazhsk, Sov. Un., 10,000 E13 27
Rybatskoye, Sov. Un., 10,000 s31 25
Rybinsk (Shcherbakov), Sov. Un., 192,000 B1 29
Rybinsk, res., Sov. Un. B1 29
Rybnik, Pol., 34,000 C5, g9 26
Rybnitsa, Sov. Un. B9 22
Rycroft, Alta., Can., 500 B1 69
Rydal, Montgomery, Pa., 1,500 *F11 114
Ryde, Eng., 19,796 D6 12
Ryder, Ward, N. Dak., 264 B4 110
Ryderwood, Cowlitz, Wash., 380 C2 122
Rye, Cleveland, Ark., 50 D4 81
Rye, Pueblo, Colo., 179 D6 83
Rye, Eng., 4,429 D8 12
Rye, Rockingham, N.H., 450 (3,244▲) D5 105
Rye, Westchester, N.Y., 14,225 D2, D3 108
Rye, lake, N.Y. A5 106
Ryeå, riv., Den. A3 24
Rye Beach, Rockingham, N.H., 165 E5 105
Ryegate, Golden Valley, Mont., 314 D7 102
Ryegate, Caledonia, Vt., 30 (894▲) C4 120
Rye Patch, res., Nev. C3 104
Ryerson, Sask., Can., 15 H5 70
Rygge, Nor. p28 25
Ryley, Alta., Can., 469 C4 69
Rylsk, Sov. Un., 10,000 F10 27
Rynda, Sov. Un. C17 25
Ryomgaard, Den., 861 B4 24
Ryozu, Jap., 12,100 G9 37
Rypin, Pol., 7,350 B5 26
Ryuku, is., Asia F10 34
Ryukyu Is. (Southern), U.S. occ., Asia, 745,194 F10 34
Rzadza, riv., Pol. k14 26
Rzeszow, Pol., 62,000 C6 26
Rzepin, Pol., 2,000 B3 26
Rzhev, Sov. Un., 54,000 C10 27

S

Saale, riv., Ger. B6 17
Saaler, bay, Ger. D6 24
Saalfield, Ger., 26,900 C6 17
Saalfelden, Aus., 8,901 B8 18
Saanen, Switz., 5,649 D3 19
Saar, state, Ger., 1,072,600 E6 15, D3 16
Saar, riv., Ger. E6 15
Saarbrücken, Ger., 130,700 (★350,000) D1 17
Saarburg, Ger., 5,600 E6 15
Saaremaa, isl., Sov. Un. B4 27
Saarlouis, Ger., 36,800 D1 17
Saas-Almagell, Switz., 359 D4 19
Saas-Fee, Switz., 739 D4 19
Saavedra, Arg., 2,130 B4 54
Saavedra, Chile B2 54
Sabá, Hond. C4 62
Sabac, Yugo., 30,231 C4 22
Sabadell, Sp., 105,152 B7 20
Sabael, Hamilton, N.Y., 100 B6 108
Sab'ah, mtn., Libya D3 43
Sabalana, is., Indon. G5 35
Sabana, P.R. B8, f12 65
Sabana de la Mar, Dom. Rep., 4,032 F9 64
Sabanagrande, Hond., 1,678 D4 62
Sabana Grande, P.R., 3,318 C3 65
Sabana Grande, mun., P.R., 15,910 C3 65
Sabanalarga, Col., 13,982 A3 70
Sabana Llana, P.R. C5 65
Sabana Seca, P.R. B6 65
Sabang, Indon, 6,855 k11 35
Sabará, Braz., 10,004 E2 57
Sabarmati, riv., India F4 40
Sabaskong, bay, Ont., Can. E4 71
Sabasţiyah, Jordan, 1,000 f11 32
Sabattus, Androscoggin, Maine, 850 D2 96
Sabattus, pond, Maine D2 96
Sabaudia, It., 6,262 D4 21
Sabbathday, pond, Maine D5 96
Sabderat, Eth. B4 47
Sabetha, Nemaha, Kans., 2,318 C8 93
Sabhah, Libya, 3,640 D2 43
Sabillasville, Frederick, Md., 300 A3 85
Sabin, Clay, Minn., 251 D2 99
Sabina, Clinton, Ohio, 2,313 C4 111
Sabinal, Uvalde, Tex., 1,747 E3 118
Sabinas, Mex., 15,953 B4 63
Sabinas Hidalgo, Mex., 11,558 B4 63
Sabine, Jefferson, Tex., 100 E6 118
Sabine, Wyoming, W. Va., 300 D3 123
Sabine, co., Tex., 7,302 D6 118
Sabine, par., La., 18,564 C2 95
Sabine, lake, La. E2 95
Sabine, lake, Tex. E2 95
Sabine, pass, La. E2 95
Sabine, peak, Ant. B29 5
Sabine, riv., La., Tex. C2 95, D6 118
Sabine Pass., Jefferson, Tex., 850 E6 118
Sabinópolis, Braz., 2,957 E2 57
Sabinoso, San Miguel, N. Mex., 10 B5 107
Sabinsville, Tioga, Pa., 300 C6 114
Sable, cape, N.S., Can. H19 67
Sable, cape, Fla. G5 86
Sable, isl., N.S., Can. H21 67
Sable Island, cape, N.S., Can. F4 74
Sable River, N.S., Can., 144 F4 74
Sablé [-sur-Sarthe], Fr., 6,885 D3 14
Sabon Birni, Nig. D6 45
Sabor, riv., Port. B2 20
Sabrina, coast, Ant. C24 5
Sabula, Jackson, Iowa, 894 B7 92
Sabula, Iron, Mo., 30 D7 101
Sabula, Clearfield, Pa., 80 D4 114
Sabzawar, Afg., 10,000 E11 41
Sabzevār, Iran, 28,151 C8 41
Sac, co., Iowa, 17,007 B2 92
Sac, riv., Mo. D4 101
Sacaba, Bol., 2,752 C2 55
Sacajawea, peak, Oreg. B9 113
Sacandaga, lake, N.Y. B6 108
Sacandaga, res., N.Y. B6 108
Sacaton, Pinal, Ariz., 700 D4 80
Sacavém, Port., 5,569 f9 20
Sacavém, riv., Port. f9 20
Sac City, Sac, Iowa, 3,354 B2 92
Sachem Head, New Haven, Conn., 200 D6 84
Sachigo, lake, Ont., Can. C5 71
Sachigo, riv., Ont., Can. B6 71
Sachse, Dallas, Tex., 359 B6 118
Sachsen, former state, Ger., 5,558,566 B8 17
Sachsen-Anhalt (Saxony-Anhalt), former state, Ger., 4,160,539 B6 17
Sacile, It., 5,340 D8 18
Sackets Harbor, Jefferson, N.Y., 1,279 B4 108
Sackville, N.B., Can., 3,038 D5 74
Saclay, Fr., 312 h9 14
Saco, Pike, Ala., 150 D4 78
Saco, York, Maine, 10,515 E2, E4 96
Saco, Phillips, Mont., 490 B9 102
Saco, riv., Maine, N.H. B4 105
Sacramento, Sacramento, Calif., 191,667 (★536,000) A6, C3 82
Sacramento, McLean, Ky., 429 C2 94
Sacramento, co., Calif., 502,778 C3 82
Sacramento, mts., N. Mex. E4 107
Sacramento, riv., Calif. C3 84
Sacramento, riv., N. Mex. E4 107
Sacramento, val., Calif. C2 84
Sacré-Coeur Saguenay, Que., Can., 1,108 A8 73
Sacred Heart, Renville, Minn., 696 F3 99
Sacrofano, It., 1,756 g8 21
Sacul, Nacogdoches, Tex., 200 D5 118
Sá da Bandeira, Ang., 13,867 D1 48
Şa'd'ah, Yemen, 25,000 B5 47
Saddle, mtn., Colo. C5 83
Saddle, mtn., Oreg. B3 113
Saddle, mts., Wash. C6 122
Saddle, riv., N.J. D5 106
Saddleback, mtn., Maine B4 96
Saddleback, mtn., Maine D2 96
Saddle Ball, mtn., Mass. A1 97
Saddle Brook, Bergen, N.J., 13,834 *A4 106
Saddle Bunch, keys, Fla. H5 86
Saddle River, Bergen, N.J., 1,776 *A4 106
Saddle Rock, Nassau, N.Y., 1,109 *G2 84
Saddlerock, mtn., Maine C3 96
Saddlestring, Johnson, Wyo., 5 A6 125
Sadieville, Scott, Ky., 276 B5 94
Sadiya, India, 5,044 C10 39
Sado, isl., Jap. G9 37
Sado, riv., Port. C1 20
Sadorus, Champaign, Ill., 384 D5 90
Sadska, Czech., 3,047 n18 26
Saeby, Den., 3,669 A4 24
Saegertown, Crawford, Pa., 1,131 C1 114
Saeki, Jap., 32,000 J5 37
Saengchon, Kor. G3 37
Safad, Isr., 10,586 B7 32
Safata, bay, W. Sam. 52
Safatulafei, W. Sam, 1,108 52
Safety Harbor, Pinellas, Fla., 1,787 D4, E2 86
Safety Valve, entrance, Fla. G6 86
Saffell, Lawrence, Ark., 150 B4 81
Safford, Dallas, Ala., 200 C2 78
Safford, Graham, Ariz., 4,648 E6 80
Saffordville, Chase, Kans., 40 D7 93
Saffron Walden, Eng., 7,810 B8 12
Safi, Mor., 81,072 C3 44
Safīdābeh, Iran F10 41
Safonovo, Sov. Un., 24,700 D9 27
Safranbolu, Tur., 7,400 B9 31
Safune, W. Sam., 476 52
Safune, bay, W. Sam. 52
Saga, Jap., 129,888 J5 37
Saga, pref., Jap., 942,874 *J5 37
Sagadahoc, co., Maine, 22,793 E3 96
Sagaing, Bur., 15,439 D10 39
Sagami, sea, Jap. n18 37
Sagamihara, Jap, 101,655 *n18 37
Sagamore, Barnstable, Mass., 987 C6 97
Sagamore, Armstong, Pa., 800 E3 114
Sagamore Hills, Summit, Ohio, 3,848 B2 111
Sagån, riv., Swe. t34 25
Saganaga, lake, Ont., Can., Minn. A6, B8 99
Sagaponack, Suffolk, N.Y., 100 F7 84
Sagara, Fiji Is. 52
Sagara, Jap., 12,000 o17 37
Sage, Lincoln, Wyo., 35 D2 125
Saggart, Ire., 426 D5 11
Sag Harbor, Suffolk, N.Y., 2,346 D4 108
Saginaw, Shelby, Ala., 200 B3 78
Saginaw, Saginaw, Mich., 98,265 (★160,900) E7 98
Saginaw, Lane, Oreg., 100 D3 113
Saginaw, Tarrant, Tex., 1,001 B5 118
Saginaw, co., Mich., 190,752 E6 98
Saginaw, bay, Mich. E7 98
Sagiz, Sov. Un. D4 29
Sagiz, riv., Sov. Un. D5 29
Saglek, bay, Newf., Can. f9 75
Sagola, Dickinson, Mich., 150 B3 98
Sagra, mtn., Sp. D4 20
Sag Sag, Bis. Arch. k12 50
Saguache, Saguache, Colo., 722 C4 83
Saguache, co., Colo., 4,473 C4 83
Saguache, creek, Colo. C4 83
Sagua de Tánamo, Cuba, 2,864 E6 64
Sagua la Grande, Cuba, 26,187 D3 64
Saguaro, nat. mon., Ariz. E5 80
Saguenay, co., Que., Can., 81,900 A8 73
Saguenay, riv., Que., Can. A7 73
Saguia el Hamra (Sekia el Hamra), Sp. overseas prov., Afr., 6,445 D2 44
Sagunto, Sp., 40,293 C5 20
Saḥāb, Jordan, 1,000 C8 32

Sahagún, Col., 5,910 B2 60
Sahara, des., Afr D5 42
Saharan Atlas, mts., Alg C5 44
Saharanpur, India, 185,213 C6 39, C6 40
Şahrajat al Kubrá, Eg., U.A.R., 1,000 D3 32
Sahuaripa, Mex., 3,836 B3 63
Sahuarita, Pima, Ariz., 250 F5 80
Sahuaro, res., Ariz G3 80
Sahuayo, Mex., 25,673 m12 63
Saïda, Alg., 20,289 G7 30
Saida (Sidon), Leb., 19,853 F10 31
Sa'īdābād, Iran, 8,074 G7 41
Saidpur, Pak., 61,369 C9 39
Saigo, Jap., 8,000 H6 37
Saigon, Viet., 1,383,000 (★1,550,000) G7 38
Saijo, Jap., 34,200 *I6 37
Saikhoa Ghat, India F4 34
Sail, rock, Vir. Is f14 63
Sailor Springs, Clay, Ill., 187 E5 90
Saimaa, lake, Fin G13 25
St. Abb's, head, Scot E6 13
St. Adelphe, Que., Can., 787 C5 73
St. Adolphe, Que., Can., 410 D6 73
St. Affrique, Fr., 5,670 F5 14
St. Agapit, Que., Can., 1,117 C6 73
St. Agatha, Aroostook, Maine, 500 (1,137▲) A4 96
Ste. Agathe, Man., Can., 298 E3 71
Ste. Agathe, Que., Can., 600 C6 73
Ste. Agathe des Monts, Que., Can., 5,173 C3 73
Ste. Agnes, Que., Can., 110 D3 73
St. Aime, Que., Can., 580 D5 73
St. Alban, Que, Can., 786 C5 73
St. Albans, Eng., 50,276 C7 12
St. Albans, Somerset, Maine, 350 (927▲) D3 96
St. Albans, Newf., Can., 1,547 E4 75
St. Albans, Franklin, Vt., 8,806 B2 120
St. Albans, Kanawha, W. Va., 15,103 C3 123
St. Albans, bay, Vt B2 120
St. Albans, head, Eng D5 12
St. Albans Bay, Franklin, Vt., 350 B2 120
St. Albert, Alta., Can., 4,059 C4 69
St. Alexandre, Que., Can., 872 B8 73
St. Alexandre, Que., Can., 425 D4 73
St. Aleix des Monts, Que., Can., 1,964 C4 73
St. Alphonse, Que., Can., 103 D5 73
St. Amand, Fr., 16,674 D3 15
St. Amand-Mont-Rond, Fr., 10,890 D5 14
St. Amant, Ascension, La., 100 B6 95
St. Amarin, Fr., 2,044 B3 18
St. Ambroise, Que., Can., 1,576 A6 73
St. Anaclet, Que., Can., 722 A9 73
St. Andre, cape, Malag g8 49
St. Andre [de Kamouraska], Que., Can., 550 B8 73
St. Andrews, bay, Fla G3 86
St. Andrews, sound, Ga F5 87
St. Andrews, N.B., Can., 1,531 D2 74
St. Andrew's, Newf., Can., 294 E2 75
St. Andrews, Scot., 9,888 D6 13
St. Andrews, Charleston, S.C., 1,500 *F2 115
St. Andrews, Franklin, Tenn., 250 B6 117
St. Andrews, bay, Scot D6 13
St. Andrews East, Que., Can., 1,183 D3 73
Ste. Angele de Rimouski, Que., Can., 655 A2 73
St. Anicet, Que., Can., 132 D3 73
St. Ann., St. Louis, Mo., 12,155 *A8 101
St. Anna, Sheboygan, Wis., 100 B6 124
Ste. Anne, Que., Can., 900 D4 73
Ste. Anne, Guad., 2,384 n16 64
St. Anne, Kankakee, Ill., 1,378 B6 90
St. Anne, lake, Alta., Can C3 69
Ste. Anne, riv., Que., Can B7 73
Ste. Anne, riv., Que., Can C6 73
Ste. Anne de Beaupré, Que., Can., 1,878 B7 73
Ste. Anne de Bellevue, Que., Can., 4,044 D8 73
Ste. Anne-de-la-Perade, Que., Can., 1,179 C5 73
St. Anne de la Pocatiere, Que., Can., 3,086 B7 73
Ste. Anne des Chenes, Man., Can., 653 E3 71
St. Anns, N.S., Can., 115 C9 74
St. Ann's Bay, Jam., 5,086 F5 64
St. Anselme, Que., Can., 1,131 C7 73
St. Ansgar, Mitchell, Iowa, 1,014 C7 94
St. Anthony, Newf., Can., 1,820 C4 75
St. Anthony, Fremont, Idaho, 2,700 F7 89
St. Anthony, Dubois, Ind., 165 H4 91
St. Anthony, Marshall, Iowa, 130 B4 92
St. Anthony, Hennepin, Minn., 5,084 *F5 99
St. Anthony, Morton, N. Dak., 88 C5 110
St. Antoine, Que., Can., 290 C6, C8 73
St. Antoine, Que., Can., 435 D4 73
St. Antoine de Kent, N.B., Can., 670 C5 74
St. Antoine des Laurentides, Que., Can, 3,005 *D3 73
St. Antonin, Que., Can., 247 B8 73
St. Apollinaire, Que., Can., 968 C6, D8 73
Ste. Apolline, Que., Can., 353 C7 73
St. Arnaud, Austl., 3,150 H4 51
St. Arsene, Que., Can., 523 B8 73
St. Athanase, Que., Can., 168 B8 73
St. Aubert, Que., Can., 735 B7 73
St. Augustine, Que., Can., 477 C2 75
St. Augustin, Que., Can., 488 C8 73
St. Augustin, Que., Can., 393 D8 73
St. Augustin, riv., Que., Can C2 75
St. Augustine, St. Johns, Fla., 14,734 C5, C7 86
St. Austell, Eng., 25,027 D3 12
St. Avold, Fr., 15,247 C7 14
Ste. Barbe, Que., Can., 181 D3 73
St. Barbe, dist., Newf., Can C3 75
St. Barbe, is., Newf., Can C4 75
St. Barnabe, Que., Can., 204 D5 73
St. Barnabe Nord, Que., Can., 541 C5 73
St. Barthelemy, Que., Can., 620 C4 73
St. Barthélemy (St. Bartholomew), isl., Guad n15 64
St. Basile, Que., Can., 1,210 D4 73
St. Basile [de Portneuf], Que., Can., 1,709 C6 73
St. Basile Station, Que., Can., 1,635 *C6 73
Ste. Beatrix, Que., Can., 360 C4 73
St. Bees, head, Eng C5 10
St. Benedict, Sask., Can., 205 E3 70
St. Benedict, Kossuth, Iowa, 100 A3 92
St. Benedict, Nemaha, Kans., 75 C7 93
St. Benedict, Marion, Oreg., 450 B2 113
St. Benoit, Que., Can., 571 D8 73
St. Benoit Labre, Que., Can., 514 C7 73
St. Bernard, Cullman, Ala., 700 A3 78
St. Bernard, Que., Can., 95 C6 73
St. Bernard, St. Bernard, La., 350 C8, E6 95
St. Bernard, Platte, Nebr., 25 C8 103
St. Bernard, Hamilton, Ohio, 6,778 D2 111
St. Bernard, par., La., 32,186 E6 95
St. Bernard, see Grand St. Bernard, pass, Switz., It.
St. Bernice, Perry, Ind., 800 E2 91
St. Bethlehem, Montgomery, Tenn., 200 A4 117
Ste. Blandine, Que., Can., 490 A9 73
St. Bonaventure, Que., Can., 384 D5 73
St. Bonaventure, Cattaraugus, N.Y., 2,000 *C2 108
St. Boniface, Man., Can., 37,600 E3 71
St. Boswells, Sask., Can., 56 G2 70
St. Brendan's, Newf., Can., 387 D5 75
St. Bride, mtn., Alta., Can D2 69
St. Bride's, Newf., Can., 397 E4 75
St. Brides, Norfolk, Va., 130 E6 121
St. Brides, bay, Wales C2 12
St. Brieuc, Fr., 43,142 C2 14
St. Brieux, Sask., Can., 364 E3 70
Ste. Brigide, Que., Can., 233 D4 73
Ste. Brigitte, Que., Can., 124 C5 73
St. Bruno, Que., Can., 1,158 B8 73
St. Calais, Fr., 3,045 D4 14
St. Calixte, Que., Can., 512 D4 73
St. Camille, Que., Can., 689 C7 73
St. Casimir, Que., Can., 1,386 C5 73
St. Catharine, Washington, Ky., 200 C4 94
St. Catharines, Ont., Can., 84,472 (★95,577) D5 72
Ste. Catherine, Que., Can., 893 C6, C8 73
St. Catherine, lake, Vt E2 120
St. Catherines, isl., Ga E5 87
St. Catherine's, pt., Bermuda E14 77
St. Catherines, pt., Eng D6 12
St. Catherines, sound, Ga E5 87
Ste. Cecile, Que., Can., 141 D7 73
St. Celestin, Que., Can., 368 C5 73
St. Cesaire, Que., Can., 2,097 D4 73
St. Chamond, Fr., 17,107 E6 14
St. Charles, Arkansas, Ark., 255 C4 81
St. Charles, Bear Lake, Idaho, 300 G4 89
St. Charles, Kane, Ill., 9,269 B5, F1 90
St. Charles, Madison, Iowa, 355 C4 92
St. Charles, Hopkins, Ky., 421 C2 94
St. Charles, Saginaw, Mich., 1,959 E6 98
St. Charles, Winona, Minn., 1,882 G6 99
St. Charles, St. Charles, Mo., 21,189 A7, C7 101
St. Charles, Gregory, S. Dak., 58 D6 116
St. Charles, Lee, Va., 368 B2 121
St. Charles, co., Mo., 52,970 C7 101
St. Charles, par., La., 21,219 E5 95
St. Charles, cape, Newf., Can B4 75
St. Charles de Bellechasse, Que., Can., 981 C7 73
St. Chély-d'Apcher, Fr., 3,900 E5 14
Ste. Christine, Que., Can., 214 D5 73
St. Chrysostome, Que., Can., 972 D4 73
St. Clair, Burke, Ga., 50 C4 87
St. Clair, St. Clair, Mich., 4,538 F8 98
St. Clair, Blue Earth, Minn., 373 F5 99
St. Clair, Franklin, Mo., 2,711 C6 101
St. Clair, Schuylkill, Pa., 5,159 E9 114
St. Clair, co., Ala., 25,388 B3 78
St. Clair, co., Ill., 262,509 E4 90
St. Clair, co., Mich., 107,201 F8 98
St. Clair, co., Mo., 8,421 C4 101
St. Clair, lake, Ont., Can., Mich E2 72, F8 98
St. Clair, riv., Ont., Can., Mich E2 72, F8 98
St. Claire, Que., Can., 1,338 C7 73
St. Clair Hills, St. Clair, Ill., 250 B8 101
St. Clair Shores, Macomb, Mich., 76,657 A8 98
St. Clairsville, Belmont, Ohio, 3,865 B7 111
St. Claude, Man., Can., 609 E2 71
St. Claude, Que., Can., 71 D6 73
St. Claude [-sur-Bienne], Fr., 12,114 D6 14
St. Clement, Que., Can., 409 B8 73
St. Cléophas, Que., Can., 227 C4 73
Ste. Clothilde, Que., Can., 359 D5 73
St. Cloud, Osceola, Fla., 4,353 D5 86
St. Cloud, Fr., 26,476 g9 14
St. Cloud, Stearns, Benton and Sherburne, Minn., 33,815 E4 99
St. Cloud, Fond du Lac, Wis., 530 B5, E5 124
St. Come, Que., Can., 598 C4 73
St. Constant, Que., Can., 2,739 D9 73
Ste. Croix, N.B., Can., 173 D2 74
Ste. Croix, Que., Can., 1,363 C6, D8 73
Ste. Croix, Perry, Ind., 100 H4 91
Ste. Croix, Switz., 6,925 C2 19
St. Croix, co., Wis., 29,164 C1 124
St. Croix, isl., Vir. Is n14 60, k17 65
St. Croix, lake, N.S., Can E5 74
St. Croix, lake, Wis D1 124
St. Croix, riv., N.B., Can D2 74
St. Croix, riv., Maine C5 96
St. Croix, riv., Minn., Wis E6 99
St. Croix, stream, Maine B4 96
St. Croix Falls, Polk, Wis., 1,249 C1 124
St. Cuthbert, Que., Can., 392 C4 73
St. Cyprien, Que., Can., 370 A8 73
St. Cyrille [de L'Islet], Que., Can., 655 B7 73
St. Cyrille [de Wendover], Que., Can., 1,138 D5 73
St. Damase, Que., Can., 879 B7 73
St. Damase, Que., Can., 277 D4 73
St. Damien, Que., Can., 431 C4 73
St. Damien, Que., Can., 1,396 C7 73
St. David, Cochise, Ariz., 650 F5 80
St. David, Que., Can., 410 C9 73
St. David, Que., Can., 277 D5 73
St. David, Fulton, Ill., 862 C3 90
St. David, Aroostook, Maine, 80 A4 96
St. David's, Newf., Can., 317 D2 75
St. Davids, Delaware, Pa., 1,200 *A10 114
St. David's, Wales, 1,505 C2 12
St. David's, head, Wales C2 12
St. David's, isl., Bermuda E14 77
St. Denis, Que., Can., 269 B8 73
St. Denis, Que., Can., 1,063 D4 73
St. Denis, Fr., 94,264 C5, g10 14
St. Denis, Reunion, 25,332 H24 3
St. Didace, Que., Can., 92 C4 73
St. Dié, Fr., 23,108 C7 14
St. Dizier, Fr., 34,407 C6 14
St. Dominique, Que., Can., 532 D5 73
St. Dominique, Que., Can., 180 D8 73
St. Donat, Que., Can., 1,190 C3 73
St. Donatus, Jackson, Iowa, 100 B7 92
St. Edouard, Que., Can., 186 D4, E9 73
St. Edward, Boone, Nebr., 777 C8 103
St. Edwidge, Que., Can., 205 D6 73
St. Eleuthere, Que., Can., 1,014 B8 73
St. Elias, cape, Alsk D11 79
St. Elias, mtn., Alsk C11 79
St. Elias, mts., Alsk., Yukon, Can C11 79
St. Elie, Fr. Gu B4 59
Ste. Elizabeth, Que., Can., 557 C4 73
St. Elizabeth, Miller, Mo., 57 C5 101
St. Elmo, Mobile, Ala., 600 E1 78
St. Elmo, Fayette, Ill., 1,503 D5 90
St. Eloi, Que., Can., 252 A8 73
St. Elzear, Que., Can., 4,150 D8 73
Ste. Emelie, Que., Can., 721 C4 73
St. Emile de Suffolk, Que., Can., 244 D3 73
St. Ephrem, Que., Can., 898 C7 73
Saintes, Fr., 25,717 E3 14
St. Esprit, Que., Can., 778 D4 73
St. Etienne, Que., Can., 235 C9 73
St. Etienne, Que., Can., 155 E8 73
St. Étienne, Fr., 201,242 (★285,000) E6 14
St.-Eugéne, Alg., 25,491 *B5 44
Ste. Eulalie, Que., Can., 193 C5 73
Ste. Euphemie, Que., Can., 158 C7 73
St. Eusebe, Que., Can., 244 B9 73
St. Eustache, Que., Can., 5,463 D4, D8 73
St. Eustache sur le Lac, Que., Can., 7,274 D8 73
St. Eustatius, isl., W.I n15 64
St. Fabien, Que., Can., 1,466 A9 73
Ste. Famille, Que., Can., 310 C7 73
St. Famille d'Aumond, Que., Can, 223 C2 73
Ste. Felicite, Que., Can., 1,057 A2 73
St. Felix de Valois, Que., Can., 1,399 C4 73
St. Ferdinand, Que., Can., 2,706 C6 73
St. Fereol, Que., Can., 268 B7 73
St. Fidele, Que., Can., 317 B7 73
Saintfield, N. Ire., 604 C6 11
St. Fintan's, Newf., Can., 107 D2 75
St. Flavien, Que., Can., 610 C6 73
Ste. Flore, Que., Can., 622 C5 73
St. Florent [-sur-Cher], Fr., 5,453 D5 14
St. Florian, Lauderdale, Ala., 100 A2 78
St. Flour, Fr., 5,846 E5 14
St. Fortunat, Que., Can., 216 D6 73
Ste. Foy, Que., Can., 29,716 C9 73
Ste. Foy-la-Grande, Fr., 3,152 E4 14
St. Francis, Clay, Ark., 224 A5 81
St. Francis, Cheyenne, Kans., 1,594 C2 93
St. Francis, Aroostook, Maine, 450 (1,058▲) A4 96
St. Francis, Anoka, Minn., 175 E5 99
St. Francis, Todd, S. Dak., 421 D4 116
St. Francis, Milwaukee, Wis., 10,065 E2 124
St. Francis, co., Ark., 33,303 B5 81
St. Francis, cape, Newf., Can E5 75
St. Francis, lake, Que., Can D6 73
St. Francis, riv., Ark C5 81
St. Francis, riv., Que., Can., Maine B9 73, A3 96
St. Francis, riv., Que., Can D5 73
St. Francis, riv., Mo E7 101
St. Francisville, Lawrence, Ill., 1,040 E6 90
St. Francisville, West Feliciana, La., 1,661 D4 95
St. Francois, Que., Can., 5,122 C7 73
St. Francois, co., Mo., 36,516 D7 101
St. Francois [-du-Lac], Que., Can., 977 C5 73
Ste. Francoise, Que., Can., 476 A8 73
St. Francois Xavier, Que., Can., 433 D5 73
St. Frederic, Que., Can., 307 C7 73
St. Froid, lake, Maine B4 96
St. Gabriel, Iberville, La., 75 B5 95
St. Gabriel [de Brandon], Que., Can., 3,425 C4 73
St. Gallen, see Sankt Gallen, Switz.
St. Gallen, see Sankt Gallen, canton, Switz.
St. Gaudens, Fr., 7,949 F4 14
St. Gedeon, Que., Can., 930 D7 73
Ste. Genevieve, Que., Can., 532 C5 73
Ste. Genevieve, Ste., Genevieve, Mo., 4,443 D7 101
Ste. Genevieve, co., Mo., 12,116 D7 101
Ste. Genevieve de Pierrefonds, Que., Can., 2,397 *D8 73
St. George, Austl., 2,209 D7 51
St. George, Bermuda, 1,869 E14 77
St. George, N.B., Can., 1,133 D3 74
St. George, Ont., Can., 791 D4 72
St. George, Charlton, Ga., 582 F4 87
St. George, Pottawatomie, Kans., 259 C7 93
St. George, St. Louis, Mo., 1,323 *C7 101
St. George, Dorchester, S.C., 1,833 E6 115
St. George, Greene, Va., 5 C4 121
St. George, Washington, Utah, 5,130 F2 119
St. George, cape, Newf., Can D2 75
St. George, cape, Fla C1 86
St. George, cape, N. Gui h13 50
St. George, isl., Alsk D6 79
St. George, isl., Fla C2 86
St. George Island, St. Marys, Md., 200 D4 85
St. Georges, Bel., 5,854 D5 15
St. Georges, Newf., Can., 1,181 D2 75
St. Georges, Que., Can., 252 A4 73
St. Georges, Que., Can., 1,775 C5 73
St. Georges, Que., Can., 385 D6 73
St. Georges, New Castle, Del., 339 A6 85
St. Georges, Fr. Gu., 1,502 B4 59
St. George's, Grenada, 19,582 p16 64
St. Georges, bay, Newf., Can D2 75
St. George's, chan., Wales E3 10
St. George's, isl., Bermuda E14 77
St. Georges Ouest, Que., Can., 4,755 *C7 73
St. Georges-Port au Port, dist., Newf., Can D2 75
St. Gerard, Que., Can., 662 D6 73
St. Germain, Que., Can., 143 B8 73
St. Germain, forest, Fr g9 14
St. Germain [de Grantham], Que., Can., 1,015 D5 73
St. Germain-en-Laye, Fr., 34,621 g9 14
Ste. Germaine Station, Que., Can., 332 C7 73
Ste. Gertrude, Que., Can., 362 C5 73
St. Gertrude, St. Tammany, La., 75 D5 95
St. Gervais, Que., Can., 576 C7 73
St. Gervais-les-Bains, Fr., 1,551 D2 18
St. Gilles, Que., Can., 822 C6 73
St. Gilles [-du-Gard], Fr., 4,791 F6 14
St. Gilles [-sur-Vie], Fr., 2,511 D3 14
St.-Gingolph, Switz., 751 D2 19
St. Girons, Fr., 7,368 F4 14
St. Goarshausen, see Sankt Goarshausen, Ger.
St. Gobain, Fr., 2,012 C5 14
St. Gotthard, tunnel, Switz C6 19
St. Goven's, head, Wales C3 12
St. Gregoire, Que., Can., 673 C5 73
St. Gregor, Sask., Can., 170 E3 70
St. Gregory, mtn., Newf., Can D2 75
St. Guillaume, Que., Can., 792 D5 73
St. Helen, lake, Mich D6 98
St. Helena, Napa, Calif., 2,722 A5, C2 82
St. Helena, Cedar, Nebr., 63 B8 103
St. Helena, Br. dep., Afr., 5,250 H5 42
St. Helena, par., La., 9,162 D5 95
St. Helena, bay, S. Afr D2 49
St. Helena, isl., Atl. O H9 6
St. Helena, isl., S.C G6 115
St. Helena, mtn., Calif A5 82
St. Helena, sound, S.C G7 115
Ste. Helene, Que., Can., 529 B8 73
St. Helens, Eng., 108,348 (★145,000) A5 12
St. Helens, Columbia, Oreg., 5,022 B4 113
St. Helens, mtn., Wash C3 122
St. Helier, Que., Can A4 73
St. Hélier, Jersey, 25,364 F5 10
Ste. Henedine, Que., Can., 518 C7 73
St. Henri, Que., Can., 782 C9 73
St. Henry, Mercer, Ohio, 978 B3 111
St. Hermas, Que., Can., 285 D8 73
St. Hermenegilde, Que., Can., 204 D6 73
St. Hilaire, Que., Can D7 73
St. Hilaire, Pennington, Minn., 270 B2 99
St. Hilaire Est, Que., Can., 2,911 D4 73
St. Hilarion, Que., Can., 493 B7 73
St. Honore, Que., Can., 891 A6 73
St. Honore, Que., Can., 528 B8 73
St. Honore, Que., Can., 943 D7 73
St. Hubert, Bel., 3,108 D5 15
St. Hubert, Que., Can., 724 B8 73
St. Hugues, Que., Can., 435 D5 73
St. Hyacinthe, Que., Can., 22,354 D5 73
St. Hyacinthe, co., Que., Can., 44,993 D4 73
St. Ignace, Mackinac, Mich., 3,334 C6 98
St. Ignatius, Lake, Mont., 940 C2 102
St. Ignatius Mission, Br. Gu B3 59
St. Imier, Switz., 6,704 B3 19
St. Inigoes, St. Marys, Md., 125 D5 85
St. Irenee, Que., Can., 701 B7 73
St. Isidore, Que., Can., 373 D6 73
St. Isidore, Que., Can., 290 E8 73
St. Isidore [de Prescott], Ont., Can., 458 B10 72
St. Ives, Eng., 4,076 B7 12
St. Ives, Eng., 9,337 D2 12
St. Jacob, Madison, Ill., 529 E4 90
St. Jacobs, Ont., Can., 669 D4 72
St. Jacques, Que., Can., 2,038 D4 73
St. Jacques, cape, Viet H7 38
St. Jacques le Mineur, Que., Can., 273 E9 73
St. James, Stone, Ark., 30 B4 81
St. James, Man., Can., 33,977 *E3 71
St. James, St. James, La., 280 C6 95
St. James, Charlevoix, Mich., 180 C5 98
St. James, Watonwan, Minn., 4,174 G4 99
St. James, Phelps, Mo., 2,384 D6 101
St. James, Suffolk, N.Y., 3,524 F4 84
St. James, par., La., 18,369 D5 95
St. James, Cedar, Nebr., 50 B8 103
St. James, cape, B.C., Can F6 66
St. James, is., Vir. Is f15 65
St. James City, Lee, Fla., 130 F4 86
St. Janvier, Que., Can., 1,811 D4, D8 73
St. Jean, Que., Can., 26,988 D4 73
St. Jean, co., Que., Can., 38,470 D4 73
St. Jean, riv., Que., Can A7 73
St. Jean Baptiste, Man., Can., 521 E3 71
St. Jean Chrysostome, Que., Can., 563 C9 73
St. Jean-d'Angély, Fr., 8,660 E3 14
St. Jean de Dieu, Que., Can., 1,177 A9 73
St. Jean-de-Luz, Fr., 10,241 F3 14
St. Jean de Matha, Que., Can., 846 C4 73
St. Jean [-de-Maurienne], Fr., 4,252 D2 18
St. Jean Eudes, Que., Can., 2,873 *A7 73
St. Jean Port Joli, Que., Can., 1,615 B7 73
St. Jerome, Que., Can., 24,546 D3 73
St. Jo, Montague, Tex., 977 C4 118
St. Joachim, Que., Can., 988 B7 73
St. Joachim, Que., Can., 25 D5 73
St. Joe, Searcy, Ark., 150 A3 81
St. Joe, Benewah, Idaho, 50 B2 89
St. Joe, DeKalb, Ind., 499 B8 91
St. Joe, riv., Idaho B3 89
Saint John, N.B., Can., 55,153 (★95,563) D3 74
St. John, Lake, Ind., 1,128 B3 91
St. John, Stafford, Kans., 1,753 E5 93
St. John, Aroostook, Maine, 400 (407▲) A4 96
St. John, St. Louis, Mo., 7,342 *A8 95
St. John, Rolette, N. Dak., 420 A6 110
St. John, Whitman, Wash., 545 B8 122
St. John, co., N.B., Can., 89,251 D4 74
St. John, bay, Newf., Can C3 75
St. John, cape, Newf., Can D4 75
St. John, isl., Newf., Can C3 75
St. John, isl., Vir. Is f16 65
St. John, lake, Newf., Can D4 75
St. John, lake, Que., Can A5 73
St. John, riv., Can., Maine B3 96, C2 74
St. Johns, Apache, Ariz., 1,310 C6 80
St. John's, Newf., Can., 63,633 (★90,838) E5 75
St. Johns, Perry, Ill., 206 E4 90
St. Johns, Clinton, Mich., 5,629 E6 98
St. Johns, Auglaize, Ohio, 220 B3 111
St. John's, Antigua, 21,637 n16 64
St. Johns, co., Fla., 30,034 C5 86
St. John's, pt., Ire C3 11
St. John's, pt., Ire C6 11
St. Johns, riv., Fla B5 86
St. Johnsbury, Caledonia, Vt., 6,809 (8,869▲) C4 120
St. Johnsbury Center, Caledonia, Vt., 300 C4 120
St. John's East, dist., Newf., Can E5 75
St. Johns River, entrance, Fla B6 86
St. Johnsville, Montgomery, N.Y., 2,196 B6 108
St. John's West, dist., Newf., Can E5 75
St. John the Baptist, par., La., 18,439 D5 95
St. Jones, riv., Del B6 85
St. Joseph, N.B., Can., 748 D5 74
St. Joseph, Champaign, Ill., 1,210 C5 90
St Joseph, Tensas, La., 1,653 C4 95
St. Joseph, Berrien, Mich., 11,755 F4 98
St. Joseph, Stearns, Minn., 1,487 E4 99
St. Joseph, Buchanan, Mo., 79,673 B3 101
St. Joseph, Lawrence, Tenn., 547 B4 117
St. Joseph, Dominica, 3,050 o16 64
St. Joseph, co., Ind., 238,614 A5 91
St. Joseph, co., Mich., 42,332 G5 98
St. Joseph, bay, Fla C1 86
St. Joseph, isl., Ont., Can B7 98
St. Joseph, isl., Mich B6 98
St. Joseph, isl., Tex E4 118
St. Joseph, lake, Ont., Can E8 72
St. Joseph, lake, Que., Can C8 73
St. Joseph, pt., Fla H3 86
St. Joseph, riv., Ind., Mich., Ohio B8 91, F5 98, A3 111
St. Joseph de Beauce, Que., Can., 2,484 C7 73
St. Joseph de Grantham, Que., Can. (part of Drummondville) *D5 73
St. Joseph de St. Hyacinthe, Que., Can., 3,799 *D5 73
St. Joseph de Sorel, Que., Can., 3,588 *C4 73
St. Joseph du Lac, Que., Can., 358 D8 73
St. Joseph's, Newf., Can., 301 E5 75
St. Josephs Hill, Clark and Floyd, Ind., 100 A4 94
St. Josephs, sound, Fla E1 86
St. Jovite, Que., Can., 2,692 C3 73
St. Jovite Station, Que., Can., 225 C3 73
St. Jude, Que., Can., 515 D5 73
Ste. Julie, Que., Can., 336 C6 73
Ste. Julienne, Que., Can., 753 D4 73
St. Junien, Fr., 8,449 E4 14
St. Just, Eng., 3,636 D2 12
St. Just-en-Chaussée, Fr., 3,575 E2 15
Ste. Justine, Que., Can., 513 D3 73
St. Keverne, Eng., 1,709 D2 12
St. Kitts, isl., St. Kitts-Nevis-Anguilla n15 64
St. Kitts-Nevis-Anguilla, Br. dep., N.A., 56,644 n15 64
St. Lambert, Que., Can., 520 C6 73
St. Lambert, Que., Can., 14,531 D9 73
St. Landry, Evangeline, La., 425 D3 95
St. Landry, par., La., 81,493 D3 95
St. Laurent, Man., Can., 869 D3 71
St. Laurent, Que., Can., 49,805 D9 73
St. Laurent, Fr. Gu., 2,095 A4 59
St. Laurent-Blangy, Fr., 3,681 D2 15
St. Laurent-de-la-Salanque, Fr., 3,300 F5 14
St. Laurent-du-Jura, Fr., 694 C1 18
St. Lawrence, Austl., 264 D8 50
St. Lawrence, Newf., Can., 2,095 E4 75
St. Lawrence, Berks, Pa., 929 *F10 114
St. Lawrence, Hand, S. Dak., 290 C7 116
St. Lawrence, co., N.Y., 111,239 A6 117
St. Lawrence, cape, N.S., Can B9 74
St. Lawrence, gulf, Can G20 67
St. Lawrence, isl., Alsk C5 79
St. Lawrence, riv., Ont., Que., Can, N.Y G19 67
St. Lazare, Man., Can., 449 D1 71
St. Lazare, Que., Can., 345 D8 73
St. Leo, Pasco, Fla., 278 D4 86
St. Leo, Yellow Medicine, Minn., 129 F2 99
St. Leon, Que., Can., 222 C5 73
St. Leon, Que., Can., 475 C7 73
St. Leon, Dearborn, Ind., 319 F8 91
St. Leonard, N.B., Can., 1,666 B2 74
St. Leonard, Que., Can., 454 C6 73
St. Leonard, Calvert, Md., 140 D4 85
St. Leonard d'Aston, Que., Can., 852 C5 73
St. Léonard [-de-Noblat], Fr., 3,671 E4 14
St. Leon [de Chicoutimi], Que., Can., 432 A6 73
St. Lewis, riv., Newf., Can B3 75
St. Lewis, sound, Newf., Can B4 75
St. Liboire, Que., Can., 577 D5 73
St. Libory, St. Clair, Ill., 346 E4 90
St. Libory, Howard, Nebr., 150 C7 103
St. Lô, Fr., 15,388 C3 14
St. Louis, P.E.I., Can., 325 C5 74
St. Louis, Que., Can., 550 D5 73
St. Louis, Sask., Can., 344 E3 70
St. Louis, Sen., 39,100 C1 45
St. Louis, Gratiot, Mich., 3,808 E6 98
St. Louis (Independent City), Mo., 750,026 (★2,050,800) B8, C7 101
St. Louis, Pottawatomie, Okla., 76 B5 102
St. Louis, co., Minn., 231,588 C6 99
St. Louis, co., Mo., 703,532 C7 101
St. Louis, bay, Miss E2 100
St. Louis, lake, Que., Can D8 73
St. Louis, riv., Minn D6 99
St. Louis de Gonzague, Que., Can., 541 D3, E8 73
St. Louis de Kent, N.B., Can., 861 C5 74
St. Louis du Ha Ha, Que., Can., 843 B8 73
Ste. Louise, Que., Can., 493 B7 74
St. Louis Park, Hennepin, Minn., 43,310 E6 99
St. Louisville, Licking, Ohio, 349 B5 111
St. Loup-sur-Semouse, Fr., 2,864 B2 18
St.-Luc, Switz., 193 D4 19
St. Lucas, Fayette, Iowa, 211 A6 92
St. Lucia, Br. dep., N.A., 86,194 p16 64
St. Lucia, cape, S. Afr C5 49
St. Lucia, chan., W.I o16 64
St. Lucia, isl., N.A p16 64

Ste. Lucie, Que., Can., 475..C3 73
Ste. Lucie, Que., Can., 297..C7 73
St. Lucie, St. Lucie, Fla., 500..E6 86
St. Lucie, co., Fla., 39,294..E6 86
St. Lucie, canal, Fla..E6 86
St. Lucie, inlet, Fla..E6 86
St. Ludger, Que., Can., 326..D7 73
Ste. Madeleine, Que., Can., 964..D4 73
St. Magloire, Que., Can., 672..C7 73
St. Magnus, bay, Scot..g10 10
St. Maixent-l'École, Fr., 7,068..D3 14
St. Malachie, Que., Can., 338..C7 73
St. Malo, Fr., 17,137..C2 14
St. Malo, gulf, Fr..C3 14
St. Mandé, Fr., 24,325..g10 14
St. Marc, Que., Can., 236..D4 73
St. Marc, Hai, 10,485..F7 64
St. Marc [des Carrieres], Que., Can., 2,622..C5 73
St. Marcel, Que., Can., 258..C7 73
St. Marcellin, Fr., 5,298..E6 14
St. Margaret, bay, Newf., Can..C3 75
St. Margarets, Anne Arundel, Md., 75..B5 85
St. Margaret's Hope, Scot..B6 13
Ste. Marguerite, Que., Can., 324..C7 73
Ste. Marguerite, Que., Can., 75..D3 73
Ste. Marguerite, riv., Que., Can..A7 73
Ste. Marie, Que., Can., 181..C5 73
Ste. Marie, Que., Can., 3,662..C6 73
Ste. Marie, Jasper, Ill., 347..E5 90
Ste. Marie, cape, Malag..K8 49
Ste. Marie, isl., Malag..g9 49
Ste. Marie-aux-Mines, Fr., 7,897..A3 18
St. Maries, Benewah, Idaho, 2,435..B2 89
Ste. Marie-Sur-Mer., 435..B5 74
St. Mark, Sedgwick, Kans..B5 93
St. Marks, Wakulla, Fla., 350..B2 86
St. Martin, Que., Can..D7 73
St. Martin, Que., Can., 6,440..*D8 73
St. Martin, par., La., 29,063..D4 95
St. Martin, isl., Mich..C4 98
St. Martin, isl., W.I..m15 64
St. Martin, lake, Man., Can..D2 71
St. Martin, riv., Md..D7 85
St. Martin-Boulogne, Fr., 10,888..D9 12
St. Martin [-de-Ré], Fr., 2,262..D3 14
Ste. Martine, Que., Can., 1,695..D4, E8 73
St. Martins, N.B., Can., 509..D4 74
St. Martin Station, Man., Can., 54..D2, D5 71
St. Martinville, St. Martin, La., 6,468..D4 95
St. Martory, Fr., 1,066..F4 14
St. Mary, Marion, Ky., 250..C4 94
St. Mary, par., La., 48,833..E4 95
St. Mary, bay, N.S., Can..E3 74
St. Mary, cape, N.S., Can..E3 74
St. Mary, is., Que., Can..C2 73
St. Mary, riv., Alta., Can..E4 69
St. Mary, riv., B.C., Can..E2 69
St. Mary-of-the-Woods, Vigo, Ind., 700..E3 91
St. Mary's, Newf., Can., 434..E5 75
St. Marys, Ont., Can., 4,482..D3 72
St. Marys, Camden, Ga., 3,272..F5 87
St. Marys, St. Joseph, Ind., 900..A5 91
St. Marys, Pottawatomie, Kans., 1,509..C7 93
St. Marys, Ste. Genevieve, Mo., 620..D8 101
St. Marys, Auglaize, Ohio, 7,737..B3 111
St. Marys, Elk, Pa., 8,065..D4 114
St. Marys, Pleasants, W. Va., 2,443..B3 123
St. Marys, co., Md., 38,915..D4 85
St. Mary's, bay, Newf., Can..E5 75
St. Mary's, cape, Newf., Can..E4 75
St. Marys, entrance, Fla..B5 86
St. Marys, riv., N.S., Can..D8 74
St. Marys, riv., Fla., Ga..B5 80, F5 87
St. Marys, riv., Ind., Ohio..C8 85, B3 111
St. Marys, riv., Md..D5 85
St. Marys, riv., Mich..B6 98
St. Matthew, isl., Alsk..C5 79
St. Matthew, isl., Bur..H3 38
St. Matthews, Jefferson, Ky., 8,738..A4, B4 94
St. Matthews, Calhoun, S.C., 2,433..D6 115
St. Mathieu, Que., Can., 545..A8 73
St. Mathieu, Que., Can., 95..E9 73
St. Maur-des-Fossés, Fr., 70,397..g10 14, F2 15
St. Maurice, Que., Can., 333..C5 73
St. Maurice, Switz., 3,196..D3 19
St. Maurice, co., Que., Can., 109,873..C4 73
St. Maurice, riv., Que., Can..B4, C5 73
St. Maxime, Que., Can., 210..C6 73
St. Meinrad, Spencer, Ind., 850..H4 91
Ste. Melanie, Que., Can., 279..C4 73
Ste. Menehould, Fr., 3,406..C6 14
St. Methode, Que., Can., 690..C6 73
St. Michael, Alsk., 205..C7 79
St. Michael, Alta., Can., 129..C4 69
St. Michael, Cambria, Pa., 1,292..*F4 114
St. Michaels, Apache, Ariz., 50..B6 80
St. Michaels, Talbot, Md., 1,484..C5 85
St. Michael, Wright, Minn., 707..E5 187
St. Michael, Benson, N. Dak., 40..B7 110
St. Michaels, bay, Newf., Can..B4 75
St. Michel, Que., Can., 675..C7 73
St. Michel, Que., Can., 145..E9 73
St. Michel, Fr., 4,502..E4 15
St. Michel-de-l'Atalaye, Hai., 2,328..F7 64
St. Michel [-de-Maurienne], Fr., 2,313..D2 18
St. Michel des Saints, Que., Can., 1,763..C4 73
St. Mihiel, Fr., 5,253..C6 14
St. Modeste, Que., Can., 73..B8 73
Ste. Monique, Que., Can., 229..C5 73
St. Moritz, see Sankt Moritz, Switz.
St. Nazaire, Que., Can., 816..A6 73
St. Nazaire, Que., Can., 229..C7 73
St. Nazaire, Que., Can., 158..D5 73
St. Nazaire, Fr., 58,286..D2 14
St. Nazianz, Manitowoc, Wis., 669..B6, D6 124
St. Neots, Eng., 5,570..B7 12
St. Neree, Que., Can., 411..C7 73
St. Nicholas, Que., Can., 424..C8 73
St. Nicolas, see Sint Niklaas, Bel.
St. Nicolas-de-Port, Fr., 5,761..F6 15
St. Noel, Que., Can., 1,124..*B3 73
St. Norbert, Man., Can., 695..E3 73
St. Norbert [d'Arthabaska], Que., Can., 291..C6 73
St. Odilon, Que., Can., 686..C7 73
St. Olaf, Clayton, Iowa, 169..B6 94
St. Olof, Swe., 468..C8 24
St. Omer, Que., Can., 470..B3 73
St. Omer, Fr., 19,283..B5 14
St. Onge, Lawrence, S. Dak., 100..C2 116
Saintonge, former prov., Fr., 286,000..E3 14
St. Ouen, Fr., 51,956..g10 14
St. Ours, Que., Can., 711..D4 73
St. Pacome, Que., Can., 1,242..B8 73
St. Pamphile, Que., Can., 1,839..C8 73
St. Paris, Champaign, Ohio, 1,460..B4 111
St. Pascal, Que., Can., 2,144..B8 73
St. Patrick, lake, Que., Can..A7 72
St. Paul, Madison, Ark., 118..B2 81
St. Paul, Alta., Can., 2,823..*B5 69
St. Paul, Que., Can., 193..B8 73
St. Paul, Que., Can., 317..C3 75
St. Paul, Que., Can., 835..C7 73
St. Paul, Decatur and Shelby, Ind., 702..F6 91
St. Paul, Neosho, Kans., 675..E8 93
St. Paul, Ramsey, Minn., 313,411..E7, F6 99
St. Paul, St. Charles, Mo., 125..A7 101
St. Paul, Howard, Nebr., 1,714..C7 103
St. Paul, Marion, Oreg., 254..B1 113
St. Paul, Clarendon, S.C., 75..D7 115
St. Paul, Wise, Va., 1,156..B2 121
St. Paul, isl., Alsk..D5 79
St. Paul, isl., N.S., Can..B9 74
St. Paul, isl., Indian O..I2 2
St. Paul, riv., Newf., Que., Can..C3 75
St. Paul, riv., Lib..E3 45
St. Paul, rocks, Atl. O..F8 6
St. Paul du Nord, Que., Can., 337..A8 73
St. Paulin, Que., Can., 920..C4 73
St. Paul Park, Washington, Minn., 3,267..F6, F7 99
St. Pauls, Blaine, Mont., 75..C8 102
St. Pauls, Robeson, N.C., 2,249..C5 109
Ste. Perpetue, Que., Can., 515..B8 73
Ste. Perpetue, Que., Can., 203..C5 73
St. Peter, Fayette, Ill., 397..E5 90
St. Peter, Graham, Kans., 60..C3 93
St. Peter, Nicollett, Minn., 8,484..F5 99
St. Peter, Cascade, Mont., 75..C5 102
St. Peter, lake, Que., Can..C5 73
St. Peter Port, Guernsey, 16,720..F5 10
St. Peters, N.S., Can., 762..D9 74
St. Peters, Franklin, Ind., 100..F7 91
St. Peters, St. Charles, Mo., 404..A7, C7 101
St. Peters Bay, P.E.I., Can., 321..C7 74
St. Petersburg, Pinellas, Fla., 181,298 (★355,200)..E2, E4 86
St. Petersburg, Clarion, Pa., 417..D2 114
St. Petersburg Beach, Pinellas, Fla., 6,268..F1 86
Ste. Petronille, Que., Can., 510..C9 73
St. Philemon, Que., Can., 539..C7 73
St. Philip, Posey, Ind., 100..I2 91
St. Philippe, Que., Can., 510..D4, D9 73
St. Philippe de Neri, Que., Can., 746..B8 73
St. Phillips, Wibaux, Mont., 30..D12 102
Ste. Philomene, Que., Can., 386..D4, E8 73
St. Pie, Que., Can., 1,434..D5 73
St. Pierre, Que., Can., 364..C7 73
St. Pierre, Mart., 6,218..o16 64
St. Pierre, St. Pierre and Miquelon..E3 75
St. Pierre, isl., St. Pierre & Miquelon..E3 75
St. Pierre & Miquelon, Fr. dep., N.A., 4,606..E3 75
St. Pierre d'Albigny, Fr., 838..D2 18
St. Pierre-en-Port, Fr., 995..E8 12
St. Pierre-Jolys, Man., Can., 856..E3 71
St. Pierre [les Becquets], Que., Can., 453..C5 73
St. Pius, Stark, N. Dak., 75..C3 110
St. Placide, Que., Can., 336..D7 73
St. Pol [-de-Léon], Fr., 6,037..C2 14
St. Pol-sur-Mer, Fr., 18,686..C2 15
St. Pol [-sur-Ternoise], Fr., 5,193..B5 14
St. Pourçain [-sur-Sioule], Fr., 3,182..D5 14
St. Prime, Que., Can., 659..A5 73
St. Prosper, Que., Can., 425..C5 73
St. Prosper, Que., Can., 1,357..C7 73
St. Quentin, N.B., Can., 2,089..B2 74
St. Quentin, Fr., 61,071..C5 14, E3 15
St. Raphael, Que., Can., 1,134..C7 73
St. Raphaël, Fr., 9,470..F7 14
St. Raymond, Que., Can., 3,931..C6 73
St. Redempteur, Que., Can., 872..C9 73
St. Redempteur, Que., Can., 1,035..D3 73
St. Regis, Mineral, Mont., 600..C1 102
St. Regis, riv., N.Y..A2 108
St. Regis Falls, Franklin, N.Y., 800..A2 108
St. Regis Park, Jefferson, Ky., 1,179..*H6 94
St. Remi, Que., Can., 2,276..D4, E9 73
St. Remi, Que., Can., 186..D6 73
St. Remi d'Amherst, Que., Can., 396..D3 73
St. Roch., Que., Can., 523..D4 73
St. Roch des Aulnaies, Que., Can., 335..B7 73
St. Romain, Que., Can., 465..D6 73
St. Romuald, Que., Can., 4,000..C6, C9 73
Ste. Rosalie, Que., Can., 1,255..*D5 73
Ste. Rose, Que., Can., 7,571..D4, D8 73
Ste. Rose, Guad., 1,288..n16 64
St. Rose, St. Charles, La., 1,099..C7 95
Ste. Rose de Lima, Que., Can., 2,965..D2 73
Ste. Rose du Degele, Que., Can., 1,943..B9 73
Ste. Rose du Lac, Man., Can., 790..D2 71
Ste. Sabine, Que., Can., 444..C7 73
St. Samuel, Que., Can., 535..D7 73
St. Sauveur [-des Montagnes], Que., Can., 1,702..D3 73
Ste. Savine, Fr., 11,864..C5 14
Ste. Scholastique, Que., Can., 838..D3, D8 73
St. Sebastien, Que., Can., 516..D7 73
St. Sebastien, cape, Malag..f9 49
St. Servan [-sur-Mer], Fr., 14,963..C3 14
St. Sever, Fr., 2,028..F3 14
St. Severin, Que., Can., 230..C6 73
St. Shotts, Newf., Can., 189..E5 75
St. Simeon, Que., Can., 1,197..B8 73
St. Siméon, Fr., 191..F3 15
St. Simon, Que., Can., 535..A8 73
St. Simons, isl., Ga..E5 87
St. Simons, sound, Ga..E5 87
St. Simons Island, Glynn, Ga., 3,199..E5 87
St. Sixte, Que., Can., 86..D2 73
Ste. Sophie, Que., Can., 398..C5 73
Ste. Sophie, Que., Can., 267..C6 73
St. Stanislas [de Champlain], Que., Can., 590..C5 73
St. Stephen, N.B., Can., 3,380..D2 74
Ste. Stephen, Berkeley, S.C., 1,462..E8 115
St. Stephen, Fremont, Wyo., 5..C4 125
St. Stephens, Washington, Ala., 150..D1 78
St. Sylvestre, Que., Can., 652..C6 73
St. Tammany, St. Tammany, La., 50..B8, D6 95
St. Tammany, par., La., 38,643..D5 95
St. Thecle, Que., Can., 2,009..C5 73
St. Theodore, Que., Can., 315..C4 73
St. Theophile, Que., Can., 510..D7 73
Ste. Therese de Blainville, Que., Can., 11,771..D4, D8 73
St. Thomas, Ont., Can., 22,469..E3 72
St. Thomas, Que., Can., 508..C4 73
St. Thomas, Cole, Mo., 180..C5 101
St. Thomas, Pembina, N. Dak., 660..A8 110
St. Thomas, Franklin, Pa., 600..G6 114
St. Thomas, isl., Vir. Is..f15 65
St. Thuribe, Que., Can., 208..C5 73
St. Timothée, Que., Can., 1,003..E8 73
St. Tite, Que., Can., 3,250..C5 73
St. Tite des Caps, Que., Can., 1,227..B7 73
St. Tropez, Fr., 3,988..F7 14
St. Ubald, Que., Can., 764..C5 73
St. Ulric, Que., Can., 1,021..*A2 73
St. Urbain, Que., Can., 878..B7 73
St. Urbain, Que., Can., 255..E8 73
St. Valere, Que., Can., 197..C5 73
St. Valerien, Que., Can., 226..D5 73
St. Valéry-en-Caux, Fr., 2,905..E8 12
St. Valéry-sur-Sommne, Fr., 3,169..B4 14
St. Vallier, Que., Can., 540..C7 73
St. Vallier, Fr., 4,124..E6 14
Ste. Veronique, Que., Can., 250..C3 73
St. Victor [de Beauce], Que., Can., 931..C7 73
St. Vincent, Kittson, Minn., 217..B1 99
St. Vincent, Br. dep., N.A., 80,005..p16 64
St. Vincent, cape, Malog..h8 49
St. Vincent, cape, Port..D1 20
St. Vincent, gulf, Austl..G2 51
St. Vincent, isl., Fla..C1 86
St. Vincent, isl., N.A..p16 64
St. Vincent, passage, Win. Is..p16 64
St. Vincent de Paul, Que., Can., 11,214..D9 73
St. Vincent's, Newf., Can., 599..E5 75
St. Vith, Bel., 2,708..D6 15
St. Vrain, Curry, N. Mex., 15..C6 107
St. Walburg, Sask., Can., 689..D1 70
St. Wendells, Posey, Ind., 160..H2 91
St. Williams, Ont., Can., 391..E4 72
St. Xavier, Big Horn., Mont., 75..E9 102
St. Yrieix-la-Perche, Fr., 4,368..E4 14
St. Yvon, Que., Can., 428..A4 73
St. Zacharie, Que., Can., 1,361..C7 73
St. Zephirin, Que., Can., 247..C5 73
Saipan, chan., Saipan.. 52
Saipan, isl., Mariana Is.. 52
Sa'ir, Jordan, 3,000..C7 32
Saitama, pref., Jap., 2,403,871..*I9 37
Saiyidabad, Afg., 10,000..D14 41
Sajama, mtn., Bol..C2 55
Saka, China, 5,000..C10 40
Saka, Ken..B6 48
Sakai, Jap., 339,863..o14 37
Sakākā, Sau. Ar., 10,000..H13 31
Sakania, Con. L., 12,100..D4 48
Sakaraha, Malag..h8 49
Sakarya (Sangarius), riv., Tur..B8 31
Sakashita, Jap., 3,400..n16 37
Sakata, Jap., 55,600 (97,671▲)..G9 37
Sakchu, Kor., 13,568..F2 37
Sakhalin, isl., Sov. Un..D17 28
Sakiai, Sov. Un..A7 26
Sakimotobu, Okinawa, 20,409.. 52
Sakishima, is., Ryukyu Is..G9 34
Sakmara, riv., Sov. Un..C5 29
Sakon Nakhon, Thai, 14,940..D6 38
Sakonnet, Newport, R.I., 100..D12 84
Sakonnet, riv., R.I..C12 84
Sak'ot'ā, Eth..C4 47
Sakripe, Lib..E3 45
Sakskōbing, Den., 2,526 (★4,035)..D5 24
Sakti, China..A9 40
Sakti, India, 8,125..F9 40
Sal, isl., C.V. Is..*E3 42
Sal, pt., Calif..E3 82
Sal, riv., Sov. Un..D2 29
Sala, Swe., 11,000..H7, t34 25
Sala Consilina, It., 6,897..D5 21
Saladas, Arg., 3,900..E4 55
Saladillo, Arg., 7,586..B5 54
Salado, riv., Arg..A3 54
Salado, riv., Arg..B5 54
Salado, riv., Arg..C3 54
Salado, riv., Arg..E3 55
Salado, riv., Mex..B4 63
Salaga, Ghana, 3,156..E4 45
Salajar, isl., Indon..G6 35
Salamá, Guat., 2,760..C4 62
Salamá, Hond., 947..C4 62
Salamanca, Chile, 2,819..A2 54
Salamanca, Mex., 32,192..m13 63
Salamanca, Cattaraugus, N.Y., 8,480..C2 108
Salamanca, Sp., 90,498..B3 20
Salamanca, prov., Sp., 405,729..*B3 20
Salamat, riv., Chad..C3 46
Salamaua, N. Gui., 270..k12 50
Salamina, Col., 7,940..B2 60
Salamis, Grc., 11,161..D4, h11 23
Salamis, isl., Grc..h10 23
Salamonia, Jay, Ind., 142..D8 91
Salamonie, riv., Ind..C7 91
Salas, Sp., 2,522..A2 20
Salaverry, Peru, 3,403..C2 58
Salavina, Arg..E3 55
Salawati, is., W. Irian..F8 35
Sala-y-Gomez, isl., Pac. O..H16 7
Salcedo, Ec..B2 58
Saldanha, S. Afr., 2,195..D2 49
Saldus, Sov. Un., 10,000..C4 27
Sale, Austl., 7,899..I6 51
Salé, Mor., 75,799..C3 44
Salealua, bay, W. Sam.. 52
Sale City, Mitchell, Ga., 275..E2 87
Sale Creek, Hamilton, Tenn., 800..D8 117
Salekhard, Sov. Un., 16,000..B22 9
Salem, Lee, Ala., 200..C4 78
Salem, Fulton, Ark., 713..A4 75
Salem, New London, Conn., 300 (925▲)..D7 84
Salem, Taylor, Fla., 200..C3 86
Salem, Marion, Ill., 6,165..E5 90
Salem, India, 249,145..F6 39
Salem, Washington, Ind., 4,546..G5 91
Salem, Henry, Iowa, 442..D6 92
Salem, Livingston, Ky., 480..A3 94
Salem, Essex, Mass., 39,211..A6, C3 97
Salem, Washtenaw, Mich., 250..A7 98
Salem, Dent, Mo., 3,870..D6 101
Salem, Richardson, Nebr., 261..D10 103
Salem, Rockingham, N.H., 950 (9,210▲)..E4 105
Salem, Salem, N.J., 8,941..D2 106
Salem, Dona Ana, N. Mex., 100..E2 107
Salem, Washington, N.Y., 1,076..B7 108
Salem, Columbiana, Ohio, 13,854..B7 111
Salem, Marion and Polk, Oreg., 49,142..C1, C4 112
Salem, Oconee, S.C., 206..B2 115
Salem, McCook, S. Dak., 1,188..D8 116
Salem, Utah, Utah, 920..C4 119
Salem, Roanoke, Va., 16,058..D2 121
Salem, Harrison, W. Va., 2,366..B4, B6 123
Salem, Kenosha, Wis., 500..F1 121
Salem, co., N.J., 58,711..D2 106
Salem, creek, Ohio..A1 123
Salem, fork, W. Va..B6 123
Salem, plat., Mo..D6 101
Salem, pond, Vt..B4 120
Salem, riv., N.J..D2 106
Salemburg, Sampson, N.C., 569..B5 109
Salem Depot, Rockingham, N.H., 2,523..E4 105
Salem Heights, Marion, Oreg., 10,770..*C4 113
Salemi, It., 13,300..F4 21
Salen, Scot..D3 13
Salerno, Martin, Fla., 900..E6 86
Salerno, It., 117,363 (★185,000)..D5 21
Salerno, gulf, It..D5 21
Sales, pt., Eng..C8 12
Salfīt, Jordan, 2,000..B7, g11 32
Salford, Eng., 154,963..A5 12
Salge, Bonner, Idaho, 100..A2 89
Salgotarjan, Hung., 26,682..A4 22
Salgueiro, Braz., 8,936..C3 57
Salida, Stanislaus, Calif., 1,109..*D3 82
Salida, Chaffee, Colo., 4,560..C5 83
Salies-de-Béarn, Fr., 2,859..F3 14
Salihli, Tur., 24,100..C7 23
Salima, Nya., 1,450..D5 48
Salin, Bur., 5,000..B1 38
Salina, Saline, Kans., 43,202..D6 93
Salina, Mayes, Okla., 972..A6 112
Salina, Sevier, Utah, 1,618..E4 119
Salina, isl., It..E5 21
Salina Cruz, Mex., 14,881..D5 63
Salinas, Braz., 5,186..E2 57
Salinas, Monterey, Calif., 28,957..C6, D3 82
Salinas, Ec., 2,868..B1 58
Salinas, P.R., 3,666..D5 65
Salinas, mun., P.R., 23,133..C5 65
Salinas, bay, P.R..D2 65
Salinas, cape, Sp..C7 20
Salinas, pampa, Arg..A3 54
Salinas, peak, N. Mex..D3 107
Salinas, pt., Ang..D1 48
Salinas, pt., Peru..D2 58
Salinas, pt., P.R..B6 65
Salinas, riv., Calif..D3 82
Salinas de Garcí Mendoza, Bol., 635..C2 55
Salinas, Grandes, salt flat, Arg..E3 55
Salina Springs, Apache, Ariz., 200..A6 80
Saline, Bienville, La., 329..B3 95
Saline, Washtenaw, Mich., 2,334..F7 98
Saline, co., Ark., 28,956..C3 81
Saline, co., Ill., 26,227..F5 90
Saline, co., Kans., 54,715..D6 93
Saline, co., Mo., 25,148..B4 101
Saline, co., Nebr., 12,542..D8 103
Saline, bayou, La..B3 95
Saline, riv., Ark..C1 81
Saline, riv., Ark..D3 81
Saline, riv., Ill..F5 90
Saline, riv., Kans..C4 93
Salineno, Starr, Tex., 175..F3 118
Salineville, Columbiana, Ohio, 1,898..B7 111
Salinópolis, Braz., 4,101..B1 57
Salins-les-Bains, Fr., 4,476..D6 14
Salisbury, N.B., Can., 589..C4 74
Salisbury, Litchfield, Conn., 368 (3,309▲)..B3 84
Salisbury, Eng., 35,471..C6 12
Salisbury, Wicomico, Md., 16,302..D6 85
Salisbury, Essex, Mass., 950 (3,154▲)..A6 97
Salisbury, Chariton, Mo., 1,787..B5 101
Salisbury, Merrimack, N.H., 100 (415▲)..D3 105
Salisbury, Rowan, N.C., 21,297..B3 109
Salisbury, Somerset, Pa., 862..G3 114
Salisbury, S. Rh., 190,000 (★299,200)..B5 49
Salisbury, Addison, Vt., 130 (575▲)..D2 120
Salisbury, isl., N.W. Ter., Can..D17 67
Salisbury, plain, Eng..C6 12
Salisbury Center, Herkimer, N.Y., 224..B6 108
Salisbury West, Rowan, N.C., 1,323..*B3 109
Salitpa, Clarke, Ala., 425..D1 78
Salix, Woodbury, Iowa., 394..B1 92
Salkehatchie, riv., S.C..E5 115
Salkum, Lewis, Wash., 200..C3 122
Salladasburg, Lycoming, Pa., 255..D7 114
Sallanches, Fr., 3,552..D2 18
Salley, Aiken, S.C., 403..D5 115
Salliqueló, Arg., 3,938..B4 54
Sallis, Attala, Miss., 223..B4 100
Sallisaw, Sequoyah, Okla., 3,351..B7 112
Salm, isl., Sov. Un..B9 4
Salmo, B.C., Can., 889..E9 68
Salmon, Lemhi, Idaho, 2,944..D5 89
Salmon, mtn., N.H..A4 105
Salmon, mts., Calif..B2 82
Salmon, peak, Tex..E2 118
Salmon, res., N.Y..B5 108
Salmon, riv., B.C., Can..B6 68
Salmon, riv., N.B., Can..B2 74
Salmon, riv., Idaho..D3 89
Salmon, riv., N.Y..A3 108
Salmon, val., B.C., Can..B6 68
Salmon Arm, B.C., Can., 1,506..D8 68
Salmon Bay, Que., Can., 86..C3, h10 75
Salmon Creek, Clark, Wash., 175..D3 122
Salmon Falls, Strafford, N.H., 1,210..D5 105
Salmon Falls, creek, Nev..B7 104
Salmon Falls, riv., Idaho..G4 89
Salmon Falls, riv., N.H., Maine..D5 105
Salmon Gums, Austl., 61..F3 50
Salmon River, mts., Idaho..D3 89
Salo, Fin., 11,000..G10 25
Saló, It., 5,568..D6 18
Salol, Roseau, Minn., 68..B3 99
Salome, Yuma, Ariz., 200..D2 80
Salon, riv., Fr..B1 18
Salon-de-Provence, Fr., 17,267 (21,393▲)..F6 14
Salonika (Thessaloniki), Grc., 216,838..B4 23
Salonika (Thessaloniki), prov., Grc., 544,394..*B4 23
Salonika, gulf, Grc..B4 23
Salonta, Rom., 16,276..B5 22
Salpi, lake, It..D6 21
Salsacate, Arg..A3 54
Salsette, isl., India..H4 40
Salsk, Sov. Un., 18,500..D2 29
Salsomaggiore, It., 8,600..B2 21
Salt (As Salt), Jordan, 15,478..F10 31
Salt, basin, Tex..E2 118
Salt, creek, Ind..G5 91
Salt, creek, Kans..A3 93
Salt, creek, Nebr..E2 103
Salt, creek, Ohio..C5 111
Salt, creek, Wyo..B6 125
Salt, fork, Okla..A3 112
Salt, fork, Okla..C2 112
Salt, isl., Vir. Is..f16 65
Salt, lake, Austl..D1 50
Salt, lake, Sask., Can..D5 70
Salt, lake, Haw..g10 88
Salt, lake, N. Mex..E5 107
Salt, marsh, Kans..D5 93
Salt, pt., Calif..C2 82
Salt, riv., Ariz..D4 80
Salt, riv., Ky..C4 94
Salt, riv., Mo..B5 101
Salta, Arg., 108,000..D2 55
Salta, prov., Arg., 412,652..D2 55
Saltair, Salt Lake, Utah, 50..C3 119
Saltash, Eng., 7,420..D3 12
Salt Ash, mtn., Vt..D3 120
Saltburn-by-the-Sea, Eng., 3,911 (Saltburn and Marske-by-the-Sea, 12,482)..F8 13
Salt Cay, isl., W.I.F..f14 65
Saltcoats, Sask., Can., 490..F4 70
Saltcoats, Scot., 14,187..E4 13
Salt Creek, pass, Oreg..D4 113
Saltee, is., Ire..E5 11
Salter Path, Carteret, N.C., 135..C7 109
Salters, Williamsburg, S.C., 100..D8 115
Saltese, Mineral, Mont., 100..C1 102
Saltfjord, fjord, Nor..D6 25
Saltfork, Grant, Okla., 35..A4 112
Saltfork, creek, Kans..E4 93
Salt Fork of Arkansas, riv., Okla..A4 112
Saltholm, isl., Den..C6 24
Saltillo, Washington, Ind., 121..G5 91
Saltillo, Mex., 99,101..B4 63
Saltillo, Lee, Miss., 536..A5 100
Saltillo, Huntingdon, Pa., 395..F5 114
Saltillo, Hardin, Tenn., 397..B3 117
Salt Lake, co., Utah, 383,035..C4 119
Salt Lake City, Salt Lake, Utah, 189,454 (★410,200)..C4 119
Salt Lick, Bath, Ky., 370..B6 94
Salto, Braz., 12,643..m8 56
Salto, Ur., 46,500..E1 56
Salto, dept., Ur., 78,600..*E1 56
Salto Grande, Braz., 3,016..C2 56
Salton, sea, Calif..F6 82
Saltonstall, lake, Conn..D5 84
Salt Peter, cave, Ga..B2 87
Saltpond, Ghana, 6,968..E4 45
Salt River, Bullitt, Ky., 400..C4 94
Salt River, Indian res., Ariz..G2 80
Salt River, mts., Ariz..G2 80
Salt River, range, Wyo..C2 125
Saltrou, Hai..F7 64
Saltsburg, Indiana, Pa., 1,054..F3 114
Saltsjöbaden, Swe., 5,300..t36 25
Salt-Spring, isl., B.C., Can..E6 68
Salt Springs, Marion, Fla., 755..C5 86
Saltville, Smyth and Washington, Va., 2,844..B3 121
Salt Wells, Churchill, Nev., 3..D3 104
Saluafata, hbr., W. Sam.. 52
Salud, Pan., 211..k10 62
Saluda, Polk, N.C., 570..D4 109
Saluda, Saluda, S.C., 2,089..D4 115
Saluda, Middlesex, Va., 300..D6 121
Saluda, co., S.C., 14,554..C4 115
Saluda, mtn., N.C..D3 109
Saluda, riv., S.C..C3 115
Saluda Gardens, Lexington, S.C., 2,000..*D5 115
Salur, India, 26,111..H9 40
Saluvia, Fulton, Pa., 25..F5 114
Saluzzo, It., 11,100..B1 21
Salvador, Braz., 630,878..D3 57
Salvador, Sask., Can., 137..E1 70
Salvador, El, see El Salvador, country, N.A.
Salvador, lake, La..E5 95
Salvage, Newf., Can., 270..D5 75
Salvage, is., Port..m14 20
Salvatierra, Mex., 14,417..m13 63
Salvisa, Mercer, Ky., 350..C5 94
Salween, riv., Bur..D10 39
Salyany, Sov. Un., 23,000..F3 29
Salyersville, Magoffin, Ky., 1,173..C6 94
Salym, marsh, Sov. Un..B8 29
Salzach, riv., Aus., Ger..A8 18
Salzburg, Aus., 108,114..E6 16, B9 18
Salzburg, state, Aus., 347,292..B9 18
Salzgitter, Ger., 110,200..A5 17
Salzwedel, Ger., 20,700..B5 16
Sama, China..C8 38
Samadan, Switz., 2,106..C5 18
Samaipata, Bol., 1,656..C3 55
Samalá, riv., Guat..C2 62
Samālūt, Eg., U.A.R., 29,300..D6 43
Samaná, Dom. Rep., 3,309..F9 64
Samaná, bay, Dom. Rep..F9 64
Samana, isl., Ba. Is..D7 64
Samaniego, Col., 2,303..C2 60
Samannūd, Eg., U.A.R., 23,300..D3 32
Samar, prov., Phil., 871,900..*C7 35
Samar, isl., Phil..C7 35
Samara, riv., Sov. Un..E19 9
Samarai, Pap..m13 50
Samarga, Sov. Un..C9 37
Samaria, Oneida, Idaho, 150..G6 89
Samarinda, Indon., 68,095..F5 35
Samarkand, Sov. Un., 209,000..H22 9
Samarra, Iraq, 8,867..D2 41
Samata, W. Sam., 469.. 52
Sambalpur, India, 38,915..G9 40
Sambar, cape, Indon..F4 35
Sambas, Indon., 12,000..E3 35
Sambava, Malag..f10 49
Sambhal, India, 68,940..C7 40
Sambhar, India, 14,139..D5 40
Sambonifacio, It., 5,470..D7 18
Sambor, Sov. Un., 41,200..G4 27
Samborombón, bay, Arg..B5 54
Sambre, riv., Bel..D4 15
Samburg, Obion, Tenn., 451..A2 117
Same, Tan., 4,428..B6 48
Samedan, Switz., 1,685..C8 19
Samit, Camb..G5 38
Sammanish, lake, Wash..D2 122
Sammylane, Stone, Mo..E4 101
Sam Neua, Laos..B6 38
Samnorwood, Collingsworth, Tex., 65..B2 118
Samnū, Libya..D2 43
Samoa, Humboldt, Calif., 600..B1 82
Samoa, American, see American Samoa, U.S. dep., Oceania
Samoa, Western, see Western Samoa, country, Pac. O.
Samokov, Bul., 12,784..D6 22
Samos, prov., Grc., 52,022..*D6 23
Samos, isl., Grc..D6 23
Samoset, Manatee, Fla., 4,824..F2 86
Samosir, isl., Indon..m11 35
Samosir, mtn., Indon..K3 38
Samothraki, isl., Grc..B5 23
Sampacho, Arg., 3,554..A4 34
Sampit, Indon..F4 35
Sampit, riv., S.C..E9 115
Sampson, co., N.C., 48,013..C5 109
Sampwe, Con. L..C4 48
Samrong, Camb..E5 38

Samsö, isl., Den C4 24
Samsö Belt, strait, Den C4 24
Samson, Geneva, Ala., 1,932 D3 78
Samsun, Tur., 87,300 B11 31
Sams Valley, Jackson, Oreg E4 113
Samtown, Rapides, La., 4,008 *C3 95
Samtredia, Sov. Un., 10,000 A14 31
Samu, Jordan, 3,000 C7 32
Samuel, hill, Ky A4 94
Samuels, Bonner, Idaho, 10 A2 89
Samui, isl., Thai H4 38
Samus, Sov. Un., 10,000 D25 9
Samutprakan, Thai., 21,607 *F4 38
Samut Sakhǒn, Thai., 27,163 F4 38
Samwari, Pak C1 40
San, Mali, 7,800 D4 45
San, riv., Camb F7 51
San, riv., Pol C7 26
Şan'ā', Yemen, 89,000 B5 47
San Acacia, Socorro, N. Mex., 80 C3 107
San Acacio, Costilla, Colo., 160 D5 83
Sanaga, riv., Cam E2 46
Sanagha, riv., Con. B E3 46
San Agustín, Arg A3 54
San Agustín, Col., 2,493 C2 60
San Agustin, cape, Phil D7 35
Sanak, isl., Alsk E7 79
San Ambrosio, isl., Pac. O H17 7
Sanana, isl., Indon F7 35
Sanandaj, Iran, 40,641 D3 41
San Andreas, Calaveras, Calif., 1,416 C3 82
San Andrés, Col., 2,139 D7 62
San Andrés, isl., Col D7 62
San Andres, mts., N. Mex E3 107
San Andrés de Giles, Arg., 5,392 g7 54
San Andrés Tetepilco, Mex., 11,266 h9 63
San Andrés Totoltepec, Mex., 1,999 h9 63
San Andrés Tuxtla, Mex., 19,830 D5 63
San Andrés y Providencia, intendencia, Col., 5,330 *D7 62
San Angelo, Tom Green, Tex., 58,815 D2 118
San Anselmo, Marin, Calif., 11,584 B5 82
San Antioco, isl., It E2 21
San Antonio, Arg D2 55
San Antonio, Chile, 27,000 A2 54
San Antonio, Chile E1 55
San Antonio, Pasco, Fla., 479 D4 86
San Antonio, Bernalillo, N. Mex., 70 G5 107
San Antonio, Socorro, N. Mex., 500 D3 107
San Antonio, Phil., 7,439 o13 35
San Antonio, P.R A2 65
San Antonio, Bexar, Tex., 587,718 (★689,700) B3, E3 118
San Antonio, bay, Tex E4 118
San Antonio, cape, Arg B5 54
San Antonio, cape, Cuba E1 64
San Antonio, peak, Calif F3 82
San Antonio, riv., Tex B4 118
San Antonio Abad, Sp., 5,377 C6 20
San Antonio de Areco, Arg., 7,436 g7 54
San Antonio de los Baños, Cuba, 17,783 D2 64
San Antonio de los Cobres, Arg D2 55
San Antonio Heights, San Bernardino, Calif., 1,100 *E5 82
San Antonio Oeste, Arg., 3,847 C4 54
San Antonio Spring, McKinley, N. Mex B1 107
Sanarate, Guat., 2,946 C2 62
San Ardo, Monterey, Calif., 500 D3 82
Sanariapo, Ven B4 60
Sanator, Custer, S. Dak., 21 D2 116
Sanatorium, Tom Green, Tex D2 118
San Augustine, San Augustine, Tex., 2,584 D5 118
San Augustine, co., Tex., 7,722 D5 118
San Bartolomeo [in Galdo], It., 8,767 D5 21
San Benedetto del Tronto, It., 22,500 (31,274▲) C4 21
San Benito, Cameron, Tex., 16,422 F4 118
San Benito, co., Calif., 15,396 D3 82
San Benito, mtn., Calif D3 82
San Benito, riv., Calif C6 82
San Bernardino, San Bernardino, Calif., 91,922 (★460,000) E5, F3 82
San Bernardino, Switz D7 19
San Bernardino, co., Calif., 503,591 E5 82
San Bernardino, mts., Calif F3 82
San Bernardino, riv., Mex G6 80
San Bernardo, Chile, 53,000 A2 54
San Blas, Mex., 1,597 m11 63
San Blas, cape, Fla H3 86
San Blas, mts., Pan F8 62
San Blas, riv., Guat B2 62
San Borja, Bol., 708 B2 55
San Borja, riv., Mex h9 63
Sanborn, O'Brien, Iowa, 1,323 A2 92
Sanborn, Redwood, Minn., 521 F3 99
Sanborn, Barnes, N. Dak., 263 C7 110
Sanborn, Ashland, Wis., 100 B3 124
Sanborn, co., S. Dak., 4,641 D7 116
Sanbornton, Belknap, N.H., 100 (857▲) D3 105
Sanbornville, Carroll, N.H., 400 C4 105
San Bruno, San Mateo, Calif., 29,063 B5, D2 82
San Candido, It., 2,782 C8 18
San Carlos, Arg A3 54
San Carlos, Gila, Ariz., 900 D5 80
San Carlos, San Mateo, Calif., 21,370 B5 82
San Carlos, Chile, 11,094 B2 54
San Carlos, Mex., 832 B4 63
San Carlos, Nic., 1,238 E5 62
San Carlos, Phil., 12,000 (74,000▲) o13 35
San Carlos, Ur., 12,400 E2 56
San Carlos, Ven., 607 C4 60
San Carlos, Ven., 11,656 B4 60
San Carlos, Indian res., Ariz D5 80
San Carlos, res., Ariz D5 80
San Carlos [de Bariloche], Arg., 6,562 C2 54
San Carlos [del Zulia], Ven., 14,478 B3 38
San Cataldo, It., 22,544 F4 21
Sancerre, Fr., 2,200 D5 14
Sanchan, China B3 37
Sánchez, Dom. Rep., 4,587 F9 64
Sanchez, San Miguel, N. Mex B5 107
Sánchez Román, Mex., 4,413 C4, m12 63
Sanchiang, China L3 36
Sanchor, India E3 40
San Clemente, Orange, Calif., 8,527 F5 82
San Clemente, Sp., 6,948 C4 20
San Clemente, isl., Calif F4 82
San Cristóbal, Arg., 9,071 A4 54
San Cristóbal, Dom. Rep., 15,525 F8 64
San Cristobal, Taos, N. Mex., 10 A4 107
San Cristóbal, Pan F7 62
San Cristóbal, Ven., 96,102 B3 60
San Cristobal (San Cristoval), isl., Sol. Is G9 7
San Cristóbal (Chatham), isl., Ec g6 58
Sancti-Spíritus, Cuba, 37,741 E4 64
Sanctuary, Sask., Can., 84 G1 70
Sand, creek, Colo A7 83
Sand, creek, Ind F6 91
Sand, creek, Kans A5 93
Sand, creek, Wyo B7 125
Sand, isl., Haw g10 88
Sand, isl., Midway Is 52
Sand, isl., Wis B3 124
Sand, islet, Midway Is 52
Sand, lake, Ont., Can D4 71
Sand, mtn., Ala A2 78
Sand, mtn., Colo A3 83
Sand, riv., Alta., Can B5 69
Sand, riv., Minn D6 99
Sanda, Jap., 16,300 o14 37
Sanda, isl., Scot E3 13
Sandakan, Mala., 28,806 D5 35
Sandani, Tan C6 48
San Daniele del Friuli, It., 7,553 C9 18
Sanday, isl., Scot A6 13
Sanday, sound, Scot A6 13
Sandborn, Knox, Ind., 547 G3 91
Sand Brook, Hunterdon, N.J., 150 C3 106
Sandcoulee, Cascade, Mont., 385 C5 102
Sand Creek, McCone, Mont., 15 C11 102
Sand Creek, Grant, Okla., 15 A3 112
Sand Creek, Dunn, Wis., 150 C2 124
Sand Cut, Palm Beach, Fla., 200 F6 86
Sande, Nor., 477 p28 25
Sandefjord, Nor., 7,000 p28 25
Sandefjord, bay, Ant C20 5
Sanders, Apache, Ariz., 250 B6 80
Sanders, Benewah, Idaho, 10 B2 89
Sanders, Carroll, Ky., 203 B5 94
Sanders, Treasure, Mont., 35 D9 102
Sanders, co., Mont., 6,880 C1 102
Sanderson, Terrell, Tex., 2,189 D1 118
Sandersville, Washington, Ga., 5,425 D4 87
Sandersville, Jones, Miss., 657 D4 100
Sandfly, lake, Sask., Can B2 70
Sandfontein, S.W. Afr B2 49
Sand Fork (Layopolis), Gilmer, W. Va., 237 C4 123
Sandhammaren, cape, Swe C8 24
Sandhill, Rankin, Miss., 150 C4 100
Sand Hill, riv., Newf., Can B3 75
Sand Hill, riv., Minn C2 99
Sandhills, Plymouth, Mass., 800 D4 97
Sandia, Peru, 1,482 E4 58
Sandia, Jim Wells, Tex., 200 F3 118
Sandia, peak, N. Mex G5 107
Sandia Park, Bernalillo, N. Mex., 150 G6 107
San Diego, San Diego, Calif., 573,224 (★1,065,000) E2, F5 82
San Diego, Duval and Jim Wells, Tex., 4,351 F3 118
San Diego, co., Calif., 1,033,011 F5 82
San Diego, cape, Arg h12 54
San Diego, riv., Calif E2 82
Sandikli, Tur., 9,400 C8 23
Sandila, India, 18,407 C7 36, D8 40
Sandilands, Man., Can., 133 E3 71
San Dimas, Los Angeles, Calif., 7,200 *F3 82
San Dimas, Mex., 190 C3 63
Sand Lake, Kent, Mich., 394 E5 98
Sandlake, Tillamook, Oreg B3 113
Sandlick, creek, W. Va D6 123
Sandnes, Nor., 4,000 H1 25
Sandoa, Con. L C3 48
Sandomierz, Pol., 8,357 C6 26
Sandon, B.C., Can., 150 E9 68
Sandoná, Col., 4,767 C2 60
San Donà di Piave, It., 11,100 B4 21
Sandoval, Marion, Ill., 1,356 E4 90
Sandoval, Sandoval, N. Mex., 600 B3, G5 107
Sandoval, co., N. Mex., 14,201 B2 107
Sandoway, Bur., 5,172 E9 39
Sandown, Eng., 14,257 D6 12
Sandown, Rockingham, N.H., 40 (366▲) E4 105
Sand Pass, Washoe, Nev., 10 C2 104
Sand Point, Alsk., 254 D7 79
Sand Point, Ont., Can., 86 B8 72
Sandpoint, Bonner, Idaho, 4,355 A2 89
Sands, San Bernardino, Calif., 5 E6 82
Sands, Marquette, Mich B3 98
Sands, key, Fla F3 86
Sands, pt., N.Y F2 84
Sandspit, B.C., Can., 466 C2 68
Sands Point, Nassau, N.Y., 2,161 F2 84
Sand Springs, Garfield, Mont., 10 C9 102
Sand Springs, Tulsa, Okla., 7,754 A5 112
Sandston, Henrico, Va., 4,500 C7 121
Sandstone, Austl., 101 E2 50
Sandstone, Pine, Minn., 1,552 D6 99
Sandstone, Summers, W. Va., 300 D4, D7 123
Sandusky, Sanilac, Mich., 2,066 E8 98
Sandusky, Cattaraugus, N.Y., 230 C2 108
Sandusky, Erie, Ohio, 31,989 A5 111
Sandusky, co., Ohio, 56,486 A4 111
Sandusky, bay, Ohio A5 111
Sandusky, riv., Ohio A4 111
Sandusky South, Erie, Ohio, 4,724 *A5 111
Sandvig, Den., 608 (part of Allinge-Sandvig) C8 24
Sandviken, Swe., 22,000 G7 25
Sandwich, DeKalb, Ill., 3,842 B5 90
Sandwich, Barnstable, Mass., 1,099 C7 97
Sandwich, Carroll, N.H., 50 (620▲) C4 105
Sandwich, bay, Newf., Can B3 75
Sandwich, range, N.H C3 105
Sandwith, Sask., Can D1 70
Sandy, Clearfield, Pa., 2,070 *D4 114
Sandy, Salt Lake, Utah, 3,322 C4 119
Sandy, brook, Conn A4 84
Sandy, cape, Austl B9 51
Sandy, creek, Ohio B6 111
Sandy, creek, W. Va C3 123
Sandy, creek, Wyo C3 125
Sandy, des., Austl C5 50
Sandy, hook, N.J C5 106
Sandy, isl., S.C D9 115
Sandy, lake, Newf., Can D3 75
Sandy, lake, Ont., Can D8 72
Sandy, lake, Minn D5 99
Sandy, neck, Mass C7 97
Sandy, pt., R.I E10 84
Sandy, pond, Mass C2 97
Sandy, ridge Va B3 121
Sandy, riv., Maine D2 96
Sandy Bay, mtn., Maine C2 96
Sandy Beach, Erie, N.Y., 1,000 *B2 108
Sandy Beach, lake, Ont., Can E5 71
Sandy Creek, Oswego, N.Y., 697 B4 108
Sandy Hook, Fairfield, Conn., 950 D3 84
Sandy Hook, Elliott, Ky., 195 B6 94
Sandy Hook, Washington, Md., 300 B2 85
Sandy Hook, Marion, Miss., 85 D4 100
Sandy Lake, Man., Can., 383 D1 71
Sandy Lake, Mercer, Pa., 838 D1 114
Sandypoint, Waldo, Maine, 200 D4 96
Sandy Ridge, Stokes, N.C., 250 A3 109
Sandy Springs, Fulton, Ga., 5,000 A5 87
Sandy Springs, Anderson, S.C., 174 B2 115
Sandyville, Warren, Iowa, 115 B7 92
Sandyville, Jackson, W. Va., 300 C3 123
San Elizario, El Paso, Tex., 1,064 F1 118
San Enrique, Arg B4 54
San Estanislao, Par., 10,948 D4 55
San Esteban, Hond., 359 C5 62
San Fabian, Phil., 2,337 n13 35
San Felice sul Panaro, It., 2,552 E7 18
San Felipe, Chile, 15,476 A2 54
San Felipe, Sandoval, N. Mex., 1,034 B3, G5 107
San Felipe, Ven., 27,774 A4 60
San Felipe, Indian res., N. Mex G5 107
San Felipe, pt., Mex A2 63
San Feliu de Guixols, Sp., 10,307 B7 20
San Felix, isl., Pac. O H17 7
San Fernando, Arg., 84,000 A5, g7 54
San Fernando, Los Angeles, Calif., 16,093 F2 82
San Fernando, Chile, 22,500 A2 54
San Fernando, Mex., 1,866 C5 63
San Fernando, Phil., 10,000 (56,700▲) o13 35
San Fernando, Phil., 4,636 B6, n13 35
San Fernando, Phil., 1,427 *C6 35
San Fernando, Sp., 52,389 D2 20
San Fernando, Trin., 41,800 A5 60
San Fernando de Apure, Ven., 21,544 B4 60
San Fernando de Atabapo, Ven., 876 C4 60
San Fernando de Henares, Sp., 2,209 p17 20
San Fidel, Valencia, N. Mex., 80 B2 107
Sanfjället, mtn., Swe F5 25
Sanford, Covington, Ala., 247 D3 78
Sanford, Man., Can., 103 E3 71
Sanford, Conejos, Colo., 679 D5 83
Sanford, Seminole, Fla., 19,175 D5 86
Sanford, Vigo, Ind., 350 E2 91
Sanford, York, Maine, 10,936 E2 96
Sanford, Midland, Mich., 450 E6 98
Sanford, Covington, Miss., 100 D4 100
Sanford, Lee, N.C., 12,253 B4 109
Sanford, Hutchinson, Tex., 400 B2 118
Sanford, mtn., Alsk f19 79
Sanford, mtn., Conn D5 84
San Francisco, Arg., 24,354 A4 54
San Francisco, San Francisco, Calif., 740,316 (★3,275,000) B5, D2 82
San Francisco, Col C2 60
San Francisco, Sal., 2,883 D3 62
San Francisco, co., Calif., 740,316 D2 82
San Francisco, bay, Calif B5 82
San Francisco, cape, Ec A1 58
San Francisco, mts., Ariz B4 80
San Francisco, pass, Arg E2 55
San Francisco, riv., Arg D2 55
San Francisco, riv., N. Mex D1 107
San Francisco Central, P.R D3 65
San Francisco de Borja, Mex., 1,015 B3 63
San Francisco del Oro, Mex., 11,317 B3 63
San Francisco del Rincón, Mex., 20,864 m13 63
San Francisco de Macorís, Dom. Rep., 26,000 F8 64
Sangabar, Afg., 5,000 D12 41
San Gabriel, Los Angeles, Calif., 22,561 F2 82
San Gabriel, Ec., 6,382 A2 58
San Gabriel Chilac, Mex., 6,131 D5, n15 63
San Gabriel, mts., Calif F2 82
San Gabriel, pt., Mex B2 63
Sangallan, isl., Peru D2 58
Sangamner, India, 21,729 H5 40
Sangamon, co., Ill., 146,539 D4 90
Sangamon, riv., Ill C3 90
Sangatte, Fr., 573 D9 12
Sangélima, Cam E2 46
Sanger, Fresno, Calif., 8,533 D4 82
Sanger, Oliver, N. Dak., 35 B4 110
Sanger, Denton, Tex., 1,190 C4 118
Sangerhausen, Ger., 23,800 B6 17
San Germán, P.R., 7,790 C2 65
San Germán, mun., P.R., 27,667 C2 65
Sangerville, Piscataquis, Maine, 600 (1,157▲) C3 96
Sanggau, Indon E4 35
Sangha, riv., Con. B A2 48
San Giacomo, pass, Switz D5 19
San Gil, Col., 10,149 B3 60
Sang-i-Masha, Afg., 10,000 E13 41
San Gimignano, It., 11,135 C3 21
San Giovanni in Fiore, It., 17,400 E6 21
San Giovanni in Persiceto, It., 7,000 E7 18
Sangju, Kor., 21,600 H4 37
Sangonera, riv., Sp D5 20
San Gorgonio, mtn., Calif E5 82
Sangre de Cristo, range, Colo D5 83
Sangre de Cristo, range, N. Mex B4 107
San Gregorio, San Mateo, Calif., 50 C5 82
San Gregorio Atlapulco, Mex., 5,555 h9 63
Sangro, riv., It D5 21
Sangrur, India, 28,344 B5 40
Sangudo, Alta, Can., 325 C3 69
Sangue, riv., Braz E3 59
Sangüesa, Sp., 4,323 A5 20
Sanguinara, riv., It h8 21
Sangwin, Lib E3 45
Sanhedrin, mtn., Calif C2 82
Sanibel, Lee, Fla., 250 F4 86
Sanibel, isl., Fla F4 86
San Ignacio, Bol., 1,757 B2 55
San Ignacio, Bol., 1,819 C3 55
San Ignacio, Mex., 898 B2 63
San Ignacio, Guadalupe, N. Mex C5 107
San Ignacio, Par., 10,766 E4 55
Sanilac, co., Mich., 32,314 E8 98
San Ildefonso, Santa Fe, N. Mex., 500 B3, F6 107
San Ildefonso, Sp., 3,245 B3 20
San Ildefonso, cape, Phil n14 35
San Ildefonso, pen., Phil B6, n14 35
San Isabel, Custer, Colo., 20 D5 83
San Isidro, Arg., 196,188 g7 54
San Isidro, Phil., 11,855 *C6 35
Sanitz, Ger., 2,184 D6 24
San Jacinto, Riverside, Calif., 2,553 F4, F5 82
San Jacinto, Elko, Nev., 15 B7 104
San Jacinto, Phil., 3,877 *C6 35
San Jacinto, co., Tex., 6,153 D5 118
San Jacinto, riv., Tex F5 118
San Jaime, Arg A5 54
San Javier, Arg., 2,961 A5 54
San Javier, Bol., 564 C3 55
San Javier, Chile, 7,006 B2 54
San Javier, riv., Arg A4 54, E4 55
San Jerónimo [Aculco Lídice], Mex., 3,009 h9 63
San Jerónimo, mts., Col B2 60
Sanjo, Jap., 48,800 H9 37
San Joao das Lampas, Port f9 20
San Joaquín, Par., 4,200 D4 55
San Joaquin, co., Calif., 249,989 C3 82
San Joaquin, riv., Calif D3 82
San Joaquin, val., Calif D3 82
San Jon, Quay, N. Mex., 411 B6 107
San Jose, Graham, Ariz., 100 E6 80
San Jose, Bol., 1,933 C3 55
San Jose, Br. Hond., 365 B3 62
San Jose, Santa Clara, Calif., 204,196 C6, D3 82
San José, C.R., 115,700 (★260,000) F5 62
San Jose, Duval, Fla., 8,000 *B5 86
San José, Guat., 2,822 D2 62
San Jose, Mason and Logan, Ill., 1,093 C4 90
San Jose, San Miguel, N. Mex., 200 B4 107
San Jose, Phil., 10,544 o13 35
San Jose, Phil., 8,183 o13 35
San Jose, Phil., 2,259 C6 35
San Jose, Phil., 5,159 *C6 35
San José, Ur., 21,200 E1 56
San José, dept., Ur., 67,200 *E1 56
San José, prov., C.R., 430,300 *F5 62
San José, isl., Mex B2 63
San Jose, riv., B.C., Can D7 68
San José, riv., Ur g8 54
San José Boquerón, Arg E3 55
San José de Feliciano, Arg., 7,643 A5 54
San José del Cabo, Mex., 1,838 C3 63
San José del Guaviare, Col C3 60
San Jose [del Monte], Phil., 1,071 o13 35
San José de los Molinos, Peru, 1,221 D2 58
San Josef Bay, B.C., Can., 25 D3 68
San Juan, Arg., 106,746 A3 54
San Juan, Las Animas, Colo., 100 D6 83
San Juan, Dom. Rep., 20,449 F8 64
San Juan, Grant, N. Mex., 95 E2 107
San Juan, Pan., 1 k11 62
San Juan, Phil., 2,121 p13 35
San Juan, Phil., 1,745 n13 35
San Juan, Phil., 2,870 *D7 35
San Juan, P.R., 432,377 (★588,805) B6 65
San Juan, Hidalgo, Tex., 4,371 F3 118
San Juan, Ven., 1,121 A4 60
San Juan, co., Colo., 849 D3 83
San Juan, co., N. Mex., 53,306 A1 107
San Juan, co., Utah, 9,040 F5 119
San Juan, co., Wash., 2,872 A3 122
San Juan, mun., P.R., 451,658 B6 65
San Juan, prov., Arg., 352,461 A3 54
San Juan, isl., Wash A2 122
San Juan, mts., Colo D3 83
San Juan, passage, P.R B8 65
San Juan, riv., Arg A3 54
San Juan, riv., B.C., Can B8 68
San Juan, riv., Col D3 83
San Juan, riv., Nic E6 62
San Juan, riv., U.S C5 76
San Juan Bautista, San Benito, Calif., 1,046 C6 82
San Juan Bautista, Par., 9,054 E4 55
San Juan Capistrano, Orange, Calif., 1,120 F5 82
San Juan de Aragón, Mex., 3,098 h9 63
San Juan de Colón, Ven., 9,210 B3 60
San Juan del Norte, Nic., 307 E6 62
San Juan de los Lagos, Mex., 22,713 m12 63
San Juan de los Morros, Ven., 25,821 B4 60
San Juan del Río, Mex., 11,179 m13 63
San Juan del Sur, Nic., 1,019 E5 62
San Juan Ixtayoapan, Mex., 2,595 h10 63
San Juan Nepomuceno, Col., 5,832 B2 63
San Juan y Martínez, Cuba, 4,142 D2 64
San Julián, Arg., 3,050 D3 54
San Justo, Arg., 6,571 A4 54
San Justo, Arg., 88,853 *A5 54
Sankarani, riv., Guinea, Mali D3 45
Sankt Anton [am Arlberg], Aus., 1,741 B6 18
Sankt Blasien, Ger., 3,200 B4 18
Sankt Gallen, Switz., 76,279 (★101,000) B7 19
Sankt Gallen, canton, Switz., 327,600 B7 19
Sankt Georgen [im Attergau], Aus., 2,659 B9 18
Sankt Goarshausen, Ger., 2,000 C2 17
Sankt Johann [in Tirol], Aus., 4,713 B8 18
S[ankt] Margrethen, Switz., 4,286 B8 19
Sankt Michaelisdonn, Ger., 3,100 E3 24
Sankt Moritz, Switz., 3,751 C8 19
Sankt Pölten, Aus., 40,112 D7 16
Sankt Veit [an der Glan], Aus., 10,950 E7 16
Sankt Wendel, Ger., 10,600 D2 17
Sankuru, riv., Con. L B3 48
San Leandro, Alameda, Calif., 65,962 B5 82
San Lorenzo, Arg., 11,109 A4 54
San Lorenzo, Alameda, Calif., 23,773 *B5 82
San Lorenzo, Ec A2 58
San Lorenzo, Hond., 2,723 D4 62
San Lorenzo, Grant, N. Mex., 200 E2 107
San Lorenzo, P.R., 5,551 C7 65
San Lorenzo, Ven., 500 B3 60
San Lorenzo, mun., P.R., 27,950 C7 65
San Lorenzo del Escorial, Sp., 7,965 B3, o16 20
San Lorenzo Tezonco, Mex., 3,208 h9 63
Sanlúcar, Sp., 34,000 D2 20
San Lucas, Bol., 925 D2 55
San Lucas, Mex., 548 C3 63
San Luis, Arg., 25,147 A3 54
San Luis, Pima, Ariz., 15 E3 80
San Luis, Yuma, Ariz., 50 E1 80
San Luis, Costilla, Colo., 800 D5 83
San Luis, Cuba h12 64
San Luis, Cuba, 11,110 E6 64
San Luis, Guat., 562 B3 62
San Luis, Mex., 28,336 A2 63
San Luis, Sandoval, N. Mex B2 107
San Luis, prov., Arg., 174,251 A3 54
San Luis, creek, Colo C5 83
San Luis, pass, Tex G5 118
San Luis, peak, Colo D4 83
San Luis, pt., Calif E3 82
San Luis, riv., Mex h9 63
San Luis, val., Colo D4 83
San Luis de la Paz, Mex., 8,268 m13 63
San Luis Jilotepeque, Guat., 4,208 C3 62
San Luis Obispo, San Luis Obispo, Calif., 20,437 E3 82
San Luis Obispo, co., Calif., 81,044 E3 82
San Luis Potosí, Mex., 159,640 C4, k13 63
San Lusi Potosí, state, Mex., 1,054,206 C4, k13 63
San Manuel, Pinal, Ariz., 4,524 E5 80
San Marcial, Socorro, N. Mex., 25 D3 107
San Marco [in Lamis], It., 19,014 D5 21
San Marcos, Col., 3,966 B2 60
San Marcos, Guat., 4,694 C2 62
San Marcos, Hays, Tex., 12,713 A4, E5 118
San Marcos, riv., Tex A4 118
San Marcos de Colón, Hond., 2,289 D4 62
San Marino, Los Angeles, Calif., 13,658 F2 82
San Marino, San Marino, 2,410 C4 21
San Marino, country, Eur., 12,950 C4 21
San Martín, Arg., 8,748 A3 54
San Martín (General San Martín), Arg., 279,213 g7 54
San Martin, Santa Clara, Calif., 1,162 C6 82
San Martín, Col., 3,094 C3 60
San Martín, dept., Peru, 120,913 C2 58
San Martín de la Vega, Sp., 2,676 p17 20
San Martín de Los Andes, Arg., 2,366 C2 54
San Martino dei Calvi, It., 2,354 D5 18
San Martino di Castrozza, It., 268 C7 18
San Mateo, San Mateo, Calif., 69,870 B5 82
San Mateo, Putnam, Fla., 850 C5 86
San Mateo, Valencia, N. Mex., 230 B2 107
San Mateo, Sp., 2,452 B6 20
San Mateo, Ven., 1,829 B5 60
San Mateo, co., Calif., 444,387 D2 82
San Mateo, cape, Ec B1 58
San Mateo, mts., N. Mex D2 107
San Mateo Xalpa, Mex., 1,253 h9 63
San Matias, Bol., 887 C4 55
San Matias, gulf, Arg C4 54
San Miguel, Pima, Ariz., 256 F4 80
San Miguel, San Luis Obispo, Calif., 500 E3 82
San Miguel, Dona Ana, N. Mex., 300 E3 107
San Miguel, Pan., 1,071 F8 62
San Miguel, Sal., 38,330 D3 62
San Miguel, co., Colo., 2,944 D2 83
San Miguel, co., N. Mex., 23,468 B4 107
San Miguel, isl., Calif E3 82
San Miguel, mts., Colo D2 83
San Miguel, riv., Colo C2 83
San Miguel de Allende, Mex., 14,853 m13 63
San Narciso, Phil., 1,341 p14 35
San Narciso, Phil., 9,908 o13 35
San Nicholas, Phil., 1,816 o13 35
San Nicolás, Arg., 25,926 A4 54
San Nicolas, Phil., 3,938 n13 35
San Nicolas, isl., Calif F4 82
San Nicolás [Totolapan], Mex., 2,971 h9 63
Sannicolaul-Mare, Rom., 9,956 B5 22
Sannois, Fr., 16,490 g10 14
Sanok, Pol., 13,800 D7 26
San Onofre, Col., 4,668 B2 60
San Pablo, Contra Costa, Calif., 19,687 B5 82
San Pablo, Costilla, Colo., 75 D5 83
San Pablo, Phil., 30,000 (71,800▲) o13 35
San Pablo, bay, Calif B5 82
San Patricio, Lincoln, N. Mex., 200 D4 107
San Patricio, co., Tex., 45,021 F4 118
San Patricio, bayou, La C2 95
San Pedro, Arg., 12,798 A5, f7 54
San Pedro, Arg., 6,105 D3 55
San Pedro, Bol., 262 B3 55
San Pedro, Bol., 1,094 C2 55
San Pedro, Los Angeles, Calif., 55,300 F2 82
San Pedro, Chile D2 55
San Pedro, Cuba k11 64
San Pedro, Par., 16,691 D4 55
San Pedro, Nueces, Tex., 7,634 *E4 118
San Pedro, dept., Par., 64,534 D4 55
San Pedro, mtn., N. Mex A3 107
San Pedro, riv., Ariz E5 80
San Pedro, riv., Mex C4 63
San Pedro, riv., Mex G1 107
San Pedro, Mex., 324 G2 107
San Pedro de Atacama, Chile D2 55
San Pedro de las Colonias, Mex., 25,183 B4 63
San Pedro de Lloc, Peru, 5,286 C2 58
San Pedro de Macorís, Dom. Rep., 22,935 F9 64
San Pedro Martir, mts., Mex A1 63
San Pedro Sula, Hond., 58,126 C3 62
Sanpete, co., Utah., 11,053 D4 119
San Pierre, Starke, Ind., 300 B4 91
San Pietro, isl., It E2 21
Sanpoil, riv., Wash A7 122
Sanquhar, Scot., 2,182 E5 13
San Quintin, Phil., 2,633 o13 35
San Rafael, Arg., 28,847 A3 54
San Rafael, Marin, Calif., 20,460 B5, D2 82
San Rafael, Valencia, N. Mex., 600 B2 107
San Rafael, Ven., 7,110 A3 61
San Rafael, mts., Calif E4 82
San Rafael, riv., Utah D5 137
San Rafael, swell, Utah E5 119
San Rafael, val., Utah E5 119
San Rafael del Norte, Nic., 810 D4 62
San Rafael Knob, mtn., Utah E5 119
San Ramón, Peru, 1,275 D2 58
San Ramon, Ur., 3,982 G9 54
San Remo, It., 43,500 C1 21
San Roque, Sp., 13,676 D3 20
San Rosendo, Chile, 3,315 B2 54
San Saba, San Saba, Tex., 2,728 D3 118
San Saba, co., Tex., 6,381 D3 118
San Salvador, Arg., 3,532 A5 54
San Salvador, Sal., 248,100 (★350,000) D3 62
San Salvador (Watling), isl., Ba. Is C6 64
San Salvador (Santiago), isl., Ec g5 58
Sansanné-Mango, Togo D5 45
San Sebastián, P.R., 4,019 B3 65
San Sebastián, Sp., 135,149 (★185,000) A4 20
San Sebastián, mun., P.R., 33,451 B3 65
San Sebastián, cape, Arg h12 54
San Sebastián de los Reyes, Sp., 3,350 o17 20
San Sepolcro, It., 7,242 C4 21
San Severo, It., 48,443 D5 21
Sanshui, China, 25,000 G7 34
San Simeon, San Luis Obispo, Calif., 35 E3 82
San Simon, Cochise, Ariz., 200 E6 80
San Simon, creek, Ariz E6 80

Sansom Park Village, Tarrant, Tex., 4,175....*B5 118
Sans Souci, Greenville, S.C., 7,000....*B3 115
Santa, Benewah, Idaho, 100..B2 89
Santa, Peru, 1,089....C2 58
Santa, riv., Peru....C2 58
Santa Amaro, isl., Braz....m8 58
Santa Ana, Bol., 2,225....B2 55
Santa Ana, Bol., 171....C2 55
Santa Ana, Orange, Calif., 100,350....F5, F3 82
Santa Ana, Ec., 3,976....B1 58
Santa Ana, Mex., 3,976....A2 63
Santa Ana, Sandoval, N. Mex., 300....G5 107
Santa Ana, Peru, 201....D3 58
Santa Ana, Sal., 73,864....D3 62
Santa Ana, Ven., 3,584....B3 60
Santa Ana, mts., Calif....F5 82
Santa Ana, riv., Calif....F3 82
Santa Ana Pueblo, Indian res., N. Mex....G5 107
Santa Anita, Mex., 4,441....h9 63
Santa Anna, Coleman, Tex., 1,320....D3 118
Santa Barbara, Braz., 13,571....m8 56
Santa Bárbara, Braz., 4,200..E2 57
Santa Barbara, Santa Barbara, Calif., 58,768....E4 82
Santa Bárbara, Chile, 2,292..B2 54
Santa Bárbara, Hond., 2,684.C3 62
Santa Bárbara, Mex., 15,892.B3 63
Santa Barbara, co., Calif., 168,962....E3 82
Santa Barbara, chan., Calif....E4 82
Santa Barbara, is., Calif....F4 82
Santa Bárbara Central, P.R...C4 65
Santa Catalina, Arg....D2 55
Santa Catalina, Chile....E2 55
Santa Catalina, gulf, Calif....F5 82
Santa Catalina, isl., Calif....F4 82
Santa Catalina, mts., Ariz....E5 80
Santa Catarina, state, Braz., 2,146,909....D2 56
Santa Catarina, isl., Braz....D3 56
Santa Clara, Santa Clara, Calif., 58,880....D3, C6 82
Santa Clara, Cuba, 77,398....D4 64
Santa Clara, Franklin, N.Y., 110....A2 108
Santa Clara, Ur., 2,499....E2 56
Santa Clara, Washington, Utah, 291....F2 119
Santa Clara, co., Calif., 642,315....D3 82
Santa Clara, riv., Calif....E4, F2 82
Santa Clara, val., Calif....C6 82
Santa Claus, Spencer, Ind., 50....H4 91
Santa Coloma de Farnés, Sp., 4,583....B7 20
Santa Croce, cape, It....F5 21
Santa Cruz, Arg....E3 54
Santa Cruz, Pinal, Ariz. D3, G2 80
Santa Cruz., Bol., 34,837....C3 55
Santa Cruz, Braz., 5,286.C3, h6 57
Santa Cruz, Santa Cruz, Calif., 25,596....D2, C5 82
Santa Cruz, Chile, 2,132....A2 54
Santa Cruz, C.R., 7,430....E5 62
Santa Cruz, Santa Fe, N. Mex., 600....B3 107
Santa Cruz, Phil., 3,851....p13 35
Santa Cruz, Phil., 2,189....o12 35
Santa Cruz, Phil., 1,093....o13 35
Santa Cruz, co., Ariz., 10,808....F5 80
Santa Cruz, co., Calif., 84,219....D2 82
Santa Cruz, dept., Bol., 286,145....C3 55
Santa Cruz, prov., Arg., 52,853....D3 54
Santa Cruz, isl., Calif....F4 82
Santa Cruz, is., Pac. O....G10 7
Santa Cruz, mts., Calif....C5 82
Santa Cruz, riv., Ariz....E4 80
Santa Cruz, Barillas, Guat., 1,296....C2 62
Santa Cruz de la Palma, Can. Is., 8,835 (11,609▲). D1 44, m13 20
Santa Cruz de la Zarza, Sp., 5,588....C4 20
Santa Cruz del Quiché, Guat., 4,211....C2 62
Santa Cruz del Sur, Cuba, 2,571....E5 64
Santa Cruz de Tenerife, Can. Is., 133,100....D1 44, m13 20
Santa Cruz de Tenerife, prov., Sp., 490,655....*m13 20
Santa Cruz do Rio Pardo, Braz., 13,789....C3 56
Santa Cruz do Sul, Braz., 18,898....D2 56
Santa Cruz (Indefatigable), isl., Ec....g5 58
Santa Elena, Ec., 2,764....B1 58
Santa Elena, Starr, Tex., 250....F3 118
Santa Elena, Ven., 620....C5 60
Santa Eugenia [de Ribeira], Sp., 4,543....A1 20
Santa Eulalia del Río, Sp., 7,564....C6 20
Santa Fe, Arg., 208,000....A4 54
Santa Fé, Cuba, 1,098....E2 64
Santa Fe, Mex., 3,706....h9 63
Santa Fe, Santa Fe, N. Mex., 33,394....B4, F6 107
Santa Fe, Auglaize and Logan, Ohio, 170....B3 111
Santa Fe, Phil., 4,061....*C6 35
Santafé, Sp., 8,387....D4 20
Santa Fe, Maury, Tenn., 125....B4 117
Santa Fe, co., N. Mex., 44,970....B3 107
Santa Fe, prov., Arg., 1,865,537....A4 54
Santa Fé, isl., Ec....g5 58
Santa Fe, lake, Fla....C4 86
Santa Fe, riv., N. Mex....F6 107
Santa Fe Baldy, mtn., N. Mex..B4 107
Santa Fe Springs, Los Angeles, Calif., 16,342....*F2 82
Santa Filomena, Braz., 652..C1 57
Santai, China, 24,000....E6 34
Santa Inés, isl., Chile....h11 54
Santa Inés Ahuatempan, Mex., 2,465....n14 63
Santa Isabel, Arg....B3 54
Santa Isabel, Chile....D2 55
Santa Isabel, Fernado Póo, 11,098....E1 46
Santa Isabel, P.R., 4,712....D5 65
Santa Isabel, see Paso de los Toros, Ur.
Santa Isabel, mun., P.R., 14,542....D5 65
Santa Isabel de Siguas, Peru, 80....E3 58
Santa Juana Central, P.R...C6 65
Santa Lucía, Cuba, 1,969....E6 64
Santa Lucía, Ur., 8,258....E1 56
Santa Lucia, range, Calif....E3 82
Santa Margarita, San Luis Obispo, Calif., 600....E3 82
Santa Margarita, isl., Mex....C2 63
Santa María, Arg., 2,052....E2 55
Santa María, Braz., 78,682..D2 57
Santa Maria, Santa Barbara, Calif., 20,027....E3 82
Santa Maria, Phil., 2,510....o13 35
Santa María, P.R....g13 65
Santa María, isl., Az. Is....h9 44
Santa María, isl., Ec....g5 58
Santa Maria, mts., Ariz....C3 80
Santa Maria, riv., Ariz....C2 80
Santa María, riv., Mex....m14 63
Santa María, riv., Mex....G2 107
Santa María, riv., Pan....F7 62
Santa Maria [Capua Vetere], It., 30,024....D5 21
Santa María del Oro, Mex., 3,246....B3 63
Santa Maria di Leuca, cape, It..E7 21
Santa Maria Madalena, Braz., 1,530....C4 56
Santa Marta, Col., 42,700....A3 60
Santa Marta, Sp., 5,142....C2 20
Santa Marta, mts., Col....A3 60
Santa Monica, Los Angeles, Calif., 83,249....F2 82
Santan, Pinal, Ariz....D4, G2 80
Santan, mtn., Ariz....G3 80
Santana, Braz., 4,357....D2 57
Santana, Port., 4,953....h12 20
Santana do Ipanema, Braz., 8,139....C3 57
Santander, Col., 5,669....C2 60
Santander, Sp., 118,435....A4 20
Santander, prov., Sp., 432,132....*A4 20
Santander, dept., Col., 804,490....B3 60
Santander, mts., Col....B3 60
Santander Jiménez, Mex., 1,358....C5 63
Sant'Angelo Romano, It., 1,878....g9 21
Santanoni, peak, N.Y....A6 108
Santanópole, Braz., 2,218....C3 57
Santañy, Sp., 6,295....C7 20
Santa Paula, Ventura, Calif., 13,279....E4 82
Santaquin, Utah, Utah, 1,183....C4 119
Santa Quitéria, Braz., 2,351.B2 57
Santarém, Braz., 24,924....C4 59
Santarém, Port., 13,114....C1 20
Santaren, chan., W.I....C4 64
Santa Rita, Braz., 4,427....k8 56
Santa Rita, Braz., 20,623.C4, h6 57
Santa Rita, Guam, 1,630.... 52
Santa Rita, Glacier, Mont., 110....B4 102
Santa Rita, Grant, N. Mex., 1,772....E1 107
Santa Rita, Ven., 11,623....A3 60
Santa Rita, mts., Pan....k11 62
Santa Rita Park, Merced, Calif., 100....D3 82
Santa Rosa, Arg., 3,564....A3 54
Santa Rosa, Arg., 2,999....A4 54
Santa Rosa, Arg., 14,623....B4 54
Santa Rosa, Bol....B2 55
Santa Rosa, Braz., 2,761....k8 56
Santa Rosa, Braz., 12,283....D2 56
Santa Rosa, Sonoma, Calif., 31,027....A5, C2 82
Santa Rosa, Col., 4,668....B2 60
Santa Rosa, Ec., 4,672....B2 58
Santa Rosa, Walton, Fla., 300....G3 86
Santa Rosa, DeKalb, Mo., 75....B3 101
Santa Rosa, Guadalupe, N. Mex., 2,220....C5 107
Santa Rosa, Phil., 3,889....o13 35
Santa Rosa, Cameron, Tex., 1,572....*F4 118
Santa Rosa, co., Fla., 29,547.G2 86
Santa Rosa, isl., Calif....F3 82
Santa Rosa, mtn., Guam.... 52
Santa Rosa, range, Nev....B4 104
Santa Rosa Beach, Wlaton, Fla., 300....G2 86
Santa Rosa de Aguán, Hond., 1,257....C5 62
Santa Rosa de Copán, Hond., 7,972....C3 62
Santa Rosa Island. nat. mon., Fla....G2 86
Santa Rosalía, Mex., 5,361....B2 63
Santa Rosalia, pt., Mex....B2 63
Santa Susana, Ventura, Calif., 2,310....*E4 82
Santa Teresa, Gallura, It., 2,570....D2 21
Santa Ursula, Mex., 3,570....h9 63
Santa Venetia, Marin, Calif., 3,000....*D2 82
Santa Vitória do Palmar, Braz., 8,224....E2 56
Santa Ynez., Santa Barbara, Calif., 400....E3 82
Santee, San Diego, Calif., 2,000....*E2 82
Santee, Knox, Nebr., 65....B8 103
Santee, dam, S.C....E7 115
Santee, Indian res., Nebr....B8 103
Santee, riv., S.C....E8 115
Sant'Eufemia, gulf, It....E6 21
Santiago, Bol., 218....C2 55
Santiago, Braz., 15,140....D2 56
Santiago, Chile, 646,000 (*2,125,000)....A2 54
Santiago, Dom. Rep., 83,523....F8 64
Santiago, Mex., 635....C3 63
Santiago, Pan., 8,746....F7 62
Santiago, Par., 7,834....E4 55
Santiago, Phil., 5,807....n13 35
Santiago, Sp., 34,000....A1 20
Santiago, prov., Chile, 2,429,539....A2 54
Santiago, cape, Phil....p13 35
Santiago, isl., Phil....n12 35
Santiago, isl., P.R....C8 65
Santiago, mts., Tex....E1 118
Santiago, peak, Calif....F3 82
Santiago, riv., Peru....B2 58
Santiago de Cao, Peru, 957..C2 58
Santiago de Cuba, Cuba, 163,237....E6 64
Santiago del Estero, Arg., 80,000 (*100,000)....E3 55
Santiago del Estero, prov., Arg., 477,156....E3 55
Santiago Ixcuintla, Mex., 10,985....C3, m11 63
Santiago Morona, prov., Ec., 22,329....B2 58
Santiago Papasquiaro, Mex., 5,317....B3 63
Santiago Tepalcatlálpan, Mex., 2,766....h9 63
Santian, riv., Oreg....C2 113
Santipur, India, 51,190....F12 40
Santis, mtn., Switz....B7 19
Santisteban del Puerto, Sp., 8,678....C4 20
Santo, Palo Pinto, Tex., 500.C3 118
Santo Amaro, Braz., 17,226.D3 57
Santo André, Braz., 230,196....*C3 56
Santo Ângelo, Braz., 25,415..D2 56
Santo Antão, isl., C.V. Is....*E2 42
Santo Antônio, Braz., 2,978....C3, h6 57
Santo Antônio de Jesus, Braz., 14,902....D3 57
Santo Antônio do Zaire, Ang., 528....C1 48
Santo Domingo, Dom. Rep., 367,053....F9 64
Santo Domingo, Nic., 3,110..D5 62
Santo Domingo de la Calzada, Sp., 5,436....A4 20
Santo Domingo de los Colorados, Ec....B2 58
Santo Domingo Pueblo, Sandoval, N. Mex., 900....B3, F5 107
San Tomas, Santa Clara, Calif., 1,500....*D3 82
Santoña, Sp., 9,082....A4 20
Santo Pueblo, Indian res., N. Mex....F5 107
Santos, Braz., 262,048 (*400,000)....C3, m8 56
Santos Dumont, Braz., 20,414....C4, g6 56
Santo Tomas, Dona Ana, N. Mex....E3 107
Santo Tomás, Peru, 877....D3 58
Santo Tomás, Ven....B5 60
Santo Tomé, Arg., 8,348....E4 55
Santuao, China....K8 36
Santuc, Union, S.C., 40....B4 115
San Valentín, mtn., Chile....D2 54
San Vicente, Col., 1,002....C3 60
San Vicente, Sal., 15,189....D3 62
San Vicente Central, P.R....B5 65
San Vicente de Alcántara, Sp., 8,212....C2 20
San Vicente de la Barquera, Sp., 3,002....A3 20
San Vito al Tagliamento, It., 5,065....B4 21
San Xavier, Indian res., Ariz....E4 80
Sanyati, riv., S. Rh....A4 49
San Ygnacio, Zapata, Tex., 900....F3 118
San Ysidro, San Diego, Calif. (part of San Diego)....E2, F5 82
Sanza Pomba, Ang., 269....C2 48
São Bento do Sul, Braz., 6,470....D3 56
São Bento do Una, Braz., 5,096....k5 57
São Bernardo [do Campo], Braz., 61,645....C3, m8 56
São Borja, Braz., 20,339....D1 56
São Carlos, Braz., 50,010....C3, m8 56
São Cristóvão, Braz., 7,624..D3 57
São Domingos, Braz., 907..D1 57
São Fidélis, Braz., 6,145....C4 56
São Francisco, Braz., 4,074..E2 57
São Francisco, Braz., 869....C5 60
São Francisco, riv., Braz....C3 57
São Francisco do Sul, Braz., 11,593....D3 56
São Gabriel, Braz., 22,967..E2 56
São Gotardo, Braz., 6,227..E1 57
São Jerônimo, Braz., 5,568..D2 56
São Jeronimo, mts., Braz....B2 56
São João, isl., Braz....B2 57
São João da Barra, Braz., 3,441....C4 56
São João da Boa Vista, Braz., 25,226....C3, k8 56
São João das Lampas, Port., 4,637....f9 20
São João del Rei, Braz., 34,654....C4, g5 56
São João do Carirí, Braz., 622....C3, h5 57
São João do Piauí, Braz., 2,688....C2 57
São João Nepomuceno, Braz., 9,436....C4, g6 56
São Joaquim, Braz., 3,811....D3 56
São Jorge, isl., Az. Is....g8 44
São José de Mipibú, Braz., 5,179....h6 57
São José do Rio Pardo, Braz., 14,186....k8 56
São José do Rio Prêto, Braz., 66,476....C3 56
São José dos Campos, Braz., 55,349....C3, m9 56
São José dos Pinhais, Braz., 7,574....D3 56
São Leopoldo, Braz., 41,023..D2 56
São Lourenço, Braz., 14,680..C3 56
São Lourenço, riv., Braz....B1 56
São Lourenço do Sul, Braz., 6,877....E2 56
São Luís, Braz., 124,606....B2 57
São Luís do Quitunde, Braz., 2,618....k6 57
São Luís Gonzaga, Braz., 12,926....D2 56
São Manuel, Braz., 10,009....C3, m7 56
São Mateus, Braz., 6,075....E3 57
São Miguel, isl., Az. Is....h9 44
São Miguel Arcanjo, Braz., 3,633....m8 56
São Miguel dos Campos, Braz., 6,511....k5 57
Saône, riv., Fr....D6 14
Saône-et-Loire, dept., Fr., 535,772....*D6 14
São Nicolau, isl., C.V. Is....*E3 42
São Paulo, Braz., 3,825,351 (*4,650,000)....C3, m8 56
São Paulo, state, Braz., 12,974,699....C3, m8 56
São Paulo de Olivença, Braz., 1,157....D4 60
São Pedro, Braz., 4,474....m8 56
São Pedro do Piauí, Braz., 2,139....C2 57
São Raimundo Nonato, Braz., 3,751....C2 57
São Roque, Braz., 12,409....m8 56
São Roque, cape, Braz....C3 57
São Salvador, Ang., 2,965....C1 48
São Sebastião, cape, Moz....B6 49
São Sebastião, isl., Braz....C3 56
São Sebastião do Paraíso, Braz., 14,451....C3 56
São Simão, Braz., 5,742..C3, k8 56
Saõ Tiago, isl., C.V. Is....*E3 42
São Tomé, São Tomé....E1 46
São Tomé, isl., Sao Tome & Principe....E1 46
São Tomé & Principe, Port. dep., Afr., 60,159....E1 46
Saoura, riv., Alg....D4 44
São Vicente, Braz., 73,578....C3 56
São Vicente, Port., 6,663....h11 20
São Vicente, isl., C.V. Is....*E3 42
Sapai, Grc., 5,698....B5 23
Sapatu, isl., Viet....H8 38
Sapele, Nig., 33,638....E6 45
Sapello, San Miguel, N. Mex., 10....B4 107
Sapelo, isl., Ga....E5 87
Sapelo, sound, Ga....E5 87
Sapinero, Gunnison, Colo., 10....C3 83
Saponac, Penobscot, Maine, 20....C4 96
Saposoa, Peru, 3,243....C2 58
Sapozhok, Sov. Un., 10,000.E13 27
Sappa, creek, Nebr., Kans....C3 93
Sappemeer, Neth., 4,565....A6 15
Sapphire, mts., Mont....D3 102
Sappho, Clallam, Wash., 100....A1 122
Sappington, St. Louis, Mo., 10,000....B8 101
Sappington, Gallatin, Mont., 17....E5 102
Sapporo, Jap., 523,839 (*615,000)....E10 37
Sapri, It., 6,188....D5 21
Sapulpa, Creek, Okla., 14,282....B5 112
Saqqara, Eg., U.A.R., 8,230.E3 32
Saqqez, Iran, 10,479....C3 41
Sāquarema, Braz., 1,467.C4, h6 56
Sara, dune, Libya....E4 43
Sara, riv., Chad, Cen. Afr. Rep.D3 44
Sara Buri, Thai., 20,245....E4 38
Sarafand, Isr., 153....h10 32
Saragosa, Reeves, Tex., 150....D1, F2 118
Saragossa, Walker, Ala., 250..B2 78
Saragossa, Pan....k11 62
Saragossa, see Zaragoza, prov. and city, Sp.
Saraguro, Ec., 1,334....B2 58
Sarah, Tate, Miss., 125....A3 100
Sarajevo, Yugo., 142,423....D4 22
Sara Kaeo, Thai....F5 38
Saraktash, Sov. Un., 10,000.E20 9
Sarala, Sov. Un., 10,000....E26 9
Saraland, Mobile, Ala., 4,595.E1 78
Saranac, Ionia, Mich., 1,081..F5 98
Saranac, Clinton, N.Y., 250..A3 108
Saranac, lakes, N.Y....A6 108
Saranac, riv., N.Y....B3 108
Saranac Inn, Franklin, N.Y., 40....B2 108
Saranac Lake, Franklin and Essex, N.Y., 6,421....B3 108
Saranap, Contra Costa, Calif., 6,450....*B5 82
Sarandë, Alb., 3,444....C2 23
Sarandí del Yí, Ur., 4,437....E1 56
Sarandí Grande, Ur., 4,539..E1 56
Sarangani, isl., Phil....D6 35
Sarangpur, India, 11,263....F6 40
Saransk, Sov. Un., 108,000..C3 29
Sarapul, Sov. Un., 76,000....B4 29
Sarasota, Sarasota, Fla., 34,083....E4, F2 86
Sarasota, co., Fla., 76,895....E4 86
Sarasota, bay, Fla....E4 86
Sarata, Sov. Un....B9 22
Saratoga, Howard, Ark., 62..D2 81
Saratoga, Santa Clara, Calif., 14,861....C2 82
Saratoga, Randolph, Ind., 363....D8 91
Saratoga, Howard, Iowa, 90..A5 92
Saratoga, Wilson, N.C., 409..B6 109
Saratoga, Hardin, Tex., 800..D5 118
Saratoga, Carbon, Wyo., 1,133....D6 125
Saratoga, co., N.Y., 89,096..B7 108
Saratoga, lake, N.Y....C7 108
Saratoga Place, Nansemond, Va., 1,478....*E6 121
Saratoga Springs, Saratoga, N.Y., 16,630....B7 108
Saratov, Sov. Un., 622,000 (*740,000)....E18 9, C3 29
Saravane, Laos, 25,000....E7 38
Sarawak, reg., Mala., 546,385....I14 33, E4 35
Saray, Tur., 5,300....B6 23
Saraykoy, Tur., 6,900....D7 23
Sarbāz, Iran....H10 41
Sarben, Keith, Nebr., 105....C4 103
Sárbogárd, Hung., 6,859....B4 22
Sarca, riv., It....D6 18
Sarcelles, Fr., 35,885....g10 14
Sarcoxie, Jasper, Mo., 1,056.D3 101
Sarda, riv., India....C8 40
Sardalas, Libya....D2 43
Sardegna, pol. dist., It., 1,419,362....*E2 21
Sardinia, Decatur, Ind., 170..F6 91
Sardinia, Brown, Ohio, 799..C4 111
Sardinia, isl., It....D2 21
Sardis, Dallas, Ala., 300....C3 78
Sardis, B.C., Can., 898....B11 68
Sardis, Burke, Ga., 829....D5 87
Sardis, Mason, Ky., 190....B6 94
Sardis, Panola, Miss., 2,098..A4 100
Sardis, Monroe, Ohio, 500..C7 111
Sardis, Henderson, Tenn., 274....B3 117
Sardis, dam, Miss....A4 100
Sardis, res., Miss....A4 100
Sardo, Eth....C5 47
Sarektjakko, mtn., Swe....D7 25
Sarepta, Webster, La., 737..B2 95
Sarepta, Calhoun, Miss., 75..A4 100
Sar-e Yazd, Iran....F7 41
Sargans, Switz., 2,571....B7 19
Sargeant, Mower, Minn., 113....G6 99
Sargent, Coweta, Ga., 900....C2 87
Sargent, Custer, Nebr., 876..C6 103
Sargent, co., N. Dak., 6,856..C8 110
Sargents, Saguache, Colo., 60....C4 83
Sargodha, Pak., 129,291....A4 40
Sarhad, Afg., 5,000....k17 41
Sārī, Iran, 23,990....C6 41
Saria, isl., Grc....E6 23
Sarikamis, Tur., 17,600....B14 31
Sarine, riv., Switz....C3 19
Sariñena, Sp., 3,389....B5 20
Sari-Pul, Afg., 5,000....C12 41
Sarita, Kenedy, Tex., 250....F4 118
Sariwon, Kor., 42,957....G2 37
Sark, isl., Guernsey....F5 10
Sarkand, Sov. Un....F24 9
Sarkisla, Tur., 3,731....C11 31
Sarkoy, Tur., 4,000....B6 23
Sarlat, Fr., 5,251....E4 14
Sarles, Cavalier and Towner, N. Dak., 225....A7 110
Sarna, Swe., 1,276....G5 25
Sarnen, Switz., 6,554....C5 19
Sarnen, lake, Switz....C5 19
Sarnia, Ont., Can., 50,976 (*61,293)....E2 72
Sarny, Sov. Un., 10,000....F6 27
Säro, Swe., 259....A5 24
Sarona, Washburn, Wis., 90.C2 124
Saronic, gulf, Grc....D4 23
Saronno, It., 25,190....B2 21
Saronville, Clay, Nebr., 71..B8 103
Saros, gulf, Tur....B6 23
Sarospatak, Hung., 9,610....A5 22
Sarova, Sov. Un., 5,000....D14 27
Sar Planina, mts., Yugo....D5 22
Sarpsborg, Nor., 13,300....p29 25
Sarpy, Big Horn, Mont....E10 102
Sarpy, co., Nebr., 31,281....C9 103
Sarrebourg, Fr., 11,080....F7 15
Sarreguemines, Fr., 17,866..C7 14
Sarre-Union, Fr., 2,645....F7 15
Sarria, Sp., 3,935....A2 20
Sarsfield, Ont., Can., 259..A10 70
Sarstedt, Ger., 10,200....A4 17
Sarstun, riv., Guat....C3 62
Sartell, Stearns, Minn., 791..E4 99
Sartène, Fr., 4,067....D2 21
Sarthe, dept., Fr., 443,019..*D4 14
Sarthe, riv., Fr....D3 14
Sartrouville, Fr., 31,267....g9 14
Sarufutsu, Jap., 2,900....D11 37
Sarvar, Hung., 11,021....B3 22
Sarvestān, Iran....G6 41
Sarybyevo, Sov. Un....o18 27
Sary-Ishikotrau, des., Sov. Un..D9 29
Sary-Ozek, Sov. Un....E9 29
Sarysu, riv., Sov. Un....D7 29
Sarzana, It., 8,600....B2 21
Sasabe, Pima, Ariz., 50....F4 80
Sasaginnigak, lake, Man., Can..D4 71
Sasakwa, Seminole, Okla., 253....C5 112
Sasalaguan, mtn., Guam.... 52
Sasaram, India, 37,782....E10 40
Sasebo, Jap., 262,484....J4 37
Saskatchewan, prov., Can., 925,181....66, 70
Saskatchewan, riv., Sask., Can..F12 66
Saskatoon, Sask., Can., 95,526....E2 70
Saskeram, riv., Sask., Can....D5 70
Sasovo, Sov. Un., 26,000....C2 29
Saspamco, Wilson, Tex., 300.B4 118
Sassafras, mtn., S.C....A2 115
Sassafras, riv., Md....B5 85
Sassandra, I.C., 2,600....E3 45
Sassandra, riv., I.C....E3 45
Sassari, It., 90,037....D2 21
Sassenberg, Ger., 3,700....B3 17
Sasser, Terrell, Ga., 382....E2 87
Sassnitz, Ger., 13,400....A6 16
Sasso Marconi, It., 1,308....E7 18
Sassuolo, It., 10,300....E6 18
Sastre, Arg., 2,308....A4 54
Sasuri, Kor....F3 37
Sasykkol, lake, Sov. Un....D10 29
Satalo, W. Sam., 181.... 52
Satan, mtn., B.C., Can....C5 68
Satanta, Haskell, Kans., 686.E3 93
Satapuala, W. Sam., 637.... 52
Satara, India, 44,353....I4 40
Satartia, Yazoo, Miss., 126..C3 100
Sataua, bay, W. Sam.... 52
Säter, Swe., 4,500....G6 25
Saticoy, Ventura, Calif., 2,283....E4 82
Satilla, riv., Ga....E4 87
Satipo, Peru....D3 58
Satka, Sov. Un., 38,900....B5 29
Satmala, range, India....H7 40
Satna, India, 38,046....E8 40
Satoraljaujhely, Hung., 16,197....A5 22
Satpura, range, India....G5 40
Sätrabrunn, Swe., 169....t34 25
Satrup, Ger., 1,750....D3 24
Satsop, Grays Harbor, Wash., 150....B2 122
Satsuma, Mobile, Ala., 1,491....E1 78
Satsuma, Putnam, Fla., 300..C5 86
Satthaip, Thai., 4,478....F4 38
Sattler, Comal, Tex., 30....A4 118
Satu-Mare, Rom., 52,099....B6 22
Satun, Thai., 5,615....I4 38
Satupaitea, Samoa, 1,047.... 52
Saturna, isl., B.C., Can....E6 68
Satus, creek, Wash....C5 122
Sauage, Scott, Minn., 1,094....F5, F6 99
Sauce, Arg., 3,017....A5 54
Sauceda, mts., Ariz....E3 80
Saucier, Harrison, Miss., 300.E4 100
Saucillo, Mex., 6,820....B3 63
Sauda, Nor., 3,055....H2 25
Saudi Arabia, country, Asia, 6,036,000....G7 33
Saugatuck, Fairfield, Conn. (part of Westport)....E3 84
Saugatuck, Allegan, Mich., 927....F4 98
Saugatuck, riv., Conn....D3 84
Saugerties, Ulster, N.Y., 4,286....C7 108
Saugor, India, 85,491....F7 40
Saugus, Los Angeles, Calif., 200....F2 82
Saugus, Essex, Mass., 20,666....B5, C3 98
Saugus, riv., Mass....C3 97
Sauk, co., Wis., 36,179....E4 124
Sauk, riv., Minn....E4 99
Sauk, riv., Wash....A4 122
Sauk Centre, Stearns, Minn., 3,573....E4 99
Sauk City, Sauk, Wis., 2,095.E4 124
Sauk Rapids, Benton, Minn., 4,038....E4 99
Sauk Village, Cook, Ill., 5,774....*B6 90
Saukville, Ozaukee, Wis., 1,038....E5 124
Saulgrub, Ger., 1,100....B7 18
Saulnierville, N.S., Can., 450.E3 74
Saulsbury, Hardeman, Tenn., 141....B2 117
Saulston, Wayne, N.C., 100..B6 109
Sault au Mouton, Que., Can., 876....A8 73
Sault Ste. Marie, Ont., Can., 43,088 (*58,460)....E9 72
Sault Ste. Marie, Chippewa, Mich., 18,722....B6 98
Saumlakki, Indon....G8 35
Saumur, Fr., 20,773....D4 14
Saunders, Alta., Can., 145..C3 69
Saunders, co., Nebr., 17,270.C9 84
Saunderstown, Washington, R.I., 400....C11 84
Saundersville, Sumner, Tenn., 100....E9 117
Saunemin, Livingston, Ill., 392....C5 90
Saurashtra, state, India, 4,137,359....*G3 40
Saurashtra, pen., India....D4 39
Sausalito, Marin, Calif., 5,331....B5, D2 82
Sauveterre, Fr., 334....E5 14
Sava, riv., Yugo....C4 22
Savage, Howard, Md., 1,341.B4 85
Savage, Scott, Minn., 1,094....F5, F6 99
Savage, Tate, Miss., 75....A3 100
Savage, Richland, Mont., 275....C12 102
Savage, riv., Md....D1 85
Savageton, Campbell, Wyo..B7 125
Savaii, isl., W. Sam.... 52
Salvalan, mtn., Iran....B3 41
Savalou, Dah., 4,100....E5 45
Savanna, Carroll, Ill., 4,950..A3 90
Savanna, Pittsburg, Okla., 620....C6 112
Savannah, Chatham, Ga., 149,245 (*189,200)....D5 87
Savannah, Andrew, Mo., 2,455....B3 101
Savannah, Wayne, N.Y., 602.B4 108
Savannah, Ashland, Ohio, 409....B5 111
Savannah, Hardin, Tenn., 4,315....B3 117
Savannah, lake, Md....D6 85
Savannah, riv., Ga., S.C....F5 115
Savannah Beach, Chatham, Ga., 1,385....D6 87
Savannakhet, Laos....D6 38
Savanna-la-Mar, Jam., 9,783....F4 64
Savé, Dah., 5,100....E5 45
Save, riv., Moz....B5 49
Sāveh, Iran, 15,365....D5 41
Saveni, Rom., 6,470....B8 22
Saverne, Fr., 9,056....C7 14
Saverton, Ralls, Mo., 135....B6 101
Savery, Carbon, Wyo., 25..D5 125
Savigliano, It., 14,800....B1 21
Savigny [-sur-Orgel] Fr., 24,316....F2 15
Savo, riv., It....h9 21
Savoie, dept., Fr., 266,678..D2 18
Savona, B.C., Can., 532....D7 68
Savona, It., 72,115 (*103,000)....B2 21
Savona, Steuben, N.Y., 904..C3 108
Savonburg Allen, Kans., 131.E8 93
Savonlinna, Fin., 14,700....G13 25
Savoonga, Alsk., 200....C5 79
Savoy, Berkshire, Mass., 100 (277▲)....A1 97
Savoy, Blaine, Mont., 10....B8 102
Savoy (Savoie), former prov., Fr., 546,000....E7 14
Savran, Sov. Un., 10,000....G8 27
Sävsjö, Swe., 5,100....A8 24
Savu, is., Indon....B3 36
Savu, sea, Indon....G6 35
Savur, Tur., 3,400....D13 31
Savusavu, bay, Fiji Is.... 52
Saw, Bur., 1,277....D9 39
Sawankhalok, Thai., 7,742..D3 38
Sawatch, range, Colo....B4 83
Sawback, range, Alta., Can....D3 69
Sawbill Landing, Lake, Minn., 100....C7 99
Sawda, mtn., Libya....D2 43
Sawdy, Alta., Can., 95....B4 69
Sawe, Indon....L2 38
Sawhāj, Eg., U.A.R., 48,300.D6 43
Sawknah, Libya....D3 43
Sawmill, Apache, Ariz., 150..B6 80
Saw Mill, riv., N.Y....D6 106
Sawtooth, mts., Idaho....E3 89
Sawtooth, ridge, Wash....A5 122
Sawyer, Pratt, Kans., 192....E5 93
Sawyer, Berrien, Mich., 1,300....G4 98
Sawyer, Carlton, Minn., 75..D6 99
Sawyer, Ward, N. Dak., 390..A4 110
Sawyer, Choctaw, Okla., 235.C6 112
Sawyer, co., Wis., 9,475....C2 124
Sawyer, lake, Wash....D2 122
Sawyerville, Que., Can., 789.D6 73
Sawyerwood, Summit, Ohio, 1,600....*A6 111
Saxe, Charlotte, Va., 125....E4 121
Saxis, Accomack, Va., 577..D7 121
Saxman, Alsk., 153....n24 79
Saxman, Nicholas, W. Va., 40....C4 123
Saxmundham, Eng., 1,538..B9 12
Saxon, Spartanburg, S.C., 3,917....B4 115
Saxon, Raleigh, W. Va., 225....D3, D6 123
Saxon, Iron, Wis., 250....B3 124
Saxonburg, Butler, Pa., 876..E2 114
Saxonville, Middlesex, Mass. (part of Framingham)....D1 97
Saxony, reg., Ger....C6 16
Saxony-Anhalt (Sachsen-Anhalt), former state, Ger., 4,160,539....B6 17
Saxton, Whitley, Ky., 650..D5 94
Saxton, Bedford, Pa., 977....F5 114
Saxtons, riv., Vt....E3 120
Saxtons River, Windham, Vt., 725....E3 120
Say, Niger, 2,200....D5 45
Sayabec, Que., Can., 2,314.G19 67
Sayaboury, Laos, 10,000....C4 38
Sayán, Peru, 1,229....D2 58
Sayan, mts., Sov. Un....D12 28
Saybrook, McLean, Ill., 859.C5 90
Saybrook Point, Middlesex, Conn., 500....D7 84
Sayle, Powder River, Mont., 7....E11 100

Saylorsburg, Monroe, Pa., 600 E11 114
Sayner, Vilas, Wis., 350 B4 124
Sayre, Jefferson, Ala., 1,200 B3 78
Sayre, Beckham, Okla., 2,913 B2 112
Sayre, Bradford, Pa., 7,917 C8 114
Sayreton, Jefferson, Ala., 1,000 C3, E4 78
Sayreville, Middlesex, N.J., 22,553 C4 106
Sayr Usa, Mong C6 34
Sayula, Mex., 11,596 D4, n12 63
Sayula, lake, Mex n12 63
Sayville, Suffolk, N.Y., 6,500 G4 84
Sazan, isl., Alb B2 23
Sazava, Czech., 1,435 o18 26
Sazava, riv., Czech o18 26
Sazliyka, riv., Bul D7 22
Sbeitla, prov., Tun., 164,395 *C6 44
Scafell Pike, mtn., Eng F5 13
Scalby, Eng., 7,251 F8 13
Scales Mound, Jo Daviess, Ill., 399 A3 90
Scalp Level, Cambria, Pa., 1,445 F4 114
Scaly, Macon, N.C., 100 D3 109
Scammon, Cherokee, Kans., 429 E9 93
Scammon Bay, Alsk., 115 C6 79
Scandia, Alta., Can., 51 D4 69
Scandia, Republic, Kans., 643 C6 93
Scandia, Washington, Minn., 150 E8 99
Scandinavia, Waupaca, Wis., 266 D4 124
Scanlon, Carlton, Minn., 1,126 D6 99
Scansano, It., 8,560 C3 21
Scanterbury, Man., Can., 75 D3 71
Scantic, Hartford, Conn., 100 B6 84
Scanzano, It., 683 D6 21
Scapa, Alta., Can D5 69
Scapa Flow, bay, Scot B5 13
Scapegoat, mtn., Mont C4 102
Scappoose, Columbia, Oreg., 923 B4 113
Scarba, isl., Scot D3 13
Scarboro, Ont., Can E7 72
Scarboro, Jenkins, Ga., 150 D5 87
Scarboro, Cumberland, Maine, 500 (6,418▲) E5 96
Scarborough, Ont., Can., 900 E7 72
Scarborough, Eng., 42,587 C6 10
Scarbro, Fayette, W. Va., 900 D7 123
Scariff, isl., Ire F1 11
Scarp, isl., Scot B1 13
Scarriff, Ire., 600 E3 11
Scarsdale, Westchester, N.Y., 17,968 D2 108
Scartaglen, Ire., 137 E2 11
Scarth, Man., Can., 80 E1 71
Scarville, Winnebago, Iowa, 105 A4 92
Scauri, It F3 21
Sceaux, Fr., 19,024 g10 14
Scenic, Pennington, S. Dak., 83 D3 116
Sceptre, Sask., Can., 257 G1 70
Sceui Ghimira, Eth D4 47
Schaal, Howard, Ark., 200 D2 81
Schaal, lake, Ger E4 24
Schaefferstown, Lebanon, Pa., 800 F9 114
Schaffer, Delta, Mich., 130 C3 98
Schaffhausen, Switz., 30,904 (*47,129) A6 19
Schaffhausen, canton, Switz., 65,981 A5 19
Schagen, Neth., 3,889 B4 15
Schaghticoke, Rensselaer, N.Y., 720 C7 108
Schaller, Sac, Iowa, 896 B2 92
Schangnau, Switz., 1,030 C4 19
Schärding, Aus., 5,710 E8 17
Scharhörn, isl., Ger E2 24
Schaumberg, Cook, Ill., 3,296 E2 90
Schefferville (Knob Lake), Que., Can., 3,178 B3 73
Schefield, Stark, N. Dak., 25 C3 110
Schellbourne, White Pine, Nev., 10 D7 104
Schell City, Vernon, Mo., 343 C3 101
Schellsburg, Bedford, Pa., 288 F4 114
Schenectady, Schenectady, N.Y., 81,682 C7 108
Schenectady, co., N.Y., 152,896 C6 108
Schenefeld, Ger., 1,500 D3 24
Schenevus, Otsego, N.Y., 493 C6 108
Schererville, Lake, Ind., 2,875 B3 91
Scherfede, Ger., 2,700 B4 17
Schertz, Guadalupe, Tex., 2,281 B4 118
Scheveningen, Neth. (part of The Hague) B4 15
Schiedam, Neth., 8,600 C4 15
Schiermonnikoog, isl., Neth A6 15
Schifferstadt, Ger., 15,600 D3 17
Schiltigheim, Fr., 25,081 C7 14
Schilpario, It., 1,751 C6 18
Schiller Park, Cook, Ill., 5,687 E2 90
Schio, It., 28,298 B3 21
Schivelbein, see Swidwin, Pol.
Schkeuditz, Ger., 18,300 B7 17
Schladming, Aus., 3,249 E6 16
Schlater, Leflore, Miss., 300 B3 100
Schlei, inlet, Ger D3 24
Schleicher, co., Tex., 2,791 D2 118
Schleiz, Ger., 7,493 C6 17
Schleswig, Ger., 33,800 A4 16
Schleswig, Crawford, Iowa, 785 B2 92
Schleswig-Holstein, reg., Ger A4 16
Schleswig-Holstein, state, Ger., 2,316,600 D4 24
Schley, Gloucester, Va., 175 D6 121
Schley, co., Ga., 3,256 D2 87
Schlieren, Switz., 10,043 B5 19
Schlitz, Ger., 5,000 C4 17
Schlüchtern, Ger., 5,800 C4 17
Schmalkalden, Ger., 14,000 C5 17
Schmölln, Ger., 13,800 C7 17
Schnackenburg, Ger., 700 E3 24
Schneeberg, Ger., 21,600 C7 17
Schneeberg, mtn., Ger C6 17
Schneidemühl, see Piła, Pol.
Schneider, Lake, Ind., 405 B3 91
Schoenchen, Ellis, Kans., 188 D4 93
Schofield, Marathon, Wis., 3,038 D4 124
Schofield Barracks, Honolulu, Haw B3, g9 88
Schoharie, Schoharie, N.Y., 1,168 C6 108
Schoharie, co., N.Y., 22,616 C6 108
Schoharie, creek, N.Y C6 108
Scholle, Socorro, N. Mex., 10 C3 107
Scholls, Washington, Oreg., 200 B2 113
Schönbach, Aus., 725 D7 16
Schönebeck, Ger., 44,400 A6 17
Schöneck, Ger., 4,521 C7 17
Schongau, Ger., 8,800 E5 16
Schöningen, Ger., 16,100 A5 17
Schoodic, lake, Maine C4 96
Schoolcraft, Kalamazoo, Mich., 1,205 F5 98
Schoolcraft, co., Mich., 8,953 B4 98
Schoonhoven, Neth., 6,300 C4 15
Schopfheim, Ger., 7,800 B3 18
Schorndorf, Ger., 18,800 E4 17
Schoten, Bel., 26,060 C4 15
Schötmar, Ger., 9,200 A3 17
Schouten, is., W. Irian F9 35
Schramberg, Ger., 18,100 D4 16
Schram City, Montgomery, Ill., 698 D4 90
Schriever, Terrebonne, La., 650 C6, E5 95
Schrobenhausen, Ger., 8,700 E6 17
Schroeder, Cook, Minn., 550 B6, C8 99
Schroon, lake, N.Y B7 108
Schroon Lake, Essex, N.Y., 800 B7 108
Schuchk, Pima, Ariz., 20 E4 80
Schuchuli, Pima, Ariz., 35 E3 80
Schulenburg, Fayette, Tex., 2,207 E4 118
Schuler, Alta., Can., 156 D5 69
Schull, Ire., 419 F2 11
Schulte, Sedgwick, Kans., 50 B5 93
Schulter, Okmulgee, Okla., 500 B6 112
Schüpfheim, Switz., 3,771 B5 19
Schurz, Mineral, Nev., 180 E3 104
Schüttorf, Ger., 8,100 A2 17
Schuyler, Colfax, Nebr., 3,096 C8 103
Schuyler, Nelson, Va., 450 D4 121
Schuyler, co., Ill., 8,746 C3 90
Schuyler, co., Mo., 5,052 A5 101
Schuyler, co., N.Y., 15,044 C4 108
Schuyler Lake, Otsego, N.Y., 200 C5 108
Schuylerville, Saratoga, N.Y., 1,361 B7 108
Schuylkill, co., Pa., 173,027 E9 114
Schuylkill, riv., Pa A11 114
Schuylkill Haven, Schuylkill, Pa., 6,470 E9 114
Schwabach, Ger., 23,700 D6 17
Schwäbische Alb, plat., Ger E4 17
Schwäbisch Gmünd, Ger., 41,100 E4 17
Schwäbisch Hall, Ger., 21,900 D4 17
Schwabmünchen, Ger., 7,300 A6 18
Schwandorf, Ger., 16,100 D7 17
Schwaner, mts., Indon F4 35
Schwarmstedt, Ger., 3,100 F3 24
Schwarze Elster, riv., Ger B8 17
Schwarzenbach [an der sächsischen Saale], Ger., 7,400 C6 17
Schwarzenbek, Ger., 7,800 E4 24
Schwarzenberg, Ger., 14,900 C7 17
Schwarzenfeld, Ger., 5,000 D7 17
Schwarzheide, Ger., 7,449 B8 17
Schwaz, Aus., 9,455 E5 16
Schwedt, Ger., 5,961 B7 16
Schweibus, see Swiebodzin, Pol.
Schweinfurt, Ger., 56,900 C5 17
Schwenningen, Ger., 31,700 D4 16
Schwerin, Ger., 92,900 B5 16
Schweriner See, lake, Ger E5 24
Schwetzingen, Ger., 15,000 D3 17
Schwyz, Switz., 11,007 B6 19
Schwyz, canton, Switz., 78,048 B6 19
Schyan, riv., Que., Can A7 72
Sciacca, It., 25,000 F4 21
Science Hill, Pulaski, Ky., 463 C5 94
Scilly, is., Eng F3 10
Scio, Allegany, N.Y., 600 C3 108
Scio, Harrison, Ohio, 1,135 B6 111
Scio, Linn, Oreg., 441 C2, C4 113
Sciota, McDonough, Ill., 120 C3 90
Scioto, co., Ohio, 84,216 D4 111
Scioto, riv., Ohio B4 111
Sciotodale, Scioto, Ohio, 800 *D5 111
Scioto Furnace, Scioto, Ohio, 300 D5 111
Scipio, Jennings, Ind., 200 F6 91
Scipio, Millard, Utah, 328 D3 119
Scitico, Hartford, Conn. (part of Hazardville) B6 84
Scituate, Plymouth, Mass., 3,229 B6, D4 97
Scituate, res., R.I B10 84
Scituate Center, Plymouth, Mass., 500 D4 97
Scobey, Daniels, Mont., 1,726 B11 102
Scofield, Carbon, Utah, 158 D4 119
Scollard, Alta., Can., 50 D4 69
Scooba, Kemper, Miss., 513 C5 100
Scopus, mtn., Jordan m14 32
Scoresbysund, Grnld., 90 B17 4
Scotch Plains, Union, N.J., 18,491 B4 106
Scotch Village, N.S., Can., 266 D6 74
Scotia, Humboldt, Calif., 1,122 B1 82
Scotia, Greeley, Nebr., 350 C7 103
Scotia, Schenectady, N.Y., 7,625 C7 108
Scotia, Hampton, S.C., 102 F5 115
Scotland, Van Buren, Ark., 150 B3 81
Scotland, Ont., Can., 612 D4 72
Scotland, Windham, Conn., 250 (684▲) C8 84
Scotland, Telfair, Ga., 236 D4 87
Scotland, Greene, Ind., 100 G4 91
Scotland, St. Marys, Md., 100 D5 85
Scotland, Cheshire, N.H., 40 E2 105
Scotland, Franklin, Pa., 500 G6 114
Scotland, Bon Homme, S. Dak., 1,077 D8 116
Scotland, reg., U.K., 5,178,490 13
Scotland, co., Mo., 6,484 A5 101
Scotland, co., N.C., 25,183 C4 109
Scotland Neck, Halifax, N.C., 2,974 A6 109
Scotlandville, East Baton Rouge, La., 10,000 A5, D4 95
Scotrun, Monroe, Pa., 150 D11 114
Scotsburn, N.S., Can., 292 D7 74
Scotsguard, Sask., Can., 71 H1 70
Scotstown, Que., Can., 1,038 D6 73
Scott, Lonoke and Pulaski, Ark., 240 C3, D6 81
Scott, Sask., Can., 281 E1 70
Scott, Johnson, Ga., 149 D4 87
Scott, Lafayette, La., 902 D3 95
Scott, Bolivar, Miss., 315 B2 100
Scott, Paulding and Van Wert, Ohio, 365 B3 111
Scott, co., Ark., 7,297 C1 81
Scott, co., Ill., 6,377 D3 90
Scott, co., Ind., 14,643 G6 91
Scott, co., Iowa, 119,067 C7 92
Scott, co., Kans., 5,228 D3 93
Scott, co., Ky., 15,376 B5 94
Scott, co., Minn., 21,909 F5 99
Scott, co., Miss., 21,187 C4 100
Scott, co., Mo., 32,748 D8 101
Scott, co., Tenn., 15,413 C9 117
Scott, co., Va., 25,813 B2 121
Scott, cape, B.C., Can D3 68
Scott, glacer, Ant A33 5
Scott, is., Ant C30 5
Scott, isl., B.C., Can D3 68
Scott, mtn., Okla C3 112
Scott, range, Ant C18 5
Scott City, Scott, Kans., 3,555 D3 93
Scott City, Scott, Mo., 1,963 D8 101
Scottdale, De Kalb, Ga., 4,000 *B5 87
Scottdale, Westmoreland, Pa., 6,244 F2 114
Scottish, sea, Scot D1 13
Scottland, Edgar, Ill., 13 D6 90
Scotts, Kalamazoo, Mich., 280 F5 98
Scotts, mtn., N.J B2 106
Scottsbluff, Scotts Bluff, Nebr., 13,377 C2 103
Scotts Bluff, co., Nebr., 33,809 C2 103
Scotts Bluff, nat. mon., Nebr C2 103
Scottsboro, Jackson, Ala., 6,449 A3 78
Scottsburg, Scott, Ind., 3,810 G6 91
Scottsburg, Douglas, Oreg., 200 D3 113
Scottsburg, Halifax, Va., 188 E4 121
Scottsdale, Maricopa, Ariz., 10,026 D4, G2 80
Scotts Hill, Pender, N.C., 150 C6 109
Scotts Hill, Henderson and Decatur, Tenn., 298 B3 117
Scotts Mills, Marion, Oreg., 155 B4, C2 113
Scottsville, Pope, Ark., 145 B3 81
Scottsville, Mitchell, Kans., 60 C6 93
Scottsville, Allen, Ky., 3,324 D3 94
Scottsville, Monroe, N.Y., 1,863 B3 108
Scottsville, Ashe, N.C., 25 A2 109
Scottsville, Albemarle and Fluvanna, Va., 353 D4 121
Scottville, Macoupin, Ill. 186 D3 90
Scottville, Mason, Mich., 1,245 E4 98
Scourie, Scot B3 13
Scout Lake, Sask., Can., 96 H3 70
Scraggly, lake, Maine B4 96
Scranton, Logan, Ark., 229 B2 81
Scranton, Greene, Iowa, 865 B3 92
Scranton, Osage, Kans., 576 D8 93
Scranton, Menifee, Ky., 127 C6 94
Scranton, Hyde, N.C., 110 B7 109
Scranton, Bowman, N. Dak., 358 C2 110
Scranton, Lackawanna, Pa., 111,443 (*215,600) A9, D10 114
Scranton, Florence, S.C., 613 D8 115
Scraper, Cherokee, Okla., 40 A7 112
Screven, Wayne, Ga., 1,010 E4 87
Screven, co., Ga., 14,919 D5 87
Scribner, Dodge, Nebr., 1,021 C9 103
Scridain, bay, Scot D2 13
Scrivia, riv., It E4 18
Scrolls Caves, Jordan G4 33
Scullville, Atlantic, N.J., 350 E3 106
Scunthorpe, Eng., 67,257 A7 12
Scuol (Schuls), Switz., 1,429 C9 19
Scurry, co., Tex., 20,369 C2 118
Scusciuban, Som C7 47
Scutari (Shkodër), Alb., 40,900 A2 23
Scutari (Shkoder), pref., Alb., 150,000 *A2 23
Scutari, lake, Alb A2 23
Scyrene, Clarke, Ala., 40 D2 78
Sdom, Isr C7 32
Seaboard, Northampton, N.C., 624 A6 109
Seabra, Braz., 2,292 C3 58
Sea Breeze, Monroe, N.Y. (part of Irondequoit) B3 108
Sea Bright, Monmouth, N.J., 1,138 C5 106
Seabrook, Prince Georges, Md., 3,000 *C4 85
Seabrook, Rockingham, N.H., 700 (2,209▲) E5 105
Seabrook, Cumberland, N.J., 1,798 *E2 106
Seabrook, Harris, Tex., 950 F5 118
Seabrook, isl., S.C F7 115
Sea Cliff, Nassau, N.Y., 5,669 F2 84
Seadrift, Calhoun, Tex., 1,082 E4 118
Seaford, Sussex, Del., 4,430 C6 85
Seaford, Nassau, N.Y., 14,718 G3 108
Seaford, York, Va., 900 A6 121
Seaforth, Ont., Can., 2,255 D3 72
Sea Girt, Monmouth, N.J., 1,798 C4 106
Seagoville, Dallas, Tex., 3,745 B6 118
Segram, lake, Sask., Can E1 70
Seagrave, Ont., Can., 76 C6 72
Seagraves, Gaines, Tex., 2,307 C1 118
Seagrove, Randolph, N.C., 323 B4 109
Seagull, Currituck, N.C. A8 109
Seaham, Eng., 26,048 F7 13
Seahurst, King, Wash., 2,500 D1 122
Sea Island, Glynn, Ga., 500 E5 87
Sea Isle City, Cape May, N.J., 1,393 E3 106
Seal, bay, Ant B11 5
Seal, lake, Newf., Can g9 75
Seal, riv., Man., Can C5 71
Seal Beach, Orange, Calif., 6,994 *F2 82
Seal Cove, N.B., Can., 549 E3 74
Seal Cove, Newf., Can., 446 D3 75
Seal Cove, Hancock, Maine, 130 D4 96
Seale, Russell, Ala., 350 C4 78
Sealevel, Carteret, N.C., 500 C7 109
Seal Harbor, Hancock, Maine, 200 D4 96
Seal Rock, Lincoln, Oreg., 240 C2 113
Sealston, King George, Va., 150 C5 121
Sealy, Austin, Tex., 2,328 E4 118
Seama, Valencia, N. Mex., 250 B2 107
Seaman, Adams, Ohio, 714 D4 111
Seanor, Somerset, Pa., 300 F4 114
Searchlight, Clark, Nev., 180 H7 104
Searcy, White, Ark., 8,215 B4 81
Searcy, co., Ark., 8,124 B3 81
Searight, Crenshaw, Ala., 64 D3 78
Searles, Tuscaloosa, Ala., 100 B2 78
Searles, Brown, Minn., 105 F4 99
Searles, lake, Calif E5 82
Sears, Osceola, Mich., 60 E5 98
Sears, falls, Nebr B5 103
Searsboro, Poweshiek, Iowa, 165 C5 92
Searsburg, Bennington, Vt., 50 (73▲) F3 120
Searsport, Waldo, Maine, 783 (1,838▲) D4 96
Searston, Newf., Can., 153 E2 75
Seaside, Monterey, Calif., 19,353 C6 82
Seaside, Clatsop, Oreg., 3,877 B3 113
Seaside Heights, Ocean, N.J., 954 D4 106
Seaside Park, Ocean, N.J., 1,054 D4 106
Seatack, Princess Anne, Va., 3,120 *E7 121
Seaton, Eng., 3,410 D4 12
Seaton, Mercer, Ill., 235 B3 90
Seat Pleasant, Prince Georges, Md., 5,365 C2, C4 85
Seattle, King, Wash., 557,087 (*938,400) B3, D1 122
Sea View, Plymouth, Mass., 400 E4 97
Seaview, Pacific, Wash., 600 C1 122
Seaville, Cape May, N.J., 100 E3 106
Seba Beach, Alta., Can., 113 C3 69
Sebáco, Nic., 1,339 D4 62
Sebago, lake, Maine E2 96
Sebago Lake, Cumberland, Maine, 350 E2 96
Sebastian, Indian River, Fla., 698 E6 86
Sebastian, co., Ark., 66,685 B1 81
Sebastian, cape, Oreg E2 113
Sebastian, inlet, Fla E6 86
Sebastián Vizcaíno, bay, Mex B2 63
Sebasticook, lake, Maine D3 96
Sebastopol, Sonoma, Calif., 2,694 A5, C2 82
Sebastopol, Scott, Miss., 343 C4 100
Sebatik, isl., Indon E5 35
Sebec, Piscataquis, Maine, 150 (384▲) C3 96
Sebec, lake, Maine C3 96
Sebeka, Wadena, Minn., 823 D3 99
Sebeș, Rom., 11,628 C6 22
Sebewaing, Huron, Mich., 2,026 E7 98
Sebinkarahisar, Tur., 8,800 B12 31
Sebnitz, Ger., 14,800 C9 17
Seboeis, Penobscot, Maine, 60 (77▲) C4 96
Seboeis, riv., Maine B4 96
Seboois, lake, Maine C4 96
Seboomook, Somerset, Maine, 10 (16▲) C3 96
Seboomook, lake, Maine C3 96
Seboruco, P.R C4 65
Sebou, riv., Mor G4 33
Seboyeta, Valencia, N. Mex., 167 B2 107
Sebree, Webster, Ky., 1,139 C2 94
Sebrell, Southampton, Va., 200 E5 121
Sebring, Highlands, Fla., 6,939 E5 86
Sebring, Mahoning, Ohio, 4,439 B6 111
Sebringville, Ont., Can., 520 D3 72
Secane, Delaware, Pa., 2,000 *B11 114
Secaucus, Hudson, N.J., 12,154 D5 106
Secchia, riv., It B3 21
Sechelt, B.C., Can., 488 E6 68
Sechura, Peru, 3,826 C1 58
Sechura, bay, Peru C1 58
Seco, Letcher, Ky., 531 C7 94
Second, lake, Maine B4 96
Second, lake, N.H A2 105
2nd Cataract (Nile River), Sud A2 47
Second Mesa, Navajo, Ariz., 110 B5 80
Secor, Woodford, Ill., 427 C4 90
Secor Gardens, Lucas, Ohio, 3,000 *A2 111
Secretan, Sask., Can., 60 G2 70
Secretary, Dorchester, Md., 351 C6 85
Secretary, isl., N.Z P11 51
Section, Jackson, Ala., 595 A4 78
Secunderabad, India, 187,471 E6 39, I7 40
Security, El Paso, Colo., 10,000 C6 83
Sedalia, Alta., Can., 50 D5 69
Sedalia, Douglas, Colo., 250 B6 83
Sedalia, Clinton, Ind., 170 D4 91
Sedalia, Graves, Ky., 258 B2 94
Sedalia, Pettis, Mo., 23,874 C4 101
Sedalia (Midway), Madison, Ohio, 341 C4 111
Sedan, Fr., 20,336 C6 14
Sedan, Chautauqua, Kans., 1,677 E7 93
Sedan, Pope, Minn., 91 E3 99
Sedan, Union, N. Mex., 45 A6 107
Sedbergh, Eng., 2,049 F6 13
Sédérog, Niger C5 45
Sedgefield, Guilford, N.C., 1,000 *A4 109
Sedgewick, Alta., Can., 655 C5 69
Sedgewick, Hancock, Maine, 150 (574▲) D4 96
Sedgewickville, Bollinger, Mo., 91 D8 101
Sedgwick, Lawrence, Ark., 206 B5 81
Sedgwick, Sedgwick, Colo., 299 A8 83
Sedgwick, Harvey, Kans., 1,095 B5, E6 93
Sedgwick, co., Colo., 4,242 A8 83
Sedgwick, co., Kans., 343,231 E6 93
Sédhiou, Sen D1 45
Sedlcany, Czech., 2,288 D9 75
Sedley, Sask., Can., 391 G3 70
Sedley, Southampton, Va., 500 E6 121
Sedona, Coconino, Ariz., 400 C4 80
Sedro Woolley, Skagit, Wash., 3,705 A3 122
Sedrun, Switz C6 19
Seebe, Alta., Can., 137 D3 69
Seeber, riv., Man., Can B5 71
Seeber, riv., Ont., Can C5 71
Seebert, Pocahontas, W. Va., 120 C4 123
Seehausen, Ger., 4,951 F5 24
Seeheim, S.W. Afr C2 49
Seeis, S.W. Afr B2 49
Seekonk, Bristol, Mass., 8,399 C5 97
Seeley, Imperial, Calif., 600 F6 82
Seeley Lake, Missoula, Mont., 300 C3 102
Seeleys Bay, Ont., Can., 344 C8 72
Seelow, Ger., 2,757 F8 24
Seely, Crook, Wyo., 5 A8 125
Seelyville, Vigo, Ind., 1,114 F3 91
Seelyville, Wayne, Pa., 400 C11 114
Sées, Fr., 3,138 C4 14
Seesen, Ger., 12,100 B5 17
Sefadu, S.L E2 45
Seferihisar, Tur., 4,400 C6 25
Seffner, Hillsborough, Fla., 500 D4 86
Sefid, riv., Iran C4 41
Segamat, Mala., 18,445 K5 38
Segesta, It., 2,340 F4 21
Segezha, Sov. Un F16 25
Seglora, Swe A6 24
Sego, Grand, Utah, 20 D6 119
Segorbe, Sp., 7,538 C5 20
Ségou, Mali, 17,400 D3 45
Segovia, Sp., 33,360 B3 20
Segovia, Kimble, Tex., 100 D3 118
Segovia, prov., Sp., 195,602 *B3 20
Segovia or Wanks, riv., Nic C5 62
Segozero, lake, Sov. Un F15 25
Segré, Fr., 5,203 D3 14
Segre, riv., Sp B6 20
Segreganset, Bristol, Mass., 200 C5 97
Seguedine (Well), Niger B7 45
Séguéla, I.C., 4,200 E3 45
Seguin, Guadalupe, Tex., 14,299 B4, E4 118
Seguine, pt., N.Y E4 106
Seguin Falls, Ont., Can., 65 B5 72
Segundo, Las Animas, Colo., 175 D6 83
Segura, riv., Sp C5 20
Sehkūheh, Iran F10 41
Sehore, India, 28,489 F6 40
Sehwan, Pak., 3,827 D1 40
Seibert, Kit Carson, Colo., 210 B8 83
Seibo, Dom. Rep., 3,164 F9 64
Seiling, Dewey, Okla., 910 A3 112
Seille, riv., Fr F6 15
Seinäjoki, Fin., 15,800 F10 25
Seine, dept., Fr., 5,646,446 *C5, g10 14
Seine, bay, Fr C3 14
Seine, riv., Fr C4 14
Seine-et-Marne, dept., Fr., 524,486 F2 15
Seine-et-Oise, dept., Fr., 2,298,931 F2 15
Seine-Maritime, dept., Fr., 1,035,844 E9 12
Seis de Septiembre, Arg., 110,344 g7 54
Seistan, reg., Iran, Afg F10 41
Seixal, Port., 4,125 f9 20
Sejerby, Den C5 24
Sejerö, Den C5 24
Sejerö, isl., Den C5 24
Sekenke, Tan B5 48
Seki, Jap., 23,500 o15 37
Sekia el Hamra (Saguia el Hamra), Sp. overseas prov. Afr., 6,445 D2 44
Sekiu, Clallam, Wash., 150 A1 122
Sekondi-Takoradi, Ghana, 26,757 (*120,793) F4 45
Selah, Yakima, Wash., 2,824 C5 122
Selangor, state, Mala., 1,012,929 K4 38
Selanovtsi, Bul., 7,259 D7 22
Selaru, isl., Indon G8 35
Selatan, cape, Indon F4 35
Selatpandjang, Indon L5 38
Selawik, Alsk., 348 B7 79
Selawik, lake, Alsk B7 79
Selb, Ger., 19,200 C7 17
Selbu, Nor F4 25
Selby, Eng., 9,869 A6 12
Selby, Walworth, S. Dak., 979 B5 116
Selbysport, Garrett, Md D1 85
Selbyville, Sussex, Del., 1,080 D7 85
Selbyville, Upshur, W. Va., 65 C4 123
Selde, Den., 411 B3 24
Selden, Sheridan, Kans., 347 C3 93
Selden, Aroostook, Maine, 25 C5 96
Seldovia, Alsk., 460 D9, h16 79
Selenge, riv., Mong B5 34
Selenicë, Alb., 3,182 B2 23
Selenter, lake, Ger D4 24
Sélestat, Fr., 13,818 C7 14
Seletytengiz, lake, Sov. Un C8 29
Selfridge, Sioux, N. Dak., 371 C5 110
Selibaby, Maur., 1,100 C2 45
Selidovka, Sov. Un., 10,000 q20 27
Seligenstadt, Ger., 9,600 C3 17
Seliger, lake, Sov. Un C9 27
Seligman, Yavapai, Ariz., 900 B3 80
Seligman, Barry, Mo., 387 E4 101
Selima (Oasis), Sud A2 47
Selinsgrove, Snyder, Pa., 3,948 E8 114
Selinunte, ruins, It F4 21
Selizharovo, Sov. Un., 5,000 C9 27
Selkirk, Man., Can., 8,576 D3 71
Selkirk, Ont., Can., 375 E5 72
Selkirk, Wichita, Kans., 50 D2 93
Selkirk, Scot., 5,634 E6 13
Selkirk, co., Scot., 21,055 E5 13
Selkirk, mts., Idaho A2 89
Selkirk, mts., B.C., Can D9 68
Selleck, King, Wash., 100 B4, D2 122
Seller, lake, Man., Can B4 71
Sellers, Montgomery, Ala., 150 C3 78
Sellers, Dillon and Marion, S.C., 431 C9 115
Sellersburg, Clark, Ind., 2,679 H6 91
Sellersville, Bucks, Pa., 2,497 F11 114
Sells, Pima, Ariz., 750 F4 80
Selm, Ger., 14,200 B2 17
Selma, Dallas, Ala., 28,385 C2 78
Selma, Drew, Ark., 300 D4 81
Selma, Fresno, Calif., 6,934 D4 82
Selma, Delaware, Ind., 562 D7 91
Selma, Grant, La C3 95
Selma, Liberty, Mont., 5 B5 102
Selma, Johnston, N.C., 3,102 B5 109
Selma, Josephine, Oreg., 30 E3 113
Selmah, N.S., Can., 135 D6 74
Selman, Harper, Okla., 60 A2 112
Selmer, McNairy, Tenn., 1,897 B3 117
Selsey, Eng., 3,889 D7 12
Selsey Bill, pt., Eng D7 12
Selukwe, S. Rh., 1,200 A5 49
Selva, Arg E3 55
Selvas, forests, Braz D4 53
Selvin, Warrick, Ind., 150 H3 91
Selway, riv., Idaho C3 89
Selwyn, Austl D7 50
Selwyn, lake, N.W. Ter., Can D12 66
Selwyn, mtn., B.C., Can B6 68
Selwyn, range, Yukon, Can D6 66
Selz, Pierce, N. Dak., 150 B6 110
Seman, Elmore, Ala., 70 G3 78
Seman, riv., Alb B2 23
Semans, Sask., Can., 386 F3 70
Semarang, Indon., 487,006 G4 35
Semenov, Sov. Un., 98,000 D17 9
Semenovka, Sov. Un., 9,000 E9 27
Seminary, Covington, Miss., 288 D4 100
Seminary Hill, Tarrant, Tex. (part of Fort Worth) B5 118
Seminoe, mts., Wyo C6 125
Seminoe, res., Wyo D6 125
Seminoe Dam, Carbon, Wyo., 55 C6 125
Seminole, Baldwin, Ala., 200 E2 78
Seminole, Pinellas, Fla., 1,500 *E4 86
Seminole, Seminole, Okla., 11,464 B5 112
Seminole, Gaines, Tex., 5,737 C1 118
Seminole, co., Fla., 54,947 D5 86
Seminole, co., Ga., 6,802 F2 87
Seminole, co., Okla., 28,066 B5 112
Seminole, Indian res., Fla F6 86
Seminole (Big Cypress), Indian res., Fla F5 86
Seminole (Brighton), Indian res., Fla E5 86
Semipalatinsk, Sov. Un., 177,000 C10 29
Semitau, Indon E4 35
Semiyarskoye, Sov. Un C9 29
Semliki, riv., Con. L A4 48
Semmens, lake, Man., Can B4 71
Semmering, pass, Aus E7 16
Semnān, Iran, 23,078 D6 41
Semora, Caswell, N.C., 350 A4 109
Semoy, riv., Bel E5 15
Sempach, Switz., 1,345 B5 19
Sempacher, lake, Switz B5 19
Semur-en-Auxois, Fr., 3,399 D6 14
Sen, riv., Camb F6 38
Sena, San Miguel, N. Mex., 45 B4 107
Senaca, La Salle, Ill., 1,719 B5 90
Senachwine, lake, Ill B4 90
Senador Pompeu, Braz., 8,210 C3 57
Sena Madureira, Braz., 1,962 C4 58
Senanga, N. Rh., 2,785 E3 48
Senate, Sask., Can., 75 H1 70
Senath, Dunklin, Mo., 1,369 E7 101
Senatobia, Tate, Miss., 3,259 A4 100
Sendai, Jap., 425,272 (*515,000) G10 37
Sendai (Kagoshima pref.), Jap., 25,600 K5 37
Sendai, bay, Jap G10 37
Seneca, Gila, Ariz., 50 D5 80
Seneca, La Salle, Ill., 1,719 B5 90
Seneca, Nemaha, Kans., 2,072 C7 93
Seneca, Newton, Mo., 1,478 E3 101
Seneca, Thomas, Nebr., 160 B5 103
Seneca, Union, N. Mex., 15 A6 107
Seneca, Grant, Oreg., 400 C8 113
Seneca, Venango, Pa., 950 D2 114
Seneca, Oconee, S.C., 5,227 B2 115
Seneca, Faulk, S. Dak., 161 B6 116
Seneca, Crawford, Wis., 180 E3 124
Seneca, co., N.Y., 31,984 C4 108
Seneca, co., Ohio, 59,326 A4 111
Seneca, caverns, W. Va C5 123
Seneca, lake, N.Y C4 108
Seneca, rocks, W. Va C5 123
Seneca Falls, Seneca, N.Y., 7,439 C4 108
Seneca Gardens, Jefferson, Ky., 928 *H6 94
Senecaville, Guernsey, Ohio, 576 C6 111
Senecaville, res., Ohio C6 111
Seneffe, Bel., 2,971 D4 15
Senegal, country, Afr., 2,973,000 D2 45, E4 42
Sénégal, riv., Maur., Sen C2 45

Senekal, S. Afr., 7,409 ...... C4 49
Seney, Schoolcraft, Mich., 80 . B5 98
Senftenberg, Ger., 21,000 ... B9 17
Senga Hill, N. Rh. C5 48
Senguerr, riv., Arg D3 54
Senhor do Bonfim, Braz., 13,958 .... D2 57
Senhoshi, Jap D10 37
Senigallia, It., 16,700 ...... C4 21
Senj, Yugo., 3,909 .... C2 22
Senja, isl., Nor C7 25
Senlac, Sask., Can., 127 ..... E1 70
Senlis, Fr., 9,371 .... C5 14
Sennār, Sud., 8,093 .... C3 47
Senneterre, Que., Can., 3,246 .... B2 73
Senneville, Que., Can., 1,262 .... *D8 73
Senoia, Coweta, Ga., 782 .... C2 87
Senorita, Sandoval, N. Mex. . B3 107
Sens, Fr., 20,015 .... C5 14
Sense, riv., Switz C3 19
Senta, Yugo., 24,987 .... C5 22
Sentery, Con. L C4 48
Sentinel, Maricopa, Ariz., 20 . E2 80
Sentinel, Washita, Okla., 1,154 .... B2 112
Sentinel, butte, N. Dak C2 110
Sentinel Butte, Golden Valley, N. Dak., 160 .... C2 110
Senzu, Jap n17 37
Seo de Urgel, Sp., 7,195 ..... A6 20
Seoni, India, 30,274 .... F7 40
Seoul (Sŏul), Kor., 2,444,900 (*2,600,000) H3 37
Separ, Grant, N. Mex., 44 ... E1 107
Sepik, riv., N. Gui h11 50
Sepolno, Pol., 4,214 .... B4 26
Sept Iles (Seven Islands), Que., Can., 14,196 .... B3 73
Sepulga, riv., Ala D3 78
Sequatchie, Marion, Tenn., 400 .... D8 117
Sequatchie, co., Tenn., 5,915 .... D8 117
Sequatchie, riv., Tenn D7 117
Sequim, Clallam, Wash., 1,164 .... A2 122
Sequoia, nat. park, Calif D4 82
Sequoia National Park, Tulare, Calif D4 82
Sequoyah, co., Okla., 18,001 .... B7 112
Serafimovich, Sov. Un., 8,800 .... D2 29
Seraing, Bel., 41,239 .... D5 15
Serakhs, Sov. Un C10 41
Serang, Indon., 11,163 ...... G3 35
Serasan, is., Indon K8 38
Serbia, reg., Yugo D5 22
Serbia, rep., Yugo., 7,629,113 .... *D5 22
Serdobsk, Sov. Un., 10,000 . E15 27
Sered, Czech., 6,208 .... D4 26
Sereflikochisar, Tur., 8,700 ... C9 31
Seremban, Mala., 52,091 ..... K4 38
Serengeti, plain, Tan C3 48
Serenje, N. Rh., 510 .... D5 48
Serenli, Som E5 47
Sergeant, McKean, Pa., 120 . C4 114
Sergeant Bluff, Woodbury, Iowa, 813 .... B1 92
Sergeantsville, Hunterdon, N.J., 165 .... C3 106
Sergipe, state, Braz., 760,273 . D3 57
Seria, Mala., 17,595 .... E4 35
Serifos, Grc., 2,372 .... D5 23
Serifos, isl., Grc D5 23
Seringapatam, India, 11,423 . F6 39
Serles, Hardeman, Tenn., 25 . B2 117
Sermaize, Fr., 2,964 .... F4 15
Sernyy Zavod, Sov. Un B9 41
Seroei, W. Irian, 2,200 ...... F9 35
Serón, Sp., 7,091 .... D4 20
Serov, Sov. Un., 102,000 ..... B6 29
Serowe, Bech., 15,935 .... B4 49
Serpa, Port., 7,273 .... D2 20
Serpentine, lakes, Austl E4 50
Serpentine, mts., N.B., Can B3 74
Serpukhov, Sov. Un., 111,000 .... C1 29
Serra dos Aimorés, disputed reg., Braz., 384,297 ...... *E2 57
Serrai, Grc., 40,063 .... B4 23
Serrai, prov., Grc., 248,041 . *B4 23
Serrana Bank, shoals, Caribbean Sea C7 62
Serra Negra, Braz., 5,221 .... m8 56
Serranilla Bank, shoals, Caribbean Sea C8 62
Serra Talhada, Braz., 5,353 .. C3 57
Serres, Fr., 770 .... E1 18
Serrezuela, Arg A3 54
Serrinha, Braz., 10,284 ...... D3 57
Sêrro, Braz., 4,594 .... E2 57
Sertã, Port., 7,281 .... C1 20
Sertânia, Braz., 7,556 .... C3 57
Serua, isl., Indon G8 35
Seruli, Bech B4 49
Servia, Grc., 3,236 .... B4 23
Servia, Wabash, Ind., 150 ... C6 91
Service, Choctaw, Ala., 100 . D1 78
Service, buttes, Oreg B7 113
Service Creek, Wheeler, Oreg C7 113
Sese, isl., Ug B5 48
Sesheke, N. Rh., 124 .... E3 48
Sésia, riv., It B2 21
Sesikinaga, lake, Ont., Can D5 71
Sesoko, isl., Okinawa 52
Sesoko, Okinawa 52
Sesser, Franklin, Ill., 1,764 .. E4 90
Sessums, Oktibbeha, Miss., 100 .... B5 100
Sesto [Fiorentino], It., 22,453 . C3 21
Sestokai, Sov. Un A7 26
Sesto San Giovanni, It., 71,384 .... B2 21
Sestriere, It., 73 .... E2 18
Sestri Levante, It., 9,100 ..... B2 21
Sestroretsk, Sov. Un., 34,000 (part of Leningrad) A8 27
Setana, Jap., 4,100 .... E9 37
Setauket, Suffolk, N.Y., 1,207 .... F4 84
Sète (Cette), Fr., 36,301 ..... F5 14
Sete Lagoas, Braz., 36,302 .... E2 57
Sete Quedas, falls, Braz C2 56
Seth, Boone, W. Va., 800 .... C3, D6 123
Seth Ward, Hale, Tex., 1,328 .... *B2 118
Sétif, Alg., 53,057 .... B6 44
Seto, Jap., 82,101 .... I8, n16 37
Seton Portage, B.C., Can., 107 .... D6 68
Settat, Mor., 29,617 .... C3 44
Setté-Cama, Gabon F1 46
Settee, lake, Sask., Can B4 72
Setting, lake, Man., Can B2 71
Settle, Eng., 2,297 .... F6 13
Setúbal, Port., 44,600 .... C1 20
Setúbal, bay, Port C1 20
Seul, lake, Ont., Can F14 66
Seul Choix, pt., Mich C5 98
Seut, lake, Ont., Can D6 72
Sevan, lake, Sov. Un G18 9
Sevastopol, Sov. Un., 163,000 . I9 27
Seven, heads, Ire F3 11
Seven Devils, mts., Idaho D2 89
Seven Harbors, Oakland, Mich., 2,748 .... F7 98
Seven Hills, Cuyahoga, Ohio, 5,708 .... *A6 111
Seven Islands, see Sept Iles, Que., Can.
Seven Mile, Butler, Ohio, 690 .... C3 111
Seven Mile, beach, N.J E3 106
Sevenoaks, Eng., 17,604 ..... C8 12
Seven Persons, Alta., Can., 76 .... E5 69
Seven Sisters, Duval, Tex., 75 .... E3 118
Seven Sisters, mtn., B.C., Can .. B3 68
Seven Springs, Wayne, N.C., 207 .... B6 109
70 Mile House, B.C., Can., 512 .... D7 68
Severance, Weld, Colo., 70 . A6 83
Severance, Doniphan, Kans., 146 .... C8 93
Severka, riv., Sov. Un n18 27
Severn, Anne Arundel, Md., 280 .... B4 85
Severn, Northampton, N.C., 310 .... A6 109
Severn, Gloucester, Va., 300 D6 121
Severn, mouth, Eng C5 12
Severn, riv., Ont., Can D8 72
Severn, riv., Eng B5 12
Severn, riv., Md B4 85
Severna Park, Anne Arundel, Md., 3,100 .... B4 85
Severnaya Zemlya, reg., Sov. Un B13 28
Severomorsk, Sov. Un., 30,000 .... C15 25
Severouralsk, Sov. Un., 23,200 .... C20 9
Severy, Greenwood, Kans., 492 .... E7 93
Sevier, Sevier, Utah, 10 ..... E3 119
Sevier, co., Ark., 10,156 ..... D1 81
Sevier, co., Tenn., 24,251 ... D10 117
Sevier, co., Utah, 10,565 .... E4 119
Sevier, des., Utah D3 119
Sevier, lake, Utah E2 119
Sevier, riv., Utah D3 119
Sevierville, Sevier, Tenn., 2,890 .... D10 117
Sevilla, Col., 17,210 .... C2 60
Sevilla (Seville), Sp., 442,300 .... D3 20
Sevilla (Seville), prov., Sp., 1,234,435 .... *D3 20
Seville, Volusia, Fla., 623 .... C5 86
Seville, Wilcox, Ga., 179 ..... E3 87
Seville, Medina, Ohio, 1,190 . A6 111
Sevlievo, Bul., 9,856 .... D7 22
Sevogle, riv., N.B., Can B3 74
Sevran, Fr., 17,969 .... g11 14
Sèvre, riv., Fr D3 14
Sèvre Niortaise, riv., Fr D3 14
Sèvres, Fr., 20,129 .... g9 14
Sewal, Wayne, Iowa, 100 .... D4 92
Sewanee, Franklin, Tenn., 1,464 .... B6 117
Seward, Alsk., 1,891 ... C10, g17 79
Seward, Winnebago, Ill., 150 . A4 90
Seward, Stafford, Kans., 92 . D5 93
Seward, Seward, Nebr., 4,208 .... D8 103
Seward, Logan, Okla., 49 ... B4 112
Seward, Westmoreland, Pa., 754 .... F3 114
Seward, co., Kans., 15,930 ... E3 93
Seward, co., Nebr., 13,581 ... D8 103
Seward, pen., Alsk B7 79
Seward Roads, chan., Midway Is 52
Sewaren, Middlesex, N.J., 1,500 .... E4 106
Sewell, Chile, 9,009 .... A2 54
Sewell, Breathitt, Ky., 265 ... C6 94
Sewell, Glouchester, N.J., 900 .... D2 106
Sewickley, Allegheny, Pa., 6,157 .... A5, E1 114
Sewickley Heights, Allegheny, Pa., 931 .... *E1 114
Sexsmith, Alta., Can., 531 .... B1 69
Sextonville, Richland, Wis., 250 .... E3 124
Seychelles, Br. dep., Afr., 34,632 .... G24 3
Seychelles, is., Indian O G24 3
Seydisehir, Tur., 6,300 .... D8 31
Seydisfjördur, Ice., 742 ...... n25 25
Seyhan (Sarus), riv., Tur C10 31
Seym, riv., Sov. Un F9 27
Seymchan, Sov. Un C17 28
Seymour, Austl., 5,103 .... H5 51
Seymour, New Haven, Conn., 10,100 .... D4 84
Seymour, Jackson, Ind., 11,629 .... G6 91
Seymour, Wayne, Iowa, 1,117 .... D4 92
Seymour, Webster, Mo., 1,046 .... D5 101
Seymour, Sevier, Tenn., 40 .... D10, E12 117
Seymour, Baylor, Tex., 3,789 . C3 118
Seymour, Outagamie, Wis., 2,045 .... A5, D5 124
Seymour, inlet, B.C., Can D4 68
Seymour, lake, Vt B4 120
Seymour, range, B.C., Can B8 68
Seymour, riv., B.C., Can D4 68
Seymourville, Iberville, La., 1,788 .... B5 95
Sézanne, Fr., 5,300 .... C5 14
Sezimbra, Port., 6,957 .... C1 20
Sezze, It., 7,544 .... D4 21
Sfantul-Gheorghe, Rom., 17,638 .... C7 22
Sfax, Tun., 65,645 . C7 44, G12 30
Sfax, prov., Tun., 338,268 .. *C7 44
Sgarbhbreae, mtn., Scot F2 13
's Gravenhage, see The Hague, Neth.
Sgurr Mor, mtn., Scot C3 13
Sha, China K7 36
Shabalē, riv., Eth D5 47
Shabani, S. Rh., 10,800 ..... B5 49
Shabbona, DeKalb, Ill., 690 . B5 90
Shabrakhit, Eg., U.A.R., 1,000 .... C2 32
Shabunda, Con. L B4 48
Shabwah, Aden B6 47
Shackelford, co., Tex., 3,990 .. C3 118
Shackleton, Sask., Can., 96 . G1 70
Shackleton, ice shelf, Ant C22 5
Shade, riv., Ohio C5 111
Shadehill, Perkins, S. Dak., 20 .... B3 116
Shadehill, dam, S. Dak B3 116
Shadehill, res., S. Dak B3 116
Shades, creek, Ala E4 78
Shades, mtn., Ala E4 78
Shadeville, Franklin, Ohio, 250 .... C2 111
Shadrinsk, Sov. Un., 59,000 . B6 29
Shady Cove, Jackson, Oreg., 875 .... E4 113
Shady Dale, Jasper, Ga., 201 . C3 87
Shady Grove, Pike, Ala., 100 .... D3 78
Shady Grove, Taylor, Fla., 300 .... B3 86
Shady Grove, Crittenden, Ky., 50 .... A3, C2 94
Shadygrove, Franklin, Pa., 500 .... G6 114
Shadypoint, LeFlore, Okla., 300 .... B7 112
Shady Side, Anne Arundel, Md., 749 .... C4 85
Shadyside, Belmont, Ohio, 5,028 .... C7 111
Shady Spring, Raleigh, W. Va., 850 .... D3 123
Shady Valley, Johnson, Tenn., 50 .... C12 117
Shafer, lake, Ind C4 91
Shaft (William Penn), Schuylkill, Pa., 850 .... *E9 114
Shafter, Kern, Calif., 4,576 .. E4 82
Shafter, Elko, Nev., 10 ...... C7 104
Shafter, Presidio, Tex., 975 .. F2 118
Shaftesbury, Eng., 3,366 .... D5 12
Shag, rocks, Atl. O J7 6
Shagamu, Nig., 30,099 ..... *E5 45
Shageluk, Alsk., 155 .... C8 79
Shag Harbour, N.S., Can., 249 .... F4 74
Shagwong, pt., N.Y E9 84
Shah, riv., Iran C5 41
Shahdād (Khabīs), Iran, 15,000 .... F8 41
Shahdadkot, Pak., 8,994 ..... D1 40
Shah Fuladi, mtn., Afg D13 41
Shahgarh, India D2 40
Shaḥḥāt, Libya, 4,149 .... C4 43
Shahi, isl., Iran C2 41
Shahjahanpur, India, 110,432 (*117,702) C6 38, D7 40
Shahjui, Afg., 5,000 .... E13 41
Shahpur, India, 11,776 ...... I6 40
Shāhpūr (Dīlmān), Iran, 13,161 .... B2 41
Shahpur, Pak C4 39
Shahpura, India, 12,165 ..... E5 40
Shahr-e Bābak, Iran, 10,000 . F7 41
Shahreẓā, Iran, 23,980 ...... E6 41
Shāhrūd, Iran, 23,132 .... C7 41
Shahsavār, Iran, 5,046 .... C5 41
Shaib al Qur, wadi, Sau. Ar ... G13 31
Shaib Hub, riv., Iraq G14 31
Shaikh Shuaib, isl., Iran H6 41
Shailerville, Middlesex, Conn., 230 .... D7 84
Shaker Heights, Cuyahoga, Ohio, 36,460 .... A6, B2 111
Shakhrisyabz, Sov. Un., 15,000 .... B13 41
Shakhty, Sov. Un., 201,000 . H13 27
Shakhunya, Sov. Un B3 29
Shaki, Nig., 22,983 .... E5 45
Shakopee, Scott, Minn., 5,201 .... F5 99
Shakotan, cape, Jap E10 37
Shaktoolik, Alsk., 187 .... C7 79
Shalalth, B.C., Can., 182 .... D6 68
Shalimar, Okaloosa, Fla., 754 .... G2 86
Shallmar, Garrett, Md., 100 . D1 85
Shallotte, Brunswick, N.C., 480 .... D5 109
Shallotte, inlet, N.C D5 109
Shallow Lake, Ont., Can., 340 .... C3 72
Shallow Water, Scott, Kans., 85 .... D3 93
Shallowater, Lubbock, Tex., 1,001 .... C2 118
Shambat, Sud., 6,611 .... B3 47
Shambaugh, Page, Iowa, 206 . D2 92
Shambe, Sud D3 47
Shamīl, Iran, 5,000 .... H8 41
Shamokin, Northumberland, Pa., 13,674 .... E8 114
Shamokin Dam, Snyder, Pa., 1,093 .... E8 114
Shamrock, Sask., Can., 126 .. G2 70
Shamrock, Dixie, Fla., 600 ... C3 86
Shamrock, Natchitoches, La., 70 .... C2 95
Shamrock, Creek, Okla., 211 . B5 112
Shamrock, Wheeler, Tex., 3,113 .... B2 118
Shamva, S. Rh A5 49
Shandaken, Ulster, N.Y., 450 .... C6 108
Shandon, San Luis Obispo, Calif., 500 .... E3 82
Shandon, Butler, Ohio, 300 . C2 111
Shanee, Park, Colo., 100 .... B5 83
Shang, China, 6,000 .... H3 36
Shangchiu, China, 134,400 ... G6 36
Shanghai, China, 6,900,000 (9,500,000*) E9 34, I9 36
Shangjao, China, 50,000 ..... J8 36
Shangnan, China H4 36
Shangssu, China A7 38
Shangtu, China D5 36
Shaniko, Wasco, Oreg., 39 ... B6 113
Shannock, Washington, R.I., 500 .... D10 84
Shannon, Floyd, Ga., 1,629 ... B1 87
Shannon, Carroll, Ill., 766 ... A4 90
Shannon, Lee, Miss., 554 .... A5 100
Shannon, Clay, Tex., 75 ..... C3 118
Shannon, co., Mo., 7,087 .... D6 101
Shannon, co., S. Dak., 6,000 . D3 116
Shannon, airport, Ire D3 11
Shannon, isl., Grnld B16 4
Shannon, lake, Wash A4 122
Shannon, riv., Ire E2 11
Shannon, riv. mouth, Ire E2 11
Shannon City, Union and Ringgold, Iowa, 127 ...... D3 92
Shannontown, Sumter, S.C., 7,064 .... *D7 115
Shanshan (Pichan), China ... C3 34
Shansi, prov., China, 15,960,000 .... D7 34
Shantar, is., Sov. Un D16 28
Shantung, prov., China, 54,030,000 .... D8 34
Shantung, pen., China .. D9 34, F9 36
Shanwa, Tan B5 48
Shaohsing, China, 130,600 .... E9 34, I9 36
Shaopo, China H8 36
Shaowu, China, 12,000 ...... K7 36
Shaoyang, China, 117,700 .... F7 34, K4 36
Shap, Eng., 1,152 .... F6 13
Shapinsay, isl., Scot A6 13
Shapio, lake, Newf., Can g9 75
Shapki, Sov. Un s32 25
Sharafkhāneh, Iran, 1,260 ... B2 41
Sharangad, Mong B2 36
Sharasume (Chenghwa), China, 25,000 .... B2 34
Sharbot Lake, Ont., Can., 481 .... C8 72
Shari, Jap., 8,100 .... E12 37
Sharita, cape, Mus. & Om H8 41
Shark, bay, Austl E1 50
Shark, pt., Fla G5 86
Sharkey, co., Miss., 10,738 ... C3 100
Sharkh, Mus. & Om D2 39
Sharon, Litchfield, Conn., 800 (2,141*) B3 84
Sharon, Taliaferro, Ga., 264 .... C4 87
Sharon, Barber, Kans., 272 .. E5 93
Sharon, Norfolk, Mass., 10,070 .... B5, E2 97
Sharon, Madison, Miss., 50 .. C4 100
Sharon, Hillsboro, N.H., 50 (78*) E3 105
Sharon, Schoharie, N.Y C6 108
Sharon, Steele, N. Dak., 251 . B8 110
Sharon, Woodward, Okla., 97 .... A2 112
Sharon, Mercer, Pa., 25,267 . D1 114
Sharon, York, S.C., 280 ...... B5 115
Sharon, Weakley, Tenn., 966 . A3 117
Sharon, Windsor, Vt., 155 (485*) D4 120
Sharon, Spokane, Wash D7 122
Sharon, Kanawha, W. Va., 100 .... D6 123
Sharon, Walworth, Wis., 1,167 .... F5 124
Sharon Grove, Todd, Ky., 100 .... D2 94
Sharon Hill, Delaware, Pa., 7,123 .... G11 114
Sharon Springs, Wallace, Kans., 966 .... D2 93
Sharon Springs, Schoharie, N.Y., 351 .... C6 108
Sharon Valley, Litchfield, Conn. (part of Sharon) B3 84
Sharonville, Hamilton, Ohio, 6,457 .... D2 111
Sharp, Burlington, N.J C3 106
Sharp, co., Ark., 6,319 ...... A4 81
Sharpe, lake, Man., Can B5 71
Sharpes, Brevard, Fla., 700 . D6 86
Sharples, Logan, W. Va., 40 .... D3, D5 123
Sharpsburg, Taylor, Iowa, 130 .... D3 92
Sharpsburg, Bath, Ky., 311 ... B6 94
Sharpsburg, Washington, Md., 861 .... B2 85
Sharpsburg, Nash, Edgecombe and Wilson, N.C., 490 ...... B5 109
Sharpsburg, Allegheny, Pa., 6,096 .... A5 114
Sharps Chapel, Union, Tenn., 25 .... C10 117
Sharpsville, Tipton, Ind., 663 .... D5 91
Sharpsville, Mercer, Pa., 6,067 .... D1 114
Sharp Top, mtn., Ark C2 81
Sharptown, Wicomico, Md., 620 .... C6 85
Sharptown, Salem, N.J., 220 .... D2 106
Sharya, Sov. Un., 21,700 .... A3 29
Shāshamani, Eth D4 47
Shashi, riv., Bech., S. Rh B4 49
Shashke, Sov. Un C7 29
Shasi, China, 85,800 .... E7 34
Shasta, co., Calif., 59,468 .... B3 82
Shasta, lake, Calif B2 82
Shasta, mtn., Calif B2 82
Shatsk, Sov. Un., 10,000 .... D13 27
Shatney, mtn., N.H A1 105
Shatra, Iraq, 9,543 .... F3 41
Shattuck, Ellis, Okla., 1,625 .. A2 112
Shattuckville, Franklin, Mass., 150 .... A2 97
Shatura, Sov. Un., 50,000 ... D12 27
Shauck, Morrow, Ohio, 200 .. B5 111
Shaunavon, Sask., Can., 2,154 .... H1 70
Shavano, mtn., Colo C4 83
Shaver, mtn., W. Va C5 123
Shavers, fork, W. Va C5 123
Shavertown, Delaware, N.Y .. C6 108
Shavertown, Luzerne, Pa., 2,000 .... *B8 114
Shaw, Lincoln, Colo B7 83
Shaw, St. Louis, Minn., 50 ... C6 99
Shaw, Bolivar, Miss., 2,062 ... B3 100
Shaw, Marion, Oreg., 75 ..... C2 113
Shaw, Mineral, W. Va., 200 .. B5 123
Shawā, reg., Eth., 2,100,000 .. D4 47
Shawanaga, Ont., Can., 75 ... B4 72
Shawangunk, mts., N.Y D6 108
Shawano, Shawano, Wis., 6,103 .... D5 124
Shawano, co., Wis., 34,351 ... D5 124
Shawano, lake, Wis D5 124
Shawatun, China D9 36
Shawboro, Currituck, N.C., 60 .... A7 109
Shawbridge, Que., Can., 1,034 .... D3 73
Shawhan, Bourbon, Ky., 250 . B5 94
Shawinigan, Que., Can., 32,169 (*63,518) C5 73
Shawinigan Est, Que., Can., 2,451 .... *C5 73
Shawmut, Chambers, Ala., 1,898 .... C4 78
Shawmut, Pike, Ark C2 81
Shawmut, Somerset, Maine, 225 .... D3 96
Shawmut, Wheatland, Mont D7 102
Shawnee, Johnson, Kans., 9,072 .... B8 93
Shawnee, Perry, Ohio, 1,000 .... C5 111
Shawnee, Pottawatomie, Okla., 24,326 .... B5 112
Shawnee, Converse, Wyo., 18 .... C7 125
Shawnee, co., Kans., 141,286 .... D8 93
Shawneetown, Gallatin, Ill., 1,399 .... F5 90
Shawnigan Lake, B.C., Can., 224 .... B9 68
Shawsheen, riv., Mass C2 97
Shawsheen Village, Essex, Mass., 3,000 .... A5 97
Shawsville, Hartford, Md., 250 .... A4 85
Shawsville, Montgomery, Va., 300 .... D2 121
Shawver Mill, Tazewell, Va B3 121
Shawville, Que., Can., 1,534 . B8 72
Shayang, China I5 36
Shchekino, Sov. Un., 30,000 .... D11 27
Shchelkovo, Sov. Un., 65,000 .... n18 27
Shcherbakov, see Rybinsk, Sov. Un.
Shchetovo, Sov. Un., 10,000 . q22 27
Shchigry, Sov. Un., 10,000 .. F11 27
Shchors, Sov. Un., 10,000 .... F8 27
Shchuchinsk, Sov. Un C8 29
Shchurovo, Sov. Un., 15,000 . n18 27
Shearer Dale, B.C., Can A7 68
Sheaville, Malheur, Oreg D9 113
Sheboygan, Sheboygan, Wis., 45,747 .... B6, E6 124
Sheboygan, co., Wis., 86,484 .... E6 124
Sheboygan, riv., Wis B5 124
Sheboygan Falls, Sheboygan, Wis., 4,061 .... B6, E6 124
Shebshi, mts., Nig E7 45
Shechichen, China H5 36
Shedd, Linn, Oreg., 150 ..... C3 113
Shedden, Ont., Can., 295 .... E3 72
Shediac, N.B., Can., 2,159 ... C5 74
Sheelin, lake, Ire D4 11
Sheenjek, riv., Alsk B11 79
Sheep, mtn., Ariz E1 80
Sheep, mtn., Wyo B2 125
Sheep, mtn., Wyo B3 125
Sheep, peak, Nev G6 104
Sheep, range, Nev G6 104
Sheep Creek, B.C., Can., 56 .... E9 68
Sheep Haven, bay, Ire B4 11
Sheep Springs, San Juan, N. Mex., 10 .... A1 107
Sheerness, Alta., Can., 93 ... D5 69
Sheerness, Eng., 14,123 ..... C8 12
Sheet Harbour, N.S., Can., 883 .... E7 74
Sheffield, Colbert, Ala., 13,491 .... A2 78
Sheffield, Eng., 493,954 (*725,000) A6 12
Sheffield, Bureau, Ill., 1,078 . B4 90
Sheffield, Franklin, Iowa, 1,156 .... B4 92
Sheffield, Berkshire, Mass., 700 (2,138*) B1 97
Sheffield, N.Z., 151 .... O14 51
Sheffield, Lorain, Ohio, 1,664 .... A5 111
Sheffield, Warren, Pa., 1,971 . C3 114
Sheffield, Pecos, Tex., 350 ... D2 118
Sheffield, Caledonia, Vt., 225 (342*) B4 120
Sheffield, lake, Newf., Can D3 75
Sheffield Lake, Lorain, Ohio, 6,884 .... *A5 111
Shefford, co., Que., Can., 54,963 .... D5 73
Shefford, mtn., Que., Can D5 73
Sheguiandah, Ont., Can., 116 .... B3 72
Shebo, Sask., Can., 391 ..... F4 70
Shehsien, China J8 36
Shehy, mts., Ire F2 11
Sheikh, Som D6 47
Shekar Dzong, China, 5,000 . C11 40
Shelbiana, Pike, Ky., 500 .... C7 94
Shelbina, Shelby, Mo., 2,067 . B5 101
Shelburn, Sullivan, Ind., 1,299 .... F3 91
Shelburn, Linn, Oreg., 40 .... C2, C4 113
Shelburne, Ont., Can., 1,239 . C4 72
Shelburne, N.S., Can., 2,408 . F4 74
Shelburne, Franklin, Mass., 100 (1,739*) A2 97
Shelburne, Coos, N.H., 50 (226*) B4 105
Shelburne, Chittenden, Vt., 250 (1,805*) C2 120
Shelburne, co., N.S., Can., 15,208 .... F4 74
Shelburne, pond, Vt C2 120
Shelburne Falls, Franklin, Mass., 2,097 .... A2 97
Shelby, Shelby, Ala., 600 .... B3 78
Shelby, Lake, Ind., 500 ...... B3 91
Shelby, Shelby, Iowa, 533 ... C2 92
Shelby, Oceana, Mich., 1,603 .... E4 98
Shelby, Bolivar, Miss., 2,384 . B3 100
Shelby, Toole, Mont., 4,017 .. B5 102
Shelby, Polk, Nebr., 613 ..... C8 103
Shelby, Cleveland, N.C., 17,698 .... B2 109
Shelby, Richland, Ohio, 9,106 .... B5 111
Shelby, co., Ala., 32,132 ..... B3 78
Shelby, co., Ill., 23,404 ...... D5 90
Shelby, co., Ind., 34,093 ..... E6 91
Shelby, co., Iowa, 15,825 .... C2 92
Shelby, co., Ky., 18,493 ..... B4 94
Shelby, co., Mo., 9,063 ..... B5 101
Shelby, co., Ohio, 33,586 .... B3 111
Shelby, co., Tenn., 627,019 .. B2 117
Shelby, co., Tex., 20,479 .... D5 118
Shelby City, Boyle, Ky., 500 . C5 94
Shelby Village, Macomb, Mich., 1,900 .... *F7 98
Shelbyville, Shelby, Ill., 4,821 .... D5 90
Shelbyville, Shelby, Ind., 14,317 .... E6 91
Shelbyville, Shelby, Ky., 4,525 .... B4 94
Shelbyville, Shelby, Mo., 657 . B5 101
Shelbyville, Bedford, Tenn., 10,466 .... B5 117
Sheldahl, Boone, Polk, and Story, Iowa, 279 .... A7 92
Sheldon, Iroquois, Ill., 1,137 . C6 90
Sheldon, O'Brien, Iowa, 4,251 .... A2 92
Sheldon, Vernon, Mo., 434 ... D3 101
Sheldon, Ransom, N. Dak., 221 .... C8 110
Sheldon, Beaufort, S.C., 200 . F6 115
Sheldon, Harris, Tex., 100 ... F5 118
Sheldon, Franklin, Vt., 300 (1,281*) B3 120
Sheldon, Rusk, Wis., 240 .... C3 124
Sheldon Springs, Franklin, Vt., 250 .... B3 120
Sheldonville, Norfolk, Mass., 200 .... B5 97
Shelekhov, gulf, Sov. Un C18 28
Shelikof, strait, Alsk D9 79
Shell, creek, Wyo A5 125
Shell, lake, Minn D3 99
Shell, lake, Wis C1 124
Shell, riv., Man., Can D1 71
Shell Beach, San Luis Obispo, Calif., 1,820 .... *E3 82
Shell Beach, St. Bernard, La., 125 .... E6 95
Shellbrook, Sask., Can., 1,042 .... D2 70
Shell Camp, Gregg, Tex., 500 .... *C5 118
Shell Creek, Carter, Tenn., 400 .... C11 117
Shell Creek, range, Nev D7 104
Shelley, Bingham, Idaho, 2,612 .... F6 89
Shell Lake, Sask., Can., 241 . D2 70
Shell Lake, Washburn, Wis., 1,016 .... C2 124
Shellman, Randolph, Ga., 1,050 .... E2 87
Shellman Bluff, McIntosh, Ga., 150 .... E5 87
Shellmouth, Man., Can., 98 . D1 71
Shell Rock, Butler, Iowa, 1,112 .... B5 92
Shellrock, riv., Iowa B4, B5 92
Shellsburg, Benton, Iowa, 625 .... B6 92
Shelly, Norman, Minn., 310 . C2 99
Shelter, isl., N.Y E7 84
Shelter Bay, Que., Can., 1,171 .... h8 75
Shelter Island, Suffolk, N.Y., 800 .... D4 108
Shelter Island Heights, Suffolk, N.Y., 600 .... E7 84
Shelton, Fairfield, Conn., 18,190 .... D4 84
Shelton, Buffalo, Nebr., 904 . D7 103
Shelton, Fairfield, S.C., 150 .. C5 115
Shelton, Mason, Wash., 5,651 .... B2 122
Shemogue, N.B., Can., 93 .... C5 74
Shenandoah, Page, Iowa, 6,567 .... D2 92
Shenandoah, Schuylkill, Pa., 11,073 .... E9 114
Shenandoah, Page, Va., 1,839 .... C4 121
Shenandoah, co., Va., 21,825 .... C4 121
Shenandoah, mtn., Va C3 121
Shenandoah, nat. park, Va C4 121
Shenandoah, riv., Va B5 121
Shenandoah, val., Va C4 121
Shenandoah Heights, Schuylkill, Pa., 1,721 .... *E9 114
Shenandoah Tower, mtn., Va., W. Va C3 121, C5 123
Shenchiu, China H6 36
Shendi, Sud., 11,031 .... B3 47
Shenipsit, lake, Conn B7 84
Shenmu, China, 10,000 ...... D7 34
Shenorock, Westchester, N.Y., 1,402 .... *D7 108
Shensi, prov., China, 18,130,000 .... E6 34
Shentsa Dzong, China, 5,000 .... B12 40
Sheopur, India, 14,591 ...... E6 40
Shepard, Alta., Can., 66 .... D3 69
Shepardsville, Vigo, Ind., 350 .... E3 91
Shepaug, riv., Conn C3 84
Shepetovka, Sov. Un., 10,000 .... F6 27
Shepherd, Isabella, Mich., 1,293 .... E6 98
Shepherd, Yellowstone, Mont., 100 .... E8 102
Shepherd, San Jacinto, Tex., 800 .... D5 118
Shepherd Brook, mtn., Maine ... B3 96
Shepherdstown, Jefferson, W. Va., 1,328 .... B7 123
Shepherdsville, Bullitt, Ky., 1,525 .... A4, C4 94
Shepp, Haywood, Tenn., 40 . B2 117
Sheppards, Buckingham, Va . D4 121
Shepparton, Austl., 13,579 ... H5 51
Sheppey, isl., Eng C8 12
Sheppton, Schuylkill, Pa., 800 .... E9 114
Shepton Mallet, Eng., 5,518 .... C5 12
Sherard, Coahoma, Miss., 60 .... A3 100
Sherborn, Middlesex, Mass., 500 (1,806*) D2 97
Sherborne, Eng., 6,062 ...... D5 12
Sherbrooke, Que., Can., 66, 554(*70,253) C2, D6 73
Sherbrooke, N.S., Can., 384 . D7 74
Sherbrooke, co., Que., Can., 80,490 .... D5 73
Sherbrooke, lake, N.S., Can E5 74
Sherburn, Martin, Minn., 1,227 .... G4 99
Sherburne, Chenango, N.Y., 1,647 .... C5 108
Sherburne, co., Minn., 12,861 .... E5 99
Sherburne Center, Rutland, Vt., 60 (266*) D3 120
Shercock, Ire., 254 .... D5 11
Shereik, Sud B3 47
Shereshevo, Sov. Un., 10,000 . B8 26
Shergarh, India D4 40
Sheridan, Grant, Ark., 1,938 .... C3 81
Sheridan, Arapahoe, Colo., 3,559 .... B6 83
Sheridan, La Salle, Ill., 704 . B5 90
Sheridan, Hamilton, Ind., 2,165 .... D5 91
Sheridan, Crittenden, Ky., 60 .... A3 94
Sheridan, Aroostook, Maine, 350 .... B4 96
Sheridan, Montcalm, Mich., 606 .... E5 98
Sheridan, Worth, Mo., 277 ... A3 101
Sheridan, Madison, Mont., 539 .... E4 102

Sheridan, Yamhill, Oreg., 1,763 ... B3 113
Sheridan, Sheridan, Wyo., 11,651 ... A6 125
Sheridan, co., Kans., 4,267 ... C3 93
Sheridan, co., Mont., 6,458 ... B12 102
Sheridan, co., Nebr., 9,049 ... B3 103
Sheridan, co., N. Dak., 4,350 ... B5 104
Sheridan, co., Wyo., 18,989 ... A5 125
Sheridan, mtn., Wyo ... A2 125
Sheridan Beach, King, Wash., 1,500 ... *B3 122
Sheridan Lake, Kiowa, Colo., 90 ... C8 83
Sheringham, Eng., 4,836 ... B9 12
Sherkaly, Sov. Un ... C22 9
Sherman, Fairfield, Conn., 250 (825▲) ... C3 84
Sherman, Aroostook, Maine, 100 (1,034▲) ... C4 96
Sherman, Pontotoc and Union, Miss., 403 ... A5 100
Sherman, St. Louis, Mo., 300 ... A7 101
Sherman, Grant, N. Mex., 10 ... E2 107
Sherman, Chautauqua, N.Y., 873 ... C1 108
Sherman, Summit, Ohio, 1,000 ... *B6 111
Sherman, Major, Okla ... A3 112
Sherman, Minnehaha, S. Dak., 116 ... D9 116
Sherman, Grayson, Tex., 24,988 ... C4 118
Sherman, co., Kans., 6,682 ... C2 93
Sherman, co., Nebr., 5,382 ... C6 103
Sherman, co., Oreg., 2,446 ... B6 113
Sherman, co., Tex., 2,605 ... A2 118
Sherman, mtn., Ark ... A2 81
Sherman, res., Nebr ... C7 103
Sherman Mills, Aroostook, Maine, 450 ... C4 96
Sherman Station, Penobscot, Maine, 375 ... C4 96
Sherpur, Pak., 19,312 ... E12 40
Sherrard, Mercer, Ill., 574 ... B3 90
Sherridon, Man., Can., 88 ... B1 71
Sherrill, Jefferson, Ark., 241 ... C4 81
Sherrill, Oneida, N.Y., 2,922 ... B5 108
Sherrodsville, Carroll, Ohio, 480 ... B6 111
's Hertogenbosch, Neth., 76,700 (★116,000) ... C5 15
Sherwood, Pulaski, Ark., 1,222 ... D5 81
Sherwood, Choctaw, Miss., 65 ... B5 100
Sherwood, Renville, N. Dak., 360 ... A4 110
Sherwood, Defiance, Ohio, 578 ... A3 111
Sherwood, McCurtain, Okla., 100 ... C7 112
Sherwood, Washington, Oreg., 680 ... B2 113
Sherwood, Franklin, Tenn., 650 ... B6 117
Sherwood, Irion, Tex., 200 ... D2 118
Sherwood, Calumet, Wis., 300 ... A5 124
Sheshebee, Aitkin, Minn., 100 ... D5 99
Shetek, lake, Minn ... F3 99
Shetland, co., Scot., 17,809 ... *g10 10
Shetland, is., Scot ... g10 10
Shetucket, riv., Conn ... C8 84
Shevlin, Clearwater, Minn., 203 ... C3 99
Shevlin, Klamath, Oreg ... D5 113
Sheyenne, Eddy, N. Dak., 423 ... B6 110
Sheyenne, riv., N. Dak ... C8 110
Shfaram, Isr., 5,029 ... B7 32
Shfayim, Isr., 713 ... g10 32
Shiant, isl., Scot ... C2 13
Shiawassee, co., Mich., 53,446 ... F6 98
Shibarghan (Shibargan), Afg., 22,464 ... C12 41
Shibetsu, Jap., 12,750 ... D11 37
Shibīn al Kawm, Eg., U.A.R., 47,100 ... D3 32
Shībin al Qanāṭir, Eg., U.A.R., 11,610 ... D3 32
Shichito, isl., Pac. O ... D8 7
Shickley, Fillmore, Nebr., 371 ... D8 103
Shickshinny, Luzerne, Pa., 1,843 ... D9 114
Shideler, Delaware, Ind., 240 ... D7 91
Shiderty, riv., Sov. Un ... C8 29
Shidler, Osage, Okla., 870 ... A5 112
Shieldaig, Scot ... C3 13
Shields, Lane, Kans., 50 ... D3 93
Shields, Harlan, Ky., 900 ... *D6 94
Shields, Grant, N. Dak., 100 ... C4 110
Shiga, pref., Jap., 842,695 ... *I8 37
Shigawake, Que., Can., 125 ... B4 73
Shihchiachuang, China, 126,000 ... E6 36
Shihchuan, China, 5,000 ... H3 36
Shihkiachwang, China, 598,000 ... D7 34
Shihmen, China, 5,000 ... J4 36
Shihshou, China, 19,000 ... J5 36
Shihtaokuo, China ... F10 36
Shikarpur, Pak., 45,376 ... C4 38, D2 40
Shikoku, isl., Jap ... J6 37
Shikuka, see Poronaysk, Sov. Un.
Shilka, Sov. Un., 23,000 ... D14 28
Shilla, peak, India ... A7 40
Shillington, Berks, Pa., 5,639 ... F10 114
Shillong, India, 72,438 ... C9 39
Shiloh, Marengo, Ala., 100 ... C2 78
Shiloh, Cleburne, Ark., 6 ... B3 81
Shiloh, Harris, Ga., 250 ... D2 87
Shiloh, St. Clair, Ill., 701 ... B9 101
Shiloh, Cumberland, N.J., 554 ... E2 106
Shiloh, Camden, N.C. (part of Asheville) ... A7 109
Shiloh, Montgomery, Ohio, 9,500 ... C3 111
Shiloh, Richland, Ohio, 724 ... B5 111
Shiloh, York, Pa., 1,500 ... *G8 114
Shiloh, Montgomery, Tenn., 40 ... A4 117
Shiloh, nat. military park and cemetery, Tenn ... B3 117
Shimabara, Jap., 28,200 ... J5 37
Shimada, Jap., 36,500 ... o17 37
Shimane, pref., Jap., 888,886 ... *I6 37
Shimanovsk, Sov. Un., 17,000 ... D15 28
Shimizu, Jap., 142,983 ... I9, n17 37
Shimo, isl., Jap ... J5 37
Shimoda, Jap., 16,200 ... o17 37
Shimodate, Jap., 23,000 ... m18 37
Shimoga, India, 63,764 ... F6 39
Shimonoseki, Jap., 246,941 ... I5 37
Shimotsuma, Jap., 9,700 ... m18 37
Shin, lake, Scot ... B4 13
Shinall, mtn., Ark ... D5 81
Shiner, Lavaca, Tex., 1,945 ... E4 118
Shinewell, McCurtain, Okla., 50 ... D7 112
Shingbwiyang, Bur., 5,000 ... C10 39
Shinglehouse, Potter, Pa., 1,298 ... C5 114
Shingleton, Alger, Mich., 450 ... B4 98
Shingu, Jap., 39,114 ... J7 37
Shinjo, Jap., 22,100 ... G10 37
Shinkolobwe, Con. L., 10,900 ... D4 48
Shinnecock, bay, N.Y ... F6 84
Shinnston, Harrison, W. Va., 2,724 ... A7, B4 123
Shin Pond, Penobscot, Maine, 40 ... B4 96
Shinshiro, Jap., 14,100 ... o16 37
Shinyanga, Tan., 2,907 ... B5 48
Shio, cape, Jap ... J7 37
Shiocton, Outagamie, Wis., 685 ... A5, D5 124
Shiogama, Jap., 55,325 ... G10 37
Shiojiri, Jap., 13,500 ... m16 37
Shioya, cape, Jap ... H10 37
Ship, isl., Miss ... E5 100
Ship Bottom, Ocean, N.J., 717 ... D4 106
Ship Cove, Newf., Can., 66 ... E4 75
Ship Harbour, N.S., Can., 212 ... E7 74
Shipiskan, lake, Newf., Can ... g9 75
Ship Island, pass., Miss ... E2 100
Shipka, pass, Bul ... D7 22
Shipki, pass, India ... B7 40
Shipman, Sask., Can., 57 ... D3 70
Shipman, Macoupin, Ill., 417 ... D3 90
Shipman, Nelson, Va., 500 ... D4 121
Shippegan, N.B., Can., 1,631 ... B5 74
Shippegan, isl., N.B., Can ... B5 74
Shippensburg, Cumberland and Franklin, Pa., 6,138 ... F6 114
Shippenville, Clarion, Pa., 599 ... D3 114
Shiprock, San Juan, N. Mex., 900 ... A1 107
Shipshewana, Lagrange, Ind., 312 ... A6 91
Shirabad, Sov. Un ... C13 41
Shiratori, Jap., 5,819 ... n15 37
Shīrāz, Iran, 170,659 ... G6 41
Shirbīn, Eg., U.A.R., 13,293 ... C8 31
Shire, riv., Nya ... A5 49
Shiremanstown, Cumberland, Pa., 1,212 ... *F8 114
Shire Nor, China ... E3 34
Shiretoko, cape, Jap ... D12 37
Shireza, Pak ... C4 39
Shiriya, cape, Jap ... C4 39
Shir Kuh, mtn., Iran ... F7 41
Shirley, Van Buren, Ark., 197 ... B3 81
Shirley, McLean, Ill., 130 ... C4 90
Shirley, Hancock and Henry, Ind., 1,038 ... E6 91
Shirley, Middlesex, Mass., 1,762 ... A4 97
Shirley, Custer, Mont., 3 ... D11 102
Shirley, Salem, N.J. ... D2 106
Shirley, Suffolk, N.Y., 1,800 ... *D4 108
Shirley, Tyler, W. Va., 150 ... A6 123
Shirley, Carbon, Wyo ... C6 125
Shirley, basin, Wyo ... C6 125
Shirley Center, Middlesex, Mass., 150 ... A4 97
Shirley Mills, Piscataquis, Maine, 200 (214▲) ... C3 96
Shirleysburg, Huntingdon, Pa., 170 ... F6 114
Shishaldin, vol., Alsk ... E7 79
Shishido, Jap., 11,018 ... m19 37
Shishmaref, Alsk., 217 ... B6 79
Shively, Humboldt, Calif., 100 ... B2 82
Shively, Jefferson, Ky., 15,155 ... A4, B4 94
Shivers, Simpson, Miss., 10 ... D4 100
Shivpuri, India, 28,681 ... E6 40
Shivwits, Washington, Utah, 40 ... F2 119
Shivwits, Indian res., Utah ... F2 119
Shizuoka, Jap., 323,819 (★485,000) ... I9, o17 37
Shizuoka, pref., Jap., 2,756,271 ... *I9 37
Shkodër (Scutari), Alb., 40,900 ... A2 23
Shkodër (Scutari), pref., Alb., 150,000 ... *A2 23
Shkotovo, Sov. Un ... E6 37
Shoal, creek, Tenn ... B4 117
Shoal, lake, Ont., Can ... B6 99
Shoal, lake, Ont., Can ... E4 71
Shoal, lakes, Man., Can ... D3 71
Shoal, riv., Man., Can ... C1 71
Shoal Harbour, Newf., Can., 544 ... D5 75
Shoal Lake, Man., Can., 751 ... D1 71
Shoals, Martin, Ind., 1,022 ... G4 91
Shoalwater, cape, Wash ... C1 122
Shobankazgan, Sov. Un ... E6 29
Shobonier, Fayette, Ill., 200 ... E4 90
Shoe, pt., Newf., Can ... D5 75
Shoe Cove, Newf., Can., 152 ... D4 75
Shoeheel, creek, S.C. ... B9 115
Shoemakersville, Berks, Pa., 1,464 ... E10 114
Shola, lake, Eth ... D4 47
Sholapur, India, 337,583 ... E6 39, I5 40
Sholes, Wayne, Nebr., 26 ... B8 103
Shona, isl., Scot ... D2 13
Shongopovi, Navajo, Ariz., 120 ... B5 80
Shonkin, Chouteau, Mont., 11 ... C6 102
Shonto, Navajo, Ariz., 15 ... A5 80
Shooks, Beltrami, Minn., 25 ... C4 99
Shooting Creek, Clay, N.C., 50 ... D2 109
Shop Spring, Wilson, Tenn., 175 ... A5 117
Shore Acres, Contra Costa, Calif., 3,093 ... *B5 82
Shoreacres, B.C., Can., 80 ... E9 68
Shore Acres, Plymouth, Mass., 980 ... *B6 97
Shoreham, Berrien, Mich., 443 ... F4 98
Shoreham, Suffolk, N.Y., 164 ... F5 84
Shoreham, Addison, Vt., 130 (786▲) ... D2 120
Shoreview, Ramsey, Minn., 7,157 ... *F5 99
Shorewood, Hennepin, Minn., 3,197 ... *F5 99
Shorewood, Milwaukee, Wis., 15,990 ... E2, E6 124
Shorewood Hills, Dane, Wis., 2,320 ... *E4 124
Short, Sequoyah, Okla ... B7 112
Short, creek, Ohio ... B1 123
Short, mtn., Tenn ... B6 117
Short, mtn., Tenn ... C10 117
Short Beach, New Haven, Conn., 950 ... D5 84
Short Creek, Mohave, Ariz., 200 ... A3 80
Short Creek, Brooke, W. Va., 400 ... B2 123
Shorter, Macon, Ala., 200 ... C4 78
Shorterville, Henry, Ala., 300 ... D4 78
Short Falls, Merrimack, N.H., 100 ... D4 105
Shorts Creek, Carroll, Va., 50 ... E2 121
Shortsville, Ontario, N.Y., 1,382 ... C3 108
Shoshone, Garfield, Colo., 15 ... B3 83
Shoshone, Lincoln, Idaho, 1,416 ... G4 89
Shoshone, co., Idaho, 20,876 ... B3 89
Shoshone, basin, Wyo ... B4 125
Shoshone, falls, Idaho ... G4 89
Shoshone, lake, Wyo ... A2 125
Shoshone, mtn., Nev ... G5 104
Shoshone, mts., Nev ... E4 104
Shoshone, peak, Nev ... D4 104
Shoshone, riv., Wyo ... A4 125
Shoshone Cavern, nat. mon., Wyo ... A3 125
Shoshoni, Fremont, Wyo., 766 ... B4 125
Shostka, Sov. Un., 30,000 ... F9 27
Shou, China ... H7 36
Shouldice, Alta., Can ... D4 69
Shouns, Johnson, Tenn., 250 ... C12 117
Shoup, Lemhi, Idaho, 10 ... D4 89
Showak, Sud., 2,171 ... C4 47
Showell, Worcester, Md., 200 ... D7 85
Show Low, Navajo, Ariz., 1,625 ... C5 80
Shpola, Sov. Un., 10,000 ... G8 27
Shreve, Wayne, Ohio, 1,617 ... B5 111
Shreveport, Caddo, La., 164,372 (★245,200) ... B2 95
Shrewsbury, Eng., 49,726 ... B5 12
Shrewsbury, Worcester, Mass., 16,622 ... B4 97
Shrewsbury, St. Louis, Mo., 4,730 ... *C7 101
Shrewsbury, Monmouth, N.J., 3,222 ... C4 106
Shrewsbury, York, Pa., 943 ... G8 114
Shrewsbury, riv., N.J ... C5 106
Shropshire, co., Eng., 297,313 ... B5 12
Shrub Oak, Westchester, N.Y., 1,874 ... *D7 108
Shrule, Ire., 250 ... D2 11
Shuangcheng, China, 81,000 ... B10 34, D3 37
Shuangchiang, China ... G4 34
Shuangshan, China, 5,000 ... E1 37
Shuangyang, China, 5,000 ... E2 37
Shuangyashan, China, 50,000 ... B11 34
Shubenacadie, N.S., Can., 579 ... D6 74
Shubert, Richardson, Nebr., 231 ... D10 103
Shubrā Khit, Eg., U.A.R., 1,000 ... C2 32
Shubuta, Clarke, Miss., 718 ... D5 100
Shucheng, China ... I7 36
Shufat, Jordan, 2,000 ... h11 32
Shuford, Panola, Miss., 60 ... A4 100
Shujabad, Pak., 14,602 ... C3 40
Shuksan, mtn., Wash ... A4 122
Shulan, China, 5,000 ... D3 37
Shulaps, peak, B.C., Can ... D6 68
Shulerville, Berkeley, S.C., 250 ... E8 115
Shullsburg, Lafayette, Wis., 1,324 ... F3 124
Shumagin, isl., Alsk ... D7 79
Shuman House, Alsk., 20 ... B11 79
Shumaykh, Libya ... I13 30
Shumen, co., Bul., 1,408,188 ... *D8 22
Shumerlya, Sov. Un., 26,800 ... B3 29
Shumikha, Sov. Un ... B6 29
Shunan, China, 6,000 ... J8 36
Shunat Nimrin, Jordan, 1,000 ... h13 32
Shunchang, China, 4,000 ... K7 36
Shungnak, Alsk., 135 ... B8 79
Shunk, Sullivan, Pa., 65 ... C8 114
Shunner Fell, mtn., Eng ... F6 13
Shunning, China ... G4 34
Shuo, China ... E5 36
Shuqrā', Aden Prot ... C6 47
Shuqualak, Noxubee, Miss., 550 ... C5 100
Shur, riv., Iran ... D5 41
Shur, riv., Iran ... F7 41
Shur, riv., Iran ... F10 41
Shur, riv., Iran ... H7 41
Shūrāb, Iran ... G10 41
Shūrāb, Iran ... G10 41
Shuri, Okinawa, 17,537 ... 52
Sūsh, Iran ... E4 41
Shushan, Washington, N.Y., 275 ... B7 108
Shushong, Bech ... B4 49
Shūshtar, Iran, 23,654 ... E4 41
Shusht el Maghara, mtn., Eg ... D5 32
Shuswap, B.C., Can., 125 ... D8 68
Shuswap, lake, B.C., Can ... D8 68
Shuswap, riv., B.C., Can ... D8 68
Shutesbury, Franklin, Mass., 150 (2,650▲) ... B3 97
Shuwaykah, Jordan, 3,000 ... f11 32
Shuya, Sov. Un., 67,000 ... B2 29
Shuzenji, Jap., 7,800 ... o17 37
Shwangliao (Chengchiatun), China, 120,100 ... C9 34
Shwebo, Bur., 17,842 ... D10 39
Shwegyin, Bur., 10,000 ... D2 38
Si, riv., China ... G7 34
Siah Band, mtn., Afg ... E11 41
Siahan, range, Iran, Pak ... H11 41
Sialkot, Pak., 135,000 (★164,346) ... B5 39, A5 40
Sialum, N. Gui ... k12 50
Siam, see Thailand, country, Asia
Siam, gulf, Asia ... G4 38
Sian (Hsian), China, 1,310,000 ... E6 34
Siangtan, China, 183,600 ... F7 34, K5 36
Siantan, isl., Indon ... K7 38
Siasconset, Nantucket, Mass., 200 ... D8 97
Siatista, Grc., 4,969 ... B3 23
Siaton, Phil., 2,364 ... *D6 35
Siau, isl., Indon ... E7 35
Siauliai, Sov. Un., 65,000 ... D4 27
Sibay, Sov. Un ... C5 29
Sibbald, Alta., Can., 75 ... D5 69
Sibenik, Yugo., 26,253 ... D2 22
Sibert, Clay, Ky., 700 ... C6 94
Siberut, isl., Indon ... F1 35
Sibi, Pak., 11,842 ... C4 39
Sibiti, Con. B ... F2 46
Sibiu, Rom., 90,478 ... C7 22
Sibley, Ford, Ill., 386 ... C5 90
Sibley, Osceola, Iowa, 2,852 ... A2 92
Sibley, Webster, La., 595 ... B2 95
Sibley, Adams, Miss., 50 ... D2 100
Sibley, Jackson, Mo., 177 ... E3 101
Sibley, co., Minn., 16,228 ... F4 99
Sibolga, Indon., 37,171 ... L3 38
Sibsagar, India, 15,106 ... C9 39
Sibu, Mala., 29,630 ... E4 35
Sibuatan, mtn., Indon ... K3 38
Sibuko, Phil., 2,136 ... *D6 35
Sibutu, isl., Phil ... E5 35
Sibuyan, sea, Phil ... C6 35
Sicamous, B.C., Can., 588 ... D8 68
Sicapoo, mtn., Phil ... B6 35
Sicard, Ouachita, La., 2,000 ... *B3 95
Sichomovi, Navajo, Ariz., 100 ... B5 80
Sicilia, pol. dist., It., 4,721,001 ... *F4 21
Sicily, isl., It ... F4 21
Sicily Island, Catahoula, La., 761 ... C4 95
Sicklerville, Camden, N.J., 350 ... D3 106
Sico, riv., Hond ... C5 62
Sicuaní, Peru, 7,036 ... D3 58
Sidādah, Libya ... C2 43
Sidāmo, reg., Eth., 1,250,000 ... E4 47
Sideling, hill, Md., W. Va ... A1 85
Sideling Hill, creek, Md ... A1 85
Sidell, Vermilion, Ill., 614 ... D6 90
Siderno Marina, It., 6,915 ... E6 21
Sidewood, Sask., Can., 58 ... G1 70
Sidheros, cape, Grc ... E6 23
Sidhirokastron, Grc., 9,022 ... B4 23
Sidi Abdalkah Ben Ali, Alg ... D5 44
Sidi Abd el Hakem, Alg ... D5 44
Sīdī Barrānī, Eg., U.A.R., 1,000 ... G6 31
Sidi-bel-Abbès, Alg., 80,632 ... B4 44
Sidi Bennour, Mor., 5,479 ... C3 44, H2 30
Sidi bou Haous, Alg ... H8 30
Sidi Hadjed Dine, Alg ... C5 44
Sidi Ifni, Ifni, 7,991 ... D2 44
Sidikalang, Indon ... m11 35
Sīdī Sālim, Eg., U.A.R., 1,000 ... C2 32
Sidlaw, hills, Scot ... D5 13
Sidley, mtn., Ant ... B36 5
Sidmouth, Eng., 11,139 ... D4 12
Sidnaw, Houghton, Mich., 200 ... B2 98
Sidney, Sharp, Ark., 97 ... A4 81
Sidney, B.C., Can., 1,558 ... B9, E6 68
Sidney, Man., Can., 154 ... E2 71
Sidney, Champaign, Ill., 686 ... C5 90
Sidney, Kosciusko, Ind., 208 ... B6 91
Sidney, Fremont, Iowa, 1,057 ... D2 92
Sidney, Kennebec, Maine, 50 (988▲) ... D3 96
Sidney, Richland, Mont., 4,564 ... C12 102
Sidney, Cheyenne, Nebr., 8,004 ... C3 103
Sidney, Delaware, N.Y., 5,157 ... C5 108
Sidney, Shelby, Ohio, 14,663 ... B3 111
Sidney, lake, Ont., Can ... D4 71
Sidney Center, Delaware, N.Y., 500 ... C5 108
Sidney Lanier, lake, Ga ... B3 87
Sidon, White, Ark., 90 ... B4 81
Sidon, Leflore, Miss., 410 ... B3 100
Sidonia, Weakley, Tenn., 120 ... A3 117
Sidorovka, Sov. Un., 10,000 ... D25 9
Sidra (Khalij Surt), gulf, Libya ... C3 43
Siedlce, Pol., 32,000 ... B7 26
Sieg, riv., Ger ... C3 16
Siegburg, Ger., 34,000 ... C2 17
Siegen, Ger., 49,400 (★120,000) ... C3 17
Sieglar, Ger., 19,500 ... C2 17
Siemianowice Slaskie, Pol., 62,000 ... g10 26
Siemiatycze, Pol., 4,106 ... B7 26
Siem Reap, Camb., 10,000 ... F5 38
Siena, It., 47,000 (61,453▲) ... C3 21
Sieper, Rapides, La., 150 ... C3 95
Sieradz, Pol., 11,700 ... C5 26
Siero, Sp., 30,931 ... A3 20
Sierpc, Pol., 10,200 ... B5 26
Sierra, co., Calif., 2,247 ... C3 82
Sierra, co., N. Mex., 6,409 ... E2 107
Sierra Ancha, mts., Ariz ... D4 80
Sierra Blanca, Hudspeth, Tex., 800 ... F1 118
Sierra Blanca, mtn., N. Mex ... D4 107
Sierra City, Sierra, Calif., 150 ... C3 82
Sierra Colorada, Arg ... C3 54
Sierra Del Carmen, mts., Tex ... E1 118
Sierra Del Pinacate, mtn., Ariz ... F2 80
Sierra Diablo, mts., Tex ... F2 118
Sierra Estrella, mts., Ariz ... G1 80
Sierra Gordo, Chile ... D2 55
Sierra Leone, country, Afr., 2,400,000 ... E2 45, F4 42
Sierra Madre, Los Angeles, Calif., 9,732 ... F2 82
Sierra Madre, mts., Wyo ... D6 125
Sierra Mojada, Mex., 954 ... B4 63
Sierra Nevada, mts., Calif ... D4 82
Sierraville, Sierra, Calif., 150 ... C3 82
Sierra Vista, Cochise, Ariz., 3,121 ... F5 80
Sierre, Switz., 8,690 ... D4 19
Siesta, key, Fla ... E4 86
Siewierz, Pol., 2,385 ... g10 26
Sifnos, isl., Grc ... D5 23
Sifton, Man., Can., 245 ... D1 71
Sigean, Fr., 2,346 ... F5 14
Sigel, Shelby, Ill., 387 ... D5 90
Sigel, Jefferson, Pa., 190 ... D3 114
Sighet, Rom., 22,361 ... B6 22
Sighişoara, Rom., 20,363 ... B7 22
Sighty Crag, mtn., Eng ... E6 13
Sigli, Indon., 3,327 ... J1 38
Siglufjördur, Ice., 2,756 ... m23 25
Sigma, Phil., 1,218 ... *C6 35
Sigmaringen, Ger., 6,578 ... A5 18
Signakhi, Sov. Un., 4,338 ... B15 31
Signal, mtn., Va ... B4 121
Signal Hill, Los Angeles, Calif., 4,627 ... *F2 82
Signal Hill, St. Clair, Ill., 1,200 ... *E3 90
Signal Mountain, Hamilton, Tenn., 3,413 ... D8, E10 117
Signau, Switz., 2,555 ... C4 19
Sigourney, Keokuk, Iowa, 2,387 ... C5 92
Sigsig, Ec., 1,632 ... B2 58
Sigtuna, Swe., 3,200 ... t35 25
Siguatepeque, Hond., 2,618 ... C4 62
Sigüenza, Sp., 4,620 ... B4 20
Siguiri, Guinea, 11,400 ... D3 45
Sigurd, Sevier, Utah, 339 ... E4 119
Siirt, Tur., 22,900 ... D13 31
Sikar, India, 50,636 ... D5 40
Sikasso, Mali, 13,600 ... D3 45
Sikes, Winn, La., 233 ... B3 95
Sikeston, Scott, Mo., 13,765 ... E8 101
Sikhote-Alin, mts., Sov. Un ... D7 37
Sikia, Grc., 2,457 ... B4 23
Sikinos, isl., Grc ... D5 23
Sikionia, Grc., 5,113 ... C4 23
Sikkim, country, Asia, 162,189 ... C8 39, D12 40
Siklos, Hung., 5,905 ... C4 22
Sil, riv., Sp ... A2 20
Silandro, It., 3,958 ... A3 21
Silang, Phil., 4,920 ... o13 35
Silao, Mex., 24,138 ... m13 63
Silas, Choctaw, Ala., 353 ... D1 78
Sīlat adh Dhahr, Jordan, 3,000 ... f11 32
Silchar, India, 41,062 ... D9 39
Sile, Sandoval, N. Mex., 60 ... F5 107
Sile, Tur., 2,700 ... B7 23
Siler City, Chatham, N.C., 4,455 ... B4 109
Silerton, Hardeman, Tenn., 84 ... B3 117
Silesia, Prince Georges, Md., 60 ... C3 85
Silesia, Carbon, Mont., 50 ... E8 102
Silesia, reg., Pol ... C4 26
Siletz, Lincoln, Oreg., 583 ... C3 113
Silex, Lincoln, Mo., 176 ... B6 101
Silgarhi, Doti, Nep., 1,461 ... C8 40
Silhuas, Peru, 1,432 ... C2 58
Silica, Lucas, Ohio, 100 ... A1 111
Silifke, Tur., 9,200 ... D9 31
Silīguri, India, 65,471 ... D12 40
Silistra, Bul., 16,180 ... C8 22
Silivri, Tur., 4,129 ... B7 23
Siljan, Nor ... p27 25
Siljan, lake, Swe ... G6 25
Silkeborg, Den., 24,465 ... B3 24
Sillery, Que., Can., 14,109 ... C9 73
Sillimans Fossil, mtn., N. W. Ter., Can ... D18 67
Silloth, Eng., 3,081 ... F5 13
Sil Nakya, Pima, Ariz., 60 ... E4 80
Siloam, Greene, Ga., 321 ... C3 87
Siloam Springs, Benton, Ark., 3,953 ... A1 81
Siloam Springs, Howell, Mo., 35 ... E5 101
Silsbee, Hardin, Tex., 6,277 ... D5 118
Silsby, lake, Man., Can ... B4 71
Sils im Domleschg, Switz., 737 ... D8 19
Silt, Garfield, Colo., 384 ... B3 83
Silton, Sask., Can., 97 ... G3 70
Siltou, well, Chad ... B3 46
Siluria, Shelby, Ala., 736 ... B3 78
Silute, Sov. Un., 5,000 ... A6 26
Silva, Wayne, Mo., 100 ... D7 101
Silva, Pierce, N. Dak., 56 ... A6 110
Silva Jardim, Braz., 1,774 ... h6 57
Silva Pôrto, Ang., 12,146 ... D2 48
Silver, Clarendon, S.C., 50 ... D7 96
Silver, creek, Ill ... B9 101
Silver, creek, Ind ... H6 91
Silver, creek, Nebr ... D2 103
Silver, creek, Oreg ... D7 113
Silver, lake, Ont., Can ... E4 71
Silver, lake, Iowa ... A3 92
Silver, lake, Maine ... C3 96
Silver, lake, Nev ... B5 104
Silver, lake, N.H. ... C4 105
Silver, lake, N.H. ... E2 105
Silver, lake, Oreg ... D6 113
Silver, lake, Oreg ... D7 113
Silver, lake, Wash ... D7 122
Silver, riv., N.S., Can ... E4 74
Silver Bank, passage, Ba. Is ... E8 64
Silver Bay, Lake, Minn., 3,723 ... C7 99
Silver Bay, Warren, N.Y., 200 ... B7 108
Silver Bell, Pima, Ariz., 700 ... E4 80
Silverbow, Silver Bow, Mont., 30 ... D4 102
Silver Bow, co., Mont., 46,454 ... E4 102
Silver Bow Park, Silver Bow, Mont., 4,798 ... *D4 102
Silver City (Rainbow City), C.Z., 3,688 ... k11 62
Silver City, Owyhee, Idaho ... F2 89
Silver City, Mills, Iowa, 281 ... C2 93
Silver City, Humphreys, Miss., 431 ... B3 100
Silver City, Lyon, Nev., 120 ... D2 104
Silver City, Grant, N. Mex., 6,972 ... E1 107
Silver City, Pennington, S. Dak., 150 ... C2 116
Silver City, Juab, Utah, 16 ... D3 119
Silver Cliff, Custer, Colo., 153 ... C5 83
Silver Creek, Floyd, Ga., 200 ... B1 87
Silver Creek, Lawrence, Miss., 229 ... D3 100
Silver Creek, Merrick, Nebr., 431 ... C8 103
Silver Creek, Chautauqua, N.Y., 3,310 ... C1 108
Silverdale, Cowley, Kans., 50 ... E7 93
Silverdale, Onslow, N.C. ... C6 109
Silverdale, Kitsap, Wash., 950 ... B3, D1 122
Silver Gate, Park, Mont., 10 ... E6 102
Silver Grove, Campbell, Ky., 1,207 ... A7 94
Silverhill, Baldwin, Ala., 417 ... E2 78
Silver Hill, Prince Georges, Md. (part of Suitland) ... C1 85
Silver Lake, Kosciusko, Ind., 514 ... B6 91
Silver Lake, Shawnee, Kans., 392 ... C8 93
Silver Lake, Middlesex, Mass., 4,654 ... C2 97
Silver Lake, McLeod, Minn., 646 ... F4 99
Silver Lake, Carroll, N.H., 150 ... C4 105
Silver Lake, Summit, Ohio, 2,655 ... *A6 111
Silver Lake, Lake, Oreg., 97 ... D5 113
Silverlake, Cowlitz, Wash., 300 ... C3 122
Silver Lake, Kenosha, Wis., 1,077 ... F1, F5 124
Silvermines, Ire., 222 ... E3 11
Silverpeak, Esmeralda, Nev., 45 ... F4 104
Silver Point, Putnam, Tenn., 150 ... C8 117
Silver River, mtn., Newf., Can ... D3 75
Silver Run, Carroll, Md., 125 ... A3 85
Silver Spring, Montgomery, Md., 66,348 ... C1, C3 85
Silver Springs, Marion, Fla., 300 ... C4 86
Silver Springs, Lyon, Nev., 60 ... D2 104
Silver Springs, Wyoming, N.Y., 726 ... C2 108
Silver Star, Madison, Mont., 50 ... E4 102
Silver Star, mtn., Wash ... A4 122
Silverstreet, Newberry, S.C., 181 ... C4 115
Silverthrone, mtn., B.C., Can ... D4 68
Silvertip, mtn., Mont ... C3 102
Silverton, B.C., Can., 285 ... E9 68
Silverton, San Juan, Colo., 822 ... D3 83
Silverton, Shoshone, Idaho, 700 ... B3 89
Silverton, Ocean, N.J., 600 ... C4 106
Silverton, Hamilton, Ohio, 6,682 ... D2 111
Silverton, Marion, Oreg., 3,081 ... B4, C2 113
Silverton, Briscoe, Tex., 1,098 ... B2 118
Silves, Port., 4,361 ... D1 20
Silvia, Col., 2,499 ... C2 60
Silvies, Grant, Oreg ... C8 113
Silvies, riv., Oreg ... D7 113
Silvis, Rock Island, Ill., 3,973 ... B3 90
Silvis Heights, Rock Island, Ill., 1,500 ... *B3 90
Silwān, Jordan, 5,000 ... h11, m14 32
Simanggang, Mala., 5,648 ... E4 35
Simav, Tur., 6,300 ... C7 23
Simav, riv., Tur ... C7 23
Simcoe, Ont., Can., 8,754 ... E4 72
Simcoe, McHenry, N. Dak., 100 ... A5 110
Simcoe, co., Ont., Can., 141,271 ... C5 72
Simcoe, creek, Wash ... C5 122
Simcoe, lake, Ont., Can ... C5 72
Simcoe, mtn., Wash ... C5 122
Simcoe, mts., Wash ... C5 122
Simdega, India, 10,438 ... F10 40
Simeulue, isl., Indon ... K1 38
Simferopol, Sov. Un., 196,000 ... I10 27
Simi, Ventura, Calif., 2,107 ... *E4 82
Simi, isl., Grc ... D6 23
Similkameen, riv., B.C., Can ... E7 68
Simití, Col., 1,742 ... B3 60
Simla, Elbert, Colo., 450 ... B6 83
Simla, India, 42,597 ... B6 39
Simleul-Silvaniei, Rom., 8,560 ... B6 22
Simme, riv., Switz ... C3 19
Simmern, Ger., 5,200 ... D2 17
Simmesport, Avoyelles, La., 2,125 ... D4 95
Simmie, Sask., Can., 136 ... H1 70
Simmons, Texas, Mo., 85 ... D5 101
Simmons, cave, Mo ... D5 101
Simms, Cascade, Mont., 100 ... C5 102
Simnasho, Wasco, Oreg ... C5 113
Simonette, riv., Alta., Can ... B1 69
Simonhouse, lake, Man., Can ... B1 71
Simonstorp, Swe., 238 ... u34 25
Simonsville, Windsor, Vt., 35 ... E3 120
Simoom Sound, B.C., Can., 250 ... D4 68
Simpang Kiri, riv., Indon ... K2 38
Simplíco Mendes, Braz., 1,682 ... C2 57
Simplon, pass, Switz ... D5 19
Simplon, tunnel, Switz., It ... D5 19
Simpson, Sask., Can., 340 ... F3 70
Simpson, Johnson, Ill., 89 ... F5 90
Simpson, Mitchell, Kans., 154 ... C6 93
Simpson, Vernon, La., 400 ... C2 95
Simpson, Olmsted, Minn., 100 ... G6 99
Simpson, Hill, Mont., 5 ... B6 102
Simpson, Pitt., N.C., 302 ... B6 109
Simpson, Lackawanna, Pa., 1,800 ... C11 114
Simpson, Taylor, W. Va., 250 ... B7 123
Simpson, co., Ky., 11,548 ... D3 94
Simpson, co., Miss., 20,454 ... D4 100
Simpson, creek, W. Va ... B7 123
Simpson, des., Austl ... E6 50
Simpson, pen., N.W. Ter., Can ... C15 66
Simpsonville, Shelby, Ky., 220 ... B4 94
Simpsonville, Greenville, S.C., 2,282 ... B3 115
Simrishamn, Swe., 7,300 ... J6 25
Sims, Wayne, Ill., 376 ... E5 90
Sims, Grant, Ind., 225 ... D6 91
Sims, Wilson, N.C., 205 ... B5 109
Sims, stream, N.H. ... A4, B1 105
Simsboro, Lincoln, La., 363 ... B3 95

Simsbury, Hartford, Conn., 2,745 B5 84
Sims Chapel, Washington, Ala., 50 D1 78
Simunjan, Mala., 1,679 E4 35
Sinabang, Indon K2 38
Sinai, Brookings, S. Dak., 166 C8 116
Sinai, pen., Eg., U.A.R. E5 32
Sinaia, Rom., 9,006 C7 22
Sinajana, Guam, 2,861 52
Sinaloa, Mex., 1,284 B3 63
Sinaloa, state, Mex., 790,679 C3 63
Sinanju, Kor., 16,493 G2 37
Sīnarū, Eg., U.A.R., 1,000 E2 32
Sīnāwan, Libya, 609 C2 43
Sinawi, Afg., 10,000 D15 41
Sincé, Col., 7,112 B2 60
Sincelejo, Col., 21,625 B2 60
Sinclair, Carbon, Wyo., 621 D5 125
Sinclair's, bay, Scot B5 12
Sinclair Station, Man., Can., 85 E1 71
Sinclairville, Chautauqua, N.Y., 726 C1 108
Sind, reg., Pak C4 39
Sind, riv., India D7 40
Sindal, Den., 1,400 A4 24
Sindara, Gabon F2 46
Sindelfingen, Ger., 26,100 E3 17
Sindirgi, Tur., 3,209 C7 23
Sinelnikovo, Sov. Un., 10,000 G10 27
Sines, Port., 4,893 D1 20
Singa, Sud., 9,436 C3 47
Singapore, Mala., 925,241 (*1,476,694) L5 38
Singapore, reg., Mala., 1,476,694 I13 33, L5 38
Singapore, strait, Asia L5 38
Singaradja, Indon., 12,345 G5 35
Singen, Ger., 33,300 B4 18
Singer, Beauregard, La., 150 D2 95
Singers Glen, Rockingham, Va., 102 C4 121
Singhampton, Ont., Can., 138 C4 72
Singida, Tan., 3,938 B5 48
Singitic, gulf, Grc B5 23
Singkawang, Indon., 7,127 L8 38
Singkep, isl., Indon F2 35
Singkil, Indon E1, m11 35
Singleton, Austl., 4,523 F8 51
Singu, Bur B1 38
Sinhai, China, 207,600 G8 36
Sinhsien, China D7 34
Sinhung, Kor., 7,583 F3 37
Sining (Hsining), China, 300,000 D5 34
Siniscola, It., 6,559 D2 21
Sinj, Yugo., 4,133 D3 22
Sinjil, Jordan, 2,000 g12 32
Sinkat, Sud., 5,175 B4 47
Sinkiang Uighur, prov., China, 5,640,000 C2 34
Sinking, creek, Ky D3 94
Sinking Springs, Highland, Ohio, 202 C4 111
Sinking Spring, Berks, Pa., 2,244 *F9 114
Sinks Grove, Monroe, W. Va., 100 D4 123
Sinnamahoning, Cameron, Pa., 400 D5 114
Sinnamahoning, creek, Pa D5 114
Sinnamary, Fr. Gu., 1,373 A4 59
Sinnuris, Eg., U.A.R., 23,537 E2 32
Sinoia, S. Rh., 4,200 A5 49
Sinop, Tur., 9,900 A10 31
Sinsiang, China, 170,500 G5 36
Sintaluta, Sask., Can., 376 G4 70
Sint-Amandsberg Bel., 24,359 C3 15
Sintang, Indon., 4,474 E4 35
Sint Jacobiparochie, Neth., 1,199 A5 15
Sint-Lenaarts, Bel., 4,301 C4 15
Sint-Niklaas, Bel., 47,819 C4 15
Sinton, San Patricio, Tex., 6,008 E4 118
Sintra, Port., 7,150 f9 20
Sint-Truiden, Bel., 20,341 D5 15
Sinú, riv., Col B2 60
Sinuiju, Kor., 118,414 F2 37
Sinyavino, Sov. Un s32 25
Sinzig, Ger., 6,100 C2 17
Sion, Switz., 16,051 D3 19
Sioux, co., Iowa, 26,375 A1 92
Sioux, co., Nebr., 2,575 B2 103
Sioux, co., N. Dak., 3,662 C4 110
Sioux Center, Sioux, Iowa, 2,275 A1 92
Sioux City, Woodbury, Iowa, 89,159 (*101,500) B1 92
Sioux Falls, Minnehaha, S. Dak., 65,466 D9 116
Sioux Lookout, Ont., Can., 2,453 E8 72
Sioux Rapids, Buena Vista, Iowa, 962 B2 92
Sipanok, chan., Sask., Can D4 70
Sipes (Midway), Seminole, Fla., 1,500 *D5 86
Sipiwesk, lake, Man., Can B3 71
Siple, mtn., Ant B36 5
Sipolilo, S. Rh A5 49
Sipsey, Walker, Ala., 900 B2 78
Sipsey, riv., Ala B2 78
Sipura, isl., Indon F1 35
Siquirres, C.R., 4,053 E6 62
Siquisique, Ven., 2,354 A4 60
Sir Abu Nuair, isl., Sau. Ar I7 41
Siracusa, It., 89,407 F5 21
Sirājganj, Pak., 37,858 E12 40
Sir Alexander, mtn., B.C., Can C7 68
Sirdar, B.C., Can., 100 E9 68
Sir Douglas, mtn., Alta., B.C., Can D3 69
Sir Edward Pellew Group, is., Austl C6 50
Siren, Burnett, Wis., 679 C1 124
Siret, Rom., 5,664 B8 22
Siretul, riv., Rom B8 22
Sir Francis Drake's, chan., Vir. Is f16 65
Sirhān, wadi, Libya G3 31
Sirik, Iran, 5,000 H8 41
Sirinhaém, Braz., 1,772 k6 57
Sir James McBrien, mtn., N.W. Ter., Can D7 66
Sirmione, It., 712 D6 18
Sirnai, isl., Grc D6 23
Sironj, India, 17,288 E6 40
Siros, Grc., 16,953 D5 23
Siros, isl., Grc D5 23
Sirpur, India, 4,466 H7 40
Sirretta, peak, Calif E4 82
Sirri, isl., Iran I7 41
Sirsa, India, 33,363 C5 40
Sir Sanford, mtn., B.C., Can D9 68
Sirte (Surt), Libya, 890 C3 43
Sirvintos, Sov. Un A8 26
Sir Wilfrid, mtn., Que C2 73
Sir Wilfrid Laurier, mtn., B.C., Can C8 68
Sisak, Yugo., 19,238 C3 22
Sisaket, Thai., 9,778 E6 38
Sisib, lake, Man., Can C2 71
Sisipuk, lake, Man., Sask., Can B5 64, B1 71
Siskiyou, co., Calif., 32,885 B2 82
Siskiyou, gap, Oreg E4 113
Siskiyou, mts., Calif., Oreg F3 113
Sisophon, Camb F5 38
Sisquoc, Santa Barbara, Calif., 40 E3 82
Sissach, Switz., 4,574 B4 19
Sisseton, Roberts, S. Dak., 3,218 B8 116
Sissiboo, riv., N.S., Can E4 74
Sissonville, Kanawha, W. Va., 900 C3 123
Sister Bay, Door, Wis., 520 C6 124
Sisterdale, Kendall, Tex., 50 E3 118
Sisteron, Fr., 3,286 E6 14
Sisters, Deschutes, Oreg., 602 C5 113
Sisterville, Tyler, W. Va., 2,331 B4 123
Sitapur, India, 53,884 D8 40
Site Six, Mohave, Ariz C1 80
Sitia, Grc., 4,393 E6 23
Sítio da Abadia, Braz., 482 A3 56
Sitionuevo, Col., 4,694 A3 60
Sitka, Alsk., 3,237 D12, m22 79
Sitka, Sharp, Ark., 50 A4 81
Sitka, Clark, Kans., 115 E4 93
Sitka, nat. mon., Alsk D12 79
Sitka, sound, Alsk m22 79
Sitkum, Coos, Oreg., 50 D3 113
Sittang, riv., Bur E10 39
Sittard, Neth., 30,700 D5 15
Sitter, riv., Switz A7 19
Siushui, China G25 9
Siutu, W. Sam., 1,284 52
Sivas, Tur., 93,800 C11 31
Siverek, Tur., 26,100 D12 31
Sivrihisar, Tur., 7,200 C8 31
Siwa, oasis, Eg., U.A.R D5 43
Sīwah, Eg., U.A.R., 878 D5 43
Siwalik, range, India, Nep C7 40
Siwana, India E4 40
Siwani, cape, Pak I10 41
Six Mile, Pickens, S.C., 218 B2 115
Sixmile, creek, Fla C6 86
Sixmile, lake, La E4 95
Sixmilecross, N. Ire., 245 C4 11
Sixteen Island Lake, Que., Can., 271 D3 73
6th. Cataract (Nile River), Sud B3 47
Sizerville, Cameron, Pa., 30 C5 114
Sjaelland, isl., Den C5 24
Sjenica, Yugo., 5,499 D5 22
Sjöbo, Swe., 2,419 C7 24
Skaelskör, Den., 2,889 C5 24
Skaerbaek, Den., 1,989 C2 24
Skagen, Den., 10,390 A4 24
Skagen, cape, Den A4 24
Skagerrak, chan., Eur I3 25
Skagit, co., Wash., 51,350 A4 122
Skagit, riv., Wash A4 122
Skagway, Alsk., 659 D12, k22 79
Skälderviken, bay, Swe B6 24
Skamania, Skamania, Wash., 325 D3 122
Skamania, co., Wash., 5,207 D4 122
Skamokawa, Wahkiakum, Wash., 200 C2 122
Skanderborg, Den., 5,482 B3 24
Skanderborg, co., Den., 136,495 C3 24
Skaneateles, Onondaga, N.Y., 2,921 C4 108
Skaneateles, lake, N.Y C4 108
Skanee, Baraga, Mich., 55 B2 98
Skanör, Swe., 900 C6 24
Skantzoura, isl., Grc C5 23
Skara, Swe., 9,100 H5 25
Skaraborg, co., Swe., 249,900 *H6 25
Skarkar, Afg., 5,000 C15 41
Skarven, lake, Swe t35 25
Skawa, riv., Pol h11 26
Skawina, Pol., 3,638 D5 26
Skedee, Pawnee, Okla., 128 A5 112
Skeena, riv., B.C., Can B3 68
Skegness, Eng., 12,843 A8 12
Skelde, Den., 330 D3 24
Skeldon, Br. Gu., 2,654 A3 59
Skellefteå, Swe., 22,700 E9 25
Skellefteälven, riv., Swe E8 25
Skellytown, Carson, Tex., 967 B2 118
Skelton, Raleigh, W. Va., 500 D7 123
Skene, Bolivar, Miss., 200 B3 100
Skerries, Ire., 2,721 D5 11
Skhimatarion, Grc., 1,369 g11 23
Skiathos, isl., Grc C4 23
Skiatook, Tulsa and Osage, Okla., 2,503 A5 112
Skibbereen, Ire., 2,028 F2 11
Skibby, Den., 1,040 C5 24
Skiddaw, mtn., Eng F5 13
Skidegate, B.C., Can., 90 C2 68
Skidegate, inlet, B.C., Can C2 68
Skidmore, Nodaway, Mo., 425 A2 101
Skidmore, Bee, Tex., 550 E4 118
Skien, Nor., 15,500 H3, p27 25
Skierniewice, Pol., 22,000 C6 26
Skiff, Alta., Can., 40 E5 69
Skihist, mtn., B.C., Can D7 68
Skiles, Bertie, N.C B7 109
Skillet, fork, Ill E5 90
Skillman, Somerset, N.J., 50 C3 106
Skime, Roseau, Minn., 5 B3 99
Skipperville, Dale, Ala., 60 D4 78
Skipton, Eng., 12,988 G7 13
Skiros, Grc., 3,395 C5 23
Skiros, isl., Grc C5 23
Skive, riv., Den B3 24
Skjern, Den., 5,349 C2 24
Skjern Å, riv., Den C2 24
Skoczow, Pol., 4,480 h9 26
Sködstrup, Den., 390 B4 24
Skofja Loka, Yugo., 3,367 B2 22
Skokie, Cook, Ill., 65,281 E3 90
Skokomish, Indian res., Wash B2 122
Skokomish, mtn., Wash B2 122
Skole, Sov. Un., 5,000 D7 26
Skönnarbo, Swe u33 25
Skopelos, isl., Grc C4 23
Skopin, Sov. Un., 10,000 E13 27
Skopje (Skoplje), Yugo., 161,983 D5 22
Skörping, Den., 1,461 B3 24
Skotovataya, Sov. Un., 6,400 q20 27
Skotselv, Nor., 183 p27 25
Skövde, Swe., 23,900 H5 25
Skovorodino, Sov. Un., 26,000 D15 28
Skowhegan, Somerset, Maine, 6,667 D3 96
Skowman, Man., Can., 61 D2 71
Skradin, Yugo., 928 D2 22
Skreen, Ire C3 11
Skull Rock, pass, Utah D2 119
Skull Valley, Yavapai, Ariz., 100 C3 80
Skull Valley, Indian res., Utah C3 119
Skuna, riv., Miss B4 100
Skunk, riv., Iowa C4 92
Skurup, Swe., 4,700 C7 24
Skvira, Sov. Un., 10,000 G7 27
Skwentna, riv., Alsk g16 79
Skwierzyna, Pol., 2,822 B3 26
Sky, hill, Mass B1 97
Skye, isl., Scot C2 13
Skykomish, King, Wash., 366 B4 122
Skykomish, riv., Wash B4 122
Skyland, De Kalb, Ga., 2,000 *C2 87
Skyland, Buncombe, N.C., 1,500 D3 109
Skylight, Oldham, Ky., 100 A4 94
Skylight, mtn., Ark B1 81
Skyline, Jackson, Ala., 100 A3 78
Skyring, lake, Chile E2 54
Slade, Powell, Ky., 200 C6 94
Sladesville, Hyde, N.C., 50 B7 109
Slagelse, Den., 20,562 C5 24
Slagle, Vernon, La., 100 C2 95
Slamet, vol., Indon G3 35
Slane, Ire., 421 D5 11
Slaney, riv., Ire E5 11
Slangerup, Den., 1,638 C6 24
Slănic-Prahova, Rom., 6,842 C7 22
Slany, Czech., 12,000 n17 26
Slapy, Czech., 704 o17 26
Slate, Wood, W. Va., 125 B3 123
Slater, Moffat, Colo., 3 A3 83
Slater, Story, Iowa, 717 A7, C4 92
Slater, Saline, Mo., 2,767 B4 101
Slater, Greenville, S.C., 900 A3 105
Slater, Platte, Wyo., 15 D8 121
Slatersville, Providence, R.I., 1,000 B10 84
Slate Run, Lycoming, Pa., 75 D6 114
Slate Spring, Calhoun, Miss., 123 B4 100
Slatina, Rom., 13,381 C7 22
Slatington, Lehigh, Pa., 4,316 E10 114
Slaton, Lubbock, Tex., 6,568 C2 118
Slätthög, Swe A8 24
Slaughter, East Feliciana, La., 403 D4 95
Slaughter Beach, Sussex, Del., 107 C7 85
Slaughters, Webster, Ky., 284 C2 94
Slave, riv., Alta., Can D10 66
Slave Lake, Alta., Can., 468 B3 69
Slavgorod, Sov. Un., 44,000 C9 29
Slavonia, reg., Yugo C3 22
Slavonska Požega, Yugo., 13,112 C3 22
Slavsk, Sov. Un A6 26
Slavyanoserbsk, Sov. Un., 5,000 q21 27
Slavyansk, Sov. Un., 86,000 G11, q20 27
Slavyanskaya, Sov. Un., 83,000 I12 27
Slawkow, Pol g10 26
Slawno, Pol., 4,845 A4 26
Slayden, Marshall, Miss., 100 A4 100
Slayton, Murray, Minn., 2,487 G3 99
Slayton, Golden Valley, Mont., 2 D7 102
Slea, head, Ire E1 11
Sleaford, Eng., 7,834 A7 12
Sleat, pt., Scot C2 13
Sleat, sound, Scot C3 13
Sled, lake, Sask., Can C2 70
Sledge, Quitman, Miss., 440 F3 100
Sleeper, Laclede, Mo., 111 D5 101
Sleeping Bear, pt., Mich D4 98
Sleeping Deer, mtn., Idaho E4 89
Sleepy Creek, Morgan, W. Va., 200 B6 123
Sleepy Eye, Brown, Minn., 3,492 F4 99
Sleepy Hollow, Marin, Calif., 1,200 *B5 82
Sleepy Hollow, Fairfax, Va., 1,200 *C5 121
Sleepy, pt., Fiji Is 52
Sleetmute, Alsk., 122 C8 79
Sliabh Gaoil, mtn., Scot E3 13
Slick, Creek, Okla., 151 B5 112
Slick Rock, San Miguel, Colo C2 83
Slickville, Westmoreland, Pa., 950 F2 114
Slide, mtn., N.Y D6 108
Slidell, St. Tammany, La., 6,356 B8, D6 95
Sliderock, mtn., Mont D3 102
Slieve Aughty, mts., Ire D3 11
Slieve Beagh, mtn., N. Ire C4 11
Slieve Bloom, mts., Ire D4 11
Slieve Callan, mtn., Ire E2 11
Slieve Car, mtn., Ire C2 11
Slieve Croob, mtn., N. Ire C5 11
Slieve Donard, mtn., N. Ire C6 11
Slieve Gamph, mts., Ire C3 11
Slieve Mish, mts., Ire E2 11
Slieve Miskish, mts., Ire F2 11
Slievenaman, mtn., Ire E4 11
Sligo, Weld, Colo A6 83
Sligo, Ire., 13,145 C3 11
Sligo, Clarion, Pa., 814 D3 114
Sligo, co., Ire., 53,561 C3 11
Slinger, Washington, Wis., 1,141 E1, E5 124
Slingerlands, Albany, N.Y., 1,500 *C7 108
Slippery Rock, Butler, Pa., 2,563 D1 114
Sliven, Bul., 35,553 D8 22
Slivenec, Czech., 1,726 n17 26
Sloan, Woodbury, Iowa, 704 B1 92
Sloan, Clark, Nev., 40 H6 104
Sloan, Erie, N.Y., 5,803 *C2 108
Sloans Valley, Pulaski, Ky., 250 D5 94
Sloat, Plumas, Calif., 200 C3 82
Sloatsburg, Rockland, N.Y., 2,565 A4 106
Slobodskoy, Sov. Un., 28,700 B4 29
Slobozia, Rom., 9,632 C8 22
Slocan, B.C., Can., 293 E9 68
Slocan, lake, B.C., Can E9 68
Slocomb, Geneva, Ala., 1,368 D4 78
Slocum, Washington, R.I., 100 C10 84
Slonim, Sov. Un., 10,000 E5 27
Slope, co., N. Dak., 1,893 C2 110
Slough, Eng., 80,503 C7 12
Slovac, Prairie, Ark., 30 C4 81
Slovakia (Slovensko), reg., Czech., 4,175,000 D5 26
Slovaktown, Prairie, Ark C4 81
Slovan, Washington, Pa., 1,018 F1 114
Slovenia, reg., Yugo C2 22
Slovenia, rep., Yugo., 1,584,368 *C2 22
Slubice, Pol., 1,689 A9 17
Sluch, riv., Sov. Un F6 27
Slunj, Yugo., 1,260 C2 22
Slupa, riv., Pol A4 26
Slupca, Pol., 5,133 B4 26
Slupsk, Pol., 53,000 A4 26
Slutsk, Sov. Un., 5,000 s31 25
Slutsk, Sov. Un., 20,000 E6 27
Slyne, head, Ire D1 11
Slyudyanka, Sov. Un., 17,500 D13 28
Smackover, Union, Ark., 2,434 D3 81
Smackover, creek, Ark D2 81
Smaalandsfarvandet, bay, Den C5 24
Smali Anadolu, mts., Tur B11 31
Small, pt., Maine E6 96
Small Point Beach, Sagadahoc, Maine, 25 E6 96
Smara, Sp. Sahara, 395 D2 44
Smarr, Monroe, Ga., 70 D3 87
Smarts, mtn., N.H C2 105
Smartt, Warren, Tenn., 125 D8 117
Smeaton, Sask., Can., 322 D3 70
Smecno, Czech., 2,446 n17 26
Smederevo, Yugo., 27,104 C5 22
Smela, Sov. Un., 45,000 G8 27
Smeltertown, El Paso, Tex. (part of El Paso) E1 118
Smelterville, Shoshone, Idaho, 1,127 B2 89
Smethport, McKean, Pa., 1,725 C5 114
Smethwick, Eng., 68,372 B6 12
Smicksburg, Indiana, Pa., 80 E3 114
Smidovich, Sov. Un., 5,000 B6 37
Smiley, Sask., Can., 232 F1 70
Smiley, Gonzales, Tex., 455 B4, E4 118
Smith, Alta., Can., 86 B3, D1 69
Smith, Lyon, Nev., 35 E2 104
Smith, co., Kans., 7,776 C5 93
Smith, co., Miss., 14,303 C4 100
Smith, co., Tenn., 12,059 C8 117
Smith, co., Tex., 86,350 C5 118
Smith, bay, Alsk A9 79
Smith, cape, Ont., Can B3 72
Smith, isl., Ant C6 5
Smith, isl., Md., Va D5 85, D6 121
Smith, isl., Va D7 121
Smith, peak, Idaho A2 89
Smith, pt., N.S., Can D6 74
Smith, pt., Mass D7 97
Smith, pt., Va D6 121
Smith, res., Ala A2 78
Smith, riv., Mont D5 102
Smith, riv., Va E3 121
Smith, sound, Arc. O B22 4
Smith, sound, B.C., Can D4 68
Smithboro, Bond, Ill., 213 E4 90
Smithburg, Doddridge, W. Va., 200 B4, B6 123
Smith Center, Smith, Kans., 2,379 C5 93
Smith Creek, Wakulla, Fla., 40 B2 86
Smithdale, Amite, Miss., 75 D3 100
Smithers, B.C., Can., 2,487 B4 68
Smithers, Fayette, W. Va., 1,696 C3, C6 123
Smithfield, Fulton, Ill., 329 C3 90
Smithfield, Somerset, Maine, 200 (382▲) D3 96
Smithfield, Jasper, Mo D3 101
Smithfield, Gosper, Nebr., 85 D6 103
Smithfield, Johnston, N.C., 6,117 B5 109
Smithfield, Jefferson, Ohio, 1,312 B7 111
Smithfield, Fayette, Pa., 939 G2 114
Smithfield, Tarrant, Tex B5 118
Smithfield, Cache, Utah, 2,512 B4 119
Smithfield, Isle of Wight, Va., 917 B6, E6 121
Smithfield, Wetzel, W. Va., 361 A6 123
Smithland, Woodbury, Iowa, 349 B2 92
Smithland, Livingston, Ky., 541 A3 94
Smithland, Marion, Tex., 100 C5 118
Smithmill, Clearfield, Pa., 600 E5 114
Smith Mills, Henderson, Ky., 300 C2 94
Smithonia, Oglethorpe, Ga., 30 B3 87
Smith River, Del Norte, Calif., 600 B1 82
Smiths, Lee, Ala., 950 C4 78
Smithsburg, Washington, Md., 586 A2 85
Smiths Cove, N.S., Can., 408 E4 74
Smiths Falls, Ont., Can., 9,603 C8 72
Smiths Ferry, Valley, Idaho, 15 E2 89
Smiths Grove, Warren, Ky., 613 C3 94
Smithshire, Warren, Ill., 140 C3 90
Smiths Lake, McKinley, N. Mex B1 107
Smithsons Valley, Comal, Tex A4 118
Smithton, Clark, Ark., 75 D2 81
Smithton, Austl., 2,671 o15 50
Smithton, St. Clair, Ill., 629 E4 90
Smithton, Pettis, Mo., 395 C4 101
Smith Town, McCreary, Ky., 300 D5 94
Smithtown, Rockingham, N.H., 150 E5 105
Smithtown, Suffolk, N.Y., 4,000 F4 84
Smithtown, bay, N.Y F4 84
Smithville, Lawrence, Ark., 75 A4 81
Smithville, Ont., Can., 947 D5 72
Smithville, Lee, Ga., 865 E2 87
Smithville, Monroe, Ind., 400 F4 91
Smithville, Monroe, Miss., 489 A5 100
Smithville, Clay, Mo., 1,254 B3, D2 101
Smithville, Atlantic, N.J E4 106
Smithville, Burlington, N.J., 200 D3 106
Smithville, Wayne, Ohio, 1,024 B6 111
Smithville, McCurtain, Okla., 110 C7 112
Smithville, DeKalb, Tenn., 2,348 D8 117
Smithville, Bastrop, Tex., 2,933 D4 118
Smithville, Ritchie, W. Va., 350 B3 123
Smithville Flats, Chenango, N.Y., 200 C5 108
Smithwick, Fall River, S. Dak., 60 D2 116
Smittle, cave, Mo D5 101
Smoaks, Colleton, S.C., 145 E6 115
Smock, Fayette, Pa., 1,012 *G2 114
Smoke Bend, Ascension, La., 450 B5 95
Smoke Creek, Washoe, Nev., 16,618 C2 104
Smoke Creek, des., Nev C2 104
Smokerun, Clearfield, Pa., 250 E5 114
Smoky, cape, Austl E9 51
Smoky, cape, N.S., Can C9 74
Smoky, mts., Idaho F4 89
Smoky, riv., Alta., Can B1 69
Smoky Hill, riv., Kans D5 93
Smoky Junction, Scott, Tenn., 200 C9 117
Smoky Lake, Alta., Can., 626 B4 69
Smöla, isl., Nor F2 25
Smolan, Saline, Kans., 210 D6 93
Smolensk, Sov. Un., 159,000 D8 27
Smolyan, Bul., 3,395 E7 22
Smoot, Lincoln, Wyo., 100 C2 121
Smooth Rock Falls, Ont., Can., 1,131 *E9 72
Smoothstone, lake, Sask., Can C2 70
Smoothstone, riv., Sask., Can B2 70
Smoots, creek, Kans B4 93
Smyadovo, Bul., 5,939 D8 22
Smyley, cape, Ant B4 5
Smyre, Gaston, N.C., 1,197 *B2 109
Smyrna, Kent, Del., 3,241 B6 85
Smyrna, Cobb, Ga., 10,157 A5, C2 87
Smyrna, Chenango, N.Y., 286 C5 108
Smyrna, York, S.C., 52 A5 115
Smyrna, Rutherford, Tenn., 3,612 B5 117
Smyrna, see Izmir, Tur.
Smyrna, riv., Del B6 85
Smyrna Mills, Aroostook, Maine, 200 (331▲) B4 96
Smyth, co., Va., 31,066 B3 121
Snaefell, mtn., I of Man F4 13
Snake, creek, Nebr B3 103
Snake, falls, Nebr B5 103
Snake, lake, Sask., Can B2 70
Snake, mtn., N.C A2 109
Snake, range, Nev E7 104
Snake, riv., Minn B1 99
Snake, riv., Minn D5 99
Snake, riv., Oreg B10 113
Snake, riv., U.S B4 76
Snake, riv., Wash C7 122
Snake Indian, riv., Alta., Can C1 69
Snake River, canyon, Idaho, Oreg D2 89, C10 113
Snake River, plain, Idaho F2 89
Snake River, range, Idaho, Wyo F7 89
Snares, is., N.Z R11 51
Sneads, Jackson, Fla., 1,399 B1 86
Sneads Ferry, Onslow, N.C., 500 C6 109
Snedsted, Den., 1,030 B2 24
Sneedville, Hancock, Tenn., 799 C10 117
Sneek, Neth., 21,100 A5 15
Sneem, Ire., 282 F2 11
Snelling, Barnwell, S.C., 100 E5 115
Snellville, Gwinnett, Ga., 468 A6 87
Snellville, Worcester, Mass B3 97
Snezhnoye, Sov. Un., 10,000 q21 27
Snezhnu, peak, Sov. Un D14 28
Sniardwy, lake, Pol B7 26
Snipe, keys, Fla H5 86
Snipe, lake, Alta., Can B2 69
Snizort, bay, Scot C2 13
Snöhetta, mtn., Nor F3 25
Snohomish, Snohomish, Wash., 3,894 B3 122
Snohomish, co., Wash., 172,199 A4 122
Snomac, Seminole, Okla., 100 B5 112
Snoqualmie, King, Wash., 1,216 B4 122
Snov, riv., Sov. Un F8 27
Snover, Sanilac, Mich., 250 E8 98
Snow, Pushmataha, Okla., 20 C6 112
Snow, mtn., Maine C2 96
Snow, peak, Wash A4 122
Snowball, Searcy, Ark., 125 B3 81
Snowbank, lake, Minn B7 99
Snow Camp, Alamance, N.C., 130 B4 109
Snowden, Sask., Can., 79 D3 70
Snowdon, mtn., Wales A3 12
Snowdon, mts., Wales A4 12
Snowdoun, Montgomery, Ala., 250 C3 78
Snowfield, peak, Wash A4 122
Snowflake, Navajo, Ariz., 982 C5 80
Snowflake, Man., Can., 73 E2 71
Snow Hill, Wilcox, Ala., 120 D2 78
Snow Hill, Ouachita, Ark., 50 D3 81
Snow Hill, Worcester, Md., 2,311 D7 85
Snow Hill, Greene, N.C., 1,043 B6 109
Snowking, mtn., Wash A4 122
Snow Lake, Desha, Ark., 119 C4 81
Snowmass, Petkin, Colo., 8 B4 83
Snowmass, mtn., Colo B3 83
Snow Road Station, Ont., Can., 60 C8 72
Snow Shoe, Centre, Pa., 714 D6 114
Snowshoe, lake, Maine B4 96
Snowshoe, peak, Mont B1 102
Snowville, Box Elder, Utah, 159 B3 119
Snowville, Pulaski, Va., 100 D2 121
Snow Water, lake, Nev C7 104
Snowyside, mtn., Idaho F4 89
Snyatyn, Sov. Un., 10,000 A7 22
Snyder, Ashley, Ark., 75 D4 81
Snyder, Morgan, Colo., 200 A7 83
Snyder, Dodge, Nebr., 325 C9 103
Snyder, Kiowa, Okla., 1,663 C3 112
Snyder, Scurry, Tex., 13,850 C2 118
Snyder, co., Pa., 25,922 E7 114
Snyder Knob, mtn., W. Va C4 123
Soai Rieng, Camb., 5,000 G6 38
Soalala, Malag., 759 g9 49
Soap Lake, Grant, Wash., 1,591 B6 122
Soar, riv., Eng B6 12
Soatá, Col., 3,116 B3 60
Soay, isl., Scot C2 13
Sobat, riv., Sud D3 47
Sobeslav, Czech., 4,299 D9 17
Sobieski, Oconto, Wis., 80 D5 124
Sobinka, Sov. Un., 10,000 D13 27
Sobota Rimavska, Czech., 10,700 D6 26
Sobral, Braz., 32,281 B2 57
Sobti (Well), Mali B4 45
Söby, Den., 728 D4 24
Sochaczew, Pol., 13,300 B6 26
Soche, see Yarkand, China
Sochi, Sov. Un., 101,000 G16 9
Social Circle, Walton, Ga., 1,780 C3 87
Social Hill, Hot Spring, Ark., 100 D6 81
Society, is., Fr. Polynesia H12 7
Society Hill, Darlington, S.C., 677 B8 115
Socmbawa, isl., Indon G5 35
Socorro, Braz., 6,402 m8 56
Socorro, Col., 11,842 B3 60
Socorro, Socorro, N. Mex., 5,271 C3 107
Socorro, El Paso, Tex., 400 *F1 118
Socorro, co., N. Mex., 10,168 D2 107
Socotra, isl., Aden H8 33
Socrum, Polk, Fla., 175 D4 86
Soc Trang, Viet., 16,890 H6 38
Socuéllamos, Sp., 14,828 C4 20
Soda Creek, B.C., Can., 113 C6 68
Sodankylä, Fin., 2,500 D12 25
Soda Springs, Caribou, Idaho, 2,424 G7 89
Sodaville, Linn, Oreg., 145 C4 113
Soddy, Hamilton, Tenn., 2,206 D8, E10 117
Söderala, Swe., 499 G7 25
Söderhamn, Swe., 13,000 G7 25
Söderköping, Swe., 3,648 H7 25
Södermanland, co., Swe., 230,400 H7 25
Södertälje, Swe., 35,300 t35 25
Sodiri, Sud., 1,804 C2 47
Sodo, Eth D4 47
Sodom, Warren, N.Y., 55 B7 108
Sodus, Berrien, Mich., 50 F4 98
Sodus, Wayne, N.Y., 1,645 B3 108
Sodus Point, Wayne, N.Y., 868 B4 108
Soela, isl., Indon F6 35
Soenda (Sunda), strait, Indon G3 35
Soest, Ger., 33,300 B3 17
Sofadhes, Grc., 4,046 C4 23
Sofia (Sofiya), Bul., 366,925 (*434,888) D6 22
Sofia, riv., Malag g9 49
Sofiya, see Sofia, Bul.
Sofiya, co., Bul., 1,457,742 *D6 22
Sofiyevka, Sov. Un., 10,000 G9 27
Sofre, Pan., 193 F7 62
Sofu-Gan, isl., Pac. O E8 7
Sogamoso, Col., 13,574 B3 60
Sögel, Ger., 2,900 B7 15
Sognefjord, fjord, Nor G1 25
Sogn og Fjordane, co., Nor., 99,900 *G1 25
Sogod, Phil., 3,344 *C7 35
Sogut, Tur., 2,900 C8 23
Sohagpur, India, 9,382 F8 40
Soham, San Miguel, N. Mex., 10 B4 107
Soignies, Bel., 10,874 D4 15
Sointula, B.C., Can., 682 D4 68
Soissons, Fr., 23,150 C5 14
Sokal, Sov. Un., 10,000 F5 27
Soke, Tur., 23,400 D6 23
Soke of Peterborough, co., Eng., 74,442 *D6 10
Sokhondo, mtn., Sov. Un E14 28
Sokhta Chinar, Afg., 5,000 D13 41
Sokol, Sov. Un., 36,000 B13 27
Sokolka, Pol., 4,879 B7 26
Sokolo, Mali D3 45
Sokolov, Czech., 17,600 C7 17
Sokolow, Pol., 7,515 B7 26
Sokoto, Nig., 47,643 D6 45
Solana, Charlotte, Fla., 1,309 *F4 86
Solana Beach, San Diego, Calif., 3,000 *F5 76
Solander, isl., N.Z Q11 51
Solano, Harding, N. Mex., 75 B5 107
Solano, Phil., 9,497 n13 35
Solano, co., Calif., 134,597 C2 82
Solbad Hall [in Tirol], Aus., 10,750 E5 16
Soldatovo, Sov. Un F26 9
Soldier, Monona, Iowa, 284 C2 92
Soldier, Jackson, Kans., 171 C8 93
Soldier, Carter, Ky., 150 B6 94
Soldier, key, Fla F3 86

Soldier, riv., Iowa C2 92
Soldier Pond, Aroostook, Maine, 500 A4 96
Soldiers Grove, Crawford, Wis., 663 E3 124
Soldier Summit, Wasatch, Utah, 33 D4 119
Soledad, Monterey, Calif., 2,837 D3 82
Soledad, Col., 20,158 A3 60
Soledad, Ven., 5,259 B5 60
Soleduck, riv., Wash B1 122
Solen, Sioux, N. Dak., 250 C5 110
Solent, chan., Eng D6 12
Solesmes, Fr., 5,722 D3 15
Solihull, Eng., 96,010 B6 12
Solikamsk, Sov. Un., 84,000 B5 29
Sol-Iletsk, Sov. Un., 19,100 C5 29
Soliman, Tun., 6,980 F12 30
Solimões (Amazon), riv., Braz., Peru B4 58
Solingen, Ger., 162,800 B2 17
Solitar, mtn., Tex B3 118
Sollas, Scot C1 13
Sollefteå, Swe., 9,900 F7 25
Sóller, Sp., 6,817 C7 20
Soller Central, P.R B3 65
Sölleröd, Den., 25,877 *C6 24
Sollihögda, Nor p28 25
Sollum, gulf, Eg., U.A.R. G5 31
Sollyu-Bong, mts., Kor G2 37
Solna, Swe., 52,300 H7 25
Solok, Indon., 6,214 F2 35
Solomea, Samoa 52
Solomon, Graham, Ariz., 500 E6 80
Solomon, Dickinson, Kans., 1,008 D6 93
Solomon Is. (Austl.), reg., N. Gui., 55,000 *G9 7
Solomon Is., British, dep., Oceania, 102,000 *G9 7
Solomon, riv., Kans C6 93
Solomons, Calvert, Md., 183 D5 85
Solon, Johnson, Iowa, 604 C6 92
Solon, Somerset, Maine, 500 (669▲) D3 96
Solon, Cuyahoga, Ohio, 6,333 A6 111
Solonika, see Thessaloniki, Grc.
Solon Mills, McHenry, Ill., 130 D2 90
Solon Springs, Douglas, Wis., 530 B2 124
Solothurn, Switz., 17,800 B4 19
Solothurn, Switz., 18,394 (★30,405) B4 19
Solothurn, canton, Switz., 200,816 B4 19
Solovyevsk, Sov. Un., 10,000 D15 28
Solsberry, Greene, Ind., 150 F4 91
Solsgirth, Man., Can., 78 D1 71
Solta, isl., Yugo D3 22
Soltau, Ger., 14,400 B4 16
Solvang, Santa Barbara, Calif., 1,325 *E3 82
Solvay, Onondaga, N.Y., 8,732 B4 108
Sölvesborg, Swe., 6,000 B8 24
Solway, Beltrami, Minn., 100 C3 99
Solway, firth, Eng., Scot F5 13
Solwezi, N. Rh C4 48
Soma, Tur., 13,100 C6 23
Somaliland, French, see French Somaliland, dep., Afr.
Somali Republic, country, Afr., 1,980,000 E10 42, E5 47
Sombor, Yugo., 37,802 C4 22
Sombra, Ont., Can., 520 E2 72
Sombrerete, Mex., 9,260 C4 63
Sombrero, chan., India G9 39
Somerdale, Camden, N.J., 4,839 D2 106
Somers, Tolland, Conn., 950 (3,702▲) B7 84
Somers, Calhoun, Iowa, 203 B3 92
Somers, Flathead, Mont., 700 B2 102
Somers, Kenosha, Wis., 200 F2, F6 124
Somerset, Man., Can., 587 E2 71
Somerset, Gunnison, Colo., 150 C3 83
Somerset, Wabash, Ind., 250 C6 91
Somerset, Miami, Kans., 100 D9 93
Somerset, Pulaski, Ky., 7,112 C5 94
Somerset, Montgomery, Md., 1,444 *C3 85
Somerset, Bristol, Mass., 12,196 C5 97
Somerset, Somerset, N.J., 12,000 C4 106
Somerset, Perry, Ohio, 1,361 C5 111
Somerset, Somerset, Pa., 6,347 F3 114
Somerset, Bexar, Tex., 700 B3 118
Somerset, St. Croix, Wis., 729 C1 124
Somerset, co., Eng., 598,556 C5 12
Somerset, co., Maine, 39,749 C2 96
Somerset, co., Md., 19,623 D6 85
Somerset, co., N.J., 143,913 B3 106
Somerset, co., Pa., 77,450 G3 114
Somerset, isl., Bermuda E13 77
Somerset, isl., N.W. Ter., Can B14 66
Somerset, res., Vt E3 120
Somerset Bridge, Bermuda E13 77
Somerset East, S. Afr., 9,779 D4 49
Somers Point, Atlantic, N.J., 4,504 E3 106
Somersville, Tolland, Conn., 500 B7 84
Somersworth, Strafford, N.H., 8,529 D5 105
Somerton, Yuma, Ariz., 1,613 E1 80
Somervell, co., Tex., 2,577 C4 118
Somerville, Morgan, Ala., 166 A3 78
Somerville, Gibson, Ind., 317 H3 91
Somerville, Middlesex, Mass., 94,697 B5, D3 97
Somerville, Somerset, N.J., 12,458 B3 106
Somerville, Butler, Ohio, 478 C3 111
Somerville, Fayette, Tenn., 1,820 B2 117
Somerville, Burleson, Tex., 1,177 D4 118
Somesul, riv., Rom B6 22
Somme, dept., Fr., 488,225 E2 15
Somme, riv., Fr B4 14
Sommepy, Fr., 588 E4 15
Sömmerda, Ger., 13,800 B6 17
Sommesous, Fr., 540 F4 15
Somogy, co., Hung., 371,783 *B3 22
Somonauk, DeKalb, Ill., 899 B5 90
Somosomo, strait, Fiji Is 52
Somoto, Nic., 2,322 D4 62
Sompeta, India, 10,588 H10 40
Somuncura, plat., Arg C3 54
Somvix, Switz., 2,004 C6 19
Son, riv., India E9 40
Soná, Pan., 3,176 F7 62
Sonchon, Kor., 22,725 G2 40
Soncino, It., 9,809 D5 15
Sönder, riv., Den D3 24
Sönderborg, Den., 20,653 D3 24
Sönderborg, co., Den., 49,604 D3 24
Sönderho, Den., 410 C2 24
Sonderhausen, Ger., 19,000 B5 17
Sönder Omme, Den., 1,308 C2 24
Söndervig, Den B2 24
Sondheimer, East Carroll, La., 350 B4 95
Sondrio, It., 18,944 A2 21
Sonestown, Sullivan, Pa., 250 D8 114
Song, Nig E7 45
Songarh, India, 2,858 G4 40
Song Cau, Viet., 5,000 F8 38
Songchon, Kor., 9,148 G3 37
Songea, Tan, 1,401 D6 48
Songkhla, Thai., 31,001 I4 38
Songololo, Con. L C1 48
Sonhat, India F9 40
Son La, Viet., 10,000 B5 38
Sonmiani, Pak C4 39
Sonmiani, bay, Pak I12 41
Sonneberg, Ger., 28,900 C6 17
Sonnette, Powder River, Mont., 5 E11 102
Sonningdale, Sask., Can., 129 E2 70
Sono, riv., Braz C1 57
Sonobe, Jap., 7,300 n14 37
Sonoita, Santa Cruz, Ariz., 5 F5 80
Sonoita, Mex., 1,275 A2 63
Sonoma, Sonoma, Calif., 3,023 B5, C2 82
Sonoma, co., Calif., 147,375 C2 82
Sonoma, peak, Nev C4 104
Sonoma, range, Nev C4 104
Sonora, Pinal, Ariz., 1,244 D4 80
Sonora, Tuolumne, Calif., 2,725 D3 82
Sonora, Hardin, Ky., 268 C4 94
Sonora, Muskingum, Ohio, 200 C6 111
Sonora, Sutton, Tex., 2,619 D2 118
Sonora, state, Mex., 771,663 B2 63
Sonora, riv., Mex G5 80
Sonoraville, Gordon, Ga., 78 B2 87
Sonpur, India, 7,108 G9 40
Sonqor, Iran, 12,126 D3 41
Sonsón, Col., 10,913 B2 60
Sonsonate, Sal., 23,137 D3 62
Sontag, Lawrence, Miss., 200 D3 100
Son Tay, Viet., 16,640 B6 38
Sonyea, Livingston, N.Y., 500 C3 108
Soo, locks, Mich B6 98
Soochow (Suchou), China, 633,000 E9 34
Soo Junction, Luce, Mich B5 98
Sooke, B.C., Can., 1,121 B9, E6 68
Sopchoppy, Wakulla, Fla., 450 B2 86
Soper, Choctaw, Okla., 309 C6 112
Soperton, Treutlen, Ga., 2,317 D4 87
Soperton, Forest, Wis. (part of Wabeno) C5 124
Sophia, Randolph, N.C., 250 B4 109
Sophia, Raleigh, W. Va., 1,284 D3 123
Sopot, Pol., 44,000 A5 26
Sopris, Las Animas, Colo., 150 D6 83
Sopron, Hung., 41,246 B3 22
Sop's Arm, Newf., Can., 296 D3 75
Soquel, Santa Cruz, Calif., 950 C6 82
Sôr, riv., Port C1 20
Sora, It., 9,000 D4 21
Sorak-San, peak, Kor G4 37
Sorata, Bol., 2,087 C2 55
Sorau, see Zary, Pol.
Sorbas, Sp., 5,961 D4 20
Sorel, Que., Can., 17,147 C4 73
Sorell, cape, Austl o15 50
Sorento, Bond, Ill., 681 E4 90
Soresina, It., 9,100 B2 21
Soria, Sp., 19,301 B4 20
Soria, prov., Sp., 147,052 *B4 20
Soriano, Ur., 1,003 f7 54
Soriano, dept., Ur., 70,500 *E1 56
Sorö, Den., 5,494 C5 24
Sorö, co., Den., 128,176 C5 24
Sorocaba, Braz., 109,258 C3, m8 56
Sorochinsk, Sov. Un., 18,400 C4 29
Soroco, P.R B8, f12 65
Soroki, Sov. Un., 15,000 G7 27
Sorong, W. Irian, 8,000 F8 35
Soroti, Ug., 6,645 A5 48
Söröy, isl., Nor B10 25
Sorraia, riv., Port C1 20
Sorrento, Lake, Fla., 500 D5 86
Sorrento, It., 7,900 D5 21
Sorrento, Ascension, La., 1,151 B6, D5 95
Sorris Sorris, S. W. Afr B1 49
Sorsogon, Phil., 14,000 (35,500▲) C6 35
Sorsogon, prov., Phil., 348,700 *C6 35
Sortavala, Sov. Un., 16,400 G14 25
Sör-Tröndelag, co., Nor., 211,700 *F4 25
Sorum, Perkins, S. Dak., 10 B3 116
Sörumsand, Nor., 726 p29 25
Sosan, Kor., 13,500 H3 37
Sosnogorsk, Sov. Un C19 9
Sosnovka, Sov. Un., 10,000 E13 27
Sosnowiec, Pol., 132,000 C5, g10 26
Soso, Jones, Miss., 150 D4 100
Sosvinskaya, Sov. Un C21 9
Sota, riv., Pol h10 26
Sottern, lake, Swe t33 25
Sotteville-lès-Rouen, Fr., 25,625 C4 14
Souanké, Con. B E2 46
Soubré, I.C., 1,300 E3 45
Soucook, riv., N.H D4 105
Soudan, St. Louis, Minn., 810 C6 99
Souderton, Montgomery, Pa., 5,381 F11 114
Souflíon, Grc., 7,435 B6 23
Soufriere, St. Lucia, 3,550 p16 64
Souhegan, riv., N.H E3 105
Souk Ahras, Alg., 17,444 F10 28
Souk el Arba, Tun., 6,469 F11 28
Souk el Arba, prov., Tun., 196,113 *C6 44
Sŏul, see Seoul, Kor.
Soulac [-sur-Mer], Fr., 1,192 E3 14
Soulanges, co., Que., Can., 10,075 D3 73
Sound Beach, Suffolk, N.Y., 1,625 *F5 84
Sound View, New London, Conn., 75 D7 84
Sources, mtn., Bas, S. Afr C4 49
Sourdnahunk, lake, Maine B3 96
Soure, Braz., 6,666 C5 59
Soure, Port., 9,317 B1 20
Souris, Man., Can., 1,759 E1 71
Souris, Bottineau, N. Dak., 213 A5 111
Souris, riv., Man., Sask., Can., N. Dak G12 66
Souris East, P.E.I., Can., 1,537 C7 74
Sourlake, Hardin, Tex., 1,602 D5 118
Sous, riv., Mor C3 44
Sousa, Braz., 12,350 C3 57
Sousse, Tun., 48,185 B7 44
Sousse, prov., Tun., 447,093 *B7 44
Soustons, Fr., 1,901 F3 14
South, div., Ice., 13,416 *n21 25
South, cape, N.Z Q11 51
South, cape, Nor B13 4
South, isl., N.Z O12 51
South, isl., S.C E9 115
South, isl., Truk 52
South, mtn., Idaho G2 89
South, mtn., Md A2 85
South, mtn., Pa C1 106
South, mts., N.C B2 109
South, pass, Kwajalein 52
South, pass, La F6 95
South, pass, Wyo C4 125
South, pt., Md D7 85
South, pt., Mich D7 98
South, riv., Ont., Can B5 72
South, riv., Ga B5 87
South, riv., Iowa C4 92
South, riv., Md C4 85
South, riv., N.C C5 109
South Acton, Middlesex, Mass., 1,700 C1 97
South Acworth, Sullivan, N.H., 100 D2 105
South Addison, Washington, Maine, 150 D5 96
South Africa, country, Afr., 15,995,312 C3 49
South Amboy, Middlesex, N.J., 8,422 C4, E4 106
South Amboy Junction, Middlesex, N.J. (part of South Amboy) E4 106
South America, cont., 118,100,000 H17 3, 53
South Amherst, Lorain, Ohio, 1,657 *A5 111
Southampton, Ont., Can., 1,640 C3 72
Southampton, N.S., Can., 188 D5 74
Southampton, Eng., 204,707 (★340,000) D6 12
Southampton, Hampshire, Mass., 400 (2,192▲) B2 97
Southampton, Suffolk, N.Y., 4,582 D4 108
Southampton, Bucks, Pa., 4,500 *F11 114
Southampton (Hampshire), Co., Eng., 1,336,084 C6 12
Southampton, co., Va., 27,195 E5 121
Southampton, cape, N.W. Ter., Can D16 67
Southampton, isl., N.W. Ter., Can D16 67
South Andaman, isl., India F9 39
South Anna, riv., Va D5 121
South Apopka, Orange, Fla., 2,484 *D5 86
South Arabia, Fed. of, reg., Asia C6 47
Southard, Monmouth, N.J., 120 C4 106
Southard, Blaine, Okla., 385 A3 112
South Ashburnham, Worcester, Mass., 700 A4 97
South Ashfield, Franklin, Mass., 100 A2 97
South Atlantic, ocean H20 3
South Attleboro, Bristol, Mass. (part of Attleboro) C5 97
South Aulatsivik, isl., Newf., Can g9 75
South Australia, state, Austl., 969,340 F6 50
South Baker, Baker, Oreg. (part of Baker) C9 113
South Bancroft, Aroostook, Maine C5 96
South Barnstead, Belknap, N.H., 45 D4 105
South Barre, Worcester, Mass., 900 B3 97
South Barre, Washington, Vt., 450 C3 120
South Bay, Palm Beach, Fla., 1,631 F6 86
South Bellingham, Norfolk, Mass., 2,300 B5 97
South Belmar, Monmouth, N.J., 1,537 C4 106
South Belmont, Gaston, N.C., 2,286 *B2 109
South Beloit, Winnebago, Ill., 3,781 A4 90
South Bend, Lincoln, Ark *C4 81
South Bend, St. Joseph, Ind., 132,445 (★265,100) A5 91
South Bend, Cass, Nebr., 86 E2 103
South Bend, Young, Tex., 130 C3 118
South Bend, Pacific, Wash., 1,671 C2 122
South Benfleet, Eng., 32,372 C8 12
South Bennettsville, Marlboro, S.C., 1,025 *B8 115
South Bentinck Arm, chan., B.C., Can C4 68
South Berwick, York, Maine, 1,773 E2 96
South Bethlehem, Albany, N.Y., 400 C7 108
South Bitter, creek, Wyo D4 125
South Bloomfield, Pickaway, Ohio, 424 C5 111
South Boardman, Kalkaska, Mich., 175 D5 98
South Bocagrande, Lee, Fla F4 86
South Boise, Ada, Idaho, 1,452 *F2 89
South Bolton, Que., Can., 167 D5 73
Southboro, Worcester, Mass., 1,114 B4, D1 97
South Boston (Independent City), Va., 5,974 E4 121
South Bound, Brook, Somerset, N.J., 3,626 B3 106
South Braintree, Norfolk, Mass. (part of Braintree) D3 97
South Branch, Newf., Can., 311 E2 75
Southbranch, Ogemaw, Mich., 80 D7 98
South Branch, lake, Maine C4 96
South Branch, mtn., W. Va B6 123
Southbridge, Worcester, Mass., 16,523 B3 97
South Bridgeview, Cook, Ill., 1,000 *B6 90
South Bristol, Lincoln, Maine, 550 (610▲) E3 96
South Britain, New Haven, Conn., 300 D4 84
South Broadway, Yakima, Wash., 3,661 *C5 122
South Brookfield, N.S., Can., 240 E5 74
South Brooksville, Hancock, Maine, 90 D4 96
South Bruny, isl., Austl o15 50
South Burlington, Chittenden, Vt., 6,903 C2 120
Southbury, New Haven, Conn., 800 (5,186▲) D4 84
South Byfield, Essex, Mass., 30 A6 97
South Byron, Genesee, N.Y., 250 B2 108
South Cabot, Washington, Vt C4 120
South Canaan, Wayne, Pa., 150 C11 114
South Carolina, state, U.S., 2,382,594 D11 77, 115
South Carver, Plymouth, Mass., 300 C6 97
South Charleston, Clark, Ohio, 1,505 C4 111
South Charleston, Kanawha, W. Va., 19,180 C6 123
South Charlestown, Sullivan, N.H., 100 D2 105
South Chatham, Barnstable, Mass., 400 C7 97
South Chatham, Carroll, N.H., 50 B4 105
South Chaves, McKinley, N. Mex., 12 B1 107
South Chelmsford, Middlesex, Mass., 1,500 C1 97
South Cheney, Spokane, Wash D7 122
South Chicago Heights, Cook, Ill., 4,043 F3 90
South China, Kennebec, Maine, 115 D3 96
South China, sea, Asia H14 33
South Cle Elum, Kittitas, Wash., 383 B5 122
South Clement, creek, Md D4 85
South Cleveland, Bradley, Tenn., 1,512 *D9 117
South Clinton, Anderson, Tenn., 1,356 *C9 117
South Coatesville, Chester, Pa., 2,032 *G10 114
South Coffeyville, Nowata, Okla., 622 A6 112
South Colby, Kitsap, Wash., 350 D1 122
South Colton, St. Lawrence, N.Y., 660 B2 108
South Connellsville, Fayette, Pa., 2,434 G2 114
South Corning, Steuben, N.Y., 1,448 *C3 108
South Covington, Va. (part of Covington) D2 121
South Dakota, state, U.S., 680,514 B7 76, 116
South Danbury, Merrimack, N.H., 35 D3 105
South Danville, Rockingham, N.H., 100 E4 105
South Dartmouth, Bristol, Mass., 6,000 C6 97
South Dayton, Cattaraugus, N.Y., 696 C1 108
South Daytona, Volusia, Fla., 1,954 C5 86
South Decatur, De Kalb, Ga., 15,000 *C2 87
South Deerfield, Franklin, Mass., 1,253 B2 97
South Deerfield, Rockingham, N.H., 50 D4 105
South Deer Isle, Hancock, Maine, 115 D4 96
South Dennis, Cape May, N.J., 365 E3 106
South Dorset, Bennington, Vt., 160 E2 120
South Downs, hills, Eng D7 12
South Durham, Que., Can., 438 D5 73
South Duxbury, Plymouth, Mass., 900 B6 97
Southeast, pass, La E6 95
South Easton, Bristol, Mass., 795 B5 97
South Effingham, Carroll, N.H., 80 C5 105
South Egremont, Berkshire, Mass., 350 B1 97
South Elgin, Kane, Ill., 2,624 E2 90
South El Monte, Los Angeles, Calif., 4,850 *E5 82
South Elwood, Madison, Ind., 400 D6 91
Southend-on-Sea, Eng., 164,976 C8 12
South English, Keokuk, Iowa, 217 C5 92
Southern, prov., Nya E6 48
Southern, prov., N. Rh E4 48
Southern, reg., Ken B6 48
Southern, reg., Tan., 1,014,265 D6 48
Southern, uplands, Scot E4 13
Southern Alps, mts., N.Z O13 51
Southern Bug, riv., Sov. Un H8 27
Southern Cross, Austl., 760 F2 50
Southern Cross, Deer Lodge, Mont., 25 D3 102
Southern Highlands, reg., Tan., 1,030,041 C5 48
Southern Indian, lake, Man., Can A2 71
Southern Pines, Moore, N.C., 5,198 B4 109
Southern Rhodesia, Br. dep., Afr., 3,150,000 A4 49
Southern Slopes (Lone Oak), Spartanburg, S.C., 1,435 *B4 115
Southern Ute, Indian res., Colo D3 83
Southern View, Sangamon, Ill., 1,485 *D4 90
South Erradale, Scot C3 13
South Esk., riv., Scot D6 13
South Essex, Essex, Mass., 700 A6 97
South Euclid, Cuyahoga, Ohio, 27,569 B2 111
Southey, Sask., Can., 483 G3 70
South Fabius, riv., Mo A6 101
South Fallsburg, Sullivan, N.Y., 1,290 D6 108
South Farmingdale, Nassau, N.Y., 16,318 *E7 108
South Farms, Middlesex, Conn. (part of Middletown) C6 84
South Fayetteville, Cumberland, N.C., 3,411 *B5 109
Southfield, Berkshire, Mass., 500 B1 97
Southfield, Oakland, Mich., 31,501 *F7 98
South Flevoland Polder, reg., Neth B5 15
South Floral Park, Nassau, N.Y., 1,090 *G2 84
Southford, New Haven, Conn D4 84
South Fork, Humboldt, Calif., 75 B2 82
South Fork, Sask., Can., 100 H1 70
South Fork, Rio Grande, Colo., 175 D4 83
South Fork, Cambria, Pa., 2,053 F4 114
South Fork, res., Wash B4 122
South Fork Republican, riv., Colo B8 83
South Fort Mitchell, Kenton, Ky., 4,086 A7 94
South Fort Smith, Sebastian, Ark. (part of Fort Smith) B1 81
South Foster, Providence, R.I., 90 B10 84
South Fox, isl., Mich C5 98
South Freeport, Cumberland, Maine, 350 E5 96
South Fulton, Obion, Tenn., 2,512 A3 117
South Gamboa, C.Z k11 62
South Gardiner, Kennebec, Maine (part of Gardiner) D3 96
South Gate, Los Angeles, Calif., 53,831 F2 82
Southgate, Campbell, Ky., 2,070 A7 94
Southgate, Wayne, Mich., 29,404 *F7 98
South Georgia, isl., Atl. O D9 5
South Gifford, Macon, Mo., 93 A5 101
South Glastonbury, Hartford, Conn., 1,000 C6 84
South Glens Falls, Saratoga, N.Y., 4,129 *B7 108
South Grafton (Fisherville), Worcester, Mass., 3,000 *B4 97
South Grand, riv., Mo C3 101
South Gray, Cumberland, Maine, 70 E5 96
South Greenfield, Dade, Mo., 179 D4 101
South Greensburg, Westmoreland, Pa., 3,058 *F2 114
South Greenwood, Greenwood, S.C., 2,520 C3 115
South Groveland, Essex, Mass., 600 A5 97
South Hackensack, Bergen, N.J., 1,841 *D5 106
South Hadley, Hampshire, Mass., 5,000 (14,956▲) B2 97
South Hadley Falls, Hampshire, Mass., 3,100 B2 97
South Hamilton, Essex, Mass., 2,000 A6, C3 97
South Hampton, Rockingham, N.H., 100 (443▲) E5 105
South Hanover, Plymouth, Mass., 400 E4 97
South Harpswell, Cumberland, Maine, 500 E5 96
South Harriman, Roane, Tenn., 2,884 *D9 117
South Harris, pen., Scot C2 13
South Haven, Sumner, Kans., 408 E6 93
South Haven, Van Buren, Mich., 6,149 F4 98
South Heart, Stark, N. Dak., 97 C2 110
South Hempstead, Nassau, N.Y., 3,000 *D2 108
South Henderson, Vance, N.C., 2,017 *A5 115
South Hero, Grand Isle, Vt., 70 (614▲) B2 120
South Hill, Mecklenburg, Va., 2,569 E4 121
South Hingham, Plymouth, Mass., 600 B6, D3 97
South Holland, Cook, Ill., 12,603 *F3 90
South Holland (Zuidholland), prov., Neth., 2,726,200 B4 15
South Holston, lake, Tenn., Va C11 117, B2 121
South Holston Lake, res., Tenn., Va C11 117, B3 121
South Hooksett, Merrimack, N.H., 1,700 D4 105
South Houston, Harris, Tex., 7,523 F5 118
South Humboldt, riv., Nev C6 104
South Huntingdon (Smithfield), Huntingdon, Pa., 2,547 *E5 114
South Hutchinson, Reno, Kans., 1,672 A4 93
Southington, Hartford, Conn., 14,000 C5 84
South International Falls, Koochiching, Minn., 2,479 B5 99
South Jacksonville, Morgan, Ill., 2,340 D3 90
South Jordan, Salt Lake, Utah, 1,354 C4 119
South Junction, Man., Can., 233 E4 71
South Junction, Wasco, Oreg., 35 B5 113
South Kent, Litchfield, Conn., 150 C3 84
Southkent, Kent, Mich., 15,000 *F5 98
South Killingly, Windham, Conn., 150 B9 84
South Kingston, Rockingham, N.H., 100 E4 105
South Klamath, Klamath, Oreg E5 113
South Laguna, Orange, Calif., 2,000 *F5 82
Southlake, Tarrant, Tex., 1,023 *C4 118
South Lancaster, Worcester, Mass., 1,891 B4 97
Southland, Jackson, Mich., 2,000 *F6 98
Southland, Garza, Tex., 153 C2 118
South Lansing, Tompkins, N.Y., 100 C4 108
Southlawn, Sangamon, Ill., 2,300 *D4 90
South Lebanon, Warren, Ohio, 2,720 C3 111
South Lee, Berkshire, Mass., 375 B1 97
South Lee, Strafford, N.H D4 105
South Liberty, Waldo, Maine, 100 D3 96
South Lincoln, Middlesex, Mass., 500 C2 97
South Lincoln, Addison, Vt., 50 C3 120
South Londonderry, Windham, Vt., 250 E3 120
South Loup, riv., Nebr C6 103
South Lubec, Washington, Maine, 230 D5 96
South Lunenburg, Essex, Vt., 100 C5 120
South Lyme, New London, Conn., 250 D8 84
South Lyndeboro, Hillsboro, N.H., 150 E3 105
South Lynnfield, Essex, Mass. (part of Lynnfield) C3 97
South Lyon, Oakland, Mich., 1,753 A6, F7 98
South Macon, Bibb, Ga., 9,000 *D3 87
Southmag, Ont., Can B4 72
South Magnetic Pole, Ant C27 5
South Manchester, Hartford, Conn. (part of Manchester) B6 84
South Manitou, isl., Mich C4 98
South Mansfield, De Soto, La., 616 B2 95
South Marsh, isl., Md D5 85
South Medford, Jackson, Oreg., 2,306 *E4 113
South Meriden, New Haven, Conn. (part of Meriden) C5 84
South Merrimack, Hillsboro, N.H., 125 E3 105
South Miami, Dade, Fla., 9,846 F3, G6 86
South Middleboro, Plymouth, Mass., 400 C6 97
South Milford, Lagrange, Ind., 350 A7 91
South Milford, Worcester, Mass. (part of Hopedale) B4, E1 97
South Milford, Hillsboro, N.H E3 105
South Mills, Camden, N.C., 479 A7 109
South Milwaukee, Milwaukee, Wis., 20,307 F2, F6 124
Southminster, Eng., 1,444 C8 12
South Modesto, Stanislaus, Calif., 9,000 *D3 82
South Molton, Eng., 2,994 C4 12
South Monroe, Monroe, Mich., 2,919 *G7 98
Southmont, Davidson, N.C., 700 B3 109
Southmont, Cambria, Pa., 2,857 *F4 114
South Montrose, Susquehanna, Pa., 250 C10 114
South Mountain, Franklin, Pa., 600 G7 114
South Naknek, Alsk., 33 D8 79
South Natick, Middlesex, Mass. (part of Natick) D2 97
South Natuna, is., Indon E3 35
South New Berlin, Chenango, N.Y., 421 C5 108
South Newburg, Penobscot, Maine D4 96
South Newbury, Merrimack, N.H., 70 D2 105
South Newbury, Orange, Vt., 60 C4 120
South Newfane, Windham, Vt., 100 F3 120
South New River, canal, Fla E3 86
South Northfield, Washington, Vt., 65 C3 120
South Nyack, Rockland, N.Y., 3,113 C6 106
South Ogden, Weber, Utah, 7,405 B4 119
Southold, Suffolk, N.Y., 950 D4 108
South Orange, Essex, N.J., 16,175 B4 106
South Orkney, is., Atl. O C8 5
South Oroville, Butte, Calif., 3,704 *C3 82
South Orrington, Penobscot, Maine, 600 D4 96
South Otselic, Chenango, N.Y., 450 C5 108
South Oyster, bay, N.Y G3 84
South Pacific, ocean H12 2
South Paris, Oxford, Maine, 2,063 D2 96
South Park, Sonoma, Calif., 3,261 *C2 82
South Park, Kane, Ill., 2,063 *B5 90
South Pasadena, Los Angeles, Calif., 19,706 F2 82
South Pass City, Fremont, Wyo., 15 C4 125
South Peabody, Essex, Mass. (part of Peabody) C3 97

South Peacham, Caledonia, Vt., 40 C4 120
South Pekin, Tazewell, Ill., 1,007 C4 90
South Penobscot, Hancock, Maine, 100 D4 96
South Pittsburg, Marion, Tenn., 4,130 D8 117
South Plainfield, Middlesex, N.J., 17,879 B4 106
South Platte, riv., Colo., Nebr. B5 84, C4 103
South Point, Lawrence, Ohio, 1,663 *D5 111
South Polar, plateau, Ant A 5
South Pole, Ant A 5
South Pomfret, Windsor, Vt., 65 D3 120
South Pond, mtn., N.Y. B6 108
South Porcupine, Ont., Can., 5,144 E9 72
Southport (South Coast), Austl., 33,716 C9 51
Southport, Fairfield, Conn. (part of Fairfield) E3 84
Southport, Eng., 81,976 A4 12
Southport, Bay, Fla., 900 G3 86
Southport, Marion, Ind., 892 E5, I8 91
Southport, Jefferson, La. (part of Jefferson) C7 95
Southport (Elmira Southeast), Chemung, N.Y., 6,698 *C4 108
Southport, Brunswick, N.C., 2,034 D5 109
South Portland, Cumberland, Maine, 22,788 E2, E5 96
South Portsmouth, Greenup, Ky., 600 B7 94
South Pottstown, Chester, Pa., 1,850 *F10 114
South Poultney, Rutland, Vt., 100 E2 120
South Prairie, Pierce, Wash., 214 B3 118
South Queensferry, Scot., 2,700 E5 13
South Range, Houghton, Mich., 760 A2 98
South Range, Douglas, Wis., 100 B2 124
South Renovo, Clinton, Pa., 777 D6 114
South Richford, Franklin, Vt B3 120
South River, Ont., Can., 1,044 B5 72
South River, Ont., Can., 995 B5 72
South River, Middlesex, N.J., 13,397 C4 106
South Robbinston, Washington, Maine C5 96
South Rockwood, Monroe, Mich., 1,337 *F7 98
South Ronaldsay, isl., Scot B6 13
South Roxana, Madison, Ill., 2,010 *E3 90
South Roxton, Que., Can., 86 D5 73
South Royalston, Worcester, Mass., 350 A3 97
South Royalton, Windsor, Vt., 450 D3 120
South Russell, Geauga, Ohio, 1,276 *A6 111
South Ryegate, Caledonia, Vt., 360 C4 120
South Sacramento, Sacramento, Calif., 10,960 *C3 82
South St. Paul, Dakota, Minn., 22,032 E7, F5 99
South Salem, Ross, Ohio, 180 C4 111
South Salisbury, Rowan, N.C., 3,065 *B3 109
South Salt Lake, Salt Lake, Utah, 9,520 C4 119
South Sandwich, is., Atl. O D10 5
South Sanford, York, Maine E2 96
South San Francisco, San Mateo, Calif., 39,418 B5 82
South San Gabriel, Los Angeles, Calif., 26,213 *F2 82
South San Leandro, Alameda, Calif., 17,150 *B5 82
South Saskatchewan, riv., Alta., Sask., Can F11 66
South Seaville, Cape May, N.J., 350 E3 106
South Shaftsbury, Bennington, Vt., 600 F2 120
South Shetland, is., Ant C6 5
South Shields, Eng., 109,533 C6 10
South Shore, St. Charles, Mo., 200 A7 101
South Shore, Codington, S. Dak., 259 B9 116
Southside, Lincoln, N.C., 645 B2 109
Southside, Montgomery, Tenn., 100 A4 117
Southside Estates, Duval, Fla., 4,000 *B5 86
South Side Place, Harris, Tex., 1,282 F5 118
South Sioux City, Dakota, Nebr., 7,200 B9 103
South Slocan, B.C., Can., 168 E9 68
South Solon, Madison, Ohio, 414 C4 111
South Spencer, Worcester, Mass., 75 B3 97
South Spring, Chaves, N. Mex D5 107
South Sterling, Wayne, Pa., 150 D11 114
South Stickney, Cook, Ill., 23,000 *B6 90
South Strafford, Orange, Vt., 75 D4 120
South Streator, Livingston, Ill., 1,923 *B5 90
South Sudbury, Middlesex, Mass B5, D1 97
South Sumatra, see Sumatra, isl., Indon.
South Superior, Sweetwater, Wyo., 401 D4 125
South Sutton, Merrimack, N.H., 100 D3 105
South Swansea, Bristol, Mass., 1,100 C5 97
South Taft, Kern, Calif., 1,910 *E4 82
South Tamworth, Carroll, N.H., 160 C4 105
South Temple, Berks, Pa., 1,500 *F10 114
South Tent, mtn., Utah D4 119
South Toms River, Ocean, N.J., 1,603 *D4 106
South Torrington, Goshen, Wyo., 950 C8 125
South Trail (Hayden), Sarasota, Fla., 5,471 *E4 86
South Tucson, Pima, Ariz., 7,004 E5 80
South Tunnel, Sumner, Tenn., 200 A5 117
South Turlock, Stanislaus, Calif., 1,577 *D3 82
South Twin, lake, Newf., Can D4 75
South Twin, mtn., N.H. B3 105
South Tyne, riv., Eng F6 13
South Uist, isl., Scot C1 13
South Uniontown, Fayette, Pa., 3,603 *G2 114
South Vernon, Franklin, Mass., 80 A3 97
South Vienna, Clark, Ohio, 400 C4 111
Southville, Worcester, Mass., 360 D1 97
South Vineland, Cumberland, N.J. (part of Vineland) E3 106
South Wabiskaw, lake, Alta., Can B4 69
South Wadesboro, Anson, N.C., 189 C3 109
South Walden, Caledonia, Vt. C4 120
South Wallingford, Rutland, Vt., 120 E3 120
South Walpole, Norfolk, Mass., 700 E2 97
South Wardsboro, Windham, Vt. E3 120
South Wareham, Plymouth, Mass., 300 C6 97
South Waterford, Oxford, Maine, 230 D2 96
South Waverly, Bradford, Pa., 1,382 C8 114
South Wayne, Lafayette, Wis., 354 F4 124
South Weare, Hillsboro, N.H., 30 D3 105
South Webster, Scioto, Ohio, 803 D5 111
South Wellfleet, Barnstable, Mass., 400 C8 97
South Wellington, B.C., Can., 409 A9 68
South Wenatchee, Chelan, Wash B5 122
Southwest, Westmoreland, Pa., 800 F2 114
South West, div., Ice., 77,976 *n22 25
Southwest, cape, Que., Can B7 74
Southwest, cape, Vir. Is. k17 65
Southwest, chan., Fla E4 86
Southwest, head, N.B., Can E3 74
Southwest, pass, La E3 95
Southwest, pass, La F6 95
Southwest, pt., R.I. E10 84
South-West Africa, S. Afr., mandate, Afr., 512,496 B2 49, I7 42
South Westbury, Nassau, N.Y., 11,977 *G2 84
South West City, McDonald, Mo., 504 E3 101
Southwest Dillon, Dillon, S.C., 1,048 *C9 115
Southwestern, mts., Va C4 121
South West Fargo, Cass., N. Dak., 3,328 C9 110
Southwest Greensburg, Westmoreland, Pa., 3,264 *F2 114
Southwest Harbor, Hancock, Maine, 900 (1,480▲) D4 96
South Westminster, B.C., Can., 260 A10 68
South Weymouth, Norfolk, Mass. (part of Weymouth) B6, D3 97
South Wheelock, Caledonia, Vt. B4 120
South Whitley, Whitley, Ind., 1,325 B6 91
Southwick, Nez Perce, Idaho, 50 C2 89
Southwick, Hampden, Mass., 1,242 B2 97
South Williamson, Pike, Ky., 1,097 C7 94
South Williamsport, Lycoming, Pa., 6,972 D7 114
South Willington, Tolland, Conn., 300 B7 84
South Wilmington, Grundy, Ill., 730 B5 90
South Windermere, Charleston, S.C., 1,500 *F8 115
South Windham, Windham, Conn., 380 C8 84
South Windham, Cumberland, Maine, 1,142 E2, E4 96
South Windsor, Hartford, Conn., 900 (9,460▲) B6 85
Southwold, Eng., 2,228 B9 12
South Wolfeboro, Carroll, N.H., 150 C4 105
Southwood Acres, Hartford, Conn., 3,000 *B6 84
South Woodbury, Washington, Vt., 35 C4 120
South Woodstock, Windham, Conn., 400 B9 84
South Woodstock, Windsor, Vt., 85 D3 120
South Woodstown, Salem, N.J (part of Woodstown) D2 106
South Worthington, Hampshire, Mass., 60 B2 97
South Yarmouth, Barnstable, Mass., 2,029 C7 97
South Zanesville, Muskingum, Ohio, 1,557 C5 111
Soverato, It., 4,750 E6 21
Sovereign, Sask., Can., 125 F2 70
Sovetsk (Tilsit), Sov. Un., 85,900 A6 26, D3 27
Sovetskaya Gavan, Sov. Un., 49,000 E16 28, B10 37
Soviet Union (U.S.S.R.), country, Eur., Asia, 208,826,000 28, D11 33
Soya, cape, Jap D10 37
Sozopol, Bul., 3,178 D8 22
Spa, Bel., 9,055 B6 15
Spadra, Johnson, Ark., 300 B2 81
Spain, country, Eur., 30,430,698 G8 8, 20
Spalding, Austl., 240 F2 51
Spalding, Sask., Can., 416 E3 70
Spalding, Eng., 14,821 B7 12
Spalding, Nez Perce, Idaho, 200 C2 89
Spalding, Greeley, Nebr., 683 C7 103
Spalding co., Ga., 35,404 C2 87
Spanaway, Pierce, Wash., 2,500 B3 122
Spangle, Spokane, Wash., 208 B8, D7 122
Spangler, Cambria, Pa., 2,658 E4 114
Spangler, hill, Ohio C2 111
Spaniard's Bay, Newf., Can., 1,289 E5 75
Spanish, Ont., Can., 1,536 A2 72
Spanish, peak, Oreg C7 113
Spanish, riv., Ont., Can A3 72
Spanishburg, Mercer, W. Va., 200 D3 123
Spanish Fork, Utah, Utah, 6,472 C4 119
Spanish Fort, Montague, Tex., 100 C4 118
Spanish Possessions in North Africa, dep., Afr., 152,768 *B3, *B4 44
Spanish Ranch, Plumas, Calif., 150 C3 82
Spanish Sahara, dep., Afr., 19,000 E2 44
Spanish Town, Jam., 14,439 G5 64
Sparenberg, Dawson, Tex., 25 C2 118
Sparkill, Rockland, N.Y., 1,100 C5 106
Sparkman, Dallas, Ark., 787 D3 81
Sparks, Cook, Ga., 1,158 E3 87
Sparks, Cherry, Nebr., 5 B5 103
Sparks, Washoe, Nev., 16,618 D2 104
Sparks, Lincoln, Okla., 186 B5 112
Sparksville, Jackson, Ind., 100 G5 91
Sparksville, Adair, Ky., 150 C4 94
Sparland, Marshall, Ill., 534 B4 90
Sparr, Marion, Fla., 400 C4 86
Sparreholm, Swe., 974 t34 25
Sparrow Lake, Ont., Can., 53 C5 72
Sparrows Point, Baltimore, Md., 3,300 B5 85
Sparta, Hancock, Ga., 1,921 C4 87
Sparta, Randolph, Ill., 3,452 E4 90
Sparta, Gallatin, Ky., 235 B5 94
Sparta, Kent, Mich., 2,749 E5 98
Sparta, Christian, Mo., 272 E4 101
Sparta, Sussex, N.J., 500 A3 106
Sparta, Alleghany, N.C., 1,047 A2 109
Sparta, Morrow, Ohio, 228 B5 111
Sparta, Baker, Oreg C9 113
Sparta, White, Tenn., 4,510 D8 117
Sparta, Monroe, Wis., 6,080 E3 124
Sparta, mts., N.J. B3 106
Spartanburg, Randolph, Ind., 200 D8 91
Spartanburg, Spartanburg, S.C., 44,352 B4 115
Spartanburg, co., S.C., 156,830 B4 115
Spartansburg, Crawford, Pa., 500 C2 114
Spárti (Sparta), Grc., 10,412 D4 23
Spas-Demensk, Sov. Un., 10,000 D10 27
Spassk-Dalniy, Sov. Un., 29,400 E16 28
Spassk-Ryazanskiy, Sov. Un., 14,500 D13 27
Spaulding, Jefferson, Ala., 300 E4 78
Spaulding, Hughes, Okla., 150 B5 112
Spavinaw, Mayes, Okla., 319 A6 112
Spavinaw, creek, Okla A7 112
Spean Bridge, Scot D4 13
Spear, Avery, N.C., 170 C4 109
Spear, cape, Newf., Can E5 75
Spearfish, Lawrence, S. Dak., 3,682 C2 116
Spearhill, Man., Can., 75 D2 71
Spearman, Hansford, Tex., 3,555 A2 118
Spearsville, Union, La., 90 B3 95
Spearville, Ford, Kans., 602 E4 93
Spectacle, pond, Maine D4 96
Speculator, Hamilton, N.Y., 372 B6 108
Spedden, Alta., Can., 123 B5 69
Spednik, lake, N.B., Can C2 74
Speed, Clark, Ind., 950 H6 91
Speed, Phillips, Kans., 75 C4 93
Speed, Edgecombe, N.C., 142 B6 109
Speedway, Marion, Ind., 9,624 E5, H7 91
Speedwell, Claiborne, Tenn., 75 C10 117
Speedwell, Wythe, Va., 200 E1 121
Speer, Laramie, Wyo D8 125
Speers, Sask., Can., 175 E2 70
Speers, Washington, Pa., 1,479 *F2 108
Speigener, Elmore, Ala., 125 C3 78
Speight, Pike, Ky., 500 C7 94
Spelter, Harrison, W. Va., 500 B7 123
Spenard, Alsk., 9,074 *g17 79
Spencer, Clark, Idaho, 100 E6 89
Spencer, Owen, Ind., 2,557 F4 91
Spencer, Clay, Iowa, 8,864 A2 92
Spencer, Somerset, Maine C2 96
Spencer, Worcester, Mass., 5,593 B4 97
Spencer, Boyd, Nebr., 671 B7 103
Spencer, Tioga, N.Y., 767 C4 108
Spencer, Rowan, N.C., 2,904 B3 109
Spencer, Medina, Ohio, 742 A5 111
Spencer, Oklahoma, Okla., 1,189 *B4 112
Spencer, McCook, S. Dak., 460 D8 116
Spencer, Van Buren, Tenn., 870 D8 117
Spencer, Henry, Va., 200 E2 121
Spencer, Roane, W. Va., 2,660 C3 123
Spencer, Marathon, Wis., 897 D3 124
Spencer, co., Ind., 16,074 H4 91
Spencer, co., Ky., 5,680 B4 94
Spencer, butte, Oreg D3 113
Spencer, cape, Alsk k21 79
Spencer, gulf, Austl F6 50
Spencer, lake, Maine C2 96
Spencer, mts., Maine C3 96
Spencer, pond, Maine C3 96
Spencerport, Monroe, N.Y., 2,461 B3 108
Spencers Island, N.S., Can., 119 D5 74
Spencerville, DeKalb, Ind., 340 B8 91
Spencerville, Montgomery, Md., 900 B4 85
Spencerville, Allen, Ohio, 2,061 B3 111
Spences Bridge, B.C., Can., 239 D7 68
Spennymoor, Eng., 19,104 F7 13
Sperkhios, riv., Grc C4 23
Sperling, Man., Can., 172 E3 71
Sperrin, mts., N. Ire C4 11
Sperry, Des Moines, Iowa, 70 D6 92
Sperry, Tulsa, Okla., 883 A6 112
Sperryville, Rappahannock, Va., 300 C4 121
Spesutie, isl., Md B5 85
Spey, riv., Scot C5 13
Speyer, Ger., 38,500 D3 17
Spiceland, Henry, Ind., 863 E7 91
Spicer, Kandiyohi, Minn., 589 E4 99
Spickard, Grundy, Mo., 450 A4 101
Spicket, hill, Mass A5 97
Spider, lake, Wis B2 124
Spiekeroog, Ger., 770 A7 15
Spiekeroog, isl., Ger B3 16
Spielman, Washington, Md., 75 A2 85
Spiess, Santa Fe, N. Mex., 150 G6 107
Spiez, Switz., 8,168 C4 19
Spigno Monferrato, It., 3,014 E4 18
Spillimacheen, riv., B.C., Can D2 69
Spillville, Winneshiek, Iowa, 389 A6 92
Spilsby, Eng., 1,486 A8 12
Spinazzola, It., 10,850 D6 21
Spindale, Rutherford, N.C., 4,082 B2, D4 109
Spink, Union, S. Dak., 25 E9 116
Spink, co., S. Dak., 11,706 C7 116
Spink Colony, Spink, S. Dak., 100 C7 116
Spirit, Price, Wis C3 124
Spirit, lake, Iowa A2 92
Spirit, lake, Wash C3 122
Spirit, res., Wis C4 124
Spirit Lake, Kootenai, Idaho, 693 B2 89
Spirit Lake, Dickinson, Iowa, 2,685 A2 92
Spirit River, Alta., Can., 890 B1 69
Spiritwood, Sask., Can., 548 D2 70
Spiritwood, Stutsman, N. Dak., 90 C7 110
Spiro, LeFlore, Okla., 1,450 B7 112
Spišská Nová Ves., Czech., 16,900 D6 26
Spithead, roadstead, Eng D6 12
Spitsbergen, see Svalbard, is., Nor.
Spittal, Aus., 10,045 E6 16
Spivey, Kingman, Kans., 98 E5 93
Split, Yugo., 99,462 D3 22
Split, cape, N.S., Can D5 74
Split, lake, Man., Can A3 71
Splügen, Switz., 346 C7 19
Splügen, pass, Switz C7 19
Spofford, Cheshire, N.H., 300 E2 105
Spofford, Kinney, Tex., 138 E2 118
Spofford, lake, N.H E2 105
Spokane, Spokane, Wash., 181,608 (*252,000) B8, D7 122
Spokane, co., Wash., 278,333 B8 122
Spokane, mtn., Wash B8 122
Spokane, riv., Wash B7 122
Spokane, val., Wash D7 122
Spoleto, It., 24,500 C4 21
Spondin, Alta., Can D5 69
Spoon, butte, Wyo C8 125
Spoon, riv., Ill C3 90
Spooner, Washburn, Wis., 2,398 C2 124
Spooner, lake, Wis C1 124
Sporades, is., Grc C5 23
Spotswood, Middlesex, N.J., 5,788 C4 106
Spotsylvania, Spotsylvania, Va., 150 C5 121
Spotsylvania, co., Va., 13,819 C5 121
Spotted, isl., Newf., Can B4 75
Spotted Horse, Campbell, Wyo., 10 A7 125
Spöttrup (Rödding), Den., 628 B2 24
Spottsville, Henderson, Ky., 465 C2 94
Spragge, Ont., Can., 430 A2 72
Sprague, Montgomery, Ala., 150 C3 78
Sprague, Man., Can., 364 E4 71
Sprague, Lancaster, Nebr., 120 D9 103
Sprague, Lincoln, Wash., 597 B8 122
Sprague, Raleigh, W. Va., 3,073 D7 123
Sprague, lake, Wash B7 122
Sprague, riv., Oreg E5 113
Sprague River, Klamath, Oreg., 150 E5 113
Spragueville, Jackson, Iowa, 100 B7 92
Spragueville, St Lawrence, N.Y., 100 A5, B1 108
Spragueville, Providence, R.I., 400 B10 84
Sprakensehl, Ger., 620 F4 24
Spray, Rockingham, N.C., 4,565 A4 109
Spray, Wheeler, Oreg., 194 C7 113
Spreckelsville, Maui, Haw., 950 C5 88
Spreckles, Monterey, Calif., 800 C6 82
Spree, riv., Ger B9 17
Spremberg, Ger., 22,900 B9 17
Sprigg, Mingo, W. Va., 250 D2 123
Spring, Harris, Tex., 950 F5 118
Spring, brook, Pa B9 114
Spring, creek, Ga F2 87
Spring, isl., S.C. G6 115
Spring, lake, Ill A3 90
Spring, lake, Maine C2 96
Spring, mts., Nev G6 104
Spring, peak, Nev B4 104
Spring, riv., Ark., Mo A4 81, D3 101
Spring Arbor, Jackson, Mich., 700 F6 98
Springbok, S. Afr., 3,111 C2 49
Springboro, Warren, Ohio, 917 C3 111
Springboro, Crawford, Pa., 583 C1 114
Spring Branch, Comal, Tex., 15 A4 118
Springbrook, Ont., Can., 140 C7 72
Springbrook, Jackson, Iowa, 139 B7 92
Spring Brook, Williams, N. Dak., 35 A2 110
Springbrook, Lackawanna, Pa B9 114
Springbrook, Washburn, Wis., 100 C2 124
Spring Canyon, Carbon, Utah, 250 D5 119
Spring City, Chester, Pa., 3,162 F10 114
Spring City, Rhea, Tenn., 1,800 D9 117
Spring City, Sanpete, Utah, 463 D4 119
Spring Coulee, Alta., Can., 76 E4 69
Spring Creek, Madison, N.C., 50 D3 109
Spring Creek, Warren, Pa., 160 C2 114
Springcreek, Madison, Tenn., 100 B3 117
Springdale, Washington, Ark., 11,895 A1 81
Springdale, Newf., Can., 1,638 D3 75
Springdale, Park, Mont., 60 E6 102
Springdale, Hamilton, Ohio, 3,550 *D2 111
Springdale, Multnomah, Oreg., 150 B4 113
Springdale, Allegheny, Pa., 5,602 A7, E2 114
Springdale, Lexington, S.C., 1,002 *D5 115
Springdale, Washington, Utah, 248 F3 119
Springdale, Stevens, Wash., 254 A8 122
Spring Dale, Fayette, W. Va., 200 D4 123
Springe, Ger., 10,200 A4 17
Springer, Colfax, N. Mex., 1,564 A5 107
Springer, Carter, Okla., 212 C4 112
Springers, Lawrence, Tenn B4 117
Springerton, White, Ill., 232 E5 90
Springerville, Apache, Ariz., 719 C4 80
Springfield, Conway, Ark., 125 B3 81
Springfield, N.B., Can., 109 D4 74
Springfield, N.S., Can., 266 E5 74
Springfield, Ont., Can., 539 E4 72
Springfield, Baca, Colo., 1,791 D8 83
Springfield, Bay, Fla., 4,628 G3 86
Springfield, Effingham, Ga., 858 D5 87
Springfield, Bingham, Idaho, 80 F6 89
Springfield, Sangamon, Ill., 83,271 (*122,700) D4 90
Springfield, Washington, Ky., 2,382 C4 94
Springfield, Livingston, La., 268 B6, D5 95
Springfield, Penobscot, Maine, 75 (426▲) C4 96
Springfield, Hampden, Mass., 174,463 (*429,400) B2 97
Springfield, Brown, Minn., 2,701 F4 99
Springfield, Greene, Mo., 95,865 (*108,700) D4 101
Springfield, Sarpy, Nebr., 506 C9, E3 103
Springfield, Sullivan, N.H., 50 (283▲) C2 105
Springfield, Union, N.J., 14,467 *B4 106
Springfield, Clark, Ohio, 82,723 (*112,100) C4 111
Springfield, Lane, Oreg., 19,616 C4 113
Springfield, Delaware, Pa., 26,733 *B11 114
Springfield, Orangeburg, S.C., 787 D5 115
Springfield, Bon Homme, S. Dak., 1,194 E8 116
Springfield, Robertson, Tenn., 9,221 A5 117
Springfield, Windsor, Vt., 6,600 E4 120
Springfield, Fairfax, Va., 10,783 B5 121
Springfield, Hampshire, W. Va., 300 B6 123
Springfield, lake, Ill D4 90
Springfield, plat., Mo D4 101
Springfield Place, Calhoun, Mich., 5,136 *F5 98
Springfontein, S. Afr., 2,850 D4 49
Spring Garden, Cherokee, Ala., 65 B4 78
Spring Glen, Duval, Fla., 5,000 *B5 86
Spring Green, Sauk, Wis., 1,146 E3 124
Spring Grove, McHenry, Ill., 301 E2 90
Spring Grove, Wayne, Ind., 471 E8 91
Spring Grove, Houston, Minn., 1,342 G7 99
Spring Grove, York, Pa., 1,675 G8 114
Springhill, Pike, Ala., 80 D4 78
Spring Hill, Hempstead, Ark., 200 D2 81
Springhill, N.S., Can., 5,836 D6 74
Spring Hill, Que., Can., 300 D6 73
Spring Hill, Warren, Iowa, 111 B7 92
Spring Hill, Johnson, Kans., 909 D9 93
Springhill, Webster, La., 6,437 A2 95
Spring Hill, Cambria, Pa., 1,127 *F4 114
Spring Hill, Maury, Tenn., 689 B5 117
Spring Hill Junction, N.S., Can., 217 D5 74
Spring Hope, Nash., N.C., 1,336 B4 109
Springhouse, B.C., Can., 81 D6 68
Spring Lake, Hernando, Fla., 90 D4 86
Spring Lake, Ottawa, Mich., 2,063 E4 98
Spring Lake, Monmouth, N.J., 2,922 C4 106
Spring Lake, Cumberland, N.C., 4,110 B5 109
Spring Lake, Klamath, Oreg E5 113
Spring Lake, Utah, Utah D4 119
Spring Lake Heights, Monmouth, N.J., 3,309 C4 106
Spring Lake Park, Anoka and Ramsey, Minn., 3,260 *E5 99
Springlee, Jefferson, Ky., 987 *H6 94
Spring Lick, Grayson, Ky., 125 C3 94
Spring Mill, Centre, Pa., 600 E6 114
Spring Mills, Lancaster, S.C., 1,069 *B6 115
Springmont, Berks, Pa., 1,000 *F10 114
Spring Place, Murray, Ga., 194 B2 87
Springport, Jackson, Mich., 693 F6 98
Springs, Suffolk, N.Y., 200 E8 84
Springs, S. Afr., 137,253 C4 49
Springside, Sask., Can., 326 F4 70
Springsure, Austl., 719 B7 51
Springton, res., Pa D2 106
Springvale, Austl A3 51
Springvale, Randolph, Ga., 57 E2 87
Springvale, York, Maine, 2,379 E2 96
Spring Valley, Colbert, Ala., 100 A2 78
Spring Valley, San Diego, Calif., 7,000 *E2 82
Spring Valley, Sask., Can., 86 H3 70
Spring Valley, Bureau, Ill., 5,371 B4 90
Spring Valley, Fillmore, Minn., 2,628 G6 99
Spring Valley, Rockland, N.Y., 6,538 D1 108
Spring Valley, Greene, Ohio, 678 C3 111
Spring Valley, Harris, Tex., 3,004 *E5 118
Spring Valley, Grayson, Va., 25 E1 121
Spring Valley, Pierce, Wis., 977 D1 124
Springview, Keya Paha, Nebr., 281 B6 103
Springville, St. Clair, Ala., 822 B3 78
Springville, Lawrence, Ind., 150 G4 91
Springville, Linn, Iowa, 785 B6 92
Springville, Livingston, La., 60 B6, D5 95
Springville, Pontotoc, Miss., 25 A4 100
Springville, Erie, N.Y., 3,852 C2 108
Springville, Susquehanna, Pa., 300 C10 114
Springville, Henry, Tenn., 40 A3 117
Springville, Utah, Utah, 7,913 C4 119
Springwater, Sask., Can., 120 F1 70
Springwater, Livingston, N.Y., 300 C3 108
Sprole, Roosevelt, Mont., 2 B11 102
Spruce, fork, W. Va D5 123
Spruce, mtn., Nev C7 104
Spruce, peak, Vt E2 120
Spruce, riv., Sask., Can D3 70
Spruce Brook, Newf., Can., 117 D2 75
Spruce Creek, Huntingdon, Pa., 200 E5 114
Sprucedale, Ont., Can., 213 B5 72
Spruce Grove, Alta., Can., 465 C4 69
Spruce Knob, mtn., W. Va C5 123
Spruce Lake, Sask., Can., 111 D1 70
Spruce Pine, Franklin, Ala., 600 A2 78
Spruce Pine, Mitchell, N.C., 2,504 C4 109
Spruga, Switz D6 19
Spry, Garfield, Utah, 25 F3 119
Spud, rock, Ariz E5 80
Spungabera, Moz B5 49
Spur, Dickens, Tex., 2,170 C2 118
Spurfield, Alta., Can., 103 B3 69
Spurgeon, Pike, Ind., 269 H3 91
Spur Lake, Catron, N. Mex D1 107
Spurr, mtn., Alsk g15 79
Spy Hill, Sask., Can., 204 G5 70
Squak, mtn., Wash D2 122
Squam, butte, Oreg D7 113
Squam, lake, N.H C4 105
Squam, mts., N.H C3 105
Squamish, B.C., Can., 1,557 E6 68
Squa Pan, Aroostook, Maine, 50 B4 96
Squapan, lake, Maine B4 96
Square, lake, Maine A4 96
Square Butte, Chouteau, Mont., 85 C6 102
Square Islands, Newf., Can B4, h11 75
Squatteck, Que., Can., 1,088 B9 73
Squaw, mtn., Maine C3 96
Squaw Cap, mtn., N.B., Can B3 74
Squaw Lake, Itasca, Minn., 129 C4 99
Squibnocket, pt., Mass D6 97
Squillace, gulf, It E6 21
Squire, McDowell, W. Va., 900 *D3 123
Squires, Douglas, Mo., 50 E5 101
Squirrel, Fremont, Idaho, 5 E7 89
Srbobran, Yugo., 14,392 C4 22
Sredne-Kolymsk, Sov. Un., 3,000 C18 28
Srem, Pol., 8,308 B4 26
Sremska Mitrovica, Yugo., 20,790 C4 22
Sremski Karlovci, Yugo., 6,383 C4 22
Sretensk, Sov. Un., 24,600 D14 28
Srinagar, India, 285,257 (*295,084) B6 39
Sroda, Pol., 11,700 B4 26
Sroda Slaska, Pol., 4,301 C4 26
Ssu, China H7 36
Ssu-chan, China B2 37
Ssumao, China, 2,000 G5 34
Ssunan, China, 8,000 K3 36
Ssushui, China G7 36

Staatsburg, Dutchess, N.Y., 450 . . . D7 108
Stab, Pulaski, Ky., 300 . . . C5 94
Stacks, mts., Ire . . . E2 11
Stacy, Carteret, N.C., 400 . . . C7 109
Stacyville, Mitchell, Iowa, 588 . . . A5 92
Stacyville, Penobscot, Maine, 130 (673▲) . . . C4 96
Stade, Ger., 30,500 . . . B4 16
Stadskanaal, Neth., 6,800 . . . A6 15
Stadthagen, Ger., 14,900 . . . A4 17
Stadtkyll, Ger., 940 . . . D6 15
Stadtlohn, Ger., 8,500 . . . B1 17
Stadtroda, Ger., 6,261 . . . C6 17
Stäfa, Switz., 6,947 . . . B6 19
Staffa, isl., Scot . . . D2 13
Staffin, Scot . . . C2 13
Stafford, Tolland, Conn., 350 (7,476▲) . . . B7 84
Stafford, Eng., 47,814 . . . B5 12
Stafford, Stafford, Kans., 1,862 . . . E5 93
Stafford, Holt, Nebr., 3 . . . B7 103
Stafford, Genesee, N.Y., 225 . . . C2 108
Stafford, Monroe, Ohio, 113 . . . C6 111
Stafford, Custer, Okla., 25 . . . B2 112
Stafford, Fort Bend, Tex., 1,485 . . . *F5 118
Stafford, Stafford, Va., 500 . . . C5 121
Stafford, co., Eng., 1,733,887 . . . B5 12
Stafford, co., Kans., 7,451 . . . D5 93
Stafford, co., Va., 16,876 . . . C5 121
Stafford Springs, Tolland, Conn., 3,322 . . . B7 84
Stafford Springs, Jasper, Miss . . . D4 100
Staffordsville, Johnson, Ky., 700 . . . C7 94
Staffordville, Tolland, Conn., 400 . . . B7 84
Staffordville, Ocean, N.J., 75 . . . D4 106
Staines, Eng., 49,259 . . . C7 12
Stains, Fr., 27,503 . . . g10 14
Stalactite, cavern, Mo . . . E6 101
Stålboga, Swe . . . t34 25
Stalden, Switz., 1,007 . . . D4 19
Staley, Randolph, N.C., 260 . . . B4 109
Stalin, see Varna, Bul.
Stalinograd, see Katowice, Pol.
Stallcup, cave, Mo . . . D5 101
Stallo, Neshoba, Miss., 200 . . . C4 100
Stalwart, Sask., Can., 53 . . . F3 70
Stalwart, Chippewa, Mich., 15 . . . B6 98
Stambaugh, Iron, Mich., 1,876 . . . B2 98
Stamford, Fairfield, Conn., 92,713 . . . E2 84
Stamford, Eng., 11,743 . . . B7 12
Stamford, Harlan, Nebr., 220 . . . D6 103
Stamford, Delaware, N.Y., 1,166 . . . C6 108
Stamford, Jackson, S. Dak., 50 . . . D4 116
Stamford, Jones, Tex., 5,259 . . . C3 118
Stamford, Bennington, Vt., 150 (600▲) . . . E2 120
Stamping Ground, Scott, Ky., 353 . . . B5 94
Stampried, S.W. Afr., 433 . . . B2 49
Stamps, Lafayette, Ark., 2,591 . . . D2 81
Stanaford, Raleigh, W. Va., 950 . . . D3, D7 123
Stanardsville, Greene, Va., 283 . . . C4 121
Stanberry, Gentry, Mo., 1,409 . . . A3 101
Stanchfield, Isanti, Minn., 150 . . . E5 99
Standale, Kent, Mich., 1,000 . . . *F5 98
Standard, Alta., Can., 266 . . . D4 69
Standard, La Salle, La., 150 . . . C3 95
Standard, Westmoreland, Pa., 700 . . . *F2 114
Standerton, S. Afr., 16,868 . . . C4 49
Standing Rock, Chambers, Ala., 450 . . . B4 78
Standing Rock, McKinley, N. Mex . . . B1 107
Standing Rock, Indian res., S. Dak . . . B4 116
Standish, Cumberland, Maine, 200 (2,095▲) . . . E2 96
Standish, Plymouth, Mass., 130 . . . E4 97
Standish, Arenac, Mich., 1,214 . . . E7 98
Standish, Clinton, N.Y., 300 . . . A3 108
Standrod, Box Elden, Utah . . . B2 119
Stanfield, Pinal, Ariz., 300 . . . E3 80
Stanfield, Umatilla, Oreg., 745 . . . B7 113
Stanford, Santa Clara, Calif., 9,000 . . . *D2 82
Stanford, McLean, Ill., 479 . . . C4 90
Stanford, Monroe, Ind., 100 . . . F4 91
Stanford, Lincoln, Ky., 2,019 . . . C5 94
Stanford, Judith Basin, Mont., 615 . . . C6 102
Stanford, range, B.C., Can . . . D3 69
Stangelville, Kewaunee, Wis., 100 . . . A6, D6 124
Stanger, S. Afr., 9,557 . . . C5 49
Stanhope, Que., Can., 167 . . . D6 73
Stanhope, Eng., 2,195 . . . F6 13
Stanhope, Hamilton, Iowa, 461 . . . B4 92
Stanhope, Sussex, N.J., 1,814 . . . B3 106
Staniard Creek, Ba. Is., 618 . . . C5 64
Stanislaus, co., Calif., 157,294 . . . D3 82
Stanislaus, riv., Calif . . . D3 82
Stanislawow, Pol., 2,000 . . . k15 26
Stanke Dimitrov, Bul., 19,239 . . . D6 22
Stanley, N.B., Can., 301 . . . C3 74
Stanley (Port Stanley), Falk. Is., 1,250 . . . I5 53
Stanley, Custer, Idaho, 35 . . . E4 89
Stanley, Johnson, Kans., 330 . . . D9 93
Stanley, Daviess, Ky., 170 . . . C2 94
Stanley, Santa Fe, N. Mex., 70 . . . B4, G6 107
Stanley, Gaston, N.C., 1,980 . . . B2 109
Stanley, Mountrail, N. Dak., 1,795 . . . A3 110
Stanley, Pushmataha, Okla., 5 . . . C6 112
Stanley, Page, Va., 1,039 . . . C4 121
Stanley, Chippewa, Wis., 2,014 . . . D3 124
Stanley, co., S. Dak., 4,085 . . . C5 116
Stanley, falls, Con L . . . A3 48
Stanleytown, Henry, Va., 500 . . . E3 121
Stanleyville, Con. L., 80,000 . . . A4 48
Stanleyville, Forsyth, N.C., 1,138 . . . *A3 109
Stanly, co., N.C., 40,873 . . . B3 109
Stanmore, Alta., Can., 75 . . . D5 69
Stann Creek, Br. Hond., 3,414 . . . B3 62
Stanovoi, mts., Sov. Un . . . D15 33
Stans, Switz., 4,337 . . . C5 19
Stansbury Estates, Baltimore, Md., 3,500 . . . *B5 85
Stansbury, Sweetwater, Wyo., 50 . . . D3 125
Stanstead, co., Que., Can., 36,095 . . . D5 73
Stanthorpe, Austl., 3,234 . . . D8 51
Stanton, Chilton, Ala., 250 . . . C3 78
Stanton, Orange, Calif., 11,163 . . . *F3 82
Stanton, New Castle, Del., 2,000 . . . *A6 85
Stanton, Montgomery, Iowa, 514 . . . D2 92
Stanton, Powell, Ky., 753 . . . C6 94
Stanton, Montcalm, Mich., 1,139 . . . E5 98
Stanton, Adams, Miss., 250 . . . D2 100
Stanton, Franklin, Mo., 163 . . . C6 101
Stanton, Stanton, Nebr., 1,317 . . . C8 103
Stanton, Hunterdon, N.J., 200 . . . B3 106
Stanton, Mercer, N. Dak., 409 . . . B4 110
Stanton, Haywood, Tenn., 458 . . . B2 117
Stanton, Martin, Tex., 2,228 . . . C3 118
Stanton, co., Kans., 2,108 . . . E2 93
Stanton, co., Nebr., 5,783 . . . C8 103
Stantonsburg, Wilson, N.C., 897 . . . B6 109
Stantonville, McNairy, Tenn., 125 . . . B3 117
Stanwood, Cedar, Iowa, 598 . . . C6 92
Stanwood, Mecosta, Mich., 205 . . . E5 98
Stanwood, Snohomish, Wash., 1,123 . . . A3 122
Staplehurst, Seward, Nebr., 240 . . . D8 103
Staples, Todd, Minn., 2,706 . . . D4 99
Staples, Guadalupe, Tex., 175 . . . A4 118
Stapleton, Baldwin, Ala., 800 . . . E2 78
Stapleton, Jefferson, Ga., 356 . . . C4 87
Stapleton, Logan, Nebr., 359 . . . C5 103
Stapp, Le Flore, Okla . . . C7 112
Star, Ada, Idaho, 400 . . . F2 89
Star, Rankin, Miss., 300 . . . C3 100
Star, Montgomery, N.C., 745 . . . B4 109
Star, Mills, Tex., 75 . . . D3 118
Star, lake, Minn . . . D3 99
Star, peak, Nev . . . C3 104
Stara Boleslav, Czech., 4,744 . . . n18 26
Starachowice, Pol., 36,000 . . . C6 26
Staraya Russa, Sov. Un., 35,400 . . . C8 27
Stara Zagora, Bul., 37,057 . . . D7 22
Stara-Zagora, co., Bul., 821,764 . . . *D7 22
Starbuck, Man., Can., 240 . . . E3 71
Starbuck, Pope, Minn., 1,099 . . . E3 99
Starbuck, Columbia, Wash., 161 . . . C7 122
Starbuck, isl., Pac. O . . . G12 7
Star City, Lincoln, Ark., 1,573 . . . D4 81
Star City, Sask., Can., 571 . . . E3 70
Star City, Pulaski, Ind., 500 . . . C4 91
Star City, Monongalia, W. Va., 1,236 . . . B5 123
Starford, Indiana, Pa., 400 . . . E4 114
Stargard, Pol., 31,000 . . . B3 26
Stargo, Greenlee, Ariz., 1,075 . . . D6 80
Staritsa, Sov. Un., 7,500 . . . C10 27
Starjunction, Fayette, Pa., 1,142 . . . *F2 114
Stark, Neosho, Kans., 96 . . . E4 93
Stark, Coos, N.H., 35 (327▲) . . . A4 105
Stark, co., Ill., 8,152 . . . B4 90
Stark, co., N. Dak., 18,451 . . . C3 110
Stark, co., Ohio, 340,345 . . . B6 111
Starke, Bradford, Fla., 4,806 . . . C4 86
Starke, co., Ind., 17,911 . . . B4 91
Starkey, Union, Oreg . . . B8 113
Starkey, Roanoke, Va., 800 . . . D3 121
Starks, Calcasieu, La., 500 . . . D2 95
Starks, Somerset, Maine, 150 (306▲) . . . D3 96
Starksboro, Addison, Vt., 150 (502▲) . . . C2 120
Starkville, Las Animas, Colo., 261 . . . D6 83
Starkville, Oktibbeha, Miss., 9,041 . . . B5 100
Starkville, Herkimer, N.Y., 100 . . . C6 108
Starkweather, Ramsey, N. Dak., 223 . . . A7 110
Starlight, Wayne, Pa., 75 . . . C11 114
Starnberg, Ger., 10,000 . . . B7 18
Starobelsk, Sov. Un., 10,000 . . . G12 27
Starodub, Sov. Un., 10,000 . . . E9 27
Starogard, Pol., 26,000 . . . B5 26
Starokonstantinov, Sov. Un., 10,000 . . . G6 27
Starominskaya, Sov. Un., 10,000 . . . F16 9, H12 27
Star Prairie, St. Croix, Wis., 331 . . . C1 124
Starr, Anderson, S.C., 243 . . . C2 115
Starr, co., Tex., 17,137 . . . F3 118
Starr, mtn., Tenn . . . D9 117
Starrking, Coos, N.H. . . . B4 105
Starrucca, Wayne, Pa., 330 . . . C11 114
Start, bay, Eng . . . D4 12
Start, pt., Eng . . . D4 12
Startex, Spartanburg, S.C., 950 . . . B3 115
Startup, Snohomish, Wash., 250 . . . B4 122
Starý Plzenec, Czech., 3,376 . . . D8 17
Stary Sacz, Pol., 4,586 . . . D6 26
Staryy Oskol, Sov. Un., 10,000 . . . F11 27
Stassfurt, Ger., 26,300 . . . B6 17
Staszow, Pol., 4,586 . . . C6 26
State Center, Marshall, Iowa, 1,142 . . . B4 92
State College, Craighead, Ark. (part of Jonesboro) . . . B5 81
State College, Oktibbeha, Miss., 5,000 . . . B5 100
State College, Centre, Pa., 22,409 . . . E6 114
State Line, Warren, Ind., 171 . . . D2 91
State Line, Berkshire, Mass., 100 . . . B1 97
State Line, Greene, Miss., 653 . . . B5 100
State Line, Cheshire, N.H., 30 . . . E2 105
State Line, Franklin, Pa., 400 . . . G6 114
Staten, isl., N.Y . . . E1 108
Staten Island, see Richmond, borough and co., N.Y.
Statenville, Echols, Ga., 400 . . . F4 87
State Road, Aroostook, Maine, 65 . . . B4 96
State Road, Surry, N.C., 600 . . . A3 109
State Sanatorium, Logan, Ark . . . B2 81
Statesboro, Bulloch, Ga., 8,356 . . . D5 87
State Schools, Drew, Ark., 500 . . . D4 81
Statesville, Iredell, N.C., 19,844 . . . B3 109
Statham, Barrow, Ga., 711 . . . C3 87
Stathelle, Nor., 700 . . . p27 25
Station No. 6, Sud . . . A3 47
Statue of Liberty, nat. mon., N.Y . . . E5 108
Staufen, Ger., 3,200 . . . B3 18
Stauffer, Lake, Oreg . . . D6 113
Staunton, Macoupin, Ill., 4,228 . . . D4 90
Staunton, Clay, Ind., 490 . . . F3 91
Staunton (Independent City), Va., 22,232 . . . C3 121
Stavanger, Nor., 52,800 (*101,000) . . . H1 25
Stave, lake, B.C., Can . . . A10 68
Staveley, Eng., 18,071 . . . A6 12
Stavelot, Bel., 4,500 . . . D5 15
Stavely, Alta., Can., 349 . . . D4 69
Stavenhagen, Ger., 6,521 . . . E6 24
Stavenisse, Neth., 1,416 . . . C4 15
Staveren, Neth., 900 . . . B5 15
Stavropol, Sov. Un., 151,000 . . . F17 9, D2 29
Stavropol, Sov. Un., 78,000 . . . E18 9
Stawell, Austl., 5,504 . . . H4 51
Stayner, Ont., Can., 1,671 . . . C4 72
Stayton, Marion, Oreg., 2,108 . . . C2, C4 113
Stead, Union, N. Mex., 5 . . . A6 107
Steamboat, Washoe, Nev., 300 . . . D2 104
Steamboat, mtn., Mont . . . C4 102
Steamboat Canyon, Apache, Ariz., 200 . . . B6 80
Steamboat Rock, Hardin, Iowa, 426 . . . B4 92
Steamboat Springs, Routt, Colo., 1,843 . . . A4 83
Stearns, McCreary, Ky., 950 . . . D5 94
Stearns, co., Minn., 80,345 . . . E4 99
Stearns, lake, Fla . . . E5 86
Stebbins, Alsk., 158 . . . C7 79
Stebbins, Aroostook, Maine, 70 . . . B5 96
Steblev, Sov. Un., 10,000 . . . G8 27
Stechovice, Czech., 1,086 . . . o17 26
Stecker, Caddo, Okla., 100 . . . C3 112
Stecoah, Graham, N.C., 100 . . . D2 109
Stedman, Cumberland, N.C., 458 . . . B5 109
Steel, mtn., Idaho . . . F3 89
Steele, St. Clair, Ala., 625 . . . B3 78
Steele, Pemiscot, Mo., 2,301 . . . E8 101
Steele, Kidder, N. Dak., 847 . . . C6 110
Steele, co., Minn, 25,029 . . . F5 99
Steele, co., N. Dak., 4,719 . . . B8 110
Steele, mtn., Wyo . . . D6 125
Steele City, Jefferson, Nebr., 173 . . . D8 103
Steeleville, Randolph, Ill., 1,569 . . . E4 90
Steelmanville, Atlantic, N.J., 200 . . . E3 106
Steelton, Dauphin, Pa., 11,266 . . . F8 114
Steelville, Crawford, Mo., 1,127 . . . D6 101
Steen, Rock, Minn., 198 . . . G2 99
Steenbergen, Neth., 5,000 . . . C4 15
Steenburg, Ont., Can., 90 . . . C7 72
Steens, Lowndes, Miss., 120 . . . B5 100
Steens, mtn., Oreg . . . E8 113
Steenwijk, Neth., 10,700 . . . B6 15
Steep Falls, Cumberland, Maine, 400 . . . E2 96
Steephill, lake, Sask., Can . . . B4 70
Steep Rock, Man., Can., 168 . . . D2 71
Steep Rock Lake, Ont., Can., 80 . . . E8 72
Stefanesti, Rom., 7,770 . . . B8 22
Stefanie, lake, Eth . . . E5 47
Steffisburg, Switz., 10,757 . . . C4 19
Stege Den., 2,620 (*3,816) . . . D6 24
Steger, Cook and Will, Ill., 6,432 . . . B6, F3 90
Stegi, Swaz . . . C5 49
Steiermark (Styria), state, Aus., 1,137,865 . . . *E7 16
Steilacoom, Pierce, Wash., 1,569 . . . E1 122
Stein, Ger., 7,500 . . . D6 17
Steinach, Aus., 2,155 . . . B7 18
Stein am Rhein, Switz., 2,588 . . . A6 19
Steinauer, Pawnee, Nebr., 124 . . . D9 103
Steinbach, Man., Can., 3,739 . . . E3 71
Steinbach-Hallenberg, Ger., 7,010 . . . C5 17
Steinfort, Lux., 2,338 . . . E5 15
Steinhatchee, Taylor, Fla., 325 . . . C3 86
Steinhatchee, riv., Fla . . . C3 86
Steinhausen, S.W. Afr . . . B2 49
Steinhuder, lake, Ger . . . F3 24
Steinkjer, Nor., 4,200 . . . E4 25
Steins, Hidalgo, N. Mex., 25 . . . E1 107
Stella, Izard, Ark . . . A4 81
Stella, Ont., Can., 96 . . . C8 72
Stella, Calloway, Ky., 100 . . . B3 94
Stella, Newton, Mo., 166 . . . E3 101
Stella, Richardson, Nebr., 262 . . . D10 103
Stella, mtn., It . . . C5 18
Stellarton, N.S., Can., 5,327 . . . D7 74
Stellenbosch, S. Afr., 22,233 . . . *D2 49
Stelvio, pass., Switz . . . C9 19
Stem, Granville, N.C., 221 . . . A5 109
Stemmers Run, Baltimore, Md. (part of Essex) . . . B7 85
Stenay, Fr., 3,829 . . . E5 15
Stendal, Ger., 37,000 . . . B5 16
Stendal, Pike, Ind., 180 . . . H3 91
Stenen, Sask., Can., 275 . . . F4 70
Stengårdshult, Swe . . . A7 24
Stennes, lake, Scot . . . A5 13
Stensele, Swe., 870 . . . E7 25
Stepanakert, Sov. Un., 22,000 . . . F3 29
Stepenitz, riv., Ger . . . E5 24
Stephan, Hyde, S. Dak., 300 . . . C6 116
Stephen, Marshall, Minn., 858 . . . B2 99
Stephens, Ouachita, Ark., 1,275 . . . D2 81
Stephens, Oglethorpe, Ga., 150 . . . C3 87
Stephens, co., Ga., 18,391 . . . B3 87
Stephens, co., Okla., 37,990 . . . C4 112
Stephens, co., Tex., 8,885 . . . C3 118
Stephens, passage, Alsk . . . m23 79
Stephensburg, Morris, N.J., 150 . . . B3 106
Stephens City, Frederick, Va., 876 . . . B4 121
Stephenson, Menominee, Mich., 820 . . . C3 98
Stephenson, co., Ill., 46,207 . . . A4 90
Stephensport, Breckinridge, Ky., 150 . . . D3 94
Stephenston, Warrick, Ind., 60 . . . H3 91
Stephenville, Newf., Can., 6,043 . . . D2 75
Stephenville, Erath, Tex., 7,359 . . . C3 118
Stephenville Crossing, Newf., Can., 2,209 . . . D2 75
Stepney Depot, Fairfield, Conn . . . D4 84
Stepnoy, Sov. Un . . . D3 29
Stepnyak, Sov. Un., 20,000 . . . C8 29
Steps, pt., Am. Sam . . . 52
Steptoe, White Pine, Nev., 5 . . . D7 104
Steptoe, Whitman, Wash., 100 . . . B8 122
Steptoe, creek, Nev . . . E7 104
Sterco, Alta., Can., 125 . . . C2 68
Sterkstroom, S. Afr., 3,744 . . . D4 49
Sterley, Floyd, Tex., 100 . . . B2 111
Sterling, Logan, Colo., 10,751 . . . A7 83
Sterling, Windham, Conn., 450 (1,397▲) . . . C9 84
Sterling, Bingham, Idaho, 70 . . . F6 89
Sterling, Whiteside, Ill., 15,688 . . . B4 90
Sterling, Fountain, Ind., 430 . . . D3 91
Sterling, Rice, Kans., 2,303 . . . D5 93
Sterling, Worcester, Mass., 700 (3,193▲) . . . B4 97
Sterling, Arenac, Mich., 470 . . . D6 98
Sterling, Johnson, Nebr., 471 . . . D9 103
Sterling, Burleigh, N. Dak., 100 . . . C5 110
Sterling, Comanche, Okla., 562 . . . C3 112
Sterling, Sanpete, Utah, 137 . . . D4 119
Sterling, Loudoun, Va., 300 . . . A4, B5 121
Sterling, co., Scot., 190,500 . . . D4 13
Sterling, co., Tex., 1,177 . . . D2 118
Sterling, res., Colo . . . A7 83
Sterling City, Sterling, Tex., 854 . . . D2 118
Sterling Run, Cameron, Pa., 150 . . . D5 114
Sterlington, Ouachita, La., 1,200 . . . B3 95
Sterlitamak, Sov. Un., 125,000 . . . C5 29
Sternberg, Ger., 4,479 . . . E5 24
Sternberk, Czech., 11,200 . . . D4 26
Sterrett, Shelby, Ala., 450 . . . B3 78
Steti, Czech., 1,176 . . . C9 17
Stetson, Penobscot, Maine, 150 (420▲) . . . D3 96
Stetson, mtn., Maine . . . C5 96
Stetsonville, Taylor, Wis., 319 . . . C3 124
Stettin, see Szczecin, Pol.
Stettin, lagoon, Ger., Pol . . . E8 24
Stettler, Alta., Can., 3,638 . . . C4 69
Steuben, Washington, Maine, 350 (673▲) . . . D5 96
Steuben, Schoolcraft, Mich., 15 . . . B4 98
Steuben, Crawford, Wis., 193 . . . E3 124
Steuben, co., Ind., 17,184 . . . A7 91
Steuben, co., N.Y., 97,691 . . . C3 108
Steubenville, Jefferson, Ohio, 32,495 (*121,300) . . . B7 111
Steve, Yell, Ark., 10 . . . C2 81
Stevenage, Eng., 42,964 . . . C7 12
Stevens, Burlington, N.J., 200 . . . C3 106
Stevens, Catron, N. Mex., 45 . . . D1 107
Stevens, co., Kans., 4,400 . . . E2 93
Stevens, co., Minn., 11,262 . . . E3 99
Stevens, co., Wash., 17,884 . . . A8 122
Stevens, peak, Idaho, Mont . . . C1 102
Stevenson, Jackson, Ala., 1,456 . . . A4 78
Stevenson, Fairfield, Conn., 200 . . . D4 84
Stevenson, Skamania, Wash., 927 . . . D4 122
Stevenson, lake, Man., Can . . . C4 71
Stevenson, mtn., Ark . . . B1 81
Stevenson, mtn., Oreg . . . C6 113
Stevenson, riv., Man., Can . . . B4 71
Stevens Point, Portage, Wis., 17,837 . . . D4 124
Stevens Pottery, Baldwin, Ga., 300 . . . D3 87
Stevens Village, Alsk., 102 . . . B10 79
Stevensville, Queen Annes, Md., 400 . . . C5 85
Stevensville, Berrien, Mich., 697 . . . F4 98
Stevensville, Ravalli, Mont., 784 . . . D2 102
Steveston, B.C., Can. 2,207 . . . A9 68
Steward, Lee, Ill., 264 . . . B4 90
Stewardson, Shelby, Ill., 656 . . . D5 90
Stewart, Hale, Ala., 50 . . . C2 78
Stewart, B.C., Can., 327 . . . B3 68
Stewart, McLeod, Minn., 676 . . . F4 99
Stewart, Montgomery, Miss., 162 . . . B4 100
Stewart, Ormsby, Nev., 900 . . . D2 104
Stewart, Athens, Ohio, 300 . . . C6 111
Stewart, Houston, Tenn., 150 . . . A4 117
Stewart, co., Ga., 7,371 . . . D2 87
Stewart, co., Tenn., 7,851 . . . A4 117
Stewart, isl., N.Z . . . Q11 51
Stewart, riv., Yukon, Can . . . D5 66
Stewart Manor, Nassau, N.Y., 2,422 . . . *E3 108
Stewarton, Scot., 3,387 . . . E4 13
Stewartstown, Coos, N.H., 125 (918▲) . . . B1 105
Stewartstown, N. Ire., 620 . . . C5 11
Stewartstown, York, Pa., 1,164 . . . G8 114
Stewartsville, Coosa, Ala., 60 . . . B3 78
Stewartsville, Posey, Ind., 235 . . . H2 91
Stewartsville, DeKalb, Mo., 466 . . . B3 101
Stewartsville, Warren, N.J., 875 . . . B2 106
Stewartsville, Bedford, Va., 150 . . . D3 121
Stewart Valley, Sask., Can., 181 . . . G2 70
Stewartville, Olmsted, Minn., 1,670 . . . G6 99
Stewiacke, N.S., Can., 1,042 . . . D6 74
Steynsburg, S. Afr., 3,365 . . . D4 49
Steyr, Aus., 38,306 . . . D7 16
Stibnite, Valley, Idaho, 25 . . . E3 89
Stickney, Cook, Ill., 6,239 . . . *F3 90
Stickney, Aurora, S. Dak., 456 . . . D7 116
Stigler, Haskell, Okla., 1,923 . . . B6 112
Stikine, mts., B.C., Can . . . E7 66
Stikine, riv., B.C., Can . . . E6 66
Stiles, Macon, N.C. . . . D2 109
Stiles, Reagan, Tex., 15 . . . D2 118
Stilesville, Hendricks, Ind., 361 . . . E4 91
Stilis, Grc., 3,606 . . . C4 23
Stillaguamish, riv., Wash . . . A4 122
Stillman Valley, Ogle, Ill., 598 . . . A4 90
Stillmore, Emanuel, Ga., 354 . . . D4 87
Still Pond, Kent, Md., 350 . . . B5 85
Still River, Worcester, Mass., 150 . . . B4 97
Stillwater, B.C., Can., 165 . . . E5 68
Stillwater, Penobscot, Maine (part of Old Town) . . . D4 96
Stillwater, Washington, Minn., 8,310 . . . E6, E8 99
Stillwater, Churchill, Nev., 20 . . . D3 104
Stillwater, Sussex, N.J., 200 . . . A3 106
Stillwater, Saratoga, N.Y., 1,398 . . . C7 108
Stillwater, Payne, Okla., 23,965 . . . A4 112
Stillwater, Columbia, Pa., 193 . . . D9 114
Stillwater, Providence, R.I., 75 . . . B10 84
Stillwater, co., Mont., 5,526 . . . E7 102
Stillwater, range, Nev . . . D3 104
Stillwell, Effingham, Ga., 50 . . . D5 87
Stillwell, LaPorte, Ind., 225 . . . A4 91
Stilson, Bullock, Ga., 160 . . . D5 87
Stilwell, Johnson, Kans., 162 . . . D9 93
Stilwell, Adair, Okla., 1,916 . . . B7 112
Stimson, mtn., Mont . . . B3 102
Stinchar, riv., Scot . . . E4 13
Stinesville, Monroe, Ind., 288 . . . F4 91
Stinnett, Hutchinson, Tex., 2,695 . . . B2 118
Stinson Lake, Grafton, N.H., 40 . . . C3 105
Štip, Yugo., 18,650 . . . E6 22
Stirling, Alta., Can., 468 . . . E4 69
Stirling, Ont., Can., 1,315 . . . C7 72
Stirling, Que., Can . . . B5 73
Stirling, Morris, N.J., 1,382 . . . B4 106
Stirling, Scot., 27,553 . . . D5 13
Stirling, co., Scot., 194,858 . . . D4 13
Stirling City, Butte, Calif., 350 . . . C3 82
Stirrat, Logan, W. Va., 900 . . . D3 123
Stirum, Sargent, N. Dak., 80 . . . C8 110
Stissing, mtn., N.Y . . . B2 84
Stites, Idaho, Idaho, 299 . . . C2 89
Stittsville, Ont., Can., 1,508 . . . B9 73
Stjördal, Nor., 6,133 . . . F4 25
Stoa Pikt, Pima, Ariz., 25 . . . E3 80
Stobnica, Pol., 2,000 . . . E8 24
Stockbridge, Henry, Ga., 1,201 . . . C2 87
Stockbridge, Berkshire, Mass., 900 (2,161▲) . . . B1 97
Stockbridge, Ingham, Mich., 1,097 . . . F6 98
Stockbridge, Windsor, Vt., 35 (392▲) . . . D3 120
Stockbridge, Calumet, Wis., 476 . . . B5, D5 124
Stockbridge, Indian res., Wis . . . D4 124
Stockdale, Pike, Ohio, 175 . . . D5 111
Stockdale, Wilson, Tex., 1,111 . . . B4, E4 118
Stockerau, Aus., 11,853 . . . D8 16
Stockertown, Northampton, Pa., 777 . . . E11 114
Stockett, Cascade, Mont., 400 . . . C5 102
Stockham, Hamilton, Nebr., 69 . . . D8 103
Stockholm, Sask., Can., 238 . . . G4 70
Stockholm, Aroostook, Maine, 500 (649▲) . . . A4 96
Stockholm, Sussex, N.J., 200 . . . A3 106
Stockholm, Grant, S. Dak., 155 . . . B9 116
Stockholm, Swe., 807,100 (*1,130,000) . . . H8, t36 25
Stockholm, Pepin, Wis., 106 . . . D1 124
Stockholm, co., Swe., 487,100 . . . *H8 25
Stockhorn, mtn., Switz . . . C4 19
Stockland, Iroquois, Ill., 150 . . . C6 90
Stockport, Eng., 142,469 . . . A5 12
Stockport, Van Buren, Iowa, 342 . . . D6 92
Stockport, Morgan, Ohio, 458 . . . C6 111
Stockton, Baldwin, Ala., 950 . . . E2 78
Stockton, San Joaquin, Calif., 86,321 (*160,000) . . . B6, D3 82
Stockton, Man., Can., 61 . . . E2 71
Stockton, Lanier, Ga., 500 . . . F4 87
Stockton, Jo Daviess, Ill., 1,800 . . . A3 90
Stockton, Rooks, Kans., 2,073 . . . C4 93
Stockton, Worcester, Md., 300 . . . D7 85
Stockton, Cedar, Mo., 838 . . . D4 101
Stockton, Hunterdon, N.J., 520 . . . C3 106
Stockton, Chautauqua, N.Y., 165 . . . C1 108
Stockton, Tooele, Utah, 362 . . . C3 119
Stockton, isl., Wis . . . B3 124
Stockton-on-Tees, Eng., 81,198 . . . C6 10
Stockton Springs, Waldo, Maine, 400 (980▲) . . . D4 96
Stockville, Frontier, Nebr., 91 . . . D5 103
Stockwell, Tippecanoe, Ind., 400 . . . D4 91
Stod, Czech., 2,502 . . . D8 17
Stoddard, Cheshire, N.H., 100 (146▲) . . . D2 105
Stoddard, Vernon, Wis., 552 . . . E2 124
Stoddard, co., Mo., 29,490 . . . E8 101
Stoeckl, mtn., B.C., Can . . . A2 68
Stoke Centre, Que., Can., 241 . . . D6 73
Stoke-on-Trent, Eng., 265,506 (*430,000) . . . A5 12
Stokes, Pitt, N.C., 195 . . . B6 109
Stokes, co., N.C., 22,314 . . . A3 109
Stokesdale, Guilford, N.C., 900 . . . A4 109
Stokesley, Eng., 1,980 . . . F7 13
Stokke, Nor . . . p28 25
Stolac, Yugo., 2,950 . . . D3 22
Stolberg, Ger., 37,500 . . . C3 16
Stolbovaya, Sov. Un . . . n17 27
Stollberg, Ger., 13,000 . . . C7 17
Stolzenau, Ger., 2,800 . . . F3 24
Stone, Eng., 8,791 . . . B5 12
Stone, Oneida, Idaho, 20 . . . G6 89
Stone, Pike, Ky., 728 . . . C7 94
Stone, co., Ark., 6,294 . . . B3 81
Stone, co., Miss., 7,013 . . . E4 100
Stone, co., Mo., 8,176 . . . E4 101
Stone, lake, Wis . . . C5 124
Stone, mtn., Ga . . . C2 89
Stone, mtn., Tenn . . . D10 117
Stone, mtn., Vt . . . B5 120
Stone, mts., Tenn . . . C12 117
Stonebluff, Fountain, Ind., 170 . . . D3 91
Stoneboro, Mercer, Pa., 1,267 . . . D1 114
Stoneboro, Kershaw, S.C., 100 . . . B6 115
Stone City, Pueblo, Colo., 35 . . . C6 83
Stone City, Jones, Iowa, 200 . . . B6 92
Stone Corral, lake, Oreg . . . E7 113
Stonefort, Saline and Williamson, Ill., 349 . . . F5 90
Stonega, Wise, Va., 800 . . . B2 121
Stoneham, Que., Can., 500 . . . C6 73
Stoneham, Weld, Colo., 80 . . . A7 83
Stoneham, Middlesex, Mass., 17,821 . . . C3 97
Stone Harbor, Cape May, N.J., 834 . . . E3 106
Stonehaven, N.B., Can., 95 . . . B4 74
Stonehaven, Scot., 4,500 . . . D6 13
Stonehenge, Austl., 38 . . . B4 51
Stone Lake, Sawyer, Wis., 175 . . . C2 124
Stoneleigh, Baltimore, Md., 8,000 . . . *B4 85
Stone Mountain, De Kalb, Ga., 1,976 . . . B5, C2 87
Stone Park, Cook, Ill., 3,038 . . . *F2 90
Stoner, Montezuma, Colo., 30 . . . D2 83
Stones, riv., Tenn . . . A5 117
Stones River, nat. military park and cemetery, Tenn . . . B5 117
Stones River Homes, Rutherford, Tenn., 1,800 . . . *B5 117
Stoneville, Washington, Miss., 300 . . . B3 100
Stoneville, Rockingham, N.C., 951 . . . A4 109
Stoneville, Meade, S. Dak., 9 . . . C3 116
Stonewall, Man., Can., 1,420 . . . D3 71
Stonewall, Fulton, Ga., 800 . . . B4 87
Stonewall, De Soto, La., 100 . . . B2 95
Stonewall, Clarke, Miss., 1,126 . . . C5 100
Stonewall, Pontotoc, Okla., 584 . . . C5 112
Stonewall, Gillespie, Tex., 170 . . . D3 118
Stonewall, co., Tex., 3,017 . . . C2 118
Stonewood, Harrison, W. Va., 2,202 . . . *B4 123
Stoney, creek, Va . . . D5 121
Stoney, isl., Newf., Can . . . B4 75
Stoney Creek, Ont., Can., 6,043 . . . *D5 72
Stonington, Baca, Colo., 36 . . . D8 83
Stonington, New London, Conn., 1,622 (13,969▲) . . . D9 84
Stonington, Christian, Ill., 1,076 . . . D4 90
Stonington, Hancock, Maine, 800 (1,408▲) . . . D4 96
Stonington, Delta, Mich., 25 . . . C4 98
Stono, inlet, S.C . . . F8 115
Stono, riv., S.C . . . F2 115
Stony, brook, N.J . . . C3 106
Stony, isl., N.Y . . . B4 108
Stony, lake, Ont., Can . . . C6 72
Stony, riv., W. Va . . . B5 123
Stony Beach, Sask., Can., 56 . . . G3 70
Stony Brook, Suffolk, N.Y., 3,548 . . . F4 84
Stony Brook, hbr., N.Y . . . F4 84
Stony Creek, New Haven, Conn., 950 . . . D5 84
Stony Creek, Warren, N.Y., 450 . . . B7 108
Stony Creek, Sussex, Va., 437 . . . E5 121
Stony Creek Mills, Berks, Pa., 1,500 . . . *F10 114
Stonyford, Colusa, Calif., 125 . . . C2 82

Stony Mountain, Man., Can., 1,130 D3 71
Stony Plain, Alta., Can., 1,311 C3 69
Stony Point, Rockland, N.Y., 3,330 A5 106
Stony Point, Alexander, N.C., 1,015 B2 109
Stony Rapids, Sask., Can., 107 E11 66
Stony Ridge, Wood, Ohio, 335 A2 111
Stony Wold, Franklin, N.Y., 50 B3 108
Stor, riv., Ger B4 16
Storaa, riv., Den B2 24
Stora Luleträsk, lake, Swe D8 25
Stora Sundby, Swe t34 25
Storavan, lake, Swe E8 25
Storden, Cottonwood, Minn., 390 F3 99
Store Belt, strait, Den C4 24
Store-Heddinge, Den., 2,082 C6 24
Storey, co., Nev., 568 D2 104
Storfjord, fjord, Nor F2 25
Storkerson, cape, N.W. Ter., Can B11 66
Storkow, Ger., 4,738 A6 17
Storla, Aurora, S. Dak., 50 D7 116
Storm, lake, Iowa B2 92
Storm Lake, Buena Vista, Iowa, 7,728 B2 92
Stormont, co., Ont., Can., 57,867 B10 72
Stornoway, Sask., Can., 89 F4 70
Stornoway, Scot., 5,221 B2 13
Storozhinets, Sov. Un., 10,000 G5 27
Storr, mtn., Scot C2 13
Storrs, Tolland, Conn., 6,054 B7 84
Storsjön, lake, Swe F6 25
Storsjö, Swe F5 25
Storthoaks, Sask., Can., 227 H5 70
Storuman, lake, Swe E7 25
Story, Sheridan, Wyo., 200 A6 125
Story, co., Iowa, 49,327 B4 92
Story City, Story, Iowa, 1,773 B4 92
Story Prairie, Sandusky, Ohio, 1,720 *A4 111
Stotesbury, Vernon, Mo., 64 D3 101
Stotts City, Lawrence, Mo., 221 D4 101
Stottville, Columbia, N.Y., 1,040 C7 108
Stouffville, Ont., Can., 3,188 D5, E7 72
Stoughton, Sask., Can., 606 H4 70
Stoughton, Norfolk, Mass., 16,328 B5, E3 97
Stoughton, Dane, Wis., 5,555 F4 124
Stour, riv., Eng C8 12
Stour, riv., Eng C9 12
Stour, riv., Eng D5 12
Stourbridge, Eng., 43,917 B5 12
Stourport-on-Severn, Eng., 11,751 B5 12
Stout, Grundy, Iowa, 145 B5 92
Stout (Rome), Adams, Ohio, 149 D4 111
Stoutland, Camden, Mo., 172 D5 101
Stoutsville, Monroe, Mo., 109 B6 101
Stoutsville, Fairfield, Ohio, 500 C5 111
Stovall, Meriwether, Ga., 250 D2 87
Stovall, Coahoma, Miss., 125 A3 100
Stovall, Granville, N.C., 570 A5 109
Stove Creek, Sask., Can E4 70
Stover, Tallahatchie, Miss., 150 A3 100
Stover, Morgan, Mo., 757 C4 101
Stövring, Den., 1,373 B3 24
Stow, Oxford, Maine., 25 (108▲) D2 96
Stow, Middlesex, Mass., 800 (2,573▲) B4, C1 97
Stow, Summit, Ohio, 12,194 A6 111
Stow, creek, N.J E2 106
Stowe [Township], Allegheny, Pa., 11,730 *B5 114
Stowe, Montgomery, Pa., 3,501 F10 114
Stowe, Lamoille, Vt., 534 (1,901▲) C3 120
Stowmarket, Eng., 7,790 B8 12
Stoy, Crawford, Ill., 185 D6 90
Stoyoma, mtn., B.C., Can E7 68
Stoystown, Somerset, Pa., 360 F4 114
Strabane, N. Ire., 7,786 C4 11
Strabane, Washington, Pa., 1,940 F1 114
Strachur, Scot., 578 D3 13
Stradbally, Ire., 792 D4 11
Strader, Tangipahoa, La B7 95
Stradone, Ire., 113 D4 11
Strafford, Greene, Mo., 300 D4 101
Strafford, Strafford, N.H., 135 (722▲) D4 105
Strafford, Chester, Pa., 2,500 *A10 114
Strafford, Orange, Vt., 100 (548▲) D4 120
Strafford, co., N.H., 59,799 D4 105
Straffordville, Ont., Can., 487 E4 72
Strakonice, Czech., 14,100 D2 26
Straldzha, Bul., 5,348 D8 22
Stralsund, Ger., 66,100 A6 16
Strandburg, Grant, S. Dak., 105 B9 116
Strandby, Den., 1,303 A4 24
Strandhill, Ire., 301 C3 11
Strandquist, Marshall, Minn., 160 B2 99
Strang, Fillmore, Nebr., 68 D8 103
Strang, Mayes, Okla., 176 A6 112
Stranger, creek, Kans B8 93
Strangford, N. Ire., 413 C6 11
Strangford, lake, N. Ire C6 11
Strängnäs, Swe., 8,300 t35 25
Strångsjö, Swe., 241 u34 25
Stranraer, Sask., Can., 78 F1 70
Stranraer, Scot., 9,249 F3 13
Strasbourg, Fr., 228,971 (★320,000) C7 14, F7 15
Strasbourg Station, Sask., Can., 636 F3 70
Strasburg, Adams and Arapahoe, Colo., 439 B6 83
Strasburg, Ger., 6,994 E7 24
Strasburg, Shelby, Ill., 467 D5 90
Strasburg, Cass, Mo., 213 C3 101
Strasburg, Emmons, N. Dak., 612 C5 110
Strasburg, Tuscarawas, Ohio, 1,687 B6 111
Strasburg, Lancaster, Pa., 1,416 G9 114
Strasburg, Shenandoah, Va., 2,428 C4 121
Strass, Aus., 495 B7 18
Strasswalchen, Aus., 4,163 B9 18
Stratford, Kings, Calif., 500 D4 82
Stratford, Ont., Can., 20,467 D4 72
Stratford, Fairfield, Conn., 45,012 E4 84
Stratford, Hamilton and Webster, Iowa, 703 B4 92
Stratford, Coos, N.H., 130 (1,029▲) A3 105
Stratford, Camden, N.J., 4,308 D2 106
Stratford, Fulton, N.Y., 200 B6 108
Stratford, N.Z., 5,273 M15 51
Stratford, Garvin, Okla., 1,058 C5 112
Stratford, Brown, S. Dak., 109 B7 116
Stratford, Sherman, Tex., 1,380 A1 118
Stratford, Marathon, Wis., 1,106 D3 124
Stratford, pt., Conn E4 84
Stratford Center, Que., Can., 466 D6 73
Stratford Hills, Chesterfield, Va., 2,500 *D5 121
Stratford-on-Avon, Eng., 16,847 B6 12
Stratham, Rockingham, N.H., 160 (1,033▲) D5 105
Strathaven, Scot., 5,867 E4 13
Strathclair, Man., Can., 465 D1 71
Strathcona, Ont., Can., 106 C8 72
Strathcona, Roseau, Minn., 64 B2 99
Strathcona, park, B.C., Can E5 68
Strathlorne, N.S., Can., 176 C8 70
Strathmere, Cape May, N.J., 100 E3 106
Strathmore, Tulare, Calif., 1,095 D4 82
Strathmore, Alta., Can., 924 D4 69
Strathnaver, B.C., Can., 191 C6 68
Strathroy, Ont., Can., 5,150 E3 72
Strattanville, Clarion, Pa., 547 D3 114
Stratton, Ont., Can., 112 E6 72
Stratton, Kit Carson, Colo., 680 B8 83
Stratton, Franklin, Maine, 500 C2 96
Stratton, Hitchcock, Nebr., 492 D4 103
Stratton, Jefferson, Ohio, 311 B7 111
Straubing, Ger., 36,300 E7 17
Straubville, Sargent, N. Dak., 40 C8 110
Straughn, Henry, Ind., 349 E7 91
Strausberg, Ger., 13,800 F7 24
Strausstown, Berks, Pa., 380 F9 114
Straw, Fergus, Mont., 20 D7 102
Strawberry, Lawrence, Ark., 200 B4 81
Strawberry, mtn., Oreg C8 113
Strawberry, mts., Oreg C8 113
Strawberry, peak, Utah C4 119
Strawberry, pt., Mass B6 97
Strawberry, res., Utah C4 119
Strawberry, riv., Ark A4 81
Strawberry, riv., Utah C5 119
Strawberry Plains, Jefferson, Tenn., 400 C10 117
Strawberry Point, Marin, Calif., 1,500 *B5 82
Strawberry Point, Clayton, Iowa, 1,303 B6 92
Strawn, Livingston, Ill., 152 B5 90
Strawn, Coffey, Kans., 100 D8 93
Strawn, Palo Pinto, Tex., 817 C3 118
Straznice, Czech., 4,989 D4 26
Streamstown, Alta., Can., 55 C5 69
Streamwood, Cook, Ill., 6,751 *A5 90
Streator, La Salle and Livingston, Ill., 16,868 B5 90
Streator East, La Salle, Ill., 1,517 *B5 90
Streeter, Stutsman, N. Dak., 491 C6 110
Streeter, Mason, Tex., 30 D3 118
Streetman, Freestone and Navarro, Tex., 300 D4 118
Streetsboro, Portage, Ohio, 1,000 *A6 111
Streetsville, Ont., Can., 5,056 E6 72
Strehaia, Rom., 8,545 C6 22
Strehlen, see Strzelin, Pol.
Strelka, Sov. Un C13 28
Strelka, Sov. Un D27 9
Stresa, It., 2,953 D4 18
Stribling, Stewart, Tenn A4 117
Stribro, Czech., 3,950 D8 17
Strichen, Scot., 1,949 C6 13
Strimon, gulf, Grc B4 23
Strimon, riv., Grc B4 23
Stringer, Jasper, Miss., 150 D4 100
Stringtown, Lake, Colo., 500 B4 83
Stringtown, Anderson, Ky., 300 B5 94
Stringtown, Bolivar, Miss., 150 B3 100
Stringtown, Atoka, Okla., 414 C5 112
Strofadhes, isl., Grc D3 23
Stroh, Lagrange, Ind., 475 A7 91
Strokestown, Ire., 707 D3 11
Stroma, isl., Scot B5 13
Stromboli, isl., It E5 21
Strome, Alta., Can., 311 C4 69
Strome Ferry, Scot C3 13
Stromness, Scot., 1,477 B5 13
Stromsburg, Polk, Nebr., 1,244 C8 103
Stronach, Manistee, Mich., 350 D4 98
Stroner, Crook, Wyo., 5 A8 125
Strong, Union, Ark., 741 D3 81
Strong, Franklin, Maine, 300 (976▲) D2 96
Strong, Monroe, Miss., 300 B5 100
Strong, riv., Miss C4 100
Strong City, Chase, Kans., 659 D7 93
Strong City, Roger Mills, Okla., 51 B2 112
Strongfield, Sask., Can., 218 F2 70
Stronghurst, Henderson, Ill., 815 C3 90
Strongs, Chippewa, Mich., 225 B6 98
Strongsville, Cuyahoga, Ohio, 8,504 A6, B2 111
Stronsay, firth, Scot A6 13
Stronsay, isl., Scot A6 13
Strontian, Scot D3 13
Stroud, Eng., 17,461 C5 12
Stroud, Lincoln, Okla., 2,456 B5 112
Stroudsburg, Monroe, Pa., 6,070 E11 114
Stroudsburg West, Monroe, Pa., 1,569 *E11 114
Struble, Plymouth, Iowa, 74 B1 92
Struer, Den., 8,335 B2 24
Struga, Yugo., 6,871 E5 22
Strule, riv., Ire C4 11
Strum, Trempealeau, Wis., 663 D2 124
Struma, riv., Bul E6 22
Strumble, head, Wales B2 12
Strumica, Yugo., 15,978 E6 22
Strunk, McCreary, Ky., 450 D5 94
Struthers, Mahoning, Ohio, 15,631 A7 111
Stryama, riv., Bul D7 22
Stryker, Lincoln, Mont., 60 B2 102
Stryker, Williams, Ohio, 1,205 A3 111
Strykersville, Wyoming, N.Y., 360 C2 108
Stryy, Sov. Un., 47,000 G4 27
Stryy, riv., Sov. Un D7 26
Strzegom, Pol., 7,137 C4 26
Strzelce, Pol., 10,300 C5 26
Strzelce. Krajenskie, Pol., 1,552 B3 26
Strzelecki, creek, Austl D3 51
Strzelin, Pol., 7,334 C4 26
Strzelno, Pol., 5,264 B5 26
Strzemieszyce, Pol g10 26
Stuart, Martin, Fla., 4,791 E6 86
Stuart, Guthrie and Adair, Iowa, 1,486 C3 92
Stuart, Holt, Nebr., 794 B6 103
Stuart, Hughes, Okla., 271 C5 112
Stuart, Patrick, Va., 974 E2 121
Stuart, lake, B.C., Can B5 68
Stuart, mtn., Wash B4 122
Stuart, riv., B.C., Can B6 68
Stuart, range, Austl E5 50
Stuarts Draft, Augusta, Va., 600 C3 121
Stub, hill, N.H A2 105
Stubbeköbing, Den., 2,097 D6 24
Studley, Sheridan. Kans., 60 C3 93
Stull, Douglas, Kans., 35 B7 93
Stull, lake, Ont., Can B5 71
Stull, riv., Man., Can B5 71
Stump, lake, N. Dak B7 110
Stump Creek, Jefferson, Pa., 500 D4 114
Stumptown, Gilmer, W. Va., 100 C4 123
Stumpy Point, Dare, N.C., 250 B8 109
Stung Treng, Camb., 10,000 F6 38
Stupart, riv., Man., Can B4 71
Stupino, Sov. Un., 36,000 D12 27
Stura, riv., It D3 18
Sturbridge, Worcester, Mass., 400 (3,604▲) B3 97
Sturgeon, Boone, Mo., 619 B5 101
Sturgeon, Allegheny, Pa., 1,000 *F1 114
Sturgeon, bay, Man., Can C3 71
Sturgeon, lake, Alta., Can B2 69
Sturgeon, riv., Sask., Can D2 70
Sturgeon Bay, Door, Wis., 7,353 D6 124
Sturgeon Falls, Ont., Can., 6,288 A5 72
Sturgeon Lake, Pine, Minn., 151 D6 99
Sturgeon Landing, Sask., Can., 104 C5 70
Sturgeon-weir, riv., Sask., Can C4 70
Sturgis, Sask., Can., 611 F4 70
Sturgis, Union, Ky., 2,209 C2 94
Sturgis, St. Joseph, Mich., 8,915 G5 98
Sturgis, Oktibbeha, Miss., 358 B4 100
Sturgis, Meade, S. Dak., 4,639 C2 116
Sturkie, Fulton, Ark A4 81
Sturtevant, Racine, Wis., 1,488 F2, F6 124
Stutsman, co., N. Dak., 25,137 B6 110
Stutterheim, S. Afr., 9,015 D4 49
Stuttgart, Arkansas, Ark., 9,661 C4 81
Stuttgart, Ger., 637,500 (★1,350,000) E4 17
Stuttgart, Phillips, Kans., 100 C4 93
Styr, riv., Sov. Un F5 27
Styria, reg., Aus E7 16
Su, China H7 36
Suakin, Sud., 4,228 B4 47
Suao, For., 5,000 G9 34
Subiaco, Logan, Ark., 290 B2 81
Subi-Besar, isl., Indon K8 38
Subic, Phil., 2,000 (13,000▲) C6, o13 35
Subic, bay, Phil o13 35
Sublette, Lee, Ill., 306 B4 90
Sublette, Haskell, Kans., 1,077 E3 93
Sublette, co., Wyo., 3,778 C2 125
Sublimity, Marion, Oreg., 490 C2, C4 113
Subotica, Yugo., 74,832 B4 22
Sucarnoochee, Kemper, Miss., 100 C5 100
Sucarnoochee, creek, Ala., Miss C5 100
Succasunna, Morris, N.J., 2,500 B3 106
Success, Clay, Ark., 226 A5 81
Success, Sask., Can., 117 G1 70
Suceava, Rom., 20,949 B8 22
Suceava, riv., Rom B7 22
Sucha, Pol., 5,866 D5 26
Suchan, Sov. Un., 47,200 E6 37
Suchdol, Czech., 3,730 n17 26
Suches, Union, Ga., 600 B2 87
Suchitoto, Sal., 4,380 D3 62
Suchocin, Pol., 500 k13 26
Suchou, see Soochow, China
Suchow, China, 676,000 E8 34
Suck, riv., Ire D3 11
Sucre, Bol., 40,128 C2 55
Sucre, state, Ven., 401,992 A5 60
Suçuapara, Braz., 1,860 B3 59
Sucuriú, riv., Braz E2 56
Sucy-en-Brie, Fr., 13,258 g11 14
Sudan, Lamb, Tex., 1,235 B1 118
Sudan, country, Afr., 11,390,000 E8 42, C2 47
Sudbury, Ont., Can., 80,120 (★110,694) A4, E9 72
Sudbury, Eng., 6,643 B8 12
Sudbury, Middlesex, Mass., 1,800 (7,447▲) D1 97
Sudbury, Rutland, Vt., 50 (249▲) D2 120
Sudbury, co., Ont., Can., 165,862 A3 72
Sudbury, res., Mass D1 97
Sudbury, riv., Mass B5 97
Sudd, swamp, Sud D2 47
Süderbrarup, Ger., 3,200 D3 24
Sudetes, mts., Pol C4 26
Sudlersville, Queen Annes, Md., 394 B6 85
Sudley, Anne Arundel, Md., 80 C4 85
Sud-Nord, canal, Ger B7 15
Sudogda, Sov. Un., 10,500 D13 27
Sudzha, Sov. Un., 10,000 F10 27
Sueca, Sp., 20,612 C5 20
Suez, Eg., U.A.R., 162,826 C6 43
Suez, canal, Eg., U.A.R C6 43
Suez, gulf, Afr D6 43
Suffern, Rockland, N.Y., 5,094 D1, D2, D6 108
Suffield, Alta., Can., 130 D5 69
Suffield, Hartford, Conn., 1,069 B6 84
Suffolk, Fergus, Mont., 10 C7 102
Suffolk (Independent City), Va., 12,609 B6, E6 121
Suffolk (East Suffolk, West Suffolk), co., Eng., 472,665 B8 12
Suffolk, co., Mass., 791,329 B5 97
Suffolk, co., N.Y., 666,784 D3 108
Sufu, see Kashgar, China
Sugar, creek, Ind F6 91
Sugar, creek, Ind E3 91
Sugar, creek, Pa C8 114
Sugar, isl., Mich B6 98
Sugar, riv., N.H D2 105
Sugar, riv., Wis F4 124
Sugar City, Crowley, Colo., 409 C7 83
Sugar City, Madison, Idaho, 584 F7 89
Sugar Creek, Jackson, Mo., 2,663 E2 101
Sugarcreek, Tuscarawas, Ohio, 982 *B6 111
Sugar Grove, Logan, Ark., 100 B2 81
Sugar Grove, Watauga, N.C., 500 A2 109
Sugar Grove, Fairfield, Ohio, 479 C5 111
Sugargrove, Warren, Pa., 636 C3 114
Sugar Grove, Smyth, Va., 800 B3, E1 121
Sugar Grove, Pendleton, W. Va., 50 C5 123
Sugar Hill, Gwinnett, Ga., 1,175 B2 87
Sugar Hill, Grafton, N.H., 100 B3 105
Sugarite, Colfax, N. Mex A5 107
Sugar Land, Fort Bend, Tex., 2,802 E5 F4 118
Sugarloaf, hill, Ohio A6 111
Sugarloaf, mtn., Maine C2 96
Sugar Loaf, mtn., Md B3 85
Sugarloaf, mtn., Mont C4 102
Sugarloaf, mtn., N.H A4, B1 105
Sugarloaf, mts., Okla B7 112
Sugar Notch, Luzerne, Pa., 1,524 B8 114
Sugarpine, mtn., Oreg D5 113
Sugar Run, Bradford, Pa., 65 C9 114
Sugar Tree, Decatur, Tenn., 40 B3 117
Sugar Valley, Gordon, Ga., 165 B1 87
Sugden, Jefferson, Okla., 68 C4 112
Suget, pass, China, India B6 39
Suggi, lake, Sask., Can C4 70
Suggsville, Clarke, Ala., 100 D2 78
Şuḥār, Mus. & Om., 5,000 D2 39
Suhl, Ger., 25,500 C5 17
Sui, China G6 36
Suiattle, riv., Wash A4 122
Suichi, China, 1,000 B9 38
Suichiang, China, 5,000 F5 34
Suichuan, China, 13,000 F7 34
Suichung, China D9 36
Suifenho, China C11 34
Suihsien, China, 31,000 E7 34
Suihua (Peilintzu), China, 40,000 B10 34
Suilai, see Manas, China
Suileng, China, 5,000 C3 37
Suilu, China A7 38
Suipacha, Arg., 3,006 g7 54
Suipacha, Bol D2 55
Suipin, China, 5,000 C5 37
Suippes, Fr., 2,738 E4 15
Suir, riv., Ire E4 11
Suisun City, Solano, Calif., 2,470 B5 82
Suita, Jap., 116,765 *o14 37
Suite, China, 15,000 D7 34
Suiter, Bland, Va D1 121
Suitland, Prince Georges, Md., 10,300 C2 85
Suitung, China C9 36
Sukabumi, Indon, 78,806 G3 35
Sukadana, Indon F3 35
Sukhaya Tunguska, Sov. Un B26 9
Sukhobuzimskoye, Sov. Un., 10,000 D27 9
Sukhona, riv., Sov. Un C17 9
Sukhumi, Sov. Un., 67,000 G17 9
Sukkur, Pak., 103,216 C4 38, D2 40
Sukunka, riv., B.C., Can B7 68
Sul, chan., Braz C5 59
Sula, Ravalli, Mont., 15 E3 102
Sula, isl., Indon F6 35
Sula, riv., Sov. Un F9 27
Sulaiman, range, Pak C2 40
Sulecin, Pol., 2,566 B3 26
Sulgen, Switz., 1,252 *A7 19
Sulina, Rom., 3,622 C9 22
Sulingen, Ger., 7,300 F2 24
Sulitjelma, mtn., Swe D7 25
Suliwula, riv., China D4 34
Sullana, Peru, 26,330 B1 58
Sulligent, Lamar, Ala., 1,346 B1 78
Sullivan, Moultrie, Ill., 3,946 D5 90
Sullivan, Sullivan, Ind., 4,979 F3 91
Sullivan, Union, Ky., 250 C2 94
Sullivan, Franklin and Crawford, Mo., 4,098 C6 101
Sullivan, Cheshire, N.H., 35 (261▲) D2 105
Sullivan, Ashland, Ohio, 340 A5 111
Sullivan, Jefferson, Wis., 418 E5 124
Sullivan, co., Ind., 21,721 F3 91
Sullivan, co., Mo., 8,783 A4 101
Sullivan, co., N.H., 28,067 D2 105
Sullivan, co., N.Y., 45,272 D6 108
Sullivan, co., Pa., 6,251 D8 114
Sullivan, co., Tenn., 114,139 C11 117
Sullivan, isl., Bur G3 38
Sullivan, lake, Alta., Can D5 69
Sullivans Island, Charleston, S.C., 1,358 F3 115
Sully, Jasper, Iowa, 508 C5 92
Sully, co., S. Dak., 2,607 C5 116
Sulmona, It., 18,400 C4 21
Sulphide, Ont., Can., 217 C7 72
Sulphur, Henry, Ky., 275 B4 94
Sulphur, Calcasieu, La., 11,429 D2 95
Sulphur, Murray, Okla., 4,737 C5 112
Sulphur, fork, Tenn A4 117
Sulphur, riv., Ark., Tex D1 75, C5 118
Sulphur, riv., Alta., Can C1 69
Sulphur Rock, Independence, Ark., 225 B4 81
Sulphur South, Calcasieu, La., 1,351 *D2 95
Sulphur Spring, val., Ariz E6 80
Sulphur Springs, Benton, Ark., 460 A1 81
Sulphur Springs, Henry, Ind., 400 D7 91
Sulphur Springs, Jefferson, Mo., 110 B8 101
Sulphur Springs, Crawford, Ohio, 350 B5 111
Sulphur Springs, Douglas, Oreg D3 113
Sulphur Springs, Hopkins, Tex., 9,160 C5 118
Sultan, Snohomish, Wash., 821 B4 122
Sultanabad, India H7 40
Sultanpur, India, 26,081 D9 40
Sulu, prov., Phil., 327,100 *D6 35
Sulu, arch., Phil D6 35
Sulu, riv., China C4 34
Sulu, sea, Phil D5 35
Sulūq, Libya, 1,000 C4 43
Sulyukta, Sov. Un., 15,000 H22 9
Sulzbach, Ger., 23,800 D2 17
Sulzbach-Rosenberg, Ger., 19,600 D6 17
Sulzberger, bay, Ant B33 5
Sumach, Yakima, Wash., 1,345 *C5 122
Sumas, Whatcom, Wash., 629 A3 122
Sumatra, Liberty, Fla., 250 B2 86
Sumatra, Rosebud, Mont., 45 D9 102
Sumatra, isl., Indon E2 35
Sumava Resorts, Newton, Ind., 200 B2 91
Sumay, Guam 52
Sumba, isl., Indon G6 35
Sumbar, riv., Sov. Un B7 41
Sumbawa. Indon G5 35
Sumbawanga. Tan, 4,590 C5 48
Sumbay, Peru E3 58
Sumburgh, pt., Scot h10 13
Sumeg, Hung., 5,941 B3 22
Sumenep, Indon., 17,824 G4 35
Sumgait, Sov. Un., 63,000 A4 41
Sumiswald, Switz., 5,525 B4 19
Sumiton, Walker, Ala., 1,287 B2 78
Summan, des., Sau. Ar H3 41
Summer, isl., Mich C4 98
Summer, lake, Oreg E6 113
Summerberry, Sask., Can., 84 G4 70
Summerberry, riv., Man., Can C1 71
Summerdale, Baldwin, Ala., 533 E2 78
Summerdale, Cumberland, Pa., 1,200 *F8 114
Summerfield, Marion, Fla., 450 C4 86
Summerfield, Marshall, Kans., 237 C7 93
Summerfield, Claiborne, La., 200 B3 95
Summerfield, Maries, Mo., 100 C6 101
Summerfield, Guilford, N.C., 700 A4 109
Summerfield, Noble, Ohio, 352 C6 111
Summerfield, Castro, Tex., 45 B1 118
Summerford, Newf., Can., 570 D4 75
Summer Hill, Pike, Ill., 150 D3 90
Summerhill, Ire., 97 D5 11
Summer Lake, Lake, Oreg., 5 E6 113
Summerland, B.C., Can., 2,500 E8 68
Summerland Key, Monroe, Fla., 350 H5 87
Summers, co., W. Va., 15,640 D4 123
Summerset, Warren, Iowa B7 92
Summer Shade, Metcalfe, Ky., 250 D4 94
Summerside, P.E.I., Can., 8,611 C6 74
Summersville, Green, Ky., 350 D4 94
Summersville, Texas, Mo., 356 D6 101
Summersville, Nicholas, W. Va., 2,008 C4, C7 123
Summerton, Clarendon, S.C., 1,504 D7 115
Summertown, Emanuel, Ga., 100 D4 87
Summertown, Lawrence, Tenn., 900 B4 117
Summerville, Chattooga, Ga., 4,706 B1 87
Summerville, Union, Oreg., 76 B8 113
Summerville, Jefferson, Pa., 895 D3 114
Summerville, Dorchester, S.C., 3,633 E2, E7 115
Summit, Marion, Ark., 239 A3 81
Summit, San Bernardino, Calif., 150 F3 82
Summit, C.Z., 48 k11 62
Summit, Cook, Ill., 10,374 F2 90
Summit, Pike, Miss., 1,663 D3 100
Summit, Union, N.J., 23,677 B4 106
Summit, Benton, Oreg., 50 C3 113
Summit, Kent, R.I., 100 C10 84
Summit, Roberts, S. Dak., 283 B8 116
Summit, Hamilton, Tenn., 200 E10 117
Summit, Iron, Utah, 150 F3 119
Summit, co., Colo., 2,073 B4 83
Summit, co., Ohio, 513,569 A6 111
Summit, co., Utah, 5,673 C4 119
Summit, lake, Iowa C3 92
Summit, mtn., Nev D5 104
Summit, mtn., N.Z N16 51
Summit, peak, Colo D4 83
Summit Hill, Carbon, Pa., 4,386 E10 114
Summit Lake, Indian res., Nev B3 104
Summit Point, Jefferson, W. Va., 150 B7 123
Summit Station, Licking, Ohio, 400 C2 111
Summitville, Rio Grande, Colo., 20 D4 83
Summitville, Madison, Ind., 1,048 D6 91
Summitville, Coffee, Tenn., 400 B6 117
Summum, Fulton, Ill., 225 C3 90
Sumner, Worth, Ga., 193 E3 87
Sumner, Lawrence, Ill., 1,035 E6 90
Sumner, Bremer, Iowa, 2,170 B5 92
Sumner, Gratiot, Mich., 85 E6 98
Sumner, Tallahatchie, Miss., 551 B3 100
Sumner, Chariton, Mo., 234 B4 101
Sumner, Dawson, Nebr., 254 D6 103
Sumner, Noble, Okla., 27 A4 112
Sumner, Pierce, Wash., 3,156 B3 122
Sumner, co., Kans., 25,316 E6 93
Sumner, co., Tenn., 36,217 A5 117
Sumner, strait, Alsk m23 79
Sumoto, Jap., 34,000 I7 37
Sumperk, Czech., 19,400 D4 26
Sumpter, Baker, Oreg., 96 C8 113
Sumrall, Lamar, Miss., 797 D4 100
Sumter, Sumter, Ga., 100 E2 87
Sumter, Sumter, S.C., 23,062 D7 115
Sumter, co., Ala., 20,041 C1 78
Sumter, co., Fla., 11,869 D4 86
Sumter, co., Ga., 24,652 D2 87
Sumter, co., S.C., 74,941 D7 115
Sumter, fort, S.C F3 115
Sumterville, Sumter, Ala., 50 C1 78
Sumurr, Maricopa, Ariz E3 80
Sumy, Sov. Un., 108,000 F10 27
Sun, St. Tammany, La., 224 D6 95
Sun, riv., Mont C4 102
Suna, Tan C5 48
Sunagawa, Jap., 24,200 E10 37
Sunapee, Sullivan, N.H., (1,164▲) D2 105
Sunapee, lake, N.H D2 105
Sunapee, mtn., N.H D2 105
Sunart, inlet, Scot D3 13
Sunbeam, Custer, Idaho, 5 E4 89
Sunbright, Morgan, Tenn., 550 C9 117
Sunbright, Scott, Va B2 121
Sunburg, Kandiyohi, Minn., 161 E3 99
Sunburst, Toole, Mont., 882 B5 102
Sunbury, Livingston, Ill B5 90
Sunbury, Gates, N.C. 450 A7 109
Sunbury, Delaware, Ohio, 1,360 B5 111
Sunbury, Northumberland, Pa., 13,687 E8 114
Sunbury, co., N.B., Can., 22,796 D3 74
Sunchales, Arg., 5,048 A4 54
Suncho Corral, Arg., 3,020 E3 55
Sunchon, Kor., 20,682 G2 37
Sunchon, Kor., 44,400 (69,500▲) I3 37
Sun City, Maricopa, Ariz., 2,000 *D3 80
Sun City, Hillsborough, Fla., 300 F2 86
Sun City, Barber, Kans., 188 E5 93
Suncook, Merrimack, N.H., 3,807 D4 105
Suncook, ponds, N.H D4 105
Suncook, riv., N.H D4 105
Sun Crest, San Diego, Calif., 1,166 *F5 82
Sunda, is., Asia G3 2
Sunda (Soenda), strait, Indon G3 35
Sundance, Crook, Wyo., 908 A8 125
Sundance, mtn., Wyo A8 125
Sunday, strait, Austl C3 50
Sundbyberg, Swe., 27,100 t35 25
Sundarbans, swamp, India G12 40
Sunderland, Ont., Can., 634 C5 72
Sunderland, Eng., 189,629 (★250,000) C6 10
Sunderland, Calvert, Md., 25 C4 85
Sunderland, Franklin, Mass., 400 (1,279▲) B2 97
Sunderland, Bennington, Vt., 40 (566▲) E2 120
Sundown, Man., Can., 196 E3 71
Sundown, Hockley, Tex., 1,186 C1 118
Sundre, Alta., Can., 853 D3 69
Sundridge, Ont., Can., 756 B5 72
Sunds, Den., 1,039 B3 24
Sundsvall, Swe., 29,800 F7 25
Sunfield, Eaton, Mich., 626 F5 98
Sunflower, Maricopa, Ariz., 5 D4 80
Sunflower, Sunflower, Miss., 662 B3 100
Sunflower, co., Miss., 45,750 B3 100
Sunflower, mtn., Kans C2 93
Sunflower, riv., Miss B3 100
Sungaiguntung, Indon E2 35
Sungari, res., China E3 37
Sungari, riv., China C5 37
Sungchiang, China, 70,000 I9 36
Sungei Patani, Mala., 22,916 J4 38
Sunghsien, China E7 34
Sungkan, China J2 36
Sungtao, China, 5,000 J3 36
Sungurlu, Tur., 10,500 B10 31
Sunlight, creek, Wyo A3 125
Sunman, Ripley, Ind., 446 F7 91
Sunne, Swe., 3,300 H5 25
Suniland, Dade, Fla., 1,500 *G6 86
Sunny Acres, Kenton, Ky., 844 *A7 94
Sunnybrae, N.S., Can., 180 D7 74

Sunnybrook, Alta., Can., 66..C3 69
Sunnydale, Sedgwick, Kans..B5 93
Sunnyland, Sarasota, Fla., 4,761..................*E4 86
Sunnyland, Tazewell, Ill., 1,000..................*C4 90
Sunnymead, Riverside, Calif., 3,404..................*F5 82
Sunnynook, Alta., Can., 76..D5 69
Sunnyside, Newf., Can., 533..E5 75
Sunnyside, Bay, Fla., 125....G3 86
Sunnyside, Leflore, Miss.....B3 100
Sunnyside, Carbon, Utah, 1,740..................D5 119
Sunnyside, Yakima, Wash., 6,208..................C5 122
Sunnyslope, Alta., Can., 61..D4 69
Sunny South, Wilcox, Ala., 200..................D2 78
Sunnyvale, Santa Clara Calif., 52,898..........B5 82
Sunnyvale, Dallas, Tex., 969..................*A6 118
Sunnyview, Brookings, S. Dak., 75..............C9 116
Sunol, Alameda, Calif., 750..B6 82
Sunol, Cheyenne, Nebr., 100.C3 103
Sun Prairie, Dane, Wis., 4,008..................E4 124
Sunray, Moore, Tex., 1,967..B2 118
Sunrise, Falls, Tex., 1,708..*D4 118
Sunrise, Platte, Wyo., 300...C8 125
Sunrise Heights, Calhoun, Mich., 1,569............*F5 98
Sun River, Cascade, Mont., 104..................C5 102
Sunset, St. Landry, La., 1,307..................D3 95
Sunset, Hancock, Maine, 150.D4 96
Sunset, Lincoln, N. Mex......D4 107
Sunset, Montague, Tex., 500.C4 118
Sunset, Davis, Utah, 4,235...B4 119
Sunset Beach, Orange, Calif., 1,300..................*F2 82
Sunset Crater, nat. mon., Ariz...B4 80
Sunset Hills, St. Louis, Mo., 3,525..................*B8 101
Sunset Park, Sedgwick, Kans., 1,000.............*B5 93
Sunshine, Los Angeles, Calif., 9,000..................*F4 82
Sunshine, Hancock, Maine, 120..................D4 96
Sunshine, Park, Wyo.........A4 125
Suntar, Sov. Un..........C14 28
Suntaug, lake, Mass.........C3 97
Suntex, Harney, Oreg.......D7 113
Suntrana, Alsk., 81.........C10 79
Sun Valley, Blaine, Idaho, 317..................F4 89
Sunwui, China, 85,000......G7 34
Sunyani, Ghana, 4,570.......E4 45
Suo, sea, Jap..................J5 37
Suoyarvi, Sov. Un..........F15 25
Supai, Coconino, Ariz., 140..A3 80
Superb, Sask., Can., 52......F1 70
Superior, Pinal, Ariz., 4,875..D4 80
Superior, Boulder, Colo., 173.B5 83
Superior, Dickinson, Iowa, 190..................A3 92
Superior, Mineral, Mont., 1,242..................C2 102
Superior, Nuckolls, Nebr., 2,935..................D7 103
Superior, Lawrence, Ohio, 5..................D5 111
Superior, Douglas, Wis., 33,563..................B1 124
Superior, Sweetwater, Wyo., 241..................D4 125
Superior, lake, U.S., Can......A10 77
Superior, McDowell, W. Va., 900..................*D3 123
Suphan Buri, Thai., 14,258..E4 38
Supi Oidak, Pima, Ariz., 75..F4 80
Suplee, Crook, Oreg.........C7 113
Supply, Randolph, Ark.......A5 81
Supply, Brunswick, N.C., 95.C5 109
Suprise, Maricopa, Ariz., 1,574..................G1 80
Suqash Shuyukh, Iraq, 7,735.F3 41
Suquamish, Kitsap, Wash., 950..................B3 122
Şūr, Mus. & Om.............D2 39
Sura, riv., Sov. Un...........C3 29
Surabaja (Soerabaja), Indon., 989,734 (*1,050,000)....G4 35
Surakarta (Soerakarta), Indon., 363,167..................G4 35
Surakhany, Sov. Un..........E4 29
Surany, Czech., 5,381........D5 26
Surat, Austl., 406............C7 51
Surat, India, 288,026.........D5 39, G4 40
Suratgarh, India, 8,330......C4 40
Surat Thani, Thai., 18,460..H3 38
Surazh, Sov. Un., 10,000....E9 27
Şūr Bāhir, Jordan...........k11 32
Surbiton, Eng., 62,940.........m11 10, C7 12
Suresnes, Fr., 39,100..........g9 14
Suretka, C.R..................F6 62
Surette Island, N.S., Can., 212..................F3 74
Surf City, Ocean, N.J., 419..D4 106
Surfside, Dade, Fla., 3,157...F3 86
Surgères, Fr., 4,839..........D3 14
Surgoinsville, Hawkins, Tenn., 1,132..............C11 117
Surgut, Sov. Un., 3,500...C23 9
Suri, India, 22,841..........F11 40
Suriapet, India, 12,443.......I7 40
Suribachi, mtn., Iwo..............52
Surigao, Phil., 12,870.........D7 35
Surigao, prov., Phil., 360,100..................*D7 35
Surinam (Netherlands Guiana), Neth. dep., S.A., 237,000..........C5 53, B3 59
Suriname, riv., Sur............B3 59
Suring, Oconto, Wis., 513...D5 124
Surkhan Darya, riv., Sov. Un..C13 41
Surprise, Butler, Nebr., 79...C8 103
Surrency, Appling, Ga., 312..E4 87
Surrey, Ward, N. Dak., 309..A4 110
Surrey, co., Eng., 1,733,036..C7 12
Surrey, Hancock, Maine, 180 (547▲)..................D4 96
Surry, Cheshire, N.H., 200 (362▲)..................D2 105
Surry, Surry, Va., 288...B5, D6 121
Surry, co., N.C., 48,205.....A3 109
Surry, co., Va., 6,220.......D6 121
Sursee, Switz., 5,324........B5 19
Surt, see Sirte, Libya
Suruc, Tur., 3,632.........D12 31
Surud Ad, mtn., Som.........C6 47
Suruga, bay, Jap............o17 37
Surveyor, Raleigh, W. Va., 125..................D6 123
Susa, It., 5,891..............B1 21
Susac, isl., Yugo.............D3 22
Susak, isl., Yugo.............C2 22
Susana Knolls, Ventrua, Calif., 900..................*E4 82
Susanino, Sov. Un..........s31 25
Susank, Barton, Kans., 87...D5 93
Susanville, Lassen, Calif., 5,598..................B3 82
Susanville, Grant, Oreg......C8 113
Susice, Czech., 6,793........D2 26
Susitna, Alsk., 42............g16 79
Susitna, riv., Alsk...........C10 79
Susquehanna, Susquehanna, Pa., 2,591..............C10 114
Susquehanna, co., Pa., 33,137..................C10 114
Susquehanna, riv., Md., N.Y., Pa.......A5 85, C4 108, E8 114
Susques, Arg..................D2 55
Sussex, N.B., Can., 3,457....D4 74
Sussex, Sussex, N.J., 1,656...A3 106
Sussex, Sussex, Va., 75......E5 121
Sussex, Waukesha, Wis., 1,087..................E1 124
Sussex, Johnson, Wyo., 5....B6 125
Sussex, co., Del., 73,195.....C6 85
Sussex, co., Eng., 1,075,893..D8 12
Sussex, co., N.J., 49,255.....A3 106
Sussex, co., Va., 12,411......E5 121
Sustut, riv., B.C., Can.......A4 68
Susua, P.R...................C3 65
Susung, China.................I7 36
Susurluk, Tur., 6,147........C7 23
Sutcliffe, Washoe, Nev., 80..D2 104
Sutersville, Westmoreland, Pa., 964..................*F2 114
Sutherland, O'Brien, Iowa 883..................B2 92
Sutherland, Lincoln, Nebr., 867..................C4 103
Sutherland, S. Afr., 1,809...D3 49
Sutherland, Dinwiddie, Va., 65..................C7, D5 121
Sutherland, co., Scot., 13,442..................B4 13
Sutherland, res. and canal system, Nebr..................D5 103
Sutherland Springs, Wilson, Tex., 300..................B4 118
Sutherlin, Douglas, Oreg., 2,452..................D3 113
Sutlej, riv., India, Pak.........B4 40
Sutter, Sutter, Calif., 1,219.*C3 82
Sutter, co., Calif., 33,380....C3 82
Sutter Creek, Amador, Calif., 1,161..................C3 82
Suttle, Perry, Ala., 250......C2 78
Sutton, Nevada, Ark., 70....D2 81
Sutton, Que., Can., 1,755....D5 73
Sutton, Eng., 78,969.......m12 10
Sutton, Clay, Nebr., 1,252...D8 103
Sutton, Merrimack, N.H., 200 (487▲)..................D3 105
Sutton, Griggs, N. Dak., 150.B7 110
Sutton, Caledonia, Vt., 125 (476▲)..................B4 120
Sutton, Braxton, W. Va., 967..................C4 123
Sutton, co., Tex., 3,738......D2 118
Sutton, res., W. Va............C4 123
Sutton Coldfield, Eng., 72,143..................B6 12
Sutton-in-Ashfield, Eng., 40,438..................A6 12
Suttons Bay, Leelanau, Mich., 421..................D5 98
Sutton West, Ont., Can., 1,470..................C5 72
Suttsu, Jap., 6,200..........E10 37
Suujiin Hudag, Mong........A2 36
Suva, Fiji Is., 37,371..............52
Suva, hbr., Fiji Is...............52
Suver, Polk, Oreg...........C1 113
Suveydiye, Tur..............D10 31
Suwa, Jap., 31,400.........m17 37
Suwalki, Pol., 20,000........A7 26
Suwanee, Gwinnett, Ga., 541.B2 87
Suwanee, mtn., Ga............B2 87
Suwannee, Dixie, Fla., 200..C3 86
Suwannee, co., Fla., 14,961...B3 86
Suwannee, riv., Fla...........C4 86
Suwannee, riv., Ga............F4 87
Suwannee, sound, Fla.........C3 86
Suwanoochee, creek, Ga........F4 87
Suwanose, isl., Jap............L4 37
Şuwaylih, Jordan, 2,000.....B7 32
Suyel, cape, Sov. Un.........G19 9
Suyo, Peru, 744..............B1 58
Suzu, cape, Jap...............H8 37
Suzuka, Jap., 44,800.........o15 37
Suzuka-Sammyaku, mts., Jap..o15 37
Suzzara, It., 6,631..........E6 18
Svalbard, Nor. dep., Eur., 1,200 (no permanent pop.)........B13 4
Svaneke, Den., 1,167.........A3 26
Svärtagård, Swe., 2,392....u35 25
Svartån, riv., Swe............t34 25
Svartisen, mtn., Nor..........C5 25
Svatovo, Sov. Un., 10,000..G12 27
Svedala, Swe., 3,114..........C7 24
Sveg, Swe., 2,300............F6 25
Svendborg, Den., 23,892......C4 24
Svendborg, co., Den., 150,365..................C4 24
Svene, Nor..................p27 25
Svenljunga, Swe., 2,600......A7 24
Svensen, Clatsop, Oreg., 50..A3 113
Svenstrup, Den., 1,254.......B3 24
Sverdlovsk, Sov. Un., 65,000.q22 27
Sverdlovsk, Sov. Un., 832,000 (*950,000).......D21 9, B6 29
Sverdrup, is., N.W. Ter., Can.B25 4
Svetlaya, Sov. Un., 10,000...C9 37
Svicha, riv., Sov. Un..........D8 26
Svilajnac, Yugo., 5,905.......C5 22
Svilengrad, Bul., 9,918.......E8 22
Svindal, Nor.................p29 25
Svinninge, Den., 1,437.......C5 24
Svir, riv., Sov. Un...........A10 27
Svirstroy, Sov. Un., 10,000..A9 27
Svishtov, Bul., 12,949........D7 22
Svitavy, Czech., 13,900......D4 26
Svobodnyy, Sov. Un., 58,000..................D15 28
Svobodnyy, cape, Sov. Un......C11 37
Svyatoy, cape, Sov. Un........B16 28
Swadlincote, Eng., 19,222....B6 12
Swaffham, Eng., 3,210........B8 12
Swain, co., N.C., 8,387.......D2 109
Swain, mtn., Ark..............A5 81
Swain, reefs, Austl...........D9 50
Swains, isl., Pac. O..........G11 7
Swainsboro, Emanuel, Ga., 5,943..................D4 87
Swainton, Cape May, N.J., 75..................E3 106
Swakopmund, S.W. Afr., 2,842..................B1 49
Swale, riv., Eng..............C6 10
Swaledale, Cerro Gordo, Iowa, 217..................B4 92
Swalwell, Alta., Can., 85....D4 69
Swamp, riv., N.Y.............C2 84
Swampers, Franklin, La., 20.B4 95
Swampscott, Essex, Mass., 13,294..................B6, C3 97
Swan, Marion, Iowa, 168.....C4 92
Swan, creek, Ohio............A2 111
Swan, falls, Idaho............F2 89
Swan, is., Caribbean Sea......B6 72
Swan, lake, Man., Can........C1 71
Swan, lake, Maine............D4 96
Swan, lake, Nebr.............C3 103
Swan, lake, Wash.............D2 122
Swan, peak, Mont.............C3 102
Swan, pt., Md................B5 85
Swan, range, Mont............B3 102
Swan, riv., Austl.............F2 50
Swan, riv., Man., Can.........C1 71
Swan, riv., Sask., Can.........E4 70
Swanage, Eng., 8,112.........D6 12
Swan Creek, Warren, Ill., 125..................C3 90
Swandale, Clay, W. Va., 200..................C7, C4 123
Swan Hill, Austl., 6,185......G4 51
Swanington, Benton, Ind., 150..................C3 91
Swan Lake, Man., Can., 307..................E2 71
Swan Lake, Bannock, Idaho, 150..................G6 89
Swan Lake, Tallahatchie, Miss., 300..................B3 100
Swan Lake, Lake, Mont., 200..................C3 102
Swanlinbar, Ire., 306.......C4 11
Swannanoa, Buncombe, N.C., 2,189..................D4 109
Swanquarter, Hyde, N.C., 200..................B7 109
Swan River, Man., Can., 3,163..................C1 71
Swan River, Itasca, Minn., 150..................C5 99
Swans, isl., Maine............D4 96
Swansboro, Onslow, N.C., 1,104..................C6 109
Swansea, Ont., Can., 9,628..E6 72
Swansea, St. Clair, Ill., 3,018..................B9 101
Swansea, Bristol, Mass., 1,000 (9,916▲)..................C5 97
Swansea, Lexington, S.C., 776..................D5 115
Swansea, Wales, 166,740 (*265,000)..................C4 12
Swansea, bay, Wales...........C4 12
Swans Island, Hancock, Maine, 300 (402▲)...........D4 96
Swanson, Sask., Can., 40.....F2 70
Swanson Lake, res., Kans.....B2 93
Swanton, Garrett, Md., 100..E1 85
Swanton, Saline, Nebr., 190..D8 103
Swanton, Fulton, Ohio, 2,306..................A4 111
Swanton, Franklin, Vt., 2,390 (3,946▲)...........B2 120
Swan Valley, Bonneville, Idaho, 217..................F7 89
Swanville, Morrison, Minn., 342..................E4 99
Swanzey, Cheshire, N.H., 150 (3,626▲)...........E2 105
Swarthmore, Delaware, Pa., 5,753..................B10 114
Swarthwood, Sussex, N.J., 100..................A3 106
Swartswood, lake, N.J........A3 106
Swartz, Ouachita, La., 300..B4 95
Swartz Creek, Genesee, Mich., 3,006..................F7 98
Swatara, Aitkin, Minn., 90..D5 99
Swatow (Shantou), China, 280,400..................G8 34
Swayzee, Grant, Ind., 863...C6 91
Swaziland, Br. dep., Afr., 267,000..................C5 49
Swea City, Kossuth, Iowa, 805..................A3 92
Sweatman, Montgomery, Miss., 25..................B4 100
Swedeborg, Pulaski, Mo., 175..................D5 101
Swedeburg, Saunders, Nebr., 65..................E2 103
Swedeland, Montgomery, Pa., 950..................A11 114
Sweden, country, Eur., 7,542,600......C12 8, F7 25
Swedesboro, Gloucester, N.J., 2,449..................D2 106
Swedesburg, Montgomery, Pa., 950..................*A11 114
Sweeny, Brazoria, Tex., 3,087..................G4 118
Sweet, Gem, Idaho, 100......F2 89
Sweet Briar, Amherst, Va., 850..................D3 121
Sweetgrass, Toole, Mont., 205..................B5 102
Sweet Grass, co., Mont., 3,290..................E7 102
Sweet Hall, King William, Va., 50..................D6 121
Sweet Home, Pulaski, Ark., 900..................C3, D6 81
Sweet Home, Linn, Oreg., 3,353..................C4 113
Sweet Home, Lavaca, Tex., 300..................E4 118
Sweetsburg, Que., Can., 958..................D5 73
Sweetsers, Grant, Ind., 896..C6 91
Sweet Springs, Saline, Mo., 1,452..................C4 101
Sweetsprings, Monroe, W. Va., 500..................D4 123
Sweet Valley, Luzerne, Pa., 250..................D9 114
Sweet Water, Marengo, Ala., 300..................C2 78
Sweetwater, B.C., Can.......B7 68
Sweetwater, Dade, Fla., 645..................*G6 86
Sweetwater, NezPerce, Idaho, 80..................C2 89
Sweetwater, Beckham and Roger Mills, Okla., 50...........B2 112
Sweetwater, Monroe, Tenn., 4,145..................D8 117
Sweetwater, Nolan, Tex., 13,914..................C2 118
Sweetwater, co., Wyo., 17,920..................D3 125
Sweet Water, canyon, Utah....D6 119
Sweetwater, riv., Wyo.........C4 125
Sweime, Jordan, 1,000......h13 32
Swenson, Stonewall, Tex., 150..................C2 118
Swepsonville, Alamance, N.C., 800..................A4 109
Swider, riv., Pol.............m14 26
Swidnica, Pol., 39,000.......C4 26
Świdwin, Pol., 6,098........B3 26
Swiebodzice, Pol., 6,078.....C4 26
Swiebodzin, Pol., 11,200.....B3 26
Swiecie, Pol., 8,358.........B5 26
Swietochlowice, Pol., 57,000..g9 26
Swift, Roseau, Minn., 30.....B3 99
Swift, Hardin, Tenn.........B3 117
Swift, co., Minn., 14,936.....E3 99
Swift, creek, N.C............B6 109
Swift, creek, Va.............C7 121
Swift, res., Wash............C3 122
Swift, riv., N.H.............C4 105
Swift, riv., N.H.............B4 105
Swift Current, Sask., Can., 12,186..................G2 70
Swiftcurrent, creek, Sask., Can..G1 70
Swift Diamond, riv., N.H......B2 105
Swifton, Jackson, Ark., 601..B4 81
Swiftown, Leflore, Miss., 125.B3 100
Swiftwater, Grafton, N.H., 60..................B3 105
Swilly, inlet, Ire.............B4 11
Swinburne, cape, N.W. Ter., Can..................B12 66
Swindle, isl., B.C., Can.......C3 68
Swindon, Eng., 91,736.......C6 12
Swinemünde, see Swinoujscie, Pol.
Swinford, Ire., 1,115........D3 11
Swink, Otero, Colo., 348.....C7 83
Swinomish, Indian res., Wash..A3 122
Swinoujscie, Pol., 10,600....B3 26
Swisher, Johnson, Iowa, 271.C6 92
Swisher, co., Tex., 10,607....B2 118
Swiss, Yancey, N.C..........C3 109
Swiss, Nicholas, W. Va., 500..................C3, C7 123
Swissvale, Allegheny, Pa., 15,089..................B6 114
Switz City, Greene, Ind., 339..................F3 91
Switzer, Logan, W. Va., 1,131..................D5 123
Switzerland, St. Johns, Fla., 150..................C6 86
Switzerland, Jasper, S.C., 50..................G5 115
Switzerland, co., Ind., 7,092.G7 91
Switzerland, country, Eur., 5,429,061..........F10 8, 19
Swords, Morgan, Ga., 200...C3 87
Swords, Ire., 1,816..........D5 11
Swoyersville, Luzerne, Pa., 6,751..................B8, D10 114
Sycamore, Talladega, Ala., 900..................B3 78
Sycamore, Turner, Ga., 501..E3 87
Sycamore, De Kalb, Ill., 6,961..................B5 90
Sycamore, Montgomery, Kans., 187..................E8 93
Sycamore, Ozark, Mo., 42....E5 101
Sycamore, Wyandot, Ohio, 998..................B4 111
Sycamore, Allendale, S.C., 401..................E5 115
Sycamore, Pittsylvania, Va..D3 121
Sycamore, creek, Tenn.........A4 117
Sycamore, creek, W. Va........C7 123
Sychevka, Sov. Un..........D10 27
Sycow, Pol., 2,108..........C4 26
Sydenham, Ont., Can., 737..C8 72
Sydney, Austl., 172,192 (*2,235,000)..................F8 51
Sydney, N.S., Can., 33,617 (*40,300)..................C9 74
Sydney Mines, N.S., Can., 9,122..................C9 74
Sykeston, Wells, N. Dak., 236..................B6 110
Sykesville, Carroll, Md., 1,196..................B4 85
Sykesville, Burlington, N.J., 100..................C3 106
Sykesville, Jefferson, Pa., 1,479..................D4 114
Syktyvkar, Sov. Un., 73,000..................C19 9
Sylacauga, Talladega, Ala., 12,857..................B3 78
Sylamore, Izard, Ark., 45....B3 81
Sylhet, Pak., 33,124.........D9 39
Sylling, Nor., 268..........p28 25
Sylt, isl., Ger................D2 24
Sylva, Jackson, N.C., 1,564..D3 109
Sylvan, Multnomah, Oreg., 600..................B2 113
Sylvan, Franklin, Pa., 20....G5 114
Sylvan, lake, Ind.............B7 91
Sylvan Beach, Oneida, N.Y., 800..................B5 108
Sylvan Grove, Lincoln, Kans., 400..................C5 93
Sylvan Hills, Pulaski, Ark., 2,000..................*C3 81
Sylvania, DeKalb, Ala., 400..A4 78
Sylvania, Sask., Can., 116...E3 70
Sylvania, Screven, Ga., 3,469..................D5 87
Sylvania, Jefferson, Ky., 1,200..................A3 94
Sylvania, Lucas, Ohio, 5,187..................A1, A4 111
Sylvania, Bradford, Pa., 243.C8 114
Sylvan Lake, Alta., Can., 1,381..................C3 69
Sylvan Lake, Oakland, Mich., 2,004..................F7 98
Sylvan Shores, Lake, Fla., 1,214..................*D5 86
Sylvarena, Smith, Miss., 69..C4 100
Sylvester, Worth, Ga., 3,610.E3 87
Sylvester, Fisher, Tex., 405..C2 118
Sylvester, mtn., Newf., Can....D4 75
Sylvia, Reno, Kans., 402.....E5 93
Sylvia, Dickson, Tenn., 100..A4 117
Sym, riv., Sov. Un............C26 9
Symmes, creek, Ohio...........D5 111
Symsonia, Graves, Ky., 400..A2 94
Syosset, Nassau, N.Y., 14,000..................F2 84
Syracuse, Kosciusko, Ind., 1,595..................B6 91
Syracuse, Hamilton, Kans., 1,888..................E2 93
Syracuse, Morgan, Mo., 180..................C5 101
Syracuse, Otoe, Nebr., 1,261..................D9, F3 103
Syracuse, Onondaga, N.Y., 216,038 (*442,300).......B4 108
Syracuse, Meigs, Ohio, 731..D6 111
Syracuse, Davis, Utah, 1,061.B3 119
Syr Darya, riv., Sov. Un......G22 9
Syria, country, Asia, 4,565,000.........E12 31, F6 33
Syriam, Bur., 15,070........D2 38
Syslādobsis, lake, Maine......C4 96
Sysola, riv., Sov. Un.........C19 9
Syzran, Sov. Un., 157,000...C3 29
Szabadszallas, Hung., 4,878..B4 22
Szabolcs-Szatmar, co., Hung., 587,257..........*B5 22
Szamos, riv., Hung............A6 22
Szamotuly, Pol., 10,800......B4 26
Szarvas, Hung., 12,277.......B5 22
Szczakowa, Pol., 4,285......g10 26
Szczebrzeszyn, Pol., 5,122...C7 26
Szczecin (Stettin), Pol., 269,000..........E8 24, B3 26
Szczecinek, Pol., 23,000.....B4 26
Szczuczyn, Pol., 2,479.......B7 26
Szczytno, Pol., 3,645........B6 26
Szechwan, prov., China, 72,160,000..................E5 34
Szeged, Hung., 99,061........B5 22
Szekesfehervar, Hung., 55,934..................B4 22
Szekszard, Hung., 16,409 (19,347▲)..................B4 22
Szengen, China..............G6 34
Szentendre, Hung., 10,307...B4 22
Szentes, Hung., 24,807 (31,175▲)..................B5 22
Szeping, China, 125,900......C9 32
Szigetvar, Hung., 7,395......B3 22
Szolnok, Hung., 45,553.......B5 22
Szolnok, co., Hung., 462,516..................*B5 22
Szombathely, Hung., 54,465..................B3 22
Szprotawa, Pol., 2,672.......C3 26
Sztum, Pol., 3,111...........B5 26
Szubin, Pol., 3,742..........B4 26
Szydlowiec, Pol., 4,010......C6 26
Szymanow, Pol., 2,000.....m12 26

T

Taal, Phil., 4,752..........p13 35
Taal, lake, Phil.............p13 35
Taasinge, isl., Den...........C4 24
Tab, Warren, Ind., 100.......D3 91
Tabaco, Phil., 8,308.........*C6 35
Ţābah, Sau. Ar...............I14 31
Tabas, Iran..................E8 41
Tabas, Iran,.................E8 41
Ţabas, Iran, 17,743.........E10 41
Tabasco, state, Mex., 471,808..................D6 63
Tabatinga, mts., Braz..........D6 57
Tabayoc, mtn., Phil...........n13 35
Tabelbala, Alg...............D4 44
Taber, Alta., Can., 3,951.....E4 69
Taberg, Oneida, N.Y., 375...B5 108
Tabernacle, Burlington, N.J., 100..................D3 106
Tabernas, Sp., 4,121.........D4 20
Tabernash, Grand, Colo., 275..................B5 83
Taberville, St. Clair, Mo., 100..................C4 101
Tabik, isl., Kwajalein...........52
Tabiona, Duchesne, Utah, 167..................C5 119
Tablas, cape, Chile............A2 54
Tablas, isl., Phil.............C6 35
Table, bay, Newf., Can........B3 75
Table, bay, S. Afr.............D2 49
Table, head, Newf., Can.......B4 75
Table, mtn., Ariz.............E5 80
Table, mtn., Newf., Can.......E2 75
Table, rock, Oreg.............C4 113
Table Grove, Fulton, Ill., 500..................C3 90
Table Rock, Pawnee, Nebr., 422..................D9 103
Table Rock, Jackson, Oreg..E4 113
Tables, isl., Scot.............C2 13
Table Top, mtn., Ariz.........E3 80
Tablónes, P.R..............C8, g12 65
Taboada, Sp., 8,162..........A2 20
Tabor, Czech., 19,600.......D3 26
Tabor, Fremont and Mills, Iowa, 909..................D2 92
Tabor, Polk, Minn., 100......B2 99
Tabor (Mount Tabor), Morris, N.J., 1,000..................*B4 106
Tabor, Bon Homme, S. Dak., 378..................E8 116
Tabor, mtn., Isr..............B7 32
Tabora, Tan., 15,361.........C5 48
Tabor City, Columbus, N.C., 2,338..................C5 109
Tabou, I.C., 2,100...........F3 45
Tabrīz, Iran, 290,195........B3 41
Tabūk, Sau. Ar., 10,000......D7 43
Tabusintac, riv., N.B., Can....B4 74
Tacámbaro de Codallos, Mex., 7,286...............n13 63
Tacheng, China..............E7 36
Tachie, riv., B.C., Can........B5 68
Tachiiwa, pt., Iwo..............52
Tachikawa, Jap., 67,949.....n18 37
Táchira, state, Ven., 399,163..................B3 60
Tachov, Czech., 5,200........D7 17
Tacloban, Phil., 37,000 (53,600▲)..................C6 35
Tacna, Yuma, Ariz., 150......E1 80
Tacna, Peru, 13,514..........E3 58
Tacna, dept., Peru, 42,874....E3 58
Tacoma, La Plata, Colo., 15.D3 83
Tacoma, Pierce, Wash., 147,979 (*298,000)......B3, D1 122
Tacoma Park, Brown, S. Dak., 25..................B7 116
Taconic, Litchfield, Conn., 200..................A3 84
Taconic, range, Mass..........A1 97
Tacoronte, Can. Is., 10,282.m13 20
Tacuarembó, Ur., 21,200.....E1 56
Tacuarembó, dept., Ur., 71,100..................*E1 56
Tacuatí, Par., 1,538.........D4 55
Tacubaya, Mex. (part of Mexico City)..............h9 63
Tad, Kanawha, W. Va., 500.C6 123
Tademait, plat., Alg...........D5 44
Tadent, riv., Alg..............E6 44
Tadjoura, Fr. Som., 1,150....C5 47
Tadoule, lake, Man., Can......C5 71
Tadoussac, Que., Can., 1,083..................A8 73
Tadzhik S.S.R., rep., Sov. Un., 2,104,000..................F9 28
Taegu, Kor., 678,300.........I4 37
Taejon, Kor., 229,400........H3 37
Taeyudong, Kor., 5,000......F2 37
Tafalla, Sp., 7,320...........A5 20
Tafí Viejo, Arg., 15,374......E2 55
Tafoya, Colfax, N Mex......A5 107
Taft, Kern, Calif., 3,822......E4 82
Taft, B.C., Can., 50..........D8 68
Taft, Orange, Fla., 1,214....D5 86
Taft, St. Charles, La., 260...B6 75
Taft, Muskogee, Okla., 386..B6 112
Taft, Lincoln, Oreg., 577....C3 113
Taft, Lincoln, Tenn., 200....B5 117
Taft, San Patricio, Tex., 3,463..................F4 118
Taft Heights, Kern, Calif., 2,661..................*E4 82
Taft Southwest, San Patricio, Texas, 1,927..............*F4 118
Taftsville, Windsor, Vt., 100..................D4 120
Taftville, New London, Conn. (part of Norwich)...........C8 84
Tagachan, pt., Guam............52
Taganrog, Sov. Un., 214,000..................H12 27
Taganrog, gulf, Sov. Un......H12 27
Tagawa, Jap., 100,071.......*J5 37
Tagbileran, Phil., 5,879......D6 35
Taghrīfat, Libya.............D3 43
Taghmaconnell, Ire..........D5 11
Taghmon, Ire., 347..........E5 11
Tagliamento, riv., It..........C8 18
Tagolo, pt., Phil..............D6 35
Tagourareт (Well), Maur....C3 45
Taguan, pt., Guam...............52
Taguatinga, Braz., 1,496....D1 57
Taguchi, Jap., 4,400........n16 37
Tagus, Mountrail, N. Dak., 72..................A4 110
Tagus, see Tejo, riv., Port.
Tagus, see Tajo, riv., Sp.
Tahan, mtn., Mala.............J5 38
Tahat, mtn., Alg..............E5 44
Tahiti, isl., Society Is........H13 7
Tahlequah, Cherokee, Okla., 5,840..................B7 112
Tahoe, lake, Calif., Nev.......C3 82
Tahoe City, Placer, Calif., 350..................C3 82
Tahoka, Lynn, Tex., 3,012...C2 118
Taholah, Grays Harbor, Wash., 400..................B1 122
Tahoma, Placer, Calif., 50...C3 82
Tahona, LeFlore, Okla., 35..................B7 112
Tahoua, Niger, 12,400.......D6 45
Tahquamenon, falls, Mich......B5 98
Tahsien, China, 26,000.......I2 36
Ţahţā, Eg., U.A.R., 41,400..D6 43
Tahtsa, lake, B.C., Can.......C4 68
Tahtsa, peak, B.C., Can.......C4 68
Tahtsa, riv., B.C., Can........C4 68
Tahuya, Mason, Wash., 150..B2 122
Tai, China...................E5 36
Tai, China, 159,800..........E8 36
Tai, I.C.....................H3 45
Tai, lake, China...............I9 36
Taian, China................D10 36
Taian, China, 80,000.........F7 36
Taiban, De Baca, N. Mex., 120..................C5 107
Taichintala, China...........B9 36
T'aichung, Taiwan, 188,000 (272,700▲)..................G9 34
Taihape, N.Z., 2,682.......M15 51
Taikang, China, 5,000........C2 37
Taiku, China.................F5 36
Tailagoin, mts., Mong.........C6 34
Tailai, China, 5,000..........C1 37
Tailem Bend, Austl., 2,049..G2 51
Taileva, pt., Fiji Is..............52
Tain, Scot., 1,699............C4 13
T'ainan, Taiwan, 239,300 (312,200▲)..................G9 34
Taining, China..............K7 36
Taipei (Taihoku), Taiwan, 813,800 (*950,000)...........G9 34
Taiping, Mala., 48,206.......J4 38
Taipu, Braz., 1,162..........g6 57
Taira, Jap., 71,115 (*103,000)..................H10 37
Taishun, China, 11,000.......K8 36
Taitao, pen., Chile............D1 54
Taitiarato, Pap..............k11 50
T'aitung, Taiwan, 8,886.....G9 34
Taivalkoski, Fin., 850.......E13 25
Taiwan (Formosa), country (Nationalist China), Asia, 8,265,000..................G9 34
Taiwan, isl., Asia.............G9 34
Taiyüan (Yangkü), China, 1,020,000...........D7 34, F5 36
Tajarhī, Libya...............E2 43
Tajimi, Jap., 53,793.......I8, n16 37
Tajique, Torrance, N. Mex., 115..................C3 107
Tajo (Tagus), riv., Sp.........C3 20
Tajumulco, peak, Guat.........C2 62
Tajuna, riv., Sp...............B4 20
Tājūrā', Libya, 2,670.........C2 43
Tak, Thai., 12,918...........D3 38
Taka Banare, isl., Okinawa.........52
Takabba, Ken...............A7 48
Takada, Jap., 46,200........H9 37
Takaka, N.Z., 769...........N14 51
Takamatsu, Jap., 228,172.....I7 37
Takaoka, Jap., 135,190.......H8 37
Takasaki, Jap., 125,000 (142,152▲)..................H9, m18 37
Takata, Jap., 23,025.........H9 37
Takato, Jap., 4,100.........n17 37
Takatsuki, Jap., 79,043......o14 37
Takaw, Bur..................B3 38
Takayama, Jap., 34,800..................H8, m16 37
Takazze, riv., Eth............C4 47
Take, isl., Jap................H5 37
Takefu, Jap., 37,100........n15 37
Takeo, Camb., 5,000.........G6 38
Tākestān, Iran, 10,543.......C4 41
Takhta-Bazar, Sov. Un.......D11 41
Takhta-Kupyr, Sov. Un......E6 29
Takht-i-Sulaiman, mtn., Iran...C5 41
Takihara, Jap., 4,457........o15 37
Takilma, Josephine, Oreg., 50..................E3 113
Takingeun, Indon., 1,411.....J2 38
Takla, lake, B.C., Can.........B5 68
Taklakhar, China............B8 40
Takla Makan, des., China.....F11 33
Takoma Park, Montgomery and Prince Georges, Md., 16,799..................C1 85

Takotna, Alsk., 40 C8 79
Takouchen, China D7 36
Taku, China E7 36
Takua Pa, Thai., 5,646 H3 38
Takut, Bur., 5,000 A3 38
Ṭalā, Eg., U.A.R., 18,570 D2 32
Tala, Mex., 12,541 m12 63
Tala, Ur., 1,957 E1 56
Talaga, Phil., 1,746 p13 35
Talagante, Chile, 7,966 A2 54
Talai, China, 24,921 D2 37
Talakan, Sov. Un B6 37
Talakmau, mtn., Indon E2 35
Talala, Rogers, Okla., 147 A6 112
Talamanca, mts., C.R. F6 62
Talanga, Hond., 2,312 C4 62
Talanquera, pt., Cuba h12 64
Talara, Peru, 12,985 B1 58
Talas, Sov. Un., 10,000 E8 29
Talaud, is., Indon E7 35
Talavera de la Reina, Sp., 31,900 B2 20
Talayan, Phil., 9,240 *D6 35
Talbert, Breathitt, Ky., 150 C6 94
Talbiya, Isr. (part of Jerusalem) m14 32
Talbot, Benton, Ind., 100 D3 89
Talbot, Marion, Oreg., 65 C1 113
Talbot, co., Ga., 7,127 D2 87
Talbot, co., Md., 21,578 C5 85
Talbot, isl., Fla B5 86
Talbot, lake, Man., Can G8 71
Talbotton, Talbot, Ga., 1,163 D2 87
Talca, Chile, 70,000 B2 54
Talca, prov., Chile, 205,448 B2 54
Talcahuano, Chile, 82,000 B2 54
Talcher, India, 8,147 G10 40
Talco, Titus, Tex., 1,024 C5 118
Talcott, Summers, W. Va., 600 D4 123
Talcottville, Tolland, Conn., 700 B7 84
Talcottville, Lewis, N.Y., 70 B5 108
Taldy-Kurgan, Sov. Un., 41,000 D9 29, F24 9
Talence, Fr., 25,874 E3 14
Talent, Jackson, Oreg., 868 E4 113
Talha, Chad A4 46
Tali, China, 9,000 F5 34
Tali, China, 15,000 G4 36
Taliabu, isl., Indon F6 35
Taliaferro, co., Ga., 3,370 C4 87
Talien, see Dairen, China
Talihina, LeFlore, Okla., 1,048 C6 112
Talim, isl., Phil o13 35
Tali Post, Sud D3 47
Talisay, Phil., 1,512 o14 35
Talisayan, Phil., 10,646 *D6 35
Talish, mts., Iran, Sov. Un B4 41
Talitsa, Sov. Un., 17,300 B6 29
Täljeån, riv., Swe t33 25
Talkeetna, Alsk., 76 C9, f16 79
Talkeetna, mts., Alsk f17 79
Ṭalkhā, Eg., U.A.R., 13,216 C3 32
Talking Rock, Pickens, Ga., 84 B2 87
Tallaboa, P.R. C4 65
Talladega, Talladega, Ala., 17,742 B3 78
Talladega, co., Ala., 65,495 B3 78
Talladega, mts., Ala B4 78
Talladega Springs, Talladega, Ala., 177 B3 78
Tallahala, creek, Miss D4 100
Tallahassee, Leon, Fla., 48,174 B2 86
Tallahatchie, co., Miss., 24,081 B3 100
Tallahatchie, riv., Miss B3 100
Tallangatta, Austl., 853 H6 51
Tallant, Osage, Okla., 25 A5 112
Tallapoosa, Haralson, Ga., 2,744 C1 87
Tallapoosa, New Madrid, Mo., 225 E8 101
Tallapoosa, co., Ala., 35,007 C4 78
Tallapoosa, riv., Ala C3 78
Tallard, Fr., 550 E2 18
Tallassee, Elmore and Tallapoosa, Ala., 4,934 C4 78
Tallevast, Manatee, Fla., 500 F2 86
Talleyville, New Castle, Del., 1,000 *A6 85
Tallieu, Assumption, La C5 95
Tallinn, Sov. Un., 298,000 B5 27
Tallmadge, Summit, Ohio, 10,246 A6 111
Tallow, Ire., 819 E3 11
Tallula, Menard, Ill., 547 D4 90
Tallula, Issaquena, Miss., 100 C2 100
Tallulah, Madison, La., 9,413 B4 95
Tallulah, mts., Ga B3 87
Tallūzā, Jordan, 2,000 f12 32
Talmage, Sask., Can., 63 H4 70
Talmage, Dickinson, Kans., 200 C6 93
Talmage, Otoe, Nebr., 361 D9 103
Talmage, Duchesne, Utah, 10 C5 119
Talnoye, Sov. Un., 10,000 G8 27
Talo, mtn., Eth C4 47
Talodi, Sud., 2,736 C3 47
Talofofo, Guam, 947 52
Talofofo, bay, Guam 52
Taloga, Dewey, Okla., 322 A3 112
Talon, lake, Ont., Can A5 72
Talowah, Lamar, Miss., 50 D4 100
Talpa, Taos, N. Mex., 500 A4 107
Talpa, Coleman, Tex., 195 D3 118
Talpa de Allende, Mex., 3,157 m11 63
Talquin, lake, Fla B2 86
Talsi, Sov. Un., 10,000 C4 27
Taltal, Chile, 4,901 E1 55
Tama, Tama, Iowa, 2,925 C5 92
Tama, co., Iowa, 21,413 B5 92
Tamaha, Haskell, Okla., 80 B7 112
Tamalameque, Col., 1,843 B3 60
Tamale, Ghana, 16,164 E4 45
Tamalpais, mtn., Calif B5 82
Tamalpais Valley, Marin, Calif., 1,500 *B5 82
Taman, Sov. Un., 10,000 I11 27
Tamanar, Mor., 783 C3 44
Tamano, Jap., 65,292 I6 37
Tamanrasset, see Fort Laperrine, Alg.
Tamanrasset, riv., Alg E5 44
Tamaqua, Shuylkill, Pa., 10,173 E10 114
Tamar, riv., Eng D3 12
Tamarack, Adams, Idaho, 50 E2 89
Tamarack, Aitkin, Minn., 112 D5 99
Tamarite, Sp., 4,272 B6 20
Tamaroa, Perry, Ill., 696 E4 90
Tamassee, Oconee, S.C., 350 B1 115
Tamatave, Malag., 36,133 g9 49
Tamatave, prov., Malag g9 49
Tamaulipas, state, Mex., 1,009,800 C5 63
Tamazula de Gordiano, Mex., 10,784 n12 63
Tamazunchale, Mex., 8,393 C5, m14 63
Tambach, Ken A6 48
Tambacounda, Sen., 4,600 D2 45
També, Braz., 4,149 C3, h6 57
Tambelan, is., Indon E3 35
Tambellaga (Well), Niger C6 45
Tambo, Austl., 404 B6 51
Tambo, riv., Peru D3 58
Tambo, riv., Peru E3 58
Tambo Grande, Peru, 4,078 B1 58
Tambov, Sov. Un., 170,000 C2 29
Tambre, riv., Sp A1 20
Tamdy-Bulak, Sov. Un E6 29
Tambura, Sud D2 47
Tamchaket, Maur C2 45
Tame, Col., 1,383 B3 60
Tamega, riv., Port B2 20
Tamel Aike, Arg D2 54
Tamgue, mtn., Guinea D2 45
Tamiahua, Mex., 4,055 C5, m15 63
Tamiahua, lagoon, Mex C5 63
Tamiami, canal, Fla G6 86
Tamiao, China C4 36
Tamina, Montgomery, Tex., 130 D5 118
Tamina, riv., Switz C7 19
Taming, China F6 36
Tamins, Switz., 881 C7 19
Tamis, riv., Yugo C5 23
Tam Ky, Viet E8 38
Tamms, Alexander, Ill., 548 F4 90
Tammūn, Jordan, 2,000 B7, f12 32
Tamney, Ire B4 11
Tamo, Jefferson, Ark., 100 C4 81
Tamola, Kemper, Miss., 50 C5 100
Tamora, Seward, Nebr., 88 D8 103
Tamora, Ponape 52
Tamoroi, Ponape 52
Tampa, Hillsborough, Fla., 274,970 (*356,200) E2, E4 86
Tampa, Marion, Kans., 145 D6 93
Tampa, bay, Fla E4 86
Tampasak, Mala D5 35
Tampere, Fin., 126,600 (*164,000) G10 25
Tampico, Whiteside, Ill., 790 B4 90
Tampico, Jackson, Ind., 100 G6 91
Tampico, Mex., 122,197 (*180,000) C5, k15 63
Tampico, Valley, Mont., 30 B10 102
Tampico, Grainger, Tenn C10 117
Tams, Raleigh, W. Va., 500 D3 123
Tamworth, Austl., 18,984 E8 51
Tamworth, Ont., Can., 375 C8 72
Tamworth, Eng., 13,555 B6 12
Tamworth, Carroll, N.H., 250 (1,016*) C4 105
Tana, Chile C2 55
Tana, lake, Eth C4 47
Tana, riv., Ken B7 48
Tanabe, Jap., 48,673 J7 37
Tanacross, Alsk., 102 C11 79
Tanaelv, riv., Fin C11 25
Tanafjord, fjord, Nor B13 25
Tanaga, isl., Alsk E4 79
Tanahbala, isl., Indon F1 35
Tanahgrogot, Indon F5 35
Tanahmasa, isl., Indon F1 35
Tanakpur, India C8 40
Tanami, Austl D4 50
Tanana, Alsk., 349 B9 79
Tanana, riv., Alsk C11 79
Tananarive, Malag., 248,000 g9 49
Tananarive, prov., Malag g9 49
Tanapag, Saipan 52
Tanapag, hbr., Saipan 52
Tanaro, riv., It B2 21
Tanauan, Phil., 4,265 o13 35
Tanaunella, It., 208 D2 21
Tanavuso, pt., Fiji Is 52
Tancheng, China G8 36
Tanchon, Kor., 32,761 F4 37
Tancook Island, N.S., Can., 323 E5 74
Tanda, India, 32,687 D9 40
Tandag, Phil., 6,735 *D7 35
Tandarei, Rom., 2,353 C8 24
Tandil, Arg., 32,309 B5 54
Tandjung, Indon F5 35
Tandjungbalai, Indon., 27,315 E1, m11 35
Tandjungpandan, Indon., 15,708 F3 35
Tandjungselor, Indon., 1,991 E5 35
Tando-Adam, Pak., 21,275 E2 40
Tandou, lake, Austl F4 51
Tanega, isl., Jap K5 37
Tanega, strait, Jap K5 37
Taney, co., Mo., 10,238 E9 101
Taneycomo, lake, Mo E5 101
Taneytown, Carroll, Md., 1,519 A3 85
Taneyville, Taney, Mo., 134 E4 101
Tanezrouft, des., Alg., Mali E4 44
Tanga, Tan., 38,053 C6 48
Tanga, reg., Tan., 687,846 C6 48
Tangancícuaro [de Arista], Mex., 9,408 n12 63
Tanganyika, country, Afr., 9,077,000 G9 42, C5 48
Tanganyika, lake, Con. L., Tan C4 48
Tangent, Linn, Oreg., 150 C3, D1 113
Tangerhütte, Ger., 6,679 A6 17
Tangermünde, Ger., 13,800 B5 16
Tangho, China H5 36
Tangier, N.S., Can., 230 E7 74
Tangier, Parke, Ind., 95 E3 91
Tangier, Mor., 141,714 B3 44
Tangier, Woodward, Okla A2 112
Tangier, Accomack, Va., 876 D7 121
Tangier, isl., Va D6 121
Tangier, sound, Md D6 85
Tangipahoa, Tangipahoa, La., 465 D5 95
Tangipahoa, par., La., 59,434 D5 95
Tangipahoa, riv., La D5 95
Tangra, lake, China F7 34
Tangshan, China, 800,000 E8 36
Tangshan, China, 65,000 G7 36
Tangtu, China I8 36
Tanguiéta, Dah D5 45
Tanguisson, pt., Guam 52
Tangwang, riv., China B10 34
Tangyang, China, 18,000 I4 36
Tangyuan, China, 5,000 C4 36
Tanhsien, China C8 38
Tanimbar, is., Indon G8 35
Taninges, Fr., 1,212 C2 18
Tanjay, Phil., 12,503 *D6 35
Tanjore, India, 111,099 F6 39
Tank, Pak., 6,899 A3 40
Tank, cape, Iran I9 41
Tännäs, Swe F5 25
Tanner, Limestone, Ala., 500 A3 78
Tanner, Gilmer, W. Va., 60 C4 123
Tannis, bay, Den A4 24
Tannu-Ola, mts., Sov. Un., Mong A3 34
Tannura, cape, Sau. Ar A5 41
Tanout, Niger C6 45
Tanque Verde, Pima, Ariz., 1,053 *C4 80
Tanshui, For., 10,000 *F9 34
Ṭanṭā, Eg., U.A.R., 151,700 G8 31, D2 32, C6 43
Tantallon, Sask., Can., 145 G5 70
Tantoyuca, Mex., 7,476 m14 63
Tanum, Swe H4 25
Tanunak, Alsk., 183 C6 79
Tanza, Phil., 2,760 o13 35
Tao, riv., China B9 34
Taoan, China B10 36
Taoershan, China B9 36
Taohsien, China F7 34
Taokou, China G6 36
Taolin, China, 4,000 D5 36
Taonan, China, 65,000 B10 36
Taopi, Mower, Minn., 92 G6 99
Taormina, It., 6,521 F5 21
Taos, Taos, N. Mex., 2,163 A4 107
Taos, co., N. Mex., 15,934 A4 107
Taos Pueblo, Taos, N. Mex., 900 A4 107
Taoudenni, Mali B4 45
Taourirt, Mor., 7,343 C4 44
Taoyuan, China J4 36
Tapacarí, Bol., 980 C2 55
Tapachula, Mex., 41,701 E6 63
Tapaga, cape, W. Sam 52
Tapajós, riv., Braz D3 59
Tapak, Ponape 52
Tapak, isl., Ponape 52
Tapaktuan, Indon m11 35
Tapalqué, Arg., 3,018 B4 54
Tapanahoni, riv., N. Gui B3 59
Tapanshang, China C8 36
Tapanui N.Z., 767 P12 51
Tapa Shan, mts., China I4 36
Tapauá, riv., Braz D4 58
Taperoá, Braz., 2,207 h5 57
Tapicitoes, Rio Arriba, N. Mex., 10 A3 107
Tapirapé, riv., Braz E4 59
Tapis, mtn., Mala K5 38
Tapotchau, mtn., Saipan 52
Tappahannock, Essex, Va., 1,086 D6 121
Tappan, Rockland, N.Y., 2,100 D1 108
Tappan, res., Ohio B6 111
Tappen, Kidder, N. Dak., 326 C6 110
Tappita, Lib E3 45
Tapti, riv., India D5 39
Tapuaenuku, mtn., N.Z. O14 51
Tapuhsing, China E4 34
Tapurucuara, Braz., 298 D4 58
Taputapu, cape, Am. Sam 52
Taquara, Braz., 11,282 D2 56
Taquara, mts., Braz B2 56
Taquaritinga do Norte, Braz., 1,042 h5 57
Taquarí, riv., Braz B1 56
Taquarí, riv., Braz E2 56
Taquaritinga, Braz., 11,624 k7 56
Tar, riv., N.C. B5 109
Tara, Ont., Can., 481 C3 72
Tara, St. Louis, Mo., 2,000 *C7 101
Tara, Sov. Un., 20,400 B8 29
Tara, hill, Ire D5 11
Tara, riv., Sov. Un B9 29
Tara, riv., Yugo D4 23
Tarabuco, Bol., 2,833 C3 55
Tarabulus (Tripoli), Leb., 110,000 E10 31
Ṭarābulus, see Tripoli, Libya
Ṭarābulus, see Tripolitania, prov., Libya
Tarague, Guam 52
Tarakan, Indon., 11,589 E5 35
Tarakliya, Sov. Un C9 22
Tarancón, Sp., 7,714 B4 20
Taransay, isl., Scot C1 13
Taranto, It., 194,609 D6 21
Taranto, gulf, It D6 21
Tarapaca, Col D4 60
Tarapacá, prov., Chile, 122,665 D1 55
Tarapoto, Peru, 9,249 C2 58
Tarar, strait, Sov. Un D17 28
Tarare, Fr., 12,131 E6 14
Tarascon [-sur-Ariège], Fr., 3,680 F4 14
Tarascon [-sur-Rhône], Fr., 6,485 F6 14
Tarasp, Switz., 396 C9 19
Tarata, Bol., 3,016 C2 55
Tarata, Peru, 2,827 E3 58
Tarauacá, riv., Braz C3 58
Tarawa, isl., Gilbert Is 52
Tarawera, N.Z., 153 M16 51
Tarazona, Sp., 12,059 B5 20
Tarazona, Sp., 6,850 C5 20
Tarbagatay, range, Sov. Un D10 29
Tarbat Ness, pt., Scot C5 13
Tarbert, Ire., 455 E2 11
Tarbert, Scot C2 13
Tarbert, Scot E3 13
Tarbert, inlet, Scot E3 13
Tarbert, East, inlet, Scot C2 13
Tarbert, West, inlet, Scot C1 13
Tarbes, Fr., 46,600 F4 14
Tarbet, Scot. (pop. incl. with Arrochar) D4 13
Tarboro, Camden, Ga., 185 F5 87
Tarboro, Edgecombe, N.C., 8,411 B6 109
Tarbū, Libya D3 43
Tarcento, It., 4,395 C9 18
Tarcoola, Austl., 225 F5 50
Taree, Austl., 10,050 E9 51
Tarentum, Pike, Ala., 100 D4 78
Tarentum, Allegheny, Pa., 8,232 A7, E2 114
Tarfaya, Mor., 18,361 D2 44
Tarfside, Scot D6 13
Targhee, pass, Idaho, Mont E7 83, F5 102
Târgoviște, Rom., 24,360 C7 22
Târgul-Frumos, Rom., 4,665 B8 22
Târgul-Jiu, Rom., 19,618 C6 22
Târgul-Neamt, Rom., 10,373 B8 22
Târgul-Ocna, Rom., 11,227 B8 22
Târgul-Săcuesc, Rom., 7,500 B8 22
Tarhūna, Libya C2 43
Tarifa, Sp., 7,737 D3 20
Tariffville, Hartford, Conn., 650 B5 84
Tarigtig, pt., Phil n14 35
Tarija, Bol., 16,869 D3 55
Tarija, dept., Bol., 126,752 D3 55
Tarik, isl., Truk 52
Tarim Darya, riv., China E11 33
Taritai, isl., Tarawa 52
Tarkiln, Providence, R.I., 100 B10 84
Tarkio, Atchison, Mo., 2,160 A2 101
Tarkio, Mineral, Mont., 10 C2 102
Tarklin, hill, Maine E4 96
Tarko-Sale, Sov. Un C24 9
Tarkwa, Ghana, 7,840 (*25,232) E4 45
Tarlac, Phil., 30,000 (98,100*) o13, B6 35
Tarlac, prov., Phil., 427,300 *B6 35
Tarland, Scot., 682 C6 13
Tarlton, Pickaway, Ohio, 377 C5 111
Tarm, Den., 2,270 C2 24
Tarma, Peru, 7,876 D2 58
Tarn, dept., Fr., 319,560 *F5 14
Tarn, riv., Fr F4 14
Tarna, riv., Hung B5 22
Tarnava, riv., Rom B7 22
Târnăveni, Rom., 14,883 B7 22
Tårnby, Den., 42,688 *C6 24
Tarn-et-Garonne, dept., Fr., 175,847 *E4 14
Tarnobrzeg, Pol., 4,140 C6 26
Tarnopol, Sask., Can., 110 E3 70
Tarnov, Platte, Nebr., 70 C8 103
Tarnow, Pol., 71,000 C6 26
Tarnowskie Gory, Pol., 28,000 C5, g9 26
Taro, riv., It B2 21
Tārom, Iran G7 41
Taroom, Austl., 468 B7 51
Taroudant, Mor., 17,141 C3 44
Tarp, Ger., 1,200 D3 24
Tarpley, Bandera, Tex., 25 E3 118
Tarpon Springs, Pinellas, Fla., 6,768 D4 86
Tarquí, Peru B2 58
Tarquinia, It., 9,776 C3 21
Tárraga, Sp., 6,059 B6 20
Tarragona, Sp., 43,519 B6 20
Tarragona, prov., Sp., 362,679 *B6 20
Tarrant, Jefferson, Ala., 7,810 C3, E4 78
Tarrant, co., Tex., 538,495 C4 118
Tarrasa, Sp., 92,234 B7 20
Tarratine, Somerset, Maine C3 96
Tarryall, Park, Colo., 15 B5 83
Tarryall, creek, Colo B5 83
Tarryall, mts., Colo B5 83
Tarrytown, Montgomery, Ga., 191 D4 87
Tarrytown, Westchester, N.Y., 11,109 D1, D3 108
Tårs, Den A4 24
Tarshiha, Isr., 639 A7 32
Tarsney, Jackson, Mo., 400 E3 101
Tarso Ahon, mtn., Chad A3 46
Tarsus, Tur., 51,300 D10 31
Tartagal, Arg., 8,539 D3 55
Tartu, Sov. Un., 76,000 B6 27
Tartuguan, pt., Guam 52
Tartūs, Syr., 15,400 E10 31
Tarut, isl., Sau. Ar H5 41
Tarutung, Indon., 3,436 K3 38
Tarutao, isl., Thai I3 38
Tarver, Echols, Ga., 100 F4 87
Tarves, Scot., 1,852 C6 13
Tarya, riv., Bol D3 55
Tasawāh, Libya D2 43
Taschereau, Que., Can., 950 B2 73
Tascosa, Oldham, Tex B1 118
Taseko, lake, B.C., Can D6 68
Taseko, mtn., B.C., Can D6 68
Taseko, riv., B.C., Can D6 68
Tashauz, Sov. Un., 45,000 E5 29
Tashigong, China, 10,000 A7 40
Tashkent, Sov. Un., 971,000 (*1,100,000) G22 9, E7 29
Tashkumyr, Sov. Un., 12,000 E8 29
Tashkurghan, Afg., 20,000 C13 41
Taskan, Sov. Un C18 28
Taskopru, Tur., 5,700 B10 31
Tasman, bay, N.Z. N14 51
Tasman, pen., Austl o15 50
Tasman, sea, Austl H8 51
Tasmania, state, Austl., 350,340 o15 50
Tassili-n-Ajjer, plat., Alg D6 44
Tassili oua-n-Ahaggar, plat., Alg E6 44
Tasso, Bradley, Tenn., 150 D9 117
Taswell, Crawford, Ind., 125 H4 91
Tata, Hung., 17,333 B4 22
Tatabánya, Hung., 52,044 B4 22
Tatamagouche, N.S., Can., 581 D6 74
Tatar, strait, Sov. Un C10 37
Tatarbunary, Sov. Un., 10,000 C9 22
Tatarsk, Sov. Un., 31,100 B9 29
Tate, Sask., Can., 30 F3 70
Tate, Pickens, Ga., 900 B2 87
Tate, co., Miss., 18,138 A4 100
Tate Cove, Evangeline, La D3 95
Tateville, Pulaski, Ky., 725 D5 94
Tateyama, Jap., 34,900 I9, o18 37
Tathlina, lake, N.W. Ter., Can D9 66
Tatien, China, 4,000 L7 36
Tatitlek, Alsk., 96 C10, g18 79
Tatlayoko Lake, B.C., Can., 57 D5 68
Tatle, lake, B.C., Can C5 68
Tatnam, cape, Man., Can C5 71
Tatra, mts., Pol., Czech D5 26
Tatta, Pak., 9,716 E1 40
Tattershall, Eng., 518 A7 12
Tattnall, co., Ga., 15,837 D4 87
Tatuí, Braz., 22,550 C3, m8 56
Tatuk, lake, B.C., Can C5 68
Tatum, Lea, N. Mex., 1,168 D6 107
Tatum, Marlboro, S.C., 132 B8 115
Tatumville, Dyer, Tenn., 300 A2 117
Tatung, China, 228,500 C7 34, D5 36
Tatung, China I7 36
Tatvan, Tur., 6,500 C14 31
Tau, isl., Thai G3 38
Tauá, Braz., 4,904 C2 57
Taubaté, Braz., 64,863 C3 56
Tauber, riv., Ger D4 17
Tauberbischofsheim, Ger., 7,000 D4 17
Taucha, Ger., 15,600 B7 17
Tauern, tunnel, Aus C9 18
Taumatawhakatangihangakoauauotamateapokaiwhenuakitanatahu, hill, N.Z. *N16 51
Taum Sauk, mtn., Mo D7 101
Taung, S. Afr., 1,496 C3 49
Taungdwingyi, Bur., 16,233 D10 39
Taunggyi, Bur., 8,652 D10 39
Taunton, Eng., 35,178 C4 12
Taunton, Bristol, Mass., 41,132 C5 97
Taunton, Lyon, Minn., 233 F2 99
Taunton, Burlington, N.J., 200 D3 106
Taunton, riv., Mass C5 97
Taunus, mts., Ger D2 17
Taupo, lake, N.Z. M15 51
Taurage, Sov. Un., 10,465 D4 27
Tauranga, N.Z., 13,468 (*24,659) L16 51
Taurianova, It., 14,300 E6 21
Taurus, mts., Tur D9 31
Tauste, Sp., 6,634 B5 20
Tavannes, Switz., 3,939 B3 19
Tavares, Lake, Fla., 2,724 D5 86
Tavas, Tur., 7,110 D7 23
Tavda, Sov. Un., 40,800 B7 29
Tavda, riv., Sov. Un D21 9
Tavernier, Monroe, Fla., 800 G6 86
Taveuni, isl., Fiji Is 52
Taviche, Mex., 1,085 D5 63
Tavira, Port., 7,496 D2 20
Tavistock, Ont., Can., 1,232 D4 72
Tavistock, Eng., 6,086 D3 12
Tavolzhan, Sov. Un C9 29
Tavoy, Bur., 40,312 F10 39
Tavoy, isl., Bur F3 38
Tavoy, pt., Bur F3 38
Tavsanli, Tur., 11,500 C7 23
Taw, riv., Eng D4 12
Tawas, lake, Mich D7 98
Tawas City, Iosco, Mich., 1,810 D7 98
Tawatinaw, Alta., Can., 100 B4 69
Tawe, riv., Wales C4 12
Tawi Tawi, isl., Phil D6 35
Tawi Tawi, is., Phil D6 35
Tāwurghā', Libya C3 43
Taxco de Alarcón, Mex., 10,025 n14 63
Tay, firth, Scot D5 13
Tay, lake, Scot D4 13
Tay, riv., Scot D5 13
Tayabamba, Peru, 1,179 C2 58
Tayabas, bay, Phil p13 35
Tayāsīr, Jordan, 1,000 f12 32
Taycheedah, Fond du Lac, Wis., 400 B5 124
Tayga, Sov. Un., 34,800 B11 29
Taylor, Columbia, Ark., 734 D2 81
Taylor, Navajo, Ariz., 400 C5 80
Taylor, Wayne, Mich., 49,658 *F7 98
Taylor, Lafayette, Miss., 122 A4 100
Taylor, Loup, Nebr., 280 C6 103
Taylor, Stark, N. Dak., 215 C3 110
Taylor, Lackawanna, Pa., 6,148 A9, D10 114
Taylor, Williamson, Tex., 9,434 D4 118
Taylor, Jackson, Wis., 334 D2 124
Taylor, co., Fla., 13,168 B3 86
Taylor, co., Ga., 8,311 D2 87
Taylor, co., Iowa, 10,288 D3 92
Taylor, co., Ky., 16,285 C4 94
Taylor, co., Tex., 101,078 C3 118
Taylor, co., W. Va., 15,010 B4 123
Taylor, co., Wis., 17,843 C3 124
Taylor, dam, Nev C3 104
Taylor, mtn., Idaho E4 89
Taylor, mtn., N.Z. O13 51
Taylor, ridge, Ga B1 87
Taylor, riv., Man., Can B2 71
Taylor, riv., Colo C4 83
Taylor Knob, mtn., Tenn E9 117
Taylor Mill, Kenton, Ky., 710 A7 94
Taylor Park, res., Colo C4 83
Taylors, Greenville, S.C., 1,071 B3 115
Taylors, isl., Md D5 85
Taylors Falls, Chisago, Minn., 546 E6 99
Taylors Island, Dorchester, Md., 50 D5 85
Taylors Island, pt., Md D5 85
Taylorsport, Boone, Ky., 200 A7 94
Taylor Springs, Montgomery, Ill., 550 D4 90
Taylorsville, Bartow, Ga., 226 B2 87
Taylorsville, Bartholomew, Ind., 350 F6 91
Taylorsville, Spencer, Ky., 937 B4 94
Taylorsville, Smith, Miss., 1,132 D4 100
Taylorsville, Alexander, N.C., 1,470 B2 109
Taylorville, Christian, Ill., 8,801 D4 90
Taymā', Sau. Ar D7 43
Taymouth, N.B., Can., 299 C3 74
Taymyr, lake, Sov. Un B13 28
Taymyr, pen., Sov. Un B12 28
Tayncha, Sov. Un C7 29
Tayport, Scot., 3,151 D6 13
Tayshet, Sov. Un., 28,900 D28 9
Taytay, Phil., 506 C5 35
Taytsy, Sov. Un s31 25
Tayu, China, 15,000 F7 34
Tayug, Phil., 5,108 n13 35
Tayung, China J4 36
Taz, riv., Sov. Un C11 28
Taza, Mor., 31,667 C4 44
Tazerbo (Oasis), Libya D4 43
Tazewell, Marion, Ga., 112 D2 87
Tazewell, Claiborne, Tenn., 1,264 C10 117
Tazewell, Tazewell, Va., 3,000 B3 121
Tazewell, co., Ill., 99,789 C4 90
Tazewell, co., Va., 44,791 B3 121
Tazovskaya, bay, Sov. Un C7 4
Tazrouk, Alg E6 44
Tbilisi, Sov. Un., 724,000 (*815,000) E2 29
Tchepone, Laos D5 38
Tchibanga, Gabon F2 46
Tchula, Holmes, Miss., 882 B3 100
Tczew, Pol., 34,000 A5 26
Te, China F7 36
Tea, Lincoln, S. Dak., 188 D9 116
Tea, creek, Sask., Can C2 70
Teacheys, Duplin, N.C., 187 C5 109
Teague, Freestone, Tex., 2,728 D4 118
Te Anau, lake, N.Z. P11 51
Teaneck, Bergen, N.J., 42,085 D5 106
Teangue, Scot C3 13
Teapa, Mex., 2,793 D6 63
Tearaght, isl., Ire E1 11
Teasdale, Wayne, Utah, 200 E4 119
Teaticket, Barnstable, Mass., 400 C6 97
Te Awamutu, N.Z., 5,425 M15 51
Tebbetts, Callaway, Mo., 211 C6 101
Tebicuary, riv., Par E4 55
Tebingtinggi, Indon., 24,826 E1, m11 35
Tecalitlán, Mex., 4,710 n12 63
Tecate, Mex., 7,074 F5 82
Teche, bayou, La D4 95
Techirghiol, Rom., 2,705 C9 22
Techny, Cook, Ill E2 90
Tecka, Arg C2 54
Teckla, Campbell, Wyo B7 125
Tecolote, Lincoln, N. Mex., 15 C4 107
Tecolotenos, San Miguel, N. Mex B4 107
Tecolula, riv., Mex m15 63
Tecomán, Mex., 14,374 n12 63
Tecopa, Inyo, Calif., 100 E5 82
Tecpan de Galeana, Mex., 5,835 D4 63
Tecuala, Mex., 10,747 C3 63
Tecuci, Rom., 23,400 C8 22
Tecumseh, Ont., Can., 4,476 E2 72
Tecumseh, Shawnee, Kans., 100 B7 93
Tecumseh, Lenawee, Mich., 7,045 F7 98
Tecumseh, Johnson, Nebr., 1,887 D9 103
Tecumseh, Pottawatomie, Okla., 2,630 B5 112
Tecumseh, mtn., N.H C3 105
Ted, Som E5 47
Tedrow, Fulton, Ohio, 240 A3 111
Tedzhen, Sov. Un C10 41
Teegarden, Marshall, Ind., 150 B5 91
Teema, Okinawa 52
Tees, Alta., Can., 63 C4 69
Tees, riv., Eng C6 12
Teeswater, Ont., Can., 919 D3 72
Tefé, Braz., 2,781 D5 60
Tefé, lake, Braz C2 59
Tefé, riv., Braz B4 58
Tefenni, Tur., 2,900 D7 23
Tefft, Jasper, Ind., 130 B4 91
Tegarden, Woods, Okla A3 112
Tegelen, Neth., 17,000 C6 15
Tegern, lake, Ger B7 18
Tegucigalpa, Hond., 133,887 C4 62
Teguise, Can. Is., 5,547 m15 20
Tehachapi, Kern, Calif., 3,161 E4 82
Tehama, co., Calif., 25,305 B2 82
Tehrān, Iran, 1,513,164 (*1,590,000) D5 41
Tehran, prov., Iran D4 41
Tehri, India, 4,508 B7 40
Tehsing, China, 5,000 J7 36
Tehtsin, China F4 34
Tehua, China, 10,000 C8 36
Tehuacán, Mex., 31,724 D5, n15 63
Tehuantepec, Mex., 13,440 D5 63
Tehuantepec, gulf, Mex D6 63
Tehuantepec, isth., Mex D6 63
Tehuitzingo, Mex., 2,930 n14 63
Teide, peak, Can. Is m13 20
Teifi, riv., Wales B3 12
Teigen, Petroleum, Mont., 2 C8 102
Teignmouth, Eng., 11,576 D4 12
Teiteiripucchi, isl., Eniwetok 52
Tejo (Tagus), riv., Port C1 20
Tekamah, Burt., Nebr., 1,788 C9 103
Tekapo, lake, N.Z. O13 51
Tekax de Alvaro Obregón, Mex., 7,797 C7 63
Tekeli, Sov. Un D9 29
Tekirdag, Tur., 23,900 B6 23
Tekkali, India, 11,636 H10 40
Tekoa, Whitman, Wash., 911 B8 122
Tekoa, mtn., Wash B8 122
Tekonsha, Calhoun, Mich., 744 F5 98
Tekro (Well), Chad B4 46
Te Kuiti, N.Z., 4,492 M15 51
Tela, Hond., 11,662 C4 62
Tel Abiad, Syr D12 31
Tel 'Afar, Iraq, 19,951 D14 31
Telavi, Sov. Un., 22,600 E3 29
Tel Aviv-Jaffa, Isr., 386,612 (*640,000) B6, g10 32
Telde, Can. Is., 14,100 m14 20
Telegraph, Kimble, Tex., 25 D3 118
Telegraph, range, B.C., Can C6 68
Telegraph Creek, B.C., Can., 132 E6 66
Tel el Farama, ruins, Eg., U.A.R. C4 32
Telemark, co., Nor., 149,900 *H3 25
Telén, Arg B3 54
Teleneshty, Sov. Un B9 22
Teleorman, riv., Rom C7 22
Telephone, Fannin, Tex., 150 C4 118
Telerhteba, mtn., Alg E6 44
Telescope, peak, Calif D5 82
Telfair, co., Ga., 11,715 E4 87
Telford, Montgomery, Pa., 2,763 F11 114
Telford, Washington, Tenn., 100 C11 117
Telfs, Aus., 5,438 B7 18
Télimélé, Guinea D2 45
Telkalakh, Syr., 4,600 E11 31
Telkwa, B.C., Can., 576 B4 68
Tell Jordan, 2,000 g11 32
Tell, Childress, Tex., 100 B2 118
Tell City, Perry, Ind., 6,609 I4 91
Teller, Alsk., 217 B6 79
Teller, co., Colo., 2,495 C5 83
Telli, lake, China B2 34
Tellico Plains, Monroe, Tenn., 794 D9 117
Telluride, San Miguel, Colo., 677 D3 83
Tel Mond, Isr., 741 B6 32
Telocaset, Union, Oreg., 45 B9 113
Telogia, Liberty, Fla., 500 B2 86

Telok Anson, Mala., 37,042...J4 38
Teloloapan, Mex., 7,297.D5, n14 63
Telos, lake, Maine...B3 96
Telsen, Arg...C3 54
Telukbetung, Indon., 132,312...G3 35
Tema, Ghana...E5 45
Temax, Mex., 3,804...C7 63
Tembo, Con. L...C2 48
Temecula, Riverside, Calif., 500...F5 82
Temerloh-Mentekab, Mala., 12,296...K5 38
Temir-Tau, Sov. Un., 20,000...E26 9
Temir-Tau, Sov. Un., 113,000...C8 29
Temiscouata, co., Que., Can., 29,079...B9 73
Temiscouata, lake, Que., Can...B9 73
Te Moak, Indian res., Nev...G6 104
Temora, Austl., 4,467...G6 51
Temósachic, Mex., 1,164...B3 63
Tempe, Maricopa, Ariz., 24,897...D4, G2 80
Tempe Downs, Austl...D5 50
Temperance, Monroe, Mich., 2,215...G7 98
Temperance, riv., Minn...C8 99
Temperanceville, Accomack, Va., 400...D7 121
Tempio Pausania, It., 8,300..D2 21
Tempiute, Lincoln, Nev., 10.F6 104
Temple, Carroll, Ga., 788...C1 87
Temple, Clare, Mich., 100...D5 98
Temple, Hillsboro, N.H., 65 (361▲)...E3 105
Temple, Williams, N. Dak., 25...A2 110
Temple, Cotton, Okla., 1,282.C3 112
Temple, Berks, Pa., 1,633...F10 114
Temple, Bell, Tex., 30,419...D4 118
Temple City, Los Angeles, Calif., 26,500...*F2 82
Temple Hill, Barren, Ky., 100...D4 94
Templemore, Ire., 1,779...E4 11
Temple Terrace, Hillsborough, Fla., 3,812...E2 86
Templeton, Benton, Ind., 130...C3 91
Templeton, Carroll, Iowa, 354...C3 92
Templeton, Worcester, Mass., 900 (5,371▲)...A3 97
Templeton, Armstrong, Pa., 700...E3 114
Templeville, Queen Annes, Md., 98...B6 85
Templin, Ger., 11,600...B6 16
Templow, Trousdale, Tenn., 50...A5 117
Tempoal, riv., Mex...m14 63
Temryuk, Sov. Un., 10,000..I11 27
Temuco, Chile, 68,000...B2 54
Temuka, N.Z., 2,431...P13 51
Temvik, Emmons, N. Dak., 45...C5 110
Tena, Ec., 331...B2 58
Tenafly, Bergen, N.J., 14,264...B5, D5 106
Tenaha, Shelby, Tex., 1,097..D5 118
Tenakee Springs, Alsk., 60...D12, m22 79
Tenakihi, range, B.C., Can...A5 68
Tenancingo [de Degollado], Mex., 14,769...D5, n14 63
Tenango del Valle, Mex., 7,536...n14 63
Tenant, mtn., N.Y...B6 108
Tenants Harbor, Knox, Maine, 400...E3 96
Tenasserim, Bur., 1,194...F10 39
Tenasserim, riv., Bur...F3 38
Tenay, Fr., 2,132...E6 14
Tenbridge, Hamilton, Tenn.E10 117
Tenby, Wales, 4,752...C3 12
Tendal, Madison, La., 75...B4 95
Tende, Fr., 1,253...E7 14
Ten Degree, chan., India...G9 39
Tendelti, Sud., 7,555...C3 47
Tendoy, Lemhi, Idaho, 20...E5 89
Tendoy, mts., Mont...F4 102
Tendre, mtn., Switz...C1 19
Ténéré, des., Niger...B7 45
Tenerife, isl., Can. Is...m13 20
Ténès, Alg., 7,266 (12,372▲).B5 44
Tenes, cape, Alg...F7 30
Teng, China...G7 36
Tengchung, China...F4 34
Tenggol, isl., Mala...J5 38
Tengiz, lake, Sov. Un...E22 9
Tengkou, China...E2 36
Tengrela, I.C...D3 45
Tengri Khan, mtn., China...G25 9
Tenino, Thurston, Wash., 836...C3 122
Tenke, Con. L...D4 48
Tenkiller Ferry, res., Okla...B6 112
Tenkodogo, Upper Volta, 3,700...D4 45
Ten Mile, Charleston, S.C., (part of Charleston Heights)...F2, F7 115
Ten Mile, Meigs, Tenn., 100.D9 117
Tenmile, creek W. Va...B6 123
Ten Mile, lake, Newf., Can...C3 75
Tenmile, lake, Minn...D4 99
Tennant, Shelby, Iowa, 95...C2 92
Tennant Creek, Austl., 567...C5 50
Tennent, Monmouth, N.J., 150...C4 106
Tennessee, McDonough, Ill., 206...C3 90
Tennessee, state, U.S., 3,567,089...C10 77, 117
Tennessee, cave, Tenn...D8 117
Tennessee, pass, Colo...B4 83
Tennessee, riv., U.S...D10 77
Tennessee City, Dickson, Tenn., 175...A4 117
Tennessee Ridge, Houston, Tenn., 324...A4 117
Tennga, Murray, Ga., 250...B2 87
Tennile, Washington, Ga., 1,837...D4 87
Tenniya, Okinawa... 52
Tennyson, Warrick, Ind., 312...H3 91
Tenosique, Mex., 6,423...D6 63
Tensas, par., La., 11,796...B4 95
Tensas, riv., La...B4 95
Tensed, Benewah, Idaho, 184...B2 89
Ten Sleep, Washakie, Wyo., 314...A5 125
Tenstrike Beltrami, Minn., 147...C4 99
Tenterfield, Austl., 3,104...D9 51
Ten Thousand, is., Fla...G5 86
Teocaltiche, Mex., 11,066...C4, m12 63
Teófilo Otoni, Braz., 41,013..E2 57
Tepa, Indon...G7 35
Tepalcatepec, Mex., 2,555..n12 63
Tepalcatepec, riv., Mex...n12 63
Tepatitán [de Morelos], Mex., 19,609...C4, m12 63
Tepelenë, Alb., 1,100...B3 23
Tepepan, Mex., 3,163...h9 63
Tepic, Mex., 53,955...C4, m11 63
Teplice, Czech., 42,900...C2 26
Tepoca, cape, Mex...A2 63
Tequila, Mex., 7,381...m12 63
Téra, Niger...D5 45
Teramo, It., 25,000...C4 21
Ter-Apel, Neth., 3,600...B7 15
Tercan, Tur., 1,551...C13 31
Terceira, isl., Az. Is...g9 44
Tercero, riv., Arg...A4 54
Terebovlya, Sov. Un...G5 27
Terek, pass, Sov. Un...G23 9
Terek, riv., Sov. Un...G18 9
Terekty, Sov. Un...D7 29
Terence, Man., Can., 50...E1 71
Terence Bay, N.S., Can., 1,055...E6 74
Tererro, San Miguel, N. Mex., 10...F6 107
Teresina, Braz., 100,006...C2 57
Teresópolis, Braz., 29,540...C4, h6 56
Terewah, lake, Austl...D6 51
Tergnier, Fr., 5,827 (★20,000)...E3 15
Terhune, Boone, Ind., 80...D5 91
Teriberka, Sov. Un...C16 25
Teri Nam, lake, China...B8 39
Terlingua, Brewster, Tex., 10...E1, G2 118
Terlingua, creek, Tex...G2 118
Terlton, Pawnee, Okla., 90...A5 112
Termet (Well), Niger...C7 45
Termez, Sov. Un., 24,000...H22 9
Termini Imerese, It., 23,690.F4 21
Términos, lagoon, Mex...D6 63
Termo, Lassen, Calif., 5...B3 82
Termoli, It., 9,686...C5 21
Termon, Ire...B4 11
Termoncarragh, Ire...C1 11
Ternate, Indon., 21,200...E7 35
Ternate, isl., Indon...E7 35
Terneuzen, Neth., 12,100...C3 15
Terney, Sov. Un., 10,000...D8 37
Terni, It., 95,072...C4 21
Ternopol, Sov. Un., 56,000...G5 27
Terpeniye, bay, Sov. Un...B11 37
Terra Alta, Preston, W. Va., 1,504...B5 123
Terrace, B.C., Can., 4,682...B3 68
Terrace, Pope, Minn., 60...E3 99
Terrace, mts., Utah...B2 119
Terra Ceia, isl., Fla...F2 86
Terrace Park, Hamilton, Ohio, 2,023...A7 94
Terracina, It., 17,300...D4 21
Terra Linda, Marin, Calif., 3,000...*D2 82
Terral, Jefferson, Okla., 585..D4 112
Terra Nova, Newf., Can., 194...D4 75
Terra Nova, nat. park, Newf., Can...D4 75
Terrasson, Fr., 2,785...E4 14
Terraville, Lawrence, S. Dak., 200...C2 116
Terrebonne, Que., Can., 6,207...D4, D9 73
Terrebonne, Deschutes, Oreg., 275...C5 113
Terrebonne, co., Que., Can., 102,275...D3 73
Terrebonne, par., La., 60,771...E5 95
Terrebonne, bay, La...E5 95
Terrebonne, bayou, La...E5 95
Terre Haute, Vigo, Ind., 72,500...F3 91
Terre Hill, Lancaster, Pa., 1,129...F9 114
Terrell, Kaufman, Tex., 13,803...C4 118
Terrell, co., Ga., 12,742...E2 87
Terrell, co., Tex., 2,600...D1 118
Terrell Hills, Bexar, Tex., 5,572...B4 118
Terrenceville, Newf., Can., 616...E4 75
Terreton, Jefferson, Idaho, 10...F6 89
Terrible, mtn., Switz...B3 19
Terrible, mtn., Vt...E3 120
Terril, Dickinson, Iowa, 382.A3 92
Terrill, mtn., Utah...E4 119
Terry, Hinds, Miss., 585...C3 100
Terry, Prairie, Mont., 1,140.D11 102
Terry, co., Tex., 16,286...C1 118
Terry, peak, S. Dak...C2 116
Terrytown, Jefferson, La., 5,000...*C7 95
Terry Town, Scotts Bluff, Nebr., 164...C2 103
Terryville, Litchfield, Conn., 5,231...C4 84
Tersakkan, riv., Sov. Un...C7 29
Terschelling, isl., Neth...A5 15
Terskey Alatau, range, Sov. Un...E9 29
Teruel, Col., 1,099...C2 60
Teruel, Sp., 19,726...B5 20
Teruel, prov., Sp., 215,183..*B5 20
Terwagne, Bel., 334...D5 15
Tes, riv., Mong...B4 36
Tešanj, Yugo., 3,150...C3 22
Tescott, Ottawa, Kans., 396..C6 93
Teshekpuk, lake, Alsk...A9 79
Teshio, Jap., 4,000...D10 37
Teslin, Yukon, Can., 231...D6 66
Teslin, lake, B.C., Yukon, Can.D6 66
Tes Nos Pes, Apache, Ariz., 50...A6 80
Tesnus, Brewster, Tex...D1 118
Tessalit, Mali...B5 45
Tessaoua, Niger, 4,000...D6 45
Tessenei, Eth...B4 47
Tessier, Sask., Can., 115...F2 70
Tessin, Ger., 4,530...D6 24
Test, riv., Eng...C6 12
Testigos, is., Ven...A4 60
Tesuque, Santa Fe, N. Mex., 500...B4, F6 107
Tesuque Pueblo, Santa Fe, N. Mex., 300...F6 107
Tetachuk, lake, B.C., Can...C5 68
Tetagouche, riv., N.B., Can...B4 74
Tetbury, Eng., 2,501...C5 12
Tete, Moz...A5 49
Tete, prov., Moz...A5 49
Tete a la Baleine, Que., Can., 311...C2 75
Tete Jaune Cache, B.C., Can., 75...C8 68
Teterow, Ger., 11,100...B6 16
Teteven, Bul., 4,701...D7 22
Tetlin, Alsk., 122...C11 79
Teton, Fremont, Idaho, 399..F7 89
Teton, co., Idaho, 2,639...F7 89
Teton, co., Mont., 7,295...C4 102
Teton, co., Wyo., 3,062...B2 125
Teton, mts., Wyo...B2 125
Teton, riv., Mont...C5 102
Tetonia, Teton, Idaho, 194...F7 89
Tetovo, Yugo., 25,203...D5 22
Tetrino, Sov. Un...D18 25
Tetu, China, 22,024...B3 37
Tetu, lake, Ont., Can...D4 71
Tetuán, Mor., 101,352...B3 44
Tetyukhe-Pristan, Sov. Un., 5,000...D7 37
Teuchern, Ger., 6,985...B7 17
Teuco, riv., Arg...D3 55
Teufen, Switz., 5,110...B7 19
Teulada, It., 5,161...E2 21
Téul de González Ortega, Mex., 1,616...m12 63
Teulon, Man., Can., 749...D3 71
Teure, isl., Jap...D10 37
Teutopolis, Effingham, Ill., 1,140...D5 90
Tevere (Tiber), riv., It...C4 21
Teviot, riv., Scot...E6 13
Tevrin, Ire...D4 11
Tevriz, Sov. Un...B8 29
Te Waewae, bay, N.Z...Q11 51
Tewkesbury, Eng., 5,814...C5 12
Tewksbury, Middlesex, Mass., 1,800...A5, C2 97
Texa, isl., Scot...E2 13
Texada. isl., B.C., Can...E5 68
Texanna, McIntosh, Okla., 10...B6 112
Texarkana, Miller, Ark., 19,788...D1 81
Texarkana, Bowie, Tex., 30,218...C5 118
Texas, Baltimore, Md., 120..B4 85
Texas, state, U.S., 9,579,677...D8 76, 118
Texas, co., Mo., 17,758...D5 101
Texas, co., Okla., 14,162...D3 112
Texas, pt., Tex...E2 95
Texas City, Galveston, Tex., 32,065...E5, G5 118
Texcaltitlán, Mex., 3,404...n14 63
Texcoco, Mex., 10,935...n14 63
Texcoco, lake, Mex...h9 63
Texel, isl., Neth...A4 15
Texhoma, Texas, Okla., 911...D2 112
Texhoma, Sherman, Tex., 350...A2 118
Texico, Curry, N. Mex., 889..C6 107
Texline, Dallam, Tex., 430...A1 118
Texola, Beckham, Okla., 202...B2 112
Texoma, lake, Okla., Tex...D5 112
Texon, Reagan, Tex., 350...D2 118
Teziutlán, Mex., 17,384...D5, n15 63
Tezpur, India, 24,159...C9 39
Tezzeron, lake, B.C., Can...B5 68
Thabeikkyin, Bur...A2 38
Thacker, Mingo, W. Va., 175...D2 123
Thackeray, Hamilton, Ill., 100...E5 90
Thackerville, Love, Okla., 185...D4 112
Tha Hin, Thai., 22,184...E4 38
Thailand (Siam), country, Asia, 25,520,638...E4 38, H13 33
Thakhek, Laos...D6 38
Thal, Pak., 5,757...B5 39
Thale, Ger., 17,300...B6 16
Thalia, Foard, Tex., 200...C3 118
Thalwil, Switz., 11,481...B6 19
Thame, Eng., 4,197...C7 12
Thames, N.Z., 5,315...L15 51
Thames, firth, N.Z...L15 51
Thames, riv., Ont., Can...E3 72
Thames, riv., Conn...D8 84
Thames, riv., Eng...C6 12
Thames, riv., mouth, Eng...C8 12
Thamesville, Ont., Can., 1,054...E3 72
Thamilat Suwaylimah (Oasis), Eg., U.A.R...D6 32
Thane, Alsk., 20...k22 79
Thanglha, mts., China...E3 34
Thanh Hoa, Viet., 25,000...C6 38
Thann, Fr., 7,736...D7 14
Thannhausen, Ger., 600...A6 18
Thano Bula Khan, Pak...E1 40
Thaon [-les-Vosges], Fr., 8,166...C7 14
Tharad, India, 7,566...E3 40
Thargomindah, Austl., 96...C4 51
Thar or Indian, des., India...D3 40
Tharptown (Uniontown), Northumberland, Pa., 1,085...*E8 114
Tharrawaddy, Bur., 8,977...E10 39
Thasos, Grc., 1,748...B5 23
Thasos, isl., Grc...B5 23
Thatch, isl., Vir. Is...f15 65
Thatcher, Graham, Ariz., 1,581...E6 80
Thatcher, Las Animas, Colo., 10...D6 83
Thatcher, Franklin, Idaho, 100...G7 89
Thatcher, Box Elder, Utah, 160...B3 119
Thaton, Bur., 38,047...E10 39
Thawville, Iroquois, Ill., 246...C5 90
Thaxton, Pontotoc, Miss., 200...A4 100
Thayer, Sangamon, Ill., 649.D4 90
Thayer, Newton, Ind., 200...B3 91
Thayer, Union, Iowa, 101...C3 92
Thayer, Neosho, Kans., 396..E8 93
Thayer, Oregon, Mo., 1,713..E6 101
Thayer, York, Nebr., 78...D8 103
Thayer, Fayette, W. Va., 200...D7 123
Thayer, co., Nebr., 9,118...D8 103
Thayer Apartments, Eddy, N. Mex...E5 107
Thayer Junction, Sweetwater, Wyo., 20...D4 125
Thayetmyo, Bur., 11,164...C1 38
Thayne, Lincoln, Wyo., 214..C2 125
Thazi, Bur., 5,000...B2 38
Theadville, Clarke, Miss., 80.D5 100
Thealka, Johnson, Ky., 600..C7 94
Theba, Maricopa, Ariz., 150.E3 80
Thebes, Alexander, Ill., 471..F4 90
Thebes, ruins, Eg., U.A.R...D6 43
Thebes, see Thivai, Grc.,
The Cheviot, mtn., Eng...E6 13
The Dalles (Dalles City), Wasco, Oreg., 10,493...B5 113
Thedford, Ont., Can., 759...D3 72
Thedford, Thomas, Nebr., 303...C5 103
The Father, mtn., N. Gui...k13 50
The Forks, Somerset, Maine, 40 (53▲)...C3 96
The Gap, Coconino, Ariz., 25.A4 80
The Hague ('s Gravenhage), Neth., 605,900 (★820,000).B4 15
The Hollow, Patrick, Va., 25...E2 121
The Hummocks, R.I...C12 172
Thelan, Golden Valley, N. Dak., 5...C2 110
Thelma, Halifax, N.C., 25...A6 109
Thelon, riv., N.W. Ter., Can..D12 66
Thenon, Fr., 696...E4 14
Theodore, Mobile, Ala., 950.E1 78
Theodore, Austl., 386...B8 51
Theodore, Sask., Can., 455...F4 70
Theodore Roosevelt, lake, Ariz..D4 80
Theodore Roosevelt, nat. mem. park, N. Dak...B2 110
Theodosia, Ozark, Mo...E5 101
The Pas, Man., Can., 4,671..C1 71
The Plains, Athens, Ohio, 1,148...C5 111
The Plains, Fauquier, Va., 484...C5 121
Thérain, riv., Fr...E2 15
The Range, N.B., Can., 100.C4 74
The Raven, pt., Ire...E5 11
Theresa, Jefferson, N.Y., 956...A5, B1 108
Theresa, Dodge, Wis., 576...E5 124
Theressa. Bradford, Fla., 35...C4 86
Therien, Alta., Can., 83...B5 69
Thermalito, Butte, Calif., 1,200...*C3 82
Thermon, Grc., 2,665...C3 23
Thermopolis, Hot Springs, Wyo., 3,955...B4 125
The Rock, Upson, Ga., 115...D2 87
Thesprotia, prov., Grc., 52,125...*C3 23
Thessalon, Ont., Can., 1,725.E9 72
Thessalon, riv., Ont., Can...B7 98
Thessaloniki (Salonika), Grc., 250,920 (★373,635)...B4 23
Thessaloniki (Salonika), prov., Grc., 544,394...*B4 23
Thessaly, reg., Grc...C4 23
Theta, Maury, Tenn., 250...B4 117
Thetford, Eng., 5,398...B8 12
Thetford, Orange, Vt., 100 (1,049▲)...D4 120
Thetford Center, Orange, Vt., 150...D4 120
Thetford Mines, Que., Can., 21,618...C6 73
The Village, Oklahoma, Okla., 12,118...*B4 112
The Weirs (Weirs Beach), Belknap, N.H...C4 105
Thibar, Tun...F11 30
Thibaudeau, Man., Can...A4 71
Thibodaux, Lafourche, La., 13,403...C6, E5 95
Thida, Independence, Ark., 50...B4 81
Thief, lake, Minn...B3 99
Thief, riv., Minn...B2 99
Thief River Falls, Pennington, Minn., 7,151...B2 99
Thielson, mtn., Oreg...D5 113
Thiene, It., 9,900...D7 18
Thiensville, Ozaukee, Wis., 2,507...E2, E6 124
Thiers, Fr., 12,418 (16,369▲).E5 14
Thiès, Sen., 42,500...D1 45
Thika, Ken...B6 48
Thimbles, The, is., Conn...E6 84
Thimbu, Bhu...D12 40
Thio, Eth...C5 47
Thionville, Fr., 31,811...C7 14
Thira, isl., Grc...D5 23
Third, lake, N.H...A2 105
3rd Cataract (Nile River), Sud..B2 47
Thirsk, Eng., 2,670...F7 13
Thirty One Mile, lake, Que., Can...C2 73
Thisted, Den., 8,768...B2 24
Thisted, co., Den., 86,703...B2 24
Thistle, Utah, Utah, 150...D3 119
Thithia, isl., Fiji Is... 52
Thivai (Thebes), Grc., 15,779...C4, g10 23
Thog Jalung, China...A8 40
Thok Daurakpa, China...A10 40
Tholen, Neth., 3,400...C4 15
Thomas, Dorchester, Md., 190...C5 85
Thomas, Custer, Okla., 1,211...B3 112
Thomas, Hamlin, S. Dak., 15.C8 116
Thomas, King, Wash., 300..D2 122
Thomas, Tucker, W. Va., 830...B5 123
Thomas, co., Ga., 34,319...F3 87
Thomas, co., Kans., 7,358...C2 93
Thomas, co., Nebr., 1,078...C5 103
Thomas, creek, Fla...B6 86
Thomas, range, Utah...D2 119
Thomasboro, Champaign, Ill., 458...C5 90
Thomaston, Marengo, Ala., 857...C2 78
Thomaston, Litchfield, Conn., 3,579...C4 84
Thomaston, Upson, Ga., 9,336...D2 87
Thomaston, Knox, Maine, 2,780...D3 96
Thomaston, Nassau, N.Y., 2,767...*E3 108
Thomaston, De Witt, Tex., 300...E4 118
Thomaston, flood control res., Conn...C4 84
Thomastown, Ire., 1,209...E4 11
Thomasville, Clarke, Ala., 3,182...D2 78
Thomasville, Thomas, Ga., 18,246...F3 87
Thomasville, Oregon, Mo., 62...E6 101
Thomasville, Davidson, N.C., 15,190...B3 109
Thomasville, Cheatham, Tenn., 30...A4 117
Thomlinson, mtn., B.C., Can...B4 68
Thompson, Man., Can., 3,418...B3 71
Thompson, Windham, Conn., 500 (6,217▲)...B9 84
Thompson, Winnebago, Iowa, 689...A4 92
Thompson, Schoolcraft, Mich., 60...C4 98
Thompson, Audrain, Mo., 82...B5 101
Thompson, Grand Forks, N. Dak., 211...B8 110
Thompson, Geauga, Ohio, 250...A6 111
Thompson, Susquehanna, Pa., 286...C10 114
Thompson, Grand, Utah, 100...E6 119
Thompson, isl., Mass...D3 97
Thompson, lake, Maine...D2 96
Thompson, peak, N. Mex...F6 107
Thompson, res., Oreg...E5 113
Thompson, riv., B.C., Can...D7 68
Thompson, riv., Iowa, Mo...D3 86, A4 101
Thompson Falls, Sanders, Mont., 1,274...C1 102
Thompsons, creek, Miss...D5 100
Thompsons Station, Williamson, Tenn., 300...B5 117
Thompsontown, Juniata, Pa., 713...E7 114
Thompsonville, Hartford, Conn., 19,000...A6 84
Thompsonville, Benzie, Mich., 243...D5 98
Thomson, McDuffie, Ga., 4,522...C4 87
Thomson, Carroll, Ill., 543...B3 90
Thomson, riv., Austl...B4 51
Thomson's Falls, Ken...A6 48
Thonburi, Thai., 370,455...*F4 38
Thonon-les-Bains, Fr., 17,080.C2 18
Thor, Humboldt, Iowa, 234..B3 92
Thorburn, N.S., Can., 1,000.D7 74
Thoreau, McKinley, N. Mex., 300...B1 107
Thorhild, Alta., Can., 312...B4 69
Thorn, see Torun, Pol.
Thornaby-on-Tees, Eng., 22,786...C6 10
Thornbury, Ont., Can., 1,097...C4 72
Thorndale, Ont., Can., 437..D3 72
Thorndale, Milam, Tex., 995...D4 118
Thorndike, Waldo, Maine, 150 (457▲)...D3 96
Thorndike, Hampden, Mass., 850...B3 97
Thorne, Mineral, Nev., 20...E3 104
Thorne, Rolette, N. Dak., 30.A6 110
Thornfield, Ozark, Mo., 50..E5 101
Thornhill, Man., Can., 105...E2 71
Thornhill, Ont., Can., 1,167...D5, E6 72
Thornhill, Scot., 1,160...E5 13
Thornton, Calhoun, Ark., 658...D3 81
Thornton, San Joaquin, Calif., 700...B6 82
Thornton, Ont., Can., 260...C5 72
Thornton, Adams, Colorado, 11,353...*B6 83
Thornton, Madison, Idaho, 200...F7 89
Thornton, Cook, Ill., 2,895..*F3 90
Thornton, Cerro Gordo, Iowa, 449...B4 92
Thornton, Holmes, Miss., 250...B3 100
Thornton, Limestone, Tex., 504...D4 118
Thornton, Whitman, Wash., 220...B8 122
Thornton, Taylor, W. Va., 200...A7 123
Thornton, Weston, Wyo...A8 125
Thornton Heights, Cumberland, Maine (part of South Portland)...E2, E5 96
Thorntown, Boone, Ind., 1,486...D4 91
Thornville, Perry, Ohio, 521.C5 111
Thornwood, Westchester, N.Y., 3,000...*D7 108
Thornwood, Pocahontas, W. Va., 45...C5 123
Thorny, mtn., Mo...D6 101
Thorofare, Gloucester, N.J., 1,100...D2 106
Thorold, Ont., Can., 8,633...D5 72
Thorp, Kittitas, Wash., 430..B5 122
Thorp, Clark, Wis., 1,496...D3 124
Thorpe, McDowell, W. Va., 1,102...D3 123
Thorsby, Chilton, Ala., 968..C3 78
Thorsby, Alta., Can., 491...C3 69
Thorshavn, Faeroe Is. (Den.), 7,447...C7 8
Thouars, Fr., 11,257...D3 14
Thousand, is., N.Y...A4, B1 108
Thousand Lake, mtn., Utah...E4 119
Thousand Oaks, Ventura, Calif., 2,934...*E4 82
Thousand Spring, creek, Nev...B7 104
Thrall, Greenwood, Kans., 60...D7 93
Thrapston, Eng., 1,798...B7 12
Three Bridges, Hunterdon, N.J., 750...B3 106
Three Fingered Jack, mtn., Oreg...C5 113
Three Forks, Gallatin, Mont., 1,161...E5 102
Three Hills, Alta., Can., 1,491...D4 69
Three Kings, is., N.Z...K14 51
Three Lakes, Oneida, Wis., 800...C4 124
Three Mile Bay, Jefferson, N.Y., 350...A4 108
Three Oaks, Berrien, Mich., 1,763...G4 98
Threepoint, lake, Man., Can...B2 71
Three Points, cape, Ghana...F4 45
Three Rivers, Hampden, Mass., 3,082...B3 97
Three Rivers, St. Joseph, Mich., 7,092...G5 98
Three Rivers, Jackson, Miss., 150...E5 100
Three Rivers, Otero, N. Mex., 10...D3 107
Three Rivers, Live Oak, Tex., 1,932...E2 118
Three Rock Cove, Newf., Can., 272...D2 75
Three Sands, Noble, Okla...A4 112
Three Sisters, mts., Oreg...C5 113
Three Springs, Huntingdon, Pa., 475...F6 114
Thrifty, Brown, Tex...D3 118
Throckmorton, Throckmorton, Tex., 1,299...C3 118
Throckmorton, co., Tex., 2,767...C3 118
Throop, Lackawanna, Pa., 4,732...A9 114
Thu Da Mot, Viet., 5,000...G7 38
Thuir, Fr., 3,097...F5 14
Thule, Grnld., 123...B21 4
Thule, mtn., N.W. Ter., Can..B17 67
Thun, Switz., 29,034 (★45,298)...C4 19
Thun, lake, Switz...C4 19
Thunder, bay, Mich...C7 98
Thunder, butte, S. Dak...B4 116
Thunder Bay, co., Ont., Can., 138,518...E8 72
Thunder Bay, riv., Mich...C7 98
Thunderbolt, Chatham, Ga., 1,925...D5 87
Thunder Butte, Ziebach, S. Dak., 50...B4 116
Thunder Butte, creek, S. Dak...B3 116
Thunder Hawk, Corson, S. Dak., 70...B4 116
Thungsong, Thai., 3,249...H3 38
Thur, riv., Switz...A6 19
Thurgau, canton, Switz., 166,420...A6 19
Thüringen (Thuringia), former state, Ger., 2,927,497...C6 17
Thuringer Wald, mts., Ger...C5 17
Thuringia (Thüringen), former state, Ger., 2,927,497...C6 17
Thuringia, reg., Ger...C5 16
Thurles, Ire., 6,421...E4 11
Thurlow, dam, Ala...C4 78
Thurman, Fremont, Iowa, 268...D2 92
Thurmont, Frederick, Md., 1,998...A3 85
Thursday, isl., Austl...B7 50
Thurso, Que., Can., 3,310...D2 73
Thurso, Scot., 8,038...B5 13
Thurso, riv., Scot...B5 13
Thurston, Thurston, Nebr., 140...B9 103
Thurston, Fairfield, Ohio, 429...C5 111
Thurston, co., Nebr., 7,237..B9 103
Thurston, co., Wash., 55,049...C3 122
Thurston, isl., Ant...B2 5
Thusis, Switz., 1,998...C7 19
Thuvu, Fiji Is... 52
Thwaites, ice tongue, Ant...B2 5
Thysville, Con. L., 11,600...C1 48
Tiahuanaco, Bol., 1,127...C2 55
Tianna, beach, N.Y...F6 84
Tiaong, Phil., 2,879...p13 35
Tiaret, Alg., 24,830...B5 44
Tiawah, Rogers, Okla...A6 112
Tibagi, Braz., 1,746...C2 56
Tibagi, riv., Braz...C2 56
Tibasti, des., Libya...E3 43
Tibati, Cam...D2 46
Tibbie, Washington, Ala., 300...D1 78
Tiber, Liberty, Mont., 50...B5 102
Tiber, res., Mont...B5 102
Tiber, see Tevere, riv., It.
Tiberias, Isr., 19,782...B7 32
Tiberias (Sea of Galilee), lake, Isr...B7 32
Tibesti, reg., Chad...A3 46
Tibesti Massif, plat., Chad...A3 46
Tibet, prov., China, 1,270,000...E3 34
Tibet, plat., China...B10 40
Tibiao, Phil., 1,994...*C6 35
Tibnine, Leb., 4,000...A7 32
Tibooburra, Austl., 135...D4 51
Tiburon, Marin, Calif., 1,500...*B5 82
Tiburón, isl., Mex...B2 63
Tice, Lee, Fla., 4,377...F5 86
Tichborne, Ont., Can., 125..C8 72
Tichfield, Sask., Can., 65...F2 70
Tichinane (Well), Maur...C2 45
Tichit, Maur...C3 45
Tichnor, Arkansas, Ark., 100.C4 81
Ticino, canton, Switz., 195,566...D6 19
Ticino, riv., It., Switz.B2 21, D6 19
Tickfaw, Tangipahoa, La., 317...D5 95
Tickfaw, riv., La...D5 95
Ticomán, Mex., 2,852...g9 63
Ticonic, Monona, Iowa, 45...B2 92
Ticonderoga, Essex, N.Y., 3,568...B7 108
Ticul, Mex., 10,809...C7 63
Tide Head, N.B., Can., 702...B3 74
Tidikelt, reg., Alg...D5 44
Tidioute, Warren, Pa., 860...C3 114
Tidjikja, Maur., 6,000...C2 45
Tiehling, China, 45,000...C9 32, C10 36
Tiel, Neth., 18,400...C5 15
Tiel, Sen...D1 45
Tielmes de Tajuña, Sp., 1,763...p18 20
Tielt, Bel., 13,455...C3 15
Tienchen, China...D6 36
Tienchiang, China, 18,000...I2 36
Tienen, Bel., 22,736...D4 15
Tienmen, China, 26,000...I5 36
Tienpai, China, 10,000...G7 34
Tienpao, China...G6 34
Tien Shan, mts., Asia...E10 33
Tienshui, China, 63,000...E6 34
Tientsin, China, 2,850,000 (3,220,000▲)...D8 34, E7 36
Tien Yen, Viet., 5,000...B7 38
Tie Plant, Pulaski, Ark. (part of North Little Rock)...D6 81
Tie Plant, Grenada, Miss., 1,491...B4 100
Tierp, Swe., 3,800...G7 25
Tierra Amarilla, Chile, 1,086...E1 55
Tierra Amarilla, Rio Arriba, N. Mex., 500...A3 107
Tierra Blanca, Mex., 16,474...D5, n15 63
Tierra del Fuego, ter., Arg., 7,064...h12 54
Tierra del Fuego, isl., Arg., Chile...h12 54
Tierra Vieja, mts., Tex...F2 118
Tie Siding, Albany, Wyo., 50...D7 125
Tietar, riv., Sp...C3 20
Tietê, Braz., 8,729...C3, m8 56

Tietê, riv., Braz...........C2 56
Tieton, Yakima, Wash., 479..C5 122
Tieton, dam, Wash..........C4 122
Tieton, peak, Wash.........C4 122
Tieton, riv., Wash.........C4 122
Tiffany, LaPlata, Colo., 80..D3 83
Tiffany, mtn., Wash........A6 122
Tift City, McDonald, Mo.,
100......................E3 101
Tiffin, Johnson, Iowa, 311...C6 92
Tiffin, Seneca, Ohio, 21,478..A4 111
Tiffin, riv., Ohio.........A3 111
Tiflis, see Tbilisi, Sov. Un.
Tift, co., Ga., 23,487.......E3 87
Tifton, Tift, Ga., 9,903.....E3 87
Tiftona, Hamilton, Tenn.,
3,520...................*D8 117
Tigard, Washington, Oreg.,
5,000...................B2 113
Tiger, Rabun, Ga., 277......B3 87
Tiger, Pend Oreille, Wash.,
10......................A8 122
Tigerton, Shawano, Wis.,
781.....................D4 124
Tigerville, Greenville, S.C.,
105.....................A3 115
Tigh, ridge, Oreg..........B5 113
Tighnabruaich, Scot........E3 13
Tigil, Sov. Un., 1,200......D18 28
Tignall, Wilkes, Ga., 556...C4 87
Tignere, Cam..............D2 45
Tignish, P.E.I., Can., 994...C6 74
Tigre, reg., Eth., 1,000,000..C2 47
Tigre, isl., Vict..........D7 38
Tigre, pt., La.............E3 95
Tigre, riv., Peru..........B2 58
Tigres, pen., Ang..........E1 48
Tigrett, Dyer, Tenn., 150...B2 117
Tigris, riv., Asia........118 9, E2 41
Tiguabos, Cuba, 1,148......E6 64
Tihuatlán, Mex., 2,636.....m15 63
Tihwa, see Urumchi, China
Tijeras, Bernalillo, N. Mex.,
150.....................G5 107
Tiji, Libya, 1,270.........C2 43
Tijuana, Mex., 151,939.....A1 63
Tijucas, Braz., 4,420......D3 56
Tikal, ruins, Guat.........
Tikamgarh, India, 20,469...E7 40
Tikhoretsk, Sov. Un., 52,000.I13 27
Tikhvin, Sov. Un., 34,300...B9 27
Tikrit, Iraq, 5,788........E14 31
Tiksi, Sov. Un., 1,000.....B15 28
Tilamuta, Indon...........E6 35
Tilburg, Neth., 138,500....C5 15
Tilbury, Ont., Can., 3,030..E2 72
Tilcara, Arg..............D2 55
Tilden, Randolph, Ill., 808..E4 90
Tilden, Itawamba, Miss., 50.A5 100
Tilden, Madison and Antelope,
Nebr., 917...............B8 103
Tilden, McMullen, Tex., 250.E3 118
Tilemsi, val., Mali........C5 45
Tilford, Meade, S. Dak., 60..C2 116
Tilghman, Talbot, Md., 800..C5 85
Tilghman, isl., Md.........C5 85
Tiline, Livingston, Ky., 150..A3 94
Till, riv., Eng............E6 13
Tillabéri, Niger, 1,300.....D5 45
Tillamook, Tillamook, Oreg.,
4,244...................B3 113
Tillamook, co., Oreg.,
18,955..................B3 113
Tillar, Drew, Ark., 232.....D4 81
Tillatoba, Yalobusha, Miss.,
102.....................B4 100
Tillberga, Swe., 1,532.....t34 25
Tiller, Douglas, Oreg., 120..D4 113
Tilley, Alta., Can., 257....D5 69
Tillicum, Pierce, Wash.,
1,500...................*E1 122
Tillman, Jasper, S.C., 250..G5 115
Tillman, co., Okla., 14,654..C3 112
Tiline, Livingston, Ky., 100..A3 94
Tillson, Ulster, N.Y., 900...D6 108
Tillsonburg, Ont., Can.,
6,600...................E4 72
Tillyfourie, Scot..........C6 13
Tilos, isl., Grc...........D6 23
Tilpa, Austl., 64..........E5 51
Tilremt, Alg.............C5 44
Tilsit, see Sovetsk, Sov. Un.
Tilston, Man., Can., 101....E1 71
Tilting, Newf., Can., 432...D4 75
Tilton, Whitefield, Ga., 80..B2 87
Tilton, Vermilion, Ill.,
2,598...................C6 90
Tilton, Belknap, N.H., 1,129.D3 105
Tiltonsville, Jefferson, Ohio,
2,454...................B7 111
Tim, Sov. Un., 10,000......F11 27
Timaná, Col., 2,439........C2 60
Timaru, N.Z., 24,821......
(*26,424)...............P13 51
Timashevskaya, Sov. Un.,
10,000..................I12 27
Timbalier, bay, La.........E5 95
Timbalier, isl., La.........E5 95
Timbaúba, Braz., 21,019....h6 57
Timbered Knob, mtn., Mo....E5 101
Timberlake, Person, N.C.,
200.....................A5 109
Timber Lake, Dewey, S. Dak.,
624.....................B4 116
Timberlake, Campbell, Va.,
2,700...................*D3 121
Timberville, Rockingham,
Va., 412................C4 121
Timblin, Jefferson, Pa., 240.E3 114
Timbo, Stone, Ark., 75.....B3 81
Timbuctu, see Tombouctou,
Fr. W. Afr.
Timerzit, Mor.............C4 44
Times Beach, St. Louis, Mo.,
986.....................*C7 101
Timewell (Mound Station),
Brown, Ill., 204.........C3 90
Timimoun, Alg., 3,038
(29,002)................D5 44
Timiskaming, co., Ont., Can.,
50,971..................E9 72
Timiskaming, co., Que., Can.,
60,288..................A6 72
Timiskaming Station, Que.,
Can., 2,517.............E9 72
Timişoara, Rom., 142,257...C5 22
Timken, Rush, Kans., 147...D4 93
Timmendorfer Strand [an die
Ostsee], Ger., 7,500.....E4 24
Timmins, Ont., Can.,
29,270 (*40,121)........E9 72
Timmonsville, Florence, S.C.,
2,178...................C8 115
Timnath, Larimer, Colo.,
150.....................A5 83
Timon, Natchitoches, La....C2 95
Timoneng, Guam, 5,380.....52
Timor, Portuguese, see Portuguese
Timor, dep., Asia

Timor, isl., Indon.........G6 35
Timor, sea, Indon..........G6 35
Timpanogos Cave, nat. mon.,
Utah....................C4 119
Timpas, Otero, Colo., 50...D7 83
Timpie, Tooele, Utah, 10...C3 119
Timpson, Shelby, Tex.,
1,120...................D5 118
Timra, Swe., 11,200.......F7 25
Timsâth, lake, Eg., U.A.R..D4 32
Tin, cape, Libya..........C4 43
Tina, Carroll, Mo., 199.....B4 101
Tina, bay., Eg., U.A.R......D4 32
Tinahely, Ire., 417........E5 11
Tin Amzi, riv., Alg........E5 44
Tinaquillo, Ven., 8,305....B4 60
Tin City, Duplin, N.C.......C6 109
Tindouf, Alg., 1,356
(22,372)................D3 44
Tineo, Sp., 21,338........A2 20
Tinghert, plat., Alg., Libya.D6 44
Tinghsien, China, 55,000...E6 36
Tinghsin, China...........C4 36
Tinghsing, China..........E6 36
Tingjegaon, Nep...........C9 40
Tinglev, Den., 1,406.......D3 24
Tingley, Ringgold, Iowa,
278.....................D3 92
Tingmerkpuk, mtn., Alsk....B7 79
Tingo Maria, Peru.........C2 58
Tingpien, China...........F2 36
Tingri Dzong, China,
5,000...................C11 40
Tingsryd, Swe., 1,510.....B8 24
Tingwick, Que., Can., 581..D6 73
Tingyuanying, China.......E1 36
Tinian, Tinian............52
Tinian, hbr., Tinian.......52
Tinian, Tinian, Mariana Is..52
Tinizong, Switz., 373......C8 19
Tinkisso, riv., Guinea.....D2 45
Tinley Park, Cook, Ill.,
6,392...................F2 90
Tinnie, Lincoln, N. Mex., 20.D4 107
Tinniswood, mtn., B.C., Can..C6 68
Tinogasta, Arg., 2,169.....C2 55
Tinos, Grc., 2,750.........D5 23
Tinos, isl., Grc...........D5 23
Tinquipaya, Bol., 766......C2 55
Tin Rerhoh (Oasis), Alg....E5 44
Tinrhert, plat., Alg.......D5 44
Tinsley, Yazoo, Miss., 250..C3 100
Tinsman, Calhoun, Ark.,
100.....................D3 81
Tinsukia, India, 28,468....C10 39
Tintah, Traverse, Minn.,
228.....................D2 99
Tintigny, Bel., 1,195......E5 15
Tintina, Arg., 2,219.......E3 55
Tintinara, Austl., 453.....G3 51
Tinton Falls, Monmouth,
N.J. (part of New Shrews-
bury)...................C4 106
Tin Zaovaten (Fort Pierre
Bordes), Alg............E5 44
Tioga, Hancock, Ill., 110...C2 90
Tioga, Rapides, La., 250....C3 95
Tioga, Williams, N. Dak.,
2,087...................A3 110
Tioga, Tioga, Pa., 597......C7 114
Tioga, Nicholas, W. Va.,
250.....................C4 123
Tioga, co., N.Y., 37,802....C4 108
Tioga, co., Pa., 36,614....C7 114
Tioga, riv., Pa............C7 114
Tiogue, res., R.I..........C10 84
Tioman, isl., Mala.........K6 38
Tionesta Creek, res., Pa...C4 114
Tinoa, Warren, Pa., 250....C3 114
Tione di Trento, It., 3,020..C6 18
Tionesta, Modoc, Calif., 20..B3 82
Tionesta, Forest, Pa., 778..C3 114
Tionesta, res., Pa.........C3 114
Tioughnioga, riv., N.Y.....C4 108
Tipler, Florence, Wis., 300..C5 124
Tiplersville, Tippah, Miss.,
105.....................A5 100
Tippah, co., Miss., 15,093..A5 100
Tipp City, Miami, Ohio,
4,267...................C3 111
Tippecanoe, San Juan, N. Mex.,
49......................A1 107
Tippecanoe, Marshall, Ind.,
350.....................B5 91
Tippecanoe, co., Ind.,
89,122..................D4 91
Tippecanoe, riv., Ind......C4 91
Tipperary, Ire., 4,684.....E3 11
Tipperary, co., Ire., 123,822.E3 11
Tippett, White Pine, Nev., 5.D7 104
Tippo, Tallahatchie, Miss.,
85......................B3 100
Tipton, Tulare, Calif., 300..D4 82
Tipton, Tipton, Ind., 5,604..D5 91
Tipton, Cedar, Iowa, 2,862..C6 92
Tipton, Mitchell, Kans.,
252.....................C5 93
Tipton, Moniteau, Mo.,
1,639...................C5 101
Tipton, Tillman, Okla.,
1,117...................C2 112
Tipton, co., Ind., 15,856...D5 91
Tipton, co., Tenn., 28,564..B2 117
Tipton, mtn., Ariz.........B1 80
Tiptonville, Lake, Tenn.,
2,068...................A2 117
Tip Top, Magoffin, Ky.......C6 94
Tip Top, Tazewell, Va., 100..B3 121
Tiran, isl., Sau. Ar.......I10 31
Tiranë (Tirana), Alb.,
119,000.................B2 23
Tiranë (Tirana), pref.,
Alb.....................*B2 23
Tirano, It., 7,517.........A3 21
Tiraque Chico, Bol., 1,390..C2 55
Tiraspol, Sov. Un., 62,000..H7 27
Tirat Carmel, Isr., 13,000..B6 32
Tirat Zvi, Isr............B7 32
Tire, Tur., 26,503.........C6 23
Tirebolu, Tur., 4,800......B12 31
Tiree, isl., Scot..........D2 13
Tirenno, well, Chad........A3 46
Tîrgu-Mureş, Rom., 65,194..B7 22
Tiriro, Guinea............D3 45
Tirlyanskiy, Sov. Un.,
10,000..................E20 27
Tírnavos, Grc., 10,805.....C4 23
Tiro, Wilson, Ohio, 334.....B5 111
Tirol, state, Aus., 462,899..B7 18
Tirol, reg., Aus...........B7 18
Tirso, riv., It............D2 21
Tirstrup, Den.............C4 24
Tiruchirappalli, India,
249,862 (*350,000)......F6 39
Tirunelveli, India, 87,988..G6 39
Tiruvannamalai, India,
(*190,048)..............G6 39
Tisa, riv., Yugo...........C5 22
Tisaren, lake, Swe.........t33 25

Tisch Mills, Manitowoc, Wis.,
200.....................A7 124
Tisdale, Sask., Can., 2,402..E3 70
Tishabee, Greene, Ala.,
72......................C2 78
Tishomingo, Tishomingo,
Miss., 415..............A5 100
Tishomingo, Johnston, Okla.,
2,381...................C5 112
Tishomingo, co., Miss.,
13,899..................A5 100
Tiskilwa Bureau, Ill., 951..B4 90
Tisnaren, lake, Swe........u3 25
Tisonia, Duval, Fla., 30....B6 86
Tista, riv., India.........D12 40
Tisvildeleje, Den., 862....B6 24
Tisza, riv., Hung..........B5 22
Tiszafured, Hung., 10,318..B5 22
Tiszakecske, Hung., 6,626..B5 22
Titagarh, India, 76,429...*F12 40
Titicaca, lake, Bol., Peru..C2 55
Titicus, res., N.Y.........D2 84
Titograd, Yugo., 29,043....D4 22
Titonka, Kossuth, Iowa, 647.A3 92
Titovo Užice, Yugo., 20,069.D4 22
Titov Veles, Yugo., 26,618..E5 22
Titule, Con., L............A4 48
Titus, co., Tex., 16,785....C5 118
Titus, mtn., Conn..........B3 84
Titusville, Brevard, Fla.,
6,410...................D6 86
Titusville, Mercer, N.J.,
1,000...................C3 106
Titusville, Crawford, Pa.,
8,356...................C2 114
Tiuggi, is., Mala.........K6 38
Tiumpan, head, Scot........B2 13
Tivaouane, Sen., 7,900.....D1 45
Tiverton, Ont., Can., 422..C3 72
Tiverton, N.S., Can., 366...E3 74
Tiverton, Eng., 12,296.....D4 12
Tiverton, Newport, R.I.,
3,000 (9,461)...........C12 84
Tiverton Four Corners, New-
port, R.I., 250.........C12 84
Tivoli, It., 34,067........D4, h9 21
Tivoli, Dutchess, N.Y., 732..C7 108
Tivoli, Refugio, Tex., 800..E4 118
Tixtla, Mex., 8,542.......o14 63
Tizapán, Mex., 5,620.......h9 63
Tizimín, Mex., 14,100.....C7 63
Tizi-Ouzou, Alg., 5,772
(55,497)................B5 44
Tiznit, Mor., 7,694........D3 44
Tjeukemeer, lake, Neth.....B5 15
Tlilatjap, Indon., 28,309..G3 35
Tjina, cape, Indon.........F2 35
Tjirebon, Indon., 153,405..G3 35
Tlacolula, Mex., 7,546....D5 63
Tlacotalpan, Mex., 6,382...D5 63
Tlacotepec, Mex., 2,433...o14 63
Tláhuac, Mex., 4,802.......h9 63
Tlahualilo de Zaragoza, Mex.,
3,201...................B4 63
Tlalnepantla, Mex., 23,886..g9 63
Tlalnepantla, riv., Mex.....g9 63
Tlalpan, Mex., 18,141......h9 63
Tlalpujahua, Mex., 2,283...n13 63
Tlaltenco, Mex., 3,950.....n9 63
Tlapa, Mex., 3,067........o14 63
Tlapacoyan, Mex., 8,511...n15 63
Tlapaneco, riv., Mex......o14 63
Tlaquepaque (San Pedro Tlaque-
paque), Mex., 37,249....m12 63
Tlaxcala, Mex., 7,646......D5 63
Tlaxcala, state, Mex., 347,334.D5 63
Tlaxco, Mex., 4,124.......n14 63
Tlaxiaco, Mex., 8,228......D5 63
Tlemcen, Alg.,
73,445..................C6 30, C4 44
Tlemcès (Well), Niger......C5 45
Tluszcz, Pol., 1,102......k14 26
Tmassah, Libya, 3,225.....D3 43
To, isl., Jap.............I9 37
Toa Alta, P.R., 1,284......B6 65
Toa Alta, mun., P.R.,
15,711..................B5 65
Toa Baja, P.R., 1,084......B5 65
Toa Baja, mun., P.R., 19,698.B6 65
Toadlena, San Juan, N. Mex.,
49......................A1 107
Toano, James City, Va., 250.D6 121
Toast, Surry, N.C., 2,023..A3 109
Toay, Arg., 2,457.........B4 54
Toba, Jap., 21,900.......o15 37
Toba, lake, Indon.........m11 35
Toba, riv., B.C., Can......D5 68
Tobacco Root, mts., Mont...A4 117
Tobaccoport, Stewart, Tenn.,A4 117
Tobago, isl., Vir. Is......A5 60
Tobago, isl., Vir. Is......f15 65
Tobaru, Okinawa............52
Tobata, Jap., 108,708.....*I5 37
Tobelo, Indon.............F6 35
Tobercurry, Ire., 878.....C3 11
Tobermory, Ont., Can., 363.B3 72
Tobermory, Scot., 668.....D2 13
Tobias, Saline, Nebr., 202..D8 103
Tobin, min., Nev..........C4 104
Tobins Harbor, Keweenaw,
Mich....................A2 98
Tobinsport, Perry, Ind., 35.I4 91
Tobique, riv., N.B., Can....B2 74
Tobol, riv., Sov. Un........C6 29
Tobol, riv., Sov. Un., 46,700.E21 9
Tobruk, see Ţubruq, Libya
Tobyhanna, Monroe, Pa.,
300.....................D11 114
Tobys Rock, min., Conn.....D4 84
Tocantinópolis, Braz., 4,927.C1 10
Tocantins, riv., Braz......C1 57
Toccoa, Stephens, Ga.,
7,303...................B3 87
Toccoa Falls, Stephens, Ga.,
300.....................B3 87
Toccopola, Pontotoc, Miss.,
198.....................A4 100
Tocoa, Hond., 582.........D1 55
Tocorpuri, hill, Bol.......D1 55
Tocuco, Wells, Ind., 175...C7 91
Tocumwal, Austl., 1,289...A5 51
Tocuyo, riv., Ven.........A4 60
Todd, co., Ky., 11,364....D2 94
Todd, co., Minn., 23,119...D4 99
Todd, co., S. Dak., 4,661..D5 116
Todd, fork, Ohio..........C4 111
Todd, mtn., B.C., Can......E6 68
Todd, mtn., N. Mex........C3 74
Todd, riv., Austl.........D6 50
Toddville, Dorchester, Md.,
325.....................D5 85

Toddy, pond, Maine........D4 96
Todi, It., 4,600..........C4 21
Tödi, mtn., Switz..........C6 19
Todoga, cape, Jap.........G11 37
Todos os Santos, bay, Braz..D3 57
Todos Santos, Bol., 408....C2 55
Todos Santos, Mex., 1,886..C2 63
Tod Park, Tooele, Utah,
700.....................C3 119
Todtnau, Ger., 3,000......B3 18
Toe, head, Ire............F2 11
Toe, head, Scot...........C1 13
Toecane, Mitchell, N.C.,
75......................C4 109
Tofield, Alta., Can., 905..C4 69
Tofino, B.C., Can., 440....E5 68
Tofte, Cook, Minn.,
175.....................B6, C8 99
Toftlund, Den., 1,814.....C3 24
Toga, Buckingham, Va.,
100.....................D4 121
Togiak, Alsk., 220........D7 79
Togian, is., Indon........F6 35
Togo, Sask., Can., 263....F5 70
Togo, country, Afr.,
1,440,000...............F6 42, E5 45
Togus, Kennebec, Maine (part of
Chelsea and Augusta)....D3 96
Tohakum, peak, Nev.......C2 104
Tohatchi, McKinley, N.Mex.,
200.....................B1 107
Tohickon, creek, Pa.......C2 106
Tohopekaliga, lake, Fla....D5 86
Toi, cape, Jap...........K5 37
Toimi, St. Louis, Minn.....C7 99
Toivola, Houghton, Mich.,
80......................B2 98
Toivola, St. Louis, Minn....C6 99
Tokaj, Hung., 5,031.......A5 22
Tokar, Sud., 16,802.......B4 47
Tokara, is., Jap..........L4 37
Tokara, strait, Jap.......K5 37
Tokat, Tur., 32,700.......B11 31
Tokeland, Pacific, Wash.,
150.....................C2 122
Tokelau Islands, N.Z., dep.,
Oceania, 1,580..........G11 7
Tokenton, Kor., 15,711....G3 37
Toki, pt., Wake Isl........52
Tokio, Hempstead, Ark.,
90......................C2 81
Tokio, Benson, N. Dak.,
100.....................B7 110
Tokio, Terry, Tex., 15....C1 118
Tokmak, Sov. Un., 30,000..E9 29
Tokoto, China, 11,000.....D4 36
Tokuchi, Okinawa..........52
Tokuno, isl., Ryukyu Is....F10 34
Tokushima, Jap.,
182,782.................I7 37
Tokushima, pref., Jap......I7 37
Tokuyama, Jap., 77,246....I6 37
Tōkyō, Jap., 8,310,027
(*14,085,000)...........I9, n18 37
Tōkyō, pref., Jap., 9,683,802.*I9 37
Tōkyō, bay, Jap...........n18 37
Tol, isl., Truk...........52
Tolbukhin (Dobrich), Bul.,
31,049..................D8 22
Toleak, pt., Wash.........B1 122
Toledo, Ont., Can., 156...C9 72
Toledo, Cuba.............k11 64
Toledo, Cumberland, Ill.,
998.....................D5 90
Toledo, Lucas, Ohio,
318,003 (*514,200).....A2, A4 111
Toledo, Lincoln, Oreg.,
3,053...................C3 113
Toledo, Sp., 40,651.......C3 20
Toledo, Lewis, Wash., 450..C3 122
Toledo, prov., Sp., 521,637.*C3 20
Toledo, mts., Sp..........C3 20
Toler, Pike, Ky., 700.....C7 94
Tolg, Swe................A8 24
Tolga, Alg., 4,949.......C6 44
Tolima, dept., Col., 788,030.C2 60
Tolima, vol., Col.........C2 60
Tolimán, Mex., 717.......m14 63
Tolken, lake, Swe.........A6 24
Tolland, Tolland, Conn.,
400 (2,950)............B7 84
Tolland, co., Conn., 68,737.B7 84
Tollense, lake, Ger.......E7 24
Tollense, riv., Ger.......E7 24
Tollesboro, Lewis, Ky., 480.B6 94
Tolleson, Maricopa, Ariz.,
3,886...................G1 80
Tollette, Howard, Ark., 350.D2 80
Tolley, Renville, N. Dak.,
189.....................A4 110
Tollhouse, Fresno, Calif.,
100.....................D4 82
Tolloche, Arg............E3 55
Tollville, Prairie, Ark....C4 81
Tolmezzo, It., 7,749......A4 21
Tolmin, Yugo., 1,638......B1 22
Tolna, Hung., 8,748......B4 22
Tolna, Nelson, N. Dak., 291.B7 110
Tolo, Jackson, Oreg.......E4 113
Tolo, gulf, Indon.........F6 35
Tolong, Phil., 2,159.....*D6 35
Tolono, Champaign, Ill.,
1,539...................D5 90
Tolosa, Sp., 16,281......A4 20
Tolovana Park, Clatsop, Oreg.,
150.....................B3 113
Tolsta, head, Scot........B2 13
Tolstoy, Potter, S. Dak., 142.B6 116
Tolstoy, Sov. Un..........
Tolt, Crittenden, Ky., 325..A3 94
Tolu, Col., 5,415.........B2 60
Toluca, Marshall, Ill., 1,352.B4 90
Toluca, Mex., 76,871......D5 63
Toluca, Big Horn, Mont., 7..E9 102
Tolun, China, 10,000.....D4 36
Tom, riv., N.Y............C2 84
Tom, riv., Sov. Un........E26 9
Tomah, Monroe, Wis., 5,321.E3 124
Tomahawk, Lincoln, Wis.,
3,348...................C4 124
Tomahawk, lake, Wis.......C4 124
Tomakomai, Jap., 62,384...E10 37
Tomakovka, Sov. Un.,
10,000..................H10 27
Tomales, Marin, Calif., 150.B4 82
Tomar, Port., 8,034.......C1 20
Tomari, Sov. Un., 16,100..C11 37

Tomaszow Lubelski, Pol.,
7,338...................C7 26
Tomaszow Mazowiecki, Pol.,
49,000..................C6 26
Tomatlán, Mex., 1,059.....D3, n11 63
Tomato, Mississippi, Ark.,
150.....................B6 81
Tomave, Bol., 201.........D2 55
Tombador, mts., Braz......D2 57
Tomball, Harris, Tex., 1,713.D5 118
Tombigbee, riv., Ala., Miss..B5 100
Tombouctou (Timbuktu),
Mali, 7,500.............C4 45
Tombstone, Cochise, Ariz.,
1,283...................F5 80
Tomé, Chile, 27,000.......B2 54
Tomelilla, Swe., 4,160....C7 24
Tomelloso, Sp., 27,815....C4 20
Tomi, Con., o., Tex.,
64,630..................D2 118
Tomichi, creek, Colo......C4 83
Tomini, gulf, Indon.......F6 35
Tomintoul, Scot..........C5 13
Tomkins Cove, Rockland,
N.Y., 800...............D2, D7 108
Tommot, Sov. Un., 4,800...D15 28
Tomnatin, Scot...........C5 13
Tomnolen, Webster, Miss.,
165.....................B4 100
Tomo, riv., Col...........C6 60
Tompkins, Newf. Can., 103..E2 75
Tompkins Sask., Can., 453..G1 70
Tompkins, co., N.Y., 66,164.C4 108
Tompkinsville, Monroe, Ky.,
2,091...................D4 94
Tompkinsville, Charles, Md.,
140.....................D4 85
Toms, riv., N.Y...........B3 84
Toms Creek, Wise, Va., 250..B2 121
Tomsk, Sov. Un., 269,000...B11 29
Toms River, Ocean, N.J.,
6,062...................D4 106
Tomy Town, Adair, Okla.,
15......................B7 112
Tonalá, Mex., 12,204.....D6 63
Tonalea, Coconino, Ariz., 10.A3 80
Tonasket, Okanogan, Wash.,
958.....................A6 122
Tonawanda, Erie, N.Y.,
21,561..................B2 108
Tonawanda, creek, N.Y......B2 108
Tonbridge, Eng., 22,141...C8 12
Tondano, Indon., 15,007...E6 35
Tönder, Den., 7,192.......D2 24
Tönder, co., Den., 42,842..D2 24
Toney, Madison, Ala., 120..A3 78
Tonga, Br. dep., Oceania,
47,000..................H11 7
Tonganoxie, Leavenworth,
Kans., 1,354............B8, C8 93
Tongatapu, isl., Pac. O....G12 7
Tongeren, Bel., 16,176....D5 15
Tongjoson, bay, Kor.......G3 37
Tongoi, Chile............F1 55
Tongue, riv., Mont., Wyo...E10 102
Tongue Kyle, bay, Scot....B4 13
Tongue River, Indian res.,
Mont....................E10 102
Tonica, La Salle, Ill., 750..B4 90
Tónichi, Mex., 315.......B3 63
Tonj, Sud., 2,071.........D2 47
Tonk, Tama, Iowa, 2,417...C5 92
Tonkawa, Kay, Okla.,
3,415...................A4 112
Tonkin, reg., Viet.,
11,000,000..............B6 38
Tonkin, gulf, Asia........C7 38
Tonk Raj (Tonk), India,
42,853..................C6, 39, D5 40
Tonle Sap, lake, Camb.....C5 38
Tonneins, Fr., 4,775......E4 14
Tonnerre, Fr., 5,595......D5 14
Tönning, Ger., 4,500......D2 24
Tono, Thurston, Wash......C3 122
Tonoloway, ridge, Md......A1 85
Tonopah, Nye, Nev., 1,679..E4 104
Tonosí, Pan., 559.........G7 62
Tonopah, Maricopa, Ariz.,30.D3 80
Tonquin, Washington, Oreg..B2 113
Tonsberg, Nor., 12,400
(*28,000)...............H4, p28 25
Tontitown, Washington, Ark.,
209.....................A1 81
Tonto, natural bridge, Ariz..C4 80
Tonto, nat. mon., Ariz.....D4 80
Tonto, mtn., Nev.........n15 63
Tonto Basin, Gila, Ariz., 30.D4 80
Tontogany, Wood, Ohio,
380.....................A1 111
Tonton, Phil., 1,489......n13 35
Tony, Rusk, Wis., 162.....C3 124
Tooele, Tooele, Utah, 9,133.C3 119
Tooele, co., Utah, 17,868..C2 119
Tooele, co., Mont., 7,904..B5 102
Toombs, co., Ga., 16,837..D4 87
Toomevara, Ire., 231.....E3 11
Toomsboro, Wilkinson, Ga.,
764.....................D3 87
Toone, Hardeman, Tenn.,
202.....................B3 117
Tooraweenah, Austl.,
1,539...................D5 90
Toora, see Tūkrah, Libya
Toormakeady, Ire.........C2 11
Toowoomba, Austl., 50,134..C8 51
Top, Grant, Oreg.........B3 113
Top, pond, Newf., Can......E3 75
Topanga, Los Angeles, Calif.,
3,500...................*F4 82
Topawa, Pima, Ariz., 200...F4 80
Topaz, Mono, Calif., 15....C4 82
Topeka, Lagrange, Ind., 600.A6 91
Topeka, Shawnee, Kans.,
119,484 (*135,800)......B6, C8 93
Topinabee, Cheboygan, Mich.,
200.....................C6 98
Topki, Sov. Un., 28,000...B11 29
Topley, B.C., Can., 138....C3 68
Toplica, riv., Yugo.......D5 22
Topličany, Czech., 10,400..D5 26
Topo, China, 10,000......*D3 36
Topocalma, cape, Chile.....C5 55
Topock, Mohave, Ariz.,
3,348...................C4 80
Topolnitsa, riv., Bul......D7 22
Topolobampo, Mex., 1,738...B3 63
Topolovgrad, Bul., 6,591..D8 22
Toppenish, Routt, Colo., 70.A4 83
Toppenish, Yakima, Wash.,
5,667...................C5 122
Toppenish, ridge, Wash....C5 122
Topsail, Pender, N.C......C6 109

Topsfield, Washington, Maine,
130.....................C5 96
Topsfield, Essex, Mass.,
2,000 (3,351)...........A6 97
Topsham, Sagadahoc, Maine,
2,240...................E3, E5 96
Topsham, Orange, Vt., 150
(638)...................C4 120
Topsy, Wayne, Tenn........B4 117
Topton, Lauderdale, Miss.,
35......................C5 100
Topton, Berks, Pa., 1,684..C5 114
Toquerville, Washington,
Utah, 197...............F2 119
Toquima, range, Nev.......E4 104
Tor, Eg., U.A.R...........H9 32
Tor, bay, Eng............D4 12
Torbali, Tur., 4,851......C6 23
Torbat-e Heydarīyeh, Iran
23,816..................D9 41
Torbat-e Jām, Iran, 8,870..D10 41
Torbay, Newf. Can., 1,445..E5 75
Torbert, mtn., Alsk.......g15 79
Torbreck, mtn., Austl......H5 51
Torcello, It., 127 (part of
Venezia)................D8 18
Torch, lake, Mich.........D5 98
Torch, riv., Sask., Can....D4 70
Torekov, Swe., 456........B6 24
Toreva, Navajo, Ariz.......B5 80
Torgau, Ger., 19,900......B7 17
Torhout, Bel., 13,465.....C3 15
Tori, isl., Jap...........J4 37
Toride, Jap., 13,300.....n19 37
Torino (Turin), It., 1,025,822
(*1,245,000)............B1 21, D3 18
Torit, Sud., 2,353........E3 47
Tórmes, riv., Sp..........B2 20
Tornado, peak, Alta. and
B.C., Can...............E3 69, E10 68
Tornälven, riv., Swe......D10 25
Torneträsk, lake, Swe.....C9 25
Torngat, mts., Que., Can...f8 75
Tornillo, El Paso, Tex., 600.F1 118
Tornio, Fin., 5,500......E11 25
Tornquist, Arg., 2,782....B4 54
Toro, Sabine, La., 35.....C2 95
Toro, Sp., 10,218.........B3 20
Toro, Swe................u35 25
Toro, lake, Que., Can......C4 73
Toro, mtn., P.R...........B7 65
Toro, peak, Calif.........F5 82
Toroi, Mong..............C6 36
Törökszentmiklós, Hung.,
17,952 (23,576).........B5 22
Toro Negro, riv., P.R......B4 65
Toronto, Ont., Can.,
672,407 (*1,824,481).D5, E6 72
Toronto, Woodson, Kans.,
524.....................E8 93
Toronto, Jefferson, Ohio,
7,780...................B7 111
Toronto, Deuel, S. Dak.,
268.....................C9 116
Toronto, res., Kans.......E8 93
Toropets, Sov. Un.,
25,000..................C8 27
Tororo, Ug., 6,365........A5 48
Torquay, Sask., Can., 462..H4 70
Torquay, Eng., 53,915.....D4 12
Torrance, Los Angeles, Calif.,
100,991.................F9 82
Torrance, Ont., Can., 239..C5 72
Torrance, Torrance, N. Mex.,
10......................C4 107
Torrance, co., N. Mex.,
6,497...................C3 107
Torre Annunziata, It.,
58,400..................D5 21
Torre de Cerredo, mtn., Sp..A3 20
Torre del Greco, It.,
77,576..................D5 21
Torredonjimeno, Sp.,
14,204..................D4 20
Torrejoncillo, Sp., 5,499..C2 20
Torrejón de Ardoz, Sp.,
10,794..................p18 20
Torrelavega, Sp., 14,800..A3 20
Torremaggiore, It., 17,318..D5 21
Torrens, lake, Austl......F6 50
Torrente, Sp., 24,042.....C5 20
Torreón, Mex., 179,955
(*300,000)..............B4 63
Torreon, Torrance, N. Mex.,
50......................C3 107
Torre Pacheco, Sp., 9,541..D5 20
Tôrres, Braz., 4,729......D3 56
Torres, strait, Pap........A7 50
Torres Novas, Port., 7,291..C1 20
Torres Vedras, Port., 5,151.C1 20
Torrevieja, Sp., 12,321...D5 20
Torrey, Wayne, Utah, 128...E4 119
Torrey, mtn., Mont........E4 102
Torridge, riv., Eng.......D3 12
Torriglia, It., 3,709.....C6 18
Torring, Den., 1,367......C3 24
Torrington, Alta., Can., 149.D4 69
Torrington, Litchfield, Conn.,
30,045..................B4 84
Torrington, Eng., 2,930...D3 12
Torrington, Goshen, Wyo.,
4,188...................C8 125
Torrinha, Braz., 1,731...m7 56
Torrowangee, Austl.......E3 51
Torrox, Sp., 7,384.......D4 20
Tors Cove, Newf., Can., 303.E5 75
Torshälla, Swe., 6,100...t34 25
Tortilla Flat, Maricopa,
Ariz., 100..............D4 80
Tortola, isl.,
Vir. Is............f16 65, m14 65
Tortona, It., 20,400.....B6 21
Tortorici, Sp., 14,500...D6 20
Tortosa, Sp., 14,500.....C6 20
Tortosa, cape, Sp........C6 20
Tortue, isl., Haiti.......E7 64
Tortuga, isl., Ven........A4 60
Tortugas, Dona Ana,
N. Mex. (part of
Mesilla Park)...........E3 107
Torud, Iran..............D7 41
Torun, Pol., 105,000.....B5 26
Torup, Swe., 546.........B3 24
Tory, isl., Ire..........A3 11
Tory, sound, Ire.........B3 11
Tory Hill, Ont., Can., 115.C6 72
Torysa, riv., Czech......C6 26
Torzhok, Sov. Un., 32,000.C10 27
Tosno, Sov. Un., 10,000..B8 27
Tosson Hill, mtn., Eng....E7 13

Tostado, Arg., 5,234...E3 55
Toston, Broadwater, Mont., 100...D5 105
Tosya, Tur., 13,700...B10 31
Toszek, Pol., 2,620...g9 26
Totagatic, riv., Wis....B2 124
Totana, Sp., 9,949...D5 20
Totavi, Santa Fe, N. Mex...F6 107
Toteng, Bech...B3 49
Totkomlos, Hung., 8,122...B5 22
Totnes, Eng., 6,064...D4 12
Toto, Ang...C1 48
Toto, Guam, 730... 52
Totomi, sea, Jap....o16 37
Totonicapán, Guat., 6,405...C2 62
Totorapalca, Bol...C2 55
Totowa, Passaic, N.J., 10,897...B4 106
Totoya, isl., Fiji Is.... 52
Tottenham, Austl., 367...F6 51
Tottenham, Ont., Can., 778...C5 72
Tottenham, Eng., 113,126...k12 10
Tottori, Jap., 73,000 (104,833▲)...I7 37
Tottori, pref., Jap., 599,135...*I7 37
Totz, Harlan, Ky., 700...D6 94
Totzke, Sask., Can., 55...E3 70
Touba, I.C., 1,200...E3 45
Touba, Sen...D1 45
Toubkal, mtn., Mor....C3 44
Touchet, Walla Walla, Wash., 250...C7 122
Touchet, riv., Wash....C8 122
Touchwood, hills, Sask., Can....F3 70
Touchwood, lake, Man., Can....B4 71
Toufourine (Oasis), Mali...B4 45
Tougaloo, Hinds, Miss., 1,000...C3 100
Tougan, Upper Volta, 3,500...D4 45
Touggourt, Alg., 17,380 (83,752▲)...C6 44
Touggourt, reg., Alg...C6 44
Tougy, Saunders, Nebr., 70...C9 103
Toul, Fr., 14,155...C6 14
Toulépleu, I.C., 2,400...E3 45
Toulon, Fr., 161,786 (*230,000)...F6 14
Toulon, Stark, Ill., 1,213...B4 90
Toulon, Pershing, Nev., 5...C3 104
Toulouse, Fr., 323,724...F4 14
Toummo (Well), Niger...B7 45
Toungo, Nig...E7 45
Toungoo, Bur., 31,589...E10 39
Tounin, Alg...F7 30
Touraine, former prov., Fr., 392,000...D4 14
Tourane (Da Nang), Viet., S., 109,000...D8 38
Tourcoing, Fr., 89,258...B5 14, D3 15
Tourlaville, Fr., 11,569...C3 14
Tournai, Bel., 33,263...D3 15
Tournon [-sur-Rhône], Fr., 5,894...E6 14
Tournus, Fr., 5,975...D6 14
Touros, Braz., 1,550...C3 57
Tours, Fr., 92,944 (*160,000)...D4 14
Tourville, Que., Can., 645...B8 73
Toussaint, creek, Ohio...A2 111
Toutle, riv., Wash....C3 122
Tova, reef, Fiji Is.... 52
Tovar, Ven., 8,827...B3 60
Tovey (Humphrey), Christian, Ill., 646...D4 90
Tow, Llano, Tex., 200...D3 118
Towaco, Morris, N.J., 2,500...B4 106
Towanda, lake, Jap....F10 37
Towanda, McLean, Ill., 586...C5 90
Towanda, Butler, Kans., 1,031...B6, E7 93
Towanda, Bradford, Pa., 4,293...C9 114
Towanda, creek, Pa....C8 114
Towaoc, Montezuma, Colo., 60...D2 83
Tower, Cheboygan, Mich., 300...C6 98
Tower, St. Louis, Minn., 878...C6 99
Tower, Crook, Wyo...A8 125
Tower, rock, Am. Sam.... 52
Tower City, Cass, N. Dak., 300...C8 110
Tower City, Schuylkill, Pa., 1,968...E8 114
Tower Hill, Shelby, Ill., 700...D5 90
Tower Junction, Yellowstone National Park, Wyo...A2 125
Town, creek, Md....D2 85
Town, fork, Ohio...A1 123
Town, hill, Md....A1 85
Town and Country (Arcade), Sacramento, Calif., 35,000...*C3 82
Town and Country, St. Louis, Mo., 1,440...*C7 101
Town Creek, Lawrence, Ala., 810...A2 78
Towner, Kiowa, Colo., 10...C8 83
Towner, McHenry, N. Dak., 948...A5 110
Towner, co., N. Dak., 5,624...A6 110
Townley, Walker, Ala., 649...B2 78
Town of Pines, Porter, Ind., 939...*A4 91
Town of Tonawanda, Erie, N.Y., 83,771...*B2 108
Town Point, Cecil, Md., 30...B6 85
Towns, Telfair, Ga., 91...D4 87
Towns, co., Ga., 4,538...B3 87
Townsend, New Castle, Del., 434...B6 85
Townsend, McIntosh, Ga., 100...E5 87
Townsend, Middlesex, Mass., 1,101 (3,650▲)...A4 97
Townsend, Broadwater, Mont., 1,528...D5 102
Townsend, Blount, Tenn., 283...D10 117
Townsend, Northampton, Va., 120...D7 121
Townsend Harbor, Middlesex, Mass...A4 97
Townshend, Windham, Vt., 170 (643▲)...E3 120
Townshend, res., Vt....E3 120
Townsville, Austl., 51,143...C8 50
Townsville, Vance, N.C., 195...A5 109
Townville, Crawford, Pa., 361...C2 114
Townville, Anderson, S.C., 200...B2 115
Towson, Baltimore, Md., 17,000...B4, C2 85
Towuti, lake, Indon...F6 35
Towyn, Wales, 4,466...B3 12
Toxey, Choctaw, Ala., 157...D1 78
Toy, Pershing, Nev., 15...C3 104
Toyah, Reeves, Tex., 294...F2 118
Toyahvale, Reeves, Tex., 20...F2 118
Toyama, Jap., 207,266...H8 37
Toyama, pref., Jap., 1,032,614...*H8 37
Toyama, bay, Jap...H8 37
Toyohashi, Jap., 215,515...I8, o16 37
Toyokawa, Jap., 44,900...*o16 37
Toyonaka, Jap., 199,065...I7, o14 37
Tozeur, Tun., 11,820...C6 44
Tozghi Koh, mtn., Pak...G11 41
Trabancos, riv., Sp....B3 20
Traben-Trarbach, Ger., 5,700...D2 17
Trabzon (Trebizond), Tur., 52,700...B12 31
Tracadie, N.B., Can., 1,651...B5 74
Tracadie, N.S., Can., 321...D8 74
Tracadie, riv., N.B., Can....B4 74
Tracy, San Joaquin, Calif., 11,289...B6, D3 82
Tracy, N.B., Can., 655...D3 74
Tracy, Que., Can., 6,542...*C4 73
Tracy, Marion, Iowa, 300...C5 92
Tracy, Barren, Ky., 50...D4 94
Tracy, Lyon, Minn., 2,862...F3 99
Tracy, Platte, Mo., 208...B3, D1 101
Tracy, brook, Md...C4 85
Tracy City, Grundy, Tenn., 1,577...D8 117
Tracy-le-Mont, Fr., 1,239...E3 15
Tracyton, Kitsap, Wash., 300...D1 122
Trade, Johnson, Tenn., 40...C12 117
Trade, lake, Sask., Can....B4 70
Tradesville, Lancaster, S.C., 75...B6 115
Tradewater, riv., Ky....C2 94
Trading Post, Linn., Kans., 50...D9 93
Traer, Tama, Iowa 1,623...B5 92
Traer, Decatur, Kans., 52...C3 93
Trafalgar, Johnson, Ind., 459...F5 91
Trafford, Westmoreland and Allegheny, Pa., 4,330...B7 114
Trafford, lake, Fla...F5 86
Traiguén, Chile, 8,806...B2 54
Trail, B.C., Can., 11,580...E9 68
Trail, Polk, Minn., 100...C3 99
Trail, Jackson, Oreg., 40...E4 113
Trail, ridge, Ga...F4 87
Trail City, Dewey, S. Dak., 100...B5 116
Trail Creek, LaPorte, Ind., 1,552...A5 91
Trailer Estates, Manatee, Fla., 1,562...*E4 86
Traill, co., N. Dak., 10,583...B8 110
Traill, isl., Grnld....B17 4
Trainer, Delaware, Pa., 2,358...*G11 114
Traipu, Braz., 2,393...C3 57
Trairí, riv., Braz....h6 57
Tralee, Ire., 10,723...E2 11
Tramelan Dessus, Switz., 5,567...B3 19
Tramore, Ire., 2,882...E4 11
Tramping, lake, Sask., Can....E1 70
Tramping Lake, Sask., Can., 288...E1 70
Tranås, Swe., 15,400...H6 25
Trancas, Arg...E2 55
Trancoso, Port., 3,537...B2 20
Tranebjerg, Den., 729...C4 24
Tranemo, Swe., 1,983...A7 24
Trang, Thai., 16,393...I3 38
Trangan, isl., Indon...G8 35
Trani, It., 38,129...D6 21
Tranquility, Sussex, N.J., 100...B3 106
Tranquillity, Fresno, Calif., 750...D3 82
Transcona, Man., Can., 14,248...E3 71
Transfer, Mercer, Pa., 300...D1 114
Trans-Ili Alatau, mts., Sov. Un....E9 29
Transilvania, prov., Rom., 3,420,859...*B6 22
Transjordan, see Jordan, country, Asia
Transvaal, prov., S. Afr., 6,273,477...B4 49
Transylvania, East Carroll, La., 50...B4 95
Transylvania, co., N.C., 16,372...D3 109
Transylvania, reg., Rom...B6 22
Transylvanian Alps, mts., Rom....C6 22
Trap, mtn., Ark....D5 81
Trapani, It., 77,139...E4 21
Trapiche, Guat...B2 62
Trappe, Talbot, Md., 358...C5 85
Trappe, Montgomery, Pa., 1,264...*F11 114
Trappe, creek, Md....D7 85
Trapper, peak, Mont....E2 102
Traralgon, Austl., 12,300...I6 51
Traryd, Swe., 687...B7 24
Trasimeno, lake, It....C4 21
Traskwood, Saline, Ark., 205...C3 81
Trás-os-Montes, reg., Port., 558,700...B2 20
Trás-os-Montes e Alto Douro, prov., Port., 639,846...*B2 20
Trat, Thai., 3,633...F5 38
Traun, lake, Aus...E6 16
Traun, riv., Aus....D7 16
Traunik, Alger, Mich., 50...B4 98
Traunstein, Ger., 14,400...E6 16
Trave, riv., Ger....E4 24
Travelers Rest, Greenville, S.C., 1,973...B3 115
Travers, Alta., Can., 50...D4 69
Travers, Switz., 1,550...C2 19
Traverse, co., Minn., 7,503...E2 99
Traverse, isl., Mich....A2 98
Traverse, lake, Minn...E2 99
Traverse City, Grand Traverse, Mich., 18,432...D5 98
Tra Vinh, Viet., 39,700...H7 38
Travis, co., Tex., 212,136...D4 118
Travnik, Yugo., 9,984...C3 22
Trawbreaga, bay, Ire...B4 11
Tray, mtn., Ga...B3 87
Treadway, Hancock, Tenn., 75...C10 117
Treadway, lake, Ill...C3 90
Treasure, co., Mont., 1,345...D9 102
Treasure Island, Pinellas, Fla., 3,506...*E4 86
Trebbia, riv., It....E5 18
Trebel, riv., Ger....E6 24
Trebic, Czech, 19,200...D3 26
Trebinje, Yugo., 4,072...D4 22
Trebisov, Czech, 9,300...D6 26
Trebizond (Trabzon), Tur., 52,700...B12 31
Treble, mtn., B.C., Can....B3 68
Trebnitz, see Trzebnica, Pol.
Trebon, Czech., 4,172...D3 26
Trechado, Valencia, N. Mex...C1 107
Treece, Cherokee, Kans., 280...E9 93
Tregaron, Wales, 1,243...B4 12
Tregarva, Sask., Can., 50...G3 70
Trego, Lincoln, Mont., 12...B2 102
Trego, Washburn, Wis., 175...C2 124
Trego, co., Kans., 5,473...D4 93
Tréguier, Fr., 2,885...C2 14
Treherne, Man., Can., 569...E2 71
Treig, lake, Scot...D4 13
Treinta y Tres, Ur., 18,600...E2 56
Treinta y Tres, dept., Ur., 41,400...*E2 56
Treis, Ger., 2,000...C2 12
Trélazé, Fr., 9,400...D3 14
Trelew, Arg., 5,880...C3 54
Trelleborg, Swe., 19,200...J5 25
Tremadoc, bay, Wales...B3 12
Tremblant, mtn., Que., Can....C3 73
Tremblay [-lès-Gonesse], Fr., 848...g11 14
Trembleur, lake, B.C., Can....B5 68
Trementina, San Miguel, N. Mex., 50...B5 107
Tremiti, is., It....C5 21
Tremont, Tazewell, Ill., 1,558...C4 90
Tremont, Itawamba, Miss., 300...A5 100
Tremont, Schuylkill, Pa., 1,893...E9 114
Tremont City, Clark, Ohio, 414...B4 111
Tremonton, Box Elder, Utah, 2,115...B3 119
Tremosna, Czech., 3,176...D8 17
Tremp, Sp., 4,466...A6 20
Trempealeau, Trempealeau, Wis., 704...D2 124
Trempealeau, co., Wis., 23,377...D2 124
Trempealeau, riv., Wis....D2 124
Trenary, Alger, Mich., 180...B4 98
Trenche, riv., Que., Can....B5 73
Trencin, Czech., 22,300...D5 26
Trengganu, state, Mala., 278,269...J7 38
Trenque, Lauquén, Arg., 10,887 (631)...B4 54
Trent, Ger., 1,872...D7 24
Trent, Lane, Oreg., 40...D4 113
Trent, Moody, S. Dak., 232...D9 116
Trent, Taylor, Tex., 298...C2 118
Trent, riv., Eng....A7 12
Trent, riv., N.C....B6 109
Trentino-Alto Adige, pol. dist., It., 785,967...C7 18
Trento, It., 75,154...A3 21, C7 18
Trenton, Jackson, Ala., 80...A3 78
Trenton, N.S., Can., 3,140...D7 74
Trenton, Ont., Can., 13,183...C7 72
Trenton, Gilchrist, Fla., 941...C4 86
Trenton, Dade, Ga., 1,301...B1 87
Trenton, Clinton, Ill., 1,866...E4 90
Trenton, Todd, Ky., 542...D2 94
Trenton, Wayne, Mich., 18,439...B7, F7 98
Trenton, Grundy, Mo., 6,262...A4 101
Trenton, Hitchcock, Nebr., 914...D4 103
Trenton, Mercer, N.J., 114,167 (*279,800)...C3 106
Trenton, Jones, N.C., 404...B6 109
Trenton, Williams, N. Dak., 125...A2 110
Trenton, Butler, Ohio, 3,064...C3 111
Trenton, Edgefield, S.C., 314...D4 115
Trenton, Gibson, Tenn., 4,225...B3 117
Trenton, Fannin, Tex., 712...C4 118
Trenton, Cache, Utah, 448...B4 119
Trentwood, Spokane, Wash., 1,387...*B8 122
Trepassey, Newf., Can., 495...E5 75
Trepassey, bay, Newf., Can....E5 75
Tres Algarrobos, Arg...B4 54
Tres Arboles, Ur...E1 56
Tres Arroyos, Arg., 29,996...B4 54
Tres Cerros, Arg...D3 54
Tresckow, Carbon, Pa., 1,145...E10 114
Três Corações, Braz., 17,498...C3 56
Tres Esquinas, Col...C2 60
Tres Forcas, cape, Mor....G5 36
Três Lagoas, Braz., 14,520...C2 56
Tres Lagunas, Catron, N. Mex., 5...C1 107
Tresle Creek, Bonner, Idaho, 50...A2 89
Tres Lomas, Arg., 3,425...B4 54
Tres Marías, is., Mex....C3 63
Três Marias, res., Braz...B3 56
Tres Piedras, Taos, N. Mex., 150...A4 107
Tres Pinos, San Benito, Calif., 150...C6 82
Tres Puntas, cape, Arg...D3 54
Tres Ritos, Taos, N. Mex...A4 107
Tre Teste, riv., It....h9 21
Treuchtlingen, Ger., 6,700...E5 17
Treuenbrietzen, Ger., 8,569...A7 17
Treutlen, co., Ga., 5,874...D4 87
Treviglio, It., 23,413...B2 21
Treviño, Sp...A4 20
Treviso, It., 75,017...B4 21, D8 18
Trevor, Kenosha, Wis., 250...F1 124
Trevorton, Northumberland, Pa., 2,597...E8 114
Trevose, Bucks, Pa., 6,000...*A12 114
Trevose, head, Eng...D2 12
Trevose Heights, Bucks, Pa., 1,500...*F12 114
Treynor, Pottawattamie, Iowa, 368...C2 92
Treysa, Ger., 7,800...C4 17
Trezevant, Carroll, Tenn., 944...A3 117
Trhove Sviny, Czech., 2,485...E9 17
Triadelphia, Ohio, W. Va., 600...A4, B2 123
Triadelphia, res., Md...B3 85
Triangle, Prince William, Va., 2,948...C5 121
Triangle Lake, Lane, Oreg. 100...C3 113
Triaucourt, Fr., 588...F5 15
Tribbet, Washington, Miss., 200...B3 100
Tribbey, Pottawatomie, Okla., 150...B4 112
Tribune, Sask., Can., 153...H4 70
Tribune, Greeley, Kans., 1,036...D2 93
Trichinopoly, see Tiruchirappalli, India
Trichur, India, 73,038...F6 39
Tridell, Uintah, Utah, 310...C6 119
Trident, Gallatin, Mont., 200...E5 102
Trident, peak, Nev....B3 104
Trier, Ger., 87,100...D1 17
Trieste, It., 272,723...B4 21
Trieste, gulf, Italy...B4 21
Trigg, co., Ky., 8,870...D2 94
Triglav, mtn., Yugo...B1 22
Trigo, mts., Ariz....D1 80
Trigueros, Sp., 6,454...D2 20
Trikkala, Grc., 27,876...C3 23
Trikkala, prov., Grc., 142,781...*C3 23
Tri Lakes, Whitley, Ind., 1,089...*B7 91
Trilby, Pasco, Fla., 700...D4 86
Trilby, Lucas, Ohio, 5,000...A2 111
Trilla, Coles and Cumberland, Ill., 225...D5 90
Trillick, N. Ire., 220...C4 11
Trim, Ire., 1,371...D5 11
Trimble, La Plata, Colo., 20...D3 83
Trimble, Clinton, Mo., 185...B3 101
Trimble, Dyer, Tenn., 581...A2 117
Trimble, co., Ky., 5,102...B4 94
Trimble, isl., Wash...D1 122
Trimont, Martin, Minn., 942...G4 99
Trimountain, Houghton, Mich., 400...A2 98
Trinchera, Las Animas, Colo. 150...D6 83
Trinchera, creek, Colo...D5 83
Trinchera, peak, Colo...D5 83
Trincomalee, Cey., 26,356...G7 39
Tring Jonction, Que., Can., 1,214...C6 73
Trinidad, Bol., 8,695...B3 65
Trinidad, Humboldt, Calif., 289...*B1 82
Trinidad, Las Animas, Colo., 10,691...D6 83
Trinidad, Cuba, 16,756...E4 64
Trinidad, Henderson, Tex., 786...C4 118
Trinidad, Ur., 13,600...E1 56
Trinidad & Tobago, country, N.A., 825,700...A5 60
Trinidad, bay, Pan...k11 62
Trinidad, isl., Arg...B4 54
Trinidad, isl., Trin....A5 60
Trinidade, isl., Braz...H8 6
Trinité, Mart., 7,732...o16 74
Trinity, Morgan, Ala., 454...A2 78
Trinity, Newf., Can., 692...D5 75
Trinity, Randolph, N.C., 881...B4 109
Trinity, Trinity, Tex., 1,787...D5 118
Trinity, co., Calif., 9,706...B2 82
Trinity, co., Tex., 7,539...D5 118
Trinity, bay, Newf., Can....D5 75
Trinity, isl., Alsk....D9 79
Trinity, mtn., Idaho...F3 89
Trinity, mtn., Nev....C3 104
Trinity, mts., Calif....B2 82
Trinity, range, Nev....C3 104
Trinity, res., Calif...B2 82
Trinity, riv., Calif....B2 82
Trinity, riv., Tex....D5 118
Trinity Center, Trinity, Calif., 100...B2 82
Trinity North, dist., Newf., Can...D4 75
Trinity South, dist., Newf., Can...E5 75
Trinity Springs, Martin, Ind., 125...G4 91
Trino, It., 8,100...B2 21
Trinway, Muskingum, Ohio, 500...B5 111
Trio, Williamsburg, S.C., 174...E8 115
Trion, Chattooga, Ga., 2,227...B1 87
Triplett, Chariton, Mo., 231...B4 101
Tripoli, Bremer, Iowa, 1,179...B5 92
Tripoli, see Tarabulus, Leb.
Tripoli (Tarābulus), Libya, 170,000...C2 43
Tripoli, Oneida, Wis., 45...C4 124
Tripolis, Grc., 18,500...D4 23
Tripolitania (Tarābulus), prov., Libya, 746,064...C2 43
Tripp, Hutchinson, S. Dak., 837...D8 116
Tripp, co., S. Dak., 8,761...D6 116
Tripura, ter., India, 1,142,005...*D9 39
Trischen, isl., Ger....A4 16
Tristan da Cunha, is., Atl. O...I9 6
Tristate Village, Du Page, Ill., 1,000...*B6 90
Triste, gulf, Ven....A4 60
Tritle, mtn., Ariz....C3 80
Triumph, Plaquemines, La., 900...E6 95
Triune, Williamson, Tenn., 50...B5 117
Triunfo, Braz., 3,123...C3 57
Trivandrum, India, 239,815 (*302,214)...G6 39
Trnava, Czech., 31,700...D4 26
Trobriand, is., Pap....k13 50
Trochu, Alta., Can., 671...D4 69
Trogir, Yugo., 4,995...D3 22
Trögstad, Nor...p29 25
Troisdorf, Ger., 16,700...C2 17
Trois Pistoles, Que., Can., 4,349...A8 73
Trois Rivières, Que., Can., 53,477 (*83,659)...C5 73
Trois Saumons, Que., Can., 60...B7 73
Troisvierges, Lux., 2,006...D6 15
Troitsk, Sov. Un., 79,000...C6 29
Troitskoye, Sov. Un., 1,000...B8 37
Trojan, Lawrence, S. Dak., 5...C2 116
Trolldalen, Nor., 141...p28 25
Trollhättan, Swe., 32,700...H5 25
Trombetas, riv., Braz....C3 59
Trombudo [Central], Braz., 732...D3 56
Tromper, bay, Ger....D7 24
Troms, co., Nor., 127,500...*C8 25
Tromsö, Nor., 12,400 (*19,500)...C8 25
Trona, San Bernardino, Calif., 1,138...E5 82
Tronador, mtn., Arg...C2 54
Trondheim, Nor., 58,600 (*109,000)...F3 25
Trondheimsfjord, fjord, Nor....F3 25
Troodos, mtn., Cyp....E9 31
Troon, Scot., 9,932...E4 13
Tropea, It., 6,702...E5 21
Trophy, mtn., B.C., Can....D8 68
Tropic, Garfield, Utah, 382...F3 119
Trosa, Swe., 1,300...u35 25
Trosky, Pipestone, Minn., 122...G2 99
Trossachs, Sask., Can., 89...H3 70
Trostan, mtn., N. Ire....B5 11
Trotters, Golden Valley, N. Dak., 2...B2 110
Trotwood, Montgomery, Ohio, 4,992...C3 111
Trough Creek, Huntingdon, Pa., 50...E5 114
Troup, Cherokee and Smith, Tex., 1,667...C5 118
Troup, co., Ga., 47,189...C1 87
Trousdale, Edwards, Kans., 88...E4 93
Trousdale, co., Ten., 4,914...A5 117
Trousers, lake, N.B., Can....B2 74
Trout, La Salle, La., 500...C3 95
Trout, creek, Fla....B6 86
Trout, lake, B.C., Can....D9 68
Trout, lake, N.W. Ter., Can....D8 66
Trout, lake, Ont., Can....A5 72
Trout, lake, Minn....B6 99
Trout, lake, Wis....B4 124
Trout, mtn., B.C., Can....D2 69
Trout, peak, Wyo....A3 125
Trout, riv., Alta., Can....A3 69
Trout, riv., Vt....B3 120
Trout Creek, Ont., Can., 510...B5 72
Trout Creek, Ontonagon, Mich., 350...A6 98
Trout Creek, Sanders, Mont., 64...C1 102
Trout Creek, Juab, Utah, 5...D2 119
Trout Creek, pass, Colo....C5 83
Troutdale, Somerset, Maine...C3 96
Trout Dale, Grayson, Va., 273...B3, E1 121
Trout Lake, Ont., Can...D8 72
Trout Lake, Chippewa, Mich., 200...B5 98
Trout Lake, Klickitat, Wash., 450...D4 122
Troutman, Stewart, Ga., 100...E2 87
Troutman, Iredell, N.C., 648...B3 109
Trout River, Newf., Can., 696...D2 75
Trout Run, Lycoming, Pa., 370...D7 114
Troutville, Botetourt, Va., 524...D3 121
Trouville, Fr., 6,622...C3 14
Trowbridge, Eng., 15,833...C5 12
Troy, Pike, Ala., 10,234...D4 78
Troy, Latah, Idaho, 555...C2 89
Troy, Madison, Ill., 1,778...E4 90
Troy, Perry, Ind., 528...H4 91
Troy, Davis, Iowa, 95...D5 92
Troy, Doniphan, Kans., 1,051...C8 93
Troy, Oakland, Mich., 19,382...*F7 98
Troy, Pontotoc, Miss., 250...A5 100
Troy, Lincoln, Mo., 1,779...C7 101
Troy, Lincoln, Mont., 855...B1 102
Troy, Cheshire, N.H., 950...E2 105
Troy, Rensselaer, N.Y., 67,492...C7 108
Troy, Montgomery, N.C., 2,346...B4 109
Troy, Miami, Ohio, 13,685...B3 111
Troy, Johnston, Okla., 100...C5 112
Troy, Wallowa, Oreg., 100...B9 113
Troy, Bradford, Pa., 1,478...C8 114
Troy, Greenwood, S.C., 260...D3 115
Troy, Grant, S. Dak., 15...B9 116
Troy, Obion, Tenn., 587...A2 117
Troy, Bell, Tex., 500...D4 118
Troy, Orleans, Vt., 150 (1,613▲)...B4 120
Troy, Gilmer, W. Va., 110...B4 123
Troy, peak, Nev....E6 104
Troy, ruins, Tur....C6 23
Troyes, Fr., 67,406 (*101,000)...C6 14
Troy Grove, La Salle, Ill., 271...B4 90
Troy Mills, Linn, Iowa, 150...B6 92
Trstenik, Yugo., 5,577...D5 22
Truax, Sask., Can., 76...H3 70
Trubchevsk, Sov. Un., 10,000...E9 27
Trubschachen, Switz., 1,665...C4 19
Truchas, Rio Arriba, N. Mex., 400...A4 107
Trucial Coast, country, Asia, 80,000...I7 41
Truckee, Nevada, Calif., 950...C3 82
Truckee, riv., Nev....D2 104
Trucksville, Luzerne, Pa., 2,300...*B8 114
Truesdell, Kenosha, Wis., 10...F2 124
Truhart, King and Queen, Va., 55...D6 121
Truitt, peak, Yukon, Can....D6 66
Trujillo, Hond., 2,957...C4 62
Trujillo, San Miguel, N. Mex., 10...B5 107
Trujillo, Peru, 45,899...C2 58
Trujillo, Sp., 13,326...C3 20
Trujillo, Ven., 19,358...B3 60
Trujillo, state, Ven., 326,634...B3 60
Trujillo Alto, P.R., 1,297...B6 65
Trujillo Alto, mun., P.R., 18,251...B6 65
Truk, is., Caroline Is.... 52
Truman, Martin, Minn., 1,256...G4 99
Trumann, Poinsett, Ark., 4,511...B5 81
Trumansburg, Tompkins, N.Y., 1,768...C4 108
Trumbull, Fairfield, Conn., 20,379...E4 84
Trumbull, Clay and Adams, Nebr., 173...D7 103
Trumbull, co., Ohio, 208,526...A7 111
Trun, Bul., 2,169...D6 22
Truro, N.S., Can., 12,421...D6 74
Truro, Eng., 13,328...D2 12
Truro, Madison, Iowa, 338...C4 92
Truro, Barnstable, Mass., 300 (1,002▲)...C7 97
Truscott, Knox, Tex., 150...C3 118
Trussville, Jefferson, Ala., 2,510...B3, E5 78
Truth or Consequences, Sierra, N. Mex., 4,269...D2 107
Trutnov, Czech., 23,000...C3 26
Truxton, Mohave, Ariz., 15...B2 80
Truxton, Lincoln, Mo., 85...B6 101
Truxton, Cortland, N.Y., 300...C4 108
Tryon, McPherson, Nebr., 150...C5 103
Tryon, Polk, N.C., 2,223...D4 109
Tryon, Lincoln, Okla., 254...B5 112
Trysil, Nor...G5 25
Trzcianka, Pol., 4,482...B4 26
Trzebiatowo, Pol., 5,995...A3 26
Trzebinia, Pol., 4,140...C5, g10 26
Trzebnica, Pol., 3,170...C4 26
Tsabong, Bech...C3 49
Tsagaan Hamar, Mong...C2 36
Tsala Apopka, lake, Fla...D4 86
Tsane, Bech...B3 49
Tsang, China...E7 36
Tsangpo (Brahmaputra), riv., China...C8 39
Tsangwu, see Wuchow, China
Tsaratanana, Malag., 1,145...g9 49
Tsau, Bech...B3 49
Tsavo, Ken...B6 48
Tschetter Colony, Hutchinson, S. Dak., 75...D8 116
Tschida, lake, N. Dak...C4 110
Tselinograd (Akmolinsk), Sov. Un., 114,000...C8 29
Tsentralnyy, Sov. Un...D26 9
Tsetserleg Suma, Mong...C1 36
Tshela, Con. L...B1 48
Tshikapa, Con. L., 6,400...C3 48
Tshilongo, Con. L...D4 48
Tshimbo, Con. L...C4 48
Tshofa...C4 48
Tshuapa, riv., Con. L....B3 48
Tsiafajavona, mtn., Malag...g9 49
Tsihombe, Malag., 600...k9 49
Tsimlyansk, res., Sov. Un....D2 29
Tsinan (Chinan), China, 862,000...D8 34
Tsinghai (Chinghai), prov., China, 2,050,000...D4 34
Tsingma, China...C8 38
Tsingtao (Chingtao), China, 1,121,000...D9 34, F9 36
Tsingyuan, see Paoting, China
Tsinling Shan, mts., China...E6 34
Tsintsabis, S.W. Afr...A2 49
Tsiroanomandidy, Malag...g9 49
Tsis, isl., Truk... 52
Tsitsihar (Ch'i-ch'i-ha-erh), China, 668,000...B9 34
Tsitsuti, peak, B.C., Can....C5 68
Tsivory, Malag., 800...h9 49
Tskhakaya, Sov. Un., 10,000...A13 31
Tskhinvali, Sov. Un., 24,000...E2 29
Tsna, riv., Sov. Un....C2 29
Tsodilo, mtn., Bech...A3 49
Tsoshui, China, 2,000...H3 36
Tsu, Jap., 110,900...I8, o15 37
Tsuchiura, Jap., 47,500...H10, m19 37
Tsugaru, strait, Jap....F10 37
Tsukan, isl., Okinawa... 52
Tsulukidze, Sov. Un., 9,770...A14 31
Tsumeb, S.W. Afr...A2 49
Tsumis, S.W. Afr...B2 49
Tsunami, Okinawa... 52
Tsunghua, China, 1,000...G7 34
Tsunhua, China, 15,000...D7 36
Tsuni, China, 97,500...K2 36
Tsurikake, Jap 6,800...E9 37
Tsuruga, Jap., 36,000...I8, n15 37
Tsuruoka, Jap., 56,300 (83,149▲)...G9 37
Tsushima, Jap., 43,198...n15 37
Tsu-Shima, isl., Jap...I4 37
Tsushima, strait, Jap....I4 37
Tsuyama, Jap., 44,800 (78,549▲)...I7 37
Tsuyung, China...F5 34
Túa, riv., Port...B2 20
Tual, Indon...G8 35
Tualatin, riv., Oreg....B1 113
Tuam, Ire., 3,500...D3 11
Tuamotu (Low), arch., Fr. Polynesia...H13 7
Tuangku, isl., Indon, 634...K2 38
Tuapse, Sov. Un., 45,000...G16 9
Tuasivi, Cape, W. Sam... 52
Tuatapere, N.Z., 872...Q11 51
Tuath, bay, Scot...D2 13
Tubac, Santa Cruz, Ariz., 100...F4 80
Tuba City, Coconino, Ariz., 500...A4 80
Tubarão, Braz., 29,615...D3 56
Tubas, Jordan, 5,000...B7, f12 32
Tuberose, Sask., Can., 80...G1 70
Tübingen, Ger., 49,600...E4 17
Tu Bong, Viet...F8 38
Tubre, It., 941...C6 18
Tubrug (Tobruk), Libya, 4,995...C4 43
Tubuai, is., Fr. Polynesia...*H13 7
Tucacas, Ven., 3,783...A4 60
Tucannon, canyon, Wash....C8 122
Tucannon, riv., Wash....C8 122
Tucano, Braz., 4,007...D3 57
Tuchola, Pol., 5,750...B4 26
Tuchuan, China...B9 36
Tuckahoe, Cape May, N.J., 500...E3 106
Tuckahoe, Westchester, N.Y., 6,423...D1 108
Tuckahoe, creek, Md....C6 85
Tuckahoe, riv., N.J....E3 106
Tucker, Jefferson, Ark., 350...C4 81
Tucker, De Kalb, Ga., 5,000...A5 87
Tucker, Jones, Miss., 25...D4 100
Tucker, Anderson, Tex., 75...D5 118
Tucker, Utah, Utah, 5...D4 119
Tucker, co., W. Va., 7,750...B5 123
Tucker, isl., N.J....D4 106
Tuckerman, Jackson, Ark., 1,539...B4 81
Tuckernuck, isl., Mass....D7 97
Tuckerton, Ocean, N.J., 1,536...D4 106
Tucson, Pima, Ariz., 212,892 (*243,000)...E5 80
Tucumán, Arg., 280,000...E2 55
Tucumán, prov., Arg., 780,348...E2 55
Tucumcari, Quay, N. Mex., 8,143...B6 107
Tucumcari, peak, N. Mex....B6 107
Tucupita, Ven., 9,575...B5 60
Tucuruí, Braz., 3,403...C5 59
Tuczna, riv., Pol....m13 26
Tucznobaby, Pol...g10 26

Tudela, Sp., 16,456...A5 20
Tufi, Pap...k12 50
Tug, fork, W. Va., Ky...C2 123
Tugaloo, riv., S.C....B1 115
Tugaske, Sask., Can., 267...G2 70
Tug Hill, mtn., N.Y....C2 108
Tuguegarao, Phil., 19,000 (43,000▲)...B6 35
Tuira, riv., Pan...F9 62
Tukangbesi, is., Indon...G6 35
Tūkrah (Toora), Libya, 4,643...C4 43
Tuktoyaktuk, N.W. Ter., Can., 409...C6 66
Tukums, Sov. Un., 10,000...C4 27
Tukuyu, Tan, 3,563...C5 48
Tukwila, King, Wash., 1,804.D1 122
Tukzar, Afg., 5,000...D13 41
Tula, Mex., 7,559...C5 63
Tula, Lafayette, Miss., 175...A4 100
Tula, Sov. Un., 333,000...C1 29
Tulalip, Indian res., Wash...A3 122
Tulancingo, Mex., 26,663...C5, m14 63
Tulare Tulare, Calif., 13,824...D4 82
Tulare, Spink, S. Dak., 225..C7 116
Tulare, co., Calif., 168,403...D4 82
Tularosa, Otero, N. Mex., 3,200...D3 107
Tularosa, val., N. Mex...E3 107
Tulcán, Ec., 10,658...A2 58
Tulcea, Rom., 24,639...C9 22
Tulchin, Sov. Un., 10,000...G7 27
Tulear, Malag., 18,648...h8 49
Tule River, Indian res., Calif...E4 82
Tuli, S. Rh...B4 49
Tulia, Swisher, Tex., 4,410...B2 118
Tūl Karm, Jordan, 21,872...B7, f11 32
Tulla, Ire., 389...E3 11
Tullahassee, Wagoner, Okla., 199...B6 112
Tullahoma, Coffee, Tenn., 12,242...B5 117
Tullamore, Ire., 6,243...D4 11
Tulle, Fr., 19,084...E4 14
Tullins, Fr., 3,680...E6 14
Tulln, Aus., 6,306...D8 16
Tullos, La Salle, La., 594...C3 95
Tullow, Ire., 1,725...E5 11
Tully, Austl., 2,678...C8 50
Tully, Onondaga, N.Y., 803...C4 108
Tullytown, Bucks, Pa., 2,452...*F2 114
Ţulmaythah, Libya, 350...F3 31
Tuloma, riv., Sov. Un...C15 25
Tulot, Poinsett, Ark., 80...B5 81
Tulsa, Tulsa, Okla., 261,685 (★360,900)...A6 112
Tulsa, co., Okla., 346,038...B6 112
Tulsk, Ire...D3 11
Tuluá, Col., 28,715...C2 60
Tulufan, see Turfan, China
Tulun, Sov. Un., 34,000...D13 28
Tulu Wallel, mtn., Eth...D3 47
Tulyehualco, Mex., 4,089...h9 63
Tuma, riv., Nic...D5 62
Tumaco, Col., 12,692...C2 60
Tumaco, bay, Col...C2 60
Tumacacori, Santa Cruz, Ariz., 80...F4 80
Tumacacori, nat. mon., Ariz...F4 80
Tumatumari, Br. Gu...A3 59
Tumba, Swe., 6,618...t35 25
Tumba, lake, Con. L...B2 48
Tumbarumba, Austl., 1,511...H7 51
Tumbaya, Arg...D2 55
Tumbes, Peru, 7,416...B1 58
Tumbes, dept., Peru, 30,035...B1 58
Tumble, mtn., Mont...E6 102
Tumbling Shoals, Cleburne, Ark., 100...B3 81
Tumen, China, 28,000...C10 34
Tumen, riv., China...E4 37
Tumeremo, Ven., 3,121...B5 60
Tumkur, India, 47,277...F6 39
Tummin, riv., Sov. Un...A10 37
Tumon, bay, Guam...52
Tump, range, Wyo...D2 125
Tumtum, Stevens, Wash., 25.B8 122
Tumu, Ghana...D4 45
Tumucumaque, mts., Braz...B4 59
Tumurisk, sand dunes, Iran...D8 41
Tumut, Austl., 3,489...G7 51
Tumwater, Thurston, Wash., 3,885...B3 122
Tuna, pt., P.R...D7 65
Tunas de Zaza, Cuba, 475...E4 64
Tunaycha, lake, Sov. Un...C11 37
Tunb, isl., Iran...H7 41
Tunbridge, Orange, Vt., 125 (743▲)...D4 120
Tunbridge Wells, Eng., 39,855...C8 12
Tunduru, Tan., 6,990...D6 48
Tundzha, riv., Bul...D8 22
Tunga, China...F7 36
Tunga, riv., India...E6 39
Tungan, China, 7,000...G8 34
Tungcheng, China...I7 36
Tungchiang, China, 11,000...I2 36
Tungchiang (Lahasusu), China, 5,000...C6 37
Tunghai, China...E8 34
Tungho, China, 5,000...D4 37
Tunghsiang, China, 5,000...J7 36
Tunghsing, China, 1,000...G6 34
Tunghwa, China, 129,100...F2 37
Tungjen, China, 11,000...K3 36
Tungkuan, China, 19,000...G4 36
Tungkuang, China...F7 36
Tungla, Nic...D5 62
Tungliao (Payintala), China, 40,000...C10 36
Tungning, China, 5,000...D5 37
Tungpei, China, 5,000...B10 34
Tungping, China...G7 36
Tungpu, see Rangsum, China
Tungsheng, China, 6,000...E3 36
Tungsten, Pershing, Nev., 30...C3 104
Tungsunitewang, China...C6 36
Tungtai, China...H9 36
Tungting, lake, China...J5 36
Tungtzu, China, 13,000...J2 36
Tunguragua (Tungurahua), prov., Ec., 204,726...B2 58
Tungurahua, prov., Ec., 204,726...B2 58
Tunhua, China, 35,000...C10 34
Tuni, India, 22,452...I9 40
Tunica, Tunica, Miss., 1,445...A3 100
Tunica, co., Miss., 16,826...A3 100
Tuninga, riv., Braz...D5 60
Tunis, Tun., 410,000...F11 30, B6 44
Tunis, gulf, Tun...F12 30
Tunis et Banlieue, prov., Tun., 747,967...*B6 44
Tunisia, country, Afr., 3,880,000...C6 42, B6 44
Tunis Mills, Talbot, Md., 90...C5 85
Tunja, Col., 45,680...B3 60
Tunk, lake, Maine...D4 96
Tunkhannock, Wyoming, Pa., 2,297...C10 114
Tunkí, Nic...D5 62
Tunnel Hill, Whitfield, Ga., 255...B1 87
Tunnel Springs, Monroe, Ala., 100...D2 80
Tunnelton, Lawrence, Ind., 150...G5 91
Tunnelton, Preston, W. Va., 359...B5 123
Tunö, isl., Den...C4 24
Tuntatuliak, Alsk., 145...C7 79
Tunungayualuk, isl., Newf., Can...g9 75
Tunuyán, Arg...A3 54
Tunuyán, riv., Arg...A3 54
Tuolumne, Tuolumne, Calif., 1,403...D3 82
Tuolumne, co., Calif., 14,404...C4 82
Tuolumnie, riv., Calif...D3 82
Tupã, Braz., 28,723...C2 56
Tupelo, Jackson, Ark., 201...B4 81
Tupelo, Lee, Miss., 17,221...A5 100
Tupelo, Coal, Okla., 261...C5 112
Tupiza, Bol., 8,248...D2 55
Tupman, Kern, Calif., 500...E4 82
Tupper, lake, N.Y., 5,200...A6 108
Tupper Creek, B.C., Can., 67...B7 68
Tupper Lake, Franklin, N.Y., 5,200...A6, B2 108
Tuppers Plains, Meigs, Ohio, 200...C6 111
Tupperville, Ont., Can., 137.E2 72
Tupungato, Arg...A3 54
Tupungato, mtn., Arg...A3 54
Túquerres, Col., 6,482...C2 60
Ţūr, Jordan...m14 32
Tura, India, 8,888...E13 40
Tura, Sov. Un., 2,000...C13 28
Tura, riv., Sov. Un...D21 9
Turabah (Ramadan), Sau. Ar...A5 47
Tūrān, Iran...D8 41
Turan, lowland, Sov. Un...E5 29
Turbaco, Col., 10,208...A2 60
Turbat, Pak., 3,549...C3 39
Turbenthal, Switz., 2,685...B6 19
Turbeville, Clarendon, S.C., 355...D7 115
Turbo, Col., 2,636...B2 60
Turbotville, Northumberland, Pa., 612...D8 114
Turciansky Svaty Martin, Czech., 22,400...D5 26
Turda, Rom., 33,610...B6 22
Turek, Pol., 7,179...B5 26
Turfan (Tulufan), China, 25,000...C2 34
Turfan, depression, China...F13 28
Turgar, riv., Sov. Un...D6 29
Turgay, Sov. Un., 5,800...D6 29
Turgovishte, Bul., 10,505...D8 22
Turgutlu, Tur., 31,700...C6 23
Turi, Sov. Un., 6,200...B5 27
Turí, pt., Braz...B2 57
Turia, riv., Sp...C5 20
Turiaçu, Braz., 1,826...B1 57
Turiaçu, riv., Braz...B1 57
Turin, Alta., Can., 99...E4 69
Turin, Coweta, Ga., 183...C2 87
Turin, Monona, Iowa, 163...B2 92
Turin, see Torino, It.
Turin, Lewis, N.Y., 323...B5 108
Turinsk, Sov. Un., 17,900...B6 29
Turiy Rog, Sov. Un...D5 37
Turiya, riv., Sov. Un...F5 27
Turka, Sov. Un., 10,000...G4 27
Turkestan, Sov. Un., 28,300...E7 29
Turkeve, Hung., 10,339 (12,505▲)...B5 22
Turkey, Sampson, N.C., 199...B5 109
Turkey, Hall, Tex., 813...B2 118
Turkey, country, Asia, Eur., 27,809,000...C10 31, F6 33
Turkey, creek, Nebr...D8 103
Turkey, creek, Okla...A4 112
Turkey, key, Fla...G5 86
Turkey, pt., Md...B5 85
Turkey, riv., Iowa...B6 92
Turkmen S.S.R., rep., Sov. Un., 1,626,000...G20 9
Turk Mine, S. Rh., 1,100...A4 49
Turks, is., Turks & Caicos Is...E8 64
Turks and Caicos Islands, Br. dep., N.A., 5,716...E8 64
Turks Island, passage, Ba. Is...E8 64
Turku, Fin., 124,200 (★162,000)...G10 25
Turku-Pori, dept., Fin., 607,073...*G10 25
Turkwel, riv., Ken...A6 48
Turley, Tulsa, Okla, 4,000...A6 112
Turlock, Stanislaus, Calif., 9,116...D3 82
Turmus 'Ayyā, Jordan, 1,000.B7 32
Turn, Valencia, N. Mex...C3 107
Turneffe, is., Br. Hond...B4 62
Turner, Phillips, Ark., 85...C4 81
Turner, Wyandotte, Kans., 1,000...B8 93
Turner, Androscoggin, Maine, 350 (1,890▲)...D2 96
Turner, Arenac, Mich., 206.D7 98
Turner, Blaine, Mont., 175...B8 102
Turner, Marion, Oreg., 770...C2, C4 113
Turner, co., Ga., 8,439...E3 87
Turner, co., S. Dak., 11,159.D8 116
Turnercrest, Campbell, Wyo...B7 125
Turners, Sumner, Tenn., 150.A5 117
Turners Falls, Franklin, Mass., 4,917...A2 97
Turnersville, Coryell, Tex., 100...D4 118
Turner Valley, Alta., Can., 702...D3 69
Turnerville, Lincoln, Wyo., 50...C2 125
Turney, Clinton, Mo., 144...B3 101
Turnhout, Bel., 36,444...C4 15
Turnor, lake, Sask., Can...E11 66
Turnov, Czech., 11,500...C3 26
Turnovo, Bul., 16,182...D7 22
Turnu-Măgurele, Rom., 18,055...D7 22
Turnu-Severin, Rom., 32,486...C6 22
Turö, isl., Den...C4 24
Turochak, Sov. Un...E26 9
Turon, Reno, Kans., 559...E5 93
Turpin, Beaver, Okla., 220...D3 112
Turrell, Crittenden, Ark., 794...B5 81
Turret, mtn., Calif...B3 82
Turret, peak, Ariz...C4 80
Turriff, Scot., 2,686...C6 13
Turtkul, Sov. Un., 23,700...E6 29
Turtle, lake, Sask., Can...D2 70
Turtle, mts., N. Dak...A5 110
Turtle, riv., Ont., Can...E5 71
Turtle Creek, N.B., Can., 101...D5 74
Turtle Creek, Allegheny, Pa., 10,607...B6 114
Turtleford, Sask., Can., 352..D1 70
Turtle Lake, McLean, N. Dak., 792...B5 110
Turtle Lake, Barron, Wis., 691...C1 124
Turtle Mountain, Indian res., N. Dak...A6 110
Turtle River, Beltrami, Minn., 48...C4 99
Turton, Spink, S. Dak., 140..B7 116
Turugart, pass, Sov. Un...E9 29
Turukhansk, Sov. Un., 5,000...C11 28
Tupungato, mtn., Arg...A3 54
Turvo, riv., Braz...m7 56
Tusas, Rio Arriba, N. Mex...A3 107
Tuscaloosa, Tuscaloosa, Ala., 63,370...B2 78
Tuscaloosa, co., Ala., 109,047...B2 78
Tuscaloosa, dam, Ala...B2 78
Tuscany, reg., It...C3 21
Tuscarawas, co., Ohio, 76,789...B6 111
Tuscarawas, riv., Ohio...B6 111
Tuscarora, Elko, Nev., 50...B5 104
Tuscarora, mts., Nev...C5 104
Tuscarora, mts., Pa...F6 114
Tuscola, Douglas, Ill., 3,875.D5 90
Tuscola, Leake, Miss., 75...C4 100
Tuscola, Taylor, Tex., 414...C3 118
Tuscola, co., Mich., 43,305..E7 98
Tusculum, ruins, It...h9 21
Tusculum College, Greene, Tenn., 1,433...*C5 117
Tuscumbia, Colbert, Ala., 8,994...A2 78
Tuscumbia, Miller, Mo., 231...C5 101
Tushan, China, 12,000...L2 36
Tushihkou, China...D6 36
Tushka, Atoka, Okla., 350...C5 112
Tuskahoma, Pushmataha, Okla., 300...C6 112
Tuskeegee, Graham, N.C...D2 109
Tuskegee, Macon, Ala., 7,240...C4 78
Tuskegee Institute, Macon, Ala., 5,380...C4 78
Tusket, N.S., Can., 331...F4 74
Tustin, Orange, Calif., 2,006.F3 82
Tustin, Osceola, Mich., 248..D5 98
Tustumena, lake, Alsk...g16 79
Tūt, Iran...E7 41
Tutamoe, mtn., N.Z...k14 51
Tutayev, Sov. Un., 17,700..C12 27
Tuthill, Bennett, S. Dak., 50.D4 116
Tuticorin, India, 124,230 (★127,356)...G6 39
Tutrakan, Bul., 7,203...C8 24
Tuttle, Gooding, Idaho, 10..G4 89
Tuttle, Kidder, N. Dak., 255.B5 110
Tuttle, Grady, Okla., 855...B4 112
Tuttle, lake, Iowa...A3 92
Tuttle Creek, dam, Kans...C7 93
Tuttle Creek, res., Kans...C7 93
Tuttlingen, Ger., 24,900...E4 16
Tutubu, Tan...C5 48
Tutuila, isl., Am. Sam...52
Tutwiler, Tallahatchie, Miss., 912...A3 100
Tuve, Swe...A5 24
Tuvu, Fiji Is., 57...52
Tuvutha, isl., Fiji Is...52
Tuxedo, Man., Can., 1,627.*E3 71
Tuxedo, Prince Georges, Md., 1,000...*C4 85
Tuxedo, Henderson, N.C., 900...D3 109
Tuxedo Park, Orange, N.Y., 723...D2, D6 108
Tuxford, Sask., Can., 141...G3 70
Tuxpan, Mex., 14,863..C3, m11 63
Tuxpan, Mex., 23,222..C5, m15 63
Tuxpan, Mex., 10,871...n12 63
Tuxpan, riv., Mex...m15 63
Tuxtepec, Mex., 8,631...n15 63
Tuxtepec, riv., Mex...n15 63
Tuxtla Gutiérrez, Mex., 41,532...D6 63
Túy, Sp., 2,779...A1 20
Tuyen Quang, Viet...B6 38
Tuymazy, Sov. Un...C4 29
Tuyun, China, 15,000...K2 36
Tuz, lake, Tur...C9 31
Tuzigoot, nat. mon., Ariz...C3 80
Tuz Khurmatli, Iraq., 6,381.D2 41
Tuzla, Yugo., 37,673...C4 22
Tuzlu, salt lake, Iran...D4 41
Tversted, Den...A4 24
Twain Harte, Tuolumne, Calif., 900...*D3 82
Twaiyya, Sau. Ar...I13 31
Twayton, Johnson, Wyo...A6 125
Tweed, Ont., Can., 1,791...C7 72
Tweed, riv., Scot...E5 13
Tweedsmuir, park, B.C., Can...C4 68
Twelve Mile, Cass, Ind., 225.C5 91
Twelve Pins, mtn., Ire...D2 11
Twelvepole, creek, W. Va...C2 123
Twentymile, creek, W. Va...C3 123
Twentynine Palms, San Bernardino, Calif., 1,000...E5 82
Twickenham, Eng., 100,822.m11 10
Twig, St. Louis, Minn., 175..D6 99
Twiggs, co., Ga., 7,935...D3 87
Twila, Harlan, Ky., 200...D6 94
Twillingate, Newf., Can., 947...D4 75
Twillingate, dist., Newf., Can...D4 75
Twillingate, is., Newf., Can...D4 75
Twin, buttes, Oreg...C4 113
Twin, creek, Ohio...C3 111
Twin, lakes, Conn...A3 84
Twin, lakes, Iowa...B3 90
Twin, lakes, Maine...C4 96
Twin, mts., Wyo...D7 125
Twin, peaks, Idaho...E4 89
Twin, peaks, Mont...E5 102
Twin Bridges, Madison, Mont., 509...E4 102
Twin Brooks, Grant, S. Dak., 86...B9 116
Twin City (Summit and Graymont), Emanuel, Ga., 1,095...D4 87
Twin Falls, Twin Falls, Idaho, 20,126...G4 89
Twin Falls, co., Idaho, 41,842...G4 89
Twining, Arenac, Mich., 199.D7 98
Twin Lake, Muskegon, Mich., 300...E4 98
Twin Lake, mtn., N.Y...B6 108
Twin Lakes, Santa Cruz, Calif., 1,849...*D2 82
Twin Lakes, Lake, Colo., 30.B4 83
Twin Lakes, Lowndes, Ga., 400...F3 87
Twin Lakes, Portage, Ohio, 1,000...*A6 111
Twin Lakes, Kenosha, Wis., 1,497...F1 124
Twin Mountain, Coos, N.H., 200...B3 105
Twin Oaks, Delaware, Pa., 1,500...*G11 114
Twin Orchards, Broome, N.Y., 2,000...*C4 108
Twin Rocks, Cambria, Pa., 950...F4 114
Twinsburg, Summit, Ohio, 4,098...A6 111
Twinton, Overton, Tenn., 125...C8 117
Twin Valley, Norman, Minn., 841...C2 99
Twisp, Okanogan, Wash., 750...A5 122
Two Butte, creek, Colo...D8 83
Two Buttes, Baca, Colo., 111...D8 83
Two Creeks, Man., Can., 110...D1 71
Twodot, Wheatland, Mont., 65...D6 102
Twoforks, riv., Sask., Can...C2 70
Two Guns, Coconino, Ariz., 5...B4 80
Two Harbors, Lake, Minn., 4,695...C7 99
Two Hills, Alta., Can., 826..C5 69
Two Mile, beach, N.J...F3 106
Two Mountains, co., Que., Can., 32,837...D3 73
Two Mountains, lake, Que., Can...D8 73
Two Prairie, bayou, Ark...D6 81
Two Rivers, Manitowoc, Wis., 12,393...B7, D6 124
Two Rivers, riv., Minn...B1 99
Tyachev, Sov. Un...A6 22
Tyaskin, Wicomico, Md., 125...D6 85
Tyborön, Den., 2,134...B2 24
Tyborön, canal, Den...B2 24
Tychy, Pol., 50,000...g9 26
Tye River, Nelson, Va., 130..D4 121
Tygart, res., W. Va...B5 123
Tygart, riv., W. Va...B4 123
Tygart River, falls, W. Va...A7 123
Tygarts, creek, Ky...B7 94
Tyger, riv., S.C...B4 115
Tygh Valley, Wasco, Oreg., 270...B5 113
Tyhee, Bannock, Idaho, 100...G6 89
Tyler, Dallas, Ala., 175...C3 78
Tyler, Lincoln, Minn., 1,138.F2 99
Tyler, Pemiscot, Mo., 75...E8 101
Tyler, Clearfield, Pa., 250...D4 114
Tyler, Smith, Tex., 51,230...C5 118
Tyler, Spokane, Wash., 30...D6 122
Tyler, co., Tex., 10,666...D5 118
Tyler, co., W. Va., 10,026...B4 123
Tyler, branch, Vt...B3 120
Tyler Heights, Kanawha, W. Va., 1,500...*C3 123
Tyler Hill, Wayne, Pa., 80..C11 114
Tyler Park, Fairfax, Va., 1,000...*C5 121
Tylersville, Clinton, Pa., 200...E7 114
Tylerton, Somerset, Md., 200...E5 85
Tylertown, Walthall, Miss., 1,532...D3 100
Tym, riv., Sov. Un...C25 9
Tymochtee, creek, Ohio...B4 111
Tynagh, Ire...D3 11
Tyndall, Man., Can., 241...D3 71
Tyndall, Bon Homme, S. Dak., 1,262...E8 116
Tyndinskiy, Sov. Un...D15 28
Tyndrum, Scot...D4 13
Tyne, riv., Eng...C5 10
Tyne, riv., Scot...E6 13
Tynec, Czech., 1,146...o18 26
Tynemouth, Eng., 70,112...C6 10
Tyner, Sask., Can., 72...G1 70
Tyner, Marshall, Ind., 200...B5 91
Tyner, Jackson, Ky., 500...C5 94
Tyner, Chowan, N.C., 325...A7 109
Tyner, Hamilton, Tenn., 800...D8, E10 117
Tyne Valley, P.E.I., Can., 248...C5 74
Tyngsboro, Middlesex, Mass., 150 (3,302▲)...A5 97
Tyn nad Vltavou, Czech., 3,523...D9 17
Tynset, Nor., 450...F4 25
Tyonek, Alsk., 187...C9, g16 79
Tyre (Sur), Leb., 10,432...A7 32
Tyringham, Bershire, Mass., 175 (197▲)...B1 97
Tyro, Lincoln, Ark., 50...D4 81
Tyro, Montgomery, Kans., 289...E8 93
Tyrol, see Tirol, reg., Aus.
Tyrone, Las Animas, Colo., 30...D6 83
Tyrone, Anderson, Ky., 240..B5 94
Tyrone, Texas, Mo., 60...D6 101
Tyrone, Grant, N. Mex., 200...E1 107
Tyrone, Texas, Okla., 456...D3 112
Tyrone, Blair, Pa., 7,792...E5 114
Tyrone, co., N. Ire., 133,930...C4 11
Tyronza, Poinsett, Ark., 601...B5 81
Tyrrell, co., N.C., 4,520...B7 109
Tyrrell, lake, Austl...G4 51
Tyrrhenian, sea, It...D3 21
Tyson, Windsor, Vt., 75...E3 120
Ty Ty, Tift, Ga., 461...E3 87
Tyukalinsk, Sov. Un...B8 29
Tyuleniy, isl., Sov. Un...B12 37
Tyumen, Sov. Un., 168,000..B7 29
Tyvan, Sask., Can., 128...G4 70
Tzekung, China, 291,300...F5 34
Tzekwei, China, 18,000...E7 34
Tzuli, China...J4 36
Tzuyuan, China...K4 36
Tzuyang, China, 25,000...G7 36
Tzuyang, China, 4,000...H3 36

U

U, cape, Pomape...52
Uardere, Eth...D6 47
Uatumã, riv., Braz...C3 59
Uaupés, Braz., 571...D4 60
Uaupês, riv., Braz...C4 60
Ubá, Braz., 21,768...C4, g6 56
Ubaíra, Braz., 2,352...D3 57
Ubaitaba, Braz., 2,581...D3 51
Ubangi, riv., Afr...F7 42
Ubaye, riv., Fr...E2 18
Ube, Jap., 166,632 (★221,824)...J5 37
Úbeda, Sp., 28,956...C4 20
Uberaba, Braz., 72,053...E1 57
Uberlândia, Braz., 70,719...E1 57
Überlingen, Ger., 10,500...E4 16
Ubiaja, Nig., 6,034...E6 45
Ubly, Huron, Mich., 819...E8 98
Ubombo, S. Afr., 273...C5 49
Uborka, Sov. Un...D7 37
Ubort, riv., Sov. Un...F6 27
Ubrique, Sp., 9,669...D3 20
Ucayali, riv., Peru...C3 58
Uccen Jargga, mtn., Nor...C8 25
Uccle, Bel., 71,725...D4 15
Uch-Aral, Sov. Un., 3,700..D10 29
Uchisa, Peru, 259...C2 58
Uchiura, bay, Jap...E10 37
Ücker, riv., Ger...B6 16
Ucluelet, B.C., Can., 782...E5 68
Ucon, Bonneville, Idaho, 532...F7 89
Ucross, Sheridan, Wyo., 25..A6 125
Udaipur, India, 111,139...D5 38, E4 40
Udall, Cowley, Kans., 600...E6 93
Udall, Ozark, Mo., 50...E5 101
Udaquiola, Arg...B5 54
Uddevalla, Swe., 34,700...H4 25
Uden, Neth., 7,200...C5 15
Udgir, India, 18,814...H6 40
Udhampur, India, 10,263...B6 39
Udine, It., 86,188...C9 19, A4 21
Udon Thani, Thai., 30,407..D5 38
Udot, isl., Truk...52
Uebi Scebeli, dist., Som...E6 47
Ueckermünde, Ger., 11,900..B7 16
Ueda, Jap., 44,900...H9, m17 37
Uehling, Dodge, Nebr., 231..C9 103
Uélé, riv., Con. L...A3 48
Uelen, Sov. Un., 800...C22 28
Uelkal, Sov. Un., 800...C21 28
Uelzen, Ger., 25,000...B5 16
Ueno, Jap., 34,200...o15 37
Uesslingen, Switz., 1,015...A6 19
Uetendorf, Switz., 2,810...C4 19
Uetersen, Ger., 16,000...E3 24
Ufa, Sov. Un., 588,000...C5 29
Ufa, riv., Sov. Un...D20 9
Uffenheim, Ger., 4,100...D5 17
Ugab, riv., S.W. Afr...B2 49
Ugalla, riv., Tan...C5 48
Uganda; country, Afr., 6,538,031...F9 42, A5 48
Ugashik, Alsk., 36...D8 79
Ugashik, lakes, Alsk...D8 79
Ugie, riv., Scot...C7 13
Ugines, Fr., 6,389...D2 18
Uglegorsk (Esutoru), Sov. Un., 20,000...B11 37
Uglezavodsk, Sov. Un...C11 37
Uglich, Sov. Un., 25,600...B1 29
Ugra, riv., Sov. Un...D10 27
Ugurchin, Bul., 8,035...D7 22
Uh, riv., Czech., Sov. Un...D7 26
Uherské Hradiště, Czech., 12,800...D4 26
Uhlava, riv., Czech...D8 17
Uhrichsville, Tuscarawas, Ohio, 6,201...B6 111
Uhrineves, Czech., 4,581...n18 26
Uig, Scot...B2 13
Uig, Scot...C2 13
Uíge, dist., Ang., 310,258...C2 48
Uijec, isl., Truk...52
Uiju (Gishu), Kor., 27,378...C9 34
Uil, Sov. Un., 2,500...D4 29
Uil, riv., Sov. Un...F20 9
Uinamarca, lake, Bol., Peru...C2 55
Uinta, co., Wyo., 7,484...D2 125
Uinta, riv., Utah...C5 119
Uintah, co., Utah, 11,852...D6 119
Uintah, mts., Utah...C5 119
Uintah and Ouray, Indian res., Utah...C5 119
Uitenhage, S. Afr., 48,755...D4 49
Uithuizen, Neth., 3,353...A6 15
Ujdomai, Okinawa...52
Ujelang, isl., Marshall Is...F9 7
Uji, isl., Jap...K4 37
Ujiji, Tan., 12,011...B4 48
Ujiyamada, Jap., 99,026.I8, o15 37
Ujjain, India, 144,161...D6 39, F5 40
Uka, Okinawa...52
Uka, Sov. Un...D19 28
Ukerewe, isl., Tan...B5 48
Ukhta, Sov. Un., 15,000...C19 9
Ukhta, Sov. Un...E14 25
Ukiah, Mendocino, Calif., 9,900...C2 82
Ukiah, Umatilla, Oreg., 200.B8 113
Ukmerge, Sov. Un., 12,292..D5 27
Ukraine (S.S.R.), rep., Sov. Un., 43,091,000...F15 8
Ulaan Goom, Mong., 5,000..B3 34
Ulalu, isl., Truk...52
Ulan Bator (Urga), Mong., 70,000...B6 34
Ulan-Ude, Sov. Un., 188,000...D13 28
Ulba, Sov. Un., 10,000...E25 9
Ulchin, Kor., 10,700...H4 37
Ulcinj, Yugo., 5,639...E4 22
Uldza, riv., Mong...B7 34
Ulen, Boone, Ind., 130...D5 91
Ulen, Clay, Minn., 481...C2 99
Ulfborg, Den., 1,174...B2 24
Ulhasnagar, India, 107,760.*H4 40
Ulifauro, pass, Truk...52
Ulindi, riv., Con. L...B4 48
Ulithi, is., Pac. O...F8 7
Ulla, Sov. Un., 10,000...D7 27
Ulla, riv., Sp...A1 20
Ulladulla, Austl., 1,458...G8 51
Ullapool, Scot...C3 13
Ullared, Swe., 418...A6 24
Ullin, Pulaski, Ill., 577...F4 90
Ullswater, lake, Eng...F6 13
Ullŭng, isl., Kor...H5 37
Ulm, Prairie, Ark., 140...C4 81
Ulm, Ger., 92,700 (★137,000)...E4 17
Ulm, Cascade, Mont., 75...C5 102
Ulm, Sheridan, Wyo., 25...A6 125
Ulman, Miller, Mo., 100...C5 101
Ulmer, mtn., Ant...B4 5
Ulmers, Allendale, S.C., 168.E5 115
Ulricehamn, Swe., 8,100...I5 25
Ulsan, Kor., 18,100...I4 37
Ülsen, Ger., 2,089...B6 15
Ulster, Bradford, Pa., 400...C8 114
Ulster, co., N.Y., 118,804...D6 108
Ulster, prov., Ire., 217,524...B3 11
Ulster, canal, Ire...C4 11
Ulu Dag, mtn., Tur...B7 23
Ulugh Muztagh, mtn., China...A8 39
Ulva, isl., Scot...D2 13
Ulverston, Eng., 10,515...C5 10
Ulverstone, Austl., 5,962...o15 50
Ulyanovka, Sov. Un., 5,000...s31 25
Ulyanovsk, Sov. Un., 226,000...C3 29
Ulysses, Grant, Kans., 3,157.E2 93
Ulysses, Butler, Nebr., 357...C8 103
Ulysses (Lewisville), Potter, Pa., 590...C6 114
Ulzburg, Ger., 2,700...E3 24
Uman, Sov. Un., 63,000...G8 27
Uman, isl., Truk...52
Umarkot, Pak., 5,142...E2 40
Umatac, bay, Guam...52
Umatilla, Lake, Fla., 1,717..D5 86
Umatilla, Umatilla, Oreg., 617...B7 113
Umatilla, co., Oreg., 44,352..B8 113
Umatilla, riv., Oreg...B7 113
Umbagog, lake, N.H...A4, B2 105
Umbria, pol. dist., It., 794,745...*C4 21
Umbria, reg., It...C4 21
Umbuzeiro, Braz., 1,223...C3, h6 57
Umcolcus, lake, Maine...B4 96
Umeå, Swe., 23,200 (★35,000)...F9 25
Umeälven, riv., Swe...E8 25
Umhausen, Aus., 1,834...B6 18
Umiat, Alsk...B9 79
Umm al Maradim, isl., Arabia..G4 41
Umm el Fahm, Isr., 4,861...B7 32
Umm el Nabayil, cape, Eg., U.A.R...C2 32
Umm Keddada, Sud...C2 47
Umm Khunān, Eg., U.A.R..E3 32
Umm Ruwāba, Sud., 7,805..C3 47
Umnak, isl., Alsk...E6 79
Umpire, Howard, Ark., 64...C1 81
Umpqua, Douglas, Oreg., 20...D3 113
Umpqua, riv., Oreg...D3 113
Um Ruaba, Sud., 7,805...E9 42
Umsaskis, lake, Maine...B3 96
Umtali, S. Rh., 36,200...A5 49
Umtata, S. Afr., 9,185...D4 49
Umuahia, Nig...E6 45
Umvuma, S. Rh., 600...A5 49
Umzinto, S. Afr., 4,095...D5 49
Una, Spartanburg, S.C., 1,500...*B4 115
Una, Davidson, Tenn., 100..E9 117
Una, riv., Braz...k6 57
Una, riv., Yugo...C2 22
Unadilla, Dooly, Ga., 1,304..D3 87
Unadilla, Otoe, Nebr., 254...F2 103
Unadilla, Otsego, N.Y., 1,586...C5 108
Unadilla, riv., N.Y...C5 108
Unalakleet, Alsk., 574...C7 79
Unalaska, Alsk., 218...E6 79
Unalaska, isl., Alsk...E6 79
Unango, Moz...D6 48
Uncas, Kay, Okla., 100...A5 114
Uncasville, New London, Conn., 1,381...D8 84
Uncía, Bol., 4,507...C2 55
Uncompahgre, mts., Colo...C3 83
Uncompahgre, peak, Colo...C3 83
Uncompahgre, plat., Colo...C2 83
Uncompahgre, riv., Colo...C3 83
Underhill, Chittenden, Vt., 225 (730▲)...B3 120
Underhill, Oconto, Wis., 80.D5 124
Underhill Center, Chittenden, Vt., 150...B3 120
Underwood, Shelby, Ala., 250...B3 78
Underwood, Clark, Ind., 200.G6 91
Underwood, Pottawattamie, Iowa, 337...C2 92
Underwood, Otter Trail, Minn., 314...D3 99
Underwood, McLean, N. Dak., 819...B4 110
Underwood, Skamania, Wash., 350...D4 122
Undu, cape, Fiji Is...52
Unecha, Sov. Un., 10,000...E9 27
Unezhma, Sov. Un...F17 25
Unga, Alsk., 43...D7 79
Ungarie, Austl., 367...F6 51
Ungava, bay, Que., Can...A3 73
Unger, Choctaw, Okla...C6 112
Unggi, Kor., 20,882...E5 37
Unhost, Czech., 3,063...n17 26
União, Braz., 4,296...B2 57
União da Vitória, Braz., 15,822...D2 56
União dos Palmares, Braz., 10,406...C3, k6 57
Unicoi, Unicoi, Tenn., 500.C11 117
Unicoi, co., Tenn., 15,082..C11 117
Unicoi, mts., N.C...D2 109
Unicoi, mts., Tenn...D9 117
Unije, isl., Yugo...C2 22
Unimak, isl., Alsk...E7 79
Unimak, pass, Alsk...E6 79
Union, Fulton, Ark., 50...A4 81
Union, McHenry, Ill., 480...A5 90
Union, Pike, Ind., 150...H3 91
Union, Hardin, Iowa, 534...B4 92
Union, Boone, Ky., 135...A7 94

Union, St. James, La., 640......B5, D5 95
Union, Knox, Maine, 300 (1,196*)......D3 96
Union, Newton and Neshoba, Miss., 1,726......C4 100
Union, Franklin, Mo., 3,937.C6 101
Union, Cass, Nebr., 303 D10, E3 103
Union, Carroll, N.H., 300..D4 105
Union, N.J., 51,499..B4 106
Union, Hertford, N.C., 306..A6 109
Union, Cavalier, N. Dak., 25.A8 110
Union, Montgomery, Ohio, 1,072......*C3 111
Union, Canadian, Okla., 329.B4 112
Union, Union, Oreg., 1,490..B9 113
Union, Union, S.C., 10,191..B4 115
Union, Mason, Wash., 500..B2 122
Union, Monroe, Wis., 411......D4 123
Union, co., Ark., 49,518....D3 81
Union, co., Fla., 6,043......B4 86
Union, co., Ga., 6,510......B2 87
Union, co., Ill., 17,645......F4 90
Union, co., Ind., 6,457......E8 91
Union, co., Iowa, 13,712....C8 92
Union, co., Ky., 14,537....C2 94
Union, co., Miss., 18,904....A4 100
Union, co., N.J., 504,255..B4 106
Union, co., N. Mex., 6,068..A6 107
Union, co., N.C., 44,670....B3 109
Union, co., Ohio, 22,853....B4 111
Union, co., Oreg., 18,180..B8 113
Union, co., Pa., 25,646....E7 114
Union, co., S.C., 30,015....B4 115
Union, co., S. Dak., 10,197..E9 116
Union, co., Tenn., 8,498....C10 117
Union, par., La., 17,624....B3 95
Union, lake, N.J.........E2 106
Union, riv., Maine.......A4 96
Union Bay, B.C., Can., 600..E5 68
Union Beach, Monmouth, N.J., 5,862......C4 106
Union Bridge, Carroll, Md., 833........A3 85
Union Center, Meade, S. Dak., 35......C3 116
Union Center, Juneau, Wis., 252........E3 124
Union City, Alameda, Calif., 6,618......B5 82
Union City, New Haven, Conn. (part of Naugatuck).C4 84
Union City, Fulton, Ga., 2,118......B4, C2 87
Union City, Randolph, Ind., 4,047........D8 91
Union City, Branch and Calhoun, Mich., 1,669..F5 98
Union City, Hudson, N.J., 52,180........D5 106
Union City, Darke, Ohio, 1,657......*B3 111
Union City, Erie, Pa., 3,819.C2 114
Union City, Obion, Tenn., 8,837........A2 117
Union Creek, Jackson, Oreg., 25........E4 113
Uniondale, Wells, Ind., 311..C7 91
Uniondale, Nassau, N.Y., 20,041......*G2 84
Union Dale, Susquehanna, Pa., 287.........C11 114
Unión de Reyes, Cuba, 5,503........B4, C2 87
Unión de Tula, Mex., 5,584.m11 63
Union Flat, creek, Wash...C8 122
Union Furnace, Hocking, Ohio, 300.......C5 111
Union Gap, Yakima, Wash., 2,100.......C5 122
Union Grove, Delaware, N.Y........C6 108
Union Grove, Racine, Wis., 1,970......F1, F5 124
Union Hall, Franklin, Va., 50........D3 121
Unionhill, Independence, Ark., 25.......B4 81
Union Lake, Oakland, Mich., 2,000......*F7 98
Union Mills, LaPorte, Ind., 450........B4 91
Union Mills, Carroll, Md., 60.........A3 85
Union of South Africa, see South Africa, country, Afr.
Union of Soviet Socialist Republics, country, Eur., Asia, 216,151,000....26, D11 33
Union Park, Orange, Fla., 1,000......*D5 86
Union Pier, Berrien, Mich., 900.........G4 98
Union Point, Man., Can., 100.........E3 71
Union Point, Greene, Ga., 1,615......C3 87
Union Springs, Bullock, Ala., 3,704......C4 78
Union Springs, Cayuga, N.Y., 1,066......C4 108
Union Star, DeKalb, Mo., 392.......B3 101
Uniontown, Perry, Ala., 1,993......C2 78
Uniontown, Perry, Ind., 20..H5 91
Uniontown, Bourbon, Kans., 211.......E9 93
Uniontown, Union, Ky., 1,255......C2 94
Uniontown, Carroll, Md., 260.......A3 96
Uniontown, Perry, Mo., 125.D8 101
Uniontown, Stark, Ohio, 1,668......B6 111
Uniontown, Fayette, Pa., 17,942......G2 114
Uniontown, Whitman, Wash., 242........C8 122
Union Village, Providence, R.I., 1,500......B10 84
Union Village, Orange and Windsor, Vt., 75......D4 120
Unionville, Ont., Can., 945.......D5, E7 72
Unionville, Hartford, Conn., 2,246......D5 84
Unionville, Bibb, Ga., 1,000.*D3 87
Unionville, Tift, Ga., 1,607.*E3 87
Unionville, Whiteside, Ill., 100.......D3 90
Unionville, Appanoose, Iowa, 185.......D5 92
Unionville, Washington, Maine.......D5 96
Unionville, Norfolk, Mass., 125.....B4, D1 97
Unionville, Tuscola, Mich., 629........E7 98
Unionville, Putnam, Mo., 1,896......A4 101
Unionville, Ashtabula and Lake, Ohio, 500......A7 111
Unionville, Bedford, Tenn., 100......C5 117
Unionville, Orange, Va., 250.C5 121
Unionville Center, Union, Ohio, 305......B4 111
Unisan, Phil., 1,890......p13 35
United, Westmoreland, Pa., 2,044......F3 114
United Arab Republic (Egypt), country, Afr., 26,059,000 ......H6 31, D5 43
United Kingdom of Great Britain and Northern Ireland, country, Eur., 52,673,221.......E9 8,10
United Nations Headquarters, N.Y........D6 106
United Provinces, see Uttar Pradesh, state, India
United Pueblos, Indian res., N. Mex........A3 107
United States, country, N.A., 179,325,100.......F11 61, 76
U.S. Air Force Academy, El Paso, Colo., 5,000.....C6 83
U.S. Naval Ammunition Depot, Mineral, Nev.......E3 104
Unity, Sask., Can., 1,902...E1 70
Unity, Alexander, Ill., 110...F4 90
Unity, Waldo, Maine, 400 (983*).......D3 96
Unity, Sullivan, N.H., 50 (708*)......D2 105
Unity, Baker, Oreg., 150....C8 113
Unity, Marathon and Clark, Wis., 386......D3 124
Unity, dam, Oreg........C8 113
Unity, pond, Maine......D4 96
Unityville, McCook, S. Dak., 70.......D8 116
Universal, Vermillion, Ind., 424.......E3 91
Universales, mts., Sp......E3 14
University, Tuscaloosa, Ala. (part of Tuscaloosa)......B2 78
University, Lafayette, Miss., 3,597......A4 100
University City, St. Louis, Mo., 51,249.....A8, C7 101
University Gardens, Prince Georges, Md., 1,000....*C4 85
University Heights, Cuyahoga, Ohio, 16,641......B2 111
University Hills, Prince Georges, Md., 1,700......*C4 85
University Park, Mahaska, Iowa, 569......C5 92
University Park, Prince Georges, Md., 3,098......*C4 85
University Park, Dona Ana, N. Mex., 2,400.....*E3 107
University Park, Dallas, Tex., 23,202......B5 118
University View, Franklin, Ohio, 1,000......*C2 111
Unlingen, Ger., 900......A5 18
Unna, Ger., 31,500......B2 17
Unnao, India......B4 38
Unnen, lake, Swe......B7 24
Unsan, Kor........G2 37
Unsernherrn, Ger., 5,900...E6 17
Unst, isl., Scot........g10 10
Unstrut, riv., Ger........B5 17
Unterwalden, canton, Switz., 45,323......C5 19
Unterwasser, Switz......B7 19
Unuk, riv., B.C., Can......A2 68
Unuwhao, mtn., N.Z......K14 51
Unwin, Sask., Can., 60....E1 70
Unye, Tur., 11,400......B11 31
Unzha, riv., Sov. Un......D17 9
Uondo, Eth........D4 47
Uorra Ilu, Eth........C4 47
Upalco, Duchesne, Utah, 10.C5 119
Upemba, lake, Con. L......C4 48
Upernavik, Grnld., 321....B20 4
Upham, McHenry, N. Dak., 333.......A5 110
Upia, riv., Col........C3 60
Upington, S. Afr., 20,366...C3 49
Upland, San Bernardino, Calif., 15,918......E5, F3 82
Upland, Grant, Ind., 1,999..D7 91
Upland, Franklin, Nebr., 237.D7 103
Upland, Delaware, Pa., 4,343......*G10 114
Upolu, isl., W. Sam......52
Upolu, pt., Haw......C6 88
Upper, reg., Ghana......D4 45
Upper, bay, N.J......E5 106
Upper Ammonoosuc, riv., N.H..A4 105
Uproyan, mts., P.R......C5 65
Urao, Col., 5,958......B2 60
Upper Anton Chico, Guadalupe, N. Mex., 120......B4 107
Upper Arlington, Franklin, Ohio, 28,486......C2, C4 111
Upper Arrow, lake, B.C., Can..D9 68
Upper Black Eddy, Bucks, Pa., 400......E11 114
Upper Blackville, N.B., Can., 438......C4 74
Upper Brookville, Nassau, N.Y., 1,045......*D3 108
Upperco, Baltimore, Md., 150......A4 85
Upper Darby, Delaware, Pa., 44,000......B11, G11 114
Upper Erne, lake, N. Ire......C4 11
Upper Fairmount, Somerset, Md., 500......D6 85
Upper Falls, Baltimore, Md., 160......A5 85
Upper Frenchville, Aroostook, Maine, 275......D3 96
Upper Gagetown, N.B., Can., 244......D3 74
Upper Ganges, canal, India......C6 40
Upper Giuba, dist., Som......D5 47
Upper Gloucester, Cumberland, Maine, 150......D5, E2 96
Upper Humber, riv., Newf., Can......D3 75
Upper Indian Pond, lake, Newf., Can........D2 75
Upper Iowa, riv., Iowa......A6 92
Upper Island Cove, Newf., Can., 1,669......E5 75
Upper Jay, Essex, N.Y., 130.B3 108
Upper Kapuas, mts., Indon., Mala........E4 34
Upper Kent, N.B., Can., 174.C2 76
Upper Klamath, lake, Oreg..E5 113

Upper Lake, Lake, Calif., 400........C2 82
Upper Marlboro, Prince Georges, Md., 673......C4 85
Upper Musquodoboit, N.S., Can., 328......D7 74
Upper Nile, reg., Sud., 888,611......D3 47
Upper Nyack, Rockland, N.Y., 1,833......*D7 108
Upper Red, lake, Minn......B4 99
Upper Sackville, N.B., Can., 254......D5 74
Upper Saddle River, Bergen, N.J., 3,570......A4 106
Upper Sandusky, Wyandot, Ohio, 4,941......B4 111
Upper Seal, lake, Que., Can..E18 67
Upper Silesia, reg., Pol......g9 26
Upper Strasburg, Franklin, Pa., 75......F6 114
Upper Tract, Pendleton, W. Va., 75......C5 123
Upperville, Fauquier, Va., 400......C5 121
Upper Volta, country, Afr., 3,567,000......E5 42, D4 45
Upper Wilson, pond, Maine..C3 96
Upper Wood Harbour, N.S., Can., 130......F4 74
Uppsala, Swe., 79,300...H7, t35 25
Uppsala, co., Swe., 170,000.*H7 25
Upsala, Morrison, Minn., 356......C4 99
Upsalquitch, N.B., Can., 165.B3 69
Upshur, co., Tex., 19,793....C5 118
Upshur, co., W. Va., 18,292.C4 123
Upson, co., Ga., 23,800....D2 87
Upson, Iron, Wis., 150....B3 124
Uptanum, ridge, Wash......C5 122
Uptergrove, Ont., Can., 90..C5 72
Upton, Que., Can., 830....D5 73
Upton, Hardin and Larue, Ky., 547......C4 94
Upton, Oxford, Maine......D1 96
Upton, Worcester, Mass., 1,000......B4, D1 97
Upton, Texas, Mo., 56....D5 101
Upton, Summit, Utah......C4 119
Upton, Weston, Wyo., 1,224.A8 125
Upton, co., Tex., 6,239....D 118
Urabá, gulf, Col........B2 60
Uracas, isl., Mariana Is......E8 7
Urakawa, Jap., 12,300....E11 37
Ural, Lincoln, Mont., 5....B1 102
Ural, riv., Sov. Un......E20 9
Ural, mts., Sov. Un......D11 9
Uralsk, Sov. Un., 111,000...C4 9
Urandi, Braz., 1,497......D2 57
Urania, La Salle, La., 1,063..C3 95
Uranium City, Sask., Can., 1,665......A1 70
Urawa, Jap., 168,757..I9, n18 37
Urbana, Union, Ark., 400...D3 81
Urbana, Champaign, Ill., 27,294......C5 90
Urbana, Wabash, Ind., 350..C6 91
Urbana, Benton, Iowa, 544..B6 92
Urbana, Frederick, Md., 100.B3 85
Urbana, Dallas, Mo., 348...D4 101
Urbana, Champaign, Ohio, 10,461......B4 111
Urbancrest, Franklin, Ohio, 1,029......C2 111
Urbandale, Polk, Iowa, 5,821......A7, C4 92
Urbank, Otter Tail, Minn., 177......D3 99
Urbanna, Middlesex, Va., 512......D6 121
Urbino, It., 6,500 (18,874*).C4 21
Urcos, Peru, 2,096......D3 58
Urdaneta, Phil., 4,474....o13 35
Urdzhar, Sov. Un., 3,500..D10 29
Ure, riv., Eng........F7 13
Ures, Mex., 3,456......B5 50
Urfa, Tur., 59,900......D12 31
Urga, Sov. Un........E6 29
Urgench, Sov. Un., 49,000..E6 29
Urgun, Afg., 5,000......E14 41
Urgut, Sov. Un........B6 37
Uri, canton, Switz., 32,021..C6 19
Uriah, Monroe, Ala., 800....D2 78
Uribia, Col., 1,101......A3 60
Urich, Henry, Mo., 408....C4 101
Urique, Mex., 256......B3 63
Urisk, Sov. Un......s31 25
Urla, Tur., 10,800......C6 23
Urlingford, Ire., 562......E4 11
Urmi, riv., Sov. Un......B6 37
Urmia, see Rezā'īyeh, Iran
Urmia, salt lake, Iran......C2 41
Uroyan, mts., P.R......C3 65
Urrao, Col., 5,958......B2 60
Ursa, Adams, Ill., 325....C2 90
Ursatyevskaya, Sov. Un., 12,300......E7 29
Ursina, Somerset, Pa., 313..G3 114
Ursine, Lincoln, Nev., 60....F7 104
Uruapan, Mex., 45,580.D4, n13 63
Urubamba, Peru, 3,481....D3 58
Urubambá, riv., Peru......D3 58
Urubici, riv., Braz......C3 59
Urubú, riv., Braz......C3 59
Urucará, Braz., 1,203......C3 59
Uruçuí, Braz., 2,253......C2 57
Urucuia, riv., Braz......E1 57
Uruguai, riv., Braz......D2 56
Uruguaiana, Braz., 48,358..D1 56
Uruguay, country, S.A., 2,163,000......G5 53, E1 56
Uruguay, riv., Arg., Ur......A5 54
Urukthapel, isl., Palau Is......52
Urumchi (Tihwa), China......B2 34
Urungu, riv., China......B2 34
Urusha, Sov. Un., 10,000..D15 28
Uryupinsk, Sov. Un., 29,900.C22 9
Urzhum, Sov. Un., 11,200...B3 29
Urziceni, Rom., 6,061....C3 22
Usa, riv., Sov. Un......B20 9
Usakos, S.W. Afr., 2,355...B2 49
Usedom, Ger., 2,562......E8 17
Usedom, isl., Ger......E8 24
Usher, Levy, Fla., 70......D7 86
Ushiro, see Orlovo, Sov. Un.
Ush-Tobe, Sov. Un., 16,300.D9 29
Ushturinan Kuh, mtn., Iran..E4 41

Ushuaia, Arg., 1,950......h12 54
Usingen, Ger., 3,800......C3 17
Usk, B.C., Can......A5 68
Usk, Pend Oreille, Wash., 300......A8 122
Uskudar, Tur., 60,722......B7 23, B7 31
Uslar, Ger., 6,400......C4 17
Uslava, riv., Czech......D8 17
Usman, Sov. Un., 50,000...E12 27
Usolye-Sibirskoye, Sov. Un., 50,000......D13 28
Ussuri, riv., China, Sov. Un...C7 37
Ussuriysk, Sov. Un......C7 37
Ust, riv., China, Sov. Un......C7 37
Ussö, Nor., 3,000......B13 25
Ust-Aldan, Sov. Un......C15 28
Ust-Bolsheretsk, Sov. Un...D18 28
Uster, Switz., 17,252......B6 19
Ust-Ishim, Sov. Un......B8 29
Ustka, Pol., 2,807......A4 26
Ust-Kamchatsk, Sov. Un...D19 28
Ust-Kamenogorsk, Sov. Un., 173,000......C10 29
Ust-Kut, Sov. Un., 21,900.D13 28
Ust-Maya, Sov. Un., 2,300.C16 28
Ust-Olenek, Sov. Un......B14 28
Uspon, co., Ga., 23,800....D2 87
Ust-Srednikan, Sov. Un., 800......C18 28
Ust-Tsilma, Sov. Un., 7,900.B19 9
Ust-Tym, Sov. Un......B9 29
Ust-Tyrma, Sov. Un......A5 37
Ust-Urt, plat., Sov. Un......G20 9
Ust-Usa, Sov. Un., 2,500..B20 9
Ustyuzhna, Sov. Un., 1,237......C7 108
Usta Barrette, Que., Can., 557......C2 73
Usulután, Sal., 12,094....D3 62
Usumbura, Burundi, 50,000......B4 48
Usuti, co., Utah, 106,991...C4 119
Utah, state, U.S., 890,627......C5 76, 119
Utah, lake, Utah......C3 119
Utajärvi, Fin., 700......E12 25
Ute, Monona, Iowa, 511....B2 92
Ute, creek, N. Mex......A6 107
Ute Mountain, Indian res., Colo..D2 83
Utete, Tan., 970......C6 48
Utica (North Utica), La Salle, Ill., 1,014......B4 90
Utica, Clark, Ind., 800....H6 91
Utica, Ness, Kans., 322....D3 93
Utica, Daviess, Ky., 300....C2 94
Utica, Macomb, Mich., 1,454......A8, F7 98
Utica, Hinds, Miss., 764....C3 100
Utica, Livingston, Mo., 450..B4 101
Utica, Judith Basin, Mont., 45......D6 102
Utica, Seward, Nebr., 564...D8 103
Utica, Oneida, N.Y., 100,410......B5 108
Utica, Licking, Ohio, 1,854..B5 111
Utica, Bryan, Okla., 100...C5 112
Utica, Venango, Pa., 274...D2 114
Utica, Oconee, S.C., 1,294..*B1 115
Utica, Yankton, S. Dak., 70.E8 116
Utica Heights, Macomb, Mich., 2,000......*F8 98
Utiel, Sp., 10,076......C5 20
Utik, lake, Man. Can......B4 71
Utikuma, lake, Alta., Can...B3 69
Utleyville, Baca, Colo., 5...D7 83
Uto, isl., Swe......u36 25
Utopia, Uvalde, Tex., 500..E3 118
Utopia, lake, N.B., Can......D3 74
Utrata, riv., Pol......m13 26
Utrecht, Neth., 256,300 (*395,000)......B5 15
Utrecht, prov., Neth., 686,600......B5 15
Utrera, Sp., 35,200......D3 20
Utsjoki, Fin., 1,200......C12 25
Utsunomiya, Jap., 193,000 (239,007*)......H9 37
Uttaradit, Thai., 9,372....D4 38
Uttar Pradesh, state, India, 73,746,401......C6 39
Utterson, Ont., Can., 184..B5 72
Uttoxeter, Eng., 8,168....I6 13
Utuado, P.R., 9,870......A6 65
Utuado, mun., P.R., 40,449.A6 65
Utulei, Am. Sam., 719....52
Utvade, Uvalde, Tex., 10,293......E3 118
Uvalde, co., Tex., 16,814...E3 118
Uvaly, Czech., 4,706......n18 26
Uvarovo, Sov. Un., 10,000.F14 27
Uvat, Sov. Un........B7 29
Uverite, isl., Fiji Is......52
Uvinza, Tan., 1,880......C5 48
Uvira, Con. L., 70,700....B4 48
Uwajima, Jap., 49,500....J6 37
Uxbridge, Ont., Can., 2,316.C5 72
Uxbridge, Eng., 63,762....C7 12
Uxbridge, Worcester, Mass., 3,377......B4 97
Uyak, Alsk., 11......D9 79
Uyar, Sov. Un., 10,000...D12 28
Uyuni, Bol......D2 55
Uyuni, salt flat, Bol......D2 55
Uzbek S.S.R., rep., Sov. Un., 8,665,000......G11 9
Uzès, Fr., 4,390......E6 14
Uzgorod, Sov. Un., 52,000.G4 27
Uzh, riv., Sov. Un......F7 27
Uzin, riv., China......B2 34
Uzlovaya, Sov. Un., 54,000.E16 9
Uzunkopru, Tur., 18,300...B6 23

Vac, Hung., 24,748......B4 22
Vaca, key, Fla......H5 86
Vaca, mtn., Calif......A5 82
Vaca, pt., P.R......f13, g12 65
Vacaville, Solano, Calif., 60,877......B5, C3 82
Vaccina, riv., It......h8 21
Vaclava, riv., Czech......D3 17
Vacha, Ger., 4,383......C5 17
Vacherie, St. James, La., 950......B6 95
Vacia Talega, pt., P.R......B7 65
Vader, Lewis, Wash., 380...C3 122
Vadis, Lewis, W. Va., 2,600.D8 29
Vadito, Taos, N. Mex., 10...A4 107
Vadnais Heights, Ramsey, Minn., 2,459......*F5 99
Vado, Dona Ana, N. Mex......E3 107
Vadsö, Nor., 3,000......B13 25
Vaduz, Liech., 3,398......B5 18
Vagay, Sov. Un......B7 29
Vagnhärad, Swe., 857....u35 25
Vagos, Port., 2,180......B1 20
Vah, riv., Czech......D4 26
Vaigai, riv., India......G6 39
Vaigalu, Sam., 802......52
Vaihingen [an der Enz], Ger., 8,000......E3 17
Vail, Pima, Ariz., 150......E5 80
Vail, Crawford, Iowa, 473...B2 92
Vail, lake, Scot......D4 13
Vail City, Pike, Ill., 109...D3 90
Vail City, Barnes, N. Dak., 7,809......C7 110
Vail Homes, Monmouth, N.J., 1,204......*C4 106
Vails, Warren, N.J., 50....C2 106
Vaitele, bay, W. Sam......52
Vakfikebir (Buyukliman), Tur., 1,341......B12 31
Vakh, riv., Sov. Un......C25 9
Vakhrushev, Sov. Un......B11 37
Valais, canton, Switz., 177,783......D3 19
Valatie, Columbia, N.Y., 1,237......C7 108
Valcartier Village, Que., Can., 163......C6, C8 73
Valcheta, Arg......C3 54
Valdagno, It., 23,000....D7 18
Val David, Que., Can., 1,118.C3 73
Valday, Sov. Un., 14,400...C9 27
Valday, hills, Sov. Un......C9 27
Valdemarsvik, Swe., 3,200..H7 25
Valdemorillo, Sp., 1,470.D6,p17 20
Valdepeñas, Sp., 25,706...C4 20
Valders, Manitowoc, Wis., 622......B6, D6 124
Valdese, Burke, N.C., 2,941..B2 109
Valdez, Alsk., 555...C10, g18 79
Valdez, Las Animas, Colo., 400......D6 83
Valdilecha, Sp., 1,539....p18 20
Valdivia, Chile, 50,000....B2 54
Valdivia, Col., 1,169......B2 60
Valdivia, prov., Chile, 255,109......B2 54
Valdobbiadene, It., 11,430..D7 18
Val d'Or, Que., Can., 10,983......E9 72
Valdosta, Lowndes, Ga., 30,652......F3 87
Vale, Malheur, Oreg., 1,491.D9 113
Vale, Butte, S. Dak., 108...C2 116
Vale, Carroll, Tenn., 125...A3 117
Valeene, Orange, Ind., 70...H5 91
Valemount, B.C., Can., 631..C8 68
Valença, Braz., 17,137....D3 57
Valença, Braz., 2,825......A1 20
Valença do Piauí, Braz., 3,046......C2 57
Valence, Fr., 52,532......E6 14
Valencia, Venez., 161,413..A4 60
Valencia, co., N. Mex., 39,085......C1 107
Valencia, prov., Sp., 1,429,708......*C5 20
Valencia, Sp., 505,066 (*660,000)......C5 20
Valencia, isl., Ire......F1 11
Valencia, Butler, Pa., 310...E2 114
Valencia, Sp., 13,159......C2 20
Valencia, Ven......A4 60
Valencia, co., N. Mex......C1 107
Valenciennes, Fr., 45,379......B5 14
Valentigney, Fr., 11,241....D7 14
Valentine, Mohave, Ariz., 50.B2 80
Valentine, Pulaski, Ark......D6 81
Valentine, Cherry, Nebr., 2,875......B5 103
Valentine, Jeff Davis, Tex., 420......F2 118
Valera, Ven., 44,566......B3 60
Valhalla, Westchester, N.Y., 2,000......D7 108
Valhalla, mts., B.C., Can......D9 68
Valhermosa Springs, Morgan, Ala., 500......A3 78
Valier, Franklin, Ill., 649...E4 90
Valier, Pondera, Mont., 724.B4 102
Valier, Jefferson, Pa., 800...E3 114
Valjevo, Yugo., 22,070....C4 22
Valka, Sov. Un., 25,000....C6 27
Valkeakoski, Fin., 14,200..G11 25
Valkenswaard, Neth., 19,300.C3 15
Valki, Sov. Un., 10,000...G10 27
Valladolid, Ec......B2 60
Valladolid, Mex., 9,306....C7 63
Valladolid, Sp., 151,807...B3 20
Valladolid, prov., Sp., 363,106......*B3 20
Vallauris, Fr., 4,337......F7 14
Valldemosa, Sp......C1 107
Vallecito, La Plata, Colo....D3 83
Vallecito, res., Colo......D3 83
Vallecitos, Rio Arriba, N. Mex., 250......A3 107
Vallecitos, Sandoval, N. Mex........*F5 107
Valle d'Aosta, It., pol. dist., 100,959......*B1 21
Valle de Bravo, Mex., 4,459.n13 63
Valle de Cauca, dept., Col., 1,396,630......C2 60
Valle de la Pascua, Ven., 24,051......B4 60
Valle de Santiago, Mex., 20,879......m13 63

Valledupar, Col., 9,011....A3 60
Vallee Jonction, Que., 1,405......C7 73
Valle Grande, Bol., 5,094...C3 55
Valle Grande, mts., N. Mex..F5 107
Vallejo, Solano, Calif., 60,877......B5, C3 82
Vallenar, Chile, 9,677......E1 55
Valles Mines, Jefferson, Mo., 225......C7 101
Valletta, Malta, 18,666 (*181,414)......G14 30
Valley, Douglas, Nebr., 1,452......C9, D2 103
Valley, Avery, N.C......A9 109
Valley, Stevens, Wash., 250..A8 122
Valley, co., Idaho, 3,663...E3 89
Valley, co., Mont., 17,080..B10 102
Valley, co., Nebr., 6,590....C6 103
Valley, creek, Ala......C4 81
Valley Bend, Randolph, W. Va., 500......C4 123
Valley Brook, Oklahoma, Okla., 1,378......*B4 112
Valley Center, Sedgwick, Kans., 857......B5, E6 93
Valley Centre, Sask., Can., 52......F2 70
Valley City, Pike, Ill., 109..D3 90
Valley City, Barnes, N. Dak., 7,809......C7 110
Valley City, Medina, Ohio, 400......A6 111
Valley Cottage, Rockland, N.Y., 2,200......D1 108
Valleydale, Los Angeles, Calif., 1,000......*F3 82
Valley Falls, Jefferson, Kans., 1,193......B7, C8 93
Valley Falls, Lake, Oreg......E6 113
Valley Farms, Pinal, Ariz., 200......E4 80
Valleyfield, Newf., Can., 509......D5 75
Valleyfield, Que., Can., 27,297 (*29,849)...D3, E8 73
Valleyford, Spokane, Wash., 100......B8, D8 122
Valley Forge, Chester, Pa., 450......A10 114
Valley Grove, Ohio, W. Va., 548......A4, B2 123
Valley Head, De Kalb, Ala., 424......A4 78
Valley Head, Randolph, W. Va., 600......C4 123
Valley Lee, St. Marys, Md., 300......D4 85
Valley Mills, Marion, Ind., 150......H7 91
Valley Mills, Bosque, Tex., 1,061......D4 118
Valley Park, Issaquena, Miss., 100......C3 100
Valley Park, St. Louis, Mo., 3,452......B7 101
Valley Ranch, San Miguel, N. Mex........F6 107
Valley Spring, Llano, Tex., 60......D3 118
Valley Springs, Boone, Ark......A3 81
Valley Springs, Minnehaha, S. Dak., 472......D9 116
Valley Station, Jefferson, Ky., 10,553......A4 94
Valley Stream, Nassau, N.Y., 38,629......E2 108
Valley view, Alta., Can......B2 69
Valley View, Kane, Ill., 1,741......*A5 91
Valley View, Madison, Ky., 200......C6 94
Valley View, Cuyahoga, Ohio, 1,221......*A6 111
Valley View, Schuylkill, Pa., 1,540......E9 114
Valley View, Cooke, Tex., 500......C4 118
Valliant, McCurtain, Okla., 477......D6 112
Valli di Comacchio, lake, It..E8 18
Vallimanca, riv., Arg......B4 54
Vallo della Lucania, It., 6,863......D5 21
Vallonia, Jackson, Ind., 500......G5 91
Vallorbe, Switz., 3,990....C1 19
Valls, Sp., 11,886......B6 20
Vallscreek, McDowell, W. Va., 500......D3 123
Val Marie, Sask., Can., 443.H2 70
Valmeyer, Monroe, Ill., 709.E3 90
Valmiera, Sov. Un., 10,000.C5 27
Valmont, Que., Can., 230...C5 73
Valmontone, It., 7,314....h9 21
Val Morin, Que., Can., 290.C3 73
Valmy, Humboldt, Nev......C4 104
Valognes, Fr., 3,938......C3 14
Valois, Que., Can., 390....D8 73
Valois, Schuyler, N.Y., 15...C4 108
Valona, see Vlonë, Alb.
Valor, Sask., Can., 65......H2 70
Valparaíso, Chile, 255,000 (*440,000)......A2 54
Valparaiso, Okaloosa, Fla., 5,975......G2 86
Valparaiso, Porter, Ind., 15,227......B3 91
Valparaiso, Mex., 5,083....C4 63
Valparaiso, Saunders, Nebr., 394......C9 103
Valparaíso, prov., Chile, 613,405......A2 54
Val Racine, Que., Can., 153.D6 73
Valréas, Fr., 4,189......E6 14
Valsch, cape, W. Irian......G9 35
Vals Platz, Switz......C7 19
Valuyki, Sov. Un., 18,596...C5 20
Valverde, Dom. Rep., 17,885......F8 65
Val Verde, co., Tex., 24,461.E2 118
Valyermo, Los Angeles, Calif., 65......F5 82
Vamdrup, Den., 2,313......C3 24
Vamori, Pima, Ariz., 75....F4 80
Van, Van Zandt, Tex., 1,103.C5 118
Van, Tur., 22,000......C14 31
Van, Boone, W. Va., 400...D6 123
Van, lake, Tur......C14 31

Van Alstyne, Grayson, Tex., 1,608....C4 118
Vananda, Rosebud, Mont., 35....D10 102
Van Bruyssel, Que., Can., 53....B5 73
Van Buren, Crawford, Ark., 6,787....B1 81
Van Buren, Grant, Ind., 929.C6 91
Vanburen, Anderson, Ky., 50....C4 94
Van Buren, Aroostook, Maine, 3,589....A5 96
Van Buren, Carter, Mo., 575....E6 101
Vanburen, Hancock, Ohio, 374....A4 111
Van Buren, co., Ark., 7,228..B3 81
Van Buren, co., Iowa, 9,778....D6 92
Van Buren, co., Mich., 48,395....F4 98
Van Buren, co., Tenn., 3,671....D8 117
Vance, Quitman, Miss., 100..A3 100
Vance, Orangeburg, S.C., 85....E7 115
Vance, Real, Tex....E2 118
Vance, co., N.C., 32,002....A5 109
Vanceboro, Washington, Maine, 450 (389▲)....C5 96
Vanceboro, Craven, N.C., 806....B6 109
Vanceburg, Lewis, Ky., 1,881....B6 94
Vancleave, Jackson, Miss., 350....E5 100
Vancouver, B.C., Can., 384,522 (*790,165)..A10, E6 68
Vancouver, Clark, Wash., 32,464....D3 122
Vancouver, dist., B.C., Can..D5 68
Vancouver, isl., B.C., Can....E4 68
Vandalia, Fayette, Ill., 5,537.E4 90
Vandalia, Audrain, Mo., 3,055....B6 101
Vandalia, Valley, Mont., 15....B10 102
Vandemere, Pamlico, N.C., 452....B7 109
Vanderbilt, Otsego, Mich., 509....C6 98
Vanderbilt, Fayette, Pa., 826....F2 114
Vanderbilt, Jackson, Tex., 750....E4 118
Vanderburgh, co., Ind., 165,794....H2 91
Vandercook, Jackson, Mich., 4,000....*F6 98
Vandergrift, Westmoreland, Pa., 8,742....E2 114
Vanderhoof, B.C., Can., 1,460....C5 68
Vanderpool, Bandera, Tex., 10....E3 118
Vandervoort, Polk, Ark., 450.C1 81
Vander Wagen, McKinley, N. Mex., 30....B1 107
Van Diemen, cape, Austl....B5 50
Van Diemen, gulf, Austl....B5 50
Vandiver, Shelby, Ala., 700..B3 78
Vandling, Lackawanna, Pa., 578....C11 114
Vanduser, Scott, Mo., 272...E8 101
Van Dyke, Macomb, Mich. (part of Warren)....A8 98
Vandyne, Fond du Lac, Wis., 200....B5 124
Vanegas, Mex., 2,246....C4 63
Vänern, (Väner), lake, Swe....H5 25
Vänersborg, Swe., 18,500....H5 25
Van Etten, Chemung, N.Y., 507....C4 108
Vanga, Ken....B6 48
Vånga, Swe....B8 24
Vangaindrano, Malag....h9 49
Vanguard, Sask., Can., 433..H2 70
Van Hill, Hawkins, Tenn...C11 117
Van Hiseville, Ocean, N.J., 200....C4 106
Van Horn, Culberson, Tex., 1,953....F2 118
Van Horne, Benton, Iowa, 554....B5 92
Van Houten, Colfax, N. Mex....A5 107
Vankleek Hill, Ont., Can., 1,735....B10 72
Van Kull, kill, N.J....E5 106
Van Lear, Johnson, Ky., 921....C7 94
Vanleer, Dickson, Tenn., 234....A4 117
Vanlue, Hancock, Ohio, 386.B4 111
Van Meter, Dallas, Iowa, 385....C4 92
Vanna, Hart, Ga., 152....B3 87
Vanndale, Cross, Ark., 300...B5 81
Vannes, Fr., 30,411....D2 14
Van Norman, Garfield, Mont., 3....C10 102
Van Raub, Bexar, Tex....A3 118
Van Rees, mts., W. Irian....F9 35
Vanrhynsdorp, S. Afr., 2,129.D2 49
Vansant, Buchanan, Va., 850.B3 121
Vanscoy, Sask., Can., 136...F2 70
Vantage, Sask., Can., 59....H2 70
Van Tassell, Niobrara, Wyo., 15....C8 125
Vanua Levu, isl., Fiji Is....52
Vanua Mbalavu, isl., Fiji Is....52
Vanua Vatu, isl., Fiji Is....52
Vanves, Fr., 25,585....g10 14
Van Vleck, Matagorda, Tex., 900....G4 118
Van Vleet, Chickasaw, Miss., 200....B5 100
Van Wert, Polk, Ga., 311....C1 87
Van Wert, Decatur, Iowa, 253....D4 92
Van Wert, Van Wert, Ohio, 11,323....B3 111
Van Wert, co., Ohio, 28,840.B3 111
Van Winkle, B.C., Can....C7 68
Van Winkle, Hinds, Miss., 2,000....C3 100
Van Wyck, Lancaster, S.C., 300....B6 115
Van Wyksvlei, S. Afr., 1,460.D3 49
Van Yen, Viet., 10,000....B6 38
Van Zandt, co., Tex., 19,091....C5 118
Vanzant, Douglas, Mo., 30..E5 101
Var, dept., Fr., 469,557....*F7 14
Varallo, It., 7,7..52....D4 18
Varangerfjord, fjord, Nor....B14 25
Varano, lake, It....D5 21
Varaždin, Yugo, 26,239....B3 22
Varazze, It., 9,200....B2 21
Varberg, Swe., 14,500....I5 25
Vardaman, Calhoun, Miss., 637....B4 100
Vardar, riv., Yugo....E5 22
Varde, Den., 9,577....C2 24
Varde Å, riv., Den....C2 24
Vardö, Nor., 3,500....B14 25
Varel, Ger., 12,400....B4 16
Varella, cape, Viet....F8 38
Varena, Sov. Un....A8 26
Varennes, Que., Can., 2,240....D4, D9 73
Vares, Yugo, 7,688....C4 22
Varese, It., 66,963..D4 18, B2 21
Varginha, Braz., 24,944....C3 56
Vari, Grc....h11 23
Variadero, San Miguel, N. Mex., 60....B5 107
Varilla, Chile....D1 55
Varina, Pocahontas, Iowa, 162....B3 92
Varina, Wake, N.C. (part of Fuquay Springs)....B5 109
Varina, Henrico, Va., 100...C7 121
Varina Grove, Henrico, Va..C7 121
Varkaus, Fin., 22,200....F12 25
Värmland, co., Swe., 290,200....*H5 25
Varna (Stalin), Bul., 119,769.D8 22
Varna, Ont., Can., 75....D3 72
Varna, Marshall, Ill., 373....B4 90
Varna (Stalin), co., Bul., 550,375....D8 22
Varnado, Washington, La., 331....D6 95
Värnamo, Swe., 13,200....I6 25
Varner, Kingman, Kans., 52.B4 93
Varney, Ont., Can., 55....C4 72
Varney, Madison, Mont....E5 102
Varney, Torrance, N. Mex...C4 107
Varnsdorf, Czech., 13,400...C3 26
Varnville, Hampton, S.C., 1,461....F5 115
Väröbacka, Swe....A6 24
Varpalota, Hung., 21,509....B4 22
Vars, Ont., Can., 342....A10 72
Vartsila, Sov. Un....F14 25
Varysburg, Wyoming, N.Y., 300....C2 108
Varzea, riv., Braz....D2 56
Varzi, It., 5,630....E5 18
Varzuga, riv., Sov. Un....D17 25
Vas, co., Hung., 283,096...*B3 22
Vashon, King, Wash., 850....B3, D1 122
Vashon, isl., Wash....D1 122
Vashon, pt., Wash....D1 122
Vasilkov, Sov. Un., 10,000...F8 27
Vaslui, Rom., 14,850....B8 22
Vasquez, Grand, Colo., 125..B5 83
Vass, Moore, N.C., 767....B4 109
Vassar, Man., Can., 243....E4 71
Vassar, Tuscola, Mich., 2,680....E7 98
Vasser, Osage, Kans., 75....D8 93
Vassouras, Braz., 6,546....h6 56
Västerås, Swe., 79,200..H7, t34 25
Västerbotten, co., Swe., 238,600....*F9 25
Västerhaninge, Swe., 4,778..t36 25
Västernörrland, co., Swe., 283,300....*F8 25
Västervik, Swe., 18,200....I7 25
Västmanland, Swe., co., 235,200....*H7 25
Vasto, It., 12,600....C5 21
Vasyugan, Sov. Un....D24 9
Vasyugan, riv., Sov. Un....D24 9
Vaternish, pt., Scot....C2 13
Vathi, Grc., 5,052....D6 23
Vatia, pt., Fiji Is....52
Vatican City, country, Eur., 890....D4, h8 21
Vaticano, cape, It....E5 21
Vatomandry, Malag., 2,323..g9 49
Vatra-Dornei, Rom., 10,822....B7 22
Vättern (Vätter), lake, Swe....H6 25
Vatu Leile, isl., Fiji Is....52
Vatu Vara, isl., Fiji Is....52
Vaucluse, Aiken, S.C., 490..D4 115
Vaucluse, dept., Fr., 303,536....*F6 14
Vaucouleurs, Fr., 3,041....F5 15
Vaud, canton, Switz., 429,512....C1 19
Vaudreuil [sur-le-Lac], Que., Can., 200....D3, D8 73
Vaudreuil, co., Que., Can., 28,681....D3 73
Vaughan, Yazoo, Miss., 200....C3 100
Vaughan, Warren, N.C., 122....A6 109
Vaughan, Nicholas, W. Va., 175....C7 123
Vaughn, Cascade, Mont., 135....C5 102
Vaughn, Guadalupe, N. Mex., 1,170....C4 107
Vaughn, Lane, Oreg., 180...C3 113
Vaughn's Gap, Davidson, Tenn., 35....E9 117
Vaughnsville, Putnam, Ohio, 300....B3 111
Vaupés, comisaría, Col., 9,750....C3 60
Vaupés, riv., Col....C3 60
Vauvillers, Fr., 244....B2 18
Vauxhall, Alta., Can., 942...D4 69
Vavoua, I.C., 3,000....E3 45
Vawn, Sask., Can., 102....D1 70
Vaxholm, Swe., 3,800....t36 25
Växjö, Swe., 24,000....I6 25
Våxtorp, Swe., 427....B7 24
Vay, Bonner, Idaho, 10....A2 89
Vaygach, isl., Sov. Un....B20 9
Vayland, Hand, S. Dak., 14.C7 116
Veach, San Augustine, Tex..D5 118
Veachland, Shelby, Ky., 700....B4 94
Veadeiros, plat., Braz....D1 59
Vealmoor, Howard, Tex., 35....C2 118
Veazie, Penobscot, Maine, 1,354....*D4 96
Veberöd, Swe., 990....C7 24
Veblen, Marshall, S. Dak., 437....B8 116
Vechta, Ger., 13,500....F2 24
Vechte, riv., Ger....B6 15
Veckholm, Swe....t35 25
Vecses, Hung., 15,537....B4 22
Vedado, Cuba....h11 64
Vedea, riv., Rom....C7 22
Vederslöv, Swe....B8 24
Vedia, Arg., 3,676....A4 54
Veedersburg, Fountain, Ind., 1,762....D3 91
Veendam, Neth., 12,300....A6 15
Vega, Oldham, Tex., 658....B1 118
Vega, isl., Nor....E4 25
Vega Alta, P.R., 3,182....B5 65
Vega Alta, mun., P.R., 17,603....B5 65
Vega Baja, P.R., 3,718....B5 65
Vega Baja, mun., P.R., 30,189....B5 65
Vegas Heights, Clark, Nev., 1,200....*G6 104
Vegreville, Alta., Can., 2,908....C4 69
Veguita, Socorro, N. Mex., 170....C3 107
Veinticinco de Mayo, Arg...A3 54
Veinticinco de Mayo, Arg...B3 54
Veinticinco de Mayo, Arg., 9,063....B4 54
Vejen, Den., 4,582....C3 24
Vejer, Sp., 10,691....D2 20
Vejle, Den., 31,362 (*39,498)....C3 24
Vejle, co., Den., 207,881....C3 24
Vejprty, Czech., 5,476....C2 26
Vela, cape, Col....A3 60
Velarde, Rio Arriba, N. Mex., 50....A4 107
Velasco, Brazoria, Tex. (part of Freeport)....E5, G5 118
Velbert, Ger., 51,500....B2 17
Velda Village Hills, St. Louis, Mo., 1,365....*A8 101
Velden, Ger., 2,000....A8 18
Velebit, mts., Yugo....C2 22
Velestinon, Grc., 2,984....C4 23
Vélez, Col., 4,305....B3 60
Vélez-Blanco, Sp., 6,335....D4 20
Vélez-Málaga, Sp., 13,000...D3 20
Vélez-Rubio, Sp., 4,484....D5 20
Velgast, Ger., 1,793....D6 24
Velhas, riv., Braz....E2 57
Velika, riv., Yugo....E5 22
Velikaya, riv., Sov. Un....C7 27
Velikiye Luki, Sov. Un., 65,000....C8 27
Velikiy Ustyug, Sov. Un., 41,300....C18 9
Vélingara, Sen....D2 45
Velizh, Sov. Un., 10,000....D8 27
Velke Mezirici, Czech., 6,217....D4 26
Velletri, It., 16,200....D4, h9 21
Vellinge, Swe., 1,426....C7 24
Vellore, India, 113,742....F6 39
Velma, Stephens, Okla., 700....C4 112
Velma, mtn., Nev....B6 104
Velpen, Pipe, Ind., 185....H3 91
Velsen, Neth., 1,232....B4 15
Velsk, Sov. Un., 14,300....C17 9
Veltrusy, Czech., 1,883....n17 26
Velva, McHenry, N. Dak., 1,330....A5 110
Velvary, Czech., 2,169....n17 26
Vemb, Den., 1,017....B2 24
Ven, isl., Swe....C6 24
Venaco, Fr., 1,571....C2 21
Venado Tuerto, Arg., 15,947.A4 54
Venango, Perkins, Nebr., 227.D3 103
Venango, Crawford, Pa., 318.C1 114
Venango, co., Pa., 65,295...D2 114
Venator, Harney, Oreg....D8 113
Vendée, dept., Fr., 408,928.*D3 14
Vendée, hills, Fr....D3 14
Vendôme, Fr., 13,556....D4 14
Vendrell, Sp., 6,124....B6 20
Venedocia, Van Wert, Ohio, 200....B3 111
Veneta, Lane, Oreg., 750....C3 113
Veneta, lagoon, It....D8 18
Venetian Alps, mts., It....A4 21
Venetian Village, Lake, Ill., 2,084....*E2 90
Venetie, Alsk., 107....B10 79
Veneto, pol., dist., It., 3,846,562....*B4 17
Veneto, reg., It....A3 21
Venev, Sov. Un., 13,800...D12 27
Venezia (Venice), It., 347,347....B4 21, D8 18
Venezuela, country, S.A., 7,583,999....B4 60, C4 53
Venezuela, gulf, Ven....A3 60
Veniaminof, vol., Alsk....D8 79
Venice, Alta., Can., 10....B4 69
Venice, Sarasota, Fla., 3,444.E4 86
Venice, Madison, Ill., 5,380..E3 90
Venice, see Venezia, It.
Venice, Plaquemines, La., 500....E6 95
Venice, Douglas, Nebr., 25..D2 103
Venice, Erie, Ohio, 350....A3 111
Venice, Sevier, Utah, 250....E4 119
Venice, gulf, It....B4 21
Vénissieux, Fr., 29,040....E6 14
Venkatapuram, India, 5,000.H8 40
Venlo, Neth., 55,200....C6 15
Venn, Sask., Can., 56....F3 70
Vennachar, Ont., Can., 70...B7 72
Venö, isl., Den....B2 24
Vent, Aus....C6 18
Venta, riv., Sov. Un....C4 27
Ventana, Pima, Ariz., 35....E3 80
Venter, riv., Aus....B6 18
Ventimiglia, It., 12,200....C1 21
Ventnor, Ont., Can., 63....C9 72
Ventnor, Eng., 6,410....D6 12
Ventnor, Atlantic, N.J., 8,688....E4 106
Venton, Somerset, Md., 60..D6 85
Ventoux, mtn., Fr....E6 14
Ventry, Ire....E1 11
Ventspils, Sov. Un., 26,200..C3 27
Ventuari, riv., Ven....C4 60
Ventura, Ventura, Calif., 29,114....E4 82
Ventura, Cerro Gordo, Iowa, 280....A4 92
Ventura, co., Calif., 199,138.E4 82
Venturia, McIntosh, N. Dak., 148....D6 110
Ventura North (Chrisman), Ventura, Calif., 3,923....*E4 82
Venus, Highlands, Fla., 250..E5 86
Venus, Knox, Nebr....B7 103
Venus, Clarion and Venango, Pa., 120....D3 114
Venus, Johnson, Tex., 324...B9 118
Venustiano Carranza, Mex., 10,649....D6 63
Vera, Arg., 7,667....E3 55
Vera, Washington, Okla., 125....A6 112
Vera, Sp., 4,889....D5 20
Vera, Knox, Tex., 150....C3 118
Vera Cruz, Wells, Ind., 176..C7 91
Veracruz, Mex., 144,232....D5, n15 63
Veracruz, state, Mex., 2,749,235....D5, n15 63
Veradale, Spokane, Wash., 2,000....*B8 122
Veranopolis, Braz., 4,827....D2 56
Veraval, India, 46,637....G3 40
Verbania, It., 29,810....D4 18
Verbena, Chilton, Ala., 400..C3 78
Vercelli, It., 50,907....B2 21
Vercheres, Que., Can., 1,768.D4 73
Verchères, co., Que., Can., 25,697....D4 73
Verda, Harlan, Ky., 950....D6 94
Verde, isl., Phil....p13 35
Verde, riv., Ariz....C4 80
Verde, riv., Braz....A1 56, E3 59
Verde, riv., Braz....B2 56
Verde, riv., Braz....C2 56
Verde, riv., Mex....m14 63
Verde, riv., Par....D4 55
Verde Grande, riv., Mex....m12 63
Verde Island Passage, strait, Phil....p13 35
Verdel, Knox, Nebr., 123....B7 103
Verden, Ger., 17,500....B4 16
Verden, Grady, Okla., 405...B3 112
Verdi, Lincoln, Minn., 112...F2 99
Verdi, Washoe, Nev., 500...D2 104
Verdi, peak, Nev....C6 104
Verdigre, Knox, Nebr., 584..B7 103
Verdigris, Rogers, Okla., 40.A6 112
Verdigris, riv., Kans., Okla....A6 112, E8 93
Verdon, Richardson, Nebr., 267....D10 103
Verdon, Brown, S. Dak., 28.B7 116
Verdun, Que., Can., 78,317.D9 73
Verdun, Fr., 21,982....C6 14
Verdun, Scott, Tenn....C9 117
Verdunville, Logan, W. Va., 2,260....*D2 123
Vereeniging, S. Afr., 78,835..C4 49
Verendrye, McHenry, N. Dak., 100....A5 110
Vereshchagino, Sov. Un., 17,500....D19 9
Verga, Gloucester, N.J., 1,000....*D2 106
Vergara, Sp., 13,162....A4 20
Vergara, Ur., 2,480....E2 56
Vergas, Otter Tail, Minn., 292....D3 99
Vergennes, Jackson, Ill., 298.F4 90
Vergennes, Addison, Vt., 1,921....C2 120
Verigin, Sask., Can., 238....F4 70
Verín, Sp., 8,137....B2 20
Veríssimo Sarmento, Ang....C3 48
Verkhnyaya Tunguska, riv., Sov. Un....D12 28
Verkhne-Imbatskoye, Sov. Un....C26 9
Verkhne-Kolymsk, Sov. Un.C18 28
Verkhne-Uralsk, Sov. Un., 11,400....E20 9
Verkhne-Usinskoye, Sov. Un....E27 9
Verkhneye, Sov. Un., 10,000....G12, q21 27
Verkhniy Baskunchak, Sov. Un....D3 29
Verkhniy Ufaley, Sov. Un., 32,700....B6 29
Verkhoyansk, Sov. Un., 1,200....C16 28
Verkhoyansk, mts., Sov. Un....C15 28
Verlo, Sask., Can., 150....G1 70
Vermilion, Alta., Can., 2,449.C5 69
Vermilion, Edgar, Ill., 317...D6 90
Vermilion, Erie, Ohio, 5,220.A5 111
Vermilion, co., Ill., 96,176...C6 90
Vermilion, par., La., 38,855..E3 95
Vermilion, bay, La....E3 95
Vermilion, iron range, Minn....C6 99
Vermilion, lake, Ont., Can....D5 71
Vermilion, lake, Minn....C6 99
Vermilion, pass, Alta., Can....D2 69
Vermilion, riv., Alta., Can....C5 69
Vermilion, riv., Ill., Ind....B5 90, D2 91
Vermilion, riv., La....E3 95
Vermilion, riv., Minn....B6 99
Vermilion, riv., Ohio....A5 111
Vermilion Bay, Ont., Can., 503....E5 71
Vermilion Heights, Vermilion, Ill., 1,568....*C6 90
Vermilion-on-the-Lake, Lorain, Ohio, 1,273....*A5 111
Vermillion, Marshall, Kans., 265....C7 93
Vermillion, Clay, S. Dak., 6,102....E9 116
Vermillion, Sevier, Utah, 120.E4 119
Vermillion, co., Ind., 17,683.E3 91
Vermillion, bluffs, Colo....A2 83
Vermillion, riv., Que., Can....B4 73
Vermont, Fulton, Ill., 903...C3 90
Vermont, state, U.S., 389,881....B13 77, 120
Vermontville, Eaton, Mich., 768....F5 98
Vernal, Uintah, Utah, 3,655.C6 119
Vernalis, San Joaquin, Calif., 5....B6 82
Verndale, Wadena, Minn., 606....D3 99
Verne, Uinta, Wyo., 20....D2 125
Verner, Ont., Can., 965....A4 72
Verneuil, Fr., 4,611....C4 14
Verneuk, lake, S. Afr....C3 49
Vernon, Lamar, Ala., 1,492..B1 78
Vernon, Apache, Ariz., 100..C6 80
Vernon, B.C., Can., 10,250..D8 68
Vernon, Yuma, Colo., 60....B8 83
Vernon, Tolland, Conn., 500 (16,961▲)....B7 84
Vernon, Kent, Del., 60....C6 85
Vernon, Washington, Fla., 624....G3 86
Vernon, Fr., 17,247....C4 14
Vernon, Marion, Ill., 235....E4 90
Vernon, Jennings, Ind., 461.G6 91
Vernon, Sussex, N.J., 150....A4 106
Vernon, Oneida, N.Y., 913.*B5 108
Vernon, McIntosh, Okla., 150....B6 112
Vernon, Hickman, Tenn., 35....B4 117
Vernon, Wilbarger, Tex., 12,141....B3 118
Vernon, Tooele, Utah, 511...C3 119
Vernon, Windham, Vt., 130 (865▲)....F3 120
Vernon, Waukesha, Wis....F1 124
Vernon, co., Mo., 20,540....D3 101
Vernon, co., Wis., 25,663....E3 124
Vernon, par., La., 18,301....C2 95
Vernon, creek, Ind....G6 91
Vernon, lake, Ont., Can....B5 72
Vernon Bridge, P.E.I., Can., 219....C7 74
Vernon Center, Blue Earth, Minn., 333....G4 99
Vernon Hill, Halifax, Va., 20....E3 121
Vernonia, Columbia, Oreg., 1,089....B3 113
Vernon River, P.E.I., Can., 171....C7 74
Vernon Valley, Suffolk, N.Y., 5,998....*D3 108
Vero Beach, Indian River, Fla., 8,849....E6 86
Veroia, Grc., 25,765....B4 23
Verona, Ont., Can., 601....C8 72
Verona, Grundy, Ill., 192....B5 90
Verona, It., 221,221.B3 21, D7 18
Verona, Boone, Ky., 225....A7 94
Verona, Lee, Miss., 824....A5 100
Verona, Lawrence, Mo., 401....E4 101
Verona, Essex, N.J., 13,782..B4 106
Verona, Onslow, N.C., 500..C6 109
Verona, La Moure, N. Dak., 162....C7 110
Verona, Montgomery and Preble, Ohio, 527....C3 111
Verona, Allegheny, Pa., 4,032....A6 114
Verona, Marshall, Tenn., 30.B5 117
Verona, Dane, Wis., 1,471...F4 124
Verona Park, Calhoun, Mich., 1,884....*F5 98
Verónica, Arg., 2,405....g8 54
Verplanck, Westchester, N.Y., 1,800....*D7 108
Verret, lake, La....E4 95
Verrières-le-Buisson, Fr., 7,063....h10 14
Versailles, New London, Conn., 300....C8 84
Versailles, Fr., 86,759....C4, g9 14, F2 15
Versailles, Brown, Ill., 427...D3 90
Versailles, Ripley, Ind., 1,158....F7 91
Versailles, Woodford, Ky., 4,060....B5 94
Versailles, Morgan, Mo., 2,047....C5 101
Versailles, Darke, Ohio, 2,159....B3 111
Versailles, Allegheny, Pa., 2,297....*F2 114
Verse, Converse, Wyo....A7 125
Versmold, Ger., 5,900....A3 17
Versoix, Switz., 3,426....D1 19
Verte, isl., Que., Can....A8 73
Verte, riv., Que., Can....B8 73
Vertedero, P.R....C6 65
Vertentes, riv., Braz....E4 59
Verton, Fr., 490....D9 12
Vertou, Fr., 2,330....D3 14
Vertus, Fr., 2,617....F4 15
Verviers, Bel., 35,453....D5 15
Vervins, Fr., 2,735....C5 14
Verwood, Sask., Can., 84....H3 70
Verzasca, riv., Switz....D6 19
Verzenay, Fr., 1,348....C6 14
Veseleyville, Walsh, N. Dak., 150....A8 110
Veseli nad Luznici, Czech., 3,922....D9 17
Veseloye, Sov. Un., 25,000.H10 27
Vesey, Red River, Tex....C5 118
Vesoul, Fr., 13,678....D7 14
Vesper, Lincoln, Kans., 100..C5 93
Vesper, Wood, Wis., 351....D4 124
Vesta, C.R....F6 62
Vesta, Redwood, Minn., 318.F3 99
Vesta, Johnson, Nebr., 75...D9 103
Vestaburg, Montcalm, Mich., 450....E6 98
Vestaburg, Washington, Pa., 950....*F2 114
Vest-Agder, co., Nor., 109,000....*H2 25
Vestal, Broome, N.Y., 7,000.*C4 108
Vestal Center, Broome, N.Y., 500....C4 108
Vestavia Hills, Jefferson, Ala., 4,029....B3 78
Vestby, Nor., 460....p29 25
Vesterålen, is., Nor....C6 25
Vesterö Havn, Den., 517....A4 24
Vesteröya, isl., Nor....p28 25
Vestervig, Den., 573....B2 24
Vestfold, co., Nor., 174,200.*H4 25
Vestmannaeyjar, Ice., 4,702.o22 25
Vestry, Jackson, Miss., 150..E5 100
Vesuvius, Rockbridge, Va., 400....D3 121
Vesuvius, vol., It....D5 21
Vesyegonsk, Sov. Un., 9,700.B11 27
Veszprem, Hung., 25,495....B3 22
Veszprem, co., Hung., 393,331....*B3 22
Veszto, Hung., 9,645....B5 22
Vetal, Bennett, S. Dak., 20..D4 116
Veteran, Alta., Can., 239....C5 69
Veteran, Goshen, Wyo., 40..D8 125
Vetlanda, Swe., 9,400....I6 25
Vetluga, riv., Sov. Un....B3 29
Vetluzhskiy, Sov. Un....B3 29
Vetovo, Bul., 4,981....D8 22
Vetren, Bul., 6,566....D7 22
Veurne, Bel., 7,330....C2 15
Vevay, Switzerland, Ind., 1,503....G7 91
Vevey, Switz., 16,269 (*26,235)....D2 19
Veynes, Fr., 3,474....E6 14
Veyo, Washington, Utah, 60.F2 119
Vézère, riv., Fr....E4 14
Vezhkurya, Sov. Un....B21 9
Vezirkopru, Tur., 8,100....B10 31
Viacha, Bol., 6,607....C2 55
Viadana, It., 5,487....B3 21
Viamonte, Arg....A4 54
Vian, Sequoyah, Okla., 930..B7 112
Viana, Braz., 5,385....B2 57
Viana del Bollo, Sp., 8,447...A2 20
Viana do Alentejo, Port., 3,566....C1 20
Viana do Castelo, Port., 14,023....B1 20
Vianen, Neth., 5,300....C5 15
Vianópolis, Braz., 2,090....E1 57
Viar, riv., Sp....D3 20
Viareggio, It., 47,323....C3 21
Vibank, Sask., Can., 308....G4 70
Viborg, Den., 23,265....B3 24
Viborg, Turner, S. Dak., 699....D8 116
Viborg, co., Den., 160,018...B3 24
Vibo Valentia, It., 12,800...E6 21
Viburnum, Iron, Mo., 590..D6 101
Viby, Den., 15,036....*B4 24
Vicco, Perry, Ky., 900....C6 94
Vic [-en-Bigorre], Fr., 3,728..F4 14
Vicente, pt., Calif....F2 82
Vicente López, Arg., 250,823....*A5 54
Vicenza, It., 98,019.B3 21, D7 18
Viceroy, Sask., Can., 225....H3 70
Vich, Sp., 12,414....B7 20
Vichada, comisaría, Col., 13,860....C3 60
Vichuga, Sov. Un., 53,000..D17 9
Vichuquén, Chile....A2 54
Vichy, Fr., 30,614 (*47,500)....D5 14
Vichy, Maries, Mo., 200....C6 101
Vici, Dewey, Okla., 601....A2 112
Vick, Avoyelles, La., 100....C3 95
Vickery, Sandusky, Ohio, 285....A5 111
Vickery, Dallas, Tex. (part of Dallas)....B5 118
Vicksburg, Yuma, Ariz., 10..D2 80
Vicksburg, Greene, Ind., 175.F3 91
Vicksburg, Kalamazoo, Mich., 2,224....F5 98
Vicksburg, Warren, Miss., 29,143....C3 100
Viçosa, Braz., 7,285....C3, k5 57
Viçosa, Braz., 9,342....F2 57
Viçosa do Ceará, Braz., 2,629.B2 57
Vicosoprano, Switz., 471....D8 19
Vicovaro, It., 3,404....g9 21
Vic [-sur-Seille], Fr., 1,213...F6 15
Victor, Teller, Colo., 434....C5 83
Victor, Teton, Idaho, 240....F7 89
Victor, Iowa and Poweshiek, Iowa, 870....C5 92
Victor, Ravalli, Mont., 375..D2 102
Victor, Ontario, N.Y., 1,180.C3 108
Victor, Roberts, S. Dak., 30..B9 116
Victor Harbor, Austl., 2,036.G2 52
Victoria, Coffee, Ala....D4 78
Victoria, Arg., 17,711....A4 54
Victoria, Cam., 8,025....F6 45
Victoria, B.C., Can., 54,941 (*154,152)....B9, E6 68
Victoria, Newf., Can., 1,506.E5 75
Victoria, P.E.I., Can., 148...C6 74
Victoria, Chile, 10,671....B2 54
Victoria, Hong Kong, 674,962 (*2,800,000)....G7 34
Victoria, Knox, Ill., 453....B3 90
Victoria, Ellis, Kans., 1,170..D4 93
Victoria, Marshall, Miss., 500....A4 100
Victoria, Jefferson, Mo., 150....C7 101
Victoria, Gloucester, N.J., 100....D3 106
Victoria, Phil, 5,672....o13 35
Victoria, P.R....B2 65
Victoria, Marion, Tenn., 600....D8 117
Victoria, Victoria, Tex., 33,047....E4 118
Victoria, Lunenburg, Va., 1,737....E4 121
Victoria, co., N.B., Can., 19,712....B2 74
Victoria, co., N.S., Can., 8,226....C9 74
Victoria, co., Ont., Can., 29,750....C6 72
Victoria, co., Tex., 46,475...E2 118
Victoria, dist., B.C., Can....E6 68
Victoria, state, Austl. 2,930,113....G7 51
Victoria, falls, S. Rh....A4 49
Victoria, isl., N.W. Ter., Can.B10 66
Victoria, lake, Austl....F3 51
Victoria, lake, Newf., Can....D3 75
Victoria, lake, Tan., Ug....B5 48
Victoria, mtn., Pap....k12 50
Victoria, riv., Austl....C5 50
Victoria, riv., Newf., Can....D3 75
Victoria, strait, N.W. Ter., Can....C13 66
Victoria Beach, Man., Can., 74....D3 71
Victoria Central, P.R....B7 65
Victoria di las Tunas, Cuba, 12,754....E5 64
Victoria Falls, S. Rh., 1,455..A4 49
Victoria Harbour, Ont., Can., 1,066....C5 64
Victoria Land, reg., Ant....B29 5
Victoria Mines, Ont., Can., 115....A3 72
Victoria Park, Los Angeles, Calif., 2,400....*F2 82
Victoria Point, Bur., 1,519..F10 39
Victoria River Downs, Austl.C5 50
Victoria Road, Ont., Can., 125....C6 72
Victoriaville, Que., Can., 18,720....C6 73
Victoria West, S. Afr., 3,745....D3 49
Victorica, Arg....B3 54
Victorino de la Plaza, Arg...B4 54
Victorio, Luna, N. Mex....F1 107
Victor Mills, Spartanburg, S.C., 2,018....*B3 115
Victorville, San Bernardino, Calif., 5,000....E5 82
Victory, Cumberland, N.C...B5 109
Victory, Vernon, Wis., 143...E2 124
Victory Gardens, Morris, N.J., 1,085....*B3 106
Victory Heights, Chemung, N.Y., 1,030....*C4 108
Vicuña, Chile, 3,415....F1 55
Vida, McCone, Mont., 85...C11 102
Vidal, San Bernardino, Calif., 100....E6 82
Vidalia, Toombs, Ga., 7,569.D4 87
Vidalia, Concordia, La., 4,313....C4 95
Vidette, Burke, Ga., 103....C4 87
Vidin, Bul., 18,580....D6 22
Vidor, Orange, Tex., 4,938..D5 118
Vidora Sask., Can., 65....H1 70
Viechtach, Ger., 3,600....D7 17
Viedma, Arg., 4,683....C4 54
Vieja, peak, Tex....F2 118
Vielsalm, Bel., 3,698....D5 15
Vienenburg, Ger., 6,700....B5 17
Vienna (Wien), Aus., 1,627,566 (*1,990,000)....D8 16
Vienna, Ont., Can., 373....E4 72
Vienna, Dooly, Ga., 2,099....D3 87
Vienna, Johnson. Ill. 1 094..F5 90

Vienna, Dorchester, Md., 420 D6 85
Vienna, Maries, Mo., 536 C6 101
Vienna, Warren, N.J., 250 B3 106
Vienna, Clark, S. Dak., 191 C8 116
Vienna, Fairfax, Va., 11,440 B4, C5 121
Vienna, Wood, W. Va., 9,381 B3 123
Vienne, Fr., 26,977 E6 14
Vienne, dept., Fr., 331,619 *D4 14
Vienne, riv., Fr. D4 14
Vientiane, Laos, 100,000 D5 38
Vieques, P.R., 2,487 g13 65
Vieques, mun., P.R., 7,210 g13 65
Vieques, isl., P.R. g13 65
Vieques, sound, P.R. g13 5
Viernheim, Ger., 19,900 D3 17
Viersen, Ger., 41,900 C6 15
Vierzon, Fr., 31,549 D5 14
Viesca, Mex., 3,043 B4 63
Vieste, It., 12,679 D6 21
Vietnam, reg., Asia E8 38, H13 33
Vietnam, North, country, Asia, 14,500,000 C6 38
Vietnam, South, country, Asia, 11,000,000 F8 38
Viewfield, Meade, S. Dak., 5 C3 116
View Park, Los Angeles, Calif., 2,500 *E4 82
Vigan, Phil., 7,424 B6 35
Viger, Que., Can., 435 B8 73
Vigevano, It., 57,069 B2 21
Vignola, It., 5,868 E7 18
Vigo, Sp., 144,914 A1 20
Vigo, co., Ind., 108,458 F3 91
Vigsö, bay, Den. A2 24
Vihowa, Pak., 2,827 B5 39, B3 40
Vijayavada, India, 230,397 E7 39
Vik, Ice., 339 o23 25
Viken, Swe., 750 B6 24
Vikesund, Nor., 1,267 p27 25
Viking, Alta., Can., 1,043 C5 69
Viking, Marshall, Minn., 128 B2 99
Vila Cabral, Moz. D6 48
Vila da Feira, Port. B1 20
Vila da Ponte, Ang., 329 D2 48
Vila de Aljustrel, Ang. D2 48
Vila de João Belo, Moz., 1,936 C5 49
Vila de Manica, Moz. A5 49
Vila de Rei, Port., 5,982 C1 20
Vila do Conde, Port., 7,772 B1 20
Vila Fontes, Moz. A6 49
Vila Franca de Xira, Port., 8,296 C1 20
Vila Gago Coutinho, Ang., 1,411 D3 48
Vila General Machado, Ang., 2,387 D2 48
Vila João de Almeida, Ang. E1 48
Vila Junqueiro, Moz. A6 49
Vilaine, riv., Fr. D2 14
Vila Luso, Ang., 2,821 D2 48
Vila Macedo de Cavaleiros, Ang. D2 48
Vila Marechal Carmona, Ang., 8,300 C2 48
Vila Mariano Machado, Ang., 349 D1 48
Vilanculos, Moz. B6 49
Vila Nova de Foz Coa, Port., 3,481 B2 20
Vila Nova de Gaia, Port., 45,700 B1 20
Vila Nova de Milfontes, Port., 2,460 D1 20
Vila Nova de Seles, Ang., 1,115 D1 48
Vila Pereira d' Eça, Ang., 416 E2 48
Vila Pery, Moz. A5 49
Vila Real, Port., 9,285 B2 20
Vila Real de Santo António, Port, 6,086 D2 20
Vila Robert Williams, Ang., 3,679 D2 48
Vila Salazar, Ang., 2,105 C1 48
Vilas, Baca, Colo., 107 D8 83
Vilas, Liberty, Fla., 15 B2 86
Vilas, Miner, S. Dak., 49 C8 116
Vilas, co., Wis., 9,332 B4 124
Vila Serpa Pinto, Ang., 387 D2 48
Vila Teixeira da Silva, Ang., 4,897 D2 48
Vila Teixeira de Sousa, Ang., 870 D3 48
Vila Vasco de Gama, Moz D5 48
Vila Viçosa, Port., 3,802 C2 20
Vila Vila, Bol., 658 C2 55
Vildbjerg, Den., 1,108 B2 24
Vildo, Hardeman, Tenn., 40 B2 117
Vileyka, Sov., Un., 10,000 D6 27
Vilhelmina, Swe., 3,000 E7 25
Viljandi, Sov. Un., 12,941 B5 27
Vilkaviskis, Sov. Un., 8,699 A7 26
Vilkovo, Sov. Un., 20,000 I7 27
Villa Acuña, Mex., 20,204 B4 63
Villa Ahumada, Mex., 2,489 A3 63
Villa Alhucemas, Mor., 11,262 G5 30
Villa Angela, Arg., 7,345 E3 55
Villa Aroma, Bol., 1,486 C2 55
Villa Bella, Bol., 88 B2 55
Villablino, Sp., 7,647 A2 20
Villacañas, Sp., 10,113 C4 20
Villacarrillo, Sp., 13,090 C4 20
Villach, Aus., 32,971 E6 16
Villacidro, It., 11,266 E2 21
Villa Cisneros, Sp. Sahara, 1,011 E1 44
Villa Colón, C.R., 4,309 F5 62
Villa Constitución, Arg., 9,183 A4 54
Villa Crespo, Arg., 4,289 A4 54
Villa Cuauhtémoc, Mex., 2,436 k15 63
Villa de Cura, Ven., 19,644 A4 60
Villa del Rosario, Arg., 4,461 A4 54
Villa Dolores, Arg., 13,835 A3 54
Villafamés, Sp., 3,652 C5 20
Villa Federal, Arg., 9,158 A5 54
Villafranca, It., 15,712 B3 21
Villafranca del Bierzo, Sp., 4,512 A2 20
Villafranca de los Barros, Sp., 16,671 C2 20
Villafranca del Panadés, Sp., 11,985 B6 20
Villafranca di Verona, It., 6,015 D6 18
Villa García, Mex., 1,877 k13 63
Villagarcía de Arosa, Sp., 4,986 A1 20
Village, Columbia, Ark., 85 D2 81
Village, Richmond, Va., 140 D6 121
Village, creek, Ala. E4 78
Village Richelieu, Que., Can., 1,612 *D4 73
Village Springs, Blount, Ala., 125 B3 78
Villaggio Duca degli Abruzzi, Som., 9,000 (15,900▲) E6 47
Villa Grove, Saguache, Colo., 100 C5 83
Villa Grove, Douglas, Ill., 2,308 D5 90
Villaguay, Arg., 17,607 A5 54
Villa Hayes, Par., 12,590 E4 55
Villa Heights, Prince Georges, Md., 1,000 *C4 85
Villahermosa, Mex., 51,611 D6 63
Villa Huidobro, Arg. A4 54
Villa Iris, Arg., 2,422 B4 54
Villajoyosa, Sp., 9,412 C5 20
Villalba, P.R., 1,892 C5 65
Villalba, Sp., 3,180 A2 20
Villalba, mun., P.R., 16,239 C5 65
Villaldama, Mex., 2,529 B4 63
Villalon, Cuba k11 64
Villalonga, Arg. B4 54
Villalpando, Sp., 2,599 B3 20
Villa María, Arg., 30,362 A4 54
Villamartín, Sp., 9,849 D3 20
Villamil, Ec. g5 58
Villa Montes, Bol., 3,105 D3 55
Villanova, Delaware, Pa., 4,000 *A10 114
Villanueva, Col., 5,830 A3 60
Villanueva, San Miguel, N. Mex., 300 B4 107
Villanueva de Córdoba, Sp., 15,719 C3 20
Villanueva del Arzobispo, Sp., 9,712 C4 20
Villanueva de la Serena, Sp., 20,812 C3 20
Villanueva [del Río y Minas], Sp., 10,982 D3 20
Villanueva y Geltrú, Sp., 25,669 B6 20
Villa Obregón, Mex., 25,908 h9 63
Villa Oliva, Par., 4,042 E4 55
Villa Park, DuPage, Ill., 23,294 F2 90
Villa Pedro Montoya, Mex., 4,443 m14 63
Villa Pérez, P.R. C3 65
Villaputzu, It., 3,731 E2 21
Villa Ranchaero, Pennington, S. Dak., 3,000 C3 116
Villard, Pope, Minn., 235 E3 99
Villard-Bonnot, Fr., 6,499 E6 14
Villa Rica, Carroll and Douglas, Ga., 3,450 C1 87
Villa Ridge, Pulaski, Ill., 550 F4 90
Villa Ridge, Franklin, Mo., 150 C7 101
Villarreal, Sp., 19,700 C5 20
Villarrica, Chile, 7,036 B2 54
Villarrica, Par., 14,680 E4 55
Villarrobledo, Sp., 21,356 C4 20
Villarrubia, Sp., 9,043 C4 20
Villars-le-Terroir, Switz., 511 D3 19
Villas, Cape May, N.J., 2,085 E3 106
Villa Unión, Arg. E2 55
Villa Unión, Mex., 4,199 C3 63
Villa Valeria, Arg. A4 54
Villavicencio, Col., 17,126 C3 60
Villaviciosa, Sp., 2,322 A3 20
Villazon, Bol., 6,261 D2 55
Villazón, Bol. D3 55
Villé, Fr., 1,326 A3 18
Ville-de-Tracy, Que., Can., 8,171 D4 73
Villefranche [-de-Rouergue], Fr., 7,969 E5 14
Villefranche [-sur-Saône], Fr., 24,516 E6 14
Villegreen, Las Animas, Colo., 10 D7 83
Villejuif, Fr., 46,116 g10 14
Ville Marie, Que., Can., 1,710 E9 72
Villemomble, Fr., 24,540 g10 14
Villena, Sp., 15,687 C5 20
Villenauxe-la-Grande, Fr., 1,925 F3 15
Villeneuve-le-Roi, Fr., 22,300 h10 14
Villeneuve-St. Georges, Fr., 28,231 h10 14
Villeneuve-sur-Lot, Fr., 15,296 (17,295▲) E4 14
Villeneuve-sur-Yonne, Fr., 4,021 C5 14
Villepinte, Fr., 483 g11 14
Ville Platte, Evangeline, La., 7,512 D3 95
Villeroy, Que., Can., 198 C6 72
Villers-Bretonneux, Fr., 3,342 E2 15
Villers-Cotterêts, Fr., 5,489 C5 14
Villers-Outréaux, Fr., 2,420 D3 15
Villerupt, Fr., 14,377 C6 14
Ville St. Georges, Que., Can., 4,082 C7 73
Ville St. Pierre, Que., Can., 6,795 *D9 73
Villeta, Par., 14,729 E4 55
Villeurbanne, Fr., 105,416 E6 14
Villia, Grc., 3,151 g10 23
Villiers-St. Georges, Fr., 839 F3 15
Villingen, Ger., 31,900 D4 16
Villisca, Montgomery, Iowa, 1,690 D3 92
Vilna, Alta., Can., 400 B5 69
Vilnius, Sov. Un., 255,000 D5 27
Vilonia, Faulkner, Ark., 234 B3 81
Vilppula, Fin., 1,800 F11 25
Vils, riv., Ger. E7 17
Vils, riv., Ger. D6 17
Vilsbiburg, Ger., 5,900 E7 17
Vilseck, Ger., 2,200 D6 17
Vilshofen, Ger., 5,800 E8 17
Vilvoorde, Bel., 31,441 D4 15
Vilyuy, riv., Sov. Un. C15 28
Vilyuysk, Sov. Un., 3,600 C15 28
Vim, Slope, N. Dak C3 110
Vimianzo, Sp., 654 A1 20
Vimmerby, Swe., 6,400 I6 25
Vimperk, Czech., 2,940 D2 26
Vina, Franklin, Ala., 184 A1 78
Vina, Tehama, Calif., 200 C2 82
Viña del Mar, Chile, 118,000 A2 54
Vinalhaven, Knox, Maine, 950 (1,273▲) D4 96
Vinalhaven, isl., Maine D4 96
Vinaroz, Sp., 10,968 B6 20
Vincennes, Fr., 50,436 g10 14
Vincennes, Knox, Ind., 18,046 G2 91
Vincennes, bay, Ant. C24 5
Vincent, Shelby, Ala., 1,402 B3 78
Vincent, Crittenden, Ark., 100 E8 81
Vincent, Webster, Iowa, 173 B3 92
Vincent, Calcasieu, La., 75 D2 95
Vincent, Lorain, Ohio, 2,100 *A5 111
Vincent, Washington, Ohio, 340 C6 111
Vincent, Howard, Tex., 50 C2 118
Vincentown, Burlington, N.J., 545 D3 106
Vinces, Ec., 4,129 B2 58
Vinchina, Arg. E2 55
Vinco, Payne, Okla., 50 B4 112
Vindelälven, riv., Swe. E8 25
Vinderup, Den., 1,910 B2 24
Vindex, Garrett, Md., 80 D1 85
Vindhya, mts., India F6 40
Vine, brook, Mass. C2 97
Vine Grove, Hardin, Ky., 2,435 C4 94
Vine Hill (Martinez East) Contra Costa, Calif., 3,958 *C2 82
Vineland, Orange, Fla., 100 D5 86
Vineland, Cumberland, N.J., 37,685 E2 106
Vinemont, Cullman, Ala., 500 A3 78
Vineyard, Lee, Ark. C5 81
Vineyard, Utah, Utah C4 119
Vineyard, sound, Mass. D6 97
Vineyard Haven (Tisbury), Dukes, Mass., 2,169 D6 97
Vingåker, Swe., 4,300 t33 25
Vinh, Viet., 30,000 C6 38
Vinhais, Port., 2,911 B2 20
Vinh Long, Viet., 30,000 G6 38
Vinh Yen, Viet., 3,820 B6 38
Vining, Clay, Kans., 128 C6 93
Vining, Otter Tail, Minn., 136 D3 99
Vinings, Cobb, Ga. A5 87
Vinita, Craig, Okla., 6,027 A6 112
Vinita Park, St. Louis, Mo., 2,204 *C7 101
Vinkovci, Yugo., 23,113 C4 22
Vinnitsa, Sov. Un., 131,000 G7 27
Vinson, Harmon, Okla., 75 C2 112
Vinton, Benton, Iowa, 4,781 B5 92
Vinton, Calcasieu, La., 2,987 D2 95
Vinton, Gallia, Ohio, 374 D5 111
Vinton, Roanoke, Va., 3,432 D3 121
Vinton, co., Ohio, 10,274 C5 111
Vintondale, Cambria, Pa., 938 F4 114
Viola, Fulton, Ark., 196 A4 81
Viola, Kent, Del., 159 B6 85
Viola, Latah, Idaho, 60 C1 89
Viola, Mercer, Ill., 812 B3 90
Viola, Sedgwick, Kans., 203 E6 93
Viola, Warren, Tenn., 206 D8 117
Viola, Richland and Vernon, Wis., 721 E3 124
Viola, Lincoln, Wyo. C2 125
Violet, St. Bernard, La., 900 C8 95
Vipiteno, It., 3,151 C7 18
Virac, Phil., 8,539 *C6 35
Virden, Man., Can., 2,708 E1 71
Virden, Macoupin, Ill., 3,309 D4 90
Virden, Hidalgo, N. Mex., 135 E1 107
Vire, Fr., 9,518 C3 14
Vire, riv., Fr. C3 14
Virgelle, Chouteau, Mont., 5 B6 102
Virgenes, cape, Arg. E3 54
Virgil, Greenwood, Kans., 229 E7 93
Virgil, Beadle, S. Dak., 81 C7 116
Virgil, Cortland, N.Y., 75 C4 107
Virgilina, Halifax, Va., 286 E4 121
Virgin, Washington, Utah, 124 F2 119
Virgin, mts., Nev G7 104
Virgin, riv., Utah F3 119
Virgin Gorda, isl., Vir. Is. m14 64
Virgin Islands, nat. park, Vir. Is. f15 65
Virgin Islands, Br. dep., N.A., 7,338 f15 65
Virgin Islands of the U.S., dep., N.A., 26,665 m14 64, f15 65
Virginia, Bannock, Idaho, 50 G6 89
Virginia, Cass, Ill., 1,669 D3 90
Virginia, Ire., 515 D4 11
Virginia, St. Louis, Minn., 14,034 C6 99
Virginia, Gage, Nebr., 88 D9 103
Virginia, S. Afr., 18,273 (*40,359) *C4 49
Virginia, state, U.S., 3,966,949 C12 77, 121
Virginia, peak, Nev D2 104
Virginia, peak, Wyo. C2 125
Virginia Beach (Independent City), Va., 8,091 B7, E7 121
Virginia City, Madison, Mont., 194 E5 102
Virginia City, Storey, Nev., 500 D2 104
Virginia Gardens, Dade, Fla., 2,159 *G6 87
Viroflay, Fr., 16,004 g9 14
Viroqua, Vernon, Wis., 3,926 E3 124
Virovitica, Yugo., 14,027 C3 22
Vir-Pazar, Yugo 323 D4 22
Virrat, Fin., 1,600 F10 25
Virserum, Swe., 2,414 I6 25
Virton, Bel., 3,421 E5 15
Virú, Peru, 2,573 C2 58
Vis, Yugo., 2,844 D3 22
Vis, isl., Yugo. D3 22
Visakhapatnam, India, 182,004 E7 39, I9 40
Visalia, Tulare, Calif., 15,791 D4 82
Visalia, Kenton, Ky., 253 A7 94
Visalia North (Crowley), Tulare, Calif., 3,950 *D4 82
Visayan, sea, Phil. C6 35
Visby, Swe., 15,600 I8 25
Visconde do Rio Branco, Braz., 12,363 C4 56
Viscount, Sask., Can., 303 F3 70
Viscount Melville, sound, N.W. Ter., Can. B11 66
Visé, Bel., 6,018 D5 15
Višegrad, Yugo., 3,316 D4 22
Viseu, Braz., 1,606 B1 57
Viseu, Port., 13,190 B2 20
Vishera, riv., Sov. Un. A8 29
Vislanda, Swe., 1,400 B8 24
Viso, mtn., It. B1 21, E3 18
Visoko, Yugo., 7,461 D4 22
Visonau, Fiji Is. 52
Visp, Switz., 3,658 D4 19
Vista, San Diego, Calif., 14,795 F5 82
Vista, Man., Can., 79 D1 65
Vista, St. Clair, Mo., 50 D4 101
Vista, Washoe, Nev., 25 D2 104
Vista Park, Kern, Calif., 3,500 *E4 82
Vistillas, Lake, Ore. E6 113
Vistula (Frisches Haff), lagoon, Sov. Un. A5 26
Vita, Man., Can., 316 E3 71
Vitanovak, Yugo., 1,127 D5 22
Vitebsk, Sov. Un., 162,000 D8 27
Viterbo, It., 29,000 (50,047▲) C4 21
Viti Levu, bay, Fiji Is. 52
Viti Levu, isl., Fiji Is. 52
Vitim, Sov. Un., 2,300 D14 28
Vitim, riv, Sov. Un. D14 28
Vitor, Peru, 2,343 E3 58
Vitória, Braz., 82,748 (*165,000) F2 57
Vitória, Braz. k5 57
Vitória, Braz. C4 61
Vitoria, Sp., 73,701 A4 20
Vitória da Conquista, Braz., 46,778 D2 57
Vitória [de Santo Antão], Braz., 27,053 C3, k6 57
Vitória do Mearim, Braz., 1,494 B2 57
Vitré, Fr., 10,380 C3 14
Vitry-le-François, Fr., 14,795 C6 14
Vitry [-sur-Seine], Fr., 65,734 g10 14
Vittel, Fr., 5,012 C6 14
Vittoria, Ont., Can., 407 E4 72
Vittoria, It., 45,035 F5 21
Vittorio Veneto, It., 21,500 B4 21
Vivero, Sp., 3,628 A2 20
Vivian, Caddo, La., 2,624 B2 95
Vivian, Lyman, S. Dak., 300 D5 116
Vivian, McDowell, W. Va., 900 D3 123
Vivoratá, Arg B5 54
Viwa, isl., Fiji Is. 52
Vizcaíno, des., Mex. B2 63
Vizcaíno, mts., Mex. B2 63
Vizcaya, prov., Sp., 754,383 *A4 20
Vize, Tur., 6,200 B6 23
Vizianagram, India, 76,808 E7 39, H9 40
Vizille, Fr., 6,493 E6 14
Viziru, Rom., 5,414 C8 22
Vizzini, It., 10,806 F5 21
Vlaardingen, Neth., 68,900 C4 15
Vladimir, Sov. Un., 167,000 B2 29
Vladimiro-Aleksandrovskoye, Sov. Un., 10,000 E6 37
Vladimirovo, Sov. Un. D3 29
Vladimir-Volynskiy, Sov. Un., 10,000 F5 27
Vladivostok, Sov. Un., 317,000 E5 37, E16 28
Vlasenica, Yugo., 3,047 C4 22
Vlasim, Czech., 5,066 D9 17
Vlasotince, Yugo., 5,924 D6 22
Vlčany, Czech., 4,555 D4 26
Vlieland, Neth., 700 A5 15
Vlieland, isl., Neth. A4 15
Vlissingen, Neth., 29,400 C3 15
Vlkava, Czech., 526 n18 26
Vlkava, riv., Czech. n18 26
Vlonë (Valona), Alb., 32,700 B2 23
Vlonë, pref., Alb., 58,000 *B2 23
Vlotho, Ger., 8,100 A3 17
Vltava, riv., Czech. n17 26
Voca, McCulloch, Tex., 100 D3 118
Vodlozero, lake, Sov. Un. F17 25
Vodnany, Czech., 4,576 D9 17
Voeune Sai, Camb., 10,000 F7 38
Vogelkop, pen., W. Irian F8 35
Voghera, It., 35,747 B2 21
Vohenstrauss, Ger., 3,700 D7 17
Vohipeno, Malag. h9 49
Voi, Ken. B6 48
Void, Fr., 1,118 F5 15
Voiotia (Boeotia), prov., Grc., 114,256 *C4 23
Voiron, Fr., 11,150 E6 14
Voitsberg, Aus., 6,353 E7 16
Voivis, lake, Grc. C4 23
Vojens, Den., 3,563 C3 24
Vokhma, Sov. Un. B3 29
Volary, Czech., 2,278 E8 17
Volborg, Custer, Mont., 8 E11 102
Volcano, isl., Phil. o13 35
Volchansk, Sov. Un., 10,000 F11 27
Volga, Clayton, Iowa, 361 B6 92
Volga, Brookings, S. Dak., 780 C9 116
Volga, Barbour, W. Va., 125 B4 123
Volga, plat., Sov. Un. C3 29
Volga, riv., Sov. Un. F18 9
Volga, riv. mouths, Sov. Un. F18 9
Volgograd, Sov. Un., 632,000 (*725,000) F17 9, D2 29
Volgograd, res., Sov. Un D3 29
Volin, Yankton, S. Dak., 171 E8 116
Volkach, Ger., 3,600 D5 17
Volkhov, Sov. Un., 16,500 B9 27
Volkhov, riv., Sov. Un. B8 27
Völklingen, Ger., 42,600 D1 17
Volkovysk, Sov. Un., 10,000 E5 27
Vollenhove, Neth., 1,918 B5 15
Volney, Grayson, Va., 40 E1 121
Volnovakha, Sov. Un., 10,000 H11 27
Volo, Lake, Ill., 150 E2 90
Volochayevka Vtoraya, Sov. Un. B7 37
Volochisk, Sov. Un., 10,000 G6 27
Volodarskiy, Sov. Un., 10,000 s31 25
Volodarskoye, Sov. Un. C7 29
Vologda, Sov. Un., 148,000 B1 29
Volokolamsk, Sov. Un., 10,000 C10 27
Volos, Grc., 49,221 (*67,424) C4 23
Volos, gulf, Grc. C4 23
Volsk, Sov. Un., 66,000 C3 29
Volta, reg., Ghana E5 45
Volta, riv., Ghana E4 45
Voltaire, McHenry, N. Dak., 70 A5 110
Volta Redonda, Braz., 83,973 (*135,000) C4, h5 56
Volterra, It., 9,300 C3 21
Volturno, riv., It. D5 21
Voluntown, New London, Conn., 500 (1,028▲) C9 84
Volusia, co., Fla., 125,319 C5 86
Volzhskiy, Sov. Un., 71,000 F17 9
Vona, Kit Carson, Colo., 130 B8 83
Vonda, Sask., Can., 238 E2 70
Von Frank, mtn., Alsk. C8 79
Vonitsa, Grc., 2,800 C3 23
Vonore, Monroe, Tenn., 525 D9 117
Von Ormy, Bexar, Tex., 350 B3 118
Voorheesville, Albany, N.Y., 1,228 *C7 108
Vopnafjordur, Ice., 388 n25 25
Vorarlberg, state, Aus., 226,323 E4 16
Vorderthal, Switz., 918 B6 19
Vordingborg, Den., 11,780 C5 24
Vorkuta, Sov. Un., 59,000 B21 9
Vorona, riv., Sov. Un. E14 27
Voronezh, Sov. Un., 496,000 C1 29
Voronezh, riv., Sov. Un. E12 27
Voronya, riv., Sov. Un. C16 25
Voroshilov, see Ussuriysk, Sov. Un.
Voroshilovgrad, see Lugansk, Sov. Un.
Vorskla, riv., Sov. Un. G10 27
Vosburg, S. Afr., 718 D3 49
Vosges, dept., Fr., 380,676 A2 18
Vosges, mts., Fr. C7 14
Voskresensk, Sov. Un., 39,000 n18 27
Voss, Nor G2 56
Voss, Walsh, N. Dak., 40 A8 110
Vossburg, Jasper, Miss., 250 D5 100
Vostochnyy, Sov. Un. B11 37
Vostok, isl., Pac. O. G12 7
Votaw, Hardin, Tex., 350 D5 118
Votice, Czech, 1,933 D9 17
Votkinsk, Sov. Un., 65,000 B4 29
Vouga, riv., Port. B1 20
Vouliagmeni, Grc., 694 h11 23
Vouvray, Fr., 1,616 D4 14
Vouziers, Fr., 3,973 C6 14
Vozhega, Sov. Un., 2,000 A13 27
Voznesensk, Sov. Un., 10,000 H8 27
Voznesenskoye, Sov. Un., 1,000 A8 37
Voznesenye, Sov. Un., 2,000 A10 27
Vrå, Den., 1,994 A3 24
Vrå, Swe. B7 24
Vranany, Czech., 659 n17 26
Vranje, Yugo., 16,457 D5 22
Vratsa, Bul., 19,448 D6 22
Vratsa, co., Bul., 771,486 D6 22
Vrbas, Yugo., 19,272 C4 22
Vrbas, riv., Yugo. C3 22
Vrchlabi, Czech., 9,900 C3 26
Vrede, S. Afr., 6,770 C4 49
Vreden, Ger., 7,200 A1 17
Vredenburgh, Monroe, Ala., 632 D2 78
Vrena, Swe., 590 u34 25
Vriezenveen, Neth., 7,200 B6 15
Vrigstad, Swe., 820 A8 24
Vrin, Switz., 393 C7 19
Vršac, Yugo., 31,551 C5 22
Vrsovice, Czech., 43,663 (part of Prague) n17 26
Vryburg, S. Afr., 14,597 C3 49
Vryheid, S. Afr., 10,753 C5 49
Vsetaty, Czech., 1,636 n18 26
Vsetin, Czech., 18,200 D5 26
Vsevidof, mtn., Alsk. E6 79
Vsevolozhskiy, Sov. Un., 5,000 r31 25
Vucha, riv., Bul. E7 22
Vukovar, Yugo., 25,826 C4 22
Vulcan, Alta., Can., 1,310 D4 69
Vulcan, Dickinson, Mich., 450 C3 98
Vulcan, Iron, Mo., 100 D7 101
Vulcano, isl., It. E5 21
Vulchedrum, Bul., 8,068 D6 22
Vuoloyarvi, Sov. Un. r31 25
Vuya, pt., Fiji Is. 52
Vyartsilya, Sov. Un. F14 25
Vyatka, riv., Sov. Un. D18 9
Vyatskiye Polyany, Sov. Un., 23,200 B4 29
Vyazemskiy, Sov. Un., 10,000 C7 37
Vyazma, Sov. Un., 26,700 D10 27
Vyazma, riv., Sov. Un. n18 27
Vyazniki, Sov. Un., 42,300 C14 27
Vyborg, Sov. Un., 56,000 G13 25, A7 27
Vychegda, riv., Sov. Un. C18 9
Vygozero, lake, Sov. Un. F16 25
Vyksa, Sov. Un., 28,600 D14 27
Vym, riv., Sov. Un. C19 9
Vyritsa, Sov. Un., 5,000 s31 25
Vyrovka, riv., Czech. n19 26
Vyshniy Volochek, Sov. Un., 70,000 C10 27
Vyskov, Czech., 12,400 D4 26
Vysoka u Melnika, Czech., 392 n18 26
Vysoke Myto, Czech., 7,983 D4 26
Vysoké Tatry, Czech., 9,000 D6 26
Vyssi Brod., Czech., 1,066 E9 17
Vytegra, Sov. Un., 11,800 A11 27

W

Wa, Ghana, 5,165 D4 45
Waal, riv., Neth. C5 15
Waalwijk, Neth., 18,200 C5 15
Wabamun, Alta., Can., 444 C3 69
Wabamun, lake, Alta., Can. C3 69
Wabana (Bell Island) Newf., Can., 8,026 E5 75
Wabash, Phillips, Ark., 115 C5 81
Wabash, Wabash, Ind., 12,621 C6 91
Wabash, Cass, Nebr., 30 E2 103
Wabash, co., Ill., 14,047 E6 90
Wabash, co., Ind., 32,605 C6 91
Wabash, riv., U.S. C10 77
Wabasha, Wabasha, Minn., 2,500 F6 99
Wabasha, co., Minn., 17,007 F6 99
Wabasso, Indian River, Fla., 500 E6 86
Wabasso, Redwood, Minn., 789 F3 99
Wabaunsee, Wabaunsee, Kans., 100 C7 93
Wabaunsee, co., Kans., 6,648 D7 93
Wabbaseka, Jefferson, Ark., 432 C4 81
Wabek, Mountrail, N. Dak., 14 B4 110
Wabeno, Forest, Wis., 800 C5 124
Wabigoon, Ont., Can., 433 E5 71
Wabigoon, lake, Ont., Can. E5 71
Wabigoon, riv., Ont., Can. E5 71
Wabiskaw, riv., Alta., Can. A3 69
Wabowden, Man., Can., 327 B2 71
Wabrzezno, Pol., 9,320 B5 26
Wabuska, Lyon, Nev., 60 D2 104
Waccamaw, riv., N.C., S.C. D5 109
Waccasassa, bay, Fla. C4 86
Wachapreague, Accomack, Va., 507 D7 121
Wachusett, mtn., Mass. B4 97
Wachusett, res., Mass. B4 97
Wacissa, Jefferson, Fla., 300 B3 86
Waco, Haralson, Ga., 381 C1 87
Waco, York, Nebr., 166 D8 103
Waco, Cleveland, N.C., 256 B2 109
Waco, McLennan, Tex., 97,808 (*129,000) D4 118
Waconia, Carver, Minn., 2,048 F5 99
Waddān, Libya D3 43
Waddān, mts., Libya D3 43
Waddell, Maricopa, Ariz., 5 D3, G1 80
Wadden, sea, Neth. A5 15
Wädenswil, Switz., 11,677 B6 19
Waddington, St. Lawrence, N.Y., 921 A2 108
Waddington, mtn., B.C., Can. D5 68
Waddy, Shelby, Ky., 300 B4 94
Waddy, lake, Sask., Can. A4 70
Wade, Cumberland, N.C., 500 B5 109
Wade, Bryan, Okla., 150 D5 112
Wade, mtn., Ant. A30 5
Wadena, Sask., Can., 1,311 F4 70
Wadena, Fayette, Iowa, 275 B6 92
Wadena, Wadena, Minn., 4,381 D3 99
Wadena, co., Minn., 12,199 D4 99
Wadesboro, Tangipahoa, La., 150 B6 95
Wadesboro, Anson, N.C., 3,744 C3 109
Wadesville, Posey, Ind., 300 H2 91
Wadham, is., Newf., Can. D5 75
Wadhams, Essex, N.Y., 150 A7, B3 108
Wadhwan, India, 27,104 F3 40
Wādī ar Ratam, Jordan m14 32
Wadi el Joz, Jordan m14 32
Wadi Halfa, Sud., 11,006 A3 47
Wading, riv., N.J. D3 106
Wading River, Burlington, N.J. D4 106
Wading River, Suffolk, N.Y., 900 D4 108
Wadi Sirhān, val., Sau. Ar D8 43
Wadley, Randolph, Ala., 605 B4 78
Wadley, Jefferson, Ga., 1,898 D4 87
Wadmalaw, isl., S.C. G2 115
Wad Medani, Sud., 47,677 C3 47
Wadowice, Pol., 7,123 D5, h10 26
Wadsworth, Autauga, Ala., 40 C3 78
Wadsworth, Lake, Ill., 150 E2 90
Wadsworth, Leavenworth, Kans. B8, C9 93
Wadsworth, Washoe, Nev., 200 D2 104
Wadsworth, Medina, Ohio, 10,635 A6 111
Waelder, Gonzales, Tex., 1,270 E4 118
Wagener, Aiken, S.C., 614 D5 115
Wager, bay, N.W. Ter., Can. C15 66
Wagga, Sud., 4,676 B4 47
Wagga Wagga, Austl., 22,087 G6 51
Waggoner, Montgomery, Ill., 219 D4 90
Waging [am See], Ger., 1,900 B8 18
Waginger See, lake, Ger. E6 16
Wagner, Phillips, Mont., 50 B8 102
Wagner, Charles Mix, S. Dak., 1,586 D7 116
Wagoner, Yavapai, Ariz., 5 C3 80
Wagoner, Wagoner, Okla., 4,469 B6 112
Wagoner, co., Okla., 15,673 B6 112
Wagon Mound, Mora, N. Mex., 760 A5 107
Wagontire, mtn., Oreg. D6 113
Wagon Wheel Gap, Mineral, Colo D4 83
Wagon Wheel Gap, res., Colo. D4 83
Wagram, Scotland, N.C., 562 C4 109
Wagrowiec, Pol., 10,800 B4 26
Wahai, Indon. F7 35
Wahak Hotrontk, Pima, Ariz. E3 80
Wahiawa, Honolulu, Haw., 15,512 B3, f9 88
Wahkiakum, co., Wash., 3,426 C2 122
Wahkon, Mille Lacs, Minn., 172 D5 99
Wahlern, Switz., 4,723 C3 19
Wahneta, Polk, Fla., 1,796 *E5 86
Wahoo, Saunders, Nebr., 3,610 C9, D2 103
Wahoo, creek, Nebr. E1 103
Wahpeton, Richland, N. Dak., 5,876 C9 110
Wahsatch, Summit, Utah, 25 B4 119
Wah Wah, mts., Utah E2 119
Wah Wa Springs, Beaver, Utah E2 119
Waia, isl., Fiji Is. 52
Waiakoa, Maui, Haw., 450 C5 88
Waialee, Honolulu, Haw., 75 B3, f9 88
Waialua (Waialua Mill), Honolulu, Haw., 2,689 B3, f9 88
Waianae, Honolulu, Haw., 4,120 B3, g9 88
Waianae, mts., Haw. g9 88
Waiawa, riv., Haw. g10 88
Waiblingen, Ger., 22,600 E4 17
Waidhofen, Aus., 3,748 D7 16
Waidhofen [an der Ybbs], Aus., 5,586 E7 16
Waigeo, isl., W. Irian E8 35
Waihee, Maui, Haw., 500 C5 88
Waikabubak, Indon. G5 35
Waikalo, Indon. G5 35
Waikane, Honolulu, Haw., 40 g10 88
Waikapu, Maui, Haw., 549 C5 88

Waikari, N.Z., 378........O14 51
Waikato, riv., N.Z.........L15 51
Waikawa, N.Z., 82........Q12 51
Waikerie, Austl., 950........G2 51
Waikii, Hawaii, Haw., 45...D6 88
Wailangilala, isl., Fiji Is......... 52
Wailea, Hawaii, Haw., 250..D6 88
Wailua (Wailua Houselots), Kauai, Haw., 1,129............A2 28
Wailuku, Maui, Haw., 6969.C5 88
Wailuku, riv., Haw.............D6 88
Waimanalo, Honolulu, Haw., 3,011..................B4, g11 88
Waimanalo, bay, Haw...........g11 88
Waimate, N.Z., 3,310......P13 51
Waimea, Honolulu, Haw., 400..........................f9 88
Waimea, Kauai, Haw., 1,312.........................B2 88
Wainganga, riv., India..........G7 40
Waingapu, Indon., 2,217....G6 35
Waini, pt., Br. Gu............A3 59
Wainola, Ontonagon, Mich..A6 98
Wainunu, bay, Fiji Is........... 52
Wainwright, Alsk., 253......A7 79
Wainwright, Alta., Can., 3,351..........................C5 69
Wainwright, East Carroll, La............................B4 95
Wainwright, Tuscarawas, Ohio, 400..................B6 111
Wainwright, Muskogee, Okla., 114.......................B6 112
Waiohinu, Hawaii, Haw., 163..............................D6 88
Waipahu, Honolulu, Haw., 7,650....................B3, g9 88
Waipara, N.Z., 349........O14 51
Waipawa, N.Z., 1,714.....M16 51
Waipio, Hawaii, Haw.........C6 88
Waipio Acres, Honolulu, Haw., 1,158.............................g9 88
Waipukurau, N.Z., 3,250...N16 51
Wairoa, N.Z., 4,303.........M16 51
Waitaki, riv., N.Z...........P13 51
Waitara, N.Z., 4,372........M15 51
Waite, Washington, Maine, 65 (73▲)........................C5 96
Waite Park, Stearns, Minn., 2,016..............................E4 99
Waiteville, Monroe, W. Va., 200..............................D4 123
Waits, riv., Vt.................C4 120
Waitsburg, Walla Walla, Wash., 1,010...................C7 122
Waitsfield, Washington, Vt., 175 (658▲)..................C3 120
Waits River, Orange, Vt., 75..................................C4 120
Waitville, Sask., Can.........E3 70
Waiwo, W. Irian..............F8 35
Wajima, Jap., 18,000........H8 37
Wajir, Ken......................A7 48
Waka, Con. L..................A3 48
Waka, Ochiltree, Tex., 100..A2 118
Wakarusa, Elkhart, Ind., 1,145..............................A5 91
Wakarusa, Shawnee, Kans., 90..................................D8 93
Wakarusa, riv., Kans...........D8 93
Wakasa, bay, Jap..................I7 37
Wakatipu, lake, N.Z...........P12 51
Wakatomika, creek, Ohio.......B5 111
Wakaw, Sask., Can., 974.....E3 70
Wakaya, isl., Fiji Is............... 52
Wakayama, Jap., 285,155.....................I7, o14 37
Wakayama, pref., Jap., 1,002,191.......................*I7 37
Wake, co., N.C., 169,082....B5 109
Wake, isl., Oceania................ 52
WaKeeney, Trego, Kans., 2,808..............................C4 93
Wakefield, Que., Can., 381..D2 73
Wakefield, Eng., 61,591......A6 12
Wakefield, Clay, Kans., 603..................................C6 93
Wakefield, Middlesex, Mass., 24,295......................A5, C3 97
Wakefield, Gogebic, Mich., 3,231..............................A5 98
Wakefield, Dixon and Wayne, Nebr., 1,068....................B9 103
Wakefield, Carroll, N.H., 100 (1,223▲)......................C5 105
Wakefield, Pike, Ohio, 135...D4 111
Wakefield, Washington, R.I., 3,570..............................D11 84
Wakefield, Sussex, Va., 1,015.E6 121
Wake Forest, Wake, N.C., 2,664..............................B5 109
Wakeham Bay, Que., Can., 112..............................D18 67
Wake Island, U.S. dep., Oceania, 1,097......................... 52
Wakeman, Huron, Ohio, 728..................................A5 111
Wakenda, Carroll, Mo., 146.B4 101
Wake Village, Bowie, Texas, 1,140..........................*C5 118
Wakita, Grant, Okla., 452....A4 112
Wakkanai, Jap., 42,100....D10 37
Wakkerstroom, S. Afr., 2,495..............................C5 49
Wakomata, lake, Ont., Can.....B7 98
Wakonda, Clay, S. Dak., 382..................................E8 116
Wakopa, Man., Can., 35.....E2 71
Wakpala, Corson, S. Dak., 200..................................B5 116
Wakuach, lake, Newf., Can......g8 75
Wakulla, Wakulla, Fla., 200.B2 87
Wakulla, co., Fla., 5,257......B2 86
Wakwekobi, lake, Ont., Can....B7 98
Walachia, reg., Rom..........C7 22
Walagā, reg., Eth., 1,000,000................................D4 47
Walbourg, Fr., 463............F7 15
Walbridge, Wood, Ohio, 2,142................................A2 111
Walbrzych, Pol., 117,000.....C4 26
Walchen, lake, Ger.............B7 18
Walcott, Greene, Ark., 75....A5 81
Walcott, Scott, Iowa, 664.....C7 92
Walcott, Richland, N. Dak., 30..................................C9 110
Walcott, Carbon, Wyo., 30..D6 125
Walcz, Pol., 13,600...........B4 26
Wald, Switz., 7,778............B6 19
Waldbröl, Ger., 13,000.......C2 17
Waldeck, Sask., Can., 237...G2 70
Walden, Jackson, Colo., 809.A4 83
Walden, Orange, N.Y., 4,851..............................D6 108
Walden, Caledonia, Vt., 30 (427▲)..........................C14 120
Walden, ridge, Tenn............D7 117
Waldenburg, Poinsett, Ark., 113..................................B4 81
Walden Heights, Caledonia, Vt..................................C4 120
Waldheim, Sask., Can., 515.E2 70
Waldheim, Ger., 11,400......B8 17
Waldheim, St. Tammany, La..................................D5 95
Waldkappel, Ger., 1,900......B4 17
Waldkirchen, Ger., 2,400.....E8 17
Waldmünchen, Ger., 4,100..D7 17
Waldo, Columbia, Ark., 1,722..............................D2 81
Waldo, B.C., Can., 117......E10 68
Waldo, Alachua, Fla., 735....C4 86
Waldo, Russell, Kans., 178..C5 93
Waldo, Santa Fe, N. Mex....G6 107
Waldo, Marion, Ohio, 374...B4 111
Waldo, co., Maine, 22,632...D3 96
Waldo, hills, Oreg.............C2 113
Waldo, lake, Mass.............E3 97
Waldo, lake, Oreg.............D4 113
Waldoboro, Lincoln, Maine, 705 (2,882▲)....................D3 96
Waldorf, Charles, Md., 1,048.C4 85
Waldorf, Waseca, Minn., 270..................................G5 99
Waldort, Charles Md., 1,048.C4 85
Waldport, Lincoln, Oreg., 667..................................C2 113
Waldron, Scott, Ark., 1,758.C1 81
Waldron, Sask., Can., 99....G4 70
Waldron, Shelby, Ind., 700..F6 91
Waldron, Harper, Kans., 38.E5 93
Waldron, Hillsdale, Mich., 454..................................G6 98
Waldron, Platte, Mo., 200...E1 101
Waldsassen, Ger., 7,600......C7 17
Waldshut, Ger., 10,900.......E4 16
Waldwick, Bergen, N.J., 10,495..............................A4 106
Wales, Alsk., 128...............B6 99
Wales, Ont., Can., 176......B10 72
Wales, Hampden, Mass., 300 (659▲)..........................B3 97
Wales, Lake, Minn., 40......C7 99
Wales, Cavalier, N. Dak., 151..................................A7 110
Wales, Giles, Tenn., 100.....B4 117
Wales, Sanpete, Utah, 130...D4 119
Wales, reg., U.K., 2,640,632.............D5 10, B4 12
Waleska, Cherokee, Ga., 479..................................B2 87
Walford Station, Ont., Can., 153..................................A2 72
Walgett, Austl.,1,726..........E7 51
Walgreen, coast, Ant............B2 5
Walhachin, B.C., Can., 65...D7 68
Walhalla, Mason, Mich., 300.E4 98
Walhalla, Pembina, N. Dak., 1,432..............................A8 110
Walhalla, Oconee, S.C., 3,431..............................B1 115
Walhonding, riv., Ohio..........B5 111
Walikale, Con. L...............B4 48
Walker, Yavapai, Ariz., 30..C3 80
Walker, Linn, Iowa, 584......B6 92
Walker, Ellis, Kans., 100.....D4 93
Walker, Livingston, La., 912.A6 95
Walker, Cass, Minn., 1,180..C4 99
Walker, Vernon, Mo., 235...D3 101
Walker, Corson, S. Dak., 20..B4 116
Walker, co., Ala., 54,211.....B2 76
Walker, co., Ga., 45,264.....B1 87
Walker, co., Tex., 21,475.....D5 118
Walker, creek, Wyo.............C7 125
Walker, lake, Man., Can........B3 71
Walker, lake, Nev...............E3 104
Walker, mtn., Ga................B3 87
Walker, mtn., Oreg..............D5 113
Walker, mtn., Va................D1 121
Walker, rim, Oreg...............D5 113
Walker Knob, mtn., Tenn........D8 117
Walker River, Indian res., Nev..E3 104
Walker Springs, Clarke, Ala., 500..................................D2 78
Walkersville, Frederick, Md., 1,020..............................B3 85
Walkersville, Lewis, W. Va., 200..................................C4 123
Walkerton, Ont., Can., 3,851..............................C3 72
Walkerton, St. Joseph, Ind., 2,044..............................B5 91
Walkertown, Forsyth, N.C., 1,240..............................A3 109
Walkerville, Oceana, Mich., 261..................................E4 98
Walkerville, Silver Bow, Mont., 1,453.....................D4 102
Wall, Allegheny, Pa., 1,493.*F2 114
Wall, Pennington, S. Dak., 629..................................D3 116
Wall, lake, Iowa...............B4 92
Wallace, Escambia, Ala., 150..................................D2 78
Wallace, N.S., Can., 276.....D6 74
Wallace, Shoshone, Idaho, 2,412..............................B3 89
Wallace, Wallace, Kans., 110..................................D2 93
Wallace, Menominee, Mich., 120..................................C3 98
Wallace, Lincoln, Nebr., 293..................................D4 103
Wallace, Steuben, N.Y., 300..................................C3 108
Wallace, Duplin, N.C., 2,285.C5 109
Wallace, Codington, S. Dak., 132..................................B8 116
Wallace, Washington, Va., 200..................................B3 121
Wallace, Harrison, W. Va., 200..............................A6, B4 123
Wallace, co., Kans., 2,069...D2 93
Wallace, creek, Wyo............C5 125
Wallaceburg, Ont., Can., 7,881..............................E2 72
Wallacetown, Ont., Can., 200..................................E3 72
Wallagrass, Aroostook, Maine, (818▲).....................A4 96
Wallal Downs, Austl..........C3 50
Walland, Blount, Tenn., 250..............................D10 117
Wallaroo, Austl., 2,237.......F6 50
Wallasey, Eng., 103,213......A4 12
Wallaston, cape, N.W. Ter., Can.............................B9 66
Walla Walla, Walla Walla, Wash., 24,536..................C7 122
Walla Walla, co., Wash., 42,195..............................C7 122
Walla Walla, riv., Wash.......C7 122
Walla Walla West (Garrett), Walla Walla, Wash., 1,641..........................*C7 122
Walldürn, Ger., 7,400.......D4 17
Walled Lake, Oakland, Mich., 3,550...................A7 98
Wallen, Allen, Ind..........B7 91
Wallen, lake, Switz.............B7 19
Wallenpaupack, lake, Pa.......D11 114
Wallenstadt, Switz., 3,296...B7 19
Waller, Waller, Tex., 900...D5 118
Waller, co., Tex., 12,071....E4 118
Wallerville, Union, Miss.....A5 100
Wallingford, New Haven, Conn., 29,920....................D5 84
Wallingford, Eng., 4,829....C6 12
Wallingford, Emmet, Iowa, 228..................................A3 92
Wallingford, Delaware, Pa., 3,000..........................*B10 114
Wallingford, Rutland, Vt., 900 (1,439▲)....................E3 120
Wallington, Eng., 32,500....m12 10
Wallington, Bergen, N.J., 9,261..............................D5 106
Wallis, Austin, Tex., 950.....E4 118
Wallis & Futuna Is., Fr. dep., Oceania, 9,000..........*G11 7
Wallkill, Ulster, N.Y., 1,215..............................D6 108
Wallkill, riv., N.Y.............D6 108
Wall Lake, Sac, Iowa, 812...B2 92
Wallo, reg., Eth., 1,000,000..C4 47
Walloomsac, riv., Vt............F2 120
Walloon, lake, Mich............C6 98
Wallowa, Wallowa, Oreg., 989..................................B9 113
Wallowa, co., Oreg., 7,102...B9 113
Wallowa, mts., Oreg............B9 113
Wallpack Center, Sussex, N.J., 25..............................A3 106
Walls, West Baton Rouge, La..................................D4 95
Walls, De Soto, Miss., 300...A3 100
Wallsburg, Wasatch, Utah, 180..................................C4 119
Wallsend, Eng., 49,785.......C6 10
Wallula, Walla Walla, Wash., 150..................................C7 122
Wallum, lake, R.I...............A9 84
Wallville, Calvert, Md.,.......D4 86
Walney, isl., Eng...............C5 10
Walnut, Los Angeles, Calif., 934..............................*F3 82
Walnut, Bureau, Ill., 1,192...B4 90
Walnut, Pottawattamie, Iowa, 777..............................C2 92
Walnut, Crawford, Kans., 381..................................E8 93
Walnut, Tippah, Miss., 390..A5 100
Walnut, Madison, N.C., 450.C3 109
Walnut, creek, Kans............D4 93
Walnut, hill, Mass..............B2 97
Walnut, riv., Kans..............E7 93
Walnut Bottom, Cumberland, Pa., 325..........................F7 114
*Walnut Canyon, nat. mon., Ariz.*B4 80
Walnut Cove, Stokes, N.C., 1,288..............................A3 109
Walnut Creek, Contra Costa, Calif., 9,903....................B5 82
Walnut Grove, Etowah, Ala., 237..............................A3 78
Walnut Grove, Sacramento, Calif., 725..........................B6 82
Walnut Grove, Redwood, Minn., 886..........................F3 99
Walnut Grove, Leake, Miss., 433..................................C4 100
Walnut Grove, Greene, Mo., 373..................................D4 101
Walnut Heights, Contra Costa, Calif., 5,080..................*B5 82
Walnut Hill, Lafayette, Ark., 25..................................D2 81
Walnut Hill, Escambia, Fla., 150..................................C1 86
Walnut Hill, Cumberland, Maine..............................E5 96
Walnut Park, Los Angeles, Calif., 7,500..........................*F2 82
Walnutport, Northampton, Pa., 1,609..........................E10 114
Walnut Ridge, Lawrence, Ark., 3,547..........................A5 81
Walpole, Sask., Can., 95....H5 70
Walpole, Norfolk, Mass., 7,000 (14,068▲).............B5, E2 97
Walpole, Cheshire, N.H., 800 (2,825▲)....................D2 105
Walsall, Eng., 117,896.......B5 12
Walsenburg, Huerfano, Colo., 5,071..........................D6 83
Walsh, Alta., Can., 97.......E5 69
Walsh, Baca, Colo., 856.....D8 83
Walsh, co., N. Dak., 17,997..A8 110
Walsingham, cape, N.W. Ter., Can...............................C20 67
Walsrode, Ger., 13,000.......F3 24
Walston, Jefferson, Pa., 350..E4 114
Walter Bathurst, cape, N.W. Ter., Can..............................B17 67
Walterboro, Colleton, S.C., 5,417..............................F6 115
Walterhill, Rutherford, Tenn., 100..........................B5 117
Walters, Mississippi, Ark.....B5 81
Walters, Faribault, Minn., 133..................................G5 99
Walters, Cotton, Okla., 2,825..............................C3 112
Walters, Isle of Wight, Va., 135..................................E6 121
Walters Falls, Ont., Can., 140..................................C4 72
Waltershausen, Ger., 13,400.C5 17
Waltersville, Saline, Miss., 400..................................C3 100
Walterville, Lane, Oreg., 170.C4 113
Walthall, Webster, Miss., 153..................................B4 100
Walthall, Chesterfield, Va., 230..................................C5 121
Walthall, co., Miss., 13,512..D3 100
Waltham, Hancock, Maine, 30 (153▲)..........................D4 96
Waltham, Middlesex, Mass., 55,413....................B5, D2 97
Waltham, Mower, Minn., 207..................................G6 99
Waltham, Chouteau, Mont., 25..................................C6 102
Waltham Station, Que., Can., 413..................................B8 102
Walthamstow, Eng., 108,788..............................k12 10
Walthill, Thurston, Nebr., 844..................................B9 103
Walthourville, Long, Ga., 600..................................E5 87
Waltman, Natrona, Wyo., 5..B5 125
Walton, N.S., Can., 393.....D6 74
Walton, Ont., Can., 62.......D3 72
Walton, St. Lucie, Fla., 300.E6 86
Walton, Cass, Ind., 1,079....C5 91
Walton, Harvey, Kans., 225..............................A5, D6 93
Walton, Boone, Ky., 1,530..............................A7, B5 94
Walton, Lancaster, Nebr., 80.E2 103
Walton, Delaware, N.Y., 3,855..............................C5 108
Walton, Roane, W. Va., 300.C3 123
Walton, co., Fla., 15,576....G3 86
Walton, co., Ga., 20,481.....C3 87
Walton Hills, Cuyahoga, Ohio, 1,776..............................*A6 111
Waltonville, Jefferson, Ill., 394..................................E4 90
Waltreak, Yell, Ark., 50......C2 81
Walum, Griggs, N. Dak., 65..B7 110
Walville, peak, Wash............C2 122
Walvis Bay, S. Afr., 12,235..............................B1 49
Walworth, Wayne, N.Y., 480..................................B3 108
Walworth, Walworth, Wis., 1,494..............................F5 124
Walworth, co., S. Dak., 8,097..................................B5 116
Walworth, co., Wis., 52,368..F5 124
Wama, Afg., 5,000..........D15 41
Wamac, Marion, Clinton and Washington, Ill., 1,394....E4 90
Wamba, Con. L................A4 48
Wamba, Nig....................E6 45
Wamba, riv., Con. L............C2 48
Wamego, Pottawatomie, Kans., 2,363..........................C7 93
Wamesit, Middlesex, Mass., 150....................A5, C2 97
Wamgumbaug, lake, Conn.......B7 84
Wami, riv., Tan.................C6 48
Wamic, Wasco, Oreg., 125...B5 113
Wampoo, Pulaski, Ark.......E6 81
Wampsville, Madison, N.Y., 564..................................B5 108
Wampum, Lawrence, Pa., 1,085..............................E1 114
Wamsutter, Sweetwater, Wyo., 110..........................D4 125
Wana, Pak......................A2 40
Wanaaring, Austl., 76........D5 51
Wanakah, Erie, N.Y., 2,000.*C2 108
Wanakena, St. Lawrence, N.Y..................................A6 108
Wanamaker, Marion, Ind., 600..................................H8 91
Wanamassa, Monmouth, N.J., 3,928..........................C4 106
Wanamie, Luzerne, Pa., 950..................................D9 114
Wanamingo, Goodhue, Minn., 540..........................F6 99
Wanan, China, 3,000.........K6 36
Wananish, Columbus, N.C. (part of Lake Waccamon).......C5 109
Wanapitei, riv., Ont., Can......A4 72
Wanaque, Passaic, N.J., 7,126..............................A4 106
Wanaque, res., N.J...............A4 106
Wanatah, La Porte, Ind., 800..................................B4 91
Wanblee, Washabaugh, S. Dak., 200..........................D4 116
Wanchese, Dare, N.C., 600..B8 109
Wandering River, Alta., Can.B4 69
Wando, Kor., 7,800............I3 37
Wando, Berkeley, S.C., 100.............................E8, F3 115
Wando, riv., S.C................F8 115
Wandoan, Austl., 370.........C7 51
Wandsworth, Eng., 347,209 (part of London)..............C7 12
Waneta, B.C., Can., 67......E9 68
Wanette, Pottawatomie, Okla., 381..............................C4 112
Wanganui, N.Z., 33,316 (*35,694)..........................M15 51
Wangaratta, Austl., 13,783...H6 51
Wangava, isl., Fiji Is............. 52
Wangching (Paitsaokou), China..............................C10 34
Wangen, Ger., 13,300........B5 18
Wangerooge, isl., Ger...........B3 16
Wangkuei, China, 5,000......C3 37
Wangyehmiao, China., 51,400..............................A10 36
Wanham, Alta., Can., 251...B1 69
Wanhsien, China...............D6 34
Wanhsien, China, 90,000.....I3 36
Wanilla, Lawrence, Miss., 100..................................D3 100
Wanipigow, riv., Man., Can....D4 71
Wankie, S. Rh., 18,200......A4 49
Wankie, nat. park, S. Rh.......A4 49
Wann, Saunders, Nebr., 35..E2 103
Wann, Nowata, Okla., 157...A6 112
Wannaska, Roseau, Minn., 50..................................B3 99
Wanne-Eickel, Ger., 107,200..........................*C7 15
Wansreck, riv., Eng............E6 13
Wantagh, Nassau, N.Y., 34,172..............................G2 84
Wantsai, China, 15,000.......J6 36
Wanup, Ont., Can., 150......A4 72
Wapakoneta, Auglaize, Ohio, 6,756..............................B3 111
Wapanucka, Johnson, Okla., 459..................................C5 112
Wapato, Yakima, Wash., 3,137..............................C5 122
Wapawekka, hills, Sask., Can...C3 70
Wapawekka, lake, Sask., Can....C3 70
Wapella, Sask., Can., 584....G5 70
Wapella, DeWitt, Ill., 526....C5 90
Wapello, Louisa, Iowa, 1,745.C6 92
Wapello, co., Iowa, 46,126...C5 92
Wapisu, lake, Man., Can........B2 71
Wapiti, pass., B.C., Can........B7 68
Wapiti, range, Wyo.............A3 125
Wapiti, riv., B.C., Alta., Can.......................B7 68, B1 69
Wappapello, Wayne, Mo., 150..................................E7 101
Wappapello, res., Mo............D7 101
Wapping, Hartford, Conn., 400..................................B6 84
Wappinger, creek, N.Y...........B1 84
Wappingers Falls, Dutchess, N.Y., 4,447....................D7 108
Wapsipinicon, riv., Iowa........B5 92
Wapske, N.B., Can., 182.....B5 96
Wapus, lake, Sask., Can........A4 70
Wapwallopen, Luzerne, Pa., 200..................................D9 114
Wāqid, Eg., U.A.R.............D2 32
Waqqāş, Jordan, 1,000.......B7 32
Waquoit, Barnstable, Mass., 336..................................C6 97
War, McDowell, W. Va., 3,006..............................D3 123
War, ridge, W. Va..............D7 123
Waramaug, lake, Conn..........C3 84
Warangal, India, 156,106..................E6 38, H7 40
Warba, Itasca, Minn., 162...C5 99
Warbowden, Man., Can., 75.D5 71
Warburg, Alta., Can., 285...C3 69
Warburg, Ger., 9,200.........B4 17
Warburton, Austl., 1,630.....H5 51
Warburton, riv., Austl...........C2 51
Ward, Sumter, Ala., 100......C1 78
Ward, Lonoke, Ark., 470......B4 81
Ward, Saluda, S.C., 162......D4 115
Ward, Moody, S. Dak., 74...C9 116
Ward, Kanawha, W. Va., 1,109..............................C6 123
Ward, co., N. Dak., 47,072..A4 110
Ward, co., Tex., 14,917......D1 118
Ward, mtn., Mont...............D2 102
Wardān, Eg., U.A.R., 3,000.D2 32
Wardell, Pemiscot, Mo., 331.E8 101
Warden, Grant, Wash., 949..C6 122
Warden Junction, Alta., Can., 60..................................C4 69
Wardensville, Hardy, W. Va., 278..................................B6 123
Warder, lake, Ger...............E4 24
Wardha, India, 49,113.......G7 40
Ward Hill, Essex, Mass. (part of Haverhill).................A5 97
Wardlow, Alta., Can., 30....D5 69
Wardner, B.C., Can., 171....E10 68
Wardner, Shoshone, Idaho, 577..................................B2 89
Ward Ridge, Gulf, Fla., 45..............................*C1 86
Wardsboro, Windham, Vt., 125 (322▲)......................E3 120
Wardville, Rapides, La., 1,086..............................*C3 95
Wardville, Atoka, Okla., 150.C5 112
Ware, Pocahontas, Iowa 25.B3 92
Ware, Hampshire, Mass., 6,650..............................B3 97
Ware, co., Ga., 34,219........E4 87
Ware, riv., Mass................B3 97
Wareagle, Benton, Ark., 40..A2 81
War Eagle, Mingo, W. Va., 180..................................D3 123
War Eagle, creek, Ark...........A2 81
Ware Center, Hampshire, Mass..................................B3 97
Wareham, Eng., 3,094........D5 12
Wareham, Plymouth, Mass., 1,739 (9,451▲)..................C6 97
Wareham Center, Plymouth, Mass..................................C6 97
Warehouse Point, Hartford, Conn., 1,936....................B6 84
Waren, Ger., 19,700...........B6 16
Waren, W. Irian................F9 35
Warendorf, Ger., 15,800......B2 17
Waresboro, Ware, Ga., 350..E4 87
Ware Shoals, Greenwood, S.C., 2,671..........................C3 115
Waretown, Ocean, N.J., 500.D4 106
Warfield, B.C., Can., 2,212.*E9 68
Warfield, Martin, Ky., 295...C7 94
Warfield, Brunswick, Va., 80.E5 121
Warfordsburg, Fulton, Pa., 100..................................G5 114
Warin, Ger., 3,652............E5 24
Waring, Kendall, Tex., 75...E3 118
Wark, Eng., 690................E6 13
Warkworth, Ont., Can., 514.C7 72
Warkworth, N.Z., 991........L15 51
Warland, Lincoln, Mont., 60.B1 102
Warman, Sask., Can., 659...E2 70
Warmbad, S. Afr., 6,351......B4 49
Warmbad, S.W. Afr., 177....C2 49
Warm Beach, Snohomish, Wash., 300..........................A3 122
Warminister, Bucks, Pa., 3,000..............................*F11 114
Warminster, Eng., 9,855.....C5 12
Warm River, Fremont, Idaho, 20..................................E7 89
Warm Springs, Randolph, Ark., 40..............................A4 81
Warm Springs, Meriwether, Ga., 538..............................D2 87
Warmsprings, Deer Lodge, Mont..................................D4 102
Warm Springs, Nye, Nev., 5.E5 104
Warm Springs, Jefferson, Oreg., 250..........................C5 113
Warm Springs, Bath, Va., 300..................................C3 121
*Warm Springs, Indian res., Oreg.*C5 113
Warm Springs, res., Oreg.......D8 113
Warner, Alta., Can., 472.....E4 69
Warner, Henry, Ill., 20........D7 92
Warner, Merrimack, N.H., 750 (1,004▲)....................D3 105
Warner, Washington, Ohio, 200..................................C6 111
Warner, Muskogee, Okla., 881..................................B6 112
Warner, Brown, S. Dak., 135.B7 116
Warner, mtn., Mass..............B1 97
Warner, mts., Calif., Oreg......F6 113
Warner Robins, Houston, Ga., 18,633..............................D3 87
Warner Springs, San Diego, Calif., 150..........................F5 82
Warnerton, Washington, La., 35..................................D5 95
Warnes, Bolivia, 1,581.......C3 55
Warnow, riv., Ger..............B6 16
Warpath, riv., Man., Can........C2 71
Warracknabeal, Austl., 3,061 H4 51
Warr Acres, Oklahoma, Okla., 7,135..............................B4 112
Warragul, Austl., 6,404.......I5 51
Warren, Bradley, Ark., 6,752..............................D3 81
Warren, Austl., 1,505.........E6 51
Warren, Ont., Can., 557.....A4 72
Warren, Litchfield, Conn., 100 (600▲)..........................C3 84
Warren, Idaho, Idaho, 30....D3 89
Warren, Jo Daviess, Ill., 1,470..............................A4 90
Warren, Huntington, Ind., 1,241..............................C7 91
Warren, Knox, Maine, 850 750 (1,678▲)....................D3 96
Warren, Worcester, Mass., 1,616 (3,383▲)..................B3 97
Warren, Macomb, Mich., 89,246..............................F7 98
Warren, Marshall, Minn., 2,007..............................B2 99
Warren, Carbon, Mont., 10..E8 102
Warren, Grafton, N.H., 400 (548▲)..........................C3 105
Warren, Herkimer, N.Y., 80..................................C6 108
Warren, Trumbull, Ohio, 59,648..............................A7 111
Warren, Warren, Pa., 14,505.C3 114
Warren, Bristol, R.I., 8,750..............................C11 84
Warren, Tyler, Tex., 360....D5 117
Warren, Washington, Vt., 200 (469▲)........................C3 120
Warren, co., Ga., 7,360......C4 87
Warren, co., Ill., 21,587.....C3 90
Warren, co., Ind., 8,545.....D3 91
Warren, co., Iowa, 20,829...C4 92
Warren, co., Ky., 45,491.....D3 94
Warren, co., Miss., 42,206...C3 100
Warren, co., Mo., 8,750.....C6 101
Warren, co., N.J., 63,220....B3 106
Warren, co., N.Y., 44,002....B7 108
Warren, co., N.C., 19,652....A5 109
Warren, co., Ohio, 65,711...C3 111
Warren, co., Pa., 45,582.....C3 114
Warren, co., Tenn., 23,102..D8 117
Warren, co., Va., 14,655.....C4 121
Warren Center, Bradford, Pa., 75..................................C9 114
Warrendale, Allegheny, Pa., 800..................................A5 114
Warren Grove, Ocean, N.J..D4 106
Warren Park, Marion, Ind., 852..................................H8 91
Warrenpoint, N. Ire., 3,238.C5 11
Warren Point, Bergen, N.J. (part of Fair Lawn).........D5 106
Warrens, Monroe, Wis., 280.D3 124
Warrensburg, Macon, Ill., 681..................................D4 90
Warrensburg, Johnson, Mo., 9,689..............................C4 101
Warrensburg, Warren, N.Y., 2,240..............................B7 108
Warrensburg, Greene, Tenn., 100..................................C10 117
Warrensville, Cuyahoga, Ohio, 10,609........................B2 111
Warrensville, Lycoming, Pa., 200..................................D8 114
Warrenton, Warren, Ga., 1,770..............................C4 87
Warrenton, Warren, Mo., 1,869..............................C6 101
Warrenton, Warren, N.C., 1,124..............................A5 109
Warrenton, Clatsop, Oreg., 1,717..............................A3 113
Warrenton, S. Afr., 5,980...C3 49
Warrenton, Fauquier, Va., 3,522..............................C5 121
Warrentown, Fayette, Ky....B5 94
Warrenville, Windham, Conn., 100..........................B8 84
Warrenville, Du Page, Ill., 3,134..............................F2 90
Warrenville, Aiken, S.C., 1,128..............................D4 115
Warri, Nig., 10,726...........E6 45
Warrick, co., Ind., 23,577...H3 91
Warrington, Eng., 75,533 (*120,000)..........................A5 12
Warrington, Escambia, Fla., 16,752..............................G2 86
Warrior, Jefferson, Ala., 2,448..............................B3 78
Warrior, mtn., Md...............D2 85
Warriors, Mark, Huntingdon, Pa., 220..........................E5 114
Warrnambool, Austl., 15,702.I4 51
Warroad, Roseau, Minn., 1,309..............................B3 99
Warsaw, Ont., Can., 258....C6 72
Warsaw, Hancock, Ill., 1,938..............................C2 90
Warsaw, Kosciusko, Ind., 7,234..............................B6 91
Warsaw, Gallatin, Ky., 981..............................B5, B6 94
Warsaw, Benton, Mo., 1,054.C4 101
Warsaw, Wyoming, N.Y., 3,653..............................C2 108
Warsaw, Duplin, N.C., 2,221..............................B5 109
Warsaw, Coshocton, Ohio, 594..................................B5 111
Warsaw (Warszawa), Pol., 1,136,000 (*1,480,000)................B6, m14 26
Warsaw, Richmond, Va., 549..................................D6 121
Warson Woods, St. Louis, Mo., 1,746..........................*C7 101
Warspite, Alta., Can., 153...B4 69
Warstein, Ger., 9,000.........B3 17
Warszawa, see Warsaw, Pol.
Warta, Pol., 2,896.............C5 26
Warta, riv., Pol.................B5 26
Wartburg, Morgan, Tenn., 800..................................C9 117
Warth, Aus., 120...............B6 18
Warthen, Washington, Ga., 275..................................C4 87
Wartime, Sask., Can., 96....F1 70
Wartrace, Bedford, Tenn., 545..................................B5 117
Warwick, Austl., 9,843........D9 51
Warwick, Ont., Can., 110....D3 72
Warwick, Que., Can., 2,487..D6 73
Warwick, Eng., 16,032.......B6 12
Warwick, Worth, Ga., 434...E3 87
Warwick, Cecil, Md., 350....B6 85
Warwick, Franklin, Mass., 150 (426▲)..........................A3 97
Warwick, Orange, N.Y., 3,218..........................D2, D6 108
Warwick, Benson, N. Dak., 204..................................B7 110
Warwick, Lincoln, Okla., 250..................................B5 112
Warwick, Chester, Pa., 120.F10 114
Warwick, Kent, R.I., 68,504..............................C10 84
Warwick, co., Eng., 2,023,289..............................B6 11
Wasatch, co., Utah, 5,308...C4 119
Wasatch, mts., Utah............C4 119
Wasco, Kern, Calif., 6,841...E4 82
Wasco, Sherman, Oreg., 348..................................B6 113
Wasco, co., Oreg., 20,205....B5 113
Waseca, Sask., Can., 103....D1 70
Waseca, Waseca, Minn., 5,898..............................F5 99
Waseca, co., Minn., 16,041...F5 99
Wash, The, bay, Eng............B8 12
Washabaugh, co., S. Dak., 1,042..............................D4 116
Washademoak, Lake, N.B., Can..................................D4 74
Washago, Ont., Can., 355....C5 72
Washakie, co., Wyo., 8,883..B5 125

Washakie Needles, mtn., Wyo...B3 125
Washburn, Woodford and Marshall, Ill., 1,064......C4 90
Washburn, Black Hawk, Iowa, 900......B5 92
Washburn, Aroostook, Maine, 1,055 (2,083▲)......B4 96
Washburn, Barry, Mo., 325..E4 101
Washburn, McLean, N. Dak., 993......B4 110
Washburn, Bayfield, Wis., 1,896......B3 124
Washburn, co., Wis., 10,301......C2 124
Washigomog, lake, Ont., Can...B6 72
Washington, Hempstead, Ark., 321......D2 81
Washington, Litchfield, Conn., 500 (2,603▲)......C3 84
Washington, D.C., 763,956 (★2,053,600)..C1, C3 85
Washington, Wilkes, Ga., 4,440......C4 87
Washington, Tazewell, Ill., 5,919......C4 90
Washington, Daviess, Ind., 10,846......G3 91
Washington, Washington, Iowa, 6,037......C6 92
Washington, Washington, Kans., 1,506......C6 93
Washington, Mason, Ky., 600......B6 94
Washington, St. Landry, La., 1,291......D3 95
Washington, Berkshire, Mass., 80 (290▲)......B1 97
Washington, Adams, Miss., 200......D2 100
Washington, Franklin, Mo., 7,961......C6 101
Washington, Washington, Nebr., 44......D2 103
Washington, Sullivan, N.H., 100 (162▲)......D2 105
Washington, Warren, N.J., 5,723......B3 106
Washington, Beaufort, N.C., 9,939......B7 109
Washington, McClain, Okla., 278......B4 112
Washington, Washington, Pa., 23,545......F1 114
Washington, Rhea, Tenn., 90......D9 117
Washington, Washington, Utah, 445......F2 119
Washington, Orange, Vt., 220 (565▲)......C4 120
Washington, Rappahannock, Va., 255......C4 121
Washington, Wood, W. Va., 50......B3 123
Washington, co., Ala., 15,372......D1 78
Washington, co., Ark., 55,797......A1 81
Washington, co., Colo., 6,625......B7 83
Washington, co., Fla., 11,249......G3 86
Washington, co., Ga., 18,903......C4 87
Washington, co., Idaho, 8,378......E2 89
Washington, co., Ill., 13,569.E4 90
Washington, co., Ind., 17,819......G5 91
Washington, co., Iowa, 19,406......C6 92
Washington, co., Kans., 10,739......C7 93
Washington, co., Ky., 11,168.C4 94
Washington, co., Maine, 32,908......D5 96
Washington, co., Md., 91,219......A2 85
Washington, co., Minn., 52,432......E6 99
Washington, co., Miss., 78,638......B3 100
Washington, co., Mo., 14,346......D2 101
Washington, co., Nebr., 12,103......C9 103
Washington, co., N.Y., 48,476......B7 108
Washington, co., N.C., 13,488......B7 109
Washington, co., Ohio, 51,489......C6 111
Washington, co., Okla., 42,347......A6 112
Washington, co., Oreg., 92,237......B3 113
Washington, co., Pa., 217,271......F1 114
Washington, co., R.I., 59,054......D10 84
Washington, co., Tenn., 64,832......C11 117
Washington, co., Tex., 19,145......D4 118
Washington, co., Utah, 10,271......F2 119
Washington, co., Vt., 42,860.C3 120
Washington, co., Va., 38,076.B3 121
Washington, co., Wis., 46,119......E5 124
Washington, par., La., 44,015......D5 95
Washington, state, U.S., 2,853,214......A3 76, 122
Washington, cape, Fiji Is......52
Washington, isl., Pac. O......F12 7
Washington, isl., Wis......C7 124
Washington, lake, Fla......D6 86
Washington, lake, Minn......E4 99
Washington, lake, Miss......B2 100
Washington, lake, Wash......D2 122
Washington, mtn., N.H......B4 105
Washington Bald, mtn., Maine..D2 74
Washington Court House, Fayette, Ohio, 12,388......C4 111
Washington Crossing, Mercer, N.J., 500......C3 106
Washington Depot, Litchfield, Conn., 400......C3 84
Washington Heights, Orange, N.Y., 1,231......*D6 108
Washington North, Washington, Pa., 2,077......*F1 122
Washington Park, St. Clair, Ill., 6,601......E3 90
Washington Place, Marion, Ind., 2,000......*E5 91
Washington Terrace, Weber, Utah, 6,441......B4 119
Washingtonville, Montour, Pa., 198......D8 114
Washington West, Washington, Pa., 3,951......*F1 114
Washir, Afg., 10,000......E11 41
Washita, co., Okla., 18,121..B2 112
Washita, riv., Okla......B3 112
Washoe, Washoe, Nev., 175..D2 104
Washoe, co., Nev., 84,743...C2 104
Washougal, Clark, Wash., 2,672......D3 122
Washow, bay, Man., Can......D3 71
Washta, Cherokee, Iowa, 310......B2 92
Washtenaw, co., Mich., 172,440......F7 98
Washtucna, Adams, Wash., 331......C7 122
Washunga, Kay, Okla., 60..A5 112
Wasigny, Fr., 535......E4 15
Wasilkow, Pol., 3,948......B7 26
Wasilla, Alsk., 112....C10, g17 79
Waskada, Man., Can., 297..E1 71
Waskaiowaka, lake, Man., Can.A3 71
Waskatenau, Alta., Can., 305......B4 69
Waskesiu, lake, Sask., Can......D2 70
Waskish, Beltrami, Minn., 35......B4 99
Waskom, Harrison, Tex., 1,336......C5 118
Wasque, pt., Mass......D7 97
Wass, lake, Man., Can......C4 71
Wassaw, sound, Ga......E6 87
Wassenaar, Neth., 25,400....B4 15
Wasseralfingen, Ger., 10,900.E5 17
Wasserburg am Inn, Ger., 6,500......A8 18
Wasserkuppe, mtn., Ger......C4 17
Wassookeag, lake, Maine......C3 96
Wasson, Saline, Ill., 100......F5 90
Wassuk, range, Nev......E3 104
Wassy-sur-Blaise, Fr., 2,818..F4 15
Wasta, Pennington, S. Dak., 196......C3 116
Wataga, Knox, Ill., 570......B3 90
Watalula, Franklin, Ark......B2 81
Watampone, Indon., 2,515...F6 35
Wataroa, N.Z., 159......O13 51
Watatic, mtn., Mass......A4 97
Watauga, Corson, S. Dak., 74......B4 116
Watauga, Carter, Tenn., 500......C11 117
Watauga, co., N.C., 17,529..A2 109
Watauga, res., Tenn......C11 117
Watauga, riv., Tenn......C12 117
Watchang, pond, R.I......D10 84
Watchet, Eng., 2,596......C4 12
Watch Hill, Washington, R.I., 300......D9 84
Watchung, Somerset, N.J., 3,312......*B4 106
Water, isl., Vir. Is......I15 65
Waterboro, York, Maine, 300 (1,059▲)......E2 96
Waterbury, New Haven, Conn., 107,130 (★190,300)......C4 84
Waterbury, Dixon, Nebr., 81......B9 103
Waterbury, Washington, Vt., 2,984 (4,303▲)......C3 120
Waterbury, riv., Vt......C3 120
Waterbury Center, Washington, Vt., 400......C3 120
Waterdown, Ont., Can., 1,844......D5 72
Wateree, Richland, S.C., 75.D6 115
Wateree, res., S.C......C6 115
Wateree, riv., S.C......D6 115
Waterflow, San Juan, N. Mex., 15......A1 107
Waterford, Stanislaus, Calif., 1,780......*D3 82
Waterford, Ont., Can., 2,221......E4 72
Waterford, New London, Conn., 5,000 (15,391▲)....D8 84
Waterford, LaPorte, Ind., 200......A4 91
Waterford, Ire., 28,216......E4 11
Waterford, Spencer, Ky., 60......A4 94
Waterford, Oakland, Mich., 1,000......*F7 98
Waterford, Marshall, Miss., 175......A4 100
Waterford, Saratoga, N.Y., 2,915......C7 108
Waterford, Washington, Ohio, 450......C6 111
Waterford, Erie, Pa., 1,390..C1 114
Waterford, Loudoun, Va., 247......B5 121
Waterford, Racine, Wis., 1,500......F1, F5 124
Waterford, co., Ire., 71,439..E4 11
Waterford, hbr., Ire......E5 11
Waterford Mills, Elkhart, Ind., 150......A6 91
Waterford Works, Camden, N.J., 700......D3 106
Watergrasshill, Ire., 143......E3 11
Waterhen, lake, Man., Can......C2 71
Waterloo, Lauderdale Ala., 215......A1 78
Waterloo, Nevada, Ark., 200......D2 81
Waterloo, Bel., 11,846......D4 15
Waterloo, Ont., Can., 21,366.D4 72
Waterloo, Que., Can., 4,543.D5 73
Waterloo, Monroe, Ill., 3,739......E3 90
Waterloo, DeKalb, Ind., 1,432......B7 91
Waterloo, Black Hawk, Iowa, 71,755 (★114,300)......B5 92
Waterloo, Kingham, Kans., 35......B4 93
Waterloo, Madison, Mont., 200......E4 102
Waterloo, Douglas, Nebr., 516......D2 103
Waterloo, Seneca, N.Y., 5,098......C4 108
Waterloo, Linn, Oreg., 151..C4 113
Waterloo, S.L., 2,312......E2 45
Waterloo, Laurens, S.C., 148......C3 115
Waterloo, Jefferson, Wis., 1,947......E5 124
Waterloo, co., Ont., Can., 176,754......D4 72
Waterman, DeKalb, Ill., 916......B5 90
Waterman, Wheeler, Oreg...C7 113
Water Mill, Suffolk, N.Y., 700......F7 84
Waterport, Orleans, N.Y., 200......B2 108
Water Proof, Tensas, La., 1,412......C4 95
Waters, Otsego, Mich., 35...D6 98
Watersmeet, Gogebic, Mich., 500......A6 98
Waterton, riv., Alta., Can......E4 69
Waterton-Glacier International Peace Park, Can., U.S......A5 76
Waterton Lakes, nat. park, Alta., Can......E3 69
Waterton Park, Alta., Can., 225......E4 69
Watertown, Litchfield, Conn., 5,500 (14,837▲)......C4 84
Watertown, Columbia, Fla., 2,109......B4 86
Watertown, Middlesex, Mass., 39,092......D2 97
Watertown, Carver, Minn., 1,046......*F5 99
Watertown, Jefferson, N.Y., 33,306......B5 108
Watertown, Washington, Ohio, 160......C6 111
Watertown, Codington, S. Dak., 14,077......C8 116
Watertown, Wilson, Tenn., 919......A5 117
Watertown, Dodge and Jefferson, Wis., 13,943......E5 124
Water Valley, Graves, Ky., 267......B2 94
Water Valley, Yalobusha, Miss., 3,206......A4 100
Water Valley, Tom Green, Tex., 150......D2 118
Water View, Middlesex, Va., 150......D6 121
Water Village, Carroll, N.H., 50......C4 105
Waterville, N.S., Can., 886..D5 74
Waterville, Que., Can., 1,330......D6 73
Waterville, Allamakee, Iowa, 184......A6 92
Waterville, Ire., 702......F1 11
Waterville, Marshall, Kans., 700......C7 93
Waterville, Kennebec, Maine, 18,695......D3 96
Waterville, LeSueur, Minn., 1,623......F5 99
Waterville, Grafton, N.H., 11 (14▲)......C3 105
Waterville, Oneida, N.Y., 1,901......C5 108
Waterville, Lucas, Ohio, 1,856......A1, A4 111
Waterville, Lycoming, Pa., 125......D7 114
Waterville, Lamoille, Vt., 250 (332▲)......B3 120
Waterville, Douglas, Wash., 1,013......B5 122
Watervliet, Bel., 1,922......C3 15
Watervliet, Berrien, Mich., 1,818......F4 98
Watervliet, Albany, N.Y., 13,917......C7 108
Waterways, Alta., Can., 250..A5 69
Watford, Ont., Can., 1,293..E3 72
Watford, Eng., 75,630......C7 12
Watford City, McKenzie, N. Dak., 1,865......B2 110
Watha, Pender, N.C., 174...C5 109
Wathena, Doniphan, Kans., 837......C9 93
Watino, Alta., Can., 93......B2 69
Watkins, Benton, Iowa, 120..C6 92
Watkins, Meeker, Minn., 744.E4 99
Watkins Glen, Schuyler, N.Y., 2,813......C4 108
Watkinsville, Oconee, Ga., 758......C3 87
Watlam, China......A9 38
Watonga, Blaine, Okla., 3,252......B3 112
Watonwan, co., Minn., 14,460......F4 99
Watonwan, riv., Minn......F4 99
Watou, Bel., 2,671......D2 15
Watova, Nowata, Okla., 80..A6 112
Watrous, Sask., Can., 1,461..F3 70
Watrous, Mora, N. Mex., 150......B5 107
Watsa, Con. L., 6,000......A4 48
Watseka, Iroquois, Ill., 5,219......C6 90
Watson, Desha, Ark., 312...D4 81
Watson, Sask., Can., 910....E3 70
Watson, Effingham, Ill., 247.D5 90
Watson, Clark, Ind., 500....A4 94
Watson, Chippewa, Minn., 267......E3 99
Watson, Atchison, Mo., 181......A2 101
Watson Lake, Yukon, Can., 597......D7 66
Watsontown, Northumberland, Pa., 2,431......D8 114
Watsonville, Santa Cruz., Calif., 13,293......C6, D3 82
Watten, Fr., 3,100......D10 12
Watten, lake, Scot......B5 13
Wattenberg, Weld, Colo., 150......A6 83
Wattensaw, bayou, Ark......C4 81
Wattenscheid, Ger., 79,200.*C7 15
Wattis, Carbon, Utah, 100...D5 119
Watton, Eng., 3,104......B8 12
Watton, Baraga, Mich., 70...B2 98
Wattrelos, Fr., 41,319......B5 14
Watts, Adair, Okla., 268....A7 112
Watts Bar, lake, Tenn......D9 117
Watts Bar Dam, Rhea, Tenn., 25......D9 117
Wattsburg, Erie, Pa., 401....C2 114
Wattsville, St. Clair, Ala., 700......B3 78
Wattsville, Laurens, S.C., 1,438......B4 115
Wattwil, Switz., 7,480......B7 19
Watu, Con. L......B3 48
Watuppa, pond, Mass......C5 97
Wau, Sud., 8,009......D2 47
Waubamick, Ont., Can., 170......B4 72
Waubaushene, Ont., Can., 597......C5 72
Waubay, Day, S. Dak., 851..B8 116
Waubay, lake, S. Dak......B8 116
Waubun, Mahnomen, Minn., 350......C3 99
Wauchope, Sask., Can., 83..H5 70
Wauchula, Hardee, Fla., 3,411......E5 86
Waucoma, Fayette, Iowa, 364......A5 92
Wauconda, Lake, Ill., 3,227..E2 90
Waugh, N.B., Can., 80......B5 74
Waukau, Winnebago, Wis., 150......E5 124
Waukee, Dallas, Iowa, 687...C4 92
Waukeenah, Jefferson, Fla., 500......B3 86
Waukegan, Lake, Ill., 55,719......A6, E2 90
Waukesha, Waukesha, Wis., 30,004......E1, F5 124
Waukesha, co., Wis., 158,249.E5 124
Waukomis, Garfield, Okla., 516......A4 112
Waukon, Allamakee, Iowa, 3,639......A6 92
Wauna, Clatsop, Oreg., 175.A3 113
Wauna, Pierce, Wash., 130..D1 122
Waunakee, Dane, Wis., 1,611......E4 124
Wauneta, Chase, Nebr., 794.D4 103
Waupaca, Waupaca, Wis., 3,984......D4 124
Waupaca, co., Wis., 35,340..D5 124
Waupun, Dodge and Fond du Lac, Wis., 7,935......E5 124
Wauregan, Windham, Conn., 950......C9 84
Waurika, Jefferson, Okla., 1,933......C4 112
Wausa, Knox, Nebr., 724....B8 103
Wausau, Washington, Fla., 300......G3 86
Wausau, Marathon, Wis., 31,943......D4 124
Wausaukee, Marinette, Wis., 608......C6 124
Wauseon, Fulton, Ohio, 4,311......A3 111
Waushara, co., Wis., 13,497.D4 124
Wautoma, Waushara, Wis., 1,466......D4 124
Wauwatosa, Milwaukee, Wis., 56,923......E1 124
Wauzeka, Crawford, Wis., 494......E3 124
Wave, Dallas, Ark......C3 81
Wave Hill, Austl., 75......C5 50
Waveland, Montgomery, Ind., 549......E3 91
Waveland, Hancock, Miss., 1,106......E1, E4 100
Waveney, riv., Eng......B9 12
Waverly, Chambers and Lee, Ala., 250......C4 78
Waverly, Crittenden, Ark....E8 117
Waverly, N.S., Can., 1,142..E6 74
Waverly, Polk, Fla., 1,160..*E5 86
Waverly, Camden, Ga., 165..E5 87
Waverly, Morgan, Ill., 1,375.D4 90
Waverly, Morgan, Ind., 150.E5 91
Waverly, Bremer, Iowa, 6,357......B5 92
Waverly, Coffey, Kans., 381.D8 93
Waverly, Union, Ky., 331....C2 94
Waverly, Madison, La., 75...B4 95
Waverly, Wright, Minn., 574......E5 99
Waverly, Lafayette, Mo., 837......B4 101
Waverly, Lancaster, Nebr., 511......D9, E2 103
Waverly, Tioga, N.Y., 5,950.C4 108
Waverly, Pike, Ohio, 3,830..C5 111
Waverly, Humphreys, Tenn., 2,891......A4 117
Waverly, Sussex, Va., 1,601..D5 121
Waverly, Wood., W. Va., 300......B3 123
Waverly Hall, Harris, Ga., 712......D2 87
Wavre, Bel., 9,706......D4 15
Wawaka, Noble, Ind., 300...B7 91
Wāw al Kabīr, Libya......D3 43
Wawanesa, Man., Can., 456.E2 71
Waw an Nāmūs (Well), Libya......E3 43
Wawasee, lake, Ind......B6 91
Wawayanda, lake, N.J......A4 106
Waweig, N.B., Can., 179....D2 74
Wawota, Sask., Can., 453...H4 70
Waxahachie, Ellis, Tex., 12,749......B5, C4 118
Waxhaw, Union, N.C., 729..C3 109
Waxia, St. Landry, La......D4 95
Waxweiler, Ger., 950......D6 15
Way, Madison, Miss., 50....C3 100
Way, is., Viet......H5 38
Wayan, Caribou, Idaho, 10..G7 89
Waygamack, lake, Que., Can....B5 73
Waycross, Ware, Ga., 20,944.E4 87
Wayland, Henry, Iowa, 597.C6 92
Wayland, Floyd, Ky., 1,340..C7 94
Wayland, Middlesex, Mass., 2,000 (10,444▲)......D2 97
Wayland, Allegan, Mich., 2,019......F5 98
Wayland, Clark, Mo., 384...A6 101
Wayland, Steuben, N.Y., 2,003......C3 108
Wayland Springs, Lawrence, Tenn., 200......B4 117
Waymart, Wayne, Pa., 1,106......C11 114
Wayne, Alta., Can., 116....D4 69
Wayne, Du Page, Ill., 373...E2 90
Wayne, Republic, Kans., 50.C6 93
Wayne, Wayne, Mich., 16,034......A7 98
Wayne, Wayne, Nebr., 4,217......B8 103
Wayne, Passaic, N.J., 29,353......*A4 106
Wayne, Schuyler, N.Y., 250.C3 108
Wayne, Wood, Ohio, 949....A4 111
Wayne, McClain, Okla., 517......C4 112
Wayne, Delaware, Pa., 10,000......A10, F11 114
Wayne, Wayne, W. Va., 1,274......C2 123
Wayne, co., Ga., 17,921......E5 87
Wayne, co., Ill., 19,008......E5 90
Wayne, co., Ind., 74,039......E7 91
Wayne, co., Iowa, 9,800......D4 92
Wayne, co., Ky., 14,700......D5 94
Wayne, co., Mich., 2,666,297......F7 98
Wayne, co., Miss., 16,258...D5 100
Wayne, co., Mo., 8,638......D7 101
Wayne, co., Nebr., 9,959....B8 103
Wayne, co., N.Y., 67,989....B3 108
Wayne, co., N.C., 82,059....B5 109
Wayne, co., Ohio, 75,497....B6 111
Wayne, co., Pa., 28,237....C11 114
Wayne, co., Tenn., 11,908...B4 117
Wayne, co., Utah, 1,728.....E4 119
Wayne, co., W. Va., 38,977..C2 123
Wayne City, Wayne, Ill., 903......E5 90
Waynesboro, Burke, Ga., 5,359......C4 87
Waynesboro, Wayne, Miss., 3,892......D5 100
Waynesboro, Franklin, Pa., 10,427......G6 114
Waynesboro, Wayne, Tenn., 1,343......B4 117
Waynesboro (Independent City), Va., 15,694......C4 121
Waynesburg, Lincoln, Ky., 450......C5 94
Waynesburg, Stark, Ohio, 1,442......B6 111
Waynesburg, Greene, Pa., 5,188......G1 114
Waynesville, Brantley, Ga., 250......E5 87
Waynesville, DeWitt, Ill., 510......C4 90
Waynesville, Pulaski, Mo., 2,377......D5 101
Waynesville, Haywood, N.C., 6,159......D3 109
Waynesville, Warren, Ohio, 1,298......C3 111
Waynetown, Montgomery, Ind., 933......D3 91
Waynoka, Woods, Okla., 1,794......A3 112
Wayside, Montgomery, Kans., 100......E8 93
Wayside, Washington, Miss., 300......B2 100
Wayside, Dawes, Nebr......B2 103
Wayside, Monmouth, N.J., 200......C4 106
Wayside, Armstrong, Tex., 35......B2 118
Wayside, Brown, Wis., 130...A6 124
Wayzata, Hennepin, Minn., 3,219......*F5 99
Wazi Khwa (Marjan), Afg..E14 41
We, isl., Indon......J1 38
Weakley, co., Tenn., 24,227..A3 117
Weare, Hillsboro, N.H., 250 (1,420▲)......D3 105
Weatherby, De Kalb, Mo., 450......B3 101
Weatherford, Custer, Okla., 4,499......B3 112
Weatherford, Parker, Tex., 9,759......C4 118
Weatherly, Carbon, Pa., 2,591......E10 114
Weathersfield, Windsor, Vt., (1,254▲) 30......E4 120
Weatogue, Hartford, Conn., 500......B5 84
Weaubleau, Hickory, Mo., 349......D4 101
Weaver, Calhoun, Ala., 1,401......B4 78
Weaver, lake, Man., Can......C3 71
Weaver, mts., Ariz......C3 80
Weaversford, Ashe, N.C., 35.A2 109
Weaverville, Trinity, Calif., 1,736......B2 82
Weaverville, Buncombe, N.C., 1,041......D3 109
Webatuck, creek, N.Y......B2 84
Webb, Houston, Ala., 331...D4 78
Webb, Sask., Can., 168......G1 70
Webb, Clay, Iowa, 236......B2 92
Webb, Tallahatchie, Miss., 686......B3 100
Webb, Webb, Tex., 15......F3 118
Webb, co., Tex., 64,791......F3 118
Webb, hill, Mass......B3 97
Webb, lake, Maine......D2 96
Webb City, Franklin, Ark., 60......B2 81
Webb City, Jasper, Mo., 6,740......D3 101
Webb City, Osage, Okla., 233......A5 112
Webber, Jewell, Kans., 58...C5 93
Webber, lake, Man., Can......B5 71
Webbers Falls, Muskogee, Okla., 441......B6 112
Webberville, Ingham, Mich., 664......F6 98
Webb Lake, Burnett, Wis., 25......B1 124
Webbville, Lawrence, Ky., 225......B7 94
Webbwood, Ont., Can., 546......A3 72
Weber, Davis, Utah......B4 119
Weber, co., Utah, 110,744...B3 119
Weber, mtn., B.C., Can......B3 68
Weber City, Curry, N. Mex., 15......C6 107
Weber City, Scott, Va., 1,274......*B2 121
Weberville, Alta., Can......A2 69
Webhannet, York, Maine....E2 96
Webi Shebeli, riv., Som......E5 44
Webster, Alta., Can......B1 69
Webster, Sumter, Fla., 366..D4 86
Webster, Keokuk, Iowa, 137.C5 92
Webster, Androscoggin, Maine, 25 (1,302▲)......D5 96
Webster, Worcester Mass., 13,680......B4 97
Webster, Rice, Minn., 140...F5 99
Webster, Fallon, Mont., 4..D12 102
Webster, Merrimack, N.H., 100 (457▲)......D3 105
Webster, Monroe, N.Y., 3,060......B3 108
Webster, Ramsey, N. Dak., 105......A7 110
Webster, Darke, Ohio, 80....B3 111
Webster, Westmoreland, Pa., 950......F2 114
Webster, Day, S. Dak., 2,409......B8 116
Webster, Harris, Tex., 329...F5 118
Webster, Burnett, Wis., 514..C1 124
Webster, co., Ga., 3,247......D2 87
Webster, co., Iowa, 47,810...B3 92
Webster, co., Ky., 14,244......C2 94
Webster, co., Miss., 10,580..B4 100
Webster, co., Mo., 13,753...D5 101
Webster, co., Nebr., 6,224..D7 103
Webster, co., W. Va., 13,719......C4 123
Webster, par., La., 39,701...B2 95
Webster, res., Kans......C4 93
Webster City, Hamilton, Iowa, 8,520......B4 92
Webster Groves, St. Louis, Mo., 28,990......B8 101
Webster Mills, Fulton, Pa., 40......G5 114
Webster Springs (Addison), Webster, W. Va., 1,132...C4 123
Websterville, Washington, Vt., 800......C4 120
Wecota, Faulk, S. Dak., 25...B6 116
Weda, Indon......E7 35
Weddell, sea, Ant......B8 5
Wedgefield, Sumter, S.C., 500......D7 115
Wedgeport, N.S., Can., 665......F3 74
Wedowee, Randolph, Ala., 917......B4 78
Wedron, La Salle, Ill., 250..B5 90
Weed, Sidkiyou, Calif., 3,223.B2 82
Weed, Otero, N. Mex., 100..E4 107
Weed Heights, Lyon, Nev., 1,092......D2 104
Weedon, Que., Can., 1,426..D6 73
Weedpatch, hill, Ind......F5 91
Weedsport, Cayuga, N.Y., 1,731......B4 108
Weedville, Elk, Pa., 500.....D5 114
Weehawken, Hudson, N.J., 13,504......D5 106
Weekapaug, Washington, R.I., 30......D9 84
Weeks (Weeks Island), Iberia, La., 1,138......E4 95
Weeksbury, Floyd, Ky., 700......C7 94
Weekstown, Atlantic, N.J., 50......D3 106
Weeksville, Pasquotank, N.C., 250......A7 109
Weems, Lancaster, Va., 250.D6 121
Weener, Ger., 5,500......A7 15
Weenusk, Ont., Can......D9 72
Weeping Water, Cass, Nebr., 1,048......D9, E3 103
Weeping Water, creek, Nebr....E3 103
Weert, Neth., 12,900......C5 15
Wessen, Switz., 1,280......B7 19
Weesp, Neth., 11,000......B5 15
Weferlingen, Ger., 4,806....A6 17
Wegdahl, Chippewa, Minn., 100......F3 99
Weggis, Switz., 2,243......B5 19
Weggs, cape, Que., Can......D18 67
Wegoborck, see Wegorzewo, Pol.
Wegorzewo, Pol., 1,184.....A6 26
Wegrow, Pol., 5,185......B7 26
Weichang, China......C8 34
Wei-Chou, isl., China......B8 38
Weida, Ger., 12,000......C7 17
Weiden, Ger., 41,700......D7 17
Weidenau, Ger., 17,200.....C3 17
Weidman, Isabella, Mich., 350......E5 98
Weifang, China, 148,900....D8 34
Weihai, China, 45,000......D9 34, F10 36
Weihsi, China......F4 34
Weilburg, Ger., 6,700......C3 17
Weilheim, Ger., 12,400......E5 16
Weimar, Ger., 63,700......C6 17
Weimar, Colorado, Tex., 2,006......E4 118
Wein, Marathon, Wis......D3 124
Weinan, China......E6 34
Weinböhla, Ger., 10,300.....B8 17
Weiner, Poinsett, Ark., 669..B5 81
Weinert, Haskell, Tex., 251..C3 112
Weinfelden, Switz., 6,954....A7 19
Weingarten, Ger., 7,700......D3 17
Weinheim, Ger., 27,900......D3 17
Weining, China, 11,000......F5 34
Weinsberg, Ger., 7,200......D4 17
Weippe, Clearwater, Idaho, 600......C3 89
Weir, Que., Can., 213......D3 73
Weir, Cherokee, Kans., 699..E9 93
Weir, Muhlenberg, Ky., 150.C2 94
Weir, Choctaw, Miss., 522...B4 100
Weir, lake, Fla......C5 86
Weir, riv., Man., Can......A4 71
Weirdale, Sask., Can., 98....D3 70
Weir River, Man., Can......A4 71
Weirsdale, Marion, Fla., 900......D5 86
Weirton, Brooke and Hancock, W. Va., 28,201...A2, A4 123
Weiser, Washington, Idaho, 4,208......E2 89
Weiser, riv., Idaho......E2 89
Weishan, lake, China......G7 36
Weisner, mtn., Ala......A4 78
Weiss, St. Helena, La......D5 95
Weiss, res., Ala......A4 78
Weisse Elster, riv., Ger......B7 17
Weissen, lake, Aus......C9 18
Weissenburg [in Bayern], Ger., 13,900......D5 17
Weissenburg, Switz......C3 19
Weissenfels, Ger., 45,900....B6 17
Weissenhorn, Ger., 6,400....A6 18
Weissert, Custer, Nebr., 20..C6 103
Weisshorn, mtn., Switz......D4 19
Weisswasser, Ger., 14,100....B9 17
Weitchpec, Humboldt, Calif., 100......B2 82
Weitra, Aus., 1,883......E9 17
Wejherowo, Pol., 25,000.....A5 26
Wekusko, lake, Man., Can......B2 71
Welaka, Putnam, Fla., 526...C5 86
Welborn, Wyandotte, Kans., 6,500......B8 93
Welch, Craig, Okla., 557....A6 112
Welch, Dawson, Tex., 500...C1 118
Welch, McDowell, W. Va., 5,313......D3 123
Welcome, St. James, La., 300......B6 95
Welcome, Martin, Minn., 733......G4 99
Welcome, Greenville, S.C., 1,500......*B3 115
Weld, Franklin, Maine, 130 (348▲)......D2 96
Weld, co., Colo., 72,344....A6 83
Welda, Anderson, Kans., 180......D8 93
Welden, Ger., 1,900......E5 17
Weldon, De Witt, Ill., 449...C5 90
Weldon, Decatur, Iowa, 202......D4 92
Weldon, Halifax, N.C., 2,165......A6 109
Weldon, Houston, Tex., 200.D5 118
Weldona, Morgan, Colo., 90......A7 83
Weldon Spring, St. Charles, Mo., 150......A7 101
Weleetka, Okfuskee, Okla., 1,231......B5 112
Welford, Austl......B4 51
Welkom, S. Afr., 48,069 (★97,614)......C4 49

Welland, Ont., Can., 36,079 . E5 72
Welland, co., Ont., Can., 164,741 . . . D5 72
Welland, riv., Eng. . . . D6 10, B7 12
Wellandport, Ont., Can., 191 . . . D5 72
Wellborn, Suwannee, Fla., 300 . . . B4 86
Wellborn, Brazos, Tex., 200 . D4 118
Wellersburg, Somerset, Pa., 303 . . . G4 114
Welles, hbr., Midway Is. . . . 52
Wellesley, Ont., Can., 649 . . D4 72
Wellesley, Norfolk, Mass., 26,071 . . . B5, D2 97
Wellesley, is., Austl. . . . C6 50
Wellesley Hills, Norfolk Mass. (part of Wellesley) . . . B5, D2 97
Wellfleet, Barnstable, Mass., 850 (1,404▲) . . . C7 97
Wellfleet, Lincoln, Nebr., 67 . D5 103
Wellford, Spartanburg, S.C., 1,040 . . . B3 115
Wellin, Bel., 1,079 . . . D5 15
Welling, Cherokee, Okla., 100 . . . B7 112
Wellingborough, Eng., 30,579 . . . D7 12
Wellington, Calhoun, Ala., 125 . . . B4 78
Wellington, Austl., 5,597 . . . F7 51
Wellington, B.C., Can., 599 . . A8 68
Wellington, Ont., Can., 1,064 . . . *D7 72
Wellington, Larimer, Colo., 532 . . . A5 83
Wellington, Eng., 13,630 . . . B5 12
Wellington, Eng., 7,523 . . . D4 12
Wellington, Iroquois, Ill., 334 . . . C6 90
Wellington, Sumner, Kans., 8,809 . . . E6 93
Wellington, Piscataquis, Maine, 100 (231▲) . . . C3 96
Wellington, Lafayette, Mo., 65 . . . B4 101
Wellington, Lyon, Nev., 50 . . E2 104
Wellington, N.Z., 123,969 (*249,532) . . . N15 51
Wellington, Lorain, Ohio, 3,599 . . . A5 111
Wellington, Collingsworth, Tex., 3,137 . . . B2 118
Wellington, Carbon, Utah, 1,066 . . . D5 119
Wellington, Fairfax, Va., 8,000 . . . *C5 121
Wellington, co., Ont., Can., 84,702 . . . D4 72
Wellington, chan., N.W., Ter., Can. . . . A14 66
Wellington, isl., Chile . . . D2 54
Wellington Station, N.S., Can., 100 . . . E6 74
Wellington Station, P.E.I., Can., 292 . . . C5 74
Wellman, Washington, Iowa, 1,085 . . . C6 92
Wellman, Terry, Tex., 150 . . C1 118
Wellpinit, Stevens, Wash., 100 . . . B8 122
Wells, B.C., Can., 740 . . . C7 68
Wells, Eng., 6,100 . . . C5 12
Wells [-next-the-sea], Eng., 2,490 . . . B8 12
Wells, Ottawa, Kans., 90 . . . C6 93
Wells, York, Maine, 500 (3,528▲) . . . E2 96
Wells, Delta, Mich., 900 . . . C3 98
Wells, Fairbault, Minn., 2,897 . . . G5 99
Wells, Elko, Nev., 1,071 . . . B7 104
Wells, Hamilton, N.Y., 400 . B6 108
Wells, Cherokee, Tex., 544 . D2 118
Wells, Rutland, Vt., 200 (419▲) . . . E2 120
Wells, co., Ind., 21,220 . . . C7 91
Wells, co., N. Dak., 9,237 . . . B6 110
Wells, lake, Austl. . . . E3 50
Wells, riv., Vt. . . . C4 120
Wells Beach, York, Maine . . E2 96
Wellsboro, La Porte, Ind., 100 . . . B4 91
Wellsboro, Tioga, Pa., 4,369 . C7 114
Wellsburg, Grundy, Iowa, 827 . . . B5 92
Wellsburg, Chemung, N.Y., 643 . . . C4 108
Wellsburg, Wells, N. Dak., 50 . . . B6 110
Wellsburg, Brooke, W. Va., 5,514 . . . A4, B2 123
Wellsdale, Benton, Oreg . . . C1 113
Wellsford, Kiowa, Kans., 24 . E4 93
Wells Gray, park, B.C., Can. . . . C8 68
Wellshire, Arapahoe, Colo., 3,000 . . . *B6 83
Wells River, Orange, Vt., 472 . . . C4 120
Wells Tannery, Fulton, Pa., 130 . . . F5 114
Wellston, Manistee, Mich., 175 . . . D5 98
Wellston, St. Louis, Mo., 7,979 . . . A8 101
Wellston, Jackson, Ohio, 5,728 . . . C5 111
Wellston, Lincoln, Okla., 630 . . . B4 112
Wellsville, Franklin, Kans., 984 . . . D8 93
Wellsville, Montgomery, Mo., 1,523 . . . B6 101
Wellsville, Allegany, N.Y., 5,967 . . . C3 108
Wellsville, Columbiana, Ohio, 7,117 . . . B7 111
Wellsville, York, Pa., 320 . . . F8 114
Wellsville, Cache, Utah, 1,106 . . . B4 119
Wellton, Yuma, Ariz., 700 . . E1 80
Wellwood, Man., Can., 73 . . D2 71
Wels, Aus., 41,060 . . . D7 16
Welsford, N.B., Can., 424 . . D3 74
Welsh, Jefferson Davis, La., 3,332 . . . D3 95
Welshpool, N.B., Can., 199 . . E3 74
Welshpool, Wales, 6,332 . . . B4 12
Welton, Clinton, Iowa, 88 . . C7 92
Welty, Okfuskee, Okla., 100 . B5 112
Welview, Concho, Tex. . . . D3 118
Welwyn, Sask., Can., 182 . . G5 70
Welzow, Ger., 7,304 . . . B9 17
Wem, Eng., 2,603 . . . B5 12
Wema, Con. L. . . . B3 48
Wembere, riv., Tan. . . . B5 48
Wembley, Alta., Can., 303 . . B1 69
Wembley, Eng., 124,843 . . . k11 10
Wemding, Ger., 4,200 . . . E5 17
Wenasaga, riv., Ont., Can. . . . D5 71
Wenasoga, Alcorn, Miss., 150 . . . A5 100
Wenatchee, Chelan, Wash., 16,726 . . . B5 122
Wenatchee, lake, Wash. . . . B5 122
Wenatchee, mts., Wash. . . . B5 122
Wenatchee, riv., Wash. . . . B5 122
Wenceslau Braz., Braz., 4,347 . . . C3 56
Wenchang, China . . . C9 39
Wenchi, Ghana, 3,812 . . . E4 45
Wenchow (Yungkia), China, 201,600 . . . F9 34, K9 36
Wenchüan, China, 5,000 . . . B8 34
Wendel, Lassen, Calif., 60 . . . B3 82
Wendel, Taylor, W. Va., 75 . . . B7 123
Wendell, Gooding, Idaho, 1,232 . . . G4 89
Wendell, Franklin, Mass., 100 (292▲) . . . A3 97
Wendell, Grant, Minn., 253 . D2 99
Wendell, Sullivan, N.H., 100 . D2 105
Wendell, Wake, N.C., 1,620 . B5 109
Wenden, Yuma, Ariz., 200 . D2 80
Wendling, Lane, Oreg . . . C4 113
Wendover, Elko, Nev., 60 . . . C7 104
Wendover, Tooele, Utah, 609 . . . C1 119
Wendover, Platte, Wyo . . . C8 125
Wenham, Essex, Mass., 2,798 . . . A6, C3 97
Wenham, lake, Mass. . . . C3 97
Wenahm, swamp, Mass. . . . C3 97
Wenlock, riv., Austl. . . . B7 50
Wennington, Eng., 145 . . . F6 13
Wenona, Marshall, Ill., 1,005 . . . B4 90
Wenona, Somerset, Md., 325 . . . D6 85
Wenonah, Gloucester, N.J., 2,100 . . . D2 106
Wenshan, China, 13,000 . . . G5 34
Wensum, riv., Eng. . . . D7 10
Wentworth, Austl., 1,154 . . . G3 51
Wentworth, N.S., Can., 128 . . . D6 74
Wentworth, Newton, Mo., 174 . . . E3 101
Wentworth, Grafton, N.H., 275 (300▲) . . . C3 105
Wentworth, Rockingham, N.C., 115 . . . A4 109
Wentworth, Lake, S. Dak., 211 . . . C8 116
Wentworth, Douglas, Wis., 35 . . . B2 124
Wentworth, co., Ont., Can., 358,837 . . . D4 72
Wentworth, lake, N.H. . . . C4 105
Wentzville, St. Charles, Mo., 2,742 . . . C7 101
Weogufka, mtn., Ala. . . . C3 78
Weohyakapka, lake, Fla. . . . E5 86
Weona, Poinsett, Ark., 100 . . . B5 81
Weott, Humboldt, Calif., 350 . . . B2 82
Wepener, S. Afr., 3,911 . . . C4 49
Werbomont, Bel., 420 . . . D5 15
Werdau, Ger., 24,500 . . . C7 17
Werder, Ger., 10,300 . . . A7 17
Werdohl, Ger., 22,100 . . . B2 17
Werfen, Aus., 3,106 . . . E6 16
Werl, Ger., 17,500 . . . B2 17
Wernberg, Ger., 1,250 . . . D7 17
Werne [an der Lippe], Ger., 19,900 . . . B2 17
Werner, Dunn, N. Dak., 59 . . B3 110
Wernersville, Berks., Pa., 1,462 . . . *F9 114
Werneuchen, Ger., 4,218 . . . B6 16
Wernigerode, Ger., 33,100 . . . B5 17
Werra, riv., Ger . . . C5 17
Wertach, riv., Ger . . . D5 16
Wertheim, Ger., 11,300 . . . D4 17
Wertingen, Ger., 3,200 . . . E5 17
Wervik, Bel., 12,442 . . . D3 15
Wesco, Crawford, Mo., 100 . . D6 101
Wesconnett, Duval, Fla., 5,000 . . . B6 86
Wesel, Ger., 32,000 . . . B1 17
Weser, canal, Ger . . . B4 16
Weser, riv., Ger . . . A4 17
Weskan, Wallace, Kans., 240 . . . D2 83
Weslaco, Hidalgo, Tex., 15,649 . . . F3 118
Weslaco North, Hidalgo, Tex., 1,049 . . . *F3 118
Weslemkoon, lake, Ont., Can. . . . C7 72
Wesley, Madison, Ark., 65 . . A2 81
Wesley, Emanuel, Ga., 18 . . D4 87
Wesley, Kassuth, Iowa, 514 . . A4 92
Wesley, Washington, Maine, 65 (145▲) . . . D5 96
Wesleyville, Newf., Can., 1,285 . . . D5 75
Wesleyville, Erie, Pa., 3,534 . . B2 114
Wessel, is., Austl. . . . B6 50
Wesselburen, Ger., 3,600 . . . D2 24
Wessington, Beadle and Hand, S. Dak., 378 . . . C7 116
Wessington Springs, Jerauld, S. Dak., 1,488 . . . C7 116
Wesson, Union, Ark., 250 . . . D3 81
Wesson, Copiah, Miss., 1,157 . . . D3 100
West, Holmes, Miss., 282 . . B4 100
West, McLennan, Tex., 2,352 . . . D4 118
West, bay, La. . . . E6 95
West, bay, Tex. . . . G5 118
West, butte, Mont. . . . B5 102
West, chan., Man., Can. . . . C2 71
West, fork, W. Va. . . . D6 123
West, hill, Mass. . . . E1 97
West, ice shelf, Ant. . . . C21 5
West, isl., Mass. . . . C6 97
West, lake, Maine . . . C4 96
West, mtn., Mass. . . . A2 97
West, mtn., N.Y. . . . B6 108
West, mtn., Vt. . . . B5 120
West, riv., N.S., Can. . . . D7 74
West, riv., Mass. . . . E1 97
West, riv., Vt. . . . E3 120
West, spit, Eniwetok . . . 52
West Acton, Middlesex, Mass., 950 . . . C1 97
West Alburgh, Grand Isle, Vt . . . B2 120
West Alexander, Washington, Pa., 468 . . . F1 114
West Alexandria, Preble, Ohio, 1,524 . . . C3 111
West Allis, Milwaukee, Wis., 68,157 . . . E2 124
West Alton, St. Charles, Mo., 350 . . . A8 101
West Alton, Belknap, N.H., 100 . . . C4 105
West Amityville, Nassau, N.Y., 4,000 . . . *E3 108
West Andover, Merrimack, N.H . . . D3 105
West Arichat, N.S., Can., 470 . . . D8 74
West Arlington, Bennington, Vt . . . E2 120
West Athens, Somerset, Maine, 150 . . . C3 96
West Auburn, Androscoggin, Maine . . . D2 96
West Auburn, Worcester, Mass. (part of Auburn) . . . B4 97
West Augusta, Augusta, Va., 10 . . . C3 121
West Ausdale, Richland, Ohio, 1,354 . . . *B5 111
West Avon, Hartford, Conn. (part of Avon) . . . B5 84
West Babylon, Suffolk, N.Y., 16,000 . . . *E3 108
West Baden Springs, Orange, Ind., 879 . . . G4 91
West Bainbridge, Decatur, Ga., 2,500 . . . *F2 87
West Baldwin, Cumberland, Maine, 75 . . . E2 96
Westbank, B.C., Can., 284 . . . E8 68
West Barnet, Caledonia, Vt., 113 . . . C4 120
West Barnstable, Barnstable, Mass., 600 . . . C7 97
West Barrington, Bristol, R.I., 4,000 . . . B11 84
West Bath, Sagadahoc, Maine, (766▲) . . . E6 96
West Baton Rouge, par., La., 14,796 . . . D4 95
West Battle, lake, Minn. . . . D3 99
West Bay, N.S., Can., 125 . . . D8 74
Westbay, Bay, Fla., 600 . . . G3 86
West Bedford, Middlesex, Mass. (part of Bedford) . . . C2 97
West Belmar, Monmouth, N.J., 2,511 . . . C4 106
West Bend, Sask., Can., 91 . . . F4 70
West Bend, Palo Alto, Iowa, 910 . . . B3 92
West Bend, Washington, Wis., 9,969 . . . E5 124
West Bengal, state, India, 34,926,279 . . . F11 40
West Benson, Douglas, Nebr., 900 . . . D3 103
West Berkshire, Franklin, Vt., 95 . . . B3 120
West Berlin, see Berlin, West, Ger.
West Berlin (Carters), Worcester, Mass., 250 . . . B4 97
West Berlin, Camden, N.J., 3,363 . . . *D3 106
West Berlin, state, Ger., 2,197,600 . . . *B6 16
West Bethel, Oxford, Maine, 160 . . . D2 96
West Billings, Yellowstone, Mont., 3,500 . . . *E8 102
West Blocton, Bibb, Ala., 1,156 . . . B2 78
West Bloomfield, Ontario, N.Y., 350 . . . C3 108
West Bolton, Chittenden, Vt., 50 . . . C3 120
Westboro, Ont., Can., 3,450 . A9 72
Westboro, Atchison, Mo., 262 . . . A2 101
Westboro, Taylor, Wis., 500 . C3 124
Westborough, Worcester, Mass., 4,011 (9,599▲) . . . D1, B4 97
West Bountiful, Davis, Utah, 945 . . . *C4 119
Westbourne, Man., Can., 123 . . . D2 71
Westbourne, Campbell, Tenn., 100 . . . C9 117
West Bowdoin, Sagadahoc, Maine, 50 . . . D5 96
West Boylston, Worcester, Mass., 2,000 (5,526▲) . . . B4 97
West Branch, Cedar, Iowa, 1,053 . . . C6 92
West Branch, Ogemaw, Mich., 2,025 . . . D6 98
West Branch, res., Conn . . . A4 84
West Branch, res., N.Y. . . . D2 84
West Brattleboro, Windham, Vt. (part of Brattleboro) . . . F3 120
West Brentwood, Rockingham, N.H., 135 . . . E4 105
West Bridgewater, Plymouth, Mass., 2,000 (5,061▲) . . . B5 97
West Bridgewater, Beaver, Pa., 1,292 . . . *E1 114
West Bridgewater, Rutland and Windsor, Vt., 80 . . . D3 120
West Brimfield, Hampden, Mass., 50 . . . B3 97
West Bromwich, Eng., 95,909 . . . B6 12
Westbrook, Middlesex, Conn., 950 (2,399▲) . . . D7 84
Westbrook, Cumberland, Maine, 13,820 . . . E2, E5 96
Westbrook, Cottonwood, Minn., 1,012 . . . F3 99
Westbrook, Mitchell, Tex., 214 . . . C2 118
West Brookfield, Worcester, Mass., 1,250 (2,053▲) . . . B3 97
West Brooklyn, Lee, Ill., 182 . B4 90
Westbrook Park, Delaware, Pa., 5,000 . . . *B11 114
Westbrookville, Sullivan, N.Y., 280 . . . D6 108
West Brownsville, Washington, Pa., 1,907 . . . *F2 114
West Brunswick, Somerset, N.J., 2,000 . . . *C4 106
West Burke, Caledonia, Vt., 369 . . . B5 120
West Burlington, Des Moines, Iowa, 2,560 . . . D6 92
Westbury, Nassau, N.Y., 14,757 . . . E3 108
West Buxton, York, Maine, 300 . . . E2 96
Westby, Sheridan, Mont., 309 . . . B12 102
Westby, Vernon, Wis., 1,544 . . . E3 124
West Cache, creek, Okla. . . . C3 112
West Caldwell, Essex, N.J., 8,314 . . . *B4 106
West Camp, Ulster, N.Y., 275 . . . C7 108
West Campton, Grafton, N.H., 100 . . . C3 105
West Canaan, Grafton, N.H., 95 . . . C2 105
West Canada, creek, N.Y. . . . B6 108
West Cape May, Cape May, N.J., 1,030 . . . F3 106
West Carroll, par., La., 14,177 . . . B4 95
West Carrollton, Montgomery, Ohio, 4,749 . . . C3 111
West Carry, pond, Maine . . . C2 96
West Carteret, Middlesex, N.J. (part of Carteret) . . . E4 106
West Carthage, Jefferson, N.Y., 2,167 . . . B5 108
West Castleton, Rutland, Vt. . D2 120
West Catasauqua, Lehigh, Pa., 900 . . . *E11 114
West Charleston, Orleans, Vt., 150 . . . B4 120
West Chazy, Clinton, N.Y., 600 . . . A3 108
West Chelmsford, Middlesex, Mass., 350 . . . C1 97
West Cheshire, New Haven, Conn. (part of Cheshire) . . C5 84
Westchester, N.S., Can., 125 . D6 74
Westchester, Cook, Ill., 18,092 . . . *F3 90
West Chester, Washington, Iowa, 253 . . . C6 92
West Chester, Butler, Ohio, 450 . . . C2 111
West Chester, Chester, Pa., 15,705 . . . G10 114
Westchester, co., N.Y., 808,891 . . . D7 108
West Chesterfield, Cheshire, N.H., 175 . . . E1 105
West Chevy Chase Heights, Montgomery, Md., 1,800 . *C3 85
West Chicago, Du Page, Ill., 6,854 . . . F2 90
West Chop, pt., Mass. . . . D6 97
West City, Franklin, Ill., 814 . . . E5 90
West Claremont, Sullivan, N.H. (part of Claremont) . D2 105
West Clarkston, Asotin, Wash., 2,851 . . . *C8 122
Westcliffe, Custer, Colo., 306 . . . C5 83
West College Corner, Union, Ind., 613 . . . E8 91
West Collingswood, Camden, N.J., 2,000 . . . *D2 106
West Collingswood Heights, Camden, N.J., 1,100 . . . *D2 106
West Columbia, Lexington, S.C., 6,410 . . . D5 115
West Columbia, Brazoria, Tex., 2,947 . . . E5, G4 118
West Concord, Middlesex, Mass., 1,556 . . . B5, C1 97
West Concord, Dodge, Minn., 810 . . . F6 99
West Concord, Cabarrus, N.C., 5,510 . . . *B3 109
Westconnaug, res., R.I. . . . C10 85
West Conshohocken, Montgomery, Pa., 2,254 . A11 114
West Corners, Broome, N.Y., 2,000 . . . *C4 108
West Cornwall, Litchfield, Conn., 200 . . . B3 84
West Cote Blanche, bay, La. . . . E4 95
West Covina, Los Angeles, Calif., 50,645 . . . *F3 82
Westcreek, Douglas, Colo., 5 . B5 83
West Creek, Ocean, N.J., 600 . . . D4 106
West Crossett, Ashley, Ark., 255 . . . D4 81
West Cumberland, Cumberland, Maine, 200 . . . E2, E5 96
West Cummington, Hampshire, Mass., 140 . . . B2 97
Westdale, Cook, Ill., 1,000 . . *F2 90
West Danby, Tompkins, N.Y., 55 . . . C4 108
West Danville, Caledonia, Vt., 85 . . . C4 120
West Decatur, Clearfield, Pa., 750 . . . E5 114
West Dennis, Barnstable, Mass., 900 . . . *C7 97
West Derry, Westmoreland, Pa., 1,000 . . . *F3 114
West Des Moines, Polk, Iowa, 11,949 . . . A7, C4 92
West Dudley, Worcester, Mass . . . B4 97
West Dummerston, Windham, Vt., 90 . . . F3 120
West Duxbury, Plymouth, Mass. 250 . . . B6 97
West Easton, Northampton, Pa., 1,228 . . . *E11 114
West Eden, Hanock, Maine . D4 96
West Elizabeth, Allegheny, Pa., 921 . . . *F2 114
West Elk, mts., Colo. . . . C3 83
West Ellicott, Chautauqua, N.Y., 1,500 . . . *C1 108
West Elmira, Chemung, N.Y., 5,763 . . . *C4 108
West Elwood, Tipton, Ind., 200 . . . D6 91
West Eminence, Shannon, Mo., 145 . . . D6 101
West Emma, creek, Kans. . . . D6 93
West End, Jefferson, Ark., 2,208 . . . *C3 81
West End, Ba. Is., 543 . . . B4 64
Westend, San Bernardino, Calif., 300 . . . E5 82
West End, Marion, Fla., 3,124 . . . *C4 86
West End, Moore, N.C., 600 . B4 109
West End, Otsega, N.Y., 1,436 . . . *C2 108
West End Anniston, Calhoun, Ala., 5,485 . . . *B4 78
West Endicott, Broome, N.Y., 2,500 . . . *C4 108
West Enfield, Penobscot, Maine, 500 . . . C4 96
West Englewood, Bergen, N.J (part of Teaneck) . . D5 106
West Enosburg, Franklin, Vt., 75 . . . B3 120
West Epping, Rockingham, N.H., 350 . . . D4 105
Westerham, Sask., Can., 89 . G1 70
Westerkappeln, Ger., 8,500 . A2 17
Westerland, Ger., 8,700 . . . A4 16
Westerly, Washington, R.I., 9,698 (14,267▲) . . . D9 84
Western, Saline, Nebr., 351 . D8 103
Western, prov., N. Rh . . . D4 48
Western, reg., Ghana . . . E4 45
Western, reg., Nig., 6,087,000 . . . E5 45
Western, reg., Tan., 955,852 . . C5 48
Western, reg., Ug . . . A5 48
Western, downs, Eng . . . D5 13
Western, isl., Newf., Can. . . . C4 75
Western Australia, state, Austl., 736,629 . . . D3 50
Western Azerbaijan, see Azerbaijan, reg., Ir
Western Ghats, range, India . . . E5 39
Western Grove, Newton, Ark., 148 . . . A3 81
Western Hills, Adams, Colo., 1,500 . . . *B6 83
Western Peninsula, div., Ice., 11,394 . . . *m21 25
Westernport, Allegany, Md., 3,559 . . . D2 85
Western Samoa, country, Oceania, 83,096 . . . 52
Western Shoshone, Indian res., Idaho, Nev. . . . B5 104
Western Springs, Cook, Ill., 10,838 . . . F2 90
Wester Schelde, chan., Neth . . . C3 15
Westerstede, Ger., 15,400 . . . A7 15
Westervelt, Shelby, Ill., 180 . D5 90
Westerville, Custer, Nebr., 40 . . . C6 103
Westerville, Franklin, Ohio, 7,011 . . . B5 111
Westerwald, mts., Ger . . . C2 17
West Fairlee, Orange, Vt., 185 (333▲) . . . D4 120
West Fairview, Cumberland, Pa., 1,718 . . . F8 114
Westfall, Lincoln, Kans., 75 . D5 93
Westfall, Malheur, Oreg., 8 . D9 113
West Falmouth, Cumberland, Maine (part of Falmouth) . E5 96
West Falmouth, Barnstable, Mass., 600 . . . C6 97
West Fargo, Cass, N. Dak., 93 . . . C9 110
West Farmington, Franklin, Maine, 500 . . . D2 96
West Farmington, Trumbull, Ohio, 614 . . . A7 111
West Feliciana, par., La., 12,395 . . . D4 95
Westfield, Jefferson, Ala., 2,000 . . . E4 78
Westfield, N.B., Can., 140 . . D3 74
Westfield, Clark, Ill., 636 . . D6 90
Westfield, Hamilton, Ind., 1,217 . . . D5 91
Westfield, Plymouth, Iowa, 187 . . . B1 92
Westfield, Aroostook, Maine, 350 (569▲) . . . B5 96
Westfield, Hampden, Mass., 26,302 . . . B2 97
Westfield, Union, N.J., 31,447 . . . B4 106
Westfield, Chautauqua, N.Y., 3,878 . . . C1 108
Westfield, Surry, N.C., 200 . A3 109
Westfield, Emmons, N. Dak., 40 . . . C5 110
Westfield, Tioga, Pa., 1,333 . C6 114
Westfield, Orleans, Vt., 90 (347▲) . . . B4 120
Westfield, Marquette, Wis., 919 . . . E4 124
Westfield, riv., Mass . . . B2 97
Westfir, Lane, Oreg., 500 . . . D4 113
West Flanders, prov., Bel., 1,068,976 . . . C3 15
Westford, Windham, Conn., 130 . . . B8 84
Westford, Middlesex, Mass., 700 (6,261▲) . . . C1 97
Westford, Chittenden, Vt., 90 (680▲) . . . B2 120
West Fork, Washington, Ark., 350 . . . B1 81
West Fork, riv., W. Va. . . . B4 123
West Forks, Somerset, Maine, (930▲) . . . C3 96
West Frankfort, Franklin, Ill., 9,027 . . . F5 90
West Friendship, Howard, Md., 50 . . . B4 85
West Frisian, is., Neth. . . . A5 15
West Gardiner, Kennebec, Maine, 35 (1,144▲) . . . D3 96
Westgate, Palm Beach, Fla., 2,500 . . . F6 86
Westgate, Fayette, Iowa, 214 . B6 92
West Glacier Park, Flathead, Mont., 300 . . . B3 102
West Glens Falls, Warren, N.Y., 2,725 . . . *B7 108
West Glocester, Providence, R.I., 50 . . . B9 84
West Gloucester, Essex, Mass. (part of Gloucester) . . A6, C4 97
West Glover, Orleans, Vt., 64 . . . B4 120
West Gorham, Cumberland, Maine, 40 . . . E4 96
West Goshen, Litchfield, Conn., 150 . . . B3 84
West Gouldsboro, Hancock, Maine, 90 . . . D4 96
West Granby, Hartford, Conn., 375 . . . B5 84
West Granville, Hampden, Mass., 150 . . . B2 97
West Gravenhurst, Ont., Can., 150 . . . C5 72
West Gray, Cumberland, Maine, 75 . . . E2, E5 96
West Green, Coffee, Ga., 300 . . . E4 87
West Greene, Greene, Ala., 60 . . . C1 96
West Groton, Middlesex, Mass., 600 . . . A4 97
West Groton, Caledonia, Vt . C4 120
West Grove, Davis, Iowa, 100 . . . D5 92
West Grove, Chester, Pa., 1,607 . . . G10 114
West Halifax, Windham, Vt., 100 . . . F3 120
West Ham, Eng., 157,186 . . . k13 10, C8 12
West Hamlin, Lincoln, W. Va., 788 . . . C2 123
West Hampstead, Rockingham, N.H., 150 . . . E4 105
Westhampton, Suffolk, N.Y., 900 . . . F6 84
Westhampton, beach, N.Y. . . . F6 84
Westhampton Beach, Suffolk, N.Y., 1,460 . . . D4 108
West Hanover, Plymouth, Mass., 350 . . . E3 97
West Harrison, Dearborn, Ind., 341 . . . F8 91
West Hartford, Hartford, Conn., 62,382 . . . B6 84
West Hartford, Windsor, Vt., 100 . . . D4 120
West Hartland, Hartford, Conn., 200 . . . A4 84
West Hartlepool, Eng., 77,073 . . . C6 10
West Hatfield, Hampshire, Mass., 100 . . . B2 97
West Haven, New Haven, Conn., 43,002 . . . D5 84
West Haven, Va. (part of Portsmouth) . . . B6 121
West Haverstraw, Rockland, N.Y., 5,020 . . . *D7 108
West Hazleton, Luzerne, Pa., 6,278 . . . E9 114
West Helena, Phillips, Ark., 8,385 . . . C5 81
West Hempstead, Nassau, N.Y., 15,800 . . . *G2 84
West Hickory, Forest, Pa., 450 . . . C3 114
West Hill, Ont., Can., 1,800 . . . D5, E7 72
Westhoff, De Witt, Tex., 400 . . . E4 118
West Hollywood, Los Angeles, Calif., 28,870 . . . *F2 82
West Hollywood, Broward, Fla., 52,000 . . . E3 86
West Homestead Allegheny, Pa., 4,155 . . . *F2 114
Westhope, Bottineau, N. Dak., 824 . . . A4 110
West Hopkinton, Merrimack, N.H., 45 . . . D3 105
West Indies, is., N.A. . . . G13 61, 64
West Irian, Indon. dep., Asia, 700,000 . . . F9 35
West Java, see Java, isl., Indon.
West Jefferson, Ashe, N.C., 1,000 . . . A2 109
West Jefferson, Madison, Ohio, 2,774 . . . C1, C4 111
West Jonesport, Washington, Maine, 540 . . . D5 96
West Jordan, Salt Lake, Utah, 3,009 . . . C4 119
West Junction, Shelby, Tenn., 4,000 . . . B1, E8 117
West Kankakee, Kankakee, Ill., 3,197 . . . B6 90
Westkapelle, Bel., 2,355 . . . C3 15
Westkapelle, Neth., 2,400 . . . C3 15
West Keansburg, Monmouth, N.J., 3,000 . . . *C4 106
West Kennebunk, York, Maine, 350 . . . E2 96
West Kildonan, Man., Can., 20,077 . . . E3 71
West Kingston, Washington, R.I., 900 . . . D10 84
West La Crosse, La Crosse, Wis., 1,440 . . . *E2 124
West Lafayette, Tippecanoe, Ind., 12,680 . . . D4 91
West Lafayette, Coshocton, Ohio, 1,476 . . . B6 111
Westlake, Calcasieu, La., 3,311 . . . D2 95
Westlake, Cuyahoga, Ohio, 12,906 . . . B1 111
Westlake, Lane, Oreg., 150 . D2 113
Westland, Tarrant, Tex., 1,000 . . . *B5 118
Westlands, Middlesex, Mass., 2,100 . . . *A5 97
West Lanham Hills, Prince Georges, Md., 1,000 . . . *C4 85
West Lawn, Berks, Pa., 2,059 . . . F10 114
West Lawn, Fairfax, Va., 1,400 . . . *C5 121
West Lebanon, Warren, Ind., 720 . . . D3 91
West Lebanon, York, Maine, 160 . . . E2 96
West Lebanon, Grafton, N.H. (part of Lebanon) . . . C2 105
West Lebanon, Lebanon, Pa., 1,054 . . . *F9 114
West Leechburg, Westmoreland, Pa., 1,323 . . . *E2 114
West Leisenring, Fayette, Pa., 800 . . . G2 114
Westley, Stanislaus, Calif., 250 . . . B6 82
West Leyden, Lewis, N.Y., 250 . . . B5 108
West Liberty, Jasper, Ill., 160 . E5 90
West Liberty, Muscatine, Iowa, 2,042 . . . C6 92
West Liberty, Morgan, Ky., 1,165 . . . C6 94
West Liberty, Logan, Ohio, 1,522 . . . B4 111
West Liberty, Ohio, W. Va., 1,500 . . . B2 123
West Lima, Richland, Wis., 140 . . . E3 124
West Lincoln, Lancaster, Nebr., 507 . . . E1 103
West Lincoln, Addison, Vt., 70 . . . C2 120
Westline, McKean, Pa., 160 . C4 114
Westlink, Sedgwick, Kans., 2,500 . . . B5 93
West Linn, Clackamas, Oreg., 3,933 . . . B2, B4 113
Westlock, Alta., Can., 1,838 . B4 69
West Long Branch, Monmouth, N.J., 5,337 . . . C4 106
West Lorne, Ont., Can., 1,070 . . . E3 72
West Lothian, co., Scot., 92,764 . . . E5 13
West Louisville, Daviess, Ky., 250 . . . C2 94
West Lubec, Washington, Maine, 150 . . . D5 96
West Manayunk, Montgomery, Pa., 1,900 . . . *F11 114
West Manchester, Preble, Ohio, 460 . . . C3 111
West Mansfield, Logan, Ohio, 791 . . . B4 111
West Marion, Marion, S.C., 45 . . . C9 115
West Mayfield, Beaver, Pa., 2,201 . . . *E1 114
Westmeath, Ont., Can., 260 . B8 72

Westmeath, co., Ire., 52,861 . D4 11
West Medway, Norfolk, Mass., 1,818 .... B5, E1 97
West Melbourne, Brevard, Fla., 2,266 .... *D6 86
West Memphis, Crittenden, Ark., 19,374 .... B5 81
Westmere, Albany, N.Y., 1,500 .... *C7 108
West Methow, riv., Wash. A5 122
West Miami, Dade, Fla., 5,296 .... F3 86
West Middlesex, Mercer, Pa., 1,301 .... D1 114
West Middleton, Howard, Ind., 280 .... D5 91
West Mifflin, Allegheny, Pa., 27,289 .... F2 114
West Milan, Coos, N.H., 100 . A4 105
West Milford, Passaic, N.J., 800 .... A4 106
West Milford, Harrison, W. Va., 367 .... B4 123
West Milton, Miami, Ohio, 2,972 .... C3 111
West Milton, Union, Pa., 750 .... D8 114
West Milwaukee, Milwaukee, Wis., 5,043 .... E2 124
West Mineral, Cherokee, Kans., 262 .... E9 93
Westminster, Orange, Calif., 25,750 .... *F3 82
Westminster, Adams, Colo., 13,850 .... B5 83
Westminster, Carroll, Md., 6,123 .... A4 85
Westminster, Worcester, Mass., 1,047 (4,022▲) A4 97
Westminster, Oconee, S.C., 2,413 .... B1 115
Westminster, Windham, Vt., 333 (1,602▲) E4 120
Westminster, dist., B.C., Can. E6 68
Westminster Station, Windham, Vt., 60 .... E4 120
Westminster West, Windham, Vt., 60 .... E3 120
West Monroe, Ouachita, La., 15,215 .... B3 95
Westmont, Du Page, Ill., 5,997 .... F2 90
Westmont, Camden, N.J., 14,000 .... D2 106
Westmont, Cambria, Pa., 6,573 .... F4 114
West Monterey, Clarion, Pa., 300 .... D2 114
Westmoreland, Pottawatomie, Kans., 460 .... C7 93
Westmoreland, Cheshire, N.H., 100 (921▲) E2 105
Westmoreland, Sumner, Tenn., 865 .... A5 117
Westmoreland, co., Pa., 352,629 .... F2 114
Westmoreland, co., Va., 11,042 .... C6 121
Westmoreland City, Westmoreland, Pa., 1,300 .... *F2 114
Westmorland, Imperial, Calif., 1,404 .... F6 82
Westmorland, co., N.B., Can., 93,679 .... C5 74
Westmorland, co., Eng., 67,222 .... *C5 10
Westmount, Que., Can., 25,012 .... *D9 73
West Muncie, Delaware, Ind., 300 .... D7 91
West Musquash, lake, Maine .. C5 96
West Mystic, New London, Conn., 3,268 .... D9 84
West Nanticoke, Luzerne, Pa., 800 .... *B7 114
West Newbury, Essex, Mass., 800 (1,844▲) A6 97
West Newbury, Orange, Vt., 50 .... C4 120
West Newfield, York, Maine, 100 .... E2 96
West New Guinea, see West Irian, Indon. dep., Asia
West Newton, Marion, Ind., 400 .... I7 91
West Newton, Westmoreland, Pa., 3,982 .... F2 114
West New York, Hudson, N.J., 35,547 .... D5 106
West Nicholson, S. Rh. B4 49
West Nishnabotna, riv., Iowa .. C2 92
West Norfolk, Norfolk, Va., 500 .... B6 121
West Nottingham, Rockingham, N.H., 50 .... D4 105
West Nyack, Rockland, N.Y., 3,000 .... *D7 108
West Olive, Ottawa, Mich., 60 .... F4 98
Weston, Ont., Can., 9,715 ... E6 72
Weston, Las Animas, Colo., 350 .... D6 83
Weston, Fairfield, Conn., 3,000 (4,039▲) E3 84
Weston, Webster, Ga., 120 .. E2 87
Weston, Franklin, Idaho, 284 .... G7 89
Weston, McLean, Ill., 150 ... C5 90
Weston, Aroostook, Maine, 50 (202▲) C5 96
Weston, Middlesex, Mass., 4,000 (8,261▲) D2 97
Weston, Platte, Mo., 1,057 .... B3, D1 101
Weston, Saunders, Nebr., 340 .... C9 103
Weston, Wood, Ohio, 1,075 . A4 111
Weston, Umatilla, Oreg., 783 .... B8 113
Weston, Windsor, Vt., 200 (442▲) E3 120
Weston, Lewis, W. Va., 8,754 .... B4 123
Weston, Dunn, Wis., 40 .... D1 124
Weston, co., Wyo., 7,929 ... B8 125
Weston-super-Mare, Eng., 43,923 .... C5 12
West Orange, Franklin, Mass., 60 .... A3 97
West Orange, Essex, N.J., 39,895 .... B4 106
West Orange, Orange, Tex., 4,848 .... *D6 118
West Ossipee, Carroll, N.H., 100 .... C4 105
Westover, Shelby, Ala., 350 .. B3 78
Westover, Somerset, Md., 250 .... D6 85
Westover, Broome, N.Y., 1,500 .... *C5 108
Westover, Clearfield, Pa., 492 .... E4 114
Westover, Monongalia, W. Va., 4,749 .... A7, B5 123
West Pakistan, prov., Pak., 42,976,261 .... *C5 39
West Palm Beach, Palm Beach, Fla., 56,208 (★157,200) F6 86
West Paris, Oxford, Maine, 600 (1,050▲) D2 96
West Park, Ulster, N.Y., 500 .... D7 108
West Paterson, Passaic, N.J., 7,602 .... *B4 106
West Pawlet, Rutland, Vt., 280 .... E2 120
West Peabody, Essex, Mass. (part of Peabody) . A6, C3 97
West Pearl, riv., La. D6 95
West Pelzer, Anderson, S.C., 687 .... B2 115
West Pembroke, Washington, Maine, 400 .... D5 96
West Pensacola, Escambia, Fla., 25,000 .... G2 86
West Peru, Oxford, Maine, 350 .... D2 96
West Peterborough, Hillsboro, N.H., 325 .... E3 105
West Petersburg, Alsk., 20 .. m23 79
Westphalia, Knox, Ind., 300 . G3 91
Westphalia, Anderson, Kans., 249 .... D8 93
Westphalia, Clinton, Mich., 560 .... F6 98
Westphalia, Osage, Mo., 316 . C5 101
Westphalia, reg., Ger. C3 16
West Pittsburg (Shell Point), Contra Costa, Calif., 5,188 . *B6 82
West Pittsburg, Lawrence, Pa., 850 .... E1 114
West Pittsfield, Berkshire, Mass. (part of Pittsfield) .. B1 97
West Pittston, Luzerne, Pa., 6,998 .... B8 114
West Plains, Howell, Mo., 5,836 .... E6 101
West Point, Cullman, Ala., 200 .... A3 78
West Point, White, Ark., 97 . B4 81
West Point, Calaveras, Calif., 900 .... C3 82
West Point, Troup, Ga., 4,610 .... D1 87
West Point, Hancock, Ill., 234 .... C2 90
Westpoint, Tippecanoe, Ind., 350 .... D3 91
West Point, Lee Iowa, 758 .. D6 92
West Point, Hardin, Ky., 2,005 .... C4 94
Westpoint, Sagadahoc, Maine, 170 .... E6 96
West Point, Clay, Miss., 8,550 .... B5 100
West Point, Cuming, Nebr., 2,921 .... C9 103
West Point, Orange, N.Y., 4,000 .... D2, D7 108
Westpoint, Lawrence, Tenn., 300 .... B4 117
West Point, Davis, Utah, 599 .... *B3 119
West Point, King William, Va., 1,678 .... D6 121
West Point, mtn., Alsk. .. C11 79
Westport, Newf., Can., 222 .... D3, k10 75
Westport, N.S., Can., 413 .... E3 74
Westport, Ont., Can., 711 .... C8 72
Westport, Fairfield, Conn., 20,955 .... E3 84
Westport, Decatur, Ind., 833 .... F6 91
Westport, Ire., 2,882 .... D2 11
Westport, Oldham, Ky., 125 .... A4, B4 94
Westport, Bristol, Mass., 600 (6,641▲) C5 97
Westport, Pope, Minn., 87 .... E3 99
Westport, Cheshire, N.H., 170 .... E2 105
Westport, Essex, N.Y., 723 .... A7, B3 108
Westport, N.Z., 5,460 .... N13 51
Westport, Clatsop, Oreg., 300 .... A3 113
Westport, Brown, S. Dak., 35 .... B7 116
Westport, Carroll, Tenn., 125 .... B3 117
Westport, Grays Harbor, Wash., 976 .... C1 122
West Portland, Multnomah, Oreg., 2,000 .... *B4 113
Westport Point, Bristol, Mass., 300 .... C5 97
West Portsmouth, Scioto, Ohio, 3,100 .... D4 111
West Prairie, riv., Alta., Can. .. B2 69
West Pubnico, N.S., Can., 247 .... F4 74
West Pueblo, Pueblo, Colo., 1,700 .... *C6 83
West Punjab, reg., Pak. B3 40
West Quoddy Head, cape, Maine . D6 96
Westray, firth, Scot. A5 13
Westray, isl., Scot. A5 13
West Reading, Berks, Pa., 4,938 .... F10 114
West Redding, Fairfield, Conn., 250 .... D3 84
Westrhauderfehn, Ger., 4,500 .... A7 15
West Richland, Benton, Wash., 1,347 .... *C6 122
West Riding, Yorkshire, see York, co., Eng.
West Rindge, Cheshire, N.H., 130 .... E2 105
West River, Anne Arundel, Md., 100 .... C4 85
West Road, riv., B.C., Can. .. C5 68
West Roag, bay, Scot. B1 13
West Roanoke Rapids, Halifax, N.C., 2,000 .... *A6 109
West Rockingham, Richmond, N.C., 1,128 .... *C4 109
West Rumney, Grafton, N.H., 40 .... C3 105
West Rupert, Bennington, Vt., 160 .... E2 120
West Rutland, Rutland, Vt., 2,302 (2,302▲) D2 120
West Rye, Rockingham, N.H., 150 .... E5 105
West Sacramento, Yolo, Calif., 10,000 .... *A6 82
West St. Modeste, Newf., Can., 141 .... C3 75
West St. Paul, Dakota, Minn., 13,101 .... E7 99
West Salem, Edwards, Ill., 956 .... E5 90
West Salem, Wayne, Ohio, 1,017 .... B5 111
West Salem, Polk, Oreg. (part of Salem) C1, C3 113
West Salem, La Crosse, Wis., 1,707 .... E2 124
West Salisbury, Addison, Vt., 85 .... D2 120
West Sayville, Suffolk, N.Y., 3,000 .... G4 84
West Scarboro, Cumberland, Maine, 900 .... E2, E4 96
West Seboois, Penobscot, Maine, 50 .... C4 96
West Seneca, Erie, N.Y., 33,644 .... *C2 108
West Shefford, Que., Can., 406 .... D5 73
West Siberian, plain, Sov. Un. .. A6 29
Westside, Crawford, Iowa, 367 .... B2 92
West Side, Lake, Oreg E6 113
West Simsbury, Hartford, Conn., 400 .... B5 84
West Slope, Washington, Oreg., 9,000 .... B2 113
West Somerset, Pulaski, Ky., 400 .... C5 94
West Spanish, peak, Colo D6 83
West Spitsbergen, isl., Nor B13 4
West Springfield, Hampden, Mass., 24,924 .... B2 97
West Springfield, Sullivan, N.H., 35 .... D2 105
West Stafford, Tolland, Conn., 300 .... B7 84
West Sterling, Whiteside, Ill., 1,430 .... *B4 90
West Stewartstown, Coos, N.H., 225 .... A1 105
West Stockbridge, Berkshire, Mass., 800 (1,244▲) B1 97
West Stoughton, Norfolk, Mass. (part of Stoughton) . E2 97
West Suffield, Hartford, Conn., 700 .... B6 84
West Sullivan, Hancock, Maine, 125 .... D4 96
West Summerland, B.C., Can., 2,500 .... E8 68
West Sumner, Oxford, Maine, 125 .... D2 96
West Sunbury, Butler, Pa., 252 .... E2 114
West Swanzey, Cheshire, N.H., 900 .... E2 105
West Tarbert, bay, Scot. C1 13
West Tavaputs, plat., Utah D5 119
West Terre Haute, Vigo, Ind., 3,006 .... F3 91
Westterschelling, Neth., 1,995 .... A5 15
West Thornton, Grafton, N.H., 130 .... C3 105
West Tisbury, Dukes, Mass., 350 (360▲) D6 97
West Tocoi, Clay, Fla., 30 .... C5 86
West Topsham, Orange, Vt., 100 .... C4 120
West Torrington, Litchfield, Conn. (part of Torrington) . B4 84
West Townsend, Middlesex, Mass., 900 .... A4 97
West Townshend, Windham, Vt., 95 .... E3 120
West Trenton, Mercer, N.J., 26,628 .... C3 106
West Union, Clark, Ill., 400 . D6 90
West Union, Fayette, Iowa, 2,551 .... B6 92
West Union, Adams, Ohio, 1,762 .... D4 111
West Union, Oconee, S.C., 443 .... B2 115
West Union, Doddridge, W. Va., 1,186 .... B4, B6 123
West Unity, Williams, Ohio, 1,192 .... A3 111
West University Place, Harris, Tex., 14,628 .... F4 118
West Upsalquitch, riv., N.B., Can. B3 74
West Upton, Worcester, Mass., 990 .... B4, D1 97
Westvale, Onondaga, N.Y., 6,000 .... *B4 108
West Valley, Cattaraugus, N.Y., 350 .... C2 108
West Van Lear, Johnson, Ky., 900 .... C7 94
West Vernon, Wilbarger, Tex. (part of Vernon) B3 118
Westview, B.C., Can., 4,000 . E5 68
Westview, Cuyahoga, Ohio, 1,303 .... B1 111
West View, Allegheny, Pa., 8,079 .... A5 114
West View Park, Sullivan, Tenn., 1,000 .... C11 117
Westville, N.S., Can., 4,159 . D7 74
Westville, Holmes, Fla., 250 . G3 86
Westville, Vermilion, Ill., 3,497 .... C6 90
Westville, LaPorte, Ind., 789 . A4 91
Westville, Rockingham, N.H., 400 .... E4 105
Westville, Gloucester, N.J., 4,951 .... D2 106
Westville, Champaign, Ohio, 180 .... B4 111
Westville, Adair, Okla., 727 . B7 112
Westville, Jefferson, Pa., 160 . D4 114
Westville, Kershaw, S.C., 200 .... C4 115
Westville Grove, Gloucester, N.J., 2,500 .... *D2 106
West Virginia, state, U.S., 1,860,421 .... C11 77, 123
West Walker, riv., Nev. E2 104
West Wareham, Plymouth, Mass., 500 .... C6 97
West Warren, Worcester, Mass, 1,124 .... B3 97
West Warwick, Kent, R.I., 21,414 .... C10 84
Westwater, Grand, Utah, 15 .... D6 119
Westway, Tolland, Conn B7 84
Westwego, Jefferson, La., 9,815 .... C7, E5 95
West Wenatchee, Chelan, Washington, 2,518 .... *B5 122
West Wickham, Eng. (part of Beckenham) m12 10
West Willington, Tolland, Conn., 300 .... B7 84
West Wilton, Hillsboro, N.H., 100 .... E3 105
West Winfield, Herkimer, N.Y., 960 .... C5 108
West Winfield, Butler, Pa., 300 .... E2 114
West Winter Haven, Polk, Fla., 5,050 .... *D5 86
Westwold, B.C., Can., 327 .. D8 68
Westwood, Lassen, Calif., 1,209 .... B3 82
Westwood, Johnson, Kans., 2,040 .... *D9 93
Westwood, Boyd, Ky., 6,000 . B7 94
Westwood, Norfolk, Mass., 5,800 (10,354▲) B5, D2 97
Westwood, Kalamazoo, Mich., 6,500 .... *F5 98
Westwood, Bergen, N.J., 9,046 .... B4, D5 106
Westwood Lakes, Dade, Fla., 22,517 .... F3 86
West Woodstock, Windham, Conn., 100 .... B8 84
West Woodstock, Windsor, Vt., 60 .... D3 120
Westworth Village, Tarrant, Tex., 3,321 .... *C4 118
West Wyalong, Austl., 2,399 .... F6 51
West Wyoming, Luzerne, Pa., 3,166 .... B8 114
West Wyomissing, Berks, Pa., 2,500 .... *F9 114
West Yarmouth, Barnstable, Mass., 1,365 .... C7 97
West Yellowstone, Gallatin, Mont., 150 .... F5 102
West York, Crawford, Ill., 250 .... D6 90
West York, York, Pa., 5,526 . G8 114
Wet, mts., Colo C5 83
Wetar, isl., Indon G7 35
Wetaskiwin, Alta., Can., 5,300 .... C4 69
Wethersfield, Hartford, Conn., 20,561 .... C6 84
Wetmore, Custer, Colo., 100 . C5 83
Wetmore, Nemaha, Kans., 390 .... C8 93
Wetmore, Alger, Mich., 200 .. B4 98
Wetonka, McPherson, S. Dak., 46 .... B7 116
Wettingen, Switz., 17,613 .... *B5 19
Wetumka, Hughes, Okla., 1,798 .... B5 112
Wetumpka, Elmore, Ala., 3,672 .... C3 78
Wetzel, co., W. Va., 19,347 . B4 123
Wetzikon, Switz., 10,421 .... B6 19
Wetzlar, Ger., 37,300 (★67,000) C3 17
Wevelgem, Bel., 12,805 .... D3 15
Wever, Lee, Iowa, 100 .... D6 92
Wewahitchka, Gulf, Fla., 1,436 .... B1 86
Wewahotee, Orange, Fla D5 86
Wewak, N. Gui., 59 .... h11 50
Wewela, Tripp, S. Dak., 22 . D6 116
Wewoka, Seminole, Okla., 5,954 .... B5 112
Wexford, Ire., 11,328 .... E5 12
Wexford, co., Ire., 83,308 .... E5 12
Wexford, co., Mich., 18,466 . D5 98
Wexford, bay, Ire. E5 12
Weyakwin, lake, Sask., Can. C3 70
Weyanoke, West Feliciana, La., 50 .... D4 95
Weyauwega, Waupaca, Wis., 1,239 .... D5 124
Weyburn, Sask., Can., 9,101 . H4 70
Weyerhauser, Rusk, Wis., 339 .... C2 124
Weyers Cave, Augusta, Va., 300 .... C4 121
Weymouth, N.S., Can., 671 .. E4 74
Weymouth, Eng., 40,962 .... D5 12
Weymouth, Norfolk, Mass., 48,177 .... B6, D3 97
Weymouth, bay, Eng D5 12
Whakatane, N.Z., 7,167 .... L16 51
Whalan, Fillmore, Minn., 146 .... G7 99
Whale, riv., Que., Can g8 75
Whaley, pond, N.Y C2 84
Whaleysville, Worcester, Md., 240 .... D7 85
Whaleyville, Nansemond, Va., 402 .... E6 121
Whangarei, N.Z., 17,880 (★21,790) K15 51
Whangarei, harbor, N.Z K15 51
Whappen Rig, mtn., Scot E5 13
Wharfe, riv., Eng D6 10
Wharton, Madison, Ark A2 81
Wharton, Morris, N.J., 5,006 .... B3 106
Wharton, Wyandot, Ohio, 463 .... B4 111
Wharton, Potter, Pa., 25 .... C5 114
Wharton, Wharton, Tex., 5,734 .... E4 118
Wharton, Boone, W. Va., 1,055 .... D3, D6 123
Wharton, co., Tex., 38,152 .... E4 118
Wharton West, Wharton, Tex., 1,609 .... *E4 118
Whatcheer, Alta., Can., 15 .. D5 69
What Cheer, Keokuk, Iowa, 956 .... C5 92
Whatcom, co., Wash., 70,317 .... A4 122
Whatcom, lake, Wash A3 122
Whatley, Franklin, Mass., 150 (1,037▲) B2 97
Whatley, Clarke, Ala., 400 .. D2 78
Wheatcroft, Webster, Ky., 317 .... C2 94
Wheatfield, Jasper, Ind., 679 . B3 91
Wheatland, Yuba, Calif., 813 .... C3 82
Wheatland, Knox, Ind., 614 . G3 91
Wheatland, Clinton, Iowa, 643 .... C7 92
Wheatland, Hickory, Mo., 305 .... D4 101
Wheatland, Cass, N. Dak., 112 .... C8 110
Wheatland, Oklahoma, Okla. (part of Oklahoma City) .. B4 112
Wheatland, Mercer, Pa., 1,813 .... *D1 114
Wheatland, Kenosha, Wis., 20 .... F1 124
Wheatland, Platte, Wyo., 2,350 .... C8 125
Wheatland, co., Mont., 3,026 .... D7 102
Wheatley, Saint Francis, Ark., 443 .... C4 81
Wheatley, Ont., Can., 1,362 . E2 72
Wheaton, Du Page, Ill., 24,312 .... B5, F2 90
Wheaton, Pottawatomie, Kans., 114 .... C7 93
Wheaton, Montgomery, Md., 54,635 .... *B3 85
Wheaton, Traverse, Minn., 2,102 .... E2 99
Wheaton, Barry, Mo., 341 .. E3 101
Wheat Ridge, Jefferson, Colo., 21,619 .... *B5 83
Wheat Road, Atlantic, N.J. (part of Buena) D3 106
Wheelbarrow, peak, Nev. F5 104
Wheeler, Lawrence, Ala., 150 .... A2 78
Wheeler, Jasper, Ill., 173 .... D5 90
Wheeler, Porter, Ind., 500 .. A3 91
Wheeler, Cheyenne, Kans., 40 .... C2 93
Wheeler, Prentiss, Miss., 250 . A5 100
Wheeler, Valley, Mont., 25 . B10 102
Wheeler, Tillamook, Oreg., 237 .... B3 113
Wheeler, Wheeler, Tex., 1,174 .... B2 118
Wheeler, Dunn, Wis., 227 .... C2 124
Wheeler, co., Ga., 5,342 .... D4 87
Wheeler, co., Nebr., 1,297 .. C7 103
Wheeler, co., Oreg., 2,722 .. C6 113
Wheeler, co., Tex., 7,947 .... B2 118
Wheeler, dam, Ala A2 78
Wheeler, nat, mon., Colo D4 83
Wheeler, peak, Nev. E7 104
Wheeler, peak, N. Mex A4 107
Wheeler, res., Ala A2 78
Wheelersburg, Scioto, Ohio, 2,682 .... D5 111
Wheeless, Cimarron, Okla., 10 .... D1 112
Wheeling, Cook, Ill., 9,627 .. E2 90
Wheeling, Livingston, Mo., 302 .... B4 101
Wheeling, Ohio, W. Va., 53,400 (★126,600) A4, B2 123
Wheeling, creek, W. Va. B2 123
Wheelock, Williams, N. Dak., 82 .... A2 110
Wheelock, Caledonia, Vt., 70 (246▲) B4 120
Wheelock, mtn., Vt. B4 120
Wheelwright, Floyd, Ky., 1,518 .... C7 94
Wheelwright, Worcester, Mass., 200 .... B3 97
Whelen Springs, Clark, Ark., 155 .... D2 81
Whernside, mtn., Eng. F6 13
Whickham, Eng., 24,791 .... F7 13
Whidbey, isl., Wash. A3 122
Whigham, Grady, Ga., 463 . F2 87
Whipholt, Cass, Minn., 25 .. C4 99
Whippany, Morris, N.J., 6,000 .... B4 106
Whipple, Fayette, W. Va., 500 .... D7 123
Whiskey Chitto, creek, La. C2 95
Whiskey Gap, Alta., Can., 75 .... E4 69
Whitacres, Hartford, Conn., 1,000 .... *A6 84
Whitaker, Allegheny, Pa., 2,130 .... *F2 114
Whitakers, Edgecombe and Nash, N.C., 1,004 .... A5 109
Whitbourne, Newf., Can., 1,085 .... E5 75
Whitby, Ont., Can., 14,685 . D6 72
Whitby, Eng., 11,662 .... C6 10
Whitchurch, Eng., 7,159 .... B5 12
White, Bartow, Ga., 439 .... B2 87
White, Brookings, S. Dak., 417 .... C9 116
White, Shelby, Tenn. (part of Memphis) E8 117
White, co., Ark., 32,745 .... B4 81
White, co., Ga., 6,935 .... B3 87
White, co., Ill., 19,373 .... E5 90
White, co., Ind., 19,709 .... C4 91
White, co., Tenn., 15,577 .. D8 117
White, bay, Newf., Can. D3 75
White, butte, N. Dak. C2 110
White, isl., Nor. A11 4
White, lake, Ont., Can. B8 72
White, lake, La. E3 95
White, mts., N.H. B3 105
White, pass, B.C., Can., U.S. .. E5 66
White, riv., Ariz. D5 80
White, riv., Ark., Mo. E4 101
White, riv., Colo., Utah C6 119
White, riv., Ind. G5 91
White, riv., Mich. E4 98
White, riv., Nebr., S. Dak. .. D5 116
White, riv., Nev. E6 104
White, riv., Tex. C2 118
White, riv., Vt. D3 120
White, riv., Wash. B4 122
White, riv., Wash. B5 122
White, rock, Oreg. D4 113
White, sea, Sov. Un. E17 25
Whiteadder, riv., Scot. E6 13
White Bay, dist., Newf., Can. D3 75
White Bear, Sask., Can., 139 . G1 70
White Bear, is., Newf., Can. .. A3 75
White Bear, lake, Newf., Can. .. A2 75
White Bear, riv., Newf., Can. B3, E3 75
White Bear Beach, Ramsey, Minn. (part of White Bear Lake) E7 99
White Bear Lake, Ramsey, Minn., 12,849 .... E5, E7 99
White Bird, Idaho, Idaho, 253 .... D2 89
White Bluff, Dickson, Tenn., 486 .... A4 117
Whitebreast, creek, Iowa C4 92
White Butte, Perkins, S. Dak., 50 .... B3 116
White Canyon, San Juan, Utah, 98 .... F5 119
White Cap, mtn., Maine C3 96
White Carpathians, mts., Czech. . D4 26
White Castle, Iberville, La., 2,253 .... B5, D4 95
White City, Gulf, Fla., 600 . C1 86
White City, Saint Lucie, Fla., 900 .... E6 86
White City, Morris, Kans., 459 .... D7 93
Whiteclay, Sheridan, Nebr., 80 .... B3 103
White Clay, creek, Del., Pa. A6 85
White Clay, creek, Nebr., S. Dak. B3 103
White Cliffs, Austl., 76 .... E4 51
White Cloud, Doniphan, Kans., 238 .... C8 93
White Cloud, Newaygo, Mich., 1,001 .... E5 98
Whiteclouds, peaks, Idaho E4 89
Whitecomb, mtn., N.H. A4 105
White Cone, Navajo, Ariz., 12 .... B5 80
White Coomb, mtn., Scot. E5 13
White Court, Alta., Can., 1,054 .... B3 69
Whiteday, creek, W. Va. A7 123
Whitedeer, Union, Pa., 300 . D8 114
White Deer, Carson, Tex., 1,057 .... B2 118
White Earth, Becker, Minn., 350 .... C3 99
White Earth, Mountrail, N. Dak., 208 .... A3 110
White Earth, Indian res., Minn. . C3 99
White Earth, lake, Minn. C3 99
Whiteface, Cochran, Tex., 535 .... C1 118
Whiteface, mtn., N.Y. B3 108
White Face, mtn., Vt. B3 120
Whiteface, riv., Minn. C6 99
Whitefield, Coos, N.H., 1,244 (1,581▲) B3 105
Whitefield, Haskell, Okla., 200 .... B6 112
Whitefish, Ont., Can., 726 .. A3 72
Whitefish, Flathead, Mont., 2,965 .... B2 102
Whitefish, bay, Ont., Can. .. A4 99
Whitefish, bay, Mich. B6 98
Whitefish, lake, Ont., Can. A8 99
Whitefish, lake, Minn. D4 99
Whitefish, pt., Wis. D6 124
Whitefish, range, Mont. B2 102
Whitefish, riv., Mich. B4 98
Whitefish Bay, Milwaukee, Wis., 18,390 .... E2 124
Whitefish Falls, Ont., Can., 211 .... A3 72
Whitefish Point, Chippewa, Mich., 100 .... B6 98
Whiteflat, Motley, Tex., 100 . B2 118
Whiteford, Hartford, Md., 120 .... A5 85
White Fox, Sask., Can., 396 . D3 70
Whitefox, riv., Sask., Can. D3 70
White Gull, creek, Sask., Can. .. D3 70
White Gull, lake, Newf., Can. .. g8 75
White Hall, Lowndes, Ala., 100 .... C3 78
White Hall, Clarke, Ga., 409 .... C3 87
White Hall, Greene, Ill., 3,012 .... D3 90
Whitehall, Ire. E4 11
Whitehall, Livingston, La., 150 .... B6, D5 95
White Hall, Baltimore, Md., 115 .... A4 85
Whitehall, Muskegon, Mich., 2,590 .... E4 98
Whitehall, Jefferson, Mont., 898 .... E4 102
Whitehall, Washington, N.Y., 4,016 .... B7 108
Whitehall, Franklin, Ohio, 20,818 .... C5 111
Whitehall, Allegheny, Pa., 16,075 .... *B5 114
White Hall, Colleton, S.C. .. F6 115
Whitehall, Trempealeau, Wis., 1,446 .... D2 124
Whitehall, pond, Mass. B4 97
Whitehaven, Eng., 27,541 .... C5 10
Whitehaven, Wicomico, Md., 60 .... D6 85
White Haven, Luzerne, Pa., 1,778 .... D10 114
Whitehaven, Shelby, Tenn., 13,894 .... E8 117
Whitehead, Lauderdale, Ala., 100 .... A2 78
Whitehead, N. Ire., 2,174 .... C5 11
White Heath, Piatt, Ill., 350 . C5 90
Whitehorn, pt., Wash. A3 122
Whitehorse, Yukon, Can., 5,031 .... D5 66
White Horse, Mercer, N.J. (part of Hamilton Township) .. C3 106
Whitehorse, Dewey, S. Dak., 50 .... B5 116
White Horse Beach, Plymouth, Mass., 150 .... C6 97
White House, Duval, Fla., 500 .... B6 86
Whitehouse, Lucas, Ohio, 1,135 .... A1, A4 111
Whitehouse, Scot. C6 13
White House Station, Hunterdon, N.J., 700 .... B3 106
White Knob, mts., Idaho F5 89
White Lake, Ont., Can., 209 . B8 72
White Lake, Aurora, S. Dak., 397 .... D7 116
White Lake, Langlade, Wis., 325 .... C5 124
White Lakes, Santa Fe, N. Mex. B4, G6 107
Whiteland, Johnson, Ind., 1,368 .... E5 91
Whitelaw, Alta., Can., 264 .. A1 69
Whitelaw, Manitowoc, Wis., 420 .... B6 124
Whiteleysburg, Kent, Del., 25 .... C6 85
White Mesa, natural bridge, Ariz. A4 80
White Mills, Wayne, Pa., 500 .... C11 114
White Mountain, Alsk., 151 . C7 79
White Mountain, peak, Calif. . D4 82
Whitemouth, Man., Can., 385 .... E4 71
Whitemouth, lake, Man., Can. . E4 71
Whitemouth, riv., Man., Can. . E4 71
Whitemud, riv., Alta., Can. .. A2 69
White Oak, Barbour, Ala. .. D4 78
White Oak, Camden, Ga., 150 .... E5 87
White Oak, Guilford, N.C. .. C5 109
White Oak, Hamilton, Ohio, 5,000 .... D2 111
White Oak, Craig, Okla., 75 .... A6 112
White Oak, Allegheny, Pa., 9,047 .... *B6 114
White Oak, Fairfield, S.C., 150 .... C5 115

Whiteoak, Williamson, Tenn. ... B4 117
White Oak, Gregg, Texas, 1,250 ... *C5 118
Whiteoak, creek, Ohio ... D4 111
Whiteoak, creek, Tenn. ... A4 111
White Oak, mtn., Ark. ... B3 81
White Oak, mtn., Ark. ... C2 81
Whiteoak, swamp, N.C. ... C6 109
White Owl, Meade, S. Dak., 10 ... C3 116
White Pigeon, St. Joseph, Mich., 1,399 ... G5 98
White Pine, Ontonagon, Mich., 950 ... A5 98
White Pine, Sanders, Mont., 50 ... C1 102
White Pine, Jefferson, Tenn., 1,035 ... C10 117
White Pine, co., Nev., 9,808 ... D6 104
White Plains, Calhoun, Ala., 150 ... B4 78
White Plains, Greene, Ga., 273 ... C3 87
White Plains, Hopkins, Ky., 359 ... C2 94
White Plains, Charles, Md., 540 ... C4 85
White Plains, Westchester, N.Y., 50,485 ... D2, D3, D7 108
White Plains, Surry, N.C., 500 ... A3 109
White Pond, Aiken, S.C., 150 ... A5 115
Whiteriver, Navajo, Ariz., 450 ... D6 80
White River, Mellette, S. Dak., 583 ... D5 116
White River, plat., Colo ... B3 83
White River Junction, Windsor, Vt., 2,546 ... D4 120
White Rock, B.C., Can., 6,453 ... B10, E6 68
White Rock, Los Alamos, N. Mex. (part of Los Alamos) F6 107
White Rock, Washington, R.I., 400 ... D9 84
White Rock, Richland, S.C., 150 ... C5 115
White Rock, Roberts, S. Dak., 76 ... B9 116
White Rock, creek, Kans ... C5 93
Whiterocks, Unitah, Utah, 170 ... C6 119
White Rocks, mtn., Ky ... D6 94
Whites, Clay, Miss., 200 ... B5 100
Whites, Grays Harbor, Wash., 55 ... B2 122
Whites, creek, Tenn ... E9 117
Whitesail, lake, B.C., Can ... C4 68
White Salmon, Klickitat, Wash., 1,590 ... D4 122
White Salmon, riv., Wash ... D4 122
Whitesand, bay, Eng ... D3 12
Whitesand, riv., Sask., Can ... F4 70
White Sands, nat. mon., N.M ... E3 107
Whitesbog, Burlington, N.J., 100 ... D3 106
Whitesboro, Cape May, N.J., 700 ... E3 106
Whitesboro, Oneida, N.Y., 4,784 ... *B5 108
Whitesboro, Grayson, Tex., 2,485 ... C4 118
Whites Brook, N.B., Can., 268 ... B2 74
Whitesburg, Carroll, Ga., 366 ... C2 87
Whitesburg, Letcher, Ky., 1,774 ... C7 94
Whites City, Eddy, N. Mex., 175 ... E5 107
Whites Creek, Davidson, Tenn., 100 ... E9 117
White Settlement, Tarrant, Tex., 11,513 ... *B5 118
Whiteshield, mtn., B.C., Alta., Can ... C8 68, C1 69
Whiteside, Lincoln, Mo., 122 ... B6 101
Whiteside, Marion, Tenn., 500 ... D8 117
Whiteside, co., Ill., 59,887 ... B3 90
White Signal, Grant, N. Mex ... E1 107
Whiteson, Yamhill, Oreg., 100 ... B1, B3 113
White Springs, Hamilton, Fla., 633 ... B4 86
Whitestone, Gilmer, Ga., 300 ... B2 87
White Stone, Lancaster, Va., 395 ... D6 121
Whitestone, lake, Man., Can ... A3 71
Whitestown, Boone, Ind., 613 ... D5 91
White Sulphur Springs, Meagher, Mont., 1,519 ... D6 102
White Sulphur Springs, Greenbrier, W. Va., 2,676 ... D4 123
Whitesville, Daviess, Ky., 713 ... C3 94
Whitesville, Ocean, N.J., 300 ... C4 106
Whitesville, Allegany, N.Y., 600 ... C3 108
Whitesville, Boone, W. Va., 774 ... D6 123
White Swan, Yakima, Wash., 300 ... C5 122
Whiteswan, lakes, Sask., Can ... C3 70
Whitetail, Daniels, Mont., 250 ... B11 102
White Tail, Otero, N. Mex., 70 ... D4 107
White Tank, mts., Ariz ... G1 80
Whiteville, Columbus, N.C., 4,683 ... C5 109
Whiteville, Hardeman, Tenn., 757 ... B2 117
White Volta, riv., Ghana ... E4 45
Whitewater, Man., Can., 110 ... E1 71
Whitewater, Mesa, Colo., 170 ... C2 83
Whitewater, Butler, Kans., 499 ... B6, E6 93
Whitewater, Cape Girardeau, Mo., 169 ... D8 101
Whitewater, Phillips, Mont., 90 ... B6 102
Whitewater, Grant, N. Mex., 10 ... E1 107
Whitewater, Walworth, Wis., 6,380 ... F5 124
Whitewater, bay, Fla ... G6 86
Whitewater, riv., Ind ... F7 91
Whitewater, riv., Kans ... B6 93
Whitewater Baldy, mtn., N. Mex ... D1 107
White Woman, creek, Kans ... D2 93
Whitewood, Sask., Can., 900 ... G4 70
Whitewood, Lawrence, S. Dak., 470 ... C2 116
Whitewright, Grayson, Tex., 1,315 ... C4 118
Whitfield, Sumter, Ala., 200 ... C1 78
Whitfield, co., Ga., 42,109 ... B2 87
Whitfield Estate, Manatee, Fla., 600 ... F2 86
Whitham, Chariton, Mo ... B4 101
Whithorn, Scot., 986 ... F4 13
Whiting, Lake, Ind., 8,137 ... A3 91
Whiting, Monona, Iowa, 595 ... B1 92
Whiting, Jackson, Kans., 233 ... A7, C8 93
Whiting, Washington, Maine, 230 (339▲) ... D5 96
Whiting, Ocean, N.J., 308 ... D4 106
Whiting, Addison, Vt., 70 (304▲) ... D2 120
Whiting, Portage, Wis., 1,193 ... D4 124
Whiting Bay, Scot ... E3 13
Whitingham, Windham, Vt., 100 (838▲) ... F3 120
Whitingham, res., Vt ... F3 120
Whitinsville, Worcester, Mass., 5,102 ... B4 97
Whitkow, Sask., Can., 56 ... E2 70
Whitla, Alta., Can., 78 ... E5 69
Whitlash, Liberty, Mont., 10 ... B5 102
Whitley, co., Ind., 20,954 ... B6 91
Whitley, co., Ky., 25,815 ... D5 94
Whitley City, McCreary, Ky., 1,034 ... D5 94
Whitlock, Henry, Tenn., 130 ... A3 117
Whitman, Plymouth, Mass., 10,485 ... B6 97
Whitman, Grant, Nebr., 150 ... B4 103
Whitman, Nelson, N. Dak., 100 ... A7 110
Whitman, Niobrara, Wyo ... C8 125
Whitman, co., Wash., 31,263 ... B8 122
Whitman, nat. mon., Wash ... C7 122
Whitman Knob, mtn., W. Va ... C4 123
Whitmans, pond, Mass ... D3 97
Whitmell, Pittsylvania, Va., 35 ... E3 121
Whitmer, Randolph, W. Va., 250 ... C5 123
Whitmire, Newberry, S.C., 2,663 ... B4 115
Whitmore, mts., Ant ... A2 5
Whitmore City, Honolulu, Haw., 1,820 ... f9 88
Whitmore Lake, Washtenaw, Mich., 950 ... A6 98
Whitnel, Caldwell, N.C., 1,232 ... B2 109
Whitney, Ont., Can., 828 ... B6 72
Whitney, Ada, Idaho, 13,603 ... F2 89
Whitney, Franklin, Idaho, 80 ... G7 89
Whitney, Menominee, Mich ... C3 98
Whitney, Dawes, Nebr., 98 ... B2 103
Whitney, Clark, Nev., 800 ... G6 104
Whitney, Baker, Oreg ... C8 113
Whitney, Westmoreland, Pa., 800 ... F3 114
Whitney, Spartanburg, S.C., 2,502 ... B4 115
Whitney, Hill, Tex., 1,050 ... D4 118
Whitney, mtn., Calif ... D4 82
Whitney Estates, Albany, N.Y., 1,000 ... *C7 108
Whitney Point, Broome, N.Y., 1,049 ... C5 108
Whitneyville, New Haven, Conn. (part of Hamden) ... D5 84
Whitneyville, Washington, Maine, 200 (229▲) ... D5 96
Whitsett, Guilford, N.C., 200 ... A4 109
Whitstable, Eng., 19,534 ... C9 12
Whitt, Parker, Tex., 200 ... C4 118
Whittaker, Washtenaw, Mich., 200 ... A6 98
Whittemore, Kossuth, Iowa, 741 ... A3 92
Whittemore, Iosco, Mich., 460 ... D7 98
Whitten, Hardin, Iowa, 184 ... B4 92
Whittier, Alsk., 809 ... C10, g17 79
Whittier, Los Angeles, Calif., 33,663 ... F2, F4 82
Whittier, Jackson, N.C., 700 ... D3 109
Whittier Downs, Los Angeles, Calif., 7,000 ... *F4 82
Whittlesey, Eng., 9,324 ... B7 12
Whittle Springs, Knox, Tenn. (part of Knoxville) ... C10, E11 117
Whitwell, Marion, Tenn., 1,857 ... D8 117
Wholdaia, lake, N.W. Ter., Can ... D12 66
Whonock, B.C., Can., 1,062 ... A10 68
Whyalla, Austl., 13,711 ... F6 50
Whycocomagh, N.S., Can., 308 ... D8 74
Wiarton, Ont., Can., 2,138 ... C3 72
Wiay, isl., Scot ... C1 13
Wibaux, Wibaux, Mont., 766 ... D12 102
Wibaux, co., Mont., 1,698 ... D12 102
Wiborg, McCreary, Ky., 500 ... D5 94
Wichita, Sedgwick, Kans., 254,698 (★346,200) ... B5, E6 93
Wichita, Clackamas, Oreg., 6,000 ... *B4 113
Wichita, co., Kans., 2,765 ... D2 93
Wichita, co., Tex., 123,528 ... B3 118
Wichita, mts., Okla ... C3 112
Wichita Falls, Wichita, Tex., 101,724 (★116,900) ... C3 118
Wick, Scot., 7,397 ... B5 13
Wickatunk, Monmouth, N.J., 100 ... C4 106
Wicked, pt., Ont., Can ... D7 72
Wickenburg, Maricopa, Ariz., 2,445 ... D3 80
Wickes, Polk, Ark., 368 ... C1 81
Wickes, Jefferson, Mont., 5 ... D4 102
Wickett, Ward, Tex., 900 ... D1 118
Wickham, Que., Can., 378 ... D5 73
Wickiup, res., Oreg ... D5 113
Wickliffe, Ballard, Ky., 917 ... A2 94
Wickliffe, Lake, Ohio, 15,760 ... A6, B2 111
Wickliffe, Mahoning, Ohio, 6,000 ... *A7 111
Wicklow, Ire., 3,125 ... E5 11
Wicklow, co., Ire., 58,473 ... E5 11
Wicklow, head, Ire ... E6 11
Wicklow, mts., Ire ... E5 11
Wicksville, Pennington, S. Dak., 25 ... C3 116
Wicomico, Charles, Md ... D4 85
Wicomico, co., Md., 49,050 ... D6 85
Wicomico, riv., Md ... D4 85
Wicomico, riv., Md ... D6 85
Wiconisco, Dauphin, Pa., 1,402 ... E8 114
Wide, passage, Eniwetok ... 52
Widemouth, Mercer, W. Va., 200 ... D3 123
Widen, Clay, W. Va., 600 ... C4, C7 123
Widener, St. Francis, Ark., 203 ... B5 81
Wide Ruin, Apache, Ariz., 15 ... B6 80
Widnes, Eng., 52,168 ... A5 12
Widtsoe, Garfield, Utah ... F4 119
Widuchowo, Pol., 1,000 ... E8 24
Wiebelskirchen, Ger., 11,000 ... D2 17
Wiedenbrück, Ger., 14,500 ... B3 17
Wieliczka, Pol., 10,200 ... C6 26
Wielka Wies, Pol., 600 ... A5 26
Wielun, Pol., 10,500 ... C5 26
Wien (Vienna), state, Aus., 1,627,566 ... D8 16
Wiener Neustadt, Aus., 33,845 ... E8 16
Wieprz, riv., Pol ... C7 26
Wieprzowka, riv., Pol ... h10 26
Wiergate, Newton, Tex., 400 ... D6 118
Wiesau, Ger., 3,900 ... D7 17
Wiesbaden, Ger., 253,300 (★500,000) ... C3 17
Wiesenburg, Ger., 2,136 ... A7 17
Wiesental, Ger., 6,500 ... D3 17
Wiesloch, Ger., 13,700 ... D3 17
Wigan, Eng., 78,702 ... A5 12
Wiggins, Morgan, Colo., 400 ... A6 83
Wiggins, Stone, Miss., 1,591 ... E4 100
Wiggins, peak, Wyo ... B3 125
Wight, isl., Eng ... D6 12
Wigton, Eng., 4,085 ... F5 13
Wigtown, Scot., 1,201 ... F4 13
Wigtown, co., Scot., 29,107 ... F4 13
Wigtown, bay, Scot ... F4 13
Wiheish, isl., Eg., U.A.R ... C2 32
Wijhe, Neth., 2,626 ... B6 15
Wikieup, Mohave, Ariz., 5 ... C2 80
Wikwemikong, Ont., Can., 500 ... B3 72
Wil, Switz., 10,927 ... B7 19
Wilbarger, co., Tex., 17,748 ... B3 118
Wilber, Saline, Nebr., 1,358 ... D9 103
Wilberforce, Ont., Can., 212 ... B6 72
Wilberforce, Greene, Ohio, 1,800 ... C4 111
Wilbraham, Hampden, Mass., 1500 (7,387▲) ... *B3 97
Wilbur, Douglas, Oreg., 350 ... D3 113
Wilbur, Lincoln, Wash., 1,138 ... B7 118
Wilburn, Cleburne, Ark., 72 ... B4 81
Wilburton, Latimer, Okla., 1,772 ... C6 112
Wilcannia, Austl., 823 ... E4 51
Wilcox, Sask., Can., 258 ... G3 70
Wilcox, Gilchrist, Fla., 50 ... C4 86
Wilcox, Kearney, Nebr., 260 ... D6 103
Wilcox, Elk, Pa., 900 ... D6, C4 114
Wilcox, co., Ala., 18,739 ... D2 78
Wilcox, co., Ga., 7,905 ... E3 87
Wild, brook, Vt ... B4 120
Wild, riv., N.H ... B4 105
Wilda, Rapides, La., 20 ... C3 95
Wildbad, Ger., 6,300 ... E3 17
Wildcat, Okmulgee, Okla., 142 ... B6 112
Wildcat, creek, Ind ... D4 91
Wildcat, hill, Sask., Can ... D4 70
Wild Cherry, Fulton, Ark., 40 ... A3 81
Wilden, Northampton, Pa., 1,787 ... *E11 114
Wilder, Canyon, Idaho 603 ... F2 89
Wilder, Fentress, Tenn., 300 ... C8 117
Wilder, Windsor, Vt., 1,322 ... D4 120
Wilder, dam, N.H ... C2 105
Wildersville, Henderson, Tenn., 300 ... B3 117
Wildervank, Neth., 4,858 ... A6 15
Wilderville, Josephine, Oreg., 100 ... E3 113
Wildeshausen, Ger., 9,200 ... B2 17
Wildhaus, Switz., 1,179 ... B7 19
Wildhorn, mtn., Switz ... D3 19
Wild Horse, Cheyenne, Colo., 30 ... C8 83
Wildhorse, creek, Okla ... C4 112
Wild Horse, creek, Wyo ... A7 125
Wild Horse, res., Nev ... B6 104
Wildnest, lake, Sask., Can ... C4 70
Wildomar, Riverside, Calif., 400 ... F3 82
Wildorado, Oldham, Tex., 160 ... B1 118
Wild Rice, Cass, N. Dak., 25 ... C9 110
Wild Rice, riv., Minn ... C2 99
Wild Rice, riv., N. Dak ... C8 110
Wildrose, Williams, N. Dak., 361 ... A2 110
Wild Rose, Waushara, Wis., 594 ... D4 124
Wildspitze, mtn., Aus ... C6 18
Wildsville, Concordia, La., 150 ... C4 95
Wildwood, Alta., Can., 479 ... C3 69
Wildwood, Sumter, Fla., 2,170 ... D4 86
Wildwood, Cape May., N.J., 4,690 ... F3 106
Wildwood Crest, Cape May, N.J., 3,011 ... F3 106
Wildwood, Allegheny, Pa., 2,500 ... *E2 114
Wildwood, Utah, Utah ... C4 119
Wiley, Prowers, Colo., 383 ... C8 83
Wilhelm, mtn., N. Gui ... k12 50
Wilhelm II, coast, Ant ... C21 5
Wilhelmina, mtn., W. Irian ... F9 35
Wilhelmina, mts., Sur ... B3 59
Wilhelm-Pieck-Stadt Guben, Ger., 22,500 ... B9 17
Wilhelmshaven, Ger., 100,200 ... B4 14, E2 24
Wilhoit, Yavapai, Ariz., 40 ... C3 80
Wilkau, Ger., 13,800 ... C7 17
Wilkes, co., Ga., 10,961 ... C4 87
Wilkes, co., N.C., 45,269 ... A2 109
Wilkes, isl., Wake Isl ... 52
Wilkes-Barre, Luzerne, Pa., 63,551 (★250,700) ... B8, D10 114
Wilkesboro, Wilkes, N.C., 1,568 ... A2 109
Wilkesboro, flood control res., N.C ... A2 109
Wilkes Land, reg., Ant ... B27 5
Wilkeson, Pierce, Wash., 412 ... B3 122
Wilkesville, Vinton, Ohio, 190 ... C5 111
Wilkie, Sask., Can., 1,612 ... E1 70
Wilkin, co., Minn., 10,650 ... D2 99
Wilkins, Elko, Nev., 60 ... B7 104
Wilkinsburg, Allegheny, Pa., 30,066 ... B6, F2 114
Wilkinson, Hancock, Ind., 388 ... E6 91
Wilkinson, Wilkinson, Miss., 50 ... D2 100
Wilkinson, Logan, W. Va., 350 ... D3 123
Wilkinson, co., Ga., 9,250 ... D3 87
Wilkinson, co., Miss., 13,235 ... D2 100
Wilkinsville, Tipton, Tenn ... B2 117
Will, co., Ill., 191,617 ... B6 90
Willacoochee, Atkinson, Ga., 1,061 ... E3 87
Willacy, co., Tex., 20,084 ... F4 118
Willamette, riv., Oreg ... C3 113
Willamina, Yamhill, Oreg., 960 ... B3 113
Willapa, Pacific, Wash., 300 ... C2 122
Willapa, bay, Wash ... C1 122
Willapa, hills, Wash ... C2 122
Willard, Logan, Colo., 5 ... A7 83
Willard, Shawnee, Kans., 94 ... C8 93
Willard, Carter, Ky., 164 ... B7 94
Willard, Greene, Mo., 357 ... D4 101
Willard, Fallon, Mont., 10 ... D12 102
Willard, Torrance, N. Mex., 294 ... C3 107
Willard, Seneca, N.Y., 625 ... C4 108
Willard, Pender, N.C., 300 ... C5 109
Willard, Huron, Ohio, 5,457 ... A5 111
Willard, Box Elder, Utah, 814 ... B3 119
Willard, Clark, Wis., 100 ... D3 124
Willard, stream, Vt ... B5 120
Willards, Wicomico, Md., 531 ... D7 85
Willcox, Cochise, Ariz., 2,441 ... E6 80
Willemstad, Neth. W.I., 44,062 ... A4 60
Willemstad, Neth., 905 ... C4 15
Willernie, Washington, Minn., 664 ... E7 99
Willesden, Eng., 170,835 ... k11 10
Willet, Cortland, N.Y., 200 ... C5 108
Willette, Macon, Tenn., 100 ... C8 117
Willetts, Concordia, La ... C4 95
Willey House, Carroll, N.H., 8 ... B4 105
William, lake, Man., Can ... C2 71
William Creek, Austl ... E6 50
Williams, Coconino, Ariz., 3,559 ... B3 80
Williams, Colusa, Calif., 1,370 ... C2 82
Williams, Lawrence, Ind., 400 ... C4 91
Williams, Hamilton, Iowa, 490 ... B4 92
Williams, Lake of the Woods, Minn., 317 ... B4 99
Williams, Pondera, Mont., 5 ... B4 102
Williams, Le Flore, Okla., 100 ... B7 112
Williams, Colleton, S.C., 194 ... E6 115
Williams, co., N. Dak., 22,051 ... A2 110
Williams, co., Ohio, 29,968 ... A3 111
Williams, cape, Ant ... C29 5
Williams, mtn., Okla ... C7 112
Williams, riv., Ariz ... C1 80
Williams, riv., Vt ... E3 120
Williams, riv., W. Va ... C4 123
Williams Bay, Walworth, Wis., 1,347 ... F5 124
Williamsburg, Ont., Can., 442 ... C9 72
Williamsburg, Wayne, Ind., 400 ... E8 91
Williamsburg, Iowa, Iowa, 1,342 ... C5 92
Williamsburg, Franklin, Kans., 255 ... D8 93
Williamsburg, Whitley, Ky., 3,478 ... D5 94
Williamsburg, Dorchester, Md., 400 ... C6 85
Williamsburg, Hampshire, Mass., 900 (2,186▲) ... B2 97
Williamsburg, Grand Traverse, Mich., 140 ... D5 98
Williamsburg, Sierra, N. Mex., 225 ... D2 107
Williamsburg, Clermont, Ohio, 1,956 ... C3 111
Williamsburg, Blair, Pa., 1,792 ... F5 114
Williamsburg (Independent City), Va., 6,832 ... D6 121
Williamsburg, Greenbrier, W. Va., 250 ... D4 123
Williamsburg, co., S.C., 40,932 ... D8 115
Williamsfield, Knox, Ill., 548 ... C3 90
Williams Lake, B.C., Can., 2,120 ... C6, E3 68
Williamson, Pike, Ga., 215 ... C2 87
Williamson, Madison, Ill., 324 ... A9 101
Williamson, Lucas, Iowa, 262 ... C4 92
Williamson, Wayne, N.Y., 1,690 ... B3 108
Williamson, Mingo, W. Va., 6,746 ... D2 123
Williamson, co., Ill., 46,117 ... F5 90
Williamson, co., Tenn., 25,267 ... B5 117
Williamson, co., Tex., 35,044 ... D4 118
Williamson, head, Ant ... C28 5
Williams Park, Lake, Ill., 900 ... *E2 90
Williamsport, Newf., Can., 151 ... C3 75
Williamsport, Warren, Ind., 1,353 ... D3 91
Williamsport, Washington, Md., 1,853 ... A2 85
Williamsport, Pickaway, Ohio, 840 ... C4 111
Williamsport, Lycoming, Pa., 41,967 ... D7 114
Williamsport, Maury, Tenn., 100 ... B4 117
Williamston, Ingham, Mich., 2,214 ... F6 98
Williamston, Martin, N.C., 6,924 ... B6 109
Williamston, Anderson, S.C., 3,721 ... B3 115
Williamstown, Ont., Can., 325 ... B10 72
Williamstown, Jefferson, Kans., 70 ... B7 93
Williamstown, Grant, Ky., 1,611 ... B5 94
Williamstown, Berkshire, Mass., 5,428 (7,322▲) ... A1 97
Williamstown, Lewis, Mo., 125 ... A6 101
Williamstown, Gloucester, N.J., 2,722 ... D3 106
Williamstown, Oswego, N.Y., 200 ... B5 108
Williamstown, Hancock, Ohio, 100 ... B4 111
Williamstown, Dauphin, Pa., 2,097 ... E8 114
Williamstown, Orange, Vt., 400 (1,553▲) ... C3 120
Williamstown, Wood, W. Va., 2,632 ... B3 123
Williamsville, Sangamon, Ill., 735 ... D4 90
Williamsville, Attala, Miss., 100 ... B4 100
Williamsville, Wayne, Mo., 412 ... E7 101
Williamsville, Erie, N.Y., 6,316 ... C2 108
Williamsville, Windham, Vt., 125 ... F3 120
Williaumez, pen., Bis. Arch ... 51
Williford, Sharp, Ark., 195 ... A4 81
Williford, Gilchrist, Fla., 75 ... C4 86
Willimantic, Windham, Conn., 13,881 ... C8 84
Willimantic, res., Conn ... B7 84
Willimantic, riv., Conn ... B7 84
Willingdon, Alta., Can., 429 ... C4 69
Willis, Brown, Kans., 109 ... C8 93
Willis, Washtenaw, Mich., 175 ... A7 98
Willis, Marshall, Okla., 10 ... D5 112
Willis, Montgomery, Tex., 975 ... D5 118
Willis, Floyd, Va., 70 ... E2 121
Willis, isl., Newf., Can ... D5 75
Willis, isl., Newf., Can ... D5 75
Willis Beach, Dakota, Nebr., 100 ... B9 103
Willisburg, Washington, Ky., 300 ... C4 94
Williston, Levy, Fla., 1,582 ... C4 86
Williston, Carteret, N.C., 300 ... C7 109
Williston, Williams, N. Dak., 11,866 ... A2 110
Williston, Ottawa, Ohio, 500 ... A2 111
Williston, Barnwell, S.C., 2,722 ... E5 115
Williston, Fayette, Tenn., 150 ... B2 117
Williston, S. Afr., 2,873 ... D3 49
Williston, Chittenden, Vt., 250 (1,484▲) ... C2 120
Williston Basin, reg., Mont., N. Dak ... B11 102, A1 110
Williston Park, Nassau, N.Y., 8,255 ... G2 84
Willisville, Nevada, Ark., 100 ... D2 81
Willisville, Perry, Ill., 532 ... F4 90
Willis Wharf, Northampton, Va., 528 ... D7 121
Willits, Mendocino, Calif., 3,410 ... C2 82
Willmar, Kandiyohi, Minn., 10,417 ... E3 99
Willmar Station, Sask., Can., 51 ... H4 70
Willoughby, Lake, Ohio, 15,058 ... A6 111
Willoughby, lake, Vt ... B4 120
Willoughby Hills, Lake, Ohio, 4,241 ... *A6 111
Willow, Alsk., 78 ... g16 79
Willow, Dallas, Ark., 60 ... C3 81
Willow, Greer, Okla., 187 ... B2 112
Willow, creek, Utah ... D6 119
Willow, creek, Wyo ... B6 125
Willow, res., Wis ... C3 124
Willow, riv., B.C., Can ... C6 68
Willow Branch, Hancock, Ind., 200 ... E6 91
Willow Brook, Los Angeles, Calif., 22,000 ... *F2 82
Willowbrook, Sask., Can., 86 ... F4 70
Willow Bunch, Sask., Can., 698 ... H3 70
Willowbunch, lake, Sask., Can ... H3 70
Willow City, Bottineau, N. Dak., 494 ... A5 110
Willow City, Gillespie, Tex., 75 ... D3 118
Willow Creek, Gallatin, Mont., 250 ... E5 102
Willowcreek, Malheur, Oreg., 50 ... C9 113
Willowdale, Ont., Can., 9,000 ... D5, E6 72
Willowgrove, Kent, Del., 100 ... B6 85
Willow Grove, Salem, N.J ... D2 106
Willow Grove, Montgomery, Pa., 10,000 ... A11, F11 114
Willow Hill, Jasper, Ill., 335 ... E5 90
Willow Hill, Franklin, Pa., 30 ... F6 114
Willowick, Lake, Ohio, 18,749 ... B2 111
Willow Island, Dawson, Nebr., 85 ... D5 93
Willow Lake, Clark, S. Dak., 467 ... C8 116
Willow Lawn, Henrice, Va., 2,500 ... *D5 121
Willowmore, S. Afr., 3,454 ... D3 49
Willow Ranch, Modoc, Calif., 300 ... B3 82
Willow River, Pine, Minn., 343 ... D6 99
Willow Run, Washtenaw, Mich., 4,100 ... *F7 98
Willows, Glenn, Calif., 4,139 ... C2 82
Willows, Sask., Can., 45 ... H3 70
Willows, Gilliam, Oreg ... B6 113
Willow Springs, Cook, Ill., 2,348 ... F2 90
Willow Springs, Howell, Mo., 1,913 ... E6 101
Wills, creek, W. Va ... C6 123
Wills, hill, Mass ... C3 97
Wills, mtn., Md ... D2 85
Wills, riv., Alta ... A3 78
Wills, val., Ala ... A4 78
Willsboro, Essex, N.Y., 800 ... B3 108
Willshire, Van Wert, Ohio, 601 ... B3 111
Wills Point, Van Zandt, Tex., 2,281 ... *C5 118
Wilmar, Drew, Ark., 718 ... D4 81
Wilmer, Mobile, Ala., 200 ... E1 78
Wilmer, Dallas, Tex., 1,785 ... B6 118
Wilmerding, Allegheny, Pa., 4,349 ... B6 114
Wilmersdorf, Ger., 398 ... E7 24
Wilmette, Cook, Ill., 28,268 ... A6, E3 90
Wilmington, New Castle, Del., 95,827 (★318,700) ... A6 85
Wilmington, Will, Ill., 4,210 ... B5 90
Wilmington, Middlesex, Mass., 2,250 (12,475▲) ... A5, C2 97
Wilmington, Essex, N.Y., 700 ... B3 108
Wilmington, New Hanover, N.C., 44,013 ... C6 109
Wilmington, Clinton, Ohio, 8,915 ... C4 111
Wilmington, Windham, Vt., 591 (1,245▲) ... F3 120
Wilmington, lake, Fla ... E6 86
Wilmont, Nobles, Minn., 473 ... G3 99
Wilmore, Comanche, Kans., 99 ... E4 93
Wilmore, Jessamine, Ky., 2,773 ... C5 94
Wilmot, Ashley, Ark., 732 ... D4 81
Wilmot, Cowley, Kans., 60 ... E7 93
Wilmot, Merrimack, N.H., 75 (391▲) ... D3 105
Wilmot, Stark, Ohio, 402 ... B6 111
Wilmot, Roberts, S. Dak., 545 ... B9 116
Wilmot Flat, Merrimack, N.H., 75 ... D3 105
Wilmot Station, N.S., Can., 363 ... E4 74
Wilmslow, Eng., 21,393 ... A5 12
Wilmurt, Herkimer, N.Y., 60 ... B6 108
Wilna, Grant, N. Mex., 10 ... E1 107
Wilno, Ont., Can., 161 ... B7 72
Wilno, see Vilnius, Sov. Un.
Wilsall, Park, Mont., 200 ... E6 102
Wilsey, Morris, Kans., 224 ... D7 93
Wilson, Mississippi, Ark., 1,191 ... B5 81
Wilson, Hartford, Conn., 2,500 ... B6 84
Wilson, Brevard, Fla., 100 ... D6 86
Wilson, Ellsworth, Kans., 905 ... D5 93
Wilson, East Feliciana, La., 300 ... D4 95
Wilson, Menominee, Mich., 70 ... C3 98
Wilson, Niagara, N.Y., 1,320 ... B2 110
Wilson, Wilson, N.C., 28,753 ... B6 109
Wilson, Carter, Okla., 1,647 ... C4 112
Wilson, Northampton, Pa., 8,465 ... E11 114
Wilson, Lynn, Tex., 403 ... C2 118
Wilson, Teton, Wyo., 35 ... B2 125
Wilson, co., Kans., 13,077 ... E8 93
Wilson, co., N.C., 57,716 ... B6 109
Wilson, co., Tenn., 27,668 ... A5 117
Wilson, co., Tex., 13,267 ... E3 118
Wilson, creek, Wash ... B5 122
Wilson, creek, Wash ... B6 122
Wilson, dam, Ala ... A2 78
Wilson, mtn., Calif ... F2 82
Wilson, mtn., Nev ... E7 104
Wilson, mtn., Oreg ... B5 113
Wilson, mtn., Vt ... C3 120
Wilson, peak, Colo ... D3 83
Wilson, pond, Maine ... C3 96
Wilson, riv., Austl ... C4 51
Wilson Creek, Grant, Wash., 252 ... B6 122
Wilson Mills (Lincoln Plantation), Oxford, Maine, 65 (99▲) ... D1 96
Wilson Mills, Johnston, N.C., 280 ... B5 109
Wilsons, Dinwiddie, Va., 150 ... D5 121
Wilson's promontory, Austl ... I6 51
Wilsons Beach, N.B., Can., 768 ... E3 74
Wilsonville, Shelby, Ala., 683 ... B3 78
Wilsonville, Windham, Conn., 150 ... A9 84
Wilsonville, Macoupin, Ill., 688 ... D4 90
Wilsonville, Spencer, Ky., 30 ... A4 94
Wilsonville, Furnas, Nebr., 289 ... D5 103
Wilsonville, Clackamas, Oreg., 185 ... B2 113
Wilstedt, Ger., 1,150 ... E3 18
Wilster, Ger., 4,900 ... E3 18
Wilton, Shelby, Ala., 428 ... B3 78
Wilton, Little River, Ark., 329 ... D1 81
Wilton, Fairfield, Conn., 3,500 (8,026▲) ... E3 84
Wilton, Franklin, Maine, 1,761 (3,274▲) ... D2 96
Wilton, Beltrami, Minn., 112 ... C3 99
Wilton, Hillsboro, N.H., 1,425 (2,025▲) ... E3 105
Wilton, Saratoga, N.Y., 200 ... B7 108
Wilton, Burleigh and McLean, N. Dak., 739 ... B5 110
Wilton, Monroe, Wis., 578 ... E3 124
Wilton Junction (Wilton), Muscatine, Iowa, 1,750 ... C6 92
Wilton Manors, Broward, Fla., 8,257 ... *F6 86
Wiltshire, co., Eng., 422,753 ... C6 12
Wiltz, Lux, 3,904 ... E5 15
Wiluna, Austl., 576 ... E3 51
Wimapedi, riv., Man., Can ... B2 71
Wimauma, Hillsborough, Fla., 583 ... E4 86
Wimbledon, Eng., 56,994 ... m12 10, C7 12
Wimbledon, Barnes, N. Dak., 402 ... B7 110

Wimborne, Alta., Can., 80..D4 69
Wimborne, Eng., 4,156.....D6 12
Wimer, Jackson, Oreg.......E3 113
Wimico, lake, Fla..........C1 86
Winagami, lake, Alta., Can.....B2 69
Winamac, Pulaski, Ind., 2,375.....B4 90
Winburg, S. Afr., 4,968......C4 49
Winburne, Clearfield, Pa., 800.....E5 114
Wincheck, pond, R.I......C9 84
Winchell, mtn., Mass.......B2 97
Winchendon, Worcester, Mass., 3,839 (6,237▲).....A3 97
Winchendon Springs, Worcester, Mass., 350.....A3 97
Winchester, Drew, Ark., 185.D4 81
Winchester, Riverside, Calif., 200.....F3 82
Winchester, Ont., Can., 1,429.....B9 72
Winchester, Eng., 28,643....C6 12
Winchester, Lewis, Idaho, 427.....C2 89
Winchester, Scott, Ill., 1,657.D3 90
Winchester, Randolph, Ind., 5,742.....D8 91
Winchester, Jefferson, Kans., 428.....B7 93
Winchester, Clark, Ky., 10,187.....C5 94
Winchester, Middlesex, Mass., 19,376.....C2 97
Winchester, Wayne, Miss., 50.....D5 100
Winchester, Clark, Mo.....A6 101
Winchester, St. Louis, Mo., 1,299.....*B7 101
Winchester, Clark, Nev., 600.G6 104
Winchester, Cheshire, N.H., 950 (2,411▲).....E2 105
Winchester, Adams, Ohio, 788.....D4 111
Winchester, Franklin, Tenn., 4,760.....B5 117
Winchester (Independent City), Va., 15,110.....B4 121
Winchester, Washakie, Wyo., 20.....B4 125
Winchester Bay, Douglas, Oreg., 500.....D2 113
Winchester Center, Litchfield, Conn., 250 (10,496▲).....B4 84
Wind, lake, Wis......F1 124
Wind, riv., Wash......D4 122
Wind, riv., Wyo......B7 125
Wind, riv., Wyo......B3 125
Windber, Somerset, Pa., 6,994.....F4 114
Wind Cave, nat. park, S. Dak...D2 116
Windemere, Ingham, Mich., 2,000.....*F6 98
Winder, Barrow, Ga., 5,555..C3 87
Windermere, B.C., Can., 391.....D10 68
Windermere, Ont., Can., 137.....B5 72
Windermere, Tolland, Conn., 75.....B7 84
Windermere, Eng., 6,556....C5 10
Windermere, lake, Eng......F6 13
Windfall, Tipton, Ind., 1,135.D6 91
Windgap, Northampton, Pa., 1,930.....E11 114
Windham, Windham, Conn., 350 (16,973▲).....C8 84
Windham, Judith Basin, Mont., 100.....C6 102
Windham, Rockingham, N.H., 30 (1,317▲).....E4 105
Windham, Greene, N.Y., 300.....C6 108
Windham, Portage, Ohio, 3,777.....A6 111
Windham, co., Conn., 68,572.B8 84
Windham, co., Vt., 29,776..F3 120
Windhoek, S.W. Afr., 36,051.....B2 49
Windigo Lake, Ont., Can....E8 72
Windigo, riv., Que., Can......B4 73
Winding Stair, mtn., Okla......C7 112
Wind Lake, Racine, Wis., 1,305.....F5 124
Windmill, pt., Va......D6 121
Windom, McPherson, Kans., 168.....D6 93
Windom, Cottonwood, Minn., 3,691.....G3 99
Wind, peak, Colo......D3 83
Windorah, Austl., 48.....B4 51
Window Rock, Apache, Ariz., 500.....B6 80
Wind Ridge, Greene, Pa., 250.....G1 114
Wind River, Fremont, Wyo. (part of Fort Washakie)...C4 125
Wind River, basin, Wyo......B4 125
Wind River, Indian res., Wyo...B3 125
Wind River, range, Wyo......B3 125
Windsbach, Ger., 2,900.....D5 17
Windsheim, Ger., 8,200.....D5 17
Windsor, Sonoma, Calif., 600.....A5 82
Windsor, Newf., Can., 5,505.D4 75
Windsor, N.S., Can., 3,823...E5 74
Windsor, Ont., Can., 114,367 (*193,365).....E1 72
Windsor, Que., Can., 6,589..D5 73
Windsor, Weld, Colo., 1,509.A6 83
Windsor, Hartford, Conn., 12,000 (19,467▲).....B6 84
Windsor, Eng., 27,126.....C7 12
Windsor, Shelby, Ill., 1,021..D5 90
Windsor, Berkshire, Mass., 100 (384▲).....A1 97
Windsor, Henry, Mo., 2,714.C4 101
Windsor, Mercer, N.J., 300..C3 106
Windsor, Broome, N.Y., 1,026.....*C5 108
Windsor, Bertie, N.C., 1,813.B7 109
Windsor, Stutsman, N. Dak., 55.....C6 110
Windsor, York, Pa., 1,029...G8 114
Windsor, Aiken, S.C., 200...E4 115
Windsor, Windsor, Vt., 3,256 (4,468▲).....E4 120
Windsor, Isle of Wight, Va., 579.....B6, E6 121
Windsor, co., Vt., 42,483.....D3 120
Windsor Heights, Polk, Iowa, 4,715.....C4 92
Windsor Hills, Los Angeles, Calif., 3,500.....*F2 82
Windsor Locks, Hartford, Conn., 11,411.....B6 84
Windsorville, Hartford, Conn., 180.....B6 84
Windthorst, Sask., Can., 202.G4 70
Windthorst, Archer, Tex., 170.....C3 118
Windward, passage, W.I......F6 64
Windward Islands, see Dominica, Grenada, St. Lucia, and St. Vincent, Br. dep., N.A...o16 64
Windy, Wirt, W. Va., 15...B3 123
Windy, lake, Sask., Can......C4 70
Windy, peak, Wash......A6 122
Windy, pt., Newf., Can......C4 75
Windy Hill, Florence, S.C., 2,201.....*C8 115
Windy Hills, Jefferson, Ky., 1,371.....*H6 94
Wine, isl., La......E5 95
Winefred, lake, Alta., Can......B5 69
Winefred, riv., Alta., Can......B5 69
Winesap, Cumberland, Tenn.D8 117
Winesap, Chelan, Wash.....B5 122
Winfall, Perquimans, N.C., 269.....A7 109
Winfield, Marion, Ala., 2,907.....B2 78
Winfield, Alta., Can., 238...C3 69
Winfield, Du Page, Ill., 1,575.....*F2 90
Winfield, Henry, Iowa, 862..C6 92
Winfield, Cowley, Kans., 11,117.....E7 93
Winfield, Carroll, Md., 100..B3 85
Winfield, Lincoln, Mo., 564.....A7, C7 101
Winfield, Union, N.J., 2,458.*E4 106
Winfield, Union, Pa., 270....E8 114
Winfield, Scott, Tenn., 200..C9 117
Winfield, Putnam, W. Va., 318.....C3 123
Winfred, Lake, S. Dak., 137.....C8 116
Wing, Covington, Ala., 82...D3 78
Wing, Yell, Ark., 15.....C2 81
Wing, Burleigh, N. Dak., 303.....B5 110
Wing, riv., Minn......D3 99
Wingate, Eng., 12,688.....F7 13
Wingate, Montgomery, Ind., 431.....D3 91
Wingate, Dorchester, Md., 400.....D5 85
Wingate, McKinley, N. Mex.B1 107
Wingate, Union, N.C., 1,304.C3 109
Wingate, Centre, Pa., 100...E6 114
Wingdale, Dutchess, N.Y., 100.....D7 108
Winger, Polk, Minn., 292...C3 99
Wingham, Ont., Can., 2,922.D3 72
Wing Lake, Oakland, Mich., 1,500.....*F7 98
Wingo, Graves, Ky., 340....B2 94
Winifred, Fergus, Mont., 220.....C7 102
Winifreda, Arg.....B4 54
Winifrede, Kanawha, W. Va., 300.....D6 123
Winigan, Sullivan, Mo., 100.A5 101
Winisk, riv., Ont., Can......D8 72
Wink, Winkler, Tex., 1,863..D1 118
Winkelman, Gila, Ariz., 1,123.....E5 80
Winkle, Highland, Ohio, 150.C4 111
Winkler, Man., Can., 2,529..E3 71
Winkler, co., Tex., 13,652...D1 118
Winlaw, B.C., Can., 392.....E9 68
Winlock, Wheeler, Oreg.....C7 113
Winlock, Lewis, Wash., 808..C3 122
Winn, Penobscot, Maine, 200 (526▲).....C4 96
Winn, Isabella, Mich., 300...E6 98
Winn, par., La., 16,034.....C3 95
Winnabow, Brunswick, N.C., 150.....C5 109
Winneba, Ghana, 15,171.....E4 45
Winnebago, Winnebago, Ill., 1,059.....A4 90
Winnebago, Faribault, Minn., 2,088.....G4 99
Winnebago, Thurston, Nebr., 682.....B9 103
Winnebago, Winnebago, Wis., 150.....B5 124
Winnebago, co., Ill., 209,765.....A4 90
Winnebago, co., Iowa, 13,099.....A4 92
Winnebago, co., Wis., 107,928.....D5 124
Winnebago, Indian res., Nebr.....B9 103
Winnebago, lake, Wis......D5 124
Winneconne, Winnebago, Wis., 1,273.....D5 124
Winnegance, Sagadahoc, Maine, 125.....E3, E6 96
Winnemucca, Humboldt, Nev., 3,453.....C4 104
Winnemucca, lake, Nev......C2 104
Winner, Tripp, S. Dak., 3,705.....D6 116
Winneshiek, co., Iowa, 21,651.....A6 92
Winnetka, Cook, Ill., 13,368.....A6, E3 90
Winnetoon, Knox, Nebr., 85.....B8 103
Winnett, Petroleum, Mont., 360.....C8 102
Winnfield, Winn, La., 7,022.C3 95
Winnibigoshish, lake, Minn.....C4 99
Winnie, Chambers, Texas, 1,114.....*E5 118
Winnifred, Alta., Can., 52...E5 69
Winning Pool, Austl.....D1 50
Winnipeg, Man., Can., 265,429 (*475,989).....E3 71
Winnipeg, lake, Man., Can......C3 71
Winnipeg, riv., Man., Can......D4 71
Winnipeg Beach, Man., Can., 807.....D3 71
Winnipegosis, Man., Can., 980.....D2 71
Winnipegosis, lake, Man., Can..C2 71
Winnipesaukee, lake, N.H......C4 105
Winnisquam, Belknap, N.H., 80.....C3 105
Winnisquam, lake, N.H......C3 105
Winnsboro, Franklin, La., 4,437.....B4 95
Winnsboro, Fairfield, S.C., 3,479.....C5 115
Winnsboro, Franklin and Wood, Tex., 2,675.....C5 118
Winokur, Charlton, Ga., 75..E4 87
Winona, Ont., Can., 294.....D5 72
Winona, Starke, Ind., 100...B4 91
Winona, Logan, Kans., 393.C2 93
Winona, Houghton, Mich., 120.....B2 98
Winona, Winona, Minn., 24,895.....F7 99
Winona, Montgomery, Miss., 4,282.....B4 100
Winona, Shannon, Mo., 562.D6 101
Winona, Whitman, Wash., 100.....C8 122
Winona, Fayette, W. Va., 650.....C4, D7 123
Winona, co., Minn., 40,937..F7 99
Winona Lake, Kosciusko, Ind., 1,928.....B6 91
Winona Lakes, Orange, N.Y., 1,655.....*D6 108
Winooski, Chittenden, Vt., 7,420.....C2 120
Winooski, riv., Vt......C3 120
Winschoten, Neth., 16,600...A7 15
Winsen, Ger., 9,700.....E4 24
Winsford, Eng., 12,738.....A5 11
Winside, Wayne, Nebr., 416.B8 103
Winslow, Navajo, Ariz., 8,862.....C5 80
Winslow, Washington, Ark., 183.....B1 81
Winslow, Stephenson, Ill., 366.....A4 90
Winslow, Pike, Ind., 1,089...H3 91
Winslow, Kennebec, Maine, 3,640 (5,891▲).....D3 96
Winslow, Dodge, Nebr., 136.C9 97
Winslow, Camden, N.J., 400.D3 106
Winslow, Kitsap, Wash., 919.D1 122
Winsted, Litchfield, Conn., 1,836.....B4 84
Winsted, McLeod, Minn., 1,163.....F4 99
Winston, Polk, Fla., 3,323...D4 86
Winston, Daviess, Mo., 236..B3 101
Winston, Broadwater, Mont., 40.....D5 102
Winston, Sierra, N. Mex., 30.....D2 107
Winston, Douglas, Oreg., 2,395.....*D3 113
Winston, co., Ala., 14,858...A2 78
Winston, co., Miss., 19,246...B4 100
Winston-Salem, Forsyth, N.C., 111,135 (*185,700)..A3 109
Winstonville, Bolivar, Miss., 413.....B3 100
Winsum, Neth., 996.....A6 15
Winter, Sask., Can., 65.....E1 70
Winter, Sawyer, Wis., 500...C2 124
Winter Beach, Indian River, Fla., 700.....E6 86
Winterberg, Ger., 3,400.....B3 17
Winter Garden, Orange, Fla., 5,513.....D5 86
Winter Gardens, San Diego, Calif., 1,000.....*E2 82
Winter Harbor, Hancock, Maine, 500 (756▲).....D4 96
Winter Harbour, B.C., Can., 103.....D3 68
Winterhaven, Imperial, Calif., 800.....F6 82
Winter Haven, Polk, Fla., 16,277.....D5 86
Wintering, lake, Man., Can.....B3 71
Winter Park, Grand, Colo., 50.....B5 83
Winter Park, Orange, Fla., 17,162.....D5 86
Winterpock, Chesterfield, Va., 130.....D5 121
Winterport, Waldo, Maine, 900 (2,088▲).....D4 96
Winter Rim, mts., Oreg......E6 113
Winters, Yolo, Calif., 1,700.....A5, C3 82
Winters, Runnels, Tex., 3,266.....D3 118
Winterset, Madison, Iowa, 3,639.....C3 92
Wintersville, Jefferson, Ohio, 3,597.....B7 111
Winterswijk, Neth., 15,000...C6 15
Winterthur, Switz., 80,352 (*95,000).....A6 19
Winterton, Newf., Can., 808.E5 75
Winterton, Sullivan, N.Y., 100.....D6 108
Winterville, Clarke, Ga., 497.C3 87
Winterville, Aroostook, Maine, 100 (215▲).....B4 96
Winterville, Washington, Miss., 300.....B2 100
Winterville, Pitt, N.C., 1,418.....B6 109
Winthrop, Little River, Ark., 225.....D1 81
Winthrop, Buchanan, Iowa, 649.....B6 92
Winthrop, Kennebec, Maine, 2,260 (3,537▲).....D3 96
Winthrop, Suffolk, Mass., 20,303.....B6, D3 97
Winthrop, Sibley, Minn., 1,381.....F4 99
Winthrop, Buchanan, Mo., 80.....B2 101
Winthrop, St. Lawrence, N.Y., 10.....A2 108
Winthrop, Okanogan, Wash., 359.....A5 122
Winthrop, lake, Mass......D1 97
Winthrop Harbor, Lake, Ill., 3,848.....A6, D2 90
Winton, Austl., 1,784.....D7 50
Winton, St. Louis, Minn., 182.....C7 99
Winton, Hertford, N.C., 835.A6 109
Winton, see Jessup, Pa.
Winton, Sweetwater, Wyo...D3 125
Wintzenheim, Fr., 2,762.....A3 18
Winyah, bay, S.C......E9 115
Wiota, Cass, Iowa, 195.....C3 92
Wirral, N.B., Can., 123.....D3 74
Wirt, Itasca, Minn., 60.....C5 99
Wirt, Carter, Okla., 500.....C4 112
Wirt, co., W. Va., 4,391.....B3 123
Wirtz, Franklin, Va., 75.....D3 121
Wisbech, Eng., 17,512.....B8 12
Wiscassett, Lincoln, Maine, 950 (1,800▲).....D3 96
Wisconsin, state, U.S., 3,951,777.....B10 77, 124
Wisconsin, riv., Wis......E3 124
Wisconsin Dells, Columbia, Wis., 2,105.....E4 124
Wisconsin Rapids, Wood, Wis., 15,042.....D4 124
Wisdom, Beaverhead, Mont., 185.....E3 102
Wise, Warren, N.C., 350.....A5 109
Wise, Wise, Va., 2,614.....B2 121
Wise, co., Tex., 17,012.....C4 118
Wise, co., Va., 43,579.....B2 121
Wiseman, Alsk., 12.....B9 79
Wise River, Beaverhead, Mont., 50.....E4 102
Wiseton, Sask., Can., 246....F2 70
Wishart, Sask., Can., 270....F3 70
Wishaw, Scot. (part of Motherwell & Wishaw)...E5 13
Wishek, McIntosh, N. Dak., 1,290.....C3 110
Wishram, Klickitat, Wash., 750.....D4 122
Wisla, riv., Pol......B5 26
Wislok, riv., Pol......C7 26
Wisloka, riv., Pol......D6 26
Wismar, Br. Gu.....A3 59
Wismar, Ger., 55,400.....B5 16
Wismar, bay, Ger......D5 24
Wisner, Franklin, La., 1,254.C4 95
Wisner, Cuming, Nebr., 1,192.....C9 103
Wissant, Fr., 951.....D9 12
Wissembourg, Fr., 5,278.....C7 14
Wissmann Pool, lake, Con. L....B2 48
Wissota, lake, Wis......D2 124
Wister, LeFlore, Okla., 592..C7 112
Wister, res. Okla......C7 112
Witbank, S. Afr., 25,881.....C4 49
Witchekan, lake, Sask., Can.....D2 70
Witch Lake, Marquette, Mich., 50.....B2 98
Witham, Eng., 9,457.....C8 12
Witham, riv., Eng......A7 12
Withamsville, Clermont, Ohio, 2,811.....A7 94
Withee, Clark, Wis., 442....D3 124
Witherbee, Essex, N.Y., 800.....A7 108
Witherbee, Berkeley, S.C., 50.....E8 115
Withernsea, Eng., 4,963.....A8 12
Witherspoon, mtn., Alsk......g18 79
Withlacoochee, riv., Fla., Ga....B3 86
Withrow, Douglas, Wash., 25.B6 122
Witless Bay, Newf., Can., 498.E5 75
Witney, Eng., 9,217.....C6 12
Witry-lès-Reims, Fr., 1,466...E4 15
Witt, Montgomery, Ill., 1,101.....D4 90
Witt, Torrance, N. Mex.....C3 107
Witt, Hamblen, Tenn., 250.C10 117
Witten, Ger., 96,500.....B2 17
Witten, Tripp, S. Dak., 146..D6 116
Witten, lake, Ger......D3 24
Wittenberg, Ger., 45,700.....B7 17
Wittenberg, Shawano, Wis., 892.....D4 124
Wittenberge, Ger., 31,600....B5 16
Wittenheim, Fr., 2,556.....B3 18
Wittingen, Ger., 5,100.....B5 16
Wittlich, Ger., 9,500.....D1 17
Wittman, Talbot, Md., 370.C5 85
Wittmann, Maricopa, Ariz., 200.....D3, F1 80
Wittmund, Ger., 4,500.....A7 15
Wittstock, Ger., 9,800.....B6 16
Witvlei, S.W. Afr., 362.....B2 49
Witzenhausen, Ger., 7,900...B4 17
Wiveliscombe, Eng., 1,218...C4 12
Wiville, Woodruff, Ark., 40..B4 81
Wiwon, Kor.....F3 37
Wixom, Oakland, Mich., 1,531.....A7 98
Wixom, lake, Mich......E6 98
Wkra, riv., Pol......B6 26
Wloclawek, Pol., 63,000.....B5 26
Wlodawa, Pol., 4,438.....C7 26
Wloszczowa, Pol., 4,683.....C5 26
Woburn, Que., Can., 412....D5 73
Woburn, Middlesex, Mass., 31,214.....B5, C2 97
Woden, Hancock, Iowa, 283.A4 92
Woerden, Neth., 13,700.....B4 15
Wohlen, Switz., 8,636.....B5 19
Woito, Ont., Can., 55.....B7 72
Wokam, isl., Indon......G8 35
Woking, Alta., Can., 157....B1 69
Woking, Eng., 67,485.....C7 12
Wokingham, Eng., 11,400...C7 12
Wolbach, Greeley, Nebr., 382.....C7 103
Wolco, Osage, Okla.....A5 112
Wolcott, Eagle, Colo., 35....B4 83
Wolcott, New Haven, Conn., 1,500 (8,889▲).....C5 84
Wolcott, White, Ind., 877....C3 91
Wolcott, Wayne, N.Y., 1,641.B4 108
Wolcott, Lamoille, Vt., 150 (633▲).....B4 120
Wolcottville, Lagrange and Noble, Ind., 720.....A7 91
Woldegk, Ger., 3,583.....E7 24
Wolf, Sheridan, Wyo., 40....A5 125
Wolf, creek, Iowa.....B5 92
Wolf, creek, Mich......D7 98
Wolf, creek, Mont......C7 102
Wolf, creek, Okla......A2 112
Wolf, creek, W. Va......D7 123
Wolf, isl., Que., Can......B8 74
Wolf, isl., Ec......f5 58
Wolf, lake, Ill......F3 90
Wolf, riv., Miss., Tenn......A4 100
Wolf, riv., Miss......E4 100
Wolf, riv., Wis......D5 124
Wolf Creek, Lewis and Clark, Mont., 175.....C4 102
Wolf Creek, Josephine, Oreg., 550.....E3 113
Wolf Creek, Cocke, Tenn...D11 117
Wolf Creek, pass, Colo......D4 83
Wolfe, Sask., Can., 40.....E1 70
Wolfe, co., Que., Can., 18,335.....D6 73
Wolfe, co., Ky., 6,534.....C6 94
Wolfeboro, Carroll, N.H., 1,557 (2,689▲).....C4 105
Wolfeboro Center, Carroll, N.H., 125.....C4 105
Wolfeboro Falls, Carroll, N.H., 500.....C4 105
Wolfe City, Hunt, Tex., 1,317.....C4 118
Wolfe Island, Ont., Can., 425.....C8 72
Wolfen, Ger., 13,700.....B7 17
Wolfenbüttel, Ger., 38,100...A5 17
Wolfenschiessen, Switz., 1,647.....C5 19
Wolfestown, Que., Can., 295.....D6 73
Wolfhagen, Ger., 5,900.....B4 17
Wolf Island, Miss., Mo., 175.E8 101
Wolf Lake, Union, Ill., 200..F4 90
Wolflake, Noble, Ind., 375...B7 91
Wolf Lake, Muskegon, Mich., 2,525.....E4 98
Wolf Lake, Becker, Minn., 83.....D3 99
Wolford, Pierce, N. Dak., 136.....A6 110
Wolf Point, Roosevelt, Mont., 3,585.....B11 102
Wolfratshausen, Ger., 7,100..B7 18
Wolf Run, Jefferson, Ohio, 200.....A1 123
Wolfsberg, Aus., 9,470.....E7 16
Wolfsburg, Ger., 64,600.....B5 16
Wolfsville, Frederick, Md., 20.....A2 85
Wolfville, N.S., Can., 2,413..D5 74
Wolgast, Ger., 14,400.....A6 16
Wolhusen, Switz., 3,446.....B5 19
Wolin, Pol., 2,369.....E8 24
Wolin, isl., Pol......E8 24
Wollaston, isl., Chile.....k12 54
Wollaston, lake, Sask., Can....E12 66
Wollongong, Austl., 131,764.G8 51
Wolmaransstad, S. Afr., 6,041.....C4 49
Wolmirstedt, Ger., 7,179.....A6 17
Wolomin, Pol., 21,000.....k14 26
Wolow, Pol., 2,902.....C4 26
Wolseley, Sask., Can., 1,031..G4 70
Wolseth, Ward, N. Dak., 20..A4 110
Wolsey, Beadle, S. Dak., 354.C7 116
Wolsztyn, Pol., 4,967.....B4 26
Wolverhampton, Eng., 150,385.....B5 12
Wolverine, Cheboygan, Mich., 292.....C6 98
Wolverine Lake, Oakland, Mich., 2,404.....*F7 98
Wolvering, riv., B.C., Can......B7 68
Wolverton, Eng., 13,116.....B7 12
Wolverton, Wilkin, Minn., 204.....D2 99
Womack Hill, Choctaw, Ala., 25.....D1 78
Womelsdorf, Berks, Pa., 1,471.....F9 114
Women, lake, Minn......D4 99
Wompi, cave, Iowa.....A6 92
Wonalancet, Carroll, N.H., 25.....C4 105
Wonder, Josephine, Oreg., 65.....E3 113
Wonder, cave, Iowa.....A6 92
Wonder Lake, McHenry, Ill., 3,543.....*E1 90
Wonewoc, Juneau, Wis., 878.....E3 124
Wonsan, Kor., 112,952.....G3 37
Wonthaggi, Austl., 4,190.....I5 51
Wood, Franklin, N.C., 94....A5 109
Wood, Huntingdon, Pa., 450.F5 114
Wood, Mellette, S. Dak., 267.....D5 116
Wood, co., Ohio, 72,596.....A4 111
Wood, co., Tex., 17,653.....C5 118
Wood, co., W. Va., 78,331...B3 123
Wood, co., Wis., 59,105.....D3 124
Wood, lake, Sask., Can......B4 70
Wood, mtn., Sask., Can......H2 70
Wood, mtn., Mont......E7 102
Wood, pond, Maine.....C2 96
Wood, riv., B.C., Can......C8 68
Wood, riv., Sask., Can......H2 70
Wood, riv., Wyo......B3 125
Wood Acres, Montgomery, Md., 1,000.....*C3 85
Woodall, mtn., Miss......A5 100
Woodard, Bertie, N.C., 25...B7 109
Woodberry, Calhoun, Ark., 44.....D3 81
Woodbine, Camden, Ga., 845.....F5 87
Woodbine, Jo Daviess, Ill., 150.....A3 90
Woodbine, Harrison, Iowa, 1,304.....C2 92
Woodbine, Dickinson, Kans., 173.....D7 93
Woodbine, Whitley, Ky., 500.....D5 94
Woodbine, Carroll, Md., 130.....B3 85
Woodbine, Cape May, N.J., 2,823.....E3 106
Woodbine, Davidson, Tenn., 11,500.....A5, E9 117
Woodboro, Oneida, Wis., 35.....C4 124
Woodbourne, Sullivan, N.Y., 900.....D6 108
Woodbridge, San Joaquin, Calif., 800.....B6 82
Woodbridge, Ont., Can., 2,315.....D5, E6 72
Woodbridge, New Haven, Conn., 5,182.....D4 84
Woodbridge, Eng., 5,927....B9 12
Woodbridge, Middlesex, N.J., 17,000.....B4, E4 106
Woodbridge, Prince William, Va., 1,100.....C5 121
Wood Buffalo, nat. park, Alta. and N.W. Ter., Can......E10 66
Woodburn, Allen, Ind., 585..B8 91
Woodburn, Clarke, Iowa, 202.....C4 92
Woodburn, Warren, Ky., 291.....D3 94
Woodburn, Marion, Oreg., 3,120.....B2, B4 113
Woodbury, Litchfield, Conn., 1,000 (3,910▲).....C4 84
Woodbury, Meriwether, Ga., 1,230.....D2 87
Woodbury, Butler, Ky., 122..C3 94
Woodbury, Gloucester, N.J., 12,453.....D2 106
Woodbury, Nassau, N.Y., 1,500.....*F3 84
Woodbury, Bedford, Pa., 280.....F5 114
Woodbury, Cannon, Tenn., 1,562.....B5 117
Woodbury, Washington, Vt., 250 (317▲).....C4 120
Woodbury, co., Iowa, 107,849.....B1 92
Woodbury Heights, Gloucester, N.J., 1,723.....*D2 106
Woodcliff, Screven, Ga., 50..D5 87
Woodcliff, lake, N.J......C5 106
Woodcliff Lake, Bergen, N.J., 2,742.....C5 106
Woodcroft, Marion, Ind., 2,000.....*E5 91
Wood Dale, Du Page, Ill., 4,424.....E2 90
Wood End, pt., Mass......B7 97
Woodfibre, B.C., Can., 524..E6 68
Woodford, Eng., 61,259.....k13 10
Woodford, Ire., 264.....D3 11
Woodford, Orangeburg, S.C., 172.....D5 115
Woodford, co., Ill., 24,579...C4 90
Woodford, co., Ky., 11,913..B5 94
Wood Green, Eng., 47,897..k12 10
Woodhull, Henry, Ill., 779...B3 90
Woodhull, Steuben, N.Y., 321.....C3 108
Woodinville, King, Wash., 650.....B3 122
Woodlake, Tulare, Calif., 2,623.....D4 82
Wood Lake, Yellow Medicine, Minn., 506.....F3 99
Wood Lake, Cherry, Nebr., 197.....B5 103
Woodlake, Trinity, Tex., 250.....D5 118
Woodland, Randolph, Ala., 200.....B4 78
Woodland, Yolo, Calif., 13,524.....A6, C3 82
Woodland, Talbot, Ga., 720.D2 87
Woodland, Iroquois, Ill., 344.....C6 90
Woodland, Aroostook, Maine, (1,372▲).....B4 96
Woodland, Washington, Maine, 1,393.....C5 96
Woodland, Barry, Mich., 374.....F5 98
Woodland, Chickasaw, Miss., 100.....B4 100
Woodland, Northampton, N.C., 651.....A6 109
Woodland, Clearfield, Pa., 900.....E5 114
Woodland, Cowlitz., Wash., 1,336.....D3 122
Woodland Beach, Monroe, Mich., 1,944.....G7 98
Woodland Mills, Obion, Tenn., 200.....A2 117
Woodland Park, Teller, Colo., 666.....C5 83
Woodlawn, Ont., Can., 75...A8 72
Woodlawn, Jefferson, Ill., 241.....E4 90
Woodlawn, McCracken, Ky., 1,688.....A2 94
Woodlawn, Baltimore, Md., 6,000.....*B4 85
Woodlawn, Prince Georges, Md., 3,000.....*C4 85
Woodlawn, Hamilton, Ohio, 3,007.....C3 111
Woodlawn, Montgomery, Tenn., 25.....A4 117
Woodlawn, Carroll, Va., 30.....E2 121
Woodlawn Beach, Erie, N.Y., 1,800.....C2 108
Woodlawn Orchards, Jackson, Mich., 2,000.....*F6 98
Woodlawn Park, Jefferson, Ky., 1,137.....*H6 94
Woodlawn Park, Anne Arundel, Md., 1,200.....*C4 85
Woodleaf, Rowan, N.C., 500.B3 109
Woodley Hills, Fairfax, Va., 2,000.....*A5 121
Woodlin, Sanders, Mont., 36.C1 102
Woodlyn, Delaware, Pa., 6,000.....*B10 114
Wood-Lynne, Camden, N.J., 3,128.....*D2 106
Woodman, Carroll, N.H., 25.C5 105
Woodmere, Nassau, N.Y., 14,011.....G2 84
Woodmont, New Haven, Conn. (part of Milford)...E4 84
Wood Mountain Station, Sask., Can., 135.....H2 70
Woodpecker, B.C., Can.....C6 68
Woodport, Morris, N.J., 150.B3 106
Woodridge, Man., Can., 289.E3 71
Wood-Ridge, Bergen, N.J., 7,964.....D5 106
Wood River, Madison, Ill., 11,694.....E3 90
Wood River, Hall, Nebr., 828.....D7 103
Wood River Junction, Washington, R.I., 90....D10 84
Woodroffe, mtn., Austl......E5 50
Woodrow, Cleburne, Ark., 25.....B3 81
Woodrow, Sask., Can., 155..H2 70
Woodruff, Navajo, Ariz., 125.C5 80
Woodruff, Phillips, Kans., 55.C4 93
Woodruff, Spartanburg, S.C., 3,679.....B3 115
Woodruff, Rich, Utah, 169..B4 119
Woodruff, Marshall, W. Va., 45.....C2 123
Woodruff, Oneida, Wis., 500.C4 124
Woodruff, co., Ark., 13,954..B4 81
Woodruff Place, Marion, Ind., 1,501.....H8 91
Woods, Newf., Can.....D2 71
Woods, co., Okla., 11,932....A3 112
Woods, lake, Austl......C5 50
Woods, lake, Ont., Can., Minn......E7 72, A4 99
Woods, mtn., Ark......B2 81
Woodsboro, Frederick, Md., 430.....A3 85
Woodsboro, Refugio, Tex., 2,081.....E4 118
Woodsburgh, Nassau, N.Y., 907.....*E2 108
Woods Cross, Davis, Utah, 1,098.....C4 119
Woodsdale, Person, N.C., 100.....A4 109
Woodsfield, Monroe, Ohio, 2,956.....C6 111
Woods Hole, Barnstable, Mass., 950.....C6 97
Woodside, Austl., 171.....I6 51
Woodside, San Mateo, Calif., 3,592.....*D2 82
Woodside, Kent, Del., 189...B6 85
Woodside, Ravalli, Mont., 75.....D2 102
Woodside, Emery, Utah, 22.D5 119
Woodson, Pulaski, Ark., 450.C3 81
Woodson, Morgan, Ill., 229.D3 90
Woodson, Throckmorton, Tex., 337.....C3 118
Woodson, co., Kans., 5,423..E8 93
Woodson Terrace, St. Louis, Mo., 6,048.....*C7 101
Woodstock, Bibb, Ala., 350..B2 78
Woodstock, N.B., Can., 4,305.....C2 74
Woodstock, Ont., Can., 20,486.....D4 72
Woodstock, Windham, Conn., 200 (3,177▲).....B9 84
Woodstock, Cherokee, Ga., 726.....B2 87
Woodstock, McHenry, Ill., 8,897.....A5 90

Woodstock, Howard and Baltimore, Md., 500......B4 85
Woodstock, Pipestone, Minn., 213......F2 99
Woodstock, Grafton, N.H., 90 (827▲)......C3 105
Woodstock, Champaign, Ohio, 310......B4 111
Woodstock, Shelby, Tenn....E7 117
Woodstock, Windsor, Vt., 1,415 (2,786▲)......D3 120
Woodstock, Shenandoah, Va., 2,083......C4 121
Woodstock Valley, Windham, Conn., 100......B8 84
Woodston, Rooks, Kans., 332......C4 93
Woodstown, Salem, N.J., 2,942......D2 106
Woodsville, Grafton, N.H., 1,596......B2 105
Woodtick, New Haven, Conn., 400......C5 84
Woodville, Tulare, Calif., 1,045......*D4 82
Woodville, Ont., Can., 399..C6 72
Woodville, Leon, Fla., 400...B2 86
Woodville, Greene, Ga., 372.C3 87
Woodville, Bingham, Idaho, 300......F6 89
Woodville, McCracken, Ky., 100......A2 94
Woodville, Middlesex, Mass., 350......D1 97
Woodville, Jackson, Mich., 2,000......*F6 98
Woodville, Wilkinson, Miss., 1,856......D2 100
Woodville, N.Z., 1,530.....N15 51
Woodville, Bertie, N.C., 344.A6 109
Woodville, Sandusky, Ohio 1,700......A2, A4 111
Woodville (New Woodville), Marshall, Okla., 98......D5 112
Woodville, Washington, R.I., 50......D9 84
Woodville, Tyler, Tex., 1,920......D5 118
Woodville, St. Croix, Wis., 430......D1 124
Woodward, Jefferson, Ala., 1,000......B3, E4 78
Woodward, Dallas, Iowa, 967......C4 92
Woodward, Woodward, Okla., 7,747......A2 112
Woodward, co., Okla., 13,902......A2 112
Woodwards Cove, N.B., Can., 188......E3 74
Woodway, Fairfield, Conn. (part of Darien)......E2 84
Woodway, McLennan, Tex., 1,244......*D4 118
Woodway, Lee, Va., 400....B2 121
Woodworth, Rapides, La., 320......C3 95
Woodworth, Stutsman, N. Dak., 221......B6 110
Woody, riv., Sask., Man., Can.......E5 70, C1 71
Wooldridge, Cooper, Mo., 100......C5 101
Wooler, Eng., 1,791......E6 13
Woolford, Dorchester, Md., 350......C5 85
Wool Market, Harrison, Miss., 350......E2, E5 100
Woolper, creek, Ky.......A6 94
Woolrich, Clinton, Pa., 500..D7 114
Woolsey, Washington, Ark...B1 81
Woolsey, peak, Ariz.......D3 80
Woolsey, Pershing, Nev., 8...C3 104
Woolstock, Wright, Iowa, 269......B4 92
Woolwich, Eng., 146,397 (part of London)......m13 10
Woolwich, Sagadahoc, Maine, 400 (1,417▲)......E6 96
Woomera, Austl., 4,808......F6 50
Woonsocket, Providence, R.I., 47,080......A10 84
Woonsocket, Sanborn, S. Dak., 1,035......C7 116
Woonsocket, hill, R.I.......B10 84
Wooramel, Austl......E1 50
Wooster, Faulkner, Ark., 161.B3 81
Wooster, Wayne, Ohio, 17,046......B6 111
Wooster, Harris, Tex., 3,000.*F5 118
Woosung, Ogle, Ill., 120.....B4 90
Worb, Switz., 5,885......C4 19
Worbis, Ger., 2,896......B5 17
Worcester, Eng., 65,865......B5 12
Worcester, Worcester, Mass., 186,587 (*316,200)......B4 97
Worcester, Otsego, N.Y., 799......C6 108
Worcester, S. Afr., 32,274...D2 49
Worcester, Washington, Vt., 140 (417▲)......C3 120
Worcester, co., Eng., 568,642.B5 12
Worcester, co., Md., 23,733..D7 85
Worcester, co., Mass., 583,228......A3 97
Worcester, mts., Vt.......C3 120
Worden, Madison, Ill., 1,060.E4 90
Worden, Yellowstone, Mont., 225......E8 102
Worden, Klamath, Oreg., 50......E5 113
Worden, pond, R.I.......D10 84
Wordsworth, Sask., Can., 80......H4 70
Workington, Eng., 29,507....C5 10
Worksop, Eng., 34,237......A6 12
Workum, Neth., 3,287......B5 15
Worland, Bates, Mo., 57....C3 101
Worland, Washakie, Wyo., 5,806......A5 125
World, 2,620,000,000...... 2
Worley, Kootenai, Idaho, 241......B2 89
Worley, Hamilton, Tenn., 300......E10 117
Wormerveer, Neth., 13,400..B4 15
Wormleysburg, Cumberland, Pa., 1,794......*F8 114
Worms, Ger., 62,400......D3 17
Worms, Merrick, Nebr., 25..C7 103
Woronoco, Hampden, Mass., 300......B2 97
Worth, Cook, Ill., 9,097......*F3 90
Worth, Worth, Mo., 135....A3 101
Worth, co., Ga., 16,682......E3 87
Worth, co., Iowa, 10,259....A4 92
Worth, co., Mo., 3,936......A3 101
Wortham, Freestone, Tex., 1,087......D4 118
Worthing, Eng., 80,143.....D7 12
Worthing, Lincoln, S. Dak., 304......D9 116
Worthington, Ont., Can., 164......A3 72
Worthington, Union, Fla., 200......C4 86
Worthington, Greene, Ind., 1,635......F4 91
Worthington, Dubuque, Iowa, 360......B6 92
Worthington, Greenup, Ky., 1,235......B7 94
Worthington, Hampshire, Mass., 150 (297▲)......B2 97
Worthington, Nobles, Minn., 9,015......G3 99
Worthington, Putnam, Mo., 175......A5 101
Worthington, Franklin, Ohio, 9,239......B4, C2 111
Worthington, Armstrong, Pa., 772......E2 114
Worthington, Marion, W. Va., 361......A7 123
Worthington, mts., Nev.......F6 104
Worthville, Carroll, Ky., 247.B4 94
Worthville, Randolph, N.C., 400......B4 109
Wortis, Sharp, Ark......A4 81
Worton, Kent, Md., 175......B5 85
Wostok, Alta., Can., 130....C4 69
Wotton, Que., Can., 726....D6 73
Wounded Knee, Shannon, S. Dak., 25......D3 116
Wowoni, isl., Indon......F6 35
Wozniki, Pol., 2,312......f10 26
Wrangel, isl., Sov. Un.......B21 28
Wrangell, Alsk., 1,315.D13, m23 79
Wrangell, isl., Alsk.......m24 79
Wrangell, mtn., Alsk.......f19 79
Wrangell, mts., Alsk.......C11 79
Wrath, cape, Scot.......B3 13
Wray, Yuma, Colo., 2,082...A8 83
Wray, Irwin, Ga., 125......E3 87
Wren, Van Wert, Ohio, 287.B3 111
Wrens, Jefferson, Ga., 1,628..C4 87
Wrenshall, Carlton, Minn., 189......D6 99
Wrentham, Alta., Can., 111 . E4 69
Wrentham, Norfolk, Mass., 1,790 (6,685▲)......B5 97
Wrexham, Wales, 35,427....A5 12
Wriezen, Ger., 4,806......B7 16
Wright, Lauderdale, Ala., 120......A2 78
Wright, Mahaska, Iowa, 90..C5 92
Wright, Ford, Kans., 90.....E4 93
Wright, Carlton, Minn., 169.D5 99
Wright, co., Iowa, 19,447....B4 92
Wright, co., Minn., 29,935...E4 99
Wright, co., Mo., 14,183....D5 101
Wright, mtn., Mont......C4 102
Wright Brothers nat. memorial, N.C.......A8 109
Wright City, Warren, Mo., 738......C6 101
Wright City, McCurtain, Okla., 1,161......C6 112
Wrightstown, Burlington, N.J., 4,846......C3 106
Wrightstown, Brown, Wis., 840......A6, D5 124
Wrightsville, Pulaski, Ark., 350......C3, D6 81
Wrightsville, Johnson, Ga., 2,056......D4 87
Wrightsville, York, Pa., 2,345.F8 114
Wrightsville, res., Vt.......C3 120
Wrightsville Beach, New Hanover, N.C., 723......C6 109
Wrightsville Sound, New Hanover, N.C......C6 109
Wrightview, Greene, Ohio, 2,500......*C3 111
Wrightwood, San Bernardino, Calif., 500......E5, F3 82
Wrigley, gulf, Ant.......B36 5
Wrigley, Hickman, Tenn., 499......B4 117
Wroclaw (Breslau), Pol., 429,000......C4 26
Wroclaw, prov., Poland, 1,799,000......B10 17
Wrong, lake, Man., Can.......C3 71
*Wrottesley, cape, N.W. Ter., Can.*B8 66
Wroxeter, Ont., Can., 292....D3 72
Wroxton, Sask., Can., 127....F5 70
Wrzesnia, Pol., 11,800......B4 26
Wschowa, Pol., 4,075......C4 26
Wu, riv., China......J3 36
Wuchan, China......A2 37
Wuchang (part of Wuhan), China
Wuchang, China, 5,000......D3 37
Wuch'i, Taiwan......G9 34
Wu Chin Shan, mtn., China....C8 38
Wuchow (Tsangwu), China, 110,800......G7 34
Wuchuan, China, 7,000.....D4 36
Wuchuan, China, 3,000......J3 36
Wufeng, China, 9,000......I4 36
Wuhan, China, 2,146,000 (incl. Hankow, Hanyang, and Wuchang)....E7 34, I6 36
Wuho, China......H7 36
Wuhsing, China, 62,700......I9 36
Wuhu, China, 242,100......I8 36
Wui, China......F6 36
Wukang, China......K4 36
Wukari, Nig......E6 45
Wümme, riv., Ger......E3 24
Wunsiedel, Ger., 9,000......C7 17
Wunstorf, Ger., 13,700......A4 17
Wuntho, Bur., 2,602......D10 39
Wupatki, nat. mon., Ariz.......B4 80
Wuppertal, Ger., 420,700 (*880,000)......B2 17
Würm, lake, Ger......E5 16
Württemberg, reg., Ger......D4 16
Wurtsboro, Sullivan, N.Y., 655......D6 108
Wurzach, Ger., 2,248......B5 18
Würzburg, Ger., 116,900....D4 17
Wurzen, Ger., 23,500......B7 17
Wushan, China, 26,000......I3 36
Wusih, China, 613,000......E9 34
Wuskwatim, lake, Man., Can...B2 71
Wusu, China......C1 34
Wusung, China......I9 36
Wutu, China......E5 34
Wuwei, China......D5 34
Wuwei, China......I7 36
Wuyuan, China, 10,000......D3 36
Wuyun, China......B10 34
Wyaconda, Clark, Mo., 402..A6 101
Wyaconda, riv., Mo.......A6 101
Wyalusing, Bradford, Pa., 685......C9 114
Wyandanch, Suffolk, N.Y., 5,000......*E7 108
Wyandot, co., Ohio, 21,648..B4 111
Wyandotte, Wayne, Mich., 43,519......A7, F7 98
Wyandotte, Ottawa, Okla., 226......A7 112
Wyandotte, co., Kans., 185,495......C9 93
Wyandotte, cove, Ind.......H5 91
Wyandra, Austl., 153......C5 51
Wyanet, Bureau, Ill., 938....B4 90
Wyarno, Sheridan, Wyo., 20.A6 125
Wyatt, St. Joseph, Ind., 300.A5 91
Wyatt, Mississippi, Mo., 711.E8 101
Wybark, Muskogee, Okla....B6 112
Wyckoff, Bergen, N.J., 11,205......*A4 106
Wyco, Wyoming, W. Va., 200......D3 123
Wye, riv., Eng.......C5 12
Wyebridge, Ont., Can., 139..C5 72
Wye Mills, Talbot, Md., 125.C5 85
Wyesocking, bay, N.C.......B7 109
Wyevale, Ont., Can., 94.....C5 72
Wyeville, Monroe, Wis., 220.D3 124
Wygietrzow, Pol......g10 26
Wykoff, Fillmore, Minn., 391......G6 99
Wylie, Collin, Tex., 1,804...*C4 118
Wylliesburg, Charlotte, Va., 150......E4 121
Wyman, Louisa, Iowa, 50....C6 92
Wyman, lake, Maine......C3 96
Wymark, Sask., Can., 257...G2 70
Wymondham, Eng., 5,896...B9 12
Wymore, Gage, Nebr., 1,975.D9 103
Wynantskill, Rensselaer, N.Y., 5,000......*C7 108
Wyncote, Montgomery, Pa., 6,000......*A11 114
Wyndham, Austl., 458......C4 50
Wyndmere, Richland, N. Dak., 644......C8 110
Wyndmoor, Montgomery, Pa., 5,800......*A11 114
Wynnburg, Lake, Tenn., 300......A2 117
Wynne, Cross, Ark., 4,922....B5 81
Wynne Wood, Garvin, Okla., 2,509......C4 112
Wynnewood, Montgomery, Pa., 7,200......*B11 114
Wynona, Osage, Okla., 652..A5 112
Wynoochee, riv., Wash.......B2 122
Wynot, Cedar, Nebr., 209....B8 103
Wynyard, Sask., Can., 1,686.F3 70
Wyocena, Columbia, Wis., 747......E4 124
Wyodak, Campbell, Wyo., 60.A7 125
Wyola, Big Horn, Mont., 87.E9 102
Wyoming, Ont., Can., 880....E2 72
Wyoming, Kent, Del., 1,172.B6 85
Wyoming, Stark, Ill., 1,559..B4 90
Wyoming, Jones, Iowa, 797..B6 92
Wyoming, Kent, Mich., 45,829......*F5 98
Wyoming, Chicago, Minn., 435......E6 99
Wyoming, Wyoming, N.Y., 526......C2 108
Wyoming, Hamilton, Ohio, 7,736......D2 111
Wyoming, Luzerne, Pa., 4,127......B8 114
Wyoming, Washington, R.I., 350......C10 84
Wyoming, co., N.Y., 34,793..C2 108
Wyoming, co., Pa., 16,813...D9 114
Wyoming, co., W. Va., 34,836......D3 123
Wyoming, state, U.S., 330,066......B6 76, 125
Wyoming, basin, Wyo.......D3 125
Wyoming, peak, Wyo.......C2 125
Wyoming, range, Wyo.......B2 125
Wyomissing, Berks, Pa., 5,044......F10 114
Wyomissing Hills, Berks, Pa., 1,644......*F10 114
Wyre, riv., Eng.......G6 13
Wyrzsk, Pol., 3,039......B4 26
Wysokie Mazowieckie, Pol., 9,748......B7 26
Wyszkow, Pol., 5,021......B6 26
Wythe, co., Va., 21,975......E1 121
Wytheville, Wythe, Va., 5,634......E1 121
Wytopitlock, Aroostook, Maine, 200......C4 96
Wyvis, mtn., Scot.......C4 13

X

Xanthi, Grc., 26,377......B5 23
Xanthi, prov., Grc., 89,594.*B5 23
Xapecó, riv., Braz.......D2 56
Xapurí, Braz., 2,000......D4 58
Xauen, Mor., 13,712......B3 44
Xavantes, mts., Braz.......E5 59
Xavier, Pima, Ariz......E5 80
Xavier, Leavenworth, Kans., 600......B8 93
Xbonil, Mex., 20......D6 63
Xcalak, Mex., 527......D7 63
Xenia, Clay, Ill., 491......E5 90
Xenia, Greene, Ohio, 20,445.C4 111
Xieng, Khouang, Laos......C5 38
Xilitla, Mex., 1,901......m14 63
Xingú, riv., Braz......C4 59
Xique-Xique, Braz., 5,467...D2 57
Xochimilco, Mex., 20,687......h9, n14 63

Y

Yaak, Lincoln, Mont., 40....B1 102
Yaan, China, 55,200......F5 34
Yabbenohr, isl., Kwajalein...... 52
Yablis, Nic......C6 62
Yablonovy, mts., Sov. Un.......D14 28
Yabucoa, P.R., 3,734......C7 65
Yabucoa, mun., P.R., 29,782.C7 65
Ya'bud, Jordan, 3,000......B7 32
Yachats, Lincoln, Oreg., 250.C2 113
Yaco, riv., Braz......C4 60
Yacolt, Clark, Wash., 375...D3 122
Yacuiba, Bol., 5,027......D3 55
Yadkin, co., N.C., 22,804....A3 109
Yadkin, riv., N.C.......B3 109
Yadkin Valley, Caldwell, N.C., 125......A2 109
Yadkinville, Yadkin, N.C., 1,644......A3 109
Yagi, Jap., 8,372......o14 37
Yaguachi, Ec., 2,879......B2 58
Yaguajay, Cuba, 4,867......D4 64
Yaguas, riv., Peru......B3 58
Yahk, B.C., Can., 243......E9 68
Yahuma, Con. L......A3 48
Yainax, butte, Oreg.......E5 113
Yaizu, Jap., 50,600......o17 37
Yakima, Yakima, Wash., 43,284......C5 122
Yakima, co., Wash., 145,112......C5 122
Yakima, ridge, Wash.......C5 122
Yakima, riv., Wash.......C5 122
Yakima, val., Wash.......C5 122
Yakoma, Con. L......A3 48
Yaku, isl., Jap......K5 37
Yakutat, Alsk., 230......D12 79
Yakutat, bay, Alsk.......D12 79
Yakutsk, Sov. Un., 77,000..C15 28
Yalaha, Lake, Fla., 650......D5 86
Yale, B.C., Can., 297......E7 68
Yale, Jasper, Ill., 123......D5 90
Yale, Guthrie, Iowa, 260......C3 92
Yale, St. Clair, Mich., 1,621..E8 98
Yale, Payne, Okla., 1,369....A5 112
Yale, Beadle, S. Dak., 171...C8 116
Yale, Sussex, Va., 150......E5 121
Yale, dist., B.C., Can......D8 68
Yale, mtn., Colo.......C4 83
Yale, res., Wash......C3 122
Yalesville, New Haven, Conn. (part of Wallingford)......D5 84
Yalinga, Cen. Afr. Rep......D4 46
Yalobusha, co., Miss., 12,502......A4 100
Yalobusha, riv., Miss.......B4 100
Yalova, Tur., 11,500......B7 23
Yalta, Sov., Un., 47,100.....I10 27
Yalu, China, 10,000......B9 34
Yalu, riv., China, Kor.......D11 36
Yalung, riv., China......E5 34
Yalutrovsk, Sov. Un., 18,700......B7 29
Yalvac, Tur., 10,400......C8 31
Yama, Sov. Un., 5,000......q21 27
Yamachiche, Que., Can., 1,186......C5 73
Yamagata, Jap., 125,000 (188,597▲)......G10 37
Yamagata, pref., Jap., 1,320,664......*G10 37
Yamaguchi, Jap., 54,400 (87,695▲)......I5 37
Yamaguchi, pref., Jap., 1,602,207......*I5 37
Yamanashi, pref., Jap., 782,062......*I9 37
Yamantau, mtn., Sov. Un.......C5 29
Yamaska, Que., Can., 454...C5 73
Yamaska, co., Que., Can., 16,058......C5 73
Yambio, Sud., 3,890......E2 47
Yambol, Bul., 30,311......D8 22
Yamdrok, lake, China......C9 39
Yamethin, Bur., 12,030....D10 39
Yamhill, Yamhill, Oreg., 407......B1 113
Yamhill, co., Oreg., 32,478..B3 113
Yamkino, Sov. Un., 1,000...n18 27
Yamma Yamma, lake, Austl....C3 51
Yampa, Routt, Colo., 312....A4 83
Yampa, mtn., Colo.......A3 83
Yampa, plat., Colo., Utah......C6 119
Yampa, riv., Colo.......A2 83
Yamparaez, Bol., 725......C2 55
Yamsey, mtn., Oreg.......E5 113
Yamsk, Sov. Un., 800......D18 28
Yana, riv., Sov. Un.......C16 28
Yanam, India, 7,032......E7 39, I9 40
Yanao, Austl., 252......H3 51
Yanaoca, Peru, 1,384......D3 58
Yanaul, Sov. Un., 10,000....D20 9
Yanbu', Sau. Ar., 10,000......E7 43
Yancey, co., N.C., 14,008....C4 109
Yanceyville, Caswell, N.C., 1,113......A4 109
Yanfolila, Mali......D3 45
Yang, China......H2 36
Yangambi, Con. L., 12,800..A3 48
Yangasa Cluster, is., Fiji Is....... 52
Yangchiang, China, 31,000..G7 34
Yangchow, China, 180,200...E8 34
Yangchun, China, 11,000....G7 34
Yangeshiri, isl., Jap......D10 37
Yangi-Yul, Sov. Un., 30,000......G22 9
Yangku, see Taiyuan, China
Yangtze, riv., China......H9 36
Yangyang, Kor., 4,900......G4 37
Yankeetown, Citrus and Levy, Fla., 425......C4 86
Yankeetown, Warrick, Ind., 250......I3 91
Yankton, Yankton, S. Dak., 9,279......E8 116
Yankton, co., S. Dak., 17,551......D8 116
Yanonge, Con. L......A3 48
Yantic, New London, Conn. (part of Norwich)......C8 84
Yantley, Chactaw, Ala., 200.C1 78
Yantra, riv., Bul.......D7 22
Yanush, Latimer, Okla., 60..C6 112
Yao, Chad, 3,679......C3 46
Yao, Jap., 122,832......*o14 37
Yaosca, Nic......D5 62
Yaoundé, Cam., 61,750......E2 46
Yap, isl., Pac. O.......F8 7
Yaphank, Suffolk, N.Y., 600..F5 84
*Yaque del Norte, riv., Dom. Rep.*F8 64
Yaqui, riv., Mex.......B3 63
Yar, Sov. Un......B4 29
Yaracuy, state, Ven., 175,291......A4 60
Yaraka, Austl., 27......B5 52
Yarbo, Washington, Ala., 125......D1 78
Yarda, well, Chad......B3 46
Yardley, Bucks, Pa., 2,271..F12 114
Yardville, Mercer, N.J., 3,000......C3 106
Yarensk, Sov. Un., 2,600...C18 9
Yarí, riv., Col.......C3 60
Yariga-Take, peak, Jap.......H8 37
Yaritagua, Ven., 6,036......A4 60
Yarkand (Soche), China, 80,000......H24 9
Yarker, Ont., Can., 314......C8 72
Yarkon, riv., Isr.......B6, g10 32
Yarkovo, Sov. Un......B7 29
Yarmouth, N.S., Can., 8,636.F3 74
Yarmouth, Des Moines, Iowa, 175......C6 92
Yarmouth, Cumberland, Maine, 2,913 (3,517▲). E2, E5 96
Yarmouth, Barnstable, Mass., 450 (5,504▲)......C7 97
Yarmouth, co., N.S., Can., 23,386......F4 74
Yarmuk, riv., Jordan and Syr....B7 32
Yarnell, Yavapai, Ariz., 500.C3 80
Yarnema, Sov. Un......F18 25
Yaroslavl, Sov. Un., 433,000......B1 29
Yarovskoye, marsh, Sov. Un....B8 29
Yarraden, Austl......B7 50
Yarram, Austl., 2,053......I6 51
Yarrowsburg, Washington, Md., 175......B2 85
Yartsevo, Sov. Un., 25,800..D9 27
Yarumal, Col., 10,349......B2 60
Yasana, Con. L......A4 48
Yasawa, isl., Fiji Is....... 52
Yasawa Group, is., Fiji Is....... 52
Yashi, Nig......D6 45
Yasinovataya, Sov. Un., 10,000......q20 27
Yasinya, Sov. Un......A7 22
Yasothon, Thai., 10,851......E6 38
Yata, Bol......B2 55
Yates, Harding, N. Mex., 25......A6 107
Yates, co., N.Y., 18,614......C3 108
Yatesboro, Armstrong, Pa., 900......E3 114
Yates Center, Woodson, Kans., 2,080......E8 93
Yates City, Knox, Ill., 802...C3 90
Yatesville, Upson, Ga., 354..D2 87
Yathata, isl., Fiji Is....... 52
Yathkyed, lake, N.W. Ter., Can.......D13 66
Yatsuga Take, peak, Jap.......I9 37
Yatsushiro, Jap., 68,000......J5 37
Yatsushiro, sea, Jap......J5 37
Yaṭṭah, Jordan, 5,000......C7 32
Yatung, China......D12 40
Yauca, Peru, 596......E3 58
Yauco, P.R., 8,996......C3 65
Yauco, mun., P.R., 34,780...C3 65
Yauco, riv., P.R.......C3 65
Yauli, Peru, 821......D2 58
Yaupi, Ec......B2 58
Yauri, Peru, 1,487......D3 58
Yautepec, Mex., 8,649......n14 63
Yauyos, Peru, 1,058......D2 58
Yavapai, co., Ariz., 28,912...D3 80
Yavarí, riv., Braz., Peru......B3 58
Yavne, Isr., 5,397......C6, h9 32
Yavneel, Isr., 1,884......B7 32
Yavorov, Sov. Un., 10,000...G4 27
Yawata, isl., Truk...... 52
Yawatahama, Jap., 35,200...J6 37
Yawhee, plat., Oreg.......E5 113
Yawngseng, Bur......A3 38
Yazd (Yezd), Iran, 63,502...F7 41
Yazdan, Iran......E10 41
Yazd-e Khvāst, Iran, 5,000..F6 41
Yazoo, co., Miss., 31,653....C3 100
Yazoo, riv., Miss.......C3 100
Yazoo City, Yazoo, Miss., 11,236......C3 100
Ybbs, riv., Aus.......D7 16
Yding Skovhöj, hill, Den......C3 24
Ye, Bur., 12,743......E10 39
Yeadon, Delaware, Pa., 11,610......B11 114
Yeager, Hughes, Okla., 129..B5 112
Yeagertown, Mifflin, Pa., 1,349......E6 114
Yebbi Bou, Chad, 146......A3 46
Yecla, Sp., 18,400......C5 20
Yeddo, Fountain, Ind., 150..D3 91
Yefremov, Sov. Un., 42,900.E12 27
Yegendybulak, Sov. Un......D9 29
Yegoryevsk, Sov. Un., 61,000......B1 29
Yeguas, pt., P.R.......C7 65
Yehpaishou, China......D8 36
Yei, Sud., 739......E3 47
Yekaterinoslavka, Sov. Un., 5,000......A4 39
Yelabuga, Sov. Un., 16,900.D19 9
Yelan, Sov. Un., 10,000....F14 27
Yelanskoye, Sov. Un......C15 28
Yelets, Sov. Un., 83,000......C1 29
Yélimané, Mali......C2 45
Yelizarovo, Sov. Un......C22 9
Yell, co., Ark., 11,940......B2 81
Yell, isl., Scot.......g10 10
Yellandalpad, India, 10,955..I8 40
Yelleq, mtn., Eg., U.A.R.......D5 32
Yellow, cliff, Ky.......D5 94
Yellow, creek, Ohio......A1 111
Yellow, creek, Tenn.......A4 117
Yellow, lake, Wis.......C1 124
Yellow, riv., see Hwang Ho, China
Yellow, riv., Fla.......G2 86
Yellow, riv., Ga.......A6 87
Yellow, riv., Ind.......B4 91
Yellow, riv., Wis.......D3 124
Yellow, sea, China......D9 34
Yellow Bluff, Wilcox, Ala., 25......D2 78
Yellow Creek, Sask., Can., 182......E3 70
Yellow Grass, Sask., Can., 527......H3 70
Yellowhead, pass, B.C., Alta., Can.......F9 66
Yellow Jacket, Montezuma, Colo., 3......D2 83
Yellow Jacket, mts., Idaho......D4 89
Yellowknife, N.W. Ter., Can., 3,141......D10 66
Yellow Medicine, co., Minn., 15,523......F2 99
Yellow Pine, Washington, Ala., 150......D1 78
Yellow Pine, Valley, Idaho, 45......E3 89
Yellow Springs, Frederick, Md., 130......B3 85
Yellow Springs, Greene, Ohio, 4,167......C4 111
Yellowstone, co., Mont., 79,015......D8 102
Yellowstone, lake, Wyo.......A2 125
Yellowstone, nat. park, Idaho, Mont., Wyo.......A1 125
Yellowstone, riv., Mont., Wyo.......C12 102, A2 125
Yellowstone National Park (part), co., Wyo., 420......A2 125
Yellowstone National Park, Yellowstone National Park, Wyo., 300..A2 125
Yellville, Marion, Ark., 636..A3 81
Yelm, Thurston, Wash., 479.C3 122
Yelnya, Sov. Un., 10,500....D9 27
Yeltes, riv., Sp.......B2 20
Yelvington, Daviess, Ky., 100......C3 94
Yelwa, Nig., 2,142......D5 45
Yemanzhelinsk, Sov. Un., 33,500......C6 29
Yemasse, Beaufort and Hampton, S. C., 473......F6 115
Yemen, country, Asia, 5,000,000......H7 33
Yen, riv., China......G8 36
Yena, Sov. Un......D14 25
Yenakiyevo, Sov. Un., 93,000......G12, q21 27
Yenan, (Fushih), China......D6 34
Yenangyat, Bur......B1 38
Yenangyaung, Bur., 11,098..B1 38
Yen Bay, Viet., 10,000......B6 38
Yencheng, China......H5 36
Yencheng, China, 105,000...H9 36
Yenchi (Chutzuchieh), China, 70,000......C10 34
Yendi, Ghana, 7,694......E4 45
Yendua, isl., Fiji Is....... 52
Yenice, riv., Tur.......B9 31
Yenisey, riv., Sov. Un.......C11 28
Yeniseysk, Sov. Un., 18,300.D12 28
Yeniseyskiy Kryazh, mts., Sov. Un.......D27 9
Yenki, see Kara Shahr, China
Yenshan, China......E7 36
Yenshih, China......G5 36
Yenshou, China, 5,000......D4 37
Yentna, riv., Alsk.......f16 79
Yeo, lake, Austl.......E3 50
Yeoman, Carroll, Ind., 172..C4 91
Yeotmal, India, 45,587......G7 40
Yeovil, Eng., 24,552......D5 12
Yerania, mtn., Grc.......g10 23
Yerba Buena, Chile......E1 55
Yerevan, Sov. Un., 558,000......G17 9, E2 29
Yerington, Lyon, Nev., 1,764......E2 104
Yerkéhida (Well), Niger......B7 45
Yermakovskoye, Sov. Un....E27 9
Yermo, San Bernardino, Calif., 800......E5 82
Yermolayevo, Sov. Un......C5 29
Yerres, Fr., 10,747......h10 14
Yerres, riv., Fr.......F2 15
Yerseke, Neth., 5,000......C4 15
Yershov, Sov. Un., 16,500...C3 29
Yerupaja, mtn., Peru......D2 58
Yesilova, Tur., 1,165......D7 23
Yeso, De Baca, N. Mex., 350......C5 107
Yessentuki, Sov. Un., 32,600......G17 9
Yessey, Sov. Un., 10,000....C13 28
Yeste, Sp., 9,997......C4 20
Yetter, Calhoun, Iowa, 85...B3 92
Yeu, isl., Fr.......D2 14
Yevlakh, Sov. Un., 15,000....E3 29
Yevpatoriya, Sov. Un., 60,000......I9 27
Yew, mtn., W. Va.......C4 123
Yeysk, Sov. Un., 60,000....H12 27
Yezd, see Yazd, Iran
Yezd, reg., Iran......F6 41
Yi, riv., Ur.......E1 56
Yiannitsa, Grc., 19,693......B4 23
Yiaros, isl., Grc.......D5 23
Yinchwan (Ningsia), China, 84,000......E2 36
Ying, China......E5 36
Yingkow, China, 131,400...D10 36
Yingshang, China......H7 36
Yingtak, China......G7 34
Yin Shan, mtn., China......D4 36
Yioura, isl., Grc.......C5 23
Yirol, Sud., 1,895......D3 47
Yithion, Grc., 7,110......D4 23
Ylig, bay, Guam...... 52
Ylikitka, lake, Fin.......D13 25
Ymer, isl., Grnld.......B17 4
Ymir, B.C., Can., 323......E9 68
Ymir, mtn., B.C., Can.......E2 69
Yoakum, DeWitt and Lavaca, Tex., 5,761......E4 118
Yoakum, co., Tex., 8,032....C1 118
Yockanookany, riv., Miss.......C4 100
Yocona, riv., Miss.......A4 100
Yoder, Arkansas, Ark......C4 81
Yoder, El Paso, Colo., 30....C6 83
Yoder, Allen, Ind., 200......C7 91
Yoder, Reno, Kans., 75......B4 93
Yoder, Goshen, Wyo., 83....D8 125
Yoho, nat. park, B.C., Can.......D9 68
Yoichi, Jap., 18,600......E10 37
Yojoa, lake, Hond......C3 62
Yokadouma, Cam......E2 46
Yokkaichi, Jap., 195,974......I8, o15 37
Yoko, Cam......D2 46
Yokoate, isl., Jap......L4 37
Yokohama, Jap., 1,375,710......I9, n18 37
Yokoshiba, Jap., 3,900......n19 37
Yokosuka, Jap., 287,309.I9, n18 37
Yokosuka, Jap., 6,000......o16 37
Yokote, Jap., 24,700......G10 37
Yokun Seat, mtn., Mass.......B1 97
Yola, Nig., 8,573......E7 45
Yolaina, mts., Nic.......E5 62
Yolo, co., Calif., 65,727......C2 82
Yolyn, Logan, W. Va., 800..D5 123
Yom, riv., Thai.......C4 38
Yomakyo, mtn., Bur.......C2 38
Yona, Guam, 1,105...... 52
Yonago, Jap., 62,900 (94,808▲)......I6 37
Yoncalla, Douglas, Oreg., 698......D3 113
Yonezawa, Jap., 60,300 (96,991▲)......H10 37
Yongan, Kor......F4 37
Yonges Island, Charleston, S.C., 250......G1, G7 115
Yonghung, Kor., 18,445......G3 37
Yongil, bay, Kor.......H4 37
Yonker, Sask., Can., 30......E1 70
Yonkers, Westchester, N.Y., 190,634......D1, D3, E7 108
Yonkers, Wagoner, Okla......A6 112
Yonne, dept., Fr., 269,826..*D5 14
Yonne, riv., Fr.......C5 14
Yorba Linda, Orange, Calif., 1,198......*F5 82
York, Sumter, Ala., 2,932....C1 78
York, Austl., 1,524......F2 50
York, Eng., 104,468......D6 10
York, York, Maine, 950 (4,663▲)......E1 96
York, York, Nebr., 6,173....D8 103
York, Benson, N. Dak., 148..A6 110

York, York, Pa., 54,504 (★146,600) G8 114
York, York, S.C., 4,758 B5 115
York, co., N.B., Can., 52,672 C3 74
York, co., Ont., Can., 1,733,108 D5 72
York, co., Eng., 4,722,661 A7 12
York, co., Maine, 99,402 E2 96
York, co., Nebr., 13,724 D8 103
York, co., Pa., 238,336 G8 114
York, co., S.C., 78,760 A5 115
York, co., Va., 21,583 D6 121
York, cape, Grnld B22 4
York, pt., Newf., Can C4 75
York, riv., Ont., Can B7 72
York, riv., Va D6 121
York Beach, York, Maine, 400 E2 96
York Center, Du Page, Ill., 1,100 *B6 90
York Corners, York, Maine (part of York) E2 96
Yorke, pen., Austl F6 50
York Factory, Man., Can A5 71
Yorkfield, Du Page, Ill., 1,200 *B6 90
York Harbor, York, Maine, 850 E2 96
York Haven, York, Pa., 736 F8 114
Yorklyn, New Castle, Del., 400 A6 85
Yorkshire, Cattaraugus, N.Y., 350 C2 108
Yorkshire, York, Pa., 1,000 *G8 114
Yorkshire, Prince William, Va., 1,500 *C5 121
York Springs, Adams, Pa., 384 F7 114
Yorkton, Sask., Can., 9,995 F4 72
Yorktown, Lincoln, Ark., 200 C4 81
Yorktown, Delaware, Ind., 1,137 D6 91
Yorktown, Page, Iowa, 150 D2 92
Yorktown, Salem, N.J., 150 D2 106
Yorktown, Westchester, N.Y., 3,576 *D7 108
Yorktown, DeWitt, Tex., 2,527 E4 118
Yorktown, York, Va., 311 A6, D6 121
Yorktown Heights, Westchester, N.Y., 2,478 A5 106
Yorktown Manor, Washington, R.I., 1,300 C11 84
Yorkville, Kendall, Ill., 1,568 B5 90
Yorkville, Oneida, N.Y., 3,749 B5 108
Yorkville, Belmont and Jefferson, Ohio, 1,801 B7 111
Yorkville, Gibson, Tenn., 250 A2 117
Yorkville, Racine, Wis., 35 F1 124
York Wolds, hills, Eng C6 10
Yoro, Hond., 1,471 C4 62
Yoseki, Con. L A3 48
Yosemite, nat. park, Calif D4 82
Yosemite National Park, Mariposa, Calif., 900 D4 82
Yosemite, Casey, Ky., 200 C5 94
Yoshiwara, Jap., 80,944 (★174,000) n17 37
Yoshkar-Ola, Sov. Un., 103,000 B3 29
Yost, Box Elder, Utah, 87 B2 119
Yosu, Kor., 76,000 (87,300▲) I3 37
Yotala, Bol., 1,554 C2 55
Youbou, B.C., Can., 1,153 B8 68
Youghal, Ire., 5,043 F4 11
Youghal, bay, Ire F4 11
Youghiogheny, riv., Md., Pa D1 85, F1 114
Youghiogheny River, res., Pa G3 114
Youkounkoun, Guinea, 700 D2 45
Young, Gila, Ariz., 30 C5 80
Young, Austl., 5,448 G7 51
Young, Sask., Can., 341 F3 70
Young, Ur., 4,923 E1 56
Young, co., Tex., 17,254 C3 118
Young America, Cass, Ind., 250 C5 91
Young Harris, Towns, Ga., 743 B3 87
Youngs Creek, Orange, Ind., 45 H4 91
Youngs Point, Ont., Can., 206 C6 72
Youngstown, Alta., Can., 321 D5 69
Youngstown, Bay, Fla., 600 B1, G3 86
Youngstown, Vigo, Ind., 200 F3 91
Youngstown, Polk, Iowa A7 92
Youngstown, Niagara, N.Y., 1,848 B1 108
Youngstown, Trumbull and Mahoning, Ohio, 166,689 (★467,600) A7 111
Youngsville, Lafayette, La., 946 D3 95
Youngsville, Rio Arriba, N. Mex., 20 A3 107
Youngsville, Franklin, N.C., 596 A5 109
Youngsville, Warren, Pa., 2,211 C3 114
Youngtown, Maricopa, Ariz., 1,559 G1 80
Youngwood, Westmoreland, Pa., 2,813 F2 114
Yountville, Napa, Calif., 600 A5 82
Yozgat, Tur., 18,300 C10 31
Ypacaraí, Par., 8,118 E4 55
Ypané, riv., Par D4 55
Ypres, see Ieper, Bel.
Ypsilanti, Washtenaw, Mich., 20,957 A6, F7 98
Ypsilanti, Stutsman, N. Dak., 110 C7 110
Yreka, Siskiyou, Calif., 4,759 B2 82
Ysleta, El Paso, Tex. (part of El Paso) F1 118
Yssingeaux, Fr., 2,702 E6 14
Ystad, Swe., 13,700 J5 25
Ythan, riv., Scot C6 13
Yü, China E6 36
Yü, riv., China G6 34
Yuan, riv., China J4 36
Yuanan, China, 13,000 I4 36
Yuanchiang, China G5 34
Yuanling, China, 28,000 J4 36
Yuanshih, China F6 36
Yuba, co., Calif., 33,859 C3 82
Yuba, riv., Calif C3 82
Yuba City, Sutter, Calif., 11,507 C3 82
Yuba City South, Sutter, Calif., 3,200 *C3 82
Yubari, Jap., 107,972 E10 37
Yubi, cape, Sp. Sahara D2 44
Yucaipa, San Bernardino, Calif., 6,000 *F3 82
Yucatán, state, Mex., 612,047 C7 63
Yucatán, chan., Mex C7 63
Yucca, Mohave, Ariz., 100 C1 80
Yucca, mtn., Nev G5 104
Yucheng, China F7 36
Yuchi, China, 4,000 K8 36
Yuehyang, China J5 36
Yug, riv., Sov. Un A3 29
Yugoslavia, country, Eur., 18,512,805 G12 8, C3 22
Yuhuan, China, 9,000 J9 36
Yukhnov, Sov. Un., 5,100 D10 27
Yukon, Duval, Fla., 2,000 B5, B6 86
Yukon, Canadian, Okla., 3,076 B4 112
Yukon, McDowell, W. Va., 400 D3 123
Yukon, ter., Can., 14,628 D5 66
Yukon, riv., Alsk., Yukon, Can B9 79
Yuksekkum, Tur D7 23
Yulee, Nassau, Fla., 500 B5, B6 86
Yuli, Nig E7 45
Yulin, China C8 38
Yulin, China E3 36
Yulin, Phil C8 35
Yuma, Yuma, Ariz., 23,974 E1 80
Yuma, Yuma, Colo., 1,919 A8 83
Yuma, Wexford, Mich., 50 D5 98
Yuma, Carroll, Tenn., 80 B3 117
Yuma, co., Ariz., 46,235 D1 80
Yuma, co., Colo., 8,912 A8 83
Yuma, des., Ariz. E1 80
Yuma, Indian res., Calif F6 82
Yumari, peak, Ven C4 60
Yumbi, Con L B4 48
Yumen, China, 50,000 C4 34
Yunabaru, Okinawa 52
Yuncheng, China G4 36
Yunes, riv., P.R B4 65
Yungan, China, 12,000 F8 34, L7 36
Yungas, mts., Bol C2 55
Yungay, Chile, 3,671 B2 54
Yungay, Peru, 2,517 C2 58
Yungchi, China, 10,000 G4 36
Yungching, China E5 34
Yungchow, see Lingling, China
Yungera, Austl G4 51
Yunghsiu, China, 7,000 J6 36
Yungkia, see Wenchow, China
Yungnien, China F6 36
Yungning (Nanning), China, 264,000 G6 34
Yungshou, China, 6,000 G2 36
Yungshun, China J3 36
Yungsui, China J3 36
Yunho, China, 3,000 J8 36
Yunhsiao, China, 18,000 G8 34
Yunhsien, China, 20,000 G5 34
Yunhsien, China H4 36
Yunnan, plat., China F5 34
Yunnan, prov., China 19,100,000 G5 34
Yunnanfu, see Kunming, China
Yunta, Austl., 111 F2 51
Yurécuaro, Mex., 12,088 m12 63
Yurga, Sov. Un., 40,000 D25 9
Yurimaguas, Peru, 5,918 C2 58
Yurino, Sov. Un., 5,000 C16 27
Yuriria, Mex., 10,221 m13 63
Yurochi, isl., Bikini 52
Yuryevets, Sov. Un., 10,000 C14 27
Yuryev-Polskiy, Sov. Un., 10,000 C12 27
Yuscarán, Hond., 1,189 D4 62
Yüshan, China, 16,000 J8 36
Yushih, China G6 36
Yushkozero, Sov. Un E15 25
Yuta, China F8 34
Yutan, Saunders, Nebr., 335 C9, D2 103
Yutien, China E7 36
Yutu, China, 16,000 F8 34
Yuty, Par., 16,262 E4 55
Yutzu, China, 60,000 F2 36
Yuwang, China F2 36
Yuyang, China, 18,000 J3 36
Yuyao, China, 26,000 I9 36
Yuyü, China D5 36
Yuzha, Sov. Un., 25,000 C13 27
Yuzhno-Sakhalinsk (Toyohara), Sov. Un., 85,000 C11 37
Yuzhnoye, Sov. Un., 10,000 C11 37
Yverdon, Switz., 16,338 C2 19
Yvetot, Fr., 7,932 C4 14
Yvonand, Switz., 1,290 C2 19
Ywathit, Bur C2 38

Z

Zaandam, Neth., 50,100 B4 15
Zabkowice, Pol., 10,127 C4 26
Zabkowice, Pol., 2,544 g10 26
Zābol, Iran F10 41
Zābolī, Iran H10 41
Zabrze, Pol., 189,000 C5, g9 26
Zacapa, Guat., 8,260 C3 62
Zacapu, Mex., 22,241 n13 63
Zacatecas, Mex., 31,851 C4 63
Zacatecas, state, Mex., 798,232 C4, m12 63
Zacatecoluca, Sal., 11,173 D3 62
Zachary, East Baton Rouge, La., 3,268 D4 95
Zachun, Ger., 273 E5 24
Zack, Searcy, Ark B3 81
Zacoalco [de Torres], Mex., 8,675 C4, m12 63
Zacualpan, Mex., 1,657 m14 63
Zacualtipán, Mex., 3,661 m14 63
Zadar, Yugo., 25,132 C2 22
Zadonsk, Sov. Un., 10,000 E12 27
Zadzbork, see Mragowo, Pol.
Za 'farānah, Eg., U.A.R H9 31
Zafra, Sp., 10,723 C2 20
Zagan, Pol., 15,300 C3 26
Zaghouan, Tun B7 44
Zagora, Grc., 3,223 C4 23
Zagora, Mor., 2,200 C3 44
Zagorsk, Sov. Un., 77,000 B1 29
Zagreb, Yugo., 427,319 C2 22
Zagros, mts., Iran E3 41
Zagyva, riv., Hung B5 22
Zāhedān, Iran, 5,000 G10 41
Zahl, Williams, N. Dak., 100 A2 110
Zahle, Leb., 30,387 F10 31
Zahna, Ger., 5,992 B7 17
Zaindeh, riv., Iran E5 41
Zaire, dist., Ang., 89,755 C1 48
Zaječar, Yugo., 18,545 D6 22
Zakataly, Sov. Un., 5,603 B16 31
Zákinthos, Grc., 9,506 D3 23
Zakinthos (Zante), prov., Grc., 35,509 *D3 23
Zakinthos (Zante), isl., Grc D3 23
Zakopane, Pol., 25,000 D6 26
Zakroczym, Pol., 3,358 k13 26
Zala, co., Hung., 274,161 *B3 22
Zala, riv., Hung B3 22
Zalaegerszeg, Hung., 23,738 B3 22
Zalamea de la Serena, Sp., 8,543 C3 20
Zalamea la Real, Sp., 6,065 D2 20
Zalău, Rom., 13,378 B6 22
Zaleski, Vinton, Ohio, 336 C5 111
Zalewo, Pol., 2,634 B5 26
Zalingei, Sud., 3,314 C1 47
Zalma, Bollinger, Mo., 141 D7 101
Zaltan, mts., Libya D3 43
Zaltbommel, Neth., 6,000 C5 15
Zama, Attala, Miss., 150 C4 96
Zambales, prov., Phil., 213,600 B6 35
Zambezi, riv., Afr H8 42
Zambézia, prov., Moz A6 49
Zamboanga, Phil., 25,000 (131,400▲) D6 35
Zamboanga del Norte, prov., Phil., 280,400 *D6 35
Zamboanga del Sur, prov., Phil., 744,500 *D6 35
Zambrano, Col., 5,863 B3 60
Zambrow, Pol., 4,150 B7 26
Zamora, Ec., 485 B2 58
Zamora, Mex., 31,991 C4, m12 63
Zamora, Sp., 42,060 B3 20
Zamora, prov., Sp., 301,129 *B3 20
Zamora, riv., Ec B2 58
Zamość, Pol., 28,000 C7 26
Zampa, pt., Okinawa 52
Zamsar, China B13 40
Zamzam, Wadi, Libya C2 43
Zanaga, Con. B F2 46
Zandvoort, Neth., 14,400 B4 15
Zanesville, Wells, Ind., 400 C7 91
Zanesville, Miskingum, Ohio, 39,077 C6 111
Zanja Blanca, P.R C5 65
Zanjān, Iran, 47,159 C4 41
Zante (Zakinthos), prov., Grc., 35,509 *D3 23
Zanzibar, Zan., 45,284 C6 48
Zanzibar, Br. dep., Afr., 299,111 G9 42, C6 48
Zanzibar, isl., Afr C6 48
Zaoviet Tahtania, Alg C4 44
Zap, Mercer, N. Dak., 339 B4 110
Zapadna Morava, riv., Yugo D5 22
Zapala, Arg., 3,387 B2 54
Zapallar, Arg E4 55
Zapata, Zapata, Tex., 2,031 F3 118
Zapata, co., Tex., 4,393 F3 118
Zapata, pen., Cuba D3 64
Zapatoca, Col., 5,629 B3 60
Zaporozhye, Sov. Un., 475,000 H10 27
Zapotillo, Ec B1 58
Zapotitlán, Mex., 3,248 h9 63
Zaqaziq, see Zagazig, Eg., U.A.R.
Zara, Tur., 7,400 C11 31
Zaragoza, Col., 1,732 B3 60
Zaragoza, Mex., 5,334 B4 63
Zaragoza, Mex., 2,464 k13 63
Zaragoza, Sp., 326,316 B5 20
Zaragoza, prov., Sp., 656,772 *B5 20
Zarand, Iran, 4,493 F8 41
Zárate, Arg., 35,197 A5, g7 54
Zaraysk, Sov. Un., 10,000 D12 27
Zaraza, Ven., 9,624 B4 60
Zardalu, Pak B1 40
Zardeh Kuh, mtn., Iran E5 41
Zarembo, isl., Alsk m23 79
Zarephath, Somerset, N.J., 250 B3 106
Zaria, Nig., 32,559 (★53,974) D6 45
Zarnekow, Ger., 434 E7 24
Zarnowiec, Pol., 700 A5 26
Zarga, Eg., U.A.R., 1,000 C3 32
Zarrentin, Ger., 3,796 E4 24
Zaruma, Ec., 3,855 B2 58
Zarumilla, Peru, 1,738 B1 58
Zary, Pol., 25,000 C3 26
Zary, Pol g9 26
Zarzal, Col., 7,395 C2 60
Zasenbeck, Ger., 420 F4 24
Zashiversk, Sov. Un., 800 C17 28
Zasieki, Pol C3 26
Zasmuky, Czech., 1,533 o19 26
Zastavna, Sov. Un A7 22
Zastron, S. Afr., 4,440 D4 49
Zatec, Czech., 14,800 C2 26
Zator, Pol., 1,844 h10 26
Zavala, co., Tex., 12,696 E3 118
Zavalla, Angelina, Tex., 700 D5 118
Zavitaya, Sov. Un., 14,500 D15 28
Zavodo-Petrovskiy, Sov. Un B7 29
Zawiercie, Pol., 33,000 C5, g10 26
Zawilah, Libya, 1,409 D3 43
Zāwiyat Masūs, Libya C4 43
Zaysan, Sov. Un., 15,600 F25 9
Zaysan Nor, lake, Sov. Un F25 9, E11 28
Zaza, riv., Cuba D4 64
Zbraslav, Czech., 4,643 o17 26
Zdice, Czech., 3,037 o16 26
Zdolbunov, Sov. Un., 10,000 F6 27
Zduńska Wola, Pol., 25,000 C5 26
Zealandia, Sask., Can., 192 F2 70
Zearing, Story, Iowa, 528 B4 92
Zeba, Baraga, Mich., 75 B2 98
Zeballos, B.C., Can., 235 D4 68
Zebdani, Syr., 8,900 F11 31
Zebrzydowice, Pol., 2,429 h9 26
Zebulon, Pike, Ga., 563 C2 87
Zebulon, Pike, Ky., 500 C7 94
Zebulon, Wake, N.C., 1,534 B5 109
Zeebrugge, Bel., 3,000 (part of Brugge) C3 15
Zeeland, Ottawa, Mich., 3,702 F4 98
Zeeland, McIntosh, N. Dak., 427 D6 110
Zeeland, prov., Neth., 283,900 C3 15
Zeerust, S. Afr., 6,907 C4 49
Zegrze, Pol., 2,000 k14 26
Zehdenick, Ger., 12,800 B6 16
Zeidab, Sud B3 47
Zeigler, Franklin, Ill., 2,133 F4 90
Zeila, Som C5 47
Zeist, Neth., 52,400 B5 15
Zeitz, Ger., 45,400 B7 17
Zekiah, swamp, Md C4 85
Zelechow, Pol., 3,892 C6 26
Zelenodolsk, Sov. Un., 64,000 B3 29
Zelenogorsk, Sov. Un., 32,000 (part of Leningrad) r30 25, A7 27
Zelezna Ruda Mestys, Czech., 1,322 D8 17
Zelienople, Butler, Pa., 3,284 E1 114
Zell, Faulk, S. Dak., 100 C7 116
Zella-Mehlis, Ger., 16,400 C5 17
Zell am Ziller, Aus., 1,510 B5 18
Zeller, lake, Aus B8 18
Zellwood, Orange, Fla., 900 D5 86
Zelma, Sask., Can., 81 F3 70
Zelzate, Bel., 10,593 C3 15
Zemetchino, Sov. Un., 10,000 E14 27
Zemio, Cen. Afr. Rep D5 46
Zena, Delaware, Okla., 150 A7 112
Zenda, Kingman, Kans., 157 E5 93
Zenia, Trinity, Calif., 160 B2 82
Zenica, Yugo., 32,552 C3 22
Zenith, Wayne, Ill., 40 E5 90
Zenith, Stafford, Kans., 50 E5 93
Zenith, King, Wash., 600 D1 122
Zenon Park, Sask., Can., 384 D4 70
Zenoria, La Salle, La., 100 C3 95
Zeona, Perkins, S. Dak., 5 B3 116
Zepče, Yugo., 2,709 C4 22
Zephyr, Brown, Tex., 250 D3 118
Zephyr Cove, Douglas, Nev., 200 E2 104
Zephyrhills, Pasco, Fla., 2,887 D4 86
Zerbst, Ger., 17,900 B7 17
Zerka, riv., Jordan B7 32
Zermatt, Switz., 2,731 D4 19
Zernez, Switz., 712 C9 19
Zero, Prairie, Mont., 10 D11 102
Zeta, riv., Yugo D4 22
Zeulenroda, Ger., 13,600 C6 17
Zeven, Ger., 7,700 E3 24
Zevenaar, Neth., 4,600 C6 15
Zevgolatio, Grc., 1,089 D3 23
Zeway, lake, Eth D4 47
Zeya, Sov. Un., 15,100 D15 28
Zeya, riv., Sov. Un D15 28
Zezere, riv., Port B2 20
Zgierz, Pol., 37,000 C5 26
Zgorzelec, Pol., 5,261 B10 17
Zhangiz-Tobe, Sov. Un D10 29
Zharkamys, Sov. Un D5 29
Zhdanov (Mariupol), Sov. Un., 310,000 H11 27
Zhidachov, Sov. Un D8 26
Zhigansk, Sov. Un C15 28
Zhikatse, China, 15,000 C8 39
Zhilaya Kosa, Sov. Un D4 29
Zhilevo, Sov. Un o18 27
Zhitomir, Sov. Un., 114,000 F7 27
Zhizdra, Sov. Un., 10,000 E10 27
Zhmerinka, Sov. Un., 25,000 G7 27
Zhob, riv., Pak B2 40
Zhukovskiy, Sov. Un., 33,000 n18 27
Zīā'ābād, Iran D4 41
Zia Pueblo, Indian res., N. Mex. F5 107
Zibā, Sau. Ar I10 31
Ziebach, co., S. Dak., 2,495 C4 116
Ziel, mtn., Austl D5 50
Zielona Gora, Pol., 51,000 C3 26
Zierikzee, Neth., 7,100 C3 15
Ziesar, Ger., 3,598 A7 17
Ziftá, Eg., U.A.R., 29,700 D3 32
Zigey, Chad C3 46
Ziguinchor, Sen., 22,400 D1 45
Zikhron Ya'aqov, Isr., 4,100 B6 32
Zile, Tur., 21,400 B10 31
Zilina, Czech., 32,500 D5 26
Ziling (Goring), lake, China B12 40
Zillah, Libya D3 43
Zillah, Yakima, Wash., 1,059 C5 122
Ziller, riv., Aus B7 18
Zilwaukee, Saginaw, Mich., 1,800 E7 98
Zima, Sov. Un., 33,200 D13 28
Zimapán, Mex., 2,343 C5, m14 63
Zimba, N. Rh., 95 E4 48
Zimmerman, Rapides, La., 500 C3 95
Zimmerman, Sherburne, Minn., 302 E5 99
Zimnicea, Rom., 12,445 D7 22
Zinal, Switz D4 19
Zinc, Boone, Ark., 68 A3 81
Zincville, Ottawa, Okla., 65 A7 112
Zindajan, Afg., 10,000 D10 41
Zinder, Niger, 13,300 D6 45
Zingst, Ger., 3,238 D6 24
Zion, Lake, Ill., 11,941 A6, D2 90
Zion, Cecil, Md., 360 A6 85
Zion, Somerset, N.J., 50 C3 106
Zion, Marion, S.C., 216 C9 115
Zion, nat. park, Utah F3 119
Zionsville, Boone, Ind., 1,822 E5, G7 91
Zionville, Watauga, N.C., 75 A2 109
Zipaquirá, Col., 12,708 B3 60
Zipp, Vanderburgh, Ind., 200 H2 91
Zippori, Isr., 281 B7 32
Zirándaro, Mex., 1,205 n13 63
Zirkel, mtn., Colo A4 83
Zirndorf, Ger., 12,000 D5 17
Zistersdorf, Aus., 3,011 D8 16
Zitácuaro, Mex., 23,410 n13 63
Zittau, Ger., 43,200 C9 17
Ziza, Jordan, 2,000 C7 32
Zlatograd, Bul., 4,169 E7 22
Zlatoust, Sov. Un., 166,000 B5 29
Zlitan, Libya, 4,000 H14 30
Zloczew, Pol., 2,948 C5 26
Zlonice, Czech., 271 n17 26
Zlotoryja, Pol., 4,613 C3 26
Zlotow, Pol., 5,275 B4 26
Zlynka, Sov. Un., 10,000 E8 27
Zmeinogorsk, Sov. Un C10 29
Znamenka, Sov. Un., 37,300 G9 27
Znin, Pol., 5,615 B4 26
Znojmo, Czech., 24,000 D4 26
Zoar, Newf., Can., 4 g9 75
Zoe, Le Flore, Okla C7 112
Zofingen, Switz., 8,779 B4 19
Zolfo Springs, Hardee, Fla., 838 E5 86
Zollikofen, Switz., 6,237 B3 19
Zollikon, Switz., 10,060 B6 19
Zolochev, Sov. Un., 28,600 G5 27
Zolotonosha, Sov. Un., 10,000 G9 27
Zolotoy, cape, Sov. Un C9 37
Zomba, Nya., 15,500 A6 49
Zomergem, Bel., 6,029 C3 15
Zona, Washington, La., 50 D5 95
Zongo, Con. L A2 48
Zonguldak, Tur., 54,000 B8 31
Zonza, riv., Pol k14 26
Zook Spur, Dallas, Iowa, 40 A7 92
Zörbig, Ger., 6,806 B7 17
Zorita, Sp., 5,718 C3 20
Zorritos, Peru B1 58
Zortman, Phillips, Mont., 120 D11 102
Zossen, Ger., 5,958 A8 17
Zottegem, Bel., 6,630 D3 15
Zouar, Chad, 272 A3 46
Zoutkamp, Neth., 1,068 A6 15
Zrenjanin (Petrovgrad), Yugo., 48,954 C5 22
Zrmanja, riv., Yugo C2 22
Zschopau, Ger., 8,983 C8 17
Zubtsov, Sov. Un., 9,100 C10 27
Zuehl, Guadalupe, Tex B4 112
Zuénoula, I.C., 4,000 E3 45
Zuera, Sp., 3,802 B5 20
Zug, Switz., 19,792 B6 19
Zug, canton, Switz., 52,489 B6 19
Zug, lake, Switz B6 19
Zugdidi, Sov. Un., 10,000 A13 31
Zugspitze, mtn., Ger B6 18
Zuhreh, riv., Iran F5 41
Zuider Zee (IJsselmeer), sea, Neth B5 15
Zuidhorn, Neth., 1,863 A6 15
Zulia, state, Ven., 919,863 B3 60
Zülpich, Ger., 4,900 C1 17
Zulueta, Cuba, 4,337 D4 64
Zumbo, Moz., 5,000 A5 49
Zumbro, riv., Minn F6 99
Zumbro Falls, Wabasha, Minn., 164 F6 99
Zumbrota, Goodhue, Minn., 1,830 F6 99
Zumpango [de Ocampo], Mex., 6,771 n14 63
Zungeru, Nig., 1,661 E6 45
Zuni, McKinley, N. Mex., 3,585 B1 107
Zuni, Isle of Wight, Va., 155 B5 121
Zuni, Indian res., N. Mex B1 107
Zuni, mts., N. Mex B1 107
Zuni, res., Ariz C6 80
Zuni, riv., Ariz., N. Mex C6 80, B1 107
Zurich, Ont., Can., 723 D3 72
Zurich, Rooks, Kans., 244 C4 93
Zürich, Switz., 440,170 (★665,000) B6 19
Zürich, canton, Switz., 952,304 A6 19
Zürich, lake, Switz B6 19
Zuru, Nig D6 45
Zusam, riv., Ger E5 17
Zutphen, Neth., 25,200 B6 15
Zuwārah, Libya, 2,380 C2 43
Zuwayzā, Jordan C7 32
Zuyevka, Sov. Un., 18,200 B4 29
Zvenigorodka, Sov. Un., 10,000 G8 27
Zvolen, Czech., 19,600 D5 26
Zvornik, Yugo., 5,438 C4 22
Zwartsluis, Neth., 3,600 B6 15
Zweibrücken, Ger., 32,900 D2 17
Zweisimmen, Switz., 2,676 C3 19
Zwenkau, Ger., 10,300 B7 17
Zwettl, Aus., 3,836 D7 16
Zwickau, Ger., 128,700 (★180,000) C6 14
Zwickauer Mulde, riv., Ger C7 17
Zwiesel, Ger., 8,100 D8 17
Zwingle, Dubuque and Jackson, Iowa, 110 B7 92
Zwolle, Sabine, La., 1,326 C2 95
Zwolle, Neth., 56,300 B6 15
Zylks, Caddo, La., 200 A1 95
Żyrardów, Pol., 30,000 B6, m12 26
Zyryanovsk, Sov. Un., 56,000 D10 29, F25 9
Zywiec, Pol., 17,000 D5 26